DEVELOPMENTAL MATHEMATICS

Published by McGraw Hill LLC, 1325 Avenue of the Americas, New York, NY 10019. Copyright ©2022 by McGraw Hill LLC. All rights reserved. Printed in the United States of America. No part of this publication may be reproduced or distributed in any form or by any means, or stored in a database or retrieval system, without the prior written consent of McGraw Hill LLC, including, but not limited to, in any network or other electronic storage or transmission, or broadcast for distance learning.

Some ancillaries, including electronic and print components, may not be available to customers outside the United States.

This book is printed on acid-free paper.

1 2 3 4 5 6 7 8 9 GPC 26 25 24 23 22 21

ISBN 978-1-264-42917-2
MHID 1-264-42917-7

Cover Image: ©*UpperCut Images/Alamy Stock Photo*

All credits appearing on page or at the end of the book are considered to be an extension of the copyright page.

The Internet addresses listed in the text were accurate at the time of publication. The inclusion of a website does not indicate an endorsement by the authors or McGraw Hill LLC, and McGraw Hill LLC does not guarantee the accuracy of the information presented at these sites.

mheducation.com/highered

Letter from the Authors

Dear Colleagues,

Across the country, Developmental Math courses are in a state of flux, and we as instructors are at the center of it all. As many of our institutions are grappling with the challenges of placement, retention, and graduation rates, we are on the front lines with our students—supporting all of them in their educational journey.

Flexibility—No Matter Your Course Format!

The three of us each teach differently, as do many of our current users. The Miller/O'Neill/Hyde series is designed for successful use in a variety of course formats, both traditional and modern—classroom lecture settings, flipped classrooms, hybrid classes, and online-only classes.

Ease of Instructor Preparation

We've all had to fill in for a colleague, pick up a last-minute section, or find ourselves running across campus to yet a different course. The Miller/O'Neill/Hyde series is carefully designed to support instructors teaching in a variety of different settings and circumstances. Experienced, senior faculty members can draw from a massive library of static and algorithmic content found in ALEKS to meticulously build assignments and assessments sharply tailored to individual student needs. Newer instructors and part-time adjunct instructors, on the other hand, will find support through a wide range of digital resources and prebuilt assignments ready to go on Day One. With these tools, instructors with limited time to prepare for class can still facilitate successful student outcomes.

Many instructors want to incorporate discovery-based learning and groupwork into their courses but don't have time to write or find quality materials. Each section of the text has numerous discovery-based activities that we have tested in our own classrooms. These are found in the text and Student Resource Manual along with other targeted worksheets for additional practice and materials for a student portfolio.

Student Success—Now and in the Future

Too often our math placement tests fail our students, which can lead to frustration, anxiety, and often withdrawal from their education journey. We encourage you to learn more about ALEKS Placement, Preparation, and Learning (ALEKS PPL), which uses adaptive learning technology to place students appropriately. No matter the skills they come in with, the Miller/O'Neill/Hyde series provides resources and support that uniquely position them for success in that course and for their next course. Whether they need a brush-up on their basic skills, ADA supportive materials, or advanced topics to help them cross the bridge to the next level, we've created a support system for them.

We hope you are as excited as we are about the series and the supporting resources and services that accompany it. Please reach out to any of us with any questions or comments you have about our texts.

Julie Miller   Molly O'Neill   Nancy Hyde

About the Authors

Photo courtesy of Molly O'Neill

Julie Miller is from Daytona State College, where she taught developmental and upper-level mathematics courses for 20 years. Prior to her work at Daytona State College, she worked as a software engineer for General Electric in the area of flight and radar simulation. Julie earned a Bachelor of Science in Applied Mathematics from Union College in Schenectady, New York, and a Master of Science in Mathematics from the University of Florida. In addition to this textbook, she has authored textbooks for college algebra, trigonometry, and precalculus, as well as several short works of fiction and nonfiction for young readers.

"My father is a medical researcher, and I got hooked on math and science when I was young and would visit his laboratory. I can remember using graph paper to plot data points for his experiments and doing simple calculations. He would then tell me what the peaks and features in the graph meant in the context of his experiment. I think that applications and hands-on experience made math come alive for me, and I'd like to see math come alive for my students."

—*Julie Miller*

Molly O'Neill is also from Daytona State College, where she taught for 22 years in the School of Mathematics. She has taught a variety of courses from developmental mathematics to calculus. Before she came to Florida, Molly taught as an adjunct instructor at the University of Michigan–Dearborn, Eastern Michigan University, Wayne State University, and Oakland Community College. Molly earned a Bachelor of Science in Mathematics and a Master of Arts and Teaching from Western Michigan University in Kalamazoo, Michigan. Besides this textbook, she has authored several course supplements for college algebra, trigonometry, and precalculus and has reviewed texts for developmental mathematics.

"I differ from many of my colleagues in that math was not always easy for me. But in seventh grade I had a teacher who taught me that if I follow the rules of mathematics, even I could solve math problems. Once I understood this, I enjoyed math to the point of choosing it for my career. I now have the greatest job because I get to do math every day and I have the opportunity to influence my students just as I was influenced. Authoring these texts has given me another avenue to reach even more students."

—*Molly O'Neill*

Nancy Hyde served as a full-time faculty member of the Mathematics Department at Broward College for 24 years. During this time she taught the full spectrum of courses from developmental math through differential equations. She received a Bachelor of Science in Math Education from Florida State University and a Master's degree in Math Education from Florida Atlantic University. She has conducted workshops and seminars for both students and teachers on the use of technology in the classroom. In addition to this textbook, she has authored a graphing calculator supplement for *College Algebra*.

"I grew up in Brevard County, Florida, where my father worked at Cape Canaveral. I was always excited by mathematics and physics in relation to the space program. As I studied higher levels of mathematics I became more intrigued by its abstract nature and infinite possibilities. It is enjoyable and rewarding to convey this perspective to students while helping them to understand mathematics."

—*Nancy Hyde*

Dedication
To Our Students

Julie Miller ⚜ Molly O'Neill ⚜ Nancy Hyde

The Miller/O'Neill/Hyde Developmental Math Series

Julie Miller, Molly O'Neill, and Nancy Hyde originally wrote their developmental math series because students were entering their College Algebra course underprepared. The students were not mathematically mature enough to understand the concepts of math, nor were they fully engaged with the material. The authors began their developmental mathematics offerings with Intermediate Algebra to help bridge that gap. This in turn evolved into several series of textbooks from Prealgebra through Precalculus to help students at all levels before Calculus.

What sets all of the Miller/O'Neill/Hyde series apart is that they address course content through an author-created digital package that maintains a consistent voice and notation throughout the program. This consistency—in videos, PowerPoints, Lecture Notes, and Integrated Video and Study Guides—coupled with the power of ALEKS, ensures that students master the skills necessary to be successful in Developmental Math through Precalculus and prepares them for the Calculus sequence.

Developmental Math Series
The Developmental Math series is traditional in approach, delivering a purposeful balance of skills and conceptual development. It places a strong emphasis on conceptual learning to prepare students for success in subsequent courses.
- Basic College Mathematics, Third Edition
- Prealgebra, Third Edition
- Prealgebra & Introductory Algebra, Second Edition
- Beginning Algebra, Sixth Edition
- Beginning & Intermediate Algebra, Sixth Edition
- Intermediate Algebra, Sixth Edition
- Developmental Mathematics: Prealgebra, Beginning Algebra, & Intermediate Algebra, Second Edition

The Miller/Gerken College Algebra/Precalculus Series
The Precalculus series serves as the bridge from Developmental Math coursework to future courses by emphasizing the skills and concepts needed for Calculus.
- College Algebra with Corequisite Support, First Edition
- College Algebra, Second Edition
- College Algebra and Trigonometry, First Edition
- Precalculus, First Edition

Acknowledgments

The author team most humbly would like to thank all the people who have contributed to this project and the Miller/O'Neill/Hyde Developmental Math series as a whole.

First and foremost, our utmost gratitude to Sarah Alamilla for her close partnership, creativity, and collaboration throughout this revision. Special thanks to our team of digital contributors for their thousands of hours of work: to Kelly Jackson, Jody Harris, Lizette Hernandez Foley, Lisa Rombes, Kelly Kohlmetz, and Leah Rineck for their devoted work. To Donna Gerken: thank you for the countless grueling hours working through spreadsheets to ensure thorough coverage of our content in ALEKS. To our digital authors, Linda Schott, Michael Larkin, and Alina Coronel: thank you for digitizing our content so it could be brought into ALEKS. We also offer our sincerest appreciation to the outstanding video talent: Alina Coronel, Didi Quesada, Tony Alfonso, and Brianna Ashley. So many students have learned from you! To Jennifer Blue, Carey Lange, John Murdzek, and Kevin Campbell: thank you so much for ensuring accuracy in our manuscripts.

We also greatly appreciate the many people behind the scenes at McGraw Hill without whom we would still be on page 1. To Megan Platt, our product developer: thank you for being our help desk and handling all things math, English, and editorial. To Brittney Merriman and Jennifer Morales, our portfolio managers and team leaders: thank you so much for leading us down this path. Your insight, creativity, and commitment to our project has made our job easier.

To the marketing team, Michele McTighe, Noah Evans, and Mary Ellen Rahn: thank you for your creative ideas in making our books come to life in the market. Thank you as well to Debbie McFarland, Justin Washington, and Sherry Bartel for continuing to drive our long-term content vision through their market development efforts. And many thanks to the team at ALEKS for creating its spectacular adaptive technology and for overseeing the quality control.

To the production team: Jane Mohr, David Hash, Lorraine Buczek, and Sandy Ludovissy—thank you for making the manuscript beautiful and for keeping the unruly authors on track. To our copyeditor Kevin Campbell and proofreader John Murdzek, who have kept a watchful eye over our manuscripts—the two of you are brilliant. To our compositor Manvir Singh and his team at Aptara, you've been a dream to work with. And finally, to Kathleen McMahon and Caroline Celano, thank you for supporting our projects for many years and for the confidence you've always shown in us.

Most importantly, we give special thanks to the students and instructors who use our series in their classes.

Julie Miller
Molly O'Neill
Nancy Hyde

Contents

Chapter 1 — Whole Numbers   1

- 1.1 Introduction to Whole Numbers   2
- 1.2 Addition and Subtraction of Whole Numbers and Perimeter   9
- 1.3 Rounding and Estimating   26
- 1.4 Multiplication of Whole Numbers and Area   33
- 1.5 Division of Whole Numbers   47
- **Problem Recognition Exercises:** Operations on Whole Numbers   58
- 1.6 Exponents, Algebraic Expressions, and the Order of Operations   59
- 1.7 Mixed Applications and Computing Mean   68
- Chapter 1   Review Exercises   75

Chapter 2 — Integers and Algebraic Expressions   79

- 2.1 Integers, Absolute Value, and Opposite   80
- 2.2 Addition of Integers   87
- 2.3 Subtraction of Integers   96
- 2.4 Multiplication and Division of Integers   103
- **Problem Recognition Exercises:** Operations on Integers   112
- 2.5 Order of Operations and Algebraic Expressions   113
- Chapter 2   Review Exercises   121

Chapter 3 — Solving Equations   125

- 3.1 Simplifying Expressions and Combining *Like Terms*   126
- 3.2 Addition and Subtraction Properties of Equality   135
- 3.3 Multiplication and Division Properties of Equality   143
- 3.4 Solving Equations with Multiple Steps   149
- **Problem Recognition Exercises:** Identifying Expressions and Equations   156
- 3.5 Applications and Problem Solving   156
- Chapter 3   Review Exercises   166

Chapter 4 — Fractions and Mixed Numbers   169

- 4.1 Introduction to Fractions and Mixed Numbers   170
- 4.2 Simplifying Fractions   180
- 4.3 Multiplication and Division of Fractions   195
- 4.4 Least Common Multiple and Equivalent Fractions   210
- 4.5 Addition and Subtraction of Fractions   220
- 4.6 Estimation and Operations on Mixed Numbers   230
- **Problem Recognition Exercises:** Operations on Fractions and Mixed Numbers   243
- 4.7 Order of Operations and Complex Fractions   244
- 4.8 Solving Equations Containing Fractions   253
- **Problem Recognition Exercises:** Comparing Expressions and Equations   261
- Chapter 4   Review Exercises   262

Chapter 5

Decimals 267

- 5.1 Decimal Notation and Rounding 268
- 5.2 Addition and Subtraction of Decimals 279
- 5.3 Multiplication of Decimals and Applications with Circles 288
- 5.4 Division of Decimals 301
  - **Problem Recognition Exercises:** Operations on Decimals 313
- 5.5 Fractions, Decimals, and the Order of Operations 314
- 5.6 Solving Equations Containing Decimals 329
  - Chapter 5 Review Exercises 336

Chapter 6

Ratio and Proportion 341

- 6.1 Ratios 342
- 6.2 Rates and Unit Cost 350
- 6.3 Proportions 358
  - **Problem Recognition Exercises:** Operations on Fractions versus Solving Proportions 366
- 6.4 Applications of Proportions and Similar Figures 367
  - Chapter 6 Review Exercises 378

Chapter 7

Percents 383

- 7.1 Percents, Fractions, and Decimals 384
- 7.2 Percent Proportions and Applications 397
- 7.3 Percent Equations and Applications 407
  - **Problem Recognition Exercises:** Percents 416
- 7.4 Applications of Sales Tax, Commission, Discount, Markup, and Percent Increase and Decrease 417
- 7.5 Simple and Compound Interest 431
  - Chapter 7 Review Exercises 440

Chapter 8

Measurement and Geometry 445

- 8.1 U.S. Customary Units of Measurement 446
- 8.2 Metric Units of Measurement 459
- 8.3 Converting Between U.S. Customary and Metric Units 473
  - **Problem Recognition Exercises:** U.S. Customary and Metric Conversions 481
- 8.4 Medical Applications Involving Measurement 482
- 8.5 Lines and Angles 485
- 8.6 Triangles and the Pythagorean Theorem 496
- 8.7 Perimeter, Circumference, and Area 508
  - **Problem Recognition Exercises:** Area, Perimeter, and Circumference 520
- 8.8 Volume and Surface Area 521
  - Chapter 8 Review Exercises 530

Chapter 9 Graphs and Statistics 535

- 9.1 Tables, Bar Graphs, Pictographs, and Line Graphs 536
- 9.2 Frequency Distributions and Histograms 548
- 9.3 Circle Graphs 556
- 9.4 Introduction to Probability 564
- 9.5 Mean, Median, and Mode 572
    - Chapter 9 Review Exercises 584

Chapter 10 Linear Equations and Inequalities 589

- 10.1 Addition, Subtraction, Multiplication, and Division Properties of Equality 590
- 10.2 Solving Linear Equations 603
- 10.3 Linear Equations: Clearing Fractions and Decimals 612
    - Problem Recognition Exercises: Equations vs. Expressions 620
- 10.4 Applications of Linear Equations: Introduction to Problem Solving 621
- 10.5 Applications Involving Percents 632
- 10.6 Formulas and Applications of Geometry 640
- 10.7 Mixture Applications and Uniform Motion 651
- 10.8 Linear Inequalities 662
    - Chapter 10 Review Exercises 678

Chapter 11 Graphing Linear Equations in Two Variables 683

- 11.1 Rectangular Coordinate System 684
- 11.2 Linear Equations in Two Variables 694
- 11.3 Slope of a Line and Rate of Change 711
- 11.4 Slope-Intercept Form of a Linear Equation 727
    - Problem Recognition Exercises: Linear Equations in Two Variables 738
- 11.5 Point-Slope Formula 740
- 11.6 Applications of Linear Equations and Modeling 750
    - Chapter 11 Review Exercises 759

Chapter 12 Systems of Linear Equations in Two Variables 765

- 12.1 Solving Systems of Equations by the Graphing Method 766
- 12.2 Solving Systems of Equations by the Substitution Method 777
- 12.3 Solving Systems of Equations by the Addition Method 788
    - Problem Recognition Exercises: Systems of Equations 799
- 12.4 Applications of Linear Equations in Two Variables 802
- 12.5 Systems of Linear Equations in Three Variables 813
- 12.6 Applications of Systems of Linear Equations in Three Variables 823
    - Chapter 12 Review Exercises 828

Chapter 13 — Polynomials and Properties of Exponents 831

- 13.1 Multiplying and Dividing Expressions With Common Bases 832
- 13.2 More Properties of Exponents 843
- 13.3 Definitions of b^0 and b^{-n} 849
- Problem Recognition Exercises: Properties of Exponents 859
- 13.4 Scientific Notation 860
- 13.5 Addition and Subtraction of Polynomials 866
- 13.6 Multiplication of Polynomials and Special Products 876
- 13.7 Division of Polynomials 887
- Problem Recognition Exercises: Operations on Polynomials 899
- Chapter 13 Review Exercises 900

Chapter 14 — Factoring Polynomials 903

- 14.1 Greatest Common Factor and Factoring by Grouping 904
- 14.2 Factoring Trinomials of the Form $x^2 + bx + c$ 915
- 14.3 Factoring Trinomials: Trial-and-Error Method 922
- 14.4 Factoring Trinomials: AC-Method 933
- 14.5 Difference of Squares and Perfect Square Trinomials 941
- 14.6 Sum and Difference of Cubes 948
- Problem Recognition Exercises: Factoring Strategy 956
- 14.7 Solving Equations Using the Zero Product Rule 957
- Problem Recognition Exercises: Polynomial Expressions versus Polynomial Equations 965
- 14.8 Applications of Quadratic Equations 966
- Chapter 14 Review Exercises 974

Chapter 15 — Rational Expressions and Equations 977

- 15.1 Introduction to Rational Expressions 978
- 15.2 Multiplication and Division of Rational Expressions 989
- 15.3 Least Common Denominator 997
- 15.4 Addition and Subtraction of Rational Expressions 1004
- Problem Recognition Exercises: Operations on Rational Expressions 1015
- 15.5 Complex Fractions 1016
- 15.6 Rational Equations 1025
- Problem Recognition Exercises: Comparing Rational Equations and Rational Expressions 1037
- 15.7 Applications of Rational Equations and Proportions 1038
- Chapter 15 Review Exercises 1051

Chapter 16 — Relations and Functions 1055

- 16.1 Introduction to Relations 1056
- 16.2 Introduction to Functions 1066
- 16.3 Graphs of Functions 1079
- Problem Recognition Exercises: Characteristics of Relations 1091
- 16.4 Algebra of Functions and Composition 1092
- 16.5 Variation 1100
- Chapter 16 Review Exercises 1110

Chapter 17 — More Equations and Inequalities 1113

- 17.1 Compound Inequalities 1114
- 17.2 Polynomial and Rational Inequalities 1128
- 17.3 Absolute Value Equations 1140
- 17.4 Absolute Value Inequalities 1147
  - Problem Recognition Exercises: Equations and Inequalities 1158
- 17.5 Linear Inequalities and Systems of Linear Inequalities in Two Variables 1159
  - Chapter 17 Review Exercises 1174

Chapter 18 — Radicals and Complex Numbers 1179

- 18.1 Definition of an nth Root 1180
- 18.2 Rational Exponents 1193
- 18.3 Simplifying Radical Expressions 1201
- 18.4 Addition and Subtraction of Radicals 1209
- 18.5 Multiplication of Radicals 1216
- 18.6 Division of Radicals and Rationalization 1225
  - Problem Recognition Exercises: Operations on Radicals 1236
- 18.7 Solving Radical Equations 1237
- 18.8 Complex Numbers 1248
  - Chapter 18 Review Exercises 1260

Chapter 19 — Quadratic Equations and Functions 1263

- 19.1 Square Root Property and Completing the Square 1264
- 19.2 Quadratic Formula 1276
- 19.3 Equations in Quadratic Form 1292
  - Problem Recognition Exercises: Quadratic and Quadratic Type Equations 1299
- 19.4 Graphs of Quadratic Functions 1299
- 19.5 Vertex of a Parabola: Applications and Modeling 1314
  - Chapter 19 Review Exercises 1326

Chapter 20 — Exponential and Logarithmic Functions and Applications 1329

- 20.1 Inverse Functions 1330
- 20.2 Exponential Functions 1341
- 20.3 Logarithmic Functions 1352
  - Problem Recognition Exercises: Identifying Graphs of Functions 1366
- 20.4 Properties of Logarithms 1367
- 20.5 The Irrational Number e and Change of Base 1376
  - Problem Recognition Exercises: Logarithmic and Exponential Forms 1390
- 20.6 Logarithmic and Exponential Equations and Applications 1391
  - Chapter 20 Review Exercises 1405

Chapter 21 — Conic Sections 1409

- 21.1 Distance Formula, Midpoint Formula, and Circles 1410
- 21.2 More on the Parabola 1422
- 21.3 The Ellipse and Hyperbola 1432
  - **Problem Recognition Exercises:** Formulas and Conic Sections 1441
- 21.4 Nonlinear Systems of Equations in Two Variables 1442
- 21.5 Nonlinear Inequalities and Systems of Inequalities in Two Variables 1449
  - **Chapter 21 Review Exercises** 1459

Chapter 22 — Binomial Expansions, Sequences, and Series 1463

- 22.1 Binomial Expansions 1464
- 22.2 Sequences and Series 1471
- 22.3 Arithmetic Sequences and Series 1481
- 22.4 Geometric Sequences and Series 1488
  - **Problem Recognition Exercises:** Identifying Arithmetic and Geometric Sequences 1498
  - **Chapter 22 Review Exercises** 1499

Chapter 23 (Online) — Transformations, Piecewise-Defined Functions, and Probability 23-1

- 23.1 Transformations of Graphs and Piecewise-Defined Functions 23-2
- 23.2 Fundamentals of Counting 23-15
- 23.3 More on Probability 23-24
  - **Chapter 23 Review Exercises** 23-33

Additional Topics Appendix A-1

- A.1 Additional Factoring A-1
- A.2 Solving Systems of Linear Equations by Using Matrices A-6
- A.3 Determinants and Cramer's Rule A-14
- A.4 Introduction to Modeling A-26

Appendix B (Online) B-1

- B.1 Review of the Set of Real Numbers B-1
- B.2 Review of Linear Equations and Linear Inequalities B-10
- B.3 Review of Graphing B-15
- B.4 Review of Systems of Linear Equations in Two Variables B-25
- B.5 Review of Polynomials and Properties of Exponents B-31
- B.6 Review of Factoring Polynomials and Solving Quadratic Equations B-38
- B.7 Review of Rational Expressions B-43

Student Answer Appendix SA-1

Index I-1

Formula Charts PC-1

To the Student

Take a deep breath and know that you aren't alone. Your instructor, fellow students, and we, your authors, are here to help you learn and master the material for this course and prepare you for future courses. You may feel like math just isn't your thing, or maybe it's been a long time since you've had a math class—that's okay!

We wrote the text and all the supporting materials with you in mind. Most of our students aren't really sure how to be successful in math, but we can help with that.

As you begin your class, we'd like to offer some specific suggestions:

1. **Attend class.** Arrive on time and be prepared. If your instructor has asked you to read prior to attending class—do it. How often have you sat in class and thought you understood the material, only to get home and realize you don't know how to get started? By reading and trying a couple of Skill Practice exercises, which follow each example, you will be able to ask questions and gain clarification from your instructor when needed.

2. **Be an *active* learner.** Whether you are at lecture, watching an author lecture or exercise video, or are reading the text, pick up a pencil and work out the examples given. Math is learned only by doing; we like to say, "Math is not a spectator sport." If you like a bit more guidance, we encourage you to use the Integrated Video and Study Guide. It was designed to provide structure and note-taking for lectures and while watching the accompanying videos.

3. **Schedule time to do some math every day.** Exercise, foreign language study, and math are three things that you must do every day to get the results you want. If you are used to cramming and doing all of your work in a few hours on a weekend, you should know that even mathematicians start making silly errors after an hour or so! Check your answers. Skill Practice exercises all have the answers at the bottom of that page. Odd-numbered exercises throughout the text have answers in the back of the text. If you didn't get it right, don't throw in the towel. Try again, revisit an example, or bring your questions to class for extra help.

4. **Prepare for quizzes and exams.** Each chapter has a set of Chapter Review Exercises at the end to help you integrate all of the important concepts. In addition, there is a detailed Chapter Summary and a Chapter Test located in the online resources. If you use ALEKS, use all of the tools available within the program to test your understanding.

5. **Use your resources.** This text comes with numerous supporting resources designed to help you succeed in this class and in your future classes. Additionally, your instructor can direct you to resources within your institution or community. Form a student study group. Teaching others is a great way to strengthen your own understanding, and they might be able to return the favor if you get stuck.

We wish you all the best in this class and in your educational journey!

Julie Miller   Molly O'Neill   Nancy Hyde

Student Guide to the Text

Clear, Precise Writing
Learning from our own students, we have written this text in simple and accessible language. Our goal is to keep you engaged and supported throughout your coursework.

Call-Outs
Just as your instructor will share tips and math advice in class, we provide call-outs throughout the text to offer tips and warn against common mistakes.

- Tip boxes offer additional insight into a concept or procedure.
- Avoiding Mistakes help fend off common student errors.
- For Review boxes positioned strategically throughout the text remind students of key skills relating to the current topic.

Examples
- Each example is step-by-step, with thorough annotation to the right explaining each step.
- Following each example is a similar **Skill Practice** exercise to give you a chance to test your understanding. You will find the answer at the bottom of the page—providing a quick check.

Exercise Sets
Each type of exercise is built so you can successfully learn the materials and show your mastery on exams.

- **Activities for discovery-based learning** appear before the exercise sets to walk students through the concepts presented in each section of the text.
- **Study Skills Exercises** integrate your studies of math concepts with strategies for helping you grow as a student overall.
- **Vocabulary and Key Concept Exercises** check your understanding of the language and ideas presented within the section.
- **Prerequisite Review** exercises keep fresh your knowledge of math content already learned by providing practice with concepts explored in previous sections.
- **Concept Exercises** assess your comprehension of the specific math concepts presented within the section.
- **Mixed Exercises** evaluate your ability to successfully complete exercises that combine multiple concepts presented within the section.
- **Expanding Your Skills** challenge you with advanced skills practice exercises around the concepts presented within the section.
- **Problem Recognition Exercises** appear in strategic locations in each chapter of the text. These will require you to distinguish between similar problem types and to determine what type of problem-solving technique to apply.
- **Technology Exercises** appear where appropriate.

End-of-Chapter Materials
The features at the end of each chapter and online are perfect for reviewing before test time.

- **Chapter Review Exercises** provide additional opportunities to practice material from the entire chapter.
- **Section-by-section summaries** provide references to key concepts, examples, and vocabulary.
- **Chapter tests** are an excellent way to test your complete understanding of the chapter concepts.

Get Better Results

How Will Miller/O'Neill/Hyde Help Your Students *Get Better Results?*

Clarity, Quality, and Accuracy

Julie Miller, Molly O'Neill, and Nancy Hyde know what students need to be successful in mathematics. Better results come from clarity in their exposition, quality of step-by-step worked examples, and accuracy of their exercise sets; but it takes more than just great authors to build a textbook series to help students achieve success in mathematics. Our authors worked with a strong team of mathematics instructors from around the country to ensure that the clarity, quality, and accuracy you expect from the Miller/O'Neill/Hyde series was included in this edition.

Exercise Sets

Comprehensive sets of exercises are available for every student level. Julie Miller, Molly O'Neill, and Nancy Hyde worked with a board of advisors from across the country to offer the appropriate depth and breadth of exercises for your students. **Problem Recognition Exercises** were created to improve student performance while testing.

Practice exercise sets help students progress from skill development to conceptual understanding. Student tested and instructor approved, the Miller/O'Neill/Hyde exercise sets will help your students *get better results*.

- ▶ **Activities for Discovery-Based Learning**
- ▶ **Prerequisite Review Exercises**
- ▶ **Problem Recognition Exercises**
- ▶ **Skill Practice Exercises**
- ▶ **Study Skills Exercises**
- ▶ **Mixed Exercises**
- ▶ **Expanding Your Skills Exercises**
- ▶ **Vocabulary and Key Concepts Exercises**
- ▶ **Technology Exercises**

Step-By-Step Pedagogy

This text provides enhanced step-by-step learning tools to help students *get better results*.

- ▶ **For Review** tips placed in the margin guide students back to related prerequisite skills needed for full understanding of course-level topics.
- ▶ **Worked Examples** provide an "easy-to-understand" approach, clearly guiding each student through a step-by-step approach to master each practice exercise for better comprehension.
- ▶ **TIPs** offer students extra cautious direction to help improve understanding through hints and further insight.
- ▶ **Avoiding Mistakes** boxes alert students to common errors and provide practical ways to avoid them. Both of these learning aids will help students get better results by showing how to work through a problem using a clearly defined step-by-step methodology that has been class tested and student approved.

Get Better Results

Formula for Student Success

Step-by-Step Worked Examples

▶ Do you get the feeling that there is a disconnect between your students' class work and homework?
▶ Do your students have trouble finding worked examples that match the practice exercises?
▶ Do you prefer that your students see examples in the textbook that match the ones you use in class?

Miller/O'Neill/Hyde's *Worked Examples* offer a clear, concise methodology that replicates the mathematical processes used in the authors' classroom lectures.

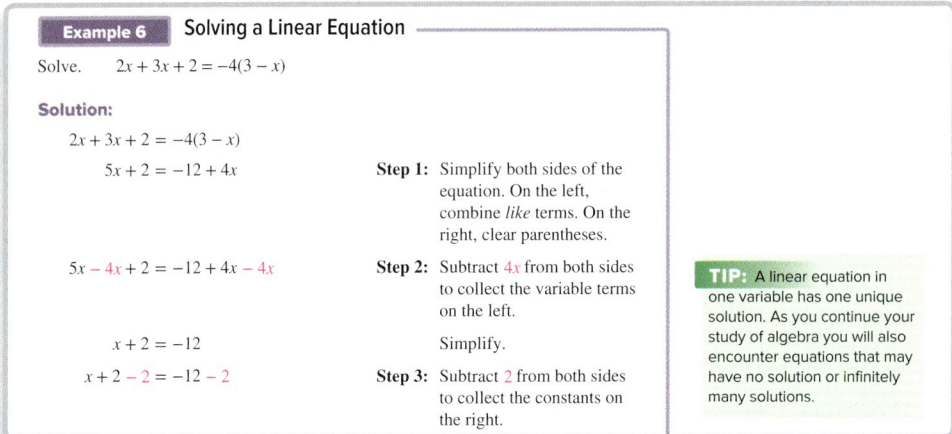

Classroom Examples

To ensure that the classroom experience also matches the examples in the text and the practice exercises, we have included references to even-numbered exercises to be used as Classroom Examples. These exercises are highlighted in the Practice Exercises at the end of each section.

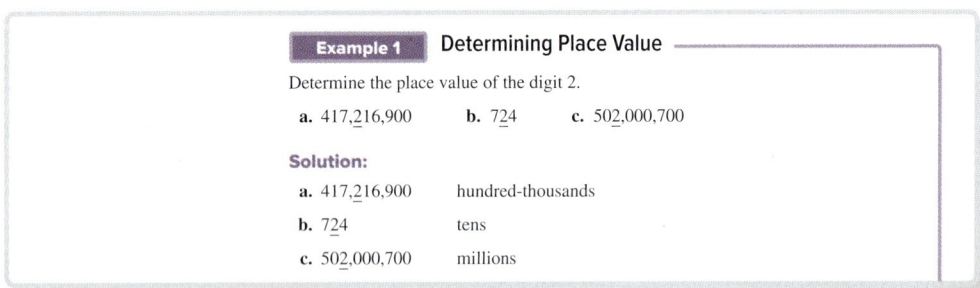

Get Better Results

Quality Learning Tools

For Review Boxes

Throughout the text, just-in-time tips and reminders of prerequisite skills appear in the margin alongside the concepts for which they are needed. References to prior sections are given for cases where more comprehensive review is available earlier in the text.

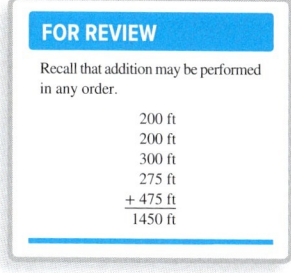

FOR REVIEW
Recall that addition may be performed in any order.

$$\begin{array}{r} 200 \text{ ft} \\ 200 \text{ ft} \\ 300 \text{ ft} \\ 275 \text{ ft} \\ + 475 \text{ ft} \\ \hline 1450 \text{ ft} \end{array}$$

TIP and Avoiding Mistakes Boxes

TIP and **Avoiding Mistakes** boxes have been created based on the authors' classroom experiences—they have also been integrated into the **Worked Examples**. These pedagogical tools will help students get better results by learning how to work through a problem using a clearly defined step-by-step methodology.

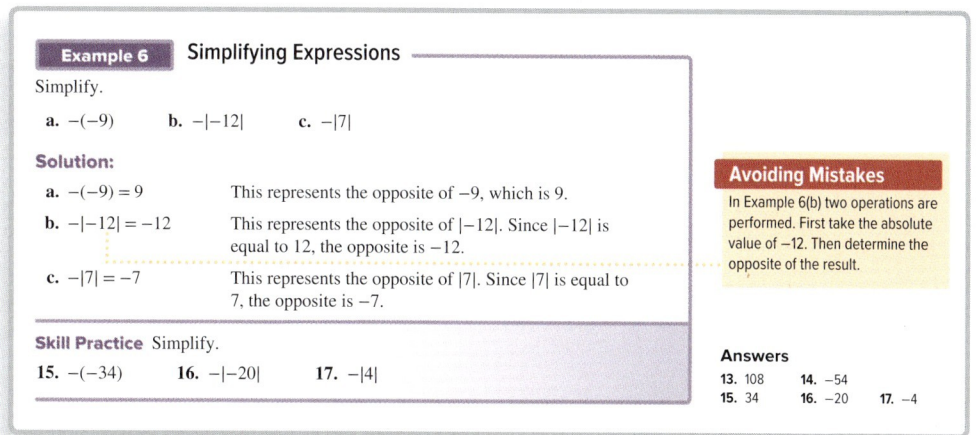

Avoiding Mistakes Boxes:
Avoiding Mistakes boxes are integrated throughout the textbook to alert students to common errors and how to avoid them.

TIP Boxes
Teaching tips are usually revealed only in the classroom. Not anymore! TIP boxes offer students helpful hints and extra direction to help improve understanding and provide further insight.

Get Better Results

Better Exercise Sets and Better Practice Yield Better Results
- ▶ Do your students have trouble with problem solving?
- ▶ Do you want to help students overcome math anxiety?
- ▶ Do you want to help your students improve performance on math assessments?

Problem Recognition Exercises
Problem Recognition Exercises present a collection of problems that look similar to a student upon first glance, but are actually quite different in the manner of their individual solutions. Students sharpen critical thinking skills and better develop their "solution recall" to help them distinguish the method needed to solve an exercise—an essential skill in mathematics.

Problem Recognition Exercises were tested in the authors' developmental mathematics classes and were created to improve student performance on tests.

Problem Recognition Exercises

Operations on Whole Numbers
For Exercises 1–14, perform the indicated operations.

1. a. $96 + 24$ b. $96 - 24$ c. 96×24 d. $24\overline{)96}$
2. a. $550 + 25$ b. $550 - 25$ c. 550×25 d. $25\overline{)550}$
3. a. $612 + 334$ b. $946 - 334$
4. a. $612 - 334$ b. $278 + 334$
5. a. $5500 - 4299$ b. $1201 + 4299$
6. a. $22{,}718 + 12{,}137$ b. $34{,}855 - 12{,}137$
7. a. $50 \cdot 400$ b. $20{,}000 \div 50$
8. a. $548 \cdot 63$ b. $34{,}524 \div 63$
9. a. $5060 \div 22$ b. $230 \cdot 22$
10. a. $1875 \div 125$ b. $125 \cdot 15$
11. a. $4\overline{)1312}$ b. $328\overline{)1312}$
12. a. $547\overline{)4376}$ b. $8\overline{)4376}$
13. a. $418 \cdot 10$ b. $418 \cdot 100$ c. $418 \cdot 1000$ d. $418 \cdot 10{,}000$
14. a. $350{,}000 \div 10$ b. $350{,}000 \div 100$ c. $350{,}000 \div 1000$ d. $350{,}000 \div 10{,}000$

Get Better Results

Student-Centered Applications

The Miller/O'Neill/Hyde Board of Advisors partnered with our authors to bring the *best applications* from every region in the country! These applications include real data and topics that are more relevant and interesting to today's student.

> **24.** Liu earned $312 on an investment of $800. How much would $1100 have earned in the same investment?
>
> **25.** A skyscraper in Chicago is 1454 ft high. If a model is made in which 1 in. represents 50 ft, how high would the building be in the model?

Activities

Each section of the text ends with an activity that steps the student through the major concepts of the section. The purpose of the activities is to promote active, discovery-based learning for the student. The implementation of the activities is flexible for a variety of delivery methods. For face-to-face classes, the activities can be used to break up lecture by covering the exercises intermittently during the class. For the flipped classroom and hybrid classes, students can watch the videos and try the activities. Then, in the classroom, the instructor can go over the activities or have the students compare their answers in groups. For online classes, the activities provide great discussion questions.

Section 1.1 Activity

A.1. In a recent presidential election, the State of Wisconsin had 1,902,505 people request an absentee ballot.
   a. Determine the place value of the underlined digit. 1,9̲02,505 _____
   b. Convert 1,902,505 to expanded form.
   c. Write 1,902,505 in words.

A.2. Of the 1,902,505 total absentee ballots requested in Wisconsin, <u>one million, eight hundred ninety-six thousand, five hundred thirty-one</u> ballots were sent to voters. Write this number in standard form.

For Exercises A.3–A.6:
   a. Write two true inequalities (one using > and one using <) for each pair of values given below.
   b. Translate one of the inequalities to words.

A.3. 210 and 201
   a. _____ or _____
   b. _____

A.4. 2233 and 2323
   a. _____ or _____
   b. _____

A.5. 79 and 76
   a. _____ or _____
   b. _____

A.6. 614 and 641
   a. _____ or _____
   b. _____

A.7. Consider the numbers 5, 9, 2, and 7.
   a. What is the greatest four-digit number that can be formed from the digits? Use each digit only once.
   b. What is the smallest four-digit number that can be formed from the digits? Use each digit only once.
   c. Write the number from part (b) in words.

Get Better Results

Additional Supplements

Lecture Videos Created by the Authors

Julie Miller began creating these lecture videos for her own students to use when they were absent from class. The student response was overwhelmingly positive, prompting the author team to create the lecture videos for their entire developmental math book series. In these videos, the authors walk students through the learning objectives using the same language and procedures outlined in the book. Students learn and review right alongside the author! Students can also access the written notes that accompany the videos.

Integrated Video and Study Workbooks

The Integrated Video and Study Workbooks were built to be used in conjunction with the Miller/O'Neill/Hyde Developmental Math series online lecture videos. These new video guides allow students to consolidate their notes as they work through the material in the book, and they provide students with an opportunity to focus their studies on particular topics that they are struggling with rather than entire chapters at a time. Each video guide contains written examples to reinforce the content students are watching in the corresponding lecture video, along with additional written exercises for extra practice. There is also space provided for students to take their own notes alongside the guided notes already provided. By the end of the academic term, the video guides will not only be a robust study resource for exams, but will serve as a portfolio showcasing the hard work of students throughout the term.

Dynamic Math Animations

The authors have constructed a series of animations to illustrate difficult concepts where static images and text fall short. The animations leverage the use of on-screen movement and morphing shapes to give students an interactive approach to conceptual learning. Some provide a virtual laboratory for which an application is simulated and where students can collect data points for analysis and modeling. Others provide interactive question-and-answer sessions to test conceptual learning.

Exercise Videos

The authors, along with a team of faculty who have used the Miller/O'Neill/Hyde textbooks for many years, have created exercise videos for designated exercises in the textbook. These videos cover a representative sample of the main objectives in each section of the text. Each presenter works through selected problems, following the solution methodology employed in the text.

The video series is available online as part of ALEKS 360. The videos are closed-captioned for the hearing impaired and meet the Americans with Disabilities Act Standards for Accessible Design.

Student Resource Manual

The *Student Resource Manual (SRM),* created by the authors, is a printable, electronic supplement available to students through ALEKS. Instructors can also choose to customize this manual and package it with their course materials. With increasing demands on faculty schedules, this resource offers a convenient means for both full-time and adjunct faculty to promote active learning and success strategies in the classroom.

This manual supports the series in a variety of different ways:

- Additional group activities developed by the authors to supplement what is already available in the text
- Discovery-based classroom activities written by the authors for each section
- Excel activities that not only provide students with numerical insights into algebraic concepts, but also teach simple computer skills to manipulate data in a spreadsheet

Get Better Results

- Worksheets for extra practice written by the authors, including Problem Recognition Exercise Worksheets
- Lecture Notes designed to help students organize and take notes on key concepts
- Materials for a student portfolio

Annotated Instructor's Edition

In the *Annotated Instructor's Edition* (*AIE*), answers to all exercises appear adjacent to each exercise in a color used *only* for annotations. The *AIE* also contains Instructor Notes that appear in the margin. These notes offer instructors assistance with lecture preparation. In addition, there are Classroom Examples referenced in the text that are highlighted in the Practice Exercises. Also found in the *AIE* are icons within the Practice Exercises that serve to guide instructors in their preparation of homework assignments and lessons.

PowerPoints

The PowerPoints present key concepts and definitions with fully editable slides that follow the textbook. An instructor may project the slides in class or post to a website in an online course.

Test Bank

Among the supplements is a computerized test bank using the algorithm-based testing software TestGen® to create customized exams quickly. Hundreds of text-specific, open-ended, and multiple-choice questions are included in the question bank.

ALEKS PPL: Pave the Path to Graduation with Placement, Preparation, and Learning

- **Success in College Begins with Appropriate Course Placement:** A student's first math course is critical to his or her success. With a unique combination of adaptive assessment and personalized learning, ALEKS Placement, Preparation, and Learning (PPL) accurately measures the student's math foundation and creates a personalized learning module to review and refresh lost knowledge. This allows the student to be placed and successful in the right course, expediting the student's path to complete their degree.
- **The Right Placement Creates Greater Value:** Students invest thousands of dollars in their education. ALEKS PPL helps students optimize course enrollment by avoiding courses they don't need to take and helping them pass the courses they do need to take. With more accurate student placement, institutions will retain the students that they recruit initially, increasing their recruitment investment and decreasing their DFW rates. Understanding where your incoming students are placing helps you to plan and develop course schedules and allocate resources efficiently.
- **See ALEKS PPL in Action:** http://bit.ly/ALEKSPPL

McGraw-Hill Create allows you to select and arrange content to match your unique teaching style, add chapters from McGraw-Hill textbooks, personalize content with your syllabus or lecture notes, create a cover design, and receive your PDF review copy in minutes! Order a print or eBook for use in your course, and update your material as often as you'd like. Additional third-party content can be selected from a number of special collections on Create. Visit **McGraw-Hill Create** to browse Create Collections: http://create.mheducation.com.

Our Commitment to Market Development and Accuracy

McGraw Hill's development process is an ongoing, market-oriented approach to building accurate and innovative print and digital products. We begin developing a series by partnering with authors who have a vision for positively impacting student success. Next, we share these ideas and the manuscript with instructors to review and provide feedback to ensure that the authors' ideas represent the needs of that discipline. Throughout multiple drafts, we help our authors to incorporate ideas and suggestions from reviewers to ensure that the series follows the pulse of today's classroom. With all editions, we commit to accuracy in the print text, supplements, and online platforms. In addition to involving instructors as we develop our content, we also perform accuracy checks throughout the various stages of development and production. Through our commitment to this process, we are confident that our series features content that has been thoughtfully developed and vetted to meet the needs of both instructors and students.

Acknowledgments and Reviewers

Paramount to the development of this series was the invaluable feedback provided by the instructors from around the country who reviewed the manuscript or attended a market development event over the course of the several years the text was in development.

Maryann Faller, *Adirondack Community College*
Albert Miller, *Ball State University*
Debra Pearson, *Ball State University*
Patricia Parkison, *Ball State University*
Robin Rufatto, *Ball State University*
Melanie Walker, *Bergen Community College*
Robert Fusco, *Bergen Community College*
Latonya Ellis, *Bishop State Community College*
Ana Leon, *Bluegrass Community College & Technical College*
Kaye Black, *Bluegrass Community College & Technical College*
Barbara Elzey, *Bluegrass Community College & Technical College*
Cheryl Grant, *Bowling Green State University*
Beth Rountree, *Brevard College*
Juliet Carl, *Broward College*
Lizette Foley, *Broward College*
Angie Matthews, *Broward College*
Mitchel Levy, *Broward College*
Jody Harris, *Broward College*
Michelle Carmel, *Broward College*
Antonnette Gibbs, *Broward College*
Kelly Jackson, *Camden Community College*
Elizabeth Valentine, *Charleston Southern University*
Adedoyin Adeyiga, *Cheyney University of Pennsylvania*
Dot French, *Community College of Philadelphia*
Brad Berger, *Copper Mountain College*

Donna Troy, *Cuyamaca College*
Brianna Kurtz, *Daytona State College–Daytona Beach*
Jennifer Walsh, *Daytona State College–Daytona Beach*
Marc Campbell, *Daytona State College–Daytona Beach*
Richard Rupp, *Del Mar College*
Joseph Hernandez, *Delta College*
Randall Nichols, *Delta College*
Thomas Wells, *Delta College*
Paul Yun, *El Camino College*
Catherine Bliss, *Empire State College–Saratoga Springs*
Laurie Davis, *Erie Community College*
Linda Kuroski, *Erie Community College*
David Usinski, *Erie Community College*
Ron Bannon, *Essex County College*
David Platt, *Front Range Community College*
Alan Dinwiddie, *Front Range Community College*
Shanna Goff, *Grand Rapids Community College*
Betsy McKinney, *Grand Rapids Community College*
Cathy Gardner, *Grand Valley State University*
Jane Mays, *Grand Valley State University*
John Greene, *Henderson State University*
Fred Worth, *Henderson State University*
Ryan Baxter, *Illinois State University*
Angela Mccombs, *Illinois State University*
Elisha Van Meenen, *Illinois State University*
Teresa Hasenauer, *Indian River State College*

Tiffany Lewis, *Indian River State College*
Deanna Voehl, *Indian River State College*
Joe Jordan, *John Tyler Community College*
Sally Copeland, *Johnson County Community College*
Nancy Carpenter, *Johnson County Community College*
Susan Yellott, *Kilgore College*
Kim Miller, *Labette Community College*
Michelle Hempton, *Lansing Community College*
Michelle Whitmer, *Lansing Community College*
Kuen Lee, *Los Angeles Trade Tech*
Nic Lahue, *MCC-Longview Community College*
Jason Pallett, *MCC-Longview Community College*
Janet Wyatt, *MCC-Longview Community College*
Rene Barrientos, *Miami Dade College—Kendall*
Nelson De La Rosa, *Miami Dade College—Kendall*
Jody Balzer, *Milwaukee Area Technical College*
Shahla Razavi, *Mt. San Jacinto College*
Shawna Bynum, *Napa Valley College*
Tammy Ford, *North Carolina A & T University*
Ebrahim Ahmadizadeh, *Northampton Community College*
Christine Wetzel-Ulrich, *Northampton Community College*
Sharon Totten, *Northeast Alabama Community College*
Rodolfo Maglio, *Northeastern Illinios University*
Christine Copple, *Northwest State Community College*
Sumitana Chatterjee, *Nova Community College*
Charbel Fahed, *Nova Community College*
Ken Hirschel, *Orange County Community College*
Linda K. Schott, *Ozarks Technical Community College*
Matthew Harris, *Ozarks Technical Community College*
Daniel Kopsas, *Ozarks Technical Community College*
Andrew Aberle, *Ozarks Technical Community College*
Alan Papen, *Ozarks Technical Community College*
Angela Shreckhise, *Ozarks Technical Community College*
Jacob Lewellen, *Ozarks Technical Community College*
Marylynne Abbott, *Ozarks Technical Community College*
Jeffrey Gervasi, *Porterville College*
Stewart Hathaway, *Porterville College*
Lauran Johnson, *Richard Bland College*
Matthew Nickodemus, *Richard Bland College*
Cameron English, *Rio Hondo College*
Lydia Gonzalez, *Rio Hondo College*
Mark Littrell, *Rio Hondo College*
Matthew Pitassi, *Rio Hondo College*
Wayne Lee, *Saint Philips College*

Paula Looney, *Saint Philips College*
Fred Bakenhus, *Saint Philips College*
Lydia Casas, *Saint Philips College*
Gloria Guerra, *Saint Philips College*
Sounny Slitine, *Saint Philips College*
Jessica Lopez, *Saint Philips College*
Lorraine Lopez, *San Antonio College*
Peter Georgakis, *Santa Barbara City College*
Sandi Nieto-Navarro, *Santa Rosa Junior College*
Steve Drucker, *Santa Rosa Junior College*
Jean-Marie Magnier, *Springfield Tech Community College*
Dave Delrossi, *Tallahassee Community College*
Natalie Johnson, *Tarrant County College South*
Marilyn Peacock, *Tidewater Community College*
Yvonne Aucoin, *Tidewater Community College*
Cynthia Harris, *Triton College*
Jennifer Burkett, *Triton College*
Christyn Senese, *Triton College*
Jennifer Dale, *Triton College*
Patricia Hussey, *Triton College*
Glenn Jablonski, *Triton College*
Myrna La Rosa, *Triton College*
Michael Maltenfort, *Truman College*
Abdallah Shuaibi, *Truman College*
Marta Hidegkuti, *Truman College*
Sandra Wilder, *University of Akron*
Sandra Jovicic, *University of Akron*
Edward Migliore, *University of California–Santa Cruz*
Kelly Kohlmetz, *University of Wisconsin—Milwaukee*
Leah Rineck, *University of Wisconsin—Milwaukee*
Carolann Van Galder, *University of Wisconsin—Rock County*
Claudia Martinez, *Valencia College*
Stephen Toner, *Victor Valley Community College*
David Cooper, *Wake Tech Community College*
Karlata Elliott, *Wake Tech Community College*
Laura Kalbaugh, *Wake Tech Community College*
Kelly Vetter, *Wake Tech Community College*
Jacqui Fields, *Wake Tech Community College*
Jennifer Smeal, *Wake Tech Community College*
Shannon Vinson, *Wake Tech Community College*
Kim Walaski, *Wake Tech Community College*
Lisa Rombes, *Washtenaw Community College*
Maziar Ouliaeinia, *Western Iowa Tech Community College*
Keith McCoy, *Wilbur Wright College*

Create More Lightbulb Moments.

Every student has different needs and enters your course with varied levels of preparation. ALEKS® pinpoints what students already know, what they don't and, most importantly, what they're ready to learn next. Optimize your class engagement by aligning your course objectives to ALEKS® topics and layer on our textbook as an additional resource for students.

ALEKS® Creates a Personalized and Dynamic Learning Path

ALEKS® creates an optimized path with an ongoing cycle of learning and assessment, celebrating students' small wins along the way with positive real-time feedback. Rooted in research and analytics, ALEKS® improves student outcomes by fostering better preparation, increased motivation, and knowledge retention.

*visit **bit.ly/whatmakesALEKSunique** to learn more about the science behind the most powerful adaptive learning tool in education!

Preparation & Retention

The more prepared your students are, the more effective your instruction is. Because ALEKS® understands the prerequisite skills necessary for mastery, students are better prepared when a topic is presented to them. ALEKS® provides personalized practice and guides students to what they need to learn next to achieve mastery. ALEKS® improves knowledge and student retention through periodic knowledge checks and personalized learning paths. This cycle of learning and assessment ensures that students remember topics they have learned, are better prepared for exams, and are ready to learn new content as they continue into their next course.

Flexible Implementation: Your Class Your Way!

ALEKS® enables you to structure your course regardless of your instruction style and format. From a traditional classroom, to various co-requisite models, to an online prep course before the start of the term, ALEKS® can supplement your instruction or play a lead role in delivering the content.

*visit **bit.ly/ALEKScasestudies** to see how your peers are delivering better outcomes across various course models!

Outcomes & Efficacy

Our commitment to improve student outcomes services a wide variety of implementation models and best practices, from lecture-based to labs and co-reqs to summer prep courses. Our case studies illustrate our commitment to help you reach your course goals, and our research demonstrates our drive to support all students, regardless of their math background and preparation level.

*visit **bit.ly/outcomesandefficacy** to review empirical data from ALEKS® users around the country

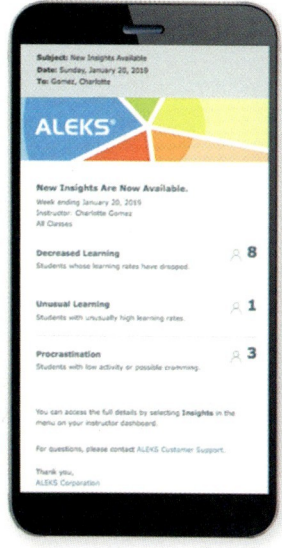

Turn Data Into Actionable Insights

ALEKS® Reports are designed to inform your instruction and create more meaningful interactions with your students when they need it the most. ALEKS® Insights alert you when students might be at risk of falling behind so that you can take immediate action. Insights summarize students exhibiting at least one of four negative behaviors that may require intervention, including Failed Topics, Decreased Learning, Unusual Learning, and Procrastination & Cramming.

Winner of 2019 Digital Edge 50 Award for Data Analytics!

bit.ly/ALEKS_MHE

Whole Numbers

CHAPTER OUTLINE

- **1.1** Introduction to Whole Numbers 2
- **1.2** Addition and Subtraction of Whole Numbers and Perimeter 9
- **1.3** Rounding and Estimating 26
- **1.4** Multiplication of Whole Numbers and Area 33
- **1.5** Division of Whole Numbers 47

  Problem Recognition Exercises: Operations on Whole Numbers 58
- **1.6** Exponents, Algebraic Expressions, and the Order of Operations 59
- **1.7** Mixed Applications and Computing Mean 68

  Chapter 1 Review Exercises 75

Numbers on Vacation

Since the beginning of human civilization, the need to communicate with one another in a precise, quantifiable language has become increasingly important. For example, to take a vacation to Disney World, a family would want to know the driving distance to the park, the time required to drive there, the cost for tickets, the number of nights for a hotel room, and the estimated amount spent on food and incidentals. Such numerical (quantifiable) information is essential for the family to determine if the vacation is affordable and to form a budget for the vacation.

Suppose the family lives 300 miles from Disney World, drives a car that gets 30 miles per gallon of gasoline, and travels 60 miles per hour. These numerical values are called whole numbers. Whole numbers include 0 and the counting numbers 1, 2, 3, and so on. Operations on whole numbers can help us solve a variety of applications. For example, dividing the whole number 300 miles by 30 miles per gallon tells us that the family will use 10 gallons of gasoline. Furthermore, dividing 300 miles by 60 miles per hour tells us that the family will arrive at Disney World in 5 hours. As you work through this chapter, reflect on how important numbers are to everyday living and how different our world would be without the precision of numerical values.

Ilene MacDonald/Alamy Stock Photo

Section 1.1 Introduction to Whole Numbers

Concepts

1. Place Value
2. Standard Notation and Expanded Notation
3. Writing Numbers in Words
4. The Number Line and Order

1. Place Value

Numbers provide the foundation that is used in mathematics. We begin this chapter by discussing how numbers are represented and named. All numbers in our numbering system are composed from the **digits** 0, 1, 2, 3, 4, 5, 6, 7, 8, and 9. In mathematics, the numbers 0, 1, 2, 3, 4, 5, 6, 7, 8, 9, 10, 11, 12, . . . are called the *whole numbers*. (The three dots are called *ellipses* and indicate that the list goes on indefinitely.)

For large numbers, commas are used to separate digits into groups of three called **periods**. For example, the number of live births in the United States in a recent year was 4,058,614. (*Source: The World Almanac*) Numbers written in this way are said to be in **standard form**. The position of each digit determines the place value of the digit. To interpret the number of births in the United States, refer to the place value chart (Figure 1-1).

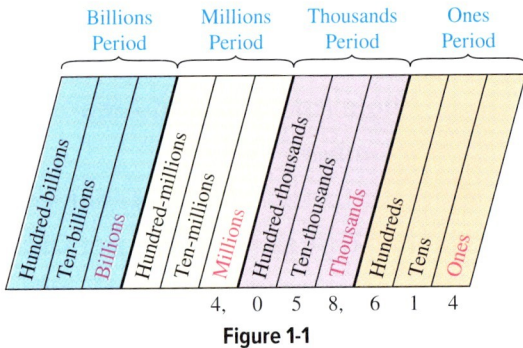

Figure 1-1

The digit 5 in 4,058,614 represents 5 ten-thousands because it is in the ten-thousands place. The digit 4 on the left represents 4 millions, whereas the digit 4 on the right represents 4 ones.

Example 1   Determining Place Value

Determine the place value of the digit 2.

    a. 417,216,900   **b.** 724   **c.** 502,000,700

Solution:

    a. 417,216,900   hundred-thousands

    b. 724   tens

    c. 502,000,700   millions

Skill Practice   Determine the place value of the digit 4.

1. 547,098,632
2. 1,659,984,036
3. 6420

Answers

1. Ten-millions
2. Thousands
3. Hundreds

Section 1.1 Introduction to Whole Numbers 3

Example 2   Determining Place Value

The altitude of Mount Everest, the highest mountain on Earth, is 29,035 feet (ft). Give the place value for each digit.

Solution:

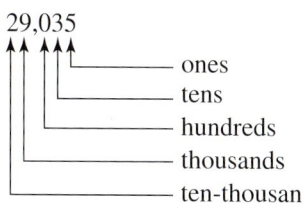
- ones
- tens
- hundreds
- thousands
- ten-thousands

Daniel Prudek/iStockphoto/Getty Images

Skill Practice

4. Alaska is the largest state geographically. Its land area is 571,962 square miles (mi^2). Give the place value for each digit.

2. Standard Notation and Expanded Notation

A number can also be written in an expanded form by writing each digit with its place value unit. For example, 287 can be written as

$$287 = 2 \text{ hundreds} + 8 \text{ tens} + 7 \text{ ones}$$
$$= 2 \times 100 + 8 \times 10 + 7 \times 1$$
$$= 200 + 80 + 7$$

This is called **expanded form**.

Example 3   Converting Standard Form to Expanded Form

Convert to expanded form.

a. 4,672   b. 257,016

Solution:

a. 4,672   4 thousands + 6 hundreds + 7 tens + 2 ones
= 4 × 1,000 + 6 × 100 + 7 × 10 + 2 × 1
= 4,000 + 600 + 70 + 2

b. 257,016   2 hundred-thousands + 5 ten-thousands +
7 thousands + 1 ten + 6 ones
= 2 × 100,000 + 5 × 10,000 + 7 × 1,000 + 1 × 10 + 6 × 1
= 200,000 + 50,000 + 7,000 + 10 + 6

Skill Practice Convert to expanded form.

5. 837   6. 4,093,062

Answers

4. 5: hundred-thousands
   7: ten-thousands
   1: thousands   9: hundreds
   6: tens   2: ones
5. 8 hundreds + 3 tens + 7 ones;
   8 × 100 + 3 × 10 + 7 × 1
6. 4 millions + 9 ten-thousands +
   3 thousands + 6 tens + 2 ones;
   4 × 1,000,000 + 9 × 10,000 +
   3 × 1,000 + 6 × 10 + 2 × 1

Example 4 — Converting Expanded Form to Standard Form

Convert to standard form.

a. 2 hundreds + 5 tens + 9 ones

b. 1 thousand + 2 tens + 5 ones

Solution:

a. 2 hundreds + 5 tens + 9 ones = 259

b. Each place position from the thousands place to the ones place must contain a digit. In this problem, there is no reference to the hundreds place digit. Therefore, we assume 0 hundreds. Thus,

$$1 \text{ thousand} + 0 \text{ hundreds} + 2 \text{ tens} + 5 \text{ ones} = 1{,}025$$

Skill Practice Convert to standard form.

7. 8 thousands + 5 hundreds + 5 tens + 1 one
8. 5 hundred-thousands + 4 thousands + 8 tens + 3 ones

3. Writing Numbers in Words

The word names of some two-digit numbers appear with a hyphen, while others do not. For example:

Number	Number Name
12	twelve
68	sixty-eight
40	forty
42	forty-two

To write a three-digit or larger number, begin at the leftmost group of digits. The number named in that group is followed by the period name, followed by a comma. Then the next period is named, and so on.

Example 5 — Writing a Number in Words

Write 621,417,325 in words.

Solution:

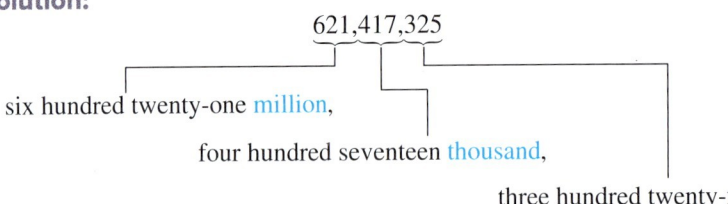

six hundred twenty-one million,
four hundred seventeen thousand,
three hundred twenty-five

Skill Practice

9. Write 1,450,327,214 in words.

Notice from Example 5 that when naming numbers, the name of the ones period is not attached to the last group of digits. Also note that for whole numbers, the word *and* should not appear in word names. For example, 405 should be written as four hundred five.

Answers

7. 8,551   8. 504,083
9. One billion, four hundred fifty million, three hundred twenty-seven thousand, two hundred fourteen

Example 6 Writing a Number in Standard Form

Write the number in standard form.

　　　Six million, forty-six thousand, nine hundred three

Solution:

　　　six million　　nine hundred three
　　　　　　6,046,903
　　　　　　forty-six thousand

Skill Practice

10. Write the number in standard form: fourteen thousand, six hundred nine.

We have seen several examples of writing a number in standard form, in expanded form, and in words. Standard form is the most concise representation. Also note that when we write a four-digit number in standard form, the comma is often omitted. For example, 4,389 is often written as 4389.

4. The Number Line and Order

Whole numbers can be visualized as equally spaced points on a line called a *number line* (Figure 1-2).

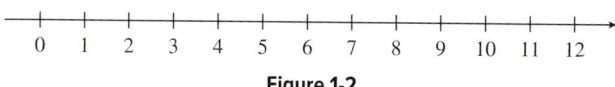

Figure 1-2

The whole numbers begin at 0 and are ordered from left to right by increasing value.

　A number is graphed on a number line by placing a dot at the corresponding point. For any two numbers graphed on a number line, the number to the left is less than the number to the right. Similarly, a number to the right is greater than the number to the left. In mathematics, the symbol < is used to denote "is less than," and the symbol > means "is greater than." Therefore,

3 < 5　means　3 is less than 5
5 > 3　means　5 is greater than 3

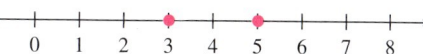

Example 7 Determining Order of Two Numbers

Fill in the blank with the symbol < or >.

　a. 7 ☐ 0　　b. 30 ☐ 82

Solution:

　a. 7 > 0

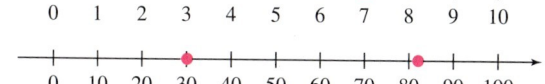

　b. 30 < 82

To visualize 82 and 30 on the number line, it may be necessary to use a different scale. Rather than setting equally spaced marks in units of 1, we can use units of 10. Then 82 must be somewhere between 80 and 90 on the number line.

Skill Practice Fill in the blank with the symbol < or >.

11. 9 ☐ 5　　12. 8 ☐ 18

Answers
10. 14,609
11. >　　12. <

Section 1.1 Activity

A.1. In a recent presidential election, the State of Wisconsin had 1,902,505 people request an absentee ballot.
  a. Determine the place value of the underlined digit. 1,9̱02,505 _____
  b. Convert 1,902,505 to expanded form.
  c. Write 1,902,505 in words.

A.2. Of the 1,902,505 total absentee ballots requested in Wisconsin, <u>one million, eight hundred ninety-six thousand, five hundred thirty-one</u> ballots were sent to voters. Write this number in standard form.

For Exercises A.3–A.6:
  a. Write two true inequalities (one using > and one using <) for each pair of values given below.
  b. Translate one of the inequalities to words.

A.3. 210 and 201
  a. _____ or _____
  b. _____

A.4. 2233 and 2323
  a. _____ or _____
  b. _____

A.5. 79 and 76
  a. _____ or _____
  b. _____

A.6. 614 and 641
  a. _____ or _____
  b. _____

A.7. Consider the numbers 5, 9, 2, and 7.
  a. What is the greatest four-digit number that can be formed from the digits? Use each digit only once.
  b. What is the smallest four-digit number that can be formed from the digits? Use each digit only once.
  c. Write the number from part (b) in words.

Section 1.1 Practice Exercises

Study Skills Exercise

To enhance your learning experience, we provide study skills throughout this textbook that focus on three main areas: mindset (ability to learn new concepts, grit, and overcoming math anxiety), study habits (managing time, taking notes, and test preparation), and mastering mathematical concepts (writing mathematically, reading comprehension, and memory techniques).

Each activity requires only a few minutes and will help you pass this course and become a better math student. Many of these skills can be carried over to other disciplines and help you become a model college student. To begin, write down the following information:

  a. Instructor's name
  b. Instructor's office number
  c. Instructor's telephone number
  d. Instructor's email address
  e. Instructor's office hours
  f. Days of the week that the class meets
  g. The room number in which the class meets
  h. Is there a lab requirement for this course? If so, where is the lab located and how often must you go?

Vocabulary and Key Concepts

1. **a.** For large numbers, commas are used to separate digits into groups called _____.
   b. The place values of the digits in the ones period are the ones, tens, and _____ places.
   c. The place values of the digits in the _____ period are the thousands, ten-thousands, and hundred-thousands places.

Concept 1: Place Value

2. Name the place value for each digit in 36,791.
3. Name the place value for each digit in 8,213,457.
4. Name the place value for each digit in 103,596.

For Exercises 5–24, determine the place value for each underlined digit. (See Example 1.)

5. 3<u>2</u>1
6. 6<u>8</u>9
7. 2<u>1</u>4
8. 73<u>8</u>
9. 8,<u>7</u>10
10. 2,<u>2</u>93
11. <u>1</u>,430
12. <u>3</u>,101
13. <u>4</u>52,723
14. <u>6</u>55,878
15. <u>1</u>,023,676,207
16. <u>3</u>,111,901,211
17. <u>2</u>2,422
18. <u>5</u>8,106
19. 51,<u>0</u>33,201
20. 93,<u>9</u>71,224

21. The number of U.S. travelers abroad in a recent year was <u>1</u>0,677,881. (See Example 2.)
22. The area of Lake Superior is 31,<u>8</u>20 square miles (mi^2).

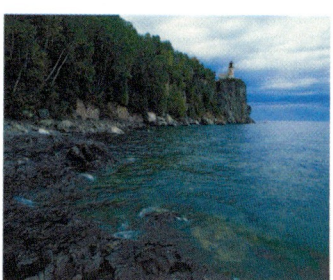
Morey Milbradt/Getty Images

23. For a recent year, the total number of U.S. $1 bills in circulation was <u>7</u>,653,468,440.
24. For a certain flight, the cruising altitude of a commercial jet is <u>3</u>1,000 ft.

Concept 2: Standard Notation and Expanded Notation

For Exercises 25–32, convert the numbers to expanded form. (See Example 3.)

25. 58
26. 71
27. 539
28. 382
29. 5,203
30. 7,089
31. 10,241
32. 20,873

For Exercises 33–40, convert the numbers to standard form. (See Example 4.)

33. 5 hundreds + 2 tens + 4 ones
34. 3 hundreds + 1 ten + 8 ones
35. 1 hundred + 5 tens
36. 6 hundreds + 2 tens
37. 1 thousand + 9 hundreds + 6 ones
38. 4 thousands + 2 hundreds + 1 one

39. 8 ten-thousands + 5 thousands + 7 ones

40. 2 ten-thousands + 6 thousands + 2 ones

41. Name the first four periods of a number (from right to left).

42. Name the first four place values of a number (from right to left).

Concept 3: Writing Numbers in Words

For Exercises 43–50, write the number in words. (See Example 5.)

43. 241

44. 327

45. 603

46. 108

47. 31,530

48. 52,160

49. 100,234

50. 400,199

51. The Shuowen jiezi dictionary, an ancient Chinese dictionary that dates back to the year 100, contained 9535 characters. Write 9535 in words.

52. Interstate I-75 is 1377 miles (mi) long. Write 1377 in words.

53. The altitude of Denali in Alaska is 20,310 ft. Write 20,320 in words.

54. There are 1800 seats in a theater. Write 1800 in words.

55. Researchers calculate that about 590,712 stone blocks were used to construct the Great Pyramid. Write 590,712 in words.

56. In the United States, there are approximately 60,000,000 cats living in households. Write 60,000,000 in words.

Photov.com/Pixtal/age fotostock

GK Hart/Vikki Hart/Getty Images

For Exercises 57–62, convert the number to standard form. (See Example 6.)

57. Six thousand, five

58. Four thousand, four

59. Six hundred seventy-two thousand

60. Two hundred forty-eight thousand

61. One million, four hundred eighty-four thousand, two hundred fifty

62. Two million, six hundred forty-seven thousand, five hundred twenty

Concept 4: The Number Line and Order

For Exercises 63 and 64, graph the numbers on the number line.

63. a. 6   **b.** 13   **c.** 8   **d.** 1

64. a. 5   **b.** 3   **c.** 11   **d.** 9

65. On a number line, what number is 4 units to the right of 6?

66. On a number line, what number is 8 units to the left of 11?

67. On a number line, what number is 3 units to the left of 7?

68. On a number line, what number is 5 units to the right of 0?

For Exercises 69–72, translate the inequality to words.

69. $8 > 2$ **70.** $6 < 11$ **71.** $3 < 7$ **72.** $14 > 12$

For Exercises 73–84, fill in the blank with the inequality symbol $<$ or $>$. (See Example 7.)

73. 6 ☐ 11 **74.** 14 ☐ 13 **75.** 21 ☐ 18 **76.** 5 ☐ 7

77. 3 ☐ 7 **78.** 14 ☐ 24 **79.** 95 ☐ 89 **80.** 28 ☐ 30

81. 0 ☐ 3 **82.** 8 ☐ 0 **83.** 90 ☐ 91 **84.** 48 ☐ 47

Expanding Your Skills

85. Answer true or false. 12 is a digit.

86. Answer true or false. 26 is a digit.

87. What is the greatest two-digit number?

88. What is the greatest three-digit number?

89. What is the greatest whole number?

90. What is the least whole number?

91. How many zeros are there in the number ten million?

92. How many zeros are there in the number one hundred billion?

93. What is the greatest three-digit number that can be formed from the digits 6, 9, and 4? Use each digit only once.

94. What is the greatest three-digit number that can be formed from the digits 0, 4, and 8? Use each digit only once.

Addition and Subtraction of Whole Numbers and Perimeter

Section 1.2

1. Addition of Whole Numbers

We use addition of whole numbers to represent an increase in quantity. For example, suppose Jonas typed 5 pages of a report before lunch. Later in the afternoon he typed 3 more pages. The total number of pages that he typed is found by adding 5 and 3.

$$5 \text{ pages} + 3 \text{ pages} = 8 \text{ pages}$$

Concepts

1. Addition of Whole Numbers
2. Properties of Addition
3. Subtraction of Whole Numbers
4. Translations and Applications Involving Addition and Subtraction
5. Perimeter

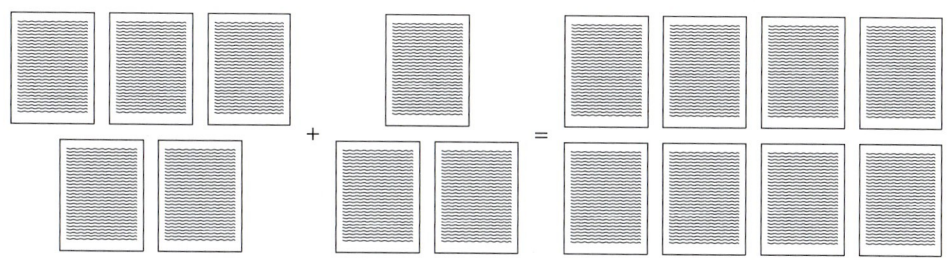

The result of an addition problem is called the **sum**, and the numbers being added are called **addends**. Thus,

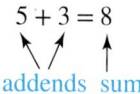

$5 + 3 = 8$
addends sum

The number line is a useful tool to visualize the operation of addition. To add 5 and 3 on a number line, begin at 5 and move 3 units to the right. The final location indicates the sum.

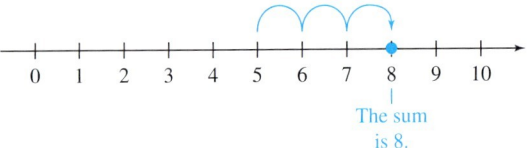

The sum is 8.

You can use a number line to find the sum of any pair of digits. The sums for all possible pairs of one-digit numbers should be memorized (see Exercise 7). Memorizing these basic addition facts will make it easier for you to add larger numbers.

To add whole numbers with several digits, line up the numbers vertically by place value. Then add the digits in the corresponding place positions.

Example 1 Adding Whole Numbers

Add. $261 + 28$

Solution:

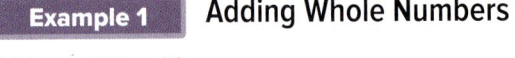

```
  261   ┌─ Add digits in
+  28   │  ones column.
  ───   │
  289   ├─ Add digits in
        │  tens column.
        │
        └─ Add digits in
           hundreds column.
```

Skill Practice Add.

1. $4135 + 210$

Sometimes when adding numbers, the sum of the digits in a given place position is greater than 9. If this occurs, we must do what is called *carrying* or *regrouping*. Example 2 illustrates this process.

Example 2 Adding Whole Numbers with Carrying

Add. $35 + 48$

Solution:

$$35 = 3 \text{ tens} + 5 \text{ ones}$$
$$+48 = 4 \text{ tens} + 8 \text{ ones}$$
$$\overline{\, 7 \text{ tens} + 13 \text{ ones}}$$

The sum of the digits in the ones place exceeds 9. But 13 ones is the same as 1 ten and 3 ones. We can *carry* 1 ten to the tens column while leaving the 3 ones in the ones column. Notice that we placed the carried digit above the tens column.

$$\ 1 \text{ ten}$$
$$35 = 3 \text{ tens} + 5 \text{ ones}$$
$$+48 = 4 \text{ tens} + 8 \text{ ones}$$
$$\overline{83 = 8 \text{ tens} + 3 \text{ ones}}$$

The sum is 83.

Skill Practice Add.

2. $43 + 29$

Answers
1. 4345
2. 72

Addition of numbers may include more than two addends.

Example 3 Adding Whole Numbers

Add. $21{,}076 + 84{,}158 + 2419$

Solution:

$$\begin{array}{r} \overset{12}{21{,}076} \\ 84{,}158 \\ +2{,}419 \\ \hline 107{,}653 \end{array}$$

In this example, the sum of the digits in the ones column is 23. Therefore, we write the 3 and carry the 2.

In the tens column, the sum is 15. Write the 5 in the tens place and carry the 1.

Skill Practice Add.

3. $57{,}296$
    $4{,}089$
    $+9{,}762$

2. Properties of Addition

A **variable** is a letter or symbol that represents a number. The following are examples of variables: a, b, and c. We will use variables to present three important properties of addition.

Most likely you have noticed that 0 added to any number is that number. For example:

$6 + 0 = 6$ $527 + 0 = 527$ $0 + 88 = 88$ $0 + 15 = 15$

In each example, the number in red can be replaced with any number that we choose, and the statement would still be true. This fact is stated as the addition property of 0.

> **Addition Property of 0**
> For any number a,
> $$a + 0 = a \quad \text{and} \quad 0 + a = a$$
> The sum of any number and 0 is that number.

The order in which we add two numbers does not affect the result. For example: $11 + 20 = 20 + 11$. This is true for any two numbers and is stated in the next property.

> **Commutative Property of Addition**
> For any numbers a and b,
> $$a + b = b + a$$
> Changing the order of two addends does not affect the sum.

In mathematics we use parentheses ( ) as grouping symbols. To add more than two numbers, we can group them and then add. For example:

$(2 + 3) + 8$ Parentheses indicate that $2 + 3$ is added first. Then 8 is
$= 5 + 8$ added to the result.
$= 13$

$2 + (3 + 8)$ Parentheses indicate that $3 + 8$ is added first. Then the
$= 2 + 11$ result is added to 2.
$= 13$

Answer

3. 71,147

> **Associative Property of Addition**
>
> For any numbers a, b, and c,
>
> $$(a + b) + c = a + (b + c)$$
>
> The manner in which addends are grouped does not affect the sum.

Example 4   **Applying the Properties of Addition**

a. Rewrite $9 + 6$, using the commutative property of addition.
b. Rewrite $(15 + 9) + 5$, using the associative property of addition.

Solution:

a. $9 + 6 = 6 + 9$ Change the order of the addends.
b. $(15 + 9) + 5 = 15 + (9 + 5)$ Change the grouping of the addends.

Skill Practice

4. Rewrite $3 + 5$, using the commutative property of addition.
5. Rewrite $(1 + 7) + 12$, using the associative property of addition.

3. Subtraction of Whole Numbers

Jeremy bought a case of 12 sodas, and on a hot afternoon he drank 3 of the sodas. We can use the operation of subtraction to find the number of sodas remaining.

12 sodas − 3 sodas   =   9 sodas
Jill Braaten/McGraw Hill

The symbol "−" between two numbers is a subtraction sign, and the result of a subtraction is called the **difference**. The number being subtracted (in this case, 3) is called the **subtrahend**. The number 12 from which 3 is subtracted is called the **minuend**.

$12 - 3 = 9$ is read as   "12 minus 3 is equal to 9"

minuend   subtrahend   difference

Subtraction is the reverse operation of addition. To find the number of sodas that remain after Jeremy takes 3 sodas away from 12 sodas, we ask the question:

"3 added to what number equals 12?"

That is,

$12 - 3 = ?$ is equivalent to $? + 3 = 12$

Subtraction can also be visualized on the number line. To evaluate $7 - 4$, start from the point on the number line corresponding to the minuend (7 in this case). Then move to the *left* 4 units. The resulting position on the number line is the difference.

Answers
4. $3 + 5 = 5 + 3$
5. $(1 + 7) + 12 = 1 + (7 + 12)$

Section 1.2 Addition and Subtraction of Whole Numbers and Perimeter 15

Example 8 **Subtracting Whole Numbers with Borrowing**

Subtract and check the result with addition.   500 − 247

Solution:

$$\begin{array}{r} 500 \\ -247 \\ \hline \end{array}$$
In the ones place, 7 is greater than 0. We try to borrow 1 ten from the tens place. However, the tens place digit is 0. Therefore we must first borrow from the hundreds place.

$$\begin{array}{r} \overset{4}{\cancel{5}}\overset{10}{\cancel{0}}0 \\ -247 \\ \hline \end{array}$$ ← 1 hundred = 10 tens

$$\begin{array}{r} \overset{4}{\cancel{5}}\overset{\overset{9}{\cancel{10}}}{\cancel{0}}\overset{10}{\cancel{0}} \\ -247 \\ \hline 253 \end{array}$$ ← Now we can borrow 1 ten to add to the ones place.

Subtract.

Check: $\begin{array}{r} \overset{1\;1}{253} \\ +247 \\ \hline 500\;\checkmark \end{array}$

Skill Practice Subtract. Check by addition.
13. 700 − 531

4. Translations and Applications Involving Addition and Subtraction

In the English language, there are many different words and phrases that imply addition. A partial list is given in Table 1-1.

Table 1-1

Word/Phrase	Example	In Symbols
Sum	The sum of 6 and x	$6 + x$
Added to	3 added to 8	$8 + 3$
Increased by	y increased by 2	$y + 2$
More than	10 more than 6	$6 + 10$
Plus	8 plus 3	$8 + 3$
Total of	The total of a and b	$a + b$

Example 9 **Translating an English Phrase to a Mathematical Statement**

Translate each phrase to an equivalent mathematical statement and simplify.

a. 12 added to 109

b. The sum of 1386 and 376

Answer
13. 169

Solution:

a. $109 + 12$

$$\begin{array}{r} \overset{1}{1}09 \\ +\ \ 12 \\ \hline 121 \end{array}$$

b. $1386 + 376$

$$\begin{array}{r} \overset{1\,1}{1}386 \\ +\ \ 376 \\ \hline 1762 \end{array}$$

Skill Practice Translate and simplify.

14. 50 more than 80   **15.** 12 increased by 14

Table 1-2 gives several key phrases that imply subtraction.

Table 1-2

Word/Phrase	Example	In Symbols
Minus	15 minus x	$15 - x$
Difference	The difference of 10 and 2	$10 - 2$
Decreased by	a decreased by 1	$a - 1$
Less than	5 less than 12	$12 - 5$
Subtract . . . from	Subtract 3 from 8	$8 - 3$
Subtracted from	6 subtracted from 10	$10 - 6$

In Table 1-2, make a note of the last three entries. The phrases *less than*, *subtract . . . from* and *subtracted from* imply a specific order in which the subtraction is performed. In all three cases, begin with the second number listed and subtract the first number listed.

Example 10 Translating an English Phrase to a Mathematical Statement

Translate the English phrase to a mathematical statement and simplify.

a. The difference of 150 and 38   b. 30 subtracted from 82

Solution:

a. From Table 1-2, the *difference* of 150 and 38 implies $150 - 38$.

$$\begin{array}{r} 1\overset{4\,10}{5}0 \\ -\ \ 38 \\ \hline 112 \end{array}$$

b. The phrase "30 subtracted from 82" implies that 30 is taken away from 82. We have $82 - 30$.

$$\begin{array}{r} 82 \\ -\ 30 \\ \hline 52 \end{array}$$

Skill Practice Translate the English phrase to a mathematical statement and simplify.

16. Twelve decreased by eight   **17.** Subtract three from nine.

We noted earlier that addition is commutative. That is, the order in which two numbers are added does not affect the sum. This is *not* true for subtraction. For example, $82 - 30$ is not equal to $30 - 82$. The symbol $\neq$ means "is not equal to." Thus, $82 - 30 \neq 30 - 82$.

Answers
14. $80 + 50$; 130   **15.** $12 + 14$; 26
16. $12 - 8$; 4   **17.** $9 - 3$; 6

In Examples 11 and 12, we use addition and subtraction of whole numbers to solve application problems.

Example 11 Solving an Application Problem Involving a Table

The table gives the number of gold, silver, and bronze medals won in a recent Winter Olympics for selected countries.

a. Find the total number of medals won by Canada.

b. Determine the total number of silver medals won by these three countries.

	Gold	Silver	Bronze
Germany	10	13	7
USA	9	15	13
Canada	14	7	5

Solution:

a. The number of medals won by Canada appears in the last row of the table. The word "total" implies addition.

$14 + 7 + 5 = 26$ Canada won 26 medals.

b. The number of silver medals is given in the middle column. The total is

$13 + 15 + 7 = 35$ There were 35 silver medals won by these countries.

Skill Practice Refer to the table in Example 11.

18. a. Find the total number of bronze medals won.
    b. Find the number of medals won by the United States.

Example 12 Solving an Application Problem

A criminal justice student did a study of the number of robberies that occurred in the United States over a period of several years. The graph shows his results for five selected years.

a. Find the increase in the number of reported robberies from year 4 to year 5.

b. Find the decrease in the number of reported robberies from year 1 to year 2.

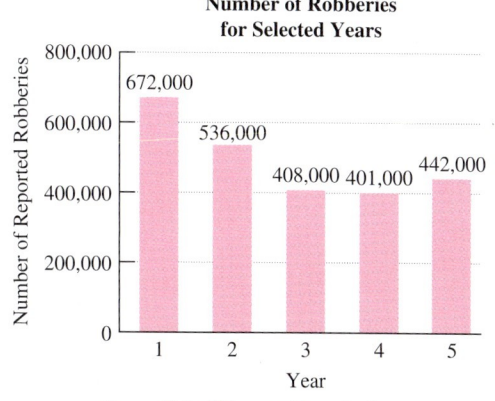

Number of Robberies for Selected Years

Source: Federal Bureau of Investigation

Solution:

For the purpose of finding an amount of increase or decrease, we will subtract the smaller number from the larger number.

a. Because the number of robberies went *up* from year 4 to year 5, there was an *increase*. To find the amount of increase, subtract the smaller number from the larger number.

$$\begin{array}{r} 442{,}000 \\ -401{,}000 \\ \hline 41{,}000 \end{array}$$

From year 4 to year 5, there was an increase of 41,000 reported robberies in the United States.

Answer

18. a. 25 medals   b. 37 medals

b. Because the number of robberies went *down* from year 1 to year 2, there was a *decrease*. To find the amount of decrease, subtract the smaller number from the larger number.

$$\begin{array}{r} 6\overset{6}{\cancel{7}}\overset{12}{\cancel{2}},000 \\ -536,000 \\ \hline 136,000 \end{array}$$

From year 1 to year 2, there was a decrease of 136,000 reported robberies in the United States.

Skill Practice Refer to the graph for Example 12.

19. a. Has the number of robberies increased or decreased from year 2 to year 5?
  b. Determine the amount of increase or decrease.

5. Perimeter

One special application of addition is to find the perimeter of a polygon. A **polygon** is a flat closed figure formed by line segments connected at their ends. Familiar figures such as triangles, rectangles, and squares are examples of polygons. See Figure 1-3.

Triangle   Rectangle   Square
Figure 1-3

The **perimeter** of any polygon is the distance around the outside of the figure. To find the perimeter, add the lengths of the sides.

Example 13 Finding Perimeter

A paving company wants to edge the perimeter of a parking lot with concrete curbing. Find the perimeter of the parking lot.

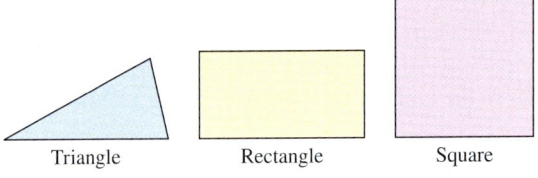

Solution:
The perimeter is the sum of the lengths of the sides.

$$\begin{array}{r} \overset{3}{1}90 \text{ ft} \\ 50 \text{ ft} \\ 60 \text{ ft} \\ 50 \text{ ft} \\ 250 \text{ ft} \\ +100 \text{ ft} \\ \hline 700 \text{ ft} \end{array}$$

The distance around the parking lot (the perimeter) is 700 ft.

Skill Practice
20. Find the perimeter of the garden.

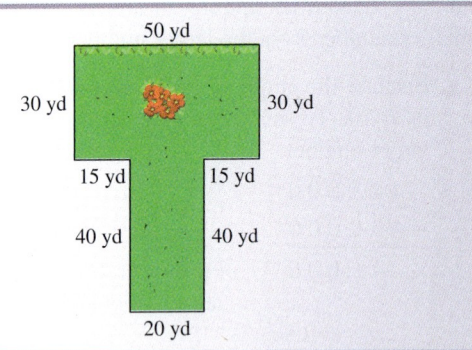

Answers
19. a. decreased **b.** 94,000 robberies
20. 240 yd

Section 1.2 Activity

A.1. There are many different words and phrases that imply addition. Some examples of these words and phrases are *sum, increased by, more than*, and *total*.
  a. Write the mathematical expression of 13 + 6 using the word *sum*.
  b. Write the mathematical expression of 2 + 8 + 10 using the word *total*.

A.2. Jennifer and Mia went shopping together. They purchased items of the following prices: $19, $5, $25, and $11.
  a. Jennifer calculated the total bill by adding:
     19 + 5 + 25 + 11
     What was the sum that Jennifer calculated?
  b. Mia calculated the total bill by adding:
     19 + 11 + 5 + 25
     What was the sum that Mia calculated?
  c. Are the sums from parts (a) and (b) equal?
  d. Which property confirms that the sums from parts (a) and (b) are equal, the associative property of addition or the commutative property of addition?

A.3. Write three words or phrases that imply subtraction.
  a.
  b.
  c.

A.4. Corey and Lucas could not agree on whether or not 12 minus 8 yields the same value as 12 subtracted from 8. Corey thinks the phrases are equal. Lucas thinks the phrases result in different answers.
  a. Who is correct, Corey or Lucas?
  b. Explain your answer to part (a)

A.5. The figure gives the height of five of the tallest buildings in the world.

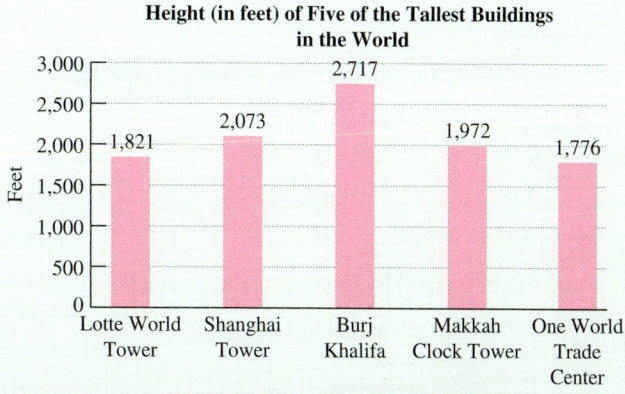

  a. What is the difference between the tallest and the shortest of the buildings?
  b. What is the difference between the Shanghai Tower and Lotte World Tower?

A.6. The perimeter of a polygon is the total distance around the figure. Find the perimeter of an optimum size soccer field.

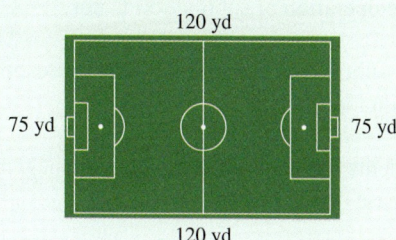

A.7. Find the length of the missing sides given the perimeter.
   a. Perimeter = 54 ft
   b. Perimeter = 372 cm

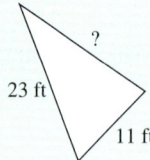
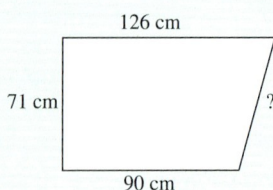

Section 1.2 Practice Exercises

Study Skills Exercise
Mindset plays an important role in your approach to learning mathematics. Mindset is our thoughts, beliefs, and attitudes about our abilities based on previous experiences throughout our lifetime. There are two types of mindsets: fixed mindsets and growth mindsets. People with a fixed mindset believe that they are born with a certain amount of intelligence that cannot be changed despite their actions. On the other hand, a person with a growth mindset believes that intelligence is dynamic and can be increased with effort and learning. What type of mindset do you have? Think about the following questions:
- Have you said to yourself, "I'm just not good at math"?
- Do you believe you have the skills required to understand math?
- Can you recall an experience that positively impacted your self-confidence in mathematics?

Vocabulary and Key Concepts

1. a. The numbers being added in an addition problem are called the _____.
    b. The result of an addition problem is called the _____.
    c. A _____ is a letter or symbol that represents a number.
    d. The _____ property of addition states that the order in which two numbers are added does not affect the sum.
    e. For any number a, the addition property of 0 states that $a + 0 = $ ___ and that $0 + a = $ ___.
    f. The associative property of addition states that $(a + b) + c = $ _____. This implies that the manner in which addends are grouped under addition does not affect the sum.
    g. Given the subtraction statement $15 - 4 = 11$, the number 15 is called the _____, the number 4 is called the _____, and the number 11 is called the _____.
    h. A _____ is a flat closed figure formed by line segments connected at their ends.
    i. The _____ of a polygon is the sum of the lengths of the sides.

2. When adding numbers, you must *carry* or *regroup* if the sum of the digits in a given place position is greater than _____.

3. Unlike addition, the operation of subtraction is not _____.

4. When a digit in the subtrahend is larger than the corresponding digit in the minuend, you must _____ a value from the column to the left.

5. The number of students enrolled in the business management program changed from 462 student to 580. Is this an example of an increase or a decrease?

6. The phrases *decreased by* and *less than* imply the operation of _____.

Concept 1: Addition of Whole Numbers

7. Fill in the table.

+	0	1	2	3	4	5	6	7	8	9
0										
1										
2										
3										
4										
5										
6										
7										
8										
9										

For Exercises 8–10, identify the addends and the sum.

8. $11 + 10 = 21$

9. $1 + 13 + 4 = 18$

10. $5 + 8 + 2 = 15$

For Exercises 11–18, add. **(See Example 1.)**

11. $\quad 42$
$\underline{+33}$

12. $\quad 21$
$\underline{+53}$

13. $\quad 12$
$\quad 15$
$\underline{+32}$

14. $\quad 10$
$\quad\; 8$
$\underline{+30}$

15. $890 + 107$

16. $444 + 354$

17. $4 + 13 + 102$

18. $11 + 221 + 5$

For Exercises 19–32, add the whole numbers with carrying. **(See Examples 2 and 3.)**

19. $\quad 76$
$\underline{+45}$

20. $\quad 25$
$\underline{+59}$

21. $\quad 87$
$\underline{+24}$

22. $\quad 38$
$\underline{+77}$

23. $\quad 658$
$\underline{+231}$

24. $\quad 642$
$\underline{+295}$

25. $\quad 152$
$\underline{+549}$

26. $\quad 462$
$\underline{+388}$

27. $79 + 112 + 12$

28. $62 + 907 + 34$

29. $4980 + 10{,}223$

30. $23{,}112 + 892$

31. $10{,}223 + 25{,}782 + 4980$

32. $92{,}377 + 5622 + 34{,}659$

Concept 2: Properties of Addition

For Exercises 33–36, rewrite the addition problem, using the commutative property of addition. **(See Example 4.)**

33. $101 + 44 = \square + \square$

34. $8 + 13 = \square + \square$

35. $x + y = \square + \square$

36. $t + q = \square + \square$

For Exercises 37–40, rewrite the addition problem using the associative property of addition, by inserting a pair of parentheses. **(See Example 4.)**

37. $(23 + 9) + 10 = 23 + 9 + 10$

38. $7 + (12 + 8) = 7 + 12 + 8$

39. $r + (s + t) = r + s + t$

40. $(c + d) + e = c + d + e$

41. Explain the difference between the commutative and associative properties of addition.

42. Explain the addition property of 0. Then simplify the expressions.
    a. $423 + 0$
    b. $0 + 25$
    c. $\begin{array}{r} 67 \\ + 0 \\ \hline \end{array}$
    d. $0 + x$

Concept 3: Subtraction of Whole Numbers

For Exercises 43 and 44, identify the minuend, subtrahend, and the difference.

43. $12 - 8 = 4$

44. $\begin{array}{r} 9 \\ - 6 \\ \hline 3 \end{array}$

For Exercises 45–48, write the subtraction problem as a related addition problem. For example, $19 - 6 = 13$ can be written as $13 + 6 = 19$.

45. $27 - 9 = 18$
46. $20 - 8 = 12$
47. $102 - 75 = 27$
48. $211 - 45 = 166$

For Exercises 49–52, subtract, then check the answer by using addition. **(See Example 5.)**

49. $8 - 3$ Check: $\square + 3 = 8$
50. $7 - 2$ Check: $\square + 2 = 7$
51. $4 - 1$ Check: $\square + 1 = 4$
52. $9 - 1$ Check: $\square + 1 = 9$

For Exercises 53–56, subtract and check the answer by using addition. **(See Example 6.)**

53. $\begin{array}{r} 1347 \\ - 221 \\ \hline \end{array}$
54. $\begin{array}{r} 4865 \\ - 713 \\ \hline \end{array}$
55. $14{,}356 - 13{,}253$
56. $34{,}550 - 31{,}450$

For Exercises 57–72, subtract the whole numbers involving borrowing. **(See Examples 7 and 8.)**

57. $\begin{array}{r} 76 \\ - 59 \\ \hline \end{array}$
58. $\begin{array}{r} 64 \\ - 48 \\ \hline \end{array}$
59. $\begin{array}{r} 710 \\ - 189 \\ \hline \end{array}$
60. $\begin{array}{r} 850 \\ - 303 \\ \hline \end{array}$

61. $\begin{array}{r} 6002 \\ - 1238 \\ \hline \end{array}$
62. $\begin{array}{r} 3000 \\ - 2356 \\ \hline \end{array}$
63. $\begin{array}{r} 10{,}425 \\ - 9{,}022 \\ \hline \end{array}$
64. $\begin{array}{r} 23{,}901 \\ - 8{,}064 \\ \hline \end{array}$

65. $\begin{array}{r} 62{,}088 \\ - 59{,}871 \\ \hline \end{array}$
66. $\begin{array}{r} 32{,}112 \\ - 28{,}334 \\ \hline \end{array}$
67. $3700 - 2987$
68. $8000 - 3788$

69. $32{,}439 - 1498$
70. $21{,}335 - 4123$
71. $8{,}007{,}234 - 2{,}345{,}115$
72. $3{,}045{,}567 - 1{,}871{,}495$

73. Use the expression $7 - 4$ to explain why subtraction is not commutative.

74. Is subtraction associative? Use the numbers 10, 6, 2 to explain.

Concept 4: Translations and Applications Involving Addition and Subtraction

For Exercises 75–92, translate the English phrase to a mathematical statement and simplify. **(See Examples 9 and 10.)**

75. The sum of 13 and 7
76. The sum of 100 and 42
77. 45 added to 7
78. 81 added to 23
79. 5 more than 18
80. 2 more than 76

81. 1523 increased by 90

82. 1320 increased by 448

83. The total of 5, 39, and 81

84. 78 decreased by 6

85. Subtract 100 from 422.

86. Subtract 42 from 89.

87. 72 less than 1090

88. 60 less than 3111

89. The difference of 50 and 13

90. The difference of 405 and 103

91. Subtract 35 from 103.

92. Subtract 14 from 91.

93. A mountain climber attempting to climb Mount Everest must climb the mountain in stages to become acclimated to the extremely high altitude. This process generally takes about 6 weeks. The climb from Base Camp to Camp II results in a gain in altitude of 3010 ft. The climb from Camp II to Camp III is a gain of 1300 ft in altitude, and the climb to Camp IV is another 1700 ft.
   a. How much altitude has the climber gained from Base Camp to Camp IV?
   b. If the climber gains another 6029 ft from Camp IV to the summit, what is the total gain in altitude from Base Camp to the summit?

94. To schedule enough drivers for an upcoming week, a local pizza shop manager recorded the number of deliveries each day from the previous week: 38, 54, 44, 61, 97, 103, 124. What was the total number of deliveries for the week?

95. A portion of Jonathan's checking account register is shown. What is the total amount of the three checks written? (See Example 11.)

Check No.	Description	Payment	Deposit	Balance
1871	Electric	$60		$180
1872	Groceries	82		98
	Payroll		$1256	1354
1874	Restaurant	58		1296
	Deposit		150	1446

96. The table gives the number of desks and chairs delivered each quarter to an office supply store. Find the total number of desks delivered for the year.

	Chairs	Desks
1st Quarter	220	115
2nd Quarter	185	104
3rd Quarter	201	93
4th Quarter	198	111

97. The altitude of White Mountain Peak in California is 14,246 ft. Denali in Alaska is 20,310 ft. How much higher is Denali than White Mountain Peak?

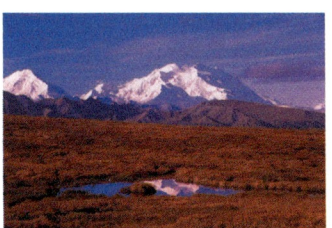

Robert Glusic/Getty Images

98. There are 55 DVDs to shelve one evening at a video rental store. If Jason puts away 39 before leaving for the day, how many are left for Patty to put away?

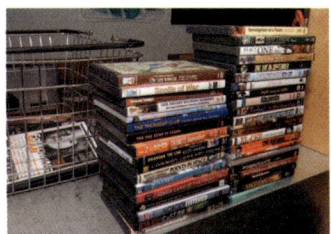

Jill Braaten/McGraw Hill

For Exercises 99–102, use the information from the graph. (See Example 12.)

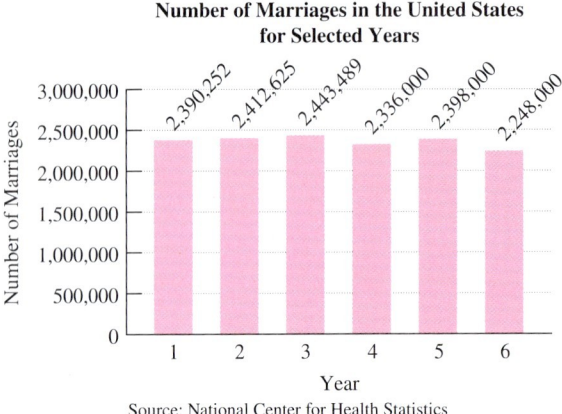

Figure for Exercises 99–102

99. What is the difference in the number of marriages between year 1 and year 5?

100. Find the decrease in the number of marriages in the United States between year 5 and year 6.

101. What is the difference in the number of marriages between the year having the greatest and the year having the least?

102. Between which two consecutive years did the greatest increase in the number of marriages occur? What is the increase?

103. The staff for U.S. public schools is categorized in the pie graph. Determine the number of staff other than teachers.

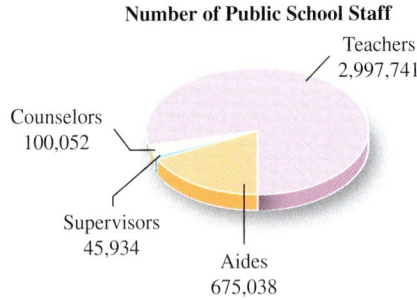

104. The pie graph shows the costs incurred in managing Sub-World sandwich shop for one month. From this information, determine the total cost for one month.

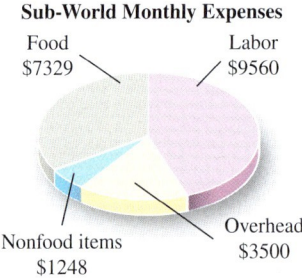

105. Pinkham Notch Visitor Center in the White Mountains of New Hampshire has an elevation of 2032 ft. The summit of nearby Mt. Washington has an elevation of 6288 ft. What is the difference in elevation?

106. Bo Jackson was a Heisman Trophy winner in college football, and then went on to play both professional football and professional baseball. He was named by ESPN as the "greatest athlete of all time." Unfortunately, his career in the National Football League (NFL) was cut short in his fourth year because of a hip injury. During his time in the NFL, he gained 2782 yd rushing and 352 yd receiving. How many more yards did Bo Jackson gain running than receiving?

107. Jeannette has two children who each attended college. Her son Ricardo attended a local community college where the yearly tuition and fees came to $4215. Her daughter Ricki attended an out-of-state university where the yearly tuition and fees totaled $22,416. If Jeannette paid the full amount for both children to go to school, what was her total expense for tuition and fees for 1 year?

108. Clyde and Mason each leave a rest area on the Florida Turnpike. Clyde travels north and Mason travels south. After 2 hr, Clyde has gone 138 mi and Mason, who ran into heavy traffic, traveled only 96 mi. How far apart are they?

Concept 5: Perimeter

For Exercises 109–112, find the perimeter. (See Example 13.)

109.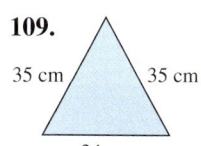

110.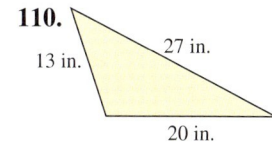

111.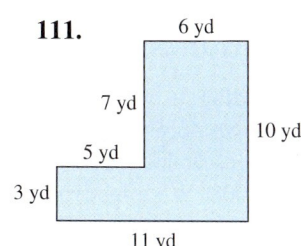

112.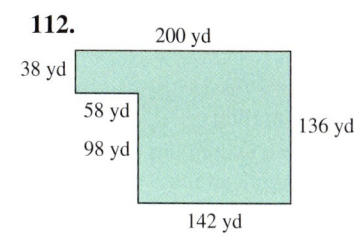

113. Find the perimeter of an NBA basketball court.

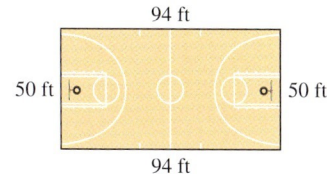

114. A major league baseball diamond is in the shape of a square. Find the distance a batter must run if he hits a home run.

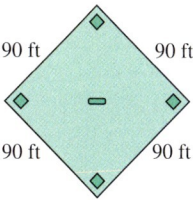

For Exercises 115 and 116, find the missing length.

115. The perimeter of the triangle is 39 m.

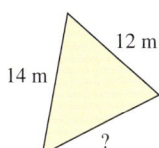

116. The perimeter of the figure is 547 cm.

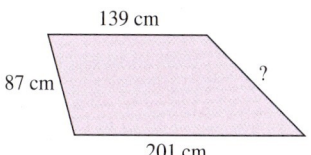

Technology Connections

For Exercises 117–122, perform the indicated operation by using a calculator.

117.
```
   45,418
   81,990
    9,063
 + 56,309
```

118.
```
  9,300,050
  7,803,513
  3,480,009
 +  907,822
```

119.
```
  3,421,019
    822,761
  1,003,721
 +    9,678
```

120.
```
  4,905,620
 -  458,318
```

121.
```
  953,400,415
 -  56,341,902
```

122.
```
   82,025,160
 - 79,118,705
```

For Exercises 123–126, refer to the table showing the land area for five states.

State	Land Area (mi²)
Rhode Island	1,045
Tennessee	41,217
West Virginia	24,078
Wisconsin	54,310
Colorado	103,718

123. Find the difference in the land area between Colorado and Wisconsin.

124. Find the difference in the land area between Tennessee and West Virginia.

125. What is the combined land area for Rhode Island, Tennessee, and Wisconsin?

126. What is the combined land area for all five states?

Section 1.3 Rounding and Estimating

Concepts
1. Rounding
2. Estimation
3. Using Estimation in Applications

1. Rounding

Rounding a whole number is a common practice when we do not require an exact value. For example, Madagascar lost 3956 mi^2 of rainforest between 1990 and 2008. We might round this number to the nearest thousand and say that approximately 4000 mi^2 was lost. In mathematics, we use the symbol $\approx$ to read "is approximately equal to." Therefore, 3956 mi$^2 \approx$ 4000 mi^2.

A number line is a helpful tool to understand rounding. For example, 48 is closer to 50 than it is to 40. Therefore, 48 rounded to the nearest ten is 50.

Courtesy of Julie Miller

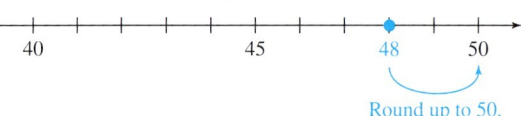

The number 43, on the other hand, is closer to 40 than to 50. Therefore, 43 rounded to the nearest ten is 40.

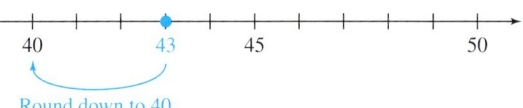

The number 45 is halfway between 40 and 50. In such a case, our convention will be to round *up* to the next-larger ten.

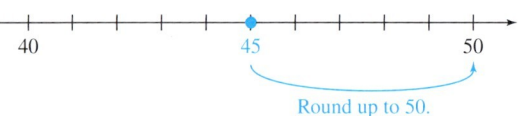

The decision to round up or down to a given place value is determined by the digit to the *right* of the given place value. The following steps outline the procedure.

Rounding Whole Numbers

Step 1 Identify the digit one position to the right of the given place value.

Step 2 If the digit in step 1 is a 5 or greater, then add 1 to the digit in the given place value. If the digit in step 1 is less than 5, leave the given place value unchanged.

Step 3 Replace each digit to the right of the given place value by 0.

Section 1.3 Rounding and Estimating 27

Example 1 Rounding a Whole Number

Round 3741 to the nearest hundred.

Solution:

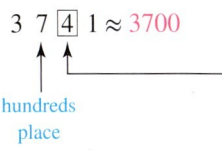

3 7 4 1 ≈ 3700

↑ hundreds place

↑ This is the digit to the right of the hundreds place. Because 4 is less than 5, leave the hundreds digit unchanged. Replace the digits to its right by zeros.

Skill Practice

1. Round 12,461 to the nearest thousand.

Example 1 could also have been solved by drawing a number line. Use the part of a number line between 3700 and 3800 because 3741 lies between these numbers.

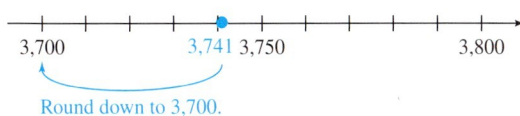

Round down to 3,700.

Example 2 Rounding a Whole Number

Round 1,790,641 to the nearest hundred-thousand.

Solution:

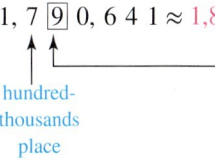

1, 7 9 0, 6 4 1 ≈ 1,800,000

↑ hundred-thousands place

↑ This is the digit to the right of the given place value. Because 9 is greater than 5, add 1 to the hundred-thousands place, add: 7 + 1 = 8. Replace the digits to the right of the hundred-thousands place by zeros.

Skill Practice

2. Round 147,316 to the nearest ten-thousand.

Example 3 Rounding a Whole Number

Round 1503 to the nearest thousand.

Solution:

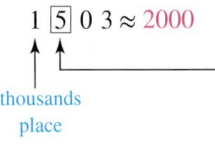

1 5 0 3 ≈ 2000

↑ thousands place

↑ This is the digit to the right of the thousands place. Because this digit is 5, we round up. We increase the thousands place digit by 1. That is, 1 + 1 = 2. Replace the digits to its right by zeros.

Skill Practice

3. Round 7,521,460 to the nearest million.

Answers

1. 12,000   2. 150,000
3. 8,000,000

Example 4 Rounding a Whole Number

Round the number 24,961 to the hundreds place.

Solution:

$$2\,4,\overset{+1}{9}\,\boxed{6}\,1 \approx 25{,}000$$

This is the digit to the right of the hundreds place. Because 6 is greater than 5, add 1 to the hundreds place digit. Replace the digits to the right of the hundreds place with 0.

Skill Practice

4. Round 39,823 to the nearest thousand.

2. Estimation

We use the process of rounding to estimate the result of numerical calculations. For example, to estimate the following sum, we can round each addend to the nearest ten.

31	rounds to	→	30
12	rounds to	→	10
+ 49	rounds to	→	+ 50
			90

The estimated sum is 90 (the actual sum is 92).

Example 5 Estimating a Sum

Estimate the sum by rounding to the nearest thousand.

$$6109 + 976 + 4842 + 11{,}619$$

Solution:

6,109	rounds to	→	$\overset{1}{6{,}000}$
976	rounds to	→	1,000
4,842	rounds to	→	5,000
+ 11,619	rounds to	→	+ 12,000
			24,000

The estimated sum is 24,000 (the actual sum is 23,546).

TIP: Rounding can be useful to mentally estimate an amount. For this purpose, we usually round so that we have one or two nonzero digits with which to work. In Example 5 we rounded to the thousands place, giving (6 + 1 + 5 + 12) thousands = 24 thousand. The estimate is 24,000.

Skill Practice

5. Estimate the sum by rounding each number to the nearest hundred.
   $3162 + 4931 + 2206$

Example 6 Estimating a Difference

Estimate the difference $4817 - 2106$ by rounding each number to the nearest hundred.

Solution:

4817	rounds to	→	4800
− 2106	rounds to	→	− 2100
			2700

The estimated difference is 2700 (the actual difference is 2711).

Skill Practice

6. Estimate the difference by rounding each number to the nearest million.
   $35{,}264{,}000 - 21{,}906{,}210$

Answers
4. 40,000   5. 10,300
6. 13,000,000

3. Using Estimation in Applications

Example 7   Estimating a Sum in an Application

A driver for a delivery service must drive from Chicago, Illinois, to Dallas, Texas, and make several stops on the way. The driver follows the route given on the map. Estimate the total mileage by rounding each distance to the nearest hundred miles.

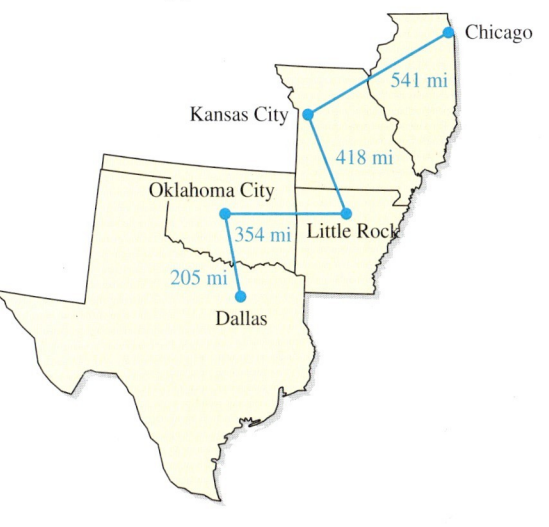

Solution:

```
    541    rounds to  →    500
    418    rounds to  →    400
    354    rounds to  →    400
  + 205    rounds to  →  + 200
                          1500
```

The driver traveled approximately 1500 mi.

Skill Practice

7. The two countries with the largest areas of rainforest are Brazil and the Democratic Republic of Congo. The rainforest areas are 4,776,980 square kilometers (km^2) and 1,336,100 km^2, respectively. Round the values to the nearest hundred-thousand. Estimate the total area of rainforest for these two countries.

Example 8   Estimating a Difference in an Application

In a recent year, the U.S. Census Bureau reported that the number of males over the age of 18 was 100,994,367. The same year, the number of females over 18 was 108,133,727. Round each value to the nearest million. Estimate how many more females over 18 there were than males over 18.

Solution:

The number of males was approximately 101,000,000. The number of females was approximately 108,000,000.

```
    108,000,000
  − 101,000,000
      7,000,000
```

There were approximately 7 million more women over age 18 in the United States than men.

Skill Practice

8. In a recent year, there were 135,073,000 persons over the age of 16 employed in the United States. During the same year, there were 6,742,000 persons over the age of 16 who were unemployed. Approximate each value to the nearest million. Use these values to approximate how many more people were employed than unemployed.

Answers

7. 6,100,000 km^2
8. 128,000,000 people

Section 1.3 Activity

A.1. Consider the number 1449.
  a. Identify the digit that is in the hundreds place.
  b. Identify the digit one position to the right of the hundreds place (digit in the tens place).
  c. Is the number identified in part (b) less than 5 or is it 5 or greater? Based on your answer, do you keep the digit the same or add 1 to the digit?
  d. Using the rules of rounding whole numbers, round 1449 to the nearest hundred based on your answer to part (c).
  e. Plot 1449 on the number line.

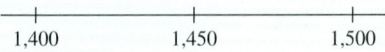

  f. Does your point on the number line in part (e) confirm your answer in part (d)?

A.2. Consider the number 495.
  a. Plot 495 on the number line.

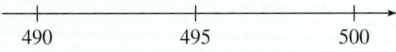

  b. What do you notice about the point 495 with respect to the numbers 490 and 500?
  c. If rounding 495 to the nearest ten, is the convention to round up or to round down?
  d. Round 495 to the nearest ten.

For Exercises A.3–A.6, round each number to the given place value.

		Tens	Hundreds	Thousands
A.3.	3456	a.	b.	c.
A.4.	9753	a.	b.	c.
A.5.	7999	a.	b.	c.
A.6.	2199	a.	b.	c.

A.7. The number of students enrolled at the local community college is 36,935. The local university has 27,156 students enrolled.
  a. Round each value to the nearest thousand.
  b. Estimate how many more students are enrolled in the community college than in the university.

Section 1.3 Practice Exercises

Study Skills Exercise

Class notes serve as a summary of key concepts that can be used later to prepare for tests. Look over the notes that you took today. Do you understand what you wrote? Identify any material or challenging problems that you need to revisit. Add any comments to make your notes clearer to you or rewrite the notes in your own words. If there were any rules, definitions, or formulas, highlight them so that they can be easily found when studying for the test.
Remember that note-taking skills are learned by practice.

Vocabulary and Key Concepts

1. A process called _____ is common practice when the exact value of a number is not required.

2. A helpful visual tool that can be drawn to understand rounding is a _____.

3. The decision to round up or down to a given place value is determined by the digit to the _____ of the given place value.

4. In mathematics, the symbol ≈ means _____.

5. Explain how to round a whole number to the hundreds place.

6. Explain how to round a whole number to the tens place.

Concept 1: Rounding

7. Determine the place value of the digit 6 in the number 1,860,432.

8. Determine the place value of the digit 4 in the number 1,860,432.

9. Round to the nearest ten by plotting the number 73 on the number line.

10. Round to the nearest hundred by plotting the number 151 on the number line.

For Exercises 11–28, round each number to the given place value. **(See Examples 1–4.)**

11. 342; tens
12. 834; tens
13. 725; tens
14. 445; tens
15. 9384; hundreds
16. 8363; hundreds
17. 8539; hundreds
18. 9817; hundreds
19. 34,992; thousands
20. 76,831; thousands
21. 2578; thousands
22. 3511; thousands
23. 9982; hundreds
24. 7974; hundreds
25. 109,337; thousands
26. 437,208; thousands
27. 489,090; ten-thousands
28. 388,725; ten-thousands

29. An automobile collector recently paid $372,533 for an Italian sports car. Round this number to the nearest ten-thousand.

30. The median per capita personal income in the United States in a recent year was $51,939. Round this number to the nearest thousand.

31. The average center-to-center distance from the Earth to the Moon is 238,863 mi. Round this to the thousands place.

32. A shopping center in Edmonton, Alberta, Canada, covers an area of 492,000 square meters (m^2). Round this number to the hundred-thousands place.

Concept 2: Estimation

For Exercises 33–36, estimate the sum or difference by first rounding each number to the nearest ten. (See Examples 5 and 6.)

33. 57
    82
    + 21

34. 33
    78
    + 41

35. 639
    − 422

36. 851
    − 399

For Exercises 37–40, estimate the sum or difference by first rounding each number to the nearest hundred. (See Examples 5 and 6.)

37. 892
    + 129

38. 347
    + 563

39. 3412
    − 1252

40. 9771
    − 4544

Concept 3: Using Estimation in Applications

For Exercises 41 and 42, refer to the table.

Brand	Manufacturer	Sales ($)
M&Ms	Mars	97,404,576
Hershey's Milk Chocolate	Hershey Chocolate	81,296,784
Reese's Peanut Butter Cups	Hershey Chocolate	54,391,268
Snickers	Mars	53,695,428
KitKat	Hershey Chocolate	38,168,580

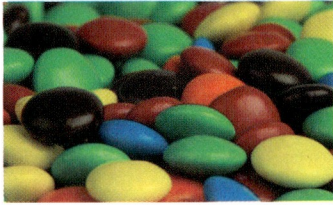
David Planchet/McGraw Hill

41. Round the individual sales to the nearest million to estimate the total sales brought in by the Mars company. (See Example 7.)

42. Round the sales to the nearest million to estimate the total sales brought in by the Hershey Chocolate Company.

43. For a presidential election, one candidate received 65,915,796 votes in the popular vote and the other candiate received 60,933,500. Round each value to the nearest million to estimate how many more votes the winner received. (See Example 8.)

44. In a recent year, the average annual salary for a public school teacher in Iowa was $47,950. The average salary for a public school teacher in California was $69,260. Round each value to the nearest thousand to estimate how much more a school teacher in California earned compared to one from Iowa.

For Exercises 45–48, use the given table.

45. Round each revenue to the nearest hundred-thousand to estimate the total revenue for the years 1 through 3.

46. Round the revenue to the nearest hundred-thousand to estimate the total revenue for the years 4 through 6.

47. a. Determine the year with the greatest revenue. Round this revenue to the nearest hundred-thousand.
    b. Determine the year with the least revenue. Round this revenue to the nearest hundred-thousand.

48. Estimate the difference between the year with the greatest revenue and the year with the least revenue.

Beach Parking Revenue for Daytona Beach, Florida

Year	Revenue
1	$3,316,897
2	3,272,028
3	3,360,289
4	3,470,295
5	3,173,050
6	1,970,380

Table for Exercises 45–48

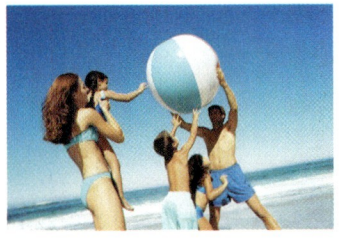
BananaStock/age fotostock

For Exercises 49–52, use the graph provided.

49. Determine the state with the greatest number of students enrolled in 9th grade. Round this number to the nearest thousand.

50. Determine the state with the least number of students enrolled in 9th grade. Round this number to the nearest thousand.

51. Use the information in Exercises 49 and 50 to estimate the difference between the number of students in the state with the greatest enrollment and that of the least enrollment.

52. Estimate the total number of students enrolled in 9th grade in the selected states by first rounding the number of students to the thousands place.

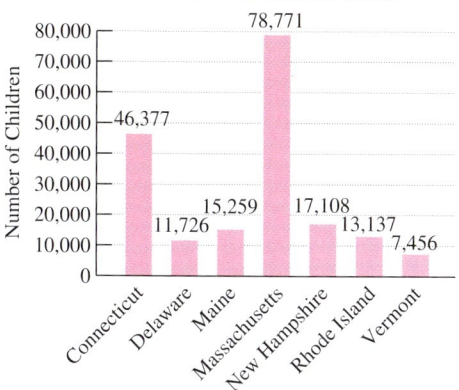

Source: National Center for Education Statistics
Figure for Exercises 49–52

Expanding Your Skills

For Exercises 53–56, round the numbers to estimate the perimeter of each figure. (Answers may vary.)

53.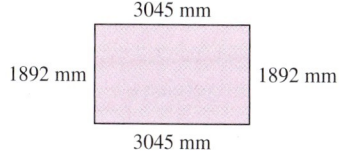

54.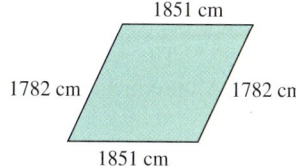

55.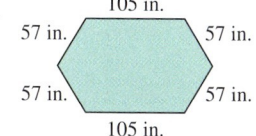

56.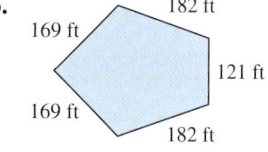

Multiplication of Whole Numbers and Area

Section 1.4

1. Introduction to Multiplication

Suppose that Carmen buys three cartons of eggs to prepare a large family brunch. If there are 12 eggs per carton, then the total number of eggs can be found by adding three 12s.

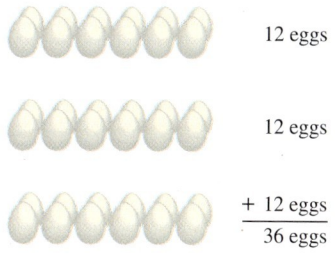

Concepts

1. Introduction to Multiplication
2. Properties of Multiplication
3. Multiplying Many-Digit Whole Numbers
4. Estimating Products by Rounding
5. Applications Involving Multiplication
6. Area of a Rectangle

When each addend in a sum is the same, we have what is called *repeated* addition. Repeated addition is also called **multiplication**. We use the multiplication sign × to express repeated addition more concisely.

$$12 + 12 + 12 \quad \text{is equal to} \quad 3 \times 12$$

The expression 3×12 is read "3 times 12" to signify that the number 12 is added 3 times. The numbers 3 and 12 are called **factors**, and the result, 36, is called the **product**.

The symbol · may also be used to denote multiplication such as in the expression $3 \cdot 12 = 36$. Two factors written adjacent to each other with no other operator between them also implies multiplication. The quantity $2y$, for example, is understood to be 2 times y. If we use this notation to multiply two numbers, parentheses are used to group one or both factors. For example:

$$3(12) = 36 \qquad (3)12 = 36 \quad \text{and} \quad (3)(12) = 36$$

all represent the product of 3 and 12.

TIP: In the expression 3(12), the parentheses are necessary because two adjacent factors written together with no grouping symbol would look like the number 312.

The products of one-digit numbers such as $4 \times 5 = 20$ and $2 \times 7 = 14$ are basic facts. All products of one-digit numbers should be memorized (see Exercise 6).

Example 1   Identifying Factors and Products

Identify the factors and the product.

a. $6 \times 3 = 18$ **b.** $5 \cdot 2 \cdot 7 = 70$

Solution:

a. Factors: 6, 3; product: 18   **b.** Factors: 5, 2, 7; product: 70

Skill Practice Identify the factors and the product.

1. $3 \times 11 = 33$
2. $2 \cdot 5 \cdot 8 = 80$

2. Properties of Multiplication

Recall that the order in which two numbers are added does not affect the sum. The same is true for multiplication. This is stated formally as the *commutative property of multiplication*.

Commutative Property of Multiplication

For any numbers, a and b,

$$a \cdot b = b \cdot a$$

Changing the order of two factors does not affect the product.

Answers
1. Factors: 3 and 11; product: 33
2. Factors: 2, 5, and 8; product: 80

These rectangular arrays help us visualize the commutative property of multiplication. For example, 2 · 5 = 5 · 2.

$2 \cdot 5 = 10$ 2 rows of 5

$5 \cdot 2 = 10$ 5 rows of 2

Multiplication is also an associative operation. For example,

$$(3 \cdot 5) \cdot 2 = 3 \cdot (5 \cdot 2)$$

Associative Property of Multiplication

For any numbers, a, b, and c,

$$(a \cdot b) \cdot c = a \cdot (b \cdot c)$$

The manner in which factors are grouped under multiplication does not affect the product.

TIP: The variable "x" is commonly used in algebra and can be confused with the multiplication symbol "×." For this reason it is customary to limit the use of "×" for multiplication.

Example 2 Applying Properties of Multiplication

a. Rewrite the expression $3 \cdot 9$, using the commutative property of multiplication. Then find the product.

b. Rewrite the expression $(4 \cdot 2) \cdot 3$, using the associative property of multiplication. Then find the product.

Solution:

a. $3 \cdot 9 = 9 \cdot 3$ The product is 27.

b. $(4 \cdot 2) \cdot 3 = 4 \cdot (2 \cdot 3)$

To find the product, we have

$$4 \cdot (2 \cdot 3)$$
$$= 4 \cdot (6)$$
$$= 24$$

The product is 24.

Skill Practice

3. Rewrite the expression $6 \cdot 5$, using the commutative property of multiplication. Then find the product.

4. Rewrite the expression $3 \cdot (1 \cdot 7)$, using the associative property of multiplication. Then find the product.

Two other important properties of multiplication involve factors of 0 and 1.

Answers

3. $5 \cdot 6$; product is 30
4. $(3 \cdot 1) \cdot 7$; product is 21

> **Multiplication Property of 0**
>
> For any number a,
>
> $$a \cdot 0 = 0 \quad \text{and} \quad 0 \cdot a = 0$$
>
> The product of any number and 0 is 0.

The product $5 \cdot 0 = 0$ can easily be understood by writing the product as repeated addition.

$$\underbrace{0 + 0 + 0 + 0 + 0}_{\text{Add 0 five times.}} = 0$$

> **Multiplication Property of 1**
>
> For any number a,
>
> $$a \cdot 1 = a \quad \text{and} \quad 1 \cdot a = a$$
>
> The product of any number and 1 is that number.

The next property of multiplication involves both addition and multiplication. First consider the expression $2(4 + 3)$. By performing the operation within parentheses first, we have

$$2(4 + 3) = 2(7) = 14$$

We get the same result by multiplying 2 times each addend within the parentheses:

$$2(4 + 3) = (2 \cdot 4) + (2 \cdot 3) = 8 + 6 = 14$$

This result illustrates the distributive property of multiplication over addition (sometimes we simply say *distributive property* for short).

> **Distributive Property of Multiplication over Addition**
>
> For any numbers a, b, and c,
>
> $$a(b + c) = a \cdot b + a \cdot c$$
>
> The product of a number and a sum can be found by multiplying the number by each addend.

Example 3 Applying the Distributive Property of Multiplication over Addition

Apply the distributive property and simplify.

a. $3(4 + 8)$ **b.** $7(3 + 0)$

Solution:

a. $3(4 + 8) = (3 \cdot 4) + (3 \cdot 8) = 12 + 24 = 36$

b. $7(3 + 0) = (7 \cdot 3) + (7 \cdot 0) = 21 + 0 = 21$

Skill Practice Apply the distributive property and simplify.

5. $2(6 + 4)$ **6.** $5(0 + 8)$

Answers
5. $(2 \cdot 6) + (2 \cdot 4)$; 20
6. $(5 \cdot 0) + (5 \cdot 8)$; 40

3. Multiplying Many-Digit Whole Numbers

When multiplying numbers with several digits, it is sometimes necessary to carry. To see why, consider the product $3 \cdot 29$. By writing the factors in expanded form, we can apply the distributive property. In this way, we see that 3 is multiplied by both 20 and 9.

$$3 \cdot 29 = 3(20 + 9) = (3 \cdot 20) + (3 \cdot 9)$$
$$= 60 + 27$$
$$= 6 \text{ tens} + 2 \text{ tens} + 7 \text{ ones}$$
$$= 8 \text{ tens} + 7 \text{ ones}$$
$$= 87$$

Now we will multiply $29 \cdot 3$ in vertical form.

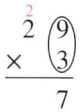   Multiply $3 \cdot 9 = 27$. Write the 7 in the ones column and carry the 2.

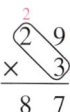   Multiply $3 \cdot 2$ tens $= 6$ tens. Add the carry: 6 tens $+ 2$ tens $= 8$ tens. Write the 8 in the tens place.

Example 4 Multiplying a Many-Digit Number by a One-Digit Number

Multiply.
$$\begin{array}{r} 368 \\ \times 5 \\ \hline \end{array}$$

Solution:

Using the distributive property, we have

$$5(300 + 60 + 8) = 1500 + 300 + 40 = 1840$$

This can be written vertically as:

$$\begin{array}{r} 368 \\ \times 5 \\ \hline 40 \\ 300 \\ +\ 1500 \\ \hline 1840 \end{array}$$

    40   Multiply $5 \cdot 8$.
   300   Multiply $5 \cdot 60$.
$+\ 1500$ Multiply $5 \cdot 300$.
 1840   Add.

The numbers 40, 300, and 1500 are called *partial products*. The product of 386 and 5 is found by adding the partial products. The product is 1840.

Skill Practice Multiply.

7. $\begin{array}{r} 247 \\ \times 3 \\ \hline \end{array}$

The solution to Example 4 can also be found by using a shorter form of multiplication. We outline the procedure:

$$\begin{array}{r} \overset{4}{3}68 \\ \times 5 \\ \hline 0 \end{array}$$ Multiply $5 \cdot 8 = 40$. Write the 0 in the ones place and carry the 4.

Answer

7. 741

$$\begin{array}{r}\overset{3\;4}{368}\\ \times\;\;\;5\\ \hline 40\end{array}$$ Multiply $5 \cdot 6$ tens = 300. Add the carry. $300 + 4$ tens = 340.
Write the 4 in the tens place and carry the 3.

$$\begin{array}{r}\overset{3\;4}{368}\\ \times\;\;\;5\\ \hline 1840\end{array}$$ Multiply $5 \cdot 3$ hundreds = 1500. Add the carry.
$1500 + 3$ hundreds = 1800. Write the 8 in the hundreds place and the 1 in the thousands place.

Example 5 demonstrates the process to multiply two factors with many digits.

Example 5 — Multiplying a Two-Digit Number by a Two-Digit Number

Multiply.
$$\begin{array}{r}72\\ \times\;83\\ \hline\end{array}$$

Solution:

Writing the problem vertically and computing the partial products, we have

$$\begin{array}{r}\overset{1}{72}\\ \times\;\;83\\ \hline 216\\ +\;5760\\ \hline 5976\end{array}$$

Multiply $3 \cdot 72$.
Multiply $80 \cdot 72$.
Add.

The product is 5976.

Skill Practice Multiply.

8. $\;\;\;59$
   $\times\;26$

Example 6 — Multiplying Two Multidigit Whole Numbers

Multiply. $\;\;368(497)$

Solution:

$$\begin{array}{r}\overset{2\;3}{\underset{4\;5}{\overset{6\;7}{368}}}\\ \times\;\;\;497\\ \hline 2{,}576\\ 33{,}120\\ +\;147{,}200\\ \hline 182{,}896\end{array}$$

Multiply $7 \cdot 368$.
Multiply $90 \cdot 368$.
Multiply $400 \cdot 368$.

Skill Practice Multiply.

9. $\;\;\;274$
   $\times\;586$

4. Estimating Products by Rounding

A special pattern occurs when one or more factors in a product ends in zero. Consider the following products:

$12 \cdot 20 = 240$	$120 \cdot 20 = 2400$
$12 \cdot 200 = 2400$	$1200 \cdot 20 = 24{,}000$
$12 \cdot 2000 = 24{,}000$	$12{,}000 \cdot 20 = 240{,}000$

Answers
8. 1534   9. 160,564

Notice in each case the product is 12 · 2 = 24 followed by the total number of zeros from each factor. Consider the product 1200 · 20.

```
  12|00      Shift the numbers 1200 and 20 so that the zeros appear to the right
×  2|0       of the multiplication process. Multiply 12 · 2 = 24.
  24|000     Write the product 24 followed by the total number of zeros from
             each factor.
```

Example 7 — Estimating a Product

Estimate the product 795(4060) by rounding 795 to the nearest hundred and 4060 to the nearest thousand.

Solution:

```
 795   rounds to →   800          8|00
4060   rounds to →  4000      ×   4|000
                               32|00000
```

The product is approximately 3,200,000.

Skill Practice

10. Estimate the product 421(869) by rounding each factor to the nearest hundred.

Example 8 — Estimating a Product in an Application

For a trip from Atlanta to Los Angeles, the average cost of a plane ticket was $495. If the plane carried 218 passengers, estimate the total revenue for the airline. (*Hint:* Round each number to the hundreds place and find the product.)

Walter Lockwood/Corbis Premium/ Alamy Stock Photo

Solution:

```
$495   rounds to →   $ 5|00
 218   rounds to →   × 2|00
                     $10|0000
```

The airline received approximately $100,000 in revenue.

Skill Practice

11. A small newspaper has 16,850 subscribers. Each subscription costs $149 per year. Estimate the revenue for the year by rounding the number of subscriptions to the nearest thousand and the cost to the nearest ten.

5. Applications Involving Multiplication

In English there are many different words that imply multiplication. A partial list is given in Table 1-3.

Table 1-3

Word/Phrase	Example	In Symbols
Product	The product of 4 and 7	4 · 7
Times	8 times 4	8 · 4
Multiply . . . by . . .	Multiply 6 by 3	6 · 3

Multiplication may also be warranted in applications involving unit rates. In Example 8, we multiplied the cost per customer ($495) by the number of customers (218). The value $495 is a unit rate because it gives the cost per one customer (per one unit).

Answers
10. 360,000   11. $2,550,000

Example 9 Solving an Application Involving Multiplication

The average weekly income for production workers is $489. How much does a production worker earn in 1 year (assume 52 weeks in 1 year)?

Solution:

The value $489 per week is a unit rate. The total for 1 year is given by $489 · 52.

$$\begin{array}{r} \overset{4\ 4}{\underset{1\ 1}{4}}89 \\ \times\ 52 \\ \hline 978 \\ +\ 24{,}450 \\ \hline 25{,}428 \end{array}$$ The yearly total is $25,428.

TIP: This product can be estimated quickly by rounding the factors.

$$\begin{array}{rcl} 489 & \text{rounds to} \longrightarrow & 5\,|00 \\ 52 & \text{rounds to} \longrightarrow & \times\ 5\,|0 \\ & & \hline 25\,|000 \end{array}$$

The total yearly income is approximately $25,000. Estimating gives a quick approximation of a product. Furthermore, it also checks for the reasonableness of our exact product. In this case $25,000 is close to our exact value of $25,428.

Skill Practice

12. Ella can type 65 words per minute. How many words can she type in 45 minutes?

6. Area of a Rectangle

Another application of multiplication of whole numbers lies in finding the area of a region. **Area** measures the amount of surface contained within the region. For example, a square that is 1 in. by 1 in. occupies an area of 1 square inch, denoted as $1\ \text{in.}^2$. Similarly, a square that is 1 centimeter (cm) by 1 cm occupies an area of 1 square centimeter. This is denoted by $1\ \text{cm}^2$.

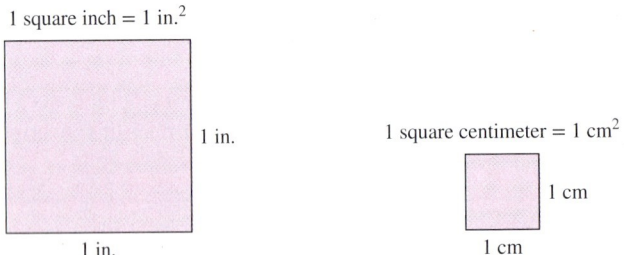

The units of square inches and square centimeters (in.^2 and cm^2) are called *square units*. To find the area of a region, measure the number of square units occupied in that region. For example, the region in Figure 1-4 occupies $6\ \text{cm}^2$.

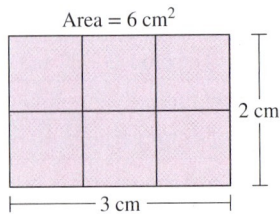

Figure 1-4

Answer
12. 2925 words

The 3-cm by 2-cm region in Figure 1-4 suggests that to find the area of a rectangle, multiply the length by the width. If the area is represented by A, the length is represented by l, and the width is represented by w, then we have

$$\textbf{Area of rectangle} = (\text{length}) \cdot (\text{width})$$
$$A = l \cdot w$$

The letters A, l, and w are variables because their values *vary* as they are replaced by different numbers.

Example 10   **Finding the Area of a Rectangle**

Find the area and perimeter of the rectangle.

Solution:

Area:

$A = l \cdot w$
$A = (7 \text{ yd}) \cdot (4 \text{ yd})$
   $= 28 \text{ yd}^2$

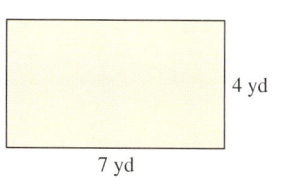

The perimeter of a polygon is the sum of the lengths of the sides. In a rectangle the opposite sides are equal in length.

Perimeter:

$P = 7 \text{ yd} + 4 \text{ yd} + 7 \text{ yd} + 4 \text{ yd}$
   $= 22 \text{ yd}$

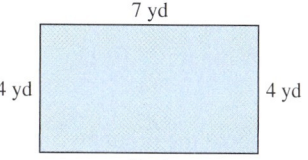

The area is 28 yd² and the perimeter is 22 yd.

Skill Practice

13. Find the area and perimeter of the rectangle.

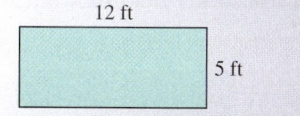

Avoiding Mistakes

Notice that area is measured in square units (such as yd²) and perimeter is measured in units of length (such as yd). It is important to apply the correct units of measurement.

Example 11   **Finding Area in an Application**

The state of Wyoming is approximately the shape of a rectangle (Figure 1-5). Its length is 355 mi and its width is 276 mi. Approximate the total area of Wyoming by rounding the length and width to the nearest ten.

Solution:

355   rounds to ⟶   36|0
276   rounds to ⟶   × 28|0
                            288
                            720
                        1008|00

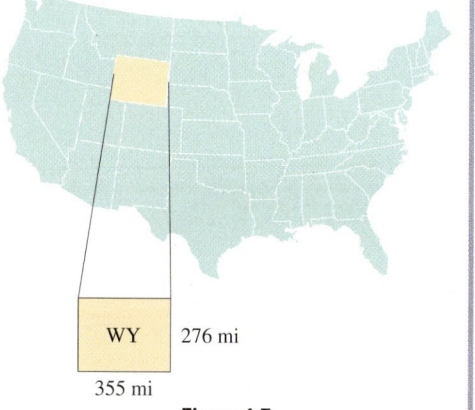

Figure 1-5

The area of Wyoming is approximately 100,800 mi².

Skill Practice

14. A house sits on a rectangular lot that is 192 ft by 96 ft. Approximate the area of the lot by rounding the length and width to the nearest hundred.

Answers

13. Area: 60 ft²; perimeter: 34 ft
14. 20,000 ft²

Section 1.4 Activity

A.1. Simplify the expression using order of operations.
$5(2 + 4)$

A.2. Simplify the expression by multiplying 5 times each addend within the parentheses.
$5(2 + 4)$

A.3. What do you observe about the results from Exercises A.1 and A.2? What property is applied in Exercise A.2?

For Exercises A.4–A.8, complete the statement and match it with the appropriate property.

A.4. $8 \cdot 1 =$ _____
A.5. $2 \cdot 5 = 5 \cdot$ _____
A.6. $6(3 + 9) = 6 \cdot 3 +$ _____
A.7. $12 \cdot$ _____ $= 0$
A.8. $7 \cdot (11 \cdot 4) = (7 \cdot 11) \cdot$ _____

a. Commutative property of multiplication
b. Multiplication property of 0
c. Associative property of multiplication
d. Multiplication property of 1
e. Distributive property of multiplication over addition

A.9. Find the product for parts (a)–(d).
   a. $4 \cdot 8 =$
   b. $40 \cdot 8 =$
   c. $400 \cdot 8 =$
   d. $400 \cdot 80 =$
   e. What observations can you make about the products?

A.10. Each student in the mathematics course is supplied with a Chromebook. If each Chromebook costs $130 and there are 24 students enrolled in the course, what is the total cost of the Chromebooks?

A.11. a. Find the area of an optimum size soccer field shown below. Include the appropriate units in your answer.

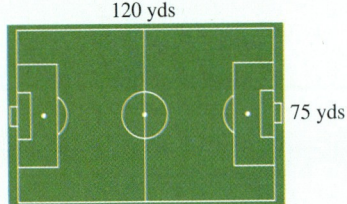

120 yds
75 yds

b. If the cost to maintain the field is $3 per square yard each year, find the annual cost of maintenance.

Section 1.4 Practice Exercises

Study Skills Exercise

Reading comprehension in a mathematics course is essential. Reading comprehension can range from understanding mathematical operations and symbols to solving lengthy word problems. When reading mathematics, it is not unusual to find multiple symbolic representations for mathematical expressions. For example,
- Write four different ways to express 5×4 using different notations.

You will also find different ways to express a mathematical statement in words. For example,
- Write three different English statements that represent $8 - 5$.

Vocabulary and Key Concepts

1. **a.** Two numbers being multiplied are called _____ and the result is called the _____.
   b. The _____ property of multiplication states that $a \cdot b = b \cdot a$. That is, the order of factors does not affect the product.
   c. The _____ property of multiplication indicates that the manner in which factors are grouped under multiplication does not affect the product. That is, $(a \cdot b) \cdot c = $ _____.
   d. The multiplication property of 0 states that $a \cdot 0 = $ _____ and $0 \cdot a = $ _____.
   e. The multiplication property of 1 states that $a \cdot 1 = $ _____ and $1 \cdot a = $ _____.
   f. The _____ property of multiplication over addition states that $a \cdot (b + c) = $ _____.
   g. The _____ of a region measures the amount of surface contained within the region.
   h. Given a rectangle of length l and width w, the area A is given by $A = $ _____.

2. The mathematical operation of multiplication is repeated _____.

3. The expression $2 \times 6 = 12$ can be written using different operators that signify multiplication. Write the expression three other ways.

Concept 1: Introduction to Multiplication

For Exercises 4–5, write the expression that represents the rectangular array shown.

4.    5.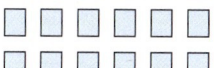

6. Fill in the table of multiplication facts.

×	0	1	2	3	4	5	6	7	8	9
0										
1										
2										
3										
4										
5										
6										
7										
8										
9										

For Exercises 7–10, write the repeated addition as multiplication and simplify.

7. $5 + 5 + 5 + 5 + 5 + 5$
8. $2 + 2 + 2 + 2 + 2 + 2 + 2 + 2 + 2$
9. $9 + 9 + 9$
10. $7 + 7 + 7 + 7$

For Exercises 11–14, identify the factors and the product. (See Example 1.)

11. $13 \times 42 = 546$

12. $26 \times 9 = 234$

13. $3 \cdot 5 \cdot 2 = 30$

14. $4 \cdot 3 \cdot 8 = 96$

15. Write the product of 5 and 12, using three different notations. (Answers may vary.)

16. Write the product of 23 and 14, using three different notations. (Answers may vary.)

Concept 2: Properties of Multiplication

For Exercises 17–22, match the property with the statement.

17. $8 \cdot 1 = 8$

18. $6 \cdot 13 = 13 \cdot 6$

19. $2(6 + 12) = 2 \cdot 6 + 2 \cdot 12$

20. $5 \cdot (3 \cdot 2) = (5 \cdot 3) \cdot 2$

21. $0 \cdot 4 = 0$

22. $7(14) = 14(7)$

a. Commutative property of multiplication

b. Associative property of multiplication

c. Multiplication property of 0

d. Multiplication property of 1

e. Distributive property of multiplication over addition

For Exercises 23–28, rewrite the expression, using the indicated property. (See Examples 2 and 3.)

23. $14 \cdot 8$; commutative property of multiplication

24. $3 \cdot 9$; commutative property of multiplication

25. $6 \cdot (2 \cdot 10)$; associative property of multiplication

26. $(4 \cdot 15) \cdot 5$; associative property of multiplication

27. $5(7 + 4)$; distributive property of multiplication over addition

28. $3(2 + 6)$; distributive property of multiplication over addition

Concept 3: Multiplying Many-Digit Whole Numbers

For Exercises 29–60, multiply. (See Examples 4–6.)

29. 24×6

30. 18×5

31. 26×2

32. 71×3

33. 131×5

34. 725×3

35. 344×4

36. 105×9

37. 1410×8

38. 2016×6

39. 3312×7

40. 4801×5

41. $42{,}014 \times 9$

42. $51{,}006 \times 8$

43. 32×14

44. 41×21

45. $68 \cdot 24$

46. $55 \cdot 41$

47. $72 \cdot 12$

48. $13 \cdot 46$

Section 1.4 Multiplication of Whole Numbers and Area   45

49. (143)(17)   **50.** (722)(28)   **51.** (349)(19)   **52.** (512)(31)

53.   151
      × 127

54.   703
      × 146

55.   222
      × 841

56.   387
      × 506

57. 3532
      × 6014

58. 2810
      × 1039

59. 4122
      × 982

60. 7026
      × 528

Concept 4: Estimating Products by Rounding

For Exercises 61–68, multiply. (See Example 7.)

61.   600
      × 40

62.   900
      × 50

63. 3000
      × 700

64. 4000
      × 400

65. 8000
      × 9000

66. 1000
      × 2000

67. 90,000
      × 400

68. 50,000
      × 6000

For Exercises 69–72, estimate the product by first rounding the number to the indicated place value.

69. 11,784 · 5201; thousands place

70. 45,046 · 7812; thousands place

71. 82,941 · 29,740; ten-thousands place

72. 630,229 · 71,907; ten-thousands place

73. Suppose a hotel room costs $189 per night. Round this number to the nearest hundred to estimate the cost for a five-night stay. (See Example 8.)

74. The science department of a high school must purchase a set of calculators for a class. If the cost of one calculator is $129, estimate the cost of 28 calculators by rounding the numbers to the tens place.

75. The price for a ticket to see a popular country singer is $137. If a concert stadium seats 10,256 fans, estimate the amount of money received during that performance by rounding the number of seats to the nearest ten-thousand.

76. A breakfast buffet at a small restaurant serves 48 people. Estimate the maximum revenue for one week (7 days) if the price of a breakfast is $12 by rounding the numbers to the tens place.

Concept 5: Applications Involving Multiplication

77. The 4-gigabyte (4-GB) music player is advertised to store approximately 1000 songs. Assuming the average length of a song is 4 minutes (min), how many minutes of music can be stored on the device? (See Example 9.)

78. One CD can hold 700 megabytes (MB) of data. How many megabytes can 15 CDs hold?

HBSS/Getty Images

79. It costs about $45 for a cat to have a medical exam. If a humane society has 37 cats, find the cost of medical exams for its cats.

80. Marina's hybrid car gets 55 miles per gal (mpg) on the highway. How many miles can it go on 20 gal of gas?

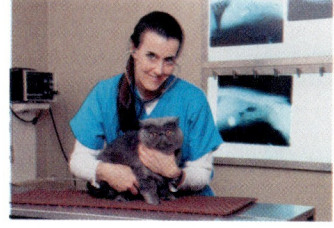
David Buffington/Photodisc/Getty Images

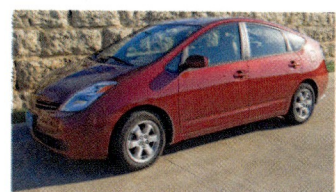
Jill Braaten/McGraw Hill

81. A can of soda contains 12 fluid ounces (fl oz). Find the number of ounces in a case of soda containing 12 cans.

82. A 3-credit-hour class at a college meets 3 hours (hr) per week. If a semester is 16 weeks long, for how many hours will the class meet during the semester?

83. PaperWorld shipped 115 cases of copy paper to a business. There are 5 reams of paper in each case and 500 sheets of paper in each ream. Find the number of sheets of paper delivered to the business.

84. A dietary supplement bar has 14 grams (g) of protein. If Kathleen eats 2 bars a day for 6 days, how many grams of protein will she get from this supplement?

85. Tylee's car gets 31 miles per gallon (mpg) on the highway. How many miles can he travel if he has a full tank of gas (12 gal)?

86. Sherica manages a small business called Pizza Express. She has 23 employees who work an average of 32 hr per week. How many hours of work does Sherica have to schedule each week?

Concept 6: Area of a Rectangle

For Exercises 87–90, find the area. (See Example 10.)

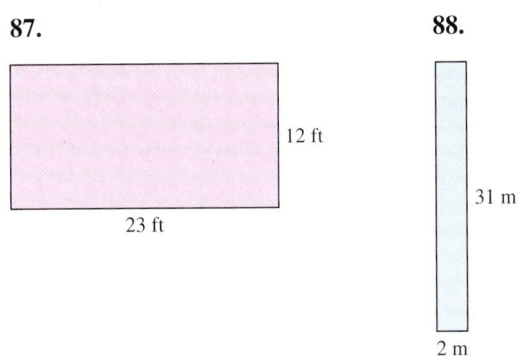

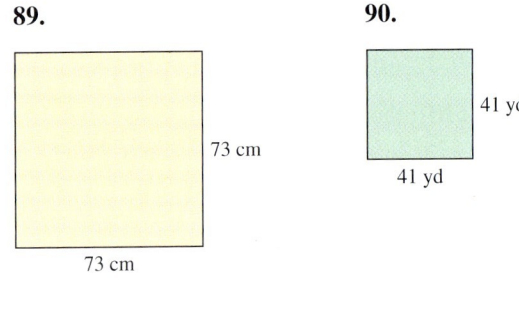

87. 23 ft by 12 ft

88. 2 m by 31 m

89. 73 cm by 73 cm

90. 41 yd by 41 yd

91. The state of Colorado is approximately the shape of a rectangle. Its length is 388 mi and its width is 269 mi. Approximate the total area of Colorado by rounding the length and width to the nearest ten. (See Example 11.)

92. A parcel of land has a width of 132 yd and a length of 149 yd. Approximate the total area by rounding each dimension to the nearest ten.

93. The front of a building has windows that are 44 in. by 58 in.

    a. Approximate the area of one window.

    b. If the building has three floors and each floor has 14 windows, how many windows are there?

    c. What is the approximate total area of all of the windows?

94. The length of a carport is 51 ft and its width is 29 ft. Approximate the area of the carport.

95. Mr. Slackman wants to paint his garage door that is 8 ft. by 16 ft. To decide how much paint to buy, he must find the area of the door. What is the area of the door?

96. To carpet a rectangular room, Erika must find the area of the floor. If the dimensions of the room are 10 yd by 15 yd, how much carpeting does she need?

Division of Whole Numbers

Section 1.5

Concepts

1. Introduction to Division
2. Properties of Division
3. Long Division
4. Dividing by a Many-Digit Divisor
5. Translations and Applications Involving Division

1. Introduction to Division

Suppose 12 pieces of pizza are to be divided evenly among 4 children (Figure 1-6). The number of pieces that each child would receive is given by 12 ÷ 4, read "12 divided by 4."

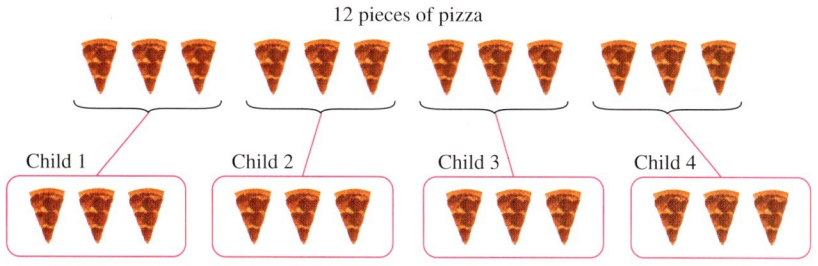

Burke/Triolo/Brand X Pictures/Getty Images

Figure 1-6

The process of separating 12 pieces of pizza evenly among 4 children is called **division**. The statement 12 ÷ 4 = 3 indicates that each child receives 3 pieces of pizza. The number 12 is called the **dividend**. It represents the number to be divided. The number 4 is called the **divisor**, and it represents the number of groups. The result of the division (in this case 3) is called the **quotient**. It represents the number of items in each group.

Division can be represented in several ways. For example, the following are all equivalent statements.

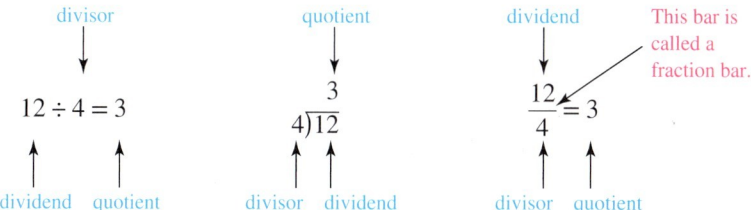

Recall that subtraction is the reverse operation of addition. In the same way, division is the reverse operation of multiplication. For example, we say 12 ÷ 4 = 3 because 3 · 4 = 12.

Example 1 Identifying the Dividend, Divisor, and Quotient

Simplify each expression. Then identify the dividend, divisor, and quotient.

a. $48 \div 6$ b. $9\overline{)36}$ c. $\dfrac{63}{7}$

Solution:

a. $48 \div 6 = 8$ because $8 \cdot 6 = 48$
   The dividend is 48, the divisor is 6, and the quotient is 8.

b. $\begin{array}{r}4\\9\overline{)36}\end{array}$ because $4 \cdot 9 = 36$
   The dividend is 36, the divisor is 9, and the quotient is 4.

c. $\dfrac{63}{7} = 9$ because $9 \cdot 7 = 63$
   The dividend is 63, the divisor is 7, and the quotient is 9.

Skill Practice Identify the dividend, divisor, and quotient.

1. $56 \div 7$ 2. $4\overline{)20}$ 3. $\dfrac{18}{2}$

Answers

1. Dividend: 56; divisor: 7; quotient: 8
2. Dividend: 20; divisor: 4; quotient: 5
3. Dividend: 18; divisor: 2; quotient: 9

2. Properties of Division

Properties of Division

1. $a \div a = 1$ for any number a, where $a \neq 0$.
   Any nonzero number divided by itself is 1.   Example: $9 \div 9 = 1$
2. $a \div 1 = a$ for any number a.
   Any number divided by 1 is the number itself.   Example: $3 \div 1 = 3$
3. $0 \div a = 0$ for any number a, where $a \neq 0$.
   Zero divided by any nonzero number is zero.   Example: $0 \div 5 = 0$
4. $a \div 0$ is undefined for any number a.
   Any number divided by zero is undefined.   Example: $9 \div 0$ is undefined.

To help remember the difference between $0 \div 2$ and $2 \div 0$, consider this application:

$\$8 \div 2 = \$4$ means that if we divide \$8 between 2 people, each person will receive \$4.
$\$0 \div 2 = \$0$ means that if we divide \$0 between 2 people, each person will receive \$0.
$\$2 \div 0$ means that we would like to divide \$2 between 0 people. This cannot be done. So $2 \div 0$ is undefined.

Example 2 illustrates the important properties of division.

Example 2   Dividing Whole Numbers

Divide.

   a. $8 \div 8$ **b.** $\dfrac{6}{6}$ **c.** $5 \div 1$ **d.** $1\overline{)7}$

   e. $0 \div 6$ **f.** $\dfrac{0}{4}$ **g.** $6 \div 0$ **h.** $\dfrac{10}{0}$

Solution:

   a. $8 \div 8 = 1$ because $1 \cdot 8 = 8$

   b. $\dfrac{6}{6} = 1$ because $1 \cdot 6 = 6$

   c. $5 \div 1 = 5$ because $5 \cdot 1 = 5$

   d. $1\overline{)7}^{\,7}$ because $7 \cdot 1 = 7$

   e. $0 \div 6 = 0$ because $0 \cdot 6 = 0$

   f. $\dfrac{0}{4} = 0$ because $0 \cdot 4 = 0$

Avoiding Mistakes

There is no mathematical symbol to describe the result when dividing by 0. We must write out the word *undefined* to denote that we cannot divide by 0.

   g. $6 \div 0$ is *undefined* because there is no number that when multiplied by 0 will produce a product of 6.

   h. $\dfrac{10}{0}$ is *undefined* because there is no number that when multiplied by 0 will produce a product of 10.

Skill Practice Divide.

   4. $3\overline{)3}$ **5.** $5 \div 5$ **6.** $\dfrac{4}{1}$ **7.** $8 \div 0$ **8.** $\dfrac{0}{7}$ **9.** $3\overline{)0}$

Answers
4. 1   **5.** 1   **6.** 4   **7.** Undefined
8. 0   **9.** 0

You should also note that unlike addition and multiplication, division is neither commutative nor associative. In other words, reversing the order of the dividend and divisor may produce a different quotient. Similarly, changing the manner in which numbers are grouped with division may affect the outcome. See Exercises 31 and 32.

3. Long Division

To divide larger numbers we use a process called **long division**. This process uses a series of estimates to find the quotient. We illustrate long division in Example 3.

Example 3 Using Long Division

Divide. $7\overline{)161}$

Solution:

Estimate $7\overline{)161}$ by first estimating $7\overline{)16}$ and writing the result above the tens place of the dividend. Since $7 \cdot 2 = 14$, there are at least 2 sevens in 16.

$$\begin{array}{r} 2 \\ 7\overline{)161} \\ -140 \\ \hline 21 \end{array}$$

The 2 in the tens place represents 20 in the quotient.
← Multiply $7 \cdot 20$ and write the result under the dividend.
Subtract 140. We see that our estimate leaves 21.

Repeat the process. Now divide $7\overline{)21}$ and write the result in the ones place of the quotient.

$$\begin{array}{r} 23 \\ 7\overline{)161} \\ -140 \\ \hline 21 \\ -21 \\ \hline 0 \end{array}$$

← Multiply $7 \cdot 3$.
Subtract.

The quotient is 23.   Check: $\begin{array}{r} 23 \\ \times\, 7 \\ \hline 161\ \checkmark \end{array}$

Skill Practice Divide.

10. $8\overline{)136}$

We can streamline the process of long division by "bringing down" digits of the dividend one at a time.

Example 4 Using Long Division

Divide. $6138 \div 9$

Solution:

$$\begin{array}{r} 682 \\ 9\overline{)6138} \\ -54 \\ \hline 73 \\ -72 \\ \hline 18 \\ -18 \\ \hline 0 \end{array}$$

$61 \div 9$ is estimated at 6. Multiply $9 \cdot 6 = 54$ and subtract.

$73 \div 9$ is estimated at 8. Multiply $9 \cdot 8 = 72$ and subtract.

$18 \div 9 = 2$. Multiply $9 \cdot 2 = 18$ and subtract.

The quotient is 682.   Check: $\begin{array}{r} 7\,1 \\ 682 \\ \times\ \ 9 \\ \hline 6138\ \checkmark \end{array}$

Skill Practice Divide.

11. $2891 \div 7$

Answers
10. 17   **11.** 413

In many instances, quotients do not come out evenly. For example, suppose we had 13 pieces of pizza to distribute among 4 children (Figure 1-7).

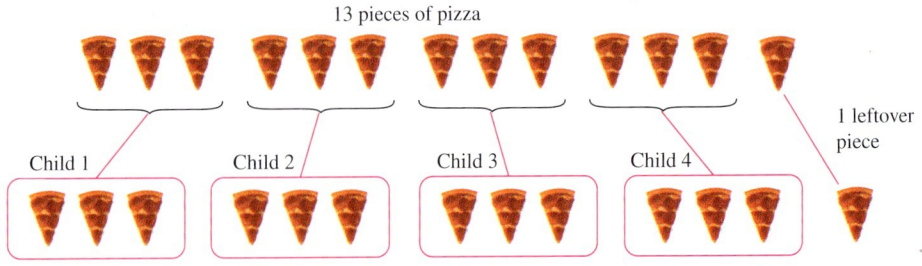

Burke/Triolo/Brand X Pictures/Getty Images   Figure 1-7

The mathematical term given to the "leftover" piece is called the **remainder**. The division process may be written as

$$4\overline{)13} \;\; \begin{array}{r} 3 \text{ R1} \\ -12 \\ \hline 1 \end{array}$$

The remainder is written next to the 3.

The **whole part of the quotient** is 3, and the remainder is 1. Notice that the remainder is written next to the whole part of the quotient.

We can check a division problem that has a remainder. To do so, multiply the divisor by the whole part of the quotient and then add the remainder. The result must equal the dividend. That is,

(Divisor)(whole part of quotient) + remainder = dividend

Thus,

$$(4)(3) + 1 \stackrel{?}{=} 13$$
$$12 + 1 \stackrel{?}{=} 13$$
$$13 = 13 \;\checkmark$$

Example 5 Using Long Division

Divide. $1253 \div 6$

Solution:

$$6\overline{)1253} \;\; \begin{array}{r} 208 \text{ R5} \\ -12 \downarrow \\ \hline 05 \\ -00 \downarrow \\ \hline 53 \\ -48 \\ \hline 5 \end{array}$$

$6 \cdot 2 = 12$ and subtract.
Bring down the 5.
Note that 6 does not divide into 5, so we put a 0 in the quotient.
Bring down the 3.
$6 \cdot 8 = 48$ and subtract.
The remainder is 5.

To check, verify that $(6)(208) + 5 = 1253$. ✓

Skill Practice Divide.

12. $5107 \div 5$

Answer

12. 1021 R2

4. Dividing by a Many-Digit Divisor

When the divisor has more than one digit, we still use a series of estimations to find the quotient.

Example 6   **Dividing by a Two-Digit Number**

Divide. $32\overline{)1259}$

Solution:

To estimate the leading digit of the quotient, estimate the number of times 30 will go into 125. Since $30 \cdot 4 = 120$, our estimate is 4.

$$\begin{array}{r} 4 \\ 32\overline{)1259} \\ -128 \end{array}$$

$32 \cdot 4 = 128$ is too big. We cannot subtract 128 from 125. Revise the estimate in the quotient to 3.

$$\begin{array}{r} 3 \\ 32\overline{)1259} \\ -96 \\ \hline 299 \end{array}$$

$32 \cdot 3 = 96$ and subtract.

Bring down the 9.

Now estimate the number of times 30 will go into 299. Because $30 \cdot 9 = 270$, our estimate is 9.

$$\begin{array}{r} 39 \text{ R}11 \\ 32\overline{)1259} \\ -96 \\ \hline 299 \\ -288 \\ \hline 11 \end{array}$$

$32 \cdot 9 = 288$ and subtract.

The remainder is 11.

To check, verify that $(32)(39) + 11 = 1259$. ✓

Skill Practice Divide.

13. $63\overline{)4516}$

Example 7   **Dividing by a Many-Digit Number**

Divide. $\dfrac{82{,}705}{602}$

Solution:

$$\begin{array}{r} 137 \text{ R}231 \\ 602\overline{)82{,}705} \\ -602 \\ \hline 2250 \\ -1806 \\ \hline 4445 \\ -4214 \\ \hline 231 \end{array}$$

$602 \cdot 1 = 602$ and subtract.

Bring down the 0.

$602 \cdot 3 = 1806$ and subtract.

Bring down the 5.

$602 \cdot 7 = 4214$ and subtract.

The remainder is 231.

To check, verify that $(602)(137) + 231 = 82{,}705$. ✓

Skill Practice Divide.

14. $304\overline{)62{,}405}$

Answers

13. 71 R43   **14.** 205 R85

5. Translations and Applications Involving Division

Several words and phrases imply division. A partial list is given in Table 1-4.

Table 1-4

Word/Phrase	Example	In Symbols
Divide	Divide 12 by 3	$12 \div 3$ or $\frac{12}{3}$ or $3\overline{)12}$
Quotient	The quotient of 20 and 2	$20 \div 2$ or $\frac{20}{2}$ or $2\overline{)20}$
Per	110 mi per 2 hr	$110 \div 2$ or $\frac{110}{2}$ or $2\overline{)110}$
Divides into	4 divides into 28	$28 \div 4$ or $\frac{28}{4}$ or $4\overline{)28}$
Divided, or shared equally among	64 shared equally among 4	$64 \div 4$ or $\frac{64}{4}$ or $4\overline{)64}$

Example 8 Solving an Application Involving Division

A painting business employs 3 painters. The business collects $1950 for painting a house. If all painters are paid equally, how much does each person make?

Solution:

This is an example where $1950 is shared equally among 3 people. Therefore, we divide.

$$\begin{array}{r} 650 \\ 3\overline{)1950} \\ -18\downarrow \\ \hline 15 \\ -15\downarrow \\ \hline 00 \\ -0 \\ \hline 0 \end{array}$$

$3 \cdot 6 = 18$ and subtract.
Bring down the 5.
$3 \cdot 5 = 15$ and subtract.
Bring down the 0.
$3 \cdot 0 = 0$ and subtract.
The remainder is 0.

Each painter makes $650.

Bojan Kontrec/E+/Getty Images

Skill Practice

15. Four people play Hearts with a standard 52-card deck of cards. If the cards are equally distributed, how many cards does each player get?

Answer

15. 13 cards

Example 9 — Solving an Application Involving Division

The graph in Figure 1-8 depicts the number of calories burned per hour for selected activities.

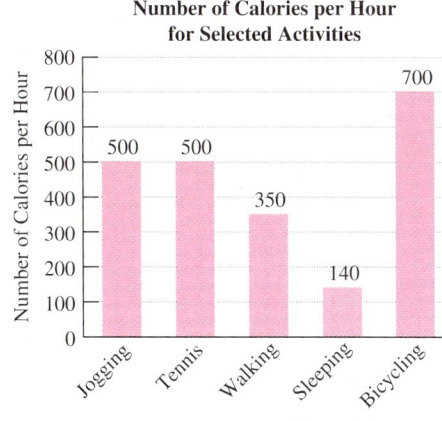

Figure 1-8

a. Janie wants to burn 3500 calories per week exercising. For how many hours must she jog?

b. For how many hours must Janie bicycle to burn 3500 calories?

Solution:

a. The total number of calories must be divided into 500-calorie increments. Thus, the number of hours required is given by 3500 ÷ 500.

$$\begin{array}{r} 7 \\ 500\overline{)3500} \\ -3500 \\ \hline 0 \end{array}$$ Janie requires 7 hr of jogging to burn 3500 calories.

b. 3500 calories must be divided into 700-calorie increments. The number of hours required is given by 3500 ÷ 700.

$$\begin{array}{r} 5 \\ 700\overline{)3500} \\ -3500 \\ \hline 0 \end{array}$$ Janie requires 5 hr of bicycling to burn 3500 calories.

Skill Practice

16. The cost for four different types of pastry at a French bakery is shown in the graph.

    Melissa has $360 to spend on desserts.

    a. If she spends all the money on chocolate éclairs, how many can she buy?

    b. If she spends all the money on apple tarts, how many can she buy?

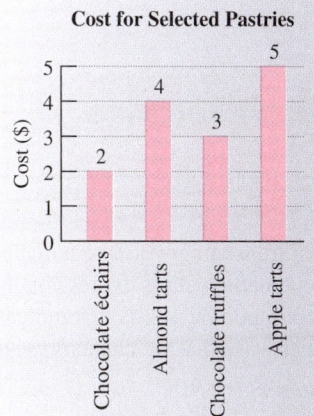

Answer

16. a. 180 chocolate éclairs
    b. 72 apple tarts

Section 1.5 Activity

A.1. Division is the process of separating a group of things into equal parts. Say I have a total of $50 to split evenly between my two children.
   a. Write 3 mathematical expressions to model this scenario using a different notation for division in each one.
   b. Identify the dividend, divisor, and quotient.
   c. Is there a remainder in this problem? Why or why not?
   d. What operation can be used to check your answer from part (a)? Verify that your answer is correct.

A.2. a. Evaluate $0 \div 8$ on a calculator. Explain your answer.
   b. Evaluate $8 \div 0$ on a calculator. Explain your answer.
   c. Do you get the same answers in parts (a) and (b)? Does the order in which you divide two numbers matter? Why or why not?

A.3. Divide.
   a. $15 \div 15$ b. $7 \div 7$ c. $x \div x$
   d. What statement can you make based on your results from parts (a)–(c)?

A.4. Divide.
   a. $15 \div 1$ b. $7 \div 1$ c. $x \div 1$
   d. What statement can you make based on your results from parts (a)–(c)?

A.5. Consider the division problem, $521 \div 6$.
   a. What is the correct way to write $521 \div 6$ as long division? Circle the correct answer.
      $521\overline{)6}$ $6\overline{)521}$
   b. Divide $521 \div 6$ using the long division process.
   c. Check your answer from part (b) by multiplying the divisor and the quotient. Do not forget to add the remainder.

A.6. José and Karin want to borrow $14,400 for several home improvements. They plan to pay off the loan with equal monthly payments. Determine the monthly payments given the number of years to pay on the loan.
   a. 3 years (36 months) _____
   b. 4 years (48 months) _____
   c. 5 years (60 months) _____
   d. If their budget allows for a maximum loan payment of $280 a month, which of these loans would they need to select?

Section 1.5 Practice Exercises

Study Skills Exercise

Writing mathematically is the skill of translating words to a mathematical expression or expressing your understanding of math in words. Translating words to a mathematical expression helps us to solve real-world application problems. *Sum, difference, product,* and *quotient* are important words that indicate the operations of addition, subtraction, multiplication and division. However, there are other ways that we can express these operations in words.
- Use three different phrases to represent $6 + 10$.
- Use three different phrases to represent $8.1 - 7.5$.
- Use three different phrases to represent $(5)(4)$.
- Use three different phrases to represent $24 \div 6$.

Vocabulary and Key Concepts

1. **a.** Given the division statement $15 \div 3 = 5$, the number 15 is called the _____, the number 3 is called the _____, and the number 5 is called the _____.
   b. $5 \div 5 =$ _____
   c. $5 \div 1 =$ _____
   d. $0 \div 5 =$ _____
   e. $5 \div 0$ is _____ because no number multiplied by 0 equals 5.
   f. If 17 is divided by 5, the whole part of the quotient is 3 and the _____ is 2.

2. The expression $15 \div 3 = 5$ can be written using different operators that signify division. Write the expression two other ways.

3. The reverse operation of division is _____.

4. Long division is the process used to find the quotient of larger numbers by using a series of _____.

5. Is the operation of division commutative; for example, does $15 \div 5 = 5 \div 15$?

6. Is the operation of division associative; for example, does $(20 \div 10) \div 2 = 20 \div (10 \div 2)$?

7. Write three words or phrases that imply division.

8. A student simplified $217 \div 5$ and found the quotient to be 43 R2. Is the student correct? Why or why not?

Concept 1: Introduction to Division

For Exercises 9–10, write the expression that represents the even division of rectangles.

9.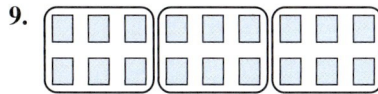

10.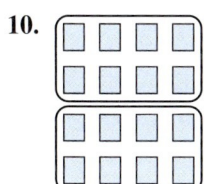

For Exercises 11–16, simplify each expression. Then identify the dividend, divisor, and quotient. **(See Example 1.)**

11. $72 \div 8$
12. $32 \div 4$
13. $8\overline{)64}$
14. $5\overline{)35}$
15. $\dfrac{45}{9}$
16. $\dfrac{20}{5}$

Concept 2: Properties of Division

17. In your own words, explain the difference between dividing a number by zero and dividing zero by a number.

18. Explain what happens when a number is either divided or multiplied by 1.

For Exercises 19–30, use the properties of division to simplify the expression, if possible. (See Example 2.)

19. $15 \div 1$
20. $21\overline{)21}$
21. $0 \div 10$
22. $\dfrac{0}{3}$

23. $0\overline{)9}$
24. $4 \div 0$
25. $\dfrac{20}{20}$
26. $1\overline{)9}$

27. $\dfrac{16}{0}$
28. $\dfrac{5}{1}$
29. $8\overline{)0}$
30. $13 \div 13$

31. Show that $6 \div 3 = 2$ but $3 \div 6 \neq 2$ by using multiplication to check.
32. Show that division is not associative, using the numbers 36, 12, and 3.

Concept 3: Long Division

33. Explain the process for checking a division problem when there is no remainder.
34. Show how checking by multiplication can help us remember that $0 \div 5 = 0$ and that $5 \div 0$ is undefined.

For Exercises 35–46, divide and check by multiplying. (See Examples 3 and 4.)

35. $78 \div 6$
    Check: $6 \cdot \square = 78$
36. $364 \div 7$
    Check: $7 \cdot \square = 364$
37. $5\overline{)205}$
    Check: $5 \cdot \square = 205$
38. $8\overline{)152}$
    Check: $8 \cdot \square = 152$

39. $\dfrac{972}{2}$
40. $\dfrac{582}{6}$
41. $1227 \div 3$
42. $236 \div 4$

43. $5\overline{)1015}$
44. $5\overline{)2035}$
45. $\dfrac{4932}{6}$
46. $\dfrac{3619}{7}$

For Exercises 47–54, check the division problems. If it is incorrect, find the correct answer.

47. $4\overline{)224}^{\,56}$
48. $7\overline{)574}^{\,82}$
49. $761 \div 3 = 253$
50. $604 \div 5 = 120$

51. $\dfrac{1021}{9} = 113 \text{ R}4$
52. $\dfrac{1311}{6} = 218 \text{ R}3$
53. $8\overline{)203}^{\,25 \text{ R}6}$
54. $7\overline{)821}^{\,117 \text{ R}5}$

For Exercises 55–70, divide and check the answer. (See Example 5.)

55. $61 \div 8$
56. $89 \div 3$
57. $9\overline{)92}$
58. $5\overline{)74}$

59. $\dfrac{55}{2}$
60. $\dfrac{49}{3}$
61. $593 \div 3$
62. $801 \div 4$

63. $\dfrac{382}{9}$
64. $\dfrac{428}{8}$
65. $3115 \div 2$
66. $4715 \div 6$

67. $6014 \div 8$
68. $9013 \div 7$
69. $6\overline{)5012}$
70. $2\overline{)1101}$

Concept 4: Dividing by a Many-Digit Divisor

For Exercises 71–86, divide. (See Examples 6 and 7.)

71. $9110 \div 19$

72. $3505 \div 13$

73. $24\overline{)1051}$

74. $41\overline{)8104}$

75. $\dfrac{8008}{26}$

76. $\dfrac{9180}{15}$

77. $68{,}012 \div 54$

78. $92{,}013 \div 35$

79. $\dfrac{1650}{75}$

80. $\dfrac{3649}{89}$

81. $520\overline{)18{,}201}$

82. $298\overline{)6278}$

83. $69{,}712 \div 304$

84. $51{,}107 \div 221$

85. $114\overline{)34{,}428}$

86. $421\overline{)87{,}989}$

Concept 5: Translations and Applications Involving Division

For Exercises 87–92, for each English sentence, write a mathematical expression and simplify.

87. Find the quotient of 497 and 71.

88. Find the quotient of 1890 and 45.

89. Divide 877 by 14.

90. Divide 722 by 53.

91. Divide 6 into 42.

92. Divide 9 into 108.

93. There are 392 students signed up for Anatomy 101. If each classroom can hold 28 students, find the number of classrooms needed. (See Example 8.)

94. A wedding reception is planned to take place in the fellowship hall of a church. The bride anticipates 120 guests, and each table will seat 8 people. How many tables should be set up for the reception to accommodate all the guests?

95. A case of tomato sauce contains 32 cans. If a grocer has 168 cans, how many cases can he fill completely? How many cans will be left over?

96. Austin has $425 to spend on dining room chairs. If each chair costs $52, does he have enough to purchase 8 chairs? If so, will he have any money left over?

Ned Frisk/Blend Images

97. At one time, a community college had 3000 students who registered for Beginning Algebra. If the average class size is 25 students, how many Beginning Algebra classes will the college have to offer?

98. Eight people are to share equally in an inheritance of $84,480. How much money will each person receive?

99. A hybrid automobile gets 45 mpg in stop-and-go traffic. How many gallons will it use in 405 mi of stop-and-go driving?

100. A couple traveled at an average speed of 52 mph for a cross-country trip. If the couple drove 1352 mi, how many hours was the trip?

101. Suppose Genny can type 1234 words in 22 min. Round each number to estimate her rate in words per minute.

102. On a trip to California from Illinois, Lavu drove 2780 mi. The gas tank in his car allows him to travel 405 mi. Round each number to the hundreds place to estimate the number of tanks of gas needed for the trip.

103. A group of 18 people goes to a concert. Ticket prices are given in the graph. If the group has $450, can they all attend the concert? If so, which type of seats can they buy? (See Example 9.)

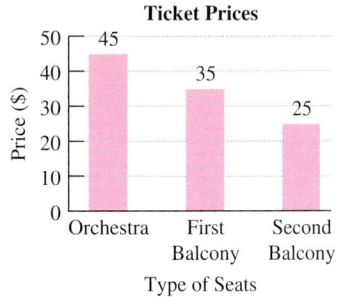

104. The graph gives the average annual starting salary for four professions: teacher, professor, accountant, and programmer. Find the monthly income for each of the four professions.

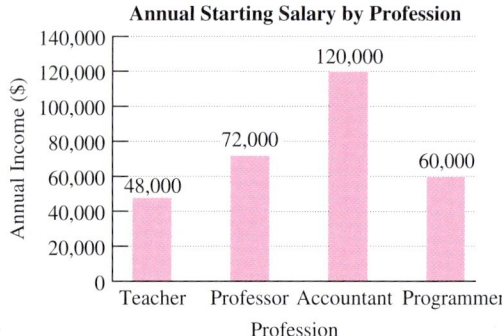

Technology Connections

For Exercises 105–108, solve the problem. Use a calculator to perform the calculations.

105. The United States consumes approximately 21,000,000 barrels (bbl) of oil per day. (*Source:* U.S. Energy Information Administration) How much does it consume in 1 year?

106. The average time to commute to work for people living in Washington State is 26 min (round trip 52 min). (*Source:* U.S. Census Bureau) How much time in minutes does a person spend commuting to and from work in 1 year if the person works 5 days a week for 50 weeks per year?

107. The budget for the U.S. federal government for a recent year was approximately $3552 billion. (*Source:* www.gpo.gov) How much could the government spend each quarter and still stay within its budget?

108. At a weigh station, a truck carrying 96 crates weighs in at 34,080 lb. If the truck weighs 9600 lb when empty, how much does each crate weigh?

Problem Recognition Exercises

Operations on Whole Numbers

For Exercises 1–14, perform the indicated operations.

1. a. 96 + 24 b. 96 − 24 c. 96 × 24 d. 24)96

2. a. 550 + 25 b. 550 − 25 c. 550 × 25 d. 25)550

3. a. 612 + 334 b. 946 − 334

4. a. 612 − 334 b. 278 + 334

5. a. 5500 − 4299 b. 1201 + 4299

6. a. 22,718 + 12,137 b. 34,855 − 12,137

7. a. 50 · 400 b. 20,000 ÷ 50

8. a. 548 · 63 b. 34,524 ÷ 63

9. a. 5060 ÷ 22 b. 230 · 22

10. a. 1875 ÷ 125 b. 125 · 15

11. a. 4)1312 b. 328)1312

12. a. 547)4376 b. 8)4376

13. a. 418 · 10 b. 418 · 100 c. 418 · 1000 d. 418 · 10,000

14. a. 350,000 ÷ 10 b. 350,000 ÷ 100 c. 350,000 ÷ 1000 d. 350,000 ÷ 10,000

Section 1.6
Exponents, Algebraic Expressions, and the Order of Operations

Concepts
1. Exponents
2. Square Roots
3. Order of Operations
4. Algebraic Expressions

1. Exponents

In this section we present the concept of an **exponent** to represent repeated multiplication. For example, the product

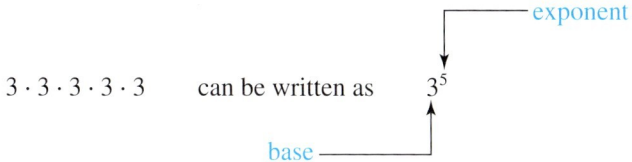

$$3 \cdot 3 \cdot 3 \cdot 3 \cdot 3 \quad \text{can be written as} \quad 3^5$$

The expression 3^5 is written in exponential form. The exponent, or **power**, is 5 and represents the number of times the **base**, 3, is used as a factor. The expression 3^5 is read as "three to the fifth power." Other expressions in exponential form are shown next.

5^2	is read as	"five squared" or "five to the second power"
5^3	is read as	"five cubed" or "five to the third power"
5^4	is read as	"five to the fourth power"
5^5	is read as	"five to the fifth power"

TIP: The expression $5^1 = 5$. Any number without an exponent explicitly written has a power of 1.

Exponential form is a shortcut notation for repeated multiplication. However, to simplify an expression in exponential form, we often write out the individual factors.

Example 1 — Evaluating Exponential Expressions

Evaluate.

a. 6^2 **b.** 5^3 **c.** 2^4

Solution:

a. $6^2 = 6 \cdot 6$ The exponent, 2, indicates the number of times the base, 6, is used as a factor.
$= 36$

b. $5^3 = 5 \cdot 5 \cdot 5$ When three factors are multiplied, we can group the first two factors and perform the multiplication.
$= (5 \cdot 5) \cdot 5$
$= (25) \cdot 5$ Then multiply the product of the first two factors by the last factor.
$= 125$

c. $2^4 = 2 \cdot 2 \cdot 2 \cdot 2$
$= (2 \cdot 2) \cdot 2 \cdot 2$ Group the first two factors.
$= 4 \cdot 2 \cdot 2$ Multiply the first two factors.
$= (4 \cdot 2) \cdot 2$ Multiply the product by the next factor to the right.
$= 8 \cdot 2$
$= 16$

Skill Practice Evaluate.
1. 8^2 2. 4^3 3. 2^5

Answers
1. 64   2. 64   3. 32

One important application of exponents lies in recognizing **powers of 10**. A power of 10 is 10 raised to a whole-number power. For example, consider the following expressions.

$$10^1 = 10$$
$$10^2 = 10 \cdot 10 = 100$$
$$10^3 = 10 \cdot 10 \cdot 10 = 1000$$
$$10^4 = 10 \cdot 10 \cdot 10 \cdot 10 = 10{,}000$$
$$10^5 = 10 \cdot 10 \cdot 10 \cdot 10 \cdot 10 = 100{,}000$$
$$10^6 = 10 \cdot 10 \cdot 10 \cdot 10 \cdot 10 \cdot 10 = 1{,}000{,}000$$

From these examples, we see that a power of 10 results in a 1 followed by several zeros. The number of zeros is the same as the exponent on the base of 10.

2. Square Roots

To square a number means that we multiply the base times itself. For example, $5^2 = 5 \cdot 5 = 25$.

To find a positive **square root** of a number means that we reverse the process of squaring. For example, finding the square root of 25 is equivalent to asking, "What positive number, when squared, equals 25?" The symbol $\sqrt{}$ (called a *radical sign*), is used to denote the positive square root of a number. Therefore, $\sqrt{25}$ is the positive number, that when squared, equals 25. Thus, $\sqrt{25} = 5$ because $(5)^2 = 25$.

Example 2 Evaluating Square Roots

Find the square roots.

a. $\sqrt{9}$ b. $\sqrt{64}$ c. $\sqrt{1}$ d. $\sqrt{0}$

Solution:

a. $\sqrt{9} = 3$ because $(3)^2 = 3 \cdot 3 = 9$
b. $\sqrt{64} = 8$ because $(8)^2 = 8 \cdot 8 = 64$
c. $\sqrt{1} = 1$ because $(1)^2 = 1 \cdot 1 = 1$
d. $\sqrt{0} = 0$ because $(0)^2 = 0 \cdot 0 = 0$

Skill Practice Find the square roots.
4. $\sqrt{4}$ 5. $\sqrt{100}$ 6. $\sqrt{400}$ 7. $\sqrt{121}$

TIP: To simplify square roots, it is advisable to become familiar with these squares and square roots.

$0^2 = 0 \longrightarrow \sqrt{0} = 0$ $7^2 = 49 \longrightarrow \sqrt{49} = 7$
$1^2 = 1 \longrightarrow \sqrt{1} = 1$ $8^2 = 64 \longrightarrow \sqrt{64} = 8$
$2^2 = 4 \longrightarrow \sqrt{4} = 2$ $9^2 = 81 \longrightarrow \sqrt{81} = 9$
$3^2 = 9 \longrightarrow \sqrt{9} = 3$ $10^2 = 100 \longrightarrow \sqrt{100} = 10$
$4^2 = 16 \longrightarrow \sqrt{16} = 4$ $11^2 = 121 \longrightarrow \sqrt{121} = 11$
$5^2 = 25 \longrightarrow \sqrt{25} = 5$ $12^2 = 144 \longrightarrow \sqrt{144} = 12$
$6^2 = 36 \longrightarrow \sqrt{36} = 6$ $13^2 = 169 \longrightarrow \sqrt{169} = 13$

Answers
4. 2 5. 10 6. 20 7. 11

3. Order of Operations

A numerical expression may contain more than one operation. For example, the following expression contains both multiplication and subtraction.

$$18 - 5(2)$$

The order in which the multiplication and subtraction are performed will affect the overall outcome.

Multiplying first yields
$18 - 5(2) = 18 - 10$
$ = 8$ (correct)

Subtracting first yields
$18 - 5(2) = 13(2)$
$ = 26$ (incorrect)

To avoid confusion, mathematicians have outlined the proper order of operations. In particular, multiplication is performed before addition or subtraction. The guidelines for the order of operations are given next. These rules must be followed in all cases.

Order of Operations

Step 1 First perform all operations inside parentheses or other grouping symbols.
Step 2 Simplify any expressions containing exponents or square roots.
Step 3 Perform multiplication or division in the order that they appear from left to right.
Step 4 Perform addition or subtraction in the order that they appear from left to right.

Example 3   Using the Order of Operations

Simplify.

a. $15 - 10 \div 2 + 3$ b. $12 - 2 \cdot (5 - 3)$ c. $\sqrt{64 + 36} - 2^3$

Solution:

a. $15 - \underline{10 \div 2} + 3$

$= \underline{15 - 5} + 3$ Perform the division $10 \div 2$ first.

$= 10 + 3$ Perform addition and subtraction from left to right.

$= 13$ Add.

b. $12 - 2 \cdot \underline{(5 - 3)}$

$= 12 - \underline{2 \cdot (2)}$ Perform the operation inside parentheses first.

$= 12 - 4$ Perform multiplication before subtraction.

$= 8$ Subtract.

c. $\sqrt{64 + 36} - 2^3$

$= \sqrt{100} - 2^3$ The radical sign is a grouping symbol. Perform the operation within the radical first.

$= 10 - 8$ Simplify any expressions with exponents or square roots. Note that $\sqrt{100} = 10$, and $2^3 = 2 \cdot 2 \cdot 2 = 8$.

$= 2$ Subtract.

Skill Practice Simplify.

8. $18 + 6 \div 2 - 4$ 9. $(20 - 4) \div 2 + 1$ 10. $2^3 - \sqrt{16 + 9}$

Example 4   Using the Order of Operations

Simplify.

a. $300 \div (7 - 2)^2 \cdot 2^2$ b. $36 + (7^2 - 3)$ c. $\dfrac{3^2 + 6 \cdot 1}{10 - 7}$

Solution:

a. $300 \div (7 - 2)^2 \cdot 2^2$ Perform the operation within parentheses first.

$= 300 \div (5)^2 \cdot 2^2$ Simplify exponents: $5^2 = 5 \cdot 5 = 25$ and $2^2 = 2 \cdot 2 = 4$.

$= 300 \div 25 \cdot 4$ From left to right, division appears before multiplication.

$= 12 \cdot 4$ Multiply.

$= 48$

b. $36 + (7^2 - 3)$

$= 36 + (49 - 3)$ Perform the operations within parentheses first. The guidelines indicate that we simplify the expression with the exponent before we subtract: $7^2 = 49$.

$= 36 + 46$ Add.

$= 82$

c. $\dfrac{3^2 + 6 \cdot 1}{10 - 7}$

$= \dfrac{9 + 6}{3}$ Simplify the expressions above and below the fraction bar by using the order of operations.

$= \dfrac{15}{3}$ Simplify.

$= 5$ Divide.

TIP: A division bar within an expression acts as a grouping symbol. In Example 4(c), we must simplify the expressions above and below the division bar first before dividing.

Skill Practice Simplify.

11. $40 \div (3 - 1)^2 \cdot 5^2$ 12. $42 - (50 - 6^2)$ 13. $\dfrac{6^2 - 3 \cdot 4}{8 \cdot 3}$

Sometimes an expression will have parentheses within parentheses. These are called *nested parentheses*. The grouping symbols ( ), [ ], or { } are all used as parentheses. The different shapes make it easier to match up the pairs of parentheses. For example,

$$\{300 - 4[4 + (5 + 2)^2] + 8\} - 31$$

When nested parentheses are present, simplify the innermost set first. Then work your way out.

Answers

8. 17   9. 9   10. 3
11. 250   12. 28   13. 1

Example 5 Using the Order of Operations

Simplify. $\{300 - 4[4 + (5 + 2)^2] + 8\} - 31$

Solution:

$\{300 - 4[4 + (5 + 2)^2] + 8\} - 31$ Simplify within the innermost parentheses first ( ).

$= \{300 - 4[4 + (7)^2] + 8\} - 31$ Simplify the exponent.

$= \{300 - 4[4 + 49] + 8\} - 31$ Simplify within the next innermost parentheses [ ].

$= \{300 - 4[53] + 8\} - 31$ Multiply before adding.

$= \{300 - 212 + 8\} - 31$ Subtract and add in order from left to right within the parentheses { }.

$= \{88 + 8\} - 31$ Simplify within the parentheses { }.

$= 96 - 31$ Simplify.

$= 65$

Skill Practice Simplify.

14. $4^2 - 2[12 - (3 + 6)]$

4. Algebraic Expressions

The formula $A = l \cdot w$ represents the area of a rectangle in terms of its length and width. The letters A, l, and w are called variables. **Variables** are used to represent quantities that are subject to change. Quantities that do not change are called **constants**. Variables and constants are used to build algebraic expressions. The following are all examples of algebraic expressions.

$$n + 30, \quad x - y, \quad 3w, \quad \frac{a}{4}$$

The value of an algebraic expression depends on the values of the variables within the expression. In Examples 6 and 7, we practice evaluating expressions for given values of the variables.

Example 6 Evaluating an Algebraic Expression

Evaluate the expression for the given values of the variables.

$5a + b$ for $a = 6$ and $b = 10$

Solution:

$5a + b$

$= 5(\ ) + (\ )$ When we substitute a number for a variable, use parentheses in place of the variable.

$= 5(6) + (10)$ Substitute 6 for a and 10 for b, by placing the values within the parentheses.

$= 30 + 10$ Apply the order of operations. Multiplication is performed before addition.

$= 40$

Skill Practice

15. Evaluate $x + 7y$ for $x = 8$ and $y = 4$.

Answers

14. 10   15. 36

> **Example 7** **Evaluating an Algebraic Expression**
>
> Evaluate the expression for the given values of the variables.
>
> $(x - y + z)^2$ for $x = 12$, $y = 9$, and $z = 4$
>
> **Solution:**
>
> $(x - y + z)^2$
>
> $= [(\ ) - (\ ) + (\ )]^2$ Use parentheses in place of the variables.
>
> $= [(12) - (9) + (4)]^2$ Substitute 12 for x, 9 for y, and 4 for z.
>
> $= [3 + 4]^2$ Subtract and add within the grouping symbols from left to right.
>
> $= (7)^2$
>
> $= 49$ The value $7^2 = 7 \cdot 7 = 49$.
>
> **Skill Practice**
>
> 16. Evaluate $(m - n)^2 + p$ for $m = 11$, $n = 5$, and $p = 2$.

Answer

16. 38

Section 1.6 Activity

A.1. Given 3^2,
    a. Identify the base.   **b.** Identify the exponent.
    c. Write as repeated multiplication.   **d.** Evaluate.

A.2. Given $\sqrt{9}$,
    a. In your own words, explain the meaning of a square root using the specific example of $\sqrt{9}$.

    b. What relationship do you see between this problem and the result obtained from 3^2?

For Exercises A.3–A.8, simplify the exponents and square roots.

A.3. 2^3 **A.4.** 5^2 **A.5.** 4^3

A.6. $\sqrt{16}$ **A.7.** $\sqrt{64}$ **A.8.** $\sqrt{100}$

A.9. Consider the expression $8 + 6 \div 2$.
    a. When you simplify the expression, is the result 7 or 11? Explain your answer.
    b. Why is it important to perform mathematical operations in a specific order?
    c. State the order of operations.

A.10. Simplify:
    a. $10 + 3^2$
    b. $(10 + 3)^2$
    c. Are the results from parts (a) and (b) different? Why or why not?

For Exercises A.11–A.14, simplify the expressions.

A.11. $12 - 3(8 - 5)$ **A.12.** $20 - 2[10 - (6 + 1)]$

A.13. $5 \cdot 3^2 + 2(3^2 + 1)$ **A.14.** $60 - \{11 + 3[20 - (8 - 6)^2]\}$

A.15. Consider the expression $z - y - x$.
  a. Substitute $x = 2$, $y = 5$, and $z = 10$ into the expression.
  b. Simplify the expression. This process is referred to as "evaluating the expression."
  c. Do any variables remain when you evaluate an expression?

For Exercises A.16–A.18, evaluate the expressions for the given values of the variables: $x = 2$, $y = 5$, and $z = 10$.

A.16. $z - (y - x)$ **A.17.** $z + y - x^2$ **A.18.** $z - (y - x)^2$

Practice Exercises

Section 1.6

Study Skills Exercise
Memorization techniques can used to move mathematical concepts, formulas, and properties from your short-term memory to your long-term memory. Students can use these techniques in a math course to retain necessary information for problem-solving. Self-testing is one such technique. Self-testing is the process where students write review questions and key concepts on study cards. On one side of the card, pose a practice problem or concept. On the other side, write the answer with an explanation. Repeated review of these cards helps build proficiency in the subject matter.
  Write down the key concept covered in this section, and create five example problems to help you review for the next test.

Vocabulary and Key Concepts
1. a. Given the expression 5^4, the _____ is 5 and the exponent is _____.
   b. The values 10^1, 10^2, 10^3, and so on are called _____ of 10.
   c. The positive _____ of 81 is denoted by $\sqrt{81}$. The value of $\sqrt{81}$ is 9 because $9 \cdot 9 =$ _____.
   d. The _____ of _____ is a process used to simplify expressions involving more than one mathematical operation.
   e. A _____ is a letter or symbol such as x, y, or z that represents a number. Quantities that do not change, such as 5 and 11, are called _____.
2. Exponential form is shortcut notation for repeated _____.
3. Another word that implies that a base is raised to the power of 2 is _____.
4. Another word that implies that a base is raised to the power of 3 is _____.
5. When evaluating 10 raised to any power, the exponent is the same as the number of _____ that follow the number 1 in the power of 10.
6. The $\sqrt{}$ symbol that is used to denote the positive square root of a number is called a _____ sign.
7. When simplifying an expression using order of operations, ( ), [ ], { }, and a division bar are all used as parentheses, and the _____ set are simplified first when nested parentheses are present.
8. When evaluating an algebraic expression, always use _____ in place of the variable when substituting a number.

Concept 1: Exponents
9. Write an exponential expression with 9 as the base and 4 as the exponent.
10. Write an exponential expression with 3 as the base and 8 as the exponent.

For Exercises 11–14, write the repeated multiplication in exponential form. Do not simplify.

11. $3 \cdot 3 \cdot 3 \cdot 3 \cdot 3 \cdot 3$ **12.** $7 \cdot 7 \cdot 7 \cdot 7$ **13.** $4 \cdot 4 \cdot 4 \cdot 4 \cdot 2 \cdot 2 \cdot 2$ **14.** $5 \cdot 5 \cdot 5 \cdot 10 \cdot 10 \cdot 10$

For Exercises 15–18, expand the exponential expression as a repeated multiplication. Do not simplify.

15. 8^4 16. 2^6 17. 4^8 18. 6^2

For Exercises 19–30, evaluate the exponential expressions. (See Example 1.)

19. 2^3 20. 4^2 21. 3^2 22. 5^2

23. 3^3 24. 11^2 25. 5^3 26. 10^3

27. 2^5 28. 6^3 29. 3^4 30. 5^4

31. Evaluate 1^2, 1^3, 1^4, and 1^5. Explain what happens when 1 is raised to any power.

For Exercises 32–35, evaluate the powers of 10.

32. 10^2 33. 10^3 34. 10^4 35. 10^5

36. Explain how to get 10^9 *without* performing the repeated multiplication. (See Exercises 32–35.)

Concept 2: Square Roots

For Exercises 37–44, evaluate the square roots. (See Example 2.)

37. $\sqrt{4}$ 38. $\sqrt{9}$ 39. $\sqrt{36}$ 40. $\sqrt{81}$

41. $\sqrt{100}$ 42. $\sqrt{49}$ 43. $\sqrt{0}$ 44. $\sqrt{16}$

Concept 3: Order of Operations

45. Does the order of operations indicate that addition is always performed before subtraction? Explain.

46. Does the order of operations indicate that multiplication is always performed before division? Explain.

For Exercises 47–87, simplify using the order of operations. (See Examples 3 and 4.)

47. $6 + 10 \cdot 2$ 48. $4 + 3 \cdot 7$ 49. $10 - 3^2$ 50. $11 - 2^2$

51. $(10 - 3)^2$ 52. $(11 - 2)^2$ 53. $36 \div 2 \div 6$ 54. $48 \div 4 \div 2$

55. $15 - (5 + 8)$ 56. $41 - (13 + 8)$ 57. $(13 - 2) \cdot 5 - 2$ 58. $(8 + 4) \cdot 6 + 8$

59. $4 + 12 \div 3$ 60. $9 + 15 \div \sqrt{25}$ 61. $30 \div 2 \cdot \sqrt{9}$ 62. $55 \div 11 \cdot 5$

63. $7^2 - 5^2$ 64. $3^3 - 2^3$ 65. $(7 - 5)^2$ 66. $(3 - 2)^3$

67. $100 \div 5 \cdot 2$ 68. $60 \div 3 \cdot 2$ 69. $20 - 5(11 - 8)$ 70. $38 - 6(10 - 5)$

71. $\sqrt{36 + 64} + 2(9 - 1)$ 72. $\sqrt{16 + 9} + 3(8 - 3)$ 73. $\dfrac{36}{2^2 + 5}$ 74. $\dfrac{42}{3^2 - 2}$

75. $80 - 20 \div 4 \cdot 6$ **76.** $300 - 48 \div 8 \cdot 40$ **77.** $\dfrac{42 - 26}{4^2 - 8}$ **78.** $\dfrac{22 + 14}{2^2 \cdot 3}$

79. $(18 - 5) - (23 - \sqrt{100})$ **80.** $(\sqrt{36} + 11) - (31 - 16)$ **81.** $80 \div (9^2 - 7 \cdot 11)^2$

82. $108 \div (3^3 - 6 \cdot 4)^2$ **83.** $22 - 4(\sqrt{25} - 3)^2$ **84.** $17 + 3(7 - \sqrt{9})^2$

85. $96 - 3(42 \div 7 \cdot 6 - 5)$ **86.** $50 - 2(36 \div 12 \cdot 2 - 4)$ **87.** $16 + 5(20 \div 4 \cdot 8 - 3)$

For Exercises 88–95, simplify the expressions with nested parentheses. (See Example 5.)

88. $3[4 + (6 - 3)^2] - 15$ **89.** $2[5(4 - 1) + 3] \div 6$

90. $8^2 - 5[12 - (8 - 6)]$ **91.** $3^3 - 2[15 - (2 + 1)^2]$

92. $3[(10 - 4) - (5 - 1)]^2$ **93.** $10[(6 + 4) - (8 - 5)^2]$

94. $5\{21 - [3^2 - (4 - 2)]\}$ **95.** $4\{18 - [(10 - 8) + 2^3]\}$

Concept 4: Algebraic Expressions

For Exercises 96–103, evaluate the expressions for the given values of the variables. $x = 12$, $y = 4$, $z = 25$, and $w = 9$. (See Examples 6 and 7.)

96. $10y - z$ **97.** $8w - 4x$ **98.** $3x + 6y + 9w$ **99.** $9y - 4w + 3z$

100. $(z - x - y)^2$ **101.** $(y + z - w)^2$ **102.** $\sqrt{z}$ **103.** $\sqrt{w}$

Technology Connections

For Exercises 104–109, use a calculator to perform the indicated operations.

104. 156^2 **105.** 418^2 **106.** 12^5 **107.** 35^4 **108.** 43^3 **109.** 71^3

For Exercises 110–115, simplify the expressions by using the order of operations. For each step use the calculator to simplify the given operation.

110. $8126 - 54{,}978 \div 561$ **111.** $92{,}168 + 6954 \times 29$

112. $(3548 - 3291)^2$ **113.** $(7500 \div 625)^3$

114. $\dfrac{89{,}880}{384 + 2184}$ *Hint:* This expression has implied grouping symbols. $\dfrac{89{,}880}{(384 + 2184)}$

115. $\dfrac{54{,}137}{3393 - 2134}$ *Hint:* This expression has implied grouping symbols. $\dfrac{54{,}137}{(3393 - 2134)}$

Section 1.7 Mixed Applications and Computing Mean

Concepts

1. Applications Involving Multiple Operations
2. Computing a Mean (Average)

1. Applications Involving Multiple Operations

Sometimes more than one operation is needed to solve an application problem.

Example 1 Solving a Consumer Application

Jorge bought a car for $18,340. He paid $2500 down and then paid the rest in equal monthly payments over a 4-year period. Find the amount of Jorge's monthly payment (not including interest).

Tom Grill/Fuse/Getty Images

Solution:

Familiarize and draw a picture.

Given: total price: $18,340
        down payment: $2500

        payment plan: 4 years
        (48 months)

Find: monthly payment

- $18,340 — Original cost of car
- −2,500 — Minus down payment
- $15,840 — Amount to be paid off
- Divide payments over 4 years (48 months)

Operations:

1. The amount of the loan to be paid off is equal to the original cost of the car minus the down payment. We use subtraction:

$$\begin{array}{r} \$18{,}340 \\ -2{,}500 \\ \hline \$15{,}840 \end{array}$$

2. This money is distributed in equal payments over a 4-year period. Because there are 12 months in 1 year, there are 4 · 12 = 48 months in a 4-year period. To distribute $15,840 among 48 equal payments, we divide.

$$\begin{array}{r} 330 \\ 48\overline{)15{,}840} \\ -144 \\ \hline 144 \\ -144 \\ \hline 00 \end{array}$$

Jorge's monthly payments will be $330.

TIP: The solution to Example 1 can be checked by multiplication. Forty-eight payments of $330 each amount to 48($330) = $15,840. This added to the down payment totals $18,340 as desired.

Skill Practice

1. Danielle buys a new entertainment center with a new television for $1680. She pays $300 down, and the rest is paid off in equal monthly payments for 1 year. Find Danielle's monthly payment.

Answer

1. $115 per month

Section 1.7 Mixed Applications and Computing Mean   69

Example 2 **Solving a Travel Application**

Linda must drive from Clayton to Oakley. She can travel directly from Clayton to Oakley on a mountain road, but will only average 40 mph. On the route through Pearson, she travels on highways and can average 60 mph. Which route will take less time?

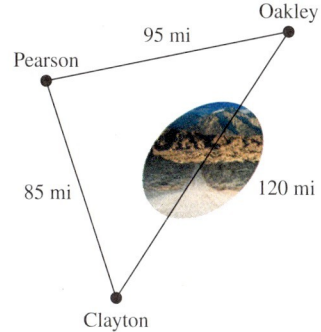

Photo 24/Brand X Pictures/Getty Images

Solution:

Read and familiarize: A map is presented in the problem.

Given: The distance for each route and the speed traveled along each route

Find: Find the time required for each route. Then compare the times to determine which will take less time.

Operations:

1. First note that the total distance of the route through Pearson is found by using addition.

   $$85 \text{ mi} + 95 \text{ mi} = 180 \text{ mi}$$

2. The speed of the vehicle gives us the distance traveled per hour. Therefore, the time of travel equals the total distance divided by the speed.

   From Clayton to Oakley through the mountains, we divide 120 mi by 40-mph increments to determine the number of hours.

   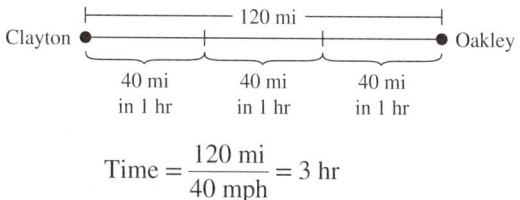

   $$\text{Time} = \frac{120 \text{ mi}}{40 \text{ mph}} = 3 \text{ hr}$$

   From Clayton to Oakley through Pearson, we divide 180 mi by 60-mph increments to determine the number of hours.

   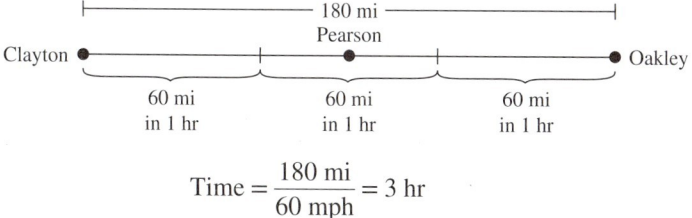

   $$\text{Time} = \frac{180 \text{ mi}}{60 \text{ mph}} = 3 \text{ hr}$$

Therefore, each route takes the same amount of time, 3 hr.

Skill Practice

2. Taylor makes $18 per hour for the first 40 hr worked each week. His overtime rate is $27 per hour for hours exceeding the normal 40-hr workweek. If his total salary for one week is $963, determine the number of hours of overtime worked.

Answer

2. 9 hr overtime

Example 3 — Solving a Construction Application

A rancher must fence the corral shown in Figure 1-9. However, no fencing is required on the side adjacent to the barn. If fencing costs $4 per foot, what is the total cost?

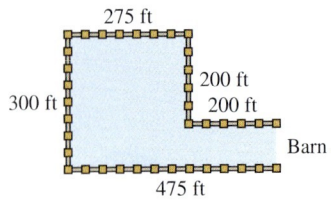

Figure 1-9

Solution:
Read and familiarize: A figure is provided.

Strategy

With some application problems, it helps to work backward from your final goal. In this case, our final goal is to find the total cost. However, to find the total cost, we must first find the total distance to be fenced. To find the total distance, we add the lengths of the sides that are being fenced.

$$\begin{array}{r} \overset{1\ 1}{275} \text{ ft} \\ 200 \text{ ft} \\ 200 \text{ ft} \\ 475 \text{ ft} \\ +\ 300 \text{ ft} \\ \hline 1450 \text{ ft} \end{array}$$

Therefore,

$$\begin{pmatrix} \text{Total cost} \\ \text{of fencing} \end{pmatrix} = \begin{pmatrix} \text{total} \\ \text{distance} \\ \text{in feet} \end{pmatrix} \begin{pmatrix} \text{cost} \\ \text{per foot} \end{pmatrix}$$

$$= (1450 \text{ ft})(\$4 \text{ per ft})$$

$$= \$5800$$

The total cost of fencing is $5800.

FOR REVIEW

Recall that addition may be performed in any order.

$$\begin{array}{r} 200 \text{ ft} \\ 200 \text{ ft} \\ 300 \text{ ft} \\ 275 \text{ ft} \\ +\ 475 \text{ ft} \\ \hline 1450 \text{ ft} \end{array}$$

Skill Practice

3. Alain wants to put molding around the base of the room shown in the figure. No molding is needed where the door, closet, and bathroom are located. Find the total cost if molding is $2 per foot.

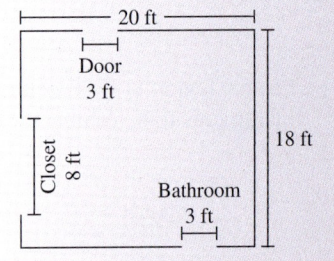

2. Computing a Mean (Average)

The order of operations must be used when we compute an average. The technical term for the average of a list of numbers is the **mean** of the numbers. To find the mean of a set of numbers, first compute the sum of the values. Then divide the sum by the number of values. This is represented by the formula

$$\text{Mean} = \frac{\text{sum of the values}}{\text{number of values}}$$

Answer

3. $124

Example 4 Computing a Mean (Average)

Ashley took 6 tests in Chemistry. Find her mean (average) score.

$$89, 91, 72, 86, 94, 96$$

Solution:

$$\text{Average score} = \frac{89 + 91 + 72 + 86 + 94 + 96}{6}$$

$$= \frac{528}{6} \quad \text{Add the values in the list first.}$$

$$= 88 \quad \text{Divide.} \quad \begin{array}{r} 88 \\ 6\overline{)528} \\ \underline{-48} \\ 48 \\ \underline{-48} \\ 0 \end{array}$$

Ashley's mean (average) score is 88.

TIP: The division bar in $\dfrac{89 + 91 + 72 + 86 + 94 + 96}{6}$ is also a grouping symbol and implies parentheses:

$$\frac{(89 + 91 + 72 + 86 + 94 + 96)}{6}$$

Skill Practice

4. The ages (in years) of 5 students in an algebra class are given here. Find the mean age.

$$22, 18, 22, 32, 46$$

Answer
4. 28 years

Practice Exercises

Section 1.7

Study Skills Exercise

Studying for a mathematics test is very different than studying for other subjects. There are a number of test-taking strategies that you can practice to be more prepared mentally and physically.

- Space out your study time and practice time so that you give your subconscious mind time to process the information. This is better than cramming information in one sitting.
- Prepare a one-page summary sheet with the most important information that you need for the test. Also include a positive statement to yourself on your summary sheet to build a growth mindset.
- On the day of the test, look at this sheet several times to refresh your memory instead of trying to memorize new information.
- When you sit down to take your test, consider writing the important formulas on the test paper or scrap paper.

Prerequisite Review

For Exercises R.1–R.8, translate the English phrase into a mathematical statement and simplify.

R.1. Fifteen subtracted from 20.

R.2. Seven subtracted from twenty.

R.3. 71 increased by 14.

R.4. 16 more than 42.

R.5. The product of 10 and 13.

R.6. Twice 14.

R.7. The quotient of 24 and 6.

R.8. Divide 12 into 60.

Vocabulary and Key Concepts

1. The _____ or average of a set of numbers is found by taking the sum of the values and then dividing by the number of values.

2. (True/False) When calculating the mean, the set of numbers can be added in any order.

Concept 1: Applications Involving Multiple Operations

3. Jackson purchased a car for $16,540. He paid $2500 down and paid the rest in equal monthly payments over a 36-month period. How much were his monthly payments?

4. Lucio purchased a refrigerator for $1170. He paid $150 at the time of purchase and then paid off the rest in equal monthly payments over 1 year. How much was his monthly payment?
   (See Example 1.)

5. Monika must drive from Watertown to Utica. She can travel directly from Watertown to Utica on a small county road, but will only average 40 mph. On the route through Syracuse, she travels on highways and can average 60 mph. Which route will take less time?

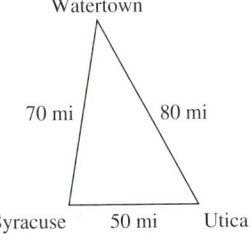

6. Rex has a choice of two routes to drive from Oklahoma City to Fort Smith. On the interstate, the distance is 220 mi and he can drive 55 mph. If he takes the back roads, he can only travel 40 mph, but the distance is 200 mi. Which route will take less time?
   (See Example 2.)

7. If you wanted to line the outside of a garden with a decorative border, would you need to know the area of the garden or the perimeter of the garden?

8. If you wanted to know how much sod to lay down within a rectangular backyard, would you need to know the area of the yard or the perimeter of the yard?

9. Arisu wants to buy molding for a room that is 12 ft by 11 ft. No molding is needed for the doorway, which measures 3 ft. See the figure. If molding costs $2 per foot, how much money will it cost?

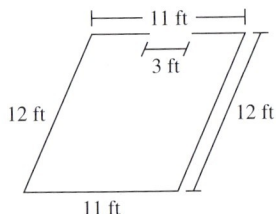

10. A homeowner wants to fence her rectangular backyard. The yard is 75 ft by 90 ft. If fencing costs $5 per foot, how much will it cost to fence the yard?
    (See Example 3.)

11. What is the cost to carpet the room whose dimensions are shown in the figure? Assume that carpeting costs $34 per square yard and that there is no waste.

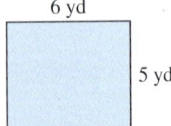

12. What is the cost to tile the room whose dimensions are shown in the figure? Assume that tile costs $3 per square foot.

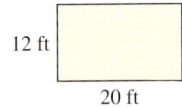

13. The balance in Gina's checking account is $278. If she writes checks for $82, $59, and $101, how much will be left over?

14. The balance in José's checking account is $3455. If he write checks for $587, $36, and $156, how much will be left over?

15. A community college bought 72 new computers and 6 new printers for a computer lab. If computers were purchased for $2118 each and the printers for $256 each, what was the total bill (not including tax)?

16. Tickets to a zoo cost $27 for children aged 3–11 and $37 for adults. How much money is required to buy tickets for a class of 33 children and 6 adult chaperones?

17. A discount music store buys used CDs from its customers for $3. Furthermore, a customer can buy any used CD in the store for $8. Latayne sells 16 CDs.

    a. How much money does she receive by selling the 16 CDs?
    b. How many CDs can she then purchase with the money?

Fuse/Getty Images

18. Shevona earns $12 per hour and works a 40-hr workweek. At the end of the week, she cashes her paycheck and then buys two tickets to a concert.

    a. How much is her paycheck?
    b. If the concert tickets cost $89 each, how much money does she have left over from her paycheck after buying the tickets?

19. During his 13-year career with the Chicago Bulls, Michael Jordan scored 12,192 field goals (worth 2 points each). He scored 581 three-point shots and 7327 free-throws (worth 1 point each). How many total points did he score during his career with the Bulls?

20. A matte is to be cut and placed over five small square pictures before framing. Each picture is 5 in. wide, and the matte frame is 37 in. wide, as shown in the figure. If the pictures are to be equally spaced (including the space on the left and right edges), how wide is the matte between them?

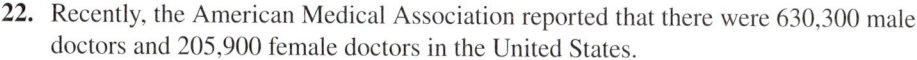

21. Winston the cat was prescribed a suspension of methimazole for hyperthyroidism. This suspension comes in a 60-milliliter bottle with instructions to give 1 milliliter twice a day. The label also shows there is one refill, but it must be called in 2 days ahead. Winston had his first two doses on September 1.

    a. For how many days will one bottle last?
    b. On what day, at the latest, should his owner order a refill to avoid running out of medicine?

Courtesy of Julie Miller

22. Recently, the American Medical Association reported that there were 630,300 male doctors and 205,900 female doctors in the United States.

    a. What is the difference between the number of male doctors and the number of female doctors?
    b. What is the total number of doctors?

23. On a map, 1 in. represents 60 mi.

    a. If Las Vegas and Salt Lake City are approximately 6 in. apart on the map, what is the actual distance between the cities?
    b. If Madison, Wisconsin, and Dallas, Texas, are approximately 840 mi apart, how many inches would this represent on the map?

24. On a map, each inch represents 40 mi.

    a. If Wichita, Kansas, and Des Moines, Iowa, are approximately 8 in. apart on the map, what is the actual distance between the cities?
    b. If Seattle, Washington, and Sacramento, California, are approximately 600 mi apart, how many inches would this represent on the map?

25. A textbook company ships books in boxes containing a maximum of 12 books. If a bookstore orders 1250 books, how many boxes can be filled completely? How many books will be left over?

26. A farmer sells eggs in containers holding a dozen eggs. If he has 4257 eggs, how many containers will be filled completely? How many eggs will be left over?

27. Marc pays for an $84 dinner with $20 bills.

    a. How many bills must he use?

    b. How much change will he receive?

28. Byron buys three CDs for a total of $54 and pays with $10 bills.

    a. How many bills must he use?

    b. How much change will he receive?

David Buffington/Getty Images

29. Ling has three jobs. He works for a lawn maintenance service 4 days a week. He also tutors math and works as a waiter on weekends. His hourly wage and the number of hours for each job are given for a 1-week period. How much money did Ling earn for the week?

	Hourly Wage	Number of Hours
Tutor	$30/hr	4
Waiter	10/hr	16
Lawn maintenance	8/hr	30

30. An electrician, a plumber, a mason, and a carpenter work at a construction site. The hourly wage and the number of hours each person worked are summarized in the table. What was the total amount paid for all four workers?

	Hourly Wage	Number of Hours
Electrician	$36/hr	18
Plumber	28/hr	15
Mason	26/hr	24
Carpenter	22/hr	48

Concept 2: Computing a Mean (Average)

For Exercises 31–33, find the mean (average) of each set of numbers. (See Example 4.)

31. 19, 21, 18, 21, 16

32. 105, 114, 123, 101, 100, 111

33. 1480, 1102, 1032, 1002

34. Neelah took six quizzes and received the following scores: 19, 20, 18, 19, 18, 14. Find her quiz average.

35. Shawn's scores on his last four tests were 83, 95, 87, and 91. What is his average for these tests?

36. Automobile companies have been striving to improve the fuel efficiency of their cars. In a recent year, four of the most efficient cars built by one company had mileage ratings of 42 mpg, 41 mpg, 31 mpg, and 30 mpg. What was the average mileage rating for these vehicles?

37. One of the most popular automobile companies has been very competitive at building fuel-efficient autos. In a recent year, five of the most efficient cars had mileage ratings of 49 mpg, 30 mpg, 34 mpg, 31 mpg, and 26 mpg. What was the average mileage rating for these vehicles?

38. The monthly rainfall for Seattle, Washington, is given in the table. All values are in millimeters (mm).

	Jan.	Feb.	Mar.	Apr.	May	Jun.	Jul.	Aug.	Sep.	Oct.	Nov.	Dec.
Rainfall	122	94	80	52	47	40	15	21	44	90	118	123

Find the average monthly rainfall for the months of November, December, and January.

39. The monthly snowfall for Alpena, Michigan, is given in the table. All values are in inches.

	Jan.	Feb.	Mar.	Apr.	May	Jun.	Jul.	Aug.	Sep.	Oct.	Nov.	Dec.
Snowfall	22	16	13	5	1	0	0	0	0	1	9	20

Find the average monthly snowfall for the months of November, December, January, February, and March.

Elena Kovalevich/Moment/Getty Images

Chapter 1 Review Exercises

Section 1.1

For Exercises 1 and 2, determine the place value for each underlined digit.

1. 1̲0,024
2. 8̲21,811

For Exercises 3 and 4, convert the numbers to standard form.

3. 9 ten-thousands + 2 thousands + 4 tens + 6 ones
4. 5 hundred-thousands + 3 thousands + 1 hundred + 6 tens

For Exercises 5 and 6, convert the numbers to expanded form.

5. 3,400,820
6. 30,554

For Exercises 7 and 8, write the numbers in words.

7. 245
8. 30,861

For Exercises 9 and 10, write the numbers in standard form.

9. Three thousand, six-hundred two
10. Eight hundred thousand, thirty-nine

For Exercises 11 and 12, place the numbers on the number line.

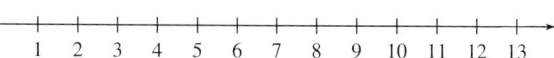

11. 2
12. 7

For Exercises 13 and 14, determine if the inequality is true or false.

13. 3 < 10
14. 10 > 12

Section 1.2

For Exercises 15 and 16, identify the addends and the sum.

15. 105 + 119 = 224
16. 53 + 21 = 74

For Exercises 17–20, add.

17. 18 + 24 + 29
18. 27 + 9 + 18
19. 8403 + 9007
20. 68,421 + 2,221

21. For each of the mathematical statements, identify the property used. Choose from the commutative property or the associative property.

    a. 6 + (8 + 2) = (8 + 2) + 6
    b. 6 + (8 + 2) = (6 + 8) + 2
    c. 6 + (8 + 2) = 6 + (2 + 8)

For Exercises 22 and 23, identify the minuend, subtrahend, and difference.

22. 14 − 8 = 6
23. 102 − 78 = 24

For Exercises 24 and 25, subtract and check your answer by addition.

24. 37 − 11   Check: ☐ + 11 = 37
25. 61 − 41   Check: ☐ + 41 = 61

For Exercises 26–29, subtract.

26. 2005 − 1884
27. 1389 − 299
28. 86,000 − 54,981
29. 67,000 − 32,812

For Exercises 30–37, translate the English phrase to a mathematical statement and simplify.

30. The sum of 403 and 79
31. 92 added to 44
32. 38 minus 31
33. 111 decreased by 15
34. 7 more than 36
35. 23 increased by 6
36. Subtract 42 from 251.
37. The difference of 90 and 52

38. The table gives the number of cars sold by three dealerships during one week.

	Honda	Ford	Toyota
Bob's Discount Auto	23	21	34
AA Auto	31	25	40
Car World	33	20	22

a. What is the total number of cars sold by AA Auto?

b. What is the total number of Fords sold by these three dealerships?

Use the bar graph to answer exercises 39–41. The graph represents the distribution of the U.S. population by age group for a recent year. (*Source:* U.S. Census Bureau)

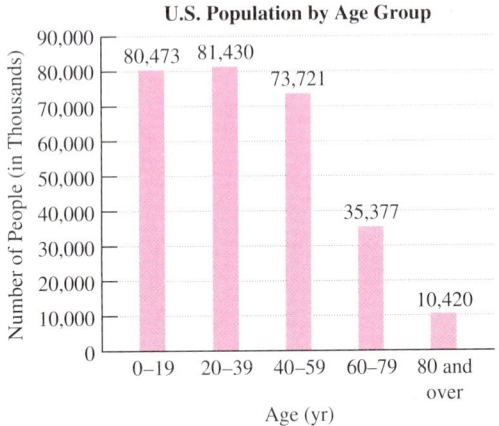

U.S. Population by Age Group

39. Determine the number of seniors (aged 60 and over).

40. Compute the difference in the number of people in the 20–39 age group and the number in the 40–59 age group.

41. How many more people are in the 0–19 age group than in the 60–79 age group?

42. For a recent year, 95,192,000 tons of watermelon and 23,299,000 tons of cantaloupe were grown. Determine the difference between the amount of watermelon grown and the amount of cantaloupe grown.

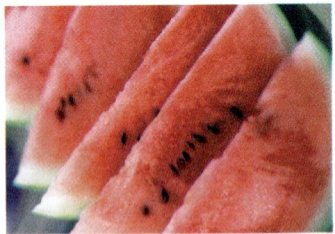

Corbis

43. For a recent year, Phil Mickelson earned $25,800,000 from both the golf tour and endorsements. Vijay Singh earned $18,600,000. Find the difference in their earnings.

44. Find the perimeter of the figure.

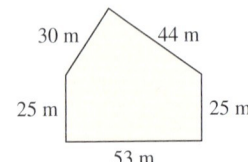

Section 1.3

For Exercises 45 and 46, round each number to the given place value.

45. 5,234,446; millions

46. 9,332,945; ten-thousands

For Exercises 47 and 48, estimate the sum or difference by rounding to the indicated place value.

47. 894,004 − 123,883; hundred-thousands

48. 330 + 489 + 123 + 571; hundreds

49. For a recent year, the population of Russia was 140,041,247, and the population of Japan was 127,078,679. Estimate the difference in their populations by rounding to the nearest million.

50. The state of Missouri has two dams: Fort Peck with a volume of 96,050 cubic meters (m^3) and Oahe with a volume of 66,517 m^3. Round the numbers to the nearest thousand to estimate the total volume of these two dams.

Section 1.4

51. Identify the factors and the product. $33 \cdot 40 = 1320$

52. Indicate whether the statement is equal to the product of 8 and 13.

   a. 8(13)　　b. (8) · 13　　c. (8) + (13)

For Exercises 53–57, for each property listed, choose an expression from the right column that demonstrates the property.

53. Associative property of multiplication　　a. 3(4) = 4(3)

54. Distributive property of multiplication over addition　　b. 19 · 1 = 19

55. Multiplication property of 0　　c. (1 · 8) · 3 = 1 · (8 · 3)

56. Commutative property of multiplication　　d. 0 · 29 = 0

57. Multiplication property of 1　　e. 4(3 + 1) = 4 · 3 + 4 · 1

For Exercises 58–60, multiply.

58. 142
× 43

59. (1024)(51)

60. 6000
× 500

61. A discussion group needs to purchase books that are accompanied by a workbook. The price of the book is $26, and the workbook costs an additional $13. If there are 11 members in the group, how much will it cost the group to purchase both the text and workbook for each student?

62. Orcas, or killer whales, eat 551 pounds (lb) of food a day. If an animal rescue park has two adult killer whales, how much food will they eat in 1 week?

ventdusud/Shutterstock

Section 1.5

For Exercises 63 and 64, perform the division. Then identify the divisor, dividend, and quotient.

63. 42 ÷ 6

64. 4)$\overline{52}$

For Exercises 65–68, use the properties of division to simplify the expression, if possible.

65. 3 ÷ 1

66. 3 ÷ 3

67. 3 ÷ 0

68. 0 ÷ 3

69. Explain how you check a division problem if there is no remainder.

70. Explain how you check a division problem if there is a remainder.

For Exercises 71–73, divide and check the answer.

71. 348 ÷ 6

72. 11)$\overline{458}$

73. $\dfrac{1043}{20}$

For Exercises 74 and 75, write the English phrase as a mathematical expression and simplify.

74. The quotient of 72 and 4

75. 108 divided by 9

76. Quinita has 105 photographs that she wants to divide equally among herself and three siblings. How many photos will each person receive? How many photos will be left over?

77. Ashley has $60 to spend on souvenirs at a surf shop. The prices of several souvenirs are given in the graph.

a. How many souvenirs can Ashley buy if she chooses all T-shirts?

b. How many souvenirs can Ashley buy if she chooses all hats?

Section 1.6

For Exercises 78 and 79, write the repeated multiplication in exponential form. Do not simplify.

78. 8 · 8 · 8 · 8 · 8

79. 2 · 2 · 2 · 2 · 5 · 5 · 5

For Exercises 80–83, evaluate the exponential expressions.

80. 5^3

81. 4^4

82. 1^7

83. 10^6

For Exercises 84 and 85, evaluate the square roots.

84. $\sqrt{64}$

85. $\sqrt{144}$

For Exercises 86–92, evaluate the expression using the order of operations.

86. 14 ÷ 7 · 4 − 1

87. $10^2 − 5^2$

88. 90 − 4 + 6 ÷ 3 · 2

89. 2 + 3 · 12 ÷ 2 − $\sqrt{25}$

90. $6^2 − [4^2 + (9 − 7)^3]$

91. 26 − 2(10 − 1) + (3 + 4 · 11)

92. $\dfrac{5 \cdot 3^2}{7 + 8}$

For Exercises 93–96, evaluate the expressions for $a = 20$, $b = 10$, and $c = 6$.

93. $a + b + 2c$
94. $5a - b^2$
95. $\sqrt{b + c}$
96. $(a - b)^2$

Section 1.7

97. Doris drives her son to extracurricular activities each week. She drives 5 mi round-trip to baseball practice 3 times a week and 6 mi round-trip to piano lessons once a week.

    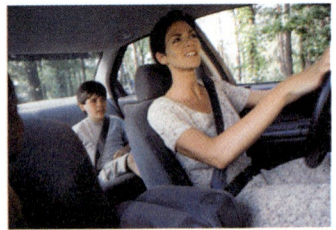
    Comstock/Alamy Stock Photo

    a. How many miles does she drive in 1 week to get her child to his activities?

    b. Approximately how many miles does she travel during a school year consisting of 10 months (there are approximately 4 weeks per month)?

98. At one point in his baseball career, Alex Rodriquez signed a contract for $252,000,000 for a 9-year period. Suppose federal taxes amount to $75,600,000 for the contract. After taxes, how much did Alex receive per year?

99. Aletha wants to buy plants for a rectangular garden in her backyard that measures 12 ft by 8 ft. She wants to divide the garden into 2-square-foot (2 ft²) areas, one for each plant.

    Ariel Skelley/Blend Images

    a. How many plants should Aletha buy?
    b. If the plants cost $3 each, how much will it cost Aletha for the plants?
    c. If she puts a fence around the perimeter of the garden that costs $2 per foot, how much will it cost for the fence?
    d. What will be Aletha's total cost for this garden?

100. Find the mean (average) for the set of numbers 7, 6, 12, 5, 7, 6, 13.

101. Carolyn's electric bills for the past 5 months have been $80, $78, $101, $92, and $94. Find her average monthly charge.

102. The table shows the number of homes sold by a realty company in the last 6 months. Determine the average number of houses sold per month for these 6 months.

Month	Number of Houses
May	6
June	9
July	11
August	13
September	5
October	4

Integers and Algebraic Expressions

2

CHAPTER OUTLINE

- **2.1** Integers, Absolute Value, and Opposite   80
- **2.2** Addition of Integers   87
- **2.3** Subtraction of Integers   96
- **2.4** Multiplication and Division of Integers   103
    - Problem Recognition Exercises: Operations on Integers   112
- **2.5** Order of Operations and Algebraic Expressions   113
    - Chapter 2 Review Exercises   121

The Need to Be Negative

Suppose that a running back in a football game runs forward 3 yd and then gets pushed back 5 yd. At the end of this play, the running back's position is behind the point where he started. In football terms, we say that he lost 2 yd. In mathematical terms, we might say that the net result is −2 yd. The number −2 is called an integer. The integers make up the set of numbers consisting of the whole numbers, 0, 1, 2, 3, and so on, along with their negative counterparts, −1, −2, −3, and so on. Thus, the integers give us our first look at negative numbers.

In this chapter, we will learn that negative numbers are far from unfavorable. Negative integers are important in many contexts, including discussions about a negative balance on a bank account, temperatures below 0°, elevations below sea level, and scoring under par in golf. Furthermore, we will begin our study of algebra by learning how to add, subtract, multiply, and divide integers.

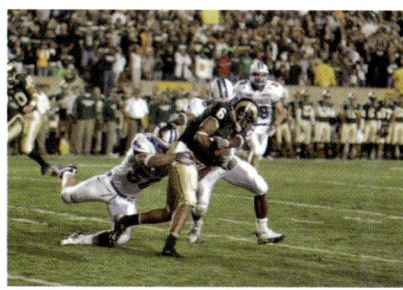

Source: U.S. Air Force photo by John Van Winkle

Section 2.1 Integers, Absolute Value, and Opposite

Concepts
1. Integers
2. Absolute Value
3. Opposite

1. Integers

The numbers 1, 2, 3, . . . are **positive numbers** because they lie to the right of zero on the number line (Figure 2-1).

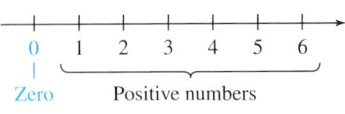

Figure 2-1

In some applications of mathematics we need to use *negative* numbers. For example:

- On a winter day in Buffalo, the low temperature was 3 degrees below zero: $-3°$
- A golfer's score in the U.S. Open was 7 below par: -7
- Carmen is $128 overdrawn on her checking account. Her balance is: $-\$128$

The values $-3°$, -7, and $-\$128$ are negative numbers. **Negative numbers** lie to the *left* of zero on a number line (Figure 2-2). The number 0 is neither negative nor positive.

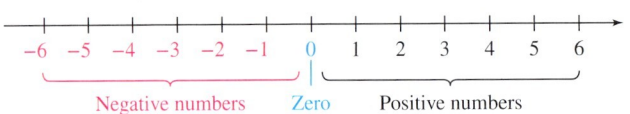

Figure 2-2

The numbers . . . $-3, -2, -1, 0, 1, 2, 3, \ldots$ and so on are called **integers**.

Example 1 — **Writing Integers**

Write an integer that denotes each number.

a. Liquid nitrogen freezes at 346°F below zero.

b. The shoreline of the Dead Sea on the border of Israel and Jordan is the lowest land area on Earth. It is 1300 ft below sea level.

c. Jenna's 10-year-old daughter weighs 14 lb more than the average child her age.

Glow Images/Getty Images

Solution:

a. $-346°F$

b. -1300 ft

c. 14 lb

Skill Practice Write an integer that denotes each number.

1. The average temperature at the South Pole in July is 65°C below zero.
2. Sylvia's checking account is overdrawn by $156.
3. The price of a new car is $2000 more than it was one year ago.

Answers
1. $-65°C$ 2. $-\$156$
3. $2000

Section 2.1 Integers, Absolute Value, and Opposite 81

Example 2 Locating Integers on the Number Line

Locate each number on the number line.

a. −4   b. −7   c. 0   d. 2

Solution:

On the number line, negative numbers lie to the left of 0, and positive numbers lie to the right of 0.

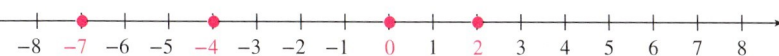

Skill Practice Locate each number on the number line.

4. −5   5. −1   6. 4

As with whole numbers, the order between two integers can be determined using the number line.

- A number a is less than b (denoted $a < b$) if a lies to the left of b on the number line (Figure 2-3).
- A number a is greater than b (denoted $a > b$) if a lies to the right of b on the number line (Figure 2-4).

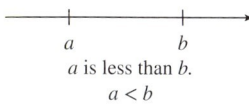

a is less than b.
$a < b$

Figure 2-3

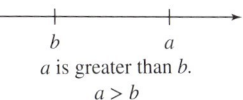

a is greater than b.
$a > b$

Figure 2-4

Example 3 Determining Order Between Two Integers

Use the number line from Example 2 to fill in the blank with $<$ or $>$ to make a true statement.

a. −7 ☐ −4   b. 0 ☐ −4   c. 2 ☐ −7

Solution:

a. −7 $<$ −4   −7 lies to the *left* of −4 on the number line.
                Therefore, −7 < −4.

b. 0 $>$ −4   0 lies to the *right* of −4 on the number line.
               Therefore, 0 > −4.

c. 2 $>$ −7   2 lies to the *right* of −7 on the number line.
               Therefore, 2 > −7.

Skill Practice Fill in the blank with $<$ or $>$.

7. −3 ☐ −8   8. −3 ☐ 8   9. 0 ☐ −11

2. Absolute Value

On the number line, pairs of numbers like 4 and −4 are the same distance from zero (Figure 2-5). The distance between a number and zero on the number line is called its **absolute value**.

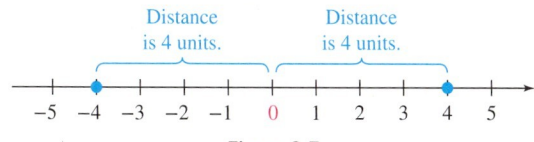

Figure 2-5

Answers

4–6
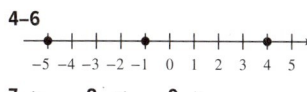

7. $>$ 8. $<$ 9. $>$

> **Absolute Value**
>
> The absolute value of a number a is denoted $|a|$. The value of $|a|$ is the distance between a and 0 on the number line.

From the number line, we see that $|-4| = 4$ and $|4| = 4$.

Example 4 Finding Absolute Value

Determine the absolute value.

a. $|-5|$ **b.** $|3|$ **c.** $|0|$

Solution:

a. $|-5| = 5$ The number -5 is 5 units from 0 on the number line.

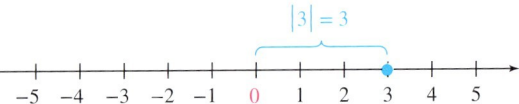

b. $|3| = 3$ The number 3 is 3 units from 0 on the number line.

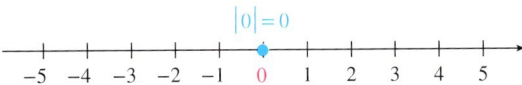

c. $|0| = 0$ The number 0 is 0 units from 0 on the number line.

TIP: The absolute value of a nonzero number is always positive. The absolute value of zero is 0.

Skill Practice Determine the absolute value.

10. $|-8|$ **11.** $|1|$ **12.** $|-16|$

3. Opposite

Two numbers that are the same distance from zero on the number line, but on opposite sides of zero, are called **opposites**. For example, the numbers -2 and 2 are opposites (see Figure 2-6).

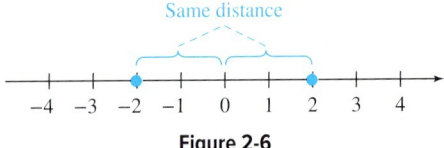

Figure 2-6

The opposite of a number, a, is denoted $-(a)$.

Original number a	Opposite $-(a)$	Simplified Form
5	$-(5)$	-5
-7	$-(-7)$	7

The opposite of a positive number is a negative number.
The opposite of a negative number is a positive number.

The opposite of a negative number is a positive number. That is, for a positive number, a, the value of $-a$ is negative and $-(-a) = a$.

Answers
10. 8   **11.** 1   **12.** 16

Example 5 Finding the Opposite of an Integer

Find the opposite.

a. 4   b. −99

Solution:

a. If a number is positive, its opposite is negative. The opposite of 4 is −4.

b. If a number is negative, its opposite is positive. The opposite of −99 is 99.

TIP: To find the opposite of a nonzero number, change the sign.

Skill Practice Find the opposite.

13. −108   14. 54

Example 6 Simplifying Expressions

Simplify.

a. −(−9)   b. −|−12| c. −|7|

Solution:

a. −(−9) = 9   This represents the opposite of −9, which is 9.

b. −|−12| = −12   This represents the opposite of |−12|. Since |−12| is equal to 12, the opposite is −12.

c. −|7| = −7   This represents the opposite of |7|. Since |7| is equal to 7, the opposite is −7.

Avoiding Mistakes
In Example 6(b) two operations are performed. First take the absolute value of −12. Then determine the opposite of the result.

Skill Practice Simplify.

15. −(−34)   16. −|−20| 17. −|4|

Answers
13. 108   14. −54
15. 34   16. −20   17. −4

Section 2.1 Activity

A.1. Write two applications in which negative numbers are used. An example has been provided for you.
  a. The yards lost from the line of scrimmage in a football game.
  b. _____
  c. _____

A.2. Consider the integers −5, 0, and 5.
  a. Locate the integers on the number line.

  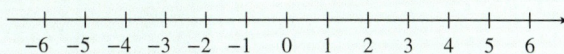

  For parts (b)–(d) fill in the blank with < or >.
  b. 5 ____ −5   c. −5 ____ 0   d. 0 ____ 5
  e. What is the distance of −5 from 0 on the number line?
  f. What is the distance of 5 from 0 on the number line?

g. The distance between a number and zero on the number line is called its **absolute value**. Do −5 and 5 have the same absolute value? Why or why not?

h. Two numbers that are the same distance from zero on the number line but on opposite sides of zero are called **opposites**. Are 5 and −5 opposites? Why or why not?

A.3. Simplify the expressions. Explain the difference between the notations used in each expression.
   a. −(−7)
   b. −|−7|

For Exercises A.4–A.10, simplify the expressions.

A.4. |−2| A.6. −|15| A.8. −(30)   A.10. −|0|
A.5. |6| A.7. −|−15| A.9. −(−8)

Section 2.1 Practice Exercises

Study Skills Exercise
When working with signed numbers, keep a simple example in your mind, such as temperature. We understand that 10 degrees below zero is colder than 2 degrees below zero, so the inequality −10 < −2 makes sense. Write down another example involving signed numbers that you can easily remember.

Prerequisite Review
For Exercises R.1–R.4 fill in the blank with < or > to make a true statement.

R.1. 12 ☐ 18   R.2. 18 ☐ 12
R.3. 198 ☐ 189   R.4. 189 ☐ 198

Vocabulary and Key Concepts
1. a. On a number line, _____ numbers lie to the right of zero and _____ numbers lie to the left of zero.

   b. The numbers . . . −3, −2, −1, 0, 1, 2, 3, . . . are called _____.

   c. The distance between a number and zero on the number line is called its _____ value.

   d. Two numbers that are the same distance from zero on the number line, but on opposite sides of zero are called _____.

Concept 1: Integers
For Exercises 2–12, write an integer that represents each numerical value. **(See Example 1.)**

2. A submarine dove to a depth of 340 ft below sea level.
3. Death Valley, California, is 86 m below sea level.

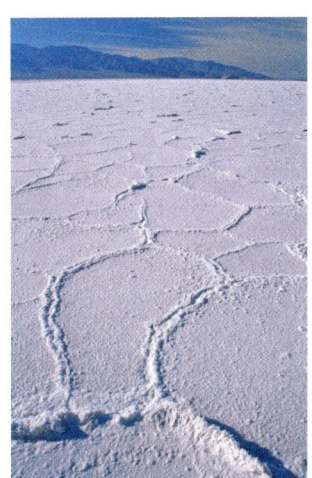
Geofile/Doug Sherman

4. In a card game, Jack lost $45.

5. Playing *Wheel of Fortune*, Sally won $3800.

6. Jim's golf score is 5 over par.

7. Rena lost $500 in the stock market in 1 month.

8. LaTonya earned $23 in interest on her savings account.

9. Patrick lost 14 lb on a diet.

10. A plane descended 2000 ft.

11. The number of Internet users rose by about 1,400,000.

12. A small business experienced a loss of $20,000 last year.

For Exercises 13 and 14, graph the numbers on the number line. (See Example 2.)

13. $-6, 0, -1, 2$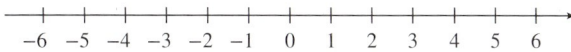

14. $-2, 4, -5, 1$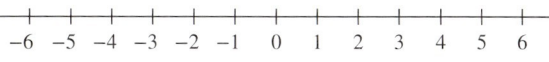

15. Which number is closer to -4 on the number line? -2 or -7

16. Which number is closer to 2 on the number line? -5 or 8

For Exercises 17–24, fill in the blank with < or > to make a true statement. (See Example 3.)

17. 0 ☐ -3
18. -1 ☐ 0
19. -8 ☐ -9
20. -5 ☐ -2
21. 8 ☐ 9
22. 5 ☐ 2
23. -226 ☐ 198
24. 408 ☐ -416

Concept 2: Absolute Value

For Exercises 25–32, determine the absolute value. (See Example 4.)

25. $|-2|$
26. $|-9|$
27. $|2|$
28. $|9|$
29. $|-427|$
30. $|-615|$
31. $|100,000|$
32. $|64,000|$

For Exercises 33–40, determine which value is greater.

33. a. -12 or -8?
    b. $|-12|$ or $|-8|$?
34. a. -14 or -20?
    b. $|-14|$ or $|-20|$?
35. a. 5 or 7?
    b. $|5|$ or $|7|$?
36. a. 3 or 4?
    b. $|3|$ or $|4|$?

37. -5 or $|-5|$?
38. -9 or $|-9|$?
39. 10 or $|10|$?
40. 256 or $|256|$?

Concept 3: Opposite

For Exercises 41–48, find the opposite. (See Example 5.)

41. 5
42. 31
43. -12
44. -25
45. 0
46. 1
47. -1
48. -612

For Exercises 49–60, simplify the expression. (See Example 6.)

49. −(−15) 50. −(−4) 51. −|−15| 52. −|−4|

53. −|15| 54. −|4| 55. |−15| 56. |−4|

57. −(−36) 58. −(−19) 59. −|−107| 60. −|−26|

Mixed Exercises

For Exercises 61–64, simplify the expression.

61. a. |−6| b. −(−6) c. −|6| d. |6| e. −|−6|

62. a. −(−12) b. |12| c. |−12| d. −|−12| e. −|12|

63. a. −|8| b. |8| c. −|−8| d. −(−8) e. |−8|

64. a. −|−1| b. −(−1) c. |1| d. |−1| e. −|1|

For Exercises 65–74, write in symbols, do not simplify.

65. The opposite of 6

66. The opposite of 23

67. The opposite of negative 2

68. The opposite of negative 9

69. The absolute value of 7

70. The absolute value of 11

71. The absolute value of negative 3

72. The absolute value of negative 10

73. The opposite of the absolute value of 14

74. The opposite of the absolute value of 42

For Exercises 75–84, fill in the blank with <, >, or =.

75. |−12| ☐ |12| 76. −(−4) ☐ −|−4| 77. |−22| ☐ −(22) 78. −8 ☐ −10

79. −44 ☐ −54 80. −|0| ☐ −|1| 81. |−55| ☐ −(−65) 82. −(82) ☐ |46|

83. −|32| ☐ |0| 84. −|22| ☐ 0

For Exercises 85–91, refer to the contour map for wind chill temperatures for a day in January. Give an *estimate* of the wind chill for the given city. For example, the wind chill in Phoenix is between 30°F and 40°F, but closer to 30°F. We might estimate the wind chill in Phoenix to be 33°F.

85. Portland 86. Atlanta

87. Bismarck 88. Denver

89. Eugene 90. Orlando

91. Dallas

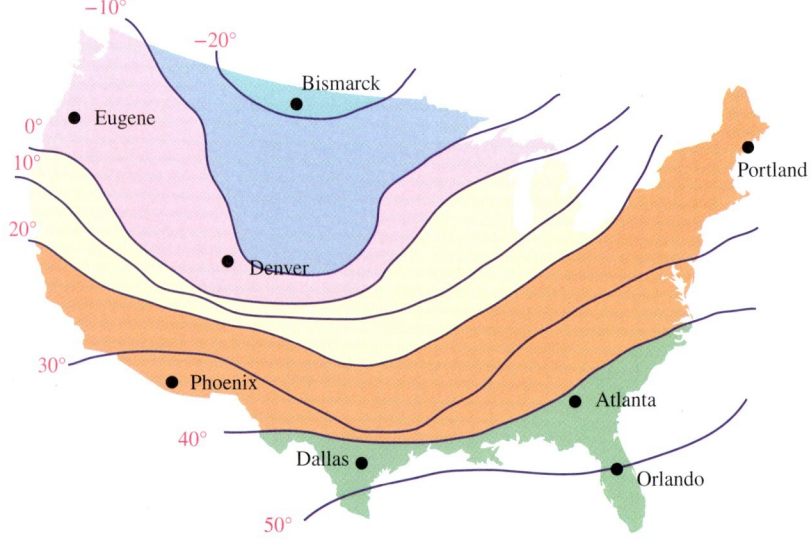

Wind Chill Temperatures, °F

Late spring and summer rainfall in the Lake Okeechobee region in Florida is important to replenish the water supply for large areas of south Florida. Each bar in the graph indicates the number of inches of rainfall above or below average for the given month. Use the graph for Exercises 92–94.

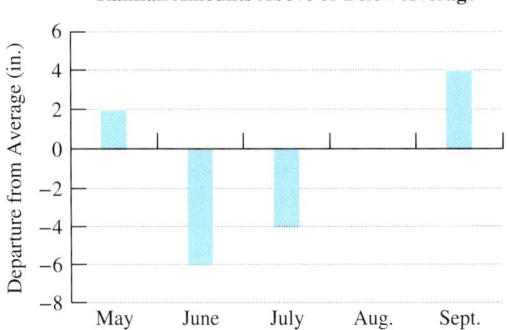
Rainfall Amounts Above or Below Average

92. Which month had the greatest amount of rainfall *below* average? What was the departure from average?

93. Which month had the greatest amount of rainfall *above* average?

94. Which month had the average amount of rainfall?

Expanding Your Skills

For Exercises 95 and 96, rank the numbers from least to greatest.

95. $-|-46|, -(-24), -60, 5^2, |-12|$

96. $-15, -(-18), -|20|, 4^2, |-3|^2$

97. If a represents a negative number, then what is the sign of $-a$?

98. If b represents a negative number, then what is the sign of $|b|$?

99. If c represents a negative number, then what is the sign of $-|c|$?

100. If d represents a negative number, then what is the sign of $-(-d)$?

Addition of Integers

Section 2.2

1. Addition of Integers by Using a Number Line

Addition of integers can be visualized on a number line. To do so, we locate the first addend on the number line. Then to add a positive number, we move to the right on the number line. To add a negative number, we move to the left on the number line. This is demonstrated in Example 1.

Concepts

1. Addition of Integers by Using a Number Line
2. Addition of Integers
3. Translations and Applications of Addition

Example 1   Using a Number Line to Add Integers

Use a number line to add.

a. $5 + 3$ b. $-5 + 3$

Solution:

a. $5 + 3 = 8$

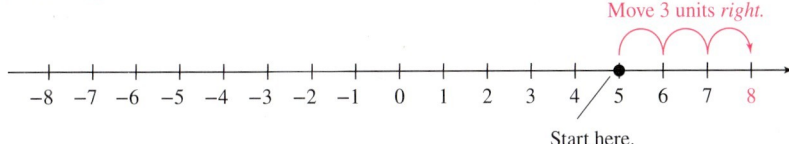

Begin at 5. Then, because we are adding *positive* 3, move to the *right* 3 units. The sum is 8.

b. $-5 + 3 = -2$

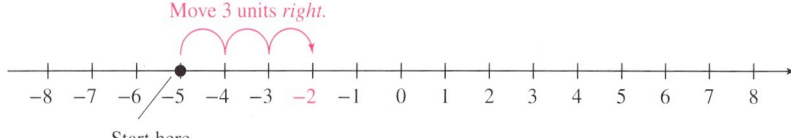

Begin at -5. Then, because we are adding *positive* 3, move to the *right* 3 units. The sum is -2.

Skill Practice Use a number line to add.

1. $3 + 2$ 2. $-3 + 2$

Example 2 Using a Number Line to Add Integers

Use a number line to add.

a. $5 + (-3)$ **b.** $-5 + (-3)$

Solution:

a. $5 + (-3) = 2$

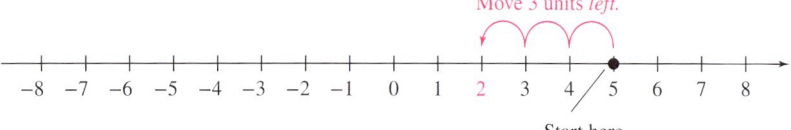

Begin at 5. Then, because we are adding *negative* 3, move to the *left* 3 units. The sum is 2.

b. $-5 + (-3) = -8$

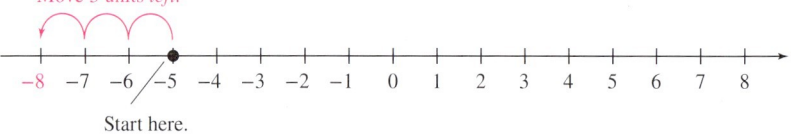

Begin at -5. Then, because we are adding *negative* 3, move to the *left* 3 units. The sum is -8.

TIP: In Example 2, parentheses are inserted for clarity. The parentheses separate the number -3 from the symbol for addition, $+$.

$5 + (-3)$ and $-5 + (-3)$

Skill Practice Use a number line to add.

3. $3 + (-2)$ 4. $-3 + (-2)$

2. Addition of Integers

It is inconvenient to draw a number line each time we want to add signed numbers. Therefore, we offer two rules for adding integers. The first rule is used when the addends have the *same* sign (that is, if the numbers are both positive or both negative).

Adding Numbers with the Same Sign

To add two numbers with the same sign, add their absolute values and apply the common sign.

Answers
1. 5 2. -1
3. 1 4. -5

Section 2.2 Addition of Integers 89

Example 3 — Adding Integers with the Same Sign

Add.

a. $-2 + (-4)$ **b.** $-12 + (-37)$ **c.** $10 + 66$

Solution:

a. $\quad -2 + (-4)$ First find the absolute value of each addend.

$|-2| = 2$ and $|-4| = 4$

$= -(2 + 4)$ Add their absolute values and apply the common sign (in this case, the common sign is negative).

Common sign is negative.

$= -6$ The sum is -6.

b. $\quad -12 + (-37)$ First find the absolute value of each addend.

$|-12| = 12$ and $|-37| = 37$

$= -(12 + 37)$ Add their absolute values and apply the common sign (in this case, the common sign is negative).

Common sign is negative.

$= -49$ The sum is -49.

c. $\quad 10 + 66$ First find the absolute value of each addend.

$|10| = 10$ and $|66| = 66$

$= +(10 + 66)$ Add their absolute values and apply the common sign (in this case, the common sign is positive).

Common sign is positive.

$= 76$ The sum is 76.

TIP: Parentheses are used to show that the absolute values are added *before* applying the common sign.

Skill Practice Add.

5. $-6 + (-8)$ **6.** $-84 + (-27)$ **7.** $14 + 31$

The next rule helps us add two numbers with different signs.

Adding Numbers with Different Signs

To add two numbers with different signs, subtract the smaller absolute value from the larger absolute value. Then apply the sign of the number having the larger absolute value.

Example 4 — Adding Integers with Different Signs

Add.

a. $2 + (-7)$ **b.** $-6 + 24$ **c.** $-8 + 8$

Solution:

a. $2 + (-7)$ First find the absolute value of each addend.

$|2| = 2$ and $|-7| = 7$

Note: The absolute value of -7 is greater than the absolute value of 2. Therefore, the sum is negative.

$= -(7 - 2)$ Next, subtract the smaller absolute value from the larger absolute value.

Apply the sign of the number with the larger absolute value.

$= -5$

Answers

5. -14 **6.** -111 **7.** 45

TIP: Parentheses are used to show that the absolute values are subtracted *before* applying the appropriate sign.

b. $-6 + 24$ First find the absolute value of each addend.
$|-6| = 6$ and $|24| = 24$

Note: The absolute value of 24 is greater than the absolute value of −6. Therefore, the sum is positive.

$= +(24 - 6)$ Next, subtract the smaller absolute value from the larger absolute value.

Apply the sign of the number with the larger absolute value.

$= 18$

c. $-8 + 8$ First find the absolute value of each addend.
$|-8| = 8$ and $|8| = 8$

$= (8 - 8)$ The absolute values are equal. Therefore, their difference is 0. The number zero is neither positive nor negative.

$= 0$

Skill Practice Add.

8. $5 + (-8)$ **9.** $-12 + 37$ **10.** $-4 + 4$

Example 4(c) illustrates that the sum of a number and its opposite is zero. For example:

$$-8 + 8 = 0 \quad -12 + 12 = 0 \quad 6 + (-6) = 0$$

Adding Opposites
For any number a,
$$a + (-a) = 0 \quad \text{and} \quad -a + a = 0$$

That is, the sum of any number and its opposite is zero. (This is also called the *additive inverse property*.)

Example 5   **Adding Several Integers**

Simplify. $-30 + (-12) + 4 + (-10) + 6$

Solution:

$-30 + (-12) + 4 + (-10) + 6$ Apply the order of operations by adding from left to right.
$= -42 + 4 + (-10) + 6$
$= -38 + (-10) + 6$
$= -48 + 6$
$= -42$

Skill Practice Simplify.

11. $-24 + (-16) + 8 + 2 + (-20)$

TIP: When several numbers are added, we can reorder and regroup the addends using the commutative property and associative property of addition. In particular, we can group all the positive addends together, and we can group all the negative addends together. This makes the arithmetic easier. For example,

$$-30 + (-12) + 4 + (-10) + 6 = \overbrace{4 + 6}^{\text{positive addends}} + \overbrace{(-30) + (-12) + (-10)}^{\text{negative addends}}$$
$$= 10 + (-52)$$
$$= -42$$

Answers
8. −3   **9.** 25
10. 0   **11.** −50

3. Translations and Applications of Addition

Example 6 — Translating an English Phrase to a Mathematical Expression

Translate to a mathematical expression and simplify.

−6 added to the sum of 2 and −11

Solution:

$[2 + (-11)] + (-6)$ Translate: −6 added to the sum of 2 and −11
Notice that the sum of 2 and −11 is written first, and then −6 is added to that value.

$= -9 + (-6)$ Find the sum of 2 and −11 first. $2 + (-11) = -9$.

$= -15$

Skill Practice Translate to a mathematical expression and simplify.

12. −2 more than the total of −8, −10, and 5

Example 7 — Applying Addition of Integers to Determine Snowfall Amount

Winter snowfall in northwest Montana is a critical source of water for forests, rivers, and lakes when the snow melts in the spring. Below average snowfall can contribute to severe forest fires in the summer. The graph indicates the number of inches of snowfall above or below average for given months.

Find the total departure from average snowfall for these months. Do the results show that the region received above average snowfall for the winter or below average?

Solution:

From the graph, we have the following departures from the average snowfall.

Month	Amount (in.)
Nov.	−8
Dec.	−12
Jan.	16
Feb.	12
Mar.	−4

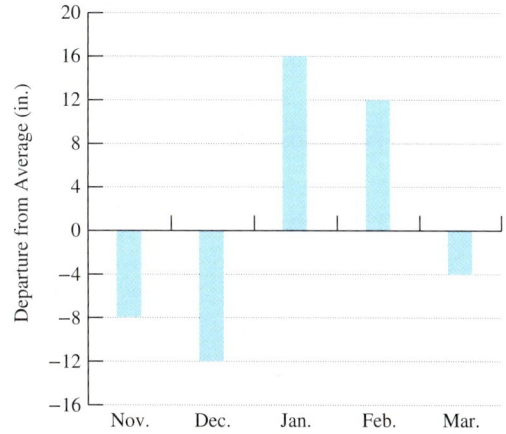

Snowfall Amounts Above or Below Average

To find the total, we can group the addends conveniently.

Total: $(-8) + (-12) + (-4) + 16 + 12$

$\; = \quad -24 \quad + \quad 28$

$\; = 4$ The total departure from average is 4 in. The region received above average snowfall.

Skill Practice

13. Jonas played 9 holes of golf and received the following scores. Positive scores indicate that Jonas was above par. Negative scores indicate that he was below par. Find Jonas's total after 9 holes. Is his score above or below par?

+2, +1, −1, 0, 0, −1, +3, +4, −1

Answers

12. $[-8 + (-10) + 5] + (-2)$; −15
13. The total score is 7. Jonas's score is above par.

Section 2.2 Activity

A.1. Use a number line to add $5 + (-3)$.

  a. Plot a point at the first addend.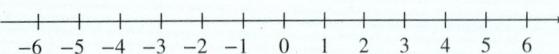

  b. Move 3 places to the left since 3 is negative. The end point is the sum.

A.2. Use a number line to add $-1 + (-3)$.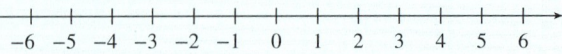

A.3. Explain how to add two real numbers with different signs.

A.4. Explain how to add two real numbers with the same signs.

For Exercises A.5–A.12, add the integers.

A.5. $-3 + 10 = $ _____

A.6. $14 + (-6) = $ _____

A.7. $-15 + (-30) = $ _____

A.8. $-189 + (-50) = $ _____

A.9. $-16 + 16 = $ _____

A.10. $-5 + 7 + (-8) = $ _____

A.11. $58 + 23 + (-45) = $ _____

A.12. $-75 + (-32) + (-3) = $ _____

A.13. Cameron is balancing his new checking account. Over the last month he paid five different bills: $125 cell phone, $110 cable, $230 electric, $86 groceries, and $750 rent. He received two deposits from his employer in the amounts of $1100 and $955.

  a. Write an expression that represents the withdrawals and deposits from Cameron's account.

  b. Assuming that Cameron started with a $0 balance, how much money is in his account at the end of the month?

A.14. Create an application problem involving sports, finances, or diets that describes the equation below. Then solve the equation.

$$10 + (-2) + 6 + (-5) + (-4) = x$$

Section 2.2 Practice Exercises

Study Skills Exercise

As a college student, you need to balance the demands of your job, your family, your personal life, and your education. The first step in achieving this balance is to efficiently manage your time. To do so, analyze your schedule and plan out your week. Keep a calendar to maintain your school, home, and work responsibilities. Insert due dates into the calendar and schedule your study time throughout the week in 20- to 30-minute sessions.

Section 2.2 Addition of Integers   93

Prerequisite Review

For Exercises R.1–R.6, place the correct symbol (>, <, or =) between the two numbers.

R.1. −6 ☐ −5   **R.2.** −33 ☐ −34
R.3. |−4| ☐ −|4| **R.4.** |6| ☐ |−6|
R.5. −(−2) ☐ −|−2| **R.6.** −|−10| ☐ 10

Vocabulary and Key Concepts

1. **a.** The sum of a number and its opposite is _____.
   b. The sum of two negative numbers is (positive/negative).
   The sum of two positive numbers is (positive/negative).
2. **a.** When using a number line to add integers, you start at the first addend and then move to the _____ when adding a *positive* number.
   b. When using a number line to add integers, you start at the first addend and then move to the _____ when adding a *negative* number.
3. For the expression 8 + (−4), the parentheses are used for clarity to separate the number −4 from the _____.
4. True or False. When adding several numbers together, the positive addends can be grouped together and the negative addends can be grouped together to make the arithmetic easier.

Concept 1: Addition of Integers by Using a Number Line

For Exercises 5–20, refer to the number line to add the integers. (See Examples 1 and 2.)

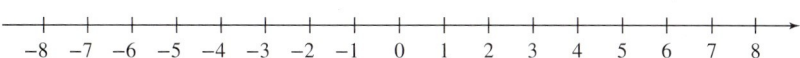

5. 1 + 6   **6.** 2 + 4   **7.** −8 + 8   **8.** −2 + 1
9. −3 + 5   **10.** −6 + 3   **11.** 2 + (−4)   **12.** 5 + (−1)
13. −4 + (−4)   **14.** −2 + (−5)   **15.** −3 + 9   **16.** −1 + 5
17. 0 + (−7)   **18.** (−5) + 0   **19.** −1 + (−3)   **20.** −4 + (−3)

Concept 2: Addition of Integers

21. Explain the process to add two numbers with the same sign.

For Exercises 22–29, add the numbers with the same sign. (See Example 3.)

22. 23 + 12   **23.** 12 + 3   **24.** −8 + (−3)   **25.** −10 + (−6)
26. −7 + (−9)   **27.** −100 + (−24)   **28.** 23 + 50   **29.** 44 + 45

30. Explain the process to add two numbers with different signs.

For Exercises 31–42, add the numbers with different signs. (See Example 4.)

31. 7 + (−10)   **32.** −8 + 2   **33.** 12 + (−7)   **34.** −3 + 9
35. −90 + 66   **36.** −23 + 49   **37.** 78 + (−33)   **38.** 10 + (−23)
39. 2 + (−2)   **40.** −6 + 6   **41.** −13 + 13   **42.** 45 + (−45)

Mixed Exercises

For Exercises 43–72, simplify. (See Example 5.)

43. $12 + (-3)$
44. $-33 + (-1)$
45. $-23 + (-3)$
46. $-5 + 15$
47. $4 + (-45)$
48. $-13 + (-12)$
49. $(-103) + (-47)$
50. $119 + (-59)$
51. $0 + (-17)$
52. $-29 + 0$
53. $-19 + (-22)$
54. $-300 + (-24)$
55. $-222 + 751$
56. $620 + (-818)$
57. $1158 + (-378)$
58. $-2022 + 997$
59. $6 + (-12) + 8$
60. $20 + (-12) + (-5)$
61. $-33 + (-15) + 18$
62. $3 + 5 + (-1)$
63. $7 + (-3) + 6$
64. $12 + (-6) + (-9)$
65. $-10 + (-3) + 5$
66. $-23 + (-4) + (-12) + (-5)$
67. $-18 + (-5) + 23$
68. $14 + (-15) + 20 + (-42)$
69. $4 + (-12) + (-30) + 16 + 10$
70. $24 + (-5) + (-19)$
71. $-79 + (-356) + 244$
72. $620 + (-949) + 758$

Concept 3: Translations and Applications of Addition

For Exercises 73–78, translate to a mathematical expression. Then simplify the expression. (See Example 6.)

73. The sum of -23 and 49
74. The sum of 89 and -11
75. The total of 3, -10, and 5
76. The total of -2, -4, 14, and 20
77. -5 added to the sum of -8 and 6
78. -15 added to the sum of -25 and 7

79. The graph gives the number of inches below or above the average snowfall for the given months for Marquette, Michigan. Find the total departure from average. Is the snowfall for Marquette above or below average? (See Example 7.)

80. The graph gives the number of inches below or above the average rainfall for the given months for Hilo, Hawaii. Find the total departure from average. Is the total rainfall for these months above or below average?

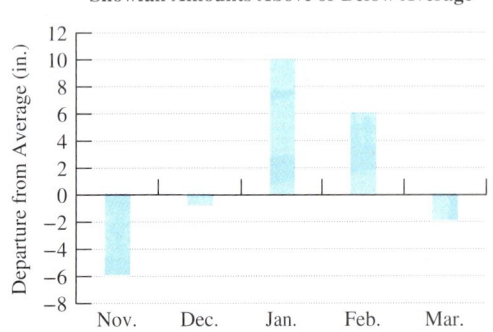

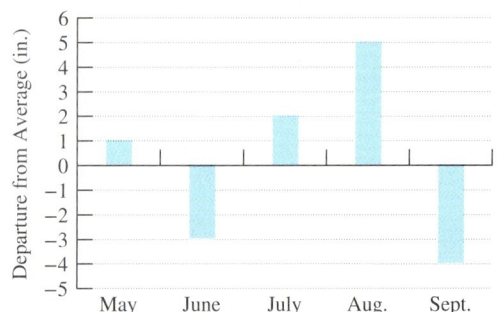

For Exercises 81 and 82, refer to the table. The table gives the scores for the top two finishers at a recent PGA Master's golf tournament.

81. Compute Phil Mickelson's total score.

82. Compute Fred Couples's total score.

	Round 1	Round 2	Round 3	Round 4
Phil Mickelson	−5	−1	−5	−5
Fred Couples	−6	3	−4	−2

83. At 6:00 A.M. the temperature was −4°F. By noon, the temperature had risen by 12°F. What was the temperature at noon?

84. At midnight the temperature was −14°F. By noon, the temperature had risen 10°F. What was the temperature at noon?

85. Jorge's checking account is overdrawn. His beginning balance was −$56. If he deposits his paycheck for $389, what is his new balance?

Source: NPS Photo by DL Coe

86. Ellen's checking account balance is $23. If she writes a check for $40, what is her new balance?

87. A contestant on a popular game show scored the following amounts for several questions he answered. Determine his total score.

   −$200,   −$400, $1000,   −$400, $600

88. The number of yards gained or lost by a running back in a football game are given. Find the total number of yards.

   3,   2,   −8,   5,   −2,   4,   21

Ilene MacDonald/Alamy Stock Photo

89. The table gives the scores for the first 9 holes in the first round for a golfer in the U.S. Open Women's Golf Championship. Find the sum of the scores.

Hole	1	2	3	4	5	6	7	8	9
Score	0	2	−1	−1	0	−1	1	0	0

90. The table gives the scores for the first 9 holes of the final round for a golfer in the PGA Golf Championship. Find the sum of the scores.

Hole	1	2	3	4	5	6	7	8	9
Score	1	1	0	0	−1	−1	0	0	2

Technology Connections

For Exercises 91–96, add using a calculator.

91. 302 + (−422)

92. −900 + 334

93. −23,991 + (−4423)

94. −1034 + (−23,291)

95. 23 + (−125) + 912 + (−99)

96. 891 + 12 + (−223) + (−341)

Expanding Your Skills

97. Find two integers whose sum is −10. Answers may vary.

98. Find two integers whose sum is −14. Answers may vary.

99. Find two integers whose sum is −2. Answers may vary.

100. Find two integers whose sum is 0. Answers may vary.

Section 2.3 Subtraction of Integers

Concepts

1. Subtraction of Integers
2. Translations and Applications of Subtraction

1. Subtraction of Integers

Subtraction of integers is defined in terms of the addition process. For example, consider the following subtraction problem. The corresponding addition problem produces the same result.

$$6 - 4 = 2 \Leftrightarrow 6 + (-4) = 2$$

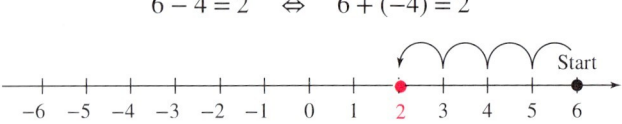

In each case, we start at 6 on the number line and move to the *left* 4 units. Adding the *opposite* of 4 produces the same result as subtracting 4. This is true in general. To subtract two integers, add the opposite of the second number to the first number.

> **Subtracting Signed Numbers**
>
> For two numbers a and b, $a - b = a + (-b)$.
>
> Therefore, to perform subtraction, follow these steps:
>
> **Step 1** Leave the first number (the minuend) unchanged.
> **Step 2** Change the subtraction sign to an addition sign.
> **Step 3** Add the opposite of the second number (the subtrahend).

For example:

$$\left. \begin{array}{l} 10 - 4 = 10 + (-4) = 6 \\ -10 - 4 = -10 + (-4) = -14 \end{array} \right\} \text{Subtracting 4 is the same as adding } -4.$$

$$\left. \begin{array}{l} 10 - (-4) = 10 + (4) = 14 \\ -10 - (-4) = -10 + (4) = -6 \end{array} \right\} \text{Subtracting } -4 \text{ is the same as adding 4.}$$

Example 1 Subtracting Integers

Subtract. **a.** $15 - 20$ **b.** $-7 - 12$ **c.** $40 - (-8)$

Solution:

a. $15 - 20 = 15 + (-20) = -5$ Rewrite the subtraction in terms of addition.
(Add the opposite of 20. Change subtraction to addition.)   Subtracting 20 is the same as adding -20.

b. $-7 - 12 = -7 + (-12)$ Rewrite the subtraction in terms of addition.
$= -19$ Subtracting 12 is the same as adding -12.

c. $40 - (-8) = 40 + (8)$ Rewrite the subtraction in terms of addition.
$= 48$ Subtracting -8 is the same as adding 8.

TIP: After subtraction is written in terms of addition, the rules of addition are applied.

Skill Practice Subtract.

1. $12 - 19$ **2.** $-8 - 14$ **3.** $30 - (-3)$

Answers

1. -7 **2.** -22 **3.** 33

Example 2 Adding and Subtracting Several Integers

Simplify. $-4 - 6 + (-3) - 5 + 8$

Solution:

$-4 - 6 + (-3) - 5 + 8$

$= -4 + (-6) + (-3) + (-5) + 8$ Rewrite all subtractions in terms of addition.
$= -10 + (-3) + (-5) + 8$ Add from left to right.
$= -13 + (-5) + 8$
$= -18 + 8$
$= -10$

Skill Practice Simplify.
4. $-8 - 10 + (-6) - (-1) + 4$

2. Translations and Applications of Subtraction

The table provides several key words that imply subtraction.

Word/Phrase	Example	In Symbols
a minus b	-15 minus 10	$-15 - 10$
The difference of a and b	The difference of 10 and -2	$10 - (-2)$
a decreased by b	9 decreased by 1	$9 - 1$
a less than b	-12 less than 5	$5 - (-12)$
Subtract a from b	Subtract -3 from 8	$8 - (-3)$
b subtracted from a	-2 subtracted from -10	$-10 - (-2)$

Example 3 Translating to a Mathematical Expression

Translate to a mathematical expression. Then simplify.

a. The difference of -52 and 10.
b. -35 decreased by -6.

Solution:

a.   *the difference of*

$-52 - 10$ Translate: The difference of -52 and 10.
$= -52 + (-10)$ Rewrite subtraction in terms of addition.
$= -62$ Add.

b.   *decreased by*

$-35 - (-6)$ Translate: -35 decreased by -6.
$= -35 + (6)$ Rewrite subtraction in terms of addition.
$= -29$ Add.

Avoiding Mistakes
Subtraction is not commutative. The order of the numbers being subtracted is important.

Skill Practice Translate to a mathematical expression. Then simplify.
5. The difference of -16 and 4
6. -8 decreased by -9

Answers
4. -19 5. $-16 - 4$; -20
6. $-8 - (-9)$; 1

Example 4: Translating to a Mathematical Expression

Translate each English phrase to a mathematical expression. Then simplify.

a. 12 less than −8 **b.** Subtract 27 from 5.

Solution:

a. To translate "12 less than −8," we must *start* with −8 and subtract 12.

$$-8 - 12 \quad \text{Translate: 12 less than } -8.$$
$$= -8 + (-12) \quad \text{Rewrite subtraction in terms of addition.}$$
$$= -20 \quad \text{Add.}$$

b. To translate "subtract 27 from 5," we must *start* with 5 and subtract 27.

$$5 - 27 \quad \text{Translate: Subtract 27 from 5.}$$
$$= 5 + (-27) \quad \text{Rewrite subtraction in terms of addition.}$$
$$= -22 \quad \text{Add.}$$

Skill Practice Translate to a mathematical expression. Then simplify.

7. 6 less than 2 **8.** Subtract −4 from −1.

Example 5: Applying Subtraction of Integers

A helicopter is hovering at a height of 200 ft above the ocean. A submarine is directly below the helicopter 125 ft below sea level. Find the difference in elevation between the helicopter and the submarine.

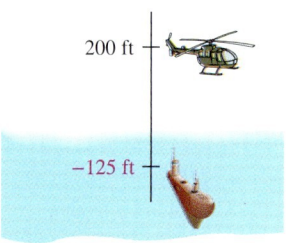

Solution:

$$\begin{pmatrix} \text{Difference between} \\ \text{elevation of helicopter} \\ \text{and submarine} \end{pmatrix} = \begin{pmatrix} \text{elevation of} \\ \text{helicopter} \end{pmatrix} - \begin{pmatrix} \text{"elevation" of} \\ \text{submarine} \end{pmatrix}$$

$$= 200 \text{ ft} - (-125 \text{ ft})$$
$$= 200 \text{ ft} + (125 \text{ ft}) \quad \text{Rewrite as addition.}$$
$$= 325 \text{ ft}$$

The helicopter and submarine are 325 ft apart.

Skill Practice

9. The highest point in California is Mt. Whitney at 14,494 ft above sea level. The lowest point in California is Death Valley, which has an "altitude" of −282 ft (282 ft below sea level). Find the difference in the elevations of the highest point and lowest point in California.

Answers
7. 2 − 6; −4 **8.** −1 − (−4); 3
9. 14,776 ft

Section 2.3 Activity

A.1. Using a calculator, add or subtract the integers.

a. 4 + 9 = _____

b. 4 − (−9) = _____

c. −8 + 12 = _____

d. −8 − (−12) = _____

e. Compare the results in parts (a) and (b), and parts (c) and (d). What can you conclude about subtracting integers?

f. State the rule for subtracting integers in terms of addition.

For Exercises A.2–A.5, write each English phrase as an algebraic expression. Then simplify.

	English Phrase	Algebraic Expression	Simplified Result
A.2.	The difference of -13 and 4		
A.3.	-21 decreased by 16		
A.4.	-18 subtracted from 25		
A.5.	4 less than -2		

A.6. The lowest recorded temperature in Alaska is $-80°F$. The highest recorded temperature in Alaska is $100°F$. What is the difference in the high and low temperatures?

A.7. Write the next three numbers in the sequence.
   a. 13, 7, 1, -5, ____, ____, ____
   b. $-4, -18, -32, -46$, ____, ____, ____

Practice Exercises

Section 2.3

Study Skills Exercise

Sometimes you may run into a problem with homework, or you may find that you are having trouble keeping up with the pace of the class. Seek support at the first sign of struggle. Mathematics labs, peer tutors, and workshops are just some of the common support services institutions offer students in the area of mathematics.

a. List the services that are available to help you complete your mathematics course successfully.

b. What costs, if any, are associated with each service?

c. How do you sign up for the support?

Prerequisite Review

For Exercises R.1–R.6, simplify.

R.1. $34 + (-13)$

R.2. $-34 + (-13)$

R.3. $-34 + 13$

R.4. $-3 + 13$

R.5. $(-9) + (-8) + 16 + (-3)$

R.6. $(-9) + 8 + (-16) + (-3)$

Vocabulary and Key Concepts

1. If a and b are real numbers, then $a - b = a +$ _____.
2. a. To write $-5 - (-4)$ as an equivalent statement using addition, we write _____.
   b. To write $-5 - 4$ as an equivalent statement using addition, we write _____.

Concept 1: Subtraction of Integers

3. Explain the process to subtract integers.

For Exercises 4–6, use the number line to prove that the subtraction problem and the corresponding addition problem produce the same result.

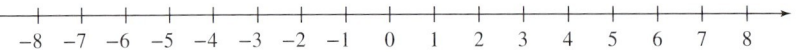

4. $7 - 3 = 4 \Leftrightarrow 7 + (-3) = 4$

5. $1 - 5 = -4 \Leftrightarrow 1 + (-5) = -4$

6. $-2 - 3 = -5 \Leftrightarrow -2 + (-3) = -5$

For Exercises 7–16, rewrite the subtraction problem as an equivalent addition problem. Then simplify. **(See Example 1.)**

7. $9 - 2 =$ ___ $+$ ___ $=$ ___
8. $11 - 5 =$ ___ $+$ ___ $=$ ___
9. $2 - 9 =$ ___ $+$ ___ $=$ ___
10. $5 - 11 =$ ___ $+$ ___ $=$ ___
11. $4 - (-3) =$ ___ $+$ ___ $=$ ___
12. $12 - (-8) =$ ___ $+$ ___ $=$ ___
13. $-3 - 15 =$ ___ $+$ ___ $=$ ___
14. $-7 - 21 =$ ___ $+$ ___ $=$ ___
15. $-11 - (-13) =$ ___ $+$ ___ $=$ ___
16. $-23 - (-9) =$ ___ $+$ ___ $=$ ___

For Exercises 17–52, simplify. **(See Examples 1 and 2.)**

17. $35 - (-17)$
18. $23 - (-12)$
19. $-24 - 9$
20. $-5 - 15$
21. $50 - 62$
22. $38 - 46$
23. $-17 - (-25)$
24. $-2 - (-66)$
25. $-8 - (-8)$
26. $-14 - (-14)$
27. $120 - (-41)$
28. $91 - (-62)$
29. $-15 - 19$
30. $-82 - 44$
31. $3 - 25$
32. $6 - 33$
33. $-13 - 13$
34. $-43 - 43$
35. $24 - 25$
36. $43 - 98$
37. $-6 - (-38)$
38. $-75 - (-21)$
39. $-48 - (-33)$
40. $-29 - (-32)$
41. $-320 - (-198)$
42. $444 - 576$
43. $-1011 - (-2020)$
44. $987 - (-337)$
45. $300 - (-386) + 575$
46. $-40 + 605 - 815$
47. $2 + 5 - (-3) - 10$
48. $4 - 8 + 12 - (-1)$

49. −5 + 6 + (−7) − 4 − (−9)

50. −2 − 1 + (−11) + 6 − (−8)

51. 25 − 13 − (−40)

52. −35 + 15 − (−28)

Concept 2: Translations and Applications of Subtraction

53. State at least two words or phrases that would indicate subtraction.

54. Is subtraction commutative; for example, does 3 − 7 = 7 − 3?

For Exercises 55–66, translate each English phrase to a mathematical expression. Then simplify. (See Examples 3 and 4.)

55. 14 minus 23

56. 27 minus 40

57. The difference of 105 and 110

58. The difference of 70 and 98

59. 320 decreased by −20

60. 150 decreased by 75

61. Subtract 12 from 5.

62. Subtract 10 from 16.

63. 21 less than −34

64. 22 less than −90

65. Subtract 24 from −35

66. Subtract 189 from 175

67. The liquid hydrogen in the space shuttle's main engine is −423°F. The temperature in the engine's combustion chamber reaches 6000°F. Find the difference between the temperature in the combustion chamber and the temperature of the liquid hydrogen. (See Example 5.)

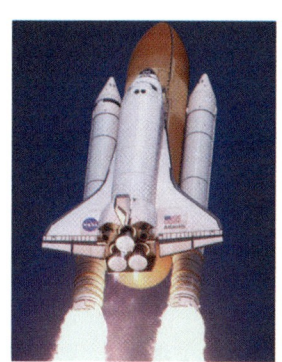
Purestock/Getty Images

68. Temperatures on the Moon range from −184°C during its night to 214°C during its day. Find the difference between the highest temperature on the Moon and the lowest temperature.

Ingram Publishing/SuperStock

69. Ivan owes $320 on his credit card; that is, his balance is −$320. If he charges $55 for a night out, what is his new balance?

70. If Justin's balance on his credit card was −$210 and he made the minimum payment of $25, what is his new balance?

71. The Campus Food Court reports its total profit or loss each day. During a 1-week period, the following profits or losses were reported. If the Campus Food Court's balance was $17,476 at the beginning of the week, what is the balance at the end of the reported week?

Monday	$1786
Tuesday	−$2342
Wednesday	−$754
Thursday	$321
Friday	$1597

72. Jeff's balance in his checking account was $2036 at the beginning of the week. During the week, he wrote two checks, made two deposits, and made one ATM withdrawal. What is his ending balance?

Check	−$150
Check	−$25
Paycheck (deposit)	$480
ATM	−$200
Cash (deposit)	$80

For Exercises 73–76, refer to the graph indicating the change in value of the Dow Jones Industrial Average for a given week.

73. What is the difference in the change in value between Tuesday and Wednesday?

74. What is the difference in the change in value between Thursday and Friday?

75. What is the total change for the week?

76. Based on the values in the graph, did the Dow gain or lose points for the given week?

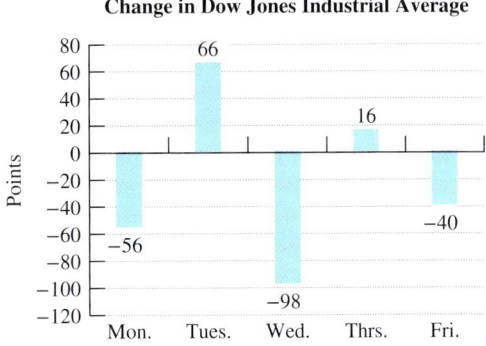

For Exercises 77–78, find the range. The *range* of a set of numbers is the difference between the highest value and the lowest value. That is, range = highest − lowest.

77. Low temperatures for 1 week in Anchorage (°C): −4°, −8°, 0°, 3°, −8°, −1°, 2°

78. Low temperatures for 1 week in Fargo (°C): −6°, −2°, −10°, −4°, −12°, −1°, −3°

79. Find two integers whose difference is −6. Answers may vary.

80. Find two integers whose difference is −20. Answers may vary.

For Exercises 81 and 82, write the next three numbers in the sequence.

81. 5, 1, −3, −7, ___, ___, ___

82. −13, −18, −23, −28, ___, ___, ___

Technology Connections

For Exercises 83–88, subtract the integers using a calculator.

83. −190 − 223

84. −288 − 145

85. −23,624 − (−9001)

86. −14,593 − (−34,499)

87. 892,904 − (−23,546)

88. 104,839 − (−24,938)

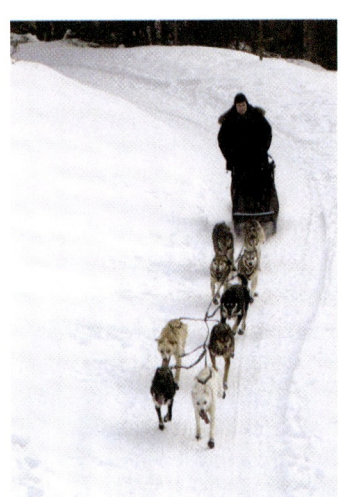

Source: U.S. Air Force photo by Tech. Sgt. Keith Brown

89. The altitude of Mount Everest is 29,029 ft. The lowest point on the surface of the Earth is −35,798 ft (that is, 35,798 ft below sea level) occurring at the Mariana Trench on the Pacific Ocean floor. What is the difference in altitude between the height of Mt. Everest and the Mariana Trench?

90. Mt. Rainier is 4392 m at its highest point. Death Valley, California, is 86 m below sea level (−86 m) at the basin, Badwater. What is the difference between the altitude of Mt. Rainier and the altitude at Badwater?

Expanding Your Skills

For Exercises 91–98, assume $a > 0$ (this means that a is positive) and $b < 0$ (this means that b is negative). Find the sign of each expression.

91. $a - b$

92. $b - a$

93. $|a| + |b|$

94. $|a + b|$

95. $-|a|$

96. $-|b|$

97. $-(a)$

98. $-(b)$

Multiplication and Division of Integers

Section 2.4

1. Multiplication of Integers

Concepts
1. Multiplication of Integers
2. Multiplying Many Factors
3. Exponential Expressions
4. Division of Integers

We know from our knowledge of arithmetic that the product of two positive numbers is a positive number. This can be shown by using repeated addition.

For example: $3(4) = 4 + 4 + 4 = 12$

Now consider a product of numbers with different signs.

For example: $3(-4) = -4 + (-4) + (-4) = -12$ (3 times -4)

This example suggests that the product of a positive number and a negative number is *negative*.

Now what if we have a product of two negative numbers? To determine the sign, consider the following pattern of products.

$$3 \cdot -4 = -12$$
$$2 \cdot -4 = -8$$
$$1 \cdot -4 = -4$$
$$0 \cdot -4 = 0$$
$$-1 \cdot -4 = 4$$
$$-2 \cdot -4 = 8$$
$$-3 \cdot -4 = 12$$

The pattern increases by 4 with each row.

The product of two negative numbers is *positive*.

From the first four rows, we see that the product increases by 4 for each row. For the pattern to continue, it follows that the product of two negative numbers must be *positive*.

Multiplying Signed Numbers

1. The product of two numbers with the *same* sign is positive.

   Examples: $(5)(6) = 30$
   $(-5)(-6) = 30$

2. The product of two numbers with *different* signs is negative.

   Examples: $4(-10) = -40$
   $-4(10) = -40$

3. The product of any number and zero is zero.

   Examples: $3(0) = 0$
   $0(-6) = 0$

Avoiding Mistakes

Try not to confuse the rule for multiplying two negative numbers with the rule for adding two negative numbers.

- The product of two negative numbers is positive.
- The sum of two negative numbers is negative.

Example 1 Multiplying Integers

Multiply.

a. $-8(-7)$ **b.** $-5 \cdot 10$ **c.** $(18)(-2)$ **d.** $16 \cdot 2$ **e.** $-3 \cdot 0$

Solution:

a. $-8(-7) = 56$ *Same* signs. Product is positive.

b. $-5 \cdot 10 = -50$ *Different* signs. Product is negative.

c. $(18)(-2) = -36$ *Different* signs. Product is negative.

d. $16 \cdot 2 = 32$ *Same* signs. Product is positive.

e. $-3 \cdot 0 = 0$ The product of any number and zero is zero.

Skill Practice Multiply.

1. $-2(-6)$ **2.** $3 \cdot (-10)$ **3.** $-14(3)$ **4.** $8 \cdot 4$ **5.** $-5 \cdot 0$

Recall that the terms *product*, *multiply*, and *times* imply multiplication.

Example 2 Translating to a Mathematical Expression

Translate each phrase to a mathematical expression. Then simplify.

a. The product of 7 and -8 **b.** -3 times -11

FOR REVIEW

Recall that multiplication is commutative. The product of 7 and -8 may be performed in any order.

$7(-8)$ or $(-8)(7)$ or $(7)(-8)$

are all equivalent.

Solution:

a. $7(-8)$ Translate: The product of 7 and -8.

   $= -56$ *Different* signs. Product is negative.

b. $(-3)(-11)$ Translate: -3 times -11.

   $= 33$ *Same* signs. Product is positive.

Skill Practice Translate each phrase to a mathematical expression. Then simplify.

6. The product of -4 and -12.   **7.** Eight times -5.

2. Multiplying Many Factors

In each of the following products, we can apply the order of operations and multiply from left to right.

two negative factors
$(-2)(-2)$
$= 4$
Product is positive.

three negative factors
$(-2)(-2)(-2)$
$= 4(-2)$
$= -8$
Product is negative.

four negative factors
$(-2)(-2)(-2)(-2)$
$= 4(-2)(-2)$
$= (-8)(-2)$
$= 16$
Product is positive.

five negative factors
$(-2)(-2)(-2)(-2)(-2)$
$= 4(-2)(-2)(-2)$
$= (-8)(-2)(-2)$
$= 16(-2)$
$= -32$
Product is negative.

These products illustrate the following rules.

- The product of an *even* number of negative factors is *positive*.
- The product of an *odd* number of negative factors is *negative*.

Answers
1. 12 **2.** -30 **3.** -42
4. 32 **5.** 0 **6.** $(-4)(-12)$; 48
7. $8(-5)$; -40

Example 3 Multiplying Several Factors

Multiply.

a. $(-2)(-5)(-7)$ b. $(-4)(2)(-1)(5)$

Solution:

a. $(-2)(-5)(-7)$ This product has an odd number of negative factors.
   $= -70$ The product is negative.

b. $(-4)(2)(-1)(5)$ This product has an even number of negative factors.
   $= 40$ The product is positive.

Skill Practice Multiply.

8. $(-3)(-4)(-8)(-1)$ 9. $(-1)(-4)(-6)(5)$

3. Exponential Expressions

Be particularly careful when evaluating exponential expressions involving negative numbers. An exponential expression with a negative base is written with parentheses around the base, such as $(-3)^4$.

To evaluate $(-3)^4$, the base -3 is multiplied 4 times:

$$(-3)^4 = (-3)(-3)(-3)(-3) = 81$$

If parentheses are *not* used, the expression -3^4 has a different meaning:

- The expression -3^4 has a base of 3 (not -3) and can be interpreted as $-1 \cdot 3^4$. Hence,

$$-3^4 = -1 \cdot (3)(3)(3)(3) = -81$$

- The expression -3^4 can also be interpreted as "the opposite of 3^4." Hence,

$$-3^4 = -(3 \cdot 3 \cdot 3 \cdot 3) = -81$$

Example 4 Simplifying Exponential Expressions

Simplify.

a. $(-4)^2$ b. -4^2 c. $(-5)^3$ d. -5^3

Solution:

a. $(-4)^2 = (-4)(-4)$ The base is -4.
   $= 16$ Multiply.

b. $-4^2 = -(4)(4)$ The base is 4. This is equal to $-1 \cdot 4^2 = -1 \cdot (4)(4)$.
   $= -16$ Multiply.

c. $(-5)^3 = (-5)(-5)(-5)$ The base is -5.
   $= -125$ Multiply.

d. $-5^3 = -(5)(5)(5)$ The base is 5. This is equal to $-1 \cdot 5^3 = -1 \cdot (5)(5)(5)$.
   $= -125$ Multiply.

Avoiding Mistakes

In Example 4(b) the base is positive because the negative sign is not enclosed in parentheses with the base.

Skill Practice Simplify.

10. $(-5)^2$ 11. -5^2 12. $(-2)^3$ 13. -2^3

Answers

8. 96   9. -120
10. 25   11. -25
12. -8 13. -8

4. Division of Integers

When we divide two numbers, we can check our result by using multiplication. Because multiplication and division are related in this way, it seems reasonable that the same sign rules that apply to multiplication also apply to division. For example:

$$24 \div 6 = 4 \qquad \text{Check:} \qquad 6 \cdot 4 = 24 \checkmark$$
$$-24 \div 6 = -4 \qquad \text{Check:} \qquad 6 \cdot (-4) = -24 \checkmark$$
$$24 \div (-6) = -4 \qquad \text{Check:} \qquad -6 \cdot (-4) = 24 \checkmark$$
$$-24 \div (-6) = 4 \qquad \text{Check:} \qquad -6 \cdot 4 = -24 \checkmark$$

We now summarize the rules for dividing signed numbers along with the properties of division involving zero.

Dividing Signed Numbers

1. The quotient of two numbers with the *same* sign is positive.

   Examples: $20 \div 5 = 4$ and $-20 \div (-5) = 4$

2. The quotient of two numbers with *different* signs is negative.

   Examples: $16 \div (-8) = -2$ and $-16 \div 8 = -2$

3. Zero divided by any nonzero number is zero.

   Examples: $0 \div 12 = 0$ and $0 \div (-3) = 0$

4. Any nonzero number divided by zero is undefined.

   Examples: $-5 \div 0$ is undefined

Example 5 Dividing Integers

Divide.

a. $50 \div (-5)$ b. $\dfrac{-42}{-7}$ c. $\dfrac{-39}{3}$ d. $0 \div (-7)$ e. $\dfrac{-8}{0}$

Solution:

a. $50 \div (-5) = -10$ *Different* signs. The quotient is negative.

b. $\dfrac{-42}{-7} = 6$ *Same* signs. The quotient is positive.

c. $\dfrac{-39}{3} = -13$ *Different* signs. The quotient is negative.

d. $0 \div (-7) = 0$ Zero divided by any nonzero number is 0.

e. $\dfrac{-8}{0}$ is undefined.   Any nonzero number divided by 0 is undefined.

TIP: In Example 5(e), $\dfrac{-8}{0}$ is undefined because there is no number that when multiplied by 0 will equal -8.

Skill Practice Divide.

14. $-40 \div 10$ 15. $\dfrac{-36}{-12}$ 16. $\dfrac{18}{-2}$ 17. $0 \div (-12)$ 18. $\dfrac{-2}{0}$

Answers
14. -4 15. 3
16. -9 17. 0
18. Undefined

Example 6 Translating to a Mathematical Expression

Translate the phrase into a mathematical expression. Then simplify.

a. The quotient of 26 and −13   b. −45 divided by 5

c. −4 divided into −24

Solution:

a. The word quotient implies division. The quotient of 26 and −13 translates to $26 \div (-13)$.

$$26 \div (-13) = -2 \quad \text{Simplify.}$$

b. −45 divided by 5 translates to $-45 \div 5$.

$$-45 \div 5 = -9 \quad \text{Simplify.}$$

c. −4 divided into −24 translates to $-24 \div (-4)$.

$$-24 \div (-4) = 6 \quad \text{Simplify.}$$

FOR REVIEW

Recall that division is not commutative.

$$26 \div (-13) \neq (-13) \div 26$$

Skill Practice Translate the phrase into a mathematical expression. Then simplify.
19. The quotient of −40 and −4
20. 60 divided by −3
21. −8 divided into −24

Example 7 Applying Division of Integers

Between midnight and 6:00 A.M., the change in temperature was −18°F. Find the average hourly change in temperature.

Solution:

In this example, we have a change of −18°F in temperature to distribute evenly over a 6-hr period (from midnight to 6:00 A.M. is 6 hr). This implies division.

$$-18 \div 6 = -3 \quad \text{Divide } -18°F \text{ by 6 hr.}$$

The temperature changed by −3°F per hour.

Skill Practice
22. A severe cold front blew through Atlanta and the temperature change over a 6-hr period was −24°F. Find the average hourly change in temperature.

Answers
19. $-40 \div (-4)$; 10
20. $60 \div (-3)$; −20
21. $-24 \div (-8)$; 3
22. −4°F

Section 2.4 Activity

A.1. Using a calculator, multiply and divide the integers.
   a. $9 \cdot 5 =$ _____
   b. $-9 \cdot (-5) =$ _____
   c. $18 \div 2 =$ _____
   d. $-18 \div (-2) =$ _____
   e. Compare the results in parts (a) and (b), and parts (c) and (d). What can you conclude about multiplying and dividing numbers of the same sign?
   f. State the rule for multiplying or dividing two numbers with the same sign.

A.2. Using a calculator, multiply and divide the integers.
 a. $-9 \cdot 5 =$ _____
 b. $9 \cdot (-5) =$ _____
 c. $-18 \div 2 =$ _____
 d. $18 \div (-2) =$ _____
 e. Compare the results in parts (a) and (b), and parts (c) and (d). What can you conclude about multiplying and dividing numbers with different signs?
 f. State the rule for multiplying or dividing two numbers with different signs.

A.3. Multiply the factors below.
 a. $(-1)(-1) =$ _____
 b. $(-1)(-1)(-1) =$ _____
 c. $(-1)(-1)(-1)(-1) =$ _____
 d. $(-1)(-1)(-1)(-1)(-1) =$ _____
 e. Is the product of an even number of negative factors positive or negative?
 f. Is the product of an odd number of negative factors positive or negative?
 g. Will the product of 891 negative factors be positive or negative? How do you know?

A.4. Multiply or divide.
 a. $0(-5) =$ _____
 b. $-2(-45) =$ _____
 c. $-60 \cdot 40 =$ _____
 d. $-2(5)(-10) =$ _____
 e. $0 \div (-8) =$ _____
 f. $82 \div (-1) =$ _____
 g. $-16\overline{)256} =$ _____
 h. $\dfrac{-42}{-2} =$ _____

A.5. Consider the exponential expression: $(-5)^2$.
 a. Identify the base in the expression.
 b. Identify the exponent in the expression.
 c. Write $(-5)^2$ in expanded form.
 d. Simplify $(-5)^2$.
 e. Evaluate $(-5)^2$ using your calculator and confirm the answer obtained in part (d).

A.6. Consider the exponential expression: -5^2.
 a. Identify the base in the expression.
 b. Identify the exponent in the expression.
 c. Write -5^2 in expanded form.
 d. Simplify -5^2.
 e. Evaluate $(-5)^2$ using your calculator and confirm the answer obtained in part (d).

For Exercises A.7–A.10, simplify the expression.

A.7. $(-2)^3 =$ _____

A.8. $-2^3 =$ _____

A.9. $(-2)^4 =$ _____

A.10. $-2^4 =$ _____

A.11. Observe your answers from Exercises A.7 and A.8. Do you get the same or opposite answers when the exponent is an odd number?

A.12. Observe your answers from Exercises A.9 and A.10. Do you get the same or opposite answers when the exponent is an even number?

A.13. Consider the expression: $(-2)^{105}$. Without the use of a calculator, will the result be positive or negative? How do you know?

Practice Exercises

Section 2.4

Study Skills Exercise

Students often learn a rule about signs that states "two negatives make a positive". This rule is incomplete and therefore not always true. Note the following combinations of two negatives:

$-2 + (-4)$ the sum of two negatives
$-(-5)$ the opposite of a negative
$-|-10|$ the opposite of an absolute value
$(-3)(-6)$ the product of two negatives

Simplify the expressions to determine which are negative and which are positive. Then write the rule for multiplying two numbers with the same sign.

$-2 + (-4)$ _____

$-(-5)$ _____

$-|-10|$ _____

$(-3)(-6)$ _____

When multiplying two numbers with the same sign, the product is _____.

Prerequisite Review

For Exercises R.1–R.4, add or subtract as indicated.

R.1. a. $14 - (-5)$ b. $(-5) + 14$

R.2. a. $-24 - 50$ b. $-24 - (-50)$

R.3. $23 - 12 + (-4) - (-10)$

R.4. $-23 + 12 - (-4) - (-10)$

Vocabulary and Key Concepts

1. a. The product of two numbers with the same sign is (positive/negative).
   The product of two numbers with different signs is (positive/negative).

   b. The quotient of two numbers with the same sign is (positive/negative).
   The quotient of two numbers with different signs is (positive/negative).

2. The product of an even number of negative factors is (positive/negative).
   The product of an odd number of negative factors is (positive/negative).

3. The base in the expression -3^2 is (positive/negative).
   The base in the expression $(-3)^2$ is (positive/negative).

4. a. $0 \div (-12) = $ _____
   b. $-12 \div 0 = $ _____ because no number multiplied by 0 equals -12.

Concept 1: Multiplication of Integers

For Exercises 5–24, multiply the integers. (See Example 1.)

5. $6(-4)$ **6.** $9(-5)$ **7.** $0(-2)$ **8.** $0(-14)$

9. $-3(5)$ **10.** $-2(13)$ **11.** $(-5)(-8)$ **12.** $(-12)(-2)$

13. $7(-3)$ **14.** $5(-12)$ **15.** $-12(-4)$ **16.** $-6(-11)$

17. $-15(3)$ **18.** $-3(25)$ **19.** $9(-8)$ **20.** $8(-3)$

21. $-14 \cdot 0$ **22.** $-8 \cdot 0$ **23.** $-95(-1)$ **24.** $-144(-1)$

For Exercises 25–30, translate to a mathematical expression. Then simplify. (See Example 2.)

25. Multiply −3 and −1
26. Multiply −12 and −4
27. The product of −5 and 3
28. The product of 9 and −2
29. 3 times −5
30. −3 times 6

Concept 2: Multiplying Many Factors

For Exercises 31–40, multiply. (See Example 3.)

31. $(-5)(-2)(-4)(-10)$
32. $(-3)(-5)(-2)(-4)$
33. $(-11)(-4)(-2)$
34. $(-20)(-3)(-1)$
35. $(24)(-2)(0)(-3)$
36. $(3)(0)(-13)(22)$
37. $(-1)(-1)(-1)(-1)(-1)(-1)$
38. $(-1)(-1)(-1)(-1)(-1)(-1)(-1)$
39. $(-2)(2)(2)(-2)(2)$
40. $(2)(-2)(2)(2)$

Concept 3: Exponential Expressions

For Exercises 41–56, simplify. (See Example 4.)

41. -10^2
42. -8^2
43. $(-10)^2$
44. $(-8)^2$
45. -10^3
46. -8^3
47. $(-10)^3$
48. $(-8)^3$
49. -5^4
50. -4^4
51. $(-5)^4$
52. $(-4)^4$
53. $(-1)^2$
54. $(-1)^3$
55. -1^4
56. -1^5

Concept 4: Division of Integers

For Exercises 57–72, divide the real numbers, if possible. (See Example 5.)

57. $60 \div (-3)$
58. $46 \div (-2)$
59. $\dfrac{-56}{-8}$
60. $\dfrac{-48}{-3}$
61. $\dfrac{-15}{5}$
62. $\dfrac{30}{-6}$
63. $-84 \div (-4)$
64. $-48 \div (-6)$
65. $\dfrac{-13}{0}$
66. $\dfrac{-41}{0}$
67. $\dfrac{0}{-18}$
68. $\dfrac{0}{-6}$
69. $(-20) \div (-5)$
70. $(-10) \div (-2)$
71. $\dfrac{204}{-6}$
72. $\dfrac{300}{-2}$

For Exercises 73–78, translate the English phrase to a mathematical expression. Then simplify. (See Example 6.)

73. The quotient of −100 and 20
74. The quotient of 46 and −23
75. −64 divided by −32
76. −108 divided by −4
77. 13 divided into −52
78. −15 divided into −45

79. During a 10-min period, a SCUBA diver's depth changed by −60 ft. Find the average change in depth per minute. (See Example 7.)

80. When a severe winter storm moved through Albany, New York, the change in temperature between 4:00 P.M. and 7:00 P.M. was −27°F. What was the average hourly change in temperature?

81. One of the most famous blizzards in the United States was the blizzard of 1888. In one part of Nebraska, the temperature plunged from 40°F to −25°F in 5 hr. What was the average change in temperature during this time?

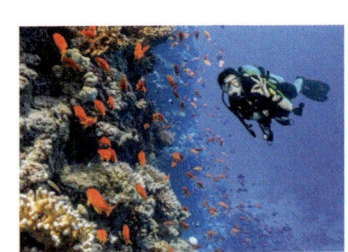

Jag_cz/Shutterstock

82. A submarine descended from a depth of −528 m to −1804 m in a 2-hr period. What was the average change in depth per hour during this time?

83. Travis wrote five checks to the employees of his business, each for $225. If the original balance in his checking account was $890, what is his new balance?

84. Jennifer's checking account had $320 when she wrote two checks for $150, and one check for $82. What is her new balance?

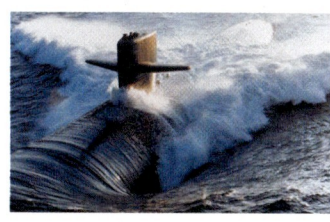
StockTrek/SuperStock

85. During a severe drought in Georgia, the water level in Lake Lanier dropped. During a 1-month period from June to July, the lake's water level changed by −3 ft. If this continued for 6 months, by how much did the water level change?

86. During a drought, the change in water level for a retention pond was −9 in. over a 1-month period. At this rate, what will the change in water level be after 5 months?

Mixed Exercises

For Exercises 87–102, perform the indicated operation.

87. $18(-6)$
88. $24(-2)$
89. $18 \div (-6)$
90. $24 \div (-2)$
91. $(-9)(-12)$
92. $-36 \div (-12)$
93. $-90 \div (-6)$
94. $(-5)(-4)$
95. $\dfrac{0}{-2}$
96. $-24 \div 0$
97. $-90 \div 0$
98. $\dfrac{0}{-5}$
99. $(-2)(-5)(4)$
100. $(10)(-2)(-3)(-5)$
101. $(-7)^2$
102. -7^2

103. a. What number must be multiplied by −5 to obtain −35?
   b. What number must be multiplied by −5 to obtain 35?

104. a. What number must be multiplied by −4 to obtain −36?
   b. What number must be multiplied by −4 to obtain 36?

Technology Connections

For Exercises 105–108, use a calculator to perform the indicated operations.

105. $(-413)(871)$
106. $-6125 \cdot (-97)$
107. $\dfrac{-576{,}828}{-10{,}682}$
108. $5{,}945{,}308 \div (-9452)$

Expanding Your Skills

The electrical charge of an atom is determined by the number of protons and number of electrons the atom has. Each proton gives a positive charge (+1) and each electron gives a negative charge (−1). An atom that has a total charge of 0 is said to be electrically neutral. An atom that is not electrically neutral is called an ion. For Exercises 109–112, determine the total charge for an atom with the given number of protons and electrons.

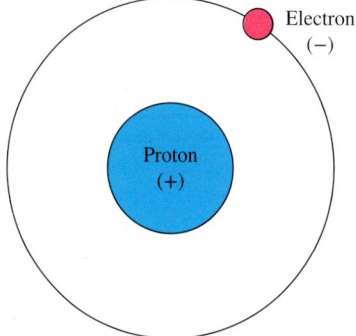
The Hydrogen Atom

109. 1 proton, 0 electrons
110. 17 protons, 18 electrons
111. 8 protons, 10 electrons
112. 20 protons, 18 electrons

For Exercises 113–118, assume $a > 0$ (this means that a is positive) and $b < 0$ (this means that b is negative). Find the sign of each expression.

113. $a \cdot b$
114. $b \div a$
115. $|a| \div b$
116. $a \cdot |b|$
117. $-a \div b$
118. $a(-b)$

Problem Recognition Exercises

Operations on Integers

1. Perform the indicated operations.
   a. $(-24)(-2)$
   b. $(-24) - (-2)$
   c. $(-24) + (-2)$
   d. $(-24) \div (-2)$

2. Perform the indicated operations.
   a. $12(-3)$
   b. $12 - (-3)$
   c. $12 + (-3)$
   d. $12 \div (-3)$

For Exercises 3–14, translate each phrase to a mathematical expression. Then simplify.

3. The sum of -5 and -3
4. The product of 9 and -5
5. The difference of -3 and -7
6. The quotient of 28 and -4
7. -23 times -2
8. 18 subtracted from -4
9. 42 divided by -2
10. -13 added to -18
11. -12 subtracted from 10
12. -7 divided into -21
13. The product of -6 and -9
14. The total of -7, 4, 8, -16, and -5

For Exercises 15–38, perform the indicated operations.

15. a. $15 - (-5)$
    b. $15(-5)$
    c. $15 + (-5)$
    d. $15 \div (-5)$

16. a. $-36(-2)$
    b. $-36 - (-2)$
    c. $\dfrac{-36}{-2}$
    d. $-36 + (-2)$

17. a. $20(-4)$
    b. $-20(-4)$
    c. $-20(4)$
    d. $20(4)$

18. a. $|-50|$
    b. $-(-50)$
    c. $|50|$
    d. $-|-50|$

19. a. $-5 - 9 - 2$
    b. $-5(-9)(-2)$

20. a. $10 + (-3) + (-12)$
    b. $10 - (-3) - (-12)$

21. a. $(-1)(-2)(-3)(-4)$
    b. $(-1)(-2)(-3)(4)$

22. a. $(5)(-2)(-6)(1)$
    b. $(-5)(-2)(6)(-1)$

23. $\dfrac{0}{-8}$

24. $-55 \div 0$

25. $-615 - (-705)$

26. $-184 - 409$

27. $420 \div (-14)$

28. $-3600 \div (-90)$

29. $-44 - (-44)$

30. $-37 - (-37)$

31. $(-9)^2$

32. $(-2)^5$

33. -9^2

34. -2^5

35. $\dfrac{-46}{0}$

36. $0 \div (-16)$

37. $-15{,}042 + 4893$

38. $-84{,}506 + (-542)$

Order of Operations and Algebraic Expressions

Section 2.5

Concepts
1. Order of Operations
2. Translations Involving Variables
3. Evaluating Algebraic Expressions

1. Order of Operations

The order of operations applies when simplifying expressions with integers.

> **Applying the Order of Operations**
> 1. First perform all operations inside parentheses and other grouping symbols.
> 2. Simplify expressions containing exponents, square roots, or absolute values.
> 3. Perform multiplication or division in the order that they appear from left to right.
> 4. Perform addition or subtraction in the order that they appear from left to right.

Example 1   Applying the Order of Operations

Simplify. $-12 - 6(7 - 5)$

Solution:

$-12 - 6(7 - 5)$
$= -12 - 6(2)$ Simplify within parentheses first.
$= -12 - 12$ Multiply before subtracting.
$= -24$ Subtract. *Note:* $-12 - 12 = -12 + (-12) = -24$

Skill Practice Simplify.
1. $8 - 2(3 - 10)$

Example 2   Applying the Order of Operations

Simplify. $6 + 48 \div 4 \cdot (-2)$

Solution:

$6 + \underline{48 \div 4} \cdot (-2)$ Perform division and multiplication from left to right before addition.
$= 6 + \underline{12 \cdot (-2)}$ Perform multiplication before addition.
$= 6 + (-24)$ Add.
$= -18$

Skill Practice Simplify.
2. $-10 + 24 \div 2 \cdot (-3)$

Answers
1. 22   2. −46

Example 3 Applying the Order of Operations

Simplify. $3^2 - 10^2 \div (-1 - 4)$

Solution:

$3^2 - 10^2 \div (-1 - 4)$

$= 3^2 - 10^2 \div (-5)$ Simplify within parentheses.
Note: $-1 - 4 = -1 + (-4) = -5$

$= 9 - 100 \div (-5)$ Simplify exponents.
Note: $3^2 = 3 \cdot 3 = 9$ and $10^2 = 10 \cdot 10 = 100$

$= 9 - (-20)$ Perform division before subtraction.
Note: $100 \div (-5) = -20$.

$= 29$ Subtract. *Note:* $9 - (-20) = 9 + (20) = 29$

Skill Practice Simplify.

3. $8^2 - 2^3 \div (-7 + 5)$

Example 4 Applying the Order of Operations

Simplify.

a. $\dfrac{|-8 + 24|}{5 - 3^2}$ b. $5 - 2[8 + (-7 - 5)]$

Solution:

a. $\dfrac{|-8 + 24|}{5 - 3^2}$ Simplify the expressions above and below the division bar by first adding within the absolute value and simplifying exponents.

$= \dfrac{|16|}{5 - 9}$

$= \dfrac{16}{-4}$ Simplify the absolute value and subtract $5 - 9$.

$= -4$ Divide.

b. $5 - 2[8 + (-7 - 5)]$

$= 5 - 2[8 + (-12)]$ First simplify the expression within the innermost parentheses.

$= 5 - 2[-4]$ Continue simplifying within parentheses.

$= 5 - (-8)$ Perform multiplication before subtraction.

$= 13$ Subtract. *Note:* $5 - (-8) = 5 + 8 = 13$

Skill Practice Simplify.

4. $\dfrac{|50 + (-10)|}{-2^2}$ 5. $1 - 3[5 - (8 - 1)]$

FOR REVIEW

Recall that the absolute value of a number is the distance between the number and zero on the number line.

$|16| = 16$
$|-16| = 16$

2. Translations Involving Variables

Recall that **variables** are used to represent quantities that are subject to change. For this reason, we can use variables and algebraic expressions to represent one or more unknowns in a word problem.

Answers
3. 68   4. −10   5. 7

Example 5 Using Algebraic Expressions in Applications

a. At a discount CD store, each CD costs $8. Suppose n is the number of CDs that a customer buys. Write an expression that represents the cost for n CDs.

b. The length of a rectangle is 5 in. longer than the width w. Write an expression that represents the length of the rectangle.

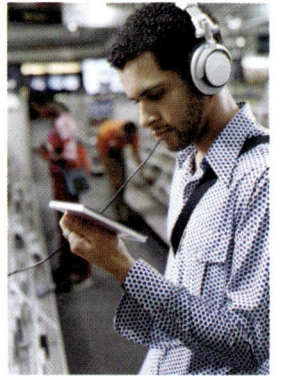
Chad Baker/Jason Reed/Ryan McVay/Getty Images

Solution:

a. The cost of 1 CD is 8(1) dollars.   From this pattern,
   The cost of 2 CDs is 8(2) dollars. we see that the total
   The cost of 3 CDs is 8(3) dollars. cost is the cost per
   The cost of n CDs is 8(n) dollars CD times the num-
   or simply $8n$ dollars.   ber of CDs.

b. The length of a rectangle is 5 in. more than the width. The phrase "more than" implies addition. Thus, the length (in inches) is represented by

$$\text{length} = w + 5$$

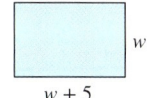
w
$w + 5$

Skill Practice

6. Smoked turkey costs $7 per pound. Write an expression that represents the cost of p pounds of turkey.

7. The width of a basketball court is 44 ft shorter than its length L. Write an expression that represents the width.

Example 6 Translating to an Algebraic Expression

Write each phrase as an algebraic expression.

a. The product of −6 and x
b. p subtracted from 7
c. The quotient of c and d
d. Twice the sum of y and 4

Solution:

a. The product of −6 and x: $-6x$ "Product" implies multiplication.

b. p subtracted from 7: $7 - p$ To subtract p from 7, we must "start" with 7 and then perform the subtraction.

c. The quotient of c and d: $\dfrac{c}{d}$ "Quotient" implies division.

d. Twice the sum of y and 4: $2(y + 4)$ The word "twice the sum" implies that we multiply the sum by 2. The sum must be enclosed in parentheses so that the entire quantity is doubled.

Skill Practice Write each phrase as an algebraic expression.

8. The quotient of w and −4
9. 11 subtracted from x
10. The product of −9 and p
11. Four times the sum of m and n

Answers

6. $7p$ 7. $L - 44$ 8. $\dfrac{w}{-4}$
9. $x - 11$ 10. $-9p$ 11. $4(m + n)$

3. Evaluating Algebraic Expressions

The value of an algebraic expression depends on the values of the variables within the expression.

Example 7   Evaluating an Algebraic Expression

Evaluate the expression for the given values of the variables.

$$3x - y \quad \text{for} \quad x = 7 \quad \text{and} \quad y = -3$$

Solution:

$3x - y$

$= 3(\ ) - (\ )$ When substituting a number for a variable, use parentheses in place of the variable.

$= 3(7) - (-3)$ Substitute 7 for x and -3 for y.

$= 21 + 3$ Apply the order of operations. Multiplication is performed before subtraction.

$= 24$

Skill Practice

12. Evaluate $4a + b$ for $a = 8$ and $b = -52$.

Example 8   Evaluating an Algebraic Expression

Evaluate the expressions for the given values of the variables.

a. $-|-z|$ for $z = -6$ **b.** x^2 for $x = -4$ **c.** $-y^2$ for $y = -6$

Solution:

a. $-|-z|$

$= -|-(\ )|$ Replace the variable with empty parentheses.

$= -|-(-6)|$ Substitute the value -6 for z. Apply the order of operations. Inside the absolute value bars, we take the opposite of -6, which is 6.

$= -|6|$

$= -6$ The opposite of $|6|$ is -6.

b. x^2

$= (\ )^2$ Replace the variable with empty parentheses.

$= (-4)^2$ Substitute -4 for x.

$= (-4)(-4)$ The expression $(-4)^2$ has a base of -4. Therefore, $(-4)^2 = (-4)(-4) = 16$.

$= 16$

c. $-y^2$

$= -(\ )^2$ Replace the variable with empty parentheses.

$= -(-6)^2$ Substitute -6 for y.

$= -(36)$ Simplify the expression with exponents first. The expression $(-6)^2 = (-6)(-6) = 36$.

$= -36$

Answer

12. -20

Skill Practice Evaluate the expressions for the given values of the variable.

13. Evaluate $-|-y|$ for $y = -12$.
14. Evaluate p^2 for $p = -5$.
15. Evaluate $-w^2$ for $w = -10$.

Example 9 — Evaluating an Algebraic Expression

Evaluate the expression for the given values of the variables.

$$-5|x - y + z|, \quad \text{for } x = -9, y = -4, \text{ and } z = 2$$

Solution:

$-5|x - y + z|$

$= -5|() - () + ()|$ Replace the variables with empty parentheses.

$= -5|(-9) - (-4) + (2)|$ Substitute -9 for x, -4 for y, and 2 for z.

$= -5|-9 + 4 + 2|$ Simplify inside absolute value bars first. Rewrite subtraction in terms of addition.

$= -5|-3|$ Add within the absolute value bars.

$= -5 \cdot 3$ Evaluate the absolute value before multiplying.

$= -15$

TIP: Absolute value bars act as grouping symbols. Therefore, you must perform the operations within the absolute value first.

Skill Practice

16. Evaluate the expression for the given value of the variable.
    $3 - |a + b + 4|$ for $a = -5$ and $b = -12$

Answers
13. -12 14. 25
15. -100 16. -10

Section 2.5 Activity

For Exercises A.1–A.4, simplify the expression. Write the mathematical steps in the first column. Write a description of each step in the right column.

Mathematical Steps	Description
A.1. $-4 - 8(12 - 15)$	
A.2. $8^2 + 4^2 \div (-8 \div 4)$	

Mathematical Steps	Description
A.3. $(-10)^2 - (15)(-8)$	
A.4. $|3 - 8| - 15 \div 3$	

A.5. Consider the expression $y^2 - x^2$ when $x = -4$ and $y = 6$.
   a. Write the expression using parentheses in place of the variables.
   b. Substitute the values of x and y into the expression.
   c. Evaluate the expression.
   d. Evaluate the expression without the use of parentheses. Do you obtain the same answer as in part (c)?
   e. Which answer is the correct answer, part (c) or part (d)? Why?

For Exercises A.6–A.9, evaluate the expression for the given values of the variables: $x = -4$, $y = 6$, and $z = -2$.

A.6. $x - y - z$

A.7. $|z - y|$

A.8. $-z + xy$

A.9. $(x - y)^2 - |z|$

Section 2.5 Practice Exercises

Study Skills Exercise

How you respond when you are faced with tough times and adversity is a personal characteristic called "grit." Grit is defined as the sustained perseverance and passion for long-term goals. Achievement and success are not the result of intelligence and IQ alone. Rather, they require a combination of both skill *and* effort.

Reflect on the following questions. Write down your answers in your math notebook. Revisit your responses at any point during the course when you find yourself challenged or inclined to give up.
- What are your long-term career goals?
- How is this course relevant to your future career?
- What are your strengths?
- How can your strengths be used to help you succeed in this course?

Prerequisite Review

For Exercises R.1–R.6, perform the indicated operation.

R.1. a. $-7 \div 0$ b. $(-7)(0)$

R.2. a. $0 \div (-7)$ b. $(0)(-7)$

R.3. a. $(-100) \div (-4)$ b. $(-100) - (-4)$

R.4. a. $(-100)(-4)$ b. $(-100) + (-4)$

R.5. a. $(-12)^2$ b. -12^2

R.6. a. -2^3 b. $(-2)^3$

Vocabulary and Key Concepts

1. The order of operations (does/does not) change when simplifying expressions with integers.
2. When evaluating an algebraic expression, replace the _____ with an empty parentheses.

Concept 1: Order of Operations

For Exercises 3–56, simplify using the order of operations. (See Examples 1–4.)

3. $-8 - 3(2)$
4. $-10 - 2(4)$
5. $8 - 3(-2)$
6. $10 - 2(-4)$
7. $-7 - 2 - 5$
8. $(-7)(-2)(-5)$
9. $-1 - 5 - 8 - 3$
10. $-2 - 6 - 3 - 10$
11. $(-1)(-5)(-8)(-3)$
12. $(-2)(-6)(-3)(-10)$
13. $5 + 2(3 - 5)$
14. $6 - 4(8 - 10)$
15. $-2(3 - 6) + 10$
16. $-4(1 - 3) - 8$
17. $-8 - 6^2$
18. $-10 - 5^2$
19. $120 \div (-4)(5)$
20. $36 \div (-2)(3)$
21. $40 - 32 \div (-4)(2)$
22. $48 - 36 \div (6)(-2)$
23. $100 - 2(3 - 8)$
24. $55 - 3(2 - 6)$
25. $|-10 + 13| - |-6|$
26. $|4 - 9| - |-10|$
27. $\sqrt{100 - 36} - 3\sqrt{16}$
28. $\sqrt{36 - 11} + 2\sqrt{9}$
29. $5^2 - (-3)^2$
30. $6^2 - (-4)^2$
31. $-3 + 2(5 - 9)^2$
32. $-5 + 4(8 - 10)^2$
33. $12 + (14 - 16)^2 \div (-4)$
34. $-7 + (1 - 5)^2 \div 4$
35. $-48 \div 12 \div (-2)$
36. $-100 \div (-5) \div (-5)$
37. $90 \div (-3) \cdot (-1) \div (-6)$
38. $64 \div (-4) \cdot 2 \div (-16)$
39. $[7^2 - 9^2] \div (-5 + 1)$
40. $[(-8)^2 - 5^2] \div (-4 + 1)$
41. $2 + 2^2 - 10 - 12$
42. $14 - 4^2 + 2 - 10$
43. $\dfrac{3^2 - 27}{-9 + 6}$
44. $\dfrac{8 + (-2)^2}{-5 + (-1)}$
45. $\dfrac{13 - (2)(4)}{-1 - 2^2}$
46. $\dfrac{10 - (-3)(5)}{-9 - 4^2}$
47. $\dfrac{|-23 + 7|}{5^2 - (-3)^2}$
48. $\dfrac{|10 - 50|}{6^2 - (-4)^2}$
49. $21 - [4 - (5 - 8)]$
50. $15 - [10 - (20 - 25)]$
51. $-17 - 2[18 \div (-3)]$
52. $-8 - 5(-45 \div 15)$
53. $4 + 2[9 + (-4 + 12)]$
54. $-13 + 3[11 + (-15 + 10)]$
55. $-36 \div (-2) \div 6(-3) \cdot 2$
56. $-48 \div (4) \div 2(-5) \cdot 2$

Concept 2: Translations Involving Variables

57. Carolyn sells homemade candles. Write an expression for her total revenue if she sells x candles for $15 each. (See Example 5.)

58. Maria needs to buy 12 wine glasses. Write an expression for the cost of 12 glasses at p dollars each.

John Foxx/Getty Images

59. Jonathan is 4 in. taller than his brother. Write an expression for Jonathan's height if his brother is t inches tall. (See Example 5.)

60. It takes Perry 1 hr longer than David to mow the lawn. If it takes David h hours to mow the lawn, write an expression for the amount of time it takes Perry to mow the lawn.

61. A sedan travels 6 mph slower than a sports car. Write an expression for the speed of the sedan if the sports car travels v mph.

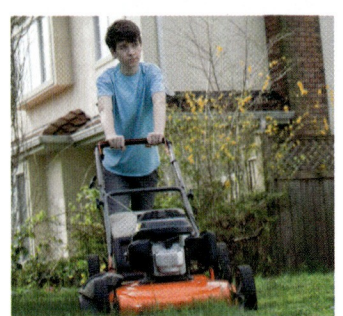

Pixel_Pig/Colleen Butler/iStock/Getty Images

62. Bill's daughter is 30 years younger than he is. Write an expression for his daughter's age if Bill is A years old.

63. The price of gas has doubled over the last 3 years. If gas cost g dollars per gallon 3 years ago, write an expression for the current price per gallon.

64. Suppose that the amount of rain that fell on Monday was twice the amount that fell on Sunday. Write an expression for the amount of rain on Monday, if Sunday's amount was t inches.

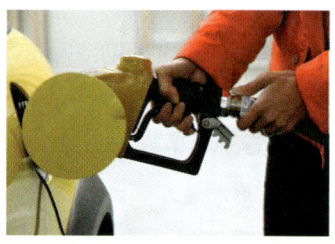

Frederic Charpentier/Alamy Stock Photo

For Exercises 65–76, write each phrase as an algebraic expression. (See Example 6.)

65. The product of -12 and n
66. The product of -3 and z
67. x subtracted from -9
68. p subtracted from -18
69. The quotient of t and -2
70. The quotient of -10 and w
71. -14 added to y
72. -150 added to c
73. Twice the sum of c and d
74. Twice the sum of a and b
75. The difference of x and -8
76. The difference of m and -5

Concept 3: Evaluating Algebraic Expressions

For Exercises 77–96, evaluate the expressions for the given values of the variables. (See Examples 7–9.)

77. $x + 9z$ for $x = -10$ and $z = -3$
78. $a + 7b$ for $a = -3$ and $b = -6$
79. $x + 5y + z$ for $x = -10$, $y = 5$, and $z = 2$
80. $9p + 4t + w$ for $p = 2$, $t = 6$, and $w = -50$
81. $a - b + 3c$ for $a = -7$, $b = -2$, and $c = 4$
82. $w + 2y - z$ for $w = -9$, $y = 10$, and $z = -3$
83. $-3mn$ for $m = -8$ and $n = -2$
84. $-5pq$ for $p = -4$ and $q = -2$
85. $|-y|$ for $y = -9$
86. $|-z|$ for $z = -18$
87. $-|-w|$ for $w = -4$
88. $-|-m|$ for $m = -15$
89. x^2 for $x = -3$
90. n^2 for $n = -9$
91. $-x^2$ for $x = -3$
92. $-n^2$ for $n = -9$
93. $-4|x + 3y|$ for $x = 5$ and $y = -6$
94. $-2|4a - b|$ for $a = -8$ and $b = -2$
95. $6 - |m - n^2|$ for $m = -2$ and $n = 3$
96. $4 - |c^2 - d^2|$ for $c = 3$ and $d = -5$

Expanding Your Skills

97. Find the average temperature: $-8°, -11°, -4°, 1°, 9°, 4°, -5°$

98. Find the average temperature: $15°, 12°, 10°, 3°, 0°, -2°, -3°$

99. Find the average score: $-8, -8, -6, -5, -2, -3, 3, 3, 0, -4$

100. Find the average score: $-6, -2, 5, 1, 0, -3, 4, 2, -7, -4$

Chapter 2 Review Exercises

Section 2.1

For Exercises 1 and 2, write an integer that represents each numerical value.

1. The plane descended 4250 ft.

2. The company's profit fell by $3,000,000.

For Exercises 3–6, graph the numbers on the number line.

3. −2   4. −5   5. 0   6. 3

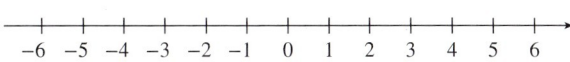

For Exercises 7 and 8, determine the opposite and the absolute value for each number.

7. −4   8. 6

For Exercises 9–16, simplify.

9. $|-3|$ 10. $|-1000|$ 11. $|74|$

12. $|0|$ 13. $-(-9)$ 14. $-(-28)$

15. $-|-20|$ 16. $-|-45|$

For Exercises 17–20, fill in the blank with $<$, $>$, or $=$, to make a true statement.

17. -7 ☐ $|-7|$ 18. -12 ☐ -5

19. $-(-4)$ ☐ $-|-4|$ 20. -20 ☐ $-|-20|$

Section 2.2

For Exercises 21–26, add the integers using the number line.

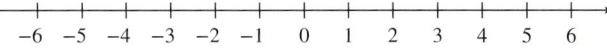

21. $6 + (-2)$ 22. $-3 + 6$

23. $-3 + (-2)$ 24. $-3 + 0$

25. $-3 + (-3)$ 26. $-5 + 4$

For Exercises 27–32, add the integers.

27. $35 + (-22)$ 28. $-105 + 90$

29. $-29 + (-41)$ 30. $-98 + (-42)$

31. $-3 + (-10) + 12 + 14 + (-10)$

32. $9 + (-15) + 2 + (-7) + (-4)$

For Exercises 33–38, translate each phrase to a mathematical expression. Then simplify the expression.

33. The sum of 23 and −35   34. 57 plus −10

35. The total of −5, −13, and 20

36. −42 increased by 12   37. 3 more than −12

38. −89 plus −22

39. The graph gives the number of inches below or above the average snowfall for the given months for Caribou, Maine. Find the total departure from average. Is the snowfall for Caribou above or below average?

Digital Vision/Getty Images

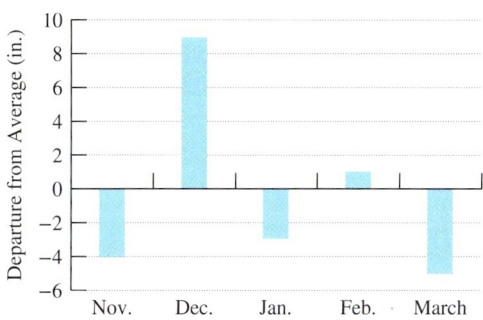

40. The table gives the scores for a golfer at a recent PGA Open golf tournament. Find the golfer's total score after all four rounds.

	Round 1	Round 2	Round 3	Round 4
Score	2	−2	−1	−4

Section 2.3

41. Write the subtraction statement as an addition statement.
    $-7 - 5$

For Exercises 42–49, simplify.

42. $4 - (-23)$ 43. $19 - 44$

44. $-2 - (-24)$ 45. $-289 - 130$

46. $2 - 7 - 3$ 47. $-45 - (-77) + 8$

48. $-16 - 4 - (-3)$ 49. $99 - (-7) - 6$

50. Translate the phrase to a mathematical expression. Then simplify.
    a. The difference of 8 and 10
    b. 8 subtracted from 10

For Exercises 51 and 52, translate the mathematical statement to an English phrase. Answers will vary.

51. $-2 - 14$

52. $-25 - (-7)$

53. The temperature in Fargo, North Dakota, rose from $-6°F$ to $-1°F$. By how many degrees did the temperature rise?

54. Sam's balance in his checking account was $-\$40$, so he deposited $\$132$. What is his new balance?

55. Find the average of the golf scores: $-3, 4, 0, 9, -2, -1, 0, 5, -3$. (These scores are the number of strokes above or below par.)

56. A missile was launched from a submarine from a depth of -1050 ft below sea level. If the maximum height of the missile is 2400 ft above sea level, find the vertical distance between its greatest height and its depth at launch.

Section 2.4

For Exercises 57–72, simplify.

57. $6(-3)$

58. $\dfrac{-12}{4}$

59. $\dfrac{-900}{-60}$

60. $(-7)(-8)$

61. $-36 \div 9$

62. $60 \div (-5)$

63. $(-12)(-4)(-1)(-2)$

64. $(-1)(-8)(2)(1)(-2)$

65. $-15 \div 0$

66. $\dfrac{0}{-5}$

67. -5^3

68. $(-5)^3$

69. $(-6)^2$

70. -6^2

71. $(-1)^{10}$

72. $(-1)^{21}$

73. What is the sign of the product of three negative factors?

74. What is the sign of the product of four negative factors?

For Exercises 75 and 76, translate the English phrase to a mathematical expression. Then simplify.

75. The quotient of -45 and -15

76. The product of -4 and 19

77. Between 8:00 P.M. and midnight, the change in temperature was $-12°F$. Find the average hourly change in temperature.

78. Suzie wrote four checks to a vendor, each for $\$160$. If the original balance in her checking account was $\$550$, what is her new balance?

Section 2.5

For Exercises 79–88, simplify using the order of operations.

79. $50 - 3(6 - 2)$

80. $48 - 8 \div (-2) + 5$

81. $28 \div (-7) \cdot 3 - (-1)$

82. $(-4)^2 \div 8 - (-6)$

83. $[10 - (-3)^2] \cdot (-11) + 4$

84. $[-9 - (-7)]^2 \cdot 3 \div (-6)$

85. $\dfrac{100 - 4^2}{(-7)(6)}$

86. $\dfrac{18 - 3(-2)}{4^2 - 8}$

87. $5 - 2[-3 + (2 - 5)]$

88. $-10 + 3[4 - (-2 + 7)]$

89. Michael is 8 years older than his sister. Write an expression for Michael's age if his sister is a years old.

90. At a movie theater, drinks are $\$3$ each. Write an expression for the cost of n drinks.

Ryan McVay/Getty Images

For Exercises 91–96, write each phrase as an algebraic expression.

91. The product of −5 and x

92. The difference of p and 12

93. Two more than the sum of a and b

94. The quotient of w and 4

95. −8 subtracted from y

96. Twice the sum of 5 and z

For Exercises 97–104, evaluate the expression for the given values of the variable.

97. $3x - 2y$ for $x = -5$ and $y = 4$

98. $5(a - 4b)$ for $a = -3$ and $b = 2$

99. $-2(x + y)^2$ for $x = 6$ and $y = -9$

100. $-3w^2 - 2z$ for $w = -4$ and $z = -9$

101. $-|x|$ for $x = -2$

102. $-|-x|$ for $x = -5$

103. $-(-x)$ for $x = -10$

104. $-(-x)$ for $x = 5$

Solving Equations

3

CHAPTER OUTLINE

3.1 Simplifying Expressions and Combining *Like* Terms   126

3.2 Addition and Subtraction Properties of Equality   135

3.3 Multiplication and Division Properties of Equality   143

3.4 Solving Equations with Multiple Steps   149

  Problem Recognition Exercises: Identifying Expressions and Equations   156

3.5 Applications and Problem Solving   156

  Chapter 3 Review Exercises   166

Mathematics and Speeding

Suppose that driving home one day after work, your speedometer "creeps up" above the posted speed limit. (Not your fault, of course!) Unfortunately for you, a police officer poised with a radar detector is parked along the same road. After pulling you over, the officer explains that the fine for speeding is $90 plus $10 for every mile per hour you were caught traveling over the speed limit. If your speeding ticket costs $260, how many miles per hour were you traveling over the posted speed limit?

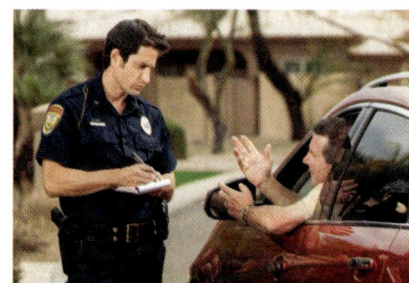

Jacom Stephens/Avid Creative, Inc./iStockphoto

To solve this riddle, you might subtract the fixed cost of $90 from the total bill of $260, leaving you with $170. Then, after dividing $170 by $10, you would know that you were traveling 17 mph over the speed limit.

This scenario can also be solved by using a mathematical equation called a **linear equation**. If x is the number of miles per hour over the speed limit, the equation $90 + 10x = 260$ can be used to solve for x. To solve an equation, we have a prescribed, methodical approach which takes out the guesswork in solving applications.

Although the equation given here is relatively simple, many complicated mathematical equations are at play in our daily lives. For example, equations are involved in transportation systems, financial systems, communications, and health and crime statistics.

Section 3.1 Simplifying Expressions and Combining *Like* Terms

Concepts

1. Identifying *Like* Terms
2. Commutative, Associative, and Distributive Properties
3. Combining *Like* Terms
4. Simplifying Expressions

1. Identifying *Like* Terms

A **term** is a number or the product or quotient of numbers and variables. An algebraic expression is the sum of one or more terms. For example, the expression

$$-8x^3 + xy - 40 \quad \text{can be written as} \quad -8x^3 + xy + (-40)$$

This expression consists of the terms $-8x^3$, xy, and -40. The terms $-8x^3$ and xy are called **variable terms** because the value of the term will change when different numbers are substituted for the variables. The term -40 is called a **constant term** because its value will never change.

It is important to distinguish between a term and the factors within a term. For example, the quantity xy is one term, and the values x and y are factors within the term. The **coefficient** of the term is the numerical factor of the term.

Term	Coefficient of the Term
$-8x^3$	-8
xy or $1xy$	1
-40	-40
$\dfrac{x}{4}$ or $\dfrac{1}{4}x$	$\dfrac{1}{4}$

TIP: Variables without a coefficient explicitly written have a coefficient of 1. Thus, the term *xy* is equal to 1*xy*. The 1 is understood.

Terms are said to be *like* **terms** if they each have the same variables, and the corresponding variables are raised to the same powers. For example:

Like Terms		Unlike Terms	
$-4x$ and $6x$		$-4x$ and $6y$	(different variables)
$18ab$ and $4ba$		$18ab$ and $4a$	(different variables)
$7m^2n^5$ and $3m^2n^5$		$7m^2n^5$ and $3mn^5$	(different powers of m)
$5p$ and $-3p$		$5p$ and 3	(different variables)
8 and 10		8 and $10x$	(different variables)

Example 1 Identifying Terms, Coefficients, and *Like* Terms

a. List the terms of the expression: $14x^3 - 6x^2 + 3x + 5$
b. Identify the coefficient of each term: $14x^3 - 6x^2 + 3x + 5$
c. Which two terms are *like* terms? $-6x, 5, -3y,$ and $4x$
d. Are the terms $3x^2y$ and $5xy^2$ *like* terms?

Solution:

a. The expression $14x^3 - 6x^2 + 3x + 5$ can be written as $14x^3 + (-6x^2) + 3x + 5$. Therefore, the terms are $14x^3, -6x^2, 3x,$ and 5.

b. The coefficients are $14, -6, 3,$ and 5.

c. The terms $-6x$ and $4x$ are *like* terms.

d. $3x^2y$ and $5xy^2$ are not *like* terms because the exponents on the x variables are different, and the exponents on the y variables are different.

Skill Practice For Exercises 1 and 2, use the expression:
$$-48y^5 + 8y^2 - y - 8$$
1. List the terms of the expression.
2. List the coefficient of each term.
3. Are the terms $5x$ and x *like* terms?
4. Are the terms $-6a^3b^2$ and a^3b^2 *like* terms?

2. Commutative, Associative, and Distributive Properties

Several important properties of whole numbers involving addition and multiplication have already been introduced. These properties also hold for integers and algebraic expressions and will be used to simplify expressions (Tables 3-1 through 3-3).

Table 3-1 Commutative Properties

Property	In Symbols/Examples	Comments/Notes
Commutative property of addition	$a + b = b + a$ ex: $-4 + 7 = 7 + (-4)$ $x + 3 = 3 + x$	The order in which two numbers are added does not affect the sum.
Commutative property of multiplication	$a \cdot b = b \cdot a$ ex: $-5 \cdot 9 = 9 \cdot (-5)$ $8y = y \cdot 8$	The order in which two numbers are multiplied does not affect the product.

Avoiding Mistakes

In an expression such as $x + 3$, the two terms x and 3 cannot be combined into one term because they are not *like* terms. Addition of *like* terms will be discussed in Examples 6 and 7.

Table 3-2 Associative Properties

Property	In Symbols/Examples	Comments/Notes
Associative property of addition	$(a + b) + c = a + (b + c)$ ex: $(5 + 8) + 1 = 5 + (8 + 1)$ $(t + n) + 3 = t + (n + 3)$	The manner in which three numbers are grouped under addition does not affect the sum.
Associative property of multiplication	$(a \cdot b) \cdot c = a \cdot (b \cdot c)$ ex: $(-2 \cdot 3) \cdot 6 = -2 \cdot (3 \cdot 6)$ $3 \cdot (m \cdot n) = (3 \cdot m) \cdot n$	The manner in which three numbers are grouped under multiplication does not affect the product.

Table 3-3 Distributive Property of Multiplication over Addition

Property	In Symbols/Examples	Comments/Notes
Distributive property of multiplication over addition*	$a \cdot (b + c) = a \cdot b + a \cdot c$ ex: $-3(2 + x) = -3(2) + (-3)(x)$ $= -6 + (-3)x$ $= -6 - 3x$	The factor outside the parentheses is multiplied by each term in the sum.

*Note that the distributive property of multiplication over addition is sometimes referred to as the *distributive property*.

Answers
1. $-48y^5, 8y^2, -y, -8$
2. $-48, 8, -1, -8$
3. Yes   4. Yes

Example 2 demonstrates the use of the commutative properties.

Example 2 **Applying the Commutative Properties**

Apply the commutative property of addition or multiplication to rewrite the expression.

a. $6 + p$ b. $-5 + n$ c. $y(7)$ d. xy

Solution:

a. $6 + p = p + 6$ Commutative property of addition
b. $-5 + n = n + (-5)$ or $n - 5$ Commutative property of addition
c. $y(7) = 7y$ Commutative property of multiplication
d. $xy = yx$ Commutative property of multiplication

Skill Practice Apply the commutative property of addition or multiplication to rewrite the expression.

5. $x + 3$ 6. $z(2)$ 7. $-12 + p$ 8. ab

We have learned that subtraction is not a commutative operation. However, if we rewrite the difference of two numbers $a - b$ as $a + (-b)$, then we can apply the commutative property of addition. For example:

$$x - 9 = x + (-9)$$ Rewrite as addition of the opposite.
$$= -9 + x$$ Apply the commutative property of addition.

Example 3 demonstrates the associative properties of addition and multiplication.

Example 3 **Applying the Associative Properties**

Use the associative property of addition or multiplication to rewrite each expression. Then simplify the expression.

a. $12 + (45 + y)$ b. $5(7w)$

Solution:

a. $12 + (45 + y) = (12 + 45) + y$ Apply the associative property of addition.
$ = 57 + y$ Simplify.

b. $5(7w) = (5 \cdot 7)w$ Apply the associative property of multiplication.
$ = 35w$ Simplify.

Skill Practice Use the associative property of addition or multiplication to rewrite the expression. Then simplify the expression.

9. $6 + (14 + x)$ 10. $2(8w)$

Note that in most cases, a detailed application of the associative properties will not be given. Instead the process will be written in one step, such as:

$$12 + (45 + y) = 57 + y, \qquad 5(7w) = 35w$$

Example 4 demonstrates the use of the distributive property.

Answers
5. $3 + x$ 6. $2z$
7. $p + (-12)$ or $p - 12$ 8. ba
9. $(6 + 14) + x$; $20 + x$
10. $(2 \cdot 8)w$; $16w$

Example 4 Applying the Distributive Property

Apply the distributive property.

a. $3(x + 4)$ **b.** $2(3y - 5z + 1)$

Solution:

a. $3(x + 4) = 3(x) + 3(4)$ Apply the distributive property.
$= 3x + 12$ Simplify.

b. $2(3y - 5z + 1) = 2[3y + (-5z) + 1]$ First write the subtraction as addition of the opposite.
$= 2[3y + (-5z) + 1]$ Apply the distributive property.
$= 2(3y) + 2(-5z) + 2(1)$
$= 6y + (-10z) + 2$ Simplify.
$= 6y - 10z + 2$

Avoiding Mistakes
Note that $6y - 10z + 2$ cannot be simplified further because $6y$, $-10z$, and 2 are not *like* terms.

TIP: In Example 4(b), we rewrote the expression by writing the subtraction as addition of the opposite. Often this step is not shown and fewer steps are shown overall. For example:

$$2(3y - 5z + 1) = 2(3y) + 2(-5z) + 2(1)$$
$$= 6y - 10z + 2$$

Skill Practice Apply the distributive property.
11. $4(2 + m)$ **12.** $6(5p - 3q + 1)$

Example 5 Applying the Distributive Property

Apply the distributive property.

a. $-8(2 - 5y)$ **b.** $-(-4a + b + 3c)$

Solution:

a. $-8(2 - 5y)$
$= -8[2 + (-5y)]$ Write the subtraction as addition of the opposite.
$= -8[2 + (-5y)]$ Apply the distributive property.
$= -8(2) + (-8)(-5y)$
$= -16 + 40y$ Simplify.

b. $-(-4a + b + 3c)$
$= -1 \cdot (-4a + b + 3c)$ The negative sign preceding the parentheses indicates that we take the opposite of the expression within parentheses. This is equivalent to multiplying the expression within parentheses by -1.

$= -1(-4a) + (-1)(b) + (-1)(3c)$ Apply the distributive property.
$= 4a - b - 3c$ Simplify.

TIP: Notice that a negative factor outside the parentheses changes the signs of all terms to which it is multiplied.

$-1 \cdot (-4a + b + 3c)$
$= +4a - b - 3c$

Skill Practice Apply the distributive property.
13. $-4(6 - 10x)$ **14.** $-(2x - 3y + 4z)$

Answers
11. $8 + 4m$ **12.** $30p - 18q + 6$
13. $-24 + 40x$ **14.** $-2x + 3y - 4z$

3. Combining *Like* Terms

Two terms may be combined if they are *like* terms. To add or subtract *like* terms, we use the distributive property as shown in Example 6.

Example 6   **Using the Distributive Property to Add and Subtract *Like* Terms**

Add or subtract as indicated.

   a. $8y + 6y$ **b.** $7x^2y - 10x^2y$ **c.** $-15w + 4w - w$

Solution:

a. $8y + 6y = (8 + 6)y$	Apply the distributive property.
$= 14y$	Simplify.
b. $7x^2y - 10x^2y = (7 - 10)x^2y$	Apply the distributive property.
$= -3x^2y$	Simplify.
c. $-15w + 4w - w = -15w + 4w - 1w$	First note that $-w = -1w$.
$= (-15 + 4 - 1)w$	Apply the distributive property.
$= (-12)w$	Simplify within parentheses.
$= -12w$	

Skill Practice   Add or subtract as indicated.

15. $-4w + 11w$ **16.** $6a^3b^2 - 11a^3b^2$ **17.** $z - 4z + 22z$

Although the distributive property is used to add and subtract *like* terms, it is tedious to write each step. Observe that adding or subtracting *like* terms is a matter of adding or subtracting the coefficients and leaving the variable factors unchanged. This can be shown in one step.

$$8y + 6y = 14y \quad \text{and} \quad -15w + 4w - 1w = -12w$$

This shortcut will be used throughout the text.

Example 7   **Adding and Subtracting *Like* Terms**

Simplify by combining *like* terms. $-3x + 8y + 4x - 19 - 10y$

Solution:

$-3x + 8y + 4x - 19 - 10y$

$= -3x + 8y + 4x + (-19) + (-10y)$	Rewrite subtraction as addition of the opposite.
$= -3x + 4x + 8y + (-10y) + (-19)$	Use the commutative and associative properties of addition to arrange terms so that *like* terms are grouped together.
$= 1x - 2y - 19$	Combine *like* terms.
$= x - 2y - 19$	Note that $1x = x$. Also note that the remaining terms cannot be combined further because they are not *like*. The variable factors are different.

Skill Practice   Simplify.

18. $4a - 10b - a + 16b + 9$

Answers

15. $7w$ **16.** $-5a^3b^2$
17. $19z$ **18.** $3a + 6b + 9$

4. Simplifying Expressions

For expressions containing parentheses, it is necessary to apply the distributive property before combining *like* terms. This is demonstrated in Examples 8 and 9. Notice that when we apply the distributive property, the parentheses are dropped. This is often called *clearing parentheses*.

Example 8   Clearing Parentheses and Combining *Like* Terms

Simplify by clearing parentheses and combining *like* terms. $6 - 3(2y + 9)$

Solution:

$6 - 3(2y + 9)$ The order of operations indicates that we must perform multiplication before subtraction.

It is also important to understand that a factor of -3 (not 3) will be multiplied to all terms within the parentheses. To see why, we can rewrite the subtraction in terms of addition of the opposite.

$6 - 3(2y + 9) = 6 + (-3)(2y + 9)$ Rewrite subtraction as addition of the opposite.

$\quad = 6 + (-3)(2y) + (-3)(9)$ Apply the distributive property.

$\quad = 6 + (-6y) + (-27)$ Simplify.

$\quad = -6y + 6 + (-27)$ Group *like* terms together.

$\quad = -6y + (-21)$ or $-6y - 21$ Combine *like* terms.

Avoiding Mistakes
Multiplication is performed before subtraction. It is incorrect to subtract $6 - 3$ first.

Skill Practice Simplify.

19. $8 - 6(w + 4)$

Example 9   Clearing Parentheses and Combining *Like* Terms

Simplify by clearing parentheses and combining *like* terms.

$$-8(x - 4) - (5x + 7)$$

Solution:

$-8(x - 4) - (5x + 7)$

$= -8[x + (-4)] + (-1)(5x + 7)$ Rewrite subtraction as addition of the opposite.

$= -8[x + (-4)] + (-1)(5x + 7)$ Apply the distributive property.

$= -8(x) + (-8)(-4) + (-1)(5x) + (-1)(7)$

$= -8x + 32 + (-5x) + (-7)$ Simplify.

$= -8x + (-5x) + 32 + (-7)$ Group *like* terms together.

$= -13x + 25$ Combine *like* terms.

FOR REVIEW

The product of two numbers with the same sign is *positive*.

$$(8)(4) = 32$$
$$(-8)(-4) = 32$$

The product of two numbers with different signs is *negative*.

$$(-8)(4) = -32$$
$$(8)(-4) = -32$$

Skill Practice Simplify.

20. $-5(10 - m) - 2(m + 1)$

Answers
19. $-6w - 16$ **20.** $3m - 52$

Section 3.1 Activity

A.1. Consider the expression $-8x^2 - x + 7$.
  a. List the terms in the expression separated by commas. Recall that a term is a number or the product or quotient of numbers and variables.
  b. List the variable term(s).
  c. Identify the constant term(s).
  d. What is the coefficient of the $-x$ term?

A.2. Now consider the expression $-8x^2 - x + 7 + 3x^2 - 5x + 3$.
  a. List the terms in the expression separated by commas.
  b. The terms $-8x^2$ and $3x^2$ are *like terms*. Explain why.
  c. Identify the other *like* terms in the expression.
  d. Simplify the expression by combining *like* terms.

A.3. Mia and Xavier are debating whether or not you can combine the following terms: $5x^2yz$ and $-zx^2y$.

Mia thinks you can combine the terms because they are *like* terms. Xavier thinks the order of the variables must be the same in order to be *like* terms. Who is correct and why?

A.4. Consider the expressions $2(9 + y)$ and $2(9y)$.
  a. How do these expressions differ from one another?
  b. Simplify $2(9 + y)$ and state the property used.
  c. Simplify $2(9y)$ and state the property used.

A.5. Simplify the expression by clearing the parentheses.
  a. $5(x - y + 3)$
  b. $-5(x - y + 3)$
  c. Compare the results from parts (a) and (b). When a negative factor is distributed throughout parentheses, what happens to the signs of the terms within parentheses?

For Exercises A.6–A.10, simplify the expressions by clearing parentheses where necessary and combining *like* terms.

A.6. $-6t - 9t$

A.7. $13a - 16b - 52 - 37a + 51b + 6$

A.8. $6x + 5(4 - 3x)$

A.9. $11 - 3(8 - w)$

A.10. $-6(t + 3) + 4(6 - t) + 7$

Section 3.1 Practice Exercises

Study Skills Exercise

After you get a test back, it is a good idea to correct the test so that you do not make the same errors again. One recommended approach is to use a clean sheet of paper and divide the paper down the middle vertically, as shown. For each problem that you missed on the test, rework the problem correctly on the left-hand side of the paper. Then write a written explanation on the right-hand side of the paper. Take the time this week to make corrections from your last test.

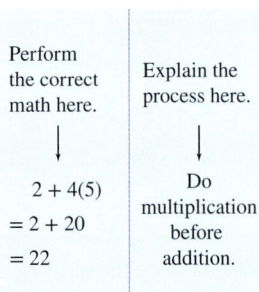

Prerequisite Review

For Exercises R.1–R.6, simplify.

R.1. $-4 - 2(9 - 3)$
R.2. $-4 + 2(-9 + 3)$
R.3. $6 - 4^2 \div (-2)$
R.4. $(6 - 4)^2 \div (-2)$
R.5. $3(10 - 8) - 2(2^2 + 1)$
R.6. $-3(10 + 8) - 2(2 + 1)$

Vocabulary and Key Concepts

1. **a.** A _____ is a number or the product or quotient of numbers and variables.

   b. The term $-5xy$ is called a (constant/variable) term, whereas the term -5 is called a (constant/variable) term.

   c. The numerical factor within a term is called the _____ of the term.

   d. Two terms are _____ terms if they each have the same variables and the corresponding variables are raised to the same powers.

   e. The commutative property of addition tells us that the expression $6 + x$ can be written as _____.
   The commutative property of multiplication tells us that the expression $x(6)$ can be written as _____.

   f. The _____ property of addition tells us that $-5 + (7 + x)$ can be written as $(-5 + 7) + x$.

   g. The _____ property of multiplication tells us that $-5(7x)$ can be written as $(-5 \cdot 7)x$.

   h. The distributive property of multiplication over addition tells us that $a \cdot (b + c) =$ _____. Using this property, the expression $4(x + 5)$ can be written as _____.

2. The coefficient of the term ab^2 is _____.

3. *Like* terms can be combined using the mathematical operations of _____ and _____.

4. The expression $10xy + 6yx + 7x$ (can/cannot) be simplified further.

5. Explain why the terms in the given sum cannot be combined.
   $3x + 7x^2$

6. To simplify the expression $-3(2x - 5) - 4(x + 3)$ you would apply the distributive property _____ time(s).

Concept 1: Identifying *Like* Terms

For Exercises 7–12, for each expression, list the terms and identify each term as a variable term or a constant term. **(See Example 1.)**

7. $a^2 + 3$
8. $y - z - 11$
9. $2a + 5b^2 + 6$
10. $-5x - 4 + 7y$
11. $8 + 9a$
12. $12 - 8k$

For Exercises 13–16, identify the coefficients for each term. **(See Example 1.)**

13. $6p - 4q$
14. $-5a^3 - 2a$
15. $-h - 12$
16. $8x - 9$

For Exercises 17–24, determine if the two terms are *like* terms or unlike terms. If they are unlike terms, explain why. **(See Example 1.)**

17. $3a, -2a$
18. $8b, 12b$
19. $4xy, 4y$
20. $-9hk, -9h$
21. $14y^2, 14y$
22. $25x, 25x^2$
23. $17, 17y$
24. $-22t, -22$

Concept 2: Commutative, Associative, and Distributive Properties

For Exercises 25–32, apply the commutative property of addition or multiplication to rewrite each expression. **(See Example 2.)**

25. $5 + w$ **26.** $t + 2$ **27.** $r(2)$ **28.** $a(-4)$

29. $t(-s)$ **30.** $d(-c)$ **31.** $-p + 7$ **32.** $-q + 8$

For Exercises 33–40, apply the associative property of addition or multiplication to rewrite each expression. Then simplify the expression. **(See Example 3.)**

33. $3 + (8 + t)$ **34.** $7 + (5 + p)$ **35.** $-2(6b)$ **36.** $-3(2c)$

37. $3(6x)$ **38.** $9(5k)$ **39.** $-9 + (-12 + h)$ **40.** $-11 + (-4 + s)$

For Exercises 41–52, apply the distributive property. **(See Examples 4 and 5.)**

41. $4(x + 8)$ **42.** $5(3 + w)$ **43.** $4(a + 4b - c)$ **44.** $2(3q - r + s)$

45. $-2(p + 4)$ **46.** $-6(k + 2)$ **47.** $-(3x + 9 - 5y)$ **48.** $-(a - 8b + 4c)$

49. $-4(3 - n^2)$ **50.** $-2(13 - t^2)$ **51.** $-3(-5q - 2s - 3t)$ **52.** $-2(-10p - 12q + 3)$

For Exercises 53–60, apply the appropriate property to simplify the expression.

53. $6(2x)$ **54.** $-3(12k)$ **55.** $6(2 + x)$ **56.** $-3(12 + k)$

57. $-8 + (4 - p)$ **58.** $3 + (25 - m)$ **59.** $-8(4 - p)$ **60.** $-3(25 - m)$

Concept 3: Combining *Like* Terms

For Exercises 61–72, combine the *like* terms. **(See Examples 6 and 7.)**

61. $6r + 8r$ **62.** $4x + 21x$ **63.** $-4h + 12h - h$

64. $9p - 13p + p$ **65.** $4a^2b - 6a^2b$ **66.** $13xy^2 + 8xy^2$

67. $10x - 12y - 4x - 3y + 9$ **68.** $14a - 5b + 3a - b - 3$ **69.** $-8 - 6k - 9k + 12k + 4$

70. $-5 - 11p + 23p - p + 4$ **71.** $-8uv + 6u + 12uv$ **72.** $9pq - 9p + 13pq$

Concept 4: Simplifying Expressions

For Exercises 73–94, clear parentheses and combine *like* terms. **(See Examples 8 and 9.)**

73. $5(t - 6) + 2$ **74.** $7(a - 4) + 8$ **75.** $-3(2x + 1) - 13$

76. $-2(4b + 3) - 10$ **77.** $4 + 6(y - 3)$ **78.** $11 + 2(p - 8)$

79. $21 - 7(3 - q)$ **80.** $10 - 5(2 - 5m)$ **81.** $-3 - (2n + 1)$

82. $-13 - (6s + 5)$ **83.** $10(x + 5) - 3(2x + 9)$ **84.** $6(y - 9) - 5(2y - 5)$

85. $-(12z + 1) + 2(7z - 5)$ **86.** $-(8w + 5) + 3(w - 15)$ **87.** $3(w + 3) - (4w + y) - 3y$

88. $2(s + 6) - (8s - t) + 6t$ **89.** $20a - 4(b + 3a) - 5b$ **90.** $16p - 3(2p - q) + 7q$

91. $6 - (3m - n) - 2(m + 8) + 5n$ **92.** $12 - (5u + v) - 4(u - 6) + 2v$

93. $15 + 2(w - 4) - (2w - 5z^2) + 7z^2$ **94.** $7 + 3(2a - 5) - (6a - 8b^2) - 2b^2$

Mixed Exercises

95. Demonstrate the commutative property of addition by evaluating the expressions for $x = -3$ and $y = 9$.

   a. $x + y$ **b.** $y + x$

96. Demonstrate the commutative property of addition by evaluating the expressions for $m = -12$ and $n = -5$.

   a. $m + n$ **b.** $n + m$

97. Demonstrate the associative property of addition by evaluating the expressions for $x = -7$, $y = 2$, and $z = 10$.

   a. $(x + y) + z$ **b.** $x + (y + z)$

98. Demonstrate the associative property of addition by evaluating the expressions for $a = -4$, $b = -6$, and $c = 18$.

   a. $(a + b) + c$ **b.** $a + (b + c)$

99. Demonstrate the commutative property of multiplication by evaluating the expressions for $x = -9$ and $y = 5$.

   a. $x \cdot y$ **b.** $y \cdot x$

100. Demonstrate the commutative property of multiplication by evaluating the expressions for $c = 12$ and $d = -4$.

   a. $c \cdot d$ **b.** $d \cdot c$

101. Demonstrate the associative property of multiplication by evaluating the expressions for $x = -2$, $y = 6$, and $z = -3$.

   a. $(x \cdot y) \cdot z$ **b.** $x \cdot (y \cdot z)$

102. Demonstrate the associative property of multiplication by evaluating the expressions for $b = -4$, $c = 2$, and $d = -5$.

   a. $(b \cdot c) \cdot d$ **b.** $b \cdot (c \cdot d)$

Addition and Subtraction Properties of Equality

Section 3.2

1. Definition of a Linear Equation in One Variable

An **equation** is a statement that indicates that two quantities are equal. The following are equations.

$$x = 7 \qquad z + 3 = 8 \qquad -6p = 18$$

All equations have an equal sign. Furthermore, notice that the equal sign separates the equation into two parts, the left-hand side and the right-hand side. A **solution to an equation** is a value of the variable that makes the equation a true statement. Substituting a solution to an equation for the variable makes the right-hand side equal to the left-hand side.

Concepts

1. Definition of a Linear Equation in One Variable
2. Addition and Subtraction Properties of Equality

Equation	Solution	Check	
$x = 7$	7	$x = 7$ $7 \stackrel{?}{=} 7$ ✓	Substitute 7 for x. The right-hand side equals the left-hand side.
$z + 3 = 8$	5	$z + 3 = 8$ $5 + 3 \stackrel{?}{=} 8$ $8 \stackrel{?}{=} 8$ ✓	Substitute 5 for z. The right-hand side equals the left-hand side.
$-6p = 18$	-3	$-6p = 18$ $-6(-3) \stackrel{?}{=} 18$ $18 \stackrel{?}{=} 18$ ✓	Substitute -3 for p. The right-hand side equals the left-hand side.

Avoiding Mistakes

It is important to distinguish between an equation and an expression. An equation has an equal sign, whereas an expression does not. For example:

$$2x + 4 = 16 \quad \text{equation}$$
$$7x - 9 \quad \text{expression}$$

Example 1 — Determining Whether a Number Is a Solution to an Equation

Determine whether the given number is a solution to the equation.

a. $2x - 9 = 3;\ 6$ **b.** $20 = 8p - 4;\ -2$

Solution:

a. $2x - 9 = 3$

$2(6) - 9 \stackrel{?}{=} 3$ Substitute 6 for x.

$12 - 9 \stackrel{?}{=} 3$ Simplify.

$3 \stackrel{?}{=} 3$ ✓   The right-hand side equals the left-hand side. Thus, 6 is a solution to the equation $2x - 9 = 3$.

b. $20 = 8p - 4$

$20 \stackrel{?}{=} 8(-2) - 4$ Substitute -2 for p.

$20 \stackrel{?}{=} -16 - 4$ Simplify.

$20 \neq -20$ The right-hand side does not equal the left-hand side. Thus, -2 is *not* a solution to the equation $20 = 8p - 4$.

FOR REVIEW

Recall that when substituting a number for a variable, use parentheses in place of the variable.

$2x - 9$
$2(\ ) - 9$
$2(6) - 9$

Skill Practice Determine whether the given number is a solution to the equation.

1. $2 + 3x = 23;\ 7$ **2.** $9 = -4x + 1;\ 2$

In the study of algebra, you will encounter a variety of equations. In this chapter, we will focus on a specific type of equation called a linear equation in one variable.

Definition of a Linear Equation in One Variable

Let a, b, and c be numbers such that $a \neq 0$. A **linear equation in one variable** is an equation that can be written in the form

$$ax + b = c$$

Note: A linear equation in one variable is often called a first-degree equation because the variable x has an implied exponent of 1.

Examples	Notes
$2x + 4 = 20$	$a = 2, b = 4, c = 20$
$-3x - 5 = 16$ can be written as $-3x + (-5) = 16$	$a = -3, b = -5, c = 16$
$5x + 9 - 4x = 1$ can be written as $x + 9 = 1$	$a = 1, b = 9, c = 1$

Answers

1. Yes **2.** No

2. Addition and Subtraction Properties of Equality

Given the equation $x = 3$, we can easily determine that the solution is 3. The solution to the equation $2x + 14 = 20$ is also 3. These two equations are called **equivalent equations** because they have the same solution. However, while the solution to $x = 3$ is obvious, the solution to $2x + 14 = 20$ is not obvious. Our goal in this chapter is to learn how to *solve* equations.

To solve an equation we use algebraic principles to write an equation like $2x + 14 = 20$ in an equivalent but simpler form, such as $x = 3$. The addition and subtraction properties of equality are the first tools we will use to solve an equation.

Addition and Subtraction Properties of Equality

Let a, b, and c represent algebraic expressions.

1. The **addition property of equality**: If $a = b$, then, $a + c = b + c$
2. The **subtraction property of equality**: If $a = b$, then, $a - c = b - c$

The addition and subtraction properties of equality indicate that adding or subtracting the same quantity to each side of an equation results in an equivalent equation. This is true because if two equal quantities are increased (or decreased) by the same amount, then the resulting quantities will also be equal (Figure 3-1).

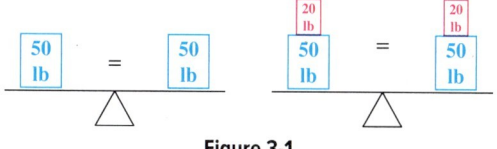

Figure 3-1

Example 2 — Applying the Addition Property of Equality

Solve the equations and check the solutions.

a. $x - 6 = 18$ **b.** $-12 = x - 7$ **c.** $-4 + y = 5$

Solution:

To solve an equation, the goal is to isolate the variable on one side of the equation. That is, we want to create an equivalent equation of the form $x = number$. To accomplish this, we can use the fact that the sum of a number and its opposite is zero.

a. $x - 6 = 18$

$x + (-6) = 18$ Rewrite subtraction as addition of the opposite.

$x + (-6) + 6 = 18 + 6$ To isolate x, add 6 to both sides because $-6 + 6 = 0$.

$x + 0 = 24$ Simplify.

$x = 24$ The variable is isolated (by itself) on the left-hand side of the equation. The solution is 24.

Check: $x - 6 = 18$ Original equation

$(24) - 6 \stackrel{?}{=} 18$ Substitute 24 for x.

$18 \stackrel{?}{=} 18$ ✓   True

TIP: Notice that the variable may be isolated on *either* side of the equal sign. In Example 2(a), the variable appears on the left. In Example 2(b), the variable appears on the right.

Avoiding Mistakes

To check the solution to an equation, always substitute the solution into the *original* equation. Then independently evaluate the expression on each side of the equation.

b.
$-12 = x - 7$
$-12 = x + (-7)$ Rewrite subtraction as addition of the opposite.
$-12 + 7 = x + (-7) + 7$ To isolate x, add 7 to both sides because $-7 + 7 = 0$.
$-5 = x + 0$ Simplify.
$-5 = x$ The variable is isolated on the right-hand side of the equation. The solution is -5.

The equation $-5 = x$ is equivalent to $x = -5$.

Check: $-12 = x - 7$ Original equation
$-12 \stackrel{?}{=} (-5) - 7$ Substitute -5 for x.
$-12 \stackrel{?}{=} -12$ ✓   True

c.
$-4 + y = 5$
$-4 + 4 + y = 5 + 4$ To isolate y, add 4 to both sides, because $-4 + 4 = 0$.
$0 + y = 9$ Simplify.
$y = 9$ The solution is 9.

Check: $-4 + y = 5$ Original equation
$-4 + (9) \stackrel{?}{=} 5$ Substitute 9 for y.
$5 \stackrel{?}{=} 5$ ✓   True

Skill Practice Solve the equation and check the solution.

3. $y - 4 = 12$ **4.** $-18 = w + 5$ **5.** $-9 + z = 2$

In Example 2, the addition property of equality was used to *add* a *positive* number to both sides of an equation to isolate the variable. Recall that addition of a *negative* number is equivalent to subtraction. This is the basis for using the subtraction property of equality for the equations in Example 3.

Example 3   Applying the Subtraction Property of Equality

Solve the equations and check.

a. $z + 11 = 14$ **b.** $-8 = 2 + q$

Solution:

a. $z + 11 = 14$ We can isolate z by adding -11 to both sides. Since adding -11 is the same as subtracting 11, we use the subtraction property of equality.

$z + 11 - 11 = 14 - 11$ To isolate z, subtract 11 from both sides, because $11 - 11 = 0$.
$z + 0 = 3$ Simplify.
$z = 3$ The solution is 3.

Check: $z + 11 = 14$ Original equation
$(3) + 11 \stackrel{?}{=} 14$ Substitute 3 for z.
$14 \stackrel{?}{=} 14$ ✓   True

Answers
3. 16   **4.** -23 **5.** 11

b. $-8 = 2 + q$

$-8 - 2 = 2 - 2 + q$ To isolate q, subtract 2 from both sides, because $2 - 2 = 0$.

$-8 + (-2) = 0 + q$ On the left side, change subtraction to addition of the opposite.

$-10 = 0 + q$ Simplify.

$-10 = q$ The solution is -10.

Check: $-8 = 2 + q$ Original equation

$-8 \stackrel{?}{=} 2 + (-10)$ Substitute -10 for q.

$-8 \stackrel{?}{=} -8$ ✓   True

Skill Practice Solve the equation and check the solution.

6. $m + 8 = 21$
7. $-16 = 1 + z$

The equations in Example 4 require that we simplify the expressions on both sides of the equation, then apply the addition or subtraction property as needed. As you read through each example, remember that you want to isolate the variable.

Example 4   **Applying the Addition and Subtraction Properties of Equality**

Solve the equations.

a. $-8 + 3x - 2x = 12 - 9$ **b.** $-5 = 2(y + 1) - y$

Solution:

a. $-8 + 3x - 2x = 12 - 9$

$-8 + 3x + (-2x) = 12 + (-9)$ Rewrite subtraction as addition of the opposite.

$-8 + x = 3$ Combine *like* terms.

$-8 + 8 + x = 3 + 8$ To isolate x, add 8 to both sides, because $-8 + 8 = 0$.

$x = 11$ The solution is 11.

Check: $-8 + 3x - 2x = 12 - 9$

$-8 + 3(11) - 2(11) \stackrel{?}{=} 12 - 9$

$-8 + 33 - 22 \stackrel{?}{=} 3$

$25 - 22 \stackrel{?}{=} 3$

$3 \stackrel{?}{=} 3$ ✓ True

Answers

6. 13   **7.** -17

b. $-5 = 2(y + 1) - y$

$-5 = 2y + 2 - y$ Apply the distributive property to clear parentheses.

$-5 = y + 2$ Combine *like* terms.

$-5 - 2 = y + 2 - 2$ To isolate y, subtract 2 from both sides, because $2 - 2 = 0$.

$-5 + (-2) = y + 0$ On the left side, rewrite subtraction as addition of the opposite.

$-7 = y$ The solution is -7.

Check: $-5 = 2(y + 1) - y$
$-5 \stackrel{?}{=} 2(-7 + 1) - (-7)$
$-5 \stackrel{?}{=} 2(-6) + 7$
$-5 \stackrel{?}{=} -12 + 7$
$-5 \stackrel{?}{=} -5$ ✓ True

Skill Practice Solve the equations.

8. $15 + 6y - 5y = 4 - 9$
9. $-1 = 3(x + 4) - 2x$

Answers
8. -20 9. -13

Section 3.2 Activity

A.1. Trey solved an equation and obtained two different solutions. His first solution was -20 and the second solution was -14. Which of the given values is the correct solution to the equation? Show your work to support your answer.
$3 + x = -17$

A.2. Determine if 0 is a solution to the equation: $2(x + 5) = x + 10$.

A.3. a. To solve the equation of $y - 7 = 2$ would you add 7 to both sides or subtract 7 from both sides? Explain.
   b. Solve the equation and check your answer.
   c. Solve the equation: $2 = y - 7$. Do the steps in solving this equation differ from those used in part (b)?

For Exercises A.4–A.8, solve the equation and check.

A.4. $x + 5 = 12$ **A.5.** $4 = q + 26$ **A.6.** $-4 + t = 8$

A.7. $x - 40 = 30$ **A.8.** $-18 = 92 + x$

A.9. Consider the equation $4x - 3x + 5 = 24 - 10$.
   a. What differences do you see in this equation in comparison to the equations given in Exercises A.4–A.8?
   b. Simplify the equation by combining *like* terms on the left and right sides of the equation.
   c. Solve the equation and check.

For Exercises A.10–A.12, first simplify each side of the equation by clearing parentheses (if necessary) and combining *like* terms. Then solve the equation and check.

A.10. $10x - 9x - 11 = 15$ **A.11.** $3(r - 2) - 2r = 6 + (-2)$ **A.12.** $4(k + 2) - 3k = -6 + 9$

Section 3.2 Addition and Subtraction Properties of Equality

Practice Exercises

Section 3.2

Study Skills Exercise

A student's motivation to learn and understand the course material can be increased by believing the course has value and relevance with the real world. As a result, you will be more likely to stay engaged, apply the mathematics, and persist when the course becomes challenging.
Answer the questions:
- Why study mathematics?
- How is this course relevant to your personal life and/or future career?

Prerequisite Review

For Exercises R.1–R.6, simplify the expression.

R.1. $-10a + 3b - 3a + 13b + a$
R.2. $2x - 15y - 12x + 6y + 10x$
R.3. $4 - 23y^2 + 11 - 16y^2$
R.4. $4y^2 - 23y^2 + 11 - 16y^2$
R.5. $6x - (5x + 2) + 10$
R.6. $6 - (5x + 2) + 10x$
R.7. Evaluate the expression $\sqrt{3x - y}$ for $x = 10$ and $y = -6$.
R.8. Evaluate the expression $(3x - y)^2$ for $x = 2$ and $y = -6$.

Vocabulary and Key Concepts

1. **a.** A _____ equation in one variable is an equation that can be written in the form $ax + b = c$, $(a \neq 0)$.
   b. A _____ to an equation is a value of the variable that makes the equation a true statement.
   c. The equations $3x = 12$ and $x = 4$ are called _____ equations because they have the same solution.
   d. The _____ property of equality tells us that adding the same quantity to both sides of an equation results in an equivalent equation.
   e. The _____ property of equality tells us that subtracting the same quantity from both sides of an equation results in an equivalent equation.

2. Considering that a linear equation in one variable can be written in the form $ax + b = c$, identify a, b, and c in the equation: $-6x + 1 = -10$.

Concept 1: Definition of a Linear Equation in One Variable

For Exercises 3–10, determine whether the given number is a solution to the equation. **(See Example 1.)**

3. $5x + 3 = -2$; -1
4. $3y - 2 = 4$; 2
5. $-z + 8 = 20$; 12

6. $-7 - w = -10$; -3
7. $13 = 13 + 6t$; 0
8. $25 = 25 - 2x$; 0

9. $15 = -2q + 9$; 3
10. $39 = -7p + 4$; 5

For Exercises 11–16, identify as an expression or an equation.

11. $8x - 9 = 7$
12. $24 - 5x = 12 + x$
13. $8x - 9 - 7$

14. $24 - 5x + 12 + x$
15. $2(x - 4) - x = 1$
16. $2(x - 4) - x + 1$

Concept 2: Addition and Subtraction Properties of Equality

For Exercises 17–22, fill in the blank with the appropriate number.

17. $13 + (-13) = $ _____

18. $6 + $ _____ $= 0$

19. _____ $+ (-7) = 0$

20. $1 + (-1) = $ _____

21. $0 = -3 + $ _____

22. $0 = 9 + $ _____

For Exercises 23–34, solve the equation using the addition property of equality. (See Example 2.)

23. $x - 23 = 14$

24. $y - 12 = 30$

25. $-4 + k = 12$

26. $-16 + m = 4$

27. $-18 = n - 3$

28. $-9 = t - 6$

29. $9 = -7 + t$

30. $-6 = -10 + z$

31. $k - 44 = -122$

32. $a - 465 = -206$

33. $13 = -21 + w$

34. $2 = -17 + p$

For Exercises 35–40, fill in the blank with the appropriate number.

35. $52 - $ _____ $= 0$

36. $2 - 2 = $ _____

37. $18 - 18 = $ _____

38. _____ $- 15 = 0$

39. _____ $- 100 = 0$

40. $21 - $ _____ $= 0$

For Exercises 41–52, solve the equation using the subtraction property of equality. (See Example 3.)

41. $x + 34 = 6$

42. $y + 12 = 4$

43. $17 + b = 20$

44. $5 + c = 14$

45. $-32 = t + 14$

46. $-23 = k + 11$

47. $82 = 21 + m$

48. $16 = 88 + n$

49. $z + 145 = 90$

50. $c + 80 = -15$

51. $52 = 10 + k$

52. $43 = 12 + p$

Mixed Exercises

For Exercises 53–72, solve the equation using the appropriate property of equality. (See Example 4.)

53. $1 + p = 0$

54. $r - 12 = 13$

55. $-34 + t = -40$

56. $7 + q = 4$

57. $-11 = w - 23$

58. $-9 = p - 10$

59. $16 = x + 21$

60. $-4 = y + 18$

61. $5h - 4h + 4 = 3$

62. $10x - 9x - 11 = 15$

63. $-9 - 4x + 5x = 3 - 1$

64. $11 - 7x + 8x = -4 - 2$

65. $3(x + 2) - 2x = 5$

66. $4(x - 1) - 3x = 2$

67. $3(r - 2) - 2r = -2 + 6$

68. $4(k + 2) - 3k = -6 + 9$

69. $9 + (-2) = 4 + t$

70. $-13 + 15 = p + 5$

71. $-2 = 2(a - 15) - a$

72. $-1 = 6(t - 4) - 5t$

Multiplication and Division Properties of Equality

Section 3.3

1. Multiplication and Division Properties of Equality

Concepts

1. Multiplication and Division Properties of Equality
2. Comparing the Properties of Equality

Adding or subtracting the same quantity on both sides of an equation results in an equivalent equation. The same is true when we multiply or divide both sides of an equation by the same nonzero quantity.

> **Multiplication and Division Properties of Equality**
>
> Let a, b, and c represent algebraic expressions, where $c \neq 0$.
>
> 1. The **multiplication property of equality:** If $a = b$, then, $c \cdot a = c \cdot b$
> 2. The **division property of equality:** If $a = b$ then, $\dfrac{a}{c} = \dfrac{b}{c}$

To understand the multiplication property of equality, suppose we start with a true equation such as $10 = 10$. If both sides of the equation are multiplied by a constant such as 3, the result is also a true statement (Figure 3-2).

$10 = 10$
$3 \cdot 10 = 3 \cdot 10$
$30 = 30$

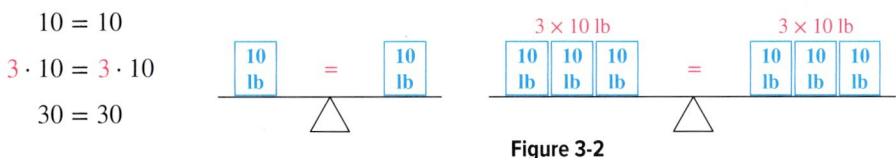

Figure 3-2

To solve an equation in the variable x, the goal is to write the equation in the form $x =$ number. In particular, notice that we want the coefficient of x to be 1. That is, we want to write the equation as $1 \cdot x =$ number. Therefore, to solve an equation such as $3x = 12$, we can *divide* both sides of the equation by 3. We do this because $\frac{3}{3} = 1$, and that leaves $1x$ on the left-hand side of the equation.

$3x = 12$

$\dfrac{3x}{3} = \dfrac{12}{3}$ Divide both sides by 3.

$1 \cdot x = 4$ The coefficient of the x term is now 1.

$x = 4$ Simplify.

> **TIP:** Recall that the quotient of a nonzero number and itself is 1. For example:
>
> $\dfrac{3}{3} = 1$ and $\dfrac{-5}{-5} = 1$

Example 1 Applying the Division Property of Equality

Solve the equations.

a. $10x = 50$ **b.** $28 = -4p$ **c.** $-y = 34$

Solution:

a. $10x = 50$

$\dfrac{10x}{10} = \dfrac{50}{10}$ To obtain a coefficient of 1 for the x term, divide both sides by 10, because $\dfrac{10}{10} = 1$.

$1x = 5$ Simplify.

$x = 5$ The solution is 5.

Check: $10x = 50$ Original equation
$10(5) \stackrel{?}{=} 50$ Substitute 5 for x.
$50 \stackrel{?}{=} 50$ ✓   True

b. $28 = -4p$

$\dfrac{28}{-4} = \dfrac{-4p}{-4}$ To obtain a coefficient of 1 for the p term, divide both sides by -4, because $\dfrac{-4}{-4} = 1$.

$-7 = 1p$ Simplify.

$-7 = p$ The solution is -7 and checks in the original equation.

c. $-y = 34$

$-1y = 34$ Note that $-y$ is the same as $-1 \cdot y$. To isolate y, we need a coefficient of *positive* 1.

$\dfrac{-1y}{-1} = \dfrac{34}{-1}$ To obtain a coefficient of 1 for the y term, divide both sides by -1.

$1y = -34$ Simplify.

$y = -34$ The solution is -34 and checks in the original equation.

Avoiding Mistakes

In Example 1(b), the operation between -4 and p is multiplication. We must divide by -4 (rather than 4) so that the resulting coefficient on p is positive 1.

TIP: In Example 1(c), we could have also multiplied both sides by -1 to obtain a coefficient of 1 for x.

$(-1)(-y) = (-1)34$
$y = -34$

Skill Practice Solve the equations.

1. $4x = 32$ 2. $18 = -2w$ 3. $19 = -m$

The multiplication property of equality indicates that multiplying both sides of an equation by the same nonzero quantity results in an equivalent equation. For example, consider the equation $\dfrac{x}{2} = 6$. The variable x is being divided by 2. To solve for x, we need to reverse this process. Therefore, we will *multiply* by 2.

$\dfrac{x}{2} = 6$

$2 \cdot \dfrac{x}{2} = 2 \cdot 6$ Multiply both sides by 2.

$\dfrac{2}{2} \cdot x = 12$ The expression $2 \cdot \dfrac{x}{2}$ can be written as $\dfrac{2}{2} \cdot x$. This process is called *regrouping factors*.

$1 \cdot x = 12$ The x coefficient is 1, because $\dfrac{2}{2}$ equals 1.

$x = 12$ The solution is 12.

Answers
1. 8   2. -9 3. -19

Section 3.3 Multiplication and Division Properties of Equality   145

Example 2 **Applying the Multiplication Property of Equality**

Solve the equations.

a. $\dfrac{x}{4} = -5$ b. $2 = \dfrac{t}{-8}$

Solution:

a. $\dfrac{x}{4} = -5$

$4 \cdot \dfrac{x}{4} = 4 \cdot (-5)$ To isolate x, multiply both sides by 4.

$\dfrac{4}{4} \cdot x = -20$ Regroup factors.

$1x = -20$ The x coefficient is now 1, because $\dfrac{4}{4}$ equals 1.

$x = -20$ The solution is -20.   Check: $\dfrac{x}{4} = -5$

$\dfrac{(-20)}{4} \stackrel{?}{=} -5$

$-5 \stackrel{?}{=} -5$ ✓ True

b. $2 = \dfrac{t}{-8}$

$-8 \cdot (2) = -8 \cdot \dfrac{t}{-8}$ To isolate t, multiply both sides by -8.

$-16 = \dfrac{-8}{-8} \cdot t$ Regroup factors.

$-16 = 1 \cdot t$ The coefficient on t is now 1.

$-16 = t$ The solution is -16 and checks in the original equation.

Skill Practice Solve the equations.

4. $\dfrac{y}{6} = -3$ 5. $5 = \dfrac{w}{-10}$

2. Comparing the Properties of Equality

It is important to determine which property of equality should be used to solve an equation. For example, compare equations:

$$4 + x = 12 \quad \text{and} \quad 4x = 12$$

In the first equation, the operation between 4 and x is addition. Therefore, we want to reverse the process by *subtracting* 4 from both sides. In the second equation, the operation between 4 and x is multiplication. To isolate x, we reverse the process by *dividing* by 4.

$$4 + x = 12 \qquad\qquad 4x = 12$$
$$4 - 4 + x = 12 - 4 \qquad \dfrac{4x}{4} = \dfrac{12}{4}$$
$$x = 8 \qquad\qquad x = 3$$

Answers
4. -18 5. -50

To solve each equation, we want to isolate the variable. If the operation between a term and the variable is addition or subtraction, we apply the subtraction or addition property of equality. If the variable is multiplied or divided by a constant, then we apply the division or multiplication property of equality.

In Example 3, we practice distinguishing which property of equality to use.

Example 3   **Solving Linear Equations**

Solve the equations.

a. $\dfrac{m}{12} = -3$ b. $x + 18 = 2$

c. $20 + 8 = -10t + 3t$ d. $-4 + 10 = 4t - 3(t + 2) - 1$

Solution:

a. $\dfrac{m}{12} = -3$ The operation between m and 12 is division. To obtain a coefficient of 1 for the m term, *multiply* both sides by 12.

$12 \cdot \dfrac{m}{12} = 12(-3)$ Multiply both sides by 12.

$\dfrac{12}{12} \cdot m = 12(-3)$ Regroup.

$m = -36$ Simplify both sides. The solution -36 checks in the original equation.

b. $x + 18 = 2$ The operation between x and 18 is addition. To isolate x, *subtract* 18 from both sides.

$x + 18 - 18 = 2 - 18$ Subtract 18 on both sides.

$x = -16$ Simplify. The solution is -16 and checks in the original equation.

c. $20 + 8 = -10t + 3t$ Begin by simplifying both sides of the equation.

$28 = -7t$ The relationship between t and -7 is multiplication. To obtain a coefficient of 1 on the t term, *divide* both sides by -7.

$\dfrac{28}{-7} = \dfrac{-7t}{-7}$ Divide both sides by -7.

$-4 = t$ Simplify. The solution is -4 and checks in the original equation.

d. $-4 + 10 = 4t - 3(t + 2) - 1$ Begin by simplifying both sides of the equation.

$6 = 4t - 3t - 6 - 1$ Apply the distributive property on the right-hand side.

$6 = t - 7$ Combine *like* terms on the right-hand side.

$6 + 7 = t - 7 + 7$ To isolate the t term, *add* 7 to both sides. This is because $-7 + 7 = 0$.

$13 = t$ The solution is 13 and checks in the original equation.

Skill Practice Solve the equations.

6. $\dfrac{t}{5} = -8$ 7. $x + 46 = 12$

8. $10 = 5p - 7p$ 9. $-5 + 20 = -3 + 6w - 5(w + 1)$

Answers
6. -40 7. -34
8. -5 9. 23

Section 3.3 Activity

A.1. Divide.

    a. $\dfrac{3}{3} =$ **b.** $\dfrac{12}{12} =$ **c.** $\dfrac{-4}{-4} =$

A.2. When solving an equation, the goal is to write the equation in the form $x =$ number.

    a. What is the numerical coefficient of x?
    b. Consider the equation: $5x = 15$. What number would you have to divide both sides of the equation by to obtain $1x$ on the left side?
    c. Solve the equation and check.
    d. Consider the equation $-5x = 15$. What number would you have to divide both sides of the equation by to obtain $1x$ on the left side?
    e. Solve the equation and check.
    f. Ernesto solved the equation in part (d), $-5x = 15$, by dividing both sides of the equation by 5. His final answer was: $-x = 3$. Do you agree with his answer? Why or why not?

A.3. Solve the equation $\dfrac{x}{3} = 12$ by multiplying both sides of the equation by 3.

A.4. Explain why multiplication was used instead of division in Exercise A.3.

For Exercises A.5–A.8, divide your paper into three columns. Solve the equation in the left column. Explain each step in the middle column. Check the answer in the right column.

Solve the Equation	Explain	Check
$5x = 20$		$5x = 20$
		$5(\ \ ) = 20$

A.5. $5x = 20$ **A.6.** $\dfrac{x}{5} = 20$

A.7. $5 + x = 20$ **A.8.** $20 = -5 + x$

For Exercises A.9–A.14, first simplify each side of the equation. Then solve the equation.

A.9. $13y - 10y = -18$ **A.10.** $3x + 4x = 25 - 4$

A.11. $-2(a + 3) - 6a + 6 = 8$ **A.12.** $5 + 4p - 2 - 3p = 0$

A.13. $5(t + 5) - 4t = -9$ **A.14.** $-5 + 10 - 5 = 2p + 3p$

Practice Exercises   Section 3.3

Study Skills Exercise

Now is a good time in the course to reflect on how your note-taking skills are progressing. Several note-taking strategies are given here. Check the activities that you routinely do and write down how the other suggestions may improve your learning.

_____ Read your notes after class and fill in details.

_____ Highlight important terms and definitions.

_____ Review your notes from the previous class.

_____ Sit in the class where you can clearly read the board and hear your instructor. Put away distractions such as cell phones.

_____ Add any Key Concepts that are not included in your notes.

Prerequisite Review

For Exercises R.1–R.6, solve the equation.

R.1. $p - 12 = 33$

R.2. $-12 = s - 3$

R.3. $-20 = 5x - 4x + 4$

R.4. $-8x + 9x - 6 = 17$

R.5. $7x - 3(2x + 1) = -2$

R.6. $2 = 6 + 3(x - 1) - 2x$

Vocabulary and Key Concepts

1. a. The _____ property of equality tells us that multiplying both sides of an equation by the same nonzero quantity results in an equivalent equation.

   b. The _____ property of equality tells us that dividing both sides of an equation by the same nonzero quantity results in an equivalent equation.

2. True or False. The expression $5 \cdot \dfrac{x}{5}$ is equivalent to both $\dfrac{5}{5}x$ and $1x$.

Concept 1: Multiplication and Division Properties of Equality

For Exercises 3–6, determine whether the given number is a solution to the equation.

3. $17x = -51$;   3

4. $60 = 12y$;   5

5. $\dfrac{a}{8} = -6$; -48

6. $\dfrac{a}{-4} = -9$; -36

For Exercises 7–34, solve the equation using the multiplication or division properties of equality. **(See Examples 1 and 2.)**

7. $22x = 22$

8. $9y = 9$

9. $10a = -150$

10. $11t = -121$

11. $14b = 42$

12. $6p = 12$

13. $-8k = 56$

14. $-5y = 25$

15. $-16 = -8z$

16. $-120 = -10p$

17. $-t = -13$

18. $-h = -17$

19. $5 = -x$

20. $30 = -a$

21. $\dfrac{b}{7} = -3$

22. $\dfrac{a}{4} = -12$

23. $\dfrac{u}{-2} = -15$

24. $\dfrac{v}{-10} = -4$

25. $-4 = \dfrac{p}{7}$

26. $-9 = \dfrac{w}{6}$

27. $-28 = -7t$

28. $-33 = -3r$

29. $5m = 0$

30. $6q = 0$

31. $\dfrac{x}{7} = 0$

32. $\dfrac{t}{8} = 0$

33. $-6 = -z$

34. $-9 = -n$

Concept 2: Comparing the Properties of Equality

For Exercises 35–38, choose the appropriate property of equality to use to solve the equation.

35. $-7x = 14$ division property or addition property?

36. $x + 7 = 14$ multiplication property or subtraction property?

37. $x - 7 = 14$ multiplication property or addition property?

38. $\dfrac{x}{7} = 14$ multiplication property or subtraction property?

Mixed Exercises

For Exercises 39–74, solve the equation. (See Example 3.)

39. $4 + x = -12$
40. $6 + z = -18$
41. $4y = -12$
42. $6p = -18$

43. $q - 4 = -12$
44. $p - 6 = -18$
45. $\dfrac{h}{4} = -12$
46. $\dfrac{w}{6} = -18$

47. $-18 = -9a$
48. $-40 = -8x$
49. $7 = r - 23$
50. $11 = s - 4$

51. $5 = \dfrac{y}{-3}$
52. $1 = \dfrac{h}{-5}$
53. $-52 = 5 + y$
54. $-47 = 12 + z$

55. $-4a = 0$
56. $-7b = 0$
57. $100 = 5k$
58. $95 = 19h$

59. $31 = -p$
60. $11 = -q$
61. $-3x + 7 + 4x = 12$
62. $6x + 7 - 5x = 10$

63. $5(x - 2) - 4x = 3$
64. $3(y - 6) - 2y = 8$

65. $3p + 4p = 25 - 4$
66. $2q + 3q = 54 - 9$

67. $-5 + 7 = 5x - 4(x - 1)$
68. $-3 + 11 = -2z + 3(z - 2)$

69. $-10 - 4 = 6m - 5(3 + m)$
70. $-15 - 5 = 9n - 8(2 + n)$

71. $5x - 2x = -15$
72. $13y - 10y = -18$

73. $-2(a + 3) - 6a + 6 = 8$
74. $-(b - 11) - 3b - 11 = -16$

Solving Equations with Multiple Steps

Section 3.4

1. Solving Equations with Multiple Steps

Some linear equations require one step to find the solution. For example, to solve the equation $x - 5 = 12$, we apply the addition property of equality. Addding 5 to both sides results in $x = 17$. We now combine the addition, subtraction, multiplication, and division properties of equality to solve equations that require multiple steps. This is shown in Example 1.

Concepts

1. Solving Equations with Multiple Steps
2. General Procedure to Solve a Linear Equation

Example 1 — Solving a Linear Equation

Solve. $2x - 3 = 15$

Solution:

Remember that our goal is to isolate x. Therefore, in this equation, we will first isolate the *term* containing x. This can be done by adding 3 to both sides.

$2x - 3 + 3 = 15 + 3$ Add 3 to both sides, because $-3 + 3 = 0$.

$\quad\quad\quad 2x = 18$ The term containing x is now isolated (by itself). The resulting equation now requires only one step to solve.

$\quad\quad\quad \dfrac{2x}{2} = \dfrac{18}{2}$ Divide both sides by 2 to make the x coefficient equal to 1.

$\quad\quad\quad\quad x = 9$ Simplify. The solution is 9.

Check: $2x - 3 = 15$ Original equation

$\quad\quad\quad 2(9) - 3 \stackrel{?}{=} 15$ Substitute 9 for x.

$\quad\quad\quad 18 - 3 \stackrel{?}{=} 15$ ✓ True

Skill Practice Solve.

1. $3x + 7 = 25$

As Example 1 shows, we will generally apply the addition (or subtraction) property of equality to isolate the variable term first. Then we will apply the multiplication (or division) property of equality to obtain a coefficient of 1 on the variable term.

Example 2 — Solving a Linear Equation

Solve. $22 = -3c + 10$

Solution:

We first isolate the term containing the variable by subtracting 10 from both sides.

$22 - 10 = -3c + 10 - 10$ Subtract 10 from both sides because $10 - 10 = 0$.

$\quad\quad 12 = -3c$ The term containing c is now isolated.

$\quad\quad \dfrac{12}{-3} = \dfrac{-3c}{-3}$ Divide both sides by -3 to make the c coefficient equal to 1.

$\quad\quad -4 = c$ Simplify. The solution is -4.

Check: $22 = -3c + 10$ Original equation

$\quad\quad\quad 22 = -3(-4) + 10$ Substitute -4 for c.

$\quad\quad\quad 22 = 12 + 10$ ✓ True

Skill Practice Solve.

2. $-12 = -5t + 13$

Answers

1. 6   **2.** 5

Example 3 — Solving a Linear Equation

Solve. $14 = \dfrac{y}{2} + 8$

Solution:

$14 = \dfrac{y}{2} + 8$

$14 - 8 = \dfrac{y}{2} + 8 - 8$ Subtract 8 from both sides. This will isolate the term containing the variable, y.

$6 = \dfrac{y}{2}$ Simplify.

$2 \cdot 6 = 2 \cdot \dfrac{y}{2}$ Multiply both sides by 2 to make the y coefficient equal to 1.

$12 = y$ Simplify. The solution is 12.

Check: $14 = \dfrac{y}{2} + 8$

$14 \stackrel{?}{=} \dfrac{(12)}{2} + 8$ Substitute 12 for y.

$14 \stackrel{?}{=} 6 + 8$ ✓   True

Skill Practice Solve.

3. $-3 = \dfrac{x}{3} + 9$

In Example 4, the variable x appears on both sides of the equation. In this case, apply the addition or subtraction properties of equality to collect the variable terms on one side of the equation and the constant terms on the other side.

Example 4 — Solving a Linear Equation with Variables on Both Sides

Solve. $4x + 5 = -2x - 13$

Solution:

To isolate x, we must first "move" all x terms to one side of the equation. For example, suppose we add $2x$ to both sides. This would "remove" the x term from the right-hand side because $-2x + 2x = 0$. The term $2x$ is then combined with $4x$ on the left-hand side.

$4x + 5 = -2x - 13$

$4x + 2x + 5 = -2x + 2x - 13$ Add $2x$ to both sides.

$6x + 5 = -13$ Simplify. Next, we want to isolate the *term* containing x.

$6x + 5 - 5 = -13 - 5$ Subtract 5 from both sides to isolate the x term.

$6x = -18$ Simplify.

$\dfrac{6x}{6} = \dfrac{-18}{6}$ Divide both sides by 6 to obtain an x coefficient of 1.

$x = -3$ The solution is -3 and checks in the original equation.

> **Avoiding Mistakes**
> Remember to rewrite subtraction as addition of the opposite to perform calculations.
> $-13 - 5$
> $= -13 + (-5)$
> $= -18$

Skill Practice Solve.

4. $8y - 3 = 6y + 17$

Answers
3. -36 4. 10

TIP: It should be noted that the variable may be isolated on either side of the equation. In Example 4 for instance, we could have isolated the x term on the right-hand side of the equation.

$4x + 5 = -2x - 13$

$4x - 4x + 5 = -2x - 4x - 13$ Subtract $4x$ from both sides. This "removes" the x term from the left-hand side.

$5 = -6x - 13$

$5 + 13 = -6x - 13 + 13$ Add 13 to both sides to isolate the x term.

$18 = -6x$ Simplify.

$\dfrac{18}{-6} = \dfrac{-6x}{-6}$ Divide both sides by -6.

$-3 = x$ This is the same solution as in Example 4.

2. General Procedure to Solve a Linear Equation

In Examples 1–4, we used multiple steps to solve equations. We also learned how to collect the variable terms on one side of the equation so that the variable could be isolated. The following procedure summarizes the steps to solve a linear equation.

Solving a Linear Equation in One Variable

Step 1 Simplify both sides of the equation.
- Clear parentheses if necessary.
- Combine *like* terms if necessary.

Step 2 Use the addition or subtraction property of equality to collect the variable terms on one side of the equation.

Step 3 Use the addition or subtraction property of equality to collect the constant terms on the *other* side of the equation.

Step 4 Use the multiplication or division property of equality to make the coefficient of the variable term equal to 1.

Step 5 Check the answer in the original equation.

Example 5   Solving a Linear Equation

Solve. $2(y + 10) = 8 - 4y$

Solution:

$2(y + 10) = 8 - 4y$

$2y + 20 = 8 - 4y$ **Step 1:** Simplify both sides of the equation. Clear parentheses.

$2y + 4y + 20 = 8 - 4y + 4y$ **Step 2:** Add $4y$ to both sides to collect the variable terms on the left.

$6y + 20 = 8$ Simplify.

$6y + 20 - 20 = 8 - 20$	**Step 3:**	Subtract 20 from both sides to collect the constants on the right.
$6y = -12$		Simplify.
$\dfrac{6y}{6} = \dfrac{-12}{6}$	**Step 4:**	Divide both sides by 6 to obtain a coefficient of 1 on the y term
$y = -2$		The solution is -2.
Check: $2(y + 10) = 8 - 4y$	**Step 5:**	Check the solution in the original equation.
$2(-2 + 10) \stackrel{?}{=} 8 - 4(-2)$		Substitute -2 for y.
$2(8) \stackrel{?}{=} 8 - (-8)$		
$16 \stackrel{?}{=} 16 \checkmark$		The solution checks.

Skill Practice Solve.

5. $6(z + 4) = -16 - 4z$

Example 6   **Solving a Linear Equation**

Solve. $2x + 3x + 2 = -4(3 - x)$

Solution:

$2x + 3x + 2 = -4(3 - x)$		
$5x + 2 = -12 + 4x$	**Step 1:**	Simplify both sides of the equation. On the left, combine *like* terms. On the right, clear parentheses.
$5x - 4x + 2 = -12 + 4x - 4x$	**Step 2:**	Subtract $4x$ from both sides to collect the variable terms on the left.
$x + 2 = -12$		Simplify.
$x + 2 - 2 = -12 - 2$	**Step 3:**	Subtract 2 from both sides to collect the constants on the right.
$x = -14$	**Step 4:**	The x coefficient is already 1. The solution is -14.
Check: $2x + 3x + 2 = -4(3 - x)$	**Step 5:**	Check in the original equation.
$2(-14) + 3(-14) + 2 \stackrel{?}{=} -4[3 - (-14)]$		Substitute -14 for x.
$-28 - 42 + 2 \stackrel{?}{=} -4(17)$		
$-70 + 2 \stackrel{?}{=} -68$		
$-68 \stackrel{?}{=} -68 \checkmark$		The solution checks.

TIP: A linear equation in one variable has one unique solution. As you continue your study of algebra you will also encounter equations that may have no solution or infinitely many solutions.

Skill Practice Solve.

6. $-3y - y - 4 = -5(y - 8)$

Answers

5. -4 6. 44

Section 3.4 Activity

A.1. Solve the equation $4x - 3 = 13$ by following these steps:
  a. Add 3 to both sides.
  b. Divide both sides by 4.
  c. Solve for x. Check your solution.

A.2. Consider the equation in Exercise A.1. If you divided both sides by 4 then added 3 to each side, would you get the same solution? Show your work to support your answer.

For Exercises A.3–A.5, divide your paper into three columns. Solve the equation in the left column. Explain each step in the middle column. Check the answer in the right column.

Solve the Equation	Explain	Check
$-8 = 2x + 6$		$-8 = 2x + 6$
		$-8 = 2(\ ) + 6$

A.3. $-8 = 2x + 6$

A.4. $106 = -52y - 206$

A.5. $\dfrac{z}{3} - 6 = 1$

A.6. Solve the equation by moving the variable terms to the left of the equal sign.
$1 + 5y = 4y - 9$

A.7. Solve the equation by moving the variable terms to the right of the equal sign.
$1 + 5y = 4y - 9$

A.8. Compare your solution from Exercises A.6 and A.7. When solving a linear equation, does it matter to which side of the equal sign the variable terms are moved?

For Exercises A.9–A.11, solve the equation.

A.9. $-6w + 8 = -3w - 19$

A.10. $-7 + 4(x + 3) = -1 - 2(x + 6)$

A.11. $3 - x + 7 = 5 - 2(x - 3)$

A.12. Summarize the five steps to solving a linear equation.

Section 3.4 Practice Exercises

Study Skills Exercise

Flashcards can be used as an effective memory technique to remember key mathematical concepts, formulas, and properties.
  1. Write down the five steps to Solving a Linear Equation in One Variable, each step on a separate 3 × 5 card.
  2. Write down two or three odd-numbered problems or examples from this section. Write the problem on one side and the section, page, problem number, and answer on the other side.

Use these cards to review and study for the next test. This study skill can be used for all sections of the textbook.

Section 3.4 Solving Equations with Multiple Steps

Prerequisite Review

For Exercises R.1–R.8, solve the equation.

R.1. $4c = 12$ **R.2.** $-5x = 15$ **R.3.** $\frac{t}{3} = -4$ **R.4.** $-\frac{t}{5} = 10$

R.5. $10x - 8x - 6 - x = 1 + (-3)$ **R.6.** $-8 - 8x + 6 + 7x = -14 + 16$

R.7. $4 - 5(y + 3) + 6y = 7$ **R.8.** $7y - 3(2y + 3) = 7 - 18$

Vocabulary and Key Concepts

1. Consider the equation $3x - 6 = 18$. According to the recommended procedure to solve a linear equation, should we add 6 to both sides first, or should we divide both sides by 3 first?

2. When checking the solution to a linear equation, substitute the answer for x into the _____ equation.

Concept 1: Solving Equations with Multiple Steps

For Exercises 3–6, determine whether the given number is a solution to the equation.

3. $-7 + 5x = 13$; 4
4. $\frac{t}{2} - 14 = 10$; 12
5. $5p - 14 = -4p + 4$; 18
6. $3(5y + 4) = 7y + 32 - 2y$; 2

For Exercises 7–26, solve the equation. (See Examples 1–3.)

7. $4a + 1 = -7$ **8.** $5n + 48 = -2$ **9.** $2x + 5 = 5$ **10.** $7y + 28 = 28$

11. $3m - 2 = 16$ **12.** $2n - 5 = 3$ **13.** $8c - 12 = 36$ **14.** $5t - 1 = -11$

15. $1 = -4z + 21$ **16.** $-4 = -3p + 14$ **17.** $9 = 12x - 15$ **18.** $7 = 5y - 8$

19. $\frac{b}{3} - 12 = -9$ **20.** $\frac{c}{5} + 2 = 4$ **21.** $-9 = \frac{w}{2} - 3$ **22.** $-16 = \frac{t}{4} - 14$

23. $\frac{m}{-3} + 40 = 60$ **24.** $\frac{c}{-4} - 3 = 5$ **25.** $-y - 7 = 14$ **26.** $-p + 8 = 20$

For Exercises 27–38, solve the equation. (See Example 4.)

27. $8 + 4b = 2 + 2b$ **28.** $2w + 10 = 5w - 5$ **29.** $7 - 5t = 3t - 9$

30. $4 - 2p = -3 + 5p$ **31.** $4 - 3d = 5d - 4$ **32.** $-3k + 14 = -4 + 3k$

33. $12p = 3p + 36$ **34.** $2x + 10 = 4x$ **35.** $4 + 2a - 7 = 3a + a + 3$

36. $4b + 2b - 7 = 2 + 4b + 5$ **37.** $-8w + 8 + 3w = 2 - 6w + 2$ **38.** $-12 + 5m + 10 = -2m - 10 - m$

Concept 2: General Procedure to Solve a Linear Equation

For Exercises 39–56, solve the equation. (See Examples 5 and 6.)

39. $5(z + 7) = 9 + 3z$ **40.** $6(w + 2) = 20 + 2w$ **41.** $2(1 - m) = 5 - 3m$

42. $3(2 - g) = 12 - g$ **43.** $3n - 4(n - 1) = 16$ **44.** $4p - 3(p + 2) = 18$

45. $4x + 2x - 9 = -3(5 - x)$ **46.** $-3x + x - 8 = -2(6 - x)$ **47.** $-4 + 2x + 1 = 3(x - 1)$

48. $-2 + 5x + 8 = 6(x + 2) - 6$ **49.** $9q - 5(q - 3) = 5q$ **50.** $6h - 2(h + 6) = 10h$

51. $-4(k - 2) + 14 = 3k - 20$ **52.** $-3(x + 4) - 9 = -2x + 12$ **53.** $3z + 9 = 3(5z - 1)$

54. $4y + 4 = 8(y - 2)$ **55.** $6(u - 1) + 5u + 1 = 5(u + 6) - u$ **56.** $2(2v + 3) + 8v = 6(v - 1) + 3v$

Problem Recognition Exercises

Identifying Expressions and Equations

Sometimes students mistake expressions for equations and vice versa. Remember that an equation has an equal sign (=) and an expression does not. Furthermore, we simplify expressions, but we *solve* equations. For example, compare the following.

$3(x - 4) - x + 2$

This is an expression and can be simplified by clearing parentheses and combining *like* terms.

$3(x - 4) - x + 2$
$= 3x - 12 - x + 2$
$= 3x - x - 12 + 2$
$= 2x - 10$

The expression is simplified.

$3(x - 4) - x + 2 = 0$

This is an equation (notice the = sign). To solve the equation, simplify both sides, then apply properties of equality to isolate x.

$3(x - 4) - x + 2 = 0$
$3x - 12 - x + 2 = 0$
$2x - 10 = 0$
$2x - 10 + 10 = 0 + 10$
$2x = 10$
$\dfrac{2x}{2} = \dfrac{10}{2}$
$x = 5$

The equation is solved.

For Exercises 1–6, identify the problem as an expression or as an equation.

1. $-5 + 4x - 6x = 7$
2. $-8(4 - 7x) + 4$
3. $4 - 6(2x - 3) + 1$
4. $10 - x = 2x + 19$
5. $6 - 3(x + 4) = 6$
6. $9 - 6(x + 1)$

For Exercises 7–30, first identify the problem as an expression or an equation. Then simplify the expression or solve the equation.

7. $5t = 20$
8. $6x = 36$
9. $5t - 20t$
10. $6x - 36x$
11. $5(w - 3)$
12. $6(x - 2)$
13. $5(w - 3) = 20$
14. $6(x - 2) = 36$
15. $5 + t = 20$
16. $6 + x = 36$
17. $5 + t + 20$
18. $6 + x + 36$
19. $5 + 3p - 2 = 0$
20. $16 - 2k + 2 = 0$
21. $23u + 2 = -12u + 72$
22. $75w - 27 = 14w + 156$
23. $23u + 2 - 12u + 72$
24. $75w - 27 - 14w + 156$
25. $-2(x - 3) + 14 + 10 - (x + 4)$
26. $26 - (3x + 12) + 11 - 4(x + 1)$
27. $-2(x - 3) + 14 = 10 - (x + 4)$
28. $26 - (3x + 12) = 11 - 4(x + 1)$
29. $2 - 3(y + 1) = -4y + 7$
30. $5(t + 5) - 3t = t - 9$

Section 3.5 Applications and Problem Solving

Concepts

1. Problem-Solving Flowchart
2. Translating Verbal Statements into Equations
3. Applications of Linear Equations

1. Problem-Solving Flowchart

Linear equations can be used to solve many real-world applications. In this section, we offer guidelines to write equations to solve applications. Consider the problem-solving flowchart.

Section 3.5 Applications and Problem Solving 157

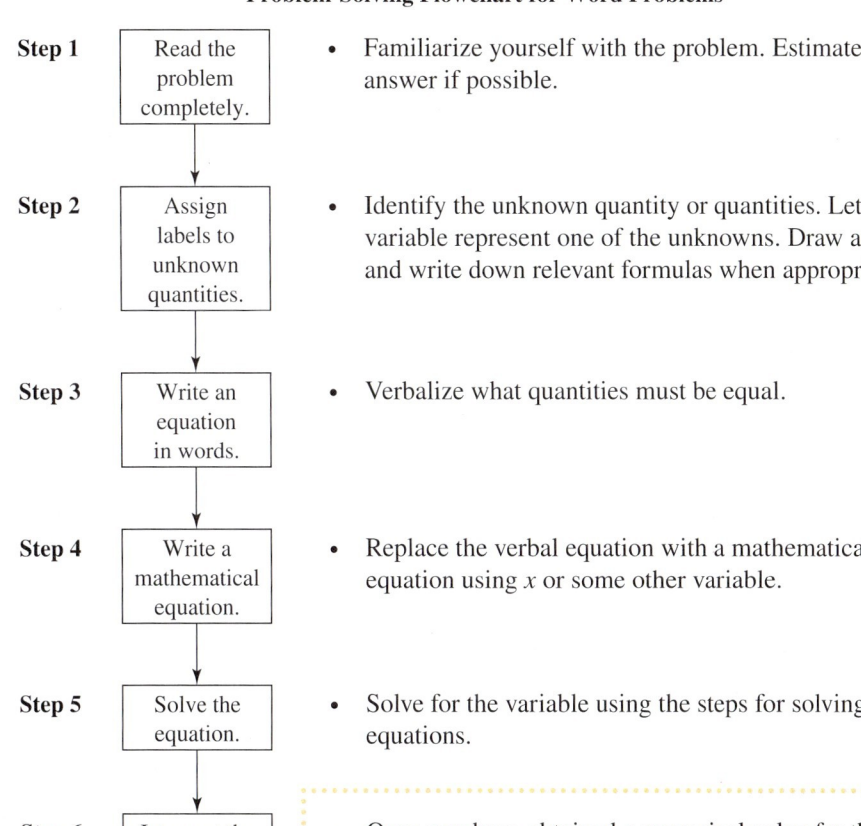

2. Translating Verbal Statements into Equations

We begin solving word problems with practice translating between an English sentence and an algebraic equation. First, spend a minute to recall some of the key words that represent addition, subtraction, multiplication, and division. See Table 3-4.

Table 3-4

Addition: $a + b$	Subtraction: $a - b$
The *sum* of a and b a plus b b added to a b more than a a increased by b The total of a and b	The *difference* of a and b a minus b b subtracted from a a decreased by b b less than a
Multiplication: $a \cdot b$	**Division:** $a \div b$
The *product* of a and b a times b a multiplied by b	The *quotient* of a and b a divided by b b divided into a The ratio of a and b a over b a per b

Avoiding Mistakes

It is always a good idea to check your answer in the context of the problem. This will help you determine if your answer is reasonable.

Example 1 Translating a Sentence to a Mathematical Equation

A number decreased by 7 is 12. Find the number.

Solution:

Let x represent the number.	**Step 1:** Read the problem completely.
	Step 2: Label the unknown.
A number decreased by 7 is 12.	**Step 3:** Write the equation in words
$\quad\quad x \quad\quad - \quad\quad 7 = 12$	**Step 4:** Translate to a mathematical equation.
$x - 7 = 12$	**Step 5:** Solve the equation.
$x - 7 + 7 = 12 + 7$	Add 7 to both sides.
$x = 19$	
The number is 19.	**Step 6:** Interpret the answer in words.

Avoiding Mistakes

To check the answer to Example 1, we see that 19 decreased by 7 is 12.

Skill Practice

1. A number minus 6 is -22. Find the number.

Example 2 Translating a Sentence to a Mathematical Equation

Seven less than 3 times a number results in 11. Find the number.

Solution:

Let x represent the number.

Step 1: Read the problem completely.
Step 2: Label the unknown.

Seven less than 3 times a number results in 11.

$$3x - 7 = 11$$
(7 less than; three times a number; results in 11)

Step 4: Translate to a mathematical equation.

$$3x - 7 + 7 = 11 + 7$$

Step 5: Solve the equation. Add 7 to both sides.

$$3x = 18$$

Simplify.

$$\frac{3x}{3} = \frac{18}{3}$$

Divide both sides by 3.

$$x = 6$$

The number is 6.

Step 6: Interpret the answer in words.

Avoiding Mistakes

To check the answer to Example 2, we have:
The value 3 times 6 is 18, and 7 less than 18 is 11.

Skill Practice

2. 4 subtracted from 8 times a number is 36. Find the number.

Answers

1. The number is -16.
2. The number is 5.

Example 3: Translating a Sentence to a Mathematical Equation

Two times the sum of a number and 8 equals 38. Find the number.

Solution:

Step 1: Read the problem completely.

Let x represent the number. Step 2: Label the unknown.

Two times the sum of a number and 8 equals 38.

$2 \cdot (x + 8) = 38$ (two times — the sum of a number and 8 — equals 38)

Step 3: Write the equation in words.
Step 4: Translate to a mathematical equation.

$2(x + 8) = 38$ Step 5: Solve the equation.
$2x + 16 = 38$ Clear parentheses.
$2x + 16 - 16 = 38 - 16$ Subtract 16 from both sides.
$2x = 22$ Simplify.
$\dfrac{2x}{2} = \dfrac{22}{2}$ Divide both sides by 2.
$x = 11$

The number is 11. Step 6: Interpret the answer in words.

Avoiding Mistakes
The sum $(x + 8)$ must be enclosed in parentheses so that the entire sum is multiplied by 2.

Avoiding Mistakes
To check the answer to Example 3, we have:
The sum of 11 and 8 is 19. Twice this amount gives 38 as expected.

Skill Practice

3. 8 added to twice a number is 22. Find the number.

3. Applications of Linear Equations

In Example 4, we practice representing quantities within a word problem by using variables.

Example 4: Representing Quantities Algebraically

a. Kathleen works twice as many hours in one week as Kevin. If Kevin works for h hours, write an expression representing the number of hours that Kathleen works.

b. At a carnival, rides cost $3 each. If Alicia takes n rides during the day, write an expression for the total cost.

c. Josie made $430 less during one week than her friend Annie made. If Annie made D dollars, write an expression for the amount that Josie made.

Purestock/Getty Images

Solution:

a. Let h represent the number of hours that Kevin works.

Kathleen works twice as many hours as Kevin.

$2h$ is the number of hours that Kathleen works.

Answer
3. The number is 7.

b. Let n represent the number of rides Alicia takes during the day.

Rides cost \$3 each.

$3n$ is the total cost.

c. Let D represent the amount of money that Annie made during the week.

Josie made \$430 less than Annie.

$D - 430$ represents the amount that Josie made.

Skill Practice

4. Tasha ate three times as many M&Ms as her friend Kate. If Kate ate x M&Ms, write an expression for the number that Tasha ate.
5. Casey bought seven books from a sale rack. If the books cost d dollars each, write an expression for the total cost.
6. One week Kim worked 8 hr more than her friend Tom. If Tom worked x hours, write an expression for the number of hours that Kim worked.

In Examples 5 and 6, we practice solving application problems using linear equations.

Example 5 — Applying a Linear Equation to Carpentry

A carpenter must cut a 10-ft board into two pieces to build a brace for a picnic table. If one piece needs to be four times longer than the other piece, how long should each piece be?

Solution:

We can let x represent the length of either piece. However, if we choose x to be the length of the shorter piece, then the longer piece has to be $4x$ (4 times as long).

Let $x =$ the length of the shorter piece.

Then $4x =$ the length of the longer piece.

Step 1: Read the problem completely.

Step 2: Label the unknowns. Draw a picture.

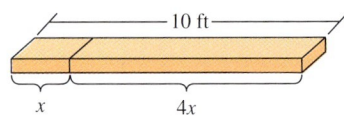

$$\begin{pmatrix}\text{Length of}\\\text{one piece}\end{pmatrix} + \begin{pmatrix}\text{length of the}\\\text{other piece}\end{pmatrix} = \begin{pmatrix}\text{total}\\\text{length}\end{pmatrix}$$

$$x \quad + \quad 4x \quad = \quad 10$$

Step 3: Write an equation in words.

Step 4: Write a mathematical equation.

$x + 4x = 10$ **Step 5:** Solve the equation.

$5x = 10$ Combine *like* terms.

$\dfrac{5x}{5} = \dfrac{10}{5}$ Divide both sides by 5.

$x = 2$ **Step 6:** Interpret the results in words.

Recall that x represents the length of the shorter piece. Therefore, the shorter piece is 2 ft. The longer piece is given by $4x$ or $4(2 \text{ ft}) = 8$ ft.

The pieces are 2 ft and 8 ft.

Avoiding Mistakes

In Example 5, the two pieces should total 10 ft. We have,
2 ft + 8 ft = 10 ft as desired.

Skill Practice

7. A piece of cable 92 ft long is to be cut into two pieces. One piece must be three times longer than the other. How long should each piece be?

Answers
4. $3x$ 5. $7d$ 6. $x + 8$
7. One piece should be 23 ft and the other should be 69 ft long.

Example 6 Applying a Linear Equation

One model of a 42″ HDTV sells for $800 more than a smaller 32″ HDTV. The combined cost for these two televisions is $2000. Find the cost for each television.

Solution:

Justin Kase/Alamy Stock Photo

Step 1: Read the problem completely.

Let x represent the cost of the 32″ TV.   Step 2: Label the unknowns.

Then $x + 800$ represents the cost of the 42″ TV.

$$\begin{pmatrix}\text{Cost of}\\32''\end{pmatrix} + \begin{pmatrix}\text{cost of}\\42''\end{pmatrix} = \begin{pmatrix}\text{total}\\\text{cost}\end{pmatrix}$$

Step 3: Write an equation in words.

$$x + x + 800 = 2000$$

Step 4: Write a mathematical equation.

$$x + x + 800 = 2000$$

Step 5: Solve the equation.

$$2x + 800 = 2000 \quad \text{Combine like terms.}$$
$$2x + 800 - 800 = 2000 - 800 \quad \text{Subtract 800 from both sides.}$$
$$2x = 1200$$
$$\frac{2x}{2} = \frac{1200}{2} \quad \text{Divide both sides by 2.}$$
$$x = 600$$

Step 6: Interpret the results in words.

Since $x = 600$, the 32″ TV costs $600.

The cost of the 42″ model is represented by $x + 800 = \$600 + \$800 = \$1400$.

Skill Practice

8. A kit of cordless 18-volt tools made by Craftsman cost $310 less than a similar kit made by DeWalt. The combined cost for both models is $690. Find the cost for each model. (*Source: Consumer Reports*)

TIP: In Example 6, we could have let x represent *either* the cost of the 32″ TV or the 42″ TV.

Suppose we had let x represent the cost of the 42″ model.
Then $x - 800$ is the cost of the 32″ model (the 32″ model is *less* expensive).

$$\begin{pmatrix}\text{Cost of}\\32''\end{pmatrix} + \begin{pmatrix}\text{cost of}\\42''\end{pmatrix} = \begin{pmatrix}\text{total}\\\text{cost}\end{pmatrix}$$
$$x - 800 + x = 2000$$
$$2x - 800 = 2000$$
$$2x - 800 + 800 = 2000 + 800$$
$$2x = 2800$$
$$x = 1400$$

Therefore, the 42″ TV costs $1400 as expected.
The 32″ TV costs $x - 800$ or $1400 - \$800 = \$600$.

Answer

8. The Craftsman model costs $190 and the DeWalt model costs $500.

Section 3.5 Activity

For Exercises A.1–A.5, solve the word problem and format your solution using the problem-solving steps:
  a. Read the problem again.
  b. Label the variable(s).
  c. Write a verbal model (that is, write an equation in *words*).
  d. Write an equation to represent the verbal model.
  e. Solve the equation.
  f. Interpret the results in words using the proper units of measurement, and verify that your solution makes sense in the context of the problem.

A.1. A number increased by 8 equals −4. Find the number.

A.2. The sum of −3 times a number and 6 is 60. Find the number.

A.3. −3 times the sum of a number and 6 results in 60. Find the number. (*Note the difference in wording between this exercise and Exercise A.2.*)

A.4. An electrician cuts a 136-ft piece of cable. One piece is 16 ft less than 3 times the length of the other piece. Find the length of each piece.

A.5. One model of a new DVD player sold for $22 more than a used DVD player. The combined cost for these two players was $68. Find the individual cost for each player.

A.6. The perimeter of a computer screen is 42 in. If the width is 3 in. less than the length, find the length and the width of the computer screen.

Section 3.5 Practice Exercises

Study Skills Exercise

A lot of anxiety that students feel while taking a math test can be minimized by adequately preparing for the test. When preparing for the test, ask yourself what material your instructor emphasizes on tests. For example, is the material from the textbook, notes, handouts, homework, or a combination of these sources?

List five types of problems that you think will be on the test for this chapter. These practices will reduce your anxiety and improve your performance on the test.

Prerequisite Review

For Exercises R.1–R.8, solve the equation.

R.1. $3t - 15 = -24$

R.2. $-6x - 4 = 20$

R.3. $\dfrac{b}{5} + 5 = -4$

R.4. $-3 + \dfrac{z}{2} = 10$

R.5. $5 = -3(x - 1) - 4$

R.6. $8(c + 1) + 1 = -15$

R.7. $4d - 3 = -2d + 9$

R.8. $10 - 9x = -20 + x$

Vocabulary and Key Concepts

For Exercises 1–6, identify and explain the step that is described in the problem-solving flowchart for word problems.

1. Verbalize what quantities must be equal.

2. Familiarize yourself with the problem and estimate the answer if possible.

3. Replace the verbal equation with a mathematical equation using a variable.

4. Check that your answer makes sense in the context of the problem. Write the answer in words.

5. Identify the unknown(s) and assign a variable to each. Draw a picture or write the formula if appropriate.

6. Solve for the unknown.

7. *Ratio* and *per* are words that imply the mathematical operation of _____.

8. Consider the verbal statement: Twice the sum of a and b. The sum of a and b must be enclosed in parentheses indicating the operation of _____ should be performed before _____.

Concept 2: Translating Verbal Statements into Equations

For Exercises 9–40,

a. write an equation that represents the given statement.

b. solve the problem. (See Examples 1–3.)

9. The sum of a number and 6 is 19. Find the number.

10. The sum of 12 and a number is 49. Find the number.

11. The difference of a number and 14 is 20. Find the number.

12. The difference of a number and 10 is 18. Find the number.

13. The quotient of a number and 3 is -8. Find the number.

14. The quotient of a number and -2 is 10. Find the number.

15. The product of a number and -6 is -60. Find the number.

16. The product of a number and -5 is -20. Find the number.

17. The difference of -2 and a number is -14. Find the number.

18. A number subtracted from -30 results in 42. Find the number.

19. 13 increased by a number results in -100. Find the number.

20. The total of 30 and a number is 13. Find the number.

21. Sixty is -5 times a number. Find the number.

22. Sixty-four is -4 times a number. Find the number.

23. Nine more than 3 times a number is 15. Find the number.

24. Eight more than twice a number is 20. Find the number.

25. Five times a number when reduced by 12 equals -27. Find the number.

26. Negative four times a number when reduced by 6 equals 14. Find the number.

27. Five less than the quotient of a number and 4 is equal to -12. Find the number.

28. Ten less than the quotient of a number and -6 is -2. Find the number.

29. Eight decreased by the product of a number and 3 is equal to 5. Find the number.

30. Four decreased by the product of a number and 7 is equal to 11. Find the number.

31. Three times the sum of a number and 4 results in −24. Find the number.

32. Twice the sum of 9 and a number is 30. Find the number.

33. Negative four times the difference of 3 and a number is −20. Find the number.

34. Negative five times the difference of 8 and a number is −55. Find the number.

35. The product of −12 and a number is the same as the sum of the number and 26. Find the number.

36. The difference of a number and 16 is the same as the product of the number and −3. Find the number.

37. Ten times the total of a number and 5 is 80. Find the number.

38. Three times the difference of a number and 5 is 15. Find the number.

39. The product of 3 and a number is the same as 10 less than twice the number.

40. Six less than a number is the same as 3 more than twice the number.

Concept 3: Applications of Linear Equations

41. A metal rod is cut into two pieces. One piece is five times as long as the other. If x represents the length of the shorter piece, write an expression for the length of the longer piece. (See Example 4a.)

42. Jackson scored three times as many points in a basketball game as Tony. If p represents the number of points that Tony scored, write an expression for the number of points that Jackson scored.

43. Tickets to a college baseball game cost $6 each. If Stan buys n tickets, write an expression for the total cost. (See Example 4b.)

44. The tuition and fees at a two-year college are $95 per credit-hour. If Tyler takes h credit-hours, write an expression for the cost of his tuition and fees.

45. Bill's daughter is 30 years younger than he is. Write an expression for his daughter's age if Bill is A years old. (See Example 4c.)

46. Carlita spent $88 less on tuition and fees than her friend Carlo did. If Carlo spent d dollars, write an expression for the amount that Carlita spent.

47. The number of prisoners at the Fort Dix Federal Correctional Facility is 1481 more than the number at Big Spring in Texas. If p represents the number of prisoners at Big Spring, write an expression for the number at Fort Dix.

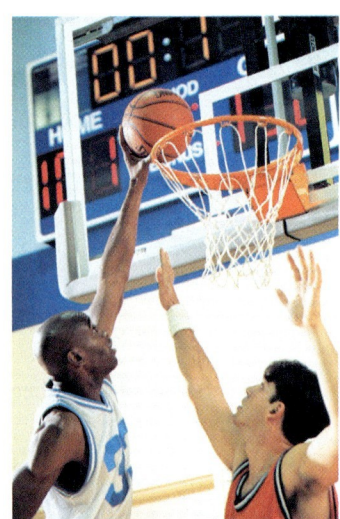

Comstock/SuperStock

48. Race car driver A. J. Foyt won 15 more races than Mario Andretti. If Mario Andretti won r races, write an expression for the number of races won by A. J. Foyt.

49. The typical cost for text messaging is 20 cents per message. Write an expression for the cost of sending t messages.

50. Sandy bought eight tomatoes for c cents each. Write an expression for the total cost.

Andrew Paterson/Alamy Stock Photo

51. The cost of a new car is about 5 times the cost of the same model that is ten years old. If x represents the cost of the ten-year-old car, write an expression for the cost of a new car.

52. Jacob scored four times as many points in his hockey game on Monday as he did in Wednesday's game. If p represents the number of points scored on Wednesday, write an expression for the number scored on Monday.

For Exercises 53–64, use the problem-solving flowchart to set up and solve an equation to solve each problem.

53. Felicia has an 8-ft piece of ribbon. She wants to cut the ribbon into two pieces so that one piece is three times the length of the other. Find the length of each piece. **(See Example 5.)**

54. Richard and Linda enjoy visiting Hilton Head Island, South Carolina. The distance from their home to Hilton Head is 954 mi, so the drive takes them 2 days. Richard and Linda travel twice as far the first day as they do the second. How many miles do they travel each day?

55. A motorist drives from Minneapolis, Minnesota, to Madison, Wisconsin, and then on to Chicago, Illinois. The total distance is 360 mi. The distance between Minneapolis and Madison is 120 mi more than the distance from Madison to Chicago. Find the distance between Minneapolis and Madison.

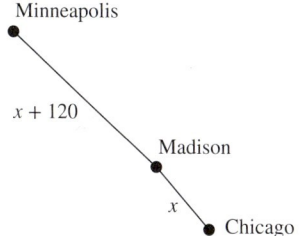

56. Two hikers on the Appalachian Trail hike from Hawk Mountain to Gooch Mountain. The next day, they hike from Gooch Mountain to Woods Hole. The total distance is 19 mi. If the distance between Gooch Mountain and Woods Hole is 5 mi more than the distance from Hawk Mountain to Gooch Mountain, find the distance they hiked each day.

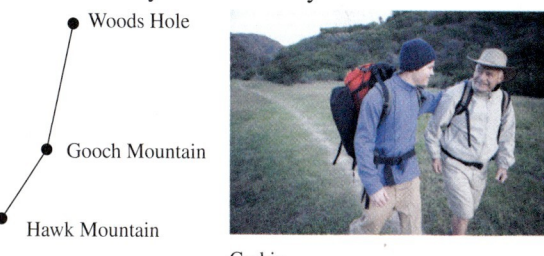

57. The Beatles had 9 more number one albums than Elvis Presley. Together, they had a total of 29 number one albums. How many number one albums did each have? **(See Example 6.)**

58. A two-piece set of luggage costs $150. If sold individually, the larger bag costs $40 more than the smaller bag. What are the individual costs for each bag?

59. In a football game, the New England Patriots scored 3 points less than the New York Giants. A total of 31 points was scored in the game. How many points did each team score?

60. The total number of goals scored in a hockey game was 21. Team A scored 3 more goals than team B. How many goals did each team score?

61. An apartment complex charges a refundable security deposit when renting an apartment. The deposit is $350 less than the monthly rent. If Charlene paid a total of $950 for her first month's rent and security deposit, how much is her monthly rent? How much is the security deposit?

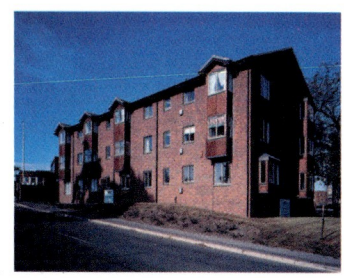

62. Becca paid a total of $695 for tuition and lab fees for fall semester. Tuition for her courses cost $605 more than lab fees. How much did her tuition cost? What was the cost of the lab fees?

63. Stefan is paid a salary of $480 a week at his job. When he works overtime, he receives $18 an hour. If his weekly paycheck came to $588, how many hours of overtime did he put in that week?

64. Mercedes is paid a salary of $480 a week at her job. She worked 8 hr of overtime during the holidays and her weekly paycheck came to $672. What is her overtime pay per hour?

Chapter 3 Review Exercises

Section 3.1

For Exercises 1 and 2, list the coefficients of the terms.

1. $3a^2 - 5a + 12$
2. $-6xy - y + 2x + 1$

For Exercises 3–6, determine if the two terms are *like* terms or unlike terms.

3. $5t^2, 5t$
4. $4h, -2h$
5. $21, -5$
6. $-8, -8k$

For Exercises 7 and 8, apply the commutative property of addition or multiplication to rewrite the expression.

7. $t - 5$
8. $h \cdot 3$

For Exercises 9 and 10, apply the associative property of addition or multiplication to rewrite the expression. Then simplify the expression.

9. $-4(2 \cdot p)$
10. $(m + 10) - 12$

For Exercises 11–14, apply the distributive property.

11. $3(2b + 5)$
12. $5(4x + 6y - 3z)$
13. $-(4c - 6d)$
14. $-(-4k + 8w - 12)$

For Exercises 15–22, combine *like* terms. Clear parentheses if necessary.

15. $-5x - x + 8x$
16. $-3y - 7y + y$
17. $6y + 8x - 2y - 2x + 10$
18. $12a - 5 + 9b - 5a + 14$
19. $5 - 3(x - 7) + x$
20. $6 - 4(z + 2) + 2z$
21. $4(u - 3v) - 5(u + v)$
22. $-5(p + 4) + 6(p + 1) - 2$

Section 3.2

For Exercises 23 and 24, determine if -3 is a solution to the equation.

23. $5x + 10 = -5$
24. $-3(x - 1) = -9 + x$

For Exercises 25–32, solve the equation using either the addition property of equality or the subtraction property of equality.

25. $r + 23 = -12$
26. $k - 3 = -15$
27. $10 = p - 4$
28. $21 = q + 3$
29. $5a + 7 - 4a = 20$
30. $-7t - 4 + 8t = 11$
31. $-4(m - 3) + 7 + 5m = 21$
32. $-2(w - 3) + 3w = -8$

Section 3.3

For Exercises 33–38, solve the equation using either the multiplication property of equality or the division property of equality.

33. $4d = -28$
34. $-3c = -12$
35. $\dfrac{t}{-2} = -13$
36. $\dfrac{p}{5} = 7$
37. $-42 = -7p$
38. $-12 = \dfrac{m}{4}$

Section 3.4

For Exercises 39–52, solve the equation.

39. $9x + 7 = -2$
40. $8y + 3 = 27$
41. $45 = 6m - 3$
42. $-25 = 2n - 1$
43. $\dfrac{p}{8} + 1 = 5$
44. $\dfrac{x}{-5} - 2 = -3$
45. $5x + 12 = 4x - 16$
46. $-4t - 2 = -3t + 5$
47. $-8 + 4y = 7y + 4$
48. $15 - 2c = 5c + 1$
49. $6(w - 2) + 15 = 3w$
50. $-4(h - 5) + h = 7h$
51. $-(5a + 3) - 3(a - 2) = 24 - a$
52. $-(4b - 7) = 2(b + 3) - 4b + 13$

Section 3.5

For Exercises 53–58,

a. write an equation that represents the statement.
b. solve the problem.

53. Four subtracted from a number is 13. Find the number.

54. The quotient of a number and −7 is −6. Find the number.

55. Three more than −4 times a number is −17. Find the number.

56. Seven less than 3 times a number is −22. Find the number.

57. Twice the sum of a number and 10 is 16. Find the number.

58. Three times the difference of 4 and a number is −9. Find the number.

59. A rack of discount CDs are on sale for $9 each. If Mario buys n CDs, write an expression for the total cost.

Stock 4B RF/Getty Images

60. Henri bought four sandwiches for x dollars each. Write an expression that represents the total cost.

61. It took Winston 2 hr longer to finish his psychology paper than it did for Gus to finish. If Gus finished in x hours, write an expression for the amount of time it took Winston.

62. Gerard is 6 in. taller than his younger brother Dwayne. If Dwayne is h inches tall, write an expression for Gerard's height.

63. Monique and Michael drove from Ormond Beach, Florida, to Knoxville, Tennessee. Michael drove three times as far as Monique drove. If the total distance is 480 mi, how far did each person drive?

64. Joel ate twice as much pizza as Angela. Together, they finished off a 12-slice pizza. How many pieces did each person have?

Dynamic Graphics Group/Getty Images

65. A city contractor needs two pieces of sewer pipe to line a ditch. One piece is 5 ft shorter than the other. Together the two pieces span 65 ft. Determine the length of each individual piece.

66. Raul signed up for his classes for the spring semester. His load was 4 credit-hours less in the spring than in the fall. If he took a total of 28 hours in the two semesters combined, how many hours did he take in the fall? How many hours did he take in the spring?

Fractions and Mixed Numbers

4

CHAPTER OUTLINE

- **4.1** Introduction to Fractions and Mixed Numbers 170
- **4.2** Simplifying Fractions 180
- **4.3** Multiplication and Division of Fractions 195
- **4.4** Least Common Multiple and Equivalent Fractions 210
- **4.5** Addition and Subtraction of Fractions 220
- **4.6** Estimation and Operations on Mixed Numbers 230

  Problem Recognition Exercises: Operations on Fractions and Mixed Numbers 243
- **4.7** Order of Operations and Complex Fractions 244
- **4.8** Solving Equations Containing Fractions 253

  Problem Recognition Exercises: Comparing Expressions and Equations 261

  Chapter 4 Review Exercises 262

Part of a Whole

Whole numbers and integers are insufficient to represent all quantifiable values we encounter in day-to-day life. For example, if a whole pizza is to feed 5 people, then the pizza must be cut into 5 pieces, and each piece is a fraction (or part) of the pie. The use of fractions is fundamentally important to a wide variety of applications. For example, a machinist might construct a bolt that is $3\frac{1}{8}$ in. wide. A plumber might cut a piece of pipe $4\frac{1}{2}$ ft in length. A carpenter might use a $\frac{5}{8}$-in. drill bit to fix a fence.

Being able to add, subtract, multiply, and divide fractions is likewise critical. For example, suppose that a book designer wants to design the layout of a novel with an overall page size of $5\frac{1}{2}$ in. by $8\frac{1}{2}$ in. The designer knows that she must leave $\frac{7}{8}$ in. margins on the top and bottom of the page. For the left and right sides of the page, the "inside" margin where the spine is located must be larger to account for the "bend" in the pages. The "outside" margin is smaller. In the figure shown, the spine of the book is on the left. Thus, the designer leaves $\frac{3}{4}$ in. for the left margin and $\frac{5}{8}$ in. for the right margin. The "live area" (area where actual text can be displayed is $5\frac{1}{2} - (\frac{3}{4} + \frac{5}{8}) = 4\frac{1}{8}$ in. wide and $8\frac{1}{2} - (\frac{7}{8} + \frac{7}{8}) = 6\frac{3}{4}$ in. high.

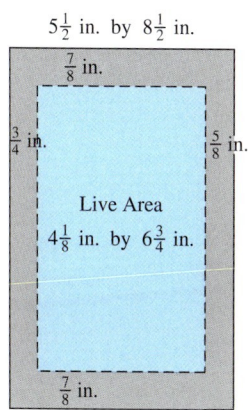

Section 4.1 Introduction to Fractions and Mixed Numbers

Concepts
1. Definition of a Fraction
2. Proper and Improper Fractions
3. Mixed Numbers
4. Fractions and the Number Line

1. Definition of a Fraction

We have already studied operations on whole numbers. In this chapter, we work with numbers that represent part of a whole. When a whole unit is divided into parts, we call the parts **fractions** of a whole. For example, the pizza in Figure 4-1 is divided into 5 parts of equal size. One-fifth $\left(\frac{1}{5}\right)$ of the pizza has been eaten, and four-fifths $\left(\frac{4}{5}\right)$ of the pizza remains.

$\frac{1}{5}$ has been eaten

$\frac{4}{5}$ remains

Mark Dierker/McGraw Hill
Figure 4-1

A fraction is written in the form $\frac{a}{b}$, where a and b are whole numbers and $b \neq 0$. In the fraction $\frac{5}{8}$, the "top" number, 5, is called the **numerator**. The "bottom" number, 8, is called the **denominator**.

$$\text{numerator} \longrightarrow \frac{5}{8} \longleftarrow \text{denominator} \qquad \frac{2x^2}{3y} \longleftarrow \text{numerator} \atop \longleftarrow \text{denominator}$$

A fraction whose numerator is an integer and whose denominator is a nonzero integer is also called a **rational number**.

The denominator of a fraction denotes the number of equal pieces into which a whole unit is divided. The numerator denotes the number of pieces being considered. For example, the garden in Figure 4-2 is divided into 10 parts of equal size. Three sections contain tomato plants. Therefore, $\frac{3}{10}$ of the garden contains tomato plants.

Avoiding Mistakes
The fraction $\frac{3}{10}$ can also be written as 3/10. However, we discourage the use of the "slanted" fraction bar. In later applications of algebra, the slanted fraction bar can cause confusion.

$\frac{3}{10}$ tomato plants

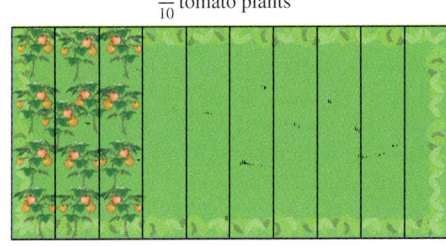

Figure 4-2

Example 1 Writing Fractions

Write a fraction for the shaded portion and a fraction for the unshaded portion of the figure.

Solution:

Shaded portion: $\dfrac{13}{16}$ ← 13 pieces are shaded.
← The triangle is divided into 16 equal pieces.

Unshaded portion: $\dfrac{3}{16}$ ← 3 pieces are not shaded.
← The triangle is divided into 16 equal pieces.

Skill Practice

1. Write a fraction for the shaded portion and a fraction for the unshaded portion.

Answer
1. Shaded portion: $\frac{3}{8}$; unshaded portion: $\frac{5}{8}$

In addition to representing a portion of a whole unit, a fraction can represent division. For example, the fraction $\frac{5}{1} = 5 \div 1 = 5$. Interpreting fractions as division leads to the following important properties.

Properties of Fractions

Suppose that a and b represent nonzero numbers.

1. $\frac{a}{1} = a$ Example: $\frac{-8}{1} = -8$
2. $\frac{0}{a} = 0$ Example: $\frac{0}{-7} = 0$
3. $\frac{a}{0}$ is undefined   Example: $\frac{11}{0}$ is undefined
4. $\frac{a}{a} = 1$ Example: $\frac{-3}{-3} = 1$
5. $\frac{-a}{-b} = \frac{a}{b}$ Example: $\frac{-2}{-5} = \frac{2}{5}$
6. $-\frac{a}{b} = \frac{-a}{b} = \frac{a}{-b}$ Example: $-\frac{2}{3} = \frac{-2}{3} = \frac{2}{-3}$

Property 6 tells us that a negative fraction can be written with the negative sign in the numerator, in the denominator, or out in front. This is because a quotient of two numbers with opposite signs is negative.

2. Proper and Improper Fractions

A positive fraction whose numerator is less than its denominator (or the opposite of such a fraction) is called a **proper fraction**. Furthermore, a positive proper fraction represents a number less than 1 whole unit. The following are proper fractions.

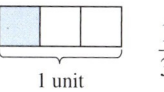
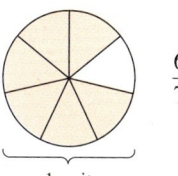

A positive fraction whose numerator is greater than or equal to its denominator (or the opposite of such a fraction) is called an **improper fraction**. For example:

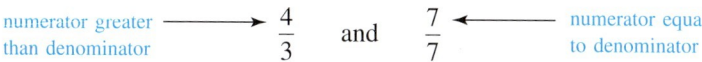

A positive improper fraction represents a quantity greater than 1 whole unit or equal to 1 whole unit.

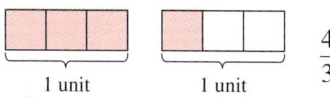
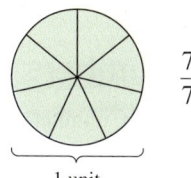

Example 2 Categorizing Fractions

Identify each fraction as proper or improper.

a. $\dfrac{12}{5}$ b. $\dfrac{5}{12}$ c. $\dfrac{12}{12}$

Solution:

a. $\dfrac{12}{5}$ Improper fraction (numerator is greater than denominator)

b. $\dfrac{5}{12}$ Proper fraction (numerator is less than denominator)

c. $\dfrac{12}{12}$ Improper fraction (numerator is equal to denominator)

Skill Practice Identify each fraction as proper or improper.

2. $\dfrac{10}{10}$ 3. $\dfrac{7}{9}$ 4. $\dfrac{9}{7}$

Example 3 Writing Improper Fractions

a. Write an improper fraction to represent the length of the screw shown in the figure.

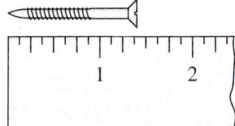

Avoiding Mistakes
Each whole unit is divided into 8 pieces. Therefore, the screw is $\tfrac{11}{8}$ in., not $\tfrac{11}{16}$ in.

b. Write an improper fraction representing the shaded area.

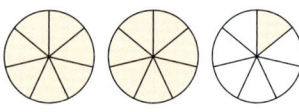

Solution:

a. Each 1-in. unit is divided into 8 parts, and the screw extends for 11 parts. Therefore, the screw is $\tfrac{11}{8}$ in.

b. Each circle is divided into 7 sections of equal size. Of these sections, 15 are shaded. Therefore, the shaded area can be represented by $\tfrac{15}{7}$.

Skill Practice

5. Write an improper fraction to represent the length of the nail shown in the figure.

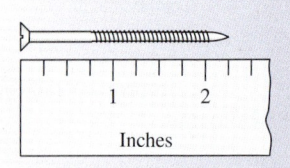

3. Mixed Numbers

Sometimes a mixed number is used instead of an improper fraction to denote a quantity greater than one whole. For example, suppose a typist typed $\tfrac{9}{4}$ pages of a report. We would be more likely to say that the typist typed $2\tfrac{1}{4}$ pages (read as "two and one-fourth pages"). The number $2\tfrac{1}{4}$ is called a *mixed number* and represents 2 wholes plus $\tfrac{1}{4}$ of a whole.

  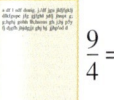

$\dfrac{9}{4} = 2\dfrac{1}{4}$

Answers
2. Improper   3. Proper
4. Improper   5. $\dfrac{9}{4}$ in.

In general, a **mixed number** is a sum of a whole number and a fractional part of a whole. However, by convention the plus sign is left out. For example:

$$3\frac{1}{2} \quad \text{means} \quad 3 + \frac{1}{2}$$

A negative mixed number implies that both the whole number part and the fraction part are negative. Therefore, we interpret the mixed number $-3\frac{1}{2}$ as

$$-3\frac{1}{2} = -\left(3 + \frac{1}{2}\right) = -3 + \left(-\frac{1}{2}\right)$$

Suppose we want to change a mixed number to an improper fraction. From Figure 4-3, we see that the mixed number $3\frac{1}{2}$ is the same as $\frac{7}{2}$.

$3\frac{1}{2} = 3 + \frac{1}{2}$ = 3 whole units + 1 half

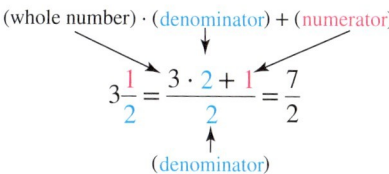

$3\frac{1}{2}$ = 3 · 2 halves + 1 half = $\frac{7}{2}$

Figure 4-3

The process to convert a mixed number to an improper fraction can be summarized as follows.

Changing a Mixed Number to an Improper Fraction
Step 1 Multiply the whole number by the denominator.
Step 2 Add the result to the numerator.
Step 3 Write the result from step 2 over the denominator.

(whole number) · (denominator) + (numerator)

$$3\frac{1}{2} = \frac{3 \cdot 2 + 1}{2} = \frac{7}{2}$$

(denominator)

Example 4 — Converting Mixed Numbers to Improper Fractions

Convert the mixed number to an improper fraction.

a. $7\frac{1}{4}$ **b.** $-8\frac{2}{5}$

Solution:

a. $7\frac{1}{4} = \frac{7 \cdot 4 + 1}{4}$

$= \frac{28 + 1}{4}$

$= \frac{29}{4}$

b. $-8\frac{2}{5} = -\left(8\frac{2}{5}\right)$ The entire mixed number is negative.

$= -\left(\frac{8 \cdot 5 + 2}{5}\right)$

$= -\left(\frac{40 + 2}{5}\right)$

$= -\frac{42}{5}$

Avoiding Mistakes
The negative sign in the mixed number applies to both the whole number and the fraction.

Skill Practice Convert the mixed number to an improper fraction.

6. $10\frac{5}{8}$ **7.** $-15\frac{1}{2}$

Answers

6. $\frac{85}{8}$ **7.** $-\frac{31}{2}$

Now suppose we want to convert an improper fraction to a mixed number. In Figure 4-4, the improper fraction $\frac{13}{5}$ represents 13 slices of pizza where each slice is $\frac{1}{5}$ of a whole pizza. If we divide the 13 pieces into groups of 5, we make 2 whole pizzas with 3 pieces left over. Thus,

$$\frac{13}{5} = 2\frac{3}{5}$$

13 pieces = 2 groups of 5 + 3 left over

$$\frac{13}{5} = 2 + \frac{3}{5}$$

Mark Dierker/McGraw Hill
Figure 4-4

> **Avoiding Mistakes**
> When writing a mixed number, the + sign between the whole number and fraction should not be written.

This process can be accomplished by division.

$$\frac{13}{5} \longrightarrow 5\overline{)13} \quad 2\frac{3}{5}$$
$$\underline{-10}$$
$$3$$

(quotient 2, remainder 3, divisor 5)

> **Changing an Improper Fraction to a Mixed Number**
> **Step 1** Divide the numerator by the denominator to obtain the quotient and remainder.
> **Step 2** The mixed number is then given by
>
> $$\text{Quotient} + \frac{\text{remainder}}{\text{divisor}}$$

Example 5 Converting Improper Fractions to Mixed Numbers

Convert to a mixed number.

a. $\dfrac{25}{6}$ b. $-\dfrac{39}{4}$

Solution:

a. $\dfrac{25}{6} \longrightarrow 6\overline{)25} \quad 4\dfrac{1}{6}$
$\phantom{\dfrac{25}{6} \longrightarrow 6}\underline{-24}$
$\phantom{\dfrac{25}{6} \longrightarrow 6xx}1$

(quotient 4, remainder 1, divisor 6)

b. $-\dfrac{39}{4} = -\left(\dfrac{39}{4}\right)$ First perform division.

$4\overline{)39} \quad 9\dfrac{3}{4}$
$\underline{-36}$
3

(quotient 9, remainder 3, divisor 4)

$= -9\dfrac{3}{4}$ Then take the opposite of the result.

Skill Practice Convert the improper fraction to a mixed number.

8. $\dfrac{14}{5}$ 9. $-\dfrac{95}{22}$

Answers
8. $2\dfrac{4}{5}$ 9. $-4\dfrac{7}{22}$

The process to convert an improper fraction to a mixed number indicates that the result of a division operation can be written as a mixed number.

Section 4.1 Introduction to Fractions and Mixed Numbers

Example 6 Writing a Quotient as a Mixed Number

Divide. Write the quotient as a mixed number.

$$25\overline{)529}$$

Solution:

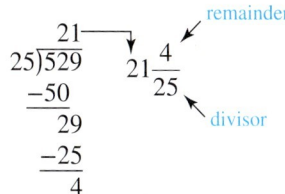

Skill Practice Divide and write the quotient as a mixed number.
10. $5967 \div 41$

4. Fractions and the Number Line

Fractions can be visualized on a number line. For example, to graph the fraction $\frac{3}{4}$, divide the distance between 0 and 1 into 4 equal parts. To plot the number $\frac{3}{4}$, start at 0 and count over 3 parts.

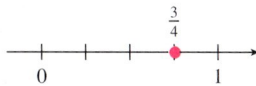

Example 7 Plotting Fractions on a Number Line

Plot the point on the number line corresponding to each fraction.

a. $\frac{1}{2}$ b. $\frac{5}{6}$ c. $-\frac{21}{5}$

Solution:

a. $\frac{1}{2}$ Divide the distance between 0 and 1 into 2 equal parts.

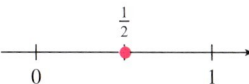

b. $\frac{5}{6}$ Divide the distance between 0 and 1 into 6 equal parts.

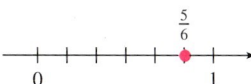

c. $-\frac{21}{5} = -4\frac{1}{5}$ Write $-\frac{21}{5}$ as a mixed number.

The value $-4\frac{1}{5}$ is located one-fifth of the way between -4 and -5 on the number line. Divide the distance between -4 and -5 into 5 equal parts. Plot the point one-fifth of the way from -4 to -5.

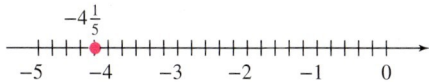

Skill Practice Plot the numbers on a number line.
11. $\frac{4}{5}$ 12. $\frac{1}{3}$ 13. $-\frac{13}{4}$

Answers
10. $145\frac{22}{41}$
11.
12.
13.
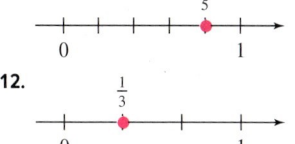

Recall that the absolute value of a number a, denoted $|a|$, is the distance between a and 0 on the number line. Two numbers are opposites if they have the same distance from 0 on the number line, but are on opposite sides of zero. For example, 2 and -2 are opposites. We now apply these concepts to fractions.

Example 8   **Determining the Absolute Value and Opposite of a Fraction**

Simplify.   **a.** $\left|-\dfrac{2}{7}\right|$ **b.** $\left|\dfrac{1}{5}\right|$ **c.** $-\left|-\dfrac{4}{9}\right|$ **d.** $-\left(-\dfrac{4}{9}\right)$

Solution:

a. $\left|-\dfrac{2}{7}\right| = \dfrac{2}{7}$ The distance between $-\dfrac{2}{7}$ and 0 on the number line is $\dfrac{2}{7}$.

b. $\left|\dfrac{1}{5}\right| = \dfrac{1}{5}$ The distance between $\dfrac{1}{5}$ and 0 on the number line is $\dfrac{1}{5}$.

c. $-\left|-\dfrac{4}{9}\right| = -\left(\dfrac{4}{9}\right)$ Take the absolute value of $-\dfrac{4}{9}$ first. This gives $\dfrac{4}{9}$. Then take the opposite of $\dfrac{4}{9}$, which is $-\dfrac{4}{9}$.

$= -\dfrac{4}{9}$

d. $-\left(-\dfrac{4}{9}\right) = \dfrac{4}{9}$ Take the opposite of $-\dfrac{4}{9}$, which is $\dfrac{4}{9}$.

Skill Practice Simplify.

14. $\left|-\dfrac{12}{5}\right|$ **15.** $\left|\dfrac{3}{7}\right|$ **16.** $-\left|-\dfrac{3}{4}\right|$ **17.** $-\left(-\dfrac{3}{4}\right)$

Answers
14. $\dfrac{12}{5}$ **15.** $\dfrac{3}{7}$
16. $-\dfrac{3}{4}$ **17.** $\dfrac{3}{4}$

Section 4.1 Practice Exercises

Study Skills Exercise

Reflecting on your performance in the course is necessary to make any adjustments to your work habits and mindset. One way to do this is to keep track of your grade throughout the semester. Take a minute to compute your grade at this point. Ask yourself if you are earning the grade that you want in the course thus far. Do you feel you are on a path of success? If you answered "no" to either of these questions, here are some suggestions to improve your performance in the course:

1. Make an appointment with your instructor to express your concerns and to develop a plan for success in the course.
2. Organize a study group with fellow students in your class.
3. Enroll in or increase your attendance in the math support center or tutoring center.

Vocabulary and Key Concepts

1. **a.** When a whole unit is divided into parts, we call the parts _____ of the whole unit.
   b. Given a fraction $\dfrac{a}{b}$, where a and b are whole numbers and $b \neq 0$, the top number a is called the _____ and the bottom number b is called the _____.
   c. A fraction whose numerator is an integer and whose denominator is a nonzero integer is called a _____ number.
   d. A positive fraction whose numerator is less than its denominator (or the opposite of such a fraction) is called a _____ fraction.
   e. A positive fraction whose numerator is greater than or equal to its denominator (or the opposite of such a fraction) is called an _____ fraction.
   f. A _____ number is a sum of a whole number and a fractional part of a whole.

Concept 1: Definition of a Fraction

For Exercises 2–5, identify the numerator and the denominator for each fraction.

2. $\dfrac{2}{3}$
3. $\dfrac{8}{9}$
4. $\dfrac{12x}{11y^2}$
5. $\dfrac{7p}{9q}$

For Exercises 6–11, write a fraction that represents the shaded area. (See Example 1.)

6.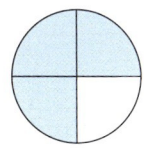
7.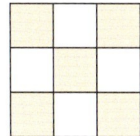
8.
9.
10.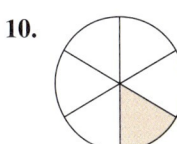
11.

12. Write a fraction to represent the portion of gas in a gas tank represented by the gauge.

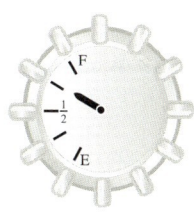

13. The scoreboard for a recent men's championship swim meet in Melbourne, Australia, shows the final standings in the event. What fraction of the finalists are from the USA?

Name	Country	Time
Maginni, Filippo	ITA	48.43
Hayden, Brent	CAN	48.43
Sullivan, Eamon	AUS	48.47
Cielo Filho, Cesar	BRA	48.51
Lezak, Jason	USA	48.52
Van Den Hoogenband, Pieter	NED	48.63
Schoeman, Roland Mark	RSA	48.72
Neethling, Ryk	RSA	48.81

14. Refer to the scoreboard from Exercise 13. What fraction of the finalists are from the Republic of South Africa (RSA)?

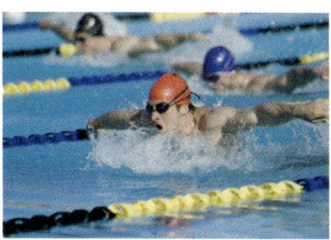

Bob Thomas/Vetta/Getty Images

15. The graph categorizes a sample of people by blood type. What fraction of the sample represents people with type O blood?

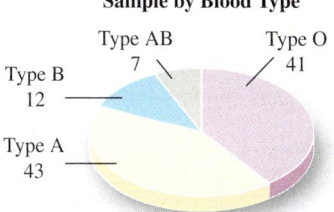

16. Refer to the graph from Exercise 15. What fraction of the sample represents people with type A blood?

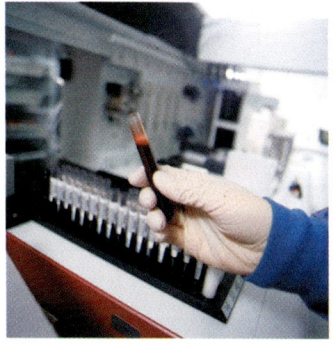

liquidlibrary/PictureQuest/Getty Images

17. A class has 21 children—11 girls and 10 boys. What fraction of the class is made up of boys?

18. In a neighborhood in Ft. Lauderdale, Florida, 10 houses are for sale and 53 are not for sale. Write a fraction representing the portion of houses that are for sale.

For Exercises 19–28, simplify if possible.

19. $\dfrac{-13}{1}$

20. $\dfrac{-14}{1}$

21. $\dfrac{2}{2}$

22. $\dfrac{8}{8}$

23. $\dfrac{0}{-3}$

24. $\dfrac{0}{7}$

25. $\dfrac{-3}{0}$

26. $\dfrac{-11}{0}$

27. $\dfrac{-9}{-10}$

28. $\dfrac{-13}{-6}$

29. Which expressions are equivalent to $\dfrac{-9}{10}$?

  a. $\dfrac{9}{10}$ b. $-\dfrac{9}{10}$ c. $\dfrac{9}{-10}$ d. $\dfrac{-9}{-10}$

30. Which expressions are equivalent to $-\dfrac{4}{5}$?

  a. $\dfrac{-4}{5}$ b. $\dfrac{-4}{-5}$ c. $\dfrac{4}{-5}$ d. $-\dfrac{5}{4}$

31. Which expressions are equivalent to $\dfrac{-4}{1}$?

  a. $\dfrac{4}{-1}$ b. -4 c. $-\dfrac{4}{1}$ d. $\dfrac{1}{-4}$

32. Which expressions are equivalent to $\dfrac{-9}{1}$?

  a. $\dfrac{-9}{-1}$ b. $\dfrac{9}{-1}$ c. -9 d. $-\dfrac{9}{1}$

Concept 2: Proper and Improper Fractions

For Exercises 33–38, label the fraction as proper or improper. (See Example 2.)

33. $\dfrac{7}{8}$

34. $\dfrac{2}{3}$

35. $\dfrac{10}{10}$

36. $\dfrac{3}{3}$

37. $\dfrac{7}{2}$

38. $\dfrac{21}{20}$

For Exercises 39–42, write an improper fraction for the shaded portion of each group of figures. (See Example 3.)

39.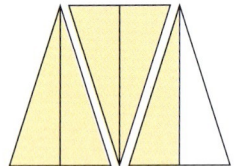

40.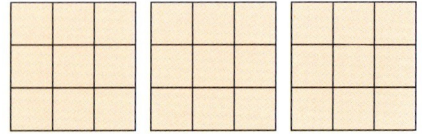

41.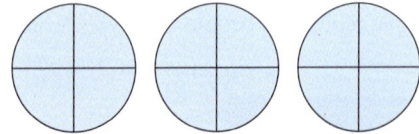

42.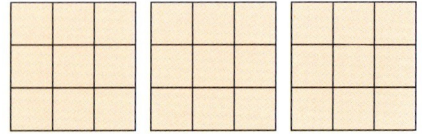

For Exercises 43 and 44, write an improper fraction to represent the length of the objects shown in the figure. (See Example 3.)

43.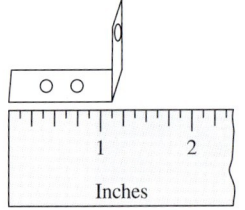

44.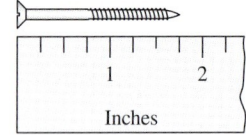

Concept 3: Mixed Numbers

For Exercises 45 and 46, write an improper fraction and a mixed number for the shaded portion of each group of figures.

45.

46.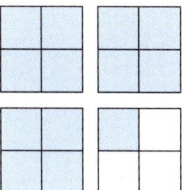

47. Write an improper fraction and a mixed number to represent the length of the nail.

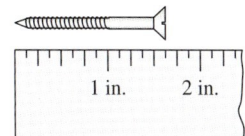

48. Write an improper fraction and a mixed number for the number of cups of sugar indicated in the figure.

Jill Braaten/McGraw Hill

For Exercises 49–60, convert the mixed number to an improper fraction. (See Example 4.)

49. $1\frac{3}{4}$
50. $6\frac{1}{3}$
51. $-4\frac{2}{9}$
52. $-3\frac{1}{5}$

53. $-3\frac{3}{7}$
54. $-8\frac{2}{3}$
55. $6\frac{3}{4}$
56. $10\frac{3}{5}$

57. $11\frac{5}{12}$
58. $12\frac{1}{6}$
59. $-21\frac{3}{8}$
60. $-15\frac{1}{2}$

61. How many thirds are in 10?
62. How many sixths are in 2?

63. How many eighths are in $2\frac{3}{8}$?
64. How many fifths are in $2\frac{3}{5}$?

65. How many fourths are in $1\frac{3}{4}$?
66. How many thirds are in $5\frac{2}{3}$?

For Exercises 67–78, convert the improper fraction to a mixed number. (See Example 5.)

67. $\frac{37}{8}$
68. $\frac{13}{7}$
69. $-\frac{39}{5}$
70. $-\frac{19}{4}$

71. $-\frac{27}{10}$
72. $-\frac{43}{18}$
73. $\frac{52}{9}$
74. $\frac{67}{12}$

75. $\frac{133}{11}$
76. $\frac{51}{10}$
77. $-\frac{23}{6}$
78. $-\frac{115}{7}$

For Exercises 79–86, divide. Write the quotient as a mixed number. (See Example 6.)

79. $7\overline{)309}$
80. $4\overline{)921}$
81. $5281 \div 5$
82. $7213 \div 8$

83. $8913 \div 11$
84. $4257 \div 23$
85. $15\overline{)187}$
86. $34\overline{)695}$

Concept 4: Fractions and the Number Line

For Exercises 87–96, plot the fraction on the number line. (See Example 7.)

87. $\dfrac{3}{4}$

88. $\dfrac{1}{4}$

89. $\dfrac{1}{5}$

90. $\dfrac{1}{3}$

91. $-\dfrac{2}{3}$

92. $-\dfrac{5}{6}$

93. $\dfrac{7}{6}$

94. $\dfrac{7}{5}$

95. $-\dfrac{4}{3}$

96. $-\dfrac{3}{2}$

For Exercises 97–104, simplify. (See Example 8.)

97. $\left|-\dfrac{3}{4}\right|$

98. $\left|-\dfrac{8}{7}\right|$

99. $\left|\dfrac{1}{10}\right|$

100. $\left|\dfrac{3}{20}\right|$

101. $-\left|-\dfrac{7}{3}\right|$

102. $-\left|-\dfrac{1}{4}\right|$

103. $-\left(-\dfrac{7}{3}\right)$

104. $-\left(-\dfrac{1}{4}\right)$

Expanding Your Skills

105. True or false? Whole numbers can be written both as proper and improper fractions.

106. True or false? Suppose m and n are nonzero numbers, where $m > n$. Then $\dfrac{m}{n}$ is an improper fraction.

107. True or false? Suppose m and n are nonzero numbers, where $m > n$. Then $\dfrac{n}{m}$ is a proper fraction.

108. True or false? Suppose m and n are nonzero numbers, where $m > n$. Then $\dfrac{n}{3m}$ is a proper fraction.

Section 4.2 Simplifying Fractions

Concepts

1. Factorizations and Divisibility
2. Prime Factorization
3. Equivalent Fractions
4. Simplifying Fractions to Lowest Terms
5. Applications of Simplifying Fractions

1. Factorizations and Divisibility

Recall that two numbers multiplied to form a product are called factors. For example, $2 \cdot 3 = 6$ indicates that 2 and 3 are factors of 6. Likewise, because $1 \cdot 6 = 6$, the numbers 1 and 6 are factors of 6. In general, a **factor** of a number n is a nonzero whole number that divides evenly into n.

The products $2 \cdot 3$ and $1 \cdot 6$ are called factorizations of 6. In general, a **factorization** of a number n is a product of factors that equals n.

Example 1 Finding Factorizations of a Number

Find four different factorizations of 12.

Solution:

$$12 = \begin{cases} 1 \cdot 12 \\ 2 \cdot 6 \\ 3 \cdot 4 \\ 2 \cdot 2 \cdot 3 \end{cases}$$

TIP: Notice that a factorization may include more than two factors.

Skill Practice

1. Find four different factorizations of 18.

A factor of a number must divide evenly into the number. There are several rules by which we can quickly determine whether a number is divisible by 2, 3, 4, 5, 6, 9, or 10. These are called divisibility rules.

Divisibility Rules for 2, 3, 4, 5, 6, 9, and 10

- *Divisibility by 2.* A whole number is divisible by 2 if it is an even number. That is, the ones-place digit is 0, 2, 4, 6, or 8.
  Examples: 26 and 384

- *Divisibility by 3.* A whole number is divisible by 3 if the sum of its digits is divisible by 3.
  Example: 312   (sum of digits is $3 + 1 + 2 = 6$, which is divisible by 3)

- *Divisibility by 4.* A whole number is divisible by 4 if the number formed by the last two digits (tens-place and ones-place digits) is divisible by 4.
  Examples: 140 and 916

- *Divisibility by 5.* A whole number is divisible by 5 if its ones-place digit is 5 or 0.
  Examples: 45 and 260

- *Divisibility by 6.* A whole number is divisible by 6 if it is divisible by both 2 and 3.
  Examples: 90 and 456

- *Divisibility by 9.* A whole number is divisible by 9 if the sum of its digits is divisible by 9.
  Examples: 72 and 882

- *Divisibility by 10.* A whole number is divisible by 10 if its ones-place digit is 0.
  Examples: 30 and 170

The divisibility rules for other numbers are harder to remember. In these cases, it is often easier simply to perform division to test for divisibility.

Answer

1. For example.
   1 · 18
   2 · 9
   3 · 6
   2 · 3 · 3

> **Example 2** Applying the Divisibility Rules
>
> Determine whether the given number is divisible by 2, 3, 4, 5, 6, 9, or 10.
>
> **a.** 720   **b.** 84
>
> **TIP:** When in doubt about divisibility, you can check by division. When we divide 84 by 2, the remainder is zero. This means that 2 divides evenly into 84.
>
> **Solution:**
>
> **Test for Divisibility**
>
> **a.** 720
> - By 2: Yes. The number 720 is even.
> - By 3: Yes. The sum $7 + 2 + 0 = 9$ is divisible by 3.
> - By 4: Yes. The number formed by the last two digits, 20, is divisible by 4.
> - By 5: Yes. The ones-place digit is 0.
> - By 6: Yes. The number is divisible by both 2 and 3.
> - By 9: Yes. The sum $7 + 2 + 0 = 9$ is divisible by 9.
> - By 10: Yes. The ones-place digit is 0.
>
> **b.** 84
> - By 2: Yes. The number 84 is even.
> - By 3: Yes. The sum $8 + 4 = 12$ is divisible by 3.
> - By 4: Yes. The number formed by the last two digits, 84, is divisible by 4.
> - By 5: No. The ones-place digit is not 5 or 0.
> - By 6: Yes. The number is divisible by both 2 and 3.
> - By 9: No. The sum $8 + 4 = 12$ is not divisible by 9.
> - By 10: No. The ones-place digit is not 0.
>
> **Skill Practice** Determine whether the given number is divisible by 2, 3, 4, 5, 6, 9, or 10.
>
> **2.** 75   **3.** 2120

2. Prime Factorization

Two important classifications of whole numbers are prime numbers and composite numbers.

> **Definition of Prime and Composite Numbers**
> - A **prime number** is a whole number greater than 1 that has only two factors (itself and 1).
> - A **composite number** is a whole number greater than 1 that is not prime. That is, a composite number will have at least one factor other than 1 and the number itself.
>
> *Note:* The whole numbers 0 and 1 are neither prime nor composite.

> **Example 3** Identifying Prime and Composite Numbers
>
> Determine whether the number is prime, composite, or neither.
>
> **a.** 19   **b.** 51   **c.** 1
>
> **Solution:**
>
> **a.** The number 19 is prime because its only factors are 1 and 19.
> **b.** The number 51 is composite because $3 \cdot 17 = 51$. That is, 51 has factors other than 1 and 51.
> **c.** The number 1 is neither prime nor composite by definition.
>
> **Skill Practice** Determine whether the number is prime, composite, or neither.
>
> **4.** 39   **5.** 0   **6.** 41

Answers
2. Divisible by 3 and 5
3. Divisible by 2, 4, 5, 10
4. Composite **5.** Neither **6.** Prime

Prime numbers are used in a variety of ways in mathematics. It is advisable to become familiar with the first several prime numbers: 2, 3, 5, 7, 11, 13, 17, 19, 23, 29, ...

In Example 1 we found four factorizations of 12.

$$1 \cdot 12$$
$$2 \cdot 6$$
$$3 \cdot 4$$
$$2 \cdot 2 \cdot 3$$

TIP: The number 2 is the only even prime number.

The last factorization $2 \cdot 2 \cdot 3$ consists of only prime-number factors. Therefore, we say $2 \cdot 2 \cdot 3$ is the prime factorization of 12.

> **Definition of Prime Factorization**
>
> The **prime factorization** of a number is the factorization in which every factor is a prime number.
>
> *Note:* The order in which the factors are written does not affect the product.

Prime factorizations of numbers will be particularly helpful when we add, subtract, multiply, divide, and simplify fractions.

Example 4 Determining the Prime Factorization of a Number

Find the prime factorization of 220.

Solution:

One method to factor a whole number is to make a factor tree. Begin by determining *any* two numbers that when multiplied equal 220. Then continue factoring each factor until the branches "end" in prime numbers.

← Branches end in prime numbers.

TIP: The prime factorization from Example 4 can also be expressed by using exponents as $2^2 \cdot 5 \cdot 11$.

Therefore, the prime factorization of 220 is $2 \cdot 2 \cdot 5 \cdot 11$.

Skill Practice

7. Find the prime factorization of 90.

TIP: In creating a factor tree, you can begin with any two factors of the number. The result will be the same. In Example 4, we could have started with the factors of 4 and 55.

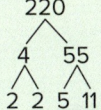 The prime factorization is $2 \cdot 2 \cdot 5 \cdot 11$ as expected.

Another technique to find the prime factorization of a number is to divide the number by the smallest known prime factor of the number. Then divide the quotient by its smallest prime factor. Continue dividing in this fashion until the quotient is a prime number. The prime factorization is the product of divisors and the final quotient. This is demonstrated in Example 5.

Answer

7. $2 \cdot 3 \cdot 3 \cdot 5$ or $2 \cdot 3^2 \cdot 5$

> **Example 5** Determining Prime Factorizations

Find the prime factorization.

a. 198   b. 153

Solution:

a. 2 is the smallest prime factor of 198. ⟶ 2)198
   3 is the smallest prime factor of 99. ⟶ 3)99
   3 is the smallest prime factor of 33. ⟶ 3)33
   The last quotient is prime.   ⟶   11

The prime factorization of 198 is $2 \cdot 3 \cdot 3 \cdot 11$ or $2 \cdot 3^2 \cdot 11$.

b. 3)153
   3)51
     17

The prime factorization of 153 is $3 \cdot 3 \cdot 17$ or $3^2 \cdot 17$

Skill Practice Find the prime factorization of the given number.

8. 168   9. 990

3. Equivalent Fractions

The fractions $\frac{3}{6}$, $\frac{2}{4}$, and $\frac{1}{2}$ all represent the same portion of a whole. See Figure 4-5. Therefore, we say that the fractions are *equivalent*.

One method to show that two fractions are equivalent is to calculate their cross products. For example, to show that $\frac{3}{6} = \frac{2}{4}$, we have

$$\frac{3}{6} \bowtie \frac{2}{4}$$

$$3 \cdot 4 \stackrel{?}{=} 6 \cdot 2$$

$$12 = 12 \qquad \text{Yes. The fractions are equivalent.}$$

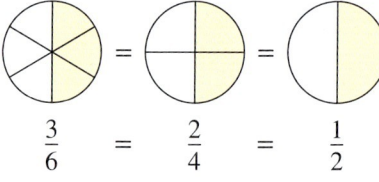

$$\frac{3}{6} = \frac{2}{4} = \frac{1}{2}$$

Figure 4-5

> **Example 6** Determining Whether Two Fractions Are Equivalent

Fill in the blank ☐ with = or ≠.   a. $\frac{18}{39} \square \frac{6}{13}$ b. $\frac{5}{7} \square \frac{7}{9}$

Solution:

a. $\frac{18}{39} \bowtie \frac{6}{13}$

   $18 \cdot 13 \stackrel{?}{=} 39 \cdot 6$

   $234 = 234$

   Therefore, $\frac{18}{39} \boxed{=} \frac{6}{13}$.

b. $\frac{5}{7} \bowtie \frac{7}{9}$

   $5 \cdot 9 \stackrel{?}{=} 7 \cdot 7$

   $45 \neq 49$

   Therefore, $\frac{5}{7} \boxed{\neq} \frac{7}{9}$.

Skill Practice Fill in the blank ☐ with = or ≠.

10. $\frac{13}{24} \square \frac{6}{11}$ 11. $\frac{9}{4} \square \frac{54}{24}$

4. Simplifying Fractions to Lowest Terms

In Figure 4-5, we see that $\frac{3}{6}$, $\frac{2}{4}$, and $\frac{1}{2}$ all represent equal quantities. However, the fraction $\frac{1}{2}$ is said to be in **lowest terms** because the numerator and denominator share no common factors other than 1.

To simplify a fraction to lowest terms, we apply the following important principle.

Answers
8. $2 \cdot 2 \cdot 2 \cdot 3 \cdot 7$ or $2^3 \cdot 3 \cdot 7$
9. $2 \cdot 3 \cdot 3 \cdot 5 \cdot 11$ or $2 \cdot 3^2 \cdot 5 \cdot 11$
10. ≠   11. =

Section 4.2 Simplifying Fractions

Fundamental Principle of Fractions

Suppose that a number, c, is a common factor in the numerator and denominator of a fraction. Then

$$\frac{a \cdot c}{b \cdot c} = \frac{a}{b} \cdot \frac{c}{c} = \frac{a}{b} \cdot 1 = \frac{a}{b} \quad \text{provided } b \neq 0.$$

To simplify a fraction, we begin by factoring the numerator and denominator into prime factors. This will help identify the common factors.

Example 7 — Simplifying a Fraction to Lowest Terms

Simplify to lowest terms.   **a.** $\dfrac{6}{10}$ **b.** $-\dfrac{170}{102}$ **c.** $\dfrac{20}{24}$

Solution:

a. $\dfrac{6}{10} = \dfrac{3 \cdot 2}{5 \cdot 2}$ Factor the numerator and denominator. Notice that 2 is a common factor.

$= \dfrac{3}{5} \cdot \dfrac{2}{2}$ Apply the fundamental principle of fractions.

$= \dfrac{3}{5} \cdot 1$ Any nonzero number divided by itself is 1.

$= \dfrac{3}{5}$

b. $-\dfrac{170}{102} = -\dfrac{5 \cdot 2 \cdot 17}{3 \cdot 2 \cdot 17}$ Factor the numerator and denominator.

$= -\dfrac{5}{3} \cdot \dfrac{2}{2} \cdot \dfrac{17}{17}$ Apply the fundamental principle of fractions.

$= -\dfrac{5}{3} \cdot 1 \cdot 1$ Any nonzero number divided by itself is 1.

$= -\dfrac{5}{3}$

c. $\dfrac{20}{24} = \dfrac{5 \cdot 2 \cdot 2}{3 \cdot 2 \cdot 2 \cdot 2}$ Factor the numerator and denominator.

$= \dfrac{5}{3 \cdot 2} \cdot \dfrac{2}{2} \cdot \dfrac{2}{2}$ Apply the fundamental principle of fractions.

$= \dfrac{5}{6} \cdot 1 \cdot 1$

$= \dfrac{5}{6}$

TIP: To check that you have simplified a fraction correctly, verify that the cross products are equal.

$\dfrac{6}{10} \times \dfrac{3}{5}$

$6 \cdot 5 \stackrel{?}{=} 10 \cdot 3$
$30 = 30$ ✓

Skill Practice Simplify to lowest terms.

12. $\dfrac{15}{35}$ **13.** $-\dfrac{26}{195}$ **14.** $\dfrac{150}{105}$

In Example 7, we show numerous steps to simplify fractions to lowest terms. However, the process is often made easier. For instance, we sometimes divide common factors, and replace them with the new common factor of 1.

$$\frac{20}{24} = \frac{5 \cdot \cancel{2} \cdot \cancel{2}}{3 \cdot 2 \cdot \cancel{2} \cdot \cancel{2}} = \frac{5}{6}$$

Answers
12. $\dfrac{3}{7}$ **13.** $-\dfrac{2}{15}$ **14.** $\dfrac{10}{7}$

The largest number that divides evenly into two or more integers is called their **greatest common factor** or **GCF**. To find the greatest common factor of 20 and 24, factor each number into its prime factors. Then identify the common factors.

$20 = \boxed{2 \cdot 2} \cdot 5$ The factors that are common to both lists are circled.

$24 = \boxed{2 \cdot 2} \cdot 2 \cdot 3$ The GCF is $2 \cdot 2 = 4$

By identifying the greatest common factor between the numerator and denominator of a fraction we can simplify the fraction.

$$\frac{20}{24} = \frac{5 \cdot \cancel{4}}{6 \cdot \cancel{4}} = \frac{5}{6}$$

Notice that "dividing out" the common factor of 4 has the same effect as dividing the numerator and denominator by 4. This is often done mentally.

$$\frac{\overset{5}{\cancel{20}}}{\underset{6}{\cancel{24}}} = \frac{5}{6} \;\;\longleftarrow\; 20 \text{ divided by 4 equals 5.} \\ \;\;\longleftarrow\; 24 \text{ divided by 4 equals 6.}$$

> **TIP:** Simplifying a fraction is also called reducing a fraction to lowest terms. For example, the simplified (or reduced) form of $\frac{20}{24}$ is $\frac{5}{6}$.

Example 8 Simplifying Fractions to Lowest Terms

Simplify the fraction. Write the answer as a fraction or integer.

a. $\dfrac{75}{25}$ b. $-\dfrac{12}{60}$

Solution:

a. $\dfrac{75}{25} = \dfrac{3 \cdot \cancel{25}}{1 \cdot \cancel{25}} = \dfrac{3}{1} = 3$ Factor the numerator and denominator to find the greatest common factor.

$75 = 3 \cdot \boxed{5 \cdot 5}$
$25 = \boxed{5 \cdot 5}$ The GCF is $5 \cdot 5 = 25$.

Or alternatively: $\dfrac{\overset{3}{\cancel{75}}}{\underset{1}{\cancel{25}}} = \dfrac{3}{1} \;\;\longleftarrow\; 75 \text{ divided by 25 equals 3.} \\ \;\;\longleftarrow\; 25 \text{ divided by 25 equals 1.}$

$= 3$

> **TIP:** Recall that any fraction of the form $\frac{n}{1} = n$. Therefore, $\frac{3}{1} = 3$.

b. $-\dfrac{12}{60} = -\dfrac{1 \cdot \cancel{12}}{5 \cdot \cancel{12}} = -\dfrac{1}{5}$ Factor the numerator and denominator to find the greatest common factor.

$12 = \boxed{2 \cdot 2 \cdot 3}$
$60 = \boxed{2 \cdot 2 \cdot 3} \cdot 5$ The GCF is $2 \cdot 2 \cdot 3 = 12$.

> **Avoiding Mistakes**
> Do not forget to write the "1" in the numerator of the fraction $-\frac{1}{5}$.

Or alternatively: $= -\dfrac{\overset{1}{\cancel{12}}}{\underset{5}{\cancel{60}}} = -\dfrac{1}{5} \;\;\longleftarrow\; 12 \text{ divided by 12 equals 1.} \\ \;\;\longleftarrow\; 60 \text{ divided by 12 equals 5.}$

Skill Practice Simplify the fraction. Write the answer as a fraction or an integer.

15. $\dfrac{39}{3}$ 16. $-\dfrac{15}{90}$

Avoiding Mistakes

Suppose that you do not recognize the *greatest* common factor in the numerator and denominator. You can still divide by *any* common factor. However, you will have to repeat this process more than once to simplify the fraction completely. For instance, consider the fraction from Example 8(b).

$-\dfrac{\overset{2}{\cancel{12}}}{\underset{10}{\cancel{60}}} = -\dfrac{2}{10}$ Dividing by the common factor of 6 leaves a fraction that can be simplified further.

$= -\dfrac{\overset{1}{\cancel{2}}}{\underset{5}{\cancel{10}}} = -\dfrac{1}{5}$ Divide again, this time by 2. The fraction is now simplified completely because the greatest common factor in the numerator and denominator is 1.

Answers
15. 13   16. $-\dfrac{1}{6}$

Section 4.2 Simplifying Fractions

Example 9 — Simplifying Fractions by 10, 100, and 1000

Simplify each fraction to lowest terms by first reducing by 10, 100, or 1000. Write the answer as a fraction.

a. $\dfrac{170}{30}$ b. $\dfrac{2500}{75{,}000}$

Solution:

a. $\dfrac{170}{30} = \dfrac{17 \cdot \cancel{10}}{3 \cdot \cancel{10}} = \dfrac{17}{3}$ Notice that dividing numerator and denominator by 10 has the effect of eliminating the 0 in the ones place from each number: $\dfrac{17\cancel{0}}{3\cancel{0}}$

Avoiding Mistakes
The "strike through" method only works for the digit 0 at the *end* of the numerator and denominator.

b. $\dfrac{2500}{75{,}000} = \dfrac{25\cancel{00}}{75{,}0\cancel{00}}$ Both 2500 and 75,000 are divisible by 100. "Strike through" two zeros. This is equivalent to dividing by 100.

$= \dfrac{\overset{1}{\cancel{25}}}{\underset{30}{\cancel{750}}}$ Simplify further. Both 25 and 750 have a common factor of 25.

$= \dfrac{1}{30}$

Skill Practice Simplify to lowest terms by first reducing by 10, 100, or 1000.

17. $\dfrac{630}{190}$ 18. $\dfrac{1300}{52{,}000}$

The process to simplify a fraction is the same for fractions that contain variables. This is shown in Example 10.

Example 10 — Simplifying a Fraction Containing Variables

Simplify.   a. $\dfrac{10xy}{6x}$ b. $\dfrac{2a^3}{4a^4}$

Solution:

a. $\dfrac{10xy}{6x} = \dfrac{2 \cdot 5 \cdot x \cdot y}{2 \cdot 3 \cdot x}$ Factor the numerator and denominator. Common factors are shown in red.

Avoiding Mistakes
Since division by 0 is undefined, we know that the value of the variable cannot make the denominator equal 0. In Example 10(a), $x \neq 0$. In Example 10(b), $a \neq 0$.

$= \dfrac{\cancel{2} \cdot 5 \cdot \cancel{x} \cdot y}{\cancel{2} \cdot 3 \cdot \cancel{x}}$ Simplify.

$= \dfrac{5y}{3}$

b. $\dfrac{2a^3}{4a^4} = \dfrac{2 \cdot a \cdot a \cdot a}{2 \cdot 2 \cdot a \cdot a \cdot a \cdot a}$ Factor the numerator and denominator. Common factors are shown in red.

$= \dfrac{\cancel{2} \cdot \cancel{a} \cdot \cancel{a} \cdot \cancel{a}}{\cancel{2} \cdot 2 \cdot \cancel{a} \cdot \cancel{a} \cdot \cancel{a} \cdot a}$ Simplify.

$= \dfrac{1}{2a}$

Skill Practice Simplify.

19. $\dfrac{18d}{15cd}$ 20. $\dfrac{9w^2}{36w^3}$

Answers
17. $\dfrac{63}{19}$ 18. $\dfrac{1}{40}$
19. $\dfrac{6}{5c}$ 20. $\dfrac{1}{4w}$

The fractions from Example 10 can also be simplified by dividing out the greatest common factor from the numerator and denominator.

a. $10xy = 2 \cdot 5 \cdot x \cdot y$
$6x = 2 \cdot 3 \cdot x$ The GCF is $2x$.

Thus, $\dfrac{10xy}{6x} = \dfrac{5y \cdot 2x}{3 \cdot 2x} = \dfrac{5y}{3}$

In Example 10(b), the numerator and denominator share a common numerical factor of 2. Furthermore, there are three factors of a in the numerator and four factors of a in the denominator. Therefore, the numerator and denominator also have a^3 in common.

b. $2a^3 = 2 \cdot a \cdot a \cdot a$
$4a^4 = 2 \cdot 2 \cdot a \cdot a \cdot a \cdot a$ The GCF is $2a^3$.

Thus, $\dfrac{2a^3}{4a^4} = \dfrac{1 \cdot 2a^3}{2a \cdot 2a^3} = \dfrac{1}{2a}$

5. Applications of Simplifying Fractions

Example 11   Simplifying Fractions in an Application

Madeleine got 28 out of 35 problems correct on an algebra exam. David got 27 out of 45 questions correct on a different algebra exam.

a. What fractional part of the exam did each student answer correctly?

b. Which student performed better?

Solution:

a. Fractional part correct for Madeleine:

$\dfrac{28}{35}$ or equivalently $\dfrac{28}{35} = \dfrac{4 \cdot 7}{5 \cdot 7} = \dfrac{4}{5}$

Fractional part correct for David:

$\dfrac{27}{45}$ or equivalently $\dfrac{27}{45} = \dfrac{3 \cdot 9}{5 \cdot 9} = \dfrac{3}{5}$

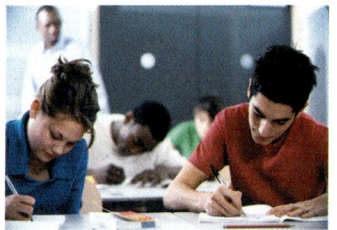
Image100 Ltd/Alamy Stock Photo

b. From the simplified form of each fraction, we see that Madeleine performed better because $\frac{4}{5} > \frac{3}{5}$. That is, 4 parts out of 5 is greater than 3 parts out of 5. This is also easily verified on a number line.

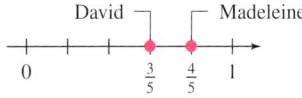

Skill Practice

21. Joanne planted 77 seeds in her garden and 55 sprouted. Geoff planted 140 seeds and 80 sprouted.

   a. What fractional part of the seeds sprouted for Joanne and what part sprouted for Geoff?

   b. For which person did a greater portion of seeds sprout?

Answer

21. a. Joanne: $\frac{5}{7}$; Geoff: $\frac{4}{7}$
    b. Joanne had a greater portion of seeds sprout.

Section 4.2 Activity

A.1. Explain how you know a number is divisible by 2, 3, 5, and 10.
   a. Divisible by 2: _____
   b. Divisible by 3: _____
   c. Divisible by 5: _____
   d. Divisible by 10: _____

A.2. Place each given number in the appropriate box. Some numbers may appear in more than one box or none of the boxes.
   a. 4892
   b. 3915
   c. 9360
   d. 1327

Divisible by 2	Divisible by 3	Divisible by 5	Divisible by 10

A.3. A **prime number** is a whole number greater than 1 that has only two factors (itself and 1).
   a. List the first 10 prime numbers starting with the number 2.

   ____, ____, ____, ____, ____, ____, ____, ____, ____, ____

   b. Write a two-digit prime number that is not included in the list from part (a).

A.4. If a number has a least one factor other than 1 and itself, then the number is a **composite number.**
   a. What is the smallest composite number?
   b. Write a three-digit composite number.

A.5. List all of the factors for each number. The first one is started for you.
   a. $4 = \begin{cases} 1 \cdot 4 \\ 2 \cdot 2 \end{cases}$
   b. 24
   c. 47
   d. 100

A.6. Find the prime factorization of 840 by following these steps:
   a. List two factors of 840 (any two numbers that when multiplied equal 840). Write the two numbers under the branches in the factor tree provided below.
   b. Continue factoring each factor until the branches "end" in only prime numbers.
   c. Write the prime factorization of 840 as the product of the prime numbers. Check the result by multiplying.

   840

   d. What is the other method that can be used to find the prime factorization of a number?

For Exercises A.7–A.10, find the prime factorization using the method of your choice.

A.7. 126

A.8. 140

A.9. 819

A.10. 595

A.11. Consider the following fractions: $\frac{2}{3}$ and $\frac{8}{12}$.

  a. Multiply the numerator of the first fraction by the denominator of the second fraction.
  b. Multiply the denominator of the first fraction by the numerator of the second fraction.
  c. In parts (a) and (b), you are calculating the *cross product*. Are the cross products from parts (a) and (b) equal? If so, then the fractions $\frac{2}{3}$ and $\frac{8}{12}$ are equivalent.

For Exercises A.12–A.14, determine if the fractions are equivalent by calculating the cross products.

A.12. $\frac{9}{48} \square \frac{3}{16}$ **A.13.** $\frac{5}{8} \square \frac{25}{64}$ **A.14.** $\frac{7}{17} \square \frac{2}{5}$

A.15. a. Write the prime factorization of 16.
  b. Write the prime factorization of 56.
  c. Simplify the fraction $\frac{16}{56}$ to lowest terms by canceling common factors.

A.16. a. Write the prime factorization of $16y^2$.
  b. Write the prime factorization of $32y^3$.
  c. Simplify the fraction $\frac{16y^2}{32y^3}$ to lowest terms by canceling common factors.

For Exercises A.17–A.20, simplify the fractions to lowest terms.

A.17. $-\frac{24}{15}$ **A.18.** $\frac{300}{900}$ **A.19.** $\frac{2xy}{6x}$ **A.20.** $\frac{30v^2z^4}{12v^5z}$

A.21. Out of 340 employees at a community college, 240 are faculty. At the university, there are 275 faculty out of 425 employees. What fraction of employees are faculty at the community college and at the university?

Section 4.2 Practice Exercises

Prerequisite Review

For Exercises R.1–R.4, simplify if possible.

R.1. $\frac{12}{-1}$ **R.2.** $\frac{-1}{-8}$

R.3. $\frac{-7}{0}$ **R.4.** $\frac{0}{-4}$

R.5. Write the fraction $\frac{38}{7}$ as a mixed number.   **R.6.** Write the fraction $\frac{53}{2}$ as a mixed number.

R.7. Write the mixed number $8\frac{4}{5}$ as a fraction.   **R.8.** Write the mixed number $12\frac{2}{3}$ as a fraction.

Vocabulary and Key Concepts

1. a. A _____ of a number n is a nonzero whole number that divides evenly into n.
   b. A _____ number is a whole number greater than 1 that has only two factors, itself and 1.
   c. A _____ number is a whole number greater than 1 that is not prime.

Section 4.2 Simplifying Fractions

  d. The _____ factorization of a number n is the product of prime numbers that equals n.

  e. A fraction is said to be in _____ terms if the numerator and denominator share no common factors other than 1.

2. $2 \cdot 12$, $3 \cdot 8$, and $4 \cdot 6$ are all _____ of 24.

3. The numbers of 50, 120, and 270 are all divisible by 10 since the digit in the _____ place is 0.

4. $3 \cdot 3 \cdot 3 \cdot 5$ or $3^3 \cdot 5$ is the _____ factorization of 135.

5. Simplifying a fraction is also called _____ a fraction to lowest terms.

6. True or False. Consider the expression $\frac{5xy}{10x}$ that can be factored to $\frac{5 \cdot x \cdot y}{2 \cdot 5 \cdot x}$. Both the common factors of 5 and x can be simplified to 1.

Concept 1: Factorizations and Divisibility

For Exercises 7–10, find two different factorizations of each number. (Answers may vary.) **(See Example 1.)**

7. 8   8. 20   9. 24   10. 14

For Exercises 11–16, state the divisibility rule for the given numbers.

11. Divisibility by 2   12. Divisibility by 5

13. Divisibility by 3   14. Divisibility by 6

15. Divisibility by 4   16. Divisibility by 9

For Exercises 17–24, determine if the given number is divisible by 2, 3, 4, 5, 6, 9, and 10. **(See Example 2.)**

17. 108   18. 40   19. 137   20. 241

21. 225   22. 1040   23. 3042   24. 2115

25. Ms. Berglund has 28 students in her class. Can she distribute a package of 84 candies evenly to her students?

26. Mr. Whalen has 22 students in an algebra class. He has 110 sheets of graph paper. Can he distribute the graph paper evenly among his students?

Concept 2: Prime Factorization

For Exercises 27–34, determine whether the number is prime, composite, or neither. **(See Example 3.)**

27. 7   28. 17   29. 10   30. 21

31. 1   32. 0   33. 97   34. 57

35. One method for finding prime numbers is the *sieve of Eratosthenes*. The natural numbers from 2 to 50 are shown in the table. Start at the number 2 (the smallest prime number). Leave the number 2 and cross out every second number after the number 2. This will eliminate all numbers that are multiples of 2. Then go back to the beginning of the chart and leave the number 3, but cross out every third number after the number 3 (thus eliminating the multiples of 3). Begin at the next open number and continue this process. The numbers that remain are prime numbers. Use this process to find the prime numbers less than 50.

	2	3	4	5	6	7	8	9	10
11	12	13	14	15	16	17	18	19	20
21	22	23	24	25	26	27	28	29	30
31	32	33	34	35	36	37	38	39	40
41	42	43	44	45	46	47	48	49	50

36. True or false? The square of any prime number is also a prime number.

37. True or false? All odd numbers are prime.

38. True or false? All even numbers are composite.

For Exercises 39–42, determine whether the factorization represents the prime factorization. If not, explain why.

39. $36 = 2 \cdot 2 \cdot 9$
40. $48 = 2 \cdot 3 \cdot 8$
41. $210 = 5 \cdot 2 \cdot 7 \cdot 3$
42. $126 = 3 \cdot 7 \cdot 3 \cdot 2$

For Exercises 43–50, find the prime factorization. **(See Examples 4 and 5.)**

43. 70
44. 495
45. 260
46. 175
47. 147
48. 231
49. 616
50. 364

Concept 3: Equivalent Fractions

For Exercises 51 and 52, shade the second figure so that it expresses a fraction equivalent to the first figure.

51.

52.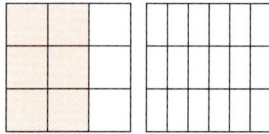

53. True or false? The fractions $\frac{4}{5}$ and $\frac{5}{4}$ are equivalent.

54. True or false? The fractions $\frac{3}{1}$ and $\frac{1}{3}$ are equivalent.

For Exercises 55–62, determine if the fractions are equivalent. Then fill in the blank with either $=$ or $\neq$. **(See Example 6.)**

55. $\frac{2}{3} \square \frac{3}{5}$
56. $\frac{1}{4} \square \frac{2}{9}$
57. $\frac{1}{2} \square \frac{3}{6}$
58. $\frac{6}{16} \square \frac{3}{8}$
59. $\frac{12}{16} \square \frac{3}{4}$
60. $\frac{4}{5} \square \frac{12}{15}$
61. $\frac{8}{9} \square \frac{20}{27}$
62. $\frac{5}{6} \square \frac{12}{18}$

Concept 4: Simplifying Fractions to Lowest Terms

For Exercises 63–70, determine the greatest common factor.

63. 12 and 18
64. 15 and 20
65. 21 and 28
66. 24 and 40
67. 75 and 25
68. 36 and 72
69. $3x^2y^3$ and $6xy^4$
70. $15c^3d^3$ and $10c^4d^2$

For Exercises 71–90, simplify the fraction to lowest terms. Write the answer as a fraction or an integer. (See Examples 7 and 8.)

71. $\dfrac{12}{24}$ 72. $\dfrac{15}{18}$ 73. $\dfrac{6}{18}$ 74. $\dfrac{21}{24}$

75. $-\dfrac{36}{20}$ 76. $-\dfrac{49}{42}$ 77. $-\dfrac{15}{12}$ 78. $-\dfrac{30}{25}$

79. $\dfrac{9}{9}$ 80. $\dfrac{2}{2}$ 81. $\dfrac{105}{140}$ 82. $\dfrac{84}{126}$

83. $-\dfrac{33}{11}$ 84. $-\dfrac{65}{5}$ 85. $\dfrac{77}{110}$ 86. $\dfrac{85}{153}$

87. $\dfrac{385}{195}$ 88. $\dfrac{39}{130}$ 89. $\dfrac{34}{85}$ 90. $\dfrac{69}{92}$

For Exercises 91–98, simplify to lowest terms by first reducing the powers of 10. (See Example 9.)

91. $\dfrac{120}{160}$ 92. $\dfrac{720}{800}$ 93. $-\dfrac{3000}{1800}$ 94. $-\dfrac{2000}{1500}$

95. $\dfrac{42{,}000}{22{,}000}$ 96. $\dfrac{50{,}000}{65{,}000}$ 97. $\dfrac{5100}{30{,}000}$ 98. $\dfrac{9800}{28{,}000}$

For Exercises 99–106, simplify the expression. (See Example 10.)

99. $\dfrac{16ab}{10a}$ 100. $\dfrac{25mn}{10n}$ 101. $\dfrac{14xyz}{7z}$ 102. $\dfrac{18pqr}{6q}$

103. $\dfrac{5x^4}{15x^3}$ 104. $\dfrac{4y^3}{20y}$ 105. $-\dfrac{6ac^2}{12ac^4}$ 106. $-\dfrac{3m^2n}{9m^2n^4}$

Concept 5: Applications of Simplifying Fractions

107. André tossed a coin 48 times and heads came up 20 times. What fractional part of the tosses came up heads? What fractional part came up tails?

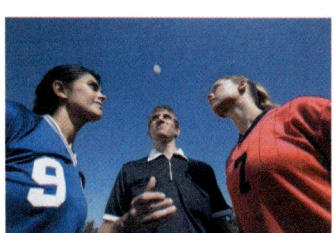

PNC/Stockbyte/Getty Images

108. At Pizza Company, Lee made 70 pizzas one day. There were 105 pizzas sold that day. What fraction of the pizzas did Lee make?

OMG/BananaStock/Getty Images

109. a. What fraction of the alphabet is made up of vowels? (Include the letter y as a vowel, not a consonant.)

   b. What fraction of the alphabet is made up of consonants?

110. Of the 88 constellations that can be seen in the night sky, 12 are associated with astrological horoscopes. The names of as many as 36 constellations are associated with animals or mythical creatures.

   a. Of the 88 constellations, what fraction is associated with horoscopes?

   b. What fraction of the constellations have names associated with animals or mythical creatures?

111. Jonathan and Jared both sold raffle tickets for a fundraiser. Jonathan sold 25 of his 35 tickets, and Jared sold 24 of his 28 tickets. (See Example 11.)

  a. What fractional part of his total number of tickets did each person sell?

  b. Which person sold the greater fractional part?

112. Lisa and Lynette are taking online courses. Lisa has completed 14 out of 16 assignments in her course while Lynette has completed 15 out of 24 assignments.

  a. What fractional part of her total number of assignments did each woman complete?

  b. Which woman has completed more of her course?

Norman Pogson/Alamy Stock Photo

113. Raymond read 720 pages of a 792-page book. His roommate, Travis, read 540 pages from a 660-page book.

  a. What fractional part of the book did each person read?

  b. Which of the roommates read a greater fraction of his book?

114. Mr. Zahnen and Ms. Waymire both gave exams today. By mid-afternoon, Mr. Zahnen had finished grading 16 out of 36 exams, and Ms. Waymire had finished grading 15 out of 27 exams.

  a. What fractional part of her total has Ms. Waymire completed?

  b. What fractional part of his total has Mr. Zahnen completed?

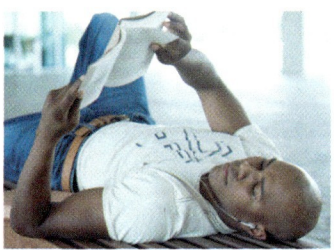
Laurence Mouton/PhotoAlto/Getty Images

Technology Connections

For Exercises 115–122, use a calculator to simplify the fractions. Write the answer as a proper or improper fraction.

115. $\dfrac{792}{891}$

116. $\dfrac{728}{784}$

117. $\dfrac{779}{969}$

118. $\dfrac{462}{220}$

119. $\dfrac{493}{510}$

120. $\dfrac{871}{469}$

121. $\dfrac{969}{646}$

122. $\dfrac{713}{437}$

Expanding Your Skills

123. Write three fractions equivalent to $\frac{3}{4}$.

124. Write three fractions equivalent to $\frac{1}{3}$.

125. Write three fractions equivalent to $-\frac{12}{18}$.

126. Write three fractions equivalent to $-\frac{80}{100}$.

Multiplication and Division of Fractions

Section 4.3

Concepts
1. Multiplication of Fractions
2. Area of a Triangle
3. Reciprocal
4. Division of Fractions
5. Applications of Multiplication and Division of Fractions

1. Multiplication of Fractions

Suppose Elija takes $\frac{1}{3}$ of a cake and then gives $\frac{1}{2}$ of this portion to his friend Max. Max gets $\frac{1}{2}$ of $\frac{1}{3}$ of the cake. This is equivalent to the expression $\frac{1}{2} \cdot \frac{1}{3}$. See Figure 4-6.

Elija takes $\frac{1}{3}$

 Max gets $\frac{1}{2}$ of $\frac{1}{3} = \frac{1}{6}$

Figure 4-6

From the illustration, the product $\frac{1}{2} \cdot \frac{1}{3} = \frac{1}{6}$. Notice that the product $\frac{1}{6}$ is found by multiplying the numerators and multiplying the denominators. This is true in general to multiply fractions.

> **Multiplying Fractions**
>
> To multiply fractions, write the product of the numerators over the product of the denominators. Then simplify the resulting fraction, if possible.
>
> $$\frac{a}{b} \cdot \frac{c}{d} = \frac{a \cdot c}{b \cdot d} \quad \text{provided } b \text{ and } d \text{ are not equal to } 0.$$

Example 1 — Multiplying Fractions

Multiply and write the answer as a fraction.

a. $\dfrac{2}{5} \cdot \dfrac{4}{7}$ b. $-\dfrac{8}{3} \cdot 5$

Solution:

a. $\dfrac{2}{5} \cdot \dfrac{4}{7} = \dfrac{2 \cdot 4}{5 \cdot 7} = \dfrac{8}{35}$ ← Multiply the numerators.
   ← Multiply the denominators.

Notice that the product $\frac{8}{35}$ is simplified completely because there are no common factors shared by 8 and 35.

b. $-\dfrac{8}{3} \cdot 5 = -\dfrac{8}{3} \cdot \dfrac{5}{1}$ First write the whole number as a fraction.

$= -\dfrac{8 \cdot 5}{3 \cdot 1}$ Multiply the numerators. Multiply the denominators.
The product of two numbers of different signs is negative.

$= -\dfrac{40}{3}$ The product cannot be simplified because there are no common factors shared by 40 and 3.

Skill Practice Multiply. Write the answer as a fraction.

1. $\dfrac{2}{3} \cdot \dfrac{5}{9}$ 2. $-\dfrac{7}{12} \cdot 11$

Answers

1. $\dfrac{10}{27}$ 2. $-\dfrac{77}{12}$

Example 2 illustrates a case where the product of fractions must be simplified.

> **Example 2** **Multiplying and Simplifying Fractions**
>
> Multiply the fractions and simplify if possible. $\quad \dfrac{4}{30} \cdot \dfrac{5}{14}$
>
> **Solution:**
>
> $\dfrac{4}{30} \cdot \dfrac{5}{14} = \dfrac{4 \cdot 5}{30 \cdot 14}$ Multiply the numerators. Multiply the denominators.
>
> $= \dfrac{20}{420}$ Simplify by first dividing 20 and 420 by 10.
>
> $= \dfrac{2}{42}$ Simplify further by dividing 2 and 42 by 2.
>
> $= \dfrac{1}{21}$
>
> **Skill Practice** Multiply and simplify.
>
> **3.** $\dfrac{7}{20} \cdot \dfrac{4}{3}$

It is often easier to simplify *before* multiplying. Consider the product from Example 2.

$$\dfrac{4}{30} \cdot \dfrac{5}{14} = \dfrac{\overset{2}{\cancel{4}}}{\underset{6}{\cancel{30}}} \cdot \dfrac{\overset{1}{\cancel{5}}}{\underset{7}{\cancel{14}}} \quad \text{4 and 14 share a common factor of 2.}$$
$$\text{30 and 5 share a common factor of 5.}$$

$$= \dfrac{\overset{1}{\cancel{\overset{2}{\cancel{4}}}}}{\underset{\underset{3}{\cancel{6}}}{\cancel{30}}} \cdot \dfrac{\overset{1}{\cancel{5}}}{\underset{7}{\cancel{14}}} \quad \text{2 and 6 share a common factor of 2.}$$

$$= \dfrac{1}{21}$$

> **Example 3** **Multiplying and Simplifying Fractions**
>
> Multiply and simplify. $\quad \left(-\dfrac{10}{18}\right)\left(-\dfrac{21}{55}\right)$
>
> **Solution:**
>
> $\left(-\dfrac{10}{18}\right)\left(-\dfrac{21}{55}\right) = +\left(\dfrac{10}{18} \cdot \dfrac{21}{55}\right)$ First note that the product will be positive. The product of two numbers with the same sign is positive.
>
> $\dfrac{10}{18} \cdot \dfrac{21}{55} = \dfrac{\overset{2}{\cancel{10}}}{\underset{6}{\cancel{18}}} \cdot \dfrac{\overset{7}{\cancel{21}}}{\underset{11}{\cancel{55}}}$ 10 and 55 share a common factor of 5.
> 18 and 21 share a common factor of 3.
>
> $= \dfrac{\overset{1}{\cancel{\overset{2}{\cancel{10}}}}}{\underset{\underset{3}{\cancel{6}}}{\cancel{18}}} \cdot \dfrac{\overset{7}{\cancel{21}}}{\underset{11}{\cancel{55}}}$ We can simplify further because 2 and 6 share a common factor of 2.
>
> $= \dfrac{7}{33}$

Skill Practice Multiply and simplify.

4. $\left(-\dfrac{6}{25}\right)\left(-\dfrac{15}{18}\right)$

Answers

3. $\dfrac{7}{15}$ **4.** $\dfrac{1}{5}$

Section 4.3 Multiplication and Division of Fractions

Example 4 — Multiplying Fractions Containing Variables

Multiply and simplify.

a. $\dfrac{5x}{7} \cdot \dfrac{2}{15x}$ **b.** $\dfrac{2a^2}{3b} \cdot \dfrac{b^3}{a}$

Avoiding Mistakes

In Example 4, $x \neq 0$, $a \neq 0$, and $b \neq 0$. The denominator of a fraction cannot equal 0.

Solution:

a. $\dfrac{5x}{7} \cdot \dfrac{2}{15x} = \dfrac{5 \cdot x \cdot 2}{7 \cdot 3 \cdot 5 \cdot x}$ Multiply fractions and factor.

$= \dfrac{\cancel{5} \cdot \cancel{x} \cdot 2}{7 \cdot 3 \cdot \cancel{5} \cdot \cancel{x}}$ Simplify. The common factors are in red.

$= \dfrac{2}{21}$

b. $\dfrac{2a^2}{3b} \cdot \dfrac{b^3}{a} = \dfrac{2 \cdot a \cdot a \cdot b \cdot b \cdot b}{3 \cdot b \cdot a}$ Multiply fractions and factor.

$= \dfrac{2 \cdot a \cdot \cancel{a} \cdot \cancel{b} \cdot b \cdot b}{3 \cdot \cancel{b} \cdot \cancel{a}}$ Simplify. The common factors are in red.

$= \dfrac{2ab^2}{3}$

Skill Practice Multiply and simplify.

5. $\dfrac{3w}{11} \cdot \dfrac{5}{6w}$ **6.** $\dfrac{5x^3}{6y} \cdot \dfrac{y^2}{x}$

2. Area of a Triangle

Recall that the area of a rectangle with length l and width w is given by

$A = l \cdot w$

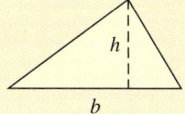

Area of a Triangle

The formula for the area of a triangle is given by $A = \tfrac{1}{2}bh$, read "one-half base times height."

The value of b is the measure of the base of the triangle. The value of h is the measure of the height of the triangle. The base b can be chosen as the length of any of the sides of the triangle. However, once you have chosen the base, the height must be measured as the shortest distance from the base to the opposite vertex (or point) of the triangle.

Figure 4-7 shows the same triangle with different orientations. Figure 4-8 shows a situation in which the height must be drawn "outside" the triangle. In such a case, notice that the height is drawn down to an imaginary extension of the base line.

Answers

5. $\dfrac{5}{22}$ **6.** $\dfrac{5x^2y}{6}$

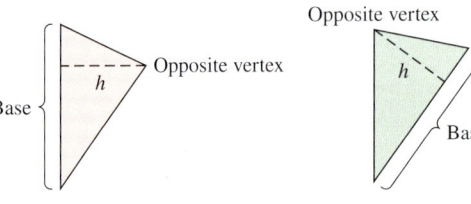
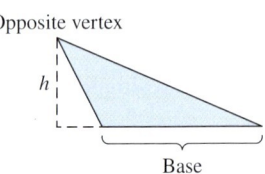

Figure 4-7   **Figure 4-8**

Example 5 — Finding the Area of a Triangle

Find the area of the triangle.

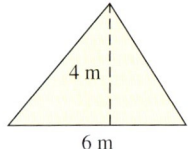

Solution:

$b = 6$ m and $h = 4$ m   Identify the measure of the base and the height.

$A = \dfrac{1}{2}bh$

$= \dfrac{1}{2}(6 \text{ m})(4 \text{ m})$ Apply the formula for the area of a triangle.

$= \dfrac{1}{2}\left(\dfrac{6}{1}\text{ m}\right)\left(\dfrac{4}{1}\text{ m}\right)$ Write the whole numbers as fractions.

$= \dfrac{1}{\cancel{2}}\left(\dfrac{\cancel{6}^{3}}{1}\text{ m}\right)\left(\dfrac{4}{1}\text{ m}\right)$ Simplify.

$= \dfrac{12}{1}\text{ m}^2$ Multiply numerators. Multiply denominators.

$= 12 \text{ m}^2$ The area of the triangle is 12 square meters (m²).

Skill Practice Find the area of the triangle.

7.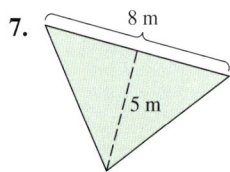

Example 6 — Finding the Area of a Triangle

Find the area of the triangle.

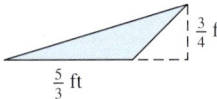

Solution:

$b = \dfrac{5}{3}$ ft and $h = \dfrac{3}{4}$ ft   Identify the measure of the base and the height.

$A = \dfrac{1}{2}bh$

$= \dfrac{1}{2}\left(\dfrac{5}{3}\text{ ft}\right)\left(\dfrac{3}{4}\text{ ft}\right)$ Apply the formula for the area of a triangle.

$= \dfrac{1}{2}\left(\dfrac{5}{\cancel{3}}\text{ ft}\right)\left(\dfrac{\cancel{3}^{1}}{4}\text{ ft}\right)$ Simplify.

$= \dfrac{5}{8}\text{ ft}^2$ The area of the triangle is $\dfrac{5}{8}$ square feet (ft²).

Answer

7. 20 m²

Skill Practice Find the area of the triangle.

8.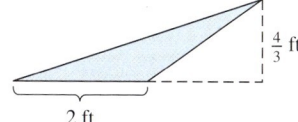

3. Reciprocal

Two numbers whose product is 1 are *reciprocals* of each other. For example, consider the product of $\frac{3}{8}$ and $\frac{8}{3}$.

$$\frac{3}{8} \cdot \frac{8}{3} = \frac{\cancel{3}}{\cancel{8}} \cdot \frac{\cancel{8}}{\cancel{3}} = 1$$

Because the product equals 1, we say that $\frac{3}{8}$ is the reciprocal of $\frac{8}{3}$ and vice versa.

To divide fractions, first we need to learn how to find the reciprocal of a fraction.

Finding the Reciprocal of a Fraction

To find the **reciprocal** of a nonzero fraction, interchange the numerator and denominator of the fraction. If a and b are nonzero numbers, then the reciprocal of $\frac{a}{b}$ is $\frac{b}{a}$. This is because $\frac{a}{b} \cdot \frac{b}{a} = 1$.

Example 7   Finding Reciprocals

Find the reciprocal.

a. $\dfrac{2}{5}$ b. $\dfrac{1}{9}$ c. 5   d. 0   e. $-\dfrac{3}{7}$

Solution:

a. The reciprocal of $\frac{2}{5}$ is $\frac{5}{2}$.
b. The reciprocal of $\frac{1}{9}$ is $\frac{9}{1}$, or 9.
c. First write the whole number 5 as the improper fraction $\frac{5}{1}$. The reciprocal of $\frac{5}{1}$ is $\frac{1}{5}$.
d. The number 0 has no reciprocal because $\frac{1}{0}$ is undefined.
e. The reciprocal of $-\frac{3}{7}$ is $-\frac{7}{3}$.   This is because $-\frac{3}{7} \cdot \left(-\frac{7}{3}\right) = 1$.

TIP: From Example 7(e) we see that a negative number will have a negative reciprocal.

Skill Practice Find the reciprocal.

9. $\dfrac{7}{10}$ 10. $\dfrac{1}{4}$ 11. 7   12. 1   13. $-\dfrac{9}{8}$

4. Division of Fractions

To understand the division of fractions, we compare it to the division of whole numbers. The statement $6 \div 2$ asks, "How many groups of 2 can be found among 6 wholes?" The answer is 3.

$$6 \div 2 = 3$$

In fractional form, the statement $6 \div 2 = 3$ can be written as $\frac{6}{2} = 3$. This result can also be found by multiplying.

$$6 \cdot \frac{1}{2} = \frac{6}{1} \cdot \frac{1}{2} = \frac{6}{2} = 3$$

Answers

8. $\dfrac{4}{3}$ ft² or $1\dfrac{1}{3}$ ft²   9. $\dfrac{10}{7}$
10. 4   11. $\dfrac{1}{7}$
12. 1   13. $-\dfrac{8}{9}$

That is, to divide by 2 is equivalent to multiplying by the reciprocal $\frac{1}{2}$.

In general, to divide two nonzero numbers we can multiply the dividend by the reciprocal of the divisor. This is how we divide by a fraction.

> **Dividing Fractions**
>
> To divide two fractions, multiply the dividend (the "first" fraction) by the reciprocal of the divisor (the "second" fraction).

The process to divide fractions can be written symbolically as

$$\frac{a}{b} \div \frac{c}{d} = \frac{a}{b} \cdot \frac{d}{c} \quad \text{provided } b, c, \text{ and } d \text{ are not } 0.$$

Change division to multiplication. Take the reciprocal of the divisor.

Example 8 — Dividing Fractions

Divide and simplify, if possible.

a. $\dfrac{2}{5} \div \dfrac{7}{4}$ b. $\dfrac{2}{27} \div \left(-\dfrac{8}{15}\right)$

Solution:

a. $\dfrac{2}{5} \div \dfrac{7}{4} = \dfrac{2}{5} \cdot \dfrac{4}{7}$ Multiply by the reciprocal of the divisor ("second" fraction).

$= \dfrac{2 \cdot 4}{5 \cdot 7}$ Multiply numerators. Multiply denominators.

$= \dfrac{8}{35}$

b. $\dfrac{2}{27} \div \left(-\dfrac{8}{15}\right) = \dfrac{2}{27} \cdot \left(-\dfrac{15}{8}\right)$ Multiply by the reciprocal of the divisor.

$= -\left(\dfrac{2}{27} \cdot \dfrac{15}{8}\right)$ The product will be negative.

$= -\left(\dfrac{\overset{1}{2}}{\underset{9}{27}} \cdot \dfrac{\overset{5}{15}}{\underset{4}{8}}\right)$ Simplify.

$= -\dfrac{5}{36}$ Multiply.

Avoiding Mistakes

Do not try to simplify until after taking the reciprocal of the divisor. In Example 8(a) it would be incorrect to "cancel" the 2 and the 4 in the expression $\frac{2}{5} \div \frac{7}{4}$.

Skill Practice Divide and simplify.

14. $\dfrac{1}{4} \div \dfrac{2}{5}$ 15. $\dfrac{3}{8} \div \left(-\dfrac{9}{10}\right)$

Answers

14. $\dfrac{5}{8}$ 15. $-\dfrac{5}{12}$

Section 4.3 Multiplication and Division of Fractions

Example 9 Dividing Fractions

Divide and simplify. Write the answer as a fraction.

a. $\dfrac{35}{14} \div 7$ b. $-12 \div \left(-\dfrac{8}{3}\right)$

Solution:

a. $\dfrac{35}{14} \div 7 = \dfrac{35}{14} \div \dfrac{7}{1}$ Write the whole number 7 as an improper fraction *before* multiplying by the reciprocal.

$= \dfrac{35}{14} \cdot \dfrac{1}{7}$ Multiply by the reciprocal of the divisor.

$= \dfrac{\overset{5}{\cancel{35}}}{14} \cdot \dfrac{1}{\underset{1}{\cancel{7}}}$ Simplify.

$= \dfrac{5}{14}$ Multiply.

b. $-12 \div \left(-\dfrac{8}{3}\right) = +\left(12 \div \dfrac{8}{3}\right)$ First note that the quotient will be positive. The quotient of two numbers with the same sign is positive.

$12 \div \dfrac{8}{3} = \dfrac{12}{1} \div \dfrac{8}{3}$ Write the whole number 12 as an improper fraction.

$= \dfrac{12}{1} \cdot \dfrac{3}{8}$ Multiply by the reciprocal of the divisor.

$= \dfrac{\overset{3}{\cancel{12}}}{1} \cdot \dfrac{3}{\underset{2}{\cancel{8}}}$ Simplify.

$= \dfrac{9}{2}$ Multiply.

Skill Practice Divide and simplify. Write the quotient as a fraction.

16. $\dfrac{15}{4} \div 10$ 17. $-20 \div \left(-\dfrac{12}{5}\right)$

Example 10 Dividing Fractions Containing Variables

Divide and simplify. $-\dfrac{10x^2}{y^2} \div \dfrac{5}{y}$

Solution:

$-\dfrac{10x^2}{y^2} \div \dfrac{5}{y} = -\dfrac{10x^2}{y^2} \cdot \dfrac{y}{5}$ Multiply by the reciprocal of the divisor.

$= -\dfrac{2 \cdot 5 \cdot x \cdot x \cdot y}{y \cdot y \cdot 5}$ Multiply fractions and factor.

$= -\dfrac{2 \cdot \overset{1}{\cancel{5}} \cdot x \cdot x \cdot \overset{1}{\cancel{y}}}{\underset{1}{\cancel{y}} \cdot y \cdot \underset{1}{\cancel{5}}}$ Simplify. The common factors are in red.

$= -\dfrac{2x^2}{y}$

Skill Practice Divide and simplify.

18. $-\dfrac{8y^3}{7} \div \dfrac{4y}{5}$

Answers

16. $\dfrac{3}{8}$ 17. $\dfrac{25}{3}$ 18. $-\dfrac{10y^2}{7}$

5. Applications of Multiplication and Division of Fractions

Sometimes it is difficult to determine whether multiplication or division is appropriate to solve an application problem. Division is generally used for a problem that requires you to separate or "split up" a quantity into pieces. Multiplication is generally used if it is necessary to take a fractional part of a quantity.

Example 11 — Using Division in an Application

A road crew must mow the grassy median along a stretch of highway I-95. If they can mow $\frac{5}{8}$ mile (mi) in 1 hr, how long will it take them to mow a 15-mi stretch?

Solution:

Read and familiarize.

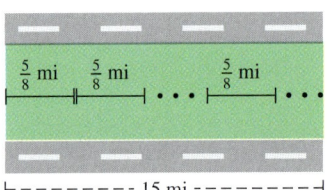

Strategy/operation: From the figure, we must separate or "split up" a 15-mi stretch of highway into pieces that are $\frac{5}{8}$ mi in length. Therefore, we must divide 15 by $\frac{5}{8}$.

$$15 \div \frac{5}{8} = \frac{15}{1} \cdot \frac{8}{5} \qquad \text{Write the whole number as a fraction. Multiply by the reciprocal of the divisor.}$$

$$= \frac{\overset{3}{\cancel{15}}}{1} \cdot \frac{8}{\underset{1}{\cancel{5}}}$$

$$= 24$$

The 15-mi stretch of highway will take 24 hr to mow.

Skill Practice

19. A cookie recipe requires $\frac{2}{5}$ package of chocolate chips for each batch of cookies. If a restaurant has 20 packages of chocolate chips, how many batches of cookies can it make?

Example 12 — Using Division in an Application

A $\frac{9}{4}$-ft length of wire must be cut into pieces of equal length that are $\frac{3}{8}$ ft long. How many pieces can be cut?

Solution:

Read and familiarize.
Operation: Here we divide the total length of wire into pieces of equal length.

$$\frac{9}{4} \div \frac{3}{8} = \frac{9}{4} \cdot \frac{8}{3} \qquad \text{Multiply by the reciprocal of the divisor.}$$

$$= \frac{\overset{3}{\cancel{9}}}{\underset{1}{\cancel{4}}} \cdot \frac{\overset{2}{\cancel{8}}}{\underset{1}{\cancel{3}}} \qquad \text{Simplify.}$$

$$= 6$$

Six pieces of wire can be cut.

Answer
19. 50 batches

Skill Practice

20. A $\frac{25}{2}$-yd ditch will be dug to put in a new water line. If piping comes in segments of $\frac{5}{4}$ yd, how many segments are needed to line the ditch?

Example 13 Using Multiplication in an Application

Carson estimates that his total cost for college for 1 year is $12,600. He has financial aid to pay $\frac{2}{3}$ of the cost.

a. How much money will be paid by financial aid?

b. How much money will Carson have to pay?

c. If Carson's parents help him by paying $\frac{1}{2}$ of the amount not paid by financial aid, how much money will be paid by Carson's parents?

Solution:

a. Carson's financial aid will pay $\frac{2}{3}$ of $12,600. Because we are looking for a fraction of a quantity, we multiply.

$$\frac{2}{3} \cdot 12{,}600 = \frac{2}{3} \cdot \frac{12{,}600}{1}$$

$$= \frac{2}{\underset{1}{\cancel{3}}} \cdot \frac{\overset{4200}{\cancel{12{,}600}}}{1}$$

$$= 8400$$

Financial aid will pay $8400.

b. Carson will have to pay the remaining portion of the cost. This can be found by subtraction.

$$\$12{,}600 - \$8400 = \$4200$$

Carson will have to pay $4200.

TIP: The answer to Example 13(b) could also have been found by noting that financial aid paid $\frac{2}{3}$ of the cost. This means that Carson must pay $\frac{1}{3}$ of the cost, or

$$\frac{1}{3} \cdot \frac{\$12{,}600}{1} = \frac{1}{\underset{1}{\cancel{3}}} \cdot \frac{\overset{4200}{\cancel{\$12{,}600}}}{1} = \$4200$$

c. Carson's parents will pay $\frac{1}{2}$ of $4200.

$$\frac{1}{\underset{1}{\cancel{2}}} \cdot \frac{\overset{2100}{\cancel{4200}}}{1}$$

Carson's parents will pay $2100.

Skill Practice

21. A new school will cost $20,000,000 to build, and the state will pay $\frac{3}{5}$ of the cost.
    a. How much will the state pay?
    b. How much will the state not pay?
    c. The county school district issues bonds to pay $\frac{4}{5}$ of the money not covered by the state. How much money will be covered by bonds?

Answers

20. 10 segments of piping
21. a. $12,000,000
    b. $8,000,000
    c. $6,400,000

Section 4.3 Activity

A.1. **a.** Multiply the fractions by writing the product of the numerators over the product of the denominators.

$$\frac{25}{6} \cdot \frac{12}{35}$$

   b. Simplify the resulting fraction to lowest terms. Write the answer as a mixed number.

A.2. Consider the multiplication problem from Exercise A.1.

$$\frac{25}{6} \cdot \frac{12}{35}$$

   a. Simplify before multiplying by dividing out common factors.
   b. Multiply the simplified numerators and the denominators.
   c. Is the answer in part (b) in lowest terms?

A.3. **a.** Multiply the fractions and write the answer as a simplified fraction.

$$\frac{4}{15} \cdot \frac{3}{2}$$

   b. Multiply the fractions and write the answer as a simplified fraction.

$$\frac{x^2}{y} \cdot \frac{y}{x}$$

   c. Considering the answers from parts (a) and (b), multiply the fractions and write the answer as a simplified fraction.

$$\frac{4x^2}{15y} \cdot \frac{3y}{2x}$$

For Exercises A.4–A.7, multiply the fractions and simplify to lowest terms. Write the answer as an improper fraction when necessary.

A.4. $\dfrac{-1}{64} \cdot \dfrac{48}{-5}$ **A.5.** $-8 \cdot \left(-\dfrac{9}{16}\right)$ **A.6.** $\dfrac{1}{3} \cdot \dfrac{16}{5} \cdot \left(-\dfrac{27}{4}\right)$ **A.7.** $\dfrac{6a^3}{5b^3} \cdot \dfrac{5b^2}{3a}$

For Exercises A.8–A.10, find the area of the figures. Leave your answer in fraction form when necessary.

A.8. 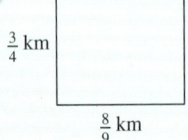   **A.9.** 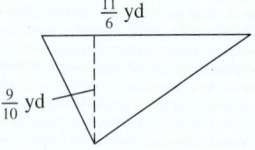   **A.10.**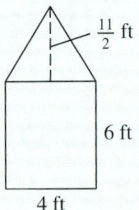

A.11. **a.** What is the reciprocal of $\dfrac{3}{4}$?

   b. What is the product of $\dfrac{3}{4}$ and its reciprocal?

   c. What is the product of $-\dfrac{2}{7}$ and its reciprocal?

   d. Do whole numbers have a reciprocal? For example, what is the reciprocal of 5?

A.12. a. Divide the fractions by multiplying the first fraction by the reciprocal of the second fraction. Simplify to lowest terms and write the answer as an improper fraction if necessary.

$$\frac{27}{16} \div \frac{3}{4}$$

b. How would the process and/or quotient change, if at all, if one of the fractions were negative?

For Exercises A.13–A.16, divide the fractions and express the answer in lowest terms.

A.13. $\frac{2}{9} \div \frac{4}{45}$ **A.14.** $-10 \div \left(-\frac{5}{6}\right)$ **A.15.** $\frac{20c^3}{9} \div \frac{10c^5}{21}$

A.16. For a recent year, Florida had $\frac{1}{9}$ of the 270 electoral votes needed to elect a president. How many votes did Florida have?

A.17. A 4-ft wooden rod needs to be cut into pieces that are each $\frac{2}{3}$ ft long. How many pieces can be cut?

Practice Exercises

Section 4.3

Prerequisite Review

R.1. The number 350 is divisible by which of the following?
  a. 2 **b.** 3 **c.** 5 **d.** 7 **e.** 10 **f.** 100

R.2. The number 126 is divisible by which of the following?
  a. 2 **b.** 3 **c.** 5 **d.** 6 **e.** 10 **f.** 12

For Exercises R.3–R.6, simplify.

R.3. $\frac{2100}{7000}$ **R.4.** $\frac{330}{2100}$ **R.5.** $\frac{12x^2}{15x}$ **R.6.** $\frac{30y^2}{12y^4}$

Vocabulary and Key Concepts

1. **a.** Given a triangle with base b and height h, the area of the triangle is given by $A = $ _____.

   b. Two numbers whose product is 1 are called _____ of each other.

2. When multiplying fractions, multiply the _____, multiply the _____, and simplify the resulting product if necessary.

3. When dividing fractions, multiply the first fraction by the _____ of the second fraction.

4. When multiplying or dividing a whole number by a fraction, write the whole number as an improper fraction with a denominator of _____.

5. The product or quotient of two negative fractions is (positive/negative).

6. When dividing two fractions, you (can/cannot) simplify fractions before taking the reciprocal of the divisor.

Concept 1: Multiplication of Fractions

7. Shade the portion of the figure that represents $\frac{1}{4}$ of $\frac{1}{4}$.

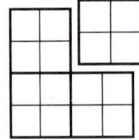

8. Shade the portion of the figure that represents $\frac{1}{3}$ of $\frac{1}{4}$.

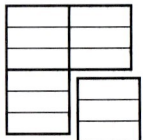

For Exercises 9–36, multiply the fractions and simplify to lowest terms. Write the answer as an improper fraction when necessary. (See Examples 1–4.)

9. $\dfrac{1}{2} \cdot \dfrac{3}{8}$

10. $\dfrac{2}{3} \cdot \dfrac{1}{3}$

11. $\left(-\dfrac{12}{7}\right)\left(-\dfrac{2}{5}\right)$

12. $\left(-\dfrac{9}{10}\right)\left(-\dfrac{7}{4}\right)$

13. $8 \cdot \left(\dfrac{1}{11}\right)$

14. $3 \cdot \left(\dfrac{2}{7}\right)$

15. $-\dfrac{4}{5} \cdot 6$

16. $-\dfrac{5}{8} \cdot 5$

17. $\dfrac{2}{9} \cdot \dfrac{3}{5}$

18. $\dfrac{1}{8} \cdot \dfrac{4}{7}$

19. $\dfrac{5}{6} \cdot \dfrac{3}{4}$

20. $\dfrac{7}{12} \cdot \dfrac{18}{5}$

21. $\dfrac{21}{5} \cdot \dfrac{25}{12}$

22. $\dfrac{16}{25} \cdot \dfrac{15}{32}$

23. $\dfrac{24}{15} \cdot \left(-\dfrac{5}{3}\right)$

24. $\dfrac{49}{24} \cdot \left(-\dfrac{6}{7}\right)$

25. $\left(\dfrac{6}{11}\right)\left(\dfrac{22}{15}\right)$

26. $\left(\dfrac{12}{45}\right)\left(\dfrac{5}{4}\right)$

27. $-12 \cdot \left(-\dfrac{15}{42}\right)$

28. $-4 \cdot \left(-\dfrac{8}{92}\right)$

29. $\dfrac{3y}{10} \cdot \dfrac{5}{y}$

30. $\dfrac{7z}{12} \cdot \dfrac{4}{z}$

31. $-\dfrac{4ab}{5} \cdot \dfrac{1}{8b}$

32. $-\dfrac{6cd}{7} \cdot \dfrac{1}{18d}$

33. $\dfrac{5x}{4y^2} \cdot \dfrac{y}{25x}$

34. $\dfrac{14w}{3z^4} \cdot \dfrac{z^2}{28w}$

35. $\left(-\dfrac{12m^3}{n}\right)\left(-\dfrac{n}{3m}\right)$

36. $\left(-\dfrac{15p^4}{t^2}\right)\left(-\dfrac{t^3}{3p}\right)$

Concept 2: Area of a Triangle

For Exercises 37–40, label the height with h and the base with b, as shown in the figure.

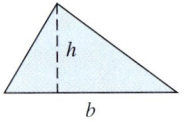

37.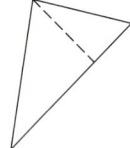

38.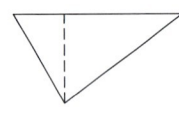

39.

40.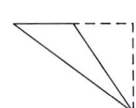

For Exercises 41–44, find the area of the triangle. (See Examples 5 and 6.)

41.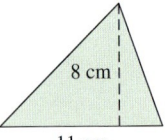

42.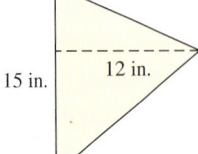

43.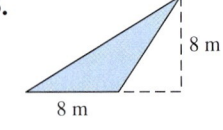

44.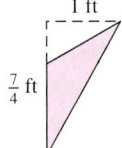

For Exercises 45–48, find the area of each figure.

45. 5 yd, $\frac{8}{5}$ yd

46. 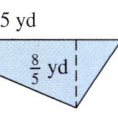 $\frac{16}{9}$ mm, 3 mm

47. 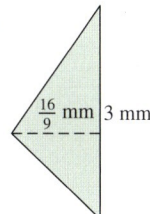 $\frac{7}{8}$ in., $\frac{7}{8}$ in.

48. 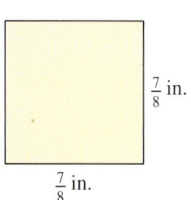 $\frac{1}{10}$ cm, $\frac{15}{16}$ cm

Concept 3: Reciprocal

For Exercises 49–56, find the reciprocal of the number, if it exists. (See Example 7.)

49. $\frac{7}{8}$

50. $\frac{5}{6}$

51. $-\frac{10}{9}$

52. $-\frac{14}{5}$

53. -4

54. -9

55. 0

56. $\frac{0}{4}$

Concept 4: Division of Fractions

For Exercises 57–60, fill in the blank.

57. Dividing by 3 is the same as multiplying by _____.

58. Dividing by 5 is the same as multiplying by _____.

59. Dividing by -8 is the same as _____ by $-\frac{1}{8}$.

60. Dividing by -12 is the same as _____ by $-\frac{1}{12}$.

For Exercises 61–80, divide and simplify the answer to lowest terms. Write the answer as a fraction or an integer. (See Examples 8–10.)

61. $\frac{2}{15} \div \frac{5}{12}$

62. $\frac{11}{3} \div \frac{6}{5}$

63. $\left(-\frac{7}{13}\right) \div \left(-\frac{2}{5}\right)$

64. $\left(-\frac{8}{7}\right) \div \left(-\frac{3}{10}\right)$

65. $\frac{14}{3} \div \frac{6}{5}$

66. $\frac{11}{2} \div \frac{3}{4}$

67. $\frac{15}{2} \div \left(-\frac{3}{2}\right)$

68. $\frac{9}{10} \div \left(-\frac{9}{2}\right)$

69. $\frac{3}{4} \div \frac{3}{4}$

70. $\frac{6}{5} \div \frac{6}{5}$

71. $-7 \div \frac{2}{3}$

72. $-4 \div \frac{3}{5}$

73. $\frac{12}{5} \div 4$

74. $\frac{20}{6} \div 2$

75. $-\frac{9}{100} \div \frac{13}{1000}$

76. $-\frac{1000}{17} \div \frac{10}{3}$

77. $\frac{4xy}{3} \div \frac{14x}{9}$

78. $\frac{3ab}{7} \div \frac{15a}{14}$

79. $-\frac{20c^3}{d^2} \div \frac{5c}{d^3}$

80. $-\frac{24w^2}{z} \div \frac{3w}{z^3}$

Mixed Exercises

For Exercises 81–96, multiply or divide as indicated. Write the answer as a fraction or an integer.

81. $\dfrac{7}{8} \div \dfrac{1}{4}$

82. $\dfrac{7}{12} \div \dfrac{5}{3}$

83. $\dfrac{5}{8} \cdot \dfrac{2}{9}$

84. $\dfrac{1}{16} \cdot \dfrac{4}{3}$

85. $6 \cdot \left(-\dfrac{4}{3}\right)$

86. $-12 \cdot \dfrac{5}{6}$

87. $\left(-\dfrac{16}{5}\right) \div (-8)$

88. $\left(-\dfrac{42}{11}\right) \div (-7)$

89. $\dfrac{1}{8} \cdot 16$

90. $\dfrac{2}{3} \cdot 9$

91. $8 \div \dfrac{16}{3}$

92. $5 \div \dfrac{15}{4}$

93. $\dfrac{13x^2}{y^2} \div \left(-\dfrac{26x}{y^3}\right)$

94. $\dfrac{11z}{w^3} \div \left(-\dfrac{33}{w}\right)$

95. $\left(-\dfrac{ad}{3}\right) \div \left(-\dfrac{ad^2}{6}\right)$

96. $\left(-\dfrac{c^2}{5}\right) \div \left(-\dfrac{c^3}{25}\right)$

Concept 5: Applications of Multiplication and Division of Fractions

97. During the month of December, a department store wraps packages free of charge. Each package requires $\frac{2}{3}$ yd of ribbon. If Li used up a 36-yd roll of ribbon, how many packages were wrapped? (See Example 11.)

98. A developer sells lots of land in increments of $\frac{3}{4}$ acre. If the developer has 60 acres, how many lots can be sold?

Brian Hagiwara/Brand X Pictures/Getty Images

99. If one cup is $\frac{1}{16}$ gal, how many cups of orange juice can be filled from $\frac{3}{2}$ gal? (See Example 12.)

100. If 1 centimeter (cm) is $\frac{1}{100}$ meter (m), how many centimeters are in a $\frac{5}{4}$-m piece of rope?

101. Dorci buys 16 sheets of plywood, each $\frac{3}{4}$ in. thick, to cover her windows in the event of a hurricane. She stacks the wood in the garage. How high will the stack be?

102. Davey built a bookshelf 36 in. long. Can the shelf hold a set of encyclopedias if there are 24 books and each book averages $\frac{5}{4}$ in. thick? Explain your answer.

103. A radio station allows 18 minutes (min) of advertising each hour. How many 40-second ($\frac{2}{3}$-min) commercials can be run in

   a. 1 hr   b. 1 day

104. A television station has 20 min of advertising each hour. How many 30-second ($\frac{1}{2}$-min) commercials can be run in

   a. 1 hr   b. 1 day

105. Ricardo wants to buy a new house for $240,000. The bank requires $\frac{1}{10}$ of the cost of the house as a down payment. As a gift, Ricardo's mother will pay $\frac{2}{3}$ of the down payment. (See Example 13.)

   a. How much money will Ricardo's mother pay toward the down payment?

   b. How much money will Ricardo have to pay toward the down payment?

   c. How much is left over for Ricardo to finance?

106. Althea wants to buy a new car for a total cost of $21,000. The dealer requires $\frac{1}{15}$ of the money as a down payment. Althea's parents have agreed to pay one-half of the down payment for her.

   a. How much money will Althea's parents pay toward the down payment?

   b. How much will Althea pay toward the down payment?

   c. How much will Althea have to finance?

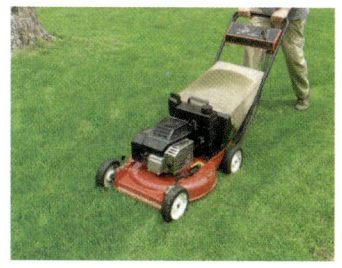
Emilio Ereza/Pixtal/age fotostock

107. Frankie's lawn measures 40 yd by 36 yd. In the morning he mowed $\frac{2}{3}$ of the lawn. How many square yards of lawn did he already mow? How much is left to be mowed?

108. Bob laid brick to make a rectangular patio in the back of his house. The patio measures 20 yd by 12 yd. On Saturday, he put down bricks for $\frac{3}{8}$ of the patio area. How many square yards is this?

C Squared Studios/Photodisc/Getty Images

109. In a certain sample of individuals, $\frac{2}{5}$ are known to have blood type O. Of the individuals with blood type O, $\frac{1}{4}$ are Rh-negative. What fraction of the individuals in the sample have O negative blood?

110. Rob has half a pizza left over from dinner. If he eats $\frac{1}{4}$ of this for breakfast, what fractional part of the whole pizza did he eat for breakfast?

111. A lab technician has $\frac{7}{4}$ liters (L) of alcohol. If she needs samples of $\frac{1}{8}$ L, how many samples can she prepare?

112. Troy has a $\frac{7}{8}$-in. nail that he must hammer into a board. Each strike of the hammer moves the nail $\frac{1}{16}$ in. into the board. How many strikes of the hammer must he make?

Antenna/fStop/Getty Images

113. At a paint store, Liu needs to make a light blue color from mixing white paint and dark blue paint. For each gallon of white paint, she mixes $\frac{1}{16}$ gal of dark blue paint. If she uses 20 gal of white paint, how many gallons of dark blue paint will she need?

114. The Bishop Gaming Center hosts a football pool. There is $1200 in prize money. The first-place winner receives $\frac{2}{3}$ of the prize money. The second-place winner receives $\frac{1}{4}$ of the prize money, and the third-place winner receives $\frac{1}{12}$ of the prize money. How much money does each person get?

115. How many eighths are in $\frac{9}{4}$?

116. How many sixths are in $\frac{4}{3}$?

117. Find $\frac{2}{5}$ of $\frac{1}{5}$.

118. Find $\frac{2}{3}$ of $\frac{1}{3}$.

Expanding Your Skills

119. The rectangle shown here has an area of 30 ft². Find the length.

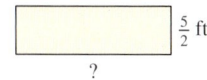

120. The rectangle shown here has an area of 8 m². Find the width.

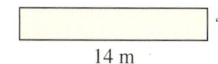

121. Find the next number in the sequence: $\frac{1}{2}, \frac{1}{4}, \frac{1}{8}, \frac{1}{16}$, _____

122. Find the next number in the sequence: $\frac{2}{3}, \frac{2}{9}, \frac{2}{27}$, _____

123. Which is greater, $\frac{1}{2}$ of $\frac{1}{8}$ or $\frac{1}{8}$ of $\frac{1}{2}$?

124. Which is greater, $\frac{2}{3}$ of $\frac{1}{4}$ or $\frac{1}{4}$ of $\frac{2}{3}$?

Section 4.4 Least Common Multiple and Equivalent Fractions

Concepts

1. Least Common Multiple
2. Applications of the Least Common Multiple
3. Writing Equivalent Fractions
4. Ordering Fractions

1. Least Common Multiple

To add or subtract fractions or to order fractions from least to greatest, the fractions must have the same denominator. To write fractions with the same denominator (a **common denominator**), we use the idea of a least common multiple of two or more numbers.

When we multiply a number by the whole numbers 1, 2, 3, and so on, we form the **multiples** of the number. For example, some of the multiples of 6 and 9 are shown below.

Multiples of 6	Multiples of 9
$6 \cdot 1 = 6$	$9 \cdot 1 = 9$
$6 \cdot 2 = 12$	$9 \cdot 2 = 18$
$6 \cdot 3 = 18$	$9 \cdot 3 = 27$
$6 \cdot 4 = 24$	$9 \cdot 4 = 36$
$6 \cdot 5 = 30$	$9 \cdot 5 = 45$
$6 \cdot 6 = 36$	$9 \cdot 6 = 54$
$6 \cdot 7 = 42$	$9 \cdot 7 = 63$
$6 \cdot 8 = 48$	$9 \cdot 8 = 72$
$6 \cdot 9 = 54$	$9 \cdot 9 = 81$

In red, we have indicated several multiples that are common to both 6 and 9.

The **least common multiple (LCM)** of two given numbers is the smallest whole number that is a multiple of each given number. For example, the LCM of 6 and 9 is 18.

Multiples of 6:   6, 12, 18, 24, 30, 36, 42, . . .

Multiples of 9:   9, 18, 27, 36, 45, 54, 63, . . .

TIP: There are infinitely many numbers that are common multiples of both 6 and 9. These include 18, 36, 54, 72, and so on. However, 18 is the smallest, and is therefore the *least* common multiple.

If one number is a multiple of another number, then the LCM is the larger of the two numbers. For example, the LCM of 4 and 8 is 8.

Multiples of 4:   4, 8, 12, 16, . . .

Multiples of 8:   8, 16, 24, 32, . . .

Example 1 Finding the LCM by Listing Multiples

Find the LCM of the given numbers by listing several multiples of each number.

a. 15 and 12   **b.** 10, 15, and 8

Solution:

a. Multiples of 15:   15, 30, 45, 60
Multiples of 12:   12, 24, 36, 48, 60

The LCM of 15 and 12 is 60.

b. Multiples of 10:   10, 20, 30, 40, 50, 60, 70, 80, 90, 100, 110, 120
Multiples of 15:   15, 30, 45, 60, 75, 90, 105, 120
Multiples of 8:   8, 16, 24, 32, 40, 48, 56, 64, 72, 80, 88, 96, 104, 112, 120

The LCM of 10, 15, and 8 is 120.

Skill Practice Find the LCM by listing several multiples of each number.

1. 15 and 25   **2.** 4, 6, and 10

Answers
1. 75   **2.** 60

Section 4.4 Least Common Multiple and Equivalent Fractions

In Example 1 we used the method of listing multiples to find the LCM of two or more numbers. As you can see, the solution to Example 1(b) required several long lists of multiples. Here we offer another method to find the LCM of two given numbers by using their prime factors.

> **Using Prime Factors to Find the LCM of Two Numbers**
> **Step 1** Write each number as a product of prime factors.
> **Step 2** The LCM is the product of unique prime factors from both numbers. Use repeated factors the maximum number of times they appear in either factorization.

This process is demonstrated in Example 2.

Example 2 Finding the LCM by Using Prime Factors

Find the LCM.

a. 14 and 12   b. 50 and 24   c. 45, 54, and 50

Solution:

a. Find the prime factorization for 14 and 12.

	2's	3's	7's
14 =	2 ·		⑦
12 =	②·②	③	

For the factors of 2, 3, and 7, we circle the greatest number of times each occurs. The LCM is the product.

FOR REVIEW
The product $2 \cdot 2 \cdot 3 \cdot 7$ can also be written as $2^2 \cdot 3 \cdot 7$.

LCM = $2 \cdot 2 \cdot 3 \cdot 7 = 84$

b. Find the prime factorization for 50 and 24.

	2's	3's	5's
50 =	2 ·		⑤·⑤
24 =	②·②·②	③	

The factor 5 is repeated twice.
The factor 2 is repeated 3 times.
The factor 3 is used only once.

LCM = $2 \cdot 2 \cdot 2 \cdot 3 \cdot 5 \cdot 5 = 600$ (The LCM can also be written as $2^3 \cdot 3 \cdot 5^2$.)

c. Find the prime factorization for 45, 54, and 50.

	2's	3's	5's
45 =		3 · 3 ·	5
54 =	2 ·	③·③·③	
50 =	②		⑤·⑤

LCM = $2 \cdot 3 \cdot 3 \cdot 3 \cdot 5 \cdot 5 = 1350$ (The LCM can also be written as $2 \cdot 3^3 \cdot 5^2$.)

Skill Practice Find the LCM by using prime factors.

3. 9 and 24
4. 16 and 9
5. 36, 42, and 30

Answers
3. 72   4. 144   5. 1260

2. Applications of the Least Common Multiple

Example 3   Using the LCM in an Application

A tile wall is to be made from 6-in., 8-in., and 12-in. square tiles. A design is made by alternating rows with different-size tiles. The first row uses only 6-in. tiles, the second row uses only 8-in. tiles, and the third row uses only 12-in. tiles. Neglecting the grout seams, what is the shortest length of wall space that can be covered using only whole tiles?

Solution:

The length of the first row must be a multiple of 6 in., the length of the second row must be a multiple of 8 in., and the length of the third row must be a multiple of 12 in. Therefore, the shortest-length wall that can be covered is given by the LCM of 6, 8, and 12.

$$6 = 2 \cdot \textcircled{3}$$
$$8 = \overparen{2 \cdot 2 \cdot 2}$$
$$12 = 2 \cdot 2 \cdot 3$$

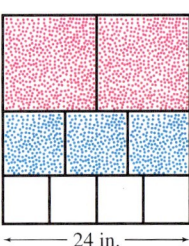

The LCM is $2 \cdot 2 \cdot 2 \cdot 3 = 24$. The shortest-length wall is 24 in.

This means that four 6-in. tiles can be placed on the first row, three 8-in. tiles can be placed on the second row, and two 12-in. tiles can be placed in the third row. See Figure 4-9.

← 24 in. →

Figure 4-9

Skill Practice

6. Three runners run on an oval track. One runner takes 60 sec to complete the loop. The second runner requires 75 sec, and the third runner requires 90 sec. Suppose the runners begin "lined up" at the same point on the track. Find the minimum amount of time required for all three runners to be lined up again.

3. Writing Equivalent Fractions

A fractional amount of a whole may be represented by many fractions. For example, the fractions $\frac{1}{2}, \frac{2}{4}, \frac{3}{6},$ and $\frac{4}{8}$ all represent the same portion of a whole. See Figure 4-10.

$$\frac{1}{2} = \frac{2}{4} = \frac{3}{6} = \frac{4}{8}$$

Figure 4-10

Expressing a fraction in an equivalent form is important for several reasons. We need this skill to order fractions and to add and subtract fractions.

Writing a fraction as an equivalent fraction is an application of the fundamental principle of fractions.

Example 4   Writing Equivalent Fractions

Write the fraction with the indicated denominator. $\dfrac{2}{9} = \dfrac{}{36}$

Solution:

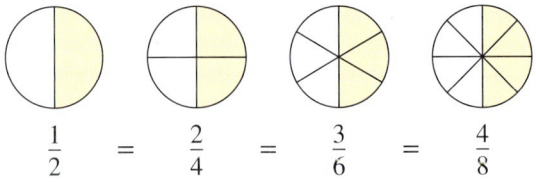

Therefore, $\dfrac{2}{9}$ is equivalent to $\dfrac{8}{36}$.

What number must we multiply 9 by to get 36?

Multiply numerator and denominator by 4.

Answer

6. After 900 sec (15 min) the runners will again be "lined up."

Skill Practice Write the fraction with the indicated denominator.

7. $\dfrac{2}{3} = \dfrac{}{15}$

TIP: In Example 4, we multiplied numerator and denominator of the fraction by 4. This is the same as multiplying the fraction by a convenient form of 1.

$$\dfrac{2}{9} = \dfrac{2}{9} \cdot 1 = \dfrac{2}{9} \cdot \dfrac{4}{4} = \dfrac{2 \cdot 4}{9 \cdot 4} = \dfrac{8}{36}$$

This is the same as multiplying numerator and denominator by 4.

Example 5 Writing Equivalent Fractions

Write the fraction with the indicated denominator. $\dfrac{11}{8} = \dfrac{}{56}$

Solution:

$\dfrac{11}{8} = \dfrac{}{56}$ $\dfrac{11 \cdot 7}{8 \cdot 7} = \dfrac{77}{56}$ Therefore, $\dfrac{11}{8}$ is equivalent to $\dfrac{77}{56}$.

What number must we multiply 8 by to get 56?   Multiply numerator and denominator by 7.

Skill Practice Write the fraction with the indicated denominator.

8. $\dfrac{5}{6} = \dfrac{}{54}$

Example 6 Writing Equivalent Fractions

Write the fractions with the indicated denominator.

a. $-\dfrac{5}{6} = -\dfrac{}{30}$ b. $\dfrac{9}{-4} = \dfrac{}{8}$

Solution:

a. $-\dfrac{5}{6} = -\dfrac{}{30}$

$-\dfrac{5 \cdot 5}{6 \cdot 5} = -\dfrac{25}{30}$ We must multiply by 5 to get a denominator of 30. Multiply numerator and denominator by 5.

Therefore, $-\dfrac{5}{6}$ is equivalent to $-\dfrac{25}{30}$.

b. $\dfrac{9 \cdot (-2)}{-4 \cdot (-2)} = \dfrac{-18}{8}$ We must multiply by -2 to get a denominator of 8. Multiply numerator and denominator by -2.

Therefore, $\dfrac{9}{-4}$ is equivalent to $\dfrac{-18}{8}$.

TIP: Note that $\dfrac{-18}{8} = \dfrac{18}{-8} = -\dfrac{18}{8}$. All are equivalent to $\dfrac{9}{-4}$.

Skill Practice Write the fractions with the indicated denominator.

9. $-\dfrac{10}{3} = -\dfrac{}{12}$ 10. $\dfrac{3}{-8} = \dfrac{}{16}$

Answers

7. $\dfrac{10}{15}$ 8. $\dfrac{45}{54}$

9. $-\dfrac{40}{12}$ 10. $\dfrac{-6}{16}$

Example 7 — Writing Equivalent Fractions

Write the fractions with the indicated denominator.

a. $\dfrac{2}{3} = \dfrac{}{9x}$ b. $\dfrac{4}{y} = \dfrac{}{y^2}$

Solution:

a. $\dfrac{2}{3} = \dfrac{}{9x}$ $\dfrac{2 \cdot 3x}{3 \cdot 3x} = \dfrac{6x}{9x}$ Therefore, $\dfrac{2}{3}$ is equivalent to $\dfrac{6x}{9x}$.

   What must we multiply 3 by to get $9x$?   Multiply numerator and denominator by $3x$.

b. $\dfrac{4}{y} = \dfrac{}{y^2}$ $\dfrac{4 \cdot y}{y \cdot y} = \dfrac{4y}{y^2}$ Therefore, $\dfrac{4}{y}$ is equivalent to $\dfrac{4y}{y^2}$.

   What must we multiply y by to get y^2?   Multiply numerator and denominator by y.

Skill Practice Write the fractions with the indicated denominator.

11. $\dfrac{7}{9} = \dfrac{}{18y}$ 12. $\dfrac{10}{w} = \dfrac{}{w^2}$

4. Ordering Fractions

Suppose we want to determine which of two fractions is larger. Comparing fractions with the same denominator, such as $\dfrac{3}{5}$ and $\dfrac{2}{5}$ is relatively easy. Clearly 3 parts out of 5 is greater than 2 parts out of 5.

Thus, $\dfrac{3}{5} > \dfrac{2}{5}$.

So how would we compare the relative size of two fractions with *different* denominators such as $\dfrac{3}{5}$ and $\dfrac{4}{7}$? Our first step is to write the fractions as equivalent fractions with the same denominator, called a common denominator. The **least common denominator (LCD)** of two fractions is the LCM of the denominators of the fractions. The LCD of $\dfrac{3}{5}$ and $\dfrac{4}{7}$ is 35, because this is the least common multiple of 5 and 7. In Example 8, we convert the fractions $\dfrac{3}{5}$ and $\dfrac{4}{7}$ to equivalent fractions having 35 as the denominator.

Example 8 — Comparing Two Fractions

Fill in the blank with $<$, $>$, or $=$. $\dfrac{3}{5} \;\square\; \dfrac{4}{7}$

Solution:

The fractions have different denominators and cannot be compared by inspection. The LCD is 35. We need to convert each fraction to an equivalent fraction with a denominator of 35.

$\dfrac{3}{5} = \dfrac{3 \cdot 7}{5 \cdot 7} = \dfrac{21}{35}$ Multiply numerator and denominator by 7 because $5 \cdot 7 = 35$.

$\dfrac{4}{7} = \dfrac{4 \cdot 5}{7 \cdot 5} = \dfrac{20}{35}$ Multiply numerator and denominator by 5 because $7 \cdot 5 = 35$.

Because $\dfrac{21}{35} > \dfrac{20}{35}$, then $\dfrac{3}{5} \boxed{>} \dfrac{4}{7}$.

Answers

11. $\dfrac{14y}{18y}$ 12. $\dfrac{10w}{w^2}$

The relationship between $\frac{3}{5}$ and $\frac{4}{7}$ is shown in Figure 4-11. The position of the two fractions is also illustrated on the number line. See Figure 4-12.

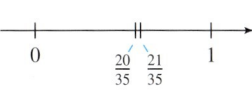

Figure 4-11

Figure 4-12

Skill Practice Fill in the blank with <, >, or =.

13. $\dfrac{3}{8} \;\square\; \dfrac{4}{9}$

Example 9   **Ranking Fractions in Order from Least to Greatest**

Rank the fractions from least to greatest. $-\dfrac{9}{20},\; -\dfrac{7}{15},\; -\dfrac{4}{9}$

Solution:

We want to convert each fraction to an equivalent fraction with a common denominator. The least common denominator is the LCM of 20, 15, and 9.

$$\left.\begin{array}{l} 20 = \boxed{2 \cdot 2} \cdot \boxed{5} \\ 15 = 3 \cdot 5 \\ 9 = \boxed{3 \cdot 3} \end{array}\right\} \quad \text{The least common denominator is } 2 \cdot 2 \cdot 3 \cdot 3 \cdot 5 = 180.$$

Now convert each fraction to an equivalent fraction with a denominator of 180.

$-\dfrac{9}{20} = -\dfrac{9 \cdot 9}{20 \cdot 9} = -\dfrac{81}{180}$ Multiply numerator and denominator by 9 because $20 \cdot 9 = 180$.

$-\dfrac{7}{15} = -\dfrac{7 \cdot 12}{15 \cdot 12} = -\dfrac{84}{180}$ Multiply numerator and denominator by 12 because $15 \cdot 12 = 180$.

$-\dfrac{4}{9} = -\dfrac{4 \cdot 20}{9 \cdot 20} = -\dfrac{80}{180}$ Multiply numerator and denominator by 20 because $9 \cdot 20 = 180$.

The relative position of these fractions is shown on the number line.

Ranking the fractions from least to greatest we have $-\dfrac{84}{180},\; -\dfrac{81}{180},$ and $-\dfrac{80}{180}.$ This is equivalent to $-\dfrac{7}{15},\; -\dfrac{9}{20},$ and $-\dfrac{4}{9}.$

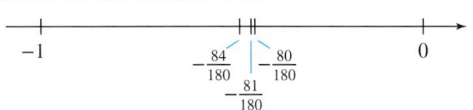

Skill Practice Rank the fractions from least to greatest.

14. $-\dfrac{5}{9},\; -\dfrac{8}{15},$ and $-\dfrac{3}{5}$

Answers

13. <   14. $-\dfrac{3}{5},\; -\dfrac{5}{9},$ and $-\dfrac{8}{15}$

Section 4.4 Activity

A.1. Consider the numbers 9, 6, and 27. Multiples of each number can be written as follows:

9: 9, 18, 27, 36, 45, 54, 63, 72, ...
6: 6, 12, 18, 24, 30, 36, 42, 48, 54, 60, ...
27: 27, 54, 81, 108, ...

a. Circle the multiples that are common to each number.

b. Write the least common multiple (LCM). The LCM of the given numbers is the smallest common multiple.

A.2. Consider the numbers 9, 6, and 27. The numbers can be written as a product of prime factors:

$9 = 3 \cdot 3$
$6 = 2 \cdot 3$
$27 = 3 \cdot 3 \cdot 3$

a. Circle the greatest number of times each factor occurs.

b. Find the LCM of the numbers 9, 6, and 27 by multiplying the factors you circled in part (a).

For Exercises A.3–A.4, find the LCM of each list of numbers by listing multiples of each number or by using prime factors.

A.3. 8, 10, 15

A.4. 15, 21, 35, 105

For Exercises A.5–A.8, write each fraction with the indicated denominator.

A.5. $\dfrac{4}{5} = \dfrac{}{30}$

A.6. $-\dfrac{2}{9} = \dfrac{}{45}$

A.7. $\dfrac{3}{7} = \dfrac{}{14x}$

A.8. $\dfrac{4}{a} = \dfrac{}{a^2}$

A.9. a. Find the LCM of 36 and 9.

b. Find the least common denominator (LCD) of $\dfrac{19}{36}$ and $\dfrac{5}{9}$.

c. Write each fraction as an equivalent fraction with the LCD from part (b).

d. Compare the numerators of each fraction to determine which fraction is smaller.

A.10. Order the following fractions from least to greatest by first finding the LCD, then writing equivalent fractions.

$$\dfrac{13}{25}, \dfrac{37}{75}, \dfrac{17}{30}, \dfrac{1}{10}$$

Section 4.4 Practice Exercises

Prerequisite Review

For Exercises R.1–R.4, simplify the fraction.

R.1. $-\dfrac{104}{36}$

R.2. $\dfrac{350}{210}$

R.3. $\dfrac{30xy}{20x^3}$

R.4. $-\dfrac{18xy^2}{15x^2y}$

For Exercises R.5–R.8, multiply or divide as indicated.

R.5. $\left(-\dfrac{22}{5}\right)(-30)$

R.6. $\left(-\dfrac{22}{5}\right) \div (-30)$

R.7. $\left(-\dfrac{14}{15}\right) \div \left(\dfrac{21}{10}\right)$

R.8. $\left(-\dfrac{14}{15}\right)\left(\dfrac{21}{10}\right)$

Vocabulary and Key Concepts

1. a. A _____ of a number is the product of the number and a nonzero whole number.

b. The _____ _____ _____ (LCM) of two numbers is the smallest whole number that is a multiple of each given number.

c. The _____ _____ _____ (LCD) of two fractions is the LCM of the denominators of the fractions.

2. There are an _____ number of common multiples between two numbers, but only one that is considered the *least* common multiple.

3. There are two methods to find the least common denominator. One method is to list multiples of each number, and the second method is to use _____ _____.

4. Consider the fractions $\frac{2}{7}$ and $\frac{8}{28}$. These are called _____ fractions because they represent the same portion of a whole.

5. Expressing fractions in equivalent form is a skill needed to perform the mathematical operations of _____ and _____ of fractions.

6. Multiplying a fraction by $\frac{x}{x}$ is the same as multiplying the fraction by the number _____.

7. The fraction $\frac{3}{8}$ is (less than/greater than) $\frac{5}{8}$.

8. When ranking fractions in order, first write all of the fractions with a _____ denominator, then compare the numerators.

Concept 1: Least Common Multiple

9. a. Circle the multiples of 24: 4, 8, 48, 72, 12, 240
   b. Circle the factors of 24: 4, 8, 48, 72, 12, 240

10. a. Circle the multiples of 30: 15, 90, 120, 3, 5, 60
    b. Circle the factors of 30: 15, 90, 120, 3, 5, 60

11. a. Circle the multiples of 36: 72, 6, 360, 12, 9, 108
    b. Circle the factors of 36: 72, 6, 360, 12, 9, 108

12. a. Circle the multiples of 28: 7, 4, 2, 56, 140, 280
    b. Circle the factors of 28: 7, 4, 2, 56, 140, 280

For Exercises 13–32, find the LCM. (See Examples 1 and 2.)

13. 10 and 25
14. 21 and 14
15. 16 and 12
16. 20 and 12
17. 18 and 24
18. 9 and 30
19. 12 and 15
20. 27 and 45
21. 42 and 70
22. 6 and 21
23. 8, 10, and 12
24. 4, 6, and 14
25. 12, 15, and 20
26. 20, 30, and 40
27. 16, 24, and 30
28. 20, 42, and 35
29. 6, 12, 18, and 20
30. 21, 35, 50, and 75
31. 5, 15, 18, and 20
32. 28, 10, 21, and 35

Concept 2: Applications of the Least Common Multiple

33. A tile floor is to be made from 10-in., 12-in., and 15-in. square tiles. A design is made by alternating rows with different-size tiles. The first row uses only 10-in. tiles, the second row uses only 12-in. tiles, and the third row uses only 15-in. tiles. Neglecting the grout seams, what is the shortest length of floor space that can be covered evenly by each row? (See Example 3.)

34. A patient admitted to the hospital was prescribed a pain medication to be given every 4 hr and an antibiotic to be given every 5 hr. Bandages applied to the patient's external injuries needed changing every 12 hr. The nurse changed the bandages and gave the patient both medications at 6:00 A.M. Monday morning.

    a. How many hours will pass before the patient is given both medications and has his bandages changed at the same time?

    b. What day and time will this be?

35. Four satellites revolve around the Earth once every 6, 8, 10, and 15 hr, respectively. If the satellites are initially "lined up," how many hours must pass before they will again be lined up?

36. Mercury, Venus, and Earth revolve around the Sun approximately once every 3 months, 7 months, and 12 months, respectively (see the figure). If the planets begin "lined up," what is the minimum number of months required for them to be aligned again? (Assume that the planets lie roughly in the same plane.)

StockTrek/Brand X Pictures/PunchStock

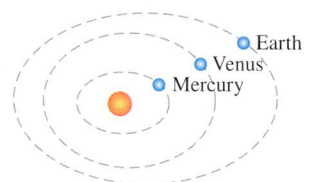

Concept 3: Writing Equivalent Fractions

For Exercises 37–66, rewrite each fraction with the indicated denominators. (See Examples 4–7.)

37. $\dfrac{2}{3} = \dfrac{}{21}$

38. $\dfrac{7}{4} = \dfrac{}{32}$

39. $\dfrac{5}{8} = \dfrac{}{16}$

40. $\dfrac{2}{9} = \dfrac{}{27}$

41. $-\dfrac{3}{4} = -\dfrac{}{16}$

42. $-\dfrac{3}{10} = -\dfrac{}{50}$

43. $\dfrac{4}{-5} = \dfrac{}{15}$

44. $\dfrac{3}{-7} = \dfrac{}{70}$

45. $\dfrac{7}{6} = \dfrac{}{42}$

46. $\dfrac{10}{3} = \dfrac{}{18}$

47. $\dfrac{11}{9} = \dfrac{}{99}$

48. $\dfrac{7}{5} = \dfrac{}{35}$

49. $5 = \dfrac{}{4}$ (Hint: $5 = \dfrac{5}{1}$)

50. $3 = \dfrac{}{12}$ (Hint: $3 = \dfrac{3}{1}$)

51. $\dfrac{11}{4} = \dfrac{}{4000}$

52. $\dfrac{18}{7} = \dfrac{}{700}$

53. $-\dfrac{11}{3} = -\dfrac{}{15}$

54. $-\dfrac{1}{6} = -\dfrac{}{60}$

55. $\dfrac{-5}{8} = \dfrac{}{24}$

56. $\dfrac{-20}{7} = \dfrac{}{35}$

57. $\dfrac{4y}{7} = \dfrac{}{28}$

58. $\dfrac{3v}{13} = \dfrac{}{26}$

59. $\dfrac{3}{8} = \dfrac{}{8y}$

60. $\dfrac{7}{13} = \dfrac{}{13u}$

61. $\dfrac{3}{5} = \dfrac{}{25p}$

62. $\dfrac{4}{9} = \dfrac{}{18v}$

63. $\dfrac{2}{x} = \dfrac{}{x^2}$

64. $\dfrac{6}{w} = \dfrac{}{w^2}$

65. $\dfrac{8}{ab} = \dfrac{}{ab^3}$

66. $\dfrac{9}{cd} = \dfrac{}{c^3d}$

Concept 4: Ordering Fractions

For Exercises 67–74, fill in the blanks with <, >, or =. (See Example 8.)

67. $\dfrac{7}{8} \;\square\; \dfrac{3}{4}$

68. $\dfrac{7}{15} \;\square\; \dfrac{11}{20}$

69. $\dfrac{13}{10} \;\square\; \dfrac{22}{15}$

70. $\dfrac{15}{4} \;\square\; \dfrac{21}{6}$

71. $-\dfrac{3}{12} \;\square\; -\dfrac{2}{8}$

72. $-\dfrac{4}{20} \;\square\; -\dfrac{6}{30}$

73. $-\dfrac{5}{18} \;\square\; -\dfrac{8}{27}$

74. $-\dfrac{9}{24} \;\square\; -\dfrac{8}{21}$

75. Which of the fractions has the greatest value? $\dfrac{2}{3}, \dfrac{7}{8}, \dfrac{5}{6}, \dfrac{1}{2}$

76. Which of the fractions has the least value? $\dfrac{1}{6}, \dfrac{1}{4}, \dfrac{2}{15}, \dfrac{2}{9}$

For Exercises 77–82, rank the fractions from least to greatest. (See Example 9.)

77. $\dfrac{7}{8}, \dfrac{2}{3}, \dfrac{3}{4}$

78. $\dfrac{5}{12}, \dfrac{3}{8}, \dfrac{2}{3}$

79. $-\dfrac{5}{16}, -\dfrac{3}{8}, -\dfrac{1}{4}$

80. $-\dfrac{2}{5}, -\dfrac{3}{10}, -\dfrac{5}{6}$

81. $-\dfrac{4}{3}, -\dfrac{13}{12}, \dfrac{17}{15}$

82. $-\dfrac{5}{7}, \dfrac{11}{21}, -\dfrac{18}{35}$

83. A patient had three cuts that needed stitches. A nurse recorded the lengths of the cuts. Where did the patient have the longest cut? Where did the patient have the shortest cut?

upper right arm ¾ in.
Right hand 11/16 in.
above left eye 7/8 in.

84. Three screws have lengths equal to $\frac{3}{4}$ in., $\frac{5}{8}$ in., and $\frac{11}{16}$ in. Which screw is the longest? Which is the shortest?

85. For a party, Aman had $\frac{3}{4}$ lb of cheddar cheese, $\frac{7}{8}$ lb of Swiss cheese, and $\frac{4}{5}$ lb of pepper jack cheese. Which type of cheese is in the least amount? Which type is in the greatest amount?

86. Susan buys $\frac{2}{3}$ lb of smoked turkey, $\frac{3}{5}$ lb of ham, and $\frac{5}{8}$ lb of roast beef. Which type of meat did she buy in the greatest amount? Which type did she buy in the least amount?

Expanding Your Skills

87. Which of the fractions is between $\frac{1}{4}$ and $\frac{5}{6}$? Identify all that apply.

    a. $\dfrac{5}{12}$ b. $\dfrac{2}{3}$ c. $\dfrac{1}{8}$

88. Which of the fractions is between $\frac{1}{3}$ and $\frac{11}{15}$? Identify all that apply.

    a. $\dfrac{2}{3}$ b. $\dfrac{4}{5}$ c. $\dfrac{2}{5}$

The LCM of two or more numbers can also be found with a method called "division of primes." For example, consider the numbers 32, 48, and 30. To find the LCM, first divide by any prime number that divides evenly into any of the numbers. Then divide and write the quotient as shown.

$$2\overline{)32 \ \ 48 \ \ 30}$$
$$16 \ \ 24 \ \ 15$$

Repeat this process and bring down any number that is not divisible by the chosen prime.

$$2\overline{)32 \ \ 48 \ \ 30}$$
$$2\overline{)16 \ \ 24 \ \ 15}$$
$$8 \ \ 12 \ \ 15 \quad \text{Bring down the 15.}$$

Continue until all quotients are 1. The LCM is the product of the prime factors on the left.

$$2\overline{)32 \ \ 48 \ \ 30}$$
$$2\overline{)16 \ \ 24 \ \ 15}$$
$$2\overline{)8 \ \ 12 \ \ 15}$$
$$2\overline{)4 \ \ 6 \ \ 15}$$
$$2\overline{)2 \ \ 3 \ \ 15}$$
$$3\overline{)1 \ \ 3 \ \ 15} \quad \longleftarrow \text{At this point, the prime number 2 does not divide evenly into any of the quotients.}$$
$$5\overline{)1 \ \ 1 \ \ 5} \quad \text{We try the next-greater prime number, 3.}$$
$$1 \ \ 1 \ \ 1$$

The LCM is $2 \cdot 2 \cdot 2 \cdot 2 \cdot 2 \cdot 3 \cdot 5 = 480$.

For Exercises 89–92, use "division of primes" to determine the LCM of the given numbers.

89. 16, 24, and 28

90. 15, 25, and 35

91. 20, 18, and 27

92. 9, 15, and 42

Section 4.5 Addition and Subtraction of Fractions

Concepts

1. Addition and Subtraction of Like Fractions
2. Addition and Subtraction of Unlike Fractions
3. Applications of Addition and Subtraction of Fractions

1. Addition and Subtraction of Like Fractions

The main focus of this section is to add and subtract fractions. The operation of addition can be thought of as combining like groups of objects. For example:

$$3 \text{ apples} + 1 \text{ apple} = 4 \text{ apples}$$

three-fifths + one-fifths = four-fifths

$$\frac{3}{5} + \frac{1}{5} = \frac{4}{5}$$

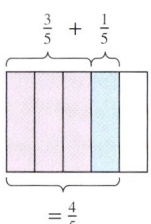

The fractions $\frac{3}{5}$ and $\frac{1}{5}$ are **like fractions** because their denominators are the same. That is, the fractions have a **common denominator**.

The following property leads to the procedure to add and subtract like fractions.

Addition and Subtraction of Like Fractions

To add or subtract like fractions, add or subtract the numerators and write the result over the common denominator. Simplify to lowest terms if possible.

$$\frac{a}{c} + \frac{b}{c} = \frac{a+b}{c} \quad \text{and} \quad \frac{a}{c} - \frac{b}{c} = \frac{a-b}{c}, \text{ provided } c \neq 0.$$

Example 1 Adding and Subtracting Like Fractions

Add. Write the answer as a fraction.

a. $\dfrac{1}{4} + \dfrac{5}{4}$ b. $\dfrac{2}{15} + \dfrac{1}{15} - \dfrac{13}{15}$

Avoiding Mistakes

Notice that when adding fractions, we do not add the denominators. We add *only* the numerators.

Solution:

a. $\dfrac{1}{4} + \dfrac{5}{4} = \dfrac{1+5}{4}$ Add the numerators.

$= \dfrac{6}{4}$ Write the sum over the common denominator.

$= \dfrac{\overset{3}{\cancel{6}}}{\underset{2}{\cancel{4}}}$ Simplify to lowest terms.

$= \dfrac{3}{2}$

b. $\dfrac{2}{15} + \dfrac{1}{15} - \dfrac{13}{15} = \dfrac{2+1-13}{15}$ Add and subtract terms in the numerator. Write the result over the common denominator.

$= \dfrac{-10}{15}$ Simplify. The answer will be a negative fraction.

$= -\dfrac{\overset{2}{\cancel{10}}}{\underset{3}{\cancel{15}}}$ Simplify to lowest terms.

$= -\dfrac{2}{3}$

Skill Practice Add. Write the answer as a fraction.

1. $\dfrac{2}{9}+\dfrac{4}{9}$ 2. $\dfrac{7}{12}-\dfrac{5}{12}-\dfrac{11}{12}$

Example 2 Subtracting Fractions

Subtract. $\dfrac{3x}{5y}-\dfrac{2}{5y}$

Solution:

The fractions $\dfrac{3x}{5y}$ and $\dfrac{2}{5y}$ are like fractions because they have the same denominator.

$\dfrac{3x}{5y}-\dfrac{2}{5y}=\dfrac{3x-2}{5y}$ Subtract the numerators. Write the result over the common denominator.

$\qquad\quad=\dfrac{3x-2}{5y}$ Notice that the numerator cannot be simplified further because $3x$ and 2 are not *like* terms.

Skill Practice Subtract.

3. $\dfrac{4a}{7b}-\dfrac{19}{7b}$

2. Addition and Subtraction of Unlike Fractions

Adding fractions can be visualized by using a diagram. For example, the sum $\tfrac{1}{2}+\tfrac{1}{3}$ is illustrated in Figure 4-13.

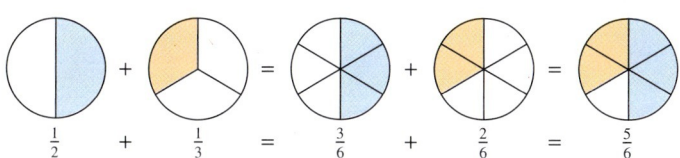

Figure 4-13

Two fractions are **unlike fractions** if they have different denominators. In Example 3 we show that the first step in adding or subtracting unlike fractions is to identify the LCD. Then we change the unlike fractions to like fractions having the LCD as the denominator.

Example 3 Adding Unlike Fractions

Add. $\dfrac{1}{6}+\dfrac{3}{4}$

Solution:

We cannot add $\tfrac{1}{6}$ and $\tfrac{3}{4}$ as they are because they have different denominators. Using the LCD of 12, we can convert each individual fraction to an equivalent fraction with 12 as the denominator.

$\dfrac{1}{6}=\dfrac{1\cdot 2}{6\cdot 2}=\dfrac{2}{12}$ Multiply numerator and denominator by 2 because $6\cdot 2=12$.

$\dfrac{3}{4}=\dfrac{3\cdot 3}{4\cdot 3}=\dfrac{9}{12}$ Multiply numerator and denominator by 3 because $4\cdot 3=12$.

Answers

1. $\dfrac{2}{3}$ 2. $-\dfrac{3}{4}$ 3. $\dfrac{4a-19}{7b}$

Thus, $\frac{1}{6} + \frac{3}{4}$ becomes $\frac{2}{12} + \frac{9}{12} = \frac{11}{12}$.

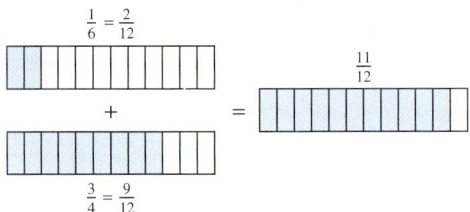

Skill Practice Add.

4. $\frac{1}{10} + \frac{1}{15}$

The general procedure to add or subtract unlike fractions is outlined as follows.

Adding and Subtracting Unlike Fractions
Step 1 Identify the LCD.
Step 2 Write each individual fraction as an equivalent fraction with the LCD.
Step 3 Add or subtract the numerators and write the result over the common denominator.
Step 4 Simplify to lowest terms, if possible.

Example 4 Subtracting Unlike Fractions

Subtract. $\frac{7}{10} - \frac{1}{5}$

Solution:

$\frac{7}{10} - \frac{1}{5}$ The LCD is 10. We must convert $\frac{1}{5}$ to an equivalent fraction with 10 as the denominator.

$= \frac{7}{10} - \frac{1 \cdot 2}{5 \cdot 2}$ Multiply numerator and denominator by 2 because $5 \cdot 2 = 10$.

$= \frac{7}{10} - \frac{2}{10}$ The fractions are now like fractions.

$= \frac{7 - 2}{10}$ Subtract the numerators.

$= \frac{5}{10}$

$= \frac{\overset{1}{\cancel{5}}}{\underset{2}{\cancel{10}}}$ Simplify to lowest terms.

$= \frac{1}{2}$

Avoiding Mistakes
When adding or subtracting fractions, we do not add or subtract the denominators.

Skill Practice Subtract. Write the answer as a fraction.

5. $\frac{9}{5} - \frac{7}{15}$

Answers
4. $\frac{1}{6}$ 5. $\frac{4}{3}$

Section 4.5 Addition and Subtraction of Fractions

Example 5 — Subtracting Unlike Fractions

Subtract. $\quad -\dfrac{4}{15} - \dfrac{1}{10}$

Solution:

$-\dfrac{4}{15} - \dfrac{1}{10}$ The LCD is 30.

$= -\dfrac{4 \cdot 2}{15 \cdot 2} - \dfrac{1 \cdot 3}{10 \cdot 3}$ Write the fractions as equivalent fractions with the denominators equal to the LCD.

$= -\dfrac{8}{30} - \dfrac{3}{30}$ The fractions are now like fractions.

$= \dfrac{-8 - 3}{30}$ The fraction $-\frac{8}{30}$ can be written as $\frac{-8}{30}$.

$= \dfrac{-11}{30}$ Subtract the numerators.

$= -\dfrac{11}{30}$ The fraction $-\frac{11}{30}$ is in lowest terms because the only common factor of 11 and 30 is 1.

Skill Practice Subtract.

6. $-\dfrac{7}{12} - \dfrac{1}{8}$

Example 6 — Adding Unlike Fractions

Add. Write the answer as a fraction. $\quad -5 + \dfrac{3}{4}$

Solution:

$-5 + \dfrac{3}{4} = -\dfrac{5}{1} + \dfrac{3}{4}$ Write the whole number as a fraction.

$= -\dfrac{5 \cdot 4}{1 \cdot 4} + \dfrac{3}{4}$ The LCD is 4.

$= -\dfrac{20}{4} + \dfrac{3}{4}$ The fractions are now like fractions.

$= \dfrac{-20 + 3}{4}$ Add the terms in the numerator.

$= \dfrac{-17}{4}$ Simplify.

$= -\dfrac{17}{4}$

Skill Practice Add. Write the answer as a fraction.

7. $-3 + \dfrac{4}{5}$

Example 7 — Adding and Subtracting Unlike Fractions

Add and subtract as indicated. $\quad -\dfrac{7}{12} - \dfrac{2}{15} + \dfrac{7}{24}$

Answers

6. $-\dfrac{17}{24}$ 7. $-\dfrac{11}{5}$

Solution:

$$-\frac{7}{12} - \frac{2}{15} + \frac{7}{24}$$ To find the LCD, factor each denominator.

$$= -\frac{7}{2 \cdot 2 \cdot 3} - \frac{2}{3 \cdot 5} + \frac{7}{2 \cdot 2 \cdot 2 \cdot 3}$$

$$\left.\begin{array}{l} 12 = 2 \cdot 2 \cdot \textcircled{3} \\ 15 = 3 \cdot \textcircled{5} \\ 24 = \textcircled{2 \cdot 2 \cdot 2} \cdot 3 \end{array}\right\}$$ The LCD is $2 \cdot 2 \cdot 2 \cdot 3 \cdot 5 = 120$.

We want to convert each fraction to an equivalent fraction having a denominator of $2 \cdot 2 \cdot 2 \cdot 3 \cdot 5 = 120$. Multiply numerator and denominator of each fraction by the factors missing from the denominator.

$$-\frac{7 \cdot (2 \cdot 5)}{2 \cdot 2 \cdot 3 \cdot (2 \cdot 5)} - \frac{2 \cdot (2 \cdot 2 \cdot 2)}{3 \cdot 5 \cdot (2 \cdot 2 \cdot 2)} + \frac{7 \cdot (5)}{2 \cdot 2 \cdot 2 \cdot 3 \cdot (5)}$$

$$= -\frac{70}{120} - \frac{16}{120} + \frac{35}{120}$$ The fractions are now like fractions.

$$= \frac{-70 - 16 + 35}{120}$$ Add and subtract terms in the numerator.

$$= \frac{-51}{120}$$

$$= -\frac{\overset{17}{\cancel{51}}}{\underset{40}{\cancel{120}}}$$ Simplify to lowest terms. The numerator and denominator share a common factor of 3.

$$= -\frac{17}{40}$$

Skill Practice Add and subtract as indicated.

8. $-\dfrac{7}{18} - \dfrac{4}{15} + \dfrac{7}{30}$

Example 8   **Adding and Subtracting Fractions with Variables**

Add or subtract as indicated.   **a.** $\dfrac{4}{5} + \dfrac{3}{x}$ **b.** $\dfrac{7}{x} - \dfrac{3}{x^2}$

Solution:

a. $\dfrac{4}{5} + \dfrac{3}{x} = \dfrac{4 \cdot x}{5 \cdot x} + \dfrac{3 \cdot 5}{x \cdot 5}$ The LCD is $5x$. Multiply by the missing factor from each denominator.

$$= \dfrac{4x}{5x} + \dfrac{15}{5x}$$ Simplify each fraction.

$$= \dfrac{4x + 15}{5x}$$ The numerator cannot be simplified further because $4x$ and 15 are not *like* terms.

b. $\dfrac{7}{x} - \dfrac{3}{x^2} = \dfrac{7 \cdot x}{x \cdot x} - \dfrac{3}{x^2}$ The LCD is x^2.

$$= \dfrac{7x}{x^2} - \dfrac{3}{x^2}$$ Simplify each fraction.

$$= \dfrac{7x - 3}{x^2}$$

> **Avoiding Mistakes**
>
> The fraction $\frac{4x+15}{5x}$ cannot be simplified because the numerator is a sum and not a product. Only *factors* can be "divided out" when simplifying a fraction.

Skill Practice Add or subtract as indicated.

9. $\dfrac{7}{8} + \dfrac{5}{z}$ 10. $\dfrac{11}{t^2} - \dfrac{4}{t}$

Answers

8. $-\dfrac{19}{45}$ 9. $\dfrac{7z + 40}{8z}$

10. $\dfrac{11 - 4t}{t^2}$

3. Applications of Addition and Subtraction of Fractions

Example 9 — Applying Operations on Unlike Fractions

A new Kelly Safari SUV tire has $\frac{7}{16}$-in. tread. After being driven 50,000 mi, the tread depth has worn down to $\frac{7}{32}$ in. By how much has the tread depth worn away?

Tread

Solution:

In this case, we are looking for the difference in the tread depth.

$$\begin{pmatrix}\text{Difference in}\\\text{tread depth}\end{pmatrix} = \begin{pmatrix}\text{original}\\\text{tread depth}\end{pmatrix} - \begin{pmatrix}\text{final}\\\text{tread depth}\end{pmatrix}$$

$= \dfrac{7}{16} - \dfrac{7}{32}$ The LCD is 32.

$= \dfrac{7 \cdot 2}{16 \cdot 2} - \dfrac{7}{32}$ Multiply numerator and denominator by 2 because $16 \cdot 2 = 32$.

$= \dfrac{14}{32} - \dfrac{7}{32}$ The fractions are now like.

$= \dfrac{7}{32}$ Subtract.

The tire lost $\frac{7}{32}$ inches in tread depth after 50,000 mi of driving.

TIP: You can check your result by adding the final tread depth to the difference in tread depth to get the original tread depth.

$$\dfrac{7}{32} + \dfrac{7}{32} = \dfrac{14}{32} = \dfrac{7}{16}$$

Skill Practice

11. On Monday, $\frac{2}{3}$ in. of rain fell on a certain town. On Tuesday, $\frac{1}{5}$ in. of rain fell. How much more rain fell on Monday than on Tuesday?

Example 10 — Finding Perimeter

A parcel of land has the following dimensions. Find the perimeter.

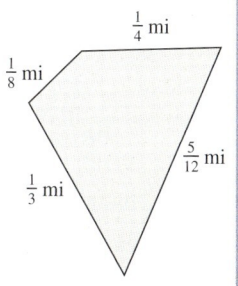

Solution:

To find the perimeter, add the lengths of the sides.

$\dfrac{1}{8} + \dfrac{1}{4} + \dfrac{5}{12} + \dfrac{1}{3}$ The LCD is 24.

$= \dfrac{1 \cdot 3}{8 \cdot 3} + \dfrac{1 \cdot 6}{4 \cdot 6} + \dfrac{5 \cdot 2}{12 \cdot 2} + \dfrac{1 \cdot 8}{3 \cdot 8}$ Convert to like fractions.

$= \dfrac{3}{24} + \dfrac{6}{24} + \dfrac{10}{24} + \dfrac{8}{24}$

$= \dfrac{27}{24}$ Add the fractions.

$= \dfrac{9}{8}$ or $1\dfrac{1}{8}$ Simplify to lowest terms.

The perimeter is $1\frac{1}{8}$ mi.

FOR REVIEW

Recall that addition is commutative. The lengths of the sides may be added in any order.

Skill Practice

12. Twelve members of a college hiking club hiked the perimeter of a canyon. How far did they hike?

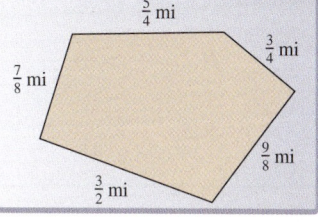

Answers

11. $\dfrac{7}{15}$ in.

12. They hiked $\dfrac{11}{2}$ mi. or equivalently $5\frac{1}{2}$ mi.

Section 4.5 Activity

A.1. a. Shade the third figure by adding the shaded regions of the first and second figure. Write the fractions and the sum illustrated.

b. What do you notice about the denominators?

For Exercises A.2–A.5, add or subtract as indicated. Write the answer as a fraction.

A.2. $\dfrac{15}{7} + \dfrac{2}{7}$ **A.3.** $\dfrac{8}{5} - \dfrac{2}{5}$ **A.4.** $-\dfrac{11}{21} + \dfrac{3}{21}$ **A.5.** $\dfrac{1}{y} - \dfrac{12}{y}$

A.6. a. Identify the least common denominator of the fractions $\dfrac{5}{8}$ and $\dfrac{3}{4}$.
b. Write each fraction as an equivalent fraction with the denominator from part (a).
c. Add the fractions and express the sum in lowest terms.
$$\dfrac{5}{8} + \dfrac{3}{4}$$

For Exercises A.7–A.11, add or subtract as indicated. Write the answer as a fraction in lowest terms.

A.7. $\dfrac{11}{14} + \left(-\dfrac{2}{7}\right)$ **A.8.** $\dfrac{1}{6} - \dfrac{9}{8}$ **A.9.** $\dfrac{5}{12} + \dfrac{4}{15}$

A.10. $\dfrac{7}{8} - \dfrac{1}{12}$ **A.11.** $-\dfrac{3}{5} + \left(-\dfrac{5}{8}\right)$

A.12. a. Identify the least common denominator of the fractions $\dfrac{6}{x}$ and $\dfrac{5}{x^2}$.
b. Write each fraction as an equivalent fraction with the denominator from part (a).
c. Add the fractions.
$$\dfrac{6}{x} + \dfrac{5}{x^2}$$

For Exercises A.13–A.14, add or subtract as indicated.

A.13. $\dfrac{2}{3} - \dfrac{5}{y}$ **A.14.** $\dfrac{10}{3z} + \dfrac{1}{z}$

A.15. On a submarine sandwich, Candice had $\dfrac{1}{10}$ lb of turkey, $\dfrac{3}{10}$ lb of ham, and $\dfrac{1}{10}$ lb of salami. How much meat was on the sandwich?

A.16. A decorated two-layer cake stands $\dfrac{17}{8}$ in. tall. If there is $\dfrac{1}{8}$ in. of frosting on the top and $\dfrac{1}{8}$ in. of frosting between the two layers of cake, how many inches thick is the cake itself?

Section 4.5 Practice Exercises

Prerequisite Review

For Exercises R.1–R.6, rewrite each fraction with the indicated denominator.

R.1. $-\dfrac{3}{5} = -\dfrac{}{15}$ **R.2.** $-\dfrac{5}{7} = -\dfrac{}{21}$ **R.3.** $\dfrac{8}{1} = \dfrac{}{12}$

R.4. $\dfrac{3}{1} = \dfrac{}{17}$ **R.5.** $\dfrac{2}{5x} = \dfrac{}{15x^2}$ **R.6.** $\dfrac{12}{7x} = \dfrac{}{14xy}$

Section 4.5 Addition and Subtraction of Fractions   227

Vocabulary and Key Concepts

1. **a.** Two fractions are _____ fractions if they have the same denominator.
   b. To add or subtract fractions, a common denominator (is/is not) needed.
   c. To multiply or divide fractions, a common denominator (is/is not) needed.

2. When adding or subtracting fractions, you (do/do not) add the denominators.

Concept 1: Addition and Subtraction of Like Fractions

For Exercises 3–7, identify whether the fractions are like or unlike fractions.

3. $\dfrac{9}{10}$ and $\dfrac{9}{20}$ 4. $-\dfrac{4}{1}$ and $-\dfrac{1}{4}$ 5. $\dfrac{3}{7}$ and $-\dfrac{5}{7}$

6. $-\dfrac{1}{xy}$ and $\dfrac{8}{xy}$ 7. $\dfrac{7}{ab^2}$ and $\dfrac{2}{ab}$

For Exercises 8 and 9, shade in the portion of the third figure that represents the addition of the first two figures.

8.    9.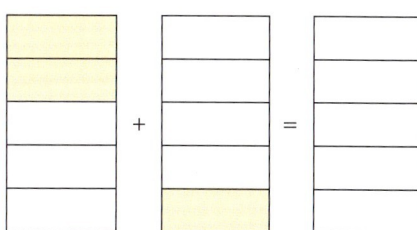

For Exercises 10 and 11, shade in the portion of the third figure that represents the subtraction of the first two figures.

10. 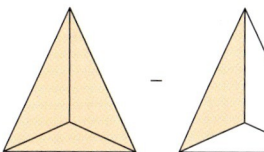   11.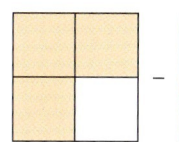

12. Explain the difference between evaluating the two expressions $\frac{2}{5} \cdot \frac{7}{5}$ and $\frac{2}{5} + \frac{7}{5}$.

For Exercises 13–28, add or subtract as indicated. Write the answer as a fraction in lowest terms or as an integer.
(See Examples 1 and 2.)

13. $\dfrac{7}{8} + \dfrac{5}{8}$ 14. $\dfrac{1}{21} + \dfrac{13}{21}$ 15. $\dfrac{23}{12} - \dfrac{15}{12}$ 16. $\dfrac{13}{6} - \dfrac{5}{6}$

17. $\dfrac{18}{14} + \dfrac{11}{14} + \dfrac{6}{14}$ 18. $\dfrac{7}{18} + \dfrac{22}{18} + \dfrac{10}{18}$ 19. $\dfrac{14}{15} + \dfrac{2}{15} - \dfrac{4}{15}$ 20. $\dfrac{19}{6} - \dfrac{11}{6} + \dfrac{5}{6}$

21. $-\dfrac{7}{2} + \dfrac{3}{2} - \dfrac{1}{2}$ 22. $-\dfrac{8}{3} - \dfrac{2}{3} + \dfrac{1}{3}$ 23. $\dfrac{5}{12} - \dfrac{19}{12} - \dfrac{7}{12}$ 24. $\dfrac{5}{18} - \dfrac{7}{18} - \dfrac{13}{18}$

25. $\dfrac{3y}{2w} + \dfrac{5}{2w}$ 26. $\dfrac{11a}{4b} + \dfrac{7}{4b}$ 27. $\dfrac{x}{5y} - \dfrac{3x}{5y}$ 28. $\dfrac{a}{9x} - \dfrac{5a}{9x}$

Concept 2: Addition and Subtraction of Unlike Fractions

For Exercises 29–64, add or subtract. Write the answer as a fraction simplified to lowest terms. (See Examples 3–8.)

29. $\dfrac{7}{8} + \dfrac{5}{16}$

30. $\dfrac{2}{9} + \dfrac{1}{18}$

31. $\dfrac{1}{15} + \dfrac{1}{10}$

32. $\dfrac{5}{6} + \dfrac{3}{8}$

33. $\dfrac{5}{6} + \dfrac{8}{7}$

34. $\dfrac{2}{11} + \dfrac{4}{5}$

35. $\dfrac{7}{8} - \dfrac{1}{2}$

36. $\dfrac{9}{10} - \dfrac{4}{5}$

37. $\dfrac{13}{12} - \dfrac{3}{4}$

38. $\dfrac{29}{30} - \dfrac{7}{10}$

39. $-\dfrac{10}{9} - \dfrac{5}{12}$

40. $-\dfrac{7}{6} - \dfrac{1}{15}$

41. $\dfrac{9}{8} - 2$

42. $\dfrac{11}{9} - 3$

43. $4 - \dfrac{4}{3}$

44. $2 - \dfrac{3}{8}$

45. $-\dfrac{16}{7} + 2$

46. $-\dfrac{15}{4} + 3$

47. $-\dfrac{1}{10} - \left(-\dfrac{9}{100}\right)$

48. $-\dfrac{3}{100} - \left(-\dfrac{21}{1000}\right)$

49. $\dfrac{3}{10} + \dfrac{9}{100} + \dfrac{1}{1000}$

50. $\dfrac{1}{10} + \dfrac{3}{100} + \dfrac{7}{1000}$

51. $\dfrac{5}{3} - \dfrac{7}{6} + \dfrac{5}{8}$

52. $\dfrac{7}{12} - \dfrac{2}{15} + \dfrac{5}{18}$

53. $-\dfrac{7}{10} - \dfrac{1}{20} - \left(-\dfrac{5}{8}\right)$

54. $\dfrac{1}{12} - \dfrac{3}{5} - \left(-\dfrac{3}{10}\right)$

55. $\dfrac{1}{2} + \dfrac{1}{4} - \dfrac{1}{8} - \dfrac{1}{16}$

56. $\dfrac{1}{3} - \dfrac{1}{9} + \dfrac{1}{27} - \dfrac{1}{81}$

57. $\dfrac{3}{4} + \dfrac{2}{x}$

58. $\dfrac{9}{5} + \dfrac{3}{y}$

59. $\dfrac{10}{x} + \dfrac{7}{y}$

60. $\dfrac{7}{a} + \dfrac{8}{b}$

61. $\dfrac{10}{x} - \dfrac{2}{x^2}$

62. $\dfrac{11}{z} - \dfrac{8}{z^2}$

63. $\dfrac{5}{3x} - \dfrac{2}{3}$

64. $\dfrac{13}{5t} - \dfrac{4}{5}$

Concept 3: Applications of Addition and Subtraction of Fractions

65. When doing her laundry, Inez added $\tfrac{3}{4}$ cup of bleach to $\tfrac{3}{8}$ cup of liquid detergent. How much total liquid is added to her wash?

66. What is the smallest possible length of screw needed to pass through two pieces of wood, one that is $\tfrac{7}{8}$ in. thick and one that is $\tfrac{1}{2}$ in. thick?

67. Before a storm, a rain gauge has $\tfrac{1}{8}$ in. of water. After the storm, the gauge has $\tfrac{9}{32}$ in. How many inches of rain did the storm deliver? (See Example 9.)

68. In one week it rained $\tfrac{5}{16}$ in. If a garden needs $\tfrac{9}{8}$ in. of water per week, how much more water does it need?

Santasan/Shutterstock

69. The information in the graph shows the distribution of a college student body by class.

   a. What fraction of the student body consists of upper classmen (juniors and seniors)?

   b. What fraction of the student body consists of freshmen and sophomores?

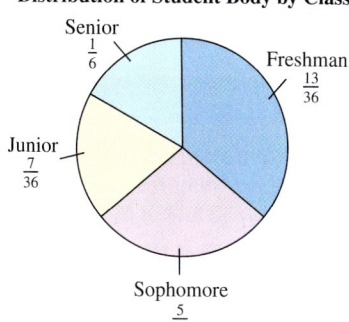
Distribution of Student Body by Class

70. A group of college students took part in a survey. One of the survey questions read:

"Do you think the government should spend more money on research to produce alternative forms of fuel?"

The results of the survey are shown in the figure.

   a. What fraction of the survey participants chose to strongly agree or agree?

   b. What fraction of the survey participants chose to strongly disagree or disagree?

For Exercises 71 and 72, find the perimeter. **(See Example 10.)**

71.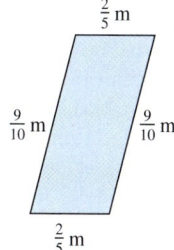

72.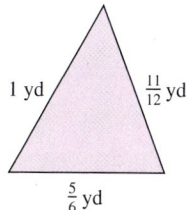

For Exercises 73 and 74, find the missing dimensions. Then calculate the perimeter.

73.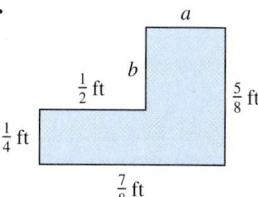

74.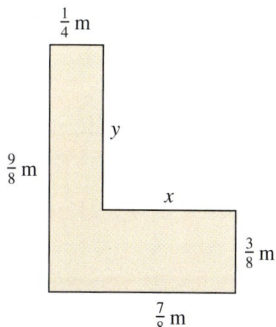

Expanding Your Skills

75. Which fraction is closest to $\frac{1}{2}$?

    a. $\frac{3}{4}$ **b.** $\frac{7}{10}$ **c.** $\frac{5}{6}$

76. Which fraction is closest to $\frac{3}{4}$?

    a. $\frac{5}{8}$ **b.** $\frac{7}{12}$ **c.** $\frac{5}{6}$

Section 4.6 Estimation and Operations on Mixed Numbers

Concepts

1. Multiplication and Division of Mixed Numbers
2. Addition of Mixed Numbers
3. Subtraction of Mixed Numbers
4. Addition and Subtraction of Negative Mixed Numbers
5. Applications of Mixed Numbers

1. Multiplication and Division of Mixed Numbers

Recall that to multiply fractions, we write the product of the numerators over the product of the denominators and then simplify. In order to apply this procedure to mixed numbers, we must first convert the mixed numbers to fractions.

> **Multiplying Mixed Numbers**
> **Step 1** Change each mixed number to an improper fraction.
> **Step 2** Multiply the improper fractions and simplify to lowest terms, if possible.
>
> An answer greater than or equal to 1 may be written as an improper fraction or as a mixed number, depending on the directions of the problem.

Example 1 Multiplying Mixed Numbers

Multiply and write the answer as a mixed number or integer.

a. $7\frac{1}{2} \cdot 4\frac{2}{3}$ **b.** $-12 \cdot \left(1\frac{7}{9}\right)$

FOR REVIEW

To write a mixed number as an improper fraction, first multiply the whole number by the denominator. Add the result to the numerator, then write the sum over the denominator.

$7\frac{1}{2} = \frac{7 \cdot 2 + 1}{2} = \frac{14 + 1}{2} = \frac{15}{2}$

$4\frac{2}{3} = \frac{4 \cdot 3 + 2}{3} = \frac{12 + 2}{3} = \frac{14}{3}$

Solution:

a. $7\frac{1}{2} \cdot 4\frac{2}{3} = \frac{15}{2} \cdot \frac{14}{3}$ Write each mixed number as an improper fraction.

$= \frac{\overset{5}{\cancel{15}}}{\underset{1}{\cancel{2}}} \cdot \frac{\overset{7}{\cancel{14}}}{\underset{1}{\cancel{3}}}$ Simplify.

$= \frac{35}{1}$ Multiply.

$= 35$

Avoiding Mistakes

Do not try to multiply mixed numbers by multiplying the whole-number parts and multiplying the fractional parts. You will not get the correct answer.

For the expression $7\frac{1}{2} \cdot 4\frac{2}{3}$, it would be incorrect to multiply $(7)(4)$ and $\frac{1}{2} \cdot \frac{2}{3}$. Notice that these values do not equal 35.

b. $-12 \cdot \left(1\frac{7}{9}\right) = -\frac{12}{1} \cdot \frac{16}{9}$ Write the integer and mixed number as improper fractions.

$= -\frac{\overset{4}{\cancel{12}}}{1} \cdot \frac{16}{\underset{3}{\cancel{9}}}$ Simplify.

$= -\frac{64}{3}$ Multiply.

$= -21\frac{1}{3}$ Write the improper fraction as a mixed number.

$$\begin{array}{r} 21 \\ 3\overline{)64} \\ \underline{-6} \\ 4 \\ \underline{-3} \\ 1 \end{array}$$

Skill Practice Multiply and write the answer as a mixed number or integer.

1. $16\frac{1}{2} \cdot 3\frac{7}{11}$ **2.** $-10 \cdot \left(7\frac{1}{6}\right)$

Answers
1. 60   **2.** $-71\frac{2}{3}$

Section 4.6 Estimation and Operations on Mixed Numbers

To divide mixed numbers, we use the following steps.

> **Dividing Mixed Numbers**
> **Step 1** Change each mixed number to an improper fraction.
> **Step 2** Divide the improper fractions and simplify to lowest terms, if possible. Recall that to divide fractions, multiply the dividend by the reciprocal of the divisor.
>
> An answer greater than or equal to 1 may be written as an improper fraction or as a mixed number, depending on the directions of the problem.

Example 2   Dividing Mixed Numbers

Divide and write the answer as a mixed number.

$$7\frac{1}{2} \div 4\frac{2}{3}$$

Solution:

$$7\frac{1}{2} \div 4\frac{2}{3} = \frac{15}{2} \div \frac{14}{3}$$ Write the mixed numbers as improper fractions.

$$= \frac{15}{2} \cdot \frac{3}{14}$$ Multiply by the reciprocal of the divisor.

$$= \frac{45}{28}$$ Multiply.

$$= 1\frac{17}{28}$$ Write the improper fraction as a mixed number.

Avoiding Mistakes
Be sure to take the reciprocal of the divisor *after* the mixed number is changed to an improper fraction.

Skill Practice Divide and write the answer as a mixed number.

3. $10\frac{1}{3} \div 2\frac{5}{6}$

Example 3   Dividing Mixed Numbers

Divide and write the answers as mixed numbers.

a. $-6 \div \left(-5\frac{1}{7}\right)$ b. $13\frac{5}{6} \div (-7)$

Solution:

a. $-6 \div \left(-5\frac{1}{7}\right) = -\frac{6}{1} \div \left(-\frac{36}{7}\right)$ Write the integer and mixed number as improper fractions.

$$= -\frac{6}{1} \cdot \left(-\frac{7}{36}\right)$$ Multiply by the reciprocal of the divisor.

$$= -\frac{\overset{1}{6}}{1} \cdot \left(-\frac{7}{\underset{6}{36}}\right)$$ Simplify.

$$= \frac{7}{6}$$ Multiply.

$$= 1\frac{1}{6}$$ Write the improper fraction as a mixed number.

Answer
3. $3\frac{11}{17}$

b. $13\frac{5}{6} \div (-7) = \frac{83}{6} \div \left(-\frac{7}{1}\right)$ Write the integer and the mixed number as improper fractions.

$= \frac{83}{6} \cdot \left(-\frac{1}{7}\right)$ Multiply by the reciprocal of the divisor.

$= -\frac{83}{42}$ Multiply. The product is negative.

$= -1\frac{41}{42}$ Write the improper fraction as a mixed number.

Skill Practice Divide and write the answers as mixed numbers.

4. $-8 \div \left(-4\frac{4}{5}\right)$ **5.** $12\frac{4}{9} \div (-8)$

2. Addition of Mixed Numbers

Now we will learn to add and subtract mixed numbers. To find the sum of two or more mixed numbers, add the whole-number parts and add the fractional parts.

Example 4 Adding Mixed Numbers

Add. $1\frac{5}{9} + 2\frac{1}{9}$

Solution:

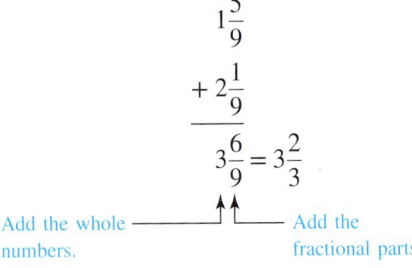

Add the whole numbers. Add the fractional parts.

The sum is $3\frac{2}{3}$.

TIP: To understand why mixed numbers can be added in this way, recall that $1\frac{5}{9} = 1 + \frac{5}{9}$ and $2\frac{1}{9} = 2 + \frac{1}{9}$. Therefore,

$1\frac{5}{9} + 2\frac{1}{9} = 1 + \frac{5}{9} + 2 + \frac{1}{9}$

$= 3 + \frac{6}{9}$

$= 3\frac{6}{9}$

$= 3\frac{2}{3}$

Skill Practice Add.

6. $7\frac{2}{15} + 2\frac{1}{15}$

When we perform operations on mixed numbers, it is often desirable to estimate the answer first. When rounding a mixed number, we offer the following convention.

Rounding Mixed Numbers

Step 1 If the fractional part of a mixed number is greater than or equal to $\frac{1}{2}$ (that is, if the numerator is half the denominator or greater), round to the next-greater whole number. For example: $6\frac{9}{16}$ and $6\frac{1}{2}$ both round to 7.

Step 2 If the fractional part of the mixed number is less than $\frac{1}{2}$ (that is, if the numerator is less than half the denominator), the mixed number rounds down to the whole number. For example: $6\frac{7}{16}$ rounds to 6.

Answers
4. $1\frac{2}{3}$ **5.** $-1\frac{5}{9}$ **6.** $9\frac{1}{5}$

Section 4.6 Estimation and Operations on Mixed Numbers   233

Example 5 Adding Mixed Numbers

Estimate the sum and then find the actual sum. $42\frac{1}{12} + 17\frac{7}{8}$

Solution:

To estimate the sum, first round the addends.

$42\frac{1}{12}$ rounds to 42

$+ 17\frac{7}{8}$ rounds to $+ 18$

$\phantom{+ 17\frac{7}{8} \text{ rounds to } +} 60$ The estimated value is 60.

To find the actual sum, we must first write the fractional parts as like fractions. The LCD is 24.

$$42\frac{1}{12} = 42\frac{1 \cdot 2}{12 \cdot 2} = 42\frac{2}{24}$$
$$+ 17\frac{7}{8} = + 17\frac{7 \cdot 3}{8 \cdot 3} = + 17\frac{21}{24}$$
$$\phantom{+ 17\frac{7}{8} = + 17\frac{7 \cdot 3}{8 \cdot 3} = +\ } 59\frac{23}{24}$$

The actual sum is $59\frac{23}{24}$. This is close to our estimate of 60.

Skill Practice Estimate the sum and then find the actual sum.

7. $6\frac{1}{11} + 3\frac{1}{2}$

Example 6 Adding Mixed Numbers with Carrying

Estimate the sum and then find the actual sum. $7\frac{5}{6} + 3\frac{3}{5}$

Solution:

$7\frac{5}{6}$ rounds to 8

$+ 3\frac{3}{5}$ rounds to $+ 4$

$\phantom{+ 3\frac{3}{5} \text{ rounds to } +} 12$ The estimated value is 12.

To find the actual sum, first write the fractional parts as like fractions. The LCD is 30.

$$7\frac{5}{6} = 7\frac{5 \cdot 5}{6 \cdot 5} = 7\frac{25}{30}$$
$$+ 3\frac{3}{5} = + 3\frac{3 \cdot 6}{5 \cdot 6} = + 3\frac{18}{30}$$
$$\phantom{+ 3\frac{3}{5} = + 3\frac{3 \cdot 6}{5 \cdot 6} = +\ } 10\frac{43}{30}$$

Notice that the number $\frac{43}{30}$ is an improper fraction. By convention, a mixed number is written as a whole number and a *proper* fraction. We have $\frac{43}{30} = 1\frac{13}{30}$. Therefore,

$$10\frac{43}{30} = 10 + 1\frac{13}{30} = 11\frac{13}{30}$$

The sum is $11\frac{13}{30}$. This is close to our estimate of 12.

Skill Practice Estimate the sum and then find the actual sum.

8. $5\frac{2}{5} + 7\frac{8}{9}$

Answers
7. Estimate: 10; actual sum: $9\frac{13}{22}$
8. Estimate: 13; actual sum: $13\frac{13}{45}$

234 Chapter 4 Fractions and Mixed Numbers

We have shown how to add mixed numbers by writing the numbers in columns. Another approach to add or subtract mixed numbers is to write the numbers first as improper fractions. Then add or subtract the fractions. To demonstrate this process, we add the mixed numbers from Example 6.

Example 7 Adding Mixed Numbers by Using Improper Fractions

Add. $7\frac{5}{6} + 3\frac{3}{5}$

Solution:

$7\frac{5}{6} + 3\frac{3}{5} = \frac{47}{6} + \frac{18}{5}$ Write each mixed number as an improper fraction.

$= \frac{47 \cdot 5}{6 \cdot 5} + \frac{18 \cdot 6}{5 \cdot 6}$ Convert the fractions to like fractions. The LCD is 30.

$= \frac{235}{30} + \frac{108}{30}$ The fractions are now like fractions.

$= \frac{343}{30}$ Add the like fractions.

$= 11\frac{13}{30}$ Convert the improper fraction to a mixed number.

$$\begin{array}{r} 11 \\ 30\overline{)343} \\ -30 \\ \hline 43 \\ -30 \\ \hline 13 \end{array} \quad 11\tfrac{13}{30}$$

The mixed number $11\frac{13}{30}$ is the same as the value obtained in Example 6.

Skill Practice Add the mixed numbers by first converting the addends to improper fractions. Write the answer as a mixed number.

9. $12\frac{1}{3} + 4\frac{3}{4}$

3. Subtraction of Mixed Numbers

To subtract mixed numbers, we subtract the fractional parts and subtract the whole-number parts.

Example 8 Subtracting Mixed Numbers

Estimate the difference and then find the actual difference.

$$15\frac{2}{3} - 4\frac{1}{6}$$

Solution:

To estimate the difference, first round to the nearest whole number.

$15\frac{2}{3}$ rounds to 16

$-4\frac{1}{6}$ rounds to -4

$\phantom{-4\frac{1}{6}\text{ rounds to }}12$ The estimated value is 12.

Answer

9. $17\frac{1}{12}$

To subtract the fractional parts, we need a common denominator. The LCD is 6.

$$15\frac{2}{3} = 15\frac{2\cdot 2}{3\cdot 2} = 15\frac{4}{6}$$
$$-4\frac{1}{6} = -4\frac{1}{6} = -4\frac{1}{6}$$
$$\phantom{-4\frac{1}{6} = -4\frac{1}{6} =} 11\frac{3}{6} = 11\frac{1}{2}$$

Subtract the whole numbers. ⟶ ⟵ Subtract the fractional parts.

The difference is $11\frac{1}{2}$. This is close to the estimate of 12.

Skill Practice Estimate the difference then find the actual difference.

10. $6\frac{3}{4} - 2\frac{1}{3}$

Borrowing is sometimes necessary when subtracting mixed numbers.

Example 9 Subtracting Mixed Numbers with Borrowing

Subtract.

a. $17\frac{2}{7} - 11\frac{5}{7}$ **b.** $14\frac{2}{9} - 9\frac{3}{5}$

Solution:

a. We will subtract $\frac{5}{7}$ from $\frac{2}{7}$ by borrowing 1 from the whole number 17. The borrowed 1 is written as $\frac{7}{7}$ because the common denominator is 7.

$$17\frac{2}{7} = \overset{16}{\cancel{17}}\frac{2}{7} + \frac{7}{7} = 16\frac{9}{7}$$
$$-11\frac{5}{7} = -11\frac{5}{7} = -11\frac{5}{7}$$
$$\phantom{-11\frac{5}{7} = -11\frac{5}{7} =} 5\frac{4}{7}$$

The difference is $5\frac{4}{7}$.

b. To subtract the fractional parts, we need a common denominator. The LCD is 45.

$$14\frac{2}{9} = 14\frac{2\cdot 5}{9\cdot 5} = 14\frac{10}{45}$$
$$-9\frac{3}{5} = -9\frac{3\cdot 9}{5\cdot 9} = -9\frac{27}{45}$$

We will subtract $\frac{27}{45}$ from $\frac{10}{45}$ by borrowing. Therefore, borrow 1 (or equivalently $\frac{45}{45}$) from 14.

$$= \overset{13}{\cancel{14}}\frac{10}{45} + \frac{45}{45} = 13\frac{55}{45}$$
$$= -9\frac{27}{45} = -9\frac{27}{45}$$
$$\phantom{= -9\frac{27}{45} =} 4\frac{28}{45}$$

The difference is $4\frac{28}{45}$.

Skill Practice Subtract.

11. $24\frac{2}{7} - 8\frac{5}{7}$ **12.** $9\frac{2}{3} - 8\frac{3}{4}$

Answers

10. $5; 4\frac{5}{12}$ **11.** $15\frac{4}{7}$ **12.** $\frac{11}{12}$

Chapter 4 Fractions and Mixed Numbers

Example 10 Subtracting Mixed Numbers with Borrowing

Subtract. $4 - 2\dfrac{5}{8}$

Solution:

$$\begin{array}{r} 4 \\ -2\dfrac{5}{8} \\ \hline \end{array}$$

In this case, we have no fractional part from which to subtract.

$$\begin{array}{r} 4\dfrac{\overset{3}{\cancel{8}}}{8} \\ -2\dfrac{5}{8} \\ \hline 1\dfrac{3}{8} \end{array}$$

We can borrow 1 or equivalently $\tfrac{8}{8}$ from the whole number 4.

TIP: The borrowed 1 is written as $\tfrac{8}{8}$ because the common denominator is 8.

The difference is $1\tfrac{3}{8}$.

TIP: The subtraction operation $4 - 2\tfrac{5}{8} = 1\tfrac{3}{8}$ can be checked by adding:

$$1\dfrac{3}{8} + 2\dfrac{5}{8} = 3\dfrac{8}{8} = 3 + 1 = 4 \checkmark$$

Skill Practice Subtract.

13. $10 - 3\dfrac{1}{6}$

In Example 11, we show the alternative approach to subtract mixed numbers by first writing each mixed number as an improper fraction.

Example 11 Subtracting Mixed Numbers by Using Improper Fractions

Subtract by first converting to improper fractions. Write the answer as a mixed number.

$$10\dfrac{2}{5} - 4\dfrac{3}{4}$$

Solution:

$$10\dfrac{2}{5} - 4\dfrac{3}{4} = \dfrac{52}{5} - \dfrac{19}{4}$$

Write each mixed number as an improper fraction.

$$= \dfrac{52 \cdot 4}{5 \cdot 4} - \dfrac{19 \cdot 5}{4 \cdot 5}$$

Convert the fractions to like fractions. The LCD is 20.

$$= \dfrac{208}{20} - \dfrac{95}{20}$$

Subtract the like fractions.

$$= \dfrac{113}{20}$$

$$= 5\dfrac{13}{20}$$

Write the result as a mixed number.

$$\begin{array}{r} 5 \\ 20\overline{\smash{)}113} \\ -100 \\ \hline 13 \end{array}$$

Skill Practice Subtract by first converting to improper fractions. Write the answer as a mixed number.

14. $8\dfrac{2}{9} - 3\dfrac{5}{6}$

Answers

13. $6\dfrac{5}{6}$ **14.** $4\dfrac{7}{18}$

As you can see from Examples 7 and 11, when we convert mixed numbers to improper fractions, the numerators of the fractions become larger numbers. Thus, we must add (or subtract) larger numerators than if we had used the method involving columns. This is one drawback. However, an advantage of converting to improper fractions first is that there is no need for carrying or borrowing.

4. Addition and Subtraction of Negative Mixed Numbers

In Examples 4–11, we have shown two methods for adding and subtracting mixed numbers. The method of changing mixed numbers to improper fractions is preferable when adding or subtracting negative mixed numbers.

Example 12   Adding and Subtracting Signed Mixed Numbers

Add or subtract as indicated. Write the answer as a fraction.

a. $-3\frac{1}{2} - \left(-6\frac{2}{3}\right)$ **b.** $-4\frac{2}{3} - 1\frac{1}{6} + 2\frac{3}{4}$

Solution:

a. $-3\frac{1}{2} - \left(-6\frac{2}{3}\right) = -3\frac{1}{2} + 6\frac{2}{3}$ Rewrite subtraction in terms of addition.

$= -\frac{7}{2} + \frac{20}{3}$ Change the mixed numbers to improper fractions.

$= -\frac{7 \cdot 3}{2 \cdot 3} + \frac{20 \cdot 2}{3 \cdot 2}$ The LCD is 6.

$= -\frac{21}{6} + \frac{40}{6}$ Simplify.

$= \frac{-21 + 40}{6}$ Add the numerators. Note that the fraction $-\frac{21}{6}$ can be written as $\frac{-21}{6}$.

$= \frac{19}{6}$ Simplify.

b. $-4\frac{2}{3} - 1\frac{1}{6} + 2\frac{3}{4} = -\frac{14}{3} - \frac{7}{6} + \frac{11}{4}$ Change the mixed numbers to improper fractions.

$= -\frac{14 \cdot 4}{3 \cdot 4} - \frac{7 \cdot 2}{6 \cdot 2} + \frac{11 \cdot 3}{4 \cdot 3}$ The LCD is 12.

$= -\frac{56}{12} - \frac{14}{12} + \frac{33}{12}$ Change each fraction to an equivalent fraction with denominator 12.

$= \frac{-56 - 14 + 33}{12}$ Add and subtract the numerators.

$= \frac{-37}{12}$ or $-\frac{37}{12}$ Simplify.

Skill Practice   Add or subtract as indicated. Write the answers as fractions.

15. $-5\frac{3}{4} - \left(-1\frac{1}{2}\right)$ **16.** $-2\frac{5}{8} + 4\frac{3}{4} - 3\frac{1}{2}$

Answers

15. $-\frac{17}{4}$ **16.** $-\frac{11}{8}$

5. Applications of Mixed Numbers

Example 13   Subtracting Mixed Numbers in an Application

The average height of a 3-year-old girl is $38\frac{1}{3}$ in. The average height of a 4-year-old girl is $41\frac{3}{4}$ in. On average, by how much does a girl grow between the ages of 3 and 4?

Solution:

We use subtraction to find the difference in heights.

$$41\frac{3}{4} = 41\frac{3 \cdot 3}{4 \cdot 3} = 41\frac{9}{12}$$
$$-38\frac{1}{3} = -38\frac{1 \cdot 4}{3 \cdot 4} = -38\frac{4}{12}$$
$$\overline{\phantom{-38\frac{1}{3}}} \phantom{= -38\frac{1 \cdot 4}{3 \cdot 4} =} \;\; 3\frac{5}{12}$$

The average amount of growth is $3\frac{5}{12}$ in.

Skill Practice

17. On December 1, the snow base at the Bear Mountain Ski Resort was $4\frac{1}{3}$ ft. By January 1, the base was $6\frac{1}{2}$ ft. By how much did the base amount of snow increase?

Example 14   Dividing Mixed Numbers in an Application

A construction site brings in $6\frac{2}{3}$ tons of soil. Each truck holds $\frac{2}{3}$ ton. How many truckloads are necessary?

Solution:

The $6\frac{2}{3}$ tons of soil must be distributed in $\frac{2}{3}$-ton increments. This will require division.

$$6\frac{2}{3} \div \frac{2}{3} = \frac{20}{3} \div \frac{2}{3} \quad \text{Write the mixed number as an improper fraction.}$$

$$= \frac{20}{3} \cdot \frac{3}{2} \quad \text{Multiply by the reciprocal of the divisor.}$$

$$= \frac{\overset{10}{\cancel{20}}}{\underset{1}{\cancel{3}}} \cdot \frac{\overset{1}{\cancel{3}}}{\underset{1}{\cancel{2}}} \quad \text{Simplify.}$$

$$= \frac{10}{1} \quad \text{Multiply.}$$

$$= 10$$

A total of 10 truckloads of soil will be required.

Skill Practice

18. A department store wraps packages for $2 each. Ribbon $2\frac{5}{8}$ ft long is used to wrap each package. How many packages can be wrapped from a roll of ribbon 168 ft long?

Answers

17. $2\frac{1}{6}$ ft   **18.** 64 packages

Practice Exercises

Section 4.6

Prerequisite Review

For Exercises R.1–R.8, perform the indicated operation. Write the answers as fractions.

R.1. $\dfrac{9}{5} + 3$ **R.2.** $8 + \dfrac{4}{7}$ **R.3.** $\dfrac{25}{8} - \dfrac{23}{24}$

R.4. $\dfrac{11}{15} - \dfrac{2}{5}$ **R.5.** $\left(-\dfrac{42}{11}\right)\left(-\dfrac{22}{7}\right)$ **R.6.** $\left(-\dfrac{8}{45}\right)\left(\dfrac{9}{4}\right)$

R.7. $\dfrac{52}{18} \div 26$ **R.8.** $42 \div \dfrac{3}{7}$

For Exercises R.9–R.10, write the mixed number as an improper fraction.

R.9. $3\dfrac{2}{5}$ **R.10.** $4\dfrac{3}{8}$

For Exercises R.11–R.12, write the improper fraction as a mixed number.

R.11. $\dfrac{77}{6}$ **R.12.** $\dfrac{41}{9}$

Concept 1: Multiplication and Division of Mixed Numbers

For Exercises 1–20, multiply or divide the mixed numbers. Write the answer as a mixed number or an integer. (See Examples 1–3.)

1. $\left(2\dfrac{2}{5}\right)\left(3\dfrac{1}{12}\right)$
2. $\left(5\dfrac{1}{5}\right)\left(3\dfrac{3}{4}\right)$
3. $-2\dfrac{1}{3} \cdot \left(-6\dfrac{3}{5}\right)$
4. $-6\dfrac{1}{8} \cdot \left(-2\dfrac{3}{4}\right)$
5. $(-9) \cdot 4\dfrac{2}{9}$
6. $(-6) \cdot 3\dfrac{1}{3}$
7. $\left(5\dfrac{3}{16}\right)\left(5\dfrac{1}{3}\right)$
8. $\left(8\dfrac{2}{3}\right)\left(2\dfrac{1}{13}\right)$
9. $\left(7\dfrac{1}{4}\right) \cdot 10$
10. $\left(2\dfrac{2}{3}\right) \cdot 3$
11. $4\dfrac{1}{2} \div 2\dfrac{1}{4}$
12. $5\dfrac{5}{6} \div 2\dfrac{1}{3}$
13. $5\dfrac{8}{9} \div \left(-1\dfrac{1}{3}\right)$
14. $12\dfrac{4}{5} \div \left(-2\dfrac{3}{5}\right)$
15. $-2\dfrac{1}{2} \div \left(-1\dfrac{1}{16}\right)$
16. $-7\dfrac{3}{5} \div \left(-1\dfrac{7}{12}\right)$
17. $2 \div 3\dfrac{1}{3}$
18. $6 \div 4\dfrac{2}{5}$
19. $8\dfrac{1}{4} \div (-3)$
20. $6\dfrac{2}{5} \div (-2)$

Concept 2: Addition of Mixed Numbers

For Exercises 21–28, add the mixed numbers. (See Examples 4 and 5.)

21. $2\dfrac{1}{11}$
    $+ 5\dfrac{3}{11}$

22. $5\dfrac{2}{7}$
    $+ 4\dfrac{3}{7}$

23. $12\dfrac{1}{14}$
    $+ 3\dfrac{5}{14}$

24. $1\dfrac{3}{20}$
    $+ 17\dfrac{7}{20}$

25. $4\dfrac{5}{16}$
    $+ 11\dfrac{1}{4}$

26. $21\dfrac{2}{9}$
    $+ 10\dfrac{1}{3}$

27. $6\dfrac{2}{3}$
    $+ 4\dfrac{1}{5}$

28. $7\dfrac{1}{6}$
    $+ 3\dfrac{5}{8}$

For Exercises 29–32, round the mixed number to the nearest whole number.

29. $5\dfrac{1}{3}$ 30. $2\dfrac{7}{8}$ 31. $1\dfrac{3}{5}$ 32. $6\dfrac{3}{7}$

For Exercises 33–36, write the mixed number in proper form (that is, as a whole number with a proper fraction that is simplified to lowest terms).

33. $2\dfrac{6}{5}$ 34. $4\dfrac{8}{7}$ 35. $7\dfrac{5}{3}$ 36. $1\dfrac{9}{5}$

For Exercises 37–42, round the numbers to estimate the answer. Then find the exact sum. In Exercise 37, the estimate is done for you. (See Examples 6 and 7.)

37. Estimate: 7, +8/15   Exact: $6\dfrac{3}{4}$, $+7\dfrac{3}{4}$

38. Estimate: ___, +___   Exact: $8\dfrac{3}{5}$, $+13\dfrac{4}{5}$

39. Estimate: ___, +___   Exact: $14\dfrac{7}{8}$, $+8\dfrac{1}{4}$

40. Estimate: ___, +___   Exact: $21\dfrac{3}{5}$, $+24\dfrac{9}{10}$

41. Estimate: ___, +___   Exact: $3\dfrac{7}{16}$, $+15\dfrac{11}{12}$

42. Estimate: ___, +___   Exact: $7\dfrac{7}{9}$, $+8\dfrac{5}{6}$

For Exercises 43–50, add the mixed numbers. Write the answer as a mixed number. (See Examples 6 and 7.)

43. $3\dfrac{3}{4} + 5\dfrac{2}{3}$ 44. $6\dfrac{5}{7} + 10\dfrac{3}{5}$ 45. $11\dfrac{5}{8} + \dfrac{7}{6}$ 46. $9\dfrac{5}{6} + \dfrac{3}{4}$

47. $3 + 6\dfrac{7}{8}$ 48. $5 + 11\dfrac{1}{13}$ 49. $124\dfrac{2}{3} + 46\dfrac{5}{6}$ 50. $345\dfrac{3}{5} + 84\dfrac{7}{10}$

Concept 3: Subtraction of Mixed Numbers

For Exercises 51–54, subtract the mixed numbers. (See Example 8.)

51. $21\dfrac{9}{10}$, $-10\dfrac{3}{10}$

52. $19\dfrac{2}{3}$, $-4\dfrac{1}{3}$

53. $18\dfrac{5}{6}$, $-6\dfrac{2}{3}$

54. $21\dfrac{17}{20}$, $-20\dfrac{1}{10}$

For Exercises 55–60, round the numbers to estimate the answer. Then find the exact difference. In Exercise 55, the estimate is done for you. (See Examples 9–11.)

55. Estimate: 25, −14/11   Exact: $25\dfrac{1}{4}$, $-13\dfrac{3}{4}$

56. Estimate: ___, −___   Exact: $36\dfrac{1}{5}$, $-12\dfrac{3}{5}$

57. Estimate: ___, −___   Exact: $17\dfrac{1}{6}$, $-15\dfrac{5}{12}$

	Estimate	Exact		Estimate	Exact		Estimate	Exact
58.	___	$22\frac{5}{18}$ $-10\frac{7}{9}$	59.	___	$46\frac{3}{7}$ $-38\frac{1}{2}$	60.	___	$23\frac{1}{2}$ $-18\frac{10}{13}$

For Exercises 61–68, subtract the mixed numbers. Write the answers as fractions or mixed numbers. (See Examples 9–11.)

61. $6 - 2\frac{5}{6}$ **62.** $9 - 4\frac{1}{2}$ **63.** $12 - 9\frac{2}{9}$ **64.** $10 - 9\frac{1}{3}$

65. $3\frac{7}{8} - 3\frac{3}{16}$ **66.** $3\frac{1}{6} - 1\frac{23}{24}$ **67.** $12\frac{1}{5} - 11\frac{2}{7}$ **68.** $10\frac{1}{8} - 2\frac{17}{18}$

Concept 4: Addition and Subtraction of Negative Mixed Numbers

For Exercises 69–76, add or subtract the mixed numbers. Write the answer as a mixed number. (See Example 12.)

69. $-4\frac{2}{7} - 3\frac{1}{2}$ **70.** $-9\frac{1}{5} - 2\frac{1}{10}$ **71.** $-4\frac{3}{8} + 7\frac{1}{4}$ **72.** $-5\frac{2}{11} + 8\frac{1}{2}$

73. $3\frac{1}{2} - \left(-2\frac{1}{4}\right)$ **74.** $6\frac{1}{6} - \left(-4\frac{1}{3}\right)$ **75.** $5 + \left(-7\frac{5}{6}\right)$ **76.** $6 + \left(-10\frac{3}{4}\right)$

Mixed Exercises

For Exercises 77–84, perform the indicated operations. Write the answers as fractions or integers.

77. $\left(3\frac{1}{5}\right)\left(-2\frac{1}{4}\right)$ **78.** $(-10)\left(-3\frac{2}{5}\right)$ **79.** $4\frac{1}{8} - 3\frac{5}{6}$ **80.** $-5\frac{5}{6} - 3\frac{1}{3}$

81. $-8\frac{1}{3} \div \left(-2\frac{1}{6}\right)$ **82.** $10\frac{2}{3} \div (-8)$ **83.** $-6\frac{3}{10} + 4\frac{5}{6} - \left(-3\frac{1}{2}\right)$ **84.** $4\frac{4}{5} - \left(-2\frac{1}{4}\right) - 1\frac{3}{10}$

Concept 5: Applications of Mixed Numbers

For Exercises 85–88, use the table to find the lengths of several common birds.

Bird	Length
Cuban Bee Hummingbird	$2\frac{1}{4}$ in.
Sedge Wren	$3\frac{1}{2}$ in.
Great Carolina Wren	$5\frac{1}{2}$ in.
Belted Kingfisher	$11\frac{1}{4}$ in.

USO/iStock/Getty Images

85. How much longer is the Belted Kingfisher than the Sedge Wren? (See Example 13.)

86. How much longer is the Great Carolina Wren than the Cuban Bee Hummingbird?

87. Estimate or measure the length of your index finger. Which is longer, your index finger or a Cuban Bee Hummingbird?

88. For a recent year, the smallest living dog in the United States was Brandy, a female Chihuahua, who measures 6 in. in length. How much longer is a Belted Kingfisher than Brandy?

89. A cellular device measures $5\frac{1}{2}$ in. by $2\frac{5}{8}$ in. Find the perimeter and area of the device.

90. Tabitha's living room is a rectangle that measures $12\frac{3}{4}$ ft by 12 ft. Find the perimeter and the area of the room.

91. According to the U.S. Census Bureau's Valentine's Day Press Release, the average American consumes $25\frac{7}{10}$ lb of chocolate in a year. Over the course of 25 years, how many pounds of chocolate would the average American consume?

92. Derrick is a middle distance runner for his college indoor track team. One day his coach suggests a workout in which Derrick runs 8 repetitions of $2\frac{1}{2}$ laps at full speed with a two-minute rest between repetitions. How many laps does he run in this workout?

93. A student has three part-time jobs. She tutors, delivers newspapers, and takes notes for a blind student. During a typical week she works $8\frac{2}{3}$ hr delivering newspapers, $4\frac{1}{2}$ hr tutoring, and $3\frac{3}{4}$ hr note-taking. What is the total number of hours worked in a typical week?

94. A contractor ordered three loads of gravel. The orders were for $2\frac{1}{2}$ tons, $3\frac{1}{8}$ tons, and $4\frac{1}{3}$ tons. What is the total amount of gravel ordered?

95. The age of a small kitten can be approximated by the following rule. The kitten's age is given as 1 week for every quarter pound of weight. **(See Example 14.)**

   a. Approximately how old is a $1\frac{3}{4}$-lb kitten?

   b. Approximately how old is a $2\frac{1}{8}$-lb kitten?

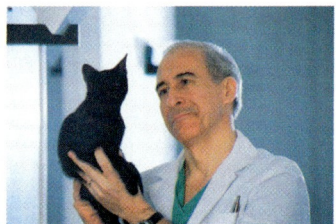

Comstock Images/Alamy Stock Photo

96. Richard's estate is to be split equally among his three children. If his estate is worth $\$1\frac{3}{4}$ million, how much will each child inherit?

97. In the summer at the South Pole, heavy equipment is used 24 hr a day. For every gallon of fuel actually used at the South Pole, it takes $3\frac{1}{2}$ gal to get it there.

   a. If in one day 130 gal of fuel was used, how many gallons of fuel did it take to transport the 130 gal?

   b. How much fuel was used in all?

98. A roll of wallpaper covers an area of 28 ft². If the roll is $1\frac{17}{24}$ ft wide, how long is the roll?

SW Productions/Brand X Pictures/Getty Images

99. A plumber fits together two pipes. Find the length of the larger piece.

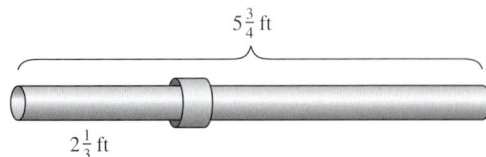

100. Find the thickness of the carpeting and pad.

101. When using a word processor, the default margins are $1\frac{1}{4}$ in. for the left and right margins. If an $8\frac{1}{2}$-in. by 11-in. piece of paper is used, what is the width of the printing area?

102. A water gauge in a pond measured $25\frac{7}{8}$ in. on Monday. After 2 days of rain and runoff, the gauge read $32\frac{1}{2}$ in. By how much did the water level rise?

103. A patient admitted to the hospital was dehydrated. In addition to intravenous (IV) fluids, the doctor told the patient that she must drink at least 4 L of an electrolyte solution within the next 12 hr. A nurse recorded the amounts the patient drank in the patient's chart.

   a. How many liters of electrolyte solution did the patient drink?

   b. How much more would the patient need to drink to reach 4 L?

Time	Amount
7 A.M.–10 A.M.	$1\frac{1}{4}$ L
10 A.M.–1 P.M.	$\frac{7}{8}$ L
1 P.M.–4 P.M.	$\frac{3}{4}$ L
4 P.M.–7 P.M.	$\frac{1}{2}$ L

104. Benjamin loves to hike in the White Mountains of New Hampshire. It takes him $4\frac{1}{2}$ hr to hike from the Pinkham Notch Visitor Center to the summit of Mt. Washington. If the round trip usually takes 8 hr, how long does it take for the return trip?

Technology Connections

For Exercises 105–116, use a calculator to perform the indicated operations and simplify. Write the answer as a mixed number.

Nico Schueler/Shutterstock

105. $12\frac{2}{3} \cdot 25\frac{1}{8}$

106. $38\frac{1}{3} \div 12\frac{1}{2}$

107. $56\frac{5}{6} \div 3\frac{1}{6}$

108. $25\frac{1}{5} \cdot 18\frac{1}{2}$

109. $\frac{23}{42} + \frac{17}{24}$

110. $\frac{14}{75} + \frac{9}{50}$

111. $\frac{31}{44} - \frac{14}{33}$

112. $\frac{29}{68} - \frac{7}{92}$

113. $32\frac{7}{18} + 14\frac{2}{27}$

114. $21\frac{3}{28} + 4\frac{31}{42}$

115. $7\frac{11}{21} - 2\frac{10}{33}$

116. $5\frac{14}{17} - 2\frac{47}{68}$

Expanding Your Skills

For Exercises 117–120, fill in the blank to complete the pattern.

117. $1, 1\frac{1}{3}, 1\frac{2}{3}, 2, 2\frac{1}{3}, \square$

118. $\frac{1}{4}, 1, 1\frac{3}{4}, 2\frac{1}{2}, 3\frac{1}{4}, \square$

119. $\frac{5}{6}, 1\frac{1}{6}, 1\frac{1}{2}, 1\frac{5}{6}, \square$

120. $\frac{1}{2}, 1\frac{1}{4}, 2, 2\frac{3}{4}, 3\frac{1}{2}, \square$

Problem Recognition Exercises

Operations on Fractions and Mixed Numbers

When performing operations on fractions, take particular care to note the operation symbol between the fractions (that is, +, −, ·, or ÷). Recall that with addition or subtraction, you need a common denominator, but with multiplication and division you do not. With division, be sure to take the reciprocal of the second fraction (that is, invert the second fraction) before multiplying. Furthermore, with operations on mixed numbers, you always have the option of converting the mixed numbers to improper fractions.

Addition

$-\frac{11}{6} + \frac{2}{3}$

$= -\frac{11}{6} + \frac{2 \cdot 2}{3 \cdot 2}$

$= -\frac{11}{6} + \frac{4}{6}$

$= \frac{-11 + 4}{6}$

$= \frac{-7}{6}$

$= -\frac{7}{6}$

Subtraction

$\frac{15}{4} - 3$

$= \frac{15}{4} - \frac{3}{1}$

$= \frac{15}{4} - \frac{3 \cdot 4}{1 \cdot 4}$

$= \frac{15}{4} - \frac{12}{4}$

$= \frac{15 - 12}{4}$

$= \frac{3}{4}$

Multiplication

$\frac{4}{5} \cdot \left(1\frac{1}{2}\right)$

$= \frac{4}{5} \cdot \frac{3}{2}$

$= \frac{\overset{2}{\cancel{4}}}{5} \cdot \frac{3}{\underset{1}{\cancel{2}}}$

$= \frac{6}{5}$ or $1\frac{1}{5}$

Division

$-6\frac{3}{8} \div \left(2\frac{1}{4}\right)$

$= -\frac{51}{8} \div \left(\frac{9}{4}\right)$

$= -\frac{51}{8} \cdot \frac{4}{9}$

$= -\frac{\overset{17}{\cancel{51}}}{\underset{2}{\cancel{8}}} \cdot \frac{\overset{1}{\cancel{4}}}{\underset{3}{\cancel{9}}}$

$= -\frac{17}{6}$ or $-2\frac{5}{6}$

For Exercises 1–10, perform the indicated operations. Estimate to check that your answer is reasonable.

1. a. $-\dfrac{7}{5} + \dfrac{2}{5}$ b. $-\dfrac{7}{5} \cdot \dfrac{2}{5}$ c. $-\dfrac{7}{5} \div \dfrac{2}{5}$ d. $-\dfrac{7}{5} - \dfrac{2}{5}$

2. a. $\dfrac{4}{3} \cdot \dfrac{5}{6}$ b. $\dfrac{4}{3} \div \dfrac{5}{6}$ c. $\dfrac{4}{3} + \dfrac{5}{6}$ d. $\dfrac{4}{3} - \dfrac{5}{6}$

3. a. $2\dfrac{3}{4} + \left(-1\dfrac{1}{2}\right)$ b. $2\dfrac{3}{4} - \left(-1\dfrac{1}{2}\right)$ c. $2\dfrac{3}{4} \div \left(-1\dfrac{1}{2}\right)$ d. $2\dfrac{3}{4} \cdot \left(-1\dfrac{1}{2}\right)$

4. a. $\left(4\dfrac{1}{3}\right) \cdot \left(2\dfrac{5}{6}\right)$ b. $\left(4\dfrac{1}{3}\right) \div \left(2\dfrac{5}{6}\right)$ c. $\left(4\dfrac{1}{3}\right) - \left(2\dfrac{5}{6}\right)$ d. $\left(4\dfrac{1}{3}\right) + \left(2\dfrac{5}{6}\right)$

5. a. $-4 - \dfrac{3}{8}$ b. $-4 \cdot \dfrac{3}{8}$ c. $-4 \div \dfrac{3}{8}$ d. $-4 + \dfrac{3}{8}$

6. a. $3\dfrac{2}{3} \div 2$ b. $3\dfrac{2}{3} - 2$ c. $3\dfrac{2}{3} + 2$ d. $3\dfrac{2}{3} \cdot 2$

7. a. $-4\dfrac{1}{5} - \left(-\dfrac{2}{3}\right)$ b. $-4\dfrac{1}{5} + \left(-\dfrac{2}{3}\right)$ c. $\left(-4\dfrac{1}{5}\right) \cdot \left(-\dfrac{2}{3}\right)$ d. $\left(-4\dfrac{1}{5}\right) \div \left(-\dfrac{2}{3}\right)$

8. a. $\dfrac{25}{9} \div 2$ b. $\dfrac{25}{9} \cdot 2$ c. $\dfrac{25}{9} - 2$ d. $\dfrac{25}{9} + 2$

9. a. $-1\dfrac{4}{5} \cdot \dfrac{5}{9}$ b. $-1\dfrac{4}{5} + \dfrac{5}{9}$ c. $-1\dfrac{4}{5} \div \dfrac{5}{9}$ d. $-1\dfrac{4}{5} - \dfrac{5}{9}$

10. a. $8 \cdot \dfrac{1}{8}$ b. $\dfrac{1}{9} \cdot 9$ c. $-\dfrac{3}{7} \cdot \left(-\dfrac{7}{3}\right)$ d. $-\dfrac{5}{13} \cdot \left(-\dfrac{13}{5}\right)$

Section 4.7 Order of Operations and Complex Fractions

Concepts

1. Order of Operations
2. Complex Fractions
3. Simplifying Algebraic Expressions

1. Order of Operations

At this point in our study, we will apply the order of operations to expressions involving fractions. We begin with a review of a fractional expression containing exponents. Recall that to square a number, multiply the number times itself. For example: $5^2 = 5 \cdot 5 = 25$. The same process is used to square a fraction.

$$\left(\dfrac{3}{7}\right)^2 = \dfrac{3}{7} \cdot \dfrac{3}{7} = \dfrac{9}{49}$$

Section 4.7 Order of Operations and Complex Fractions

Example 1 — Simplifying Expressions with Exponents

Simplify. a. $\left(\dfrac{2}{5}\right)^2$ b. $\left(-\dfrac{2}{5}\right)^2$ c. $-\left(\dfrac{2}{5}\right)^2$ d. $\left(-\dfrac{1}{4}\right)^3$

Solution:

a. $\left(\dfrac{2}{5}\right)^2 = \dfrac{2}{5} \cdot \dfrac{2}{5} = \dfrac{4}{25}$

b. $\left(-\dfrac{2}{5}\right)^2 = \left(-\dfrac{2}{5}\right)\left(-\dfrac{2}{5}\right) = \dfrac{4}{25}$ The base is negative. The product of two negatives is positive.

c. $-\left(\dfrac{2}{5}\right)^2 = -\left(\dfrac{2}{5}\right)\left(\dfrac{2}{5}\right) = -\dfrac{4}{25}$ First square the base. Then take the opposite.

d. $\left(-\dfrac{1}{4}\right)^3 = \left(-\dfrac{1}{4}\right)\left(-\dfrac{1}{4}\right)\left(-\dfrac{1}{4}\right) = -\dfrac{1}{64}$ The base is negative. The product of *three* negatives is negative.

Skill Practice Simplify.

1. $\left(\dfrac{5}{4}\right)^2$ 2. $\left(-\dfrac{5}{4}\right)^2$ 3. $-\left(\dfrac{5}{4}\right)^2$ 4. $\left(-\dfrac{1}{3}\right)^3$

From our study of exponents, we learned to recognize powers of 10. These are $10^1 = 10$, $10^2 = 100$, and so on. In this section, we learn to recognize the **powers of one-tenth**. That is, $\dfrac{1}{10}$ raised to a whole number power.

$$\left(\dfrac{1}{10}\right)^1 = \dfrac{1}{10}$$

$$\left(\dfrac{1}{10}\right)^2 = \dfrac{1}{10} \cdot \dfrac{1}{10} = \dfrac{1}{100}$$

$$\left(\dfrac{1}{10}\right)^3 = \dfrac{1}{10} \cdot \dfrac{1}{10} \cdot \dfrac{1}{10} = \dfrac{1}{1000}$$

$$\left(\dfrac{1}{10}\right)^4 = \dfrac{1}{10} \cdot \dfrac{1}{10} \cdot \dfrac{1}{10} \cdot \dfrac{1}{10} = \dfrac{1}{10{,}000}$$

$$\left(\dfrac{1}{10}\right)^5 = \dfrac{1}{10} \cdot \dfrac{1}{10} \cdot \dfrac{1}{10} \cdot \dfrac{1}{10} \cdot \dfrac{1}{10} = \dfrac{1}{100{,}000}$$

From these examples, we see that a power of one-tenth results in a fraction with a 1 in the numerator. The denominator has a 1 followed by the same number of zeros as the exponent.

Example 2 — Applying the Order of Operations

Simplify. $\left(-\dfrac{2}{15} \cdot \dfrac{3}{4}\right)^2$

Solution:

$\left(-\dfrac{2}{15} \cdot \dfrac{3}{4}\right)^2 = -\left(\dfrac{\cancel{2}}{\cancel{15}} \cdot \dfrac{\cancel{3}}{\cancel{4}}\right)^2 = \left(-\dfrac{1}{10}\right)^2$ Multiply fractions within parentheses.

$= \left(-\dfrac{1}{10}\right)\left(-\dfrac{1}{10}\right)$ Square the fraction $-\dfrac{1}{10}$.

$= \dfrac{1}{100}$

TIP: The instruction "simplify" means to perform all indicated operations. Fractional answers should always be written in lowest terms.

Skill Practice Simplify.

5. $\left(-\dfrac{7}{8} \cdot \dfrac{4}{35}\right)^3$

Answers

1. $\dfrac{25}{16}$ 2. $\dfrac{25}{16}$ 3. $-\dfrac{25}{16}$ 4. $-\dfrac{1}{27}$ 5. $-\dfrac{1}{1000}$

Chapter 4 Fractions and Mixed Numbers

FOR REVIEW

Recall that we may simplify common factors in multiplication but *not* in addition.

$$\frac{1}{2} + \left(\frac{18}{5}\right) \cdot \left(-\frac{10}{9}\right)$$

Thus, we may not use the factor of 2 in the first denominator to simplify.

Example 3 Applying the Order of Operations

Simplify. Write the answer as a fraction. $\frac{1}{2} + \left(3\frac{3}{5}\right) \cdot \left(-\frac{10}{9}\right)$

Solution:

$$\frac{1}{2} + \left(3\frac{3}{5}\right) \cdot \left(-\frac{10}{9}\right)$$ We must perform multiplication before addition.

$$= \frac{1}{2} + \left(\frac{18}{5}\right) \cdot \left(-\frac{10}{9}\right)$$ Write the mixed number $3\frac{3}{5}$ as $\frac{18}{5}$.

$$= \frac{1}{2} + \left(\frac{\overset{2}{\cancel{18}}}{\underset{1}{\cancel{5}}}\right) \cdot \left(-\frac{\overset{2}{\cancel{10}}}{\underset{1}{\cancel{9}}}\right)$$ Multiply fractions. The product will be negative.

$$= \frac{1}{2} - \frac{4}{1}$$

$$= \frac{1}{2} - \frac{4 \cdot 2}{1 \cdot 2}$$ Write the whole number over 1 and obtain a common denominator.

$$= \frac{1}{2} - \frac{8}{2}$$

$$= \frac{1 - 8}{2}$$ Subtract the fractions.

$$= -\frac{7}{2}$$ Simplify.

Skill Practice Simplify. Write the answer as a fraction.

6. $\frac{2}{3} + \left(3\frac{3}{4}\right) \cdot \left(-\frac{8}{5}\right)$

Example 4 Evaluating an Algebraic Expression

Evaluate the expression. $2 \div y \cdot z$ for $y = -\frac{14}{3}$ and $z = -\frac{1}{3}$

Solution:

$$2 \div y \cdot z$$

$$= 2 \div \left(-\frac{14}{3}\right) \cdot \left(-\frac{1}{3}\right)$$ Substitute $-\frac{14}{3}$ for y and $-\frac{1}{3}$ for z.

$$= \frac{2}{1} \cdot \left(-\frac{3}{14}\right) \cdot \left(-\frac{1}{3}\right)$$ Write the whole number over 1. Multiply by the reciprocal of $-\frac{14}{3}$.

$$= \frac{\overset{1}{\cancel{2}}}{1} \cdot \left(-\frac{\overset{1}{\cancel{3}}}{\underset{7}{\cancel{14}}}\right) \cdot \left(-\frac{1}{\underset{1}{\cancel{3}}}\right)$$ Simplify by dividing out common factors. The product is positive.

$$= \frac{1}{7}$$

Avoiding Mistakes

Do not forget to write the "1" in the numerator of the fraction.

Skill Practice Evaluate the expression for $a = -\frac{7}{15}$ and $b = \frac{21}{10}$.

7. $3 \cdot a \div b$

Answers

6. $-\frac{16}{3}$ 7. $-\frac{2}{3}$

2. Complex Fractions

A **complex fraction** is a fraction in which the numerator and denominator contain one or more terms with fractions. We will simplify complex fractions in Examples 5 and 6.

Example 5 Simplifying a Complex Fraction

Simplify. $\dfrac{\frac{3}{5}}{\frac{m}{10}}$

Solution:

a. $\dfrac{\frac{3}{5}}{\frac{m}{10}}$ ← This fraction bar denotes division.

$= \dfrac{3}{5} \div \dfrac{m}{10}$ Write the expression as a division of fractions.

$= \dfrac{3}{5} \cdot \dfrac{10}{m}$ Multiply by the reciprocal of $\dfrac{m}{10}$.

$= \dfrac{3}{\underset{1}{\cancel{5}}} \cdot \dfrac{\overset{2}{\cancel{10}}}{m}$ Reduce common factors.

$= \dfrac{6}{m}$ Simplify.

Skill Practice Simplify.

8. $\dfrac{\frac{5}{6}}{\frac{y}{21}}$

Example 6 Simplifying a Complex Fraction

Simplify. $\dfrac{\frac{2}{5} - \frac{1}{3}}{1 - \frac{3}{5}}$

Solution:

$\dfrac{\frac{2}{5} - \frac{1}{3}}{1 - \frac{3}{5}}$ This expression can be simplified using the order of operations. Subtract the fractions in the numerator and denominator separately. Then divide the results.

$= \dfrac{\frac{2 \cdot 3}{5 \cdot 3} - \frac{1 \cdot 5}{3 \cdot 5}}{\frac{1 \cdot 5}{1 \cdot 5} - \frac{3}{5}}$ The LCD in the numerator is 15.
In the denominator, the LCD is 5.

$= \dfrac{\frac{6}{15} - \frac{5}{15}}{\frac{5}{5} - \frac{3}{5}}$

Answer

8. $\dfrac{35}{2y}$

Avoiding Mistakes

Perform all operations in the numerator and denominator before performing division.

$$= \frac{\frac{1}{15}}{\frac{2}{5}} \quad \leftarrow \text{This fraction bar represents division.}$$

$$= \frac{1}{\underset{3}{\cancel{15}}} \cdot \frac{\overset{1}{\cancel{5}}}{2} \quad \text{Multiply by the reciprocal of } \frac{2}{5} \text{ and reduce common factors.}$$

$$= \frac{1}{6}$$

Skill Practice Simplify.

9. $\dfrac{\frac{3}{10} - \frac{1}{2}}{2 + \frac{1}{5}}$

As you can see from Example 6, simplifying a complex fraction can be a tedious process. For this reason, we offer an alternative approach, as demonstrated in Example 7.

Example 7 Simplifying a Complex Fraction by Multiplying by the LCD

Simplify. $\dfrac{\frac{2}{5} - \frac{1}{3}}{1 - \frac{3}{5}}$

Solution:

$\dfrac{\frac{2}{5} - \frac{1}{3}}{1 - \frac{3}{5}}$ — First identify the LCD of all four terms in the expression. The LCD of $\frac{2}{5}, \frac{1}{3}, 1,$ and $\frac{3}{5}$ is 15.

$= \dfrac{15 \cdot \left(\frac{2}{5} - \frac{1}{3}\right)}{15 \cdot \left(1 - \frac{3}{5}\right)}$ — Group the terms in the numerator and denominator within parentheses. Then multiply each term in the numerator and denominator by the LCD, 15.

$= \dfrac{\left(\overset{3}{\cancel{15}} \cdot \frac{2}{\cancel{5}_1}\right) - \left(\overset{5}{\cancel{15}} \cdot \frac{1}{\cancel{3}_1}\right)}{\left(15 \cdot 1\right) - \left(\overset{3}{\cancel{15}} \cdot \frac{3}{\cancel{5}_1}\right)}$ — Apply the distributive property to multiply each term by 15. Then simplify each product.

$= \dfrac{6 - 5}{15 - 9} = \dfrac{1}{6}$ — Simplify the fraction. This is the same result we obtained in Example 6.

Skill Practice Simplify the complex fraction by multiplying numerator and denominator by the LCD of all four terms in the fraction.

10. $\dfrac{\frac{3}{10} - \frac{1}{2}}{2 + \frac{1}{5}}$

Answers

9. $-\dfrac{1}{11}$ 10. $-\dfrac{1}{11}$

3. Simplifying Algebraic Expressions

To simplify algebraic expressions, we combine *like* terms by applying the distributive property. For example,

$$2x - 3x + 8x = (2 - 3 + 8)x \quad \text{Apply the distributive property.}$$
$$= 7x$$

In a similar way, we can combine *like* terms when the coefficients are fractions. This is demonstrated in Example 8.

TIP: Recall that *like* terms can also be combined by combining the coefficients and keeping the variable factor unchanged.

$$2x - 3x + 8x = 7x$$

Example 8 — Simplifying an Algebraic Expression

Simplify. $\quad \dfrac{2}{5}x - \dfrac{1}{3}x$

Solution:

$\dfrac{2}{5}x - \dfrac{1}{3}x \qquad$ The two terms are *like* terms.

$= \left(\dfrac{2}{5} - \dfrac{1}{3}\right)x \qquad$ Combine *like* terms by applying the distributive property.

$= \left(\dfrac{2 \cdot 3}{5 \cdot 3} - \dfrac{1 \cdot 5}{3 \cdot 5}\right)x \qquad$ The LCD is 15.

$= \left(\dfrac{6}{15} - \dfrac{5}{15}\right)x \qquad$ Simplify.

$= \dfrac{1}{15}x$

TIP: $\dfrac{1}{15}x$ can also be written as $\dfrac{x}{15}$.

$$\dfrac{1}{15}x = \dfrac{1}{15} \cdot x = \dfrac{1}{15} \cdot \dfrac{x}{1} = \dfrac{x}{15}$$

Skill Practice Simplify.

11. $\dfrac{7}{4}y + \dfrac{2}{3}y$

Answer

11. $\dfrac{29}{12}y$

Section 4.7 Activity

A.1. Consider the expression $-\dfrac{2}{5} + \dfrac{3}{4}\left(\dfrac{1}{6}\right)$.

  a. What operation would you perform first, addition or multiplication? Perform that operation.

  b. What operation would you perform second? Perform that operation and write the answer as a fraction in lowest terms.

A.2. Consider the expression $\left(-\dfrac{2}{5} + \dfrac{3}{4}\right)\dfrac{1}{6}$.

  a. Compare the expression in Exercise A.2 to the expression given in Exercise A.1. How do they differ?

  b. What operation would you perform first, addition or multiplication? Perform that operation.

  c. What operation would you perform second? Perform that operation and write the answer as a fraction in lowest terms.

For Exercises A.3–A.6, simplify. Show each step by performing only one operation at a time.

A.3. $\dfrac{18}{7} \div \dfrac{3}{14} - \dfrac{3}{2}$

A.4. $\dfrac{18}{7} \div \left(\dfrac{3}{14} - \dfrac{3}{2}\right)$

A.5. $\dfrac{7}{2} + \dfrac{9}{8} \div \dfrac{3}{4}$

A.6. $1\dfrac{2}{5} \cdot \left(\dfrac{7}{8} + 2\dfrac{1}{4}\right)$

A.7. **a.** Identify the base that is raised to the second power in the expression $\left(\frac{1}{9} - \frac{2}{3}\right)^2$.
   b. Write the expression in expanded form.
   c. Simplify.

A.8. **a.** Identify the base that is raised to the second power in the expression $\frac{1}{9} - \left(\frac{2}{3}\right)^2$.
   b. Write the expression in expanded form.
   c. Simplify.

For Exercises A.9–A.10, simplify.

A.9. $\left(-\frac{3}{4}\right)^2$

A.10. $-\left(\frac{3}{4}\right)^2$

A.11. **a.** Add the fractions: $\frac{2}{3} + \frac{1}{2}$
   b. Subtract the fractions: $\frac{3}{4} - \frac{1}{3}$
   c. Simplify the complex fraction by dividing the sum from part (a) by the difference from part (b).

$$\frac{\frac{2}{3} + \frac{1}{2}}{\frac{3}{4} - \frac{1}{3}}$$

A.12. **a.** Find the LCD of the fractions: $\frac{2}{3}, \frac{1}{2}, \frac{3}{4}, \frac{1}{3}$.
   b. Apply the distributive property to multiply each term by the LCD, then simplify each product: $\frac{2}{3} + \frac{1}{2}$
   c. Apply the distributive property to multiply each term by the LCD, then simplify each product: $\frac{3}{4} - \frac{1}{3}$
   d. Simplify the fraction.

$$\frac{\frac{2}{3} + \frac{1}{2}}{\frac{3}{4} - \frac{1}{3}}$$

For Exercises A.13–A.14, simplify the complex fractions by using order of operations (see Exercise A.11) or by multiplying by the LCD (see Exercise A.12).

A.13. $\dfrac{\frac{1}{10} - \frac{2}{5}}{\frac{4}{5} + \frac{1}{10}}$

A.14. $\dfrac{\frac{2}{w}}{\frac{1}{w^2}}$

For Exercises A.15–A.18, simplify the expression by combining *like* terms.

A.15. $2x - 3x$

A.16. $\frac{2}{5}x - \frac{3}{5}x$

A.17. $\frac{2}{5}x - \frac{3}{4}x$

A.18. $\frac{10}{9}x - \left(-\frac{5}{6}y\right) + \frac{4}{3}y - \frac{11}{6}x$

Section 4.7 Order of Operations and Complex Fractions

Practice Exercises

Section 4.7

Prerequisite Review

For Exercises R.1–R.6, perform the indicated operations. Write the answers as fractions

R.1. $-\dfrac{2}{5} + \dfrac{7}{6}$

R.2. $\dfrac{7}{8} - \dfrac{2}{5}$

R.3. $-\dfrac{2}{5}\left(\dfrac{7}{6}\right)$

R.4. $\left(\dfrac{7}{8}\right)\left(-\dfrac{2}{5}\right)$

R.5. $-\dfrac{2}{5} \div \dfrac{7}{6}$

R.6. $\dfrac{7}{8} \div \dfrac{2}{5}$

For Exercises R.7–R.10, perform the indicated operations. Write the answers as mixed or whole numbers.

R.7. $3\dfrac{1}{4} + 2\dfrac{5}{6}$

R.8. $2\dfrac{2}{5} + 5\dfrac{5}{6}$

R.9. $\left(3\dfrac{1}{4}\right)\left(2\dfrac{5}{6}\right)$

R.10. $\left(2\dfrac{2}{5}\right)\left(5\dfrac{5}{6}\right)$

Vocabulary and Key Concepts

1. **a.** The values $\left(\tfrac{1}{10}\right)^1$, $\left(\tfrac{1}{10}\right)^2$, $\left(\tfrac{1}{10}\right)^3$, and so on are called powers of _____.

   In simplified form these equal $\tfrac{1}{10}$, $\tfrac{1}{100}$, $\tfrac{1}{1000}$, and so on.

   b. A _____ fraction is an expression containing one or more fractions in the numerator, denominator, or both.

2. Consider the expression $\left(-\dfrac{5}{8}\right)^2$. The base raised to the second power is _____.

3. Consider the expression $-\left(\dfrac{5}{8}\right)^2$. The base raised to the second power is _____.

4. The fraction bar of a complex fraction functions as a _____ symbol, so perform all operations in the numerator and denominator before dividing.

5. The fraction $\dfrac{1}{5}x$ can also be written as _____.

6. When simplifying an algebraic expression with fractional coefficients, *like* terms can be combined as long as the coefficients have a common _____.

7. In the expression $\left(\dfrac{1}{2}\right)^2 \div \left(\dfrac{1}{5} + \dfrac{2}{3}\right)$, the first operation to perform is to (simplify the exponent/divide).

8. When simplifying the complex fraction $\dfrac{2 + \tfrac{3}{4}}{\tfrac{1}{3}}$, the first operation to perform is (addition/division).

Concept 1: Order of Operations

For Exercises 9–44, simplify. (See Examples 1–3.)

9. $\left(\dfrac{1}{9}\right)^2$

10. $\left(\dfrac{1}{4}\right)^2$

11. $\left(-\dfrac{1}{9}\right)^2$

12. $\left(-\dfrac{1}{4}\right)^2$

13. $-\left(\dfrac{3}{2}\right)^3$

14. $-\left(\dfrac{4}{3}\right)^3$

15. $\left(-\dfrac{3}{2}\right)^3$

16. $\left(-\dfrac{4}{3}\right)^3$

17. $\left(\dfrac{1}{10}\right)^3$

18. $\left(\dfrac{1}{10}\right)^4$

19. $\left(-\dfrac{1}{10}\right)^6$

20. $\left(-\dfrac{1}{10}\right)^5$

21. $\left(-\dfrac{10}{3}\cdot\dfrac{3}{100}\right)^3$
22. $\left(-\dfrac{1}{6}\cdot\dfrac{3}{5}\right)^2$
23. $-\left(4\cdot\dfrac{3}{4}\right)^3$
24. $-\left(5\cdot\dfrac{2}{5}\right)^2$

25. $\dfrac{1}{6}+\left(2\dfrac{1}{3}\right)\cdot\left(1\dfrac{3}{4}\right)$
26. $\dfrac{7}{9}+\left(2\dfrac{1}{6}\right)\cdot\left(3\dfrac{1}{3}\right)$
27. $6-5\dfrac{1}{7}\div\left(-\dfrac{1}{7}\right)$
28. $11-6\dfrac{1}{3}\div\left(-1\dfrac{1}{6}\right)$

29. $-\dfrac{1}{3}\cdot\left|-\dfrac{21}{4}\cdot\dfrac{8}{7}\right|$
30. $-\dfrac{1}{6}\cdot\left|-\dfrac{24}{5}\cdot\dfrac{30}{8}\right|$
31. $\dfrac{16}{9}\cdot\left(\dfrac{1}{2}\right)^3$
32. $\dfrac{28}{6}\cdot\left(\dfrac{3}{2}\right)^2$

33. $\dfrac{54}{21}\div\dfrac{2}{3}\cdot\dfrac{7}{9}$
34. $\dfrac{48}{56}\div\dfrac{3}{8}\cdot\dfrac{7}{8}$
35. $7\dfrac{1}{8}\div\left(-1\dfrac{1}{3}\right)\div\left(-2\dfrac{1}{4}\right)$
36. $\left(-3\dfrac{1}{8}\right)\div 5\dfrac{5}{7}\div 1\dfrac{5}{16}$

37. $-\dfrac{5}{4}\div\dfrac{3}{2}-\left(-\dfrac{5}{6}\right)$
38. $\dfrac{1}{7}\div\dfrac{2}{21}-\left(-\dfrac{5}{2}\right)$
39. $\left(\dfrac{1}{3}-\dfrac{1}{2}\right)^2$
40. $\left(-\dfrac{2}{3}+\dfrac{1}{6}\right)^2$

41. $\left(\dfrac{1}{4}\right)^2\div\left(\dfrac{5}{6}-\dfrac{2}{3}\right)+\dfrac{7}{12}$
42. $\left(\dfrac{1}{2}+\dfrac{1}{3}\right)\cdot\left(\dfrac{2}{5}\right)^2+\dfrac{3}{10}$
43. $\left(5-1\dfrac{7}{8}\right)\div\left(3-\dfrac{13}{16}\right)$
44. $\left(4+2\dfrac{1}{9}\right)\div\left(2-1\dfrac{11}{36}\right)$

For Exercises 45–52, evaluate the expression for the given values of the variables. (See Example 4.)

45. $-3\cdot a\div b$ for $a=-\dfrac{5}{6}$ and $b=\dfrac{3}{10}$
46. $4\div w\cdot z$ for $w=\dfrac{2}{7}$ and $z=-\dfrac{1}{5}$

47. xy^2 for $x=2\dfrac{1}{3}$ and $y=\dfrac{3}{2}$
48. $c^3 d$ for $c=-1\dfrac{1}{2}$ and $d=\dfrac{1}{3}$

49. $4x+6y$ for $x=\dfrac{1}{2}$ and $y=-\dfrac{3}{2}$
50. $2m-3n$ for $m=-\dfrac{3}{4}$ and $n=\dfrac{1}{6}$

51. $(4-w)(3+z)$ for $w=2\dfrac{1}{3}$ and $z=1\dfrac{2}{3}$
52. $(2+a)(7-b)$ for $a=-1\dfrac{1}{2}$ and $b=5\dfrac{1}{4}$

Concept 2: Complex Fractions

For Exercises 53–68, simplify the complex fractions. (See Examples 5–7.)

53. $\dfrac{\frac{5}{8}}{\frac{3}{4}}$
54. $\dfrac{\frac{8}{9}}{\frac{7}{12}}$
55. $\dfrac{-\frac{21}{10}}{\frac{6}{5}}$
56. $\dfrac{\frac{20}{3}}{-\frac{5}{8}}$

57. $\dfrac{\frac{3}{7}}{\frac{12}{x}}$
58. $\dfrac{\frac{5}{p}}{\frac{30}{11}}$
59. $\dfrac{\frac{-15}{w}}{\frac{25}{w}}$
60. $\dfrac{-\frac{18}{t}}{\frac{15}{t}}$

61. $\dfrac{\frac{4}{3}-\frac{1}{6}}{1-\frac{1}{3}}$
62. $\dfrac{\frac{6}{5}+\frac{3}{10}}{2+\frac{1}{2}}$
63. $\dfrac{\frac{1}{2}+3}{\frac{9}{8}+\frac{1}{4}}$
64. $\dfrac{\frac{5}{2}+1}{\frac{3}{4}+\frac{1}{3}}$

65. $\dfrac{\frac{5}{7}-\frac{1}{14}}{\frac{1}{2}-\frac{3}{7}}$
66. $\dfrac{-\frac{1}{4}+\frac{1}{6}}{-\frac{4}{3}-\frac{5}{6}}$
67. $\dfrac{-\frac{7}{4}+\frac{3}{2}}{\frac{7}{8}-\frac{1}{4}}$
68. $\dfrac{\frac{8}{3}-\frac{5}{6}}{\frac{1}{4}-\frac{4}{3}}$

Concept 3: Simplifying Algebraic Expressions

For Exercises 69–76, simplify the expressions. (See Example 8.)

69. $\frac{1}{2}y + \frac{3}{2}y$

70. $-\frac{4}{5}p + \frac{2}{5}p$

71. $\frac{3}{4}a - \frac{1}{8}a$

72. $\frac{1}{3}b + \frac{2}{9}b$

73. $\frac{4}{5}x - \frac{3}{10}x + \frac{1}{15}x$

74. $-\frac{3}{2}y - \frac{4}{3}y + \frac{3}{4}y$

75. $\frac{3}{2}y + \frac{1}{4}z - \frac{1}{6}y - 2z$

76. $-\frac{5}{8}a + 3b + \frac{1}{4}a - \frac{7}{2}b$

Expanding Your Skills

77. Evaluate

    a. $\left(\frac{1}{6}\right)^2$
    b. $\sqrt{\frac{1}{36}}$

78. Evaluate

    a. $\left(\frac{2}{7}\right)^2$
    b. $\sqrt{\frac{4}{49}}$

For Exercises 79–82, evaluate the square roots.

79. $\sqrt{\frac{1}{25}}$

80. $\sqrt{\frac{1}{100}}$

81. $\sqrt{\frac{64}{81}}$

82. $\sqrt{\frac{9}{4}}$

Solving Equations Containing Fractions

Section 4.8

1. Solving Equations Containing Fractions

In this section we will solve linear equations that contain fractions. To begin, review the addition, subtraction, multiplication, and division properties of equality.

Concepts

1. Solving Equations Containing Fractions
2. Solving Equations by Clearing Fractions

Property	Example
Addition Property of Equality If $a = b$, then $a + c = b + c$	Solve. $\quad x - 4 = 6$ $\qquad x - 4 + 4 = 6 + 4$ $\qquad x = 10$
Subtraction Property of Equality If $a = b$, then $a - c = b - c$	Solve. $\quad x + 3 = 5$ $\qquad x + 3 - 3 = 5 - 3$ $\qquad x = 2$
Multiplication Property of Equality If $a = b$, then $c \cdot a = c \cdot b$	Solve. $\quad \frac{x}{5} = 3$ $\qquad 5 \cdot \frac{x}{5} = 5 \cdot 3$ $\qquad x = 15$
Division Property of Equality If $a = b$, then $\frac{a}{c} = \frac{b}{c}$ (provided that $c \neq 0$)	Solve. $\quad 9x = 27$ $\qquad \frac{9x}{9} = \frac{27}{9}$ $\qquad x = 3$

We will use the same properties to solve equations containing fractions. In Examples 1 and 2, we use the addition and subtraction properties of equality.

Example 1 — Using the Addition Property of Equality

Solve. $x - \dfrac{3}{10} = \dfrac{9}{20}$

Solution:

$$x - \dfrac{3}{10} = \dfrac{9}{20}$$

$$x - \dfrac{3}{10} + \dfrac{3}{10} = \dfrac{9}{20} + \dfrac{3}{10} \qquad \text{Add } \dfrac{3}{10} \text{ to both sides to isolate } x.$$

$$x = \dfrac{9}{20} + \dfrac{3 \cdot 2}{10 \cdot 2} \qquad \text{To add the fractions on the right, the LCD is 20.}$$

$$x = \dfrac{9}{20} + \dfrac{6}{20}$$

$$x = \dfrac{15}{20} \qquad \text{Now simplify the fraction.}$$

$$x = \dfrac{3}{4} \qquad \text{The solution is } \dfrac{3}{4}.$$

Check: $x - \dfrac{3}{10} = \dfrac{9}{20}$

$$\dfrac{3}{4} - \dfrac{3}{10} \stackrel{?}{=} \dfrac{9}{20} \qquad \text{Substitute } \dfrac{3}{4} \text{ for } x.$$

$$\dfrac{3 \cdot 5}{4 \cdot 5} - \dfrac{3 \cdot 2}{10 \cdot 2} \stackrel{?}{=} \dfrac{9}{20}$$

$$\dfrac{15}{20} - \dfrac{6}{20} \stackrel{?}{=} \dfrac{9}{20} \quad \checkmark \text{ True}$$

Skill Practice Solve.

1. $w - \dfrac{1}{3} = \dfrac{4}{15}$

Example 2 — Using the Subtraction Property of Equality

Solve. $-\dfrac{5}{4} = \dfrac{1}{2} + t$

Solution:

$$-\dfrac{5}{4} = \dfrac{1}{2} + t$$

$$-\dfrac{5}{4} - \dfrac{1}{2} = \dfrac{1}{2} - \dfrac{1}{2} + t \qquad \text{Subtract } \dfrac{1}{2} \text{ from both sides to isolate } t.$$

$$-\dfrac{5}{4} - \dfrac{1 \cdot 2}{2 \cdot 2} = t \qquad \text{To subtract the fractions on the left, the LCD is 4.}$$

$$-\dfrac{5}{4} - \dfrac{2}{4} = t$$

$$-\dfrac{7}{4} = t \qquad \text{The solution is } -\tfrac{7}{4} \text{ and checks in the original equation.}$$

Skill Practice Solve.

2. $-\dfrac{7}{9} = \dfrac{1}{3} + y$

Answers

1. $\dfrac{3}{5}$ 2. $-\dfrac{10}{9}$

Section 4.8 Solving Equations Containing Fractions

Recall that the product of a number and its reciprocal is 1. For example:

$$\frac{2}{7} \cdot \frac{7}{2} = 1, \qquad -\frac{3}{8} \cdot \left(-\frac{8}{3}\right) = 1, \qquad 5 \cdot \frac{1}{5} = 1$$

We will use this fact and the multiplication property of equality to solve the equations in Examples 3–5.

Example 3 Using the Multiplication Property of Equality

Solve. $\frac{4}{5}x = \frac{2}{7}$

Solution:

$\frac{4}{5}x = \frac{2}{7}$ In this equation, we will apply the multiplication property of equality. Multiply both sides of the equation by the reciprocal of $\frac{4}{5}$. Do this because $\frac{5}{4} \cdot \frac{4}{5}x$ is equal to $1x$. This isolates the variable x.

$\frac{5}{4} \cdot \frac{4}{5}x = \frac{5}{4} \cdot \frac{2}{7}$ Multiply both sides of the equation by the reciprocal of $\frac{4}{5}$.

$1x = \frac{10}{28}$

$x = \frac{\overset{5}{\cancel{10}}}{\underset{14}{\cancel{28}}}$ Simplify.

$x = \frac{5}{14}$ The solution is $\frac{5}{14}$ and checks in the original equation.

Skill Practice Solve.

3. $\frac{3}{4}x = \frac{5}{6}$

Example 4 Using the Multiplication Property of Equality

Solve. $8 = -\frac{1}{6}y$

Solution:

$8 = -\frac{1}{6}y$

$-6 \cdot (8) = -6 \cdot \left(-\frac{1}{6}y\right)$ Multiply both sides of the equation by the reciprocal of $-\frac{1}{6}$. Do this because $-6 \cdot \left(-\frac{1}{6}y\right)$ is equal to $1y$. This isolates the variable y.

$-48 = 1y$

$-48 = y$ The solution is -48 and checks in the original equation.

Skill Practice Solve.

4. $7 = -\frac{1}{4}y$

Answers

3. $\frac{10}{9}$ 4. -28

Example 5 Using the Multiplication Property of Equality

Solve. $\quad -\dfrac{2}{9} = -4w$

Solution:

$$-\dfrac{2}{9} = -4w$$

$$-\dfrac{1}{4}\left(-\dfrac{2}{9}\right) = -\dfrac{1}{4}(-4w)$$

Multiply both sides of the equation by the reciprocal of -4. Do this because $-\dfrac{1}{4} \cdot (-4w)$ is equal to $1w$. This isolates the variable w.

$$-\dfrac{1}{\overset{}{\underset{2}{4}}}\left(-\dfrac{\overset{1}{2}}{9}\right) = 1w$$

$$\dfrac{1}{18} = w$$

The solution is $\dfrac{1}{18}$ and checks in the original equation.

Skill Practice Solve.

5. $-\dfrac{2}{5} = -4x$

In Example 6, we solve an equation in which we must apply both the addition property of equality and the multiplication property of equality.

Example 6 Using Multiple Steps to Solve an Equation with Fractions

Solve. $\quad \dfrac{3}{4}x - \dfrac{1}{3} = \dfrac{5}{6}$

Solution:

$$\dfrac{3}{4}x - \dfrac{1}{3} = \dfrac{5}{6}$$

$$\dfrac{3}{4}x - \dfrac{1}{3} + \dfrac{1}{3} = \dfrac{5}{6} + \dfrac{1}{3}$$

To isolate the x term, add $\dfrac{1}{3}$ to both sides.

$$\dfrac{3}{4}x = \dfrac{5}{6} + \dfrac{1 \cdot 2}{3 \cdot 2}$$

To add the fractions on the right, the LCD is 6.

$$\dfrac{3}{4}x = \dfrac{5}{6} + \dfrac{2}{6}$$

$$\dfrac{3}{4}x = \dfrac{7}{6}$$

$$\dfrac{4}{3} \cdot \dfrac{3}{4}x = \dfrac{4}{3} \cdot \dfrac{7}{6}$$

Multiply both sides of the equation by the reciprocal of $\dfrac{3}{4}$. Do this because $\dfrac{4}{3} \cdot \left(\dfrac{3}{4}x\right)$ is equal to $1x$. This isolates the variable x.

$$x = \dfrac{\overset{2}{4}}{3} \cdot \dfrac{7}{\underset{3}{6}}$$

Simplify the product.

$$x = \dfrac{14}{9}$$

The solution is $\dfrac{14}{9}$ and checks in the original equation.

Skill Practice Solve.

6. $\dfrac{2}{3}x - \dfrac{1}{4} = \dfrac{3}{2}$

Answers

5. $\dfrac{1}{10}$ 6. $\dfrac{21}{8}$

2. Solving Equations by Clearing Fractions

As you probably noticed, Example 6 required tedious manipulation of fractions to isolate the variable. Therefore, we will now show you an alternative technique that eliminates the fractions immediately. This technique is called **clearing fractions**. Its basis is to multiply both sides of the equation by the LCD of all terms in the equation. This is demonstrated in Examples 7 and 8.

Example 7 — Solving an Equation by First Clearing Fractions

Solve. $\dfrac{y}{8} + \dfrac{3}{2} = \dfrac{y}{4}$

Solution:

$\dfrac{y}{8} + \dfrac{3}{2} = \dfrac{y}{4}$ The LCD of $\dfrac{y}{8}, \dfrac{3}{2},$ and $\dfrac{y}{4}$ is 8.

$8 \cdot \left(\dfrac{y}{8} + \dfrac{3}{2}\right) = 8 \cdot \left(\dfrac{y}{4}\right)$ Apply the multiplication property of equality. Multiply both sides of the equation by 8.

$8 \cdot \left(\dfrac{y}{8}\right) + 8 \cdot \left(\dfrac{3}{2}\right) = 8 \cdot \left(\dfrac{y}{4}\right)$ Use the distributive property to multiply each term by 8.

$\overset{1}{8} \cdot \left(\dfrac{y}{\underset{1}{8}}\right) + \overset{4}{8} \cdot \left(\dfrac{3}{\underset{1}{2}}\right) = \overset{2}{8} \cdot \left(\dfrac{y}{\underset{1}{4}}\right)$ Simplify each term.

$y + 12 = 2y$ The fractions have been "cleared." Now isolate the variable.

$y - y + 12 = 2y - y$ Subtract y from both sides to collect the variable terms on one side.

$12 = y$ Simplify.

The solution is 12 and checks in the original equation.

Skill Practice Solve.

7. $\dfrac{x}{10} + \dfrac{1}{2} = \dfrac{x}{5}$

Clearing fractions is a technique that "removes" fractions from an equation and produces a simpler equation. Because this technique is so powerful, we will add it to step 1 of our procedure box for solving a linear equation in one variable.

Solving a Linear Equation in One Variable

Step 1 Simplify both sides of the equation.
- Clear parentheses if necessary.
- Combine *like* terms if necessary.
- Consider clearing fractions if necessary.

Step 2 Use the addition or subtraction property of equality to collect the variable terms on one side of the equation.

Step 3 Use the addition or subtraction property of equality to collect the constant terms on the *other* side of the equation.

Step 4 Use the multiplication or division property of equality to make the coefficient of the variable term equal to 1.

Step 5 Check the answer in the original equation.

Answer

7. 5

Example 8 Solving an Equation by First Clearing Fractions

Solve. $-\dfrac{4}{7}x - 2 = \dfrac{3}{14}$

Solution:

$-\dfrac{4}{7}x - 2 = \dfrac{3}{14}$ — The LCD of $-\dfrac{4}{7}x$, -2, and $\dfrac{3}{14}$ is 14.

$14 \cdot \left(-\dfrac{4}{7}x - 2\right) = 14 \cdot \left(\dfrac{3}{14}\right)$ — Multiply both sides by 14.

$14 \cdot \left(-\dfrac{4}{7}x\right) + 14 \cdot (-2) = 14 \cdot \left(\dfrac{3}{14}\right)$ — Use the distributive property to multiply each term by 14.

$\overset{2}{14} \cdot \left(-\dfrac{4}{7}x\right) + 14 \cdot (-2) = \overset{1}{14} \cdot \left(\dfrac{3}{\underset{1}{14}}\right)$ — Simplify each term.

$-8x - 28 = 3$ — The fractions have been "cleared."

$-8x - 28 + 28 = 3 + 28$ — Add 28 to both sides to isolate the x term.

$-8x = 31$

$\dfrac{-8x}{-8} = \dfrac{31}{-8}$ — Apply the division property of equality to isolate x.

$x = -\dfrac{31}{8}$ — The solution is $-\dfrac{31}{8}$ and checks in the original equation.

Avoiding Mistakes

Be sure to multiply each term by 14. This includes the constant term, -2.

Skill Practice Solve.

8. $-\dfrac{3}{8}x - 3 = \dfrac{1}{4}$

Answer

8. $-\dfrac{26}{3}$

Section 4.8 Activity

A.1. a. Solve the equation: $y - 4 = 2$
  b. Explain the step(s) used to solve the equation in part (a).
  c. Solve the equation: $y - \dfrac{4}{7} = \dfrac{2}{7}$
  d. Explain the step(s) used to solve the equation in part (c).
  e. Observe the similarities between the steps used to solve each equation. Do the steps change because fractions appear in the equation?

A.2. a. What is the reciprocal of $\dfrac{7}{4}$?
  b. What product results when you multiply $\dfrac{7}{4}$ by the reciprocal?
  c. Solve the equation by multiplying both sides of the equation by the reciprocal from part (a): $\dfrac{7}{4}x = -\dfrac{2}{3}$
  d. Check the solution by substituting the answer back into the original equation.

For Exercises A.3–A.4, solve the equations.

A.3. $\dfrac{1}{2}t - \dfrac{3}{4} = \dfrac{11}{8}$

A.4. $-\dfrac{4}{3}x - \dfrac{7}{6} = \dfrac{1}{2}x + \dfrac{2}{3}$

A.5. **a.** Find the LCD of the fractions: $\dfrac{5}{2}, \dfrac{3}{4}, \dfrac{7}{3}$.

  b. Multiply each term in the equation by the LCD from part (a). If done correctly, the resulting equation should not contain fractions.

$$\dfrac{5}{2}x + \dfrac{3}{4} = -\dfrac{7}{3}$$

  c. Solve the equation.

  d. Check the solution by substituting the answer back into the original equation.

For Exercises A.6–A.7, determine the LCD, then solve the equations by clearing the fractions.

A.6. $-2 + \dfrac{3}{5}y = \dfrac{1}{2}y + \dfrac{3}{10}$

A.7. $p - \dfrac{5}{6} = \dfrac{1}{8}p - 2$

Practice Exercises

Section 4.8

Study Skills Exercise

When you solve equations involving several steps, it is recommended that you write an explanation for each step. In the example below, an equation is solved with each step shown. Your job is to write an explanation for each step for the following equation.

$$\dfrac{3}{5}x - \dfrac{7}{10} = \dfrac{1}{2}$$

Explanation

$$10\left(\dfrac{3}{5}x - \dfrac{7}{10}\right) = 10\left(\dfrac{1}{2}\right)$$

$$6x - 7 = 5$$

$$6x - 7 + 7 = 5 + 7$$

$$6x = 12$$

$$\dfrac{6x}{6} = \dfrac{12}{6}$$

$$x = 2 \qquad \text{The solution is 2.}$$

Prerequisite Review

For Exercises R.1–R.8, perform the indicated operation. Write the answers as fractions or integers.

R.1. $\dfrac{3}{4} - \dfrac{5}{6}$

R.2. $\dfrac{5}{8} - \dfrac{5}{12}$

R.3. $\dfrac{7}{12} + \dfrac{2}{3}$

R.4. $\dfrac{6}{15} + \dfrac{1}{10}$

R.5. $\dfrac{1}{3} + \dfrac{4}{5} - 1$

R.6. $3 - \dfrac{5}{8} + \dfrac{1}{4}$

R.7. $-18\left(\dfrac{5}{9} + \dfrac{1}{3}\right)$

R.8. $20\left(\dfrac{1}{5} - \dfrac{3}{4}\right)$

Concept 1: Solving Equations Containing Fractions

For Exercises 1–34, solve the equations. Write the answers as fractions or integers. (See Examples 1–6.)

1. $p - \dfrac{5}{6} = \dfrac{1}{3}$
2. $q - \dfrac{3}{4} = \dfrac{3}{2}$
3. $-\dfrac{7}{10} = \dfrac{3}{5} + a$
4. $-\dfrac{3}{8} = \dfrac{1}{4} + b$

5. $\dfrac{2}{3} = y - \dfrac{5}{12}$
6. $\dfrac{7}{11} = z + \dfrac{3}{11}$
7. $t + \dfrac{3}{8} = 2$
8. $r - \dfrac{4}{7} = -1$

9. $\dfrac{1}{6} = -\dfrac{11}{6} + m$
10. $n + \dfrac{1}{2} = -\dfrac{2}{3}$
11. $\dfrac{3}{5}y = \dfrac{7}{10}$
12. $\dfrac{7}{2}x = \dfrac{5}{4}$

13. $\dfrac{5}{4}k = -\dfrac{1}{2}$
14. $-\dfrac{11}{12}h = -\dfrac{1}{6}$
15. $6 = -\dfrac{1}{4}x$
16. $3 = -\dfrac{1}{5}y$

17. $\dfrac{2}{3}m = 14$
18. $\dfrac{5}{9}n = 40$
19. $\dfrac{b}{7} = -3$
20. $\dfrac{a}{4} = 12$

21. $-\dfrac{u}{2} = -15$
22. $-\dfrac{v}{10} = -4$
23. $0 = \dfrac{3}{8}m$
24. $0 = \dfrac{1}{10}n$

25. $6x = \dfrac{12}{5}$
26. $7t = \dfrac{14}{3}$
27. $-\dfrac{5}{9} = -10x$
28. $-\dfrac{4}{3} = -6y$

29. $\dfrac{2}{5}x - \dfrac{1}{4} = \dfrac{3}{2}$
30. $\dfrac{5}{9}y - \dfrac{1}{3} = \dfrac{5}{6}$
31. $-\dfrac{4}{7} = \dfrac{1}{2} + \dfrac{3}{14}w$
32. $-\dfrac{1}{8} = \dfrac{3}{4} + \dfrac{5}{2}z$

33. $3p + \dfrac{1}{2} = \dfrac{5}{4}$
34. $2t - \dfrac{3}{8} = \dfrac{9}{16}$

Concept 2: Solving Equations by Clearing Fractions

For Exercises 35–46, solve the equations by first clearing fractions. (See Examples 7 and 8.)

35. $\dfrac{x}{5} + \dfrac{1}{2} = \dfrac{7}{10}$
36. $\dfrac{p}{4} + \dfrac{1}{2} = \dfrac{5}{8}$
37. $-\dfrac{5}{7}y - 1 = \dfrac{3}{2}$
38. $-\dfrac{2}{3}w - 3 = -\dfrac{1}{2}$

39. $\dfrac{2}{3} = \dfrac{5}{9} + \dfrac{1}{6}t$
40. $\dfrac{4}{5} = \dfrac{9}{15} + \dfrac{1}{3}n$
41. $\dfrac{m}{3} + \dfrac{m}{6} = \dfrac{5}{9}$
42. $\dfrac{4}{5} = \dfrac{n}{15} - \dfrac{n}{3}$

43. $\dfrac{x}{3} + \dfrac{7}{6} = \dfrac{x}{2}$
44. $\dfrac{p}{4} = \dfrac{p}{8} + \dfrac{1}{2}$
45. $\dfrac{3}{2}y + 3 = 2y + \dfrac{1}{2}$
46. $\dfrac{1}{4}x - 1 = 2 - \dfrac{1}{2}x$

Mixed Exercises

For Exercises 47–64, solve the equations.

47. $\dfrac{h}{4} = -12$
48. $\dfrac{w}{6} = -18$
49. $\dfrac{2}{3} + t = 1$
50. $\dfrac{3}{4} + q = 1$

51. $-\dfrac{3}{7}x = \dfrac{9}{10}$
52. $-\dfrac{2}{11}y = \dfrac{4}{15}$
53. $4c = -\dfrac{1}{3}$
54. $\dfrac{1}{3}b = -4$

55. $-p = -\dfrac{7}{10}$
56. $-8h = 0$
57. $-9 = \dfrac{w}{2} - 3$
58. $-16 = \dfrac{t}{4} - 14$

59. $2x - \dfrac{1}{2} = \dfrac{1}{6}$
60. $3z - \dfrac{3}{4} = \dfrac{1}{2}$
61. $\dfrac{5}{4}x = \dfrac{5}{6}x + \dfrac{2}{3}$
62. $\dfrac{3}{4}y = \dfrac{3}{2}y + \dfrac{1}{5}$

63. $-4 - \dfrac{3}{2}d = \dfrac{2}{5}$
64. $-2 - \dfrac{5}{4}z = \dfrac{5}{8}$

Expanding Your Skills

For Exercises 65–68, solve the equations by clearing fractions.

65. $p - 1 + \dfrac{1}{4}p = 2 + \dfrac{3}{4}p$
66. $\dfrac{4}{3} + \dfrac{2}{3}q = -\dfrac{5}{3} - q - \dfrac{1}{3}$

67. $\dfrac{5}{3}x - \dfrac{4}{5} = \dfrac{2}{3}x + 1$
68. $\dfrac{3}{4}y + \dfrac{9}{7} = -\dfrac{1}{4}y + 2$

Problem Recognition Exercises

Comparing Expressions and Equations

For Exercises 1–24, first identify the problem as an expression or as an equation. Then simplify the expression or solve the equation. Two examples are given for you.

Example: $\dfrac{1}{3} + \dfrac{1}{6} - \dfrac{5}{6}$

This is an expression. Combine *like* terms.

$$\dfrac{1}{3} + \dfrac{1}{6} - \dfrac{5}{6}$$
$$= \dfrac{2 \cdot 1}{2 \cdot 3} + \dfrac{1}{6} - \dfrac{5}{6}$$
$$= \dfrac{2}{6} + \dfrac{1}{6} - \dfrac{5}{6}$$
$$= \dfrac{-2}{6}$$
$$= -\dfrac{1}{3}$$

Example: $\dfrac{1}{3}x + \dfrac{1}{6} = \dfrac{5}{6}$

This is an equation. Solve the equation.

$$\dfrac{1}{3}x + \dfrac{1}{6} = \dfrac{5}{6}$$
$$6 \cdot \left(\dfrac{1}{3}x + \dfrac{1}{6}\right) = 6 \cdot \left(\dfrac{5}{6}\right)$$
$$2x + 1 = 5$$
$$2x + 1 - 1 = 5 - 1$$
$$2x = 4$$
$$\dfrac{2x}{2} = \dfrac{4}{2}$$
$$x = 2 \quad \text{The solution is 2.}$$

1. $\dfrac{5}{3}x = \dfrac{1}{6}$
2. $\dfrac{7}{8}y = \dfrac{3}{4}$
3. $\dfrac{5}{3} - \dfrac{1}{6}$

4. $\dfrac{7}{8} - \dfrac{3}{4}$
5. $1 + \dfrac{2}{5}$
6. $1 + \dfrac{4}{5}$

7. $z + \dfrac{2}{5} = 0$
8. $w + \dfrac{4}{5} = 0$
9. $\dfrac{2}{9}x - \dfrac{1}{3} = \dfrac{5}{9}$

10. $\dfrac{3}{10}x - \dfrac{2}{5} = \dfrac{1}{10}$
11. $\dfrac{2}{9} - \dfrac{1}{3} + \dfrac{5}{9}$
12. $\dfrac{3}{10} - \dfrac{2}{5} + \dfrac{1}{10}$

13. $3(x - 4) + 2x = 12 + x$
14. $5(x + 4) - 2x = 10 - x$
15. $3(x - 4) + 2x - 12 + x$

16. $5(x + 4) - 2x + 10 - x$
17. $\dfrac{3}{4}x = 2$
18. $\dfrac{5}{7}y = 5$

19. $\dfrac{3}{4} \cdot 2$
20. $\dfrac{5}{7} \cdot 5$
21. $\dfrac{4}{7} = 2c$

22. $\dfrac{9}{4} = 3d$
23. $\dfrac{4}{7}c + 2c$
24. $\dfrac{9}{4}d + 3d$

Chapter 4 Review Exercises

Section 4.1

For Exercises 1 and 2, write a fraction that represents the shaded area.

1.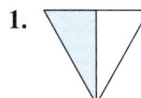
2.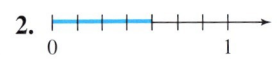

3. a. Write a fraction that has denominator 3 and numerator 5.
   b. Label this fraction as proper or improper.

4. a. Write a fraction that has numerator 1 and denominator 6.
   b. Label this fraction as proper or improper.

5. Simplify.
   a. $\left|-\dfrac{3}{8}\right|$
   b. $\left|\dfrac{2}{3}\right|$
   c. $-\left(-\dfrac{4}{9}\right)$

For Exercises 6 and 7, write a fraction and a mixed number that represent the shaded area.

6.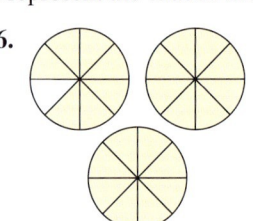
7.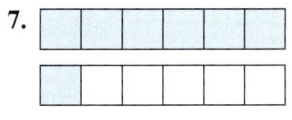

For Exercises 8 and 9, convert the mixed number to a fraction.

8. $6\dfrac{1}{7}$
9. $11\dfrac{2}{5}$

For Exercises 10 and 11, convert the improper fraction to a mixed number.

10. $\dfrac{47}{9}$
11. $\dfrac{23}{21}$

For Exercises 12–15, locate the numbers on the number line.

12. $-\dfrac{10}{5}$ 13. $-\dfrac{7}{8}$ 14. $\dfrac{13}{8}$ 15. $-1\dfrac{3}{8}$

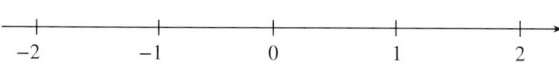

For Exercises 16 and 17, divide. Write the answer as a mixed number.

16. $7\overline{)941}$ 17. $26\overline{)1582}$

Section 4.2

For Exercises 18 and 19, refer to this list of numbers:
21, 43, 51, 55, 58, 124, 140, 260, 1200.

18. List all the numbers that are divisible by 3.

19. List all the numbers that are divisible by 5.

20. Identify the prime numbers in the list.
    2, 39, 53, 54, 81, 99, 112, 113

21. Identify the composite numbers in the list.
    1, 12, 27, 51, 63, 97, 130

For Exercises 22 and 23, find the prime factorization.

22. 330   23. 900

For Exercises 24 and 25, determine if the fractions are equivalent. Fill in the blank with = or ≠.

24. $\dfrac{3}{6} \square \dfrac{5}{9}$ 25. $\dfrac{15}{21} \square \dfrac{10}{14}$

For Exercises 26–33, simplify the fraction to lowest terms. Write the answer as a fraction.

26. $\dfrac{5}{20}$ 27. $\dfrac{7}{35}$

28. $-\dfrac{24}{16}$ 29. $-\dfrac{63}{27}$

30. $\dfrac{120}{1500}$ 31. $\dfrac{140}{20{,}000}$

32. $\dfrac{4ac}{10c^2}$ 33. $\dfrac{24t^3}{30t}$

34. On his final exam, Gareth got 42 out of 45 questions correct. What fraction of the test represents correct answers? What fraction represents incorrect answers?

35. Isaac proofread 6 pages of his 10-page term paper. Yulisa proofread 6 pages of her 15-page term paper.

    a. What fraction of his paper did Isaac proofread?

    b. What fraction of her paper did Yulisa proofread?

Section 4.3

For Exercises 36–41, multiply the fractions and simplify to lowest terms. Write the answer as a fraction or an integer.

36. $-\dfrac{2}{5} \cdot \dfrac{15}{14}$ 37. $-\dfrac{4}{3} \cdot \dfrac{9}{8}$

38. $-14 \cdot \left(-\dfrac{9}{2}\right)$ 39. $-33 \cdot \left(-\dfrac{5}{11}\right)$

40. $\dfrac{2x}{5} \cdot \dfrac{10}{x^2}$ 41. $\dfrac{3y^3}{14} \cdot \dfrac{7}{y}$

42. Write the formula for the area of a triangle.

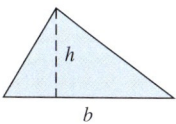

43. Find the area of the shaded region.

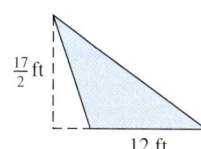

For Exercises 44 and 45, multiply.

44. $-\dfrac{3}{4} \cdot \left(-\dfrac{4}{3}\right)$ 45. $\dfrac{1}{12} \cdot 12$

For Exercises 46 and 47, find the reciprocal of the number, if it exists.

46. $\dfrac{7}{2}$ 47. -7

For Exercises 48–53, divide and simplify the answer to lowest terms. Write the answer as a fraction or an integer.

48. $\dfrac{28}{15} \div \dfrac{21}{20}$ 49. $\dfrac{7}{9} \div \dfrac{35}{63}$

50. $-\dfrac{6}{7} \div 18$ 51. $12 \div \left(-\dfrac{6}{7}\right)$

52. $\dfrac{4a^2}{7} \div \dfrac{a}{14}$ 53. $\dfrac{11}{3y} \div \dfrac{22}{9y^3}$

54. How many $\tfrac{2}{3}$-lb bags of candy can be filled from a 24-lb sack of candy?

55. Chuck is an elementary school teacher and needs 22 pieces of wood, $\tfrac{3}{8}$ ft long, for a class project. If he has a 9-ft board from which to cut the pieces, will he have enough $\tfrac{3}{8}$-ft pieces for his class? Explain.

For Exercises 56 and 57, refer to the graph. The graph represents the distribution of the students at a college by race/ethnicity.

Distribution of Student Body by Race/Ethnicity

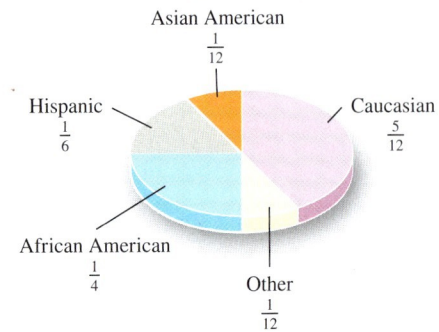

Asian American $\frac{1}{12}$
Hispanic $\frac{1}{6}$
Caucasian $\frac{5}{12}$
African American $\frac{1}{4}$
Other $\frac{1}{12}$

56. If the college has 3600 students, how many are African American?

57. If the college has 3600 students, how many are Asian American?

58. Amelia worked only $\frac{4}{5}$ of her normal 40-hr workweek. If she makes $18 per hour, how much money did she earn for the week?

Section 4.4

59. Find the prime factorization.

    a. 100 b. 65 c. 70

For Exercises 60 and 61, find the LCM by using any method.

60. 105 and 28 61. 16, 24, and 32

62. Sharon and Tony signed up at a gym on the same day. Sharon will be able to go to the gym every third day and Tony will go to the gym every fourth day. In how many days will they meet again at the gym?

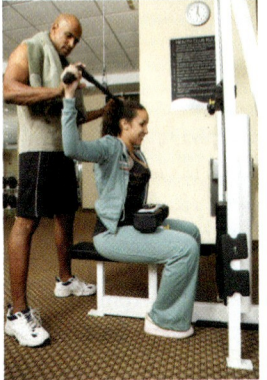

Tanya Constantine/Blend Images/Getty Images

For Exercises 63–66, rewrite each fraction with the indicated denominator.

63. $\frac{5}{16} = \frac{}{48}$

64. $\frac{9}{5} = \frac{}{35}$

65. $\frac{7}{12} = \frac{}{60y}$

66. $\frac{-7}{x} = \frac{}{4x}$

For Exercises 67–69, fill in the blanks with <, >, or =.

67. $\frac{11}{24} \square \frac{7}{12}$

68. $\frac{5}{6} \square \frac{7}{9}$

69. $-\frac{5}{6} \square -\frac{15}{18}$

70. Rank the numbers from least to greatest.

$-\frac{7}{10}, -\frac{72}{105}, -\frac{8}{15}, -\frac{27}{35}$

Section 4.5

For Exercises 71–80, add or subtract. Write the answer as a fraction simplified to lowest terms.

71. $\frac{5}{6} + \frac{4}{6}$

72. $\frac{4}{15} + \frac{6}{15}$

73. $\frac{9}{10} - \frac{61}{100}$

74. $\frac{11}{25} - \frac{2}{5}$

75. $-\frac{25}{11} - 2$

76. $-4 - \frac{37}{20}$

77. $\frac{2}{15} - \left(-\frac{5}{8}\right) - \frac{1}{3}$

78. $\frac{11}{14} - \frac{4}{7} - \left(-\frac{3}{2}\right)$

79. $\frac{7}{5w} + \frac{2}{w}$

80. $\frac{11}{a} + \frac{4}{b}$

For Exercises 81 and 82, find (a) the perimeter and (b) the area.

81. Triangle with sides $\frac{63}{16}$ m, $\frac{25}{16}$ m, $\frac{13}{4}$ m, and height $1\frac{5}{4}$ m.

82. Rectangle with sides $\frac{3}{2}$ yd and $\frac{7}{3}$ yd.

Section 4.6

For Exercises 83–88, multiply or divide as indicated.

83. $\left(3\frac{2}{3}\right)\left(6\frac{2}{5}\right)$

84. $\left(11\frac{1}{3}\right)\left(2\frac{3}{34}\right)$

85. $-3\frac{5}{11} \div \left(-3\frac{4}{5}\right)$

86. $-7 \div \left(-1\frac{5}{9}\right)$

87. $-4\frac{6}{11} \div 2$

88. $10\frac{1}{5} \div (-17)$

For Exercises 89 and 90, round the numbers to estimate the answer. Then find the exact sum or difference.

89. $65\frac{1}{8} - 14\frac{9}{10}$

   Estimate: _____
   Exact: _____

90. $43\frac{13}{15} - 20\frac{23}{25}$

   Estimate: _____
   Exact: _____

For Exercises 91–100, add or subtract the mixed numbers.

91. $9\frac{8}{9}$
    $+1\frac{2}{7}$

92. $10\frac{1}{2}$
    $+3\frac{15}{16}$

93. $7\frac{5}{24}$
    $-4\frac{7}{12}$

94. $5\frac{1}{6}$
    $-3\frac{1}{4}$

95. 6
    $-2\frac{3}{5}$

96. 8
    $-4\frac{11}{14}$

97. $42\frac{1}{8} - \left(-21\frac{13}{16}\right)$

98. $38\frac{9}{10} - \left(-11\frac{3}{5}\right)$

99. $-4\frac{2}{3} + 1\frac{5}{6}$

100. $6\frac{3}{8} + \left(-10\frac{1}{4}\right)$

101. Corry drove for $4\frac{1}{2}$ hr in the morning and $3\frac{2}{3}$ hr in the afternoon. Find the total number of hours he drove.

George Doyle/Stockbyte/SuperStock

102. Denise owned $2\frac{1}{8}$ acres of land. If she sells $1\frac{1}{4}$ acres, how much will she have left?

103. It takes $1\frac{1}{4}$ gal of paint for Neva to paint her living room. If her great room is $2\frac{1}{2}$ times larger than the living room, how many gallons will it take to paint the great room?

SW Productions/Brand X Pictures/Getty Images

104. A roll of ribbon contains $12\frac{1}{2}$ yd. How many pieces of length $1\frac{1}{4}$ yd can be cut from this roll?

Section 4.7

For Exercises 105–116, simplify.

105. $\left(\frac{3}{8}\right)^2$

106. $\left(-\frac{3}{8}\right)^2$

107. $\left(-\frac{3}{8} \cdot \frac{4}{15}\right)^5$

108. $\left(\frac{1}{25} \cdot \frac{15}{6}\right)^4$

109. $-\frac{2}{5} - \left(1\frac{2}{3}\right) \cdot \frac{3}{2}$

110. $\frac{7}{5} - \left(-2\frac{1}{3}\right) \div \frac{7}{2}$

111. $\left(\frac{2}{3} - \frac{5}{6}\right)^2 + \frac{5}{36}$

112. $\left(-\frac{1}{4} - \frac{1}{2}\right)^2 - \frac{1}{8}$

113. $\dfrac{\frac{8}{5}}{\frac{4}{7}}$

114. $\dfrac{\frac{14}{9}}{\frac{7}{x}}$

115. $\dfrac{-\frac{2}{3} - \frac{5}{6}}{3 + \frac{1}{2}}$

116. $\dfrac{\frac{3}{5} - 1}{-\frac{1}{2} - \frac{3}{10}}$

For Exercises 117–120, evaluate the expressions for the given values of the variables.

117. $x \div y \div z$ for $x = \frac{2}{3}$, $y = \frac{5}{6}$, and $z = -\frac{3}{5}$

118. a^2b for $a = -\frac{3}{5}$ and $b = 1\frac{2}{3}$

119. $t^2 + v^2$ for $t = \frac{1}{2}$ and $v = -\frac{1}{4}$

120. $2(w + z)$ for $w = 3\frac{1}{3}$ and $z = 2\frac{1}{2}$

For Exercises 121–124, simplify the expressions.

121. $-\frac{3}{4}x - \frac{2}{3}x$

122. $\frac{1}{5}y - \frac{3}{2}y$

123. $-\frac{4}{3}a + \frac{1}{2}c + 2a - \frac{1}{3}c$

124. $\frac{4}{5}w + \frac{2}{3}y + \frac{1}{10}w + y$

Section 4.8

For Exercises 125–134, solve the equations.

125. $x - \frac{3}{5} = \frac{2}{3}$

126. $y + \frac{4}{7} = \frac{3}{14}$

127. $\frac{3}{5}x = \frac{2}{3}$

128. $\frac{4}{7}y = \frac{3}{14}$

129. $-\frac{6}{5} = -2c$

130. $-\frac{9}{4} = -3d$

131. $-\frac{2}{5} + \frac{y}{10} = \frac{y}{2}$

132. $-\frac{3}{4} + \frac{w}{2} = \frac{w}{8}$

133. $2 = \frac{1}{2} - \frac{x}{10}$

134. $1 = \frac{7}{3} - \frac{t}{9}$

Decimals

5

CHAPTER OUTLINE

- **5.1** Decimal Notation and Rounding 268
- **5.2** Addition and Subtraction of Decimals 279
- **5.3** Multiplication of Decimals and Applications with Circles 288
- **5.4** Division of Decimals 301
  - **Problem Recognition Exercises:** Operations on Decimals 313
- **5.5** Fractions, Decimals, and the Order of Operations 314
- **5.6** Solving Equations Containing Decimals 329
  - **Chapter 5** Review Exercises 336

Decimals Have a Point

Suppose that the temperature rises from 70° in the morning to 85° in the mid-afternoon. Intuitively, it makes sense that the temperature must attain each degree measure between 70° and 85° at some point during the day. That is to say, at some point, the temperature is 71°, at some point the temperature is 72°, and so on. But the temperature must also take on the intermediate values between the whole numbers. In this chapter, we introduce decimal notation to represent both whole numbers and the intermediate values between them.

SketchbookDesign/iStockphoto/Getty Images

A decimal number has two parts separated by a decimal point. For example, using our temperature scenario, at some point during the day the temperature must be 70.1°, 70.2°, 70.3°, and so on. Likewise, at some point, the temperature must be 70.01°, 70.02°, 70.03°, and so on. These values are written in decimal form. The number to the left of the decimal point is called the *whole number part*, and the number to the right of the decimal point is called the *fractional part*. The positions of the digits to the right of the decimal point represent powers of one-tenth. For example, the number 70.1° represents $70\frac{1}{10}$° and the number 70.2° represents $70\frac{2}{10}$°. The number 70.01° represents $70\frac{1}{100}$°.

In this chapter, we study operations on decimals and learn how they are related to fractions.

Section 5.1 Decimal Notation and Rounding

Concepts

1. Decimal Notation
2. Writing Decimals as Mixed Numbers or Fractions
3. Ordering Decimal Numbers
4. Rounding Decimals

1. Decimal Notation

We learned that fraction notation denotes equal parts of a whole. In this chapter, we introduce decimal notation to denote parts of a whole. We first introduce the concept of a decimal fraction. A **decimal fraction** is a fraction whose denominator is a power of 10. The following are examples of decimal fractions.

$$\frac{3}{10} \text{ is read as "three-tenths"}$$

$$\frac{7}{100} \text{ is read as "seven-hundredths"}$$

$$-\frac{9}{1000} \text{ is read as "negative nine-thousandths"}$$

We now want to write these fractions in **decimal notation**. This means that we will write the numbers by using place values, as we did with whole numbers. The place value chart can be extended as shown in Figure 5-1.

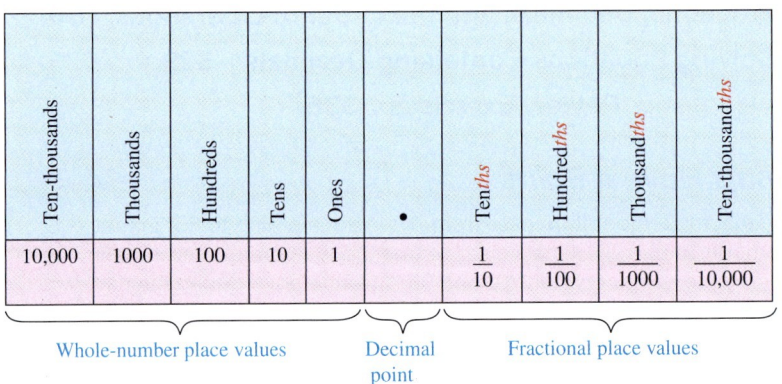

Figure 5-1

From Figure 5-1, we see that the decimal point separates the whole number part from the fractional part. The place values for decimal fractions are located to the right of the decimal point. Their place value names are similar to those for whole numbers, but end in *ths*. Notice the correspondence between the tens place and the ten*ths* place. Similarly notice the hundreds place and the hundred*ths* place. Each place value on the left has a corresponding place value on the right, with the exception of the ones place. There is no "one*ths*" place.

Example 1 Identifying Place Values

Identify the place value of each underlined digit.

a. 30,804.0<u>9</u>   b. −0.8469<u>2</u>0   c. 2<u>9</u>3.604

Solution:

a. 30,804.0<u>9</u>   The digit 9 is in the hundredths place.

b. −0.8469<u>2</u>0   The digit 2 is in the hundred-thousandths place.

c. 2<u>9</u>3.604   The digit 9 is in the tens place.

Skill Practice Identify the place value of each underlined digit.
1. 24.38<u>6</u>   2. −218.02168<u>4</u>   3. 1<u>3</u>16.42

For a whole number, the decimal point is understood to be after the ones place, and is usually not written. For example:

$$42. = 42$$

Using Figure 5-1, we can write the numbers $\frac{3}{10}$, $\frac{7}{100}$, and $-\frac{9}{1000}$ in decimal notation.

Fraction	Word name	Decimal notation
$\frac{3}{10}$	Three-tenths	0.3 ↑ tenths place
$\frac{7}{100}$	Seven-hundredths	0.07 ↑ hundredths place
$-\frac{9}{1000}$	Negative nine-thousandths	−0.009 ↑ thousandths place

TIP: The 0 to the left of the decimal point is a placeholder so that the position of the decimal point can be easily identified. It does not contribute to the value of the number. Thus, 0.3 and .3 are equal.

Now consider the number $15\frac{7}{10}$. This value represents 1 ten + 5 ones + 7 tenths. In decimal form we have 15.7.

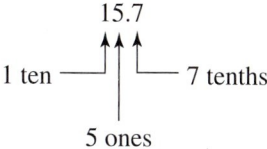

1 ten — 5 ones — 7 tenths

The decimal point is interpreted as the word *and*. Thus, 15.7 is read as "fifteen *and* seven tenths." The number 356.29 can be represented as

$$356 + 2 \text{ tenths} + 9 \text{ hundredths} = 356 + \frac{2}{10} + \frac{9}{100}$$

$$= 356 + \frac{20}{100} + \frac{9}{100} \quad \text{We can use the LCD of 100 to add the fractions.}$$

$$= 356\frac{29}{100}$$

We can read the number 356.29 as "three hundred fifty-six *and* twenty-nine hundredths."
This discussion leads to a quicker method to read decimal numbers.

Answers
1. Thousandths
2. Millionths
3. Hundreds

> **Reading a Decimal Number**
>
> **Step 1** The part of the number to the left of the decimal point is read as a whole number. *Note:* If there is no whole-number part, skip to step 3.
> **Step 2** The decimal point is read *and*.
> **Step 3** The part of the number to the right of the decimal point is read as a whole number but is followed by the name of the place position of the digit farthest to the right.

Example 2 Reading Decimal Numbers

Write the word name for each number.

a. 1028.4   **b.** 2.0736   **c.** −0.478

Solution:

a. 1028.4   is written as "one thousand, twenty-eight and four-tenths."
b. 2.0736   is written as "two and seven hundred thirty-six ten-thousandths."
c. −0.478   is written as "negative four hundred seventy-eight thousandths."

Skill Practice Write a word name for each number.

4. 1004.6   **5.** 3.042   **6.** −0.0063

Example 3 Writing a Numeral from a Word Name

Write the word name as a numeral.

a. Four hundred eight and fifteen ten-thousandths
b. Negative five thousand eight hundred and twenty-three hundredths

Solution:

a. Four hundred eight *and* fifteen ten-thousandths: 408.0015
b. Negative five thousand eight hundred *and* twenty-three hundredths: −5800.23

Skill Practice Write the word name as a numeral.

7. Two hundred and two hundredths
8. Negative seventy-nine and sixteen thousandths

2. Writing Decimals as Mixed Numbers or Fractions

A fractional part of a whole may be written as a fraction or as a decimal. To convert a decimal to an equivalent fraction, it is helpful to think of the decimal in words. For example:

Decimal	Word name	Fraction	
0.3	Three tenths	$\frac{3}{10}$	
0.67	Sixty-seven hundredths	$\frac{67}{100}$	
0.048	Forty-eight thousandths	$\frac{48}{1000} = \frac{6}{125}$	(simplified)
6.8	Six and eight-tenths	$6\frac{8}{10} = 6\frac{4}{5}$	(simplified)

From the list, we notice several patterns that can be summarized as follows.

Answers
4. One thousand, four and six-tenths
5. Three and forty-two thousandths
6. Negative sixty-three ten-thousandths
7. 200.02   **8.** −79.016

Section 5.1 Decimal Notation and Rounding

Converting a Decimal to a Mixed Number or Proper Fraction

Step 1 The digits to the right of the decimal point are written as the numerator of the fraction.
Step 2 The place value of the digit farthest to the right of the decimal point determines the denominator.
Step 3 The whole-number part of the number is left unchanged.
Step 4 Once the number is converted to a fraction or mixed number, simplify the fraction to lowest terms, if possible.

Example 4 Writing Decimals as Proper Fractions or Mixed Numbers

Write the decimals as proper fractions or mixed numbers and simplify.

a. 0.847   **b.** −0.0025   **c.** 4.16

Solution:

a. $0.847 = \dfrac{847}{1000}$

↑ thousandths place

b. $-0.0025 = -\dfrac{25}{10{,}000} = -\dfrac{\overset{1}{\cancel{25}}}{\underset{400}{\cancel{10{,}000}}} = -\dfrac{1}{400}$

↑ ten-thousandths place

c. $4.16 = 4\dfrac{16}{100} = 4\dfrac{\overset{4}{\cancel{16}}}{\underset{25}{\cancel{100}}} = 4\dfrac{4}{25}$

↑ hundredths place

FOR REVIEW

Recall that we may simplify with any common factors, not just the GCF.

$$-\dfrac{25}{10{,}000} = -\dfrac{\overset{5}{\cancel{25}}}{\underset{2{,}000}{\cancel{10{,}000}}} = -\dfrac{\overset{1}{\cancel{5}}}{\underset{400}{\cancel{2{,}000}}} = -\dfrac{1}{400}$$

Skill Practice Write the decimals as proper fractions or mixed numbers.
9. 0.034   **10.** −0.00086   **11.** 3.184

A decimal number greater than 1 can be written as a mixed number or as an improper fraction. The number 4.16 from Example 4(c) can be expressed as follows.

$$4.16 = 4\dfrac{16}{100} = 4\dfrac{4}{25} \quad \text{or} \quad \dfrac{104}{25}$$

A quick way to obtain an improper fraction for a decimal number greater than 1 is outlined here.

Writing a Decimal Number Greater Than 1 as an Improper Fraction

Step 1 The denominator is determined by the place position of the digit farthest to the right of the decimal point.
Step 2 The numerator is obtained by removing the decimal point of the original number. The resulting whole number is then written over the denominator.
Step 3 Simplify the improper fraction to lowest terms, if possible.

Answers
9. $\dfrac{17}{500}$ **10.** $-\dfrac{43}{50{,}000}$ **11.** $3\dfrac{23}{125}$

For example:

$$4.16 = \frac{416}{100} = \frac{104}{25} \quad \text{(simplified)}$$

(Remove decimal point. hundredths place)

Example 5 — Writing Decimals as Improper Fractions

Write the decimals as improper fractions and simplify.

a. 40.2   b. −2.113

Solution:

a. $40.2 = \frac{402}{10} = \frac{\cancel{402}^{201}}{\cancel{10}_{5}} = \frac{201}{5}$

b. $-2.113 = -\frac{2113}{1000}$ Note that the fraction is already in lowest terms.

Skill Practice Write the decimals as improper fractions and simplify.

12. 6.38   13. −15.1

3. Ordering Decimal Numbers

It is often necessary to compare the values of two decimal numbers.

Comparing Two Positive Decimal Numbers

Step 1 Starting at the left (and moving toward the right), compare the digits in each corresponding place position.

Step 2 As we move from left to right, the first instance in which the digits differ determines the order of the numbers. The number having the greater digit is greater overall.

Example 6 — Ordering Decimals

Fill in the blank with < or >.

a. 0.68 ☐ 0.7   b. 3.462 ☐ 3.4619

Solution:

a. 0.68 $<$ 0.7   (different 6 < 7)

b. 3.462 $>$ 3.4619   (different 2 > 1, same)

TIP: Decimal numbers can also be ordered by comparing their fractional forms:

$0.68 = \frac{68}{100}$ and $0.7 = \frac{7}{10} = \frac{70}{100}$

Therefore, 0.68 < 0.7.

Skill Practice Fill in the blank with < or >.

14. 4.163 ☐ 4.159   15. 218.38 ☐ 218.41

Answers

12. $\frac{319}{50}$ 13. $-\frac{151}{10}$
14. >   15. <

When ordering two negative numbers, you must be careful to visualize their relative positions on the number line. For example, 1.4 < 1.5 but −1.4 > −1.5. See Figure 5-2.

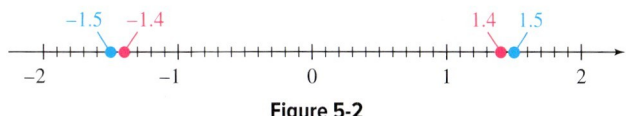

Figure 5-2

Also note that when ordering two decimal numbers, sometimes it is necessary to insert additional zeros after the rightmost digit in a number. This does not change the value of the number. For example:

$$0.7 = 0.70 \text{ because } \frac{7}{10} = \frac{70}{100}$$

Example 7 — Ordering Negative Decimals

Fill in the blank with < or >.

a. −0.2 ☐ −0.05   b. −0.04591 ☐ −0.0459

Solution:

a. Insert a zero in the hundredths place for the number on the left. The digits in the tenths place are different.

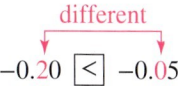   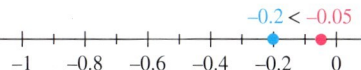

b. Insert a zero in the hundred-thousandths place for the number on the right. The digits in the hundred-thousandths place are different.

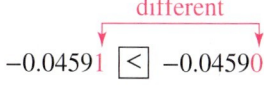   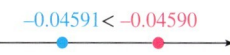

Skill Practice Fill in the blank with < or >.

16. −0.32 ☐ −0.062   **17.** −0.873 ☐ −0.8731

4. Rounding Decimals

The process to round the decimal part of a number is nearly the same as rounding whole numbers. The main difference is that the digits to the right of the rounding place position are dropped instead of being replaced by zeros.

Rounding Decimals to a Place Value to the Right of the Decimal Point

Step 1 Identify the digit one position to the right of the given place value.
Step 2 If the digit in step 1 is 5 or greater, add 1 to the given digit. If the digit in step 1 is less than 5, leave the given digit unchanged.
Step 3 Discard all digits to the right of the given digit.

Answers
16. <   **17.** >

Example 8 — Rounding Decimal Numbers

a. Round 4.81542 to the thousandths place.

b. Round 52.9999 to the hundredths place.

Solution:

a. $4.81542 \approx 4.815$ — remaining digits discarded

The digit in the thousandths place — this digit is less than 5. Discard it and all digits to the right.

b. $5\overset{1}{2}.\overset{1}{9}\overset{+1}{9}\overline{99}$ — discard remaining digits

The hundredths place — this digit is greater than 5. Add 1 to the hundredths-place digit.

- Since the hundredths-place digit is 9, adding 1 requires us to carry 1 to the tenths-place digit.
- Since the tenths-place digit is 9, adding 1 requires us to carry 1 to the ones-place digit.

≈ 53.00

Skill Practice

18. Round 45.372 to the hundredths place.
19. Round 134.9996 to the thousandths place.

Example 9 — Rounding Decimal Numbers

Round 14.795 to the indicated place value.

a. Tenths   b. Hundredths

Solution:

a. $14.\overset{+1}{7}95 \approx 14.8$ — remaining digits discarded

tenths place — this digit is 5 or greater. Add 1 to the tenths place.

b. $14.7\overset{+1}{9}5 \approx 14.80$ — remaining digit discarded

hundredths place — this digit is 5 or greater. Add 1 to the hundredths place.

- Since the hundredths-place digit is 9, adding 1 requires us to carry 1 to the tenths-place digit.

Skill Practice Round 187.26498 to the indicated place value.

20. Hundredths   21. Ten-thousandths

In Example 9(b) the 0 in 14.80 indicates that the number was rounded to the hundredths place. It would be incorrect to drop the zero. Even though 14.8 has the same numerical value as 14.80, it implies a different level of accuracy. For example, when measurements are taken using some instrument such as a ruler or scale, the measured values are not exact. The place position to which a number is rounded reflects the accuracy of the measuring device. Thus, the value 14.8 lb indicates that the scale is accurate to the nearest tenth of a pound. The value 14.80 lb indicates that the scale is accurate to the nearest hundredth of a pound.

Answers
18. 45.37
19. 135.000
20. 187.26
21. 187.2650

Section 5.1 Activity

A.1. The diameter of a half dollar coin is 1.205 inches.
  a. Identify the place value of the 0.
  b. Identify the place value of the 1.
  c. Identify the place value of the 5.
  d. Identify the place value of the 2.

For Exercises A.2–A.7, write the decimal as a word name or write the word name as a decimal.

	Decimal	Word Name
A.2.		Three and six thousandths
A.3.		Thirteen hundredths
A.4.		Twenty and twelve ten-thousandths
A.5.	7.64	
A.6.	205	
A.7.	0.065	

A.8. For the decimal 0.15,
  a. Write the word name for the decimal.
  b. Use the word name from part (a) to help you write the given decimal as a fraction, with the number as the numerator and the place value as the denominator.
  c. What do you observe about the number of digits to the right of the decimal point and the number of zeros in the denominator?
  d. Simplify the fraction to lowest terms.

A.9. Now consider the number 4.15.
  a. What is the whole number part of 4.15?
  b. What is the decimal part of 4.15?
  c. Write the decimal part as a fraction in lowest terms (see Exercise A.8).
  d. Include the whole number part with the fraction in part (d). The answer should be a mixed fraction.
  e. How would the answer change if the number is 4238.15?

A.10. Rank the decimals from least to greatest. Before comparing the decimals, make sure all numbers have the same number of digits after the decimal point by adding zeros where necessary.

   0.59   −0.6   −0.568   0.594   0.568   −0.64

For Exercises A.11–A.14, round each number to the given place value.

		Tenths	Hundredths	Thousandths
A.11.	1.8275	a.	b.	c.
A.12.	6.1529	a.	b.	c.
A.13.	30.3999	a.	b.	c.
A.14.	21.6959	a.	b.	c.

Section 5.1 Practice Exercises

Vocabulary and Key Concepts

1. a. A _____ fraction is a fraction whose denominator is a power of 10.

    b. The first three place values to the right of the decimal point are the _____ place, the _____ place, and the _____ place.

Concept 1: Decimal Notation

2. State the first five place values to the right of the decimal point in order from left to right.

For Exercises 3–10, expand the powers of 10 or $\frac{1}{10}$.

3. 10^2

4. 10^3

5. 10^4

6. 10^5

7. $\left(\frac{1}{10}\right)^2$

8. $\left(\frac{1}{10}\right)^3$

9. $\left(\frac{1}{10}\right)^4$

10. $\left(\frac{1}{10}\right)^5$

For Exercises 11–22, identify the place value of each underlined digit. (See Example 1.)

11. 3.9̲83

12. 34.8̲2

13. 440.39̲

14. 24̲8.94

15. 48̲9.02

16. 4.092̲84

17. −9.2834̲5

18. −0.321̲

19. 0.489̲

20. 58̲.211

21. −93̲.834

22. −5.000001̲

For Exercises 23–30, write the word name for each decimal fraction.

23. $\frac{9}{10}$

24. $\frac{7}{10}$

25. $\frac{23}{100}$

26. $\frac{19}{100}$

27. $-\frac{33}{1000}$

28. $-\frac{51}{1000}$

29. $\frac{407}{10,000}$

30. $\frac{20}{10,000}$

For Exercises 31–38, write the word name for the decimal. (See Example 2.)

31. 3.24

32. 4.26

33. −5.9

34. −3.4

35. 52.3

36. 21.5

37. 6.219

38. 7.338

For Exercises 39–44, write the word name as a numeral. (See Example 3.)

39. Negative eight thousand, four hundred seventy-two and fourteen thousandths

40. Negative sixty thousand, twenty-five and four hundred one ten-thousandths

41. Seven hundred and seven hundredths

42. Nine thousand and nine thousandths

43. Negative two million, four hundred sixty-nine thousand and five hundred six thousandths

44. Negative eighty-two million, six hundred fourteen and ninety-seven ten-thousandths

Concept 2: Writing Decimals as Mixed Numbers or Fractions

For Exercises 45–56, write the decimal as a proper fraction or as a mixed number and simplify. (See Example 4.)

45. 3.7   **46.** 1.9   **47.** 2.8   **48.** 4.2

49. 0.25   **50.** 0.75   **51.** −0.55   **52.** −0.45

53. 20.812   **54.** 32.905   **55.** −15.0005   **56.** −4.0015

For Exercises 57–64, write the decimal as an improper fraction and simplify. (See Example 5.)

57. 8.4   **58.** 2.5   **59.** 3.14   **60.** 5.65

61. −23.5   **62.** −14.6   **63.** 11.91   **64.** 21.33

Concept 3: Ordering Decimal Numbers

For Exercises 65–72, fill in the blank with < or >. (See Examples 6 and 7.)

65. 6.312 ☐ 6.321   **66.** 8.503 ☐ 8.530   **67.** 11.21 ☐ 11.2099   **68.** 10.51 ☐ 10.5098

69. −0.762 ☐ −0.76   **70.** −0.1291 ☐ −0.129   **71.** −51.72 ☐ −51.721   **72.** −49.06 ☐ −49.062

73. Which number is between 3.12 and 3.13? Circle all that apply.

   a. 3.127   **b.** 3.129   **c.** 3.134   **d.** 3.139

74. Which number is between 42.73 and 42.86? Circle all that apply.

   a. 42.81   **b.** 42.64   **c.** 42.79   **d.** 42.85

75. The batting averages for five baseball legends are given in the table. Rank the players' batting averages from lowest to highest. (*Source:* Baseball Almanac)

Player	Average
Joe Jackson	0.3558
Ty Cobb	0.3664
Lefty O'Doul	0.3493
Ted Williams	0.3444
Rogers Hornsby	0.3585

76. The average speed, in miles per hour (mph), of the Daytona 500 for selected years is given in the table. Rank the speeds from slowest to fastest. (*Source:* NASCAR)

Year	Driver	Speed (mph)
1989	Darrell Waltrip	148.466
1991	Ernie Irvan	148.148
1997	Jeff Gordon	148.295
2007	Kevin Harvick	149.333

Source: Library of Congress Prints and Photographs Division [LC-USZ62-97880]

U.S. Air Force photo by Master Sgt Michael A. Kaplan

Concept 4: Rounding Decimals

77. The numbers given all have equivalent value. However, suppose they represent measured values from a scale. Explain the difference in the interpretation of these numbers.

$$0.25, \quad 0.250, \quad 0.2500, \quad 0.25000$$

78. Which number properly represents 3.499999 rounded to the thousandths place?

   a. 3.500   **b.** 3.5   **c.** 3.500000   **d.** 3.499

79. Which value is rounded to the nearest tenth, 7.1 or 7.10?

80. Which value is rounded to the nearest hundredth, 34.50 or 34.5?

For Exercises 81–92, round the decimals to the indicated place values. (See Examples 8 and 9.)

81. 49.943; tenths
82. 12.7483; tenths
83. 33.416; hundredths

84. 4.359; hundredths
85. −9.0955; thousandths
86. −2.9592; thousandths

87. 21.0239; tenths
88. 16.804; hundredths
89. 6.9995; thousandths

90. 21.9997; thousandths
91. 0.0079499; ten-thousandths
92. 0.00084985; ten-thousandths

93. A snail moves at a rate of about 0.00362005 miles per hour. Round the decimal value to the ten-thousandths place.

Photodisc/Getty Images

For Exercises 94–97, round the number to the indicated place value.

	Number	Hundreds	Tens	Tenths	Hundredths	Thousandths
94.	349.2395					
95.	971.0948					
96.	79.0046					
97.	21.9754					

Expanding Your Skills

98. What is the least number with three places to the right of the decimal that can be created with the digits 2, 9, and 7? Assume that the digits cannot be repeated.

99. What is the greatest number with three places to the right of the decimal that can be created from the digits 2, 9, and 7? Assume that the digits cannot be repeated.

Addition and Subtraction of Decimals

Section 5.2

Concepts

1. Addition and Subtraction of Decimals
2. Applications of Addition and Subtraction of Decimals
3. Algebraic Expressions

1. Addition and Subtraction of Decimals

In this section, we learn to add and subtract decimals. To begin, consider the sum $5.67 + 3.12$.

$$5.67 = 5 + \frac{6}{10} + \frac{7}{100}$$

$$+\,3.12 = +3 + \frac{1}{10} + \frac{2}{100}$$

$$8 + \frac{7}{10} + \frac{9}{100} = 8.79$$

Notice that the decimal points and place positions are lined up to add the numbers. In this way, we can add digits with the same place values because we are effectively adding decimal fractions with like denominators. The intermediate step of using fraction notation is often skipped. We can get the same result more quickly by adding digits in like place positions.

Adding and Subtracting Decimals

Step 1 Write the numbers in a column with the decimal points and corresponding place values lined up. (You may insert additional zeros as placeholders after the last digit to the right of the decimal point.)

Step 2 Add or subtract the digits in columns from right to left, as you would whole numbers. The decimal point in the answer should be lined up with the decimal points from the original numbers.

Example 1   Adding Decimals

Add. $27.486 + 6.37$

Solution:

$$\begin{array}{r} 27.486 \\ +\ 6.370 \end{array}$$ Line up the decimal points.
Insert an extra zero as a placeholder.

$$\begin{array}{r} \overset{1\ \ 1}{2}7.486 \\ +\ \ 6.370 \\ \hline 33.856 \end{array}$$ Add digits with common place values.

Line up the decimal point in the answer.

Skill Practice Add.

1. $184.218 + 14.12$

With operations on decimals it is important to locate the correct position of the decimal point. A quick estimate can help you determine whether your answer is reasonable. From Example 1, we have

$$\begin{array}{rll} 27.486 & \text{rounds to} & 27 \\ 6.370 & \text{rounds to} & +6 \\ & & \overline{33} \end{array}$$

The estimated value, 33, is close to the actual value of 33.856.

Answer
1. 198.338

Example 2 — Adding Decimals

Add. $3.7026 + 43 + 816.3$

Solution:

$$\begin{array}{r} 3.7026 \\ 43.0000 \\ + 816.3000 \end{array}$$

Line up the decimal points.
Insert a decimal point and four zeros after it.
Insert three zeros.

$$\begin{array}{r} \overset{1\,1}{3.7026} \\ 43.0000 \\ + 816.3000 \\ \hline 863.0026 \end{array}$$

Add digits with common place values.

Line up the decimal point in the answer.

The sum is 863.0026.

> **TIP:** To check that the answer is reasonable, round each addend.
>
> | | | |
> |---|---|---|
> | 3.7026 | rounds to | 4 |
> | 43 | rounds to | $\overset{1}{43}$ |
> | 816.3 | rounds to | + 816 |
> | | | 863 |
>
> which is close to the actual sum, 863.0026.

Skill Practice   Add.

2. $2.90741 + 15.13 + 3$

Example 3 — Subtracting Decimals

Subtract.

a. $0.2868 - 0.056$ **b.** $139 - 28.63$ **c.** $192.4 - 89.387$

Solution:

a.
$$\begin{array}{r} 0.2868 \\ -0.0560 \\ \hline 0.2308 \end{array}$$

Line up the decimal points.
Insert an extra zero as a placeholder.
Subtract digits with common place values.
The decimal point in the answer is lined up.

b.
$$\begin{array}{r} 139.00 \\ -28.63 \end{array}$$

Insert extra zeros as placeholders.
Line up the decimal points.

$$\begin{array}{r} \overset{8\,\overset{9}{\cancel{10}}\,10}{13\cancel{9}.\cancel{0}\cancel{0}} \\ -28.63 \\ \hline 110.37 \end{array}$$

Subtract digits with common place values.
Borrow where necessary.
Line up the decimal point in the answer.

c.
$$\begin{array}{r} 192.400 \\ -89.387 \end{array}$$

Insert extra zeros as placeholders.
Line up the decimal points.

$$\begin{array}{r} \overset{8\,12\;\;3\,\overset{9}{\cancel{10}}\,10}{19\cancel{2}.4\cancel{0}\cancel{0}} \\ -89.387 \\ \hline 103.013 \end{array}$$

Subtract digits with common place values.
Borrow where necessary.
Line up the decimal point in the answer.

Skill Practice   Subtract.

3. $3.194 - 0.512$ **4.** $0.397 - 0.1584$ **5.** $566.4 - 414.231$

Answers
2. 21.03741   **3.** 2.682
4. 0.2386   **5.** 152.169

In Example 4, we will use the rules for adding and subtracting signed numbers.

Example 4 Adding and Subtracting Signed Decimal Numbers

Simplify.

a. $-23.9 + 45.8$ b. $-0.694 - 0.482$ c. $2.61 - 3.79 - (-6.29)$

Solution:

a. $-23.9 + 45.8$ The sum will be *positive* because $|45.8|$ is greater than $|-23.9|$.

$ = +(45.8 - 23.9)$ Because the numbers have different signs, subtract the smaller absolute value from the larger absolute value. Apply the sign from the number with the larger absolute value.

$$\begin{array}{r} \overset{4\ 18}{4\cancel{5}.8} \\ -23.9 \\ \hline 21.9 \end{array}$$

$ = 21.9$ The result is positive.

b. $-0.694 - 0.482$

$ = -0.694 + (-0.482)$ Change subtraction to addition of the opposite.

$ = -(0.694 + 0.482)$ Because the numbers have the same signs, add their absolute values and apply the common sign.

$$\begin{array}{r} \overset{1}{}0.694 \\ +\ 0.482 \\ \hline 1.176 \end{array}$$

$ = -1.176$ The result is negative.

c. $2.61 - 3.79 - (-6.29)$

$= \underbrace{2.61 + (-3.79)}_{} + (6.29)$ Change subtraction to addition of the opposite. Add from left to right.

$= -1.18 + 6.29$

$$\begin{array}{r} 3.79 \\ -2.61 \\ \hline 1.18 \end{array}$$ Subtract the smaller absolute value from the larger absolute value.

$= 5.11$

$$\begin{array}{r} 6.29 \\ -1.18 \\ \hline 5.11 \end{array}$$ Subtract the smaller absolute value from the larger absolute value.

Skill Practice Simplify.

6. $-39.46 + 29.005$ 7. $-0.345 - 6.51$ 8. $-3.79 - (-6.2974)$

2. Applications of Addition and Subtraction of Decimals

Decimals are used often in measurements and in day-to-day applications.

Answers
6. -10.455 7. -6.855 8. 2.5074

Example 5: Applying Addition and Subtraction of Decimals in a Checkbook

Fill in the balance for each line in the checkbook register, shown in Figure 5-3. What is the ending balance?

Check No.	Description	Payment	Deposit	Balance
				$684.60
2409	Doctor	$75.50		
2410	Mechanic	215.19		
2411	Groceries	94.56		
	Paycheck		$981.46	
2412	Veterinarian	49.90		

Figure 5-3

Solution:

We begin with $684.60 in the checking account. For each debit, we subtract. For each credit, we add.

Check No.	Description	Payment	Deposit	Balance	
				$684.60	
2409	Doctor	$75.50		609.10	= $684.60 − $75.50
2410	Mechanic	215.19		393.91	= $609.10 − $215.19
2411	Groceries	94.56		299.35	= $393.91 − $94.56
	Paycheck		$981.46	1280.81	= $299.35 + $981.46
2412	Veterinarian	49.90		1230.91	= $1280.81 − $49.90

The ending balance is $1230.91.

Skill Practice

9. Fill in the balance for each line in the checkbook register.

Payment	Deposit	Balance
		$437.80
$82.50		
	$514.02	
26.04		

Example 6: Applying Decimals to Perimeter

a. Find the length of the sides labeled x.
b. Find the length of the side labeled y.
c. Find the perimeter of the figure.

Solution:

a. If we extend the line segment labeled with the dashed line as shown below, we see that the sum of side x and the dashed line must equal 14 m. Therefore, subtract $14 - 2.9$ to find the length of side x.

Length of side x:

$$\begin{array}{r} \overset{3\ 10}{1\cancel{4}.\cancel{0}} \\ -2.9 \\ \hline 11.1 \end{array}$$

Side x is 11.1 m long.

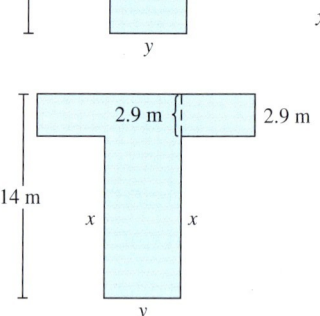

Answer

9.

Payment	Deposit	Balance
		$437.80
$82.50		355.30
	$514.02	869.32
26.04		843.28

b. The dashed line in the figure below has the same length as side y. We also know that 4.8 + 5.2 + y must equal 15.4. Since 4.8 + 5.2 = 10.0,

$$y = 15.4 - 10.0$$
$$= 5.4$$

The length of side y is 5.4 m.

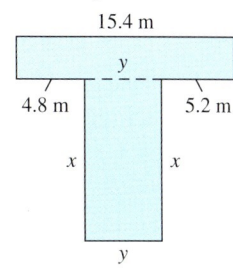

c. Now that we have the lengths of all sides, add them to get the perimeter.

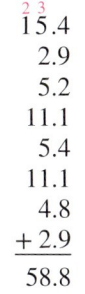

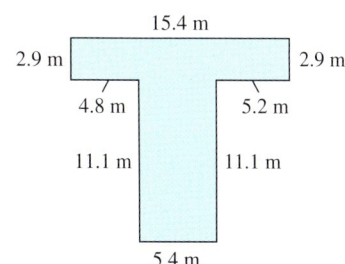

The perimeter is 58.8 m.

Skill Practice

10. Consider the figure.
  a. Find the length of side x.
  b. Find the length of side y.
  c. Find the perimeter.

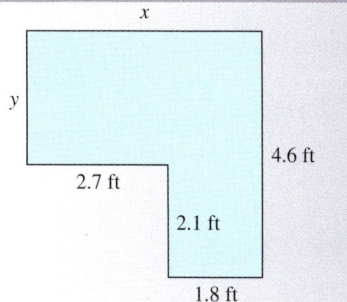

3. Algebraic Expressions

In Example 7, we practice combining *like* terms. In this case, the terms have decimal coefficients.

Example 7   Combining *Like* Terms

Simplify by combining *like* terms.

a. $-2.3x + 8.6x$ **b.** $0.042x + 0.539y + 0.65x - 0.21y$

Solution:

a. $-2.3x + 8.6x = (-2.3 + 8.6)x$ Apply the distributive property.

$= (6.3)x$
$= 6.3x$

$\begin{array}{r} 8.6 \\ -2.3 \\ \hline 6.3 \end{array}$ Subtract the smaller absolute value from the larger.

b. $0.042x + 0.539y + 0.65x - 0.21y$

$= 0.042x + 0.65x + 0.539y - 0.21y$ Group *like* terms together.

$= 0.692x - 0.329y$ Combine *like* terms by adding or subtracting the coefficients.

TIP: To combine *like* terms, we can simply add or subtract the coefficients and leave the variable unchanged.

$0.042x + 0.65x = 0.692x$

Skill Practice Simplify by combining *like* terms.

11. $-9.15t - 2.34t$ **12.** $0.31w + 0.46z + 0.45w - 0.2z$

Answers

10. a. Side x is 4.5 ft.
   b. Side y is 2.5 ft.
   c. The perimeter is 18.2 ft.
11. $-11.49t$ **12.** $0.76w + 0.26z$

Section 5.2 Activity

A.1. **a.** Tatianna is shopping for a new cell phone case. One retailer is advertising a price of $28, while another retailer is advertising a price of $28.00. Are the retailers selling the cell phone case for the same price? Why or why not?

b. Based on the answer from part (a), does the number 28 have a decimal point? If so, where is it located?

c. Does adding zeros after the decimal point change the value of the number?

A.2. Donnell purchased two items online, one for $82.55 and the other for $36.90. What was the total of his purchases before tax and shipping?

A.3. Jacqueline had $421.88 in her checking account before paying her electric bill for $176.35. How much was left in her account after paying the electric bill?

A.4. Looking at your work for Exercises A.2 and A.3, explain how to add and subtract decimals.

For Exercises A.5–A.12, add or subtract the decimal numbers. In some cases it may be helpful to first change subtraction to addition of the opposite.

A.5. $6.002 + 42.3$

A.6. $0.931 + 0.0901$

A.7. $4.1 - 1.012$

A.8. $5 - 3.289$

A.9. $-10.2 + 21.31$

A.10. $7.252 - (-4)$

A.11. $11.89 - 23.67$

A.12. $12 - (-0.23) + 103.2 - 0.098$

For Exercises A.13–A.15, simplify by combining *like* terms.

A.13. $-7.2x + 9.5x$

A.14. $14.3y - 8.8y + 2y$

A.15. $0.21c - c$

Section 5.2 Practice Exercises

Prerequisite Review

R.1. Which number is equal to 5.03?
 a. 5.30 **b.** 5.030 **c.** 5.3 **d.** 5.0300

R.2. Which number is equal to -0.024?
 a. -0.0024 **b.** -0.0240 **c.** -0.2040 **d.** -0.2400

R.3. Which number is equal to $2\frac{7}{100}$?
 a. 2.07 **b.** 2.70 **c.** 2.070 **d.** 0.027

R.4. Which number is equal to $3\frac{1}{10}$?
 a. 3.01 **b.** 3.10 **c.** 0.31 **d.** 3.1

For Exercises R.5–R.8, round the decimals to the indicated place value.

R.5. 23.489; tenths

R.6. 18.948; tenths

R.7. 25.99952; thousandths

R.8. 25.99912; thousandths

Vocabulary and Key Concepts

1. When adding or subtracting decimals, line up the _____ and corresponding place values.

2. When adding and subtracting decimals, it may be necessary to add additional _____ as placeholders after the last digit to the right of the decimal point.

3. _____ can help determine if the sum or difference is reasonable.

4. The placement of the decimal point for a whole number is after the _____ place.

Section 5.2 Addition and Subtraction of Decimals

5. To calculate the end balance of a bank statement, _____ are added and _____ are subtracted.

6. To combine *like* terms, we add or subtract the coefficients and (change/do not change) the variable.

Concept 1: Addition and Subtraction of Decimals

For Exercises 7–14, add the decimal numbers. Then round the numbers and find the sum to determine if your answer is reasonable. The first estimate is done for you. **(See Examples 1 and 2.)**

	Expression	Estimate		Expression	Estimate
7.	12.99 + 19.25	13 + 19	8.	6.75 + 22.88	
9.	44.6 + 18.6		10.	28.2 + 23.2	
11.	5.306 + 3.645		12.	3.451 + 7.339	
13.	12.9 + 3.091		14.	4.125 + 5.9	

For Exercises 15–26, add the decimals. **(See Examples 1 and 2.)**

15. 78.9 + 0.9005
16. 44.2 + 0.7802
17. 23 + 8.0148
18. 7.9302 + 34
19. 34 + 23.0032 + 5.6
20. 23 + 8.01 + 1.0067
21. 68.394 + 32.02
22. 2.904 + 34.229

23. 103.94
    + 24.5

24. 93.2
    + 43.336

25. 54.2
    23.993
    + 3.87

26. 13.9001
    72.4
    + 34.13

For Exercises 27–32, subtract the decimal numbers. Then round the numbers and find the difference to determine if your answer is reasonable. The first estimate is done for you. **(See Example 3.)**

	Expression	Estimate		Expression	Estimate
27.	35.36 − 21.12	35 − 21 = 14	28.	53.9 − 22.4	
29.	7.24 − 3.56		30.	23.3 − 20.8	
31.	45.02 − 32.7		32.	66.15 − 42.9	

For Exercises 33–44, subtract the decimals. **(See Example 3.)**

33. 14.5 − 8.823
34. 33.2 − 21.932
35. 2 − 0.123
36. 4 − 0.42
37. 55.9 − 34.2354
38. 49.1 − 24.481
39. 18.003 − 3.238
40. 21.03 − 16.446
41. 183.01 − 23.452
42. 164.23 − 44.3893
43. 1.001 − 0.0998
44. 2.0007 − 0.0689

Mixed Exercises: Addition and Subtraction of Signed Decimals

For Exercises 45–60, add or subtract as indicated. **(See Example 4.)**

45. −506.34 + 83.4
46. −89.041 + 76.43
47. −0.489 − 0.87
48. −0.78 − 0.439
49. 47.82 − (−3.159)
50. 4.2 − (−9.8458)
51. 55.3 − 68.4 − (−9.83)
52. 3.45 − 8.7 − (−10.14)

53. 5 − 9.432

54. 7 − 11.468

55. −6.8 − (−8.2)

56. −10.3 − (−5.1)

57. −28.3 + (−82.9)

58. −92.6 + (−103.8)

59. 2.6 − 2.06 − 2.006 + 2.0006

60. 5.84 − 5.084 − 5.0084 + 58.4

Concept 2: Applications of Addition and Subtraction of Decimals

61. Fill in the balance for each line in the checkbook register shown in the figure. What is the ending balance?
    (See Example 5.)

Check No.	Description	Payment	Deposit	Balance
				$ 245.62
2409	Electric bill	$ 52.48		
2410	Groceries	72.44		
2411	Department store	108.34		
	Paycheck		$1084.90	
2412	Restaurant	23.87		
	Transfer from savings		200	

62. A section of a bank statement is shown in the figure. Find the mistake that was made by the bank.

Date	Action	Payment	Deposit	Balance
				$1124.35
Jan. 2	Check #4214	$749.32		375.03
Jan. 3	Check #4215	37.29		337.74
Jan. 4	Transfer from savings		$ 400.00	737.74
Jan. 5	Paycheck		1451.21	2188.95
Jan. 6	Cash withdrawal	150.00		688.95

63. A normal human red blood cell count is between 4.2 and 6.9 million cells per microliter (μL). A cancer patient undergoing chemotherapy has a red blood cell count of 2.85 million cells per microliter. How far below the lower normal limit is this?

64. A laptop computer was originally priced at $1299.99 and was discounted to $998.95. By how much was it marked down?

65. Water flows into a pool at a constant rate. The water level is recorded at several 1-hr intervals.

    a. From the table, how many inches is the water level rising each hour?

    b. At this rate, what will the water level be at 1:00 P.M.?

    c. At this rate, what will the water level be at 3:00 P.M.?

Time	Water Level
9:00 A.M.	4.2 in.
10:00 A.M.	5.9 in.
11:00 A.M.	7.6 in.
12:00 P.M.	9.3 in.

66. The amount of time that it takes Mercury, Venus, Earth, and Mars to revolve about the Sun is given in the graph.

    a. How much longer does it take Mars to complete a revolution around the Sun than the Earth?

    b. How much longer does it take Venus than Mercury to revolve around the Sun?

Section 5.2 Addition and Subtraction of Decimals   287

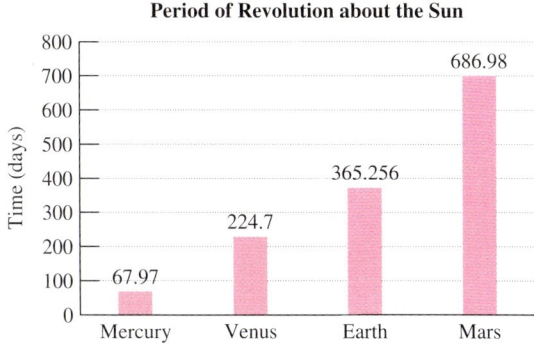

Source: National Aeronautics and Space Administration

rwarnick/iStockphoto/Getty Images

67. The table shows the thickness of four U.S. coins. If you stacked three quarters and a dime in one pile and two nickels and two pennies in another pile, which pile would be higher?

Coin	Thickness
Quarter	1.75 mm
Dime	1.35 mm
Nickel	1.95 mm
Penny	1.55 mm

Source: U.S. Department of the Treasury

68. Refer to Exercise 67. How much thicker is a nickel than a quarter?

Skip ODonnell/E+/Getty Images

For Exercises 69–72, find the lengths of the sides labeled x and y. Then find the perimeter. **(See Example 6.)**

69.

27.3 in.
6.7 in.
x
22.1 in.
y
18.4 in.

70.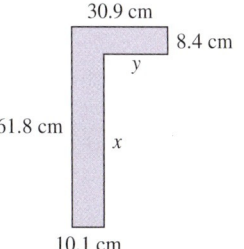

30.9 cm
8.4 cm
y
61.8 cm
x
10.1 cm

71.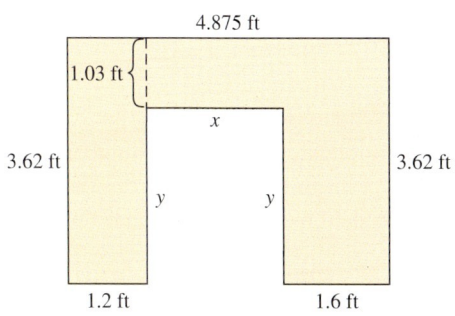

4.875 ft
1.03 ft
x
3.62 ft
3.62 ft
y y
1.2 ft   1.6 ft

72.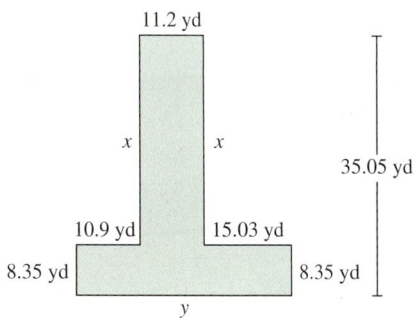

11.2 yd
x x
35.05 yd
10.9 yd   15.03 yd
8.35 yd   8.35 yd
y

73. A city bus follows the route shown in the map. How far does it travel in one circuit?

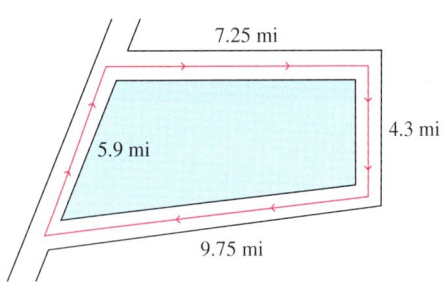

74. Santos built a new deck and needs to put a railing around the sides. He does not need railing where the deck is against the house. How much railing should he purchase?

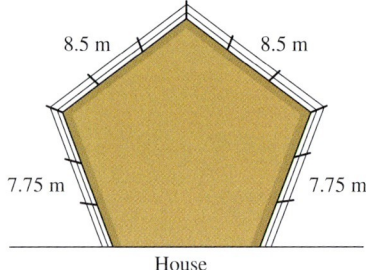

Concept 3: Algebraic Expressions

For Exercises 75–84, simplify by combining *like* terms. (See Example 7.)

75. $-5.83t + 9.7t$
76. $-2.158w + 10.4w$
77. $-4.5p - 8.7p$
78. $-98.46a - 12.04a$
79. $y - 0.6y$
80. $0.18x - x$
81. $0.025x + 0.83y + 0.82x - 0.31y$
82. $0.008m - 0.06n - 0.0043m + 0.092n$
83. $0.92c - 4.78d + 0.08c - 0.22d$
84. $3.62x - 80.09y - 0.62x + 10.09y$

Technology Connections

For Exercises 85–90, refer to the table. The table gives the closing stock prices (in dollars per share) for the first day of trading for the given month.

Stock	January	February	March	April	May
IBM	132.45	125.53	128.57	128.25	130.46
FedEx	83.45	80.67	85.81	92.17	90.01

85. By how much did the IBM stock decrease between January and May?
86. By how much did the FedEx stock increase between January and May?
87. Between which two consecutive months did the FedEx stock increase the most? What was the amount of increase?
88. Between which two consecutive months did the IBM stock increase the most? What was the amount of increase?
89. Between which two consecutive months did the FedEx stock decrease the most? What was the amount of decrease?
90. Between which two consecutive months did the IBM stock decrease the most? What was the amount of decrease?

Section 5.3 Multiplication of Decimals and Applications with Circles

Concepts

1. Multiplication of Decimals
2. Multiplication by a Power of 10 and by a Power of 0.1
3. Applications Involving Multiplication of Decimals
4. Circumference and Area of a Circle

1. Multiplication of Decimals

Multiplication of decimals is much like multiplication of whole numbers. However, we need to know where to place the decimal point in the product. Consider the product $(0.3)(0.41)$. One way to multiply these numbers is to write them first as decimal fractions.

$$(0.3)(0.41) = \frac{3}{10} \cdot \frac{41}{100} = \frac{123}{1000} \text{ or } 0.123$$

Another method multiplies the factors vertically. First we multiply the numbers as though they were whole numbers. We temporarily disregard the decimal point in the product because it will be placed later.

$$\begin{array}{r} 0.41 \\ \times\, 0.3 \\ \hline 123 \end{array} \leftarrow \text{decimal point not yet placed}$$

From the first method, we know that the correct answer to this problem is 0.123. Notice that 0.123 contains the same number of decimal places as the two factors combined. That is,

$$\begin{array}{r} 0.41 \\ \times\, 0.3 \\ \hline .123 \end{array} \begin{array}{l} \leftarrow\ 2 \text{ decimal places} \\ \leftarrow\ 1 \text{ decimal place} \\ \leftarrow\ 3 \text{ decimal places} \end{array}$$

TIP: When multiplying decimals, it is *not* necessary to line up the decimal points as we do when we add or subtract decimals. Instead, we line up the right-most digits.

The process to multiply decimals is summarized as follows.

> **Multiplying Two Decimals**
>
> **Step 1** Multiply as you would integers.
>
> **Step 2** Place the decimal point in the product so that the number of decimal places equals the combined number of decimal places of both factors.
>
> *Note:* You may need to insert zeros to the left of the whole-number product to get the correct number of decimal places in the answer.

In Example 1, we multiply decimals by using this process.

Example 1 Multiplying Decimals

Multiply. $\begin{array}{r} 11.914 \\ \times\ 0.8 \end{array}$

Solution:

$$\begin{array}{r} \overset{\scriptscriptstyle 17\ 13}{11.914} \\ \times\ \ 0.8 \\ \hline 9.5312 \end{array} \quad \begin{array}{l} 3 \text{ decimal places} \\ +\ 1 \text{ decimal place} \\ 4 \text{ decimal places} \end{array}$$

The product is 9.5312.

Skill Practice Multiply.

1. $(19.7)(4.1)$

Example 2 Multiplying Decimals

Multiply. Then use estimation to check the location of the decimal point.

$$(29.3)(2.8)$$

Solution:

$$\begin{array}{r} \overset{\scriptscriptstyle 1}{2}\overset{\scriptscriptstyle 7\ 2}{9}.3 \\ \times\ 2.8 \\ \hline 2344 \\ 5860 \\ \hline 82.04 \end{array} \quad \begin{array}{l} 1 \text{ decimal place} \\ +\ 1 \text{ decimal place} \\ \\ \\ 2 \text{ decimal places} \end{array} \quad \text{The product is 82.04.}$$

Answer

1. 80.77

To check the answer, we can round the factors and estimate the product. The purpose of the estimate is primarily to determine whether we have placed the decimal point correctly. Therefore, it is usually sufficient to round each factor to the left-most nonzero digit. This is called **front-end rounding**. Thus,

$$\begin{array}{rcr} 29.3 & \text{rounds to} & 30 \\ 2.8 & \text{rounds to} & \times\ 3 \\ \hline & & 90 \end{array}$$

The first digit for the actual product 8̲2.04 and the first digit for the estimate 9̲0 is the tens place. Therefore, we are reasonably sure that we have located the decimal point correctly. The estimate 90 is close to 82.04.

Skill Practice

2. Multiply (1.9)(29.1) and check your answer using estimation.

Example 3 — Multiplying Decimals

Multiply. Then use estimation to check the location of the decimal point.

$$-2.79 \times 0.0003$$

Solution:

The product will be *negative* because the factors have opposite signs.

Actual product:

$$\begin{array}{r} \overset{2\ 2}{-2.79} \\ \times\ 0.0003 \\ \hline -.000837 \end{array}$$

2 decimal places
+ 4 decimal places
6 decimal places
(insert 3 zeros to the left)

Estimate:

$$\begin{array}{rcr} -2.79 & \text{rounds to} & -3 \\ \times\ 0.0003 & \text{rounds to} & \times\ 0.0003 \\ \hline & & -0.0009 \end{array}$$

The first digit for both the actual product and the estimate is in the ten-thousandths place. We are reasonably sure the decimal point is positioned correctly.

The product is -0.000837.

Skill Practice

3. Multiply -4.6×0.00008, and check your answer using estimation.

2. Multiplication by a Power of 10 and by a Power of 0.1

Consider the number 2.7 multiplied by 10, 100, 1000 . . .

$$\begin{array}{ccc} 10 & 100 & 1000 \\ \times\ 2.7 & \times\ 2.7 & \times\ 2.7 \\ \hline 70 & 700 & 7000 \\ 200 & 2000 & 20000 \\ \hline 27.0 & 270.0 & 2700.0 \end{array}$$

Multiplying 2.7 by 10 moves the decimal point 1 place to the right.
Multiplying 2.7 by 100 moves the decimal point 2 places to the right.
Multiplying 2.7 by 1000 moves the decimal point 3 places to the right.

This leads us to the following generalization.

Answers

2. 55.29; ≈ 2 · 30 = 60, which is close to 55.29.
3. −0.000368; ≈ −5 × 0.00008 = −0.0004, which is close to −0.000368.

Multiplying a Decimal by a Power of 10

Move the decimal point to the right the same number of decimal places as the number of zeros in the power of 10.

Example 4 Multiplying by Powers of 10

Multiply.

a. $14.78 \times 10{,}000$ **b.** 0.0064×100 **c.** $-8.271 \times (-1{,}000{,}000)$

Solution:

a. $14.78 \times 10{,}000 = 147{,}800$ Move the decimal point 4 places to the right.

b. $0.0064 \times 100 = 00.64 = 0.64$ Move the decimal point 2 places to the right.

c. The product will be positive because the factors have the same sign.

$-8.271 \times (-1{,}000{,}000) = 8{,}271{,}000$ Move the decimal point 6 places to the right.

Skill Practice Multiply.

4. 81.6×1000 **5.** $0.0000085 \times 10{,}000$ **6.** $-2.396 \times (-10{,}000{,}000)$

Multiplying a positive decimal by 10, 100, 1000, and so on increases its value. Therefore, it makes sense to move the decimal point to the *right*. Now suppose we multiply a decimal by 0.1, 0.01, and 0.001. These numbers represent the decimal fractions $\frac{1}{10}$, $\frac{1}{100}$, and $\frac{1}{1000}$, respectively, and are easily recognized as powers of 0.1. Taking one-tenth of a positive number or one-hundredth of a positive number makes the number smaller. To multiply by 0.1, 0.01, 0.001, and so on (powers of 0.1), move the decimal point to the *left*.

$$\begin{array}{ccc} 3.6 & 3.6 & 3.6 \\ \times\ 0.1 & \times\ 0.01 & \times\ 0.001 \\ \hline .36 & .036 & .0036 \end{array}$$

Multiplying a Decimal by Powers of 0.1

Move the decimal point to the left the same number of places as there are decimal places in the power of 0.1.

Example 5 Multiplying by Powers of 0.1

Multiply.

a. 62.074×0.0001 **b.** 7965.3×0.1 **c.** -0.0057×0.00001

Solution:

a. $62.074 \times 0.0001 = 0.0062074$ Move the decimal point 4 places to the left. Insert extra zeros.

b. $7965.3 \times 0.1 = 796.53$ Move the decimal point 1 place to the left.

c. $-0.0057 \times 0.00001 = -0.000000057$ Move the decimal point 5 places to the left.

Skill Practice Multiply.

7. 471.034×0.01 **8.** $9{,}437{,}214.5 \times 0.00001$ **9.** -0.0004×0.001

Answers
4. 81,600   **5.** 0.085
6. 23,960,000   **7.** 4.71034
8. 94.372145   **9.** −0.0000004

Sometimes people prefer to use number names to express very large numbers. For example, we might say that the U.S. population in a recent year was approximately 310 million. To write this in decimal form, we note that 1 million = 1,000,000. In this case, we have 310 of this quantity. Thus,

$$310 \text{ million} = 310 \times 1,000,000 \text{ or } 310,000,000$$

Example 6 Naming Large Numbers

Write the decimal number representing each word name.

a. The distance between the Earth and Sun is approximately 92.9 million miles.

b. Recently the number of deaths in the United States due to heart disease was projected to be 8 hundred thousand.

c. A recent estimate claimed that collectively Americans throw away 472 billion pounds of garbage each year.

Arthur Tilley/Nextdoor Images/ Stockbyte/Creatas Images/Getty Images

Solution:

a. 92.9 million = 92.9 × 1,000,000 = 92,900,000

b. 8 hundred thousand = 8 × 100,000 = 800,000

c. 472 billion = 472 × 1,000,000,000 = 472,000,000,000

Skill Practice Write a decimal number representing the word name.

10. The population in Bexar County, Texas, is approximately 1.7 million.
11. Light travels approximately 5.9 trillion miles in 1 year.
12. The legislative branch of the federal government employs approximately 31 thousand employees.

3. Applications Involving Multiplication of Decimals

Example 7 Applying Decimal Multiplication

Jane Marie bought eight cans of tennis balls for $1.98 each. She paid $1.03 in tax. What was the total bill?

Solution:

The cost of the tennis balls before tax is

$$8(\$1.98) = \$15.84$$

$$\begin{array}{r} \overset{7\ 6}{1.98} \\ \times\ \ \ 8 \\ \hline 15.84 \end{array}$$

Adding the tax to this value, we have

$$\binom{\text{Total}}{\text{cost}} = \binom{\text{cost of}}{\text{tennis balls}} + (\text{tax})$$

$$= \$15.84$$
$$+\ \ 1.03$$
$$\$16.87 \quad \text{The total cost is } \$16.87.$$

Ken Karp/McGraw Hill

Skill Practice

13. A book club ordered 12 books for $8.99 each. The shipping cost was $4.95. What was the total bill?

Answers
10. 1,700,000
11. 5,900,000,000,000
12. 31,000
13. The total bill was $112.83.

Example 8 Finding the Area of a Rectangle

The *Mona Lisa* is perhaps the most famous painting in the world. It was painted by Leonardo da Vinci somewhere between 1503 and 1506 and now hangs in the Louvre in Paris, France. The dimensions of the painting are 30 in. by 20.875 in. What is the total area?

Solution:

Recall that the area of a rectangle is given by

$$A = l \cdot w$$
$$A = (30 \text{ in.})(20.875 \text{ in.})$$

$$\begin{array}{r} 20.875 \\ \times\ 30 \\ \hline 0 \\ 626250 \\ \hline 626.250 \end{array}$$

$$= 626.25 \text{ in.}^2$$

The area of the *Mona Lisa* is 626.25 in.^{2}

Skill Practice

14. The IMAX movie screen at the Museum of Science and Discovery is 18 m by 24.4 m. What is the area of the screen?

4. Circumference and Area of a Circle

A **circle** is a figure consisting of all points in a flat surface located the same distance from a fixed point called the center. The **radius** of a circle is the length of any line segment from the center to a point on the circle. The **diameter** of a circle is the length of any line segment connecting two points on the circle and passing through the center. See Figure 5-4.

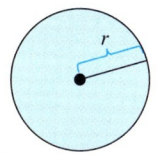
r is the length of the radius.

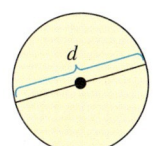
d is the length of the diameter.

Figure 5-4

Notice that the length of a diameter is twice the radius. Therefore, we have

$$d = 2r \quad \text{or equivalently} \quad r = \frac{1}{2}d$$

Example 9 Finding Diameter and Radius

a. Find the length of a radius.

b. Find the length of a diameter.

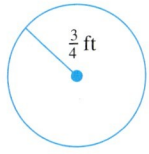

Solution:

a. $r = \frac{1}{2}d = \frac{1}{2}(46 \text{ cm}) = 23 \text{ cm}$

b. $d = 2r = \overset{1}{2}\left(\frac{3}{\underset{2}{4}}\text{ ft}\right) = \frac{3}{2}\text{ ft}$

Answer

14. The screen area is 439.2 m^2.

Skill Practice

15. Find the length of a radius.

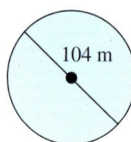

16. Find the length of a diameter.

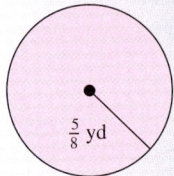

The distance around a circle is called the **circumference**. In any circle, if the circumference C is divided by the diameter, the result is equal to the number π (read "pi"). The number π in decimal form is 3.1415926535...., which goes on forever without a repeating pattern. We approximate π by 3.14 or $\frac{22}{7}$ to make it easier to use in calculations. The relationship between the circumference and the diameter of a circle is $\frac{C}{d} = \pi$. This gives us the following formulas.

> **Circumference of a Circle**
>
> The circumference C of a circle is given by
>
> $$C = \pi d \quad \text{or} \quad C = 2\pi r$$
>
> where π is approximately 3.14 or $\frac{22}{7}$.

Example 10 — Determining Circumference of a Circle

Determine the circumference. Use 3.14 for π.

Solution:

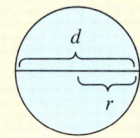

The radius is given, $r = 4$ cm.

$$\begin{aligned} C &= 2\pi r \\ &= 2(\pi)(4 \text{ cm}) &&\text{Substitute 4 cm for } r. \\ &= 8\pi \text{ cm} \\ &\approx 8(3.14) \text{ cm} &&\text{Approximate } \pi \text{ by 3.14.} \\ &= 25.12 \text{ cm} &&\text{The distance around the circle is approximately 25.12 cm.} \end{aligned}$$

TIP: The value 3.14 is an approximation for π. Therefore, using 3.14 in a calculation results in an approximate answer.

Skill Practice Find the circumference. Use 3.14 for π.

17.

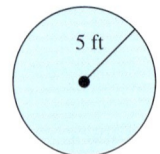

TIP: The circumference from Example 10 can also be found from the formula $C = \pi d$. In this case, the diameter is 8 cm.

$$\begin{aligned} C &= \pi d \\ C &= \pi(8 \text{ cm}) \\ &= 8\pi \text{ cm} \\ &\approx 8(3.14) \text{ cm} \\ &= 25.12 \text{ cm} \end{aligned}$$

Answers

15. 52 m **16.** $\frac{5}{4}$ yd

17. ≈ 31.4 ft

The formula for the area of a circle also involves the number π.

Area of a Circle

The **area of a circle** is given by

$$A = \pi r^2$$

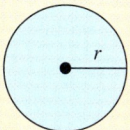

Avoiding Mistakes

To express the formula for the circumference of a circle, we can use either the radius ($C = 2\pi r$) or the diameter ($C = \pi d$).

To find the area of a circle, we will always use the radius ($A = \pi r^2$).

Example 11   Determining the Area of a Circle

Determine the area of a circle that has radius 0.3 ft. Approximate the answer by using 3.14 for π. Round to two decimal places.

Solution:

$A = \pi r^2$
$ = \pi(0.3 \text{ ft})^2$ Substitute $r = 0.3$ ft.
$ = \pi(0.09 \text{ ft}^2)$ Square the radius. $(0.3 \text{ ft})^2 = (0.3 \text{ ft})(0.3 \text{ ft}) = 0.09 \text{ ft}^2$
$ = 0.09\pi \text{ ft}^2$
$ \approx 0.09(3.14) \text{ ft}^2$ Approximate π by 3.14.
$ = 0.2826 \text{ ft}^2$ Multiply decimals.
$ \approx 0.28 \text{ ft}^2$ The area is approximately 0.28 ft².

Avoiding Mistakes

$(0.3)^2 \neq (0.9)$. Be sure to count the number of places needed to the right of the decimal point. $(0.3)^2 = (0.3)(0.3) = 0.09$

TIP: Using the approximation 3.14 for π results in approximate answers for area and circumference. To express the exact answer, we must leave the result written in terms of π. In Example 11, the exact area is given by $0.09\pi \text{ ft}^2$.

Skill Practice

18. Find the area of a circular clock having a radius of 6 in. Approximate the answer by using 3.14 for π. Round to the nearest whole unit.

Answer
18. $\approx 113 \text{ in.}^2$

Practice Exercises   Section 5.3

Study Skills Exercise

To help you remember the formulas for circumference and area of a circle, list them together and note the similarities and differences in the formulas.

Circumference: _____   Area: _____

Similarities: Differences:

Chapter 5 Decimals

Prerequisite Review

For Exercises R.1–R.2, round the decimals to the indicated place value.

R.1. −0.399; hundredths

R.2. −8.09499; hundredths

For Exercises R.3–R.4, fill in the blank with < or >.

R.3. −51.4382 _____ −51.4389

R.4. −1.508 _____ −1.5

For Exercises R.5–R.6, multiply.

R.5. 258 × 1000

R.6. −18 × 10,000

Vocabulary and Key Concepts

1. a. Rounding a number to the left-most nonzero digit is called _____ -end rounding.

   b. A _____ is a figure consisting of all points in a flat surface located the same distance from a fixed point called the center.

   c. The _____ of a circle is the length of the line segment from the center to a point on the circle.

   d. The _____ of a circle is the length of a line segment connecting two points on the circle and passing through the center of the circle.

   e. The distance around a circle is called its _____.

   f. If the circumference of a circle is divided by its diameter, then the result is equal to the number _____.

   g. The number π is often approximated by the decimal number _____ or by the fraction _____.

   h. Which formula can be used to find the circumference C of a circle with radius r and diameter d? $C = 2\pi r$ or $C = \pi d$

   i. A formula for the area A of a circle with radius r is given by $A =$ _____.

2. When multiplying a decimal by a power of 10, move the decimal point to the _____ the same number of decimal places as the number of zeros in the power of 10.

3. When multiplying a decimal by a power of 0.1, move the decimal point to the _____ the same number of decimal places as the number of zeros in the power of 0.1.

4. The product of (26.3)(5.25) would have _____ digits after the decimal point.

Concept 1: Multiplication of Decimals

For Exercises 5–16, multiply the decimals. (See Examples 1–3.)

5. 8 × 0.5

6. 0.6 × 5

7. (9)(0.4)

8. (2)(0.9)

9. 0.8 × 0.5

10. 0.6 × 0.5

11. (0.9)(4)

12. (0.2)(9)

13. (−60)(−0.003)

14. (−40)(−0.005)

15. 22.38 × 0.8

16. 31.67 × 0.4

For Exercises 17–30, multiply the decimals. Then estimate the answer by rounding. The first estimate is done for you. (See Examples 2 and 3.)

17. Exact: 8.3 × 4.5 Estimate: 8 × 5 = 40

18. Exact: 4.3 × 9.2 Estimate:

19. Exact: 0.58 × 7.2 Estimate:

For Exercises 73–76, find the circumference of the circle. Approximate the answer by using 3.14 for π. (See Example 10.)

73.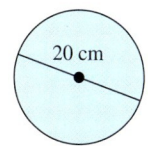

74.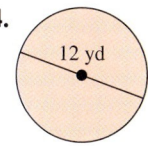

75.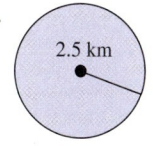

76.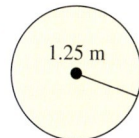

For Exercises 77–80, use 3.14 for π.

77. Find the circumference of a can of soda.

Mark Dierker/McGraw Hill

78. Find the circumference of the can of tuna.

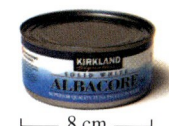
Mark Dierker/McGraw Hill

79. Find the circumference of a compact disk.

80. Find the outer circumference of a pipe with 1.75-in. radius.

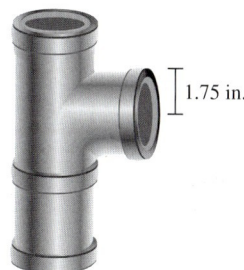

For Exercises 81–84, find the area. Approximate the answer by using 3.14 for π. Round to the nearest whole unit. (See Example 11.)

81.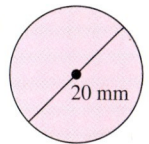

82.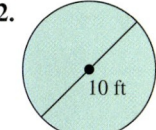

83.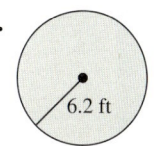

84.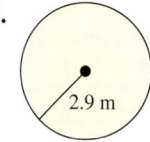

85. The Large Hadron Collider (LHC) is a particle accelerator and collider built to detect subatomic particles. The accelerator is in a huge circular tunnel that straddles the border between France and Switzerland. The diameter of the tunnel is 5.3 mi. Find the circumference of the tunnel. (*Source:* European Organization for Nuclear Research)

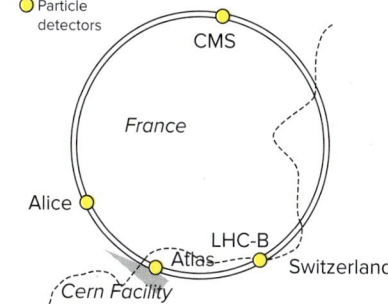

86. A ceiling fan blade rotates in a full circle. If the fan blades are 2 ft long, what is the area covered by the fan blades?

87. An outdoor torch lamp shines light a distance of 30 ft in all directions. What is the total ground area lighted?

88. Hurricane Katrina's eye was 32 mi wide. The eye of a storm of similar intensity is usually only 10 mi wide. (*Source:* Associated Press 10/8/05 "Mapping Katrina's Storm Surge")

   a. What area was covered by the eye of Katrina? Round to the nearest square mile.

   b. What is the usual area of the eye of a similar storm?

Technology Connections

For Exercises 89–92, find the area and circumference rounded to four decimals places. Use the π key on your calculator.

89.

90.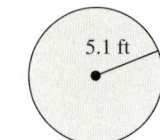

91.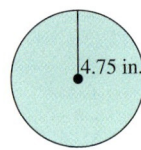

92.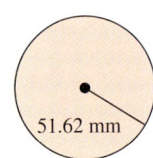

Expanding Your Skills

93. A hula hoop has a 30-in. diameter.

    a. Find the circumference. Use 3.14 for π.

    b. How far will the hula hoop roll if it turns 16 times? Will it reach the end of a driveway that is 40 yd long (1440 in.)?

94. A bicycle wheel has a 26-in. diameter.

    a. Find the circumference. Use 3.14 for π.

    b. How much distance will the wheel cover if it turns 147 times? Is this enough for the rider to travel a distance of 1000 ft (12,000 in.)?

Art Vandalay/Digital Vision/Getty Images

Stockbyte/PunchStock/Getty Images

95. Latasha has a bicycle, and the wheel has a 22-in. diameter. If the wheels of the bike turned 1000 times, how far did she travel? Use 3.14 for π. Give the answer to the nearest inch and to the nearest foot (1 ft = 12 in.).

96. The exercise wheel for Al's dwarf hamster has a diameter of 6.75 in.

    a. Find the circumference. Use 3.14 for π and round to the nearest inch.

    b. How far does Al's hamster travel if he completes 25 revolutions? Write the answer in feet and round to the nearest foot.

Division of Decimals

Section 5.4

Concepts
1. Division of Decimals
2. Rounding a Quotient
3. Applications of Decimal Division

1. Division of Decimals

Dividing decimals is much the same as dividing whole numbers. However, we must determine where to place the decimal point in the quotient.

First consider the quotient $3.5 \div 7$. We can write the numbers in fractional form and then divide.

$$3.5 \div 7 = \frac{35}{10} \div \frac{7}{1} = \frac{35}{10} \cdot \frac{1}{7} = \frac{35}{70} = \frac{5}{10} = 0.5$$

Now consider the same problem by using the efficient method of long division: $7\overline{)3.5}$.

When the divisor is a whole number, we place the decimal point directly above the decimal point in the dividend. Then we divide as we would whole numbers.

decimal point placed above the decimal point in the dividend

$$7\overline{)3.5} \;\; .5$$

Dividing a Decimal by a Whole Number
To divide by a whole number:
Step 1 Place the decimal point in the quotient directly above the decimal point in the dividend.
Step 2 Divide as you would whole numbers.

Example 1 — Dividing by a Whole Number

Divide and check the answer by multiplying.

$$30.55 \div 13$$

Solution:

Locate the decimal point in the quotient.

$$13\overline{)30.55}$$

$$\begin{array}{r} 2.35 \\ 13\overline{)30.55} \\ -26 \\ \hline 45 \\ -39 \\ \hline 65 \\ -65 \\ \hline 0 \end{array}$$

Divide as you would whole numbers.

Check by multiplying:

$$\begin{array}{r} 2.35 \\ \times\; 13 \\ \hline 705 \\ 2350 \\ \hline 30.55 \checkmark \end{array}$$

Skill Practice Divide. Check by using multiplication.
1. $502.96 \div 8$

When dividing decimals, we do not use a remainder. Instead we insert zeros to the right of the dividend and continue dividing. This is demonstrated in Example 2.

Answer
1. 62.87

Example 2 Dividing by an Integer

Divide and check the answer by multiplying.

$$\frac{-3.5}{-4}$$

Solution:

The quotient will be *positive* because the divisor and dividend have the same sign. We will perform the division without regard to sign, and then write the quotient as a positive number.

$$4\overline{)3.5}$$ ← Locate the decimal point in the quotient.

$$\begin{array}{r} .8 \\ 4\overline{)3.5} \\ -32 \\ \hline 3 \end{array}$$ ← Rather than using a remainder, we insert zeros in the dividend and continue dividing.

$$\begin{array}{r} .875 \\ 4\overline{)3.500} \\ -32 \\ \hline 30 \\ -28 \\ \hline 20 \\ -20 \\ \hline 0 \end{array}$$

Check by multiplying:

$$\begin{array}{r} 32 \\ 0.875 \\ \times4 \\ \hline 3.500 \checkmark \end{array}$$

The quotient is 0.875.

FOR REVIEW

Recall decimal multiplication.

0.875	3 decimal places
× 4	+ 0 decimal places
3.500	3 decimal places

Skill Practice Divide.

2. $\dfrac{-6.8}{-5}$

Example 3 Dividing by an Integer

Divide and check the answer by multiplying. $40\overline{)5}$

Solution:

$40\overline{)5.}$ ← The dividend is a whole number, and the decimal point is understood to be to its right. Insert the decimal point above it in the quotient.

$$\begin{array}{r} .125 \\ 40\overline{)5.000} \\ -40 \\ \hline 100 \\ -80 \\ \hline 200 \\ -200 \\ \hline 0 \end{array}$$

Since 40 is greater than 5, we need to insert zeros to the right of the dividend.

Check by multiplying.

$$\begin{array}{r} 1\,2 \\ 0.125 \\ \times40 \\ \hline 000 \\ 5000 \\ \hline 5.000 \checkmark \end{array}$$

The quotient is 0.125.

Skill Practice Divide.

3. $20\overline{)3}$

Answers

2. 1.36 **3.** 0.15

Section 5.4 Division of Decimals

Sometimes when dividing decimals, the quotient follows a repeated pattern. The result is called a **repeating decimal**.

Example 4   Dividing Where the Quotient Is a Repeating Decimal

Divide. $1.7 \div 30$

Solution:

$$
\begin{array}{r}
.05666\ldots \\
30\overline{)1.70000} \\
\underline{-150} \\
200 \\
\underline{-180} \\
200 \\
\underline{-180} \\
200
\end{array}
$$

Notice that as we continue to divide, we get the same values for each successive step. This causes a pattern of repeated digits in the quotient. Therefore, the quotient is a *repeating decimal*.

The quotient is $0.05666\ldots$. To denote the repeated pattern, we often use a bar over the first occurrence of the repeat cycle to the right of the decimal point. That is,

$0.05666\ldots = 0.05\overline{6}$ ← repeat bar

Avoiding Mistakes

In Example 4, notice that the repeat bar goes over only the 6. The 5 is not being repeated.

Skill Practice   Divide.

4. $2.4 \div 9$

Example 5   Dividing Where the Quotient Is a Repeating Decimal

Divide. $11\overline{)68}$

Solution:

$$
\begin{array}{r}
6.1818\ldots \\
11\overline{)68.0000} \\
\underline{-66} \\
20 \\
\underline{-11} \\
90 \\
\underline{-88} \\
20 \\
\underline{-11} \\
90 \\
\underline{-88} \\
20
\end{array}
$$

Could have stopped here →

Once again, we see a repeated pattern. The quotient is a repeating decimal. Notice that we could have stopped dividing when we obtained the second value of 20.

The quotient is $6.\overline{18}$.

Avoiding Mistakes

Be sure to put the repeating bar over the entire block of numbers that is being repeated. In Example 5, the bar extends over both the 1 and the 8. We have $6.\overline{18}$.

Skill Practice   Divide.

5. $11\overline{)57}$

Answers

4. $0.2\overline{6}$ 5. $5.\overline{18}$

The numbers $0.05\overline{6}$ and $6.\overline{18}$ are examples of repeating decimals. A decimal that "stops" is called a **terminating decimal**. For example, 6.18 is a terminating decimal, whereas $6.\overline{18}$ is a repeating decimal.

In Examples 1–5, we performed division where the divisor was an integer. Suppose now that we have a divisor that is *not* an integer, for example, $0.56 \div 0.7$. Because division can also be expressed in fraction notation, we have

$$0.56 \div 0.7 = \frac{0.56}{0.7}$$

If we multiply the numerator and denominator by 10, the denominator (divisor) becomes the whole number 7.

$$\frac{0.56}{0.7} = \frac{0.56 \times 10}{0.7 \times 10} = \frac{5.6}{7} \longrightarrow 7\overline{)5.6}$$

Recall that multiplying decimal numbers by 10 (or any power of 10, such as 100, 1000, etc.) moves the decimal point to the right. We use this idea to divide decimal numbers when the divisor is not a whole number.

Dividing When the Divisor Is Not a Whole Number

Step 1 Move the decimal point in the divisor to the right to make it a whole number.

Step 2 Move the decimal point in the dividend to the right the same number of places as in step 1.

Step 3 Place the decimal point in the quotient directly above the decimal point in the dividend.

Step 4 Divide as you would whole numbers. Then apply the correct sign to the quotient.

Example 6 Dividing Decimals

Divide.

a. $0.56 \div (-0.7)$ **b.** $0.005\overline{)3.1}$

Solution:

a. The quotient will be *negative* because the divisor and dividend have different signs. We will perform the division without regard to sign and then apply the negative sign to the quotient.

$.7\overline{).56}$ Move the decimal point in the divisor and dividend one place to the right.

$7\overline{)5.6}$ Line up the decimal point in the quotient.

$$\begin{array}{r} 0.8 \\ 7\overline{)5.6} \\ \underline{-5\ 6} \\ 0 \end{array}$$

The quotient is -0.8.

b. $.005\overline{)3.100}$ Move the decimal point in the divisor and dividend three places to the right. Insert additional zeros in the dividend if necessary. Line up the decimal point in the quotient.

Section 5.4 Division of Decimals

$$\begin{array}{r} 620. \\ 5\overline{)3100.} \\ \underline{-30} \\ 10 \\ \underline{-10} \\ 00 \end{array}$$

The quotient is 620.

Skill Practice Divide.

6. $0.64 \div (-0.4)$ 7. $5.4 \div 0.03$

Example 7   Dividing Decimals

Divide. $\dfrac{-50}{-1.1}$

Solution:

The quotient will be *positive* because the divisor and dividend have the same sign. We will perform the division without regard to sign, and then write the quotient as a positive number.

$1.1\overline{)50.0}$ Move the decimal point in the divisor and dividend one place to the right. Insert an additional zero in the dividend. Line up the decimal point in the quotient.

$$\begin{array}{r} 45.45\ldots \\ 11\overline{)500.000} \\ \underline{-44} \\ 60 \\ \underline{-55} \\ 50 \\ \underline{-44} \\ 60 \\ \underline{-55} \\ 50 \end{array}$$

The quotient is the repeating decimal, $45.\overline{45}$.

The quotient is $45.\overline{45}$.

> **Avoiding Mistakes**
> The quotient in Example 7 is a repeating decimal. The repeat cycle actually begins to the left of the decimal point. However, the repeat bar is placed on the first repeated block of digits to the *right* of the decimal point. Therefore, we write the quotient as $45.\overline{45}$.

Skill Practice Divide.

8. $\dfrac{-70}{-0.6}$

When we multiply a number by 10, 100, 1000, and so on, we move the decimal point to the right. However, dividing a positive number by 10, 100, or 1000 decreases its value. Therefore, we move the decimal point to the *left*.

For example, suppose 3.6 is divided by 10, 100, and 1000.

$$\begin{array}{r} .36 \\ 10\overline{)3.60} \\ \underline{-30} \\ 60 \\ \underline{-60} \\ 0 \end{array} \qquad \begin{array}{r} .036 \\ 100\overline{)3.600} \\ \underline{-300} \\ 600 \\ \underline{-600} \\ 0 \end{array} \qquad \begin{array}{r} .0036 \\ 1000\overline{)3.6000} \\ \underline{-3000} \\ 6000 \\ \underline{-6000} \\ 0 \end{array}$$

Answers

6. -1.6 7. 180   8. $116.\overline{6}$

> **Dividing by a Power of 10**
>
> To divide a number by a power of 10, move the decimal point to the *left* the same number of places as there are zeros in the power of 10.

Example 8 Dividing by a Power of 10

Divide.

a. $214.3 \div 10{,}000$ b. $0.03 \div 100$

Solution:

a. $214.3 \div 10{,}000 = 0.02143$ Move the decimal point four places to the left. Insert an additional zero.

b. $0.03 \div 100 = 0.0003$ Move the decimal point two places to the left. Insert two additional zeros.

Skill Practice Divide.

9. $162.8 \div 1000$ 10. $0.0039 \div 10$

2. Rounding a Quotient

In Example 7, we found that $50 \div 1.1 = 45.\overline{45}$. To check this result, we could multiply $45.\overline{45} \times 1.1$ and show that the product equals 50. However, at this point we do not have the tools to multiply repeating decimals. What we can do is round the quotient and then multiply to see if the product is *close* to 50.

Example 9 Rounding a Repeating Decimal

Round $45.\overline{45}$ to the hundredths place. Then use the rounded value to estimate whether the product $45.\overline{45} \times 1.1$ is close to 50. (This will serve as a check to the division problem in Example 7.)

Solution:

To round the number $45.\overline{45}$, we must write out enough of the repeated pattern so that we can view the digit to the right of the rounding place. In this case, we must write out the number to the thousandths place.

$$45.\overline{45} = 45.454\ldots \approx 45.45$$

hundredths place — This digit is less than 5. Discard it and all others to its right.

Now multiply the rounded value by 1.1.

```
      45.45
   ×   1.1
      4545
     45450
     49.995
```

This value is close to 50. We are reasonably sure that we divided correctly in Example 7.

Skill Practice Round to the indicated place value.

11. $2.3\overline{85}$; thousandths

Answers
9. 0.1628   10. 0.00039
11. 2.386

Sometimes we may want to round a quotient to a given place value. To do so, divide until you get a digit in the quotient one place value to the right of the rounding place. At this point, you can stop dividing and round the quotient.

Example 10 — Rounding a Quotient

Round the quotient to the tenths place. $\dfrac{47.3}{5.4}$

Solution:

$5.4\overline{)47.3}$ — Move the decimal point in the divisor and dividend one place to the right. Line up the decimal point in the quotient.

```
       8.75  ← hundredths place
   54)473.00
      −432
       410
      −378
        320
       −270
         50
```
← tenths place

To round the quotient to the tenths place, we must determine the hundredths-place digit and use it to base our decision on rounding. The hundredths-place digit is 5. Therefore, we round the quotient to the tenths place by increasing the tenths-place digit by 1 and discarding all digits to its right.

The quotient is approximately 8.8.

Skill Practice Round the quotient to the indicated place value.

12. $42.68 \div 5.1$; hundredths

3. Applications of Decimal Division

Examples 11 and 12 show how decimal division can be used in applications. Remember that division is used when we need to distribute a quantity into equal parts.

Example 11 — Applying Division of Decimals

Suppose a lunch costs $45.80, and the bill is to be split equally among 5 people. How much would each person pay?

Solution:

We want to distribute $45.80 equally among 5 people, so we must divide $45.80 ÷ 5.

```
      9.16
   5)45.80      Each person would pay $9.16.
    −45
      08
     −5
      30
     −30
       0
```

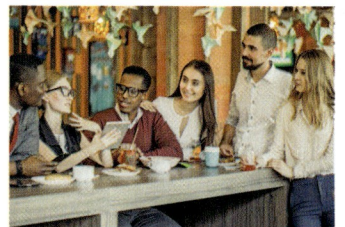
UfaBizPhoto/Shutterstock

TIP: To check if your answer is reasonable, divide $45 among 5 people.

$$5\overline{)45} = 9$$

The estimated value of $9 suggests that the answer to Example 11 is correct.

Skill Practice

13. A 46.5-ft cable is to be cut into 6 pieces of equal length. How long is each piece?

Answers

12. 8.37   13. Each piece is 7.75 ft.

Division is also used in practical applications to express rates. In Example 12, we find the rate of speed in meters per second (m/sec) for the world record time in the men's 400-m run.

Example 12 Using Division to Find a Rate of Speed

In a recent year, the world-record time in the men's 400-m run was 43.2 sec. What is the speed in meters per second? Round to one decimal place. (*Source:* International Association of Athletics Federations)

Solution:

To find the rate of speed in meters per second, we must divide the distance in meters by the time in seconds.

$$43.2\overline{)400.0}$$

$$9.25$$
$$432\overline{)4000.00}$$
$$\underline{-3888}$$
$$1120$$
$$\underline{-864}$$
$$2560$$
$$\underline{-2160}$$
$$400$$

TIP: In Example 12, we had to find speed in meters per second. The units of measurement required in the answer give a hint as to the order of the division. The word *per* implies division. So to obtain meters *per* second implies 400 m ÷ 43.2 sec.

To round the quotient to the tenths place, determine the hundredths-place digit and use it to make the decision on rounding. The hundredths-place digit is 5, which is 5 or greater. Therefore, add 1 to the tenths-place digit and discard all digits to its right.

The speed is approximately 9.3 m/sec.

Skill Practice

14. For many years, the world-record time in the women's 200-m run was held by the late Florence Griffith-Joyner. She ran the race in 21.3 sec. Find the speed in meters per second. Round to the nearest tenth.

Answer
14. The speed was 9.4 m/sec.

Section 5.4 Activity

A.1. **a.** Divide using long division.

$$6\overline{)3888}$$

b. Divide using long division. Place the decimal point in the quotient directly above the decimal point in the dividend.

$$6\overline{)38.88}$$

c. What similarities do you observe in the quotients from parts (a) and (b)?
d. What operation can be used to check your answers from parts (a) and (b)? Check the answer in part (a).

A.2. **a.** Divide using long division. If there is a remainder, insert zeros to the right of the dividend and continue to divide until the remainder is zero or repeats.

$$-4.9 \div (-21)$$

b. What do you notice about the remainder in each successive step as you continue to divide?
c. Write the quotient using a repeat bar.

For Exercises A.3–A.5, divide.

A.3. $3.15 \div 5$

A.4. $\dfrac{412.8}{-4}$

A.5. $3\overline{)0.921}$

A.6. a. What number could you multiply 5.7 by to get a whole number 57?
   b. What number could you multiply 2.1 by to get a whole number 21?
   c. Write an equivalent fraction by multiplying the numerator and denominator by the factor found in parts (a) and (b).

$$\frac{5.7}{2.1} = \frac{}{}$$

   d. Is the quotient of $2.1\overline{)5.7}$ the same as the quotient of $21\overline{)57}$?

For Exercises A.7–A.9, divide.

A.7. $0.4\overline{)0.0236}$

A.8. $5.1\overline{)31.62}$

A.9. $3.69\overline{)8.2}$

A.10. a. If $500.00 is divided equally between 10 people, how much money would each person receive?
   b. How many places did the decimal point move and in what direction?
   c. How many zeros are in the number 10 (10 is referred to as a *power of 10*)? How does that correspond to the number of places the decimal point moves?
   d. If $500.00 is divided equally between 100 people, how much money would each person receive?

A.11. Divide $500 \div 0.01$ (notice that the power of 10 is less than 1 compared to Exercise A.10 where the power of 10 was greater than 1).

For Exercises A.12–A.15, divide by powers of 10 and 0.1 by moving the decimal point accordingly.

A.12. $321.038 \div 1000$

A.13. $321.038 \div 0.1$

A.14. $321.038 \div 100$

A.15. $321.038 \div 0.001$

Practice Exercises

Section 5.4

Prerequisite Review

For Exercises R.1–R.2, round the decimals to the indicated place value.

R.1. 17.060606
   a. hundredths   b. thousandths

R.2. 3.939393
   a. hundredths   b. thousandths

For Exercises R.3–R.6, multiply.

R.3. $4.23 \times 10{,}000$

R.4. 17.0185×1000

R.5. 4.23×0.001

R.6. 17.0185×0.01

Vocabulary and Key Concepts

1. a. If a decimal number has an infinite number of digits with a repeated pattern, then the number is called a _____ decimal.

   b. If a decimal number has a finite number of digits after the decimal point, then the number is called a _____ decimal.

2. When dividing decimals with a remainder, insert _____ to the right of the dividend and continue dividing.

3. When dividing a decimal by a power of 10, move the decimal point to the _____ the same number of decimal places as the number of zeros in the power of 10.

4. The decimal point in the quotient is placed _____ the decimal point in the dividend.

5. Check your answer when dividing by multiplying the _____ by the _____.

6. The repeating decimal 0.818181 . . . , would be written with a repeating bar as _____.

7. Before dividing 0.256 ÷ 0.5, the divisor and the dividend must be multiplied by _____.

8. If rounding a quotient to the hundredths place, divide until there is a digit in the _____ place.

Concept 1: Division of Decimals

For Exercises 9–18, divide. Check the answer by using multiplication. (See Example 1.)

9. 7.2 ÷ 4   Check: _____ × 4 = 7.2
10. 8.4 ÷ 7   Check: _____ × 7 = 8.4
11. $\frac{8.1}{9}$ Check: _____ × 9 = 8.1
12. $\frac{4.8}{6}$ Check: _____ × 6 = 4.8
13. $6\overline{)1.08}$ Check: _____ × 6 = 1.08
14. $4\overline{)2.08}$ Check: _____ × 4 = 2.08
15. −4.24 ÷ (−8)
16. −5.75 ÷ (−25)
17. $5\overline{)105.5}$
18. $7\overline{)221.2}$

For Exercises 19–50, divide. (See Examples 2–7.)

19. $5\overline{)9.8}$
20. $3\overline{)2.07}$
21. 0.28 ÷ 8
22. 0.54 ÷ 8
23. $5\overline{)84.2}$
24. $2\overline{)89.1}$
25. $50\overline{)6}$
26. $80\overline{)6}$
27. −4 ÷ 25
28. −12 ÷ 60
29. 16 ÷ 3
30. 52 ÷ 9
31. −19 ÷ (−6)
32. −9.1 ÷ (−3)
33. $33\overline{)71}$
34. $11\overline{)42}$
35. 5.03 ÷ 0.01
36. 3.2 ÷ 0.001
37. 0.992 ÷ 0.1
38. 123.4 ÷ 0.01
39. $\frac{57.12}{-1.02}$
40. $\frac{95.89}{-2.23}$
41. $\frac{-2.38}{-0.8}$
42. $\frac{-5.51}{-0.2}$
43. $0.3\overline{)62.5}$
44. $1.05\overline{)22.4}$
45. −6.305 ÷ (−0.13)
46. −42.9 ÷ (−0.25)
47. 1.1 ÷ 0.001
48. 4.44 ÷ 0.01
49. 420.6 ÷ 0.01
50. 0.31 ÷ 0.1

51. If 45.62 is divided by 100, will the decimal point move to the right or to the left? By how many places?

52. If 5689.233 is divided by 100,000, will the decimal point move to the right or to the left? By how many places?

For Exercises 53–60, divide by the powers of 10. (See Example 8.)

53. 3.923 ÷ 100
54. 5.32 ÷ 100
55. −98.02 ÷ 10
56. −11.033 ÷ 10
57. −0.027 ÷ (−100)
58. −0.665 ÷ (−100)
59. 1.02 ÷ 1000
60. 8.1 ÷ 1000

Concept 2: Rounding a Quotient

For Exercises 61–66, round to the indicated place value. (See Example 9.)

61. Round $2.\overline{4}$ to the
    a. Tenths place
    b. Hundredths place
    c. Thousandths place

62. Round $5.\overline{2}$ to the
    a. Tenths place
    b. Hundredths place
    c. Thousandths place

63. Round $1.7\overline{8}$ to the
    a. Tenths place
    b. Hundredths place
    c. Thousandths place

64. Round $4.2\overline{7}$ to the
   a. Tenths place
   b. Hundredths place
   c. Thousandths place

65. Round $3.\overline{62}$ to the
   a. Tenths place
   b. Hundredths place
   c. Thousandths place

66. Round $9.\overline{38}$ to the
   a. Tenths place
   b. Hundredths place
   c. Thousandths place

For Exercises 67–75, divide. Round the answer to the indicated place value. Use the rounded quotient to check. **(See Example 10.)**

67. $7\overline{)1.8}$ hundredths

68. $2.1\overline{)75.3}$ hundredths

69. $-54.9 \div 3.7$ tenths

70. $-94.3 \div 21$ tenths

71. $0.24\overline{)4.96}$ thousandths

72. $2.46\overline{)27.88}$ thousandths

73. $0.9\overline{)32.1}$ hundredths

74. $0.6\overline{)81.4}$ hundredths

75. $2.13\overline{)237.1}$ tenths

Concept 3: Applications of Decimal Division

When multiplying or dividing decimals, it is important to place the decimal point correctly. For Exercises 76–79, determine whether you think the number is reasonable or unreasonable. If the number is unreasonable, move the decimal point to a position that makes more sense.

76. Steve computed the gas mileage for his compact car to be 3.2 miles per gallon.

77. The sale price of a new kitchen refrigerator is $96.0.

78. Mickey makes $18.50 per hour. He estimates his weekly paycheck to be $7400.

79. Jason works in a legal office. He computes the average annual income for the attorneys in his office to be $1400 per year.

For Exercises 80–88, solve the application. Check to see if your answers are reasonable.

80. The amount that Brooke owes on her mortgage including interest is $40,540.08. If her monthly payment is $965.24, how many months does she still need to pay? How many years is this?

81. A membership at a health club costs $560 per year. The club has a payment plan in which a member can pay $50 down and the rest in 12 equal monthly payments. How much is each payment? **(See Example 11.)**

82. It is reported that on average 42,000 tennis balls are used and 650 matches are played at the Wimbledon tennis tournament each year. On average, how many tennis balls are used per match? Round to the nearest whole unit.

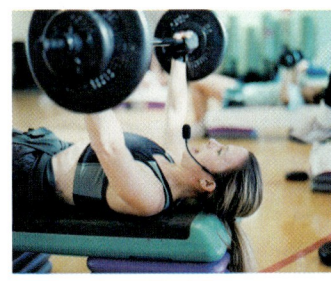

Jack Hollingsworth/Photodisc/ Getty Images

83. A standard 75-watt lightbulb costs $0.75 and lasts about 800 hr. An energy efficient fluorescent bulb that gives off the same amount of light costs $5.00 and lasts about 10,000 hr.
   a. How many standard lightbulbs would be needed to provide 10,000 hr of light?
   b. How much would it cost using standard lightbulbs to provide 10,000 hr of light?
   c. Which is more cost effective long term?

Luther Caverly/Brownstock/Alamy Stock Photo

84. According to government tests, a late model hybrid car gets 45 miles per gallon of gas. If gasoline sells at $3.20 per gallon, how many miles will the car travel on $20.00 of gas?

85. In baseball, the batting average is found by dividing the number of hits by the number of times a batter was at bat. Babe Ruth had 2873 hits in 8399 times at bat. What was his batting average? Round to the thousandths place.

86. Baseball legend Ty Cobb was at bat 11,434 times and had 4189 hits, giving him the all-time best batting average. Find his average. Round to the thousandths place. (Refer to Exercise 85.)

87. Manny hikes 12 mi in 5.5 hr. What is his speed in miles per hour? Round to one decimal place. **(See Example 12.)**

88. Alicia rides her bike 33.2 mi in 2.5 hr. What is her speed in miles per hour? Round to one decimal place.

Technology Connections

For Exercises 89–94, multiply or divide as indicated.

89. (2749.13)(418.2)

90. (139.241)(24.5)

91. $(43.75)^2$

92. $(9.3)^5$

93. $21.5 \overline{)2056.75}$

94. $14.2 \overline{)4167.8}$

95. A large SUV uses 1260 gal of gas to travel 12,000 mi per year. A smaller hybrid SUV uses 375 gal of gas to go the same distance. Use the current cost of gasoline in your area, to determine the amount saved per year by driving the hybrid SUV rather than the large SUV.

96. A small truck gets 16.5 mpg and a small compact car averages 32 mpg. Suppose a driver drives 15,000 mi per year. Use the current cost of gasoline in your area to determine the amount saved per year by driving the car rather than the truck.

97. Recently, the U.S. capacity to generate wind power was 25,369 megawatts (MW). Texas generates approximately 9,410 MW of this power. (*Source:* American Wind Energy Association)

    a. What fraction of the U.S. wind power is generated in Texas? Express this fraction as a decimal number rounded to the nearest hundredth of a megawatt.

    b. Suppose there is a claim in a news article that Texas generates about one-third of all wind power in the United States. Is this claim accurate? Explain using your answer from part (a).

98. Population density is defined to be the number of people per square mile of land area. If California has 37,253,956 people with a land area of 155,779 square miles, what is the population density? Round to the nearest whole unit. (*Source:* U.S. Census Bureau)

99. The Earth travels approximately 584,000,000 mi around the Sun each year.

    a. How many miles does the Earth travel in one day?

    b. Find the speed of the Earth in miles per hour.

100. Although we say the time for the Earth to revolve about the Sun is 365 days, the actual time is 365.256 days. Multiply the fractional amount (0.256) by 4 to explain why we have a leap year every 4 years. (A leap year is a year in which February has an extra day, February 29.)

Expanding Your Skills

101. What number is halfway between −47.26 and −47.27?

102. What number is halfway between −22.4 and −22.5?

103. Which numbers when divided by 8.6 will produce a quotient less than 12.4? Circle all that apply.

    a. 111.8   b. 103.2   c. 107.5   d. 105.78

104. Which numbers when divided by 5.3 will produce a quotient greater than 15.8? Circle all that apply.

    a. 84.8   b. 84.27   c. 83.21   d. 79.5

Problem Recognition Exercises

Operations on Decimals

Remember the following rules when applying operations on decimal numbers:

- When adding or subtracting decimal numbers, write the numbers in a column with the decimal points aligned. Then add or subtract digits in corresponding place positions from right to left.

Add.

$$\begin{array}{r} 407.38 \\ +\ 28.0516 \end{array} \xrightarrow{\text{Insert zeros as needed}} \begin{array}{r} 1\ \ 1 \\ 407.3800 \\ +\ 28.0516 \\ \hline 435.4316 \end{array}$$

Subtract.

$$\begin{array}{r} 60.721 \\ -\ 48.48 \end{array} \xrightarrow{\text{Insert zeros as needed}} \begin{array}{r} 5\ 10\ \ 6\ 12 \\ \cancel{6}\cancel{0}.\cancel{7}\cancel{2}1 \\ -48.480 \\ \hline 12.241 \end{array}$$

- To multiply decimals, write the numbers right-justified in a column. Multiply as you would whole numbers. Then place the decimal point in the answer so that the number of decimal places equals the combined number of decimal places of both factors.
- To perform long division of decimals, move the decimal point in the divisor to the right to make it a whole number. Then move the decimal point in the dividend the same number of positions to the right. Place the decimal point in the quotient directly above the decimal point in the dividend. Divide as you would whole numbers and apply the correct sign to the quotient.

Multiply.

$$\begin{array}{r} -3.261 \\ \times 4.2 \end{array} \longrightarrow \begin{array}{r} 1\ 2 \\ 1 \\ 3.261\quad\text{3 decimal places} \\ \times4.2\quad\text{1 decimal place} \\ \hline 6522 \\ +\ 130440 \\ \hline 13.6962\quad\text{4 decimal places} \end{array}$$

The product is negative because the factors have different signs.

Product: -13.6962

Divide.

$$.34\overline{)27.54} \xrightarrow{\text{Move decimal points}} \begin{array}{r} 81. \\ 34\overline{)2754.} \\ \underline{-272} \\ 34 \\ \underline{-34} \\ 0 \end{array}$$

For Exercises 1–24, perform the indicated operations.

1. **a.** $123.04 + 100$
   b. 123.04×100
   c. $123.04 - 100$
   d. $123.04 \div 100$
   e. $123.04 + 0.01$
   f. 123.04×0.01
   g. $123.04 \div 0.01$
   h. $123.04 - 0.01$

2. **a.** $5078.3 + 1000$
   b. 5078.3×1000
   c. $5078.3 - 1000$
   d. $5078.3 \div 1000$
   e. $5078.3 + 0.001$
   f. 5078.3×0.001
   g. $5078.3 \div 0.001$
   h. $5078.3 - 0.001$

3. **a.** $-4.8 + (-2.391)$
   b. $2.391 - (-4.8)$

4. **a.** $-632.46 + (-98.0034)$
   b. $98.0034 - (-632.46)$

5. **a.** $(32.9)(1.6)$
   b. $(-1.6)(-32.9)$

6. **a.** $(74.23)(0.8)$
   b. $(-0.8)(-74.23)$

7. a. $-3.47-(-9.2)$

   b. $-3.47-9.2$

9. a. $-448 \div 5.6$

   b. $(5.6)(-80)$

11. $8(0.125)$

13. $280 \div 0.07$

15. $490\overline{)98{,}000{,}000}$

17. $(-4500)(-300{,}000)$

19. 83.4
    $\underline{-78.9999}$

21. $4(21.6)$

23. $2(92.5)$

8. a. $-0.042-(-0.097)$

   b. $-0.042-0.097$

10. a. $-496.8 \div 9.2$

    b. $(-54)(9.2)$

12. $20(0.05)$

14. $6400 \div 0.001$

16. $2000\overline{)5{,}400{,}000}$

18. $(-340)(-5000)$

20. 124.7
    $\underline{-47.9999}$

22. $-0.25(21.6)$

24. $-0.5(92.5)$

Section 5.5 Fractions, Decimals, and the Order of Operations

Concepts

1. Writing Fractions as Decimals
2. Writing Decimals as Fractions
3. Decimals and the Number Line
4. Order of Operations Involving Decimals and Fractions
5. Applications of Decimals and Fractions

1. Writing Fractions as Decimals

Sometimes it is possible to convert a fraction to its equivalent decimal form by rewriting the fraction as a decimal fraction. That is, try to multiply the numerator and denominator by a number that will make the denominator a power of 10.

For example, the fraction $\frac{3}{5}$ can easily be written as an equivalent fraction with a denominator of 10.

$$\frac{3}{5} = \frac{3 \cdot 2}{5 \cdot 2} = \frac{6}{10} = 0.6$$

The fraction $\frac{3}{25}$ can easily be converted to a fraction with a denominator of 100.

$$\frac{3}{25} = \frac{3 \cdot 4}{25 \cdot 4} = \frac{12}{100} = 0.12$$

This technique is useful in some cases. However, some fractions such as $\frac{1}{3}$ cannot be converted to a fraction with a denominator that is a power of 10. This is because 3 is not a factor of any power of 10. For this reason we recommend the following procedure for converting fractions to decimals. Every fraction can be expressed as a terminating or repeating decimal.

> **Writing Fractions as Decimals**
> **Step 1** Divide the numerator by the denominator.
> **Step 2** Continue the division process until the quotient is a terminating decimal or a repeating pattern is recognized.

Section 5.5 Fractions, Decimals, and the Order of Operations

Example 1 Writing Fractions as Decimals

Write each fraction or mixed number as a decimal.

a. $\dfrac{3}{5}$ b. $-\dfrac{68}{25}$ c. $3\dfrac{5}{8}$

Solution:

a. $\dfrac{3}{5}$ means $3 \div 5$.

$\dfrac{3}{5} = 0.6$

```
    .6
5)3.0
   -30
     0
```
Divide the numerator by the denominator.

b. $-\dfrac{68}{25}$ means $-(68 \div 25)$.

$-\dfrac{68}{25} = -2.72$

```
       2.72
 25)68.00
    -50
     180
    -175
      50
     -50
       0
```
Divide the numerator by the denominator.

c. $3\dfrac{5}{8} = 3 + \dfrac{5}{8} = 3 + (5 \div 8)$

$= 3 + 0.625$

$= 3.625$

```
      .625
 8)5.000
   -48
     20
    -16
     40
    -40
      0
```
Divide the numerator by the denominator.

Skill Practice Write each fraction or mixed number as a decimal.

1. $\dfrac{3}{8}$ 2. $-\dfrac{43}{20}$ 3. $12\dfrac{5}{16}$

The fractions in Example 1 are represented by terminating decimals. The fractions in Example 2 are represented by repeating decimals.

Example 2 Converting Fractions to Repeating Decimals

Write each fraction as a decimal.

a. $\dfrac{4}{9}$ b. $-\dfrac{3}{22}$

Solution:

a. $\dfrac{4}{9}$ means $4 \div 9$.

$\dfrac{4}{9} = 0.\overline{4}$

```
     .44 . . .
 9)4.00
   -36
     40
    -36
     40
```
The quotient is a repeating decimal.

Answers

1. 0.375   2. −2.15   3. 12.3125

b. $-\dfrac{3}{22}$ means $-(3 \div 22)$

$$22\overline{)3.0000}^{\,.1363\ldots}$$

The quotient is a repeating decimal.

$-\dfrac{3}{22} = -0.1\overline{36}$

Avoiding Mistakes

Be sure to carry out the division far enough to see the repeating digits. In Example 2(b), the next digit in the quotient is 3. The 1 does not repeat.

$$-\dfrac{3}{22} = -0.1363636\ldots$$

Skill Practice Write each fraction as a decimal.

4. $\dfrac{8}{9}$ **5.** $-\dfrac{17}{11}$

Several fractions are used quite often. Their decimal forms are worth memorizing and are presented in Table 5-1.

Table 5-1

$\dfrac{1}{4} = 0.25$	$\dfrac{2}{4} = \dfrac{1}{2} = 0.5$	$\dfrac{3}{4} = 0.75$	
$\dfrac{1}{9} = 0.\overline{1}$	$\dfrac{2}{9} = 0.\overline{2}$	$\dfrac{3}{9} = \dfrac{1}{3} = 0.\overline{3}$	$\dfrac{4}{9} = 0.\overline{4}$
$\dfrac{5}{9} = 0.\overline{5}$	$\dfrac{6}{9} = \dfrac{2}{3} = 0.\overline{6}$	$\dfrac{7}{9} = 0.\overline{7}$	$\dfrac{8}{9} = 0.\overline{8}$

Example 3 Converting Fractions to Decimals with Rounding

Convert the fraction to a decimal rounded to the indicated place value.

a. $\dfrac{162}{7}$; tenths place   **b.** $-\dfrac{21}{31}$; hundredths place

Solution:

a. $\dfrac{162}{7}$

$$7\overline{)162.00}^{\,23.14}$$

To round to the tenths place, we must determine the hundredths-place digit and use it to base our decision on rounding.

$23.14 \approx 23.1$

The fraction $\dfrac{162}{7}$ is approximately 23.1.

Answers
4. $0.\overline{8}$ **5.** $-1.\overline{54}$

Section 5.5 Fractions, Decimals, and the Order of Operations 317

b. $-\dfrac{21}{31}$

```
           .677  ← hundredths place
       _____      ← thousandths place
   31)21.000
      −186
       240
      −217
        230
       −217
         13
```

To round to the hundredths place, we must determine the thousandths-place digit and use it to base our decision on rounding.

$0.67\overset{1}{7} \approx 0.68$

The fraction $-\dfrac{21}{31}$ is approximately -0.68.

Skill Practice Convert the fraction to a decimal rounded to the indicated place value.

6. $\dfrac{9}{7}$; tenths 7. $-\dfrac{17}{37}$; hundredths

2. Writing Decimals as Fractions

To convert a terminating decimal to a fraction, we write the decimal as a decimal fraction and then reduce the fraction to lowest terms. For example:

$$0.46 = \dfrac{46}{100} = \dfrac{\overset{23}{\cancel{46}}}{\underset{50}{\cancel{100}}} = \dfrac{23}{50}$$

We do not yet have the tools to convert a repeating decimal to its equivalent fraction form. However, we can make use of our knowledge of the common fractions and their repeating decimal forms from Table 5-1.

Example 4 Writing Decimals as Fractions

Write the decimals as fractions.

a. 0.475 b. $0.\overline{6}$ c. -1.25

Solution:

a. $0.475 = \dfrac{475}{1000} = \dfrac{19 \cdot 25}{40 \cdot 25} = \dfrac{19}{40}$

b. From Table 5-1, the decimal $0.\overline{6} = \dfrac{2}{3}$.

c. $-1.25 = -\dfrac{125}{100} = -\dfrac{5 \cdot 25}{4 \cdot 25} = -\dfrac{5}{4}$

FOR REVIEW

Recall that the place value of the farthest right digit is the denominator of the fraction.

$0.475 = \dfrac{475}{1000}$

thousandths

Skill Practice Write the decimals as fractions.

8. 0.875 9. $0.\overline{7}$ 10. -1.55

3. Decimals and the Number Line

Recall that a **rational number** is a number that can be written as a fraction whose numerator is an integer and whose denominator is a nonzero integer. The following are all rational numbers.

Answers

6. 1.3 7. -0.46 8. $\dfrac{7}{8}$
9. $\dfrac{7}{9}$ 10. $-\dfrac{31}{20}$

$$\frac{2}{3}$$

$-\frac{5}{7}$ can be written as $\frac{-5}{7}$ or as $\frac{5}{-7}$.

6 can be written as $\frac{6}{1}$.

0.37 can be written as $\frac{37}{100}$.

$0.\overline{3}$ can be written as $\frac{1}{3}$.

Rational numbers consist of all numbers that can be expressed as terminating decimals or as repeating decimals.

- All numbers that can be expressed as *repeating decimals* are rational numbers.

  For example, $0.\overline{3} = 0.3333\ldots$ is a rational number.

- All numbers that can be expressed as *terminating decimals* are rational numbers.

  For example, 0.25 is a rational number.

- A number that *cannot* be expressed as a repeating or terminating decimal is not a rational number. These are called **irrational numbers**. An example of an irrational number is $\sqrt{2}$.

  For example, $\sqrt{2} \approx 1.41421356237\ldots$

  The digits never repeat and never terminate

TIP: The number π is also an irrational number. We approximate π with 3.14 or $\frac{22}{7}$ when calculating area and circumference of a circle.

The rational numbers and the irrational numbers together make up the set of **real numbers**. Furthermore, every real number corresponds to a point on the number line.

In Example 5, we rank the numbers from least to greatest and visualize the position of the numbers on the number line.

Example 5 Ordering Decimals and Fractions

Rank the numbers from least to greatest. Then approximate the position of the points on the number line.

$$0.\overline{45}, \quad 0.45, \quad \frac{1}{2}$$

Solution:

First note that $\frac{1}{2} = 0.5$ and that $0.\overline{45} = 0.454545\ldots$. By writing each number in decimal form, we can compare the place values of each digit.

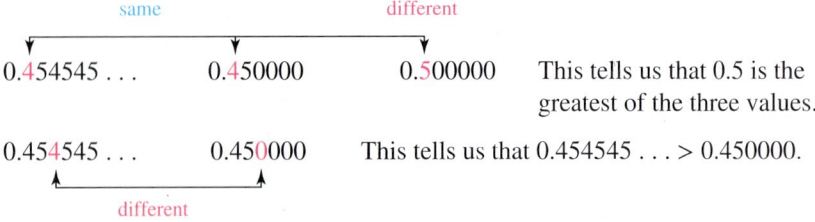

Ranking the numbers from least to greatest we have: $0.45, \quad 0.\overline{45}, \quad 0.5$

The position of these numbers can be seen on the number line. Note that we have expanded the segment of the number line between 0.4 and 0.5 to see more place values to the right of the decimal point.

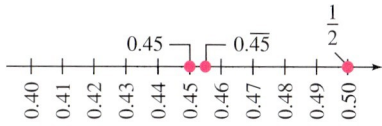

Recall that numbers that lie to the left on the number line have lesser value than numbers that lie to the right.

Skill Practice

11. Rank the numbers from least to greatest. Then approximate the position of the points on the number line.

    0.161, $\frac{1}{6}$, 0.16

    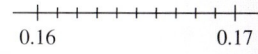

Avoiding Mistakes

Be careful when ordering negative numbers. If the numbers in Example 5 had been negative, the order would be reversed. $-0.5 < -0.\overline{45} < -0.45$

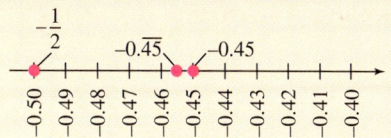

4. Order of Operations Involving Decimals and Fractions

In Example 6, we perform the order of operations with an expression involving decimal numbers.

Applying the Order of Operations

Step 1 First perform all operations inside parentheses or other grouping symbols.
Step 2 Simplify expressions containing exponents, square roots, or absolute values.
Step 3 Perform multiplication or division in the order that they appear from left to right.
Step 4 Perform addition or subtraction in the order that they appear from left to right.

Example 6 Applying the Order of Operations

Simplify. $16.4 - (6.7 - 3.5)^2$

Solution:

$16.4 - (6.7 - 3.5)^2$	Perform the subtraction within parentheses first.	6.7 − 3.5 3.2
$= 16.4 - (3.2)^2$		
$= 16.4 - 10.24$	Perform the operation involving the exponent.	3.2 × 3.2 64 960 10.24
$= 6.16$	Subtract.	16.4 0 − 1 0. 2 4 6. 1 6

Skill Practice Simplify.
12. $(5.8 - 4.3)^2 - 2$

Answers
11. 0.16, 0.161, $\frac{1}{6}$

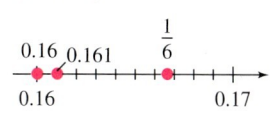

12. 0.25

Chapter 5 Decimals

In Example 7, we apply the order of operations on fractions and decimals combined.

Example 7   **Applying the Order of Operations**

Simplify. $(-6.4) \cdot 2\dfrac{5}{8} \div \left(\dfrac{3}{5}\right)^2$

Solution:

Approach 1
Convert all numbers to fractional form.

$(-6.4) \cdot 2\dfrac{5}{8} \div \left(\dfrac{3}{5}\right)^2 = -\dfrac{64}{10} \cdot \dfrac{21}{8} \div \left(\dfrac{3}{5}\right)^2$ Convert the decimal and mixed number to fractions.

$= -\dfrac{64}{10} \cdot \dfrac{21}{8} \div \dfrac{9}{25}$ Square the quantity $\dfrac{3}{5}$.

$= -\dfrac{64}{10} \cdot \dfrac{21}{8} \cdot \dfrac{25}{9}$ Multiply by the reciprocal of $\dfrac{9}{25}$.

$= -\dfrac{\cancel{64}^{4}}{\cancel{10}_{2}} \cdot \dfrac{\cancel{21}^{7}}{\cancel{8}_{1}} \cdot \dfrac{\cancel{25}^{5}}{\cancel{9}_{3}}$ Simplify common factors.

$= -\dfrac{140}{3}$ or $-46\dfrac{2}{3}$ or $-46.\overline{6}$ Multiply.

Approach 2
Convert all numbers to decimal form.

$(-6.4) \cdot 2\dfrac{5}{8} \div \left(\dfrac{3}{5}\right)^2 = (-6.4)(2.625) \div (0.6)^2$ The fraction $\dfrac{5}{8} = 0.625$ and $\dfrac{3}{5} = 0.6$.

$= (-6.4)(2.625) \div 0.36$ Square the quantity 0.6. That is, $(0.6)(0.6) = 0.36$.

Multiply $(-6.4)(2.625)$.
$$\begin{array}{r} 2.625 \\ \times\ 6.4 \\ \hline 10500 \\ 157500 \\ \hline 16.8000 \end{array}$$

$= -16.8 \div 0.36$ Divide $-16.8 \div 0.36$.

$= -46.\overline{6}$

$$.36\overline{)16.80}$$

$$\begin{array}{r} 46.6\ldots \\ 36\overline{)1680} \\ -144 \\ \hline 240 \\ -216 \\ \hline 240 \end{array}$$

Skill Practice Simplify.

13. $(-2.6) \cdot \dfrac{3}{13} \div \left(1\dfrac{1}{2}\right)^2$

Answer

13. $-\dfrac{4}{15}$ or $-0.2\overline{6}$

When performing operations on fractions, sometimes it is desirable to write the answer as a decimal. For example, suppose that Sabina receives $\frac{1}{4}$ of a $6245 inheritance. She would want to express this answer as a decimal. Sabina should receive:

$$\frac{1}{4}(\$6245) = (0.25)(\$6245) = \$1561.25$$

If Sabina had received $\frac{1}{3}$ of the inheritance, then the decimal form of $\frac{1}{3}$ would have to be rounded to some desired place value. This would cause *round-off error*. Any calculation performed on a rounded number will compound the error. For this reason, we recommend that fractions be kept in fractional form as long as possible as you simplify an expression. Then perform division and rounding in the last step. This is demonstrated in Example 8.

Example 8 — Dividing a Fraction and Decimal

Divide $\frac{4}{7} \div 3.6$. Round the answer to the nearest hundredth.

Solution:

If we attempt to write $\frac{4}{7}$ as a decimal, we find that it is the repeating decimal $0.\overline{571428}$. Rather than rounding this number, we choose to change 3.6 to fractional form: $3.6 = \frac{36}{10}$.

$$\frac{4}{7} \div 3.6 = \frac{4}{7} \div \frac{36}{10} \quad \text{Write 3.6 as a fraction.}$$

$$= \frac{\overset{1}{4}}{7} \cdot \frac{10}{\underset{9}{36}} \quad \text{Multiply by the reciprocal of the divisor.}$$

$$= \frac{10}{63} \quad \text{Multiply and reduce to lowest terms.}$$

We must write the answer in decimal form, rounded to the nearest hundredth.

```
      .158
63)10.000
   -63
    370
   -315
    550
   -504
     46
```

Divide. To round to the hundredths place, divide until we find the thousandths-place digit in the quotient. Use that digit to make a decision for rounding.

≈ 0.16 Round to the nearest hundredth.

Skill Practice Divide. Round the answer to the nearest hundredth.

14. $4.1 \div \frac{12}{5}$

Answer

14. 1.71

Example 9 — Evaluating an Algebraic Expression

Determine the area of the triangle.

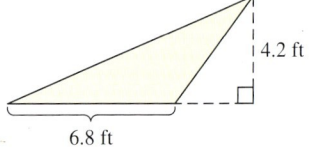

Solution:

In this triangle, the base is 6.8 ft. The height is 4.2 ft and is drawn outside the triangle.

$$A = \frac{1}{2}bh \qquad \text{Formula for the area of a triangle}$$

$$A = \frac{1}{2}(6.8 \text{ ft})(4.2 \text{ ft}) \qquad \text{Substitute } b = 6.8 \text{ ft and } h = 4.2 \text{ ft.}$$

$$= 0.5(6.8 \text{ ft})(4.2 \text{ ft}) \qquad \text{Write } \tfrac{1}{2} \text{ as the terminating decimal, 0.5.}$$

$$= 14.28 \text{ ft}^2 \qquad \text{Multiply from left to right.}$$

The area is 14.28 ft^2.

Skill Practice Determine the area of the triangle.

15.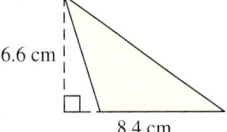

5. Applications of Decimals and Fractions

Example 10 — Using Decimals and Fractions in a Consumer Application

Joanne filled the gas tank in her car and noted that the odometer read 22,341.9 mi. Ten days later she filled the tank again with $11\frac{1}{2}$ gal of gas. Her odometer reading at that time was 22,622.5 mi.

a. How many miles had she driven between fill-ups?

b. How many miles per gallon did she get?

Gary He/McGraw Hill

Solution:

a. To find the number of miles driven, we need to subtract the initial odometer reading from the final reading.

$$\begin{array}{r} 2\,2,\!6\,2\,2.\,5 \\ -2\,2,\!3\,4\,1.\,9 \\ \hline 2\,8\,0.\,6 \end{array}$$

Recall that to add or subtract decimals, line up the decimal points.

Joanne had driven 280.6 mi between fill-ups.

Answer

15. 27.72 cm^2

Section 5.5 Fractions, Decimals, and the Order of Operations

b. To find the number of miles per gallon (mi/gal), we divide the number of miles driven by the number of gallons.

$$280.6 \div 11\frac{1}{2} = 280.6 \div 11.5 \quad \text{We convert to decimal form because the fraction } 11\frac{1}{2} \text{ is recognized as } 11.5.$$

$$= 24.4$$

```
            24.4
11.5)280.6    115)2806.0
              −230
              506
              −460
               460
              −460
                 0
```

Joanne got 24.4 mi/gal.

Skill Practice

16. The odometer on a car read 46,125.9 mi. After a $13\frac{1}{4}$-hr trip, the odometer read 46,947.4 mi.

  a. Find the total distance traveled on the trip.
  b. Find the average speed in miles per hour (mph).

Answer
16. a. 821.5 mi   **b.** 62 mph

Practice Exercises   Section 5.5

Prerequisite Review

For Exercises R.1–R.2, reduce the fractions to lowest terms.

R.1. $\dfrac{15}{1000}$ **R.2.** $\dfrac{24}{100}$

For Exercises R.3–R.8, perform the indicated operation.

R.3. $(-0.04)^2$ **R.4.** $(-0.02)^3$

R.5. $1.04 - 2.1$ **R.6.** $-3.4 - 1.002$

R.7. $3.4 \div 0.17$ **R.8.** $25.4 \div 0.04$

For Exercises R.9–R.10, simplify by using the order of operations.

R.9. $-5^2 + 2(14 \div 7)$ **R.10.** $(-5)^2 + 2(14) \div 7$

Vocabulary and Key Concepts

1. a. A _____ number is a number that can be written as a fraction whose numerator is an integer and whose denominator is a nonzero integer.

  b. All rational numbers can be expressed as terminating or _____ decimals.

  c. An _____ number is a number that cannot be expressed as a terminating decimal or as a repeating decimal.

  d. The set of _____ numbers includes the rational numbers and irrational numbers.

2. To write a fraction as a decimal, use the mathematical operation of _____.

324 Chapter 5 Decimals

3. The decimal 0.625 would be written as the fraction $\frac{625}{\square}$ before reducing.

4. The decimal -0.78 is (less than/greater than) the decimal -0.87.

Concept 1: Writing Fractions as Decimals

For Exercises 5–12, write each fraction as a decimal fraction, that is, a fraction whose denominator is a power of 10. Then write the number in decimal form.

5. $\frac{1}{2}$
6. $\frac{1}{4}$
7. $\frac{21}{25}$
8. $\frac{4}{25}$

9. $\frac{2}{5}$
10. $\frac{4}{5}$
11. $\frac{49}{50}$
12. $\frac{3}{50}$

For Exercises 13–24, write each fraction or mixed number as a decimal. (See Example 1.)

13. $\frac{7}{25}$
14. $\frac{4}{25}$
15. $-\frac{316}{500}$
16. $-\frac{19}{500}$

17. $-\frac{16}{5}$
18. $-\frac{68}{25}$
19. $-5\frac{3}{12}$
20. $-6\frac{5}{8}$

21. $\frac{18}{24}$
22. $\frac{24}{40}$
23. $7\frac{9}{20}$
24. $3\frac{11}{25}$

For Exercises 25–32, write each fraction or mixed number as a repeating decimal. (See Example 2.)

25. $3\frac{8}{9}$
26. $4\frac{7}{9}$
27. $\frac{19}{36}$
28. $\frac{7}{12}$

29. $-\frac{14}{111}$
30. $-\frac{58}{111}$
31. $\frac{25}{22}$
32. $\frac{45}{22}$

For Exercises 33–40, convert the fraction to a decimal and round to the indicated place value. (See Example 3.)

33. $\frac{15}{16}$; tenths
34. $\frac{3}{11}$; tenths
35. $\frac{1}{7}$; thousandths
36. $\frac{2}{7}$; thousandths

37. $\frac{1}{13}$; hundredths
38. $\frac{9}{13}$; hundredths
39. $-\frac{5}{7}$; hundredths
40. $-\frac{1}{8}$; hundredths

41. Write the fractions as decimals. Explain how to memorize the decimal form for these fractions with a denominator of 9.

    a. $\frac{1}{9}$ b. $\frac{2}{9}$ c. $\frac{4}{9}$ d. $\frac{5}{9}$

42. Write the fractions as decimals. Explain how to memorize the decimal forms for these fractions with a denominator of 3.

    a. $\frac{1}{3}$ b. $\frac{2}{3}$

Concept 2: Writing Decimals as Fractions

For Exercises 43–46, complete the table. (See Example 4.)

43.

	Decimal Form	Fraction Form
a.	0.45	
b.		$1\frac{5}{8}$ or $\frac{13}{8}$
c.	$-0.\overline{7}$	
d.		$-\frac{5}{11}$

44.

	Decimal Form	Fraction Form
a.		$\frac{4}{9}$
b.	1.6	
c.		$-\frac{152}{25}$
d.	$-0.\overline{2}$	

45.

	Decimal Form	Fraction Form
a.	$0.\overline{3}$	
b.	-2.125	
c.		$-\dfrac{19}{22}$
d.		$\dfrac{42}{25}$

46.

	Decimal Form	Fraction Form
a.	0.75	
b.		$-\dfrac{7}{11}$
c.	$-1.\overline{8}$	
d.		$\dfrac{74}{25}$

Concept 3: Decimals and the Number Line

For Exercises 47–58, identify the number as rational or irrational.

47. $-\dfrac{2}{5}$ **48.** $-\dfrac{1}{9}$ **49.** 5 **50.** 3

51. 3.5 **52.** 1.1 **53.** $\sqrt{7}$ **54.** $\sqrt{11}$

55. π **56.** 2π **57.** $0.\overline{4}$ **58.** $0.\overline{9}$

For Exercises 59–66, insert the appropriate symbol. Choose from <, >, or =

59. $0.2 \square \dfrac{1}{5}$ **60.** $1.5 \square \dfrac{3}{2}$ **61.** $0.2 \square 0.\overline{2}$ **62.** $\dfrac{3}{5} \square 0.\overline{6}$

63. $\dfrac{1}{3} \square 0.3$ **64.** $\dfrac{2}{3} \square 0.66$ **65.** $-4\dfrac{1}{4} \square -4.\overline{25}$ **66.** $-2.12 \square -2.\overline{12}$

For Exercises 67–70, rank the numbers from least to greatest. Then approximate the position of the points on the number line. **(See Example 5.)**

67. $1.8, 1.75, 1.\overline{7}$

68. $5\dfrac{1}{6}, 5.\overline{6}, 5.0\overline{6}$

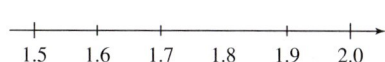

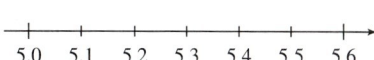

69. $-0.\overline{1}, -\dfrac{1}{10}, -\dfrac{1}{5}$

70. $-3\dfrac{1}{4}, -3\dfrac{1}{3}, -3.3$

Concept 4: Order of Operations Involving Decimals and Fractions

For Exercises 71–86, simplify by using the order of operations. **(See Example 6.)**

71. $(3.7 - 1.2)^2$ **72.** $(6.8 - 4.7)^2$ **73.** $16.25 - (18.2 - 15.7)^2$ **74.** $11.38 - (10.42 - 7.52)^2$

75. $12.46 - 3.05 - 0.8^2$ **76.** $15.06 - 1.92 - 0.4^2$ **77.** $63.75 - 9.5(4)$ **78.** $6.84 + (3.6)(9)$

79. $-6.8 \div 2 \div 1.7$ **80.** $-8.4 \div 2 \div 2.1$ **81.** $2.2 - [9.34 + (1.2)^2]$ **82.** $4.2 \div 2.1 - (3.1)^2$

83. $16.04 \div [(2.2)^2 - 0.83]$ **84.** $6[(3.1)(4) - 8.1]$ **85.** $42.82 - 3(4.8 - 1.6)^2$ **86.** $14.28 \div [(1.1)^2 + 5.79]$

For Exercises 87–92, simplify by using the order of operations. Express the answer in decimal form. (See Example 7.)

87. $89.8 \div 1\frac{1}{3}$ **88.** $-30.12 \div \left(-1\frac{3}{5}\right)$ **89.** $-20.04 \div \left(-\frac{4}{5}\right)$

90. $(78.2 - 60.2) \div \frac{9}{13}$ **91.** $14.4\left(\frac{7}{4} - \frac{1}{8}\right)$ **92.** $6.5 + \frac{1}{8}\left(\frac{1}{5}\right)^2$

For Exercises 93–98, perform the indicated operations. Round the answer to the nearest hundredth when necessary. (See Example 8.)

93. $(2.3)\left(\frac{5}{9}\right)$ **94.** $(4.6)\left(\frac{7}{6}\right)$ **95.** $6.5 \div \left(-\frac{3}{5}\right)$

96. $\left(\frac{1}{12}\right)(6.24) \div (-2.1)$ **97.** $(42.81 - 30.01) \div \frac{9}{2}$ **98.** $\left(\frac{2}{7}\right)(5.1)\left(\frac{1}{10}\right)$

For Exercises 99 and 100, find the area of the triangle. (See Example 9.)

99.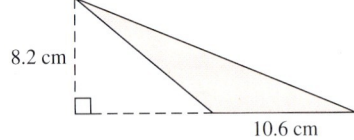

100.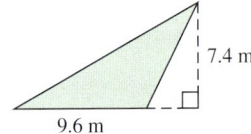

For Exercises 101–104, evaluate the expression for the given value(s) of the variables. Use 3.14 for π.

101. $\frac{1}{3}\pi r^2$ for $r = 9$ **102.** $\frac{4}{3}\pi r^3$ for $r = 3$

103. $-\frac{1}{2}gt^2$ for $g = 9.8$ and $t = 4$ **104.** $\frac{1}{2}mv^2$ for $m = 2.5$ and $v = 6$

Concept 5: Applications of Decimals and Fractions

105. Oprah and Gayle traveled for $7\frac{1}{2}$ hr between Atlanta, Georgia, and Orlando, Florida. At the beginning of the trip, the car's odometer read 21,345.6 mi. When they arrived in Orlando, the odometer read 21,816.6 mi. (See Example 10.)

   a. How many miles had they driven on the trip?

   b. Find the average speed in miles per hour (mph).

106. Jennifer earns $720.00 per week. Alex earns $\frac{3}{4}$ of what Jennifer earns per week.

   a. How much does Alex earn per week?

   b. If they both work 40 hr per week, how much do they each make per hour?

107. A cell phone plan has a $39.95 monthly fee and includes 450 min. For time on the phone over 450 min, the charge is $0.40 per minute. How much is Jorge charged for a month in which he talks for 597 min?

108. A night at a hotel in Dallas costs $185.95 with a nightly room tax of $24.17. If John stays for 5 nights, how much is his total bill?

109. Olivia's diet allows her 60 grams (g) of fat per day. If she has $\frac{1}{4}$ of her total fat grams for breakfast and a hamburger (20.7 g of fat) for lunch, how many grams does she have left for dinner?

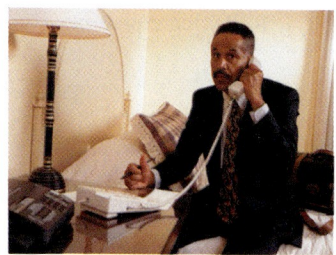

John A. Rizzo/Photodisc/Getty Images

110. Todd is establishing his beneficiaries for his life insurance policy. The policy is for $150,000.00 and $\frac{1}{2}$ will go to his daughter, $\frac{3}{8}$ will go to his stepson, and the rest will go to his grandson. What dollar amount will go to the grandson?

111. Hannah bought three packages of printer paper for $4.79 each. The sales tax for the merchandise was $0.86. If Hannah paid with a $20 bill, how much change should she receive?

112. Mr. Timpson bought dinner for $28.42. He left a tip of $6.00. He paid the bill with two $20 bills. How much change should he receive?

Technology Connections

113. Suppose that Deanna owns 50 shares of stock in Company A, valued at $132.05 per share. She decides to sell these shares and use the money to buy stock in Company B, valued at $27.80 per share. Assume there are no fees for either transaction.

    a. How many full shares of Company B stock can she buy?

    b. How much money will be left after she buys the Company B stock?

114. One megawatt (MW) of wind power produces enough electricity to supply approximately 275 homes. For a recent year, the state of Texas produced 3352 MW of wind power. (*Source:* American Wind Energy Association)

    a. About how many homes can be supplied with electricity using wind power produced in Texas?

    b. The given table outlines new proposed wind power projects in Texas. If these projects are completed, approximately how many additional homes could be supplied with electricity?

Project	MW
JD Wind IV	79.8
Buffalo Gap, Phase II	232.5
Lone Star I (3Q)	128
Sand Bluff	90
Roscoe	209
Barton Chapel	120
Stanton Wind Energy Center	120
Whirlwind Energy Center	59.8
Sweetwater V	80.5
Champion	126.5

115. Health-care providers use *body mass index* (BMI) as one way to assess a person's risk of developing diabetes and heart disease. BMI is calculated with the following equation:

$$\text{BMI} = \frac{703w}{h^2}$$

where w is the person's weight in pounds and h is the person's height in inches. A person whose BMI is between 18.5 and 24.9 has a low risk of developing diabetes and heart disease. A BMI of 25.0 to 29.9 indicates moderate risk. A BMI above 30.0 indicates high risk.

a. What is the risk level for a person whose height is 67.5 in. and whose weight is 195 lb?

b. What is the risk level for a person whose height is 62.5 in. and whose weight is 110 lb?

116. Marty bought a home for $145,000. He paid $25,000 as a down payment and then financed the rest with a 30-year mortgage. His monthly payments are $798.36 and go toward paying off the loan and interest on the loan.

a. How much money does Marty have to finance?

b. How many months are in a 30-year period?

c. How much money will Marty pay over a 30-year period to pay off the loan?

d. How much money did Marty pay in interest over the 30-year period?

117. An inheritance for $80,460.60 is to be divided equally among four heirs. However, before the money can be distributed, approximately one-third of the money must go to the government for taxes. How much does each person get after the taxes have been taken?

118. Kiah performed in four events at a state gymnastics meet. She received the following scores from the judges:

Event	Score
Beam	7.425
Floor	8.875
Vault	8.725
Uneven Bars	9.300

a. If her all-around score is determined by the mean of her top three events, what was Kiah's all-around score at the meet? Round the answer to the nearest thousandth.

b. If an all-around score of 8.950 is required to compete at the regional gymnastics competition, did Kiah qualify?

Expanding Your Skills

For Exercises 119–122, perform the indicated operations.

119. $0.\overline{3} \cdot 0.3 + 3.375$

120. $0.\overline{5} \div 0.\overline{2} - 0.75$

121. $(0.\overline{8} + 0.\overline{4}) \cdot 0.39$

122. $(0.\overline{7} - 0.\overline{6}) \cdot 5.4$

Solving Equations Containing Decimals

Section 5.6

1. Solving Equations Containing Decimals

In this section, we will practice solving linear equations. In Example 1, we will apply the addition and subtraction properties of equality.

Concepts
1. Solving Equations Containing Decimals
2. Solving Equations by Clearing Decimals
3. Applications and Problem Solving

Example 1   Applying the Addition and Subtraction Properties of Equality

Solve.   **a.** $x + 3.7 = 5.42$ **b.** $-35.4 = -6.1 + y$

Solution:

a.
$$x + 3.7 = 5.42$$
$$x + 3.7 - 3.7 = 5.42 - 3.7$$
$$x = 1.72$$

The value 3.7 is added to x. To isolate x, we must reverse this process. Therefore, *subtract* 3.7 from both sides.

Check: $x + 3.7 = 5.42$
$$(1.72) + 3.7 \stackrel{?}{=} 5.42$$
$$5.42 = 5.42 \checkmark$$

The solution is 1.72.

b.
$$-35.4 = -6.1 + y$$
$$-35.4 + 6.1 = -6.1 + 6.1 + y$$
$$-29.3 = y$$

To isolate y, add 6.1 to both sides.

Check: $-35.4 = -6.1 + y$
$$-35.4 \stackrel{?}{=} -6.1 + (-29.3)$$
$$-35.4 = -35.4 \checkmark$$

The solution is -29.3.

Skill Practice   Solve.
1. $t + 2.4 = 9.68$ **2.** $-97.5 = -31.2 + w$

In Example 2, we will apply the multiplication and division properties of equality.

Example 2   Applying the Multiplication and Division Properties of Equality

Solve.   **a.** $\dfrac{t}{10.2} = -4.5$ **b.** $-25.2 = -4.2z$

Solution:

a.
$$\frac{t}{10.2} = -4.5$$
$$10.2\left(\frac{t}{10.2}\right) = 10.2(-4.5)$$
$$t = -45.9$$

The variable t is being divided by 10.2. To isolate t, *multiply* both sides by 10.2.

Check: $\dfrac{t}{10.2} = -4.5 \longrightarrow \dfrac{-45.9}{10.2} \stackrel{?}{=} -4.5$
$$-4.5 = -4.5 \checkmark$$

The solution is -45.9.

b.
$$-25.2 = -4.2z$$
$$\frac{-25.2}{-4.2} = \frac{-4.2z}{-4.2}$$

The variable z is being multiplied by -4.2. To isolate z, *divide* by -4.2. The quotient will be positive.

Answers
1. 7.28   **2.** -66.3

$6 = z$ Check: $-25.2 = -4.2z$

$-25.2 \stackrel{?}{=} -4.2(6)$

The solution is 6. $-25.2 = -25.2$ ✓

Skill Practice Solve.

3. $\dfrac{z}{11.5} = -6.8$ **4.** $-8.37 = -2.7p$

In Examples 3 and 4, multiple steps are required to solve the equation. As you read through Example 3, remember that we first isolate the variable term. Then we apply the multiplication or division property of equality to make the coefficient on the variable term equal to 1.

Example 3   Solving Equations Involving Multiple Steps

Solve. $2.4x - 3.85 = 8.63$

Solution:

$2.4x - 3.85 = 8.63$

$2.4x - 3.85 + 3.85 = 8.63 + 3.85$ To isolate the x term, add 3.85 to both sides.

$24x = 12.48$ Divide both sides by 2.4.

$\dfrac{2.4x}{2.4} = \dfrac{12.48}{2.4}$ This makes the coefficient on the x term equal to 1.

$x = 5.2$ The solution 5.2 checks in the original equation.

The solution is 5.2.

Skill Practice Solve.

5. $5.8y - 14.4 = 55.2$

Example 4   Solving Equations Involving Multiple Steps

Solve. $0.03(x + 4) = 0.01x + 2.8$

Solution:

$0.03(x + 4) = 0.01x + 2.8$ Apply the distributive property to clear parentheses.

$0.03x + 0.12 = 0.01x + 2.8$

$0.03x - 0.01x + 0.12 = 0.01x - 0.01x + 2.8$ Subtract $0.01x$ from both sides. This places the variable terms all on one side.

$0.02x + 0.12 = 2.8$

$0.02x + 0.12 - 0.12 = 2.8 - 0.12$ Subtract 0.12 from both sides. This places the constant terms all on one side.

$0.02x = 2.68$

$\dfrac{0.02x}{0.02} = \dfrac{2.68}{0.02}$ Divide both sides by 0.02. This makes the coefficient on the x term equal to 1.

$x = 134$ The value 134 checks in the original equation.

The solution is 134.

Skill Practice Solve.

6. $0.05(x + 6) = 0.02x - 0.18$

Answers

3. -78.2 **4.** 3.1
5. 12   **6.** -16

2. Solving Equations by Clearing Decimals

We have already learned that an equation containing fractions can be easier to solve if we clear the fractions. Similarly, when solving an equation with decimals, students may find it easier to clear the decimals first. To do this, we can multiply both sides of the equation by a power of 10 (10, 100, 1000, etc.). This will move the decimal point to the right in the coefficient on each term in the equation. This process is demonstrated in Example 5.

Example 5   Solving an Equation by Clearing Decimals

Solve. $0.05x - 1.45 = 2.8$

Solution:

To determine a power of 10 to use to clear decimals, identify the term with the most digits to the right of the decimal point. In this case, the terms $0.05x$ and 1.45 each have two digits to the right of the decimal point. Therefore, to clear decimals, we must multiply by 100. This will move the decimal point to the right two places.

$$0.05x - 1.45 = 2.8$$
$$100(0.05x - 1.45) = 100(2.8) \quad \text{Multiply by 100 to clear decimals.}$$
$$100(0.05x) - 100(1.45) = 100(2.80) \quad \text{The decimal point will move to the right two places.}$$
$$5x - 145 = 280$$
$$5x - 145 + 145 = 280 + 145 \quad \text{Add 145 to both sides.}$$
$$5x = 425$$
$$\frac{5x}{5} = \frac{425}{5} \quad \text{Divide both sides by 5.}$$
$$x = 85 \quad \text{The value 85 checks in the original equation.}$$

The solution is 85.

Skill Practice   Solve.

7. $0.02x - 3.42 = 1.6$

3. Applications and Problem Solving

In Examples 6–8, we will practice using linear equations to solve application problems.

Example 6   Translating to an Algebraic Expression

The sum of a number and 7.5 is three times the number. Find the number.

Solution:

Let x represent the number.

    the sum   3 times the
      of   number

$$x + 7.5 = 3x$$

a number   7.5

Step 1: Read the problem completely.
Step 2: Label the unknown.
Step 3: Write the equation in words.
Step 4: Translate to a mathematical equation.

Answer

7. 251

$$x - x + 7.5 = 3x - x$$
$$7.5 = 2x$$
$$\frac{7.5}{2} = \frac{2x}{2}$$
$$3.75 = x$$

The number is 3.75.

Step 5: Solve the equation. Subtract x from both sides. This will bring all variable terms to one side.

Divide by 2.

Step 6: Interpret the answer in words.

> **Avoiding Mistakes**
> Check the answer to Example 6. The sum of 3.75 and 7.5 is 11.25. The product 3(3.75) is also equal to 11.25.

Skill Practice

8. The sum of a number and 15.6 is the same as four times the number. Find the number.

Example 7 Solving an Application Involving Geometry

The perimeter of a triangle is 22.8 cm. The longest side is 8.4 cm more than the shortest side. The middle side is twice the shortest side. Find the lengths of the three sides.

Solution:

Let x represent the length of the shortest side.

Then the longest side is $(x + 8.4)$.

The middle side is $2x$.

Step 1: Read the problem completely.

Step 2: Label the unknowns and draw a figure.

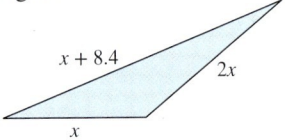

The sum of the lengths of the sides is 22.8 cm.

$$x + (x + 8.4) + 2x = 22.8$$
$$x + x + 8.4 + 2x = 22.8$$
$$4x + 8.4 = 22.8$$
$$4x + 8.4 - 8.4 = 22.8 - 8.4$$
$$4x = 14.4$$
$$\frac{4x}{4} = \frac{14.4}{4}$$
$$x = 3.6$$

Step 3: Write the equation in words.

Step 4: Write a mathematical equation.

Step 5: Solve the equation.
Combine *like* terms.
Subtract 8.4 from both sides.

Divide by 4 on both sides.

Step 6: Interpret the answer in words.

The shortest side is 3.6 cm.
The longest side is $(x + 8.4)$ cm = $(3.6 + 8.4)$ cm = 12 cm.
The middle side is $(2x)$ cm = $2(3.6)$ cm = 7.2 cm.

The lengths of the sides are 3.6 cm, 12 cm, and 7.2 cm.

> **Avoiding Mistakes**
> To check Example 7, notice that the sum of the lengths of the sides is 22.8 cm as expected.
> 3.6 cm + 12 cm + 7.2 cm
> = 22.8 cm

Skill Practice

9. The perimeter of a triangle is 16.6 ft. The longest side is 6.1 ft more than the shortest side. The middle side is three times the shortest side. Find the lengths of the three sides.

Answers
8. The number is 5.2.
9. The lengths are 2.1 ft, 6.3 ft, and 8.2 ft.

Example 8 Using a Linear Equation in a Consumer Application

The cost to buy business cards is $18.99 for the design of the card. In addition, there is a printing charge of $9.99 per 100 cards. If cards must be purchased in increments of 100, and Joanne's bill comes to $98.91, how many cards did she buy?

Solution:

Let x represent the number of business cards in increments of 100.

Step 1: Read the problem.
Step 2: Label the unknown.

$$\begin{pmatrix} \text{Cost of} \\ x \text{ hundred} \\ \text{cards} \end{pmatrix} + \begin{pmatrix} \text{cost of} \\ \text{the design} \end{pmatrix} = \begin{pmatrix} \text{total} \\ \text{cost} \end{pmatrix}$$

Step 3: Write an equation in words.

$$9.99x + 18.99 = 98.91$$

Step 4: Write a mathematical equation.

$$9.99x + 18.99 - 18.99 = 98.91 - 18.99$$

Step 5: Solve the equation.

$$9.99x = 79.92 \qquad \text{Subtract } 18.99.$$

$$\frac{9.99x}{9.99} = \frac{79.92}{9.99} \qquad \text{Divide by } 9.99.$$

$$x = 8$$

Joanne purchased 800 cards.

Skill Practice

10. D.J. signs up for a new credit card that earns travel miles with a certain airline. She initially earns 15,000 travel miles by signing up for the new card. Then for each dollar spent she earns 2.5 travel miles. If at the end of one year she has 38,500 travel miles, how many dollars did she charge on the credit card?

Answer
10. D.J. charged $9400.

Section 5.6 Activity

A.1.
  a. Solve the equation $t + 86 = 401$
  b. Solve the equation $t + 0.86 = 4.01$
  c. Does the solving process change when an equation contains decimals (part b) as opposed to whole numbers (part a)?

A.2.
  a. Consider the equation $-52x = 312$
     What step should be performed to solve for x?
  b. Consider the equation $-5.2x = 31.2$
     What step should be performed to solve for x?
  c. Did you use the same mathematical operation to solve the equations in parts (a) and (b)?
  d. Do you get the same solution for the equations in part (a) as part (b)?
  e. What power of 10 was used to multiply each term in the equation in part (b) to obtain the equation in part (a)?

A.3.
  a. Solve the equation $t + 0.86 = 4.01$
  b. Solve the equation $100t + 86 = 401$
  c. Do you get the same solution for the equation in part (a) as in part (b)? Why?

For Exercises A.4–A.7, solve the equation. If you prefer to clear the decimals first before solving, determine the power of 10 that can be used to clear the decimals, then solve.

A.4. $-0.1x + 0.02 = -0.16$

A.5. $2.2y - 5.2 = -3(0.1y - 0.3)$

A.6. $1.2x + 35 = 20.6$

A.7. $2y - 0.03 = 0.4y + 3.97$

For Exercises A.8–A.9, solve the word problem and format your solution using the problem-solving steps:

a. Read the problem again.
b. Label the variable(s).
c. Write a verbal model (that is, write an equation in *words*).
d. Write an equation to represent the verbal model.
e. Solve the equation.
f. Interpret the results in words using the proper units of measurement and verify that your solution makes sense in the context of the problem.

A.8. The sum of a number and 8.55 is 2.99. Find the number.

A.9. Jenna plans to attend college in the fall semester. The college has a one-time admissions fee of $35, plus an additional cost of $92.50 per credit hour. The equation $C = 35 + 92.50x$ represents the cost C (in dollars) for a new student to take x credit hours. How many credit hours is Jenna taking if her cost is $1145?

Section 5.6 Practice Exercises

Prerequisite Review

For Exercises R.1–R.6, solve the equations.

R.1. $13x - 6 = 2x + 60$

R.2. $8x + 5 = 10x - 3$

R.3. $2x + \dfrac{5}{6} = \dfrac{1}{3}x - 2$

R.4. $\dfrac{3}{2}x - \dfrac{1}{4} = 3x + 1$

R.5. $-4(3x - 2) = 4x + 8$

R.6. $6(4 - x) = 4x - 6$

Concept 1: Solving Equations Containing Decimals

For Exercises 1–24, solve the equations. **(See Examples 1–4.)**

1. $y + 8.4 = 9.26$

2. $z + 1.9 = 12.41$

3. $t - 3.92 = -8.7$

4. $w - 12.69 = -15.4$

5. $-141.2 = -91.3 + p$

6. $-413.7 = -210.6 + m$

7. $-0.07 + n = 0.025$

8. $-0.016 + k = 0.08$

9. $\dfrac{x}{-4.6} = -9.3$

10. $\dfrac{y}{-8.1} = -1.5$

11. $6 = \dfrac{z}{-0.02}$

12. $7 = \dfrac{a}{-0.05}$

13. $19.43 = -6.7n$

14. $94.08 = -8.4q$

15. $-6.2y = -117.8$

16. $-4.1w = -73.8$

17. $8.4x + 6 = 48$

18. $9.2n + 6.4 = 43.2$

19. $-3.1x - 2 = -29.9$

20. $-5.2y - 7 = -22.6$

21. $0.04(p - 2) = 0.05p + 0.16$

22. $0.06(t - 9) = 0.07t + 0.27$

23. $-2.5x + 5.76 = 0.4(6 - 5x)$

24. $-1.5m + 14.26 = 0.2(18 - m)$

Concept 2: Solving Equations by Clearing Decimals

For Exercises 25–32, solve by first clearing decimals. (See Example 5.)

25. $0.04x - 1.9 = 0.1$

26. $0.03y - 2.3 = 0.7$

27. $-4.4 = -2 + 0.6x$

28. $-3.7 = -4 + 0.5x$

29. $4.2 = 3 - 0.002m$

30. $3.8 = 7 - 0.016t$

31. $6.2x - 4.1 = 5.94x - 1.5$

32. $1.32x + 5.2 = 0.12x + 0.4$

Concept 3: Applications and Problem Solving

33. Nine times a number is equal to 36 more than the number. Find the number. (See Example 6.)

34. Six times a number is equal to 30.5 more than the number. Find the number.

35. The difference of 13 and a number is 2.2 more than three times the number. Find the number.

36. The difference of 8 and a number is 1.7 more than two times the number. Find the number.

37. The quotient of a number and 5 is -1.88. Find the number.

38. The quotient of a number and -2.5 is 2.72. Find the number.

39. The product of 2.1 and a number is 8.36 more than the number. Find the number.

40. The product of -3.6 and a number is 48.3 more than the number. Find the number.

41. The perimeter of a triangle is 21.5 yd. The longest side is twice the shortest side. The middle side is 3.1 yd longer than the shortest side. Find the lengths of the sides. (See Example 7.)

42. The perimeter of a triangle is 2.5 m. The longest side is 2.4 times the shortest side, and the middle side is 0.3 m more than the shortest side. Find the lengths of the sides.

43. Toni, Rafa, and Henri are all servers at the Chez Joëlle Restaurant. The tips collected for the night amount to $167.80. Toni made $22.05 less in tips than Rafa. Henri made $5.90 less than Rafa. How much did each person make?

DreamPictures/Blend Images

44. Bob bought a popcorn, a soda, and a hotdog at the movies for $8.25. Popcorn costs $1 more than a hotdog. A soda costs $0.25 less than a hotdog. How much is each item?

45. The U-Rent-It home supply store rents pressure cleaners for $4.95, plus $4 per hour. A painter rents a pressure cleaner and the bill comes to $18.95. For how many hours did he rent the pressure cleaner? (See Example 8.)

46. A dog kennel charges $23.50 per day to board a dog plus a non-refundable deposit of $35.50. If Kelly has $200 budgeted for boarding her dog, for how many days can she board the dog?

47. Karla's credit card bill is $420.90. Part of the bill is from the previous month's balance, part is in new charges, and part is from a late fee of $39. The previous balance is $172.40 less than the new charges. How much is the previous balance and how much is in new charges?

48. Thayne's credit card bill is $879.10. This includes his charges and interest. If the new charges are $794.10 more than the interest, find the amount in charges and the amount in interest.

49. Madeline and Kim each rode 15 miles in a bicycle relay. Madeline's time was 8.25 min less than Kim's time. If the total time was 1 hr, 56.75 min, for how long did each person ride?

50. The two-night attendance for a Friday/Saturday basketball tournament at a small college was 2570. There were 522 more people on Saturday for the finals than on Friday. How many people attended each night?

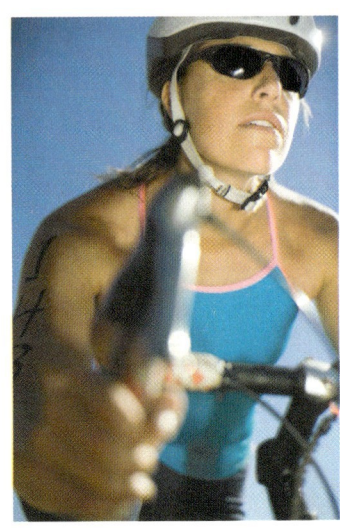

Fuse/Corbis/Getty Images

Chapter 5 Review Exercises

Section 5.1

1. Identify the place value for each digit in the number 32.16.

2. Identify the place value for each digit in the number 2.079.

For Exercises 3–6, write the word name for the decimal.

3. 5.7
4. 10.21
5. −51.008
6. −109.01

For Exercises 7 and 8, write the word name as a numeral.

7. Thirty-three thousand, fifteen and forty-seven thousandths.

8. Negative one hundred and one hundredth.

For Exercises 9 and 10, write the decimal as a proper fraction or mixed number.

9. −4.8
10. 0.025

For Exercises 11 and 12, write the decimal as an improper fraction.

11. 1.3
12. 6.75

For Exercises 13 and 14, fill in the blank with either < or >.

13. −15.032 ☐ −15.03
14. 7.209 ☐ 7.22

15. The batting average for five members of the American League for a recent season is given in the table. Rank the averages from least to greatest.

Player	AVG
Robinson Cano	.330
Josh Hamilton	.354
Joe Mauer	.325
Adrian Beltre	.333
Miguel Cabrera	.338

For Exercises 16 and 17, round the decimal to the indicated place value.

16. 89.9245; hundredths

17. 34.8895; thousandths

18. A quality control manager tests the amount of cereal in several brands of breakfast cereal against the amount advertised on the box. She selects one box at random. She measures the contents of one 12.5-oz box and finds that the box has 12.46 oz.

   a. Is the amount in the box less than or greater than the advertised amount?

b. If the quality control manager rounds the measured value to the tenths place, what is the value?

19. Which numbers are equal to 571.24? Circle all that apply.

a. 571.240 b. 571.2400

c. 571.024 d. 571.0024

20. Which numbers are equal to 3.709? Circle all that apply.

a. 3.7 b. 3.7090

c. 3.709000 d. 3.907

Section 5.2

For Exercises 21–28, add or subtract as indicated.

21. $45.03 + 4.713$ **22.** $239.3 + 33.92$

23. $34.89 - 29.44$ **24.** $5.002 - 3.1$

25. $-221 - 23.04$ **26.** $34 + (-4.993)$

27. $17.3 + 3.109 - 12.6$

28. $189.22 - (-13.1) - 120.055$

For Exercises 29 and 30, simplify by combining *like* terms.

29. $-5.1y - 4.6y + 10.2y$

30. $-2(12.5x - 3) + 11.5x$

31. The closing prices for a mutual fund are given in the graph for a 5-day period.

a. Determine the difference in price between the two consecutive days for which the price increased the most.

b. Determine the difference in price between the two consecutive days for which the price decreased the most.

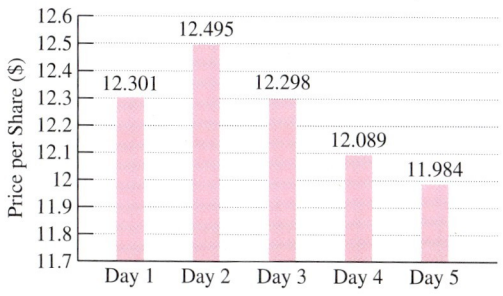

Mutual Fund Price

Section 5.3

For Exercises 32–39, multiply the decimals.

32. 3.9
   × 2.1

33. 57.01
   × 1.3

34. $(60.1)(-4.4)$ **35.** $(-7.7)(45)$

36. 85.49×1000 **37.** 1.0034×100

38. $-92.01 \times (-0.01)$ **39.** $-104.22 \times (-0.01)$

For Exercises 40 and 41, write the decimal number representing each word name.

40. The population of Guadeloupe is approximately 4.32 hundred thousand.

41. A recent season premier of a popular television show had 33.8 million viewers.

42. A store advertises a package of two 9-volt batteries on sale for $3.99.

a. What is the cost of buying 8 batteries?

b. If another store has an 8-pack for the regular price of $17.99, how much can a customer save by buying batteries at the sale price?

43. If the monthly bill for cable television is $124.49, what is the total cost for a year?

44. Find the area and perimeter of the rectangle.

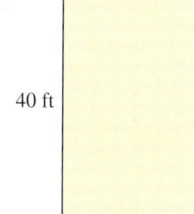

45. Population density gives the approximate number of people per square mile. The population density for Texas is given in the graph for selected years.

a. Approximately how many people would have been located in a 200-mi² area in 1960?

b. Approximately how many people would have been located in a 200-mi² area in 2010?

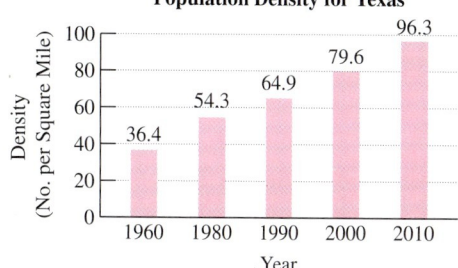

Population Density for Texas

Source: U.S. Census Bureau

46. Determine the diameter of a circle with radius 3.75 m.

47. Determine the radius of a circle with diameter 27.2 ft.

48. Find the area and circumference of a circular fountain with diameter 24 ft. Use 3.14 for π.

49. Find the area and circumference of a circular garden with radius 30 yd. Use 3.14 for π.

Section 5.4

For Exercises 50–59, divide. Write the answer in decimal form.

50. $8.55 \div 0.5$
51. $64.2 \div 1.5$
52. $0.06\overline{)0.248}$
53. $0.3\overline{)2.63}$
54. $-18.9 \div 0.7$
55. $-0.036 \div 1.2$
56. $493.93 \div 100$
57. $90.234 \div 10$
58. $-553.8 \div (-0.001)$
59. $-2.6 \div (-0.01)$

60. For each number, round to the indicated place.

	$8.\overline{6}$	$52.\overline{52}$	$0.\overline{409}$
Tenths			
Hundredths			
Thousandths			
Ten-thousandths			

For Exercises 61 and 62, divide and round the answer to the nearest hundredth.

61. $104.6 \div (-9)$
62. $71.8 \div (-6)$

63. a. A generic package of toilet paper costs $5.99 for 12 rolls. What is the cost per roll? (Round the answer to the nearest cent, that is, the nearest hundredth of a dollar.)

b. A package of four rolls costs $2.29. What is the cost per roll?

c. Which of the two packages offers the better buy?

Section 5.5

For Exercises 64–67, write the fraction or mixed number as a decimal.

64. $2\frac{2}{5}$
65. $3\frac{13}{25}$
66. $-\frac{24}{125}$
67. $-\frac{7}{16}$

For Exercises 68–70, write the fraction as a repeating decimal.

68. $\frac{7}{12}$
69. $\frac{55}{36}$
70. $-4\frac{7}{22}$

For Exercises 71–73, write the fraction as a decimal rounded to the indicated place value.

71. $\frac{5}{17}$; hundredths
72. $\frac{20}{23}$; tenths
73. $-\frac{11}{3}$; thousandths

74. Identify the numbers as rational or irrational.

a. $6.\overline{4}$
b. π
c. $\sqrt{10}$
d. $-\frac{8}{5}$

For Exercises 75 and 76, write a fraction or mixed number for the repeating decimal.

75. $0.\overline{2}$
76. $3.\overline{3}$

77. Complete the table, giving the closing value of stocks.

Stock	Closing Price ($) (Decimal)	Closing Price ($) (Mixed Number)
Ford		$13\frac{1}{50}$
Microsoft	30.50	
Citibank	29.37	

For Exercises 78 and 79, insert the appropriate symbol. Choose from <, >, or =.

78. $1\frac{1}{3} \square 1.33$
79. $-0.14 \square -\frac{1}{7}$

For Exercises 80–85, perform the indicated operations. Write the answers in decimal form.

80. $-7.5 \div \frac{3}{2}$
81. $\frac{1}{2}(4.6)(-2.4)$
82. $(-5.46 - 2.24)^2 - 0.29$
83. $[-3.46 - (-2.16)]^2 - 0.09$
84. $\left(\frac{1}{4}\right)^2 \left(\frac{4}{5}\right) - 3.05$
85. $-1.25 - \frac{1}{3} \cdot \left(\frac{3}{2}\right)^2$

86. An audio Spanish course is available online at one website for the following prices. How much money is saved by buying the combo package versus the three levels individually?

Level	Price
Spanish I	$189.95
Spanish II	199.95
Spanish III	219.95
Combo (Spanish I, II, III combined)	519.95

87. Marvin drives the route shown in the figure each day, making deliveries. He completes one-third of the route before lunch. How many more miles does he still have to drive after lunch?

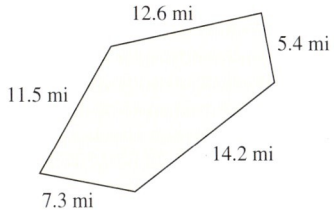

Section 5.6

For Exercises 88–96, solve the equation.

88. $x + 4.78 = 2.2$ **89.** $6.2 + y = 4.67$

90. $11 = -10.4 + w$ **91.** $-20.4 = -9.08 + z$

92. $4.6 = 16.7 + 2.2m$ **93.** $21.8 = 12.05 + 3.9x$

94. $-0.2(x - 6) = 0.3x + 3.8$

95. $-0.5(9 + y) = -0.6y + 6.5$

96. $0.04z - 3.5 = 0.06z + 4.8$

For Exercises 97 and 98, determine a number that can be used to clear decimals in the equation. Then solve the equation.

97. $2p + 3.1 = 0.14$ **98.** $2.5w - 6 = 0.9$

99. Four times a number is equal to 19.5 more than the number. Find the number.

100. The sum of a number and 4.8 is three times the number. Find the number.

101. The product of a number and 0.05 is equal to 80. Find the number.

102. The quotient of a number and 0.05 is equal to 80. Find the number.

103. The perimeter of a triangle is 24.8 yd. The longest side is twice the shortest side. The middle side is 2.4 yd longer than the shortest side. Find the lengths of the sides.

104. For a temporary job out of town, Deanna rents an apartment and a car for a weekly total of $861.35. The weekly cost for the apartment is $602.35 more than the weekly cost of the car. Find the weekly cost to rent each item.

105. The U-Store-It Company rents a 10-ft by 12-ft storage space for $59, plus $89 per month. Mykeshia wrote a check for $593. How many months of storage will this cover?

Ratio and Proportion

CHAPTER OUTLINE

6.1 Ratios 342

6.2 Rates and Unit Cost 350

6.3 Proportions 358

    Problem Recognition Exercises: Operations on Fractions versus Solving Proportions 366

6.4 Applications of Proportions and Similar Figures 367

    Chapter 6 Review Exercises 378

Let's Compare

In this chapter, we study ratios and rates. These are two fundamentally important concepts that you use every day. A **ratio** is the quotient of two numbers used to compare the relative values of the numbers. For example, a daycare might advertise that the ratio of children to adults is 3 to 1 (also written as 3 : 1, or 3/1). A builder might mix cement and a sand/gravel combination in a 1 : 6 ratio to make concrete.

A rate is a ratio that usually compares two numbers with different units. For example, you might drive to the grocery store at 35 miles per hour in a car that gets 25 miles per gallon. While at the store, you might buy sliced turkey for $3.99 per pound and socks at a rate of 3 pairs for $10. A unit rate is a rate in which a quantity is compared to 1 unit of another quantity. For example, the rate 35 miles per hour (35 miles per 1 hour) is a unit rate.

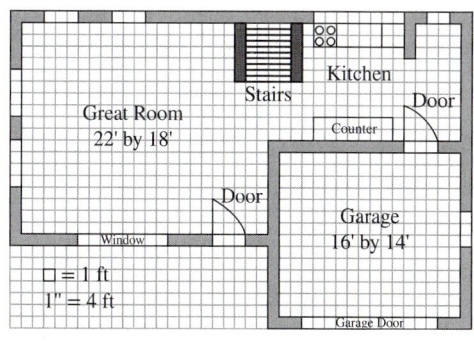

A blueprint or a map offers a scaled-down paper drawing of a much larger physical object or location. Although the paper version is smaller, it is said to be proportional to (or similar to) the real object. That is, corresponding lengths between the drawing and the real object are in proportion. For example, 1 in. on a map might correspond to 120 mi in true physical distance. On an architectural drawing for a house plan, 1 in. might represent 4 ft.

Section 6.1 Ratios

Concepts
1. Writing a Ratio
2. Writing Ratios of Mixed Numbers and Decimals
3. Applications of Ratios

1. Writing a Ratio

Thus far, we have seen two interpretations of fractions.

- The fraction $\frac{5}{8}$ represents 5 parts of a whole that has been divided evenly into 8 pieces.
- The fraction $\frac{5}{8}$ represents $5 \div 8$.

Now we consider a third interpretation.

- The fraction $\frac{5}{8}$ represents the ratio of 5 to 8.

A **ratio** is a comparison of two quantities. There are three different ways to write a ratio.

Writing a Ratio

The ratio of a to b can be written as follows, provided $b \neq 0$.

1. a to b
2. $a : b$
3. $\dfrac{a}{b}$

The colon means "to."   The fraction bar means "to."

Although there are three ways to write a ratio, we primarily use the fraction form.

Example 1 Writing a Ratio

In an algebra class there are 15 women and 17 men.

a. Write the ratio of women to men.
b. Write the ratio of men to women.
c. Write the ratio of women to the total number of people in the class.

Mike Watson Images Limited/Glow Images

Solution:

It is important to observe the *order* of the quantities mentioned in a ratio. The first quantity mentioned is the numerator. The second quantity is the denominator.

a. The ratio of women to men is

$$\frac{15}{17}$$

b. The ratio of men to women is

$$\frac{17}{15}$$

c. First find the total number of people in the class.

Total = number of women + number of men
$= 15 + 17$
$= 32$

Therefore, the ratio of women to the total number of people in the class is

$$\frac{15}{32}$$

Skill Practice

1. For a recent flight from Atlanta to San Diego, 291 seats were occupied and 29 were unoccupied. Write the ratio of:
   a. The number of occupied seats to unoccupied seats
   b. The number of unoccupied seats to occupied seats
   c. The number of occupied seats to the total number of seats

Answer

1. a. $\dfrac{291}{29}$ b. $\dfrac{29}{291}$ c. $\dfrac{291}{320}$

It is often desirable to write a ratio in lowest terms. The process is similar to simplifying fractions to lowest terms.

Example 2 — Writing Ratios in Lowest Terms

Write each ratio in lowest terms.

a. 15 ft to 10 ft **b.** $20 to $10

Solution:

In part (a) we are comparing feet to feet. In part (b) we are comparing dollars to dollars. We can divide out the like units in the numerator and denominator as we would common factors.

a. $\dfrac{15 \text{ ft}}{10 \text{ ft}} = \dfrac{3 \cdot 5 \text{ ft}}{2 \cdot 5 \text{ ft}}$

$= \dfrac{3 \cdot \cancel{5} \text{ ft}}{2 \cdot \cancel{5} \text{ ft}}$ Simplify common factors.

$= \dfrac{3}{2}$ The ratio is 3 to 2.

TIP: Even though the number $\frac{3}{2}$ is equivalent to $1\frac{1}{2}$, we do not write the ratio as a mixed number. Remember that a ratio is a comparison of *two* quantities. If you did convert $\frac{3}{2}$ to the mixed number $1\frac{1}{2}$, you would write the ratio as $\dfrac{1\frac{1}{2}}{1}$. This would imply that the numerator is one and one-half times as large as the denominator.

b. $\dfrac{\$20}{\$10} = \dfrac{\cancel{\$20}^{2}}{\cancel{\$10}_{1}}$ Simplify common factors.

$= \dfrac{2}{1}$

Although the fraction $\frac{2}{1}$ is equivalent to 2, we do not generally write ratios as whole numbers. Again, a ratio compares *two* quantities. In this case, we say that there is a 2-to-1 ratio between the original dollar amounts.

Skill Practice Write the ratios in lowest terms.

2. 72 m to 16 m   **3.** 30 gal to 5 gal

2. Writing Ratios of Mixed Numbers and Decimals

It is often desirable to express a ratio in lowest terms by using whole numbers in the numerator and denominator. This is demonstrated in Examples 3 and 4.

Example 3 — Writing a Ratio as a Ratio of Whole Numbers

The length of a rectangular picture frame is 7.5 in., and the width is 6.25 in. Express the ratio of the length to the width. Then rewrite the ratio as a ratio of whole numbers simplified to lowest terms.

Solution:

The ratio of length to width is $\dfrac{7.5}{6.25}$. We now want to rewrite the ratio, using whole numbers in the

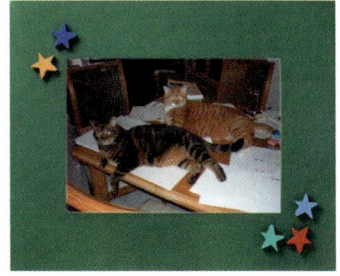
Courtesy of Molly O'Neill

Answers

2. $\dfrac{9}{2}$ **3.** $\dfrac{6}{1}$

numerator and denominator. If we multiply 7.5 by 10, the decimal point will move to the right one place, resulting in a whole number. If we multiply 6.25 by 100, the decimal point will move to the right two places, resulting in a whole number. Because we want to multiply the numerator and denominator by the *same* number, we choose the greater number, 100.

$$\frac{7.5}{6.25} = \frac{7.5 \times 100}{6.25 \times 100}$$ Multiply numerator and denominator by 100.

$$= \frac{750}{625}$$ Because the numerator and denominator are large numbers, we write the prime factorization of each. The common factors are now easy to identify.

$$= \frac{2 \cdot 3 \cdot \cancel{5} \cdot \cancel{5} \cdot \cancel{5}}{5 \cdot \cancel{5} \cdot \cancel{5} \cdot \cancel{5}}$$ Simplify common factors to lowest terms.

$$= \frac{6}{5}$$ The ratio of length to width is $\frac{6}{5}$.

Skill Practice Write the ratio as a ratio of whole numbers expressed in lowest terms.

4. $4.20 to $2.88

In Example 3, we multiplied by 100 to move the decimal point *two* places to the right. Multiplying by 10 would not have been sufficient, because $6.25 \times 10 = 62.5$, which is not a whole number.

Example 4   Writing a Ratio as a Ratio of Whole Numbers

Ling walked $2\frac{1}{4}$ mi on Monday and $3\frac{1}{2}$ mi on Tuesday. Write the ratio of miles walked Monday to miles walked Tuesday. Then rewrite the ratio as a ratio of whole numbers reduced to lowest terms.

Lars A. Niki

Solution:

The ratio of miles walked on Monday to miles walked on Tuesday is $\dfrac{2\frac{1}{4}}{3\frac{1}{2}}$.

To convert this to a ratio of whole numbers, first we rewrite each mixed number as an improper fraction. Then we can divide the fractions and simplify.

$$\frac{2\frac{1}{4}}{3\frac{1}{2}} = \frac{\frac{9}{4}}{\frac{7}{2}}$$ ← Write the mixed numbers as improper fractions. Recall that a fraction bar also implies division.

$$= \frac{9}{4} \div \frac{7}{2}$$

$$= \frac{9}{4} \cdot \frac{2}{7}$$ Multiply by the reciprocal of the divisor.

$$= \frac{9}{\cancel{4}} \cdot \frac{\cancel{2}}{7}$$ Simplify common factors to lowest terms.

$$= \frac{9}{14}$$ This is a ratio of whole numbers in lowest terms.

Skill Practice

5. A recipe calls for $2\frac{1}{2}$ cups of flour and $\frac{3}{4}$ cup of sugar. Write the ratio of flour to sugar. Then rewrite the ratio as a ratio of whole numbers in lowest terms.

Answers

4. $\dfrac{35}{24}$ 5. $\dfrac{2\frac{1}{2}}{\frac{3}{4}}, \dfrac{10}{3}$

3. Applications of Ratios

Ratios are used in a number of applications.

Example 5   Using Ratios to Express Population Increase

After the tragedy of Hurricane Katrina, New Orleans showed signs of recovery as more people moved back into the city. Three years after the storm, New Orleans had the fastest growth rate of any city in the United States. During a 1-year period, its population rose from 210,000 to 239,000. (*Source:* U.S. Census Bureau) Write a ratio expressing the increase in population to the original population for that year.

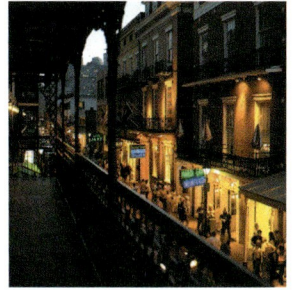
Steve Allen/Getty Images

FOR REVIEW

Recall that whole numbers ending in 0 are divisible by a power of 10.

$$\frac{29,000}{210,000} = \frac{29 \times 1000}{210 \times 1000}$$
$$= \frac{29}{210}$$

"Strike through" 3 zeros.

$$\frac{29,\cancel{000}}{210,\cancel{000}} = \frac{29}{210}$$

Solution:

To write this ratio, we need to know the increase in population.

$$\text{Increase} = 239,000 - 210,000 = 29,000$$

The ratio of the increase in population to the original number is

increase in population ⟶ $\dfrac{29,000}{210,000}$ ⟵ original population

$= \dfrac{29}{210}$ Simplify to lowest terms.

Skill Practice

6. The U.S. Department of Transportation reported that more airports will have to expand to meet the growing demand of air travel. During a 5-year period, the Phoenix Sky Harbor Airport went from servicing 20 million passengers to 26 million passengers. Write a ratio of the increase in passengers to the original number.

Example 6   Applying Ratios to Unit Conversion

A fence is 12 yd long and 1 ft high.

a. Write the ratio of length to height with all units measured in yards.

b. Write the ratio of length to height with all units measured in feet.

1 ft = $\frac{1}{3}$ yd

12 yd

Solution:

a. 3 ft = 1 yd, therefore, 1 ft = $\frac{1}{3}$ yd.

Measuring in yards, we see that the ratio of length to height is

$$\frac{12 \text{ yd}}{\frac{1}{3} \text{ yd}} = \frac{12}{1} \cdot \frac{3}{1} = \frac{36}{1}$$

b. The length is 12 yd = 36 ft. (Since 1 yd = 3 ft, then 12 yd = 12 · 3 ft = 36 ft.)

Measuring in feet, we see that the ratio of length to height is $\dfrac{36 \text{ ft}}{1 \text{ ft}} = \dfrac{36}{1}$

Notice that regardless of the units used, the ratio is the same, 36 to 1. This means that the length is 36 times the height.

Skill Practice

7. A painting is 2 yd in length by 2 ft wide.
  a. Write the ratio of length to width with all units measured in feet.
  b. Write the ratio of length to width with all units measured in yards.

Answers

6. $\dfrac{3}{10}$ 7. a. $\dfrac{3}{1}$ b. $\dfrac{3}{1}$

Section 6.1 Activity

A.1. There are 16 business management students and 22 marketing students in the business law course. Write the ratios described below as a fraction in lowest terms.
  a. Write the ratio of business management students to marketing students.
  b. Write the ratio of marketing students to business management students.
  c. Write the ratio of business management students to the total number of students in the class.

A.2. Sean makes $15.50 an hour during the weekdays and $22.00 an hour on the weekend. Write the ratio of Sean's weekly hourly pay to his weekend hourly pay as a ratio of whole numbers. Simplify to lowest terms.

A.3. To comply with state regulations, the ratio of infants to daycare teachers is 4:1. If there are 26 infants scheduled to attend the daycare and 6 teachers staffed, is the daycare meeting state regulations?

A.4. A homeowner mixes $1\frac{1}{8}$ cups bleach with $11\frac{1}{4}$ cups of water to make an antibacterial solution. Write the ratio of water to bleach as a ratio of whole numbers. Simplify to lowest terms.

A.5. Vivian and her mom raced each other down the street. Vivian reached the end of the street in 45 seconds. Her mom reached the end of the street in 1 minute and 5 seconds. Write a ratio in lowest terms of Vivian's time to her mom's time.

Section 6.1 Practice Exercises

Prerequisite Review

For Exercises R.1–R.4, reduce to lowest terms. Write the answer as a fraction.

R.1. $\dfrac{3600}{6000}$

R.2. $\dfrac{450}{300}$

R.3. $\dfrac{92}{4}$

R.4. $\dfrac{136}{8}$

Vocabulary and Key Concepts

1. The statement "a to b" or "$a : b$" or $\dfrac{a}{b}$ represents the _____ of a to b.

Concept 1: Writing a Ratio

2. Write a ratio of the number of females to males in your class.

For Exercises 3–8, write the ratio in two other ways.

3. 5 to 6

4. 3 to 7

5. 11 : 4

6. 8 : 13

7. $\dfrac{1}{2}$

8. $\dfrac{1}{8}$

For Exercises 9–14, write the ratios in fraction form. (See Examples 1 and 2.)

9. For a recent year, there were 10 named tropical storms and 6 hurricanes in the Atlantic. (*Source:* NOAA)

StockTrek/Getty Images

   a. Write a ratio of the number of tropical storms to the number of hurricanes.
   b. Write a ratio of the number of hurricanes to the number of tropical storms.
   c. Write a ratio of the number of hurricanes to the total number of named storms.

10. For a recent year, 250 million persons were covered by health insurance in the United States and 45 million were not covered. (*Source:* U.S. Census Bureau)

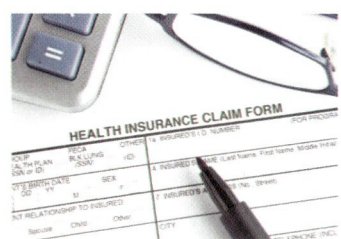

Numbeos/E+/Getty Images

   a. Write a ratio of the number of insured persons to the number of uninsured persons.
   b. Write a ratio of the number of uninsured persons to the number of insured persons.
   c. Write a ratio of the number of uninsured persons to the total number of persons.

11. In a certain neighborhood, 60 houses were on the market to be sold. During a 1-year period during a housing crisis, only 8 of these houses actually sold.

   a. Write a ratio of the number of houses that sold to the total number that had been on the market.
   b. Write a ratio of the number of houses that sold to the number that did not sell.

12. There are 52 cars in a parking lot, of which 21 are silver.

   a. Write a ratio of the number of silver cars to the total number of cars.
   b. Write a ratio of the number of silver cars to the number of cars that are not silver.

13. In a recent survey of a group of computer users, 21 were MAC users and 54 were PC users.

   a. Write a ratio of the number of MAC users to PC users.
   b. Write a ratio for the number of MAC users to the total number of people surveyed.

14. At a school sporting event, the concession stand sold 450 bottles of water, 200 cans of soda, and 125 cans of iced tea.

   a. Write a ratio of the number of bottles of water sold to the number of cans of soda sold.
   b. Write a ratio of the number of cans of iced tea sold to the total number of drinks sold.

For Exercises 15–26, write the ratio in lowest terms. (See Example 2.)

15. 4 yr to 6 yr
16. 10 lb to 14 lb
17. 5 mi to 25 mi
18. 20 ft to 12 ft

19. 8 m to 2 m
20. 14 oz to 7 oz
21. 33 cm to 15 cm
22. 21 days to 30 days

23. $60 to $50
24. 75¢ to 100¢
25. 18 in. to 36 in.
26. 3 cups to 9 cups

Concept 2: Writing Ratios of Mixed Numbers and Decimals

For Exercises 27–38, write the ratio in lowest terms with whole numbers in the numerator and denominator. (See Examples 3 and 4.)

27. 3.6 ft to 2.4 ft **28.** 10.15 hr to 8.12 hr **29.** 8 gal to $9\frac{1}{3}$ gal **30.** 24 yd to $13\frac{1}{3}$ yd

31. $16\frac{4}{5}$ m to $18\frac{9}{10}$ m **32.** $1\frac{1}{4}$ in. to $1\frac{3}{8}$ in. **33.** $16.80 to $2.40 **34.** $18.50 to $3.70

35. $\frac{1}{2}$ day to 4 days **36.** $\frac{1}{4}$ mi to $1\frac{1}{2}$ mi **37.** 10.25 L to 8.2 L **38.** 11.55 km to 6.6 km

Concept 3: Applications of Ratios

39. In 1981, a giant sinkhole in Winter Park, Florida, "swallowed" a home, a public pool, and a car dealership (including five Porsches). One witness said that in a single day, the hole widened from 5 ft in diameter to 320 ft.

  a. Find the increase in the diameter of the sinkhole.

  b. Write a ratio representing the increase in diameter to the original diameter of 5 ft. (See Example 5.)

40. The temperature at 8:00 A.M. in Los Angeles was 66°F. By 2:00 P.M., the temperature had risen to 90°F.

  a. Find the increase in temperature from 8:00 A.M. to 2:00 P.M.

  b. Write a ratio representing the increase in temperature to the temperature at 8:00 A.M.

U.S. Geological Survey

41. A window is 2 ft wide and 3 yd in length (2 ft is $\frac{2}{3}$ yd). (See Example 6.)

  a. Find the ratio of width to length with all units in yards.

  b. Find the ratio of width to length with all units in feet.

42. A construction company needs 2 weeks to construct a family room and 3 days to add a porch.

  a. Find the ratio of the time it takes for constructing the porch to the time constructing the family room, with all units in weeks.

  b. Find the ratio of the time it takes for constructing the porch to the time constructing the family room, with all units in days.

For Exercises 43–46, refer to the table showing Alex Rodriguez's salary (rounded to the nearest $100,000) for selected years during his career. Write each ratio in lowest terms.

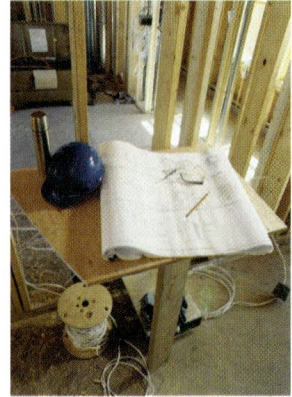
Ryan McVay/Photodisc/Getty Images

Year	Team	Salary	Position
1994	Seattle Mariners	$400,000	Shortstop
2000	Seattle Mariners	$4,400,000	Shortstop
2003	Texas Rangers	$22,000,000	Shortstop
2009	New York Yankees	$33,000,000	Third baseman
2012	New York Yankees	$29,000,000	Third baseman

43. Write the ratio of Alex's salary for the year 1994 to the year 2000.

44. Write a ratio of Alex's salary for the year 2009 to the year 2003.

45. Write a ratio of the increase in Alex's salary between the years 1994 and 2000 to his salary in 1994.

46. Write a ratio of the increase in Alex's salary between the years 2003 and 2009 to his salary in 2003.

For Exercises 47–50, refer to the table that shows the average spending per person for reading (books, newspapers, magazines, etc.) by age group. Write each ratio in lowest terms.

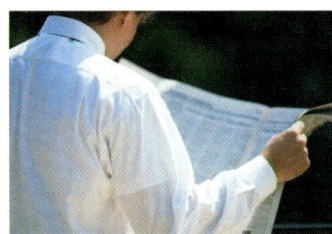
MIXA/Getty Images

Age Group	Annual Average ($)
Under 25 years	60
25 to 34 years	111
35 to 44 years	136
45 to 54 years	172
55 to 64 years	183
65 to 74 years	159
75 years and over	128

Source: Mediamark Research Inc.

47. Find the ratio of spending for the group under 25 years old to the spending for the group 75 years and over.

48. Find the ratio of spending for the group 25 to 34 years old to the spending for the group of 65 to 74 years old.

49. Find the ratio of spending for the group under 25 years old to the spending for the group of 55 to 64 years old.

50. Find the ratio of spending for the group 35 to 44 years old to the spending for the group 45 to 54 years old.

For Exercises 51–54, find the ratio of the shortest side to the longest side. Write each ratio in lowest terms with whole numbers in the numerator and denominator.

51.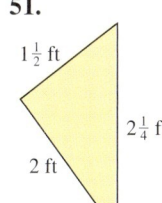

52.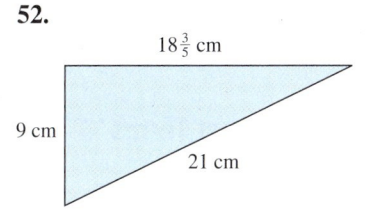

53.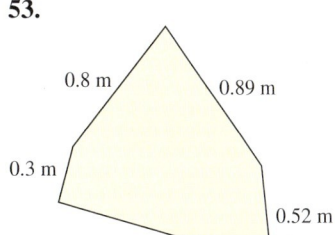

54.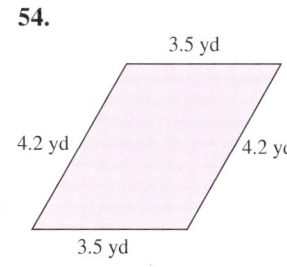

Expanding Your Skills

For Exercises 55–57, refer to the figure. The lengths of the sides for squares A, B, C, and D are given.

55. What are the lengths of the sides of square E?

56. Find the ratio of the lengths of the sides for the given pairs of squares.

  a. Square B to square A
  b. Square C to square B
  c. Square D to square C
  d. Square E to square D

57. Write the decimal equivalents for each ratio in Exercise 56. Do these values seem to be approaching a number close to 1.618 (this is an approximation for the *golden ratio*, which is equal to $\frac{1+\sqrt{5}}{2}$)? Applications of the golden ratio are found throughout nature. In particular, as a result of the geometrically pleasing pattern, artists and architects have proportioned their work to approximate the golden ratio.

58. The ratio of a person's height to the length of the person's lower arm (from elbow to wrist) is approximately 6.5 to 1. Measure your own height and lower arm length. Is the ratio you get close to the average of 6.5 to 1?

59. The ratio of a person's height to the person's shoulder width (measured from outside shoulder to outside shoulder) is approximately 4 to 1. Measure your own height and shoulder width. Is the ratio you get close to the average of 4 to 1?

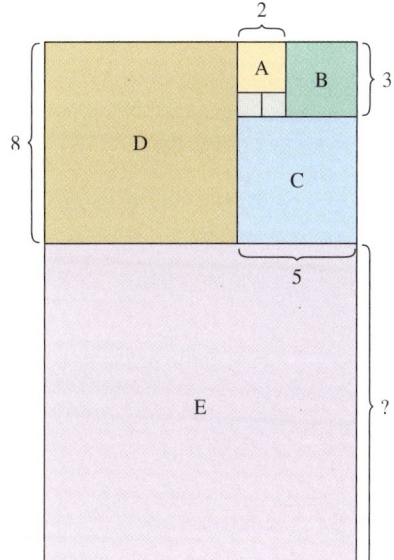

Section 6.2 Rates and Unit Cost

Concepts

1. Definition of a Rate
2. Unit Rates
3. Unit Cost
4. Applications of Rates

1. Definition of a Rate

A **rate** is a type of ratio used to compare different types of quantities, for example:

$$\frac{270 \text{ mi}}{13 \text{ gal}} \quad \text{and} \quad \frac{\$8.55}{1 \text{ hr}}$$

Several key words imply rates. These are given in Table 6-1.

Table 6-1

Key Word	Example	Rate
Per	117 miles per 2 hours	$\frac{117 \text{ mi}}{2 \text{ hr}}$
For	$12 for 3 lb	$\frac{\$12}{3 \text{ lb}}$
In	400 meters in 43.5 seconds	$\frac{400 \text{ m}}{43.5 \text{ sec}}$
On	270 miles on 12 gallons of gas	$\frac{270 \text{ mi}}{12 \text{ gal}}$

Because a rate compares two different quantities it is important to include the units in both the numerator and the denominator. It is also desirable to write rates in lowest terms.

Example 1 Writing Rates in Lowest Terms

Write each rate in lowest terms.

a. In one region, there are approximately 640 trees on 12 acres.

b. Latonya drove 138 mi on 6 gal of gas.

c. After a cold front, the temperature changed by −10°F in 4 hr.

Solution:

a. The rate of 640 trees on 12 acres can be expressed as $\frac{640 \text{ trees}}{12 \text{ acres}}$.

Now write this rate in lowest terms. $\frac{\overset{160}{\cancel{640}} \text{ trees}}{\underset{3}{\cancel{12}} \text{ acres}} = \frac{160 \text{ trees}}{3 \text{ acres}}$

b. The rate of 138 mi on 6 gal of gas can be expressed as $\frac{138 \text{ mi}}{6 \text{ gal}}$.

Now write this rate in lowest terms. $\frac{\overset{23}{\cancel{138}} \text{ mi}}{\underset{1}{\cancel{6}} \text{ gal}} = \frac{23 \text{ mi}}{1 \text{ gal}}$

c. −10°F in 4 hr can be represented by $\frac{-10°F}{4 \text{ hr}}$.

Writing this in lowest terms, we have: $\frac{\overset{5}{\cancel{-10}}°F}{\underset{2}{\cancel{4}} \text{ hr}} = -\frac{5°F}{2 \text{ hr}}$

Skill Practice Write each rate in lowest terms.
1. Maria reads 15 pages in 10 min.
2. A Chevrolet Corvette Z06 got 163.4 mi on 8.6 gal of gas.
3. Marty's balance in his investment account changed by −$254 in 4 months.

2. Unit Rates

A rate having a denominator of 1 unit is called a **unit rate**. Furthermore, the number 1 is often omitted in the denominator.

$$\frac{23 \text{ mi}}{1 \text{ gal}} = 23 \text{ mi/gal} \quad \text{is read as "twenty-three miles per gallon."}$$

$$\frac{52 \text{ ft}}{1 \text{ sec}} = 52 \text{ ft/sec} \quad \text{is read as "fifty-two feet per second."}$$

$$\frac{\$15}{1 \text{ hr}} = \$15/\text{hr} \quad \text{is read as "fifteen dollars per hour."}$$

> **Converting a Rate to a Unit Rate**
> To convert a rate to a unit rate, divide the numerator by the denominator and maintain the units of measurement.

Example 2 Finding Unit Rates

Write each rate as a unit rate. Round to three decimal places if necessary.

a. A health club charges $125 for 20 visits. Find the unit rate in dollars per visit.

b. In 1960, Wilma Rudolph won the women's 200-m run in 24 sec. Find her speed in meters per second.

c. During one baseball season, Barry Bonds got 149 hits in 403 at bats. Find his batting average. (*Hint:* Batting average is defined as the number of hits per the number of at bats.)

Solution:

a. The rate of $125 for 20 visits can be expressed as $\frac{\$125}{20 \text{ visits}}$.

To convert this to a unit rate, divide $125 by 20 visits.

$$\frac{\$125}{20 \text{ visits}} = \frac{\$6.25}{1 \text{ visit}} \text{ or } \$6.25 \text{ per visit}$$

```
        6.25
   20)125.00
      −120
        50
       −40
        100
       −100
          0
```

b. The rate of 200 m per 24 sec can be expressed as $\frac{200 \text{ m}}{24 \text{ sec}}$.

To convert this to a unit rate, divide 200 m by 24 sec.

Answers
1. $\frac{3 \text{ pages}}{2 \text{ min}}$
2. $\frac{19 \text{ mi}}{1 \text{ gal}}$ or 19 mi/gal
3. $-\frac{\$127}{2 \text{ months}}$

$$\frac{200 \text{ m}}{24 \text{ sec}} \approx \frac{8.333 \text{ m}}{1 \text{ sec}} \quad \text{or approximately } 8.333 \text{ m/sec}$$

Wilma Rudolph's speed was approximately 8.333 m/sec.

$$\begin{array}{r} 8.\overline{3} \\ 24\overline{)200.00} \\ -192 \\ \hline 80 \\ -72 \\ \hline 80 \end{array}$$

The quotient repeats.

Avoiding Mistakes

Units of measurement must be included for the answer to be complete.

c. The rate of 149 hits in 403 at bats can be expressed as $\frac{149 \text{ hits}}{403 \text{ at bats}}$.

To convert this to a unit rate, divide 149 hits by 403 at bats.

$$\frac{149 \text{ hits}}{403 \text{ at bats}} \approx \frac{0.370 \text{ hit}}{1 \text{ at bat}} \quad \text{or } 0.370 \text{ hit/at bat}$$

Skill Practice Write a unit rate.

4. It costs $3.90 for 12 oranges.
5. A flight from Dallas to Des Moines travels a distance of 646 mi in 1.4 hr. Round the unit rate to the nearest mile per hour.
6. Under normal conditions, 98 in. of snow is equivalent to about 10 in. of rain.

3. Unit Cost

A **unit cost** or unit price is the cost per 1 unit of something. At the grocery store, for example, you might purchase meat for $3.79/lb ($3.79 per 1 lb). Unit cost is useful in day-to-day life when we compare prices. Example 3 compares the prices of three different sizes of apple juice.

Example 3 Finding Unit Costs

Apple juice comes in a variety of sizes and packaging options. Find the unit cost per ounce and determine which is the best buy.

a. $4.19 — Apple Juice 46 oz
Jill Braaten/McGraw Hill

b. $4.89 — Apple Juice 64 oz
Jill Braaten/McGraw Hill

c. $3.55 — Apple Juice 8-pack 4 oz each
Mark Dierker/McGraw Hill

Solution:

When we compute a unit cost, the cost is always placed in the numerator of the rate. Furthermore, when we divide the cost by the amount, we need to obtain enough digits in the quotient to see the variation in unit cost. In this example, we have rounded to the nearest thousandth of a dollar (nearest tenth of a cent). This means that we use the ten-thousandths-place digit in the quotient on which to base our decision on rounding.

Answers

4. $0.325 per orange
5. 461 mi/hr
6. 9.8 in. snow per 1 in. of rain

	Rate	Quotient	Unit Rate (Rounded)
a.	$\dfrac{\$4.19}{46 \text{ oz}}$	$\$4.19 \div 46 \text{ oz} = \$0.0911/\text{oz}$	$\$0.091/\text{oz}$ or $9.1¢/\text{oz}$
b.	$\dfrac{\$4.89}{64 \text{ oz}}$	$\$4.89 \div 64 \text{ oz} = \$0.0764/\text{oz}$	$\$0.076/\text{oz}$ or $7.6¢/\text{oz}$
c.	$\dfrac{\$3.55}{32 \text{ oz}}$ (4 oz × 8 = 32 oz)	$\$3.55 \div 32 \text{ oz} = \$0.1109/\text{oz}$	$\$0.111/\text{oz}$ or $11.1¢/\text{oz}$

From the table, we see that the most economical buy is the 64-oz size because its unit rate is the least expensive.

Skill Practice

7. A sport's drink comes in several size packages. Compute the unit cost per ounce for each option (round to the nearest thousandth of a dollar). Then determine which is the best buy.
   a. $3.25 for a 64-oz bottle
   b. $9.59 for eight 20-oz bottles
   c. $1.95 for a 32-oz bottle

4. Applications of Rates

Example 4 uses a unit rate for comparison in an application.

Example 4 Computing Mortality Rates

Mortality rate is defined to be the total number of people who die due to some risk behavior divided by the total number of people who engage in the risk behavior. Based on the following statistics, compare the mortality rate for undergoing heart bypass surgery to the mortality rate of flying on the Space Shuttle.

   a. Roughly 28 people will die for every 1000 who undergo heart bypass surgery. (*Source:* The Society of Thoracic Surgeons)

   b. When the Space Shuttle program came to a close, there had been 14 astronauts killed in shuttle missions out of 912 astronauts who had flown.

Solution:

   a. Mortality rate for heart bypass surgery: $\dfrac{28}{1000} = 0.028$ death/surgery

   b. Mortality rate for flying on the shuttle: $\dfrac{14}{912} \approx 0.015$ death/flight

Comparing these rates shows that it is riskier to have heart bypass surgery than to fly on the Space Shuttle.

Skill Practice

8. In Ecuador roughly 450 out of 500 adults can read. In Brazil, approximately 1700 out of 2000 adults can read.
   a. Compute the unit rate for Ecuador (this unit rate is called the *literacy rate*).
   b. Compute the unit rate for Brazil.
   c. Which country has the greater literacy rate?

Answers

7. a. $0.051/oz
   b. $0.060/oz
   c. $0.061/oz
   The 64-oz bottle is the best buy.
8. a. 0.9 reader per adult
   b. 0.85 reader per adult
   c. Ecuador

Section 6.2 Activity

For Exercises A.1–A.3, express each rate as a unit rate.

A.1. A Jeep Cherokee gets 210 miles on 14 gallons of gas.

A.2. Sonja travelled 165 miles in 3 hours.

A.3. Damian earned $970 for 40 hours of work.

A.4. A cell phone company offered the following plans for 1500 minutes and 1000 text messages:

Service (Days)	Cost ($)	Unit Cost ($/day)
Plan A: 365	$125	
Plan B: 90	$50	
Plan C: 60	$20	

  a. Calculate the unit cost in dollars per day for each plan. Round to the nearest cent if necessary.

  b. What service plan has the lowest cost per day?

A.5. a. Traveling at 55 mph, a car travels 250 miles on 10 gallons of gas. Find the unit rate in miles per gallon.

  b. Traveling at 70 mph, a car travels 321 miles on 15 gallons of gas. Find the unit rate in miles per gallon.

  c. Compare your answers from parts (a) and (b). Do you save more gas by driving slower or faster?

  d. How much gas would be saved if a person drove 1000 miles at 55 mph versus 70 mph? Round to the nearest tenth of a gallon.

  e. Assume gas costs $2.15 per gallon. How much money would the driver save by slowing down based on the answer in part (d)? Round to the nearest cent.

A.6. Gus is trying to determine what package of diapers is the better buy. The jumbo box of diapers is $35.99 for 102 diapers. The large pack is $21.99 for 60 diapers. If Gus has a $1.00 off coupon for the large pack, which is more cost effective?

Section 6.2 Practice Exercises

Prerequisite Review

For Exercises R.1–R.6, write the ratio in lowest terms.

R.1. $6\frac{3}{4}$ ft to $8\frac{1}{4}$ ft

R.2. $3\frac{1}{3}$ cm to $2\frac{1}{2}$ cm

R.3. 1.08 mi to 2.04 mi

R.4. 0.45 in. to 0.0035 in.

R.5. $15.50 to $3.10

R.6. $34.50 to $0.15

Vocabulary and Key Concepts

1. a. A _____ is a type of ratio used to compare different types of quantities—for example: 200 miles per 4 hours.

  b. A _____ rate is a rate that has a denominator of 1 unit.

2. Always include the _____ in both the numerator and denominator of a rate.

3. To convert a rate to a unit rate, _____ the numerator by the denominator.

4. When computing a unit cost, the cost is always in the _____ of the rate.

Concept 1: Definition of a Rate

For Exercises 5–18, write each rate in lowest terms. (See Example 1.)

5. Tanya can bike 66 mi in 8 hr.

6. The cost of tuition is $1650 for 9 credits.

7. A type of laminate flooring sells for $32 for 5 ft^2.

8. A remote control car can go up to 44 ft in 5 sec.

9. Elaine drives 234 mi in 4 hr.

10. Travis has 14 blooms on 6 of his plants.

11. Tyler earned $58 in 8 hr.

12. Neil can type only 336 words in 15 min.

13. A printer can print 13 pages in 26 sec.

14. During a bad storm there was 2 in. of rain in 6 hr.

15. There are 130 calories in 8 snack crackers.

16. There are 50 students assigned to 4 advisers.

17. The temperature changed by −18°F in 8 hr.

18. Jill's portfolio changed by −$2600 in 6 months.

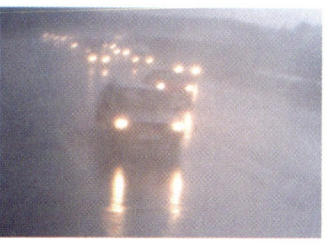

Steve Allen/Getty Images

Bildagentur-online.com/th-foto/
Alamy Stock Photo

Concept 2: Unit Rates

19. Of the following rates, identify those that are unit rates.

    a. $\dfrac{\$0.37}{1 \text{ oz}}$ b. $\dfrac{333.2 \text{ mi}}{14 \text{ gal}}$ c. 16 ft/sec   d. $\dfrac{59 \text{ mi}}{1 \text{ hr}}$

20. Of the following rates, identify those that are unit rates.

    a. $\dfrac{3 \text{ lb}}{\$1.00}$ b. $\dfrac{21 \text{ ft}}{1 \text{ sec}}$ c. 50 mi/hr   d. $\dfrac{232 \text{ words}}{2 \text{ min}}$

For Exercises 21–32, write each rate as a unit rate and round to the nearest hundredth when necessary. (See Example 2.)

21. The Osborne family drove 452 mi in 4 days.

22. The book of poetry *The Prophet* by Kahlil Gibran has estimated sales of $6,000,000 over an 80-year period.

23. A submarine practicing evasive maneuvers descended 100 m in $\frac{1}{4}$ hr. Find a unit rate representing the change in "altitude" per hour.

24. The ground dropped 74.4 ft in 24 hr in the famous Winter Park sinkhole of 1981. Find the unit rate representing the change in "altitude" per hour.

25. If Oscar bought an easy chair for $660 and plans to make 12 payments, what is the amount per payment?

26. The jockey David Gall had 7396 wins in 43 years of riding.

27. At the market, bananas cost $2.76 for 4 lb.

28. Ceramic tile sells for $13.08 for a box of 12 tiles. Find the price per tile.

29. Lottery prize money of $1,792,000 is for seven people.

30. One WeightWatchers group lost 123 lb for its 11 members.

31. A male speed skater skated 500 m in 35 sec. Find the rate in meters per second.

32. A female speed skater skated 500 m in 38 sec. Find the rate in meters per second.

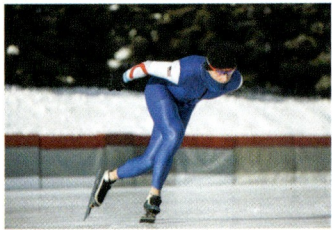

Jmichl/iStock/Getty Images

Concept 3: Unit Cost

For Exercises 33–42, find the unit costs (that is, dollars per unit). Round the answers to three decimal places when necessary. (See Example 3.)

33. A laundry detergent costs $14.99 for 100 oz.

34. A liquid body wash costs $10.25 for 24 oz.

35. Soda costs $1.99 for a 2-L bottle.

36. Four chairs cost $221.00.

37. A set of four tires costs $210.

38. A package of three shirts costs $64.80.

39. A package of six newborn bodysuits costs $32.50.

40. A package of eight AAA batteries costs $9.84.

41. a. 25 oz of shampoo for $8.35
    b. 15 oz of shampoo for $5.01
    c. Which is the better buy?

42. a. 10 lb of potting soil for $2.99
    b. 30 lb of potting soil for $8.97
    c. Which is the better buy?

43. Creamed corn comes in two size cans, 15 oz and 8.5 oz. The larger can costs $1.85 and the smaller can costs $1.39. Find the unit cost of each can. Which is the better buy? (Round to three decimal places.)

44. Napkins come in a variety of packages. A package of 400 napkins sells for $5.95, and a package of 100 napkins sells for $1.95. Find the unit cost of each package. Which is the better buy? (Round to three decimal places.)

Concept 4: Applications of Rates

45. Carbonated beverages come in different sizes and contain different amounts of sugar. Compute the amount of sugar (in grams) per fluid ounce for each soda. Then determine which has the greatest amount of sugar per fluid ounce. (*Source*: Coca-Cola Product Facts) (See Example 4.)

Soda	Amount	Sugar
Coca-Cola	20 fl oz	65 g
Mello Yello	12 fl oz	47 g
Canada Dry Ginger Ale	8 fl oz	24 g

46. Compute the amount of sodium per fluid ounce for each soda. Then determine which has the greatest amount of sodium per fluid ounce. (*Source*: Coca-Cola Product Facts)

Soda	Amount	Sodium
Coca-Cola	20 fl oz	75 mg
Mello Yello	12 fl oz	45 mg
Canada Dry Ginger Ale	8 fl oz	33 mg

47. Carbonated beverages come in different sizes and have a different number of calories. Compute the number of calories per fluid ounce for each soda. Then determine which has the least number of calories per fluid ounce. (*Source*: Coca-Cola Product Facts)

Soda	Amount	Calories
Coca-Cola	20 fl oz	240
Mello Yello	12 fl oz	170
Canada Dry Ginger Ale	8 fl oz	90

48. According to the National Institutes of Health, a platelet count below 20,000 per microliter of blood is considered a life-threatening condition. Suppose a patient's test results yield a platelet count of 13,000,000 for 100 microliters. Write this as a unit rate (number of platelets per microliter). Does the patient have a life-threatening condition?

49. The number of motor vehicles produced in the United States increased steadily by a total of 5,310,000 in an 18-year period. Compute the rate representing the increase in the number of vehicles produced per year during this time period. (*Source:* American Automobile Manufacturers Association)

50. The total number of prisoners in the United States increased steadily by a total of 344,000 in an 8-year period. Compute the rate representing the increase in the number of prisoners per year.

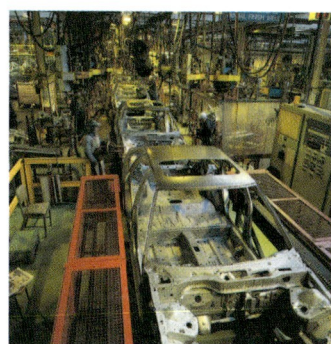

Digital Vision/Getty Images

51. a. The population of Mexico increased steadily by 22 million people in a 10-year period. Compute the rate representing the increase in the population per year.

    b. The population of Brazil increased steadily by 10.2 million in a 5-year period. Compute the rate representing the increase in the population per year.

    c. Which country has a greater rate of increase in population per year?

52. a. The price per share of Microsoft stock rose $18.24 in a 24-month period. Compute the rate representing the increase in the price per month.

    b. The price per share of IBM stock rose $22.80 in a 12-month period. Compute the rate representing the increase in the price per month.

    c. Which stock had a greater rate of increase per month?

53. A cheetah can run 120 m in 4.1 sec. An antelope can run 50 m in 2.1 sec. Compare their unit speeds to determine which animal is faster. Round to the nearest whole unit.

Jeff Vanuga/Corbis Super RF/Alamy Stock Photo

Technology Connections

Don Shula coached football for 33 years. He had 328 wins and 156 losses. Tom Landry coached football for 29 years. He had 250 wins and 162 losses. Use this information to answer Exercises 54 and 55.

54. a. Compute a unit rate representing the average number of wins per year for Don Shula. Round to one decimal place.

    b. Compute a unit rate representing the average number of wins per year for Tom Landry. Round to one decimal place.

    c. Which coach had a better rate of wins per year?

55. a. Compute a unit rate representing the number of wins to the number of losses for Don Shula. Round to one decimal place.

b. Compute a unit rate representing the number of wins to the number of losses for Tom Landry. Round to one decimal place.

c. Which coach had a better win/loss rate?

56. Compare three packages of soap. Find the price per ounce and determine the best buy. (Round to two decimal places.)

a. $9.59 for a 6-bar pack of 4.25-oz bars

b. $6.39 for a 8-bar pack of 4.5-oz bars

c. $2.55 for a 3-bar pack of 4.5-oz bars

57. Mayonnaise comes in 48-, 22-, and 18-oz jars. They are priced at $8.69, $6.15, and $4.59, respectively. Find the unit cost of each size jar to find the best buy. (Round to three decimal places.)

58. Tuna fish comes in different size cans. Find the unit cost of each package to find the best buy. (Round to three decimal places.)

a. 2-pack of 4.5-oz cans for $3.61

b. One 12-oz can for $2.00

c. 3-pack of 3.3-oz cans for $3.19

59. Root beer is sold in a variety of different packages. Find the unit cost of each package to find the better buy. (Round to three decimal places.)

a. 6-pack of 8-oz cans for $2.99

b. 12-pack of 12-oz cans for $3.33

Section 6.3 Proportions

Concepts

1. Definition of a Proportion
2. Determining Whether Two Ratios Form a Proportion
3. Solving Proportions

1. Definition of a Proportion

Recall that a statement indicating that two quantities are equal is called an equation. In this section, we are interested in a special type of equation called a proportion. A **proportion** states that two ratios or rates are equal. For example:

$$\frac{1}{4} = \frac{10}{40} \text{ is a proportion.}$$

We know that the fractions $\frac{1}{4}$ and $\frac{10}{40}$ are equal because $\frac{10}{40}$ reduces to $\frac{1}{4}$.

We read the proportion $\frac{1}{4} = \frac{10}{40}$ as follows: "1 is to 4 as 10 is to 40."

We also say that the numbers 1 and 4 are *proportional to* the numbers 10 and 40.

Example 1 — Writing Proportions

Write a proportion for each statement.

a. 5 is to 12 as 30 is to 72.
b. 240 mi is to 4 hr as 300 mi is to 5 hr.
c. The numbers −3 and 7 are proportional to the numbers −12 and 28.

Solution:

a. $\dfrac{5}{12} = \dfrac{30}{72}$ 5 is to 12 as 30 is to 72.

b. $\dfrac{240 \text{ mi}}{4 \text{ hr}} = \dfrac{300 \text{ mi}}{5 \text{ hr}}$ 240 mi is to 4 hr as 300 mi is to 5 hr.

c. $\dfrac{-3}{7} = \dfrac{-12}{28}$ −3 and 7 are proportional to −12 and 28.

TIP: A proportion is an equation and must have an equal sign.

Skill Practice Write a proportion for each statement.

1. 7 is to 28 as 13 is to 52.
2. $17 is to 2 hr as $102 is to 12 hr.
3. The numbers 5 and −11 are proportional to the numbers 15 and −33.

2. Determining Whether Two Ratios Form a Proportion

To determine whether two ratios form a proportion, we must determine whether the ratios are equal. Recall that two fractions are equal whenever their cross products are equal. That is,

$$\dfrac{a}{b} = \dfrac{c}{d} \quad \text{implies} \quad a \cdot d = b \cdot c \quad \text{(and vice versa)}.$$

Example 2 — Determining Whether Two Ratios Form a Proportion

Determine whether the ratios form a proportion.

a. $\dfrac{3}{5} \stackrel{?}{=} \dfrac{9}{15}$ b. $\dfrac{8}{4} \stackrel{?}{=} \dfrac{10}{5\frac{1}{2}}$

Avoiding Mistakes

An equation from cross products can be formed only when there are two fractions separated by an equal sign.

Solution:

a. $\dfrac{3}{5} \stackrel{?}{=} \dfrac{9}{15}$

$(3)(15) \stackrel{?}{=} (5)(9)$ Form an equation from the cross products.
$45 = 45$ ✓

The cross products are equal. Therefore, the ratios form a proportion.

b. $\dfrac{8}{4} \stackrel{?}{=} \dfrac{10}{5\frac{1}{2}}$

$(8)\left(5\dfrac{1}{2}\right) \stackrel{?}{=} (4)(10)$ Form an equation from the cross products.

$\dfrac{\overset{4}{\cancel{8}}}{1} \cdot \dfrac{11}{\underset{1}{\cancel{2}}} \stackrel{?}{=} 40$ Write the mixed number as an improper fraction.

$44 \neq 40$ Multiply fractions.

The cross products are not equal. The ratios do not form a proportion.

Skill Practice Determine whether the ratios form a proportion.

4. $\dfrac{4}{9} \stackrel{?}{=} \dfrac{12}{27}$ 5. $\dfrac{3\frac{1}{4}}{5} \stackrel{?}{=} \dfrac{8}{12}$

Answers

1. $\dfrac{7}{28} = \dfrac{13}{52}$ 2. $\dfrac{\$17}{2 \text{ hr}} = \dfrac{\$102}{12 \text{ hr}}$
3. $\dfrac{5}{-11} = \dfrac{15}{-33}$ 4. Yes   5. No

Example 3 — Determining Whether Pairs of Numbers Are Proportional

Determine whether the numbers 2.7 and −5.3 are proportional to the numbers 8.1 and −15.9.

Solution:

Two pairs of numbers are proportional if their ratios are equal.

$$\frac{2.7}{-5.3} \stackrel{?}{=} \frac{8.1}{-15.9}$$

$(2.7)(-15.9) \stackrel{?}{=} (-5.3)(8.1)$ Form an equation from the cross products.

$-42.93 = -42.93$ ✓   Multiply decimals.

The cross products are equal. The pairs of numbers are proportional.

Skill Practice

6. Determine whether the numbers −1.2 and 2.5 are proportional to the numbers −2 and 5.

3. Solving Proportions

A proportion is made up of four values. If three of the four values are known, we can solve for the fourth.

Consider the proportion $\frac{x}{20} = \frac{3}{4}$. We let the variable x represent the unknown value in the proportion. To solve for x, we can equate the cross products to form an equivalent equation.

$$\frac{x}{20} \stackrel{?}{=} \frac{3}{4}$$

$4x = 3 \cdot 20$ Form an equation from the cross products.

$4x = 60$ Simplify.

$\dfrac{4x}{4} = \dfrac{60}{4}$ Divide both sides by 4 to isolate x.

$x = 15$

We can check the value of x in the original proportion.

Check: $\dfrac{x}{20} = \dfrac{3}{4}$ $\xrightarrow{\text{substitute 15 for } x}$ $\dfrac{15}{20} \stackrel{?}{=} \dfrac{3}{4}$

$(15)(4) \stackrel{?}{=} (3)(20)$

$60 = 60$ ✓   The solution 15 checks.

The steps to solve a proportion are summarized next.

Solving a Proportion

Step 1 Set the cross products equal to each other.
Step 2 Solve the equation.
Step 3 Check the solution in the original proportion.

Answer

6. No

Section 6.3 Proportions

Example 4 — Solving a Proportion

Solve the proportion. $\quad \dfrac{x}{13} = \dfrac{6}{39}$

Solution:

$\dfrac{x}{13} = \dfrac{6}{39}$

$39x = (6)(13)\quad$ Set the cross products equal.

$39x = 78 \quad$ Simplify.

$\dfrac{39x}{39} = \dfrac{78}{39} \quad$ Divide both sides by 39 to isolate x.

$x = 2 \quad$ Check: $\dfrac{x}{13} = \dfrac{6}{39} \xrightarrow{\text{substitute } x=2} \dfrac{2}{13} \stackrel{?}{=} \dfrac{6}{39}$

$(2)(39) \stackrel{?}{=} (6)(13)$

The solution 2 checks in the original proportion. $\quad 78 = 78$ ✓

Skill Practice Solve the proportion. Be sure to check your answer.

7. $\dfrac{2}{9} = \dfrac{n}{81}$

Example 5 — Solving a Proportion

Solve the proportion. $\quad \dfrac{4}{15} = \dfrac{9}{n}$

Solution:

$\dfrac{4}{15} = \dfrac{9}{n} \quad$ The variable can be represented by any letter.

$4n = (9)(15) \quad$ Set the cross products equal.

$4n = 135 \quad$ Simplify.

$\dfrac{4n}{4} = \dfrac{135}{4} \quad$ Divide both sides by 4 to isolate n.

$n = \dfrac{135}{4} \quad$ The fraction $\dfrac{135}{4}$ is in lowest terms.

The solution may be written as $n = \dfrac{135}{4}$ or $n = 33\dfrac{3}{4}$ or $n = 33.75$.

To check the solution in the original proportion, we may use any of the three forms of the answer. We will use the decimal form.

Check: $\dfrac{4}{15} = \dfrac{9}{n} \xrightarrow{\text{substitute } n = 33.75} \dfrac{4}{15} \stackrel{?}{=} \dfrac{9}{33.75}$

$(4)(33.75) \stackrel{?}{=} (9)(15)$

$135 = 135$ ✓

The solution 33.75 checks in the original proportion.

> **Avoiding Mistakes**
>
> When solving a proportion, do not try to "cancel" like factors on opposite sides of the equal sign. The proportion, $\dfrac{4}{15} = \dfrac{9}{n}$ cannot be simplified.

Skill Practice Solve the proportion. Be sure to check your answer.

8. $\dfrac{3}{w} = \dfrac{21}{77}$

Answers

7. 18   8. 11

Example 6 Solving a Proportion

Solve the proportion. $\dfrac{0.8}{-3.1} = \dfrac{4}{p}$

Solution:

$$\dfrac{0.8}{-3.1} = \dfrac{4}{p}$$

$0.8p = 4(-3.1)$ Set the cross products equal.

$0.8p = -12.4$ Simplify.

$\dfrac{0.8p}{0.8} = \dfrac{-12.4}{0.8}$ Divide both sides by 0.8.

$p = -15.5$ The value -15.5 checks in the original equation.

The solution is -15.5.

Skill Practice Solve the proportion. Be sure to check your answer.

9. $\dfrac{0.6}{x} = \dfrac{1.5}{-2}$

We chose to give the solution to Example 6 in decimal form because the values in the original proportion are decimal numbers. However, it would be correct to give the solution as a mixed number or fraction. The solution -15.5 is also equivalent to $-15\frac{1}{2}$ or $-\frac{31}{2}$.

Example 7 Solving a Proportion

Solve the proportion. $\dfrac{-12}{-8} = \dfrac{x}{\frac{2}{3}}$

Solution:

$$\dfrac{-12}{-8} = \dfrac{x}{\frac{2}{3}}$$

$(-12)\left(\dfrac{2}{3}\right) = -8x$ Set the cross products equal.

$\left(-\dfrac{\overset{4}{\cancel{12}}}{1}\right)\left(\dfrac{2}{\underset{1}{\cancel{3}}}\right) = -8x$ Write the whole number as an improper fraction.

$-8 = -8x$ Simplify.

$\dfrac{-8}{-8} = \dfrac{-8x}{-8}$ Divide both sides by -8.

$1 = x$ The value 1 checks in the original equation.

The solution is 1.

Skill Practice Solve the proportion. Be sure to check your answer.

10. $\dfrac{\frac{1}{2}}{3.5} = \dfrac{x}{14}$

Answers

9. -0.8 10. 2

Section 6.3 Activity

A.1. 4 women are to 12 men as 16 women are to 48 men.
  a. Write a proportion for the statement.
  b. Show these ratios form a proportion. Support your answer using two different methods to prove equality.

For Exercises A.2–A.5, determine whether the ratios form a proportion.

A.2. Is 2 DVDs for $29.98 proportional to 5 DVDs for $74.95?

A.3. $\dfrac{4}{-11} = \dfrac{14}{-36}$ **A.4.** $\dfrac{3.8}{17.1} = \dfrac{2.5}{11.7}$ **A.5.** $\dfrac{3\frac{1}{4}}{10} = \dfrac{19\frac{1}{2}}{60}$

A.6. Determine the missing value by inspection. That is, find the numerator by writing equivalent fractions.

$$\dfrac{1}{4} = \dfrac{}{8}$$

A.7. Form an equation by setting the cross products equal to each other, then solve.

$$\dfrac{1}{4} = \dfrac{x}{8}$$

A.8. Are the answers the same in Exercises A.6 and A.7? Describe the process used in Exercise A.7 to solve an equation that contains two ratios set equal to each other.

For Exercises A.9–A.11, solve the proportion by cross multiplying and solving the resulting equation.

A.9. $\dfrac{6}{-8} = \dfrac{n}{12}$ **A.10.** $\dfrac{-3}{20} = \dfrac{7.5}{w}$ **A.11.** $\dfrac{15}{p} = \dfrac{3}{\frac{3}{4}}$

A.12. a. Simplify: $5(x-4)$.
  b. Solve the proportion. $\dfrac{-2}{5} = \dfrac{(x-4)}{20}$

Practice Exercises

Section 6.3

Prerequisite Review

For Exercises R.1–R.2, write as a ratio in lowest terms.

  R.1. 25 ft to 8.5 ft   **R.2.** 18.3 hr to 6.9 hr

For Exercises R.3–R.4, write as a rate in lowest terms.

  R.3. 18 apples for 4 pies   **R.4.** 35 people for 70 dogs

For Exercises R.5–R.6, write as a unit rate.

  R.5. 820 revolutions in 40 sec   **R.6.** 337.2 mi on 12 gal of gas

Vocabulary and Key Concepts

1. A _____ is a statement indicating that two ratios or rates are equal.

2. A proportion must be written with an equal sign because it is a(n) _____.

3. Two ratios form a proportion if their _____ _____ are equal.

4. A proportion is made up of four values; _____ are known and _____ is unknown.

5. The numbers −2 and 7 (are/are not) proportional to the numbers −16 and 56.

6. You (can/cannot) cancel like factors on opposite sides of the equal sign in the proportion $\frac{x}{12} = \frac{4}{13}$.

Concept 1: Definition of a Proportion

For Exercises 7–20, write a proportion for each statement. (See Example 1.)

7. 2 is to 48 as 1 is to 24.

8. 1 is to 11 as 9 is to 99.

9. 4 is to 16 as 5 is to 20.

10. 3 is to 18 as 4 is to 24.

11. −25 is to 15 as −10 is to 6.

12. 35 is to −14 as 20 is to −8.

13. The numbers 2 and 3 are proportional to the numbers 4 and 6.

14. The numbers 2 and 1 are proportional to the numbers 26 and 13.

15. The numbers −30 and −25 are proportional to the numbers 12 and 10.

16. The numbers −24 and −18 are proportional to the numbers 8 and 6.

17. $6.25 per hour is proportional to $187.50 per 30 hr.

18. $115 per week is proportional to $460 per 4 weeks.

19. 1 in. is to 7 mi as 5 in. is to 35 mi.

20. 16 flowers is to 5 plants as 32 flowers is to 10 plants.

Concept 2: Determining Whether Two Ratios Form a Proportion

For Exercises 21–28, determine whether the ratios form a proportion. (See Example 2.)

21. $\frac{5}{18} \stackrel{?}{=} \frac{4}{16}$

22. $\frac{9}{10} \stackrel{?}{=} \frac{8}{9}$

23. $\frac{16}{24} \stackrel{?}{=} \frac{2}{3}$

24. $\frac{4}{5} \stackrel{?}{=} \frac{24}{30}$

25. $\frac{2\frac{1}{2}}{3\frac{2}{3}} \stackrel{?}{=} \frac{15}{22}$

26. $\frac{1\frac{3}{4}}{3} \stackrel{?}{=} \frac{7}{12}$

27. $\frac{2}{-3.2} \stackrel{?}{=} \frac{10}{-16}$

28. $\frac{4.7}{-7} \stackrel{?}{=} \frac{23.5}{-35}$

For Exercises 29–34, determine whether the pairs of numbers are proportional. (See Example 3.)

29. Are the numbers 48 and 18 proportional to the numbers 24 and 9?

30. Are the numbers 35 and 14 proportional to the numbers 5 and 2?

31. Are the numbers $2\frac{3}{8}$ and $1\frac{1}{2}$ proportional to the numbers $9\frac{1}{2}$ and 6?

32. Are the numbers $1\frac{2}{3}$ and $\frac{5}{6}$ proportional to the numbers 5 and $2\frac{1}{2}$?

33. Are the numbers -6.3 and 9 proportional to the numbers -12.6 and 16?

34. Are the numbers -7.1 and 2.4 proportional to the numbers -35.5 and 10?

Concept 3: Solving Proportions

For Exercises 35–38, determine whether the given value is a solution to the proportion.

35. $\dfrac{x}{40} = \dfrac{1}{-8}$; $x = -5$

36. $\dfrac{14}{x} = \dfrac{12}{-18}$; $x = -21$

37. $\dfrac{12.4}{31} = \dfrac{8.2}{y}$; $y = 20$

38. $\dfrac{4.2}{9.8} = \dfrac{z}{36.4}$; $z = 15.2$

For Exercises 39–58, solve the proportion. Be sure to check your answers. (See Examples 4–7.)

39. $\dfrac{12}{16} = \dfrac{3}{x}$

40. $\dfrac{20}{28} = \dfrac{5}{x}$

41. $\dfrac{9}{21} = \dfrac{x}{7}$

42. $\dfrac{15}{10} = \dfrac{3}{x}$

43. $\dfrac{p}{12} = \dfrac{-25}{4}$

44. $\dfrac{p}{8} = \dfrac{-30}{24}$

45. $\dfrac{3}{40} = \dfrac{w}{10}$

46. $\dfrac{5}{60} = \dfrac{z}{8}$

47. $\dfrac{16}{-13} = \dfrac{20}{t}$

48. $\dfrac{12}{b} = \dfrac{8}{-9}$

49. $\dfrac{m}{12} = \dfrac{5}{8}$

50. $\dfrac{16}{12} = \dfrac{21}{a}$

51. $\dfrac{17}{12} = \dfrac{4\frac{1}{4}}{x}$

52. $\dfrac{26}{30} = \dfrac{5\frac{1}{5}}{x}$

53. $\dfrac{0.5}{h} = \dfrac{1.8}{9}$

54. $\dfrac{2.6}{h} = \dfrac{1.3}{0.5}$

55. $\dfrac{\frac{3}{8}}{6.75} = \dfrac{x}{72}$

56. $\dfrac{12.5}{\frac{1}{4}} = \dfrac{120}{y}$

57. $\dfrac{4}{\frac{1}{10}} = \dfrac{-\frac{1}{2}}{z}$

58. $\dfrac{6}{\frac{1}{3}} = \dfrac{-\frac{1}{2}}{t}$

For Exercises 59–64, write a proportion for each statement. Then solve for the variable.

59. 25 is to 100 as 9 is to y.

60. 65 is to 15 as 26 is to y.

61. 15 is to 20 as t is to 10.

62. 9 is to 12 as w is to 30.

63. The numbers -3.125 and 5 are proportional to the numbers -18.75 and k.

64. The numbers -4.75 and 8 are proportional to the numbers -9.5 and k.

Expanding Your Skills

For Exercises 65–72, solve the proportions.

65. $\dfrac{x+1}{3} = \dfrac{5}{7}$

66. $\dfrac{2}{5} = \dfrac{x-2}{4}$

67. $\dfrac{x-3}{3x} = \dfrac{2}{3}$

68. $\dfrac{x-2}{4x} = \dfrac{3}{8}$

69. $\dfrac{x+3}{x} = \dfrac{5}{4}$

70. $\dfrac{x-5}{x} = \dfrac{3}{2}$

71. $\dfrac{x}{3} = \dfrac{x-2}{4}$

72. $\dfrac{x}{6} = \dfrac{x-1}{3}$

Problem Recognition Exercises

Operations on Fractions versus Solving Proportions

Recall that a proportion is a statement that indicates that two ratios are equal. A proportion looks like two fractions with an equal sign separating 304. To solve a proportion, we cross multiply and solve the resulting equation. However, don't confuse solving a proportion with the operation of multiplying fractions. When we multiply two fractions, we multiply "straight across." That is, we multiply the numerators and multiply the denominators.

Solve the proportion. *(Notice the equal sign.)*

$$\frac{3}{20} = \frac{x}{48} \quad \text{Cross multiply.}$$

$3 \cdot (48) = 20 \cdot x$

$144 = 20x$

$\dfrac{144}{20} = \dfrac{20x}{20}$

$7.2 = x$

Multiply the fractions. *(Notice the multiplication sign.)*

$$\frac{3}{20} \cdot \frac{5}{48} \quad \begin{matrix}\text{Multiply numerators.}\\ \text{Multiply denominators.}\end{matrix}$$

$= \dfrac{3 \cdot 5}{20 \cdot 48}$

$= \dfrac{\overset{1}{3} \cdot \overset{1}{5}}{\underset{4}{20} \cdot \underset{16}{48}}$

$= \dfrac{1}{64}$

For Exercises 1–6, identify the problem as a proportion or as a product of fractions. Then solve the proportion or multiply the fractions.

1. a. $\dfrac{x}{4} = \dfrac{15}{8}$ b. $\dfrac{1}{4} \cdot \dfrac{15}{8}$
2. a. $\dfrac{2}{5} \cdot \dfrac{3}{10}$ b. $\dfrac{2}{5} = \dfrac{y}{10}$
3. a. $\dfrac{2}{7} \times \dfrac{3}{14}$ b. $\dfrac{2}{7} = \dfrac{n}{14}$
4. a. $\dfrac{m}{5} = \dfrac{6}{15}$ b. $\dfrac{3}{5} \times \dfrac{6}{15}$
5. a. $\dfrac{48}{p} = \dfrac{16}{3}$ b. $\dfrac{48}{8} \cdot \dfrac{16}{3}$
6. a. $\dfrac{10}{7} \cdot \dfrac{28}{5}$ b. $\dfrac{10}{7} = \dfrac{28}{t}$

For Exercises 7–10, solve the proportion or perform the indicated operation on fractions.

7. a. $\dfrac{3}{7} = \dfrac{6}{z}$ b. $\dfrac{3}{7} \div \dfrac{6}{35}$ c. $\dfrac{3}{7} + \dfrac{6}{35}$ d. $\dfrac{3}{7} \cdot \dfrac{6}{35}$

8. a. $\dfrac{4}{5} \div \dfrac{20}{3}$ b. $\dfrac{4}{v} = \dfrac{20}{3}$ c. $\dfrac{4}{5} \times \dfrac{20}{3}$ d. $\dfrac{4}{5} - \dfrac{20}{3}$

9. a. $\dfrac{14}{5} \cdot \dfrac{10}{7}$ b. $\dfrac{14}{5} = \dfrac{x}{7}$ c. $\dfrac{14}{5} - \dfrac{10}{7}$ d. $\dfrac{14}{5} \div \dfrac{10}{7}$

10. a. $\dfrac{11}{3} = \dfrac{66}{y}$ b. $\dfrac{11}{3} + \dfrac{66}{11}$ c. $\dfrac{11}{3} \div \dfrac{66}{11}$ d. $\dfrac{11}{3} \times \dfrac{66}{11}$

Applications of Proportions and Similar Figures

Section 6.4

Concepts
1. Applications of Proportions
2. Similar Figures

1. Applications of Proportions

Proportions are used in a variety of applications. In Examples 1 through 4, we take information from the wording of a problem and form a proportion.

Example 1 — Using a Proportion in a Consumer Application

Linda drove 145 mi on 5 gal of gas. At this rate, how far can she drive on 12 gal?

Solution:

Let x represent the distance Linda can go on 12 gal.

This problem involves two rates. We can translate this to a proportion. Equate the two rates.

$$\underset{\text{number of gallons}}{\overset{\text{distance}}{\longrightarrow}} \frac{145 \text{ mi}}{5 \text{ gal}} = \frac{x \text{ mi}}{12 \text{ gal}} \underset{\longleftarrow \text{ number of gallons}}{\overset{\longleftarrow \text{ distance}}{}}$$

Solve the proportion.

$(145)(12) = 5x$ Form an equation from the cross products.

$1740 = 5x$

$\dfrac{1740}{5} = \dfrac{5x}{5}$ Divide both sides by 5.

$348 = x$ Divide. $1740 \div 5 = 348$

Linda can drive 348 mi on 12 gal of gas.

TIP: Notice that the two rates have the same units in the numerator (miles) and the same units in the denominator (gallons).

Skill Practice

1. Jacques bought 3 lb of tomatoes for $5.55. At this rate, how much would 7 lb cost?

In Example 1 we could have set up a proportion in many different ways.

$$\frac{145 \text{ mi}}{5 \text{ gal}} = \frac{x \text{ mi}}{12 \text{ gal}} \quad \text{or} \quad \frac{5 \text{ gal}}{145 \text{ mi}} = \frac{12 \text{ gal}}{x \text{ mi}} \quad \text{or}$$

$$\frac{5 \text{ gal}}{12 \text{ gal}} = \frac{145 \text{ mi}}{x \text{ mi}} \quad \text{or} \quad \frac{12 \text{ gal}}{5 \text{ gal}} = \frac{x \text{ mi}}{145 \text{ mi}}$$

Notice that in each case, the cross products produce the same equation. We will generally set up the proportions so that the units in the numerators are the same and the units in the denominators are the same.

Answer

1. The price for 7 lb would be $12.95.

Example 2 Using a Proportion in a Construction Application

If a cable 25 ft long weighs 1.2 lb, how much will a 120-ft cable weigh?

Solution:

Let w represent the weight of the 120-ft cable.   Label the unknown.

$$\underset{\text{weight}}{\overset{\text{length}}{}}\ \frac{25 \text{ ft}}{1.2 \text{ lb}} = \frac{120 \text{ ft}}{w \text{ lb}}\ \underset{\text{weight}}{\overset{\text{length}}{}}$$ Translate to a proportion.

$25w = (1.2)(120)$ Equate the cross products.

$25w = 144$

$\dfrac{25w}{25} = \dfrac{144}{25}$ Divide both sides by 25.

$w = 5.76$ Divide. $144 \div 25 = 5.76$

The 120-ft cable weighs 5.76 lb.

Skill Practice

2. It takes 2.5 gal of paint to cover 900 ft² of wall. How much area could be painted with 4 gal of the same paint?

Example 3 Using a Proportion in a Geography Application

The distance between Phoenix and Los Angeles is 348 mi. On a certain map, this is represented by 6 in. On the same map, the distance between San Antonio and Little Rock is $8\frac{1}{2}$ in. What is the actual distance between San Antonio and Little Rock?

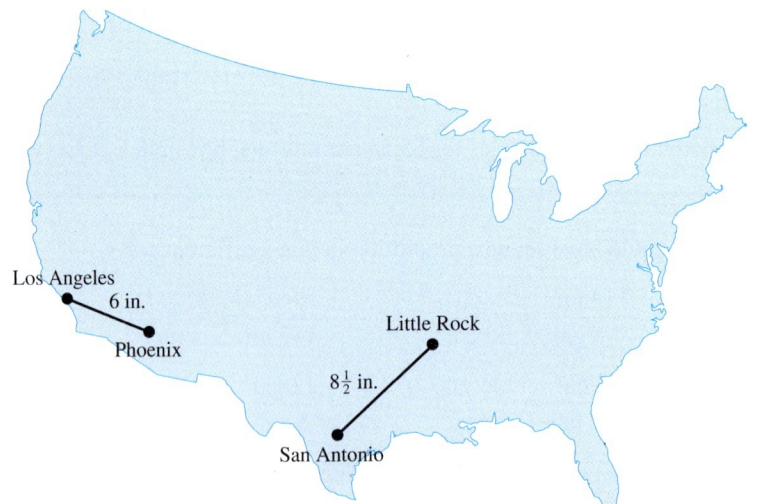

Solution:

Let d represent the distance between San Antonio and Little Rock.

Answer

2. An area of 1440 ft² could be painted.

$$\text{actual distance} \longrightarrow \frac{348 \text{ mi}}{6 \text{ in.}} = \frac{d \text{ mi}}{8\frac{1}{2} \text{ in.}} \longleftarrow \text{actual distance}$$
$$\text{distance on map} \longrightarrow \phantom{\frac{348 \text{ mi}}{6 \text{ in.}}} \phantom{\frac{d \text{ mi}}{8\frac{1}{2} \text{ in.}}} \longleftarrow \text{distance on map}$$

Translate to a proportion.

$(348)(8\frac{1}{2}) = 6d$ Equate the cross products.

$(348)(8.5) = 6d$ Convert the values to decimal.

$2958 = 6d$

$\dfrac{2958}{6} = \dfrac{6d}{6}$ Divide both sides by 6.

$493 = d$ Divide. $2958 \div 6 = 493$

The distance between San Antonio and Little Rock is 493 mi.

Skill Practice

3. On a certain map, the distance between Shreveport, Louisiana, and Memphis, Tennessee, is represented as $4\frac{1}{2}$ in. The actual distance is 288 mi. The map distance between Seattle, Washington, and San Francisco, California, is represented as 11 in. Find the actual distance.

Example 4   Applying a Proportion to Environmental Science

A biologist wants to estimate the number of elk in a wildlife preserve. She sedates 25 elk and clips a small radio transmitter onto the ear of each animal. The elk return to the wild, and after 6 months, the biologist studies a sample of 120 elk in the preserve. Of the 120 elk sampled, 4 have radio transmitters. Approximately how many elk are in the whole preserve?

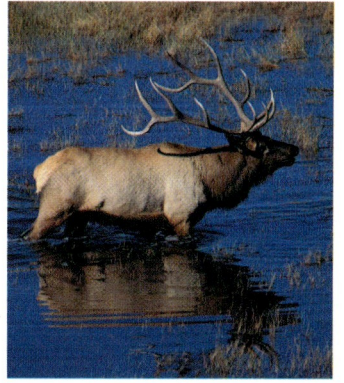
Jeff Vanuga/Getty Images

Solution:

Let n represent the number of elk in the whole preserve.

                Sample   Population

number of elk in the sample with radio transmitters $\longrightarrow \dfrac{4}{120} = \dfrac{25}{n} \longleftarrow$ number of elk in the population with radio transmitters

total elk in the sample $\longrightarrow \phantom{\dfrac{4}{120}} \phantom{\dfrac{25}{n}} \longleftarrow$ total elk in the population

$4n = (120)(25)$ Equate the cross products.

$4n = 3000$

$\dfrac{4n}{4} = \dfrac{3000}{4}$ Divide both sides by 4.

$n = 750$ Divide. $3000 \div 4 = 750$

There are approximately 750 elk in the wildlife preserve.

Skill Practice

4. To estimate the number of fish in a lake, the park service catches 50 fish and tags them. After several months the park service catches a sample of 100 fish and finds that 6 are tagged. Approximately how many fish are in the lake?

Answers

3. The distance between Seattle and San Francisco is 704 mi.
4. There are approximately 833 fish in the lake.

2. Similar Figures

Two triangles whose corresponding sides are proportional are called **similar triangles**. This means that the corresponding angles have equal measures, but the triangles may be different sizes. The following pairs of triangles are similar (Figure 6-1).

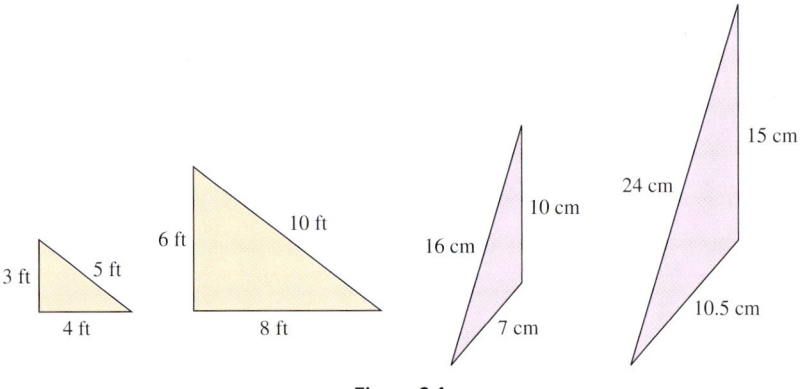

Figure 6-1

Consider the triangles:

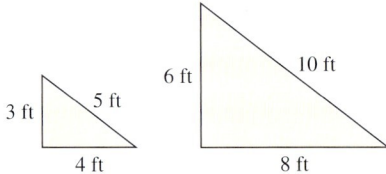

Notice that the ratio formed by the "left" sides of the triangles is $\frac{3}{6}$. This is the same as the ratio formed by the "bottom" sides, $\frac{4}{8}$, and is the same as the ratio formed by the "right" sides, $\frac{5}{10}$. Each ratio simplifies to $\frac{1}{2}$. Because all ratios are equal, the corresponding sides are proportional.

Example 5 — Finding the Unknown Sides in Similar Triangles

The triangles in Figure 6-2 are similar.

a. Solve for x. **b.** Solve for y.

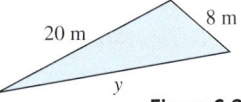 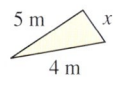

Figure 6-2

Solution:

a. Notice that the lengths of the "left" sides of both triangles are known. This forms a known ratio of $\frac{20}{5}$. Because the triangles are similar, the ratio of the other corresponding sides must be equal to $\frac{20}{5}$. To solve for x, we have

large triangle "left" side ⟶ $\dfrac{20}{5} = \dfrac{8}{x}$ ⟵ large triangle "right" side
small triangle "left" side ⟶ ⟵ small triangle "right" side

$20x = (5)(8)$ Equate the cross products.

$20x = 40$

$\dfrac{20x}{20} = \dfrac{40}{20}$ Divide both sides of the equation by 20.

$x = 2$ Divide. $40 \div 20 = 2$

b. To solve for y, we have

large triangle "left" side → $\dfrac{20}{5} = \dfrac{y}{4}$ ← large triangle "bottom" side
small triangle "left" side   small triangle "bottom" side

$(20)(4) = 5y$ Equate the cross products.

$80 = 5y$

$\dfrac{80}{5} = \dfrac{5y}{5}$ Divide both sides by 5.

$16 = y$ Divide. $80 \div 5 = 16$

The values for x and y are $x = 2$ m and $y = 16$ m.

> **TIP:** In Example 5, notice that the corresponding sides of the triangles form a 4 to 1 ratio.
>
>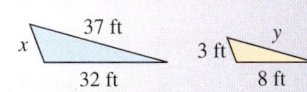

Skill Practice

5. The triangles are similar.
  a. Solve for x.   **b.** Solve for y.

Example 6 — Finding the Unknown Side in a Similar Triangle

The triangles are similar. Solve for x.

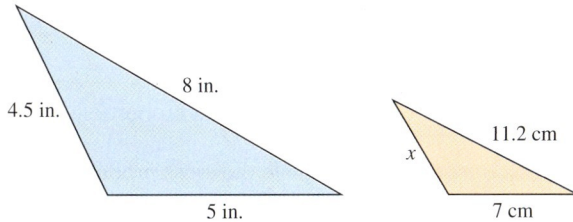

Solution:

The ratio $\dfrac{5}{7}$ can be formed from the lengths of the bottom sides of the triangles. Then this ratio and the ratio of the left sides of the triangles form a proportion.

large triangle bottom → $\dfrac{5}{7} = \dfrac{4.5}{x}$ ← large triangle "left" side
small triangle bottom   small triangle "left" side

Notice that the proportion is still true even though we are comparing inches to centimeters.

$5x = (4.5)(7)$ Equate the cross products.

$5x = 31.5$

$\dfrac{5x}{5} = \dfrac{31.5}{5}$ Divide both sides of the equation by 5.

$x = 6.3$ Divide. $31.5 \div 5 = 6.3$

The value of x will be in centimeters. The length of side x is 6.3 cm.

Skill Practice

6. The given triangles are similar. Solve for y.

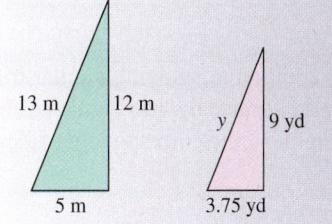

Answers
5. a. $x = 12$ ft   **b.** $y = 9.25$ ft
6. $y = 9.75$ yd

Example 7 — Using Similar Triangles in an Application

On a sunny day, a 6-ft man casts a 3.2-ft shadow on the ground. At the same time, a building casts an 80-ft shadow. How tall is the building?

Solution:

We illustrate the scenario in Figure 6-3. Note, however, that the distances are not drawn to scale.

Let h represent the height of the building in feet.

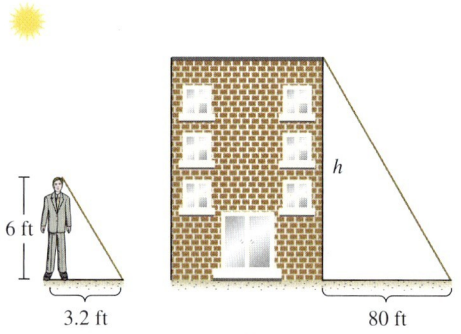

Figure 6-3

We will assume that the triangles are similar because the readings were taken at the same time of day.

$$\frac{6}{h} = \frac{3.2}{80}$$ Translate to a proportion.

$(6)(80) = (3.2) \cdot h$ Equate the cross products.

$480 = 3.2h$

$$\frac{480}{3.2} = \frac{3.2h}{3.2}$$ Divide both sides by 3.2.

$150 = h$ Divide. $480 \div 3.2 = 150$

The height of the building is 150 ft.

Skill Practice

7. Donna has drawn a pattern from which to make a scarf. Assuming that the triangles are similar (the pattern and scarf have the same shape), find the length of side x (in yards).

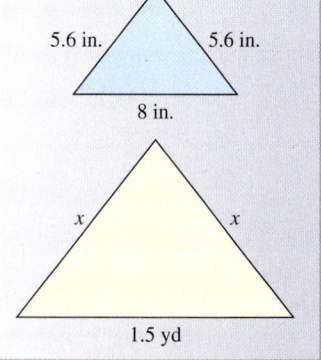

In addition to studying similar triangles, we present similar polygons. A polygon is a flat figure formed by line segments connected at their ends. For a pair of **similar polygons**, the lengths of the corresponding sides of the two polygons are proportional.

Answer

7. Side x is 1.05 yd long.

Example 8 Using Similar Polygons in an Application

A negative for a photograph is 3.5 cm by 2.5 cm. If the width of the resulting picture is 4 in., what is the length of the picture? See Figure 6-4.

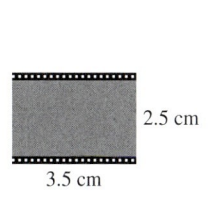

Figure 6-4

Solution:

Let x represent the length of the photo.

The photo and its negative are similar polygons.

$$\frac{3.5 \text{ cm}}{x \text{ in.}} = \frac{2.5 \text{ cm}}{4 \text{ in.}} \quad \text{Translate to a proportion.}$$

$$(3.5)(4) = (2.5) \cdot x \quad \text{Equate the cross products.}$$

$$14 = 2.5x$$

$$\frac{14}{2.5} = \frac{2.5x}{2.5} \quad \text{Divide both sides by 2.5.}$$

$$5.6 = x \quad \text{Divide. } 14 \div 2.5 = 5.6$$

The picture is 5.6 in. long.

Skill Practice

8. The first- and second-place plaques for a softball tournament are similar figures. Find the length of side x.

Answer

8. Side x is 6.25 in.

Section 6.4 Activity

A.1. In Kansas City, one motorist has an 18-min commute to travel 7.5 mi. In similar traffic, how long would it take another motorist to commute 5.5 mi?

A.2. A pet groomer can groom 24 dogs in an 8-hr day. At the same rate, how many dogs can be scheduled for a 36-hr work week?

A.3. A long-distance runner burns 315 calories when running a 5k race (3.1 mi). Assuming the runner keeps the same pace and burns calories at the same rate, how many calories will they burn when running a half-marathon (13.1 mi)? Round to the nearest whole calorie.

A.4. A 1.75-oz protein bar contains 17 g of protein. How many grams of protein are in a 2.25-oz bar? Round to the nearest tenth of a gram.

A.5. It takes Orlando 45 min to mow a lawn that is $\frac{1}{8}$ acre. At the same rate, how long will it take Orlando to cut a lawn that is $\frac{3}{4}$ acre?

A.6. A 40-ft tall tree casts a shadow of 18 ft. How long will the shadow be for a 5-ft tall teenager?

A.7. Shelia is enlarging a photograph. Assuming the pair of rectangles are similar, solve for the indicated variable.

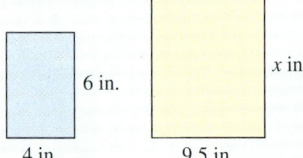

Section 6.4 Practice Exercises

Prerequisite Review

For Exercises R.1–R.6, solve the proportion.

R.1. $\dfrac{12}{x} = \dfrac{15}{24}$

R.2. $\dfrac{10}{4} = \dfrac{x}{15}$

R.3. $\dfrac{y}{2.1} = \dfrac{7}{3}$

R.4. $\dfrac{0.25}{1.2} = \dfrac{y}{24}$

R.5. $\dfrac{3\frac{1}{2}}{k} = \dfrac{2\frac{1}{3}}{4}$

R.6. $\dfrac{k}{3\frac{1}{3}} = \dfrac{1\frac{4}{5}}{2}$

Vocabulary and Key Concepts

1. **a.** Two triangles are called _____ triangles if the lengths of their corresponding sides are proportional.

   b. If two polygons are _____, then the lengths of their corresponding sides are proportional.

2. The two proportions: $\dfrac{1 \text{ ft}}{12 \text{ in.}} = \dfrac{15 \text{ ft}}{x}$ and $\dfrac{12 \text{ in.}}{1 \text{ ft}} = \dfrac{x}{15 \text{ ft}}$ will have the (same/different) result.

Concept 1: Applications of Proportions

3. Pam drives her hybrid car 244 mi in city driving on 4 gal of gas. At this rate how many miles can she drive on 10 gal of gas? (See Example 1.)

4. Didi takes her pulse for 10 sec and counts 13 beats. How many beats per minute is this?

5. To cement a garden path, it takes crushed rock and cement in a ratio of 3.25 kg of rock to 1 kg of cement. If a 24-kg bag of cement is purchased, how much crushed rock will be needed? (See Example 2.)

6. Suppose two adults produce 63.4 lb of garbage in one week. At this rate, how many pounds will 50 adults produce in one week?

Fuse/Getty Images

7. On a map, the distance from Sacramento, California, to San Francisco, California, is 8 cm. The legend gives the actual distance at 91 mi. On the same map, Faythe measured 7 cm from Sacramento to Modesto, California. What is the actual distance? (Round to the nearest mile.) **(See Example 3.)**

8. On a map, the distance from Nashville, Tennessee, to Atlanta, Georgia, is 3.5 in., and the actual distance is 210 mi. If the map distance between Dallas, Texas, and Little Rock, Arkansas, is 4.75 in., what is the actual distance?

9. At Central Community College, the ratio of female students to male students is 31 to 19. If there are 6200 female students, how many male students are there?

10. Evelyn won an election by a ratio of 6 to 5. If she received 7230 votes, how many votes did her opponent receive?

11. If you flip a coin many times, the coin should come up heads about 1 time out of every 2 times it is flipped. If a coin is flipped 630 times, about how many heads do you expect to come up?

12. A die is a small cube used in games of chance. It has six sides, and each side has 1, 2, 3, 4, 5, or 6 dots painted on it. If you roll a die, the number 4 should come up about 1 time out of every 6 times the die is rolled. If you roll a die 366 times, about how many times do you expect the number 4 to come up?

Mark Dierker/McGraw Hill

13. A pitcher gave up 42 earned runs in 126 innings. Approximately how many earned runs will he give up in one game (9 innings)? This value is called the earned run average.

14. During a football game, a quarterback completed 34 passes for 357 yd. At this rate how many yards would be gained for 22 passes?

Don Farrall/Photodisc/Getty Images

15. Pierre bought 600 Euros with $750 American. At this rate, how many Euros can he buy with $900 American?

16. Erik bought $624 Canadian with $600 American. At this rate, how many Canadian dollars can he buy with $250 American?

17. Each gram of fat consumed has 9 calories. If a $\frac{1}{2}$-cup serving of gelato has 81 calories from fat, how many grams of fat are in this serving?

18. Approximately 24 out of 100 Americans over the age of 12 smoke. How many smokers would you expect in a group of 850 Americans over the age of 12?

19. A recipe from the National Audubon Society for making hummingbird nectar recommends using a ratio of one part table sugar to four parts water. How much water should be added to one-sixth cup of sugar to create nectar for a hummingbird feeder?

20. A new car depreciates by a total of $8000 over a 5-year period. At this rate, how much will it depreciate in $3\frac{1}{2}$ years?

21. Professor Fusco had 14 students in his online course and received 91 emails in 1 week. He spends about 3 min per email with his response.

    a. If he had a class of 30 students, how many emails would he expect to receive?

    b. With a class of 30 students, how much time would he expect to spend answering email?

22. A serving of 6 crackers has 225 mg of sodium.

    a. How much sodium would be in 14 crackers?

    b. The FDA suggests that the maximum sodium intake per day is 2300 mg. If Donna ate 20 crackers, how much sodium would be allowed that day from other foods?

23. A computer animation that is captured at 30 frames per second takes 2.45 megabytes of memory. If the animation is captured at 12 frames per second, how much memory would be used?

24. Liu earned $312 on an investment of $800. How much would $1100 have earned in the same investment?

25. A skyscraper in Chicago is 1454 ft high. If a model is made in which 1 in. represents 50 ft, how high would the building be in the model?

26. The Golden Gate Bridge is 8981 ft long. If a model is made in which 1 in. represents 100 ft, how long would the bridge be in the model?

27. A chemist mixes 3 parts acid to 5 parts water. How much acid would the chemist need if she used 420 mL of water?

28. Mitch mixes 5 parts white paint to 7 parts blue paint. If he has 4 qt of white paint, how much blue paint would he need?

29. Park officials stocked a man-made lake with bass last year. To approximate the number of bass this year, a sample of 75 bass is taken out of the lake and tagged. Then later a different sample is taken, and it is found that 21 of 100 bass are tagged. Approximately how many bass are in the lake? Round to the nearest whole unit. **(See Example 4.)**

30. Laws have been instituted in Florida to help save the manatee. To establish the number in Florida, a sample of 150 manatees was marked and let free. A new sample was taken and found that there were 3 marked manatees out of 40 captured. What is the approximate population of manatees in Florida?

James R.D. Scott/Moment/Getty Images

31. Yellowstone National Park in Wyoming has the largest population of free-roaming bison. To approximate the number of bison, 200 are captured and tagged and then let free to roam. Later, a sample of 120 bison is observed and 6 have tags. Approximate the population of bison in the park.

32. In Cass County, Michigan, there are about 20 white-tailed deer per square mile. If the county covers 492 mi^2, about how many white-tailed deer are in the county?

Concept 2: Similar Figures

For Exercises 33–38, the pairs of triangles are similar. Solve for x and y. **(See Examples 5 and 6.)**

33.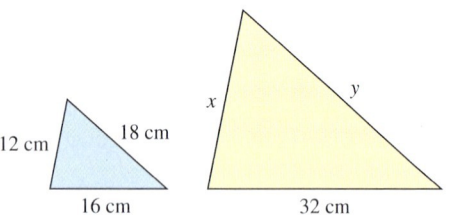

34.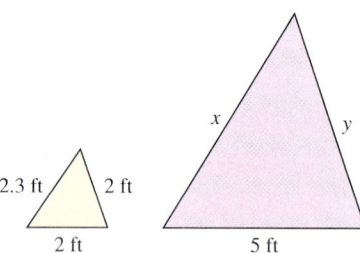

35.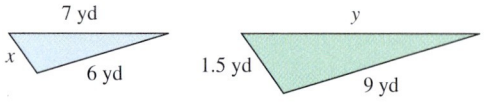

36.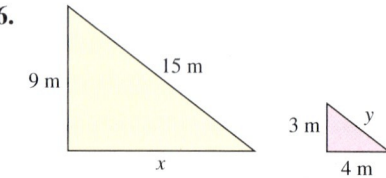

37.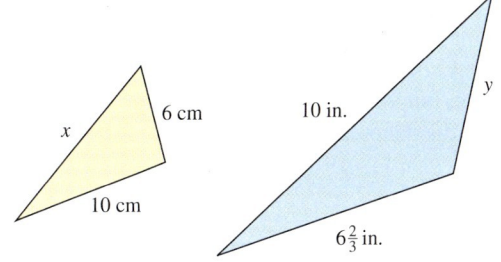

38.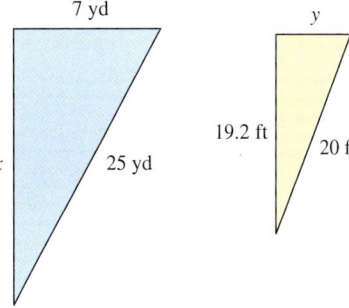

39. The height of a flagpole can be determined by comparing the shadow of the flagpole and the shadow of a yardstick. From the figure, determine the height of the flagpole. (See Example 7.)

40. A 15-ft flagpole casts a 4-ft shadow. How long will the shadow be for a 90-ft building?

41. A person 1.6 m tall stands next to a lifeguard observation platform. If the person casts a shadow of 1 m and the lifeguard platform casts a shadow of 1.5 m, how high is the platform?

42. A 32-ft tree casts a shadow of 18 ft. How long will the shadow be for a 22-ft tree?

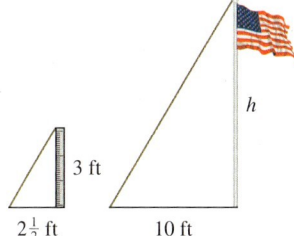

Figure for Exercise 39

For Exercises 43–46, the pairs of polygons are similar. Solve for the indicated variables. (See Example 8.)

43.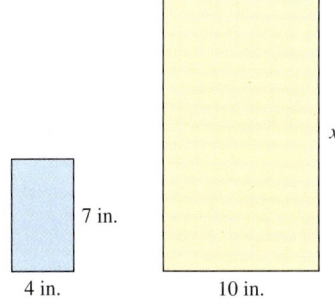

44.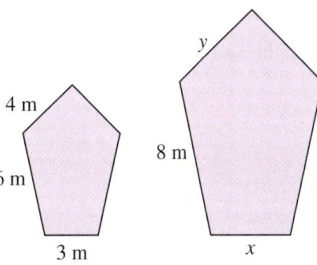

45.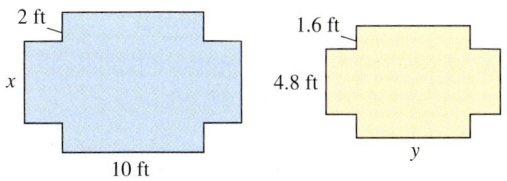

46.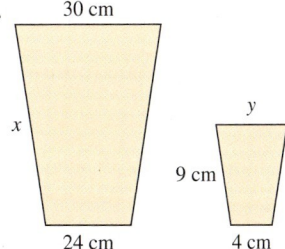

47. A carpenter makes a schematic drawing of a porch he plans to build. On the drawing, 2 in. represents 7 ft. Find the lengths denoted by x, y, and z.

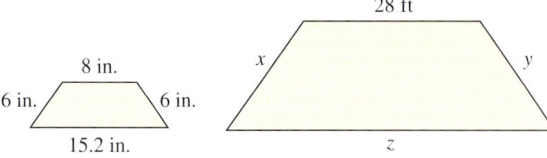

48. The Great Cookie Company has a sign on the front of its store as shown in the figure. The company would like to put a sign of the same shape in the back, but with dimensions $\frac{1}{3}$ as large. Find the lengths denoted by x and y.

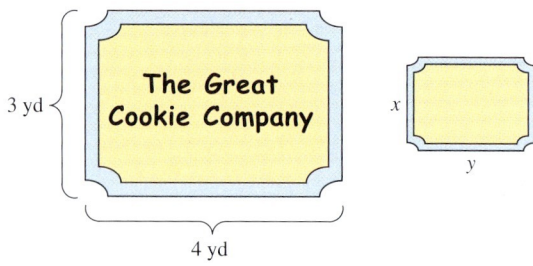

Technology Connections

49. In a recent year, the annual crime rate for Oklahoma was 4743 crimes per 100,000 people. If Oklahoma had approximately 3,500,000 people at that time, approximately how many crimes were committed? (*Source:* Oklahoma Department of Corrections)

50. To measure the height of the Washington Monument, a student 5.5 ft tall measures his shadow to be 3.25 ft. At the same time of day, he measured the shadow of the Washington Monument to be 328 ft long. Estimate the height of the monument to the nearest foot.

51. In a recent year, the rate of breast cancer in women was 110 cases per 100,000 women. At that time, the state of California had approximately 14,000,000 women. How many women in California would be expected to have breast cancer? (*Source:* Centers for Disease Control)

52. In a recent year, the rate of prostate disease in U.S. men was 118 cases per 1000 men. At that time, the state of Massachusetts had approximately 2,500,000 men. How many men in Massachusetts would be expected to have prostate disease? (*Source:* National Center for Health Statistics)

Chapter 6 Review Exercises

Section 6.1

For Exercises 1–3, write the ratios in two other ways.

1. $5:4$ **2.** 3 to 1   **3.** $\dfrac{8}{7}$

For Exercises 4–6, write the ratios in fraction form.

4. Saul had three daughters and two sons.

  a. Write a ratio of the number of sons to the number of daughters.

  b. Write a ratio of the number of daughters to the number of sons.

  c. Write a ratio of the number of daughters to the total number of children.

5. In his refrigerator, Jonathan has four bottles of soda and five bottles of juice.

  a. Write a ratio of the number of bottles of soda to the number of bottles of juice.

  b. Write a ratio of the number of bottles of juice to the number of bottles of soda.

  c. Write a ratio of the number of bottles of juice to the total number of bottles.

6. There are 12 face cards in a regular deck of 52 cards.

David A. Tietz/Editorial Image, LLC

a. Write a ratio of the number of face cards to the total number of cards.

b. Write a ratio of the number of face cards to the number of cards that are not face cards.

For Exercises 7–10, write the ratio in lowest terms.

7. 52 cards to 13 cards
8. $21 to $15
9. 80 ft to 200 ft
10. 7 days to 28 days

For Exercises 11–14, write the ratio in lowest terms with whole numbers in the numerator and denominator.

11. $1\frac{1}{2}$ hr to $\frac{1}{3}$ hr
12. $\frac{2}{3}$ yd to $2\frac{1}{6}$ yd
13. $2.56 to $1.92
14. 42.5 mi to 3.25 mi

15. This year a high school had an increase of 320 students. The enrollment last year was 1200 students.

a. How many students will be attending this year?

b. Write a ratio of the increase in the number of students to the total enrollment of students this year. Simplify to lowest terms.

16. A living room has dimensions of 3.8 m by 2.4 m. Find the ratio of length to width and reduce to lowest terms.

For Exercises 17 and 18, refer to the table that shows the number of personnel who smoke in a particular workplace.

	Smokers	Nonsmokers	Totals
Office personnel	12	20	32
Shop personnel	60	55	115

17. Find the ratio of the number of office personnel who smoke to the number of shop personnel who smoke.

18. Find the ratio of the total number of personnel who smoke to the total number of personnel.

Section 6.2

For Exercises 19–22, write each rate in lowest terms.

19. A concession stand sold 20 hot dogs in 45 min.

20. Mike can skate 4 mi in 34 min.

21. During a period in which the economy was weak, Evelyn's balance on her investment account changed by −$3400 in 6 months.

22. A submarine's "elevation" changed by −75 m in 45 min.

23. What is the difference between rates in lowest terms and unit rates?

For Exercises 24–27, write each rate as a unit rate.

24. A pheasant can fly 44 mi in $1\frac{1}{3}$ hr.

25. The temperature changed −14° in 3.5 hr.

26. A hummingbird can flap its wings 2700 times in 30 sec.

27. It takes David's lawn company 66 min to cut six lawns.

For Exercises 28 and 29, find the unit costs. Round the answers to three decimal places when necessary.

28. Body lotion costs $5.99 for 10 oz.

29. Three towels cost $20.00.

For Exercises 30 and 31, compute the unit cost (round to three decimal places). Then determine the best buy.

30. a. 32 oz of detergent for $8.39

b. 45 oz of detergent for $12.59

31. a. 24 oz of spaghetti sauce for $4.19

b. 44 oz of spaghetti sauce for $6.99

32. Suntan lotion costs $5.99 for 8 oz. If Jody has a coupon for $2.00 off, what will be the unit cost of the lotion after the coupon has been applied?

33. A 24-roll pack of bathroom tissue costs $8.99 without a discount card. The package is advertised at 29¢ per roll if the buyer uses the discount card. What is the difference in price per roll when the buyer uses the discount card? Round to the nearest cent.

34. In Wilmington, North Carolina, Hurricane Floyd dropped 15.06 in. of rain during a 24-hr period. What was the average rainfall per hour? (*Source:* National Weather Service)

35. For a recent year, a car manufacturer steadily increased the number of hybrid vehicles for sale in the United States from 130,000 to 250,000.

a. What was the increase in the number of hybrid vehicles?

b. How many additional hybrid vehicles will be available each month?

36. A nutrition student studied histotical records and found that the average per capita amount of vegetables that Americans consumed per year increased from 386 lb to 449 lb over an 18-year period.

John Foxx/Getty Images

a. What was the increase in the number of pounds of vegetables?

b. How many additional pounds of vegetables did Americans add to their diet per year?

Section 6.3

For Exercises 37–42, write a proportion for each statement.

37. 16 is to 14 as 12 is to $10\frac{1}{2}$.

38. 8 is to 20 as 6 is to 15.

39. The numbers -5 and 3 are proportional to the numbers -10 and 6.

40. The numbers 4 and -3 are proportional to the numbers 20 and -15.

41. $11 is to 1 hr as $88 is to 8 hr.

42. 2 in. is to 5 mi as 6 in. is to 15 mi.

For Exercises 43–46, determine whether the ratios form a proportion.

43. $\frac{64}{81} \stackrel{?}{=} \frac{8}{9}$

44. $\frac{3\frac{1}{2}}{7} \stackrel{?}{=} \frac{7}{14}$

45. $\frac{5.2}{3} \stackrel{?}{=} \frac{15.6}{9}$

46. $\frac{6}{10} \stackrel{?}{=} \frac{6.3}{10.3}$

For Exercises 47–50, determine whether the pairs of numbers are proportional.

47. Are the numbers $2\frac{1}{8}$ and $4\frac{3}{4}$ proportional to the numbers $3\frac{2}{5}$ and $7\frac{3}{5}$?

48. Are the numbers $5\frac{1}{2}$ and 6 proportional to the numbers $6\frac{1}{2}$ and 7?

49. Are the numbers -4.25 and -8 proportional to the numbers 5.25 and 10?

50. Are the numbers 12.4 and 9.2 proportional to the numbers -3.1 and -2.3?

For Exercises 51–56, solve the proportion.

51. $\frac{100}{16} = \frac{25}{x}$

52. $\frac{y}{6} = \frac{45}{10}$

53. $\frac{1\frac{6}{7}}{b} = \frac{13}{21}$

54. $\frac{p}{6\frac{1}{3}} = \frac{3}{9\frac{1}{2}}$

55. $\frac{2.5}{-6.8} = \frac{5}{h}$

56. $\frac{0.3}{1.2} = \frac{k}{-3.6}$

Section 6.4

57. One year of a dog's life is about the same as 7 years of a human life. If a dog is 12 years old in dog years, how does that equate to human years?

Jarvell Jardey/Alamy Stock Photo

58. Lavu bought 9500 Japanese yen with $100 American. At this rate, how many yen can he buy with $450 American?

59. The number of births in Alabama in a recent year was approximately 59,800. If the birthrate was about 13 per 1000, what was the approximate population of Alabama? (Round to the nearest person.)

60. If the tax on a $25.00 item is $1.20, what would be the tax on an item that costs $145.00?

61. The triangles shown in the figure are similar. Solve for x and y.

62. The height of a building can be approximated by comparing the shadows of the building and of a meterstick. From the figure, find the height of the building.

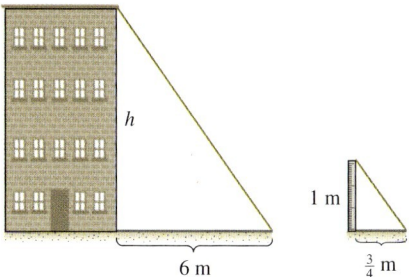

63. The polygons shown in the figure are similar. Solve for x and y.

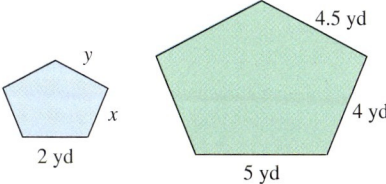

64. The figure shows two picture frames that are similar. Use this information to solve for x and y.

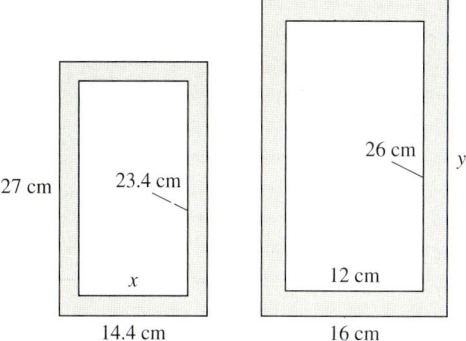

Percents

CHAPTER OUTLINE

7.1 Percents, Fractions, and Decimals   384
7.2 Percent Proportions and Applications   397
7.3 Percent Equations and Applications   407
   Problem Recognition Exercises: Percents   416
7.4 Applications of Sales Tax, Commission, Discount, Markup, and Percent Increase and Decrease   417
7.5 Simple and Compound Interest   431
   Chapter 7 Review Exercises   440

Numbers Per 100

In this chapter, we study **percent**. The word "percent" comes from the Latin phrase "per centum" and refers to a ratio of parts per 100. For example, if 8 percent (8%) of the population gets a winter cold, this means that 8 people out of 100 people on average will get a cold this winter.

Percent is an important concept because it gives us an intuitive sense for the portion of a whole unit. For example, suppose that a business considers advertising its new skateboard line during a television show called *Space Invasion*. Advertising on television is very expensive, so the business wants to make sure that the show's audience is part of the target market for skateboards (youth between 10 and 24 years old). Therefore, the marketing director conducts a survey of the viewing audience and finds that 612 viewers out of the 900 surveyed are between 10 and 24 years old. The number 612 by itself does not tell us much. But the ratio of 612 to the number of viewers sampled, 900, tells us more: the ratio 612/900 is just over 2/3. Extrapolating to the whole viewing audience, we conclude that roughly 2/3 of the viewing audience is in the targeted age range.

Galina Barskaya/Alamy Stock Photo

To further our intuitive understanding of this ratio, we can interpret the ratio as a portion taken out of 100.

$\dfrac{612}{900} = \dfrac{x}{100}$ Solving the proportion gives us $x = 68$.

Thus, 612/900 is the same ratio as 68/100, and we say that 68 percent (68%) of the viewers are between 10 and 24 years old. Thus, *Space Invasion* is a good venue for advertising new skateboards!

Section 7.1 Percents, Fractions, and Decimals

Concepts

1. Definition of Percent
2. Converting Percents to Fractions
3. Converting Percents to Decimals
4. Converting Fractions and Decimals to Percents
5. Percents, Fractions, and Decimals: A Summary

1. Definition of Percent

In this chapter we study the concept of percent. Literally, the word **percent** means *per one hundred*. To indicate percent, we use the percent symbol %. For example, 45% (read as "45 percent") of the land area in South America is rainforest (shaded in green). This means that if South America were divided into 100 squares of equal size, 45 of the 100 squares would cover rainforest. See Figures 7-1 and 7-2.

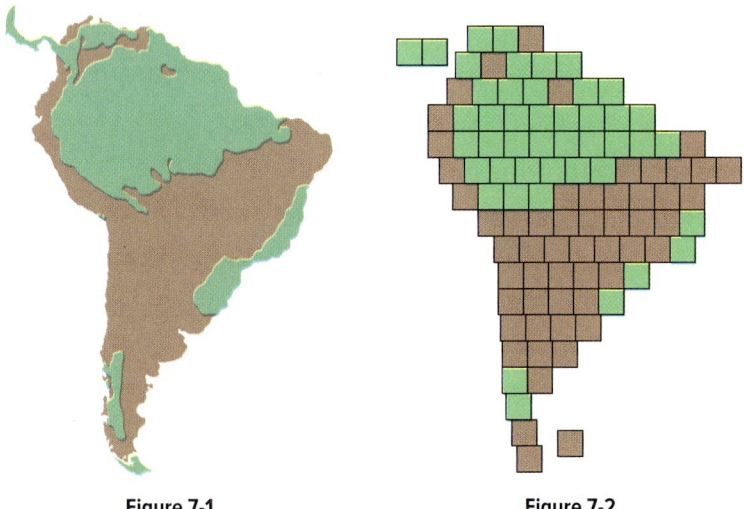

Figure 7-1   Figure 7-2

Consider another example. For a recent year, the population of Virginia could be described as follows.

21%	African American	21 out of 100 Virginians are African American
72%	Caucasian (non-Hispanic)	72 out of 100 Virginians are Caucasian (non-Hispanic)
3%	Asian American	3 out of 100 Virginians are Asian American
3%	Hispanic	3 out of 100 Virginians are Hispanic
1%	Other	1 out of 100 Virginians have other backgrounds

Figure 7-3 represents a sample of 100 residents of Virginia.

AA African American
C   Caucasian (non-Hispanic)
A   Asian American
H   Hispanic
O   Other

Figure 7-3

2. Converting Percents to Fractions

By definition, a percent represents a ratio of parts per 100. Therefore, we can write percents as fractions.

Percent		Fraction	Example/Interpretation
7%	=	$\dfrac{7}{100}$	A sales tax of 7% means that 7 cents in tax is charged for every 100 cents spent.
35%	=	$\dfrac{35}{100}$	To say that 35% of households own a cat means that 35 per every 100 households own a cat.

Notice that $35\% = \dfrac{35}{100} = 35 \times \dfrac{1}{100} = 35 \div 100$.

From this discussion we have the following rule for converting percents to fractions.

> **Converting Percents to Fractions**
> **Step 1** Replace the symbol % by $\times \dfrac{1}{100}$ (or by $\div\, 100$).
> **Step 2** Simplify the fraction to lowest terms, if possible.

Example 1 Converting Percents to Fractions

Convert each percent to a fraction.

a. 56%   b. 125%   c. 0.4%

Solution:

a. $56\% = 56 \times \dfrac{1}{100}$ Replace the % symbol by $\times \dfrac{1}{100}$.

 $= \dfrac{56}{100}$ Multiply.

 $= \dfrac{14}{25}$ Simplify to lowest terms.

b. $125\% = 125 \times \dfrac{1}{100}$ Replace the % symbol by $\times \dfrac{1}{100}$.

 $= \dfrac{125}{100}$ Multiply.

 $= \dfrac{5}{4}$ or $1\dfrac{1}{4}$ Simplify to lowest terms.

c. $0.4\% = 0.4 \times \dfrac{1}{100}$ Replace the % symbol by $\times \dfrac{1}{100}$.

 $= \dfrac{4}{10} \times \dfrac{1}{100}$ Write 0.4 in fraction form.

 $= \dfrac{4}{1000}$ Multiply.

 $= \dfrac{1}{250}$ Simplify to lowest terms.

Skill Practice Convert each percent to a fraction.
1. 55%   2. 175%   3. 0.06%

Answers
1. $\dfrac{11}{20}$ 2. $\dfrac{7}{4}$ 3. $\dfrac{3}{5000}$

Note that $100\% = 100 \times \frac{1}{100} = 1$. That is, 100% represents 1 whole unit. In Example 1(b), $125\% = \frac{5}{4}$ or $1\frac{1}{4}$. This illustrates that any percent greater than 100% represents a quantity greater than 1 whole. Therefore, its fractional form may be expressed as an improper fraction or as a mixed number.

Note that $1\% = 1 \times \frac{1}{100} = \frac{1}{100}$. In Example 1(c), the value 0.4% represents a quantity less than 1%. Its fractional form is less than one-hundredth.

3. Converting Percents to Decimals

To express part of a whole unit, we can use a percent, a fraction, or a decimal. We would like to be able to convert from one form to another. The procedure for converting a percent to a decimal is the same as that for converting a percent to a fraction. We replace the % symbol by $\times \frac{1}{100}$. However, when converting to a decimal, it is usually more convenient to use the form $\times 0.01$.

> **Converting Percents to Decimals**
> Replace the % symbol by $\times 0.01$. (This is equivalent to $\times \frac{1}{100}$ and $\div 100$.)
> *Note:* Multiplying a decimal by 0.01 (or dividing by 100) is the same as moving the decimal point 2 places to the left.

Example 2   Converting Percents to Decimals

Convert each percent to its decimal form.

a. 31%   **b.** 6.5%   **c.** 428%   **d.** $1\frac{3}{5}\%$ **e.** 0.05%

Solution:

a. $31\% = 31 \times 0.01$ Replace the % symbol by $\times 0.01$.

$ = 0.31$ Move the decimal point 2 places to the left.

b. $6.5\% = 6.5 \times 0.01$ Replace the % symbol by $\times 0.01$.

$ = 0.065$ Move the decimal point 2 places to the left.

c. $428\% = 428 \times 0.01$ Because 428% is greater than 100% we expect the decimal form to be a number greater than 1.

$ = 4.28$

d. $1\frac{3}{5}\% = 1.6 \times 0.01$ Convert the mixed number to decimal form.

$\phantom{1\frac{3}{5}\%} = 0.016$ Because the percent is just over 1%, we expect the decimal form to be just slightly greater than 0.01.

e. $0.05\% = 0.05 \times 0.01$ The value 0.05% is less than 1%. We expect the decimal form to be less than 0.01.

$ = 0.0005$

> **TIP:** In Example 2(d), convert $\frac{3}{5}$ to decimal form by dividing the numerator by the denominator.
> $\frac{3}{5} = 0.6$, therefore, $1\frac{3}{5} = 1.6$.

Skill Practice Convert each percent to its decimal form.

4. 67%   **5.** 8.6%   **6.** 321%

7. $6\frac{1}{4}\%$ **8.** 0.7%

Answers
4. 0.67   **5.** 0.086
6. 3.21   **7.** 0.0625
8. 0.007

4. Converting Fractions and Decimals to Percents

To convert a percent to a decimal or fraction, we replace the % symbol by $\times\,0.01$ (or $\times\,\frac{1}{100}$). We will now reverse this process. To convert a decimal or fraction to a percent, multiply by 100 and apply the % symbol.

> **Converting Fractions and Decimals to Percent Form**
> Multiply the fraction or decimal by 100%.
> *Note:* Multiplying a decimal by 100 moves the decimal point 2 places to the right.

Example 3   Converting Decimals to Percents

Convert each decimal to its equivalent percent form.

  a. 0.62   **b.** 1.75   **c.** 1   **d.** 0.004   **e.** 8.9

Solution:

a. $0.62 = 0.62 \times 100\%$ Multiply by 100%.
         $= 62\%$ Multiplying by 100 moves the decimal point 2 places to the right.

b. $1.75 = 1.75 \times 100\%$ Multiply by 100%.
         $= 175\%$ The decimal number 1.75 is greater than 1. Therefore, we expect a percent greater than 100%.

c. $1 = 1 \times 100\%$ Multiply by 100%.
     $= 100\%$ Recall that 1 whole is equal to 100%.

d. $0.004 = 0.004 \times 100\%$ Multiply by 100%.
           $= 00.4\%$ Move the decimal point to the right 2 places.

e. $8.9 = 8.90 \times 100\%$ Multiply by 100%.
       $= 890\%$

Skill Practice Convert each decimal to its percent form.

  9. 0.46   **10.** 3.25   **11.** 2
12. 0.0006   **13.** 2.5

> **TIP:** Multiplying a number by 100% is equivalent to multiplying the number by 1. Thus, the value of the number is not changed.

Answers
9. 46%   **10.** 325%   **11.** 200%
12. 0.06%   **13.** 250%

Example 4 Converting a Fraction to Percent Notation

Convert the fraction to percent notation. $\dfrac{3}{5}$

Solution:

$$\dfrac{3}{5} = \dfrac{3}{5} \times 100\% \qquad \text{Multiply by 100\%.}$$

$$= \dfrac{3}{5} \times \dfrac{100}{1}\% \qquad \text{Convert the whole number to an improper fraction.}$$

$$= \dfrac{3}{\cancel{5}} \times \dfrac{\cancel{100}^{20}}{1}\% \qquad \text{Multiply fractions and simplify to lowest terms.}$$

$$= 60\%$$

TIP: We could also have converted $\dfrac{3}{5}$ to decimal form first (by dividing the numerator by the denominator) and then converted the decimal to a percent.

$$\dfrac{3}{5} = 0.60 \quad = \quad 0.60 \times 100\% = 60\%$$

Skill Practice Convert the fraction to percent notation.

14. $\dfrac{7}{10}$

Example 5 Converting a Fraction to Percent Notation

Convert the fraction to percent notation. $\dfrac{2}{3}$

Solution:

$$\dfrac{2}{3} = \dfrac{2}{3} \times 100\% \qquad \text{Multiply by 100\%.}$$

$$= \dfrac{2}{3} \times \dfrac{100}{1}\% \qquad \text{Convert the whole number to an improper fraction.}$$

$$= \dfrac{200}{3}\%$$

The number $\dfrac{200}{3}\%$ can be written as $66\dfrac{2}{3}\%$ or as $66.\overline{6}\%$.

TIP: First converting $\dfrac{2}{3}$ to a decimal before converting to percent notation is an alternative approach.

$$\dfrac{2}{3} = 0.\overline{6} \quad = \quad 0.666\ldots \times 100\% = 66.\overline{6}\%$$

Skill Practice Convert the fraction to percent notation.

15. $\dfrac{1}{9}$

Answers

14. 70%
15. $\dfrac{100}{9}\%$ or $11\dfrac{1}{9}\%$ or $11.\overline{1}\%$

In Example 6, we convert an improper fraction and a mixed number to percent form.

Section 7.1 Percents, Fractions, and Decimals

Example 6 — Converting Improper Fractions and Mixed Numbers to Percents

Convert to percent notation.

a. $2\frac{1}{4}$ **b.** $\frac{13}{10}$

Solution:

a. $2\frac{1}{4} = 2\frac{1}{4} \times 100\%$ Multiply by 100%.

$= \frac{9}{4} \times \frac{100}{1}\%$ Convert to improper fractions.

$= \frac{9}{\cancel{4}_{1}} \times \frac{\cancel{100}^{25}}{1}\%$ Multiply and simplify to lowest terms.

$= 225\%$

b. $\frac{13}{10} = \frac{13}{10} \times 100\%$ Multiply by 100%.

$= \frac{13}{10} \times \frac{100}{1}\%$ Convert the whole number to an improper fraction.

$= \frac{13}{\cancel{10}_{1}} \times \frac{\cancel{100}^{10}}{1}\%$ Multiply and simplify to lowest terms.

$= 130\%$

Skill Practice Convert to percent notation.

16. $1\frac{7}{10}$ **17.** $\frac{11}{4}$

Notice that both answers in Example 6 are greater than 100%. This is reasonable because any number greater than 1 whole unit represents a percent greater than 100%.

In Example 7, we approximate a percent from its fraction form.

Example 7 — Approximating a Percent by Rounding

Convert the fraction $\frac{5}{13}$ to percent notation rounded to the nearest tenth of a percent.

Solution:

$\frac{5}{13} = \frac{5}{13} \times 100\%$ Multiply by 100%.

$= \frac{5}{13} \times \frac{100}{1}\%$ Write the whole number as an improper fraction.

$= \frac{500}{13}\%$

To round to the nearest tenth of a percent, we must divide. We will obtain the hundredths-place digit in the quotient on which to base the decision on rounding.

$38.4\overset{1}{6} \approx 38.5$

Thus, $\frac{5}{13} \approx 38.5\%$

```
     38.46
13)500.00
   -39
    110
   -104
     60
    -52
     80
    -78
      2
```

Avoiding Mistakes

In Example 7, we converted a fraction to a percent where rounding was necessary. We converted the fraction to percent form *before* dividing and rounding. If you try to convert to decimal form first, you might round too soon.

Skill Practice

18. Write the fraction $\frac{3}{7}$ in percent notation to the nearest tenth of a percent.

Answers

16. 170%   **17.** 275%   **18.** 42.9%

5. Percents, Fractions, and Decimals: A Summary

The diagram in Figure 7-4 summarizes the methods for converting fractions, decimals, and percents.

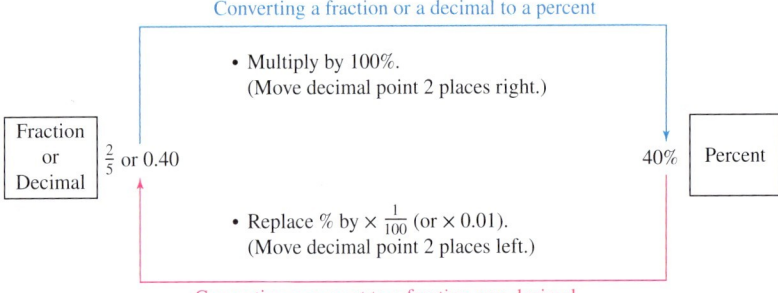

Converting a fraction or a decimal to a percent

- Multiply by 100%.
  (Move decimal point 2 places right.)

Fraction or Decimal $\frac{2}{5}$ or 0.40 → 40% Percent

- Replace % by $\times \frac{1}{100}$ (or $\times 0.01$).
  (Move decimal point 2 places left.)

Converting a percent to a fraction or a decimal

Figure 7-4

Table 7-1 shows some common percents and their equivalent fraction and decimal forms.

Table 7-1

Percent	Fraction	Decimal	Example/Interpretation
100%	1	1.00	Of people who give birth, 100% are female.
50%	$\frac{1}{2}$	0.50	Of the population, 50% is male. That is, one-half of the population is male.
25%	$\frac{1}{4}$	0.25	Approximately 25% of the U.S. population smokes. That is, one-quarter of the population smokes.
75%	$\frac{3}{4}$	0.75	Approximately 75% of homes have computers. That is, three-quarters of homes have computers.
10%	$\frac{1}{10}$	0.10	Of the population, 10% is left-handed. That is, one-tenth of the population is left-handed.
20%	$\frac{1}{5}$	0.20	In a recent year, approximately 20% of the population of the United States was under age 15. That is, one-fifth of the population was under age 15.
1%	$\frac{1}{100}$	0.01	Approximately 1% of babies are born underweight. That is, about 1 in 100 babies is born underweight.
$33\frac{1}{3}\%$	$\frac{1}{3}$	$0.\overline{3}$	A basketball player made $33\frac{1}{3}\%$ of her shots. That is, she made about 1 basket for every 3 shots attempted.
$66\frac{2}{3}\%$	$\frac{2}{3}$	$0.\overline{6}$	Of the population, $66\frac{2}{3}\%$ prefers chocolate ice cream to other flavors. That is, 2 out of 3 people prefer chocolate ice cream.

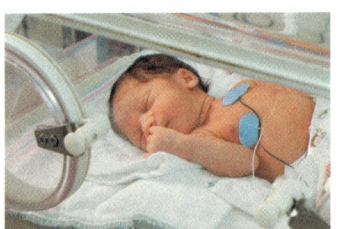

Larry Mulvehill/Corbis/age fotostock

C Squared Studios/Photodisc/Getty Images

Example 8   Converting Fractions, Decimals, and Percents

Complete the table.

	Fraction	Decimal	Percent
a.		0.55	
b.	$\frac{1}{200}$		
c.			160%
d.		2.4	
e.			$66\frac{2}{3}\%$
f.	$\frac{2}{9}$		

Solution:

a. 0.55 to fraction: $0.55 = \dfrac{55}{100} = \dfrac{11}{20}$

0.55 to percent: $0.55 \times 100\% = 55\%$

b. $\dfrac{1}{200}$ to decimal: $1 \div 200 = 0.005$

$\dfrac{1}{200}$ to percent: $\dfrac{1}{200} \times 100\% = \dfrac{100}{200}\% = 0.5\%$

c. 160% to fraction: $160 \times \dfrac{1}{100} = \dfrac{160}{100} = \dfrac{8}{5}$ or $1\dfrac{3}{5}$

160% to decimal: $160 \times 0.01 = 1.6$

d. 2.4 to fraction: $\dfrac{24}{10} = \dfrac{12}{5}$ or $2\dfrac{2}{5}$

2.4 to percent: $2.4 \times 100\% = 240\%$

e. $66\dfrac{2}{3}\%$ to fraction: $66\dfrac{2}{3} \times \dfrac{1}{100} = \dfrac{200}{3} \times \dfrac{1}{100} = \dfrac{2}{3}$

$66\dfrac{2}{3}\%$ to decimal: $66\dfrac{2}{3} \times 0.01 = 66.\overline{6} \times 0.01 = 0.\overline{6}$

f. $\dfrac{2}{9}$ to decimal: $2 \div 9 = 0.\overline{2}$

$\dfrac{2}{9}$ to percent: $\dfrac{2}{9} \times 100\% = \dfrac{2}{9} \times \dfrac{100}{1}\% = \dfrac{200}{9}\% = 22\dfrac{2}{9}\%$ or $22.\overline{2}\%$

The completed table is as follows.

	Fraction	Decimal	Percent
a.	$\dfrac{11}{20}$	0.55	55%
b.	$\dfrac{1}{200}$	0.005	0.5%
c.	$\dfrac{8}{5}$ or $1\dfrac{3}{5}$	1.6	160%
d.	$\dfrac{12}{5}$ or $2\dfrac{2}{5}$	2.4	240%
e.	$\dfrac{2}{3}$	$0.\overline{6}$	$66\dfrac{2}{3}\%$
f.	$\dfrac{2}{9}$	$0.\overline{2}$	$22\dfrac{2}{9}\%$ or $22.\overline{2}\%$

Skill Practice Complete the table.

	Fraction	Decimal	Percent
19.		1.41	
20.	$\dfrac{1}{50}$		
21.			18%
22.		0.58	
23.			$33\dfrac{1}{3}\%$
24.	$\dfrac{7}{9}$		

Answers

	Fraction	Decimal	Percent
19.	$\dfrac{141}{100}$ or $1\dfrac{41}{100}$	1.41	141%
20.	$\dfrac{1}{50}$	0.02	2%
21.	$\dfrac{9}{50}$	0.18	18%
22.	$\dfrac{29}{50}$	0.58	58%
23.	$\dfrac{1}{3}$	$0.\overline{3}$	$33\dfrac{1}{3}\%$
24.	$\dfrac{7}{9}$	$0.\overline{7}$	$77.\overline{7}\%$

Section 7.1 Activity

A.1.
  a. What mathematical operation is implied by the word *per*?
  b. How many cents are there in one dollar?
  c. Define the word *percent* in your own words.

A.2. The figure below illustrates the number of freshman (F), sophomores (S), juniors (J), and seniors (Sr) out of 100 total students enrolled in a college psychology class. Determine the number, fraction, and percent of students in each grade level. Do not simplify the fraction to lowest terms.

	Number of Students	Fraction of Students	Percent of Students
Freshman			
Sophomores			
Juniors			
Seniors			

A.3. Your two favorite retailers advertise that everything in the store is on sale! Retailer A is selling everything 40% off the original price. Retailer B is selling everything at 0.75% off the original price. Which sale is worth shopping and why?

A.4. Match each percent with the equivalent fraction or whole number.

  a. 25%   A. 1
  b. 90%   B. $\frac{2}{3}$
  c. 1%   C. $\frac{1}{8}$
  d. 100%   D. $\frac{9}{10}$
  e. $66\frac{2}{3}\%$ E. $1\frac{7}{25}$
  f. 0.8%   F. $\frac{1}{4}$
  g. 128%   G. $\frac{1}{100}$
  h. 12.5%   H. $\frac{1}{125}$

A.5. Consider the percent to decimal conversions:

Percent →	Decimal
60%	0.60
6%	0.06
0.6%	0.006

  a. Based on the examples above, what number was the percent divided by to obtain the decimal equivalent?
  b. How does the placement of the decimal point change when converting a percent to a decimal?

A.6. Consider the decimal to percent conversions:

Decimal →	Percent
2.25	225%
0.25	25%
0.025	2.5%

a. Based on the examples above, what number was the decimal multiplied by to obtain the percent equivalent?
b. How does the placement of the decimal point change when converting a decimal to percent?

A.7. Convert $\frac{4}{5}$ to percent notation by:
a. Write an expression showing the fraction multiplied by 100% in the form of $\frac{100}{1}$%.
b. Multiply the fractions and simplify to lowest terms.

A.8. Convert $\frac{4}{5}$ to percent notation by:
a. Converting the fraction to decimal form by dividing the numerator by the denominator.
b. Multiply the decimal by 100%.
c. Is the answer from Exercise A.7(b) the same as the answer from Exercise A.8(b)?
d. What approach for converting a fraction to percent notation do you prefer?

A.9. a. Convert $\frac{7}{13}$ to a decimal. Does the decimal terminate?
b. Convert the decimal obtained in part (a) to percent notation. Round the percent to the nearest tenth of a percent.
c. Round the decimal obtained in part (a) to the nearest tenth. Convert it to percent notation.
d. Are the answers obtained in parts (b) and (c) the same? Why or why not?
e. Convert $\frac{2}{9}$ to percent notation and round to the nearest tenth of a percent.

A.10. Complete the table.

Fraction	Decimal	Percent
$\frac{3}{8}$		
	0.36	
		0.4%
	1.5	
$\frac{1}{3}$		
		200%

For Exercises A.11–A.16, support your answer to each question.

A.11. Is the number 1.40 less than or greater than 100%?

A.12. Does the fraction $\frac{7}{11}$ represent more or less than 50%?

A.13. Is the number 0.02 less than or greater than 1%?

A.14. Does the fraction $\frac{19}{16}$ represent more or less than 100%?

A.15. Is the number 600% more or less than 0.6?

A.16. Is the number 0.057 less than or greater than 50%

Section 7.1 Practice Exercises

Prerequisite Review

For Exercises R.1–R.2, change to a fraction.

R.1. 0.024   **R.2.** 0.15

For Exercises R.3–R.6, change to a decimal.

R.3. $\dfrac{12}{25}$ **R.4.** $\dfrac{23}{8}$ **R.5.** $\dfrac{17}{12}$ **R.6.** $\dfrac{5}{22}$

Vocabulary and Key Concepts

1. The word _____ means per hundred.

Concept 1: Definition of Percent

For Exercises 2–6, use a percent to express the shaded portion of each drawing.

2.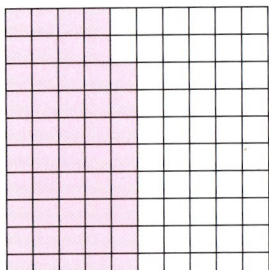

3.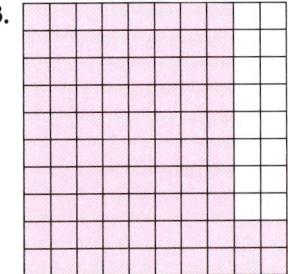

4.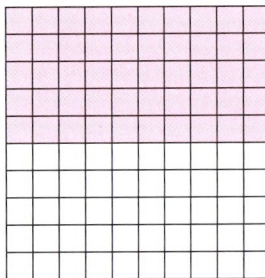

5.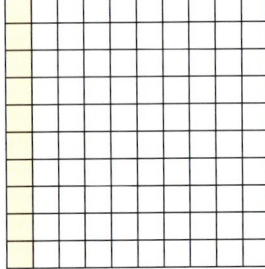

6.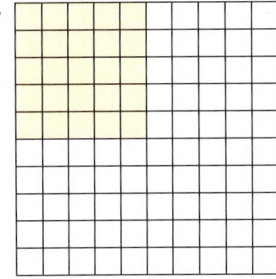

For Exercises 7–10, write a percent for each statement.

7. A bank pays $2 in interest for every $100 deposited.

8. In South Dakota, 5 out of every 100 people work in construction.

9. Out of 100 acres, 70 acres were planted with corn.

10. On TV, 26 out of every 100 minutes are filled with commercials.

Concept 2: Converting Percents to Fractions

11. Explain the procedure to change a percent to a fraction.

12. What fraction represents 50%?

For Exercises 13–28, change the percent to a simplified fraction or mixed number. (See Example 1.)

13. 3%
14. 7%
15. 84%
16. 32%
17. 3.4%
18. 5.2%
19. 115%
20. 150%
21. 0.5%
22. 0.2%
23. 0.25%
24. 0.75%
25. $5\frac{1}{6}\%$
26. $66\frac{2}{3}\%$
27. $124\frac{1}{2}\%$
28. $110\frac{1}{2}\%$

Concept 3: Converting Percents to Decimals

29. Explain the procedure to change a percent to a decimal.

30. To change 26.8% to decimal form, which direction would you move the decimal point, and by how many places?

For Exercises 31–42, change the percent to a decimal. (See Example 2.)

31. 58%
32. 72%
33. 8.5%
34. 12.9%
35. 142%
36. 201%
37. 0.55%
38. 0.75%
39. $26\frac{2}{5}\%$
40. $16\frac{1}{4}\%$
41. $55\frac{1}{20}\%$
42. $62\frac{1}{5}\%$

Concept 4: Converting Fractions and Decimals to Percents

For Exercises 43–54, convert the decimal to a percent. (See Example 3.)

43. 0.27
44. 0.51
45. 0.19
46. 0.33
47. 1.75
48. 2.8
49. 0.124
50. 0.277
51. 0.006
52. 0.0008
53. 1.014
54. 2.203

For Exercises 55–66, convert the fraction to a percent. (See Examples 4–6.)

55. $\frac{71}{100}$
56. $\frac{89}{100}$
57. $\frac{7}{8}$
58. $\frac{5}{8}$
59. $\frac{5}{6}$
60. $\frac{5}{12}$
61. $1\frac{3}{4}$
62. $2\frac{1}{8}$
63. $\frac{11}{9}$
64. $\frac{14}{9}$
65. $1\frac{2}{3}$
66. $1\frac{1}{6}$

For Exercises 67–74, write the fraction in percent notation to the nearest tenth of a percent. (See Example 7.)

67. $\frac{3}{7}$
68. $\frac{6}{7}$
69. $\frac{1}{13}$
70. $\frac{3}{13}$
71. $\frac{5}{11}$
72. $\frac{8}{11}$
73. $\frac{13}{15}$
74. $\frac{1}{15}$

Concept 5: Percents, Fractions, and Decimals: A Summary (Mixed Exercises)

For Exercises 75–80, match the percent with its fraction form.

75. $66\frac{2}{3}\%$
76. 10%
77. 90%
78. 75%
79. 25%
80. 150%

a. $\frac{3}{2}$
b. $\frac{3}{4}$
c. $\frac{2}{3}$
d. $\frac{1}{10}$
e. $\frac{9}{10}$
f. $\frac{1}{4}$

For Exercises 81–86, match the decimal with its percent form.

81. 0.30
82. $0.\overline{3}$
83. 5
84. 0.5
85. 0.05
86. 0.8

a. 5%
b. 50%
c. 80%
d. $33\frac{1}{3}\%$
e. 30%
f. 500%

For Exercises 87–90, complete the table. (See Example 8.)

87.

	Fraction	Decimal	Percent
a.	$\frac{1}{4}$		
b.		0.92	
c.			15%
d.		1.6	
e.	$\frac{1}{100}$		
f.			0.8%

88.

	Fraction	Decimal	Percent
a.			0.6%
b.	$\frac{2}{5}$		
c.		2	
d.	$\frac{1}{2}$		
e.		0.12	
f.			45%

89.

	Fraction	Decimal	Percent
a.			14%
b.		0.87	
c.		1	
d.	$\frac{1}{3}$		
e.			0.2%
f.	$\frac{19}{20}$		

90.

	Fraction	Decimal	Percent
a.		1.3	
b.			22%
c.	$\frac{3}{4}$		
d.		0.73	
e.			$22.\overline{2}\%$
f.	$\frac{1}{20}$		

For Exercises 91–94, write the fraction as a percent.

91. One-quarter of Americans say they entertain at home two or more times a month. (*Source: USA TODAY*)

92. According to the Centers for Disease Control (CDC), $\frac{37}{100}$ of U.S. teenage boys say they rarely or never wear their seat belts.

93. According to the Centers for Disease Control, $\frac{1}{10}$ of teenage girls in the United States say they rarely or never wear their seat belts.

94. In a recent year, $\frac{2}{3}$ of the beds in U.S. hospitals were occupied.

Buccina Studios/Photodisc/Getty Images

For Exercises 95–98, find the decimal and fraction equivalents of the percent given in the sentence.

95. For a recent year, the unemployment rate in the United States was 9.6%.

96. During a slow travel season, one airline cut its seating capacity by 15.8%.

97. During a period of a weak economy and rising fuel prices, a low-fare airline raised its average fare during the second quarter by 8.4%. (*Source: USA TODAY*)

98. One year, the average U.S. income tax rate was 18.2%.

99. Explain the difference between $\frac{1}{2}$ and $\frac{1}{2}\%$.

100. Explain the difference between $\frac{3}{4}$ and $\frac{3}{4}\%$.

101. Explain the difference between 25% and 0.25%.

102. Explain the difference between 10% and 0.10%.

103. Which of the numbers represent 125%?
   a. 1.25
   b. 0.125
   c. $\frac{5}{4}$
   d. $\frac{5}{4}\%$

104. Which of the numbers represent 60%?
   a. 6.0
   b. 0.60%
   c. 0.6
   d. $\frac{3}{5}$

105. Which of the numbers represent 30%?
   a. $\frac{3}{10}$
   b. $\frac{1}{3}$
   c. 0.3
   d. 0.03%

106. Which of the numbers represent 180%?
   a. 18
   b. 1.8
   c. $\frac{9}{5}$
   d. $\frac{9}{5}\%$

Expanding Your Skills

107. Is the number 1.4 less than or greater than 100%?

108. Is the number 0.0087 less than or greater than 1%?

109. Is the number 0.052 less than or greater than 50%?

110. Is the number 25 less than or greater than 25%?

Section 7.2 Percent Proportions and Applications

1. Introduction to Percent Proportions

Recall that a percent is a ratio in parts per 100. For example, $50\% = \frac{50}{100}$. However, a percent can be represented by infinitely many equivalent fractions. Thus,

$$50\% = \frac{50}{100} = \frac{1}{2} = \frac{2}{4} = \frac{3}{6} \quad \text{and infinitely many more.}$$

Equating a percent to an equivalent ratio forms a proportion that we call a **percent proportion**. A percent proportion is a proportion in which one ratio is written with a denominator of 100. For example:

$$\frac{50}{100} = \frac{3}{6} \quad \text{is a percent proportion.}$$

Concepts

1. Introduction to Percent Proportions
2. Solving Percent Proportions
3. Applications of Percent Proportions

We will be using percent proportions to solve a variety of application problems. But first we need to identify and label the parts of a percent proportion.

A percent proportion can be written in the form:

$$\frac{\text{Amount}}{\text{Base}} = p\% \quad \text{or} \quad \frac{\text{Amount}}{\text{Base}} = \frac{p}{100}$$

For example:

4 L out of 8 L is 50% $\quad \frac{4}{8} = 50\% \quad$ or $\quad \frac{4}{8} = \frac{50}{100}$

amount   base   p

In this example, 8 L is some total (or base) quantity and 4 L is some part (or amount) of that whole. The ratio $\frac{4}{8}$ represents a fraction of the whole equal to 50%. In general, we offer the following guidelines for identifying the parts of a percent proportion.

Identifying the Parts of a Percent Proportion

A percent proportion can be written as

$$\frac{\text{Amount}}{\text{Base}} = p\% \quad \text{or} \quad \frac{\text{Amount}}{\text{Base}} = \frac{p}{100}$$

- The **base** is the total or whole amount being considered. It often appears after the word *of* within a word problem.
- The **amount** is the part being compared to the base. It sometimes appears with the word *is* within a word problem.

Example 1 Identifying Amount, Base, and *p* for a Percent Proportion

Identify the amount, base, and *p* value, and then set up a percent proportion.

a. 25% of 60 students is 15 students.
b. $32 is 50% of $64.
c. 5 of 1000 employees is 0.5%.

Solution:

For each problem, we recommend that you identify *p* first. It is the number in front of the symbol %. Then identify the base. In most cases it follows the word *of*. Then, by the process of elimination, find the amount.

a. 25% of 60 students is 15 students.

  p (before % symbol)   base (after the word *of*)   amount

  amount → $\frac{15}{60} = \frac{25}{100}$ ← *p*
  base →   ←100

b. $32 is 50% of $64.

  amount   *p*   base

  amount → $\frac{32}{64} = \frac{50}{100}$ ← *p*
  base →   ←100

c. 5 of 1000 employees is 0.5%

  amount   base   *p*

  amount → $\frac{5}{1000} = \frac{0.5}{100}$ ← *p*
  base →   ←100

Skill Practice Identify the amount, base, and *p* value. Then set up a percent proportion.

1. 12 mi is 20% of 60 mi.
2. 14% of $600 is $84.
3. 32 books out of 2000 books is 1.6%.

Answers

1. Amount = 12; base = 60; $p = 20$; $\frac{12}{60} = \frac{20}{100}$
2. Amount = 84; base = 600; $p = 14$; $\frac{84}{600} = \frac{14}{100}$
3. Amount = 32; base = 2000; $p = 1.6$; $\frac{32}{2000} = \frac{1.6}{100}$

2. Solving Percent Proportions

In Example 1, we practiced identifying the parts of a percent proportion. Now we consider percent proportions in which one of these numbers is unknown. Furthermore, we will see that the examples come in three types:
- Amount is unknown.
- Base is unknown.
- Value p is unknown.

However, the process for solving in each case is the same.

Example 2 Solving Percent Proportions—Amount Unknown

What is 30% of 180?

Solution:

What is 30% of 180?
amount (x) p base

The base and value for p are known.

Let x represent the unknown amount.

$$\frac{x}{180} = \frac{30}{100}$$

Set up a percent proportion.

$$100x = (30)(180)$$

Equate the cross products.

$$100x = 5400$$

$$\frac{100x}{100} = \frac{5400}{100}$$

Divide both sides of the equation by 100.

$$x = 54$$

Simplify to lowest terms.

Therefore, 54 is 30% of 180.

TIP: We can check the answer to Example 2 as follows. Ten percent of a number is $\frac{1}{10}$ of the number. Furthermore, $\frac{1}{10}$ of 180 is 18. Thirty percent of 180 must be 3 times this amount.

$3 \times 18 = 54$ ✓

Skill Practice

4. What is 82% of 250?

Example 3 Solving Percent Proportions—Base Unknown

40% of what number is 25?

Solution:

40% of what number is 25?
p base (x)   amount

The amount and value of p are known.

Let x represent the unknown base.

$$\frac{25}{x} = \frac{40}{100}$$

Set up a percent proportion.

$$(25)(100) = 40x$$

Equate the cross products.

$$2500 = 40x$$

$$\frac{2500}{40} = \frac{40x}{40}$$

Divide both sides 40.

$$62.5 = x$$

Therefore, 40% of 62.5 is 25.

Skill Practice

5. 21% of what number is 42?

Answers

4. 205   5. 200

Example 4 — Solving Percent Proportions—p Unknown

12.4 mi is what percent of 80 mi?

Solution:

12.4 mi is what percent of 80 mi?
- amount: 12.4
- p
- base: 80

The amount and base are known.
The value of p is unknown.

$$\frac{12.4}{80} = \frac{p}{100}$$ Set up a percent equation.

$(12.4)(100) = 80p$ Equate the cross products.

$1240 = 80p$

$$\frac{1240}{80} = \frac{80p}{80}$$ Divide both sides 80.

$15.5 = p$

Therefore, 12.4 mi is 15.5% of 80.

Avoiding Mistakes

Remember that p represents the number of *parts* per 100. However, Example 4 asked us to find the value of p%. Therefore, it was necessary to attach the % symbol to our value of p. For Example 4, we have $p = 15.5$. Therefore, p% is 15.5%.

Skill Practice

6. What percent of $48 is $15?

3. Applications of Percent Proportions

We now use percent proportions to solve application problems involving percents.

Example 5 — Using Percents in Meteorology

Buffalo, New York, receives an average of 94 in. of snow each year. One year it had 120% of the normal annual snowfall. How much snow did Buffalo get that year?

Solution:

This situation can be translated as:

"The amount of snow Buffalo received is 120% of 94 in."
- amount (x)
- p
- base

$$\frac{x}{94} = \frac{120}{100}$$ Set up a percent equation.

$100x = (120)(94)$ Equate the cross products.

$100x = 11{,}280$

$$\frac{100x}{100} = \frac{11{,}280}{100}$$ Divide both sides 100.

$x = 112.8$

Buffalo had 112.8 in. of snow.

Skill Practice

7. In a recent year it was estimated that 24.7% of U.S. adults smoked tobacco products regularly. In a group of 2000 adults, how many would be expected to be smokers?

TIP: In a word problem, it is always helpful to check the reasonableness of your answer. In Example 5, the percent is greater than 100%. This means that the amount must be greater than the base. Therefore, we suspect that our solution is reasonable.

Answers
6. 31.25%   7. 494 people

Example 6 Using Percents in Statistics

In a recent year, the freshman class of a prestigious university had 18% Asian American students. If this represented 380 students, how many students were admitted to the freshman class? Round to the nearest student.

Solution:

This situation can be translated as:

"380 is 18% of what number?"
amount p   base (x)

$$\frac{380}{x} = \frac{18}{100}$$ Set up a percent proportion.

$(380)(100) = 18x$ Equate the cross products.

$38,000 = 18x$

$$\frac{38,000}{18} = \frac{18x}{18}$$ Note that $38,000 \div 18 \approx 2111.1$. Rounded to the nearest whole unit (whole person), this is 2111.

$2111 \approx x$

The freshman class had approximately 2111 students.

Skill Practice

8. Eight students in a statistics class received a grade of "A" in the class. If this represents about 19% of the class, how many students are in the class? Round to the nearest student.

TIP: We can check the answer to Example 6 by substituting $x = 2111$ back into the original proportion. The cross products will not be exactly the same because we had to round the value of x. However, the cross products should be *close*.

$$\frac{380}{2111} \stackrel{?}{\approx} \frac{18}{100}$$ Substitute $x = 2111$ into the proportion.

$(380)(100) \stackrel{?}{\approx} (18)(2111)$

$38,000 \approx 37,998$ ✓   The values are very close.

Example 7 Using Percents in Business

Suppose a tennis pro who is ranked 90th in the world on the men's professional tour earns $280,000 per year in tournament winnings and endorsements. He pays his coach $100,000 per year. What percent of his income goes toward his coach? Round to the nearest tenth of a percent.

Stockbyte/Getty Images

Answer

8. 42 students are in the class.

Solution:

This can be translated as:

"What percent of $280,000 is $100,000?"

- The value of p is unknown.
- base (x)
- amount

$$\frac{100{,}000}{280{,}000} = \frac{p}{100}$$ Set up a percent proportion.

$$\frac{100{,}\cancel{000}}{280{,}\cancel{000}} = \frac{p}{100}$$ The ratio on the left side of the equation can be simplified by a factor of 10,000. "Strike through" four zeros in the numerator and denominator.

$$\frac{10}{28} = \frac{p}{100}$$

$(10)(100) = 28p$ Equate the cross products.

$1000 = 28p$

$$\frac{1000}{28} = \frac{28p}{28}$$ Divide both sides by 28.

$$\frac{1000}{28} = p$$

$35.7 \approx p$ Dividing $1000 \div 28$, we get approximately 35.7.

The tennis pro spends about 35.7% of his income on his coach.

Skill Practice

9. There were 425 donations made for various dollar amounts at an animal sanctuary. If 60 donations were made in the $100–$199 range, what percent does this represent? Round to the nearest tenth of a percent.

Answer

9. Of the donations, approximately 14.1% are in the $100–$199 range.

Section 7.2 Activity

A.1. List the key word or symbol that helps to identify:
   a. Amount   **b.** Percent   **c.** Base

A.2. Paris is purchasing a used car that costs $8000. Her parents have agreed to give her 20% of the total cost of the car, which equates to $1600.

Identify: (a) the percent, p, (b) the amount, and (c) the base in the problem.

For Exercises A.3–A.5,
  a. Identify the amount, base, and the value of p, and set up a percent proportion in the form of: $\frac{amount}{base} = \frac{p}{100}$
  b. Solve the proportion.

A.3. What is 38% of 950?

A.4. 288 is 120% of what number?

A.5. 5 is what percent of 250?

A.6. In a class of forty students, sixteen earned an "A" on the latest assessment. What percent of students in the class earned a grade other than an "A"? Round to the nearest whole percent.
  a. Identify the amount, base, and the value of p, and set up a percent proportion.
  b. Solve the proportion set up in part (a). Considering that the amount of students who earned an "A" on the test was used in the proportion, what does the resulting percent represent?
  c. What percent of students earned a grade other than an "A"?

A.7. One day, the Dow Jones Industrial Average fell 2% from 29,600.
  a. How many points did the Dow lose?
  b. What was the value of the Dow at the end of the day?

A.8. For a recent year, the freshman class at MIT was 49% women. If there were 1076 women, how large was the freshman class? Round to the nearest whole number.

A.9. The cost of a textbook at the college bookstore is 135% of the publisher's price. If the publisher sells the book for $60, what is the bookstore's price?

A.10. At one time there were 20,000 Humvees in the Middle East. Of these, approximately 5900 were the M-114 model, which is armored top to bottom. Determine the percent of M-114 models in the Middle East.

Practice Exercises

Section 7.2

Prerequisite Review

For Exercises R.1–R.6, solve the proportion.

R.1. $\dfrac{x}{100} = \dfrac{16}{25}$

R.2. $\dfrac{x}{100} = \dfrac{3}{20}$

R.3. $\dfrac{24}{100} = \dfrac{3}{n}$

R.4. $\dfrac{33}{100} = \dfrac{22}{n}$

R.5. $\dfrac{7.8}{100} = \dfrac{x}{30}$

R.6. $\dfrac{14.5}{100} = \dfrac{x}{8}$

Vocabulary and Key Concepts

1. **a.** A _____ proportion is a proportion in which one ratio is written with a denominator of 100.

   b. The first step to solve a percent proportion is to equate the _____ products.

2. The base in a percent proportion often appears after the word _____.

3. Other words for the base are _____ or _____.

4. The amount in a percent proportion sometimes appears with the word _____.

5. Another word for the amount is _____.

6. The percent in a percent proportion appears in front of the symbol _____.

7. If you are given the percent and the amount in a word problem, you are solving for the _____.

8. If you are given the amount and the base in a word problem, you are solving for the _____.

9. If you are given the percent and the base in a word problem, you are solving for the _____.

10. When the percent is greater than 100%, that means the _____ is greater than the _____.

Concept 1: Introduction to Percent Proportions

For Exercises 11–14, identify if the proportion given is a percent proportion.

11. $\dfrac{7}{100} = \dfrac{x}{32}$ 12. $\dfrac{42}{x} = \dfrac{15}{100}$ 13. $\dfrac{2}{3} = \dfrac{x}{9}$ 14. $\dfrac{7}{28} = \dfrac{10}{x}$

For Exercises 15–20, identify the amount, base, and *p* value. (See Example 1.)

15. 12 balloons is 60% of 20 balloons.

16. 25% of 400 cars is 100 cars.

17. $99 of $200 is 49.5%.

18. 45 of 50 children is 90%.

19. 50 hr is 125% of 40 hr.

20. 175% of 2 in. of rainfall is 3.5 in.

For Exercises 21–26, write the percent proportion.

21. 10% of 120 trees is 12 trees.

22. 15% of 20 pictures is 3 pictures.

23. 72 children is 80% of 90 children.

24. 21 dogs is 20% of 105 dogs.

25. 21,684 college students is 104% of 20,850 college students.

26. 103% of $40,000 is $41,200.

Concept 2: Solving Percent Proportions

For Exercises 27–36, solve the percent problems with an unknown amount. (See Example 2.)

27. Compute 54% of 200 employees.

28. Find 35% of 412.

29. What is $\tfrac{1}{2}$% of 40?

30. What is 1.8% of 900 g?

31. Find 112% of 500.

32. Compute 106% of 1050.

33. Pedro pays 28% of his salary in income tax. If he makes $72,000 in taxable income, how much income tax does he pay?

34. A car dealer sets the sticker price of a car by taking 115% of the wholesale price. If a car sells wholesale at $19,000, what is the sticker price?

35. A recent study in Missouri showed that over a 2-year period, 72% of the teens (age 15–19) killed in traffic accidents were not wearing seat belts. If a total of 304 teens were killed, approximately how many were not wearing seat belts? (Round to the nearest whole number.)

Fotog/Tetra images/Getty Images

36. In a psychology class, 61.9% of the class consists of freshmen. If there are 42 students, how many are freshmen? Round to the nearest whole unit.

For Exercises 37–46, solve the percent problems with an unknown base. (See Example 3.)

37. 18 is 50% of what number?

38. 22% of what length is 44 ft?

39. 30% of what weight is 69 lb?

40. 70% of what number is 28?

41. 9 is $\frac{2}{3}$% of what number?

42. 9.5 is 200% of what number?

43. Albert saves $120 per month. If this is 7.5% of his monthly income, how much does he make per month?

44. Janie and Don left their house in South Bend, Indiana, to visit friends in Chicago. They drove 80% of the distance before stopping for lunch. If they had driven 56 mi before lunch, what would be the total distance from their house to their friends' house in Chicago?

45. Aimee read 14 emails, which was only 40% of her total emails. What is her total number of emails?

46. A recent survey found that 10% of the population of Charlotte, North Carolina, was unemployed. If Charlotte had 75,000 unemployed, what was the population of Charlotte at this time?

For Exercises 47–54, solve the percent problems with p unknown. **(See Example 4.)**

47. What percent of $120 is $42?

48. 112 is what percent of 400?

49. 84 is what percent of 70?

50. What percent of 12 letters is 4 letters?

51. What percent of 320 mi is 280 mi?

52. 54¢ is what percent of 48¢?

53. A student answered 29 problems correctly on a final exam of 40 problems. What percent of the questions did she answer correctly?

54. During his college basketball season, Jeff made 520 baskets out of 1280 attempts. What was his shooting percentage? Round to the nearest whole percent.

For Exercises 55–58, use the table given. The data represent 600 police officers broken down by gender and by the number of officers promoted.

	Promoted	Not Promoted	Total
Male	140	340	480
Female	20	100	120
Total	160	440	600

55. What percent of the officers are female?

56. What percent of the officers are male?

57. What percent of the officers were promoted? Round to the nearest tenth of a percent.

58. What percent of the officers were not promoted? Round to the nearest tenth of a percent.

Darrin Klimek/Digital Vision/Alamy Stock Photo

Mixed Exercises

For Exercises 59–70, solve the problem using a percent proportion.

59. What is 15% of 50?

60. What is 28% of 70?

61. What percent of 240 is 96?

62. What percent of 600 is 432?

63. 85% of what number is 78.2?

64. 23% of what number is 27.6?

65. What is $3\frac{1}{2}$% of 2200?

66. What is $6\frac{3}{4}$% of 800?

67. 0.5% of what number is 44?

68. 0.8% of what number is 192?

69. 80 is what percent of 50?

70. 20 is what percent of 8?

Concept 3: Applications of Percent Proportions

71. The rainfall at Birmingham Airport in the United Kingdom averages 56 mm per month. In August the amount of rain that fell was 125% of the average monthly rainfall. How much rain fell in August?
(See Example 5.)

72. In a recent survey, 38% of people in the United States say that gas prices have affected the type of vehicle they will buy. In a sample of 500 people who are in the market for a new vehicle, how many would you expect to be influenced by gas prices?

73. In a recent year, a university in Florida accepted 6400 students to its freshman class. If this represented approximately 23% of all applicants, how many students applied for admission? (See Example 6.)

74. Yellowstone National Park has 3366 mi² of undeveloped land. If this represents 99% of the total area, find the total area of the park.

75. At a hospital, 546 beds out of 650 are occupied. What is the percent occupancy?
(See Example 7.)

76. For a recent year, 6 hurricanes struck the United States coastline out of 16 named storms to make landfall. What percent does this represent?

77. The risk of breast cancer relapse after surviving 10 years is shown in the graph according to the stage of the cancer at the time of diagnosis. (*Source: Journal of the National Cancer Institute*)

   a. How many women out of 200 diagnosed with Stage II breast cancer would be expected to relapse after having survived 10 years?

   b. How many women out of 500 diagnosed with Stage I breast cancer would *not* be expected to relapse after having survived 10 years?

78. A computer has 74.4 GB (gigabytes) of memory available. If 7.56 GB is used, what percent of the memory is used? Round to the nearest percent.

Medioimages/Photodisc/Getty Images

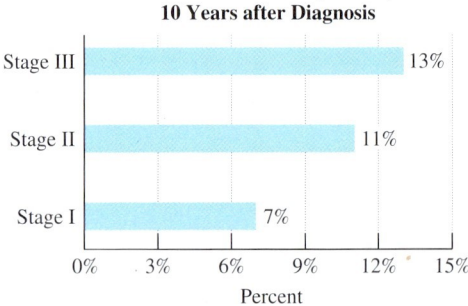
Figure for Exercise 77

A used car dealership sells several makes of vehicles. For Exercises 79–82, refer to the graph. Round the answers to the nearest whole unit.

79. If the dealership sold 215 vehicles in 1 month, how many were Chevys?

80. If the dealership sold 182 vehicles in 1 month, how many were Fords?

81. If the dealership sold 27 Hondas in 1 month, how many total vehicles were sold?

82. If the dealership sold 10 cars in the "Other" category, how many total vehicles were sold?

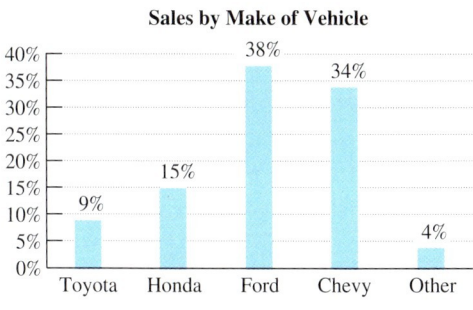

Expanding Your Skills

83. Carson had $600 and spent 44% of it on clothes. Then he spent 20% of the remaining money on dinner. How much did he spend altogether?

84. Melissa took $52 to the mall and spent 24% on makeup. Then she spent one-half of the remaining money on lunch. How much did she spend altogether?

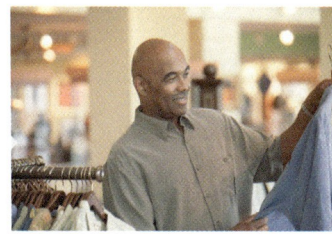

Ariel Skelley/Blend Images/Getty Images

It is customary to leave a 15–20% tip for the server in a restaurant. However, when you are at a restaurant in a social setting, you probably do not want to take out a pencil and piece of paper to figure out the tip. It is more socially acceptable to compute the tip mentally. Try this method.

Step 1: First, if the bill is not a whole dollar amount, simplify the calculations by rounding the bill to the next-higher whole dollar.

Step 2: Take 10% of the bill. This is the same as taking one-tenth of the bill. Move the decimal point to the left 1 place.

Step 3: If you want to leave a 20% tip, double the value found in step 2.

Step 4: If you want to leave a 15% tip, first note that 15% is 5% + 10%. Therefore, add one-half of the value found in step 2 to the number in step 2.

85. Estimate a 20% tip on a bill of $57.65. (*Hint:* Round up to $58 first.)

86. Estimate a 20% tip on a bill of $18.79.

87. Estimate a 15% tip on a dinner bill of $42.00.

88. Estimate a 15% tip on a luncheon bill of $12.00.

Percent Equations and Applications

Section 7.3

1. Solving Percent Equations—Amount Unknown

In this section, we investigate an alternative method to solve applications involving percents. We use percent equations. A **percent equation** represents a percent proportion in an alternative form. For example, recall that we can write a percent proportion as follows:

$$\frac{\text{amount}}{\text{base}} = p\% \qquad \text{percent proportion}$$

This is equivalent to writing   Amount = $(p\%) \cdot$ (base)   percent equation

Concepts

1. Solving Percent Equations—Amount Unknown
2. Solving Percent Equations—Base Unknown
3. Solving Percent Equations—Percent Unknown
4. Applications of Percent Equations

To set up a percent equation, it is necessary to translate an English sentence into a mathematical equation. As you read through the examples in this section, you will notice several key words. In the phrase *percent of*, the word *of* implies multiplication. The verb *to be* (am, is, are, was, were, been) often implies =.

Example 1 — Solving a Percent Equation—Amount Unknown

What is 30% of 60?

Solution:

We translate the words to mathematical symbols.

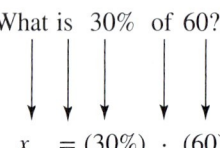

$x = (30\%) \cdot (60)$ Let x represent the unknown amount.

In this context, the word *of* means to multiply.

To find x, we must multiply 30% by 60. However, 30% means $\frac{30}{100}$ or 0.30. For the purpose of calculation, we *must* convert 30% to its equivalent decimal or fraction form. The equation becomes

$x = (0.30)(60)$
$= 18$

The value 18 is 30% of 60.

TIP: The solution to Example 1 can be checked by noting that 10% of 60 is 6. Therefore, 30% is equal to $(3)(6) = 18$.

Skill Practice Use a percent equation to solve.

1. What is 40% of 90?

Example 2 — Solving a Percent Equation—Amount Unknown

142% of 75 amounts to what number?

Solution:

142% of 75 amounts to what number?

$(142\%) \cdot (75) = x$

$(1.42)(75) = x$

$106.5 = x$

Let x represent the unknown amount.

The word *of* implies multiplication.

The phrase *amounts to* implies =.

Convert 142% to its decimal form (1.42).

Multiply.

Therefore, 142% of 75 amounts to 106.5.

Skill Practice Use a percent equation to solve.

2. 235% of 60 amounts to what number?

FOR REVIEW

Recall that to convert a percent to a decimal, we remove the % symbol and divide by 100.

Examples 1 and 2 illustrate that the percent equation gives us a quick way to find an unknown amount. For example, because $(p\%) \cdot (\text{base}) = \text{amount}$, we have

$50\% \text{ of } 80 = 0.50(80) = 40$

$25\% \text{ of } 20 = 0.25(20) = 5$

$250\% \text{ of } 90 = 2.50(90) = 225$

2. Solving Percent Equations—Base Unknown

Examples 3 and 4 illustrate the case in which the base is unknown.

Answers
1. 36 **2.** 141

Section 7.3 Percent Equations and Applications 409

Example 3 Solving a Percent Equation—Base Unknown

40% of what number is 225?

Solution:

40% of what number is 225?   Let x represent the base number.

$(0.40) \cdot x = 225$ Notice that we immediately converted 40% to its decimal form 0.40 so that we would not forget.

$0.40x = 225$

$\dfrac{0.40x}{0.40} = \dfrac{225}{0.40}$ Divide both sides by 0.40.

$x = 562.5$ 40% of 562.5 is 225.

Skill Practice Use a percent equation to solve.
3. 80% of what number is 94?

Example 4 Solving a Percent Equation—Base Unknown

0.19 is 0.2% of what number?

Solution:

0.19 is 0.2% of what number?   Let x represent the base number.

$0.19 = (0.002) \cdot x$ Convert 0.2% to its decimal form 0.002.

$0.19 = 0.002x$

$\dfrac{0.19}{0.002} = \dfrac{0.002x}{0.002}$ Divide both sides by 0.002.

$95 = x$ Therefore, 0.19 is 0.2% of 95.

Skill Practice Use a percent equation to solve.
4. 5.6 is 0.8% of what number?

3. Solving Percent Equations—Percent Unknown

Examples 5 and 6 demonstrate the process to find an unknown percent.

Example 5 Solving a Percent Equation—Percent Unknown

75 is what percent of 250?

Solution:

75 is what percent of 250?

$75 = x \cdot (250)$ Let x represent the unknown percent.

$75 = 250x$

$\dfrac{75}{250} = \dfrac{250x}{250}$ Divide both sides by 250.

$0.3 = x$

Answers
3. 117.5   4. 700

At this point, we have $x = 0.3$. To write the value of x in percent form, multiply by 100%.

$$x = 0.3$$
$$= 0.3 \times 100\%$$
$$= 30\%$$

> **Avoiding Mistakes**
>
> When solving for an unknown percent using a percent equation, it is necessary to convert x to its percent form.

Thus, 75 is 30% of 250.

Skill Practice Use a percent equation to solve.

5. 16.1 is what percent of 46?

Example 6 Solving a Percent Equation—Percent Unknown

What percent of $60 is $92? Round to the nearest tenth of a percent.

Solution:

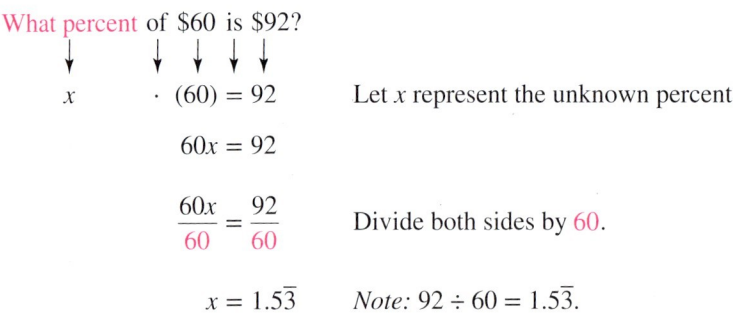

$$60x = 92$$

$$\frac{60x}{60} = \frac{92}{60} \qquad \text{Divide both sides by 60.}$$

$$x = 1.5\overline{3} \qquad \text{Note: } 92 \div 60 = 1.5\overline{3}.$$

At this point, we have $x = 1.5\overline{3}$. To convert x to its percent form, multiply by 100%.

$$x = 1.5\overline{3}$$
$$= 1.5\overline{3} \times 100\% \qquad \text{Convert from decimal form to percent form.}$$
$$= (1.53333\ldots) \times 100\%$$
$$= 153.333\ldots\%$$

The hundredths-place digit is less than 5. Discard it and the digits to its right.

Round to the nearest tenth of a percent.

$$\approx 153.3\%$$

> **Avoiding Mistakes**
>
> Notice that in Example 6 we converted the final answer to percent form first *before* rounding. With the number written in percent form, we are sure to round to the nearest tenth of a percent.

Therefore, $92 is approximately 153.3% of $60. (Notice that $92 is just over $1\frac{1}{2}$ times $60, so our answer seems reasonable.)

Skill Practice Use a percent equation to solve.

6. What percent of $150 is $213?

4. Applications of Percent Equations

In Examples 7, 8, and 9, we use percent equations in applications. An important part of this process is to extract the base, amount, and percent from the wording of the problem.

Answers
5. 35%   6. 142%

Example 7 — Using a Percent Equation in Ecology

Forty-six panthers are thought to live in Florida's Big Cypress National Preserve. This represents 53% of the panthers living in Florida. How many panthers are there in Florida? Round to the nearest whole unit. (*Source:* U.S. Fish and Wildlife Services)

Solution:

This problem translates to

"46 is 53% of the number of panthers living in Florida."

$46 = (0.53) \cdot x$ — Let x represent the total number of panthers.
$46 = 0.53x$

$\dfrac{46}{0.53} = \dfrac{0.53x}{0.53}$ — Divide both sides by 0.53.

$87 \approx x$ — *Note:* $46 \div 0.53 \approx 87$ (rounded to the nearest whole number).

There are approximately 87 panthers in Florida.

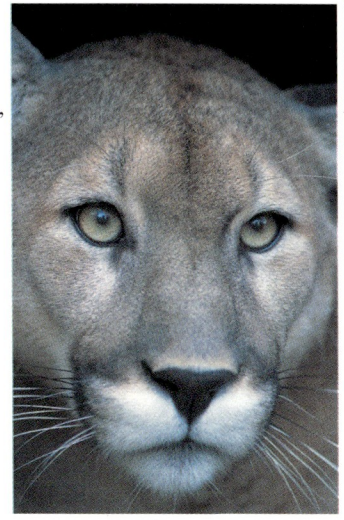

Digital Vision/Getty Images

Skill Practice

7. Brianna read 143 pages in a book. If this represents 22% of the book, how many pages are in the book?

Example 8 — Using a Percent Equation in Sports Statistics

Football player Tom Brady of the New England Patriots led his team to numerous Super Bowl victories. For one game during the season, Brady completed 23 of 30 passes. What percent of passes did he complete? Round to the nearest tenth of a percent.

Solution:

This problem translates to

"23 is what percent of 30?"

$23 = x \cdot 30$ — Let x represent the unknown.
$23 = 30x$

$\dfrac{23}{30} = \dfrac{30x}{30}$ — Divide both sides by 30.

$0.767 \approx x$ — Divide. $23 \div 30 \approx 0.767$

The decimal value 0.767 has been rounded to 3 decimal places. We did this because the next step is to convert the decimal to a percent. Move the decimal point to the right 2 places and attach the % symbol. We have 76.7% which is rounded to the nearest tenth of a percent.

$x \approx 0.767$
$= 0.767 \times 100\%$
$= 76.7\%$

Tom Brady completed approximately 76.7% of his passes.

Answer

7. The book has 650 pages.

Skill Practice

8. Hector had $60 in his wallet to take himself and a date to dinner and a movie. If he spent $28 on dinner and $19 on the movie, what percent of his money did he spend? Round to the nearest tenth of a percent.

Example 9 Using a Percent Equation in Ecology

On April 20, 2010, an explosion occurred on an offshore oil rig in the Gulf of Mexico. The explosion left 11 crewmembers dead and an uncontained oil leak that threatened the ecosystem in the Gulf of Mexico and surrounding beaches.

The United States consumes approximately 20 million barrels of oil per day. If the oil obtained from the Gulf of Mexico represents 8% of U.S. daily consumption, how much oil does the United States produce from the Gulf of Mexico?

Aaron Roeth Photography

Solution:

This situation translates to

"What number is 8% of 20?"

$x = 0.08 \cdot 20$ Write 8% in decimal form.

$x = (0.08)(20)$ Let x represent the number of millions of barrels produced by the United States per day in the Gulf of Mexico.

$x = 1.6$ Multiply.

The United States produces approximately 1.6 million barrels of oil per day from the Gulf of Mexico.

Skill Practice

9. In a science class, 85% of the students passed the class. If there were 40 people in the class, how many passed?

Answers
8. Hector spent about 78.3% of his money.
9. 34 students passed the class.

Section 7.3 Practice Exercises

Prerequisite Review

For Exercises R.1–R.6, solve the equation.

R.1. $4x = 640$

R.2. $225 = 15x$

R.3. $\dfrac{924}{x} = \dfrac{132}{100}$

R.4. $\dfrac{693}{x} = \dfrac{165}{100}$

R.5. $2.04 = 1.2x$

R.6. $1.15 = 2.3x$

Concept 1: Solving Percent Equations—Amount Unknown

For Exercises 1–6, write the percent equation. Then solve for the unknown amount. (See Examples 1 and 2.)

1. What is 35% of 700?
2. Find 12% of 625.
3. 0.55% of 900 is what number?
4. What is 0.4% of 75?
5. Find 133% of 600.
6. 120% of 40.4 is what number?
7. What is a quick way to find 50% of a number?
8. What is a quick way to find 10% of a number?
9. Compute 200% of 14 mentally.
10. Compute 75% of 80 mentally.
11. Compute 50% of 40 mentally.
12. Compute 10% of 32 mentally.
13. Household bleach is 6% sodium hypochlorite (active ingredient). In a 64-oz bottle, how much is active ingredient?
14. One antifreeze solution is 40% alcohol. How much alcohol is in a 12.5-L mixture?
15. In football, a quarterback completed 60% of his passes. If he attempted 8358 passes, how many did he complete? Round to the nearest whole unit.

Jill Braaten/McGraw Hill

16. To pass an exit exam, a student must pass a 60-question test with a score of 80% or better. What is the minimum number of questions she must answer correctly?

Concept 2: Solving Percent Equations—Base Unknown

For Exercises 17–22, write the percent equation. Then solve for the unknown base. (See Examples 3 and 4.)

17. 18 is 40% of what number?
18. 72 is 30% of what number?
19. 92% of what number is 41.4?
20. 84% of what number is 100.8?
21. 3.09 is 103% of what number?
22. 189 is 105% of what number?

23. In tests of a new anti-inflammatory drug, it was found that 47 subjects experienced nausea. If this represents 4% of the sample, how many subjects were tested?

24. Ted typed 80% of his research paper before taking a break.
    a. If he typed 8 pages, how many total pages are in the paper?
    b. How many pages does he have left to type?

25. A recent report stated that approximately 80 million Americans had some form of cardiovascular disease. If this represents 26% of the population, approximate the total population of the United States.

26. A city has a population of 245,300 which is 110% of the population from the previous year. What was the population the previous year?

Concept 3: Solving Percent Equations—Percent Unknown

For Exercises 27–32 write the percent equation. Then solve for the unknown percent. Round to the nearest tenth of a percent if necessary. (See Examples 5 and 6.)

27. What percent of 480 is 120?
28. 180 is what percent of 2000?
29. 666 is what percent of 740?
30. What percent of 60 is 2.88?
31. What percent of 300 is 400?
32. 28 is what percent of 24?

33. Of the 8079 Americans serving in the Peace Corps for a recent year, 406 were over 50 years old. What percent is this? Round to the nearest whole percent.

34. At a softball game, the concession stand had 120 hot dogs and sold 84 of them. What percent was sold?

For Exercises 35 and 36, refer to the table that shows the 1-year absentee record for a business.

35. a. Determine the total number of employees.
    b. What percent missed exactly 3 days of work?
    c. What percent missed between 1 and 5 days, inclusive?

36. a. What percent missed at least 4 days?
    b. What percent did not miss any days?

Number of Days Missed	Number of Employees
0	4
1	2
2	14
3	10
4	16
5	18
6	10
7	6

Mixed Exercises

For Exercises 37–48, solve the problem using a percent equation.

37. What is 45% of 62?
38. What is 32% of 30?
39. What percent of 140 is 28?
40. What percent of 25 is 18?
41. 23% of what number is 34.5?
42. 12% of what number is 26.4?
43. What is $18\frac{1}{2}$% of 3000?
44. What is $\frac{1}{4}$% of 460?
45. 350% of what number is 2100?
46. 225% of what number is 18?
47. 1.2 is what percent of 600?
48. 10 is what percent of 2000?

Concept 4: Applications of Percent Equations

49. In a recent year, children and adolescents comprised 6.3 million hospital stays. If this represents 18% of all hospital stays, what was the total number of hospital stays? (See Example 7.)

50. One fruit drink advertised that it contained "10% real fruit juice." In one bottle, this was found to be 4.8 oz of real juice.
    a. How many ounces of drink does the bottle contain?
    b. How many ounces is something other than fruit juice?

51. Of the 87 panthers living in the wild in Florida, 11 are thought to live in Everglades National Park. To the nearest tenth of a percent, what percent is this? (*Source:* U.S. Fish and Wildlife Services) (See Example 8.)

52. Earth is covered by approximately 360 million km² of water. If the total surface area is 510 million km², what percent is water? (Round to the nearest tenth of a percent.)

Art Wolfe/The Image Bank/Stone/Getty Images

NASA

53. Fifty-two percent of American parents have started to put money away for their children's college education. In a survey of 800 parents, how many would be expected to have started saving for their children's education? (*Source:* USA TODAY) **(See Example 9.)**

54. Forty-four percent of Americans used online travel sites to book hotel or airline reservations. If 400 people need to make airline or hotel reservations, how many would be expected to use online travel sites?

55. Brian has been saving money to buy a 55-in. television. He has saved $1440 so far, but this is only 60% of the total cost of the television. What is the total cost?

56. Recently the number of females that were home-schooled for grades K–12 was 875 thousand. This is 202% of the number of females home-schooled in 1999. How many females were home-schooled in 1999? Round to the nearest thousand. (*Source:* National Center for Educational Statistics)

57. Mr. Asher made $49,000 as a teacher in Virginia in 2010, and he spent $8,800 on food that year. In 2011, he received a 4% increase in his salary, but his food costs increased by 6.2%.

   a. How much money was left from Mr. Asher's 2010 salary after subtracting the cost of food?

   b. How much money was left from his 2011 salary after subtracting the cost of food? Round to the nearest dollar.

58. The human body is 65% water. Mrs. Wright weighed 180 lb. After 1 year on a diet, her weight decreased by 15%.

   a. Before the diet, how much of Mrs. Wright's weight was water?

   b. After the diet, how much of Mrs. Wright's weight was water?

For Exercises 59–62, refer to the graph showing the distribution of fatal traffic accidents in the United States according to the age of the driver. (*Source:* National Safety Council)

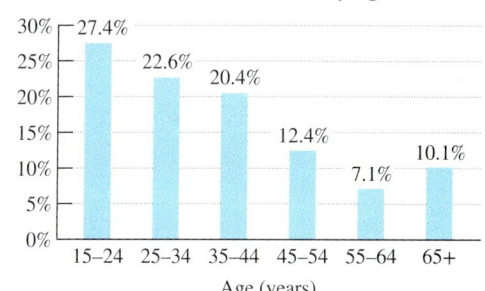

Traffic Fatalities Distributed by Age of Driver

59. If there were 60,000 fatal traffic accidents during a given year, how many would be expected to involve drivers in the 35–44 age group?

60. If there were 60,000 fatal traffic accidents, how many would be expected to involve drivers in the 15–24 age group?

61. If there were 9040 fatal accidents involving drivers in the 25–34 age group, how many total traffic fatalities were there for that year?

62. If there were 3550 traffic fatalities involving drivers in the 55–64 age group, how many total traffic fatalities were there for that year?

Expanding Your Skills

The maximum recommended heart rate (in beats per minute) is given by 220 minus a person's age. For aerobic activity, it is recommended that individuals exercise at 60%–85% of their maximum recommended heart rate. This is called the aerobic range. Use this information for Exercises 63 and 64.

63. a. Find the maximum recommended heart rate for a 20-year-old.

   b. Find the aerobic range for a 20-year-old.

64. a. Find the maximum recommended heart rate for a 42-year-old.

   b. Find the aerobic range for a 42-year-old.

Problem Recognition Exercises

Percents

Common percentages and their equivalent fractional forms can often be helpful when you want to estimate a percentage of a whole unit.

Percent	Example
10% is one-tenth of a whole unit. Move the decimal point to the left one place to find 10% of a number.	10% of 62 is 6.2.
20% is twice 10% of a whole unit. To calculate 20%, first calculate 10% and double the result.	20% of 62 is 12.4.
25% is one-quarter of a whole unit.	25% of 120 is 30.
33% is approximately one-third of a whole unit.	33% of 120 is approximately 40.
50% is half of a whole unit.	50% of 70 is 35.
75% is three-quarters of a whole unit.	75% of 200 is 150.
100% of a whole unit is the unit itself.	100% of 64 is 64.
200% is twice the value of the whole unit.	200% of 64 is 128.

For Exercises 1–6, perform the calculations mentally.

1. What is 10% of 82?
2. What is 5% of 82?
3. What is 20% of 82?
4. What is 50% of 82?
5. What is 200% of 82?
6. What is 15% of 82?
7. Is 104% of 80 less than or greater than 80?
8. Is 8% of 50 less than or greater than 5?
9. Is 11% of 90 less than or greater than 9?
10. Is 52% of 200 less than or greater than 100?

For Exercises 11–34, solve the problem by using a percent proportion or a percent equation.

11. 6 is 0.2% of what number?
12. What percent of 500 is 120?
13. 12% of 40 is what number?
14. 27 is what percent of 180?
15. 150% of what number is 105?
16. What number is 30% of 120?
17. What is 7% of 90?
18. 100 is 40% of what number?
19. 180 is what percent of 60?
20. 0.5% of 140 is what number?
21. 75 is 0.1% of what number?
22. 27 is what percent of 72?
23. What number is 50% of 50?
24. What number is 15% of 900?
25. 50 is 50% of what number?
26. 900 is 15% of what number?
27. What percent of 250 is 2?
28. 75 is what percent of 60?
29. What number is 10% of 26?
30. 11 is 55% of what number?
31. 186 is what percent of 248?
32. What number is 55% of 11?
33. 248 is what percent of 186?
34. 20 is what percent of 5?

Applications of Sales Tax, Commission, Discount, Markup, and Percent Increase and Decrease

Section 7.4

Percents are used in an abundance of applications in day-to-day life. In this section, we investigate six common applications of percents:

- Sales tax
- Discount
- Percent increase
- Commission
- Markup
- Percent decrease

Concepts

1. Applications Involving Sales Tax
2. Applications Involving Commission
3. Applications Involving Discount and Markup
4. Applications Involving Percent Increase and Decrease

1. Applications Involving Sales Tax

The first application involves computing sales tax. **Sales tax** is a tax based on a percent of the cost of merchandise.

Sales Tax Formula

$$\begin{pmatrix} \text{Amount of} \\ \text{sales tax} \end{pmatrix} = \begin{pmatrix} \text{tax} \\ \text{rate} \end{pmatrix} \cdot \begin{pmatrix} \text{cost of} \\ \text{merchandise} \end{pmatrix}$$

In this formula, the tax rate is usually given by a percent. Also note that there are three parts to the formula, just as there are in the general percent equation. The sales tax formula is a special case of a percent equation.

Example 1 — Computing Sales Tax

Suppose a new mid-size car sells for $26,000.

a. Compute the sales tax for a tax rate of 5.5%.

b. What is the total price of the car?

Solution:

a. Let x represent the amount of sales tax.   Label the unknown.

    Tax rate = 5.5%   Identify the parts of the formula.

    Cost of merchandise = $26,000

$$\begin{pmatrix} \text{Amount of} \\ \text{sales tax} \end{pmatrix} = \begin{pmatrix} \text{tax} \\ \text{rate} \end{pmatrix} \cdot \begin{pmatrix} \text{cost of} \\ \text{merchandise} \end{pmatrix}$$

$$x = (5.5\%) \cdot (\$26{,}000)$$

Substitute values into the sales tax formula.

$$x = (0.055)(\$26{,}000)$$

Convert the percent to its decimal form.

$$x = \$1430$$

The sales tax on the vehicle is $1430.

b. The total price is $26,000 + $1430 = $27,430.

> **Avoiding Mistakes**
>
> Notice that we must use the decimal form of the sales tax rate in calculations as shown in Example 1(a).

Skill Practice

1. A graphing calculator costs $110. The sales tax rate is 4.5%.
   a. Compute the amount of sales tax.
   b. Compute the total cost.

> **TIP:** Example 1(a) can also be solved by using a percent proportion.
>
> $$\frac{5.5}{100} = \frac{x}{26{,}000} \quad \text{What is 5.5\% of 26,000?}$$
>
> $$(5.5)(26{,}000) = 100x$$
>
> $$143{,}000 = 100x$$
>
> $$\frac{143{,}000}{100} = \frac{100x}{100} \quad \text{Divide both sides by 100.}$$
>
> $$1430 = x \quad \text{The sales tax is \$1430.}$$

Example 2 — Computing a Sales Tax Rate

Lindsay has just moved and must buy a new refrigerator for her home. The refrigerator costs $1200 and the sales tax is $48. Because she is new to the area, she does not know the sales tax rate. Compute the tax rate.

Solution:

Let x represent the sales tax rate.   Label the unknown.

Cost of merchandise = $1200   Identify the parts of the formula.

Amount of sales tax = $48

$$\begin{pmatrix}\text{Amount of}\\ \text{sales tax}\end{pmatrix} = \begin{pmatrix}\text{tax}\\ \text{rate}\end{pmatrix} \cdot \begin{pmatrix}\text{cost of}\\ \text{merchandise}\end{pmatrix}$$

$$48 = x \cdot (1200) \quad \text{Substitute values into the sales tax formula.}$$

$$48 = 1200x$$

$$\frac{48}{1200} = \frac{1200x}{1200} \quad \text{Divide both sides by 1200.}$$

$$0.04 = x$$

The question asks for the tax rate, which is given in percent form.

$$x = 0.04$$
$$= 0.04 \times 100\%$$
$$= 4\% \quad \text{The sales tax rate is 4\%.}$$

> **TIP:** Example 2 can also be solved by using a percent proportion.
>
> $$\frac{48}{1200} = \frac{p}{100}$$

Skill Practice

2. A DVD sells for $15. The sales tax is $0.90. What is the tax rate?

Answers
1. a. The tax is $4.95.
   b. The total cost is $114.95.
2. The tax rate is 6%.

Section 7.4 Applications of Sales Tax, Commission, Discount, Markup, and Percent Increase and Decrease

Example 3 — Computing Cost of Merchandise

The tax on a new CD comes to $1.05. If the tax rate is 6%, find the cost of the CD before tax.

Solution:

Let x represent the cost of the CD. Label the unknown.

Tax rate = 6% Identify the parts of the formula.

Amount of tax = $1.05

$$\begin{pmatrix}\text{Amount of}\\ \text{sales tax}\end{pmatrix} = \begin{pmatrix}\text{tax}\\ \text{rate}\end{pmatrix} \cdot \begin{pmatrix}\text{cost of}\\ \text{merchandise}\end{pmatrix}$$

$$1.05 = (0.06) \cdot x$$

Notice that we immediately converted 6% to its decimal form.

$$1.05 = 0.06x$$

$$\frac{1.05}{0.06} = \frac{0.06x}{0.06}$$

Divide both sides by 0.06.

$$17.5 = x$$

The CD costs $17.50.

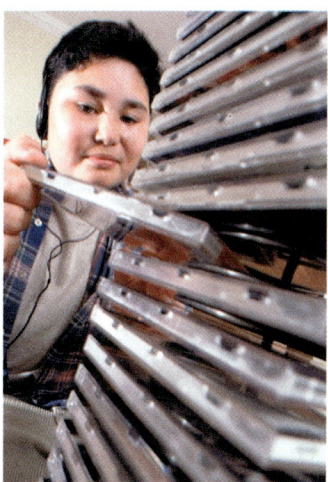

Corbis/VCG/image100/Getty Images

Skill Practice

3. Sales tax on a new lawn mower is $21.50. If the tax rate is 5%, compute the price of the lawn mower before tax.

2. Applications Involving Commission

Salespeople often receive all or part of their salary in commission. **Commission** is a form of income based on a percent of sales.

Commission Formula

$$\begin{pmatrix}\text{Amount of}\\ \text{commission}\end{pmatrix} = \begin{pmatrix}\text{commission}\\ \text{rate}\end{pmatrix} \cdot \begin{pmatrix}\text{total}\\ \text{sales}\end{pmatrix}$$

For example, if a realtor gets a 6% commission on the sale of a $200,000 home, then

$$\text{Commission} = (0.06)(\$200{,}000)$$
$$= \$12{,}000$$

Answer

3. The mower costs $430 before tax.

Example 4 Computing Commission Rate

Alexis works in real estate sales.

a. If she sells a $150,000 house and earns a commission of $10,500, what is her commission rate?

b. At this rate, how much will she earn by selling a $200,000 house?

Solution:

a. Let x represent the commission rate. Label the unknown.

Total sales = $150,000 Identify the parts of the formula.

Amount of commission = $10,500

$$\begin{pmatrix}\text{Amount of}\\\text{commission}\end{pmatrix} = \begin{pmatrix}\text{commission}\\\text{rate}\end{pmatrix} \cdot \begin{pmatrix}\text{total}\\\text{sales}\end{pmatrix}$$

$$10,500 = x \cdot (150,000)$$ Substitute values into the commission formula.

$$10,500 = 150,000x$$

$$\frac{10,500}{150,000} = \frac{150,000x}{150,000}$$ Divide both sides by 150,000.

$$0.07 = x$$

$$x = 0.07 \times 100\%$$ Convert to percent form.

$$= 7\%$$

The commission rate is 7%.

b. The commission on a $200,000 house is given by

Amount of commission = (0.07)($200,000) = $14,000

Alexis will earn $14,000 by selling a $200,000 house.

> **TIP:** A percent proportion can be used to find the commission rate in Example 4.
> $$\frac{10,500}{150,000} = \frac{p}{100}$$

Skill Practice

4. Trevor sold a home for $160,000 and earned an $8000 commission. What is his commission rate?

Example 5 Computing Sales Base

Tonya is a real estate agent. She makes $10,000 as her annual base salary for the work she does in the office. In addition, she makes 4% commission on her total sales. If her salary for the year amounts to $106,000, what was her total in sales?

Solution:

First note that her commission is her total salary minus the $10,000 for working in the office. Thus,

Amount of commission = $106,000 − $10,000 = $96,000

Let x represent Tonya's total sales. Label the unknown.

Amount of commission = $96,000 Identify the parts of the formula.

Commission rate = 4%

Answer

4. His commission rate is 5%.

$$\begin{pmatrix}\text{Amount of}\\\text{commission}\end{pmatrix} = \begin{pmatrix}\text{commission}\\\text{rate}\end{pmatrix} \cdot \begin{pmatrix}\text{total}\\\text{sales}\end{pmatrix}$$

$$96{,}000 = (0.04) \cdot x \quad \text{Substitute values into the commission formula.}$$

$$96{,}000 = 0.04x$$

$$\frac{96{,}000}{0.04} = \frac{0.04x}{0.04} \quad \text{Divide both sides by } 0.04.$$

$$\$2{,}400{,}000 = x$$

Tonya's sales totaled $2,400,000 ($2.4 million).

TIP: Example 5 can also be solved by using a percent proportion.

$$\frac{4}{100} = \frac{96{,}000}{x}$$

Skill Practice

5. A sales rep for a pharmaceutical firm makes $50,000 as his base salary. In addition, he makes 6% commission on sales. If his salary for the year amounts to $98,000, what were his total sales?

3. Applications Involving Discount and Markup

When we go to the store, we often find items discounted or on sale. For example, a printer might be discounted 20%, or a blouse might be on sale for 30% off. We compute the amount of the **discount** (the savings) as follows.

Discount Formulas

$$\begin{pmatrix}\text{Amount of}\\\text{discount}\end{pmatrix} = \begin{pmatrix}\text{discount}\\\text{rate}\end{pmatrix} \cdot \begin{pmatrix}\text{original}\\\text{price}\end{pmatrix}$$

Sale price = original price − amount of discount

Example 6 Computing Discount Rate

A gold chain originally priced $500 is marked down to $375. What is the discount rate?

Solution:

First note that the amount of the discount is given by

Discount = original price − sale price
= $500 − $375
= $125

Let x represent the discount rate.   Label the unknown.

Original price = $500   Identify the parts of the formula.

Amount of discount = $125

$$\begin{pmatrix}\text{Amount of}\\\text{discount}\end{pmatrix} = \begin{pmatrix}\text{discount}\\\text{rate}\end{pmatrix} \cdot \begin{pmatrix}\text{original}\\\text{price}\end{pmatrix}$$

$$125 = x \cdot 500 \quad \text{Substitute values into discount formula.}$$

$$125 = 500x$$

$$\frac{125}{500} = \frac{500x}{500} \quad \text{Divide both sides by } 500.$$

$$0.25 = x$$

Converting $x = 0.25$ to percent form, we have $x = 25\%$. The chain has been discounted 25%.

Answer

5. He made $800,000 in sales.

Skill Practice

6. Find the discount rate.

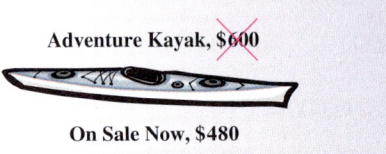

Adventure Kayak, $600

On Sale Now, $480

Retailers often buy goods from manufacturers or wholesalers. To make a profit, the retailer must increase the cost of the merchandise before reselling it. This is called **markup**.

Markup Formulas

$$\begin{pmatrix}\text{Amount of}\\ \text{markup}\end{pmatrix} = \begin{pmatrix}\text{markup}\\ \text{rate}\end{pmatrix} \cdot \begin{pmatrix}\text{original}\\ \text{price}\end{pmatrix}$$

Retail price = original price + amount of markup

Example 7 — Computing Markup

A college bookstore marks up the price of books 40%.

a. What is the markup for a textbook that has a manufacturer price of $66?
b. What is the retail price of the book?
c. If there is a 6% sales tax, how much will the book cost to take home?

Solution:

a. Let x represent the amount of markup.   Label the unknown.

Markup rate = 40%   Identify parts of the formula.

Original price = $66

$$\begin{pmatrix}\text{Amount of}\\ \text{markup}\end{pmatrix} = \begin{pmatrix}\text{markup}\\ \text{rate}\end{pmatrix} \cdot \begin{pmatrix}\text{original}\\ \text{price}\end{pmatrix}$$

$\quad\quad x \quad = \quad (0.40) \quad \cdot \quad (\$66)$ Use the decimal form of 40%.

$x = (0.40)(\$66)$

$\quad = \$26.40$

The amount of markup is $26.40.

b. Retail price = original price + markup

$\quad = \$66 + \$26.40$

$\quad = \$92.40$

The retail price is $92.40.

c. Next we must find the amount of the sales tax. This value is added to the cost of the book. The sales tax rate is 6%.

$$\begin{pmatrix}\text{Amount of}\\ \text{sales tax}\end{pmatrix} = \begin{pmatrix}\text{tax}\\ \text{rate}\end{pmatrix} \cdot \begin{pmatrix}\text{cost of}\\ \text{merchandise}\end{pmatrix}$$

Tax $\quad = (0.06) \cdot \quad (\$92.40)$

$\quad\quad \approx \$5.54$ Round the tax to the nearest cent.

The total cost of the book is $92.40 + $5.54 = $97.94.

Answer

6. The kayak is discounted 20%.

Skill Practice

7. An importer uses a markup rate of 35%.
   a. What is the markup on a wicker chair that has a manufacturer price of $100?
   b. What is the retail price?
   c. If there is a 5% sales tax, what is the total cost to buy the chair retail?

It is important to note the similarities in the formulas presented in this section. To find the amount of sales tax, commission, discount, or markup, we multiply a rate (percent) by some base value.

Formulas for Sales Tax, Commission, Discount, and Markup

$$\text{Amount} = \text{rate} \times \text{base value}$$

Sales tax: $\begin{pmatrix}\text{Amount of}\\ \text{sales tax}\end{pmatrix} = \begin{pmatrix}\text{tax}\\ \text{rate}\end{pmatrix} \cdot \begin{pmatrix}\text{cost of}\\ \text{merchandise}\end{pmatrix}$

Commission: $\begin{pmatrix}\text{Amount of}\\ \text{commission}\end{pmatrix} = \begin{pmatrix}\text{commission}\\ \text{rate}\end{pmatrix} \cdot \begin{pmatrix}\text{total}\\ \text{sales}\end{pmatrix}$

Discount: $\begin{pmatrix}\text{Amount of}\\ \text{discount}\end{pmatrix} = \begin{pmatrix}\text{discount}\\ \text{rate}\end{pmatrix} \cdot \begin{pmatrix}\text{original}\\ \text{price}\end{pmatrix}$

Markup: $\begin{pmatrix}\text{Amount of}\\ \text{markup}\end{pmatrix} = \begin{pmatrix}\text{markup}\\ \text{rate}\end{pmatrix} \cdot \begin{pmatrix}\text{original}\\ \text{price}\end{pmatrix}$

4. Applications Involving Percent Increase and Decrease

Two other important applications of percents are finding percent increase and percent decrease. For example:

- The price of gas increased 40% in 4 years.
- After taking a new drug for 3 months, a patient's cholesterol decreased by 35%.

When we compute **percent increase** or **percent decrease**, we are comparing the *change* between two given amounts to the *original value*. The change (amount of increase or decrease) is found by subtraction. To compute the percent increase or decrease, we use the following formulas.

Percent Increase or Percent Decrease

$$\begin{pmatrix}\text{Percent}\\ \text{increase}\end{pmatrix} = \begin{pmatrix}\dfrac{\text{amount of increase}}{\text{original value}}\end{pmatrix} \times 100\%$$

$$\begin{pmatrix}\text{Percent}\\ \text{decrease}\end{pmatrix} = \begin{pmatrix}\dfrac{\text{amount of decrease}}{\text{original value}}\end{pmatrix} \times 100\%$$

In Example 8, we apply the percent increase formula.

Example 8 Computing Percent Increase

The price of heating oil rose from an average of $2.20 per gallon to $2.75 per gallon in a one-year period. Compute the percent increase.

Solution:

The original price was $2.20 per gallon.   Identify the parts of the formula.

The final price after the increase is $2.75.

Answer

7. a. The markup is $35.
   b. The retail price is $135.
   c. The total cost is $141.75.

TIP: Example 8 could also have been solved by using a percent proportion.

$$\frac{p}{100} = \frac{\text{amount of increase}}{\text{original value}}$$

$$\frac{p}{100} = \frac{0.55}{2.20}$$

The amount of increase is determined by subtraction.

Amount of increase = $2.75 − $2.20
= $0.55   There was a $0.55 increase in price.

$$\begin{pmatrix}\text{Percent}\\\text{increase}\end{pmatrix} = \begin{pmatrix}\dfrac{\text{amount of increase}}{\text{original value}}\end{pmatrix} \times 100\%$$

$$= \frac{0.55}{2.20} \times 100\%$$ Apply the percent increase formula.

$$= 0.25 \times 100\%$$

$$= 25\%$$

There was a 25% increase in the price of heating oil.

Skill Practice

8. After a raise, Denisha's salary increased from $42,500 to $44,200. What is the percent increase?

Example 9   **Finding Percent Decrease in an Application**

The graph in Figure 7-5 represents the closing price of a cable company stock for a 5-day period. Compute the percent decrease between the first day and the fifth day.

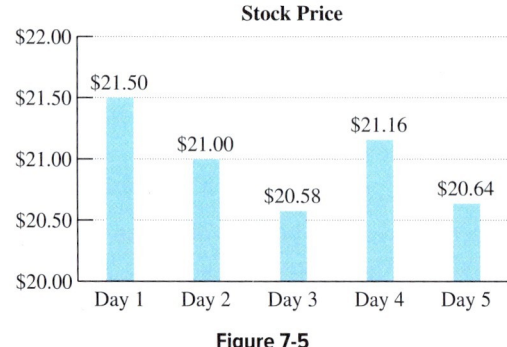

Figure 7-5

Solution:

The amount of decrease is given by: $21.50 − $20.64 = $0.86

The original amount is the closing price on Day 1: $21.50

$$\begin{pmatrix}\text{Percent}\\\text{decrease}\end{pmatrix} = \begin{pmatrix}\dfrac{\text{amount of decrease}}{\text{original value}}\end{pmatrix} \times 100\%$$

$$= \left(\frac{\$0.86}{\$21.50}\right) \times 100\%$$

$$= (0.04) \times 100\%$$

$$= 4\%$$

The stock fell by 4%.

Skill Practice

9. Refer to the graph in Example 9. Compute the percent decrease between Day 2 and Day 3.

Answers
8. Denisha's salary increased by 4%.
9. 2%

Section 7.4 Activity

A.1. The sales tax formula is:

$$(\text{Amount of sales tax}) = (\text{tax rate}) \cdot (\text{cost of merchandise})$$

    a. Identify what variable corresponds to the amount.
    b. Identify what variable corresponds to the percent, p.
    c. Identify what variable corresponds to the base.

A.2. **a.** Using the sales tax formula from Exercise A.1, calculate the amount of sales tax on a tablet that costs $120 with a tax rate of 6%. Express the answer to the nearest cent.
    b. What is the tax rate in your county?
    c. Calculate the amount of sales tax for the tablet in part (a) using the tax rate in your county.
    d. What is the total price that you would pay for the tablet if purchased in your county?

A.3. Moses paid $67.15 in sales tax on his new smartphone. If the phone costs $1199, what is the tax rate? Round to the nearest tenth of a percent.

A.4. Taylor is a real estate agent who earns a 4.5% commission on every home she sells. If she sold a home for $315,000, what was the amount of commission Taylor received?

A.5. A car dealership is having a sales competition for the month of January. The salesperson with the greatest amount of total sales will earn a $2000 bonus. Based on the sales statistics below, determine the salesperson that will receive the bonus.

	Amount of Commission	Commission Rate	Total Sales
Michael	$4,000	2.5%	
Jorge	$4,950	3%	
Dominique	$4,743.75	2.75%	

A.6. A furniture store marks up a dining room table 20%.
    a. What is the amount of markup if the wholesale price is $650?
    b. What is the retail price of the dining room table?
    c. If the sales tax rate is 6.5%, what is the total cost of the dining room table after taxes?

A.7. Retailer A is selling everything 40% off the original price. Retailer B is selling everything at 0.75% off the original price. If you are interested in purchasing a gaming system that sells for $299.00,
    a. How much of a discount do you receive at Retailer A?
    b. How much of a discount do you receive at retailer B?
    c. What retailer has the greater discount?

A.8. A cell phone charger is on sale for $16.99. The original price was $24.99.
    a. What was the amount of the discount?
    b. What is the discount rate to the nearest whole percent?

A.9. Yazmine purchased a new pair of winter boots. The boots were advertised as 35% off.
    a. If Yazmine saved $47.25, what was the original price of the boots?
    b. How much did Yazmine pay for the boots before taxes?
    c. If the salesperson receives a 12% commission on every pair of boots sold, how much commission did the salesperson receive on this sale?

A.10. The yearly tuition at a college is presently $13,500. Next year it is expected to be $15,000.
    a. Is this an increase or a decrease in tuition?
    b. What is the amount of change in tuition?

c. What was tuition before it changed (the original value)?

d. What is the percent increase or decrease? Round to the nearest tenth of a percent, if applicable.

A.11. The graph below depicts the gross movie sales in millions for two different movies over the course of two weeks. Refer to this graph to answer parts (a)–(c).

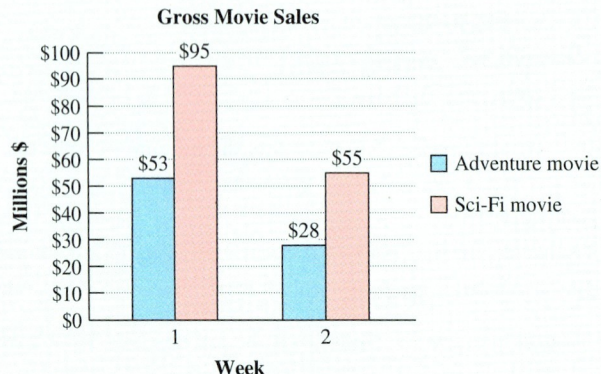

a. Calculate the amount of decrease and the percent decrease (to the nearest tenth of a percent) for the adventure movie and the Sci-Fi movie.

Movie	Amount of Decrease	Percent Decrease
Adventure		
Sci-Fi		

b. What movie showed the greater amount of decrease?

c. Which movie had the greater percent decrease in sales from week 1 to week 2? Is this consistent with the movie that had the greater amount of decrease from part (b)?

Section 7.4 Practice Exercises

Prerequisite Review

For Exercises R.1–R.6, solve for the unknown quantity.

R.1. What is 15% of 80?

R.2. 16% of 550 is what number?

R.3. What percent of 6 is 12?

R.4. 60 is what percent of 25?

R.5. 30 is 12% of what number?

R.6. 16% of what number is 72?

Vocabulary and Key Concepts

1. a. Write a formula to compute the amount of sales tax.

   b. The tax rate in the sales tax formula is given by a _____.

2. a. Write a formula to compute the amount of commission.

   b. The commission that salespeople earn is the (amount/base) in a commission problem.

Section 7.4 Applications of Sales Tax, Commission, Discount, Markup, and Percent Increase and Decrease

3. **a.** Write a formula to compute the amount of discount.

   b. The original price of an item is the (amount/base) in a discount problem.

4. **a.** Write a formula to compute the amount of markup.

   b. To find the retail price, add the markup and the _____ _____ of an item.

5. **a.** Write a formula to compute percent increase.

   b. The amount of increase in the percent increase formula is found by _____ the new amount from the old amount.

6. **a.** Write a formula to compute percent decrease.

   b. The original value in the percent decrease formula is the (smaller value/larger value) given in the problem.

7. All of the percent application formulas are special cases of the _____ equation.

8. Percent proportions (can/cannot) be used to solve percent application problems.

Concept 1: Applications Involving Sales Tax

For Exercises 9–12, write the sales tax formula. Do not solve for the unknown quantity.

9. A laptop computer sells for $1299 with a tax rate of 5.6%. Calculate the amount of sales tax.

10. Antonio paid $1.82 in tax for his video game that sold for $34.99. Find the sales tax rate.

11. The tax on a new coat is $11.72. If the tax rate is 6.2%, find the cost of the coat.

12. A pair of soccer cleats cost $235 before tax. If the sales tax was $10.58, find the tax rate.

For Exercises 13 and 14, complete the table.

13.	Cost of Item	Sales Tax Rate	Amount of Tax	Total Cost
a.	$20	5%		
b.	$12.50		$0.50	
c.		2.5%	$2.75	
d.	$55			$58.30

14.	Cost of Item	Sales Tax Rate	Amount of Tax	Total Cost
a.	$56	6%		
b.	$212		$14.84	
c.		3%	$18.00	
d.	$214			$220.42

15. A new coat costs $68.25. If the sales tax rate is 5%, what is the total bill? **(See Example 1.)**

16. Sales tax for a county in Oklahoma is 4.5%. Compute the amount of tax on a new personal music player that sells for $64.

17. The sales tax on a set of luggage is $16.80. If the luggage cost before tax is $240.00, what is the sales tax rate? **(See Example 2.)**

18. A new shirt is labeled at $42.00. Jon purchased the shirt and paid $44.10.

    a. How much was the sales tax?

    b. What is the sales tax rate?

19. The 6% sales tax on a fruit basket came to $2.67. What is the price of the fruit basket? **(See Example 3.)**

20. The sales tax on a bag of groceries came to $1.50. If the sales tax rate is 6% what was the price of the groceries before tax?

C Squared Studios/Photodisc/Getty Images

Arthur Tilley/Creatas Images/Getty Images

Concept 2: Applications Involving Commission

For Exercises 21 and 22, complete the table.

21.	Total Sales	Commission Rate	Amount of Commission
a.	$20,000	5%	
b.	$125,000		$10,000
c.		10%	$540

22.	Total Sales	Commission Rate	Amount of Commission
a.	$540	16%	
b.	$800		$24
c.		15%	$159

23. Zach works in an insurance office. He receives a commission of 7% on new policies. How much did he make last month in commission if he sold $48,000 in new policies?

24. Marisa makes a commission of 15% on sales over $400. One day she sells $750 worth of merchandise.
    a. How much over $400 did Marisa sell?
    b. How much did she make in commission that day?

25. In one week, Rodney sold $2000 worth of sports equipment. He received $300 in commission. What is his commission rate? (See Example 4.)

26. A realtor sold a townhouse for $95,000. If he received a commission of $7600, what is his commission rate?

27. A realtor makes an annual salary of $25,000 plus a 3% commission on sales. If a realtor's salary is $67,000, what was the amount of her sales? (See Example 5.)

28. A salesperson receives a weekly salary of $100, plus a 5.5% commission on sales. Her salary last week was $1090. What were her sales that week?

Concept 3: Applications Involving Discount and Markup

Richard Ransier/Corbis/VCG/Digital Stock/Getty Images

For Exercises 29–32, complete the table.

29.	Original Price	Discount Rate	Amount of Discount	Sale Price
a.	$56	20%		
b.	$900			$600
c.			$8.50	$76.50
d.		50%	$38	

30.	Original Price	Discount Rate	Amount of Discount	Sale Price
a.	$175	15%		
b.	$900			$630
c.			$33	$77
d.		40%	$23.36	

31.	Original Price	Markup Rate	Amount of Markup	Retail Price
a.	$92	5%		
b.	$110			$118.80
c.			$97.50	$422.50
d.		20%	$9	

32.	Original Price	Markup Rate	Amount of Markup	Retail Price
a.	$25	10%		
b.	$50			$57.50
c.			$175	$875
d.		18%	$31.50	

33. Hospital employees get a 15% discount at the hospital cafeteria. If the lunch bill originally comes to $5.60, what is the price after the discount?

34. A health club membership costs $550 for 1 year. If a member pays up front in a lump sum, the member will receive a 10% discount.
    a. How much money is discounted?
    b. How much will the yearly membership cost with the discount?

Digital Vision/Getty Images

Section 7.4 Applications of Sales Tax, Commission, Discount, Markup, and Percent Increase and Decrease

35. A bathing suit is on sale for $45. If the regular price is $60, what is the discount rate? **(See Example 6.)**

36. A printer that sells for $229 is on sale for $183.20. What is the discount rate?

37. A business suit has a wholesale price of $150.00. A department store's markup rate is 18%. **(See Example 7.)**
   a. What is the markup for this suit?
   b. What is the retail price?
   c. If Antonio buys this suit including a 7% sales tax, how much will he pay?

38. An import/export business marks up imported merchandise by 110%. If a wicker chair imported from Singapore originally costs $84 from the manufacturer, what is the retail price?

39. A table is purchased from the manufacturer for $300 and is sold retail at $375. What is the markup rate?

40. A $60 hairdryer is sold for $69. What is the markup rate?

41. Find the discount and the sale price of the tent in the given advertisement.

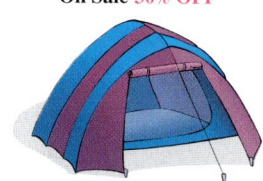

Explorer 4-person tent
On Sale 30% OFF
Was $269

42. A set of dishes had an original price of $112. Then it was discounted 50%. A week later, the new sale price was discounted another 50%. At that time, was the set of dishes free? Explain why or why not.

43. A campus bookstore adds $43.20 to the cost of a science text. If the final cost is $123.20, what is the markup rate?

44. The retail price of a golf club is $420.00. If the golf store has marked up the price by $70, what is the markup rate?

45. Find the discount and the sale price of the bike in the given advertisement.

Huffy Chopper Bike $109.99
Now 10% off.

46. Find the discount and the discount rate of the chair from the given advertisement.

Accent Chair was $235.00
and is now $188.00

Concept 4: Applications Involving Percent Increase and Decrease

47. Select the correct percent increase for a price that is double the original amount. For example, a book that originally cost $30 now costs $60.
   a. 200%   b. 2%
   c. 100%   d. 150%

48. Select the correct percent increase for a price that is greater by $\frac{1}{2}$ of the original amount. For example, an employee made $20 per hour and now makes $30 per hour.
   a. 150%   b. 50%
   c. $\frac{1}{2}$%   d. 200%

49. The number of accidents from all-terrain vehicles that required emergency room visits for children under 16 increased from 21,000 to 42,000 in an 10-year period. What was the percent increase? **(See Example 8.)**

50. The number of deaths from alcohol-induced causes rose from approximately 20,200 to approximately 20,700 in a 10-year period. (*Source:* Centers for Disease Control) What is the percent increase? Round to the nearest tenth of a percent.

51. Robin's health-care premium increased from $5000 per year to $5500 per year. What is the percent increase?

52. The yearly deductible for Diane's health-care plan rose from $800 to $1000. What is the percent increase?

53. Joel's yearly salary went from $42,000 to $45,000. What is the percent increase? Round to the nearest percent.

54. For a recent year, 67.5 million people participated in recreational boating. Sixteen years later, that number increased to 72.6 million. Determine the percent increase. Round to one decimal place.

55. During a recent housing slump, the median price of homes decreased in the United States. If James bought his house for $360,000 and the value 1 year later was $253,800, compute the percent decrease in the value of the house.

56. The U.S. government classified 8 million documents as secret in 2001. By 2003 (2 years after the attacks on 9-11), this number had increased to 14 million. What is the percent increase?

57. A stock closed at $12.60 per share on Monday. By Friday, the closing price was $11.97 per share. What was the percent decrease? **(See Example 9.)**

58. During a 5-year period, the number of participants collecting food stamps went from 27 million to 17 million. What is the percent decrease? Round to the nearest whole percent. (*Source:* U.S. Department of Agriculture)

59. Shanti bought a new water-efficient toilet for her house. Her old toilet used 5 gal of water per flush. The new toilet uses only 1.6 gal of water per flush. What is the percent decrease in water per flush?

60. Rafu put new insulation in his attic and discovered that his heating bill for December decreased from $160 to $140. What is the percent decrease?

61. Gus, the cat, originally weighed 12 lb. He was diagnosed with a thyroid disorder, and Dr. Smith the veterinarian found that his weight had decreased to 10.2 lb. What percent of his body weight did Gus lose?

62. To lose weight, Kelly reduced her Calories from 3000 per day to 1800 per day. What is the percent decrease in Calories?

Technology Connections

For Exercises 63–68, determine the percent increase in the price for the given stock funds over a two-year period. Round to the nearest tenth of a percent.

Courtesy of Julie Miller

	Stock Fund	Price Jan. Year 1 ($ per share)	Price Jan. Year 3 ($ per share)	Change ($)	Percent Increase
63.	Healthcare	$16.96	$104.47		
64.	Real Estate	$16.84	$23.90		
65.	Foreign Markets	$6.06	$23.05		
66.	Precious Metals	$28.66	$214.01		
67.	Technology	$118.37	$132.45		
68.	Industrial	$134.62	$212.70		

Expanding Your Skills

69. The retail price of four tickets to a concert is $648. The wholesale price of each ticket is $113.

    a. What is the markup amount per ticket?

    b. What is the markup rate? Round to one decimal place.

Simple and Compound Interest

Section 7.5

Concepts

1. Simple Interest
2. Compound Interest
3. Using the Compound Interest Formula

1. Simple Interest

In this section, we use percents to compute simple and compound interest on an investment or a loan.

Banks hold large quantities of money for their customers. They keep some cash for day-to-day transactions, but invest the remaining portion of the money. As a result, banks often pay customers interest.

When making an investment, **simple interest** is the money that is earned on principal (**principal** is the original amount of money). When people take out a loan, the amount borrowed is the principal. The interest is a percent of the amount borrowed that you must pay back in addition to the principal.

The following formulas can be used to compute simple interest for an investment or a loan and to compute the total amount in the account.

Simple Interest Formulas

Simple interest = principal × rate × time $I = Prt$

Total amount = principal + interest $A = P + I$

where I = amount of interest
    P = amount of principal
    r = annual interest rate (in decimal form)
    t = time (in years)
    A = total amount in an account

The time, t, is expressed in years because the rate, r, is an *annual* interest rate. If we are given a monthly interest rate, then the time, t, should be expressed in months.

Example 1   Computing Simple Interest

Suppose $2000 is invested in an account that earns 7% simple interest.

a. How much interest is earned after 3 years?

b. What is the total value of the account after 3 years?

Solution:

a. Principal: $P = \$2000$ Identify the parts of the formula.

  Annual interest rate: $r = 7\%$

  Time (in years): $t = 3$

  Let I represent the amount of interest.   Label the unknown.

  $I = Prt$

  $= (2000)(0.07)(3)$ Convert 7% to decimal form and substitute values into the formula.

  $= 420$ Multiply from left to right.

The amount of interest earned is $420.

Avoiding Mistakes

It is important to use the decimal form of the interest rate when calculating interest.

b. The total amount in the account is given by

$A = P + I$

$= \$2000 + \$420$

$= \$2420$

The total amount in the account is $2420.

Skill Practice

1. Suppose $1500 is invested in an account that earns 6% simple interest.
   a. How much interest is earned in 5 years?
   b. What is the total value of the account after 5 years?

When applying the simple interest formula, it is important that time be expressed in years. This is demonstrated in Example 2.

Example 2   Computing Simple Interest

Clyde takes out a loan for $3500. He pays simple interest at a rate of 6% for 4 years 3 months.

a. How much money does he pay in interest?
b. How much total money must he pay to pay off the loan?

Solution:

a. $P = \$3500$ Identify parts of the formula.

$r = 6\%$

$t = 4\frac{1}{4}$ years or 4.25 years   3 months $= \frac{3}{12}$ year $= \frac{1}{4}$ year

$I = Prt$

$= (\$3500)(0.06)(4.25)$ Substitute values into the interest formula. Convert 6% to decimal form 0.06.

$= \$892.50$ Multiply.

The interest paid is $892.50.

b. To find the total amount that must be paid, we have

$A = P + I$

$= \$3500 + \$892.50$

$= \$4392.50$ The total amount that must be paid is $4392.50.

Skill Practice

2. Morris takes out a loan for $10,000. He pays simple interest at 7% for 66 months.
   a. Write 66 months in terms of years.
   b. How much money does he pay in interest?
   c. How much total money must he repay to pay off the loan?

Answers

1. a. $450 in interest is earned.
   b. The total account value is $1950.
2. a. 66 months = 5.5 years
   b. Morris pays $3850 in interest.
   c. Morris must pay a total of $13,850 to pay off the loan.

2. Compound Interest

Simple interest is based only on a percent of the original principal. However, many day-to-day applications involve compound interest. **Compound interest** is based on both the original principal and the interest earned.

To compare the difference between simple and compound interest, consider this scenario in Example 3.

Example 3 Comparing Simple Interest and Compound Interest

Suppose $1000 is invested at 8% interest for 3 years.

a. Compute the total amount in the account after 3 years, using simple interest.
b. Compute the total amount in the account after 3 years of compounding interest annually.

Solution:

a. $P = \$1000$
$r = 8\%$
$t = 3$ years
$I = Prt$
$= (\$1000)(0.08)(3)$
$= \$240$

The amount of simple interest earned is $240. The total amount in the account is $1000 + $240 = $1240.

b. To compute interest compounded annually over a period of 3 years, compute the interest earned in the first year. Then add the principal plus the interest earned in the first year. This value then becomes the principal on which to base the interest earned in the second year. We repeat this process, finding the interest for the second and third years based on the principal and interest earned in the preceding years. This process is outlined in a table.

Year	Interest Earned $I = Prt$	Total Amount in Account
First year	$I = (\$1000)(0.08)(1) = \$80$	$\$1000 + \$80 = \$1080$
Second year	$I = (\$1080)(0.08)(1) = \$86.40$	$\$1080 + \$86.40 = \$1166.40$
Third year	$I = (\$1166.40)(0.08)(1) = \$93.31$	$\$1166.40 + 93.31 = \mathbf{\$1259.71}$

The total amount in the account by compounding interest annually is $1259.71.

Skill Practice

3. Suppose $2000 is invested at 5% interest for 3 years.
   a. Compute the total amount in the account after 3 years, using simple interest.
   b. Compute the total amount in the account after 3 years of compounding interest annually.

Notice that in Example 3 the final amount in the account is greater for the situation where interest is compounded. The difference is $1259.71 − $1240 = $19.71. By compounding interest we earn more money.

Interest may be compounded more than once per year.

Annually	1 time per year
Semiannually	2 times per year
Quarterly	4 times per year
Monthly	12 times per year
Daily	365 times per year

To compute compound interest, the calculations become very tedious. Banks use computers to perform the calculations quickly. You may want to use a calculator if calculators are allowed in your class.

Answer
3. a. $2300   b. $2315.25

3. Using the Compound Interest Formula

As you can see from Example 3, computing compound interest by hand is a cumbersome process. Can you imagine computing daily compound interest (365 times a year) by hand!

We now use a formula to compute compound interest. This formula requires the use of a scientific or graphing calculator. In particular, the calculator must have an exponent key y^x, x^y, or $\wedge$.

Let A = total amount in an account

P = principal

r = annual interest rate

t = time in years

n = number of compounding periods per year

Then $A = P \cdot \left(1 + \dfrac{r}{n}\right)^{n \cdot t}$ computes the total amount in an account.

To use this formula, note the following guidelines:

- Rate r must be expressed in decimal form.
- Time t must be the total time of the investment in *years*.
- Number n is the number of compounding periods per year.

Annual	$n = 1$
Semiannual	$n = 2$
Quarterly	$n = 4$
Monthly	$n = 12$
Daily	$n = 365$

Example 4 Computing Compound Interest by Using the Compound Interest Formula

Suppose $1000 is invested at 8% interest compounded annually for 3 years. Use the compound interest formula to find the total amount in the account after 3 years. Compare the result to the answer from Example 3(b).

FOR REVIEW

Recall the order of operations:

1. Operations inside parentheses and other grouping symbols
2. Simplify exponents, square roots, or absolute values
3. Multiplication or division in order from left to right
4. Addition or subtraction in order from left to right

Solution:

$P = \$1000$ Identify the parts of the formula.

$r = 8\%$ (0.08 in decimal form)

$t = 3$ years

$n = 1$ (annual compound interest is compounded 1 time per year)

$A = P \cdot \left(1 + \dfrac{r}{n}\right)^{n \cdot t}$

$= 1000 \cdot \left(1 + \dfrac{0.08}{1}\right)^{1 \cdot 3}$ Substitute values into the formula.

$= 1000(1 + 0.08)^3$ Apply the order of operations. Divide within parentheses. Simplify the exponent.

$= 1000(1.08)^3$ Add within parentheses.

$= 1000(1.259712)$ Evaluate $(1.08)^3 = (1.08)(1.08)(1.08) = 1.259712$.

$= 1259.712$ Multiply.

≈ 1259.71 Round to the nearest cent.

The total amount in the account after 3 years is $1259.71. This is the same value obtained in Example 3(b).

Skill Practice

4. Suppose $2000 is invested at 5% interest compounded annually for 3 years. Use the formula.

$$A = P \cdot \left(1 + \frac{r}{n}\right)^{n \cdot t}$$

to find the total amount after 3 years. Compare the answer to Skill Practice exercise 3(b).

Example 5 Computing Compound Interest by Using the Compound Interest Formula

Suppose $8000 is invested in an account that earns 5% interest compounded quarterly for $1\frac{1}{2}$ years. Use the compound interest formula to compute the total amount in the account after $1\frac{1}{2}$ years.

Solution:

$P = \$8000$ Identify the parts of the formula.
$r = 5\%$ (0.05 in decimal form)
$t = 1.5$ years
$n = 4$ (quarterly interest is compounded 4 times per year)

$$A = P \cdot \left(1 + \frac{r}{n}\right)^{n \cdot t}$$

$= 8000 \cdot \left(1 + \dfrac{0.05}{4}\right)^{(4)(1.5)}$ Substitute values into the formula.

$= 8000(1 + 0.0125)^6$ Apply the order of operations. Divide within parentheses. Simplify the exponent.

$= 8000(1.0125)^6$ Add within parentheses.

$\approx 8000(1.077383181)$ Evaluate $(1.0125)^6$. If your teacher allows the use of a calculator, consider using the exponent key.

≈ 8619.07 Multiply and round to the nearest cent.

The total amount in the account after $1\frac{1}{2}$ years is $8619.07.

Skill Practice

5. Suppose $5000 is invested at 9% interest compounded monthly for 30 years. Use the formula for compound interest to find the total amount in the account after 30 years.

Answers

4. $2315.25; this is the same as the result in Skill Practice exercise 3(b).
5. The account is worth $73,652.88 after 30 years.

Section 7.5 Activity

A.1. There are two different methods of calculating interest.
    a. Simple interest is based on _____ only.
    b. Compound interest is based on _____ and _____.

A.2. The formula for computing simple interest is:

$$I = Prt$$

    a. Define each variable: I, P, r, and t.
    b. What form should the interest rate, r, be expressed in?
    c. Time, t, is measured in _____.

A.3. Demetri took out a loan for $5000 for 4 years at a 7% interest rate.
    a. Calculate the amount of simple interest Demetri will pay at the end of 4 years.
    b. Calculate the total amount due at the end of 4 years.
    c. If the interest rate was 5% versus 7%, how much money would Demetri have saved in interest on the loan?

A.4. Liza made a $10,000 investment for 18 months that earns 4.25% in simple interest. Liza determined that she will earn $7650 in interest. Her sister calculated that Liza will earn $637.50 in interest. Who is correct and why?

A.5. Nathan invested $2000 for 3 years in an account that earns 7% simple interest.
    a. Calculate the amount of interest that Nathan earns.
    b. Calculate the total amount of the investment at the end of 3 years.

A.6. a. Consider Nathan's investment from Exercise A.5. Calculate the total amount in the account at the end of 3 years if the interest compounds annually. Use the simple interest formula to fill in the table below. The first row has been done for you.

Compound Period	Principal (P)	Interest = P*r*t	Amount in the Account
1st Year	$2000	($2000)(0.07)(1) = $140	$2000 + $140 = $2140
2nd Year	$2140		
3rd Year			

    b. How much interest was made after 3 years?
    c. Compare the amount of interest earned in Exercise A.5(a) to that in Exercise A.6(b). Do you earn less or more interest when interest compounds? Why?

A.7. The formula for computing compound interest is:

$$A = P \cdot \left(1 + \frac{r}{n}\right)^{n \cdot t}$$

    a. Identify the meaning of each variable: I, P, r, and t.
    b. What form should the interest rate, r, be expressed in?
    c. Time, t, is measured in _____.
    d. Verify the total amount in the account from Exercise A.6(a) using the compound interest formula.

A.8. Identify the value of n for interest that is compounded:
  a. annually
  b. semiannually
  c. quarterly
  d. monthly
  e. daily

A.9. If you are taking out a loan and had a choice between your loan compounding annually or monthly, what compounding period would you select and why?

A.10. Jeral and Yemisa plan to buy a new SUV. The car's sticker price is $32,000.
  a. If the sales tax rate is 5%, compute the total price of the car.
  b. If Jeral and Yemisa pay 20% down, how much is their down payment?
  c. How much do they need to finance? That is, how much do they have to borrow in a loan?
  d. They plan to pay off the car in 5 years. Interest on the loan is 8% and is compounded monthly. How much will they have to pay in total to pay off the loan?
  e. How much interest do they have to pay?
  f. What is the total price of the car that Jeral and Yemisa will pay? Include the down payment, amount of the loan, and the interest paid.

Practice Exercises — Section 7.5

Prerequisite Review

For Exercises R.1–R.2, write the percent in decimal form.

R.1. $8\frac{1}{2}\%$
R.2. 3.04%

For Exercises R.3–R.6, simplify.

R.3. $(4200)(0.03)(12)$
R.4. $(10{,}000)(0.003)(5)$
R.5. $8000(1 + 0.04)^2$
R.6. $200(1 + 0.02)^2$

Vocabulary and Key Concepts

1. a. _____ interest is the money that is earned (or owed) on an original amount of money invested (or borrowed). The original amount of money invested (or borrowed) is called the _____.

  b. Write a formula that represents the amount of simple interest I earned on an investment of P dollars at an annual interest rate r for t years.

  c. _____ interest is interest earned on both the original principal and interest already earned.

  d. Write a formula that represents the amount of money A in an account based on P dollars of principal compounded n times per year at an annual interest rate r for t years.

2. Write a formula to calculate the total amount in a simple interest account.

3. If the time of a loan or investment is given in months, divide by _____ to convert the time to years.

4. Compound interest is based on _____ and _____.

5. Compounding semiannually means interest is compounded _____ times a year, while compounding quarterly means interest is compounded _____ times a year.

6. Consider a $5000 investment that earns 6% interest compounding annually for 5 years. A (simple interest/compound interest) loan would earn more interest.

Concept 1: Simple Interest

For Exercises 7–14, find the simple interest and the total amount including interest. **(See Examples 1 and 2.)**

	Principal	Annual Interest Rate	Time, Years	Interest	Total Amount
7.	$6,000	5%	3	_____	_____
8.	$4,000	3%	2	_____	_____
9.	$5,050	6%	4	_____	_____
10.	$4,800	4%	3	_____	_____
11.	$12,000	4%	$4\frac{1}{2}$	_____	_____
12.	$6,230	7%	$6\frac{1}{3}$	_____	_____
13.	$10,500	4.5%	4	_____	_____
14.	$9,220	8%	4	_____	_____

15. Dale deposited $2500 in an account that pays $3\frac{1}{2}$% simple interest for 4 years. **(See Example 2.)**
    a. How much interest will he earn in 4 years?
    b. What will be the total value of the account after 4 years?

16. Charlene invested $3400 at 4% simple interest for 5 years.
    a. How much interest will she earn in 5 years?
    b. What will be the total value of the account after 5 years?

17. Gloria borrowed $400 for 18 months at 8% simple interest.
    a. How much interest will Gloria have to pay?
    b. What will be the total amount that she has to pay back?

18. Floyd borrowed $1000 for 2 years 3 months at 8% simple interest.
    a. How much interest will Floyd have to pay?
    b. What will be the total amount that he has to pay back?

19. Jozef deposited $10,300 into an account paying 4% simple interest 5 years ago. If he withdraws the entire amount of money, how much will he have?

20. Heather invested $20,000 in an account that pays 6% simple interest. If she invests the money for 10 years, how much will she have?

21. Anne borrowed $4500 from a bank that charges 10% simple interest. If she repays the loan in $2\frac{1}{2}$ years, how much will she have to pay back?

22. Dan borrowed $750 from his brother who is charging 8% simple interest. If Dan pays his brother back in 6 months, how much does he have to pay back?

Ryan McVay/Photodisc/Getty Images

Concept 2: Compound Interest

23. If a bank compounds interest semiannually for 3 years, how many total compounding periods are there?

24. If a bank compounds interest quarterly for 2 years, how many total compounding periods are there?

25. If a bank compounds interest monthly for 2 years, how many total compounding periods are there?

26. If a bank compounds interest monthly for $1\frac{1}{2}$ years, how many total compounding periods are there?

27. Mary Ellen deposited $500 in a bank. **(See Example 3.)**

   a. If the bank offers 4% simple interest, compute the amount in the account after 3 years.

   b. Now suppose the bank offers 4% interest compounded annually. Complete the table to determine the amount in the account after 3 years.

Year	Interest Earned	Total Amount in Account
1		
2		
3		

28. The amount of $8000 is invested at 4% for 3 years.

   a. Compute the ending balance if the bank calculates simple interest.

   b. Compute the ending balance if the bank calculates interest compounded annually.

   c. How much more interest is earned in the account with compound interest?

Year	Interest Earned	Total Amount in Account
1		
2		
3		

29. Fatima deposited $24,000 in an account.

   a. If the bank offers 5% simple interest, compute the amount in the account after 2 years.

   b. Now suppose the bank offers 5% compounded semiannually (twice per year). Complete the table to determine the amount in the account after 2 years.

Period	Interest Earned	Total Amount in Account
1st		
2nd		
3rd		
4th		

30. The amount of $12,000 is invested at 8% for 1 year.

   a. Compute the ending balance if the bank calculates simple interest.

   b. Compute the ending balance if the bank calculates interest compounded quarterly.

   c. How much more interest is earned in the account with compound interest?

Period	Interest Earned	Total Amount in Account
1st		
2nd		
3rd		
4th		

Concept 3: Using the Compound Interest Formula

31. For the formula $A = P \cdot \left(1 + \dfrac{r}{n}\right)^{n \cdot t}$, identify what each variable means.

32. If $1000 is deposited in an account paying 8% interest compounded monthly for 3 years, label the variables: P, r, n, and t.

Technology Connections

For Exercises 33–40, find the total amount in an account for an investment subject to the following conditions. (See Examples 4 and 5.)

	Principal	Annual Interest Rate	Time, Years	Compounded	Total Amount
33.	$5,000	4.5%	5	Annually	_____
34.	$12,000	5.25%	4	Annually	_____
35.	$6,000	5%	2	Semiannually	_____
36.	$4,000	3%	3	Semiannually	_____
37.	$10,000	6%	$1\frac{1}{2}$	Quarterly	_____
38.	$9,000	4%	$2\frac{1}{2}$	Quarterly	_____
39.	$14,000	4.5%	3	Monthly	_____
40.	$9,000	8%	2	Monthly	_____

Chapter 7 Review Exercises

Section 7.1

For Exercises 1–4, use a percent to express the shaded portion of each drawing.

1.

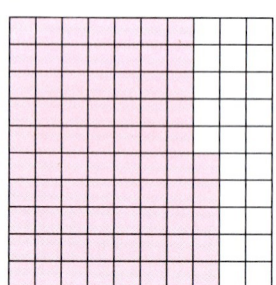

2.

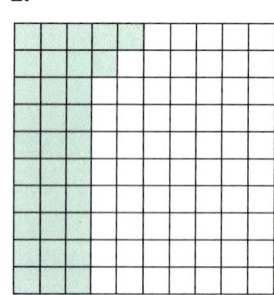

3.

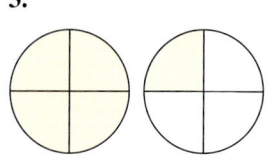

4.

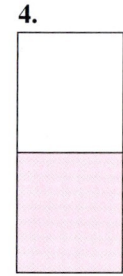

5. 68% can be expressed as which of the following forms? Identify all that apply.
   a. $\dfrac{68}{1000}$
   b. $\dfrac{68}{100}$
   c. 0.68
   d. 0.068

6. 0.4% can be expressed as which of the following forms? Identify all that apply.
   a. $\dfrac{4}{100}$
   b. 0.04
   c. $\dfrac{0.4}{100}$
   d. 0.004

For Exercises 7–14, write the percent as a fraction and as a decimal.

7. 30%

8. 95%

9. 135%

10. 212%

11. 0.2%

12. 0.6%

13. $66\frac{2}{3}\%$

14. $33\frac{1}{3}\%$

For Exercises 15–24, write the fraction or decimal as a percent. Round to the nearest tenth of a percent if necessary.

15. $\dfrac{5}{8}$ 16. $\dfrac{7}{20}$

17. $\dfrac{7}{4}$ 18. $\dfrac{11}{5}$

19. 0.006   20. 0.001

21. 4   22. 6

23. $\dfrac{3}{7}$ 24. $\dfrac{9}{11}$

Section 7.2

For Exercises 25–28, write the percent proportion.

25. 6 books of 8 books is 75%.

26. 15% of 180 lb is 27 lb.

27. 200% of $420 is $840.

28. 6 pine trees out of 2000 pine trees is 0.3%.

For Exercises 29–34, solve the percent problems, using proportions.

29. What is 12% of 50?

30. $5\tfrac{3}{4}$% of 64 is what number?

31. 11 is what percent of 88?

32. 8 is what percent of 2500?

33. 13 is $33\tfrac{1}{3}$% of what number?

34. 24 is 120% of what number?

35. Based on recent statistics, one airline expects that 4.2% of its customers will be "no-shows." If the airline sold 260 seats, how many people would the airline expect as no-shows? Round to the nearest whole unit.

36. In a survey of college students, 58% said that they wore their seat belts regularly. If this represents 493 people, how many people were surveyed?

37. Victoria spends $720 per month on rent. If her monthly take-home pay is $1800, what percent does she pay in rent?

38. Of the rental cars at the U-Rent-It company, 40% are compact cars. If this represents 26 cars, how many cars are on the lot?

Section 7.3

For Exercises 39–44, write as a percent equation and solve.

39. 18% of 900 is what number?

40. What number is 29% of 404?

41. 18.90 is what percent of 63?

42. What percent of 250 is 86?

43. 30 is 25% of what number?

44. 26 is 130% of what number?

45. A student buys a used book for $54.40. This is 80% of the original price. What was the original price?

46. Veronica has read 330 pages of a 600-page novel. What percent of the novel has she read?

47. Elaine tries to keep her fat intake to no more than 30% of her total calories. If she has a 2400-calorie diet, how many fat calories can she consume to stay within her goal?

48. In 2010, 13% of Americans were over the age of 65. By 2050 that number could rise to 20%. Suppose that the U.S. population grows from 300,000,000 in 2010 to 404,000,000 in 2050.

Big Cheese Photo/Getty Images

a. Find the number of Americans over 65 in the year 2010.

b. Find the number of Americans over 65 in the year 2050.

Section 7.4

For Exercises 49–52, solve the problem involving sales tax.

49. A TV costs $1279. Find the sales tax if the rate is 6%.

50. The sales tax on a sofa is $47.95. If the sofa costs $685.00 before tax, what is the sales tax rate?

51. Kim has a digital camera. She decides to make prints for 40 photos. If the sales tax at a rate of 5% comes to $0.44, what was the price of the photos before tax? What is the cost per photo?

52. A resort hotel charges an 11% resort tax along with the 6% sales tax. If the hotel's one-night accommodation is $225.00, what will a guest pay for 4 nights, including tax?

For Exercises 53–56, solve the problems involving commission.

53. At a recent auction, *Boy with a Pipe,* an early work by Pablo Picasso, sold for $104 million. The commission for the sale of the work was $11 million. What was the rate of commission? Round to the nearest tenth of a percent. (*Source: The New York Times*)

54. Andre earns a commission of 12% on sales of restaurant supplies. If he sells $4075 in one week, how much commission will he earn?

55. Sela sells sportswear at a department store. She earns an hourly wage of $15, and she gets a 5% commission on all merchandise that she sells over $200. If Sela works an 8-hr day and sells $420 of merchandise, how much will she earn that day?

56. A real estate agent earned $9600 for the sale of a house. If her commission rate is 4%, for how much did the house sell?

For Exercises 57–60, solve the problems involving discount and markup.

57. Find the discount and the sale price of the movie if the regular price is $28.95.

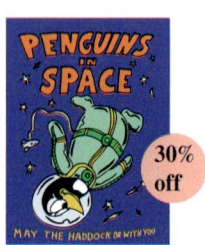

58. This laptop computer was originally priced at $1299. How much is the discount? After the $50 rebate, how much will a person pay for this computer?

59. A rug manufacturer sells a rug to a retail store for $160. The store then marks up the rug to $208. What is the markup rate?

60. Peg sold some homemade baskets to a store for $50 each. The store marks up all merchandise by 18%. What will be the retail price of the baskets after the markup?

For Exercises 61 and 62, solve the problem involving percent increase or decrease.

61. The number of species of animals on the endangered species list went from 263 in 1990 to 574 in 2010. Find the percent increase. Round to the nearest tenth of a percent. (*Source:* U.S. Fish and Wildlife Services)

62. During a weak period in the economy, the stock price for Hershey Foods fell from $50.62 per share to $32.68 per share. Compute the percent decrease. Round to the nearest tenth of a percent.

Section 7.5

For Exercises 63 and 64, find the simple interest and the total amount including interest.

	Principal	Annual Interest Rate	Time, Years	Interest	Total Amount
63.	$10,200	3%	4	_____	_____
64.	$7000	4%	5	_____	_____

65. Jean-Luc borrowed $2500 at 5% simple interest. What is the total amount that he will pay back at the end of 18 months (1.5 years)?

66. Kyle loaned his brother Steve $800 and charged 2.5% simple interest. If Steve pays all the money back (principal plus interest) at the end of 2 years, how much will Steve pay his brother?

67. Sydney deposited $6000 in a certificate of deposit that pays 4% interest compounded annually. Complete the table to determine her balance after 3 years.

Year	Interest	Total
1		
2		
3		

68. Nell deposited $10,000 in a money market account that pays 3% interest compounded semiannually. Complete the table to find her balance after 2 years.

Compound Periods	Interest	Total
Period 1 (end of first 6 months)		
Period 2 (end of year 1)		
Period 3 (end of 18 months)		
Period 4 (end of year 2)		

For Exercises 69–72, find the total amount for the investment, using compound interest. Use the formula $A = P \cdot \left(1 + \dfrac{r}{n}\right)^{n \cdot t}$.

	Principal	Annual Interest Rate	Time in Years	Compounded	Total Amount
69.	$850	8%	2	Quarterly	_____
70.	$2050	5%	5	Semiannually	_____
71.	$11,000	7.5%	6	Annually	_____
72.	$8200	4.5%	4	Monthly	_____

Measurement and Geometry

8

CHAPTER OUTLINE

- 8.1 U.S. Customary Units of Measurement 446
- 8.2 Metric Units of Measurement 459
- 8.3 Converting Between U.S. Customary and Metric Units 473

  Problem Recognition Exercises: U.S. Customary and Metric Conversions 481
- 8.4 Medical Applications Involving Measurement 482
- 8.5 Lines and Angles 485
- 8.6 Triangles and the Pythagorean Theorem 496
- 8.7 Perimeter, Circumference, and Area 508

  Problem Recognition Exercises: Area, Perimeter, and Circumference 520
- 8.8 Volume and Surface Area 521

  Chapter 8 Review Exercises 530

Mathematics in Shapes

Anyone who has done home improvement knows that a working knowledge of simple geometry can be critical to an efficient and cost-effective home project. For example, suppose that a contractor wants to build a one-room addition to a house. The room is rectangular with a length of 20 ft and a width of 16 ft. The first step is to pour a 9-in. concrete slab ($\frac{3}{4}$ ft thick) as the foundation of the room. Thus, the amount of concrete needed is given by the volume of the slab. To determine the amount of carpeting to buy, the contractor computes the area of the slab. Baseboards are to go around the perimeter of the room, with 6 ft subtracted to account for two doorways. The walls are 10 ft high, and the contractor computes the area of the four walls to estimate how much paint to buy.

Corbis/SuperStock

Volume of concrete slab: $V = lwh = (20 \text{ ft})(16 \text{ ft})(0.75 \text{ ft}) = 240 \text{ ft}^3$

Area of floor: $A = lw = (20 \text{ ft})(16 \text{ ft}) = 320 \text{ ft}^2$

Perimeter of floor: $P = 2l + 2w = 2(20 \text{ ft}) + 2(16 \text{ ft}) = 72 \text{ ft}$

Perimeter minus doors: 72 ft − 6 ft = 66 ft

Area of four walls: $A = 2 \cdot (20 \text{ ft})(10 \text{ ft}) + 2 \cdot (16 \text{ ft})(10 \text{ ft}) = 720 \text{ ft}^2$

Section 8.1 U.S. Customary Units of Measurement

Concepts

1. U.S. Customary Units
2. U.S. Customary Units of Length
3. Units of Time
4. U.S. Customary Units of Weight
5. U.S. Customary Units of Capacity

TIP: Sometimes units of feet are denoted with the ′ symbol. That is, 3 ft = 3′. Similarly, sometimes units of inches are denoted with the ″ symbol. That is, 4 in. = 4″.

1. U.S. Customary Units

In many applications in day-to-day life, we need to measure things. To measure an object means to assign it a number and a **unit of measure**. In this section, we present units of measure used in the United States for length, time, weight, and capacity. Table 8-1 gives several commonly used units and their equivalents.

Table 8-1 Summary of U.S. Customary Units of Length, Time, Weight, and Capacity

Length	Time
1 foot (ft) = 12 inches (in.)	1 year (yr) = 365 days
1 yard (yd) = 3 feet (ft)	1 week (wk) = 7 days
1 mile (mi) = 5280 feet (ft)	1 day = 24 hours (hr)
1 mile (mi) = 1760 yards (yd)	1 hour (hr) = 60 minutes (min)
	1 minute (min) = 60 seconds (sec)
Capacity	**Weight**
1 tablespoon (T) = 3 teaspoons (tsp)	1 pound (lb) = 16 ounces (oz)
1 cup (c) = 8 fluid ounces (fl oz)	1 ton = 2000 pounds (lb)
1 pint (pt) = 2 cups (c)	
1 quart (qt) = 2 pints (pt)	
1 quart (qt) = 4 cups (c)	
1 gallon (gal) = 4 quarts (qt)	

2. U.S. Customary Units of Length

In Example 1, we will demonstrate how to convert between two units of measure by multiplying by a conversion factor. A **conversion factor** is a ratio of equivalent measures.

For example, note that 1 yd = 3 ft. Therefore, $\frac{1 \text{ yd}}{3 \text{ ft}} = 1$ and $\frac{3 \text{ ft}}{1 \text{ yd}} = 1$.

These conversion factors are unit ratios or unit fractions because the quotient is 1. To convert from one unit of measure to another, we can multiply by an appropriate conversion factor. We offer these guidelines to determine the proper conversion factor to use.

Choosing a Conversion Factor

In a conversion factor,
- The unit of measure in the numerator should be the new unit you want to convert *to*.
- The unit of measure in the denominator should be the original unit you want to convert *from*.

Example 1 — Converting Units of Length by Using Conversion Factors

Convert the units of length.

a. 1500 ft = _____ yd   b. 9240 yd = _____ mi   c. 8.2 mi = _____ ft

Solution:

a. From Table 8-1, we have 1 yd = 3 ft, therefore, $\dfrac{1 \text{ yd}}{3 \text{ ft}} = 1$.

$$1500 \text{ ft} = 1500 \text{ ft} \cdot \dfrac{1 \text{ yd}}{3 \text{ ft}} \quad \leftarrow \text{new unit to convert to} \\ \leftarrow \text{unit to convert from}$$

$$= \dfrac{1500 \text{ ft}}{1} \cdot \dfrac{1 \text{ yd}}{3 \text{ ft}}$$

$$= \dfrac{1500}{3} \text{ yd}$$

$$= 500 \text{ yd}$$

Notice that the original units of ft divide out in the same way as simplifying common factors. The unit yd remains in the final answer.

b. From Table 8-1, we have 1 mi = 1760 yd, therefore, $\dfrac{1 \text{ mi}}{1760 \text{ yd}} = 1$.

$$9240 \text{ yd} = 9240 \text{ yd} \cdot \dfrac{1 \text{ mi}}{1760 \text{ yd}} \quad \leftarrow \text{new unit to convert to} \\ \leftarrow \text{unit to convert from}$$

$$= \dfrac{9240 \text{ yd}}{1} \cdot \dfrac{1 \text{ mi}}{1760 \text{ yd}} \quad \text{The units of yd divide out, leaving the answer in miles.}$$

$$= \dfrac{9240}{1760} \text{ mi} \quad \text{Multiply fractions.}$$

$$= 5.25 \text{ mi} \quad \text{Simplify.}$$

c. From Table 8-1, we have 1 mi = 5280 ft, therefore, $\dfrac{5280 \text{ ft}}{1 \text{ mi}} = 1$.

$$8.2 \text{ mi} = 8.2 \text{ mi} \cdot \dfrac{5280 \text{ ft}}{1 \text{ mi}} \quad \leftarrow \text{new unit to convert to} \\ \leftarrow \text{unit to convert from}$$

$$= \dfrac{8.2 \text{ mi}}{1} \cdot \dfrac{5280 \text{ ft}}{1 \text{ mi}} \quad \text{The units of mi divide out, leaving the answer in feet.}$$

$$= 43{,}296 \text{ ft}$$

TIP: It is important to write the units associated with the numbers. The units can help you select the correct conversion factor.

Skill Practice Convert, using conversion factors.

1. 720 in. = _____ ft
2. 4224 ft = _____ mi
3. 8 mi = _____ yd

Answers
1. 60 ft   2. 0.8 mi   3. 14,080 yd

Example 2 Making Multiple Conversions of Length

Convert the units of length.

a. 0.25 mi = _____ in. **b.** 22 in. = _____ yd

Solution:

a. To convert miles to inches, we use two conversion factors. The first converts miles to feet. The second converts feet to inches.

$$0.25 \text{ mi} = 0.25 \text{ mi} \cdot \frac{5280 \text{ ft}}{1 \text{ mi}} \cdot \frac{12 \text{ in.}}{1 \text{ ft}}$$

$$= \frac{0.25 \text{ mi}}{1} \cdot \frac{5280 \text{ ft}}{1 \text{ mi}} \cdot \frac{12 \text{ in.}}{1 \text{ ft}}$$

The units mi and ft divide out, leaving the answer in inches.

$$= 15{,}840 \text{ in.}$$

b. $22 \text{ in.} = 22 \text{ in.} \cdot \frac{1 \text{ ft}}{12 \text{ in.}} \cdot \frac{1 \text{ yd}}{3 \text{ ft}}$

Multiply by two conversion factors. The first converts inches to feet. The second converts feet to yards.

$$= \frac{22 \text{ in.}}{1} \cdot \frac{1 \text{ ft}}{12 \text{ in.}} \cdot \frac{1 \text{ yd}}{3 \text{ ft}}$$

The units in. and ft divide out, leaving the answer in yards.

$$= \frac{22}{36} \text{ yd}$$

Multiply fractions.

$$= \frac{11}{18} \text{ yd} \quad \text{or} \quad 0.6\overline{1} \text{ yd}$$

Simplify.

Skill Practice Convert, using conversion factors.

4. 6.2 yd = _____ in. **5.** 6336 in. = _____ mi

To add and subtract measurements, we must have like units. For example:

$$3 \text{ ft} + 8 \text{ ft} = 11 \text{ ft}$$

Sometimes, however, measurements have mixed units. For example, a drainpipe might be 4 ft 6 in. long or symbolically 4′6″. Measurements and calculations with mixed units can be handled in much the same way as mixed numbers.

Example 3 Adding and Subtracting Mixed Units of Measurement

a. Add. 4′6″ + 2′9″ **b.** Subtract. 8′2″ − 3′6″

Solution:

a. 4′6″ + 2′9″ =   4 ft + 6 in.
                  + 2 ft + 9 in.
                  ─────────────────
                    6 ft + 15 in.   Add like units.

= 6 ft + 1 ft + 3 in.   Because 15 in. is more than 1 ft, we can write 15 in. = 12 in. + 3 in.

= 7 ft 3 in.   or   7′3″   = 1 ft + 3 in.

Answers
4. 223.2 in.   **5.** 0.1 mi

b. $8'2'' - 3'6'' = $ $8 \text{ ft} + 2 \text{ in.} = \overset{7}{\cancel{8}} \text{ ft} + \overset{12 \text{ in.}}{2} \text{ in.}$ Borrow 1 ft = 12 in.
$\underline{-(3 \text{ ft} + 6 \text{ in.})} \quad \underline{-(3 \text{ ft} + 6 \text{ in.})}$

$$ = \begin{array}{r} 7 \text{ ft} + 14 \text{ in.} \\ \underline{-(3 \text{ ft} + 6 \text{ in.})} \\ 4 \text{ ft} + 8 \text{ in.} \quad \text{or} \quad 4'8'' \end{array}$$

Skill Practice

6. Add. $8'4'' + 4'10''$ **7.** Subtract. $6'3'' - 4'9''$

3. Units of Time

In Examples 4 and 5, we convert between two units of time.

Example 4   **Converting Units of Time**

Convert the units of time.

a. 32 hr = _____ days   **b.** 36 hr = _____ sec

Solution:

a. $32 \text{ hr} = \dfrac{32 \text{ hr}}{1} \cdot \dfrac{1 \text{ day}}{24 \text{ hr}}$ ← new unit to convert to   Recall that
      ← unit to convert from   1 day = 24 hr.

$\phantom{32 \text{ hr}} = \dfrac{32}{24} \text{ days}$ Multiply fractions.

$\phantom{32 \text{ hr}} = \dfrac{4}{3} \text{ days or } 1\dfrac{1}{3} \text{ days}$ Simplify.

                     converts   converts
                     hr to min   min to sec

b. $36 \text{ hr} = \dfrac{36 \text{ hr}}{1} \cdot \dfrac{60 \text{ min}}{1 \text{ hr}} \cdot \dfrac{60 \text{ sec}}{1 \text{ min}}$ Multiply by two conversion factors.

$\phantom{36 \text{ hr}} = 129{,}600 \text{ sec}$ Simplify.

Skill Practice Convert.

8. 16 hr = _____ days   **9.** 24 hr = _____ sec

Example 5   **Converting Units of Time**

After running a marathon, Dave crossed the finish line and noticed that the race clock read 2:20:30. Convert this time to minutes.

Solution:

The notation 2:20:30 means 2 hr 20 min 30 sec. We must convert 2 hr to minutes and 30 sec to minutes. Then we add the total number of minutes.

$2 \text{ hr} = \dfrac{2 \text{ hr}}{1} \cdot \dfrac{60 \text{ min}}{1 \text{ hr}} = 120 \text{ min}$

$30 \text{ sec} = \dfrac{30 \text{ sec}}{1} \cdot \dfrac{1 \text{ min}}{60 \text{ sec}} = \dfrac{30}{60} \text{ min} = \dfrac{1}{2} \text{ min} \quad \text{or} \quad 0.5 \text{ min}$

The total number of minutes is 120 min + 20 min + 0.5 min = 140.5 min. Dave finished the race in 140.5 min.

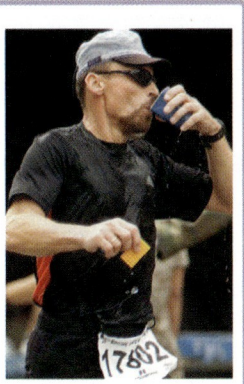

LCPL Casey N. Thurston, USMC/DoD Media

Answers

6. $13'2''$ **7.** $1'6''$

8. $\dfrac{2}{3}$ day   **9.** 86,400 sec

Skill Practice

10. When Ward Burton won the Daytona 500 his time was 2:22:45. Convert this time to minutes.

4. U.S. Customary Units of Weight

Measurements of weight record the force of an object subject to gravity. In Example 6, we convert from one unit of weight to another.

Example 6 — Converting Units of Weight

a. The average weight of an adult male African elephant is 12,400 lb. Convert this value to tons.

b. Convert the weight of a 7-lb 3-oz baby to ounces.

Solution:

a. Recall that 1 ton = 2000 lb.

$$12{,}400 \text{ lb} = \frac{12{,}400 \text{ lb}}{1} \cdot \frac{1 \text{ ton}}{2000 \text{ lb}}$$

$$= \frac{12{,}400}{2000} \text{ tons} \qquad \text{Multiply fractions.}$$

$$= \frac{31}{5} \text{ tons} \quad \text{or} \quad 6.2 \text{ tons}$$

David Fettes/Cultura/Getty Images

An adult male African elephant weighs 6.2 tons.

b. To convert 7 lb 3 oz to ounces, we must convert 7 lb to ounces.

$$7 \text{ lb} = \frac{7 \text{ lb}}{1} \cdot \frac{16 \text{ oz}}{1 \text{ lb}} \qquad \text{Recall that 1 lb = 16 oz.}$$

$$= 112 \text{ oz}$$

The baby's total weight is 112 oz + 3 oz = 115 oz.

Skill Practice

11. The blue whale is the largest animal on Earth. It is so heavy that it would be crushed under its own weight if it were taken from the water. An average adult blue whale weighs 120 tons. Convert this to pounds.

12. A box of apples weighs 5 lb 12 oz. Convert this to ounces.

Example 7 — Applying U.S. Customary Units of Weight

Jessica lifts four boxes of books. The boxes have the following weights: 16 lb 4 oz, 18 lb 8 oz, 12 lb 5 oz, and 22 lb 9 oz. How much weight did she lift altogether?

Solution:

```
  16 lb   4 oz            Add like units in columns.
  18 lb   8 oz
  12 lb   5 oz
+ 22 lb   9 oz
  68 lb  26 oz  = 68 lb + 26 oz
               = 68 lb + 1 lb + 10 oz     Recall that 1 lb = 16 oz.
               = 69 lb 10 oz
```

Jessica lifted 69 lb 10 oz of books.

Answers
10. 142.75 min 11. 240,000 lb
12. 92 oz

Skill Practice

13. A set of triplets weighed 4 lb 3 oz, 3 lb 9 oz, and 4 lb 5 oz. What is the total weight of all three babies?

5. U.S. Customary Units of Capacity

A typical can of soda contains 12 fl oz. This is a measure of capacity. Capacity is the volume or amount that a container can hold. The U.S. Customary units of capacity are fluid ounces (fl oz), cup (c), pint (pt), quart (qt), and gallon (gal).

One fluid ounce is approximately the amount of liquid that two large spoonfuls will hold. One cup is the amount in an average-size cup of tea. While Table 8-1 summarizes the relationships among units of capacity, we also offer an illustration (Figure 8-1).

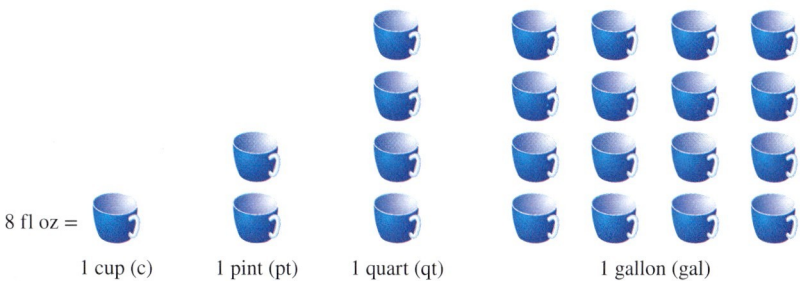

8 fl oz =   1 cup (c)   1 pint (pt)   1 quart (qt)   1 gallon (gal)

Figure 8-1

Example 8 — Converting Units of Capacity

Convert the units of capacity.

a. 1.25 pt = _____ qt **b.** 2 gal = _____ c **c.** 48 fl oz = _____ gal

Solution:

a. $1.25 \text{ pt} = \dfrac{1.25 \text{ pt}}{1} \cdot \dfrac{1 \text{ qt}}{2 \text{ pt}}$ Recall that 1 qt = 2 pt.

$= \dfrac{1.25}{2} \text{ qt}$ Multiply fractions.

$= 0.625 \text{ qt}$ Simplify.

b. $2 \text{ gal} = 2 \text{ gal} \cdot \dfrac{4 \text{ qt}}{1 \text{ gal}} \cdot \dfrac{4 \text{ c}}{1 \text{ qt}}$ Use two conversion factors. The first converts gallons to quarts. The second converts quarts to cups.

$= \dfrac{2 \text{ gal}}{1} \cdot \dfrac{4 \text{ qt}}{1 \text{ gal}} \cdot \dfrac{4 \text{ c}}{1 \text{ qt}}$

$= 32 \text{ c}$ Multiply.

c. $48 \text{ fl oz} = \dfrac{48 \text{ fl oz}}{1} \cdot \dfrac{1 \text{ c}}{8 \text{ fl oz}} \cdot \dfrac{1 \text{ qt}}{4 \text{ c}} \cdot \dfrac{1 \text{ gal}}{4 \text{ qt}}$ Convert from fluid ounces to cups, from cups to quarts, and from quarts to gallons.

$= \dfrac{48}{128} \text{ gal}$

$= \dfrac{3}{8} \text{ gal}$ or 0.375 gal

Avoiding Mistakes

It is important to note that ounces (oz) and fluid ounces (fl oz) are different quantities. An ounce (oz) is a measure of weight, and a fluid ounce (fl oz) is a measure of capacity. Furthermore,

16 oz = 1 lb
8 fl oz = 1 c

Skill Practice Convert.

14. 8.5 gal = _____ qt **15.** 2.25 qt = _____ c **16.** 40 fl oz = _____ qt

Answers

13. 12 lb 1 oz **14.** 34 qt
15. 9 c **16.** 1.25 qt

Example 9 Applying Units of Capacity

A recipe calls for $1\frac{3}{4}$ c of chicken broth. A can of chicken broth holds 14.5 fl oz. Is there enough chicken broth in the can for the recipe?

Solution:

We need to convert each measurement to the same unit of measure for comparison. Converting $1\frac{3}{4}$ c to fluid ounces, we have

$$1\frac{3}{4}c = \frac{7}{4}\cancel{c} \cdot \frac{8 \text{ fl oz}}{1 \cancel{c}} \quad \text{Recall that 1 c = 8 fl oz.}$$

$$= \frac{56}{4} \text{ fl oz} \quad \text{Multiply fractions.}$$

$$= 14 \text{ fl oz} \quad \text{Simplify.}$$

The recipe calls for $1\frac{3}{4}$ c or 14 fl oz of chicken broth. The can of chicken broth holds 14.5 fl oz, which is enough.

Jill Braaten/McGraw Hill

Skill Practice

17. A recipe calls for $3\frac{1}{2}$ c of tomato sauce. A jar of sauce holds 24 fl oz. Is there enough sauce in the jar for the recipe?

Answer

17. No, $3\frac{1}{2}$ c is equal to 28 fl oz. The jar only holds 24 fl oz.

Section 8.1 Activity

A.1. Consider the multiplication of two fractions:

$$\frac{5}{7} \cdot \frac{2}{5}$$

  a. Simplify the product. What common factor divides out?
  b. What quotient results when you divide common factors as in $\frac{5}{5}$?
  c. What is the product of any number and 1?

A.2. a. The image below illustrates a keypad that is 1 foot in length. If the ruler below the images is in increments of inches, how many inches is the keypad?

  b. Based on the relationship in part (a), complete the unit equivalent.

$$\boxed{} \text{ foot (ft)} = \boxed{} \text{ inches (in.)}$$

c. Write the unit equivalent as a conversion factor:

$$\frac{\square \text{ ft}}{\square \text{ in.}} = \frac{\square \text{ in.}}{\square \text{ ft}}$$

d. What is the quotient of each conversion factor from part (c)?
e. What is the product of any number and 1?

A.3. Consider the multiplication of a measurement of length times a conversion factor:

$$\frac{1 \text{ ft}}{1} \cdot \frac{12 \text{ in.}}{1 \text{ ft}}$$

a. Simplify the fraction. What common *unit* divides out?
b. What *unit* remains in the product?

A.4. Convert 15 ft to yards.
  a. Write 15 ft as a ratio with a denominator of 1. Make sure to include the units in the ratio.
  b. Multiply the ratio from part (a) by the conversion factor of feet and yards. Write the conversion factor as a ratio so that the unit of yards is in the numerator and the unit of feet is in the denominator.
  c. Multiply the fractions and divide out the units of feet. Be sure to include the unit of yards in the answer.
  d. Based on this exercise (circle the correct answer):
     The new unit to *convert to* is the unit of measure in the: numerator/denominator
  e. Based on this exercise (circle the correct answer):
     The unit to *convert from* is the unit of measure in the: numerator/denominator

For Exercises A.5–A.12, convert to the new unit of measure using conversion factors. Refer to Table 8-1 in the text if necessary.

A.5. 1.25 mi = _____ yd

A.6. 880 yd = _____ ft

A.7. 36 min = _____ hr

A.8. 4.2 yr = _____ days

A.9. 40.75 tons = _____ lb

A.10. 3 lb 6 oz = _____ lb

A.11. 1.75 pt = _____ qt

A.12. 6 gal = _____ c

For Exercises A.13–A.15, add or subtract.

A.13. 2.5 yd + 2.5 ft = _____ ft

A.14. 19′ − 2′9″ = _____ in.

A.15. 22 min + 330 sec _____ min

A.16. As of 2018, the world record time for the men's mile run was just under 4 minutes. The official time was 3 minutes 43.13 seconds. Convert this value to seconds.

A.17. A cat weighing 11 lb 4 oz is diagnosed with a kidney disease. Three months later, the cat weighs 9 lb 10 oz. How much weight did the cat lose? Express the answer in pounds and ounces.

A.18. A recipe for spaghetti sauce calls for 3½ c of tomato sauce. One can of tomato sauce holds 24.5 fl oz. Is there enough tomato sauce to make the recipe?

Section 8.1 Practice Exercises

Prerequisite Review

For Exercises R.1–R.6, simplify.

R.1. $24{,}300 \cdot \dfrac{1}{5} \cdot \dfrac{1}{90}$

R.2. $300 \cdot \dfrac{1}{8} \cdot \dfrac{24}{1}$

R.3. $38.4 \times \dfrac{1}{3} \times \dfrac{10}{1}$

R.4. $25 \times \dfrac{10}{1} \times \dfrac{1}{5}$

R.5. $(12)\left(\dfrac{3}{1}\right)\left(\dfrac{1}{60}\right)$

R.6. $(18)\left(\dfrac{1}{300}\right)\left(\dfrac{6}{1}\right)$

Vocabulary and Key Concepts

1. To convert between two units of measure, we multiply by a ratio of equivalent measures called a _____ factor.

Concept 2: U.S. Customary Units of Length

2. Identify an appropriate conversion factor to convert feet to inches by using multiplication.

   a. $\dfrac{12 \text{ ft}}{1 \text{ in.}}$ **b.** $\dfrac{12 \text{ in.}}{1 \text{ ft}}$ **c.** $\dfrac{1 \text{ ft}}{12 \text{ in.}}$ **d.** $\dfrac{1 \text{ in.}}{12 \text{ ft}}$

For Exercises 3–20, convert the units of length by using conversion factors. **(See Examples 1 and 2.)**

3. 9 ft = _____ yd

4. $2\dfrac{1}{3}$ yd = _____ ft

5. 3.5 ft = ___ in.

6. $4\dfrac{1}{2}$ in. = _____ ft

7. 11,880 ft = _____ mi

8. 0.75 mi = _____ ft

9. 14 ft = _____ yd

10. 75 in. = _____ ft

11. 320 mi = _____ yd

12. $3\dfrac{1}{4}$ ft = _____ in.

13. 171 in. = _____ yd

14. 0.3 mi = _____ in.

15. 2 yd = _____ in.

16. 12,672 in. = _____ mi

17. 0.8 mi = _____ in.

18. 900 in. = _____ yd

19. 31,680 in. = _____ mi

20. 6 yd = _____ in.

21. **a.** Convert 6′4″ to inches.
    b. Convert 6′4″ to feet.

22. **a.** Convert 10 ft 8 in. to inches.
    b. Convert 10 ft 8 in. to feet.

23. **a.** Convert 2 yd 2 ft to feet.
    b. Convert 2 yd 2 ft to yards.

24. **a.** Convert 3′6″ to feet.
    b. Convert 3′6″ to inches.

For Exercises 25–30, add or subtract as indicated. (See Example 3.)

25. 2 ft 8 in. + 3 ft 4 in. **26.** 5 ft 2 in. + 6 ft 10 in. **27.** 8 ft 8 in. − 5 ft 4 in.

28. 3 ft 2 in. − 1 ft 5 in. **29.** 9′2″ − 4′10″ **30.** 4′10″ + 6′4″

Concept 3: Units of Time

For Exercises 31–42, convert the units of time. (See Example 4.)

31. 2 yr = _____ days **32.** $1\frac{1}{2}$ days = _____ hr **33.** 90 min = _____ hr **34.** 3 wk = _____ days

35. 180 sec = _____ min **36.** $3\frac{1}{2}$ hr = _____ min **37.** 72 hr = _____ days **38.** 28 days = _____ wk

39. 3600 sec = _____ hr **40.** 168 hr = _____ wk **41.** 9 wk = _____ hr **42.** 1680 hr = _____ wk

For Exercises 43–46, convert the time given as hr:min:sec to minutes. (See Example 5.)

43. 1:20:30 **44.** 3:10:45 **45.** 2:55:15 **46.** 1:40:30

Concept 4: U.S. Customary Units of Weight

For Exercises 47–52, convert the units of weight. (See Example 6.)

47. 32 oz = _____ lb **48.** 2500 lb = _____ tons **49.** 2 tons = _____ lb

50. 4 lb = _____ oz **51.** $3\frac{1}{4}$ tons = _____ lb **52.** 3000 lb = _____ tons

For Exercises 53–58, add or subtract as indicated. (See Example 7.)

53. 6 lb 10 oz + 3 lb 14 oz **54.** 12 lb 11 oz + 13 lb 7 oz **55.** 30 lb 10 oz − 22 lb 8 oz

56. 5 lb − 2 lb 5 oz **57.** 10 lb − 3 lb 8 oz **58.** 20 lb 3 oz + 15 oz

Concept 5: U.S. Customary Units of Capacity

For Exercises 59–70, convert the units of capacity. (See Example 8.)

59. 16 fl oz = _____ c **60.** 5 pt = _____ c **61.** 6 gal = _____ qt **62.** 8 pt = _____ qt

63. 1 gal = _____ c **64.** 1 T = _____ tsp **65.** 2 qt = _____ gal **66.** 2 qt = _____ c

67. 1 pt = _____ fl oz **68.** 32 fl oz = _____ qt **69.** 2 T = _____ tsp **70.** 2 gal = _____ pt

Mixed Exercises

For Exercises 71–80, select the most reasonable measurement.

71. A shoebox is approximately _____ long.
   a. 6 in.   b. 14 in.
   c. 2 ft   d. 1 yd

72. A dining room table is _____ long.
   a. 2 ft   b. 20 in.
   c. 10 yd   d. $1\frac{1}{2}$ yd

73. A healthy newborn baby weighs _____.
   a. 25 lb   b. 50 oz
   c. 25 oz   d. 112 oz

74. A car weighs _____.
   a. 100 lb   b. 100 tons
   c. 2000 lb d. 500 lb

75. The height of a grown man is _____.
   a. 72 in.   b. 18 in.
   c. 18 ft   d. 3 yd

76. A typical ceiling height is _____.
   a. 5 ft   b. 7 yd
   c. 8 ft   d. 65 in.

77. The height of a one-story house is approximately _____.
   a. 30 ft   b. 60 in.
   c. 15 ft   d. 20 yd

78. The length of typical computer mouse is _____.
   a. 5 in.   b. 10 in.
   c. 2 in.   d. $\frac{2}{3}$ yd

79. A bowl of soup contains _____.
   a. 1 qt   b. 12 fl oz
   c. $\frac{1}{2}$ gal   d. 2 fl oz

80. A glass of water contains _____.
   a. 30 fl oz   b. 42 fl oz
   c. 3 fl oz   d. 10 fl oz

81. A recipe for minestrone soup calls for 3 c of spaghetti sauce. If a jar of sauce has 48 fl oz, is there enough for the recipe? **(See Example 9.)**

82. A recipe for punch calls for 6 c of apple juice. A bottle of juice has 2 qt. Is there enough juice in the bottle for the recipe?

83. A plumber used two pieces of pipe for a job. One piece was 4′6″ and the other was 2′8″. How much pipe was used?

84. A carpenter needs to put wood molding around three sides of a room. Two sides are 6′8″ long, and the third side is 10′ long. How much molding should the carpenter purchase?

85. If you have 4 yd of rope and you use 5 ft, how much is left over? Express the answer in feet.

86. The Blaisdell Arena football field, home of the Hawaiian Islanders football team, did not have the proper dimensions required by the arena football league. The width was measured to be 82 ft 10 in. Regulation width is 85 ft. What is the difference between the widths of a regulation field and the field at Blaisdell Arena?

87. Find the perimeter in feet.

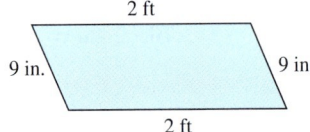

88. Find the perimeter in yards.

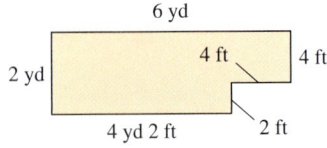

89. A 24-fl-oz jar of spaghetti sauce sells for $2.69. Another jar that holds 1 qt of sauce sells for $3.29. Which is a better buy? Explain.

90. Tatiana went to purchase bottled water. She found the following options: a 12-pack of 1-pt bottles; a 1-gal jug; and a 6-pack of 24-fl-oz bottles. Which option should Tatiana choose to get the most water? Explain.

Mark Dierker/McGraw Hill

Mark Dierker/McGraw Hill

91. A picnic table requires 5 boards that are 6′ long, 4 boards that are 3′3″ long, and 2 boards that are 18″ long. Find the total length of lumber required.

92. A roll of ribbon is 60 yd. If you wrap 12 packages that each use 2.5 ft of ribbon, how much ribbon is left over?

93. Byron lays sod in his backyard. Each piece of sod weighs 6 lb 4 oz. If he puts down 50 pieces, find the total weight.

94. A can of paint weighs 2 lb 4 oz. How much would 6 cans weigh?

95. The garden pictured needs a decorative border. The border comes in pieces that are 1.5 ft long. How many pieces of border are needed to surround the garden?

96. Monte fences all sides of a field with panels of fencing that are 2 yd long. How many panels of fencing does he need?

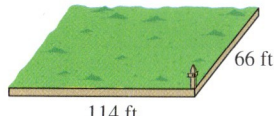

For Exercises 97 and 98, refer to the table.

97. Gil is a distance runner. The durations of his training runs for one week are given in the table. Find the total time that Gil ran that week, and express the answer in mixed units.

98. Find the difference between the amount of time Gil trained on Monday and the amount of time he trained on Friday.

Day	Time
Mon.	1 hr 10 min
Tues.	45 min
Wed.	1 hr 20 min
Thur.	30 min
Fri.	50 min
Sat.	Rest
Sun.	1 hr

Diane Collins and Jordan Hollender/Digital Vision/Getty Images

99. In a team triathlon, Torie swims $\frac{1}{2}$ mi in 15 min 30 sec. David rides his bicycle 20 mi in 50 min 20 sec. Emilie runs 4 mi in 28 min 10 sec. Find the total time for the team.

100. Joe competes in a biathlon. He runs 5 mi in 32 min 8 sec. He rides his bike 25 mi in 1 hr 2 min 40 sec. Find the total time for his race.

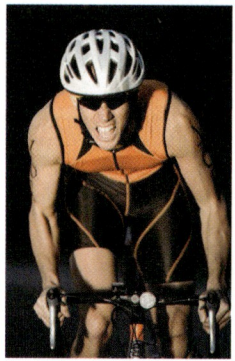

Fuse/Corbis/Getty Images

Expanding Your Skills

Area is measured in square units such as in.², ft², yd², and mi². Converting square units involves a different set of conversion factors. For example, 1 yd = 3 ft, but 1 yd² = 9 ft². To understand why, recall that the formula for the area of a rectangle is $A = l \times w$. In a square, the length and the width are the same distance s. Therefore, the area of a square is given by the formula $A = s \times s$, where s is the length of a side.

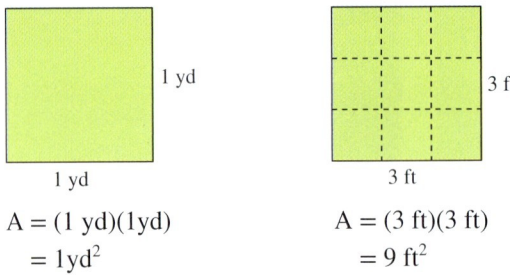

$A = (1 \text{ yd})(1 \text{ yd})$
$= 1 \text{ yd}^2$

$A = (3 \text{ ft})(3 \text{ ft})$
$= 9 \text{ ft}^2$

Instead of learning a new set of conversion factors, we can use multiples of the conversion factors that we mastered in this section.

Example: Converting Area

Convert 4 yd² to square feet.

Solution: We will use the conversion factor of $\dfrac{3 \text{ ft}}{1 \text{ yd}}$ twice.

$$\dfrac{4 \text{ yd}^2}{1} \cdot \dfrac{3 \text{ ft}}{1 \text{ yd}} \cdot \dfrac{3 \text{ ft}}{1 \text{ yd}} = \dfrac{4 \text{ yd}^2}{1} \cdot \dfrac{9 \text{ ft}^2}{1 \text{ yd}^2} = 36 \text{ ft}^2$$

Multiply first.

TIP: We use the conversion factor $\dfrac{3 \text{ ft}}{1 \text{ yd}}$ twice because there are two dimensions. Length and width must both be converted to feet.

For Exercises 101–108, convert the units of area by using multiple factors of the given conversion factor.

101. 54 ft² = _____ yd² $\left(\text{Use two factors of } \dfrac{1 \text{ yd}}{3 \text{ ft}}.\right)$

102. 108 ft² = _____ yd²

103. 432 in.² = _____ ft² $\left(\text{Use two factors of } \dfrac{1 \text{ ft}}{12 \text{ in.}}.\right)$

104. 720 in.² = _____ ft²

105. 5 ft² = _____ in.² $\left(\text{Use two factors of } \dfrac{12 \text{ in.}}{1 \text{ ft}}.\right)$

106. 7 ft² = _____ in.²

107. 3 yd² = _____ ft² $\left(\text{Use two factors of } \dfrac{3 \text{ ft}}{1 \text{ yd}}.\right)$

108. 10 yd² = _____ ft²

Metric Units of Measurement

Section 8.2

1. Introduction to the Metric System

Throughout history the lack of standard units of measure led to much confusion in trade between countries. In 1790 the French Academy of Sciences adopted a simple, decimal-based system of units. This system is known today as the **metric system**. The metric system is the predominant system of measurement used in science.

The simplicity of the metric system is a result of having one basic unit of measure for each type of quantity (length, mass, and capacity). The base units are the *meter* for length, the *gram* for mass, and the *liter* for capacity. Other units of length, mass, and capacity in the metric system are products of the base unit and a power of 10.

Concepts

1. Introduction to the Metric System
2. Metric Units of Length
3. Metric Units of Mass
4. Metric Units of Capacity
5. Summary of Metric Conversions

2. Metric Units of Length

The **meter** (m) is the basic unit of length in the metric system. A meter is slightly longer than a yard.

| 1 meter | 1 m ≈ 39 in. |
| 1 yard | 1 yd = 36 in. |

The meter was defined in the late 1700s as one ten-millionth of the distance along the Earth's surface from the North Pole to the Equator through Paris, France. Today the meter is defined as the distance traveled by light in a vacuum during $\frac{1}{299,792,458}$ sec.

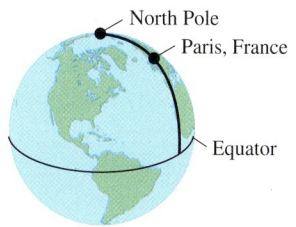

Six other common metric units of length are given in Table 8-2. Notice that each unit is related to the meter by a power of 10. This makes it particularly easy to convert from one unit to another.

TIP: The units of hectometer, dekameter, and decimeter are not frequently used.

Table 8-2 Metric Units of Length and Their Equivalents

1 kilometer (km) = 1000 m	
1 hectometer (hm) = 100 m	
1 dekameter (dam) = 10 m	
1 meter (m) = 1 m	
1 decimeter (dm) = 0.1 m	$\left(\frac{1}{10}\text{ m}\right)$
1 centimeter (cm) = 0.01 m	$\left(\frac{1}{100}\text{ m}\right)$
1 millimeter (mm) = 0.001 m	$\left(\frac{1}{1000}\text{ m}\right)$

TIP: In addition to the key facts presented in Table 8-2, the following equivalences are useful.

100 cm = 1 m
1000 mm = 1 m

Notice that each unit of length has a prefix followed by the word *meter* (kilometer, for example). You should memorize these prefixes along with their multiples of the basic unit, the meter. Furthermore, it is generally easiest to memorize the prefixes in order.

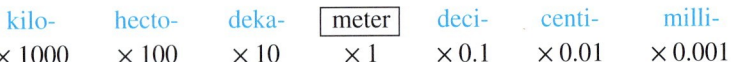

As you familiarize yourself with the metric units of length, it is helpful to have a sense of the distance represented by each unit (Figure 8-2).

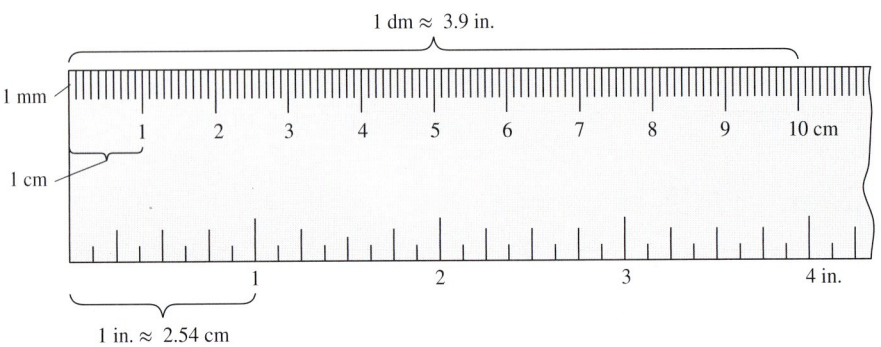

Figure 8-2

- 1 *millimeter* is approximately the thickness of five sheets of paper.
- 1 *centimeter* is approximately the width of a key on a calculator.
- 1 *decimeter* is approximately 4 in.
- 1 *meter* is just over 1 yd.
- A *kilometer* is used to express longer distances in much the same way we use miles. The distance between Los Angeles and San Diego is about 193 km.

Example 1 — Measuring Distances in Metric Units

Approximate the distance in centimeters and in millimeters.

Sven Kahns/McGraw Hill

Solution:

The numbered lines on the ruler are units of centimeters. Each centimeter is divided into 10 mm. We see that the width of the penny is not quite 2 cm. We can approximate this distance as 1.8 cm or equivalently 18 mm.

Skill Practice

1. Approximate the length of the pin in centimeters and in millimeters.

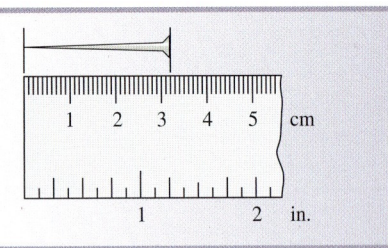

Answer

1. 3.2 cm or 32 mm

In Example 2, we convert metric units of length by using conversion factors.

Example 2 — **Converting Metric Units of Length**

a. 10.4 km = _____ m   b. 88 mm = _____ m

Solution:

From Table 8-2, 1 km = 1000 m.

a. $10.4 \text{ km} = \dfrac{10.4 \text{ km}}{1} \cdot \dfrac{1000 \text{ m}}{1 \text{ km}}$ ← new unit to convert to
   ← unit to convert from

   $= 10{,}400 \text{ m}$ Multiply.

b. $88 \text{ mm} = \dfrac{88 \text{ mm}}{1} \cdot \dfrac{1 \text{ m}}{1000 \text{ mm}}$ ← new unit to convert to
   ← unit to convert from

   $= \dfrac{88}{1000} \text{ m}$

   $= 0.088 \text{ m}$

Skill Practice Convert.

2. 8.4 km = _____ m   3. 64,000 cm = _____ m

Recall that the place positions in our numbering system are based on powers of 10. For this reason, when we multiply a number by 10, 100, or 1000, we move the decimal point 1, 2, or 3 places, respectively, to the right. Similarly, when we multiply by 0.1, 0.01, or 0.001, we move the decimal point to the left 1, 2, or 3 places, respectively.

Since the metric system is also based on powers of 10, we can convert between two metric units of length by moving the decimal point. The direction and number of place positions to move are based on the metric **prefix line**, shown in Figure 8-3.

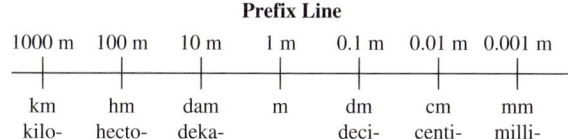

Figure 8-3

TIP: To use the prefix line effectively, you must know the order of the metric prefixes. Sometimes a mnemonic (memory device) can help. Consider the following sentence. The first letter of each word represents one of the metric prefixes.

kids	**h**ave	**d**oughnuts	until	**d**ad	**c**alls	**m**om.
kilo-	hecto-	deka-	unit	deci-	centi-	milli-

unit ↑ represents the main unit of measurement (meter, liter, or gram)

Answers

2. 8400 m   3. 640 m

> **Using the Prefix Line to Convert Metric Units**
>
> **Step 1** To use the prefix line, begin at the point on the line corresponding to the original unit you are given.
> **Step 2** Then count the number of positions you need to move to reach the new unit of measurement.
> **Step 3** Move the decimal point in the original measured value the same direction and same number of places as on the prefix line.
> **Step 4** Replace the original unit with the new unit of measure.

Example 3 Using the Prefix Line to Convert Metric Units of Length

Use the prefix line for each conversion.

a. 0.0413 m = _____ cm   b. 4700 cm = _____ km

Solution:

a.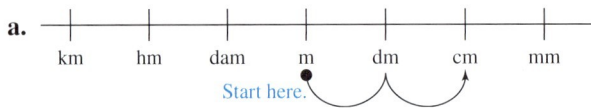

 0.0413 m = 4.13 cm   From the prefix line, move the decimal point two places to the right.

b.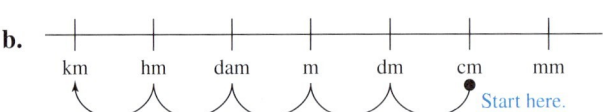

 4700 cm = 04700. cm = 0.047 km   From the prefix line, move the decimal point five places to the left.

Skill Practice Convert.

4. 864 cm = _____ m   5. 8.2 km = _____ m

3. Metric Units of Mass

We have learned that the pound and ton are two measures of weight in the U.S. Customary System. Measurements of weight give the force of an object under the influence of gravity. The mass of an object is related to its weight. However, mass is not affected by gravity. Thus, the weight of an object will be different on Earth than on the Moon because the effect of gravity is different. The mass of the object will stay the same.

The fundamental unit of mass in the metric system is the **gram** (g). A penny is approximately 2.5 g (Figure 8-4). A paper clip is approximately 1 g (Figure 8-5).

≈ 2.5 g

Figure 8-4

≈ 1 g

Figure 8-5

Answers
4. 8.64 m   5. 8200 m

Other common metric units of mass are given in Table 8-3. Once again, notice that the metric units of mass are related to the gram by powers of 10.

Table 8-3 Metric Units of Mass and Their Equivalents

1 kilogram (kg) = 1000 g	
1 hectogram (hg) = 100 g	
1 dekagram (dag) = 10 g	
1 gram (g) = 1 g	
1 decigram (dg) = 0.1 g	$\left(\frac{1}{10} g\right)$
1 centigram (cg) = 0.01 g	$\left(\frac{1}{100} g\right)$
1 milligram (mg) = 0.001 g	$\left(\frac{1}{1000} g\right)$

TIP: In addition to the key facts presented in Table 8-3, the following equivalences are useful.

100 cg = 1 g
1000 mg = 1 g

On the surface of Earth, 1 kg of mass is equivalent to approximately 2.2 lb of weight. Therefore, a 180-lb man has a mass of approximately 81.8 kg.

$$180 \text{ lb} \approx 81.8 \text{ kg}$$

The metric prefix line for mass is shown in Figure 8-6. This can be used to convert from one unit of mass to another.

Prefix Line

```
1000 g   100 g   10 g    1 g    0.1 g  0.01 g  0.001 g
  +--------+-------+-------+-------+-------+-------+
  kg       hg      dag     g      dg      cg      mg
  kilo-    hecto-  deka-          deci-   centi-  milli-
```

Figure 8-6

Example 4 Converting Metric Units of Mass

a. 1.6 kg = _____ g   **b.** 1400 mg = _____ g

Solution:

a. $1.6 \text{ kg} = \dfrac{1.6 \text{ kg}}{1} \cdot \dfrac{1000 \text{ g}}{1 \text{ kg}}$ ← new unit to convert to
← unit to convert from

$= 1600 \text{ g}$

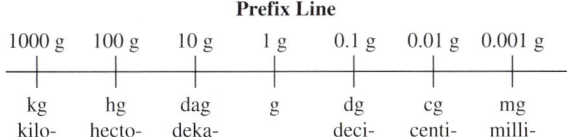

1.6 kg = 1.600 kg = 1600 g

b. $1400 \text{ mg} = \dfrac{1400 \text{ mg}}{1} \cdot \dfrac{1 \text{ g}}{1000 \text{ mg}}$ ← new unit to convert to
← unit to convert from

$= \dfrac{1400}{1000} \text{ g}$

$= 1.4 \text{ g}$

1400 mg = 1400 mg = 1.4 g

Skill Practice Convert.

6. 80 kg = _____ g   **7.** 49 cg = _____ g

Answers
6. 80,000 g   **7.** 0.49 g

4. Metric Units of Capacity

The basic unit of capacity in the metric system is the **liter** (L). One liter is slightly more than 1 qt. Other common units of capacity are given in Table 8-4.

Table 8-4 Metric Units of Capacity and Their Equivalents

1 **kilo**liter (kL) = 1000 L	
1 **hecto**liter (hL) = 100 L	
1 **deka**liter (daL) = 10 L	
1 liter (L) = 1 L	
1 **deci**liter (dL) = 0.1 L	$(\frac{1}{10} L)$
1 **centi**liter (cL) = 0.01 L	$(\frac{1}{100} L)$
1 **milli**liter (mL) = 0.001 L	$(\frac{1}{1000} L)$

TIP: In addition to the key facts presented in Table 8-4, the following equivalences are useful.

$$100 \text{ cL} = 1 \text{ L}$$
$$1000 \text{ mL} = 1 \text{ L}$$

1 mL is also equivalent to a **cubic centimeter** (**cc** or **cm³**). The unit cc is often used to measure dosages of medicine. For example, after having an allergic reaction to a bee sting, a patient might be given 1 cc of adrenaline.

The metric prefix line for capacity is similar to that of length and mass (Figure 8-7). It can be used to convert between metric units of capacity.

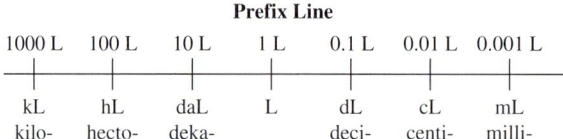
1 cc = 1 mL

Prefix Line

```
1000 L   100 L   10 L    1 L    0.1 L   0.01 L  0.001 L
  +       +       +       +       +       +       +
  kL      hL      daL     L       dL      cL      mL
 kilo-   hecto-  deka-                  centi-   milli-
                                  deci-
```

Figure 8-7

Example 5 Converting Metric Units of Capacity

a. 5.5 L = _____ mL   **b.** 150 cL = _____ L

Solution:

a. $5.5 \text{ L} = \frac{5.5 \text{ L}}{1} \cdot \frac{1000 \text{ mL}}{1 \text{ L}}$ ← new unit to convert to
← unit to convert from

$= 5500 \text{ mL}$

5.5 L = 5.500 L = 5500 mL

b. $150 \text{ cL} = \frac{150 \text{ cL}}{1} \cdot \frac{1 \text{ L}}{100 \text{ cL}}$ ← new unit to convert to
← unit to convert from

$= \frac{150}{100} \text{ L}$

$= 1.5 \text{ L}$

150 cL = 150. cL = 1.5 L

Skill Practice Convert.

8. 10.2 L = _____ cL   **9.** 150,000 mL = _____ kL

Answers
8. 1020 cL   **9.** 0.15 kL

Example 6 Converting Metric Units of Capacity

a. 15 cc = _____ mL   b. 0.8 cL = _____ cc

Solution:

a. Recall that 1 cc = 1 mL. Therefore, 15 cc = 15 mL.

b. We must convert from centiliters to milliliters, and then from milliliters to cubic centimeters.

$$0.8 \text{ cL} = \frac{0.8 \text{ cL}}{1} \cdot \frac{10 \text{ mL}}{1 \text{ cL}}$$

$$= 8 \text{ mL}$$

$$= 8 \text{ cc} \quad \text{Recall that 1 mL = 1 cc.}$$

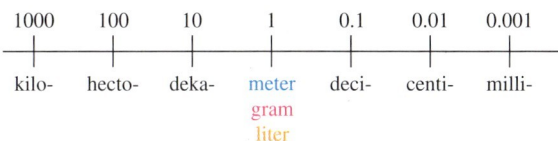

0.8 cL = 8 mL = 8 cc

Skill Practice Convert.

10. 0.5 cc = _____ mL   11. 0.04 L = _____ cc

5. Summary of Metric Conversions

The prefix line in Figure 8-8 summarizes the relationships learned thus far.

```
1000    100    10     1      0.1    0.01   0.001
 +       +     +      +       +      +      +
kilo-  hecto- deka- meter   deci- centi- milli-
                    gram
                    liter
```

Figure 8-8

Example 7 Converting Metric Units

a. The distance between San Jose and Santa Clara is 26 km. Convert this to meters.

b. A bottle of canola oil holds 946 mL. Convert this to liters.

c. The mass of a bag of rice is 90,700 cg. Convert this to grams.

d. A dose of an antiviral medicine is 0.5 cc. Convert this to milliliters.

Solution:

a. 26 km = 26.000 km = 26,000 m

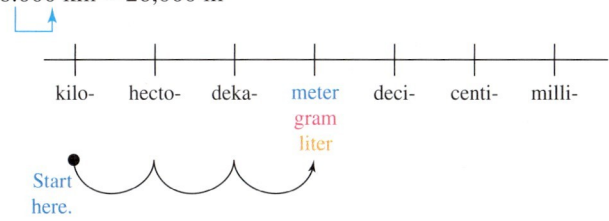

b. 946 mL = 946. mL = 0.946 L

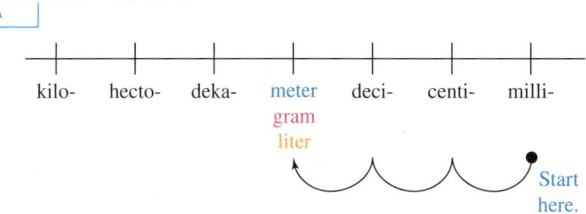

Answers

10. 0.5 mL   11. 40 cc

c. 90,700 cg = 90700. cg = 907 g

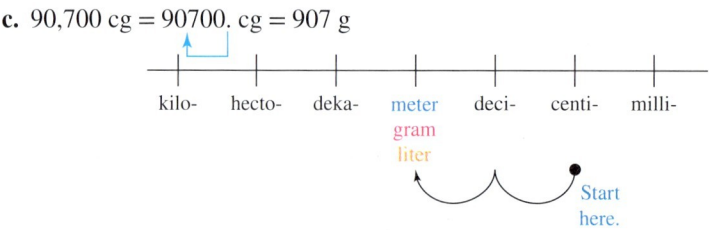

d. Recall that 1 cc = 1 mL. Therefore, 0.5 cc = 0.5 mL.

Skill Practice

12. The distance between Savannah and Hinesville is 64 km. How many meters is this?
13. A bottle of water holds 1420 mL. How many liters is this?
14. The mass of a box of cereal is 680 g. Convert this to kilograms.
15. A cat receives 1 mL of an antibiotic solution. Convert this to cc.

Answers
12. 64,000 m
13. 1.42 L
14. 0.68 kg
15. 1 cc

Section 8.2 Activity

A.1. The text has a mnemonic (memory device) to help you remember the order of the metric prefixes:

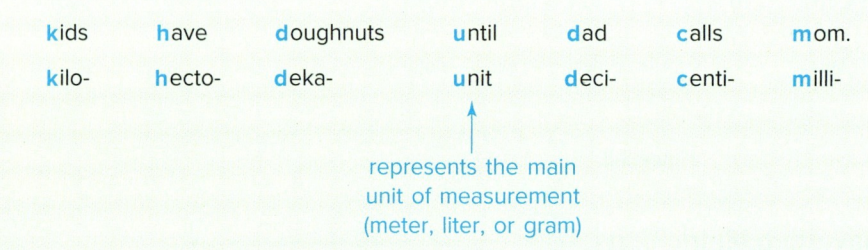

Develop your own memory device for remembering the prefix line.

A.2. Fill in the blanks with the proper prefix and measurement abbreviation for length. The first is done for you.

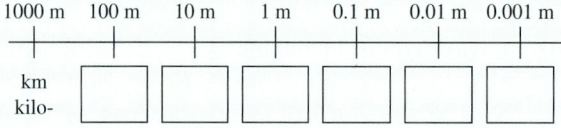

A.3. Convert 144 m to km using conversion factors from Exercise A.2.
  a. Write 144 meters as a ratio with a denominator of 1. Make sure to include the units in the ratio.
  b. Multiply the ratio from part (a) by the conversion factor of meters to kilometers. Write the conversion factor as a ratio so that the unit of kilometers is in the numerator and the unit of meters is in the denominator.
  c. Multiply the fractions and divide out the units of meters. Be sure to include the unit of kilometers in the answer.

A.4. Convert 144 m to km using the prefix line from Exercise A.2.
  a. Begin at meters on the prefix line. Count the number of positions you need to move to reach kilometers.
  b. Move the decimal point in 144 the same direction and same number of places as on the prefix line from part (a). (Recall that the decimal point is located to the right of the last digit of a whole number.)

c. Replace the original unit with the new unit of measure.
d. Compare the answer in part (c) to the answer obtained in Exercise A.3, part (c). Are the answers the same? Why or why not?

For Exercises A.5–A.9, convert by using conversion factors (see Exercise A.3) or the prefix line (see Exercise A.4).

A.5. 6.4 hm = _____ m

A.6. 0.044 cm = _____ mm

A.7. 1900 cm = _____ m

A.8. 0.00192 mm = _____ dm

A.9. 182 dm = _____ dam

A.10. In a certain neighborhood, a city block is 140 m long. How many 4-m water pipes must be connected to stretch a total distance of 9 city blocks?

A.11. Fill in the blanks with the proper prefix and measurement abbreviation for mass. The first is done for you.

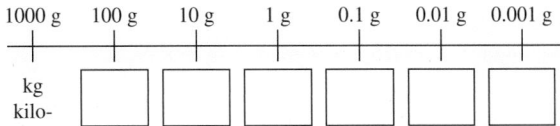

For Exercises A.12–A.16, convert using the method of your choice of multiplying by a conversion, or using the prefix line.

A.12. 382 dag = _____ kg

A.13. 9.16 g = _____ kg

A.14. 0.0003 g = _____ mg

A.15. 8.0 hg = _____ dg

A.16. 187 cg = _____ g

A.17. In Bayer Extra Strength Aspirin, each capsule contains 500 mg of aspirin. How many grams of aspirin would a person be ingesting if the person took four capsules in one day?

A.18. Fill in the blanks with the proper prefix and measurement abbreviation for capacity. The first is done for you.

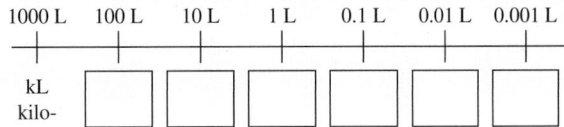

For Exercises A.19–A.24, convert using the method of your choice of multiplying by a conversion, or using the prefix line.

A.19. 60,200 L = _____ kL

A.20. 240 L = _____ mL

A.21. 100 cc = _____ mL

A.22. 150 cc = _____ L

A.23. 3 dL = _____ L

A.24. 800 cL = _____ kL

A.25. Match each description with the measurement that seems most reasonable.
    a. Distance between Dallas and Atlanta   i. 0.1 mm
    b. Mass of a dime   ii. 2 g
    c. Volume of water in a hot tub   iii. 0.710 L
    d. Width of a human hair   iv. 1258 km
    e. Length of a human toe   v. 2.27 kg
    f. Mass of a couch   vi. 5 cm
    g. Volume of a bottle of water   vii. 1.2 kL
    h. Mass of a bag of flour   viii. 120 kg

Section 8.2 Practice Exercises

Prerequisite Review

For Exercises R.1–R.6, convert the units of measurement.

R.1. 2200 yd = _____ mi

R.2. 65 mi = _____ yd

R.3. 3 pt = _____ qt

R.4. 3 pt = _____ gal

R.5. 3.5 lb = _____ oz

R.6. 48 oz = _____ lb

Vocabulary and Key Concepts

1. **a.** A decimal-based system of units called the _____ system was first proposed by the French Academy of Sciences in 1790 and is now the predominant system of measurement used in the sciences.

   b. A metric _____ line is a figure used to order metric units for the purpose of counting the number of place positions needed to convert between two metric units of measurement.

   c. The _____ is the fundamental unit of length in the metric system and is denoted by _____.

   d. The _____ is the fundamental unit of mass in the metric system and is denoted by _____.

   e. The _____ is the fundamental unit of capacity in the metric system and is denoted by _____.

   f. The _____ centimeter or cc is a unit of capacity equivalent to 1 mL.

2. Units in the metric system are products of the base unit and a power of _____.

3. When multiplying by powers of 10, the decimal point is moved to the _____ the same number of places as there are zeros in the power of 10.

4. When multiplying by powers of 0.1, the decimal point is moved to the _____ the same number of places as there are decimal places in the power of 0.1.

5. To convert a measurement in millimeters to meters, move the decimal point in the original measurement _____ places to the _____.

6. To convert a measurement in kilograms to centigrams, move the decimal point in the original measurement _____ places to the _____.

7. To convert a measurement of liters to hectoliters, move the decimal point in the original measurement _____ places to the _____.

Concept 1: Introduction to the Metric System

8. Identify the units that apply to length. Circle all that apply.

   a. Yard **b.** Ounce **c.** Fluid ounce **d.** Meter **e.** Quart
   f. Gram **g.** Pound **h.** Liter **i.** Mile **j.** Inch

9. Identify the units that apply to weight or mass. Circle all that apply.

   a. Yard **b.** Ounce **c.** Fluid ounce **d.** Meter **e.** Quart
   f. Gram **g.** Pound **h.** Liter **i.** Mile **j.** Inch

10. Identify the units that apply to capacity. Circle all that apply.

    a. Yard **b.** Ounce **c.** Fluid ounce **d.** Meter **e.** Quart
    f. Gram **g.** Pound **h.** Liter **i.** Mile **j.** Inch

Section 8.2 Metric Units of Measurement   469

Concept 2: Metric Units of Length

For Exercises 11 and 12, approximate each distance in centimeters and millimeters. (See Example 1.)

11.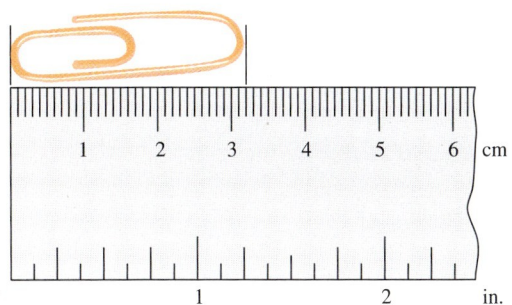

12.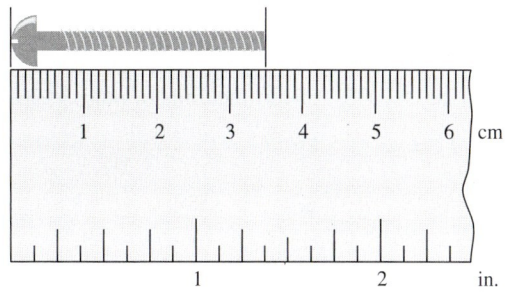

For Exercises 13–18, select the most reasonable measurement.

13. A table is _____ long.
    a. 2 m
    b. 2 cm
    c. 2 km
    d. 2 hm

14. A picture frame is _____ wide.
    a. 22 cm
    b. 22 mm
    c. 22 m
    d. 22 km

15. The distance between Albany, New York, and Buffalo, New York, is _____.
    a. 210 m
    b. 2100 cm
    c. 2.1 km
    d. 210 km

16. The distance between Denver and Colorado Springs is _____.
    a. 110 cm
    b. 110 km
    c. 11,000 km
    d. 1100 mm

17. The height of a full-grown giraffe is approximately _____.
    a. 50 m
    b. 0.05 m
    c. 0.5 m
    d. 5 m

18. The length of a canoe is _____.
    a. 5 m
    b. 0.5 m
    c. 0.05 m
    d. 500 m

Ingram Publishing/Alamy Stock Photo

Medioimages/SuperStock

For Exercises 19–30, convert metric units of length by using conversion factors or the prefix line. (See Examples 2 and 3.)

Prefix Line for Length

1000 m	100 m	10 m	1 m	0.1 m	0.01 m	0.001 m
km	hm	dam	m	dm	cm	mm
kilo-	hecto-	deka-		deci-	centi-	milli-

19. 2430 m = _____ km

20. 1251 mm = _____ m

21. 50 m = _____ mm

22. 1.3 m = _____ mm

23. 4 km = _____ m

24. 5 m = _____ cm

25. 4.31 cm = _____ mm 26. 18 cm = _____ mm 27. 3328 dm = _____ km

28. 128 hm = _____ km 29. 3 hm = _____ m 30. 450 mm = _____ dm

Concept 3: Metric Units of Mass

For Exercises 31–38, convert the units of mass. (See Example 4.)

31. 539 g = _____ kg
32. 328 mg = _____ g

33. 2.5 kg = _____ g
34. 2011 g = _____ kg

35. 0.0334 g = _____ mg
36. 0.38 dag = _____ dg

37. 409 cg = _____ g
38. 0.003 kg = _____ g

Concept 4: Metric Units of Capacity

For Exercises 39–42, fill in the blank with >, <, or =.

39. 1 cL _____ 1 L
40. 1 L _____ 1 mL

41. 1 mL _____ 1 cc
42. 1 L _____ 1 cc

43. What does the abbreviation cc represent?

44. Which of the following are measures of capacity? Circle all that apply.

    a. cm   b. cc   c. cL   d. cg

For Exercises 45–54, convert the units of capacity. (See Examples 5 and 6.)

45. 3200 mL = _____ L
46. 280 L = _____ kL

47. 7 L = _____ cL
48. 0.52 L = _____ mL

49. 42 mL = _____ dL
50. 0.88 L = _____ hL

51. 64 cc = _____ mL
52. 125 mL = _____ cc

53. 0.04 L = _____ cc
54. 38 cc = _____ L

Concept 5: Summary of Metric Conversions (Mixed Exercises)

55. Identify the units that apply to length.

    a. mL   b. mm   c. hg   d. cc   e. kg   f. hm   g. cL

56. Identify the units that apply to capacity.

    a. kg   b. km   c. cL   d. cc   e. hm   f. dag   g. mm

57. Identify the units that apply to mass.

    a. dg   b. hm   c. kL   d. cc   e. dm   f. kg   g. cL

For Exercises 58–69, complete the table.

	Object	mm	cm	m	km
58.	Distance between Orlando and Miami				402
59.	Length of the Mississippi River				3766
60.	Thickness of a dime	1.35			
61.	Diameter of a quarter	24.3			

	Object	mg	cg	g	kg
62.	Can of tuna			170	
63.	Bag of rice			907	
64.	Box of raisins		42,500		
65.	Hockey puck		17,000		

	Object	mL	cL	L	kL
66.	1 Tablespoon	15			
67.	Bottle of vanilla extract	59			
68.	Capacity of a cooler				0.0377
69.	Capacity of a gasoline tank				0.0757

For Exercises 70–75, convert the metric units as indicated. (See Example 7.)

70. The height of the tallest living tree is 112.014 m. Convert this to dekameters.

71. The Congo River is 4669 km long. Convert this to meters.

72. A multivitamin tablet contains 60 mg of vitamin C. How many tablets must be taken to consume a total of 3 g of vitamin C?

73. A can contains 305 g of soup. Convert this to milligrams.

74. A gasoline can has a capacity of 19 L. Convert this to kiloliters.

75. The capacity of a coffee cup is 0.25 L. Convert this to milliliters.

76. In one day, Stacy gets 600 mg of calcium in her daily vitamin, 500 mg in her calcium supplement, and 250 mg in the dairy products she eats. How many grams of calcium will she get in one week?

77. Cliff drives his children to their sports activities outside of school. When he drives his son to baseball practice, it is a 6-km round trip. When he drives his daughter to basketball practice, it is a 1800-m round trip. If basketball practice is 3 times a week and baseball practice is twice a week, how many kilometers does Cliff drive?

78. A gas tank holds 45 L. If it costs $74.25 to fill up the tank, what is the price per liter?

79. A can of paint holds 120 L. How many kiloliters are contained in 8 cans?

80. A bottle of water holds 710 mL. How many liters are in a 6-pack?

81. A bottle of olive oil has 33 servings of 15 mL each. How many centiliters of oil does the bottle contain?

82. A quart of milk has 130 mg of sodium per cup. How much sodium is in the whole bottle?

83. A ½-c serving of cereal has 180 mg of potassium. This is 5% of the recommended daily allowance of potassium. How many grams is the recommended daily allowance?

Tanya Constantine/Blend Images

84. Rosanna has material 1.5 m long for a window curtain. If the window is 90 cm and she needs 10 cm for a hem at the bottom and 12 cm for finishing the top, does Rosanna have enough material?

85. Veronique has a piece of framing 1 m long. Does she have enough to cut four pieces to frame the picture shown in the figure?

86. A square tile is 110 mm in length. If they are placed side by side, how many tiles will it take to cover a length of wall 1.43 m long?

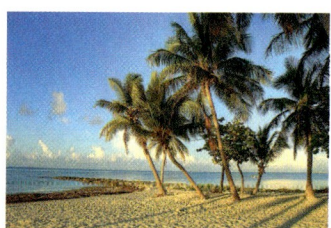
12 cm
40 cm
Crawford A. Wilson III/Moment/Flickr/Getty Images

87. Two Olympic speed skating races for women are 500 m and 5 km. What is the difference (in meters) between the lengths of these races?

Expanding Your Skills

For Exercises 88–91, convert the units of area by using the given conversion factor twice as shown in the example.

Example: Converting area

Convert 1000 mm² to square centimeters.

Solution: $\dfrac{1000 \text{ mm}^2}{1} \cdot \underbrace{\dfrac{1 \text{ cm}}{10 \text{ mm}} \cdot \dfrac{1 \text{ cm}}{10 \text{ mm}}}_{\text{Multiply first.}} = \dfrac{1000 \text{ mm}^2}{1} \cdot \dfrac{1 \text{ cm}^2}{100 \text{ mm}^2} = \dfrac{1000 \text{ cm}^2}{100} = 10 \text{ cm}^2$

88. $30{,}000 \text{ mm}^2 = $ _____ cm^2 $\left(\text{Use } \dfrac{1 \text{ cm}}{10 \text{ mm}}.\right)$

89. $65{,}000{,}000 \text{ m}^2 = $ _____ km^2 $\left(\text{Use } \dfrac{1 \text{ km}}{1000 \text{ m}}.\right)$

90. $4.1 \text{ m}^2 = $ _____ cm^2 $\left(\text{Use } \dfrac{100 \text{ cm}}{1 \text{ m}}.\right)$

91. $5600 \text{ cm}^2 = $ _____ m^2 $\left(\text{Use } \dfrac{1 \text{ m}}{100 \text{ cm}}.\right)$

In the U.S. Customary System of measurement, 1 ton = 2000 lb. In the metric system, 1 metric ton = 1000 kg. Use this information to answer Exercises 92–95.

92. Convert 3300 kg to metric tons.

93. Convert 5780 kg to metric tons.

94. Convert 10.9 metric tons to kilograms.

95. Convert 8.5 metric tons to kilograms.

Converting Between U.S. Customary and Metric Units

Section 8.3

Concepts
1. Summary of U.S. Customary and Metric Unit Equivalents
2. Converting U.S. Customary and Metric Units
3. Applications
4. Units of Temperature

1. Summary of U.S. Customary and Metric Unit Equivalents

In this section, we learn how to convert between U.S. Customary and metric units of measure. Suppose, for example, that you take a trip to Europe. A street sign indicates that the distance to Paris is 45 km (Figure 8-9). This distance may be unfamiliar to you until you convert to miles.

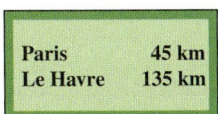

Figure 8-9

Example 1 — Converting Metric Units to U.S. Customary Units

Use the fact that 1 mi ≈ 1.61 km to convert 45 km to miles. Round to the nearest mile.

Solution:

$$45 \text{ km} \approx \frac{45 \text{ km}}{1} \cdot \frac{1 \text{ mi}}{1.61 \text{ km}} \qquad \text{Set up a conversion factor to convert kilometers to miles.}$$

$$= \frac{45}{1.61} \text{ mi} \qquad \text{Multiply fractions.}$$

$$\approx 28 \text{ mi} \qquad \text{Divide and round to the nearest mile.}$$

The distance of 45 km to Paris is approximately 28 mi.

Skill Practice

1. Use the fact that 1 mi ≈ 1.61 km to convert 184 km to miles. Round to the nearest mile.

Table 8-5 summarizes some common metric and U.S. Customary equivalents.

Table 8-5

Length	Weight/Mass (on Earth)	Capacity
1 in. = 2.54 cm	1 lb ≈ 0.45 kg	1 qt ≈ 0.95 L
1 ft ≈ 0.305 m	1 oz ≈ 28 g	1 fl oz ≈ 30 mL = 30 cc
1 yd ≈ 0.914 m		
1 mi ≈ 1.61 km		

2. Converting U.S. Customary and Metric Units

Using the U.S. Customary and metric equivalents given in Table 8-5, we can create conversion factors to convert between units.

Answer

1. 114 mi

> **Example 2** Converting Units of Length
>
> Fill in the blank. Round to two decimal places, if necessary.
>
> **a.** 18 cm = _____ in.   **b.** 15 yd ≈ _____ m   **c.** 8.2 m ≈ _____ ft
>
> **Solution:**
>
> **a.** $18 \text{ cm} = \dfrac{18 \text{ cm}}{1} \cdot \dfrac{1 \text{ in.}}{2.54 \text{ cm}}$ From Table 8-5, we know 1 in. = 2.54 cm.
>
> $\quad = \dfrac{18}{2.54} \text{ in.}$ Multiply fractions.
>
> $\quad \approx 7.09 \text{ in.}$ Divide and round to two decimal places.
>
> **b.** $15 \text{ yd} \approx \dfrac{15 \text{ yd}}{1} \cdot \dfrac{0.914 \text{ m}}{1 \text{ yd}}$ From Table 8-5, we know 1 yd ≈ 0.914 m.
>
> $\quad = 13.71 \text{ m}$ Multiply.
>
> **c.** $8.2 \text{ m} \approx \dfrac{8.2 \text{ m}}{1} \cdot \dfrac{1 \text{ ft}}{0.305 \text{ m}}$ From Table 8-5, we know 1 ft ≈ 0.305 m.
>
> $\quad = \dfrac{8.2}{0.305} \text{ ft}$ Multiply.
>
> $\quad \approx 26.89 \text{ ft}$ Divide and round to two decimal places.
>
> ---
>
> **Skill Practice** Convert. Round to one decimal place.
>
> **2.** 4 m ≈ _____ ft   **3.** 3 in. ≈ _____ cm   **4.** 6500 yd ≈ _____ m

> **Example 3** Converting Units of Weight and Mass
>
> Fill in the blank. Round to one decimal place, if necessary.
>
> **a.** 180 g ≈ _____ oz   **b.** 5.25 tons ≈ _____ kg
>
> **Solution:**
>
> **a.** $180 \text{ g} \approx \dfrac{180 \text{ g}}{1} \cdot \dfrac{1 \text{ oz}}{28 \text{ g}}$ From Table 8-5, we know 1 oz ≈ 28 g.
>
> $\quad = \dfrac{180}{28} \text{ oz}$
>
> $\quad \approx 6.4 \text{ oz}$ Divide and round to one decimal place.
>
> **b.** We can first convert 5.25 tons to pounds. Then we can use the fact that 1 lb ≈ 0.45 kg.
>
> $5.25 \text{ tons} = \dfrac{5.25 \text{ tons}}{1} \cdot \dfrac{2000 \text{ lb}}{1 \text{ ton}}$ Convert tons to pounds.
>
> $\quad = 10{,}500 \text{ lb}$
>
> $\quad \approx 10{,}500 \text{ lb} \cdot \dfrac{0.45 \text{ kg}}{1 \text{ lb}}$ Convert pounds to kilograms.
>
> $\quad = 4725 \text{ kg}$
>
> ---
>
> **Skill Practice** Convert. Round to one decimal place, if necessary.
>
> **5.** 8 kg ≈ _____ lb   **6.** 4 tons ≈ _____ kg

Answers

2. 13.1 ft   **3.** 7.6 cm   **4.** 5941 m
5. 17.8 lb   **6.** 3600 kg

Example 4 Converting Units of Capacity

Fill in the blank. Round to two decimal places, if necessary.

a. 75 mL ≈ _____ fl oz   b. 3 qt ≈ _____ L

Solution:

a. $75 \text{ mL} \approx \dfrac{75 \text{ mL}}{1} \cdot \dfrac{1 \text{ fl oz}}{30 \text{ mL}}$ From Table 8-5, we know 1 fl oz ≈ 30 mL.

$= \dfrac{75}{30} \text{ fl oz}$ Multiply fractions.

$= 2.5 \text{ fl oz}$ Divide.

b. $3 \text{ qt} \approx \dfrac{3 \text{ qt}}{1} \cdot \dfrac{0.95 \text{ L}}{1 \text{ qt}}$ From Table 8-5, we know 1 qt ≈ 0.95 L.

$= 2.85 \text{ L}$ Multiply.

Skill Practice Convert. Round to one decimal place, if necessary.

7. 120 mL ≈ _____ fl oz   8. 4 qt ≈ _____ L

3. Applications

Example 5 Converting Units in an Application

A 2-L bottle of soda sells for $2.19. A 32-fl-oz bottle of soda sells for $1.59. Compare the price per quart of each bottle to determine the better buy.

Solution:

Note that 1 qt = 2 pt = 4 c = 32 fl oz. So a 32-fl-oz bottle of soda costs $1.59 per quart. Next, if we can convert 2 L to quarts, we can compute the unit cost per quart and compare the results.

$2 \text{ L} \approx \dfrac{2 \text{ L}}{1} \cdot \dfrac{1 \text{ qt}}{0.95 \text{ L}}$ Recall that 1 qt = 0.95 L.

$\approx \dfrac{2}{0.95} \text{ qt}$ Multiply fractions.

$\approx 2.11 \text{ qt}$ Divide and round to two decimal places.

Mark Dierker/McGraw Hill

Now find the cost per quart. $\dfrac{\$2.19}{2.11 \text{ qt}} \approx \$1.04$ per quart

The cost for the 2-L bottle is $1.04 per quart, whereas the cost for 32-fl-oz is $1.59 per quart. Therefore, the 2-L bottle is the better buy.

Skill Practice

9. A 720-mL bottle of water sells for $0.79. A 32-fl-oz bottle of water sells for $1.29. Compare the price per ounce to determine the better buy.

Answers
7. 4 fl oz   8. 3.8 L
9. 720 mL is 24 oz. The cost per ounce is $0.033. The unit price for the 32-fl-oz bottle is $0.040 per ounce. The 720-mL bottle is the better buy.

> **Example 6** Converting Units in an Application
>
> In track and field, the 1500-m race is slightly less than 1 mi. How many yards less is it? Round to the nearest yard.
>
> **Solution:**
>
> We know that 1 mi = 1760 yd. If we can convert 1500 m to yards, then we can subtract the results.
>
> $$1500 \text{ m} \approx \frac{1500 \text{ m}}{1} \cdot \frac{1 \text{ yd}}{0.914 \text{ m}} \qquad \text{Recall that } 1 \text{ yd} = 0.914 \text{ m}.$$
>
> $$\approx \frac{1500}{0.914} \text{ yd} \qquad \text{Multiply fractions.}$$
>
> $$\approx 1641 \text{ yd} \qquad \text{Divide and round to the nearest yard.}$$
>
> Therefore, the difference between 1 mi and 1500 m is:
>
> $$\underset{(1 \text{ mi})}{1760 \text{ yd}} - \underset{(1500 \text{ m})}{1641 \text{ yd}} = 119 \text{ yd}$$
>
> **Skill Practice**
>
> 10. In track and field, the 800-m race is slightly shorter than a half-mile race. How many yards less is it? Round to the nearest yard.

4. Units of Temperature

In the United States, the **Fahrenheit** scale is used most often to measure temperature. On this scale, water freezes at 32°F and boils at 212°F. The symbol ° stands for "degrees," and °F means "degrees Fahrenheit."

Another scale used to measure temperature is the **Celsius** temperature scale. On this scale, water freezes at 0°C and boils at 100°C. The symbol °C stands for "degrees Celsius."

Figure 8-10 shows the relationship between the Celsius scale and the Fahrenheit scale.

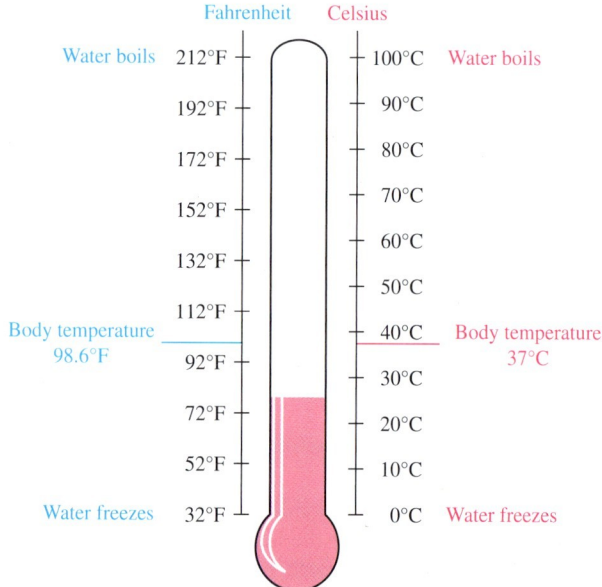

Figure 8-10

To convert back and forth between the Fahrenheit and Celsius scales, we use the following formulas.

Answer
10. 5 yd

Section 8.3 Converting Between U.S. Customary and Metric Units

> **Conversions for Temperature Scale**
>
> To convert from °C to °F:
> $$F = \frac{9}{5}C + 32$$
>
> To convert from °F to °C:
> $$C = \frac{5}{9}(F - 32)$$
>
> *Note:* Using decimal notation we can write the formulas as
> $$F = 1.8C + 32 \qquad C = \frac{F - 32}{1.8}$$

Example 7 Converting Units of Temperature

Convert a body temperature of 98.6°F to degrees Celsius.

Solution:

Because we want to convert degrees Fahrenheit to degrees Celsius, we use the formula $C = \frac{5}{9}(F - 32)$.

$$C = \frac{5}{9}(F - 32)$$

$$= \frac{5}{9}(98.6 - 32) \qquad \text{Substitute } F = 98.6.$$

$$= \frac{5}{9}(66.6) \qquad \text{Perform the operation inside parentheses first.}$$

$$= \frac{(5)(66.6)}{9}$$

$$= 37 \qquad \text{Body temperature is 37°C.}$$

Skill Practice

11. The ocean temperature in the Caribbean in August averages 84°F. Convert this to degrees Celsius and round to one decimal place.

Example 8 Converting Units of Temperature

Convert the temperature inside a refrigerator, 5°C, to degrees Fahrenheit.

Solution:

Because we want to convert degrees Celsius to degrees Fahrenheit, we use the formula $F = \frac{9}{5}C + 32$.

$$F = \frac{9}{5}C + 32$$

$$= \frac{9}{5} \cdot 5 + 32 \qquad \text{Substitute } C = 5.$$

$$= \frac{9}{\cancel{5}} \cdot \frac{\cancel{5}}{1} + 32$$

$$= 9 + 32$$

$$= 41 \qquad \text{The temperature inside the refrigerator is 41°F.}$$

Skill Practice

12. The high temperature on a day in March for Raleigh, North Carolina, was 10°C. Convert this to degrees Fahrenheit.

Answers

11. 28.9°C   12. 50°F

Section 8.3 Practice Exercises

Prerequisite Review

For Exercises R.1–R.6, convert the units of measurement.

R.1. 500 g = _____ kg

R.2. 500 mL = _____ L

R.3. 38 L = _____ mL

R.4. 23 cg = _____ mg

R.5. 0.55 mi = _____ in.

R.6. 18 hr = _____ sec

Vocabulary and Key Concepts

1. **a.** In the United States, the _____ scale is used most often to measure temperature. Using this scale, water freezes at _____°F and boils at _____°F.

   b. In the sciences the _____ scale is used most often to measure temperature. Using this scale, water freezes at _____°C and boils at _____°C.

2. When converting units, write the given measurement as a fraction with a denominator of _____ before multiplying by the conversion factor.

3. Conversion factors are unit ratios because the quotient is equal to _____.

4. The unit that you want to *convert to* should be in the (numerator/denominator) of the conversion factor.

5. The unit that you want to *convert from* should be in the (numerator/denominator) of the conversion factor.

6. The symbol ≈ means _____ equal.

Concept 1: Summary of U.S. Customary and Metric Unit Equivalents

For Exercises 7–10, write each unit equivalent as a unit ratio two different ways.

7. 1 in. = 2.54 cm   □/□ or □/□

8. 1 mi ≈ 1.61 km   □/□ or □/□

9. 1 lb ≈ 0.45 kg   □/□ or □/□

10. 1 fl oz ≈ 30 mL   □/□ or □/□

Length	Weight/Mass (on Earth)	Capacity
1 in. = 2.54 cm	1 lb ≈ 0.45 kg	1 qt ≈ 0.95 L
1 ft ≈ 0.305 m	1 oz ≈ 28 g	1 fl oz ≈ 30 mL = 30 cc
1 yd ≈ 0.914 m		
1 mi ≈ 1.61 km		

11. Identify an appropriate conversion factor to convert 5 yards to meters by using multiplication.

    a. $\dfrac{1 \text{ yd}}{0.914 \text{ m}}$ **b.** $\dfrac{0.914 \text{ m}}{1 \text{ yd}}$ **c.** $\dfrac{0.914 \text{ yd}}{1 \text{ m}}$ **d.** $\dfrac{1 \text{ m}}{0.914 \text{ yd}}$

12. Identify an appropriate conversion factor to convert 3 pounds to kilograms by using multiplication.

    a. $\dfrac{0.45 \text{ lb}}{1 \text{ kg}}$ **b.** $\dfrac{1 \text{ kg}}{0.45 \text{ lb}}$ **c.** $\dfrac{1 \text{ lb}}{0.45 \text{ kg}}$ **d.** $\dfrac{0.45 \text{ kg}}{1 \text{ lb}}$

13. Identify an appropriate conversion factor to convert 2 quarts to liters by using multiplication.

    a. $\dfrac{0.95 \text{ L}}{1 \text{ qt}}$ **b.** $\dfrac{1 \text{ qt}}{0.95 \text{ L}}$ **c.** $\dfrac{0.95 \text{ qt}}{1 \text{ L}}$ **d.** $\dfrac{1 \text{ L}}{0.95 \text{ qt}}$

14. Identify an appropriate conversion factor to convert 10 miles to kilometers by using multiplication.

    a. $\dfrac{1 \text{ mi}}{1.61 \text{ km}}$ **b.** $\dfrac{1 \text{ km}}{1.61 \text{ mi}}$ **c.** $\dfrac{1.61 \text{ km}}{1 \text{ mi}}$ **d.** $\dfrac{1.61 \text{ mi}}{1 \text{ km}}$

Concept 2: Converting U.S. Customary and Metric Units

For Exercises 15–23, convert the units of length. Round the answer to one decimal place, if necessary. (See Examples 1 and 2.)

15. 2 in. ≈ _____ cm
16. 120 km ≈ _____ mi
17. 8 m ≈ _____ yd
18. 4 ft ≈ _____ m
19. 400 ft ≈ _____ m
20. 0.75 m ≈ _____ yd
21. 45 in ≈ _____ m
22. 150 cm ≈ _____ ft
23. 0.5 ft ≈ _____ cm

For Exercises 24–32, convert the units of weight and mass. Round the answer to one decimal place, if necessary. (See Example 3.)

24. 6 oz ≈ _____ g
25. 6 lb ≈ _____ kg
26. 4 kg ≈ _____ lb
27. 10 g ≈ _____ oz
28. 14 g ≈ _____ oz
29. 0.54 kg ≈ _____ lb
30. 0.3 lb ≈ _____ kg
31. 2.2 tons ≈ _____ kg
32. 4500 kg ≈ _____ tons

For Exercises 33–38, convert the units of capacity. Round the answer to one decimal place, if necessary. (See Example 4.)

33. 6 qt ≈ _____ L
34. 5 fl oz ≈ _____ mL
35. 120 mL ≈ _____ fl oz
36. 19 L ≈ _____ qt
37. 960 cc ≈ _____ fl oz
38. 0.5 fl oz ≈ _____ cc

Concept 3: Applications (Mixed Exercises)

For Exercises 39–52, refer to the table of conversion facors.

Length	Weight/Mass (on Earth)	Capacity
1 in. = 2.54 cm	1 lb ≈ 0.45 kg	1 qt ≈ 0.95 L
1 ft ≈ 0.305 m	1 oz ≈ 28 g	1 fl oz ≈ 30 mL = 30 cc
1 yd ≈ 0.914 m		
1 mi ≈ 1.61 km		

39. A 2-lb box of sugar costs $3.19. A box that contains single-serving packets contains 354 g and costs $1.49. Find the unit costs in dollars per ounce to determine the better buy. (See Example 5.)

40. At the grocery store, Debbie compares the prices of a 2-L bottle of water and a 6-pack of bottled water. The 2-L bottle is priced at $1.59. The 6-pack costs $3.60 and each bottle in the package contains 24 fl oz. Compare the cost of water per quart to determine which is a better buy.

Leonard Zhukovsky/Shutterstock

41. A cross-country skiing race is 30 km long. Is this length more or less than 18 mi? (See Example 6.)

42. A can of cat food is 85 g. How many ounces is this? Round to the nearest ounce.

43. For a recent year, the Olympic gymnast who won the Olympic All-Around event weighed 97 lb. How many kilograms is this?

44. Warren's dad ran the 100-yd dash when he was in high school. Suppose Warren runs the 100-*meter* dash on his track team.
    a. Who runs the longer race?
    b. Find the difference between the lengths in meters.

45. In a recent year, the price of gas in Germany was $1.90 per liter. What is the price per gallon?

46. A jar of spaghetti sauce is 2 lb 8 oz. How many kilograms is this? Round to two decimal places.

47. The thickness of a hockey puck is 2.54 cm. How many inches is this?

48. A bottle of grape juice contains 1.9 L of juice. Is there enough juice to fill 10 glasses that hold 6 fl oz each? If yes, how many ounces will be left over?

49. If a football player weighs 99,790 g, how many pounds is this? Round to the nearest pound.

50. The distance between two lightposts is 6 m. How many feet is this? Round to the nearest foot.

Stock Foundry/Design Pics

51. A nurse gives a patient 45 cc of saline solution. How many fluid ounces does this represent?

52. Cough syrup comes in a bottle that contains 4 fl oz. How many milliliters is this?

53. Suppose this figure is a drawing of a room where 1 cm represents 2 ft. If you were to install molding along the edge of the floor, how many feet would you need?

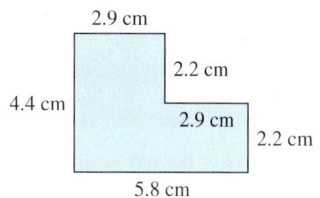

54. Suppose this figure is a drawing of a room where 1 cm represents 3 ft. If you were to install molding along the edge of the floor, how many feet would you need?

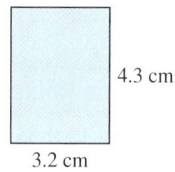

Concept 4: Units of Temperature

For Exercises 55–60, convert the temperatures by using the appropriate formula: $F = \frac{9}{5}C + 32$ or $C = \frac{5}{9}(F - 32)$.
(See Examples 7 and 8.)

55. $25°C = $ _____ $°F$

56. $113°F = $ _____ $°C$

57. $68°F = $ _____ $°C$

58. $15°C = $ _____ $°F$

59. $30°C = $ _____ $°F$

60. $104°F = $ _____ $°C$

61. The boiling point of the element boron is 4000°C. Find the boiling point in degrees Fahrenheit.

62. The melting point of the element copper is 1085°C. Find the melting point in degrees Fahrenheit.

63. If the outdoor temperature is 35°C, is it a hot day or a cold day?

64. If the temperature is 25° in Miami Beach, Florida, is it a warm or a cold day? Explain.

65. If the temperature is 25° in Florence, Italy, is it a warm or a cold day? Explain.

66. The high temperature in London, England, on a typical September day was 18°C, and the low was 13°C. Convert these temperatures to degrees Fahrenheit.

67. Use the fact that water boils at 100°C to show that the boiling point is 212°F.

68. Use the fact that water freezes at 0°C to show that the temperature at which water freezes is 32°F.

David Toase/Getty Images

Expanding Your Skills

The USDA recommends that an adult woman get 46 g of protein per day. Peanuts are 25% protein. Use this information to answer Exercises 69 and 70.

69. How many grams of peanuts would an adult woman need to satisfy a day's protein requirement?

70. How many ounces of peanuts would be required for a woman's daily protein need? Round to the nearest tenth.

In the U.S. Customary System of measurement, 1 ton = 2000 lb. In the metric system, 1 metric ton = 1000 kg. Use this information to answer Exercises 71–74.

71. A large SUV weighs 5700 lb. How many metric tons is this?

72. An elevator has a maximum capacity of 1200 lb. How many metric tons is this?

73. The average mass of a blue whale (the largest mammal in the world) is approximately 108 metric tons. How many pounds is this?

74. The mass of a small compact car is 1.25 metric tons. How many pounds is this? Round to the nearest pound.

Problem Recognition Exercises

U.S. Customary and Metric Conversions

For Exercises 1–32, use the table of conversion factors and the prefix line to convert the units.

Length	Capacity	Weight and Mass (on the Earth)
1 ft = 12 in.	3 tsp = 1 Tbl	1 lb = 16 oz
1 yd = 3 ft	1 cup = 8 fl oz	1 ton = 2000 lb
1 yd = 36 in.	1 pt = 2 cups	
1 mi = 5280 ft	1 qt = 2 pt	
1 mi = 1760 yd	1 qt = 4 cups	
	1 gal = 4 qt	

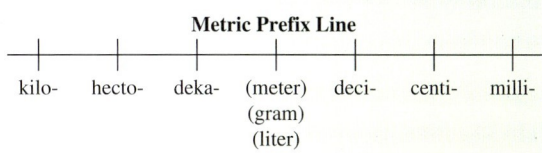

Metric Prefix Line

kilo-   hecto-   deka-   (meter) (gram) (liter)   deci-   centi-   milli-

1. 36 c = _____ qt
2. 220 cm = _____ m
3. $\frac{3}{4}$ lb = _____ oz
4. 0.3 L = _____ mL

5. 12 ft = _____ yd
6. 6.03 kg = _____ g
7. 45 dm = _____ m
8. 9 in. = _____ ft

9. $\frac{1}{2}$ mi = _____ ft
10. 6000 lb = _____ tons
11. 8 pt = _____ qt
12. 1.5 tsp = _____ T

13. 21 m = _____ km
14. 68 mg = _____ cg
15. 36 mL = _____ cc
16. 64 oz = _____ lb

17. 4322 g = _____ kg
18. 5 m = _____ mm
19. 20 fl oz = _____ c
20. 510 sec = _____ min

21. 4 pt = _____ gal
22. 26 fl oz = _____ c
23. 5.46 kg = _____ g
24. 9.02 L = _____ cL

25. 9.1 mi = _____ yd
26. 48 oz = _____ lb
27. 1.62 tons = _____ lb
28. 4.6 km = _____ m

29. 60 hr = _____ days
30. 8 cc = _____ mL
31. 8:32:24 = _____ min
32. 2 wk = _____ hr

Section 8.4 Medical Applications Involving Measurement

Concepts
1. Additional Metric Units of Mass
2. Medical Applications

1. Additional Metric Units of Mass

Scientists and people in the medical community almost exclusively use the metric system. For example:

A patient's mass may be recorded in kilograms.
A dosage of cough syrup may be measured in milliliters.
The active ingredient in a pain reliever is given in grams.

Sometimes doctors prescribe medicines in very small amounts. In these cases, it is sometimes more convenient to use units of **micrograms**. The abbreviation for microgram is mcg or sometimes μg. Furthermore,

1000 mcg $= 1$ mg   It takes 1 thousand micrograms to equal 1 milligram.

$1{,}000{,}000$ mcg $= 1$ g   It takes 1 million micrograms to equal 1 gram.

The prefix micro- comes from Greek meaning "small." In the metric system, it is one-millionth of a fundamental unit.

- $1 \text{ μg} = \dfrac{1}{1{,}000{,}000} \text{ g}$
- $1 \text{ μL} = \dfrac{1}{1{,}000{,}000} \text{ L}$
- $1 \text{ μm} = \dfrac{1}{1{,}000{,}000} \text{ m}$

Example 1 Converting Units of Micrograms

a. Convert. 0.85 mg $=$ _____ mcg
b. A doctor gives a heart patient an initial dose of 200 μg of nitroglycerin. How many milligrams is this?

Solution:

a. $0.85 \text{ mg} = \dfrac{0.85 \text{ mg}}{1} \cdot \dfrac{1000 \text{ mcg}}{1 \text{ mg}}$ ← new unit to convert to
← unit to convert from

$= 850$ mcg

b. $200 \text{ μg} = \dfrac{200 \text{ μg}}{1} \cdot \dfrac{1 \text{ mg}}{1000 \text{ μg}}$ ← new unit to convert to
← unit to convert from

$= 0.2$ mg

Skill Practice Convert.

1. 0.04 mg $=$ _____ mcg
2. $95{,}000$ μg $=$ _____ mg

2. Medical Applications

Example 2 Applying Metric Units of Measure to Medicine

A doctor orders the antibiotic oxacillin for a child. The dosage is 12.5 mg of the drug per kilogram of the child's body mass. This dosage is given 4 times a day.

a. How much of the drug should a 24-kg child get in one dose?
b. How much of the drug would the child get if she were on a 10-day course of the antibiotic?

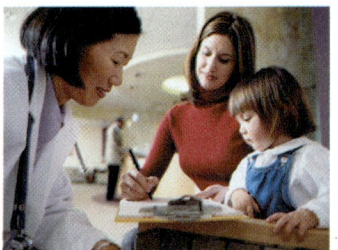
Ryan McVay/Photodisc/Getty Images

Answers
1. 40 mcg   2. 95 mg

Solution:

a. We need to multiply the unit rate of 12.5 mg per kilogram times the child's body mass.

$$\text{Single dose} = \left(12.5 \, \frac{\text{mg}}{\text{kg}}\right)(24 \, \text{kg}) = 300 \, \text{mg}$$

b. For a 10-day course, we need to multiply 300 mg by the number of doses per day (4), and the total number of days (10).

$$\text{Total amount of drug} = \left(300 \, \frac{\text{mg}}{\text{dose}}\right) \cdot \left(4 \, \frac{\text{doses}}{\text{day}}\right) \cdot (10 \, \text{days})$$

$$= 12{,}000 \, \text{mg} \quad \text{or equivalently 12 g.}$$

Skill Practice

3. A child is to receive 0.5 mg of a drug per kilogram of the child's body mass. This dosage is given two times each day.

  a. How much of the drug should a 27-kg child get in one dose?

  b. How much of the drug would the child get if he were on a 12-day course of the drug?

Example 3 Applying Metric Units of Measure to Medicine

A drug used for high blood pressure comes in a concentration of 50 mg per milliliter of solution. If a doctor prescribes 225 mg per dose, how many milliliters should be administered for a single dose?

Solution:

The statement, 50 mg per milliliter, represents the ratio, $\frac{50 \, \text{mg}}{1 \, \text{mL}}$.

$\frac{50 \, \text{mg}}{1 \, \text{mL}} = \frac{225 \, \text{mg}}{x}$ Use this ratio to set up a proportion.

$50x = 225(1)$ Equate the cross products.

$\frac{50x}{50} = \frac{225}{50}$ Divide by 50.

$x = 4.5$ A single dose is 4.5 mL.

Skill Practice

4. A doctor orders amoxicillin (an antibiotic) in a liquid suspension form. The liquid contains 75 mg of amoxicillin per 5 mL of liquid. If the doctor orders 300 mg per day for a child, how much of the suspension should be given?

Answers

3. a. 13.5 mg **b.** 324 mg
4. 20 mL

Practice Exercises

Section 8.4

Prerequisite Review

For Exercises R.1–R.4, perform the conversions.

R.1. 9.84 m = _____ mm

R.2. 80 mg = _____ cg

R.3. 4 kL = _____ cL

R.4. 0.009 kL = _____ mL

For Exercises R.5–R.6, perform the conversions. Round the answer to the tenths.

R.5. 9 kg = _____ lb

R.6. 150 g = _____ oz

Vocabulary and Key Concepts

1. The unit of capacity equivalent to $\frac{1}{1,000,000}$ of a gram is called the _____ and is denoted by mcg or µg.
2. 1000 micrograms (mcg) is equal to _____ milligram (mg).

Concept 1: Additional Metric Units of Mass

For Exercises 3–10, perform the conversions. (See Example 1.)

3. 0.01 mg = _____ µg
4. 0.005 mg = _____ µg
5. 7500 mcg = _____ mg
6. 50 mcg = _____ mg

7. 500 µg = _____ cg
8. 1000 µg = _____ cg
9. 0.05 cg = _____ mcg
10. 0.001 cg = _____ mcg

Concept 2: Medical Applications

11. A doctor orders 0.2 mg of the drug atropine given by injection. How many micrograms is this?

12. A doctor prescribes a drug used to treat thyroid disease. A patient is sometimes started on a dose of 0.05 mg/day. How many micrograms is this?

13. The drug cyanocobalamin is prescribed by a doctor in the amount of 1000 mcg. How many milligrams is this?

14. An injection of naloxone is given in the amount of 800 mcg. How many milligrams is this?

15. The drug Zovirax is sometimes used to treat chicken pox in children. One doctor recommended 20 mg of the drug per kilogram of the child's body mass, 4 times daily. (See Example 2.)

    a. How much of the drug should a 20-kg child receive for one dose?

    b. How much of the drug would be given over a 5-day period?

16. The drug Amoxil is sometimes used to treat children with bacterial infections. A doctor prescribed 40 mg of the drug per kilogram of the child's body mass, 3 times daily.

    a. How much of the drug should a 15-kg child receive for one dose?

    b. How much of the drug would be given to the child over a 10-day period?

17. A nurse must administer 45 mg of a drug. The drug is available in a liquid form with a concentration of 15 mg per milliliter of the solution. How many milliliters of the solution should the nurse give? (See Example 3.)

18. A patient must receive 500 mg of medication in a solution that has a strength of 250 mg per 5 milliliter of solution. How many milliliters of solution should be given?

19. The drug amoxicillin is an antibiotic used to treat bacterial infections. A doctor orders 250 mg every 8 hr. How many grams of the drug would be given in 1 wk?

20. A 25-lb dog suffering with arthritis is given 250 mg of glucosamine twice per day for 10 days. How much glucosamine would be given in 10 days?

21. Dr. Boyd gives a patient 2 cc of an ulcer medication. How many milliliters is this?

22. If a nurse mixed 11.5 mL of sterile water with 1.5 mL of oxacillin, how many cubic centimeters will this produce?

23. A tetanus vaccine was purchased by a group of family practice doctors. They purchased 1 L of the vaccine. How many patients can be vaccinated if the normal dose is 2 cc?

24. A pharmacist has a 1-L bottle of cough syrup. How many 20-cL bottles can she make?

25. A doctor orders 0.2 mg of a drug per kilogram of a patient's body mass. How much of the drug should be given to a patient who is 48 kg?

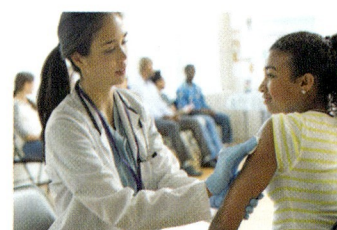

KidStock/Blend Images

26. The dosage for a painkiller is 0.05 mg per kilogram of a patient's body mass. How much of the drug should be administered to a patient who is 90 kg?

Expanding Your Skills

27. A normal value of hemoglobin in the blood for an adult male is 18 g/dL (that is, 18 grams per deciliter). How much hemoglobin would be expected in 20 mL of a males's blood?

28. A normal value of hemoglobin in the blood for an adult female is 15 g/dL (that is, 15 grams per deciliter). How much hemoglobin would be expected in 40 mL of a female's blood?

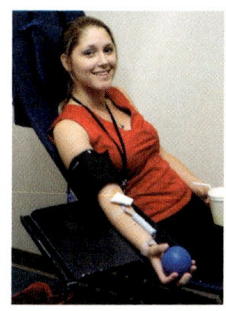

David H. Lewis/E+/Getty Images

Lines and Angles

Section 8.5

1. Basic Definitions

Concepts

1. Basic Definitions
2. Naming and Measuring Angles
3. Complementary and Supplementary Angles
4. Parallel and Perpendicular Lines

In this chapter, we will introduce some basic concepts of geometry.

A **point** is a specific location in space. We often symbolize a point by a dot and label it with a capital letter such as P.

A **line** consists of infinitely many points that follow a straight path. A line extends forever in both directions. This is illustrated by arrowheads at both ends. Figure 8-11 shows a line through points A and B. The line can be represented as $\overleftrightarrow{AB}$ or as $\overleftrightarrow{BA}$.

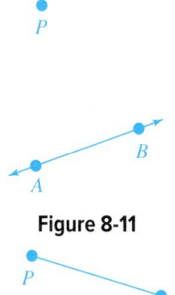

Figure 8-11

A **line segment** is a part of a line between and including two distinct endpoints. A line segment with endpoints P and Q can be denoted $\overline{PQ}$ or $\overline{QP}$. See Figure 8-12.

Figure 8-12

A **ray** is the part of a line that includes an endpoint and all points on one side of the endpoint. In Figure 8-13, ray $\overrightarrow{PQ}$ is named by using the endpoint P and another point Q on the ray. Notice that the rays $\overrightarrow{PQ}$ and $\overrightarrow{QP}$ are different because they extend in different directions. See Figure 8-13.

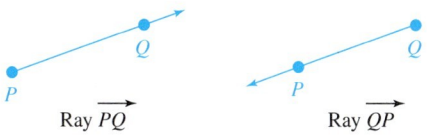

Figure 8-13

TIP: A ray has only one endpoint, which is always written first.

Example 1 Identifying Points, Lines, Line Segments, and Rays

Identify each as a point, line, line segment, or ray.

a. $\overleftrightarrow{MN}$ b. $\overrightarrow{NM}$ c. $\overline{MN}$ d. $\bullet S$

Solution:

a. The double arrowheads indicate that $\overleftrightarrow{MN}$ is a line.
b. The single arrowhead indicates that $\overrightarrow{NM}$ is a ray with endpoint N.
c. The bar drawn above the letters $\overline{MN}$ indicates a line segment with endpoints M and N.
d. The dot represents a point.

Skill Practice Identify each as a point, line, line segment, or ray.

1. $\overline{RS}$ 2. $\bullet Q$ 3. $\overrightarrow{XY}$ 4. $\overleftrightarrow{TV}$

Answers

1. Line segment
2. Point   3. Ray   4. Line

2. Naming and Measuring Angles

An **angle** is a geometric figure formed by two rays that share a common endpoint. The common endpoint is called the **vertex** of the angle. In Figure 8-14, the rays $\overrightarrow{PR}$ and $\overrightarrow{PQ}$ share the endpoint P. These rays form the sides of the angle and the angle is denoted ∠QPR or ∠RPQ. Notice that when we name an angle, the vertex must be the middle letter. Sometimes a small arc ⌒ is drawn to illustrate the location of an angle. In such a case, the angle may be named by using the symbol ∠ along with the letter of the vertex.

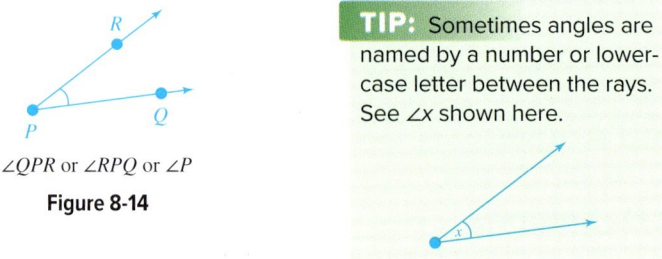

∠QPR or ∠RPQ or ∠P

Figure 8-14

TIP: Sometimes angles are named by a number or lower-case letter between the rays. See ∠x shown here.

Avoiding Mistakes

The ° symbol can be used for temperature or for measuring angles. The context of the problem tells us whether temperature or angle measure is implied by the symbol.

The most common unit to measure an angle is the degree, denoted by °. To become familiar with the measure of angles, consider the following benchmarks. Two rays that form a quarter turn of a circle make a 90° angle. A 90° angle is called a **right angle** and is often depicted with a □ symbol. Two rays that form a half turn of a circle make a 180° angle. A 180° angle is called a **straight angle** because it appears as a straight line. A full circle has 360°. For example, the second hand of a clock sweeps out an angle of 360° in 1 minute. See Figure 8-15.

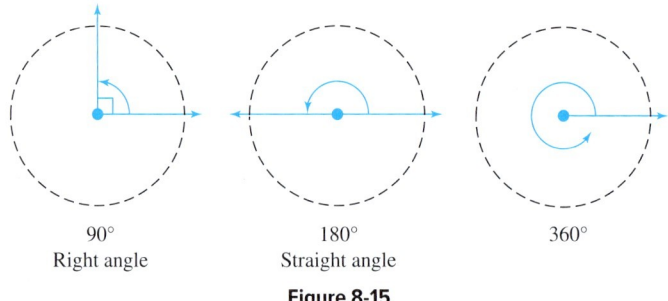

90°
Right angle

180°
Straight angle

360°

Figure 8-15

We can approximate the measure of an angle by using a tool called a *protractor*, shown in Figure 8-16. A protractor uses equally spaced tick marks around a semicircle to measure angles from 0° to 180°.

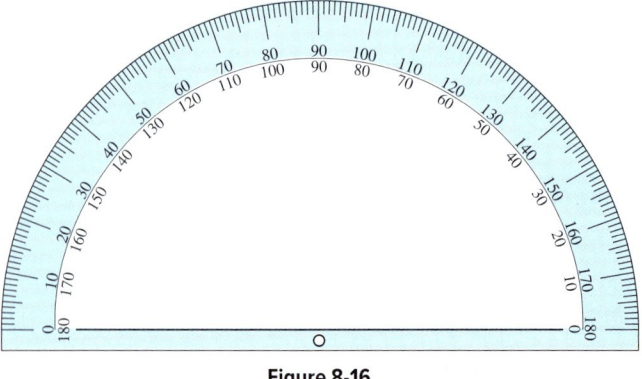

Figure 8-16

Example 2 shows how we can use a protractor to measure several angles. To denote the measure of an angle, we use the symbol *m*, written in front of the name of the angle. For example, if the measure of angle A is 30°, we write $m(\angle A) = 30°$.

Example 2 Measuring Angles

Read the protractor to determine the measure of each angle.

a. ∠AOB   b. ∠AOC   c. ∠AOD   d. ∠AOE

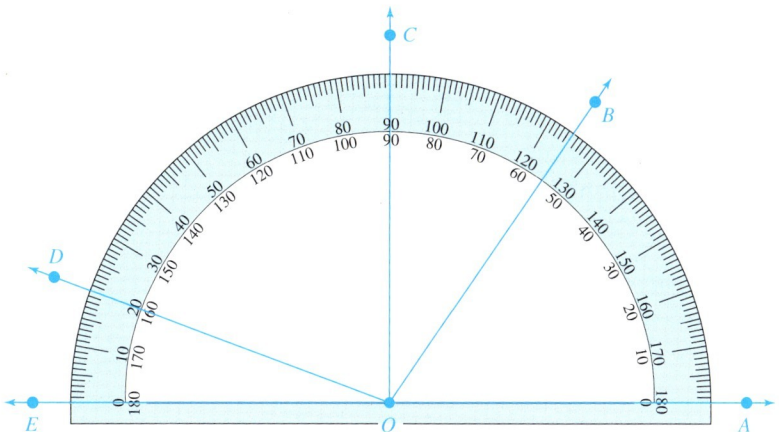

Solution:

We will use the inner scale on the protractor. This is done because we are measuring the angles in a counterclockwise direction, beginning at 0° along ray $\overrightarrow{OA}$.

a. $m(\angle AOB) = 55°$ On the inner scale, ray $\overrightarrow{OA}$ is aligned with 0° and ray $\overrightarrow{OB}$ passes through 55°. Therefore, $m(\angle AOB) = 55°$.

b. $m(\angle AOC) = 90°$ ∠AOC is a right angle.

c. $m(\angle AOD) = 160°$

d. $m(\angle AOE) = 180°$ ∠AOE is a straight angle.

Skill Practice Read the protractor to determine the measure of each angle.

5. ∠POQ
6. ∠POR
7. ∠POS
8. ∠POT

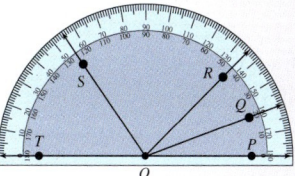

An angle is said to be an **acute angle** if its measure is between 0° and 90°. An angle is said to be an **obtuse angle** if its measure is between 90° and 180°. See Figure 8-17.

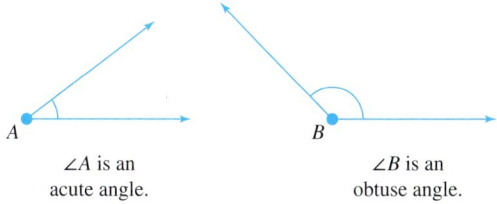

∠A is an acute angle.   ∠B is an obtuse angle.

Figure 8-17

3. Complementary and Supplementary Angles

- Two angles are said to be equal or **congruent** if they have the same measure.
- Two angles are said to be **complementary** if the sum of their measures is 90°. For example, the complement of a 60° angle is a 30° angle, and vice versa. (Figure 8-18).
- Two angles are said to be **supplementary** if the sum of their measures is 180°. For example, the supplement of a 60° angle is a 120° angle, and vice versa. (Figure 8-19).

Answers
5. 20°   6. 45°   7. 125°
8. 180°

TIP: To remember the difference between complementary and supplementary, think

Complementary ⇔ **C**orner,
Supplementary ⇔ **S**traight.

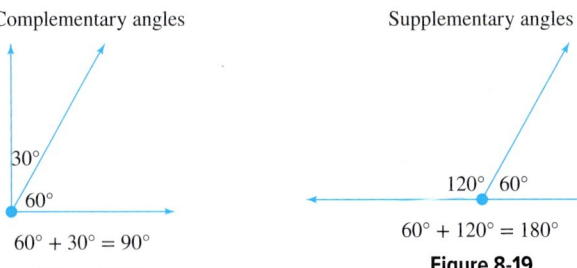

Complementary angles
$60° + 30° = 90°$
Figure 8-18

Supplementary angles
$60° + 120° = 180°$
Figure 8-19

Example 3   **Identifying Supplementary and Complementary Angles**

a. What is the supplement of a 105° angle?
b. What is the complement of a 12° angle?

Solution:

a. Let x represent the measure of the supplement of a 105° angle.

$$x + 105 = 180$$ The sum of a 105° angle and its supplement must equal 180°.

$$x + 105 - 105 = 180 - 105$$ Subtract 105 from both sides to solve for x.

$$x = 75$$ The supplement is a 75° angle.

b. Let y represent the measure of the complement of a 12° angle.

$$y + 12 = 90$$ The sum of a 12° angle and its complement must equal 90°.

$$y + 12 - 12 = 90 - 12$$ Subtract 12 from both sides to solve for y.

$$y = 78$$ The complement is a 78° angle.

Skill Practice

9. What is the supplement of a 35° angle?
10. What is the complement of a 52° angle?

4. Parallel and Perpendicular Lines

Two lines may intersect (cross) or may be parallel. **Parallel lines** lie in the same plane (on the same flat surface), but never intersect. See Figure 8-20.

TIP: Sometimes we use the symbol ∥ to denote parallel lines.

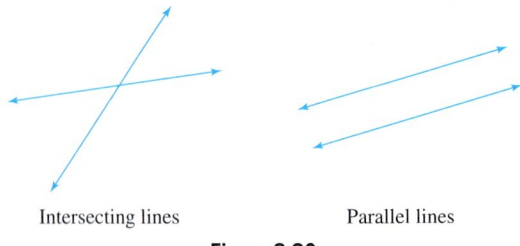

Intersecting lines   Parallel lines
Figure 8-20

Notice that two intersecting lines form four angles. In Figure 8-21, ∠a and ∠c are **vertical angles**. They appear on opposite sides of the vertex. Likewise, ∠b and ∠d are vertical angles. Vertical angles are equal in measure. That is $m(\angle a) = m(\angle c)$ and $m(\angle b) = m(\angle d)$.

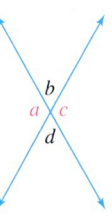

Figure 8-21

Answers
9. 145°   10. 38°

Angles that share a side are called *adjacent* angles. One pair of *adjacent* angles in Figure 8-21 is ∠a and ∠b.

If two lines intersect at a right angle, they are **perpendicular lines**. See Figure 8-22.

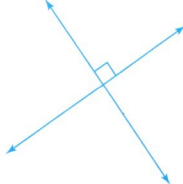

Figure 8-22

> **TIP:** Sometimes we use the symbol ⊥ to denote perpendicular lines.

In Figure 8-23, lines L_1 and L_2 are parallel lines. If a third line m intersects the two parallel lines, eight angles are formed. Suppose we label the eight angles formed by lines L_1, L_2, and m with the numbers 1–8.

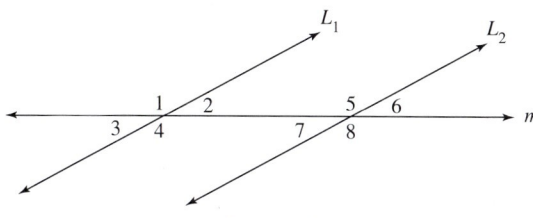

Figure 8-23

These angles have the special properties found in Table 8-6.

Table 8-6

Lines L_1 and L_2 are Parallel; Line m is an Intersecting Line	Name of Angles	Property
	The following pairs of angles are called **alternate interior angles**: ∠2 and ∠7 ∠4 and ∠5	Alternate interior angles are equal in measure. $m(∠2) = m(∠7)$ $m(∠4) = m(∠5)$
	The following pairs of angles are called **alternate exterior angles**: ∠1 and ∠8 ∠3 and ∠6	Alternate exterior angles are equal in measure. $m(∠1) = m(∠8)$ $m(∠3) = m(∠6)$
	The following pairs of angles are called **corresponding angles**: ∠1 and ∠5 ∠2 and ∠6 ∠3 and ∠7 ∠4 and ∠8	Corresponding angles are equal in measure. $m(∠1) = m(∠5)$ $m(∠2) = m(∠6)$ $m(∠3) = m(∠7)$ $m(∠4) = m(∠8)$

Example 4 — Finding the Measure of Angles in a Diagram

Assume that lines L_1 and L_2 are parallel. Find the measure of each angle, and explain how the angle is related to the given angle of 65°.

a. $\angle a$
b. $\angle b$
c. $\angle c$
d. $\angle d$

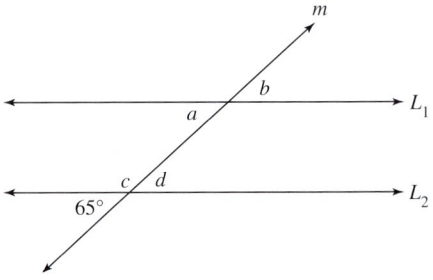

Solution:

a. $m(\angle a) = 65°$ $\angle a$ is a corresponding angle to the given angle.
b. $m(\angle b) = 65°$ $\angle b$ is an alternate exterior angle to the given angle.
c. $m(\angle c) = 115°$ $\angle c$ is the supplement to the given angle.
d. $m(\angle d) = 65°$ $\angle d$ and the given angle are vertical angles.

Skill Practice Assume that lines L_1 and L_2 are parallel. Find the measure of each angle. Explain how the angle is related to the given angle.

11. $\angle a$
12. $\angle b$
13. $\angle c$
14. $\angle d$

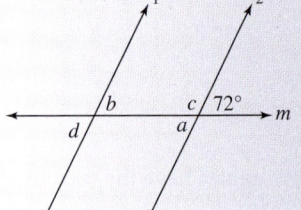

Answers
11. 72°; vertical angles
12. 72°; corresponding angles
13. 108°; supplementary angles
14. 72°; alternate exterior angles

Section 8.5 Activity

A.1. Match each symbol with the appropriate definition:

a. $\overrightarrow{AB}$ i. Point A
b. $\overline{AB}$ ii. Ray BA
c. $\overleftrightarrow{AB}$ iii. Line segment AB
d. $\overrightarrow{BA}$ iv. Line AB
e. •A   v. Ray AB

For Exercises A.2–A.5, determine the following:
a. Read the protractor to determine the measure of each angle.
b. Provide an alternative name for the angle.
c. Classify the angle as acute, right, obtuse, or straight.

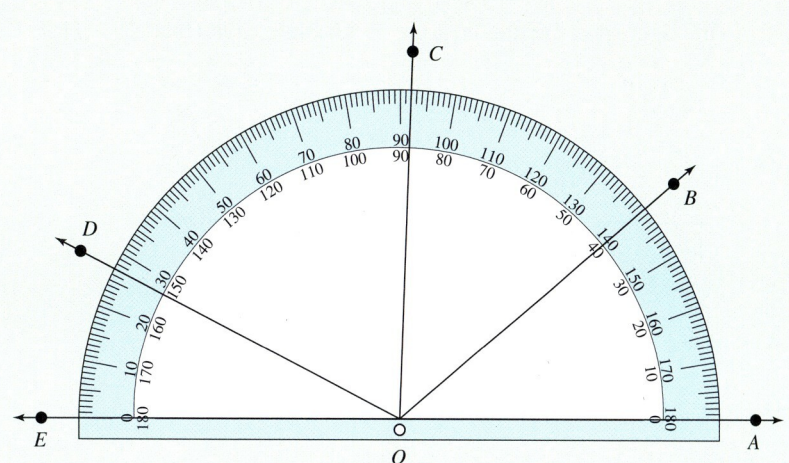

A.2. ∠AOB
   a. $m(\angle AOB) = $ _____   **b.** _____   **c.** _____

A.3. ∠AOC
   a. $m(\angle AOC) = $ _____   **b.** _____   **c.** _____

A.4. ∠AOD
   a. $m(\angle AOD) = $ _____   **b.** _____   **c.** _____

A.5. ∠AOE
   a. $m(\angle AOE) = $ _____   **b.** _____   **c.** _____

A.6. What is the vertex of each angle in Exercises A.2–A.5?

A.7. What does it mean for two angles to be complementary? Calculate the complement of 57°.

A.8. What does it mean for two angles to be supplementary? Calculate the supplement of 9°.

A.9.
  a. Two lines that lie in the same plane but never intersect are called _____ lines.
  b. Draw two lines that are parallel.
  c. Give an example of parallel lines in the real world.

A.10.
  a. Two lines that intersect at a right angle are called _____ lines.
  b. Draw two lines that are perpendicular.
  c. Give an example of perpendicular lines in the real world.

A.11. Refer to the figure and determine what angle pairs are vertical angles. What do you know about the measure of vertical angles?

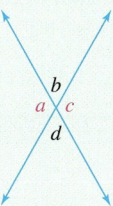

For Exercises A.12–A.18, refer to the figure and find the measure of the angle, given that ∠1 = 118°. Lines L_1 and L_2 are parallel lines.

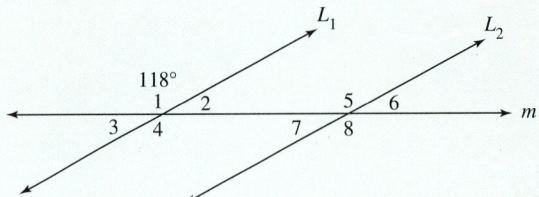

A.12. $m(\angle 2) =$ _____

A.13. $m(\angle 3) =$ _____

A.14. $m(\angle 4) =$ _____

A.15. $m(\angle 5) =$ _____

A.16. $m(\angle 6) =$ _____

A.17. $m(\angle 7) =$ _____

A.18. $m(\angle 8) =$ _____

Section 8.5   Practice Exercises

Vocabulary and Key Concepts

1. **a.** A _____ is a specific location in space.

   b. A _____ consists of infinitely many points that follow a straight path.

   c. A line _____ is part of a line between and including two distinct endpoints.

   d. The ray $\overrightarrow{PQ}$ is the set of points beginning at point _____ and extending along the line in the direction of point _____.

   e. An _____ is a geometric figure formed by two rays that share a common endpoint. The common endpoint is called the _____.

   f. A _____ angle has a measure of 90°. A straight angle has a measure of _____ °.

   g. A _____ is a tool used to measure an angle.

   h. An _____ angle measures between 0° and 90°, and an _____ angle measures between 90° and 180°.

   i. Two angles are called _____ angles if the sum of their measures is 90°. Two angles are called _____ angles if the sum of their measures is 180°.

   j. Two lines that lie on the same flat surface but never intersect are called _____ lines.

   k. Two lines that intersect at a 90° angle are called _____ lines.

Section 8.5 Lines and Angles 493

Concept 1: Basic Definitions

2. Explain the difference between a line and a line segment.

3. Explain the difference between a line and a ray.

4. Is the ray $\overrightarrow{AB}$ the same as $\overrightarrow{BA}$? Explain.

For Exercises 5–10, identify each figure as a line, line segment, ray, or point. (See Example 1.)

5. K   H

6. K   H

7. H

8. H   K

9. H   K

10. K

For Exercises 11–14, draw a figure that represents the expression.

11. $\overline{XY}$

12. A point named X

13. $\overleftrightarrow{YX}$

14. $\overrightarrow{XY}$

Concept 2: Naming and Measuring Angles

15. Sketch a right angle.

16. Sketch a straight angle.

17. Sketch an acute angle.

18. Sketch an obtuse angle.

For Exercises 19–24, use the protractor to determine the measure of each angle. (See Example 2.)

19. $m(\angle AOB)$

20. $m(\angle AOC)$

21. $m(\angle AOD)$

22. $m(\angle AOE)$

23. $m(\angle AOF)$

24. $m(\angle AOG)$

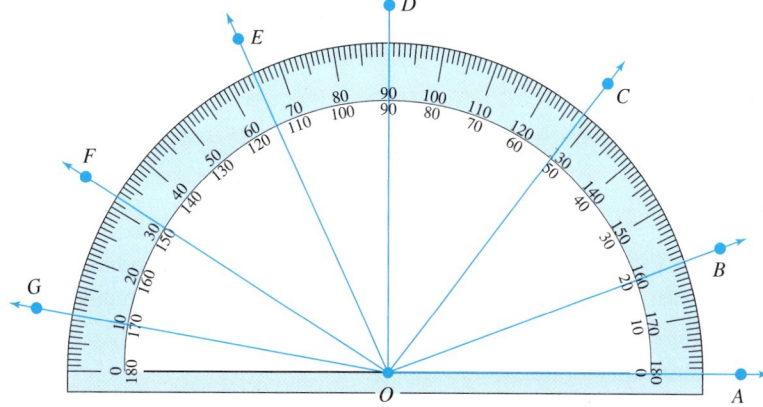

For Exercises 25–32, label each as an obtuse angle, acute angle, right angle, or straight angle.

25. $m(\angle A) = 90°$

26. $m(\angle E) = 91°$

27. $m(\angle B) = 98°$

28. $m(\angle F) = 30°$

29. $m(\angle C) = 2°$

30. $m(\angle G) = 130°$

31. $m(\angle D) = 180°$

32. $m(\angle H) = 45°$

Concept 3: Complementary and Supplementary Angles

For Exercises 33–40, the measure of an angle is given. Find the measure of the complement. (See Example 3.)

33. 80°

34. 5°

35. 27°

36. 64°

37. 29.5°

38. 13.2°

39. 89°

40. 1°

For Exercises 41–48, the measure of an angle is given. Find the measure of the supplement. (See Example 3.)

41. 80° 42. 5° 43. 127° 44. 124°

45. 37.4° 46. 173.9° 47. 179° 48. 1°

49. Can two supplementary angles both be obtuse? Why or why not?

50. Can two supplementary angles both be acute? Why or why not?

51. Can two complementary angles both be acute? Why or why not?

52. Can two complementary angles both be obtuse? Why or why not?

53. What measure angle is its own supplement?

54. What measure angle is its own complement?

Concept 4: Parallel and Perpendicular Lines

55. Sketch two lines that are parallel.

56. Sketch two lines that are *not* parallel.

57. Sketch two lines that are perpendicular.

58. Sketch two lines that are *not* perpendicular.

For Exercises 59–62, find the measure of angles *a*, *b*, *c*, and *d*.

59.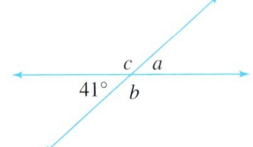

60.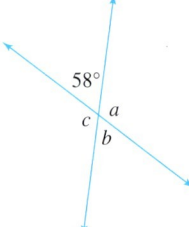

61.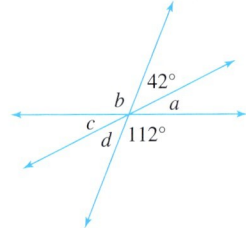

62.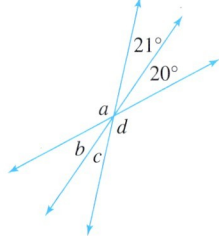

63. If two intersecting lines form vertical angles and each angle measures 90°, what can you say about the lines?

64. Can two adjacent angles formed by two intersecting lines be complementary, supplementary, or neither?

For Exercises 65 and 66, refer to the figure.

65. Describe the pair of angles, ∠a and ∠c, as complementary, vertical, or supplementary angles.

66. Describe the pair of angles, ∠b and ∠c, as complementary, vertical, or supplementary angles.

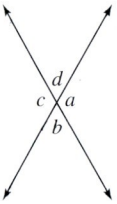

For Exercises 67–70, refer to the figure.

67. Identify a pair of vertical angles.

68. Identify a pair of alternate interior angles.

69. Identify a pair of alternate exterior angles.

70. Identify a pair of corresponding angles.

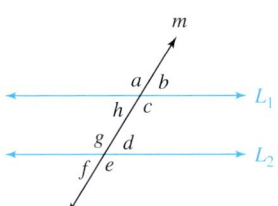

For Exercises 71–74, find the measure of angles *a–g* in the figure. Assume that L_1 and L_2 are parallel and that *m* is an intersecting line. (See Example 4.)

71.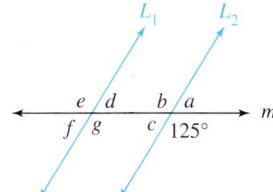

72.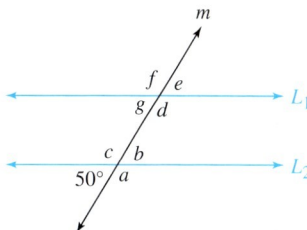

73.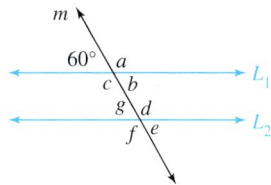

74.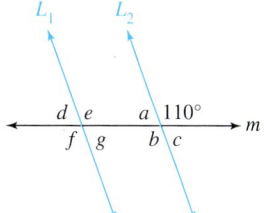

For Exercises 75–84, refer to the figure and answer true or false.

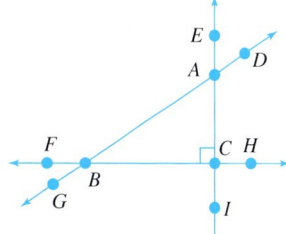

75. $\overleftrightarrow{AC}$ and $\overleftrightarrow{BC}$ are perpendicular lines.

76. $\overleftrightarrow{AB}$ and $\overleftrightarrow{AC}$ are perpendicular lines.

77. ∠GBF is an acute angle.

78. ∠EAD is an acute angle.

79. ∠EAD and ∠DAC are complementary angles.

80. ∠GBF and ∠FBA are complementary angles.

81. ∠EAD and ∠CAB are vertical angles.

82. ∠ABC and ∠FBG are vertical angles.

83. The point B is on $\overline{GA}$.

84. The point C is on $\overline{BH}$.

For Exercises 85–88, find the measure of ∠XYZ.

85.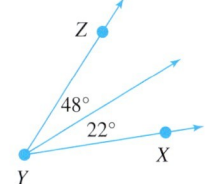

86.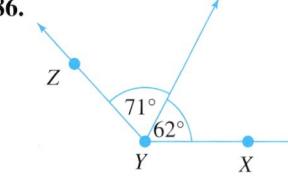

87.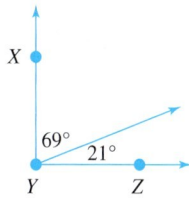

88.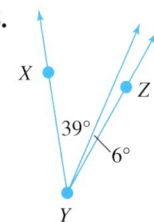

Expanding Your Skills

89. Use the figure to find the measure of each angle.

    a. ∠AOB

    b. ∠EOD

    c. ∠AOE

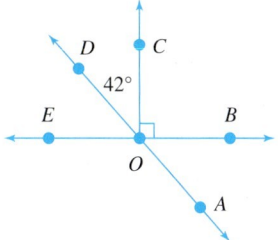

90. Use the figure to find the measure of each angle.

    a. ∠AOB

    b. ∠EOD

    c. ∠AOE

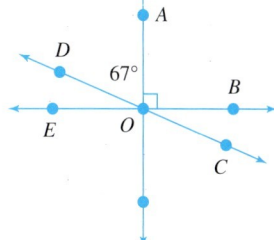

The second hand on a clock sweeps out a complete circle in 1 min. A circle forms a 360° arc. Use this information for Exercises 91–94.

91. How many degrees does a second hand on a clock move in 30 sec?

92. How many degrees does a second hand on a clock move in 15 sec?

93. How many degrees does a second hand on a clock move in 20 sec?

94. How many degrees does a second hand on a clock move in 45 sec?

Marko Kovacevic/iStockphoto/ Getty Images

Section 8.6 Triangles and the Pythagorean Theorem

Concepts

1. Triangles
2. Square Roots
3. Pythagorean Theorem

1. Triangles

Recall that a **polygon** is a flat figure formed by line segments connected at their ends. A triangle is a three-sided polygon. Furthermore, the sum of the measures of the angles within a triangle is 180°. Teachers often demonstrate this fact by tearing a triangular sheet of paper as shown in Figure 8-24. Then they align the **vertices** (points) of the triangle to form a straight angle.

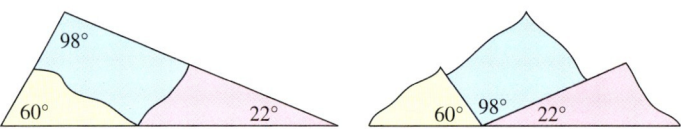

Figure 8-24

> **Angles of a Triangle**
> The sum of the measures of the angles of a triangle equals 180°.

Example 1 Finding the Measure of Angles within a Triangle

Find the measures of angles a and b.

a.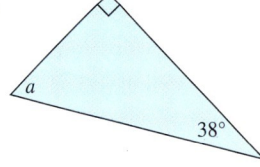

b.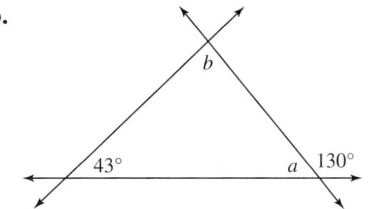

Solution:

a. Recall that the □ symbol represents a 90° angle.

$38° + 90° + m(\angle a) = 180°$ The sum of the angles within a triangle is 180°.

$128° + m(\angle a) = 180°$ Add the measures of the two known angles.

$128° - 128° + m(\angle a) = 180° - 128°$ Solve for $m(\angle a)$.

$m(\angle a) = 52°$

b. ∠*a* is the supplement of the 130° angle. Thus, *m*(∠*a*) = 50°.

$43° + 50° + m(\angle b) = 180°$ The sum of the angles within a triangle is 180°.

$93° + m(\angle b) = 180°$ Add the measures of the two known angles.

$93° - 93° + m(\angle b) = 180° - 93°$ Solve for *m*(∠*b*).
$m(\angle b) = 87°$

Skill Practice Find the measures of angles *a* and *b*.

1.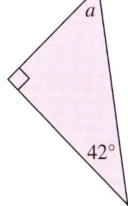
2.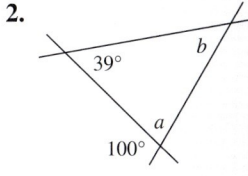

Triangles may be categorized by the measures of their angles and by the number of equal sides or angles (Figures 8-25 and 8-26).

- An **acute triangle** is a triangle in which all three angles are acute.
- A **right triangle** is a triangle in which one angle is a right angle.
- An **obtuse triangle** is a triangle in which one angle is obtuse.

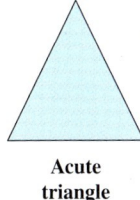 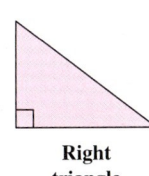 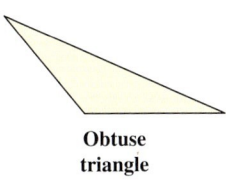

Acute triangle   Right triangle   Obtuse triangle

Figure 8-25

- An **equilateral triangle** is a triangle in which all three sides (and all three angles) are equal in measure.
- An **isosceles triangle** is a triangle in which two sides are equal in length (the angles opposite the equal sides are also equal in measure).
- A **scalene triangle** is a triangle in which no sides (or angles) are equal in measure.

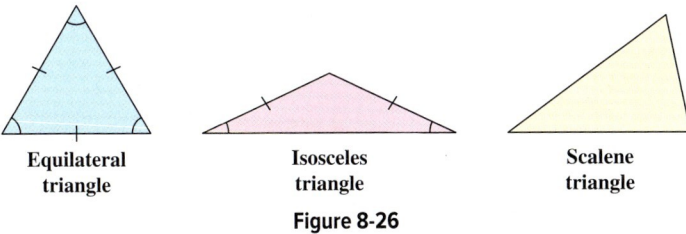

Equilateral triangle   Isosceles triangle   Scalene triangle

Figure 8-26

TIP: Sometimes we use tick marks / to denote segments of equal length. Similarly, we sometimes use a small arc) to denote angles of equal measure.

Answers
1. $m(\angle a) = 48°$
2. $m(\angle a) = 80°$
   $m(\angle b) = 61°$

2. Square Roots

In this section we present an important theorem called the Pythagorean theorem. To understand this theorem, we first need some background definitions.

Recall that to square a number means to find the product of the number and itself. Thus, $b^2 = b \cdot b$. For example:

$$6^2 = 6 \cdot 6 = 36$$

We now want to reverse this process by finding a square root of a number. Recall that this is denoted by the radical sign $\sqrt{}$. For example, $\sqrt{36}$ reads as "the positive square root of 36." Thus,

$$\sqrt{36} = 6 \quad \text{because } 6^2 = 6 \cdot 6 = 36.$$

Example 2   **Evaluating Squares and Square Roots**

Simplify.

  a. $\sqrt{64}$ **b.** $\sqrt{100}$ **c.** 100^2 **d.** $\sqrt{\dfrac{1}{4}}$

Solution:

a. $\sqrt{64} = 8$ because $8 \cdot 8 = 64$

b. $\sqrt{100} = 10$ because $10 \cdot 10 = 100$

c. $100^2 = 100 \cdot 100$
       $= 10,000$

d. $\sqrt{\dfrac{1}{4}} = \dfrac{1}{2}$ because $\dfrac{1}{2} \cdot \dfrac{1}{2} = \dfrac{1}{4}$

TIP: The numbers 0, 1, 4, 9, 16, and so on are called perfect squares, because in each case, the principal square root is a whole number.

Skill Practice Simplify.

3. $\sqrt{49}$ **4.** $\sqrt{9}$ **5.** 9^2 **6.** $\sqrt{\dfrac{1}{9}}$

3. Pythagorean Theorem

Recall that a right triangle is a triangle with a 90° angle. The two sides forming the right angle are called the **legs**. The side opposite the right angle is called the **hypotenuse**. Note that the hypotenuse is always the longest side. See Figure 8-27. We often use the letters a and b to represent the legs of a right triangle. The letter c is used to label the hypotenuse.

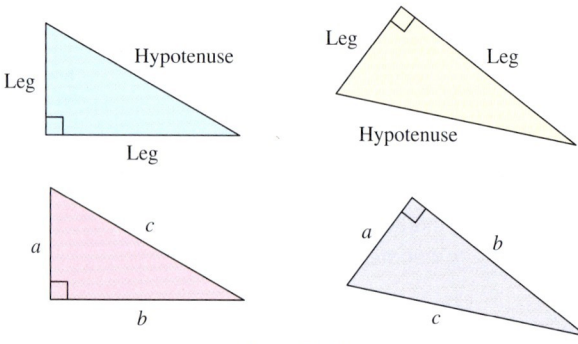

Figure 8-27

Answers
3. 7   **4.** 3   **5.** 81   **6.** $\dfrac{1}{3}$

Section 8.6 Triangles and the Pythagorean Theorem

For any right triangle, the **Pythagorean theorem** gives us the following important relationship among the lengths of the sides.

> ### Pythagorean Theorem
> For any right triangle,
> $$(\text{Leg})^2 + (\text{Leg})^2 = (\text{Hypotenuse})^2$$
>
>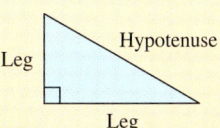
>
> Using the letters a, b, and c to represent the legs and hypotenuse, respectively, we have
> $$a^2 + b^2 = c^2$$
>
>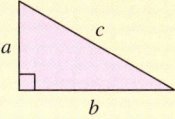

Example 3 — Finding the Length of the Hypotenuse of a Right Triangle

Find the length of the hypotenuse of the right triangle.

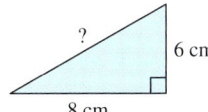

Solution:

The lengths of the legs are given.

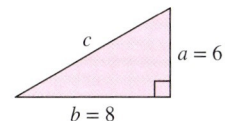

Label the triangle, using a, b, and c. It does not matter which leg is labeled a and which is labeled b.

$a^2 + b^2 = c^2$	Apply the Pythagorean theorem.
$(6)^2 + (8)^2 = c^2$	Substitute $a = 6$ and $b = 8$.
$36 + 64 = c^2$	Simplify.
$100 = c^2$	The solution to this equation is the positive number, c, that when squared equals 100.
$\sqrt{100} = c$	
$10 = c$	Simplify the square root of 100.

TIP: The value $\sqrt{100} = 10$ because $10^2 = 100$.

The solution may be checked using the Pythagorean theorem.
$$a^2 + b^2 = c^2$$
$$(6)^2 + (8)^2 \stackrel{?}{=} (10)^2$$
$$36 + 64 = 100 \checkmark$$

The length of the hypotenuse is 10 cm.

Skill Practice

7. Find the length of the hypotenuse.

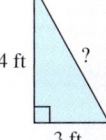

In Example 4, we solve for one of the legs of a right triangle when the other leg and the hypotenuse are known.

Answer
7. 5 ft

Example 4 — Finding the Length of a Leg in a Right Triangle

Find the length of the unknown side of the right triangle.

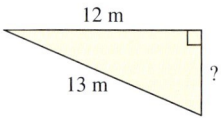

Solution:

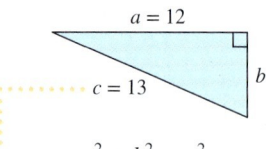

Label the triangle, using a, b, and c. One of the legs is unknown. It doesn't matter whether we call the unknown leg a or b.

$$a^2 + b^2 = c^2$$ Apply the Pythagorean theorem.

$$(12)^2 + b^2 = (13)^2$$ Substitute $a = 12$ and $c = 13$.

$$144 + b^2 = 169$$ Simplify.

$$144 - 144 + b^2 = 169 - 144$$ Solve for b^2.

$$b^2 = 25$$ The solution to this equation is the positive number b that when squared equals 25.

$$b = \sqrt{25}$$

$$b = 5$$ Simplify the square root of 25.

Avoiding Mistakes

Always remember that the hypotenuse (longest side) is given the letter "c" when applying the Pythagorean theorem.

TIP: The value $\sqrt{25} = 5$ because $5^2 = 25$.

The solution may be checked by using the Pythagorean theorem.

$$a^2 + b^2 = c^2$$

$$(12)^2 + (5)^2 \stackrel{?}{=} (13)^2$$

$$144 + 25 = 169 \checkmark$$

The length of the unknown side is 5 m.

Skill Practice

8. Find the length of the unknown side.

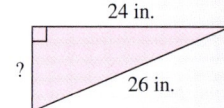

In Example 5 we use the Pythagorean theorem in an application.

Example 5 — Using the Pythagorean Theorem in an Application

When Barb swam across a river, the current carried her 300 yd downstream from her starting point. If the river is 400 yd wide, how far did Barb swim?

Solution:

We first familiarize ourselves with the problem and draw a diagram (Figure 8-28). The distance Barb actually swims is the hypotenuse of the right triangle. Therefore, we label this distance c.

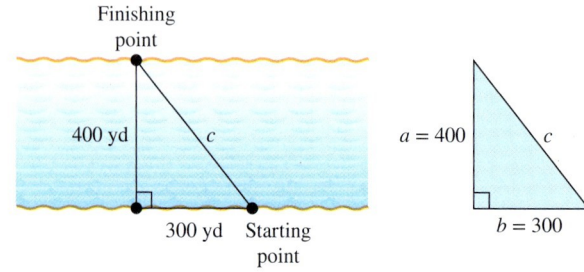

Figure 8-28

Answer

8. 10 in.

$$a^2 + b^2 = c^2$$ Apply the Pythagorean theorem.

$$(400)^2 + (300)^2 = c^2$$ Substitute $a = 400$ and $b = 300$.

$$160{,}000 + 90{,}000 = c^2$$ Simplify.

$$250{,}000 = c^2$$ Add. The solution to this equation is the positive number c that when squared equals 250,000.

$$\sqrt{250{,}000} = c$$

$$500 = c$$ Simplify.

Barb swam 500 yd.

Skill Practice

9. The bottom of a 17-ft ladder is placed 8 ft from the base of a building. How far up the building is the top of the ladder?

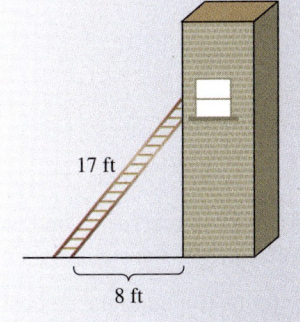

Answer
9. 15 ft

Section 8.6 Activity

A.1. The sum of the measures of $\angle a + \angle b + \angle c = $ _____ °.

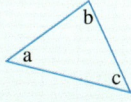

For Exercises A.2–A.7, match each triangle with its definition and figure(s).

Triangle		Definition	Figure
A.2.	Acute triangle	a. Two sides are equal in length.	i.
A.3.	Right triangle	b. No sides or angles are equal in measure.	ii.
A.4.	Obtuse triangle	c. One angle is a right angle.	iii.
A.5.	Equilateral triangle	d. All three sides and angles are equal in measure.	iv.
A.6.	Isosceles triangle	e. All three angles are acute.	
A.7.	Scalene triangle	f. One angle is obtuse.	

For Exercises A.8–A.10, refer to the general triangle below (not to scale). Calculate the measure of the missing angles.

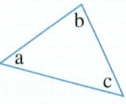

A.8. If $\angle a = 52°$ and $\angle b = 21°$, find the measure of $\angle c$.
A.9. If $\angle b = 25°$ and $\angle c = 89°$, find the measure of $\angle a$.
A.10. If $\angle a = 110°$ and $\angle c = 32°$, find the measure of $\angle b$.

For Exercises A.11–A.15, find the measures of the angles.

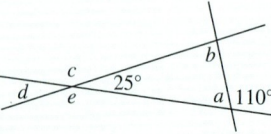

A.11. $m(\angle a) = $ _____
A.12. $m(\angle b) = $ _____
A.13. $m(\angle c) = $ _____
A.14. $m(\angle d) = $ _____
A.15. $m(\angle e) = $ _____

A.16. Recall that finding a square root of a number is to find what positive number, when squared, equals the number inside the radical. It is the reverse process of squaring a number.

Simplify the following square roots. Round to the nearest hundredth when necessary.
  a. $\sqrt{36}$
  b. $\sqrt{49}$
  c. $\sqrt{37}$
  d. What do you notice about the answer in part (c) compared to the results in parts (a) – (b)?

A.17. Solve the following equation by taking the principal square root of both sides to isolate x. Check your result.
$$x^2 = 144$$

A.18. Refer to the right triangle below.

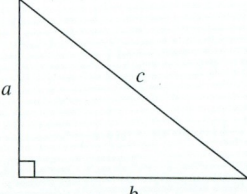

  a. What two sides are the legs of the triangle?
  b. What side is the hypotenuse of the triangle?
  c. What is always true about the length of the hypotenuse?
  d. Write the Pythagorean theorem for any right triangle.

A.19. Refer to the right triangle below.

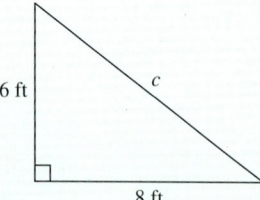

  a. Label the side that is 6 ft in length as side a.
  b. Label the side that is 8 ft in length as side b.
  c. Find the value of the hypotenuse, side c, by using the Pythagorean theorem.
  d. Is the answer from part (c) the longest side in the triangle?

A.20. Refer to the right triangle below.

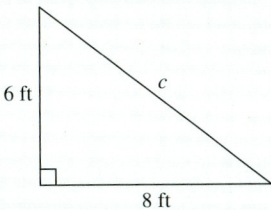

   a. Label the side that is 8 ft in length as side *a*.
   b. Label the side that is 6 ft in length as side *b*.
   c. Find the value of the hypotenuse, side *c*, by using the Pythagorean theorem.
   d. Compare your answer obtained in part (c) to that obtained in Exercise A.19, part (c). Are they equal? Why or why not?

A.21. Find the value of the unknown length.

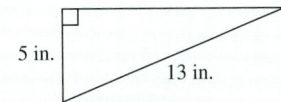

A.22. Find the length of the supporting cable.

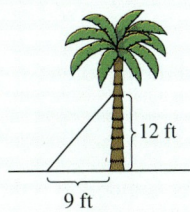

Practice Exercises

Section 8.6

Study Skills Exercise

When solving an application involving geometry, draw an appropriate figure and label the known quantities with numbers and the unknown quantities with variables. This will help you solve the problem. After reading this section, what geometric figure do you think you will be drawing most often in this section?

Prerequisite Review

   R.1. What is the measure of the supplement of a 34° angle?
   R.2. What is the measure of the supplement of a 120° angle?
   R.3. What is the measure of the complement of a 12° angle?
   R.4. What is the measure of the complement of a 60° angle?

For Exercises R.5–R.6, evaluate the square root.

   R.5. $\sqrt{81}$ **R.6.** $\sqrt{100}$

For Exercises R.7–R.8, simplify.

   R.7. $12^2 + 9^2$ **R.8.** $5^2 + 12^2$

504 Chapter 8 Measurement and Geometry

Vocabulary and Key Concepts

1. **a.** The sum of the measures of a triangle is _____°.

    b. A(n) _____ triangle is a triangle in which all three angles measure less than 90°. A(n) _____ triangle is a triangle in which one angle measures 90°. A(n) _____ triangle is a triangle in which one angle measures more than 90°.

    c. A(n) _____ triangle is a triangle in which all three sides have equal length.

    d. A(n) _____ triangle is a triangle in which two sides have equal length.

    e. A(n) _____ triangle is a triangle in which no sides have equal length.

    f. In a right triangle, the _____ is the name given to the longest side. The two shorter sides are called the _____ of the right triangle.

    g. Given a triangle with legs of lengths a and b and hypotenuse of length c, the _____ theorem states that $a^2 + b^2 =$ _____.

2. Tick marks on the segments of a triangle indicate equal _____.

3. Small arcs are used to denote _____ of equal measurement.

4. The ☐ symbol within a triangle represents a _____° angle.

5. The numbers 25, 49, 81, and 100 are examples of _____ squares because the principal square root is a whole number.

6. If given a right triangle with the length of a leg and the hypotenuse given, you can solve for the _____.

7. If given a right triangle with the length of both legs given, you can solve for the _____.

8. True or False. The Pythagorean theorem can only be applied to right triangles.

Concept 1: Triangles

For Exercises 9–16, find the measures of angles a and b. **(See Example 1.)**

9.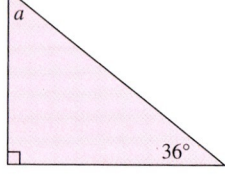

10.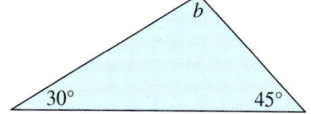

11.

12.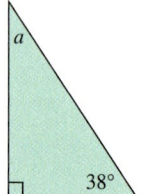

13.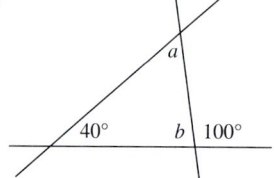

14.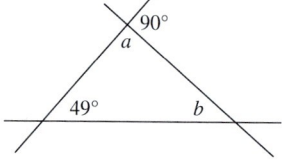

15.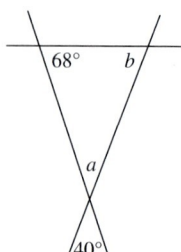

16.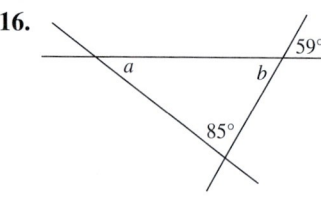

Section 8.6 Triangles and the Pythagorean Theorem   505

For Exercises 17–22, choose all figures that apply. The tick marks / denote segments of equal length, and small arcs) denote angles of equal measure.

17. Acute triangle

18. Obtuse triangle

19. Right triangle

20. Scalene triangle

21. Isosceles triangle

22. Equilateral triangle

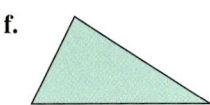

Concept 2: Square Roots

For Exercises 23–42, simplify the squares and square roots. (See Example 2.)

23. $\sqrt{49}$

24. $\sqrt{64}$

25. 7^2

26. 8^2

27. 4^2

28. 5^2

29. $\sqrt{16}$

30. $\sqrt{25}$

31. $\sqrt{36}$

32. $\sqrt{100}$

33. 6^2

34. 10^2

35. 9^2

36. 3^2

37. $\sqrt{81}$

38. $\sqrt{9}$

39. $\sqrt{\dfrac{1}{16}}$

40. $\sqrt{\dfrac{1}{144}}$

41. $\sqrt{0.04}$

42. $\sqrt{0.09}$

Concept 3: Pythagorean Theorem

For Exercises 43–46, find the length of the unknown side. (See Examples 3 and 4.)

43.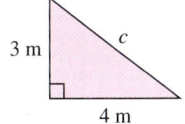

44.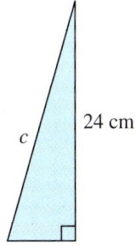

45.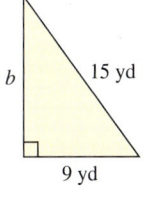

46.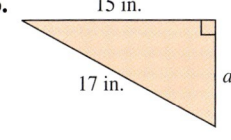

For Exercises 47–50, find the length of the unknown leg or hypotenuse.

47. Leg = 24 ft, hypotenuse = 26 ft

48. Leg = 9 km, hypotenuse = 41 km

49. Leg = 32 in., leg = 24 in.

50. Leg = 16 m, leg = 30 m

51. Find the length of the supporting brace. **(See Example 5.)**

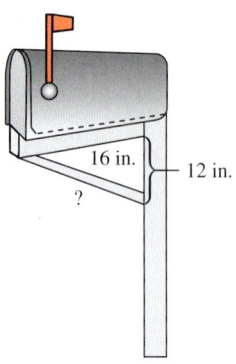

52. Find the length of the ramp.

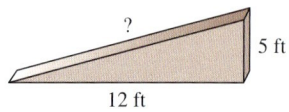

53. Find the height of the airplane above the ground.

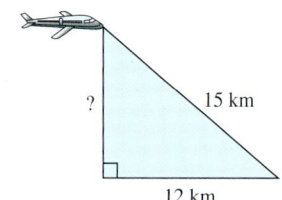

54. A 25-in. monitor measures 25 in. across the diagonal. If the width is 20 in., find the height.

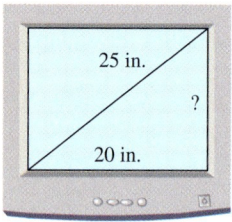

55. A car travels east 24 mi and then south 7 mi. How far is the car from its starting point?

56. A 26-ft-long wire is to be tied from a stake in the ground to the top of a 24-ft pole. How far from the bottom of the pole should the stake be placed?

Technology Connections

For Exercises 57–62, complete the table. For the estimate, find two consecutive whole numbers between which the square root lies. The first row is done for you.

	Square Root	Estimate	Calculator Approximation (Round to 3 Decimal Places)
	$\sqrt{50}$	is between <u>7</u> and <u>8</u>	7.071
57.	$\sqrt{10}$	is between ___ and ___	
58.	$\sqrt{90}$	is between ___ and ___	
59.	$\sqrt{116}$	is between ___ and ___	
60.	$\sqrt{65}$	is between ___ and ___	
61.	$\sqrt{5}$	is between ___ and ___	
62.	$\sqrt{48}$	is between ___ and ___	

For Exercises 63–70, use a calculator to approximate the square root to three decimal places.

63. $\sqrt{427.75}$ **64.** $\sqrt{3184.75}$ **65.** $\sqrt{1,246,000}$ **66.** $\sqrt{50,416,000}$

67. $\sqrt{0.49}$ **68.** $\sqrt{0.25}$ **69.** $\sqrt{0.56}$ **70.** $\sqrt{0.82}$

For Exercises 71–76, find the length of the unknown side. Round to three decimal places if necessary.

71.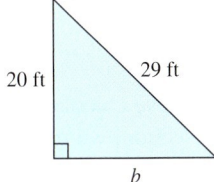

72.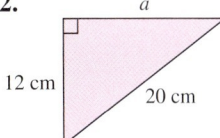

73. Leg = 5 mi, leg = 10 mi

74. Leg = 2 m, leg = 8 m

75. Leg = 12 in., hypotenuse = 22 in.

76. Leg = 15 ft, hypotenuse = 18 ft

77. A square tile is 1 ft on each side. What is the length of the diagonal? Round to the nearest hundredth of a foot.

78. A tennis court is 120 ft long and 60 ft wide. What is the length of the diagonal? Round to the nearest hundredth of a foot.

79. A contractor plans to construct a cement patio for one of the houses that he is building. The patio will be a square, 25 ft by 25 ft. After the contractor builds the frame for the cement, he checks to make sure that it is square by measuring the diagonals. Use the Pythagorean theorem to determine what the length of the diagonals should be if the contractor has constructed the frame correctly. Round to the nearest hundredth of a foot.

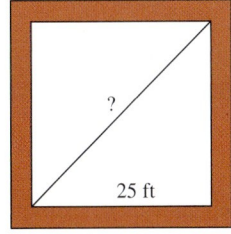

80. A rectangular intersection measures 110 ft by 126 ft. If a person crosses the intersection diagonally, what is the distance that the person would travel? Round to the nearest hundredth of a foot.

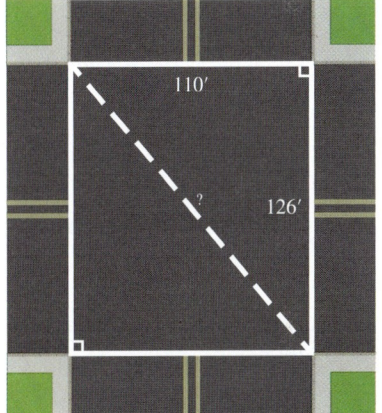

Expanding Your Skills

For Exercises 81–84, find the perimeter.

81.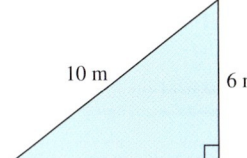

82.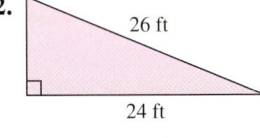

83.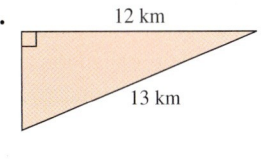

84.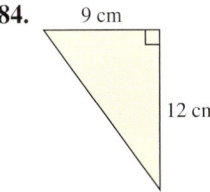

Section 8.7 Perimeter, Circumference, and Area

Concepts
1. Quadrilaterals
2. Perimeter and Circumference
3. Area

1. Quadrilaterals

At this point, you are familiar with several geometric figures. We have calculated perimeter and area of squares, rectangles, triangles, and circles. In this section, we revisit these concepts and also define some additional geometric figures.

A four-sided polygon is called a **quadrilateral**. Some quadrilaterals fall in the following categories.

A **parallelogram** is a quadrilateral with opposite sides parallel. It follows that opposite sides must be equal in length.

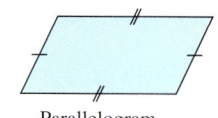
Parallelogram

A **rectangle** is a parallelogram with four right angles.

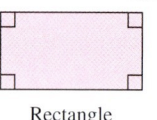
Rectangle

A **square** is a rectangle with sides of equal length.

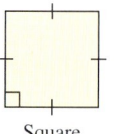
Square

A **rhombus** is a parallelogram with sides of equal length. The angles are not necessarily equal.

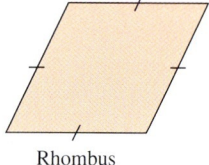
Rhombus

A **trapezoid** is a quadrilateral with exactly one pair of parallel sides.

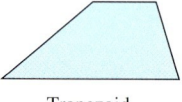
Trapezoid

Notice that some figures belong to more than one category. For example, a square is also a rectangle, a parallelogram, and a rhombus.

2. Perimeter and Circumference

Recall that the **perimeter** of a polygon is the distance around the figure. For example, we use perimeter to find the amount of fencing needed to enclose a yard. The perimeter of a polygon is found by adding the lengths of the sides. Also recall that the "perimeter" of a circle is called the **circumference**.

We summarize some of the formulas presented earlier in the text.

TIP: The approximate circumference of a circle can be calculated by using either 3.14 or $\frac{22}{7}$ for the value of π. To express the exact circumference, the answer must be written in terms of π.

Perimeter and Circumference Formulas

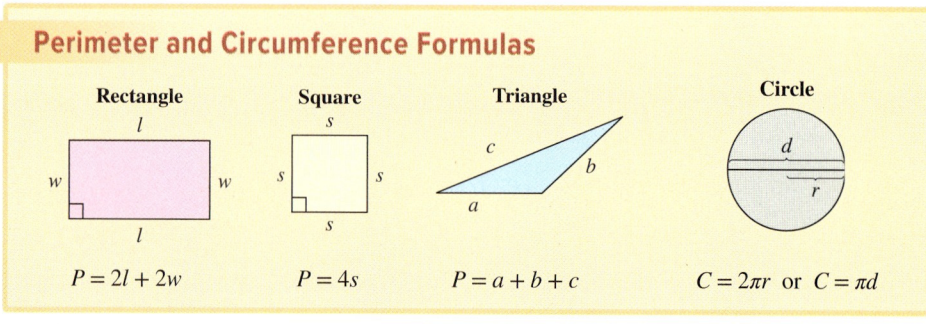

Section 8.7 Perimeter, Circumference, and Area

Example 1 Finding Perimeter

Use an appropriate formula to find the perimeter.

a.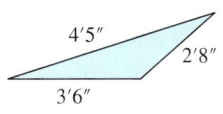

b.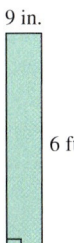
9 in.
6 ft

Solution:

a. To find the perimeter, add the lengths of the sides.

$P = a + b + c$
$P = 4'5'' + 2'8'' + 3'6''$ or
$P = 4$ ft 5 in.
     2 ft 8 in.
   + 3 ft 6 in.
   ─────────────
   9 ft 19 in. = 9 ft + (1 ft + 7 in.)   Note that 19 in. = 1 ft + 3 in.
              = 10 ft 7 in.   or   10'7''

b. First note that to add the lengths of the sides, we must have like units.

$$9 \text{ in.} = \frac{9 \text{ in.}}{1} \cdot \frac{1 \text{ ft}}{12 \text{ in.}} = \frac{9}{12} \text{ ft} = \frac{3}{4} \text{ ft or } 0.75 \text{ ft}$$

$P = 2l + 2w$ The figure is a rectangle. Use $P = 2l + 2w$.
$= 2(6 \text{ ft}) + 2(0.75 \text{ ft})$ Substitute $l = 6$ ft and $w = 0.75$ ft.
$= 12 \text{ ft} + 1.5 \text{ ft}$ Simplify.
$= 13.5 \text{ ft}$

Skill Practice

1. Find the perimeter.

   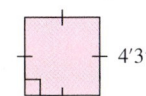
   4'3''

2. Find the perimeter in feet.

   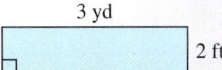
   3 yd
   2 ft

Example 2 Finding Circumference

Find the circumference. Use 3.14 for π.

1.6 m

Solution:

From the figure, $r = 1.6$ m.

$C = 2\pi(1.6 \text{ m})$ Use the formula, $C = 2\pi r$.
$= 3.2\pi$ m   This is the exact value of the circumference.
$\approx 3.2(3.14)$ m   Substitute 3.14 for π.
$= 10.048$ m   The circumference is approximately 10.048 m.

Skill Practice

3. Find the circumference. Use $\frac{22}{7}$ for π.

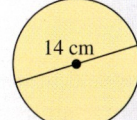
14 cm

Answers

1. 17'   **2.** 22 ft   **3.** 44 cm

3. Area

Recall that the **area** of a region is the number of square units that can be enclosed within the region. For example, the rectangle shown in Figure 8-29 encloses 6 square inches (in.²). We would compute area in such applications as finding the amount of sod needed to cover a yard or the amount of carpeting to cover a floor.

We have already presented the formulas to compute the area of a square, a rectangle, a triangle, and a circle. We summarize these along with the area formulas for a parallelogram and trapezoid. These should be memorized as common knowledge.

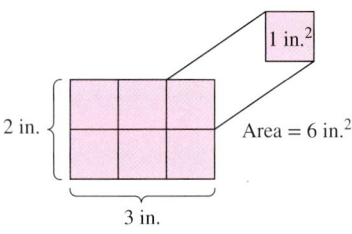

Figure 8-29

TIP: Because the height of a parallelogram is perpendicular to the base, sometimes it must be drawn *outside* the parallelogram.

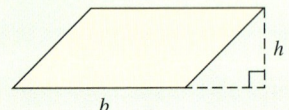

Area Formulas

Rectangle
$A = lw$

Square
$A = s^2$

Parallelogram
$A = bh$

Triangle
$A = \frac{1}{2}bh$

Trapezoid
$A = \frac{1}{2}(a+b)h$

Circle
$A = \pi r^2$

Example 3 — Finding Area

Determine the area of the field.

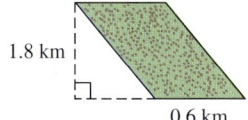

Solution:
The field is in the shape of a parallelogram. The base is 0.6 km and the height is 1.8 km.

$A = bh$ Area formula for a parallelogram.
$ = (0.6 \text{ km})(1.8 \text{ km})$ Substitute $b = 0.6$ km and $h = 1.8$ km.
$ = 1.08 \text{ km}^2$

The field is 1.08 km².

TIP: When two common units are multiplied, such as km · km, the resulting units are square units, such as km².

Skill Practice

4. Determine the area.

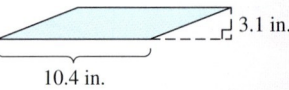

Answer
4. 32.24 in.²

Example 4 Finding Area

Determine the area of the matting.

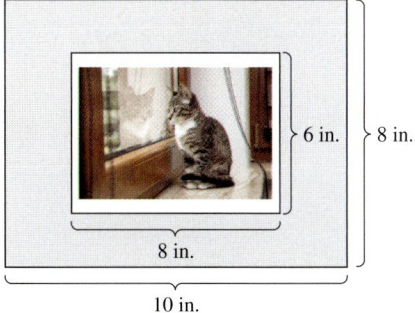
AlexandCo Studio/Shutterstock

Solution:

To find the area of the matting only, we can subtract the inner 6-in. by 8-in. area from the outer 8-in. by 10-in. area. In each case, apply the formula, $A = lw$.

$$\text{Area of matting} = \overset{\text{outer area}}{(10 \text{ in.})(8 \text{ in.})} - \overset{\text{inner area}}{(8 \text{ in.})(6 \text{ in.})}$$
$$= 80 \text{ in.}^2 - 48 \text{ in.}^2$$
$$= 32 \text{ in.}^2$$

The matting is 32 in.²

Skill Practice

5. A side of a house must be painted. Exclude the windows to find the area that must be painted.

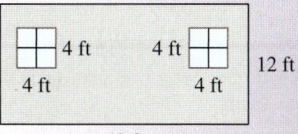

Example 5 Finding Area

Determine the area.

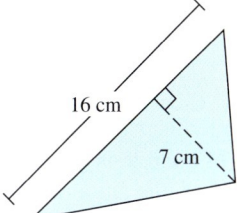

Solution:

$A = \dfrac{1}{2}bh$ Apply the formula for the area of a triangle.

$= \dfrac{1}{2}(16 \text{ cm})(7 \text{ cm})$ Substitute $b = 16$ cm and $h = 7$ cm.

$= \dfrac{1}{\underset{1}{2}}\left(\dfrac{\overset{8}{16}}{1} \text{ cm}\right)\left(\dfrac{7}{1} \text{ cm}\right)$ Multiply fractions.

$= 56 \text{ cm}^2$

Skill Practice Determine the area.

6.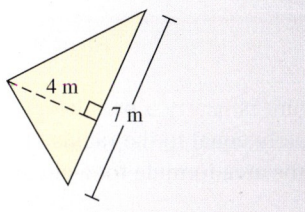

Answers

5. 448 ft²   **6.** 14 m²

Example 6 Finding Area

Determine the area.

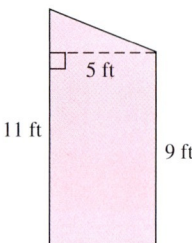

Solution:

$A = \dfrac{1}{2}(a+b)h$ Apply the formula for the area of a trapezoid.

In this case, the two parallel sides are the left-hand side and the right-hand side. Therefore, these sides are the two bases, a and b.

The "height" is the distance between the two parallel sides.

$A = \dfrac{1}{2}(11 \text{ ft} + 9 \text{ ft})(5 \text{ ft})$ Substitute $a = 11$ ft, $b = 9$ ft, and $h = 5$ ft.

$= \dfrac{1}{2}(20 \text{ ft})(5 \text{ ft})$ Simplify within parentheses first.

$= \dfrac{1}{2}\left(\dfrac{\overset{10}{\cancel{20}}}{1}\text{ ft}\right)\left(\dfrac{5}{1}\text{ ft}\right)$ Multiply fractions.

$= 50 \text{ ft}^2$

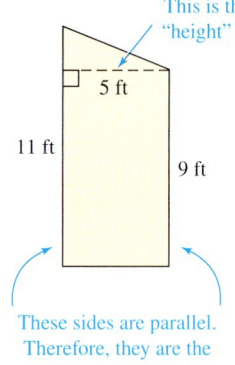

These sides are parallel. Therefore, they are the bases a and b.

Skill Practice

7. Determine the area.

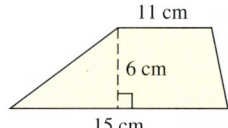

The formula for the area A of a circle with radius r is given by $A = \pi r^2$. Here we show the basis for the formula. The circumference of a circle is given by $C = 2\pi r$. The length of a *semicircle* (one-half of a circle) is one-half of this amount: $\frac{1}{2} \cdot 2\pi r = \pi r$. To visualize the formula for the area of a circle, consider the bottom half and top half of a circle cut into pie-shaped wedges. Unfold the figure as shown (Figure 8-30).

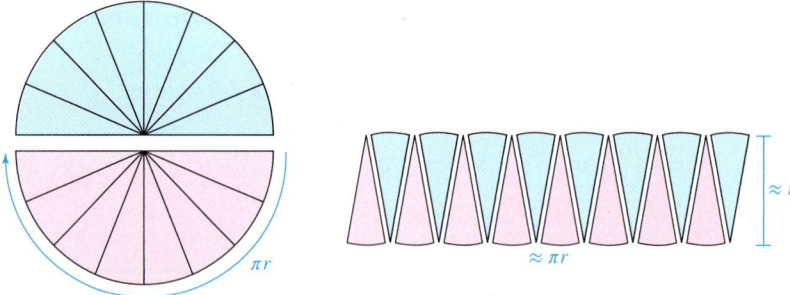

Figure 8-30

The resulting figure is nearly a parallelogram, with base approximately equal to πr and height approximately equal to the radius of the circle. The area is (base) · (height) ≈ $(\pi r) \cdot r = \pi r^2$. This is the area formula for a circle.

Answer

7. 78 cm^2

Example 7 Computing the Area of a Circle

Determine the area of the gold medal. Use $\frac{22}{7}$ for π.

7 cm

Solution:

First note that the radius is found by $r = \frac{1}{2}d$. Therefore, $r = \frac{1}{2}(7 \text{ cm}) = \frac{7}{2}$ cm.

$A = \pi \left(\frac{7}{2} \text{ cm}\right)^2$ Substitute $r = \frac{7}{2}$ cm into the formula $A = \pi r^2$.

$= \pi \left(\frac{49}{4} \text{ cm}^2\right)$ Simplify the factor with the exponent.

$= \frac{49}{4}\pi \text{ cm}^2$ This is the exact area, which is written in terms of π.

$\approx \left(\frac{49}{4} \text{ cm}^2\right)\left(\frac{22}{7}\right)$ Substitute $\frac{22}{7}$ for π.

$= \frac{77}{2} \text{ cm}^2$ The medal has an area of approximately $\frac{77}{2}$ cm² or $38\frac{1}{2}$ cm².

Skill Practice

8. Determine the area of a 9-in. diameter plate. Use 3.14 for π and round to the nearest tenth.

> **TIP:** The approximate area of a circle can be calculated using either 3.14 or $\frac{22}{7}$ for π. In Example 7 we could have used $\pi = 3.14$ and $r = \frac{7}{2} = 3.5$ cm.
>
> $A = \pi r^2$
> $A \approx (3.14)(3.5 \text{ cm})^2$
> $A = 38.465 \text{ cm}^2$

Example 8 Finding Area for a Landscaping Application

Sod can be purchased in palettes for $225. If a palette contains 240 ft² of sod, how much will it cost to cover the area in Figure 8-31?

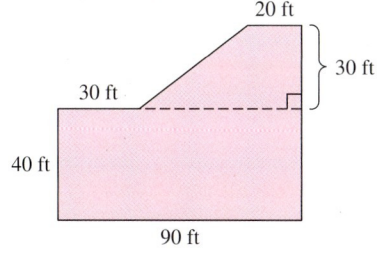

Figure 8-31

Answer

8. 63.6 in.²

Solution:

To find the total cost, we need to know the total number of square feet. Then we can determine how many 240-ft² palettes are required.

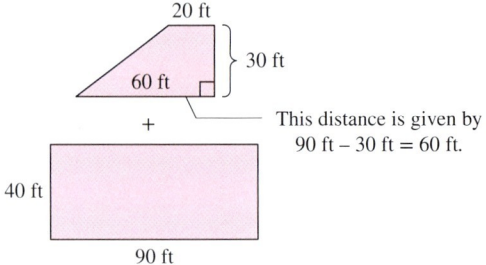

This distance is given by 90 ft − 30 ft = 60 ft.

The total area is given by

$$A = \underbrace{\tfrac{1}{2}(a+b)h}_{\text{area of trapezoid}} + \underbrace{lw}_{\text{area of rectangle}}$$

$$= \tfrac{1}{2}(60 \text{ ft} + 20 \text{ ft})(30 \text{ ft}) + (90 \text{ ft})(40 \text{ ft})$$

$$= \tfrac{1}{2}(80 \text{ ft})(30 \text{ ft}) + 3600 \text{ ft}^2$$

$$= 1200 \text{ ft}^2 + 3600 \text{ ft}^2$$

$$= 4800 \text{ ft}^2 \qquad \text{The total area is } 4800 \text{ ft}^2.$$

To determine how many 240-ft² palettes of sod are required, divide the total area by 240 ft².

Number of palettes: $\quad 4800 \text{ ft}^2 \div 240 \text{ ft}^2 = 20$

The total cost for 20 palettes is ($225 per palette) × 20 palettes = $4500.

The cost for the sod is $4500.

Skill Practice

9. A homeowner wants to apply water sealant to a pier that extends from the backyard to a lake. One gallon of sealant covers 160 ft² and sells for $11.95. How much will it cost to cover the pier?

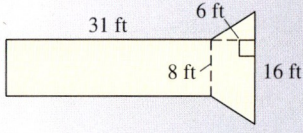

Answer

9. $23.90

Section 8.7 Practice Exercises

Study Skills Exercise

It may help to remember formulas if you understand how they were derived. For example, the perimeter of a square has the formula $P = 4s$. It was derived from the fact that perimeter measures the distance around a figure.

Explain how the formula for the perimeter of a rectangle ($P = 2l + 2w$) was derived.

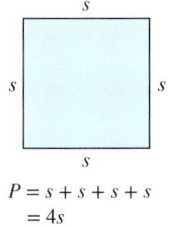

$P = s + s + s + s$
$\quad = 4s$

Prerequisite Review

For Exercises R.1–R.8, simplify.

R.1. $2(4.3) + 2(8.2)$

R.2. $2(9.1) + 2(6)$

R.3. $(7.1)^2$

R.4. $(3.5)^2$

R.5. $\frac{1}{2}(12)\left(1\frac{2}{3}\right)$

R.6. $\frac{1}{2}\left(1\frac{3}{4}\right)(24)$

R.7. $2(3.14)(12)$

R.8. $2(3.14)(1.2)$

Vocabulary and Key Concepts

1. **a.** The _____ of a polygon is the distance around the polygon. The _____ of a circle is the distance around the circle.

   b. The _____ of a region is the number of square units that can be enclosed within the region.

2. A rectangular lot was measured in feet.

   a. The perimeter of the lot is in the unit of _____.

   b. The area of the lot is in the unit of _____.

3. A four-sided figure is called a _____.

4. A _____ is a rectangle with four equal sides.

5. A parallelogram with sides of equal length and angles that are not necessarily equal is a _____.

6. A quadrilateral with exactly one pair of parallel sides is a _____.

7. The height of a parallelogram is _____ to the base, therefore, sometimes the height must be drawn outside of the parallelogram.

8. The area of a trapezoid is defined by the formula $A = $ _____.

Concept 1: Quadrilaterals

For Exercises 9–14, select all terms that apply to each figure. There may be more than one selection.

    a. quadrilateral   **b.** polygon   **c.** square   **d.** rectangle

    e. parallelogram   **f.** triangle   **g.** trapezoid   **h.** rhombus

9.

10.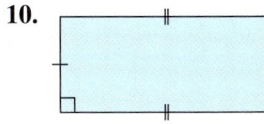

11.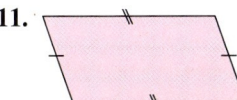

12.  13. 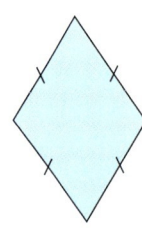 14.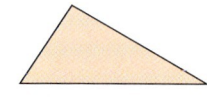

Concept 2: Perimeter and Circumference

For Exercises 15–24, determine the perimeter or circumference. Use 3.14 for π. **(See Examples 1 and 2.)**

15. Rectangle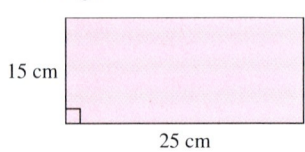

16. Square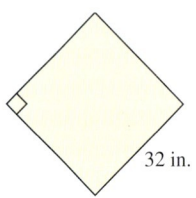

17. Square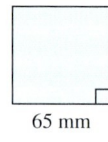

18. Rectangle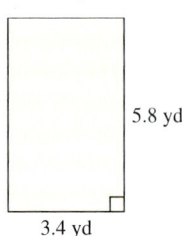

19. Trapezoid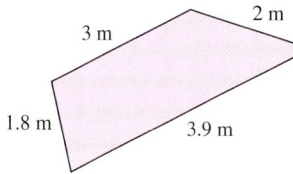

20. Trapezoid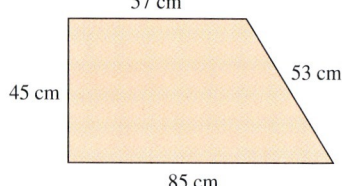

21.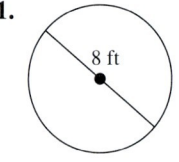

22.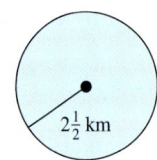

23.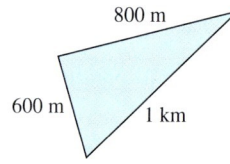

24.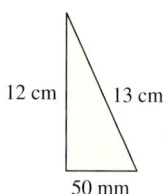

25. Find the perimeter of a triangle with sides 3 ft 8 in., 2 ft 10 in., and 4 ft.

26. Find the perimeter of a triangle with sides 4 ft 2 in., 3 ft, and 2 ft 9 in.

27. Find the perimeter of a rectangle with length 2 ft and width 6 in.

28. Find the perimeter of a rectangle with length 4 m and width 85 cm.

29. Find the lengths of the two missing sides labeled x and y. Then find the perimeter of the figure.

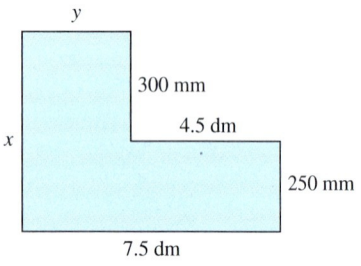

30. Find the lengths of the two missing sides labeled a and b. Then find the perimeter of the figure.

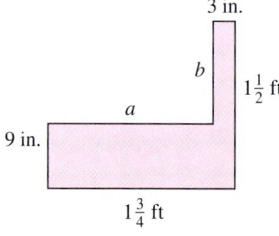

31. Rain gutters are going to be installed around the perimeter of a house. What is the total length needed?

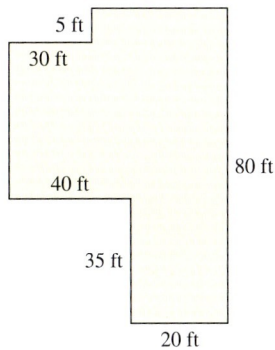

32. Wood molding needs to be installed around the perimeter of a living room floor. With no molding needed in the doorway, how much molding is needed?

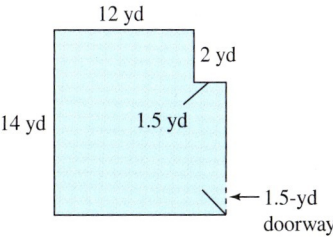

Concept 3: Area

For Exercises 33–44, determine the area of the shaded region. (See Examples 3–6.)

33.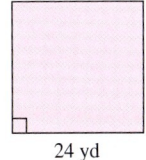

34.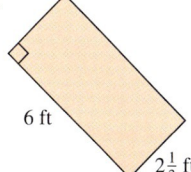

35.

36.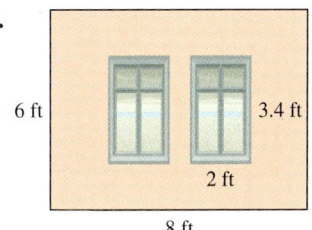

37.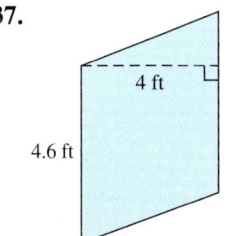

38.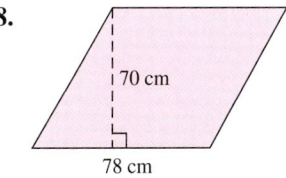

39.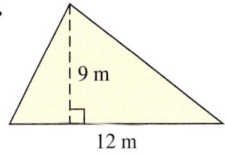

40.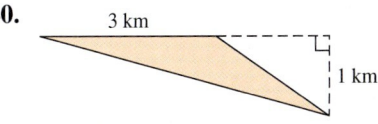

41.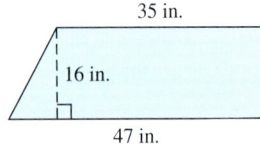

42.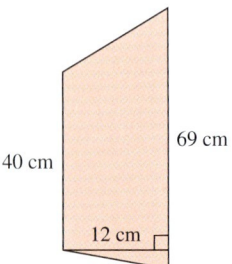

43.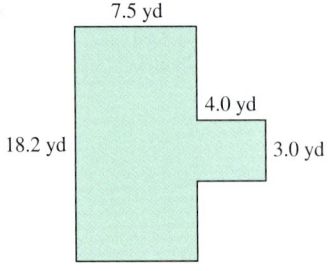

44.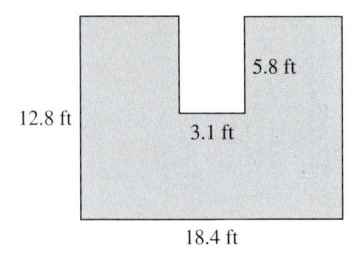

For Exercises 45–48, determine the area of the circle, using $\frac{22}{7}$ for π. **(See Example 7.)**

45.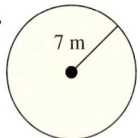

46.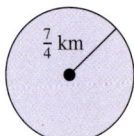

47.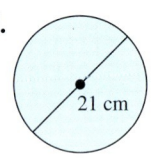

48.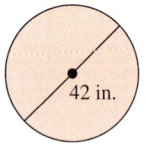

For Exercises 49–52, determine the area of the circle, using 3.14 for π. Round to the nearest whole unit.

49.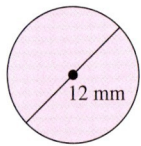

50.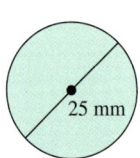

51.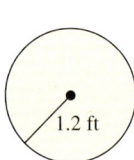

52.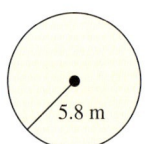

53. Determine the area of the sign.

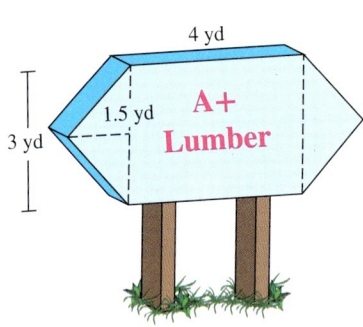

54. Determine the area of the kite.

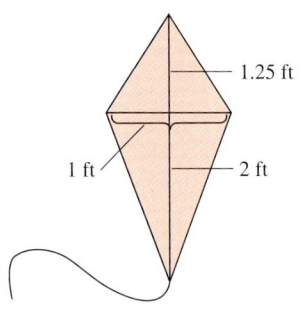

55. A rectangular living room is all to be carpeted except for the tiled portion in front of the fireplace. If carpeting is $2.50 per square foot (including installation), how much will the carpeting cost?
(See Example 8.)

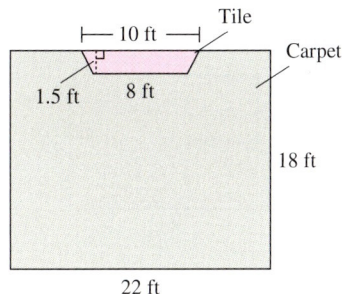

56. A patio area is to be covered with outdoor tile. If tile costs $8 per square foot (including installation), how much will it cost to tile the whole patio?

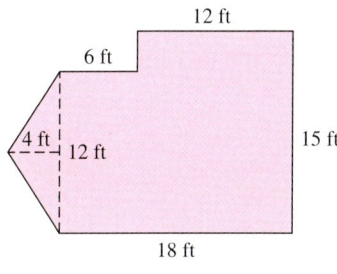

57. Lizette's backyard is a rectangle measuring 150 ft by 200 ft. She plans to seed it with grass that is environmentally friendly. Each 25-lb bag will cover 5000 square feet.

  a. What is the area of Lizette's backyard?

  b. How many 25-lb bags of seed will she need?

58. The Baker family plans to paint their garage floor with paint that resists gas, oil, and dirt from tires. The garage is 21 ft wide and 23 ft long. The paint kit they plan to use will cover approximately 250 square feet.

  a. What is the area of the garage floor?

  b. How many kits will be needed to paint the entire garage floor?

Expanding Your Skills

For Exercises 59–62, determine the area of the shaded region. Use 3.14 for π.

59.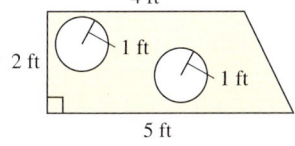

60.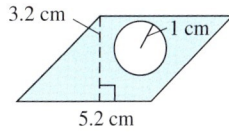

61.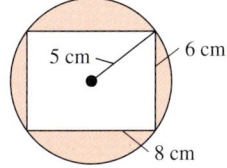

62.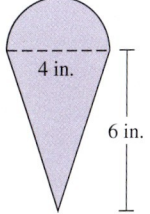

63. Find the length of side c by dividing this figure into a rectangle and a right triangle. Then find the perimeter of the figure.

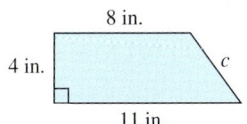

64. Find the perimeter of the figure.

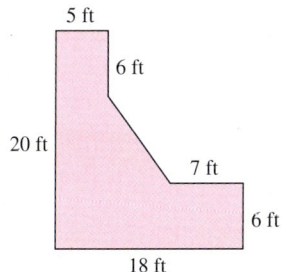

Problem Recognition Exercises

Area, Perimeter, and Circumference

For Exercises 1–14, determine the area and the perimeter or circumference for each figure.

1.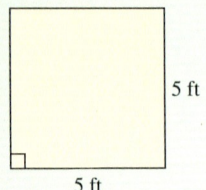

2.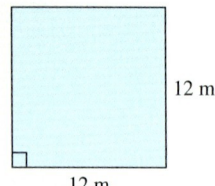

3.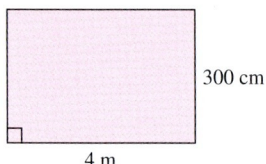

4.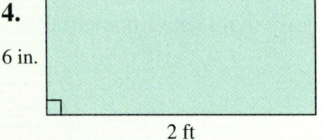

5.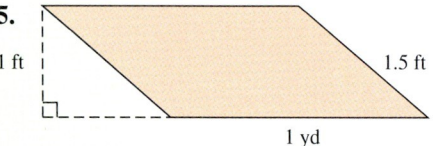

6.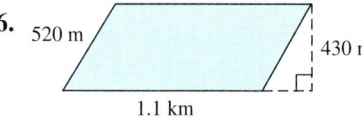

7.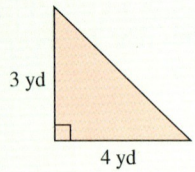

8.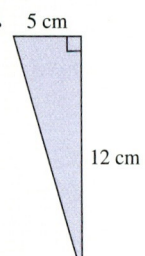

9.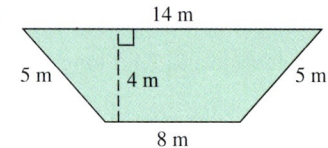

10.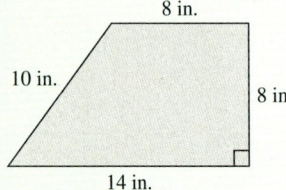

11. Use 3.14 for π.

    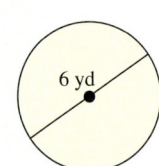

12. Use 3.14 for π.

    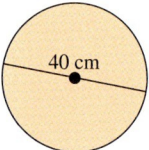

13. Use $\dfrac{22}{7}$ for π.

    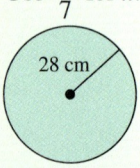

14. Use $\dfrac{22}{7}$ for π.

    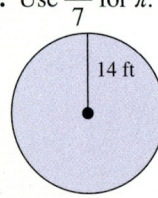

For Exercises 15–18,
   a. Choose the type of formula (perimeter, circumference, or area) needed to solve the application.
   b. Solve the application.

15. Find the amount of fencing needed to enclose a square field whose side measures 16 yd.

16. Find the amount of wood trim needed to frame a circular window with diameter $1\tfrac{3}{4}$ ft. Use $\tfrac{22}{7}$ for π.

17. Find the amount of carpeting needed to cover the floor of a rectangular room that is $12\tfrac{1}{2}$ ft by 10 ft.

18. Find the amount of sod needed to cover a circular garden with diameter 20 m. Use 3.14 for π.

Volume and Surface Area

Section 8.8

Concepts

1. Volume
2. Surface Area

1. Volume

The amount of liquid that can be held in a coffee cup or the amount of sand that can be held in a dump truck are each measures of volume. Volume is another word for capacity and is a measure of how much material can be filled within a three-dimensional object.

A cube that is 1 cm on a side has a volume of 1 cubic centimeter (1 cm³ or cc). A cube that is 1 in. on a side has a volume of 1 cubic inch (1 in.³). See Figure 8-32. Additional units of volume include cubic feet (ft³), cubic yards (yd³), cubic meters (m³), and so on.

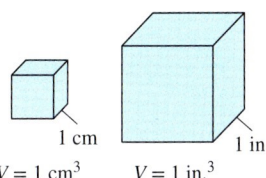

TIP: Recall that 1 cubic centimeter can also be denoted as 1 cc. Furthermore, 1 cc = 1 mL.

Figure 8-32

The formulas used to compute the volume of several common solids are given.

Volume Formulas

Rectangular Solid   **Cube**   **Right Circular Cylinder**

$V = lwh$ $V = s^3$ $V = \pi r^2 h$

Notice that the volume formulas for these three figures are given by the product of the area of the base and the height of the figure:

$V = lw\mathbf{h}$ (area of rectangular base)

$V = s \cdot s \cdot s$ (area of square base)

$V = \pi r^2 \mathbf{h}$ (area of circular base)

A right circular cone has the shape of a party hat. A sphere has the shape of a ball. To compute the volume of a cone and a sphere, we use the following formulas.

Volume Formulas

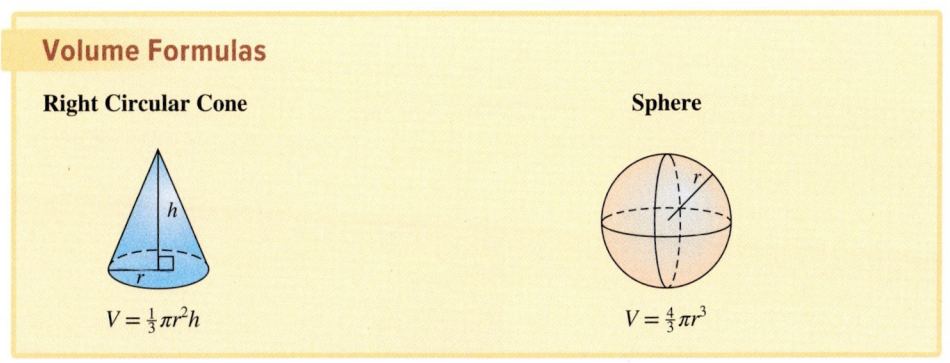

Right Circular Cone   **Sphere**

$V = \frac{1}{3}\pi r^2 h$ $V = \frac{4}{3}\pi r^3$

TIP: Notice that the formula for the volume of a right circular cone is $\frac{1}{3}$ that of a right circular cylinder.

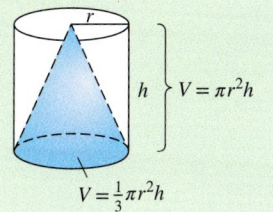

$V = \pi r^2 h$

$V = \frac{1}{3}\pi r^2 h$

Example 1 Finding Volume

Find the volume.

Solution:

$V = lwh$ — Use the volume formula for a rectangular solid. Identify the length, width, and height.

$= (4 \text{ in.})(3 \text{ in.})(5 \text{ in.})$ — $l = 4$ in., $w = 3$ in., and $h = 5$ in.

$= 60 \text{ in.}^3$

We can visualize the volume by "layering" cubes that are each 1 in. high (Figure 8-33). The number of cubes in each layer is equal to $4 \times 3 = 12$. Each layer has 12 cubes, and there are 5 layers. Thus, the total number of cubes is $12 \times 5 = 60$ for a volume of 60 in.^{3}

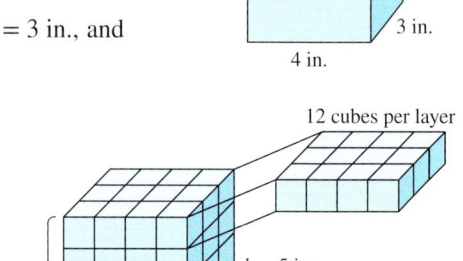

Figure 8-33

Skill Practice

1. Find the volume.

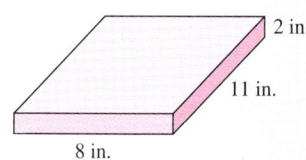

Example 2 Finding the Volume of a Cylinder

Find the volume. Use 3.14 for π. Round to the nearest whole unit.

Solution:

$V = \pi r^2 h$ — Use the formula for the volume of a right circular cylinder.

$\approx (3.14)(3.7 \text{ cm})^2(11.2 \text{ cm})$ — Substitute 3.14 for π, $r = 3.7$ cm, and $h = 11.2$ cm.

$= (3.14)(13.69 \text{ cm}^2)(11.2 \text{ cm})$ — Simplify exponents first.

$= 481.44992 \text{ cm}^3$ — Multiply from left to right.

$\approx 481 \text{ cm}^3$ — Round to the nearest whole unit.

Studiohio/McGraw Hill

Skill Practice

2. Find the volume. Use 3.14 for π.

Answers

1. 176 in.^{3} 2. $\approx$ 56.52 in.^{3}

Section 8.8 Volume and Surface Area 523

Example 3 — Finding the Volume of a Sphere

Find the volume. Use 3.14 for π. Round to one decimal place.

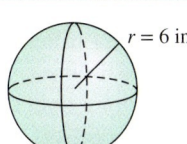
$r = 6$ in.

Solution:

$V = \dfrac{4}{3}\pi r^3$ Use the formula for the volume of a sphere.

$\approx \dfrac{4}{3}(3.14)(6 \text{ in.})^3$ Substitute 3.14 for π and $r = 6$ in.

$= \dfrac{4}{3}(3.14)(216 \text{ in.}^3)$ Simplify exponents first.
$(6 \text{ in.})^3 = (6 \text{ in.})(6 \text{ in.})(6 \text{ in.}) = 216 \text{ in.}^3$

$= \dfrac{4}{3}\left(\dfrac{3.14}{1}\right)\left(\dfrac{216 \text{ in.}^3}{1}\right)$ Multiply fractions.

$= \dfrac{4}{\underset{1}{\cancel{3}}}\left(\dfrac{3.14}{1}\right)\left(\dfrac{\overset{72}{\cancel{216}} \text{ in.}^3}{1}\right)$ Simplify to lowest terms.

$= 904.32 \text{ in.}^3$ Multiply from left to right.

$\approx 904.3 \text{ in.}^3$ Round to one decimal place.

Skill Practice

3. Find the volume. Use 3.14 for π.

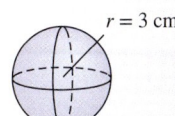
$r = 3$ cm

Example 4 — Finding the Volume of a Cone

Find the volume. Use 3.14 for π. Round to one decimal place.

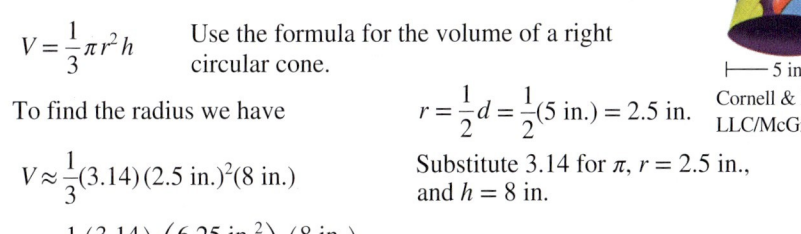
8 in.
5 in.
Cornell & McCarthy, LLC/McGraw Hill

Solution:

$V = \dfrac{1}{3}\pi r^2 h$ Use the formula for the volume of a right circular cone.

To find the radius we have $r = \dfrac{1}{2}d = \dfrac{1}{2}(5 \text{ in.}) = 2.5 \text{ in.}$

$V \approx \dfrac{1}{3}(3.14)(2.5 \text{ in.})^2(8 \text{ in.})$ Substitute 3.14 for π, $r = 2.5$ in., and $h = 8$ in.

$= \dfrac{1}{3}\left(\dfrac{3.14}{1}\right)\left(\dfrac{6.25 \text{ in.}^2}{1}\right)\left(\dfrac{8 \text{ in.}}{1}\right)$ Simplify exponents first.

$= \dfrac{157}{3} \text{ in.}^3$ Multiply fractions.

$\approx 52.3 \text{ in.}^3$ Round to one decimal place.

Skill Practice

4. Find the volume. Use 3.14 for π.

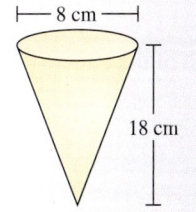
8 cm
18 cm

Answers
3. 113.04 cm^3 4. 301.44 cm^3

2. Surface Area

Surface area (often abbreviated SA) is the area of the surface of a three-dimensional object. To illustrate, consider the surface area of a soup can in the shape of a right circular cylinder.

If we peel off the label, we see that the label forms a rectangle whose length is the circumference of the base, $2\pi r$. The width of the rectangle is the height of the can, h. The area of the label is determined by multiplying the length and the width: $l \times w = 2\pi rh$. To determine the total surface area, add the areas of the top and bottom of the can to this rectangular area.

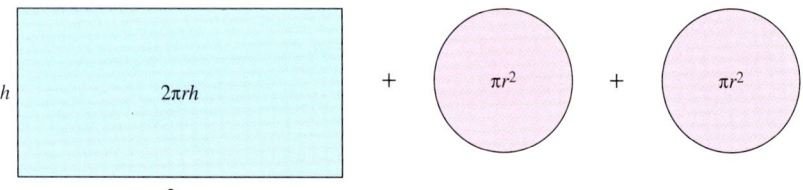

$$\text{Surface area} = 2\pi rh + \pi r^2 + \pi r^2 \text{ or}$$
$$SA = 2\pi rh + 2\pi r^2$$

Table 8-7 gives the formulas for the surface areas of four common solids.

Table 8-7 Surface Area

Cube	Rectangular Solid	Cylinder	Sphere
$SA = 6s^2$	$SA = 2lh + 2lw + 2hw$	$SA = 2\pi rh + 2\pi r^2$	$SA = 4\pi r^2$

Example 5 Determining Surface Area

Determine the surface area of the rectangular solid.

Solution:

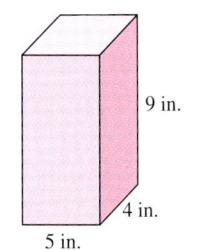

Use the formula $SA = 2lh + 2lw + 2hw$, where $l = 5$ in., $w = 4$ in., and $h = 9$ in.

$SA = 2(5 \text{ in.})(9 \text{ in.}) + 2(5 \text{ in.})(4 \text{ in.}) + 2(9 \text{ in.})(4 \text{ in.})$

$= 90 \text{ in.}^2 + 40 \text{ in.}^2 + 72 \text{ in.}^2$ Multiply.

$= 202 \text{ in.}^2$ Add.

The surface area of the rectangular solid is 202 in.^{2}

Avoiding Mistakes

Although we are working with a three-dimensional figure, we are finding area. The answer will be in square units, not cubic units, as in calculating volume.

Skill Practice

5. Determine the surface area of the cube with the side length of 3 ft.

Answer

5. 54 ft^2

Example 6 Determining Surface Area

Determine the surface area of a sphere with radius 6 m. Use 3.14 for π.

Solution:

Use the formula SA = $4\pi r^2$, where $r = 6$ m.

$SA = 4\pi (6 \text{ m})^2$
$\approx 4(3.14)(36 \text{ m}^2)$ Substitute 3.14 for π.
$\approx 452.16 \text{ m}^2$

Skill Practice

6. Determine the surface area of a cylinder with radius 15 cm and height 20 cm. Use 3.14 for π.

Answer

6. 3297 cm²

Practice Exercises   Section 8.8

Study Skills Exercise

For your next test, prepare a one-page summary sheet with the most important information. On the day of the test, look at this sheet several times to refresh your memory. Then when you sit down to take your test, write down all the information that you can remember on the test or on scrap paper. This process, called a "memory dump," allows you to take the test without worrying that you will forget something important. Always include a positive statement to yourself to build a growth mindset.

Prerequisite Review

For Exercises R.1–R.8, simplify.

R.1. $2(3)(5) + 2(3)(8) + 2(5)(8)$

R.2. $2(6)(2) + 2(6)(3) + 2(2)(3)$

R.3. $\frac{1}{3}(3.14)(6)^2(12)$

R.4. $\frac{1}{3}(3.14)(8)^2(1.5)$

R.5. $\frac{4}{3}(3.14)(6)^3$

R.6. $\frac{4}{3}(3.14)(9)^3$

Vocabulary and Key Concepts

1. The volume of a cube with sides of length s is given by $V =$ _____.

2. The volume of a rectangular solid with sides of lengths l, w, and h is given by $V =$ _____.

3. The volume of a right circular cylinder with radius r and height h is given by $V =$ _____.

4. The formula $V = \frac{1}{3}\pi r^2 h$ gives the volume of a (sphere/cone) whereas the formula $V = \frac{4}{3}\pi r^3$ gives the volume of a (sphere/cone).

5. The _____ _____ of a rectangular solid is the sum of the areas of the six sides.

6. The surface area of a cube with sides of length s is given by $SA =$ _____.

Concept 1: Volume

For Exercises 7–18, find the volume. Use 3.14 for π where necessary. (See Examples 1–4.)

7.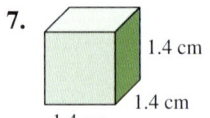

8.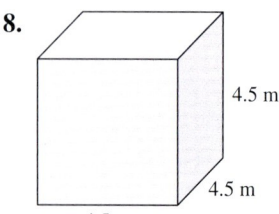

9.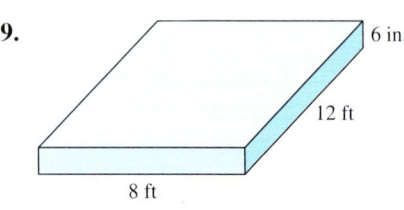

10.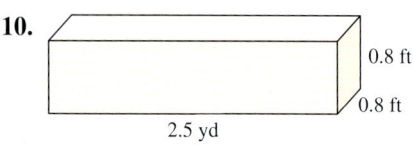

11.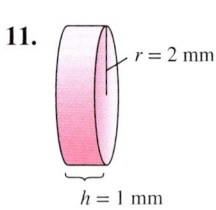

12.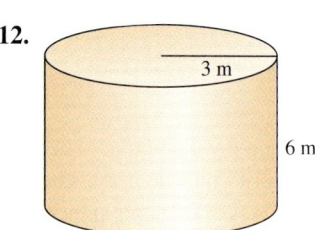

13.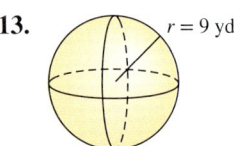

14.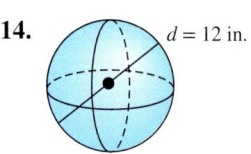

15.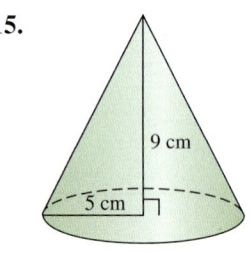

16.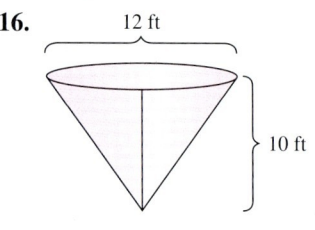

17.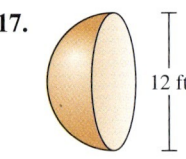

18.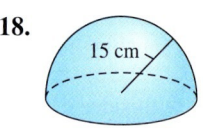

For Exercises 19–26, use 3.14 for π. Round each value to the nearest whole unit.

19. The diameter of a volleyball is 8.2 in. Find the volume.

20. The diameter of a basketball is 9 in. Find the volume.

21. Find the volume of the sand pile.

22. In decorating cakes, many people use an icing bag that has the shape of a cone. Find the volume of the icing bag.

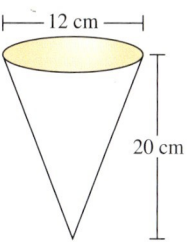

23. Find the volume of water (in cubic feet) that the pipe can hold.

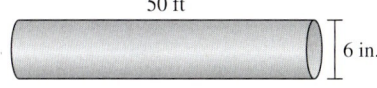

24. Find the volume of the wastebasket that has the shape of a cylinder with the height of 3 ft and diameter of 2 ft.

25. Sam bought an aboveground circular swimming pool with diameter 27 ft and height 54 in.

   a. Approximate the volume of the pool in cubic feet using 3.14 for π.

   b. How many gallons of water will it take to fill the pool? (*Hint:* 1 gal $\approx$ 0.1337 ft^3.)

26. Richard needs 3 in. of topsoil for his vegetable garden that is in the shape of a rectangle, 15 ft by 20 ft.

   a. Find the amount of topsoil needed in cubic feet.

   b. If topsoil can be purchased in bags containing 2 ft^3, how many bags must Richard purchase?

Concept 2: Surface Area

For Exercises 27–34, determine the surface area to the nearest tenth of a unit. Use 3.14 for π when necessary. (See Examples 5 and 6.)

27. Determine the amount of cardboard for the cereal box.

Jacques Cornell/McGraw Hill

28. Determine the amount of cardboard for the box of macaroni.

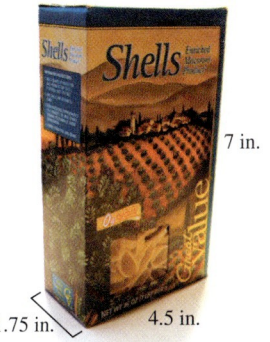

Mark Dierker/McGraw Hill

29. Determine the surface area for the sugar cube.

Jacques Cornell/McGraw Hill

30. Determine the surface area of the Sudoku cube.

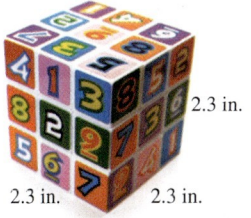

Mark Dierker/McGraw Hill

31. Determine the amount of cardboard for the container of oatmeal.

Mark Dierker/McGraw Hill

32. Determine the amount of steel needed for the can of tomato paste.

Mark Dierker/McGraw Hill

33. Determine the surface area of the volleyball.

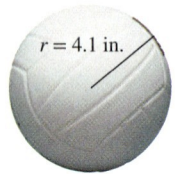

Comstock Images/Alamy Stock Photo

34. Determine the surface area of the golf ball.

Photodisc/Getty Images

For Exercises 35–42, determine the surface area of the object described. Use 3.14 for π when necessary. (See Examples 5 and 6.)

35. A rectangular solid with dimensions 12 ft by 14 ft by 3 ft

36. A rectangular solid with dimensions 8 m by 7 m by 5 m

37. A cube with each side 4 cm long

38. A cube with each side 10 yd long

39. A cylinder with radius 9 in. and height 15 in.

40. A cylinder with radius 90 mm and height 75 mm

41. A sphere with radius 10 mm

42. A sphere with radius 9 m

Expanding Your Skills

For Exercises 43–46, find the volume of the shaded region. Use 3.14 for π if necessary.

43. The height of the interior portion is 1 ft.

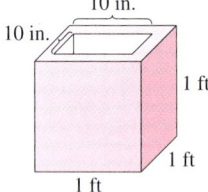

44. The height of the interior portion is 1 ft.

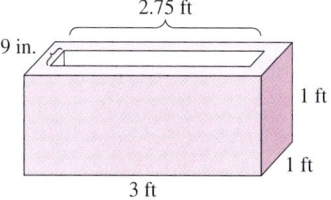

45.

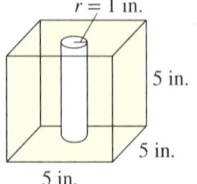

46.

47. A machine part is in the shape of a cylinder with a hole drilled through the center. Find the volume of the machine part.

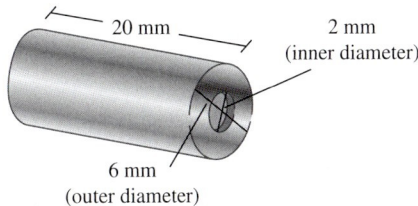

48. To insulate pipes, a cylinder of Styrofoam has a hole drilled through it to fit around a pipe. What is the volume of this piece of insulation?

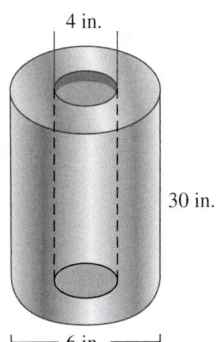

49. A silo is in the shape of a cylinder with a hemisphere on the top. Find the volume.

Bryan Mullennix/Pixtal/age fotostock

50. An ice cream cone is in the shape of a cone with a sphere on top. Assuming that ice cream is packed inside the cone, find the volume of the ice cream.

C Squared Studios/Photodisc/Getty Images

The volume formulas for right circular cylinders and right circular cones are the same for slanted cylinders and cones. For Exercises 51–54, find the volume. Use 3.14 for π.

51.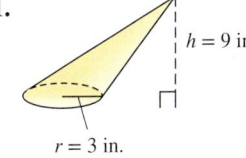

52.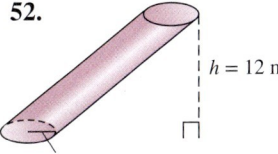

53.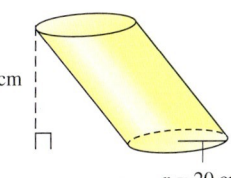

54.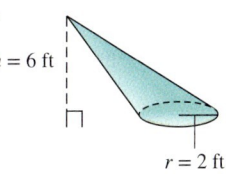

A square-based pyramid is a solid figure with a square base and triangular faces that meet at a common point called the apex. The formula for the volume is below.

$$V = \frac{1}{3}s^2 h$$

The length of the sides of the square base is s, and h is the perpendicular height from the apex to the base.

For Exercises 55 and 56, determine the volume of the pyramid.

55.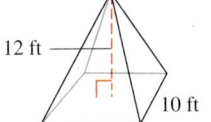

56.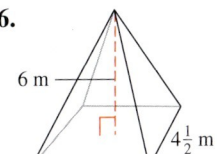

Chapter 8 Review Exercises

Section 8.1

For Exercises 1–6, convert the units of length.

1. 48 in. = _____ ft
2. $3\frac{1}{4}$ ft = _____ in.
3. 2 mi = _____ yd
4. 7040 ft = _____ mi
5. $\frac{1}{2}$ mi = _____ ft
6. 2 yd = _____ in.

For Exercises 7–10, perform the indicated operations.

7. 3 ft 9 in. + 5 ft 6 in.
8. 4′11″ + 1′5″
9. 5′3″ − 2′5″
10. 12 ft 7 in. − 8 ft 10 in.

11. Find the perimeter in feet.

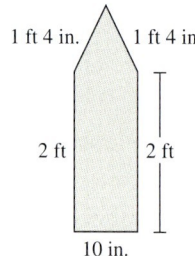

12. A roll of wire contains 50 yd of wire. If Ivan uses 48 ft, how much wire is left?

For Exercises 13–22, convert the units of time, weight, and capacity.

13. 72 hr = _____ days
14. 6 min = _____ sec
15. 5 lb = _____ oz
16. 1 wk = _____ hr
17. 12 fl oz = _____ c
18. 0.25 ton = _____ lb
19. 3500 lb = _____ tons
20. 2 gal = _____ pt
21. 12 oz = _____ lb
22. 16 qt = _____ gal

23. A runner finished a race with a time of 2:24:30. Convert the time to minutes.

24. A mother gave birth to triplets that weighed 3 lb 10 oz, 4 lb 2 oz, and 4 lb 1 oz. What was the total weight of the triplets?

Section 8.2

For Exercises 25 and 26, select the most reasonable measurement.

25. A pencil is _____ long.
    a. 16 mm
    b. 16 cm
    c. 16 m
    d. 16 km

26. The distance between Houston and Dallas is _____.
    a. 362 mm
    b. 362 cm
    c. 362 m
    d. 362 km

For Exercises 27–32, convert the metric units of length.

27. 52 cm = _____ mm
28. 93 m = _____ km
29. 34 dm = _____ m
30. 2.1 m = _____ dam
31. 4 cm = _____ m
32. 1.2 m = _____ mm

For Exercises 33–36, convert the metric units of mass.

33. 6.1 g = _____ cg
34. 420 g = _____ kg
35. 3212 mg = _____ g
36. 0.7 hg = _____ g

For Exercises 37–40, convert the metric units of capacity.

37. 830 cL = _____ L
38. 124 mL = _____ cc
39. 225 cc = _____ cL
40. 0.49 kL = _____ L

41. The dimensions of a dining room table are 2 m by 125 cm. Convert the units to meters and find the perimeter and area of the tabletop.

42. A bottle of apple juice contains 1.2 L of juice. If a glass holds 24 cL, how many glasses can be filled from this bottle?

43. An adult has a mass of 68 kg. A baby has a mass of 3200 g. What is the difference in their masses, in kilograms?

Image Source/Alamy Stock Photo

44. From a wooden board 2 m long, Jesse needs to cut 3 pieces that are each 75 cm long. Is the 2-m length of board long enough for the 3 pieces?

Section 8.3

For Exercises 45–54, refer to the table of conversion factors. Convert the units as indicated. Round to the nearest hundredth, if necessary.

Length	Weight/Mass (on Earth)
1 in. = 2.54 cm	1 lb ≈ 0.45 kg
1 ft ≈ 0.305 m	1 oz ≈ 28 g
1 yd ≈ 0.914 m	**Capacity**
1 mi ≈ 1.61 km	1 qt ≈ 0.95 L
	1 fl oz ≈ 30 mL = 33 cc

45. 6.2 in. ≈ _____ cm
46. 75 mL ≈ _____ fl oz
47. 140 g ≈ _____ oz
48. 5 L ≈ _____ qt
49. 3.4 ft ≈ _____ m
50. 100 lb ≈ _____ kg
51. 120 km ≈ _____ mi
52. 6 qt ≈ _____ L
53. 1.5 fl oz ≈ _____ cc
54. 12.5 tons ≈ _____ kg

55. The height of a computer desk is 30 in. The height of the chair is 38 cm. What is the difference in height between the desk and chair, in centimeters?

56. A bag of snack crackers contains 7.2 oz. If one serving is 30 g, approximately how many servings are in one bag?

57. The Boston Marathon is 42.195 km long. Convert this distance to miles. Round to the tenths place.

58. Write the formula to convert degrees Fahrenheit to degrees Celsius.

59. When roasting a turkey, the meat thermometer should register between 180°F and 185°F to indicate that the turkey is done. Convert these temperatures to degrees Celsius. Round to the nearest tenth, if necessary.

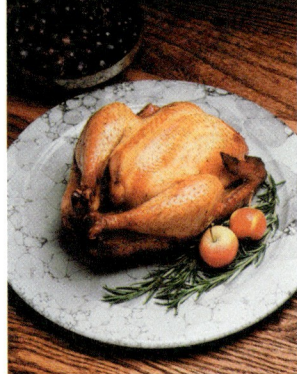

Ernie Friedlander/Cole Group/Photodisc/Getty Images

60. Write the formula to convert degrees Celsius to degrees Fahrenheit.

61. The average October temperature for Toronto, Ontario, Canada, is 8°C. Convert this temperature to degrees Fahrenheit.

Section 8.4

For Exercises 62–65, convert the metric units.

62. 0.45 mg = _____ μg
63. 1.5 mg = _____ mcg
64. 400 mcg = _____ mg
65. 5000 μg = _____ cg

66. A nasal spray delivers 2.5 mg of active ingredient per milliliter of solution. Determine the amount of active ingredient per cc.

67. A physician prescribes a drug based on a patient's mass. The dosage is given as 0.04 mg of the drug per kilogram of the patient's mass.

   a. How much would the physician prescribe for an 80-kg patient?

   b. If the dosage was to be given twice a day, how much of the drug would the patient take in a week?

68. A prescription for cough syrup indicates that 30 mL should be taken twice a day for 7 days. What is the total amount, in liters, of cough syrup to be taken?

Photodisc/Getty Images

69. A standard hypodermic syringe holds 3 cc of fluid. If a nurse uses 1.8 mL of the fluid, how much is left in the syringe?

70. A medication comes in 250-mg capsules. If Clayton took 3 capsules a day for 10 days, how many grams of the medication did he take?

Section 8.5

For Exercises 71–74, match the symbol with a description.

71. $\overleftrightarrow{AB}$
72. $\overrightarrow{AB}$
73. $\overrightarrow{BA}$
74. $\overline{AB}$

a. Ray AB
b. Line segment AB
c. Ray BA
d. Line AB

75. Describe the measure of an acute angle.

76. Describe the measure of an obtuse angle.

77. Describe the measure of a right angle.

78. Let $m(\angle X) = 33°$.
    a. Find the complement of $\angle X$.
    b. Find the supplement of $\angle X$.

79. Let $m(\angle T) = 20°$.
    a. Find the complement of $\angle T$.
    b. Find the supplement of $\angle T$.

For Exercises 80–84, refer to the figure. Find the measures of the angles.

80. $m(\angle a)$
81. $m(\angle b)$
82. $m(\angle c)$
83. $m(\angle d)$
84. $m(\angle e)$

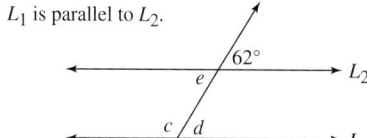

L_1 is parallel to L_2.

Section 8.6

For Exercises 85 and 86, find the measures of the angles x and y.

85.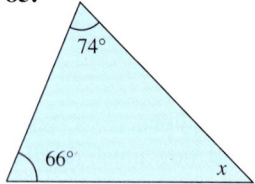

86.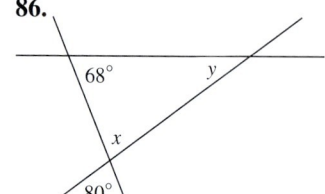

For Exercises 87 and 88, describe the characteristics of each type of triangle.

87. Equilateral triangle

88. Isosceles triangle

For Exercises 89 and 90, simplify the square roots.

89. $\sqrt{25}$

90. $\sqrt{49}$

For Exercises 91 and 92, find the length of the unknown side.

91.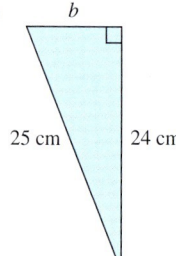

92.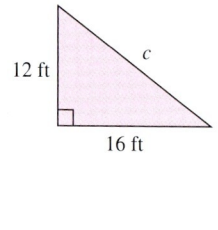

93. Kayla is flying a kite. At one point the kite is 5 m from Kayla horizontally and 12 m above her (see figure). How much string is extended? (Assume there is no slack.)

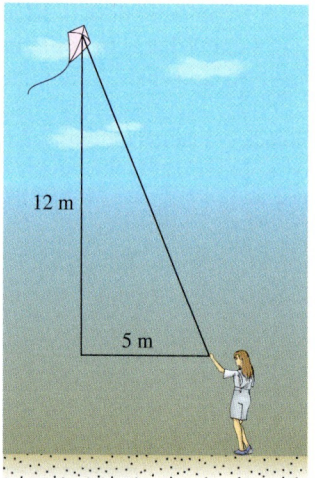

Section 8.7

94. Find the perimeter of the figure.

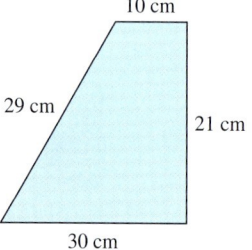

95. Find the perimeter of a triangle with sides of 4.2 m, 6.1 m, and 7.0 m.

96. How much fencing is required to put up a chain link fence around a 120-yd by 80-yd playground?

97. The perimeter of a square is 62 ft. What is the length of each side?

98. Find the circumference of a circle with a 20-cm diameter. Use 3.14 for π.

99. Find the area of the triangle.

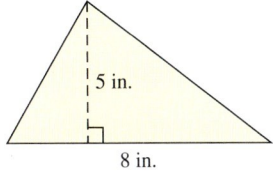

100. Fatima has a Persian rug 8.5 ft by 6 ft. What is the area?

DAJ/Getty Images

101. A lot is 150 ft by 80 ft. Within the lot, there is a 12-ft easement along all edges. An easement is the portion of the lot on which nothing may be built. What is the area of the portion that may be used for building?

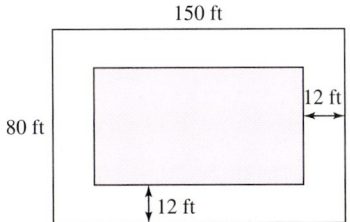

102. Find the area.

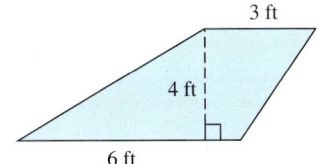

103. Find the area of a circle with a 20-cm diameter. Use 3.14 for π.

104. Find the area.

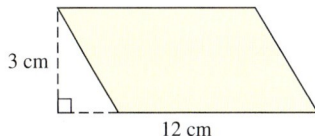

Section 8.8

For Exercises 105–108, find the volume and surface area. Use 3.14 for π.

105. 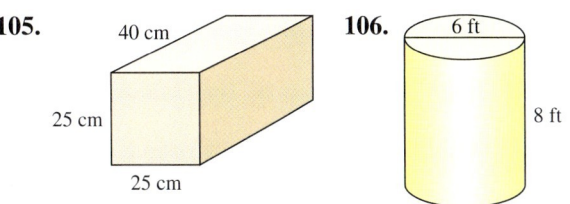 106.

107. 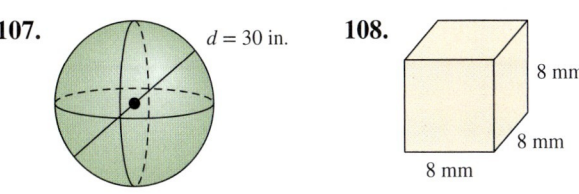 108.

109. Find the volume. Use 3.14 for π.

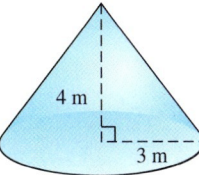

110. Find the volume of a can of paint if the can is a cylinder with diameter 6.5 in. and height 7.5 in. Round to the nearest whole unit.

Lawrence Manning/
Fuse/Corbis/Getty
Images

111. Find the volume of a ball if the diameter of the ball is approximately 6 in. Round to the nearest whole unit.

George Doyle/Stockbyte Platinum/Alamy Stock Photo

112. A microwave oven is a rectangular solid with dimensions 1 ft by 1 ft 9 in. by 1 ft 4 in. Find the volume in cubic feet.

For Exercises 113 and 114, find the volume of the shaded region. Use 3.14 for π if necessary. Round to the nearest whole unit.

113. **114.**

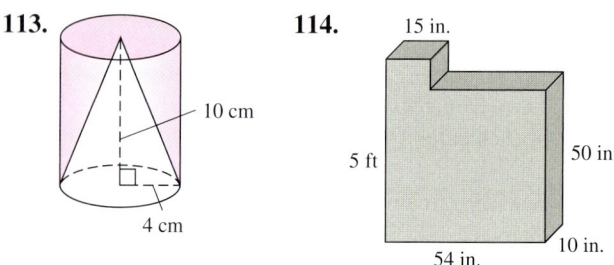

Graphs and Statistics

9

CHAPTER OUTLINE

- **9.1** Tables, Bar Graphs, Pictographs, and Line Graphs 536
- **9.2** Frequency Distributions and Histograms 548
- **9.3** Circle Graphs 556
- **9.4** Introduction to Probability 564
- **9.5** Mean, Median, and Mode 572
- Chapter 9 Review Exercises 584

Statistics in Life

In this chapter, we introduce the study of **statistics**—the science of describing data and drawing conclusions from data. This will include the concept of **probability**, which is used to estimate the likelihood of an event to occur. Probability is used in many fields, including the insurance industry. To set the premium for a life insurance policy, an insurance company would need to know the probability that an individual will survive the term of the policy (usually a period of 1 year). For example, according to the Social Security Administration, a 30-year-old male has a probability of 0.99891 to survive the year. This also means that the probability that he will not survive the year is 0.00109. Thus, out of 100,000 men of age 30, approximately 109 will die within the year. In such a case, the insurance company will need to pay his family a large sum. The premiums set by the insurance company paid over a large number of customers are set to offset this cost and leave room for company profit.

Hero/Corbis/Fancy Photography/Glow Images

We will also examine other statistical measures such as the mean, median, and mode. For example, you might be interested to know the mean (average) starting salary of a college graduate in your major. To do so, you might begin by collecting a sample of salaries of such graduates and computing the average salary. Although this may not give the exact salary you would earn when you are out of school, it will give you a reasonable estimate. You may also want to record stock prices of a company in which you want to invest. Collecting data over time can often illuminate trends in company performance.

Section 9.1 Tables, Bar Graphs, Pictographs, and Line Graphs

Concepts

1. Introduction to Data and Tables
2. Bar Graphs
3. Pictographs
4. Line Graphs

1. Introduction to Data and Tables

Statistics is the branch of mathematics that involves collecting, organizing, and analyzing **data** (information). One method to organize data is by using tables. A **table** uses rows and columns to reference information. The individual entries within a table are called **cells**.

Example 1 Interpreting Data in a Table

Table 9-1 summarizes the maximum wind speed, number of reported deaths, and estimated cost for recent hurricanes that made landfall in the United States. (*Source:* National Oceanic and Atmospheric Administration)

Table 9-1

Hurricane	Date	Landfall	Maximum Sustained Winds at Landfall (mph)	Number of Reported Deaths	Estimated Cost ($ Billions)
Katrina	2005	Louisiana, Mississippi	125	1836	81.2
Fran	1996	North Carolina, Virginia	115	37	5.8
Andrew	1992	Florida, Louisiana	145	61	35.6
Hugo	1989	South Carolina	130	57	10.8
Alicia	1983	Texas	115	19	5.9

a. Which hurricane caused the greatest number of deaths?
b. Which hurricane was the most costly?
c. What was the difference in the maximum sustained winds for hurricane Andrew and hurricane Katrina?
d. How many times greater was the death toll for Hugo than for Alicia?

Photo by Master Sgt. Michael A. Kaplan/U.S. Air Force

Solution:

a. The death toll is reported in the 5th column. The death toll for hurricane Katrina, 1836, is the greatest value.
b. The estimated cost is reported in the 6th column. Hurricane Katrina was also the costliest hurricane at $81.2 billion.
c. The wind speeds are given in the 4th column. The difference in the wind speed for hurricane Andrew and hurricane Katrina is 145 mph − 125 mph = 20 mph.
d. There were 57 deaths from Hugo and 19 from Alicia. The ratio of deaths from Hugo to deaths from Alicia is given by

$$\frac{57}{19} = 3$$

There were 3 times as many deaths from Hugo as from Alicia.

Skill Practice For Exercises 1–4, use the information in Table 9-1.

1. Which hurricane had the highest sustained wind speed at landfall?
2. Which hurricane was the most recent?
3. What was the difference in the death toll for Fran and Alicia?
4. How many times greater was the cost for Katrina than for Fran?

Example 2 **Constructing a Table from Observed Data**

The following data were taken by a student conducting a study for a statistics class. The student observed the type of vehicle and gender of the driver for 18 vehicles in the school parking lot. Complete the table.

Male–car	Female–truck	Male–truck
Male–truck	Female–car	Male–truck
Female–car	Male–truck	Male–motorcycle
Female–car	Male–car	Female–car
Male–motorcycle	Female–car	Female–car
Female–motorcycle	Male–car	Male–car

Driver \ Vehicle	Car	Truck	Motorcycle
Male			
Female			

Solution:

We need to count the number of data values that fall in each of the six cells. One method is to go through the list of data one by one. For each value place a tally mark | in the appropriate cell. For example, the first data value male–car would go in the cell in the first row, first column.

Driver \ Vehicle	Car	Truck	Motorcycle
Male	||||	||||	||
Female	⧸⧹⧸⧹|	|	|

To form the completed table, count the number of tally marks in each cell. See Table 9-2.

Table 9-2

Driver \ Vehicle	Car	Truck	Motorcycle
Male	4	4	2
Female	6	1	1

Answers

1. Andrew 2. Katrina
3. 18 deaths 4. 14 times greater

Skill Practice

5. A political poll was taken to determine the political party and gender of several registered voters. The following codes were used.

M = male
F = female
dem = Democrat
rep = Republican
ind = Independent

Complete the table given the following results.

F–dem	F–rep	M–dem
M–rep	M–ind	M–rep
F–dem	M–rep	F–dem
F–dem	M–dem	F–ind
M–dem	M–ind	M–rep
M–rep	F–rep	F–dem

	Male	Female
Democrat		
Republican		
Independent		

2. Bar Graphs

In Table 9-1, we see that hurricane Andrew had the greatest wind speed of those listed. This can be visualized in a graph. Figure 9-1 shows a bar graph of the wind speed for the hurricanes listed in Table 9-1. Notice that the bar showing wind speed for Andrew is the highest.

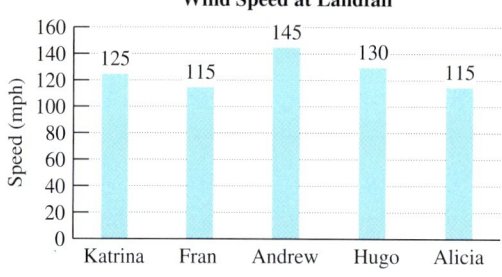

Super Nova Images/Alamy Stock Photo

Figure 9-1

Notice that the **bar graph** compares the data values through the height of each bar. The bars in a bar graph may also be presented horizontally. For example, the double bar graph in Figure 9-2 illustrates the data from Example 2.

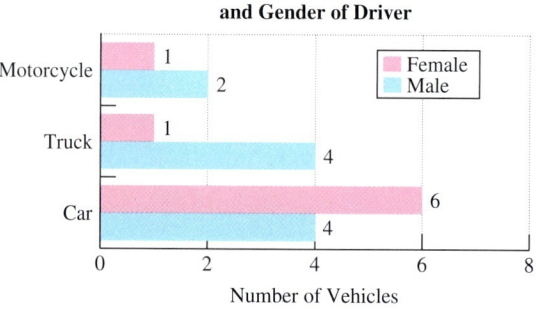

Figure 9-2

When constructing a graph or chart, be sure to include:
- A title.
- Labels on the vertical and horizontal axes.
- An appropriate range and scale.

Answer

5.

	Male	Female
Democrat	3	5
Republican	5	2
Independent	2	1

Example 3   Constructing a Bar Graph

The number of fat grams for five different ice cream brands and flavors is given in Table 9-3. Each value is based on a $\frac{1}{2}$-c serving. Construct a bar graph with vertical bars to depict this information.

Table 9-3

Brand/Flavor	Number of Fat Grams (g) per $\frac{1}{2}$-c Serving
Breyers Strawberry	6
Edy's Grand Light Mint Chocolate Chip	4.5
Healthy Choice Chocolate Fudge Brownie	2
Ben and Jerry's Chocolate Chip Cookie Dough	15
Häagen-Dazs Vanilla Swiss Almond	20

(*Source*: Caloriecount.about.com)

Solution:

First draw a horizontal line and label the different food categories. Then draw a vertical line on the left-hand side of the graph as in Figure 9-3. The vertical line represents the number of grams of fat. The vertical scale must extend to at least 20 to accommodate the largest value in the table. In Figure 9-3, the vertical scale ranges from 0 to 24 in multiples (or steps) of 4.

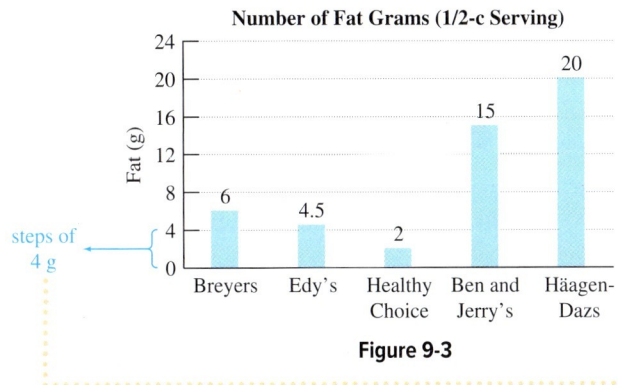

Figure 9-3

Avoiding Mistakes

The scale on the *y*-axis should begin at 0 and increase by equal intervals. We use the range of values in the data to choose a suitable scale.

Skill Practice

6. The amount of sodium in milligrams (mg) per $\frac{1}{2}$-c serving for three different brands of cereal is given in the table. Construct a bar graph with horizontal bars.

Brand/Flavor	Amount of Sodium (mg)
Post Grape Nuts	290
Post Cranberry Almond	70
Post Bran Flakes	120

(*Source:* Post Cereals)

Answer

6.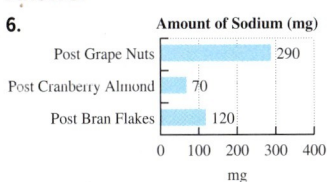

3. Pictographs

Sometimes a bar graph might use an icon or small image to convey a unit of measurement. This type of bar graph is called a **pictograph**.

Example 4 Interpreting a Pictograph

For a recent year, California led the United States in hybrid vehicle sales (83,000 sold). The graph displays the hybrid vehicle sales for five other states for the same year (Figure 9-4). (*Source:* HybridCars.com)

a. What is the value of each car icon in the graph?

b. From the graph, estimate the number of hybrids sold in the state of Washington.

c. For which state were approximately 10,000 hybrid vehicles sold?

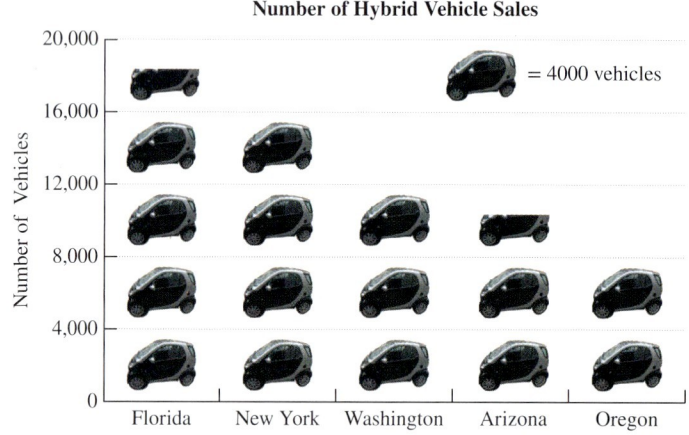

Erica Simone Leeds   Figure 9-4

Solution:

a. The legend indicates that 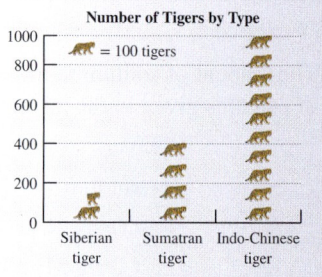 = 4000 vehicles sold.

b. The height of the "bar" for Washington is given by 3 car icons. This represents 3×4000 vehicles. Therefore, 12,000 hybrid vehicles were sold in Washington.

c. The bar containing $2\frac{1}{2}$ icons represents $2.5 \cdot (4000 \text{ vehicles}) = 10,000$ hybrids sold. This corresponds to the state of Arizona.

Skill Practice For Exercises 7–9, refer to the pictograph showing the number of tigers living in the wild.

7. What is the value of each tiger icon?

8. From the graph, estimate the number of Sumatran tigers.

9. Estimate the number of Siberian tigers.

4. Line Graphs

Line graphs are often used to track how one variable changes with respect to a second variable. For example, a line graph may illustrate a pattern or trend of a variable over time and allow us to make predictions.

Answers
7. 100 tigers   8. 400 Sumatran tigers
9. 150 Siberian tigers

Section 9.1 Tables, Bar Graphs, Pictographs, and Line Graphs   541

Example 5 Interpreting a Line Graph

Figure 9-5 shows the number of bachelor's degrees earned by men and women for selected years. (*Source:* U.S. Census Bureau)

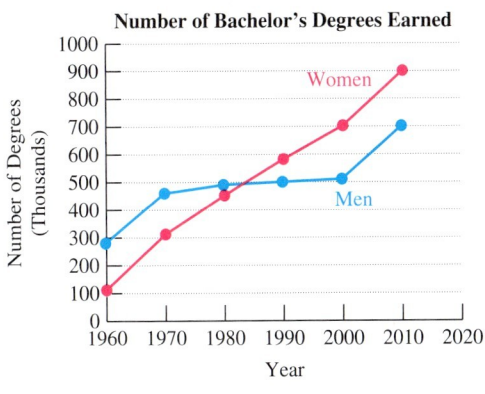

Figure 9-5

Mark Scott/Photographer's Choice/ Getty Images

a. In 1960, who earned more bachelor's degrees, men or women?

b. In 2000, who earned more bachelor's degrees, men or women?

c. If the trends from the year 2000 to the year 2010 continue, predict the number of bachelor's degrees earned by men and women in the year 2020.

Solution:

a. The blue graph represents men, and the red graph represents women. In 1960, men earned approximately 280 thousand bachelor's degrees. Women earned approximately 110 thousand. Men earned more bachelor's degrees in 1960.

b. In 2000, women earned more bachelor's degrees.

c. To predict the number of bachelor's degrees earned by men and women in the year 2020, we need to extend both line graphs. See the dashed lines in Figure 9-6. The number of bachelor's degrees earned by women is predicted to be approximately 1100 thousand. The number of degrees earned by men will be approximately 900 thousand.

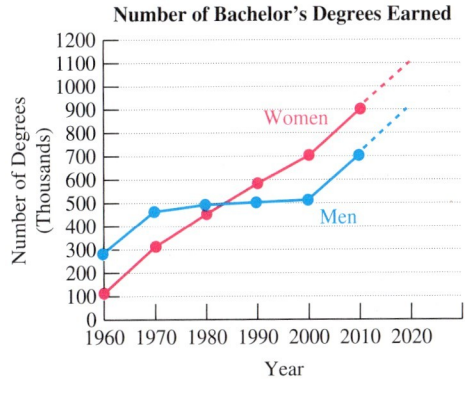

Figure 9-6

Skill Practice For Exercises 10–12, refer to the line graph showing the number of cats and dogs received by an animal shelter for the first four years since it opened.

10. Which animal did the shelter receive most?

11. Approximate the number of cats received the third year after the shelter opened.

12. Use the graph to predict the number of dogs the shelter can expect for its fifth year.

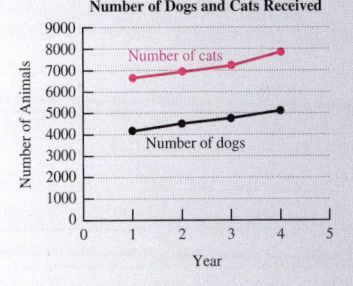

Answers

10. Cats
11. Approximately 7200 cats
12. Approximately 5400 dogs

Example 6 Constructing a Line Graph

Table 9-4 represents the amount (in millions of gallons) of oil or fuel spilled in major oil spills in the United States for selected years. Use the data given in the table to create a line graph.

Table 9-4

Year	Number of Gallons of Oil or Fuel (in millions)
1	7.00
2	2.98
3	0.50
4	0.42
5	0.05
6	4.90
7	0.04

Solution:

First draw a horizontal line and label the year. Then draw a vertical line on the left-side of the graph, as in Figure 9-7. The vertical line represents the number of gallons of oil or fuel (in millions). In Figure 9-7, the vertical scale ranges from 0 to 8 million in steps of 1.0 million. For each year, plot a point corresponding to the number of millions of gallons for that year. Then connect the points.

Figure 9-7

In Figure 9-7, we labeled the value at each data point because the exact values are difficult to read from the graph.

Skill Practice

13. The table gives the number of cars registered in the United States (in millions) for selected years. (*Source:* U.S. Department of Transportation) Create a line graph for this information.

Year	Number of Cars (millions)
1970	89
1980	122
1990	134
2000	134
2010	138

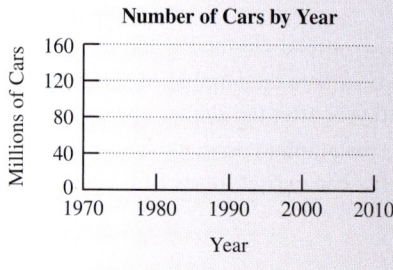

Answer

13.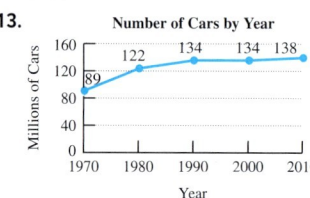

Practice Exercises

Section 9.1

Vocabulary and Key Concepts

1. a. _____ is the branch of mathematics that involves collecting, organizing, and analyzing information.

   b. A _____ uses rows and columns to reference information. The individual entries within a table are called _____.

   c. A _____ is a type of bar graph that uses an icon or small image to convey a unit of measurement.

Concept 1: Introduction to Data and Tables

For Exercises 2–6, refer to the table. The table represents the high and low temperatures for January for five selected cities of different latitudes in the United States. (The given cities are all north of the equator. The greater the latitude, the farther north of the equator the city is located.)

City	Latitude	Low Temperature (°F)	High Temperature (°F)
Miami, FL	25.8°N	60°	76°
Atlanta, GA	33.7°N	35°	51°
Kansas City, MO	39.1°N	18°	38°
Bismark, ND	46.8°N	2°	23°
Fairbanks, AK	64.8°N	−17°	1°

2. What is the latitude of the city located farthest north?

3. What is the difference between the high and low temperatures for Kansas City?

4. What is the difference between the high and low temperatures for Atlanta?

5. What is the difference between the low temperature in Miami and the low temperature in Fairbanks?

6. What is the difference between the high temperature in Atlanta and the high temperature in Bismark?

For Exercises 7–10, refer to the table. The table represents the Seven Summits (the highest peaks from each continent). **(See Example 1.)**

Mountain	Continent	Altitude (ft)
Mt. Kilimanjaro	Africa	19,340
Elbrus	Europe	18,510
Aconcagua	South America	22,834
Denali (Mt. McKinley)	North America	20,320
Vinson Massif	Antarctica	16,864
Mt. Kosciusko	Australia	7,310
Mt. Everest	Asia	29,035

DLILLC/Corbis Super/Alamy Stock Photo

7. In which continent does the highest mountain lie?

8. Which mountain among those listed is the lowest? In which continent does it lie?

9. How much higher is Aconcagua than Denali?

10. What is the difference in altitude between the highest mountain in Europe and the highest mountain in Australia?

For Exercises 11–16, refer to the table. The table gives the average ages (in years) for U.S. women and men married for the first time for selected years. (*Source:* U.S. Census Bureau)

	Men	Women
1940	24.3	21.5
1960	22.8	20.3
1980	24.7	22.0
2000	26.8	25.1

11. By how much has the average age for women increased between 1940 and 2000?

12. By how much has the average age for men increased between 1940 and 2000?

13. What is the difference between the men's and women's average age at first marriage in 1940?

14. What is the difference between the men's and women's average age at first marriage in 2000?

15. Which group, men or women, had the consistently higher age at first marriage?

16. Which group, men or women, had a greater increase in age between 1940 and 2000?

17. The following data were taken from a survey of a third-grade class. The survey denotes the gender of a student and whether the student owned a dog, a cat, or neither. Complete the table. Be sure to label the rows and columns. (See Example 2.)

Boy–dog	Boy–dog	Boy–cat	Boy–neither
Girl–dog	Girl–neither	Boy–dog	Girl–cat
Girl–neither	Girl–neither	Girl–dog	Girl–cat
Boy–dog	Girl–cat	Boy–neither	Girl–dog
Boy–neither	Girl–neither	Girl–cat	Girl–neither

	Dog	Cat	Neither
Boy			
Girl			

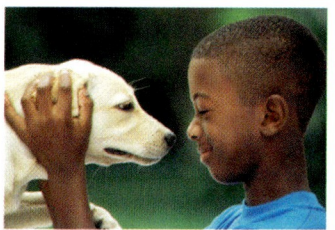

Arthur Tilley/Creatas Images/Getty Images

18. In a group of 20 women, 10 were given an experimental drug to lower cholesterol. The other 10 were given a placebo. The letter "D" indicates that the person got the drug, and the letter "P" indicates that the person received the placebo. The values "yes" or "no" indicate whether the person's cholesterol was lowered. Complete the table.

	Yes	No
Drug (D)		
Placebo (P)		

| D–yes | D–yes | P–no | D–no | P–yes | D–yes | P–no | P–no | D–yes | D–no |
| P–yes | P–no | D–yes | P–yes | P–yes | D–no | D–yes | P–no | D–yes | P–no |

Concept 2: Bar Graphs

19. A study done in a suburban area reported the percentage of Internet users in various age groups. The data are given in the table. (See Example 3.)

   a. For which age group is the percentage of Internet users the greatest?

   b. Draw a bar graph with vertical bars to illustrate these data.

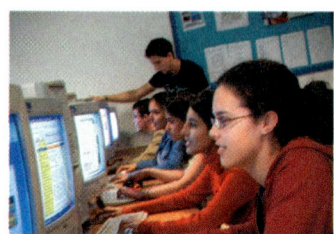

Ian Shaw/Alamy Stock Photo

Age Group (in years)	Percentage of Internet Users
18–29	93%
30–49	51%
50–64	70%
65+	38%

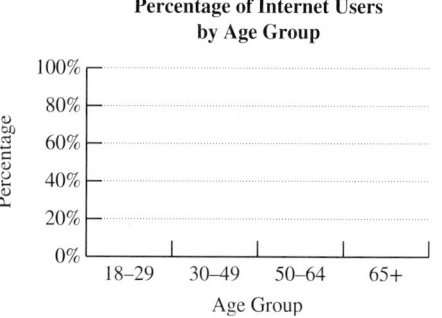

20. The number of new jobs for selected industries are given in the table. (*Source:* Bureau of Labor Statistics)

a. Which category has the greatest number of new jobs? How many new jobs is this?

b. Draw a bar graph with vertical bars to illustrate these data.

Digital Vision/Getty Images

Industry	Number of New Jobs
Health care	219,400
Temporary help	212,000
Construction	173,000
Food service	167,600
Retail	78,600

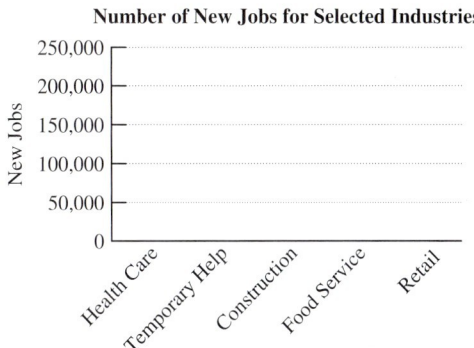

21. The table shows the percentage of broadband subscribers in various countries. Construct a bar graph with horizontal bars. The length of each bar should represent the percentage of broadband subscribers for the corresponding country. (*Source:* International Telecommunications Union)

Country	Percent of Subscribers
Sweden	37.3%
Canada	29.0%
United Kingdom	28.3%
United States	25.6%
Japan	23.5%

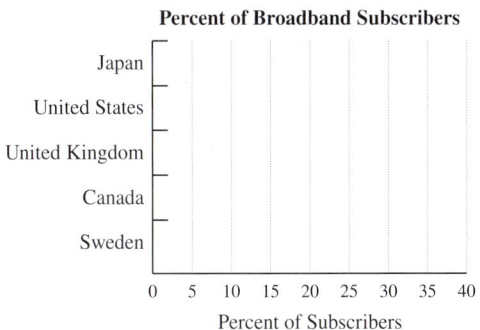

22. The table represents the world's major consumers of primary energy for a recent year. All measurements are in quadrillions of Btu. *Note:* 1 quadrillion = 1,000,000,000,000,000. (*Source:* Energy Information Administration, U.S. Department of Energy) Construct a bar graph using horizontal bars. The length of each bar gives the amount of energy consumed for that country.

Country	Amount of Energy Consumed (Quadrillions of Btu)
Germany	14
Japan	22
Russia	28
China	37
United States	99

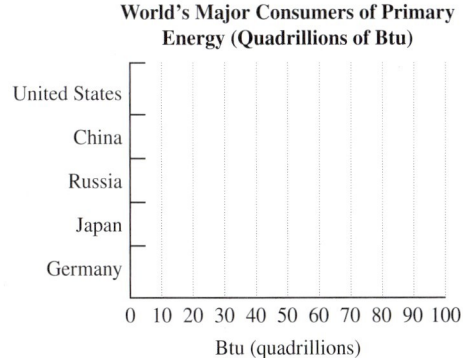

Concept 3: Pictographs

23. A local ice cream stand kept track of its ice cream sales for one weekend, as shown in the figure. (See Example 4.)

    a. What does each ice cream icon represent?

    b. From the graph, estimate the number of servings of ice cream sold on Saturday.

    c. Which day had approximately 275 servings of ice cream sold?

24. Adults access the Internet to see weather updates and check on current news. The pictograph displays the percent of adult Internet users who access these topics.

    a. What does each computer icon represent?

    b. From the graph, estimate the percent of adult users that access the Internet for weather.

    c. Which type of news is accessed about 45% of the time?

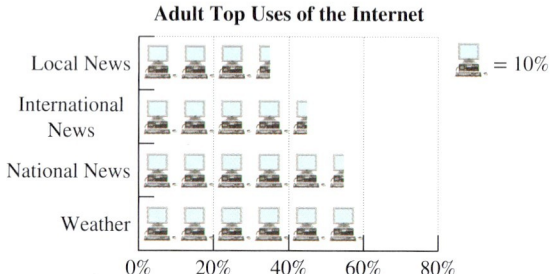

25. The figure displays the 1-day box office gross earnings for selected top-earning films.

    a. Estimate the earnings for the adventure movie.

    b. Which film grossed approximately $60 million?

    c. Estimate the total earnings for all four films.

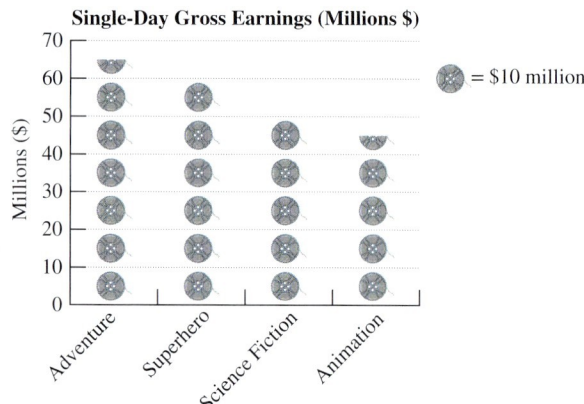

26. Recently the largest populations of senior citizens were in California, Florida, New York, Texas, and Pennsylvania, as shown in the figure. (*Source:* U.S. Bureau of the Census)

    a. Estimate the number of senior citizens living in Texas.

    b. Approximately how many more senior citizens are living in California than in Pennsylvania?

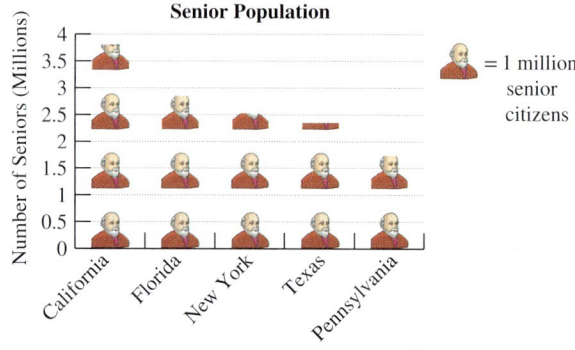

Concept 4: Line Graphs

For Exercises 27–32, use the graph provided. The graph shows the trend depicted by the percent of men and women over 65 years old in the labor force in a certain metropolitan area. (See Example 5.)

27. What was the difference in the percent of men and the percent of women over 65 in the labor force in the year 1930?

28. What was the difference in the percent of men and the percent of women over 65 in the labor force in the year 2010?

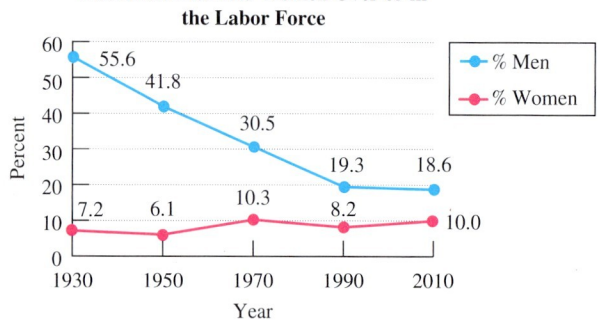

29. What was the overall trend in the percent of women over 65 in the labor force for the years shown in the graph?

30. What was the overall trend in the percent of men over 65 in the labor force for the years shown in the graph?

31. Use the graph to predict the number of men over 65 in the labor force in the year 2020. Answers will vary.

32. Use the graph to predict the number of women over 65 in the labor force in the year 2020. Answers will vary.

For Exercises 33–38, refer to the graph representing the number of hybrid cars sold in the United States over a 6-year period.

33. In which year were the most hybrid cars sold? How many were sold?

34. In which year were the fewest number of hybrid cars sold? How many were sold?

35. What is the difference between the sales in year 2 and year 1?

36. What is the difference between the sales in year 6 and year 5?

37. Between which two consecutive years was the increase in sales the greatest?

38. Between which two consecutive years did the sales decrease? How much was the decrease?

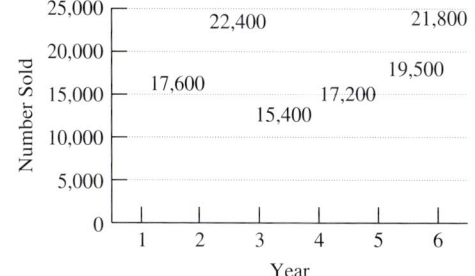

39. The data shown here give the average height for girls based on age. (*Source:* National Parenting Council)
    (See Example 6.)
    a. Make a line graph to illustrate these data.
    b. Use the line graph from part (a) to predict the average height of a 10-year-old girl. (Answers may vary.)

Age, x	Height (in.), y
2	35
3	38.5
4	41.5
5	44
6	46
7	48
8	50.5
9	53

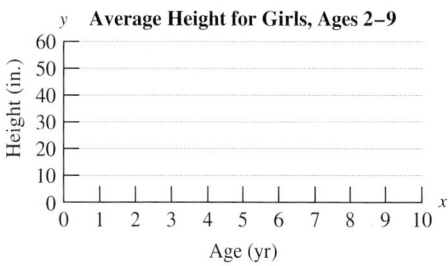

40. The data shown here give the average height for boys based on age. (*Source:* National Parenting Council)

   a. Make a line graph to illustrate these data.

   b. Use the line graph from part (a) to predict the average height of a 10-year-old boy. (Answers may vary.)

Age, x	Height (in.), y
2	36
3	39
4	42
5	44
6	46.75
7	49
8	51
9	53.5

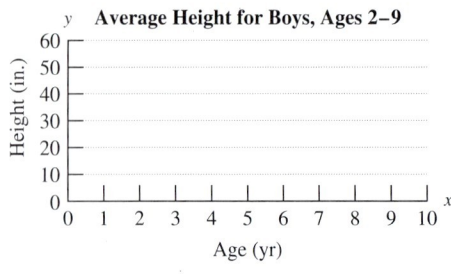

Expanding Your Skills

All packaged food items have to display nutritional facts so that the consumer can make informed choices. For Exercises 41–44, refer to the nutritional chart for Breyers French Vanilla ice cream.

41. How many servings are there per container? How much total fat is in one container of this ice cream?

42. How much total sodium is in one container of this ice cream?

43. If 8 g of fat is 13% of the daily value, what is the daily value of fat? Round to 1 decimal place.

44. If 50 mg of cholesterol is 17% of the daily value, what is the daily value of cholesterol? Round to the nearest whole unit.

Nutrition Facts	
Serving Size $\frac{1}{2}$ cup (68 g)	
Servings per Container 14	
Amount per Serving	
Calories 150	**Calories from Fat** 80
	% Daily Value
Total Fat 8 g	13%
Saturated fat 5 g	25%
Cholesterol 50 mg	17%
Sodium 45 mg	2%
Total Carbohydrate	
Dietary fiber 0 g	
Sugars 15 g	
Protein 3 g	

Section 9.2 Frequency Distributions and Histograms

Concepts

1. Frequency Distributions
2. Histograms

1. Frequency Distributions

The ages for 36 players in the NBA (National Basketball Association) are given for a recent year.

23	32	26	25	30	30	29	38	21	25	34	22
33	26	39	22	32	23	30	27	22	24	31	33
21	24	28	26	26	20	33	27	29	31	27	28

The youngest player is 20 years old and the oldest is 39 years old. Suppose we wanted to organize this information further by age groups. One way is to create a frequency distribution. A **frequency distribution** is a table displaying the number of values that fall within specific categories. When the categories represent a range of numerical values we call the categories **class intervals**. This is demonstrated in Example 1.

Example 1 Creating a Frequency Distribution

Complete the table to form a frequency distribution for the ages of the NBA players listed.

23	32	26	25	30	30	29	38	21	25	34	22
33	26	39	22	32	23	30	27	22	24	31	33
21	24	28	26	26	20	33	27	29	31	27	28

Class Intervals, Age (yr)	Tally	Frequency (Number of Players)
20–23		
24–27		
28–31		
32–35		
36–39		

Solution:

The classes represent different age groups. Go through the list of ages, and use tally marks to track the number of players that fall within each class. Tally marks are shown in red for the first column of data: 23, 33, and 21.

Table 9-5

Class Intervals, Age (yr)	Tally	Frequency (Number of Players)
20–23	|||| |||	8
24–27	|||| ||||		11
28–31	|||| ||||	9
32–35	||||		6
36–39	||	2

The frequency is a count of the tally marks within each class. See Table 9-5.

Skill Practice

1. The ages (in years) of individuals arrested on a certain day in Galveston, Texas, are listed. Complete the table to form a frequency distribution.

18	20	35	46
19	26	24	32
28	25	30	34
22	29	39	19
18	19	26	40

Class (Age)	Tally	Frequency
18–23		
24–29		
30–35		
36–41		
42–47		

Example 2 Interpreting a Frequency Distribution

Consider the frequency distribution in Table 9-5.

a. Which class has the most values?
b. How many values are represented in the table?
c. What percent of the players were 32 years old or older at the time the data were recorded?

Answer

1.

Class (Age)	Tally	Frequency
18–23	|||| ||	7
24–29	||||		6
30–35	||||	4
36–41	||	2
42–47			1

Solution:

a. The 24–27 year class has the highest frequency (greatest number of values).

b. The number of data values is given by the sum of the frequencies.

$$\text{Total number of values} = 8 + 11 + 9 + 6 + 2$$
$$= 36$$

c. The number of players 32 years old or older is given by the sum of the frequencies in the 32–35 category and the 36–39 category. This is $6 + 2 = 8$.

There are 8 players 32 years old or older.

Therefore, $\frac{8}{36} = 0.\overline{2} \approx 22.2\%$ of the players are 32 years old or older.

Skill Practice For Exercises 2–4, consider the frequency distribution from Skill Practice Exercise 1.

2. Which class (age group) has the most values?
3. How many values are represented in the table?
4. What percent of the people arrested are in the 42–47 age group?

When creating a frequency distribution, keep these important guidelines in mind.

- The classes should be equally spaced. For instance, in Example 1, we would not want one class to represent a 4-year interval and another to represent a 10-year interval.
- The classes should not overlap. That is, a value should belong to one and only one class.
- In general, we usually create a frequency distribution with between 5 and 15 classes, inclusive.

2. Histograms

A **histogram** is a special bar graph that illustrates data given in a frequency distribution. The class intervals are given on the horizontal scale. The height of each bar in a histogram represents the frequency for each class. Furthermore, the bars of a histogram touch with no space between the bars.

Example 3   Constructing a Histogram

Construct a histogram for the frequency distribution given in Example 1.

Solution:

Class Intervals, Age (yr)	Tally	Frequency (Number of Players)
20–23	⧸⧸⧸⧸ ⦀⦀⦀	8
24–27	⧸⧸⧸⧸ ⧸⧸⧸⧸ ⦀	11
28–31	⧸⧸⧸⧸ ⦀⦀⦀⦀	9
32–35	⧸⧸⧸⧸ ⦀	6
36–39	⦀⦀	2

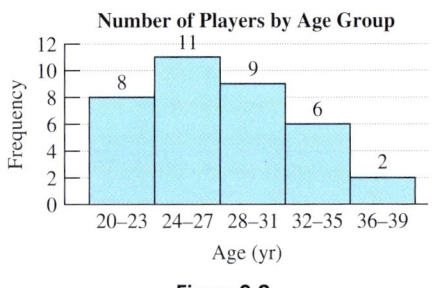

Figure 9-8

To create a histogram of these data, we list the classes (ages of players) on the horizontal scale. On the vertical scale we measure the frequency (Figure 9-8).

Skill Practice

5. Construct a histogram for the frequency distribution in Skill Practice Exercise 1.

Answers

2. 18–23 years   3. 20   4. 5%
5.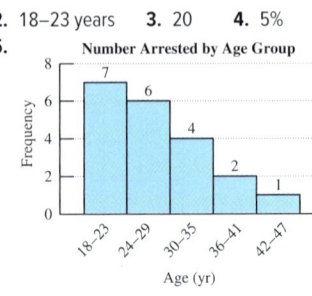

Section 9.2 Activity

For Exercises A.1–A.6, use the given data. The data represent the ages of 24 students in a prealgebra class for a recent semester.

22 24 19 31 29 33 18 26 25 42 41 29
32 19 24 21 20 38 26 28 35 40 32 21

A.1. Complete the frequency distribution.

a. Place a tally mark for each value that falls within the class intervals (age group) in the Tally column.

b. Count the tally marks within each class and place the total in the Frequency column.

Age Group	Tally	Frequency
18–21		
22–26		
27–30		
31–34		
35–38		
39–42		

A.2. Draw a histogram based on the data.

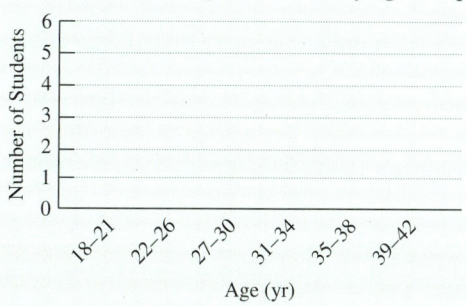

Number of Prealgebra Students by Age Group

A.3. What age group(s) is most represented in the class?

A.4. What age group(s) is represented the least in the class?

A.5. What percent of the class is between the ages of 27 and 34? Round to the nearest tenth of a percent.

A.6. What percent of the class is 21 years old or younger?

For Exercises A.7–A.10, use the given data. These data represent the ages of finalists for two seasons of a singing competition.

19 25 29 21 29 19 20 17 28 22 28 21
25 24 22 23 18 21 23 22 24 24 22 25

A.7. Complete the frequency distribution.

Age Group	Tally	Frequency
17–19		
20–22		
23–25		
26–28		
29–31		

A.8. Draw a histogram based on the data.

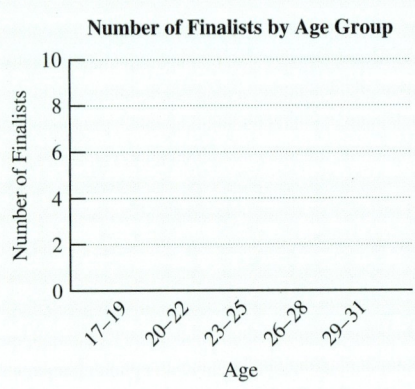

Number of Finalists by Age Group

A.9. What age group(s) is most represented in the competition?

A.10. What percentage of contestants are 25 years old or younger? Round your answer to the nearest tenth of a percent.

Section 9.2 Practice Exercises

Vocabulary and Key Concepts

1. a. A _____ distribution is a table displaying the number of values that fall within specific categories.

   b. A _____ is a bar graph that illustrates data given in a frequency distribution.

Concept 1: Frequency Distributions

For Exercises 2–6, refer to the frequency distribution.

Class Intervals	Frequency
10–19	14
20–29	25
30–39	36
49–49	24
50–59	9

2. Determine the total number of data.

3. Which class contains the most data?

4. Which class contains the least data?

5. What percent of data is contained in the 20–29 class? Round to the nearest tenth of a percent.

6. What percent of data is contained in the 10–19 class? Round to the nearest tenth of a percent.

7. From the frequency distribution, determine the total number of data.

Class Intervals	Frequency
1–4	14
5–8	18
9–12	24
13–16	10
17–20	6

8. From the frequency distribution, determine the total number of data.

Class Intervals	Frequency
1–50	29
51–100	12
101–150	6
151–200	22
201–250	56
251–300	60

9. For the table in Exercise 7, which class contains the most data?

10. For the table in Exercise 8, which class contains the most data?

11. The retirement age (in years) for 20 college professors is given. Complete the frequency distribution. (See Examples 1 and 2.)

67 56 68 70 60 65 73 72 56 65
71 66 72 69 65 65 63 65 68 70

Class Intervals (Age in Years)	Tally	Frequency (Number of Professors)
56–58		
59–61		
62–64		
65–67		
68–70		
71–73		

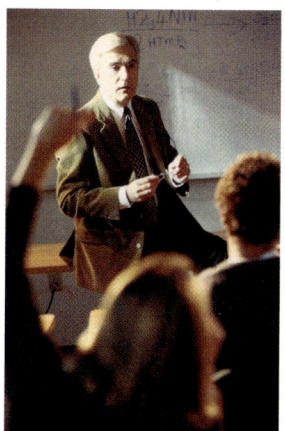
Photodisc/Getty Images

a. Which class has the most values?
b. How many data values are represented in the table?
c. What percent of the professors retire when they are 68 to 70 years old?

12. The number of miles run in one day by 16 selected runners is given. Complete the frequency distribution.

2 4 7 3 8 4 5 7
4 6 4 3 4 2 4 10

Class Intervals (Number of Miles)	Tally	Frequency (Number of Runners)
1–2		
3–4		
5–6		
7–8		
9–10		

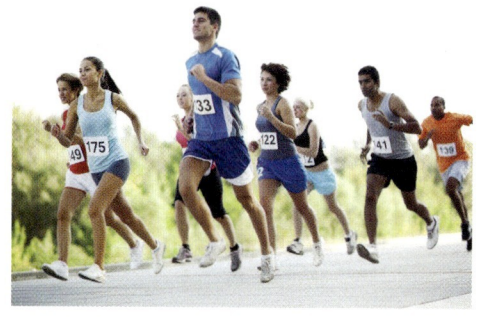
skynesher/E+/Getty Images

a. Which class contains the greatest number of values?

b. How many data values are represented in the table?

c. What percent of the runners ran 3 to 4 mi/day?

13. The number of gallons of gas purchased by 16 customers at a certain gas station is given. Complete the frequency distribution.

| 12.7 | 13.1 | 9.8 | 12.0 | 10.4 | 9.8 | 14.2 | 8.6 |
| 19.2 | 8.1 | 14.0 | 15.4 | 12.8 | 18.2 | 15.1 | 13.0 |

Class Intervals (Amount in Gal)	Tally	Frequency (Number of Customers)
8.0–9.9		
10.0–11.9		
12.0–13.9		
14.0–15.9		
16.0–17.9		
18.0–19.9		

NithidPhoto/iStock/Getty Images

a. Which class contains the greatest number of values?

b. How many data values are represented in the table?

c. What percent of the customers purchased 18 to 19.9 gal of gas?

14. The hourly wages (in dollars) for 15 employees at a department store are given. Complete the frequency distribution.

9.50	12.00	14.75	9.50	17.00
11.50	12.50	11.75	18.25	9.75
12.75	14.75	9.50	15.75	9.75

Class Intervals (Hourly Wage, $)	Tally	Frequency (Number of Employees)
9.00–10.99		
11.00–12.99		
13.00–14.99		
15.00–16.99		
17.00–18.99		

a. Which class contains the least number of values?

b. How many data values are represented in the table?

c. What percent of the employees has an hourly wage of $11.00 to $12.99?

15. Explain what is wrong with the following class intervals.

Class	Tally	Frequency
0–4		
5–10		
11–17		
18–25		
26–34		

16. Explain what is wrong with the following class intervals.

Class	Tally	Frequency
1–6		
7–11		
12–17		
18–23		
24–28		

17. Explain what is wrong with the following class intervals.

Class	Tally	Frequency
1–20		
21–40		

18. Explain what is wrong with the following class intervals.

Class	Tally	Frequency
1–33		
34–66		
67–99		

19. Explain what is wrong with the following class intervals.

Class	Tally	Frequency
10–12		
12–14		
14–16		
16–18		
18–20		

20. Explain what is wrong with the following class intervals.

Class	Tally	Frequency
1–5		
5–10		
10–15		
15–20		
20–25		

21. The heights of 20 students in a math class are given. Complete the frequency distribution.

70	71	73	62	65	70	69	70
64	66	73	63	68	67	69	72
64	66	67	69				

Class Interval (Height, in.)	Frequency (Number of Students)
62–63	
64–65	
66–67	
68–69	
70–71	
72–73	

22. The amount withdrawn in dollars from a certain ATM is given for 20 customers. Construct a frequency distribution.

40	50	200	200	100	120	200
50	100	60	100	100	30	40
100	100	50	200	150	200	

Class Interval (Amount, $)	Frequency (Number of Customers)
0–49	
50–99	
100–149	
150–199	
200–249	

Image Source/Getty Images

Concept 2: Histograms

23. Construct a histogram for the frequency table in Exercise 21. (See Example 3.)

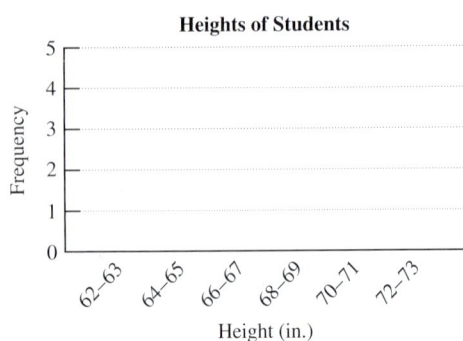

24. Construct a histogram for the frequency table in Exercise 22.

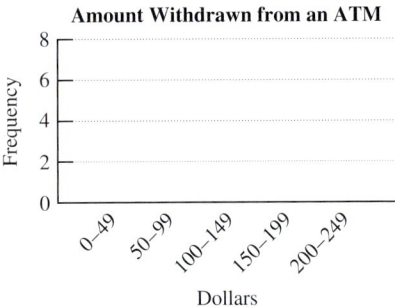

25. Construct a histogram, using the given data. Each number represents the number of Calories in a 100-g serving for selected fruits.

59	65	48	49	105	47	92	43
52	59	56	49	35	55	72	30
67	44	32	32	61	29	30	56

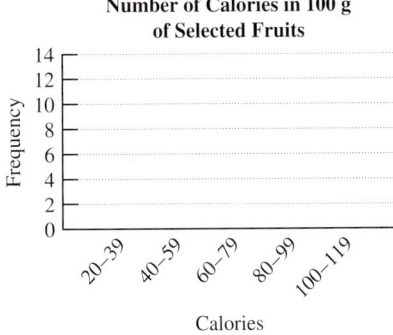

26. The list of data gives the number of children of the presidents of the United States (in no particular order). Construct a histogram.

0	4	3	3	2	2	2	10	0	5	2
4	5	7	4	3	6	4	0	7	5	6
1	4	3	0	4	3	2	6	4	6	8
3	2	1	0	2	6	0	2	2	2	8

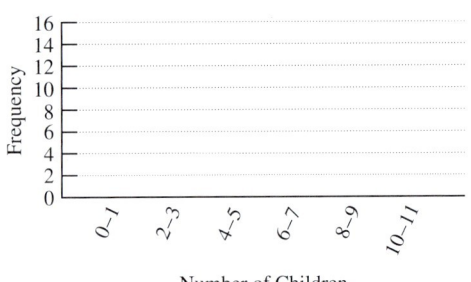

Section 9.3 Circle Graphs

Concepts

1. Interpreting Circle Graphs
2. Circle Graphs and Percents
3. Constructing Circle Graphs

1. Interpreting Circle Graphs

Thus far we have used bar graphs, line graphs, and histograms to visualize data. A **circle graph** (or pie graph) is another type of graph used to show how a whole amount is divided into parts. Each part of the circle, called a **sector,** is like a slice of pie. The size of each piece relates to the fraction of the whole it represents.

Section 9.3 Circle Graphs   557

Example 1 Interpreting a Circle Graph

The grade distribution for a math test is shown in the circle graph (Figure 9-9).

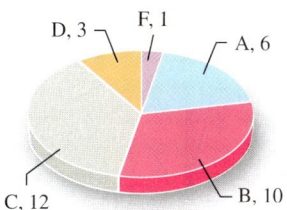

Figure 9-9

a. How many total grades are represented?
b. How many grades are "B"s?
c. How many times more "C"s are there than "D"s?
d. What percent of the grades were "A"s?

Solution:

a. The total number of grades is equal to the sum of the number of grades from each category.

$$\text{Total number of grades} = 6 + 10 + 12 + 3 + 1$$
$$= 32$$

b. The number of "B"s is represented by the red portion of the graph. There are 10 "B"s.

c. There are 12 "C"s and 3 "D"s. The ratio of "C"s to "D"s is $\frac{12}{3} = 4$. Therefore, there are 4 times as many "C"s as "D"s.

d. There are 6 "A"s. The percent of "A"s is given by

$$\frac{6}{32} = 0.1875$$

Therefore, the percent of "A"s is 18.75%.

Skill Practice A used car dealership sells cars and trucks. Use the circle graph to answer Exercises 1–4.

1. How many total vehicles are represented?
2. How many are Toyotas?
3. How many times more Hondas are there than Dodges?
4. What percent are Fords?

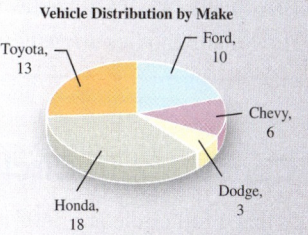

2. Circle Graphs and Percents

Sometimes circle graphs show data in percent form. This is illustrated in Example 2.

Answers
1. 50   2. 13
3. There are 6 times more Hondas than Dodges.
4. 20% are Fords.

> **Example 2** Calculating Amounts by Using a Circle Graph
>
> A certain online movie rental site carries 20,000 different titles. It groups its movie collection by the categories shown in the graph (Figure 9-10).
>
>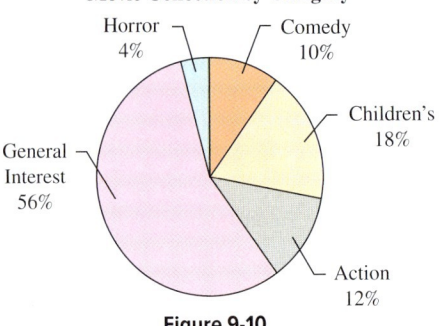
>
> Figure 9-10
>
> a. How many movies are comedy?
>
> b. How many movies are action or horror?
>
> **Solution:**
>
> a. First note that the site carries 20,000 different movies. From the graph we know that 10% are comedies. Therefore, this question can be interpreted as
>
> What is 10% of 20,000?
> $$x = (0.10) \cdot (20,000)$$
> $$= 2000 \qquad \text{There are 2000 comedies.}$$
>
> b. From the graph we know that 12% of the movies are action and 4% are horror. This accounts for 16% of the total collection. Therefore, this question asks
>
> What is 16% of 20,000?
> $$x = (0.16) \cdot (20,000)$$
> $$= 3200 \qquad \text{There are 3200 movies that are action or horror.}$$
>
> **Skill Practice** For Exercises 5 and 6, refer to Figure 9-10.
>
> 5. How many movies are action?
> 6. How many movies are general interest or children's?

FOR REVIEW

Recall that a percent equation may also be solved with a proportion.

What is 10% of 20,000?

$$\frac{x}{20,000} = \frac{10}{100}$$
$$100x = (10)(20,000)$$
$$x = 2000$$

3. Constructing Circle Graphs

Recall that a full circle is a 360° arc. To draw a circle graph, we must compute the number of degrees of arc for each sector. In Example 2, 10% of the videos are comedies. To draw the sector for this category, we must determine 10% of 360°.

$$10\% \text{ of } 360° = 0.10(360°) = 36°$$

The sector representing comedies should be drawn with a 36° angle. To do this, we can use a protractor (Figure 9-11).

Answers

5. There are 2400 action movies.
6. There are 14,800 general interest or children's movies.

To draw a sector with a 36° arc, first draw a circle. Place the hole in the protractor over the center of the circle. Using the inner scale on the protractor, place a tick mark at 0° and at 36°. Use a straightedge to draw two line segments from the center of the circle to each tick mark. See Figure 9-11.

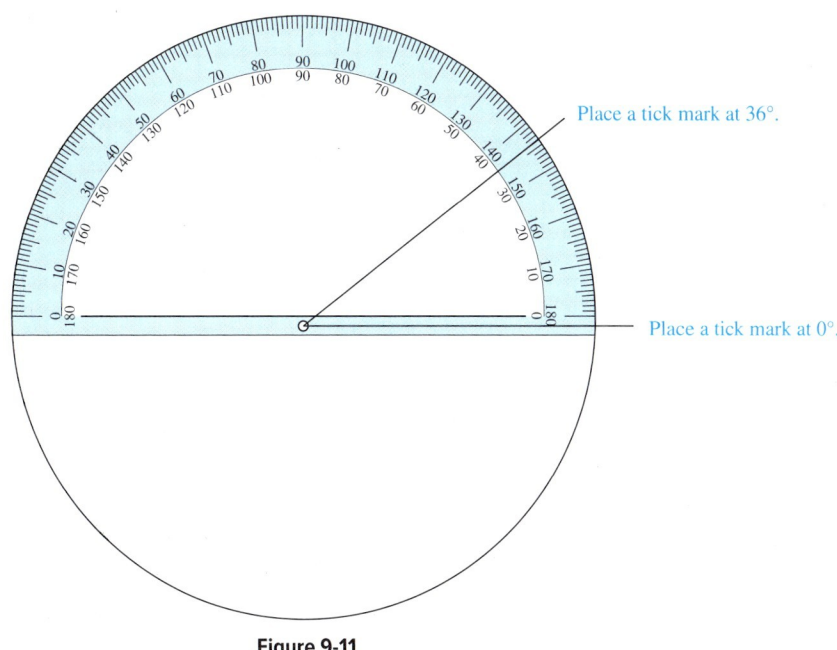

Figure 9-11

In Example 3, we use this technique to construct a circle graph.

Example 3 Constructing a Circle Graph

A teacher earns a monthly salary of $3600 after taxes. Her monthly budget is broken down in Table 9-6.

Terry Vine/Blend Images/Getty Images

Table 9-6

Budget Item	Monthly Value ($)
Rent	1260
Utilities	315
Car expenses	765
Groceries	540
Savings	450
Other	270

Construct a circle graph illustrating the information in this table. Label each sector of the graph with the percent that it represents.

Solution:

This example calls for two types of calculations: (1) For each budget item, we must compute the percent of the whole that it represents. (2) We must determine the number of degrees for each category. We can use a table to help organize our calculations.

Budget Item	Monthly Value ($)	Percent	Number of Degrees
Rent	1260	$=\dfrac{1260}{3600} = 0.35$ or 35%	35% of 360° $= 0.35(360°)$ $= 126°$
Utilities	315	$=\dfrac{315}{3600} = 0.0875$ or 8.75%	8.75% of 360° $= 0.0875(360°)$ $= 31.5°$
Car	765	$=\dfrac{765}{3600} = 0.2125$ or 21.25%	21.25% of 360° $= 0.2125(360°)$ $= 76.5°$
Groceries	540	$=\dfrac{540}{3600} = 0.15$ or 15%	15% of 360° $= 0.15(360°)$ $= 54°$
Savings	450	$=\dfrac{450}{3600} = 0.125$ or 12.5%	12.5% of 360° $= 0.125(360°)$ $= 45°$
Other	270	$=\dfrac{270}{3600} = 0.075$ or 7.5%	7.5% of 360° $= 0.075(360°)$ $= 27°$

Now construct the circle graph. Use the degree measures found in the table for each sector. Label the graph with the percent for each sector (Figure 9-12).

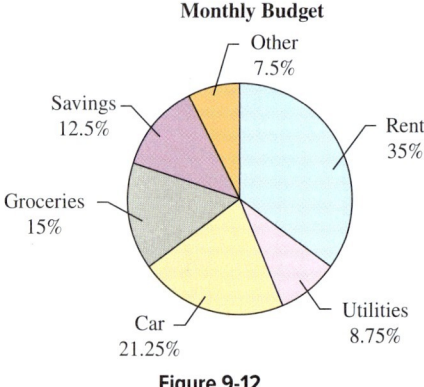

Figure 9-12

Answer

7.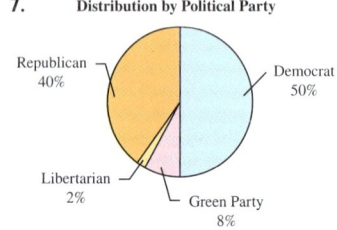

Skill Practice

7. A sample of voters in Oregon was asked to identify the political party to which they belonged. Construct a circle graph. Label each sector of the graph with the percent that it represents.

Political Affiliation	Number
Democrat	900
Republican	720
Libertarian	36
Green Party	144

Practice Exercises — Section 9.3

Prerequisite Review

R.1. What percent of 360 is 72?

R.2. 162 is what percent of 360?

R.3. What is 65% of 360?

R.4. 28% of 360 is what number?

Vocabulary and Key Concepts

1. A _____ graph or pie graph illustrates how a whole amount is divided into parts.
2. Each part of a circle graph is called a _____.
3. The size of each sector in a circle graph relates to the _____ of the whole it represents.
4. The sum of the percentages in a circle graph always equals _____ %.

Concept 1: Interpreting Circle Graphs

For Exercises 5–12, refer to the graph. The graph represents the number of traffic fatalities by age group in the United States. (*Source:* U.S. Bureau of the Census) **(See Example 1.)**

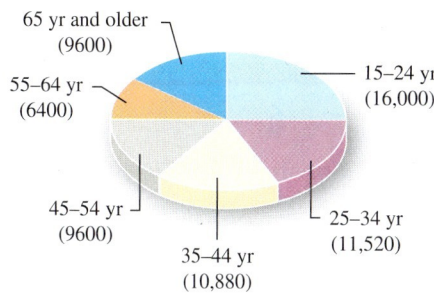

Number of U.S. Traffic Fatalities

65 yr and older (9600); 15–24 yr (16,000); 55–64 yr (6400); 45–54 yr (9600); 35–44 yr (10,880); 25–34 yr (11,520)

5. What is the total number of traffic fatalities?
6. Which of the age groups has the most fatalities?
7. How many more people died in the 25–34 age group than in the 35–44 age group?
8. How many more people died in the 45–54 age group than in the 55–64 age group?
9. What percent of the deaths were from the 15–24 age group?
10. What percent of the deaths were from the 65 and older age group?
11. How many times more deaths were from the 15–24 age group than the 55–64 age group?
12. How many times more deaths were from the 25–34 age group than from the 65 and older age group?

For Exercises 13–18, refer to the figure. The figure represents the number of different types of bicycles sold one year by a popular sporting goods store.

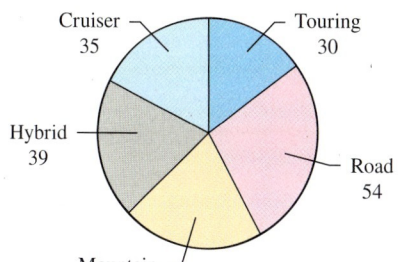

Number of Bicycles Sold

Cruiser 35; Touring 30; Hybrid 39; Road 54; Mountain 42

13. How many bicycles are represented?
14. How many bicycles were sold in the most popular category?
15. How many times more road bikes were sold than touring bikes?
16. How many times more mountain bikes were sold than cruisers?
17. What percentage of the bicycles sold were touring bikes?
18. What percentage of the bicycles sold were mountain bikes?

Concept 2: Circle Graphs and Percents

For Exercises 19–22, use the graph representing the type of music CDs found in a store containing approximately 8000 CDs. **(See Example 2.)**

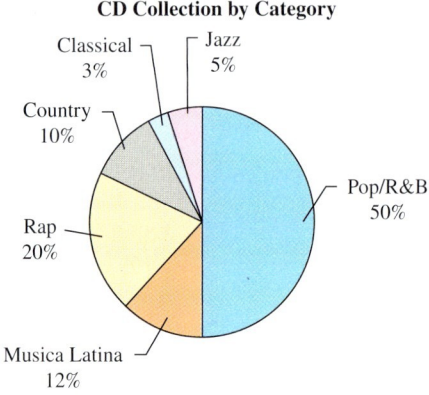

CD Collection by Category

19. How many CDs are musica Latina?

20. How many CDs are rap?

21. How many CDs are jazz or classical?

22. How many CDs are *not* Pop/R&B?

For Exercises 23–26, use the graph representing the type of health insurance coverage for individuals in a selected city with 75,000 people.

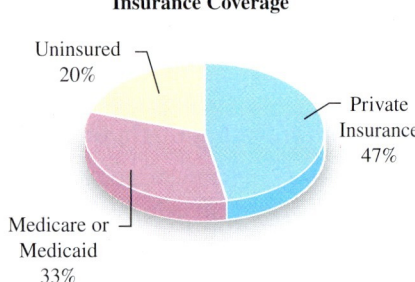

Insurance Coverage

23. How many people have private insurance? Round to the nearest thousand.

24. How many people are covered by Medicare or Medicaid? Round to the nearest thousand.

25. How many people are uninsured? Round to the nearest thousand.

26. How many people are insured? Round to the nearest thousand.

Concept 3: Constructing Circle Graphs

For Exercises 27–34, use a protractor to construct an angle of the given measure.

27. 20°   28. 70°   29. 125°

30. 270°   31. 195°   32. 5°

33. 300°   34. 90°

35. Draw a circle and divide it into sectors of 30°, 60°, 100°, and 170°.

36. Draw a circle and divide it into sectors of 125°, 180°, and 55°.

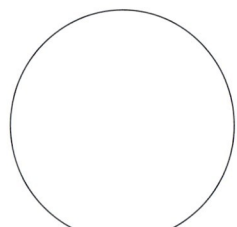

37. The Sunshine Nursery sells flowering plants, shrubs, ground cover, trees, and assorted flower pots. Construct a circle graph to show the distribution of the types of purchases.

Types of Purchases	Percent of Distribution
Flowering plants	45%
Shrubs	13%
Ground cover	18%
Trees	20%
Flower pots	4%

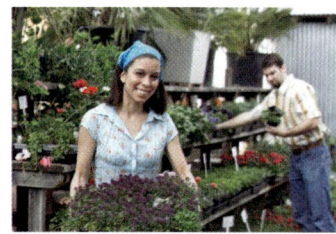
Jack Hollingsworth/Photodisc/Getty Images

Sunshine Nursery Distribution of Sales

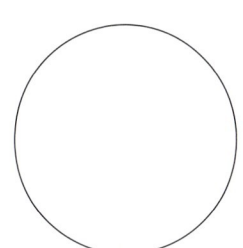

38. The party affiliation of registered Latino voters for a recent year is as follows:

45% Democrat   20% Republican

13% Other   22% Independent

Construct a circle graph from this information.

Party Affiliation of Latino Voters

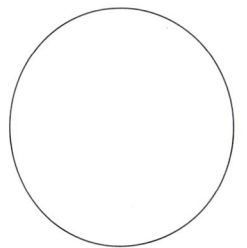

39. The table provided gives the expenses for one semester at college. (See Example 3.)

a. Complete the table.

	Expenses	Percent	Number of Degrees
Tuition	$9000		
Books	600		
Housing	2400		

b. Construct a circle graph to display the college expenses. Label the graph with percents.

College Expenses for a Semester

40. The table provided gives the number of establishments for three large pizza chains.

a. Complete the table.

	Number of Stores	Percent	Number of Degrees
Pizza Hut	8100		
Domino's	7200		
Papa Johns	2700		

Barry Barker/McGraw Hill

b. Construct a circle graph. Label the graph with percents.

Percent of Pizza Establishments

Section 9.4 Introduction to Probability

Concepts

1. Basic Definitions
2. Probability of an Event
3. Estimating Probabilities from Observed Data
4. Complementary Events

1. Basic Definitions

The probability of an event measures the likelihood of the event to occur. It is of particular interest because of its application to everyday life.

- The probability of picking the winning six-number combination for the New York lotto grand prize is $\frac{1}{45,057,474}$.
- Genetic DNA analysis can be used to determine the risk that a child will be born with cystic fibrosis. If both parents test positive, the probability is 25% that a child will be born with cystic fibrosis.

To begin our discussion, we must first understand some basic definitions.

An activity with observable outcomes such as flipping a coin or rolling a die is called an **experiment**. The collection (or set) of all possible outcomes of an experiment is called the **sample space** of the experiment.

Example 1 Determining the Sample Space of an Experiment

a. Suppose a single die is rolled. Determine the sample space of the experiment.

b. Suppose a coin is flipped. Determine the sample space of the experiment.

Solution:

a. A die is a single six-sided cube on which each side has between 1 and 6 dots painted on it. When the die is rolled, any of the six sides may come up.

The sample space is {1, 2, 3, 4, 5, 6}.   Notice that the symbols { } (called *set braces*) are used to enclose the elements.

Mark Dierker/ McGraw Hill

b. The coin may land as a head H or as a tail T. The sample space is {H, T}.

Skill Practice

1. Suppose one ball is selected from those shown and the color is recorded. Write the sample space for this experiment.

2. For an individual birth, the gender of the baby is recorded. Determine the sample space for this experiment.

Answers
1. {red, green, blue, yellow}
2. {male, female}

2. Probability of an Event

Any part of a sample space is called an **event**. For example, if we roll a die, the event of rolling number 5 or a greater number consists of the outcomes 5 and 6. In mathematics, we measure the likelihood of an event to occur by its probability.

> **Probability of an Event**
>
> $$\text{Probability of an event} = \frac{\text{number of elements in event}}{\text{number of elements in sample space}}$$

Avoiding Mistakes

From the definition, a probability value can never be negative or greater than 1.

Example 2 — Finding the Probabilities of Events

a. Find the probability of rolling a 5 or greater on a die.

b. Find the probability of flipping a coin and having it land as heads.

Solution:

a. The event can occur in 2 ways: The die lands as a 5 or 6. The sample space has 6 elements: {1, 2, 3, 4, 5, 6}.

The probability of rolling a 5 or greater: $\frac{2}{6}$ ← number of ways to roll a 5 or greater
← number of elements in the sample space

$= \frac{1}{3}$ Simplify to lowest terms.

Brian Hagiwara/Brand X Pictures/PunchStock

b. The event can occur in 1 way (the coin lands head side up). The sample space has 2 outcomes, heads or tails: {H, T}.

The probability of flipping a head on a coin: $\frac{1}{2}$ ← number of ways to get heads
← number of elements in the sample space

Skill Practice

3. Find the probability of selecting a yellow ball from those shown.
4. Find the probability of a random birth resulting in a girl.

The value of a probability can be written as a fraction, as a decimal, or as a percent. For example, the probability of a coin landing as heads is $\frac{1}{2}$ or 0.5 or 50%. In words, this means that if we flip a coin many times, theoretically we expect one-half (50%) of the outcomes to land as heads.

Example 3 — Finding Probabilities

A class has 4 freshmen, 12 sophomores, and 6 juniors. If one individual is selected at random from the class, find the probability of selecting

a. A sophomore
b. A junior
c. A senior

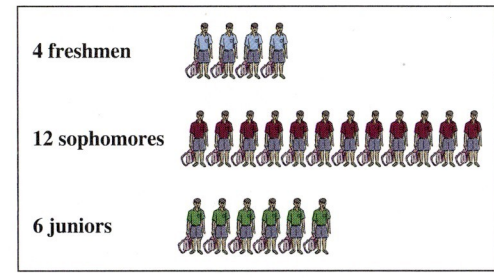

Answers

3. $\frac{1}{4}$ or 0.25   4. $\frac{1}{2}$ or 0.5

Solution:

In this case, there are 22 members of the class (4 freshmen + 12 sophomores + 6 juniors). This means that the sample space has 22 elements.

a. There are 12 sophomores in the class. The probability of selecting a sophomore is

$$\frac{12}{22}$$ There are 12 sophomores out of 22 people in the sample space.

$$=\frac{6}{11}$$ Simplify to lowest terms.

b. There are 6 juniors out of 22 people in the sample space. The probability of selecting a junior is

$$\frac{6}{22} \quad \text{or} \quad \frac{3}{11}$$

c. There are no seniors in the class. The probability of selecting a senior is

$$\frac{0}{22} \quad \text{or} \quad 0$$

A probability of 0 indicates that the event is impossible. It is impossible to select a senior from a class that has no seniors.

Skill Practice A group of registered voters has 9 Republicans, 8 Democrats, and 3 Independents. Suppose one person from the group is selected at random.

5. What is the probability that the person is a Democrat?
6. What is the probability that the person is an Independent?
7. What is the probability that the person is registered with the Libertarian Party?

From the definition of the probability of an event, it follows that the value of a probability must be between 0 and 1, inclusive. An event with a probability of 0 is called an *impossible event*. An event with a probability of 1 is called a *certain event*.

3. Estimating Probabilities from Observed Data

We were able to compute the probabilities in Examples 2 and 3 because the sample space was known. Sometimes we need to collect information to help us estimate probabilities.

Example 4 Estimating Probabilities from Observed Data

A dental hygienist records the number of times a day her patients say that they brush their teeth. Table 9-7 displays the results.

Table 9-7

Number of Times of Brushing Teeth per Day	Frequency
1	6
2	10
3	4
More than 3	1

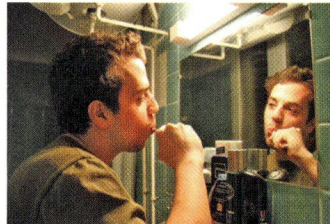
Erica Simone Leeds

If one of her patients is selected at random,

a. What is the probability of selecting a patient who brushes only one time a day?
b. What is the probability of selecting a patient who brushes more than once a day?

Answers

5. $\frac{2}{5}$ or 0.4 **6.** $\frac{3}{20}$ or 0.15 **7.** 0

Solution:

a. The table shows that there are 6 patients who brush once a day. To get the total number of patients we add all of the frequencies $(6 + 10 + 4 + 1 = 21)$. The probability of selecting a patient who brushes only once a day is

$$\frac{6}{21} \quad \text{or} \quad \frac{2}{7}$$

b. To find the number of patients who brush more than once a day, we add the frequencies for the patients who brush 2 times, 3 times, and more than 3 times $(10 + 4 + 1 = 15)$. The probability of selecting a patient who brushes more than once a day is

$$\frac{15}{21} \quad \text{or} \quad \frac{5}{7}$$

Skill Practice Refer to Table 9-7 in Example 4.

8. What is the probability of selecting a patient who brushes twice a day?
9. What is the probability of selecting a patient who brushes more than twice a day?

4. Complementary Events

The events in Example 4(a) and 4(b) are called complementary events. The **complement of an event** is the set of all elements in the sample space that are not in the event. In this case, the number of patients who brush once a day and the number of patients who brush more than once a day make up the entire sample space, yet do not overlap. For this reason, the probability of an event plus the probability of its complement is 1. For Example 4, we have $\frac{2}{7} + \frac{5}{7} = \frac{7}{7} = 1$.

Example 5 Finding the Probability of Complementary Events

Find the indicated probability.

a. The probability of getting a winter cold is $\frac{3}{10}$. What is the probability of *not* getting a winter cold?

b. If the probability that a washing machine will break before the end of the warranty period is 0.0042, what is the probability that a washing machine will *not* break before the end of the warranty period?

Solution:

a. The sum of the probability of an event and the probability of its complement must equal 1. Therefore, we have an equation with the probability of the complement, c, unknown.

$$\frac{3}{10} + c = 1$$

$$c = 1 - \frac{3}{10} \quad \text{Solve for } c.$$

$$= \frac{7}{10}$$

There is a $\frac{7}{10}$ chance (70%) of *not* getting a winter cold.

Answers

8. $\frac{10}{21}$ 9. $\frac{5}{21}$

b. The probability that a washing machine will break before the end of the warranty period is 0.0042. Then the probability that a machine will *not* break before the end of the warranty period is the probability of the complement, c.

$$0.0042 + c = 1$$
$$c = 1 - 0.0042 \quad \text{Solve for } c.$$
$$= 0.9958$$

The probability that the machine will *not* break before the end of the warranty period is 0.9958 or 99.58%.

Skill Practice

10. For one particular medicine, the probability that a patient will experience side effects is $\frac{1}{20}$. What is the probability that a patient will *not* experience side effects?

11. The probability that a flight arrives on time is 0.18. What is the probability that a flight will *not* arrive on time?

Answers
10. $\frac{19}{20}$ **11.** 0.82

Section 9.4 Activity

A.1. Flipping a coin is an *experiment*, an activity with an observable outcome.
  a. What is the sample space (possible outcomes) of flipping a coin? Use "H" for heads and "T" for tails, and use set braces to enclose the elements.
  b. How many elements are in the sample space of flipping a coin?
  c. How many ways can the event of flipping a coin and having it land on heads happen?
  d. Find the probability of flipping a coin and having it land on heads. Write the answer as a fraction and a percentage.

$$\text{Probability of an event} = \frac{\text{number of elements in an event}}{\text{number of elements in sample space}}$$

A.2. Maximum and Shelia are playing dice with a 10-sided number die (1-10).
  a. What is the sample space of rolling the die? Use set braces to enclose the elements.
  b. How many ways can the event of rolling a 2 or less occur?
  c. Calculate the probability of rolling a 2 or less on a 10-sided die. Write the answer as a fraction in simplest form.

A.3. Twenty students in a Business Law class were given the opportunity to vote on the day that their next exam would take place. The table below displays the results.

Day of the Week	Number of Votes
Monday	2
Wednesday	5
Friday	13

If one of the votes is selected at random,
  a. What is the probability of selecting Monday as the exam date? Write the answer as a percent.
  b. What is the probability of selecting Friday as the exam date? Write the answer as a percent.
  c. What is the probability of *not* selecting Friday as the exam date? Write the answer as a percent.
  d. What is the sum of the probability of selecting Friday and the probability of *not* selecting Friday as the exam date? Write the answer as a percent and as a decimal number.
  e. The probabilities of an event and its complement add up to _____.

The table displays the number of days for 125 patients to receive their test results from the date of testing. Express the answers as a percent rounded to the nearest tenth of a percent if necessary.

A.4.

Number of Days to Receive Test Result	Frequency
2	12
3	26
4	31
5	34
6	14
7	8

a. What is the probability that a patient will receive their test results in less than 5 days?
b. What is the probability that a patient will receive their test results 7 days after being tested?
c. What is the probability of a patient receiving their test results before 7 days after being tested?
d. What is the probability that a patient will receive their test result in 8 or more days?

Practice Exercises

Section 9.4

Prerequisite Review

For Exercises R.1–R.2, which values are between 0 and 1, inclusive?

R.1. a. 212%   b. $\dfrac{37}{36}$ c. 5.16%   **R.2.** a. 46%   b. $\dfrac{36}{37}$ c. 100%

R.3. If 9 out of 36 people in a class earn an "A," what percent is this?

R.4. If 114 people out of 300 people have type O-positive blood, what percent is this?

Vocabulary and Key Concepts

1. a. An activity with observable outcomes is called an _____.
   b. The set of all possible outcomes of an experiment is called the _____ space.
   c. The _____ of an event is the ratio of the number of elements in the event to the number of elements in the sample space.
   d. The _____ of an event is the set of all elements in the sample space that are not in the event.
   e. The sum of the probability of an event and the probability of its complement is _____.
   f. An event with probability 0 is called an _____ event.
   g. A certain event has a probability of _____.

2. The sample space for a six-sided die when rolled once is _____.

3. The sample space for a coin flipped once is _____.

4. The probability of an event (can/cannot) be negative.

5. The probability of an event (can/cannot) be greater than 1.

6. If the probability of an event is $\dfrac{1}{2}$, then the probability of the complement of the event is _____.

Concept 1: Basic Definitions

7. A card is chosen from a deck consisting of 10 cards numbered 1–10. Determine the sample space of this experiment. (See Example 1.)

8. A marble is chosen from a jar containing a yellow marble, a red marble, a blue marble, a green marble, and a white marble. Determine the sample space of this experiment.

9. Two dice are thrown, and the sum of the top sides is observed. Determine the sample space of this experiment.

10. A coin is tossed twice. Determine the sample space of this experiment.

11. If a die is rolled, in how many ways can an odd number come up?

12. If a die is rolled, in how many ways can a number less than 6 come up?

Concept 2: Probability of an Event

13. Which of the values can represent the probability of an event?

    a. 1.62   b. $-\dfrac{7}{5}$ c. 0   d. 1   e. 200%   f. 4.5   g. 4.5%   h. 0.87

14. Which of the values can represent the probability of an event?

    a. 1.5   b. 0   c. $\dfrac{2}{3}$ d. 1   e. 150%   f. 3.7   g. 3.7%   h. 0.92

15. If a single die is rolled, what is the probability that it will come up as a number less than 3? (See Example 2.)

16. If a single die is rolled, what is the probability that it will come up as a number greater than 5?

17. If a single die is rolled, what is the probability that it will come up with an even number?

18. If a single die is rolled, what is the probability that it will come up as an odd number?

For Exercises 19–22, refer to the figure. A sock drawer contains 2 white socks, 5 black socks, and 1 blue sock. (See Example 3.)

19. What is the probability of choosing a black sock from the drawer?

20. What is the probability of choosing a white sock from the drawer?

21. What is the probability of choosing a blue sock from the drawer?

22. What is the probability of choosing a purple sock from the drawer?

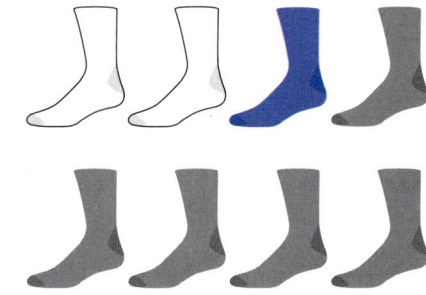

23. If a die is tossed, what is the probability that a number from 1 to 6 will come up?

24. If a die is tossed, what is the probability of getting a 7?

25. What is an impossible event?

26. What is the sum of the probabilities of an event and its complement?

27. In a deck of cards there are 12 face cards and 40 cards with numbers. What is the probability of selecting a face card from the deck?

28. In a deck of cards, 13 are diamonds, 13 are spades, 13 are clubs, and 13 are hearts. Find the probability of selecting a diamond from the deck.

29. A jar contains 7 yellow marbles, 5 red marbles, and 4 green marbles. What is the probability of selecting a red marble or a yellow marble?

30. A jar contains 10 black marbles, 12 white marbles, and 4 blue marbles. What is the probability of selecting a blue marble or a black marble?

Concept 3: Estimating Probabilities from Observed Data

31. The table displays the length of stay for vacationers at a small motel. (See Example 4.)

Kim Steele/Getty Images

Length of Stay in Days	Frequency
2	14
3	13
4	18
5	28
6	11
7	30
8	6

  a. What is the probability that a vacationer will stay for 4 days?

  b. What is the probability that a vacationer will stay for less than 4 days?

  c. Based on the information from the table, what percent of vacationers stay for more than 6 days?

32. A number of students at a large university were asked if they owned a car and if they lived in a dorm or off campus. The table shows the results.

  a. What is the probability that a student selected at random lives in a dorm?

  b. What is the probability that a student selected at random does not own a car?

	Number of Car Owners	Number Who Do Not Own a Car
Dorm resident	32	88
Lives off campus	59	26

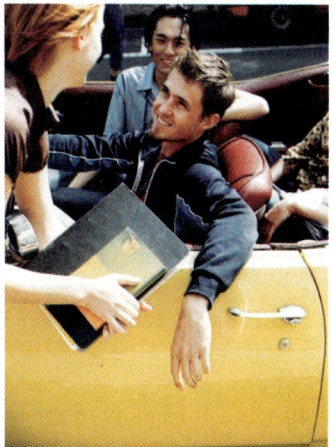

Stockbyte/Getty Images

33. A survey was made of 60 participants, asking if they drive an American-made car, a Japanese car, or a car manufactured in another foreign country. The table displays the results.

  a. What is the probability that a randomly selected car is manufactured in America?

  b. What percent of cars is manufactured in some country other than Japan?

	Frequency
American	21
Japanese	30
Other	9

34. The number of customer complaints for service representatives at a small company is given in the table. If one representative is picked at random, find the probability that the representative received

  a. Exactly 3 complaints.

  b. Between 1 and 5 complaints, inclusive.

  c. At least 4 complaints.

  d. More than 5 complaints or fewer than 2 complaints.

Number of Complaints	Number of Representatives
0	4
1	2
2	14
3	10
4	16
5	18
6	10
7	6

35. Mr. Gutierrez noted the times in which his students entered his classroom and constructed the following chart.

    a. What is the probability that a student will be early to class?

    b. What is the probability that a student will be late to class?

    c. What percent of students arrive on time or early? Round to the nearest whole percent.

Time	Number of Students
About 10 min early	1
About 5 min early	6
On time	11
About 5 min late	7
About 10 min late	3
About 15 min late	1

36. Each person at an office party purchased a raffle ticket. The graph shows the results of the raffle.

    a. How many people bought raffle tickets?

    b. What is the probability of winning the TV set?

    c. What percent of people won some type of prize?

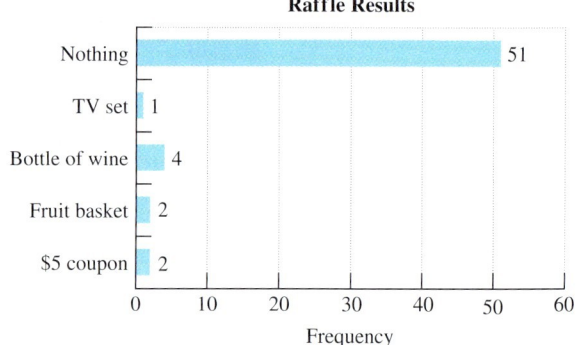

Raffle Results

- Nothing: 51
- TV set: 1
- Bottle of wine: 4
- Fruit basket: 2
- $5 coupon: 2

Frequency

Concept 4: Complementary Events

37. If the probability of the horse Lightning Bolt to win a race is $\frac{2}{11}$, what is the probability that he will not win? (See Example 5.)

38. If the probability of being hit by lighting is $\frac{1}{1,000,000}$, what is the probability of not getting hit by lighting?

39. If the probability of having twins is 1.2%, what is the probability of not having twins?

40. The probability of a woman's surviving breast cancer is 88%. What is the probability that a woman would not survive breast cancer?

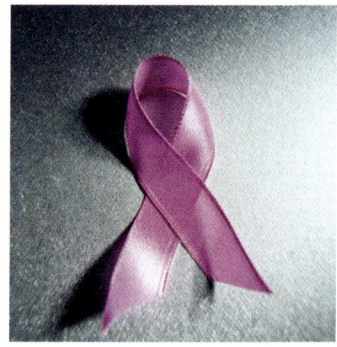

Photodisc/Getty Images

Section 9.5 Mean, Median, and Mode

Concepts

1. Mean
2. Median
3. Mode
4. Weighted Mean

1. Mean

When given a list of numerical data, it is often desirable to obtain a single number that represents the central value of the data. In this section, we discuss three such values called the mean, median, and mode. We have already introduced the concept of the mean of a set of numbers and will review the definition here.

> **Definition of Mean**
>
> The **mean** (or average) of a set of numbers is the sum of the values divided by the number of values. We can write this as a formula.
>
> $$\text{Mean} = \frac{\text{sum of the values}}{\text{number of values}}$$

Section 9.5 Mean, Median, and Mode   573

Example 1   Finding the Mean of a Data Set

A small business employs five workers. Their yearly salaries are

$42,000 $36,000 $45,000 $35,000 $38,000

a. Find the mean yearly salary for the five employees.
b. Suppose the owner of the business makes $218,000 per year. Find the mean salary for all six individuals (that is, include the owner's salary).

Solution:

a. Mean salary of five employees

$$= \frac{42{,}000 + 36{,}000 + 45{,}000 + 35{,}000 + 38{,}000}{5}$$

$$= \frac{196{,}000}{5} \quad \text{Add the data values.}$$

$$= 39{,}200 \quad \text{Divide.}$$

The mean salary for employees is $39,200.

Avoiding Mistakes

When computing a mean remember that the data are added first before dividing.

b. Mean of all six individuals

$$= \frac{42{,}000 + 36{,}000 + 45{,}000 + 35{,}000 + 38{,}000 + 218{,}000}{6}$$

$$= \frac{414{,}000}{6}$$

$$= 69{,}000$$

The mean salary with the owner's salary included is $69,000.

Skill Practice Housing prices for five homes in one neighborhood are given.

$108,000 $149,000 $164,000 $118,000 $144,000

1. Find the mean price of these five houses.
2. Suppose a new home is built in the neighborhood for $1.3 million ($1,300,000). Find the mean price of all six homes.

2. Median

In Example 1, you may have noticed that the mean salary was greatly affected by the unusually high value of $218,000. For this reason, you may want to use a different measure of "center" called the median. The **median** is the "middle" number in an ordered list of numbers.

Finding the Median

To compute the median of a list of numbers, first arrange the numbers in order from least to greatest.
- If the number of data values in the list is *odd*, then the median is the middle number in the list.
- If the number of data values is *even*, there is no single middle number. Therefore, the median is the mean (average) of the two middle numbers in the list.

Answers
1. $136,600   2. $330,500

Example 2 Finding the Median of a Data Set

Consider the salaries of the five workers from Example 1.

$42,000 $36,000 $45,000 $35,000 $38,000

a. Find the median salary for the five workers.
b. Find the median salary including the owner's salary of $218,000.

Solution:

a.   35,000   36,000   **38,000**   42,000   45,000   Arrange the data in order.

Because there are five data values (an *odd* number), the median is the middle number.

The median is $38,000.

b. Now consider the scores of all six individuals (including the owner). Arrange the data in order.

   35,000   36,000   **38,000**   **42,000**   45,000   218,000

$$\frac{38,000 + 42,000}{2}$$

There are six data values (an *even* number). The median is the average of the two middle numbers.

$$= \frac{80,000}{2}$$

Add the two middle numbers.

$$= 40,000$$

Divide.

The median of all six salaries is $40,000.

Avoiding Mistakes
The data must be arranged in order before determining the median.

Skill Practice

3. Find the median of the five housing prices given in Skill Practice 1.

$108,000 $149,000 $164,000 $118,000 $144,000

4. Find the median of the six housing prices given in Skill Practice 2.

$108,000 $149,000 $164,000
$118,000 $144,000 $1,300,000

In Examples 1 and 2, the mean of all six salaries is $69,000, whereas the median is $40,000. These examples show that the median is a better representation for a central value when the data list has an unusually high (or low) value.

Answers
3. $144,000   **4.** $146,500

Example 3 — Finding the Median of a Data Set

The average temperatures (in °C) for the South Pole are given by month. Find the median temperature. (*Source:* NOAA)

Frank Krahmer/Photographer's Choice/Getty Images

Jan.	Feb.	Mar.	April	May	June	July	Aug.	Sept.	Oct.	Nov.	Dec.
−2.9	−9.5	−19.2	−20.7	−21.7	−23.0	−25.7	−26.1	−24.6	−18.9	−9.7	−3.4

Solution:

First arrange the numbers in order from least to greatest.

−26.1 −25.7 −24.6 −23.0 −21.7 −20.7 −19.2 −18.9 −9.7 −9.5 −3.4 −2.9

$$\text{Median} = \frac{-20.7 + (-19.2)}{2} = -19.95$$

There are 12 data values (an even number). Therefore, the median is the average of the two middle numbers. The median temperature at the South Pole is −19.95°C.

Skill Practice

5. The gain or loss for a stock is given for an 8-day period. Find the median gain or loss.

   −2.4   −2.0   1.25   0.6   −1.8   −0.4   0.6   −0.9

Note: The median may not be one of the original data values. This was true in Example 3.

3. Mode

A third representative value for a list of data is called the mode.

> **Definition of Mode**
>
> The **mode** of a set of data is the value or values that occur most often.
> - If two values occur most often we say the data are **bimodal**.
> - If more than two values occur most often, we say there is no mode.

Answer

5. −0.65

Example 4 Finding the Mode of a Data Set

The student-to-teacher ratio is given for elementary schools for ten selected states. For example, California has a student-to-teacher ratio of 20.6. This means that there are approximately 20.6 students per teacher in California elementary schools. (*Source:* National Center for Education Statistics)

David Buffington/Blend Images/Getty Images

ME	ND	WI	NH	RI	IL	IN	MS	CA	UT
12.5	13.4	14.1	14.5	14.8	16.1	16.1	16.1	20.6	21.9

Find the mode of the student-to-teacher ratio for these states.

Solution:

The data value 16.1 appears the most often. Therefore, the mode is 16.1 students per teacher.

Skill Practice

6. The monthly rainfall amounts for Houston, Texas, are given in inches. Find the mode. (*Source:* NOAA)

   4.5   3.0   3.2   3.5   5.1   6.8   4.3   4.5   5.6   5.3   4.5   3.8

Example 5 Finding the Mode of a Data Set

Find the mode of the list of average monthly temperatures for Albany, New York. Values are in °F.

Jan.	Feb.	March	April	May	June	July	Aug.	Sept.	Oct.	Nov.	Dec.
22	25	35	47	58	66	71	69	61	49	39	26

Solution:

No data value occurs most often. There is no mode for this set of data.

Skill Practice

7. Find the mode of the weights of babies (in pounds) born one day at Brackenridge Hospital in Austin, Texas.

   7.2   8.1   6.9   9.3   8.3   7.7   7.9   6.4   7.5

Answers

6. 4.5 in.   **7.** No mode

Example 6 — Finding the Mode of a Data Set

The grades for a quiz in college algebra are as follows. The scores are out of a possible 10 points. Find the mode of the scores.

9	4	6	9	9	8	2	1	4	9
5	10	10	5	7	7	9	8	7	3
9	7	10	7	10	1	7	4	5	6

Solution:

Sometimes arranging the data in order makes it easier to find the repeated values.

1	1	2	3	4	4	4	5	5	5
6	6	7	7	7	7	7	7	8	8
9	9	9	9	9	9	10	10	10	10

The score of 9 occurs 6 times. The score of 7 occurs 6 times. There are two modes, 9 and 7, because these scores both occur more than any other score. We say that these data are *bimodal*.

Skill Practice

8. The ages of children participating in an afterschool sports program are given. Find the mode(s).

13	15	17	15	14	15	16	16	15	16	12	13
15	14	16	15	15	16	16	13	16	13	14	18

TIP: To remember the difference between median and mode, think of the *median* of a highway that goes down the *middle*. Think of the word *mode* as sounding similar to the word *most*.

4. Weighted Mean

Sometimes data values in a list appear multiple times. In such a case, we can compute a **weighted mean**. In Example 7, we demonstrate how to use a weighted mean to compute a grade point average (GPA). To compute GPA, each grade is assigned a numerical value. For example, an "A" is worth 4 points, a "B" is worth 3 points, and so on. Then each grade for a course is "weighted" by the number of credit-hours that the course is worth.

Example 7 — Using a Weighted Mean to Compute GPA

At a certain college, the grades A–F are assigned numerical values as follows.

$$A = 4.0 \quad B+ = 3.5 \quad B = 3.0 \quad C+ = 2.5$$
$$C = 2.0 \quad D+ = 1.5 \quad D = 1.0 \quad F = 0.0$$

Elmer takes the following classes with the grades as shown. Determine Elmer's GPA.

Course	Grade	Number of Credit-Hours
Prealgebra	A = 4 pts	3
Study Skills	C = 2 pts	1
First Aid	B+ = 3.5 pts	2
English I	D = 1 pt	4

Solution:

The data in the table can be visualized as follows.

4 pts	4 pts	4 pts	2 pts	3.5 pts	3.5 pts	1 pt	1 pt	1 pt	1 pt
A	A	A	C	B+	B+	D	D	D	D
3 of these			1 of these	2 of these		4 of these			

Answer

8. There are two modes, 15 and 16.

The number of grade points earned for each course is the product of the grade for the course and the number of credit-hours for the course. For example:

Grade points for Prealgebra: (4 pts)(3 credit-hours) = 12 points.

Course	Grade	Number of Credit-Hours (Weights)	Product Number of Grade Points
Prealgebra	A = 4 pts	3	(4 pts)(3 credit-hours) = 12 pts
Study Skills	C = 2 pts	1	(2 pts)(1 credit-hour) = 2 pts
First Aid	B+ = 3.5 pts	2	(3.5 pts)(2 credit-hours) = 7 pts
English I	D = 1 pt	4	(1 pt)(4 credit-hours) = 4 pt
		Total hours: 10	Total grade points: 25 pts

To determine GPA, we will add the number of grade points earned for each course and then divide by the total number of credit-hours taken.

$$\text{Mean} = \frac{25}{10} = 2.5 \quad \text{Elmer's GPA for this term is 2.5.}$$

Skill Practice

9. Clyde received the following grades for the semester. Use the numerical values assigned to grades from Example 7 to find Clyde's GPA.

Course	Grade	Credit-Hours
Math	B+	4
Science	C	3
Speech	A	3

In Example 7, notice that the value of each grade is "weighted" by the number of credit-hours. The grade of "A" for Prealgebra is weighted 3 times. The grade of "C" for the study skills course is weighted 1 time. The grade that hurt Elmer's GPA was the "D" in English. Not only did he receive a low grade, but the course was weighted heavily (4 credit-hours). In Exercise 47, we recompute Elmer's GPA with a "B" in English to see how this grade affects his GPA.

Answer
9. Clyde's GPA is 3.2.

Section 9.5 Activity

A.1. Amira recorded the number of calories she burned during her workouts as shown in the table below:

	Monday	Tuesday	Wednesday	Thursday	Friday
Calories Burned	274	309	362	378	237

a. Calculate the total number of calories Amira burned.
b. How many days did she work out?
c. Calculate the mean number of calories Amira burns by dividing the total number of calories by the number of days she worked out.
d. Write the formula for calculating the mean of a set of numbers.

A.2. The Red Raider basketball team is having its best season in school history. They have beaten their opponents by 23, 20, 18, 21, and 22 points in their last five games.

   a. Order the numbers from least to greatest.
   b. Find the number in the middle of the list. This is the median of the data values.
   c. In the sixth game of the season, the team won by 14 points. Order the numbers for the six games from least to greatest.
   d. Since there is an even number of data values, find the mean of the two middle numbers in the list. This is the median of the data values.

A.3. The number of touchdowns of an NFL quarterback for the last 10 years are:

$$28, 30, 28, 45, 39, 17, 38, 31, 40, 16$$

   a. Is there a value that occurs more often than the other values? If so, what is that value? The mode of the data set is the value or values that occur most often.
   b. Assume in his eleventh year, the quarterback has 38 touchdowns. Is the mode the same as in part (a)? Why or why not?

A.4. Hibeto is taking the following classes with the grades as shown. Determine Hibeto's GPA.

Course	Grade	Number of Credit-Hours	Product: Number of Grade Points
Prealgebra	A = 4 points	3	
College Writing	C = 2 points	3	
Intro to Marketing	B = 3 points	4	
		Total Hours:	Total Grade Pts:

   a. Calculate the total number of credit-hours in column 3.
   b. Calculate the number of grade points for each class in column 4 by multiplying the grade points times the number of credit hours.
   c. Calculate the total grade points by adding the number of grade points earned for each class in column 4.
   d. Divide the total grade points from part (c) by the total hours from part (a). This weighted mean is the resulting GPA.

A.5. The Humane Society has 5 cats available for adoption. While in the care of the Humane Society, they need to determine how much food is necessary to feed the cats. This is based on their weight.

Cat	Weight (lb)
Vader	8
Tinkerbell	11
Raven	9
Snickers	10
Levi	8

   a. Find the mean weight of the cats.
   b. Find the median weight of the cats.
   c. Find the mode of the cats' weight.

Section 9.5 Practice Exercises

Prerequisite Review

For Exercises R.1–R.4, simplify.

R.1. $(24.7 + 36.5) \div 2$

R.2. $(200 + 160) \div 2$

R.3. $(125 + 300 + 120 + 79) \div 4$

R.4. $(-12 - 82 - 200) \div 3$

Vocabulary and Key Concepts

1. **a.** The _____ or average of a set of numbers is the sum of the values divided by the number of values.

   b. The _____ of an *odd* number of data values ranked in order from least to greatest is the "middle" number in the list.

   c. To find the median of an *even* number of data values ranked in order from least to greatest, find the _____ (or average) of the two middle numbers in the list.

   d. The _____ of a list of data values is the value that occurs most often.

   e. A _____ mean is a mean where each data value is weighted according to the number of times it appears in the list.

Concept 1: Mean

For Exercises 2–7, find the mean of each set of numbers. (See Example 1.)

2. 4, 6, 5, 10, 4, 5, 8

3. 3, 8, 5, 7, 4, 2, 7, 4

4. 0, 5, 7, 4, 7, 2, 4, 3

5. 7, 6, 5, 10, 8, 4, 8, 6, 0

6. −10, −13, −18, −20, −15

7. −22, −14, −12, −16, −15

8. Compute the mean of your test scores for this class up to this point.

9. The flight times in hours for six flights between New York and Los Angeles are given. Find the mean flight time. Round to the nearest tenth of an hour.

   5.5, 6.0, 5.8, 5.8, 6.0, 5.6

   Digital Vision/Getty Images

10. A nurse takes the temperature of a patient every hour and records the temperatures as follows: 98°F, 98.4°F, 98.9°F, 100.1°F, and 99.2°F. Find the patient's mean temperature.

    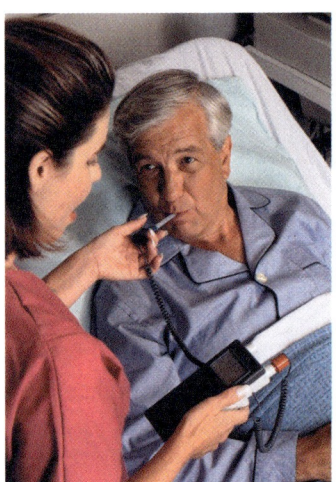
    Fuse/Digital Stock/Getty Images

11. The number of Calories for six different chicken sandwiches and chicken salads is given in the table.

    a. What is the mean number of Calories for a chicken sandwich? Round to the nearest whole unit.

    b. What is the mean number of Calories for a salad with chicken? Round to the nearest whole unit.

    c. What is the difference in the means?

Chicken Sandwiches	Salads with Chicken
360	310
370	325
380	350
400	390
400	440
470	500

12. The heights of the players from two NBA teams are given in the table. All heights are in inches.

    a. Find the mean height for the players on the Philadelphia 76ers.

    b. Find the mean height for the players on the Milwaukee Bucks.

    c. What is the difference in the mean heights?

Philadelphia 76ers' Height (in.)	Milwaukee Bucks' Height (in.)
83	70
83	83
72	82
79	72
77	82
84	85
75	75
76	75
82	78
79	77

13. Zach received the following scores for his first four tests: 98%, 80%, 78%, 90%.

    a. Find Zach's mean test score.

    b. Zach got a 59% on his fifth test. Find the mean of all five tests.

    c. How did the low score of 59% affect the overall mean of five tests?

14. The prices of four phones are $149, $179, $249, and $199.

    a. Find the mean of these prices.

    b. A model that costs $59 is added to the list. What is the mean price of all 5 phones?

    c. How does including the price of the inexpensive phone affect the mean?

Concept 2: Median

For Exercises 15–20, find the median for each set of numbers. (See Examples 2 and 3.)

15. 16, 14, 22, 13, 20, 19, 17

16. 32, 35, 22, 36, 30, 31, 38

17. 109, 118, 111, 110, 123, 100

18. 134, 132, 120, 135, 140, 118

19. −58, −55, −50, −40, −40, −55

20. −82, −90, −99, −82, −88, −87

21. The infant mortality rates for five countries are given in the table. Find the median.

Country	Infant Mortality Rate (Deaths per 1000)
Sweden	3.93
Japan	4.10
Finland	3.82
Andorra	4.09
Singapore	3.87

22. The snowfall amounts for five winter months in Burlington, Vermont, are given in the table. Find the median.

Month	Snowfall (in.)
November	6.6
December	18.1
January	18.8
February	16.8
March	12.4

23. Jonas played eight golf tournaments, each with 72 holes of golf. His scores for the tournaments are given. Find the median score.

    −3, −5, 1, 4, −8, 2, 8, −1

24. Andrew recorded the daily low temperature (in °C) at his home in Virginia for 8 days in January. Find the median temperature.

    5, 6, −5, 1, −4, −11, −8, −5

25. The number of passengers (in millions) on nine leading airlines for a recent year is listed. Find the median number of passengers. (*Source:* International Airline Transport Association)

    48.3, 42.4, 91.6, 86.8, 46.5, 71.2, 45.4, 56.4, 51.7

26. For a recent year the number of albums sold (in millions) is listed for the 10 best sellers. Find the median number of albums sold.

    2.7, 3.0, 4.8, 7.4, 3.4, 2.6, 3.0, 3.0, 3.9, 3.2

Concept 3: Mode

For Exercises 27–32, find the mode(s) for each set of numbers. **(See Examples 4 and 5.)**

27. 4, 5, 3, 8, 4, 9, 4, 2, 1, 4

28. 12, 14, 13, 17, 19, 18, 19, 17, 17

29. −28, −21, −24, −23, −24, −30, −21

30. −45, −42, −40, −41, −49, −49, −42

31. 90, 89, 91, 77, 88

32. 132, 253, 553, 255, 552, 234

33. The table gives the monthly cost of six different health insurance plans. Find the mode.

Plan	Cost ($)
Plan A	200
Plan B	300
Plan C	350
Plan D	250
Plan E	300
Plan F	500

34. The table gives the number of hazardous waste sites for selected states. Find the mode.

State	Number of Sites
Florida	51
New Jersey	112
Michigan	67
Wisconsin	39
California	96
Pennsylvania	94
Illinois	39
New York	90

35. The unemployment rates for nine countries are given. Find the mode. **(See Example 6.)**

    6.3%, 7.0%, 5.8%, 9.1%, 5.2%, 8.8%, 8.4%, 5.8%, 5.2%

36. The list gives the number of children who were absent from class for an 11-day period. Find the mode.

    4, 1, 6, 2, 4, 4, 4, 2, 2, 3, 2

Mixed Exercises

37. Six test scores for Jonathan's history class are listed. Find the mean and median. Round to the nearest tenth if necessary. Did the mean or median give a better overall score for Jonathan's performance?

    92%, 98%, 43%, 98%, 97%, 85%

38. Nora's math test results are listed. Find the mean and median. Round to the nearest tenth if necessary. Did the mean or median give a better overall score for Nora's performance?

    52%, 85%, 89%, 90%, 83%, 89%

39. Listed below are monthly costs for leasing seven different new cars. Find the mean, median, and mode (if one exists). Round to the nearest dollar.

$312, $225, $221, $256, $308, $280, $147

40. The salaries for seven Associate Professors at a university in the midwest are listed. These are salaries for 9-month contracts for a recent year. Find the mean, median, and mode (if one exists). Round to the nearest dollar.

$104,000, $107,000, $67,750, $82,500, $73,500, $88,300, $104,000

41. The prices of 10 single-family, three-bedroom homes for sale in Santa Rosa, California, are listed for a recent year. Find the mean, median, and mode (if one exists).

$850,000, $835,000, $839,000, $829,000, $850,000, $850,000, $850,000, $847,000, $1,850,000, $825,000

42. The prices of 10 single-family, three-bedroom homes for sale in Boston, Massachusetts, are listed for a recent year. Find the mean, median, and mode (if one exists).

$300,000, $2,495,000, $2,120,000, $220,000, $194,000, $391,000, $315,000, $330,000, $435,000, $250,000

Concept 4: Weighted Mean

For Exercises 43–46, use the numerical values assigned to grades to compute GPA. Round each GPA to the hundredths place. **(See Example 7.)**

$$A = 4.0 \quad B+ = 3.5 \quad B = 3.0 \quad C+ = 2.5$$
$$C = 2.0 \quad D+ = 1.5 \quad D = 1.0 \quad F = 0.0$$

43. Compute the GPA for the following grades. Round to the nearest hundredth.

Course	Grade	Number of Credit-Hours (Weights)
Intermediate Algebra	B	4
Theater	C	1
Music Appreciation	A	3
World History	D	5

44. Compute the GPA for the following grades. Round to the nearest hundredth.

Course	Grade	Number of Credit-Hours (Weights)
General Psychology	B+	3
Beginning Algebra	A	4
Student Success	A	1
Freshman English	B	3

45. Compute the GPA for the following grades. Round to the nearest hundredth.

Course	Grade	Number of Credit-Hours (Weights)
Business Calculus	B+	3
Biology	C	4
Library Research	F	1
American Literature	A	3

46. Compute the GPA for the following grades. Round to the nearest hundredth.

Course	Grade	Number of Credit-Hours (Weights)
University Physics	C+	5
Calculus I	A	4
Computer Programming	D	3
Swimming	A	1

47. Refer to the table given in Example 7. Replace the grade of "D" in English I with a grade of "B" and compute the GPA. How did Elmer's GPA differ with a better grade in the 4-hr English class?

Expanding Your Skills

48. There are 20 students enrolled in a 12th-grade math class. The graph displays the number of students by age. First complete the table, and then find the mean.

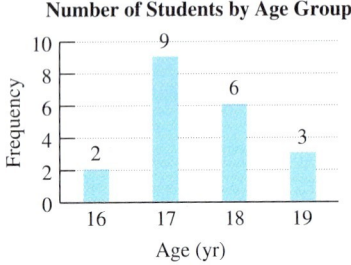

Age (yr)	Number of Students	Product
16		
17		
18		
19		
Total:		

49. A survey was made in a neighborhood of 37 houses. The graph represents the number of residents who live in each house. Complete the table and determine the mean number of residents per house.

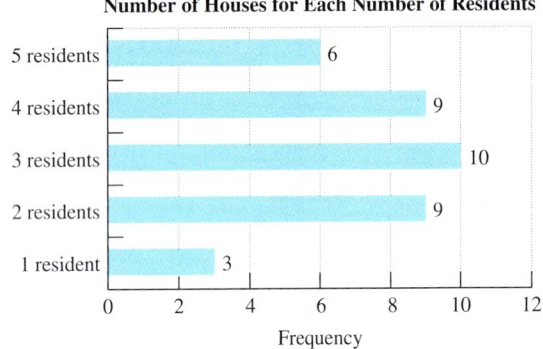

Number of Residents in Each House	Number of Houses	Product
1		
2		
3		
4		
5		
Total:		

Chapter 9 Review Exercises

Section 9.1

For Exercises 1–4, refer to the table. The table gives the number of Calories and the amount of fat, cholesterol, sodium, and total carbohydrates for a single $\frac{1}{2}$-c serving of chocolate ice cream.

Ice Cream	Calories	Fat(g)	Cholesterol (mg)	Sodium (mg)	Carbohydrate (g)
Breyers	150	8	20	35	17
Häagen-Dazs	270	18	115	60	22
Edy's Grand	150	8	25	35	17
Blue Bell	160	8	35	70	18
Godiva	290	18	65	50	28

1. Which ice cream has the most calories?

2. Which ice cream has the least amount of cholesterol?

3. How many more times the sodium does Blue Bell have per serving than Edy's Grand?

4. What is the difference in the amount of carbohydrate for Godiva and Blue Bell?

For Exercises 5–8, refer to the pictograph. The graph represents the number of tornadoes during four months with active weather.

Comstock Images/age fotostock

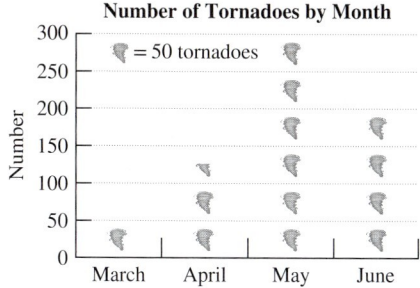

5. What does each icon represent?

6. From the graph, estimate the number of tornadoes in May.

7. Which month had approximately 200 tornadoes?

8. Estimate the difference in the number of tornadoes in April and the number in March.

For Exercises 9–12, refer to the graph. The graph represents the number of liver transplants in the United States for selected years. (*Source:* U.S. Department of Health and Human Services)

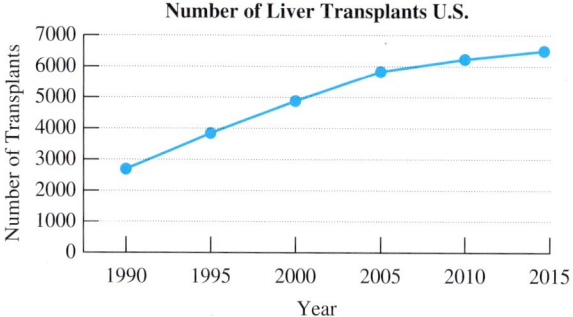

9. In which year did the greatest number of liver transplants occur?

10. Approximate the number of liver transplants for the year 2000.

11. Does the trend appear to be increasing or decreasing?

12. Extend the graph to predict the number of liver transplants for the year 2020.

13. The table shows the number of households that own four different types of pets. Construct a bar graph using horizontal bars. The length of each bar should represent the number of each type of pet represented in millions. (*Source:* American Veterinary Medical Association)

Type of Pet	Number of Households (millions)
Dog	43
Cat	36
Bird	4
Horse	2

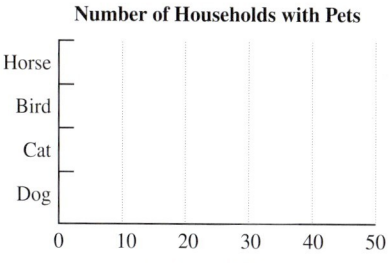

Section 9.2

Use these data for Exercises 14 and 15.

The ages of students in a Spanish class are given.

18 22 19 26 31 20 40 24 43 22
29 28 35 42 29 30 24 31 23 21

14. Complete the frequency table.

Class Intervals (Age)	Frequency
18–21	
22–25	
26–29	
30–33	
34–37	
38–41	
42–45	

15. Construct a histogram of the data in Exercise 14.

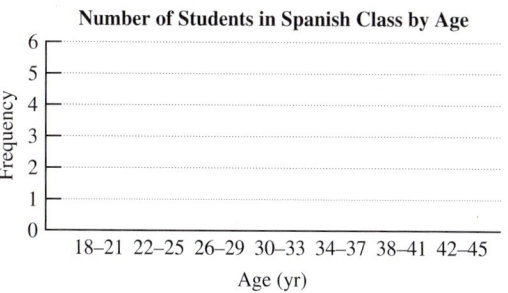

Section 9.3

The pie graph describes the types of subs offered at Larry's Sub Shop. Use the information in the graph for Exercises 16–18.

Number of Certain Types of Subs

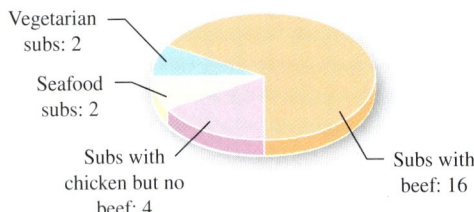

- Vegetarian subs: 2
- Seafood subs: 2
- Subs with chicken but no beef: 4
- Subs with beef: 16

16. How many types of subs are offered at Larry's?

17. What fraction of the subs at Larry's is made with beef?

18. What fraction of the subs at Larry's is not made with beef?

19. A survey was conducted with 200 people, and they were asked their highest level of education. The results of the survey are given in the table.

    a. Complete the table.

Education Level	Number of People	Percent	Number of Degrees
Grade school	10		
High school	50		
Some college	60		
Four-year degree	40		
Postgraduate	40		

   b. Construct a circle graph using percents from the information in the table.

Percent by Education Level

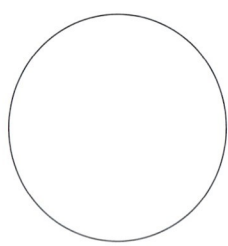

Section 9.4

20. Roberto has six pairs of socks, each a different color: blue, green, brown, black, gray, and white. If Roberto randomly chooses a pair of socks, write the sample space for this event.

21. Refer to Exercise 20. What is the probability that Roberto will select a pair of gray socks?

22. Refer to Exercise 20. What is the probability that Roberto will *not* select a blue pair of socks?

23. Which of the numbers could represent a probability?

    a. $\dfrac{1}{2}$ b. $\dfrac{5}{4}$ c. 0 d. 1

    e. 25% f. 2.5 g. 6% h. 6

24. A bicycle shop sells a child's tricycle in three colors: red, blue, and pink. In the warehouse there are 8 red tricycles, 6 blue tricycles, and 2 pink tricycles.

    a. If Kevin selects a tricycle at random, what is the probability that he will pick a red tricycle?

    b. What is the probability that he will not pick a red tricycle?

    c. What is the probability of Kevin's selecting a green tricycle?

Ingram Publishing/Alamy Stock Photo

25. Torie's math teacher, Ms. Coronel, gave a particularly difficult math test. As a result, Torie had to guess on a multiple choice question that had four possible answers.

    a. What is the probability that she would guess correctly?

    b. What is the probability that she would guess incorrectly?

Section 9.5

26. For the list of quiz scores, find the mean, median, and mode(s).

20, 20, 18, 16, 18, 17, 16, 10, 20, 20, 15, 20

27. Juanita kept track of the number of milligrams of calcium she took each day through vitamins and dairy products. Determine the mean, median, and mode for the amount of calcium she took per day. Round to the nearest 10 mg if necessary.

Pixtal/SuperStock

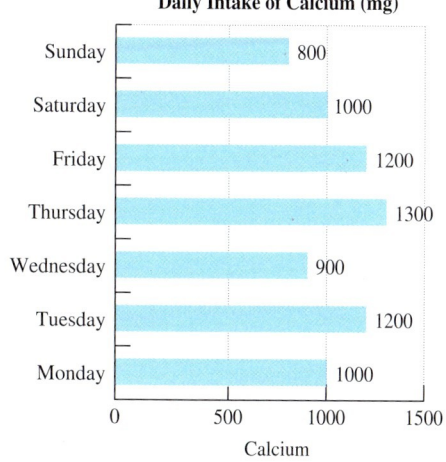

28. The seating capacity for five arenas used by the NBA is given in the table. Find the median number of seats.

Arena	Number of Seats
Philips Arena, Atlanta	20,000
TD Garden Arena, Boston	18,624
Time Warner Cable Area, Charlotte	23,799
United Center Arena, Chicago	21,500
Quicken Loans Arena, Cleveland	20,562

29. The manager of a restaurant had his customers fill out evaluations on the service that they received. A scale of 1 to 5 was used, where 1 represents very poor service and 5 represents excellent service. Given the list of responses, determine the mode(s).

4 5 3 4 4 3 2 5 5 1 4 3 4 4 5
2 5 4 4 3 2 5 5 1 4

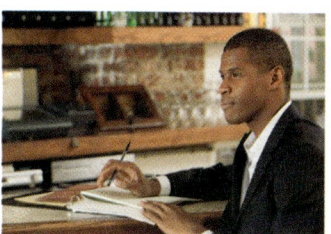

Isadora Getty Buyou/Image Source

30. Compute the GPA for the following grades.

A = 4 pts   B = 3 pts   C = 2 pts
D = 1 pt   F = 0 pts

Course	Grade	Credit-Hours
History	B	3
Reading	D	2
Biology	A	4

Linear Equations and Inequalities

10

CHAPTER OUTLINE

10.1 Addition, Subtraction, Multiplication, and Division Properties of Equality   590

10.2 Solving Linear Equations   603

10.3 Linear Equations: Clearing Fractions and Decimals   612

   Problem Recognition Exercises: Equations vs. Expressions   620

10.4 Applications of Linear Equations: Introduction to Problem Solving   621

10.5 Applications Involving Percents   632

10.6 Formulas and Applications of Geometry   640

10.7 Mixture Applications and Uniform Motion   651

10.8 Linear Inequalities   662

   Chapter 10   Review Exercises   678

Mathematics as a Language

Languages make use of symbols to represent sounds and other conventions used in speech and writing. The letters of the alphabet and punctuation marks are all examples of these symbols. Mathematicians cleverly adopted the use of symbols as a way to give a temporary nickname to *unknowns* in order to simplify the problem-solving process.

Suppose that the maximum recommended heart rate for an adult is given by 220 minus the age (in years) of the adult. If we let a be the age of an adult in years, then the expression $220 - a$ represents the adult's maximum recommended heart rate.

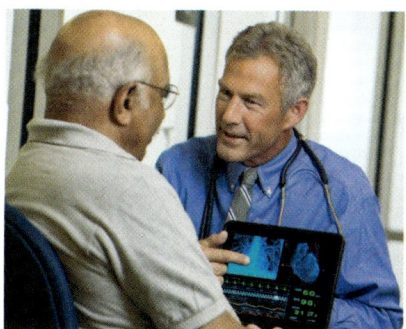
Ariel Skelley/Blend Images/Vetta/Getty Images

If Alan is 60 years old, his maximum heart rate is found by substituting 60 for a. Thus, Alan's maximum recommended heart rate is $220 - 60$, which is 160 beats per minute.

If we know that Ben's recommended heart rate is 178, then we can solve an equation to determine his age, a.

$$220 - a = 178$$
$$-a = 178 - 220$$
$$-a = -42$$
$$a = 42 \quad \text{Ben is 42 years old.}$$

Section 10.1 Addition, Subtraction, Multiplication, and Division Properties of Equality

Concepts

1. Definition of a Linear Equation in One Variable
2. Addition and Subtraction Properties of Equality
3. Multiplication and Division Properties of Equality
4. Translations

1. Definition of a Linear Equation in One Variable

An **equation** is a statement that indicates that two expressions are equal. The following are equations.

$$x = 5 \qquad y + 2 = 12 \qquad -4z = 28$$

All equations have an equal sign. Furthermore, notice that the equal sign separates the equation into two parts, the left-hand side and the right-hand side. A **solution to an equation** is a value of the variable that makes the equation a true statement. Substituting a solution into an equation for the variable makes the right-hand side equal to the left-hand side.

Equation	Solution	Check	
$x = 5$	5	$x = 5$ $\downarrow$ $5 = 5$ ✓	Substitute 5 for x. Right-hand side equals left-hand side.
$y + 2 = 12$	10	$y + 2 = 12$ $\downarrow$ $10 + 2 = 12$ ✓	Substitute 10 for y. Right-hand side equals left-hand side.
$-4z = 28$	-7	$-4z = 28$ $\downarrow$ $-4(-7) = 28$ ✓	Substitute -7 for z. Right-hand side equals left-hand side.

Avoiding Mistakes

Be sure to notice the difference between solving an equation and simplifying an expression. For example, $2x + 1 = 7$ is an equation, whose solution is 3, while $2x + 1 + 7$ is an expression that simplifies to $2x + 8$.

Example 1 Determining Whether a Number Is a Solution to an Equation

Determine whether the given number is a solution to the equation.

a. $4x + 7 = 5$; $-\frac{1}{2}$ **b.** $-4 = 6w - 14$; 3

Solution:

a.
$$4x + 7 = 5$$
$$4(-\tfrac{1}{2}) + 7 \stackrel{?}{=} 5 \qquad \text{Substitute } -\tfrac{1}{2} \text{ for } x.$$
$$-2 + 7 \stackrel{?}{=} 5 \qquad \text{Simplify.}$$
$$5 \stackrel{?}{=} 5 \checkmark \qquad \text{Right-hand side equals the left-hand side.}$$
Thus, $-\tfrac{1}{2}$ *is a solution* to the equation $4x + 7 = 5$.

b. $-4 = 6w - 14$
$$-4 \stackrel{?}{=} 6(3) - 14 \qquad \text{Substitute 3 for } w.$$
$$-4 \stackrel{?}{=} 18 - 14 \qquad \text{Simplify.}$$
$$-4 \neq 4 \qquad \text{Right-hand side does not equal left-hand side.}$$
Thus, 3 *is not a solution* to the equation $-4 = 6w - 14$.

Skill Practice Determine whether the given number is a solution to the equation.

1. $4x - 1 = 7$; 3 **2.** $9 = -2y + 5$; -2

Answers
1. No **2.** Yes

The set of all solutions to an equation is called the **solution set** and is written with set braces. For example, the solution set for Example 1(a) is $\{-\frac{1}{2}\}$.

In the study of algebra, you will encounter a variety of equations. In this chapter, we will focus on a specific type of equation called a linear equation in one variable.

Definition of a Linear Equation in One Variable

Let a, b, and c be real numbers such that $a \neq 0$. A **linear equation in one variable** is an equation that can be written in the form

$$ax + b = c$$

Note: A linear equation in one variable is often called a first-degree equation because the variable x has an implied exponent of 1.

Examples	Notes
$2x + 4 = 20$	$a = 2, b = 4, c = 20$
$-3x - 5 = 16$ can be written as $-3x + (-5) = 16$	$a = -3, b = -5, c = 16$
$5x + 9 - 4x = 1$ can be written as $x + 9 = 1$	$a = 1, b = 9, c = 1$

2. Addition and Subtraction Properties of Equality

If two equations have the same solution set, then the equations are equivalent. For example, the following equations are equivalent because the solution set for each equation is $\{6\}$.

Equivalent Equations	Check the Solution 6	
$2x - 5 = 7$	$2(6) - 5 \stackrel{?}{=} 7$	$\Rightarrow 12 - 5 \stackrel{?}{=} 7$ ✓
$2x = 12$	$2(6) \stackrel{?}{=} 12$	$\Rightarrow\ \ \ 12 \stackrel{?}{=} 12$ ✓
$x = 6$	$6 \stackrel{?}{=} 6$	$\Rightarrow\ \ \ \ \ 6 \stackrel{?}{=} 6$ ✓

To solve a linear equation, $ax + b = c$, the goal is to find *all* values of x that make the equation true. One general strategy for solving an equation is to rewrite it as an equivalent but simpler equation. This process is repeated until the equation can be written in the form x = number. We call this "isolating the variable." The addition and subtraction properties of equality help us isolate the variable.

Addition and Subtraction Properties of Equality

Let a, b, and c represent algebraic expressions.

1. **Addition property of equality:** If $a = b$,
   then $a + c = b + c$
2. ***Subtraction property of equality:** If $a = b$,
   then $a - c = b - c$

*The subtraction property of equality follows directly from the addition property, because subtraction is defined in terms of addition.

If $a + (-c) = b + (-c)$
then, $a - c = b - c$

The addition and subtraction properties of equality indicate that adding or subtracting the same quantity on each side of an equation results in an equivalent equation. This means that if two equal quantities are increased or decreased by the same amount, then the resulting quantities will also be equal (Figure 10-1).

$$50 = 50$$
$$50 + 20 = 50 + 20$$
$$70 = 70$$

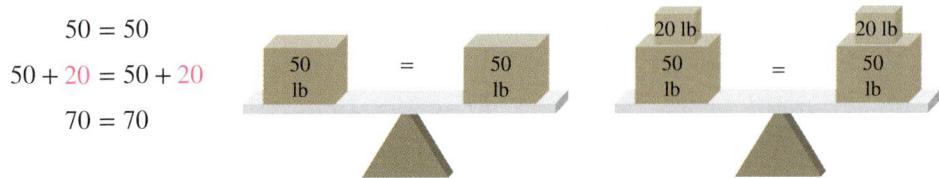

Figure 10-1

Example 2 — Applying the Addition and Subtraction Properties of Equality

Solve the equations.

a. $p - 4 = 11$ **b.** $w + 5 = -2$

Solution:

In each equation, the goal is to isolate the variable on one side of the equation. To accomplish this, we use the fact that the sum of a number and its opposite is zero and the difference of a number and itself is zero.

FOR REVIEW

Recall that the sum of a number and its opposite is zero.

$$-4 + 4 = 0$$

a.
$$p - 4 = 11$$
$$p - 4 + 4 = 11 + 4 \qquad \text{To isolate } p, \text{ add } 4 \text{ to both sides } (-4 + 4 = 0).$$
$$p + 0 = 15 \qquad \text{Simplify.}$$
$$p = 15 \qquad \text{Check by substituting } p = 15 \text{ into the original equation.}$$

Check: $p - 4 = 11$
$15 - 4 \stackrel{?}{=} 11$
$11 \stackrel{?}{=} 11 \checkmark$ True

The solution set is $\{15\}$.

b.
$$w + 5 = -2$$
$$w + 5 - 5 = -2 - 5 \qquad \text{To isolate } w, \text{ subtract } 5 \text{ from both sides. } (5 - 5 = 0).$$
$$w + 0 = -7 \qquad \text{Simplify.}$$
$$w = -7 \qquad \text{Check by substituting } w = -7 \text{ into the original equation.}$$

Check: $w + 5 = -2$
$-7 + 5 \stackrel{?}{=} -2$
$-2 \stackrel{?}{=} -2 \checkmark$ True

The solution set is $\{-7\}$.

Skill Practice Solve the equations.

3. $v - 7 = 2$ **4.** $x + 4 = 4$

Answers

3. $\{9\}$ **4.** $\{0\}$

Example 3 Applying the Addition and Subtraction Properties of Equality

Solve the equations.

a. $\dfrac{9}{4} = q - \dfrac{3}{4}$ **b.** $-1.2 + z = 4.6$

Solution:

a.
$$\dfrac{9}{4} = q - \dfrac{3}{4}$$

$$\dfrac{9}{4} + \dfrac{3}{4} = q - \dfrac{3}{4} + \dfrac{3}{4} \quad \text{To isolate } q, \text{ add } \tfrac{3}{4} \text{ to both sides } \left(-\tfrac{3}{4} + \tfrac{3}{4} = 0\right).$$

$$\dfrac{12}{4} = q + 0 \quad \text{Simplify.}$$

$$3 = q \quad \text{or equivalently,} \quad q = 3$$

Check: $\dfrac{9}{4} = q - \dfrac{3}{4}$

$\dfrac{9}{4} \stackrel{?}{=} 3 - \dfrac{3}{4}$ Substitute $q = 3$.

$\dfrac{9}{4} \stackrel{?}{=} \dfrac{12}{4} - \dfrac{3}{4}$ Common denominator

$\dfrac{9}{4} \stackrel{?}{=} \dfrac{9}{4}$ ✓ True

The solution set is $\{3\}$.

TIP: The variable may be isolated on either side of the equation.

b.
$$-1.2 + z = 4.6$$
$$-1.2 + 1.2 + z = 4.6 + 1.2 \quad \text{To isolate } z, \text{ add } 1.2 \text{ to both sides.}$$
$$0 + z = 5.8$$
$$z = 5.8$$

Check: $-1.2 + z = 4.6$

$-1.2 + 5.8 \stackrel{?}{=} 4.6$ Substitute $z = 5.8$.

$4.6 \stackrel{?}{=} 4.6$ ✓ True

The solution set is $\{5.8\}$.

Skill Practice Solve the equations.

5. $\dfrac{1}{4} = a - \dfrac{2}{3}$ **6.** $-8.1 + w = 11.5$

3. Multiplication and Division Properties of Equality

Adding or subtracting the same quantity to both sides of an equation results in an equivalent equation. In a similar way, multiplying or dividing both sides of an equation by the same nonzero quantity also results in an equivalent equation. This is stated formally as the multiplication and division properties of equality.

Answers

5. $\left\{\dfrac{11}{12}\right\}$ **6.** $\{19.6\}$

> **Multiplication and Division Properties of Equality**
>
> Let a, b, and c represent algebraic expressions, $c \neq 0$.
>
> 1. **Multiplication property of equality:** If $a = b$, then $ac = bc$
>
> 2. ***Division property of equality:** If $a = b$ then $\dfrac{a}{c} = \dfrac{b}{c}$
>
> *The division property of equality follows directly from the multiplication property because division is defined as multiplication by the reciprocal.
>
> If $a \cdot \dfrac{1}{c} = b \cdot \dfrac{1}{c}$
>
> then, $\dfrac{a}{c} = \dfrac{b}{c}$

To understand the multiplication property of equality, suppose we start with a true equation such as $10 = 10$. If both sides of the equation are multiplied by a constant such as 3, the result is also a true statement (Figure 10-2).

$10 = 10$
$3 \cdot 10 = 3 \cdot 10$
$30 = 30$

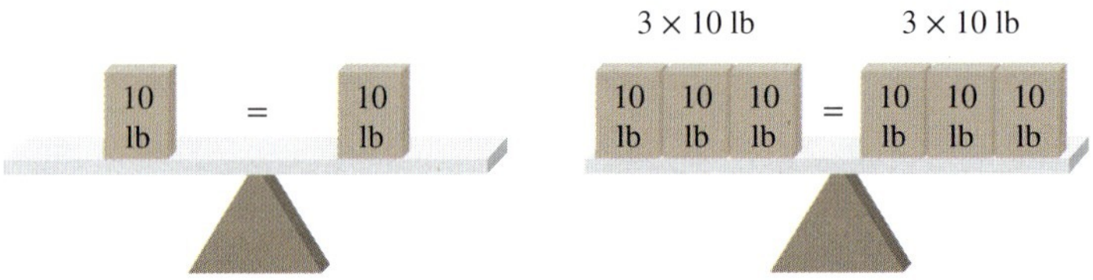

Figure 10-2

Similarly, if both sides of the equation are divided by a nonzero real number such as 2, the result is also a true statement (Figure 10-3).

$10 = 10$
$\dfrac{10}{2} = \dfrac{10}{2}$
$5 = 5$

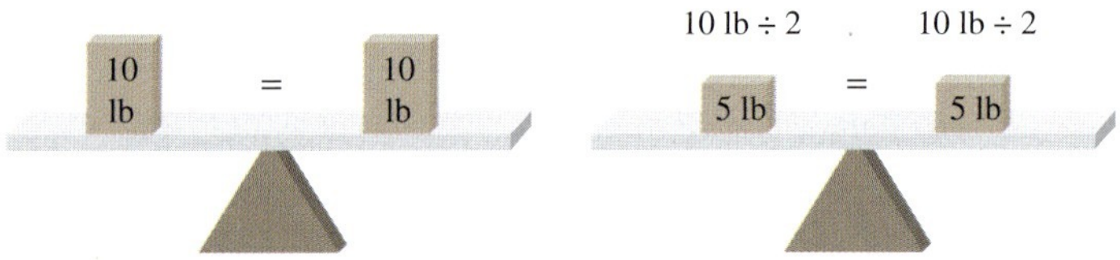

Figure 10-3

FOR REVIEW

Recall that the product of a number and its reciprocal is always 1. For example:

$\dfrac{1}{5}(5) = 1$

$-\dfrac{7}{2}\left(-\dfrac{2}{7}\right) = 1$

To solve an equation in the variable x, the goal is to write the equation in the form $x = $ number. In particular, notice that we want the coefficient of x to be 1. That is, we want to write the equation as $1x = $ number. Therefore, to solve an equation such as $5x = 15$, we can multiply both sides of the equation by the reciprocal of the x-term coefficient. In this case, multiply both sides by the reciprocal of 5, which is $\frac{1}{5}$.

$5x = 15$

$\dfrac{1}{5}(5x) = \dfrac{1}{5}(15)$ Multiply by $\dfrac{1}{5}$.

$1x = 3$ The coefficient of the x-term is now 1.

$x = 3$

Section 10.1 Addition, Subtraction, Multiplication, and Division Properties of Equality

The division property of equality can also be used to solve the equation $5x = 15$ by dividing both sides by the coefficient of the x-term. In this case, divide both sides by 5 to make the coefficient of x equal to 1.

$$5x = 15$$

$$\frac{5x}{5} = \frac{15}{5} \quad \text{Divide by 5.}$$

$$1x = 3 \quad \text{The coefficient of the } x\text{-term is now 1.}$$

$$x = 3$$

TIP: The quotient of a nonzero real number and itself is always 1. For example:

$$\frac{5}{5} = 1$$

$$\frac{-3.5}{-3.5} = 1$$

Example 4   **Applying the Division Property of Equality**

Solve the equations using the division property of equality.

   a. $12x = 60$ **b.** $48 = -8w$ **c.** $-x = 8$

Solution:

a. $12x = 60$

$$\frac{12x}{12} = \frac{60}{12} \quad \text{To obtain a coefficient of 1 for the } x\text{-term, divide both sides by 12.}$$

$$1x = 5 \quad \text{Simplify.}$$

$$x = 5$$

Check: $12x = 60$

$$12(5) \stackrel{?}{=} 60$$

$$60 \stackrel{?}{=} 60 \checkmark \quad \text{True}$$

The solution set is $\{5\}$.

b. $48 = -8w$

$$\frac{48}{-8} = \frac{-8w}{-8} \quad \text{To obtain a coefficient of 1 for the } w\text{-term, divide both sides by } -8.$$

$$-6 = 1w \quad \text{Simplify.}$$

$$-6 = w$$

Check: $48 = -8w$

$$48 \stackrel{?}{=} -8(-6)$$

$$48 \stackrel{?}{=} 48 \checkmark \quad \text{True}$$

The solution set is $\{-6\}$.

c. $-x = 8$ Note that $-x$ is equivalent to $-1 \cdot x$.

$$-1x = 8$$

$$\frac{-1x}{-1} = \frac{8}{-1} \quad \text{To obtain a coefficient of 1 for the } x\text{-term, divide by } -1.$$

$$x = -8$$

Check: $-x = 8$

$$-(-8) \stackrel{?}{=} 8$$

$$8 \stackrel{?}{=} 8 \checkmark \quad \text{True}$$

The solution set is $\{-8\}$.

TIP: In Example 4(c), we could also have *multiplied* both sides by -1 to create a coefficient of 1 on the x-term.

$$-x = 8$$

$$(-1)(-x) = (-1)8$$

$$x = -8$$

Skill Practice   Solve the equations.

   7. $4x = -20$ **8.** $100 = -4p$ **9.** $-y = -11$

Answers

7. $\{-5\}$ **8.** $\{-25\}$ **9.** $\{11\}$

Example 5 — Applying the Multiplication Property of Equality

Solve the equation by using the multiplication property of equality.

$$-\frac{2}{9}q = \frac{1}{3}$$

Solution:

$$-\frac{2}{9}q = \frac{1}{3}$$

$$\left(-\frac{9}{2}\right)\left(-\frac{2}{9}q\right) = \frac{1}{3}\left(-\frac{9}{2}\right)$$ To obtain a coefficient of 1 for the q-term, multiply by the reciprocal of $-\frac{2}{9}$, which is $-\frac{9}{2}$.

$$1q = -\frac{3}{2}$$ Simplify. The product of a number and its reciprocal is 1.

$$q = -\frac{3}{2}$$

Check: $-\frac{2}{9}q = \frac{1}{3}$

$-\frac{2}{9}\left(-\frac{3}{2}\right) \stackrel{?}{=} \frac{1}{3}$

The solution set is $\left\{-\frac{3}{2}\right\}$.

$\frac{1}{3} \stackrel{?}{=} \frac{1}{3}$ ✓ True

Skill Practice Solve the equation.

10. $-\frac{2}{3}a = \frac{1}{4}$

TIP: When applying the multiplication or division property of equality to obtain a coefficient of 1 for the variable term, we will generally use the following convention:

- If the coefficient of the variable term is expressed as a fraction, we will usually multiply both sides by its reciprocal, as in Example 5.
- If the coefficient of the variable term is an integer or decimal, we will divide both sides by the coefficient itself, as in Example 6.

Example 6 — Applying the Division Property of Equality

Solve the equation by using the division property of equality.

$$-3.43 = -0.7z$$

Solution:

$$-3.43 = -0.7z$$

$$\frac{-3.43}{-0.7} = \frac{-0.7z}{-0.7}$$ To obtain a coefficient of 1 for the z-term, divide by -0.7.

$$4.9 = 1z$$ Simplify.

$$4.9 = z$$

$$z = 4.9$$

Check: $-3.43 = -0.7z$

$-3.43 \stackrel{?}{=} -0.7(4.9)$

The solution set is $\{4.9\}$. $-3.43 \stackrel{?}{=} -3.43$ ✓ True

Answer

10. $\left\{-\frac{3}{8}\right\}$

Skill Practice Solve the equation.

11. $6.82 = 2.2w$

Example 7   **Applying the Multiplication Property of Equality**

Solve the equation by using the multiplication property of equality.

$$\frac{d}{6} = -4$$

Solution:

$\dfrac{d}{6} = -4$

$\dfrac{1}{6}d = -4$ $\dfrac{d}{6}$ is equivalent to $\dfrac{1}{6}d$.

$\dfrac{6}{1} \cdot \dfrac{1}{6}d = -4 \cdot \dfrac{6}{1}$ To obtain a coefficient of 1 for the d-term, multiply by the reciprocal of $\frac{1}{6}$, which is $\frac{6}{1}$.

$1d = -24$ Simplify.

$d = -24$ Check: $\dfrac{d}{6} = -4$

$\dfrac{-24}{6} \stackrel{?}{=} -4$

The solution set is $\{-24\}$. $-4 \stackrel{?}{=} -4$ ✓ True

---

Skill Practice Solve the equation.

12. $\dfrac{x}{5} = -8$

It is important to distinguish between cases where the addition or subtraction properties of equality should be used to isolate a variable and those in which the multiplication or division property of equality should be used. Remember the goal is to isolate the variable term and obtain a coefficient of 1. Compare the equations:

$$5 + x = 20 \quad \text{and} \quad 5x = 20$$

In the first equation, the relationship between 5 and x is addition. Therefore, we want to reverse the process by subtracting 5 from both sides. In the second equation, the relationship between 5 and x is multiplication. To isolate x, we reverse the process by dividing by 5 or equivalently, multiplying by the reciprocal, $\frac{1}{5}$.

$$5 + x = 20 \quad \text{and} \quad 5x = 20$$

$$5 - 5 + x = 20 - 5 \quad\quad\quad \dfrac{5x}{5} = \dfrac{20}{5}$$

$$x = 15 \quad\quad\quad\quad\quad x = 4$$

Answers

11. $\{3.1\}$ **12.** $\{-40\}$

4. Translations

We have already practiced writing an English sentence as a mathematical equation. Recall that several key words translate to the algebraic operations of addition, subtraction, multiplication, and division.

Example 8 — Translating to a Linear Equation

Write an algebraic equation to represent each English sentence. Then solve the equation.

a. The quotient of a number and 4 is 6.

b. The product of a number and 4 is 6.

c. Negative twelve is equal to the sum of -5 and a number.

d. The value 1.4 subtracted from a number is 5.7.

Solution:

For each case we will let x represent the unknown number.

a. The quotient of a number and 4 is 6.

$$\frac{x}{4} = 6$$

$$4 \cdot \frac{x}{4} = 4 \cdot 6 \qquad \text{Multiply both sides by } 4.$$

$$\frac{4}{1} \cdot \frac{x}{4} = 4 \cdot 6$$

$$x = 24 \qquad \underline{\text{Check:}} \ \frac{24}{4} \stackrel{?}{=} 6 \checkmark \quad \text{True}$$

The number is 24.

b. The product of a number and 4 is 6.

$$4x = 6$$

$$\frac{4x}{4} = \frac{6}{4} \qquad \text{Divide both sides by } 4.$$

$$x = \frac{3}{2} \qquad \underline{\text{Check:}} \ 4\left(\frac{3}{2}\right) \stackrel{?}{=} 6 \checkmark \quad \text{True}$$

The number is $\frac{3}{2}$.

c. Negative twelve is equal to the sum of -5 and a number.

$$-12 = -5 + x$$

$$-12 + 5 = -5 + 5 + x \qquad \text{Add } 5 \text{ to both sides.}$$

$$-7 = x \qquad \underline{\text{Check:}} \ -12 \stackrel{?}{=} -5 + (-7) \checkmark \quad \text{True}$$

The number is -7.

d. The value 1.4 subtracted from a number is 5.7.

$$x - 1.4 = 5.7$$
$$x - 1.4 + 1.4 = 5.7 + 1.4 \qquad \text{Add } 1.4 \text{ to both sides.}$$
$$x = 7.1 \qquad \text{Check: } 7.1 - 1.4 \stackrel{?}{=} 5.7 \checkmark \text{ True}$$

The number is 7.1.

Skill Practice Write an algebraic equation to represent each English sentence. Then solve the equation.

13. The quotient of a number and -2 is 8.
14. The product of a number and -3 is -24.
15. The sum of a number and 6 is -20.
16. 13 is equal to 5 subtracted from a number.

Answers
13. $\frac{x}{-2} = 8$; The number is -16.
14. $-3x = -24$; The number is 8.
15. $y + 6 = -20$; The number is -26.
16. $13 = x - 5$; The number is 18.

Section 10.1 Activity

A.1. Explain how to check whether -4 is a solution to the equation $x - 6 = -10$.

A.2. Determine if 12 is a solution to $4 - x = 16$.

A.3. Determine if $\frac{2}{3}$ is a solution to $8 = 12w$.

For Exercises A.4–A.5, solve the equations and state the property used to isolate the variable.

A.4. **a.** $x - 5 = 20$ **b.** $x + 5 = 20$ **c.** $5x = 20$ **d.** $\frac{x}{5} = 20$

A.5. **a.** $-\frac{1}{5}w = 2$ **b.** $-\frac{1}{5} + w = 2$

  c. $\frac{1}{5} + w = 2$ **d.** $-5w = 2$

A.6. **a.** Solve the equation $-3x = 18$ by dividing both sides by -3.

  b. Solve the equation $-3x = 18$ by multiplying both sides by $-\frac{1}{3}$.

  c. Which approach do you prefer to solve this equation?

A.7. **a.** Solve the equation $11 = 8 + y$ by adding -8 to both sides.

  b. Solve the equation $11 = 8 + y$ by subtracting 8 from both sides.

  c. Which approach do you prefer to solve this equation?

A.8. **a.** To solve the equation $-\frac{3}{10} = -\frac{6}{5}t$, would you multiply both sides of the equation by the reciprocal of $-\frac{3}{10}$ or by the reciprocal of $-\frac{6}{5}$?

  b. Solve the equation.

Section 10.1 Practice Exercises

Study Skills Exercise

Math anxiety is a feeling of nervousness, tension, fear, and worry that interferes with a student's ability to concentrate and solve math problems. Math anxiety causes students to believe that they are not capable of learning mathematics or performing computations, even if they know how. If you can relate to these feelings, the good news is that there are strategies to overcome math anxiety. Complete the following lists and reflect on your responses during times of angst.
- List the people who support you and your educational goals. Rely on these individuals for encouragement.
- List one or more successes that you have had in this course. Reframe your thoughts to focus on these positive moments and times of achievement.
- List your study habits that are beneficial to your success in this course. Are there other study skills or habits that you practice that would help you understand the content? Give these a try!

Prerequisite Review

For Exercises R.1–R.8, simplify the expression.

R.1. $-\dfrac{1}{3} \cdot 3y$

R.2. $5 \cdot \dfrac{1}{5}x$

R.3. $m - 6 + 6$

R.4. $t + 4 - 4$

R.5. $-4.4p + 6 + 5.4p$

R.6. $1.2q + 6.3 - 0.2q$

R.7. $-\dfrac{2}{5} \cdot \dfrac{15}{4}$

R.8. $\dfrac{9}{8} \cdot \left(-\dfrac{2}{3}\right)$

For Exercises R.9–R.12, evaluate the expression for the given value of x.

R.9. $3x - 4$ for $x = -6$

R.10. $-7x + 2$ for $x = 2$

R.11. $\dfrac{2}{15}x$ for $x = 25$

R.12. $-\dfrac{5}{6}x$ for $x = 3$

Vocabulary and Key Concepts

1. a. An _____ is a statement that indicates that two expressions are equal.
   b. A _____ to an equation is a value of the variable that makes the equation a true statement.
   c. An equation that can be written in the form $ax + b = c$ $(a \neq 0)$ is called a _____ equation in one variable.
   d. The set of all solutions to an equation is called the _____.

Concept 1: Definition of a Linear Equation in One Variable

For Exercises 2–6, identify the following as either an expression or an equation.

2. $2 - 8x + 10$ **3.** $x - 4 + 5x$ **4.** $8x + 2 = 7$ **5.** $9 = 2x - 4$ **6.** $3x^2 + x = -3$

7. Explain how to determine if a number is a solution to an equation.

8. Explain why the equations $6x = 12$ and $x = 2$ are *equivalent equations*.

For Exercises 9–14, determine whether the given number is a solution to the equation. **(See Example 1.)**

9. $x - 1 = 5$; 4
10. $x - 2 = 1$; -1
11. $5x = -10$; -2
12. $3x = 21$; 7
13. $3x + 9 = 3$; -2
14. $2x - 1 = -3$; -1

Concept 2: Addition and Subtraction Properties of Equality

For Exercises 15–34, solve each equation using the addition or subtraction property of equality. Be sure to check your answers. **(See Examples 2–3.)**

15. $x + 6 = 5$
16. $x - 2 = 10$
17. $q - 14 = 6$
18. $w + 3 = -5$
19. $2 + m = -15$
20. $-6 + n = 10$
21. $-23 = y - 7$
22. $-9 = -21 + b$
23. $4 + c = 4$
24. $-13 + b = -13$
25. $4.1 = 2.8 + a$
26. $5.1 = -2.5 + y$
27. $5 = z - \frac{1}{2}$
28. $-7 = p + \frac{2}{3}$
29. $x + \frac{5}{2} = \frac{1}{2}$
30. $\frac{7}{3} = x - \frac{2}{3}$
31. $-6.02 + c = -8.15$
32. $p + 0.035 = -1.12$
33. $3.245 + t = -0.0225$
34. $-1.004 + k = 3.0589$

Concept 3: Multiplication and Division Properties of Equality

For Exercises 35–54, solve each equation using the multiplication or division property of equality. Be sure to check your answers. **(See Examples 4–7.)**

35. $6x = 54$
36. $2w = 8$
37. $12 = -3p$
38. $6 = -2q$
39. $-5y = 0$
40. $-3k = 0$
41. $-\frac{y}{5} = 3$
42. $-\frac{z}{7} = 1$
43. $\frac{4}{5} = -t$
44. $-\frac{3}{7} = -h$
45. $\frac{2}{5}a = -4$
46. $\frac{3}{8}b = -9$
47. $-\frac{1}{5}b = -\frac{4}{5}$
48. $-\frac{3}{10}w = \frac{2}{5}$
49. $-41 = -x$
50. $32 = -y$
51. $3.81 = -0.03p$
52. $2.75 = -0.5q$
53. $5.82y = -15.132$
54. $-32.3x = -0.4522$

Concept 4: Translations

For Exercises 55–66, write an algebraic equation to represent each English sentence. (Let x represent the unknown number.) Then solve the equation. **(See Example 8.)**

55. The sum of negative eight and a number is forty-two.
56. The sum of thirty-one and a number is thirteen.
57. The difference of a number and negative six is eighteen.
58. The sum of negative twelve and a number is negative fifteen.
59. The product of a number and seven is the same as negative sixty-three.
60. The product of negative three and a number is the same as twenty-four.

61. The value 3.2 subtracted from a number is 2.1.

62. The value −3 subtracted from a number is 4.

63. The quotient of a number and twelve is one-third.

64. Eighteen is equal to the quotient of a number and two.

65. The sum of a number and $\frac{5}{8}$ is $\frac{13}{8}$.

66. The difference of a number and $\frac{2}{3}$ is $\frac{1}{3}$.

Mixed Exercises

For Exercises 67–90, solve each equation using the appropriate property of equality.

67. a. $x - 9 = 1$
   b. $-9x = 1$

68. a. $k - 2 = -4$
   b. $-2k = -4$

69. a. $-\frac{2}{3}h = 8$
   b. $\frac{2}{3} + h = 8$

70. a. $\frac{3}{4}p = 15$
   b. $\frac{3}{4} + p = 15$

71. $\frac{r}{3} = -12$

72. $\frac{d}{-4} = 5$

73. $k + 16 = 32$

74. $-18 = -9 + t$

75. $16k = 32$

76. $-18 = -9t$

77. $7 = -4q$

78. $-3s = 10$

79. $-4 + q = 7$

80. $s - 3 = 10$

81. $-\frac{1}{3}d = 12$

82. $-\frac{2}{5}m = 10$

83. $4 = \frac{1}{2} + z$

84. $3 = \frac{1}{4} + p$

85. $1.2y = 4.8$

86. $4.3w = 8.6$

87. $4.8 = 1.2 + y$

88. $8.6 = w - 4.3$

89. $0.0034 = y - 0.405$

90. $-0.98 = m + 1.0034$

For Exercises 91–98, determine if the equation is a linear equation in one variable. Answer yes or no.

91. $4p + 5 = 0$

92. $3x - 5y = 0$

93. $4 + 2a^2 = 5$

94. $-8t = 7$

95. $x - 4 = 9$

96. $2x^3 + y = 0$

97. $19b = -3$

98. $13 + x = 19$

Expanding Your Skills

For Exercises 99–104, construct an equation with the given solution set. Answers will vary.

99. $\{6\}$

100. $\{2\}$

101. $\{-4\}$

102. $\{-10\}$

103. $\{0\}$

104. $\{1\}$

For Exercises 105–108, simplify by collecting the *like* terms. Then solve the equation.

105. $5x - 4x + 7 = 8 - 2$

106. $2 + 3 = 2y + 1 - y$

107. $6p - 3p = 15 + 6$

108. $12 - 20 = 2t + 2t$

Solving Linear Equations

Section 10.2

Concepts

1. Linear Equations Involving Multiple Steps
2. Procedure for Solving a Linear Equation in One Variable
3. Conditional Equations, Identities, and Contradictions

1. Linear Equations Involving Multiple Steps

Previously we studied a one-step process to solve linear equations by using the addition, subtraction, multiplication, and division properties of equality. In Example 1, we solve the equation $-2w - 7 = 11$. Solving this equation will require multiple steps. To understand the proper steps, always remember that the ultimate goal is to isolate the variable. Therefore, we will first isolate the *term* containing the variable before dividing both sides by -2.

Example 1 Solving a Linear Equation

Solve the equation. $-2w - 7 = 11$

Solution:

$-2w - 7 = 11$

$-2w - 7 + 7 = 11 + 7$ Add 7 to both sides of the equation. This isolates the w-term.

$-2w = 18$

$\dfrac{-2w}{-2} = \dfrac{18}{-2}$ Next, apply the division property of equality to obtain a coefficient of 1 for w. Divide by -2 on both sides.

$w = -9$

Check:

$-2w - 7 = 11$

$-2(-9) - 7 \stackrel{?}{=} 11$ Substitute $w = -9$ in the original equation.

$18 - 7 \stackrel{?}{=} 11$

$11 \stackrel{?}{=} 11$ ✓   True

The solution set is $\{-9\}$.

Skill Practice   Solve the equation.
1. $-5y - 5 = 10$

Example 2 Solving a Linear Equation

Solve the equation. $2 = \dfrac{1}{5}x + 3$

Solution:

$2 = \dfrac{1}{5}x + 3$

$2 - 3 = \dfrac{1}{5}x + 3 - 3$ Subtract 3 from both sides. This isolates the x-term.

$-1 = \dfrac{1}{5}x$ Simplify.

Answer
1. $\{-3\}$

Chapter 10 Linear Equations and Inequalities

$$5(-1) = 5 \cdot \left(\frac{1}{5}x\right)$$

Next, apply the multiplication property of equality to obtain a coefficient of 1 for x.

$$-5 = 1x$$

$$-5 = x$$

Simplify. The answer checks in the original equation.

The solution set is $\{-5\}$.

Skill Practice Solve the equation.

2. $2 = \dfrac{1}{2}a - 7$

In Example 3, the variable x appears on both sides of the equation. In this case, apply the addition or subtraction property of equality to collect the variable terms on one side of the equation and the constant terms on the other side. Then use the multiplication or division property of equality to get a coefficient equal to 1 on the variable term.

Example 3 Solving a Linear Equation

Solve the equation. $\quad 6x - 4 = 2x - 8$

Solution:

$$6x - 4 = 2x - 8$$

$$6x - 2x - 4 = 2x - 2x - 8 \quad \text{Subtract } 2x \text{ from both sides, leaving } 0x \text{ on the right-hand side.}$$

$$4x - 4 = 0x - 8 \quad \text{Simplify.}$$

$$4x - 4 = -8 \quad \text{The } x\text{-terms have now been combined on one side of the equation.}$$

$$4x - 4 + 4 = -8 + 4 \quad \text{Add 4 to both sides of the equation. This combines the constant terms on the } other \text{ side of the equation.}$$

$$4x = -4$$

$$\frac{4x}{4} = \frac{-4}{4} \quad \text{To obtain a coefficient of 1 for } x, \text{ divide both sides of the equation by 4.}$$

$$x = -1 \quad \text{The answer checks in the original equation.}$$

The solution set is $\{-1\}$.

Skill Practice Solve the equation.

3. $10x - 3 = 4x - 2$

Answers

2. $\{18\}$ **3.** $\left\{\dfrac{1}{6}\right\}$

TIP: It is important to note that the variable may be isolated on either side of the equation. We will solve the equation from Example 3 again, this time isolating the variable on the right side.

$$6x - 4 = 2x - 8$$
$$6x - 6x - 4 = 2x - 6x - 8 \quad \text{Subtract } 6x \text{ on both sides.}$$
$$0x - 4 = -4x - 8$$
$$-4 = -4x - 8$$
$$-4 + 8 = -4x - 8 + 8 \quad \text{Add } 8 \text{ to both sides.}$$
$$4 = -4x$$
$$\frac{4}{-4} = \frac{-4x}{-4} \quad \text{Divide both sides by } -4.$$
$$-1 = x \quad \text{or equivalently } x = -1$$

2. Procedure for Solving a Linear Equation in One Variable

In some cases, it is necessary to simplify both sides of a linear equation before applying the properties of equality. Therefore, we offer the following steps to solve a linear equation in one variable.

Solving a Linear Equation in One Variable

Step 1 Simplify both sides of the equation.
- Clear parentheses
- Combine *like* terms

Step 2 Use the addition or subtraction property of equality to collect the variable terms on one side of the equation.

Step 3 Use the addition or subtraction property of equality to collect the constant terms on the other side of the equation.

Step 4 Use the multiplication or division property of equality to make the coefficient of the variable term equal to 1.

Step 5 Check your answer.

Example 4 — Solving a Linear Equation

Solve the equation. $7 + 3 = 2(p - 3)$

Solution:

$7 + 3 = 2(p - 3)$

$10 = 2p - 6$ **Step 1:** Simplify both sides of the equation by clearing parentheses and combining *like* terms.

    Step 2: The variable terms are already on one side.

$10 + 6 = 2p - 6 + 6$ **Step 3:** Add 6 to both sides to collect the constant terms on the other side.

$16 = 2p$

Chapter 10 Linear Equations and Inequalities

$$\frac{16}{2} = \frac{2p}{2}$$

Step 4: Divide both sides by 2 to obtain a coefficient of 1 for p.

$8 = p$

Step 5: Check:

$7 + 3 = 2(p - 3)$

$10 \stackrel{?}{=} 2(8 - 3)$

$10 \stackrel{?}{=} 2(5)$

The solution set is $\{8\}$.

$10 \stackrel{?}{=} 10$ ✓ True

Skill Practice Solve the equation.

4. $12 + 2 = 7(3 - y)$

Example 5 Solving a Linear Equation

Solve the equation. $2.2y - 8.3 = 6.2y + 12.1$

Solution:

TIP: In Examples 5 and 6 we collected the variable terms on the right side to avoid negative coefficients on the variable term.

$2.2y - 8.3 = 6.2y + 12.1$

Step 1: The right- and left-hand sides are already simplified.

$2.2y - 2.2y - 8.3 = 6.2y - 2.2y + 12.1$
$-8.3 = 4y + 12.1$

Step 2: Subtract $2.2y$ from both sides to collect the variable terms on one side of the equation.

$-8.3 - 12.1 = 4y + 12.1 - 12.1$
$-20.4 = 4y$

Step 3: Subtract 12.1 from both sides to collect the constant terms on the other side.

$$\frac{-20.4}{4} = \frac{4y}{4}$$
$-5.1 = y$
$y = -5.1$

Step 4: To obtain a coefficient of 1 for the y-term, divide both sides of the equation by 4.

Step 5: Check:

$2.2y - 8.3 = 6.2y + 12.1$

$2.2(-5.1) - 8.3 \stackrel{?}{=} 6.2(-5.1) + 12.1$

$-11.22 - 8.3 \stackrel{?}{=} -31.62 + 12.1$

The solution set is $\{-5.1\}$.

$-19.52 \stackrel{?}{=} -19.52$ ✓ True

Skill Practice Solve the equation.

5. $1.5t + 2.3 = 3.5t - 1.9$

Answers

4. $\{1\}$ 5. $\{2.1\}$

Example 6 Solving a Linear Equation

Solve the equation. $2 + 7x - 5 = 6(x + 3) + 2x$

Solution:

$2 + 7x - 5 = 6(x + 3) + 2x$

$-3 + 7x = 6x + 18 + 2x$ **Step 1:** Add *like* terms on the left. Clear parentheses on the right.

$-3 + 7x = 8x + 18$ Combine *like* terms.

$-3 + 7x - 7x = 8x - 7x + 18$ **Step 2:** Subtract $7x$ from both sides.

$-3 = x + 18$ Simplify.

$-3 - 18 = x + 18 - 8$ **Step 3:** Subtract 18 from both sides.

$-21 = x$ **Step 4:** Because the coefficient of the x term is already 1, there is no need to apply the multiplication or division property of equality.

$x = -21$

The solution set is $\{-21\}$.   **Step 5:** The check is left to the reader.

Skill Practice Solve the equation.

6. $4(2y - 1) + y = 6y + 3 - y$

Example 7 Solving a Linear Equation

Solve the equation. $9 - (z - 3) + 4z = 4z - 5(z + 2) - 6$

Solution:

$9 - (z - 3) + 4z = 4z - 5(z + 2) - 6$

$9 - z + 3 + 4z = 4z - 5z - 10 - 6$ **Step 1:** Clear parentheses.

$12 + 3z = -z - 16$ Combine *like* terms.

$12 + 3z + z = -z + z - 16$ **Step 2:** Add z to both sides.

$12 + 4z = -16$

$12 - 12 + 4z = -16 - 12$ **Step 3:** Subtract 12 from both sides.

$4z = -28$

$\dfrac{4z}{4} = \dfrac{-28}{4}$ **Step 4:** Divide both sides by 4.

$z = -7$ **Step 5:** The check is left for the reader.

The solution set is $\{-7\}$.

> **Avoiding Mistakes**
>
> When distributing a negative number through a set of parentheses, be sure to change the signs of every term within the parentheses.

Skill Practice Solve the equation.

7. $10 - (x + 5) + 3x = 6x - 5(x - 1) - 3$

Answers

6. $\left\{\dfrac{7}{4}\right\}$ 7. $\{-3\}$

3. Conditional Equations, Identities, and Contradictions

The solutions to an equation are the values of x that make the equation a true statement. A linear equation in one variable has one unique solution. Some types of equations, however, have no solution while others have infinitely many solutions.

I. Conditional Equations

An equation that is true for some values of the variable but false for other values is called a **conditional equation**. The equation $x + 4 = 6$, for example, is true on the condition that $x = 2$. For other values of x, the statement $x + 4 = 6$ is false.

$$x + 4 = 6$$
$$x + 4 - 4 = 6 - 4$$
$$x = 2 \quad \text{(Conditional equation)} \quad \text{Solution set: } \{2\}$$

II. Contradictions

TIP: The empty set is also called the null set and can be expressed by the symbol ∅.

Some equations have no solution, such as $x + 1 = x + 2$. There is no value of x, that when increased by 1 will equal the same value increased by 2. If we try to solve the equation by subtracting x from both sides, we get the contradiction $1 = 2$. This indicates that the equation has no solution. An equation that has no solution is called a **contradiction**. The solution set is the empty set. The **empty set** is the set with no elements and is denoted by $\{\ \}$.

$$x + 1 = x + 2$$
$$x - x + 1 = x - x + 2$$
$$1 = 2 \quad \text{(Contradiction)} \quad \text{Solution set: } \{\ \}$$

Avoiding Mistakes

There are two ways to express the empty set: $\{\ \}$ or ∅. Be sure that you do not use them together. It would be incorrect to write $\{∅\}$.

III. Identities

An equation that has all real numbers as its solution set is called an **identity**. For example, consider the equation $x + 4 = x + 4$. Because the left- and right-hand sides are *identical*, any real number substituted for x will result in equal quantities on both sides. If we subtract x from both sides of the equation, we get the identity $4 = 4$. In such a case, the solution is the set of all real numbers.

$$x + 4 = x + 4$$
$$x - x + 4 = x - x + 4$$
$$4 = 4 \quad \text{(Identity)} \quad \text{Solution set: The set of real numbers.}$$

Example 8   Identifying Conditional Equations, Contradictions, and Identities

Solve the equation. Identify each equation as a conditional equation, a contradiction, or an identity.

a. $4k - 5 = 2(2k - 3) + 1$ **b.** $2(b - 4) = 2b - 7$ **c.** $3x + 7 = 2x - 5$

Solution:

a.
$$4k - 5 = 2(2k - 3) + 1$$
$$4k - 5 = 4k - 6 + 1 \quad \text{Clear parentheses.}$$
$$4k - 5 = 4k - 5 \quad \text{Combine } like \text{ terms.}$$
$$4k - 4k - 5 = 4k - 4k - 5 \quad \text{Subtract } 4k \text{ from both sides.}$$
$$-5 = -5 \quad \text{(Identity)}$$

This is an identity. Solution set: The set of real numbers.

b. $2(b - 4) = 2b - 7$
  $2b - 8 = 2b - 7$ Clear parentheses.
  $2b - 2b - 8 = 2b - 2b - 7$ Subtract $2b$ from both sides.
  $-8 = -7$ (Contradiction)

This is a contradiction. Solution set: { }

c. $3x + 7 = 2x - 5$
  $3x - 2x + 7 = 2x - 2x - 5$ Subtract $2x$ from both sides.
  $x + 7 = -5$ Simplify.
  $x + 7 - 7 = -5 - 7$ Subtract 7 from both sides.
  $x = -12$ (Conditional equation)

This is a conditional equation. The solution set is $\{-12\}$. (The equation is true only on the condition that $x = -12$.)

Skill Practice Solve the equation. Identify the equation as a conditional equation, a contradiction, or an identity.

8. $4(2t + 1) - 1 = 8t + 3$ **9.** $3x - 5 = 4x + 1 - x$ **10.** $6(v - 2) = 2v - 4$

Answers
8. The set of real numbers; identity
9. { }; contradiction
10. {2}; conditional equation

Section 10.2 Activity

A.1. Write the steps to solve a linear equation in one variable.

A.2. Consider the equation $1 + 4x = 3x + 6$.
  a. Solve the equation by moving the variable terms to the left side of the equation.
  b. Solve the equation by moving the variable terms to the right side of the equation.
  c. When solving a linear equation, does it matter to which side of the equation the variable is isolated?

For Exercises A.3–A.5, divide your paper into three columns. Solve the equation in the left column. Explain each step in the middle column. Check the answer and write the solution set in the right column.

Solve the equation	Explain	Check
$-8x + 2 = -14$		$-8x + 2 = -14$
		$-8() + 2 = -14$

A.3. $-8x + 2 = -14$

A.4. $-5z + 8 = -2z + 14$

A.5. $4(p - 3) = -9 - 2(p + 6)$

A.6. Explain why the equation $x + 1 = x + 2$ is a contradiction. Write the solution set.

A.7. Explain why the equation $2x + 6 = 2(x + 3)$ is an identity. Write the solution set.

A.8. The equation $x + 4 = 10$ is true under the condition that $x =$ _____. Therefore, this equation is a conditional equation. Write the solution set.

Section 10.2 Practice Exercises

Prerequisite Review

For Exercises R.1–R.4, simplify the expression.

R.1. $5z + 2 - 7z - 3z$

R.2. $10 - 4w + 7w - 2 + w$

R.3. $-(-7p + 9) + (3p - 1)$

R.4. $8y - (2y + 3) - 19$

For Exercises R.5–R.10, solve the equation.

R.5. $7 = p - 12$

R.6. $x + 8 = -15$

R.7. $-7y = 21$

R.8. $5w = -30$

R.9. $-\dfrac{9}{8} = -\dfrac{3}{4}k$

R.10. $\dfrac{2}{5} = -\dfrac{4}{9}y$

Vocabulary and Key Concepts

1. **a.** A _____ equation is true for some values of the variable but false for other values.

   b. An equation that has no solution is called a _____.

   c. The set containing no elements is called the _____ set.

   d. An equation that has all real numbers as its solution set is called an _____.

2. By inspection, determine whether the equation has no solution, one solution, or infinitely many solutions.

   a. $3x - 1 = 2x + x - 1$
   b. $x + 6 = x + 2$
   c. $x - 1 = 5$

Concept 1: Linear Equations Involving Multiple Steps

For Exercises 3–26, solve each equation using the steps outlined in the text. (See Examples 1–3.)

3. $6z + 1 = 13$

4. $5x + 2 = -13$

5. $3y - 4 = 14$

6. $-7w - 5 = -19$

7. $-2p + 8 = 3$

8. $2b - \dfrac{1}{4} = 5$

9. $0.2x + 3.1 = -5.3$

10. $-1.8 + 2.4a = -6.6$

11. $\dfrac{5}{8} = \dfrac{1}{4} - \dfrac{1}{2}p$

12. $\dfrac{6}{7} = \dfrac{1}{7} + \dfrac{5}{3}r$

13. $7w - 6w + 1 = 10 - 4$

14. $5v - 3 - 4v = 13$

15. $11h - 8 - 9h = -16$

16. $6u - 5 - 8u = -7$

17. $3a + 7 = 2a - 19$

18. $6b - 20 = 14 + 5b$

19. $-4r - 28 = -58 - r$

20. $-6x - 7 = -3 - 8x$

21. $-2z - 8 = -z$

22. $-7t + 4 = -6t$

23. $\dfrac{5}{6}x + \dfrac{2}{3} = -\dfrac{1}{6}x - \dfrac{5}{3}$

24. $\dfrac{3}{7}x - \dfrac{1}{4} = -\dfrac{4}{7}x - \dfrac{5}{4}$

25. $3y - 2 = 5y - 2$

26. $4 + 10t = -8t + 4$

Concept 2: Procedure for Solving a Linear Equation in One Variable

For Exercises 27–48, solve each equation using the steps outlined in the text. **(See Examples 4–7.)**

27. $4q + 14 = 2$

28. $6 = 7m - 1$

29. $-9 = 4n - 1$

30. $-\dfrac{1}{2} - 4x = 8$

31. $3(2p - 4) = 15$

32. $4(t + 15) = 20$

33. $6(3x + 2) - 10 = -4$

34. $4(2k + 1) - 1 = 5$

35. $3.4x - 2.5 = 2.8x + 3.5$

36. $5.8w + 1.1 = 6.3w + 5.6$

37. $17(s + 3) = 4(s - 10) + 13$

38. $5(4 + p) = 3(3p - 1) - 9$

39. $6(3t - 4) + 10 = 5(t - 2) - (3t + 4)$

40. $-5y + 2(2y + 1) = 2(5y - 1) - 7$

41. $5 - 3(x + 2) = 5$

42. $1 - 6(2 - h) = 7$

43. $3(2z - 6) - 4(3z + 1) = 5 - 2(z + 1)$

44. $-2(4a + 3) - 5(2 - a) = 3(2a + 3) - 7$

45. $-2[(4p + 1) - (3p - 1)] = 5(3 - p) - 9$

46. $5 - (6k + 1) = 2[(5k - 3) - (k - 2)]$

47. $3(-0.9n + 0.5) = -3.5n + 1.3$

48. $7(0.4m - 0.1) = 5.2m + 0.86$

Concept 3: Conditional Equations, Identities, and Contradictions

For Exercises 49–54, solve each equation. Identify as a conditional equation, an identity, or a contradiction. **(See Example 8.)**

49. $2(k - 7) = 2k - 13$

50. $5h + 4 = 5(h + 1) - 1$

51. $7x + 3 = 6(x - 2)$

52. $3y - 1 = 1 + 3y$

53. $3 - 5.2p = -5.2p + 3$

54. $2(q + 3) = 4q + q - 9$

55. A conditional linear equation has (choose one): one solution, no solution, or infinitely many solutions.

56. An equation that is a contradiction has (choose one): one solution, no solution, or infinitely many solutions.

57. An equation that is an identity has (choose one): one solution, no solution, or infinitely many solutions.

58. If the only solution to a linear equation is 5, then is the equation a conditional equation, an identity, or a contradiction?

Mixed Exercises

For Exercises 59–82, solve each equation.

59. $4p - 6 = 8 + 2p$

60. $\dfrac{1}{2}t - 2 = 3$

61. $2k - 9 = -8$

62. $3(y - 2) + 5 = 5$

63. $7(w - 2) = -14 - 3w$

64. $0.24 = 0.4m$

65. $2(x + 2) - 3 = 2x + 1$

66. $n + \dfrac{1}{4} = -\dfrac{1}{2}$

67. $0.5b = -23$

68. $3(2r + 1) = 6(r + 2) - 6$
69. $8 - 2q = 4$
70. $\dfrac{x}{7} - 3 = 1$

71. $2 - 4(y - 5) = -4$
72. $4 - 3(4p - 1) = -8$
73. $0.4(a + 20) = 6$

74. $2.2r - 12 = 3.4$
75. $10(2n + 1) - 6 = 20(n - 1) + 12$
76. $\dfrac{2}{5}y + 5 = -3$

77. $c + 0.123 = 2.328$
78. $4(2z + 3) = 8(z - 3) + 36$
79. $\dfrac{4}{5}t - 1 = \dfrac{1}{5}t + 5$

80. $6g - 8 = 4 - 3g$
81. $8 - (3q + 4) = 6 - q$
82. $6w - (8 + 2w) = 2(w - 4)$

Expanding Your Skills

83. Suppose the solution set for x in the equation $x + a = 10$ is $\{-5\}$. Find the value of a.

84. Suppose the solution set for x in the equation $x + a = -12$ is $\{6\}$. Find the value of a.

85. Suppose the solution set for x in the equation $ax = 12$ is $\{3\}$. Find the value of a.

86. Suppose the solution set for x in the equation $ax = 49.5$ is $\{11\}$. Find the value of a.

87. Write an equation that is an identity. Answers may vary.

88. Write an equation that is a contradiction. Answers may vary.

Section 10.3 Linear Equations: Clearing Fractions and Decimals

Concepts

1. Linear Equations Containing Fractions
2. Linear Equations Containing Decimals

1. Linear Equations Containing Fractions

Linear equations that contain fractions can be solved in different ways. The first procedure, illustrated here, uses the method previously outlined.

$$\dfrac{5}{6}x - \dfrac{3}{4} = \dfrac{1}{3}$$

$$\dfrac{5}{6}x - \dfrac{3}{4} + \dfrac{3}{4} = \dfrac{1}{3} + \dfrac{3}{4}$$ To isolate the variable term, add $\dfrac{3}{4}$ to both sides.

$$\dfrac{5}{6}x = \dfrac{4}{12} + \dfrac{9}{12}$$ Find the common denominator on the right-hand side.

$$\dfrac{5}{6}x = \dfrac{13}{12}$$ Simplify.

$$\dfrac{6}{5}\left(\dfrac{5}{6}x\right) = \dfrac{6}{5}\left(\dfrac{13}{12}\right)$$ Multiply by the reciprocal of $\dfrac{5}{6}$, which is $\dfrac{6}{5}$.

$$x = \dfrac{13}{10}$$ The solution set is $\left\{\dfrac{13}{10}\right\}$.

FOR REVIEW

Recall the process to add two fractions with different denominators. First rewrite each fraction as an equivalent fraction using the least common denominator.

$$\dfrac{1}{3} + \dfrac{3}{4} = \dfrac{1 \cdot 4}{3 \cdot 4} + \dfrac{3 \cdot 3}{4 \cdot 3}$$

$$= \dfrac{4}{12} + \dfrac{9}{12}$$

$$= \dfrac{13}{12}$$

Sometimes it is simpler to solve an equation with fractions by eliminating the fractions first by using a process called **clearing fractions**. To clear fractions in the equation $\dfrac{5}{6}x - \dfrac{3}{4} = \dfrac{1}{3}$, we can apply the multiplication property of equality to multiply both sides of the equation by the least common denominator (LCD). In this case, the LCD of $\dfrac{5}{6}x$, $-\dfrac{3}{4}$, and $\dfrac{1}{3}$ is 12. Because each denominator in the equation is a factor of 12, we can simplify common factors to leave integer coefficients for each term.

Section 10.3 Linear Equations: Clearing Fractions and Decimals

Example 1 Solving a Linear Equation by Clearing Fractions

Solve the equation by clearing fractions first. $\frac{5}{6}x - \frac{3}{4} = \frac{1}{3}$

Solution:

$\frac{5}{6}x - \frac{3}{4} = \frac{1}{3}$ The LCD of $\frac{5}{6}x$, $-\frac{3}{4}$, and $\frac{1}{3}$ is 12.

$12\left(\frac{5}{6}x - \frac{3}{4}\right) = 12\left(\frac{1}{3}\right)$ Multiply both sides of the equation by the LCD, 12.

$\frac{\cancel{12}^2}{1}\left(\frac{5}{6}x\right) - \frac{\cancel{12}^3}{1}\left(\frac{3}{4}\right) = \frac{\cancel{12}^4}{1}\left(\frac{1}{3}\right)$ Apply the distributive property (recall that $12 = \frac{12}{1}$).

$2(5x) - 3(3) = 4(1)$ Simplify common factors to clear the fractions.

$10x - 9 = 4$

$10x - 9 + 9 = 4 + 9$ Add 9 to both sides.

$10x = 13$

$\frac{10x}{10} = \frac{13}{10}$ Divide both sides by 10.

$x = \frac{13}{10}$ The solution set is $\left\{\frac{13}{10}\right\}$.

TIP: Recall that the multiplication property of equality indicates that multiplying both sides of an equation by a nonzero constant results in an equivalent equation.

TIP: The fractions in this equation can be eliminated by multiplying both sides of the equation by *any* common multiple of the denominators. These include 12, 24, 36, 48, and so on. We chose 12 because it is the *least* common multiple.

Skill Practice Solve the equation by clearing fractions.

1. $\frac{2}{5}y + \frac{1}{2} = -\frac{7}{10}$

In this section, we combine the process for clearing fractions and decimals with the general strategies for solving linear equations. To solve a linear equation, it is important to follow these steps.

Solving a Linear Equation in One Variable

Step 1 Simplify both sides of the equation.
- Clear parentheses
- Consider clearing fractions and decimals (if any are present) by multiplying both sides of the equation by a common denominator of all terms
- Combine *like* terms

Step 2 Use the addition or subtraction property of equality to collect the variable terms on one side of the equation.

Step 3 Use the addition or subtraction property of equality to collect the constant terms on the other side of the equation.

Step 4 Use the multiplication or division property of equality to make the coefficient of the variable term equal to 1.

Step 5 Check your answer.

Answer

1. $\{-3\}$

Chapter 10 Linear Equations and Inequalities

FOR REVIEW

To find the least common multiple (LCM) of two or more natural numbers, first factor each number into a product of prime factors. The LCM is the product of unique factors, each raised to the highest power to which it appears.

$6 = 2 \cdot 3$
$3 = 3$
$5 = 5$

The LCM is $2 \cdot 3 \cdot 5 = 30$.

Example 2 Solving a Linear Equation Containing Fractions

Solve the equation. $\quad \dfrac{1}{6}x - \dfrac{2}{3} = \dfrac{1}{5}x - 1$

Solution:

$\dfrac{1}{6}x - \dfrac{2}{3} = \dfrac{1}{5}x - 1 \qquad$ The LCD of $\tfrac{1}{6}x$, $-\tfrac{2}{3}$, $\tfrac{1}{5}x$, and $\tfrac{-1}{1}$ is 30.

$30\left(\dfrac{1}{6}x - \dfrac{2}{3}\right) = 30\left(\dfrac{1}{5}x - 1\right) \qquad$ Multiply by the LCD, 30.

$\dfrac{\overset{5}{\cancel{30}}}{1} \cdot \dfrac{1}{\cancel{6}}x - \dfrac{\overset{10}{\cancel{30}}}{1} \cdot \dfrac{2}{\cancel{3}} = \dfrac{\overset{6}{\cancel{30}}}{1} \cdot \dfrac{1}{\cancel{5}}x - 30(1) \qquad$ Apply the distributive property (recall $30 = \tfrac{30}{1}$).

$5x - 20 = 6x - 30 \qquad$ Clear fractions.

$5x - 6x - 20 = 6x - 6x - 30 \qquad$ Subtract $6x$ from both sides.

$-x - 20 = -30$

$-x - 20 + 20 = -30 + 20 \qquad$ Add 20 to both sides.

$-x = -10$

$\dfrac{-x}{-1} = \dfrac{-10}{-1} \qquad$ Divide both sides by -1.

$x = 10 \qquad$ The check is left to the reader.

The solution set is $\{10\}$.

Skill Practice Solve the equation.

2. $\dfrac{2}{5}x - \dfrac{1}{2} = \dfrac{7}{4} + \dfrac{3}{10}x$

Example 3 Solving a Linear Equation Containing Fractions

Solve the equation. $\quad \dfrac{1}{3}(x + 7) - \dfrac{1}{2}(x + 1) = 4$

TIP: In Example 3, both parentheses and fractions are present within the equation. In such a case, we recommend that you clear parentheses first. Then clear the fractions.

Solution:

$\dfrac{1}{3}(x + 7) - \dfrac{1}{2}(x + 1) = 4$

$\dfrac{1}{3}x + \dfrac{7}{3} - \dfrac{1}{2}x - \dfrac{1}{2} = 4 \qquad$ Clear parentheses.

$6\left(\dfrac{1}{3}x + \dfrac{7}{3} - \dfrac{1}{2}x - \dfrac{1}{2}\right) = 6(4) \qquad$ The LCD of $\tfrac{1}{3}x$, $\tfrac{7}{3}$, $-\tfrac{1}{2}x$, $-\tfrac{1}{2}$, and $\tfrac{4}{1}$ is 6.

$\dfrac{\overset{2}{\cancel{6}}}{1} \cdot \dfrac{1}{\cancel{3}}x + \dfrac{\overset{2}{\cancel{6}}}{1} \cdot \dfrac{7}{\cancel{3}} + \dfrac{\overset{3}{\cancel{6}}}{1}\left(-\dfrac{1}{\cancel{2}}x\right) + \dfrac{\overset{3}{\cancel{6}}}{1}\left(-\dfrac{1}{\cancel{2}}\right) = 6(4) \qquad$ Apply the distributive property.

Avoiding Mistakes

When multiplying an equation by the LCD, be sure to multiply all terms on both sides of the equation, including terms that are not fractions.

$2x + 14 - 3x - 3 = 24$

$-x + 11 = 24 \qquad$ Combine *like* terms.

$-x + 11 - 11 = 24 - 11 \qquad$ Subtract 11.

$-x = 13$

$\dfrac{-x}{-1} = \dfrac{13}{-1} \qquad$ Divide by -1.

$x = -13 \qquad$ The check is left to the reader.

The solution set is $\{-13\}$.

Answer

2. $\left\{\dfrac{45}{2}\right\}$

Skill Practice Solve the equation.

3. $\frac{1}{5}(z+1) + \frac{1}{4}(z+3) = 2$

Example 4   Solving a Linear Equation Containing Fractions

Solve the equation. $\dfrac{x-2}{5} - \dfrac{x-4}{2} = 2$

Solution:

$$\frac{x-2}{5} - \frac{x-4}{2} = \frac{2}{1}$$

The LCD of $\frac{x-2}{5}$, $\frac{x-4}{2}$, and $\frac{2}{1}$ is 10.

$$10\left(\frac{x-2}{5} - \frac{x-4}{2}\right) = 10\left(\frac{2}{1}\right)$$

Multiply both sides by 10.

$$\frac{\overset{2}{\cancel{10}}}{1} \cdot \left(\frac{x-2}{\cancel{5}}\right) - \frac{\overset{5}{\cancel{10}}}{1} \cdot \left(\frac{x-4}{\cancel{2}}\right) = \frac{10}{1} \cdot \left(\frac{2}{1}\right)$$

Apply the distributive property.

$$2(x-2) - 5(x-4) = 20$$ Clear fractions.

$$2x - 4 - 5x + 20 = 20$$ Apply the distributive property.

$$-3x + 16 = 20$$ Simplify both sides of the equation.

$$-3x + 16 - 16 = 20 - 16$$ Subtract 16 from both sides.

$$-3x = 4$$

$$\frac{-3x}{-3} = \frac{4}{-3}$$ Divide both sides by -3.

$$x = -\frac{4}{3}$$ The check is left to the reader.

The solution set is $\left\{-\dfrac{4}{3}\right\}$.

> **Avoiding Mistakes**
>
> In Example 4, several of the fractions in the equation have two terms in the numerator. It is important to enclose these fractions in parentheses when clearing fractions. In this way, we will remember to use the distributive property to multiply the factors shown in blue with both terms from the numerator of the fractions.

Skill Practice Solve the equation.

4. $\dfrac{x+1}{4} + \dfrac{x+2}{6} = 1$

---

2. Linear Equations Containing Decimals

The same procedure used to clear fractions in an equation can be used to **clear decimals**. For example, consider the equation

$$2.5x + 3 = 1.7x - 6.6$$

Recall that any terminating decimal can be written as a fraction. Therefore, the equation can be interpreted as

$$\frac{25}{10}x + 3 = \frac{17}{10}x - \frac{66}{10}$$

A convenient common denominator of all terms is 10. Therefore, we can multiply the original equation by 10 to clear decimals.

$$10(2.5x + 3) = 10(1.7x - 6.6)$$
$$25x + 30 = 17x - 66$$

Answers

3. $\left\{\dfrac{7}{3}\right\}$ 4. $\{1\}$

Multiplying by the appropriate power of 10 moves the decimal points so that all coefficients become integers.

Example 5 — Solving a Linear Equation Containing Decimals

Solve the equation by clearing decimals. $\quad 2.5x + 3 = 1.7x - 6.6$

Solution:

$$2.5x + 3 = 1.7x - 6.6$$

$$10(2.5x + 3) = 10(1.7x - 6.6) \quad \text{Multiply both sides of the equation by 10.}$$

$$25x + 30 = 17x - 66 \quad \text{Apply the distributive property.}$$

$$25x - 17x + 30 = 17x - 17x - 66 \quad \text{Subtract } 17x \text{ from both sides.}$$

$$8x + 30 = -66$$

$$8x + 30 - 30 = -66 - 30 \quad \text{Subtract 30 from both sides.}$$

$$8x = -96$$

$$\frac{8x}{8} = \frac{-96}{8} \quad \text{Divide both sides by 8.}$$

$$x = -12 \quad \text{The check is left to the reader.}$$

The solution set is $\{-12\}$.

TIP: Notice that multiplying a decimal number by 10 has the effect of moving the decimal point one place to the right. Similarly, multiplying by 100 moves the decimal point two places to the right, and so on.

Skill Practice Solve the equation.

5. $1.2w + 3.5 = 2.1 + w$

Example 6 — Solving a Linear Equation Containing Decimals

Solve the equation by clearing decimals. $\quad 0.2(x + 4) - 0.45(x + 9) = 12$

Solution:

$$0.2(x + 4) - 0.45(x + 9) = 12$$

$$0.2x + 0.8 - 0.45x - 4.05 = 12 \quad \text{Clear parentheses first.}$$

$$100(0.2x + 0.8 - 0.45x - 4.05) = 100(12) \quad \text{Multiply both sides by 100.}$$

$$20x + 80 - 45x - 405 = 1200 \quad \text{Apply the distributive property.}$$

$$-25x - 325 = 1200 \quad \text{Simplify both sides.}$$

$$-25x - 325 + 325 = 1200 + 325 \quad \text{Add 325 to both sides.}$$

$$-25x = 1525$$

$$\frac{-25x}{-25} = \frac{1525}{-25} \quad \text{Divide both sides by } -25.$$

$$x = -61 \quad \text{The check is left to the reader.}$$

The solution set is $\{-61\}$.

TIP: The terms with the most digits following the decimal point are $-0.45x$ and -4.05. Each of these is written to the hundredths place. Therefore, we multiply both sides by 100.

Skill Practice Solve the equation.

6. $0.25(x + 2) - 0.15(x + 3) = 4$

Answers
5. $\{-7\}$ **6.** $\{39.5\}$

Section 10.3 Activity

A.1. Consider the equation $\frac{1}{3}x - \frac{5}{6} = 2 + \frac{1}{4}x$.

a. Which of the following value(s) can be used to clear the fractions?
2, 3, 4, 6, 12, 15, 18, 21, 24, 28, 36, 40, 48, 60

b. Multiply. $12 \cdot \left(\frac{1}{3}x\right)$

c. Multiply. $12 \cdot \left(-\frac{5}{6}\right)$

d. Multiply. $12 \cdot (2)$

e. Multiply. $12 \cdot \left(\frac{1}{4}x\right)$

f. Solve the equation.

A.2. Consider the equation $-4.15 - 2.5x = 17.2 + x$.

a. Which of the following value(s) can be used to clear the decimals?
10, 100, 1000, 10,000, 100,000

b. Multiply. $100 \cdot (-4.15)$

c. Multiply. $100 \cdot (-2.5x)$

d. Multiply. $100 \cdot (17.2)$

e. Multiply. $100 \cdot (x)$

f. Solve the equation.

For Exercises A.3–A.5, divide your paper into three columns. Solve the equation in the left column, explain each step in the middle column, and check the result in the right column.

Solve the equation	Explain	Check
$\frac{2}{3}(y - 5) = \frac{1}{6}(2y + 2)$		$\frac{2}{3}[(\ ) - 5] = \frac{1}{6}[2(\ ) + 2]$

A.3. $\frac{2}{3}(y - 5) = \frac{1}{6}(2y + 2)$

A.4. $0.05(2x - 1) - 0.12(x + 3) = 0.91$

A.5. $\frac{2y + 3}{4} - \frac{y - 1}{5} = 2$

Practice Exercises

Section 10.3

Prerequisite Review

For Exercises R.1–R.4, identify the least common denominator of the fractions.

R.1. $\frac{1}{2}, -\frac{2}{3},$ and $\frac{9}{4}$

R.2. $-\frac{4}{15}, \frac{1}{3}, \frac{6}{5},$ and $\frac{3}{20}$

R.3. $\frac{3}{10}, -\frac{53}{100},$ and $\frac{7}{1000}$

R.4. $6, \frac{9}{100},$ and $\frac{551}{10,000}$

For Exercises R.5–R.10, solve the equation.

R.5. $5(x + 2) - 3 = 4x + 7$

R.6. $-2(2x + 3) = -2(3 - 4x)$

R.7. $7x + 2 = 7(x - 12)$

R.8. $-4(3 - x) - x = 3x + 6$

R.9. $2(3x - 6) = 3(2x - 4)$

R.10. $-4(5 - x) + 3 = 4(x - 4) - 1$

Vocabulary and Key Concepts

1. a. The process of eliminating fractions in an equation by multiplying both sides of the equation by the LCD is called _____ _____.

    b. The process of eliminating decimals in an equation by multiplying both sides of the equation by a power of 10 is called _____ _____.

Concept 1: Linear Equations Containing Fractions

For Exercises 2–8, clear parentheses by applying the distributive property.

2. $12\left(\dfrac{2}{3}x + \dfrac{3}{4}\right)$

3. $15\left(\dfrac{3}{5}w + \dfrac{2}{3}\right)$

4. $10\left(\dfrac{2c-3}{5} - \dfrac{c+1}{2}\right)$

5. $8\left(\dfrac{4d-3}{8} - \dfrac{3d+2}{4}\right)$

6. $100(0.09x + 1.2)$

7. $1000(0.025t - 0.04)$

8. $10(5.5y - 3)$

For Exercises 9–14, determine which of the values could be used to clear fractions or decimals in the given equation.

9. $\dfrac{2}{3}x - \dfrac{1}{6} = \dfrac{x}{9}$

    Values: 6, 9, 12, 18, 24, 36

10. $\dfrac{1}{4}x - \dfrac{2}{7} = \dfrac{1}{2}x + 2$

    Values: 4, 7, 14, 21, 28, 42

11. $0.02x + 0.5 = 0.35x + 1.2$

    Values: 10; 100; 1000; 10,000

12. $0.003 - 0.002x = 0.1x$

    Values: 10; 100; 1000; 10,000

13. $\dfrac{1}{6}x + \dfrac{7}{10} = x$

    Values: 3, 6, 10, 30, 60

14. $2x - \dfrac{5}{2} = \dfrac{x}{3} - \dfrac{1}{4}$

    Values: 2, 3, 4, 6, 12, 24

For Exercises 15–36, solve each equation. **(See Examples 1–4.)**

15. $\dfrac{1}{2}x + 3 = 5$

16. $\dfrac{1}{3}y - 4 = 9$

17. $\dfrac{2}{15}z + 3 = \dfrac{7}{5}$

18. $\dfrac{1}{6}y + 2 = \dfrac{5}{12}$

19. $\dfrac{1}{3}q + \dfrac{3}{5} = \dfrac{1}{15}q - \dfrac{2}{5}$

20. $\dfrac{3}{7}x - 5 = \dfrac{24}{7}x + 7$

21. $\dfrac{12}{5}w + 7 = 31 - \dfrac{3}{5}w$

22. $-\dfrac{1}{9}p - \dfrac{5}{18} = -\dfrac{1}{6}p + \dfrac{1}{3}$

23. $\dfrac{1}{4}(3m - 4) - \dfrac{1}{5} = \dfrac{1}{4}m + \dfrac{3}{10}$

24. $\dfrac{1}{25}(20 - t) = \dfrac{4}{25}t - \dfrac{3}{5}$

25. $\dfrac{1}{6}(5s + 3) = \dfrac{1}{2}(s + 11)$

26. $\dfrac{1}{12}(4n - 3) = \dfrac{1}{4}(2n + 1)$

27. $\dfrac{2}{3}x + 4 = \dfrac{2}{3}x - 6$

28. $-\dfrac{1}{9}a + \dfrac{2}{9} = \dfrac{1}{3} - \dfrac{1}{9}a$

29. $\dfrac{1}{6}(2c - 1) = \dfrac{1}{3}c - \dfrac{1}{6}$

30. $\dfrac{3}{2}b - 1 = \dfrac{1}{8}(12b - 8)$

31. $\dfrac{2x+1}{3} + \dfrac{x-1}{3} = 5$

32. $\dfrac{4y-2}{5} - \dfrac{y+4}{5} = -3$

33. $\dfrac{3w-2}{6} = 1 - \dfrac{w-1}{3}$

34. $\dfrac{z-7}{4} = \dfrac{6z-1}{8} - 2$

35. $\dfrac{x+3}{3} - \dfrac{x-1}{2} = 4$

36. $\dfrac{5y-1}{2} - \dfrac{y+4}{5} = 1$

Concept 2: Linear Equations Containing Decimals

For Exercises 37–54, solve each equation. (See Examples 5–6.)

37. $9.2y - 4.3 = 50.9$
38. $-6.3x + 1.5 = -4.8$
39. $0.05z + 0.2 = 0.15z - 10.5$
40. $21.1w + 4.6 = 10.9w + 35.2$
41. $0.2p - 1.4 = 0.2(p - 7)$
42. $0.5(3q + 87) = 1.5q + 43.5$
43. $0.20x + 53.60 = x$
44. $z + 0.06z = 3816$
45. $0.15(90) + 0.05p = 0.1(90 + p)$
46. $0.25(60) + 0.10x = 0.15(60 + x)$
47. $0.40(y + 10) - 0.60(y + 2) = 2$
48. $0.75(x - 2) + 0.25(x + 4) = 0.5$
49. $0.12x + 3 - 0.8x = 0.22x - 0.6$
50. $0.4x + 0.2 = -3.6 - 0.6x$
51. $0.06(x - 0.5) = 0.06x + 0.01$
52. $0.125x = 0.025(5x + 1)$
53. $-3.5x + 1.3 = -0.3(9x - 5)$
54. $x + 4 = 2(0.4x + 1.3)$

Mixed Exercises

For Exercises 55–64, solve each equation.

55. $0.2x - 1.8 = -3$
56. $9.8h + 2 = 3.8h + 20$
57. $\frac{1}{4}(x + 4) = \frac{1}{5}(2x + 3)$
58. $\frac{2}{3}(y - 1) = \frac{3}{4}(3y - 2)$
59. $0.05(2t - 1) - 0.03(4t - 1) = 0.2$
60. $0.3(x + 6) - 0.7(x + 2) = 4$
61. $\frac{2k + 5}{4} = 2 - \frac{k + 2}{3}$
62. $\frac{3d - 4}{6} + 1 = \frac{d + 1}{8}$
63. $\frac{1}{8}v + \frac{2}{3} = \frac{1}{6}v + \frac{3}{4}$
64. $\frac{2}{5}z - \frac{1}{4} = \frac{3}{10}z + \frac{1}{2}$

Expanding Your Skills

For Exercises 65–68, solve each equation.

65. $\frac{1}{2}a + 0.4 = -0.7 - \frac{3}{5}a$
66. $\frac{3}{4}c - 0.11 = 0.23(c - 5)$
67. $0.8 + \frac{7}{10}b = \frac{3}{2}b - 0.8$
68. $0.78 - \frac{1}{25}h = \frac{3}{5}h - 0.5$

Problem Recognition Exercises

Equations vs. Expressions

In this set of exercises, we review the difference between expressions that can be *simplified* and equations that can be *solved*. Remember that an equation has an equal sign (=) and an expression does not. For example, compare the following.

This is an expression and can be simplified by clearing parentheses and combining *like* terms.

$$\frac{7}{2}(6x - 4) + 8 + \frac{1}{3}(9 - 3x)$$

$$= \frac{7}{2} \cdot (6x) - \frac{7}{2} \cdot (4) + 8 + \frac{1}{3} \cdot (9) - \frac{1}{3} \cdot (3x)$$

$$= 21x - 14 + 8 + 3 - x$$

$$= 20x - 3 \quad \text{The expression is simplified.}$$

This is an equation (notice the = sign). To solve the equation, simplify both sides, then apply properties of equality to isolate x.

$$\frac{7}{2}(6x - 4) + 8 = \frac{1}{3}(9 - 3x)$$

$$\frac{7}{2} \cdot (6x) - \frac{7}{2} \cdot (4) + 8 = \frac{1}{3} \cdot (9) - \frac{1}{3} \cdot (3x)$$

$$21x - 14 + 8 = 3 - x$$
$$21x - 6 = 3 - x$$
$$22x - 6 = 3$$
$$22x = 9$$
$$x = \frac{9}{22}$$

The solution set is $\left\{\frac{9}{22}\right\}$.

For Exercises 1–32, identify each exercise as an expression or an equation. Then simplify the expression or solve the equation.

1. $2b + 23 - 6b - 5$

2. $10p - 9 + 2p - 3 + 8p - 18$

3. $\frac{y}{4} = -2$

4. $-\frac{x}{2} = 7$

5. $3(4h - 2) - (5h - 8) = 8 - (2h + 3)$

6. $7y - 3(2y + 5) = 7 - (10 - 10y)$

7. $3(8z - 1) + 10 - 6(5 + 3z)$

8. $-5(1 - x) - 3(2x + 3) + 5$

9. $6c + 3(c + 1) = 10$

10. $-9 + 5(2y + 3) = -7$

11. $0.5(2a - 3) - 0.1 = 0.4(6 + 2a)$

12. $0.07(2v - 4) = 0.1(v - 4)$

13. $-\frac{5}{9}w + \frac{11}{12} = \frac{23}{36}$

14. $\frac{3}{8}t - \frac{5}{8} = \frac{1}{2}t + \frac{1}{8}$

15. $\frac{3}{4}x + \frac{1}{2} - \frac{1}{8}x + \frac{5}{4}$

16. $\frac{7}{3}(6 - 12t) + \frac{1}{2}(4t + 8)$

17. $2z - 7 = 2(z - 13)$

18. $-6x + 2(x + 1) = -2(2x + 3)$

19. $\frac{2x - 1}{4} + \frac{3x + 2}{6} = 2$

20. $\frac{w - 4}{6} - \frac{3w - 1}{2} = -1$

21. $4b - 8 - b = -3b + 2(3b - 4)$

22. $-k - 41 - 2 - k = -2(20 + k) - 3$

23. $\frac{4}{3}(6y - 3) = 0$

24. $\frac{1}{2}(2c - 4) + 3 = \frac{1}{3}(6c + 3)$

25. $3(x + 6) - 7(x + 2) - 4(1 - x)$

26. $-10(2k + 1) - 4(4 - 5k) + 25$

27. $3 - 2[4a - 5(a + 1)]$

28. $-9 - 4[3 - 2(q + 3)]$

29. $4 + 2[8 - (6 + x)] = -2(x - 1) - 4 + x$

30. $-1 - 5[2 + 3(w - 2)] = 5(w + 4)$

31. $\frac{1}{6}y + y - \frac{1}{3}(4y - 1)$

32. $\frac{1}{2} - \frac{1}{5}\left(x + \frac{1}{2}\right) + \frac{9}{10}x$

Applications of Linear Equations: Introduction to Problem Solving

Section 10.4

Concepts
1. Problem-Solving Strategies
2. Translations Involving Linear Equations
3. Consecutive Integer Problems
4. Applications of Linear Equations

1. Problem-Solving Strategies

Linear equations can be used to solve many real-world applications. However, with "word problems," students often do not know where to start. To help organize the problem-solving process, we offer the following guidelines:

Problem-Solving Flowchart for Word Problems

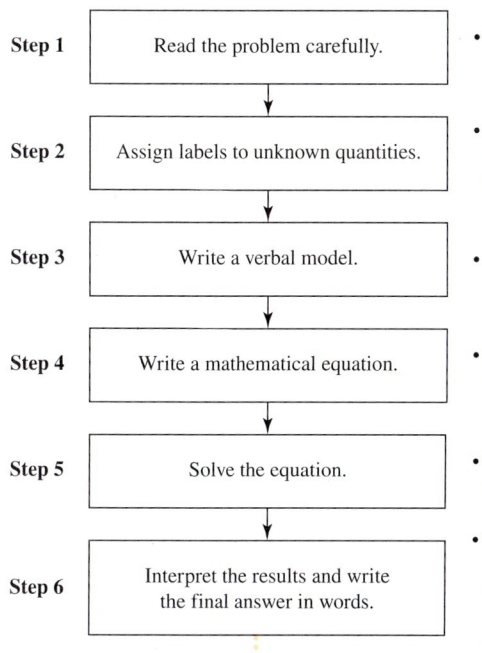

- **Step 1** — Read the problem carefully.
  - Familiarize yourself with the problem. Identify the unknown, and if possible, estimate the answer.
- **Step 2** — Assign labels to unknown quantities.
  - Identify the unknown quantity or quantities. Let x or another variable represent one of the unknowns. Draw a picture and write down relevant formulas.
- **Step 3** — Write a verbal model.
  - Write an equation in *words*.
- **Step 4** — Write a mathematical equation.
  - Replace the verbal model with a mathematical equation using x or another variable.
- **Step 5** — Solve the equation.
  - Solve for the variable using the steps for solving linear equations.
- **Step 6** — Interpret the results and write the final answer in words.
  - Once you have obtained a numerical value for the variable, recall what it represents in the context of the problem. Can this value be used to determine other unknowns in the problem? Write an answer to the word problem in *words*.

Avoiding Mistakes

Once you have reached a solution to a word problem, verify that it is reasonable in the context of the problem.

2. Translations Involving Linear Equations

We have already practiced translating an English sentence to a mathematical equation. Recall that several key words translate to the algebraic operations of addition, subtraction, multiplication, and division.

Example 1 — Translating to a Linear Equation

The sum of a number and negative eleven is negative fifteen. Find the number.

Solution:

Let x represent the unknown number.

$$\underbrace{(\text{a number})}_{\text{the sum of}} + (-11) \underbrace{=}_{\text{is}} (-15)$$

$$x + (-11) = -15$$

$$x + (-11) + 11 = -15 + 11$$

$$x = -4$$

The number is -4.

- **Step 1:** Read the problem.
- **Step 2:** Label the unknown.
- **Step 3:** Write a verbal model.
- **Step 4:** Write an equation.
- **Step 5:** Solve the equation.
- **Step 6:** Write the final answer in words.

Skill Practice

1. The sum of a number and negative seven is 12. Find the number.

Example 2 — Translating to a Linear Equation

Forty less than five times a number is fifty-two less than the number. Find the number.

Solution:

Step 1: Read the problem.

Let x represent the unknown number.

Step 2: Label the unknown.

Step 3: Write a verbal model.

$$\begin{pmatrix}5\text{ times}\\ \text{a number}\end{pmatrix} \underset{\text{less}}{-} (40) \underset{\text{is}}{=} \begin{pmatrix}\text{the}\\ \text{number}\end{pmatrix} \underset{\text{less}}{-} (52)$$

$$5x - 40 = x - 52$$

Step 4: Write an equation.

$$5x - 40 = x - 52$$

Step 5: Solve the equation.

$$5x - x - 40 = x - x - 52$$
$$4x - 40 = -52$$
$$4x - 40 + 40 = -52 + 40$$
$$4x = -12$$
$$\frac{4x}{4} = \frac{-12}{4}$$
$$x = -3$$

The number is -3.

Step 6: Write the final answer in words.

Avoiding Mistakes

It is important to remember that subtraction is not a commutative operation. Therefore, the order in which two real numbers are subtracted affects the outcome. The expression "forty less than five times a number" must be translated as: $5x - 40$ (not $40 - 5x$). Similarly, "fifty-two less than the number" must be translated as: $x - 52$ (not $52 - x$).

Skill Practice

2. Thirteen more than twice a number is 5 more than the number. Find the number.

Example 3 — Translating to a Linear Equation

Twice the sum of a number and six is two more than three times the number. Find the number.

Solution:

Step 1: Read the problem.

Let x represent the unknown number.

Step 2: Label the unknown.

$$\underset{\text{twice}}{2} \cdot \underset{\text{the sum}}{(x + 6)} \underset{\text{is}}{=} \underset{\substack{\text{2 more than} \\ \text{three times} \\ \text{a number}}}{3x + 2}$$

Step 3: Write a verbal model.

Step 4: Write an equation.

Answers

1. The number is 19.
2. The number is -8.

Section 10.4 Applications of Linear Equations: Introduction to Problem Solving   623

$$2(x + 6) = 3x + 2$$
$$2x + 12 = 3x + 2$$
$$2x - 2x + 12 = 3x - 2x + 2$$
$$12 = x + 2$$
$$12 - 2 = x + 2 - 2$$
$$10 = x$$

The number is 10.

Step 5: Solve the equation.

Step 6: Write the final answer in words.

> **Avoiding Mistakes**
>
> It is important to enclose "the sum of a number and six" within parentheses so that the entire quantity is multiplied by 2. Forgetting the parentheses would imply that only the x-term is multiplied by 2.
>
> Correct: $2(x + 6)$
> Incorrect: $2x + 6$

Skill Practice

3. Three times the sum of a number and eight is 4 more than the number. Find the number.

3. Consecutive Integer Problems

The word *consecutive* means "following one after the other in order without gaps." The numbers 6, 7, and 8 are examples of three **consecutive integers**. The numbers $-4, -2, 0,$ and 2 are examples of **consecutive even integers**. The numbers 23, 25, and 27 are examples of **consecutive odd integers**.

Notice that any two consecutive integers differ by 1. Therefore, if x represents an integer, then $(x + 1)$ represents the next larger consecutive integer (Figure 10-4).

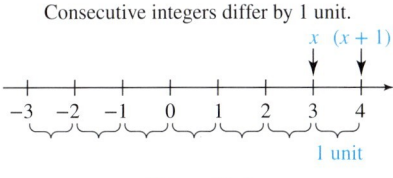

Figure 10-4

Any two consecutive even integers differ by 2. Therefore, if x represents an even integer, then $(x + 2)$ represents the next consecutive larger even integer (Figure 10-5).

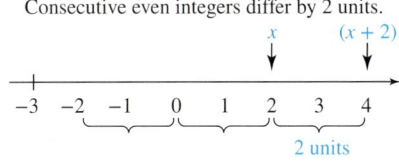

Figure 10-5

Likewise, any two consecutive odd integers differ by 2. If x represents an odd integer, then $(x + 2)$ is the next larger odd integer (Figure 10-6).

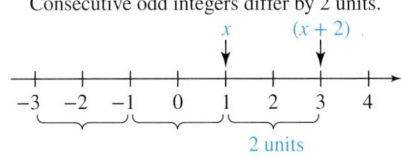

Figure 10-6

Answer

3. The number is -10.

Example 4 Solving an Application Involving Consecutive Integers

The sum of two consecutive odd integers is −188. Find the integers.

Solution:

In this example we have two unknown integers. We can let x represent either of the unknowns.

Step 1: Read the problem.

Suppose x represents the first odd integer.   **Step 2:** Label the unknowns.

Then $(x + 2)$ represents the second odd integer.

$$\begin{pmatrix}\text{First}\\\text{integer}\end{pmatrix} + \begin{pmatrix}\text{second}\\\text{integer}\end{pmatrix} = (\text{total})$$ **Step 3:** Write a verbal model.

$$x + (x+2) = -188$$ **Step 4:** Write a mathematical equation.

$$x + (x+2) = -188$$
$$2x + 2 = -188$$ **Step 5:** Solve for x.
$$2x + 2 - 2 = -188 - 2$$
$$2x = -190$$
$$\frac{2x}{2} = \frac{-190}{2}$$
$$x = -95$$

The first integer is $x = -95$.   **Step 6:** Interpret the results and write the answer in words.

The second integer is $x + 2 = -95 + 2 = -93$.

The two integers are −95 and −93.

TIP: With word problems, it is advisable to check that the answer is reasonable.
The numbers −95 and −93 are consecutive odd integers. Furthermore, their sum is −188 as desired.

Skill Practice

4. The sum of two consecutive even integers is 66. Find the integers.

Example 5 Solving an Application Involving Consecutive Integers

Ten times the smallest of three consecutive integers is twenty-two more than three times the sum of the integers. Find the integers.

Solution:

Step 1: Read the problem.

Let x represent the first integer.   **Step 2:** Label the unknowns.

$x + 1$ represents the second consecutive integer.

$x + 2$ represents the third consecutive integer.

Answer

4. The integers are 32 and 34.

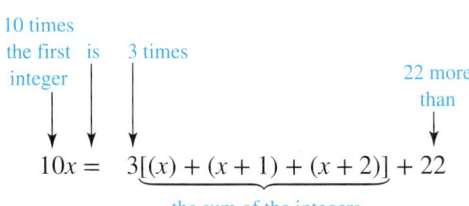

Step 3: Write a verbal model.

$10x = 3[(x) + (x+1) + (x+2)] + 22$ Step 4: Write a mathematical equation.

$10x = 3(x + x + 1 + x + 2) + 22$ Step 5: Solve the equation.

$10x = 3(3x + 3) + 22$ Clear parentheses.

$10x = 9x + 9 + 22$ Combine *like* terms.

$10x = 9x + 31$

$10x - 9x = 9x - 9x + 31$ Isolate the x-terms on one side.

$x = 31$

The first integer is $x = 31$.   Step 6: Interpret the results and write the answer in words.
The second integer is $x + 1 = 31 + 1 = 32$.
The third integer is $x + 2 = 31 + 2 = 33$.

The three integers are 31, 32, and 33.

Skill Practice

5. Five times the smallest of three consecutive integers is 17 less than twice the sum of the integers. Find the integers.

4. Applications of Linear Equations

Example 6 Using a Linear Equation in an Application

A carpenter cuts a 6-ft board in two pieces. One piece must be three times as long as the other. Find the length of each piece.

Solution:

In this problem, one piece must be three times as long as the other. Thus, if x represents the length of one piece, then $3x$ can represent the length of the other.

Step 1: Read the problem completely.

x represents the length of the smaller piece.
$3x$ represents the length of the longer piece.

Step 2: Label the unknowns. Draw a figure.

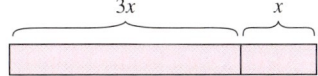

Answer

5. The integers are 11, 12, and 13.

$$\begin{pmatrix}\text{Length of}\\ \text{one piece}\end{pmatrix} + \begin{pmatrix}\text{length of}\\ \text{other piece}\end{pmatrix} = \begin{pmatrix}\text{total length}\\ \text{of the board}\end{pmatrix}$$

Step 3: Write a verbal model.

$$x + 3x = 6$$

Step 4: Write an equation.

$$4x = 6$$

Step 5: Solve the equation.

$$\frac{4x}{4} = \frac{6}{4}$$

$$x = 1.5$$

The smaller piece is $x = 1.5$ ft.

Step 6: Interpret the results.

The longer piece is $3x$ or $3(1.5 \text{ ft}) = 4.5$ ft.

TIP: The variable can represent either unknown. In Example 6, if we let x represent the length of the longer piece of board, then $\frac{1}{3}x$ would represent the length of the smaller piece. The equation would become $x + \frac{1}{3}x = 6$. Try solving this equation and interpreting the result.

Skill Practice

6. A plumber cuts a 96-in. piece of pipe into two pieces. One piece is five times longer than the other piece. How long is each piece?

Example 7 Using a Linear Equation in an Application

In a recent Olympics, the United States won the greatest number of overall medals, followed by China. The United States won 16 more medals than China, and together they brought home a total of 192 medals. How many medals did each country win?

Master Sgt. Lono Kollars/U.S. Department of Defense

Solution:

In this example, we have two unknowns. The variable x can represent either quantity. However, the number of medals won by the United States is given in terms of the number won by China.

Step 1: Read the problem.

Let x represent the number of medals won by China.

Then let $x + 16$ represent the number of medals won by the United States.

Step 2: Label the variables.

$$\begin{pmatrix}\text{Number of}\\ \text{medals won}\\ \text{by China}\end{pmatrix} + \begin{pmatrix}\text{Number of medals}\\ \text{won by the}\\ \text{United States}\end{pmatrix} = \begin{pmatrix}\text{Total}\\ \text{number}\\ \text{of medals}\end{pmatrix}$$

Step 3: Write a verbal model.

$$x + (x + 16) = 192$$

Step 4: Write an equation.

$$2x + 16 = 192$$

Step 5: Solve the equation.

$$2x = 176$$

$$x = 88$$

- Medals won by China, $x = 88$
- Medals won by the United States, $x + 16 = (88) + 16 = 104$

China won 88 medals and the United States won 104 medals.

Answer

6. One piece is 80 in. and the other is 16 in.

Skill Practice

7. There are 40 students in an algebra class. There are 4 more women than men. How many women and how many men are in the class?

Answer
7. There are 22 women and 18 men.

Section 10.4 Activity

A.1. An electrician has 400 ft of electrical wire.
   a. If she uses 56 ft to start the job, how much wire is left?
   b. Is she uses x feet to start the job, how much wire is left?

A.2. Apples cost $1.29 per pound.
   a. If a chef buys 24 lb, what is the cost?
   b. If a chef buys p pounds, what is the cost?

A.3. a. If x represents the smallest of three consecutive integers, write expressions to represent the next two consecutive integers.
   b. If x represents the largest of three consecutive integers, write expressions to represent the previous two consecutive integers.

A.4. a. If x represents the smallest of three consecutive odd integers, write expressions to represent the next two consecutive odd integers.
   b. If x represents the largest of three consecutive odd integers, write expressions to represent the previous two consecutive odd integers.

A.5. Twenty-one more than three times a number is twelve less than the sum of the number and five. To find the number, follow these steps.
   a. If x represents the unknown number, write an equation in terms of x to represent this scenario.
   b. Solve the equation from part (a).
   c. Interpret your answer from part (b).

For Exercises A.6–A.7, format your solution to the word problem as follows:
   a. Read the problem again.
   b. Label the variable(s).
   c. Write a verbal model (that is, write an equation in *words*).
   d. Write an equation to represent the verbal model.
   e. Solve the equation.
   f. Interpret the results in words using the proper units of measurement and verify that your solution makes sense in the context of the problem.

A.6. The perimeter of a triangle is 54 in. The lengths of the sides are represented by three consecutive even integers. Find the lengths of the sides of the triangle.

A.7. A manager wants to separate his 24 employees into two teams. One team has six fewer members than twice the number on the other team. How many employees are on each team?

Section 10.4 Practice Exercises

Prerequisite Review

For Exercises R.1–R.8, translate the English phrase into an algebraic expression. Let x represent the unknown number.

R.1. Sixteen more than a number

R.2. The difference of a number and eleven

R.3. Twelve subtracted from a number

R.4. A number added to negative five

R.5. The product of negative twenty-three and a number

R.6. The quotient of a number and ten

R.7. Negative four times the difference of a number and two

R.8. The sum of a number and six all divided by ten

Vocabulary and Key Concepts

1. **a.** Integers that follow one after the other without "gaps" are called _____ integers.

   b. The integers −2, 0, 2, and 4 are examples of consecutive _____ integers.

   c. The integers −3, −1, 1, and 3 are examples of consecutive _____ integers.

   d. Two consecutive integers differ by _____.

   e. Two consecutive odd integers differ by _____.

   f. Two consecutive even integers differ by _____.

Concept 2: Translations Involving Linear Equations

For Exercises 2–8, write an expression representing the unknown quantity.

2. In a math class, the number of students who received an "A" in the class was 5 more than the number of students who received a "B." If x represents the number of "B" students, write an expression for the number of "A" students.

3. There are 5,682,080 fewer men than women on a particular social media site. If x represents the number of women using that site, write an expression for the number of men using that site.

4. At a recent motorcycle rally, the number of men exceeded the number of women by 216. If x represents the number of women, write an expression for the number of men.

5. There are 10 times as many users of a social media site as there are of a social news site. If x represents the number of users of the news site, write an expression for the number of users of the social media site.

6. Rebecca downloaded twice as many songs as Nigel. If x represents the number of songs downloaded by Nigel, write an expression for the number downloaded by Rebecca.

7. Sidney made $20 less than three times Casey's weekly salary. If x represents Casey's weekly salary, write an expression for Sidney's weekly salary.

8. David scored 26 points less than twice the number of points Rich scored in a video game. If x represents the number of points scored by Rich, write an expression representing the number of points scored by David.

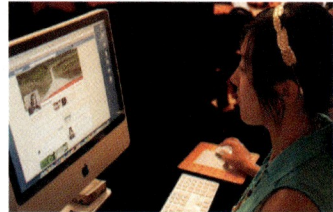
Mark Dierker/McGraw-Hill

For Exercises 9–18, use the Problem-Solving Flowchart for Word Problems. (See Examples 1–3.)

9. Six less than a number is -10. Find the number.

10. Fifteen less than a number is 41. Find the number.

11. Twice the sum of a number and seven is eight. Find the number.

12. Twice the sum of a number and negative two is sixteen. Find the number.

13. A number added to five is the same as twice the number. Find the number.

14. Three times a number is the same as the difference of twice the number and seven. Find the number.

15. The sum of six times a number and ten is equal to the difference of the number and fifteen. Find the number.

16. The difference of fourteen and three times a number is the same as the sum of the number and negative ten. Find the number.

17. If the difference of a number and four is tripled, the result is six less than the number. Find the number.

18. Twice the sum of a number and eleven is twenty-two less than three times the number. Find the number.

Concept 3: Consecutive Integer Problems

19. **a.** If x represents the smallest of three consecutive integers, write an expression to represent each of the next two consecutive integers.

    b. If x represents the largest of three consecutive integers, write an expression to represent each of the previous two consecutive integers.

20. **a.** If x represents the smallest of three consecutive odd integers, write an expression to represent each of the next two consecutive odd integers.

    b. If x represents the largest of three consecutive odd integers, write an expression to represent each of the previous two consecutive odd integers.

For Exercises 21–30, use the Problem-Solving Flowchart for Word Problems. (See Examples 4–5.)

21. The sum of two consecutive integers is -67. Find the integers.

22. The sum of two consecutive odd integers is 52. Find the integers.

23. The sum of two consecutive odd integers is 28. Find the integers.

24. The sum of three consecutive even integers is 66. Find the integers.

25. The perimeter of a pentagon (a five-sided polygon) is 80 in. The five sides are represented by consecutive integers. Find the lengths of the sides.

26. The perimeter of a triangle is 96 in. The lengths of the sides are represented by consecutive integers. Find the lengths of the sides.

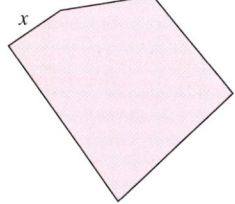

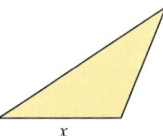

27. The sum of three consecutive even integers is 48 less than twice the smallest of the three integers. Find the integers.

28. The sum of three consecutive odd integers is 89 less than twice the largest integer. Find the integers.

29. Eight times the sum of three consecutive odd integers is 210 more than ten times the middle integer. Find the integers.

30. Five times the sum of three consecutive even integers is 140 more than ten times the smallest of the integers. Find the integers.

Concept 4: Applications of Linear Equations

For Exercises 31–42, use the Problem-Solving Flowchart for Word Problems to solve the problems.

31. A board is 86 cm in length and must be cut so that one piece is 20 cm longer than the other piece. Find the length of each piece. (See Example 6.)

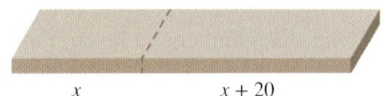

32. A rope is 54 in. in length and must be cut into two pieces. If one piece must be twice as long as the other, find the length of each piece.

33. Karen's music library contains 12 fewer playlists than Clarann's music library. The total number of playlists for both music libraries is 58. Find the number of playlists in each person's music library.

34. Maria has 15 fewer apps on her phone than Orlando. If the total number of apps on both phones is 29, how many apps are on each phone?

35. For a recent year, 31 more Democrats than Republicans were in the U.S. House of Representatives. If the total number of representatives in the House from these two parties was 433, find the number of representatives from each party.

36. For a recent year, the number of men in the U.S. Senate totaled 4 more than five times the number of women. Find the number of men and the number of women in the Senate given that the Senate has 100 members.

37. A car dealership sells SUVs and passenger cars. For a recent year, 40 more SUVs were sold than passenger cars. If a total of 420 vehicles were sold, determine the number of each type of vehicle sold. (See Example 7.)

38. Two of the largest Internet retailers are eBay and Amazon. In one quarter the estimated U.S. sales of eBay were $0.1 billion less than twice the sales of Amazon. Given the total sales of $5.6 billion, determine the sales of eBay and Amazon.

39. The longest river in Africa is the Nile. It is 2455 km longer than the Congo River, also in Africa. The sum of the lengths of these rivers is 11,195 km. What is the length of each river?

40. The average depth of the Gulf of Mexico is three times the depth of the Red Sea. The difference between the average depths is 1078 m. What is the average depth of the Gulf of Mexico and the average depth of the Red Sea?

41. Asia and Africa are the two largest continents in the world. The land area of Asia is approximately 14,514,000 km² larger than the land area of Africa. Together their total area is 74,644,000 km². Find the land area of Asia and the land area of Africa.

42. Mt. Everest, the highest mountain in the world, is 2654 m higher than Denali, the highest mountain in the United States. If the sum of their heights is 15,042 m, find the height of each mountain.

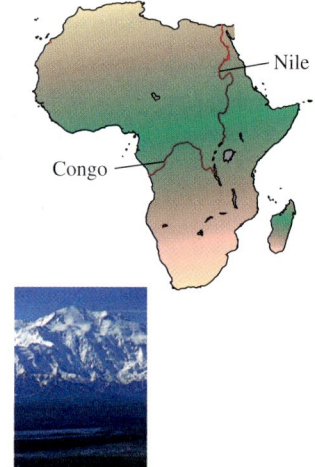

Steven Prorak/Wildnerdpix/123RF

Mixed Exercises

43. A group of hikers walked from Hawk Mountain Shelter to Blood Mountain Shelter along the Appalachian Trail, a total distance of 20.5 mi. It took 2 days for the walk. The second day the hikers walked 4.1 mi less than they did on the first day. How far did they walk each day?

Redlink/Corbis/age fotostock

44. $120 is to be divided among three restaurant servers. Angie made $10 more than Marie. Gwen, who went home sick, made $25 less than Marie. How much money should each server get?

45. A 4-ft piece of PVC pipe is cut into three pieces. The longest piece is 5 in. shorter than three times the shortest piece. The middle piece is 8 in. longer than the shortest piece. How long is each piece?

46. A 6-ft piece of copper wire must be cut into three pieces. The shortest piece is 16 in. less than the middle piece. The longest piece is twice as long as the middle piece. How long is each piece?

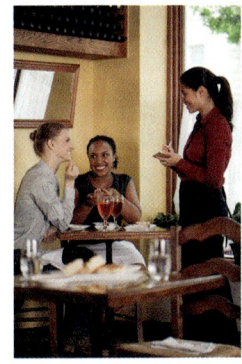

Image Source/Isadora Getty Buyou/Image Source/Getty Images

47. Three consecutive integers are such that three times the largest exceeds the sum of the two smaller integers by 47. Find the integers.

48. Four times the smallest of three consecutive odd integers is 236 more than the sum of the other two integers. Find the integers.

49. The winner and runner-up of a TV music contest had lucrative earnings immediately after the show's finale. The runner-up earned $2 million less than half of the winner's earnings. If their combined earnings totaled $19 million, how much did each person make?

50. One TV series ran 97 fewer episodes than twice the number of a second TV series. If the total number of episodes is 998, determine the number of each show produced.

51. Five times the difference of a number and three is four less than four times the number. Find the number.

52. Three times the difference of a number and seven is one less than twice the number. Find the number.

53. The sum of the page numbers on two facing pages in a book is 941. What are the page numbers?

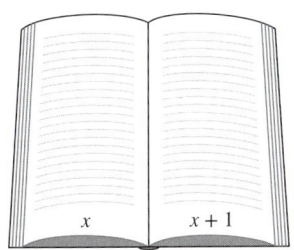

54. Three raffle tickets are represented by three consecutive integers. If the sum of the three integers is 2,666,031, find the numbers.

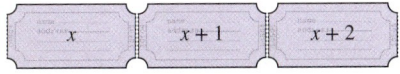

55. If three is added to five times a number, the result is forty-three more than the number. Find the number.

56. If seven is added to three times a number, the result is thirty-one more than the number. Find the number.

57. The deepest point in the Pacific Ocean is 676 m more than twice the deepest point in the Arctic Ocean. If the deepest point in the Pacific is 10,920 m, how many meters is the deepest point in the Arctic Ocean?

58. The area of Greenland is 201,900 km² less than three times the area of New Guinea. What is the area of New Guinea if the area of Greenland is 2,175,600 km²?

59. The sum of twice a number and $\frac{3}{4}$ is the same as the difference of four times the number and $\frac{1}{8}$. Find the number.

60. The difference of a number and $-\frac{11}{12}$ is the same as the difference of three times the number and $\frac{1}{6}$. Find the number.

61. The product of a number and 3.86 is equal to 7.15 more than the number. Find the number.

62. The product of a number and 4.6 is 33.12 less than the number. Find the number.

Section 10.5 Applications Involving Percents

Concepts

1. Basic Percent Equations
2. Applications Involving Simple Interest
3. Applications Involving Discount and Markup

1. Basic Percent Equations

Recall that the word *percent* means "per hundred."

Percent	Interpretation
63% of homes have a computer	63 out of 100 homes have a computer.
5% sales tax	5¢ in tax is charged for every 100¢ in merchandise.
15% commission	$15 is earned in commission for every $100 sold.

Percents come up in a variety of applications in day-to-day life. Many such applications follow the basic percent equation:

$$\text{Amount} = (\text{percent})(\text{base}) \quad \text{Basic percent equation}$$

In Example 1, we apply the basic percent equation to compute sales tax.

Example 1 **Computing Sales Tax**

A new digital camera costs $429.95.

a. Compute the sales tax if the tax rate is 4%.

b. Determine the total cost, including tax.

Solution:

a. Let x represent the amount of tax.

$$\text{Amount} = (\text{percent}) \cdot (\text{base})$$
$$\text{Sales tax} = (\text{tax rate})(\text{price of merchandise})$$

$x = (0.04)(\$429.95)$

$x = \$17.198$

$x = \$17.20$

The tax on the merchandise is $17.20.

Step 1: Read the problem.
Step 2: Label the variable.
Step 3: Write a verbal model. Apply the percent equation to compute sales tax.
Step 4: Write a mathematical equation.
Step 5: Solve the equation. Round to the nearest cent.
Step 6: Interpret the results.

Avoiding Mistakes
Be sure to use the decimal form of a percent within an equation.
$4\% = 0.04$

b. The total cost is found by:

total cost = cost of merchandise + amount of tax

Therefore, the total cost is $429.95 + $17.20 = $447.15.

Skill Practice

1. Find the amount of tax on a portable audio player that sells for $89. Assume the tax rate is 6%.
2. Find the total cost including tax.

In Example 2, we solve a problem in which the percent is unknown.

Example 2 Finding an Unknown Percent

A group of 240 college men were asked what intramural sport they most enjoyed playing. The results are in the graph. What percent of the men surveyed preferred tennis?

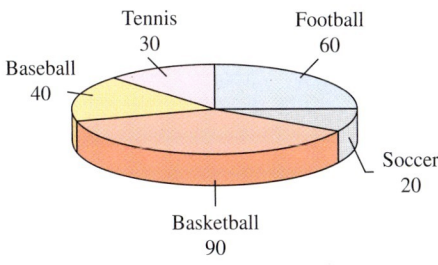

Solution:

Let x represent the unknown percent (in decimal form).

Step 1: Read the problem.
Step 2: Label the variable.

The problem can be rephrased as:

30 is what percent of 240?
↓   ↓   ↓
30 =   x   · 240

Step 3: Write a verbal model.

Step 4: Write a mathematical equation.

$30 = 240x$

Step 5: Solve the equation.

$\dfrac{30}{240} = \dfrac{240x}{240}$

Divide both sides by 240.

$0.125 = x$

$0.125 \times 100\% = 12.5\%$

Step 6: Interpret the results. Change the value of x to a percent form by multiplying by 100%.

In this survey, 12.5% of men prefer tennis.

Skill Practice   Refer to the graph in Example 2.

3. What percent of the men surveyed prefer basketball as their favorite intramural sport?

Answers
1. The amount of tax is $5.34.
2. The total cost is $94.34.
3. 37.5% of the men surveyed prefer basketball.

Example 3 Solving a Percent Equation With an Unknown Base

Andrea spends 20% of her monthly paycheck on rent each month. If her rent payment is $950, what is her monthly paycheck?

Solution:

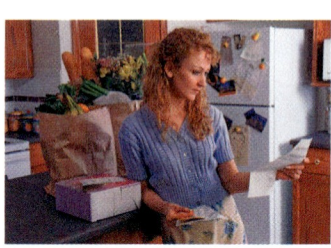

EyeWire Collection/Stockbyte/Getty Images

Let x represent the amount of Andrea's monthly paycheck.

The problem can be rephrased as:

$950 is 20% of what number?

$950 = 0.20 \cdot x$

$950 = 0.20x$

$$\frac{950}{0.20} = \frac{0.20x}{0.20}$$

$4750 = x$

Andrea's monthly paycheck is $4750.

Step 1: Read the problem.
Step 2: Label the variables.

Step 3: Write a verbal model.

Step 4: Write a mathematical equation.

Step 5: Solve the equation.

Divide both sides by 0.20.

Step 6: Interpret the results.

Skill Practice

4. To pass an exam, a student must answer 70% of the questions correctly. If answering 42 questions correctly results in a 70% score, how many questions are on the test?

2. Applications Involving Simple Interest

One important application of percents is in computing simple interest on a loan or on an investment.

Simple interest is interest that is earned or owed on principal (the original amount of money invested or borrowed). The following formula is used to compute simple interest.

$$\begin{pmatrix}\text{Simple} \\ \text{interest}\end{pmatrix} = (\text{principal})\begin{pmatrix}\text{annual} \\ \text{interest rate}\end{pmatrix}\begin{pmatrix}\text{time} \\ \text{in years}\end{pmatrix}$$

This formula is often written symbolically as $I = Prt$. In this formula, I represents the simple interest, P represents the principal, r represents the annual interest rate, and t is the time of the investment in years.

For example, to find the simple interest earned on $2000 invested at 7.5% interest for 3 years, we have $P = \$2000$, $r = 0.075$, and $t = 3$. Thus,

$$I = Prt$$
$$\text{Interest} = (\$2000)(0.075)(3)$$
$$= \$450$$

Answer

4. There are 60 questions on the test.

Example 4 Applying Simple Interest

Jorge wants to save money to buy a car in 5 years. If Jorge needs to have $20,250 at the end of 5 years, how much money would he need to invest in a certificate of deposit (CD) at a 2.5% interest rate?

Solution:

Let P represent the original amount invested.

$$\begin{pmatrix}\text{Original}\\ \text{principal}\end{pmatrix} + (\text{interest}) = (\text{total})$$

$$P + Prt = \text{total}$$
$$P + P(0.025)(5) = 20{,}250$$
$$P + 0.125P = 20{,}250$$
$$1.125P = 20{,}250$$
$$\frac{1.125P}{1.125} = \frac{20{,}250}{1.125}$$
$$P = 18{,}000$$

The original investment should be $18,000.

Step 1: Read the problem.
Step 2: Label the variables.
Step 3: Write a verbal model.
Step 4: Write a mathematical equation.
Step 5: Solve the equation.
Step 6: Interpret the results and write the answer in words.

Freeograph/Shutterstock

Avoiding Mistakes

The interest is computed on the original principal, P, not on the total amount $20,250. That is, the interest is $P(0.025)(5)$, not $(\$20{,}250)(0.025)(5)$.

Skill Practice

5. Cassandra invested some money in her bank account, and after 10 years at 4% simple interest, it has grown to $7700. What was the initial amount invested?

3. Applications Involving Discount and Markup

Applications involving percent increase and percent decrease are abundant in many real-world settings. Sales tax, for example, is essentially a markup by a state or local government. It is important to understand that percent increase or decrease is always computed on the original amount given.

In Example 5, we illustrate an example of percent decrease in an application where merchandise is discounted.

Example 5 Applying Percents to a Discount Problem

After a 38% discount, a used treadmill costs $868. What was the original cost of the treadmill?

Solution:

Let x be the original cost of the treadmill.

Step 1: Read the problem.
Step 2: Label the variables.

Answer
5. The initial investment was $5500.

Comstock Images/Alamy Stock Photo

$$\begin{pmatrix}\text{Original}\\\text{cost}\end{pmatrix} - (\text{discount}) = \begin{pmatrix}\text{sale}\\\text{price}\end{pmatrix}$$

$$x \quad - \quad 0.38(x) \quad = \quad 868$$

Step 3: Write a verbal model.

Step 4: Write a mathematical equation. The discount is a percent of the *original* amount.

$$x - 0.38x = 868$$
$$0.62x = 868$$

Step 5: Solve the equation. Combine *like* terms.

$$\frac{0.62x}{0.62} = \frac{868}{0.62}$$

Divide by 0.62.

$$x = 1400$$

The original cost of the treadmill was $1400.

Step 6: Interpret the result.

Skill Practice

6. A camera is on sale for $151.20. This is after a 20% discount. What was the original cost?

Answer
6. The camera originally cost $189.

Section 10.5 Activity

For Exercises A.1–A.3, translate each statement to an equation. Then solve the equation.

A.1. What is 38% of 81?

A.2. 33.6 is what percent of 28?

A.3. 81 is 6% of what number?

A.4. At a kitchen store, a shopper has a coupon for 15% off an item of any price and a second coupon for $20 off any item of $50 or more. However, only one coupon can be used on a given item.
   a. If the shopper buys a blender for $89, which coupon is better?
   b. If the shopper buys a comforter for $139, which coupon is better?

A.5. Jonathan pays $13,847.20 in federal income tax on taxable income of $72,880. What tax rate is Jonathan paying? Round to the nearest percent.

A.6. Michael borrowed $1700 from his father. He promised to pay the money back with interest in 9 months $\left(\frac{3}{4}\text{ yr}\right)$. If Michael pays his father 6% simple interest, how much does he owe his father at the end of 9 months?

For Exercises A.7–A.9, follow these steps to solve the application.
   a. Read the problem again.
   b. Label the variable(s).
   c. Write a verbal model (that is, write an equation in *words*).
   d. Write an equation to represent the verbal model.
   e. Solve the equation.
   f. Interpret the results in words using the proper units of measurement and verify that your solution makes sense in the context of the problem.

A.7. A furniture store purchases kitchen furniture directly from the manufacturer. The store then sells the furniture at a 40% markup. If the store sells a kitchen table for $232.40, what was the original price from the manufacturer?

A.8. The total cost (including tax) of a pair of shoes is $65.10. If the sales tax rate is 5%, what is the retail price of the shoes before tax?

A.9. A store employee makes $60 for an 8-hr shift plus a 10% commission on sales. If the employee made $148 for an 8-hr shift, what amount of merchandise did she sell?

Practice Exercises   Section 10.5

Prerequisite Review

For Exercises R.1–R.8, match the given statement with one of the values a–h.

R.1. 25% _____   a. three-quarters of a whole

R.2. 90% _____   b. one-hundredth of a whole

R.3. 1%   _____   c. half of a whole

R.4. 50% _____   d. one whole

R.5. $66.\overline{6}\%$ _____   e. nine-tenths of a whole

R.6. 100% _____   f. two-thirds of a whole

R.7. 75% _____   g. one-quarter of a whole

R.8. 10% _____   h. one-tenth of a whole

R.9. Compute a 6% sales tax on a blouse that costs $20.

R.10. Compute the commission if a salesperson makes a 20% commission on a piece of furniture that costs $820.

R.11. How would you estimate a 15% tip on a meal that costs $60?

R.12. How would you estimate a 40% discount on a $250 kitchen table?

Vocabulary and Key Concepts

1. a. Interest that is earned on principal is called _____ interest.

   b. 82% means 82 out of _____.

2. If a pair of binoculars normally costs $99 but is on sale for 10% off, mentally compute the amount of the discount.

3. If a meal costs $40 and you want to leave a 15% tip, compute the tip mentally. (*Hint:* Determine 10% of the bill and then add half of that value to get 15%.)

4. Suppose that 2400 students apply to a college each year and 25% of students are accepted. Mentally compute the number of students accepted.

Concept 1: Basic Percent Equations

For Exercises 5–16, find the missing values.

5. 45 is what percent of 360?

6. 338 is what percent of 520?

7. 544 is what percent of 640?

8. 576 is what percent of 800?

9. What is 0.5% of 150?

10. What is 9.5% of 616?

11. What is 142% of 740?

12. What is 156% of 280?

13. 177 is 20% of what number?

14. 126 is 15% of what number?

15. 275 is 12.5% of what number?

16. 594 is 45% of what number?

17. A drill is on sale for $99.99. If the sales tax rate is 7%, how much will Molly have to pay for the drill? **(See Example 1.)**

18. Patrick purchased four new tires that were regularly priced at $94.99 each, but are on sale for $20 off per tire. If the sales tax rate is 6%, how much will be charged to Patrick's VISA card?

For Exercises 19–22, use the graph showing the distribution for leading forms of cancer in men. (*Source:* Centers for Disease Control)

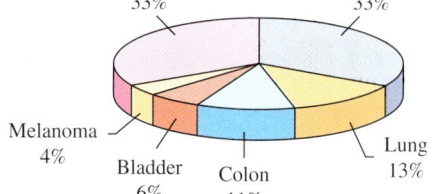

Percent of Cancer Cases by Type (Men)

19. If there are 700,000 cases of cancer in men in the United States, approximately how many are prostate cancer?

20. Approximately how many cases of lung cancer would be expected in 700,000 cancer cases among men in the United States?

21. There were 14,000 cases of cancer of the pancreas diagnosed out of 700,000 cancer cases. What percent is this? **(See Example 2.)**

22. There were 21,000 cases of leukemia diagnosed out of 700,000 cancer cases. What percent is this?

23. Javon is in a 28% tax bracket for his federal income tax. If the amount of money that he paid for federal income tax was $23,520, what was his taxable income? **(See Example 3.)**

24. In a recent survey of college-educated adults, 155 indicated that they regularly work more than 50 hr a week. If this represents 31% of those surveyed, how many people were in the survey?

Concept 2: Applications Involving Simple Interest

25. Aidan is trying to save money and has $1800 to set aside in some type of savings account. He checked his bank one day, and found that the rate for a 12-month CD had an annual percentage yield (APY) of 1.25%. The interest rate on his savings account was 0.75% APY. How much more simple interest would Aidan earn if he invested in a CD for 12 months rather than leaving the $1800 in a regular savings account?

26. How much interest will Roxanne have to pay if she borrows $2000 for 2 years at a simple interest rate of 4%?

27. Bob borrowed money for 1 year at 5% simple interest. If he had to pay back a total of $1260, how much did he originally borrow? **(See Example 4.)**

28. Andrea borrowed some money for 2 years at 6% simple interest. If she had to pay back a total of $3640, how much did she originally borrow?

29. If $1500 grows to $1950 after 5 years, find the simple interest rate.

30. If $9000 grows to $10,440 in 2 years, find the simple interest rate.

31. Perry is planning a vacation to Europe in 2 years. How much should he invest in an account that pays 3% simple interest to get the $3500 that he needs for the trip? Round to the nearest dollar.

32. Sherica invested in a mutual fund, and at the end of 20 years she has $14,300 in her account. If the mutual fund returned an average yield of 8%, how much did she originally invest?

Concept 3: Applications Involving Discount and Markup

33. A hands-free kit for a car costs $62. An electronics store has it on sale for 12% off with free installation.
    a. What is the discount on the hands-free kit?
    b. What is the sale price?

34. A tablet originally selling for $550 is on sale for 10% off.
    a. What is the discount on the tablet?
    b. What is the sale price?

35. A digital camera is on sale for $400. This price is 15% off the original price. What was the original price? Round to the nearest cent. (See Example 5.)

36. A sun hat with UV protection is on sale for $18. If this represents an 18% discount rate, what was the original price?

37. The original price of an Audio Jukebox was $250. It is on sale for $220. What percent discount does this represent?

38. During its first year, a gaming console sold for $425 in stores. This product was in such demand that it sold for $800 online. What percent markup does this represent? (Round to the nearest whole percent.)

39. A doctor ordered a dosage of medicine for a patient. After 2 days, she increased the dosage by 20% and the new dosage came to 18 cc. What was the original dosage?

40. In one area, the cable company marked up the monthly cost by 6%. The new cost is $63.60 per month. What was the cost before the increase?

Mixed Exercises

41. Sun Lei bought a laptop computer for $1800. The total cost, including tax, came to $1890. What is the tax rate?

Creatas/Getty Images

42. Jamie purchased a beach umbrella and paid $32.04, including tax. If the price before tax is $29.80, what is the sales tax rate (round to the nearest tenth of a percent)?

Tero Hakala/123RF

43. To discourage tobacco use and to increase state revenue, several states tax tobacco products. One year, the state of New York increased taxes on tobacco, resulting in an 11% increase in the retail price of a pack of cigarettes. If the new price of a pack of cigarettes is $12.85, what was the cost before the increase in tax?

44. A hotel room rented for 5 nights costs $706.25 including 13% in taxes. Find the original price of the room (before tax) for the 5 nights. Then find the price per night.

45. Deon purchased a house and sold it for a 24% profit. If he sold the house for $260,400, what was the original purchase price?

46. To meet the rising cost of energy, the yearly membership at a YMCA had to be increased by 12.5% from the past year. The yearly membership fee is currently $450. What was the cost of membership last year?

Brand X Pictures/Stockbyte/Getty Images

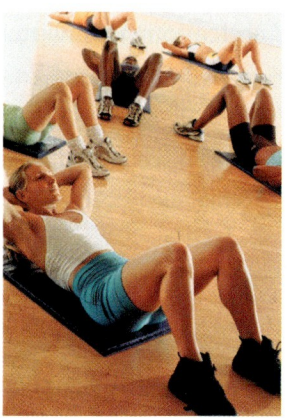

Corbis/VCG/Getty Images

47. Alina earns $1600 per month plus a 12% commission on pharmaceutical sales. If she sold $25,000 in pharmaceuticals one month, what was her salary that month?

48. Dan sold a beachfront home for $650,000. If his commission rate is 4%, what did he earn in commission?

49. Diane sells women's sportswear at a department store. She earns a regular salary and, as a bonus, she receives a commission of 4% on all sales over $200. If Diane earned an extra $25.80 last week in commission, how much merchandise did she sell over $200?

50. For selling software, Tom received a bonus commission based on sales over $500. If he received $180 in commission for selling a total of $2300 worth of software, what is his commission rate?

Section 10.6 Formulas and Applications of Geometry

Concepts

1. Literal Equations and Formulas
2. Geometry Applications

1. Literal Equations and Formulas

A *literal equation* is an equation that has more than one variable. A formula is a literal equation with a specific application. For example, the perimeter of a triangle (distance around the triangle) can be found by the formula $P = a + b + c$, where a, b, and c are the lengths of the sides (Figure 10-7).

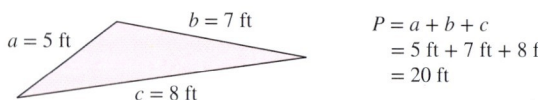

$$P = a + b + c$$
$$= 5 \text{ ft} + 7 \text{ ft} + 8 \text{ ft}$$
$$= 20 \text{ ft}$$

Figure 10-7

In this section, we will learn how to rewrite formulas to solve for a different variable within the formula. Suppose, for example, that the perimeter of a triangle is known and two of the sides are known (say, sides a and b). Then the third side, c, can be found by subtracting the lengths of the known sides from the perimeter (Figure 10-8).

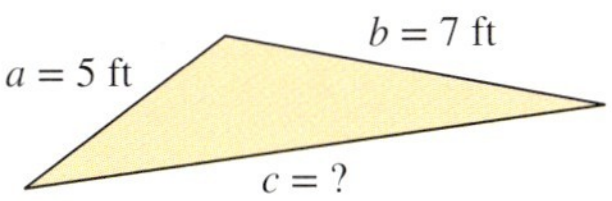

If the perimeter is 20 ft, then
$c = P - a - b$
$= 20 \text{ ft} - 5 \text{ ft} - 7 \text{ ft}$
$= 8 \text{ ft}$

Figure 10-8

To solve a formula for a different variable, we use the same properties of equality outlined in the earlier sections of this chapter. For example, consider the two equations $2x + 3 = 11$ and $wx + y = z$. Suppose we want to solve for x in each case:

$2x + 3 = 11$	$wx + y = z$
$2x + 3 - 3 = 11 - 3$ Subtract 3.	$wx + y - y = z - y$ Subtract y.
$2x = 8$	$wx = z - y$
$\dfrac{2x}{2} = \dfrac{8}{2}$ Divide by 2.	$\dfrac{wx}{w} = \dfrac{z-y}{w}$ Divide by w.
$x = 4$	$x = \dfrac{z-y}{w}$

The equation on the left has only one variable, and we can simplify the equation to find a numerical value for x. The equation on the right has multiple variables. Because we do not know the values of w, y, and z, we cannot simplify further. The value of x is left as a formula in terms of w, y, and z.

Example 1   Solving for an Indicated Variable

Solve for the indicated variable.

a. $d = rt$ for t **b.** $5x + 2y = 12$ for y

Solution:

a. $d = rt$ for t The goal is to isolate the variable t.

$\dfrac{d}{r} = \dfrac{rt}{r}$ Because the relationship between r and t is multiplication, we reverse the process by dividing both sides by r.

$\dfrac{d}{r} = t$, or equivalently $t = \dfrac{d}{r}$

b. $5x + 2y = 12$ for y The goal is to solve for y.

$5x - 5x + 2y = 12 - 5x$ Subtract $5x$ from both sides to isolate the y term.

$2y = -5x + 12$ $-5x + 12$ is the same as $12 - 5x$.

$\dfrac{2y}{2} = \dfrac{-5x + 12}{2}$ Divide both sides by 2 to isolate y.

$y = \dfrac{-5x + 12}{2}$

> **Avoiding Mistakes**
>
> In the expression $\dfrac{-5x + 12}{2}$ do not try to divide the 2 into the 12. The divisor of 2 is dividing the entire quantity, $-5x + 12$ (not just the 12).
>
> We may, however, apply the divisor to each term individually in the numerator. That is, $y = \dfrac{-5x + 12}{2}$ can be written in several different forms. Each is correct.
>
> $$y = \dfrac{-5x + 12}{2} \quad \text{or} \quad y = \dfrac{-5x}{2} + \dfrac{12}{2} \Rightarrow y = -\dfrac{5}{2}x + 6$$

Skill Practice Solve for the indicated variable.

1. $A = lw$ for l
2. $-2a + 4b = 7$ for a

Example 2 Solving for an Indicated Variable

The formula $C = \tfrac{5}{9}(F - 32)$ is used to find the temperature, C, in degrees Celsius for a given temperature expressed in degrees Fahrenheit, F. Solve the formula $C = \tfrac{5}{9}(F - 32)$ for F.

Solution:

$$C = \dfrac{5}{9}(F - 32)$$

$$C = \dfrac{5}{9}F - \dfrac{5}{9} \cdot 32 \qquad \text{Clear parentheses.}$$

$$C = \dfrac{5}{9}F - \dfrac{160}{9} \qquad \text{Multiply: } \dfrac{5}{9} \cdot \dfrac{32}{1} = \dfrac{160}{9}$$

$$9(C) = 9\left(\dfrac{5}{9}F - \dfrac{160}{9}\right) \qquad \text{Multiply by the LCD to clear fractions.}$$

$$9C = \dfrac{9}{1} \cdot \dfrac{5}{9}F - \dfrac{9}{1} \cdot \dfrac{160}{9} \qquad \text{Apply the distributive property.}$$

$$9C = 5F - 160 \qquad \text{Simplify.}$$

$$9C + 160 = 5F - 160 + 160 \qquad \text{Add 160 to both sides.}$$

$$9C + 160 = 5F$$

$$\dfrac{9C + 160}{5} = \dfrac{5F}{5} \qquad \text{Divide both sides by 5.}$$

$$\dfrac{9C + 160}{5} = F$$

The answer may be written in several forms:

$$F = \dfrac{9C + 160}{5} \quad \text{or} \quad F = \dfrac{9C}{5} + \dfrac{160}{5} \Rightarrow F = \dfrac{9}{5}C + 32$$

Skill Practice Solve for the indicated variable.

3. $y = \dfrac{1}{3}(x - 7)$ for x.

Answers

1. $l = \dfrac{A}{w}$
2. $a = \dfrac{7 - 4b}{-2}$ or $a = \dfrac{4b - 7}{2}$ or $a = 2b - \dfrac{7}{2}$
3. $x = 3y + 7$

Section 10.6 Formulas and Applications of Geometry 643

2. Geometry Applications

In Examples 3–6 and the related exercises, we use facts and formulas from geometry.

Example 3   **Solving a Geometry Application Involving Perimeter**

The length of a rectangular lot is 1 m less than twice the width. If the perimeter is 190 m, find the length and width.

Solution:

Step 1: Read the problem.

Let x represent the width of the rectangle.   Step 2: Label the variables.

Then $2x - 1$ represents the length.

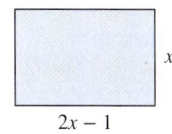

$$P = 2l + 2w$$ Step 3: Write the formula for perimeter.

$$190 = 2(2x - 1) + 2(x)$$ Step 4: Write an equation in terms of x.

$$190 = 4x - 2 + 2x$$ Step 5: Solve for x.

$$190 = 6x - 2$$

$$192 = 6x$$

$$\frac{192}{6} = \frac{6x}{6}$$

$$32 = x$$

The width is $x = 32$.   Step 6: Interpret the results and write the answer in words.

The length is $2x - 1 = 2(32) - 1 = 63$.

The width of the rectangular lot is 32 m and the length is 63 m.

Skill Practice

4. The length of a rectangle is 10 ft less than twice the width. If the perimeter is 178 ft, find the length and width.

Recall some facts about angles.

- Two angles are complementary if the sum of their measures is 90°.
- Two angles are supplementary if the sum of their measures is 180°.
- The sum of the measures of the angles within a triangle is 180°.
- The measures of vertical angles are equal.

Answer

4. The length is 56 ft, and the width is 33 ft.

Example 4 Solving a Geometry Application Involving Complementary Angles

Two complementary angles are drawn such that one angle is 4° more than seven times the other angle. Find the measure of each angle.

Solution:

Step 1: Read the problem.

Let x represent the measure of one angle. Step 2: Label the variables.

Then $7x + 4$ represents the measure of the other angle.

The angles are complementary, so their sum must be 90°.

$$\begin{pmatrix}\text{Measure of}\\ \text{first angle}\end{pmatrix} + \begin{pmatrix}\text{measure of}\\ \text{second angle}\end{pmatrix} = 90°$$

Step 3: Write a verbal model.

$$x + 7x + 4 = 90$$

Step 4: Write a mathematical equation.

$$8x + 4 = 90$$
$$8x = 86$$
$$\frac{8x}{8} = \frac{86}{8}$$
$$x = 10.75$$

Step 5: Solve for x.

One angle is $x = 10.75$.

Step 6: Interpret the results and write the answer in words.

The other angle is $7x + 4 = 7(10.75) + 4 = 79.25$.

The angles are 10.75° and 79.25°.

Skill Practice

5. Two complementary angles are constructed so that one measures 1° less than six times the other. Find the measures of the angles.

Example 5 Solving a Geometry Application Involving Angles in a Triangle

One angle in a triangle is twice as large as the smallest angle. The third angle is 10° more than seven times the smallest angle. Find the measure of each angle.

Solution:

Step 1: Read the problem.

Let x represent the measure of the smallest angle. Step 2: Label the variables.

Then $2x$ and $7x + 10$ represent the measures of the other two angles.

The sum of the angles must be 180°.

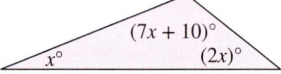

Answer

5. The angles are 13° and 77°.

$$\begin{pmatrix}\text{Measure of}\\\text{first angle}\end{pmatrix} + \begin{pmatrix}\text{measure of}\\\text{second angle}\end{pmatrix} + \begin{pmatrix}\text{measure of}\\\text{third angle}\end{pmatrix} = 180°$$

Step 3: Write a verbal model.

$$x \;+\; 2x \;+\; (7x + 10) = 180$$

Step 4: Write a mathematical equation.

$$x + 2x + 7x + 10 = 180$$
$$10x + 10 = 180$$
$$10x = 170$$
$$x = 17$$

Step 5: Solve for x.

Step 6: Interpret the results and write the answer in words.

The smallest angle is $x = 17$.

The other angles are $2x = 2(17) = 34$

$$7x + 10 = 7(17) + 10 = 129$$

The angles are 17°, 34°, and 129°.

Skill Practice

6. In a triangle, the measure of the first angle is 80° greater than the measure of the second angle. The measure of the third angle is twice that of the second. Find the measures of the angles.

Example 6 Solving a Geometry Application Involving Circumference

The distance around a circular garden is 188.4 ft. Find the radius to the nearest tenth of a foot. Use 3.14 for π.

Solution:

$$C = 2\pi r \qquad \text{Use the formula for the circumference of a circle.}$$
$$188.4 = 2\pi r \qquad \text{Substitute 188.4 for } C.$$
$$\frac{188.4}{2\pi} = \frac{2\pi r}{2\pi} \qquad \text{Divide both sides by } 2\pi.$$
$$\frac{188.4}{2\pi} = r$$
$$r \approx \frac{188.4}{2(3.14)}$$
$$= 30.0$$

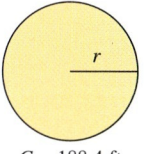

$C = 188.4$ ft

The radius is approximately 30.0 ft.

Skill Practice

7. The circumference of a drain pipe is 12.5 cm. Find the radius. Round to the nearest tenth of a centimeter.

Answers

6. The angles are 25°, 50°, and 105°.
7. The radius is 2.0 cm.

Section 10.6 Activity

A.1. The City of Ormond Beach requires that a tree removal permit be obtained before cutting down a tree with more than a 6-in. diameter. However, because we can't measure "through" the center of a tree without cutting the tree down, we can indirectly measure the diameter by using the circumference of (distance around) the tree. To measure circumference, wrap a string once around the tree. Then unwind the string and measure its length.

    a. Recall that the circumference C of a circle is related to its diameter d by the formula $C = \pi d$. Solve this equation for d.

    b. If you measure the circumference of the tree to be approximately 22 in., calculate the diameter. Use 3.14 for π. Round to the nearest inch.

    c. The measurement you made for circumference is not exact, and therefore, the calculation for diameter is also not exact. However, if you round the diameter to the nearest inch, does it appear that you would need a permit to remove the tree?

    d. Use this example to explain why a formula such as $C = \pi d$ might be valuable when it is written in a different form with a different variable isolated.

A.2. **a.** Solve the equation for y. $3y - 5 = 16$
    b. Solve the equation for y. $ay - z = w$
    c. Explain the similarities for solving each equation.

A.3. **a.** Solve the equation for y. $3x + 2y = 6$
    b. Is there more than one way to express the final answer? Explain.

A.4. How are complementary angles related?

A.5. How are supplementary angles related?

A.6. What is the sum of the measures of the angles of a triangle?

A.7. Write a formula for the perimeter P of a rectangle of length l and width w.

A.8. How are the angles a and b related?

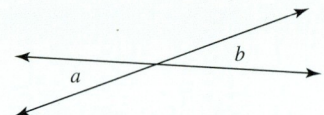

A.9. Find the measures of the vertical angles by first solving for x.

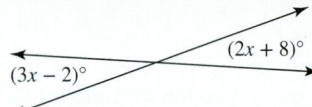

For Exercises A.10–A.12, follow these steps to solve the application.
    a. Read the problem.
    b. Label the variable(s).
    c. Write a verbal model or appropriate formula to represent this scenario.
    d. Write an equation to represent the verbal model.
    e. Solve the equation.
    f. Interpret the results in words using the proper units of measurement, and verify that your solution makes sense in the context of the problem.

A.10. The width of a rectangular picture is five inches shorter than the length. The total amount of framing required to go around the edge of the picture is 54 in. Find the length and width of the picture.

A.11. Two angles are complementary. The measure of one angle is six degrees less than five times the measure of the other angle. Find the measures of the angles.

A.12. The largest angle in a triangle is five times the measure of the smallest angle. The middle angle is 4° more than twice the measure of the smallest angle. Find the measure of each angle.

Practice Exercises

Section 10.6

Study Skills Exercise

Spaced practice is an effective memorization technique in mathematics. To become proficient in mathematics, you should work on it often and in spaced sessions. Repetition of similar problems should also occur. This will increase your understanding of the material and help you retain it for longer periods of time.

Evaluate your schedule to see how you are spending your time outside of the classroom. Ask yourself the following questions:
- How many hours are being spent outside of class studying math?
- Is my study time spread out over the course of a week, or do I cram leading up to the test?
- Am I retaining material or am I forgetting most of it after a day or two?

If you have not already done so, plan frequent, short intervals of study time into your weekly schedule.

Prerequisite Review

R.1. **a.** Give a formula for the perimeter P of a rectangle of length l and width w.
  b. Find the perimeter of a rectangle of length 6 cm and width 4 cm.

R.2. **a.** Give a formula for the area A of a rectangle of length l and width w.
  b. Find the area of a rectangle of length 12 ft and width 5 ft.

R.3. **a.** Determine the complement of a 43° angle.
  b. Determine the supplement of a 43° angle.

R.4. **a.** Determine the complement of a 77° angle.
  b. Determine the supplement of a 77° angle.

R.5. Two angles in a triangle are 56° and 82°. What is the measure of the third angle?

R.6. In a right triangle, one of the acute angles is 21°. What is the measure of the other acute angle?

Concept 1: Literal Equations and Formulas

For Exercises 1–32, solve for the indicated variable. **(See Examples 1–2.)**

1. $P = a + b + c$ for a
2. $P = a + b + c$ for b
3. $x = y - z$ for y
4. $c + d = e$ for d
5. $p = 250 + q$ for q
6. $y = 35 + x$ for x
7. $A = bh$ for b
8. $d = rt$ for r
9. $PV = nrt$ for t
10. $P_1V_1 = P_2V_2$ for V_1
11. $x - y = 5$ for x
12. $x + y = -2$ for y
13. $3x + y = -19$ for y
14. $x - 6y = -10$ for x
15. $2x + 3y = 6$ for y
16. $7x + 3y = 1$ for y
17. $-2x - y = 9$ for x
18. $3x - y = -13$ for x
19. $4x - 3y = 12$ for y
20. $6x - 3y = 4$ for y
21. $ax + by = c$ for y
22. $ax + by = c$ for x
23. $A = P(1 + rt)$ for t
24. $P = 2(L + w)$ for L

648 Chapter 10 Linear Equations and Inequalities

25. $a = 2(b + c)$ for c

26. $3(x + y) = z$ for x

27. $Q = \dfrac{x + y}{2}$ for y

28. $Q = \dfrac{a - b}{2}$ for a

29. $M = \dfrac{a}{S}$ for a

30. $A = \dfrac{1}{3}(a + b + c)$ for c

31. $P = I^2 R$ for R

32. $F = \dfrac{GMm}{d^2}$ for m

Concept 2: Geometry Applications

For Exercises 33–54, use the Problem-Solving Flowchart for Word Problems.

33. The perimeter of a rectangular garden is 24 ft. The length is 2 ft more than the width. Find the length and the width of the garden. (See Example 3.)

34. In a small rectangular wallet photo, the width is 7 cm less than the length. If the perimeter of the photo is 34 cm, find the length and width.

35. The length of a rectangular parking area is four times the width. The perimeter is 300 yd. Find the length and width of the parking area.

36. The width of Jason's workbench is $\frac{1}{2}$ the length. The perimeter is 240 in. Find the length and the width of the workbench.

37. A builder buys a rectangular lot of land such that the length is 5 m less than two times the width. If the perimeter is 590 m, find the length and the width.

38. The perimeter of a rectangular pool is 140 yd. If the length is 20 yd less than twice the width, find the length and the width.

$2w - 5$

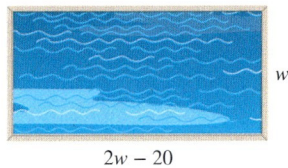
$2w - 20$

39. A triangular parking lot has two sides that are the same length, and the third side is 5 m longer. If the perimeter is 71 m, find the lengths of the sides.

40. The perimeter of a triangle is 16 ft. One side is 3 ft longer than the shortest side. The third side is 1 ft longer than the shortest side. Find the lengths of the sides.

41. Sometimes memory devices are helpful for remembering mathematical facts. Recall that the sum of two complementary angles is 90°. That is, two complementary angles when added together form a right angle or "corner." The words *Complementary* and *Corner* both start with the letter "*C*." Derive your own memory device for remembering that the sum of two supplementary angles is 180°.

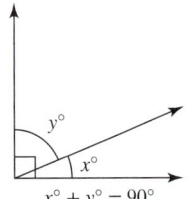
$x° + y° = 90°$
Complementary angles form a "Corner"

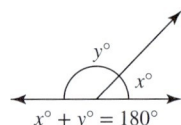
$x° + y° = 180°$
Supplementary angles . . .

42. Two angles are complementary. One angle is 20° less than the other angle. Find the measures of the angles.

43. Two angles are complementary. One angle is 4° less than three times the other angle. Find the measures of the angles. (See Example 4.)

44. Two angles are supplementary. One angle is three times as large as the other angle. Find the measures of the angles.

45. Two angles are supplementary. One angle is 6° more than four times the other. Find the measures of the angles.

46. Refer to the figure. The angles, ∠a and ∠b, are vertical angles.

   a. If the measure of ∠a is 32°, what is the measure of ∠b?

   b. What is the measure of the supplement of ∠a?

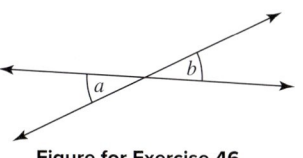

Figure for Exercise 46

47. Find the measures of the vertical angles labeled in the figure by first solving for x.

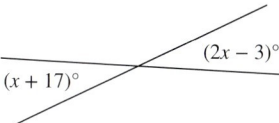

48. Find the measures of the vertical angles labeled in the figure by first solving for y.

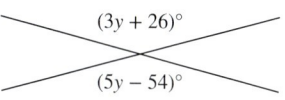

49. The largest angle in a triangle is three times the smallest angle. The middle angle is two times the smallest angle. Given that the sum of the angles in a triangle is 180°, find the measure of each angle. **(See Example 5.)**

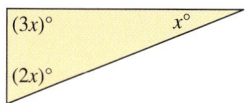

50. The smallest angle in a triangle measures 90° less than the largest angle. The middle angle measures 60° less than the largest angle. Find the measure of each angle.

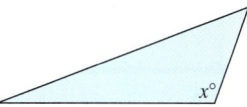

51. The smallest angle in a triangle is half the largest angle. The middle angle measures 30° less than the largest angle. Find the measure of each angle.

52. The largest angle of a triangle is three times the middle angle. The smallest angle measures 10° less than the middle angle. Find the measure of each angle.

53. Find the value of x and the measure of each angle labeled in the figure.

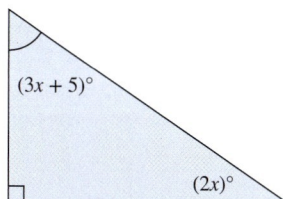

54. Find the value of y and the measure of each angle labeled in the figure.

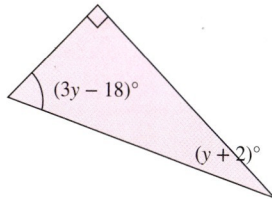

55. a. A rectangle has length l and width w. Write a formula for the area.

   b. Solve the formula for the width, w.

   c. The area of a rectangular volleyball court is 1740.5 ft² and the length is 59 ft. Find the width.

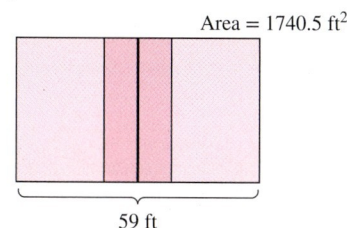

56. a. A parallelogram has height h and base b. Write a formula for the area.

   b. Solve the formula for the base, b.

   c. Find the base of the parallelogram pictured if the area is 40 m².

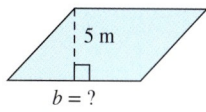

57.
a. A rectangle has length l and width w. Write a formula for the perimeter.
b. Solve the formula for the length, l.
c. The perimeter of the soccer field at Giants Stadium is 338 m. If the width is 66 m, find the length.

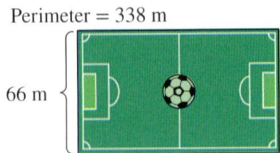
Perimeter = 338 m
66 m

58.
a. A triangle has height h and base b. Write a formula for the area.
b. Solve the formula for the height, h.
c. Find the height of the triangle pictured if the area is 12 km².

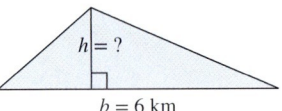
$h = ?$
$b = 6$ km

59.
a. A circle has a radius of r. Write a formula for the circumference. **(See Example 6.)**
b. Solve the formula for the radius, r.
c. The circumference of the circular Buckingham Fountain in Chicago is approximately 880 ft. Find the radius. Round to the nearest foot.

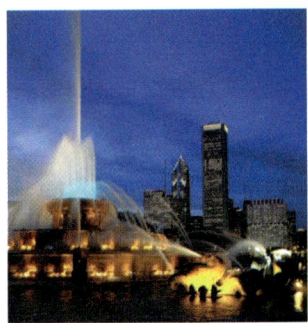
Steve Allen/Brand X Pictures/Getty Images

60.
a. The length of each side of a square is s. Write a formula for the perimeter of the square.
b. Solve the formula for the length of a side, s.
c. The Pyramid of Khufu (known as the Great Pyramid) at Giza has a square base. If the distance around the bottom is 921.6 m, find the length of the sides at the bottom of the pyramid.

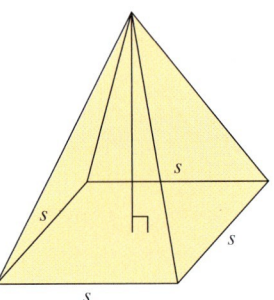

Technology Connections

For Exercises 61–64, approximate the expressions with a calculator. Round to three decimal places if necessary.

61. $\dfrac{880}{2\pi}$ **62.** $\dfrac{1600}{\pi(4)^2}$ **63.** $\dfrac{20}{5\pi}$ **64.** $\dfrac{10}{7\pi}$

Expanding Your Skills

For Exercises 65–66, find the indicated area or volume. Be sure to include the proper units, and round each answer to two decimal places if necessary.

65.
a. Find the area of a circle with radius 11.5 m. Use the π key on the calculator.
b. Find the volume of a right circular cylinder with radius 11.5 m and height 25 m.

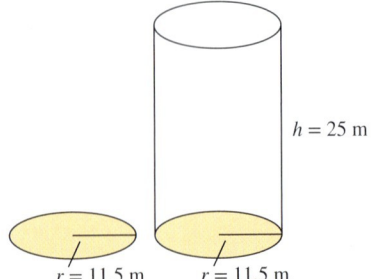
$h = 25$ m
$r = 11.5$ m $r = 11.5$ m

66.
a. Find the area of a parallelogram with base 30 in. and height 12 in.
b. Find the area of a triangle with base 30 in. and height 12 in.
c. Compare the areas found in parts (a) and (b).

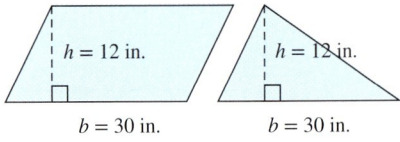
$h = 12$ in. $h = 12$ in.
$b = 30$ in. $b = 30$ in.

Mixture Applications and Uniform Motion

Section 10.7

Concepts

1. Applications Involving Cost
2. Applications Involving Mixtures
3. Applications Involving Uniform Motion

1. Applications Involving Cost

In Examples 1 and 2, we will look at different kinds of mixture problems. The first example "mixes" two types of theater tickets, adult tickets that sell for $12 and children's tickets that sell for $9. Furthermore, there were 300 tickets sold for a total revenue of $3060. Before attempting the problem, we should try to gain some familiarity. Let's try a few combinations to see how many of each type of ticket might have been sold.

Suppose 100 adult tickets were sold and 200 children's tickets were sold (a total of 300 tickets).

- 100 adult tickets at $12 each gives   100($12) = $1200
- 200 children's tickets at $9 each gives   200($9) = $1800

                       Total revenue: $3000 (not enough)

Suppose 150 adult tickets were sold and 150 children's tickets were sold (a total of 300 tickets).

- 150 adult tickets at $12 each gives   150($12) = $1800
- 150 children's tickets at $9 each gives   150($9) = $1350

                       Total revenue: $3150 (too much)

As you can see, the trial-and-error process can be tedious and time consuming. Therefore, we will use algebra to determine the correct combination of each type of ticket.

Suppose we let x represent the number of adult tickets. Then the number of children's tickets is the *total minus* x. That is,

$$\begin{pmatrix} \text{Number of} \\ \text{children's tickets} \end{pmatrix} = \begin{pmatrix} \text{total number} \\ \text{of tickets} \end{pmatrix} - \begin{pmatrix} \text{number of} \\ \text{adult tickets, } x \end{pmatrix}$$

Number of children's tickets = $300 - x$.

Notice that the number of tickets sold times the price per ticket gives the revenue:

- x adult tickets at $12 each gives a revenue of: $x(\$12)$ or simply $12x$.
- $300 - x$ children's tickets at $9 each gives: $(300 - x)(\$9)$ or $9(300 - x)$.

This will help us set up an equation in Example 1.

Example 1   Solving a Mixture Problem Involving Ticket Sales

At a community theater, 300 tickets were sold. Adult tickets cost $12 and tickets for children cost $9. If the total revenue from ticket sales was $3060, determine the number of each type of ticket sold.

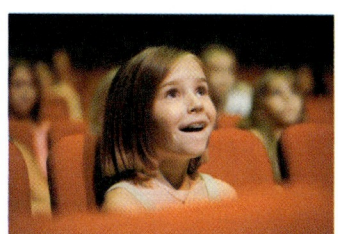

Image Studios/Uppercut/Getty Images

Solution:

Step 1: Read the problem.

Let x represent the number of adult tickets sold.
$300 - x$ is the number of children's tickets.

Step 2: Label the variables.

	$12 Tickets	$9 Tickets	Total
Number of tickets	x	$300 - x$	300
Revenue	$12x$	$9(300 - x)$	3060

$$\begin{pmatrix}\text{Revenue from}\\ \text{adult tickets}\end{pmatrix} + \begin{pmatrix}\text{revenue from}\\ \text{children's tickets}\end{pmatrix} = \begin{pmatrix}\text{total}\\ \text{revenue}\end{pmatrix}$$

$$12x + 9(300 - x) = 3060$$

Step 3: Write a verbal model.

Step 4: Write a mathematical equation.

$$12x + 9(300 - x) = 3060$$

$$12x + 2700 - 9x = 3060$$

$$3x + 2700 = 3060$$

$$3x = 360$$

$$x = 120$$

Step 5: Solve the equation.

Step 6: Interpret the results.

There were 120 adult tickets sold.
The number of children's tickets is $300 - x$ which is 180.

Avoiding Mistakes
Check that the answer is reasonable. 120 adult tickets and 180 children's tickets makes 300 total tickets.
Furthermore, 120 adult tickets at $12 each amounts to $1440, and 180 children's tickets at $9 each amounts to $1620. The total revenue is $3060 as expected.

Skill Practice

1. At a Performing Arts Center, seats in the orchestra section cost $18 and seats in the balcony cost $12. If there were 120 seats sold for one performance, for a total revenue of $1920, how many of each type of seat were sold?

2. Applications Involving Mixtures

Example 2 — Solving a Mixture Application

How many liters (L) of a 60% antifreeze solution must be added to 8 L of a 10% antifreeze solution to produce a 20% antifreeze solution?

Solution:

The information can be organized in a table. Notice that an algebraic equation is derived from the second row of the table. This relates the number of liters of pure antifreeze in each container.

Step 1: Read the problem.

Step 2: Label the variables.

	60% Antifreeze	10% Antifreeze	Final Mixture: 20% Antifreeze
Number of liters of solution	x	8	$(8 + x)$
Number of liters of pure antifreeze	$0.60x$	$0.10(8)$	$0.20(8 + x)$

TIP: To determine the amount of pure antifreeze in a given solution, multiply the concentration rate by the amount of solution. For example, 8 L of a 10% solutions means that

$$(0.10)(8\text{ L}) = 0.8\text{ L}$$

of the solution is pure antifreeze. The rest is something else such as water.

Answer
1. There were 80 seats in the orchestra section, and there were 40 in the balcony.

The amount of pure antifreeze in the final solution equals the sum of the amounts of antifreeze in the first two solutions.

$$\begin{pmatrix} \text{Pure antifreeze} \\ \text{from solution 1} \end{pmatrix} + \begin{pmatrix} \text{pure antifreeze} \\ \text{from solution 2} \end{pmatrix} = \begin{pmatrix} \text{pure antifreeze} \\ \text{in the final solution} \end{pmatrix}$$

Step 3: Write a verbal model.

$$0.60x + 0.10(8) = 0.20(8+x)$$

Step 4: Write a mathematical equation.

$$0.60x + 0.10(8) = 0.20(8+x)$$

Step 5: Solve the equation.

$$0.6x + 0.8 = 1.6 + 0.2x$$

$$0.6x - 0.2x + 0.8 = 1.6 + 0.2x - 0.2x \quad \text{Subtract } 0.2x.$$

$$0.4x + 0.8 = 1.6$$

$$0.4x + 0.8 - 0.8 = 1.6 - 0.8 \quad \text{Subtract } 0.8.$$

$$0.4x = 0.8$$

$$\frac{0.4x}{0.4} = \frac{0.8}{0.4} \quad \text{Divide by } 0.4.$$

$$x = 2$$

Step 6: Interpret the result.

Therefore, 2 L of 60% antifreeze solution is necessary to make a final solution that is 20% antifreeze.

Skill Practice

2. How many gallons of a 5% bleach solution must be added to 10 gallons (gal) of a 20% bleach solution to produce a solution that is 15% bleach?

3. Applications Involving Uniform Motion

The formula (distance) = (rate)(time) or simply, $d = rt$, relates the distance traveled to the rate of travel and the time of travel.

For example, if a car travels at 60 mph for 3 hours, then

$$d = (60 \text{ mph})(3 \text{ hours})$$
$$= 180 \text{ miles}$$

If a car travels at 60 mph for x hours, then

$$d = (60 \text{ mph})(x \text{ hours})$$
$$= 60x \text{ miles}$$

Adam Gault/Getty Images

Answer

2. 5 gal is needed.

Example 3 Solving an Application Involving Distance, Rate, and Time

One bicyclist rides 4 mph faster than another bicyclist. The faster rider takes 3 hr to complete a race, while the slower rider takes 4 hr. Find the speed for each rider.

Solution:

Step 1: Read the problem.

The problem is asking us to find the speed of each rider.

Let x represent the speed of the slower rider. Then $(x + 4)$ is the speed of the faster rider.

Step 2: Label the variables and organize the information given in the problem. A distance-rate-time chart may be helpful.

	Distance	Rate	Time
Faster rider	$3(x + 4)$	$x + 4$	3
Slower rider	$4(x)$	x	4

To complete the first column, we can use the relationship, $d = rt$.

Because the riders are riding in the same race, their distances are equal.

$$\begin{pmatrix} \text{Distance} \\ \text{by faster rider} \end{pmatrix} = \begin{pmatrix} \text{distance} \\ \text{by slower rider} \end{pmatrix}$$

Step 3: Write a verbal model.

$$3(x + 4) = 4(x)$$

Step 4: Write a mathematical equation.

$$3x + 12 = 4x$$

Step 5: Solve the equation.

$$12 = x$$

Subtract $3x$ from both sides.

The variable x represents the slower rider's rate. The quantity $x + 4$ is the faster rider's rate. Thus, if $x = 12$, then $x + 4 = 16$.

The slower rider travels 12 mph and the faster rider travels 16 mph.

Avoiding Mistakes

Check that the answer is reasonable. If the slower rider rides at 12 mph for 4 hr, he travels 48 mi. If the faster rider rides at 16 mph for 3 hr, he also travels 48 mi as expected.

Skill Practice

3. An express train travels 25 mph faster than a cargo train. It takes the express train 6 hr to travel a route, and it takes 9 hr for the cargo train to travel the same route. Find the speed of each train.

Example 4 Solving an Application Involving Distance, Rate, and Time

Two families that live 270 mi apart plan to meet for an afternoon picnic at a park that is located between their two homes. Both families leave at 9:00 A.M., but one family averages 12 mph faster than the other family. If the families meet at the designated spot $2\frac{1}{2}$ hr later, determine

Rawpixel.com/Shutterstock

a. The average rate of speed for each family.

b. The distance each family traveled to the picnic.

Answer

3. The express train travels 75 mph, and the cargo train travels 50 mph.

Solution:

For simplicity, we will call the two families Family A and Family B. Let Family A be the family that travels at the slower rate (Figure 10-9).

Step 1: Read the problem and draw a sketch.

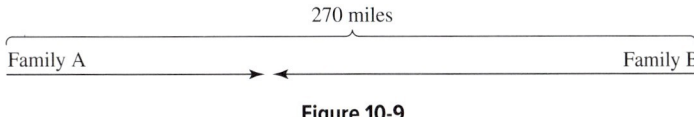

Figure 10-9

Let x represent the rate of Family A. Then $(x + 12)$ is the rate of Family B.

Step 2: Label the variables.

	Distance	Rate	Time
Family A	$2.5x$	x	2.5
Family B	$2.5(x + 12)$	$x + 12$	2.5

To complete the first column, we can multiply rate and time: $d = rt$.

To set up an equation, recall that the total distance between the two families is given as 270 mi.

$$\begin{pmatrix} \text{Distance} \\ \text{traveled by} \\ \text{Family A} \end{pmatrix} + \begin{pmatrix} \text{distance} \\ \text{traveled by} \\ \text{Family B} \end{pmatrix} = \begin{pmatrix} \text{total} \\ \text{distance} \end{pmatrix}$$

Step 3: Write a verbal model.

$$2.5x + 2.5(x+12) = 270$$

Step 4: Write a mathematical equation.

$$2.5x + 2.5(x + 12) = 270$$

Step 5: Solve for x.

$$2.5x + 2.5x + 30 = 270$$
$$5.0x + 30 = 270$$
$$5x = 240$$
$$x = 48$$

a. Family A traveled 48 mph.

Step 6: Interpret the results and write the answer in words.

Family B traveled $x + 12 = 48 + 12 = 60$ mph.

b. To compute the distance each family traveled, use $d = rt$.

Family A traveled $(48 \text{ mph})(2.5 \text{ hr}) = 120$ mi.

Family B traveled $(60 \text{ mph})(2.5 \text{ hr}) = 150$ mi.

Skill Practice

4. A Piper airplane has an average air speed that is 10 mph faster than a Cessna 150 airplane. If the combined distance traveled by these two small planes is 690 mi after 3 hr, what is the average speed of each plane?

Answer

4. The Cessna's speed is 110 mph, and the Piper's speed is 120 mph.

Section 10.7 Activity

A.1. A group of 16 college students go to see a rock concert. Some students have seats near the stage that cost $85 each. The other students have seats in the balcony that cost $50 each. If the total amount paid for all 16 tickets comes to $975, how many students sit near the stage and how many sit in the balcony?

  a. Let x represent the number of seats close to the stage. Write an expression representing the number of seats in the balcony. (*Hint:* There are 16 seats in total, and x of them are near the stage. The "leftovers" are in the balcony.)

  b. Write an equation representing $\begin{pmatrix} \text{The cost for} \\ x \text{ seats close} \\ \text{to the stage} \end{pmatrix} + \begin{pmatrix} \text{the cost for} \\ \text{seats in} \\ \text{the balcony} \end{pmatrix} = \begin{pmatrix} \text{the total} \\ \text{cost} \end{pmatrix}$

  c. Solve the equation from part (b).

  d. Interpret the results in words using proper units of measurement, and verify that your solution makes sense in the context of the problem.

A.2. A chemist has 10 liters (L) of an acid and water solution that is 20% acid. How many liters of the solution is acid and how many liters is water?

A.3. **a.** If a chemist has 5 oz of a solution that is 6% alcohol, how much is pure alcohol?

  b. If the chemist has x ounces of a solution that is 15% alcohol, write an expression that represents the amount of pure alcohol.

  c. If the chemist has $(x + 5)$ ounces of a solution that is 10% alcohol, write an expression that represents the amount of pure alcohol.

A.4. How many ounces of a 15% alcohol solution must be mixed with 5 oz of a 6% alcohol solution to produce a 10% alcohol solution?

  a. Let x represent _____.

  b. Complete the table.

	15% Alcohol Solution	6% Alcohol Solution	10% Alcohol Solution
Amount of solution (oz)			
Amount of pure alcohol (oz)			

  c. Write an equation that represents

  $\begin{pmatrix} \text{The amount of} \\ \text{pure alcohol in} \\ \text{the 15\% solution} \end{pmatrix} + \begin{pmatrix} \text{the amount of} \\ \text{pure alcohol in} \\ \text{the 6\% solution} \end{pmatrix} = \begin{pmatrix} \text{the amount of} \\ \text{pure alcohol in} \\ \text{the 10\% solution} \end{pmatrix}$

  d. Solve the equation from part (c).

  e. Interpret the results in words using proper units of measurement and verify that your solution makes sense in the context of the problem.

A.5. Complete the formula using d (for distance), r (for rate), and t (for time). $d = $ _____.

A.6. **a.** If a walker walks 4 mph, how far does she walk in 30 min ($\frac{1}{2}$ hr)?

  b. If a walker walks 3 mph, how far does she walk in 40 min ($\frac{2}{3}$ hr)?

A.7. Jennifer takes a walk every day. She walks 1 mph slower when she is pushing her son in a stroller than she does without the stroller. She can complete her circuit in 40 min ($\frac{2}{3}$ hr) when pushing her son. Without pushing the stroller, it only takes her $\frac{1}{2}$ hr. Find Jennifer's speed with and without pushing the stroller.

a. Let x represent _____

b. Complete the table.

	Distance (mi)	Rate (mph)	Time (hr)
With the stroller			
Without the stroller			

c. We know that Jennifer walks the same circuit (distance) with the stroller as she does without the stroller. Write a mathematical equation that represents this statement.

d. Solve the equation from part (c).

e. Interpret the results in words using proper units of measurement, and verify that your solution makes sense in the context of the problem.

Practice Exercises

Section 10.7

Study Skills Exercise

It is always helpful to read the material in a section and make notes before it is presented in class. That will help reduce anxiety about new material. Refer to your class syllabus and identify the section that will be covered in the next class. Read the section in the text before the class meets to familiarize yourself with the material and terminology. As you preview the material, respond to the following in your own words.

- What is the section about?
- What concepts do you find challenging or confusing?
- Write down at least one question to ask your instructor.

Prerequisite Review

For Exercises R.1–R.2, solve the equation.

R.1. $0.02x + 0.04(10 - x) = 1.26$

R.2. $0.15(x - 3) + 0.05x = 0.55$

R.3. Tickets to a play cost $12 each.
  a. How much do four tickets cost?
  b. How much do x tickets cost?

R.4. Tacos from a food truck cost $3.50 each.
  a. How much do eight tacos cost?
  b. How much do y tacos cost?

R.5. A solution of bleach and water is 25% bleach.
  a. How much pure bleach is in 2 L of the solution?
  b. How much pure bleach is in x liters of the solution?

R.6. A solution of liquid fertilizer and water is 10% fertilizer.
  a. How much fertilizer is in 5 gal of the solution?
  b. How much fertilizer is in x gallons of the solution?

R.7. A plane flies at 400 mph.
  a. If the plane travels for 4 hr, how far will it go?
  b. If the plane travels for t hours, how far will it go?

R.8. A bicyclist rides 18 mph.
  a. If he rides for 1.5 hr, how far will he go?
  b. If he rides for t hours, how far will he go?

Concept 1: Applications Involving Cost

For Exercises 1–6, write an algebraic expression as indicated.

1. Two numbers total 200. Let t represent one of the numbers. Write an algebraic expression for the other number.

2. The total of two numbers is 43. Let s represent one of the numbers. Write an algebraic expression for the other number.

3. Olivia needs to bring 100 cookies to her friend's party. She has already baked x cookies. Write an algebraic expression for the number of cookies Olivia still needs to bake.

4. Rachel needs a mixture of 55 pounds (lb) of nuts consisting of peanuts and cashews. Let p represent the number of pounds of peanuts in the mixture. Write an algebraic expression for the number of pounds of cashews that she needs to add.

Comstock/Getty Images

Jill Braaten/McGraw-Hill

5. Max has a total of $3000 in two bank accounts. Let y represent the amount in one account. Write an algebraic expression for the amount in the other account.

6. Roberto has a total of $7500 in two savings accounts. Let z represent the amount in one account. Write an algebraic expression for the amount in the other account.

7. A church had an ice cream social and sold tickets for $3 and $2. When the social was over, 81 tickets had been sold totaling $215. How many of each type of ticket did the church sell? **(See Example 1.)**

8. Sarah is a teacher at an elementary school. She purchased 72 tickets to take the first-grade children and some parents on a field trip to the zoo. She purchased children's tickets for $10 each and adult tickets for $18 each. She spent a total of $856. How many of each type of ticket did she buy?

	$3 Tickets	$2 Tickets	Total
Number of tickets			
Cost of tickets			

	Adults	Children	Total
Number of tickets			
Cost of tickets			

9. Josh downloaded 25 songs from an online site. Some songs cost $1.29 and some cost $1.49. He spent a total of $33.85. How many songs at each price were purchased?

10. During the past year, Brittney purchased 30 books at a wholesale club store. She purchased softcover books for $4.50 each and hardcover books for $13.50 each. The total cost of the books was $216. How many of each type of book did she purchase?

11. During the past year, Jolynn purchased 32 books for her e-reader. She purchased some books for $6.99 and others for $9.99. If she spent a total of $256.68, how many books from each price category did she buy?

Siede Preis/Photodisc/Getty Images

12. Steven wants to buy some candy with his birthday money. He can choose from jelly beans that sell for $6.99 per pound and a variety mix that sells for $3.99. He likes to have twice the amount of jelly beans as the variety mix. If he spent a total of $53.91, how many pounds of each type of candy did he buy?

Concept 2: Applications Involving Mixtures

For Exercises 13–16, write an algebraic expression as indicated.

13. A container holds 7 ounces (oz) of liquid. Let x represent the number of ounces of liquid in another container. Write an expression for the total amount of liquid.

14. A bucket contains 2.5 L of a bleach solution. Let n represent the number of liters of bleach solution in a second bucket. Write an expression for the total amount of bleach solution.

15. If Charlene invests $2000 in a certificate of deposit and d dollars in a stock, write an expression for the total amount she invested.

16. James has $5000 in one savings account. Let y represent the amount he has in another savings account. Write an expression for the total amount of money in both accounts.

17. How much of a 5% ethanol fuel mixture should be mixed with 2000 gal of 10% ethanol fuel mixture to get a mixture that is 9% ethanol?
(See Example 2.)

	5% Ethanol	10% Ethanol	Final Mixture: 9% Ethanol
Number of gallons of fuel mixture			
Number of gallons of pure ethanol			

18. How many ounces of a 50% antifreeze solution must be mixed with 10 oz of an 80% antifreeze solution to produce a 60% antifreeze solution?

	50% Antifreeze	80% Antifreeze	Final Mixture: 60% Antifreeze
Number of ounces of solution			
Number of ounces of pure antifreeze			

19. A pharmacist needs to mix a 1% saline (salt) solution with 24 milliliters (mL) of a 16% saline solution to obtain a 9% saline solution. How many milliliters of the 1% solution must she use?

20. A landscaper needs to mix a 75% pesticide solution with 30 gal of a 25% pesticide solution to obtain a 60% pesticide solution. How many gallons of the 75% solution must he use?

21. To clean a concrete driveway, a contractor needs a solution that is 30% acid. How many ounces of a 50% acid solution must be mixed with 15 oz of a 21% solution to obtain a 30% acid solution?

22. A veterinarian needs a mixture that contains 12% of a certain medication to treat an injured bird. How many milliliters of a 16% solution should be mixed with 6 mL of a 7% solution to obtain a solution that is 12% medication?

Concept 3: Applications Involving Uniform Motion

23. a. If a car travels 60 mph for 5 hr, find the distance traveled.

    b. If a car travels at x miles per hour for 5 hr, write an expression that represents the distance traveled.

    c. If a car travels at $x + 12$ mph for 5 hr, write an expression that represents the distance traveled.

24. a. If a plane travels 550 mph for 2.5 hr, find the distance traveled.

    b. If a plane travels at x miles per hour for 2.5 hr, write an expression that represents the distance traveled.

    c. If a plane travels at $x - 100$ mph for 2.5 hr, write an expression that represents the distance traveled.

25. A woman can walk 2 mph faster down a trail to Cochita Lake than she can on the return trip uphill. It takes her 2 hr to get to the lake and 4 hr to return. What is her speed walking down to the lake? (See Example 3.)

	Distance	Rate	Time
Downhill to the lake			
Uphill from the lake			

26. A car travels 20 mph slower in a bad rainstorm than in sunny weather. The car travels the same distance in 2 hr in sunny weather as it does in 3 hr in rainy weather. Find the speed of the car in sunny weather.

	Distance	Rate	Time
Rainstorm			
Sunny weather			

27. Bryan hiked up to the top of City Creek in 3 hr and then returned down the canyon to the trailhead in another 2 hr. His speed downhill was 1 mph faster than his speed uphill. How far up the canyon did he hike?

28. Laura hiked up Lamb's Canyon in 2 hr and then ran back down in 1 hr. Her speed running downhill was 2.5 mph faster than her speed hiking uphill. How far up the canyon did she hike?

29. Hazel and Emilie fly from Atlanta to San Diego. The flight from Atlanta to San Diego is against the wind and takes 4 hr. The return flight with the wind takes 3.5 hr. If the wind speed is 40 mph, find the speed of the plane in still air.

30. A boat on the Potomac River travels the same distance downstream with the current in $\frac{2}{3}$ hr as it does going upstream against the current in 1 hr. If the speed of the current is 3 mph, find the speed of the boat in still water.

31. Two cars are 200 mi apart and travel toward each other on the same road. They meet in 2 hr. One car travels 4 mph faster than the other. What is the speed of each car? (See Example 4.)

32. Two cars are 238 mi apart and travel toward each other along the same road. They meet in 2 hr. One car travels 5 mph slower than the other. What is the speed of each car?

33. After Hurricane Katrina, a rescue vehicle leaves a station at noon and heads for New Orleans. An hour later a second vehicle traveling 10 mph faster leaves the same station. By 4:00 P.M., the first vehicle reaches its destination, and the second is still 10 mi away. How fast is each vehicle?

34. A truck leaves a truck stop at 9:00 A.M. and travels toward Sturgis, Wyoming. At 10:00 A.M., a motorcycle leaves the same truck stop and travels the same route. The motorcycle travels 15 mph faster than the truck. By noon, the truck has traveled 20 mi farther than the motorcycle. How fast is each vehicle?

35. In a recent marathon, Jeanette's speed running is twice Sarah's speed walking. After 2 hr, Jeanette is 7 mi ahead of Sarah. Find Jeanette's speed and Sarah's speed.

36. Two canoes travel down a river, starting at 9:00 A.M. One canoe travels twice as fast as the other. After 3.5 hr, the canoes are 5.25 mi apart. Find the speed of each canoe.

Mixed Exercises

37. A certain granola mixture is 10% peanuts.

  a. If a container has 20 lb of granola, how many pounds of peanuts are there?

  b. If a container has x pounds of granola, write an expression that represents the number of pounds of peanuts in the granola.

  c. If a container has $x + 3$ lb of granola, write an expression that represents the number of pounds of peanuts.

38. A certain blend of coffee sells for $9.00 per pound.

  a. If a container has 20 lb of coffee, how much will it cost?

  b. If a container has x pounds of coffee, write an expression that represents the cost.

  c. If a container has $40 - x$ pounds of this coffee, write an expression that represents the cost.

39. The Coffee Company mixes coffee worth $12 per pound with coffee worth $8 per pound to produce 50 lb of coffee worth $8.80 per pound. How many pounds of the $12 coffee and how many pounds of the $8 coffee must be used?

	$12 Coffee	$8 Coffee	Total
Number of pounds			
Value of coffee			

40. The Nut House sells pecans worth $4 per pound and cashews worth $6 per pound. How many pounds of pecans and how many pounds of cashews must be mixed to form 16 lb of a nut mixture worth $4.50 per pound?

	$4 Pecans	$6 Cashews	Total
Number of pounds			
Value of nuts			

41. A boat in distress, 21 nautical miles from a marina, travels toward the marina at 3 knots (nautical miles per hour). A coast guard cruiser leaves the marina and travels toward the boat at 25 knots. How long will it take for the boats to reach each other?

42. An air traffic controller observes a plane heading from New York to San Francisco traveling at 450 mph. At the same time, another plane leaves San Francisco and travels 500 mph to New York. If the distance between the airports is 2850 mi, how long will it take for the planes to pass each other?

Ilene MacDonald/Dennis MacDonald/Alamy Stock Photo

43. Surfer Sam purchased a total of 21 items at the surf shop. He bought wax for $3 per package and sunscreen for $8 per bottle. He spent a total of $88. How many of each item did he purchase?

44. Tonya owns a small coffee and gift shop. She purchased 30 jars of gourmet jam for $178.50 from a vendor and plans to resell the jam in her shop. She bought raspberry jam for $6.25 per jar and strawberry jam for $5.50 per jar. How many jars of each did she purchase?

45. How many quarts of 85% chlorine solution must be mixed with 5 qt of 25% chlorine solution to obtain a 45% chlorine solution?

46. How many liters of a 58% sugar solution must be added to 14 L of a 40% sugar solution to obtain a 50% sugar solution?

Expanding Your Skills

47. How much pure water must be mixed with 12 L of a 40% alcohol solution to obtain a 15% alcohol solution? (*Hint:* Pure water is 0% alcohol.)

48. How much pure water must be mixed with 10 oz of a 60% alcohol solution to obtain a 25% alcohol solution?

49. Amtrak Acela Express is a high-speed train that runs in the United States between Washington, D.C., and Boston. In Japan, a bullet train along the Sanyo line operates at an average speed of 60 km/hr faster than the Amtrak Acela Express. It takes the Japanese bullet train 2.7 hr to travel the same distance as the Acela Express can travel in 3.375 hr. Find the speed of each train.

50. Amtrak Acela Express is a high-speed train along the northeast corridor between Washington, D.C., and Boston. Since its debut, it has cut the travel time from 4 hr 10 min to 3 hr 20 min. On average, if the Acela Express is 30 mph faster than the old train, find the speed of the Acela Express. (*Hint:* 4 hr 10 min = $4\frac{1}{6}$ hr.)

Section 10.8 Linear Inequalities

Concepts

1. Graphing Linear Inequalities
2. Set-Builder Notation and Interval Notation
3. Addition and Subtraction Properties of Inequality
4. Multiplication and Division Properties of Inequality
5. Inequalities of the Form $a < x < b$
6. Applications of Linear Inequalities

1. Graphing Linear Inequalities

Consider the following two statements.

$$2x + 7 = 11 \quad \text{and} \quad 2x + 7 < 11$$

The first statement is an equation (it has an = sign). The second statement is an inequality (it has an inequality symbol, <). In this section, we will learn how to solve linear *inequalities*, such as $2x + 7 < 11$.

> **A Linear Inequality in One Variable**
>
> A **linear inequality in one variable**, x, is any inequality that can be written in the form:
>
> $ax + b < c$, $ax + b \leq c$, $ax + b > c$, or $ax + b \geq c$, where $a \neq 0$.

The following inequalities are linear equalities in one variable.

$$2x - 3 < 11 \qquad -4z - 3 > 0 \qquad a \leq 4 \qquad 5.2y \geq 10.4$$

The number line is a useful tool to visualize the solution set of an equation or inequality. For example, the solution set to the equation $x = 2$ is $\{2\}$ and may be graphed as a single point on the number line.

$x = 2$
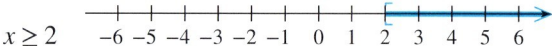

The solution set to an inequality is the set of real numbers that make the inequality a true statement. For example, the solution set to the inequality $x \geq 2$ is all real numbers 2 or greater. Because the solution set has an infinite number of values, we cannot list all of the individual solutions. However, we can graph the solution set on the number line.

$x \geq 2$
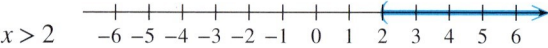

The square bracket symbol, [, is used on the graph to indicate that the point $x = 2$ is included in the solution set. By convention, square brackets, either [or], are used to *include* a point on a number line. Parentheses, (or), are used to *exclude* a point on a number line.

The solution set of the inequality $x > 2$ includes the real numbers greater than 2 but not equal to 2. Therefore, a "(" symbol is used on the graph to indicate that $x = 2$ is not included.

$x > 2$

In Example 1, we demonstrate how to graph linear inequalities. To graph an inequality means that we graph its solution set. That is, we graph all of the values on the number line that make the inequality true.

Section 10.8 Linear Inequalities 663

Example 1   Graphing Linear Inequalities

Graph the solution sets.

a. $x > -1$ **b.** $c \leq \dfrac{7}{3}$ **c.** $3 > y$

Solution:

a. $x > -1$

The solution set is the set of all real numbers strictly greater than -1. Therefore, we graph the region on the number line to the right of -1. Because $x = -1$ is not included in the solution set, we use the "(" symbol at $x = -1$.

b. $c \leq \dfrac{7}{3}$ is equivalent to $c \leq 2\dfrac{1}{3}$.

The solution set is the set of all real numbers less than or equal to $2\dfrac{1}{3}$. Therefore, graph the region on the number line to the left of and including $2\dfrac{1}{3}$. Use the symbol] to indicate that $c = 2\dfrac{1}{3}$ is included in the solution set.

c. $3 > y$ This inequality reads "3 is greater than y." This is equivalent to saying, "y is less than 3." The inequality $3 > y$ can also be written as $y < 3$.

$y < 3$

The solution set is the set of real numbers less than 3. Therefore, graph the region on the number line to the left of 3. Use the symbol ")" to denote that the endpoint, 3, is not included in the solution.

Skill Practice Graph the solution sets.

1. $y < 0$ **2.** $x \geq -\dfrac{5}{4}$ **3.** $5 \geq a$

TIP: Some textbooks use a closed circle or an open circle (● or ○) rather than a bracket or parenthesis to denote inclusion or exclusion of a value on the real number line. For example, the solution sets for the inequalities $x > -1$ and $c \leq \dfrac{7}{3}$ are graphed here.

$x > -1$

$c \leq \dfrac{7}{3}$

A statement that involves more than one inequality is called a **compound inequality**. One type of compound inequality is used to indicate that one number is between two others. For example, the inequality $-2 < x < 5$ means that $-2 < x$ and $x < 5$. In words, this is easiest to understand if we read the variable first: x is greater than -2 and x is less than 5. The numbers satisfied by these two conditions are those between -2 and 5.

Answers

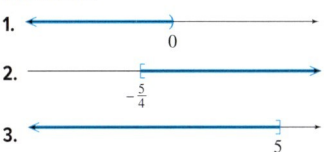

> **Example 2** Graphing a Compound Inequality
>
> Graph the solution set of the inequality: $-4.1 < y \leq -1.7$
>
> **Solution:**
>
> $-4.1 < y \leq -1.7$ means that
>
> $-4.1 < y$ and $y \leq -1.7$
>
>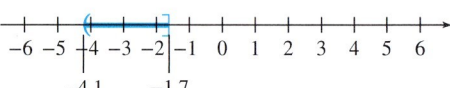
>
> Shade the region of the number line greater than -4.1 and less than or equal to -1.7.
>
> ---
>
> **Skill Practice** Graph the solution set.
>
> **4.** $0 \leq y \leq 8.5$

2. Set-Builder Notation and Interval Notation

Graphing the solution set to an inequality is one way to define the set. Two other methods are to use **set-builder notation** or **interval notation**.

Set-Builder Notation

The solution to the inequality $x \geq 2$ can be expressed in set-builder notation as follows:

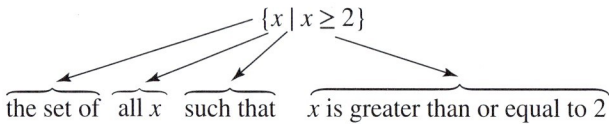

Interval Notation

To understand interval notation, first think of a number line extending infinitely far to the right and infinitely far to the left. Sometimes we use the infinity symbol, ∞, or negative infinity symbol, $-\infty$, to label the far right and far left ends of the number line (Figure 10-10).

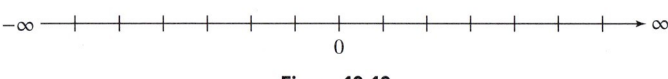

Figure 10-10

To express the solution set of an inequality in interval notation, sketch the graph first. Then use the endpoints to define the interval.

Inequality	Graph	Interval Notation
$x \geq 2$		$[2, \infty)$

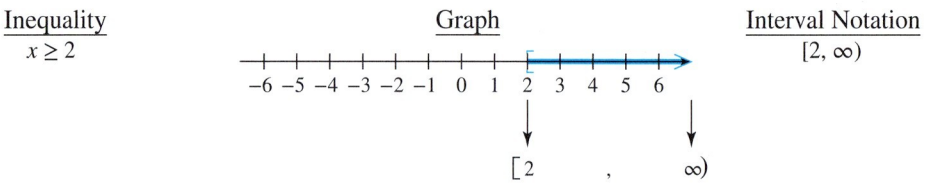

The graph of the solution set $x \geq 2$ begins at 2 and extends infinitely far to the right. The corresponding interval notation begins at 2 and extends to ∞. Notice that a square bracket [is used at 2 for both the graph and the interval notation. A parenthesis is always used at ∞ and for $-\infty$, because there is no endpoint.

Answer

4.

Using Interval Notation

- The endpoints used in interval notation are always written from left to right. That is, the smaller number is written first, followed by a comma, followed by the larger number.
- A parenthesis, (or), indicates that an endpoint is *excluded* from the set.
- A square bracket, [or], indicates that an endpoint is *included* in the set.
- A parenthesis, (or), is always used with $-\infty$ or ∞, respectively.

In Table 10-1, we present examples of eight different scenarios for interval notation and the corresponding graph.

Table 10-1

Interval Notation	Graph	Interval Notation	Graph
(a, ∞)		$[a, \infty)$	
$(-\infty, a)$		$(-\infty, a]$	
(a, b)		$[a, b]$	
$(a, b]$		$[a, b)$	

Example 3 Using Set-Builder Notation and Interval Notation

Complete the chart.

Set-Builder Notation	Graph	Interval Notation
	$-6\ -5\ -4\ -3\ -2\ -1\ 0\ 1\ 2\ 3\ 4\ 5\ 6$	
		$[-\frac{1}{2}, \infty)$
$\{y \mid -2 \leq y < 4\}$		

Solution:

Set-Builder Notation	Graph	Interval Notation
$\{x \mid x < -3\}$	$-6\ -5\ -4\ -3\ -2\ -1\ 0\ 1\ 2\ 3\ 4\ 5\ 6$	$(-\infty, -3)$
$\{x \mid x \geq -\frac{1}{2}\}$	$-6\ -5\ -4\ -3\ -2\ -1\ 0\ 1\ 2\ 3\ 4\ 5\ 6$	$[-\frac{1}{2}, \infty)$
$\{y \mid -2 \leq y < 4\}$	$-6\ -5\ -4\ -3\ -2\ -1\ 0\ 1\ 2\ 3\ 4\ 5\ 6$	$[-2, 4)$

Skill Practice Express each of the following in set-builder notation and interval notation.

5. ⟶ [at -2

6. $x < \dfrac{3}{2}$

7. (at -3] at 1

Answers

5. $\{x \mid x \geq -2\}$; $[-2, \infty)$
6. $\left\{x \mid x < \dfrac{3}{2}\right\}$; $\left(-\infty, \dfrac{3}{2}\right)$
7. $\{x \mid -3 < x \leq 1\}$; $(-3, 1]$

3. Addition and Subtraction Properties of Inequality

The process to solve a linear inequality is very similar to the method used to solve linear equations. Recall that adding or subtracting the same quantity from both sides of an equation results in an equivalent equation. The addition and subtraction properties of inequality state that the same is true for an inequality.

> **Addition and Subtraction Properties of Inequality**
>
> Let a, b, and c represent real numbers.
>
> 1. *Addition Property of Inequality: If $a < b$, then $a + c < b + c$
>
> 2. *Subtraction Property of Inequality: If $a < b$, then $a - c < b - c$
>
> *These properties may also be stated for $a \leq b$, $a > b$, and $a \geq b$.

To illustrate the addition and subtraction properties of inequality, consider the inequality $5 > 3$. If we add or subtract a real number such as 4 from both sides, the left-hand side will still be greater than the right-hand side. (See Figure 10-11.)

$5 > 3$ $5 + 4 > 3 + 4$

Figure 10-11

Example 4   Solving a Linear Inequality

Solve the inequality and graph the solution set. Express the solution set in set-builder notation and in interval notation.

$$-2p + 5 < -3p + 6$$

Solution:

$$-2p + 5 < -3p + 6$$

$-2p + 3p + 5 < -3p + 3p + 6$ Addition property of inequality (add $3p$ to both sides).

$p + 5 < 6$ Simplify.

$p + 5 - 5 < 6 - 5$ Subtraction property of inequality.

$p < 1$

Graph: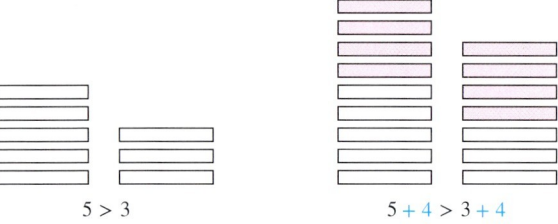

Set-builder notation: $\{p \mid p < 1\}$

Interval notation: $(-\infty, 1)$

Skill Practice Solve the inequality and graph the solution set. Express the solution set in set-builder notation and interval notation.

8. $2y - 5 < y - 11$

Answer

8. ←———•———
           -6

$\{y \mid y < -6\}$; $(-\infty, -6)$

TIP: The solution to an inequality gives a set of values that make the original inequality true. Therefore, you can test your final answer by using *test points*. That is, pick a value in the proposed solution set and verify that it makes the original inequality true. Furthermore, any test point picked outside the solution set should make the original inequality false. For example,

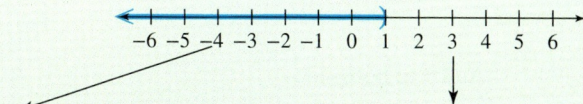

Pick $p = -4$ as an arbitrary test point within the proposed solution set.

$-2p + 5 < -3p + 6$
$-2(-4) + 5 \stackrel{?}{<} -3(-4) + 6$
$8 + 5 \stackrel{?}{<} 12 + 6$
$13 < 18 \checkmark$ True

Pick $p = 3$ as an arbitrary test point outside the proposed solution set.

$-2p + 5 < -3p + 6$
$-2(3) + 5 \stackrel{?}{<} -3(3) + 6$
$-6 + 5 \stackrel{?}{<} -9 + 6$
$-1 \stackrel{?}{<} -3$ False

4. Multiplication and Division Properties of Inequality

Multiplying both sides of an equation by the same quantity results in an equivalent equation. However, the same is not always true for an inequality. If you multiply or divide an inequality by a negative quantity, the direction of the inequality symbol must be reversed.

For example, consider multiplying or dividing the inequality, $4 < 5$ by -1.

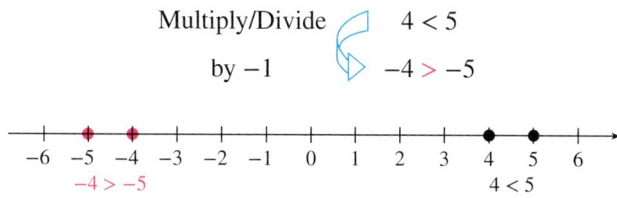

Figure 10-12

The number 4 lies to the left of 5 on the number line. However, -4 lies to the right of -5 (Figure 10-12). Changing the sign of two numbers changes their relative position on the number line. This is stated formally in the multiplication and division properties of inequality.

Multiplication and Division Properties of Inequality

Let a, b, and c represent real numbers, $c \neq 0$.

*If c is positive and $a < b$, then $ac < bc$ and $\dfrac{a}{c} < \dfrac{b}{c}$

*If c is negative and $a < b$, then $ac > bc$ and $\dfrac{a}{c} > \dfrac{b}{c}$

The second statement indicates that if both sides of an inequality are multiplied or divided by a negative quantity, the inequality sign must be reversed.

*These properties may also be stated for $a \leq b$, $a > b$, and $a \geq b$.

Example 5 Solving a Linear Inequality

Solve the inequality and graph the solution set. Express the solution set in set-builder notation and in interval notation.

$$-5x - 3 \leq 12$$

Solution:

$$-5x - 3 \leq 12$$
$$-5x - 3 + 3 \leq 12 + 3 \quad \text{Add 3 to both sides.}$$
$$-5x \leq 15$$
$$\frac{-5x}{-5} \geq \frac{15}{-5} \quad \text{Divide by } -5. \text{ Reverse the direction of the inequality sign.}$$
$$x \geq -3$$

Set-builder notation: $\{x \mid x \geq -3\}$

Interval notation: $[-3, \infty)$

TIP: The inequality $-5x - 3 \leq 12$ could have been solved by isolating x on the right-hand side of the inequality. This would create a positive coefficient on the variable term and eliminate the need to divide by a negative number.

$$-5x - 3 \leq 12$$
$$-3 \leq 5x + 12$$
$$-15 \leq 5x \quad \text{Notice that the coefficient of } x \text{ is positive.}$$
$$\frac{-15}{5} \leq \frac{5x}{5} \quad \text{Do not reverse the inequality sign because we are dividing by a positive number.}$$
$$-3 \leq x, \text{ or equivalently, } x \geq -3$$

Skill Practice Solve the inequality and graph the solution set. Express the solution set in set-builder notation and in interval notation.

9. $-5p + 2 > 22$

Example 6 Solving a Linear Inequality

Solve the inequality and graph the solution set. Express the solution set in set-builder notation and in interval notation.

$$12 - 2(y + 3) < -3(2y - 1) + 2y$$

Solution:

$$12 - 2(y + 3) < -3(2y - 1) + 2y$$
$$12 - 2y - 6 < -6y + 3 + 2y \quad \text{Clear parentheses.}$$
$$-2y + 6 < -4y + 3 \quad \text{Combine } like \text{ terms.}$$
$$-2y + 4y + 6 < -4y + 4y + 3 \quad \text{Add } 4y \text{ to both sides.}$$
$$2y + 6 < 3 \quad \text{Simplify.}$$
$$2y + 6 + (-6) < 3 + (-6) \quad \text{Add } -6 \text{ to both sides.}$$
$$2y < -3 \quad \text{Simplify.}$$
$$\frac{2y}{2} < \frac{-3}{2} \quad \text{Divide by 2. The direction of the inequality sign is } not \text{ reversed because we divided by a positive number.}$$
$$y < -\frac{3}{2}$$

Answer

9. $\{p \mid p < -4\}; (-\infty, -4)$

Set-builder notation: $\left\{y \mid y < -\dfrac{3}{2}\right\}$

Interval notation: $\left(-\infty, -\dfrac{3}{2}\right)$

Skill Practice Solve the inequality and graph the solution set. Express the solution set in set-builder notation and in interval notation.

10. $-8 - 8(2x - 4) > -5(4x - 5) - 21$

Example 7 Solving a Linear Inequality

Solve the inequality and graph the solution set. Express the solution set in set-builder notation and in interval notation.

$$-\dfrac{1}{4}k + \dfrac{1}{6} \leq 2 + \dfrac{2}{3}k$$

Solution:

$$-\dfrac{1}{4}k + \dfrac{1}{6} \leq 2 + \dfrac{2}{3}k$$

$$12\left(-\dfrac{1}{4}k + \dfrac{1}{6}\right) \leq 12\left(2 + \dfrac{2}{3}k\right)$$

Multiply both sides by 12 to clear fractions. (Because we multiplied by a positive number, the inequality sign is not reversed.)

$$\dfrac{\overset{3}{\cancel{12}}}{1}\left(-\dfrac{1}{4}k\right) + \dfrac{\overset{2}{\cancel{12}}}{1}\left(\dfrac{1}{6}\right) \leq 12(2) + \dfrac{\overset{4}{\cancel{12}}}{1}\left(\dfrac{2}{3}k\right)$$

Apply the distributive property.

$$-3k + 2 \leq 24 + 8k$$ Simplify.

$$-3k - 8k + 2 \leq 24 + 8k - 8k$$ Subtract $8k$ from both sides.

$$-11k + 2 \leq 24$$

$$-11k + 2 - 2 \leq 24 - 2$$ Subtract 2 from both sides.

$$-11k \leq 22$$

$$\dfrac{-11k}{-11} \geq \dfrac{22}{-11}$$ Divide both sides by -11. Reverse the inequality sign.

$$k \geq -2$$

Set-builder notation: $\{k \mid k \geq -2\}$

Interval notation: $[-2, \infty)$

Skill Practice Solve the inequality and graph the solution set. Express the solution set in set-builder notation and in interval notation.

11. $\dfrac{1}{5}t + 7 \leq \dfrac{1}{2}t - 2$

Answers

10.
$\{x \mid x > -5\}; (-5, \infty)$

11.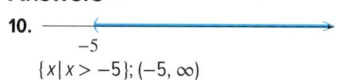
$\{t \mid t \geq 30\}; [30, \infty)$

5. Inequalities of the Form $a < x < b$

To solve a compound inequality of the form $a < x < b$, we can work with the inequality as a three-part inequality and isolate the variable, x, as demonstrated in Example 8.

Example 8   Solving a Compound Inequality of the Form $a < x < b$

Solve the inequality and graph the solution set. Express the solution set in set-builder notation and in interval notation.

$$-3 \leq 2x + 1 < 7$$

Solution:

To solve the compound inequality $-3 \leq 2x + 1 < 7$, isolate the variable x in the middle. The operations performed on the middle portion of the inequality must also be performed on the left and right sides.

$$-3 \leq 2x + 1 < 7$$

$-3 - 1 \leq 2x + 1 - 1 < 7 - 1$ Subtract 1 from all three parts of the inequality.

$-4 \leq 2x < 6$ Simplify.

$\dfrac{-4}{2} \leq \dfrac{2x}{2} < \dfrac{6}{2}$ Divide by 2 in all three parts of the inequality.

$-2 \leq x < 3$

Set-builder notation: $\{x \mid -2 \leq x < 3\}$

Interval notation: $[-2, 3)$

Skill Practice Solve the inequality and graph the solution set. Express the solution set in set-builder notation and in interval notation.

12. $-3 \leq -5 + 2y < 11$

6. Applications of Linear Inequalities

Table 10-2 provides several commonly used translations to express inequalities.

Table 10-2

English Phrase	Mathematical Inequality
a is less than b	$a < b$
a is greater than b a exceeds b	$a > b$
a is less than or equal to b a is at most b a is no more than b	$a \leq b$
a is greater than or equal to b a is at least b a is no less than b	$a \geq b$

Answer

12.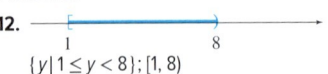
$\{y \mid 1 \leq y < 8\}; [1, 8)$

Example 9 — Translating Expressions Involving Inequalities

Write the English phrases as mathematical inequalities.

a. Claude's annual salary, s, is no more than $75,000.
b. A citizen must be at least 18 years old to vote. (Let a represent a citizen's age.)
c. An amusement park ride has a height requirement between 48 in. and 70 in. (Let h represent height in inches.)

Solution:

a. $s \leq 75,000$ Claude's annual salary, s, is no more than $75,000.
b. $a \geq 18$ A citizen must be at least 18 years old to vote.
c. $48 < h < 70$ An amusement park ride has a height requirement between 48 in. and 70 in.

Skill Practice Write the English phrase as a mathematical inequality.

13. Bill needs a score of at least 92 on the final exam. Let x represent Bill's score.
14. Fewer than 19 cars are in the parking lot. Let c represent the number of cars.
15. The heights, h, of women who wear petite size clothing are typically between 58 in. and 63 in., inclusive.

Linear inequalities are found in a variety of applications. Example 10 can help you determine the minimum grade you need on an exam to get an "A" in your math course.

Example 10 — Solving an Application With Linear Inequalities

To earn an "A" in a math class, Alsha must average at least 90 on all of her tests. Suppose Alsha has scored 79, 86, 93, 90, and 95 on her first five math tests. Determine the minimum score she needs on her sixth test to get an "A" in the class.

Solution:

Let x represent the score on the sixth exam.   Label the variable.

$$\left(\begin{array}{c}\text{Average of}\\\text{all tests}\end{array}\right) \geq 90$$ Create a verbal model.

$$\frac{79 + 86 + 93 + 90 + 95 + x}{6} \geq 90$$ The average score is found by taking the sum of the test scores and dividing by the number of scores.

$$\frac{443 + x}{6} \geq 90$$ Simplify.

$$6\left(\frac{443 + x}{6}\right) \geq (90)6$$ Multiply both sides by 6 to clear fractions.

$$443 + x \geq 540$$ Solve the inequality.

$$x \geq 540 - 443$$ Subtract 443 from both sides.

$$x \geq 97$$ Interpret the results.

Alsha must score at least 97 on her sixth exam to receive an "A" in the course.

Skill Practice

16. To get at least a "B" in math, Simon must average at least 80 on all tests. Suppose Simon has scored 60, 72, 98, and 85 on the first four tests. Determine the minimum score he needs on the fifth test to receive a "B."

Answers
13. $x \geq 92$
14. $c < 19$
15. $58 \leq h \leq 63$
16. Simon needs at least 85.

Section 10.8 Activity

Inequality statements surround us in day-to-day life. In Exercises A.1–A.4, let x represent the unknown quantity, and write a mathematical inequality to represent the given statement.

A.1. In the United States, a person must be at least 18 years old to vote.

A.2. A child younger than 2 years old can fly free on most airlines.

A.3. The normal range for hemoglobin for an adult female is between 12.0 grams per deciliter (g/dL) and 15.2 g/dL, inclusive.

A.4. To receive a discount at the zoo, a child must be at most 6 years old.

A.5. a. Graph the inequality $x > -2$. ———

    b. Graph the inequality $x \geq -2$. ———

    c. Explain when to use a parenthesis, (or), versus a bracket, [or], when graphing an inequality.

    d. Write the inequality $x > -2$ in interval notation.

    e. Write the inequality $x \geq -2$ in interval notation.

    f. Explain when to use a parenthesis, (or), versus a bracket, [or], when using interval notation.

A.6. a. Graph the inequality $x < 3$. ———

    b. Graph the inequality $3 > x$. ———

    c. What conclusion can you draw about the inequalities $x < 3$ and $3 > x$ based on their graphs?

    d. Write the inequality $x < 3$ in interval notation.

    e. Write the inequality $3 > x$ in interval notation.

A.7. Refer to the number line to fill in the blanks with $<$ or $>$.

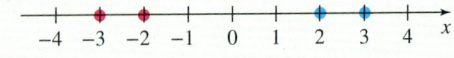

    a. $2 \square 3$ **b.** $-2 \square -3$

A.8. The following inequalities can each be solved in one step. Solve the inequality. Graph the solution set and write the set in interval notation.

    a. $5x > 10$ ———   **b.** $x + 5 > 10$ ———

    c. $\dfrac{x}{-5} > 10$ ———   **d.** $-5x > 10$ ———

    e. $x - 5 > 10$ ———

A.9. Which inequalities in Exercise A.8 required that the direction of the inequality sign be reversed? Why?

For Exercises A.10–A.13, solve the inequality. Write the solution set in interval notation.

A.10. $-4x - 3 < 17$ **A.11.** $-2(3x + 1) \geq -2(x + 4) - (x - 3)$

A.12. $\dfrac{4 - 2x}{-6} > \dfrac{1}{3}$ **A.13.** $-7 \leq 2x - 3 < 9$

Practice Exercises

Section 10.8

Prerequisite Review

For Exercises R.1–R.4, fill in the blank with <, >, or =.

R.1. $-4.5 \ \square \ -4.4$

R.2. $-12.36 \ \square \ -12.37$

R.3. $\dfrac{7}{5} \ \square \ \dfrac{4}{3}$

R.4. $\dfrac{2}{5} \ \square \ \dfrac{3}{7}$

For Exercises R.5–R.10, solve the equation.

R.5. $\dfrac{1}{4}x + \dfrac{2}{3} = \dfrac{1}{2}x - 1$

R.6. $\dfrac{1}{5} - \dfrac{3}{10}x = \dfrac{2}{3}x + 1$

R.7. $2(3 - x) + 4 = -2x + 8$

R.8. $5x - 4 = 5(x + 3) - 6$

R.9. $3x + 1 = 4x - (x - 2) - 1$

R.10. $5 - 2(3 - x) = 2(x - 1) + 1$

Vocabulary and Key Concepts

1. **a.** A relationship of the form $ax + b > c$ or $ax + b < c$ ($a \neq 0$) is called a _____ in one variable.
   b. A statement that involves more than one _____ is called a compound inequality.
   c. The notation $\{x \mid x > -9\}$ is an example of _____ notation, whereas $(-9, \infty)$ is an example of _____ notation.

2. Which two inequalities are equivalent?
   a. $t < 4$ **b.** $t \leq 4$ **c.** $4 > t$ **d.** $4 < t$

3. Which two inequalities are equivalent?
   a. $c > 6$ **b.** $c \geq 6$ **c.** $6 \geq c$ **d.** $6 \leq c$

4. Which statement describes a real number x that is between 5 and 9?
   a. $5 < x > 9$ **b.** $5 \geq x \geq 9$ **c.** $5 \leq x \leq 9$ **d.** $5 < x < 9$

Concept 1: Graphing Linear Inequalities

For Exercises 5–16, graph the solution set of each inequality. (See Examples 1–2.)

5. $x > 5$

6. $x \geq -7.2$

7. $x \leq \dfrac{5}{2}$

8. $x < -1$

9. $13 > p$

10. $-12 \geq t$

11. $2 \leq y \leq 6.5$

12. $-3 \leq m \leq \dfrac{8}{9}$

13. $0 < x < 4$

14. $-4 < y < 1$

15. $1 < p \leq 8$

16. $-3 \leq t < 3$

Concept 2: Set-Builder Notation and Interval Notation

For Exercises 17–22, graph each inequality and write the solution set in interval notation. (See Example 3.)

Set-Builder Notation	Graph	Interval Notation
17. $\{x \mid x \geq 6\}$		
18. $\{x \mid \frac{1}{2} < x \leq 4\}$		
19. $\{x \mid x \leq 2.1\}$		
20. $\{x \mid x > \frac{7}{3}\}$		
21. $\{x \mid -2 < x \leq 7\}$		
22. $\{x \mid x < -5\}$		

For Exercises 23–28, write each set in set-builder notation and in interval notation. (See Example 3.)

Set-Builder Notation	Graph	Interval Notation
23.	(at $\frac{3}{4}$, extending right	
24.	] at -0.3, extending left	
25.	(at -1,) at 8	
26.	[at 0, extending right	
27.	] at -14, extending left	
28.	(at 0,] at 9	

For Exercises 29–34, graph each set and write the set in set-builder notation. (See Example 3.)

Set-Builder Notation	Graph	Interval Notation
29.		$[18, \infty)$
30.		$[-10, -2]$
31.		$(-\infty, -0.6)$
32.		$\left(-\infty, \frac{5}{3}\right)$
33.		$[-3.5, 7.1]$
34.		$[-10, \infty)$

Concepts 3–4: Properties of Inequality

For Exercises 35–42, solve the equation in part (a). For part (b), solve the inequality and graph the solution set. Write the answer in set-builder notation and interval notation. **(See Examples 4–7.)**

35. a. $x + 3 = 6$
    b. $x + 3 > 6$

36. a. $y - 6 = 12$
    b. $y - 6 \geq 12$

37. a. $p - 4 = 9$
    b. $p - 4 \leq 9$

38. a. $k + 8 = 10$
    b. $k + 8 < 10$

39. a. $4c = -12$
    b. $4c < -12$

40. a. $5d = -35$
    b. $5d > -35$

41. a. $-10z = 15$
    b. $-10z \leq 15$

42. a. $-2w = 14$
    b. $-2w < 14$

Concept 5: Inequalities of the Form $a < x < b$

For Exercises 43–48, graph the solution and write the set in interval notation. **(See Example 8.)**

43. $-1 < y \leq 4$

44. $2.5 \leq t < 5.7$

45. $0 < x + 3 < 8$

46. $-2 \leq x - 4 \leq 3$

47. $8 \leq 4x \leq 24$

48. $-9 < 3x < 12$

Mixed Exercises

For Exercises 49–96, solve the inequality and graph the solution set. Write the solution set in (a) set-builder notation and (b) interval notation. **(See Exercises 4–8.)**

49. $x + 5 \leq 6$

50. $y - 7 < 6$

51. $3q - 7 > 2q + 3$

52. $5r + 4 \geq 4r - 1$

53. $4 < 1 + x$

54. $3 > z - 6$

55. $3c > 6$

56. $4d \leq 12$

57. $-3c > 6$

58. $-4d \leq 12$

59. $-h \leq -14$

60. $-q > -7$

61. $12 \geq -\dfrac{x}{2}$

62. $6 < -\dfrac{m}{3}$

63. $-2 \leq p + 1 < 4$

64. $0 < k + 7 < 6$

65. $-3 < 6h - 3 < 12$

66. $-6 \leq 4a - 2 \leq 12$

67. $-24 < -2x < -20$

68. $-12 \leq -3x \leq 6$

69. $-3 \leq \dfrac{1}{4}x - 1 < 5$

70. $-2 < \frac{1}{3}x - 2 \leq 2$

71. $-\frac{2}{3}y < 6$

72. $\frac{3}{4}x \leq -12$

73. $-2x - 4 \leq 11$

74. $-3x + 1 > 0$

75. $-12 > 7x + 9$

76. $8 < 2x - 10$

77. $-7b - 3 \leq 2b$

78. $3t \geq 7t - 35$

79. $4n + 2 < 6n + 8$

80. $2w - 1 \leq 5w + 8$

81. $8 - 6(x - 3) > -4x + 12$

82. $3 - 4(h - 2) > -5h + 6$

83. $3(x + 1) - 2 \leq \frac{1}{2}(4x - 8)$

84. $8 - (2x - 5) \geq \frac{1}{3}(9x - 6)$

85. $4(z - 1) - 6 \geq 6(2z + 3) - 12$

86. $3(2x + 5) + 2 < 5(2x + 2) + 3$

87. $2a + 3(a + 5) > -4a - (3a - 1) + 6$

88. $13 + 7(2y - 3) \leq 12 + 3(3y - 1)$

89. $\frac{7}{6}p + \frac{4}{3} \geq \frac{11}{6}p - \frac{7}{6}$

90. $\frac{1}{3}w - \frac{1}{2} \leq \frac{5}{6}w + \frac{1}{2}$

91. $\frac{y - 6}{3} > y + 4$

92. $\frac{5t + 7}{2} < t - 4$

93. $-1.2a - 0.4 < -0.4a + 2$

94. $-0.4c + 1.2 > -2c - 0.4$

95. $-2x + 5 \geq -x + 5$

96. $4x - 6 < 5x - 6$

For Exercises 97–100, determine whether the given number is a solution to the inequality.

97. $-2x + 5 < 4$; $x = -2$

98. $-3y - 7 > 5$; $y = 6$

99. $4(p + 7) - 1 > 2 + p$; $p = 1$

100. $3 - k < 2(-1 + k)$; $k = 4$

Concept 6: Applications of Linear Inequalities

For Exercises 101–110, write each English phrase as a mathematical inequality. (See Example 9.)

101. The length of a fish, L, was at least 10 in.

102. Tasha's average test score, t, exceeded 90.

103. The wind speed, w, exceeded 75 mph.

104. The height of a cave, h, was no more than 2 ft.

105. The temperature of the water in Blue Spring, t, is no more than 72°F.

106. The temperature on the tennis court, t, was no less than 100°F.

107. The length of the hike, L, was no less than 8 km.

108. The depth, d, of a certain pool was at most 10 ft.

109. The snowfall, h, in Monroe County is between 2 in. and 5 in.

110. The cost, c, of carpeting a room is between $300 and $400.

111. The average summer rainfall for Miami, Florida, for June, July, and August is 7.4 in. per month. If Miami receives 5.9 in. of rain in June and 6.1 in. in July, how much rain is required in August to exceed the 3-month summer average? **(See Example 10.)**

Dave Moyer

112. The average winter snowfall for Burlington, Vermont, for December, January, and February is 18.7 in. per month. If Burlington receives 22 in. of snow in December and 24 in. in January, how much snow is required in February to exceed the 3-month winter average?

113. To earn a "B" in chemistry, Trevor's average on his five tests must be at least 80. Suppose that Trevor has scored 85, 75, 72, and 82 on his first four chemistry tests. Determine the minimum score needed on his fifth test to get a "B" in the class.

114. In speech class, Carolyn needs at least a "B+" to keep her financial aid. To earn a "B+," the average of her four speeches must be at least an 85. On the first three speeches she scored 87, 75, and 82. Determine the minimum score on her fourth speech to get a "B+."

115. An artist paints wooden birdhouses. She buys the birdhouses for $9 each. However, for large orders, the price per birdhouse is discounted by a percentage off the original price. Let x represent the number of birdhouses ordered. The corresponding discount is given in the table.

Size of Order	Discount
$x \leq 49$	0%
$50 \leq x \leq 99$	5%
$100 \leq x \leq 199$	10%
$x \geq 200$	20%

   a. If the artist places an order for 190 birdhouses, compute the total cost.

   b. Which costs more: 190 birdhouses or 200 birdhouses? Explain your answer.

116. A wholesaler sells T-shirts to a surf shop at $8 per shirt. However, for large orders, the price per shirt is discounted by a percentage off the original price. Let x represent the number of shirts ordered. The corresponding discount is given in the table.

Number of Shirts Ordered	Discount
$x \leq 24$	0%
$25 \leq x \leq 49$	2%
$50 \leq x \leq 99$	4%
$100 \leq x \leq 149$	6%
$x \geq 150$	8%

   a. If the surf shop orders 50 shirts, compute the total cost.

   b. Which costs more: 148 shirts or 150 shirts? Explain your answer.

117. To print a flyer for a new business, Company A charges $39.99 for the design plus $0.50 per flyer. Company B charges $0.60 per flyer but has no design fee. For how many flyers would Company A be a better deal?

118. Melissa runs a landscaping business. She has equipment and fuel expenses of $313 per month. If she charges $45 for each lawn, how many lawns must she service to make a profit of at least $600 a month?

119. Madison is planning a 5-night trip to Cancun, Mexico, with her friends. The airfare is $475, her share of the hotel room is $54 per night, and her budget for food and entertainment is $350. She has $700 in savings and has a job earning $10 per hour babysitting. What is the minimum number of hours of babysitting that Madison needs so that she will have enough money to take the trip?

120. Luke and Landon are both tutors. Luke charges $50 for an initial assessment and $25 per hour for each hour he tutors. Landon charges $100 for an initial assessment and $20 per hour for tutoring. After how many hours of tutoring will Luke surpass Landon in earnings?

Chapter 10 Review Exercises

Section 10.1

1. Label the following as either an expression or an equation:

   a. $3x + y = 10$ b. $9x + 10x - 2xy$

   c. $4(x + 3) = 12$ d. $-5x = 7$

2. Explain how to determine whether an equation is linear in one variable.

3. Determine if the given equation is a linear equation in one variable. Answer yes or no.

   a. $4x^2 + 8 = -10$ b. $x + 18 = 72$

   c. $-3 + 2y^2 = 0$ d. $-4p - 5 = 6p$

4. For the equation $4y + 9 = -3$, determine if the given numbers are solutions.

   a. $y = 3$ b. $y = -3$

For Exercises 5–12, solve each equation using the addition property, subtraction property, multiplication property, or division property of equality.

5. $a + 6 = -2$ 6. $6 = z - 9$

7. $-\dfrac{3}{4} + k = \dfrac{9}{2}$ 8. $0.1r = 7$

9. $-5x = 21$ 10. $\dfrac{t}{3} = -20$

11. $-\dfrac{2}{5}k = \dfrac{4}{7}$ 12. $-m = -27$

13. The quotient of a number and negative six is equal to negative ten. Find the number.

14. The difference of a number and $-\tfrac{1}{8}$ is $\tfrac{5}{12}$. Find the number.

15. Four subtracted from a number is negative twelve. Find the number.

16. The product of a number and $\tfrac{1}{4}$ is $-\tfrac{1}{2}$. Find the number.

Section 10.2

For Exercises 17–28, solve each equation.

17. $4d + 2 = 6$ 18. $5c - 6 = -9$

19. $-7c = -3c - 8$ 20. $-28 = 5w + 2$

21. $\dfrac{b}{3} + 1 = 0$ 22. $\dfrac{2}{3}h - 5 = 7$

23. $-3p + 7 = 5p + 1$ 24. $4t - 6 = 12t + 18$

25. $4a - 9 = 3(a - 3)$ 26. $3(2c + 5) = -2(c - 8)$

27. $7b + 3(b - 1) + 3 = 2(b + 8)$

28. $2 + (18 - x) + 2(x - 1) = 4(x + 2) - 8$

29. Explain the difference between an equation that is a contradiction and an equation that is an identity.

For Exercises 30–35, solve the equation. Then identify the equation as a conditional equation, an identity, or a contradiction.

30. $x + 3 = 3 + x$ 31. $3x - 19 = 2x + 1$

32. $5x + 6 = 5x - 28$ 33. $2x - 8 = 2(x - 4)$

34. $-8x - 9 = -8(x - 9)$ 35. $4x - 4 = 3x - 2$

Section 10.3

For Exercises 36–55, solve each equation.

36. $\dfrac{x}{8} - \dfrac{1}{4} = \dfrac{1}{2}$ 37. $\dfrac{y}{15} - \dfrac{2}{3} = \dfrac{4}{5}$

38. $\dfrac{x + 5}{2} - \dfrac{2x + 10}{9} = 5$

39. $\dfrac{x - 6}{3} - \dfrac{2x + 8}{2} = 12$

40. $\dfrac{1}{10}p - 3 = \dfrac{2}{5}p$ 41. $\dfrac{1}{4}y - \dfrac{3}{4} = \dfrac{1}{2}y + 1$

42. $-\dfrac{1}{4}(2 - 3t) = \dfrac{3}{4}$ 43. $\dfrac{2}{7}(w + 4) = \dfrac{1}{2}$

44. $17.3 - 2.7q = 10.55$

45. $4.9z + 4.6 = 3.2z - 2.2$

46. $5.74a + 9.28 = 2.24a - 5.42$

47. $62.84t - 123.66 = 4(2.36 + 2.4t)$

48. $0.05x + 0.10(24 - x) = 0.75(24)$

49. $0.20(x + 4) + 0.65x = 0.20(854)$

50. $100 - (t - 6) = -(t - 1)$

51. $3 - (x + 4) + 5 = 3x + 10 - 4x$

52. $5t - (2t + 14) = 3t - 14$

53. $9 - 6(2x + 1) = -3(4z - 1)$

54. $5x - 3(x - 4) = x + 12$

55. $8 - 2(x + 5) = -3(x + 1) + 1$

Section 10.4

56. Twelve added to the sum of a number and two is forty-four. Find the number.

57. Twenty added to the sum of a number and six is thirty-seven. Find the number.

58. Three times a number is the same as the difference of twice the number and seven. Find the number.

59. Eight less than five times a number is forty-eight less than the number. Find the number.

60. Three times the largest of three consecutive even integers is 76 more than the sum of the other two integers. Find the integers.

61. Ten times the smallest of three consecutive integers is 213 more than the sum of the other two integers. Find the integers.

62. The perimeter of a triangle is 78 in. The lengths of the sides are represented by three consecutive integers. Find the lengths of the sides of the triangle.

63. The perimeter of a pentagon (a five-sided polygon) is 190 cm. The lengths of the sides are represented by consecutive integers. Find the lengths of the sides.

Section 10.5

For Exercises 64–69, solve each problem involving percents.

64. What is 35% of 68?

65. What is 4% of 720?

66. 53.5 is what percent of 428?

67. 68.4 is what percent of 72?

68. 24 is 15% of what number?

69. 8.75 is 0.5% of what number?

70. A couple spent a total of $50.40 for dinner. This included a 20% tip and 6% sales tax on the price of the meal. What was the price of the dinner before tax and tip?

71. Anna invested $3000 in an account paying 8% simple interest.

   a. How much interest will she earn in $3\frac{1}{2}$ years?

   b. What will her balance be at that time?

72. Eduardo invested money in an account earning 4% simple interest. At the end of 5 years, he had a total of $14,400. How much money did he originally invest?

73. A novel is discounted 30%. The sale price is $20.65. What was the original price?

Section 10.6

For Exercises 74–81, solve for the indicated variable.

74. $C = K - 273$ for K

75. $K = C + 273$ for C

76. $P = 4s$ for s

77. $P = 3s$ for s

78. $y = mx + b$ for x

79. $a + bx = c$ for x

80. $2x + 5y = -2$ for y

81. $4(a + b) = Q$ for b

For Exercises 82–88, use an appropriate geometry formula to solve the problem.

82. Find the height of a parallelogram whose area is 42 m² and whose base is 6 m.

83. The volume of a cone is given by the formula $V = \frac{1}{3}\pi r^2 h$.

    a. Solve the formula for h.

    b. Find the height of a right circular cone whose volume is 47.8 in.^{3} and whose radius is 3 in. Round to the nearest tenth of an inch.

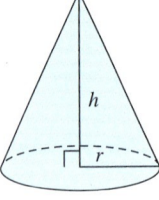

84. The smallest angle of a triangle is 2° more than $\frac{1}{4}$ of the largest angle. The middle angle is 2° less than the largest angle. Find the measure of each angle.

85. A carpenter uses a special saw to cut an angle on a piece of framing. If the angles are complementary and one angle is 10° more than the other, find the measure of each angle.

86. A rectangular window has width 1 ft less than its length. The perimeter is 18 ft. Find the length and the width of the window.

87. Find the measure of the vertical angles by first solving for x.

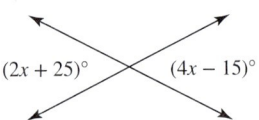

$(2x + 25)°$ $(4x - 15)°$

88. Find the measure of angle y.

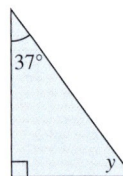

Section 10.7

89. In stormy conditions, a delivery truck can travel a route in 14 hr. In good weather, the same trip can be made in 10.5 hr because the truck travels 15 km/hr faster. Find the speed of the truck in stormy weather and the speed in good weather.

90. Winston and Gus ride their bicycles in a relay. Each person rides the same distance. Winston rides 3 mph faster than Gus and finishes the course in 2.5 hr. Gus finishes in 3 hr. How fast does each person ride?

91. Two cars leave a rest stop on Interstate I-10 at the same time. One heads east and the other heads west. One car travels 55 mph and the other 62 mph. How long will it take for them to be 327.6 mi apart?

92. Two hikers begin at the same time at opposite ends of a 9-mi trail and walk toward each other. One hiker walks 2.5 mph and the other walks 1.5 mph. How long will it be before they meet?

93. How much ground beef with 24% fat should be mixed with 8 lb of ground sirloin that is 6% fat to make a mixture that is 9.6% fat?

94. A soldering compound with 40% lead (the rest is tin) must be combined with 80 lb of solder that is 75% lead to make a compound that is 68% lead. How much solder with 40% lead should be used?

Section 10.8

For Exercises 95–97, graph each inequality and write the set in interval notation.

95. $\{x \mid x > -2\}$

96. $\{x \mid x \leq \frac{1}{2}\}$

97. $\{x \mid -1 < x \leq 4\}$

98. A landscaper buys potted geraniums from a nursery at a price of $5 per plant. However, for large orders, the price per plant is discounted by a percentage off the original price. Let x represent the number of potted plants ordered. The corresponding discount is given in the table.

Number of Plants	Discount
$x \leq 99$	0%
$100 \leq x \leq 199$	2%
$200 \leq x \leq 299$	4%
$x \geq 300$	6%

    a. Find the cost to purchase 130 plants.

    b. Which costs more, 300 plants or 295 plants? Explain your answer.

For Exercises 99–109, solve the inequality. Graph the solution set and write the answer in set-builder notation and interval notation.

99. $c + 6 < 23$

100. $3w - 4 > -5$

101. $-2x - 7 \geq 5$

102. $5(y+2) \leq -4$ _____

103. $-\dfrac{3}{7}a \leq -21$ _____

104. $1.3 > 0.4t - 12.5$ _____

105. $4k + 23 < 7k - 31$ _____

106. $\dfrac{6}{5}h - \dfrac{1}{5} \leq \dfrac{3}{10} + h$ _____

107. $-5x - 2(4x - 3) + 6 > 17 - 4(x - 1)$ _____

108. $-6 < 2b \leq 14$ _____

109. $-2 \leq z + 4 \leq 9$ _____

110. The summer average rainfall for Bermuda for June, July, and August is 5.3 in. per month. If Bermuda receives 6.3 in. of rain in June and 7.1 in. in July, how much rain is required in August to exceed the 3-month summer average?

111. Collette has $15.00 to spend on dinner. Of this, 25% will cover the tax and tip, resulting in $11.25 for her to spend on food. If Collette wants veggies and blue cheese, fries, and a drink, what is the maximum number of chicken wings she can get?

Wing Special
25¢ each
5:00–7:00 P.M.

*Add veggies and blue cheese for $2.50
*Add fries for $2.50
*Add a drink for $1.75

Graphing Linear Equations in Two Variables

11

CHAPTER OUTLINE

11.1 Rectangular Coordinate System   684
11.2 Linear Equations in Two Variables   694
11.3 Slope of a Line and Rate of Change   711
11.4 Slope-Intercept Form of a Linear Equation   727
   Problem Recognition Exercises: Linear Equations in Two Variables   738
11.5 Point-Slope Formula   740
11.6 Applications of Linear Equations and Modeling   750
   Chapter 11 Review Exercises   759

Mathematics in Gaming

Imagine that you are playing a game of *Battleship*. You say "E-5" to your opponent and his face saddens as he realizes you have sunk his ship. In this game, the "E" and the "5" represent a row or column in a grid, thus defining a *location* on a map. In mathematics we use **ordered pairs**, two numbers of the form (x, y), to determine the location of a point. Furthermore, an equation involving x and y defines a line or curve in a plane that might be the path of a battleship or other object in a computer game.

For example, suppose that a battleship follows the path defined by $y = 2x + 1$. The value of y depends on the value of x, where y is 1 more than double the value of x. We can determine the value of y for any given value of x by using substitution.

ERproductions Ltd/Getty Images

If $x = 1$, then $y = 2(1) + 1 = 3$. Thus, the ordered pair (1, 3) represents a point on the path of the ship.

If $x = 2$, then $y = 2(2) + 1 = 5$. Thus, the ordered pair (2, 5) represents a point on the path of the ship.

In this chapter, we will study equations in x and y and their related graphs. This is important content for a variety of applications, including computer gaming.

Section 11.1 Rectangular Coordinate System

Concepts

1. Interpreting Graphs
2. Plotting Points in a Rectangular Coordinate System
3. Applications of Plotting and Identifying Points

1. Interpreting Graphs

Mathematics is a powerful tool used by scientists and has directly contributed to the highly technical world in which we live. Applications of mathematics have led to advances in the sciences, business, computer technology, and medicine.

One fundamental application of mathematics is the graphical representation of numerical information (or data). For example, Table 11-1 represents the number of clients admitted to a drug and alcohol rehabilitation program over a 12-month period.

Table 11-1

Month		Number of Clients
Jan.	1	55
Feb.	2	62
March	3	64
April	4	60
May	5	70
June	6	73
July	7	77
Aug.	8	80
Sept.	9	80
Oct.	10	74
Nov.	11	85
Dec.	12	90

In table form, the information is difficult to picture and interpret. It appears that on a monthly basis, the number of clients fluctuates. However, when the data are represented in a graph, an upward trend is clear (Figure 11-1).

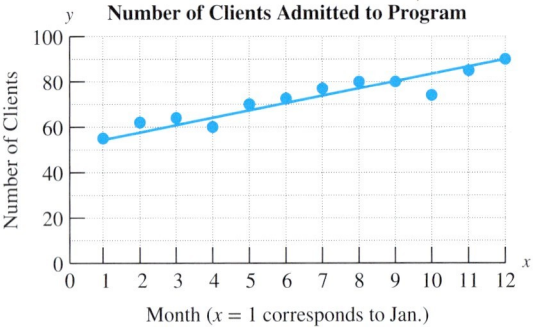

Figure 11-1

From the increase in clients shown in this graph, management for the rehabilitation center might make plans for the future. If the trend continues, management might consider expanding its facilities and increasing its staff to accommodate the expected increase in clients.

Example 1 Interpreting a Graph

Refer to Figure 11-1 and Table 11-1.

a. For which month was the number of clients the greatest?
b. How many clients were served in the first month (January)?
c. Which month corresponds to 60 clients served?
d. Between which two consecutive months did the number of clients decrease?
e. Between which two consecutive months did the number of clients remain the same?

Solution:

a. Month 12 (December) corresponds to the highest point on the graph. This represents the greatest number of clients, 90.
b. In month 1 (January), 55 clients were served.
c. Month 4 (April).
d. The number of clients decreased between months 3 and 4 and between months 9 and 10.
e. The number of clients remained the same between months 8 and 9.

Skill Practice Refer to Figure 11-1 and Table 11-1.
1. How many clients were served in October?
2. Which month corresponds to 70 clients?
3. What is the difference between the number of clients in month 12 and in month 1?
4. For which month was the number of clients the least?

2. Plotting Points in a Rectangular Coordinate System

The data in Table 11-1 represent a relationship between two variables—the month number and the number of clients. The graph in Figure 11-1 enables us to visualize this relationship. In picturing the relationship between two quantities, we often use a graph with two number lines drawn at right angles to each other (Figure 11-2). This forms a **rectangular coordinate system**. The horizontal line is called the **x-axis**, and the vertical line is called the **y-axis**. The point where the lines intersect is called the **origin**. On the x-axis, the numbers to the right of the origin are positive and the numbers to the left are negative. On the y-axis, the numbers above the origin are positive and the numbers below are negative. The x- and y-axes divide the graphing area into four regions called **quadrants**.

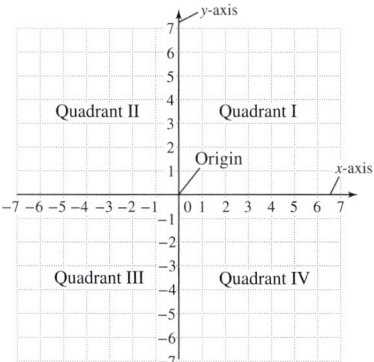

Figure 11-2

Answers
1. 74 clients
2. Month 5 (May)
3. 35 clients
4. Month 1 (January)

Points graphed in a rectangular coordinate system are defined by two numbers as an **ordered pair**, (x, y). The first number (called the **x-coordinate**, or the abscissa) is the horizontal position from the origin. The second number (called the **y-coordinate**, or the ordinate) is the vertical position from the origin. Example 2 shows how points are plotted in a rectangular coordinate system.

Example 2 — Plotting Points in a Rectangular Coordinate System

Plot the points.

a. $(4, 5)$ b. $(-4, -5)$ c. $(-1, 3)$ d. $(3, -1)$

e. $\left(\dfrac{1}{2}, -\dfrac{7}{3}\right)$ f. $(-2, 0)$ g. $(0, 0)$ h. $(\pi, 1.1)$

Solution:

See Figure 11-3.

a. The ordered pair $(4, 5)$ indicates that $x = 4$ and $y = 5$. Beginning at the origin, move 4 units in the positive x-direction (4 units to the right), and from there move 5 units in the positive y-direction (5 units up). Then plot the point. The point is in Quadrant I.

b. The ordered pair $(-4, -5)$ indicates that $x = -4$ and $y = -5$. Move 4 units in the negative x-direction (4 units to the left), and from there move 5 units in the negative y-direction (5 units down). Then plot the point. The point is in Quadrant III.

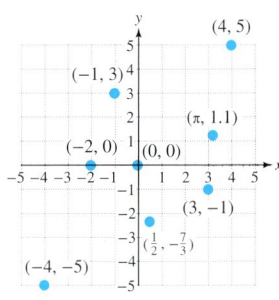

Figure 11-3

c. The ordered pair $(-1, 3)$ indicates that $x = -1$ and $y = 3$. Move 1 unit to the left and 3 units up. The point is in Quadrant II.

d. The ordered pair $(3, -1)$ indicates that $x = 3$ and $y = -1$. Move 3 units to the right and 1 unit down. The point is in Quadrant IV.

e. The improper fraction $-\dfrac{7}{3}$ can be written as the mixed number $-2\dfrac{1}{3}$. Therefore, to plot the point $\left(\dfrac{1}{2}, -\dfrac{7}{3}\right)$ move to the right $\dfrac{1}{2}$ unit, and down $2\dfrac{1}{3}$ units. The point is in Quadrant IV.

f. The point $(-2, 0)$ indicates $y = 0$. Therefore, the point is on the x-axis.

g. The point $(0, 0)$ is at the origin.

h. The irrational number π can be approximated as 3.14. Thus, the point $(\pi, 1.1)$ is located approximately 3.14 units to the right and 1.1 units up. The point is in Quadrant I.

TIP: Notice that changing the order of the x- and y-coordinates changes the location of the point. For example, the point $(-1, 3)$ is in Quadrant II, whereas $(3, -1)$ is in Quadrant IV (Figure 11-3). This is why points are represented by *ordered* pairs. The order of the coordinates is important.

Avoiding Mistakes

Points that lie on either of the axes do not lie in any quadrant.

Skill Practice

5. Plot the points.

$A(3, 4)$ $B(-2, 2)$ $C(4, 0)$ $D\left(\dfrac{5}{2}, -\dfrac{1}{2}\right)$ $E(-5, -2)$

Answer

5.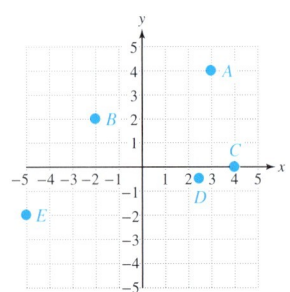

3. Applications of Plotting and Identifying Points

The effective use of graphs for mathematical models requires skill in identifying points and interpreting graphs.

Example 3 Determining Points from a Graph

A map of a national park is drawn so that the origin is placed at the ranger station (Figure 11-4). Four fire observation towers are located at points A, B, C, and D. Estimate the coordinates of the fire towers relative to the ranger station (all distances are in miles).

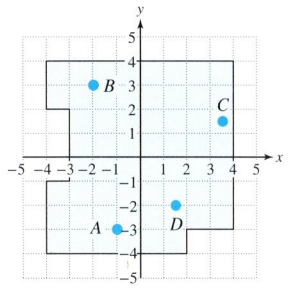

Figure 11-4

Solution:

Point A: $(-1, -3)$

Point B: $(-2, 3)$

Point C: $(3\frac{1}{2}, 1\frac{1}{2})$ or $(\frac{7}{2}, \frac{3}{2})$ or $(3.5, 1.5)$

Point D: $(1\frac{1}{2}, -2)$ or $(\frac{3}{2}, -2)$ or $(1.5, -2)$

Skill Practice

6. Towers are located at points A, B, C, and D. Estimate the coordinates of the towers.

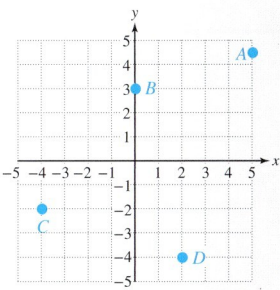

Example 4 Plotting Points in an Application

The daily low temperaures (in degrees Fahrenheit) for one week in January for Sudbury, Ontario, Canada, are given in Table 11-2.

a. Write an ordered pair for each row in the table using the day number as the x-coordinate and the temperature as the y-coordinate.

b. Plot the ordered pairs from part (a) on a rectangular coordinate system.

Table 11-2

Day Number, x	Temperature (°F), y
1	−3
2	−5
3	1
4	6
5	5
6	0
7	−4

Solution:

a. Each ordered pair represents the day number and the corresponding low temperature for that day.

$(1, -3)$ $(2, -5)$ $(3, 1)$ $(4, 6)$ $(5, 5)$ $(6, 0)$ $(7, -4)$

Answer

6. $A(5, 4\frac{1}{2})$
$B(0, 3)$
$C(-4, -2)$
$D(2, -4)$

b.

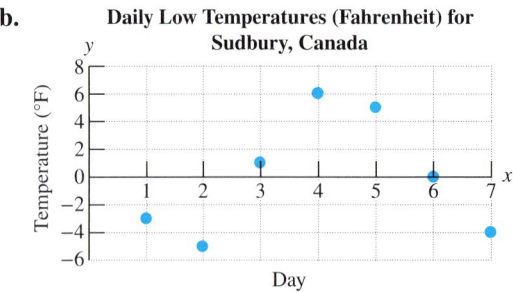

Daily Low Temperatures (Fahrenheit) for Sudbury, Canada

TIP: The graph in Example 4(b) shows only Quadrants I and IV because all *x*-coordinates are positive.

Skill Practice

7. The table shows the number of homes sold in a certain town for a 6-month period. Plot the ordered pairs.

Month, x	Number Sold, y
1	20
2	25
3	28
4	40
5	45
6	30

Answer

7.

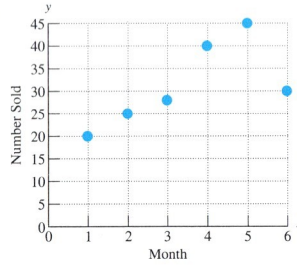

Section 11.1 Activity

A.1. a. Plot the points *A* through *E*. For each point, indicate the location of the point (choose from Quadrant I, Quadrant II, Quadrant III, Quadrant IV, the *x*-axis, the *y*-axis, or the origin).

Point	Ordered Pair	Location
A	(−2, −3)	
B	(0, 1)	
C	(2, −3)	
D	(−2, 0)	
E	(2, 0)	

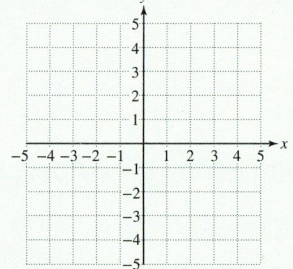

b. Connect the points using a straightedge in the order from *A* to *B*, from *B* to *C*, from *C* to *D*, from *D* to *E*, and from *E* to *A*.

c. The figure drawn in part (b) illustrates that some shapes are easily drawn by plotting a few key points and connecting the dots. Now consider a computer screen or tablet. The screen is made up of a rectangular array of dots called pixels. Each pixel can be assigned an ordered pair to indicate its location on the screen. Explain how a rectangular coordinate system might be helpful to video game programmers.

Practice Exercises

Section 11.1

Prerequisite Review

For Exercises R.1–R.8, plot the points on the number line.

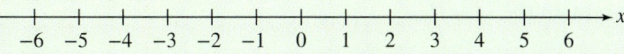

R.1. 4 R.2. -3 R.3. $-\dfrac{5}{4}$ R.4. $\dfrac{5}{3}$

R.5. π R.6. $-0.\overline{4}$ R.7. $-4\dfrac{2}{3}$ R.8. $2\dfrac{1}{4}$

Vocabulary and Key Concepts

1. **a.** In a rectangular coordinate system, two number lines are drawn at right angles to each other. The horizontal line is called the _____-axis, and the vertical line is called the _____.

   b. A point in a rectangular coordinate system is defined by an _____ pair, (x, y).

   c. In a rectangular coordinate system, the point where the x- and y-axes intersect is called the _____ and is represented by the ordered pair _____.

   d. The x- and y-axes divide the coordinate plane into four regions called _____.

   e. A point with a positive x-coordinate and a _____ y-coordinate is in Quadrant IV.

   f. In Quadrant _____, both the x- and y-coordinates are negative.

Concept 1: Interpreting Graphs

For Exercises 2–6, refer to the graphs to answer the questions. (See Example 1.)

2. The number of Botox® injection procedures (in millions) in the United States over a 6-yr period is shown in the graph.

   a. For which year was the number of procedures the greatest?

   b. Approximately how many procedures were performed in year 5?

   c. Which year corresponds to 2.3 million procedures?

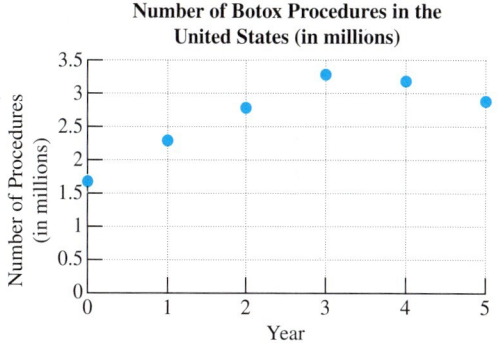

3. The number of patients served by a certain hospice care center for the first 12 months after it opened is shown in the graph.

   a. For which month was the number of patients greatest?

   b. How many patients did the center serve in the first month?

   c. Between which months did the number of patients decrease?

   d. Between which two months did the number of patients remain the same?

   e. Which month corresponds to 40 patients served?

   f. Approximately how many patients were served during the 10th month?

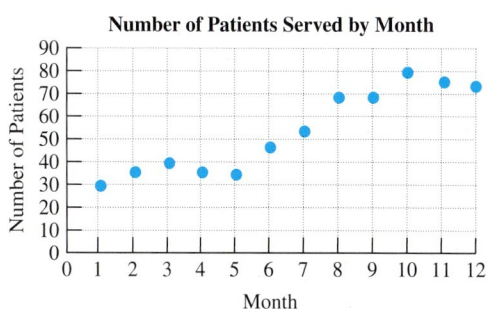

4. The number of housing permits (in thousands) issued by a county in Texas between year 0 and year 6 is shown in the graph.

   a. For which year was the number of permits greatest?
   b. How many permits did the county issue in year 0?
   c. Between which years did the number of permits decrease?
   d. Between which two years did the number of permits remain the same?
   e. Which year corresponds to 7000 permits issued?

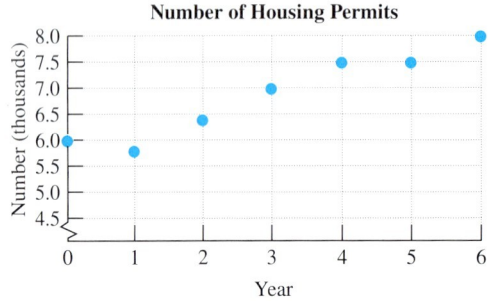

5. The price per share of a stock (in dollars) over a period of 5 days is shown in the graph.

   a. Interpret the meaning of the ordered pair (1, 89.25).
   b. What was the change in price between day 3 and day 4?
   c. What was the change in price between day 4 and day 5?

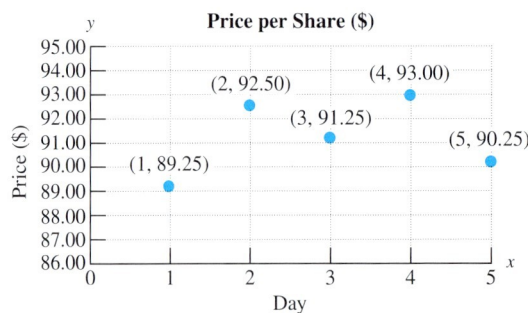

6. The price per share of a stock (in dollars) over a period of 5 days is shown in the graph.

   a. Interpret the meaning of the ordered pair (1, 10.125).
   b. What was the change between day 4 and day 5?
   c. What is the change between day 1 and day 5?

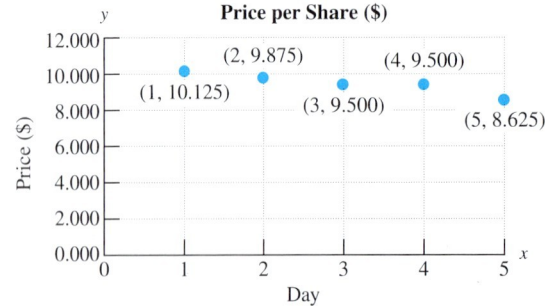

Concept 2: Plotting Points in a Rectangular Coordinate System

7. Plot the points on a rectangular coordinate system.
(See Example 2.)

a. (2, 6) b. (6, 2) c. (−7, 3)
d. (−7, −3) e. (0, −3) f. (−3, 0)
g. (6, −4) h. (0, 5)

8. Plot the points on a rectangular coordinate system.

a. (4, 5) b. (−4, 5) c. (−6, 0)
d. (6, 0) e. (4, −5) f. (−4, −5)
g. (0, −2) h. (0, 0)

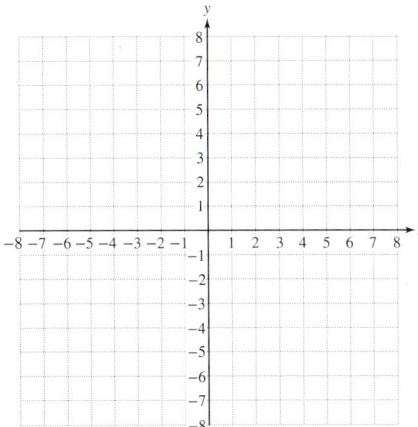

9. Plot the points on a rectangular coordinate system.

a. (−1, 5) b. (0, 4) c. $\left(-2, -\dfrac{3}{2}\right)$
d. (2, −1.75) e. (4, 2) f. (−6, 0)

10. Plot the points on a rectangular coordinate system.

a. (7, 0) b. (−3, −2) c. $\left(6\dfrac{3}{5}, 1\right)$
d. (0, 1.5) e. $\left(\dfrac{7}{2}, -4\right)$ f. $\left(-\dfrac{7}{2}, 4\right)$

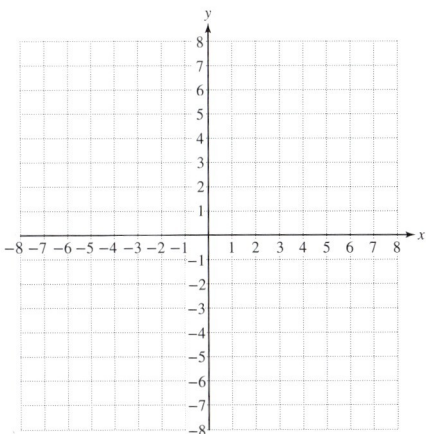
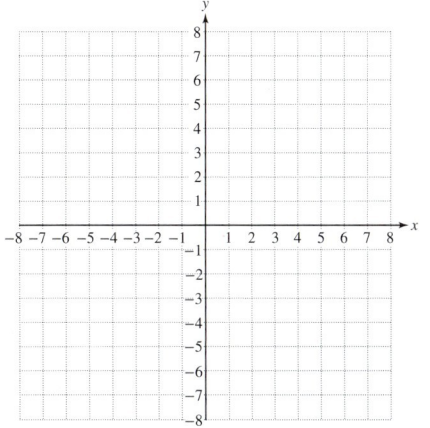

For Exercises 11–18, identify the quadrant in which the given point is located.

11. (13, −2) **12.** (25, 16) **13.** (−8, 14) **14.** (−82, −71)

15. (−5, −19) **16.** (−31, 6) **17.** $\left(\dfrac{5}{2}, \dfrac{7}{4}\right)$ **18.** (9, −40)

19. Explain why the point (0, −5) is *not* located in Quadrant IV.

20. Explain why the point (−1, 0) is *not* located in Quadrant II.

21. Where is the point $\left(\dfrac{7}{8}, 0\right)$ located?

22. Where is the point $\left(0, \dfrac{6}{5}\right)$ located?

Concept 3: Applications of Plotting and Identifying Points

For Exercises 23–24, refer to the graph. (See Example 3.)

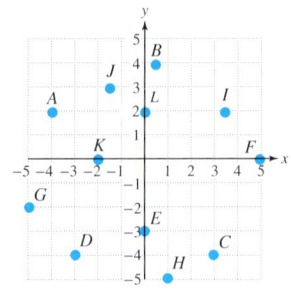

23. Estimate the coordinates of the points A, B, C, D, E, and F.

24. Estimate the coordinates of the points G, H, I, J, K, and L.

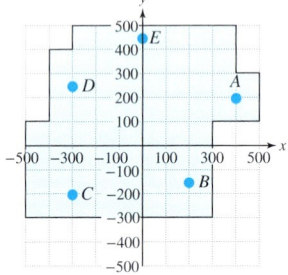

25. A map of a park is laid out with the visitor center located at the origin. Five visitors are in the park located at points A, B, C, D, and E. All distances are in meters.

    a. Estimate the coordinates of each visitor. (See Example 3.)

    b. How far apart are visitors C and D?

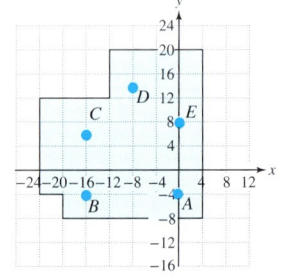

26. A townhouse has a sprinkler system in the backyard. With the water source at the origin, the sprinkler heads are located at points A, B, C, D, and E. All distances are in feet.

    a. Estimate the coordinates of each sprinkler head.

    b. How far is the distance from sprinkler head B to C?

Ryan McVay/Getty Images

27. A movie theater has kept records of popcorn sales versus movie attendance.

    a. Use the table to write the corresponding ordered pairs using the movie attendance as the x variable and sales of popcorn as the y variable. Interpret the meaning of the first ordered pair. (See Example 4.)

    b. Plot the data points on a rectangular coordinate system.

Movie Attendance (Number of People)	Sales of Popcorn ($)
250	225
175	193
315	330
220	209
450	570
400	480
190	185

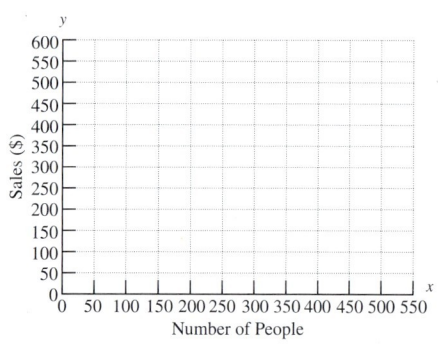

28. The age and systolic blood pressure (in millimeters of mercury, mm Hg) for eight different women are given in the table.

   a. Write the corresponding ordered pairs using the woman's age as the x variable and the systolic blood pressure as the y variable. Interpret the meaning of the first ordered pair.

   b. Plot the data points on a rectangular coordinate system.

Age (Years)	Systolic Blood Pressure (mm Hg)
57	149
41	120
71	158
36	115
64	151
25	110
40	118
77	165

29. The following table shows the average temperature in degrees Celsius for Montreal, Quebec, Canada, by month.

   a. Write the corresponding ordered pairs, letting $x = 1$ correspond to the month of January.

   b. Plot the ordered pairs on a rectangular coordinate system.

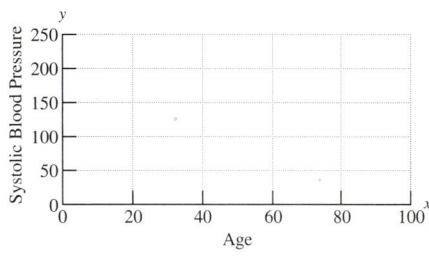

Month, x		Temperature (°C), y
Jan.	1	−10.2
Feb.	2	−9.0
March	3	−2.5
April	4	5.7
May	5	13.0
June	6	18.3
July	7	20.9
Aug.	8	19.6
Sept.	9	14.8
Oct.	10	8.7
Nov.	11	2.0
Dec.	12	−6.9

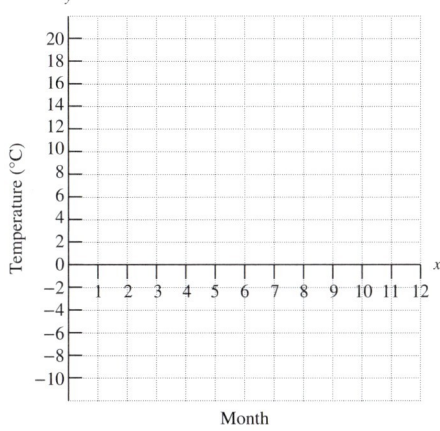

30. The table shows the average temperature in degrees Fahrenheit for Fairbanks, Alaska, by month.

   a. Write the corresponding ordered pairs, letting $x = 1$ correspond to the month of January.

   b. Plot the ordered pairs on a rectangular coordinate system.

Month, x		Temperature (°F), y
Jan.	1	−12.8
Feb.	2	−4.0
March	3	8.4
April	4	30.2
May	5	48.2
June	6	59.4
July	7	61.5
Aug.	8	56.7
Sept.	9	45.0
Oct.	10	25.0
Nov.	11	6.1
Dec.	12	−10.1

Expanding Your Skills

31. The data in the table give the percent of males and females who have completed 4 or more years of college education for selected years. Let x represent the number of years since 1960. Let y represent the percent of men and the percent of women that completed 4 or more years of college.

Year	x	Percent, y Men	Percent, y Women
1960	0	9.7	5.8
1970	10	13.5	8.1
1980	20	20.1	12.8
1990	30	24.4	18.4
2000	40	27.8	23.6
2005	45	28.9	26.5
2010	50	29.9	28.8
2020	60	35.4	36.6

a. Plot the data points for men and for women on the same graph.

b. Is the percentage of men with 4 or more years of college increasing or decreasing?

c. Is the percentage of women with 4 or more years of college increasing or decreasing?

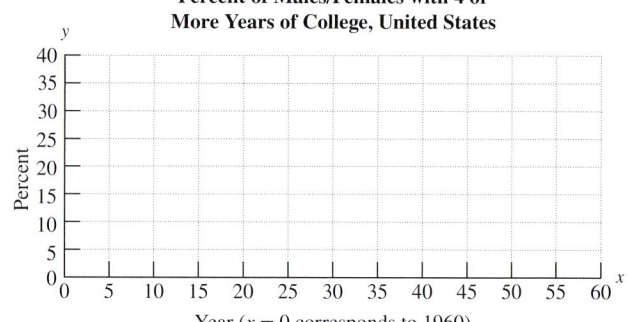

Percent of Males/Females with 4 or More Years of College, United States

32. Use the data and graph from Exercise 31 to answer the questions.

a. In which year was the difference in percentages between men and women with 4 or more years of college the greatest?

b. In which year was the difference in percentages between men and women the least?

c. In which 10-year period did the percentage of women with 4-year degrees overtake the percentage of men with 4-year degrees?

Section 11.2 Linear Equations in Two Variables

Concepts

1. Definition of a Linear Equation in Two Variables
2. Graphing Linear Equations in Two Variables by Plotting Points
3. x- and y-Intercepts
4. Horizontal and Vertical Lines

1. Definition of a Linear Equation in Two Variables

Recall that an equation in the form $ax + b = c$, where $a \neq 0$, is called a linear equation in one variable. A solution to such an equation is a value of x that makes the equation a true statement. For example, $3x + 5 = -1$ has a solution of -2.

In this section, we will look at linear equations in *two* variables.

> **Linear Equation in Two Variables**
>
> Let A, B, and C be real numbers such that A and B are not both zero. Then, an equation that can be written in the form:
>
> $$Ax + By = C$$
>
> is called a **linear equation in two variables**.

The equation $x + y = 4$ is a linear equation in two variables. A solution to such an equation is an ordered pair (x, y) that makes the equation a true statement. Several solutions to the equation $x + y = 4$ are listed here:

Solution:	Check:
(x, y)	$x + y = 4$
$(2, 2)$	$(2) + (2) = 4$ ✓
$(1, 3)$	$(1) + (3) = 4$ ✓
$(4, 0)$	$(4) + (0) = 4$ ✓
$(-1, 5)$	$(-1) + (5) = 4$ ✓

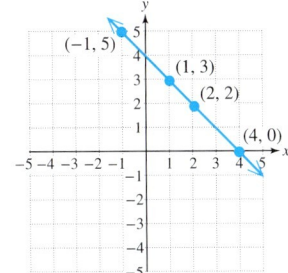

Figure 11-5

By graphing these ordered pairs, we see that the solution points line up (Figure 11-5).

Notice that there are infinitely many solutions to the equation $x + y = 4$, so they cannot all be listed. Therefore, to visualize all solutions to the equation $x + y = 4$, we draw the line through the points in the graph. Every point on the line represents an ordered pair solution to the equation $x + y = 4$, and the line represents the set of *all* solutions to the equation.

Example 1 Determining Solutions to a Linear Equation

For the linear equation $6x - 5y = 12$, determine whether the given ordered pair is a solution.

a. $(2, 0)$ **b.** $(3, 1)$ **c.** $\left(1, -\dfrac{6}{5}\right)$

Solution:

a. $6x - 5y = 12$

$6(2) - 5(0) \stackrel{?}{=} 12$ Substitute $x = 2$ and $y = 0$.

$12 - 0 \stackrel{?}{=} 12$ ✓ True   The ordered pair $(2, 0)$ is a solution.

b. $6x - 5y = 12$

$6(3) - 5(1) \stackrel{?}{=} 12$ Substitute $x = 3$ and $y = 1$.

$18 - 5 \neq 12$ The ordered pair $(3, 1)$ is *not* a solution.

c. $6x - 5y = 12$

$6(1) - 5\left(-\dfrac{6}{5}\right) \stackrel{?}{=} 12$ Substitute $x = 1$ and $y = -\dfrac{6}{5}$.

$6 + 6 \stackrel{?}{=} 12$ ✓ True   The ordered pair $\left(1, -\dfrac{6}{5}\right)$ is a solution.

FOR REVIEW

Remember to use parentheses when substituting numerical values for variables within an expression or an equation.

$6x - 5y = 12$
$6(\ ) - 5(\ ) = 12$
$6(2) - 5(0) = 12$
$12 - 0 = 12$ ✓

Skill Practice Given the equation $3x - 2y = -12$, determine whether the given ordered pair is a solution.

1. $(4, 0)$ **2.** $(-2, 3)$ **3.** $\left(1, \dfrac{15}{2}\right)$

2. Graphing Linear Equations in Two Variables by Plotting Points

In this section, we will graph linear equations in two variables.

Answers

1. No   **2.** Yes   **3.** Yes

The Graph of an Equation in Two Variables

The graph of an equation in two variables is the graph of all ordered pair solutions to the equation.

The word *linear* means "relating to or resembling a line." It is not surprising then that the solution set for any linear equation in two variables forms a line in a rectangular coordinate system. Because two points determine a line, to graph a linear equation it is sufficient to find two solution points and draw the line between them. We will find three solution points and use the third point as a check point. This process is demonstrated in Example 2.

Example 2 Graphing a Linear Equation

Graph the equation $x - 2y = 8$.

Solution:

We will find three ordered pairs that are solutions to $x - 2y = 8$. To find the ordered pairs, choose an arbitrary value of x or y. Three choices are recorded in the table. To complete the table, individually substitute each choice into the equation and solve for the missing variable. The substituted value and the solution to the equation form an ordered pair.

x	y
2	
	−1
0	

→ (2,)
→ (, −1)
→ (0,)

TIP: Usually we try to choose arbitrary values that will be convenient to graph.

From the first row, substitute $x = 2$:

$x - 2y = 8$
$(2) - 2y = 8$
$-2y = 6$
$y = -3$

From the second row, substitute $y = -1$:

$x - 2y = 8$
$x - 2(-1) = 8$
$x + 2 = 8$
$x = 6$

From the third row, substitute $x = 0$:

$x - 2y = 8$
$(0) - 2y = 8$
$-2y = 8$
$y = -4$

Avoiding Mistakes

Only two points are needed to graph a line. However, in Example 2, we found a third ordered pair, (0, −4). Notice that this point "lines up" with the other two points. If the three points do not line up, then we know that a mistake was made in solving for at least one of the ordered pairs.

The completed table is shown with the corresponding ordered pairs.

x	y
2	−3
6	−1
0	−4

→ (2, −3)
→ (6, −1)
→ (0, −4)

To graph the equation, plot the three solutions and draw the line through the points (Figure 11-6).

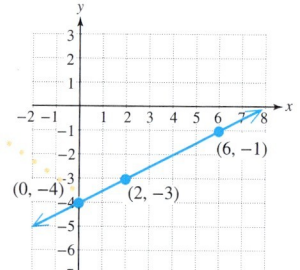

Figure 11-6

Skill Practice

4. Graph the equation $2x + y = 6$.

Answer

4.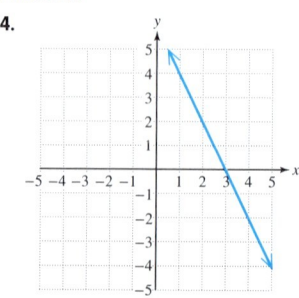

In Example 2, the original values for x and y given in the table were chosen arbitrarily by the authors. It is important to note, however, that once you choose an arbitrary value for x, the corresponding y value is determined by the equation. Similarly, once you choose an arbitrary value for y, the x value is determined by the equation.

Example 3 — Graphing a Linear Equation

Graph the equation $4x + 3y = 15$.

Solution:

We will find three ordered pairs that are solutions to the equation $4x + 3y = 15$. In the table, we have selected arbitrary values for x and y and must complete the ordered pairs. Notice that in this case, we are choosing zero for x and zero for y to illustrate that the resulting equation is often easy to solve.

x	y
0		→ (0,)
	0	→ (, 0)
3		→ (3,)

From the first row, substitute $x = 0$:

$4x + 3y = 15$
$4(0) + 3y = 15$
$3y = 15$
$y = 5$

From the second row, substitute $y = 0$:

$4x + 3y = 15$
$4x + 3(0) = 15$
$4x = 15$
$x = \dfrac{15}{4}$ or $3\dfrac{3}{4}$

From the third row, substitute $x = 3$:

$4x + 3y = 15$
$4(3) + 3y = 15$
$12 + 3y = 15$
$3y = 3$
$y = 1$

The completed table is shown with the corresponding ordered pairs.

x	y
0	5	→ (0, 5)
$3\tfrac{3}{4}$	0	→ ($3\tfrac{3}{4}$, 0)
3	1	→ (3, 1)

To graph the equation, plot the three solutions and draw the line through the points (Figure 11-7).

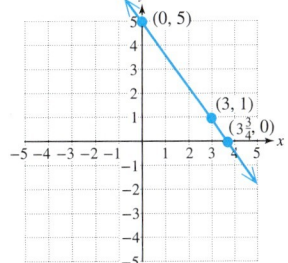

Figure 11-7

Skill Practice

5. Graph the equation $2x + 3y = 12$.

Example 4 — Graphing a Linear Equation

Graph the equation $y = -\dfrac{1}{3}x + 1$.

Solution:

Because the y variable is isolated in the equation, it is easy to substitute a value for x and simplify the right-hand side to find y. Since any number for x can be

Answer

5.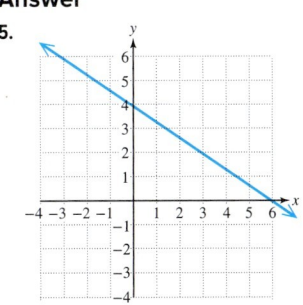

chosen, select numbers that are multiples of 3. These will simplify easily when multiplied by $-\frac{1}{3}$.

x	y
3	
0	
-3	

$y = -\frac{1}{3}x + 1$

Let $x = 3$:

$y = -\frac{1}{3}(3) + 1$

$y = -1 + 1$

$y = 0$

Let $x = 0$:

$y = -\frac{1}{3}(0) + 1$

$y = 0 + 1$

$y = 1$

Let $x = -3$:

$y = -\frac{1}{3}(-3) + 1$

$y = 1 + 1$

$y = 2$

x	y	
3	0	→ (3, 0)
0	1	→ (0, 1)
-3	2	→ (-3, 2)

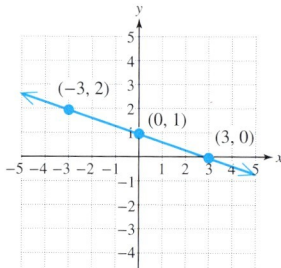

The line through the three ordered pairs (3, 0), (0, 1), and (-3, 2) is shown in Figure 11-8. The line represents the set of all solutions to the equation $y = -\frac{1}{3}x + 1$.

Figure 11-8

Skill Practice

6. Graph the equation $y = \frac{1}{2}x + 3$.

3. x- and y-Intercepts

The x- and y-intercepts are the points where the graph intersects the x- and y-axes, respectively. From Example 4, we see that the x-intercept is at the point (3, 0) and the y-intercept is at the point (0, 1). See Figure 11-8. Notice that a y-intercept is a point on the y-axis and must have an x-coordinate of 0. Likewise, an x-intercept is a point on the x-axis and must have a y-coordinate of 0.

> **Definitions of x- and y-Intercepts**
>
> An **x-intercept** of a graph is a point $(a, 0)$ where the graph intersects the x-axis.
>
> A **y-intercept** of a graph is a point $(0, b)$ where the graph intersects the y-axis.

In some applications, an x-intercept is defined as the x-coordinate of a point of intersection that a graph makes with the x-axis. For example, if an x-intercept is at the point (3, 0), it is sometimes stated simply as 3 (the y-coordinate is assumed to be 0). Similarly, a y-intercept is sometimes defined as the y-coordinate of a point of intersection that a graph makes with the y-axis. For example, if a y-intercept is at the point (0, 7), it may be stated simply as 7 (the x-coordinate is assumed to be 0).

Although any two points may be used to graph a line, in some cases it is convenient to use the x- and y-intercepts of the line. To find the x- and y-intercepts of any two-variable equation in x and y, follow these steps:

Answer

6.
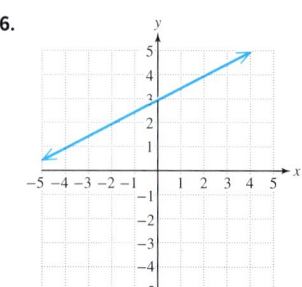

> **Finding x- and y-Intercepts**
>
> - Find the x-intercept(s) by substituting $y = 0$ into the equation and solving for x.
> - Find the y-intercept(s) by substituting $x = 0$ into the equation and solving for y.

Example 5 Finding the x- and y-Intercepts of a Line

Given the equation $-3x + 2y = 8$,

a. Find the x-intercept.

b. Find the y-intercept.

c. Graph the equation.

Solution:

a. To find the x-intercept, substitute $y = 0$.

$$-3x + 2y = 8$$
$$-3x + 2(0) = 8$$
$$-3x = 8$$
$$\frac{-3x}{-3} = \frac{8}{-3}$$
$$x = -\frac{8}{3}$$

The x-intercept is $\left(-\frac{8}{3}, 0\right)$.

b. To find the y-intercept, substitute $x = 0$.

$$-3x + 2y = 8$$
$$-3(0) + 2y = 8$$
$$2y = 8$$
$$y = 4$$

The y-intercept is $(0, 4)$.

Avoiding Mistakes

Be sure to write the x- and y-intercepts as two separate ordered pairs: $\left(-\frac{8}{3}, 0\right)$ and $(0, 4)$.

c. The line through the ordered pairs $\left(-\frac{8}{3}, 0\right)$ and $(0, 4)$ is shown in Figure 11-9. Note that the point $\left(-\frac{8}{3}, 0\right)$ can be written as $\left(-2\frac{2}{3}, 0\right)$.

The line represents the set of all solutions to the equation $-3x + 2y = 8$.

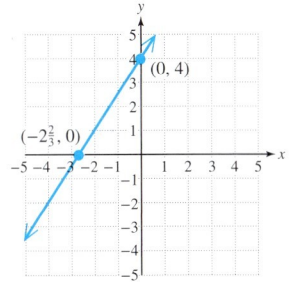

Figure 11-9

Skill Practice Given the equation $x - 3y = -4$,

7. Find the x-intercept.   8. Find the y-intercept.   9. Graph the equation.

Example 6 Finding the x- and y-Intercepts of a Line

Given the equation $4x + 5y = 0$,

a. Find the x-intercept.

b. Find the y-intercept.

c. Graph the equation.

Solution:

a. To find the x-intercept, substitute $y = 0$.

$$4x + 5y = 0$$
$$4x + 5(0) = 0$$
$$4x = 0$$
$$x = 0$$

The x-intercept is $(0, 0)$.

b. To find the y-intercept, substitute $x = 0$.

$$4x + 5y = 0$$
$$4(0) + 5y = 0$$
$$5y = 0$$
$$y = 0$$

The y-intercept is $(0, 0)$.

Answers

7. $(-4, 0)$ 8. $\left(0, \frac{4}{3}\right)$

9.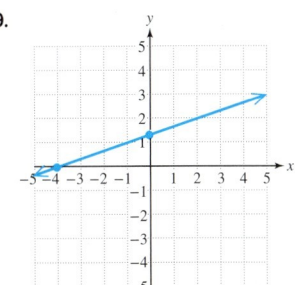

Avoiding Mistakes

Do not try to graph a line given only one point. There are infinitely many lines that pass through a single point.

c. Because the x-intercept and the y-intercept are the same point (the origin), one or more additional points are needed to graph the line. In the table, we have arbitrarily selected additional values for x and y to find two more points on the line.

x	y
-5	
	2

Let $x = -5$: $\quad 4x + 5y = 0$ $\qquad$ Let $y = 2$: $\quad 4x + 5y = 0$

$\qquad\qquad\quad 4(-5) + 5y = 0 \qquad\qquad\qquad\qquad 4x + 5(2) = 0$

$\qquad\qquad\quad -20 + 5y = 0 \qquad\qquad\qquad\qquad\; 4x + 10 = 0$

$\qquad\qquad\qquad\qquad 5y = 20 \qquad\qquad\qquad\qquad\qquad\; 4x = -10$

$\qquad\qquad\qquad\qquad\; y = 4 \qquad\qquad\qquad\qquad\qquad\quad x = -\dfrac{10}{4}$

$(-5, 4)$ is a solution.

$\qquad\qquad\qquad\qquad\qquad\qquad\qquad\qquad\qquad\qquad\qquad x = -\dfrac{5}{2}$

$\qquad\qquad\qquad\qquad\qquad\qquad\qquad\qquad\qquad\; \left(-\dfrac{5}{2}, 2\right)$ is a solution.

The line through the ordered pairs $(0, 0)$, $(-5, 4)$, and $\left(-\dfrac{5}{2}, 2\right)$ is shown in Figure 11-10. Note that the point $\left(-\dfrac{5}{2}, 2\right)$ can be written as $\left(-2\dfrac{1}{2}, 2\right)$.

The line represents the set of all solutions to the equation $4x + 5y = 0$.

x	y
-5	4
$-\dfrac{5}{2}$	2
0	0

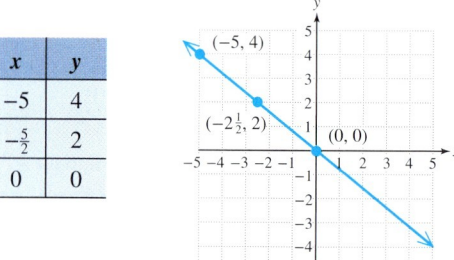

Figure 11-10

Skill Practice Given the equation $2x - 3y = 0$,

10. Find the x-intercept. $\qquad$ 11. Find the y-intercept.

12. Graph the equation. (*Hint:* You may need to find an additional point.)

4. Horizontal and Vertical Lines

Recall that a linear equation can be written in the form $Ax + By = C$, where A and B are not both zero. However, if A or B is 0, then the line is either horizontal or vertical. A horizontal line either lies on the x-axis or is parallel to the x-axis. A vertical line either lies on the y-axis or is parallel to the y-axis.

Answers

10. $(0, 0)$ **11.** $(0, 0)$
12.

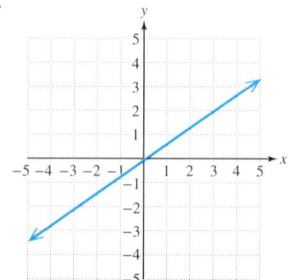

Equations of Vertical and Horizontal Lines

1. A **vertical line** can be represented by an equation of the form $x = k$, where k is a constant.

2. A **horizontal line** can be represented by an equation of the form $y = k$, where k is a constant.

Section 11.2 Linear Equations in Two Variables

Example 7 — Graphing a Horizontal Line

Graph the equation $y = 3$.

Solution:

Because this equation is in the form $y = k$, the line is horizontal and must cross the y-axis at $y = 3$ (Figure 11-11).

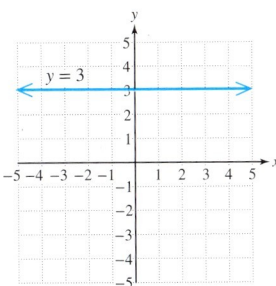

Figure 11-11

TIP: Notice that a horizontal line has a y-intercept but does not have an x-intercept (unless the horizontal line is the x-axis itself).

Alternative Solution:

Create a table of values for the equation $y = 3$. The choice for the y-coordinate must be 3, but x can be any real number.

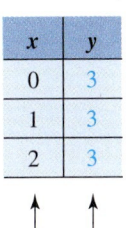

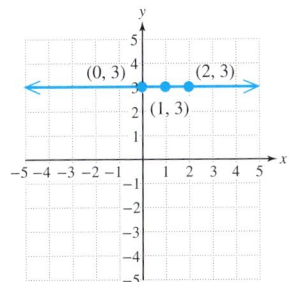

x can be any number. y must be 3.

Skill Practice

13. Graph the equation. $y = -2$

Example 8 — Graphing a Vertical Line

Graph the equation $7x = -14$.

Solution:

Because the equation does not have a y variable, we can solve the equation for x.

$7x = -14$ is equivalent to $x = -2$

This equation is in the form $x = k$, indicating that the line is vertical and must cross the x-axis at $x = -2$ (Figure 11-12).

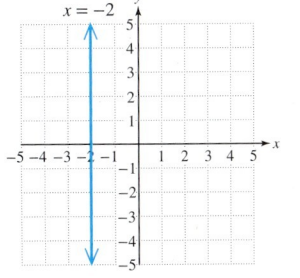

Figure 11-12

Answer

13.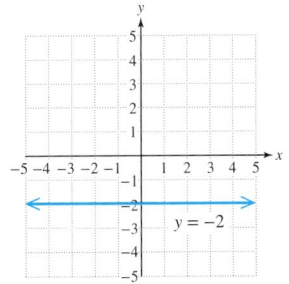

TIP: Notice that a vertical line has an *x*-intercept but does not have a *y*-intercept (unless the vertical line is the *y*-axis itself).

Answer

14.

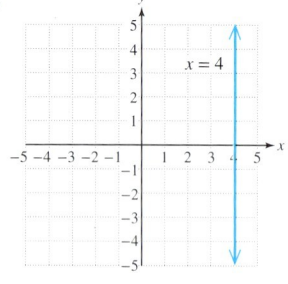

Alternative Solution:

Create a table of values for the equation $x = -2$. The choice for the *x*-coordinate must be -2, but *y* can be any real number.

x	y
-2	0
-2	3
-2	-4

x must be -2.   *y* can be any number.

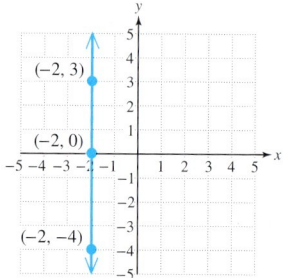

Skill Practice

14. Graph the equation. $3x = 12$

Section 11.2 Activity

A.1. a. Complete the table for the equation $2x + y = -4$.

x	y
-2	
	4
0	
	2
1	

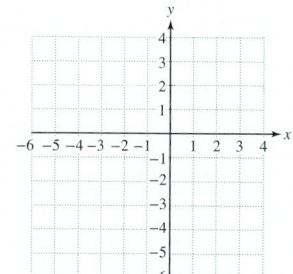

b. Graph the ordered pairs from part (a). Do the points all line up? If not, revisit part (a) to review your calculations.
c. Draw a line through the points to represent *all* solutions to the equation $2x + y = -4$.
d. The equation $2x + y = -4$ is a linear equation in two variables.
Based on the graph in part (c), why is the equation categorized as linear?

A.2. A linear equation in two variables is an equation of the form $Ax + By = C$, where *A* and *B* are not both zero. From this definition, determine whether the equation is linear. If not, explain why it is not linear.
  a. $3x + 4y = 12$
  b. $-3x - 4y = 12$
  c. $3x^2 + 4y = 12$
  d. $\dfrac{3}{x} + \dfrac{4}{y} = 12$
  e. $\dfrac{x}{3} + \dfrac{y}{4} = 12$

A.3. What are the *x*- and *y*-intercepts of the graph of an equation?

A.4. Given an equation such as $3x + 2y = 6$, how do you find the x- and y-intercepts?

For Exercises A.5–A.8,
  a. Find the x-intercept (if it exists).
  b. Find the y-intercept (if it exists).
  c. Find another point on the line.
  d. Plot the points and sketch the line.

A.5. $-3x + y = 3$

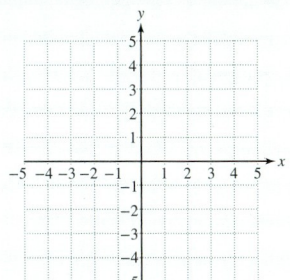

A.6. $x - y = 0$

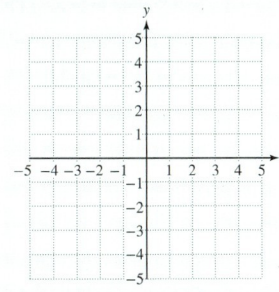

A.7. $y + 2 = 0$

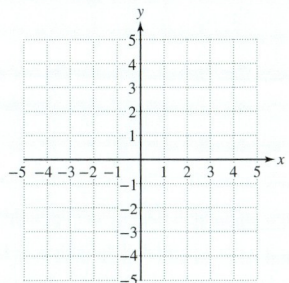

A.8. $x - 4 = -5$

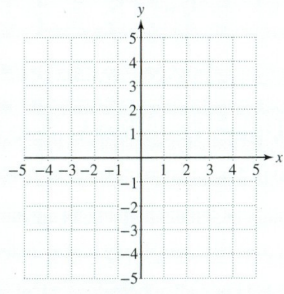

For Exercises A.9–A.11, refer to Exercises A.5–A.8.

A.9. By looking at an equation of a line, how can you tell if the line is horizontal?

A.10. By looking at an equation of a line, how can you tell if the line is vertical?

A.11. By looking at an equation of a line, how can you tell if the line is a slanted line passing through the origin?

Practice Exercises Section 11.2

Prerequisite Review

For Exercises R.1–R.4, solve for x or y.

R.1. $3x + 2 = 17$

R.2. $-4x - 3 = 21$

R.3. $8 - 5y = -2$

R.4. $12 + 2y = 8$

For Exercises R.5–R.8, evaluate the expression for the given values of x and y.

R.5. $-4x + 3y$ for $x = -2$ and $y = 7$

R.6. $2x - 7y$ for $x = 3$ and $y = -5$

R.7. $7x - 2y$ for $x = 0$ and $y = -7$

R.8. $5x + 8y$ for $x = -6$ and $y = 0$

For Exercises R.9–R.14, refer to the figure to give the coordinates of the labeled points, and state the quadrant or axis where the point is located.

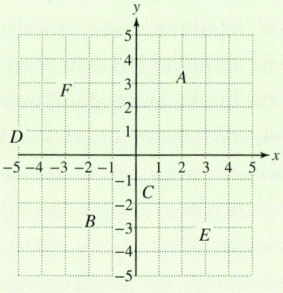

R.9. A

R.10. B

R.11. C

R.12. D

R.13. E

R.14. F

Vocabulary and Key Concepts

1. **a.** A linear equation in two variables is an equation that can be written in the form _____ where A and B are not both zero.

b. A point where a graph intersects the x-axis is called a(n) _____.

c. A point where a graph intersects the y-axis is called a(n) _____.

d. A _____ line can be represented by an equation of the form $x = k$, where k is a constant.

e. A _____ line can be represented by an equation of the form $y = k$, where k is a constant.

Concept 1: Definition of a Linear Equation in Two Variables

For Exercises 2–8, determine whether the equation is linear or nonlinear.

2. $4x + 5y = 10$

3. $-3x + 6y = 4$

4. $8x - y = 0$

5. $\dfrac{-3}{x} + \dfrac{6}{y} = 4$

6. $\dfrac{4}{x} + \dfrac{5}{y} = 10$

7. $x^2 + 2y = 6$

8. $9x + y^2 = 1$

For Exercises 9–17, determine whether the given ordered pair is a solution to the equation. **(See Example 1.)**

9. $x - y = 6$; $(8, 2)$

10. $y = 3x - 2$; $(1, 1)$

11. $y = -\dfrac{1}{3}x + 3$; $(-3, 4)$

12. $y = -\dfrac{5}{2}x + 5$; $\left(\dfrac{4}{5}, -3\right)$

13. $4x + 5y = 1$; $\left(\dfrac{1}{4}, -\dfrac{2}{5}\right)$

14. $y = 7$; $(0, 7)$

15. $y = -2$; $(-2, 6)$

16. $x = 1$; $(0, 1)$

17. $x = -5$; $(-5, 6)$

Concept 2: Graphing Linear Equations in Two Variables by Plotting Points

For Exercises 18–31, complete each table, and graph the corresponding ordered pairs. Draw the line defined by the points to represent all solutions to the equation. **(See Examples 2–4.)**

18. $x + y = 3$

x	y
2	
	3
-1	
	0

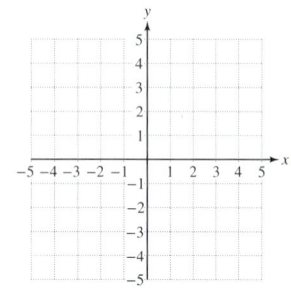

19. $x + y = -2$

x	y
1	
	0
-3	
	2

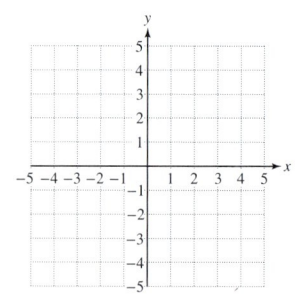

Section 11.2 Linear Equations in Two Variables

20. $y = 5x + 1$

x	y
1	
	1
-1	

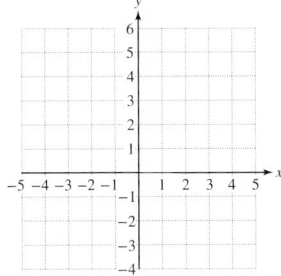

21. $y = -3x - 3$

x	y
-2	
	0
-4	

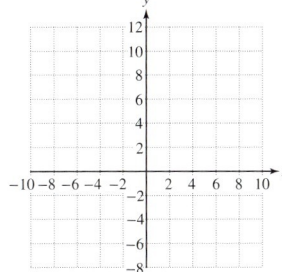

22. $2x - 3y = 6$

x	y
0	
	0
2	

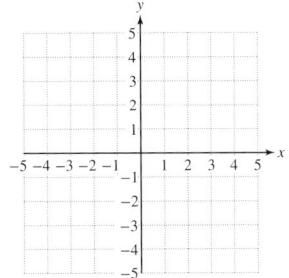

23. $4x + 2y = 8$

x	y
0	
	0
3	

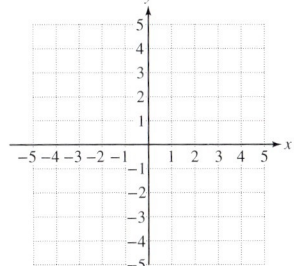

24. $y = \dfrac{2}{7}x - 5$

x	y
7	
-7	
0	

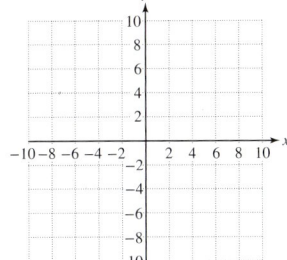

25. $y = -\dfrac{3}{5}x - 2$

x	y
0	
5	
10	

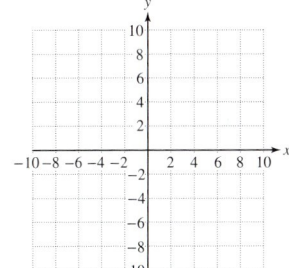

26. $y = 3$

x	y
2	
0	
-1	

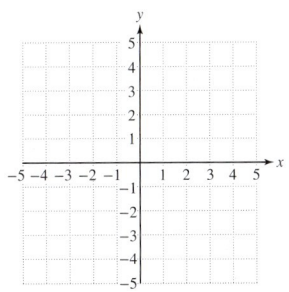

27. $y = -2$

x	y
0	
-3	
5	

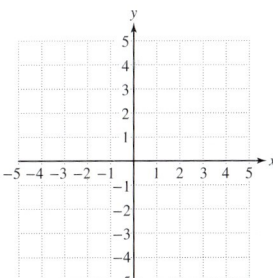

28. $x = -4$

x	y
	1
	-2
	4

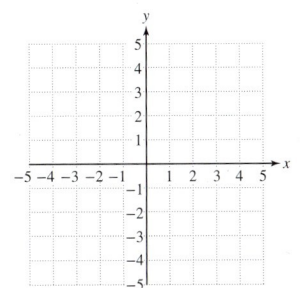

29. $x = \dfrac{3}{2}$

x	y
	-1
	2
	-3

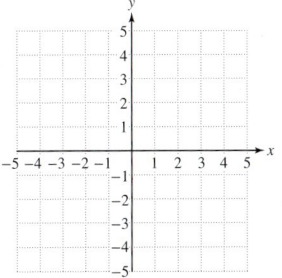

30. $y = -3.4x + 5.8$

x	y
0	
1	
2	

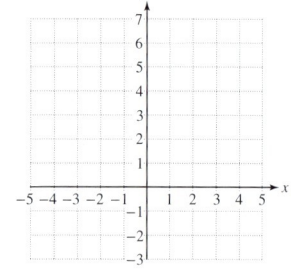

31. $y = -1.2x + 4.6$

x	y
0	
1	
2	

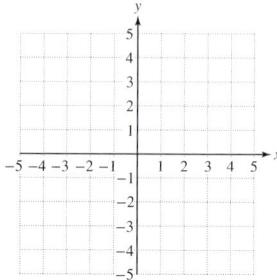

For Exercises 32–43, graph each line by making a table of at least three ordered pairs and plotting the points. **(See Example 4.)**

32. $x - y = 2$

33. $x - y = 4$

34. $-3x + y = -6$

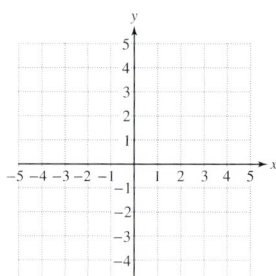
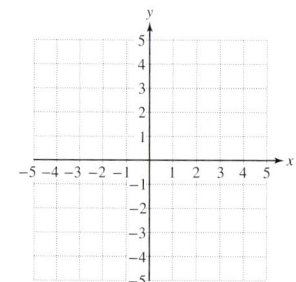
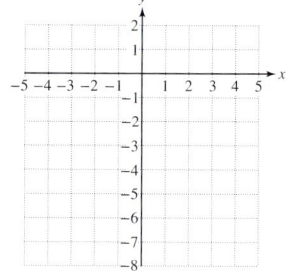

35. $2x - 5y = 10$

36. $y = 4x$

37. $y = -2x$

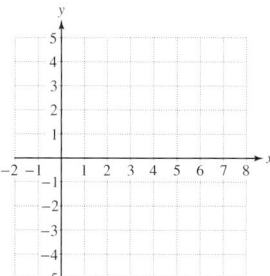
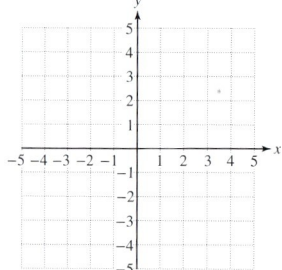
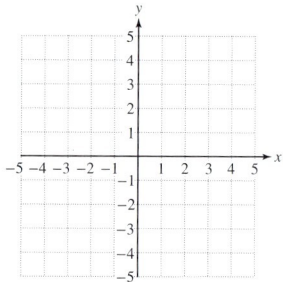

38. $y = -\dfrac{1}{2}x + 3$

39. $y = \dfrac{1}{4}x - 2$

40. $x + y = 0$

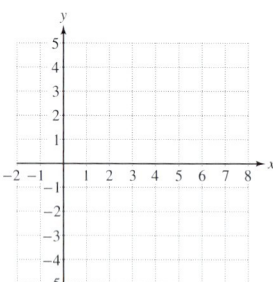
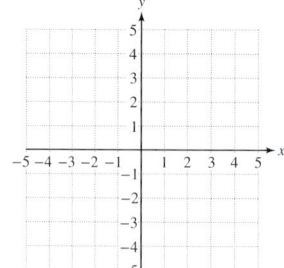
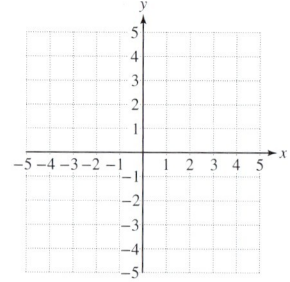

41. $-x + y = 0$

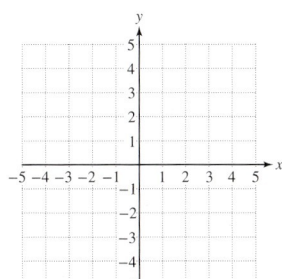

42. $50x - 40y = 200$

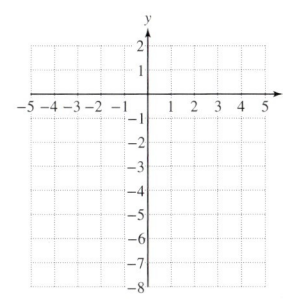

43. $-30x - 20y = 60$

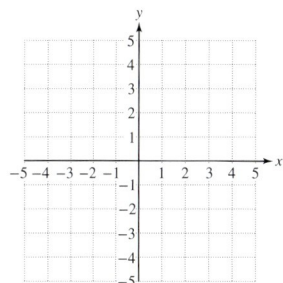

Concept 3: *x*- and *y*-Intercepts

44. The *x*-intercept is on which axis?

45. The *y*-intercept is on which axis?

For Exercises 46–49, estimate the coordinates of the *x*- and *y*-intercepts.

46.

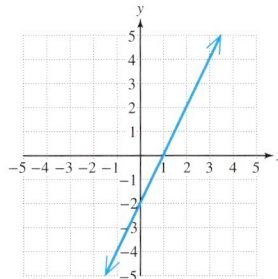

47.

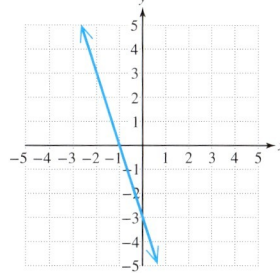

48.

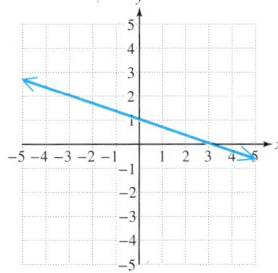

49.

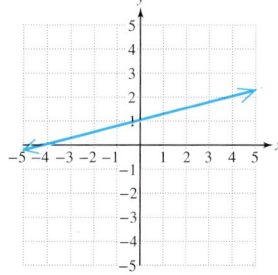

For Exercises 50–61, find the *x*- and *y*-intercepts (if they exist), and graph the line. **(See Examples 5–6.)**

50. $5x + 2y = 5$

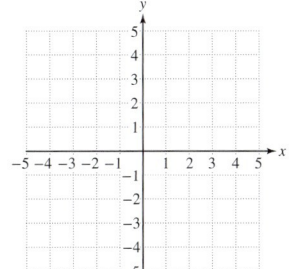

51. $4x - 3y = -9$

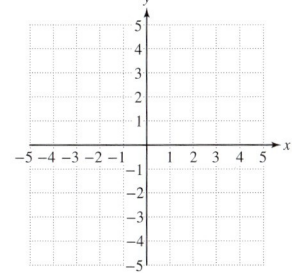

52. $y = \dfrac{2}{3}x - 1$

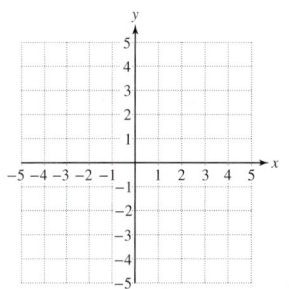

53. $y = -\dfrac{3}{4}x + 2$

54. $x - 3 = y$

55. $2x + 8 = y$

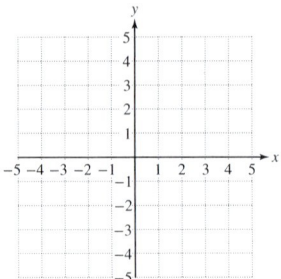
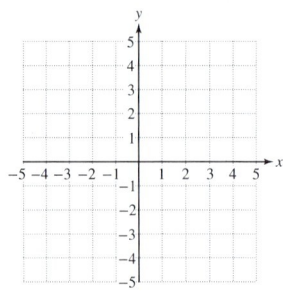
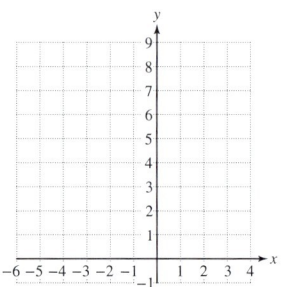

56. $-3x + y = 0$

57. $2x - 2y = 0$

58. $25y = 10x + 100$

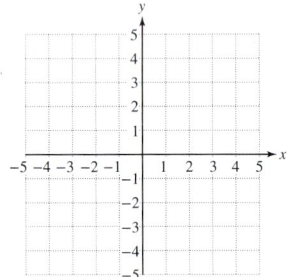
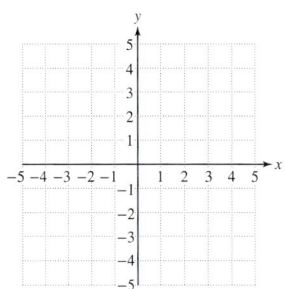
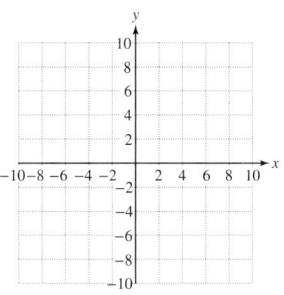

59. $20x = -40y + 200$

60. $x = 2y$

61. $x = -5y$

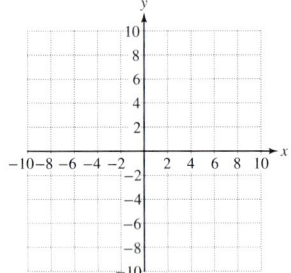
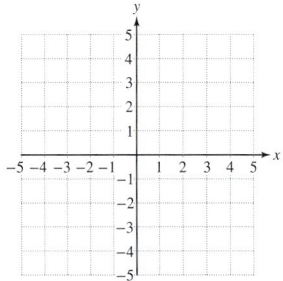
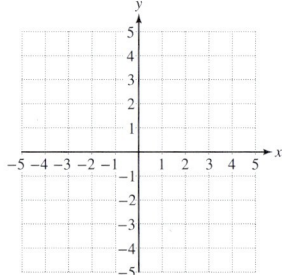

Concept 4: Horizontal and Vertical Lines

For Exercises 62–65, answer true or false. If the statement is false, rewrite it to be true.

62. The line defined by $x = 3$ is horizontal.

63. The line defined by $y = -4$ is horizontal.

64. A line parallel to the *y*-axis is vertical.

65. A line parallel to the *x*-axis is horizontal.

For Exercises 66–74,

a. Identify the equation as representing a horizontal or vertical line.

b. Graph the line.

c. Identify the *x*- and *y*-intercepts if they exist. **(See Examples 7–8.)**

66. $x = 3$

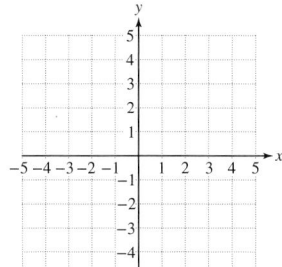

67. $y = -1$

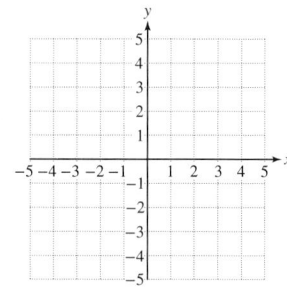

68. $-2y = 8$

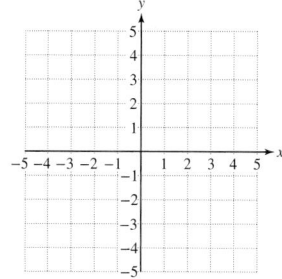

69. $5x = 20$

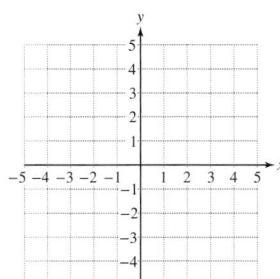

70. $x - 3 = -7$

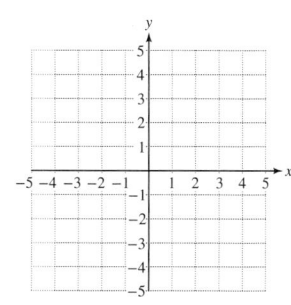

71. $y + 8 = 11$

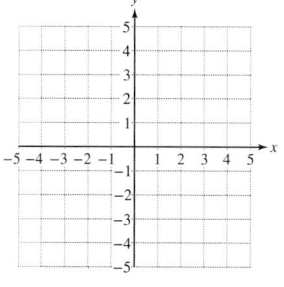

72. $3y = 0$

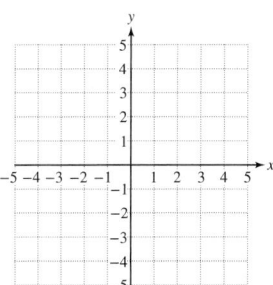

73. $5x = 0$

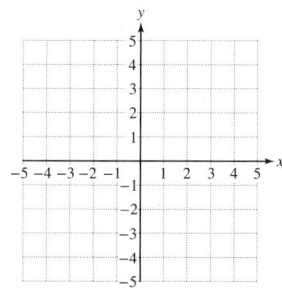

74. $2x + 7 = 10$

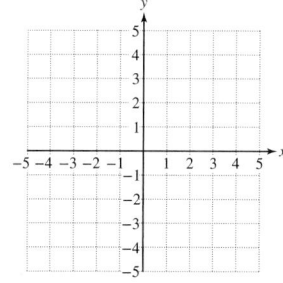

75. Explain why not every line has both an *x*- and a *y*-intercept.

76. Which of the lines has an *x*-intercept?

    a. $2x - 3y = 6$ **b.** $x = 5$ **c.** $2y = 8$ **d.** $-x + y = 0$

77. Which of the lines has a *y*-intercept?

    a. $y = 2$ **b.** $x + y = 0$ **c.** $2x - 10 = 2$ **d.** $x + 4y = 8$

Technology Connections

For Exercises 78–85, graph the equations on the standard viewing window.

78. $y = -2x + 5$

79. $y = 3x - 1$

80. $y = \frac{1}{2}x - \frac{7}{2}$

81. $y = -\frac{3}{4}x + \frac{5}{3}$

82. $4x - 7y = 21$

83. $2x + 3y = 12$

84. $-3x - 4y = 6$

85. $-5x + 4y = 10$

For Exercises 86–89, graph the equations on the given viewing window.

86. $y = 3x + 15$ Window: $-10 \le x \le 10$
$-5 \le y \le 20$

87. $y = -2x - 25$ Window: $-30 \le x \le 30$
$-30 \le y \le 30$

Xscl = 3 (sets the x-axis tick marks to increments of 3)

Yscl = 3 (sets the y-axis tick marks to increments of 3)

88. $y = -0.2x + 0.04$ Window: $-0.1 \le x \le 0.3$
$-0.1 \le y \le 0.1$

Xscl = 0.01 (sets the x-axis tick marks to increments of 0.01)

Yscl = 0.01 (sets the y-axis tick marks to increments of 0.01)

89. $y = 0.3x - 0.5$ Window: $-1 \le x \le 3$
$-1 \le y \le 1$

Xscl = 0.1 (sets the x-axis tick marks to increments of 0.1)

Yscl = 0.1 (sets the y-axis tick marks to increments of 0.1)

Expanding Your Skills

90. The store "DVDs R US" sells all DVDs for $13.99. The equation $y = 13.99x$ represents the revenue, y, (in dollars) generated by selling x DVDs.

a. Find y when $x = 13$.

b. Find x when $y = 279.80$.

c. Write the ordered pairs from parts (a) and (b), and interpret their meaning in the context of the problem.

d. Graph the ordered pairs and the line defined by the points.

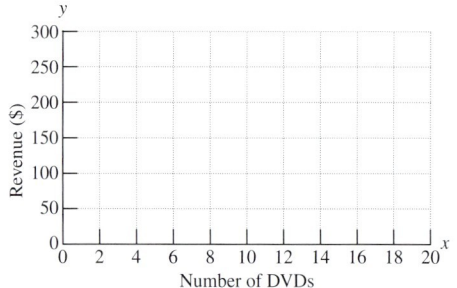

91. The value of a car depreciates once it is driven off of the dealer's lot. For a certain subcompact car, the value of the car is given by the equation $y = -1025x + 12{,}215$ ($x \ge 0$) where y is the value of the car in dollars x years after its purchase.

a. Find y when $x = 1$.

b. Find x when $y = 9140$.

c. Write the ordered pairs from parts (a) and (b), and interpret their meaning in the context of the problem.

Slope of a Line and Rate of Change

Section 11.3

Concepts

1. Introduction to Slope
2. Slope Formula
3. Parallel and Perpendicular Lines
4. Applications of Slope: Rate of Change

1. Introduction to Slope

The x- and y-intercepts represent the points where a line crosses the x- and y-axes. Another important feature of a line is its slope. Geometrically, the slope of a line measures the "steepness" of the line. For example, two hiking trails are depicted by the lines in Figure 11-13.

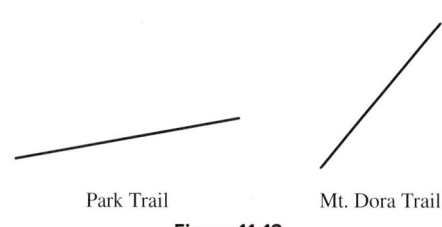

Figure 11-13

By visual inspection, Mt. Dora Trail is "steeper" than Park Trail. To measure the slope of a line quantitatively, consider two points on the line. The **slope** of the line is the ratio of the vertical change (change in y) between the two points and the horizontal change (change in x). As a memory device, we might think of the slope of a line as "rise over run." See Figure 11-14.

$$\text{Slope} = \frac{\text{change in } y}{\text{change in } x} = \frac{\text{rise}}{\text{run}}$$

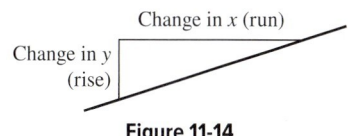

Figure 11-14

To move from point A to point B on Park Trail, rise 2 ft and move to the right 6 ft (Figure 11-15).

To move from point A to point B on Mt. Dora Trail, rise 5 ft and move to the right 4 ft (Figure 11-16).

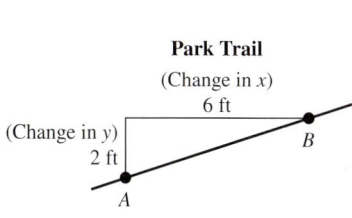

Figure 11-15

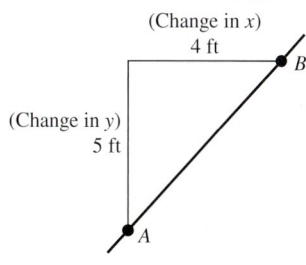

Figure 11-16

TIP: To find the slope, you can use any two points on the line. The ratio of rise to run will be the same.

$$\text{Slope} = \frac{\text{change in } y}{\text{change in } x} = \frac{2 \text{ ft}}{6 \text{ ft}} = \frac{1}{3}$$

$$\text{Slope} = \frac{\text{change in } y}{\text{change in } x} = \frac{5 \text{ ft}}{4 \text{ ft}} = \frac{5}{4}$$

Example 1 — Finding Slope in an Application

Determine the slope of the ramp up the stairs.

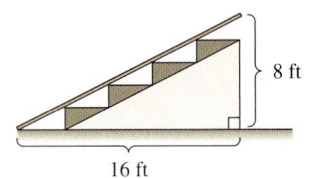

Solution:

$$\text{Slope} = \frac{\text{change in } y}{\text{change in } x} = \frac{8 \text{ ft}}{16 \text{ ft}}$$

$$\frac{8}{16} = \frac{1}{2} \quad \text{Write the ratio for the slope and simplify.}$$

The slope is $\frac{1}{2}$.

FOR REVIEW

Recall that to simplify a fraction, divide out the greatest common factor from the numerator and denominator.

$$\frac{8}{16} = \frac{1 \cdot \cancel{8}}{2 \cdot \cancel{8}} = \frac{1}{2}$$

Skill Practice

1. Determine the slope of the aircraft's takeoff path. (Figure is not drawn to scale.)

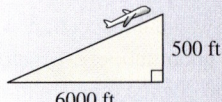

2. Slope Formula

The slope of a line may be found using any two points on the line—call these points (x_1, y_1) and (x_2, y_2). The numbers to the right and below the variables are called *subscripts*. In this instance, the subscript 1 indicates the coordinates of the first point, and the subscript 2 indicates the coordinates of the second point. The change in y between the points can be found by taking the difference of the y values: $y_2 - y_1$. The change in x can be found by taking the difference of the x values in the same order: $x_2 - x_1$ (Figure 11-17).

The slope of a line is often symbolized by the letter m and is given by the following formula.

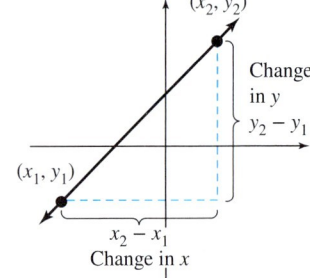

Figure 11-17

> **Slope Formula**
>
> The slope of a line passing through the distinct points (x_1, y_1) and (x_2, y_2) is
>
> $$m = \frac{y_2 - y_1}{x_2 - x_1} \quad \text{provided } x_2 - x_1 \neq 0.$$
>
> *Note:* If $x_2 - x_1 = 0$, the slope is undefined.

Answer

1. $\dfrac{500}{6000} = \dfrac{1}{12}$

Example 2 — Finding the Slope of a Line Given Two Points

Find the slope of the line through the points $(-1, 3)$ and $(-4, -2)$.

Solution:

To use the slope formula, first label the coordinates of each point and then substitute the coordinates into the slope formula.

$$\underset{(x_1, y_1)}{(-1, 3)} \text{ and } \underset{(x_2, y_2)}{(-4, -2)} \quad \text{Label the points.}$$

$$m = \frac{y_2 - y_1}{x_2 - x_1} = \frac{(-2) - (3)}{(-4) - (-1)} \quad \text{Apply the slope formula.}$$

$$= \frac{-5}{-3}$$

$$= \frac{5}{3} \quad \text{Simplify to lowest terms.}$$

The slope of the line can be verified from the graph (Figure 11-18).

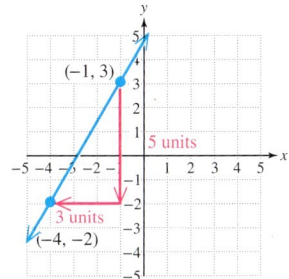

Figure 11-18

> **Avoiding Mistakes**
> When calculating slope, always write the change in y in the numerator.

Skill Practice Find the slope of the line through the given points.

2. $(-5, 2)$ and $(1, 3)$

> **TIP:** The slope formula is not dependent on which point is labeled (x_1, y_1) and which point is labeled (x_2, y_2). In Example 2, reversing the order in which the points are labeled results in the same slope.
>
> $$\underset{(x_2, y_2)}{(-1, 3)} \text{ and } \underset{(x_1, y_1)}{(-4, -2)} \quad \text{Label the points.}$$
>
> $$m = \frac{(3) - (-2)}{(-1) - (-4)} = \frac{5}{3} \quad \text{Apply the slope formula.}$$

Answer

2. $\dfrac{1}{6}$

When you apply the slope formula, you will see that the slope of a line may be positive, negative, zero, or undefined.
- Lines that increase, or rise, from left to right have a positive slope.
- Lines that decrease, or fall, from left to right have a negative slope.
- Horizontal lines have a slope of zero.
- Vertical lines have an undefined slope.

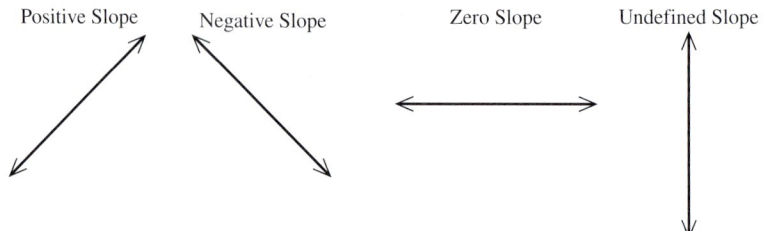

Example 3 Finding the Slope of a Line Given Two Points

Find the slope of the line passing through the points $(-5, \frac{1}{2})$ and $(2, -\frac{3}{2})$.

Solution:

$$\underset{(x_1, y_1)}{\left(-5, \frac{1}{2}\right)} \quad \text{and} \quad \underset{(x_2, y_2)}{\left(2, -\frac{3}{2}\right)} \quad \text{Label the points.}$$

$$m = \frac{y_2 - y_1}{x_2 - x_1} = \frac{\left(-\frac{3}{2}\right) - \left(\frac{1}{2}\right)}{(2) - (-5)} \quad \text{Apply the slope formula.}$$

$$= \frac{-\frac{4}{2}}{2 + 5} \quad \text{Simplify.}$$

$$= \frac{-2}{7} \quad \text{or} \quad -\frac{2}{7}$$

Avoiding Mistakes

When applying the slope formula, y_2 and x_2 are taken from the same ordered pair. Likewise y_1 and x_1 are taken from the same ordered pair.

By graphing the points $(-5, \frac{1}{2})$ and $(2, -\frac{3}{2})$, we can verify that the slope is $-\frac{2}{7}$ (Figure 11-19). Notice that the line slopes downward from left to right.

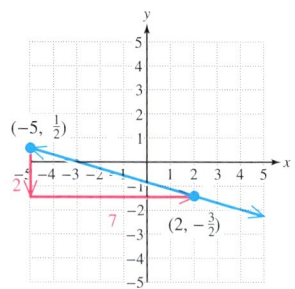

Figure 11-19

Skill Practice Find the slope of the line through the given points.

3. $\left(\frac{2}{3}, 0\right)$ and $\left(-\frac{1}{6}, 5\right)$

Answer

3. -6

Section 11.3 Slope of a Line and Rate of Change 715

Example 4 Determining the Slope of a Vertical Line

Find the slope of the line passing through the points $(2, -1)$ and $(2, 4)$.

Solution:

$(2, -1)$ and $(2, 4)$
(x_1, y_1) (x_2, y_2) Label the points.

$m = \dfrac{y_2 - y_1}{x_2 - x_1} = \dfrac{(4) - (-1)}{(2) - (2)}$ Apply the slope formula.

$m = \dfrac{5}{0}$ Undefined

Because the slope, m, is undefined, we expect the points to form a vertical line as shown in Figure 11-20.

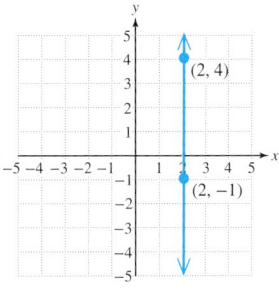

Figure 11-20

FOR REVIEW

Recall that division by zero is undefined.

$\dfrac{5}{0}$ is undefined

because no number when multiplied by 0 will equal 5.

Skill Practice Find the slope of the line through the given points.

4. $(5, 6)$ and $(5, -2)$

Example 5 Determining the Slope of a Horizontal Line

Find the slope of the line passing through the points $(3.4, -2)$ and $(-3.5, -2)$.

Solution:

$(3.4, -2)$ and $(-3.5, -2)$
(x_1, y_1) (x_2, y_2) Label the points.

$m = \dfrac{y_2 - y_1}{x_2 - x_1} = \dfrac{(-2) - (-2)}{(-3.5) - (3.4)}$ Apply the slope formula.

$= \dfrac{-2 + 2}{-3.5 - 3.4} = \dfrac{0}{-6.9} = 0$ Simplify.

Because the slope is 0, we expect the points to form a horizontal line, as shown in Figure 11-21.

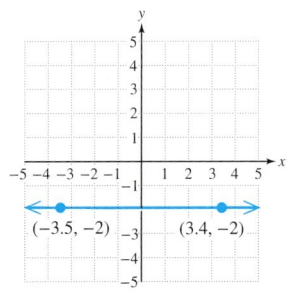

Figure 11-21

Skill Practice Find the slope of the line through the given points.

5. $(3, 8)$ and $(-5, 8)$

Answers

4. Undefined   **5.** 0

3. Parallel and Perpendicular Lines

Lines in the same plane that do not intersect are called **parallel lines**. Parallel lines have the same slope and different y-intercepts (Figure 11-22).

Lines that intersect at a right angle are **perpendicular lines**. If two lines are perpendicular then the slope of one line is the opposite of the reciprocal of the slope of the other line (provided neither line is vertical) (Figure 11-23).

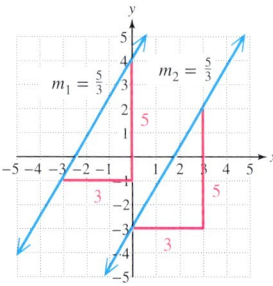

Figure 11-22

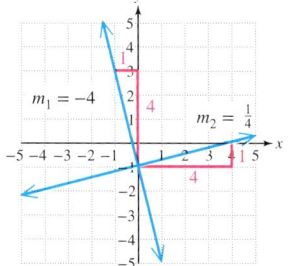

Figure 11-23

> **Slopes of Parallel Lines**
>
> If m_1 and m_2 represent the slopes of two parallel (nonvertical) lines, then
>
> $$m_1 = m_2.$$
>
> See Figure 11-22.

> **Slopes of Perpendicular Lines**
>
> If $m_1 \neq 0$ and $m_2 \neq 0$ represent the slopes of two perpendicular lines, then
>
> $$m_1 = -\frac{1}{m_2} \text{ or equivalently, } m_1 m_2 = -1. \text{ See Figure 11-23.}$$

Example 6   **Determining the Slope of Parallel and Perpendicular Lines**

Suppose a given line has a slope of -6.

  a. Find the slope of a line parallel to the line with the given slope.

  b. Find the slope of a line perpendicular to the line with the given slope.

Solution:

  a. Parallel lines must have the same slope. The slope of a line parallel to the given line is $m = -6$.

  b. For perpendicular lines, the slope of one line must be the opposite of the reciprocal of the other. The slope of a line perpendicular to the given line is $m = -\left(\frac{1}{-6}\right) = \frac{1}{6}$.

Skill Practice   A given line has a slope of $\frac{5}{3}$.

6. Find the slope of a line parallel to the given line.

7. Find the slope of a line perpendicular to the given line.

Answers

6. $\frac{5}{3}$ 7. $-\frac{3}{5}$

If the slopes of two lines are known, then we can compare the slopes to determine if the lines are parallel, perpendicular, or neither.

Example 7 — Determining If Lines Are Parallel, Perpendicular, or Neither

Lines l_1 and l_2 pass through the given points. Determine if l_1 and l_2 are parallel, perpendicular, or neither.

l_1: (2, −7) and (4, 1) l_2: (−3, 1) and (1, 0)

Solution:

Find the slope of each line.

l_1: (2, −7)   and   (4, 1)
       (x_1, y_1) (x_2, y_2)

$$m_1 = \frac{1 - (-7)}{4 - 2}$$

$$m_1 = \frac{8}{2}$$

$$m_1 = 4$$

l_2: (−3, 1)   and   (1, 0)
       (x_1, y_1) (x_2, y_2)

$$m_2 = \frac{0 - 1}{1 - (-3)}$$

$$m_2 = \frac{-1}{4}$$

$$m_2 = -\frac{1}{4}$$

TIP: You can check that two lines are perpendicular by checking that the product of their slopes is −1.

$$4\left(-\frac{1}{4}\right) = -1$$

One slope is the opposite of the reciprocal of the other slope. Therefore, the lines are perpendicular.

Skill Practice Determine if lines l_1 and l_2 are parallel, perpendicular, or neither.

8. l_1: (−2, −3) and (4, −1)
   l_2: (0, 2) and (−3, 1)

4. Applications of Slope: Rate of Change

In many applications, the interpretation of slope refers to the *rate of change* of the y variable in relation to the x variable.

Example 8 — Interpreting Slope in an Application

In late 2019, the Chinese government announced the presence of a new dangerous virus, later named COVID-19. During the first three months of 2020, the virus had infected more than 80,000 people in China and had spread to other parts of the globe. By early March the United States had detected several hundred cases, and it began testing patients experiencing symptoms of the virus. The number of confirmed cases (in 1000s) in the United States for March and April 2020 is shown in Figure 11-24.

Answer

8. Parallel

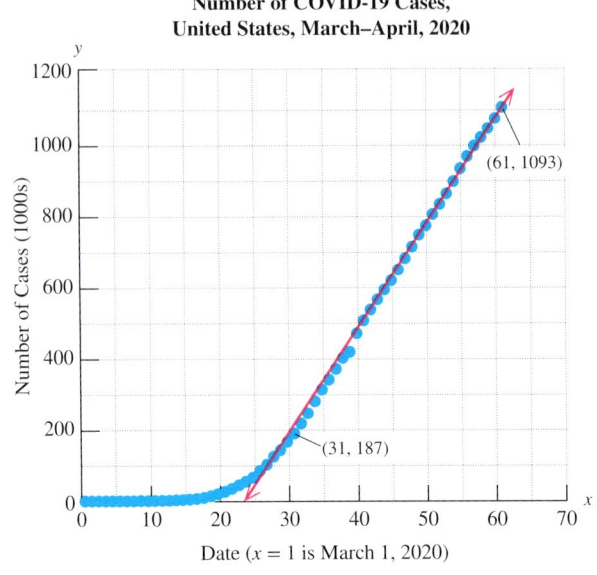

Figure 11-24

a. Interpret the meaning of the ordered pair (61, 1093).
b. The number of cases appears to increase linearly during the month of April. Find the slope of the line between the points (31, 187) and (61, 1093).
c. Interpret the meaning of the slope in the context of this problem.

Solution:

a. The ordered pair (61, 1093) means that on day 61 (April 30), the number of confirmed cases in the United States was 1093 thousand, or equivalently, 1,093,000.

b. Label (31, 187) as (x_1, y_1) and (61, 1093) as (x_2, y_2).

$$m = \frac{y_2 - y_1}{x_2 - x_1} = \frac{1093 - 187}{61 - 31} \quad \text{Apply the slope formula.}$$

$$= \frac{906}{30} \quad \text{Simplify.}$$

$$= 30.2$$

c. The slope is 30.2 and means that the average number of cases rose by 30.2 thousand (30,200) per day during April 2020.

Skill Practice

9. In the year 2000, the population of Alaska was approximately 630,000. By 2010, it had grown to 700,000. Use the ordered pairs (2000, 630,000) and (2010, 700,000) to determine the slope of the line through the points. Then interpret the meaning in the context of this problem.

Answer

9. $m = 7000$; The population of Alaska increased at a rate of 7000 people per year.

Section 11.3 Activity

A.1. A ramp in front of a movie theater has the dimensions shown. By law, a wheelchair ramp must have a slope of no more than 1 to 12 or equivalently $\frac{1}{12}$. Does the ramp meet the criteria required by law?

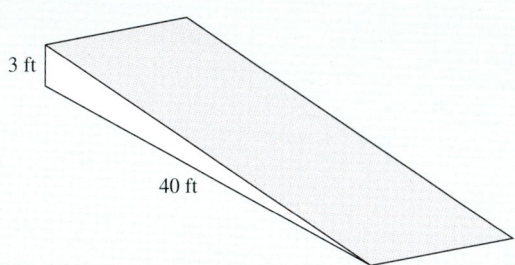

A.2. Write a formula for the slope m of a line passing through (x_1, y_1) and (x_2, y_2).

For Exercises A.3–A.6,
  a. Find the slope m of the line between the two points.
  b. Match the slope with the appropriate graph: i, ii, iii, or iv.

  i.    ii.   iii.   iv.

A.3. $(1, -2)$ and $(-8, -2)$ **A.4.** $(-1, -3)$ and $(1, 5)$

A.5. $(-5, 0)$ and $(-5, 2)$ **A.6.** $(-3, 2)$ and $(4, 1)$

For Exercises A.7–A.8, two points are given for each of two lines l_1 and l_2.
  a. Find the slope of each line.
  b. Based on the slopes, are the lines parallel, perpendicular, or neither?
  c. Verify your results from part (b) by graphing l_1 and l_2.

A.7. l_1: $(0, 3)$ and $(-1, -1)$ **A.8.** l_1: $(0, 2)$ and $(-1, -1)$
     l_2: $(4, 0)$ and $(0, 1)$ l_2: $(2, 5)$ and $(-1, -4)$

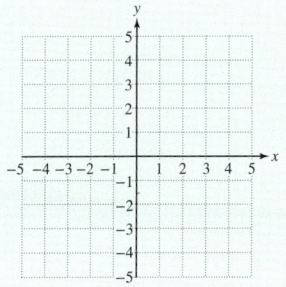 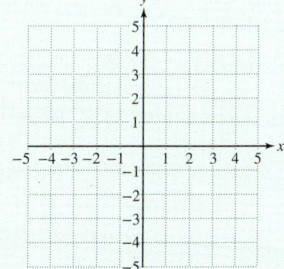

A.9. The graph depicts the amount of money per capita (per person) that Americans spent per year on prescription drugs above what was covered by insurance. Data are given for a recent study over several years.
  a. If the dollar amount is approximated by a linear trend, compute the slope of the line through the points (1, 143) and (7, 269).
  b. Interpret the meaning of the slope in the context of these data.

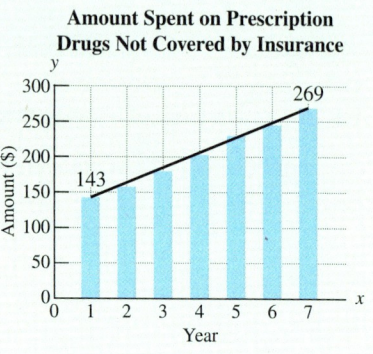

Amount Spent on Prescription Drugs Not Covered by Insurance

Section 11.3 Practice Exercises

Study Skills Exercise

Mnemonics are a great way to remember things in mathematics. A **mnemonic** is a statement or sentence that uses the first letter of each word that you are trying to remember. A well-known mnemonic in mathematics is "**P**lease **E**xcuse **M**y **D**ear **A**unt **S**ally" (PEMDAS), which helps students remember the order of operations. A mnemonic can be created for almost any content with some creativity.

Create a mnemonic to help you remember whether the slope of a line is positive, negative, zero, or undefined.

Prerequisite Review

For Exercises R.1–R.6, simplify the expression, if possible.

R.1. $\dfrac{7-(-3)}{-1-5}$

R.2. $\dfrac{-2-6}{3-(-1)}$

R.3. $\dfrac{-3-(-3)}{4-7}$

R.4. $\dfrac{7-2}{5-5}$

R.5. $\dfrac{\frac{3}{2}}{\frac{5}{6}}$

R.6. $\dfrac{-\frac{4}{5}}{\frac{3}{10}}$

R.7. a. What is the reciprocal of $\dfrac{2}{3}$?

  b. What is the opposite of the reciprocal of $\dfrac{2}{3}$?

R.8. a. What is the reciprocal of -4?

  b. What is the opposite of the reciprocal of -4?

R.9. A 12-month subscription to a channel for streaming television is $71.88. What is the monthly rate?

R.10. A discount car rental company rents a compact car for $111.93 per week. What is the daily rate?

Section 11.3 Slope of a Line and Rate of Change

Vocabulary and Key Concepts

1. **a.** The ratio of the vertical change and the horizontal change between two distinct points (x_1, y_1) and (x_2, y_2) on a line is called the _____ of the line. The slope can be computed from the formula $m = $ _____.

   b. Lines in the same plane that do not intersect are called _____ lines.

   c. Two lines are perpendicular if they intersect at a _____ angle.

   d. If m_1 and m_2 represent the slopes of two nonvertical perpendicular lines then $m_1 \cdot m_2 = $ _____.

   e. The slope of a vertical line is _____. The slope of a _____ line is 0.

2. Four lines are shown and labeled as l_1, l_2, l_3, and l_4. Rank the slopes of the four lines in order from least to greatest. (*Hint:* Remember that a line sloping downward has a negative slope.)

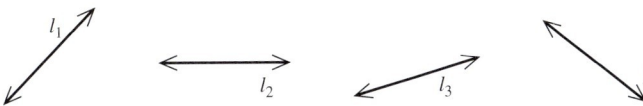

Concept 1: Introduction to Slope

3. Determine the slope of the roof.
   (See Example 1.)

   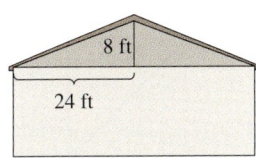

4. Determine the slope of the stairs.

   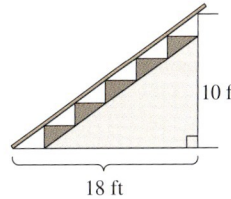

5. Calculate the slope of the handrail.

   Ryan McVay/Getty Images

6. Determine the slope of the treadmill

   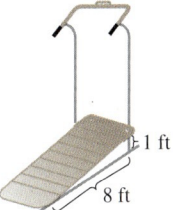

Concept 2: Slope Formula

For Exercises 7–10, fill in the blank with the appropriate term: *zero, negative, positive,* or *undefined.*

7. The slope of a line parallel to the *y*-axis is _____.

8. The slope of a horizontal line is _____.

9. The slope of a line that rises from left to right is _____.

10. The slope of a line that falls from left to right is _____.

722 Chapter 11 Graphing Linear Equations in Two Variables

For Exercises 11–19, determine if the slope is positive, negative, zero, or undefined.

11.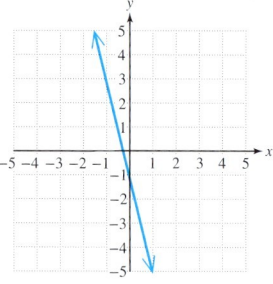

12.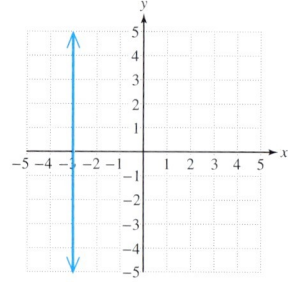

13.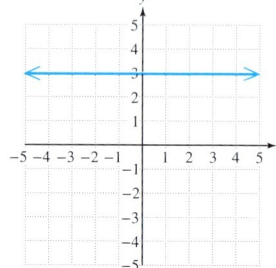

14.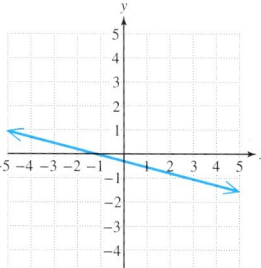

15.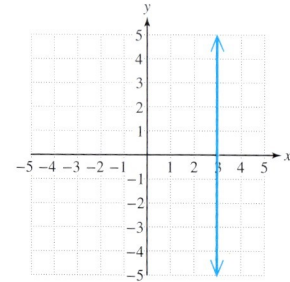

16.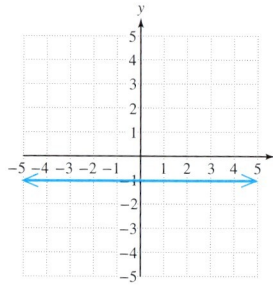

17.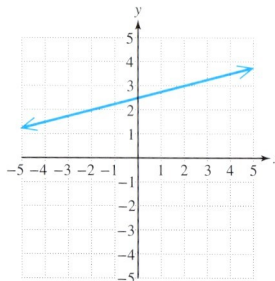

18.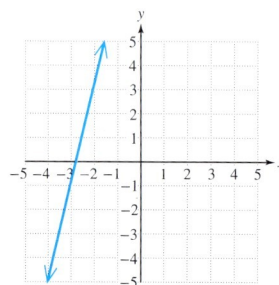

19.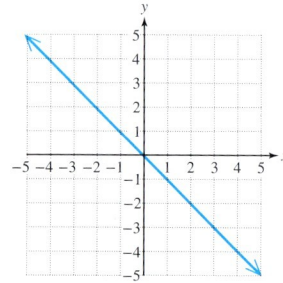

For Exercises 20–28, determine the slope by using the slope formula and any two points on the line. Check your answer by drawing a right triangle, where appropriate, and labeling the "rise" and "run."

20.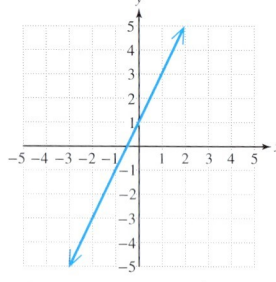

21.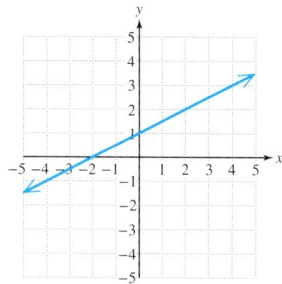

22.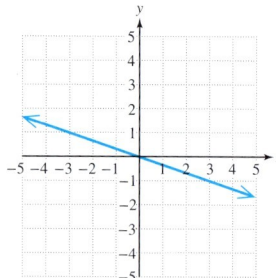

23.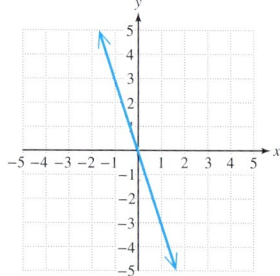

24.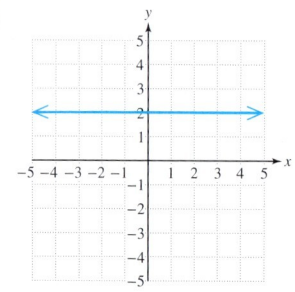

25.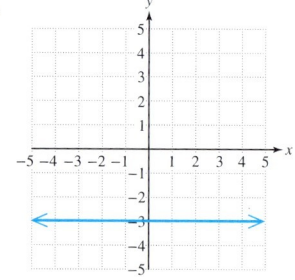

26.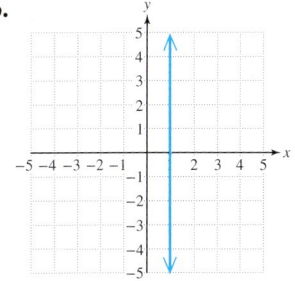
27.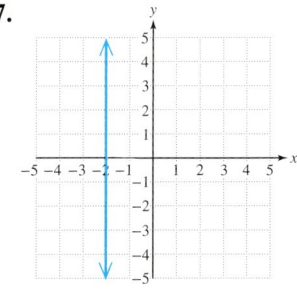
28.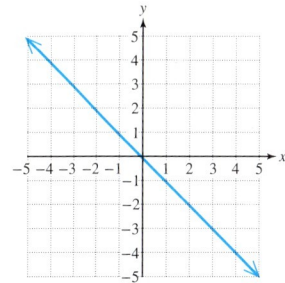

For Exercises 29–46, find the slope of the line that passes through the two points. (See Example 2–5.)

29. (2, 4) and (−4, 2)
30. (−5, 4) and (−11, 12)
31. (−2, 3) and (1, −6)
32. (−3, −4) and (1, −5)
33. (1, 5) and (−4, 2)
34. (−6, −1) and (−2, −3)
35. (5, 3) and (−2, 3)
36. (0, −1) and (−4, −1)
37. (2, −7) and (2, 5)
38. (−4, 3) and (−4, −4)
39. $\left(\frac{1}{2}, \frac{3}{5}\right)$ and $\left(\frac{1}{4}, -\frac{4}{5}\right)$
40. $\left(-\frac{2}{7}, \frac{1}{3}\right)$ and $\left(\frac{8}{7}, -\frac{5}{6}\right)$
41. (3, −1) and (−5, 6)
42. (−6, 5) and (−10, 4)
43. (6.8, −3.4) and (−3.2, 1.1)
44. (−3.15, 8.25) and (6.85, −4.25)
45. (1994, 3.5) and (2000, 2.6)
46. (1988, 4.65) and (1998, 9.25)

Concept 3: Parallel and Perpendicular Lines

For Exercises 47–52, the slope of a line is given. (See Example 6.)

a. Determine the slope of a line parallel to the line with the given slope.

b. Determine the slope of a line perpendicular to the line with the given slope.

47. $m = -2$
48. $m = \frac{2}{3}$
49. $m = 0$
50. The slope is undefined.
51. $m = \frac{4}{5}$
52. $m = -4$

For Exercises 53–58, let m_1 and m_2 represent the slopes of two lines. Determine if the lines are parallel, perpendicular, or neither. (See Example 6.)

53. $m_1 = -2$, $m_2 = \frac{1}{2}$
54. $m_1 = \frac{2}{3}$, $m_2 = \frac{3}{2}$
55. $m_1 = 1$, $m_2 = \frac{4}{4}$
56. $m_1 = \frac{3}{4}$, $m_2 = -\frac{8}{6}$
57. $m_1 = \frac{2}{7}$, $m_2 = -\frac{2}{7}$
58. $m_1 = 5$, $m_2 = 5$

For Exercises 59–64, find the slopes of the lines l_1 and l_2 defined by the two given points. Then determine whether l_1 and l_2 are parallel, perpendicular, or neither. (See Example 7.)

59. l_1: (2, 4) and (−1, −2)
    l_2: (1, 7) and (0, 5)
60. l_1: (0, 0) and (−2, 4)
    l_2: (1, −5) and (−1, −1)
61. l_1: (1, 9) and (0, 4)
    l_2: (5, 2) and (10, 1)
62. l_1: (3, −4) and (−1, −8)
    l_2: (5, −5) and (−2, 2)
63. l_1: (4, 4) and (0, 3)
    l_2: (1, 7) and (−1, −1)
64. l_1: (3, 5) and (−2, −5)
    l_2: (2, 0) and (−4, −3)

Concept 4: Applications of Slope: Rate of Change

65. For a recent year, the average earnings for male workers between the ages of 25 and 34 with a high school diploma were $32,000. Comparing this value in constant dollars to the average earnings 15 yr later showed that the average earnings have decreased to $29,600. Find the average rate of change in dollars per year for this time period. [*Hint:* Use the ordered pairs (0, 32,000) and (15, 29,600).]

66. In 1985, the U.S. Postal Service charged $0.22 for first class letters and cards up to 1 oz. By 2015, the price had increased to $0.49. Let x represent the year, and y represent the cost for 1 oz of first class postage. Find the average rate of change of the cost per year.

67. The number of deaths (in 1000s) in the United States attributed to COVID-19 for March and April 2020 is shown in the graph.
    a. Interpret the meaning of the ordered pair (31, 3.833) in the context of this problem.
    b. Find the slope of the line containing the points (31, 3.833) and (38, 12.758) and interpret its meaning in the context of this problem.
    c. From the graph, does it appear that the death rate increased or decreased the following week? Find the slope of the line containing the points (38, 12.758) and (45, 25.786) to support your answer.

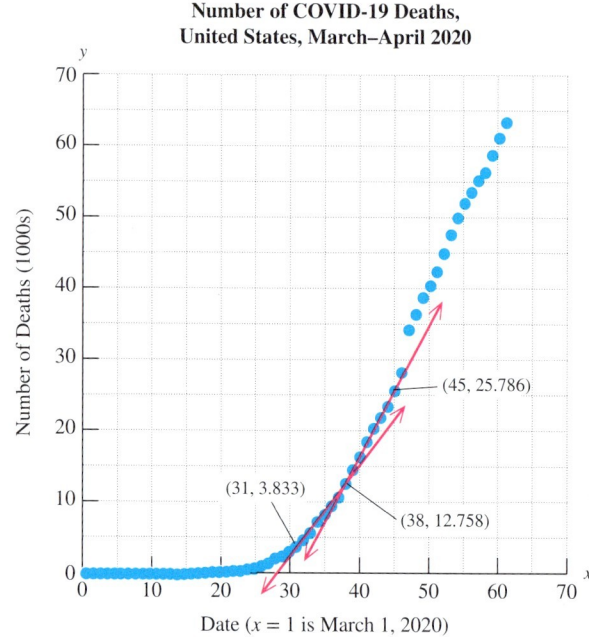

68. In Example 8, we saw that the number of confirmed COVID-19 cases rose by 30,200 per day during the month of April 2020. Find the slope of the line containing the points (21, 26) and (31, 187) and interpret its meaning in the context of this problem.

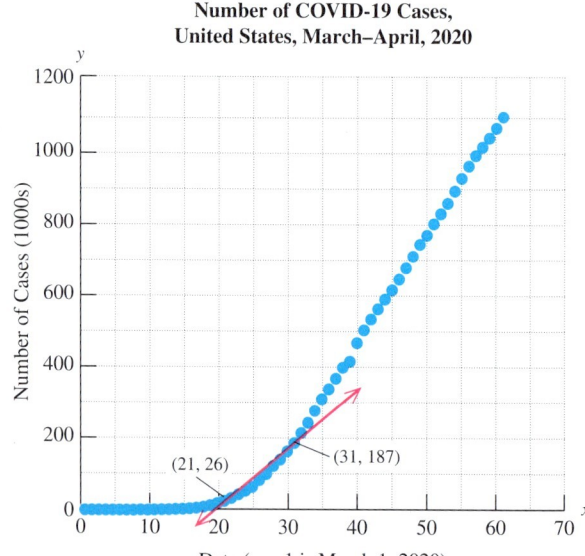

69. Michael wants to buy a used car and selects one that costs $12,600. His dad agreed to purchase the car if Michael would pay it off in equal monthly payments for the next 60 months. The equation $y = -210x + 12{,}600$ represents the amount, y (in dollars), that Michael owes his father after x months.
    a. How much does Michael owe his dad after 5 months?
    b. Determine the slope of the line and interpret its meaning in the context of this problem.

Erica Simone Leeds

70. The distance, d (in miles), between a lightning strike and an observer is given by the equation $d = 0.2t$, where t is the time (in seconds) between seeing lightning and hearing thunder.

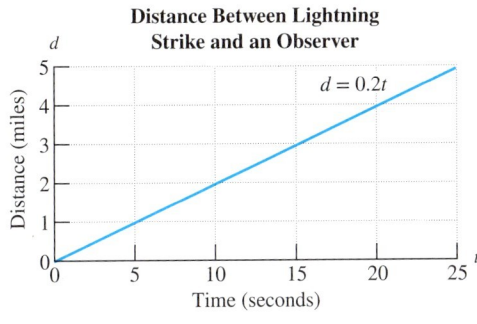

Visual Intermezzo/Shutterstock

a. If an observer counts 5 sec between seeing lightning and hearing thunder, how far away was the lightning strike?

b. If an observer counts 10 sec between seeing lightning and hearing thunder, how far away was the lightning strike?

c. If an observer counts 15 sec between seeing lightning and hearing thunder, how far away was the lightning strike?

d. What is the slope of the line? Interpret the meaning of the slope in the context of this problem.

Mixed Exercises

For Exercises 71–74, determine the slope of the line passing through points A and B.

71. Point A is located 3 units up and 4 units to the right of point B.

72. Point A is located 2 units up and 5 units to the left of point B.

73. Point A is located 5 units to the right of point B.

74. Point A is located 3 units down from point B.

75. Graph the line through the point $(1, -2)$ having slope $-\frac{2}{3}$. Then give two other points on the line.

76. Graph the line through the point $(1, 2)$ having slope $-\frac{3}{4}$. Then give two other points on the line.

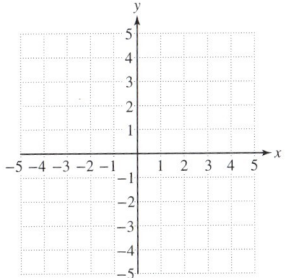

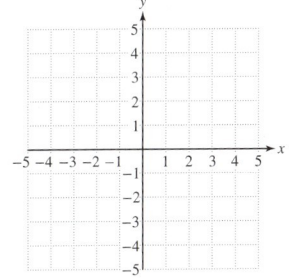

77. Graph the line through the point $(2, 2)$ having slope 3. Then give two other points on the line.

78. Graph the line through the point $(-1, 3)$ having slope 2. Then give two other points on the line.

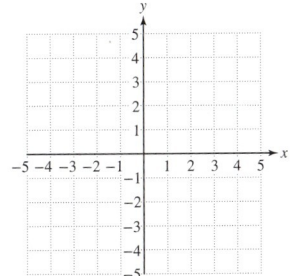

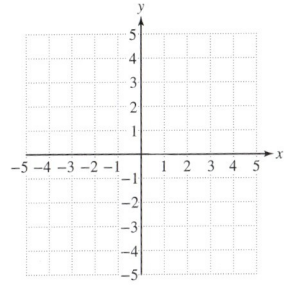

79. Graph the line through $(-3, -2)$ with an undefined slope. Then give two other points on the line.

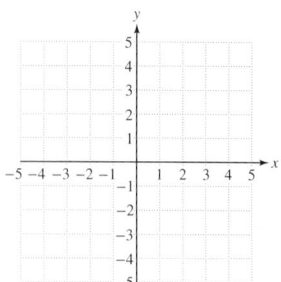

80. Graph the line through $(3, 3)$ with a slope of 0. Then give two other points on the line.

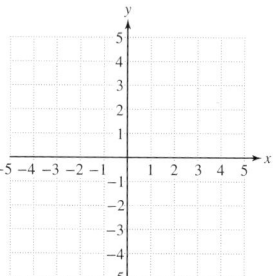

For Exercises 81–86, draw a line as indicated. Answers may vary.

81. Draw a line with a positive slope and a positive y-intercept.

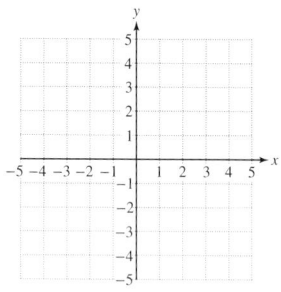

82. Draw a line with a positive slope and a negative y-intercept.

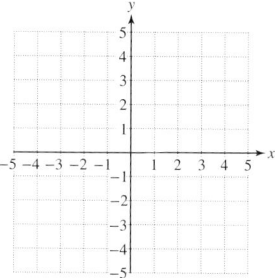

83. Draw a line with a negative slope and a negative y-intercept.

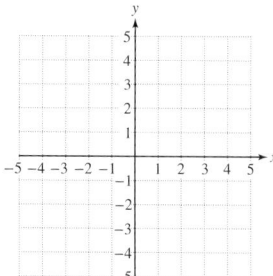

84. Draw a line with a negative slope and positive y-intercept.

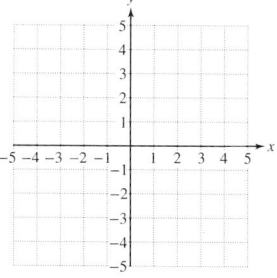

85. Draw a line with a zero slope and a positive y-intercept.

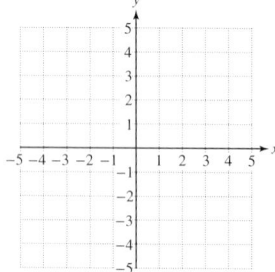

86. Draw a line with undefined slope and a negative x-intercept.

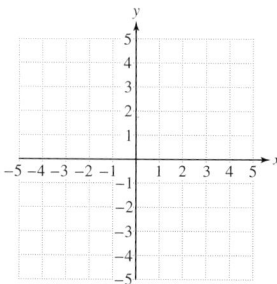

Expanding Your Skills

87. Determine the slope between the points $(a + b, 4m - n)$ and $(a - b, m + 2n)$.

88. Determine the slope between the points $(3c - d, s + t)$ and $(c - 2d, s - t)$.

89. Determine the x-intercept of the line $ax + by = c$.

90. Determine the y-intercept of the line $ax + by = c$.

91. Find another point on the line that contains the point $(2, -1)$ and has a slope of $\frac{2}{5}$.

92. Find another point on the line that contains the point $(-3, 4)$ and has a slope of $\frac{1}{4}$.

Slope-Intercept Form of a Linear Equation

Section 11.4

1. Slope-Intercept Form of a Linear Equation

We learned that the solutions to an equation of the form $Ax + By = C$ (where A and B are not both zero) represent a line in a rectangular coordinate system. An equation of a line written in this way is said to be in **standard form**. In this section, we will learn a new form, called **slope-intercept form**, that can be used to determine the slope and y-intercept of a line.

Let $(0, b)$ represent the y-intercept of a line. Let (x, y) represent any other point on the line. See Figure 11-25. Then the slope, m, of the line can be found as follows:

Let $(0, b)$ represent (x_1, y_1), and let (x, y) represent (x_2, y_2). Apply the slope formula.

$m = \dfrac{y_2 - y_1}{x_2 - x_1} \rightarrow m = \dfrac{y - b}{x - 0}$ Apply the slope formula.

$m = \dfrac{y - b}{x}$ Simplify.

$mx = \left(\dfrac{y - b}{x}\right)x$ Multiply by x to clear fractions.

$mx = y - b$

$mx + b = y - b + b$ To isolate y, add b to both sides.

$mx + b = y$ or $y = mx + b$ The equation is in slope-intercept form.

Concepts

1. Slope-Intercept Form of a Linear Equation
2. Graphing a Line from Its Slope and y-Intercept
3. Determining Whether Two Lines Are Parallel, Perpendicular, or Neither
4. Writing an Equation of a Line Using Slope-Intercept Form

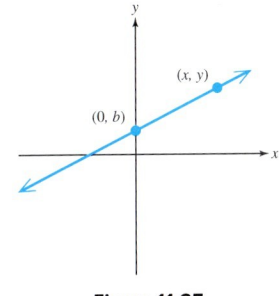

Figure 11-25

Slope-Intercept Form of a Linear Equation

$y = mx + b$ is the slope-intercept form of a linear equation.
m is the slope and the point $(0, b)$ is the y-intercept.

Example 1   Identifying the Slope and y-Intercept From a Linear Equation

For each equation, identify the slope and y-intercept.

a. $y = 3x - 1$ b. $y = -2.7x + 5$ c. $y = 4x$

Solution:

Each equation is written in slope-intercept form, $y = mx + b$. The slope is the coefficient of x, and the y-intercept is determined by the constant term.

a. $y = 3x - 1$ The slope is 3.   The y-intercept is $(0, -1)$.

b. $y = -2.7x + 5$ The slope is -2.7.   The y-intercept is $(0, 5)$.

c. $y = 4x$ can be written as $y = 4x + 0$.   The slope is 4.
The y-intercept is $(0, 0)$.

Skill Practice Identify the slope and the y-intercept.

1. $y = 4x + 6$ 2. $y = 3.5x - 4.2$ 3. $y = -7$

Given an equation of a line, we can write the equation in slope-intercept form by solving the equation for the y variable. This is demonstrated in Example 2.

Example 2   Identifying the Slope and *y*-Intercept From a Linear Equation

Given the equation $-5x - 2y = 6$,

a. Write the equation in slope-intercept form.

b. Identify the slope and y-intercept.

Solution:

a. Write the equation in slope-intercept form, $y = mx + b$, by solving for y.

$$-5x - 2y = 6$$

$-2y = 5x + 6$ Add $5x$ to both sides.

$\dfrac{-2y}{-2} = \dfrac{5x + 6}{-2}$ Divide both sides by -2.

$y = \dfrac{5x}{-2} + \dfrac{6}{-2}$ Divide each term by -2 and simplify.

$y = -\dfrac{5}{2}x - 3$ Slope-intercept form

b. The slope is $-\dfrac{5}{2}$, and the y-intercept is $(0, -3)$.

Skill Practice Given the equation $2x - 6y = -3$.

4. Write the equation in slope-intercept form.

5. Identify the slope and the y-intercept.

2. Graphing a Line from Its Slope and *y*-Intercept

Slope-intercept form is a useful tool to graph a line. The y-intercept is a known point on the line. The slope indicates the direction of the line and can be used to find a second point. Using slope-intercept form to graph a line is demonstrated in Examples 3 and 4.

Answers

1. slope: 4; y-intercept: $(0, 6)$
2. slope: 3.5; y-intercept: $(0, -4.2)$
3. slope: 0; y-intercept: $(0, -7)$
4. $y = \dfrac{1}{3}x + \dfrac{1}{2}$
5. slope is $\dfrac{1}{3}$; y-intercept is $\left(0, \dfrac{1}{2}\right)$

Example 3 Graphing a Line Using the Slope and y-Intercept

Graph the equation $y = -\frac{5}{2}x - 3$ by using the slope and y-intercept.

Solution:

First plot the y-intercept, $(0, -3)$.

The slope $m = -\frac{5}{2}$ can be written as

$m = \dfrac{-5}{2}$ ⟵ The change in y is −5.
⟵ The change in x is 2.

To find a second point on the line, start at the y-intercept and move down 5 units and to the right 2 units. Then draw the line through the two points (Figure 11-26).

Similarly, the slope can be written as

$m = \dfrac{5}{-2}$ ⟵ The change in y is 5.
⟵ The change in x is −2.

To find a second point, start at the y-intercept and move up 5 units and to the left 2 units. Then draw the line through the two points (Figure 11-27).

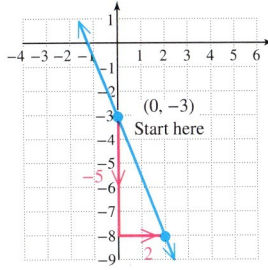

Figure 11-26

FOR REVIEW

Recall that the negative sign in a fraction can be written in the numerator, denominator, or out in front.

$$\dfrac{-5}{2} = \dfrac{5}{-2} = -\dfrac{5}{2}$$

Figure 11-27

Skill Practice

6. Graph the equation by using the slope and y-intercept. $y = 2x - 3$

Example 4 Graphing a Line Using the Slope and y-Intercept

Graph the equation $y = 4x$ by using the slope and y-intercept.

Solution:

The equation can be written as $y = 4x + 0$. Therefore, we can plot the y-intercept at $(0, 0)$.
The slope $m = 4$ can be written as

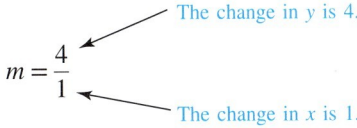

$m = \dfrac{4}{1}$

The change in y is 4.
The change in x is 1.

To find a second point on the line, start at the y-intercept and move up 4 units and to the right 1 unit. Then draw the line through the two points (Figure 11-28).

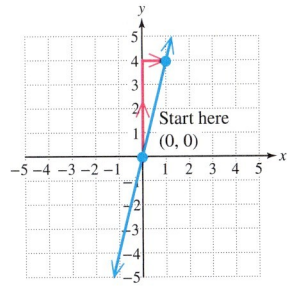

Figure 11-28

Answers

6–7.

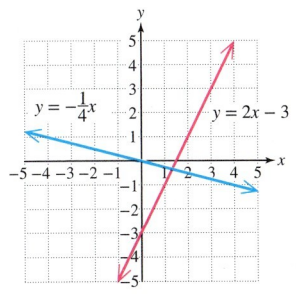

Skill Practice

7. Graph the equation by using the slope and y-intercept. $y = -\dfrac{1}{4}x$

3. Determining Whether Two Lines Are Parallel, Perpendicular, or Neither

The slope-intercept form provides a means to find the slope of a line by inspection. Recall that if the slopes of two lines are known, then we can compare the slopes to determine if the lines are parallel, perpendicular, or neither parallel nor perpendicular. (Two distinct nonvertical lines are parallel if their slopes are equal. Two lines are perpendicular if the slope of one line is the opposite of the reciprocal of the slope of the other line.)

Example 5   **Determining If Two Lines Are Parallel, Perpendicular, or Neither**

For each pair of lines, determine if they are parallel, perpendicular, or neither.

  a. l_1: $y = 3x - 5$ **b.** l_1: $y = \frac{3}{2}x + 2$
     l_2: $y = 3x + 1$ l_2: $y = \frac{2}{3}x + 1$

Solution:

  a. l_1: $y = 3x - 5$ The slope of l_1 is 3.
     l_2: $y = 3x + 1$ The slope of l_2 is 3.

Because the slopes are the same, the lines are parallel.

  b. l_1: $y = \frac{3}{2}x + 2$ The slope of l_1 is $\frac{3}{2}$.
     l_2: $y = \frac{2}{3}x + 1$ The slope of l_2 is $\frac{2}{3}$.

The slopes are not the same. Therefore, the lines are not parallel. The values of the slopes are reciprocals, but they are not opposite in sign. Therefore, the lines are not perpendicular. The lines are neither parallel nor perpendicular.

Skill Practice   For each pair of lines determine if they are parallel, perpendicular, or neither.

  8. $y = 3x - 5$ **9.** $y = \frac{5}{6}x - \frac{1}{2}$
     $y = -3x - 15$ $y = \frac{5}{6}x + \frac{1}{2}$

Example 6   **Determining If Two Lines Are Parallel, Perpendicular, or Neither**

For each pair of lines, determine if they are parallel, perpendicular, or neither.

  a. l_1: $x - 3y = -9$ **b.** l_1: $x = 2$
     l_2: $3x = -y + 4$ l_2: $2y = 8$

Answers

8. Neither   **9.** Parallel

Solution:

a. First write the equation of each line in slope-intercept form.

l_1: $x - 3y = -9$ $\qquad$ l_2: $3x = -y + 4$

$\qquad -3y = -x - 9 \qquad\qquad\quad 3x + y = 4$

$\qquad \dfrac{-3y}{-3} = \dfrac{-x}{-3} - \dfrac{9}{-3} \qquad\quad y = -3x + 4$

$\qquad y = \dfrac{1}{3}x + 3$

l_1: $y = \frac{1}{3}x + 3$ $\qquad$ The slope of l_1 is $\frac{1}{3}$.
l_2: $y = -3x + 4$ $\qquad$ The slope of l_2 is -3.

The slope of $\frac{1}{3}$ is the opposite of the reciprocal of -3. Therefore, the lines are perpendicular.

b. The equation $x = 2$ represents a vertical line because the equation is in the form $x = k$.

The equation $2y = 8$ can be simplified to $y = 4$, which represents a horizontal line.

In this example, we do not need to analyze the slopes because vertical lines and horizontal lines are perpendicular.

Skill Practice For each pair of lines, determine if they are parallel, perpendicular, or neither.

10. $x - 5y = 10$
$\quad\;\, 5x - 1 = -y$

11. $y = -5$
$\quad\;\, x = 6$

---

4. Writing an Equation of a Line Using Slope-Intercept Form

The slope-intercept form of a linear equation can be used to write an equation of a line when the slope is known and the y-intercept is known.

Example 7 — Writing an Equation of a Line Using Slope-Intercept Form

Write an equation of the line whose slope is $\frac{2}{3}$ and whose y-intercept is $(0, 8)$.

Solution:

The slope is given as $m = \frac{2}{3}$, and the y-intercept $(0, b)$ is given as $(0, 8)$. Substitute the values $m = \frac{2}{3}$ and $b = 8$ into the slope-intercept form of a line.

$$y = mx + b$$
$$y = \dfrac{2}{3}x + 8$$

Skill Practice

12. Write an equation of the line whose slope is -4 and y-intercept is $(0, -10)$.

Answers

10. Perpendicular $\qquad$ **11.** Perpendicular
12. $y = -4x - 10$

Example 8: Writing an Equation of a Line Using Slope-Intercept Form

Write an equation of the line having a slope of 2 and passing through the point $(-3, 1)$.

Solution:

To find an equation of a line using slope-intercept form, it is necessary to find the value of m and b. The slope is given in the problem as $m = 2$. Therefore, the slope-intercept form becomes

$$y = mx + b$$
$$\downarrow$$
$$y = 2x + b$$

Because the point $(-3, 1)$ is on the line, it is a solution to the equation. Therefore, to find b, substitute the values of x and y from the ordered pair $(-3, 1)$ and solve the resulting equation for b.

$y = 2x + b$
$1 = 2(-3) + b$ Substitute $y = 1$ and $x = -3$.
$1 = -6 + b$ Simplify and solve for b.
$7 = b$

Now with m and b known, the slope-intercept form is $y = 2x + 7$.

TIP: The equation from Example 8 can be checked by graphing the line $y = 2x + 7$. The slope $m = 2$ can be written as $m = \frac{2}{1}$. Therefore, to graph the line, start at the y-intercept $(0, 7)$ and move up 2 units and to the right 1 unit.

The graph verifies that the line passes through the point $(-3, 1)$ as it should.

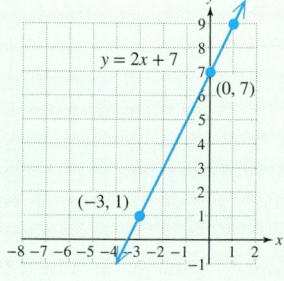

Skill Practice

13. Write an equation of the line having a slope of -3 and passing through the point $(-2, -5)$.

Answer
13. $y = -3x - 11$

Section 11.4 Activity

A.1. a. Solve the equation $2x + y = 4$ for y.
  b. Identify the slope and y-intercept.
  c. Graph the line by first graphing the y-intercept and then using the slope to find a second and third point on the line.

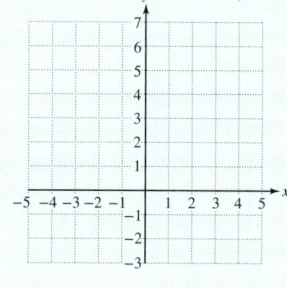

A.2. **a.** Given the equation $3x + y = 4$, write the equation in slope-intercept form.
 b. What is the slope of the line given in part (a)?
 c. Given the equation $x - 3y = -3$, write the equation in slope-intercept form.
 d. What is the slope of the line from part (c)?
 e. Are the two lines parallel, perpendicular, or neither? How did you come to that conclusion?
 f. Graph the two lines to verify your answer from part (e).

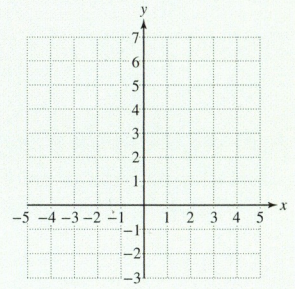

A.3. Given the linear equations $4x - y = 2$ and $y = 4x + 1$, are the lines parallel, perpendicular, or neither? How did you come to that conclusion?

A.4. Write an equation in slope-intercept form that has a slope of $\frac{2}{3}$ and a y-intercept of $(0, -4)$.

A.5. A line passes through the point $(2, -3)$ and has a slope of $-\frac{1}{2}$.

 a. Use the slope-intercept form $y = mx + b$ and substitute $-\frac{1}{2}$ for m, 2 for x, and -3 for y. Then solve the equation for b.
 b. Using the values of m and b, write an equation of the line in slope-intercept form.
 c. Graph the equation from part (b) to confirm that the line passes through $(2, -3)$ with slope $-\frac{1}{2}$.

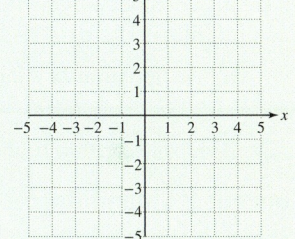

A.6. An engineering firm sets up a pump to drain water from a retention pond. The depth of the pond, y (in inches), can be approximated by the equation $y = -0.5x + 48$ where x represents the time in hours after the pump has started.

 a. How deep is the pond after 12 hr?
 b. How deep is the pond after 24 hr?
 c. Determine the slope of the line represented by this equation. Interpret its meaning in the context of this problem.
 d. Determine the y-intercept and interpret its meaning in the context of this problem.
 e. Determine the x-intercept and interpret its meaning in the context of this problem.

Practice Exercises

Section 11.4

Study Skills Exercise

Research shows that writing mathematically has significant advantages to learning. When you can communicate a mathematical concept in words, you are demonstrating a full understanding of that concept. You are also communicating your understanding and reasoning to yourself and others. This skill will help you move mathematical concepts into long-term memory.

- Suppose you are given two linear equations in slope-intercept form. Explain how you know whether the lines are parallel, perpendicular, or neither. Describe how you might check that your answers are correct.

Prerequisite Review

For Exercises R.1–R.4, solve the equation for y.

R.1. $4x + 2y = 8$ **R.2.** $5x - 3y = 15$ **R.3.** $x - \frac{1}{4}y = 0$ **R.4.** $-\frac{1}{3}x + \frac{1}{6}y = 0$

For Exercises R.5–R.8, the slope of a line is given.
  a. Determine the slope of a line parallel to the given line.
  b. Determine the slope of a line perpendicular to the given line.

R.5. $m = -4$ **R.6.** $m = 6$ **R.7.** $m = \frac{5}{7}$ **R.8.** $m = -\frac{2}{5}$

For Exercises R.9–R.10, solve the equation for b.

R.9. $5 = -2 + b$ **R.10.** $8 = 2 + b$

Vocabulary and Key Terms

1. **a.** Consider a line with slope m and y-intercept $(0, b)$. The slope-intercept form of an equation of the line is _____.
   b. An equation of a line written in the form $Ax + By = C$ where A and B are not both zero is said to be in _____ form.
2. Write an equation of the line having a slope of $\frac{3}{4}$ and y-intercept $(0, -5)$.

Concept 1: Slope-Intercept Form of a Linear Equation

For Exercises 3–22, identify the slope and y-intercept, if they exist. **(See Examples 1–2.)**

3. $y = -2x + 3$ 4. $y = \frac{2}{3}x + 5$ 5. $y = x - 2$

6. $y = -x + 6$ 7. $y = -x$ 8. $y = -5x$

9. $y = \frac{3}{4}x - 1$ 10. $y = x - \frac{5}{3}$ 11. $2x - 5y = 4$

12. $3x + 2y = 9$ 13. $3x - y = 5$ 14. $7x - 3y = -6$

15. $x + y = 6$ 16. $x - y = 1$ 17. $x + 6 = 8$

18. $-4 + x = 1$ 19. $-8y = 2$ 20. $1 - y = 9$

21. $3y - 2x = 0$ 22. $5x = 6y$

Concept 2: Graphing a Line from Its Slope and y-Intercept

For Exercises 23–26, graph the line using the slope and y-intercept. **(See Examples 3–4.)**

23. Graph the line through the point $(0, 2)$, having a slope of -4.

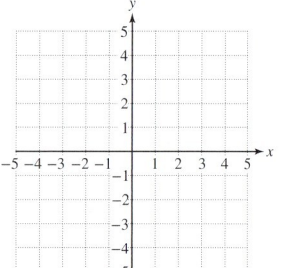

24. Graph the line through the point $(0, -1)$, having a slope of -3.

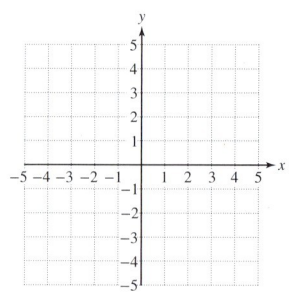

25. Graph the line through the point (0, −5), having a slope of $\frac{3}{2}$.

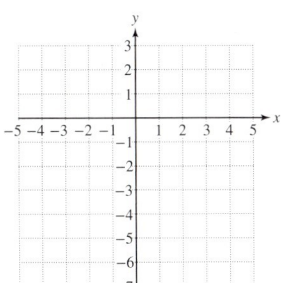

26. Graph the line through the point (0, 3), having a slope of $-\frac{1}{4}$.

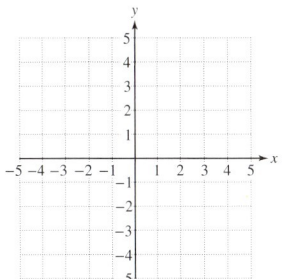

For Exercises 27–32, match the equation with the graph (a–f) by identifying if the slope is positive or negative and if the y-intercept is positive, negative, or zero.

27. $y = 2x + 3$

28. $y = -3x - 2$

29. $y = -\frac{1}{3}x + 3$

30. $y = \frac{1}{2}x - 2$

31. $y = x$

32. $y = -2x$

a.
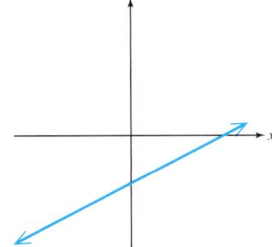

b.
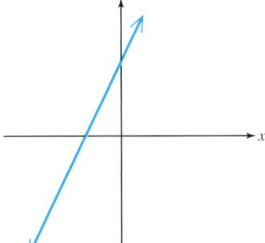

c.
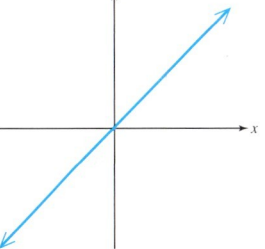

d.
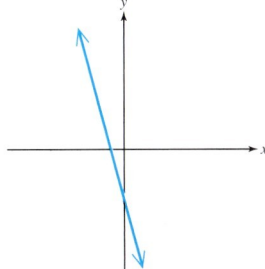

e.
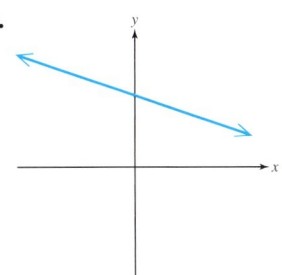

f.
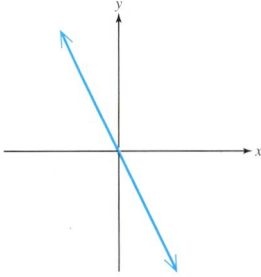

For Exercises 33–44, write each equation in slope-intercept form (if possible) and graph the line. **(See Examples 3–4.)**

33. $2x + y = 9$

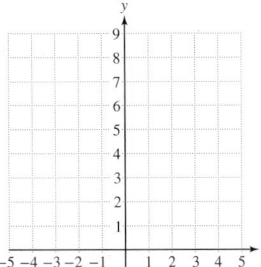

34. $-6x + y = 8$

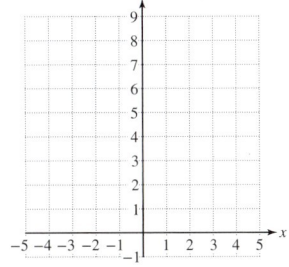

35. $x - 2y = 6$

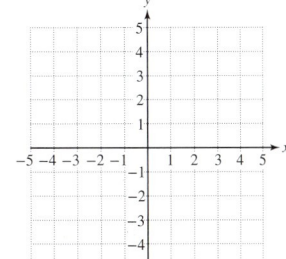

36. $5x - 2y = 2$

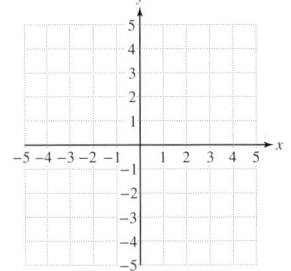

37. $2x = -4y + 6$ **38.** $6x = 2y - 14$ **39.** $x + y = 0$ **40.** $x - y = 0$

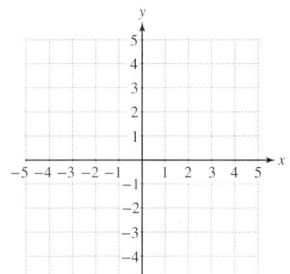 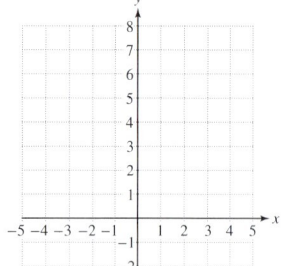 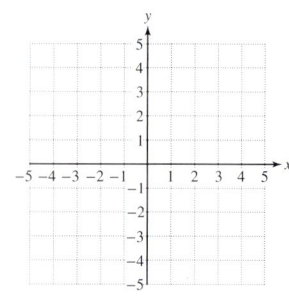 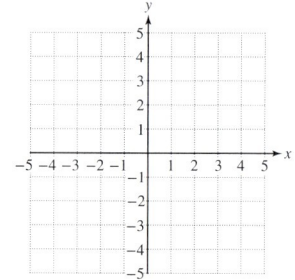

41. $5y = 4x$ **42.** $-2x = 5y$ **43.** $3y + 2 = 0$ **44.** $1 + 5y = 6$

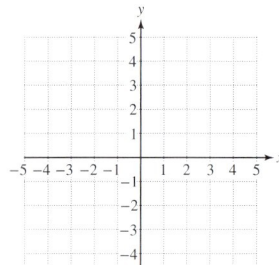 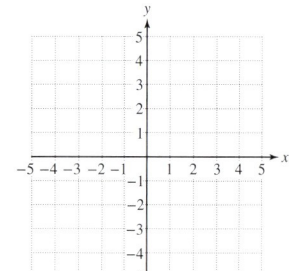 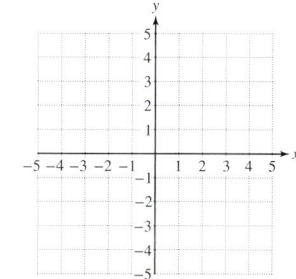 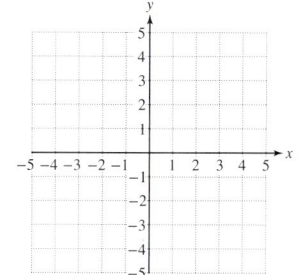

Concept 3: Determining Whether Two Lines Are Parallel, Perpendicular, or Neither

For Exercises 45–60, determine if the equations represent parallel lines, perpendicular lines, or neither. (See Examples 5–6.)

45. l_1: $y = -2x - 3$
   l_2: $y = \frac{1}{2}x + 4$

46. l_1: $y = \frac{4}{3}x - 2$
   l_2: $y = -\frac{3}{4}x + 6$

47. l_1: $y = \frac{4}{5}x - \frac{1}{2}$
   l_2: $y = \frac{5}{4}x - \frac{2}{3}$

48. l_1: $y = \frac{1}{5}x + 1$
   l_2: $y = 5x - 3$

49. l_1: $y = -9x + 6$
   l_2: $y = -9x - 1$

50. l_1: $y = 4x - 1$
   l_2: $y = 4x + \frac{1}{2}$

51. l_1: $x = 3$
   l_2: $y = \frac{7}{4}$

52. l_1: $y = \frac{2}{3}$
   l_2: $x = 6$

53. l_1: $2x = 4$
   l_2: $6 = x$

54. l_1: $2y = 7$
   l_2: $y = 4$

55. l_1: $2x + 3y = 6$
   l_2: $3x - 2y = 12$

56. l_1: $4x + 5y = 20$
   l_2: $5x - 4y = 60$

57. l_1: $4x + 2y = 6$
   l_2: $4x + 8y = 16$

58. l_1: $3x + y = 5$
   l_2: $x + 3y = 18$

59. l_1: $y = \frac{1}{5}x - 3$
   l_2: $2x - 10y = 20$

60. l_1: $y = \frac{1}{3}x + 2$
   l_2: $-x + 3y = 12$

Concept 4: Writing an Equation of a Line Using Slope-Intercept Form

For Exercises 61–72, write an equation of the line given the following information. Write the answer in slope-intercept form if possible. (See Examples 7–8.)

61. The slope is $-\frac{1}{3}$, and the y-intercept is $(0, 2)$.

62. The slope is $\frac{2}{3}$, and the y-intercept is $(0, -1)$.

63. The slope is 10, and the y-intercept is $(0, -19)$.

64. The slope is -14, and the y-intercept is $(0, 2)$.

65. The slope is 6, and the line passes through the point (1, −2).

66. The slope is −4, and the line passes through the point (4, −3).

67. The slope is $\frac{1}{2}$, and the line passes through the point (−4, −5).

68. The slope is $-\frac{2}{3}$, and the line passes through the point (3, −1).

69. The slope is 0, and the y-intercept is −11.

70. The slope is 0, and the y-intercept is $\frac{6}{7}$.

71. The slope is 5, and the line passes through the origin.

72. The slope is −3, and the line passes through the origin.

Technology Connections

For Exercises 73–75, determine if the lines are parallel, perpendicular, or neither. Then use a square viewing window to graph the lines on a graphing calculator to verify your results.

73. $x + y = 1$
    $x - y = -3$

74. $3x + y = -2$
    $6x + 2y = 6$

75. $2x - y = 4$
    $3x + 2y = 4$

76. Graph the lines defined by $y = x + 1$ and $y = 0.99x + 3$. Are these lines parallel? Explain.

77. Graph the lines defined by $y = -2x - 1$ and $y = -2x - 0.99$. Are these lines the same? Explain.

78. Graph the line defined by $y = 0.001x + 3$. Is this line horizontal? Explain.

Expanding Your Skills

For Exercises 79–84, write an equation of the line that passes through two points by following these steps:

Step 1: Find the slope of the line using the slope formula, $m = \frac{y_2 - y_1}{x_2 - x_1}$.

Step 2: Using the slope from Step 1 and either given point, follow the procedure given in Example 8 to find an equation of the line in slope-intercept form.

79. (2, −1) and (0, 3)

80. (4, −8) and (0, −4)

81. (3, 1) and (−3, 3)

82. (2, −3) and (4, −2)

83. (1, 3) and (−2, −9)

84. (1, 7) and (−2, 4)

85. A propane company charges a customer a yearly fee of $49 to rent a propane tank plus $3.50 per gallon of propane. The equation $C = 3.50x + 49$ represents the total yearly cost, C (in dollars), for renting a tank and buying x gallons of propane.

   a. Identify the slope and interpret its meaning in the context of this problem.

   b. Identify the C-intercept and interpret its meaning in the context of this problem.

   c. Use the equation to determine the total yearly cost for a homeowner to rent a tank and buy 60 gal of propane.

   d. If a homeowner's yearly bill is $196, how much propane did the homeowner purchase?

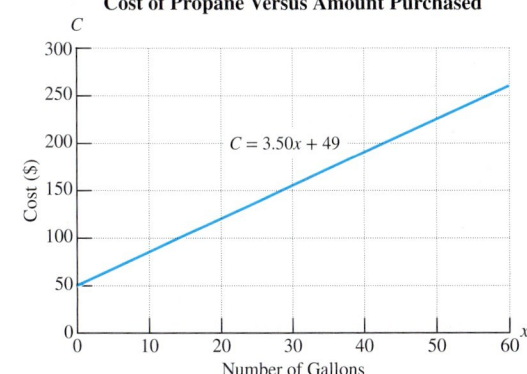
Cost of Propane Versus Amount Purchased

86. A linear equation is written in standard form if it can be written as $Ax + By = C$, where A and B are not both zero. Write the equation $Ax + By = C$ in slope-intercept form to show that the slope is given by the ratio $-\frac{A}{B}$. ($B \neq 0$.)

For Exercises 87–90, use the result of Exercise 86 to find the slope of the line.

87. $2x + 5y = 8$
88. $6x + 7y = -9$
89. $4x - 3y = -5$
90. $11x - 8y = 4$

Problem Recognition Exercises

Linear Equations in Two Variables

As you become more familiar with linear equations in two variables, you will often be able to determine characteristics of the corresponding lines by simple inspection of the equations. We review several cases here.

• A linear equation that has x (but no y or other variable) can be written in the form $x = k$ and represents a vertical line. • The x-intercept is $(k, 0)$. • The slope of a vertical line is undefined.	**Example:** $3x - 4 = 11$ This equation can be written with x isolated. $3x - 4 = 11$ $3x = 15$ $x = 5$ The graph of this equation is a vertical line with x-intercept $(5, 0)$. The slope is undefined. 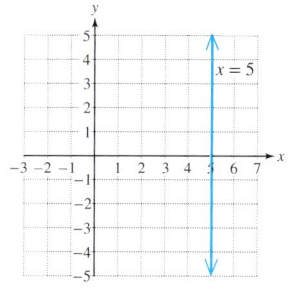
• A linear equation that has y (but no x or other variable) can be written in the form $y = k$ and represents a horizontal line. • The y-intercept is $(0, k)$. • The slope of a horizontal line is 0.	**Example:** $2 = 8 + 2y$ This equation can be written with y isolated. $2 = 8 + 2y$ $-6 = 2y$ $y = -3$ The graph of this equation is a horizontal line with y-intercept $(0, -3)$. The slope is 0. 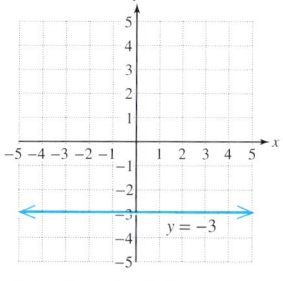
• A linear equation in two variables x and y can be written in **slope-intercept form** $y = mx + b$. This form of the equation is helpful because we can determine the slope and y-intercept by inspection. • The slope is m. • The y-intercept is $(0, b)$. • The x-intercept can be found by substituting 0 for y and solving for x. • *Note:* The graph of a linear equation in two variables with both x and y appearing in the equation is a *slanted* line.	**Example:** $y = 3x - 2$ The slope is 3. The y-intercept is $(0, -2)$. To find the x-intercept, solve the equation $0 = 3x - 2$ $2 = 3x$ $x = \dfrac{2}{3}$ The x-intercept is 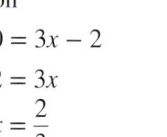. 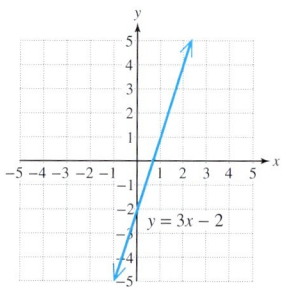

- A linear equation in two variables x and y with *no constant term* represents a line passing through the origin. The slope-intercept form of the equation is $y = -mx$.
- The slope is m.
- The x-intercept is $(0, 0)$.
- The y-intercept is $(0, 0)$.

Example: $4x + 3y = 0$

Write the equation in slope-intercept form:
$$3y = -4x$$
$$y = \frac{-4x}{3}$$
$$y = -\frac{4}{3}x$$

The slope is $-\frac{4}{3}$.

The x- and y-intercepts are $(0, 0)$.

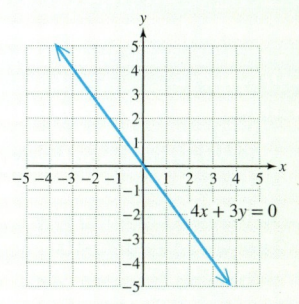

For Exercises 1–20, choose the equation(s) from the column on the right whose graph satisfies the condition described. Give all possible answers.

1. Line whose slope is positive.
2. Line whose slope is negative.
3. Line that passes through the origin.
4. Line that contains the point $(3, -2)$.
5. Line whose y-intercept is $(0, 4)$.
6. Line whose y-intercept is $(0, -5)$.
7. Line whose slope is $\frac{1}{2}$.
8. Line whose slope is -2.
9. Line whose slope is 0.
10. Line whose slope is undefined.
11. Line that is parallel to the line with equation $y = -\frac{2}{3}x + 4$.
12. Line perpendicular to the line with equation $y = 2x + 9$.
13. Line that is vertical.
14. Line that is horizontal.
15. Line whose x-intercept is $(10, 0)$.
16. Line whose x-intercept is $(6, 0)$.
17. Line that is parallel to the x-axis.
18. Line that is perpendicular to the y-axis.
19. Line with a negative slope and positive y-intercept.
20. Line with a positive slope and negative y-intercept.

a. $y = 5x$

b. $2x + 3y = 12$

c. $y = \frac{1}{2}x - 5$

d. $3x - 6y = 10$

e. $2y = -8$

f. $y = -2x + 4$

g. $3x = 1$

h. $x + 2y = 6$

Section 11.5 Point-Slope Formula

Concepts

1. Writing an Equation of a Line Using the Point-Slope Formula
2. Writing an Equation of a Line Given Two Points
3. Writing an Equation of a Line Parallel or Perpendicular to Another Line
4. Different Forms of Linear Equations: A Summary

1. Writing an Equation of a Line Using the Point-Slope Formula

The slope-intercept form of a line can be used as a tool to construct an equation of a line. Another useful tool to determine an equation of a line is the point-slope formula. The point-slope formula can be derived from the slope formula as follows:

Suppose a line passes through a given point (x_1, y_1) and has slope m. If (x, y) is any other point on the line, then the slope is given by

$$m = \frac{y - y_1}{x - x_1} \qquad \text{Slope formula}$$

$$m(x - x_1) = \frac{y - y_1}{x - x_1}(x - x_1) \qquad \text{Clear fractions.}$$

$$m(x - x_1) = y - y_1$$

$$y - y_1 = m(x - x_1) \qquad \text{Point-slope formula}$$

> **Point-Slope Formula**
>
> The **point-slope formula** is given by
>
> $$y - y_1 = m(x - x_1)$$
>
> where m is the slope of the line and (x_1, y_1) is any known point on the line.

Example 1 demonstrates how to use the point-slope formula to find an equation of a line when a point on the line and slope are given.

Example 1   Writing an Equation of a Line Using the Point-Slope Formula

Use the point-slope formula to write an equation of the line having a slope of 3 and passing through the point $(-2, -4)$. Write the answer in slope-intercept form.

Solution:

The slope of the line is given: $m = 3$

A point on the line is given: $(x_1, y_1) = (-2, -4)$

The point-slope formula:

$y - y_1 = m(x - x_1)$

$y - (-4) = 3[x - (-2)]$ Substitute $m = 3$, $x_1 = -2$, and $y_1 = -4$.

$y + 4 = 3(x + 2)$ Simplify. Because the final answer is required in slope-intercept form, simplify the equation and solve for y.

$y + 4 = 3x + 6$ Apply the distributive property.

$y = 3x + 2$ Slope-intercept form

Skill Practice

1. Use the point-slope formula to write an equation of the line having a slope of -4 and passing through $(-1, 5)$. Write the answer in slope-intercept form.

The equation $y = 3x + 2$ from Example 1 is graphed in Figure 11-29. Notice that the line does indeed pass through the point $(-2, -4)$.

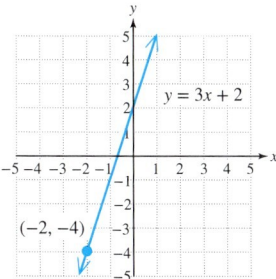

Figure 11-29

2. Writing an Equation of a Line Given Two Points

Example 2 is similar to Example 1; however, the slope must first be found from two given points.

Example 2 — Writing an Equation of a Line Given Two Points

Use the point-slope formula to find an equation of the line passing through the points $(-2, 5)$ and $(4, -1)$. Write the final answer in slope-intercept form.

Solution:

Given two points on a line, the slope can be found with the slope formula.

$(-2, 5)$ and $(4, -1)$
(x_1, y_1) (x_2, y_2) Label the points.

$$m = \frac{y_2 - y_1}{x_2 - x_1} = \frac{(-1) - (5)}{(4) - (-2)} = \frac{-6}{6} = -1$$

To apply the point-slope formula, use the slope, $m = -1$ and either given point. We will choose the point $(-2, 5)$ as (x_1, y_1).

$y - y_1 = m(x - x_1)$
$y - 5 = -1[x - (-2)]$ Substitute $m = -1$, $x_1 = -2$, and $y_1 = 5$.
$y - 5 = -1(x + 2)$ Simplify.
$y - 5 = -x - 2$
$y = -x + 3$

TIP: The point-slope formula can be applied using either given point for (x_1, y_1). In Example 2, using the point $(4, -1)$ for (x_1, y_1) produces the same result.

$y - y_1 = m(x - x_1)$
$y - (-1) = -1(x - 4)$
$y + 1 = -x + 4$
$y = -x + 3$

Skill Practice

2. Use the point-slope formula to write an equation of the line passing through the points $(1, -1)$ and $(-1, -5)$.

Answers
1. $y = -4x + 1$ 2. $y = 2x - 3$

The solution to Example 2 can be checked by graphing the line $y = -x + 3$ using the slope and y-intercept. Notice that the line passes through the points $(-2, 5)$ and $(4, -1)$ as expected. See Figure 11-30.

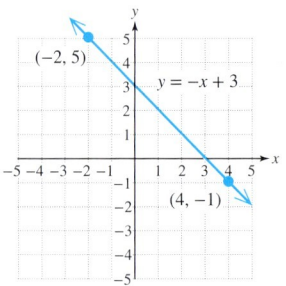

Figure 11-30

3. Writing an Equation of a Line Parallel or Perpendicular to Another Line

To write an equation of a line using the point-slope formula, the slope must be known. If the slope is not explicitly given, then other information must be used to determine the slope. In Example 2, the slope was found using the slope formula. Examples 3 and 4 show other situations in which we might find the slope.

Example 3   Writing an Equation of a Line Parallel to Another Line

Use the point-slope formula to find an equation of the line passing through the point $(-1, 0)$ and parallel to the line $y = -4x + 3$. Write the final answer in slope-intercept form.

Solution:

Figure 11-31 shows the line $y = -4x + 3$ (pictured in black) and a line parallel to it (pictured in blue) that passes through the point $(-1, 0)$. The equation of the given line, $y = -4x + 3$, is written in slope-intercept form, and its slope is easily identified as -4. The line parallel to the given line must also have a slope of -4.

Apply the point-slope formula using $m = -4$ and the point $(x_1, y_1) = (-1, 0)$.

$$y - y_1 = m(x - x_1)$$
$$y - 0 = -4[x - (-1)]$$
$$y = -4(x + 1)$$
$$y = -4x - 4$$

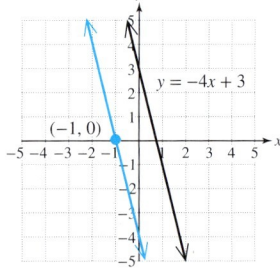

Figure 11-31

TIP: When writing an equation of a line, slope-intercept form or standard form is usually preferred. For instance, the solution to Example 3 can be written as follows.

Slope-intercept form:
$y = -4x - 4$

Standard form:
$4x + y = -4$

Skill Practice

3. Use the point-slope formula to write an equation of the line passing through $(8, 2)$ and parallel to the line $y = \frac{3}{4}x - \frac{1}{2}$.

Answer

3. $y = \frac{3}{4}x - 4$

Example 4: Writing an Equation of a Line Perpendicular to Another Line

Use the point-slope formula to find an equation of the line passing through the point $(-3, 1)$ and perpendicular to the line $3x + y = -2$. Write the final answer in slope-intercept form.

Solution:

The given line can be written in slope-intercept form as $y = -3x - 2$. The slope of this line is -3. Therefore, the slope of a line perpendicular to the given line is $\frac{1}{3}$.

Apply the point-slope formula with $m = \frac{1}{3}$, and $(x_1, y_1) = (-3, 1)$.

$y - y_1 = m(x - x_1)$	Point-slope formula
$y - (1) = \frac{1}{3}[x - (-3)]$	Substitute $m = \frac{1}{3}$, $x_1 = -3$, and $y_1 = 1$.
$y - 1 = \frac{1}{3}(x + 3)$	To write the final answer in slope-intercept form, simplify the equation and solve for y.
$y - 1 = \frac{1}{3}x + 1$	Apply the distributive property.
$y = \frac{1}{3}x + 2$	Add 1 to both sides.

A sketch of the perpendicular lines $y = \frac{1}{3}x + 2$ and $y = -3x - 2$ is shown in Figure 11-32. Notice that the line $y = \frac{1}{3}x + 2$ passes through the point $(-3, 1)$ as expected.

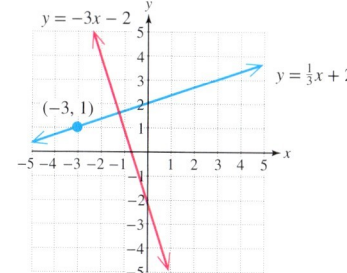

Figure 11-32

Skill Practice

4. Write an equation of the line passing through the point $(10, 4)$ and perpendicular to the line $x + 2y = 1$.

4. Different Forms of Linear Equations: A Summary

A linear equation can be written in several different forms, as summarized in Table 11-3.

Table 11-3

Form	Example	Comments
Standard Form $Ax + By = C$	$4x + 2y = 8$	A and B must not both be zero.
Horizontal Line $y = k$ (k is constant)	$y = 4$	The slope is zero, and the y-intercept is $(0, k)$.
Vertical Line $x = k$ (k is constant)	$x = -1$	The slope is undefined, and the x-intercept is $(k, 0)$.
Slope-Intercept Form $y = mx + b$ the slope is m y-intercept is $(0, b)$	$y = -3x + 7$ Slope $= -3$ y-intercept is $(0, 7)$	Solving a linear equation for y results in slope-intercept form. The coefficient of the x-term is the slope, and the constant defines the location of the y-intercept.
Point-Slope Formula $y - y_1 = m(x - x_1)$	$m = -3$ $(x_1, y_1) = (4, 2)$ $y - 2 = -3(x - 4)$	This formula is typically used to build an equation of a line when a point on the line is known and the slope of the line is known.

Answer

4. $y = 2x - 16$

Although standard form and slope-intercept form can be used to express an equation of a line, often the slope-intercept form is used to give a *unique* representation of the line. For example, the following linear equations are all written in standard form, yet they each define the same line.

$$2x + 5y = 10$$
$$-4x - 10y = -20$$
$$6x + 15y = 30$$
$$\frac{2}{5}x + y = 2$$

The line can be written uniquely in slope-intercept form as: $y = -\frac{2}{5}x + 2$.

Although it is important to understand and apply slope-intercept form and the point-slope formula, they are not necessarily applicable to all problems, particularly when dealing with a horizontal or vertical line.

Example 5   Writing an Equation of a Line

Find an equation of the line passing through the point $(2, -4)$ and parallel to the *x*-axis.

Solution:

Because the line is parallel to the *x*-axis, the line must be horizontal. Recall that all horizontal lines can be written in the form $y = k$, where k is a constant. A quick sketch can help find the value of the constant. See Figure 11-33.

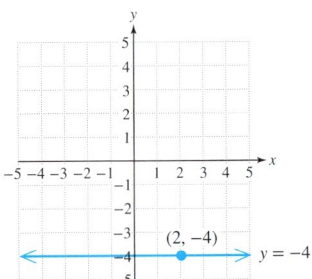

Figure 11-33

Because the line must pass through a point whose *y*-coordinate is -4, then the equation of the line must be $y = -4$.

Skill Practice

5. Write an equation for the vertical line that passes through the point $(-7, 2)$.

Answer

5. $x = -7$

Section 11.5 Activity

A.1. Write the point-slope formula for an equation of a line with slope m and known point on the line (x_1, y_1).

A.2. A line passes through the point $(-2, 4)$ and has a slope of -5.
  a. Use the point-slope formula to write an equation of the line. Write the final answer in slope-intercept form.
  b. Use the equation from part (a) to graph the line. Verify that the line passes through $(-2, 4)$ and has a slope of -5.

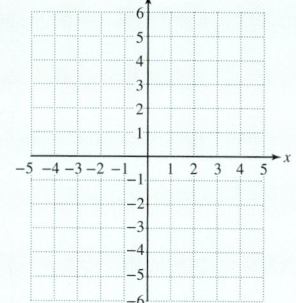

A.3. Write an equation of the line passing through $(3, 0)$ and perpendicular to the line $3x + y = 4$ by following these steps.

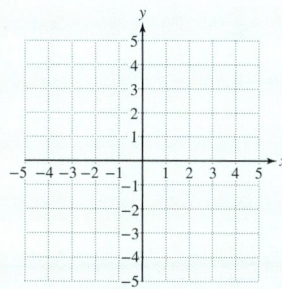

  a. Write the equation $3x + y = 4$ in slope-intercept form.
  b. What is the slope of a line perpendicular to $3x + y = 4$?
  c. Use the point-slope formula with the point $(3, 0)$ and the slope from part (b) to write an equation of the line. Write the answer in slope-intercept form.
  d. Graph the line $3x + y = 4$ and the line from part (c). Does the line from part (c) pass through the point $(3, 0)$ as expected? Do the two lines appear to be perpendicular?

A.4. How would the procedure have changed in Exercise A.3 if the two lines were to be parallel rather than perpendicular?

A.5. Write an equation of the line passing through $(-1, -2)$ and $(-3, -4)$ by following these steps.
  a. Find the slope of the line between the two given points.
  b. Apply the point-slope formula with the slope found in part (a) and the point $(-1, -2)$ to find an equation of the line through the points. Write the answer in slope-intercept form.
  c. Apply the point-slope formula with the slope found in part (a) and the point $(-3, -4)$ to find an equation of the line through the points. Write the answer in slope-intercept form.
  d. Are the equations in parts (b) and (c) the same? If not, revisit your work and look for errors.

A.6. a. Does an equation of the form $x = k$ represent a horizontal, vertical, or slanted line? What is the slope of the line?
  b. Does an equation of the form $y = k$ represent a horizontal, vertical, or slanted line? What is the slope of the line?

A.7. a. Graph the line perpendicular to the *y*-axis that passes through the point (−2, 3).

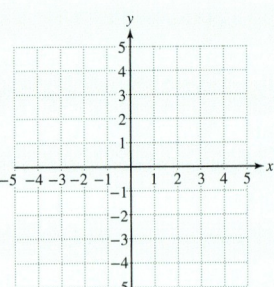

  b. Write an equation of the line perpendicular to the *y*-axis that passes through the point (−2, 3).

A.8. a. Graph the line $x = -3$.

  b. On the same coordinate system, graph the line parallel to $x = -3$ that passes through the point (1, 4).

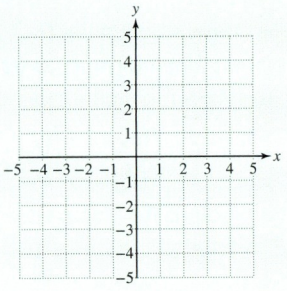

  c. Write an equation of the line parallel to $x = -3$ that passes through the point (1, 4).

Section 11.5 Practice Exercises

Prerequisite Review

For Exercises R.1–R.4, indicate whether the equation represents a horizontal line, vertical line, or slanted line.

R.1. $3x = 6$ **R.2.** $y + 3 = 6$ **R.3.** $2x = 7y$ **R.4.** $x + y = 3$

For Exercises R.5–R.8, graph each equation.

R.5. $-5x - 15 = 0$

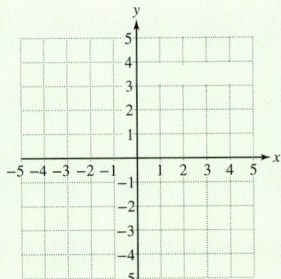

R.6. $2x - 3y = -3$

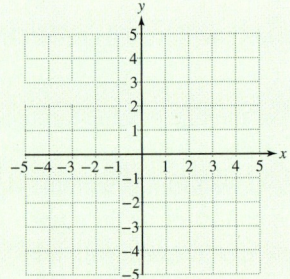

R.7. $y = -2x$

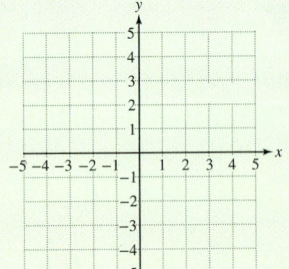

R.8. $3 - y = 9$

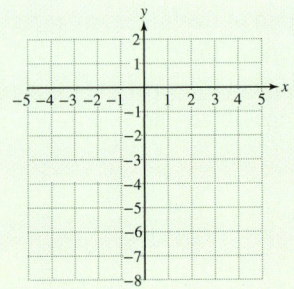

For Exercises R.9–R.12, find the slope of the line that passes through the given points.

R.9. $(1, -3)$ and $(2, 6)$

R.10. $(2, -4)$ and $(-2, 4)$

R.11. $(-2, 5)$ and $(5, 5)$

R.12. $(6.1, 2.5)$ and $(6.1, -1.5)$

For Exercises R.13–R.16, the slope of a line is given.
  a. Find the slope of a line parallel to the line with the given slope.
  b. Find the slope of a line perpendicular to the line with the given slope, if possible.

R.13. $m = -\dfrac{4}{3}$ **R.14.** $m = \dfrac{3}{8}$ **R.15.** $m = 0$ **R.16.** $m = -1$

Vocabulary and Key Concepts

1. **a.** The standard form of an equation of a line is _____, where A and B are not both zero and C is a constant.

   b. A line defined by an equation $y = k$, where k is a constant, is a (horizontal/vertical) line.

   c. A line defined by an equation $x = k$, where k is a constant, is a (horizontal/vertical) line.

   d. Given the slope-intercept form of an equation of a line, $y = mx + b$, the value of m is the _____ and b is the _____.

   e. Given a point (x_1, y_1) on a line with slope m, the point-slope formula is given by _____.

2. Suppose that a line passes through the point $(-1, 4)$ and has a slope of 2. What values of m, x_1, and y_1 would be used in the point-slope formula $y - y_1 = m(x - x_1)$ to find an equation of the line?

Concept 1: Writing an Equation of a Line Using the Point-Slope Formula

For Exercises 3–8, use the point-slope formula (if possible) to write an equation of the line given the following information. **(See Example 1.)**

3. The slope is 3, and the line passes through the point $(-2, 1)$.

4. The slope is -2, and the line passes through the point $(1, -5)$.

5. The slope is -4, and the line passes through the point $(-3, -2)$.

6. The slope is 5, and the line passes through the point $(-1, -3)$.

7. The slope is $-\frac{1}{2}$, and the line passes through $(-1, 0)$.

8. The slope is $-\frac{3}{4}$, and the line passes through $(2, 0)$.

Concept 2: Writing an Equation of a Line Given Two Points

For Exercises 9–14, use the point-slope formula to write an equation of the line given the following information. **(See Example 2.)**

9. The line passes through the points $(-2, -6)$ and $(1, 0)$.

10. The line passes through the points $(-2, 5)$ and $(0, 1)$.

11. The line passes through the points (0, −4) and (−1, −3).

12. The line passes through the points (1, −3) and (−7, 2).

13. The line passes through the points (2.2, −3.3) and (12.2, −5.3).

14. The line passes through the points (4.7, −2.2) and (−0.3, 6.8).

For Exercises 15–20, find an equation of the line through the given points. Write the final answer in slope-intercept form.

15.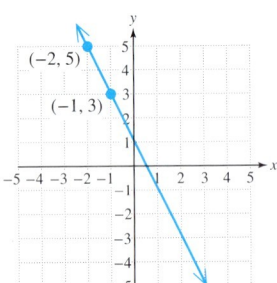

16.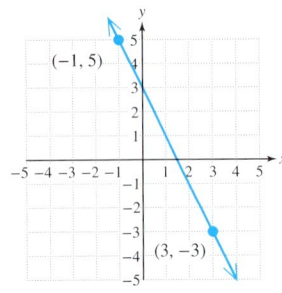

17.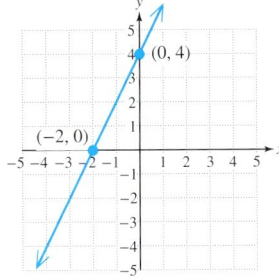

18.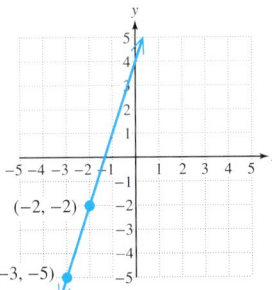

19.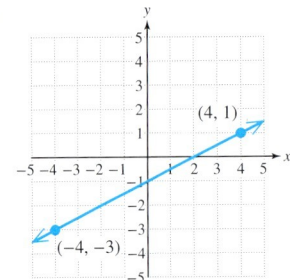

20.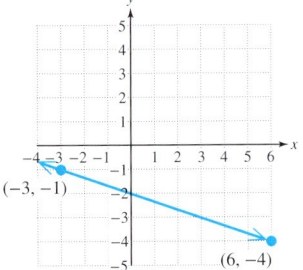

Concept 3: Writing an Equation of a Line Parallel or Perpendicular to Another Line

For Exercises 21–28, use the point-slope formula to write an equation of the line given the following information. (See Examples 3–4.)

21. The line passes through the point (−3, 1) and is parallel to the line $y = 4x + 3$.

22. The line passes through the point (4, −1) and is parallel to the line $y = 3x + 1$.

23. The line passes through the point (4, 0) and is parallel to the line $3x + 2y = 8$.

24. The line passes through the point (2, 0) and is parallel to the line $5x + 3y = 6$.

25. The line passes through the point (−5, 2) and is perpendicular to the line $y = \frac{1}{2}x + 3$.

26. The line passes through the point (−2, −2) and is perpendicular to the line $y = \frac{1}{3}x - 5$.

27. The line passes through the point (0, −6) and is perpendicular to the line $-5x + y = 4$.

28. The line passes through the point (0, −8) and is perpendicular to the line $2x - y = 5$.

Concept 4: Different Forms of Linear Equations: A Summary

For Exercises 29–34, match the form or formula on the left with its name on the right.

29. $x = k$ i. Standard form

30. $y = mx + b$ ii. Point-slope formula

31. $m = \dfrac{y_2 - y_1}{x_2 - x_1}$ iii. Horizontal line

32. $y - y_1 = m(x - x_1)$ iv. Vertical line

33. $y = k$ v. Slope-intercept form

34. $Ax + By = C$ vi. Slope formula

For Exercises 35–40, find an equation for the line given the following information. **(See Example 5.)**

35. The line passes through the point (3, 1) and is parallel to the line $y = -4$. See the figure.

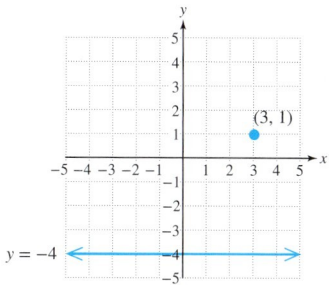

36. The line passes through the point (−1, 1) and is parallel to the line $y = 2$. See the figure.

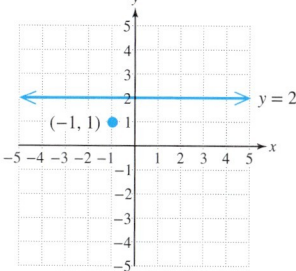

37. The line passes through the point (2, 6) and is perpendicular to the line $y = 1$. (*Hint:* Sketch the line first.)

38. The line passes through the point (0, 3) and is perpendicular to the line $y = -5$. (*Hint:* Sketch the line first.)

39. The line passes through the point (2, 2) and is perpendicular to the line $x = 0$.

40. The line passes through the point (5, −2) and is perpendicular to the line $x = 0$.

Mixed Exercises

For Exercises 41–52, write an equation of the line given the following information.

41. The slope is $\frac{1}{4}$, and the line passes through the point (−8, 6).

42. The slope is $\frac{2}{5}$, and the line passes through the point (−5, 4).

43. The line passes through the point (4, 4) and is parallel to the line $3x - y = 6$.

44. The line passes through the point (−1, −7) and is parallel to the line $5x + y = -5$.

45. The slope is 4.5, and the line passes through the point (5.2, −2.2).

46. The slope is −3.6, and the line passes through the point (10.0, 8.2).

47. The slope is undefined, and the line passes through the point (−6, −3).

48. The slope is undefined, and the line passes through the point (2, −1).

49. The slope is 0, and the line passes through the point (3, −2).

50. The slope is 0, and the line passes through the point (0, 5).

51. The line passes through the points (−4, 0) and (−4, 3).

52. The line passes through the points (1, 3) and (1, −4).

Expanding Your Skills

For Exercises 53–56, write an equation in slope-intercept form for the line shown.

53.
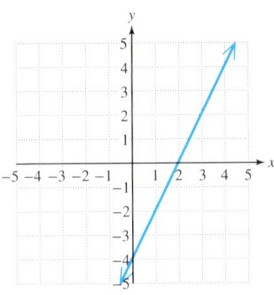

54.
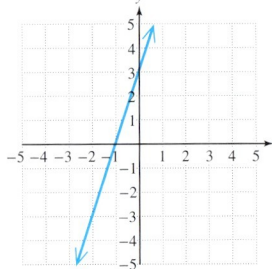

55.
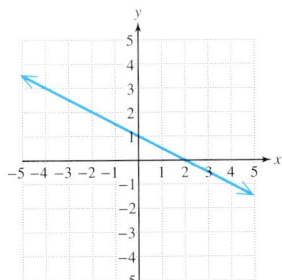

56.
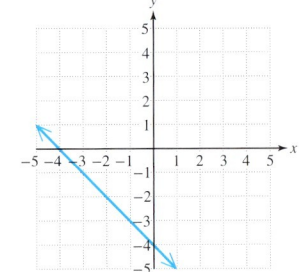

Section 11.6 Applications of Linear Equations and Modeling

Concepts

1. Interpreting a Linear Equation in Two Variables
2. Writing a Linear Model Using Observed Data Points
3. Writing a Linear Model Given a Fixed Value and a Rate of Change

1. Interpreting a Linear Equation in Two Variables

Linear equations can often be used to describe (or model) the relationship between two variables in a real-world event.

Example 1 Interpreting a Linear Equation in Context

A grocery store with long wait times at checkout wants to improve its efficiency. Management looks at data to determine times of peak customer activity. As a result of the data, the manager increases the number of checkout lanes and adjusts the schedules of the personnel within the store. Over a period of several months the average wait time for customers in checkout lanes decreases. The equation $y = -0.8x + 6.4$ approximates the wait time y (in minutes) for month x.

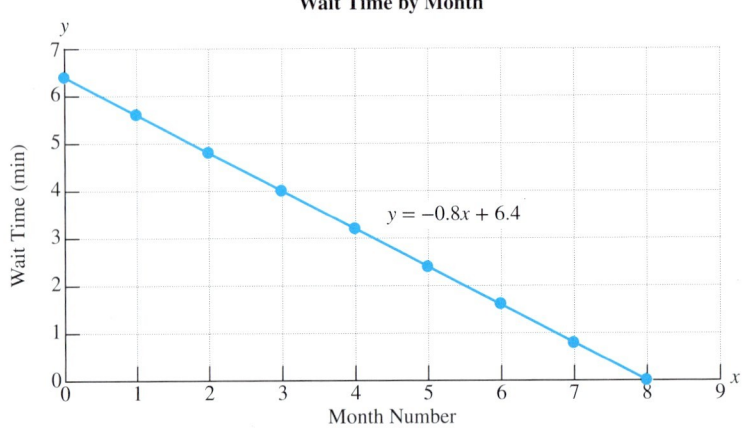

a. Use the equation to estimate the wait time in month 1.
b. Use the equation to estimate the wait time in month 4.
c. Determine the slope of the line and interpret its meaning in the context of this problem.
d. Determine the y-intercept and interpret its meaning.
e. Determine the x-intercept and interpret its meaning.

Solution:

a. Substitute $x = 1$ into the equation.

$y = -0.8(1) + 6.4$
$= 5.6$

The average wait time in month 1 is 5.6 min.

b. Substitute $x = 4$ into the equation.

$y = -0.8(4) + 6.4$
$= 3.2$

The average wait time in month 4 is 3.2 min.

c. The equation is written in slope-intercept form. The slope is -0.8.

The slope means that the average customer wait time decreased by 0.8 minutes per month.

d. The y-intercept is 6.4.

The average wait time in month 0 (before management implemented the changes) was 6.4 min.

e. To find the x-intercept, substitute $y = 0$.

$0 = -0.8x + 6.4$
$-6.4 = -0.8x$
$8 = x$

The x-intercept is $(8, 0)$. This means that 8 months after management implemented the new changes, the wait time decreased to 0 minutes.

Skill Practice

1. The cost y (in dollars) for a local move by a small moving company is given by $y = 60x + 100$, where x is the number of hours required for the move.
   a. How much would be charged for a move that requires 3 hr?
   b. How much would be charged for a move that requires 8 hr?
   c. What is the slope of the line and what does it mean in the context of this problem?
   d. Determine the y-intercept and interpret its meaning in the context of this problem.

Answers

1. a. $280   b. $580
   c. 60; This means that for each additional hour of service, the cost of the move goes up by $60.
   d. (0, 100); The $100 charge is a fixed fee in addition to the hourly rate.

2. Writing a Linear Model Using Observed Data Points

Example 2   Writing a Linear Model from Observed Data Points

The monthly sales of hybrid cars sold by a certain company are given for a recent year. The sales for the first 8 months of the year are shown in Figure 11-34. The value $x = 0$ represents January, $x = 1$ represents February, and so on.

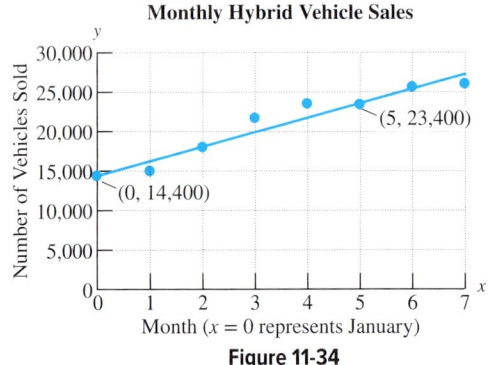

Figure 11-34

a. Use the data points from Figure 11-34 to find a linear equation that represents the monthly sales of hybrid cars. Let x represent the month number and let y represent the number of vehicles sold.

b. Use the linear equation in part (a) to estimate the number of hybrid vehicles sold in month 7 (August).

Solution:

a. The ordered pairs (0, 14,400) and (5, 23,400) are given in the graph. Use these points to find the slope.

$$\underset{(x_1, y_1)}{(0, 14{,}400)} \quad \text{and} \quad \underset{(x_2, y_2)}{(5, 23{,}400)} \quad\quad \text{Label the points.}$$

$$m = \frac{y_2 - y_1}{x_2 - x_1} = \frac{23{,}400 - 14{,}400}{5 - 0}$$

$$= \frac{9000}{5}$$

$$= 1800 \quad\quad \text{The slope is 1800. This indicates that sales increased by approximately 1800 per month during this time period.}$$

With $m = 1800$, and the y-intercept given as (0, 14,400), we have the following linear equation in slope-intercept form.

$$y = 1800x + 14{,}400$$

b. To approximate the sales in month number 7, substitute $x = 7$ into the equation from part (a).

$$y = 1800(7) + 14{,}400 \quad\quad \text{Substitute } x = 7.$$

$$= 27{,}000$$

The monthly sales for August (month 7) would be 27,000 vehicles.

Skill Practice

2. Soft drink sales at a concession stand at a softball stadium have increased linearly over the course of the summer softball season.

   a. Use the given data points to find a linear equation that relates the sales, y, to week number, x.

   b. Use the equation to predict the number of soft drinks sold in week 10.

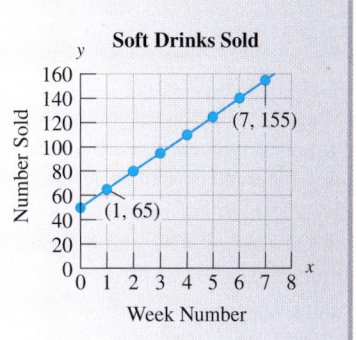

3. Writing a Linear Model Given a Fixed Value and a Rate of Change

Another way to look at the equation $y = mx + b$ is to identify the term mx as the variable term and the term b as the constant term. The value of the term mx will change with the value of x (this is why the slope, m, is called a *rate of change*). However, the term b will remain constant regardless of the value of x. With these ideas in mind, we can write a linear equation if the rate of change and the constant are known.

Example 3 Writing a Linear Model

A stack of posters to advertise a production by the theater department costs $19.95 plus $1.50 per poster at the printer.

a. Write a linear equation to compute the cost, c, of buying x posters.

b. Use the equation to compute the cost of 125 posters.

Solution:

a. The constant cost is $19.95. The variable cost is $1.50 per poster. If m is replaced with 1.50 and b is replaced with 19.95, the equation is

   $c = 1.50x + 19.95$ where c is the cost (in dollars) of buying x posters.

b. Because x represents the number of posters, substitute $x = 125$.

$$c = 1.50(125) + 19.95$$
$$= 187.5 + 19.95$$
$$= 207.45$$

The total cost of buying 125 posters is $207.45.

Skill Practice

3. The monthly cost for a "minimum use" cellular phone is $19.95 plus $0.10 per minute for all calls.

   a. Write a linear equation to compute the cost, c, of using t minutes.

   b. Use the equation to determine the cost of using 150 minutes.

Answers

2. a. $y = 15x + 50$ b. 200 soft drinks
3. a. $c = 0.10t + 19.95$ b. $34.95

Section 11.6 Activity

A.1. A fund-raising walk-a-thon raises money from cash donations as well as from pledges for the number of miles walked by each participant. The amount of money received, y, can be approximated by the equation $y = 25.5x + 350$, where x represents the number of miles walked by each participant. The graph shows the amount raised versus the number of miles walked.

a. Which variable is the dependent variable?
b. Which variable is the independent variable?
c. Use the equation to determine the amount of money raised if each participant walks 4 mi.
d. What is the slope of the line? Interpret the slope in the context of this problem.
e. What is the y-intercept and what does it represent in the context of this scenario?

A.2. The graph depicts the average selling price of a house (in $1000) sold in the United States over a 15-year period.

a. In year 0, the average price of a house was $150,000, and 15 years later the average price was $285,000. The trend upward appears to be linear. Use the ordered pairs (0, 150) and (15, 285) to compute the slope of the line.
b. What does the slope represent in the context of this problem?
c. Use the slope found in part (a) and the ordered pair (0, 150) to find an equation of a line that roughly approximates the linear trend. Write the final answer in slope-intercept form.
d. If this trend continues, use the equation found in part (c) to predict the price of a house in year 20.

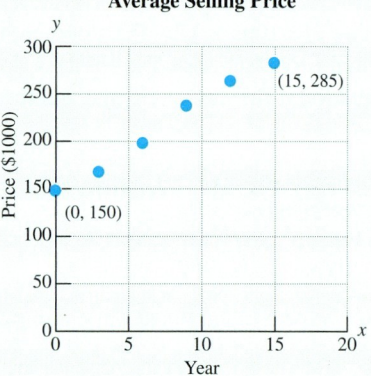

A.3. To rent a water softener, there is an installation fee of $60 plus a $25 monthly fee.
a. Write a linear equation to represent the total cost, y, of renting a water softener for x months.
b. Use the linear equation to compute the cost of renting the water softener for 6 months.
c. Use the linear equation to compute the cost of renting the water softener for 2 years.
d. What is the slope of the line, and what does it mean in the context of this problem?

Section 11.6 Practice Exercises

Prerequisite Review

For Exercises R.1–R.2, find the slope and y-intercept.

R.1. $3x - 7y = 14$ **R.2.** $-4x - y = 6$

For Exercises R.3–R.4, find the slope of the line containing the given points.

R.3. $(4, -7)$ and $(-2, -4)$ **R.4.** $(-1, -6)$ and $(5, 3)$

For Exercises R.5–R.8, find an equation of the line from the given information. Write the answer in slope-intercept form.

R.5. The line passes through the point $(-3, 4)$ with slope $-\dfrac{1}{3}$.

R.6. The line passes through the point $(5, -2)$ with slope $\dfrac{1}{5}$.

R.7. The line passes through the points $(-4, 1)$ and $(-2, -5)$.

R.8. The line passes through the points $(3, -7)$ and $(4, -3)$.

Concept 1: Interpreting a Linear Equation in Two Variables

1. The minimum hourly wage, y (in dollars per hour), in the United States can be approximated by the equation $y = 0.113x + 1.6$. In this equation, x represents the number of years since 1970 ($x = 0$ represents 1970, $x = 5$ represents 1975, and so on). **(See Example 1.)**

   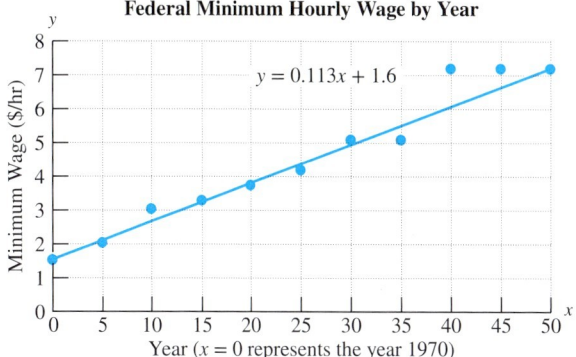

   a. Use the equation to approximate the minimum hourly wage in the year 1980.
   b. Use the equation to approximate the minimum hourly wage in 2015.
   c. Determine the y-intercept. Interpret the meaning of the y-intercept in the context of this problem.
   d. Determine the slope. Interpret the meaning of the slope in the context of this problem.

2. Veterinarians keep records of the weights of animals that are brought in for examination. Sine, the cat, weighed 42 oz when he was 70 days old. He weighed 46 oz when he was 84 days old. His brother, George, weighed 40 oz when he was 70 days old and 48 oz at 84 days old.
   a. Compute the slope of the line representing Sine's weight.

   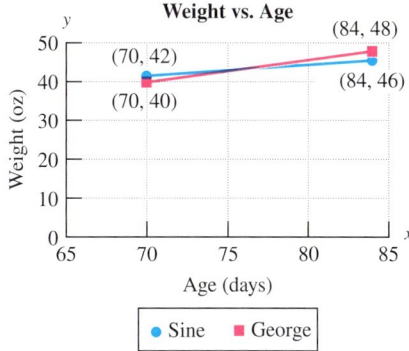
   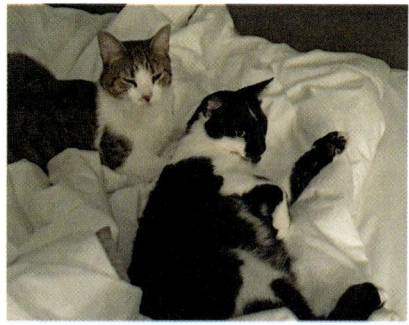

   Courtesy of Julie Miller

   b. Compute the slope of the line representing George's weight.
   c. Interpret the meaning of each slope in the context of this problem.
   d. Which cat gained weight more rapidly during this time period?

3. The average daily temperature in January for cities along the eastern seaboard of the United States and Canada generally decreases for cities farther north. A city's latitude in the northern hemisphere is a measure of how far north it is on the globe.

   The average temperature, y (measured in degrees Fahrenheit), can be described by the equation

   $y = -2.333x + 124.0$ where x is the latitude of the city.

   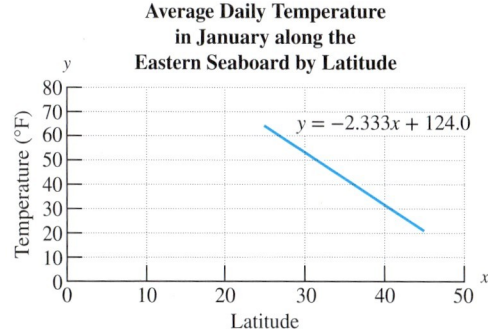

   (*Source:* U.S. National Oceanic and Atmospheric Administration)

   a. Use the equation to predict the average daily temperature in January for Philadelphia, Pennsylvania, whose latitude is 40.0°N. Round to one decimal place.

b. Use the equation to predict the average daily temperature in January for Edmundston, New Brunswick, Canada, whose latitude is 47.4°N. Round to one decimal place.

c. What is the slope of the line? Interpret the meaning of the slope in terms of latitude and temperature.

d. From the equation, determine the value of the x-intercept. Round to one decimal place. Interpret the meaning of the x-intercept in terms of latitude and temperature.

4. The graph depicts the rise in the number of jail inmates in the United States since 2000. Two linear equations are given: one to describe the number of female inmates and one to describe the number of male inmates by year.

Let y represent the number of inmates (in thousands). Let x represent the number of years since 2000.

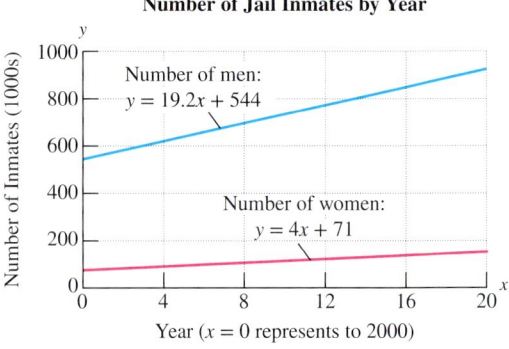

(*Source:* U.S. Bureau of Justice Statistics)

a. What is the slope of the line representing the number of female inmates? Interpret the meaning of the slope in the context of this problem.

b. What is the slope of the line representing the number of male inmates? Interpret the meaning of the slope in the context of this problem.

c. Which group, males or females, has the larger slope? What does this imply about the rise in the number of male and female prisoners?

d. Assuming this trend continues, use the equation to predict the number of female inmates in 2025.

5. The electric bill charge for a certain utility company is $0.095 per kilowatt-hour plus a fixed monthly tax of $11.95. The total cost, y, depends on the number of kilowatt-hours, x, according to the equation $y = 0.095x + 11.95$, $x \geq 0$.

a. Determine the cost of using 1000 kilowatt-hours.

b. Determine the cost of using 2000 kilowatt-hours.

c. Determine the y-intercept. Interpret the meaning of the y-intercept in the context of this problem.

d. Determine the slope. Interpret the meaning of the slope in the context of this problem.

6. For a recent year, children's admission to a State Fair was $8. Ride tickets were $0.75 each. The equation $y = 0.75x + 8$ represented the cost, y, in dollars to be admitted to the fair and to purchase x ride tickets.

a. Determine the slope of the line represented by $y = 0.75x + 8$. Interpret the meaning of the slope in the context of this problem.

b. Determine the y-intercept. Interpret its meaning in the context of this problem.

c. Use the equation to determine how much money a child needs for admission and to ride 10 rides.

Concept 2: Writing a Linear Model Using Observed Data Points

7. Meteorologists often measure the intensity of a tropical storm or hurricane by the maximum sustained wind speed and the minimum pressure. The relationship between these two quantities is approximately linear. Hurricane Katrina had a maximum sustained wind speed of 150 knots and a minimum pressure of 902 mb (millibars). Hurricane Ophelia had maximum sustained winds of 75 knots and a pressure of 976 mb. **(See Example 2.)**

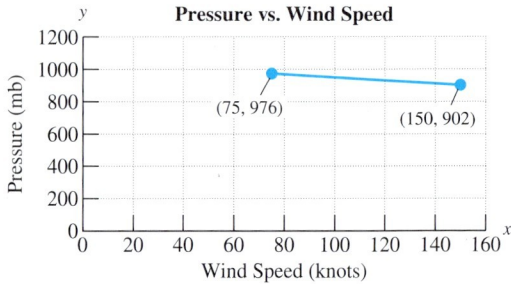

a. Find the slope of the line between these two points. Round to one decimal place.

b. Using the slope found in part (a) and the point (75, 976), find a linear model that represents the minimum pressure of a hurricane, y, versus its maximum sustained wind speed, x.

c. Hurricane Dennis had a maximum wind speed of 130 knots. Using the equation found in part (b), predict the minimum pressure.

8. The figure depicts a relationship between a person's height, y (in inches), and the length of the person's arm, x (measured in inches from shoulder to wrist).

   a. Use the points (17, 57.75) and (24, 82.25) to find a linear equation relating height to arm length.

   b. What is the slope of the line? Interpret the slope in the context of this problem.

   c. Use the equation from part (a) to estimate the height of a person whose arm length is 21.5 in.

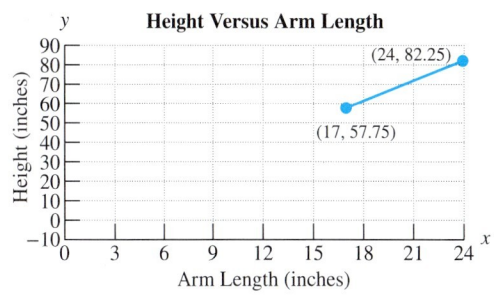

9. Wind energy is one type of renewable energy that does not produce dangerous greenhouse gases as a by-product. The graph shows the consumption of wind energy in the United States over a 6-year period. The variable y represents the amount of wind energy in trillions of Btu, and the variable x represents the year number.

   a. Use the points (0, 57) and (4, 143) to determine the slope of the line.

   b. Interpret the slope in the context of this problem.

   c. Use the points (0, 57) and (4, 143) to find a linear equation relating the consumption of wind energy, y, to the year number, x.

   d. If this linear trend continues beyond the observed data values, use the equation in part (c) to estimate the consumption of wind energy in year 10.

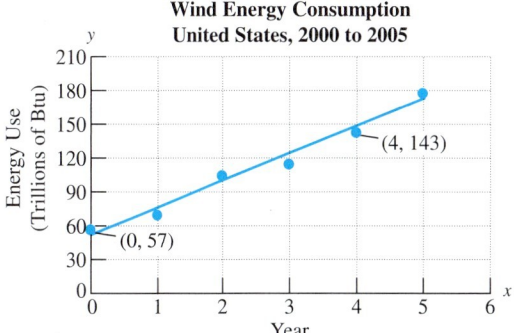

(*Source*: United States Department of Energy)

10. The graph shows the average height for boys based on age. Let x represent a boy's age, and let y represent his height (in inches).

    a. Find a linear equation that represents the height of a boy versus his age.

    b. Use the linear equation found in part (a) to predict the average height of a 5-year-old boy.

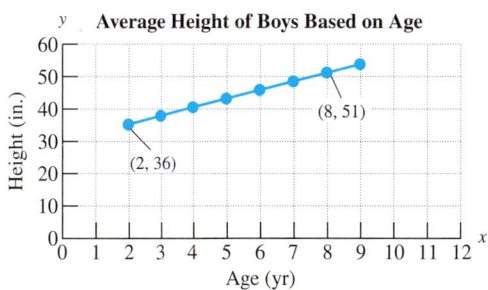

(*Source:* National Parenting Council)

Concept 3: Writing a Linear Model Given a Fixed Value and a Rate of Change

11. The owner of a restaurant franchise pays the parent company a monthly fee of $5000 plus 10% of sales. (See Example 3.)

    a. Write a linear model to compute the monthly payment, y, that the restaurant owner must pay the parent company for x dollars in sales.

    b. Use the equation to compute the amount that the owner has to pay if monthly sales are $11,300.

12. To lease a mid-size sedan, a car dealership charges $3000 up front plus $350 per month.

    a. Write an equation that represents the cost C (in dollars) of leasing a sedan for x months.

    b. Use the equation from part (a) to compute the cost of leasing the car for 4 years.

13. The cost to rent a 10 ft by 10 ft storage space is $90 per month plus a nonrefundable deposit of $105.

    a. Write a linear equation to compute the cost, y, of renting a 10 ft by 10 ft space for x months.

    b. What is the cost of renting such a storage space for 1 year (12 months)?

14. An air-conditioning and heating company has a fixed monthly cost of $5000. Furthermore, each service call costs the company $25.

    a. Write a linear equation to compute the total cost, y, for 1 month if x service calls are made.

    b. Use the equation to compute the cost for 1 month if 150 service calls are made.

Nick Gunderson/Getty Images

15. A bakery that specializes in bread rents a booth at a flea market. The daily cost to rent the booth is $100. Each loaf of bread costs the bakery $0.80 to produce.

    a. Write a linear equation to compute the total cost, y, for 1 day if x loaves of bread are produced.

    b. Use the equation to compute the cost for 1 day if 200 loaves of bread are produced.

16. A beverage company rents a booth at an art show to sell lemonade. The daily cost to rent a booth is $35. Each lemonade costs $0.50 to produce.

    a. Write a linear equation to compute the total cost, y, for 1 day if x lemonades are produced.

    b. Use the equation to compute the cost for 1 day if 350 lemonades are produced.

Technology Connections

For Exercises 17–20, use a graphing calculator to graph the lines on an appropriate viewing window. Evaluate the equation at the given values of x.

17. $y = -4.6x + 27.1$ at $x = 3$

18. $y = -3.6x - 42.3$ at $x = 0$

19. $y = 40x + 105$ at $x = 6$

20. $y = 20x - 65$ at $x = 8$

Chapter 11 Review Exercises

Section 11.1

1. Graph the points on a rectangular coordinate system.

   a. $\left(\dfrac{1}{2}, 5\right)$
   b. $(-1, 4)$
   c. $(2, -1)$
   d. $(0, 3)$
   e. $(0, 0)$
   f. $\left(-\dfrac{8}{5}, 0\right)$
   g. $(-2, -5)$
   h. $(3, 1)$

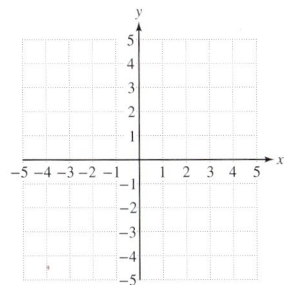

2. Estimate the coordinates of the points A, B, C, D, E, and F.

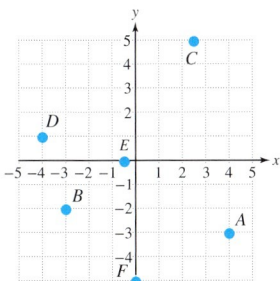

For Exercises 3–8, determine the quadrant in which the given point is located.

3. $(-2, -10)$
4. $(-4, 6)$
5. $(3, -5)$
6. $\left(\dfrac{1}{2}, \dfrac{7}{5}\right)$
7. $(\pi, -2.7)$
8. $(-1.2, -6.8)$

9. On which axis is the point $(2, 0)$ located?

10. On which axis is the point $(0, -3)$ located?

11. The price per share of a stock (in dollars) over a period of 5 days is shown in the graph.

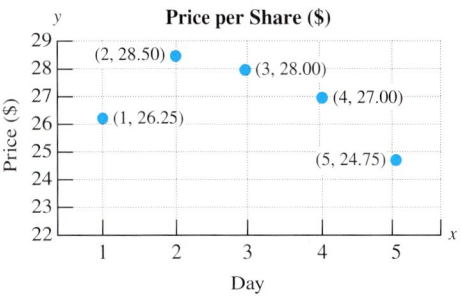

   a. Interpret the meaning of the ordered pair $(1, 26.25)$.
   b. On which day was the price the highest?
   c. What was the increase in price between day 1 and day 2?

12. The number of classes that a student missed, x, along with the student's class average, y, are shown in the given ordered pairs.

   $(3, 88)$ $(7, 67)$ $(1, 96)$ $(11, 62)$
   $(2, 90)$ $(14, 56)$ $(2, 97)$ $(5, 82)$

   a. Interpret the meaning of the ordered pairs $(7, 67)$ and $(1, 96)$.
   b. Plot the points on a rectangular coordinate system.

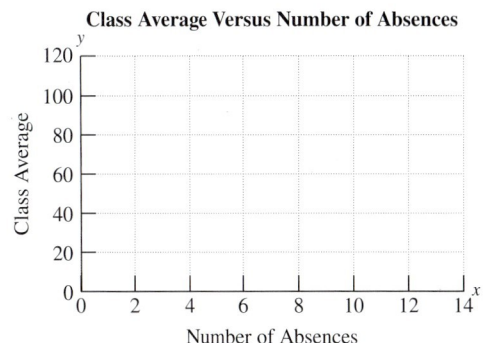

Section 11.2

For Exercises 13–16, determine if the given ordered pair is a solution to the equation.

13. $5x - 3y = 12$; (0, 4)

14. $2x - 4y = -6$; (3, 0)

15. $y = \frac{1}{3}x - 2$; (9, 1)

16. $y = -\frac{2}{5}x + 1$; (−10, 5)

For Exercises 17–20, complete the table and graph the corresponding ordered pairs. Graph the line through the points to represent all solutions to the equation.

17. $3x - y = 5$

x	y
2	
	4
1	

18. $\frac{1}{2}x + 3y = 6$

x	y
	2
−2	
	3

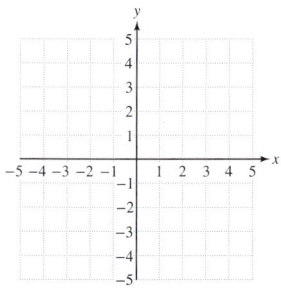

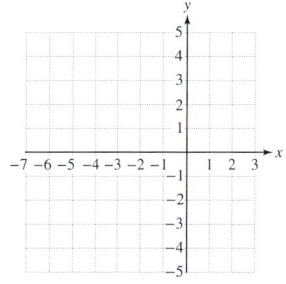

19. $y = \frac{2}{3}x - 1$

x	y
0	
3	
−6	

20. $y = -2x - 3$

x	y
0	
−3	
	1

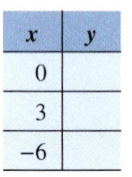

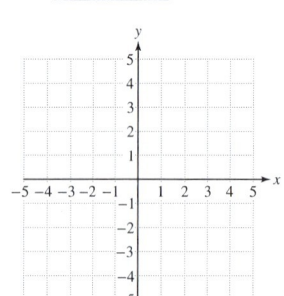

For Exercises 21–24, graph the equation.

21. $x + 2y = 4$

22. $x - y = 5$

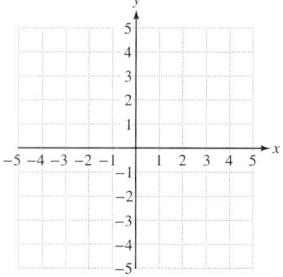

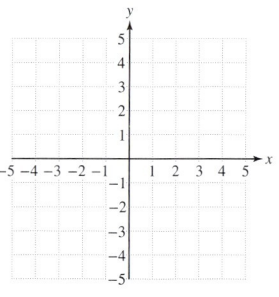

23. $y = 3x$

24. $y = \frac{1}{4}x$

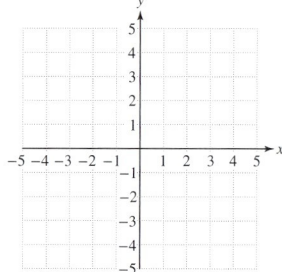

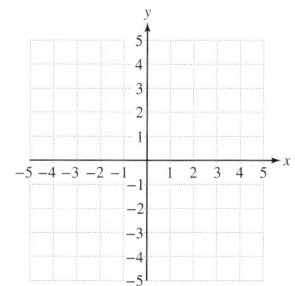

For Exercises 25–28, identify the line as horizontal or vertical. Then graph the equation.

25. $3x - 2 = 10$

26. $2x + 1 = -2$

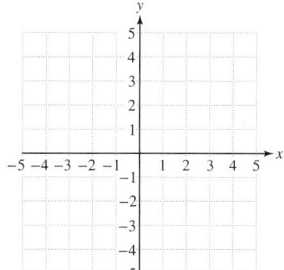

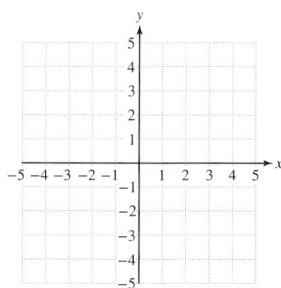

27. $6y + 1 = 13$

28. $5y - 1 = 14$

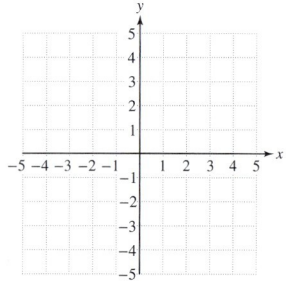

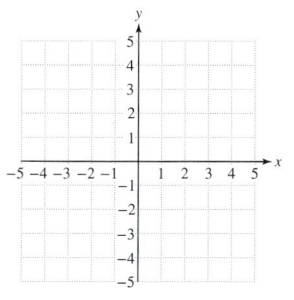

For Exercises 29–36, find the x- and y-intercepts if they exist.

29. $-4x + 8y = 12$
30. $2x + y = 6$
31. $y = 8x$
32. $5x - y = 0$
33. $6y = -24$
34. $2y - 3 = 1$
35. $2x + 5 = 0$
36. $-3x + 1 = 0$

Section 11.3

37. What is the slope of the ladder leaning up against the wall?

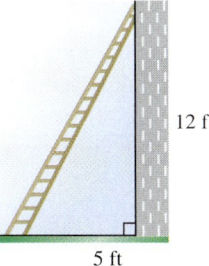

38. Point A is located 4 units down and 2 units to the right of point B. What is the slope of the line through points A and B?

39. Determine the slope of the line that passes through the points $(7, -9)$ and $(-5, -1)$.

40. Determine the slope of the line that has x- and y-intercepts of $(-1, 0)$ and $(0, 8)$.

41. Determine the slope of the line that passes through the points $(3, 0)$ and $(3, -7)$.

42. Determine the slope of the line given by $y = -1$.

43. A given line has a slope of -5.
    a. What is the slope of a line parallel to the given line?
    b. What is the slope of a line perpendicular to the given line?

44. A given line has a slope of 0.
    a. What is the slope of a line parallel to the given line?
    b. What is the slope of a line perpendicular to the given line?

For Exercises 45–48, find the slopes of the lines l_1 and l_2 from the two given points. Then determine whether l_1 and l_2 are parallel, perpendicular, or neither.

45. l_1: $(3, 7)$ and $(0, 5)$
    l_2: $(6, 3)$ and $(-3, -3)$

46. l_1: $(-2, 1)$ and $(-1, 9)$
    l_2: $(0, -6)$ and $(2, 10)$

47. l_1: $(0, \frac{5}{6})$ and $(2, 0)$
    l_2: $(0, \frac{6}{5})$ and $(-\frac{1}{2}, 0)$

48. l_1: $(1, 1)$ and $(1, -8)$
    l_2: $(4, -5)$ and $(7, -5)$

49. Carol's electric bill had an initial reading of 35,955 kilowatt-hours at the beginning of the month. At the end of the month the reading was 37,005 kilowatt-hours. Let x represent the day of the month and y represent the reading on the meter in kilowatt-hours.

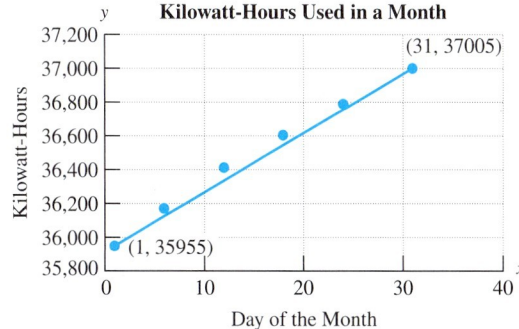

a. Using the ordered pairs $(1, 35955)$ and $(31, 37005)$, find the slope of the line.

b. Interpret the slope in the context of this problem.

50. A car company recorded new car sales over a four-year period. Let x represent the year number and y represent the number of new cars sold.

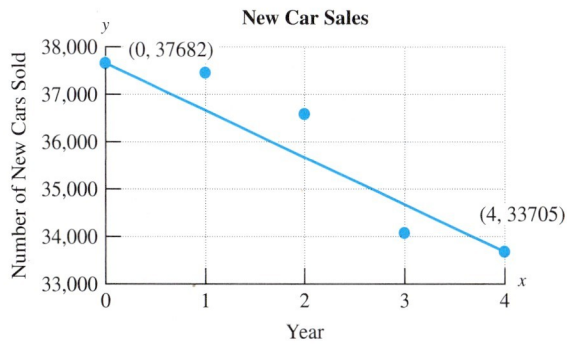

a. Using the ordered pairs $(0, 37682)$ and $(4, 33705)$, find the slope of the line. Round to the nearest whole unit.

b. Interpret the slope in the context of this problem.

Section 11.4

For Exercises 51–56, write each equation in slope-intercept form. Identify the slope and the y-intercept.

51. $5x - 2y = 10$
52. $3x + 4y = 12$
53. $x - 3y = 0$
54. $5y - 8 = 4$
55. $2y = -5$
56. $y - x = 0$

For Exercises 57–62, determine whether the equations represent parallel lines, perpendicular lines, or neither.

57. l_1: $y = \dfrac{3}{5}x + 3$
    l_2: $y = \dfrac{5}{3}x + 1$

58. l_1: $2x - 5y = 10$
    l_2: $5x + 2y = 20$

59. l_1: $3x + 2y = 6$
    l_2: $-6x - 4y = 4$

60. l_1: $y = \dfrac{1}{4}x - 3$
    l_2: $-x + 4y = 8$

61. l_1: $2x = 4$
    l_2: $y = 6$

62. l_1: $y = \dfrac{2}{9}x + 4$
    l_2: $y = \dfrac{9}{2}x - 3$

63. Write an equation of the line whose slope is $-\dfrac{4}{3}$ and whose y-intercept is $(0, -1)$.

64. Write an equation of the line that passes through the origin and has a slope of 5.

65. Write an equation of the line with slope $-\dfrac{4}{3}$ that passes through the point $(-6, 2)$.

66. Write an equation of the line with slope 5 that passes through the point $(-1, -8)$.

Section 11.5

67. Write a linear equation in two variables in slope-intercept form. (Answers may vary.)

68. Write a linear equation in two variables in standard form. (Answers may vary.)

69. Write the slope formula to find the slope of the line between the points (x_1, y_1) and (x_2, y_2).

70. Write the point-slope formula.

71. Write an equation of a vertical line (answers may vary).

72. Write an equation of a horizontal line (answers may vary).

For Exercises 73–78, write an equation of a line given the following information.

73. The slope is -6, and the line passes through the point $(-1, 8)$.

74. The slope is $\dfrac{2}{3}$, and the line passes through the point $(5, 5)$.

75. The line passes through the points $(0, -4)$ and $(8, -2)$.

76. The line passes through the points $(2, -5)$ and $(8, -5)$.

77. The line passes through the point $(5, 12)$ and is perpendicular to the line $y = -\dfrac{5}{6}x - 3$.

78. The line passes through the point $(-6, 7)$ and is parallel to the line $4x - y = 0$.

Section 11.6

79. The graph shows the average height for girls based on age (*Source:* National Parenting Council). Let x represent a girl's age, and let y represent her height (in inches).

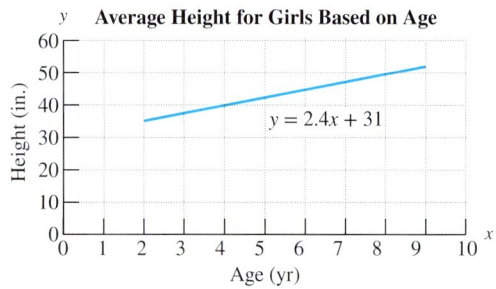

a. Use the equation to estimate the average height of a 7-year-old girl.

b. What is the slope of the line? Interpret the meaning of the slope in the context of the problem.

80. The number of drug prescriptions in the United States increased over a 16-year period. Let x represent the year number and let y represent the number of prescriptions (in millions).

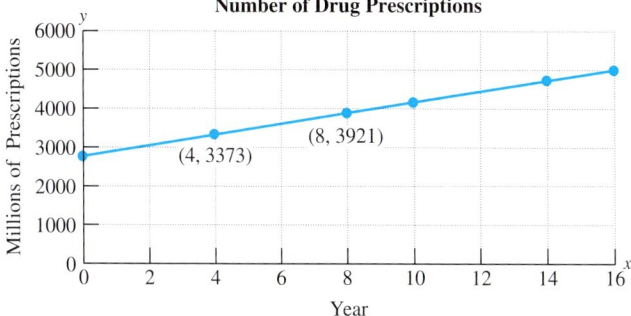

a. Using the ordered pairs (4, 3373) and (8, 3921) find the slope of the line.

b. Interpret the meaning of the slope in the context of this problem.

c. Find a linear equation that represents the number of prescriptions, y, versus the year number, x.

d. Use the equation from part (c) to estimate the number of prescriptions for year 15.

81. A water purification company charges $20 per month and a $55 installation fee.

a. Write a linear equation to compute the total cost, y, of renting this system for x months.

b. Use the equation from part (a) to determine the total cost to rent the system for 9 months.

82. A small cleaning company has a fixed monthly cost of $700 and a variable cost of $8 per service call.

a. Write a linear equation to compute the total cost, y, of making x service calls in one month.

b. Use the equation from part (a) to determine the total cost of making 80 service calls.

Systems of Linear Equations in Two Variables

12

CHAPTER OUTLINE

- **12.1** Solving Systems of Equations by the Graphing Method 766
- **12.2** Solving Systems of Equations by the Substitution Method 777
- **12.3** Solving Systems of Equations by the Addition Method 788
  - Problem Recognition Exercises: Systems of Equations 799
- **12.4** Applications of Linear Equations in Two Variables 802
- **12.5** Systems of Linear Equations in Three Variables 813
- **12.6** Applications of Systems of Linear Equations in Three Variables 823
  - Chapter 12 Review Exercises 828

Mathematics in Business

Suppose that you have been invited to an end-of-semester party. You ask your friend for directions and she says, "It's on Earl Street and 10th Avenue." If we think of Earl Street and 10th Avenue as lines, then we know that the house is located where these lines intersect. In mathematics, we call these intersections **solutions** to **systems of linear equations**. Furthermore, the applications of systems of linear equations are numerous.

Imagine that the total price of buying a shirt and a tie is normally $42. If the items are on sale and priced at 60% and 90% of the original values, respectively, then the total cost is $30. To determine the original cost s of a single shirt and the original cost t of a single tie, we can set up a system of two equations.

Lissa Harrison

Original Price: $s + t = 42$
Discounted Price: $0.60s + 0.90t = 30$

With techniques you will learn in this chapter, you can determine that the solution to this system is (26, 16). This means that the ordered pair (26, 16) satisfies both equations and that the point (26, 16) is a point of intersection of the lines defined by the equations in the system. The solution also tells us that the original cost of a shirt is $26 and the original cost of a tie is $16.

In this chapter, you will learn that some systems of linear equations have no solution, indicating that the related lines never intersect. This is similar to parallel streets that never meet. Other systems of linear equations may have infinitely many solutions. This occurs if the equations represent the same line.

Section 12.1 Solving Systems of Equations by the Graphing Method

Concepts

1. Solutions to a System of Linear Equations
2. Solving Systems of Linear Equations by Graphing

1. Solutions to a System of Linear Equations

Recall that a linear equation in two variables has an infinite number of solutions. The set of all solutions to a linear equation forms a line in a rectangular coordinate system. Two or more linear equations form a **system of linear equations**. For example, here are three systems of equations:

$$x - 3y = -5 \qquad y = \tfrac{1}{4}x - \tfrac{3}{4} \qquad 5a + b = 4$$
$$2x + 4y = 10 \qquad -2x + 8y = -6 \qquad -10a - 2b = 8$$

A **solution to a system of linear equations** is an ordered pair that is a solution to *both* individual linear equations.

Example 1   Determining Solutions to a System of Linear Equations

Determine whether the ordered pairs are solutions to the system.

$$x + y = 4$$
$$-2x + y = -5$$

a. (3, 1)   **b.** (0, 4)

Solution:

a. Substitute the ordered pair (3, 1) into both equations:

$$x + y = 4 \longrightarrow (3) + (1) \stackrel{?}{=} 4 \checkmark \quad \text{True}$$
$$-2x + y = -5 \longrightarrow -2(3) + (1) \stackrel{?}{=} -5 \checkmark \quad \text{True}$$

Because the ordered pair (3, 1) is a solution to both equations, it is a solution to the *system* of equations.

b. Substitute the ordered pair (0, 4) into both equations.

$$x + y = 4 \longrightarrow (0) + (4) \stackrel{?}{=} 4 \checkmark \quad \text{True}$$
$$-2x + y = -5 \longrightarrow -2(0) + (4) \stackrel{?}{=} -5 \quad \text{False}$$

Because the ordered pair (0, 4) is not a solution to the second equation, it is *not* a solution to the system of equations.

Avoiding Mistakes

It is important to test an ordered pair in *both* equations to determine if the ordered pair is a solution.

Skill Practice Determine whether the ordered pair is a solution to the system.

$$5x - 2y = 24$$
$$2x + y = 6$$

1. (6, 3)   **2.** (4, −2)

A solution to a system of two linear equations can be interpreted graphically as a point of intersection of the two lines. Using slope-intercept form to graph the lines from Example 1, we have

$$l_1: \quad x + y = 4 \longrightarrow y = -x + 4$$
$$l_2: \quad -2x + y = -5 \longrightarrow y = 2x - 5$$

Answers
1. No
2. Yes

All points on l_1 are solutions to the equation $y = -x + 4$.
All points on l_2 are solutions to the equation $y = 2x - 5$.
The point of intersection (3, 1) is the only point that is a solution to both equations. (See Figure 12-1.)

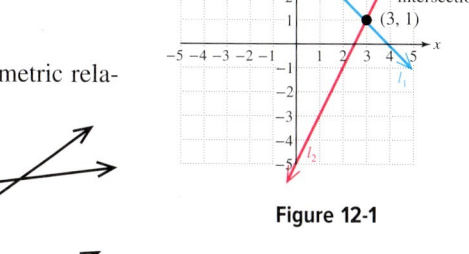

Figure 12-1

When two lines are drawn in a rectangular coordinate system, three geometric relationships are possible:

1. Two lines may intersect at *exactly one point*.

2. Two lines may intersect at *no point*. This occurs if the lines are parallel.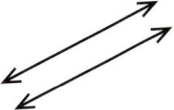

3. Two lines may intersect at *infinitely many points* along the line. This occurs if the equations represent the same line (the lines coincide).

If a system of linear equations has one or more solutions, the system is said to be **consistent**. If a system of linear equations has no solution, it is said to be **inconsistent**.

If two equations represent the same line, the equations are said to be **dependent equations**. In this case, all points on the line are solutions to the system. If two equations represent two different lines, the equations are said to be **independent equations**. In this case, the lines either intersect at one point or are parallel.

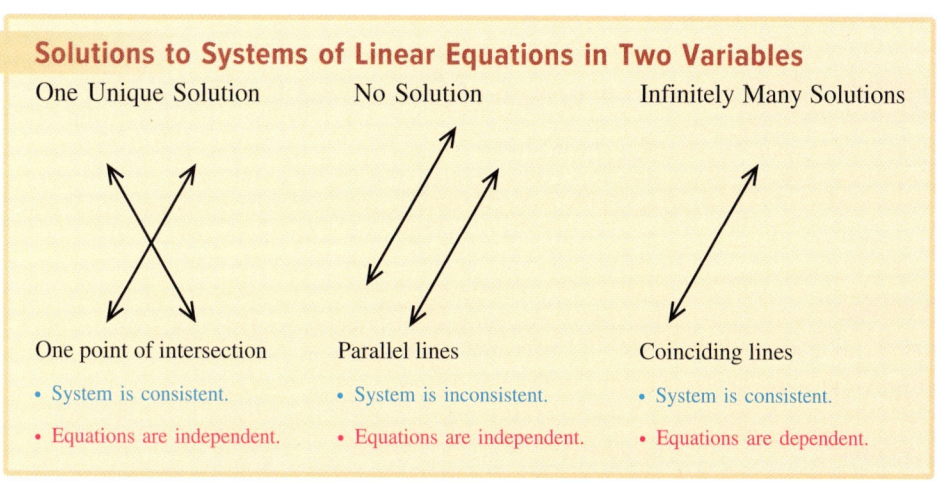

2. Solving Systems of Linear Equations by Graphing

One way to find a solution to a system of equations is to graph the equations and find the point (or points) of intersection. This is called the *graphing method* to solve a system of equations.

Example 2 Solving a System of Linear Equations by Graphing

Solve the system by the graphing method.
$y = 2x$
$y = 2$

Solution:

The equation $y = 2x$ is written in slope-intercept form as $y = 2x + 0$. The line passes through the origin, with a slope of 2.

The line $y = 2$ is a horizontal line and has a slope of 0.

Because the lines have different slopes, the lines must be different and nonparallel. From this, we know that the lines must intersect at exactly one point. Graph the lines to find the point of intersection (Figure 12-2).

The point (1, 2) appears to be the point of intersection. This can be confirmed by substituting $x = 1$ and $y = 2$ into both original equations.

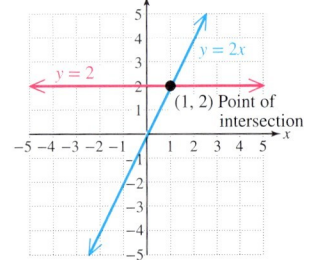

Figure 12-2

$y = 2x \quad (2) \stackrel{?}{=} 2(1)$ ✓ True
$y = 2 \quad (2) \stackrel{?}{=} 2$ ✓ True

The solution set is $\{(1, 2)\}$.

Skill Practice Solve the system by the graphing method.

3. $y = -3x$
   $x = -1$

Example 3 Solving a System of Linear Equations by Graphing

Solve the system by the graphing method.

$$x - 2y = -2$$
$$-3x + 2y = 6$$

Solution:

One method to graph the lines is to write each equation in slope-intercept form, $y = mx + b$.

Equation 1
$x - 2y = -2$
$-2y = -x - 2$
$\dfrac{-2y}{-2} = \dfrac{-x}{-2} - \dfrac{2}{-2}$
$y = \dfrac{1}{2}x + 1$

Equation 2
$-3x + 2y = 6$
$2y = 3x + 6$
$\dfrac{2y}{2} = \dfrac{3x}{2} + \dfrac{6}{2}$
$y = \dfrac{3}{2}x + 3$

Figure 12-3

From their slope-intercept forms, we see that the lines have different slopes, indicating that the lines are different and nonparallel. Therefore, the lines must intersect at exactly one point. Graph the lines to find that point (Figure 12-3).

Answer
3. $\{(-1, 3)\}$

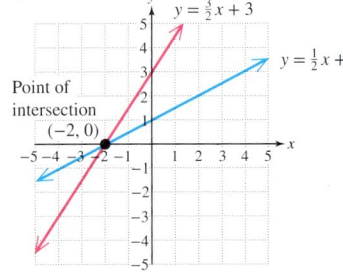

The point $(-2, 0)$ appears to be the point of intersection. This can be confirmed by substituting $x = -2$ and $y = 0$ into both equations.

$$x - 2y = -2 \longrightarrow (-2) - 2(0) \stackrel{?}{=} -2 \checkmark \quad \text{True}$$
$$-3x + 2y = 6 \longrightarrow -3(-2) + 2(0) \stackrel{?}{=} 6 \checkmark \quad \text{True}$$

The solution set is $\{(-2, 0)\}$.

Skill Practice Solve the system by the graphing method.

4. $y = 2x - 3$
   $6x + 2y = 4$

> **TIP:** In Examples 2 and 3, the lines could also have been graphed by using the x- and y-intercepts or by using a table of points. However, the advantage of writing the equations in slope-intercept form is that we can compare the slopes and y-intercepts of the two lines.
>
> 1. If the slopes differ, the lines are different and nonparallel and must intersect at exactly one point.
> 2. If the slopes are the same and the y-intercepts are different, the lines are parallel and will not intersect.
> 3. If the slopes are the same and the y-intercepts are the same, the two equations represent the same line.

Example 4 Graphing an Inconsistent System

Solve the system by graphing.

$$-x + 3y = -6$$
$$6y = 2x + 6$$

Solution:

To graph the lines, write each equation in slope-intercept form.

Equation 1
$-x + 3y = -6$
$3y = x - 6$
$\dfrac{3y}{3} = \dfrac{x}{3} - \dfrac{6}{3}$
$y = \dfrac{1}{3}x - 2$

Equation 2
$6y = 2x + 6$
$\dfrac{6y}{6} = \dfrac{2x}{6} + \dfrac{6}{6}$
$y = \dfrac{1}{3}x + 1$

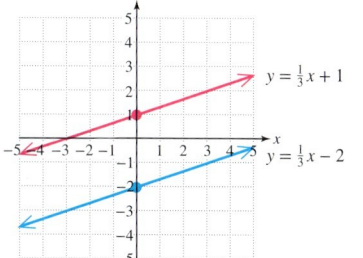

Figure 12-4

Because the lines have the same slope but different y-intercepts, they are parallel (Figure 12-4). Two parallel lines do not intersect, which implies that the system has no solution. Therefore, the solution set is the empty set, { }. The system is inconsistent.

Skill Practice Solve the system by graphing.

5. $4x + y = 8$
   $y = -4x + 3$

Answers

4. $\{(1, -1)\}$

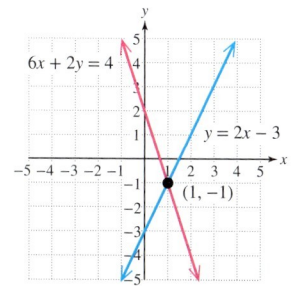

5.

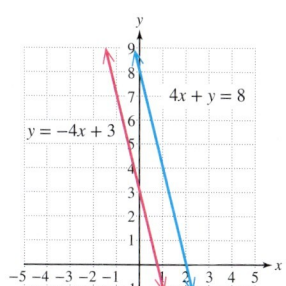

{ } The lines are parallel. The system is inconsistent.

Example 5 — Graphing a System of Dependent Equations

Solve the system by graphing.
$$x + 4y = 8$$
$$y = -\tfrac{1}{4}x + 2$$

Solution:

Write the first equation in slope-intercept form. The second equation is already in slope-intercept form.

Equation 1

$x + 4y = 8$

$4y = -x + 8$

$\dfrac{4y}{4} = \dfrac{-x}{4} + \dfrac{8}{4}$

$y = -\tfrac{1}{4}x + 2$

Equation 2

$y = -\tfrac{1}{4}x + 2$

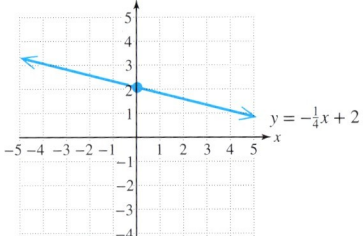

Figure 12-5

> **TIP:** The solution set to a system of dependent equations uses set-builder notation to describe the common line of intersection. Any form of the equation can be used. For example, in Example 5, we show the equation written in slope-intercept form and in standard form.
>
> $\{(x, y)\,|\,y = -\tfrac{1}{4}x + 2\}$
>
> or
>
> $\{(x, y)\,|\,x + 4y = 8\}$

Notice that the slope-intercept forms of the two lines are identical. Therefore, the equations represent the same line (Figure 12-5). The equations are dependent, and the solution to the system of equations is the set of all points on the line.

Because there are infinitely many points on the line, the ordered pairs in the solution set cannot all be listed. Therefore, we can write the solution in set-builder notation: $\{(x, y)\,|\,y = -\tfrac{1}{4}x + 2\}$. This can be read as "the set of all ordered pairs (x, y) such that the ordered pairs satisfy the equation $y = -\tfrac{1}{4}x + 2$."

In summary:

- There are infinitely many solutions to the system of equations.
- The solution set is $\{(x, y)\,|\,y = -\tfrac{1}{4}x + 2\}$, or equivalently $\{(x, y)\,|\,x + 4y = 8\}$.
- The equations are dependent.

Skill Practice Solve the system by graphing.

6. $x - 3y = 6$
$y = \tfrac{1}{3}x - 2$

Answer

6.

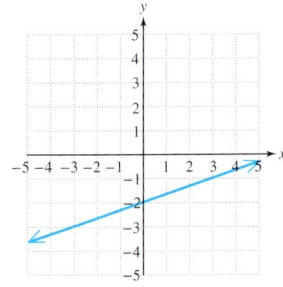

$\{(x, y)\,|\,y = \tfrac{1}{3}x - 2\}$

The equations are dependent.

Section 12.1 Activity

A.1. Consider the system.
$$x + y = 2$$
$$x - y = 4$$

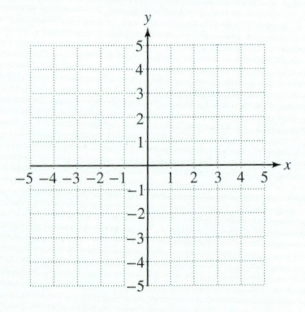

a. Write each equation in slope-intercept form.
b. Identify the slope of each line.
c. Based on the slope-intercept form of each line, do the lines intersect in a single point, are the lines parallel, or do the equations represent the same line (coinciding lines)?
d. Graph each line on the same coordinate system.
e. If the lines intersect, estimate the point of intersection.
f. The point of intersection is the solution to the system of equations. To verify that your answer is correct, check that the ordered pair satisfies both equations.

A.2. Consider the system. $\quad 2x - y = 1$
$\quad\quad\quad\quad\quad\quad\quad\quad\quad\;\; 2x - y = -2$

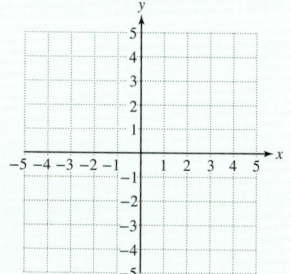

a. Write each equation in slope-intercept form.
b. Identify the slope of each line.
c. Based on the slope-intercept form of each line, do the lines intersect in a single point, are the lines parallel, or do the equations represent the same line (coinciding lines)?
d. Graph each line on the same coordinate system.
e. Is there a solution to this system? Write the solution set.
f. Is the system consistent or inconsistent?

A.3. Consider the system. $\quad x - 3y = -6$
$\quad\quad\quad\quad\quad\quad\quad\quad\quad\;\; -\dfrac{1}{3}x + y = 2$

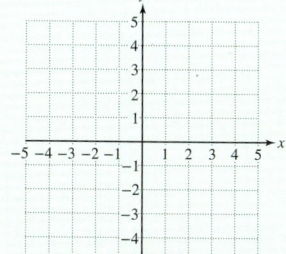

a. Write each equation in slope-intercept form.
b. Identify the slope of each line.
c. Based on the slope-intercept form of each line, do the lines intersect in a single point, are the lines parallel, or do the equations represent the same line (coinciding lines)?
d. Graph each line on the same coordinate system.
e. How many solutions are there to this system?
f. Write the solution set.
g. Are the equations in this system dependent or independent?

Practice Exercises

Section 12.1

Study Skills Exercise

Note-taking is a skill that is more challenging than it may initially appear. There is a delicate balance between writing notes and listening to your instructor. But with practice, taking notes in class will help you stay engaged and will enable you to ask and answer meaningful questions.

The following questions are key concepts from this section. Using only your notes, complete the following sentences.

- An ordered pair that satisfies both equations in a system is a _____.
- Suppose that the graphs of the equations in a system represent two parallel lines. This means that the slopes are _____ and the _____ are different.
- Suppose that a system of equations when graphed results in lines that coincide. This indicates that the system has _____ solution(s).
- A system of equations is consistent if _____.

Prerequisite Review

For Exercises R.1–R.4, identify the slope and x- and y-intercepts, if possible.

R.1. $3x - 5y = 15$

R.2. $-x - 2y = 4$

R.3. $3y - 1 = 8$

R.4. $\frac{1}{2}x = -1$

For Exercises R.5–R.8, graph the equations.

R.5. $3x - 5y = 15$

R.6. $-x - 2y = 4$

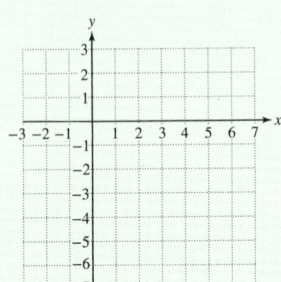 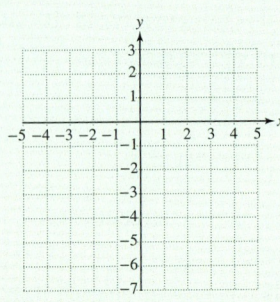

R.7. $3y - 1 = 8$

R.8. $\frac{1}{2}x = -1$

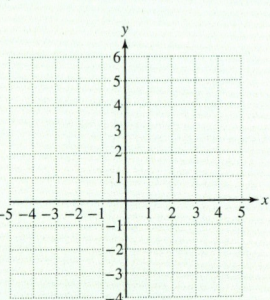 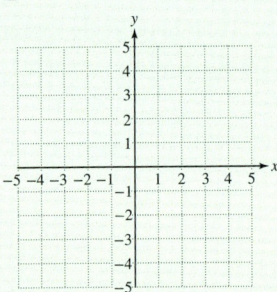

Vocabulary and Key Concepts

1. **a.** A _____ of linear equations consists of two or more linear equations.

   b. A _____ to a system of linear equations must be a solution to both individual equations in the system.

   c. Graphically, a solution to a system of linear equations in two variables is a point where the lines _____.

   d. A system of equations that has one or more solutions is said to be _____.

   e. The solution set to an inconsistent system of equations is _____.

   f. Two equations in a system of linear equations in two variables are said to be _____ if they represent the same line.

   g. Two equations in a system of linear equations in two variables are said to be _____ if they represent different lines.

Concept 1: Solutions to a System of Linear Equations

For Exercises 2–10, determine if the given point is a solution to the system. (See Example 1.)

2. $6x - y = -9$ $(-1, 3)$
    $x + 2y = 5$

3. $3x - y = 7$ $(2, -1)$
    $x - 2y = 4$

4. $x - y = 3$ $(4, 1)$
    $x + y = 5$

5. $4y = -3x + 12$ $(0, 4)$
  $y = \dfrac{2}{3}x - 4$

6. $y = -\dfrac{1}{3}x + 2$ $(9, -1)$
  $x = 2y + 6$

7. $3x - 6y = 9$ $\left(4, \dfrac{1}{2}\right)$
  $x - 2y = 3$

8. $x - y = 4$ $(6, 2)$
  $3x - 3y = 12$

9. $\dfrac{1}{3}x = \dfrac{2}{5}y - \dfrac{4}{5}$ $(0, 2)$
  $\dfrac{3}{4}x + \dfrac{1}{2}y = 2$

10. $\dfrac{1}{4}x + \dfrac{1}{2}y = \dfrac{3}{2}$ $(4, 1)$
  $y = \dfrac{3}{2}x - 6$

For Exercises 11–14, match the graph of the system of equations with the appropriate description of the solution.

11.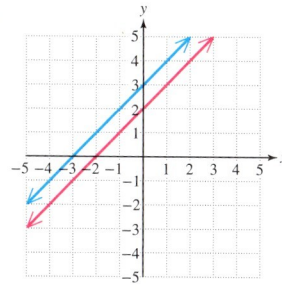

a. The solution set is $\{(1, 3)\}$.

b. $\{\ \}$

c. There are infinitely many solutions.

d. The solution set is $\{(0, 0)\}$.

12.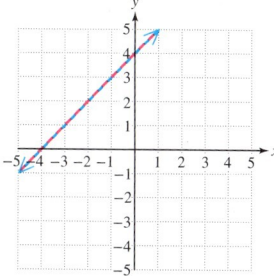

13.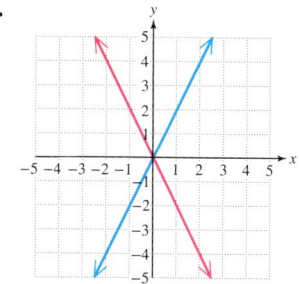

14.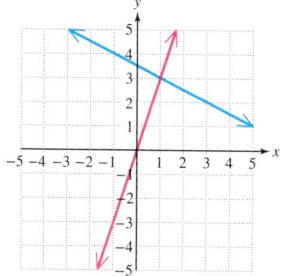

15. Graph each system of equations.

a. $y = 2x - 3$
  $y = 2x + 5$

b. $y = 2x + 1$
  $y = 4x - 1$

c. $y = 3x - 5$
  $y = 3x - 5$

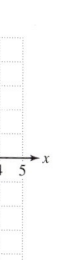
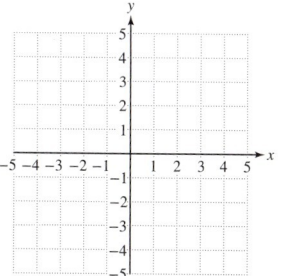
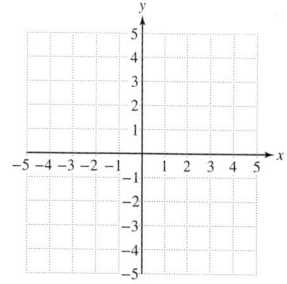

For Exercises 16–26, determine which system of equations (a, b, or c) makes the statement true. (*Hint:* Refer to the graphs from Exercise 15.)

a. $y = 2x - 3$
   $y = 2x + 5$

b. $y = 2x + 1$
   $y = 4x - 1$

c. $y = 3x - 5$
   $y = 3x - 5$

16. The lines are parallel.

17. The lines coincide.

18. The lines intersect at exactly one point.

19. The system is inconsistent.

20. The equations are dependent.

21. The lines have the same slope but different y-intercepts.

22. The lines have the same slope and same y-intercept.

23. The lines have different slopes.

24. The system has exactly one solution.

25. The system has infinitely many solutions.

26. The system has no solution.

Concept 2: Solving Systems of Linear Equations by Graphing

For Exercises 27–50, solve the system by graphing. For systems that do not have one unique solution, also state the number of solutions and whether the system is inconsistent or the equations are dependent. **(See Examples 2–5.)**

27. $y = -x + 4$
    $y = x - 2$

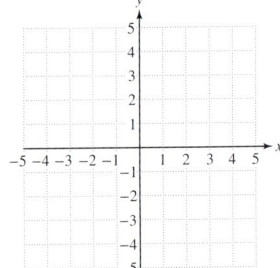

28. $y = 3x + 2$
    $y = 2x$

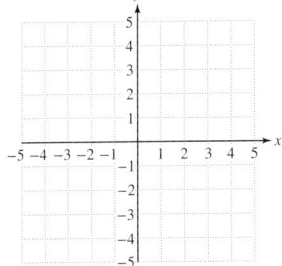

29. $2x + y = 0$
    $3x + y = 1$

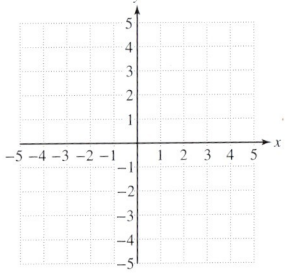

30. $x + y = -1$
    $2x - y = -5$

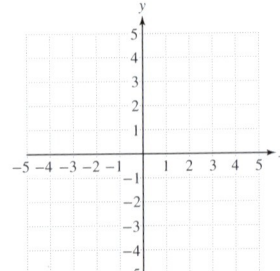

31. $2x + y = 6$
    $x = 1$

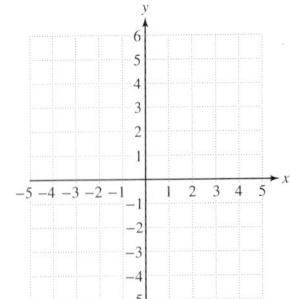

32. $4x + 3y = 9$
    $x = 3$

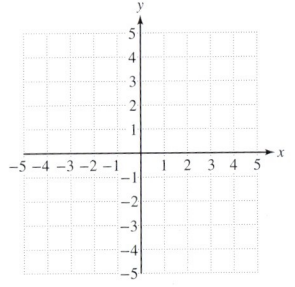

33. $-6x - 3y = 0$
$4x + 2y = 4$

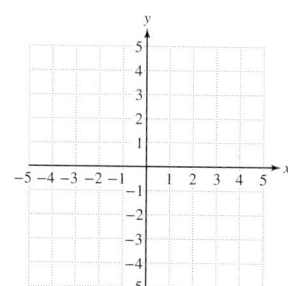

34. $2x - 6y = 12$
$-3x + 9y = 12$

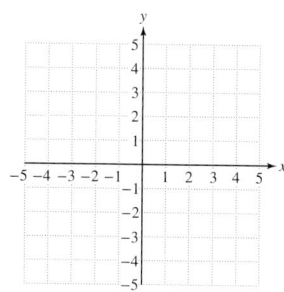

35. $-2x + y = 3$
$6x - 3y = -9$

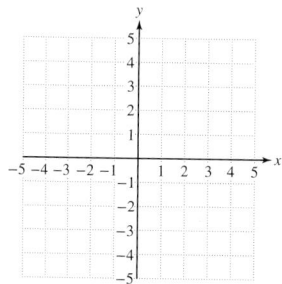

36. $x + 3y = 0$
$-2x - 6y = 0$

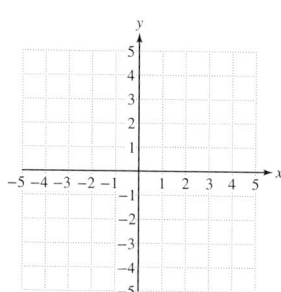

37. $y = 6$
$2x + 3y = 12$

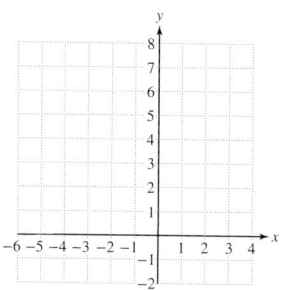

38. $y = -2$
$x - 2y = 10$

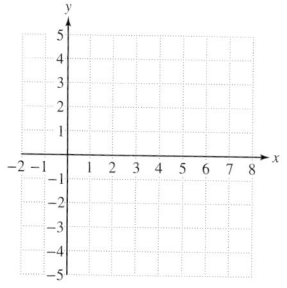

39. $x = 4 + y$
$3y = -3x$

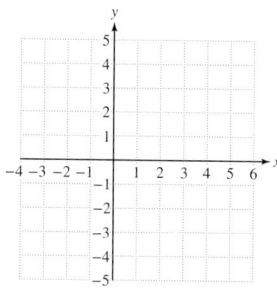

40. $3y = 4x$
$x - y = -1$

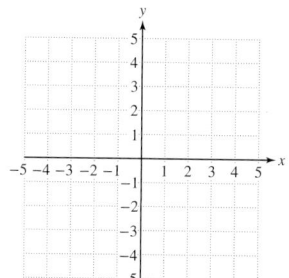

41. $-x + y = 3$
$4y = 4x + 6$

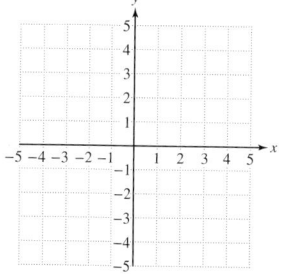

42. $x - y = 4$
$3y = 3x + 6$

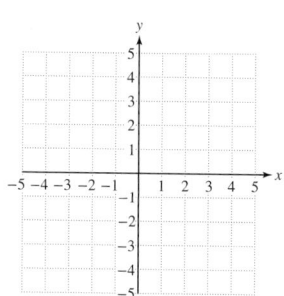

43. $x = 4$
$2y = 4$

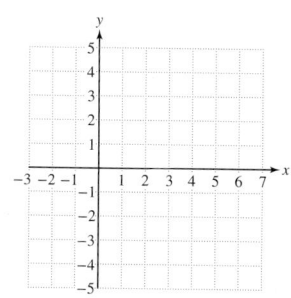

44. $-3x = 6$
$y = 2$

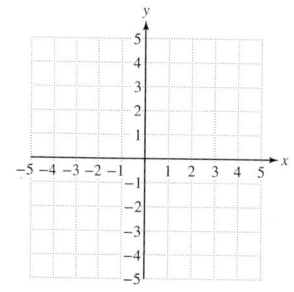

45. 2x + y = 4
4x − 2y = 0

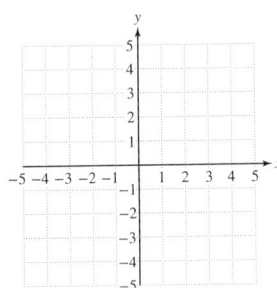

46. 3x + 3y = 3
2x − y = 5

47. y = 0.5x + 2
−x + 2y = 4

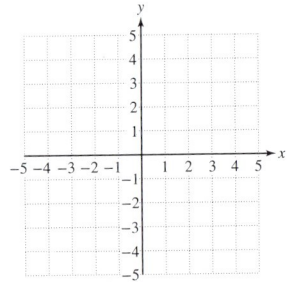

48. 3x − 4y = 6
−6x + 8y = −12

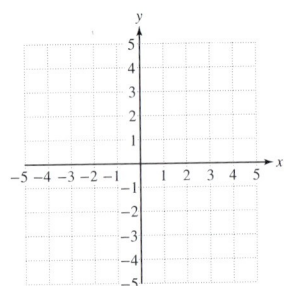

49. x − 3y = 0
y = −x − 4

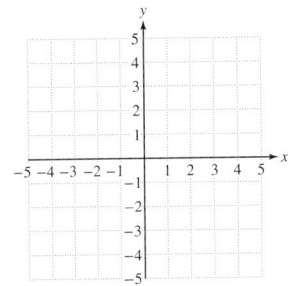

50. −6x + 3y = −6
4x + y = −2

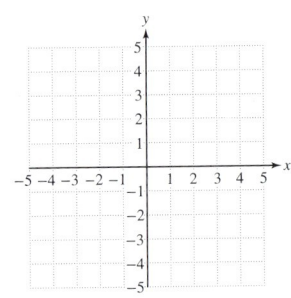

51. A wholesale club offers two types of memberships. The Executive Membership is $100 per year, including an annual 2% reward. The Business Membership is $50 per year without a reward. The total cost for membership, y, depends on the amount of money spent on merchandise, x, and can be represented by the following equations:

Executive Membership: y = 100 − 0.02x
Business Membership: y = 50

According to the graph, how much money spent on merchandise would result in the same cost for each membership?

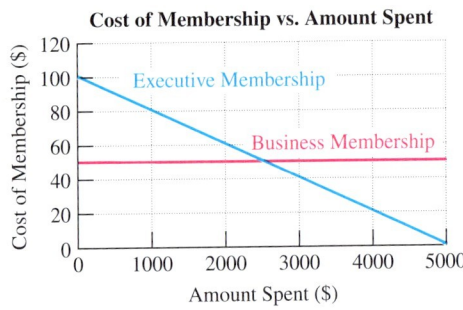

52. The cost to rent a 10 ft by 10 ft storage space is different for two different storage companies. The Storage Bin charges $90 per month plus a nonrefundable deposit of $120. AAA Storage charges $110 per month with no deposit. The total cost, y, to rent a 10 ft by 10 ft space depends on the number of months, x, according to the equations

The Storage Bin: y = 90x + 120
AAA Storage: y = 110x

From the graph, determine the number of months required for which the cost to rent space is equal for both companies.

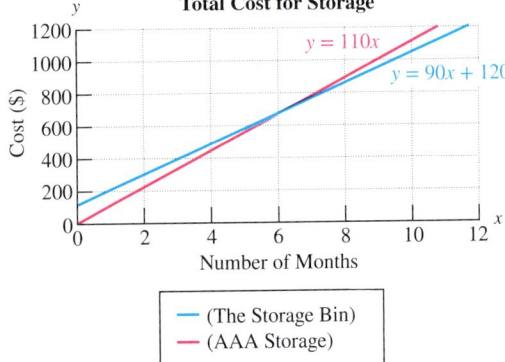

For the systems graphed in Exercises 53–54, explain why the ordered pair cannot be a solution to the system of equations.

53. $(-3, 1)$

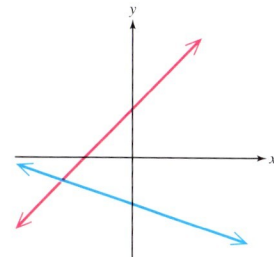

54. $(-1, -4)$

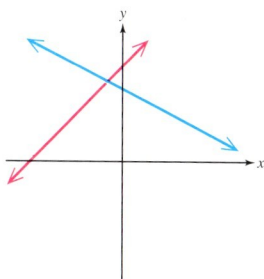

Technology Connections

For Exercises 55–60, use a graphing calculator to graph each pair of linear equations on the same viewing window. Use a *Trace* or *Intersect* feature to find the point(s) of intersection. Then write the solution set.

55. $y = 2x - 3$
$y = -4x + 9$

56. $y = -\frac{1}{2}x + 2$
$y = \frac{1}{3}x - 3$

57. $x + y = 4$ (Example 1)
$-2x + y = -5$

58. $x - 2y = -2$ (Example 3)
$-3x + 2y = 6$

59. $-x + 3y = -6$ (Example 4)
$6y = 2x + 6$

60. $x + 4y = 8$ (Example 5)
$y = -\frac{1}{4}x + 2$

Expanding Your Skills

61. Write a system of linear equations whose solution set is $\{(2, 1)\}$.

62. Write a system of linear equations whose solution set is $\{(1, 4)\}$.

63. One equation in a system of linear equations is $x + y = 4$. Write a second equation such that the system will have no solution. (Answers may vary.)

64. One equation in a system of linear equations is $x - y = 3$. Write a second equation such that the system will have infinitely many solutions. (Answers may vary.)

Solving Systems of Equations by the Substitution Method

Section 12.2

1. Solving Systems of Linear Equations by Using the Substitution Method

We have used the graphing method to find the solution set to a system of equations. However, sometimes it is difficult to determine the solution using this method because of limitations in the accuracy of the graph. This is particularly true when the coordinates of a solution are not integer values or when the solution is a point not sufficiently close to the origin. Identifying the coordinates of the point $\left(\frac{3}{17}, -\frac{23}{9}\right)$ or $(-251, 8349)$, for example, might be difficult from a graph.

In this section, we will cover the *substitution method* to solve systems of equations. This is an algebraic method that does not require graphing the individual equations. We demonstrate the substitution method in Examples 1 through 5.

Concepts

1. Solving Systems of Linear Equations by Using the Substitution Method
2. Applications of the Substitution Method

Chapter 12 Systems of Linear Equations in Two Variables

Example 1 Solving a System of Linear Equations by Using the Substitution Method

Solve the system by using the substitution method.

$$x = 2y - 3$$
$$-4x + 3y = 2$$

Solution:

The variable x has been isolated in the first equation. The quantity $2y - 3$ is equal to x and therefore can be substituted for x in the second equation. This leaves the second equation in terms of y only.

First equation: $x = \underline{2y - 3}$

Second equation: $-4x + 3y = 2$

$-4(2y - 3) + 3y = 2$ This equation now contains only one variable.

$-8y + 12 + 3y = 2$ Solve the resulting equation.

$-5y + 12 = 2$

$-5y = -10$

$y = 2$

To find x, substitute $y = 2$ back into the first equation.

$$x = 2y - 3$$
$$x = 2(2) - 3$$
$$x = 1$$

Check the ordered pair (1, 2) in both original equations.

$x = 2y - 3 \longrightarrow 1 \stackrel{?}{=} 2(2) - 3$ ✓   True

$-4x + 3y = 2 \longrightarrow -4(1) + 3(2) \stackrel{?}{=} 2$ ✓   True

The solution set is {(1, 2)}.

Skill Practice Solve the system by using the substitution method.

1. $2x + 3y = -2$
   $y = x + 1$

FOR REVIEW

Recall that when making a substitution for a variable, always use parentheses. For example, to substitute $2y - 3$ for x in the following equation, we have

$-4x + 3y = 2$
$-4() + 3y = 2$
$-4(2y - 3) + 3y = 2$

Avoiding Mistakes

Remember to solve for *both* variables in the system.

Answer

1. {(−1, 0)}

In Example 1, we eliminated the x variable from the second equation by substituting an equivalent expression for x. The resulting equation was relatively simple to solve because it had only one variable. This is the premise of the substitution method.

The substitution method can be summarized as follows.

Solving a System of Equations by Using the Substitution Method
Step 1 Isolate one of the variables from one equation.
Step 2 Substitute the expression found in step 1 into the other equation.
Step 3 Solve the resulting equation.
Step 4 Substitute the value found in step 3 back into the equation in step 1 to find the value of the remaining variable.
Step 5 Check the ordered pair in both original equations.

Example 2 Solving a System of Linear Equations by Using the Substitution Method

Solve the system by using the substitution method.

$$x + y = 4$$
$$-5x + 3y = -12$$

Solution:

The x or y variable in the first equation is easy to isolate because the coefficients are both 1. While either variable can be isolated, we arbitrarily choose to solve for the x variable.

$x + y = 4 \longrightarrow x = \underbrace{4 - y}$ **Step 1:** Isolate x in the first equation.

$-5(4 - y) + 3y = -12$ **Step 2:** Substitute $4 - y$ for x in the other equation.

$-20 + 5y + 3y = -12$ **Step 3:** Solve for y.
$-20 + 8y = -12$
$8y = 8$
$y = 1$

$x = 4 - y$ **Step 4:** Substitute $y = 1$ into the equation $x = 4 - y$.
$x = 4 - 1$
$x = 3$

Step 5: Check the ordered pair (3, 1) in both original equations.

$x + y = 4 \qquad (3) + (1) \stackrel{?}{=} 4$ ✓ True
$-5x + 3y = -12 \quad -5(3) + 3(1) \stackrel{?}{=} -12$ ✓ True

The solution set is $\{(3, 1)\}$.

Avoiding Mistakes
Although we solved for y first, be sure to write the x-coordinate first in the ordered pair. Remember that (1, 3) is not the same as (3, 1).

TIP:
The solution to a system of linear equations can be confirmed by graphing. The system from Example 2 is graphed here.

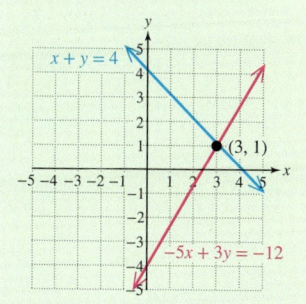

Skill Practice Solve the system by using the substitution method.
2. $\quad x + y = 3$
$\quad -2x + 3y = 9$

Answer
2. $\{(0, 3)\}$

Example 3   **Solving a System of Linear Equations by Using the Substitution Method**

Solve the system by using the substitution method.
$$3x + 5y = 17$$
$$2x - y = -6$$

Solution:

The y variable in the second equation is the easiest variable to isolate because its coefficient is -1.

$$3x + 5y = 17$$
$$2x - y = -6 \longrightarrow -y = -2x - 6$$
$$y = \underline{2x + 6}$$

Step 1: Isolate y in the second equation.

$$3x + 5(2x + 6) = 17$$

Step 2: Substitute the quantity $2x + 6$ for y in the other equation.

$$3x + 10x + 30 = 17$$
$$13x + 30 = 17$$
$$13x = 17 - 30$$
$$13x = -13$$
$$x = -1$$

Step 3: Solve for x.

$$y = 2x + 6$$
$$y = 2(-1) + 6$$
$$y = -2 + 6$$
$$y = 4$$

Step 4: Substitute $x = -1$ into the equation $y = 2x + 6$.

Step 5: The ordered pair $(-1, 4)$ can be checked in the original equations to verify the answer.

$$3x + 5y = 17 \longrightarrow 3(-1) + 5(4) \stackrel{?}{=} 17 \longrightarrow -3 + 20 \stackrel{?}{=} 17 \checkmark \quad \text{True}$$
$$2x - y = -6 \longrightarrow 2(-1) - (4) \stackrel{?}{=} -6 \longrightarrow -2 - 4 \stackrel{?}{=} -6 \checkmark \quad \text{True}$$

The solution set is $\{(-1, 4)\}$.

Avoiding Mistakes

Do not substitute $y = 2x + 6$ into the same equation from which it came. This mistake will result in an identity:
$$2x - y = -6$$
$$2x - (2x + 6) = -6$$
$$2x - 2x - 6 = -6$$
$$-6 = -6$$

Skill Practice Solve the system by using the substitution method.

3. $x + 4y = 11$
     $2x - 5y = -4$

Answer

3. $\{(3, 2)\}$

Recall that a system of linear equations may represent two parallel lines. In such a case, there is no solution to the system.

Example 4 — Solving an Inconsistent System Using Substitution

Solve the system by using the substitution method.

$$2x + 3y = 6$$
$$y = -\tfrac{2}{3}x + 4$$

Solution:

$2x + 3y = 6$

$y = \underbrace{-\tfrac{2}{3}x + 4}$

Step 1: The variable y is already isolated in the second equation.

$2x + 3(-\tfrac{2}{3}x + 4) = 6$

Step 2: Substitute $y = -\tfrac{2}{3}x + 4$ from the second equation into the first equation.

$2x - 2x + 12 = 6$

Step 3: Solve the resulting equation.

$12 = 6$ (contradiction)

The equation results in a contradiction. There are no values of x and y that will make 12 equal to 6. Therefore, the solution set is { }, and the system is inconsistent.

FOR REVIEW

Recall that an equation that reduces to a contradiction such as $12 = 6$ or $6 = 0$ has no solution. On the other hand, an equation that reduces to an identity such as $12 = 12$ or $0 = 0$ has infinitely many solutions.

Skill Practice Solve the system by using the substitution method.

4. $y = -\tfrac{1}{2}x + 3$
   $2x + 4y = 5$

TIP: The answer to Example 4 can be verified by writing each equation in slope-intercept form and graphing the lines.

Equation 1

$2x + 3y = 6$

$3y = -2x + 6$

$\dfrac{3y}{3} = \dfrac{-2x}{3} + \dfrac{6}{3}$

$y = -\tfrac{2}{3}x + 2$

Equation 2

$y = -\tfrac{2}{3}x + 4$

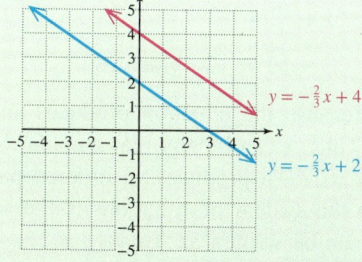

The equations indicate that the lines have the same slope but different y-intercepts. Therefore, the lines must be parallel. There is no point of intersection, indicating that the system has no solution, { }.

Answer

4. { }

Recall that a system of two linear equations may represent the same line. In such a case, the solution is the set of all points on the line.

Example 5   **Solving a System of Dependent Equations Using Substitution**

Solve the system by using the substitution method.

$$\frac{1}{2}x - \frac{1}{4}y = 1$$
$$6x - 3y = 12$$

Solution:

$\frac{1}{2}x - \frac{1}{4}y = 1$ To make the first equation easier to work with, we
$6x - 3y = 12$ have the option of clearing fractions.

$\frac{1}{2}x - \frac{1}{4}y = 1$ $\xrightarrow{\text{Multiply by 4.}}$ $4\left(\frac{1}{2}x\right) - 4\left(\frac{1}{4}y\right) = 4(1)$ $\rightarrow$ $2x - y = 4$

Now the system becomes:

$2x - y = 4$ The y variable in the first equation is the easiest to
$6x - 3y = 12$ isolate because its coefficient is -1.

$2x - y = 4$ $\xrightarrow{\text{Solve for } y.}$ $-y = -2x + 4$ $\rightarrow$ $y = \underline{2x - 4}$ **Step 1:** Isolate one of the variables.

$6x - 3y = 12$

$6x - 3(2x - 4) = 12$ **Step 2:** Substitute $y = 2x - 4$ from the first equation into the second equation.

$6x - 6x + 12 = 12$ **Step 3:** Solve the resulting equation.

$12 = 12$ (identity)

Because the equation produces an identity, all values of x make this equation true. Thus, x can be any real number. Substituting any real number, x, into the equation $y = 2x - 4$ produces an ordered pair on the line $y = 2x - 4$. Hence, the solution set to the system of equations is the set of all ordered pairs on the line $y = 2x - 4$. This can be written as $\{(x, y) \mid y = 2x - 4\}$. The equations are dependent.

Skill Practice Solve the system by using the substitution method.

5. $2x + \frac{1}{3}y = -\frac{1}{3}$
    $12x + 2y = -2$

Answer

5. Infinitely many solutions;
    $\{(x, y) \mid 12x + 2y = -2\}$; dependent equations

TIP: The solution to Example 5 can be verified by writing each equation in slope-intercept form and graphing the lines.

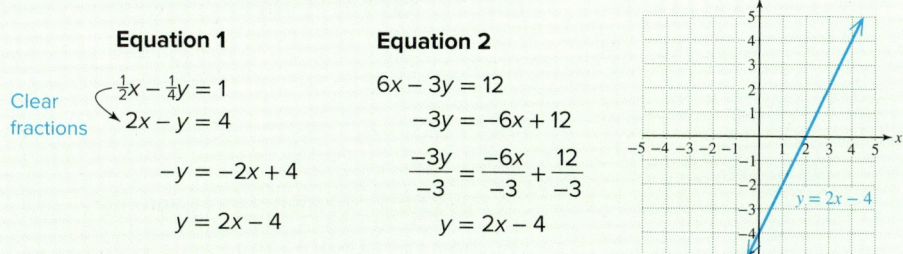

Equation 1
Clear fractions
$\frac{1}{2}x - \frac{1}{4}y = 1$
$2x - y = 4$
$-y = -2x + 4$
$y = 2x - 4$

Equation 2
$6x - 3y = 12$
$-3y = -6x + 12$
$\frac{-3y}{-3} = \frac{-6x}{-3} + \frac{12}{-3}$
$y = 2x - 4$

Notice that the slope-intercept forms for both equations are identical. The equations represent the same line, indicating that they are dependent. Each point on the line is a solution to the system of equations.

The following summary reviews the three different geometric relationships between two lines and the solutions to the corresponding systems of equations.

Interpreting Solutions to a System of Two Linear Equations

- The lines may intersect at one point (yielding one unique solution).
- The lines may be parallel and have no point of intersection (yielding no solution). This is detected algebraically when a contradiction (false statement) is obtained (e.g., $0 = -3$ or $12 = 6$).
- The lines may be the same and intersect at all points on the line (yielding an infinite number of solutions). This is detected algebraically when an identity is obtained (e.g., $0 = 0$ or $12 = 12$).

2. Applications of the Substitution Method

We have already encountered word problems using one linear equation and one variable. In this chapter, we investigate application problems with two unknowns. In such a case, we can use two variables to represent the unknown quantities. However, if two variables are used, we must write a system of *two* distinct equations.

Example 6 — Applying the Substitution Method

One number is 3 more than 4 times another. Their sum is 133. Find the numbers.

Solution:

We can use two variables to represent the two unknown numbers.

Let x represent one number.
Let y represent the other number. Label the variables.

We must now write two equations. Each of the first two sentences gives a relationship between x and y:

One number is 3 more than 4 times another. ⟶ $x = 4y + 3$ (first equation)

Their sum is 133. ⟶ $x + y = 133$ (second equation)

Avoiding Mistakes

Notice that the answer for an application is not represented by an ordered pair.

$(4y + 3) + y = 133$ Substitute $x = 4y + 3$ into the second equation, $x + y = 133$.

$5y + 3 = 133$ Solve the resulting equation.

$5y = 130$

$y = 26$

$x = 4y + 3$

$x = 4(26) + 3$ To solve for x, substitute $y = 26$ into the equation $x = 4y + 3$.

$x = 104 + 3$

$x = 107$

One number is 26, and the other is 107.

TIP: Check that the numbers 26 and 107 meet the conditions of Example 6.
- 4 times 26 is 104. Three more than 104 is 107. ✓
- The sum of the numbers should be 133: $26 + 107 = 133$ ✓

Skill Practice

6. One number is 16 more than another. Their sum is 92. Use a system of equations to find the numbers.

Example 7 Using the Substitution Method in a Geometry Application

Two angles are supplementary. The measure of one angle is 15° more than twice the measure of the other angle. Find the measures of the two angles.

Solution:

Let x represent the measure of one angle.
Let y represent the measure of the other angle.

FOR REVIEW

Recall that two angles are **supplementary** if their sum is 180°. Two angles are **complementary** if their sum is 90°.

The sum of the measures of supplementary angles is 180°. ⟶ $x + y = 180$

The measure of one angle is 15° more than twice the other angle. ⟶ $x = 2y + 15$

$x + y = 180$

$x = 2y + 15$ The x variable in the second equation is already isolated.

$(2y + 15) + y = 180$ Substitute $2y + 15$ into the first equation for x.

$2y + 15 + y = 180$ Solve the resulting equation.

$3y + 15 = 180$

$3y = 165$

$y = 55$

TIP: Check that the angles 55° and 125° meet the conditions of Example 7.
- Because $55° + 125° = 180°$, the angles are supplementary. ✓
- The angle 125° is 15° more than twice 55°: $125° = 2(55°) + 15°$ ✓

$x = 2y + 15$ Substitute $y = 55$ into the equation $x = 2y + 15$.

$x = 2(55) + 15$

$x = 110 + 15$

$x = 125$

One angle is 55°, and the other is 125°.

Skill Practice

7. The measure of one angle is 2° less than 3 times the measure of another angle. The angles are complementary. Use a system of equations to find the measures of the two angles.

Answers

6. One number is 38, and the other number is 54.
7. The measures of the angles are 23° and 67°.

Section 12.2 Activity

A.1. Consider the system. $y = x + 6$
$3x + 2y = 2$

a. The first equation tells us that the variable y is equal to the quantity $x + 6$. Therefore, the expression $x + 6$ and the variable y are interchangeable. Write the second equation with y replaced by the quantity $x + 6$.

b. After making the substitution in part (a), the equation should now have only one variable. Solve the equation for x.

c. Substitute the value of x you found in part (b) back into the first equation to solve for y. With the values of x and y now known, write the solution set.

d. Check the ordered pair in both equations.

A.2. Consider the system. $3x - 2y = -10$
$x + 4y = 6$

a. Of the four variable terms, which is the easiest to isolate?

b. Solve the system by using the substitution method.

c. Check the ordered pair from part (b) in both equations.

A.3. Consider the system. $\frac{2}{3}x - y = -2$
$4x - 6y = 6$

a. Of the four variable terms, which is the easiest to isolate?

b. Solve the system by using the substitution method.

c. How many solutions are there to this system?

d. Is the system consistent or inconsistent?

A.4. Consider the system. $5x - 3y = 15$
$x - \frac{3}{5}y = 3$

a. Of the four variable terms, which is the easiest to isolate?

b. Solve the system by using the substitution method.

c. How many solutions are there to this system?

d. Are the equations in this system dependent or independent?

To solve an application with two unknowns, it is often helpful to use two variables such as x and y. When two variables are used to solve an application, we need two independent equations that relate the variables. That is, for two unknowns, set up a system of two equations. In Exercise A.5, we walk through this process.

A.5. Two angles are complementary. One angle is 8° more than the other. What are the measures of the two angles?
a. Let x represent _____.
   Let y represent _____.
b. Set up an equation that indicates that the two angles are complementary (the sum of their measures is 90°).
   Set up an equation that indicates that one angle is 8° more than the other angle.
c. Solve the system of equations made up by the two equations from part (b).
d. Interpret the solution to the system you found in part (c).
e. Verify that your answer makes sense by asking whether the two angles are complementary. Also verify that the measure of one angle is indeed 8° more than the measure of the other angle.

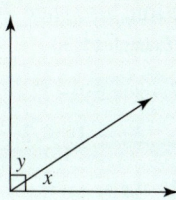

Section 12.2 Practice Exercises

Prerequisite Review

For Exercises R.1–R.6, write each pair of lines in slope-intercept form. Then identify whether the lines intersect in exactly one point or if the lines are parallel or coinciding.

R.1. $2x - y = 4$
$-2y = -4x + 8$

R.2. $x - 2y = 5$
$3x = 6y + 15$

R.3. $2x + 3y = 6$
$x - y = 5$

R.4. $x - y = -1$
$x + 2y = 4$

R.5. $2x = \frac{1}{2}y + 2$
$4x - y = 13$

R.6. $4y = 3x$
$3x - 4y = 15$

R.7. Given $3x + 2y = 11$, if $y = 4$, solve for x.

R.8. Given $2x - 4y = -2$, if $y = 2$, solve for x.

R.9. Given $y = -\frac{1}{3}x + 8$, if $x = -3$, solve for y.

R.10. Given $y = \frac{2}{5}x - 4$, if $x = 5$, solve for y.

For Exercises R.11–R.16,
a. Identify the equation as a conditional equation, an identity, or a contradiction.
b. Solve the equation.

R.11. $2(x + 4) - 5 = 2x + 6$

R.12. $5y + 8 - 8y = 3(4 - y)$

R.13. $t - (3 - t) + 2 = 2t - 1$

R.14. $5(2 - m) + 6m = 3m + 10 - 2m$

R.15. $4x + 3(2x - 1) = 7$

R.16. $3(y - 4) - 2y = -6$

Concept 1: Solving Systems of Linear Equations by Using the Substitution Method

For Exercises 1–4, solve each system by using the substitution method. (See Example 1.)

1. $3x + 2y = -3$
$y = 2x - 12$

2. $4x - 3y = -19$
$y = -2x + 13$

3. $x = -4y + 16$
$3x + 5y = 20$

4. $x = -y + 3$
$-2x + y = 6$

5. Given the system: $4x - 2y = -6$
$3x + y = 8$

a. Which variable from which equation is easiest to isolate and why?

b. Solve the system by using the substitution method.

6. Given the system: $x - 5y = 2$
$11x + 13y = 22$

a. Which variable from which equation is easiest to isolate and why?

b. Solve the system by using the substitution method.

For Exercises 7–42, solve the system by using the substitution method. For systems that do not have one unique solution, also state the number of solutions and whether the system is inconsistent or the equations are dependent. (See Examples 1–5.)

7. $x = 3y - 1$
$2x - 4y = 2$

8. $2y = x + 9$
$y = -3x + 1$

9. $-2x + 5y = 5$
$x = 4y - 10$

10. $y = -2x + 27$
$3x - 7y = -2$

11. $4x - y = -1$
$2x + 4y = 13$

12. $5x - 3y = -2$
$10x - y = 1$

13. $4x - 3y = 11$
$x = 5$

14. $y = -3x - 9$
$y = 12$

15. $4x = 8y + 4$
$5x - 3y = 5$

16. $3y = 6x - 6$
$-3x + y = -4$

17. $x - 3y = -11$
$6x - y = 2$

18. $-2x - y = 9$
$x + 7y = 15$

19. $3x + 2y = -1$	20. $5x - 2y = 6$	21. $10x - 30y = -10$	22. $3x + 6y = 6$
$\frac{3}{2}x + y = 4$	$-\frac{5}{2}x + y = 5$	$2x - 6y = -2$	$-6x - 12y = -12$
23. $2x + y = 3$	24. $-3x = 2y + 23$	25. $x + 2y = -2$	26. $x + y = 1$
$y = -7$	$x = -1$	$4x = -2y - 17$	$2x - y = -2$
27. $y = -\frac{1}{2}x - 4$	28. $y = \frac{2}{3}x - 3$	29. $y = 6$	30. $x = 9$
$y = 4x - 13$	$y = 6x - 19$	$y - 4 = -2x - 6$	$x - 3 = 6y + 12$
31. $3x + 2y = 4$	32. $4x + 3y = 4$	33. $y = 0.25x + 1$	34. $y = 0.75x - 3$
$2x - 3y = -6$	$-2x + 5y = -2$	$-x + 4y = 4$	$-3x + 4y = -12$
35. $11x + 6y = 17$	36. $3x - 8y = 7$	37. $x + 2y = 4$	38. $-y = x - 6$
$5x - 4y = 1$	$10x - 5y = 45$	$4y = -2x - 8$	$2x + 2y = 4$
39. $2x = 3 - y$	40. $2x = 4 + 2y$	41. $\frac{x}{3} + \frac{y}{2} = -4$	42. $x - 2y = -5$
$x + y = 4$	$3x + y = 10$	$x - 3y = 6$	$\frac{2x}{3} + \frac{y}{3} = 0$

Concept 2: Applications of the Substitution Method

For Exercises 43–52, set up a system of linear equations and solve for the indicated quantities. (See Examples 6–7.)

43. Two numbers have a sum of 106. One number is 10 less than the other. Find the numbers.

44. Two positive numbers have a difference of 8. The larger number is 2 less than 3 times the smaller number. Find the numbers.

45. The difference between two positive numbers is 26. The larger number is 3 times the smaller. Find the numbers.

46. The sum of two numbers is 956. One number is 94 less than 6 times the other. Find the numbers.

47. Two angles are supplementary. One angle is 15° more than 10 times the other angle. Find the measure of each angle.

48. Two angles are complementary. One angle is 1° less than 6 times the other angle. Find the measure of each angle.

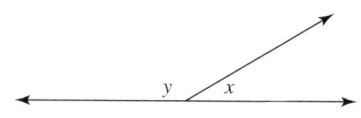

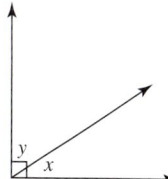

49. Two angles are complementary. One angle is 10° more than 3 times the other angle. Find the measure of each angle.

50. Two angles are supplementary. One angle is 5° less than twice the other angle. Find the measure of each angle.

51. In a right triangle, one of the acute angles is 6° less than the other acute angle. Find the measure of each acute angle.

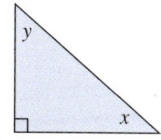

52. In a right triangle, one of the acute angles is 9° less than twice the other acute angle. Find the measure of each acute angle.

Expanding Your Skills

53. The following system consists of dependent equations and therefore has infinitely many solutions. Find three ordered pairs that are solutions to the system of equations.

$$y = 2x + 3$$
$$-4x + 2y = 6$$

54. The following system consists of dependent equations and therefore has infinitely many solutions. Find three ordered pairs that are solutions to the system of equations.

$$y = -x + 1$$
$$2x + 2y = 2$$

Section 12.3 Solving Systems of Equations by the Addition Method

Concepts

1. Solving Systems of Linear Equations by Using the Addition Method
2. Summary of Methods for Solving Systems of Linear Equations in Two Variables

1. Solving Systems of Linear Equations by Using the Addition Method

In this section, we present another algebraic method to solve a system of linear equations. This method is called the *addition method* and its underlying principle is to add multiples of the given equations to eliminate a variable from the system. For this reason, the addition method is sometimes called the *elimination method*.

Example 1   Solving a System of Linear Equations by Using the Addition Method

Solve the system by using the addition method.

$$x + y = -2$$
$$x - y = -6$$

Solution:

Notice that the coefficients of the y variables are opposites:

$$x + 1y = -2 \quad \text{Coefficient is 1.}$$
$$x - 1y = -6 \quad \text{Coefficient is } -1.$$

FOR REVIEW

Recall that the sum of an expression and its opposite is zero. For example:

$$1 + (-1) = 0$$
$$y + (-y) = 0$$
$$-10b + 10b = 0$$

Because the coefficients of the y variables are opposites, we can add the two equations to eliminate the y variable.

$$x + y = -2$$
$$\underline{x - y = -6}$$
$$2x = -8 \quad \leftarrow \text{After adding the equations, we have one equation and one variable.}$$

$$2x = -8 \quad \text{Solve the resulting equation.}$$
$$x = -4$$

To find the value of y, substitute $x = -4$ into *either* of the original equations.

$$x + y = -2 \quad \text{First equation}$$
$$(-4) + y = -2$$
$$y = -2 + 4$$
$$y = 2 \quad \text{The ordered pair is } (-4, 2).$$

Check:

$x + y = -2 \longrightarrow (-4) + (2) \stackrel{?}{=} -2 \longrightarrow -2 \stackrel{?}{=} -2 \checkmark$ True

$x - y = -6 \longrightarrow (-4) - (2) \stackrel{?}{=} -6 \longrightarrow -6 \stackrel{?}{=} -6 \checkmark$ True

The solution set is $\{(-4, 2)\}$.

Skill Practice Solve the system by using the addition method.

1. $x + y = 13$
   $2x - y = 2$

TIP: In Example 1, notice that the value $x = -4$ could have been substituted into the second equation to obtain the same value for y.

$$x - y = -6$$
$$(-4) - y = -6$$
$$-y = -6 + 4$$
$$-y = -2$$
$$y = 2$$

It is important to note that the addition method works on the premise that the two equations have *opposite* values for the coefficients of one of the variables. Sometimes it is necessary to manipulate the original equations to create two coefficients that are opposites. This is accomplished by multiplying one or both equations by an appropriate constant. The process is outlined as follows:

Solving a System of Equations by Using the Addition Method

Step 1 Write both equations in standard form: $Ax + By = C$
Step 2 Clear fractions or decimals (optional).
Step 3 Multiply one or both equations by nonzero constants to create opposite coefficients for one of the variables.
Step 4 Add the equations from step 3 to eliminate one variable.
Step 5 Solve for the remaining variable.
Step 6 Substitute the known value from step 5 into one of the original equations to solve for the other variable.
Step 7 Check the ordered pair in both equations.

Answer

1. $\{(5, 8)\}$

Example 2: Solving a System of Linear Equations by Using the Addition Method

Solve the system by using the addition method.

$$3x + 5y = 17$$
$$2x - y = -6$$

Solution:

$3x + 5y = 17$ **Step 1:** Both equations are already written in standard form.

$2x - y = -6$ **Step 2:** There are no fractions or decimals.

Notice that neither the coefficients of x nor the coefficients of y are opposites. However, multiplying the second equation by 5 creates the term $-5y$ in the second equation. This is the opposite of the term $+5y$ in the first equation.

$3x + 5y = 17$ $\qquad\qquad 3x + 5y = 17$ **Step 3:** Multiply the second equation by 5.

$2x - y = -6 \xrightarrow{\text{Multiply by 5.}} \underline{10x - 5y = -30}$

$\qquad\qquad\qquad\qquad\qquad 13x \quad\;\; = -13$ **Step 4:** Add the equations.

$\qquad\qquad\qquad\qquad\qquad 13x \quad\;\; = -13$ **Step 5:** Solve the equation.

$\qquad\qquad\qquad\qquad\qquad\; x = -1$

$3x + 5y = 17$ $\quad$ First equation $\qquad\qquad$ **Step 6:** Substitute $x = -1$ into one of the original equations.

$3(-1) + 5y = 17$

$-3 + 5y = 17$

$5y = 20$

$y = 4$ $\qquad\qquad\qquad\qquad\qquad\qquad\qquad$ **Step 7:** Check $(-1, 4)$ in both original equations.

Check:

$3x + 5y = 17 \longrightarrow 3(-1) + 5(4) \stackrel{?}{=} 17 \longrightarrow -3 + 20 \stackrel{?}{=} 17$ ✓ True

$2x - y = -6 \longrightarrow 2(-1) - (4) \stackrel{?}{=} -6 \longrightarrow -2 - 4 \stackrel{?}{=} -6$ ✓ True

The solution set is $\{(-1, 4)\}$.

Avoiding Mistakes
Remember to multiply the chosen constant on *both* sides of the equation.

TIP: In Example 2, we could have eliminated the x variable by multiplying the first equation by 2 and the second equation by -3.

Skill Practice Solve the system by using the addition method.

2. $4x + 3y = 3$
$\;\;\; x - 2y = 9$

In Example 3, the system of equations uses the variables a and b instead of x and y. In such a case, we will write the solution as an ordered pair with the variables written in alphabetical order, such as (a, b).

Answer
2. $\{(3, -3)\}$

Example 3: Solving a System of Linear Equations by Using the Addition Method

Solve the system by using the addition method.

$$5b = 7a + 8$$
$$-4a - 2b = -10$$

Solution:

Step 1: Write the equations in standard form.

The first equation becomes: $5b = 7a + 8 \longrightarrow -7a + 5b = 8$

The system becomes:
$$-7a + 5b = 8$$
$$-4a - 2b = -10$$

Step 2: There are no fractions or decimals.

Step 3: We need to obtain opposite coefficients on either the a or b term.

Notice that neither the coefficients of a nor the coefficients of b are opposites. However, it is possible to change the coefficients of b to 10 and -10 (this is because the LCM of 5 and 2 is 10). This is accomplished by multiplying the first equation by 2 and the second equation by 5.

$-7a + 5b = 8$ $\xrightarrow{\text{Multiply by 2.}}$ $-14a + 10b = 16$
$-4a - 2b = -10$ $\xrightarrow{\text{Multiply by 5.}}$ $\underline{-20a - 10b = -50}$
$\phantom{-4a - 2b = -10 \xrightarrow{\text{Multiply by 5.}}} -34a = -34$ **Step 4:** Add the equations.

$$-34a = -34$$
$$\frac{-34a}{-34} = \frac{-34}{-34}$$
$$a = 1$$

Step 5: Solve the resulting equation.

$5b = 7a + 8$ First equation   **Step 6:** Substitute $a = 1$ into one of the original equations.

$5b = 7(1) + 8$

$5b = 15$

$b = 3$ **Step 7:** Check $(1, 3)$ in the original equations.

Check:

$5b = 7a + 8 \longrightarrow 5(3) \stackrel{?}{=} 7(1) + 8 \longrightarrow 15 \stackrel{?}{=} 7 + 8$ ✓   True

$-4a - 2b = -10 \longrightarrow -4(1) - 2(3) \stackrel{?}{=} -10 \longrightarrow -4 - 6 \stackrel{?}{=} -10$ ✓   True

The solution set is $\{(1, 3)\}$.

Skill Practice Solve the system by using the addition method.

3. $8n = 4 - 5m$
   $7m + 6n = -10$

Answer

3. $\{(-4, 3)\}$

Chapter 12 Systems of Linear Equations in Two Variables

> **Example 4** **Solving a System of Linear Equations by Using the Addition Method**
>
> Solve the system by using the addition method.
>
> $$34x - 22y = 4$$
> $$17x - 88y = -19$$
>
> **Solution:**
>
> The equations are already in standard form. There are no fractions or decimals to clear.
>
> $$34x - 22y = 4 \longrightarrow 34x - 22y = 4$$
> $$17x - 88y = -19 \xrightarrow{\text{Multiply by } -2.} \underline{-34x + 176y = 38}$$
> $$154y = 42$$
>
> Solve for y. $154y = 42$
>
> $$\frac{154y}{154} = \frac{42}{154}$$
>
> Simplify. $y = \dfrac{3}{11}$
>
> To find the value of x, we normally substitute y into one of the original equations and solve for x. In this example, we will show an alternative method for finding x. By repeating the addition method, this time eliminating y, we can solve for x. This approach enables us to avoid substitution of the fractional value for y.
>
> $$34x - 22y = 4 \xrightarrow{\text{Multiply by } -4.} -136x + 88y = -16$$
> $$17x - 88y = -19 \longrightarrow \underline{17x - 88y = -19}$$
> $$-119x = -35$$
>
> Solve for x. $-119x = -35$
>
> $$\frac{-119x}{-119} = \frac{-35}{-119}$$
>
> Simplify. $x = \dfrac{5}{17}$
>
> The ordered pair $\left(\dfrac{5}{17}, \dfrac{3}{11}\right)$ can be checked in the original equations.
>
> $$34x - 22y = 4 \qquad\qquad 17x - 88y = -19$$
> $$34\left(\dfrac{5}{17}\right) - 22\left(\dfrac{3}{11}\right) \stackrel{?}{=} 4 \qquad 17\left(\dfrac{5}{17}\right) - 88\left(\dfrac{3}{11}\right) \stackrel{?}{=} -19$$
> $$10 - 6 \stackrel{?}{=} 4 \checkmark \ \text{True} \qquad\qquad 5 - 24 \stackrel{?}{=} -19 \checkmark \ \text{True}$$
>
> The solution set is $\left\{\left(\dfrac{5}{17}, \dfrac{3}{11}\right)\right\}$.
>
> **Skill Practice** Solve the system by using the addition method.
>
> 4. $15x - 16y = 1$
> $45x + 4y = 16$

Answer

4. $\left\{\left(\dfrac{1}{3}, \dfrac{1}{4}\right)\right\}$

Example 5 — Solving an Inconsistent System by the Addition Method

Solve the system by using the addition method.

$$2x - 5y = 10$$
$$\frac{1}{2}x - \frac{5}{4}y = 1$$

Solution:

$$2x - 5y = 10$$
$$\frac{1}{2}x - \frac{5}{4}y = 1$$

Step 1: The equations are in standard form.

Step 2: Multiply both sides of the second equation by 4 to clear fractions.

$$\frac{1}{2}x - \frac{5}{4}y = 1 \longrightarrow 4\left(\frac{1}{2}x - \frac{5}{4}y\right) = 4(1) \longrightarrow 2x - 5y = 4$$

Now the system becomes
$$2x - 5y = 10$$
$$2x - 5y = 4$$

To make either the x coefficients or y coefficients opposites, multiply either equation by -1.

$$2x - 5y = 10 \xrightarrow{\text{Multiply by } -1.} -2x + 5y = -10$$
$$2x - 5y = 4 \longrightarrow \underline{2x - 5y = 4}$$
$$0 = -6$$

Step 3: Create opposite coefficients.

Step 4: Add the equations.

Because the result is a contradiction, the solution set is { }, and the system of equations is inconsistent. Writing each equation in slope-intercept form verifies that the lines are parallel (Figure 12-6).

$$2x - 5y = 10 \xrightarrow{\text{slope-intercept form}} y = \frac{2}{5}x - 2$$

$$\frac{1}{2}x - \frac{5}{4}y = 1 \xrightarrow{\text{slope-intercept form}} y = \frac{2}{5}x - \frac{4}{5}$$

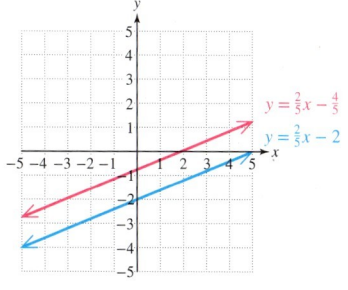

Figure 12-6

Skill Practice Solve the system by using the addition method.

5. $\frac{2}{3}x - \frac{3}{4}y = 2$
   $8x - 9y = 6$

Answer

5. { }

> **Example 6** Solving a System of Dependent Equations by the Addition Method
>
> Solve the system by using the addition method.
> $$3x - y = 4$$
> $$2y = 6x - 8$$
>
> **Solution:**
>
> $3x - y = 4 \longrightarrow 3x - y = 4$ **Step 1:** Write the equations in standard form.
>
> $2y = 6x - 8 \longrightarrow -6x + 2y = -8$ **Step 2:** There are no fractions or decimals.
>
> Notice that the equations differ exactly by a factor of -2, which indicates that these two equations represent the same line. Multiply the first equation by 2 to create opposite coefficients for the variables.
>
> $$3x - y = 4 \xrightarrow{\text{Multiply by 2.}} 6x - 2y = 8$$
> $$-6x + 2y = -8 \qquad\qquad\qquad -6x + 2y = -8$$
> $$\qquad\qquad\qquad\qquad\qquad\qquad\qquad 0 = 0$$
>
> **Step 3:** Create opposite coefficients.
>
> **Step 4:** Add the equations.
>
> Because the resulting equation is an identity, the original equations represent the same line. This can be confirmed by writing each equation in slope-intercept form.
>
> $$3x - y = 4 \longrightarrow -y = -3x + 4 \longrightarrow y = 3x - 4$$
> $$-6x + 2y = -8 \longrightarrow 2y = 6x - 8 \longrightarrow y = 3x - 4$$
>
> The solution is the set of all points on the line, or equivalently, $\{(x, y) \mid y = 3x - 4\}$.
>
> ---
>
> **Skill Practice** Solve the system by using the addition method.
>
> 6. $3x = 3y + 15$
> $2x - 2y = 10$

2. Summary of Methods for Solving Systems of Linear Equations in Two Variables

If no method of solving a system of linear equations is specified, you may use the method of your choice. However, we recommend the following guidelines:

1. If one of the equations is written with a variable isolated, the substitution method is a good choice. For example:

$$2x + 5y = 2 \qquad \text{or} \qquad y = \frac{1}{3}x - 2$$
$$x = y - 6 \qquad\qquad\qquad\qquad x - 6y = 9$$

2. If both equations are written in standard form, $Ax + By = C$, where none of the variables has coefficients of 1 or -1, then the addition method is a good choice.

$$4x + 5y = 12$$
$$5x + 3y = 15$$

3. If both equations are written in standard form, $Ax + By = C$, and at least one variable has a coefficient of 1 or -1, then either the substitution method or the addition method is a good choice.

Answer
6. $\{(x, y) \mid 2x - 2y = 10\}$

Section 12.3 Activity

A.1. a. Add the equations.

$3x - 5y = -7$
$4x + 5y = 14$

b. Add the equations.

$2x + 5y = 3$
$-4x + 3y = 7$

c. In which system was a variable eliminated? Why?

A.2. a. Solve the system by using the addition method.

$3x - 5y = -7$
$4x + 5y = 14$

b. Check the ordered pair in both equations.

A.3. a. To solve the system by using the addition method, which variable is easier to eliminate and why?

$2x + 5y = 3$
$-4x + 3y = 7$

b. Solve the system by using the addition method.
c. Check the ordered pair in both equations.

A.4. a. Solve the system by using the addition method.

$4x - 5y = 2$
$-3x + 3y = -3$

b. Check the ordered pair in both equations.

A.5. a. Given the equation $\frac{5}{2}x - 2y = 2$, what is the smallest positive number that you could use to clear the fractions in the equation?

b. Solve the system by using the addition method.

$-5x + 4y = -4$
$\frac{5}{2}x - 2y = 2$

c. How many solutions are there to the system?

Practice Exercises   Section 12.3

Study Skills Exercise

Often in mathematics there is more than one method that can be used to solve a problem. Solving a system of linear equations with two variables is one example. Now that you have learned three different ways to solve a system of linear equations, solve the given system using the graphing method, the substitution method, and the addition method. There are two advantages to this exercise. One advantage is to check your answer. You should get the same answer using all three methods. The second advantage is to determine which method is preferable for a given system.

- Solve the system by using the graphing method, the substitution method, and the addition method.

$2x + y = -7$
$x - 10 = 4y$

Prerequisite Review

For Exercises R.1–R.2, check whether the given ordered pair is a solution to the system.

R.1. $-\dfrac{3}{4}x + 2y = -10$ $(8, -2)$

$x - \dfrac{1}{2}y = 7$

R.2. $x + y = 8$ $(5, 3)$

$y = x - 2$

For Exercises R.3–R.8, solve the equation.

R.3. $3(x - 4) + 6 = 3x + 8$

R.4. $5 - 7(2 - x) - x = 6x - 3$

R.5. $\dfrac{2}{5}x - \dfrac{2}{3} = x - \dfrac{1}{15}$

R.6. $\dfrac{1}{2} + \dfrac{3}{4}y = \dfrac{2}{3} - \dfrac{1}{6}y$

R.7. $4t - 5 = 3(t - 1) + t - 2$

R.8. $n + 3(n + 2) = 7 + 4n - 1$

Concept 1: Solving Systems of Linear Equations by Using the Addition Method

For Exercises 1–5, answer the question regarding the application of the addition method. Recall that the general principle is to create opposite coefficients on either the x or y variable.

1. How would you eliminate the y variable from the given system?

    $2x - y = 2$
    $3x + 2y = 17$

2. How would you eliminate the x variable from the given system?

    $3x + 5y = 13$
    $x - 4y = -7$

3. Consider the system. $-2x + 7y = 5$
    $3x - 4y = -1$

    a. What is the least common multiple of 2 and 3?

    b. How would you eliminate the x variable?

4. Consider the system. $-2x + 7y = 5$
    $3x - 4y = -1$

    a. What is the least common multiple of 7 and 4?

    b. How would you eliminate the y variable?

5. Consider the system. $4x - 5y = 3$
    $3x - 8y = -2$

    a. What is the least common multiple of 4 and 3?

    b. How would you eliminate the x variable?

For Exercises 6–7, answer as true or false.

6. Given the system $5x - 4y = 1$
    $7x - 2y = 5$

    a. To eliminate the y variable using the addition method, multiply the second equation by 2.

    b. To eliminate the x variable using the addition method, multiply the first equation by 7 and the second equation by -5.

7. Given the system $3x + 5y = -1$
    $9x - 8y = -26$

    a. To eliminate the x variable using the addition method, multiply the first equation by -3.

    b. To eliminate the y variable using the addition method, multiply the first equation by 8 and the second equation by -5.

8. Given the system $3x - 4y = 2$
$17x + y = 35$

  a. Which variable, x or y, is easier to eliminate using the addition method?

  b. Solve the system using the addition method.

9. Given the system $-2x + 5y = -15$
$6x - 7y = 21$

  a. Which variable, x or y, is easier to eliminate using the addition method?

  b. Solve the system using the addition method.

For Exercises 10–24, solve each system using the addition method. (See Examples 1–4.)

10. $x + 2y = 8$
$5x - 2y = 4$

11. $2x - 3y = 11$
$-4x + 3y = -19$

12. $a + b = 3$
$3a + b = 13$

13. $-2u + 6v = 10$
$-2u + v = -5$

14. $-3x + y = 1$
$-6x - 2y = -2$

15. $5m - 2n = 4$
$3m + n = 9$

16. $3x - 5y = 13$
$x - 2y = 5$

17. $7a + 2b = -1$
$3a - 4b = 19$

18. $6c - 2d = -2$
$5c = -3d + 17$

19. $2s + 3t = -1$
$5s = 2t + 7$

20. $6y - 4z = -2$
$4y + 6z = 42$

21. $4k - 2r = -4$
$3k - 5r = 18$

22. $2x + 3y = 6$
$x - y = 5$

23. $6x + 6y = 8$
$9x - 18y = -3$

24. $2x - 5y = 4$
$3x - 3y = 4$

25. In solving a system of equations, suppose you get the statement $0 = 5$. How many solutions will the system have? What can you say about the graphs of these equations?

26. In solving a system of equations, suppose you get the statement $0 = 0$. How many solutions will the system have? What can you say about the graphs of these equations?

27. In solving a system of equations, suppose you get the statement $3 = 3$. How many solutions will the system have? What can you say about the graphs of these equations?

28. In solving a system of equations, suppose you get the statement $2 = -5$. How many solutions will the system have? What can you say about the graphs of these equations?

29. Suppose in solving a system of linear equations, you get the statement $x = 0$. How many solutions will the system have? What can you say about the graphs of these equations?

30. Suppose in solving a system of linear equations, you get the statement $y = 0$. How many solutions will the system have? What can you say about the graphs of these equations?

For Exercises 31–42, solve the system by using the addition method. For systems that do not have one unique solution, also state the number of solutions and whether the system is inconsistent or the equations are dependent. (See Examples 5–6.)

31. $-2x + y = -5$
$8x - 4y = 12$

32. $x - 3y = 2$
$-5x + 15y = 10$

33. $x + 2y = 2$
$-3x - 6y = -6$

34. $4x - 3y = 6$
$-12x + 9y = -18$

35. $3a + 2b = 11$
$7a - 3b = -5$

36. $4y + 5z = -2$
$5y - 3z = 16$

37. $3x - 5y = 7$
$5x - 2y = -1$

38. $4s + 3t = 9$
$3s + 4t = 12$

39. $2x + 2 = -3y + 9$
$3x - 10 = -4y$

40. $-3x + 6 + 7y = 5$
    $5y = 2x$

41. $4x - 5y = 0$
    $8(x - 1) = 10y$

42. $y = 2x + 1$
    $-3(2x - y) = 0$

Concept 2: Summary of Methods for Solving Systems of Linear Equations in Two Variables

For Exercises 43–63, solve each system of equations by either the addition method or the substitution method.

43. $5x - 2y = 4$
    $y = -3x + 9$

44. $-x = 8y + 5$
    $4x - 3y = -20$

45. $0.1x + 0.1y = 0.6$
    $0.1x - 0.1y = 0.1$

46. $0.1x + 0.1y = 0.2$
    $0.1x - 0.1y = 0.3$

47. $3x = 5y - 9$
    $2y = 3x + 3$

48. $10x - 5 = 3y$
    $4x + 5y = 2$

49. $\frac{1}{10}y = -\frac{1}{2}x - \frac{1}{2}$
    $\frac{3}{2}x - \frac{3}{4} = -\frac{3}{4}y$

50. $x + \frac{5}{4}y = -\frac{1}{2}$
    $\frac{3}{4}x = -\frac{1}{2}y - \frac{5}{4}$

51. $x = -\frac{1}{2}$
    $6x - 5y = -8$

52. $4x - 2y = 1$
    $y = 3$

53. $0.02x + 0.04y = 0.12$
    $0.03x - 0.05y = -0.15$

54. $-0.04x + 0.03y = 0.03$
    $-0.06x - 0.02y = -0.02$

55. $8x - 16y = 24$
    $2x - 4y = 0$

56. $y = -\frac{1}{2}x - 5$
    $2x + 4y = -8$

57. $\frac{m}{2} + \frac{n}{5} = \frac{13}{10}$
    $3m - 3n = m - 10$

58. $\frac{a}{4} - \frac{3b}{2} = \frac{15}{2}$
    $a + 2b = -10$

59. $2m - 6n = m + 4$
    $3m + 8 = 5m - n$

60. $m - 3n = 10$
    $3m + 12n = -12$

61. $9a - 2b = 8$
    $18a + 6 = 4b + 22$

62. $a = 5 + 2b$
    $3a - 6b = 15$

63. $6x - 5y = 7$
    $4x - 6y = 7$

For Exercises 64–69, use a system of linear equations, and solve for the indicated quantities.

64. The sum of two positive numbers is 26. Their difference is 14. Find the numbers.

65. The difference of two positive numbers is 2. The sum of the numbers is 36. Find the numbers.

66. Eight times the smaller of two numbers plus 2 times the larger number is 44. Three times the smaller number minus 2 times the larger number is zero. Find the numbers.

67. Six times the smaller of two numbers minus the larger number is −9. Ten times the smaller number plus 5 times the larger number is 5. Find the numbers.

68. Twice the difference of two angles is 64°. If the angles are complementary, find the measures of the angles.

69. The difference of an angle and twice another angle is 42°. If the angles are supplementary, find the measures of the angles.

For Exercises 70–72, solve the system by using each of the three methods: (a) the graphing method, (b) the substitution method, and (c) the addition method.

70. $2x + y = 1$
    $-4x - 2y = -2$

71. $3x + y = 6$
    $-2x + 2y = 4$

72. $2x - 2y = 6$
    $5y = 5x + 5$

Expanding Your Skills

73. Explain why a system of linear equations cannot have exactly two solutions.

74. The solution to the system of linear equations is $\{(1, 2)\}$. Find A and B.

$$Ax + 3y = 8$$
$$x + By = -7$$

75. The solution to the system of linear equations is $\{(-3, 4)\}$. Find A and B.

$$4x + Ay = -32$$
$$Bx + 6y = 18$$

Problem Recognition Exercises

Systems of Equations

When solving a system of linear equations in two variables, it is often helpful to analyze the equations to identify the number of solutions to the system. Slope-intercept form enables us to identify the slope of each line in the system and determine if the lines are

1) Intersecting (different slopes)
2) Parallel (same slope and different y-intercepts)
3) Coinciding (same slope and same y-intercept)

Example: $2x + y = 4$ ← Slanted line
$x = 3$ ← Vertical line

The first equation has both the x and y variables, indicating that it represents a slanted line. The second equation represents a vertical line. A slanted line and a vertical line intersect at exactly one point.

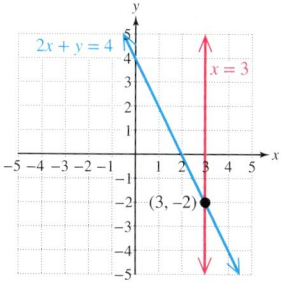

Solution: In this system, x is already isolated. Therefore, the substitution method is particularly easy to use. Substitute $x = 3$ into the first equation and solve for y.

$$2(3) + y = 4$$
$$6 + y = 4$$
$$y = -2$$

The solution set is $\{(3, -2)\}$.

Example: $3x + 2y = 6$
$5x + 3y = 7$

Both lines are slanted because the equations have both the x and y variables. Writing each equation in slope-intercept form, we have:

$$y = -\frac{3}{2}x + 3$$
$$y = -\frac{5}{3}x + \frac{7}{3}$$

The slopes are different, indicating that the lines intersect. There is exactly one point of intersection and thus, one solution.

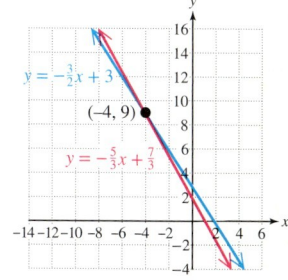

Solution: $3x + 2y = 6 \longrightarrow -9x - 6y = -18$
$5x + 3y = 7 \longrightarrow \underline{10x + 6y = 14}$
$ x = -4$

Substitute $x = -4$ into either original equation.

$$3(-4) + 2y = 6$$
$$-12 + 2y = 6$$
$$2y = 18$$
$$y = 9$$

The solution set is $\{(-4, 9)\}$.

We used the addition method here because none of the variable terms has a coefficient of 1. Thus, using the substitution method would have involved the use of fractional coefficients to solve the system.

Example: $x - 3y = 6$ $\frac{1}{3}x = y + 2$ Both lines are slanted because the equations have both the x and y variables. Writing each equation in slope-intercept form, we have: $y = \frac{1}{3}x - 2$ $y = \frac{1}{3}x - 2$ The slopes are the same ($\frac{1}{3}$) and the y-intercepts are the same $(0, -2)$. This indicates that the equations represent the same line. 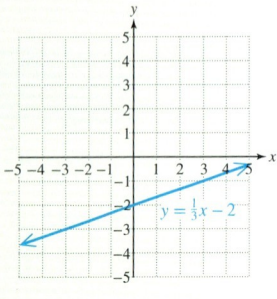	**Solution:** We can easily apply the substitution method by solving for x in the first equation or y in the second equation. $x - 3y = 6 \qquad x = 3y + 6$ $\frac{1}{3}x = y + 2 \longrightarrow \frac{1}{3}(3y + 6) = y + 2$ $\frac{1}{3}(3y + 6) = y + 2$ $y + 2 = y + 2$ $2 = 2$ The system of equations reduces to an identity. The equations represent the same line. There are infinitely many solutions.
Example: $3x + y = 2$ $6x + 2y = 0$ Both lines are slanted because the equations have both the x and y variables. Writing each equation in slope-intercept form, we have: $y = -3x + 2$ $y = -3x$ The slopes are the same (-3) but the y-intercepts are different. For the first equation, the y-intercept is $(0, 2)$, and for the second, the y-intercept is $(0, 0)$. Two lines with the same slope but different y-intercepts are parallel. 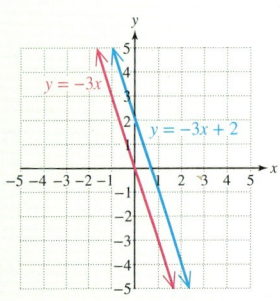	**Solution:** $3x + y = 2 \longrightarrow -6x - 2y = -4$ $6x + 2y = 0 \longrightarrow \underline{6x + 2y = 0}$ $0 = -4$ Using the addition method, we find that the system of equations reduces to a contradiction. Therefore, the lines are parallel and the system has no solution. The solution set is $\{\ \}$.

For Exercises 1–6, determine the number of solutions to the system without solving the system. Explain your answers.

1. $y = -4x + 2$
   $y = -4x + 2$

2. $y = -4x + 6$
   $y = -4x + 1$

3. $y = 4x - 3$
   $y = -4x + 5$

4. $y = 7$
   $2x + 3y = 1$

5. $2x + 3y = 1$
   $2x + 3y = 8$

6. $8x - 2y = 6$
   $12x - 3y = 9$

For Exercises 7–10, a method of solving has been suggested for each system of equations. Explain why that method was suggested for the system and then solve the system using the method given.

7. $2x - 5y = -11$ Addition Method
   $7x + 5y = -16$

8. $4x + 11y = 56$ Addition Method
   $-2x - 5y = -26$

9. $x = -3y + 4$ Substitution Method
   $5x + 4y = -2$

10. $2x + 3y = 16$ Substitution Method
    $y = x - 8$

For Exercises 11–30, solve each system using the method of your choice. For systems that do not have one unique solution, also state the number of solutions and whether the system is inconsistent or the equations are dependent.

11. $x = -2y + 5$
    $2x - 4y = 10$

12. $y = -3x - 4$
    $2x - y = 9$

13. $3x - 2y = 22$
    $5x + 2y = 10$

14. $-4x + 2y = -2$
    $4x - 5y = -7$

15. $\frac{1}{3}x + \frac{1}{2}y = \frac{2}{3}$
    $-\frac{2}{3}x + y = -\frac{4}{3}$

16. $\frac{1}{4}x + \frac{2}{5}y = 6$
    $\frac{1}{2}x - \frac{1}{10}y = 3$

17. $2c + 7d = -1$
    $c = 2$

18. $-3w + 5z = -6$
    $z = -4$

19. $y = 0.4x - 0.3$
    $-4x + 10y = 20$

20. $x = -0.5y + 0.1$
    $-10x - 5y = 2$

21. $3a + 7b = -3$
    $-11a + 3b = 11$

22. $2v - 5w = 10$
    $9v + 7w = 45$

23. $y = 2x - 14$
    $4x - 2y = 28$

24. $x = 5y - 9$
    $-2x + 10y = 18$

25. $x + y = 3200$
    $0.06x + 0.04y = 172$

26. $x + y = 4500$
    $0.07x + 0.05y = 291$

27. $3x + y - 7 = x - 4$
    $3x - 4y + 4 = -6y + 5$

28. $7y - 8y - 3 = -3x + 4$
    $10x - 5y - 12 = 13$

29. $3x - 6y = -1$
    $9x + 4y = 8$

30. $8x - 2y = 5$
    $12x + 4y = -3$

Section 12.4 Applications of Linear Equations in Two Variables

Concepts

1. Applications Involving Cost
2. Applications Involving Principal and Interest
3. Applications Involving Mixtures
4. Applications Involving Distance, Rate, and Time

1. Applications Involving Cost

We have solved several applied problems by setting up a linear equation in one variable. When solving an application that involves two unknowns, sometimes it is convenient to use a system of linear equations in two variables.

Example 1   Using a System of Linear Equations Involving Cost

At a movie theater, a couple buys one large popcorn and two small drinks for $12.50. A group of teenagers buys two large popcorns and five small drinks for $28.50. Find the cost of one large popcorn and the cost of one small drink.

Burke/Triolo/Brand X Pictures/Getty Images

Solution:

In this application we have two unknowns, which we can represent by x and y.

Let x represent the cost of one large popcorn.
Let y represent the cost of one small drink.

We must now write two equations. Each of the first two sentences in the problem gives a relationship between x and y:

$$\begin{pmatrix}\text{Cost of 1}\\\text{popcorn}\end{pmatrix} + \begin{pmatrix}\text{cost of 2}\\\text{drinks}\end{pmatrix} = \begin{pmatrix}\text{total}\\\text{cost}\end{pmatrix} \longrightarrow x + 2y = 12.50$$

$$\begin{pmatrix}\text{Cost of 2}\\\text{popcorns}\end{pmatrix} + \begin{pmatrix}\text{cost of 5}\\\text{drinks}\end{pmatrix} = \begin{pmatrix}\text{total}\\\text{cost}\end{pmatrix} \longrightarrow 2x + 5y = 28.50$$

To solve this system, we may use either the substitution method or the addition method. We will use the substitution method by solving for x in the first equation.

$x + 2y = 12.50 \longrightarrow x = -2y + 12.50$ Isolate x in the first equation.

$2x + 5y = 28.50$

$2(-2y + 12.50) + 5y = 28.50$ Substitute $x = -2y + 12.50$ into the other equation.

$-4y + 25.00 + 5y = 28.50$ Solve for y.

$y + 25.00 = 28.50$

$y = 3.50$

$x = -2y + 12.50$

$x = -2(3.50) + 12.50$ Substitute $y = 3.50$ into the equation $x = -2y + 12.50$.

$x = -7.00 + 12.50$

$x = 5.50$

The cost of one large popcorn is $5.50 and the cost of one small drink is $3.50.

Check by verifying that the solution meets the specified conditions.

1 popcorn + 2 drinks = 1($5.50) + 2($3.50) = $12.50 ✓   True

2 popcorns + 5 drinks = 2($5.50) + 5($3.50) = $28.50 ✓   True

Skill Practice

1. Lynn went to a fast-food restaurant and spent $20.00. She purchased 4 hamburgers and 5 orders of fries. The next day, Ricardo went to the same restaurant and purchased 10 hamburgers and 7 orders of fries. He spent $41.20. Use a system of equations to determine the cost of a burger and the cost of an order of fries.

2. Applications Involving Principal and Interest

Simple interest is interest computed on the principal amount of money invested (or borrowed). Simple interest, I, is found by using the formula

$$I = Prt \quad \text{where } P \text{ is the principal,}$$
$$r \text{ is the annual interest rate, and}$$
$$t \text{ is the time in years.}$$

In Example 2, we apply the concept of simple interest to two accounts to produce a desired amount of interest after 1 year.

Example 2 — Using a System of Linear Equations Involving Investments

Joanne has a total of $6000 to deposit in two accounts. One account earns 3.5% simple interest and the other earns 2.5% simple interest. If the total amount of interest at the end of 1 year is $195, find the amount she deposited in each account.

Solution:

Let x represent the principal deposited in the 2.5% account.
Let y represent the principal deposited in the 3.5% account.

	2.5% Account	3.5% Account	Total
Principal	x	y	6000
Interest ($I = Prt$)	$0.025x(1)$	$0.035y(1)$	195

Each row of the table yields an equation in x and y:

$$\begin{pmatrix}\text{Principal} \\ \text{invested} \\ \text{at } 2.5\%\end{pmatrix} + \begin{pmatrix}\text{principal} \\ \text{invested} \\ \text{at } 3.5\%\end{pmatrix} = \begin{pmatrix}\text{total} \\ \text{principal}\end{pmatrix} \longrightarrow x + y = 6000$$

$$\begin{pmatrix}\text{Interest} \\ \text{earned} \\ \text{at } 2.5\%\end{pmatrix} + \begin{pmatrix}\text{interest} \\ \text{earned} \\ \text{at } 3.5\%\end{pmatrix} = \begin{pmatrix}\text{total} \\ \text{interest}\end{pmatrix} \longrightarrow 0.025x + 0.035y = 195$$

We will choose the addition method to solve the system of equations. First multiply the second equation by 1000 to clear decimals.

Answer

1. The cost of a burger is $3.00, and the cost of an order of fries is $1.60.

$$x + y = 6000 \longrightarrow \quad x + y = 6000 \xrightarrow{\text{Multiply by } -25.} -25x - 25y = -150{,}000$$
$$0.025x + 0.035y = 195 \longrightarrow 25x + 35y = 195{,}000 \longrightarrow \underline{25x + 35y = 195{,}000}$$
$$\text{Multiply by 1000.} \qquad\qquad 10y = 45{,}000$$

$10y = 45{,}000$ After eliminating the x variable, solve for y.

$\dfrac{10y}{10} = \dfrac{45{,}000}{10}$

$y = 4500$ The amount invested in the 3.5% account is $4500.

$x + y = 6000$ Substitute $y = 4500$ into the equation $x + y = 6000$.

$x + 4500 = 6000$

$x = 1500$ The amount invested in the 2.5% account is $1500.

Joanne deposited $1500 in the 2.5% account and $4500 in the 3.5% account.

To check, verify that the conditions of the problem have been met.

1. The sum of $1500 and $4500 is $6000 as desired. ✓ True

2. The interest earned on $1500 at 2.5% is: $0.025(\$1500) = \$37.50$
   The interest earned on $4500 at 3.5% is: $\underline{0.035(\$4500) = \$157.50}$
                                           Total interest: $195.00 ✓ True

Skill Practice

2. Addie has a total of $8000 in two accounts. One pays 5% interest, and the other pays 6.5% interest. At the end of one year, she earned $475 interest. Use a system of equations to determine the amount invested in each account.

3. Applications Involving Mixtures

Example 3   **Using a System of Linear Equations in a Mixture Application**

According to new hospital standards, a certain disinfectant solution needs to be 20% alcohol instead of 10% alcohol. There is a 40% alcohol disinfectant available to adjust the mixture. Determine the amount of 10% solution and the amount of 40% solution to produce 30 L of a 20% solution.

Solution:

Each solution contains a percentage of alcohol plus some other mixing agent such as water. Before we set up a system of equations to model this situation, it is helpful to have background understanding of the problem. In Figure 12-7, the liquid depicted in blue is pure alcohol and the liquid shown in gray is the mixing agent (such as water). Together these liquids form a solution. (Realistically the mixture may not separate as shown, but this image may be helpful for your understanding.)

Let x represent the number of liters of 10% solution.

Let y represent the number of liters of 40% solution.

FOR REVIEW

Recall that the amount of a pure substance in a mixture is equal to the amount of mixture times the concentration rate of the substance. For example, in 30 L of a 20% alcohol mixture, the amount of pure alcohol is $(30 \text{ L})(0.20) = 6 \text{ L}$.

Answer

2. $3000 is invested at 5%, and $5000 is invested at 6.5%.

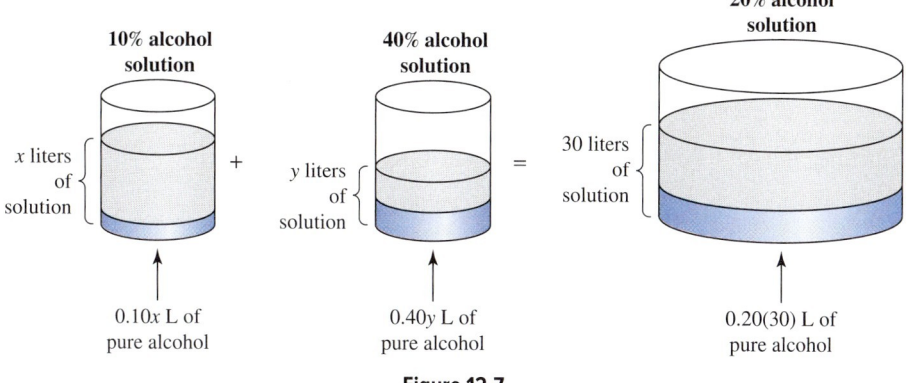

Figure 12-7

The information given in the statement of the problem can be organized in a chart.

	10% Alcohol	40% Alcohol	20% Alcohol
Number of liters of solution	x	y	30
Number of liters of pure alcohol	$0.10x$	$0.40y$	$0.20(30) = 6$

From the first row, we have

$$\begin{pmatrix} \text{Amount of} \\ \text{10\% solution} \end{pmatrix} + \begin{pmatrix} \text{amount of} \\ \text{40\% solution} \end{pmatrix} = \begin{pmatrix} \text{total amount} \\ \text{of 20\% solution} \end{pmatrix} \rightarrow x + y = 30$$

From the second row, we have

$$\begin{pmatrix} \text{Amount of} \\ \text{alcohol in} \\ \text{10\% solution} \end{pmatrix} + \begin{pmatrix} \text{amount of} \\ \text{alcohol in} \\ \text{40\% solution} \end{pmatrix} = \begin{pmatrix} \text{total amount of} \\ \text{alcohol in} \\ \text{20\% solution} \end{pmatrix} \rightarrow 0.10x + 0.40y = 6$$

We will solve the system with the addition method by first clearing decimals.

$$\begin{array}{l} x + y = 30 \\ 0.10x + 0.40y = 6 \end{array} \xrightarrow{\text{Multiply by 10.}} \begin{array}{l} x + y = 30 \\ x + 4y = 60 \end{array} \xrightarrow{\text{Multiply by } -1.} \begin{array}{r} -x - y = -30 \\ x + 4y = 60 \\ \hline 3y = 30 \end{array}$$

$3y = 30$ After eliminating the x variable, solve for y.

$y = 10$ 10 L of 40% solution is needed.

$x + y = 30$ Substitute $y = 10$ into either of the original equations.

$x + (10) = 30$

$x = 20$ 20 L of 10% solution is needed.

10 L of 40% solution must be mixed with 20 L of 10% solution.

Skill Practice

3. How many ounces of 20% and 35% acid solution should be mixed together to obtain 15 oz of 30% acid solution?

Answer

3. 10 oz of the 35% solution, and 5 oz of the 20% solution.

4. Applications Involving Distance, Rate, and Time

The following formula relates the distance traveled to the rate and time of travel.

$$d = rt \quad \text{distance} = \text{rate} \cdot \text{time}$$

For example, if a car travels 60 mph for 3 hr, then

$$d = (60 \text{ mph})(3 \text{ hr})$$
$$= 180 \text{ mi}$$

If a car travels 60 mph for x hr, then

$$d = (60 \text{ mph})(x \text{ hr})$$
$$= 60x \text{ mi}$$

The relationship $d = rt$ is used in Example 4.

Example 4: Using a System of Linear Equations in a Distance, Rate, and Time Application

A plane travels with the wind from Kansas City, Missouri, to Denver, Colorado, a distance of 600 mi in 2 hr. The return trip against the same wind takes 3 hr. Find the speed of the plane in still air, and find the speed of the wind.

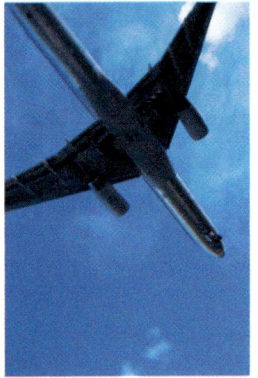
Stockbyte/Getty Images

Solution:

Let p represent the speed of the plane in still air.

Let w represent the speed of the wind.

Notice that when the plane travels with the wind, the net speed is $p + w$. When the plane travels against the wind, the net speed is $p - w$.

The information given in the problem can be organized in a chart.

	Distance	Rate	Time
With the wind	600	$p + w$	2
Against the wind	600	$p - w$	3

To set up two equations in p and w, recall that $d = rt$.

From the first row, we have

$$\begin{pmatrix} \text{Distance} \\ \text{with the wind} \end{pmatrix} = \begin{pmatrix} \text{rate with} \\ \text{the wind} \end{pmatrix} \begin{pmatrix} \text{time traveled} \\ \text{with the wind} \end{pmatrix} \longrightarrow 600 = (p + w) \cdot 2$$

From the second row, we have

$$\begin{pmatrix} \text{Distance} \\ \text{against the wind} \end{pmatrix} = \begin{pmatrix} \text{rate against} \\ \text{the wind} \end{pmatrix} \begin{pmatrix} \text{time traveled} \\ \text{against the wind} \end{pmatrix} \longrightarrow 600 = (p - w) \cdot 3$$

Using the distributive property to clear parentheses produces the following system:

$$2p + 2w = 600$$
$$3p - 3w = 600$$

The coefficients of the w variable can be changed to 6 and -6 by multiplying the first equation by 3 and the second equation by 2.

$2p + 2w = 600$ $\xrightarrow{\text{Multiply by 3.}}$ $6p + 6w = 1800$
$3p - 3w = 600$ $\xrightarrow{\text{Multiply by 2.}}$ $\underline{6p - 6w = 1200}$
$\phantom{3p - 3w = 600 \xrightarrow{\text{Multiply by 2.}}}$ $12p = 3000$

$12p = 3000$

$\dfrac{12p}{12} = \dfrac{3000}{12}$

$p = 250$ The speed of the plane in still air is 250 mph.

TIP: To create opposite coefficients on the w variables, we could have divided the first equation by 2 and divided the second equation by 3:

$2p + 2w = 600$ $\xrightarrow{\text{Divide by 2.}}$ $p + w = 300$
$3p - 3w = 600$ $\xrightarrow{\text{Divide by 3.}}$ $\underline{p - w = 200}$
$\phantom{3p - 3w = 600 \xrightarrow{\text{Divide by 3.}}}$ $2p = 500$
$\phantom{3p - 3w = 600 \xrightarrow{\text{Divide by 3.}}}$ $p = 250$

$2p + 2w = 600$ Substitute $p = 250$ into the first equation.
$2(250) + 2w = 600$
$500 + 2w = 600$
$2w = 100$
$w = 50$ The speed of the wind is 50 mph.

The speed of the plane in still air is 250 mph. The speed of the wind is 50 mph.

Skill Practice

4. Dan and Cheryl paddled their canoe 40 mi in 5 hr with the current and 16 mi in 8 hr against the current. Find the speed of the current and the speed of the canoe in still water.

Answer

4. The speed of the canoe in still water is 5 mph. The speed of the current is 3 mph.

Section 12.4 Activity

A.1. A benefit dinner included two raffles. One prize was for a gift certificate to a restaurant, and the tickets sold for $1 each. The other prize was for a TV, and the tickets sold for $2 each. The total amount of money received from both raffles was $359, and the total number of tickets sold was 248. How many of each type of ticket were sold?
   a. Set up an equation that represents the fact that the number of one-dollar tickets, x, plus the number of two-dollar tickets, y, equals 248.

   Set up an equation that indicates that the revenue generated by the one-dollar tickets plus the revenue generated by the two-dollar tickets equals $359.
   b. Solve the system of equations from part (a).
   c. Interpret the meaning of the solution from part (b) and verify that the solution meets the conditions of the problem.

A.2. a. If P dollars is invested at an annual interest rate r for t years, write a formula for the simple interest I earned.
  b. If $7000 is invested at 3% simple interest for 1 year, how much interest is earned?
  c. If x dollars is invested at 3% simple interest for 1 year, write an expression for the amount of interest earned.
  d. If y dollars is invested at 4% simple interest for 1 year, write an expression for the amount of interest earned.

A.3. Fred invests $3000 more in an account that pays 4% simple interest than he invests in an account that pays 3% simple interest. If the total interest earned for the first year is $610, determine how much did Fred invested in each account.
  a. Let x represent the amount invested in the account paying 3% interest.
  What should y represent?
  b. In this scenario, there are two "types" of money: principal (amount invested) and interest. Refer to Exercises A.2(c) and A.2(d) and fill in the table with expressions for the principal and interest in each account.

	3% Account	4% Account
Amount invested ($)		
Interest earned ($)		

  c. Set up an equation that indicates that there is $3000 more invested in the 4% account than in the 3% account.
  Set up an equation that indicates that the total annual interest generated by the 3% account plus the interest generated by the 4% account is $610.
  d. Solve the system of equations from part (c).
  e. Interpret the meaning of the solution from part (d).

A.4. Suppose that water is mixed with acid to make a diluted acid solution.
  a. If a chemist has 16 L of a 12% acid solution, how much is pure acid?
  b. If a chemist has x liters of a 12% acid solution, write an expression representing the amount of pure acid.
  c. If a chemist has y liters of a 27% acid solution, write an expression representing the amount of pure acid.

A.5. A 12% acid solution is to be mixed with a 27% acid solution to get 20 L of a 15% acid solution. How much 12% solution and how much 27% solution should be mixed?
  a. Let x represent the amount of the 12% acid solution. What should y represent?
  b. Refer to Exercises A.4(b) and A.4(c) and fill in the table.

	12% Mixture	27% Mixture	15% Mixture
Amount of solution (L)			
Amount of pure acid (L)			

  c. Set up an equation that indicates that the amount of 12% solution plus the amount of 27% solution equals a final mixture of 20 L.
  Set up an equation that indicates that the amount of pure acid from the 12% solution plus the amount of pure acid from the 27% solution equals the amount of pure acid from 20 L of the 15% solution.
  d. Solve the system of equations from part (c).
  e. Interpret the meaning of the solution from part (d).

A.6. a. Suppose that a boat normally travels 10 mph in still water (without a current). If the boat travels *with* a current of 2 mph, what is the net speed? In such a case, how far would the boat travel in 2 hr?
  b. Suppose that a boat normally travels 10 mph in still water. If the boat travels *against* a current of 2 mph, what is the net speed? In such a case, how far would the boat travel in 2 hr?

c. Suppose that a boat normally travels b mph in still water. If the boat travels *with* a current of c mph, write an expression for the net speed. Write an expression representing the distance the boat would travel in 2 hr.

d. Suppose that a boat normally travels b mph in still water. If the boat travels *against* a current of c mph, write an expression for the net speed. Write an expression representing the distance the boat would travel in 3 hr.

A.7. A cruise ship travels 48 mi in 2 hr with the current. Against the current, the same trip takes 3 hr. What is the speed of the current and the speed of the ship in still water?

a. Let b represent the speed of the ship in still water. What should c represent?

b. Refer to Exercises A.6(c) and A.6(d) and complete the table.

	Distance (mi)	Rate (mph)	Time (hr)
With the current			
Against the current			

c. Set up an equation that indicates that the distance traveled with the current equals the rate of speed with the current times the time of travel with the current.

Set up an equation that indicates that the distance traveled against the current equals the rate of speed against the current times the time of travel against the current.

d. Solve the system of equations from part (c).

e. Interpret the meaning of the solution from part (d).

Practice Exercises

Section 12.4

Study Skills Exercise

With word problems, important mathematical relationships are often hidden within the text of the problem. This requires that you read the narrative and understand the vocabulary and key ideas. For example, consider the following word problem:
One acute angle in a right triangle is 20° more than the other acute angle. Find the measures of the acute angles.

- To extract the information necessary to solve this problem, you need to be familiar with key vocabulary. In this case, what is meant by an "acute angle?"
- As you read a word problem, be thinking of key mathematical formulas or theorems that might apply to the given problem. In this case, how are three angles within a triangle related? How are the two acute angles related?

Prerequisite Review

For Exercises R.1–R.4, solve each system of equations by three different methods:

    a. Graphing method   **b.** Substitution method   **c.** Addition method

R.1. $-2x + y = 6$
       $2x + y = 2$

R.2. $x - y = 2$
       $x + y = 6$

R.3. $y = -2x + 6$
       $4x - 2y = 8$

R.4. $3x + y = 2$
       $6x - 2y = 0$

For Exercises R.5–R.12, set up a system of linear equations in two variables and solve for the unknown quantities.

R.5. The sum of two numbers is 41. One number is 5 more than the other. Find the numbers.

R.6. The sum of two numbers is 76. One number is 6 less than the other. Find the numbers.

R.7. The two acute angles in a right triangle differ by 10°. Find the measures of the angles.

R.8. One acute angle in a right triangle is 15° more than twice the measure of the other acute angle. Find the measures of the angles.

R.9. One number is eight more than twice another. Their sum is 20. Find the numbers.

R.10. The difference of two positive numbers is 264. The larger number is three times the smaller number. Find the numbers.

R.11. Two angles are complementary. The measure of one angle is 10° less than nine times the measure of the other. Find the measure of each angle.

R.12. Two angles are supplementary. The measure of one angle is 9° more than twice the measure of the other angle. Find the measure of each angle.

Concept 1: Applications Involving Cost

1. An online store sells old video games and collector DVDs as a bundle. A bundle of two video games and three DVDs can be purchased for $88. A bundle of one video game and two DVDs can be purchased for $51.50. Find the cost of one video game and the cost of one DVD in the bundle. (See Example 1.)

2. Tanya bought three adult tickets and one children's ticket to a movie for $32.00. Li bought two adult tickets and five children's tickets for $49.50. Find the cost of one adult ticket and the cost of one children's ticket.

3. Nora bought 100 shares of a technology stock and 200 shares of a mutual fund for $3800. Her sister, Erin, bought 300 shares of the technology stock and 50 shares of the same mutual fund for $5350. Find the cost per share of the technology stock, and the cost per share of the mutual fund.

4. Eight students in Ms. Reese's class decided to purchase their textbooks from two different sources. Some students purchased the textbook from the college bookstore for $95.50. The other students purchased the textbook from an online discount store for $65 per book. If the total amount spent by the eight students is $611.50, how many students purchased the book online?

5. During the COVID-19 pandemic, Armondo helped his community by making tacos to give to hospital personnel. Some tacos had pork filling and some had beef filling. He made a total of 200 tacos, and he made 30 more beef tacos than pork tacos. How many of each type of taco did he make?

6. Zoey purchased some beef and some chicken for a family barbeque. The beef cost $6.00 per pound and the chicken cost $4.50 per pound. She bought a total of 18 lb of meat and spent $96. How much of each type of meat did she purchase?

Concept 2: Applications Involving Principal and Interest

7. Shanelle invested $10,000, and at the end of 1 year, she received $805 in interest. She invested part of the money in an account earning 10% simple interest and the remaining money in an account earning 7% simple interest. How much did she invest in each account? (See Example 2.)

8. $12,000 was borrowed from two sources, one that charges 12% simple interest and the other that charges 8% simple interest. If the total interest at the end of 1 year was $1240, how much money was borrowed from each source?

	10% Account	7% Account	Total
Principal invested			
Interest earned			

	12% Account	8% Account	Total
Principal borrowed			
Interest earned			

9. Troy borrowed a total of $12,000 in two different loans to help pay for his new truck. One loan charges 9% simple interest, and the other charges 6% simple interest. If he is charged $810 in interest after 1 year, find the amount borrowed at each rate.

10. Blake has a total of $4000 to invest in two accounts. One account earns 2% simple interest, and the other earns 5% simple interest. How much should be invested in each account to earn exactly $155 at the end of 1 year?

11. Suppose a rich uncle dies and leaves you an inheritance of $30,000. You decide to invest part of the money in a relatively safe bond fund that returns 8%. You invest the rest of the money in a riskier stock fund that you hope will return 12% at the end of 1 year. If you need $3120 at the end of 1 year to make a down payment on a car, how much should you invest at each rate?

12. As part of his retirement strategy, John plans to invest $200,000 in two different funds. He projects that the moderately high-risk investments should return, over time, about 9% per year, while the low-risk investments should return about 4% per year. If he wants a supplemental income of $12,000 a year, how should he divide his investments?

Concept 3: Applications Involving Mixtures

13. How much 50% disinfectant solution must be mixed with a 40% disinfectant solution to produce 25 gal of a 46% disinfectant solution? (See Example 3.)

	50% Mixture	40% Mixture	46% Mixture
Amount of solution			
Amount of disinfectant			

14. How many gallons of 20% antifreeze solution and a 10% antifreeze solution must be mixed to obtain 40 gal of a 16% antifreeze solution?

	20% Mixture	10% Mixture	16% Mixture
Amount of solution			
Amount of antifreeze			

15. How much 45% disinfectant solution must be mixed with a 30% disinfectant solution to produce 20 gal of a 39% disinfectant solution?

16. How many gallons of a 25% antifreeze solution and a 15% antifreeze solution must be mixed to obtain 15 gal of a 23% antifreeze solution?

17. A chemist needs 50 mL of a 16% salt solution for an experiment. She can only find a 13% salt solution and an 18% salt solution in the supply room. How many milliliters of the 13% solution should be mixed with the 18% solution to produce the desired amount of the 16% solution?

18. Meadowsilver Dairy keeps two kinds of milk on hand, skim milk that has 0.3% butterfat and whole milk that contains 3.3% butterfat. How many gallons of each type of milk does the company need to produce 300 gal of 1% milk for the P&A grocery store?

19. The cooling system in most cars requires a mixture that is 50% antifreeze. How many liters of pure antifreeze and how many liters of 40% antifreeze solution should Chad mix to obtain 6 L of 50% antifreeze solution? (*Hint:* Pure antifreeze is 100% antifreeze.)

20. Silvia wants to mix a 40% apple juice drink with pure apple juice to make 2 L of a juice drink that is 80% apple juice. How much pure apple juice should she use?

Concept 4: Applications Involving Distance, Rate, and Time

21. It takes a boat 2 hr to go 16 mi downstream with the current and 4 hr to return against the current. Find the speed of the boat in still water and the speed of the current. (See Example 4.)

	Distance	Rate	Time
Downstream			
Upstream			

22. A boat takes 1.5 hr to go 12 mi upstream against the current. It can go 24 mi downstream with the current in the same amount of time. Find the speed of the current and the speed of the boat in still water.

	Distance	Rate	Time
Upstream			
Downstream			

23. A plane can fly 960 mi with the wind in 3 hr. It takes the same amount of time to fly 840 mi against the wind. What is the speed of the plane in still air and the speed of the wind?

24. A plane flies 720 mi with the wind in 3 hr. The return trip against the wind takes 4 hr. What is the speed of the wind and the speed of the plane in still air?

25. Tony flew from JFK Airport to London. It took him 6 hr to fly with the wind, and 8 hr on the return flight against the wind. If the distance is approximately 3600 mi, determine the speed of the plane in still air and the speed of the wind.

26. A riverboat cruise upstream on the Mississippi River from New Orleans, Louisiana, to Natchez, Mississippi, takes 10 hr and covers 140 mi. The return trip downstream with the current takes only 7 hr. Find the speed of the riverboat in still water and the speed of the current.

PictureNet/Getty Images

David Planchet

Mixed Exercises

27. Debi has $2.80 in a collection of dimes and nickels. The number of nickels is five more than the number of dimes. Find the number of each type of coin.

28. A child collects state quarters and new $1 coins. If she has a total of 25 coins, and the number of quarters is nine more than the number of dollar coins, how many of each type of coin does she have?

29. How many quarts of water should be mixed with a 30% vinegar solution to obtain 12 qt of a 25% vinegar solution? (*Hint:* Water is 0% vinegar.)

30. How much water should be mixed with an 8% plant fertilizer solution to make a half gallon of a 6% plant fertilizer solution?

31. In the 1961–1962 NBA basketball season, Wilt Chamberlain of the Philadelphia Warriors made 2432 baskets. Some of the baskets were free throws (worth 1 point each) and some were field goals (worth 2 points each). The number of field goals was 762 more than the number of free throws.

   a. How many field goals did he make and how many free throws did he make?

   b. What was the total number of points scored?

   c. If Wilt Chamberlain played 80 games during this season, what was the average number of points per game?

32. In the 1971–1972 NBA basketball season, Kareem Abdul-Jabbar of the Milwaukee Bucks made 1663 baskets. Some of the baskets were free throws (worth 1 point each) and some were field goals (worth 2 points each). The number of field goals he scored was 151 more than twice the number of free throws.

   a. How many field goals did he make and how many free throws did he make?

   b. What was the total number of points scored?

   c. If Kareem Abdul-Jabbar played 81 games during this season, what was the average number of points per game?

33. A small plane can fly 350 mi with the wind in $1\frac{3}{4}$ hr. In the same amount of time, the same plane can travel only 210 mi against the wind. What is the speed of the plane in still air and the speed of the wind?

34. A plane takes 2 hr to travel 1000 mi with the wind. It can travel only 880 mi against the wind in the same amount of time. Find the speed of the wind and the speed of the plane in still air.

35. A total of $60,000 is invested in two accounts, one that earns 5.5% simple interest, and one that earns 6.5% simple interest. If the total interest at the end of 1 year is $3750, find the amount invested in each account.

36. Jacques borrows a total of $15,000. Part of the money is borrowed from a bank that charges 12% simple interest per year. Jacques borrows the remaining part of the money from his sister and promises to pay her 7% simple interest per year. If Jacques' total interest for the year is $1475, find the amount he borrowed from each source.

37. At the holidays, Erica likes to sell a candy/nut mixture to her neighbors. She wants to combine candy that costs $1.80 per pound with nuts that cost $1.20 per pound. If Erica needs 20 lb of mixture that will sell for $1.56 per pound, how many pounds of candy and how many pounds of nuts should she use?

38. Mary Lee's natural food store sells a combination of teas. The most popular is a mixture of a tea that sells for $3.00 per pound with one that sells for $4.00 per pound. If she needs 40 lb of tea that will sell for $3.65 per pound, how many pounds of each tea should she use?

Jill Braaten/McGraw Hill

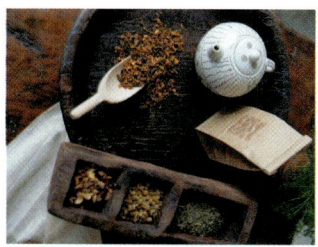

Steve Mason/Getty Images

39. In a football game, the Dallas Cowboys scored four more points than twice the number of points scored by the Buffalo Bills. If the total number of points scored by both teams was 43, find the number of points scored by each team.

40. In the 1973 Super Bowl, the undefeated Miami Dolphins football team scored twice as many points as the Washington Redskins. If the total number of points scored by both teams was 21, find the number of points scored by each team.

Mark Herreid/Getty Images

Expanding Your Skills

41. In a survey conducted among 500 college students, 340 said that the campus lacked adequate lighting. If $\frac{4}{5}$ of the women and $\frac{1}{2}$ of the men said that they thought the campus lacked adequate lighting, how many men and how many women were in the survey?

42. During a 1-hr television program, there were 22 commercials. Some commercials were 15 sec and some were 30 sec long. Find the number of 15-sec commercials and the number of 30-sec commercials if the total playing time for commercials was 9.5 min.

Section 12.5
Systems of Linear Equations in Three Variables

1. Solutions to Systems of Linear Equations in Three Variables

In this section, we will expand the discussion of solving systems of linear equations in two variables to solving systems involving three variables.

A **linear equation in three variables** can be written in the form $Ax + By + Cz = D$, where A, B, and C are not all zero. For example, the equation $2x + 3y + z = 6$ is a linear

Concepts

1. Solutions to Systems of Linear Equations in Three Variables
2. Solving Systems of Linear Equations in Three Variables

equation in three variables. Solutions to this equation are **ordered triples** of the form (x, y, z) that satisfy the equation. Some solutions to the equation $2x + 3y + z = 6$ are

<u>Solution:</u>   <u>Check:</u>

$(1, 1, 1) \longrightarrow 2(1) + 3(1) + (1) = 6$ ✓ True

$(2, 0, 2) \longrightarrow 2(2) + 3(0) + (2) = 6$ ✓ True

$(0, 1, 3) \longrightarrow 2(0) + 3(1) + (3) = 6$ ✓ True

Infinitely many ordered triples serve as solutions to the equation $2x + 3y + z = 6$.

The set of all ordered triples that are solutions to a linear equation in three variables may be represented graphically by a plane in space. Figure 12-8 shows a portion of the plane $2x + 3y + z = 6$ in a 3-dimensional coordinate system.

A solution to a system of linear equations in three variables is an ordered triple that satisfies *each* equation. Geometrically, a solution is a point of intersection of the planes represented by the equations in the system.

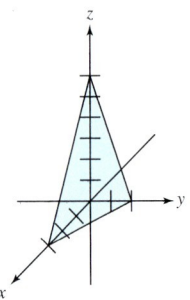

Figure 12-8

A system of linear equations in three variables may have *one unique solution, infinitely many solutions,* or *no solution*.

One unique solution (planes intersect at one point)
- The system is consistent.
- The equations are independent.

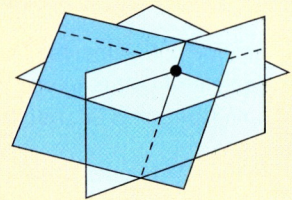

No solution (the three planes do not all intersect)
- The system is inconsistent.
- The equations are independent.

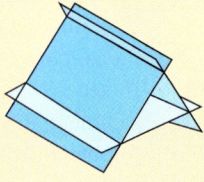 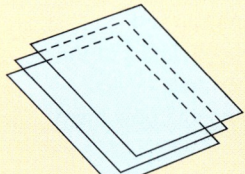 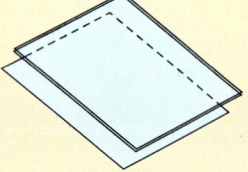

Infinitely many solutions (planes intersect at infinitely many points)
- The system is consistent.
- The equations are dependent.

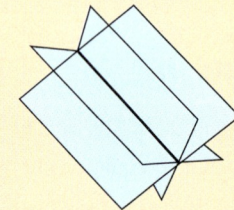 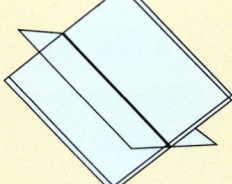 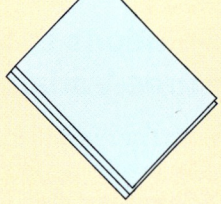

2. Solving Systems of Linear Equations in Three Variables

To solve a system involving three variables, the goal is to eliminate one variable. This reduces the system to two equations in two variables. One strategy for eliminating a variable is to pair up the original equations two at a time.

Solving a System of Three Linear Equations in Three Variables

Step 1 Write each equation in standard form $Ax + By + Cz = D$.
Step 2 Choose a pair of equations, and eliminate one of the variables by using the addition method.
Step 3 Choose a different pair of equations and eliminate the *same* variable.
Step 4 Once steps 2 and 3 are complete, you should have two equations in two variables. Solve this system by using the substitution or the addition method.
Step 5 Substitute the values of the variables found in step 4 into any of the three original equations that contain the third variable. Solve for the third variable.
Step 6 Check the ordered triple in each of the original equations.

Example 1 Solving a System of Linear Equations in Three Variables

Solve the system.
$$2x + y - 3z = -7$$
$$3x - 2y + z = 11$$
$$-2x - 3y - 2z = 3$$

Solution:

$\boxed{A}$ $\;2x + y - 3z = -7$
$\boxed{B}$ $\;3x - 2y + z = 11$
$\boxed{C}$ $-2x - 3y - 2z = 3$

Step 1: The equations are already in standard form.
- It is often helpful to label the equations.
- The y variable can be easily eliminated from equations $\boxed{A}$ and $\boxed{B}$ and from equations $\boxed{A}$ and $\boxed{C}$. This is accomplished by creating opposite coefficients for the y terms and then adding the equations.

Step 2: Eliminate the y variable from equations $\boxed{A}$ and $\boxed{B}$.

$\boxed{A}$ $\;2x + y - 3z = -7$ $\xrightarrow{\text{Multiply by 2.}}$ $\;4x + 2y - 6z = -14$
$\boxed{B}$ $\;3x - 2y + z = 11$ $\longrightarrow$ $\;3x - 2y + z = 11$
$\phantom{\boxed{B}\;3x - 2y + z = 11 \longrightarrow}$ $\;7x - 5z = -3\;\boxed{D}$

Chapter 12 Systems of Linear Equations in Two Variables

TIP: It is important to note that in steps 2 and 3, the *same* variable is eliminated.

Step 3: Eliminate the y variable again, this time from equations $\boxed{A}$ and $\boxed{C}$.

$\boxed{A}$ $2x + y - 3z = -7$ $\xrightarrow{\text{Multiply by 3.}}$ $6x + 3y - 9z = -21$
$\boxed{C}$ $-2x - 3y - 2z = 3$ $\longrightarrow$ $\underline{-2x - 3y - 2z = 3}$
$4x - 11z = -18$ $\boxed{E}$

Step 4: Now equations $\boxed{D}$ and $\boxed{E}$ can be paired up to form a linear system in two variables. Solve this system.

$\boxed{D}$ $7x - 5z = -3$ $\xrightarrow{\text{Multiply by } -4.}$ $-28x + 20z = 12$
$\boxed{E}$ $4x - 11z = -18$ $\xrightarrow{}$ $\underline{28x - 77z = -126}$
 Multiply by 7.
$-57z = -114$
$z = 2$

Once one variable has been found, substitute this value into either equation in the two-variable system, that is, either equation $\boxed{D}$ or $\boxed{E}$.

$\boxed{D}$ $7x - 5z = -3$
$7x - 5(2) = -3$ Substitute $z = 2$ into equation $\boxed{D}$.
$7x - 10 = -3$
$7x = 7$
$x = 1$

$\boxed{A}$ $2x + y - 3z = -7$
$2(1) + y - 3(2) = -7$
$2 + y - 6 = -7$
$y - 4 = -7$
$y = -3$

Step 5: Now that two variables are known, substitute these values for x and z into any of the original three equations to find the remaining variable y. Substitute $x = 1$ and $z = 2$ into equation $\boxed{A}$.

Step 6: Check the ordered triple, $(1, -3, 2)$, in the three original equations.

Check: $2x + y - 3z = -7 \rightarrow 2(1) + (-3) - 3(2) = -7$ ✓ True
$3x - 2y + z = 11 \rightarrow 3(1) - 2(-3) + (2) = 11$ ✓ True
$-2x - 3y - 2z = 3 \rightarrow -2(1) - 3(-3) - 2(2) = 3$ ✓ True

The solution set is $\{(1, -3, 2)\}$.

Skill Practice Solve the system.

1. $x + 2y + z = 1$
$3x - y + 2z = 13$
$2x + 3y - z = -8$

Answer

1. $\{(1, -2, 4)\}$

Example 2 Solving a System of Linear Equations in Three Variables

Solve the system.
$$3x + y - 5 = 0$$
$$z = 7 + 2y$$
$$x + 3z = 17$$

Solution:

A	$3x + y = 5$
B	$-2y + z = 7$
C	$x + 3z = 17$

Step 1: Write equations in standard form, $Ax + By + Cz = D$.

Steps 2 and 3: Note equation C is missing the y variable. Therefore, we can eliminate the y variable by pairing up equations A and B.

A $3x + y = 5$ $\xrightarrow{\text{Multiply by 2.}}$ $6x + 2y = 10$
B $-2y + z = 7$ $\longrightarrow$ $\underline{-2y + z = 7}$
$6x + z = 17$ D

Step 4: Now we pair up equations C and D to eliminate the z variable.

C $x + 3z = 17$ $\longrightarrow$ $x + 3y = 17$
D $6x + z = 17$ $\xrightarrow{\text{Multiply by } -3.}$ $\underline{-18x - 3z = -51}$
$-17x = -34$
$x = 2$

Step 5: Now substitute $x = 2$ into equations A and C to find the remaining variables.

A $3x + y = 5$ C $x + 3z = 17$
$3(2) + y = 5$ $(2) + 3z = 17$
$6 + y = 5$ $3z = 15$
$y = -1$ $z = 5$

Step 6: Check $(2, -1, 5)$ in the three original equations.

Check: $3x + y - 5 = 0$ $\longrightarrow$ $3(2) + (-1) - 5 = 0$ ✓ True
$z = 7 + 2y$ $\longrightarrow$ $5 = 7 + 2(-1)$ ✓ True
$x + 3z = 17$ $\longrightarrow$ $(2) + 3(5) = 17$ ✓ True

The solution set is $\{(2, -1, 5)\}$.

Skill Practice Solve the system.

2. $4y = x + 4$
$2x + z = 5$
$y - 14 = 4z$

Answer

2. $\{(4, 2, -3)\}$

Example 3 Solving a System of Dependent Equations

Solve the system. If there is not a unique solution, state the number of solutions and whether the system is inconsistent or the equations are dependent.

$$\boxed{A} \quad 3x + y - z = 8$$
$$\boxed{B} \quad 2x - y + 2z = 3$$
$$\boxed{C} \quad x + 2y - 3z = 5$$

Solution:

First, make a decision regarding the variable to eliminate. The y variable is particularly easy to eliminate because the coefficients of y in equations $\boxed{A}$ and $\boxed{B}$ are already opposites. The y variable can be eliminated from equations $\boxed{B}$ and $\boxed{C}$ by multiplying equation $\boxed{B}$ by 2.

$$\boxed{A} \quad 3x + y - z = 8$$
$$\boxed{B} \quad \underline{2x - y + 2z = 3}$$
$$\phantom{\boxed{B}} \quad 5x + z = 11 \quad \boxed{D}$$

Pair up equations $\boxed{A}$ and $\boxed{B}$ to eliminate y.

$$\boxed{B} \quad 2x - y + 2z = 3 \xrightarrow{\text{Multiply by 2.}} 4x - 2y + 4z = 6$$
$$\boxed{C} \quad x + 2y - 3z = 5 \xrightarrow{\phantom{\text{Multiply by 2.}}} \underline{x + 2y - 3z = 5}$$
$$\phantom{\boxed{C} \quad x + 2y - 3z = 5 \xrightarrow{\phantom{\text{Multiply by 2.}}}} 5x + z = 11 \quad \boxed{E}$$

Pair up equations $\boxed{B}$ and $\boxed{C}$ to eliminate y.

Because equations $\boxed{D}$ and $\boxed{E}$ are equivalent equations, this is a dependent system. By eliminating variables we obtain the identity $0 = 0$.

$$\boxed{D} \quad 5x + z = 11 \xrightarrow{\text{Multiply by } -1.} -5x - z = -11$$
$$\boxed{E} \quad 5x + z = 11 \xrightarrow{\phantom{\text{Multiply by } -1.}} \underline{5x + z = 11}$$
$$\phantom{\boxed{E} \quad 5x + z = 11 \xrightarrow{\phantom{\text{Multiply by } -1.}}} 0 = 0$$

The result $0 = 0$ indicates that there are infinitely many solutions and that the equations are dependent.

Skill Practice Solve the system. If the system does not have a unique solution, state the number of solutions and whether the system is inconsistent or the equations are dependent.

3. $x + y + z = 8$
    $2x - y + z = 6$
    $-5x - 2y - 4z = -30$

Answer

3. Infinitely many solutions; dependent equations

Example 4 Solving an Inconsistent System of Linear Equations

Solve the system. If there is not a unique solution, state the number of solutions and whether the system is inconsistent or the equations are dependent.

$$2x + 3y - 7z = 4$$
$$-4x - 6y + 14z = 1$$
$$5x + y - 3z = 6$$

Solution:

We will eliminate the x variable.

|A| $2x + 3y - 7z = 4$ $\xrightarrow{\text{Multiply by 2.}}$ $4x + 6y - 14z = 8$
|B| $-4x - 6y + 14z = 1$ $\longrightarrow$ $-4x - 6y + 14z = 1$
|C| $5x + y - 3z = 6$ $0 = 9$ (contradiction)

The result $0 = 9$ is a contradiction, indicating that the system has no solution. The system is inconsistent.

Skill Practice Solve the system. If the system does not have a unique solution, state the number of solutions and whether the system is inconsistent or the equations are dependent.

4. $x - 2y + z = 5$
    $x - 3y + 2z = -7$
    $-2x + 4y - 2z = 6$

Answer
4. No solution; inconsistent system

Section 12.5 Activity

A.1. **a.** The graph of a linear equation in two variables $Ax + By = C$ is a line, whereas the graph of a linear equation in three variables $Ax + By + Cz = D$ is a _____ in space.
   b. A solution to a linear equation in two variables $Ax + By = C$ is an ordered _____ of the form (x, y). A solution to a linear equation in three variables $Ax + By + Cz = D$ is an ordered _____ of the form (x, y, z).

A.2. Determine if the ordered triple is a solution to the system.

$$x + 2y + 3z = 1$$
$$3x - y + z = -1$$
$$2x - 3y - 4z = 4$$

   a. $(1, -3, 2)$ **b.** $(2, 4, -3)$

The basic premise to solve a system of linear equations in three variables is to eliminate one of the variables to create a system of two-variable equations. In Exercise A.3, we walk through this process.

A.3. For the given system, we must first choose a variable to eliminate. We will arbitrarily choose variable z to eliminate.

$$x + 2y + 3z = 1 \quad \boxed{A}$$
$$3x - y + z = -1 \quad \boxed{B}$$
$$2x - 3y - 4z = 4 \quad \boxed{C}$$

a. Choose two equations from the original system and eliminate the z variable. For example, pair up equations $\boxed{A}$ and $\boxed{B}$. Multiply equation $\boxed{B}$ by -3, then add the equations. Label the resulting equation as $\boxed{D}$.

$$x + 2y + 3z = 1 \quad \boxed{A}$$
$$3x - y + z = -1 \quad \boxed{B}$$

b. Repeat part (a) with a different pair of equations from the original system. For example, pair up equations $\boxed{B}$ and $\boxed{C}$. Then eliminate the z variable. Label the resulting equation as $\boxed{E}$.

$$3x - y + z = -1 \quad \boxed{B}$$
$$2x - 3y - 4z = 4 \quad \boxed{C}$$

c. Solve for the remaining variables x and y by solving the system given by equations $\boxed{D}$ and $\boxed{E}$.

d. Substitute the values of x and y found in part (c) into one of the original equations in the system. Then solve the equation for z.

e. Write the solution set as an ordered triple. Be sure to check the ordered triple in all three original equations.

A.4. The following system has missing variables within the individual equations. This makes the system slightly easier to solve.

$$3x + 4y + z = -3 \quad \boxed{A}$$
$$5x \phantom{{}+ 4y} + 4z = -10 \quad \boxed{B}$$
$$\phantom{5x +{}} 2y - 3z = 13 \quad \boxed{C}$$

a. Equations $\boxed{B}$ and $\boxed{C}$ each have a missing variable. For example, y is missing from equation $\boxed{B}$. Therefore, eliminate y from equations $\boxed{A}$ and $\boxed{C}$. Label the resulting equation as $\boxed{D}$.

b. Pair up equation $\boxed{B}$ from the original system and equation $\boxed{D}$ to form a system of equations involving the variables x and z. Solve this system for x and z.

c. Substitute the values of x and z found in part (c) into one of the original equations in the system. Then solve the equation for y.

d. Write the solution set as an ordered triple. Be sure to check the ordered triple in all three original equations.

A system of linear equations in three variables may have no solution. In such a case, you will encounter a contradiction such as $0 = 4$ and will identify the system as inconsistent. Likewise, a system of three variables may have dependent equations. In such a case, you will encounter an identity such as $0 = 0$ and will note that there are infinitely many solutions to the system. In Exercises A.5 and A.6, we investigate these two scenarios.

A.5. Consider the system.

$$4x \phantom{{}- 9y} + 6z = 6 \quad \boxed{A}$$
$$8x - 9y + 15z = 9 \quad \boxed{B}$$
$$4x - 3y + 7z = 5 \quad \boxed{C}$$

a. Eliminate y from equations $\boxed{B}$ and $\boxed{C}$. Label the resulting equation as $\boxed{D}$.

b. As you solve the system given by equations $\boxed{A}$ and $\boxed{D}$, do you encounter a contradiction or an identity? From this finding, does the original system have no solution or infinitely many solutions?

A.6. Consider the system.

$$x + 2y - 3z = -1 \quad \boxed{A}$$
$$3x - 9y + 3z = 3 \quad \boxed{B}$$
$$2x - y - 2z = 2 \quad \boxed{C}$$

a. Eliminate x from equations $\boxed{A}$ and $\boxed{B}$. Label the resulting equation as $\boxed{D}$.

b. Eliminate x from equations $\boxed{A}$ and $\boxed{C}$. Label the resulting equation as $\boxed{E}$.

c. As you solve the system given by equations $\boxed{D}$ and $\boxed{E}$, do you encounter a contradiction or an identity? From this finding, does the original system have no solution or infinitely many solutions?

d. Write the solution set.

Practice Exercises
Section 12.5

Prerequisite Review

For Exercises R.1–R.6, solve the systems by using two methods: (a) the substitution method and (b) the addition method.

R.1. $3x + y = 4$
$4x + y = 5$

R.2. $2x - 3y = 16$
$x + 2y = -6$

R.3. $2x - 5y = 3$
$-4x + 10y = 3$

R.4. $4x - y = 2$
$8x = 2y$

R.5. $4x - 6y = 5$
$x = \frac{3}{2}y + \frac{5}{4}$

R.6. $x + 3y = 3$
$y = -\frac{1}{3}x + 1$

Vocabulary and Key Concepts

1. **a.** An equation written in the form $Ax + By + Cz = D$, where A, B, and C are not all zero, is called a _____ equation in three variables.

   b. Solutions to a linear equation in three variables are of the form (x, y, z) and are called _____.

Concept 1: Solutions to Systems of Linear Equations in Three Variables

2. How many solutions are possible when solving a system of three equations with three variables?

3. Which of the following points are solutions to the system?

   $(2, 1, 7)$, $(3, -10, -6)$, $(4, 0, 2)$

   $2x - y + z = 10$
   $4x + 2y - 3z = 10$
   $x - 3y + 2z = 8$

4. Which of the following points are solutions to the system?

   $(1, 1, 3)$, $(0, 0, 4)$, $(4, 2, 1)$

   $-3x - 3y - 6z = -24$
   $-9x - 6y + 3z = -45$
   $9x + 3y - 9z = 33$

5. Which of the following points are solutions to the system?

   $(12, 2, -2)$, $(4, 2, 1)$, $(1, 1, 1)$

   $-x - y - 4z = -6$
   $x - 3y + z = -1$
   $4x + y - z = 4$

6. Which of the following points are solutions to the system?

   $(0, 4, 3)$, $(3, 6, 10)$, $(3, 3, 1)$

   $x + 2y - z = 5$
   $x - 3y + z = -5$
   $-2x + y - z = -4$

Concept 2: Solving Systems of Linear Equations in Three Variables

For Exercises 7–30, solve the system of equations. (See Examples 1–2.)

7. $2x + y - 3z = -12$
$3x - 2y - z = 3$
$-x + 5y + 2z = -3$

8. $-3x - 2y + 4z = -15$
$2x + 5y - 3z = 3$
$4x - y + 7z = 15$

9. $x - 3y - 4z = -7$
$5x + 2y + 2z = -1$
$4x - y - 5z = -6$

10. $6x - 5y + z = 7$
$5x + 3y + 2z = 0$
$-2x + y - 3z = 11$

11. $-3x + y - z = 8$
$-4x + 2y + 3z = -3$
$2x + 3y - 2z = -1$

12. $2x + 3y + 3z = 15$
$3x - 6y - 6z = -23$
$-9x - 3y + 6z = 8$

13. $4x + 2z = 12 + 3y$
$2x = 3x + 3z - 5$
$y = 2x + 7z + 8$

14. $y = 2x + z + 1$
$-3x - 1 = -2y + 2z$
$5x + 3z = 16 - 3y$

15. $x + y + z = 6$
$-x + y - z = -2$
$2x + 3y + z = 11$

16. $x - y - z = -11$
$x + y - z = 15$
$2x - y + z = -9$

17. $2x - 3y + 2z = -1$
$x + 2y = -4$
$x + z = 1$

18. $x + y + z = 2$
$2x - z = 5$
$3y + z = 2$

19. $4x + 9y = 8$
$8x + 6z = -1$
$6y + 6z = -1$

20. $3x + 2z = 11$
$y - 7z = 4$
$x - 6y = 1$

21. $2x + 3y - 2z = 8$
$x - 4y + z = -8$
$-4x + 3y + 2z = 3$

22. $5x - 3y + z = 4$
$-x + 6y + 4z = 1$
$2x + 3y - z = 3$

23. $5y = -7 + 4z$
$x + 3y + 3 = 0$
$2x = 14 + z$

24. $4(x - z) = -1$
$3y + z = 1$
$4x = 3 + y$

25. $x + z = -16 + 3y$
$2(y + z) = 12 - 5y$
$3x + 8 = y - z$

26. $4(x - 2y) = 8 - 3z$
$4y + 13 = 3x + z$
$x - 3y = -2(z + 2)$

27. $\frac{1}{3}x + \frac{1}{2}y + \frac{1}{6}z = \frac{1}{6}$
$\frac{1}{5}x + y - \frac{3}{5}z = \frac{4}{5}$
$x - \frac{4}{9}y + \frac{2}{3}z = \frac{1}{3}$

28. $\frac{1}{2}x + \frac{5}{2}z = -1$
$\frac{1}{4}x - \frac{1}{6}y - \frac{2}{3}z = 3$
$-\frac{1}{2}x + \frac{5}{8}y + \frac{1}{8}z = -3$

29. $0.1x - 0.6y + z = -6.8$
$0.7x - 0.3z = -14$
$y + 0.5z = 8$

30. $0.5x + y = 10$
$-0.2y + 0.5z = 0.5$
$-0.3x - z = -6$

Mixed Exercises

For Exercises 31–42, solve the system. If there is not a unique solution, state whether the system is inconsistent or the equations are dependent. **(See Examples 1–4.)**

31. $2x + y + 3z = 2$
$x - y + 2z = -4$
$-2x + 2y - 4z = 8$

32. $x + y = z$
$2x + 4y - 2z = 6$
$3x + 6y - 3z = 9$

33. $6x - 2y + 2z = 2$
$4x + 8y - 2z = 5$
$-2x - 4y + z = -2$

34. $3x + 2y + z = 3$
$x - 3y + z = 4$
$-6x - 4y - 2z = 1$

35. $\frac{1}{2}x + \frac{2}{3}y = \frac{5}{2}$
$\frac{1}{5}x - \frac{1}{2}z = -\frac{3}{10}$
$\frac{1}{3}y - \frac{1}{4}z = \frac{3}{4}$

36. $\frac{1}{2}x + \frac{1}{4}y + z = 3$
$\frac{1}{8}x + \frac{1}{4}y + \frac{1}{2}z = \frac{9}{8}$
$x - y - \frac{2}{3}z = \frac{1}{3}$

37. $3(2x + y) = 4(3z - 1)$
$y + 4z = 2(3 + 5x)$
$2(x + z) = y - 1$

38. $5(y - 2z) = 9x - 11$
$2(3x + z) = 5 + y$
$3x = z + 2(y + 1)$

39. $2x + y = 3(z - 1)$
$3x - 2(y - 2z) = 1$
$2(2x - 3z) = -6 - 2y$

40. $2x + y = -3$
$2y + 16z = -10$
$-7x - 3y + 4z = 8$

41. $-0.1y + 0.2z = 0.2$
$0.1x + 0.1y + 0.1z = 0.2$
$-0.1x + 0.3z = 0.2$

42. $0.1x - 0.2y = 0$
$0.3y + 0.1z = -0.1$
$0.4x - 0.1z = 1.2$

Expanding Your Skills

The systems in Exercises 43–46 are called homogeneous systems because each system has (0, 0, 0) as a solution. However, if the equations are dependent, there will be infinitely many more solutions. For each system, determine whether (0, 0, 0) is the only solution or if the equations are dependent.

43. $2x - 4y + 8z = 0$
    $-x - 3y + z = 0$
    $x - 2y + 5z = 0$

44. $2x - 4y + z = 0$
    $x - 3y - z = 0$
    $3x - y + 2z = 0$

45. $4x - 2y - 3z = 0$
    $-8x - y + z = 0$
    $2x - y - \frac{3}{2}z = 0$

46. $5x + y = 0$
    $4y - z = 0$
    $5x + 5y - z = 0$

Section 12.6
Applications of Systems of Linear Equations in Three Variables

1. Applications Involving Geometry

In this section we solve several applications of linear equations in three variables.

Concepts
1. Applications Involving Geometry
2. Applications Involving Mixtures

Example 1 Applying Systems of Linear Equations in Three Variables

In a triangle, the smallest angle measures 10° more than one-half of the largest angle. The middle angle measures 12° more than the smallest angle. Find the measure of each angle.

Solution:

Let x represent the measure of the smallest angle.

Let y represent the measure of the middle angle.

Let z represent the measure of the largest angle.

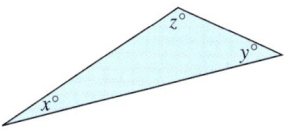

To solve for three variables, we need to establish three independent relationships among x, y, and z.

[A] $x = \frac{1}{2}z + 10$ The smallest angle measures 10° more than one-half the measure of the largest angle.

[B] $y = x + 12$ The middle angle measures 12° more than the measure of the smallest angle.

[C] $x + y + z = 180$ The sum of the interior angles of a triangle measures 180°.

Clear fractions and write each equation in standard form.

Standard Form

[A] $x = \frac{1}{2}z + 10$ $\xrightarrow{\text{Multiply by 2.}}$ $2x = z + 20$ $\longrightarrow$ $2x - z = 20$

[B] $y = x + 12$ $\longrightarrow$ $-x + y = 12$

[C] $x + y + z = 180$ $\longrightarrow$ $x + y + z = 180$

Notice equation [B] is missing the z variable. Therefore, we can eliminate z again by pairing up equations [A] and [C].

[A] $2x - z = 20$
[C] $x + y + z = 180$
      $\overline{3x + y = 200}$ [D]

$$\boxed{B}\ -x+y=12 \xrightarrow{\text{Multiply by }-1.} x-y=-12$$
$$\boxed{D}\ 3x+y=200 \longrightarrow \underline{3x+y=200}$$
$$4x\ \ \ \ =188$$
$$x=47$$

Pair up equations $\boxed{B}$ and $\boxed{D}$ to form a system of two variables.

Solve for x.

From equation $\boxed{B}$ we have $-x+y=12 \longrightarrow -47+y=12 \rightarrow y=59$

From equation $\boxed{C}$ we have $x+y+z=180 \rightarrow 47+59+z=180 \rightarrow z=74$

The smallest angle measures 47°, the middle angle measures 59°, and the largest angle measures 74°.

Skill Practice

1. The perimeter of a triangle is 30 in. The shortest side is 4 in. shorter than the longest side. The longest side is 6 in. less than the sum of the other two sides. Find the length of each side.

2. Applications Involving Mixtures

Example 2 Applying Systems of Linear Equations to Nutrition

Doctors have become increasingly concerned about the sodium intake in the U.S. diet. Recommendations by the American Medical Association indicate that most individuals should not exceed 2400 mg of sodium per day.

Liz ate 1 slice of pizza, 1 serving of ice cream, and 1 glass of soda for a total of 1030 mg of sodium. David ate 3 slices of pizza, no ice cream, and 2 glasses of soda for a total of 2420 mg of sodium. Melinda ate 2 slices of pizza, 1 serving of ice cream, and 2 glasses of soda for a total of 1910 mg of sodium. How much sodium is in one serving of each item?

Solution:

Let x represent the sodium content of 1 slice of pizza.

Let y represent the sodium content of 1 serving of ice cream.

Let z represent the sodium content of 1 glass of soda.

From Liz's meal we have: $\boxed{A}\ x+y+z=1030$

From David's meal we have: $\boxed{B}\ 3x\ \ \ \ +2z=2420$

From Melinda's meal we have: $\boxed{C}\ 2x+y+2z=1910$

Equation $\boxed{B}$ is missing the y variable. Eliminating y from equations $\boxed{A}$ and $\boxed{C}$, we have

$$\boxed{A}\ x+y+z=1030 \xrightarrow{\text{Multiply by }-1.} -x-y-z=-1030$$
$$\boxed{C}\ 2x+y+2z=1910 \longrightarrow \underline{2x+y+2z=\ 1910}$$
$$\boxed{D}\ x\ \ \ \ +z=\ 880$$

Solve the system formed by equations $\boxed{B}$ and $\boxed{D}$.

$$\boxed{B}\ 3x+2z=2420 \longrightarrow 3x+2z=\ 2420$$
$$\boxed{D}\ x+z=880 \xrightarrow{\text{Multiply by }-2.} \underline{-2x-2z=-1760}$$
$$x\ \ \ \ =\ 660$$

Answer

1. The sides are 8 in., 10 in., and 12 in.

From equation $\boxed{D}$ we have $x + z = 800 \rightarrow 660 + z = 880 \rightarrow z = 220$

From equation $\boxed{A}$ we have $x + y + z = 1030 \rightarrow 660 + y + 220 = 1030 \rightarrow y = 150$

Therefore, 1 slice of pizza has 660 mg of sodium, 1 serving of ice cream has 150 mg of sodium, and 1 glass of soda has 220 mg of sodium.

Skill Practice

2. Annette, Barb, and Carlita work in a clothing shop. One day the three had combined sales of $1480. Annette sold $120 more than Barb. Barb and Carlita combined sold $280 more than Annette. How much did each person sell?

Answer

2. Annette sold $600, Barb sold $480, and Carlita sold $400.

Section 12.6 Activity

To solve an application with three unknowns, it is often helpful to use three variables such as x, y, and z. When three variables are used to solve an application, we need three independent equations that relate the variables. That is, for three unknowns, set up a system of three equations.

A.1. The average of Randall's three test grades in economics is 92. His first test score is 6 points higher than his second test score. His third test score is 3 points lower than his first test score. Find his test scores by following these steps.

  a. Let x represent Randall's first test score.

  Let y represent _____.

  Let z represent _____.

  b. Write an equation in terms of x, y, and z that represents the fact that the average of Randall's three test scores is 92.

  c. Write an equation that represents the fact that Randall's first test score is 6 points higher than his second test score.

  d. Write an equation that represents the fact that Randall's third test score is 3 points lower than his first test score.

  e. Solve the system made up of the equations from parts (b)–(d).

  f. Interpret the answer from part (e).

A.2. Alicia invested a total of $14,500 in three different mutual funds. After 1 year, the first fund *lost* 2%, the second fund gained 3%, and the third gained 5%. She invested $500 more in the fund that earned 5% than the other two funds combined. At the end of the year, Alicia earned $485.

  a. Let x represent the amount invested in the first fund.

  Let y represent _____.

  Let z represent _____.

  b. Write an equation that represents the fact that the total amount invested is $14,500.

  c. Write an equation that represents the fact that Alicia invested $500 more in the 5% fund than in the other two funds combined.

  d. Write an equation that represents the fact that the total earnings for the year is $485.

  e. Solve the system made up of the equations from parts (b)–(d).

  f. Interpret the answer from part (e).

Section 12.6 Practice Exercises

Prerequisite Review

For Exercises R.1–R.2, solve the system.

R.1. $\quad -5y - z = -8$
$\quad\quad\; x + 10y + 2z = 9$
$\quad\quad -3x + y = 21$

R.2. $\quad 12x - y + z = -6$
$\quad\quad -6x + 2y - 3z = 5$
$\quad\quad\;\; 6x + y - z = 3$

R.3. The cost to download a song on a popular music site is $1.29.
  a. What is the cost to download 5 songs?
  b. What is the cost to download y songs?

R.4. The cost for organic apples is $1.50 per pound.
  a. What is the cost for 8 lb?
  b. What is the cost for x pounds?

R.5. Jose invests money in an account that earns 2.5% simple interest.
  a. How much interest will he earn if he invests $5000 for 2 years?
  b. How much interest will he earn if he invests x dollars for 1 year?

R.6. Eliza owes money and pays simple interest at a rate of 4%.
  a. How much interest will she owe if she borrows $8000 for 3 years?
  b. How much interest will she owe if she borrows y dollars for 1 year?

Concept 1: Applications Involving Geometry

1. A triangle has one angle that measures 5° more than twice the smallest angle, and the largest angle measures 11° less than 3 times the measure of the smallest angle. Find the measures of the three angles. **(See Example 1.)**

2. The largest angle of a triangle measures 4° less than 5 times the measure of the smallest angle. The middle angle measures twice that of the smallest angle. Find the measures of the three angles.

3. One angle of a triangle measures 6° more than twice the measure of the smallest angle. The third angle measures 1° less than four times the smallest. Find the measures of the three angles.

4. In a triangle the smallest angle measures 12° less than the middle angle. The measure of the largest angle is equal to the sum of the other two. Find the measures of the three angles.

5. The perimeter of a triangle is 55 cm. The measure of the shortest side is 8 cm less than the middle side. The measure of the longest side is 1 cm less than the sum of the other two sides. Find the lengths of the sides.

6. The perimeter of a triangle is 5 ft. The longest side of the triangle measures 20 in. more than the shortest side. The middle side is 3 times the measure of the shortest side. Find the lengths of the three sides in *inches*.

7. The perimeter of a triangle is 4.5 ft. The shortest side measures 2 in. less than half the longest side. The longest side measures 2 in. less than the sum of the other two sides. Find the lengths of the sides in *inches*.

8. The perimeter of a triangle is 7 m. The length of the longest side is twice the length of the shortest side. The sum of the lengths of the shortest side and the middle side is 1 m more than the length of the longest side. Find the lengths of the sides.

Concept 2: Applications Involving Mixtures

9. Sean kept track of his fiber intake from three sources for 3 weeks. The first week he had 3 servings of a fiber supplement, 1 serving of oatmeal, and 4 servings of cereal, which totaled 19 g of fiber. The second week he had 2 servings of the fiber supplement, 4 servings of oatmeal, and 2 servings of cereal totaling 25 g. The third week he had 5 servings of the fiber supplement, 3 servings of oatmeal, and 2 servings of cereal for a total of 30 g. Find the amount of fiber in one serving of each of the following: the fiber supplement, the oatmeal, and the cereal. **(See Example 2.)**

10. Natalie kept track of her calcium intake from three sources for 3 days. The first day she had 1 glass of milk, 1 serving of ice cream, and 1 calcium supplement in pill form which totaled 1180 mg of calcium. The second day she had 2 glasses of milk, 1 serving of ice cream, and 1 calcium supplement totaling 1680 mg. The third day she had 1 glass of milk, 2 servings of ice cream, and 1 calcium supplement for a total of 1260 mg. Find the amount of calcium in one glass of milk, in one serving of ice cream, and in one calcium supplement.

11. Kyoki invested a total of $10,000 into bonds, mutual funds, and a Treasury note. He put the same amount of money in the money market account as he did in bonds. For 1 year, the bonds paid 5% interest, the mutual funds paid 8%, and the Treasury note paid 4%. At the end of the year, Kyoki earned $660 in interest. How much did he invest in each account?

12. Walter had $25,000 to invest. He split the money into three types of investment: small caps earning 6%, global market investments earning 10%, and a balanced fund earning 9%. He put twice as much money in the global account as he did in the balanced fund. If his earnings for the first year totaled $2160, how much did he invest in each account?

13. Winston deposited $4500 into three accounts: one pays 5% interest, another pays 5.5% interest, and the third pays 4% interest. He deposits $1000 more in the 5.5% account than he does in the 4% account. After the first year, he earned a total of $225 interest. How much did he deposit into each account?

14. Raeann deposited $8000 into three accounts at her credit union: a checking account that pays 1.2% interest, a savings account that pays 2.5% interest, and a money market account that pays 3% interest. If she put 3 times more money in the 3% account than she did in the 1.2% account, and her total interest for 1 year was $202, how much did she deposit into each account?

15. A basketball player scored 29 points in one game. In basketball, some baskets are worth 3 points, some are worth 2 points, and free throws are worth 1 point. He scored four more 2-point baskets than he did 3-point baskets. The number of free throws is one less than the number of 2-point baskets. How many free throws, 2-point shots, and 3-point shots did he make?

16. Goofie Golf has 18 holes that are par 3, par 4, or par 5. Most of the holes are par 4. In fact, there are 3 times as many par 4s as par 3s. There are 3 more par 5s than par 3s. How many of each type are there?

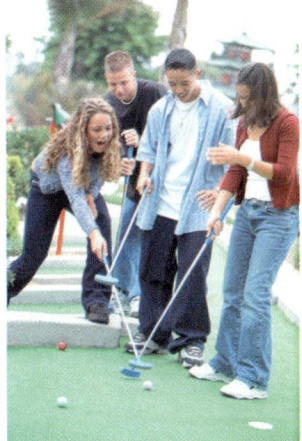

SW Productions/Getty Images

17. Combining peanuts, pecans, and cashews makes a party mixture of nuts. If the amount of peanuts equals the amount of pecans and cashews combined, and if there are twice as many cashews as pecans, how many ounces of each nut is used to make 48 oz of party mixture?

18. Souvenir hats, T-shirts, and jackets are sold at a rock concert. Three hats, two T-shirts, and one jacket cost $140. Two hats, two T-shirts, and two jackets cost $170. One hat, three T-shirts, and two jackets cost $180. Find the prices of the individual items.

19. A theater charges $70 per seat for seats in section A, $65 per seat for seats in section B, and $44 per seat for seats in section C. 2400 tickets are sold for a total of $153,000 in revenue. If there are 600 more seats sold in section A than the other two sections combined, how many seats in each section were sold?

20. Annie and Maria traveled overseas for seven days and stayed in three different hotels in three different cities: Stockholm, Sweden; Oslo, Norway; and Paris, France.

The total bill for all seven nights (not including tax) was $1040. The total tax was $106. The nightly cost (excluding tax) to stay at the hotel in Paris was $80 more than the nightly cost (excluding tax) to stay in Oslo. Find the cost per night for each hotel excluding tax.

City	Number of Nights	Cost/Night ($)	Tax Rate
Paris, France	1	x	8%
Stockholm, Sweden	4	y	11%
Oslo, Norway	2	z	10%

Chapter 12 Review Exercises

Section 12.1

For Exercises 1–4, determine if the ordered pair is a solution to the system.

1. $x - 4y = -4$ (4, 2)
   $x + 2y = 8$

2. $x - 6y = 6$ (12, 1)
   $-x + y = 4$

3. $3x + y = 9$ (1, 3)
   $y = 3$

4. $2x - y = 8$ (2, −4)
   $x = 2$

For Exercises 5–10, identify whether the system represents intersecting lines, parallel lines, or coinciding lines by comparing slopes and y-intercepts.

5. $y = -\frac{1}{2}x + 4$
   $y = x - 1$

6. $y = -3x + 4$
   $y = 3x + 4$

7. $y = -\frac{4}{7}x + 3$
   $y = -\frac{4}{7}x - 5$

8. $y = 5x - 3$
   $y = \frac{1}{5}x - 3$

9. $y = 9x - 2$
   $9x - y = 2$

10. $x = -5$
    $y = 2$

For Exercises 11–18, solve the system by graphing. For systems that do not have one unique solution, also state the number of solutions and whether the system is inconsistent or the equations are dependent.

11. $y = -\frac{2}{3}x - 2$
    $-x + 3y = -6$

12. $y = -2x - 1$
    $x + 2y = 4$

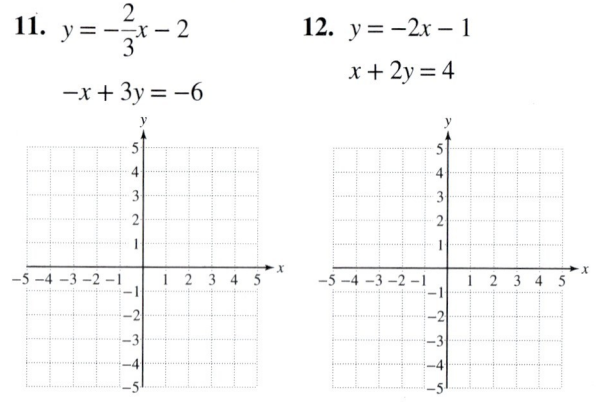

13. $4x = -2y + 10$
    $2x + y = 5$

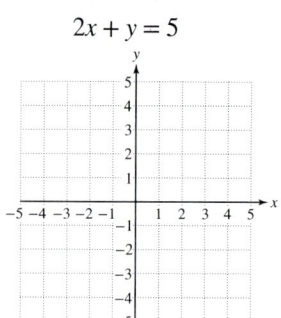

14. $10y = 2x - 10$
    $-x + 5y = -5$

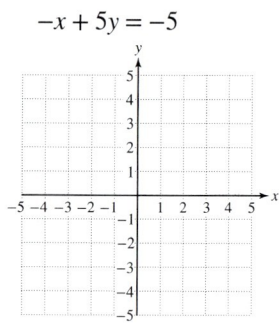

15. $6x - 3y = 9$
    $y = -1$

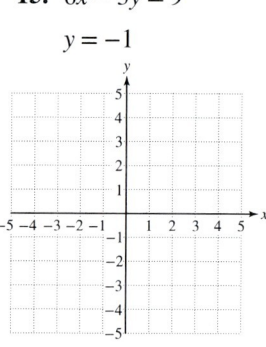

16. $5x + y = -3$
    $x = -1$

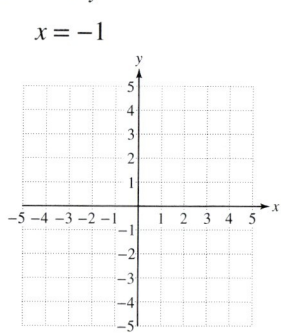

17. $x - 7y = 14$
    $-2x + 14y = 14$

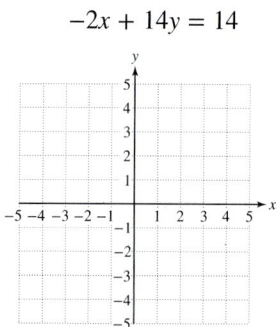

18. $y = -5x + 4$
    $10x + 2y = -4$

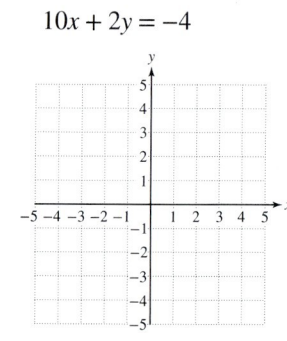

Section 12.2

19. One salesperson makes a 20% commission on sales along with a yearly base salary of $15,000. A second salesperson makes a 25% commission, but does not receive a base salary. The yearly compensation (in dollars) for x dollars in sales is given for each sales person.

    First salesperson: $y = 0.20x + 15{,}000$

    Second salesperson: $y = 0.25x$

    Find the amount in sales at which the yearly compensation for each salesperson is the same.

For Exercises 20–23, solve each system using the substitution method.

20. $6x + y = 2$
    $y = 3x - 4$

21. $2x + 3y = -5$
    $x = y - 5$

22. $2x + 6y = 10$
    $x = -3y + 6$

23. $4x + 2y = 4$
    $y = -2x + 2$

24. Given the system: $x + 2y = 11$
    $5x + 4y = 40$

    a. Which variable from which equation is easiest to isolate and why?

    b. Solve the system using the substitution method.

25. Given the system: $4x - 3y = 9$
    $2x + y = 12$

    a. Which variable from which equation is easiest to isolate and why?

    b. Solve the system using the substitution method.

For Exercises 26–29, solve each system using the substitution method.

26. $3x - 2y = 23$
    $x + 5y = -15$

27. $x + 5y = 20$
    $3x + 2y = 8$

28. $x - 3y = 9$
    $5x - 15y = 45$

29. $-3x + y = 15$
    $6x - 2y = 12$

30. The difference of two positive numbers is 42. The larger number is 2 more than 6 times the smaller number. Find the numbers.

31. In a right triangle, one of the acute angles is 8° less than the other acute angle. Find the measure of each acute angle.

32. Two angles are supplementary. One angle measures 14° less than two times the other angle. Find the measure of each angle.

Section 12.3

33. Given the system: $-2x + 7y = 30$
    $4x + 5y = 16$

    a. Which variable, x or y, is easier to eliminate using the addition method? (Answers may vary.)

    b. Solve the system using the addition method.

34. Given the system: $3x - 5y = 1$
    $2x - y = -4$

    a. Which variable, x or y, is easier to eliminate using the addition method? (Answers may vary.)

    b. Solve the system using the addition method.

35. Given the system: $9x - 2y = 14$
    $4x + 3y = 14$

    a. Which variable, x or y, is easier to eliminate using the addition method? (Answers may vary.)

    b. Solve the system using the addition method.

For Exercises 36–43, solve each system using the addition method.

36. $2x + 3y = 1$
    $x - 2y = 4$

37. $x + 3y = 0$
    $-3x - 10y = -2$

38. $8x + 8 = -6y + 6$
    $10x = 9y - 8$

39. $12x = 5y + 5$
    $5y = -1 - 4x$

40. $-4x - 6y = -2$
    $6x + 9y = 3$

41. $-8x - 4y = 16$
    $10x + 5y = 5$

42. $\frac{1}{2}x - \frac{3}{4}y = -\frac{1}{2}$
    $\frac{1}{3}x + y = -\frac{10}{3}$

43. $0.5x - 0.2y = 0.5$
    $0.4x + 0.7y = 0.4$

44. Given the system: $4x + 9y = -7$
    $y = 2x - 13$

    a. Which method would you choose to solve the system, the substitution method or the addition method? Explain your choice.

    b. Solve the system.

45. Given the system: $5x - 8y = -2$
    $3x - 7y = 1$

    a. Which method would you choose to solve the system, the substitution method or the addition method? Explain your choice.

    b. Solve the system.

Section 12.4

46. A zoo charges $19.95 for adult admission and $15.95 for children under 13. The total bill before tax for a school group of 60 people is $989. How many adults and how many children were admitted?

47. As part of his retirement strategy Winston plans to invest $600,000 in two different funds. He projects that the high-risk investments should return, over time, about 12% per year, while the low-risk investments should return about 4% per year. If he wants a supplemental income of $30,000 a year, how should he divide his investments?

48. Suppose that whole milk with 4% fat is mixed with 1% low-fat milk to make a 2% reduced fat milk. How much of the whole milk should be mixed with the low-fat milk to make 60 gal of 2% reduced fat milk?

49. A boat travels 80 mi downstream with the current in 4 hr and 80 mi upstream against the current in 5 hr. Find the speed of the current and the speed of the boat in still water.

50. A plane travels 870 mi against the wind in 3 hr. Traveling with the wind, the plane travels 700 mi in 2 hr. Find the speed of the plane in still air and the speed of the wind.

51. At a sports arena, the total cost of a soft drink and a hot dog is $8.00. The price of the hot dog is $1.00 more than the cost of the soft drink. Find the cost of a soft drink and the cost of a hot dog.

52. Carlos played two rounds of golf at Pebble Beach for a total score of 154. If his score in the second round is 10 more than his score in the first round, find the scores for each round.

Ivan Mikhaylov/123RF

Section 12.5

For Exercises 53–58, solve the systems of equations. If a system does not have a unique solution, state whether the system is inconsistent or the equations are dependent.

53.
$5x + 5y + 5z = 30$
$-x + y + z = 2$
$10x + 6y - 2z = 4$

54.
$5x + 3y - z = 5$
$x + 2y + z = 6$
$-x - 2y - z = 8$

55.
$x + y + z = 4$
$-x - 2y - 3z = -6$
$2x + 4y + 6z = 12$

56.
$3x + 4z = 5$
$ 2y + 3z = 2$
$2x - 5y = 8$

57.
$3(x - y) = -1 - 5z - y$
$3y + 6 = 2z + x$
$5x + 1 = 3(z - y) - y$

58.
$2a = -3c - 2$
$2b - 8 = 5c$
$7a + 3b = 5$

Section 12.6

59. The perimeter of a right triangle is 30 ft. One leg is 2 ft longer than twice the shortest leg. The hypotenuse is 2 ft less than 3 times the shortest leg. Find the lengths of the sides of this triangle.

60. Three pumps are working to drain a construction site. Working together, the pumps can drain 950 gal/hr of water. The slowest pump drains 150 gal/hr less than the fastest pump. The fastest pump drains 150 gal/hr less than the sum of the other two pumps. How many gallons can each pump drain per hour?

61. Theresa had $12,000 to invest among three mutual funds. Fund A had a 5% yield for the year. Fund B had a 3.5% yield, and Fund C lost 2% for the year. She invested three times as much in Fund A as in Fund C. If she gained $400 in 1 year, how much was invested in each fund?

62. In a triangle, the largest angle measures 76° more than the sum of the other two angles. The middle angle measures 3 times the smallest angle. Find the measure of each angle.

Polynomials and Properties of Exponents

CHAPTER OUTLINE

13.1 Multiplying and Dividing Expressions With Common Bases   832
13.2 More Properties of Exponents   843
13.3 Definitions of b^0 and b^{-n} 849
   Problem Recognition Exercises: Properties of Exponents   859
13.4 Scientific Notation   860
13.5 Addition and Subtraction of Polynomials   866
13.6 Multiplication of Polynomials and Special Products   876
13.7 Division of Polynomials   887
   Problem Recognition Exercises: Operations on Polynomials   899
   Chapter 13 Review Exercises   900

Mathematics to Compute Cost

Trevor is a dance instructor and wants to host a one-day dance event. He has a fixed cost of $200 to rent the dance studio. In addition, to host the event, he has the following variable costs, which depend on the number of participants.

- $1.00 per person for coffee and breakfast snacks
- $7.00 per person for lunch
- $1.20 per person for a step-sheet booklet

If n represents the number of participants, then Trevor's cost for the event is given by

$$\text{Cost} = 1.00n + 7.00n + 1.20n + 200$$

This expression is called a **polynomial**. Terms of a polynomial are separated by addition, and sometimes a polynomial can be simplified by adding *like terms*. In this case, the terms containing the variable n are *like* terms. Thus, the polynomial representing cost can be simplified as

$$\text{Cost} = (1.00 + 7.00 + 1.20)n + 200$$
$$= 9.20n + 200$$

The terms $1.00n$, $7.00n$, and $1.20n$ are combined to form $9.20n$ because we're effectively consolidating the costs for breakfast, lunch, and the booklet into one overall cost per person.

As you study polynomials in this chapter, you will encounter other applications, including those involving profit and cost.

John Flournoy/McGraw Hill

Section 13.1 Multiplying and Dividing Expressions With Common Bases

Concepts

1. Review of Exponential Notation
2. Evaluating Expressions With Exponents
3. Multiplying and Dividing Expressions With Common Bases
4. Simplifying Expressions With Exponents
5. Applications of Exponents

1. Review of Exponential Notation

Recall that an **exponent** is used to show repeated multiplication of the **base**.

> **Definition of b^n**
>
> Let b represent any real number and n represent a positive integer. Then,
>
> $$b^n = \underbrace{b \cdot b \cdot b \cdot b \cdot \ldots b}_{n \text{ factors of } b}$$

Example 1 Evaluating Expressions With Exponents

For each expression, identify the exponent and base. Then evaluate the expression.

a. 6^2 b. $\left(-\dfrac{1}{2}\right)^3$ c. 0.8^4

Solution:

Expression	Base	Exponent	Result
a. 6^2	6	2	$(6)(6) = 36$
b. $\left(-\dfrac{1}{2}\right)^3$	$-\dfrac{1}{2}$	3	$\left(-\dfrac{1}{2}\right)\left(-\dfrac{1}{2}\right)\left(-\dfrac{1}{2}\right) = -\dfrac{1}{8}$
c. 0.8^4	0.8	4	$(0.8)(0.8)(0.8)(0.8) = 0.4096$

Skill Practice For each expression, identify the base and exponent.

1. 8^3 2. $\left(-\dfrac{1}{4}\right)^2$ 3. 0.2^4

Note that if no exponent is explicitly written for an expression, then the expression has an implied exponent of 1. For example, $x = x^1$.

Consider an expression such as $3y^6$. The factor 3 has an exponent of 1, and the factor y has an exponent of 6. That is, the expression $3y^6$ is interpreted as $3^1 y^6$.

2. Evaluating Expressions With Exponents

Particular care must be taken when evaluating exponential expressions involving negative numbers. An exponential expression with a negative base is written with parentheses around the base, such as $(-3)^2$.

To evaluate $(-3)^2$, we have: $(-3)^2 = (-3)(-3) = 9$

If no parentheses are present, the expression -3^2 is the *opposite* of 3^2, or equivalently, $-1 \cdot 3^2$.

$$-3^2 = -1(3^2) = -1(3)(3) = -9$$

Answers

1. Base 8; exponent 3
2. Base $-\dfrac{1}{4}$; exponent 2
3. Base 0.2; exponent 4

Example 2 Evaluating Expressions With Exponents

Evaluate each expression.

a. -5^4 b. $(-5)^4$ c. $(-0.2)^3$ d. -0.2^3

Solution:

a. -5^4
$\quad = -1 \cdot 5^4$ 5 is the base with exponent 4.
$\quad = -1 \cdot 5 \cdot 5 \cdot 5 \cdot 5$ Multiply -1 by four factors of 5.
$\quad = -625$

b. $(-5)^4$
$\quad = (-5)(-5)(-5)(-5)$ Parentheses indicate that -5 is the base with exponent 4.
$\quad = 625$ Multiply four factors of -5.

c. $(-0.2)^3$ Parentheses indicate that -0.2 is the base with exponent 3.
$\quad = (-0.2)(-0.2)(-0.2)$ Multiply three factors of -0.2.
$\quad = -0.008$

d. -0.2^3
$\quad = -1 \cdot 0.2^3$ 0.2 is the base with exponent 3.
$\quad = -1(0.2)(0.2)(0.2)$ Multiply -1 by three factors of 0.2.
$\quad = -0.008$

Skill Practice Evaluate each expression.

4. -2^4 5. $(-2)^4$ 6. $(-0.1)^3$ 7. -0.1^3

Example 3 Evaluating Expressions With Exponents

Evaluate each expression for $a = 2$ and $b = -3$.

a. $5a^2$ b. $(5a)^2$ c. $5ab^2$ d. $(b + a)^2$

Solution:

a. $5a^2$
$\quad = 5()^2$ Use parentheses to substitute a number for a variable.
$\quad = 5(2)^2$ Substitute $a = 2$.
$\quad = 5(4)$ Simplify exponents before multiplying.
$\quad = 20$

b. $(5a)^2$
$\quad = [5()]^2$ Use parentheses to substitute a number for a variable. The original parentheses are replaced with brackets.
$\quad = [5(2)]^2$ Substitute $a = 2$.
$\quad = (10)^2$ Simplify inside the parentheses first.
$\quad = 100$

Answers
4. -16 5. 16
6. -0.001 7. -0.001

Avoiding Mistakes

In the expression $5ab^2$, the exponent, 2, applies only to the variable b. The constant 5 and the variable a both have an implied exponent of 1.

c. $5ab^2$

$= 5(2)(-3)^2$ Substitute $a = 2$, $b = -3$.

$= 5(2)(9)$ Simplify exponents before multiplying.

$= 90$ Multiply.

d. $(b + a)^2$

$= [(-3) + (2)]^2$ Substitute $b = -3$ and $a = 2$.

$= (-1)^2$ Simplify within the parentheses first.

$= 1$

Avoiding Mistakes

Be sure to follow the order of operations. In Example 3(d), it would be incorrect to square the terms within the parentheses before adding.

Skill Practice Evaluate each expression for $x = 2$ and $y = -5$.

8. $6x^2$ 9. $(6x)^2$ 10. $2xy^2$ 11. $(y - x)^2$

3. Multiplying and Dividing Expressions With Common Bases

In this section, we investigate the effect of multiplying or dividing two quantities with the same base. For example, consider the expressions: x^5x^2 and $\frac{x^5}{x^2}$. Simplifying each expression, we have:

$$x^5 x^2 = (x \cdot x \cdot x \cdot x \cdot x)(x \cdot x) = \overbrace{x \cdot x \cdot x \cdot x \cdot x \cdot x \cdot x}^{7 \text{ factors of } x} = x^7$$

$$\frac{x^5}{x^2} = \frac{x \cdot x \cdot x \cdot \overset{\downarrow}{x} \cdot \overset{\downarrow}{x}}{x \cdot x} = \frac{x \cdot x \cdot x}{1} = x^3$$

These examples suggest that to multiply two quantities with the same base, we add the exponents. To divide two quantities with the same base, we subtract the exponent in the denominator from the exponent in the numerator. These rules are stated formally in the following two properties.

> **Multiplication of Expressions With Like Bases**
>
> Assume that b is a real number and that m and n represent positive integers. Then,
>
> $$b^m b^n = b^{m+n}$$

> **Division of Expressions With Like Bases**
>
> Assume that $b \neq 0$ is a real number and that m and n represent positive integers. Then,
>
> $$\frac{b^m}{b^n} = b^{m-n}$$

Answers

8. 24   9. 144
10. 100   11. 49

Section 13.1 Multiplying and Dividing Expressions With Common Bases

Example 4 — Simplifying Expressions With Exponents

Simplify the expressions.   **a.** $w^3 w^4$ **b.** $2^3 \cdot 2^4$

Solution:

a. $w^3 w^4$ $(w \cdot w \cdot w)(w \cdot w \cdot w \cdot w)$

$\quad = w^{3+4}$ To multiply expressions with like bases, add the exponents.

$\quad = w^7$

b. $2^3 \cdot 2^4$ $(2 \cdot 2 \cdot 2)(2 \cdot 2 \cdot 2 \cdot 2)$

$\quad = 2^{3+4}$ To multiply expressions with like bases, add the exponents (the base is unchanged).

$\quad = 2^7$ or 128

> **Avoiding Mistakes**
>
> When we multiply expressions with like bases, we add the exponents. The base does not change. In Example 4(b), notice that the base 2 does not change. $2^3 \cdot 2^4 = 2^7$.

Skill Practice Simplify the expressions.
12. $q^4 q^8$ **13.** $8^4 \cdot 8^8$

Example 5 — Simplifying Expressions With Exponents

Simplify the expressions.   **a.** $\dfrac{t^6}{t^4}$ **b.** $\dfrac{5^6}{5^4}$

Solution:

a. $\dfrac{t^6}{t^4}$ $\dfrac{t \cdot t \cdot t \cdot t \cdot t \cdot t}{t \cdot t \cdot t \cdot t}$

$\quad = t^{6-4}$ To divide expressions with like bases, subtract the exponents.

$\quad = t^2$

b. $\dfrac{5^6}{5^4}$ $\dfrac{5 \cdot 5 \cdot 5 \cdot 5 \cdot 5 \cdot 5}{5 \cdot 5 \cdot 5 \cdot 5}$

$\quad = 5^{6-4}$ To divide expressions with like bases, subtract the exponents (the base is unchanged).

$\quad = 5^2$ or 25

Skill Practice Simplify the expressions.
14. $\dfrac{y^{15}}{y^8}$ **15.** $\dfrac{3^{15}}{3^8}$

Answers
12. q^{12} **13.** 8^{12}
14. y^7 **15.** 3^7

Chapter 13 Polynomials and Properties of Exponents

Example 6 Simplifying Expressions With Exponents

Simplify the expressions.   **a.** $\dfrac{z^4 z^5}{z^3}$ **b.** $\dfrac{10^7}{10^2 \cdot 10}$

Solution:

a. $\dfrac{z^4 z^5}{z^3}$

$= \dfrac{z^{4+5}}{z^3}$ Add the exponents in the numerator (the base is unchanged).

$= \dfrac{z^9}{z^3}$

$= z^{9-3}$ Subtract the exponents.

$= z^6$

b. $\dfrac{10^7}{10^2 \cdot 10}$

$= \dfrac{10^7}{10^2 \cdot 10^1}$ Note that 10 is equivalent to 10^1.

$= \dfrac{10^7}{10^{2+1}}$ Add the exponents in the denominator (the base is unchanged).

$= \dfrac{10^7}{10^3}$

$= 10^{7-3}$ Subtract the exponents.

$= 10^4$ or 10,000   Simplify.

Skill Practice Simplify the expressions.

16. $\dfrac{a^3 a^8}{a^7}$ **17.** $\dfrac{5^9}{5^2 \cdot 5^5}$

---

4. Simplifying Expressions With Exponents

Example 7 Simplifying Expressions With Exponents

Use the commutative and associative properties of real numbers and the properties of exponents to simplify the expressions.

a. $(-3p^2 q^4)(2pq^5)$ **b.** $\dfrac{16w^9 z^3}{4w^8 z}$

Solution:

a. $(-3p^2 q^4)(2pq^5)$

$= (-3 \cdot 2)(p^2 p)(q^4 q^5)$ Apply the associative and commutative properties of multiplication to group coefficients and like bases.

$= (-3 \cdot 2) p^{2+1} q^{4+5}$ Add the exponents when multiplying expressions with like bases.

$= -6 p^3 q^9$ Simplify.

Avoiding Mistakes

To simplify the expression in Example 7(a) we multiply the coefficients. However, to multiply expressions with like bases, we add the exponents.

Answers
16. a^4 **17.** 5^2 or 25

b. $\dfrac{16w^9z^3}{4w^8z}$

$= \left(\dfrac{16}{4}\right)\left(\dfrac{w^9}{w^8}\right)\left(\dfrac{z^3}{z}\right)$ Group coefficients and like bases.

$= 4w^{9-8}z^{3-1}$ Subtract the exponents when dividing expressions with like bases.

$= 4wz^2$ Simplify.

Avoiding Mistakes

In Example 7(b) we divide the coefficients. However, to divide expressions with like bases, we subtract the exponents.

Skill Practice Simplify the expressions.

18. $(-4x^2y^3)(3x^5y^7)$ 19. $\dfrac{81x^4y^7}{9xy^3}$

5. Applications of Exponents

Simple interest on an investment or loan is computed by the formula $I = Prt$, where P is the amount of principal, r is the annual interest rate, and t is the time in years. Simple interest is based only on the original principal. However, in most day-to-day applications, the interest computed on money invested or borrowed is compound interest. **Compound interest** is computed on the original principal and on the interest already accrued.

Suppose $1000 is invested at 8% interest for 3 years. Compare the total amount in the account if the money earns simple interest versus if the interest is compounded annually.

Simple Interest

The simple interest earned is given by $I = Prt$

$= (\$1000)(0.08)(3)$

$= \$240$

The total amount, A, at the end of 3 years is $A = P + I$

$= \$1000 + \$240$

$= \$1240$

Compound Annual Interest

The total amount, A, in an account earning compound annual interest may be computed using the following formula:

$A = P(1 + r)^t$ where P is the amount of principal, r is the annual interest rate (expressed in decimal form), and t is the number of years.

For example, for $1000 invested at 8% interest compounded annually for 3 years, we have $P = 1000$, $r = 0.08$, and $t = 3$.

$A = P(1 + r)^t$

$A = 1000(1 + 0.08)^3$

$= 1000(1.08)^3$

$= 1000(1.259712)$

$= 1259.712$

Rounding to the nearest cent, we have $A = \$1259.71$.

Answers

18. $-12x^7y^{10}$ 19. $9x^3y^4$

Example 8 Using Exponents in an Application

Find the amount in an account after 8 years if the initial investment is $7000, invested at 2.25% interest compounded annually.

Solution:

Identify the values for each variable.

$P = 7000$

$r = 0.0225$ — Note that the decimal form of a percent is used for calculations.

$t = 8$

$A = P(1 + r)^t$

$= 7000(1 + 0.0225)^8$ — Substitute.

$= 7000(1.0225)^8$ — Simplify inside the parentheses.

$\approx 7000(1.194831142)$ — Approximate $(1.0225)^8$.

≈ 8363.82 — Multiply (round to the nearest cent).

The amount in the account after 8 years is $8363.82.

Skill Practice

20. Find the amount in an account after 3 years if the initial investment is $4000 invested at 5% interest compounded annually.

Answer
20. $4630.50

Section 13.1 Activity

A.1. **a.** Write the expression x^3 in expanded form.
  b. Write the expression x^5 in expanded form.
  c. Write the expression $x^3 \cdot x^5$ in expanded form.
  d. Write the simplified form by filling in the blank. $x^3 \cdot x^5 = x^{\square}$
  e. Write a rule to simplify an expression of the form $x^m \cdot x^n$.

A.2. **a.** Simplify $x^{12} \cdot x^7$.
  b. Simplify $2^{12} \cdot 2^7$. (Write the answer in exponential form.)
  c. Simplify $y^3 \cdot y \cdot y^8$.

A.3. **a.** Write the expression t^6 in expanded form.
  b. Write the expression t^2 in expanded form.
  c. Write the expression $\dfrac{t^6}{t^2}$ in expanded form.
  d. Write the simplified form $\dfrac{t^6}{t^2}$ by filling in the blank. $\dfrac{t^6}{t^2} = t^{\square}$
  e. Write a rule to simplify an expression of the form $\dfrac{x^m}{x^n}$.

A.4. a. Simplify $\dfrac{p^{19}}{p^{11}}$.

b. Simplify $\dfrac{4^{19}}{4^{11}}$. (Write the answer in exponential form.)

c. Simplify $\dfrac{a^7 \cdot a}{a^4}$.

A.5. Simplify the expression $(-10c^4d^7)\left(\dfrac{1}{2}c^2d\right)$ by using the following steps.

    a. Apply the commutative and associative properties of multiplication to group the coefficients and factors with the same base.

    b. Simplify the expression from part (a).

A.6. Simplify the expression $\dfrac{-5c^{10}d^8}{-10c^4d^7}$ by using the following steps.

    a. Regroup factors so that the coefficients are grouped and factors with the same base are grouped.

    b. Simplify the expression from part (a).

Practice Exercises   Section 13.1

Study Skills Exercise

Talking out loud to yourself and/or teaching the material to another person is a great way to retain information. As you are speaking the information or describing a concept to another individual, you are increasing your understanding of the content and your ability to recall it.

- Pick a key concept that was presented in this section. Develop a lesson for teaching this concept to another student. Include important definitions or rules and example problems.

Prerequisite Review

For Exercises R.1–R.8, simplify the expression.

R.1. 6^2 **R.3.** $(-6)^2$ **R.5.** 4^3 **R.7.** $(-4)^3$

R.2. 8^2 **R.4.** $(-8)^2$ **R.6.** 2^5 **R.8.** $(-2)^5$

For this exercise set, assume all variables represent nonzero real numbers.

Vocabulary and Key Concepts

1. **a.** A(n) _____ is used to show repeated multiplication of the base.

    b. Given the expression b^n, the value b is the _____ and n is the _____.

    c. Given the expression x, the value of the exponent on x is understood to be _____.

    d. The formula to compute simple interest is _____.

    e. Interest that is computed on the original principal and on the accrued interest is called _____.

Concept 1: Review of Exponential Notation

For Exercises 2–13, identify the base and the exponent. (See Example 1.)

2. c^3
3. x^4
4. 5^2
5. 3^5
6. $(-4)^8$
7. $(-1)^4$
8. x
9. 13
10. -4^2
11. -10^3
12. $-y^5$
13. $-t^6$

14. What base corresponds to the exponent 5 in the expression $x^3y^5z^2$?

15. What base corresponds to the exponent 2 in the expression w^3v^2?

16. What is the exponent for the factor of 2 in the expression $2x^3$?

17. What is the exponent for the factor of p in the expression pq^7?

For Exercises 18–26, write the expression using exponents.

18. $(4n)(4n)(4n)$
19. $(-6b)(-6b)$
20. $4 \cdot n \cdot n \cdot n$
21. $-6 \cdot b \cdot b$
22. $(x-5)(x-5)(x-5)$
23. $(y+2)(y+2)(y+2)(y+2)$
24. $\dfrac{4}{x \cdot x \cdot x \cdot x \cdot x}$
25. $\dfrac{-2}{t \cdot t \cdot t}$
26. $\dfrac{5 \cdot x \cdot x \cdot x}{(y-7)(y-7)}$

Concept 2: Evaluating Expressions With Exponents

For Exercises 27–34, evaluate the two expressions and compare the answers. Do the expressions have the same value? (See Example 2.)

27. -5^2 and $(-5)^2$
28. -3^4 and $(-3)^4$
29. -2^5 and $(-2)^5$
30. -5^3 and $(-5)^3$
31. $\left(\dfrac{1}{2}\right)^3$ and $\dfrac{1}{2^3}$
32. $\left(\dfrac{1}{5}\right)^2$ and $\dfrac{1}{5^2}$
33. $-(-2)^4$ and $-(2)^4$
34. $-(-3)^3$ and $-(3)^3$

For Exercises 35–42, evaluate each expression. (See Example 2.)

35. 16^1
36. 20^1
37. $(-1)^{21}$
38. $(-1)^{30}$
39. $\left(-\dfrac{1}{3}\right)^2$
40. $\left(-\dfrac{1}{4}\right)^3$
41. $-\left(\dfrac{2}{5}\right)^2$
42. $-\left(\dfrac{3}{5}\right)^2$

For Exercises 43–50, simplify using the order of operations.

43. $3 \cdot 2^4$
44. $2 \cdot 0^5$
45. $-4(-1)^7$
46. $-3(-1)^4$
47. $6^2 - 3^3$
48. $4^3 + 2^3$
49. $2 \cdot 3^2 + 4 \cdot 2^3$
50. $6^2 - 3 \cdot 1^3$

For Exercises 51–62, evaluate each expression for $a = -4$ and $b = 5$. (See Example 3.)

51. $-4b^2$
52. $3a^2$
53. $(-4b)^2$
54. $(3a)^2$
55. $(a+b)^2$
56. $(a-b)^2$
57. $a^2 + 2ab + b^2$
58. $a^2 - 2ab + b^2$
59. $-10ab^2$
60. $-6a^3b$
61. $-10a^2b$
62. $-a^2b$

Concept 3: Multiplying and Dividing Expressions With Common Bases

63. Expand the following expressions first. Then simplify using exponents.
   a. $x^4 \cdot x^3$
   b. $5^4 \cdot 5^3$

64. Expand the following expressions first. Then simplify using exponents.
   a. $y^2 \cdot y^4$
   b. $3^2 \cdot 3^4$

For Exercises 65–76, simplify each expression. Write the answers in exponent form. (See Example 4.)

65. $z^5 z^3$
66. $w^4 w^7$
67. $a \cdot a^8$
68. $p^4 p$

69. $4^5 \cdot 4^9$
70. $6^7 \cdot 6^5$
71. $\left(\dfrac{2}{3}\right)^3 \left(\dfrac{2}{3}\right)$
72. $\left(\dfrac{1}{x}\right)\left(\dfrac{1}{x}\right)^2$

73. $c^5 c^2 c^7$
74. $b^7 b^2 b^8$
75. $x \cdot x^4 \cdot x^{10} \cdot x^3$
76. $z^7 \cdot z^{11} \cdot z^{60} \cdot z$

77. Expand the expressions. Then simplify.
   a. $\dfrac{p^8}{p^3}$
   b. $\dfrac{8^8}{8^3}$

78. Expand the expressions. Then simplify.
   a. $\dfrac{w^5}{w^2}$
   b. $\dfrac{4^5}{4^2}$

For Exercises 79–94, simplify each expression. Write the answers in exponent form. (See Examples 5–6.)

79. $\dfrac{x^8}{x^6}$
80. $\dfrac{z^5}{z^4}$
81. $\dfrac{a^{10}}{a}$
82. $\dfrac{b^{12}}{b}$

83. $\dfrac{7^{13}}{7^6}$
84. $\dfrac{2^6}{2^4}$
85. $\dfrac{5^8}{5}$
86. $\dfrac{3^5}{3}$

87. $\dfrac{y^{13}}{y^{12}}$
88. $\dfrac{w^7}{w^6}$
89. $\dfrac{h^3 h^8}{h^7}$
90. $\dfrac{n^5 n^4}{n^2}$

91. $\dfrac{7^2 \cdot 7^6}{7}$
92. $\dfrac{5^3 \cdot 5^8}{5}$
93. $\dfrac{10^{20}}{10^3 \cdot 10^8}$
94. $\dfrac{3^{15}}{3^2 \cdot 3^{10}}$

Concept 4: Simplifying Expressions With Exponents (Mixed Exercises)

For Exercises 95–114, use the commutative and associative properties of real numbers and the properties of exponents to simplify. (See Example 7.)

95. $(2x^3)(3x^4)$
96. $(10y)(2y^3)$
97. $(5a^2 b)(8a^3 b^4)$
98. $(10xy^3)(3x^4 y)$

99. $s^3 \cdot t^5 \cdot t \cdot t^{10} \cdot s^6$
100. $c \cdot c^4 \cdot d^2 \cdot c^3 \cdot d^3$
101. $(-2v^2)(3v)(5v^5)$
102. $(10q^5)(-3q^8)(q)$

103. $\left(\dfrac{2}{3}m^{13}n^8\right)(24m^7 n^2)$
104. $\left(\dfrac{1}{4}c^6 d^6\right)(28c^2 d^7)$
105. $\dfrac{14c^4 d^5}{7c^3 d}$
106. $\dfrac{36h^5 k^2}{9h^3 k}$

107. $\dfrac{z^3 z^{11}}{z^4 z^6}$
108. $\dfrac{w^{12} w^2}{w^4 w^5}$
109. $\dfrac{25h^3 jk^5}{12h^2 k}$
110. $\dfrac{15m^5 np^{12}}{4mp^9}$

111. $(-4p^6 q^8 r^4)(2pqr^2)$
112. $(-5a^4 bc)(-10a^2 b)$
113. $\dfrac{-12s^2 tu^3}{4su^2}$
114. $\dfrac{15w^5 x^{10} y^3}{-15w^4 x}$

Concept 5: Applications of Exponents

Use the formula $A = P(1 + r)^t$ for Exercises 115–118. **(See Example 8.)**

115. Find the amount in an account after 2 years if the initial investment is $5000, invested at 7% interest compounded annually.

116. Find the amount in an account after 5 years if the initial investment is $2000, invested at 4% interest compounded annually.

117. Find the amount in an account after 3 years if the initial investment is $4000, invested at 6% interest compounded annually.

118. Find the amount in an account after 4 years if the initial investment is $10,000, invested at 5% interest compounded annually.

For Exercises 119–122, use appropriate geometry formulas.

119. Find the area of the pizza shown in the figure. Round to the nearest square inch.

Burke/Triolo Productions/Brand X Pictures/Alamy Stock Photo

120. Find the volume of the sphere shown in the figure. Round to the nearest cubic centimeter.

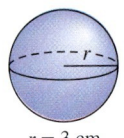

$r = 3$ cm

121. Find the volume of a spherical balloon that is 8 in. in diameter. Round to the nearest cubic inch.

122. Find the area of a circular pool 50 ft in diameter. Round to the nearest square foot.

Technology Connections

For Exercises 123–128, use a calculator to evaluate the expressions.

123. $(1.06)^5$

124. $(1.02)^{40}$

125. $5000(1.06)^5$

126. $2000(1.02)^{40}$

127. $3000(1 + 0.06)^2$

128. $1000(1 + 0.05)^3$

Expanding Your Skills

For Exercises 129–136, simplify each expression using the addition or subtraction rules of exponents. Assume that a, b, m, and n represent positive integers.

129. $x^n x^{n+1}$

130. $y^a y^{2a}$

131. $p^{3m+5} p^{-m-2}$

132. $q^{4b-3} q^{-4b+4}$

133. $\dfrac{z^{b+1}}{z^b}$

134. $\dfrac{w^{5n+3}}{w^{2n}}$

135. $\dfrac{r^{3a+3}}{r^{3a}}$

136. $\dfrac{t^{3+2m}}{t^{2m}}$

More Properties of Exponents

Section 13.2

Concepts

1. Power Rule for Exponents
2. The Properties $(ab)^m = a^m b^m$ and $\left(\dfrac{a}{b}\right)^m = \dfrac{a^m}{b^m}$

1. Power Rule for Exponents

The expression $(x^2)^3$ indicates that the quantity x^2 is cubed.

$$(x^2)^3 = (x^2)(x^2)(x^2) = (x \cdot x)(x \cdot x)(x \cdot x) = x^6$$

From this example, it appears that to raise a base to successive powers, we multiply the exponents and leave the base unchanged. This is stated formally as the power rule for exponents.

> **Power Rule for Exponents**
>
> Assume that b is a real number and that m and n represent positive integers. Then,
>
> $$(b^m)^n = b^{m \cdot n}$$

Example 1   Simplifying Expressions With Exponents

Simplify the expressions.

a. $(s^4)^2$ b. $(3^4)^2$ c. $(x^2 x^5)^4$

Solution:

a. $(s^4)^2$

    $= s^{4 \cdot 2}$ Multiply exponents (the base is unchanged).

    $= s^8$

b. $(3^4)^2$

    $= 3^{4 \cdot 2}$ Multiply exponents (the base is unchanged).

    $= 3^8$ or 6561

c. $(x^2 x^5)^4$

    $= (x^7)^4$ Simplify inside the parentheses by adding exponents.

    $= x^{7 \cdot 4}$ Multiply exponents (the base is unchanged).

    $= x^{28}$

Skill Practice   Simplify the expressions.

1. $(y^3)^5$ 2. $(2^8)^{10}$ 3. $(q^5 q^4)^3$

2. The Properties $(ab)^m = a^m b^m$ and $\left(\dfrac{a}{b}\right)^m = \dfrac{a^m}{b^m}$

Consider the following expressions and their simplified forms:

$$(xy)^3 = (xy)(xy)(xy) = (x \cdot x \cdot x)(y \cdot y \cdot y) = x^3 y^3$$

$$\left(\frac{x}{y}\right)^3 = \left(\frac{x}{y}\right)\left(\frac{x}{y}\right)\left(\frac{x}{y}\right) = \left(\frac{x \cdot x \cdot x}{y \cdot y \cdot y}\right) = \frac{x^3}{y^3}$$

The expressions are simplified using the commutative and associative properties of multiplication. The simplified forms for each expression could have been reached in one step by applying the exponent to each factor inside the parentheses.

Answers

1. y^{15} 2. 2^{80} 3. q^{27}

Chapter 13 Polynomials and Properties of Exponents

Avoiding Mistakes

The power rule of exponents can be applied to a product of bases but in general cannot be applied to a sum or difference of bases.

$(ab)^n = a^n b^n$
but $(a+b)^n \neq a^n + b^n$

Power of a Product and Power of a Quotient

Assume that a and b are real numbers. Let m represent a positive integer. Then,

$$(ab)^m = a^m b^m$$
$$\left(\frac{a}{b}\right)^m = \frac{a^m}{b^m}, \quad b \neq 0$$

Applying these properties of exponents, we have
$$(xy)^3 = x^3 y^3 \quad \text{and} \quad \left(\frac{x}{y}\right)^3 = \frac{x^3}{y^3}$$

Example 2 Simplifying Expressions With Exponents

Simplify the expressions.

a. $(-2xyz)^4$ b. $(5x^2 y^7)^3$ c. $\left(\frac{2}{5}\right)^3$ d. $\left(\frac{1}{3xy^4}\right)^2$

Solution:

a. $(-2xyz)^4$
$= (-2)^4 x^4 y^4 z^4$ Raise each factor within parentheses to the fourth power.
$= 16 x^4 y^4 z^4$

b. $(5x^2 y^7)^3$
$= 5^3 (x^2)^3 (y^7)^3$ Raise each factor within parentheses to the third power.
$= 125 x^6 y^{21}$ Multiply exponents and simplify.

c. $\left(\frac{2}{5}\right)^3$
$= \frac{(2)^3}{(5)^3}$ Raise each factor within parentheses to the third power.
$= \frac{8}{125}$ Simplify.

d. $\left(\frac{1}{3xy^4}\right)^2$
$= \frac{1^2}{3^2 x^2 (y^4)^2}$ Square each factor within parentheses.
$= \frac{1}{9 x^2 y^8}$ Multiply exponents and simplify.

Skill Practice Simplify the expressions.

4. $(3abc)^5$ 5. $(-2t^2 w^4)^3$ 6. $\left(\frac{3}{4}\right)^3$ 7. $\left(\frac{2x^3}{y^5}\right)^2$

Answers
4. $3^5 a^5 b^5 c^5$ or $243 a^5 b^5 c^5$
5. $-8 t^6 w^{12}$ 6. $\frac{27}{64}$ 7. $\frac{4x^6}{y^{10}}$

Definitions of b^0 and b^{-n}

Section 13.3

We have learned several rules that enable us to manipulate expressions containing *positive* integer exponents. In this section, we present definitions that can be used to simplify expressions with negative exponents or with an exponent of zero.

Concepts

1. Definition of b^0
2. Definition of b^{-n}
3. Properties of Integer Exponents: A Summary

1. Definition of b^0

To begin, consider the following pattern.

$3^3 = 27$ — Divide by 3.
$3^2 = 9$ — Divide by 3.
$3^1 = 3$ — Divide by 3.
$3^0 = 1$

As the exponents decrease by 1, the resulting expressions are divided by 3.

For the pattern to continue, we define $3^0 = 1$.

This pattern suggests that we should define an expression with a zero exponent as follows:

Definition of b^0

Let b be a nonzero real number. Then, $b^0 = 1$.

Note: The value of 0^0 is not defined by this definition because the base b must not equal zero.

Example 1 — Simplifying Expressions With a Zero Exponent

Simplify. Assume that $z \neq 0$.

a. 4^0 b. $(-4)^0$ c. -4^0
d. z^0 e. $-4z^0$ f. $(4z)^0$

Solution:

a. $4^0 = 1$ — By definition

b. $(-4)^0 = 1$ — By definition

c. $-4^0 = -1 \cdot 4^0 = -1 \cdot 1 = -1$ — The exponent 0 applies only to 4.

d. $z^0 = 1$ — By definition

e. $-4z^0 = -4 \cdot z^0 = -4 \cdot 1 = -4$ — The exponent 0 applies only to z.

f. $(4z)^0 = 1$ — The parentheses indicate that the exponent, 0, applies to both factors 4 and z.

Skill Practice Evaluate the expressions. Assume that $x \neq 0$ and $y \neq 0$.

1. 7^0 2. $(-7)^0$ 3. -5^0
4. y^0 5. $-2x^0$ 6. $(2x)^0$

Answers
1. 1   2. 1   3. −1
4. 1   5. −2   6. 1

The definition of b^0 is consistent with the other properties of exponents learned thus far. For example, we know that $1 = \frac{5^3}{5^3}$. If we subtract exponents, the result is 5^0.

$$1 = \frac{5^3}{5^3} = 5^{3-3} = 5^0 \quad \text{Therefore, } 5^0 \text{ must be defined as 1.}$$

Subtract exponents.

2. Definition of b^{-n}

To understand the concept of a *negative* exponent, consider the following pattern.

$3^3 = 27$
$3^2 = 9$ — Divide by 3.
$3^1 = 3$ — Divide by 3.   As the exponents decrease by 1,
$3^0 = 1$ — Divide by 3.   the resulting expressions are
$3^{-1} = \frac{1}{3}$ — Divide by 3.   divided by 3.

$3^{-1} = \frac{1}{3}$ ← For the pattern to continue, we define $3^{-1} = \frac{1}{3^1} = \frac{1}{3}$.

$3^{-2} = \frac{1}{9}$ ← For the pattern to continue, we define $3^{-2} = \frac{1}{3^2} = \frac{1}{9}$.

$3^{-3} = \frac{1}{27}$ ← For the pattern to continue, we define $3^{-3} = \frac{1}{3^3} = \frac{1}{27}$.

This pattern suggests that $3^{-n} = \frac{1}{3^n}$ for all integers, n. In general, we have the following definition involving negative exponents.

> **Definition of b^{-n}**
>
> Let n be an integer and b be a nonzero real number. Then,
>
> $$b^{-n} = \left(\frac{1}{b}\right)^n \quad \text{or} \quad \frac{1}{b^n}$$

The definition of b^{-n} implies that to evaluate b^{-n}, take the reciprocal of the base and change the sign of the exponent.

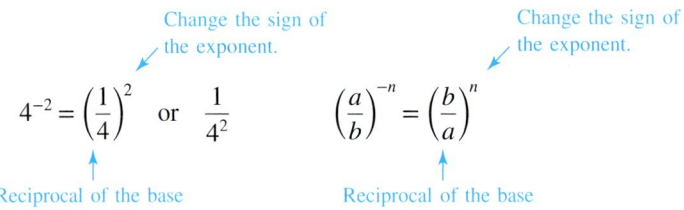

$$4^{-2} = \left(\frac{1}{4}\right)^2 \quad \text{or} \quad \frac{1}{4^2} \qquad \left(\frac{a}{b}\right)^{-n} = \left(\frac{b}{a}\right)^n$$

Reciprocal of the base — Change the sign of the exponent.

Reciprocal of the base — Change the sign of the exponent.

Example 2 Simplifying Expressions With Negative Exponents

Simplify. Assume that $c \neq 0$.

a. c^{-3} b. 5^{-1} c. $(-3)^{-4}$

Solution:

a. $c^{-3} = \dfrac{1}{c^3}$ By definition

b. $5^{-1} = \dfrac{1}{5^1}$ By definition

$= \dfrac{1}{5}$ Simplify.

c. $(-3)^{-4} = \dfrac{1}{(-3)^4}$ The base is -3 and must be enclosed in parentheses.

$= \dfrac{1}{81}$ Simplify. Note that $(-3)^4 = (-3)(-3)(-3)(-3) = 81$.

Avoiding Mistakes

A negative exponent does *not* affect the sign of the base.

Skill Practice Simplify. Assume that $p \neq 0$.

7. p^{-4} 8. 3^{-3} 9. $(-5)^{-2}$

Example 3 Simplifying Expressions With Negative Exponents

Simplify. Assume that $y \neq 0$.

a. $\left(\dfrac{1}{6}\right)^{-2}$ b. $\left(-\dfrac{3}{5}\right)^{-3}$ c. $\dfrac{1}{y^{-5}}$

Solution:

a. $\left(\dfrac{1}{6}\right)^{-2} = 6^2$ Take the reciprocal of the base, and change the sign of the exponent.

$= 36$ Simplify.

b. $\left(-\dfrac{3}{5}\right)^{-3} = \left(-\dfrac{5}{3}\right)^3$ Take the reciprocal of the base, and change the sign of the exponent.

$= -\dfrac{125}{27}$ Simplify.

c. $\dfrac{1}{y^{-5}} = \left(\dfrac{1}{y}\right)^{-5}$ Apply the power of a quotient rule.

$= (y)^5$ Take the reciprocal of the base, and change the sign of the exponent.

$= y^5$

TIP: Example 3(c) illustrates that $\dfrac{1}{b^{-n}} = b^n$, for $b \neq 0$.

Skill Practice Simplify. Assume that $w \neq 0$.

10. $\left(\dfrac{1}{3}\right)^{-1}$ 11. $\left(-\dfrac{2}{5}\right)^{-2}$ 12. $\dfrac{1}{w^{-7}}$

Answers

7. $\dfrac{1}{p^4}$ 8. $\dfrac{1}{3^3}$ or $\dfrac{1}{27}$

9. $\dfrac{1}{(-5)^2}$ or $\dfrac{1}{25}$ 10. 3

11. $\dfrac{25}{4}$ 12. w^7

> **Example 4** **Simplifying Expressions With Negative Exponents**
>
> Simplify. Assume that $x \neq 0$.
>
> **a.** $(5x)^{-3}$ **b.** $5x^{-3}$ **c.** $-5x^{-3}$
>
> **Solution:**
>
> **a.** $(5x)^{-3} = \left(\dfrac{1}{5x}\right)^3$ Take the reciprocal of the base, and change the sign of the exponent.
>
> $= \dfrac{(1)^3}{(5x)^3}$ Apply the exponent of 3 to each factor within parentheses.
>
> $= \dfrac{1}{125x^3}$ Simplify.
>
> **b.** $5x^{-3} = 5 \cdot x^{-3}$ Note that the exponent, -3, applies only to x.
>
> $= 5 \cdot \dfrac{1}{x^3}$ Rewrite x^{-3} as $\dfrac{1}{x^3}$.
>
> $= \dfrac{5}{x^3}$ Multiply.
>
> **c.** $-5x^{-3} = -5 \cdot x^{-3}$ Note that the exponent, -3, applies only to x, and that -5 is a coefficient.
>
> $= -5 \cdot \dfrac{1}{x^3}$ Rewrite x^{-3} as $\dfrac{1}{x^3}$.
>
> $= -\dfrac{5}{x^3}$ Multiply.
>
> **Skill Practice** Simplify. Assume that $w \neq 0$.
>
> **13.** $(2w)^{-4}$ **14.** $2w^{-4}$ **15.** $-2w^{-4}$

It is important to note that the definition of b^{-n} is consistent with the other properties of exponents learned thus far. For example, consider the expression

$$\dfrac{x^4}{x^7} = \dfrac{\cancel{x}\cdot\cancel{x}\cdot\cancel{x}\cdot\cancel{x}}{\cancel{x}\cdot\cancel{x}\cdot\cancel{x}\cdot\cancel{x}\cdot x \cdot x \cdot x} = \dfrac{1}{x^3}$$

Subtract exponents. Hence, $x^{-3} = \dfrac{1}{x^3}$.

By subtracting exponents, we have $\dfrac{x^4}{x^7} = x^{4-7} = x^{-3}$.

3. Properties of Integer Exponents: A Summary

The definitions of b^0 and b^{-n} enable us to extend the properties of exponents. These are summarized in Table 13-1.

Answers

13. $\dfrac{1}{16w^4}$ **14.** $\dfrac{2}{w^4}$

15. $-\dfrac{2}{w^4}$

Table 13-1

Properties of Integer Exponents

Assume that a and b are real numbers ($b \neq 0$) and that m and n represent integers.

Property	Example	Details/Notes
Multiplication of Expressions With Like Bases $b^m b^n = b^{m+n}$	$b^2 b^4 = b^{2+4} = b^6$	$b^2 b^4 = (b \cdot b)(b \cdot b \cdot b \cdot b) = b^6$
Division of Expressions With Like Bases $\dfrac{b^m}{b^n} = b^{m-n}$	$\dfrac{b^5}{b^2} = b^{5-2} = b^3$	$\dfrac{b^5}{b^2} = \dfrac{\not{b} \cdot \not{b} \cdot b \cdot b \cdot b}{\not{b} \cdot \not{b}} = b^3$
The Power Rule $(b^m)^n = b^{m \cdot n}$	$(b^4)^2 = b^{4 \cdot 2} = b^8$	$(b^4)^2 = (b \cdot b \cdot b \cdot b)(b \cdot b \cdot b \cdot b) = b^8$
Power of a Product $(ab)^m = a^m b^m$	$(ab)^3 = a^3 b^3$	$(ab)^3 = (ab)(ab)(ab) = (a \cdot a \cdot a)(b \cdot b \cdot b) = a^3 b^3$
Power of a Quotient $\left(\dfrac{a}{b}\right)^m = \dfrac{a^m}{b^m}$	$\left(\dfrac{a}{b}\right)^3 = \dfrac{a^3}{b^3}$	$\left(\dfrac{a}{b}\right)^3 = \left(\dfrac{a}{b}\right)\left(\dfrac{a}{b}\right)\left(\dfrac{a}{b}\right) = \dfrac{a \cdot a \cdot a}{b \cdot b \cdot b} = \dfrac{a^3}{b^3}$

Definitions

Assume that b is a real number ($b \neq 0$) and that n represents an integer.

Definition	Example	Details/Notes
$b^0 = 1$	$(4)^0 = 1$	Any nonzero quantity raised to the zero power equals 1.
$b^{-n} = \left(\dfrac{1}{b}\right)^n = \dfrac{1}{b^n}$	$b^{-5} = \left(\dfrac{1}{b}\right)^5 = \dfrac{1}{b^5}$	To simplify a negative exponent, take the reciprocal of the base and make the exponent positive.

Example 5 Simplifying Expressions With Exponents

Simplify the expressions. Write the answers with positive exponents only. Assume all variables are nonzero.

a. $\dfrac{a^3 b^{-2}}{c^{-5}}$ b. $\dfrac{x^2 x^{-7}}{x^3}$ c. $\dfrac{z^2}{w^{-4} w^4 z^{-8}}$

Solution:

a. $\dfrac{a^3 b^{-2}}{c^{-5}}$

$= \dfrac{a^3}{1} \cdot \dfrac{b^{-2}}{1} \cdot \dfrac{1}{c^{-5}}$

$= \dfrac{a^3}{1} \cdot \dfrac{1}{b^2} \cdot \dfrac{c^5}{1}$ Simplify negative exponents.

$= \dfrac{a^3 c^5}{b^2}$ Multiply.

b. $\dfrac{x^2 x^{-7}}{x^3}$

$= \dfrac{x^{2+(-7)}}{x^3}$ Add the exponents in the numerator.

$= \dfrac{x^{-5}}{x^3}$ Simplify.

$= x^{-5-3}$ Subtract the exponents.

$= x^{-8}$

$= \dfrac{1}{x^8}$ Simplify the negative exponent.

c. $\dfrac{z^2}{w^{-4} w^4 z^{-8}}$

$= \dfrac{z^2}{w^{-4+4} z^{-8}}$ Add the exponents in the denominator.

$= \dfrac{z^2}{w^0 z^{-8}}$

$= \dfrac{z^2}{(1) z^{-8}}$ Recall that $w^0 = 1$.

$= z^{2-(-8)}$ Subtract the exponents.

$= z^{10}$ Simplify.

Skill Practice Simplify the expressions. Assume all variables are nonzero.

16. $\dfrac{x^{-6}}{y^4 z^{-8}}$ 17. $\dfrac{x^3 x^{-8}}{x^4}$ 18. $\dfrac{p^3}{w^7 w^{-7} z^{-2}}$

Example 6   Simplifying Expressions With Exponents

Simplify the expressions. Write the answers with positive exponents only. Assume that all variables are nonzero.

a. $(-4ab^{-2})^{-3}$ b. $\left(\dfrac{2p^{-4} q^3}{5p^2 q}\right)^{-2}$

Solution:

a. $(-4ab^{-2})^{-3}$

$= (-4)^{-3} a^{-3} (b^{-2})^{-3}$ Apply the power rule of exponents.

$= (-4)^{-3} a^{-3} b^6$

$= \dfrac{1}{(-4)^3} \cdot \dfrac{1}{a^3} \cdot b^6$ Simplify the negative exponents.

$= \dfrac{1}{-64} \cdot \dfrac{1}{a^3} \cdot b^6$ Simplify.

$= -\dfrac{b^6}{64 a^3}$ Multiply fractions.

Answers

16. $\dfrac{z^8}{y^4 x^6}$ 17. $\dfrac{1}{x^9}$ 18. $p^3 z^2$

b. $\left(\dfrac{2p^{-4}q^3}{5p^2q}\right)^{-2}$ First simplify within the parentheses.

$= \left(\dfrac{2p^{-4-2}q^{3-1}}{5}\right)^{-2}$ Divide expressions with like bases by subtracting exponents.

$= \left(\dfrac{2p^{-6}q^2}{5}\right)^{-2}$ Simplify.

$= \dfrac{(2p^{-6}q^2)^{-2}}{(5)^{-2}}$ Apply the power rule of a quotient.

$= \dfrac{2^{-2}(p^{-6})^{-2}(q^2)^{-2}}{5^{-2}}$ Apply the power rule of a product.

$= \dfrac{2^{-2}p^{12}q^{-4}}{5^{-2}}$ Simplify.

$= \dfrac{5^2 p^{12}}{2^2 q^4}$ Simplify the negative exponents.

$= \dfrac{25p^{12}}{4q^4}$ Simplify.

TIP: For Example 6(b), the power rule of exponents can be performed first. In that case, the second step would be

$$\dfrac{2^{-2}p^8 q^{-6}}{5^{-2}p^{-4}q^{-2}}$$

Skill Practice Simplify the expressions. Assume all variables are nonzero.

19. $(-5x^{-2}y^3)^{-2}$ **20.** $\left(\dfrac{3x^{-3}y^{-2}}{4xy^{-3}}\right)^{-2}$

Example 7   Simplifying an Expression With Exponents

Simplify the expression $2^{-1} + 3^{-1} + 5^0$. Write the answer with positive exponents only.

Solution:

$2^{-1} + 3^{-1} + 5^0$

$= \dfrac{1}{2} + \dfrac{1}{3} + 1$ Simplify negative exponents. Simplify $5^0 = 1$.

$= \dfrac{3}{6} + \dfrac{2}{6} + \dfrac{6}{6}$ The least common denominator is 6.

$= \dfrac{11}{6}$ Simplify.

Skill Practice Simplify the expressions.

21. $2^{-1} + 4^{-2} + 3^0$

Answers

19. $\dfrac{x^4}{25y^6}$ **20.** $\dfrac{16x^8}{9y^2}$ **21.** $\dfrac{25}{16}$

Section 13.3 Activity

A.1. Simplify the expressions.

    a. 2^5 **b.** 2^4 **c.** 2^3 **d.** 2^2 **e.** 2^1

A.2. Look at the pattern of the simplified forms of the expressions from Exercise A.1. Based on this pattern, what do you think the next value 2^0 would be?

A.3. **a.** For the nonzero base b, simplify $\dfrac{b^4}{b^4}$ by expanding factors in the numerator and denominator.

    b. Now consider the expression $\dfrac{b^4}{b^4} = b^{4-4} = b^0$. How should the expression b^0 be defined so that it is consistent with the result in part (a)?

A.4. **a.** Simplify the expression $\dfrac{b^2}{b^5}$ by expanding factors in the numerator and denominator.

    b. Now consider the expression $\dfrac{b^2}{b^5} = b^{2-5} = b^{-3}$. How should the expression b^{-3} be defined so that it is consistent with the result in part (a)?

    c. For a nonzero base b and nonzero exponent n, simplify the expression b^{-n}.

A.5. **a.** Evaluate the expression 4^{-2} using a calculator.

    b. Evaluate the expression $\dfrac{1}{4^2}$ using a calculator.

    c. What do you observe about the results from parts (a) and (b)?

    d. How can a base with a negative exponent be written so that the exponent is positive?

For Exercises A.6–A.9, simplify the expression.

A.6. x^{-4} **A.7.** $\dfrac{1}{x^{-4}}$ **A.8.** $\dfrac{a^6}{b^{-7}}$ **A.9.** $\dfrac{a^{-6}}{b^7}$

A.10. **a.** Explain the difference between the process to simplify $(-6)^0$ versus -6^0.

    b. Explain the difference between the process to simplify $(-3)^{-2}$ and -3^{-2}.

    c. Explain the difference between the process to simplify x^{-4} versus $-x^{-4}$.

    d. Explain the difference between the process to simplify $\left(\dfrac{3}{4}\right)^0$ versus $\dfrac{3^0}{4}$.

A.11. Simplify the expression $\left(\dfrac{c^{-6}d^4}{4c^{-8}d^{-2}}\right)^{-2}(cd^{-3})$ by following these steps.

    a. Simplify the expression within the first parentheses.

    b. Apply the exponent of -2 to the factors in the first parentheses and simplify.

    c. Multiply the factors from the first parentheses and the factors from the second parentheses.

    d. Simplify the expression from part (c).

Practice Exercises

Section 13.3

For this exercise set, assume all variables represent nonzero real numbers.

Prerequisite Review

For Exercises R.1–R.2, simplify the expressions.

R.1. a. $\dfrac{9}{9}$ b. $\dfrac{3^2}{3^2}$ c. $\dfrac{t}{t}$ d. $\dfrac{t^2}{t^2}$

R.2. a. $\dfrac{16}{16}$ b. $\dfrac{2^4}{2^4}$ c. $\dfrac{x}{x}$ d. $\dfrac{x^4}{x^4}$

For Exercises R.3–R.6, simplify the expression by reducing common factors.

R.3. $\dfrac{125}{25} = \dfrac{5 \cdot 5 \cdot 5}{5 \cdot 5} =$

R.4. $\dfrac{32}{4} = \dfrac{2 \cdot 2 \cdot 2 \cdot 2 \cdot 2}{2 \cdot 2} =$

R.5. $\dfrac{100}{1000} = \dfrac{10 \cdot 10}{10 \cdot 10 \cdot 10} =$

R.6. $\dfrac{25}{625} = \dfrac{5 \cdot 5}{5 \cdot 5 \cdot 5 \cdot 5} =$

For Exercises R.7–R.14, simplify the expression.

R.7. $b^3 b^8$

R.8. $c^7 c^2$

R.9. $\dfrac{x^6}{x^2}$

R.10. $\dfrac{y^9}{y^8}$

R.11. $\dfrac{9^4 \cdot 9^8}{9}$

R.12. $\dfrac{3^{14}}{3^3 \cdot 3^5}$

R.13. $(6ab^3c^2)^5$

R.14. $(7w^7z^2)^4$

Vocabulary and Key Concepts

1. a. The expression b^0 is defined to be _____ provided that $b \neq 0$.
   b. The expression b^{-n} is defined as _____ provided that $b \neq 0$.

Concept 1: Definition of b^0

2. Simplify.
   a. 8^0 b. $\dfrac{8^4}{8^4}$

3. Simplify.
   a. d^0 b. $\dfrac{d^3}{d^3}$

4. Simplify.
   a. m^0 b. $\dfrac{m^5}{m^5}$

For Exercises 5–16, simplify. (See Example 1.)

5. p^0

6. k^0

7. 5^0

8. 2^0

9. -4^0

10. -1^0

11. $(-6)^0$

12. $(-2)^0$

13. $(8x)^0$

14. $(-3y^3)^0$

15. $-7x^0$

16. $6y^0$

Concept 2: Definition of b^{-n}

17. Simplify and write the answers with positive exponents.
    a. t^{-5} b. $\dfrac{t^3}{t^8}$

18. Simplify and write the answers with positive exponents.
    a. 4^{-3} b. $\dfrac{4^2}{4^5}$

For Exercises 19–38, simplify. (See Examples 2–4.)

19. $\left(\dfrac{2}{7}\right)^{-3}$

20. $\left(\dfrac{5}{4}\right)^{-1}$

21. $\left(-\dfrac{1}{5}\right)^{-2}$

22. $\left(-\dfrac{1}{3}\right)^{-3}$

23. a^{-3}

24. c^{-5}

25. 12^{-1}

26. 4^{-2}

27. $(4b)^{-2}$

28. $(3z)^{-1}$

29. $6x^{-2}$

30. $7y^{-1}$

31. $(-8)^{-2}$

32. -8^{-2}

33. $-3y^{-4}$

34. $-6a^{-2}$

35. $(-t)^{-3}$

36. $(-r)^{-5}$

37. $\dfrac{1}{a^{-5}}$

38. $\dfrac{1}{b^{-6}}$

Concept 3: Properties of Integer Exponents: A Summary

For Exercises 39–42, correct the statement.

39. $\dfrac{x^4}{x^{-6}} = x^{4-6} = x^{-2}$

40. $\dfrac{y^5}{y^{-3}} = y^{5-3} = y^2$

41. $2a^{-3} = \dfrac{1}{2a^3}$

42. $5b^{-2} = \dfrac{1}{5b^2}$

Mixed Exercises

For Exercises 43–86, simplify each expression. Write the answer with positive exponents only. (See Examples 5–6.)

43. $x^{-8}x^4$

44. s^5s^{-6}

45. $a^{-8}a^8$

46. q^3q^{-3}

47. $y^{17}y^{-13}$

48. $b^{20}b^{-14}$

49. $(m^{-6}n^9)^3$

50. $(c^4d^{-5})^{-2}$

51. $(-3j^{-5}k^6)^4$

52. $(6xy^{-11})^{-3}$

53. $\dfrac{p^3}{p^9}$

54. $\dfrac{q^2}{q^{10}}$

55. $\dfrac{r^{-5}}{r^{-2}}$

56. $\dfrac{u^{-2}}{u^{-6}}$

57. $\dfrac{a^2}{a^{-6}}$

58. $\dfrac{p^3}{p^{-5}}$

59. $\dfrac{y^{-2}}{y^6}$

60. $\dfrac{s^{-4}}{s^3}$

61. $\dfrac{7^3}{7^2 \cdot 7^8}$

62. $\dfrac{3^4 \cdot 3}{3^7}$

63. $\dfrac{a^2a}{a^3}$

64. $\dfrac{t^5}{t^2t^3}$

65. $\dfrac{a^{-1}b^2}{a^3b^8}$

66. $\dfrac{k^{-4}h^{-1}}{k^6h}$

67. $\dfrac{w^{-8}(w^2)^{-5}}{w^3}$

68. $\dfrac{p^2p^{-7}}{(p^2)^3}$

69. $\dfrac{3^{-2}}{3}$

70. $\dfrac{5^{-1}}{5}$

71. $\left(\dfrac{p^{-1}q^5}{p^{-6}}\right)^0$

72. $\left(\dfrac{ab^{-4}}{a^{-5}}\right)^0$

73. $(8x^3y^0)^{-2}$

74. $(3u^2v^0)^{-3}$

75. $(-8y^{-12})(2y^{16}z^{-2})$

76. $(5p^{-2}q^5)(-2p^{-4}q^{-1})$

77. $\dfrac{-18a^{10}b^6}{108a^{-2}b^6}$

78. $\dfrac{-35x^{-4}y^{-3}}{-21x^2y^{-3}}$

79. $\dfrac{(-4c^{12}d^7)^2}{(5c^{-3}d^{10})^{-1}}$

80. $\dfrac{(s^3t^{-2})^4}{(3s^{-4}t^6)^{-2}}$

81. $\dfrac{(2x^3y^2)^{-3}}{(3x^2y^4)^{-2}}$

82. $\dfrac{(5p^4q)^{-3}}{(p^3q^5)^{-4}}$

83. $\left(\dfrac{5cd^{-3}}{10d^5}\right)^{-2}$

84. $\left(\dfrac{4m^{10}n^4}{2m^{12}n^{-2}}\right)^{-1}$

85. $(2xy^3)\left(\dfrac{9xy}{4x^3y^2}\right)$

86. $(-3a^3)\left(\dfrac{ab}{27a^4b^2}\right)$

For Exercises 87–94, simplify. (See Example 7.)

87. $5^{-1} + 2^{-2}$
88. $4^{-2} + 8^{-1}$
89. $10^0 - 10^{-1}$
90. $3^0 - 3^{-2}$
91. $2^{-2} + 1^{-2}$
92. $4^{-1} + 8^{-1}$
93. $4 \cdot 5^0 - 2 \cdot 3^{-1}$
94. $2 \cdot 4^0 - 3 \cdot 4^{-1}$

Expanding Your Skills

For Exercises 95–98, determine the missing exponent.

95. $\dfrac{y^4 y^\square}{y^{-2}} = y^8$
96. $\dfrac{x^4 x^\square}{x^{-1}} = x^9$
97. $\dfrac{w^{-9}}{w^\square} = w^2$
98. $\dfrac{a^{-2}}{a^\square} = a^6$

Problem Recognition Exercises

Properties of Exponents

For Exercises 1–40, simplify completely. Assume that all variables represent nonzero real numbers.

1. $t^3 t^5$
2. $2^3 2^5$
3. $\dfrac{y^7}{y^2}$
4. $\dfrac{p^9}{p^3}$

5. $(r^2 s^4)^2$
6. $(ab^3 c^2)^3$
7. $\dfrac{w^4}{w^{-2}}$
8. $\dfrac{m^{-14}}{m^2}$

9. $\dfrac{y^{-7} x^4}{z^{-3}}$
10. $\dfrac{a^3 b^{-6}}{c^{-8}}$
11. $\dfrac{x^4 x^{-3}}{x^{-5}}$
12. $\dfrac{y^{-4}}{y^7 y^{-1}}$

13. $\dfrac{t^{-2} t^4}{t^8 t^{-1}}$
14. $\dfrac{w^8 w^{-5}}{w^{-2} w^{-2}}$
15. $\dfrac{1}{p^{-6} p^{-8} p^{-1}}$
16. $p^6 p^8 p$

17. $\dfrac{v^9}{v^{11}}$
18. $(c^5 d^4)^{10}$
19. $\left(\dfrac{1}{2}\right)^{-1} + \left(\dfrac{1}{3}\right)^0$
20. $\left(\dfrac{1}{4}\right)^0 - \left(\dfrac{1}{5}\right)^{-1}$

21. $(2^5 b^{-3})^{-3}$
22. $(3^{-2} y^3)^{-2}$
23. $\left(\dfrac{3x}{2y}\right)^{-4}$
24. $\left(\dfrac{6c}{5d^3}\right)^{-2}$

25. $(3ab^2)(a^2 b)^3$
26. $(4x^2 y^3)^3 (xy^2)$
27. $\left(\dfrac{xy^2}{x^3 y}\right)^4$
28. $\left(\dfrac{a^3 b}{a^5 b^3}\right)^5$

29. $\dfrac{(t^{-2})^3}{t^{-4}}$
30. $\dfrac{(p^3)^{-4}}{p^{-5}}$
31. $\left(\dfrac{2w^2 x^3}{3y^0}\right)^3$
32. $\left(\dfrac{5a^0 b^4}{4c^3}\right)^2$

33. $\dfrac{q^3 r^{-2}}{s^{-1} t^5}$
34. $\dfrac{n^{-3} m^2}{p^{-3} q^{-1}}$
35. $\dfrac{(y^{-3})^2 (y^5)}{(y^{-3})^{-4}}$
36. $\dfrac{(w^2)^{-4}(w^{-2})}{(w^5)^{-4}}$

37. $\left(\dfrac{-2a^2 b^{-3}}{a^{-4} b^{-5}}\right)^{-3}$
38. $\left(\dfrac{-3x^{-4} y^3}{2x^5 y^{-2}}\right)^{-2}$
39. $(5h^{-2} k^0)^3 (5k^{-2})^{-4}$
40. $(6m^3 n^{-5})^{-4} (6m^0 n^{-2})^5$

Section 13.4 Scientific Notation

Concepts

1. Writing Numbers in Scientific Notation
2. Writing Numbers in Standard Form
3. Multiplying and Dividing Numbers in Scientific Notation

1. Writing Numbers in Scientific Notation

In many applications in mathematics, it is necessary to work with very large or very small numbers. For example, the number of movie tickets sold in the United States recently is estimated to be 1,500,000,000. The weight of a flea is approximately 0.00066 lb. To avoid writing numerous zeros in very large or small numbers, scientific notation was devised as a shortcut.

The principle behind scientific notation is to use a power of 10 to express the magnitude of the number. For example, the numbers 4000 and 0.07 can be written as:

$$4000 = 4 \times 1000 = 4 \times 10^3$$

$$0.07 = 7.0 \times 0.01 = 7.0 \times 10^{-2} \qquad \text{Note that } 10^{-2} = \frac{1}{100} = 0.01$$

> **Definition of Scientific Notation**
>
> A positive number expressed in the form $a \times 10^n$, where $1 \leq a < 10$ and n is an integer, is said to be written in **scientific notation**.

To write a positive number in scientific notation, we apply the following guidelines:

1. Move the decimal point so that its new location is to the right of the first nonzero digit. The number should now be greater than or equal to 1 but less than 10. Count the number of places that the decimal point is moved.

2. If the original number is *large* (greater than or equal to 10), use the number of places the decimal point was moved as a *positive* power of 10.

$$450{,}000 = 4.5 \times 100{,}000 = 4.5 \times 10^5$$

5 places

3. If the original number is *small* (between 0 and 1), use the number of places the decimal point was moved as a *negative* power of 10.

$$0.0002 = 2.0 \times 0.0001 = 2.0 \times 10^{-4}$$

4 places

4. If the original number is greater than or equal to 1 but less than 10, use 0 as the power of 10.

$$7.592 = 7.592 \times 10^0 \qquad \textit{Note: } \text{A number between 1 and 10 is seldom written in scientific notation.}$$

5. If the original number is negative, then $-10 < a \leq -1$.

$$-450{,}000 = -4.5 \times 100{,}000 = -4.5 \times 10^5$$

5 places

Section 13.4 Scientific Notation   861

Example 1 Writing Numbers in Scientific Notation

Write the numbers in scientific notation.

a. 53,000   b. 0.00053

Solution:

a. $53,000. = 5.3 \times 10^4$ To write 53,000 in scientific notation, the decimal point must be moved four places to the left. Because 53,000 is larger than 10, a *positive* power of 10 is used.

b. $0.00053 = 5.3 \times 10^{-4}$ To write 0.00053 in scientific notation, the decimal point must be moved four places to the right. Because 0.00053 is between 0 and 1, a *negative* power of 10 is used.

Skill Practice Write the numbers in scientific notation.

1. 175,000,000   2. 0.000005

Example 2 Writing Numbers in Scientific Notation

Write the numbers in scientific notation.

a. The number of movie tickets sold in the United States for a recent year is estimated to be 1,500,000,000.
b. The weight of a flea is approximately 0.00066 lb.
c. The temperature on a January day in Fargo dropped to −43°F.
d. A bench is 8.2 ft long.

Solution:

a. $1,500,000,000 = 1.5 \times 10^9$ b. $0.00066 \text{ lb} = 6.6 \times 10^{-4} \text{ lb}$

c. $-43°F = -4.3 \times 10^1 \text{ °F}$ d. $8.2 \text{ ft} = 8.2 \times 10^0 \text{ ft}$

Skill Practice Write the numbers in scientific notation.

3. In the year 2011, the population of Earth was approximately 7,000,000,000.
4. The weight of a grain of salt is approximately 0.000002 oz.

2. Writing Numbers in Standard Form

Example 3 Writing Numbers in Standard Form

Write the numbers in standard form.

a. The mass of a proton is approximately 1.67×10^{-24} g.
b. The "nearby" star Vega is approximately 1.552×10^{14} mi from Earth.

Solution:

a. 1.67×10^{-24} g = 0.000 000 000 000 000 000 000 001 67 g

Because the power of 10 is negative, the value of 1.67×10^{-24} is a decimal number between 0 and 1. Move the decimal point 24 places to the *left*.

b. 1.552×10^{14} mi = 155,200,000,000,000 mi

Because the power of 10 is a positive integer, the value of 1.552×10^{14} is a large number greater than 10. Move the decimal point 14 places to the *right*.

Answers
1. 1.75×10^8 2. 5×10^{-6}
3. 7×10^9 4. 2×10^{-6} oz

Skill Practice Write the numbers in standard form.

5. The probability of winning the California Super Lotto Jackpot is 5.5×10^{-8}.
6. The Sun's mass is 2×10^{30} kg.

3. Multiplying and Dividing Numbers in Scientific Notation

To multiply or divide two numbers in scientific notation, use the commutative and associative properties of multiplication to group the powers of 10. For example:

$$400 \times 2000 = (4 \times 10^2)(2 \times 10^3) = (4 \cdot 2) \times (10^2 \cdot 10^3) = 8 \times 10^5$$

$$\frac{0.00054}{150} = \frac{5.4 \times 10^{-4}}{1.5 \times 10^2} = \left(\frac{5.4}{1.5}\right) \times \left(\frac{10^{-4}}{10^2}\right) = 3.6 \times 10^{-6}$$

Example 4   Multiplying and Dividing Numbers in Scientific Notation

Multiply or divide as indicated.

a. $(8.7 \times 10^4)(2.5 \times 10^{-12})$ b. $\dfrac{4.25 \times 10^{13}}{8.5 \times 10^{-2}}$

Solution:

a. $(8.7 \times 10^4)(2.5 \times 10^{-12})$

$= (8.7 \cdot 2.5) \times (10^4 \cdot 10^{-12})$	Regroup factors using the commutative and associative properties of multiplication.
$= 21.75 \times 10^{-8}$	The number 21.75 is not in proper scientific notation because 21.75 is not between 1 and 10.
$= (2.175 \times 10^1) \times 10^{-8}$	Rewrite 21.75 as 2.175×10^1.
$= 2.175 \times (10^1 \times 10^{-8})$	Associative property of multiplication
$= 2.175 \times 10^{-7}$	Simplify.

b. $\dfrac{4.25 \times 10^{13}}{8.5 \times 10^{-2}}$

$= \left(\dfrac{4.25}{8.5}\right) \times \left(\dfrac{10^{13}}{10^{-2}}\right)$	Regroup factors using the commutative and associative properties.
$= 0.5 \times 10^{15}$	The number 0.5×10^{15} is not in proper scientific notation because 0.5 is not between 1 and 10.
$= (5.0 \times 10^{-1}) \times 10^{15}$	Rewrite 0.5 as 5.0×10^{-1}.
$= 5.0 \times (10^{-1} \times 10^{15})$	Associative property of multiplication
$= 5.0 \times 10^{14}$	Simplify.

Skill Practice Multiply or divide as indicated.

7. $(7 \times 10^5)(5 \times 10^3)$ 8. $\dfrac{1 \times 10^{-2}}{4 \times 10^{-7}}$

Answers
5. 0.000 000 055
6. 2,000,000,000,000,000,000,000,000,000,000
7. 3.5×10^9
8. 2.5×10^4

Section 13.4 Activity

A.1. Complete the table. The first row is done for you.

Standard Form	Expanded Form	Scientific Notation
5000	5×1000	5×10^3
40,000,000		
		8×10^5
		2×10^1

A.2. Complete the table. The first row is done for you.

Standard Form	Expanded Form	Scientific Notation
0.05	$5 \times \dfrac{1}{100}$ or $5 \times \dfrac{1}{10^2}$	5×10^{-2}
0.00002		
		7×10^{-1}
		6×10^{-8}

A.3. When using scientific notation, how do you determine whether to use a positive or negative power of 10?

A.4. Suppose that the number 524,900,000 is to be written in scientific notation.
  a. Where should the decimal point be placed for the number that is multiplied by the power of 10?
  b. In general when using scientific notation, how do you determine where to place the decimal point in the number that is multiplied by the power of 10?
  c. Write 524,900,000 in scientific notation.

A.5. a. Write 0.0000458 in scientific notation.
  b. Write 85,700 in scientific notation.
  c. Write 7.49×10^{-3} in standard form.
  d. Write 2.683×10^{13} in standard form.

A.6. The mean (average) distance between Venus and the Sun is 6.726×10^7 mi.
  a. If the orbit of Venus around the Sun is approximately circular, find the distance Venus travels around the Sun in one orbit. (*Hint:* The circumference of a circle is given by $C = 2\pi r$.) Express the answer in scientific notation. Use 3.14 for π and round the nearest million.
  b. Venus makes one complete trip around the Sun in approximately 5.393×10^3 hr (224.7 days). Use the answer from part (a) to determine how fast (in miles per hour) Venus travels around the Sun. Write the answer in scientific notation rounded to the nearest thousand miles per hour.

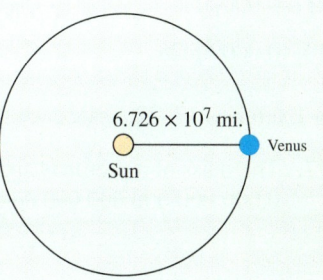

Section 13.4 Practice Exercises

Prerequisite Review

R.1. a. Simplify 10^3.
b. Write 100,000 as a power of 10.

R.2. a. Simplify 10^8.
b. Write 1,000,000 as a power of 10.

R.3. a. Simplify 10^{-3}. Write the answer as a fraction.
b. Write $\dfrac{1}{100,000}$ as a power of 10.
c. Write 0.01 as a power of 10.

R.4. a. Simplify 10^{-8}. Write the answer as a fraction.
b. Write $\dfrac{1}{1,000,000}$ as a power of 10.
c. Write 0.1 as a power of 10.

For Exercises R.5–R.12, simplify the expression.

R.5. $(3t^2)(5t^4)$

R.6. $(4m^3)(2m^7)$

R.7. $\dfrac{16n^4}{8n^7}$

R.8. $\dfrac{24x^4}{6x^5}$

R.9. $\dfrac{10^3 \cdot 10^{-4}}{10^5}$

R.10. $\dfrac{10^2}{10^5 \cdot 10^{-8}}$

R.11. $(c^4 d^2)^3$

R.12. $(x^5 y^{-3})^4$

Vocabulary and Key Concepts

1. A positive number expressed in the form $a \times 10^n$, where $1 \leq a < 10$ and n is an integer, is said to be written in _____ _____.

Concept 1: Writing Numbers in Scientific Notation

2. Explain how scientific notation might be valuable in studying astronomy. Answers may vary.

3. Explain how you would write the number 0.000 000 000 23 in scientific notation.

4. Explain how you would write the number 23,000,000,000,000 in scientific notation.

For Exercises 5–16, write the number in scientific notation. (See Example 1.)

5. 50,000
6. 900,000
7. 208,000
8. 420,000,000
9. 6,010,000
10. 75,000
11. 0.000008
12. 0.003
13. 0.000125
14. 0.00000025
15. 0.006708
16. 0.02004

For Exercises 17–22, write each number in scientific notation. (See Example 2.)

17. The mass of a proton is approximately 0.000 000 000 000 000 000 000 0017 g.

18. The total combined salaries of the president, vice president, senators, and representatives of the United States federal government is approximately $85,000,000.

19. A renowned foundation has over $27,000,000,000 from which it makes contributions to global charities.

20. One gram is equivalent to 0.0035 oz.

21. One of the world's largest tanker disasters spilled 68,000,000 gal of oil, causing widespread environmental damage over 100 mi of coastline.

22. The human heart pumps about 2100 gal of blood per day. That means that it pumps approximately 767,000 gal per year.

Concept 2: Writing Numbers in Standard Form

23. Explain how you would write the number 3.1×10^{-9} in standard form.

24. Explain how you would write the number 3.1×10^9 in standard form.

For Exercises 25–40, write each number in standard form. (See Example 3.)

25. 5×10^{-5}
26. 2×10^{-7}
27. 2.8×10^3
28. 9.1×10^6
29. 6.03×10^{-4}
30. 7.01×10^{-3}
31. 2.4×10^6
32. 3.1×10^4
33. 1.9×10^{-2}
34. 2.8×10^{-6}
35. 7.032×10^3
36. 8.205×10^2

37. One picogram (pg) is equal to 1×10^{-12} g.

38. A nanometer (nm) is approximately 3.94×10^{-8} in.

39. A normal diet contains between 1.6×10^3 Cal and 2.8×10^3 Cal per day.

40. The total land area of Texas is approximately 2.62×10^5 mi².

Ryan McVay/Getty Images

Concept 3: Multiplying and Dividing Numbers in Scientific Notation

For Exercises 41–60, multiply or divide as indicated. Write the answers in scientific notation. (See Example 4.)

41. $(2.5 \times 10^6)(2 \times 10^{-2})$
42. $(2 \times 10^{-7})(3 \times 10^{13})$
43. $(1.2 \times 10^4)(3 \times 10^7)$
44. $(3.2 \times 10^{-3})(2.5 \times 10^8)$
45. $\dfrac{7.7 \times 10^6}{3.5 \times 10^2}$
46. $\dfrac{9.5 \times 10^{11}}{1.9 \times 10^3}$
47. $\dfrac{9 \times 10^{-6}}{4 \times 10^7}$
48. $\dfrac{7 \times 10^{-2}}{5 \times 10^9}$
49. $80{,}000{,}000{,}000 \times 4000$
50. 0.0006×0.03
51. $(3.2 \times 10^{-4})(7.6 \times 10^{-7})$
52. $(5.9 \times 10^{12})(3.6 \times 10^9)$
53. $\dfrac{210{,}000{,}000{,}000}{0.007}$
54. $\dfrac{160{,}000{,}000{,}000{,}000}{0.00008}$
55. $\dfrac{5.7 \times 10^{-2}}{9.5 \times 10^{-8}}$
56. $\dfrac{2.72 \times 10^{-6}}{6.8 \times 10^{-4}}$
57. $6{,}000{,}000{,}000 \times 0.0000000023$
58. $0.000055 \times 40{,}000$
59. $\dfrac{0.0000000003}{6000}$
60. $\dfrac{420{,}000}{0.0000021}$

Mixed Exercises

61. If a piece of paper is 3×10^{-3} in. thick, how thick is a stack of 1.25×10^3 pieces of paper?

62. A box of staples contains 5×10^3 staples and weighs 15 oz. How much does one staple weigh? Write your answer in scientific notation.

63. A wealthy investor has 125,000 shares of a technology stock. If the stock price is $118 per share, how much is the investor's stock worth?

64. A state lottery had a jackpot of 5.2×10^7. This week the winner was a group of office employees that included 13 people. How much would each person receive?

65. Dinosaurs became extinct about 65 million years ago.

   a. Write the number 65 million in scientific notation.

   b. How many days is 65 million years?

   c. How many hours is 65 million years?

   d. How many seconds is 65 million years?

66. Earth is 150,000,000 km from the Sun.

   a. Write the number 150,000,000 in scientific notation.

   b. There are 1000 m in a kilometer. How many meters is Earth from the Sun?

   c. There are 100 cm in a meter. How many centimeters is Earth from the Sun?

Technology Connections

For Exercises 67–72, use a calculator to perform the indicated operations:

67. $(5.2 \times 10^6)(4.6 \times 10^{-3})$

68. $(2.19 \times 10^{-8})(7.84 \times 10^{-4})$

69. $\dfrac{4.76 \times 10^{-5}}{2.38 \times 10^9}$

70. $\dfrac{8.5 \times 10^4}{4.0 \times 10^{-1}}$

71. $\dfrac{(9.6 \times 10^7)(4.0 \times 10^{-3})}{2.0 \times 10^{-2}}$

72. $\dfrac{(5.0 \times 10^{-12})(6.4 \times 10^{-5})}{(1.6 \times 10^{-8})(4.0 \times 10^2)}$

Section 13.5 Addition and Subtraction of Polynomials

Concepts

1. Introduction to Polynomials
2. Addition of Polynomials
3. Subtraction of Polynomials
4. Polynomials and Applications to Geometry

1. Introduction to Polynomials

One commonly used algebraic expression is called a polynomial. A **polynomial** in one variable, x, is defined as a single term or a sum of terms of the form ax^n, where a is a real number and the exponent, n, is a nonnegative integer. For each term, a is called the **coefficient**, and n is called the **degree of the term**. For example:

Term (Expressed in the Form ax^n)	Coefficient	Degree
$-12z^7$	-12	7
$x^3 \rightarrow$ rewrite as $1x^3$	1	3
$10w \rightarrow$ rewrite as $10w^1$	10	1
$7 \rightarrow$ rewrite as $7x^0$	7	0

If a polynomial has exactly one term, it is categorized as a **monomial**. A two-term polynomial is called a **binomial**, and a three-term polynomial is called a **trinomial**. Usually the terms of a polynomial are written in descending order according to degree. The term with highest degree is called the **leading term**, and its coefficient is called the **leading coefficient**. The **degree of a polynomial** is the greatest degree of all of its terms. A polynomial in one variable is written in **descending order** if the term with highest degree is written first, followed by the term of next highest degree and so on. Thus, for a polynomial written in descending order, the leading term determines the degree of the polynomial.

	Expression	Descending Order	Leading Coefficient	Degree of Polynomial
Monomials	$-3x^4$	$-3x^4$	-3	4
	17	17	17	0
Binomials	$4y^3 - 6y^5$	$-6y^5 + 4y^3$	-6	5
	$\dfrac{1}{2} - \dfrac{1}{4}c$	$-\dfrac{1}{4}c + \dfrac{1}{2}$	$-\dfrac{1}{4}$	1
Trinomials	$4p - 3p^3 + 8p^6$	$8p^6 - 3p^3 + 4p$	8	6
	$7a^4 - 1.2a^8 + 3a^3$	$-1.2a^8 + 7a^4 + 3a^3$	-1.2	8

Example 1 — Identifying the Parts of a Polynomial

Given the polynomial: $3a - 2a^4 + 6 - a^3$

a. List the terms of the polynomial, and state the coefficient and degree of each term.

b. Write the polynomial in descending order.

c. State the degree of the polynomial and the leading coefficient.

d. Evaluate the polynomial for $a = -2$.

Solution:

a. term: $3a$ coefficient: 3 degree: 1

   term: $-2a^4$ coefficient: -2 degree: 4

   term: 6 coefficient: 6 degree: 0

   term: $-a^3$ coefficient: -1 degree: 3

b. $-2a^4 - a^3 + 3a + 6$

c. The degree of the polynomial is 4 and the leading coefficient is -2.

d. $-2a^4 - a^3 + 3a + 6$

$= -2(-2)^4 - (-2)^3 + 3(-2) + 6$ Substitute -2 for the variable, a. Remember to use parentheses when replacing a variable.

$= -2(16) - (-8) + 3(-2) + 6$ Simplify exponents first.

$= -32 - (-8) + (-6) + 6$ Perform multiplication before addition and subtraction.

$= -24$ Simplify.

Skill Practice

1. Given the polynomial: $5x^3 - x + 8x^4 + 3x^2$
    a. Write the polynomial in descending order.
    b. State the degree of the polynomial.
    c. State the coefficient of the leading term.
    d. Evaluate the polynomial for $x = -1$.

Polynomials may have more than one variable. In such a case, the degree of a term is the sum of the exponents of the variables contained in the term. For example, the term, $32x^2y^5z$, has degree 8 because the exponents applied to x, y, and z are 2, 5, and 1, respectively. The following polynomial has a degree of 11 because the highest degree of its terms is 11.

$$32x^2y^5z \;-\; 2x^3y \;+\; 2x^2yz^8 \;+\; 7$$

$\quad\quad$ degree 8 $\quad$ degree 4 $\quad$ degree 11 $\quad$ degree 0

Answers

1. a. $8x^4 + 5x^3 + 3x^2 - x$ b. 4   c. 8   d. 7

2. Addition of Polynomials

Recall that two terms are *like* terms if they each have the same variables, and the corresponding variables are raised to the same powers.

Like terms: $3x^2, -7x^2$ (Same exponents, Same variables) $-5yz^3, yz^3$ (Same exponents, Same variables)

Unlike terms: $9z^2, 12z^6$ (Different exponents) $\frac{1}{3}w^6, \frac{2}{5}p^6$ (Different variables) $4y, 7$ (Different variables)

Avoiding Mistakes

Note that when adding *like* terms, the exponents do *not* change.
$2x^3 + 3x^3 \neq 5x^6$

Recall that the distributive property is used to add or subtract *like* terms. For example,

$3x^2 + 9x^2 - 2x^2$
$= (3 + 9 - 2)x^2$ Apply the distributive property.
$= (10)x^2$ Simplify.
$= 10x^2$

Example 2 Adding Polynomials

Add the polynomials. $3x^2y + 5x^2y$

Solution:

$3x^2y + 5x^2y$ The terms are *like* terms.
$= (3 + 5)x^2y$ Apply the distributive property.
$= (8)x^2y$
$= 8x^2y$ Simplify.

Skill Practice Add the polynomials.

2. $13a^2b^3 + 2a^2b^3$

It is the distributive property that enables us to add *like* terms. We shorten the process by adding the coefficients of *like* terms.

Example 3 Adding Polynomials

Add the polynomials. $(-3c^3 + 5c^2 - 7c) + (11c^3 + 6c^2 + 3)$

Solution:

$(-3c^3 + 5c^2 - 7c) + (11c^3 + 6c^2 + 3)$
$= -3c^3 + 11c^3 + 5c^2 + 6c^2 - 7c + 3$ Clear parentheses, and group *like* terms.
$= 8c^3 + 11c^2 - 7c + 3$ Combine *like* terms.

Answer

2. $15a^2b^3$

> **TIP:** Polynomials can also be added by combining *like* terms in columns. The sum of the polynomials from Example 3 is shown here.
>
> $$\begin{array}{r} -3c^3 + 5c^2 - 7c + 0 \\ + 11c^3 + 6c^2 + 0c + 3 \\ \hline 8c^3 + 11c^2 - 7c + 3 \end{array}$$
>
> Placeholders such as 0 and 0c may be used to help line up *like* terms.

Skill Practice Add the polynomials.

3. $(7q^2 - 2q + 4) + (5q^2 + 6q - 9)$

Example 4   Adding Polynomials

Add the polynomials. $(4w^2 - 2x) + (3w^2 - 4x^2 + 6x)$

Solution:

$(4w^2 - 2x) + (3w^2 - 4x^2 + 6x)$

$= 4w^2 + 3w^2 - 4x^2 - 2x + 6x$ Clear parentheses and group *like* terms.

$= 7w^2 - 4x^2 + 4x$

Skill Practice Add the polynomials.

4. $(5x^2 - 4xy + y^2) + (-3x^2 - 5y^2)$

3. Subtraction of Polynomials

Subtraction of two polynomials requires us to find the opposite of the polynomial being subtracted. To find the opposite of a polynomial, take the opposite of each term. This is equivalent to multiplying the polynomial by -1.

Example 5   Finding the Opposite of a Polynomial

Find the opposite of the polynomials.

a. $5x$ b. $3a - 4b - c$ c. $5.5y^4 - 2.4y^3 + 1.1y$

Solution:

Expression	Opposite	Simplified Form
a. $5x$	$-(5x)$	$-5x$
b. $3a - 4b - c$	$-(3a - 4b - c)$	$-3a + 4b + c$
c. $5.5y^4 - 2.4y^3 + 1.1y$	$-(5.5y^4 - 2.4y^3 + 1.1y)$	$-5.5y^4 + 2.4y^3 - 1.1y$

TIP: Notice that the sign of each term is changed when finding the opposite of a polynomial.

Skill Practice Find the opposite of the polynomials.

5. $x - 3$ 6. $3y^2 - 2xy + 6x + 2$ 7. $-2.1w^3 + 4.9w^2 - 1.9w$

Subtraction of two polynomials is similar to subtracting real numbers. Add the opposite of the second polynomial to the first polynomial.

Answers
3. $12q^2 + 4q - 5$
4. $2x^2 - 4xy - 4y^2$
5. $-x + 3$
6. $-3y^2 + 2xy - 6x - 2$
7. $2.1w^3 - 4.9w^2 + 1.9w$

Chapter 13 Polynomials and Properties of Exponents

> **Subtraction of Polynomials**
> If A and B are polynomials, then $A - B = A + (-B)$.

Example 6 Subtracting Polynomials

Subtract the polynomials. $\quad (-4p^4 + 5p^2 - 3) - (11p^2 + 4p - 6)$

Solution:

$(-4p^4 + 5p^2 - 3) - (11p^2 + 4p - 6)$

$= (-4p^4 + 5p^2 - 3) + (-11p^2 - 4p + 6) \quad$ Add the opposite of the second polynomial.

$= -4p^4 + 5p^2 - 11p^2 - 4p - 3 + 6 \quad$ Group *like* terms.

$= -4p^4 - 6p^2 - 4p + 3 \quad$ Combine *like* terms.

FOR REVIEW

Recall that to subtract two real numbers, we add the opposite of the second number (the subtrahend) to the first number (the minuend). For example, $3 - 5 = 3 + (-5)$. The same is true for subtracting polynomials.

> **TIP:** Two polynomials can also be subtracted in columns by adding the opposite of the second polynomial to the first polynomial. Placeholders (shown in red) may be used to help line up *like* terms.
>
> $\quad -4p^4 + 0p^3 + 5p^2 + 0p - 3$
> $-(0p^4 + 0p^3 + 11p^2 + 4p - 6)$
>
> Add the opposite $\longrightarrow$
>
> $\quad -4p^4 + 0p^3 + 5p^2 + 0p - 3$
> $+\underline{-0p^4 - 0p^3 - 11p^2 - 4p + 6}$
> $\quad -4p^4 - 6p^2 - 4p + 3$
>
> The difference of the polynomials is $-4p^4 - 6p^2 - 4p + 3$.

Skill Practice Subtract the polynomials.

8. $(x^2 + 3x - 2) - (4x^2 + 6x + 1)$

Example 7 Subtracting Polynomials

Subtract the polynomials. $\quad (a^2 - 2ab + 7b^2) - (-8a^2 - 6ab + 2b^2)$

Solution:

$(a^2 - 2ab + 7b^2) - (-8a^2 - 6ab + 2b^2)$

$= (a^2 - 2ab + 7b^2) + (8a^2 + 6ab - 2b^2) \quad$ Add the opposite of the second polynomial.

$= a^2 + 8a^2 - 2ab + 6ab + 7b^2 - 2b^2 \quad$ Group *like* terms.

$= 9a^2 + 4ab + 5b^2 \quad$ Combine *like* terms.

Skill Practice Subtract the polynomials.

9. $(-3y^2 + xy + 2x^2) - (-2y^2 - 3xy - 8x^2)$

In Example 8, we illustrate the subtraction of polynomials by first clearing parentheses and then combining *like* terms.

Answers
8. $-3x^2 - 3x - 3$
9. $-y^2 + 4xy + 10x^2$

Section 13.5 Addition and Subtraction of Polynomials

Example 8   Subtracting Polynomials

Subtract $\frac{1}{3}t^4 + \frac{1}{2}t^2$ from $t^2 - 4$, and simplify the result.

Solution:

To subtract a from b, we write $b - a$. Thus, to subtract $\overset{a}{\overbrace{\frac{1}{3}t^4 + \frac{1}{2}t^2}}$ from $\overset{b}{\overbrace{t^2 - 4}}$, we have

$$\overset{b}{(t^2 - 4)} - \overset{a}{\left(\frac{1}{3}t^4 + \frac{1}{2}t^2\right)}$$

$$= t^2 - 4 - \frac{1}{3}t^4 - \frac{1}{2}t^2 \quad \text{Apply the distributive property to clear parentheses.}$$

$$= -\frac{1}{3}t^4 + t^2 - \frac{1}{2}t^2 - 4 \quad \text{Group } like \text{ terms in descending order.}$$

$$= -\frac{1}{3}t^4 + \frac{2}{2}t^2 - \frac{1}{2}t^2 - 4 \quad \text{The } t^2 \text{ terms are the only } like \text{ terms.}$$
$$\phantom{= -\frac{1}{3}t^4 + \frac{2}{2}t^2 - \frac{1}{2}t^2 - 4 \quad} \text{Get a common denominator for the } t^2 \text{ terms.}$$

$$= -\frac{1}{3}t^4 + \frac{1}{2}t^2 - 4 \quad \text{Add } like \text{ terms.}$$

> **Avoiding Mistakes**
>
> Example 8 involves subtracting two *expressions*. This is not an equation. Therefore, we cannot clear fractions.

Skill Practice

10. Subtract $\frac{3}{4}x^2 + \frac{2}{5}$ from $x^2 + 3x$.

4. Polynomials and Applications to Geometry

Example 9   Subtracting Polynomials in Geometry

If the perimeter of the triangle in Figure 13-1 can be represented by the polynomial $2x^2 + 5x + 6$, find a polynomial that represents the length of the missing side.

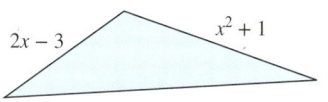

Figure 13-1

Solution:

The missing side of the triangle can be found by subtracting the sum of the two known sides from the perimeter.

$$\begin{pmatrix} \text{Length} \\ \text{of missing} \\ \text{side} \end{pmatrix} = (\text{perimeter}) - \begin{pmatrix} \text{sum of the} \\ \text{two known sides} \end{pmatrix}$$

$$\begin{pmatrix} \text{Length} \\ \text{of missing} \\ \text{side} \end{pmatrix} = (2x^2 + 5x + 6) - [(2x - 3) + (x^2 + 1)]$$

Answer

10. $\frac{1}{4}x^2 + 3x - \frac{2}{5}$

$$= 2x^2 + 5x + 6 - [2x - 3 + x^2 + 1] \quad \text{Clear inner parentheses.}$$
$$= 2x^2 + 5x + 6 - (x^2 + 2x - 2) \quad \text{Combine } like \text{ terms within [ ].}$$
$$= 2x^2 + 5x + 6 - x^2 - 2x + 2 \quad \text{Apply the distributive property.}$$
$$= 2x^2 - x^2 + 5x - 2x + 6 + 2 \quad \text{Group } like \text{ terms.}$$
$$= x^2 + 3x + 8 \quad \text{Combine } like \text{ terms.}$$

The polynomial $x^2 + 3x + 8$ represents the length of the missing side.

Skill Practice

11. If the perimeter of the triangle is represented by the polynomial $6x - 9$, find the polynomial that represents the missing side.

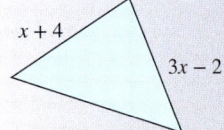

Answer

11. $2x - 11$

Section 13.5 Activity

A.1. a. Write a monomial with variable p, coefficient -5, and degree 4.

 b. Write a trinomial of degree 2 in descending order. Use the variable y and coefficients 7, -1, and -2, respectively.

A.2. a. Write a monomial of degree zero. (Answers may vary.)

 b. Write a binomial of degree 3 and leading coefficient $\frac{3}{4}$. (Answers may vary.)

A.3. a. Find the sum of $(x^2 + x) + (3x^2 - 2x + 5)$.

 b. Explain how to add two polynomials.

A.4. a. Simplify $-(-4w^3 + 2w^2 - 5)$.

 b. Explain how to find the opposite of a polynomial.

A.5. a. Find the difference of $\left(3y^2 - y + \frac{7}{3}\right)$ and $\left(-2y^2 - 2y + \frac{4}{3}\right)$.

 b. Subtract $(4a^2b - 2a^2 - 5b)$ from $(6a^2b + a^2)$.

 c. Summarize how the difference in wording in parts (a) and (b) affects the order in which the polynomials are subtracted.

Practice Exercises

Section 13.5

Study Skills

Good reading comprehension in mathematics requires the skill of understanding mathematical operations and symbols in various expressions. Let's apply good reading comprehension to the skill of adding polynomials.

- One technique that might help your understanding is to read each term of the polynomial as if it were a word in a sentence. For example, read the following expression aloud. $(4x^2 + 5x) + (-3x^2 - 2x)$
- Recall that only *like* terms can be combined within a polynomial. As you read the expression $(4x^2 + 5x) + (-3x^2 - 2x)$, ask yourself what the parentheses imply and how you might regroup terms to add the polynomials.
- As you read the expression $(4x^2 + 5x) + (-3x^2 - 2x)$, you recognize that the terms $4x^2$ and $-3x^2$ can be combined. Ask yourself whether the exponents change when combining *like* terms. If necessary, expand each term to visualize the similarities between the terms.

Prerequisite Review

For Exercises R.1–R.4, combine *like* terms.

R.1. $4x + 3x - 11x$

R.2. $5y - 7y + 8y$

R.3. $-\frac{3}{4}m + \frac{1}{3}m + \frac{1}{6}m$

R.4. $\frac{4}{15}p - \frac{2}{3}p + \frac{1}{5}p$

For Exercises R.5–R.6, apply the distributive property.

R.5. $3(5x + 7y - 3)$

R.6. $5(4m - 3n + 2)$

For Exercises R.7–R.8, clear parentheses and combine *like* terms.

R.7. $(3x + 5) - (4x - 3)$

R.8. $(-5y + 8) - (-9y - 2)$

R.9. a. Find the difference of 8 and -10.
   b. Subtract 8 from -10.

R.10. a. Subtract -7 from -5.
   b. Find the difference of -7 and -5.

Vocabulary and Key Concepts

1. **a.** A _____ is a single term or a sum of terms.
   b. For the term ax^n, a is called the _____ and n is called the _____ of the term.
   c. Given the term x, the coefficient of the term is _____.
   d. A monomial is a polynomial with exactly _____ term(s).
   e. A _____ is a polynomial with exactly two terms.
   f. A _____ is a polynomial with exactly three terms.
   g. The term with the highest degree is called the _____ term and its coefficient is called the _____ _____.
   h. The degree of a polynomial is the _____ degree of all of its terms.
   i. The degree of a nonzero constant such as 5 is _____.

Concept 1: Introduction to Polynomials

For Exercises 2–4,

    a. write the polynomial in descending order.

    b. identify the leading coefficient.

    c. identify the degree of the polynomial. (See Example 1.)

2. $10 - 8a - a^3 + 2a^2 + a^5$

3. $6 + 7x^2 - 7x^5 + 9x$

4. $\frac{1}{2}y + y^2 - 12y^4 + y^3 - 6$

For Exercises 5–14, categorize each expression as a monomial, a binomial, or a trinomial. Then evaluate the polynomial given $x = -3$, $y = 2$, and $z = -1$. (See Example 1.)

5. $10x^2 + 5x$

6. $7z + 13z^2 - 15$

7. $6x^2$

8. 9

9. $2y - y^4$

10. $7x + 2$

11. $2y^4 - 3y + 1$

12. 23

13. $-32xyz$

14. $y^4 - x^2$

Concept 2: Addition of Polynomials

15. Explain why the terms $3x$ and $3x^2$ are not *like* terms.

16. Explain why the terms $4w^3$ and $4z^3$ are not *like* terms.

For Exercises 17–34, add the polynomials. (See Examples 2–4.)

17. $23x^2y + 12x^2y$

18. $-5ab^3 + 17ab^3$

19. $3b^5d^2 + (5b^5d^2 - 9d)$

20. $4c^2d^3 + (3cd - 10c^2d^3)$

21. $(7y^2 + 2y - 9) + (-3y^2 - y)$

22. $(-3w^2 + 4w - 6) + (5w^2 + 2)$

23. $(5x + 3x^2 - x^3) + (2x^2 + 4x - 10)$

24. $(t^2 - 4t + t^4) + (3t^4 + 2t + 6)$

25. $(6.1y + 3.2x) + (4.8y - 3.2x)$

26. $(2.7m - 0.5h) + (-3.2m + 0.2h)$

27. $6a + 2b - 5c$
    $+ \underline{-2a - 2b - 3c}$

28. $-13x + 5y + 10z$
    $+ \underline{-3x - 3y + 2z}$

29. $\left(\frac{2}{5}a + \frac{1}{4}b - \frac{5}{6}\right) + \left(\frac{3}{5}a - \frac{3}{4}b - \frac{7}{6}\right)$

30. $\left(\frac{5}{9}x + \frac{1}{10}y\right) + \left(-\frac{4}{9}x + \frac{3}{10}y\right)$

31. $\left(z - \frac{8}{3}\right) + \left(\frac{4}{3}z^2 - z + 1\right)$

32. $\left(-\frac{7}{5}r + 1\right) + \left(-\frac{3}{5}r^2 + \frac{7}{5}r + 1\right)$

33. $7.9t^3 + 2.6t - 1.1$
    $+ \underline{-3.4t^2 + 3.4t - 3.1}$

34. $0.34y^2 + 1.23$
    $+ \underline{ 3.42y - 7.56}$

Concept 3: Subtraction of Polynomials

For Exercises 35–40, find the opposite of each polynomial. (See Example 5.)

35. $4h - 5$
36. $5k - 12$
37. $-2.3m^2 + 3.1m - 1.5$
38. $-11.8n^2 - 6.7n + 9.3$
39. $3v^3 + 5v^2 + 10v + 22$
40. $7u^4 + 3v^2 + 17$

For Exercises 41–60, subtract the polynomials. (See Examples 6–7.)

41. $4a^3b^2 - 12a^3b^2$
42. $5yz^4 - 14yz^4$
43. $-32x^3 - 21x^3$
44. $-23c^5 - 12c^5$
45. $(7a - 7) - (12a - 4)$
46. $(4x + 3v) - (-3x + v)$

47. $\quad\quad 4k + 3$
    $-(-12k - 6)$

48. $\quad\quad 3h - 15$
    $-(8h + 13)$

49. $25m^4 - (23m^4 + 14m)$
50. $3x^2 - (-x^2 - 12)$
51. $(5s^2 - 3st - 2t^2) - (2s^2 + st + t^2)$
52. $(6k^2 + 2kp + p^2) - (3k^2 - 6kp + 2p^2)$

53. $\quad\quad 10r - 6s + 2t$
    $-(12r - 3s - t)$

54. $\quad\quad a - 14b + 7c$
    $-(-3a - 8b + 2c)$

55. $\left(\frac{7}{8}x + \frac{2}{3}y - \frac{3}{10}\right) - \left(\frac{1}{8}x + \frac{1}{3}y\right)$
56. $\left(r - \frac{1}{12}s\right) - \left(\frac{1}{2}r - \frac{5}{12}s - \frac{4}{11}\right)$
57. $\left(\frac{2}{3}h^2 - \frac{1}{5}h - \frac{3}{4}\right) - \left(\frac{4}{3}h^2 - \frac{4}{5}h + \frac{7}{4}\right)$
58. $\left(\frac{3}{8}p^3 - \frac{5}{7}p^2 - \frac{2}{5}\right) - \left(\frac{5}{8}p^3 - \frac{2}{7}p^2 + \frac{7}{5}\right)$

59. $\quad\quad 4.5x^4 - 3.1x^2 \quad\quad - 6.7$
    $-(2.1x^4 \quad\quad + 4.4x + 1.2)$

60. $\quad\quad 1.3c^2 \quad\quad + 4.8$
    $- (4.3c^2 - 2c - 2.2)$

61. Find the difference of $(4b^3 + 6b - 7)$ and $(-12b^2 + 11b + 5)$.
62. Find the difference of $(-5y^2 + 3y - 21)$ and $(-4y^2 - 5y + 23)$.

63. Subtract $\left(\frac{3}{2}x^2 - 5x\right)$ from $(-2x^2 - 11)$.
    (See Example 8.)

64. Subtract $\left(a^5 - \frac{1}{3}a^3 + 5a\right)$ from $\left(\frac{3}{4}a^5 + \frac{1}{2}a^4 + 6a\right)$.

Concept 4: Polynomials and Applications to Geometry

65. Find a polynomial that represents the perimeter of the figure.

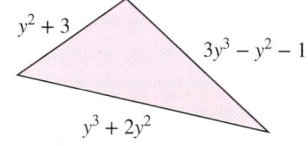

66. Find a polynomial that represents the perimeter of the figure.

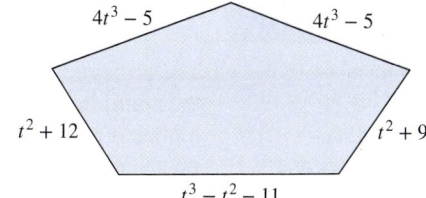

67. If the perimeter of the figure can be represented by the polynomial $6a^2 - 3a + 1$, find a polynomial that represents the length of the missing side.
(See Example 9.)

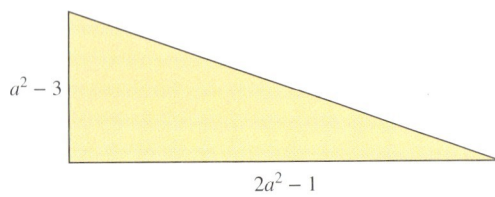

68. If the perimeter of the figure can be represented by the polynomial $6w^3 - 2w - 3$, find a polynomial that represents the length of the missing side.

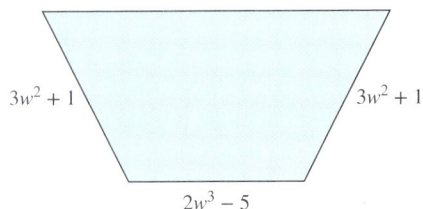

Mixed Exercises

For Exercises 69–84, perform the indicated operation.

69. $(2ab^2 + 9a^2b) + (7ab^2 - 3ab + 7a^2b)$

70. $(8x^2y - 3xy - 6xy^2) + (3x^2y - 12xy)$

71. $\quad 4z^5 \quad\quad + z^3 - 3z + 13$
$\quad -(\quad\quad - z^4 - 8z^3 \quad\quad + 15)$

72. $\quad -15t^4 \quad\quad - 23t^2 + 16t$
$\quad -(\quad\quad 21t^3 + 18t^2 + \quad t)$

73. $(9x^4 + 2x^3 - x + 5) + (9x^3 - 3x^2 + 8x + 3) - (7x^4 - x + 12)$

74. $(-6y^3 - 9y^2 + 23) - (7y^2 + 2y - 11) + (3y^3 - 25)$

75. $(0.2w^2 + 3w + 1.3) - (w^3 - 0.7w + 2)$

76. $(8.1u^3 - 5.2u^2 + 4) + (2.8u^3 + 6.3u - 7)$

77. $(7p^2q - 3pq^2) - (8p^2q + pq) + (4pq - pq^2)$

78. $(12c^2d - 2cd + 8cd^2) - (-c^2d + 4cd) - (5cd - 2cd^2)$

79. $(5x - 2x^3) + (2x^3 - 5x)$

80. $(p^2 - 4p + 2) - (2 + p^2 - 4p)$

81. $\quad 2a^2b - 4ab + \;ab^2$
$\quad -(2a^2b + \;ab - 5ab^2)$

82. $\quad -3xy + 7xy^2 + 5x^2y$
$\quad + (-8xy - 11xy^2 + 3x^2y)$

83. $[(3y^2 - 5y) - (2y^2 + y - 1)] + (10y^2 - 4y - 5)$

84. $(12c^3 - 5c^2 - 2c) + [(7c^3 - 2c^2 + c) - (4c^3 + 4c)]$

Expanding Your Skills

85. Write a binomial of degree 3. (Answers may vary.)

86. Write a trinomial of degree 6. (Answers may vary.)

87. Write a monomial of degree 5. (Answers may vary.)

88. Write a monomial of degree 1. (Answers may vary.)

89. Write a trinomial with the leading coefficient −6. (Answers may vary.)

90. Write a binomial with the leading coefficient 13. (Answers may vary.)

Section 13.6 Multiplication of Polynomials and Special Products

Concepts

1. Multiplication of Polynomials
2. Special Case Products: Difference of Squares and Perfect Square Trinomials
3. Applications to Geometry

1. Multiplication of Polynomials

The properties of exponents can be used to simplify many algebraic expressions including the multiplication of monomials. To multiply monomials, first use the associative and commutative properties of multiplication to group coefficients and like bases. Then simplify the result by using the properties of exponents.

Section 13.6 Multiplication of Polynomials and Special Products

Example 1 Multiplying Monomials

Multiply the monomials.

a. $(3x^4)(4x^2)$ b. $(-4c^5d)(2c^2d^3e)$ c. $\left(\frac{1}{3}a^4b^3\right)\left(\frac{3}{4}b^7\right)$

Solution:

a. $(3x^4)(4x^2)$
   $= (3 \cdot 4)(x^4 x^2)$ Group coefficients and like bases.
   $= 12x^6$ Multiply the coefficients and add the exponents on x.

b. $(-4c^5d)(2c^2d^3e)$
   $= (-4 \cdot 2)(c^5c^2)(dd^3)(e)$ Group coefficients and like bases.
   $= -8c^7d^4e$ Simplify.

c. $\left(\frac{1}{3}a^4b^3\right)\left(\frac{3}{4}b^7\right)$
   $= \left(\frac{1}{3} \cdot \frac{3}{4}\right)(a^4)(b^3b^7)$ Group coefficients and like bases.
   $= \frac{1}{4}a^4b^{10}$ Simplify.

> **FOR REVIEW**
> Recall that the commutative property of multiplication enables us to reorder factors: $ab = ba$. The associative property of multiplication enables us to regroup factors: $a(bc) = (ab)c$.

Skill Practice Multiply the monomials.

1. $(-5y)(6y^3)$ 2. $(7x^2y)(-2x^3y^4)$ 3. $\left(\frac{2}{5}w^5z^3\right)\left(\frac{15}{4}w^4\right)$

The distributive property is used to multiply polynomials: $a(b + c) = ab + ac$.

Example 2 Multiplying a Polynomial by a Monomial

Multiply the polynomials.

a. $2t(4t - 3)$ b. $-3a^2\left(-4a^2 + 2a - \frac{1}{3}\right)$

Solution:

a. $2t(4t - 3)$
   $= (2t)(4t) + (2t)(-3)$ Apply the distributive property by multiplying each term by $2t$.
   $= 8t^2 - 6t$ Simplify each term.

b. $-3a^2\left(-4a^2 + 2a - \frac{1}{3}\right)$
   $= (-3a^2)(-4a^2) + (-3a^2)(2a) + (-3a^2)\left(-\frac{1}{3}\right)$ Apply the distributive property by multiplying each term by $-3a^2$.
   $= 12a^4 - 6a^3 + a^2$ Simplify each term.

Answers
1. $-30y^4$
2. $-14x^5y^5$
3. $\frac{3}{2}w^9z^3$

Skill Practice Multiply the polynomials.

4. $-4a(5a - 3)$ **5.** $-4p\left(2p^2 - 6p + \dfrac{1}{4}\right)$

Thus far, we have illustrated polynomial multiplication involving monomials. Next, the distributive property will be used to multiply polynomials with more than one term.

$$(x + 3)(x + 5) = x(x + 5) + 3(x + 5) \quad \text{Apply the distributive property.}$$
$$= x(x + 5) + 3(x + 5) \quad \text{Apply the distributive property again.}$$
$$= (x)(x) + (x)(5) + (3)(x) + (3)(5)$$
$$= x^2 + 5x + 3x + 15$$
$$= x^2 + 8x + 15 \quad \text{Combine } like \text{ terms.}$$

Note: Using the distributive property results in multiplying each term of the first polynomial by each term of the second polynomial.

$$(x + 3)(x + 5) = (x)(x) + (x)(5) + (3)(x) + (3)(5)$$
$$= x^2 + 5x + 3x + 15$$
$$= x^2 + 8x + 15$$

Example 3   Multiplying a Polynomial by a Polynomial

Multiply the polynomials. $(c - 7)(c + 2)$

Solution:

$(c - 7)(c + 2)$ Multiply each term in the first polynomial by each term in the second. That is, apply the distributive property.

$$= (c)(c) + (c)(2) + (-7)(c) + (-7)(2)$$
$$= c^2 + 2c - 7c - 14 \quad \text{Simplify.}$$
$$= c^2 - 5c - 14 \quad \text{Combine } like \text{ terms.}$$

TIP: Notice that the product of two *binomials* equals the sum of the products of the **F**irst terms, the **O**uter terms, the **I**nner terms, and the **L**ast terms. The acronym **FOIL** (First Outer Inner Last) can be used as a memory device to multiply two binomials.

$$(c - 7)(c + 2) = (c)(c) + (c)(2) + (-7)(c) + (-7)(2)$$
$$= c^2 + 2c - 7c - 14$$
$$= c^2 - 5c - 14$$

Skill Practice Multiply the polynomials.

6. $(x + 2)(x + 8)$

Answers
4. $-20a^2 + 12a$
5. $-8p^3 + 24p^2 - p$
6. $x^2 + 10x + 16$

Section 13.6 Multiplication of Polynomials and Special Products

Example 4   Multiplying a Polynomial by a Polynomial

Multiply the polynomials. $(y-2)(3y^2 + y - 5)$

Solution:

$(y-2)(3y^2 + y - 5)$ Multiply each term in the first polynomial by each term in the second.

$= (y)(3y^2) + (y)(y) + (y)(-5) + (-2)(3y^2) + (-2)(y) + (-2)(-5)$

$= 3y^3 + y^2 - 5y - 6y^2 - 2y + 10$ Simplify each term.

$= 3y^3 - 5y^2 - 7y + 10$ Combine *like* terms.

Avoiding Mistakes

It is important to note that the acronym FOIL does not apply to Example 4 because the product does not involve two binomials.

TIP: Multiplication of polynomials can be performed vertically by a process similar to column multiplication of real numbers. For example,

```
      235              3y² +  y - 5
    ×  21           ×         y - 2
      235             -6y² - 2y + 10
     4700        3y³ +  y² - 5y + 0
     4935        3y³ - 5y² - 7y + 10
```

Note: When multiplying by the column method, it is important to *align like* terms vertically before adding terms.

Skill Practice Multiply the polynomials.

7. $(2y + 4)(3y^2 - 5y + 2)$

Example 5   Multiplying Polynomials

Multiply the polynomials. $2(10x + 3y)(2x - 4y)$

Solution:

$2(10x + 3y)(2x - 4y)$ In this case we are multiplying three polynomials—a monomial times two binomials. The associative property of multiplication enables us to choose which two polynomials to multiply first.

$= 2[(10x + 3y)(2x - 4y)]$ First we will multiply the binomials. Multiply each term in the first binomial by each term in the second binomial. That is, apply the distributive property.

$= 2[(10x)(2x) + (10x)(-4y) + (3y)(2x) + (3y)(-4y)]$

$= 2[20x^2 - 40xy + 6xy - 12y^2]$ Simplify each term.

$= 2(20x^2 - 34xy - 12y^2)$ Combine *like* terms.

$= 40x^2 - 68xy - 24y^2$ Multiply by 2 using the distributive property.

TIP: When multiplying three polynomials, first multiply two of the polynomials. Then multiply the result by the third polynomial.

Skill Practice Multiply.

8. $3(4a - 3c)(5a - 2c)$

Answers

7. $6y^3 + 2y^2 - 16y + 8$
8. $60a^2 - 69ac + 18c^2$

2. Special Case Products: Difference of Squares and Perfect Square Trinomials

In some cases the product of two binomials takes on a special pattern.

I. The first special case occurs when multiplying the sum and difference of the same two terms. For example:

$(2x + 3)(2x - 3)$
$= 4x^2 - 6x + 6x - 9$
$= 4x^2 - 9$

Notice that the middle terms are opposites. This leaves only the difference between the square of the first term and the square of the second term. For this reason, the product is called a *difference of squares*.

Note: The binomials $2x + 3$ and $2x - 3$ are called **conjugates**. In one expression, $2x$ and 3 are added, and in the other, $2x$ and 3 are subtracted.

II. The second special case involves the square of a binomial. For example:

$(3x + 7)^2$
$= (3x + 7)(3x + 7)$
$= 9x^2 + 21x + 21x + 49$
$= 9x^2 + 42x + 49$
$= (3x)^2 + 2(3x)(7) + (7)^2$

When squaring a binomial, the product will be a trinomial called a *perfect square trinomial*. The first and third terms are formed by squaring each term of the binomial. The middle term equals twice the product of the terms in the binomial.

Note: The expression $(3x - 7)^2$ also expands to a perfect square trinomial, but the middle term will be negative:

$(3x - 7)(3x - 7) = 9x^2 - 21x - 21x + 49 = 9x^2 - 42x + 49$

Special Case Product Formulas

1. $(a + b)(a - b) = a^2 - b^2$ The product is called a **difference of squares**.

2. $(a + b)^2 = a^2 + 2ab + b^2$
   $(a - b)^2 = a^2 - 2ab + b^2$ The product is called a **perfect square trinomial**.

You should become familiar with these special case products because they will be used again to factor polynomials.

Example 6 Multiplying Conjugates

Multiply the conjugates.

a. $(x - 9)(x + 9)$ b. $\left(\frac{1}{2}p + 6\right)\left(\frac{1}{2}p - 6\right)$

Solution:

a. $(x - 9)(x + 9)$ Apply the formula: $(a + b)(a - b) = a^2 - b^2$

$= (x)^2 - (9)^2$ Substitute $a = x$ and $b = 9$.

$= x^2 - 81$

TIP: The product of two conjugates can be checked by applying the distributive property:

$(x - 9)(x + 9)$
$= x^2 + 9x - 9x - 81$
$= x^2 - 81$

b. $\left(\dfrac{1}{2}p + 6\right)\left(\dfrac{1}{2}p - 6\right)$ Apply the formula: $(a+b)(a-b) = a^2 - b^2$

$$= \left(\dfrac{1}{2}p\right)^2 - (6)^2$$ Substitute $a = \dfrac{1}{2}p$ and $b = 6$.

$$= \dfrac{1}{4}p^2 - 36$$ Simplify each term.

Skill Practice Multiply the conjugates.

9. $(a+7)(a-7)$ 10. $\left(\dfrac{4}{5}x - 10\right)\left(\dfrac{4}{5}x + 10\right)$

Example 7   Squaring Binomials

Square the binomials.

   a. $(3w - 4)^2$ **b.** $(5x^2 + 2)^2$

Solution:

a. $(3w - 4)^2$ Apply the formula: $(a-b)^2 = a^2 - 2ab + b^2$

$$= (3w)^2 - 2(3w)(4) + (4)^2$$ Substitute $a = 3w$, $b = 4$.

$$= 9w^2 - 24w + 16$$ Simplify each term.

> **TIP:** The square of a binomial can be checked by explicitly writing the product of the two binomials and applying the distributive property:
>
> $$(3w - 4)^2 = (3w - 4)(3w - 4) = 9w^2 - 12w - 12w + 16$$
> $$= 9w^2 - 24w + 16$$

b. $(5x^2 + 2)^2$ Apply the formula: $(a+b)^2 = a^2 + 2ab + b^2$

$$= (5x^2)^2 + 2(5x^2)(2) + (2)^2$$ Substitute $a = 5x^2$, $b = 2$.

$$= 25x^4 + 20x^2 + 4$$ Simplify each term.

Avoiding Mistakes

The property for squaring two factors is different than the property for squaring two terms: $(ab)^2 = a^2b^2$ but $(a+b)^2 = a^2 + 2ab + b^2$

Skill Practice Square the binomials.

11. $(2x + 3)^2$ 12. $(5c^2 - 6)^2$

Answers

9. $a^2 - 49$
10. $\dfrac{16}{25}x^2 - 100$
11. $4x^2 + 12x + 9$
12. $25c^4 - 60c^2 + 36$

3. Applications to Geometry

Example 8 — Using Special Case Products in an Application of Geometry

Find a polynomial that represents the volume of the cube (Figure 13-2).

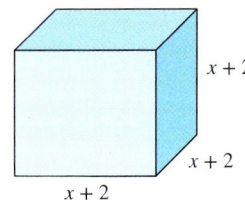

Figure 13-2

Solution:

$$\text{Volume} = (\text{length})(\text{width})(\text{height})$$

$$V = (x+2)(x+2)(x+2) \quad \text{or} \quad V = (x+2)^3$$

To expand $(x+2)(x+2)(x+2)$, multiply the first two factors. Then multiply the result by the last factor.

$V = \underline{(x+2)(x+2)}(x+2)$

$= (x^2 + 4x + 4)(x+2)$

TIP: $(x+2)(x+2) = (x+2)^2$ and results in a perfect square trinomial.
$(x+2)^2 = (x)^2 + 2(x)(2) + (2)^2$
$= x^2 + 4x + 4$

$= (x^2)(x) + (x^2)(2) + (4x)(x) + (4x)(2) + (4)(x) + (4)(2)$ Apply the distributive property.

$= x^3 + 2x^2 + 4x^2 + 8x + 4x + 8$ Group *like* terms.

$= x^3 + 6x^2 + 12x + 8$ Combine *like* terms.

The volume of the cube can be represented by

$$V = (x+2)^3 = x^3 + 6x^2 + 12x + 8$$

Skill Practice

13. Find the polynomial that represents the volume of the cube.

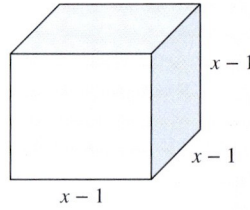

Answer

13. The volume of the cube can be represented by $x^3 - 3x^2 + 3x - 1$.

Section 13.6 Activity

A.1. Multiply the monomials by applying the properties of exponents.

$$\left(\frac{3}{5}u^2v\right)(15uv^4)$$

A.2. To multiply polynomials, multiply each term in the first polynomial by each term in the second polynomial.

  a. Given the product $(3t - 5)(t^2 - 26t + 13)$, how many terms are in the first polynomial? How many terms are in the second polynomial? How many terms should be in the product before combining *like* terms?

  b. Multiply the polynomials. $(3t - 5)(t^2 - 26t + 13)$

A.3. a. The polynomials $(6p - 1)$ and $(6p + 1)$ are called _____ of each other.

  b. Multiply the polynomials by multiplying each term in the first binomial by each term in the second binomial.

$$(6p - 1)(6p + 1) = \underline{\hspace{2cm}}$$

  c. Multiply the polynomials by applying the formula $(a - b)(a + b) = a^2 - b^2$.

$$(6p - 1)(6p + 1) = \underline{\hspace{2cm}}$$

A.4. a. Square the binomial by expanding the product of binomials and multiplying each term in the first binomial by each term in the second.

$$(2w - 5)^2 = (2w - 5)(2w - 5)$$

  b. Square the binomial by applying the formula $(a - b)^2 = a^2 - 2ab + b^2$.

$$(2w - 5)^2 = \underline{\hspace{3cm}}$$

A.5. The expression $(x + 4)^3$ can be written as $(x + 4)^3 = (x + 4)(x + 4)(x + 4)$. Use the order of operations to find the product. That is, multiply the polynomials from left to right.

Practice Exercises

Section 13.6

Prerequisite Review

For Exercises R.1–R.14, simplify the expression.

R.1. $4x + 5x$

R.2. $2y^2 - 4y^2$

R.3. $(4x)(5x)$

R.4. $(2y^2)(-4y^2)$

R.5. $-5a^3b - 2a^3b$

R.6. $7uvw^2 + uvw^2$

R.7. $(-5a^3b)(-2a^3b)$

R.8. $(7uvw^2)(uvw^2)$

R.9. $3w^2 + 4w - 7w + 2$

R.10. $11t^2 - 4t + 6t - 12$

R.11. $-3(3x - 7y - 8z)$

R.12. $-4(-a + 2b - 7c)$

R.13. $\left(\frac{1}{3}x^2 - \frac{5}{2}x - \frac{1}{10}\right) + \left(\frac{5}{3}x^2 + \frac{3}{4}x + \frac{2}{5}\right)$

R.14. $\left(\frac{1}{4}y^3 + \frac{1}{6}y^2 - \frac{5}{12}y\right) + \left(\frac{3}{4}y^3 - \frac{2}{3}y^2 + \frac{1}{2}y\right)$

Vocabulary and Key Concepts

1. a. The conjugate of $5 + 2x$ is _____.

   b. When two conjugates are multiplied, the resulting binomial is a difference of _____. This is given by the formula $(a + b)(a - b) =$ _____.

   c. When a binomial is squared, the resulting trinomial is a _____ square trinomial. This is given by the formula $(a + b)^2 =$ _____.

Concept 1: Multiplication of Polynomials

For Exercises 2–10, multiply the expressions. (See Example 1.)

2. $8(4x)$
3. $-2(6y)$
4. $-10(5z)$
5. $7(3p)$
6. $(x^{10})(4x^3)$
7. $(a^{13}b^4)(12ab^4)$
8. $(4m^3n^7)(-3m^6n)$
9. $(2c^7d)(-c^3d^{11})$
10. $(-5u^2v)(-8u^3v^2)$

For Exercises 11–46, multiply the polynomials. (See Examples 2–5.)

11. $8pq(2pq - 3p + 5q)$
12. $5ab(2ab + 6a - 3b)$
13. $(k^2 - 13k - 6)(-4k)$
14. $(h^2 + 5h - 12)(-2h)$
15. $-15pq(3p^2 + p^3q^2 - 2q)$
16. $-4u^2v(2u - 5uv^3 + v)$
17. $(y + 10)(y + 9)$
18. $(x + 5)(x + 6)$
19. $(m - 12)(m - 2)$
20. $(n - 7)(n - 2)$
21. $(3p - 2)(4p + 1)$
22. $(7q + 11)(q - 5)$
23. $(8 - 4w)(-3w + 2)$
24. $(-6z + 10)(4 - 2z)$
25. $(p - 3w)(p - 11w)$
26. $(y - 7x)(y - 10x)$
27. $(6x - 1)(2x + 5)$
28. $(3x + 7)(x - 8)$
29. $(4a - 9)(1.5a - 2)$
30. $(2.1y - 0.5)(y + 3)$
31. $(3t - 7)(1 + 3t^2)$
32. $(2 - 5w)(2w^2 - 5)$
33. $3(3m + 4n)(m + 2n)$
34. $2(7y + z)(3y + 5z)$
35. $(5s + 3)(s^2 + s - 2)$
36. $(t - 4)(2t^2 - t + 6)$
37. $(3w - 2)(9w^2 + 6w + 4)$
38. $(z + 5)(z^2 - 5z + 25)$
39. $(p^2 + p - 5)(p^2 + 4p - 1)$
40. $(-x^2 - 2x + 4)(x^2 + 2x - 6)$

41. $\quad 3a^2 - 4a + 9$
    $\times \underline{\quad\quad 2a - 5}$

42. $\quad 7x^2 - 3x - 4$
    $\times \underline{\quad\quad 5x + 1}$

43. $\quad 4x^2 - 12xy + 9y^2$
    $\times \underline{\quad\quad 2x - 3y}$

44. $\quad 25a^2 + 10ab + b^2$
    $\times \underline{\quad\quad\quad 5a + b}$

45. $\quad\quad 6x + 2y$
    $\times \underline{0.2x + 1.2y}$

46. $\quad\quad 4.5a + 2b$
    $\times \underline{\quad 2a - 1.8b}$

Concept 2: Special Case Products: Difference of Squares and Perfect Square Trinomials

For Exercises 47–58, multiply the conjugates. (See Example 6.)

47. $(y - 6)(y + 6)$
48. $(x + 3)(x - 3)$
49. $(3a - 4b)(3a + 4b)$
50. $(5y + 7x)(5y - 7x)$
51. $(9k + 6)(9k - 6)$
52. $(2h - 5)(2h + 5)$

53. $\left(\dfrac{2}{3}t - 3\right)\left(\dfrac{2}{3}t + 3\right)$

54. $\left(\dfrac{1}{4}r - 1\right)\left(\dfrac{1}{4}r + 1\right)$

55. $(u^3 + 5v)(u^3 - 5v)$

56. $(8w^2 - x)(8w^2 + x)$

57. $\left(\dfrac{2}{3} - p\right)\left(\dfrac{2}{3} + p\right)$

58. $\left(\dfrac{1}{8} - q\right)\left(\dfrac{1}{8} + q\right)$

For Exercises 59–70, square the binomials. **(See Example 7.)**

59. $(a + 5)^2$
60. $(a - 3)^2$
61. $(x - y)^2$

62. $(x + y)^2$
63. $(2c + 5)^2$
64. $(5d - 9)^2$

65. $(3t^2 - 4s)^2$
66. $(u^2 + 4v)^2$
67. $(7 - t)^2$

68. $(4 + w)^2$
69. $(3 + 4q)^2$
70. $(2 - 3b)^2$

71. **a.** Evaluate $(2 + 4)^2$ by working within the parentheses first.
    b. Evaluate $2^2 + 4^2$.
    c. Compare the answers to parts (a) and (b) and make a conjecture about $(a + b)^2$ and $a^2 + b^2$.

72. **a.** Evaluate $(6 - 5)^2$ by working within the parentheses first.
    b. Evaluate $6^2 - 5^2$.
    c. Compare the answers to parts (a) and (b) and make a conjecture about $(a - b)^2$ and $a^2 - b^2$.

Concept 3: Applications to Geometry

73. Find a polynomial expression that represents the area of the rectangle shown in the figure.

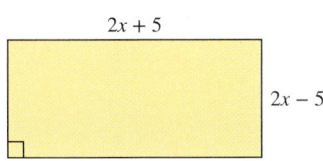

74. Find a polynomial expression that represents the area of the rectangle shown in the figure.

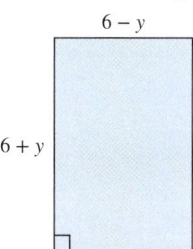

75. Find a polynomial expression that represents the area of the square shown in the figure.

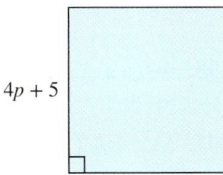

76. Find a polynomial expression that represents the area of the square shown in the figure.

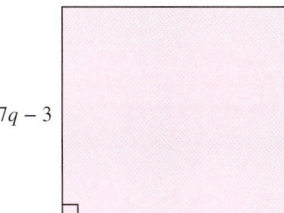

77. Find a polynomial that represents the volume of the cube shown in the figure. (See Example 8.)
(Recall: $V = s^3$)

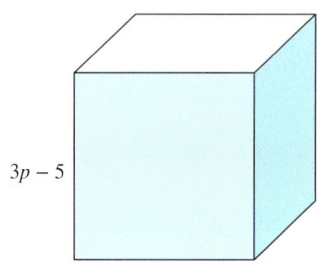

78. Find a polynomial that represents the volume of the rectangular solid shown in the figure.
(Recall: $V = lwh$)

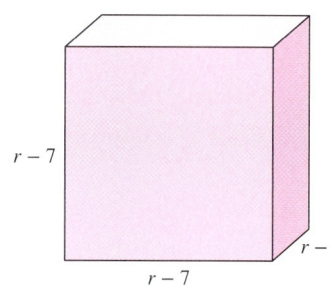

79. Find a polynomial that represents the area of the triangle shown in the figure.
(Recall: $A = \frac{1}{2}bh$)

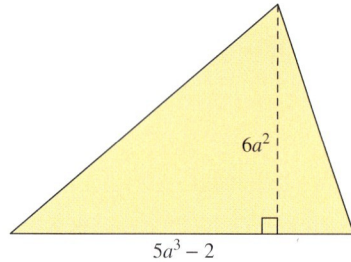

80. Find a polynomial that represents the area of the triangle shown in the figure.

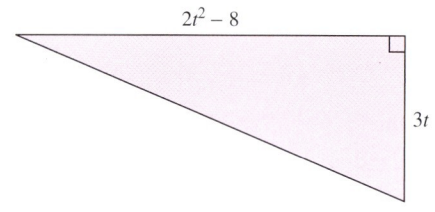

Mixed Exercises

For Exercises 81–110, multiply the expressions.

81. $(7x + y)(7x - y)$

82. $(9w - 4z)(9w + 4z)$

83. $(5s + 3t)^2$

84. $(5s - 3t)^2$

85. $(7x - 3y)(3x - 8y)$

86. $(5a - 4b)(2a - b)$

87. $\left(\frac{2}{3}t + 2\right)(3t + 4)$

88. $\left(\frac{1}{5}s + 6\right)(5s - 3)$

89. $-5(3x + 5)(2x - 1)$

90. $-4(2k - 5)(4k - 3)$

91. $(3a - 2)(5a + 1 + 2a^2)$

92. $(u + 4)(2 - 3u + u^2)$

93. $(y^2 + 2y + 4)(y - 5)$

94. $(w^2 - w + 6)(w + 2)$

95. $\left(\frac{1}{3}m - n\right)^2$

96. $\left(\frac{2}{5}p - q\right)^2$

97. $6w^2(7w - 14)$

98. $4v^3(v + 12)$

99. $(4y - 8.1)(4y + 8.1)$

100. $(2h + 2.7)(2h - 2.7)$

101. $(3c^2 + 4)(7c^2 - 8)$

102. $(5k^3 - 9)(k^3 - 2)$

103. $(3.1x + 4.5)^2$

104. $(2.5y + 1.1)^2$

105. $(k - 4)^3$

106. $(h + 3)^3$

107. $(5x + 3)^3$

108. $(2a - 4)^3$

109. $(y^2 + 2y + 1)(2y^2 - y + 3)$

110. $(2w^2 - w - 5)(3w^2 + 2w + 1)$

Expanding Your Skills

For Exercises 111–114, multiply the expressions containing more than two factors.

111. $2a(3a - 4)(a + 5)$

112. $5x(x + 2)(6x - 1)$

113. $(x - 3)(2x + 1)(x - 4)$

114. $(y - 2)(2y - 3)(y + 3)$

115. What binomial when multiplied by $(3x + 5)$ will produce a product of $6x^2 - 11x - 35$?
[*Hint:* Let the quantity $(a + b)$ represent the unknown binomial.] Then find a and b such that $(3x + 5)(a + b) = 6x^2 - 11x - 35$.

116. What binomial when multiplied by $(2x - 4)$ will produce a product of $2x^2 + 8x - 24$?

For Exercises 117–119, determine the values of k that would create a perfect square trinomial.

117. $x^2 + kx + 25$

118. $w^2 + kw + 9$

119. $a^2 + ka + 16$

Division of Polynomials

Section 13.7

Division of polynomials will be presented in this section as two separate cases: The first case illustrates division by a monomial divisor. The second case illustrates long division by a polynomial with two or more terms.

Concepts

1. Division by a Monomial
2. Long Division
3. Synthetic Division

1. Division by a Monomial

To divide a polynomial by a monomial, divide each individual term in the polynomial by the divisor and simplify the result.

Dividing a Polynomial by a Monomial

If a, b, and c are polynomials such that $c \neq 0$, then

$$\frac{a + b}{c} = \frac{a}{c} + \frac{b}{c} \quad \text{Similarly,} \quad \frac{a - b}{c} = \frac{a}{c} - \frac{b}{c}$$

Example 1 Dividing a Polynomial by a Monomial

Divide the polynomials.

a. $\dfrac{5a^3 - 10a^2 + 20}{5a}$

b. $(12y^2z^3 - 15yz^2 + 6y^2z) \div (-6y^2z)$

Solution:

a. $\dfrac{5a^3 - 10a^2 + 20}{5a}$

$= \dfrac{5a^3}{5a} - \dfrac{10a^2}{5a} + \dfrac{20}{5a}$ Divide each term in the numerator by $5a$.

$= a^2 - 2a + \dfrac{4}{a}$ Simplify each term using the properties of exponents.

b. $(12y^2z^3 - 15yz^2 + 6y^2z) \div (-6y^2z)$

$$= \frac{12y^2z^3 - 15yz^2 + 6y^2z}{-6y^2z}$$

$$= \frac{12y^2z^3}{-6y^2z} - \frac{15yz^2}{-6y^2z} + \frac{6y^2z}{-6y^2z} \quad \text{Divide each term by } -6y^2z.$$

$$= -2z^2 + \frac{5z}{2y} - 1 \quad \text{Simplify each term.}$$

Skill Practice Divide the polynomials.

1. $(36a^4 - 48a^3 + 12a^2) \div (6a^3)$

2. $\dfrac{-15x^3y^4 + 25x^2y^3 - 5xy^2}{-5xy^2}$

---

2. Long Division

FOR REVIEW

Recall the terminology associated with division. For the given division statement we have:

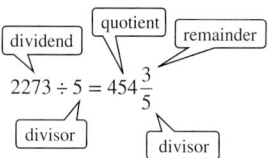

$2273 \div 5 = 454\frac{3}{5}$ (dividend ÷ divisor = quotient remainder/divisor)

If the divisor has two or more terms, a *long division* process similar to the division of real numbers is used. Take a minute to review the long division process for real numbers by dividing 2273 by 5.

```
        454  ← Quotient
     ┌──────
   5 )2273
     -20↓
       27
      -25↓
        23
       -20
         3  ← Remainder    Therefore, 2273 ÷ 5 = 454 3/5.
```

A similar procedure is used for long division of polynomials as shown in Example 2.

Example 2   Using Long Division to Divide Polynomials

Divide the polynomials using long division. $(2x^2 - x + 3) \div (x - 3)$

Solution:

$x - 3 \overline{) 2x^2 - x + 3}$ Divide the leading term in the dividend by the leading term in the divisor.

$$\frac{2x^2}{x} = 2x$$

This is the first term in the quotient.

TIP: Recall that taking the opposite of a polynomial changes the sign of each term of the polynomial.

$\begin{array}{r} 2x \\ x - 3 \overline{) 2x^2 - x + 3} \\ -(2x^2 - 6x) \end{array}$ Multiply $2x$ by the divisor: $2x(x-3) = 2x^2 - 6x$
Then subtract the result.

Answers

1. $6a - 8 + \dfrac{2}{a}$

2. $3x^2y^2 - 5xy + 1$

$$\begin{array}{r}
2x \\
x-3{\overline{\smash{\big)}\,2x^2 - x + 3}} \\
\underline{-2x^2 + 6x} \\
5x
\end{array}$$ Subtract the quantity $2x^2 - 6x$. To do this, add the opposite.

$$\begin{array}{r}
2x + 5 \\
x-3{\overline{\smash{\big)}\,2x^2 - x + 3}} \\
\underline{-2x^2 + 6x}\downarrow \\
5x + 3
\end{array}$$ Bring down the next column, and repeat the process.

Divide the leading term by x: $(5x)/x = 5$
Place 5 in the quotient.

$$\begin{array}{r}
2x + 5 \\
x-3{\overline{\smash{\big)}\,2x^2 - x + 3}} \\
\underline{-2x^2 + 6x} \\
5x + 3 \\
-(5x - 15)
\end{array}$$ Multiply the divisor by 5: $5(x-3) = 5x - 15$
Subtract the result.

$$\begin{array}{r}
2x + 5 \\
x-3{\overline{\smash{\big)}\,2x^2 - x + 3}} \\
\underline{-2x^2 + 6x} \\
5x + 3 \\
\underline{-5x + 15} \\
18
\end{array}$$ Subtract the quantity $5x - 15$ by adding the opposite.

The remainder is 18.

Summary:

The quotient is $\quad 2x + 5$
The remainder is $\quad 18$
The divisor is $\quad x - 3$
The dividend is $\quad 2x^2 - x + 3$

The solution to a long division problem is usually written in the form:

$$\text{quotient} + \frac{\text{remainder}}{\text{divisor}}$$

Hence,

$$(2x^2 - x + 3) \div (x - 3) = 2x + 5 + \frac{18}{x - 3}$$

Skill Practice Divide the polynomials using long division.

3. $(3x^2 + 2x - 5) \div (x + 2)$

The division of polynomials can be checked in the same fashion as the division of real numbers. To check Example 2, we use the **division algorithm**:

$$\text{Dividend} = (\text{divisor})(\text{quotient}) + \text{remainder}$$

$$2x^2 - x + 3 \stackrel{?}{=} (x - 3)(2x + 5) + (18)$$

$$\stackrel{?}{=} 2x^2 + 5x - 6x - 15 + (18)$$

$$= 2x^2 - x + 3 \checkmark$$

Answer

3. $3x - 4 + \dfrac{3}{x + 2}$

Chapter 13 Polynomials and Properties of Exponents

> **Example 3** Using Long Division to Divide Polynomials
>
> Divide the polynomials using long division: $(3w^3 + 26w^2 - 3) \div (3w - 1)$
>
> **Solution:**
>
> First note that the dividend has a missing power of w and can be written as $3w^3 + 26w^2 + 0w - 3$. The term $0w$ is a placeholder for the missing term. It is helpful to use the placeholder to keep the powers of w lined up.
>
> $$\begin{array}{r} w^2 \\ 3w - 1 \overline{\smash{)} 3w^3 + 26w^2 + 0w - 3} \\ -(3w^3 - w^2) \end{array}$$
>
> Divide $3w^3 \div 3w = w^2$. This is the first term of the quotient.
> Then multiply $w^2(3w - 1) = 3w^3 - w^2$.
>
> $$\begin{array}{r} w^2 \\ 3w - 1 \overline{\smash{)} 3w^3 + 26w^2 + 0w - 3} \\ \underline{-3w^3 + w^2} \\ 27w^2 + 0w \end{array}$$
>
> Subtract by adding the opposite.
>
> Bring down the next column, and repeat the process.
>
> $$\begin{array}{r} w^2 + 9w \\ 3w - 1 \overline{\smash{)} 3w^3 + 26w^2 + 0w - 3} \\ \underline{-3w^3 + w^2} \\ 27w^2 + 0w \\ -(27w^2 - 9w) \end{array}$$
>
> Divide $27w^2$ by the leading term in the divisor. $27w^2 \div 3w = 9w$.
> Place $9w$ in the quotient.
> Multiply $9w(3w - 1) = 27w^2 - 9w$.
>
> $$\begin{array}{r} w^2 + 9w \\ 3w - 1 \overline{\smash{)} 3w^3 + 26w^2 + 0w - 3} \\ \underline{-3w^3 + w^2} \\ 27w^2 + 0w \\ \underline{-27w^2 + 9w} \\ 9w - 3 \end{array}$$
>
> Subtract by adding the opposite.
>
> Bring down the next column, and repeat the process.
>
> $$\begin{array}{r} w^2 + 9w + 3 \\ 3w - 1 \overline{\smash{)} 3w^3 + 26w^2 + 0w - 3} \\ \underline{-3w^3 + w^2} \\ 27w^2 + 0w \\ \underline{-27w^2 + 9w} \\ 9w - 3 \\ -(9w - 3) \end{array}$$
>
> Divide $9w$ by the leading term in the divisor. $9w \div 3w = 3$.
> Place 3 in the quotient.
> Multiply $3(3w - 1) = 9w - 3$.
>
> $$\begin{array}{r} w^2 + 9w + 3 \\ 3w - 1 \overline{\smash{)} 3w^3 + 26w^2 + 0w - 3} \\ \underline{-3w^3 + w^2} \\ 27w^2 + 0w \\ \underline{-27w^2 + 9w} \\ 9w - 3 \\ \underline{-9w + 3} \\ 0 \end{array}$$
>
> Subtract by adding the opposite.
>
> The remainder is 0.
>
> The quotient is $w^2 + 9w + 3$, and the remainder is 0.
>
> ---
>
> **Skill Practice** Divide the polynomials using long division.
>
> 4. $\dfrac{9x^3 + 11x + 10}{3x + 2}$

Answer

4. $3x^2 - 2x + 5$

In Example 3, the remainder is zero. Therefore, we say that $3w - 1$ divides evenly into $3w^3 + 26w^2 - 3$. For this reason, the divisor and quotient are factors of $3w^3 + 26w^2 - 3$. To check, we have

$$\text{Dividend} = (\text{divisor})(\text{quotient}) + \text{remainder}$$

$$3w^3 + 26w^2 - 3 \stackrel{?}{=} (3w - 1)(w^2 + 9w + 3) + 0$$

$$\stackrel{?}{=} 3w^3 + 27w^2 + 9w - w^2 - 9w - 3$$

$$= 3w^3 + 26w^2 - 3 \checkmark$$

Example 4   Using Long Division to Divide Polynomials

Divide the polynomials using long division.

$$\frac{2y + y^4 - 5}{1 + y^2}$$

Solution:

First note that both the dividend and divisor should be written in descending order:

$$\frac{y^4 + 2y - 5}{y^2 + 1}$$

Also note that the dividend and the divisor have missing powers of y. Leave placeholders.

$$y^2 + 0y + 1 \overline{\smash{\big)}\, y^4 + 0y^3 + 0y^2 + 2y - 5}$$

$$\begin{array}{r} y^2 \\ y^2 + 0y + 1 \overline{\smash{\big)}\, y^4 + 0y^3 + 0y^2 + 2y - 5} \\ -(y^4 + 0y^3 + y^2) \end{array}$$

Divide $y^4 \div y^2 = y^2$. This is the first term of the quotient.

Multiply $y^2(y^2 + 0y + 1) = y^4 + 0y^3 + y^2$.

$$\begin{array}{r} y^2 \\ y^2 + 0y + 1 \overline{\smash{\big)}\, y^4 + 0y^3 + 0y^2 + 2y - 5} \\ -y^4 - 0y^3 - y^2 \\ \hline -y^2 + 2y - 5 \end{array}$$

Subtract by adding the opposite.

Bring down the next columns.

$$\begin{array}{r} y^2 -1 \\ y^2 + 0y + 1 \overline{\smash{\big)}\, y^4 + 0y^3 + 0y^2 + 2y - 5} \\ -y^4 - 0y^3 - y^2 \\ \hline -y^2 + 2y - 5 \\ -(-y^2 - 0y - 1) \end{array}$$

Divide $-y^2 \div y^2 = -1$.

Multiply $-1(y^2 + 0y + 1) = -y^2 - 0y - 1$.

$$\begin{array}{r} y^2 -1 \\ y^2 + 0y + 1 \overline{\smash{\big)}\, y^4 + 0y^3 + 0y^2 + 2y - 5} \\ -y^4 - 0y^3 - y^2 \\ \hline -y^2 + 2y - 5 \\ y^2 + 0y + 1 \\ \hline 2y - 4 \end{array}$$

Subtract by adding the opposite.

Remainder

Therefore, $\dfrac{y^4 + 2y - 5}{y^2 + 1} = y^2 - 1 + \dfrac{2y - 4}{y^2 + 1}$.

Skill Practice   Divide the polynomials using long division.

5. $(4 - x^2 + x^3) \div (2 + x^2)$

Answer

5. $x - 1 + \dfrac{-2x + 6}{x^2 + 2}$

Example 5   **Determining Whether Long Division Is Necessary**

Determine whether long division is necessary for each division of polynomials.

a. $\dfrac{2p^5 - 8p^4 + 4p - 16}{p^2 - 2p + 1}$ b. $\dfrac{2p^5 - 8p^4 + 4p - 16}{2p^2}$

c. $(3z^3 - 5z^2 + 10) \div (15z^3)$ d. $(3z^3 - 5z^2 + 10) \div (3z + 1)$

Solution:

- Long division is used when the divisor has *two or more terms*.
- If the divisor has *one term*, then divide each term in the dividend by the monomial divisor.

a. $\dfrac{2p^5 - 8p^4 + 4p - 16}{p^2 - 2p + 1}$ The divisor has three terms. Use long division.

b. $\dfrac{2p^5 - 8p^4 + 4p - 16}{2p^2}$ The divisor has one term. Long division is not necessary.

c. $(3z^3 - 5z^2 + 10) \div (15z^3)$ The divisor has one term. Long division is not necessary.

d. $(3z^3 - 5z^2 + 10) \div (3z + 1)$ The divisor has two terms. Use long division.

Skill Practice   Divide the polynomials using the appropriate method of division.

6. $\dfrac{6x^3 - x^2 + 3x - 5}{2x + 3}$ 7. $\dfrac{9w^3 - 18w^2 + 6w + 12}{3w}$

3. Synthetic Division

In this section, we introduced the process of long division to divide two polynomials. Next, we will learn another technique called **synthetic division** to divide two polynomials. Synthetic division can be used when dividing a polynomial by a first-degree divisor of the form $x - r$, where r is a constant. Synthetic division is considered a "shortcut" because it uses the coefficients of the divisor and dividend without writing the variables.

Consider dividing the polynomials $(3x^2 - 14x - 10) \div (x - 2)$.

$$\begin{array}{r} 3x - 8 \\ x-2 \overline{\smash{)}3x^2 - 14x - 10} \\ -(3x^2 - 6x) \\ \hline -8x - 10 \\ -(-8x + 16) \\ \hline -26 \end{array}$$

Answers

6. $3x^2 - 5x + 9 + \dfrac{-32}{2x + 3}$

7. $3w^2 - 6w + 2 + \dfrac{4}{w}$

First note that the divisor $x - 2$ is in the form $x - r$, where $r = 2$. Therefore, synthetic division can be used to find the quotient and remainder.

Step 1: Write the value of r in a box.

Step 2: Write the coefficients of the dividend to the right of the box.

Step 3: Skip a line and draw a horizontal line below the list of coefficients.

Step 4: Bring down the leading coefficient from the dividend and write it below the line.

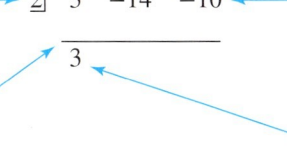

Step 5: Multiply the value of r by the number below the line $(2 \times 3 = 6)$. Write the result in the next column above the line.

Step 6: Add the numbers in the column above the line $(-14 + 6)$, and write the result below the line.

Repeat steps 5 and 6 until all columns have been completed.

Step 7: To get the final result, we use the numbers below the line. The number in the last column is the remainder. The other numbers are the coefficients of the quotient.

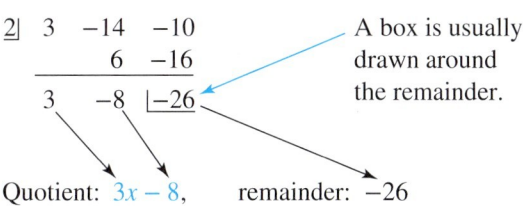

Quotient: $3x - 8$,   remainder: -26

A box is usually drawn around the remainder.

The degree of the quotient will always be 1 less than that of the dividend. Because the dividend is a second-degree polynomial, the quotient will be a first-degree polynomial. In this case, the quotient is $3x - 8$ and the remainder is -26.

Example 6 Using Synthetic Division to Divide Polynomials

Divide the polynomials $(5x + 4x^3 - 6 + x^4) \div (x + 3)$ by using synthetic division.

Solution:

As with long division, the terms of the dividend and divisor should be written in descending order. Furthermore, missing powers must be accounted for by using placeholders (shown here in red).

$$5x + 4x^3 - 6 + x^4$$
$$= x^4 + 4x^3 + 0x^2 + 5x - 6$$

To use synthetic division, the divisor must be in the form $(x - r)$. The divisor $x + 3$ can be written as $x - (-3)$. Hence, $r = -3$.

Avoiding Mistakes

It is important to check that the divisor is in the form $(x - r)$ before applying synthetic division. The variable x in the divisor must be of first degree, and its coefficient must be 1.

Step 1: Write the value of r in a box.

Step 2: Write the coefficients of the dividend to the right of the box.

$$\underline{-3|} \; 1 \quad 4 \quad 0 \quad 5 \quad -6$$
$$ 1$$

Step 3: Skip a line and draw a horizontal line below the list of coefficients.

Step 4: Bring down the leading coefficient from the dividend and write it below the line.

Step 5: Multiply the value of r by the number below the line ($-3 \times 1 = -3$). Write the result in the next column above the line.

$$\underline{-3|} \; 1 \quad 4 \quad 0 \quad 5 \quad -6$$
$$-3$$
$$1 \quad 1$$

Step 6: Add the numbers in the column above the line: $4 + (-3) = 1$.

Repeat steps 5 and 6:

$$\underline{-3|} \; 1 \quad 4 \quad 0 \quad 5 \quad -6$$
$$-3 \; -3 \quad 9 \; -42$$
$$1 \quad 1 \; -3 \; 14 \; \underline{|-48}$$

The quotient is $x^3 + x^2 - 3x + 14$.

The remainder is -48.

- remainder
- constant
- x term coefficient
- x^2 term coefficient
- x^3 term coefficient

The answer is $x^3 + x^2 - 3x + 14 + \dfrac{-48}{x+3}$.

Skill Practice Divide the polynomials by using synthetic division.

8. $(5y^2 + 2y^3 - 5) \div (y + 2)$

TIP: It is interesting to compare the long division process to the synthetic division process. For Example 5, long division is shown on the left, and synthetic division is shown on the right. Notice that the same pattern of coefficients used in long division appears in the synthetic division process.

$$\begin{array}{r}
x^3 + x^2 - 3x + 14 \\
x+3\overline{)x^4 + 4x^3 + 0x^2 + 5x - 6} \\
\underline{-(x^4 + 3x^3)} \\
x^3 + 0x^2 \\
\underline{-(x^3 + 3x^2)} \\
-3x^2 + 5x \\
\underline{-(-3x^2 - 9x)} \\
14x - 6 \\
\underline{-(14x + 42)} \\
-48
\end{array}$$

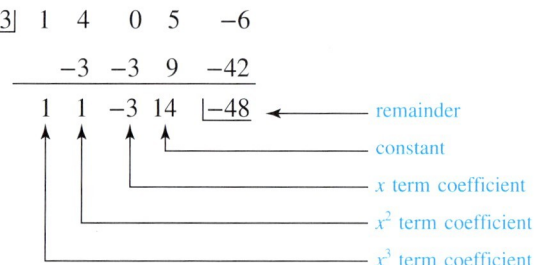

Quotient: $x^3 + x^2 - 3x + 14$
Remainder: -48

Answer

8. $2y^2 + y - 2 + \dfrac{-1}{y+2}$

Example 7 — Using Synthetic Division to Divide Polynomials

Divide the polynomials by using synthetic division. $(p^4 - 81) \div (p - 3)$

Solution:

$(p^4 - 81) \div (p - 3)$

$(p^4 + 0p^3 + 0p^2 + 0p - 81) \div (p - 3)$ Insert placeholders (red) for missing powers of p.

```
3|  1   0   0   0  -81
        3   9  27   81
    1   3   9  27   |0
```

Quotient: $p^3 + 3p^2 + 9p + 27$

Remainder: 0

The answer is $p^3 + 3p^2 + 9p + 27$.

Skill Practice Divide the polynomials by using synthetic division.

9. $(x^3 + 1) \div (x + 1)$

Answer

9. $x^2 - x + 1$

Section 13.7 Activity

A.1. **a.** Write the division expression as a sum.

$$\frac{12a^3 - 6a^2 + 18a}{-6a} = \frac{}{-6a} + \frac{}{-6a} + \frac{}{-6a}$$

    b. Simplify the expression from part (a).

A.2. Divide. $(12k^5 - 2k^3 + 8k^2 - 3) \div (2k^2)$

A.3. How do you determine when to use long division when dividing polynomials?

A.4. Divide by using long division. $31\overline{)50,417}$

Write the answer in the form quotient + $\dfrac{\text{remainder}}{\text{divisor}}$.

A.5. **a.** Divide the polynomials. $\dfrac{2x^4 - 9x^3 + 10x^2 - 7x + 10}{x - 3}$

    b. Identify the quotient.
    c. Identify the divisor.
    d. Identify the dividend.
    e. Identify the remainder.
    f. Check the result of the division by verifying that
       Dividend = (Divisor)(Quotient) + Remainder.

A.6. Consider the division expression. $(27x^3 + 8) \div (3x + 2)$

    a. Rewrite the expression with the dividend written in descending order and with placeholders for missing powers of x.
    b. Divide the polynomials.
    c. Identify the quotient.
    d. Identify the divisor.
    e. Identify the dividend.
    f. Identify the remainder.
    g. Check the result of the division.

A.7. Consider the division expression. $\dfrac{2y^3 - 6y + 3y^4 + 1}{y^2 - 2}$

  a. Rewrite the expression with the dividend written in descending order and with placeholders for missing powers of y in the dividend and divisor.
  b. Divide the polynomials.
  c. Check the result of the division.

A.8. Consider the division expression from Exercise A.5.

$$\dfrac{2x^4 - 9x^3 + 10x^2 - 7x + 10}{x - 3}$$

  a. Perform synthetic division.

   $\underline{|}\,2 \;\; -9 \;\; 10 \;\; -7 \;\; 10$

  b. From the results of part (a), identify the quotient.
  c. Identify the remainder.
  d. Write the result of the division. Is this the same result as in Exercise A.5?
  e. Was long division or synthetic division easier?

A.9. a. Divide using synthetic division. $(t^3 - 125) \div (t - 5)$
   (*Hint:* Be sure to leave 0's for missing powers of t in the dividend.)
  b. Check the result of the division.

A.10. Given two polynomials in the variable x, explain when synthetic division can be used to divide two polynomials.

For Exercises A.11–A.14, determine which technique of division could be used to perform the indicated operation. Choose from monomial division, long division, and synthetic division.

A.11. $(8y^3 + y^2 - 5y + 3) \div (2y + 1)$

A.12. $(4t^3 - 12t^2 - 6t + 3) \div (2t^2)$

A.13. $(5p^4 - 3p^2 + p - 6) \div (p + 4)$

A.14. $(2t^5 - 3t^4 + 5t + 8) \div (t^2 + 2t - 3)$

Section 13.7 Practice Exercises

Prerequisite Review

R.1. Divide $8945 \div 9$ using long division. Identify the dividend, divisor, quotient, and remainder.

R.2. Divide $25{,}461 \div 24$ using long division. Identify the dividend, divisor, quotient, and remainder.

For Exercises R.3–R.10, perform the indicated operations.

R.3. $(4x^2 - 2x + 3)(2x - 5) + 7$

R.4. $(5x^2 - 3x - 2)(7x - 1) - 9$

R.5. $8b^2(2b^2 - 5b + 12)$

R.6. $-4x^2(3x^2 + 6x - 7)$

R.7. $\begin{array}{r} 5x^3 - 4x^2 \\ -(5x^3 - 15x^2) \end{array}$

R.8. $\begin{array}{r} 2x^5 - 3x^4 \\ -(2x^5 + 4x^4) \end{array}$

R.9. $\begin{array}{r} x^3 + 4x^2 - 5x \\ -(x^3 + 2x^2 - 3x) \end{array}$

R.10. $\begin{array}{r} x^4 - 3x^3 + 9x^2 \\ -(x^4 - 7x^3 + 2x^2) \end{array}$

Vocabulary and Key Concepts

1. a. The _____ algorithm states that: Dividend = (divisor)(_____) + (_____).

   b. _____ division or long division can be used when dividing a polynomial by a divisor of the form $x - r$, where r is a constant.

2. Divide 313 by 6 and identify the dividend, divisor, quotient, and remainder.

Concept 1: Division by a Monomial

For Exercises 3–24, divide the polynomials. Check your answer by multiplication. **(See Example 1.)**

3. $\dfrac{p^{12}}{p^7}$
4. $\dfrac{c^{15}}{c^{12}}$
5. $\dfrac{14a^5}{2a}$
6. $\dfrac{12b^7}{3b}$
7. $\dfrac{24x^3y^4}{36xy^5}$
8. $\dfrac{15t^2v^3}{25t^8v^2}$

9. $\dfrac{16t^4 - 4t^2 + 20t}{-4t}$
10. $\dfrac{2x^3 + 8x^2 - 2x}{-2x}$

11. $(36y + 24y^2 + 6y^3) \div (3y)$
12. $(6p^2 - 18p^4 + 30p^5) \div (6p)$

13. $(4x^3y + 12x^2y^2 - 4xy^3) \div (4xy)$
14. $(25m^5n - 10m^4n + m^3n) \div (5m^3n)$

15. $(-8y^4 - 12y^3 + 32y^2) \div (-4y^2)$
16. $(12y^5 - 8y^6 + 16y^4 - 10y^3) \div (2y^3)$

17. $(3p^4 - 6p^3 + 2p^2 - p) \div (-6p)$
18. $(-4q^3 + 8q^2 - q) \div (-12q)$

19. $(a^3 + 5a^2 + a - 5) \div (a)$
20. $(2m^5 - 3m^4 + m^3 - m^2 + 9m) \div (m^2)$

21. $\dfrac{6s^3t^5 - 8s^2t^4 + 10st^2}{-2st^4}$
22. $\dfrac{-8r^4w^2 - 4r^3w + 2w^3}{-4r^3w}$

23. $(8p^4q^7 - 9p^5q^6 - 11p^3q - 4) \div (p^2q)$
24. $(20a^5b^5 - 20a^3b^2 + 5a^2b + 6) \div (a^2b)$

Concept 2: Long Division

25. a. Divide $(2x^3 - 7x^2 + 5x - 1) \div (x - 2)$ and identify the divisor, quotient, and remainder.

    b. Explain how to check by using multiplication.

26. a. Divide $(x^3 + 4x^2 + 7x - 3) \div (x + 3)$ and identify the divisor, quotient, and remainder.

    b. Explain how to check by using multiplication.

For Exercises 27–48, divide the polynomials by using long division. Check your answer by multiplication. **(See Examples 2–4.)**

27. $(x^2 + 11x + 19) \div (x + 4)$
28. $(x^3 - 7x^2 - 13x + 3) \div (x + 2)$

29. $(3y^3 - 7y^2 - 4y + 3) \div (y - 3)$
30. $(z^3 - 2z^2 + 2z - 5) \div (z - 4)$

31. $(-12a^2 + 77a - 121) \div (3a - 11)$
32. $(28x^2 - 29x + 6) \div (4x - 3)$

33. $(9y + 18y^2 - 20) \div (3y + 4)$
34. $(-2y + 3y^2 - 1) \div (y - 1)$

35. $(18x^3 + 7x + 12) \div (3x - 2)$
36. $(8x^3 - 6x + 22) \div (2x - 1)$

37. $(8a^3 + 1) \div (2a + 1)$
38. $(81x^4 - 1) \div (3x + 1)$

39. $(x^4 - x^3 - x^2 + 4x - 2) \div (x^2 + x - 1)$
40. $(2a^5 - 7a^4 + 11a^3 - 22a^2 + 29a - 10) \div (2a^2 - 5a + 2)$

41. $(2x^3 - 10x + x^4 - 25) \div (x^2 - 5)$
42. $(-5x^3 + x^4 - 4 - 10x) \div (x^2 + 2)$

43. $(x^4 - 3x^2 + 10) \div (x^2 - 2)$
44. $(3y^4 - 25y^2 - 18) \div (y^2 - 3)$

45. $(n^4 - 16) \div (n - 2)$

46. $(m^3 + 27) \div (m + 3)$

47. $\dfrac{3y^4 + 2y + 3}{1 + y^2}$

48. $\dfrac{2x^4 + 6x + 4}{2 + x^2}$

Concept 3: Synthetic Division

49. Explain the conditions under which you may use synthetic division to divide polynomials.

50. Can synthetic division be used directly to divide $(4x^4 + 3x^3 - 7x + 9)$ by $(2x + 5)$? Explain why or why not.

51. Can synthetic division be used to divide $(6x^5 - 3x^2 + 2x - 14)$ by $(x^2 - 3)$? Explain why or why not.

52. Can synthetic division be used to divide $(3x^4 - x + 1)$ by $(x - 5)$? Explain why or why not.

53. The following table represents the result of a synthetic division.

$$\begin{array}{r|rrrr} 5 & 1 & -2 & -4 & 3 \\ & & 5 & 15 & 55 \\ \hline & 1 & 3 & 11 & \underline{|58} \end{array}$$

Use x as the variable.

  a. Identify the divisor.
  b. Identify the quotient.
  c. Identify the remainder.

54. The following table represents the result of a synthetic division.

$$\begin{array}{r|rrrrr} -2 & 2 & 3 & 0 & -1 & 6 \\ & & -4 & 2 & -4 & 10 \\ \hline & 2 & -1 & 2 & -5 & \underline{|16} \end{array}$$

Use x as the variable.

  a. Identify the divisor.
  b. Identify the quotient.
  c. Identify the remainder.

For Exercises 55–70, divide by using synthetic division. Check your answer by multiplication. **(See Examples 6–7.)**

55. $(x^2 - 2x - 48) \div (x - 8)$

56. $(x^2 - 4x - 12) \div (x - 6)$

57. $(t^2 - 3t - 4) \div (t + 1)$

58. $(h^2 + 7h + 12) \div (h + 3)$

59. $(5y^2 + 5y + 1) \div (y - 1)$

60. $(3w^2 + w - 5) \div (w + 2)$

61. $(3 + 7y^2 - 4y + 3y^3) \div (y + 3)$

62. $(2z - 2z^2 + z^3 - 5) \div (z + 3)$

63. $(x^3 - 3x^2 + 4) \div (x - 2)$

64. $(3y^4 - 25y^2 - 18) \div (y - 3)$

65. $(a^5 - 32) \div (a - 2)$

66. $(b^3 + 27) \div (b + 3)$

67. $(x^3 - 216) \div (x - 6)$

68. $(y^4 - 16) \div (y + 2)$

69. $(4w^4 - w^2 + 6w - 3) \div \left(w - \dfrac{1}{2}\right)$

70. $(-12y^4 - 5y^3 - y^2 + y + 3) \div \left(y + \dfrac{3}{4}\right)$

Mixed Exercises

For Exercises 71–82, divide the polynomials by using an appropriate method. **(See Examples 5.)**

71. $(-x^3 - 8x^2 - 3x - 2) \div (x + 4)$

72. $(8xy^2 - 9x^2y + 6x^2y^2) \div (x^2y^2)$

73. $(22x^2 - 11x + 33) \div (11x)$

74. $(2m^3 - 4m^2 + 5m - 33) \div (m - 3)$

75. $(12y^3 - 17y^2 + 30y - 10) \div (3y^2 - 2y + 5)$

76. $(90h^{12} - 63h^9 + 45h^8 - 36h^7) \div (9h^9)$

77. $(4x^4 + 6x^3 + 3x - 1) \div (2x^2 + 1)$

78. $(y^4 - 3y^3 - 5y^2 - 2y + 5) \div (y + 2)$

79. $(16k^{11} - 32k^{10} + 8k^8 - 40k^4) \div (8k^8)$

80. $(4m^3 - 18m^2 + 22m - 10) \div (2m^2 - 4m + 3)$

81. $(5x^3 + 9x^2 + 10x) \div (5x^2)$

82. $(15k^4 + 3k^3 + 4k^2 + 4) \div (3k^2 - 1)$

Problem Recognition Exercises

Operations on Polynomials

For Exercises 1–40, perform the indicated operations and simplify.

1. **a.** $6x^2 + 2x^2$
   b. $(6x^2)(2x^2)$
2. **a.** $8y^3 + y^3$
   b. $(8y^3)(y^3)$
3. **a.** $(4x + y)^2$
   b. $(4xy)^2$
4. **a.** $(2a + b)^2$
   b. $(2ab)^2$

5. **a.** $(2x + 3) + (4x - 2)$
   b. $(2x + 3)(4x - 2)$
6. **a.** $(5m^2 + 1) + (m^2 + m)$
   b. $(5m^2 + 1)(m^2 + m)$
7. **a.** $(3z + 2)^2$
   b. $(3z + 2)(3z - 2)$
8. **a.** $(6y - 7)^2$
   b. $(6y - 7)(6y + 7)$

9. **a.** $(2x - 4)(x^2 - 2x + 3)$
   b. $(2x - 4) + (x^2 - 2x + 3)$
10. **a.** $(3y^2 + 8)(-y^2 - 4)$
    b. $(3y^2 + 8) - (-y^2 - 4)$
11. **a.** $x + x$
    b. $x \cdot x$
12. **a.** $2c + 2c$
    b. $2c \cdot 2c$

13. $(7mn)^2$
14. $(8pq)^2$

15. $(-2x^4 - 6x^3 + 8x^2) \div (2x^2)$
16. $(-15m^3 + 12m^2 - 3m) \div (-3m)$

17. $(m^3 - 4m^2 - 6) - (3m^2 + 7m) + (-m^3 - 9m + 6)$
18. $(n^4 + 2n^2 - 3n) + (4n^2 + 2n - 1) - (4n^5 + 6n - 3)$

19. $(8x^3 + 2x + 6) \div (x - 2)$
20. $(-4x^3 + 2x^2 - 5) \div (x - 3)$

21. $(2x - y)(3x^2 + 4xy - y^2)$
22. $(3a + b)(2a^2 - ab + 2b^2)$

23. $(x + y^2)(x^2 - xy^2 + y^4)$
24. $(m^2 + 1)(m^4 - m^2 + 1)$

25. $(a^2 + 2b) - (a^2 - 2b)$
26. $(y^3 - 6z) - (y^3 + 6z)$
27. $(a^3 + 2b)(a^3 - 2b)$
28. $(y^3 - 6z)(y^3 + 6z)$

29. $\dfrac{8p^2 + 4p - 6}{2p - 1}$
30. $\dfrac{4v^2 - 8v + 8}{2v + 3}$
31. $\dfrac{12x^3y^7}{3xy^5}$
32. $\dfrac{-18p^2q^4}{2pq^3}$

33. $\left(\dfrac{3}{7}x - \dfrac{1}{2}\right)\left(\dfrac{3}{7}x + \dfrac{1}{2}\right)$
34. $\left(\dfrac{2}{5}y + \dfrac{4}{3}\right)\left(\dfrac{2}{5}y - \dfrac{4}{3}\right)$
35. $\left(\dfrac{1}{9}x^3 + \dfrac{2}{3}x^2 + \dfrac{1}{6}x - 3\right) - \left(\dfrac{4}{3}x^3 + \dfrac{1}{9}x^2 + \dfrac{2}{3}x + 1\right)$

36. $\left(\dfrac{1}{10}y^2 - \dfrac{3}{5}y - \dfrac{1}{15}\right) - \left(\dfrac{7}{5}y^2 + \dfrac{3}{10}y - \dfrac{1}{3}\right)$
37. $(0.05x^2 - 0.16x - 0.75) + (1.25x^2 - 0.14x + 0.25)$

38. $(1.6w^3 + 2.8w + 6.1) + (3.4w^3 - 4.1w^2 - 7.3)$

39. $(3x^2y)(-2xy^5)$
40. $(10ab^4)(5a^3b^2)$

Chapter 13 Review Exercises

Section 13.1

For Exercises 1–4, identify the base and the exponent.

1. 5^3
2. x^4
3. $(-2)^0$
4. y

5. Evaluate the expressions.
   a. 6^2
   b. $(-6)^2$
   c. -6^2

6. Evaluate the expressions.
   a. 4^3
   b. $(-4)^3$
   c. -4^3

For Exercises 7–18, simplify and write the answers in exponent form. Assume that all variables represent nonzero real numbers.

7. $5^3 \cdot 5^{10}$
8. $a^7 a^4$
9. $x \cdot x^6 \cdot x^2$
10. $6^3 \cdot 6 \cdot 6^5$
11. $\dfrac{10^7}{10^4}$
12. $\dfrac{y^{14}}{y^8}$
13. $\dfrac{b^9}{b}$
14. $\dfrac{7^8}{7}$
15. $\dfrac{k^2 k^3}{k^4}$
16. $\dfrac{8^4 \cdot 8^7}{8^{11}}$
17. $\dfrac{2^8 \cdot 2^{10}}{2^3 \cdot 2^7}$
18. $\dfrac{q^3 q^{12}}{q q^8}$

19. Explain why $2^2 \cdot 4^4$ does *not* equal 8^6.

20. Explain why $\dfrac{10^5}{5^2}$ does *not* equal 2^3.

For Exercises 21–22, use the formula
$$A = P(1 + r)^t$$

21. Find the amount in an account after 3 years if the initial investment is $6000, invested at 6% interest compounded annually.

22. Find the amount in an account after 2 years if the initial investment is $20,000, invested at 5% interest compounded annually.

Section 13.2

For Exercises 23–40, simplify each expression. Write the answer in exponent form. Assume all variables represent nonzero real numbers.

23. $(7^3)^4$
24. $(c^2)^6$
25. $(p^4 p^2)^3$
26. $(9^5 \cdot 9^2)^4$
27. $\left(\dfrac{a}{b}\right)^2$
28. $\left(\dfrac{1}{3}\right)^4$
29. $\left(\dfrac{5}{c^2 d^5}\right)^2$
30. $\left(-\dfrac{m^2}{4n^6}\right)^5$
31. $(2ab^2)^4$
32. $(-x^7 y)^2$
33. $\left(\dfrac{-3x^3}{5y^2 z}\right)^3$
34. $\left(\dfrac{r^3}{s^2 t^6}\right)^5$
35. $\dfrac{a^4 (a^2)^8}{(a^3)^3}$
36. $\dfrac{(8^3)^4 \cdot 8^{10}}{(8^4)^5}$
37. $\dfrac{(4h^2 k)^2 (h^3 k)^4}{(2hk^3)^2}$
38. $\dfrac{(p^3 q)^3 (2p^2 q^4)^4}{(8p)(pq^3)^2}$
39. $\left(\dfrac{2x^4 y^3}{4xy^2}\right)^2$
40. $\left(\dfrac{a^4 b^6}{ab^4}\right)^3$

Section 13.3

For Exercises 41–62, simplify each expression. Write the answers with positive exponents. Assume all variables represent nonzero real numbers.

41. 8^0
42. $(-b)^0$
43. $-x^0$
44. 1^0
45. $2y^0$
46. $(2y)^0$
47. z^{-5}
48. 10^{-4}
49. $(6a)^{-2}$
50. $6a^{-2}$
51. $4^0 + 4^{-2}$
52. $9^{-1} + 9^0$
53. $t^{-6} t^{-2}$
54. $r^8 r^{-9}$
55. $\dfrac{12x^{-2} y^3}{6x^4 y^{-4}}$
56. $\dfrac{8ab^{-3} c^0}{10a^{-5} b^{-4} c^{-1}}$
57. $(-2m^2 n^{-4})^{-4}$
58. $(3u^{-5} v^2)^{-3}$
59. $\dfrac{(k^{-6})^{-2} (k^3)}{5k^{-6} k^0}$
60. $\dfrac{(3h)^{-2} (h^{-5})^{-3}}{h^{-4} h^8}$
61. $2 \cdot 3^{-1} - 6^{-1}$
62. $2^{-1} - 2^{-2} + 2^0$

Section 13.4

63. Write the numbers in scientific notation.
   a. In a recent year there were 97,400,000 packages of M&Ms sold in the United States.
   b. The thickness of a piece of paper is 0.0042 in.

64. Write the numbers in standard form.
   a. A pH of 10 means the hydrogen ion concentration is 1×10^{-10} units.
   b. A fundraising event for neurospinal research raised 2.56×10^5.

For Exercises 65–68, perform the indicated operations. Write the answers in scientific notation.

65. $(41 \times 10^{-6})(2.3 \times 10^{11})$

66. $\dfrac{9.3 \times 10^3}{6 \times 10^{-7}}$

67. $\dfrac{2000}{0.000008}$

68. $(0.000078)(21,000,000)$

69. Use your calculator to evaluate 5^{20}. Why is scientific notation necessary on your calculator to express the answer?

70. Use your calculator to evaluate $(0.4)^{30}$. Why is scientific notation necessary on your calculator to express the answer?

71. The average distance between Earth and the Sun is 9.3×10^7 mi.

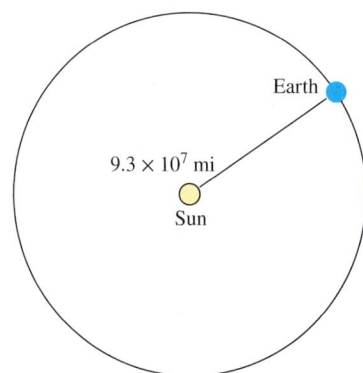

   a. If Earth's orbit is approximated by a circle, find the total distance Earth travels around the Sun in one orbit. (*Hint:* The circumference of a circle is given by $C = 2\pi r$.) Express the answer in scientific notation.

   b. If Earth makes one complete trip around the Sun in 1 year (365 days = 8.76×10^3 hr), find the average speed that Earth travels around the Sun in miles per hour. Express the answer in scientific notation.

72. The average distance between the planet Mercury and the Sun is 3.6×10^7 mi.

   a. If Mercury's orbit is approximated by a circle, find the total distance Mercury travels around the Sun in one orbit. (*Hint:* The circumference of a circle is given by $C = 2\pi r$.) Express the answer in scientific notation.

   b. If Mercury makes one complete trip around the Sun in 88 days (2.112×10^3 hr), find the average speed that Mercury travels around the Sun in miles per hour. Express the answer in scientific notation.

Section 13.5

73. For the polynomial $7x^4 - x + 6$
   a. Classify as a monomial, a binomial, or a trinomial.
   b. Identify the degree of the polynomial.
   c. Identify the leading coefficient.

74. For the polynomial $2y^3 - 5y^7$
   a. Classify as a monomial, a binomial, or a trinomial.
   b. Identify the degree of the polynomial.
   c. Identify the leading coefficient.

For Exercises 75–80, add or subtract as indicated.

75. $(4x + 2) + (3x - 5)$

76. $(7y^2 - 11y - 6) - (8y^2 + 3y - 4)$

77. $(9a^2 - 6) - (-5a^2 + 2a)$

78. $\left(5x^3 - \dfrac{1}{4}x^2 + \dfrac{5}{8}x + 2\right) + \left(\dfrac{5}{2}x^3 + \dfrac{1}{2}x^2 - \dfrac{1}{8}x\right)$

79. $\begin{array}{r} 8w^4 - 6w + 3 \\ + 2w^4 + 2w^3 - w + 1 \end{array}$

80. $\begin{array}{r} -0.02b^5 + b^4 - 0.7b + 0.3 \\ + \underline{0.03b^5 - 0.1b^3 + b + 0.03} \end{array}$

81. Subtract $(9x^2 + 4x + 6)$ from $(7x^2 - 5x)$.

82. Find the difference of $(x^2 - 5x - 3)$ and $(6x^2 + 4x + 9)$.

83. Write a trinomial of degree 2 with a leading coefficient of -5. (Answers may vary.)

84. Write a binomial of degree 5 with leading coefficient 6. (Answers may vary.)

85. Find a polynomial that represents the perimeter of the given rectangle.

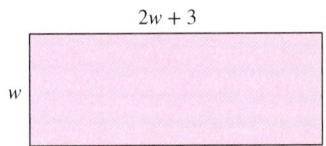

Section 13.6

For Exercises 86–103, multiply the expressions.

86. $(25x^4y^3)(-3x^2y)$

87. $(9a^6)(2a^2b^4)$

88. $5c(3c^3 - 7c + 5)$

89. $(x^2 + 5x - 3)(-2x)$

90. $(5k - 4)(k + 1)$

91. $(4t - 1)(5t + 2)$

92. $(q + 8)(6q - 1)$

93. $(2a - 6)(a + 5)$

94. $\left(7a + \dfrac{1}{2}\right)^2$

95. $(b - 4)^2$

96. $(4p^2 + 6p + 9)(2p - 3)$

97. $(2w - 1)(-w^2 - 3w - 4)$

98. $\begin{array}{r} 2x^2 - 3x + 4 \\ \times \quad\quad 2x - 1 \\ \hline \end{array}$

99. $\begin{array}{r} 4a^2 + a - 5 \\ \times \quad\quad 3a + 2 \\ \hline \end{array}$

100. $(b - 4)(b + 4)$

101. $\left(\dfrac{1}{3}r^4 - s^2\right)\left(\dfrac{1}{3}r^4 + s^2\right)$

102. $(-7z^2 + 6)^2$

103. $(2h + 3)(h^4 - h^3 + h^2 - h + 1)$

104. Find a polynomial that represents the area of the given rectangle.

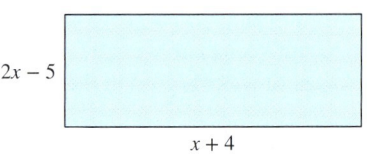

Section 13.7

For Exercises 105–117, divide the polynomials.

105. $\dfrac{20y^3 - 10y^2}{5y}$

106. $(18a^3b^2 - 9a^2b - 27ab^2) \div 9ab$

107. $(12x^4 - 8x^3 + 4x^2) \div (-4x^2)$

108. $\dfrac{10z^7w^4 - 15z^3w^2 - 20zw}{-20z^2w}$

109. $\dfrac{x^2 + 7x + 10}{x + 5}$

110. $(2t^2 + t - 10) \div (t - 2)$

111. $(2p^2 + p - 16) \div (2p + 7)$

112. $\dfrac{5a^2 + 27a - 22}{5a - 3}$

113. $\dfrac{b^3 - 125}{b - 5}$

114. $(z^3 + 4z^2 + 5z + 20) \div (5 + z^2)$

115. $(y^4 - 4y^3 + 5y^2 - 3y + 2) \div (y^2 + 3)$

116. $(3t^4 - 8t^3 + t^2 - 4t - 5) \div (3t^2 + t + 1)$

117. $\dfrac{2w^4 + w^3 + 4w - 3}{2w^2 - w + 3}$

Factoring Polynomials

14

CHAPTER OUTLINE

14.1 Greatest Common Factor and Factoring by Grouping 904
14.2 Factoring Trinomials of the Form $x^2 + bx + c$ 915
14.3 Factoring Trinomials: Trial-and-Error Method 922
14.4 Factoring Trinomials: AC-Method 933
14.5 Difference of Squares and Perfect Square Trinomials 941
14.6 Sum and Difference of Cubes 948
   Problem Recognition Exercises: Factoring Strategy 956
14.7 Solving Equations Using the Zero Product Rule 957
   Problem Recognition Exercises: Polynomial Expressions versus Polynomial Equations 965
14.8 Applications of Quadratic Equations 966
   Chapter 14 Review Exercises 974

Mathematics in the Workplace

Suppose that you are the manager of a laboratory that produces prototype electronics. This week you are tasked with producing 6 laptops over a 6-day period. With these constraints, you could produce 1 laptop per day for 6 days, 2 laptops per day for 3 days, 3 laptops per day for 2 days, or 6 laptops all in 1 day.

Any of these combinations will accomplish your task.

Laptops per Day	Number of Days	Total Number of Laptops
1	6	6
2	3	6
3	2	6
6	1	6

Henrik Jonsson/Getty Images

The word **factor** is Latin for "maker." In this example, the numbers 1, 2, 3, and 6 are **factors** of 6 because they *make* a 6 by using multiplication. In other words, we can multiply each of the numbers 1, 2, 3, and 6 by another factor to produce 6. In essence, *factoring* is the process of breaking numbers or expressions into the elements that made them via multiplication.

Factoring (finding the factors) is used extensively in algebra to make expressions easier to use. In this chapter you will learn how to factor both numerical values and polynomials.

Chapter 14 Factoring Polynomials

Section 14.1 Greatest Common Factor and Factoring by Grouping

Concepts

1. Identifying the Greatest Common Factor
2. Factoring Out the Greatest Common Factor
3. Factoring Out a Negative Factor
4. Factoring Out a Binomial Factor
5. Factoring by Grouping

1. Identifying the Greatest Common Factor

We have already learned how to multiply two or more polynomials. We now devote our study to a related operation called **factoring**. To factor an integer means to write the integer as a product of two or more integers. To factor a polynomial means to express the polynomial as a product of two or more polynomials.

In the product $2 \cdot 5 = 10$, for example, 2 and 5 are factors of 10.

In the product $(3x + 4)(2x - 1) = 6x^2 + 5x - 4$, the quantities $(3x + 4)$ and $(2x - 1)$ are factors of $6x^2 + 5x - 4$.

We begin our study of factoring by factoring integers. The number 20, for example, can be factored as $1 \cdot 20$ or $2 \cdot 10$ or $4 \cdot 5$ or $2 \cdot 2 \cdot 5$. The product $2 \cdot 2 \cdot 5$ (or equivalently $2^2 \cdot 5$) consists only of prime numbers and is called the **prime factorization**.

The **greatest common factor** (denoted **GCF**) of two or more integers is the largest factor common to each integer. To find the greatest common factor of two or more integers, it is often helpful to express the numbers as a product of prime factors as shown in the next example.

Example 1   Identifying the Greatest Common Factor

Find the greatest common factor.

  a. 24 and 36   **b.** 105, 40, and 60

Solution:

First find the prime factorization of each number. Then find the product of common factors.

a. 2|24   2|36   Factors of $24 = 2 \cdot 2 \cdot 2 \cdot 3$ Common
    2|12   2|18   factors are
    2|6   3|9   Factors of $36 = 2 \cdot 2 \cdot 3 \cdot 3$ circled.
    3   3

The numbers 24 and 36 share two factors of 2 and one factor of 3. Therefore, the greatest common factor is $2 \cdot 2 \cdot 3 = 12$.

b. 5|105   5|40   5|60   Factors of $105 = 3 \cdot 7 \cdot 5$
    3|21   2|8   3|12
    7   2|4   2|4   Factors of $40 = 2 \cdot 2 \cdot 2 \cdot 5$
                 2   2
                                     Factors of $60 = 2 \cdot 2 \cdot 3 \cdot 5$

The greatest common factor is 5.

Skill Practice Find the GCF.

  1. 12 and 20   **2.** 45, 75, and 30

Answers
1. 4   **2.** 15

In Example 2, we find the greatest common factor of two or more variable terms.

Example 2 Identifying the Greatest Common Factor

Find the GCF among each group of terms.

a. $7x^3, 14x^2, 21x^4$ b. $15a^4b, 25a^3b^2$

Solution:

List the factors of each term.

a. $7x^3 = \boxed{7 \cdot x \cdot x} \cdot x$
$14x^2 = 2 \cdot \boxed{7 \cdot x \cdot x}$
$21x^4 = 3 \cdot \boxed{7 \cdot x \cdot x} \cdot x \cdot x$
The GCF is $7x^2$.

b. $15a^4b = 3 \cdot \boxed{5 \cdot a \cdot a \cdot a} \cdot a \cdot \boxed{b}$
$25a^3b^2 = 5 \cdot \boxed{5 \cdot a \cdot a \cdot a} \cdot \boxed{b} \cdot b$
The GCF is $5a^3b$.

TIP: Notice in Example 2(b) the expressions $15a^4b$ and $25a^3b^2$ share factors of 5, a, and b. The GCF is the product of the common factors, where each factor is raised to the lowest power to which it occurs in all the original expressions.

$\left.\begin{array}{l} 15a^4b = 3^1 \cdot 5^1 a^4 b^1 \\ 25a^3b^2 = 5^2 a^3 b^2 \end{array}\right\}$

Lowest power of 5 is 1: 5^1
Lowest power of a is 3: a^3 The GCF is $5a^3b$.
Lowest power of b is 1: b^1

Skill Practice Find the GCF.

3. $10z^3, 15z^5, 40z$ 4. $6w^3y^5, 21w^4y^2$

Example 3 Identifying the Greatest Common Factor

Find the GCF of the terms $8c^2d^7e$ and $6c^3d^4$.

Solution:

$\left.\begin{array}{l} 8c^2d^7e = 2^3c^2d^7e \\ 6c^3d^4 = 2 \cdot 3c^3d^4 \end{array}\right\}$ The common factors are 2, c, and d.

The lowest power of 2 is 1: 2^1
The lowest power of c is 2: c^2 The GCF is $2c^2d^4$.
The lowest power of d is 4: d^4

Skill Practice Find the GCF.

5. $9m^2np^8, 15n^4p^5$

Sometimes polynomials share a common binomial factor, as shown in Example 4.

Answers

3. $5z$ 4. $3w^3y^2$
5. $3np^5$

Example 4   **Identifying the Greatest Common Binomial Factor**

Find the greatest common factor of the terms $3x(a + b)$ and $2y(a + b)$.

Solution:

$\left. \begin{array}{l} 3x(a + b) \\ 2y(a + b) \end{array} \right\}$ The only common factor is the binomial $(a + b)$.
              The GCF is $(a + b)$.

Skill Practice   Find the GCF.

6. $a(x + 2)$ and $b(x + 2)$

2. Factoring Out the Greatest Common Factor

The process of factoring a polynomial is the reverse process of multiplying polynomials. Both operations use the distributive property: $ab + ac = a(b + c)$.

Multiply

$5y(y^2 + 3y + 1) = 5y(y^2) + 5y(3y) + 5y(1)$
$ = 5y^3 + 15y^2 + 5y$

Factor

$5y^3 + 15y^2 + 5y = 5y(y^2) + 5y(3y) + 5y(1)$
$ = 5y(y^2 + 3y + 1)$

> **Factoring Out the Greatest Common Factor**
> **Step 1**   Identify the GCF of all terms of the polynomial.
> **Step 2**   Write each term as the product of the GCF and another factor.
> **Step 3**   Use the distributive property to remove the GCF.
> *Note:* To check the factorization, multiply the polynomials to remove parentheses.

Example 5   **Factoring Out the Greatest Common Factor**

Factor out the GCF.

   a. $4x - 20$ **b.** $6w^2 + 3w$

Solution:

   a. $4x - 20$ The GCF is 4.

        $= 4(x) - 4(5)$ Write each term as the product of the GCF and another factor.

        $= 4(x - 5)$ Use the distributive property to factor out the GCF.

> **TIP:** Any factoring problem can be checked by multiplying the factors:
>
> Check: $4(x - 5) = 4x - 20$ ✓

Answer

6. $(x + 2)$

b. $6w^2 + 3w$ The GCF is $3w$.

$= 3w(2w) + 3w(1)$ Write each term as the product of $3w$ and another factor.

$= 3w(2w + 1)$ Use the distributive property to factor out the GCF.

Check: $3w(2w + 1) = 6w^2 + 3w$ ✓

> **Avoiding Mistakes**
>
> In Example 5(b), the GCF, $3w$, is equal to one of the terms of the polynomial. In such a case, you must leave a 1 in place of that term after the GCF is factored out.

Skill Practice Factor out the GCF.

7. $6w + 18$ **8.** $21m^3 - 7m$

Example 6   Factoring Out the Greatest Common Factor

Factor out the GCF.

a. $15y^3 + 12y^4$ **b.** $9a^4b - 18a^5b + 27a^6b$

Solution:

a. $15y^3 + 12y^4$ The GCF is $3y^3$.

$= 3y^3(5) + 3y^3(4y)$ Write each term as the product of $3y^3$ and another factor.

$= 3y^3(5 + 4y)$ Use the distributive property to factor out the GCF.

Check: $3y^3(5 + 4y) = 15y^3 + 12y^4$ ✓

> **TIP:** When factoring out the GCF from a polynomial, the terms within parentheses are found by dividing the original terms by the GCF. For example:
>
> $15y^3 + 12y^4$ The GCF is $3y^3$.
>
> $\dfrac{15y^3}{3y^3} = 5$ and $\dfrac{12y^4}{3y^3} = 4y$
>
> Thus, $15y^3 + 12y^4 = 3y^3(5 + 4y)$

b. $9a^4b - 18a^5b + 27a^6b$ The GCF is $9a^4b$.

$= 9a^4b(1) - 9a^4b(2a) + 9a^4b(3a^2)$ Write each term as the product of $9a^4b$ and another factor.

$= 9a^4b(1 - 2a + 3a^2)$ Use the distributive property to factor out the GCF.

Check: $9a^4b(1 - 2a + 3a^2)$

$= 9a^4b - 18a^5b + 27a^6b$ ✓

> **Avoiding Mistakes**
>
> The GCF is $9a^4b$, not $3a^4b$. The expression $3a^4b(3 - 6a + 9a^2)$ is not factored completely.

Skill Practice Factor out the GCF.

9. $9y^2 - 6y^5$ **10.** $50s^3t - 40st^2 + 10st$

The greatest common factor of the polynomial $2x + 5y$ is 1. If we factor out the GCF, we have $1(2x + 5y)$. A polynomial whose only factors are itself and 1 is called a **prime polynomial**.

Answers

7. $6(w + 3)$
8. $7m(3m^2 - 1)$
9. $3y^2(3 - 2y^3)$
10. $10st(5s^2 - 4t + 1)$

3. Factoring Out a Negative Factor

Usually it is advantageous to factor out the *opposite* of the GCF when the leading coefficient of the polynomial is negative. This is demonstrated in Example 7. Notice that this *changes the signs* of the remaining terms inside the parentheses.

Example 7   Factoring Out a Negative Factor

Factor out -3 from the polynomial $-3x^2 + 6x - 33$.

Solution:

$-3x^2 + 6x - 33$ The GCF is 3. However, in this case, we will factor out the *opposite* of the GCF, -3.

$= -3(x^2) + (-3)(-2x) + (-3)(11)$ Write each term as the product of -3 and another factor.

$= -3[x^2 + (-2x) + 11]$ Factor out -3.

$= -3(x^2 - 2x + 11)$ Simplify. Notice that each sign within the trinomial has changed.

Check: $-3(x^2 - 2x + 11) = -3x^2 + 6x - 33$ ✓

Skill Practice   Factor out -2 from the polynomial.
11. $-2x^2 - 10x + 16$

Example 8   Factoring Out a Negative Factor

Factor out the quantity $-4pq$ from the polynomial $-12p^3q - 8p^2q^2 + 4pq^3$.

Solution:

$-12p^3q - 8p^2q^2 + 4pq^3$ The GCF is $4pq$. However, in this case, we will factor out the *opposite* of the GCF, $-4pq$.

$= -4pq(3p^2) + (-4pq)(2pq) + (-4pq)(-q^2)$ Write each term as the product of $-4pq$ and another factor.

$= -4pq[3p^2 + 2pq + (-q^2)]$ Factor out $-4pq$. Notice that each sign within the trinomial has changed.

$= -4pq(3p^2 + 2pq - q^2)$ To verify that this is the correct factorization and that the signs are correct, multiply the factors.

Check: $-4pq(3p^2 + 2pq - q^2) = -12p^3q - 8p^2q^2 + 4pq^3$ ✓

Skill Practice   Factor out $-5xy$ from the polynomial.
12. $-10x^2y + 5xy - 15xy^2$

4. Factoring Out a Binomial Factor

The distributive property can also be used to factor out a common factor that consists of more than one term, as shown in Example 9.

Answers
11. $-2(x^2 + 5x - 8)$
12. $-5xy(2x - 1 + 3y)$

Example 9 — Factoring Out a Binomial Factor

Factor out the GCF. $2w(x + 3) - 5(x + 3)$

Solution:

$2w(x + 3) - 5(x + 3)$ The greatest common factor is the quantity $(x + 3)$.

$= (x + 3)(2w - 5)$ Use the distributive property to factor out the GCF.

Skill Practice Factor out the GCF.

13. $8y(a + b) + 9(a + b)$

5. Factoring by Grouping

When two binomials are multiplied, the product before simplifying contains four terms. For example:

$$(x + 4)(3a + 2b) = (x + 4)(3a) + (x + 4)(2b)$$
$$= (x + 4)(3a) + (x + 4)(2b)$$
$$= 3ax + 12a + 2bx + 8b$$

In Example 10, we learn how to reverse this process. That is, given a four-term polynomial, we will factor it as a product of two binomials. The process is called *factoring by grouping*.

> **Factoring by Grouping**
>
> To factor a four-term polynomial by grouping:
>
> **Step 1** Identify and factor out the GCF from all four terms.
> **Step 2** Factor out the GCF from the first pair of terms. Factor out the GCF from the second pair of terms. (Sometimes it is necessary to factor out the opposite of the GCF.)
> **Step 3** If the two terms share a common binomial factor, factor out the binomial factor.

Example 10 — Factoring by Grouping

Factor by grouping. $3ax + 12a + 2bx + 8b$

Solution:

$3ax + 12a + 2bx + 8b$ **Step 1:** Identify and factor out the GCF from all four terms. In this case, the GCF is 1.

$= 3ax + 12a \mid + 2bx + 8b$ Group the first pair of terms and the second pair of terms.

Answer

13. $(a + b)(8y + 9)$

$$= 3a(x+4) + 2b(x+4)$$

Step 2: Factor out the GCF from each pair of terms. *Note:* The two terms now share a common binomial factor of $(x+4)$.

$$= (x+4)(3a+2b)$$

Step 3: Factor out the common binomial factor.

Check: $(x+4)(3a+2b) = 3ax + 2bx + 12a + 8b$ ✓

Note: Step 2 results in two terms with a common binomial factor. If the two binomials are different, step 3 cannot be performed. In such a case, the original polynomial may not be factorable by grouping, or different pairs of terms may need to be grouped and inspected.

Skill Practice Factor by grouping.

14. $5x + 10y + ax + 2ay$

TIP: One frequently asked question when factoring is whether the order can be switched between the factors. The answer is yes. Because multiplication is commutative, the order in which the factors are written does not matter.

$$(x+4)(3a+2b) = (3a+2b)(x+4)$$

Example 11 — Factoring by Grouping

Factor by grouping. $ax + ay - x - y$

Solution:

$ax + ay - x - y$

Step 1: Identify and factor out the GCF from all four terms. In this case, the GCF is 1.

$= ax + ay \mid -x - y$

Group the first pair of terms and the second pair of terms.

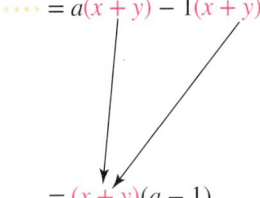

$= a(x+y) - 1(x+y)$

Step 2: Factor out a from the first pair of terms. Factor out -1 from the second pair of terms. (This causes sign changes within the second parentheses.) The terms in parentheses now match.

$= (x+y)(a-1)$

Step 3: Factor out the common binomial factor.

Check: $(x+y)(a-1) = x(a) + x(-1) + y(a) + y(-1)$
$ = ax - x + ay - y$ ✓

Skill Practice Factor by grouping.

15. $tu - tv - u + v$

Avoiding Mistakes

In step 2, the expression $a(x+y) - (x+y)$ is not yet factored completely because it is a *difference*, not a product. To factor the expression, you must carry it one step further.

$a(x+y) - 1(x+y)$
$= (x+y)(a-1)$

The factored form must be represented as a product.

Answers

14. $(x+2y)(5+a)$
15. $(u-v)(t-1)$

Example 12 Factoring by Grouping

Factor by grouping. $16w^4 - 40w^3 - 12w^2 + 30w$

Solution:

$16w^4 - 40w^3 - 12w^2 + 30w$

$= 2w[8w^3 - 20w^2 - 6w + 15]$

Step 1: Identify and factor out the GCF from all four terms. In this case, the GCF is $2w$.

$= 2w[8w^3 - 20w^2 \mid -6w + 15]$

Group the first pair of terms and the second pair of terms.

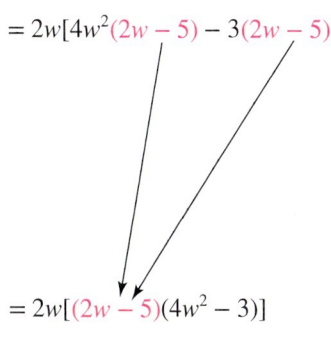

$= 2w[4w^2(2w - 5) - 3(2w - 5)]$

Step 2: Factor out $4w^2$ from the first pair of terms.

Factor out -3 from the second pair of terms. (This causes sign changes within the second parentheses.) The terms in parentheses now match.

$= 2w[(2w - 5)(4w^2 - 3)]$

Step 3: Factor out the common binomial factor.

$= 2w(2w - 5)(4w^2 - 3)$

Skill Practice Factor by grouping.

16. $3ab^2 + 6b^2 - 12ab - 24b$

Answer

16. $3b(a + 2)(b - 4)$

Section 14.1 Activity

A.1. Consider the terms $10a^2b^3$, $15a^3b^4$, and $-35a^4b$. The terms can be written in expanded form as:

$10a^2b^3 = 2 \cdot 5 \cdot a \cdot a \cdot b \cdot b \cdot b$
$15a^3b^4 = 3 \cdot 5 \cdot a \cdot a \cdot a \cdot b \cdot b \cdot b \cdot b$
$-35a^4b = -1 \cdot 5 \cdot 7 \cdot a \cdot a \cdot a \cdot a \cdot b$

  a. Circle the common factors that appear in all three lists.
  b. Identify the greatest common factor of the terms $10a^2b^3$, $15a^3b^4$, and $-35a^4b$ by multiplying the common factors identified in part (a).
  c. Factor out the greatest common factor. $10a^2b^3 + 15a^3b^4 - 35a^4b = $ ____ (____ ____ ____)

A.2. a. Identify the greatest common factor of the term $5x(c + d)$ and $7y(c + d)$.
  b. Factor out the greatest common factor. $5x(c + d) + 7y(c + d) = $ ____ (____ ____)

A.3. a. Factor out $5m$ from the polynomial. $-10m^2 + 15m = 5m$ (____ ____)
  b. Factor out $-5m$ from the polynomial. $-10m^2 + 15m = -5m$ (____ ____)
  c. When a negative factor is factored out of a polynomial, what is the effect on the signs of the terms?

A.4. Multiply. $(a - b)(a + 4)$

A.5. To factor $a^2 - ab + 4a - 4b$ by grouping, follow these steps.
  a. Factor out the greatest common factor from the first two terms.
     Factor out the greatest common factor from the last two terms.
  b. Factor out the greatest common factor from the expression. $a(a - b) + 4(a - b)$
  c. Compare your answer to the result from Exercise A.4.

A.6. a. Factor $6p^2 - pq - 30p + 5q$.
  b. Check the result by multiplying factors.

Section 14.1 Practice Exercises

Study Skills Exercise

As you study mathematics you must believe that you are capable of achieving success despite any past difficulties. This requires the right frame of mind. It requires a growth mindset. One strategy for developing a growth mindset is to make connections between existing knowledge and new concepts. For example, as you learn about factoring polynomials, recall what you already know about greatest common factor. Notice the similarities between the two examples.

- Find the greatest common factor between 16 and 24:   8

$$16 = 8 \cdot 2 \qquad 24 = 8 \cdot 3$$

- Find the greatest common factor between $16x^3$ and $24x^2$: $8x^2$

$$16x^3 = 8 \cdot 2 \cdot x \cdot x \cdot x \qquad 24x^2 = 8 \cdot 3 \cdot x \cdot x$$

Practice making connections between skills you already know and those you are learning. Your overall understanding will grow.

Prerequisite Review

For Exercises R.1–R.2, write the prime factorization for the given real number.

R.1. 120

R.2. 700

For Exercises R.3–R.10, perform the indicated operations.

R.3. $-(4x^2 - 5x + 3)$

R.4. $-(-5x + 7)$

R.5. $-10c^2(4c - 5)$

R.6. $-8m^3(2m + 7)$

R.7. $3x^2y(2x^2 - 5x + 1)$

R.8. $5ab^4(2a - 4ab + b^2)$

R.9. $(4a + 3b)(2x - 7)$

R.10. $(c - 4d)(9y - 3)$

Vocabulary and Key Concepts

1. a. Factoring a polynomial means to write it as a _____ of two or more polynomials.
   b. The prime factorization of a number consists of only _____ factors.
   c. The _____ _____ _____ (GCF) of two or more integers is the largest whole number that is a factor of each integer.
   d. A polynomial whose only factors are 1 and itself is called a _____ polynomial.
   e. The first step toward factoring a polynomial is to factor out the _____ _____ _____.
   f. To factor a four-term polynomial, we try the process of factoring by _____.

Concept 1: Identifying the Greatest Common Factor

2. List all the factors of 24.

For Exercises 3–14, identify the greatest common factor. (See Examples 1–4.)

3. 28, 63

4. 24, 40

5. 42, 30, 60

6. 20, 52, 32

7. $3xy, 7y$

8. $10mn, 11n$

9. $12w^3z, 16w^2z$

10. $20cd, 15c^3d$

11. $8x^3y^4z^2, 12xy^5z^4, 6x^2y^8z^3$

12. $15r^2s^2t^5, 5r^3s^4t^3, 30r^4s^3t^2$

13. $7(x-y), 9(x-y)$

14. $(2a-b), 3(2a-b)$

Concept 2: Factoring Out the Greatest Common Factor

15. a. Use the distributive property to multiply $3(x-2y)$.

    b. Use the distributive property to factor $3x - 6y$.

16. a. Use the distributive property to multiply $a^2(5a+b)$.

    b. Use the distributive property to factor $5a^3 + a^2b$.

For Exercises 17–36, factor out the GCF. (See Examples 5–6.)

17. $4p + 12$

18. $3q - 15$

19. $5c^2 - 10c + 15$

20. $16d^3 + 24d^2 + 32d$

21. $x^5 + x^3$

22. $y^2 - y^3$

23. $t^4 - 4t + 8t^2$

24. $7r^3 - r^5 + r^4$

25. $2ab + 4a^3b$

26. $5u^3v^2 - 5uv$

27. $38x^2y - 19x^2y^4$

28. $100a^5b^3 + 16a^2b$

29. $6x^3y^5 - 18xy^9z$

30. $15mp^7q^4 + 12m^4q^3$

31. $5 + 7y^3$

32. $w^3 - 5u^3v^2$

33. $42p^3q^2 + 14pq^2 - 7p^4q^4$

34. $8m^2n^3 - 24m^2n^2 + 4m^3n$

35. $t^5 + 2rt^3 - 3t^4 + 4r^2t^2$

36. $u^2v + 5u^3v^2 - 2u^2 + 8uv$

Concept 3: Factoring Out a Negative Factor

37. For the polynomial $-2x^3 - 4x^2 + 8x$

    a. Factor out $-2x$.

    b. Factor out $2x$.

38. For the polynomial $-9y^5 + 3y^3 - 12y$

    a. Factor out $-3y$.

    b. Factor out $3y$.

39. Factor out -1 from the polynomial.
    $-8t^2 - 9t - 2$

40. Factor out -1 from the polynomial.
    $-6x^3 - 2x - 5$

For Exercises 41–46, factor out the opposite of the greatest common factor. (See Examples 7–8.)

41. $-15p^3 - 30p^2$

42. $-24m^3 - 12m^4$

43. $-3m^4n^2 + 6m^2n - 9mn^2$

44. $-12p^3t + 2p^2t^3 + 6pt^2$

45. $-7x - 6y - 2z$

46. $-4a + 5b - c$

Concept 4: Factoring Out a Binomial Factor

For Exercises 47–52, factor out the GCF. (See Example 9.)

47. $13(a+6) - 4b(a+6)$

48. $7(x^2+1) - y(x^2+1)$

49. $8v(w^2-2) + (w^2-2)$

50. $t(r+2) + (r+2)$

51. $21x(x+3) + 7x^2(x+3)$

52. $5y^3(y-2) - 15y(y-2)$

Concept 5: Factoring by Grouping

For Exercises 53–72, factor by grouping. (See Examples 10–11.)

53. $8a^2 - 4ab + 6ac - 3bc$
54. $4x^3 + 3x^2y + 4xy^2 + 3y^3$
55. $3q + 3p + qr + pr$
56. $xy - xz + 7y - 7z$
57. $6x^2 + 3x + 4x + 2$
58. $4y^2 + 8y + 7y + 14$
59. $2t^2 + 6t - t - 3$
60. $2p^2 - p - 2p + 1$
61. $6y^2 - 2y - 9y + 3$
62. $5a^2 + 30a - 2a - 12$
63. $b^4 + b^3 - 4b - 4$
64. $8w^5 + 12w^2 - 10w^3 - 15$
65. $3j^2k + 15k + j^2 + 5$
66. $2ab^2 - 6ac + b^2 - 3c$
67. $14w^6x^6 + 7w^6 - 2x^6 - 1$
68. $18p^4x - 4x - 9p^5 + 2p$
69. $ay + bx + by + ax$ (*Hint:* Rearrange the terms.)
70. $2c + 3ay + ac + 6y$
71. $vw^2 - 3 + w - 3wv$
72. $2x^2 + 6m + 12 + x^2m$

Mixed Exercises

For Exercises 73–78, factor out the GCF first. Then factor by grouping. (See Example 12.)

73. $15x^4 + 15x^2y^2 + 10x^3y + 10xy^3$
74. $2a^3b - 4a^2b + 32ab - 64b$
75. $4abx - 4b^2x - 4ab + 4b^2$
76. $p^2q - pq^2 - rp^2q + rpq^2$
77. $6st^2 - 18st - 6t^4 + 18t^3$
78. $15j^3 - 10j^2k - 15j^2k^2 + 10jk^3$

79. The formula $P = 2l + 2w$ represents the perimeter, P, of a rectangle given the length, l, and the width, w. Factor out the GCF and write an equivalent formula in factored form.

80. The formula $P = 2a + 2b$ represents the perimeter, P, of a parallelogram given the base, b, and an adjacent side, a. Factor out the GCF and write an equivalent formula in factored form.

81. The formula $S = 2\pi r^2 + 2\pi rh$ represents the surface area, S, of a cylinder with radius, r, and height, h. Factor out the GCF and write an equivalent formula in factored form.

82. The formula $A = P + Prt$ represents the total amount of money, A, in an account that earns simple interest at a rate, r, for t years. Factor out the GCF and write an equivalent formula in factored form.

Expanding Your Skills

83. Factor out $\frac{1}{7}$ from $\frac{1}{7}x^2 + \frac{3}{7}x - \frac{5}{7}$.

84. Factor out $\frac{1}{5}$ from $\frac{6}{5}y^2 - \frac{4}{5}y + \frac{1}{5}$.

85. Factor out $\frac{1}{4}$ from $\frac{5}{4}w^2 + \frac{3}{4}w + \frac{9}{4}$.

86. Factor out $\frac{1}{6}$ from $\frac{1}{6}p^2 - \frac{3}{6}p + \frac{5}{6}$.

87. Write a polynomial that has a GCF of $3x$. (Answers may vary.)

88. Write a polynomial that has a GCF of $7y$. (Answers may vary.)

89. Write a polynomial that has a GCF of $4p^2q$. (Answers may vary.)

90. Write a polynomial that has a GCF of $2ab^2$. (Answers may vary.)

Factoring Trinomials of the Form $x^2 + bx + c$

Section 14.2

1. Factoring Trinomials With a Leading Coefficient of 1

Concept

1. Factoring Trinomials With a Leading Coefficient of 1

We have already learned how to multiply two binomials. We also saw that such a product often results in a trinomial. For example:

$$(x + 3)(x + 7) = x^2 + 7x + 3x + 21 = x^2 + 10x + 21$$

where x^2 is the product of first terms, 21 is the product of last terms, and $7x + 3x$ is the sum of products of inner terms and outer terms.

In this section, we want to reverse the process. That is, given a trinomial, we want to *factor* it as a product of two binomials. In particular, we begin our study with the case in which a trinomial has a leading coefficient of 1.

Consider the quadratic trinomial $x^2 + bx + c$. To produce a leading term of x^2, we can construct binomials of the form $(x +)(x +)$. The remaining terms can be obtained from two integers, p and q, whose product is c and whose sum is b.

$$x^2 + bx + c = (x + p)(x + q) = x^2 + qx + px + pq$$
$$= x^2 + (q + p)x + pq$$

where $q + p =$ Sum $= b$ and $pq =$ Product $= c$, and p, q are Factors of c.

This process is demonstrated in Example 1.

Example 1   Factoring a Trinomial of the Form $x^2 + bx + c$

Factor. $x^2 + 4x - 45$

Solution:

$x^2 + 4x - 45 = (x + \square)(x + \square)$ The product of the first terms in the binomials must equal the leading term of the trinomial $x \cdot x = x^2$.

We must fill in the blanks with two integers whose product is -45 and whose sum is 4. The factors must have opposite signs to produce a negative product. The possible factorizations of -45 are:

Product = −45	Sum
$-1 \cdot 45$	44
$-3 \cdot 15$	12
$-5 \cdot 9$	4
$-9 \cdot 5$	−4
$-15 \cdot 3$	−12
$-45 \cdot 1$	−44

$$x^2 + 4x - 45 = (x + \square)(x + \square)$$
$$= [x + (-5)](x + 9) \quad \text{Fill in the blanks with } -5 \text{ and } 9.$$
$$= (x - 5)(x + 9) \quad \text{Factored form}$$

Check:
$$(x - 5)(x + 9) = x^2 + 9x - 5x - 45$$
$$= x^2 + 4x - 45 \checkmark$$

Skill Practice Factor.

1. $x^2 - 5x - 14$

Multiplication of polynomials is a commutative operation. Therefore, in Example 1, we can express the factorization as $(x - 5)(x + 9)$ or as $(x + 9)(x - 5)$.

Example 2 Factoring a Trinomial of the Form $x^2 + bx + c$

Factor. $w^2 - 15w + 50$

Solution:

$w^2 - 15w + 50 = (w + \square)(w + \square)$ The product $w \cdot w = w^2$.

Find two integers whose product is 50 and whose sum is -15. To form a positive product, the factors must be either both positive or both negative. The sum must be negative, so we will choose negative factors of 50.

Product = 50	Sum
$(-1)(-50)$	-51
$(-2)(-25)$	-27
$(-5)(-10)$	-15

$$w^2 - 15w + 50 = (w + \square)(w + \square)$$
$$= [w + (-5)][w + (-10)]$$
$$= (w - 5)(w - 10) \quad \text{Factored form}$$

Check:
$$(w - 5)(w - 10) = w^2 - 10w - 5w + 50$$
$$= w^2 - 15w + 50 \checkmark$$

Skill Practice Factor.

2. $z^2 - 16z + 48$

TIP: Practice will help you become proficient in factoring polynomials. As you do your homework, keep these important guidelines in mind:
- To factor a trinomial, write the trinomial in descending order such as $x^2 + bx + c$.
- For all factoring problems, always factor out the GCF from all terms first.

Answers
1. $(x - 7)(x + 2)$
2. $(z - 4)(z - 12)$

Sign Rules for Factoring Trinomials

Given the trinomial $x^2 + bx + c$, the signs within the binomial factors are determined as follows:

Case 1 If c is *positive*, then the signs in the binomials must be the same (either both positive or both negative). The correct choice is determined by the middle term. If the middle term is positive, then both signs must be positive. If the middle term is negative, then both signs must be negative.

c is positive.
$$x^2 + 6x + 8$$
$$(x + 2)(x + 4)$$
Same signs

c is positive.
$$x^2 - 6x + 8$$
$$(x - 2)(x - 4)$$
Same signs

Case 2 If c is *negative*, then the signs in the binomials must be different.

c is negative.
$$x^2 + 2x - 35$$
$$(x + 7)(x - 5)$$
Different signs

c is negative.
$$x^2 - 2x - 35$$
$$(x - 7)(x + 5)$$
Different signs

Example 3 Factoring Trinomials

Factor. **a.** $-8p - 48 + p^2$ **b.** $-40t - 30t^2 + 10t^3$

Solution:

a. $-8p - 48 + p^2$

$= p^2 - 8p - 48$ Write in descending order.

$= (p \ \square)(p \ \square)$ Find two integers whose product is -48 and whose sum is -8. The numbers are -12 and 4.

$= (p - 12)(p + 4)$ Factored form

b. $-40t - 30t^2 + 10t^3$

$= 10t^3 - 30t^2 - 40t$ Write in descending order.

$= 10t(t^2 - 3t - 4)$ Factor out the GCF.

$= 10t(t \ \square)(t \ \square)$ Find two integers whose product is -4 and whose sum is -3. The numbers are -4 and 1.

$= 10t(t - 4)(t + 1)$ Factored form

Skill Practice Factor.

3. $-5w + w^2 - 6$ **4.** $30y^3 + 2y^4 + 112y^2$

FOR REVIEW

Given a polynomial in one variable, the **degree** of an individual term is the exponent of the variable factor within the term. A polynomial is written in **descending order** if the terms are written in order by degree, with the term of highest degree written first.

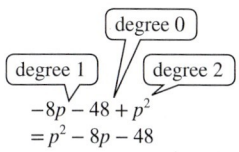

Answers

3. $(w - 6)(w + 1)$
4. $2y^2(y + 8)(y + 7)$

Example 4 Factoring Trinomials

Factor. **a.** $-a^2 + 6a - 8$ **b.** $-2c^2 - 22cd - 60d^2$

Solution:

a. $-a^2 + 6a - 8$

$= -1(a^2 - 6a + 8)$ It is generally easier to factor a trinomial with a *positive* leading coefficient. Therefore, we will factor out -1 from all terms.

$= -1(a \ \square)(a \ \square)$ Find two integers whose product is 8 and whose sum is -6. The numbers are -4 and -2.

$= -1(a - 4)(a - 2)$

$= -(a - 4)(a - 2)$

> **Avoiding Mistakes**
>
> Recall that factoring out -1 from a polynomial changes the signs of all terms within parentheses.

b. $-2c^2 - 22cd - 60d^2$

$= -2(c^2 + 11cd + 30d^2)$ Factor out -2.

$= -2(c \ \square d)(c \ \square d)$ Notice that the second pair of terms has a factor of d. This will produce a product of d^2.

$= -2(c + 5d)(c + 6d)$ Find two integers whose product is 30 and whose sum is 11. The numbers are 5 and 6.

Skill Practice Factor.

5. $-x^2 + x + 12$ **6.** $-3a^2 + 15ab - 12b^2$

To factor a trinomial of the form $x^2 + bx + c$, we must find two integers whose product is c and whose sum is b. If no such integers exist, then the trinomial is prime.

Example 5 Factoring Trinomials

Factor. $x^2 - 13x + 14$

Solution:

$x^2 - 13x + 14$ The trinomial is in descending order. The GCF is 1.

$= (x \ \square)(x \ \square)$ Find two integers whose product is 14 and whose sum is -13. No such integers exist.

The trinomial $x^2 - 13x + 14$ is prime.

Skill Practice Factor.

7. $x^2 - 7x + 28$

Answers

5. $-(x - 4)(x + 3)$
6. $-3(a - b)(a - 4b)$
7. Prime

Section 14.2 Activity

A.1. a. Find two numbers whose product is 12 and whose sum is 7.
  b. Factor. $x^2 + 7x + 12 = (x\quad)(x\quad)$
  c. Check your answer by multiplying factors.

A.2. a. Find two numbers whose product is -48 and whose sum is -8.
  b. Factor $y^2 - 8y - 48$.
  c. Check your answer by multiplying factors.

A.3. a. To factor the trinomial $t^2 + t - 12$, would the signs in the binomials be both positive, both negative, or different?
  $t^2 + t - 12 = (t\quad)(t\quad)$
  b. Factor $t^2 + t - 12$.

A.4. a. To factor the trinomial $p^2 - 9p + 14$, would the signs in the binomials be both positive, both negative, or different?
  $p^2 - 9p + 14 = (p\quad)(p\quad)$
  b. Factor $p^2 - 9p + 14$.

A.5. a. To factor the trinomial $w^2 + 11w + 24$, would the signs in the binomials be both positive, both negative, or different?
  $w^2 + 11w + 24 = (w\quad)(w\quad)$
  b. Factor $w^2 + 11w + 24$.

A.6. a. Given $x^4 + 10x^2 + 9$, fill in the blanks for the first terms of the binomials.
  $x^4 + 10x^2 + 9 = (\square\quad)(\square\quad)$
  b. Factor $x^4 + 10x^2 + 9$.
  c. Check your answer by multiplying factors.

A.7. Consider the trinomial $5y^4 + 10y^3 - 75y^2$.
  a. What is the greatest common factor?
  b. Is the polynomial written in descending order? If not, write it in descending order.
  c. Factor out the greatest common factor.
  d. Factor the polynomial completely.
  e. Check your answer by multiplying factors.

Practice Exercises — Section 14.2

Prerequisite Review

R.1. Find two integers whose product is -45 and whose sum is -4.

R.2. Find two integers whose product is 44 and whose sum is -15.

For Exercises R.3–R.4, write the trinomial in descending order.

R.3. $15y + 5y^2 + 10$ **R.4.** $24 - 18a + 6a^2$

For Exercises R.5–R.6, factor out the greatest common factor.

R.5. $5y^3 + 20y^2 + 10y$ **R.6.** $3b^4 - 18b^3 + 6b^2$

For Exercises R.7–R.12, perform the indicated operations.

R.7. $(x + 4)(x - 12)$ **R.8.** $(t - 2)(t - 11)$
R.9. $(m + 8)^2$ **R.10.** $(n - 4)^2$
R.11. $4a^2b(a - 5)(a + 6)$ **R.12.** $6cd(d - 3)(d - 2)$

Vocabulary and Key Concepts

1. **a.** Given a trinomial $x^2 + bx + c$, if c is positive, then the signs in the binomial factors are either both _____ or both negative.

   b. Given a trinomial $x^2 + bx + c$, if c is negative, then the signs in the binomial factors are (choose one: both positive, both negative, different).

2. **a.** Which is the correct factored form of $x^2 - 7x - 44$? The product $(x + 4)(x - 11)$ or $(x - 11)(x + 4)$?

   b. Which is the complete factorization of $3x^2 + 24x + 36$? The product $(3x + 6)(x + 6)$ or $3(x + 6)(x + 2)$?

Concept 1: Factoring Trinomials With a Leading Coefficient of 1

For Exercises 3–20, factor completely. **(See Examples 1, 2, and 5.)**

3. $x^2 + 10x + 16$
4. $y^2 + 18y + 80$
5. $z^2 - 11z + 18$
6. $w^2 - 7w + 12$
7. $z^2 - 3z - 18$
8. $w^2 + 4w - 12$
9. $p^2 - 3p - 40$
10. $a^2 - 10a + 9$
11. $t^2 + 6t - 40$
12. $m^2 - 12m + 11$
13. $x^2 - 3x + 20$
14. $y^2 + 6y + 18$
15. $n^2 + 8n + 16$
16. $v^2 + 10v + 25$
17. $x^4 + 10x^2 + 9$
18. $y^4 + 4y^2 - 21$
19. $w^4 + 2w^2 - 15$
20. $p^4 - 13p^2 + 40$

For Exercises 21–24, assume that b and c represent positive integers.

21. When factoring a polynomial of the form $x^2 + bx + c$, pick an appropriate combination of signs.

    a. $(\ +\ )(\ +\ )$ **b.** $(\ -\ )(\ -\ )$ **c.** $(\ +\ )(\ -\ )$

22. When factoring a polynomial of the form $x^2 + bx - c$, pick an appropriate combination of signs.

    a. $(\ +\ )(\ +\ )$ **b.** $(\ -\ )(\ -\ )$ **c.** $(\ +\ )(\ -\ )$

23. When factoring a polynomial of the form $x^2 - bx - c$, pick an appropriate combination of signs.

    a. $(\ +\ )(\ +\ )$ **b.** $(\ -\ )(\ -\ )$ **c.** $(\ +\ )(\ -\ )$

24. When factoring a polynomial of the form $x^2 - bx + c$, pick an appropriate combination of signs.

    a. $(\ +\ )(\ +\ )$ **b.** $(\ -\ )(\ -\ )$ **c.** $(\ +\ )(\ -\ )$

25. Which is the correct factorization of $y^2 - y - 12$, the product $(y - 4)(y + 3)$ or $(y + 3)(y - 4)$? Explain.

26. Which is the correct factorization of $x^2 + 14x + 13$, the product $(x + 13)(x + 1)$ or $(x + 1)(x + 13)$? Explain.

27. Which is the correct factorization of $w^2 + 2w + 1$, the product $(w + 1)(w + 1)$ or $(w + 1)^2$? Explain.

28. Which is the correct factorization of $z^2 - 4z + 4$, the product $(z - 2)(z - 2)$ or $(z - 2)^2$? Explain.

29. In what order should a trinomial be written before attempting to factor it?

30. Once a polynomial is written in descending order, what is the next step to factor the polynomial?

For Exercises 31–66, factor completely. Be sure to factor out the GCF when necessary. (See Examples 3–4.)

31. $-13x + x^2 - 30$
32. $12y - 160 + y^2$
33. $-18w + 65 + w^2$
34. $17t + t^2 + 72$
35. $22t + t^2 + 72$
36. $10q - 1200 + q^2$
37. $3x^2 - 30x - 72$
38. $2z^2 + 4z - 198$
39. $8p^3 - 40p^2 + 32p$
40. $5w^4 - 35w^3 + 50w^2$
41. $y^4z^2 - 12y^3z^2 + 36y^2z^2$
42. $t^4u^2 + 6t^3u^2 + 9t^2u^2$
43. $-x^2 + 10x - 24$
44. $-y^2 - 12y - 35$
45. $-5a^2 + 5ax + 30x^2$
46. $-2m^2 + 10mn + 12n^2$
47. $-4 - 2c^2 - 6c$
48. $-40d - 30 - 10d^2$
49. $x^3y^3 - 19x^2y^3 + 60xy^3$
50. $y^2z^5 + 17yz^5 + 60z^5$
51. $12p^2 - 96p + 84$
52. $5w^2 - 40w - 45$
53. $-2m^2 + 22m - 20$
54. $-3x^2 - 36x - 81$
55. $c^2 + 6cd + 5d^2$
56. $x^2 + 8xy + 12y^2$
57. $a^2 - 9ab + 14b^2$
58. $m^2 - 15mn + 44n^2$
59. $a^2 + 4a + 18$
60. $b^2 - 6a + 15$
61. $2q + q^2 - 63$
62. $-32 - 4t + t^2$
63. $x^2 + 20x + 100$
64. $z^2 - 24z + 144$
65. $t^2 + 18t - 40$
66. $d^2 + 2d - 99$

67. A student factored a trinomial as $(2x - 4)(x - 3)$. The instructor did not give full credit. Why?

68. A student factored a trinomial as $(y + 2)(5y - 15)$. The instructor did not give full credit. Why?

69. What polynomial factors as $(x - 4)(x + 13)$?

70. What polynomial factors as $(q - 7)(q + 10)$?

71. Raul purchased a parcel of land in the country. The given expressions represent the lengths of the boundary lines of his property.

    a. Write the perimeter of the land as a polynomial in simplified form.

    b. Write the polynomial from part (a) in factored form.

    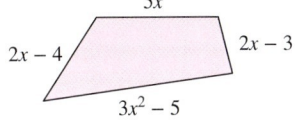

72. Jamison painted a mural in the shape of a triangle on the wall of a building. The given expressions represent the lengths of the sides of the triangle.

    a. Write the perimeter of the triangle as a polynomial in simplified form.

    b. Write the polynomial in factored form.

    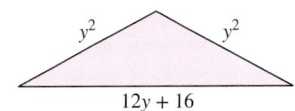

Expanding Your Skills

73. Find all integers, b, that make the trinomial $x^2 + bx + 6$ factorable.

74. Find all integers, b, that make the trinomial $x^2 + bx + 10$ factorable.

75. Find a value of c that makes the trinomial $x^2 + 6x + c$ factorable.

76. Find a value of c that makes the trinomial $x^2 + 8x + c$ factorable.

Section 14.3 Factoring Trinomials: Trial-and-Error Method

Concept

1. Factoring Trinomials by the Trial-and-Error Method

In this section we will learn how to factor a trinomial of the form $ax^2 + bx + c = 0$ (where $a \neq 0$). The method presented here is called the trial-and-error method.

1. Factoring Trinomials by the Trial-and-Error Method

To understand the basis of factoring trinomials of the form $ax^2 + bx + c$, first consider the multiplication of two binomials:

$$(2x + 3)(1x + 2) = 2x^2 + \underline{\mathbf{4x + 3x}} + 6 = 2x^2 + 7x + 6$$

Product of $2 \cdot 1$; Product of $3 \cdot 2$

Sum of products of inner terms and outer terms

To factor the trinomial, $2x^2 + 7x + 6$, this operation is reversed.

$$2x^2 + 7x + 6 = (\square x \; \square)(\square x \; \square)$$

Factors of 2 ; Factors of 6

We need to fill in the blanks so that the product of the first terms in the binomials is $2x^2$ and the product of the last terms in the binomials is 6. Furthermore, the factors of $2x^2$ and 6 must be chosen so that the sum of the products of the inner terms and outer terms equals $7x$.

To produce the product $2x^2$, we might try the factors $2x$ and x within the binomials:

$$(2x \; \square)(x \; \square)$$

To produce a product of 6, the remaining terms in the binomials must either both be positive or both be negative. To produce a positive middle term, we will try positive factors of 6 in the remaining blanks until the correct product is found. The possibilities are $1 \cdot 6, 2 \cdot 3, 6 \cdot 1$, and $3 \cdot 2$.

$(2x + 1)(x + 6) = 2x^2 + 12x + 1x + 6 = 2x^2 + 13x + 6$	Wrong middle term
$(2x + 2)(x + 3) = 2x^2 + 6x + 2x + 6 = 2x^2 + 8x + 6$	Wrong middle term
$(2x + 6)(x + 1) = 2x^2 + 2x + 6x + 6 = 2x^2 + 8x + 6$	Wrong middle term
$(2x + 3)(x + 2) = 2x^2 + 4x + 3x + 6 = 2x^2 + 7x + 6$	Correct!

The correct factorization of $2x^2 + 7x + 6$ is $(2x + 3)(x + 2)$. ✓

As this example shows, we factor a trinomial of the form $ax^2 + bx + c$ by shuffling the factors of a and c within the binomials until the correct product is obtained. However, sometimes it is not necessary to test all the possible combinations of factors. In the previous example, the GCF of the original trinomial is 1. Therefore, any binomial factor whose terms share a common factor *greater than 1* does not need to be considered. In this case, the possibilities $(2x + 2)(x + 3)$ and $(2x + 6)(x + 1)$ cannot work.

$$\underbrace{(2x + 2)}_{\text{Common factor of 2}}(x + 3) \qquad \underbrace{(2x + 6)}_{\text{Common factor of 2}}(x + 1)$$

> **Trial-and-Error Method to Factor $ax^2 + bx + c$**
>
> **Step 1** Factor out the GCF.
> **Step 2** List all pairs of positive factors of a and pairs of positive factors of c. Consider the reverse order for one of the lists of factors.
> **Step 3** Construct two binomials of the form:

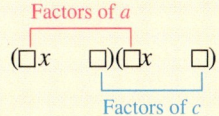

> **Step 4** Test each combination of factors and signs until the correct product is found.
> **Step 5** If no combination of factors produces the correct product, the trinomial cannot be factored further and is a *prime polynomial*.

Before we begin Example 1, keep these two important guidelines in mind:
- For any factoring problem you encounter, always factor out the GCF from all terms first.
- To factor a trinomial, write the trinomial in the form $ax^2 + bx + c$.

Example 1   Factoring a Trinomial by the Trial-and-Error Method

Factor the trinomial by the trial-and-error method. $10x^2 + 11x + 1$

Solution:

$10x^2 + 11x + 1$ **Step 1:** Factor out the GCF from all terms. In this case, the GCF is 1.

The trinomial is written in the form $ax^2 + bx + c$.

To factor $10x^2 + 11x + 1$, two binomials must be constructed in the form:

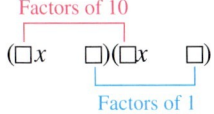

Step 2: To produce the product $10x^2$, we might try $5x$ and $2x$, or $10x$ and $1x$. To produce a product of 1, we will try the factors $(1)(1)$ and $(-1)(-1)$.

Step 3: Construct all possible binomial factors using different combinations of the factors of $10x^2$ and 1.

$(5x + 1)(2x + 1) = 10x^2 + 5x + 2x + 1 = 10x^2 + 7x + 1$ Wrong middle term

$(5x - 1)(2x - 1) = 10x^2 - 5x - 2x + 1 = 10x^2 - 7x + 1$ Wrong middle term

Because the numbers 1 and -1 did not produce the correct trinomial when coupled with $5x$ and $2x$, try using $10x$ and $1x$.

$(10x - 1)(1x - 1) = 10x^2 - 10x - 1x + 1 = 10x^2 - 11x + 1$ Wrong sign on the middle term

$(10x + 1)(1x + 1) = 10x^2 + 10x + 1x + 1 = 10x^2 + 11x + 1$ Correct!

Therefore, $10x^2 + 11x + 1 = (10x + 1)(x + 1)$.

Skill Practice Factor using the trial-and-error method.

1. $3b^2 + 8b + 4$

In Example 1, the factors of 1 must have the same signs to produce a positive product. Therefore, the binomial factors must both be sums or both be differences. Determining the correct signs is an important aspect of factoring trinomials. We suggest the following guidelines:

Sign Rules for the Trial-and-Error Method

Given the trinomial $ax^2 + bx + c$, $(a > 0)$, the signs can be determined as follows:

- If c *is positive*, then the signs in the binomials must be the same (either both positive or both negative). The correct choice is determined by the middle term. If the middle term is positive, then both signs must be positive. If the middle term is negative, then both signs must be negative.

c is positive
$20x^2 + 43x + 21$
$(4x + 3)(5x + 7)$
Same signs

c is positive
$20x^2 - 43x + 21$
$(4x - 3)(5x - 7)$
Same signs

- If c *is negative*, then the signs in the binomial must be different. The middle term in the trinomial determines which factor gets the positive sign and which gets the negative sign.

c is negative
$x^2 + 3x - 28$
$(x + 7)(x - 4)$
Different signs

c is negative
$x^2 - 3x - 28$
$(x - 7)(x + 4)$
Different signs

TIP: Look at the sign on the third term. If it is a sum, the signs will be the same in the two binomials. If it is a difference, the signs in the two binomials will be different: sum–same sign; difference–different signs.

Answer

1. $(3b + 2)(b + 2)$

Example 2 Factoring a Trinomial by the Trial-and-Error Method

Factor the trinomial. $13y - 6 + 8y^2$

Solution:

$13y - 6 + 8y^2$

$= 8y^2 + 13y - 6$ Write the polynomial in descending order.

$(\Box y \ \ \Box)(\Box y \ \ \Box)$ **Step 1:** The GCF is 1.

Factors of 8	Factors of 6
$1 \cdot 8$	$1 \cdot 6$
$2 \cdot 4$	$2 \cdot 3$
	$3 \cdot 2$ ⎫ (reverse order)
	$6 \cdot 1$ ⎭

Step 2: List the positive factors of 8 and positive factors of 6. Consider the reverse order in one list of factors.

Step 3: Construct all possible binomial factors using different combinations of the factors of 8 and 6.

$(1y \ \ 1)(8y \ \ 6)$
$(1y \ \ 3)(8y \ \ 2)$
$(2y \ \ 1)(4y \ \ 6)$ Without regard to signs, these factorizations cannot work because the red terms in the binomials share a common factor greater than 1.
$(2y \ \ 2)(4y \ \ 3)$
$(2y \ \ 3)(4y \ \ 2)$
$(2y \ \ 6)(4y \ \ 1)$

Test the remaining factorizations. Keep in mind that to produce a product of -6, the signs within the parentheses must be opposite (one positive and one negative). Also, the sum of the products of the inner terms and outer terms must be combined to form $13y$.

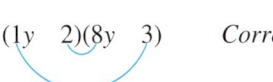   *Incorrect.*   Wrong middle term. Regardless of the signs, the product of inner terms, $48y$, and the product of outer terms, $1y$, cannot be combined to form the middle term $13y$.

$(1y \ \ 2)(8y \ \ 3)$ *Correct.*   The terms $16y$ and $3y$ can be combined to form the middle term $13y$, provided the signs are applied correctly. We require $+16y$ and $-3y$.

The correct factorization of $8y^2 + 13y - 6$ is $(y + 2)(8y - 3)$.

Skill Practice Factor.

2. $-25w + 6w^2 + 4$

Remember that the first step in any factoring problem is to remove the GCF. By removing the GCF, the remaining terms of the trinomial will have smaller coefficients.

FOR REVIEW

Recall that the factored form of a polynomial can be checked by multiplying factors. The product should equal the original polynomial.

$(y + 2)(8y - 3)$
$= 8y^2 - 3y + 16y - 6$
$= 8y^2 + 13y - 6$ ✓

Answer

2. $(6w - 1)(w - 4)$

Example 3 Factoring a Trinomial by the Trial-and-Error Method

Factor the trinomial by the trial-and-error method. $40x^3 - 104x^2 + 10x$

Solution:

$40x^3 - 104x^2 + 10x$

$= 2x(20x^2 - 52x + 5)$ **Step 1:** The GCF is $2x$.

$= 2x(\Box x \quad \Box)(\Box x \quad \Box)$ **Step 2:** List the factors of 20 and factors of 5. Consider the reverse order in one list of factors.

Factors of 20   **Factors of 5**

$1 \cdot 20$ $1 \cdot 5$

$2 \cdot 10$ $5 \cdot 1$

$4 \cdot 5$

Step 3: Construct all possible binomial factors using different combinations of the factors of 20 and factors of 5. The signs in the parentheses must both be negative.

$= 2x(1x - 1)(20x - 5)$
$= 2x(2x - 1)(10x - 5)$ *Incorrect.*   Once the GCF has been removed from the original polynomial, the binomial factors cannot contain a GCF greater than 1.
$= 2x(4x - 1)(5x - 5)$

$= 2x(1x - 5)(20x - 1)$ *Incorrect.*   Wrong middle term.

$2x(x - 5)(20x - 1)$
$= 2x(20x^2 - 1x - 100x + 5)$
$= 2x(20x^2 - 101x + 5)$

$= 2x(4x - 5)(5x - 1)$ *Incorrect.*   Wrong middle term.

$2x(4x - 5)(5x - 1)$
$= 2x(20x^2 - 4x - 25x + 5)$
$= 2x(20x^2 - 29x + 5)$

$= 2x(2x - 5)(10x - 1)$ *Correct.* $2x(2x - 5)(10x - 1)$
$= 2x(20x^2 - 2x - 50x + 5)$
$= 2x(20x^2 - 52x + 5)$
$= 40x^3 - 104x^2 + 10x$

The correct factorization is $2x(2x - 5)(10x - 1)$.

TIP: Notice that when the GCF, $2x$, is removed from the original trinomial, the new trinomial has smaller coefficients. This makes the factoring process simpler. It is easier to list the factors of 20 and 5 than the factors of 40 and 10.

Skill Practice Factor.

3. $8t^3 + 38t^2 + 24t$

Often it is easier to factor a trinomial when the leading coefficient is positive. If the leading coefficient is negative, consider factoring out the opposite of the GCF.

Answer

3. $2t(4t + 3)(t + 4)$

Example 4 Factoring a Trinomial by the Trial-and-Error Method

Factor. $-45x^2 - 3xy + 18y^2$

Solution:

$-45x^2 - 3xy + 18y^2$

$= -3(15x^2 + xy - 6y^2)$ **Step 1:** Factor out -3 to make the leading coefficient positive.

$= -3(\square x \ \square y)(\square x \ \square y)$ **Step 2:** List the factors of 15 and 6.

Factors of 15	Factors of 6
$1 \cdot 15$	$1 \cdot 6$
$3 \cdot 5$	$2 \cdot 3$
	$3 \cdot 2$
	$6 \cdot 1$

Step 3: We will construct all binomial combinations, without regard to signs first.

$-3(x \ \ y)(15x \ \ 6y)$
$-3(x \ \ 2y)(15x \ \ 3y)$
$-3(3x \ \ 3y)(5x \ \ 2y)$
$-3(3x \ \ 6y)(5x \ \ y)$

Incorrect. These combinations cannot work because the binomials in red each contain a common factor.

Test the remaining factorizations. The signs within parentheses must be opposite to produce a product of $-6y^2$. Also, the sum of the products of the inner terms and outer terms must be combined to form $1xy$.

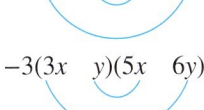

$-3(x \ \ 3y)(15x \ \ 2y)$ *Incorrect.* Regardless of signs, $45xy$ and $2xy$ cannot be combined to equal xy.

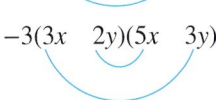

$-3(x \ \ 6y)(15x \ \ y)$ *Incorrect.* Regardless of signs, $90xy$ and xy cannot be combined to equal xy.

$-3(3x \ \ y)(5x \ \ 6y)$ *Incorrect.* Regardless of signs, $5xy$ and $18xy$ cannot be combined to equal xy.

$-3(3x \ \ 2y)(5x \ \ 3y)$ *Correct.* The terms $10xy$ and $9xy$ can be combined to form xy provided that the signs are applied correctly. We require $10xy$ and $-9xy$.

$-3(3x + 2y)(5x - 3y)$ Factored form

Avoiding Mistakes
Do not forget to write the GCF in the final answer.

Skill Practice Factor.
4. $-4x^2 + 26xy - 40y^2$

Recall that a prime polynomial is a polynomial whose only factors are itself and 1. Not every trinomial is factorable by the methods presented in this text.

Answer
4. $-2(2x - 5y)(x - 4y)$

Example 5 Factoring a Trinomial by the Trial-and-Error Method

Factor the trinomial by the trial-and-error method. $\quad 2p^2 - 8p + 3$

Solution:

$2p^2 - 8p + 3$ **Step 1:** The GCF is 1.

$= (1p \ \square)(2p \ \square)$ **Step 2:** List the factors of 2 and the factors of 3.

Factors of 2   **Factors of 3**   **Step 3:** Construct all possible binomial factors using different combinations of the factors of 2 and 3. Because the third term in the trinomial is positive, both signs in the binomial must be the same. Because the middle term coefficient is negative, both signs will be negative.

$1 \cdot 2$ $1 \cdot 3$
      $3 \cdot 1$

$(p - 1)(2p - 3) = 2p^2 - 3p - 2p + 3$
$ = 2p^2 - 5p + 3$ *Incorrect.*   Wrong middle term.

$(p - 3)(2p - 1) = 2p^2 - p - 6p + 3$
$ = 2p^2 - 7p + 3$ *Incorrect.*   Wrong middle term.

None of the combinations of factors results in the correct product. Therefore, the polynomial $2p^2 - 8p + 3$ is prime and cannot be factored further.

Skill Practice Factor.

5. $3a^2 + a + 4$

In Example 6, we use the trial-and-error method to factor a higher degree trinomial into two binomial factors.

Example 6 Factoring a Higher Degree Trinomial

Factor the trinomial. $\quad 3x^4 + 8x^2 + 5$

Solution:

$3x^4 + 8x^2 + 5$ **Step 1:** The GCF is 1.

$= (\square x^2 + \square)(\square x^2 + \square)$ **Step 2:** To produce the product $3x^4$, we must use $3x^2$ and $1x^2$. To produce a product of 5, we will try the factors $(1)(5)$ and $(5)(1)$.

          Step 3: Construct all possible binomial factors using the combinations of factors of $3x^4$ and 5.

FOR REVIEW

Recall that to multiply expressions with the same base, keep the base the same and add the exponents.

$x^2 \cdot x^2 = x^4$

$(3x^2 + 1)(x^2 + 5) = 3x^4 + 15x^2 + 1x^2 + 5 = 3x^4 + 16x^2 + 5$ Wrong middle term.

$(3x^2 + 5)(x^2 + 1) = 3x^4 + 3x^2 + 5x^2 + 5 = 3x^4 + 8x^2 + 5$ Correct!

Therefore, $3x^4 + 8x^2 + 5 = (3x^2 + 5)(x^2 + 1)$.

Skill Practice Factor.

6. $2y^4 - y^2 - 15$

Answers

5. Prime   **6.** $(y^2 - 3)(2y^2 + 5)$

Section 14.3 Activity

A.1. Consider the trinomial $2x^2 + 13x + 15$.
  a. What is the greatest common factor?
  b. Is the polynomial written in descending order? If not, write it in descending order.
  c. What is the leading coefficient?
  d. To factor the trinomial, we can set up the binomials as either $(2x \quad)(x \quad)$ or $(x \quad)(2x \quad)$. In each case, what is the product of the leading terms in the binomials?
  e. Using the binomials $(2x \quad)(x \quad)$, complete the factorization by filling in the second term of each binomial. This will require some trial and error. The two numbers you insert into the binomials must have a product of 15. But, the overall product of the binomials should also produce a middle term of $13x$.
  f. Check your answer by multiplying factors.

A.2. Consider the trinomial $5t^2 - 18t + 9$.
  a. What is the greatest common factor?
  b. Is the polynomial written in descending order? If not, write it in descending order.
  c. What is the leading coefficient?
  d. Set up two binomials so that the product of leading terms equals $5t^2$.

  $5t^2 - 18t + 9 = (\square \quad)(\square \quad)$

  e. Complete the factorization from part (d) by filling in the second term of each binomial.
  f. Check your answer by multiplying factors.

A.3. Consider the trinomial $4a^2 - 11ab + 6b^2$.
  a. What is the greatest common factor?
  b. Which product will produce a leading term of $4a^2$? There may be more than one answer.

  $4a^2 - 11ab + 6b^2 = (4a \quad)(a \quad)$
  $4a^2 - 11ab + 6b^2 = (2a \quad)(2a \quad)$
  $4a^2 - 11ab + 6b^2 = (a \quad)(4a \quad)$

  c. To factor the trinomial $4a^2 - 11ab + 6b^2$, would the signs in the binomials be both positive, both negative, or different?
  d. Choose one of the products of binomials from part (b). Then try different pairs of factors of $6b^2$ until you find the correct factorization. If you do not find the correct factorization, try using a different combination from part (b).
  e. Check your answer by multiplying factors.

Section 14.3 Practice Exercises

Prerequisite Review

For Exercises R.1–R.2, write the trinomial in descending order.

R.1. $-11y^2 + 3 + 6y^4$

R.2. $-36 + t^2 + 2t^4$

For Exercises R.3–R.8, perform the indicated operations.

R.3. $(5p - 4)(3p + 7)$

R.4. $(2n - 9)(4n - 3)$

R.5. $(3x^2 - 4)^2$

R.6. $(9y^2 + 2)^2$

R.7. $4a(2a - 3)(a + 6)$

R.8. $5b^2(3b + 1)(4b - 2)$

For Exercises R.9–R.12, factor completely.

R.9. $mn - m - 2n + 2$

R.10. $5x - 10 - xy + 2y$

R.11. $6a^2 - 30a - 84$

R.12. $10b^2 + 20b - 240$

Vocabulary and Key Concepts

1. **a.** Which is the correct factored form of $2x^2 - 5x - 12$, the product $(2x + 3)(x - 4)$ or $(x - 4)(2x + 3)$?

   b. Which is the complete factorization of $6x^2 - 4x - 10$, the product $(3x - 5)(2x + 2)$ or $2(3x - 5)(x + 1)$?

2. Consider the trinomial $8x^2 + 26x + 15$ and its factored form $(ax + b)(cx + d)$.

   a. The product ac must equal _____.

   b. The product bd must equal _____.

   c. The sum $ad + bc$ must equal _____.

Concept 1: Factoring Trinomials by the Trial-and-Error Method

For Exercises 3–6, assume a, b, and c represent positive integers.

3. When factoring a polynomial of the form $ax^2 + bx + c$, pick an appropriate combination of signs.

   a. (+)(+)

   b. (−)(−)

   c. (+)(−)

4. When factoring a polynomial of the form $ax^2 - bx - c$, pick an appropriate combination of signs.

   a. (+)(+)

   b. (−)(−)

   c. (+)(−)

5. When factoring a polynomial of the form $ax^2 - bx + c$, pick an appropriate combination of signs.

   a. (+)(+)
   b. (−)(−)
   c. (+)(−)

6. When factoring a polynomial of the form $ax^2 + bx - c$, pick an appropriate combination of signs.

   a. (+)(+)
   b. (−)(−)
   c. (+)(−)

For Exercises 7–10, a trinomial is partially factored.

a. Determine whether the signs of the remaining terms in the binomials should be the same or different.
b. Fill in the blanks (including the signs) to complete the factored form.

7. $5x^2 + 8x + 3 = (5x \ \square)(x \ \square)$

8. $7x^2 - 26x + 15 = (x \ \square)(7x \ \square)$

9. $2x^2 - 5x - 7 = (x \ \square)(2x \ \square)$

10. $3x^2 + 13x - 10 = (3x \ \square)(x \ \square)$

For Exercises 11–28, factor completely by using the trial-and-error method. (See Examples 1, 2, and 5.)

11. $3n^2 + 13n + 4$

12. $2w^2 + 5w - 3$

13. $2y^2 - 3y - 2$

14. $2a^2 + 7a + 6$

15. $5x^2 - 14x - 3$

16. $7y^2 + 9y - 10$

17. $12c^2 - 5c - 2$

18. $6z^2 + z - 12$

19. $-12 + 10w^2 + 37w$

20. $-10 + 10p^2 + 21p$

21. $-5q - 6 + 6q^2$

22. $17a - 2 + 3a^2$

23. $6b - 23 + 4b^2$

24. $8 + 7x^2 - 18x$

25. $-8 + 25m^2 - 10m$

26. $8q^2 + 31q - 4$

27. $6y^2 + 19xy - 20x^2$

28. $12y^2 - 73yz + 6z^2$

For Exercises 29–36, factor completely. Be sure to factor out the GCF first. (See Examples 3–4.)

29. $2m^2 - 12m - 80$

30. $3c^2 - 33c + 72$

31. $2y^5 + 13y^4 + 6y^3$

32. $3u^8 - 13u^7 + 4u^6$

33. $-a^2 - 15a + 34$

34. $-x^2 - 7x - 10$

35. $-80m^2 + 100mp + 30p^2$

36. $-60w^2 - 550wz + 500z^2$

For Exercises 37–42, factor the higher degree polynomial. (See Example 6.)

37. $x^4 + 10x^2 + 9$

38. $y^4 + 4y^2 - 21$

39. $w^4 + 2w^2 - 15$

40. $p^4 - 13p^2 + 40$

41. $2x^4 - 7x^2 - 15$

42. $5y^4 + 11y^2 + 2$

Mixed Exercises

For Exercises 43–82, factor each trinomial completely.

43. $20z - 18 - 2z^2$
44. $25t - 5t^2 - 30$
45. $42 - 13q + q^2$
46. $-5w - 24 + w^2$

47. $6t^2 + 7t - 3$
48. $4p^2 - 9p + 2$
49. $4m^2 - 20m + 25$
50. $16r^2 + 24r + 9$

51. $5c^2 - c + 2$
52. $7s^2 + 2s + 9$
53. $6x^2 - 19xy + 10y^2$
54. $15p^2 + pq - 2q^2$

55. $12m^2 + 11mn - 5n^2$
56. $4a^2 + 5ab - 6b^2$
57. $30r^2 + 5r - 10$
58. $36x^2 - 18x - 4$

59. $4s^2 - 8st + t^2$
60. $6u^2 - 10uv + 5v^2$
61. $10t^2 - 23t - 5$
62. $16n^2 + 14n + 3$

63. $14w^2 + 13w - 12$
64. $12x^2 - 16x + 5$
65. $a^2 - 10a - 24$
66. $b^2 + 6b - 7$

67. $x^2 + 9xy + 20y^2$
68. $p^2 - 13pq + 36q^2$
69. $a^2 + 21ab + 20b^2$
70. $x^2 - 17xy - 18y^2$

71. $t^2 - 10t + 21$
72. $z^2 - 15z + 36$
73. $5d^3 + 3d^2 - 10d$
74. $3y^3 - y^2 + 12y$

75. $4b^3 - 4b^2 - 80b$
76. $2w^2 + 20w + 42$
77. $x^2y^2 - 13xy^2 + 30y^2$
78. $p^2q^2 - 14pq^2 + 33q^2$

79. $-12u^3 - 22u^2 + 20u$
80. $-18z^4 + 15z^3 + 12z^2$
81. $8x^4 + 14x^2 + 3$
82. $6y^4 - 5y^2 - 4$

83. A rock is thrown straight upward from the top of a 40-ft building. Its height in feet after t seconds is given by the polynomial $-16t^2 + 12t + 40$.
    a. Calculate the height of the rock after 1 sec. ($t = 1$)
    b. Write $-16t^2 + 12t + 40$ in factored form. Then evaluate the factored form of the polynomial for $t = 1$. Is the result the same as from part (a)?

84. A baseball is thrown straight downward from the top of a 120-ft building. Its height in feet after t seconds is given by $-16t^2 - 8t + 120$.
    a. Calculate the height of the ball after 2 sec. ($t = 2$)
    b. Write $-16t^2 - 8t + 120$ in factored form. Then evaluate the factored form of the polynomial for $t = 2$. Is the result the same as from part (a)?

Expanding Your Skills

For Exercises 85–88, the two trinomials look similar but differ by one sign. Factor each trinomial and see how their factored forms differ.

85. a. $x^2 - 10x - 24$
    b. $x^2 - 10x + 24$

86. a. $x^2 - 13x - 30$
    b. $x^2 - 13x + 30$

87. a. $x^2 - 5x - 6$
    b. $x^2 - 5x + 6$

88. a. $x^2 - 10x + 9$
    b. $x^2 + 10x + 9$

Factoring Trinomials: AC-Method

Section 14.4

We have already learned how to factor a trinomial of the form $ax^2 + bx + c = 0$ with a leading coefficient of 1. Then we learned the trial-and-error method to factor the more general case in which the leading coefficient is any integer. In this section, we provide an alternative method to factor trinomials, called the ac-method.

Concept

1. Factoring Trinomials by the AC-Method

1. Factoring Trinomials by the AC-Method

The product of two binomials results in a four-term expression that can sometimes be simplified to a trinomial. To factor the trinomial, we want to reverse the process.

Multiply:

$$(2x + 3)(x + 2) = \xrightarrow{\text{Multiply the binomials.}} 2x^2 + 4x + 3x + 6 = \xrightarrow{\text{Add the middle terms.}} 2x^2 + 7x + 6$$

Factor:

$$2x^2 + 7x + 6 = \xrightarrow{\text{Rewrite the middle term as a sum or difference of terms.}} 2x^2 + 4x + 3x + 6 = \xrightarrow{\text{Factor by grouping.}} (2x + 3)(x + 2)$$

To factor a quadratic trinomial, $ax^2 + bx + c$, by the ac-method, we rewrite the middle term, bx, as a sum or difference of terms. The goal is to produce a four-term polynomial that can be factored by grouping. The process is outlined as follows:

AC-Method: Factoring $ax^2 + bx + c$ ($a \neq 0$)

Step 1 Factor out the GCF from all terms.
Step 2 Multiply the coefficients of the first and last terms (ac).
Step 3 Find two integers whose product is ac and whose sum is b. (If no pair of integers can be found, then the trinomial cannot be factored further and is a *prime polynomial*.)
Step 4 Rewrite the middle term, bx, as the sum of two terms whose coefficients are the integers found in step 3.
Step 5 Factor the polynomial by grouping.

The ac-method for factoring trinomials is illustrated in Example 1. However, before we begin, keep these two important guidelines in mind:
- For any factoring problem you encounter, always factor out the GCF from all terms first.
- To factor a trinomial, write the trinomial in the form $ax^2 + bx + c$.

Chapter 14 Factoring Polynomials

Example 1 Factoring a Trinomial by the AC-Method

Factor the trinomial by the ac-method. $2x^2 + 7x + 6$

Solution:

$2x^2 + 7x + 6$

Step 1: Factor out the GCF from all terms. In this case, the GCF is 1. The trinomial is written in the form $ax^2 + bx + c$.

$a = 2, b = 7, c = 6$

Step 2: Find the product $ac = (2)(6) = 12$.

12	12
$1 \cdot 12$	$(-1)(-12)$
$2 \cdot 6$	$(-2)(-6)$
$3 \cdot 4$	$(-3)(-4)$

Step 3: List all factors of ac and search for the pair whose sum equals the value of b. That is, list the factors of 12 and find the pair whose sum equals 7.

The numbers 3 and 4 satisfy both conditions: $3 \cdot 4 = 12$ and $3 + 4 = 7$.

$2x^2 + 7x + 6$

$= 2x^2 + 3x + 4x + 6$

Step 4: Write the middle term of the trinomial as the sum of two terms whose coefficients are the selected pair of numbers: 3 and 4.

$= 2x^2 + 3x + 4x + 6$

$= x(2x + 3) + 2(2x + 3)$

$= (2x + 3)(x + 2)$

Step 5: Factor by grouping.

Check: $(2x + 3)(x + 2) = 2x^2 + 4x + 3x + 6$
$= 2x^2 + 7x + 6$ ✓

FOR REVIEW

Recall that for factoring by grouping to work, the two binomial factors must be the same.

$2x^2 + 7x + 6$
$= 2x^2 + 3x + 4x + 6$
$= x(2x + 3) + 2(2x + 3)$
$= (2x + 3)(x + 2)$ ✓

Skill Practice Factor by the ac-method.

1. $2x^2 + 5x + 3$

TIP: One frequently asked question is whether the order matters when we rewrite the middle term of the trinomial as two terms (step 4). The answer is no. From the previous example, the two middle terms in step 4 could have been reversed to obtain the same result:

$2x^2 + 7x + 6$
$= 2x^2 + 4x + 3x + 6$
$= 2x(x + 2) + 3(x + 2)$
$= (x + 2)(2x + 3)$

This example also points out that the order in which two factors are written does not matter. The expression $(x + 2)(2x + 3)$ is equivalent to $(2x + 3)(x + 2)$ because multiplication is a commutative operation.

Answer

1. $(x + 1)(2x + 3)$

Example 2 Factoring a Trinomial by the AC-Method

Factor the trinomial by the ac-method. $-2x + 8x^2 - 3$

Solution:

$-2x + 8x^2 - 3$ First rewrite the polynomial in the form $ax^2 + bx + c$.

$= 8x^2 - 2x - 3$ **Step 1:** The GCF is 1.

$a = 8, b = -2, c = -3$ **Step 2:** Find the product $ac = (8)(-3) = -24$.

-24	-24
$-1 \cdot 24$	$-24 \cdot 1$
$-2 \cdot 12$	$-12 \cdot 2$
$-3 \cdot 8$	$-8 \cdot 3$
$-4 \cdot 6$	$-6 \cdot 4$

Step 3: List all the factors of -24 and find the pair of factors whose sum equals -2.

The numbers -6 and 4 satisfy both conditions: $(-6)(4) = -24$ and $-6 + 4 = -2$.

$= 8x^2 - 2x - 3$ **Step 4:** Write the middle term of the trinomial as two terms whose coefficients are the selected pair of numbers, -6 and 4.

$= 8x^2 - 6x + 4x - 3$

$= 8x^2 - 6x \mid + 4x - 3$ **Step 5:** Factor by grouping.

$= 2x(4x - 3) + 1(4x - 3)$

$= (4x - 3)(2x + 1)$

Check: $(4x - 3)(2x + 1) = 8x^2 + 4x - 6x - 3$
$= 8x^2 - 2x - 3$ ✓

Skill Practice Factor by the ac-method.

2. $13w + 6w^2 + 6$

Example 3 Factoring a Trinomial by the AC-Method

Factor the trinomial by the ac-method. $10x^3 - 85x^2 + 105x$

Solution:

$10x^3 - 85x^2 + 105x$ **Step 1:** Factor out the GCF of $5x$.

$= 5x(2x^2 - 17x + 21)$ The trinomial is in the form $ax^2 + bx + c$.

$a = 2, b = -17, c = 21$ **Step 2:** Find the product $ac = (2)(21) = 42$.

42	42
$1 \cdot 42$	$(-1)(-42)$
$2 \cdot 21$	$(-2)(-21)$
$3 \cdot 14$	$(-3)(-14)$
$6 \cdot 7$	$(-6)(-7)$

Step 3: List all the factors of 42 and find the pair whose sum equals -17.

The numbers -3 and -14 satisfy both conditions: $(-3)(-14) = 42$ and $-3 + (-14) = -17$.

Answer

2. $(2w + 3)(3w + 2)$

$$= 5x(2x^2 - 17x + 21)$$

Step 4: Write the middle term of the trinomial as two terms whose coefficients are the selected pair of numbers, -3 and -14.

$$= 5x(2x^2 - 3x - 14x + 21)$$

$$= 5x(2x^2 - 3x \mid -14x + 21)$$

Step 5: Factor by grouping.

$$= 5x[x(2x - 3) - 7(2x - 3)]$$

$$= 5x(2x - 3)(x - 7)$$

Avoiding Mistakes
Be sure to bring down the GCF in each successive step as you factor.

TIP: Notice when the GCF is removed from the original trinomial, the new trinomial has smaller coefficients. This makes the factoring process simpler because the product ac is smaller. It is much easier to list the factors of 42 than the factors of 1050.

Original trinomial	With the GCF factored out
$10x^3 - 85x^2 + 105x$	$5x(2x^2 - 17x + 21)$
$ac = (10)(105) = 1050$	$ac = (2)(21) = 42$

Skill Practice Factor by the ac-method.

3. $9y^3 - 30y^2 + 24y$

In most cases, it is easier to factor a trinomial with a positive leading coefficient.

Example 4   Factoring a Trinomial by the AC-Method

Factor the trinomial by the ac-method. $-18x^2 + 21xy + 15y^2$

Solution:

$-18x^2 + 21xy + 15y^2$

$= -3(6x^2 - 7xy - 5y^2)$

Step 1: Factor out the GCF.
Factor out -3 to make the leading term positive.

Step 2: The product $ac = (6)(-5) = -30$.

Step 3: The numbers -10 and 3 have a product of -30 and a sum of -7.

$= -3[6x^2 - 10xy + 3xy - 5y^2]$

Step 4: Rewrite the middle term, $-7xy$, as $-10xy + 3xy$.

$= -3[6x^2 - 10xy \mid + 3xy - 5y^2]$

Step 5: Factor by grouping.

$= -3[2x(3x - 5y) + y(3x - 5y)]$

$= -3(3x - 5y)(2x + y)$ Factored form

Skill Practice Factor.

4. $-8x^2 - 8xy + 30y^2$

Answers
3. $3y(3y - 4)(y - 2)$
4. $-2(2x - 3y)(2x + 5y)$

Recall that a prime polynomial is a polynomial whose only factors are itself and 1. It also should be noted that not every trinomial is factorable by the methods presented in this text.

Example 5   **Factoring a Trinomial by the AC-Method**

Factor the trinomial by the ac-method. $2p^2 - 8p + 3$

Solution:

$2p^2 - 8p + 3$ **Step 1:** The GCF is 1.

Step 2: The product $ac = 6$.

6	6
$1 \cdot 6$	$(-1)(-6)$
$2 \cdot 3$	$(-2)(-3)$

Step 3: List the factors of 6. Notice that no pair of factors has a sum of -8. Therefore, the trinomial cannot be factored.

The trinomial $2p^2 - 8p + 3$ is a prime polynomial.

Skill Practice Factor.

5. $4x^2 + 5x + 2$

In Example 6, we use the ac-method to factor a higher degree trinomial.

Example 6   **Factoring a Higher Degree Trinomial**

Factor the trinomial by the ac-method. $2x^4 + 5x^2 + 2$

Solution:

$2x^4 + 5x^2 + 2$ **Step 1:** The GCF is 1.

$a = 2, b = 5, c = 2$ **Step 2:** Find the product $ac = (2)(2) = 4$.

Step 3: The numbers 1 and 4 have a product of 4 and a sum of 5.

$2x^4 + x^2 + 4x^2 + 2$ **Step 4:** Rewrite the middle term, $5x^2$, as $x^2 + 4x^2$.

$2x^4 + x^2 \mid + 4x^2 + 2$ **Step 5:** Factor by grouping.

$x^2(2x^2 + 1) + 2(2x^2 + 1)$

$(2x^2 + 1)(x^2 + 2)$ Factored form

Skill Practice Factor.

6. $3y^4 + 2y^2 - 8$

Answers

5. Prime
6. $(3y^2 - 4)(y^2 + 2)$

Section 14.4 Activity

A.1. To use the ac-method to factor the polynomial $2x^2 + 7x + 5$, follow these steps.
  a. What is the greatest common factor?
  b. Is the polynomial written in descending order? If not, write it in descending order.
  c. The trinomial is in the form $ax^2 + bx + c$. Identify a, b, and c.
  d. What is the product of a and c? $ac =$ _____
  e. Find two numbers whose product is ac and whose sum is b. That is, the product is 10 and the sum is 7.
  f. Write the trinomial as a four-term polynomial by splitting the middle term into the sum of two terms. The coefficients of the two terms should be the numbers you found in part (e).

$$2x^2 + 7x + 5$$
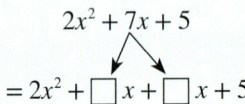
$$= 2x^2 + \Box x + \Box x + 5$$

  g. Factor by grouping.
  h. Switch the order of the middle terms found in part (f). Then factor by grouping. Compare your result to that found in part (g). Does the order of the middle terms matter?

A.2. To use the ac-method to factor the polynomial $6t^2 - 31t + 40$, follow these steps.
  a. What is the greatest common factor?
  b. Is the polynomial written in descending order? If not, write it in descending order.
  c. The trinomial is in the form $at^2 + bt + c$. Identify a, b, and c.
  d. Find two numbers whose product is ac and whose sum is b.
  e. Write the trinomial as a four-term polynomial by splitting the middle term into the sum of two terms. The coefficients of the two terms should be the numbers you found in part (d).

$$6t^2 - 31t + 40$$
$$= 6t^2 + \Box t + \Box t + 40$$

  f. Factor by grouping.
  g. Check your answer by multiplying factors.

A.3. Use the ac-method to factor the polynomial $28a^2 + 80ab - 12b^2$.
  a. What is the greatest common factor?
  b. Factor out the GCF.
  c. Factor the result from part (b) by using the ac-method.
  d. Check your answer by multiplying factors.

Section 14.4 Practice Exercises

Prerequisite Review

For Exercises R.1–R.4, factor out the greatest common factor.

R.1. $7y^2 - 14y$
R.2. $15b^2 + 5b$
R.3. $9y(y + 5) + 8(y + 5)$
R.4. $5x(x - 2) - 3(x - 2)$
R.5. Factor out $-2x$. $-2x^2 - 10x$
R.6. Factor out $-3y$. $-3y^2 + 6y$

For Exercises R.7–R.10, factor completely.

R.7. $6xy - 15x + 4y - 10$
R.8. $10ab + 25a - 12b - 30$
R.9. $6ab + 24b - 12a - 48$
R.10. $10xy + 80x - 50y - 400$

Vocabulary and Key Concepts

1. **a.** Which is the correct factored form of $10x^2 - 13x - 3$? The product $(5x + 1)(2x - 3)$ or $(2x - 3)(5x + 1)$?

   b. Which is the complete factorization of $12x^2 - 15x - 18$? The product $(4x + 3)(3x - 6)$ or $3(4x + 3)(x - 2)$?

2. To factor the trinomial $6x^2 + 19x + 10$ by using the ac-method, first rewrite the middle term of the trinomial as the sum of two terms.
$$6x^2 + mx + nx + 10$$
   a. The product mn must equal _____ and the sum $m + n$ must equal _____.

   b. Once the values of m and n have been determined, factor the resulting four-term polynomial by the _____ method.

Concept 1: Factoring Trinomials by the AC-Method

For Exercises 3–12, find the pair of integers whose product and sum are given.

3. Product: 12   Sum: 13
4. Product: 12   Sum: 7
5. Product: 8   Sum: −9
6. Product: −4   Sum: −3
7. Product: 100   Sum: −20
8. Product: 36   Sum: 12
9. Product: −20   Sum: 1
10. Product: −6   Sum: −1
11. Product: −18 Sum: 7
12. Product: −72 Sum: −6

For Exercises 13–30, factor the trinomials using the ac-method. **(See Examples 1, 2, and 5.)**

13. $3x^2 + 13x + 4$
14. $2y^2 + 7y + 6$
15. $4w^2 - 9w + 2$
16. $2p^2 - 3p - 2$
17. $x^2 + 7x - 18$
18. $y^2 - 6y - 40$
19. $2m^2 + 5m - 3$
20. $6n^2 + 7n - 3$
21. $8k^2 - 6k - 9$
22. $9h^2 - 3h - 2$
23. $4k^2 - 20k + 25$
24. $16h^2 + 24h + 9$
25. $5x^2 + x + 7$
26. $4y^2 - y + 2$
27. $10 + 9z^2 - 21z$
28. $13x + 4x^2 - 12$
29. $12y^2 + 8yz - 15z^2$
30. $20a^2 + 3ab - 9b^2$

For Exercises 31–38, factor completely. Be sure to factor out the GCF first. **(See Examples 3–4.)**

31. $50y + 24 + 14y^2$
32. $-24 + 10w + 4w^2$
33. $-15w^2 + 22w + 5$
34. $-16z^2 + 34z + 15$
35. $-12x^2 + 20xy - 8y^2$
36. $-6p^2 - 21pq - 9q^2$
37. $18y^3 + 60y^2 + 42y$
38. $8t^3 - 4t^2 - 40t$

For Exercises 39–44, factor the higher degree polynomial. **(See Example 6.)**

39. $a^4 + 5a^2 + 6$
40. $y^4 - 2y^2 - 35$
41. $6x^4 - x^2 - 15$
42. $8t^4 + 2t^2 - 3$
43. $8p^4 + 37p^2 - 15$
44. $2a^4 + 11a^2 + 14$

Mixed Exercises

For Exercises 45–80, factor completely.

45. $20p^2 - 19p + 3$
46. $4p^2 + 5pq - 6q^2$
47. $6u^2 - 19uv + 10v^2$
48. $15m^2 + mn - 2n^2$
49. $12a^2 + 11ab - 5b^2$
50. $3r^2 - rs - 14s^2$
51. $3h^2 + 19hk - 14k^2$
52. $2u^2 + uv - 15v^2$
53. $2x^2 - 13xy + y^2$
54. $3p^2 + 20pq - q^2$
55. $3 - 14z + 16z^2$
56. $10w + 1 + 16w^2$
57. $b^2 + 16 - 8b$
58. $1 + q^2 - 2q$
59. $25x - 5x^2 - 30$
60. $20a - 18 - 2a^2$
61. $-6 - t + t^2$
62. $-6 + m + m^2$
63. $v^2 + 2v + 15$
64. $x^2 - x - 1$
65. $72x^2 + 18x - 2$
66. $20y^2 - 78y - 8$
67. $p^3 - 6p^2 - 27p$
68. $w^5 - 11w^4 + 28w^3$
69. $3x^3 + 10x^2 + 7x$
70. $4r^3 + 3r^2 - 10r$
71. $2p^3 - 38p^2 + 120p$
72. $4q^3 - 4q^2 - 80q$
73. $x^2y^2 + 14x^2y + 33x^2$
74. $a^2b^2 + 13ab^2 + 30b^2$
75. $-k^2 - 7k - 10$
76. $-m^2 - 15m + 34$
77. $-3n^2 - 3n + 90$
78. $-2h^2 + 28h - 90$
79. $x^4 - 7x^2 + 10$
80. $m^4 + 10m^2 + 21$

81. Is the expression $(2x + 4)(x - 7)$ factored completely? Explain why or why not.

82. Is the expression $(3x + 1)(5x - 10)$ factored completely? Explain why or why not.

83. Colleen noticed that the number of tables placed in her restaurant affects the number of customers who eat at the restaurant. The number of customers each night is given by $-2x^2 + 40x - 72$, where x is the number of tables set up in the room, and $2 \leq x \leq 18$.

    a. Calculate the number of customers when there are 10 tables set up. ($x = 10$)

    b. Write $-2x^2 + 40x - 72$ in factored form. Then evaluate the factored form of the polynomial for $x = 10$. Is the result the same as from part (a)?

84. Roland sells cases for smartphones online. He noticed that for every dollar he discounts the price, he sells two more cases per week. His income in dollars each week is given by $-2d^2 + 30d + 200$, where d is the dollar amount of the discount in price.

    a. Calculate his income if the discount is $2. ($d = 2$)

    b. Write $-2d^2 + 30d + 200$ in factored form. Then evaluate the factored form of the polynomial for $d = 2$. Is the result the same as from part (a)?

85. A formula for finding the sum of the first n even integers is given by $n^2 + n$.

    a. Find the sum of the first 6 even integers $(2 + 4 + 6 + 8 + 10 + 12)$ by evaluating the expression for $n = 6$.

    b. Write the polynomial in factored form. Then evaluate the factored form of the expression for $n = 6$. Is the result the same as part (a)?

86. A formula for finding the sum of the squares of the first n integers is given by $\dfrac{2n^3 + 3n^2 + n}{6}$.

    a. Find the sum of the squares of the first 4 integers $(1^2 + 2^2 + 3^2 + 4^2)$ by evaluating the expression for $n = 4$.

    b. Write the polynomial in the numerator of the expression in factored form. Then evaluate the factored form of the expression for $n = 4$. Is the result the same as part (a)?

Difference of Squares and Perfect Square Trinomials

Section 14.5

1. Factoring a Difference of Squares

Up to this point, we have learned several methods of factoring, including:

- Factoring out the greatest common factor from a polynomial
- Factoring a four-term polynomial by grouping
- Factoring trinomials by the ac-method or by the trial-and-error method

In this section, we will learn to factor polynomials that fit two special case patterns: a difference of squares, and a perfect square trinomial. First recall that the product of two conjugates results in a **difference of squares**:

$$(a + b)(a - b) = a^2 - b^2$$

Therefore, to factor a difference of squares, the process is reversed. Identify a and b and construct the conjugate factors.

Concepts

1. Factoring a Difference of Squares
2. Factoring Perfect Square Trinomials

Factored Form of a Difference of Squares

$$a^2 - b^2 = (a + b)(a - b)$$

To help recognize a difference of squares, we recommend that you become familiar with the first several perfect squares.

Perfect Squares	Perfect Squares	Perfect Squares
$1 = (1)^2$	$36 = (6)^2$	$121 = (11)^2$
$4 = (2)^2$	$49 = (7)^2$	$144 = (12)^2$
$9 = (3)^2$	$64 = (8)^2$	$169 = (13)^2$
$16 = (4)^2$	$81 = (9)^2$	$196 = (14)^2$
$25 = (5)^2$	$100 = (10)^2$	$225 = (15)^2$

FOR REVIEW

Recall that when multiplying the product of conjugates, the middle terms "cancel." The result is a difference of squares.

$$(y - 5)(y + 5)$$
$$= y^2 + 5y - 5y - 25$$
$$= y^2 - 25$$

It is also important to recognize that a variable expression is a perfect square if its exponent is a multiple of 2. For example:

Perfect Squares

$$x^2 = (x)^2$$
$$x^4 = (x^2)^2$$
$$x^6 = (x^3)^2$$
$$x^8 = (x^4)^2$$
$$x^{10} = (x^5)^2$$

Example 1 Factoring Differences of Squares

Factor the binomials.

a. $y^2 - 25$ **b.** $49s^2 - 4t^4$ **c.** $18w^2z - 2z$

Solution:

a. $y^2 - 25$ The binomial is a difference of squares.
$= (y)^2 - (5)^2$ Write in the form: $a^2 - b^2$, where $a = y$, $b = 5$.
$= (y + 5)(y - 5)$ Factor as $(a + b)(a - b)$.

b. $49s^2 - 4t^4$ The binomial is a difference of squares.
$= (7s)^2 - (2t^2)^2$ Write in the form $a^2 - b^2$, where $a = 7s$ and $b = 2t^2$.
$= (7s + 2t^2)(7s - 2t^2)$ Factor as $(a + b)(a - b)$.

c. $18w^2z - 2z$ The GCF is $2z$.
$= 2z(9w^2 - 1)$ $(9w^2 - 1)$ is a difference of squares.
$= 2z[(3w)^2 - (1)^2]$ Write in the form: $a^2 - b^2$, where $a = 3w$, $b = 1$.
$= 2z(3w - 1)(3w + 1)$ Factor as $(a - b)(a + b)$.

TIP: Recall that multiplication is commutative. Therefore,
$a^2 - b^2 = (a + b)(a - b)$ or $(a - b)(a + b)$.

Skill Practice Factor the binomials.

1. $a^2 - 64$ 2. $25q^2 - 49w^2$ 3. $98m^3n - 50mn$

The difference of squares $a^2 - b^2$ factors as $(a + b)(a - b)$. However, the *sum* of squares is not factorable.

Sum of Squares

Suppose a and b have no common factors. Then the **sum of squares** $a^2 + b^2$ is *not* factorable over the real numbers.

That is, $a^2 + b^2$ is prime over the real numbers.

To see why $a^2 + b^2$ is not factorable, consider the product of binomials:

$(a + b)(a - b) = a^2 - b^2$ Wrong sign
$(a + b)(a + b) = a^2 + 2ab + b^2$ Wrong middle term
$(a - b)(a - b) = a^2 - 2ab + b^2$ Wrong middle term

After exhausting all possibilities, we see that if a and b share no common factors, then the sum of squares $a^2 + b^2$ is a prime polynomial.

Example 2 Factoring Binomials

Factor the binomials, if possible.   **a.** $p^2 - 9$ **b.** $p^2 + 9$

Solution:

a. $p^2 - 9$ Difference of squares
$= (p - 3)(p + 3)$ Factor as $a^2 - b^2 = (a - b)(a + b)$.

b. $p^2 + 9$ Sum of squares
Prime (cannot be factored)

Answers
1. $(a + 8)(a - 8)$
2. $(5q + 7w)(5q - 7w)$
3. $2mn(7m + 5)(7m - 5)$

Skill Practice Factor the binomials, if possible.
4. $t^2 - 144$ 5. $t^2 + 144$

Some factoring problems require several steps. Always be sure to factor completely.

Example 3   Factoring a Difference of Squares

Factor completely. $w^4 - 81$

Solution:

$w^4 - 81$ — The GCF is 1. $w^4 - 81$ is a difference of squares.

$= (w^2)^2 - (9)^2$ — Write in the form: $a^2 - b^2$, where $a = w^2$, $b = 9$.

$= (w^2 + 9)(w^2 - 9)$ — Factor as $(a + b)(a - b)$.

$= (w^2 + 9)(w + 3)(w - 3)$ — Note that $w^2 - 9$ can be factored further as a difference of squares. (The binomial $w^2 + 9$ is a sum of squares and cannot be factored further.)

Skill Practice Factor completely.
6. $y^4 - 1$

Example 4   Factoring a Polynomial

Factor completely. $y^3 - 5y^2 - 4y + 20$

Solution:

$y^3 - 5y^2 - 4y + 20$ — The GCF is 1. The polynomial has four terms. Factor by grouping.

$= y^3 - 5y^2 \mid -4y + 20$

$= y^2(y - 5) - 4(y - 5)$

$= (y - 5)(y^2 - 4)$ — The expression $y^2 - 4$ is a difference of squares and can be factored further as $(y - 2)(y + 2)$.

$= (y - 5)(y - 2)(y + 2)$

Check: $(y - 5)(y - 2)(y + 2) = (y - 5)(y^2 - 2y + 2y - 4)$
$= (y - 5)(y^2 - 4)$
$= (y^3 - 4y - 5y^2 + 20)$
$= y^3 - 5y^2 - 4y + 20$ ✓

Skill Practice Factor completely.
7. $p^3 + 7p^2 - 9p - 63$

2. Factoring Perfect Square Trinomials

Recall that the square of a binomial always results in a **perfect square trinomial**.

$(a + b)^2 = (a + b)(a + b) \xrightarrow{\text{Multiply.}} = a^2 + 2ab + b^2$

$(a - b)^2 = (a - b)(a - b) \xrightarrow{\text{Multiply.}} = a^2 - 2ab + b^2$

Answers
4. $(t - 12)(t + 12)$
5. Prime
6. $(y + 1)(y - 1)(y^2 + 1)$
7. $(p - 3)(p + 3)(p + 7)$

For example, $(3x + 5)^2 = (3x)^2 + 2(3x)(5) + (5)^2$
$= 9x^2 + 30x + 25$ (perfect square trinomial)

We now want to reverse this process by factoring a perfect square trinomial. The trial-and-error method or the ac-method can always be used; however, if we recognize the pattern for a perfect square trinomial, we can use one of the following formulas to reach a quick solution.

> **Factored Form of a Perfect Square Trinomial**
> $$a^2 + 2ab + b^2 = (a + b)^2$$
> $$a^2 - 2ab + b^2 = (a - b)^2$$

For example, $4x^2 + 36x + 81$ is a perfect square trinomial with $a = 2x$ and $b = 9$. Therefore, it factors as

$$4x^2 + 36x + 81 = (2x)^2 + 2(2x)(9) + (9)^2 = (2x + 9)^2$$
$$a^2 \;+\; 2(a)(b) + (b)^2 = (a + b)^2$$

To apply the formula to factor a perfect square trinomial, we must first be sure that the trinomial is indeed a perfect square trinomial.

> **Checking for a Perfect Square Trinomial**
> **Step 1** Determine whether the first and third terms are both perfect squares and have positive coefficients.
> **Step 2** If this is the case, identify a and b and determine if the middle term equals $2ab$ or $-2ab$.

Example 5   **Factoring Perfect Square Trinomials**

Factor the trinomials completely.

  a. $x^2 + 14x + 49$ **b.** $25y^2 - 20y + 4$

Solution:

a. $x^2 + 14x + 49$ The GCF is 1.

- The first and third terms are positive.
- The first term is a perfect square: $x^2 = (x)^2$.
- The third term is a perfect square: $49 = (7)^2$.

Perfect squares

$x^2 + 14x + 49$

- The middle term is twice the product of x and 7: $14x = 2(x)(7)$.

$= (x)^2 + 2(x)(7) + (7)^2$ The trinomial is in the form $a^2 + 2ab + b^2$, where $a = x$ and $b = 7$.

$= (x + 7)^2$ Factor as $(a + b)^2$.

TIP: The sign of the middle term in a perfect square trinomial determines the sign within the binomial of the factored form.
$a^2 + 2ab + b^2 = (a + b)^2$
$a^2 - 2ab + b^2 = (a - b)^2$

b. $25y^2 - 20y + 4$ The GCF is 1.

Perfect squares

- The first and third terms are positive.
- The first term is a perfect square: $25y^2 = (5y)^2$.

$25y^2 - 20y + 4$

- The third term is a perfect square: $4 = (2)^2$.

$= (5y)^2 - 2(5y)(2) + (2)^2$

- In the middle: $-20y = -2(5y)(2)$

$= (5y - 2)^2$ Factor as $(a - b)^2$.

Skill Practice Factor completely.

8. $x^2 - 6x + 9$ **9.** $81w^2 + 72w + 16$

Example 6 **Factoring Perfect Square Trinomials**

Factor the trinomials completely.

a. $18c^3 - 48c^2d + 32cd^2$ **b.** $5w^2 + 50w + 45$

Solution:

a. $18c^3 - 48c^2d + 32cd^2$

$= 2c(9c^2 - 24cd + 16d^2)$ The GCF is $2c$.

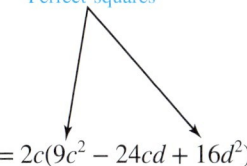

- The first and third terms are positive.
- The first term is a perfect square: $9c^2 = (3c)^2$.

$= 2c(9c^2 - 24cd + 16d^2)$

- The third term is a perfect square: $16d^2 = (4d)^2$.

$= 2c[(3c)^2 - 2(3c)(4d) + (4d)^2]$

- In the middle: $-24cd = -2(3c)(4d)$

$= 2c(3c - 4d)^2$ Factor as $(a - b)^2$.

b. $5w^2 + 50w + 45$

$= 5(w^2 + 10w + 9)$ The GCF is 5.

The first and third terms are perfect squares.

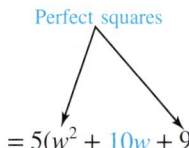

$$w^2 = (w)^2 \quad \text{and} \quad 9 = (3)^2$$

$= 5(w^2 + 10w + 9)$ However, the middle term is not 2 times the product of w and 3.

$$10w \neq 2(w)(3)$$

Therefore, this is not a perfect square trinomial.

$= 5(w + 9)(w + 1)$ To factor, use the trial-and-error method.

TIP: If you do not recognize that a trinomial is a perfect square trinomial, you can still use the trial-and-error method or ac-method to factor it.

Skill Practice Factor completely.

10. $5z^3 + 20z^2w + 20zw^2$ **11.** $40x^2 + 130x + 90$

Answers

8. $(x - 3)^2$
9. $(9w + 4)^2$
10. $5z(z + 2w)^2$
11. $10(4x + 9)(x + 1)$

Section 14.5 Activity

A.1. **a.** Identify the values that are perfect squares. 1, 2, 4, 6, 8, 9, 12, 16, 20, 25, 36, 42, 49, 64, 81, 100, 120, 121

b. Identify the expressions that are perfect squares. $x, x^2, x^3, x^4, x^6, x^8, x^9, x^{10}, x^{12}$
c. For what positive values of n is the expression x^n a perfect square?

A.2. **a.** Multiply. $(a-b)(a+b)$ **b.** Factor. $a^2 - b^2$
c. Multiply. $(4x-3)(4x+3)$ **d.** Factor. $16x^2 - 9$
e. How do you recognize a difference of squares?

For Exercises A.3–A.5, factor completely. Recall that you can check your answers by multiplying factors.

A.3. $2y^3 - 98y$

A.4. $2y^3 + 98y$

A.5. $16t^4 - 81$

A.6. **a.** Multiply. $(a+b)^2$ **b.** Factor. $a^2 + 2ab + b^2$
c. Multiply. $(5w+6)^2$ **d.** Factor. $25w^2 + 60w + 36$
e. How do you recognize a perfect square trinomial?

A.7. Factor completely. $81c^2 - 36cd + 4d^2$

A.8. **a.** What is the first step in factoring a polynomial?
b. Factor completely. $18x^2 - 24xy + 8y^2$

A.9. **a.** Factor completely. $x^2 + 17x + 16$
b. Factor completely. $x^2 + 8x + 16$
c. Compare the trinomials in parts (a) and (b). Which is a perfect square trinomial?

Section 14.5 Practice Exercises

Prerequisite Review

For Exercises R.1–R.8, perform the indicated operations.

R.1. $(2x-7)(2x+7)$ **R.2.** $(2y+9)(2y-9)$ **R.3.** $(x-2)(x+2)(x^2+4)$

R.4. $(b+3)(b-3)(b^2+9)$ **R.5.** $(3x-2)^2$ **R.6.** $(4m+3)^2$

R.7. $(6c^2+1)^2$ **R.8.** $(7-2b^2)^2$

For Exercises R.9–R.18, factor completely.

R.9. $3x^2 + x - 10$ **R.10.** $6x^2 - 17x + 5$ **R.11.** $6a^2b + 3a^3b$

R.12. $15x^2y^5 - 10xy^6$ **R.13.** $5p^2q + 20p^2 - 3pq - 12p$ **R.14.** $ax + ab - 6x - 6b$

R.15. $-6x + 5 + x^2$ **R.16.** $6y - 40 + y^2$ **R.17.** $a^2 + 7a + 1$

R.18. $t^2 + 5t - 7$

Section 14.5 Difference of Squares and Perfect Square Trinomials

Vocabulary and Key Concepts

1. **a.** The binomial $x^2 - 16$ is an example of a _____ of squares. After factoring out the GCF, we factor a difference of squares $a^2 - b^2$ as _____.
   b. The binomial $y^2 + 121$ is an example of a _____ of squares.
   c. A sum of squares with greatest common factor 1 (is/is not) factorable over the real numbers.
   d. The square of a binomial always results in a perfect _____ trinomial.
   e. A perfect square trinomial $a^2 + 2ab + b^2$ factors as _____. Likewise, $a^2 - 2ab + b^2$ factors as _____.

2. Determine whether the trinomial is a perfect square trinomial.
   a. $25x^2 - 10x - 1$ **b.** $36x^2 + 13x + 1$
   c. $25x^2 - 10x + 1$ **d.** $36x^2 + 12x + 1$

Concept 1: Factoring a Difference of Squares

3. What binomial factors as $(x - 5)(x + 5)$?
4. What binomial factors as $(n - 3)(n + 3)$?
5. What binomial factors as $(2p - 3q)(2p + 3q)$?
6. What binomial factors as $(7x - 4y)(7x + 4y)$?

For Exercises 7–30, factor each binomial completely. **(See Examples 1–3.)**

7. $x^2 - 36$
8. $r^2 - 81$
9. $3w^2 - 300$
10. $t^3 - 49t$
11. $4a^2 - 121b^2$
12. $9x^2 - y^2$
13. $49m^2 - 16n^2$
14. $100a^2 - 49b^2$
15. $9q^2 + 16$
16. $36 + s^2$
17. $y^2 - 4z^2$
18. $b^2 - 144c^2$
19. $a^2 - b^4$
20. $y^4 - x^2$
21. $25p^2q^2 - 1$
22. $81s^2t^2 - 1$
23. $c^2 - \dfrac{1}{25}$
24. $z^2 - \dfrac{1}{4}$
25. $50 - 32t^2$
26. $63 - 7h^2$
27. $x^4 - 256$
28. $y^4 - 625$
29. $16 - z^4$
30. $81 - a^4$

For Exercises 31–38, factor each polynomial completely. **(See Example 4.)**

31. $x^3 + 5x^2 - 9x - 45$
32. $y^3 + 6y^2 - 4y - 24$
33. $c^3 - c^2 - 25c + 25$
34. $t^3 + 2t^2 - 16t - 32$
35. $2x^2 - 18 + x^2y - 9y$
36. $5a^2 - 5 + a^2b - b$
37. $x^2y^2 - 9x^2 - 4y^2 + 36$
38. $w^2z^2 - w^2 - 25z^2 + 25$

Concept 2: Factoring Perfect Square Trinomials

39. Multiply. $(3x + 5)^2$
40. Multiply. $(2y - 7)^2$

41. **a.** Which trinomial is a perfect square trinomial?
    $x^2 + 4x + 4$ or $x^2 + 5x + 4$
    b. Factor the trinomials from part (a).

42. **a.** Which trinomial is a perfect square trinomial?
    $x^2 + 13x + 36$ or $x^2 + 12x + 36$
    b. Factor the trinomials from part (a).

For Exercises 43–60, factor completely. (*Hint:* Look for the pattern of a perfect square trinomial.) **(See Examples 5–6.)**

43. $x^2 + 18x + 81$
44. $y^2 - 8y + 16$
45. $25z^2 - 20z + 4$
46. $36p^2 + 60p + 25$
47. $49a^2 + 42ab + 9b^2$
48. $25m^2 - 30mn + 9n^2$
49. $-2y + y^2 + 1$
50. $4 + w^2 - 4w$
51. $80z^2 + 120zw + 45w^2$
52. $36p^2 - 24pq + 4q^2$
53. $9y^2 + 78y + 25$
54. $4y^2 + 20y + 9$
55. $2a^2 - 20a + 50$
56. $3t^2 + 18t + 27$
57. $4x^2 + x + 9$
58. $c^2 - 4c + 16$
59. $4x^2 + 4xy + y^2$
60. $100y^2 + 20yz + z^2$

61. The volume of the box shown is given as $3x^3 - 6x^2 + 3x$. Write the polynomial in factored form.

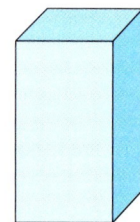

62. The volume of the box shown is given as $20y^3 + 20y^2 + 5y$. Write the polynomial in factored form.

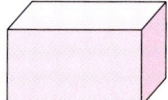

Expanding Your Skills

For Exercises 63–70, factor the difference of squares.

63. $(y - 3)^2 - 9$
64. $(x - 2)^2 - 4$
65. $(2p + 1)^2 - 36$
66. $(4q + 3)^2 - 25$
67. $16 - (t + 2)^2$
68. $81 - (a + 5)^2$
69. $(2a - 5)^2 - 100b^2$
70. $(3k + 7)^2 - 49m^2$

71. **a.** Write a polynomial that represents the area of the shaded region in the figure.

    b. Factor the expression from part (a).

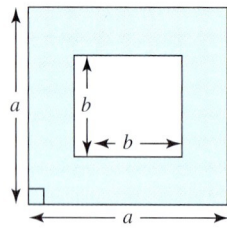

72. **a.** Write a polynomial that represents the area of the shaded region in the figure.

    b. Factor the expression from part (a).

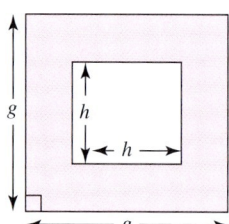

Section 14.6 Sum and Difference of Cubes

Concepts

1. Factoring a Sum or Difference of Cubes
2. Factoring Binomials: A Summary

1. Factoring a Sum or Difference of Cubes

A binomial $a^2 - b^2$ is a difference of squares and can be factored as $(a - b)(a + b)$. Furthermore, if a and b share no common factors, then a sum of squares $a^2 + b^2$ is not factorable over the real numbers. In this section, we will learn that both a difference of cubes, $a^3 - b^3$, and a sum of cubes, $a^3 + b^3$, are factorable.

Factored Form of a Sum or Difference of Cubes

Sum of Cubes: $a^3 + b^3 = (a + b)(a^2 - ab + b^2)$

Difference of Cubes: $a^3 - b^3 = (a - b)(a^2 + ab + b^2)$

Multiplication can be used to confirm the formulas for factoring a sum or difference of cubes:

$$(a + b)(a^2 - ab + b^2) = a^3 - a^2b + ab^2 + a^2b - ab^2 + b^3 = a^3 + b^3 \checkmark$$

$$(a - b)(a^2 + ab + b^2) = a^3 + a^2b + ab^2 - a^2b - ab^2 - b^3 = a^3 - b^3 \checkmark$$

To help you remember the formulas for factoring a sum or difference of cubes, keep the following guidelines in mind:

- The factored form is the product of a binomial and a trinomial.
- The first and third terms in the trinomial are the squares of the terms within the binomial factor.
- Without regard to signs, the middle term in the trinomial is the product of terms in the binomial factor.

$$x^3 + 8 = (x)^3 + (2)^3 = (x + 2)[(x)^2 - (x)(2) + (2)^2]$$

Square the first term of the binomial. Product of terms in the binomial. Square the last term of the binomial.

- The sign within the binomial factor is the same as the sign of the original binomial.
- The first and third terms in the trinomial are always positive.
- The sign of the middle term in the trinomial is opposite the sign within the binomial.

$$x^3 + 8 = (x)^3 + (2)^3 = (x + 2)[(x)^2 - (x)(2) + (2)^2]$$

Same sign — Positive — Opposite signs

TIP: To help you remember the placement of the signs in factoring the sum or difference of cubes, remember SOAP: **S**ame sign, **O**pposite signs, **A**lways **P**ositive.

To help you recognize a sum or difference of cubes, we recommend that you familiarize yourself with the first several perfect cubes:

Perfect Cubes	Perfect Cubes
$1 = (1)^3$	$216 = (6)^3$
$8 = (2)^3$	$343 = (7)^3$
$27 = (3)^3$	$512 = (8)^3$
$64 = (4)^3$	$729 = (9)^3$
$125 = (5)^3$	$1000 = (10)^3$

It is also helpful to recognize that a variable expression is a perfect cube if its exponent is a multiple of 3. For example:

Perfect Cubes

$x^3 = (x)^3$

$x^6 = (x^2)^3$

$x^9 = (x^3)^3$

$x^{12} = (x^4)^3$

Example 1 Factoring a Sum of Cubes

Factor. $w^3 + 64$

Solution:

$w^3 + 64$ w^3 and 64 are perfect cubes.

$= (w)^3 + (4)^3$ Write as $a^3 + b^3$, where $a = w$, $b = 4$.

$a^3 + b^3 = (a + b)(a^2 - ab + b^2)$ Apply the formula for a sum of cubes.

$(w)^3 + (4)^3 = (w + 4)[(w)^2 - (w)(4) + (4)^2]$

$= (w + 4)(w^2 - 4w + 16)$ Simplify.

Skill Practice Factor.

1. $p^3 + 125$

Example 2 Factoring a Difference of Cubes

Factor. $27p^3 - 1000q^3$

Solution:

$27p^3 - 1000q^3$ $27p^3$ and $1000q^3$ are perfect cubes.

$= (3p)^3 - (10q)^3$ Write as $a^3 - b^3$, where $a = 3p$, $b = 10q$.

$a^3 - b^3 = (a - b)(a^2 + ab + b^2)$ Apply the formula for a difference of cubes.

$(3p)^3 - (10q)^3 = (3p - 10q)[(3p)^2 + (3p)(10q) + (10q)^2]$

$= (3p - 10q)(9p^2 + 30pq + 100q^2)$ Simplify.

Skill Practice Factor.

2. $8y^3 - 27z^3$

2. Factoring Binomials: A Summary

After removing the GCF, the next step in any factoring problem is to recognize what type of pattern it follows. Exponents that are divisible by 2 are perfect squares and those divisible by 3 are perfect cubes. The formulas for factoring binomials are summarized in the following box:

Factored Forms of Binomials

Difference of Squares: $a^2 - b^2 = (a + b)(a - b)$
Difference of Cubes: $a^3 - b^3 = (a - b)(a^2 + ab + b^2)$
Sum of Cubes: $a^3 + b^3 = (a + b)(a^2 - ab + b^2)$

Example 3 Factoring Binomials

Factor completely.

a. $27y^3 + 1$ b. $\dfrac{1}{25}m^2 - \dfrac{1}{4}$ c. $z^6 - 8w^3$

Answers

1. $(p + 5)(p^2 - 5p + 25)$
2. $(2y - 3z)(4y^2 + 6yz + 9z^2)$

Solution:

a. $27y^3 + 1$ Sum of cubes: $27y^3 = (3y)^3$ and $1 = (1)^3$.

$= (3y)^3 + (1)^3$ Write as $a^3 + b^3$, where $a = 3y$ and $b = 1$.

$= (3y + 1)[(3y)^2 - (3y)(1) + (1)^2]$ Apply the formula
$a^3 + b^3 = (a + b)(a^2 - ab + b^2)$.

$= (3y + 1)(9y^2 - 3y + 1)$ Simplify.

b. $\dfrac{1}{25}m^2 - \dfrac{1}{4}$ Difference of squares

$= \left(\dfrac{1}{5}m\right)^2 - \left(\dfrac{1}{2}\right)^2$ Write as $a^2 - b^2$, where $a = \tfrac{1}{5}m$ and $b = \tfrac{1}{2}$.

$= \left(\dfrac{1}{5}m + \dfrac{1}{2}\right)\left(\dfrac{1}{5}m - \dfrac{1}{2}\right)$ Apply the formula $a^2 - b^2 = (a + b)(a - b)$.

c. $z^6 - 8w^3$ Difference of cubes: $z^6 = (z^2)^3$ and $8w^3 = (2w)^3$

$= (z^2)^3 - (2w)^3$ Write as $a^3 - b^3$, where $a = z^2$ and $b = 2w$.

$= (z^2 - 2w)[(z^2)^2 + (z^2)(2w) + (2w)^2]$ Apply the formula
$a^3 - b^3 = (a - b)(a^2 + ab + b^2)$.

$= (z^2 - 2w)(z^4 + 2z^2w + 4w^2)$ Simplify.

Each factorization in this example can be checked by multiplying.

Skill Practice Factor completely.

3. $1000x^3 + 1$ **4.** $25p^2 - \dfrac{1}{9}$ **5.** $27a^6 - b^3$

Some factoring problems require more than one method of factoring. In general, when factoring a polynomial, be sure to factor completely.

Example 4 Factoring a Polynomial

Factor completely. $3y^4 - 48$

Solution:

$3y^4 - 48$

$= 3(y^4 - 16)$ Factor out the GCF. The binomial is a difference of squares.

$= 3[(y^2)^2 - (4)^2]$ Write as $a^2 - b^2$, where $a = y^2$ and $b = 4$.

$= 3(y^2 + 4)(y^2 - 4)$ Apply the formula
$a^2 - b^2 = (a + b)(a - b)$.

$y^2 + 4$ is a sum of squares and cannot be factored.

$= 3(y^2 + 4)(y + 2)(y - 2)$ $y^2 - 4$ is a difference of squares and can be factored further.

Skill Practice Factor completely.

6. $2x^4 - 2$

FOR REVIEW

Remember that the first step to factoring any polynomial is to factor out the greatest common factor, GCF.

$3y^4 - 48$
$= 3(y^4 - 16)$

Answers

3. $(10x + 1)(100x^2 - 10x + 1)$

4. $\left(5p - \dfrac{1}{3}\right)\left(5p + \dfrac{1}{3}\right)$

5. $(3a^2 - b)(9a^4 + 3a^2b + b^2)$

6. $2(x^2 + 1)(x - 1)(x + 1)$

Example 5 Factoring a Polynomial

Factor completely. $4x^3 + 4x^2 - 25x - 25$

Solution:

$4x^3 + 4x^2 - 25x - 25$ The GCF is 1.

$= 4x^3 + 4x^2 \mid -25x - 25$ The polynomial has four terms. Factor by grouping.

$= 4x^2(x + 1) - 25(x + 1)$

$= (x + 1)(4x^2 - 25)$ $4x^2 - 25$ is a difference of squares.

$= (x + 1)(2x + 5)(2x - 5)$

Skill Practice Factor completely.

7. $x^3 + 6x^2 - 4x - 24$

Example 6 Factoring a Binomial

Factor the binomial $x^6 - y^6$ as

a. A difference of cubes   **b.** A difference of squares

Notice that the expressions x^6 and y^6 are both perfect squares and perfect cubes because both exponents are multiples of 2 and of 3. Consequently, $x^6 - y^6$ can be factored initially as either the difference of squares or as the difference of cubes.

Solution:

a. $x^6 - y^6$

$\qquad$ Difference of cubes

$= (x^2)^3 - (y^2)^3$ Write as $a^3 - b^3$, where $a = x^2$ and $b = y^2$.

$= (x^2 - y^2)[(x^2)^2 + (x^2)(y^2) + (y^2)^2]$ Apply the formula $a^3 - b^3 = (a - b)(a^2 + ab + b^2)$.

$= (x^2 - y^2)(x^4 + x^2y^2 + y^4)$ Factor $x^2 - y^2$ as a difference of squares.

$= (x + y)(x - y)(x^4 + x^2y^2 + y^4)$

Avoiding Mistakes

The trinomial $x^4 + x^2y^2 + y^4$ cannot be factored further with the techniques presented in this chapter.

b. $x^6 - y^6$

$\qquad$ Difference of squares

$= (x^3)^2 - (y^3)^2$ Write as $a^2 - b^2$, where $a = x^3$ and $b = y^3$.

$= (x^3 + y^3)(x^3 - y^3)$ Apply the formula $a^2 - b^2 = (a + b)(a - b)$.

Sum of cubes   Difference of cubes

Factor $x^3 + y^3$ as a sum of cubes.

Factor $x^3 - y^3$ as a difference of cubes.

$= (x + y)(x^2 - xy + y^2)(x - y)(x^2 + xy + y^2)$

Answer

7. $(x + 6)(x + 2)(x - 2)$

In a case such as this, it is recommended that you factor the expression as a difference of squares first because it factors more completely into polynomials of lower degree.

$$x^6 - y^6 = (x + y)(x^2 - xy + y^2)(x - y)(x^2 + xy + y^2)$$

Skill Practice Factor completely.

8. $z^6 - 64$

Answer

8. $(z + 2)(z - 2)(z^2 + 2z + 4)$
   $(z^2 - 2z + 4)$

Section 14.6 Activity

A.1. a. Identify the values that are perfect cubes. 1, 4, 8, 9, 12, 16, 20, 27, 36, 48, 64, 100, 125
   b. Identify the expressions that are perfect cubes. $x, x^2, x^3, x^4, x^6, x^8, x^9, x^{10}, x^{12}$
   c. For what positive values of n is the expression x^n a perfect cube?

A.2. a. Multiply. $(a - b)(a^2 + ab + b^2)$
   b. Factor. $a^3 - b^3$

A.3. Factor $8y^3 + 27$ as a sum of cubes by using the following steps.
   a. Write the formula to factor a sum of cubes. $a^3 + b^3 = (\quad)(\quad)$
   b. Write $8y^3$ as a perfect cube. $8y^3 = (\quad)^3$
   c. Write 27 as a perfect cube. $27 = (\quad)^3$
   d. To factor the polynomial $8y^3 + 27$ as a sum of cubes, $a^3 + b^3$, identify the value of a and the value of b.
   e. Factor $8y^3 + 27$ using the formula in part (a).
   f. Check the factored form by multiplying the factors.

A.4. Factor $p^6 - 64q^3$ as a difference of cubes by using the following steps.
   a. Write the formula to factor a difference of cubes. $a^3 - b^3 = (\quad)(\quad)$
   b. Write p^6 as a perfect cube. $p^6 = (\quad)^3$
   c. Write $64q^3$ as a perfect cube. $64q^3 = (\quad)^3$
   d. To factor the polynomial $p^6 - 64q^3$ as a difference of cubes, $a^3 - b^3$, identify the value of a and the value of b.
   e. Factor $p^6 - 64q^3$ using the formula in part (a).
   f. Check the factored form by multiplying the factors.

Perhaps one of the most important aspects of factoring a polynomial is to be able to identify whether the polynomial fits a special pattern. For binomials, these include: a difference of squares, a sum of squares (not factorable), a difference of cubes, and a sum of cubes. For Exercises A.5–A.9,
   a. Factor out the GCF.
   b. Identify the binomial from part (a) as a difference of squares, a sum of squares, a difference of cubes, a sum of cubes, or none of these.
   c. Factor the binomial completely.

A.5. $3p^2 + 48$

A.6. $16x^3 - 54y^6$

A.7. $125a^4 + ab^3$

A.8. $625 - w^4$

A.9. $16y^3 - 2y^2$

Section 14.6 Practice Exercises

Study Skills Exercise

A valuable skill in mathematics is the ability to clearly communicate your understanding of the steps required to solve a problem. This skill is especially important when factoring polynomials. To factor a polynomial successfully, you must learn to recognize the strategy that is appropriate for the problem and how to properly apply that strategy. At this point in the text, we have covered all major techniques for factoring a polynomial. Suppose that you wanted to tutor a friend in how to factor a polynomial.

- Write a step-by-step approach that you would follow to assess and factor a polynomial. Consider all different scenarios such as factoring two-term, three-term, and four-term polynomials. These guidelines will help your overall understanding (and your friend's understanding) and will help you commit these strategies to long-term memory. Refer to this list when completing the Practice Exercises and commit these strategies to memory.

Prerequisite Review

For Exercises R.1–R.2, multiply the polynomials.

R.1. $(y-5)(y^2+5y+25)$

R.2. $(t+6)(t^2-6t+36)$

For Exercises R.3–R.6, fill in the blank.

R.3. $27p^3 = (\_\_)^3$

R.4. $64x^3 = (\_\_)^3$

R.5. $-125m^6 = (\_\_)^3$

R.6. $-1000y^9 = (\_\_)^3$

For Exercises R.7–R.14, factor completely.

R.7. $600 - 6x^2$

R.8. $20 - 5t^2$

R.9. $ax + bx + 5a + 5b$

R.10. $2t + 2u + st + su$

R.11. $5y^2 + 13y - 6$

R.12. $3v^2 + 5v - 12$

R.13. $-c^2 - 10c - 25$

R.14. $-z^2 + 6z - 9$

Vocabulary and Key Concepts

1. **a.** The binomial $x^3 + 27$ is an example of a _____ of _____.

   b. The binomial $c^3 - 8$ is an example of a _____ of _____.

   c. A difference of cubes $a^3 - b^3$ factors as ( )( ).

   d. A sum of cubes $a^3 + b^3$ factors as ( )( ).

2. Determine whether the binomial is a sum or difference of cubes.

   a. $125x^3 - 8x$

   b. $125a^3 - b^4$

   c. $125x^6 - 8$

   d. $125a^3 + 4b^6$

Concept 1: Factoring a Sum or Difference of Cubes

3. Identify the expressions that are perfect cubes:
   $x^3, 8, 9, y^6, a^4, b^2, 3p^3, 27q^3, w^{12}, r^3s^6$

4. Identify the expressions that are perfect cubes:
   $z^9, -81, 30, 8, 6x^3, y^{15}, 27a^3, b^2, p^3q^2, -1$

5. Identify the expressions that are perfect cubes:
   $36, t^3, -1, 27, a^3b^6, -9, 125, -8x^2, y^6, 25$

6. Identify the expressions that are perfect cubes:
   $343, 15b^3, z^3, w^{12}, -p^9, -1000, a^2b^3, 3x^3, -8, 60$

For Exercises 7–22, factor the sums or differences of cubes. **(See Examples 1–2.)**

7. $y^3 - 8$
8. $x^3 + 27$
9. $1 - p^3$
10. $q^3 + 1$
11. $w^3 + 64$
12. $8 - t^3$
13. $x^3 - 1000y^3$
14. $8r^3 - 27t^3$
15. $64t^3 + 1$
16. $125r^3 + 1$
17. $1000a^3 + 27$
18. $216b^3 - 125$
19. $n^3 - \dfrac{1}{8}$
20. $\dfrac{8}{27} + m^3$
21. $125x^3 + 8y^3$
22. $27t^3 + 64u^3$

Concept 2: Factoring Binomials: A Summary

For Exercises 23–58, factor completely. **(See Examples 3–6.)**

23. $x^4 - 4$
24. $b^4 - 25$
25. $a^2 + 9$
26. $w^2 + 36$
27. $t^3 + 64$
28. $u^3 + 27$
29. $g^3 - 4$
30. $h^3 - 25$
31. $4b^3 + 108$
32. $3c^3 - 24$
33. $5p^2 - 125$
34. $2q^4 - 8$
35. $\dfrac{1}{64} - 8h^3$
36. $\dfrac{1}{125} + k^6$
37. $x^4 - 16$
38. $p^4 - 81$
39. $\dfrac{4}{9}x^2 - w^2$
40. $\dfrac{16}{25}y^2 - x^2$
41. $q^6 - 64$
42. $a^6 - 1$

(*Hint:* Factor using the difference of squares first.)

43. $x^9 + 64y^3$
44. $125w^3 - z^9$
45. $2x^3 + 3x^2 - 2x - 3$
46. $3x^3 + x^2 - 12x - 4$
47. $16x^4 - y^4$
48. $1 - t^4$
49. $81y^4 - 16$
50. $u^5 - 256u$
51. $a^3 + b^6$
52. $u^6 - v^3$
53. $x^4 - y^4$
54. $a^4 - b^4$
55. $k^3 + 4k^2 - 9k - 36$
56. $w^3 - 2w^2 - 4w + 8$
57. $2t^3 - 10t^2 - 2t + 10$
58. $9a^3 + 27a^2 - 4a - 12$

Expanding Your Skills

For Exercises 59–62, factor completely.

59. $\dfrac{64}{125}p^3 - \dfrac{1}{8}q^3$
60. $\dfrac{1}{1000}r^3 + \dfrac{8}{27}s^3$
61. $a^{12} + b^{12}$
62. $a^9 - b^9$

Use Exercises 63–64 to investigate the relationship between division and factoring.

63. **a.** Use long division to divide $x^3 - 8$ by $(x - 2)$.
    b. Factor $x^3 - 8$.

64. **a.** Use long division to divide $y^3 + 27$ by $(y + 3)$.
    b. Factor $y^3 + 27$.

65. What trinomial multiplied by $(x - 4)$ gives a difference of cubes?

66. What trinomial multiplied by $(p + 5)$ gives a sum of cubes?

67. Write a binomial that when multiplied by $(4x^2 - 2x + 1)$ produces a sum of cubes.

68. Write a binomial that when multiplied by $(9y^2 + 15y + 25)$ produces a difference of cubes.

Problem Recognition Exercises

Factoring Strategy

> **Factoring Strategy**
> **Step 1** Factor out the GCF.
> **Step 2** Identify whether the polynomial has two terms, three terms, or more than three terms.
> **Step 3** If the polynomial has more than three terms, try factoring by grouping.
> **Step 4** If the polynomial has three terms, check first for a perfect square trinomial. Otherwise, factor the trinomial with the trial-and-error method or the ac-method.
> **Step 5** If the polynomial has two terms, determine if it fits the pattern for
> - A difference of squares: $a^2 - b^2 = (a - b)(a + b)$
> - A sum of squares: $a^2 + b^2$ is prime.
> - A difference of cubes: $a^3 - b^3 = (a - b)(a^2 + ab + b^2)$
> - A sum of cubes: $a^3 + b^3 = (a + b)(a^2 - ab + b^2)$
>
> **Step 6** Be sure to factor the polynomial completely.
> **Step 7** Check by multiplying.

1. What is meant by a prime polynomial?

2. What is the first step in factoring any polynomial?

3. When factoring a binomial, what patterns can you look for?

4. What technique should be considered when factoring a four-term polynomial?

For Exercises 5–73,

a. Factor out the GCF from each polynomial. Then identify the category in which the remaining polynomial best fits. Choose from
- difference of squares
- sum of squares
- difference of cubes
- sum of cubes
- trinomial (perfect square trinomial)
- trinomial (nonperfect square trinomial)
- four terms-grouping
- none of these

b. Factor the polynomial completely.

5. $2a^2 - 162$

6. $y^2 + 4y + 3$

7. $6w^2 - 6w$

8. $16z^4 - 81$

9. $3t^2 + 13t + 4$

10. $5r^3 + 5$

11. $3ac + ad - 3bc - bd$

12. $x^3 - 125$

13. $y^3 + 8$

14. $7p^2 - 29p + 4$

15. $3q^2 - 9q - 12$

16. $-2x^2 + 8x - 8$

17. $18a^2 + 12a$
18. $54 - 2y^3$
19. $4t^2 - 100$
20. $4t^2 - 31t - 8$
21. $10c^2 + 10c + 10$
22. $2xw - 10x + 3yw - 15y$
23. $x^3 + 0.001$
24. $4q^2 - 9$
25. $64 + 16k + k^2$
26. $s^2t + 5t + 6s^2 + 30$
27. $2x^2 + 2x - xy - y$
28. $w^3 + y^3$
29. $a^3 - c^3$
30. $3y^2 + y + 1$
31. $c^2 + 8c + 9$
32. $a^2 + 2a + 1$
33. $b^2 + 10b + 25$
34. $-t^2 - 4t + 32$
35. $-p^3 - 5p^2 - 4p$
36. $x^2y^2 - 49$
37. $6x^2 - 21x - 45$
38. $20y^2 - 14y + 2$
39. $5a^2bc^3 - 7abc^2$
40. $8a^2 - 50$
41. $t^2 + 2t - 63$
42. $b^2 + 2b - 80$
43. $ab + ay - b^2 - by$
44. $6x^3y^4 + 3x^2y^5$
45. $14u^2 - 11uv + 2v^2$
46. $9p^2 - 36pq + 4q^2$
47. $4q^2 - 8q - 6$
48. $9w^2 + 3w - 15$
49. $9m^2 + 16n^2$
50. $5b^2 - 30b + 45$
51. $6r^2 + 11r + 3$
52. $4s^2 + 4s - 15$
53. $16a^4 - 1$
54. $p^3 + p^2c - 9p - 9c$
55. $81u^2 - 90uv + 25v^2$
56. $4x^2 + 16$
57. $x^2 - 5x - 6$
58. $q^2 + q - 7$
59. $2ax - 6ay + 4bx - 12by$
60. $8m^3 - 10m^2 - 3m$
61. $21x^4y + 41x^3y + 10x^2y$
62. $2m^4 - 128$
63. $8uv - 6u + 12v - 9$
64. $4t^2 - 20t + st - 5s$
65. $12x^2 - 12x + 3$
66. $p^2 + 2pq + q^2$
67. $6n^3 + 5n^2 - 4n$
68. $4k^3 + 4k^2 - 3k$
69. $64 - y^2$
70. $36b - b^3$
71. $b^2 - 4b + 10$
72. $y^2 + 6y + 8$
73. $c^4 - 12c^2 + 20$

Solving Equations Using the Zero Product Rule

Section 14.7

1. Definition of a Quadratic Equation

We have already learned to solve linear equations in one variable. These are equations of the form $ax + b = c$ ($a \neq 0$). A linear equation in one variable is sometimes called a first-degree polynomial equation because the highest degree of all its terms is 1. A second-degree polynomial equation in one variable is called a quadratic equation.

Concepts

1. Definition of a Quadratic Equation
2. Zero Product Rule
3. Solving Equations by Factoring

> **A Quadratic Equation in One Variable**
> If a, b, and c are real numbers such that $a \neq 0$, then a **quadratic equation** is an equation that can be written in the form
> $$ax^2 + bx + c = 0$$

The following equations are quadratic because they can each be written in the form $ax^2 + bx + c = 0$, $(a \neq 0)$.

$$-4x^2 + 4x = 1 \qquad x(x-2) = 3 \qquad (x-4)(x+4) = 9$$
$$-4x^2 + 4x - 1 = 0 \qquad x^2 - 2x = 3 \qquad x^2 - 16 = 9$$
$$\qquad\qquad x^2 - 2x - 3 = 0 \qquad x^2 - 25 = 0$$
$$\qquad\qquad\qquad\qquad x^2 + 0x - 25 = 0$$

2. Zero Product Rule

One method for solving a quadratic equation is to factor and apply the zero product rule. The **zero product rule** states that if the product of two factors is zero, then one or both of its factors is zero.

> **Zero Product Rule**
> If $ab = 0$, then $a = 0$ or $b = 0$.

Example 1 Applying the Zero Product Rule

Solve the equation by using the zero product rule. $\quad (x-4)(x+3) = 0$

Solution:

$(x-4)(x+3) = 0$ Apply the zero product rule.

$x - 4 = 0 \quad$ or $\quad x + 3 = 0$ Set each factor equal to zero.

$x = 4 \quad$ or $\quad x = -3$ Solve each equation for x.

Check: $x = 4$ Check: $x = -3$

$(4-4)(4+3) \stackrel{?}{=} 0 \qquad (-3-4)(-3+3) \stackrel{?}{=} 0$

$\qquad (0)(7) \stackrel{?}{=} 0 \checkmark \qquad\qquad (-7)(0) \stackrel{?}{=} 0 \checkmark$

The solution set is $\{4, -3\}$.

Skill Practice Solve.
1. $(x+1)(x-8) = 0$

Answer
1. $\{-1, 8\}$

Example 2 Applying the Zero Product Rule

Solve the equation by using the zero product rule. $(x + 8)(4x + 1) = 0$

Solution:

$(x + 8)(4x + 1) = 0$ Apply the zero product rule.

$x + 8 = 0$ or $4x + 1 = 0$ Set each factor equal to zero.

$x = -8$ or $4x = -1$ Solve each equation for x.

$x = -8$ or $x = -\dfrac{1}{4}$ The solutions check in the original equation.

The solution set is $\left\{-8, -\dfrac{1}{4}\right\}$.

Skill Practice Solve.

2. $(4x - 5)(x + 6) = 0$

Example 3 Applying the Zero Product Rule

Solve the equation using the zero product rule. $x(3x - 7) = 0$

Solution:

$x(3x - 7) = 0$ Apply the zero product rule.

$x = 0$ or $3x - 7 = 0$ Set each factor equal to zero.

$x = 0$ or $3x = 7$ Solve each equation for x.

$x = 0$ or $x = \dfrac{7}{3}$ The solutions check in the original equation.

The solution set is $\left\{0, \dfrac{7}{3}\right\}$.

Skill Practice Solve.

3. $x(4x + 9) = 0$

3. Solving Equations by Factoring

Quadratic equations, like linear equations, arise in many applications in mathematics, science, and business. The following steps summarize the factoring method for solving a quadratic equation.

> **Solving a Quadratic Equation by Factoring**
>
> **Step 1** Write the equation in the form: $ax^2 + bx + c = 0$.
> **Step 2** Factor the quadratic expression completely.
> **Step 3** Apply the zero product rule. That is, set each factor equal to zero and solve the resulting equations.
>
> *Note:* The solution(s) found in step 3 may be checked by substitution in the original equation.

Answers

2. $\left\{\dfrac{5}{4}, -6\right\}$ 3. $\left\{0, -\dfrac{9}{4}\right\}$

Chapter 14 Factoring Polynomials

> **Example 4** Solving a Quadratic Equation
>
> Solve the quadratic equation. $2x^2 - 9x = 5$
>
> **Solution:**
>
> $2x^2 - 9x = 5$
>
> $2x^2 - 9x - 5 = 0$ Write the equation in the form $ax^2 + bx + c = 0$.
>
> $(2x + 1)(x - 5) = 0$ Factor the polynomial completely.
>
> $2x + 1 = 0$ or $x - 5 = 0$ Set each factor equal to zero.
>
> $2x = -1$ or $x = 5$ Solve each equation.
>
> $x = -\dfrac{1}{2}$ or $x = 5$
>
> Check: $x = -\dfrac{1}{2}$ Check: $x = 5$
>
> $2x^2 - 9x = 5$ $2x^2 - 9x = 5$
>
> $2\left(-\dfrac{1}{2}\right)^2 - 9\left(-\dfrac{1}{2}\right) \stackrel{?}{=} 5$ $2(5)^2 - 9(5) \stackrel{?}{=} 5$
>
> $2\left(\dfrac{1}{4}\right) + \dfrac{9}{2} \stackrel{?}{=} 5$ $2(25) - 45 \stackrel{?}{=} 5$
>
> $\dfrac{1}{2} + \dfrac{9}{2} \stackrel{?}{=} 5$ $50 - 45 \stackrel{?}{=} 5$ ✓
>
> $\dfrac{10}{2} \stackrel{?}{=} 5$ ✓
>
> The solution set is $\left\{-\dfrac{1}{2}, 5\right\}$.
>
> ---
>
> **Skill Practice** Solve the quadratic equation.
>
> **4.** $2y^2 + 19y = -24$

FOR REVIEW

The purpose of factoring a quadratic equation is to rewrite the quadratic equation as two linear equations. Recall that linear equations are *first-degree* equations.

$2x^2 - 9x - 5 = 0$
$(2x + 1)(x - 5) = 0$
$2x + 1 = 0 \quad x - 5 = 0$

two linear equations

> **Example 5** Solving a Quadratic Equation
>
> Solve the quadratic equation. $4x^2 + 24x = 0$
>
> **Solution:**
>
> $4x^2 + 24x = 0$ The equation is already in the form $ax^2 + bx + c = 0$. (Note that $c = 0$.)
>
> $4x(x + 6) = 0$ Factor completely.
>
> $4x = 0$ or $x + 6 = 0$ Set each factor equal to zero.
>
> $x = 0$ or $x = -6$ The solutions check in the original equation.
>
> The solution set is $\{0, -6\}$.
>
> ---
>
> **Skill Practice** Solve the quadratic equation.
>
> **5.** $5s^2 = 45$

Answers

4. $\left\{-8, -\dfrac{3}{2}\right\}$ **5.** $\{3, -3\}$

Example 6 Solving a Quadratic Equation

Solve the quadratic equation. $5x(5x + 2) = 10x + 9$

Solution:

$5x(5x + 2) = 10x + 9$

$25x^2 + 10x = 10x + 9$ Clear parentheses.

$25x^2 + 10x - 10x - 9 = 0$ Set the equation equal to zero.

$25x^2 - 9 = 0$ The equation is in the form $ax^2 + bx + c = 0$. (Note that $b = 0$.)

$(5x - 3)(5x + 3) = 0$ Factor completely.

$5x - 3 = 0$ or $5x + 3 = 0$ Set each factor equal to zero.

$5x = 3$ or $5x = -3$ Solve each equation.

$\dfrac{5x}{5} = \dfrac{3}{5}$ or $\dfrac{5x}{5} = \dfrac{-3}{5}$

$x = \dfrac{3}{5}$ or $x = -\dfrac{3}{5}$ The solutions check in the original equation.

The solution set is $\left\{\dfrac{3}{5}, -\dfrac{3}{5}\right\}$.

Skill Practice Solve the quadratic equation.

6. $4z(z + 3) = 4z + 5$

The zero product rule can be used to solve higher degree polynomial equations provided the equations can be set to zero and written in factored form.

Example 7 Solving a Higher Degree Polynomial Equation

Solve the equation. $-6(y + 3)(y - 5)(2y + 7) = 0$

Solution:

$-6(y + 3)(y - 5)(2y + 7) = 0$ The equation is already in factored form and equal to zero.

Set each factor equal to zero.

Solve each equation for y.

$-6 \neq 0$ or $y + 3 = 0$ or $y - 5 = 0$ or $2y + 7 = 0$

No solution, $y = -3$ or $y = 5$ or $y = -\dfrac{7}{2}$

Notice that when the constant factor is set equal to zero, the result is a contradiction, $-6 = 0$. The constant factor does not produce a solution to the equation. Therefore, the solution set is $\{-3, 5, -\tfrac{7}{2}\}$. Each solution can be checked in the original equation.

Answer

6. $\left\{-\dfrac{5}{2}, \dfrac{1}{2}\right\}$

Skill Practice Solve the equation.

7. $5(p - 4)(p + 7)(2p - 9) = 0$

Example 8   **Solving a Higher Degree Polynomial Equation**

Solve the equation. $w^3 + 5w^2 - 9w - 45 = 0$

Solution:

$w^3 + 5w^2 - 9w - 45 = 0$	This is a higher degree polynomial equation.
$w^3 + 5w^2 \mid -9w - 45 = 0$	The equation is already set equal to zero. Now factor.
$w^2(w + 5) - 9(w + 5) = 0$	Because there are four terms, try factoring by grouping.
$(w + 5)(w^2 - 9) = 0$	
$(w + 5)(w - 3)(w + 3) = 0$	$w^2 - 9$ is a difference of squares and can be factored further.
$w + 5 = 0$ or $w - 3 = 0$ or $w + 3 = 0$	Set each factor equal to zero.
$w = -5$ or $w = 3$ or $w = -3$	Solve each equation.

The solution set is $\{-5, 3, -3\}$. Each solution checks in the original equation.

Skill Practice Solve the equation.

8. $x^3 + 3x^2 - 4x - 12 = 0$

Answers

7. $\left\{4, -7, \dfrac{9}{2}\right\}$ 8. $\{-2, -3, 2\}$

Section 14.7 Activity

A.1. **a.** The equation $2x + 5 = 0$ is (choose one: linear or quadratic) because the term of highest degree is (choose one: 1 or 2).

    b. The equation $2x^2 + 5x - 12 = 0$ is (choose one: linear or quadratic) because the term of highest degree is (choose one: 1 or 2).

A.2. **a.** Given the equation $ab = 0$, what do you know about a or b?

    b. Given the equation $(x + 1)(x - 4) = 0$, then either _____ = 0 or _____ = 0.

    c. Write the solution set to the equation $(x + 1)(x - 4) = 0$.

A.3. Solve the equation $(2y - 5)(y + 7) = 0$ by following these steps.

    a. First determine whether one side of the equation is zero and the other side is factored.

    b. Set each factor equal to 0.

    c. Solve the individual equations from part (b) and write the solution set for the original equation.

    d. Do the solutions check in the original equation?

A.4. Solve the equation $4x^2 = 12x$ by following these steps.
  a. Write the equation with one side equal to zero and the other side factored.
  b. Set each factor equal to 0.
  c. Solve the individual equations from part (b) and write the solution set for the original equation.
  d. Do the solutions check in the original equation?

A.5. Solve the equation $2x(x - 10) = -50$ by following these steps.
  a. Write the equation with one side equal to zero and the other side factored.
  b. Set each factor equal to 0.
  c. Solve the individual equations from part (b) and write the solution set for the original equation.
  e. Does the solution check in the original equation?

For Exercises A.6–A.11,
  a. Identify the equation as linear, quadratic, or neither.
  b. Solve the equation.

A.6. $x(x - 1) = 12$

A.7. $10x - 18 = 5x + 12$

A.8. $2(x - 6) + 5x = 16$

A.9. $14x = 3 - 5x^2$

A.10. $2(x - 3)(2x + 1)(5 - x) = 0$

A.11. $50w^3 - 2w = 0$

Practice Exercises   Section 14.7

Prerequisite Review

For Exercises R.1–R.4, solve the equation.

R.1. $y + 9 = 0$

R.2. $w - 14 = 0$

R.3. $2x - 7 = 0$

R.4. $5x + 1 = 0$

For Exercises R.5–R.10, factor completely.

R.5. $x^2 - 3x - 70$

R.6. $y^2 + 16y + 55$

R.7. $2t^2 + 9t - 5$

R.8. $3w^2 - 14w - 24$

R.9. $3h^3 - 75h$

R.10. $2x^3 - 32x$

Vocabulary and Key Concepts

1. a. An equation that can be written in the form $ax^2 + bx + c = 0$, $(a \neq 0)$, is called a _____ equation.
   b. The zero product rule states that if $ab = 0$, then $a =$ _____ or $b =$ _____.

Concept 1: Definition of a Quadratic Equation

For Exercises 2–13, identify the equations as linear, quadratic, or neither.

2. $4 - 5x = 0$
3. $5x^3 + 2 = 0$
4. $3x - 6x^2 = 0$
5. $1 - x + 2x^2 = 0$
6. $7x^4 + 8 = 0$
7. $3x + 2 = 0$
8. $\dfrac{3}{x} + \dfrac{1}{2} = 2$
9. $4x - 3 = 3x - 2$
10. $2x = 9x - 14$
11. $(4x - 3)(3x - 2) = 0$
12. $2x(9x - 4) = 0$
13. $x^2(x - 4) = 0$

Concept 2: Zero Product Rule

For Exercises 14–22, solve each equation using the zero product rule. (See Examples 1–3.)

14. $(x - 5)(x + 1) = 0$
15. $(x + 3)(x - 1) = 0$
16. $(3x - 2)(3x + 2) = 0$
17. $(2x - 7)(2x + 7) = 0$
18. $2(x - 7)(x - 7) = 0$
19. $3(x + 5)(x + 5) = 0$
20. $(3x - 2)(2x - 3) = 0$
21. $x(5x - 1) = 0$
22. $x(3x + 8) = 0$

23. For a quadratic equation of the form $ax^2 + bx + c = 0$, what must be done before applying the zero product rule?

24. What are the requirements needed to use the zero product rule to solve a quadratic equation or higher degree polynomial equation?

Concept 3: Solving Equations by Factoring

For Exercises 25–72, solve each equation. (See Examples 4–8.)

25. $p^2 - 2p - 15 = 0$
26. $y^2 - 7y - 8 = 0$
27. $z^2 + 10z - 24 = 0$
28. $w^2 - 10w + 16 = 0$
29. $2q^2 - 7q = 4$
30. $4x^2 - 11x = 3$
31. $0 = 9x^2 - 4$
32. $4a^2 - 49 = 0$
33. $2k^2 - 28k + 96 = 0$
34. $0 = 2t^2 + 20t + 50$
35. $0 = 2m^3 - 5m^2 - 12m$
36. $3n^3 + 4n^2 + n = 0$
37. $5(3p + 1)(p - 3)(p + 6) = 0$
38. $4(2x - 1)(x - 10)(x + 7) = 0$
39. $x(x - 4)(2x + 3) = 0$
40. $x(3x + 1)(x + 1) = 0$
41. $-5x(2x + 9)(x - 11) = 0$
42. $-3x(x + 7)(3x - 5) = 0$
43. $x^3 - 16x = 0$
44. $t^3 - 36t = 0$
45. $3x^2 + 18x = 0$
46. $2y^2 - 20y = 0$
47. $16m^2 = 9$
48. $9n^2 = 1$
49. $2y^3 + 14y^2 = -20y$
50. $3d^3 - 6d^2 = 24d$
51. $5t - 2(t - 7) = 0$
52. $8h = 5(h - 9) + 6$
53. $2c(c - 8) = -30$
54. $3q(q - 3) = 12$
55. $b^3 = -4b^2 - 4b$
56. $x^3 + 36x = 12x^2$
57. $3(a^2 + 2a) = 2a^2 - 9$
58. $9(k - 1) = -4k^2$
59. $2n(n + 2) = 6$
60. $3p(p - 1) = 18$
61. $x(2x + 5) - 1 = 2x^2 + 3x + 2$
62. $3z(z - 2) - z = 3z^2 + 4$
63. $27q^2 = 9q$

64. $21w^2 = 14w$
65. $3(c^2 - 2c) = 0$
66. $2(4d^2 + d) = 0$
67. $y^3 - 3y^2 - 4y + 12 = 0$
68. $t^3 + 2t^2 - 16t - 32 = 0$
69. $(x - 1)(x + 2) = 18$
70. $(w + 5)(w - 3) = 20$
71. $(p + 2)(p + 3) = 1 - p$
72. $(k - 6)(k - 1) = -k - 2$

Problem Recognition Exercises

Polynomial Expressions versus Polynomial Equations

For Exercises 1–36, factor the expressions and solve the equations.

1. a. $x^2 + 6x - 7$
   b. $x^2 + 6x - 7 = 0$
2. a. $c^2 + 8c + 12$
   b. $c^2 + 8c + 12 = 0$
3. a. $2y^2 + 7y + 3$
   b. $2y^2 + 7y + 3 = 0$
4. a. $3x^2 - 8x + 5$
   b. $3x^2 - 8x + 5 = 0$
5. a. $5q^2 + q - 4 = 0$
   b. $5q^2 + q - 4$
6. a. $6a^2 - 7a - 3 = 0$
   b. $6a^2 - 7a - 3$
7. a. $a^2 - 64 = 0$
   b. $a^2 - 64$
8. a. $v^2 - 100 = 0$
   b. $v^2 - 100$
9. a. $4b^2 - 81$
   b. $4b^2 - 81 = 0$
10. a. $36t^2 - 49$
    b. $36t^2 - 49 = 0$
11. a. $8x^2 + 16x + 6 = 0$
    b. $8x^2 + 16x + 6$
12. a. $12y^2 + 40y + 32 = 0$
    b. $12y^2 + 40y + 32$
13. a. $x^3 - 8x^2 - 20x$
    b. $x^3 - 8x^2 - 20x = 0$
14. a. $k^3 + 5k^2 - 14k$
    b. $k^3 + 5k^2 - 14k = 0$
15. a. $b^3 + b^2 - 9b - 9 = 0$
    b. $b^3 + b^2 - 9b - 9$
16. a. $x^3 - 8x^2 - 4x + 32 = 0$
    b. $x^3 - 8x^2 - 4x + 32$
17. $2s^2 - 6s + rs - 3r$
18. $6t^2 + 3t + 10tu + 5u$
19. $8x^3 - 2x = 0$
20. $2b^3 - 50b = 0$
21. $2x^3 - 4x^2 + 2x = 0$
22. $3t^3 + 18t^2 + 27t = 0$
23. $7c^2 - 2c + 3 = 7(c^2 + c)$
24. $3z(2z + 4) = -7 + 6z^2$
25. $8w^3 + 27$
26. $1000q^3 - 1$
27. $5z^2 + 2z = 7$
28. $4h^2 + 25h = -6$
29. $3b(b + 6) = b - 10$
30. $3y^2 + 1 = y(y - 3)$
31. $5(2x - 3) - 2(3x + 1) = 4 - 3x$
32. $11 - 6a = -4(2a - 3) - 1$
33. $4s^2 = 64$
34. $81v^2 = 36$
35. $(x - 3)(x - 4) = 6$
36. $(x + 5)(x + 9) = 21$

Section 14.8 Applications of Quadratic Equations

Concepts

1. Applications of Quadratic Equations
2. Pythagorean Theorem

1. Applications of Quadratic Equations

In this section we solve applications using the Problem-Solving Strategies for Word Problems.

Example 1 — Translating to a Quadratic Equation

The product of two consecutive integers is 14 more than 6 times the smaller integer.

Solution:

Let x represent the first (smaller) integer.

Then $x + 1$ represents the second (larger) integer.   Label the variables.

(Smaller integer)(larger integer) = 6 · (smaller integer) + 14   Verbal model

$$x(x + 1) = 6(x) + 14 \quad \text{Algebraic equation}$$
$$x^2 + x = 6x + 14 \quad \text{Simplify.}$$
$$x^2 + x - 6x - 14 = 0 \quad \text{Set one side of the equation equal to zero.}$$
$$x^2 - 5x - 14 = 0$$
$$(x - 7)(x + 2) = 0 \quad \text{Factor.}$$
$$x - 7 = 0 \quad \text{or} \quad x + 2 = 0 \quad \text{Set each factor equal to zero.}$$
$$x = 7 \quad \text{or} \quad x = -2 \quad \text{Solve for } x.$$

Recall that x represents the smaller integer. Therefore, there are two possibilities for the pairs of consecutive integers.

If $x = 7$, then the larger integer is $x + 1$ or $7 + 1 = 8$.

If $x = -2$, then the larger integer is $x + 1$ or $-2 + 1 = -1$.

The integers are 7 and 8, or -2 and -1.

Skill Practice

1. The product of two consecutive odd integers is 9 more than 10 times the smaller integer. Find the pair of integers.

Example 2 — Using a Quadratic Equation in a Geometry Application

A rectangular sign has an area of 40 ft². If the width is 3 ft shorter than the length, what are the dimensions of the sign?

Solution:

Let x represent the length of the sign. Then $x - 3$ represents the width (Figure 14-1).

The problem gives information about the length of the sides and about the area. Therefore, we can form a relationship by using the formula for the area of a rectangle.

Label the variables.

Figure 14-1

Jill Braaten/McGraw-Hill

Answer

1. The integers are 9 and 11 or -1 and 1.

$A = l \cdot w$	Area equals length times width.
$40 = x(x - 3)$	Set up an algebraic equation.
$40 = x^2 - 3x$	Clear parentheses.
$0 = x^2 - 3x - 40$	Write the equation in the form, $ax^2 + bx + c = 0$.
$0 = (x - 8)(x + 5)$	Factor.
$0 = x - 8$ or $0 = x + 5$	Set each factor equal to zero.
$8 = x$ or $-5 \neq x$	Because x represents the length of a rectangle, reject the negative solution.

FOR REVIEW

Recall that the area A of a rectangle of length l and width w is $A = lw$.
The area A of a triangle with base b and height h is $A = \frac{1}{2}bh$.

The variable x represents the length of the sign. The length is 8 ft.

The expression $x - 3$ represents the width. The width is 8 ft − 3 ft, or 5 ft.

Skill Practice

2. The length of a rectangle is 5 ft more than the width. The area is 36 ft². Find the length and width.

Example 3 — Using a Quadratic Equation in an Application

A stone is dropped off a 64-ft cliff and falls into the ocean below. The height of the stone above sea level is given by the equation

$h = -16t^2 + 64$ where h is the stone's height in feet, and t is the time in seconds.

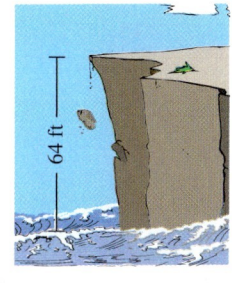

Find the time required for the stone to hit the water.

Solution:

When the stone hits the water, its height is zero. Therefore, substitute $h = 0$ into the equation.

$h = -16t^2 + 64$	The equation is quadratic.
$0 = -16t^2 + 64$	Substitute $h = 0$.
$0 = -16(t^2 - 4)$	Factor out the GCF.
$0 = -16(t - 2)(t + 2)$	Factor as a difference of squares.
$-16 \neq 0$ or $t - 2 = 0$ or $t + 2 = 0$	Set each factor to zero.
No solution, $t = 2$ or $t \neq -2$	Solve for t.

The negative value of t is rejected because the stone cannot fall for a negative time. Therefore, the stone hits the water after 2 sec.

Skill Practice

3. An object is launched into the air from the ground and its height is given by $h = -16t^2 + 144t$, where h is the height in feet after t seconds. Find the time required for the object to hit the ground.

Answers

2. The width is 4 ft, and the length is 9 ft.
3. The object hits the ground in 9 sec.

2. Pythagorean Theorem

Recall that a right triangle is a triangle that contains a 90° angle. Furthermore, the sum of the squares of the two legs (the shorter sides) of a right triangle equals the square of the hypotenuse (the longest side). This important fact is known as the Pythagorean theorem. The Pythagorean theorem is an enduring landmark of mathematical history from which many mathematical ideas have been built. Although the theorem is named after Pythagoras (sixth century B.C.E.), a Greek mathematician and philosopher, it is thought that the ancient Babylonians were familiar with the principle more than a thousand years earlier.

For the right triangle shown in Figure 14-2, the **Pythagorean theorem** is stated as:

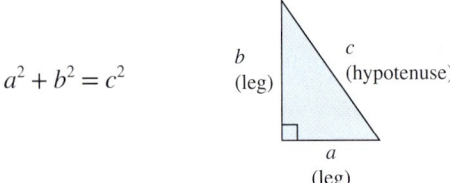

$$a^2 + b^2 = c^2$$

Figure 14-2

In this formula, a and b are the legs of the right triangle and c is the hypotenuse. Notice that the hypotenuse is the longest side of the right triangle and is opposite the 90° angle.

The triangle shown below is a right triangle. Notice that the lengths of the sides satisfy the Pythagorean theorem.

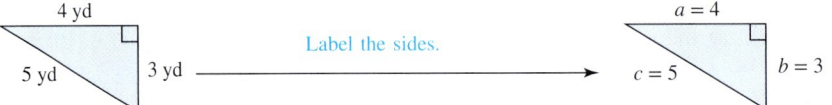

The shorter sides are labeled as a and b.

The longest side is the hypotenuse. It is always labeled as c.

$a^2 + b^2 = c^2$	Apply the Pythagorean theorem.
$(4)^2 + (3)^2 = (5)^2$	Substitute $a = 4$, $b = 3$, and $c = 5$.
$16 + 9 = 25$	
$25 = 25$ ✓	

Example 4   Applying the Pythagorean Theorem

Find the length of the missing side of the right triangle.

Solution:

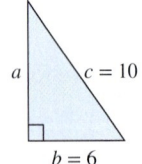

	Label the triangle.
$a^2 + b^2 = c^2$	Apply the Pythagorean theorem.
$a^2 + 6^2 = 10^2$	Substitute $b = 6$ and $c = 10$.
$a^2 + 36 = 100$	Simplify. The equation is quadratic.

$$a^2 + 36 - 100 = 100 - 100$$ Subtract 100 from both sides.

$$a^2 - 64 = 0$$ One side is now equal to zero.

$$(a+8)(a-8) = 0$$ Factor.

$$a + 8 = 0 \text{ or } a - 8 = 0$$ Set each factor equal to zero.

$a \cancel{=} -8$ or $a = 8$ Because x represents the length of a side of a triangle, reject the negative solution.

The third side is 8 ft.

Skill Practice

4. Find the length of the missing side.

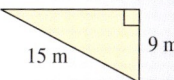

Example 5 — Using a Quadratic Equation in an Application

A 13-ft board is used as a ramp to unload furniture off a loading platform. If the distance between the top of the board and the ground is 7 ft less than the distance between the bottom of the board and the base of the platform, find both distances.

Solution:

Let x represent the distance between the bottom of the board and the base of the platform. Then $x - 7$ represents the distance between the top of the board and the ground (Figure 14-3).

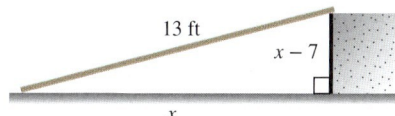

Figure 14-3

$$a^2 + b^2 = c^2$$ Pythagorean theorem

$$x^2 + (x-7)^2 = (13)^2$$

$$x^2 + (x)^2 - 2(x)(7) + (7)^2 = 169$$

$$x^2 + x^2 - 14x + 49 = 169$$

$$2x^2 - 14x + 49 = 169$$ Combine *like* terms.

$$2x^2 - 14x + 49 - 169 = 169 - 169$$ Set the equation equal to zero.

$$2x^2 - 14x - 120 = 0$$ Write the equation in the form $ax^2 + bx + c = 0$.

$$2(x^2 - 7x - 60) = 0$$ Factor.

$$2(x - 12)(x + 5) = 0$$

$2 \cancel{=} 0$ or $x - 12 = 0$ or $x + 5 = 0$ Set each factor equal to zero.

$x = 12$ or $x \cancel{=} -5$ Solve both equations for x.

> **Avoiding Mistakes**
>
> Recall that the square of a binomial results in a perfect square trinomial.
> $$(a - b)^2 = a^2 - 2ab + b^2$$
> $$(x - 7)^2 = x^2 - 2(x)(7) + 7^2$$
> $$= x^2 - 14x + 49$$
> Don't forget the middle term.

Answer

4. The length of the third side is 12 m.

Recall that x represents the distance between the bottom of the board and the base of the platform. We reject the negative value of x because a distance cannot be negative. Therefore, the distance between the bottom of the board and the base of the platform is 12 ft. The distance between the top of the board and the ground is $x - 7 = 5$ ft.

Skill Practice

5. A 5-yd ladder leans against a wall. The distance from the bottom of the wall to the top of the ladder is 1 yd more than the distance from the bottom of the wall to the bottom of the ladder. Find both distances.

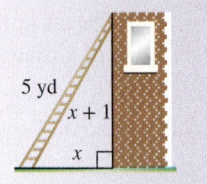

Answer

5. The distance along the wall to the top of the ladder is 4 yd. The distance on the ground from the ladder to the wall is 3 yd.

Section 14.8 Activity

A.1. The product of two consecutive odd integers is 23 more than their sum. Find the two integers by following these steps.
   a. Let x represent the first integer. Write an expression representing the second integer.
   b. Write an expression that represents the product of the two integers.
   c. Write an expression that represents 23 more than the sum of the two integers.
   d. Write an equation that indicates that the product of the two integers is 23 more than their sum.
   e. Solve the equation.
   f. Interpret your solution(s) and verify that the solutions meet the criteria of the problem.

A.2. The area of a parallelogram is 180 m² (square meters). If the base is 3 m more than the height, find the base and height of the parallelogram by following these steps.
   a. Let h represent the height of the parallelogram. Write an expression in terms of h that represents the base.
   b. Write an equation in terms of h that indicates that the area is 180 m².
   c. Solve the equation.
   d. The equation has two solutions. Is either solution unreasonable in the context of this problem?
   e. Interpret your solution and verify that the solution meets the criteria of the problem.

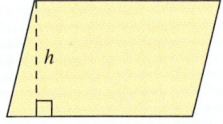

A.3. a. State the Pythagorean theorem for a right triangle with legs a and b, and hypotenuse c.
   b. Determine the length of the unknown side of the given right triangle.

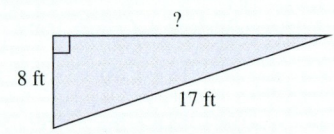

A.4. One leg of a right triangle is 24 cm. The hypotenuse is 4 cm more than 3 times the length of the other leg. Find the length of the unknown leg and the length of the hypotenuse by following these steps.
   a. Draw a right triangle and label the sides using variable expressions where necessary.
   b. Write an equation that relates the lengths of the sides.
   c. Solve the equation.
   d. Interpret your solution and verify that the solution meets the criteria of the problem.

A.5. A toy rocket is shot straight upward from ground level. The height h of the rocket (in feet) is given by the equation $h = -16t^2 + 128t$, where t is the number of seconds after launch. Find the time at which the rocket hits the ground by following these steps.

  a. What is the value of h when the rocket is at ground level?

  b. Substitue the value of h from part (a) into the equation $h = -16t^2 + 128t$ and solve the resulting quadratic equation for t.

  c. The rocket will hit the ground on the way down. Based on your answer from part (b), after how many seconds will the rocket hit the ground?

Practice Exercises — Section 14.8

Prerequisite Review

For Exercises R.1–R.8, solve the equation.

R.1. $(6x + 1)(x - 4) = 0$
R.2. $(5y - 2)(y + 8) = 0$
R.3. $4x^2 - 1 = 0$
R.4. $9w^2 - 49 = 0$
R.5. $m(m - 5) = 6$
R.6. $t(t - 1) = 12$
R.7. $3a(a - 3) = a + 8$
R.8. $2y(y + 6) = y - 14$

For Exercises R.9–R.14, write a mathematical expression for the given English phrase.

R.9. 8 less than x
R.10. y more than 7
R.11. Negative 9 added to the square of t
R.12. Twenty-five subtracted from the square of w
R.13. c squared subtracted from 6 squared
R.14. The difference of the square of p and the square of 3

Vocabulary and Key Concepts

1. **a.** If x is the smaller of two consecutive integers, then _____ represents the next greater integer.

  b. If x is the smaller of two consecutive odd integers, then _____ represents the next greater odd integer.

  c. If x is the smaller of two consecutive even integers, then _____ represents the next greater even integer.

2. **a.** The area of a rectangle of length L and width W is given by $A =$ _____.

  b. The area of a triangle with base b and height h is given by the formula $A =$ _____.

  c. Given a right triangle with legs a and b and hypotenuse c, the Pythagorean theorem is stated as _____.

Concept 1: Applications of Quadratic Equations

3. If eleven is added to the square of a number, the result is sixty. Find all such numbers.

4. If a number is added to two times its square, the result is thirty-six. Find all such numbers.

5. If twelve is added to six times a number, the result is twenty-eight less than the square of the number. Find all such numbers.

6. The square of a number is equal to twenty more than the number. Find all such numbers.

7. The product of two consecutive odd integers is sixty-three. Find all such integers. **(See Example 1.)**

8. The product of two consecutive even integers is forty-eight. Find all such integers.

9. The sum of the squares of two consecutive integers is sixty-one. Find all such integers.

10. The sum of the squares of two consecutive even integers is fifty-two. Find all such integers.

11. *Las Meninas* (Spanish for *The Maids of Honor*) is a famous painting by Spanish painter Diego Velázquez. This work is regarded as one of the most important paintings in Western art history. The height of the painting is approximately 2 ft more than its width. If the total area is 99 ft², determine the dimensions of the painting. (See Example 2.)

12. The width of a rectangular painting is 2 in. less than the length. The area is 120 in.² Find the length and width.

Simon Murrell/IT Stock Free/Alamy Stock Photo

13. The width of a rectangular slab of concrete is 3 m less than the length. The area is 28 m².

    a. What are the dimensions of the rectangle?

    b. What is the perimeter of the rectangle?

14. The width of a rectangular picture is 7 in. less than the length. The area of the picture is 78 in.²

    a. What are the dimensions of the picture?

    b. What is the perimeter of the picture?

15. The base of a triangle is 3 ft more than the height. If the area is 14 ft², find the base and the height.

16. The height of a triangle is 15 cm more than the base. If the area is 125 cm², find the base and the height.

17. The height of a triangle is 7 cm less than 3 times the base. If the area is 20 cm², find the base and the height.

18. The base of a triangle is 2 ft less than 4 times the height. If the area is 6 ft², find the base and the height.

19. In a physics experiment, a ball is dropped off a 144-ft platform. The height of the ball above the ground is given by the equation

    $h = -16t^2 + 144$ where h is the ball's height in feet, and t is the time in seconds after the ball is dropped ($t \geq 0$).

    Find the time required for the ball to hit the ground. (*Hint:* Let $h = 0$.) (See Example 3.)

20. A stone is dropped off a 256-ft cliff. The height of the stone above the ground is given by the equation

    $h = -16t^2 + 256$ where h is the stone's height in feet, and t is the time in seconds after the stone is dropped ($t \geq 0$).

    Find the time required for the stone to hit the ground.

21. An object is shot straight up into the air from ground level with an initial speed of 24 ft/sec. The height of the object (in feet) is given by the equation

    $h = -16t^2 + 24t$ where t is the time in seconds after launch ($t \geq 0$).

    Find the time(s) when the object is at ground level.

22. A rocket is launched straight up into the air from the ground with initial speed of 64 ft/sec. The height of the rocket (in feet) is given by the equation

    $h = -16t^2 + 64t$ where t is the time in seconds after launch ($t \geq 0$).

    Find the time(s) when the rocket is at ground level.

Concept 2: Pythagorean Theorem

23. Sketch a right triangle and label the sides with the words *leg* and *hypotenuse*.

24. State the Pythagorean theorem.

For Exercises 25–28, find the length of the missing side of the right triangle. (See Example 4.)

25.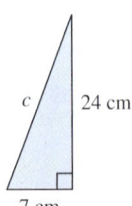

26.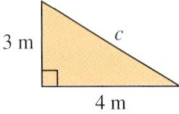

27.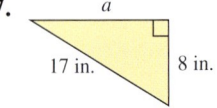

28.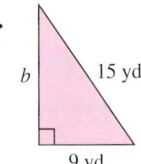

29. Find the length of the supporting brace.

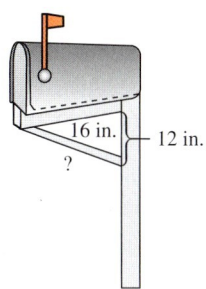

30. Find the height of the airplane above the ground.

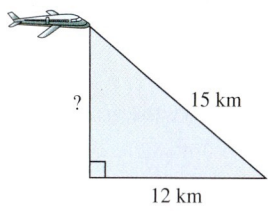

31. Darcy holds the end of a kite string 3 ft (1 yd) off the ground and wants to estimate the height of the kite. Her friend Jenna is 24 yd away from her, standing directly under the kite as shown in the figure. If Darcy has 30 yd of string out, find the height of the kite (ignore the sag in the string).

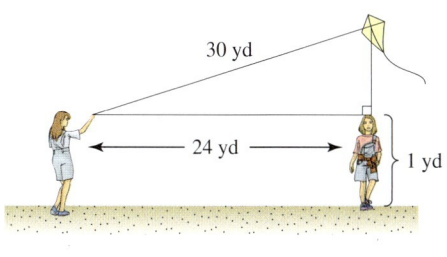

32. Two cars leave the same point at the same time, one traveling north and the other traveling east. After an hour, one car has traveled 48 mi and the other has traveled 64 mi. How many miles apart were they at that time?

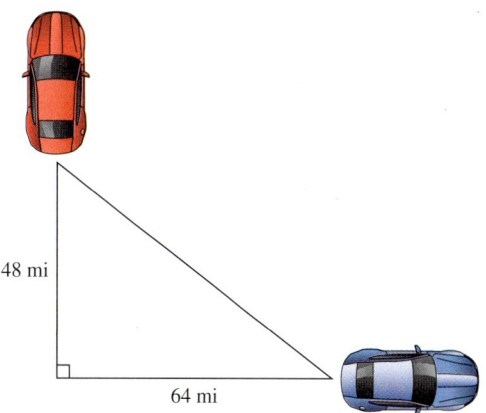

33. A 17-ft ladder rests against the side of a house. The distance between the top of the ladder and the ground is 7 ft more than the distance between the base of the ladder and the bottom of the house. Find both distances. (See Example 5.)

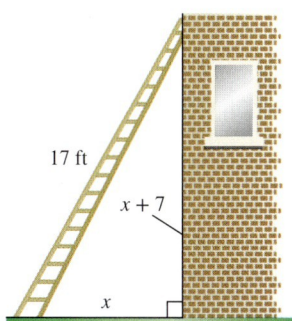

34. Two boats leave a marina. One travels east, and the other travels south. After 30 min, the second boat has traveled 1 mi farther than the first boat and the distance between the boats is 5 mi. Find the distance each boat traveled.

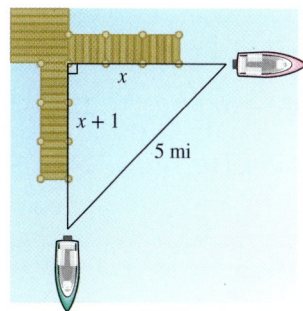

35. One leg of a right triangle is 4 m less than the hypotenuse. The other leg is 2 m less than the hypotenuse. Find the length of the hypotenuse.

36. The longer leg of a right triangle is 1 cm less than twice the shorter leg. The hypotenuse is 1 cm greater than twice the shorter leg. Find the length of the shorter leg.

Chapter 14 Review Exercises

Section 14.1

For Exercises 1–4, identify the greatest common factor for each group of terms.

1. $15a^2b^4, 30a^3b, 9a^5b^3$
2. $3(x+5), x(x+5)$
3. $2c^3(3c-5), 4c(3c-5)$
4. $-2wyz, -4xyz$

For Exercises 5–10, factor out the greatest common factor.

5. $6x^2 + 2x^4 - 8x$
6. $11w^3y^3 - 44w^2y^5$
7. $-t^2 + 5t$
8. $-6u^2 - u$
9. $3b(b+2) - 7(b+2)$
10. $2(5x+9) + 8x(5x+9)$

For Exercises 11–14, factor by grouping.

11. $7w^2 + 14w + wb + 2b$
12. $b^2 - 2b + yb - 2y$
13. $60y^2 - 45y - 12y + 9$
14. $6a - 3a^2 - 2ab + a^2b$

Section 14.2

For Exercises 15–24, factor completely.

15. $x^2 - 10x + 21$
16. $y^2 - 19y + 88$
17. $-6z + z^2 - 72$
18. $-39 + q^2 - 10q$
19. $3p^2w + 36pw + 60w$
20. $2m^4 + 26m^3 + 80m^2$
21. $-t^2 + 10t - 16$
22. $-w^2 - w + 20$
23. $a^2 + 12ab + 11b^2$
24. $c^2 - 3cd - 18d^2$

Section 14.3

For Exercises 25–28, assume that a, b, and c represent positive integers.

25. When factoring a polynomial of the form $ax^2 - bx - c$, should the signs of the binomials be both positive, both negative, or different?

26. When factoring a polynomial of the form $ax^2 - bx + c$, should the signs of the binomials be both positive, both negative, or different?

27. When factoring a polynomial of the form $ax^2 + bx + c$, should the signs of the binomials be both positive, both negative, or different?

28. When factoring a polynomial of the form $ax^2 + bx - c$, should the signs of the binomials be both positive, both negative, or different?

For Exercises 29–42, factor each trinomial using the trial-and-error method.

29. $2y^2 - 5y - 12$
30. $4w^2 - 5w - 6$
31. $10z^2 + 29z + 10$
32. $8z^2 + 6z - 9$
33. $2p^2 - 5p + 1$
34. $5r^2 - 3r + 7$
35. $10w^2 - 60w - 270$
36. $-3y^2 + 18y + 48$
37. $9c^2 - 30cd + 25d^2$
38. $x^2 + 12x + 36$
39. $6g^2 + 7gh + 2h^2$
40. $12m^2 - 32mn + 5n^2$
41. $v^4 - 2v^2 - 3$
42. $x^4 + 7x^2 + 10$

Section 14.4

For Exercises 43–44, find a pair of integers whose product and sum are given.

43. Product: -5 sum: 4
44. Product: 15 sum: -8

For Exercises 45–58, factor each trinomial using the ac-method.

45. $3c^2 - 5c - 2$
46. $4y^2 + 13y + 3$
47. $t^2 + 13tw + 12w^2$
48. $4x^4 + 17x^2 - 15$
49. $w^4 + 7w^2 + 10$
50. $p^2 - 8pq + 15q^2$
51. $-40v^2 - 22v + 6$
52. $40s^2 + 30s - 100$
53. $a^3b - 10a^2b^2 + 24ab^3$
54. $2z^6 + 8z^5 - 42z^4$

55. $m + 9m^2 - 2$ **56.** $2 + 6p^2 + 19p$

57. $49x^2 + 140x + 100$ **58.** $9w^2 - 6wz + z^2$

Section 14.5

For Exercises 59–60, write the formula to factor each binomial, if possible.

59. $a^2 - b^2$ **60.** $a^2 + b^2$

For Exercises 61–76, factor completely.

61. $a^2 - 49$ **62.** $d^2 - 64$

63. $100 - 81t^2$ **64.** $4 - 25k^2$

65. $x^2 + 16$ **66.** $y^2 + 121$

67. $y^2 + 12y + 36$ **68.** $t^2 + 16t + 64$

69. $9a^2 - 12a + 4$ **70.** $25x^2 - 40x + 16$

71. $-3v^2 - 12v - 12$ **72.** $-2x^2 + 20x - 50$

73. $2c^4 - 18$ **74.** $72x^2 - 2y^2$

75. $p^3 + 3p^2 - 16p - 48$ **76.** $4k - 8 - k^3 + 2k^2$

Section 14.6

For Exercises 77–78, write the formula to factor each binomial, if possible.

77. $a^3 + b^3$ **78.** $a^3 - b^3$

For Exercises 79–92, factor completely.

79. $64 + a^3$ **80.** $125 - b^3$

81. $p^6 + 8$ **82.** $q^6 - \dfrac{1}{27}$

83. $6x^3 - 48$ **84.** $7y^3 + 7$

85. $x^3 - 36x$ **86.** $q^4 - 64q$

87. $8h^2 + 20$ **88.** $m^2 - 8m$

89. $x^3 + 4x^2 - x - 4$ **90.** $5p^4q - 20q^3$

91. $8n + n^4$ **92.** $14m^3 - 14$

Section 14.7

93. For which of the following equations can the zero product rule be applied directly? Explain. $(x - 3)(2x + 1) = 0$ or $(x - 3)(2x + 1) = 6$

For Exercises 94–109, solve each equation using the zero product rule.

94. $(4x - 1)(3x + 2) = 0$

95. $(a - 9)(2a - 1) = 0$

96. $3w(w + 3)(5w + 2) = 0$

97. $6u(u - 7)(4u - 9) = 0$

98. $7k^2 - 9k - 10 = 0$

99. $4h^2 - 23h - 6 = 0$

100. $q^2 - 144 = 0$ **101.** $r^2 = 25$

102. $5v^2 - v = 0$ **103.** $x(x - 6) = -8$

104. $36t^2 + 60t = -25$ **105.** $9s^2 + 12s = -4$

106. $3(y^2 + 4) = 20y$ **107.** $2(p^2 - 66) = -13p$

108. $2y^3 - 18y^2 = -28y$ **109.** $x^3 - 4x = 0$

Section 14.8

110. The base of a parallelogram is 1 ft longer than twice the height. If the area is 78 ft², find the base and height of the parallelogram.

111. A ball is tossed into the air from ground level with initial speed of 16 ft/sec. The height of the ball is given by the equation

$$h = -16t^2 + 16t \quad (t \geq 0)$$

where h is the ball's height in feet, and t is the time in seconds.

Find the time(s) when the ball is at ground level.

112. Find the length of the ramp.

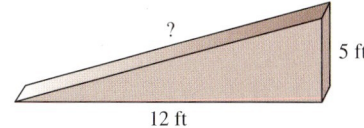

113. A right triangle has one leg that is 2 ft longer than the other leg. The hypotenuse is 2 ft less than twice the shorter leg. Find the lengths of all sides of the triangle.

114. If the square of a number is subtracted from 60, the result is −4. Find all such numbers.

115. The product of two consecutive integers is 44 more than 14 times their sum. Find all such integers.

116. The base of a triangle is 1 m longer than twice the height. If the area of the triangle is 18 m^2, find the base and height.

Rational Expressions and Equations

15

CHAPTER OUTLINE

- **15.1** Introduction to Rational Expressions 978
- **15.2** Multiplication and Division of Rational Expressions 989
- **15.3** Least Common Denominator 997
- **15.4** Addition and Subtraction of Rational Expressions 1004

  Problem Recognition Exercises: Operations on Rational Expressions 1015

- **15.5** Complex Fractions 1016
- **15.6** Rational Equations 1025

  Problem Recognition Exercises: Comparing Rational Equations and Rational Expressions 1037

- **15.7** Applications of Rational Equations and Proportions 1038

  Chapter 15 Review Exercises 1051

Mathematics in Purchasing Land

Suppose you want to purchase a parcel of land in a remote area. The current owner has a large amount of acreage that she plans to divide into n smaller lots for sale. Her total acreage is worth $120,000 and the smaller lots are of equal size. Thus, the cost (in dollars) of an individual lot is given by:

$$\frac{120{,}000}{n}$$

This fraction is called a **rational expression**, because it consists of a polynomial in the numerator and a nonzero polynomial in the denominator.

The cost to purchase an individual lot also includes closing costs. If closing costs are $5600, then the total cost (in dollars) to purchase one lot is given by:

$$\frac{120{,}000}{n} + 5600$$

AlinaMD/Shutterstock

This expression can be simplified by combining the two terms into a single rational expression. In this chapter we will learn how to perform operations on rational expressions and see numerous applications of their use.

Section 15.1 Introduction to Rational Expressions

Concepts

1. Definition of a Rational Expression
2. Evaluating Rational Expressions
3. Restricted Values of a Rational Expression
4. Simplifying Rational Expressions
5. Simplifying a Ratio of −1

1. Definition of a Rational Expression

We define a rational number as the ratio of two integers, $\frac{p}{q}$, where $q \neq 0$.

Examples of rational numbers: $\frac{2}{3}, -\frac{1}{5}, 9$

In a similar way, we define a **rational expression** as the ratio of two polynomials, $\frac{p}{q}$, where $q \neq 0$.

Examples of rational expressions: $\frac{3x-6}{x^2-4}, \frac{3}{4}, \frac{6r^5+2r}{7r^3}$

2. Evaluating Rational Expressions

Example 1 Evaluating Rational Expressions

Evaluate the rational expression (if possible) for the given values of x. $\dfrac{x}{x-3}$

a. $x = 6$ b. $x = -3$ c. $x = 0$ d. $x = 3$

Solution:

Substitute the given value for the variable. Use the order of operations to simplify.

a. $\dfrac{x}{x-3}$

$\dfrac{(6)}{(6)-3}$ Substitute $x = 6$.

$= \dfrac{6}{3}$

$= 2$

b. $\dfrac{x}{x-3}$

$\dfrac{(-3)}{(-3)-3}$ Substitute $x = -3$.

$= \dfrac{-3}{-6}$

$= \dfrac{1}{2}$

c. $\dfrac{x}{x-3}$

$\dfrac{(0)}{(0)-3}$ Substitute $x = 0$.

$= \dfrac{0}{-3}$

$= 0$

d. $\dfrac{x}{x-3}$

$\dfrac{(3)}{(3)-3}$ Substitute $x = 3$.

$= \dfrac{3}{0}$ Undefined.

Recall that division by zero is undefined.

Avoiding Mistakes

The numerator is 0 and the denominator is nonzero. Therefore, the fraction is equal to 0.

Avoiding Mistakes

The denominator is 0, therefore the fraction is undefined.

Skill Practice Evaluate the expression for the given values of x. $\dfrac{x-3}{x+5}$

1. $x = 2$ 2. $x = 0$ 3. $x = 3$ 4. $x = -5$

Answers

1. $-\dfrac{1}{7}$ 2. $-\dfrac{3}{5}$
3. 0   4. Undefined

3. Restricted Values of a Rational Expression

From Example 1 we see that not all values of x can be substituted into a rational expression. The values that make the denominator zero must be restricted.

The expression $\dfrac{x}{x-3}$ is undefined for $x = 3$, so we call $x = 3$ a restricted value.

Restricted values of a rational expression are all values that make the expression undefined, that is, make the denominator equal to zero.

> **Finding the Restricted Values of a Rational Expression**
> - Set the denominator equal to zero and solve the resulting equation.
> - The restricted values are the solutions to the equation.

Example 2 Finding the Restricted Values of Rational Expressions

Identify the restricted values for each expression.

a. $\dfrac{y-3}{2y+7}$ b. $\dfrac{-5}{x}$

Solution:

a. $\dfrac{y-3}{2y+7}$

$2y + 7 = 0$ Set the denominator equal to zero.
$2y = -7$ Solve the equation.
$\dfrac{2y}{2} = \dfrac{-7}{2}$
$y = -\dfrac{7}{2}$ The restricted value is $y = -\dfrac{7}{2}$.

b. $\dfrac{-5}{x}$

$x = 0$ Set the denominator equal to zero.
The restricted value is $x = 0$.

Skill Practice Identify the restricted values.

5. $\dfrac{a+2}{2a-8}$ 6. $\dfrac{2}{t}$

Answers
5. $a = 4$ 6. $t = 0$

Example 3 Finding the Restricted Values of Rational Expressions

Identify the restricted values for each expression.

a. $\dfrac{a + 10}{a^2 - 25}$ b. $\dfrac{2x^3 + 5}{x^2 + 9}$

Solution:

a. $\dfrac{a + 10}{a^2 - 25}$

$a^2 - 25 = 0$ Set the denominator equal to zero. The equation is quadratic.

$(a - 5)(a + 5) = 0$ Factor.

$a - 5 = 0$ or $a + 5 = 0$ Set each factor equal to zero.

$a = 5$ or $a = -5$

The restricted values are $a = 5$ and $a = -5$.

b. $\dfrac{2x^3 + 5}{x^2 + 9}$

$x^2 + 9 = 0$

$x^2 = -9$

The quantity x^2 cannot be negative for any real number, x. The denominator $x^2 + 9$ is the sum of two nonnegative values, and thus cannot equal zero. Therefore, there are no restricted values.

FOR REVIEW

Recall that the sum of squares is not factorable. The expression $x^2 + 9$ is a prime polynomial.

Skill Practice Identify the restricted values.

7. $\dfrac{w - 4}{w^2 - 9}$ 8. $\dfrac{8}{z^4 + 1}$

4. Simplifying Rational Expressions

In many cases, it is advantageous to simplify or reduce a fraction to lowest terms. The same is true for rational expressions.

The method for simplifying rational expressions mirrors the process for simplifying fractions. In each case, factor the numerator and denominator. Common factors in the numerator and denominator form a ratio of 1 and can be reduced.

Simplifying a fraction: $\dfrac{21}{35} \xrightarrow{\text{Factor}} \dfrac{3 \cdot 7}{5 \cdot 7} = \dfrac{3}{5} \cdot \dfrac{\cancel{7}}{\cancel{7}} = \dfrac{3}{5} \cdot (1) = \dfrac{3}{5}$

Simplifying a rational expression: $\dfrac{2x - 6}{x^2 - 9} \xrightarrow{\text{Factor}} \dfrac{2(x - 3)}{(x + 3)(x - 3)} = \dfrac{2}{(x + 3)} \cdot \dfrac{\cancel{(x - 3)}}{\cancel{(x - 3)}} = \dfrac{2}{(x + 3)} (1)$

$= \dfrac{2}{x + 3}$

Informally, to simplify a rational expression, we simplify the ratio of common factors to 1. Formally, this is accomplished by applying the fundamental principle of rational expressions.

Answers

7. $w = 3, w = -3$
8. There are no restricted values.

Section 15.1 Introduction to Rational Expressions

Fundamental Principle of Rational Expressions

Let p, q, and r represent polynomials where $q \neq 0$ and $r \neq 0$. Then

$$\frac{pr}{qr} = \frac{p}{q} \cdot \frac{r}{r} = \frac{p}{q} \cdot 1 = \frac{p}{q}$$

TIP: In practice we often shorten the process to reduce a rational expression by dividing out common factors.

$$\frac{p \cdot \cancel{r}}{q \cdot \cancel{r}} = \frac{p}{q}$$

Example 4 — Simplifying a Rational Expression

Given the expression $\dfrac{2p - 14}{p^2 - 49}$

a. Factor the numerator and denominator.

b. Identify the restricted values.

c. Simplify the rational expression.

Solution:

a. $\dfrac{2p - 14}{p^2 - 49}$ Factor out the GCF in the numerator.

$= \dfrac{2(p - 7)}{(p + 7)(p - 7)}$ Factor the denominator as a difference of squares.

b. $(p + 7)(p - 7) = 0$ To find the restricted values, set the denominator equal to zero. The equation is quadratic.

$p + 7 = 0$ or $p - 7 = 0$ Set each factor equal to 0.

$p = -7$ or $p = 7$ The restricted values are -7 and 7.

Avoiding Mistakes
The restricted values of a rational expression are always determined *before* simplifying the expression.

c. $\dfrac{2(\cancel{p - 7})}{(p + 7)(\cancel{p - 7})}$ Simplify the ratio of common factors to 1.

$= \dfrac{2}{p + 7}$ (provided $p \neq 7$ and $p \neq -7$)

Skill Practice Given $\dfrac{5z + 25}{z^2 + 3z - 10}$

9. Factor the numerator and the denominator.
10. Identify the restricted values.
11. Simplify the rational expression.

In Example 4, it is important to note that the expressions

$$\frac{2p - 14}{p^2 - 49} \quad \text{and} \quad \frac{2}{p + 7}$$

are equal for all values of p that make each expression a real number. Therefore,

$$\frac{2p - 14}{p^2 - 49} = \frac{2}{p + 7}$$

for all values of p except $p = 7$ and $p = -7$. (At $p = 7$ and $p = -7$, the original expression is undefined.) This is why the restricted values are determined before the expression is simplified.

Answers
9. $\dfrac{5(z + 5)}{(z + 5)(z - 2)}$
10. $z = -5, z = 2$
11. $\dfrac{5}{z - 2}$ ($z \neq 2, z \neq -5$)

982 Chapter 15 Rational Expressions and Equations

From this point forward, we will write statements of equality between two rational expressions with the assumption that they are equal for all values of the variable for which each expression is defined.

Example 5 Simplifying a Rational Expression

Simplify the rational expression. $\dfrac{18a^4}{9a^5}$

Solution:

$\dfrac{18a^4}{9a^5}$

$= \dfrac{2 \cdot 3 \cdot 3 \cdot a \cdot a \cdot a \cdot a}{3 \cdot 3 \cdot a \cdot a \cdot a \cdot a \cdot a}$ Factor the numerator and denominator.

$= \dfrac{2 \cdot (3 \cdot 3 \cdot a \cdot a \cdot a \cdot a)}{(3 \cdot 3 \cdot a \cdot a \cdot a \cdot a) \cdot a}$ Simplify common factors.

$= \dfrac{2}{a}$

> **FOR REVIEW**
>
> The expression in Example 5 can be simplified using the properties of exponents.
>
> • $\dfrac{b^m}{b^n} = b^{m-n}$
>
> • $b^{-n} = \dfrac{1}{b^n}$
>
> Thus,
>
> $\dfrac{18a^4}{9a^5} = 2a^{4-5} = 2a^{-1} = \dfrac{2}{a}$

Skill Practice Simplify the rational expression.

12. $\dfrac{15q^3}{9q^2}$

Example 6 Simplifying a Rational Expression

Simplify the rational expression. $\dfrac{2c - 8}{10c^2 - 80c + 160}$

Solution:

$\dfrac{2c - 8}{10c^2 - 80c + 160}$

$= \dfrac{2(c - 4)}{10(c^2 - 8c + 16)}$ Factor out the GCF.

$= \dfrac{2(c - 4)}{10(c - 4)^2}$ Factor the denominator.

$= \dfrac{2(c - 4)}{2 \cdot 5(c - 4)(c - 4)}$ Simplify the ratio of common factors to 1.

$= \dfrac{1}{5(c - 4)}$

> **Avoiding Mistakes**
>
> Given the expression
>
> $\dfrac{2c - 8}{10c^2 - 80c + 160}$
>
> do not be tempted to reduce before factoring. The terms $2c$ and $10c^2$ cannot be "canceled" because they are *terms*, not factors.
> The numerator and denominator must be in factored form before simplifying.

Skill Practice Simplify the rational expression.

13. $\dfrac{x^2 - 1}{2x^2 - x - 3}$

Answers

12. $\dfrac{5q}{3}$ 13. $\dfrac{x - 1}{2x - 3}$

The process to simplify a rational expression is based on the identity property of multiplication. Therefore, this process applies only to factors (remember that factors are multiplied). For example:

$$\frac{3x}{3y} = \frac{3 \cdot x}{3 \cdot y} = \underset{\text{Simplify}}{\frac{3}{3}} \cdot \frac{x}{y} = 1 \cdot \frac{x}{y} = \frac{x}{y}$$

Terms that are added or subtracted cannot be reduced to lowest terms. For example:

$$\underset{\text{Cannot be simplified}}{\frac{x+3}{y+3}}$$

The objective of simplifying a rational expression is to create an equivalent expression that is simpler to use. Consider the rational expression from Example 6 in its original form and in its reduced form. If we substitute a value c into each expression, we see that the reduced form is easier to evaluate. For example, substitute $c = 3$:

Original Expression   **Simplified Expression**

$$\frac{2c - 8}{10c^2 - 80c + 160} \qquad \frac{1}{5(c - 4)}$$

Substitute $c = 3$

$$= \frac{2(3) - 8}{10(3)^2 - 80(3) + 160} \qquad = \frac{1}{5(3 - 4)}$$

$$= \frac{6 - 8}{10(9) - 240 + 160} \qquad = \frac{1}{5(-1)}$$

$$= \frac{-2}{90 - 240 + 160} \qquad = -\frac{1}{5}$$

$$= \frac{-2}{10} \quad \text{or} \quad -\frac{1}{5}$$

5. Simplifying a Ratio of −1

When two factors are identical in the numerator and denominator, they form a ratio of 1 and can be reduced. Sometimes we encounter two factors that are *opposites* and form a ratio of −1. For example:

Simplified Form   **Details/Notes**

$\dfrac{-5}{5} = -1$ The ratio of a number and its opposite is −1.

$\dfrac{100}{-100} = -1$ The ratio of a number and its opposite is −1.

$\dfrac{x + 7}{-x - 7} = -1 \qquad \dfrac{x + 7}{-x - 7} = \dfrac{x + 7}{-1(x + 7)} = \dfrac{\overset{1}{\cancel{x+7}}}{-1(\cancel{x+7})} = \dfrac{1}{-1} = -1$

factor out −1

$\dfrac{2 - x}{x - 2} = -1 \qquad \dfrac{2 - x}{x - 2} = \dfrac{-1(-2 + x)}{x - 2} = \dfrac{-1(\cancel{x - 2})}{\cancel{x - 2}} = \dfrac{-1}{1} = -1$

Avoiding Mistakes

While the expression $2 - x$ and $x - 2$ are opposites, the expressions $2 - x$ and $2 + x$ are *not*.

Therefore, $\dfrac{2 - x}{2 + x}$ does not simplify to −1.

Recognizing factors that are opposites is useful when simplifying rational expressions.

Example 7 — Simplifying a Rational Expression

Simplify the rational expression. $\dfrac{3c - 3d}{d - c}$

Solution:

$$\dfrac{3c - 3d}{d - c}$$

$$= \dfrac{3(c - d)}{d - c} \qquad \text{Factor the numerator and denominator.}$$

Notice that $(c - d)$ and $(d - c)$ are opposites and form a ratio of -1.

$$= \dfrac{3(c - d)^{-1}}{d - c} \qquad \text{Details: } \dfrac{3(c - d)}{d - c} = \dfrac{3(-1)(-c + d)}{d - c} = \dfrac{-3(d - c)}{d - c} = -3$$

$$= 3(-1)$$

$$= -3$$

Skill Practice Simplify the rational expression.

14. $\dfrac{2t - 12}{6 - t}$

TIP: It is important to recognize that a rational expression can be written in several equivalent forms. In particular, two numbers with opposite signs form a negative quotient. Therefore, a number such as $-\tfrac{3}{4}$ can be written as:

$$-\dfrac{3}{4} \quad \text{or} \quad \dfrac{-3}{4} \quad \text{or} \quad \dfrac{3}{-4}$$

The negative sign can be written in the numerator, in the denominator, or out in front of the fraction. We demonstrate this concept in Example 8.

Example 8 — Simplifying a Rational Expression

Simplify the rational expression. $\dfrac{5 - y}{y^2 - 25}$

Solution:

$$\dfrac{5 - y}{y^2 - 25}$$

$$= \dfrac{5 - y}{(y - 5)(y + 5)} \qquad \text{Factor the numerator and denominator.}$$

Notice that $5 - y$ and $y - 5$ are opposites and form a ratio of -1.

Answer

14. -2

$$= \frac{\overset{-1}{5-y}}{(y-5)(y+5)} \qquad \text{Details: } \frac{5-y}{(y-5)(y+5)} = \frac{-1(-5+y)}{(y-5)(y+5)}$$

$$= \frac{-1(y-5)}{(y-5)(y+5)} = \frac{-1}{y+5}$$

$$= \frac{-1}{y+5} \quad \text{or} \quad \frac{1}{-(y+5)} \quad \text{or} \quad -\frac{1}{y+5}$$

Skill Practice Simplify the rational expression.

15. $\dfrac{b-a}{a^2-b^2}$

Answer

15. $-\dfrac{1}{a+b}$

Section 15.1 Activity

A.1. Explain why x cannot be 5 in the expression $\dfrac{x+2}{x-5}$.

A.2. Given $\dfrac{5d+15}{d^2-d-12}$,

   a. Factor the numerator and denominator.

   b. Identify the restricted values of d.

   c. Simplify the expression.

For Exercises A.3–A.4, simplify the expression and identify the restricted values of the variables.

A.3. $\dfrac{15z^3}{5z^5}$

A.4. $\dfrac{t^2-16}{t^2+7t+12}$

A.5. a. Write the opposite of $x-12$.

   b. Simplify $\dfrac{x-12}{12-x}$.

A.6. a. Write the opposite of $-2b-5c$.

   b. Simplify $\dfrac{2b+5c}{-2b-5c}$.

A.7. Simplify the expression and identify the restricted values of the variable.

$$\dfrac{9-w^2}{2w^2-12w+18}$$

Practice Exercises

Section 15.1

Study Skills Exercise

As the content in this course grows in complexity, it is important to remind yourself of your long-term goals. Times of challenge are great opportunities to strengthen your character. Strategies that will help you persist include self-awareness of your math skills, advocating for yourself when you need help, and putting aside fears about making mistakes, viewing them instead as learning opportunities. With persistence you will find success.

Prerequisite Review

For Exercises R.1–R.2, factor the number into a product of prime factors.

R.1. 252   **R.2.** 675

For Exercises R.3–R.12, simplify the expression.

R.3. $\dfrac{42}{120}$ **R.4.** $\dfrac{150}{360}$

R.5. $-\dfrac{108}{24}$ **R.6.** $-\dfrac{147}{63}$

R.7. $\dfrac{-25}{25}$ **R.8.** $\dfrac{32}{-32}$

R.9. $\dfrac{a^{11}}{a^3}$ **R.10.** $\dfrac{b^{13}}{b^{10}}$

R.11. $\dfrac{x^5 \cdot x^4}{x^{10}}$ **R.12.** $\dfrac{y^3 \cdot y^9}{y^{16}}$

For Exercises R.13–R.24, factor completely.

R.13. $105x^3 - 63x^2$ **R.14.** $45y^2 + 60y$

R.15. $25y^2 - 4$ **R.16.** $64z^2 - 9w^2$

R.17. $a^2 - 4a - 21$ **R.18.** $t^2 + 5t - 24$

R.19. $10x^2 - 11x + 3$ **R.20.** $8x^2 - 2x - 15$

R.21. $6ac + 8a - 15bc - 20b$ **R.22.** $18x^2 + 48x + 3xy + 8y$

R.23. $9x^2 - 30xy + 25y^2$ **R.24.** $4c^2 + 28cd + 49d^2$

Vocabulary and Key Concepts

1. a. A _____ expression is the ratio of two polynomials, $\dfrac{p}{q}$, where $q \neq 0$.

b. Restricted values of a rational expression are all values that make the _____ equal to _____.

c. For polynomials p, q, and r, where ($q \neq 0$ and $r \neq 0$), $\dfrac{pr}{qr} = $ _____.

d. The ratio $\dfrac{a-b}{a-b} = $ _____ whereas the ratio $\dfrac{a-b}{b-a} = $ _____ provided that $a \neq b$.

Concept 2: Evaluating Rational Expressions

For Exercises 2–10, substitute the given number into the expression and simplify (if possible). **(See Example 1.)**

2. $\dfrac{5}{y-4}$; $y = 6$ **3.** $\dfrac{t-2}{t^2 - 4t + 8}$; $t = 2$ **4.** $\dfrac{4x}{x-7}$; $x = 8$

5. $\dfrac{1}{x-6}$; $x = -2$ **6.** $\dfrac{w-10}{w+6}$; $w = 0$ **7.** $\dfrac{w-4}{2w+8}$; $w = 0$

8. $\dfrac{y-8}{2y^2 + y - 1}$; $y = 8$ **9.** $\dfrac{(a-7)(a+1)}{(a-2)(a+5)}$; $a = 2$ **10.** $\dfrac{(a+4)(a+1)}{(a-4)(a-1)}$; $a = 1$

11. A bicyclist rides 24 mi against a wind and returns 24 mi with the same wind. His average speed for the return trip traveling with the wind is 8 mph faster than his speed going out against the wind. If x represents the bicyclist's speed going out against the wind, then the total time, t, required for the round trip is given by

$$t = \frac{24}{x} + \frac{24}{x+8}$$ where t is measured in hours.

LittlePerfectStock/Shutterstock

a. Find the time required for the round trip if the cyclist rides 12 mph against the wind.

b. Find the time required for the round trip if the cyclist rides 24 mph against the wind.

12. The manufacturer of mountain bikes has a fixed cost of $56,000, plus a variable cost of $140 per bike. The average cost per bike, y (in dollars), is given by the equation:

$$y = \frac{56{,}000 + 140x}{x}$$ where x represents the number of bikes produced.

a. Find the average cost per bike if the manufacturer produces 1000 bikes.

b. Find the average cost per bike if the manufacturer produces 2000 bikes.

c. Find the average cost per bike if the manufacturer produces 10,000 bikes.

Jeff Maloney/Photodisc/Getty Images

Concept 3: Restricted Values of a Rational Expression

For Exercises 13–24, identify the restricted values. (See Examples 2–3.)

13. $\dfrac{5}{k+2}$

14. $\dfrac{-3}{h-4}$

15. $\dfrac{x+5}{(2x-5)(x+8)}$

16. $\dfrac{4y+1}{(3y+7)(y+3)}$

17. $\dfrac{m+12}{m^2+5m+6}$

18. $\dfrac{c-11}{c^2-5c-6}$

19. $\dfrac{x-4}{x^2+9}$

20. $\dfrac{x+1}{x^2+4}$

21. $\dfrac{y^2-y-12}{12}$

22. $\dfrac{z^2+10z+9}{9}$

23. $\dfrac{t-5}{t}$

24. $\dfrac{2w+7}{w}$

25. Construct a rational expression that is undefined for $x = 2$. (Answers will vary.)

26. Construct a rational expression that is undefined for $x = 5$. (Answers will vary.)

27. Construct a rational expression that is undefined for $x = -3$ and $x = 7$. (Answers will vary.)

28. Construct a rational expression that is undefined for $x = -1$ and $x = 4$. (Answers will vary.)

29. Evaluate the expressions for $x = -1$.

   a. $\dfrac{3x^2 - 2x - 1}{6x^2 - 7x - 3}$
   b. $\dfrac{x-1}{2x-3}$

30. Evaluate the expressions for $x = 4$.

   a. $\dfrac{(x+5)^2}{x^2+6x+5}$
   b. $\dfrac{x+5}{x+1}$

31. Evaluate the expressions for $x = 1$.

   a. $\dfrac{5x+5}{x^2-1}$
   b. $\dfrac{5}{x-1}$

32. Evaluate the expressions for $x = -3$.

   a. $\dfrac{2x^2-4x-6}{2x^2-18}$
   b. $\dfrac{x+1}{x+3}$

Concept 4: Simplifying Rational Expressions

For Exercises 33–42,

    a. Identify the restricted values.   **b.** Simplify the rational expression. (See Example 4.)

33. $\dfrac{3y+6}{6y+12}$

34. $\dfrac{8x-8}{4x-4}$

35. $\dfrac{t^2-1}{t+1}$

36. $\dfrac{r^2-4}{r-2}$

37. $\dfrac{7w}{21w^2-35w}$

38. $\dfrac{12a^2}{24a^2-18a}$

39. $\dfrac{9x^2-4}{6x+4}$

40. $\dfrac{8n-20}{4n^2-25}$

41. $\dfrac{a^2+3a-10}{a^2+a-6}$

42. $\dfrac{t^2+3t-10}{t^2+t-20}$

For Exercises 43–72, simplify the rational expression. (See Examples 5–6.)

43. $\dfrac{7b^2}{21b}$

44. $\dfrac{15c^3}{3c^5}$

45. $\dfrac{-24x^2y^5z}{8xy^4z^3}$

46. $\dfrac{60rs^4t^2}{-12r^4s^2t^3}$

47. $\dfrac{(p-3)(p+5)}{(p+5)(p+4)}$

48. $\dfrac{(c+4)(c-1)}{(c+4)(c+2)}$

49. $\dfrac{m+11}{4(m+11)(m-11)}$

50. $\dfrac{n-7}{9(n+2)(n-7)}$

51. $\dfrac{x(2x+1)^2}{4x^3(2x+1)}$

52. $\dfrac{(p+2)(p-3)^4}{(p+2)^2(p-3)^2}$

53. $\dfrac{5}{20a-25}$

54. $\dfrac{7}{14c-21}$

55. $\dfrac{3x^2-6x}{9xy+18x}$

56. $\dfrac{6p^2+12p}{2pq-4p}$

57. $\dfrac{2x+4}{x^2-3x-10}$

58. $\dfrac{5z+15}{z^2-4z-21}$

59. $\dfrac{a^2-49}{a-7}$

60. $\dfrac{b^2-64}{b-8}$

61. $\dfrac{q^2+25}{q+5}$

62. $\dfrac{r^2+36}{r+6}$

63. $\dfrac{y^2+6y+9}{2y^2+y-15}$

64. $\dfrac{h^2+h-6}{h^2+2h-8}$

65. $\dfrac{5q^2+5}{q^4-1}$

66. $\dfrac{4t^2+16}{t^4-16}$

67. $\dfrac{ac-ad+2bc-2bd}{2ac+ad+4bc+2bd}$ (*Hint:* Factor by grouping.)

68. $\dfrac{3pr-ps-3qr+qs}{3pr-ps+3qr-qs}$ (*Hint:* Factor by grouping.)

69. $\dfrac{5x^3+4x^2-45x-36}{x^2-9}$

70. $\dfrac{x^2-1}{ax^3-bx^2-ax+b}$

71. $\dfrac{2x^2-xy-3y^2}{2x^2-11xy+12y^2}$

72. $\dfrac{2c^2+cd-d^2}{5c^2+3cd-2d^2}$

Concept 5: Simplifying a Ratio of −1

73. What is the relationship between $x-2$ and $2-x$?

74. What is the relationship between $w+p$ and $-w-p$?

For Exercises 75–86, simplify the rational expressions. (See Examples 7–8.)

75. $\dfrac{x-5}{5-x}$

76. $\dfrac{8-p}{p-8}$

77. $\dfrac{-4-y}{4+y}$

78. $\dfrac{z+10}{-z-10}$

79. $\dfrac{3y-6}{12-6y}$

80. $\dfrac{4q-4}{12-12q}$

81. $\dfrac{k+5}{5-k}$

82. $\dfrac{2+n}{2-n}$

83. $\dfrac{10x-12}{10x+12}$

84. $\dfrac{4t-16}{16+4t}$

85. $\dfrac{x^2-x-12}{16-x^2}$

86. $\dfrac{49-b^2}{b^2-10b+21}$

Mixed Exercises

For Exercises 87–100, simplify the rational expressions.

87. $\dfrac{3x^2 + 7x - 6}{x^2 + 7x + 12}$

88. $\dfrac{y^2 - 5y - 14}{2y^2 - y - 10}$

89. $\dfrac{3(m - 2)}{6(2 - m)}$

90. $\dfrac{8(1 - x)}{4(x - 1)}$

91. $\dfrac{w^2 - 4}{8 - 4w}$

92. $\dfrac{15 - 3x}{x^2 - 25}$

93. $\dfrac{18st^5}{12st^3}$

94. $\dfrac{20a^4b^2}{25ab^2}$

95. $\dfrac{4r^2 - 4rs + s^2}{s^2 - 4r^2}$

96. $\dfrac{y^2 - 9z^2}{3z^2 + 2yz - y^2}$

97. $\dfrac{3y - 3x}{2x^2 - 4xy + 2y^2}$

98. $\dfrac{49p^2 - 28pq + 4q^2}{4q - 14p}$

99. $\dfrac{2t^2 - 3t}{2t^4 - 13t^3 + 15t^2}$

100. $\dfrac{4m^3 + 3m^2}{4m^3 + 7m^2 + 3m}$

Expanding Your Skills

For Exercises 101–104, factor and simplify.

101. $\dfrac{w^3 - 8}{w^2 + 2w + 4}$

102. $\dfrac{y^3 + 27}{y^2 - 3y + 9}$

103. $\dfrac{z^2 - 16}{z^3 - 64}$

104. $\dfrac{x^2 - 25}{x^3 + 125}$

Section 15.2 Multiplication and Division of Rational Expressions

Concepts
1. Multiplication of Rational Expressions
2. Division of Rational Expressions

1. Multiplication of Rational Expressions

Recall that to multiply fractions, we multiply the numerators and multiply the denominators. The same is true for multiplying rational expressions.

> **Multiplication of Rational Expressions**
>
> Let p, q, r, and s represent polynomials, such that $q \neq 0$, $s \neq 0$. Then,
>
> $$\dfrac{p}{q} \cdot \dfrac{r}{s} = \dfrac{pr}{qs}$$

For example:

Multiply the Fractions

$$\dfrac{2}{3} \cdot \dfrac{5}{7} = \dfrac{10}{21}$$

Multiply the Rational Expressions

$$\dfrac{2x}{3y} \cdot \dfrac{5z}{7} = \dfrac{10xz}{21y}$$

Sometimes it is possible to simplify a ratio of common factors to 1 *before* multiplying. To do so, we must first factor the numerators and denominators of each fraction.

$$\dfrac{15}{14} \cdot \dfrac{21}{10} = \dfrac{3 \cdot \cancel{5}}{2 \cdot \cancel{7}} \cdot \dfrac{3 \cdot \cancel{7}}{2 \cdot \cancel{5}} = \dfrac{9}{4}$$

The same process is also used to multiply rational expressions.

Multiplying Rational Expressions

Step 1 Factor the numerators and denominators of all rational expressions.
Step 2 Simplify the ratios of common factors to 1 and opposite factors to −1.
Step 3 Multiply the remaining factors in the numerator, and multiply the remaining factors in the denominator.

Example 1 — Multiplying Rational Expressions

Multiply. $\dfrac{5a^2b}{2} \cdot \dfrac{6a}{10b}$

Solution:

$$\dfrac{5a^2b}{2} \cdot \dfrac{6a}{10b}$$

$$= \dfrac{5 \cdot a \cdot a \cdot b}{2} \cdot \dfrac{2 \cdot 3 \cdot a}{2 \cdot 5 \cdot b} \quad \text{Factor into prime factors.}$$

$$= \dfrac{\cancel{5} \cdot a \cdot a \cdot \cancel{b}}{2} \cdot \dfrac{\cancel{2} \cdot 3 \cdot a}{\cancel{2} \cdot \cancel{5} \cdot \cancel{b}} \quad \text{Simplify.}$$

$$= \dfrac{3a^3}{2} \quad \text{Multiply remaining factors.}$$

Skill Practice Multiply.

1. $\dfrac{7a}{3b} \cdot \dfrac{15b}{14a^2}$

Example 2 — Multiplying Rational Expressions

Multiply. $\dfrac{3c - 3d}{6c} \cdot \dfrac{2}{c^2 - d^2}$

Solution:

$$\dfrac{3c - 3d}{6c} \cdot \dfrac{2}{c^2 - d^2}$$

$$= \dfrac{3(c - d)}{2 \cdot 3 \cdot c} \cdot \dfrac{2}{(c - d)(c + d)} \quad \text{Factor.}$$

$$= \dfrac{\cancel{3}(\cancel{c - d})}{\cancel{2} \cdot \cancel{3} \cdot c} \cdot \dfrac{\cancel{2}}{(\cancel{c - d})(c + d)} \quad \text{Simplify.}$$

$$= \dfrac{1}{c(c + d)} \quad \text{Multiply remaining factors.}$$

Avoiding Mistakes
If all the factors in the numerator reduce to a ratio of 1, a factor of 1 is left in the numerator.

Skill Practice Multiply.

2. $\dfrac{4x - 8}{x + 6} \cdot \dfrac{x^2 + 6x}{2x}$

Answers

1. $\dfrac{5}{2a}$ 2. $2(x - 2)$

Example 3 Multiplying Rational Expressions

Multiply. $\dfrac{35 - 5x}{5x + 5} \cdot \dfrac{x^2 + 5x + 4}{x^2 - 49}$

Solution:

$\dfrac{35 - 5x}{5x + 5} \cdot \dfrac{x^2 + 5x + 4}{x^2 - 49}$

$= \dfrac{5(7 - x)}{5(x + 1)} \cdot \dfrac{(x + 4)(x + 1)}{(x - 7)(x + 7)}$ Factor the numerators and denominators completely.

$= \dfrac{\overset{1}{\cancel{5}}\overset{-1}{\cancel{(7-x)}}}{\cancel{5}\cancel{(x+1)}} \cdot \dfrac{(x+4)\overset{1}{\cancel{(x+1)}}}{\cancel{(x-7)}(x+7)}$ Simplify the ratios of common factors to 1 or −1.

$= \dfrac{-1(x + 4)}{x + 7}$ Multiply remaining factors.

$= \dfrac{-(x + 4)}{x + 7}$ or $\dfrac{x + 4}{-(x + 7)}$ or $-\dfrac{x + 4}{x + 7}$

TIP: The ratio $\dfrac{7-x}{x-7} = -1$ because $7 - x$ and $x - 7$ are opposites.

Skill Practice Multiply.

3. $\dfrac{p^2 + 4p + 3}{5p + 10} \cdot \dfrac{p^2 - p - 6}{9 - p^2}$

2. Division of Rational Expressions

Recall that to divide two fractions, multiply the first fraction by the reciprocal of the second.

$\dfrac{21}{10} \div \dfrac{49}{15} \xrightarrow{\text{multiply by the reciprocal of the second fraction}} \dfrac{21}{10} \cdot \dfrac{15}{49} \xrightarrow{\text{factor}} \dfrac{3 \cdot \overset{1}{\cancel{7}}}{2 \cdot \cancel{5}} \cdot \dfrac{3 \cdot \overset{1}{\cancel{5}}}{\cancel{7} \cdot 7} = \dfrac{9}{14}$

The same process is used to divide rational expressions.

Division of Rational Expressions

Let p, q, r, and s represent polynomials, such that $q \neq 0$, $r \neq 0$, $s \neq 0$. Then,

$$\dfrac{p}{q} \div \dfrac{r}{s} = \dfrac{p}{q} \cdot \dfrac{s}{r} = \dfrac{ps}{qr}$$

Example 4 Dividing Rational Expressions

Divide. $\dfrac{5t - 15}{2} \div \dfrac{t^2 - 9}{10}$

Solution:

$\dfrac{5t - 15}{2} \div \dfrac{t^2 - 9}{10}$

$= \dfrac{5t - 15}{2} \cdot \dfrac{10}{t^2 - 9}$ Multiply the first fraction by the reciprocal of the second.

Avoiding Mistakes

When dividing rational expressions, take the reciprocal of the second fraction and change to multiplication *before* reducing like factors.

Answer

3. $\dfrac{-(p + 1)}{5}$ or $\dfrac{p + 1}{-5}$ or $-\dfrac{p + 1}{5}$

$$= \frac{5(t-3)}{2} \cdot \frac{2 \cdot 5}{(t-3)(t+3)}$$ Factor each polynomial.

$$= \frac{5(t-3)}{2} \cdot \frac{2 \cdot 5}{(t-3)(t+3)}$$ Simplify the ratio of common factors to 1.

$$= \frac{25}{t+3}$$

Skill Practice Divide.

4. $\dfrac{7y-14}{y+1} \div \dfrac{y^2+2y-8}{2y+2}$

Example 5   Dividing Rational Expressions

Divide. $\quad \dfrac{p^2 - 11p + 30}{10p^2 - 250} \div \dfrac{30p - 5p^2}{2p+4}$

Solution:

$$\frac{p^2 - 11p + 30}{10p^2 - 250} \div \frac{30p - 5p^2}{2p+4}$$

$$= \frac{p^2 - 11p + 30}{10p^2 - 250} \cdot \frac{2p+4}{30p - 5p^2}$$

Multiply the first fraction by the reciprocal of the second.

Factor the trinomial.
$p^2 - 11p + 30 = (p-5)(p-6)$

$$= \frac{(p-5)(p-6)}{2 \cdot 5(p-5)(p+5)} \cdot \frac{2(p+2)}{5p(6-p)}$$

Factor out the GCF.
$2p + 4 = 2(p+2)$

Factor out the GCF. Then factor the difference of squares.
$10p^2 - 250 = 10(p^2 - 25)$
$\qquad = 2 \cdot 5(p-5)(p+5)$

Factor out the GCF.
$30p - 5p^2 = 5p(6-p)$

$$= \frac{(p-5)(p-6)}{2 \cdot 5(p-5)(p+5)} \cdot \frac{2(p+2)}{5p(6-p)}$$

Simplify the ratio of common factors to 1 or −1.

$$= -\frac{(p+2)}{25p(p+5)}$$

Skill Practice Divide.

5. $\dfrac{4x^2 - 9}{2x^2 - x - 3} \div \dfrac{20x + 30}{x^2 + 7x + 6}$

Answers

4. $\dfrac{14}{y+4}$ 5. $\dfrac{x+6}{10}$

Example 6 Dividing Rational Expressions

Divide. $\dfrac{\frac{3x}{4y}}{\frac{5x}{6y}}$

Solution:

$$\dfrac{\frac{3x}{4y}}{\frac{5x}{6y}} \quad \longleftarrow \text{This fraction bar denotes division } (\div).$$

$$= \frac{3x}{4y} \div \frac{5x}{6y}$$

$$= \frac{3x}{4y} \cdot \frac{6y}{5x} \qquad \text{Multiply by the reciprocal of the second fraction.}$$

$$= \frac{3 \cdot \cancel{x}}{\cancel{2} \cdot 2 \cdot \cancel{y}} \cdot \frac{\cancel{2} \cdot 3 \cdot \cancel{y}}{5 \cdot \cancel{x}} \qquad \text{Simplify the ratio of common factors to 1.}$$

$$= \frac{9}{10}$$

Skill Practice Divide.

6. $\dfrac{\frac{a^3 b}{9c}}{\frac{4ab}{3c^3}}$

Sometimes multiplication and division of rational expressions appear in the same problem. In such a case, apply the order of operations by multiplying or dividing in order from left to right.

Example 7 Multiplying and Dividing Rational Expressions

Perform the indicated operations. $\dfrac{4}{c^2 - 9} \div \dfrac{6}{c - 3} \cdot \dfrac{3c}{8}$

Solution:

In this example, division occurs first, before multiplication. Parentheses may be inserted to reinforce the proper order.

$$\left(\frac{4}{c^2 - 9} \div \frac{6}{c - 3} \right) \cdot \frac{3c}{8}$$

$$= \left(\frac{4}{c^2 - 9} \cdot \frac{c - 3}{6} \right) \cdot \frac{3c}{8} \qquad \text{Multiply the first fraction by the reciprocal of the second.}$$

Answer

6. $\dfrac{a^2 c^2}{12}$

$$= \left(\frac{2 \cdot 2}{(c-3)(c+3)} \cdot \frac{c-3}{2 \cdot 3}\right) \cdot \frac{3 \cdot c}{2 \cdot 2 \cdot 2}$$

Now that each operation is written as multiplication, factor the polynomials and reduce the common factors.

$$= \frac{\overset{1}{2} \cdot \overset{1}{2}}{(c-3)(c+3)} \cdot \frac{(c-3)^{1}}{2 \cdot \cancel{3}} \cdot \frac{\cancel{3} \cdot c}{\cancel{2} \cdot \cancel{2} \cdot 2}$$

$$= \frac{c}{4(c+3)}$$

Simplify.

Skill Practice Perform the indicated operations.

7. $\dfrac{v}{v+2} \div \dfrac{5v^2}{v^2-4} \cdot \dfrac{v}{10}$

Answer

7. $\dfrac{v-2}{50}$

Section 15.2 Activity

A.1. Multiply the fractions. $\dfrac{6}{35} \cdot \dfrac{7}{12}$

A.2. Multiply the rational expressions by simplifying common factors whose ratio is 1 or −1.

$$\frac{(2p+1)(p-4)}{(p-8)(2p+1)} \cdot \frac{5(8-p)}{p(2p+1)}$$

A.3. Multiply the rational expressions $\dfrac{4t^2-20t-24}{9t^2-6t+1} \cdot \dfrac{9t^2-1}{6t^2-34t-12}$ by following these steps.

  a. Factor the numerator and denominator of both rational expressions.
  b. Multiply and write the simplified expression.

A.4. Divide the fractions. $-\dfrac{3}{2} \div \dfrac{9}{8}$

A.5. Divide the rational expressions $\dfrac{4-w}{2w-3} \div \dfrac{w-4}{2(2w-3)}$ by following these steps.

  a. Write the division expression as a related multiplication expression. That is, multiply the first fraction (the dividend) by the reciprocal of the second fraction (the divisor).
  b. Simplify common factors whose ratio is 1 or −1.

A.6. Divide the rational expressions $\dfrac{100-y^2}{24y-3} \div \dfrac{y^2-9y-10}{12y+12}$ by following these steps.

  a. Multiply the first expression by the reciprocal of the second expression.
  b. Factor the numerator and denominator of each expression.
  c. Simplify the expression.

A.7. Multiply and divide as indicated.

$$\frac{10x-5}{25-x^2} \div \frac{20x^2-10x}{x^2-x-20} \cdot \frac{x^3+5x^2}{x^2+2x-8}$$

Practice Exercises

Section 15.2

Study Skills Exercise

Math anxiety can stem from negative past experiences, fear of failure in a current course, or experiencing failure in a specific content area. However, advocating for yourself will help you understand mathematics and build confidence. Ask questions in class or privately with your instructor. Learning new concepts requires productive struggle, so don't be afraid to make mistakes, take risks, and seek assistance when needed.

- Review the key concepts of this section and identify one to three concepts or problems that you do not understand. Continue to seek help until you are satisfied that your questions have been answered and that you are confident in your understanding.

Prerequisite Review

For Exercises R.1–R.4, identify the reciprocal of the given expressions. Assume all variable expressions represent nonzero real numbers.

R.1. a. $-\dfrac{1}{2}$ b. $\dfrac{3}{7}$ c. 8 d. x

R.2. a. $\dfrac{5}{12}$ b. -6 c. $\dfrac{1}{5}$ d. $\dfrac{4}{y}$

R.3. $\dfrac{x-3}{x^2+10x+9}$

R.4. $\dfrac{2x^2+7x+5}{x+4}$

For Exercises R.5–R.12, perform the indicated operation.

R.5. $\dfrac{3}{5} \cdot \dfrac{1}{2}$

R.6. $\dfrac{6}{7} \cdot \dfrac{5}{12}$

R.7. $\dfrac{3}{4} \div \dfrac{3}{8}$

R.8. $\dfrac{18}{5} \div \dfrac{2}{5}$

R.9. $6 \cdot \dfrac{5}{12}$

R.10. $\dfrac{7}{25} \cdot 5$

R.11. $\dfrac{\frac{21}{4}}{\frac{7}{5}}$

R.12. $\dfrac{\frac{9}{2}}{\frac{3}{4}}$

Concept 1: Multiplication of Rational Expressions

For Exercises 1–16, multiply. (See Examples 1–3.)

1. $\dfrac{2xy}{5x^2} \cdot \dfrac{15}{4y}$

2. $\dfrac{7s}{t^2} \cdot \dfrac{t^2}{14s^2}$

3. $\dfrac{6x^3}{9x^6y^2} \cdot \dfrac{18x^4y^7}{4y}$

4. $\dfrac{10a^2b}{15b^2} \cdot \dfrac{30b}{2a^3}$

5. $\dfrac{4x-24}{20x} \cdot \dfrac{5x}{8}$

6. $\dfrac{5a+20}{a} \cdot \dfrac{3a}{10}$

7. $\dfrac{3y+18}{y^2} \cdot \dfrac{4y}{6y+36}$

8. $\dfrac{2p-4}{6p} \cdot \dfrac{4p^2}{8p-16}$

9. $\dfrac{10}{2-a} \cdot \dfrac{a-2}{16}$

10. $\dfrac{w-3}{6} \cdot \dfrac{20}{3-w}$

11. $\dfrac{b^2-a^2}{a-b} \cdot \dfrac{a}{a^2-ab}$

12. $\dfrac{(x-y)^2}{x^2+xy} \cdot \dfrac{x}{y-x}$

13. $\dfrac{y^2+2y+1}{5y-10} \cdot \dfrac{y^2-3y+2}{y^2-1}$

14. $\dfrac{6a^2-6}{a^2+6a+5} \cdot \dfrac{a^2+5a}{12a}$

15. $\dfrac{10x}{2x^2+3x+1} \cdot \dfrac{x^2+7x+6}{5x}$

16. $\dfrac{p-3}{p^2+p-12} \cdot \dfrac{4p+16}{p+1}$

Concept 2: Division of Rational Expressions

For Exercises 17–30, divide. (See Examples 4–6.)

17. $\dfrac{4x}{7y} \div \dfrac{2x^2}{21xy}$

18. $\dfrac{6cd}{5d^2} \div \dfrac{8c^3}{10d}$

19. $\dfrac{\dfrac{8m^4n^5}{5n^6}}{\dfrac{24mn}{15m^3}}$

20. $\dfrac{\dfrac{10a^3b}{3a}}{\dfrac{5b}{9ab}}$

21. $\dfrac{4a+12}{6a-18} \div \dfrac{3a+9}{5a-15}$

22. $\dfrac{8m-16}{3m+3} \div \dfrac{5m-10}{2m+2}$

23. $\dfrac{3x-21}{6x^2-42x} \div \dfrac{7}{12x}$

24. $\dfrac{4a^2-4a}{9a-9} \div \dfrac{5}{12a}$

25. $\dfrac{m^2-n^2}{9} \div \dfrac{3n-3m}{27m}$

26. $\dfrac{9-t^2}{15t+15} \div \dfrac{t-3}{5t}$

27. $\dfrac{3p+4q}{p^2+4pq+4q^2} \div \dfrac{4}{p+2q}$

28. $\dfrac{x^2+2xy+y^2}{2x-y} \div \dfrac{x+y}{5}$

29. $\dfrac{p^2-2p-3}{p^2-p-6} \div \dfrac{p^2-1}{p^2+2p}$

30. $\dfrac{4t^2-1}{t^2-5t} \div \dfrac{2t^2+5t+2}{t^2-3t-10}$

Mixed Exercises

For Exercises 31–56, multiply or divide as indicated.

31. $(w+3) \cdot \dfrac{w}{2w^2+5w-3}$

32. $\dfrac{5t+1}{5t^2-31t+6} \cdot (t-6)$

33. $(r-5) \cdot \dfrac{4r}{2r^2-7r-15}$

34. $\dfrac{q+1}{5q^2-28q-12} \cdot (5q+2)$

35. $\dfrac{\dfrac{5t-10}{12}}{\dfrac{4t-8}{8}}$

36. $\dfrac{\dfrac{6m+6}{5}}{\dfrac{3m+3}{10}}$

37. $\dfrac{2a^2+13a-24}{8a-12} \div (a+8)$

38. $\dfrac{3y^2+20y-7}{5y+35} \div (3y-1)$

39. $\dfrac{y^2+5y-36}{y^2-2y-8} \cdot \dfrac{y+2}{y-6}$

40. $\dfrac{z^2-11z+28}{z-1} \cdot \dfrac{z+1}{z^2-6z-7}$

41. $\dfrac{2t^2+t-1}{t^2+3t+2} \cdot \dfrac{t+2}{2t-1}$

42. $\dfrac{3p^2-2p-8}{3p^2-5p-12} \cdot \dfrac{p-3}{p-2}$

43. $(5t-1) \div \dfrac{5t^2+9t-2}{3t+8}$

44. $(2q-3) \div \dfrac{2q^2+5q-12}{q-7}$

45. $\dfrac{x^2+2x-3}{x^2-3x+2} \cdot \dfrac{x^2+2x-8}{x^2+4x+3}$

46. $\dfrac{y^2+y-12}{y^2-y-20} \cdot \dfrac{y^2+y-30}{y^2-2y-3}$

47. $\dfrac{\dfrac{w^2-6w+9}{8}}{\dfrac{9-w^2}{4w+12}}$

48. $\dfrac{\dfrac{p^2-6p+8}{24}}{\dfrac{16-p^2}{6p+6}}$

49. $\dfrac{5k^2+7k+2}{k^2+5k+4} \div \dfrac{5k^2+17k+6}{k^2+10k+24}$

50. $\dfrac{4h^2-5h+1}{h^2+h-2} \div \dfrac{6h^2-7h+2}{2h^2+3h-2}$

51. $\dfrac{ax+a+bx+b}{2x^2+4x+2} \cdot \dfrac{4x+4}{a^2+ab}$

52. $\dfrac{3my+9m+ny+3n}{9m^2+6mn+n^2} \cdot \dfrac{30m+10n}{5y^2+15y}$

53. $\dfrac{y^4 - 1}{2y^2 - 3y + 1} \div \dfrac{2y^2 + 2}{8y^2 - 4y}$

54. $\dfrac{x^4 - 16}{6x^2 + 24} \div \dfrac{x^2 - 2x}{3x}$

55. $\dfrac{x^2 - xy - 2y^2}{x + 2y} \div \dfrac{x^2 - 4xy + 4y^2}{x^2 - 4y^2}$

56. $\dfrac{4m^2 - 4mn - 3n^2}{8m^2 - 18n^2} \div \dfrac{3m + 3n}{6m^2 + 15mn + 9n^2}$

For Exercises 57–62, multiply or divide as indicated. (See Example 7.)

57. $\dfrac{y^3 - 3y^2 + 4y - 12}{y^4 - 16} \cdot \dfrac{3y^2 + 5y - 2}{3y^2 - 10y + 3} \div \dfrac{3}{6y - 12}$

58. $\dfrac{x^2 - 25}{3x^2 + 3xy} \cdot \dfrac{x^2 + 4x + xy + 4y}{x^2 + 9x + 20} \div \dfrac{x - 5}{x}$

59. $\dfrac{a^2 - 5a}{a^2 + 7a + 12} \div \dfrac{a^3 - 7a^2 + 10a}{a^2 + 9a + 18} \div \dfrac{a + 6}{a + 4}$

60. $\dfrac{t^2 + t - 2}{t^2 + 5t + 6} \div \dfrac{t - 1}{t} \div \dfrac{5t - 5}{t + 3}$

61. $\dfrac{p^3 - q^3}{p - q} \cdot \dfrac{p + q}{2p^2 + 2pq + 2q^2}$

62. $\dfrac{r^3 + s^3}{r - s} \div \dfrac{r^2 + 2rs + s^2}{r^2 - s^2}$

Least Common Denominator

Section 15.3

1. Least Common Denominator

We have already learned how to simplify, multiply, and divide rational expressions. Our next goal is to add and subtract rational expressions. As with fractions, rational expressions may be added or subtracted only if they have the same denominator.

The **least common denominator (LCD)** of two or more rational expressions is defined as the least common multiple of the denominators. For example, consider the fractions $\frac{1}{20}$ and $\frac{1}{8}$. By inspection, you can probably see that the least common denominator is 40. To understand why, find the prime factorization of both denominators:

$$20 = 2^2 \cdot 5 \quad \text{and} \quad 8 = 2^3$$

A common multiple of 20 and 8 must be a multiple of 5, a multiple of 2^2, and a multiple of 2^3. However, any number that is a multiple of $2^3 = 8$ is automatically a multiple of $2^2 = 4$. Therefore, it is sufficient to construct the least common denominator as the product of unique prime factors, in which each factor is raised to its highest power.

The LCD of $\dfrac{1}{20}$ and $\dfrac{1}{8}$ is $2^3 \cdot 5 = 40$.

Concepts

1. Least Common Denominator
2. Writing Rational Expressions With the Least Common Denominator

> **Finding the Least Common Denominator of Two or More Rational Expressions**
>
> **Step 1** Factor all denominators completely.
>
> **Step 2** The LCD is the product of unique prime factors from the denominators, in which each factor is raised to the highest power to which it appears in any denominator.

Example 1 Finding the Least Common Denominator

Find the LCD of the rational expressions.

a. $\dfrac{5}{14}; \dfrac{3}{49}; \dfrac{1}{8}$ b. $\dfrac{5}{3x^2z}; \dfrac{7}{x^5y^3}$

Solution:

a. $\dfrac{5}{14}; \dfrac{3}{49}; \dfrac{1}{8}$

$= \dfrac{5}{2 \cdot 7}; \dfrac{3}{7^2}; \dfrac{1}{2^3}$ Factor the denominators.

The LCD is $7^2 \cdot 2^3 = 392$.   The LCD is the product of unique factors, each raised to its highest power.

b. $\dfrac{5}{3x^2z}; \dfrac{7}{x^5y^3}$

$= \dfrac{5}{3 \cdot x^2 \cdot z^1}; \dfrac{7}{x^5 \cdot y^3}$ The denominators are in factored form.

The LCD is the product of $3 \cdot x^5 \cdot y^3 \cdot z^1$ or simply $3x^5y^3z$.

Skill Practice Find the LCD for each set of expressions.

1. $\dfrac{3}{8}; \dfrac{7}{10}; \dfrac{1}{15}$ 2. $\dfrac{1}{5a^3b^2}; \dfrac{1}{10a^4b}$

Example 2 Finding the Least Common Denominator

Find the LCD for each pair of rational expressions.

a. $\dfrac{a+b}{a^2-25}; \dfrac{1}{2a-10}$ b. $\dfrac{x-5}{x^2-2x}; \dfrac{1}{x^2-4x+4}$

Solution:

a. $\dfrac{a+b}{a^2-25}; \dfrac{1}{2a-10}$

$= \dfrac{a+b}{(a-5)(a+5)}; \dfrac{1}{2(a-5)}$ Factor the denominators.

The LCD is $2(a-5)(a+5)$.   The LCD is the product of unique factors, each raised to its highest power.

b. $\dfrac{x-5}{x^2-2x}; \dfrac{1}{x^2-4x+4}$

$= \dfrac{x-5}{x(x-2)}; \dfrac{1}{(x-2)^2}$ Factor the denominators.

The LCD is $x(x-2)^2$.   The LCD is the product of unique factors, each raised to its highest power.

Answers
1. 120
2. $10a^4b^2$

Skill Practice Find the LCD.

3. $\dfrac{x}{x^2-16}; \dfrac{2}{3x+12}$ 4. $\dfrac{6}{t^2+5t-14}; \dfrac{8}{t^2-3t+2}$

2. Writing Rational Expressions With the Least Common Denominator

To add or subtract two rational expressions, the expressions must have the same denominator. Therefore, we must first practice the skill of converting each rational expression into an equivalent expression with the LCD as its denominator.

Writing Equivalent Fractions With Common Denominators
Step 1 Identify the LCD for the expressions.
Step 2 Multiply the numerator and denominator of each fraction by the factors from the LCD that are missing from the original denominators.

Example 3   Converting to the Least Common Denominator

Find the LCD of each pair of rational expressions. Then convert each expression to an equivalent fraction with the denominator equal to the LCD.

a. $\dfrac{3}{2ab}; \dfrac{6}{5a^2}$ b. $\dfrac{4}{x+1}; \dfrac{7}{x-4}$

Solution:

a. $\dfrac{3}{2ab}; \dfrac{6}{5a^2}$ The LCD is $10a^2b$.

$\dfrac{3}{2ab} = \dfrac{3 \cdot 5a}{2ab \cdot 5a} = \dfrac{15a}{10a^2b}$ The first expression is missing the factor $5a$ from the denominator.

$\dfrac{6}{5a^2} = \dfrac{6 \cdot 2b}{5a^2 \cdot 2b} = \dfrac{12b}{10a^2b}$ The second expression is missing the factor $2b$ from the denominator.

b. $\dfrac{4}{x+1}; \dfrac{7}{x-4}$ The LCD is $(x+1)(x-4)$.

$\dfrac{4}{x+1} = \dfrac{4(x-4)}{(x+1)(x-4)} = \dfrac{4x-16}{(x+1)(x-4)}$ The first expression is missing the factor $(x-4)$ from the denominator.

$\dfrac{7}{x-4} = \dfrac{7(x+1)}{(x-4)(x+1)} = \dfrac{7x+7}{(x-4)(x+1)}$ The second expression is missing the factor $(x+1)$ from the denominator.

Skill Practice For each pair of expressions, find the LCD, and then convert each expression to an equivalent fraction with the denominator equal to the LCD.

5. $\dfrac{2}{rs^2}; \dfrac{-1}{r^3s}$ 6. $\dfrac{5}{x-3}; \dfrac{x}{x+1}$

Answers
3. $3(x-4)(x+4)$
4. $(t+7)(t-2)(t-1)$
5. $\dfrac{2}{rs^2} = \dfrac{2r^2}{r^3s^2}; \dfrac{-1}{r^3s} = \dfrac{-s}{r^3s^2}$
6. $\dfrac{5}{x-3} = \dfrac{5x+5}{(x-3)(x+1)}$

$\dfrac{x}{x+1} = \dfrac{x^2-3x}{(x+1)(x-3)}$

Example 4 — Converting to the Least Common Denominator

Find the LCD of the pair of rational expressions. Then convert each expression to an equivalent fraction with the denominator equal to the LCD.

$$\frac{w+2}{w^2 - w - 12}; \quad \frac{1}{w^2 - 9}$$

Solution:

$\dfrac{w+2}{w^2 - w - 12}; \dfrac{1}{w^2 - 9}$ To find the LCD, factor each denominator.

$\dfrac{w+2}{(w-4)(w+3)}; \dfrac{1}{(w-3)(w+3)}$ The LCD is $(w-4)(w+3)(w-3)$.

$\dfrac{w+2}{(w-4)(w+3)} = \dfrac{(w+2)(w-3)}{(w-4)(w+3)(w-3)}$ The first expression is missing the factor $(w-3)$ from the denominator.

$ = \dfrac{w^2 - w - 6}{(w-4)(w+3)(w-3)}$

$\dfrac{1}{(w-3)(w+3)} = \dfrac{1(w-4)}{(w-3)(w+3)(w-4)}$ The second expression is missing the factor $(w-4)$ from the denominator.

$ = \dfrac{w-4}{(w-3)(w+3)(w-4)}$

Skill Practice Find the LCD. Then convert each expression to an equivalent expression with the denominator equal to the LCD.

7. $\dfrac{z}{z^2 - 4}; \quad \dfrac{-3}{z^2 - z - 2}$

Example 5 — Converting to the Least Common Denominator

Convert each expression to an equivalent expression with the denominator equal to the LCD.

$$\frac{3}{x-7} \quad \text{and} \quad \frac{1}{7-x}$$

Solution:

Notice that the expressions $x - 7$ and $7 - x$ are opposites and differ by a factor of -1. Therefore, we may use either $x - 7$ or $7 - x$ as a common denominator. Each case is shown below.

Converting to the Denominator $x - 7$

$\dfrac{3}{x-7}; \dfrac{1}{7-x}$ Leave the first fraction unchanged because it has the desired LCD.

$\dfrac{1}{7-x} = \dfrac{(-1)1}{(-1)(7-x)}$ Multiply the *second* rational expression by the ratio $\dfrac{-1}{-1}$ to change its denominator to $x - 7$.

$ = \dfrac{-1}{-7 + x}$ Apply the distributive property.

$ = \dfrac{-1}{x - 7}$

TIP: In Example 5, the expressions

$\dfrac{3}{x-7}$ and $\dfrac{1}{7-x}$

have opposite factors in the denominators. In such a case, you do not need to include *both* factors in the LCD.

Answer

7. $\dfrac{z^2 + z}{(z-2)(z+2)(z+1)};$

$\dfrac{-3z - 6}{(z-2)(z+2)(z+1)}$

Converting to the Denominator $7 - x$

$\dfrac{3}{x-7}; \dfrac{1}{7-x}$ Leave the second fraction unchanged because it has the desired LCD.

$\dfrac{3}{x-7} = \dfrac{(-1)3}{(-1)(x-7)};$ Multiply the *first* rational expression by the ratio $\dfrac{-1}{-1}$ to change its denominator to $7 - x$.

$= \dfrac{-3}{-x+7}$ Apply the distributive property.

$= \dfrac{-3}{7-x}$

Skill Practice

8. a. Find the LCD of the expressions. $\dfrac{9}{w-2}; \dfrac{11}{2-w}$

b. Convert each expression to an equivalent fraction with denominator equal to the LCD.

Answers

8. a. The LCD is $(w-2)$ or $(2-w)$.

b. $\dfrac{9}{w-2} = \dfrac{9}{w-2};$

$\dfrac{11}{2-w} = \dfrac{-11}{w-2}$

or

$\dfrac{9}{w-2} = \dfrac{-9}{2-w};$

$\dfrac{11}{2-w} = \dfrac{11}{2-w}$

Section 15.3 Activity

A.1. a. List the first six multiples of 8.
  b. List the first six multiples of 12.
  c. Which values from parts (a) and (b) are multiples of both 8 and 12?
  d. Which value from parts (a) and (b) is the least common multiple of 8 and 12?

A.2. a. Factor 8 as a product of prime factors.
  b. Factor 12 as a product of prime factors.
  c. Use the prime factors of 8 and 12 to construct the least common multiple of 8 and 12. Explain the process you used.
  d. Identify the least common denominator of the fractions $\dfrac{3}{8}$ and $\dfrac{5}{12}$.
  e. Write each fraction with a denominator of 24.

$\dfrac{3}{8} = \dfrac{}{24}$ and $\dfrac{5}{12} = \dfrac{}{24}$

A.3. a. Identify the least common denominator of the fractions $\dfrac{11}{5p^2q}$ and $\dfrac{-1}{10pq^3}$.

  b. Write each fraction with the denominator $10p^2q^3$.

$\dfrac{11}{5p^2q} = \dfrac{}{10p^2q^3}; \quad \dfrac{-1}{10pq^3} = \dfrac{}{10p^2q^3}$

For Exercises A.4–A.5,

a. Identify the least common denominator of the expressions. (*Hint:* You may have to factor the denominators first.)

b. Write each expression as an equivalent expression with the common denominator.

A.4. $\dfrac{y-3}{(2y+1)^2}; \dfrac{4y}{(2y+1)(y-7)}$

A.5. $\dfrac{-1}{x^2-2x+15}; \dfrac{x+1}{x^2-25}$

A.6. a. Write the expression with the indicated denominator. $\dfrac{3}{5-y} = \dfrac{}{y-5}$

b. Write the expression with the indicated denominator. $\dfrac{x-2}{y-5} = \dfrac{}{5-y}$

Section 15.3 Practice Exercises

Prerequisite Review

For Exercises R.1–R.2, factor the number into a product of prime factors.

R.1. 480

R.2. 210

For Exercises R.3–R.4, find the least common multiple of the pair of numbers.

R.3. 12 and 28

R.4. 9 and 15

For Exercises R.5–R.6, evaluate the expression for the given values of the variable.

R.5. $\dfrac{2x}{x+5}$

  a. $x = -3$ **b.** $x = 1$ **c.** $x = -5$

R.6. $\dfrac{y-1}{y-3}$

  a. $y = 4$ **b.** $y = 1$ **c.** $y = 3$

For Exercises R.7–R.8, identify the restricted values. Then simplify the expression.

R.7. $\dfrac{3x+3}{5x^2-5}$

R.8. $\dfrac{x+2}{x^2-3x-10}$

For Exercises R.9–R.12, multiply or divide as indicated.

R.9. $\dfrac{a+3}{a+7} \cdot \dfrac{a^2+3a-10}{a^2+a-6}$

R.10. $\dfrac{6(a+2b)}{2(a-3b)} \cdot \dfrac{4(a+3b)(a-3b)}{9(a+2b)(a-2b)}$

R.11. $\dfrac{16y^2}{9y+36} \div \dfrac{8y^3}{3y+12}$

R.12. $\dfrac{5w^2+6w+1}{w^2+5w+6} \div (5w+1)$

R.13. Which of the expressions are equivalent to $-\dfrac{5}{x-3}$? Circle all that apply.

  a. $\dfrac{-5}{x-3}$ **b.** $\dfrac{5}{-x+3}$ **c.** $\dfrac{5}{3-x}$ **d.** $\dfrac{5}{-(x-3)}$

R.14. Which of the expressions are equivalent to $\dfrac{4-a}{6}$? Circle all that apply.

  a. $\dfrac{a-4}{-6}$ **b.** $\dfrac{a-4}{6}$ **c.** $\dfrac{-(4-a)}{-6}$ **d.** $-\dfrac{a-4}{6}$

Concept 1: Least Common Denominator

1. Explain why the least common denominator of $\dfrac{1}{x^3}$, $\dfrac{1}{x^5}$, and $\dfrac{1}{x^4}$ is x^5.

2. Explain why the least common denominator of $\dfrac{2}{y^3}$, $\dfrac{9}{y^6}$, and $\dfrac{4}{y^5}$ is y^6.

For Exercises 3–20, identify the LCD. (See Examples 1–2.)

3. $\dfrac{4}{15}; \dfrac{5}{9}$

4. $\dfrac{7}{12}; \dfrac{1}{18}$

5. $\dfrac{1}{16}; \dfrac{1}{4}; \dfrac{1}{6}$

6. $\dfrac{1}{2}; \dfrac{11}{12}; \dfrac{3}{8}$

7. $\dfrac{1}{7}; \dfrac{2}{9}$

8. $\dfrac{2}{3}; \dfrac{5}{8}$

9. $\dfrac{1}{3x^2y}; \dfrac{8}{9xy^3}$

10. $\dfrac{5}{2a^4b^2}; \dfrac{1}{8ab^3}$

11. $\dfrac{6}{w^2}; \dfrac{7}{y}$

12. $\dfrac{2}{r}; \dfrac{3}{s^2}$

13. $\dfrac{p}{(p+3)(p-1)}; \dfrac{2}{(p+3)(p+2)}$

14. $\dfrac{6}{(q+4)(q-4)}; \dfrac{q^2}{(q+1)(q+4)}$

15. $\dfrac{7}{3t(t+1)}; \dfrac{10t}{9(t+1)^2}$

16. $\dfrac{13x}{15(x-1)^2}; \dfrac{5}{3x(x-1)}$

17. $\dfrac{y}{y^2-4}; \dfrac{3y}{y^2+5y+6}$

18. $\dfrac{4}{w^2-3w+2}; \dfrac{w}{w^2-4}$

19. $\dfrac{5}{3-x}; \dfrac{7}{x-3}$

20. $\dfrac{4}{x-6}; \dfrac{9}{6-x}$

21. Explain why a common denominator of

$$\dfrac{b+1}{b-1} \text{ and } \dfrac{b}{1-b}$$

could be either $(b-1)$ or $(1-b)$.

22. Explain why a common denominator of

$$\dfrac{1}{6-t} \text{ and } \dfrac{t}{t-6}$$

could be either $(6-t)$ or $(t-6)$.

Concept 2: Writing Rational Expressions With the Least Common Denominator

For Exercises 23–46, find the LCD. Then convert each expression to an equivalent expression with the denominator equal to the LCD. (See Examples 3–5.)

23. $\dfrac{6}{5x^2}; \dfrac{1}{x}$

24. $\dfrac{3}{y}; \dfrac{7}{9y^2}$

25. $\dfrac{4}{5x^2}; \dfrac{y}{6x^3}$

26. $\dfrac{3}{15b^2}; \dfrac{c}{3b^2}$

27. $\dfrac{5}{6a^2b}; \dfrac{a}{12b}$

28. $\dfrac{x}{15y^2}; \dfrac{y}{5xy}$

29. $\dfrac{6}{m+4}; \dfrac{3}{m-1}$

30. $\dfrac{3}{n-5}; \dfrac{7}{n+2}$

31. $\dfrac{6}{2x-5}; \dfrac{1}{x+3}$

32. $\dfrac{4}{m+3}; \dfrac{-3}{5m+1}$

33. $\dfrac{6}{(w+3)(w-8)}; \dfrac{w}{(w-8)(w+1)}$

34. $\dfrac{t}{(t+2)(t+12)}; \dfrac{18}{(t-2)(t+2)}$

35. $\dfrac{6p}{p^2-4}; \dfrac{3}{p^2+4p+4}$

36. $\dfrac{5}{t^2-6t+9}; \dfrac{t}{t^2-9}$

37. $\dfrac{1}{a-4}; \dfrac{a}{4-a}$

38. $\dfrac{3b}{2b-5}; \dfrac{2b}{5-2b}$

39. $\dfrac{4}{x-7}; \dfrac{y}{14-2x}$

40. $\dfrac{4}{3x-15}; \dfrac{z}{5-x}$

41. $\dfrac{1}{a+b}; \dfrac{6}{-a-b}$

42. $\dfrac{p}{-q-8}; \dfrac{1}{q+8}$

43. $\dfrac{-3}{24y+8}; \dfrac{5}{18y+6}$

44. $\dfrac{r}{10r+5}; \dfrac{2}{16r+8}$

45. $\dfrac{3}{5z}; \dfrac{1}{z+4}$

46. $\dfrac{-1}{4a-8}; \dfrac{5}{4a}$

Expanding Your Skills

For Exercises 47–50, find the LCD. Then convert each expression to an equivalent expression with the denominator equal to the LCD.

47. $\dfrac{z}{z^2+9z+14}; \dfrac{-3z}{z^2+10z+21}; \dfrac{5}{z^2+5z+6}$

48. $\dfrac{6}{w^2-3w-4}; \dfrac{1}{w^2+6w+5}; \dfrac{-9w}{w^2+w-20}$

49. $\dfrac{3}{p^3-8}; \dfrac{p}{p^2-4}; \dfrac{5p}{p^2+2p+4}$

50. $\dfrac{7}{n^3+125}; \dfrac{n}{n^2-25}; \dfrac{12}{n^2-5n+25}$

Section 15.4 Addition and Subtraction of Rational Expressions

Concepts

1. Addition and Subtraction of Rational Expressions With the Same Denominator
2. Addition and Subtraction of Rational Expressions With Different Denominators
3. Using Rational Expressions in Translations

1. Addition and Subtraction of Rational Expressions With the Same Denominator

To add or subtract rational expressions, the expressions must have the same denominator. As with fractions, we add or subtract rational expressions with the same denominator by combining the terms in the numerator and then writing the result over the common denominator. Then, if possible, simplify the expression.

> **Addition and Subtraction of Rational Expressions**
>
> Let p, q, and r represent polynomials where $q \neq 0$. Then,
>
> **1.** $\dfrac{p}{q} + \dfrac{r}{q} = \dfrac{p+r}{q}$ **2.** $\dfrac{p}{q} - \dfrac{r}{q} = \dfrac{p-r}{q}$

Section 15.4 Addition and Subtraction of Rational Expressions

Example 1   **Adding and Subtracting Rational Expressions With the Same Denominator**

Add or subtract as indicated.   **a.** $\dfrac{1}{12} + \dfrac{7}{12}$ **b.** $\dfrac{2}{5p} - \dfrac{7}{5p}$

Solution:

a. $\dfrac{1}{12} + \dfrac{7}{12}$ The fractions have the same denominator.

$= \dfrac{1+7}{12}$ Add the terms in the numerators, and write the result over the common denominator.

$= \dfrac{\cancel{8}^{2}}{\cancel{12}_{3}}$

$= \dfrac{2}{3}$ Simplify.

b. $\dfrac{2}{5p} - \dfrac{7}{5p}$ The rational expressions have the same denominator.

$= \dfrac{2-7}{5p}$ Subtract the terms in the numerators, and write the result over the common denominator.

$= \dfrac{-5}{5p}$

$= \dfrac{\cancel{-5}^{-1}}{\cancel{5}p}$ Simplify.

$= -\dfrac{1}{p}$

Skill Practice   Add or subtract as indicated.

1. $\dfrac{3}{14} + \dfrac{4}{14}$ **2.** $\dfrac{2}{7d} - \dfrac{9}{7d}$

Example 2   **Adding and Subtracting Rational Expressions With the Same Denominator**

Add or subtract as indicated.

a. $\dfrac{2}{3d+5} + \dfrac{7d}{3d+5}$ **b.** $\dfrac{x^2}{x-3} - \dfrac{-5x+24}{x-3}$

Solution:

a. $\dfrac{2}{3d+5} + \dfrac{7d}{3d+5}$ The rational expressions have the same denominator.

$= \dfrac{2+7d}{3d+5}$ Add the terms in the numerators, and write the result over the common denominator.

$= \dfrac{7d+2}{3d+5}$ Because the numerator and denominator share no common factors, the expression is in lowest terms.

Answers

1. $\dfrac{1}{2}$ **2.** $-\dfrac{1}{d}$

Avoiding Mistakes

When subtracting rational expressions, use parentheses to group the terms in the numerator that follow the subtraction sign. This will help you remember to apply the distributive property.

b. $\dfrac{x^2}{x-3} - \dfrac{-5x+24}{x-3}$ The rational expressions have the same denominator.

$= \dfrac{x^2 - (-5x+24)}{x-3}$ Subtract the terms in the numerators, and write the result over the common denominator.

$= \dfrac{x^2 + 5x - 24}{x-3}$ Simplify the numerator.

$= \dfrac{(x+8)(x-3)}{(x-3)}$ Factor the numerator and denominator to determine if the rational expression can be simplified.

$= \dfrac{(x+8)\cancel{(x-3)}}{\cancel{(x-3)}}$ Simplify.

$= x + 8$

Skill Practice Add or subtract as indicated.

3. $\dfrac{x^2+2}{x+3} + \dfrac{4x+1}{x+3}$ 4. $\dfrac{4t-9}{2t+1} - \dfrac{t-5}{2t+1}$

2. Addition and Subtraction of Rational Expressions With Different Denominators

To add or subtract two rational expressions with unlike denominators, we must convert the expressions to equivalent expressions with the same denominator. For example, consider adding

$$\dfrac{1}{10} + \dfrac{12}{5y}$$

The LCD is $10y$. For each expression, identify the factors from the LCD that are missing from the denominator. Then multiply the numerator and denominator of the expression by the missing factor(s).

The LCD is $10y$: $\dfrac{1}{10} + \dfrac{12}{5y}$

 Missing y Missing 2

$= \dfrac{1 \cdot y}{10 \cdot y} + \dfrac{12 \cdot 2}{5y \cdot 2}$

$= \dfrac{y}{10y} + \dfrac{24}{10y}$ The rational expressions now have the same denominators.

$= \dfrac{y + 24}{10y}$ Add the numerators.

Avoiding Mistakes

In the expression $\dfrac{y+24}{10y}$, notice that you cannot reduce the 24 and 10 because 24 is not a factor in the numerator, it is a term. Only factors can be reduced—not terms.

After successfully adding or subtracting two rational expressions, always check to see if the final answer is simplified. If necessary, factor the numerator and denominator, and reduce common factors. The expression

$$\dfrac{y+24}{10y}$$

is in lowest terms because the numerator and denominator do not share any common factors.

Answers

3. $x + 1$ 4. $\dfrac{3t-4}{2t+1}$

Section 15.4 Addition and Subtraction of Rational Expressions

Adding or Subtracting Rational Expressions
Step 1 Factor the denominators of each rational expression.
Step 2 Identify the LCD.
Step 3 Rewrite each rational expression as an equivalent expression with the LCD as its denominator.
Step 4 Add or subtract the numerators, and write the result over the common denominator.
Step 5 Simplify.

Example 3 Subtracting Rational Expressions With Different Denominators

Subtract. $\dfrac{4}{7k} - \dfrac{3}{k^2}$

Solution:

$\dfrac{4}{7k} - \dfrac{3}{k^2}$

Step 1: The denominators are already factored.

Step 2: The LCD is $7k^2$.

$= \dfrac{4 \cdot k}{7k \cdot k} - \dfrac{3 \cdot 7}{k^2 \cdot 7}$ **Step 3:** Write each expression with the LCD.

$= \dfrac{4k}{7k^2} - \dfrac{21}{7k^2}$

$= \dfrac{4k - 21}{7k^2}$ **Step 4:** Subtract the numerators, and write the result over the LCD.

Step 5: The expression is in lowest terms because the numerator and denominator share no common factors.

Avoiding Mistakes
Do not reduce after rewriting the individual fractions with the LCD. You will revert back to the original expression.

Skill Practice Subtract.

5. $\dfrac{4}{3x} - \dfrac{1}{2x^2}$

Example 4 Subtracting Rational Expressions With Different Denominators

Subtract. $\dfrac{2q - 4}{3} - \dfrac{q + 1}{2}$

Solution:

$\dfrac{2q - 4}{3} - \dfrac{q + 1}{2}$

Step 1: The denominators are already factored.

Step 2: The LCD is 6.

$= \dfrac{2(2q - 4)}{2 \cdot 3} - \dfrac{3(q + 1)}{3 \cdot 2}$ **Step 3:** Write each expression with the LCD.

Answer
5. $\dfrac{8x - 3}{6x^2}$

$$= \frac{2(2q-4) - 3(q+1)}{6}$$

Step 4: Subtract the numerators, and write the result over the LCD.

$$= \frac{4q - 8 - 3q - 3}{6}$$

$$= \frac{q - 11}{6}$$

Step 5: The expression is in lowest terms because the numerator and denominator share no common factors.

Skill Practice Subtract.

6. $\dfrac{t}{12} - \dfrac{t-2}{4}$

Example 5 Adding Rational Expressions With Different Denominators

Add. $\quad \dfrac{1}{x-5} + \dfrac{-10}{x^2 - 25}$

Solution:

$\dfrac{1}{x-5} + \dfrac{-10}{x^2 - 25}$

$= \dfrac{1}{x-5} + \dfrac{-10}{(x-5)(x+5)}$

Step 1: Factor the denominators.

Step 2: The LCD is $(x-5)(x+5)$.

$= \dfrac{1(x+5)}{(x-5)(x+5)} + \dfrac{-10}{(x-5)(x+5)}$

Step 3: Write each expression with the LCD.

$= \dfrac{1(x+5) + (-10)}{(x-5)(x+5)}$

Step 4: Add the numerators, and write the result over the LCD.

$= \dfrac{x + 5 - 10}{(x-5)(x+5)}$

$= \dfrac{\overset{1}{\cancel{x-5}}}{\cancel{(x-5)}(x+5)}$

Step 5: Simplify.

$= \dfrac{1}{x+5}$

Skill Practice Add.

7. $\dfrac{1}{x-4} + \dfrac{-8}{x^2 - 16}$

Answers

6. $\dfrac{-t+3}{6}$ 7. $\dfrac{1}{x+4}$

Example 6 — Subtracting Rational Expressions With Different Denominators

Subtract. $\dfrac{p+2}{p-1} - \dfrac{2}{p+6} - \dfrac{14}{p^2+5p-6}$

Solution:

$\dfrac{p+2}{p-1} - \dfrac{2}{p+6} - \dfrac{14}{p^2+5p-6}$

$= \dfrac{p+2}{p-1} - \dfrac{2}{p+6} - \dfrac{14}{(p-1)(p+6)}$ **Step 1:** Factor the denominators.

Step 2: The LCD is $(p-1)(p+6)$.

Step 3: Write each expression with the LCD.

$= \dfrac{(p+2)(p+6)}{(p-1)(p+6)} - \dfrac{2(p-1)}{(p+6)(p-1)} - \dfrac{14}{(p-1)(p+6)}$

$= \dfrac{(p+2)(p+6) - 2(p-1) - 14}{(p-1)(p+6)}$ **Step 4:** Combine the numerators, and write the result over the LCD.

$= \dfrac{p^2 + 6p + 2p + 12 - 2p + 2 - 14}{(p-1)(p+6)}$ **Step 5:** Clear parentheses in the numerator.

$= \dfrac{p^2 + 6p}{(p-1)(p+6)}$ Combine *like* terms.

$= \dfrac{p(p+6)}{(p-1)(p+6)}$ Factor the numerator to determine if the expression is in lowest terms.

$= \dfrac{p\cancel{(p+6)}}{(p-1)\cancel{(p+6)}}$ Simplify.

$= \dfrac{p}{p-1}$

Skill Practice Subtract.

8. $\dfrac{2y}{y-1} - \dfrac{1}{y} - \dfrac{2y+1}{y^2-y}$

When the denominators of two rational expressions are opposites, we can produce identical denominators by multiplying one of the expressions by the ratio $\dfrac{-1}{-1}$. This is demonstrated in Example 7.

Example 7 — Adding Rational Expressions With Different Denominators

Add the rational expressions. $\dfrac{1}{d-7} + \dfrac{5}{7-d}$

Answer

8. $\dfrac{2y-3}{y-1}$

Solution:

$$\frac{1}{d-7} + \frac{5}{7-d}$$

The expressions $d-7$ and $7-d$ are opposites and differ by a factor of -1. Therefore, multiply the numerator and denominator of *either* expression by -1 to obtain a common denominator.

$$= \frac{1}{d-7} + \frac{(-1)5}{(-1)(7-d)}$$

Note that $-1(7-d) = -7 + d$ or $d - 7$.

$$= \frac{1}{d-7} + \frac{-5}{d-7}$$

Simplify.

$$= \frac{1+(-5)}{d-7}$$

Add the terms in the numerators, and write the result over the common denominator.

$$= \frac{-4}{d-7}$$

Skill Practice Add.

9. $\dfrac{3}{p-8} + \dfrac{1}{8-p}$

3. Using Rational Expressions in Translations

Example 8   Using Rational Expressions in Translations

Write the English phrase as a mathematical expression. Then simplify by combining the rational expressions.

The difference of the reciprocal of n and the quotient of n and 3

FOR REVIEW

Recall key terms representing arithmetic operations.

- The *difference* of a and b implies subtraction in the order $a - b$.
- The *reciprocal* of a real number n is $\dfrac{1}{n}$.
- The *quotient* of a and b is $\dfrac{a}{b}$.

Solution:

The difference of the reciprocal of n and the quotient of n and 3

$$\left(\frac{1}{n}\right) - \left(\frac{n}{3}\right)$$

The reciprocal of n The quotient of n and 3   The difference of

$$\frac{1}{n} - \frac{n}{3}$$

The LCD is $3n$.

$$= \frac{3 \cdot 1}{3 \cdot n} - \frac{n \cdot n}{3 \cdot n}$$

Write each expression with the LCD.

$$= \frac{3 - n^2}{3n}$$

Subtract the numerators.

Skill Practice Write the English phrase as a mathematical expression. Then simplify by combining the rational expressions.

10. The sum of 1 and the quotient of 2 and a

Answers

9. $\dfrac{2}{p-8}$ or $\dfrac{-2}{8-p}$

10. $1 + \dfrac{2}{a}$; $\dfrac{a+2}{a}$

Section 15.4 Addition and Subtraction of Rational Expressions

Section 15.4 Activity

For Exercises A.1–A.3, add or subtract the expressions by following these steps.
   a. Factor the denominators.
   b. Identify the LCD.
   c. Write each term as an equivalent expression with the LCD as the denominator.
   d. Add or subtract the numerators and write the result over the common denominator.
   e. Simplify the result and write the expression in lowest terms.

A.1. $\dfrac{6}{5x} + \dfrac{7}{10}$

A.2. $\dfrac{3}{2x-4} - \dfrac{6}{x^2-4}$

A.3. $\dfrac{1}{y+1} + \dfrac{1}{y-3} - \dfrac{4}{y^2-2y-3}$

A.4. Consider the expression $\dfrac{2}{t-5} + \dfrac{8}{5-t}$.
   a. Explain why the LCD can be taken as $t-5$ or as $5-t$.
   b. Add the fractions using $t-5$ as the LCD.
   c. Add the fractions using $5-t$ as the LCD.
   d. Show that the expressions $\dfrac{-6}{t-5}$ and $\dfrac{6}{5-t}$ are equivalent.

Practice Exercises   Section 15.4

Prerequisite Review

For Exercises R.1–R.8, identify the least common denominator of the given expressions.

R.1. $\dfrac{5}{18}$ and $\dfrac{7}{30}$

R.2. $\dfrac{7}{12}$ and $\dfrac{11}{42}$

R.3. $\dfrac{2}{5ab^2}$ and $-\dfrac{4}{15a^2b}$

R.4. $-\dfrac{5}{21c^3}$ and $\dfrac{6}{7c^2d}$

R.5. $\dfrac{x-3}{(x+2)^3(x-5)}$ and $\dfrac{4}{(x+2)^2(x+1)}$

R.6. $\dfrac{y}{(y-3)(y+7)^4}$ and $\dfrac{y+8}{(y-3)^2(y+7)^3}$

R.7. $\dfrac{3}{t^2-6t+9}$ and $\dfrac{t+1}{5t^2-15t}$

R.8. $\dfrac{7}{4p^2+24p}$ and $\dfrac{p+4}{p^2+12p+36}$

For Exercises R.9–R.16, simplify the expression and identify restricted values of the variables.

R.9. $\dfrac{4a^3}{8ab}$

R.10. $\dfrac{21c^5d}{7c^6}$

R.11. $\dfrac{2x^2-32}{x^2-8x+16}$

R.12. $\dfrac{2y^2+4y+2}{10y^2-10}$

R.13. $\dfrac{8w-28}{4w^2-12w-7}$

R.14. $\dfrac{14x+8}{21x^2+5x-4}$

R.15. $\dfrac{8-2x}{x-4}$

R.16. $\dfrac{3x-6}{2-x}$

For Exercises R.17–R.20, simplify the expression.

R.17. $-3x(x-2)$

R.18. $-5x(x-9)$

R.19. $(2x-5)(4x+3)$

R.20. $(8y+1)(2y-3)$

Concept 1: Addition and Subtraction of Rational Expressions With the Same Denominator

For Exercises 1–26, add or subtract the expressions with the same denominators as indicated. **(See Examples 1–2.)**

1. $\dfrac{7}{8}+\dfrac{3}{8}$

2. $\dfrac{1}{3}+\dfrac{7}{3}$

3. $\dfrac{9}{16}-\dfrac{3}{16}$

4. $\dfrac{14}{15}-\dfrac{4}{15}$

5. $\dfrac{3}{x}-\dfrac{4}{x}$

6. $\dfrac{9}{y}-\dfrac{12}{y}$

7. $\dfrac{t}{t+4}+\dfrac{5}{t+4}$

8. $\dfrac{m}{m-7}+\dfrac{3}{m-7}$

9. $\dfrac{5a}{a+2}-\dfrac{3a-4}{a+2}$

10. $\dfrac{2b}{b-3}-\dfrac{b-9}{b-3}$

11. $\dfrac{5c}{c+6}+\dfrac{30}{c+6}$

12. $\dfrac{12}{2+d}+\dfrac{6d}{2+d}$

13. $\dfrac{5}{t-8}-\dfrac{2t+1}{t-8}$

14. $\dfrac{7p+1}{2p+1}-\dfrac{p-4}{2p+1}$

15. $\dfrac{9x^2}{3x-7}-\dfrac{49}{3x-7}$

16. $\dfrac{4w^2}{2w-1}-\dfrac{1}{2w-1}$

17. $\dfrac{m^2}{m+5}+\dfrac{10m+25}{m+5}$

18. $\dfrac{k^2}{k-3}-\dfrac{6k-9}{k-3}$

19. $\dfrac{2a}{a+2}+\dfrac{4}{a+2}$

20. $\dfrac{5b}{b+4}+\dfrac{20}{b+4}$

21. $\dfrac{x^2}{x+5}-\dfrac{25}{x+5}$

22. $\dfrac{y^2}{y-7}-\dfrac{49}{y-7}$

23. $\dfrac{r}{r^2+3r+2}+\dfrac{2}{r^2+3r+2}$

24. $\dfrac{x}{x^2-x-12}-\dfrac{4}{x^2-x-12}$

25. $\dfrac{1}{3y^2+22y+7}-\dfrac{-3y}{3y^2+22y+7}$

26. $\dfrac{5}{2x^2+13x+20}+\dfrac{2x}{2x^2+13x+20}$

For Exercises 27–28, find an expression that represents the perimeter of the figure (assume that $x>0$, $y>0$, and $t>0$).

27.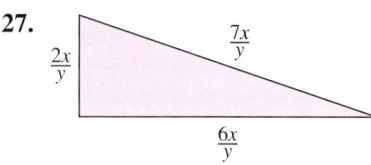

28.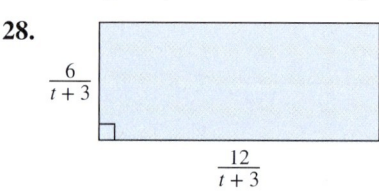

Concept 2: Addition and Subtraction of Rational Expressions With Different Denominators

For Exercises 29–70, add or subtract the expressions with different denominators as indicated. (See Examples 3–7.)

29. $\dfrac{5}{4} + \dfrac{3}{2a}$

30. $\dfrac{11}{6p} + \dfrac{-7}{4p}$

31. $\dfrac{4}{5xy^3} + \dfrac{2x}{15y^2}$

32. $\dfrac{5}{3a^2 b} - \dfrac{7}{6b^2}$

33. $\dfrac{2}{s^3 t^3} - \dfrac{3}{s^4 t}$

34. $\dfrac{1}{p^2 q} - \dfrac{2}{pq^3}$

35. $\dfrac{z}{3z-9} - \dfrac{z-2}{z-3}$

36. $\dfrac{3w-8}{2w-4} - \dfrac{w-3}{w-2}$

37. $\dfrac{5}{a+1} + \dfrac{4}{3a+3}$

38. $\dfrac{2}{c-4} + \dfrac{1}{5c-20}$

39. $\dfrac{k}{k^2-9} - \dfrac{4}{k-3}$

40. $\dfrac{7}{h+2} + \dfrac{2h-3}{h^2-4}$

41. $\dfrac{3a-7}{6a+10} - \dfrac{10}{3a^2+5a}$

42. $\dfrac{k+2}{8k} - \dfrac{3-k}{12k}$

43. $\dfrac{x}{x-4} + \dfrac{3}{x+1}$

44. $\dfrac{4}{y-3} + \dfrac{y}{y-5}$

45. $\dfrac{3x}{x^2+6x+9} + \dfrac{x}{x^2+5x+6}$

46. $\dfrac{7x}{x^2+2xy+y^2} + \dfrac{3x}{x^2+xy}$

47. $\dfrac{p}{3} - \dfrac{4p-1}{-3}$

48. $\dfrac{r}{7} - \dfrac{r-5}{-7}$

49. $\dfrac{8}{x-3} - \dfrac{1}{3-x}$

50. $\dfrac{5y}{y-1} - \dfrac{3y}{1-y}$

51. $\dfrac{4n}{n-8} - \dfrac{2n-1}{8-n}$

52. $\dfrac{m}{m-2} - \dfrac{3m+1}{2-m}$

53. $\dfrac{5}{x} + \dfrac{3}{x+2}$

54. $\dfrac{6}{y-1} + \dfrac{9}{y}$

55. $\dfrac{y}{4y+2} + \dfrac{3y}{6y+3}$

56. $\dfrac{4}{q^2-2q} - \dfrac{5}{3q-6}$

57. $\dfrac{4w}{w^2+2w-3} + \dfrac{2}{1-w}$

58. $\dfrac{z-23}{z^2-z-20} - \dfrac{2}{5-z}$

59. $\dfrac{3a-8}{a^2-5a+6} + \dfrac{a+2}{a^2-6a+8}$

60. $\dfrac{3b+5}{b^2+4b+3} + \dfrac{-b+5}{b^2+2b-3}$

61. $\dfrac{4x}{x^2+4x-5} - \dfrac{x}{x^2+10x+25}$

62. $\dfrac{x}{x^2+5x+4} - \dfrac{2x}{x^2+8x+16}$

63. $\dfrac{3y}{2y^2-y-1} - \dfrac{4y}{2y^2-7y-4}$

64. $\dfrac{5}{6y^2-7y-3} + \dfrac{4y}{3y^2+4y+1}$

65. $\dfrac{3}{2p-1} - \dfrac{4p+4}{4p^2-1}$

66. $\dfrac{1}{3q-2} - \dfrac{6q+4}{9q^2-4}$

67. $\dfrac{m}{m+n} - \dfrac{m}{m-n} + \dfrac{1}{m^2-n^2}$

68. $\dfrac{x}{x+y} - \dfrac{2xy}{x^2-y^2} + \dfrac{y}{x-y}$

69. $\dfrac{2}{a+b} + \dfrac{2}{a-b} - \dfrac{4a}{a^2-b^2}$

70. $\dfrac{-2x}{x^2-y^2} + \dfrac{1}{x+y} - \dfrac{1}{x-y}$

For Exercises 71–72, find an expression that represents the perimeter of the figure (assume that $x > 0$ and $t > 0$).

71.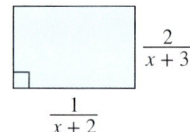

72.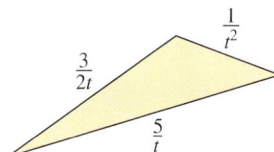

Concept 3: Using Rational Expressions in Translations

73. Let a number be represented by n. Write the reciprocal of n.

74. Write the reciprocal of the sum of a number and 6.

75. Write the quotient of 5 and the sum of a number and 2.

76. Write the quotient of 12 and p.

For Exercises 77–80, translate the English phrases into algebraic expressions. Then simplify by combining the rational expressions. **(See Example 8.)**

77. The sum of a number and the quantity seven times the reciprocal of the number.

78. The sum of a number and the quantity five times the reciprocal of the number.

79. The difference of the reciprocal of n and the quotient of 2 and n.

80. The difference of the reciprocal of m and the quotient of $3m$ and 7.

Expanding Your Skills

For Exercises 81–86, perform the indicated operations.

81. $\dfrac{-3}{w^3 + 27} - \dfrac{1}{w^2 - 9}$

82. $\dfrac{m}{m^3 - 1} + \dfrac{1}{(m - 1)^2}$

83. $\dfrac{2p}{p^2 + 5p + 6} - \dfrac{p + 1}{p^2 + 2p - 3} + \dfrac{3}{p^2 + p - 2}$

84. $\dfrac{3t}{8t^2 + 2t - 1} - \dfrac{5t}{2t^2 - 9t - 5} + \dfrac{2}{4t^2 - 21t + 5}$

85. $\dfrac{3m}{m^2 + 3m - 10} + \dfrac{5}{4 - 2m} - \dfrac{1}{m + 5}$

86. $\dfrac{2n}{3n^2 - 8n - 3} + \dfrac{1}{6 - 2n} - \dfrac{3}{3n + 1}$

For Exercises 87–90, simplify by applying the order of operations.

87. $\left(\dfrac{2}{k + 1} + 3\right)\left(\dfrac{k + 1}{4k + 7}\right)$

88. $\left(\dfrac{p + 1}{3p + 4}\right)\left(\dfrac{1}{p + 1} + 2\right)$

89. $\left(\dfrac{1}{10a} - \dfrac{b}{10a^2}\right) \div \left(\dfrac{1}{10} - \dfrac{b}{10a}\right)$

90. $\left(\dfrac{1}{2m} + \dfrac{n}{2m^2}\right) \div \left(\dfrac{1}{4} + \dfrac{n}{4m}\right)$

Problem Recognition Exercises

Operations on Rational Expressions

We have learned how to simplify, add, subtract, multiply, and divide rational expressions. The procedure for each operation is different, and it takes considerable practice to determine the correct method to apply for a given problem. The following review exercises give you the opportunity to practice the specific techniques for simplifying rational expressions.

For Exercises 1–20, perform any indicated operations, and simplify the expression.

1. $\dfrac{5}{3x+1} - \dfrac{2x-4}{3x+1}$

2. $\dfrac{\dfrac{w+1}{w^2-16}}{\dfrac{w+1}{w+4}}$

3. $\dfrac{3}{y} \cdot \dfrac{y^2-5y}{6y-9}$

4. $\dfrac{-1}{x+3} + \dfrac{2}{2x-1}$

5. $\dfrac{x-9}{9x-x^2}$

6. $\dfrac{1}{p} - \dfrac{3}{p^2+3p} + \dfrac{p}{3p+9}$

7. $\dfrac{c^2+5c+6}{c^2+c-2} \div \dfrac{c}{c-1}$

8. $\dfrac{2x^2-5x-3}{x^2-9} \cdot \dfrac{x^2+6x+9}{10x+5}$

9. $\dfrac{6a^2b^3}{72ab^7c}$

10. $\dfrac{2a}{a+b} - \dfrac{b}{a-b} - \dfrac{-4ab}{a^2-b^2}$

11. $\dfrac{p^2+10pq+25q^2}{p^2+6pq+5q^2} \div \dfrac{10p+50q}{2p^2-2q^2}$

12. $\dfrac{3k-8}{k-5} + \dfrac{k-12}{k-5}$

13. $\dfrac{20x^2+10x}{4x^3+4x^2+x}$

14. $\dfrac{w^2-81}{w^2+10w+9} \cdot \dfrac{w^2+w+2zw+2z}{w^2-9w+zw-9z}$

15. $\dfrac{8x^2-18x-5}{4x^2-25} \div \dfrac{4x^2-11x-3}{3x-9}$

16. $\dfrac{xy+7x+5y+35}{x^2+ax+5x+5a}$

17. $\dfrac{a}{a^2-9} - \dfrac{3}{6a-18}$

18. $\dfrac{4}{y^2-36} + \dfrac{2}{y^2-4y-12}$

19. $(t^2+5t-24)\left(\dfrac{t+8}{t-3}\right)$

20. $\dfrac{6b^2-7b-10}{b-2}$

Section 15.5 Complex Fractions

Concepts

1. Simplifying Complex Fractions (Method I)
2. Simplifying Complex Fractions (Method II)

1. Simplifying Complex Fractions (Method I)

A **complex fraction** is an expression containing one or more fractions in the numerator, denominator, or both. For example,

$$\frac{\frac{1}{ab}}{\frac{2}{b}} \quad \text{and} \quad \frac{1 + \frac{3}{4} - \frac{1}{6}}{\frac{1}{2} + \frac{1}{3}}$$

are complex fractions.

Two methods will be presented to simplify complex fractions. The first method (Method I) follows the order of operations to simplify the numerator and denominator separately before dividing. The process is summarized as follows:

Simplifying a Complex Fraction (Method I)

Step 1 Add or subtract expressions in the numerator to form a single fraction. Add or subtract expressions in the denominator to form a single fraction.

Step 2 Divide the rational expressions from step 1 by multiplying the numerator of the complex fraction by the reciprocal of the denominator of the complex fraction.

Step 3 Simplify to lowest terms if possible.

Example 1 Simplifying a Complex Fraction (Method I)

Simplify the expression. $\dfrac{\frac{1}{ab}}{\frac{2}{b}}$

Solution:

Step 1: The numerator and denominator of the complex fraction are already single fractions.

$$\frac{\frac{1}{ab}}{\frac{2}{b}} \quad \longleftarrow \text{This fraction bar denotes division } (\div).$$

$$= \frac{1}{ab} \div \frac{2}{b}$$

$$= \frac{1}{ab} \cdot \frac{b}{2} \quad \textbf{Step 2:} \text{ Multiply the numerator of the complex fraction by the reciprocal of } \tfrac{2}{b}, \text{ which is } \tfrac{b}{2}.$$

$$= \frac{1}{a\cancel{b}} \cdot \frac{\cancel{b}}{2} \quad \textbf{Step 3:} \text{ Reduce common factors and simplify.}$$

$$= \frac{1}{2a}$$

FOR REVIEW

The expression in Example 1 is a complex fraction with variables. Before attempting Example 1, consider a similar-looking numerical fraction.

$$\frac{\frac{1}{10}}{\frac{2}{5}} = \frac{1}{10} \div \frac{2}{5}$$

$$= \frac{1}{\underset{2}{\cancel{10}}} \cdot \frac{\overset{1}{\cancel{5}}}{2} = \frac{1}{4}$$

Skill Practice Simplify the expression.

1. $\dfrac{\dfrac{6x}{y}}{\dfrac{9}{2y}}$

Sometimes it is necessary to simplify the numerator and denominator of a complex fraction before the division can be performed. This is illustrated in Example 2.

Example 2 — Simplifying a Complex Fraction (Method I)

Simplify the expression. $\dfrac{1+\dfrac{3}{4}-\dfrac{1}{6}}{\dfrac{1}{2}+\dfrac{1}{3}}$

Solution:

$\dfrac{1+\dfrac{3}{4}-\dfrac{1}{6}}{\dfrac{1}{2}+\dfrac{1}{3}}$

Step 1: Combine fractions in the numerator and denominator separately.

$= \dfrac{1\cdot\dfrac{12}{12}+\dfrac{3}{4}\cdot\dfrac{3}{3}-\dfrac{1}{6}\cdot\dfrac{2}{2}}{\dfrac{1}{2}\cdot\dfrac{3}{3}+\dfrac{1}{3}\cdot\dfrac{2}{2}}$

The LCD in the numerator is 12.
The LCD in the denominator is 6.

$= \dfrac{\dfrac{12}{12}+\dfrac{9}{12}-\dfrac{2}{12}}{\dfrac{3}{6}+\dfrac{2}{6}}$

$= \dfrac{\dfrac{19}{12}}{\dfrac{5}{6}}$

Form a single fraction in the numerator and in the denominator.

$= \dfrac{19}{\cancel{12}_{2}}\cdot\dfrac{\cancel{6}^{1}}{5}$

Step 2: Multiply the numerator by the reciprocal of the denominator.

$= \dfrac{19}{10}$

Step 3: Simplify.

Skill Practice Simplify the expression.

2. $\dfrac{\dfrac{3}{4}-\dfrac{1}{6}+2}{\dfrac{1}{3}+\dfrac{1}{2}}$

Answers

1. $\dfrac{4x}{3}$ 2. $\dfrac{31}{10}$

Example 3 — Simplifying a Complex Fraction (Method I)

Simplify the expression.

$$\frac{\dfrac{1}{x}+\dfrac{1}{y}}{x-\dfrac{y^2}{x}}$$

Solution:

$$\frac{\dfrac{1}{x}+\dfrac{1}{y}}{x-\dfrac{y^2}{x}}$$

The LCD in the numerator is xy.
The LCD in the denominator is x.

$$=\frac{\dfrac{1\cdot y}{x\cdot y}+\dfrac{1\cdot x}{y\cdot x}}{\dfrac{x\cdot x}{1\cdot x}-\dfrac{y^2}{x}}$$

Rewrite the expressions using common denominators.

$$=\frac{\dfrac{y}{xy}+\dfrac{x}{xy}}{\dfrac{x^2}{x}-\dfrac{y^2}{x}}$$

$$=\frac{\dfrac{y+x}{xy}}{\dfrac{x^2-y^2}{x}}$$

Form single fractions in the numerator and denominator.

$$=\frac{y+x}{xy}\cdot\frac{x}{x^2-y^2}$$

Multiply the numerator by the reciprocal of the denominator.

$$=\frac{\cancel{y+x}}{\cancel{x}y}\cdot\frac{\cancel{x}}{(\cancel{x+y})(x-y)}$$

Factor and reduce. Note that $(y+x)=(x+y)$.

$$=\frac{1}{y(x-y)}$$

Simplify.

Skill Practice Simplify the expression.

3. $\dfrac{1-\dfrac{1}{p}}{\dfrac{p}{w}+\dfrac{w}{p}}$

2. Simplifying Complex Fractions (Method II)

We will now simplify the expressions from Examples 2 and 3 again using a second method to simplify complex fractions (Method II). Recall that multiplying the numerator and denominator of a rational expression by the same quantity does not change the value of the expression because we are multiplying by a number equivalent to 1. This is the basis for Method II.

Answer
3. $\dfrac{w(p-1)}{p^2+w^2}$

Section 15.5 Complex Fractions   1019

> **Simplifying a Complex Fraction (Method II)**
> **Step 1** Multiply the numerator and denominator of the complex fraction by the LCD of *all* individual fractions within the expression.
> **Step 2** Apply the distributive property, and simplify the numerator and denominator.
> **Step 3** Simplify to lowest terms if possible.

Example 4   Simplifying a Complex Fraction (Method II)

Simplify the expression.
$$\frac{1 + \frac{3}{4} - \frac{1}{6}}{\frac{1}{2} + \frac{1}{3}}$$

Solution:

$$\frac{1 + \frac{3}{4} - \frac{1}{6}}{\frac{1}{2} + \frac{1}{3}}$$

The LCD of the expressions $1, \frac{3}{4}, \frac{1}{6}, \frac{1}{2},$ and $\frac{1}{3}$ is 12.

$$= \frac{12\left(1 + \frac{3}{4} - \frac{1}{6}\right)}{12\left(\frac{1}{2} + \frac{1}{3}\right)}$$

Step 1: Multiply the numerator and denominator of the complex fraction by 12.

TIP: In step 1, we multiply the original expression by $\frac{12}{12}$, which equals 1.

$$= \frac{12 \cdot 1 + 12 \cdot \frac{3}{4} - 12 \cdot \frac{1}{6}}{12 \cdot \frac{1}{2} + 12 \cdot \frac{1}{3}}$$

Step 2: Apply the distributive property.

FOR REVIEW

Recall that to multiply a whole number by a fraction, we can write the whole number over 1.

$$12 \cdot \left(\frac{3}{4}\right) = \frac{\overset{3}{\cancel{12}}}{1} \cdot \left(\frac{3}{\cancel{4}_1}\right) = 9$$

$$= \frac{12 \cdot 1 + \overset{3}{\cancel{12}} \cdot \frac{3}{\cancel{4}} - \overset{2}{\cancel{12}} \cdot \frac{1}{\cancel{6}}}{\overset{6}{\cancel{12}} \cdot \frac{1}{\cancel{2}} + \overset{4}{\cancel{12}} \cdot \frac{1}{\cancel{3}}}$$

Simplify each term.

$$= \frac{12 + 9 - 2}{6 + 4}$$

$$= \frac{19}{10}$$

Step 3: Simplify. This is the same result as in Example 2.

Skill Practice Simplify the expression.

4. $\dfrac{1 - \frac{3}{5}}{\frac{1}{4} - \frac{7}{10} + 1}$

Answer
4. $\dfrac{8}{11}$

Chapter 15 Rational Expressions and Equations

Example 5 Simplifying a Complex Fraction (Method II)

Simplify the expression. $\dfrac{\dfrac{1}{x}+\dfrac{1}{y}}{x-\dfrac{y^2}{x}}$

Solution:

$\dfrac{\dfrac{1}{x}+\dfrac{1}{y}}{x-\dfrac{y^2}{x}}$ The LCD of the expressions $\tfrac{1}{x}, \tfrac{1}{y}, x,$ and $\tfrac{y^2}{x}$ is xy.

$=\dfrac{xy\left(\dfrac{1}{x}+\dfrac{1}{y}\right)}{xy\left(x-\dfrac{y^2}{x}\right)}$ **Step 1:** Multiply numerator and denominator of the complex fraction by xy.

$=\dfrac{xy\cdot\dfrac{1}{x}+xy\cdot\dfrac{1}{y}}{xy\cdot x-xy\cdot\dfrac{y^2}{x}}$ **Step 2:** Apply the distributive property, and simplify each term.

$=\dfrac{y+x}{x^2y-y^3}$

$=\dfrac{y+x}{y(x^2-y^2)}$ **Step 3:** Factor completely, and reduce common factors.

$=\dfrac{\cancel{y+x}}{y\cancel{(x+y)}(x-y)}$ Note that $(y+x)=(x+y)$.

$=\dfrac{1}{y(x-y)}$ This is the same result as in Example 3.

Skill Practice Simplify the expression.

5. $\dfrac{\dfrac{z}{3}-\dfrac{3}{z}}{1+\dfrac{3}{z}}$

Answer

5. $\dfrac{z-3}{3}$

Example 6 Simplifying a Complex Fraction (Method II)

Simplify the expression. $\dfrac{\dfrac{1}{k+1} - 1}{\dfrac{1}{k+1} + 1}$

Solution:

$\dfrac{\dfrac{1}{k+1} - 1}{\dfrac{1}{k+1} + 1}$ The LCD of $\dfrac{1}{k+1}$ and 1 is $(k+1)$.

$= \dfrac{(k+1)\left(\dfrac{1}{k+1} - 1\right)}{(k+1)\left(\dfrac{1}{k+1} + 1\right)}$ **Step 1:** Multiply numerator and denominator of the complex fraction by $(k+1)$.

$= \dfrac{(k+1) \cdot \dfrac{1}{(k+1)} - (k+1) \cdot 1}{(k+1) \cdot \dfrac{1}{(k+1)} + (k+1) \cdot 1}$ **Step 2:** Apply the distributive property.

$= \dfrac{1 - (k+1)}{1 + (k+1)}$ Simplify.

$= \dfrac{1 - k - 1}{1 + k + 1}$

$= \dfrac{-k}{k+2}$ **Step 3:** The expression is already in lowest terms.

Skill Practice Simplify the expression.

6. $\dfrac{\dfrac{4}{p-3} + 1}{1 + \dfrac{2}{p-3}}$

Answer

6. $\dfrac{p+1}{p-1}$

Section 15.5 Activity

A.1. a. Add the fractions. $\dfrac{2}{3} + \dfrac{1}{2}$

    b. Subtract the fractions. $\dfrac{3}{4} - \dfrac{1}{3}$

    c. Simplify the complex fraction using the order of operations (Method I). That is, simplify the numerator, simplify the denominator, and then divide the results.

$$\dfrac{\dfrac{2}{3} + \dfrac{1}{2}}{\dfrac{3}{4} - \dfrac{1}{3}}$$

A.2. **a.** Add the fractions. $\dfrac{1}{ab} + \dfrac{2}{a}$

b. Subtract the fractions. $\dfrac{3}{b} - \dfrac{1}{ab}$

c. Use Method I to simplify the complex fraction.

$$\dfrac{\dfrac{1}{ab} + \dfrac{2}{a}}{\dfrac{3}{b} - \dfrac{1}{ab}}$$

A.3. **a.** Given the fractions $\dfrac{2}{3}, \dfrac{1}{2}, \dfrac{3}{4}$, and $\dfrac{1}{3}$, identify the LCD.

b. Multiply. $12 \cdot \left(\dfrac{2}{3} + \dfrac{1}{2}\right)$

c. Multiply. $12 \cdot \left(\dfrac{3}{4} - \dfrac{1}{3}\right)$

d. Simplify the complex fraction $\dfrac{\dfrac{2}{3} + \dfrac{1}{2}}{\dfrac{3}{4} - \dfrac{1}{3}}$ by using Method II. To begin, multiply the numerator and denominator by the LCD of all terms in the fraction. Compare your answer to the answer to Exercise A.1(c).

$$\dfrac{12 \cdot \left(\dfrac{2}{3} + \dfrac{1}{2}\right)}{12 \cdot \left(\dfrac{3}{4} - \dfrac{1}{3}\right)} =$$

A.4. **a.** Given the expressions $\dfrac{1}{ab}, \dfrac{2}{a}, \dfrac{3}{b}$, and $\dfrac{1}{ab}$, identify the LCD.

b. Multiply. $ab \cdot \left(\dfrac{1}{ab} + \dfrac{2}{a}\right)$

c. Multiply. $ab \cdot \left(\dfrac{3}{b} - \dfrac{1}{ab}\right)$

d. Simplify the complex fraction by using Method II. That is, multiply the numerator and denominator by the LCD of all individual terms in the complex fraction. Compare your answer to the answer to Exercise A.2(c).

$$\dfrac{ab \cdot \left(\dfrac{1}{ab} + \dfrac{2}{a}\right)}{ab \cdot \left(\dfrac{3}{b} - \dfrac{1}{ab}\right)} =$$

A.5. Simplify the complex fraction by using either Method I or Method II.

$$\dfrac{2 - \dfrac{1}{w}}{4 - \dfrac{1}{w^2}}$$

Practice Exercises

Section 15.5

Prerequisite Review

For Exercises R.1–R.6, apply the distributive property.

R.1. $12\left(\dfrac{3}{4}x - \dfrac{1}{3}\right)$ **R.2.** $15\left(\dfrac{2}{3}y + \dfrac{3}{5}\right)$ **R.3.** $10\left(\dfrac{1}{2}w^2 - \dfrac{2}{5}w - 2\right)$

R.4. $9\left(t^2 - \dfrac{1}{3}t + \dfrac{2}{9}\right)$ **R.5.** $ab\left(\dfrac{1}{a} - \dfrac{1}{b}\right)$ **R.6.** $x^2y^2\left(\dfrac{1}{x^2y} - \dfrac{1}{xy^2}\right)$

For Exercises R.7–R.10, perform the indicated operation.

R.7. $\dfrac{2}{3} + \dfrac{5}{8}$ **R.8.** $\dfrac{3}{4} - \dfrac{5}{7}$ **R.9.** $\dfrac{5}{24} \div \dfrac{15}{28}$ **R.10.** $-\dfrac{5}{14} \div \dfrac{20}{21}$

For Exercises R.11–R.12, write the expression with positive exponents only.

R.11. $5t^{-2}$ **R.12.** $-3x^{-1}$

Vocabulary and Key Concepts

1. A _____ fraction is an expression containing one or more fractions in the numerator, denominator, or both.

For Exercises 2–6,
 a. Write the division expression as an equivalent multiplication expression.
 b. Simplify the expression.

2. $\dfrac{2}{3} \div \dfrac{10}{21}$ **3.** $\dfrac{7}{10} \div \dfrac{14}{3}$ **4.** $\dfrac{4}{15} \div \dfrac{20}{3}$

5. $\dfrac{2x}{15} \div \dfrac{4x}{25}$ **6.** $\dfrac{3t}{10} \div \dfrac{7t}{20}$

Concepts 1–2: Simplifying Complex Fractions (Methods I and II)

For Exercises 7–34, simplify the complex fractions using either method. (See Examples 1–6.)

7. $\dfrac{\dfrac{7}{18y}}{\dfrac{2}{9}}$ **8.** $\dfrac{\dfrac{a^2}{2a-3}}{\dfrac{5a}{8a-12}}$ **9.** $\dfrac{\dfrac{3x+2y}{2y}}{\dfrac{6x+4y}{2}}$ **10.** $\dfrac{\dfrac{2x-10}{4}}{\dfrac{x^2-5x}{3x}}$

11. $\dfrac{\dfrac{8a^4b^3}{3c}}{\dfrac{a^7b^2}{9c}}$ **12.** $\dfrac{\dfrac{12x^2}{5y}}{\dfrac{8x^6}{9y^2}}$ **13.** $\dfrac{\dfrac{4r^3s}{t^5}}{\dfrac{2s^7}{r^2t^9}}$ **14.** $\dfrac{\dfrac{5p^4q}{w^4}}{\dfrac{10p^2}{qw^2}}$

15. $\dfrac{\dfrac{1}{8}+\dfrac{4}{3}}{\dfrac{1}{2}-\dfrac{5}{12}}$

16. $\dfrac{\dfrac{8}{9}-\dfrac{1}{3}}{\dfrac{7}{6}+\dfrac{1}{9}}$

17. $\dfrac{\dfrac{1}{h}+\dfrac{1}{k}}{\dfrac{1}{hk}}$

18. $\dfrac{\dfrac{1}{b}+1}{\dfrac{1}{b}}$

19. $\dfrac{\dfrac{n+1}{n^2-9}}{\dfrac{2}{n+3}}$

20. $\dfrac{\dfrac{5}{k-5}}{\dfrac{k+1}{k^2-25}}$

21. $\dfrac{2+\dfrac{1}{x}}{4+\dfrac{1}{x}}$

22. $\dfrac{6+\dfrac{6}{k}}{1+\dfrac{1}{k}}$

23. $\dfrac{\dfrac{m}{7}-\dfrac{7}{m}}{\dfrac{1}{7}+\dfrac{1}{m}}$

24. $\dfrac{\dfrac{2}{p}+\dfrac{p}{2}}{\dfrac{p}{3}-\dfrac{3}{p}}$

25. $\dfrac{\dfrac{1}{5}-\dfrac{1}{y}}{\dfrac{7}{10}+\dfrac{1}{y^2}}$

26. $\dfrac{\dfrac{1}{m^2}+\dfrac{2}{3}}{\dfrac{1}{m}-\dfrac{5}{6}}$

27. $\dfrac{\dfrac{8}{a+4}+2}{\dfrac{12}{a+4}-2}$

28. $\dfrac{\dfrac{2}{w+1}+3}{\dfrac{3}{w+1}+4}$

29. $\dfrac{1-\dfrac{4}{t^2}}{1-\dfrac{2}{t}-\dfrac{8}{t^2}}$

30. $\dfrac{1-\dfrac{9}{p^2}}{1-\dfrac{1}{p}-\dfrac{6}{p^2}}$

31. $\dfrac{t+4+\dfrac{3}{t}}{t-4-\dfrac{5}{t}}$

32. $\dfrac{\dfrac{9}{4m}+\dfrac{9}{2m^2}}{\dfrac{3}{2}+\dfrac{3}{m}}$

33. $\dfrac{\dfrac{1}{k-6}-1}{\dfrac{2}{k-6}-2}$

34. $\dfrac{\dfrac{3}{y-3}+4}{8+\dfrac{6}{y-3}}$

For Exercises 35–38, write the English phrases as algebraic expressions. Then simplify the expressions.

35. The sum of one-half and two-thirds, divided by five

36. The quotient of ten and the difference of two-fifths and one-fourth

37. The quotient of three and the sum of two-thirds and three-fourths

38. The difference of three-fifths and one-half, divided by four

39. In electronics, resistors oppose the flow of current. For two resistors in parallel, the total resistance is given by

$$R = \dfrac{1}{\dfrac{1}{R_1}+\dfrac{1}{R_2}}$$

a. Find the total resistance if $R_1 = 2\ \Omega$ (ohms) and $R_2 = 3\ \Omega$.

b. Find the total resistance if $R_1 = 10\ \Omega$ and $R_2 = 15\ \Omega$.

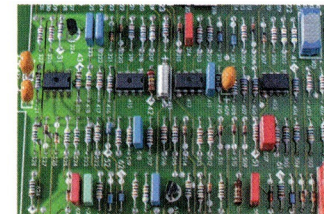

Igor Batenev/Shutterstock

40. Suppose that Joëlle makes a round trip to a location that is d miles away. If the average rate going to the location is r_1 and the average rate on the return trip is given by r_2, the average rate of the entire trip, R, is given by

$$R = \dfrac{2d}{\dfrac{d}{r_1}+\dfrac{d}{r_2}}$$

a. Find the average rate of a trip to a destination 30 mi away when the average rate going there is 60 mph and the average rate returning home is 45 mph. (Round to the nearest tenth of a mile per hour.)

b. Find the average rate of a trip to a destination that is 50 mi away if the driver travels at the same rates as in part (a). (Round to the nearest tenth of a mile per hour.)

c. Compare your answers from parts (a) and (b) and explain the results in the context of the problem.

Expanding Your Skills

For Exercises 41–50, simplify the complex fractions using either method.

41. $\dfrac{x^{-1} - y^{-1}}{x^{-2} - y^{-2}}$

42. $\dfrac{a^{-2} - 1}{a^{-1} - 1}$

43. $\dfrac{2x^{-1} + 8y^{-1}}{4x^{-1}}$

$\left(\text{Hint: } 2x^{-1} = \dfrac{2}{x}\right)$

44. $\dfrac{6a^{-1} + 4b^{-1}}{8b^{-1}}$

45. $\dfrac{(mn)^{-2}}{m^{-2} + n^{-2}}$

46. $\dfrac{(xy)^{-1}}{2x^{-1} + 3y^{-1}}$

47. $\dfrac{\dfrac{1}{z^2 - 9} + \dfrac{2}{z + 3}}{\dfrac{3}{z - 3}}$

48. $\dfrac{\dfrac{5}{w^2 - 25} - \dfrac{3}{w + 5}}{\dfrac{4}{w - 5}}$

49. $\dfrac{\dfrac{2}{x - 1} + 2}{\dfrac{2}{x + 1} - 2}$

50. $\dfrac{\dfrac{1}{y - 3} + 1}{\dfrac{2}{y + 3} - 1}$

For Exercises 51–52, simplify the complex fractions. (*Hint:* Use the order of operations and begin with the fraction on the lower right.)

51. $1 + \dfrac{1}{1 + 1}$

52. $1 + \dfrac{1}{1 + \dfrac{1}{1 + 1}}$

Rational Equations

Section 15.6

1. Introduction to Rational Equations

Thus far we have studied two specific types of equations in one variable: linear equations and quadratic equations. Recall,

$$ax + b = c, \text{ where } a \neq 0, \text{ is a } \textbf{linear equation}.$$

$$ax^2 + bx + c = 0, \text{ where } a \neq 0, \text{ is a } \textbf{quadratic equation}.$$

We will now study another type of equation called a rational equation.

Concepts

1. Introduction to Rational Equations
2. Solving Rational Equations
3. Solving Formulas Involving Rational Expressions

> **Definition of a Rational Equation**
>
> An equation with one or more rational expressions is called a **rational equation**.

The following equations are rational equations:

$$\frac{1}{x} + \frac{1}{3} = \frac{5}{6} \qquad \frac{6}{t^2 - 7t + 12} + \frac{2t}{t - 3} = \frac{3t}{t - 4}$$

To understand the process of solving a rational equation, first review the process of clearing fractions. We can clear the fractions in an equation by multiplying both sides of the equation by the LCD of all terms.

Example 1 — Solving an Equation Containing Fractions

Solve. $\quad \dfrac{y}{2} + \dfrac{y}{4} = 6$

Solution:

$\dfrac{y}{2} + \dfrac{y}{4} = 6$ \qquad The LCD of all terms in the equation is 4.

$4\left(\dfrac{y}{2} + \dfrac{y}{4}\right) = 4(6)$ \qquad Multiply both sides of the equation by 4 to clear fractions.

$\overset{2}{\cancel{4}} \cdot \dfrac{y}{2} + \overset{1}{\cancel{4}} \cdot \dfrac{y}{4} = 4(6)$ \qquad Apply the distributive property.

$2y + y = 24$ \qquad Clear fractions.

$3y = 24$ \qquad Solve the resulting equation (linear).

$y = 8$

Check: $\quad \dfrac{y}{2} + \dfrac{y}{4} = 6$

$\dfrac{(8)}{2} + \dfrac{(8)}{4} \overset{?}{=} 6$

$4 + 2 \overset{?}{=} 6$

$6 \overset{?}{=} 6 \checkmark$ (True)

The solution set is $\{8\}$.

Skill Practice Solve the equation.

1. $\dfrac{t}{5} - \dfrac{t}{4} = 2$

FOR REVIEW

The technique of clearing fractions utilizes the multiplication property of equality. Multiplying both sides of an equation by the same nonzero real number results in an equivalent equation.

Answer

1. $\{-40\}$

2. Solving Rational Equations

The same process of clearing fractions is used to solve rational equations when variables are present in the denominator. However, variables in the denominator make it necessary to take note of the restricted values.

Example 2 — Solving a Rational Equation

Solve the equation. $\dfrac{x+1}{x} + \dfrac{1}{3} = \dfrac{5}{6}$

Solution:

$$\dfrac{x+1}{x} + \dfrac{1}{3} = \dfrac{5}{6}$$

The LCD of all the expressions is $6x$. The restricted value is $x = 0$.

$$6x \cdot \left(\dfrac{x+1}{x} + \dfrac{1}{3}\right) = 6x \cdot \left(\dfrac{5}{6}\right)$$

Multiply by the LCD.

$$\overset{1}{6x} \cdot \left(\dfrac{x+1}{\cancel{x}}\right) + \overset{2}{\cancel{6x}} \cdot \left(\dfrac{1}{\cancel{3}}\right) = \overset{1}{\cancel{6x}} \cdot \left(\dfrac{5}{\cancel{6}}\right)$$

Apply the distributive property.

$$6(x+1) + 2x = 5x$$ Clear fractions.
$$6x + 6 + 2x = 5x$$ Solve the resulting equation.
$$8x + 6 = 5x$$
$$3x = -6$$
$$x = -2 \qquad \text{-2 is not a restricted value.}$$

Check: $\dfrac{x+1}{x} + \dfrac{1}{3} = \dfrac{5}{6}$

$$\dfrac{(-2)+1}{(-2)} + \dfrac{1}{3} \overset{?}{=} \dfrac{5}{6}$$

$$\dfrac{-1}{-2} + \dfrac{1}{3} \overset{?}{=} \dfrac{5}{6}$$

$$\dfrac{1}{2} + \dfrac{1}{3} \overset{?}{=} \dfrac{5}{6}$$

$$\dfrac{3}{6} + \dfrac{2}{6} \overset{?}{=} \dfrac{5}{6}$$

$$\dfrac{5}{6} \overset{?}{=} \dfrac{5}{6} \checkmark \text{ (True)}$$

The solution set is $\{-2\}$.

TIP: The restricted value tells us that $x = 0$ is *not* a possible solution to the equation.

Skill Practice Solve the equation.

2. $\dfrac{3}{4} + \dfrac{5+a}{a} = \dfrac{1}{2}$

Answer

2. $\{-4\}$

Example 3   Solving a Rational Equation

Solve the equation. $1 + \dfrac{3a}{a-2} = \dfrac{6}{a-2}$

Solution:

$$1 + \dfrac{3a}{a-2} = \dfrac{6}{a-2}$$

The LCD of all the expressions is $a - 2$. The restricted value is $a = 2$.

$$(a-2)\left(1 + \dfrac{3a}{a-2}\right) = (a-2)\left(\dfrac{6}{a-2}\right)$$

Multiply by the LCD.

$$(a-2)1 + (a-2)\left(\dfrac{3a}{a-2}\right) = (a-2)\left(\dfrac{6}{a-2}\right)$$

Apply the distributive property.

$$a - 2 + 3a = 6$$
$$4a - 2 = 6$$
$$4a = 8$$
$$a = 2$$

Solve the resulting equation (linear).

2 is a restricted value.

Check: $1 + \dfrac{3a}{a-2} = \dfrac{6}{a-2}$

$$1 + \dfrac{3(2)}{(2)-2} \stackrel{?}{=} \dfrac{6}{(2)-2}$$

$$1 + \dfrac{6}{0} \stackrel{?}{=} \dfrac{6}{0}$$

The denominator is 0 when $a = 2$.

Because the value $a = 2$ makes the denominator zero in one (or more) of the rational expressions within the equation, the equation is undefined for $a = 2$. No other potential solutions exist for the equation; therefore, the solution set is { }.

Skill Practice   Solve the equation.

3. $\dfrac{x}{x+1} - 2 = \dfrac{-1}{x+1}$

Examples 1–3 show that the steps to solve a rational equation mirror the process of clearing fractions. However, there is one significant difference. The solutions of a rational equation must not make the denominator equal to zero for any expression within the equation.

Answer

3. { } (The value −1 does not check.)

The steps to solve a rational equation are summarized as follows:

> **Solving a Rational Equation**
> **Step 1** Factor the denominators of all rational expressions. Identify the restricted values.
> **Step 2** Identify the LCD of all expressions in the equation.
> **Step 3** Multiply both sides of the equation by the LCD.
> **Step 4** Solve the resulting equation.
> **Step 5** Check potential solutions in the original equation.

After multiplying by the LCD and then simplifying, the rational equation will be either a linear equation or a higher degree equation.

Example 4 Solving a Rational Equation

Solve the equation. $\quad 1 - \dfrac{4}{p} = -\dfrac{3}{p^2}$

Solution:

$1 - \dfrac{4}{p} = -\dfrac{3}{p^2}$

Step 1: The denominators are already factored. The restricted value is $p = 0$.

Step 2: The LCD of all expressions is p^2.

$p^2\left(1 - \dfrac{4}{p}\right) = p^2\left(-\dfrac{3}{p^2}\right)$

Step 3: Multiply by the LCD.

$p^2(1) - p^2\left(\dfrac{4}{p}\right) = p^2\left(-\dfrac{3}{p^2}\right)$

Apply the distributive property.

$p^2 - 4p = -3$

Step 4: Solve the resulting quadratic equation.

$p^2 - 4p + 3 = 0$

Set the equation equal to zero and factor.

$(p - 3)(p - 1) = 0$

$p - 3 = 0 \quad \text{or} \quad p - 1 = 0$

Set each factor equal to zero.

$p = 3 \quad \text{or} \quad p = 1$

Step 5: Check: $p = 3$ $\quad$ Check: $p = 1$

3 and 1 are not restricted values.

$1 - \dfrac{4}{p} = -\dfrac{3}{p^2} \qquad 1 - \dfrac{4}{p} = -\dfrac{3}{p^2}$

$1 - \dfrac{4}{(3)} \stackrel{?}{=} -\dfrac{3}{(3)^2} \qquad 1 - \dfrac{4}{(1)} \stackrel{?}{=} -\dfrac{3}{(1)^2}$

$\dfrac{3}{3} - \dfrac{4}{3} \stackrel{?}{=} -\dfrac{3}{9} \qquad 1 - 4 \stackrel{?}{=} -3$

$-\dfrac{1}{3} \stackrel{?}{=} -\dfrac{1}{3} \checkmark \qquad -3 \stackrel{?}{=} -3 \checkmark$

The solution set is $\{3, 1\}$.

> **FOR REVIEW**
> Recall that to solve a quadratic equation, set one side equal to zero and factor the other side. If the product of factors equals zero, then one or both factors must be zero.
> Thus, if
> $(p - 3)(p - 1) = 0$, then
> $p - 3 = 0$ or $p - 1 = 0$.

Skill Practice Solve the equation.

4. $\dfrac{z}{2} - \dfrac{1}{2z} = \dfrac{12}{z}$

Answer
4. $\{5, -5\}$

Example 5   **Solving a Rational Equation**

Solve the equation. $\quad \dfrac{6}{t^2 - 7t + 12} + \dfrac{2t}{t-3} = \dfrac{3t}{t-4}$

Solution:

$$\dfrac{6}{t^2 - 7t + 12} + \dfrac{2t}{t-3} = \dfrac{3t}{t-4}$$

$$\dfrac{6}{(t-3)(t-4)} + \dfrac{2t}{t-3} = \dfrac{3t}{t-4}$$

Step 1: Factor the denominators. The restricted values are $t = 3$ and $t = 4$.

Step 2: The LCD is $(t-3)(t-4)$.

Step 3: Multiply by the LCD on both sides.

$$(t-3)(t-4)\left[\dfrac{6}{(t-3)(t-4)} + \dfrac{2t}{t-3}\right] = (t-3)(t-4)\left(\dfrac{3t}{t-4}\right)$$

$$\cancel{(t-3)}\cancel{(t-4)}\left[\dfrac{6}{\cancel{(t-3)}\cancel{(t-4)}}\right] + \cancel{(t-3)}(t-4)\left(\dfrac{2t}{\cancel{t-3}}\right) = (t-3)\cancel{(t-4)}\left(\dfrac{3t}{\cancel{t-4}}\right)$$

$6 + 2t(t-4) = 3t(t-3)$

$6 + 2t^2 - 8t = 3t^2 - 9t$ **Step 4:** Solve the resulting equation.

$0 = 3t^2 - 2t^2 - 9t + 8t - 6$ Because the resulting equation is quadratic, set the equation equal to zero and factor.

$0 = t^2 - t - 6$

$0 = (t-3)(t+2)$

$t - 3 = 0 \quad \text{or} \quad t + 2 = 0$ Set each factor equal to zero.

$t = 3 \quad \text{or} \quad t = -2$

3 is a restricted value, but -2 is not restricted.

Step 5: Check the potential solutions in the original equation.

<u>Check</u>: $t = 3$

3 cannot be a solution to the equation because it will make the denominator zero in the original equation.

$$\dfrac{6}{t^2 - 7t + 12} + \dfrac{2t}{t-3} = \dfrac{3t}{t-4}$$

$$\dfrac{6}{(3)^2 - 7(3) + 12} + \dfrac{2(3)}{(3) - 3} \stackrel{?}{=} \dfrac{3(3)}{(3) - 4}$$

$$\dfrac{6}{0} + \dfrac{6}{0} \stackrel{?}{=} \dfrac{9}{-1}$$

Zero in the denominator

The solution set is $\{-2\}$.

<u>Check</u>: $t = -2$

$$\dfrac{6}{t^2 - 7t + 12} + \dfrac{2t}{t-3} = \dfrac{3t}{t-4}$$

$$\dfrac{6}{(-2)^2 - 7(-2) + 12} + \dfrac{2(-2)}{(-2) - 3} \stackrel{?}{=} \dfrac{3(-2)}{(-2) - 4}$$

$$\dfrac{6}{4 + 14 + 12} + \dfrac{-4}{-5} \stackrel{?}{=} \dfrac{-6}{-6}$$

$$\dfrac{6}{30} + \dfrac{4}{5} \stackrel{?}{=} 1$$

$$\dfrac{1}{5} + \dfrac{4}{5} \stackrel{?}{=} 1 \;\checkmark\; \text{(True)}$$

$t = -2$ is a solution.

Skill Practice Solve the equation.

5. $\dfrac{-8}{x^2 + 6x + 8} + \dfrac{x}{x+4} = \dfrac{2}{x+2}$

Answer

5. $\{4\}$ (The value -4 does not check.)

Example 6 Translating to a Rational Equation

Ten times the reciprocal of a number is added to four. The result is equal to the quotient of twenty-two and the number. Find the number.

Solution:

Let x represent the number.

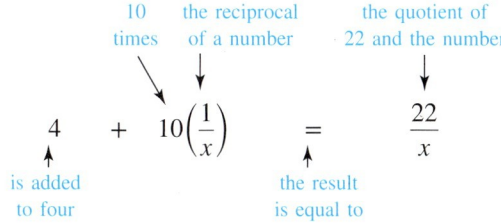

$$4 + \frac{10}{x} = \frac{22}{x}$$

Step 1: The denominators are already factored. The restricted value is $x = 0$.

Step 2: The LCD is x.

$$x\left(4 + \frac{10}{x}\right) = x\left(\frac{22}{x}\right)$$

Step 3: Multiply both sides by the LCD.

$$4x + 10 = 22$$

Apply the distributive property.

$$4x = 12$$

Step 4: Solve the resulting linear equation.

$x = 3$ is a potential solution.

Step 5: 3 is not a restricted value. Substituting $x = 3$ into the original equation verifies that it is a solution.

The number is 3.

Skill Practice

6. The quotient of ten and a number is two less than four times the reciprocal of the number. Find the number.

3. Solving Formulas Involving Rational Expressions

A rational equation may have more than one variable. To solve for a specific variable within a rational equation, we can still apply the principle of clearing fractions.

Answer

6. The number is -3.

Example 7 — Solving Formulas Involving Rational Equations

Solve for k. $F = \dfrac{ma}{k}$

Solution:

To solve for k, we must clear fractions so that k no longer appears in the denominator.

$$F = \dfrac{ma}{k} \qquad \text{The LCD is } k.$$

$$k \cdot (F) = k \cdot \left(\dfrac{ma}{k}\right) \qquad \text{Multiply both sides of the equation by the LCD.}$$

$$kF = ma \qquad \text{Clear fractions.}$$

$$\dfrac{kF}{F} = \dfrac{ma}{F} \qquad \text{Divide both sides by } F.$$

$$k = \dfrac{ma}{F}$$

Skill Practice

7. Solve for t. $C = \dfrac{rt}{d}$

Example 8 — Solving Formulas Involving Rational Equations

A formula to find the height of a trapezoid given its area and the lengths of the two parallel sides is $h = \dfrac{2A}{B + b}$. Solve for b, the length of one of the parallel sides.

Solution:

To solve for b, we must clear fractions so that b no longer appears in the denominator.

$$h = \dfrac{2A}{B + b} \qquad \text{The LCD is } (B + b).$$

$$h(B + b) = \left(\dfrac{2A}{B + b}\right) \cdot (B + b) \qquad \text{Multiply both sides of the equation by the LCD.}$$

$$hB + hb = 2A \qquad \text{Apply the distributive property.}$$

$$hb = 2A - hB \qquad \text{Subtract } hB \text{ from both sides to isolate the } b \text{ term.}$$

$$\dfrac{hb}{h} = \dfrac{2A - hB}{h} \qquad \text{Divide by } h.$$

$$b = \dfrac{2A - hB}{h}$$

Avoiding Mistakes

Algebra is case-sensitive. The variables B and b represent different values.

Skill Practice

8. Solve the formula for x. $y = \dfrac{3}{x - 2}$

Answers

7. $t = \dfrac{Cd}{r}$

8. $x = \dfrac{3 + 2y}{y}$ or $x = \dfrac{3}{y} + 2$

> **TIP:** The solution to Example 8 can be written in several forms. The quantity
> $$\frac{2A - hB}{h}$$
> can be left as a single rational expression or can be split into two fractions and simplified.
> $$b = \frac{2A - hB}{h} = \frac{2A}{h} - \frac{hB}{h} = \frac{2A}{h} - B$$

Example 9 — Solving Formulas Involving Rational Equations

Solve for z. $y = \dfrac{x - z}{x + z}$

Solution:

To solve for z, we must clear fractions so that z no longer appears in the denominator.

$y = \dfrac{x - z}{x + z}$	LCD is $(x + z)$.
$y(x + z) = \left(\dfrac{x - z}{x + z}\right)(x + z)$	Multiply both sides of the equation by the LCD.
$yx + yz = x - z$	Apply the distributive property.
$yz + z = x - yx$	Collect z terms on one side of the equation and collect terms not containing z on the other side.
$z(y + 1) = x - yx$	Factor out z.
$z = \dfrac{x - yx}{y + 1}$	Divide by $y + 1$ to solve for z.

Skill Practice

9. Solve for h. $\dfrac{b}{x} = \dfrac{a}{h} + 1$

Answer

9. $h = \dfrac{ax}{b - x}$ or $\dfrac{-ax}{x - b}$

Section 15.6 Activity

A.1. a. What is the LCD of the expressions $\dfrac{1}{3}y$, $\dfrac{1}{2}$, and $\dfrac{5}{6}y$?

b. Multiply. $6 \cdot \left(\dfrac{1}{3}y - \dfrac{1}{2}\right)$

c. Multiply. $6 \cdot \left(\dfrac{5}{6}y + \dfrac{1}{2}\right)$

d. Solve the equation by first multiplying both sides by the LCD of all terms in the equation.

$$\dfrac{1}{3}y - \dfrac{1}{2} = \dfrac{5}{6}y + \dfrac{1}{2}$$

A.2. a. Factor the denominators of the terms in the equation
$$\frac{2x}{x+2} = \frac{5}{x^2-x-6} - \frac{1}{x-3}.$$

b. What are the restrictions on the variable?

c. What is the LCD of all terms in the equation
$$\frac{2x}{x+2} = \frac{5}{x^2-x-6} - \frac{1}{x-3}?$$

d. Solve the equation by first multiplying both sides by the LCD of all terms in the equation.
$$\frac{2x}{x+2} = \frac{5}{x^2-x-6} - \frac{1}{x-3}$$

A.3. a. Solve the equation for c. $\quad d = \frac{Ka-b}{c}$

b. Solve the equation for a. $\quad d = \frac{Ka-b}{c}$

A.4. The equation $\dfrac{P_1}{V_2} = \dfrac{P_2}{V_1}$ is used in chemistry, and we want to solve the equation for V_1.

a. Is the variable V_1 located in the numerator or denominator?

b. To solve an equation for a variable, we want the variable to be in the numerator. The technique of clearing fractions enables us to do this. What is the common denominator of the two fractions?

c. Clear fractions in this equation by multiplying both sides by V_1V_2.

d. Solve the resulting equation for V_1.

Section 15.6 Practice Exercises

Prerequisite Review

For Exercises R.1–R.6, determine the least common denominator.

R.1. $\dfrac{7}{4a}$ and $\dfrac{3}{a-5}$

R.2. $\dfrac{2}{x+4}$ and $\dfrac{5}{3x}$

R.3. $\dfrac{2}{m+3}, \dfrac{5}{4m+12}$, and $-\dfrac{3}{8}$

R.4. $\dfrac{2}{4n-4}, -\dfrac{7}{4}$, and $-\dfrac{3}{n-1}$

R.5. $-\dfrac{2}{m}$ and $\dfrac{8}{m^2}$

R.6. $\dfrac{3}{y^2}$ and $\dfrac{5}{y}$

For Exercises R.7–R.10, apply the distributive property.

R.7. $4a(a-5)\left(\dfrac{7}{4a} + \dfrac{3}{a-5}\right)$

R.8. $3x(x+4)\left(\dfrac{2}{x+4} + \dfrac{5}{3x}\right)$

R.9. $m^2\left(\dfrac{8}{m^2} - \dfrac{2}{m}\right)$

R.10. $y^2\left(\dfrac{3}{y^2} + \dfrac{5}{y}\right)$

For Exercises R.11–R.12, identify the restricted values of the expression.

R.11. $\dfrac{7}{4a} + \dfrac{3}{a-5}$

R.12. $\dfrac{2}{x+4} + \dfrac{5}{3x}$

For Exercises R.13–R.16, solve for the indicated variable.

R.13. $V = lwh$ for h

R.14. $C = 2\pi r$ for r

R.15. $5x - 3y = 6$ for y

R.16. $-2x + 5y = 20$ for y

Vocabulary and Key Concepts

1. **a.** The equation $4x + 7 = -18$ is an example of a _____ equation, whereas $3y^2 - 4y - 7 = 0$ is an example of a _____ equation.

   b. The equation $\dfrac{6}{x+2} + \dfrac{1}{4} = \dfrac{2}{3}$ is an example of a _____ equation.

   c. After solving a rational equation, check each potential solution to determine if it makes the _____ equal to zero in one or more of the rational expressions. If so, that potential solution is not part of the solution set.

Concept 1: Introduction to Rational Equations

For Exercises 2–13, solve the equations by first clearing the fractions. **(See Example 1.)**

2. $\dfrac{3}{5}t = \dfrac{2}{3}$

3. $\dfrac{3}{4}m = \dfrac{6}{5}$

4. $\dfrac{1}{2}x + \dfrac{1}{3} = 2$

5. $\dfrac{1}{3}x - \dfrac{3}{4} = 1$

6. $\dfrac{1}{3}z + \dfrac{2}{3} = -2z + 10$

7. $\dfrac{5}{2} + \dfrac{1}{2}b = 5 - \dfrac{1}{3}b$

8. $\dfrac{3}{2}p + \dfrac{1}{3} = \dfrac{2p - 3}{4}$

9. $\dfrac{5}{3} - \dfrac{1}{6}k = \dfrac{3k + 5}{4}$

10. $\dfrac{2x - 3}{4} + \dfrac{9}{10} = \dfrac{x}{5}$

11. $\dfrac{4y + 2}{3} - \dfrac{7}{6} = -\dfrac{y}{6}$

12. $\dfrac{x - 5}{2} - \dfrac{2x + 1}{5} = 1$

13. $\dfrac{2y + 1}{3} - \dfrac{y - 4}{2} = 1$

Concept 2: Solving Rational Equations

14. For the equation
    $$\dfrac{1}{w} - \dfrac{1}{2} = -\dfrac{1}{4}$$
    a. Identify the restricted values.
    b. Identify the LCD of the fractions in the equation.
    c. Solve the equation.

15. For the equation
    $$\dfrac{3}{z} - \dfrac{4}{5} = -\dfrac{1}{5}$$
    a. Identify the restricted values.
    b. Identify the LCD of the fractions in the equation.
    c. Solve the equation.

16. Identify the LCD of all the denominators in the equation.
    $$\dfrac{x + 1}{x^2 + 2x - 3} = \dfrac{1}{x + 3} - \dfrac{1}{x - 1}$$

For Exercises 17–46, solve the equations. **(See Examples 2–5.)**

17. $\dfrac{1}{8} = \dfrac{3}{5} + \dfrac{5}{y}$

18. $\dfrac{2}{7} - \dfrac{1}{x} = \dfrac{2}{3}$

19. $\dfrac{7}{4a} = \dfrac{3}{a - 5}$

20. $\dfrac{2}{x + 4} = \dfrac{5}{3x}$

21. $\dfrac{5}{6x} + \dfrac{7}{x} = 1$

22. $\dfrac{14}{3x} - \dfrac{5}{x} = 2$

23. $1 - \dfrac{2}{y} = \dfrac{3}{y^2}$

24. $1 - \dfrac{2}{m} = \dfrac{8}{m^2}$

25. $\dfrac{a + 1}{a} = 1 + \dfrac{a - 2}{2a}$

26. $\dfrac{7b - 4}{5b} = \dfrac{9}{5} - \dfrac{4}{b}$

27. $\dfrac{w}{5} - \dfrac{w + 3}{w} = -\dfrac{3}{w}$

28. $\dfrac{t}{12} + \dfrac{t + 3}{3t} = \dfrac{1}{t}$

29. $\dfrac{2}{m + 3} = \dfrac{5}{4m + 12} - \dfrac{3}{8}$

30. $\dfrac{2}{4n - 4} - \dfrac{7}{4} = \dfrac{-3}{n - 1}$

31. $\dfrac{p}{p - 4} - 5 = \dfrac{4}{p - 4}$

32. $\dfrac{-5}{q+5} = \dfrac{q}{q+5} + 2$

33. $\dfrac{2t}{t+2} - 2 = \dfrac{t-8}{t+2}$

34. $\dfrac{4w}{w-3} - 3 = \dfrac{3w-1}{w-3}$

35. $\dfrac{x^2 - x}{x-2} = \dfrac{12}{x-2}$

36. $\dfrac{x^2 + 9}{x+4} = \dfrac{-10x}{x+4}$

37. $\dfrac{x^2 + 3x}{x-1} = \dfrac{4}{x-1}$

38. $\dfrac{2x^2 - 21}{2x-3} = \dfrac{-11x}{2x-3}$

39. $\dfrac{2x}{x+4} - \dfrac{8}{x-4} = \dfrac{2x^2 + 32}{x^2 - 16}$

40. $\dfrac{4x}{x+3} - \dfrac{12}{x-3} = \dfrac{4x^2 + 36}{x^2 - 9}$

41. $\dfrac{x}{x+6} = \dfrac{72}{x^2 - 36} + 4$

42. $\dfrac{y}{y+4} = \dfrac{32}{y^2 - 16} + 3$

43. $\dfrac{5}{3x-3} - \dfrac{2}{x-2} = \dfrac{7}{x^2 - 3x + 2}$

44. $\dfrac{6}{5a+10} - \dfrac{1}{a-5} = \dfrac{4}{a^2 - 3a - 10}$

45. $\dfrac{w}{w-3} = \dfrac{17}{w^2 - 7w + 12} + \dfrac{1}{w-4}$

46. $\dfrac{y}{y+6} = \dfrac{-6}{y^2 + 7y + 6} + \dfrac{2}{y+1}$

For Exercises 47–50, translate to a rational equation and solve. (See Example 6.)

47. The reciprocal of a number is added to three. The result is the quotient of 25 and the number. Find the number.

48. The difference of three and the reciprocal of a number is equal to the quotient of 20 and the number. Find the number.

49. If a number added to five is divided by the difference of the number and two, the result is three-fourths. Find the number.

50. If twice a number added to three is divided by the number plus one, the result is three-halves. Find the number.

Concept 3: Solving Formulas Involving Rational Expressions

For Exercises 51–68, solve for the indicated variable. (See Examples 7–9.)

51. $K = \dfrac{ma}{F}$ for m

52. $K = \dfrac{ma}{F}$ for a

53. $K = \dfrac{IR}{E}$ for E

54. $K = \dfrac{IR}{E}$ for R

55. $I = \dfrac{E}{R+r}$ for R

56. $I = \dfrac{E}{R+r}$ for r

57. $h = \dfrac{2A}{B+b}$ for B

58. $\dfrac{C}{\pi r} = 2$ for r

59. $\dfrac{V}{\pi h} = r^2$ for h

60. $\dfrac{V}{lw} = h$ for w

61. $x = \dfrac{at + b}{t}$ for t

62. $\dfrac{T + mf}{m} = g$ for m

63. $\dfrac{x - y}{xy} = z$ for x

64. $\dfrac{w - n}{wn} = P$ for w

65. $a + b = \dfrac{2A}{h}$ for h

66. $1 + rt = \dfrac{A}{P}$ for P

67. $\dfrac{1}{R} = \dfrac{1}{R_1} + \dfrac{1}{R_2}$ for R

68. $\dfrac{b + a}{ab} = \dfrac{1}{f}$ for b

Problem Recognition Exercises

Comparing Rational Equations and Rational Expressions

Often adding or subtracting rational expressions is confused with solving rational equations. When adding rational expressions, we combine the terms to simplify the expression. When solving an equation, we clear the fractions and find numerical solutions, if possible. Both processes begin with finding the LCD, but the LCD is used differently in each process. Compare these two examples.

Example 1:

Add. $\dfrac{4}{x} + \dfrac{x}{3}$ (The LCD is $3x$.)

$= \dfrac{3}{3} \cdot \left(\dfrac{4}{x}\right) + \left(\dfrac{x}{3}\right) \cdot \dfrac{x}{x}$

$= \dfrac{12}{3x} + \dfrac{x^2}{3x}$

$= \dfrac{12 + x^2}{3x}$ The answer is a rational expression.

Example 2:

Solve. $\dfrac{4}{x} + \dfrac{x}{3} = -\dfrac{8}{3}$ (The LCD is $3x$.)

$\dfrac{3x}{1}\left(\dfrac{4}{x} + \dfrac{x}{3}\right) = \dfrac{3x}{1}\left(-\dfrac{8}{3}\right)$

$12 + x^2 = -8x$

$x^2 + 8x + 12 = 0$

$(x + 2)(x + 6) = 0$

$x + 2 = 0$ or $x + 6 = 0$

$x = -2$ or $x = -6$ The answer is the set $\{-2, -6\}$.

For Exercises 1–20, solve the equations and simplify the expressions.

1. $\dfrac{y}{2y + 4} - \dfrac{2}{y^2 + 2y}$

2. $\dfrac{1}{x + 2} + 2 = \dfrac{x + 11}{x + 2}$

3. $\dfrac{5t}{2} - \dfrac{t - 2}{3} = 5$

4. $3 - \dfrac{2}{a - 5}$

5. $\dfrac{7}{6p^2} + \dfrac{2}{9p} + \dfrac{1}{3p^2}$

6. $\dfrac{3b}{b + 1} - \dfrac{2b}{b - 1}$

7. $4 + \dfrac{2}{h - 3} = 5$

8. $\dfrac{2}{w + 1} + \dfrac{3}{(w + 1)^2}$

9. $\dfrac{1}{x - 6} - \dfrac{3}{x^2 - 6x} = \dfrac{4}{x}$

10. $\dfrac{3}{m} - \dfrac{6}{5} = -\dfrac{3}{m}$

11. $\dfrac{7}{2x + 2} + \dfrac{3x}{4x + 4}$

12. $\dfrac{10}{2t - 1} - 1 = \dfrac{t}{2t - 1}$

13. $\dfrac{3}{5x} + \dfrac{7}{2x} = 1$

14. $\dfrac{7}{t^2 - 5t} - \dfrac{3}{t - 5}$

15. $\dfrac{5}{2a - 1} + 4$

16. $p - \dfrac{5p}{p - 2} = -\dfrac{10}{p - 2}$

17. $\dfrac{3}{u} + \dfrac{12}{u^2 - 3u} = \dfrac{u + 1}{u - 3}$

18. $\dfrac{5}{4k} - \dfrac{2}{6k}$

19. $\dfrac{-2h}{h^2 - 9} + \dfrac{3}{h - 3}$

20. $\dfrac{3y}{y^2 - 5y + 4} = \dfrac{2}{y - 4} + \dfrac{3}{y - 1}$

Section 15.7 Applications of Rational Equations and Proportions

Concepts

1. Solving Proportions
2. Applications of Proportions and Similar Triangles
3. Distance, Rate, and Time Applications
4. Work Applications

1. Solving Proportions

In this section, we look at how rational equations can be used to solve a variety of applications. The first type of rational equation that will be applied is called a proportion.

> **Definition of Ratio and Proportion**
>
> 1. The **ratio** of a to b is $\dfrac{a}{b}$ ($b \neq 0$) and can also be expressed as $a:b$ or $a \div b$.
>
> 2. An equation that equates two ratios or rates is called a **proportion**. Therefore, if $b \neq 0$ and $d \neq 0$, then $\dfrac{a}{b} = \dfrac{c}{d}$ is a proportion.

A proportion can be solved by multiplying both sides of the equation by the LCD and clearing fractions.

Example 1   Solving a Proportion

Solve the proportion. $\dfrac{3}{11} = \dfrac{123}{w}$

Solution:

$\dfrac{3}{11} = \dfrac{123}{w}$ The LCD is $11w$.

$\cancel{11w}\left(\dfrac{3}{\cancel{11}}\right) = 11\cancel{w}\left(\dfrac{123}{\cancel{w}}\right)$ Multiply by the LCD and clear fractions.

$3w = 11 \cdot 123$ Solve the resulting equation (linear).

$3w = 1353$

$\dfrac{3w}{3} = \dfrac{1353}{3}$

$w = 451$

Check: $w = 451$

$\dfrac{3}{11} = \dfrac{123}{w}$

$\dfrac{3}{11} \stackrel{?}{=} \dfrac{123}{(451)}$

The solution set is $\{451\}$. $\dfrac{3}{11} \stackrel{?}{=} \dfrac{3}{11}$ ✓ (True)   Simplify to lowest terms.

Skill Practice Solve the proportion.

1. $\dfrac{10}{b} = \dfrac{2}{33}$

Answer

1. $\{165\}$

2. Applications of Proportions and Similar Triangles

Example 2   Using a Proportion in an Application

For a recent year, the population of Alabama was approximately 4.2 million. At that time, Alabama had seven representatives in the U.S. House of Representatives. In the same year, North Carolina had a population of approximately 7.2 million. If representation in the House is based on population in equal proportions for each state, how many representatives did North Carolina have?

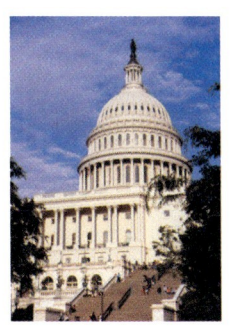

Steve Allen/Brand X Pictures/Getty Images

Solution:

Let x represent the number of representatives for North Carolina.

Set up a proportion by writing two equivalent ratios.

$$\frac{\text{Population of Alabama}}{\text{number of representatives}} \rightarrow \frac{4.2}{7} = \frac{7.2}{x} \leftarrow \frac{\text{Population of North Carolina}}{\text{number of representatives}}$$

$$\frac{4.2}{7} = \frac{7.2}{x}$$

$$7x \cdot \frac{4.2}{7} = 7x \cdot \frac{7.2}{x} \qquad \text{Multiply by the LCD, } 7x.$$

$$4.2x = (7.2)(7) \qquad \text{Solve the resulting linear equation.}$$

$$4.2x = 50.4$$

$$\frac{4.2x}{4.2} = \frac{50.4}{4.2}$$

$$x = 12 \qquad \text{North Carolina had 12 representatives.}$$

TIP: The equation from Example 2 could have been solved by first equating the cross products:

$$\frac{4.2}{7} \bowtie \frac{7.2}{x}$$

$$4.2x = (7.2)(7)$$

$$4.2x = 50.4$$

$$x = 12$$

Skill Practice

2. A university has a ratio of students to faculty of 105 to 2. If the student population at the university is 15,750, how many faculty members are needed?

Proportions are used in geometry with **similar triangles**. Two triangles are similar if their angles have equal measure. In such a case, the lengths of the corresponding sides are proportional. In Figure 15-1, triangle ABC is similar to triangle XYZ. Therefore, the following ratios are equivalent.

$$\frac{a}{x} = \frac{b}{y} = \frac{c}{z}$$

Figure 15-1

Answer

2. 300 faculty members are needed.

Example 3 Using Similar Triangles to Find an Unknown Side in a Triangle

In Figure 15-2, triangle XYZ is similar to triangle ABC.

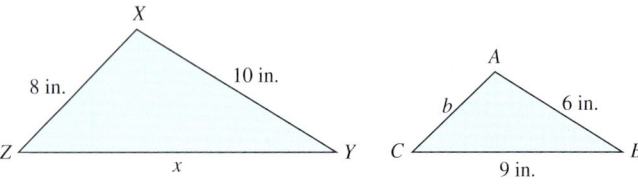

Figure 15-2

a. Solve for x.   b. Solve for b.

Solution:

a. The lengths of the upper right sides of the triangles are given. These form a known ratio of $\frac{10}{6}$. Because the triangles are similar, the ratio of the other corresponding sides must also be equal to $\frac{10}{6}$. To solve for x, we have

$$\frac{\text{Bottom side from large triangle}}{\text{bottom side from small triangle}} \rightarrow \frac{x}{9 \text{ in.}} = \frac{10 \text{ in.}}{6 \text{ in.}} \leftarrow \frac{\text{Right side from large triangle}}{\text{right side from small triangle}}$$

$\frac{x}{9} = \frac{10}{6}$ The LCD is 18.

$\overset{2}{\cancel{18}} \cdot \left(\frac{x}{\cancel{9}}\right) = \overset{3}{\cancel{18}} \cdot \left(\frac{10}{\cancel{6}}\right)$ Multiply by the LCD.

$2x = 30$ Clear fractions.

$x = 15$ Divide by 2.

The length of side x is 15 in.

b. To solve for b, the ratio of the upper left sides of the triangles must equal $\frac{10}{6}$.

$$\frac{\text{Left side from large triangle}}{\text{left side from small triangle}} \rightarrow \frac{8 \text{ in.}}{b} = \frac{10 \text{ in.}}{6 \text{ in.}} \leftarrow \frac{\text{Right side from large triangle}}{\text{right side from small triangle}}$$

$\frac{8}{b} = \frac{10}{6}$ The LCD is $6b$.

$\cancel{6b} \cdot \left(\frac{8}{\cancel{b}}\right) = 6\cancel{b} \cdot \left(\frac{10}{\cancel{6}}\right)$ Multiply by the LCD.

$48 = 10b$ Clear fractions.

$\frac{48}{10} = \frac{10b}{10}$

$4.8 = b$

The length of side b is 4.8 in.

Skill Practice

3. Triangle ABC is similar to triangle XYZ. Solve for the lengths of the missing sides.

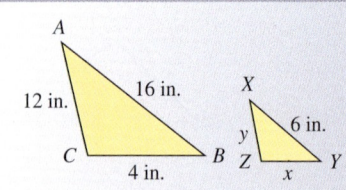

Answer

3. $x = 1.5$ in., and $y = 4.5$ in.

Example 4   Using Similar Triangles in an Application

A tree that is 20 ft from a house is to be cut down. Use the following information and similar triangles to find the height of the tree to determine if it will hit the house.

The shadow cast by a yardstick is 2 ft long. The shadow cast by the tree is 11 ft long.

Solution:

| | Step 1: | Read the problem. |

Let x represent the height of the tree.   Step 2: Label the variables.

We will assume that the measurements were taken at the same time of day. Therefore, the angle of the Sun is the same on both objects, and we can set up similar triangles (Figure 15-3).

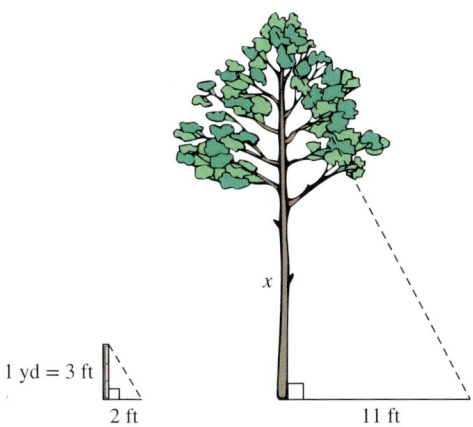

Figure 15-3

Step 3: Create a verbal model.

$$\boxed{\frac{\text{Height of yardstick}}{\text{height of tree}}} \rightarrow \frac{3 \text{ ft}}{x} = \frac{2 \text{ ft}}{11 \text{ ft}} \leftarrow \boxed{\frac{\text{Length of yardstick's shadow}}{\text{length of tree's shadow}}}$$

$\dfrac{3}{x} = \dfrac{2}{11}$ Step 4: Write a mathematical equation.

$11x\left(\dfrac{3}{x}\right) = \left(\dfrac{2}{11}\right)11x$ Step 5: Multiply by the LCD.

$33 = 2x$ Solve the equation.

$\dfrac{33}{2} = \dfrac{2x}{2}$

$16.5 = x$ Step 6: Interpret the results, and write the answer in words.

The tree is 16.5 ft high. The tree is less than 20 ft high, so it will not hit the house.

Skill Practice

4. The Sun casts a 3.2-ft shadow of a 6-ft man. At the same time, the Sun casts an 80-ft shadow of a building. How tall is the building?

Answer

4. The building is 150 ft tall.

3. Distance, Rate, and Time Applications

In Examples 5 and 6, we use the familiar relationship among the variables distance, rate, and time. Recall that $d = rt$.

Example 5: Using a Rational Equation in a Distance, Rate, and Time Application

A small plane flies 440 mi with the wind from Memphis, TN, to Oklahoma City, OK. In the same amount of time, the plane flies 340 mi against the wind from Oklahoma City to Little Rock, AR (see Figure 15-4). If the wind speed is 30 mph, find the speed of the plane in still air.

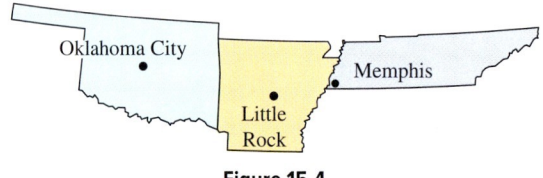

Figure 15-4

Solution:

Let x represent the speed of the plane in still air.

Then $x + 30$ is the speed of the plane with the wind.

$x - 30$ is the speed of the plane against the wind.

Organize the given information in a chart.

	Distance	Rate	Time
With the wind	440	$x + 30$	$\dfrac{440}{x + 30}$
Against the wind	340	$x - 30$	$\dfrac{340}{x - 30}$

Because $d = rt$, then $t = \dfrac{d}{r}$

The plane travels with the wind for the same amount of time as it travels against the wind, so we can equate the two expressions for time.

$$\begin{pmatrix}\text{Time with}\\\text{the wind}\end{pmatrix} = \begin{pmatrix}\text{time against}\\\text{the wind}\end{pmatrix}$$

$$\dfrac{440}{x + 30} = \dfrac{340}{x - 30} \qquad \text{The LCD is } (x + 30)(x - 30).$$

$$(x+30)(x-30) \cdot \dfrac{440}{x+30} = (x+30)(x-30) \cdot \dfrac{340}{x-30}$$

$$440(x - 30) = 340(x + 30)$$

$$440x - 13{,}200 = 340x + 10{,}200 \qquad \text{Solve the resulting linear equation.}$$

$$100x = 23{,}400$$

$$x = 234$$

The plane's speed in still air is 234 mph.

TIP: The equation

$$\dfrac{440}{x + 30} = \dfrac{340}{x - 30}$$

is a proportion. The fractions can also be cleared by equating the cross products.

$$\dfrac{440}{x + 30} \times \dfrac{340}{x - 30}$$

$$440(x - 30) = 340(x + 30)$$

Skill Practice

5. Alison paddles her kayak in a river where the current is 2 mph. She can paddle 20 mi with the current in the same amount of time that she can paddle 10 mi against the current. Find the speed of the kayak in still water.

Example 6   Using a Rational Equation in a Distance, Rate, and Time Application

A motorist drives 100 mi between two cities in a bad rainstorm. For the return trip in sunny weather, she averages 10 mph faster and takes $\frac{1}{2}$ hr less time. Find the average speed of the motorist in the rainstorm and in sunny weather.

Solution:

Let x represent the motorist's speed during the rain.

Then $x + 10$ represents the speed in sunny weather.

	Distance	Rate	Time
Trip during rainstorm	100	x	$\dfrac{100}{x}$
Trip during sunny weather	100	$x + 10$	$\dfrac{100}{x + 10}$

Because $d = rt$, then $t = \dfrac{d}{r}$

Because the same distance is traveled in $\frac{1}{2}$ hr less time, the difference between the time of the trip during the rainstorm and the time during sunny weather is $\frac{1}{2}$ hr.

$\left(\begin{array}{c}\text{Time during}\\\text{the rainstorm}\end{array}\right) - \left(\begin{array}{c}\text{time during}\\\text{sunny weather}\end{array}\right) = \left(\dfrac{1}{2}\text{ hr}\right)$ Verbal model

$\dfrac{100}{x} - \dfrac{100}{x + 10} = \dfrac{1}{2}$ Mathematical equation

$2x(x + 10)\left(\dfrac{100}{x} - \dfrac{100}{x + 10}\right) = 2x(x + 10)\left(\dfrac{1}{2}\right)$ Multiply by the LCD.

$2x(x + 10)\left(\dfrac{100}{x}\right) - 2x(x + 10)\left(\dfrac{100}{x + 10}\right) = 2x(x + 10)\left(\dfrac{1}{2}\right)$ Apply the distributive property.

$200(x + 10) - 200x = x(x + 10)$ Clear fractions.

$200x + 2000 - 200x = x^2 + 10x$ Solve the resulting equation (quadratic).

$2000 = x^2 + 10x$

$0 = x^2 + 10x - 2000$ Set the equation equal to zero.

$0 = (x - 40)(x + 50)$ Factor.

$x = 40$ or $x \ne -50$

Avoiding Mistakes

The equation

$$\dfrac{100}{x} - \dfrac{100}{x + 10} = \dfrac{1}{2}$$

is not a proportion because the left-hand side has more than one fraction. Do not try to multiply the cross products. Instead, multiply by the LCD to clear fractions.

Answer

5. The speed of the kayak is 6 mph.

Because a rate of speed cannot be negative, reject $x = -50$. Therefore, the speed of the motorist in the rainstorm is 40 mph. Because $x + 10 = 40 + 10 = 50$, the average speed for the return trip in sunny weather is 50 mph.

Skill Practice

6. Harley rode his mountain bike 12 mi to the top of a mountain and the same distance back down. His speed going up was 8 mph slower than coming down. The ride up took 2 hr longer than the ride coming down. Find his speed in each direction.

4. Work Applications

Example 7 demonstrates how work rates are related to a portion of a job that can be completed in one unit of time.

Example 7 — Using a Rational Equation in a Work Problem

A new printing press can print the morning edition in 2 hr, whereas the old printer requires 4 hr. How long would it take to print the morning edition if both printers work together?

Solution:

One method to solve this problem is to add rates.

Let x represent the time required for both printers working together to complete the job.

$$\begin{pmatrix}\text{Rate}\\\text{of old printer}\end{pmatrix} + \begin{pmatrix}\text{rate}\\\text{of new printer}\end{pmatrix} = \begin{pmatrix}\text{rate of}\\\text{both working together}\end{pmatrix}$$

$$\frac{1 \text{ job}}{4 \text{ hr}} + \frac{1 \text{ job}}{2 \text{ hr}} = \frac{1 \text{ job}}{x \text{ hr}}$$

$$\frac{1}{4} + \frac{1}{2} = \frac{1}{x}$$

$4x\left(\frac{1}{4} + \frac{1}{2}\right) = 4x\left(\frac{1}{x}\right)$ The LCD is $4x$.

$4x \cdot \frac{1}{4} + 4x \cdot \frac{1}{2} = 4x \cdot \frac{1}{x}$ Apply the distributive property.

$x + 2x = 4$ Solve the resulting linear equation.

$3x = 4$

$x = \frac{4}{3}$ The time required to print the morning edition using both printers is $1\frac{1}{3}$ hr.

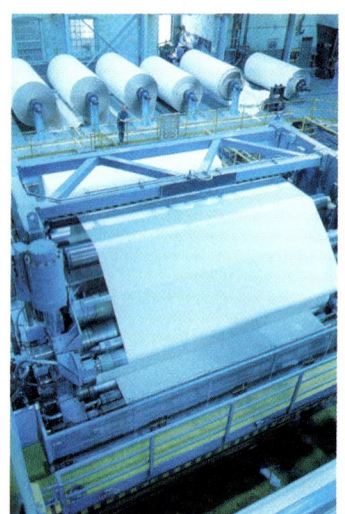
Digital Vision/Getty Images

Skill Practice

7. The computer at a bank can process and prepare the bank statements in 30 hr. A new faster computer can do the job in 20 hr. If the bank uses both computers together, how long will it take to process the statements?

Answers

6. Uphill speed was 4 mph; downhill speed was 12 mph.
7. 12 hr

An alternative approach to Example 7 is to determine the portion of the job that each printer can complete in 1 hr and extend that rate to the portion of the job completed in x hours.

- The old printer can perform the job in 4 hr. Therefore, it completes $\frac{1}{4}$ of the job in 1 hr and $\frac{1}{4}x$ jobs in x hours.
- The new printer can perform the job in 2 hr. Therefore, it completes $\frac{1}{2}$ of the job in 1 hr and $\frac{1}{2}x$ jobs in x hours.

The sum of the portions of the job completed by each printer must equal one whole job.

$$\begin{pmatrix} \text{Portion of job} \\ \text{completed by} \\ \text{old printer} \end{pmatrix} + \begin{pmatrix} \text{portion of job} \\ \text{completed by} \\ \text{new printer} \end{pmatrix} = \begin{pmatrix} 1 \\ \text{whole} \\ \text{job} \end{pmatrix}$$

$\frac{1}{4}x + \frac{1}{2}x = 1$ The LCD is 4.

$4\left(\frac{1}{4}x + \frac{1}{2}x\right) = 4(1)$ Multiply by the LCD.

$x + 2x = 4$ Solve the resulting linear equation.

$3x = 4$

$x = \frac{4}{3}$ The time required using both printers is $1\frac{1}{3}$ hr.

Section 15.7 Activity

A proportion is an equation that equates two ratios. For Exercises A.1–A.2, fill in the blanks to complete the proportion.

A.1. **a.** One inch on a map represents 50 mi. Therefore, 2 in. on the map represents _____ mi.

   b. Written as an equation, the statement in part (a) is:

   $\dfrac{1 \text{ in.}}{50 \text{ mi}} = \dfrac{2 \text{ in.}}{\Box}$

A.2. **a.** Eight apples cost $4. Therefore _____ apples can be purchased for $24.

   b. Written as an equation, the statement in part (a) is:

   $\dfrac{8 \text{ apples}}{\$4.00} = \dfrac{\Box}{\$24.00}$

A.3. It takes 2 lb of ground coffee to make 75 cups. How many pounds of ground coffee would be required to make 300 cups?

   a. In this scenario, what is the unknown quantity? Let x be this value.

   b. Write a proportion that represents this scenario.

   c. Solve the proportion and interpret your answer.

A.4. Sean runs 12 mi in the same amount of time that it takes him to ride his bicycle 72 mi. If he rides 20 mph faster than he runs, find the rate at which he runs and the rate at which he rides.

   a. The unknowns to be found are Sean's rate of speed running and his rate of speed bicycling. Let x represent one of these two unknowns.

b. Complete the Distance and Rate columns of the table.

	Distance (mi)	Rate (mph)	Time (hr)
Sean running			
Sean bicycling			

c. Solve the equation $d = rt$ for t.

d. From part (c), we know that the time of travel is the ratio of the distance to the rate. Fill in the Time column in the table.

e. Write an equation that indicates that Sean's time running and his time bicycling are the same.

f. Solve the equation.

g. Interpret the solution to the equation and verify that the solution makes sense in the context of this problem.

A.5. Charlene cleans swimming pools. She can clean a large pool in 4 hr. When she goes on vacation, Ralph takes over. It takes Ralph 6 hr to clean the same pool. When Charlene comes back from vacation, she and Ralph clean the pool together. How long does it take them to do the job together?

a. Let x represent the time required for Charlene and Ralph to clean the pool together.

b. This scenario can be modeled by adding the rates of speed at which Charlene and Ralph work. Fill in the blanks.

$$\begin{pmatrix}\text{The rate at which}\\ \text{Charlene works}\end{pmatrix} + \begin{pmatrix}\text{The rate at which}\\ \text{Ralph works}\end{pmatrix} = \begin{pmatrix}\text{The rate at which}\\ \text{they work together}\end{pmatrix}$$

$$\frac{1 \text{ job}}{4 \text{ hr}} + \frac{1 \text{ job}}{\Box \text{ hr}} = \frac{1 \text{ job}}{\Box \text{ hr}} \quad \text{or simply} \quad \frac{1}{4} + \frac{1}{\Box} = \frac{1}{\Box}$$

c. Solve the equation from part (b).

d. Interpret the solution to the equation and verify that the solution makes sense in the context of this problem.

Section 15.7 Practice Exercises

Prerequisite Review

For Exercises R.1–R.8, determine whether the exercise is an equation or an expression. If it is an equation, solve it. If it is an expression, perform the indicated operation.

R.1. $5 - \dfrac{x}{4} = -6$

R.2. $\dfrac{b}{5} + 3 = 9$

R.3. $\dfrac{m}{m-1} - \dfrac{2}{m+3}$

R.4. $\dfrac{2}{a+5} + \dfrac{5}{a^2-25}$

R.5. $\dfrac{3y+6}{20} \div \dfrac{4y+8}{8}$

R.6. $\dfrac{z^2+z}{24} \cdot \dfrac{8}{z+1}$

R.7. $\dfrac{3}{p+3} = \dfrac{12p+19}{p^2+7p+12} - \dfrac{5}{p+4}$

R.8. $\dfrac{3}{x+2} = \dfrac{x+16}{x^2-3x-10} - \dfrac{3}{x-5}$

Vocabulary and Key Concepts

1. An equation that equates two ratios is called a _____.
2. Given similar triangles, the lengths of corresponding sides are _____.

Concept 1: Solving Proportions

For Exercises 3–16, solve the proportions. (See Example 1.)

3. $\dfrac{8}{5} = \dfrac{152}{p}$

4. $\dfrac{6}{7} = \dfrac{96}{y}$

5. $\dfrac{19}{76} = \dfrac{z}{4}$

6. $\dfrac{15}{135} = \dfrac{w}{9}$

7. $\dfrac{5}{3} = \dfrac{a}{8}$

8. $\dfrac{b}{14} = \dfrac{3}{8}$

9. $\dfrac{2}{1.9} = \dfrac{x}{38}$

10. $\dfrac{16}{1.3} = \dfrac{30}{p}$

11. $\dfrac{y+1}{2y} = \dfrac{2}{3}$

12. $\dfrac{w-2}{4w} = \dfrac{1}{6}$

13. $\dfrac{9}{2z-1} = \dfrac{3}{z}$

14. $\dfrac{1}{t} = \dfrac{1}{4-t}$

15. $\dfrac{8}{9a-1} = \dfrac{5}{3a+2}$

16. $\dfrac{4p+1}{3} = \dfrac{2p-5}{6}$

17. Charles' law describes the relationship between the initial and final temperature and volume of a gas held at a constant pressure.

$$\dfrac{V_i}{V_f} = \dfrac{T_i}{T_f}$$

   a. Solve the equation for V_f.
   b. Solve the equation for T_f.

18. The relationship between the area, height, and base of a triangle is given by the proportion

$$\dfrac{A}{b} = \dfrac{h}{2}$$

   a. Solve the equation for A.
   b. Solve the equation for b.

Concept 2: Applications of Proportions and Similar Triangles

For Exercises 19–26, solve using proportions.

19. Toni drives her car 132 mi on the highway on 4 gal of gas. At this rate how many miles can she drive on 9 gal of gas? (See Example 2.)

20. Tim takes his pulse for 10 sec and counts 12 beats. How many beats per minute is this?

21. It is recommended that 7.8 mL of Grow-It-Right plant food be mixed with 2 L of water for feeding houseplants. How much plant food should be mixed with 1 gal of water to maintain the same concentration? (1 gal ≈ 3.8 L.) Express the answer in milliliters.

22. According to the website for the state of Virginia, 0.8 million tons of clothing is reused or recycled out of 7 million tons of clothing discarded. If 17.5 million tons of clothing is discarded, how many tons will be reused or recycled?

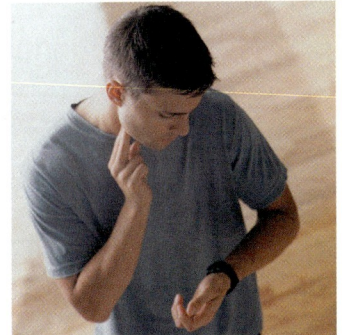

Duncan Smith/Photodisc/Getty Images

23. Andrew is on a low-carbohydrate diet. If his diet book tells him that an 8-oz serving of pineapple contains 19.2 g of carbohydrate, how many grams of carbohydrate does a 5-oz serving contain?

24. Cooking oatmeal requires 1 c of water for every $\frac{1}{2}$ c of oats. How many cups of water will be required for $\frac{3}{4}$ c of oats?

25. According to a building code, a wheelchair ramp must be at least 12 ft long for each foot of height. If the height of a newly constructed ramp is to be $1\frac{2}{3}$ ft, find the minimum acceptable length.

26. A map has a scale of 50 mi/in. If two cities measure 6.5 in. apart, how many miles does this represent?

For Exercises 27–30, triangle ABC is similar to triangle XYZ. Solve for x and y. (See Example 3.)

27.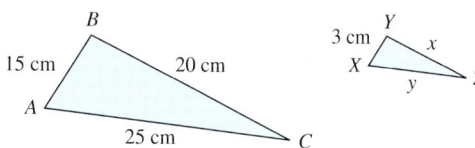

28.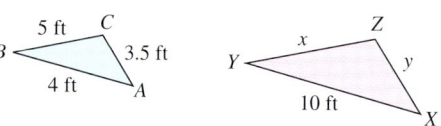

29.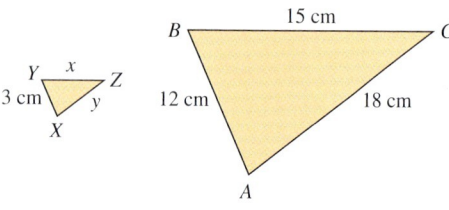

30.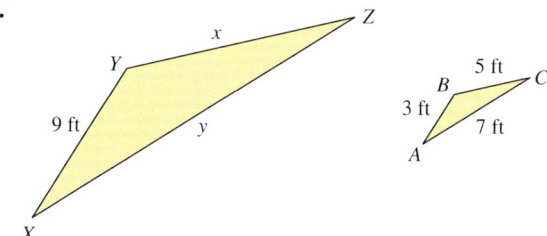

31. To estimate the height of a light pole, a mathematics student measures the length of a shadow cast by a meterstick and the length of the shadow cast by the light pole. Find the height of the light pole. (See Example 4.)

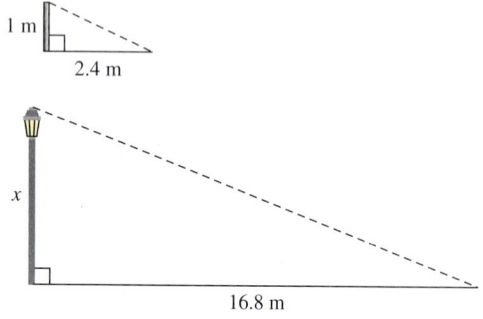

32. To estimate the height of a building, a student measures the length of a shadow cast by a yardstick and the length of the shadow cast by the building. Find the height of the building.

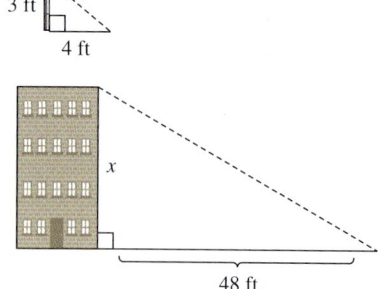

33. A 6-ft-tall man standing 54 ft from a light post casts an 18-ft shadow. What is the height of the light post?

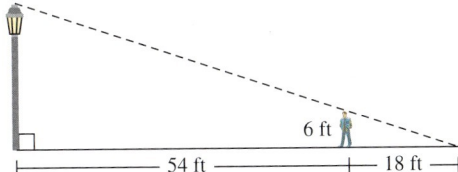

34. For a science project at school, a student must measure the height of a tree. The student measures the length of the shadow of the tree and then measures the length of the shadow cast by a yardstick. Find the height of the tree.

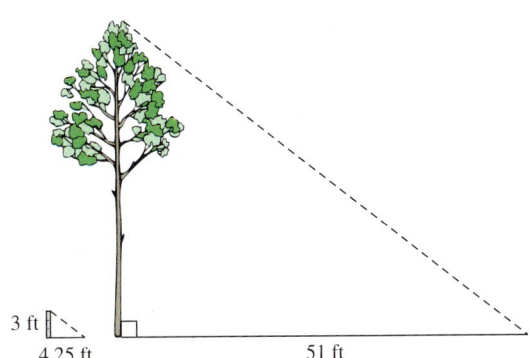

Concept 3: Distance, Rate, and Time Applications

35. A boat travels 54 mi upstream against the current in the same amount of time it takes to travel 66 mi downstream with the current. If the current is 2 mph, what is the speed of the boat in still water? (Use $t = \frac{d}{r}$ to complete the table.)
(See Example 5.)

	Distance	Rate	Time
With the current (downstream)			
Against the current (upstream)			

36. A plane flies 630 mi with the wind in the same amount of time that it takes to fly 455 mi against the wind. If this plane flies at the rate of 217 mph in still air, what is the speed of the wind? (Use $t = \frac{d}{r}$ to complete the table.)

	Distance	Rate	Time
With the wind			
Against the wind			

37. The jet stream is a fast-flowing air current found in the atmosphere at around 36,000 ft above the surface of the Earth. During one summer day, the speed of the jet stream is 35 mph. A plane flying with the jet stream can fly 700 mi in the same amount of time that it would take to fly 500 mi against the jet stream. What is the speed of the plane in still air?

38. A fishing boat travels 9 mi downstream with the current in the same amount of time that it travels 3 mi upstream against the current. If the speed of the current is 6 mph, what is the speed at which the boat travels in still water?

39. An athlete in training rides his bike 20 mi and then immediately follows with a 10-mi run. The total workout takes him 2.5 hr. He also knows that he bikes about twice as fast as he runs. Determine his biking speed and his running speed.

40. Devon can cross-country ski 5 km/hr faster than his sister Shanelle. Devon skis 45 km in the same amount of time Shanelle skis 30 km. Find their speeds.

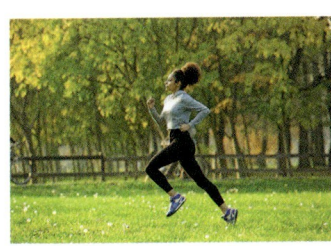

Michael Simons/123RF

41. Floyd can walk 2 mph faster than his wife, Rachel. It takes Rachel 3 hr longer than Floyd to hike a 12-mi trail through a park. Find their speeds.
(See Example 6.)

42. Janine bikes 3 mph faster than her sister, Jessica. Janine can ride 36 mi in 1 hr less time than Jessica can ride the same distance. Find each of their speeds.

43. Sergio rode his bike 4 mi. Then he got a flat tire and had to walk back 4 mi. It took him 1 hr longer to walk than it did to ride. If his rate walking was 9 mph less than his rate riding, find the two rates.

44. Amber jogs 10 km in $\frac{3}{4}$ hr less time than she can walk the same distance. If her walking rate is 3 km/hr less than her jogging rate, find her rates jogging and walking (in km/hr).

Concept 4: Work Applications

45. If the cold-water faucet is left on, the sink will fill in 10 min. If the hot-water faucet is left on, the sink will fill in 12 min. How long would it take to fill the sink if both faucets are left on?
(See Example 7.)

46. The CUT-IT-OUT lawn mowing company consists of two people: Tina and Bill. If Tina cuts a lawn by herself, she can do it in 4 hr. If Bill cuts the same lawn himself, it takes him an hour longer than Tina. How long would it take them if they worked together?

47. A manuscript needs to be printed. One printer can do the job in 50 min, and another printer can do the job in 40 min. How long would it take if both printers were used?

48. A pump can empty a small pond in 4 hr. Another more efficient pump can do the job in 3 hr. How long would it take to empty the pond if both pumps were used?

49. A pipe can fill a reservoir in 16 hr. A drainage pipe can drain the reservoir in 24 hr. How long would it take to fill the reservoir if the drainage pipe were left open by mistake? (*Hint:* The rate at which water drains should be negative.)

50. A hole in the bottom of a child's plastic swimming pool can drain the pool in 60 min. If the pool had no hole, a hose could fill the pool in 40 min. How long would it take the hose to fill the pool with the hole?

51. Tim and Al are bricklayers. Tim can construct an outdoor grill in 5 days. If Al helps Tim, they can build it in only 2 days. How long would it take Al to build the grill alone?

52. Norma is a new and inexperienced secretary. It takes her 3 hr to prepare a mailing. If her boss helps her, the mailing can be completed in 1 hr. How long would it take the boss to do the job by herself?

Expanding Your Skills

For Exercises 53–56, solve using proportions.

53. The ratio of smokers to nonsmokers in a restaurant is 2 to 7. There are 100 more nonsmokers than smokers. How many smokers and nonsmokers are in the restaurant?

54. The ratio of fiction to nonfiction books sold in a bookstore is 5 to 3. One week there were 180 more fiction books sold than nonfiction. Find the number of fiction and nonfiction books sold during that week.

55. There are 440 students attending a biology lecture. The ratio of male to female students at the lecture is 6 to 5. How many men and women are attending the lecture?

56. The ratio of dogs to cats at the humane society is 5 to 8. The total number of dogs and cats is 650. How many dogs and how many cats are at the humane society?

Chapter 15 Review Exercises

Section 15.1

1. For the rational expression $\dfrac{t-2}{t+9}$

   a. Evaluate the expression (if possible) for $t = 0, 1, 2, -3, -9$

   b. Identify the restricted values.

2. For the rational expression $\dfrac{k+1}{k-5}$

   a. Evaluate the expression for $k = 0, 1, 5, -1, -2$

   b. Identify the restricted values.

3. Which of the rational expressions are equal to -1?

   a. $\dfrac{2-x}{x-2}$ b. $\dfrac{x-5}{x+5}$

   c. $\dfrac{-x-7}{x+7}$ d. $\dfrac{x^2-4}{4-x^2}$

For Exercises 4–13, identify the restricted values. Then simplify the expressions.

4. $\dfrac{x-3}{(2x-5)(x-3)}$ 5. $\dfrac{h+7}{(3h+1)(h+7)}$

6. $\dfrac{4a^2+7a-2}{a^2-4}$ 7. $\dfrac{2w^2+11w+12}{w^2-16}$

8. $\dfrac{z^2-4z}{8-2z}$ 9. $\dfrac{15-3k}{2k^2-10k}$

10. $\dfrac{2b^2+4b-6}{4b+12}$ 11. $\dfrac{3m^2-12m-15}{9m+9}$

12. $\dfrac{n+3}{n^2+6n+9}$ 13. $\dfrac{p+7}{p^2+14p+49}$

Section 15.2

For Exercises 14–27, multiply or divide as indicated.

14. $\dfrac{3y^3}{3y-6} \cdot \dfrac{y-2}{y}$ 15. $\dfrac{2u+10}{u} \cdot \dfrac{u^3}{4u+20}$

16. $\dfrac{11}{v-2} \cdot \dfrac{2v^2-8}{22}$ 17. $\dfrac{8}{x^2-25} \cdot \dfrac{3x+15}{16}$

18. $\dfrac{4c^2+4c}{c^2-25} \div \dfrac{8c}{c^2-5c}$ 19. $\dfrac{q^2-5q+6}{2q+4} \div \dfrac{2q-6}{q+2}$

20. $\left(\dfrac{-2t}{t+1}\right)(t^2-4t-5)$ 21. $(s^2-6s+8)\left(\dfrac{4s}{s-2}\right)$

22. $\dfrac{\dfrac{a^2+5a+1}{7a-7}}{\dfrac{a^2+5a+1}{a-1}}$ 23. $\dfrac{\dfrac{n^2+n+1}{n^2-4}}{\dfrac{n^2+n+1}{n+2}}$

24. $\dfrac{5h^2-6h+1}{h^2-1} \div \dfrac{16h^2-9}{4h^2+7h+3} \cdot \dfrac{3-4h}{30h-6}$

25. $\dfrac{3m-3}{6m^2+18m+12} \cdot \dfrac{2m^2-8}{m^2-3m+2} \div \dfrac{m+3}{m+1}$

26. $\dfrac{x-2}{x^2-3x-18} \cdot \dfrac{6-x}{x^2-4}$

27. $\dfrac{4y^2-1}{1+2y} \div \dfrac{y^2-4y-5}{5-y}$

Section 15.3

28. Determine the LCD.

    $\dfrac{6}{n^2-9}; \dfrac{5}{n^2-n-6}$

29. Determine the LCD.

    $\dfrac{8}{m^2-16}; \dfrac{7}{m^2-m-12}$

30. State two possible LCDs that could be used to add the fractions.

    $\dfrac{7}{c-2} + \dfrac{4}{2-c}$

31. State two possible LCDs that could be used to subtract the fractions.

    $\dfrac{10}{3-x} - \dfrac{5}{x-3}$

For Exercises 32–37, write each fraction as an equivalent fraction with the LCD as its denominator.

32. $\dfrac{2}{5a}; \dfrac{3}{10b}$

33. $\dfrac{7}{4x}; \dfrac{11}{6y}$

34. $\dfrac{1}{x^2y^4}; \dfrac{3}{xy^5}$

35. $\dfrac{5}{ab^3}; \dfrac{3}{ac^2}$

36. $\dfrac{5}{p+2}; \dfrac{p}{p-4}$

37. $\dfrac{6}{q}; \dfrac{1}{q+8}$

Section 15.4

For Exercises 38–49, add or subtract as indicated.

38. $\dfrac{h+3}{h+1} + \dfrac{h-1}{h+1}$

39. $\dfrac{b-6}{b-2} + \dfrac{b+2}{b-2}$

40. $\dfrac{a^2}{a-5} - \dfrac{25}{a-5}$

41. $\dfrac{x^2}{x+7} - \dfrac{49}{x+7}$

42. $\dfrac{y}{y^2-81} + \dfrac{2}{9-y}$

43. $\dfrac{3}{4-t^2} + \dfrac{t}{2-t}$

44. $\dfrac{4}{3m} - \dfrac{1}{m+2}$

45. $\dfrac{5}{2r+12} - \dfrac{1}{r}$

46. $\dfrac{4p}{p^2+6p+5} - \dfrac{3p}{p^2+5p+4}$

47. $\dfrac{3q}{q^2+7q+10} - \dfrac{2q}{q^2+6q+8}$

48. $\dfrac{1}{h} + \dfrac{h}{2h+4} - \dfrac{2}{h^2+2h}$

49. $\dfrac{x}{3x+9} - \dfrac{3}{x^2+3x} + \dfrac{1}{x}$

Section 15.5

For Exercises 50–57, simplify the complex fractions.

50. $\dfrac{\frac{a-4}{3}}{\frac{a-2}{3}}$

51. $\dfrac{\frac{z+5}{z}}{\frac{z-5}{3}}$

52. $\dfrac{\frac{2-3w}{2}}{\frac{2}{w}-3}$

53. $\dfrac{\frac{2}{y}+6}{\frac{3y+1}{4}}$

54. $\dfrac{\frac{y}{x}-\frac{x}{y}}{\frac{1}{x}+\frac{1}{y}}$

55. $\dfrac{\frac{b}{a}-\frac{a}{b}}{\frac{1}{b}-\frac{1}{a}}$

56. $\dfrac{\frac{6}{p+2}+4}{\frac{8}{p+2}-4}$

57. $\dfrac{\frac{25}{k+5}+5}{\frac{5}{k+5}-5}$

Section 15.6

For Exercises 58–65, solve the equations.

58. $\dfrac{2}{x} + \dfrac{1}{2} = \dfrac{1}{4}$

59. $\dfrac{1}{y} + \dfrac{3}{4} = \dfrac{1}{4}$

60. $\dfrac{2}{h-2} + 1 = \dfrac{h}{h+2}$

61. $\dfrac{w}{w-1} = \dfrac{3}{w+1} + 1$

62. $\dfrac{t+1}{3} - \dfrac{t-1}{6} = \dfrac{1}{6}$

63. $\dfrac{w+1}{w-3} - \dfrac{3}{w} = \dfrac{12}{w^2-3w}$

64. $\dfrac{1}{z+2} = \dfrac{4}{z^2-4} - \dfrac{1}{z-2}$

65. $\dfrac{y+1}{y+3} = \dfrac{y^2-11y}{y^2+y-6} - \dfrac{y-3}{y-2}$

66. Four times a number is added to 5. The sum is then divided by 6. The result is $\tfrac{7}{2}$. Find the number.

67. Solve the formula $\dfrac{V}{h} = \dfrac{\pi r^2}{3}$ for h.

68. Solve the formula $\dfrac{A}{b} = \dfrac{h}{2}$ for b.

Section 15.7

For Exercises 69–70, solve the proportions.

69. $\dfrac{m+2}{8} = \dfrac{m}{3}$

70. $\dfrac{12}{a} = \dfrac{5}{8}$

71. A bag of popcorn states that it contains 4 g of fat per serving. If a serving is 2 oz, how many grams of fat are in a 5-oz bag?

72. Bud goes 10 mph faster on his motorcycle than Ed goes on his motorcycle. If Bud travels 105 mi in the same amount of time that Ed travels 90 mi, what are the rates of the two bikers?

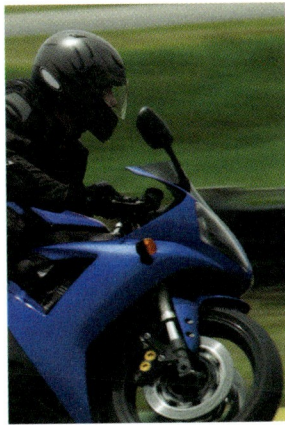

Glow Images

73. There are two pumps set up to fill a small swimming pool. One pump takes 24 min by itself to fill the pool, but the other takes 56 min by itself. How long would it take if both pumps worked together?

74. Triangle XYZ is similar to triangle ABC. Find the values of x and b.

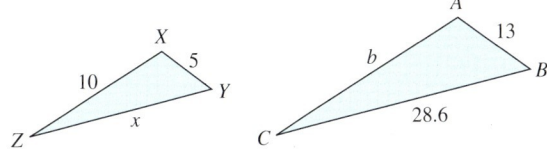

Relations and Functions

16

CHAPTER OUTLINE

16.1 Introduction to Relations 1056
16.2 Introduction to Functions 1066
16.3 Graphs of Functions 1079
   Problem Recognition Exercises: Characteristics of Relations 1091
16.4 Algebra of Functions and Composition 1092
16.5 Variation 1100
   Chapter 16 Review Exercises 1110

Mathematics in Communication

Imagine that every time you call your friend's cell phone your call is sent to a wrong device, and a different person answers the call. Suppose that one time you reach a single mother in Alabama, another time perhaps a small-business owner in Georgia, and the next time a person in Kansas.

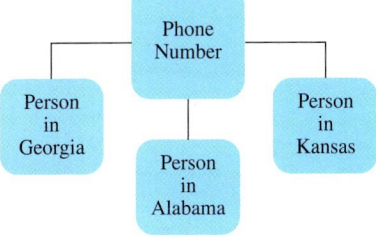

This sort of haphazard relationship between a phone number and a target would create chaos and seriously compromise the future of your cell phone provider. Instead, when we dial 10 digits on our phone we know that the call will be routed to *only one* person's device—the person to whom the phone number is registered.

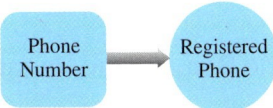

This example illustrates the importance of a **function**. That is, every item in a first set of items (in this case, a phone number being dialed) is associated with one and only one element in a second set of items (in this case, the phone of the proper recipient). Functions are relationships that take an input value and perform an operation on that value to produce a unique and predictable output value. The study of relations and functions begins this chapter and provides the springboard for mathematics at the next level.

Section 16.1 Introduction to Relations

Concepts

1. Definition of a Relation
2. Domain and Range of a Relation
3. Applications Involving Relations

1. Definition of a Relation

In many naturally occurring phenomena, two variables may be linked by some other type of relationship. Table 16-1 shows a correspondence between the length of a woman's femur and her height. (The femur is the large bone in the thigh attached to the knee and hip.)

Table 16-1

Length of Femur (cm) x	Height (in.) y	Ordered Pair
45.5	65.5	(45.5, 65.5)
48.2	68.0	(48.2, 68.0)
41.8	62.2	(41.8, 62.2)
46.0	66.0	(46.0, 66.0)
50.4	70.0	(50.4, 70.0)

Each data point from Table 16-1 may be represented as an ordered pair. In this case, the first value represents the length of a woman's femur and the second, the woman's height. The set of ordered pairs {(45.5, 65.5), (48.2, 68.0), (41.8, 62.2), (46.0, 66.0), (50.4, 70.0)} defines a relation between femur length and height.

2. Domain and Range of a Relation

> **Definition of Relation in x and y**
>
> A set of ordered pairs (x, y) is called a **relation in x and y**. Furthermore,
>
> - The set of first components in the ordered pairs is called the **domain of the relation**.
> - The set of second components in the ordered pairs is called the **range of the relation**.

Example 1 Finding the Domain and Range of a Relation

Find the domain and range of the relation linking the length of a woman's femur to her height {(45.5, 65.5), (48.2, 68.0), (41.8, 62.2), (46.0, 66.0), (50.4, 70.0)}.

Solution:

Domain: {45.5, 48.2, 41.8, 46.0, 50.4}   Set of first components

Range: {65.5, 68.0, 62.2, 66.0, 70.0}   Set of second components

Skill Practice Find the domain and range of the relation.

1. $\left\{(0, 0), (-8, 4), \left(\dfrac{1}{2}, 1\right), (-3, 4), (-8, 0)\right\}$

The x and y components that constitute the ordered pairs in a relation do not need to be numerical. This is demonstrated in Example 2.

Answer

1. Domain: $\left\{0, -8, \dfrac{1}{2}, -3\right\}$; range: {0, 4, 1}

Example 2 Writing a Relation and Finding Its Domain and Range

Table 16-2 gives five states in the United States and the corresponding number of representatives in the House of Representatives for a recent year.

a. The data in the table define a relation. Write a list of ordered pairs for this relation.

b. Write the domain and range.

Table 16-2

State x	Number of Representatives y
Alabama	7
California	53
Colorado	7
Florida	27
Kansas	4

Solution:

a. {(Alabama, 7), (California, 53), (Colorado, 7), (Florida, 27), (Kansas, 4)}

b. Domain: {Alabama, California, Colorado, Florida, Kansas}

   Range: {7, 53, 27, 4} (*Note:* The element 7 is not listed twice.)

Skill Practice The table depicts six types of animals and their average longevity.

2. Write the ordered pairs indicated by the relation in the table.
3. Find the domain and range of the relation.

Animal x	Longevity (years) y
Bear	22.5
Cat	11
Cow	20.5
Deer	12.5
Dog	11
Elephant	35

A relation may consist of a finite number of ordered pairs or an infinite number of ordered pairs. Furthermore, a relation may be defined by several different methods.

- A relation may be defined as a set of ordered pairs.

  {(1, 2), (−3, 4), (1, −4), (3, 4)}

- A relation may be defined by a correspondence (Figure 16-1). The corresponding ordered pairs are {(1, 2), (1, −4), (−3, 4), (3, 4)}.

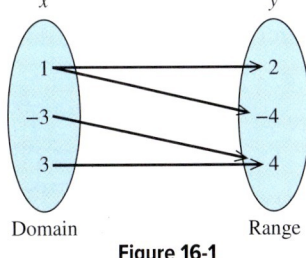

Figure 16-1

- A relation may be defined by a graph (Figure 16-2). The corresponding ordered pairs are {(1, 2), (−3, 4), (1, −4), (3, 4)}.

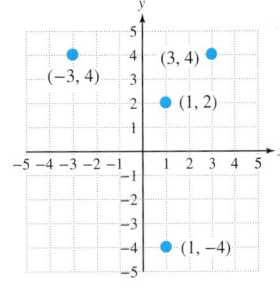

Figure 16-2

Answers

2. {(Bear, 22.5), (Cat, 11), (Cow, 20.5), (Deer, 12.5), (Dog, 11), (Elephant, 35)}
3. Domain: {Bear, Cat, Cow, Deer, Dog, Elephant}; range: {22.5, 11, 20.5, 12.5, 35}

- A relation may be expressed by an equation such as $x = y^2$. The solutions to this equation define an infinite set of ordered pairs of the form $\{(x, y) \mid x = y^2\}$. The solutions can also be represented by a graph in a rectangular coordinate system (Figure 16-3).

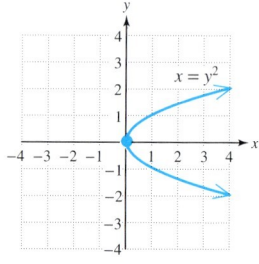

Figure 16-3

Example 3 Finding the Domain and Range of a Relation

Write the relation as a set of ordered pairs. Then find the domain and range.

a.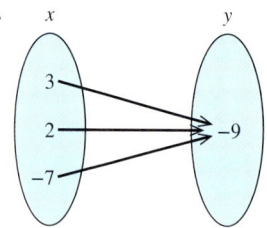

b.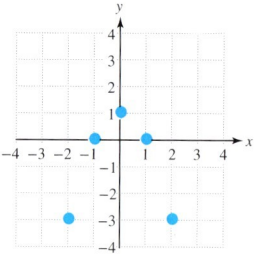

Solution:

a. From the figure, the relation defines the set of ordered pairs:
$\{(3, -9), (2, -9), (-7, -9)\}$

Domain: $\{3, 2, -7\}$

Range: $\{-9\}$

b. The points in the graph make up the set of ordered pairs:
$\{(-2, -3), (-1, 0), (0, 1), (1, 0), (2, -3)\}$

Domain: $\{-2, -1, 0, 1, 2\}$

Range: $\{-3, 0, 1\}$

Skill Practice Find the domain and range of the relations.

4.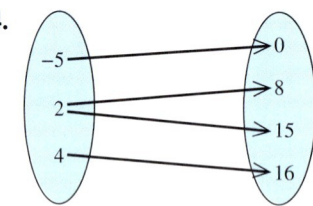

5.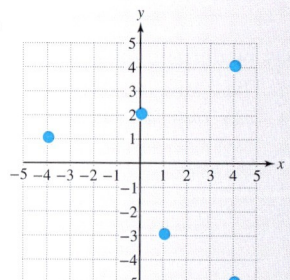

Answers

4. Domain: $\{-5, 2, 4\}$;
   range: $\{0, 8, 15, 16\}$
5. Domain: $\{-4, 0, 1, 4\}$;
   range: $\{-5, -3, 1, 2, 4\}$

Section 16.1 Introduction to Relations   1059

Example 4 — Finding the Domain and Range of a Relation

Use interval notation to express the domain and range of the relation.

a.
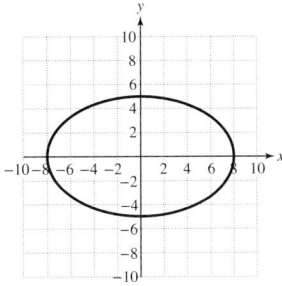

b.
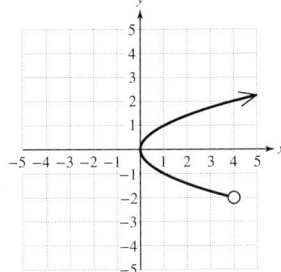

Solution:

a.
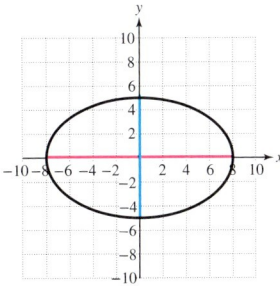

The domain consists of an infinite number of x values extending from -8 to 8 (shown in red). The range consists of all y values from -5 to 5 (shown in blue). Thus, the domain and range must be expressed in set-builder notation or in interval notation.

Domain: $[-8, 8]$

Range: $[-5, 5]$

b.
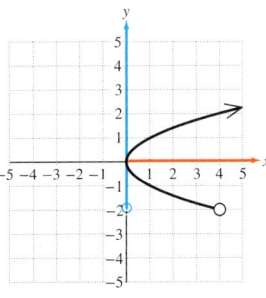

The arrow on the curve indicates that the graph extends infinitely far up and to the right. The open circle means that the graph will end at the point $(4, -2)$, but not include that point.

Domain: $[0, \infty)$

Range: $(-2, \infty)$

Skill Practice Use interval notation to express the domain and range of the relations.

6.
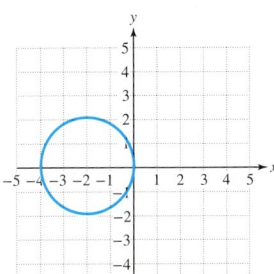

7.
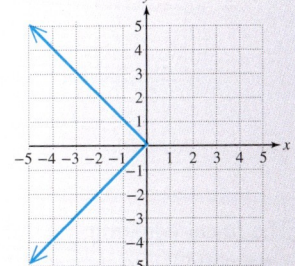

Answers

6. Domain: $[-4, 0]$
   Range: $[-2, 2]$
7. Domain: $(-\infty, 0]$
   Range: $(-\infty, \infty)$

3. Applications Involving Relations

Example 5 — Analyzing a Relation

The data in Table 16-3 depict the length of a woman's femur and her corresponding height. Based on these data, a forensics specialist can find a linear relationship between height y (in inches) and femur length x (in centimeters):

$$y = 0.91x + 24 \qquad 40 \leq x \leq 55$$

From this type of relationship, the height of a woman can be inferred based on skeletal remains.

Table 16-3

Length of Femur (cm) x	Height (in.) y
45.5	65.5
48.2	68.0
41.8	62.2
46.0	66.0
50.4	70.0

a. Find the height of a woman whose femur is 46.0 cm.

b. Find the height of a woman whose femur is 51.0 cm.

c. Why is the domain restricted to $40 \leq x \leq 55$?

Solution:

a. $y = 0.91x + 24$

$ = 0.91(46.0) + 24$ \qquad Substitute $x = 46.0$ cm.

$ = 65.86$ \qquad The woman is approximately 65.9 in. tall.

b. $y = 0.91x + 24$

$ = 0.91(51.0) + 24$ \qquad Substitute $x = 51.0$ cm.

$ = 70.41$ \qquad The woman is approximately 70.4 in. tall.

c. The domain restricts femur length to values between 40 cm and 55 cm inclusive. These values are within the normal lengths for an adult female and are in the proximity of the observed data (Figure 16-4).

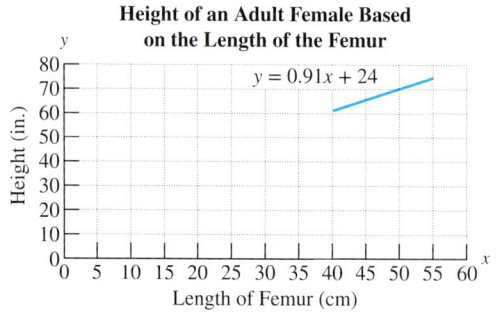

Figure 16-4

Skill Practice The linear equation, $y = -0.014x + 64.5$, for $1500 \leq x \leq 4000$, relates the weight of a car, x (in pounds), to its gas mileage, y (in mpg).

8. Find the gas mileage in miles per gallon for a car weighing 2550 lb.

9. Find the gas mileage for a car weighing 2850 lb.

10. Why is the domain restricted to $1500 \leq x \leq 4000$?

Answers

8. 28.8 mpg **9.** 24.6 mpg

10. The relation is valid only for cars weighing between 1500 lb and 4000 lb, inclusive.

Section 16.1 Activity

A.1. A set of ordered pairs of the form (x, y) is called a relation in x and y. Consider the relation {(−3, 1), (0, 4), (2, −5), (−1, −4)}.

  a. The domain of a relation is the set of _____ (choose one: x values, y values).
  b. Write the domain of the given relation.
  c. The set of y values from a relation is called the _____ of the relation.
  d. Write the range of the given relation.

A relation may be expressed as a figure, a table, or a graph. For Exercises A.2–A.3,

  a. Write the domain of the relation.
  b. Write the range of the relation.

A.2.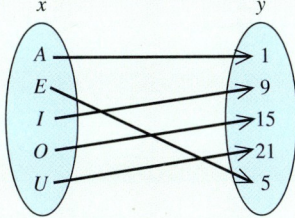

A.3.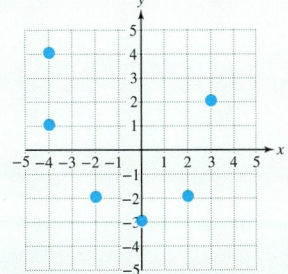

A.4. Consider the relation shown in the graph.

  a. What is the leftmost x value in the graph? What is the rightmost x value?
  b. Because the graph has no "gaps" or "breaks," every x value between the leftmost and rightmost value is accounted for. Write the domain in interval notation.
  c. What is the lowermost y value in the graph? What is the uppermost y value in the graph?
  d. Write the range of the relation in interval notation.

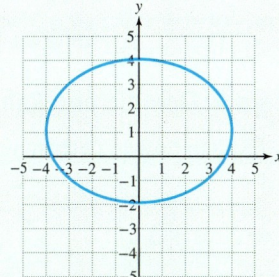

A.5. Consider the relation shown in the graph.

  a. What does the open dot at the point (1, −1) represent?
  b. How far to the left does the graph extend? How far to the right does the graph extend?
  c. Write the domain in interval notation.
  d. What is the lowermost y value in the graph? Does the graph have an uppermost y value?
  e. Write the range of the relation in interval notation.

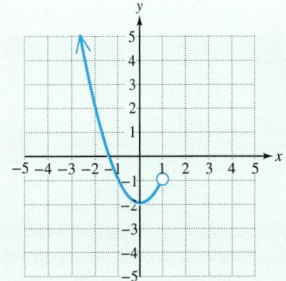

Section 16.1 Practice Exercises

Prerequisite Review

For Exercises R.1–R.6, write the set in interval notation.

R.1. $\{x \mid x > -6\}$ **R.2.** $\{x \mid x < 2\}$ **R.3.** $\left\{x \mid \frac{3}{4} \geq x\right\}$

R.4. $\left\{x \mid -\frac{10}{3} \leq x\right\}$ **R.5.** $\{x \mid 0.5 < x \leq 1.2\}$ **R.6.** $\{x \mid 0 \leq x < 4.6\}$

For Exercises R.7–R.12, estimate the coordinates of the given point.

R.7. A **R.8.** B

R.9. C **R.10.** D

R.11. E **R.12.** F

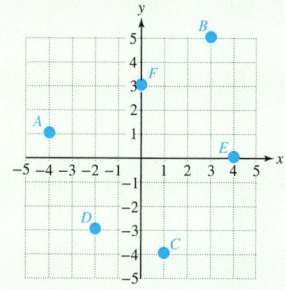

Vocabulary and Key Concepts

1. **a.** A set of ordered pairs (x, y) is called a _____ in x and y.

   b. The _____ of a relation is the set of first components in the ordered pairs.

   c. The _____ of a relation is the set of second components in the ordered pairs.

Concept 2: Domain and Range of a Relation

2. Explain how to determine the domain and range of a relation represented by a set of ordered pairs.

For Exercises 3–14,

   a. Write the relation as a set of ordered pairs.

   b. Determine the domain and range. (See Examples 1–3.)

3.

Region	Number Living in Poverty (thousands)
Northeast	54.1
Midwest	65.6
South	110.7
West	70.7

4.

x	y
0	3
−2	$\frac{1}{2}$
−7	1
−2	8
5	1

5.

Country	Year of First Man or Woman in Space
USSR	1961
USA	1962
Poland	1978
Vietnam	1980
Cuba	1980

6.

State, x	Year of Statehood, y
Maine	1820
Nebraska	1823
Utah	1847
Hawaii	1959
Alaska	1959

7.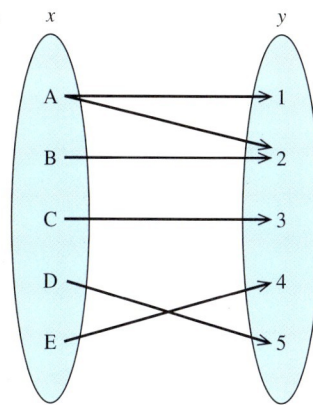

8.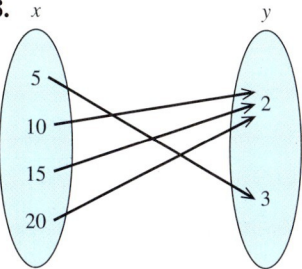

9.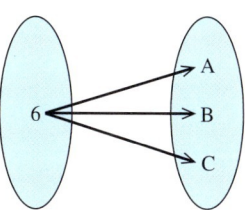

10.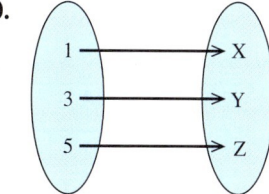

11.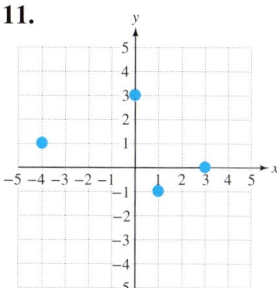

12.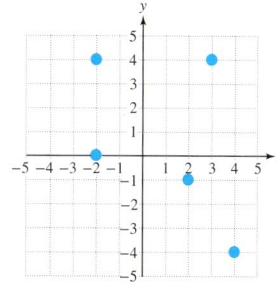

13.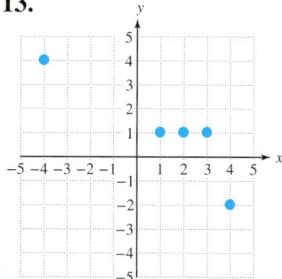

14.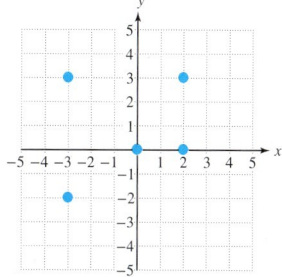

For Exercises 15–30, find the domain and range of the relations. Use interval notation where appropriate.
(See Example 4.)

15.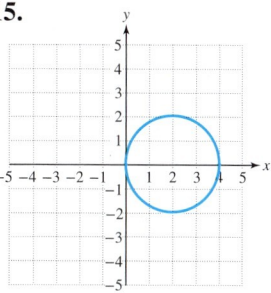

16.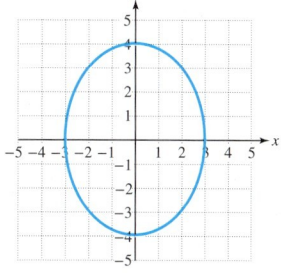

17.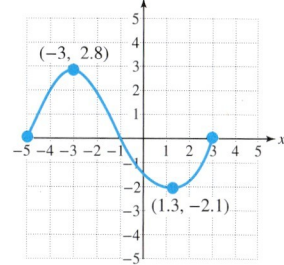

18.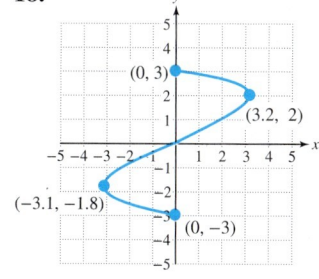

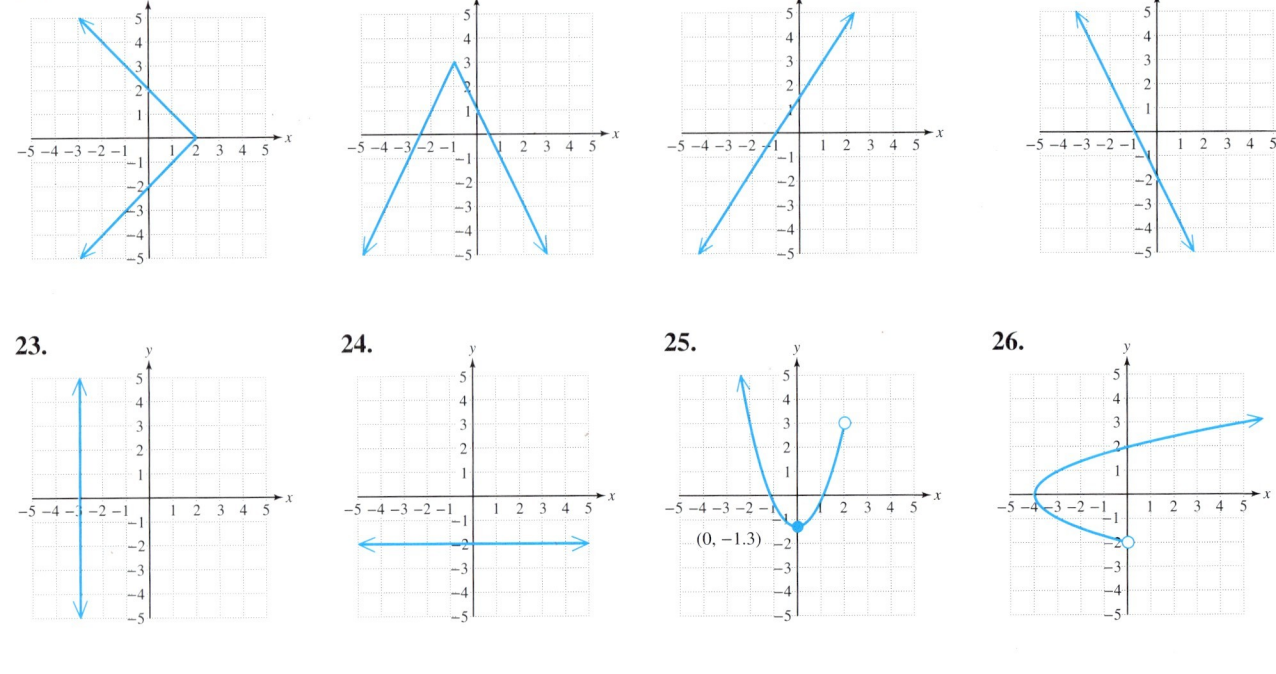

Concept 3: Applications Involving Relations

31. The table gives a relation between the month of the year and the average precipitation for that month for Miami, Florida.
(See Example 5.)

    a. What is the range element corresponding to April?

    b. What is the range element corresponding to June?

    c. Which element in the domain corresponds to the least value in the range?

    d. Complete the ordered pair: (, 2.66)

    e. Complete the ordered pair: (Sept.,)

    f. What is the domain of this relation?

Month x	Precipitation (in.) y	Month x	Precipitation (in.) y
Jan.	2.01	July	5.70
Feb.	2.08	Aug.	7.58
Mar.	2.39	Sept.	7.63
Apr.	2.85	Oct.	5.64
May	6.21	Nov.	2.66
June	9.33	Dec.	1.83

Source: U.S. National Oceanic and Atmospheric Administration

32. The table gives a relation between a person's age and the person's maximum recommended heart rate.

Age (years) x	Maximum Recommended Heart Rate (Beats per Minute) y
20	200
30	190
40	180
50	170
60	160

   a. What is the domain?

   b. What is the range?

   c. The range element 200 corresponds to what element in the domain?

   d. Complete the ordered pair: (50,   )

   e. Complete the ordered pair: (, 190)

33. The population of Canada y (in millions) can be approximated by the relation $y = 0.35x + 30.7$, where x represents the number of years since 2000.

   a. Approximate the population of Canada in the year 2010.

   b. In what year did the population of Canada reach approximately 37,350,000?

34. The world record times for women's track and field events are shown in the table. The women's world record time y (in seconds) required to run x meters can be approximated by the relation $y = 0.159x - 10.79$.

Digital Vision/Getty Images

Distance (m)	Time (sec)	Winner's Name and Country
100	10.49	Florence Griffith Joyner (United States)
200	21.34	Florence Griffith Joyner (United States)
400	47.60	Marita Koch (East Germany)
800	113.28	Jarmila Kratochvilova (Czechoslovakia)
1000	148.98	Svetlana Masterkova (Russia)
1500	230.07	Genzebe Dibaba (Ethiopia)

   a. Predict the time required for a 500-m race.

   b. Use this model to predict the time for a 1000-m race. Is this value exactly the same as the data value given in the table? Explain.

Expanding Your Skills

35. a. Define a relation with four ordered pairs such that the first element of the ordered pair is the name of a friend and the second element is your friend's place of birth.

    b. State the domain and range of this relation.

36. a. Define a relation with four ordered pairs such that the first element is a state and the second element is its capital.
    b. State the domain and range of this relation.

37. Use a mathematical equation to define a relation whose second component y is 1 less than 2 times the first component x.

38. Use a mathematical equation to define a relation whose second component y is 3 more than the first component x.

39. Use a mathematical equation to define a relation whose second component y is the square of the first component x.

40. Use a mathematical equation to define a relation whose second component y is one-fourth the first component x.

Section 16.2 Introduction to Functions

Concepts
1. Definition of a Function
2. Vertical Line Test
3. Function Notation
4. Finding Function Values From a Graph
5. Domain of a Function

1. Definition of a Function

In this section, we introduce a special type of relation called a function.

> **Definition of a Function**
>
> Given a relation in x and y, we say "y is a **function** of x" if, for each element x in the domain, there is exactly one value of y in the range.
>
> *Note:* This means that no two ordered pairs may have the same first coordinate and different second coordinates.

To understand the difference between a relation that is a function and a relation that is not a function, consider Example 1.

Example 1 — Determining Whether a Relation Is a Function

Determine which of the relations define y as a function of x.

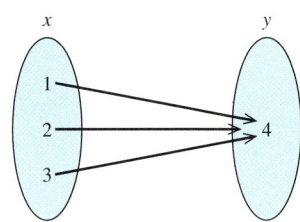

Solution:

a. This relation is defined by the set of ordered pairs

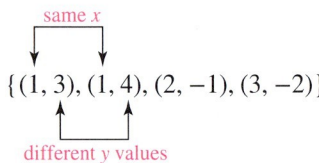

$\{(1, 3), (1, 4), (2, -1), (3, -2)\}$

When $x = 1$, there are *two* possible range elements: $y = 3$ and $y = 4$. Therefore, this relation is *not* a function.

b. This relation is defined by the set of ordered pairs $\{(1, 4), (2, -1), (3, 2)\}$.

Notice that no two ordered pairs have the same value of x but different values of y. Therefore, this relation *is* a function.

c. This relation is defined by the set of ordered pairs $\{(1, 4), (2, 4), (3, 4)\}$.

Notice that no two ordered pairs have the same value of x but different values of y. Therefore, this relation *is* a function.

Skill Practice Determine if the relation defines y as a function of x.

1.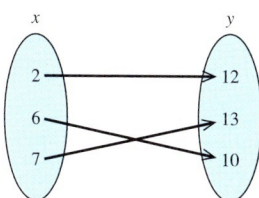

2. {(4, 2), (−5, 4), (0, 0), (8, 4)}

3. {(−1, 6), (8, 9), (−1, 4), (−3, 10)}

2. Vertical Line Test

A relation that is not a function has at least one domain element x paired with more than one range value y. For example, the set {(4, 2), (4, −2)} does not define a function because two different y values correspond to the same x. These two points are aligned vertically in the xy-plane, and a vertical line drawn through one point also intersects the other point (see Figure 16-5). If a vertical line drawn through a graph of a relation intersects the graph in more than one point, the relation cannot be a function. This idea is stated formally as the **vertical line test**.

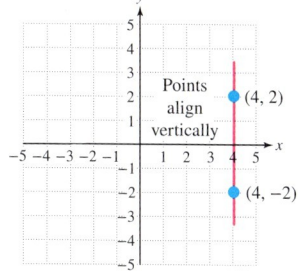

Figure 16-5

The Vertical Line Test

Consider a relation defined by a set of points (x, y) in a rectangular coordinate system. The graph defines y as a function of x if no vertical line intersects the graph in more than one point.

The vertical line test can be demonstrated by graphing the ordered pairs from the relations in Example 1.

a. {(1, 3), (1, 4), (2, −1), (3, −2)}

b. {(1, 4), (2, −1), (3, 2)}

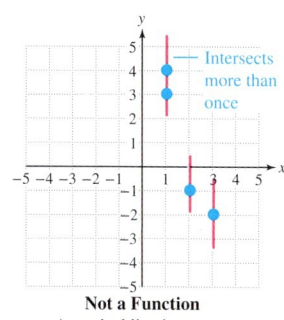

Not a Function
A vertical line intersects in more than one point.

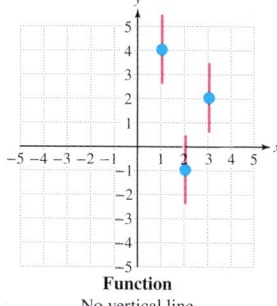

Function
No vertical line intersects more than once.

Example 2 — Using the Vertical Line Test

Use the vertical line test to determine whether the relations define y as a function of x.

a.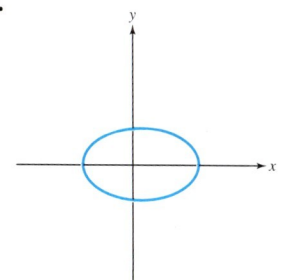

b.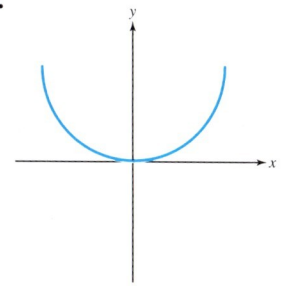

Answers
1. Yes 2. Yes 3. No

Solution:

a.
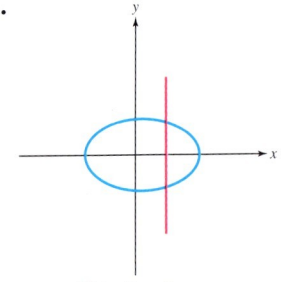
Not a Function
A vertical line intersects
in more than one point.

b.
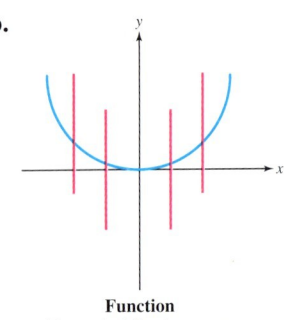
Function
No vertical line intersects
in more than one point.

Skill Practice Use the vertical line test to determine whether the relations define y as a function of x.

4.
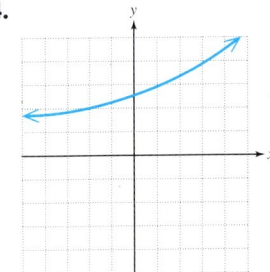

5.
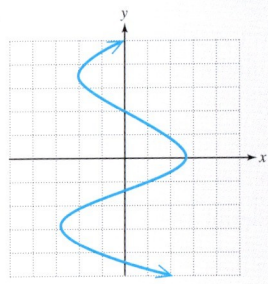

3. Function Notation

A function is defined as a relation with the added restriction that each value in the domain must have only one corresponding y value in the range. In mathematics, functions are often given by rules or equations to define the relationship between two or more variables. For example, the equation $y = 2x$ defines the set of ordered pairs such that the y value is twice the x value.

When a function is defined by an equation, we often use **function notation**. For example, the equation $y = 2x$ may be written in function notation as

$f(x) = 2x$
- f is the name of the function.
- x is an input value from the domain of the function.
- $f(x)$ is the function value (y value) corresponding to x.

Avoiding Mistakes
Be sure to note that $f(x)$ is *not* $f \cdot x$.

The notation $f(x)$ is read as "f of x" or "the value of the function f at x."

A function may be evaluated at different values of x by substituting x values from the domain into the function. For example, to evaluate the function defined by $f(x) = 2x$ at $x = 5$, substitute $x = 5$ into the function.

$$f(x) = 2x$$
$$f(5) = 2(5)$$
$$f(5) = 10$$

TIP: $f(5) = 10$ can be interpreted as the ordered pair (5, 10).

Answers
4. Yes 5. No

Thus, when $x = 5$, the corresponding function value is 10. We say:
- f of 5 is 10.
- f at 5 is 10.
- f evaluated at 5 is 10.

The names of functions are often given by either lowercase or uppercase letters, such as f, g, h, p, K, and M. The input variable may also be a letter other than x. For example, $y = P(t)$ might represent population as a function of time.

Example 3 — Evaluating a Function

Given the function defined by $g(x) = \frac{1}{2}x - 1$, find the function values.

a. $g(0)$ **b.** $g(2)$ **c.** $g(4)$ **d.** $g(-2)$

Solution:

a. $g(x) = \frac{1}{2}x - 1$

$g(0) = \frac{1}{2}(0) - 1$

$= 0 - 1$

$= -1$

We say that "g of 0 is -1."
This is equivalent to the ordered pair $(0, -1)$.

b. $g(x) = \frac{1}{2}x - 1$

$g(2) = \frac{1}{2}(2) - 1$

$= 1 - 1$

$= 0$

We say that "g of 2 is 0."
This is equivalent to the ordered pair $(2, 0)$.

c. $g(x) = \frac{1}{2}x - 1$

$g(4) = \frac{1}{2}(4) - 1$

$= 2 - 1$

$= 1$

We say that "g of 4 is 1."
This is equivalent to the ordered pair $(4, 1)$.

d. $g(x) = \frac{1}{2}x - 1$

$g(-2) = \frac{1}{2}(-2) - 1$

$= -1 - 1$

$= -2$

We say that "g of -2 is -2."
This is equivalent to the ordered pair $(-2, -2)$.

Skill Practice Given the function defined by $f(x) = -2x - 3$, find the function values.

6. $f(1)$ **7.** $f(0)$ **8.** $f(-3)$ **9.** $f\left(\dfrac{1}{2}\right)$

In Example 3, notice that $g(0)$, $g(2)$, $g(4)$, and $g(-2)$ correspond to the ordered pairs $(0, -1)$, $(2, 0)$, $(4, 1)$, and $(-2, -2)$. In the graph, these points "line up." The graph of *all* ordered pairs defined by this function is a line with a slope of $\frac{1}{2}$ and y-intercept of $(0, -1)$ (Figure 16-6). This should not be surprising because the function defined by $g(x) = \frac{1}{2}x - 1$ is equivalent to $y = \frac{1}{2}x - 1$.

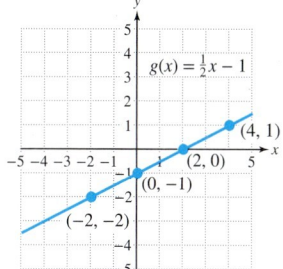

Figure 16-6

Answers

6. -5 **7.** -3 **8.** 3 **9.** -4

Example 4 Evaluating Functions

Given the functions defined by $f(x) = x^2 - 2x$ and $g(x) = 3x + 5$, find the function values.

a. $f(t)$ **b.** $g(w + 4)$ **c.** $f(-t)$

Solution:

a. $f(x) = x^2 - 2x$

$f(t) = (t)^2 - 2(t)$ Substitute $x = t$ for all values of x in the function.

$= t^2 - 2t$ Simplify.

b. $g(x) = 3x + 5$

$g(w + 4) = 3(w + 4) + 5$ Substitute $x = w + 4$ for all values of x in the function.

$= 3w + 12 + 5$

$= 3w + 17$ Simplify.

c. $f(x) = x^2 - 2x$ Substitute $-t$ for x.

$f(-t) = (-t)^2 - 2(-t)$

$= t^2 + 2t$ Simplify.

> **FOR REVIEW**
>
> Remember to use parentheses when substituting values into an expression. For example, the use of parentheses in the expression $3(w + 4) + 5$ reminds us to multiply 3 by both w and 4.

Skill Practice Given the function defined by $g(x) = 4x - 3$, find the function values.

10. $g(a)$ **11.** $g(x + h)$ **12.** $g(-x)$

4. Finding Function Values From a Graph

We can find function values by looking at the graph of a function. The value of $f(a)$ refers to the y-coordinate of a point with x-coordinate a.

Example 5 Finding Function Values From a Graph

Consider the function pictured in Figure 16-7.

a. Find $h(-1)$.

b. Find $h(2)$.

c. Find $h(5)$.

d. For what value of x is $h(x) = 3$?

e. For what values of x is $h(x) = 0$?

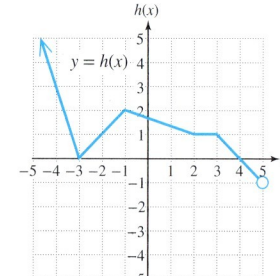

Figure 16-7

Solution:

a. $h(-1) = 2$ From the graph, when $x = -1$, the function value (y value) is 2. This corresponds to the ordered pair $(-1, 2)$.

b. $h(2) = 1$ From the graph, when $x = 2$, the function value (y value) is 1. This corresponds to the ordered pair $(2, 1)$.

Answers
10. $4a - 3$ **11.** $4x + 4h - 3$
12. $-4x - 3$

c. $h(5)$ is not defined. The open dot at $(5, -1)$ indicates that 5 is not in the domain.

d. $h(x) = 3$ for $x = -4$ We want to find the x value(s) that correspond to a function value (y value) of 3. This corresponds to the ordered pair $(-4, 3)$.

e. $h(x) = 0$ for $x = -3$ and $x = 4$ We want to find the x value(s) that correspond to a function value (y value) of 0. These correspond to the ordered pairs $(-3, 0)$ and $(4, 0)$.

Skill Practice Refer to the function graphed here.

13. Find $f(0)$.
14. Find $f(-2)$.
15. Find $f(5)$.
16. For what value(s) of x is $f(x) = 0$?
17. For what value(s) of x is $f(x) = -4$?

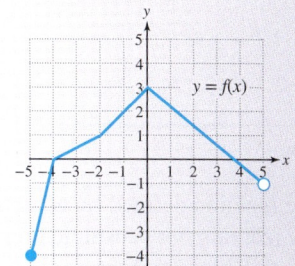

5. Domain of a Function

A function is a relation, and it is often necessary to determine its domain and range. Consider a function defined by the equation $y = f(x)$. The **domain** of f is the set of all x values that when substituted into the function produce a real number. The **range** of f is the set of all y values corresponding to the values of x in the domain.

To find the domain of a function defined by $y = f(x)$, keep these guidelines in mind.

- Exclude values of x that make the denominator of a fraction zero.
- Exclude values of x that make the expression within a square root negative.

Example 6 Finding the Domain of a Function

Write the domain in interval notation.

a. $f(x) = \dfrac{x+7}{2x-1}$ **b.** $h(x) = \dfrac{x-4}{x^2+9}$

c. $k(t) = \sqrt{t+4}$ **d.** $g(t) = t^2 - 3t$

Solution:

a. $f(x) = \dfrac{x+7}{2x-1}$ will not be a real number when the denominator is zero, that is, when

$$2x - 1 = 0$$
$$2x = 1$$
$$x = \dfrac{1}{2}$$

The value $x = \dfrac{1}{2}$ must be *excluded* from the domain.

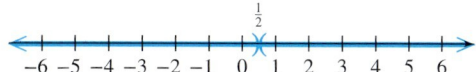

The number line indicates that the domain consists of two intervals, $(-\infty, \frac{1}{2})$ and $(\frac{1}{2}, \infty)$. We use the notation $\cup$ to combine the two intervals. The symbol $\cup$ stands for union. The union of two intervals consists of all the numbers in the first interval, along with all the numbers in the second interval.

Interval notation: $\left(-\infty, \dfrac{1}{2}\right) \cup \left(\dfrac{1}{2}, \infty\right)$

Answers
13. 3   14. 1
15. not defined
16. $x = -4$ and $x = 4$
17. $x = -5$

Chapter 16 Relations and Functions

FOR REVIEW

Recall that with interval notation, curved parentheses, (or), indicate that the endpoints *are not* included in the interval. On the other hand, square brackets, [or], indicate that the endpoints *are* included.

b. For $h(x) = \dfrac{x-4}{x^2+9}$ the quantity x^2 is greater than or equal to 0 for all real numbers x, and the number 9 is positive. The sum $x^2 + 9$ must be *positive* for all real numbers x. The denominator will never be zero; therefore, the domain is the set of all real numbers.

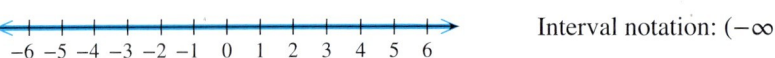

Interval notation: $(-\infty, \infty)$

c. The value of the function defined by $k(t) = \sqrt{t+4}$ will not be a real number when the expression within the square root is negative. Therefore, the domain is the set of all t values that make $t + 4$ greater than or equal to zero.

$$t + 4 \geq 0$$
$$t \geq -4$$

Interval notation: $[-4, \infty)$

d. The function defined by $g(t) = t^2 - 3t$ has no restrictions on its domain because any real number substituted for t will produce a real number. The domain is the set of all real numbers.

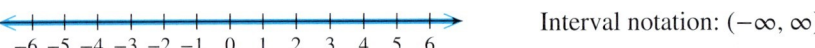

Interval notation: $(-\infty, \infty)$

Skill Practice Write the domain in interval notation.

18. $f(x) = \dfrac{2x+1}{x-9}$

19. $k(x) = \dfrac{-5}{4x^2+1}$

20. $g(x) = \sqrt{x-2}$

21. $h(x) = x + 6$

Answers
18. $(-\infty, 9) \cup (9, \infty)$ 19. $(-\infty, \infty)$
20. $[2, \infty)$ 21. $(-\infty, \infty)$

Section 16.2 Activity

A.1. Given a set of ordered pairs, how can you determine whether the relation defines y as a function of x?

For Exercises A.2–A.3, consider the given relation.
  a. Do any two ordered pairs have the same x value but different y values?
  b. Is the relation a function?

A.2. $\{(-5, 1), (3, 4), (-2, 6), (-5, 2), (0, -3)\}$

A.3. $\{(-1, 6), (2, 11), (8, 6), (-3, 1), (0.4, -0.5)\}$

A.4. a. For the graph given, draw a vertical line through the point (4, 2). Does the vertical line intersect the graph at any other point?

  b. Does this graph define y as a function of x?
  c. Using this example, explain how the vertical line test is used to determine if a graph defines y as a function of x.

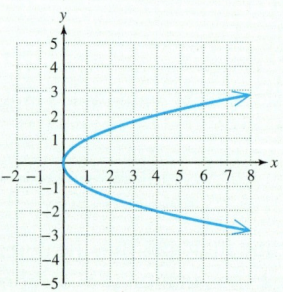

A.5. Consider the equation $y = 2x + 1$.
  a. If $x = 3$, what is the corresponding y value?
  b. Write the result of part (a) as an ordered pair (x, y).

A.6. Consider the function defined by $f(x) = 2x + 1$.
   a. Find $f(3)$. That is, evaluate the function for $x = 3$ by substituting 3 for x.
   b. Write the result of part (a) as an ordered pair (x, y).
   c. Refer to Exercise A.5 and compare the results.

A.7. Given $g(x) = \dfrac{4}{x - 1}$, find the function values if possible.
   a. $g(2)$
   b. $g(-3)$
   c. $g(0)$
   d. $g(1)$

A.8. Refer to $g(x) = \dfrac{4}{x - 1}$ from Exercise A.7.
   a. What value(s) of x must be excluded from the domain of g? Why?
   b. Write the domain of g in interval notation.

A.9. Given $h(x) = \sqrt{x + 2}$, find the function values if possible.
   a. $h(-1)$
   b. $h(2)$
   c. $h(7)$
   d. $h(-6)$

A.10. Refer to $h(x) = \sqrt{x + 2}$ from Exercise A.9.
   a. The square root of a negative number is not a real number. Therefore, what restrictions must be imposed on x?
   b. Write the domain of h in interval notation.

A.11. Consider the function defined by $f(x) = x^2 - 3x + 4$. Find the function values.
   a. $f(-2)$
   b. $f(t)$
   c. $f(a + 2)$

Practice Exercises — Section 16.2

Prerequisite Review

For Exercises R.1–R.4, substitute the given value of x. Then solve for y.

R.1. $y = -2x - 4;\ x = 6$
R.2. $y = 5x - 10;\ x = -2$
R.3. $y = 2x^2 - 4x + 1;\ x = -2$
R.4. $y = 3x^2 + x - 4;\ x = -1$

For Exercises R.5–R.8, determine the values of x for which the expression is undefined.

R.5. $\dfrac{3}{x - 5}$
R.6. $\dfrac{-4}{x + 1}$
R.7. $\dfrac{x + 4}{x^2 - 9}$
R.8. $\dfrac{x - 10}{x^2 - 25}$

For Exercises R.9–R.12, express the set in interval notation.

R.9. ⟶ [from 4
R.10. ⟵ ⟶ to −3
R.11. (−6 to −5)
R.12. [−2 to −7

For Exercises R.13–R.16, solve the inequality. Write the solution set in interval notation.

R.13. $2x - 3 \geq 0$
R.14. $5x + 1 \geq 0$
R.15. $4 - x \geq 0$
R.16. $-3 - x \geq 0$

Vocabulary and Key Concepts

1. a. Given a relation in x and y, we say that y is a _____ of x if for each element x in the domain, there is exactly one value of y in the range.

   b. If a _____ line intersects the graph of a relation in more than one point, the relation is not a function.

   c. Function notation for the equation $y = 2x + 1$ is $f(x) = $ _____.

   d. Given a function defined by $y = f(x)$, the set of all _____ that produce a real number when substituted into a function is called the domain of the function.

   e. The set of all _____ corresponding to x values in the domain is called the range of a function.

   f. To find the domain of a function defined by $y = f(x)$, exclude any values of x that make the _____ of a fraction equal to zero.

   g. To find the domain of a function defined by $y = f(x)$, exclude any values of x that make the expression within a square root _____.

For Exercises 2–4, consider the functions defined by $f(x) = x + 2$, $g(x) = 2x$, and $h(x) = x^2$. Fill in the blank with f, g, or h based on the description of the function.

2. Function _____ doubles the values of x from its domain.

3. Function _____ squares the values of x from its domain.

4. Function _____ increases by two the values of x from its domain.

Concept 1: Definition of a Function

For Exercises 5–10, determine if the relation defines y as a function of x. **(See Example 1.)**

5.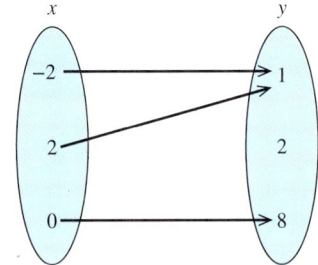

6.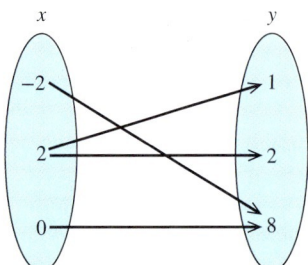

7.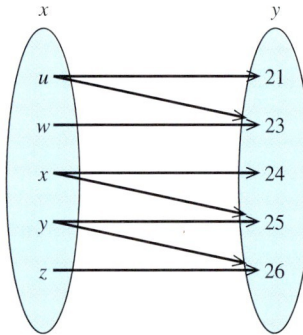

8.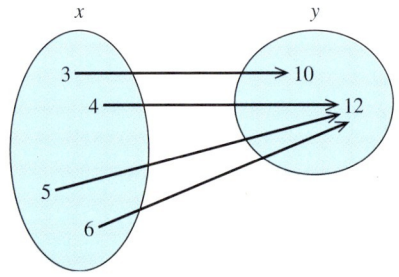

9. $\{(1, 2), (3, 4), (5, 4), (-9, 3)\}$

10. $\left\{(0, -1.1), \left(\dfrac{1}{2}, 8\right), (1.1, 8), \left(4, \dfrac{1}{2}\right)\right\}$

Concept 2: Vertical Line Test

For Exercises 11–16, use the vertical line test to determine whether the relation defines y as a function of x. **(See Example 2.)**

11.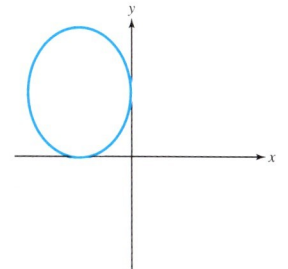

12.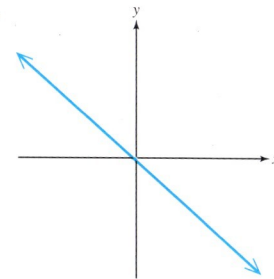

13.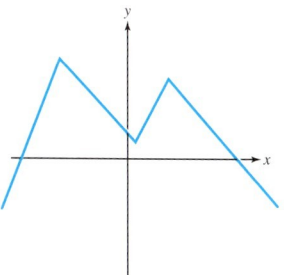

14.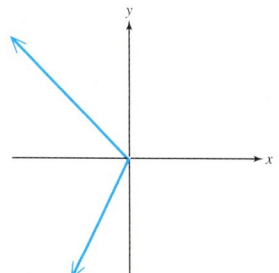

15.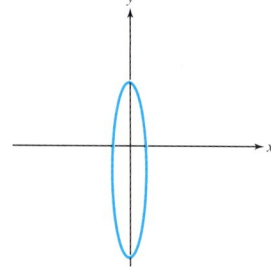

16.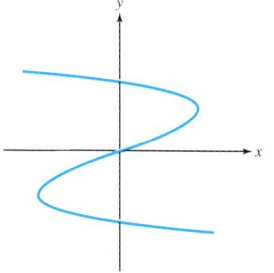

Concept 3: Function Notation

For Exercises 17–20, consider a function defined by $y = f(x)$.

17. Interpret the meaning of $f(2) = 5$.

18. Interpret the meaning of $f(-7) = 8$.

19. Write the ordered pair represented by $f(0) = -2$.

20. Write the ordered pair represented by $f(-10) = 11$.

Consider the functions defined by $f(x) = 6x - 2$, $g(x) = -x^2 - 4x + 1$, $h(x) = 7$, and $k(x) = |x - 2|$. For Exercises 21–52, find the following. **(See Examples 3–4.)**

21. $g(2)$
22. $k(2)$
23. $g(0)$
24. $h(0)$

25. $k(0)$
26. $f(0)$
27. $f(t)$
28. $g(a)$

29. $h(u)$
30. $k(v)$
31. $g(-3)$
32. $h(-5)$

33. $k(-2)$
34. $f(-6)$
35. $f(x+1)$
36. $h(x+1)$

37. $g(2x)$
38. $k(x-3)$
39. $g(-\pi)$
40. $g(a^2)$

41. $h(a+b)$
42. $f(x+h)$
43. $f(-a)$
44. $g(-b)$

45. $k(-c)$
46. $h(-x)$
47. $f\left(\frac{1}{2}\right)$
48. $g\left(\frac{1}{4}\right)$

49. $h\left(\frac{1}{7}\right)$
50. $k\left(\frac{3}{2}\right)$
51. $f(-2.8)$
52. $k(-5.4)$

Consider the functions $p = \{(\frac{1}{2}, 6), (2, -7), (1, 0), (3, 2\pi)\}$ and $q = \{(6, 4), (2, -5), (\frac{3}{4}, \frac{1}{5}), (0, 9)\}$. For Exercises 53–60, find the function values.

53. $p(2)$
54. $p(1)$
55. $p(3)$
56. $p\left(\frac{1}{2}\right)$

57. $q(2)$
58. $q\left(\frac{3}{4}\right)$
59. $q(6)$
60. $q(0)$

Concept 4: Finding Function Values From a Graph

61. The graph of $y = f(x)$ is given. **(See Example 5.)**

    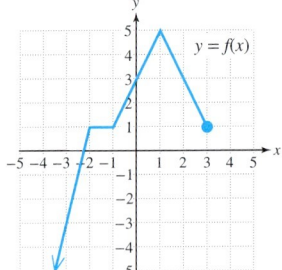

    a. Find $f(0)$.
    b. Find $f(3)$.
    c. Find $f(-2)$.
    d. For what value(s) of x is $f(x) = -3$?
    e. For what value(s) of x is $f(x) = 3$?
    f. Write the domain of f.
    g. Write the range of f.

62. The graph of $y = g(x)$ is given.

    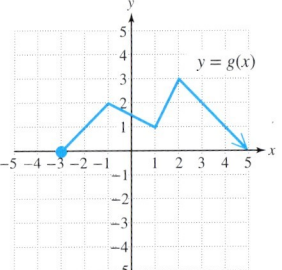

    a. Find $g(-1)$.
    b. Find $g(1)$.
    c. Find $g(4)$.
    d. For what value(s) of x is $g(x) = 3$?
    e. For what value(s) of x is $g(x) = 0$?
    f. Write the domain of g.
    g. Write the range of g.

63. The graph of $y = H(x)$ is given.

    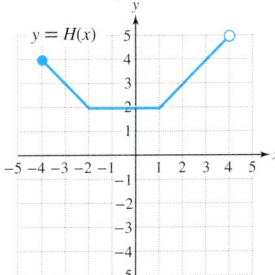

    a. Find $H(-3)$.
    b. Find $H(4)$.
    c. Find $H(3)$.
    d. For what value(s) of x is $H(x) = 3$?
    e. For what value(s) of x is $H(x) = 2$?
    f. Write the domain of H.
    g. Write the range of H.

64. The graph of $y = K(x)$ is given.

    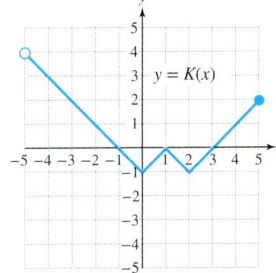

    a. Find $K(0)$.
    b. Find $K(-5)$.
    c. Find $K(1)$.
    d. For what value(s) of x is $K(x) = 0$?
    e. For what value(s) of x is $K(x) = 3$?
    f. Write the domain of K.
    g. Write the range of K.

65. The graph of $y = p(x)$ is given.

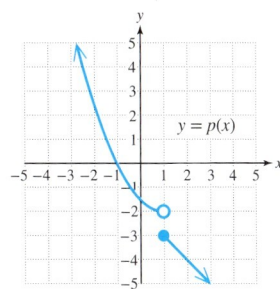

  a. Find $p(2)$.
  b. Find $p(-1)$.
  c. Find $p(1)$.
  d. For what value(s) of x is $p(x) = 0$?
  e. For what value(s) of x is $p(x) = -2$?
  f. Write the domain of p.
  g. Write the range of p.

66. The graph of $y = q(x)$ is given.

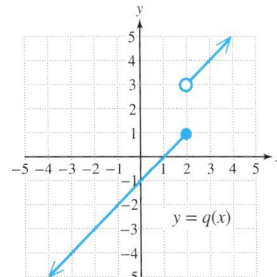

  a. Find $q(3)$.
  b. Find $q(-1)$.
  c. Find $q(2)$.
  d. For what value(s) of x is $q(x) = -4$?
  e. For what value(s) of x is $q(x) = 3$?
  f. Write the domain of q.
  g. Write the range of q.

For Exercises 67–76, refer to the functions $y = f(x)$ and $y = g(x)$, defined as follows:

$$f = \{(-3, 5), (-7, -3), (-\tfrac{3}{2}, 4), (1.2, 5)\}$$
$$g = \{(0, 6), (2, 6), (6, 0), (1, 0)\}$$

67. Identify the domain of f.

68. Identify the range of f.

69. Identify the range of g.

70. Identify the domain of g.

71. For what value(s) of x is $f(x) = 5$?

72. For what value(s) of x is $f(x) = -3$?

73. For what value(s) of x is $g(x) = 0$?

74. For what value(s) of x is $g(x) = 6$?

75. Find $f(-7)$.

76. Find $g(0)$.

Concept 5: Domain of a Function

77. Explain how to determine the domain of the function defined by $f(x) = \dfrac{x+6}{x-2}$.

78. Explain how to determine the domain of the function defined by $g(x) = \sqrt{x-3}$.

For Exercises 79–94, find the domain. Write the answer in interval notation. **(See Example 6.)**

79. $k(x) = \dfrac{x-3}{x+6}$

80. $m(x) = \dfrac{x-1}{x-4}$

81. $f(t) = \dfrac{5}{t}$

82. $g(t) = \dfrac{t-7}{t}$

83. $h(p) = \dfrac{p-4}{p^2+1}$

84. $n(p) = \dfrac{p+8}{p^2+2}$

85. $h(t) = \sqrt{t+7}$

86. $k(t) = \sqrt{t-5}$

87. $f(a) = \sqrt{a-3}$
88. $g(a) = \sqrt{a+2}$
89. $m(x) = \sqrt{1-2x}$
90. $n(x) = \sqrt{12-6x}$

91. $p(t) = 2t^2 + t - 1$
92. $q(t) = t^3 + t - 1$
93. $f(x) = x + 6$
94. $g(x) = 8x - \pi$

Mixed Exercises

95. The height (in feet) of a ball that is dropped from an 80-ft building is given by $h(t) = -16t^2 + 80$, where t is the time in seconds after the ball is dropped.

    a. Find $h(1)$ and $h(1.5)$.

    b. Interpret the meaning of the function values found in part (a).

96. A ball is dropped from a 50-m building. The height (in meters) after t seconds is given by $h(t) = -4.9t^2 + 50$.

    a. Find $h(1)$ and $h(1.5)$.

    b. Interpret the meaning of the function values found in part (a).

97. If Alicia rides a bike at an average speed of 11.5 mph, the distance that she rides can be represented by $d(t) = 11.5t$, where t is the time in hours.

    a. Find $d(1)$ and $d(1.5)$.

    b. Interpret the meaning of the function values found in part (a).

98. Brian's score on an exam is a function of the number of hours he spends studying. The function defined by

    $P(x) = \dfrac{100x^2}{50 + x^2}$ $(x \geq 0)$ indicates that he will achieve a score of $P\%$ if he studies for x hours.

    Evaluate $P(0)$, $P(5)$, $P(10)$, $P(15)$, $P(20)$, and $P(25)$ and confirm the values on the graph. (Round to one decimal place.) Interpret $P(25)$ in the context of this problem.

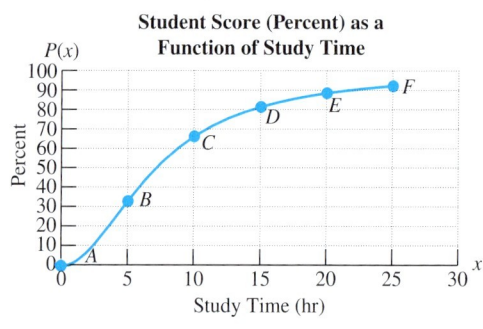

Student Score (Percent) as a Function of Study Time

For Exercises 99–102, write a function defined by $y = f(x)$ subject to the conditions given.

99. The value of $f(x)$ is three more than two times x.

100. The value of $f(x)$ is four less than the square of x.

101. The value of $f(x)$ is ten less than the absolute value of x.

102. The value of $f(x)$ is sixteen times the square root of x.

Technology Connections

103. Graph $k(t) = \sqrt{t-5}$. Use the graph to support your answer to Exercise 86.

104. Graph $h(t) = \sqrt{t+7}$. Use the graph to support your answer to Exercise 85.

Expanding Your Skills

For Exercises 105–106, write the domain in interval notation.

105. $q(x) = \dfrac{2}{\sqrt{x+2}}$

106. $p(x) = \dfrac{8}{\sqrt{x-4}}$

Graphs of Functions

Section 16.3

Concepts

1. Linear and Constant Functions
2. Graphs of Basic Functions
3. Definition of a Quadratic Function
4. Finding the x- and y-Intercepts of a Graph Defined by $y = f(x)$

1. Linear and Constant Functions

A function may be expressed as a mathematical equation that relates two or more variables. In this section, we will look at several elementary functions.

Recall that the graph of an equation $y = k$, where k is a constant, is a horizontal line. In function notation, this can be written as $f(x) = k$. For example, the graph of the function defined by $f(x) = 3$ is a horizontal line, as shown in Figure 16-8.

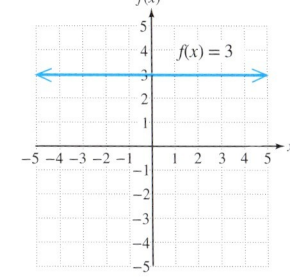

Figure 16-8

We say that a function defined by $f(x) = k$ is a constant function because for any value of x, the function value is constant.

An equation of the form $y = mx + b$ is represented graphically by a line with slope m and y-intercept $(0, b)$. In function notation, this may be written as $f(x) = mx + b$. A function in this form is called a linear function. For example, the function defined by $f(x) = 2x - 3$ is a linear function with slope $m = 2$ and y-intercept $(0, -3)$ (Figure 16-9).

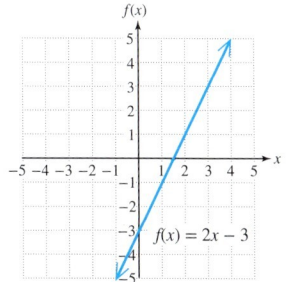

Figure 16-9

> **Linear Functions and Constant Functions**
>
> Let m and b represent real numbers such that $m \neq 0$. Then
>
> A function that can be written in the form $f(x) = mx + b$ is a **linear function**.
> A function that can be written in the form $f(x) = b$ is a **constant function**.
>
> *Note:* The graphs of linear and constant functions are lines.

2. Graphs of Basic Functions

At this point, we are able to recognize the equations and graphs of linear and constant functions. In addition to linear and constant functions, the following equations define six basic functions that will be encountered in the study of algebra:

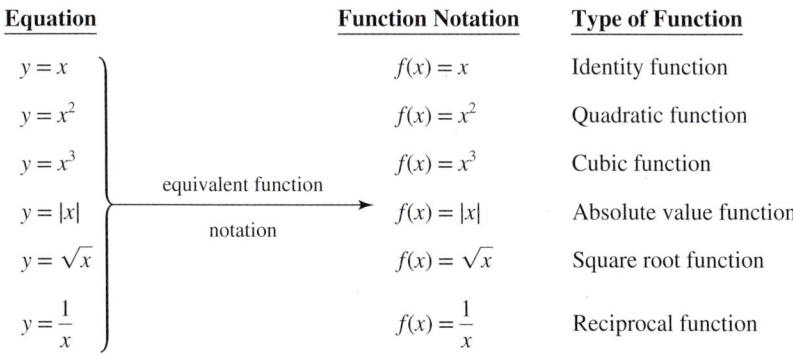

Equation	Function Notation	Type of Function
$y = x$	$f(x) = x$	Identity function
$y = x^2$	$f(x) = x^2$	Quadratic function
$y = x^3$	$f(x) = x^3$	Cubic function
$y = |x|$	$f(x) = |x|$	Absolute value function
$y = \sqrt{x}$	$f(x) = \sqrt{x}$	Square root function
$y = \dfrac{1}{x}$	$f(x) = \dfrac{1}{x}$	Reciprocal function

The graph of the function defined by $f(x) = x$ is linear, with slope $m = 1$ and y-intercept $(0, 0)$ (Figure 16-10).

To determine the shapes of the other basic functions, we can plot several points to establish the pattern of the graph. Analyzing the equation itself may also provide insight to the domain, range, and shape of the graph. To demonstrate this, we will graph $f(x) = x^2$ and $g(x) = \frac{1}{x}$.

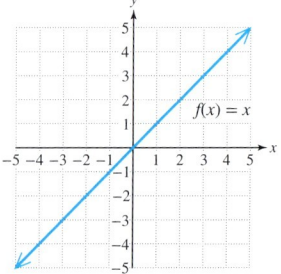

Figure 16-10

Example 1 Graphing Basic Functions

Graph the function defined by $f(x) = x^2$.

Solution:

The domain of the function given by $f(x) = x^2$ (or equivalently $y = x^2$) is all real numbers.

To graph the function, choose arbitrary values of x within the domain of the function. Be sure to choose values of x that are positive and values that are negative to determine the behavior of the function to the right and left of the origin (Table 16-4). The graph of $f(x) = x^2$ is shown in Figure 16-11.

The function values are equated to the square of x, so $f(x)$ will always be greater than or equal to zero. Hence, the y-coordinates on the graph will never be negative. The range of the function is $[0, \infty)$. The arrows on each branch of the graph imply that the pattern continues indefinitely.

Table 16-4

x	$f(x) = x^2$
0	0
1	1
2	4
3	9
−1	1
−2	4
−3	9

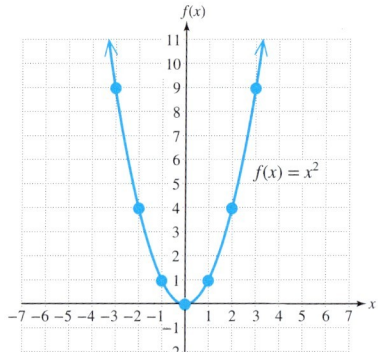

Figure 16-11

Skill Practice

1. Graph $f(x) = -x^2$ by first making a table of points.

Answer

1.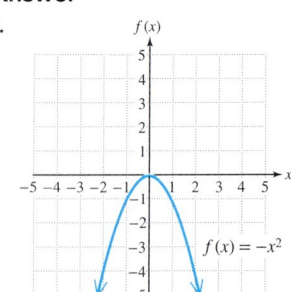

Example 2 Graphing Basic Functions

Graph the function defined by $g(x) = \dfrac{1}{x}$.

Solution:

$g(x) = \dfrac{1}{x}$ Notice that $x = 0$ is not in the domain of the function. From the equation $y = \dfrac{1}{x}$, the y values will be the reciprocal of the x values. The graph defined by $g(x) = \dfrac{1}{x}$ is shown in Figure 16-12.

x	$g(x) = \dfrac{1}{x}$
1	1
2	$\tfrac{1}{2}$
3	$\tfrac{1}{3}$
-1	-1
-2	$-\tfrac{1}{2}$
-3	$-\tfrac{1}{3}$

x	$g(x) = \dfrac{1}{x}$
$\tfrac{1}{2}$	2
$\tfrac{1}{3}$	3
$\tfrac{1}{4}$	4
$-\tfrac{1}{2}$	-2
$-\tfrac{1}{3}$	-3
$-\tfrac{1}{4}$	-4

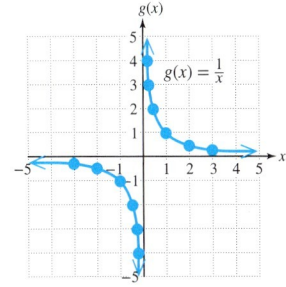

Figure 16-12

Notice that as x approaches ∞ and $-\infty$, the y values approach zero, and the graph approaches the x-axis. In this case, the x-axis is called a *horizontal asymptote*. Similarly, the graph of the function approaches the y-axis as x gets close to zero. In this case, the y-axis is called a *vertical asymptote*.

Skill Practice

2. Graph $h(x) = |x| - 1$ by first making a table of points.

Summary of Six Basic Functions and Their Graphs

Function	Graph	Domain and Range
1. $f(x) = x$ Identity function		Domain $(-\infty, \infty)$ Range $(-\infty, \infty)$
2. $f(x) = x^2$ Quadratic function	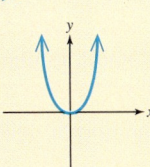	Domain $(-\infty, \infty)$ Range $[0, \infty)$
3. $f(x) = x^3$ Cubic function	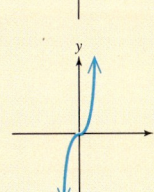	Domain $(-\infty, \infty)$ Range $(-\infty, \infty)$

Answer

2.

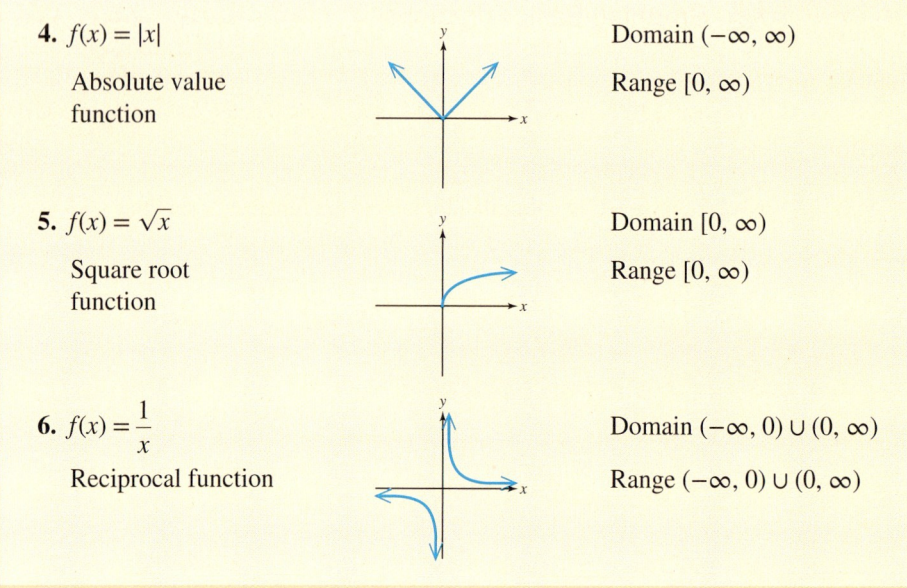

TIP: Recall that the symbol ∪ means the combination (or union) of the two intervals.

4. $f(x) = |x|$

   Absolute value function

   Domain $(-\infty, \infty)$
   Range $[0, \infty)$

5. $f(x) = \sqrt{x}$

   Square root function

   Domain $[0, \infty)$
   Range $[0, \infty)$

6. $f(x) = \dfrac{1}{x}$

   Reciprocal function

   Domain $(-\infty, 0) \cup (0, \infty)$
   Range $(-\infty, 0) \cup (0, \infty)$

The shapes of these six graphs will be developed in the homework exercises. These functions are used often in the study of algebra. Therefore, we recommend that you associate an equation with its graph and commit each to memory.

3. Definition of a Quadratic Function

In Example 1 we graphed the function defined by $f(x) = x^2$ by plotting points. This function belongs to a special category called quadratic functions.

Definition of a Quadratic Function

A **quadratic function** is a function defined by

$$f(x) = ax^2 + bx + c \qquad \text{where } a, b, \text{ and } c \text{ are real numbers and } a \neq 0.$$

The graph of a quadratic function is in the shape of a **parabola**. The leading coefficient, a, determines the direction of the parabola.

- If $a > 0$, then the parabola opens upward, and the vertex is the minimum point on the parabola. For example, the graph of $f(x) = x^2$ is shown in Figure 16-13.
- If $a < 0$, then the parabola opens downward, and the vertex is the maximum point on the parabola. For example, the graph of $f(x) = -x^2$ is shown in Figure 16-14.

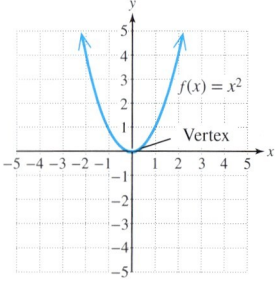

Figure 16-13

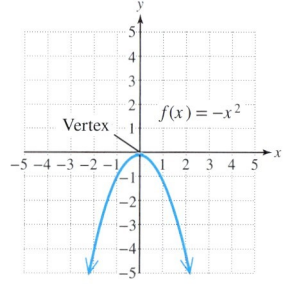

Figure 16-14

Example 3 — Identifying Functions

Identify each function as linear, constant, quadratic, or none of these.

a. $f(x) = -4$
b. $f(x) = x^2 + 3x + 2$
c. $f(x) = 7 - 2x$
d. $f(x) = \dfrac{4x + 8}{8}$
e. $f(x) = \dfrac{6}{x} + 2$

Solution:

a. $f(x) = -4$ is a constant function. It is in the form $f(x) = b$, where $b = -4$.

b. $f(x) = x^2 + 3x + 2$ is a quadratic function. It is in the form $f(x) = ax^2 + bx + c$, where $a \neq 0$.

c. $f(x) = 7 - 2x$ is linear. Writing it in the form $f(x) = mx + b$, we get $f(x) = -2x + 7$, where $m = -2$ and $b = 7$.

d. $f(x) = \dfrac{4x + 8}{8}$ is linear. Writing it in the form $f(x) = mx + b$, we get

$$f(x) = \dfrac{4x}{8} + \dfrac{8}{8}$$

$$= \dfrac{1}{2}x + 1, \text{ where } m = \dfrac{1}{2} \text{ and } b = 1.$$

e. $f(x) = \dfrac{6}{x} + 2$ fits none of these categories because the variable is in the denominator.

Skill Practice Identify each function as constant, linear, quadratic, or none of these.

3. $m(x) = -2x^2 - 3x + 7$
4. $n(x) = -6$
5. $W(x) = \dfrac{4}{3}x - \dfrac{1}{2}$
6. $R(x) = \dfrac{4}{3x} - \dfrac{1}{2}$

4. Finding the x- and y-Intercepts of a Graph Defined by y = f(x)

We have already learned that to find an x-intercept, we substitute $y = 0$ and solve the equation for x. Using function notation, this is equivalent to finding the real solutions of the equation $f(x) = 0$. To find the y-intercept, substitute $x = 0$ and solve the equation for y. In function notation, this is equivalent to finding $f(0)$.

Finding Intercepts Using Function Notation

Given a function defined by $y = f(x)$,

Step 1 The x-intercepts are the real solutions to the equation $f(x) = 0$.
Step 2 The y-intercept is given by $f(0)$.

Answers
3. Quadratic
4. Constant
5. Linear
6. None of these

> **Example 4** Finding *x*- and *y*-Intercepts

Given the function defined by $f(x) = 2x - 4$:

a. Find the *x*-intercept(s).

b. Find the *y*-intercept.

c. Graph the function.

Solution:

a. To find the *x*-intercept(s), find the real solutions to the equation $f(x) = 0$.

$f(x) = 2x - 4$

$0 = 2x - 4$ Substitute $f(x) = 0$.

$4 = 2x$

$2 = x$ The *x*-intercept is (2, 0).

b. To find the *y*-intercept, evaluate $f(0)$.

$f(0) = 2(0) - 4$ Substitute $x = 0$.

$f(0) = -4$ The *y*-intercept is (0, −4).

c. This function is linear, with a *y*-intercept of (0, −4), an *x*-intercept of (2, 0), and a slope of 2 (Figure 16-15).

Figure 16-15

Skill Practice Consider $f(x) = -5x + 1$.

7. Find the *x*-intercept.
8. Find the *y*-intercept.
9. Graph the function.

> **Example 5** Finding *x*- and *y*-Intercepts

For the function pictured in Figure 16-16, estimate

a. The real values of *x* for which $f(x) = 0$.

b. The value of $f(0)$.

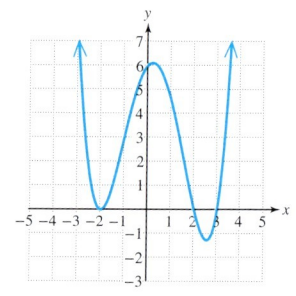

Figure 16-16

Answers

7. $(\frac{1}{5}, 0)$ 8. (0, 1)

9.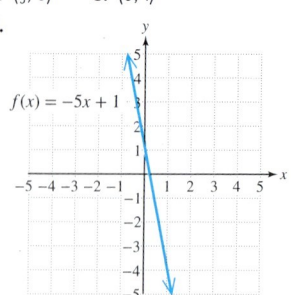

Solution:

a. The real values of *x* for which $f(x) = 0$ are the *x*-intercepts of the function. For this graph, the *x*-intercepts are located at $x = -2$, $x = 2$, and $x = 3$.

b. The value of $f(0)$ is the value of *y* at $x = 0$. That is, $f(0)$ is the *y*-intercept, $f(0) = 6$.

Skill Practice Use the function pictured.

10. a. Estimate the real value(s) of x for which $f(x) = 0$.
    b. Estimate the value of $f(0)$.

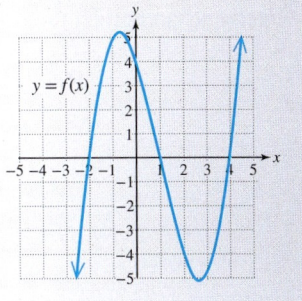

Answer
10. a. $x = -2$, $x = 1$, and $x = 4$
    b. $f(0) = 4$

Section 16.3 Activity

For Exercises A.1–A.3,
a. Identify the graph of the equation as representing a slanted line, a horizontal line, a vertical line, or a parabola.
b. Write the equation using function notation $y = f(x)$.
c. Identify the function as being a constant function (characterized by a horizontal line), a linear function (slanted line), or a quadratic function (parabola).

A.1. $y = -3x - 2$

A.2. $y = 5$

A.3. $y = x^2 - 4x + 3$

A.4. Given the function defined by $f(x) = -3x - 2$,
a. Explain how to find the x-intercept.
b. Find the x-intercept.
c. Explain how to find the y-intercept.
d. Find the y-intercept.

For Exercises A.5–A.6, identify the x- and y-intercepts for the function defined by $y = f(x)$.

A.5. $f(x) = 5$

A.6. $f(x) = x^2 - 4x + 3$

Practice Exercises — Section 16.3

Prerequisite Review

For Exercises R.1–R.4, state whether the equation represents a slanted line, a horizontal line, or a vertical line.

R.1. $3x = 5$

R.2. $y - 3 = 5$

R.3. $3x - 7y = 14$

R.4. $y = -\dfrac{1}{2}x - 4$

For Exercises R.5–R.8, find the *x*- and *y*-intercepts and graph the equation.

R.5. $y = -\frac{3}{5}x + 3$

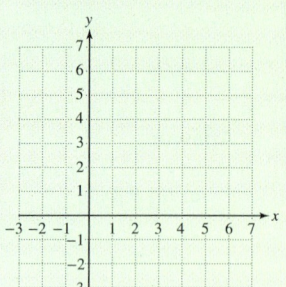

R.6. $y = -\frac{1}{4}x - 1$

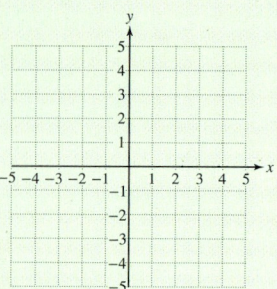

R.7. $y = -4$

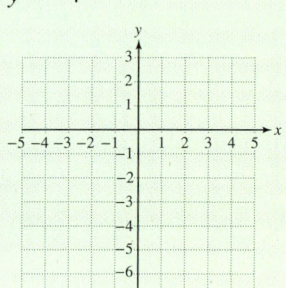

R.8. $x = -2$

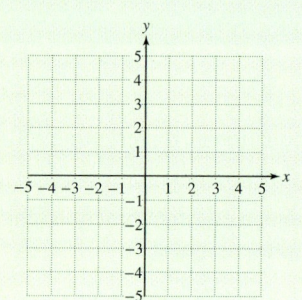

Vocabulary and Key Concepts

1. **a.** A function that can be written in the form $f(x) = mx + b$, $m \neq 0$, is a _____ function.

   b. A function that can be written in the form $f(x) = b$ is a _____ function.

   c. A function that can be written in the form $f(x) = ax^2 + bx + c$, $a \neq 0$, is a _____ function.

   d. The graph of a quadratic function is in the shape of a _____.

   e. To find the *x*-intercept(s) of a graph defined by $y = f(x)$, find the real solutions to the equation $f(x) = $ _____.

   f. To find the _____-intercept of a graph defined by $y = f(x)$, evaluate $f(0)$.

Concept 1: Linear and Constant Functions

2. Given $f(x) = 4$, find the function values.

   a. $f(1)$ **b.** $f(2)$ **c.** $f(3)$

3. Given $g(x) = -3$, find the function values.

   a. $g(1)$ **b.** $g(2)$ **c.** $g(3)$

4. Given $h(x) = 4x$, find the function values.

   a. $h(1)$ **b.** $h(2)$ **c.** $h(3)$

5. Given $k(x) = -3x$, find the function values.

   a. $k(1)$ **b.** $k(2)$ **c.** $k(3)$

6. Refer to functions *f*, *g*, *h*, and *k* from Exercises 2–5.

   a. Which functions are linear?

   b. Which functions are constant?

Section 16.3 Graphs of Functions

7. Fill in the blank with the word *vertical* or *horizontal*. The graph of a constant function is a _____ line.

8. For the linear function defined by $f(x) = mx + b$, identify the slope and y-intercept.

9. Graph the constant function $f(x) = 2$. Then use the graph to identify the domain and range of f.

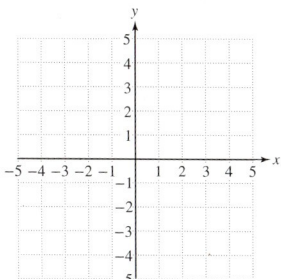

10. Graph the linear function $g(x) = -2x + 1$. Then use the graph to identify the domain and range of g.

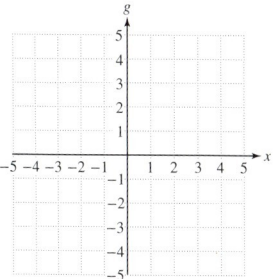

Concept 2: Graphs of Basic Functions

For Exercises 11–16, sketch a graph by completing the table and plotting the points. (See Examples 1–2.)

11. $f(x) = \dfrac{1}{x}$

x	f(x)
−2	
−1	
−½	
−¼	

x	f(x)
¼	
½	
1	
2	

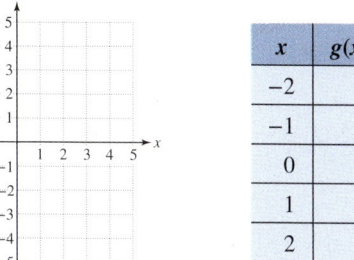

12. $g(x) = |x|$

x	g(x)
−2	
−1	
0	
1	
2	

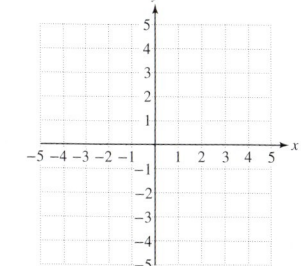

13. $h(x) = x^3$

x	h(x)
−2	
−1	
0	
1	
2	

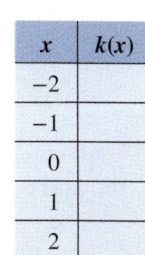

14. $k(x) = x$

x	k(x)
−2	
−1	
0	
1	
2	

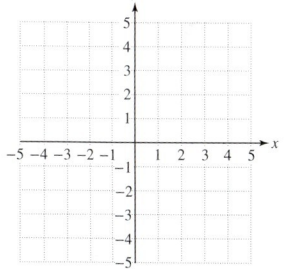

15. $q(x) = x^2$

x	q(x)
−2	
−1	
0	
1	
2	

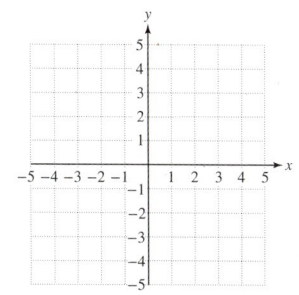

16. $p(x) = \sqrt{x}$

x	p(x)
0	
1	
4	
9	
16	

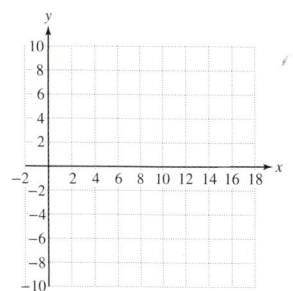

Concept 3: Definition of a Quadratic Function

For Exercises 17–28, determine if the function is constant, linear, quadratic, or none of these. (See Example 3.)

17. $f(x) = 2x^2 + 3x + 1$
18. $g(x) = -x^2 + 4x + 12$
19. $k(x) = -3x - 7$
20. $h(x) = -x - 3$

21. $m(x) = \dfrac{4}{3}$
22. $n(x) = 0.8$
23. $p(x) = \dfrac{2}{3x} + \dfrac{1}{4}$
24. $Q(x) = \dfrac{1}{5x} - 3$

25. $t(x) = \dfrac{2}{3}x + \dfrac{1}{4}$
26. $r(x) = \dfrac{1}{5}x - 3$
27. $w(x) = \sqrt{4 - x}$
28. $T(x) = -|x + 10|$

Concept 4: Finding the x- and y-Intercepts of a Graph Defined by $y = f(x)$

For Exercises 29–36, find the x- and y-intercepts, and graph the function. (See Example 4.)

29. $f(x) = 5x - 10$
30. $f(x) = -3x + 12$
31. $g(x) = -6x + 5$

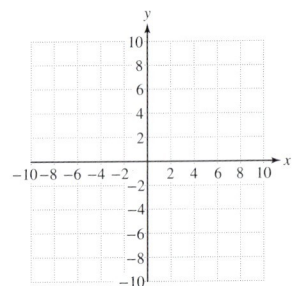
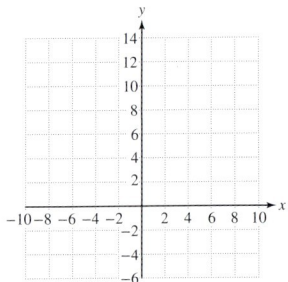
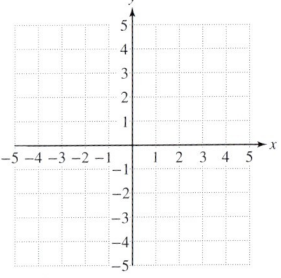

32. $h(x) = 2x + 9$
33. $f(x) = 18$
34. $g(x) = -7$

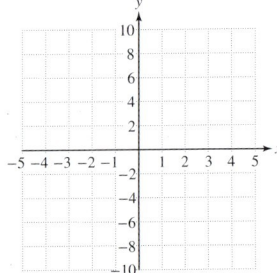
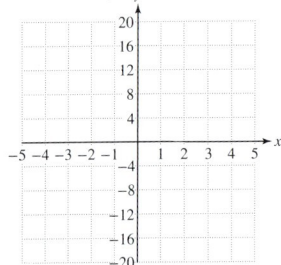
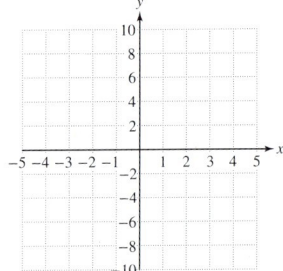

35. $g(x) = \dfrac{2}{3}x + 2$
36. $h(x) = -\dfrac{3}{5}x - 3$

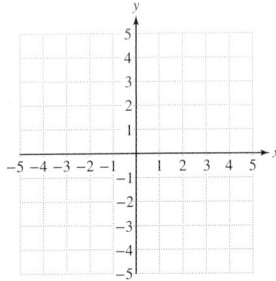
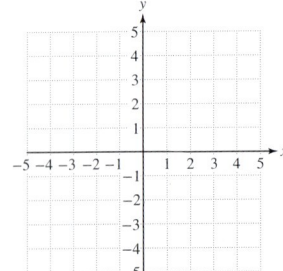

For Exercises 37–42, use the graph to estimate

    a. The real values of x for which $f(x) = 0$.   **b.** The value of $f(0)$. (See Example 5.)

37.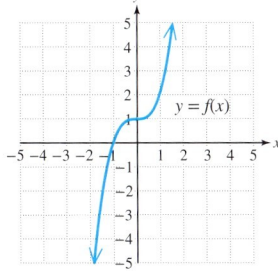

38.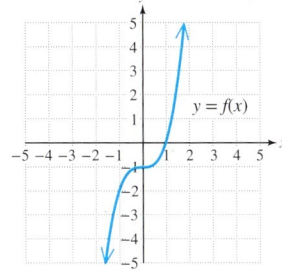

39.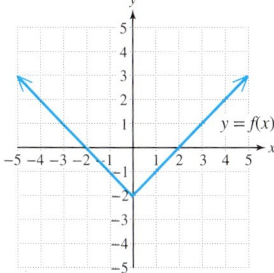

40.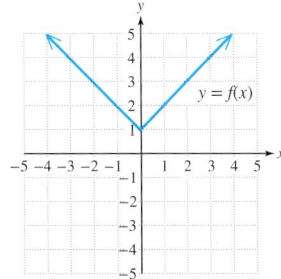

41.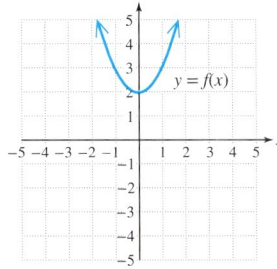

42.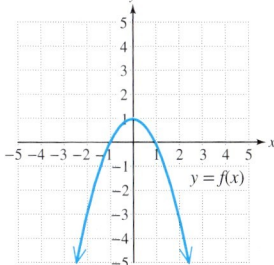

Mixed Exercises

For Exercises 43–52,

    a. Identify the domain of the function.

    b. Identify the y-intercept.

    c. Match the function with its graph by recognizing the basic shape of the graph and using the results from parts (a) and (b). Plot additional points if necessary.

43. $q(x) = 2x^2$

44. $p(x) = -2x^2 + 1$

45. $h(x) = x^3 + 1$

46. $k(x) = x^3 - 2$

47. $r(x) = \sqrt{x+1}$

48. $s(x) = \sqrt{x+4}$

49. $f(x) = \dfrac{1}{x-3}$

50. $g(x) = \dfrac{1}{x+1}$

51. $k(x) = |x+2|$

52. $h(x) = |x-1| + 2$

i.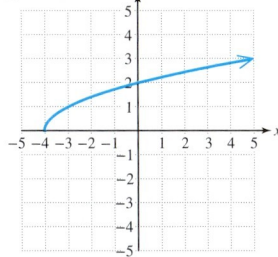

ii.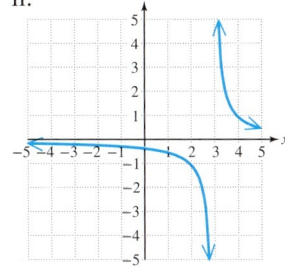

iii.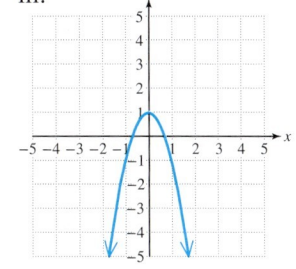

iv.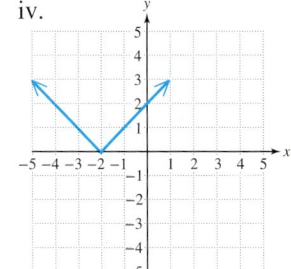

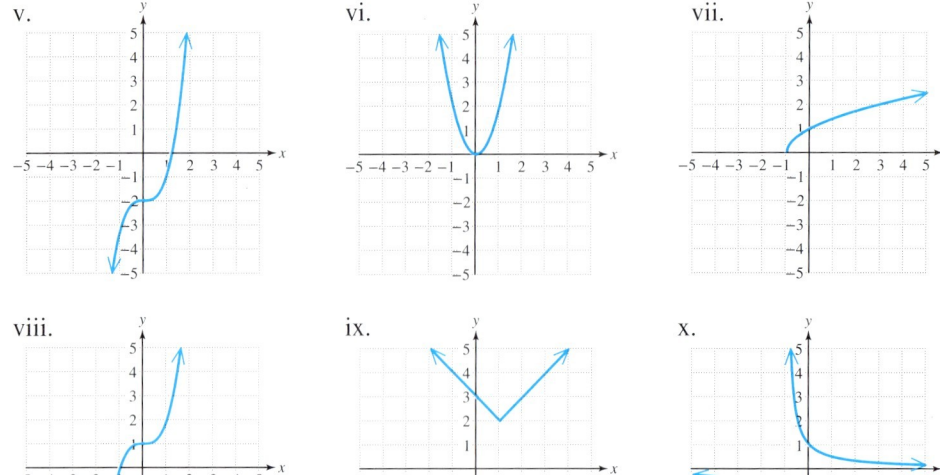

53. Suppose that a student has an 80% average on all of her chapter tests in her Intermediate Algebra class. This counts as $\frac{3}{4}$ of the student's overall grade. The final exam grade counts as the remaining $\frac{1}{4}$ of the student's overall grade. The student's overall course grade, $G(x)$, can be computed by

$$G(x) = \frac{3}{4}(80) + \frac{1}{4}x,\text{ where } x \text{ is the student's grade on the final exam.}$$

   a. Is this function linear, quadratic, or neither?

   b. Evaluate $G(90)$ and interpret its meaning in the context of the problem.

   c. Evaluate $G(50)$ and interpret its meaning in the context of the problem.

54. The median weekly earnings, $E(x)$ in dollars, for women 16 years and older working full time can be approximated by $E(x) = 0.14x^2 + 7.8x + 540$. For this function, x represents the number of years since 2000. (*Source:* U.S. Department of Labor)

   a. Is this function linear, quadratic, or neither?

   b. Evaluate $E(10)$ and interpret its meaning in the context of this problem.

   c. Evaluate $E(20)$ and interpret its meaning in the context of this problem.

For Exercises 55–60, write a function defined by $y = f(x)$ under the given conditions.

55. The value of $f(x)$ is two more than the square of x.

56. The value of $f(x)$ is the square of the sum of two and x.

57. The function f is a constant function passing through the point $(4, 3)$.

58. The function f is a constant function with y-intercept $(0, 5)$.

59. The function f is a linear function with slope $\frac{1}{2}$ and y-intercept $(0, -2)$.

60. The function f is a linear function with slope -4 and y-intercept $(0, \frac{1}{3})$.

Technology Connections

For Exercises 61–66, use a graphing calculator to graph the basic functions. Verify your answers from the Summary of Six Basic Functions and Their Graphs.

61. $f(x) = x$

62. $f(x) = x^2$

63. $f(x) = x^3$

64. $f(x) = |x|$

65. $f(x) = \sqrt{x}$

66. $f(x) = \dfrac{1}{x}$

For Exercises 67–70, find the x- and y-intercepts using a graphing calculator and the *Value* and *Zero* features.

67. $y = -\frac{1}{8}x + 1$
68. $y = -\frac{1}{2}x - 3$
69. $y = \frac{4}{5}x + 4$
70. $y = \frac{7}{2}x - 7$

Problem Recognition Exercises

Characteristics of Relations

Exercises 1–15 refer to the sets of points (x, y) described in a–h.

a. $\{(0, 8), (1, 4), (\frac{1}{2}, 4), (-3, 5), (2, 1)\}$

b. $\{(-6, 4), (2, 3), (-9, 6), (2, 1), (0, 10)\}$

c. $c(x) = 3x^2 - 2x - 1$

d. $d(x) = 5x - 9$

e.
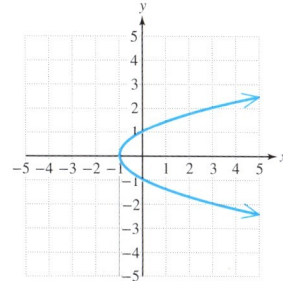

f.
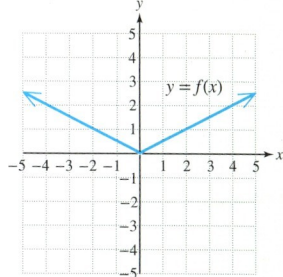

g.
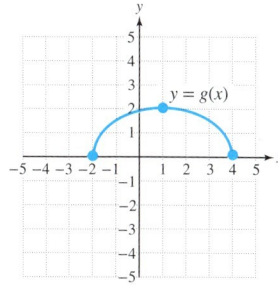

h.
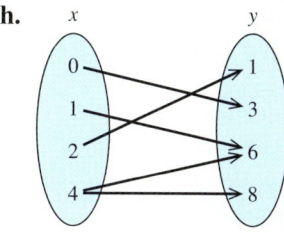

1. Which relations define y as a function of x?
2. Which relations contain the point $(2, 1)$?
3. Use relation (c) to find $c(-1)$.
4. Use relation (f) to find $f(-4)$.
5. Find the domain of relation (a).
6. Find the range of relation (b).
7. Find the domain of relation (g).
8. Find the range of relation (f).
9. Use relation (h) to complete the ordered pair (__, 3).
10. Find the x-intercept(s) of the graph of relation (d).
11. Find the y-intercept(s) of the graph shown in relation (e).
12. Use relation (g) to determine the value of x such that $g(x) = 2$.
13. Which relation describes a quadratic function?
14. Which relation describes a linear function?
15. Use relation (d) to find the value of x such that $d(x) = 6$.

Section 16.4 Algebra of Functions and Composition

Concepts

1. Algebra of Functions
2. Composition of Functions
3. Operations on Functions

1. Algebra of Functions

Addition, subtraction, multiplication, and division can be used to create a new function from two or more functions. The domain of the new function will be the intersection of the domains of the original functions.

Sum, Difference, Product, and Quotient of Functions

Given two functions f and g, the functions $f+g, f-g, f \cdot g$, and $\frac{f}{g}$ are defined as

$$(f+g)(x) = f(x) + g(x)$$
$$(f-g)(x) = f(x) - g(x)$$
$$(f \cdot g)(x) = f(x) \cdot g(x)$$
$$\left(\frac{f}{g}\right)(x) = \frac{f(x)}{g(x)} \quad \text{provided } g(x) \neq 0$$

For example, suppose $f(x) = |x|$ and $g(x) = 3$. Taking the sum of the functions produces a new function denoted by $f+g$. In this case, $(f+g)(x) = |x| + 3$. Graphically, the y values of the function $f+g$ are given by the sum of the corresponding y values of f and g. This is depicted in Figure 16-17. The function $f+g$ appears in red. In particular, notice that $(f+g)(2) = f(2) + g(2) = 2 + 3 = 5$.

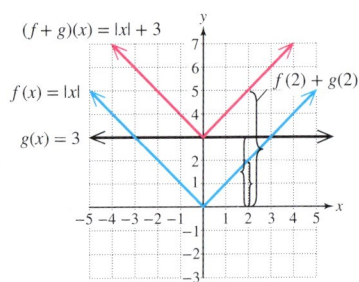

Figure 16-17

Example 1 Adding, Subtracting, Multiplying and Dividing Functions

Given: $g(x) = 4x \qquad h(x) = x^2 - 3x \qquad k(x) = x - 2$

a. Find $(g+h)(x)$. **b.** Find $(h-g)(x)$.

c. Find $(g \cdot k)(x)$. **d.** Find $\left(\dfrac{k}{h}\right)(x)$.

Solution:

a. $(g+h)(x) = g(x) + h(x)$
$\qquad\qquad = (4x) + (x^2 - 3x)$
$\qquad\qquad = 4x + x^2 - 3x$
$\qquad\qquad = x^2 + x \qquad$ Simplify.

b. $(h-g)(x) = h(x) - g(x)$
$\qquad\qquad = (x^2 - 3x) - (4x)$
$\qquad\qquad = x^2 - 3x - 4x$
$\qquad\qquad = x^2 - 7x \qquad$ Simplify.

c. $(g \cdot k)(x) = g(x) \cdot k(x)$

$= (4x)(x-2)$

$= 4x^2 - 8x$ Simplify.

d. $\left(\dfrac{k}{h}\right)(x) = \dfrac{k(x)}{h(x)}$

$= \dfrac{x-2}{x^2 - 3x}$ From the denominator we have $x^2 - 3x \neq 0$ or, equivalently, $x(x-3) \neq 0$. Hence, $x \neq 3$ and $x \neq 0$.

Skill Practice Given $f(x) = x - 1$, $g(x) = 5x^2 + x$, and $h(x) = x^2$, find

1. $(f+g)(x)$
2. $(g-f)(x)$
3. $(g \cdot h)(x)$
4. $\left(\dfrac{f}{h}\right)(x)$

2. Composition of Functions

Composition of Functions

The **composition** of f and g, denoted $f \circ g$, is defined by the rule

$(f \circ g)(x) = f(g(x))$ provided that $g(x)$ is in the domain of f

Note: $f \circ g$ is read as "f of g" or "f compose g."

For example, given $f(x) = 2x - 3$ and $g(x) = x + 5$, we have

$(f \circ g)(x) = f(g(x))$

$= f(x+5)$ Substitute $g(x) = x + 5$ into the function f.

$= 2(x+5) - 3$

$= 2x + 10 - 3$

$= 2x + 7$

In this composition, the function g is the innermost operation and acts on x first. Then the output value of function g becomes the domain element of the function f, as shown in the figure.

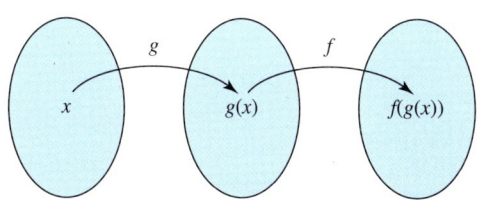

Example 2 Composing Functions

Given: $f(x) = x - 5$, $g(x) = x^2$, and $n(x) = \sqrt{x+2}$, find

a. $(f \circ g)(x)$

b. $(g \circ f)(x)$

c. $(n \circ f)(x)$

Answers

1. $5x^2 + 2x - 1$
2. $5x^2 + 1$
3. $5x^4 + x^3$
4. $\dfrac{x-1}{x^2}$, $x \neq 0$

Solution:

a. $(f \circ g)(x) = f(g(x))$

$= f(x^2)$ Evaluate the function f at x^2.

$= (x^2) - 5$ Replace x with x^2 in function f.

$= x^2 - 5$

b. $(g \circ f)(x) = g(f(x))$

$= g(x - 5)$ Evaluate the function g at $(x - 5)$.

$= (x - 5)^2$ Replace x with $(x - 5)$ in function g.

$= x^2 - 10x + 25$ Simplify.

c. $(n \circ f)(x) = n(f(x))$

$= n(x - 5)$ Evaluate the function n at $x - 5$.

$= \sqrt{(x - 5) + 2}$ Replace x with the quantity $(x - 5)$ in function n.

$= \sqrt{x - 3}$

> **TIP:** Examples 2(a) and 2(b) illustrate that the order in which two functions are composed may result in different functions. That is, $f \circ g$ does not necessarily equal $g \circ f$.

> **FOR REVIEW**
> Recall that to square a binomial $(a - b)^2$ we have:
> $(a - b)^2 = a^2 - 2ab + b^2$.
> Thus,
> $(x - 5)^2$
> $= x^2 - 2(x)(5) + (5)^2$
> $= x^2 - 10x + 25$

Skill Practice Given $f(x) = 2x^2$, $g(x) = x + 3$, and $h(x) = \sqrt{x - 1}$, find

5. $(f \circ g)(x)$ **6.** $(g \circ f)(x)$ **7.** $(h \circ g)(x)$

3. Operations on Functions

Example 3   **Combining Functions**

Given the functions defined by $f(x) = x - 7$ and $h(x) = 2x^3$, find the function values, if possible.

a. $(f \cdot h)(3)$ **b.** $\left(\dfrac{h}{f}\right)(7)$ **c.** $(h \circ f)(2)$

Solution:

a. $(f \cdot h)(3) = f(3) \cdot h(3)$ $(f \cdot h)(3)$ is a product (not a composition).

$= (3 - 7) \cdot 2(3)^3$

$= (-4) \cdot 2(27)$

$= -216$

b. $\left(\dfrac{h}{f}\right)(7) = \dfrac{h(7)}{f(7)} = \dfrac{2(7)^3}{7 - 7}$

The value $x = 7$ will make the denominator equal to 0. Therefore, $\left(\dfrac{h}{f}\right)(7)$ is undefined.

c. $(h \circ f)(2) = h(f(2))$ Evaluate $f(2)$ first. $f(2) = 2 - 7 = -5$

$= h(-5)$ Substitute the result into function h.

$= 2(-5)^3$

$= 2(-125)$

$= -250$

Answers
5. $2x^2 + 12x + 18$ **6.** $2x^2 + 3$
7. $\sqrt{x + 2}$

Skill Practice Given $h(x) = x + 4$ and $k(x) = x^2 - 3$, find

8. $(h \cdot k)(-2)$ 9. $\left(\dfrac{k}{h}\right)(-4)$ 10. $(k \circ h)(1)$

Example 4 — Finding Function Values From a Graph

For the functions f and g pictured, find the function values if possible.

a. $g(2)$

b. $(f - g)(-3)$

c. $\left(\dfrac{g}{f}\right)(5)$

d. $(f \circ g)(4)$

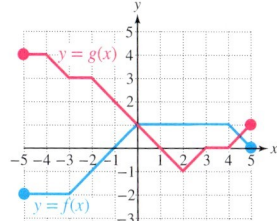

Solution:

a. $g(2) = -1$ — The value $g(2)$ represents the y value of $y = g(x)$ (the red graph) when $x = 2$. Because the point $(2, -1)$ lies on the graph, $g(2) = -1$.

b. $(f - g)(-3) = f(-3) - g(-3)$ — Evaluate the difference of $f(-3)$ and $g(-3)$.

$\qquad = -2 - (3)$ — Estimate function values from the graph.

$\qquad = -5$

c. $\left(\dfrac{g}{f}\right)(5) = \dfrac{g(5)}{f(5)}$ — Evaluate the quotient of $g(5)$ and $f(5)$.

$\qquad = \dfrac{1}{0}$ (undefined)

The function $\dfrac{g}{f}$ is undefined at 5 because the denominator is zero.

d. $(f \circ g)(4) = f(g(4))$ — From the red graph, find the value of $g(4)$ first. We see that $g(4) = 0$.

$\qquad = f(0)$ — From the blue graph, find the value of f at $x = 0$.

$\qquad = 1$

Skill Practice Find the values from the graph.

11. $g(3)$
12. $(f + g)(-4)$
13. $\left(\dfrac{f}{g}\right)(2)$
14. $(g \circ f)(-2)$

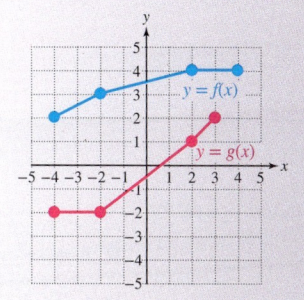

Answers

8. 2   9. undefined   10. 22
11. 2   12. 0   13. 4   14. 2

Section 16.4 Activity

A.1. Consider the graphs of f and g given in the figure.
  a. Find $f(0)$.
  b. Find $g(0)$.
  c. Find $f(0) + g(0)$.
  d. Now suppose that a new function, $f + g$, is defined as the sum of functions f and g for all real numbers for which f and g are defined. That is, $(f + g)(x) = f(x) + g(x)$. Find $(f + g)(1)$.
  e. Find $(f + g)(-2)$.

A.2. Given two functions f and g, write the definitions of the functions $f - g$, $f \cdot g$, and $\dfrac{f}{g}$.
  a. $(f - g)(x) = $ _____
  b. $(f \cdot g)(x) = $ _____
  c. $\left(\dfrac{f}{g}\right)(x) = $ _____

A.3. Use the functions defined by $f(x) = x^2 - x - 12$ and $g(x) = x - 4$.
  a. Explain the process to find $f(x - 4)$.
  b. Find $f(x - 4)$.
  c. Find $g(x^2 - x - 12)$.

A.4.
  a. The composition of functions f and g is denoted by $f \circ g$ and is defined by $(f \circ g)(x) = $ _____.
  The composition of functions g and f is denoted by $g \circ f$ and is defined by $(g \circ f)(x) = $ _____.
  b. Refer to the functions defined in Exercise A.3: $f(x) = x^2 - x - 12$ and $g(x) = x - 4$. Find $(f \circ g)(x)$.
  c. Find $(g \circ f)(x)$.

A.5. Given $h(x) = 3x - 4$ and $k(x) = \dfrac{1}{x - 5}$, fill in the blanks to find the function value.
  a. $(h \circ k)(4) = h[k(\square)] = h(\square) = $ _____
  b. $(k \circ h)(0) = k[h(\square)] = k(\square) = $ _____

A.6. Given $h(x) = 3x - 4$ and $k(x) = \dfrac{1}{x - 5}$, find the function values if possible. If a function value does not exist, explain why.
  a. $(h \circ k)(6)$
  b. $(h \circ k)(5)$
  c. $(k \circ h)(1)$
  d. $(k \circ h)(3)$

A.7. Refer to the graphs in Exercise A.1.
  a. Find $(f \cdot g)(-1)$.
  b. Find $(g - f)(-3)$.
  c. Find $\left(\dfrac{f}{g}\right)(0)$.
  d. Explain why $\left(\dfrac{g}{f}\right)(-2)$ is undefined.
  e. Find $(f \circ g)(-1)$.
  f. Find $(g \circ f)(-2)$.

Section 16.4 Practice Exercises

Prerequisite Review

For Exercises R.1–R.8, perform the indicated operations.

R.1. $(3x^2 - 4x - 5) + (6x^2 + x - 3)$

R.2. $(-x^3 + 6x - 7) + (x^3 - 7x + 3)$

R.3. $(-x^4 + 2x^2 - 1) - (x^4 - 3x^2 + 5)$

R.4. $(3x^2 + 5x - 7) - (-x^2 - 4x - 1)$

R.5. $\dfrac{1}{x-3} \cdot \dfrac{4x-12}{x-1}$ **R.6.** $\dfrac{2x-3}{x} \cdot \dfrac{x^2}{16x-24}$

R.7. $\dfrac{x^2-25}{5-x} \div \dfrac{x^2+5x}{x^2}$ **R.8.** $\dfrac{3-x}{x^2} \div \dfrac{x^2-9}{2x^2+6x}$

For Exercises R.9–R.16, write the domain in interval notation.

R.9. $f(x) = 3x + 2$ **R.10.** $g(x) = -5x - 1$

R.11. $h(x) = \dfrac{1}{3x+2}$ **R.12.** $k(x) = \dfrac{x}{-5x-1}$

R.13. $m(x) = \dfrac{3-x}{x^2+9}$ **R.14.** $n(x) = \dfrac{2+x}{x^2+4}$

R.15. $p(x) = \sqrt{4-2x}$ **R.16.** $q(x) = \sqrt{8-x}$

For Exercises R.17–R.20, given $f(x) = -3x - 4$ and $g(x) = -x^2 + 4x + 6$, evaluate the functions.

R.17. $f(-4)$ **R.18.** $f(6)$

R.19. $g(-2)$ **R.20.** $g(-3)$

Vocabulary and Key Concepts

1. **a.** Given the functions f and g, the function $(f+g)(x) =$ _____ + _____.

    b. Given the functions f and g, the function $\left(\dfrac{f}{g}\right)(x) = \dfrac{f(x)}{\square}$, provided _____ $\neq 0$.

    c. The composition of functions f and g is defined by the rule $(f \circ g)(x) =$ _____.

Concept 1: Algebra of Functions

2. Given $f(x) = x^2$ and $g(x) = 2x - 3$, find

   a. $f(-2)$ **b.** $g(-2)$ **c.** $(f+g)(-2)$

For Exercises 3–14, refer to the functions defined below.

$$f(x) = x + 4 \qquad g(x) = 2x^2 + 4x \qquad h(x) = x^2 + 1 \qquad k(x) = \dfrac{1}{x}$$

Find the indicated functions. **(See Example 1.)**

3. $(f+g)(x)$ **4.** $(f-g)(x)$ **5.** $(g-f)(x)$

6. $(f+h)(x)$ **7.** $(f \cdot h)(x)$ **8.** $(h \cdot k)(x)$

9. $(g \cdot f)(x)$ **10.** $(f \cdot k)(x)$ **11.** $\left(\dfrac{h}{f}\right)(x)$

12. $\left(\dfrac{g}{f}\right)(x)$ **13.** $\left(\dfrac{f}{g}\right)(x)$ **14.** $\left(\dfrac{f}{h}\right)(x)$

Concept 2: Composition of Functions

For Exercises 15–22, find the indicated functions. Use f, g, h, and k as defined in Exercises 3–14. (See Example 2.)

15. $(f \circ g)(x)$
16. $(f \circ k)(x)$
17. $(g \circ f)(x)$
18. $(k \circ f)(x)$
19. $(k \circ h)(x)$
20. $(h \circ k)(x)$
21. $(k \circ g)(x)$
22. $(g \circ k)(x)$

23. Based on your answers to Exercises 15 and 17, is it true in general that $(f \circ g)(x) = (g \circ f)(x)$?

24. Based on your answers to Exercises 16 and 18, is it true in general that $(f \circ k)(x) = (k \circ f)(x)$?

For Exercises 25–28, find $(f \circ g)(x)$ and $(g \circ f)(x)$.

25. $f(x) = x^2 - 3x + 1$, $g(x) = 5x$
26. $f(x) = 3x^2 + 8$, $g(x) = 2x - 4$
27. $f(x) = |x|$, $g(x) = x^3 - 1$
28. $f(x) = \dfrac{1}{x+2}$, $g(x) = |x+2|$
29. For $h(x) = 5x - 4$, find $(h \circ h)(x)$.
30. For $k(x) = -x^2 + 1$, find $(k \circ k)(x)$.

Concept 3: Operations on Functions

For Exercises 31–46, refer to the functions defined below.

$$m(x) = x^3 \qquad n(x) = x - 3 \qquad r(x) = \sqrt{x+4} \qquad p(x) = \dfrac{1}{x+2}$$

Find each function value if possible. (See Example 3.)

31. $(m \cdot r)(0)$
32. $(n \cdot p)(0)$
33. $(m + r)(-4)$
34. $(n - m)(4)$
35. $(r \circ n)(3)$
36. $(n \circ r)(5)$
37. $(p \circ m)(-1)$
38. $(m \circ n)(5)$
39. $(m \circ p)(2)$
40. $(r \circ m)(2)$
41. $(r + p)(-3)$
42. $(n + p)(-2)$
43. $(m \circ p)(-2)$
44. $(r \circ m)(-2)$
45. $\left(\dfrac{r}{n}\right)(12)$
46. $\left(\dfrac{n}{m}\right)(2)$

For Exercises 47–64, approximate each function value from the graph, if possible. (See Example 4.)

47. $f(-4)$
48. $f(1)$
49. $g(-2)$
50. $g(3)$
51. $(f + g)(2)$
52. $(g - f)(3)$
53. $(f \cdot g)(-1)$
54. $(g \cdot f)(-4)$
55. $\left(\dfrac{g}{f}\right)(0)$
56. $\left(\dfrac{f}{g}\right)(-2)$
57. $\left(\dfrac{f}{g}\right)(0)$
58. $\left(\dfrac{g}{f}\right)(-2)$
59. $(g \circ f)(-1)$
60. $(f \circ g)(0)$
61. $(f \circ g)(-4)$
62. $(g \circ f)(-4)$
63. $(g \circ g)(2)$
64. $(f \circ f)(-2)$

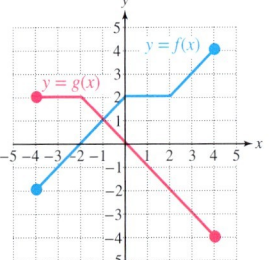

For Exercises 65–80, approximate each function value from the graph, if possible. (See Example 4.)

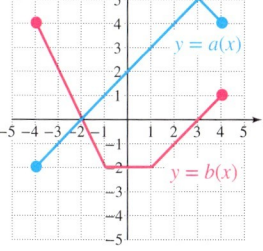

65. $a(-3)$
66. $a(1)$
67. $b(-1)$
68. $b(3)$
69. $(a-b)(-1)$
70. $(a+b)(0)$
71. $(b \cdot a)(1)$
72. $(a \cdot b)(2)$
73. $(b \circ a)(0)$
74. $(a \circ b)(-2)$
75. $(a \circ b)(-4)$
76. $(b \circ a)(-3)$
77. $\left(\dfrac{b}{a}\right)(3)$
78. $\left(\dfrac{a}{b}\right)(4)$
79. $(a \circ a)(-2)$
80. $(b \circ b)(1)$

81. The cost in dollars of producing x toy cars is $C(x) = 2.2x + 1$. The revenue for selling x toy cars is $R(x) = 5.98x$. To calculate profit, subtract the cost from the revenue.

    a. Write and simplify a function P that represents profit in terms of x.

    b. Find the profit of producing 50 toy cars.

82. The cost in dollars of producing x lawn chairs is $C(x) = 2.5x + 10.1$. The revenue for selling x lawn chairs is $R(x) = 6.99x$. To calculate profit, subtract the cost from the revenue.

    a. Write and simplify a function P that represents profit in terms of x.

    b. Find the profit in producing 100 lawn chairs.

83. The functions defined by $D(t) = 0.925t + 26.958$ and $R(t) = 0.725t + 20.558$ approximate the amount of child support (in billions of dollars) that was due $D(t)$ and the amount of child support actually received $R(t)$ in the United States for a selected number of years. The value $t = 0$ represents the first year of the study.

    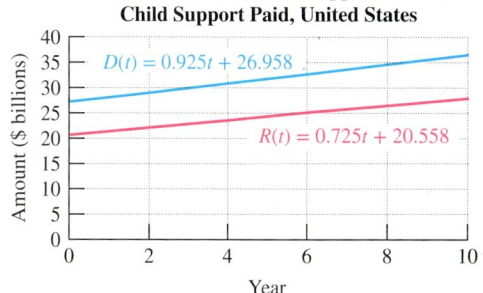

    Difference Between Child Support Due and Child Support Paid, United States

    Source: U.S. Bureau of the Census

    a. Find the function defined by $F(t) = D(t) - R(t)$. What does $F(t)$ represent in the context of this problem?

    b. Find $F(4)$. What does this function value represent in the context of this problem?

84. The rural and urban populations in the South (in the United States) between the years 1900 and 1970 can be modeled by the following:

    $r(t) = -3.497t^2 + 266.2t + 20{,}220$

    $u(t) = 0.0566t^3 + 0.952t^2 + 177.8t + 4593$

    The variable t represents the number of years since 1900, and $0 \le t \le 70$. The variable $r(t)$ represents the rural population in thousands, and $u(t)$ represents the urban population in thousands.

    Rural and Urban Populations in the South, United States, 1900–1970

    Source: Historical Abstract of the United States

    a. Find the function defined by $T(t) = r(t) + u(t)$. What does $T(t)$ represent in the context of this problem?

    b. Use the function T to approximate the total population in the South for the year 1940.

85. Joe rides a bicycle and his wheels revolve at 80 revolutions per minute (rpm). Therefore, the total number of revolutions, r, is given by $r(t) = 80t$, where t is the time in minutes. For each revolution of the wheels of the bike, he travels approximately 7 ft. Therefore, the total distance he travels $D(r)$ (in feet) depends on the total number of revolutions, r, according to the function $D(r) = 7r$.

   a. Find $(D \circ r)(t)$ and interpret its meaning in the context of this problem.

   b. Find Joe's total distance in feet after 10 min.

86. The area of a square is given by the function $a(x) = x^2$, where x is the length of the sides of the square. If carpeting costs \$9.95 per square yard, then the cost $C(a)$ (in dollars) to carpet a square room is given by $C(a) = 9.95a$, where a is the area of the room in square yards.

   a. Find $(C \circ a)(x)$ and interpret its meaning in the context of this problem.

   b. Find the cost to carpet a square room if its floor dimensions are 15 yd by 15 yd.

Section 16.5 Variation

Concepts

1. Definition of Direct and Inverse Variation
2. Translations Involving Variation
3. Applications of Variation

1. Definition of Direct and Inverse Variation

In this section, we introduce the concept of variation. Direct and inverse variation models can show how one quantity varies in proportion to another.

Definition of Direct and Inverse Variation

Let k be a nonzero constant real number. Then,

1. y varies **directly** as x.
   y is directly proportional to x. $\Big\}$ $y = kx$

2. y varies **inversely** as x.
   y is inversely proportional to x. $\Big\}$ $y = \dfrac{k}{x}$

Note: The value of k is called the constant of variation.

For a car traveling 30 mph, the equation $d = 30t$ indicates that the distance traveled is *directly proportional* to the time of travel. For positive values of k, when two variables are directly related, as one variable increases, the other variable will also increase. Likewise, if one variable decreases, the other will decrease. In the equation $d = 30t$, the longer the time of the trip, the greater the distance traveled. The shorter the time of the trip, the shorter the distance traveled.

For positive values of k, when two positive variables are *inversely related,* as one variable increases, the other will decrease, and vice versa. Consider a car traveling between Toronto and Montreal, a distance of 500 km. The time required to make the trip is inversely proportional to the speed of travel: $t = 500/r$. As the rate of speed, r, increases, the quotient $500/r$ will decrease. Thus, the time will decrease. Similarly, as the rate of speed decreases, the trip will take longer.

2. Translations Involving Variation

The first step in using a variation model is to write an English phrase as an equivalent mathematical equation.

Example 1 Translating to a Variation Model

Write each expression as an equivalent mathematical model.

a. The circumference of a circle varies directly as the radius.

b. At a constant temperature, the volume of a gas varies inversely as the pressure.

c. The length of time of a meeting is directly proportional to the *square* of the number of people present.

Solution:

a. Let C represent circumference and r represent radius. The variables are directly related, so use the model $C = kr$.

b. Let V represent volume and P represent pressure. Because the variables are inversely related, use the model $V = \frac{k}{P}$.

c. Let t represent time, and let N be the number of people present at a meeting. Because t is directly related to N^2, use the model $t = kN^2$.

Skill Practice Write each expression as an equivalent mathematical model.

1. The distance, d, driven in a particular time varies directly with the speed of the car, s.
2. The weight of an individual kitten, w, varies inversely with the number of kittens in the litter, n.
3. The value of v varies inversely as the square root of b.

Sometimes a variable varies directly as the product of two or more other variables. In this case, we have joint variation.

> **Definition of Joint Variation**
>
> Let k be a nonzero constant real number. Then the following statements are equivalent:
>
> y varies **jointly** as w and z.
> y is jointly proportional to w and z. $\Big\}$ $y = kwz$

Example 2 Translating to a Variation Model

Write each expression as an equivalent mathematical model.

a. y varies jointly as u and the square root of v.

b. The gravitational force of attraction between two planets varies jointly as the product of their masses and inversely as the square of the distance between them.

Solution:

a. $y = ku\sqrt{v}$

b. Let m_1 and m_2 represent the masses of the two planets. Let F represent the gravitational force of attraction and d represent the distance between the planets.

The variation model is: $F = \dfrac{km_1m_2}{d^2}$

Answers

1. $d = ks$ 2. $w = \dfrac{k}{n}$

3. $v = \dfrac{k}{\sqrt{b}}$

Skill Practice Write each expression as an equivalent mathematical model.

4. The value of q varies jointly as u and v.
5. The value of x varies directly as the square of y and inversely as z.

3. Applications of Variation

Consider the variation models $y = kx$ and $y = \frac{k}{x}$. In either case, if values for x and y are known, we can solve for k. Once k is known, we can use the variation equation to find y if x is known, or to find x if y is known. This concept is the basis for solving many applications involving variation.

> **Finding a Variation Model**
> **Step 1** Write a general variation model that relates the variables given in the problem. Let k represent the constant of variation.
> **Step 2** Solve for k by substituting known values of the variables into the model from step 1.
> **Step 3** Substitute the value of k into the original variation model from step 1.

Example 3 Solving an Application Involving Direct Variation

The variable z varies directly as w. When w is 16, z is 56.

a. Write a variation model for this situation. Use k as the constant of variation.
b. Solve for the constant of variation.
c. Find the value of z when w is 84.

Solution:

a. $z = kw$

b. $z = kw$

$56 = k(16)$ Substitute known values for z and w. Then solve for the unknown value of k.

$\frac{56}{16} = \frac{k(16)}{16}$ To isolate k, divide both sides by 16.

$\frac{7}{2} = k$ Simplify $\frac{56}{16}$ to $\frac{7}{2}$.

c. With the value of k known, the variation model can now be written as:

$z = \frac{7}{2}w$

$z = \frac{7}{2}(84)$ To find z when $w = 84$, substitute $w = 84$ into the equation.

$z = 294$

Skill Practice The variable t varies directly as the square of v. When v is 8, t is 32.

6. Write a variation model for this relationship.
7. Solve for the constant of variation.
8. Find t when $v = 10$.

Answers

4. $q = kuv$ 5. $x = \frac{ky^2}{z}$
6. $t = kv^2$ 7. $\frac{1}{2}$ 8. 50

Example 4 Solving an Application Involving Direct Variation

The speed of a racing canoe in still water varies directly as the square root of the length of the canoe.

a. If a 16-ft canoe can travel 6.2 mph in still water, find a variation model that relates the speed of a canoe to its length.

b. Find the speed of a 25-ft canoe.

Solution:

a. Let s represent the speed of the canoe and L represent the length. The general variation model is $s = k\sqrt{L}$. To solve for k, substitute the known values for s and L.

$$s = k\sqrt{L}$$
$$6.2 = k\sqrt{16} \quad \text{Substitute } s = 6.2 \text{ mph and } L = 16 \text{ ft.}$$
$$6.2 = k \cdot 4$$
$$\frac{6.2}{4} = \frac{4k}{4} \quad \text{Solve for } k.$$
$$k = 1.55$$
$$s = 1.55\sqrt{L} \quad \text{Substitute } k = 1.55 \text{ into the model } s = k\sqrt{L}.$$

b. $s = 1.55\sqrt{L}$
$ = 1.55\sqrt{25} \quad$ Find the speed when $L = 25$ ft.
$ = 7.75$ mph $\quad$ The speed is 7.75 mph.

Skill Practice

9. The amount of water needed by a mountain hiker varies directly as the time spent hiking. The hiker needs 2.4 L for a 4-hr hike. How much water will be needed for a 5-hr hike?

Example 5 Solving an Application Involving Inverse Variation

The loudness of sound measured in decibels (dB) varies inversely as the square of the distance between the listener and the source of the sound. If the loudness of sound is 17.92 dB at a distance of 10 ft from a home theater speaker, what is the decibel level 20 ft from the speaker?

Solution:

Let L represent the loudness of sound in decibels and d represent the distance in feet. The inverse relationship between decibel level and the square of the distance is modeled by

$$L = \frac{k}{d^2}$$

$$17.92 = \frac{k}{(10)^2} \quad \text{Substitute } L = 17.92 \text{ dB and } d = 10 \text{ ft.}$$

Answer

9. 3 L

$$17.92 = \frac{k}{100}$$

$$(17.92)100 = \frac{k}{100} \cdot 100 \qquad \text{Solve for } k \text{ (clear fractions).}$$

$$k = 1792$$

$$L = \frac{1792}{d^2} \qquad \text{Substitute } k = 1792 \text{ into the original model } L = \frac{k}{d^2}.$$

With the value of k known, we can find L for any value of d.

$$L = \frac{1792}{(20)^2} \qquad \text{Find the loudness when } d = 20 \text{ ft.}$$

$$= 4.48 \text{ dB} \qquad \text{The loudness is 4.48 dB.}$$

Notice that the loudness of sound is 17.92 dB at a distance 10 ft from the speaker. When the distance from the speaker is increased to 20 ft, the decibel level decreases to 4.48 dB. This is consistent with an inverse relationship. For $k > 0$, as one variable is increased, the other is decreased. It also seems reasonable that the farther one moves away from the source of a sound, the softer the sound becomes.

Skill Practice

10. The yield on a bond varies inversely as the price. The yield on a particular bond is 5% when the price is $100. Find the yield when the price is $125.

Example 6 Solving an Application Involving Joint Variation

The kinetic energy of an object varies jointly as the weight of the object at sea level and as the square of its velocity. During a hurricane, a 0.5-lb stone traveling at 60 mph has 81 J (joules) of kinetic energy. Suppose the wind speed doubles to 120 mph. Find the kinetic energy.

Solution:

Let E represent the kinetic energy, let w represent the weight, and let v represent the velocity of the stone. The variation model is

$$E = kwv^2$$

$$81 = k(0.5)(60)^2 \qquad \text{Substitute } E = 81 \text{ J}, w = 0.5 \text{ lb, and } v = 60 \text{ mph.}$$

$$81 = k(0.5)(3600) \qquad \text{Simplify exponents.}$$

$$81 = k(1800)$$

$$\frac{81}{1800} = \frac{k(1800)}{1800} \qquad \text{Divide by 1800.}$$

$$0.045 = k \qquad \text{Solve for } k.$$

Answer

10. 4%

With the value of k known, the model $E = kwv^2$ can now be written as $E = 0.045wv^2$. We now find the kinetic energy of a 0.5-lb stone traveling at 120 mph.

$E = 0.045(0.5)(120)^2$

$= 324$

The kinetic energy of a 0.5-lb stone traveling 120 mph is 324 J.

Skill Practice

11. The amount of simple interest earned in an account varies jointly as the interest rate and time of the investment. An account earns $72 in 4 years at 2% interest. How much interest would be earned in 3 years at a rate of 5%?

In Example 6, when the velocity increased by 2 times, the kinetic energy increased by 4 times (note that 324 J = 4 · 81 J). This factor of 4 occurs because the kinetic energy is proportional to the *square* of the velocity. When the velocity increased by 2 times, the kinetic energy increased by 2^2 times.

Answer

11. $135

Section 16.5 Activity

A.1. Which equation represents a direct relationship between z and p and which represents an inverse relationship?

$$z = kp \quad \text{and} \quad z = \frac{k}{p}$$

A.2. a. Consider the equation $z = 6p$, where z and p are positive real numbers. If p increases, then z will (choose one: increase or decrease) proportionally.

b. Consider the equation $z = \frac{6}{p}$, where z and p are positive real numbers. If p increases, then z will (choose one: increase or decrease) proportionally.

A.3. The amount of medicine that a physician prescribes for a patient varies directly as the weight of the patient. A physician prescribes 3 g (grams) of a medicine for a 150-lb person.
   a. Write a variation equation using A for the amount of medicine, w for the weight of the patient, and k as the constant of variation.
   b. Solve for k by using the fact that 3 g of medicine is given to a 150-lb person. Write the resulting variation model.
   c. Based on the variation model, as the weight of the patient increases, will more or less medicine be prescribed?
   d. How many grams should the physician prescribe for a 180-lb person?
   e. How many grams should the physician prescribe for a 225-lb person?
   f. How many grams should the physician prescribe for a 120-lb person?

A.4. The average cost of producing DVDs is inversely proportional to the number of DVDs produced. If 5000 DVDs are produced, the average cost per DVD is $0.48.
   a. Write a variation equation using C for the average cost, n for the number of DVDs produced, and k as the constant of variation.
   b. Solve for k by using the known average cost of $0.48 for 5000 DVDs. Write the resulting variation model.
   c. Based on the variation model, as the number of DVDs produced increases, will the average cost per DVD go up or down?
   d. What would the average cost be if 6000 DVDs are produced?
   e. What would the average cost be if 8000 DVDs are produced?
   f. What would the average cost be if 2400 DVDs are produced?

A.5. The variable y varies jointly as b and the square root of c.
    a. Write a variation model using k as the constant of variation.
    b. If y is 150 when b is 10 and c is 25, find the value of k.
    c. Write the variation model using the numerical value of k.
    d. Determine the value of y when b is 7 and c is 36.

Section 16.5 Practice Exercises

Prerequisite Review

For Exercises R.1–R.4, solve the equation.

R.1. $40 = k \cdot 8$

R.2. $75 = k \cdot 12$

R.3. $0.16 = \dfrac{k}{4}$

R.4. $1.25 = \dfrac{k}{5}$

R.5. Given $x = \dfrac{ky^2}{z}$ and $k = 4$, $y = 6$, and $z = 3$, solve for x.

R.6. Given $y = \dfrac{ka}{b^3}$ and $k = 10$, $a = 12$, and $b = 2$, solve for y.

R.7. Given $p = ka\sqrt{c}$ and $k = 12$, $c = 9$, and $p = 72$, solve for a.

R.8. Given $t = kxy^3$ and $k = 5$, $y = 3$, and $t = 54$, solve for x.

Vocabulary and Key Concepts

1. **a.** Let k be a nonzero constant. If y varies directly as x, then $y =$ _____, where k is the constant of variation.

   b. Let k be a nonzero constant. If y varies inversely as x, then $y =$ _____, where k is the constant of variation.

   c. Let k be a nonzero constant. If y varies jointly as x and w, then $y =$ _____, where k is the constant of variation.

Concept 1: Definition of Direct and Inverse Variation

2. In the equation $r = kt$, does r vary directly or inversely as t?

3. In the equation $w = \dfrac{k}{v}$, does w vary directly or inversely as v?

4. In the equation $P = \dfrac{k \cdot c}{v}$, does P vary directly or inversely as v?

Concept 2: Translations Involving Variation

For Exercises 5–16, write a variation model. Use k as the constant of variation. (See Examples 1–2.)

5. T varies directly as q.

6. W varies directly as z.

7. b varies inversely as c.

8. m varies inversely as t.

9. Q is directly proportional to x and inversely proportional to y.

10. d is directly proportional to p and inversely proportional to n.

11. c varies jointly as s and t.

12. w varies jointly as p and f.

13. L varies jointly as w and the square root of v.

14. q varies jointly as v and the cube root of w.

15. x varies directly as the square of y and inversely as z.

16. a varies directly as n and inversely as the square of d.

Concept 3: Applications of Variation

For Exercises 17–22, find the constant of variation, k. (See Example 3.)

17. y varies directly as x and when x is 4, y is 18.

18. m varies directly as x and when x is 8, m is 22.

19. p varies inversely as q and when q is 16, p is 32.

20. T varies inversely as x and when x is 40, T is 200.

21. y varies jointly as w and v. When w is 50 and v is 0.1, y is 8.75.

22. N varies jointly as t and p. When t is 1 and p is 7.5, N is 330.

For Exercises 23–34, solve for the indicated variable. (See Example 3.)

23. x varies directly as p. If $x = 50$ when $p = 10$, find x when p is 14.

24. y is directly proportional to z. If $y = 12$ when $z = 36$, find y when z is 21.

25. b is inversely proportional to c. If b is 4 when c is 3, find b when $c = 2$.

26. q varies inversely as w. If q is 8 when w is 50, find q when w is 125.

27. Z varies directly as the square of w. If $Z = 14$ when $w = 4$, find Z when $w = 8$.

28. m varies directly as the square of x. If $m = 200$ when $x = 20$, find m when x is 32.

29. Q varies inversely as the square of p. If $Q = 4$ when $p = 3$, find Q when $p = 2$.

30. z is inversely proportional to the square of t. If $z = 15$ when $t = 4$, find z when $t = 10$.

31. L varies jointly as a and the square root of b. If $L = 72$ when $a = 8$ and $b = 9$, find L when $a = \frac{1}{2}$ and $b = 36$.

32. Y varies jointly as the cube of x and the square root of w. $Y = 128$ when $x = 2$ and $w = 16$. Find Y when $x = \frac{1}{2}$ and $w = 64$.

33. B varies directly as m and inversely as n. $B = 20$ when $m = 10$ and $n = 3$. Find B when $m = 15$ and $n = 12$.

34. R varies directly as s and inversely as t. $R = 14$ when $s = 2$ and $t = 9$. Find R when $s = 4$ and $t = 3$.

For Exercises 35–50, use a variation model to solve for the unknown value. (See Examples 4–6.)

35. The weight of a person's heart varies directly as the person's actual weight. For a 150-lb man, his heart would weigh 0.75 lb.

    a. Approximate the weight of a 184-lb man's heart.

    b. How much does your heart weigh?

36. The number of calories, C, in beer varies directly with the number of ounces, n. If 12 oz of beer contains 153 calories, how many calories are in 40 oz of beer?

37. The number of turkeys needed for a banquet is directly proportional to the number of guests that must be fed. Master Chef Rico knows that he needs to cook 3 turkeys to feed 42 guests.

   a. How many turkeys should he cook to feed 70 guests?

   b. How many turkeys should he cook to feed 140 guests?

   c. How many turkeys should be cooked to feed 700 guests at an inaugural ball?

   d. How many turkeys should be cooked for a wedding with 100 guests?

38. An author self-publishes a book and finds that the number of books she can sell per month varies inversely as the price of the book. The author can sell 1500 books per month when the price is set at $8 per book.

   a. How many books would she expect to sell if the price were $12?

   b. How many books would she expect to sell if the price were $15?

   c. How many books would she expect to sell if the price were $6?

39. The amount of pollution entering the atmosphere over a given time varies directly as the number of people living in an area. If 80,000 people create 56,800 tons of pollutants, how many tons enter the atmosphere in a city with a population of 500,000?

Patrick Clark/Photodisc/Getty Images

40. The area of a picture projected on a wall varies directly as the square of the distance from the projector to the wall. If a 10-ft distance produces a 16-ft^2 picture, what is the area of a picture produced when the projection unit is moved to a distance 20 ft from the wall?

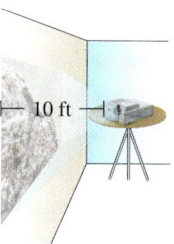

41. The stopping distance of a car varies directly as the square of the speed of the car. If a car traveling 40 mph has a stopping distance of 109 ft, find the stopping distance of a car that travels 25 mph. (Round the answer to one decimal place.)

42. The intensity of a light source varies inversely as the square of the distance from the source. If the intensity of a light bulb is 400 lumen/m^2 (lux) at a distance of 5 m, determine the intensity at 8 m.

43. The power in an electric circuit varies jointly as the current and the square of the resistance. If the power is 144 W (watts) when the current is 4 A (amperes) and the resistance is 6 Ω (ohms), find the power when the current is 3 A and the resistance is 10 Ω.

44. Some bodybuilders claim that, within safe limits, the number of repetitions that a person can complete on a given weight-lifting exercise is inversely proportional to the amount of weight lifted. Roxanne can bench press 45 lb for 15 repetitions.

   a. How many repetitions can Roxanne bench with 60 lb of weight?

   b. How many repetitions can Roxanne bench with 75 lb of weight?

   c. How many repetitions can Roxanne bench with 100 lb of weight?

45. The current in a wire varies directly as the voltage and inversely as the resistance. If the current is 9 A when the voltage is 90 V (volts) and the resistance is 10 Ω, find the current when the voltage is 185 V and the resistance is 10 Ω.

46. The resistance of a wire varies directly as its length and inversely as the square of its diameter. A 40-ft wire 0.1 in. in diameter has a resistance of 4 Ω. What is the resistance of a 50-ft wire with a diameter of 0.2 in.?

47. The weight of a medicine ball varies directly as the cube of its radius. A ball with a radius of 3 in. weighs 4.32 lb. How much does a medicine ball weigh if its radius is 5 in.?

48. The area of an equilateral triangle varies directly as the square of the length of the sides. For an equilateral triangle with 7-cm sides, the area is 21.22 cm^2. What is the area of an equilateral triangle with 17-cm sides? Round to the nearest whole unit.

49. The amount of simple interest earned in an account varies jointly as the amount of principal invested and the amount of time the money is invested. If $2500 in principal earns $500 in interest after 4 years, then how much interest will be earned on $7000 invested for 10 years?

50. The amount of simple interest earned in an account varies jointly as the amount of principal invested and the amount of time the money is invested. If $6000 in principal earns $840 in interest after 2 years, then how much interest will be earned on $4500 invested for 8 years?

Chapter 16 Review Exercises

Section 16.1

For Exercises 1–4, find the domain and range.

1. $\left\{\left(\frac{1}{3}, 10\right), \left(6, -\frac{1}{2}\right), \left(\frac{1}{4}, 4\right), \left(7, \frac{2}{5}\right)\right\}$

2.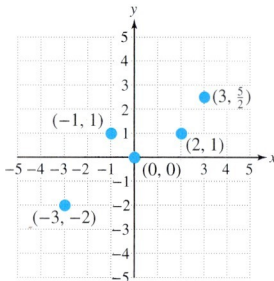

3.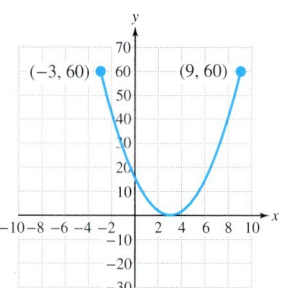

4.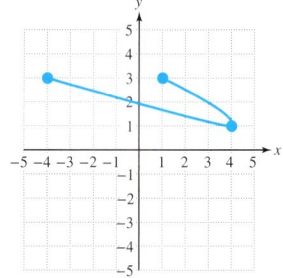

Section 16.2

5. Sketch a relation that is *not* a function. (Answers may vary.)

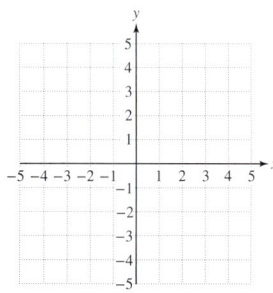

6. Sketch a relation that *is* a function. (Answers may vary.)

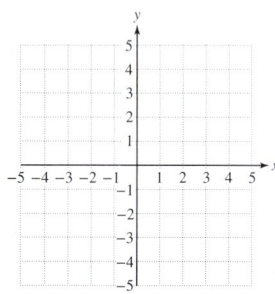

For Exercises 7–12:

a. Determine whether the relation defines y as a function of x.

b. Find the domain.

c. Find the range.

7.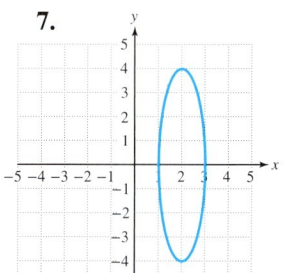

8.

9. $\{(1, 3), (2, 3), (3, 3), (4, 3)\}$

10. $\{(0, 2), (0, 3), (4, 4), (0, 5)\}$

11.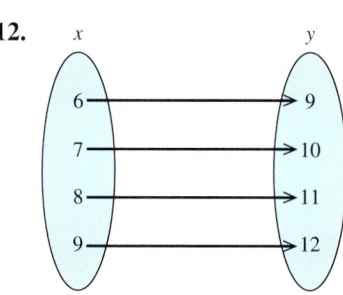

12.

For Exercises 13–20, find the function values given $f(x) = 6x^2 - 4$.

13. $f(0)$

14. $f(1)$

15. $f(-1)$

16. $f(t)$

17. $f(b)$

18. $f(\pi)$

19. $f(a)$

20. $f(-2)$

For Exercises 21–24, write the domain of each function in interval notation.

21. $g(x) = 7x^3 + 1$

22. $h(x) = \dfrac{x+10}{x-11}$

23. $k(x) = \sqrt{x-8}$

24. $w(x) = \sqrt{x+2}$

25. Anita is a waitress and makes \$12.50 per hour plus tips. Her tips average \$5 per table. In one 8-hr shift, Anita's total pay $p(x)$ can be described by $p(x) = 100 + 5x$, where x represents the number of tables she waits on. Determine how much Anita will earn if she waits on

   a. 10 tables **b.** 15 tables **c.** 20 tables

Section 16.3

For Exercises 26–31, sketch the functions from memory.

26. $h(x) = x$

27. $f(x) = x^2$

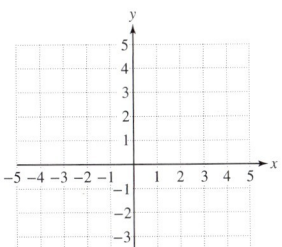
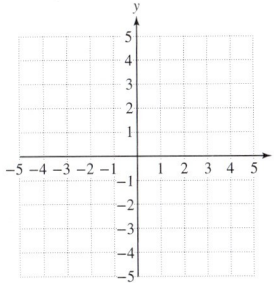

28. $g(x) = x^3$

29. $w(x) = |x|$

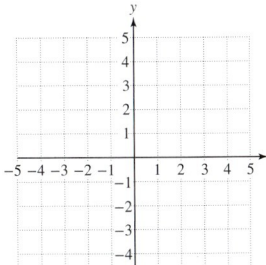
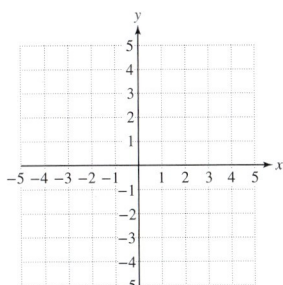

30. $s(x) = \sqrt{x}$

31. $r(x) = \dfrac{1}{x}$

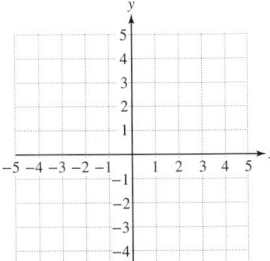
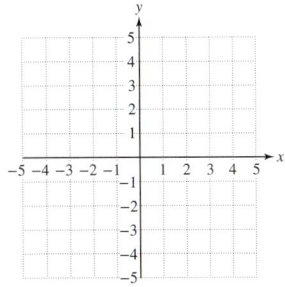

For Exercises 32–33, sketch the functions.

32. $q(x) = 3$

33. $k(x) = 2x + 1$

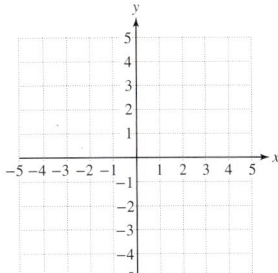
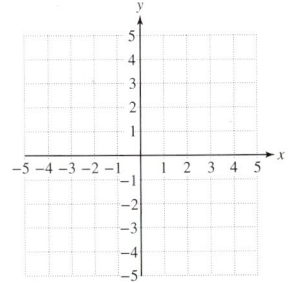

34. Given: $s(x) = (x-2)^2$

   a. Find $s(4)$, $s(-3)$, $s(2)$, $s(1)$, and $s(0)$.

   b. What is the domain of s?

35. Given: $r(x) = 2\sqrt{x-4}$

   a. Find $r(2)$, $r(4)$, $r(5)$, and $r(8)$.

   b. What is the domain of r?

36. Given: $h(x) = \dfrac{3}{x-3}$

   a. Find $h(-3)$, $h(0)$, $h(2)$, and $h(5)$.

   b. What is the domain of h?

37. Given: $k(x) = -|x+3|$

   a. Find $k(-5)$, $k(-4)$, $k(-3)$, and $k(2)$.

   b. What is the domain of k?

For Exercises 38–39, find the x- and y-intercepts.

38. $p(x) = 4x - 7$

39. $q(x) = -2x + 9$

40. The function defined by $b(t) = 1.64t + 28.3$ represents the per capita consumption of bottled water in the United States since 2010. The values of $b(t)$ are measured in gallons, and $t = 0$ corresponds to the year 2010. (*Source:* U.S. Department of Agriculture)

   a. Evaluate $b(0)$ and $b(5)$ and interpret the results in the context of this problem.

   b. Determine the slope and interpret its meaning in the context of this problem.

For Exercises 41–46, refer to the graph.

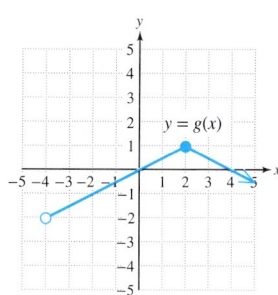

41. Find $g(-2)$.   42. Find $g(4)$.

43. For what value(s) of x is $g(x) = 0$?

44. For what value(s) of x is $g(x) = 4$?

45. Write the domain of g.

46. Write the range of g.

Section 16.4

For Exercises 47–54, refer to the functions defined here.

$$f(x) = x - 7 \qquad g(x) = -2x^3 - 8x$$

$$m(x) = x^2 \qquad n(x) = \frac{1}{x - 2}$$

Find the indicated functions.

47. $(f - g)(x)$ 48. $(f + g)(x)$

49. $(f \cdot n)(x)$ 50. $(f \cdot m)(x)$

51. $\left(\dfrac{f}{g}\right)(x)$ 52. $\left(\dfrac{g}{f}\right)(x)$

53. $(m \circ f)(x)$ 54. $(n \circ f)(x)$

For Exercises 55–58, refer to the functions defined for Exercises 47–54. Find the function values, if possible.

55. $(m \circ g)(-1)$ 56. $(n \circ g)(-1)$

57. $(f \circ g)(4)$ 58. $(g \circ f)(8)$

59. Given: $f(x) = 2x + 1$ and $g(x) = x^2$
    a. Find $(g \circ f)(x)$.
    b. Find $(f \circ g)(x)$.
    c. Based on your answers to part (a), is $f \circ g$ equal to $g \circ f$?

For Exercises 60–65, refer to the graph. Approximate the function values, if possible.

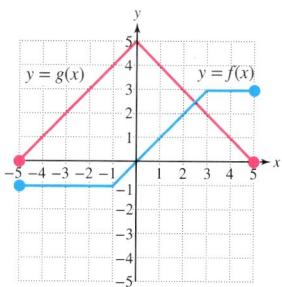

60. $\left(\dfrac{f}{g}\right)(1)$ 61. $(f \cdot g)(-2)$

62. $(f + g)(-4)$ 63. $(f - g)(2)$

64. $(g \circ f)(-3)$ 65. $(f \circ g)(4)$

Section 16.5

66. The force F applied to a spring varies directly with the distance d that the spring is stretched.
    a. Write a variation model using k as the constant of variation.
    b. When 6 lb of force is applied, the spring stretches 2 ft. Find k.
    c. How much force is required to stretch the spring 4.2 ft?

67. Suppose y varies inversely with the cube of x, and $y = 9$ when $x = 2$. Find y when $x = 3$.

68. Suppose y varies jointly with x and the square root of z, and $y = 3$ when $x = 3$ and $z = 4$. Find y when $x = 8$ and $z = 9$.

69. The distance, d, that one can see to the horizon varies directly as the square root of the height above sea level. If a person 25 m above sea level can see 30 km, how far can a person see if she is 64 m above sea level?

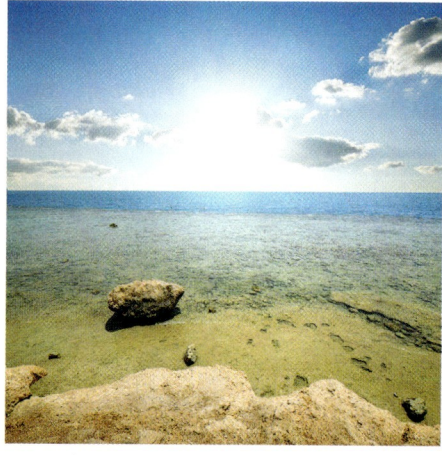

Givaga/Shutterstock

More Equations and Inequalities

17

CHAPTER OUTLINE

17.1 Compound Inequalities   1114
17.2 Polynomial and Rational Inequalities   1128
17.3 Absolute Value Equations   1140
17.4 Absolute Value Inequalities   1147
   Problem Recognition Exercises: Equations and Inequalities   1158
17.5 Linear Inequalities and Systems of Linear Inequalities in Two Variables   1159
   Chapter 17 Review Exercises   1174

Mathematics in Manufacturing

When a company manufactures a product, the goal is to obtain maximum profit at minimum cost. However, the production process is often limited by certain constraints such as the amount of labor available, the capacity of machinery, and the amount of money available for the company to invest in the process. To determine the optimal production parameters for a manufacturing process, mathematicians use **linear inequalities in two variables** and a process called linear programming.

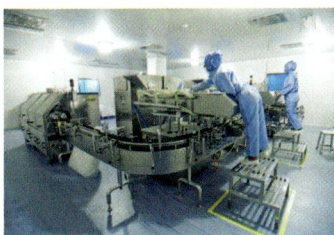
Lianxun Zhang/iStockphoto/Getty Images

To understand the concept of a linear inequality in two variables, suppose that Morgan tutors algebra and physics and that he has exactly 15 hr of time available. Let x represent the amount of time he spends tutoring algebra, and let y represent the amount of time he spends tutoring physics. Then the inequality $x + y \leq 15$ represents the distribution of his time tutoring. Furthermore, since Morgan cannot tutor for a negative period of time, we also have $x \geq 0$ and $y \geq 0$. The solution set to the three inequalities is the set of points in the first quadrant on and below the line $x + y = 15$. For example, the point (5, 8) means that Morgan tutored 5 hr of algebra and 8 hr of physics.

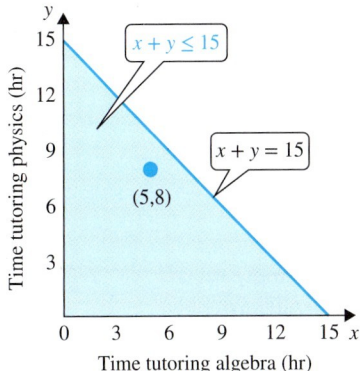

Section 17.1 Compound Inequalities

Concepts

1. Union and Intersection of Sets
2. Solving Compound Inequalities: And
3. Solving Inequalities of the Form $a < x < b$
4. Solving Compound Inequalities: Or
5. Applications of Compound Inequalities

1. Union and Intersection of Sets

We have already learned how to graph linear inequalities and to express the solution sets in interval notation and in set-builder notation. Now we will solve **compound inequalities** that involve the union or intersection of two or more inequalities.

> **Definition of A Union B and A Intersection B**
>
> The **union** of sets A and B, denoted $A \cup B$, is the set of elements that belong to set A or to set B or to both sets A and B.
>
> The **intersection** of two sets A and B, denoted $A \cap B$, is the set of elements common to both A and B.

The concepts of the union and intersection of two sets are illustrated in Figures 17-1 and 17-2.

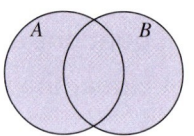
$A \cup B$
A union B
The elements in A or B or both

Figure 17-1

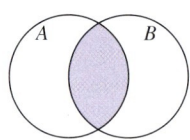
$A \cap B$
A intersection B
The elements in A and B

Figure 17-2

Example 1 Finding the Union and Intersection of Sets

Given the sets: $A = \{a, b, c, d, e, f\}$ $B = \{a, c, e, g, i, k\}$ $C = \{g, h, i, j, k\}$

Find: **a.** $A \cup B$ **b.** $A \cap B$ **c.** $A \cap C$

Solution:

a. $A \cup B = \{a, b, c, d, e, f, g, i, k\}$ — The union of A and B includes all the elements of A along with all the elements of B. Notice that the elements a, c, and e are not listed twice.

b. $A \cap B = \{a, c, e\}$ — The intersection of A and B includes only those elements that are common to both sets.

c. $A \cap C = \{\ \}$ (the empty set) — Because A and C share no common elements, the intersection of A and C is the empty set (also called the null set).

TIP: The empty set is denoted by the symbol $\{\ \}$ or by the symbol $\emptyset$.

Skill Practice Given: $A = \{r, s, t, u, v, w\}$ $B = \{s, v, w, y, z\}$ $C = \{x, y, z\}$

Find: **1.** $B \cup C$ **2.** $A \cap B$ **3.** $A \cap C$

Answers
1. $\{s, v, w, x, y, z\}$ **2.** $\{s, v, w\}$ **3.** $\{\ \}$

Example 2 Finding the Union and Intersection of Sets

Given the sets: $A = \{x \mid x < 3\}$ $B = \{x \mid x \geq -2\}$ $C = \{x \mid x \geq 5\}$

Graph the following sets. Then express each set in interval notation.

a. $A \cap B$ **b.** $A \cup C$

Solution:

It is helpful to visualize the graphs of individual sets on the number line before taking the union or intersection.

a. Graph of $A = \{x \mid x < 3\}$

Graph of $B = \{x \mid x \geq -2\}$

Graph of $A \cap B$ (the "overlap")

Interval notation: $[-2, 3)$

> **FOR REVIEW**
>
> When writing a set of real numbers using interval notation, recall the following rules.
> - A parenthesis (or) means that an endpoint is *not* included in the set.
> - A square bracket [or] means that an endpoint *is* included.
> - Use parentheses for ∞ and −∞.

Note that the set $A \cap B$ represents the real numbers greater than or equal to -2 and less than 3. This relationship can be written more concisely as a compound inequality: $-2 \leq x < 3$. We can interpret this inequality as "x is between -2 and 3, including $x = -2$."

b. Graph of $A = \{x \mid x < 3\}$

Graph of $C = \{x \mid x \geq 5\}$

Graph of $A \cup C$

Interval notation: $(-\infty, 3) \cup [5, \infty)$

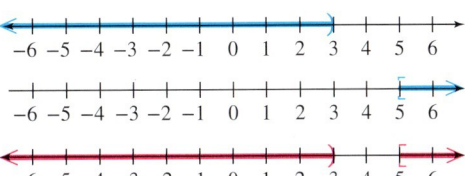

$A \cup C$ includes all elements from set A along with the elements from set C.

Skill Practice Given the sets: $A = \{x \mid x \geq -1\}$, $B = \{x \mid x < 4\}$, and $C = \{x \mid x \geq 9\}$, determine the union or intersection and express the answer in interval notation.

4. $A \cap B$ **5.** $B \cup C$

In Example 3, we find the union and intersection of sets expressed in interval notation.

Example 3 Finding the Union and Intersection of Two Intervals

Find the union or intersection as indicated. Write the answer in interval notation.

a. $(-\infty, -2) \cup [-4, 3)$ **b.** $(-\infty, -2) \cap [-4, 3)$

Solution:

a. $(-\infty, -2) \cup [-4, 3)$ To find the union, graph each interval separately. The union is the collection of real numbers that lie in the first interval, the second interval, or both intervals.

Answers

4. $[-1, 4)$ **5.** $(-\infty, 4) \cup [9, \infty)$

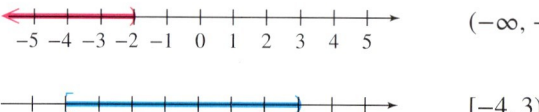 $(-\infty, -2)$

[−4, 3)

The union consists of all real numbers in the red interval along with the real numbers in the blue interval: $(-\infty, 3)$

The union is $(-\infty, 3)$.

b. $(-\infty, -2) \cap [-4, 3)$

$(-\infty, -2)$

[−4, 3)

The intersection is the "overlap" of the two intervals: [−4, −2).

The intersection is [−4, −2).

Skill Practice Find the union or intersection. Write the answer in interval notation.

6. $(-\infty, -5] \cup (-7, 0)$ **7.** $(-\infty, -5] \cap (-7, 0)$

FOR REVIEW

Multiplying or dividing an inequality by a negative factor changes the signs within the inequality. As a result, the direction of the inequality sign must be reversed. For example, $2 < 5$, but $-2 > -5$.

2. Solving Compound Inequalities: And

The solution to two inequalities joined by the word *and* is the intersection of their solution sets. For example, to play in a golf tournament for juniors, a child's age x must be at least 8 years and not more than 16 years. This is translated as $x \geq 8$ and $x \leq 16$. The word "and" joins the two inequalities and implies that we want the intersection of the individual solution sets.

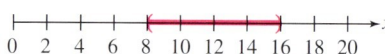

Solving a Compound Inequality: And
Step 1 Solve and graph each inequality separately.
Step 2 If the inequalities are joined by the word *and*, find the intersection of the two solution sets.
Step 3 Express the solution set in interval notation or in set-builder notation.

As you work through the examples in this section, remember that multiplying or dividing an inequality by a negative factor reverses the direction of the inequality sign.

Example 4 Solving a Compound Inequality: And

Solve the compound inequality.

$$-2x < 6 \quad \text{and} \quad x + 5 \leq 7$$

Solution:

$-2x < 6$ and $x + 5 \leq 7$ Solve each inequality separately.

$\dfrac{-2x}{-2} > \dfrac{6}{-2}$ and $x \leq 2$ Reverse the first inequality sign.

$x > -3$ and $x \leq 2$

Answers
6. $(-\infty, 0)$ **7.** $(-7, -5]$

$\{x \mid x > -3\}$

$\{x \mid x \leq 2\}$

Identify the intersection of the solution sets: $\{x \mid -3 < x \leq 2\}$

The solution is $\{x \mid -3 < x \leq 2\}$, or equivalently in interval notation, $(-3, 2]$.

Skill Practice Solve the compound inequality.

8. $5x + 2 \geq -8$ and $-4x > -24$

Example 5 Solving a Compound Inequality: And

Solve the compound inequality.

$$4.4a + 3.1 < -12.3 \quad \text{and} \quad -2.8a + 9.1 < -6.3$$

Solution:

$4.4a + 3.1 < -12.3$	and	$-2.8a + 9.1 < -6.3$	
$4.4a < -15.4$	and	$-2.8a < -15.4$	Solve each inequality separately.
$\dfrac{4.4a}{4.4} < \dfrac{-15.4}{4.4}$	and	$\dfrac{-2.8a}{-2.8} > \dfrac{-15.4}{-2.8}$	Reverse the second inequality sign.
$a < -3.5$	and	$a > 5.5$	

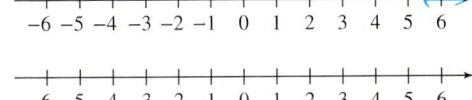

$\{a \mid a < -3.5\}$

$\{a \mid a > 5.5\}$

The intersection of the solution sets is the empty set: { }

There are no real numbers that are simultaneously less than -3.5 and greater than 5.5. There is no solution.

The solution set is { }.

Skill Practice Solve the compound inequality.

9. $3.2y - 2.4 > 16.8$ and $-4.1y \geq 8.2$

Answers

8. $\{x \mid -2 \leq x < 6\}; [-2, 6)$
9. { }

Example 6 Solving a Compound Inequality: And

Solve the compound inequality.

$$-\frac{2}{3}x \leq 6 \quad \text{and} \quad -\frac{1}{2}x < 1$$

Solution:

$$-\frac{2}{3}x \leq 6 \quad \text{and} \quad -\frac{1}{2}x < 1$$

$$-\frac{3}{2}\left(-\frac{2}{3}x\right) \geq -\frac{3}{2}(6) \quad \text{and} \quad -2\left(-\frac{1}{2}x\right) > -2(1) \quad \text{Solve each inequality separately.}$$

$$x \geq -9 \quad \text{and} \quad x > -2$$

$\{x \mid x \geq -9\}$

$\{x \mid x > -2\}$

Identify the intersection of the solution sets: $\{x \mid x > -2\}$

The solution set is $\{x \mid x > -2\}$, or in interval notation, $(-2, \infty)$.

Skill Practice Solve the compound inequality.

10. $-\frac{1}{4}z < \frac{5}{8}$ and $\frac{1}{2}z + 1 \geq 3$

FOR REVIEW

Recall that the expression $a < x$ is equivalent to $x > a$. For example, $3 < x$ is equivalent to $x > 3$.

3. Solving Inequalities of the Form $a < x < b$

An inequality of the form $a < x < b$ is a type of compound inequality, one that defines two simultaneous conditions on x.

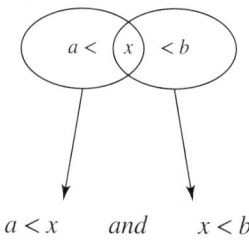

$a < x \quad \text{and} \quad x < b$

The solution set to the compound inequality $a < x < b$ is the *intersection* of the solution sets to the inequalities $a < x$ and $x < b$.

Example 7 Solving an Inequality of the Form $a < x < b$

Solve the inequality. $\quad -4 < 3x + 5 \leq 10$

Solution:

$$-4 < 3x + 5 \leq 10$$

$-4 < 3x + 5 \quad$ and $\quad 3x + 5 \leq 10 \quad$ Set up the intersection of two inequalities.

Answer
10. $\{z \mid z \geq 4\}$; $[4, \infty)$

$-9 < 3x$	and	$3x \leq 5$	Solve each inequality.
$\dfrac{-9}{3} < \dfrac{3x}{3}$	and	$\dfrac{3x}{3} \leq \dfrac{5}{3}$	
$-3 < x$	and	$x \leq \dfrac{5}{3}$	Recall $-3 < x$ is the same as $x > -3$.

$$-3 < x \leq \dfrac{5}{3}$$

Identify the intersection of the solution sets.

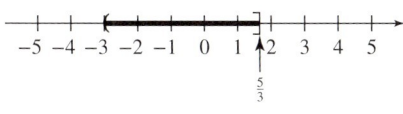

The solution is $\{x \mid -3 < x \leq \tfrac{5}{3}\}$, or equivalently in interval notation, $(-3, \tfrac{5}{3}]$.

Skill Practice Solve the inequality.

11. $-6 \leq 2x - 5 < 1$

To solve an inequality of the form $a < x < b$, we can also work with the inequality as a "three-part" inequality and isolate x. This is demonstrated in Example 8.

Example 8 Solving an Inequality of the Form $a < x < b$

Solve the inequality. $\quad 2 \geq \dfrac{p-2}{-3} \geq -1$

Solution:

$2 \geq \dfrac{p-2}{-3} \geq -1$ Isolate the variable in the middle part.

$-3(2) \leq -3\left(\dfrac{p-2}{-3}\right) \leq -3(-1)$ Multiply all three parts by -3. Remember to reverse the inequality signs.

$-6 \leq p - 2 \leq 3$ Simplify.

$-6 + 2 \leq p - 2 + 2 \leq 3 + 2$ Add 2 to all three parts to isolate p.

$-4 \leq p \leq 5$

The solution is $\{p \mid -4 \leq p \leq 5\}$, or equivalently in interval notation, $[-4, 5]$.

Skill Practice Solve the inequality.

12. $8 > \dfrac{t+4}{-2} > -5$

4. Solving Compound Inequalities: Or

In Examples 9 and 10, we solve compound inequalities that involve inequalities joined by the word "or." In such a case, the solution to the compound inequality is the union of the solution sets of the individual inequalities.

Answers

11. $\{x \mid -\tfrac{1}{2} \leq x < 3\}$; $[-\tfrac{1}{2}, 3)$
12. $\{t \mid -20 < t < 6\}$; $(-20, 6)$

For example, a resting heart rate x is potentially abnormal if it is below 50 beats per minute or above 100 beats per minute.

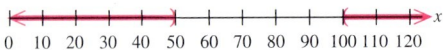

Solving a Compound Inequality: Or

Step 1 Solve and graph each inequality separately.
Step 2 If the inequalities are joined by the word *or*, find the union of the two solution sets.
Step 3 Express the solution set in interval notation or in set-builder notation.

Example 9 Solving a Compound Inequality: Or

Solve the compound inequality. $-3y - 5 > 4$ or $4 - y \leq 6$

Solution:

$$-3y - 5 > 4 \quad \text{or} \quad 4 - y \leq 6$$
$$-3y > 9 \quad \text{or} \quad -y \leq 2 \qquad \text{Solve each inequality separately.}$$
$$\frac{-3y}{-3} < \frac{9}{-3} \quad \text{or} \quad \frac{-y}{-1} \geq \frac{2}{-1} \qquad \text{Reverse the inequality signs.}$$
$$y < -3 \quad \text{or} \quad y \geq -2$$

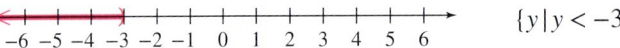 $\{y \mid y < -3\}$

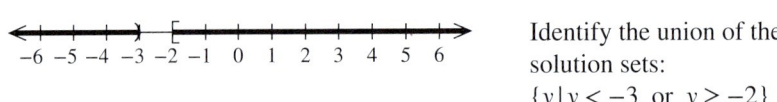 $\{y \mid y \geq -2\}$

Identify the union of the solution sets:
$\{y \mid y < -3 \text{ or } y \geq -2\}$

The solution is $\{y \mid y < -3 \text{ or } y \geq -2\}$, or equivalently in interval notation, $(-\infty, -3) \cup [-2, \infty)$.

Skill Practice Solve the compound inequality.

13. $-10t - 8 \geq 12$ or $3t - 6 > 3$

Answer

13. $\{t \mid t \leq -2 \text{ or } t > 3\}$; $(-\infty, -2] \cup (3, \infty)$

Example 10 Solving a Compound Inequality: Or

Solve the compound inequality. $4x + 3 < 16$ or $-2x < 3$

Solution:

$4x + 3 < 16$ or $-2x < 3$

$4x < 13$ or $x > -\dfrac{3}{2}$ Solve each inequality separately.

$x < \dfrac{13}{4}$ or $x > -\dfrac{3}{2}$

$\{x \mid x < \tfrac{13}{4}\}$

$\{x \mid x > -\tfrac{3}{2}\}$

Identify the union of the solution sets.

The union of the solution sets is $\{x \mid x \text{ is a real number}\}$, or equivalently, $(-\infty, \infty)$.

Skill Practice Solve the compound inequality.

14. $x - 7 > -2$ or $-6x > -48$

5. Applications of Compound Inequalities

Compound inequalities are used in many applications, as shown in Examples 11 and 12.

Example 11 Translating Compound Inequalities

The normal level of thyroid-stimulating hormone (TSH) for adults ranges from 0.4 to 4.8 microunits per milliliter (μU/mL), inclusive. Let x represent the amount of TSH measured in microunits per milliliter.

a. Write an inequality representing the normal range of TSH.
b. Write a compound inequality representing abnormal TSH levels.

Solution:

a. $0.4 \leq x \leq 4.8$ b. $x < 0.4$ or $x > 4.8$

Skill Practice The length of a normal human pregnancy, w, is from 37 to 41 weeks, inclusive.

15. Write an inequality representing the normal length of a pregnancy.
16. Write a compound inequality representing an abnormal length for a pregnancy.

Answers
14. $\{x \mid x \text{ is a real number}\}$; $(-\infty, \infty)$
15. $37 \leq w \leq 41$
16. $w < 37$ or $w > 41$

> **TIP:**
> - In mathematics, the word "between" means strictly between two values. That is, the endpoints are *excluded*.
>   Example: x is between 4 and 10 $\Rightarrow$ (4, 10).
>
> - If the word "inclusive" is added to the statement, then we *include* the endpoints.
>   Example: x is between 4 and 10, inclusive $\Rightarrow$ [4, 10].

Example 12 Translating and Solving a Compound Inequality

The sum of a number and 4 is between -5 and 12. Find all such numbers.

Solution:

Let x represent a number.

$-5 < x + 4 < 12$	Translate the inequality.
$-5 - 4 < x + 4 - 4 < 12 - 4$	Subtract 4 from all three parts of the inequality.
$-9 < x < 8$	

The number may be any real number between -9 and 8: $\{x \mid -9 < x < 8\}$.

Skill Practice

17. The sum of twice a number and 11 is between 21 and 31. Find all such numbers.

Answer

17. Any real number between 5 and 10: $\{x \mid 5 < x < 10\}$

Section 17.1 Activity

A.1. Consider sets A and B defined as $A = \{1, 2, 3, 4, 5, 6\}$ and $B = \{1, 3, 5, 7, 9\}$. Place the elements of sets A and B in the Venn diagram.

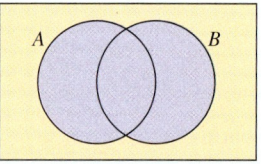

A.2. a. The _____ of set A and set B, denoted by _____, is the set of elements that belong to set A, or to set B, or to both sets A and B.

 b. Given $A = \{1, 2, 3, 4, 5, 6\}$ and $B = \{1, 3, 5, 7, 9\}$, find $A \cup B$. Refer to the Venn diagram in Exercise A.1.

A.3. a. The _____ of set A and set B, denoted by _____, is the set of elements common to both A and B.

 b. Given $A = \{1, 2, 3, 4, 5, 6\}$ and $B = \{1, 3, 5, 7, 9\}$, find $A \cap B$. Refer to the Venn diagram in Exercise A.1.

A.4. Consider the sets $C = \{x \mid x < 2\}$ and $D = \{x \mid x \geq -3\}$.

   a. Graph C.

   b. Graph D.

   c. Graph $C \cup D$.

   d. Write $C \cup D$ in interval notation.

   e. Graph $C \cap D$.

   f. Write $C \cap D$ in interval notation.

A.5. a. Solve the inequality $-4x - 1 < 19$ and write the solution set in interval notation.
   b. Solve the inequality $2x + 5 \geq 3$ and write the solution set in interval notation.
   c. Solve the compound inequality and write the solution set in interval notation.

   $$-4x - 1 < 19 \quad \text{and} \quad 2x + 5 \geq 3$$

   d. Solve the compound inequality and write the solution set in interval notation.

   $$-4x - 1 < 19 \quad \text{or} \quad 2x + 5 \geq 3$$

   e. The solution set to a compound inequality joined by the conjunction *and* is the (choose one: union/intersection) of the solution sets to the individual inequalities. (Refer to part [c].)

   f. The solution set to a compound inequality joined by the conjunction *or* is the (choose one: union/intersection) of the solution sets to the individual inequalities. (Refer to part [d].)

A.6. a. Write the inequality $-3 \leq \frac{1}{2}x - 5 < 1$ as two separate inequalities.

   b. Solve the inequality $-3 \leq \frac{1}{2}x - 5 < 1$ and write the solution set in interval notation.

A.7. The normal number of red blood cells for human blood is between 4,200,000 and 5,900,000 cells per cubic millimeter (per mm³), inclusive. Let x represent the number of red blood cells per cubic millimeter.
   a. Write a compound inequality representing the normal range of red blood cells.
   b. Write a compound inequality representing abnormal levels of red blood cells.

Practice Exercises

Section 17.1

Prerequisite Review

For Exercises R.1–R.4, write the set in interval notation.

R.1. $\{x \mid 8 > x\}$ **R.2.** $\{x \mid -4 \leq x\}$

R.3. $\{x \mid -9 < x \leq -6\}$ **R.4.** $\{x \mid 0 \leq x < 2\}$

For Exercises R.5–R.10, solve the inequality. Write the solution set in interval notation.

R.5. $-5(x - 3) - 2 \leq 43$ **R.6.** $-3(x - 4) + 8 < -10$

R.7. $-9 \leq 2x - 5 < 7$ **R.8.** $-15 < 3x + 6 \leq 12$

R.9. $\frac{2}{3}y - 1 \geq 11$ **R.10.** $4 - \frac{3}{4}t > 16$

Vocabulary and Key Concepts

1. a. The _____ of two sets A and B, denoted by _____, is the set of elements that belong to A or B or both A and B.
   b. The _____ of two sets A and B, denoted by _____, is the set of elements common to both A and B.
   c. The solution set to the compound inequality $x < c$ and $x > d$ is the (union/intersection) of the solution sets of the individual inequalities.
   d. The compound inequality $a < x$ and $x < b$ can be written as the three-part inequality _____.
   e. The solution set to the compound inequality $x < a$ or $x > b$ is the (union/intersection) of the solution sets of the individual inequalities.

Concept 1: Union and Intersection of Sets

For Exercises 2–6, refer to the Venn diagram and sets A, B, and C. Find the union or intersection as indicated.

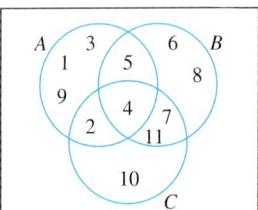

2. $A \cap B$
3. $A \cap C$
4. $B \cap C$
5. $B \cup C$
6. $A \cup C$

7. Given $M = \{-3, -1, 1, 3, 5\}$ and $N = \{-4, -3, -2, -1, 0\}$ **(See Example 1.)**

   List the elements of the following sets.
   a. $M \cap N$ b. $M \cup N$

8. Given $P = \{a, b, c, d, e, f, g, h, i\}$ and $Q = \{a, e, i, o, u\}$

   List the elements of the following sets.
   a. $P \cap Q$ b. $P \cup Q$

For Exercises 9–20, refer to the sets A, B, C, and D. Determine the union or intersection as indicated. Express the answer in interval notation, if possible. **(See Example 2.)**

$$A = \{x \mid x < -4\}, \quad B = \{x \mid x > 2\}, \quad C = \{x \mid x \geq -7\}, \quad D = \{x \mid 0 \leq x < 5\}$$

9. $A \cap C$
10. $B \cap C$
11. $A \cup B$
12. $A \cup D$
13. $A \cap B$
14. $A \cap D$
15. $B \cup C$
16. $B \cup D$
17. $C \cap D$
18. $B \cap D$
19. $C \cup D$
20. $A \cup C$

For Exercises 21–26, find the intersection and union of sets as indicated. Write the answers in interval notation. **(See Example 3.)**

21. a. $(-2, 5) \cap [-1, \infty)$
    b. $(-2, 5) \cup [-1, \infty)$

22. a. $(-\infty, 4) \cap [-1, 5)$
    b. $(-\infty, 4) \cup [-1, 5)$

23. a. $\left(-\frac{5}{2}, 3\right) \cap \left(-1, \frac{9}{2}\right)$
    b. $\left(-\frac{5}{2}, 3\right) \cup \left(-1, \frac{9}{2}\right)$

24. a. $(-3.4, 1.6) \cap (-2.2, 4.1)$

    b. $(-3.4, 1.6) \cup (-2.2, 4.1)$

25. a. $(-4, 5] \cap (0, 2]$

    b. $(-4, 5] \cup (0, 2]$

26. a. $[-1, 5) \cap (0, 3)$

    b. $[-1, 5) \cup (0, 3)$

Concept 2: Solving Compound Inequalities: And

For Exercises 27–36, solve the compound inequality and graph the solution. Write the answer in interval notation.
(See Examples 4–6.)

27. $y - 7 \geq -9$ and $y + 2 \leq 5$

28. $a + 6 > -2$ and $5a < 30$

29. $2t + 7 < 19$ and $5t + 13 > 28$

30. $5p + 2p \geq -21$ and $-9p + 3p \geq -24$

31. $2.1k - 1.1 \leq 0.6k + 1.9$ and $0.3k - 1.1 < -0.1k + 0.9$

32. $0.6w + 0.1 > 0.3w - 1.1$ and $2.3w + 1.5 \geq 0.3w + 6.5$

33. $\frac{2}{3}(2p - 1) \geq 10$ and $\frac{4}{5}(3p + 4) \geq 20$

34. $\frac{5}{2}(a + 2) < -6$ and $\frac{3}{4}(a - 2) < 1$

35. $-2 < -x - 12$ and $-14 < 5(x - 3) + 6x$

36. $-8 \geq -3y - 2$ and $3(y - 7) + 16 > 4y$

Concept 3: Solving Inequalities of the Form $a < x < b$

37. Write $-4 \leq t < \frac{3}{4}$ as two separate inequalities.

38. Write $-2.8 < y \leq 15$ as two separate inequalities.

39. Explain why $6 < x < 2$ has no solution.

40. Explain why $4 < t < 1$ has no solution.

41. Explain why $-5 > y > -2$ has no solution.

42. Explain why $-3 > w > -1$ has no solution.

For Exercises 43–54, solve the compound inequality and graph the solution set. Write the answer in interval notation.
(See Examples 7–8.)

43. $0 \leq 2b - 5 < 9$

44. $-6 < 3k - 9 \leq 0$

45. $-1 < \frac{a}{6} \leq 1$

46. $-3 \leq \dfrac{1}{2}x < 0$

47. $-\dfrac{2}{3} < \dfrac{y-4}{-6} < \dfrac{1}{3}$

48. $\dfrac{1}{3} > \dfrac{t-4}{-3} > -2$

49. $5 \leq -3x - 2 \leq 8$

50. $-1 < -2x + 4 \leq 5$

51. $12 > 6x + 3 \geq 0$

52. $-4 \geq 2x - 5 > -7$

53. $-0.2 < 2.6 + 7t < 4$

54. $-1.5 < 0.1x \leq 8.1$

Concept 4: Solving Compound Inequalities: Or

For Exercises 55–64, solve the compound inequality and graph the solution set. Write the answer in interval notation. (See Examples 9–10.)

55. $2y - 1 \geq 3$ or $y < -2$

56. $x < 0$ or $3x + 1 \geq 7$

57. $1 > 6z - 8$ or $8z - 6 \leq 10$

58. $22 > 4t - 10$ or $7 > 2t - 5$

59. $5(x - 1) \geq -5$ or $5 - x \leq 11$

60. $-p + 7 \geq 10$ or $3(p - 1) \leq 12$

61. $\dfrac{5}{3}v \leq 5$ or $-v - 6 < 1$

62. $\dfrac{3}{8}u + 1 > 0$ or $-2u \geq -4$

63. $0.5w + 5 < 2.5w - 4$ or $0.3w \leq -0.1w - 1.6$

64. $1.25a + 3 \leq 0.5a - 6$ or $2.5a - 1 \geq 9 - 1.5a$

Mixed Exercises

For Exercises 65–74, solve the compound inequality. Write the answer in interval notation.

65. **a.** $3x - 5 < 19$ and $-2x + 3 < 23$

 b. $3x - 5 < 19$ or $-2x + 3 < 23$

66. **a.** $0.5(6x + 8) > 0.8x - 7$ and $4(x + 1) < 7.2$

 b. $0.5(6x + 8) > 0.8x - 7$ or $4(x + 1) < 7.2$

67. **a.** $8x - 4 \geq 6.4$ or $0.3(x + 6) \leq -0.6$

 b. $8x - 4 \geq 6.4$ and $0.3(x + 6) \leq -0.6$

68. **a.** $-2r + 4 \leq -8$ or $3r + 5 \leq 8$

 b. $-2r + 4 \leq -8$ and $3r + 5 \leq 8$

69. $-4 \leq \dfrac{2-4x}{3} < 8$

70. $-1 < \dfrac{3-x}{2} \leq 0$

71. $5 \geq -4(t-3) + 3t$ or $6 < 12t + 8(4-t)$

72. $3 > -(w-3) + 4w$ or $-5 \geq -3(w-5) + 6w$

73. $\dfrac{-x+3}{2} > \dfrac{4+x}{5}$ or $\dfrac{1-x}{4} > \dfrac{2-x}{3}$

74. $\dfrac{y-7}{-3} < \dfrac{1}{4}$ or $\dfrac{y+1}{-2} > -\dfrac{1}{3}$

Concept 5: Applications of Compound Inequalities

75. The normal number of white blood cells for human blood is between 4800 and 10,800 cells per cubic millimeter, inclusive. Let x represent the number of white blood cells per cubic millimeter. (See Example 11.)

    a. Write an inequality representing the normal range of white blood cells per cubic millimeter.

    b. Write a compound inequality representing abnormal levels of white blood cells per cubic millimeter.

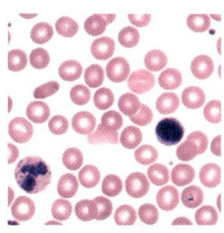

Al Telser/McGraw Hill

76. Normal hemoglobin levels in human blood for adult males are between 13 and 16 grams per deciliter (g/dL), inclusive. Let x represent the level of hemoglobin measured in grams per deciliter.

    a. Write an inequality representing normal hemoglobin levels for adult males.

    b. Write a compound inequality representing abnormal levels of hemoglobin for adult males.

77. A polling company estimates that a certain candidate running for office will receive between 44% and 48% of the votes. Let x represent the percentage of votes for this candidate.

    a. Write a strict inequality representing the expected percentage of votes for this candidate.

    b. Write a compound inequality representing the percentage of votes that would fall outside the polling company's prediction.

78. A machine is calibrated to cut a piece of wood between 2.4 in. thick and 2.6 in. thick. Let x represent the thickness of the wood after it is cut.

    a. Write a strict inequality representing the expected range of thickness of the wood after it has been cut.

    b. Write a compound inequality representing the thickness of wood that would fall outside the normal range for this machine.

79. Twice a number is between -3 and 12. Find all such numbers. (See Example 12.)

80. The difference of a number and 6 is between 0 and 8. Find all such numbers.

81. One plus twice a number is either greater than 5 or less than -1. Find all such numbers.

82. One-third of a number is either less than -2 or greater than 5. Find all such numbers.

83. Amy knows from reading her syllabus in intermediate algebra that the average of her chapter tests accounts for 80% (0.8) of her overall course grade. She also knows that the final exam counts as 20% (0.2) of her grade. Suppose that the average of Amy's chapter tests is 92%.

   a. Determine the range of grades that she would need on her final exam to get an "A" in the class. (Assume that a grade of "A" is obtained if Amy's overall average is 90% or better.)

   b. Determine the range of grades that Amy would need on her final exam to get a "B" in the class. (Assume that a grade of "B" is obtained if Amy's overall average is at least 80% but less than 90%.)

84. Robert knows from reading his syllabus in intermediate algebra that the average of his chapter tests accounts for 60% (0.6) of his overall course grade. He also knows that the final exam counts as 40% (0.4) of his grade. Suppose that the average of Robert's chapter tests is 89%.

   a. Determine the range of grades that he would need on his final exam to get an "A" in the class. (Assume that a grade of "A" is obtained if Robert's overall average is 90% or better.)

   b. Determine the range of grades that Robert would need on his final exam to get a "B" in the class. (Assume that a grade of "B" is obtained if Robert's overall average is at least 80% but less than 90%.)

85. The average high and low temperatures for Vancouver, British Columbia, in January are 5.6°C and 0°C, respectively. The formula relating Celsius temperatures to Fahrenheit temperatures is given by $C = \frac{5}{9}(F - 32)$. Convert the inequality $0.0° \leq C \leq 5.6°$ to an equivalent inequality using Fahrenheit temperatures.

86. For a day in July, the temperature in Austin, Texas, ranged from 20°C to 29°C. The formula relating Celsius temperatures to Fahrenheit temperatures is given by $C = \frac{5}{9}(F - 32)$. Convert the inequality $20° \leq C \leq 29°$ to an equivalent inequality using Fahrenheit temperatures.

Steve Allen/Brand X Pictures/Getty Images

Section 17.2 Polynomial and Rational Inequalities

Concepts

1. Solving Quadratic and Polynomial Inequalities
2. Solving Rational Inequalities

1. Solving Quadratic and Polynomial Inequalities

In this section, we will expand our study of solving inequalities.

Quadratic inequalities are inequalities that can be written in one of the following forms:

$$ax^2 + bx + c \geq 0 \qquad ax^2 + bx + c \leq 0$$
$$ax^2 + bx + c > 0 \qquad ax^2 + bx + c < 0 \qquad \text{where } a \neq 0$$

Recall that the graph of a quadratic function defined by $f(x) = ax^2 + bx + c$ is a parabola that opens upward or downward.

- The inequality $ax^2 + bx + c > 0$ is asking "For what values of x is the function positive (above the x-axis)?"
- The inequality $ax^2 + bx + c < 0$ is asking "For what values of x is the function negative (below the x-axis)?"

The graph of a quadratic function can be used to answer these questions.

Example 1 Using a Graph to Solve a Quadratic Inequality

Use the graph of $f(x) = x^2 - 6x + 5$ in Figure 17-3 to solve the inequalities.

a. $x^2 - 6x + 5 < 0$ **b.** $x^2 - 6x + 5 > 0$

Solution:

From Figure 17-3, we see that the graph of $f(x) = x^2 - 6x + 5$ is a parabola opening upward. The function factors as $f(x) = (x - 1)(x - 5)$. The x-intercepts are $(1, 0)$ and $(5, 0)$, and the y-intercept is $(0, 5)$.

FOR REVIEW

Given a quadratic function defined by $f(x) = ax^2 + bx + c$, if $a > 0$, the parabola opens upward. If $a < 0$, the parabola opens downward.

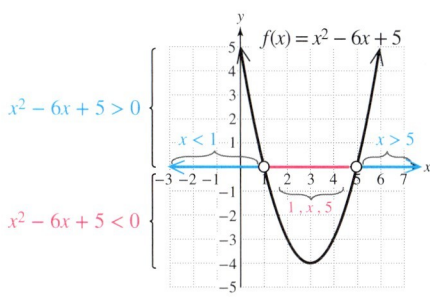

Figure 17-3

a. The solution to $x^2 - 6x + 5 < 0$ is the set of real numbers x for which $f(x) < 0$. Graphically, this is the set of all x values corresponding to the points where the parabola is below the x-axis (shown in red).

$x^2 - 6x + 5 < 0$ for $\{x \mid 1 < x < 5\}$ or in interval notation, $(1, 5)$

b. The solution to $x^2 - 6x + 5 > 0$ is the set of real numbers x for which $f(x) > 0$. This is the set of x values where the parabola is above the x-axis (shown in blue).

$x^2 - 6x + 5 > 0$ for $\{x \mid x < 1 \text{ or } x > 5\}$ or $(-\infty, 1) \cup (5, \infty)$

Skill Practice Refer to the graph of $f(x) = x^2 + 3x - 4$ to solve the inequalities.

1. $x^2 + 3x - 4 > 0$
2. $x^2 + 3x - 4 < 0$

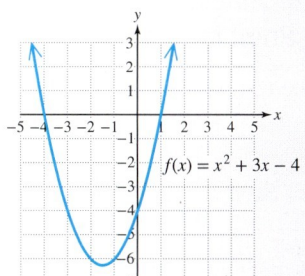

TIP: The inequalities in Example 1 are strict inequalities. Therefore, $x = 1$ and $x = 5$ (where $f(x) = 0$) are not included in the solution set. However, the corresponding inequalities using the symbols $\leq$ and $\geq$ do include the values where $f(x) = 0$.

The solution to $x^2 - 6x + 5 \leq 0$ is $\{x \mid 1 \leq x \leq 5\}$, or equivalently, $[1, 5]$.
The solution to $x^2 - 6x + 5 \geq 0$ is $\{x \mid x \leq 1 \text{ or } x \geq 5\}$ or $(-\infty, 1] \cup [5, \infty)$.

Answers

1. $\{x \mid x < -4 \text{ or } x > 1\}$; $(-\infty, -4) \cup (1, \infty)$
2. $\{x \mid -4 < x < 1\}$; $(-4, 1)$

Notice that $x = 1$ and $x = 5$ are the boundaries of the solution set to the inequalities in Example 1. These values are the solutions to the related equation $x^2 - 6x + 5 = 0$.

> **Definition of Boundary Points**
>
> The **boundary points** of an inequality consist of the real solutions to the related equation and the points where the inequality is undefined.

Testing points on intervals bounded by these points is the basis of the **test point method** to solve inequalities.

> **Solving Inequalities by Using the Test Point Method**
>
> **Step 1** Find the boundary points of the inequality.
> **Step 2** Plot the boundary points on the number line. This divides the number line into intervals.
> **Step 3** Select a test point from each interval and substitute it into the original inequality.
> - If a test point makes the original inequality true, then that interval is part of the solution set.
>
> **Step 4** Test the boundary points in the original inequality.
> - If the original inequality is strict ($<$ or $>$), do not include the boundary points in the solution set.
> - If the original inequality is defined using $\leq$ or $\geq$, then include the boundary points that are defined within the inequality.
>
> *Note:* Any boundary point that makes an expression within the inequality undefined must *always* be excluded from the solution set.

Example 2   Solving a Quadratic Inequality by Using the Test Point Method

Solve the inequality by using the test point method. $2x^2 + 5x < 12$

Solution:

$2x^2 + 5x < 12$ **Step 1:** Find the boundary points. Because polynomials are defined for all values of x, the only boundary points are the real solutions to the related equation.

$2x^2 + 5x = 12$ Solve the related equation.

$2x^2 + 5x - 12 = 0$

$(2x - 3)(x + 4) = 0$

$x = \dfrac{3}{2} \quad x = -4$ The boundary points are $\dfrac{3}{2}$ and -4.

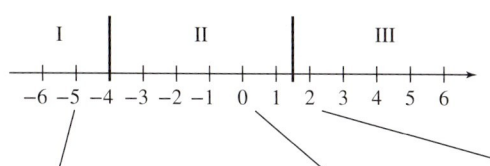

Test $x = -5$
$2x^2 + 5x < 12$
$2(-5)^2 + 5(-5) \stackrel{?}{<} 12$
$50 - 25 \stackrel{?}{<} 12$
$25 \stackrel{?}{<} 12$ False

Test $x = 0$
$2x^2 + 5x < 12$
$2(0)^2 + 5(0) \stackrel{?}{<} 12$
$0 + 0 \stackrel{?}{<} 12$
$0 \stackrel{?}{<} 12$ True

Test $x = 2$
$2x^2 + 5x < 12$
$2(2)^2 + 5(2) \stackrel{?}{<} 12$
$8 + 10 \stackrel{?}{<} 12$
$18 \stackrel{?}{<} 12$ False

Step 2: Plot the boundary points.

Step 3: Select a test point from each interval.

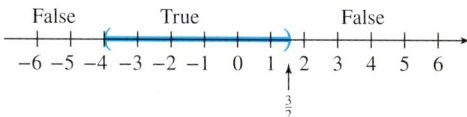

Step 4: Test the boundary points. The strict inequality excludes values of x for which $2x^2 + 5x = 12$. Therefore, the boundary points are *not* included in the solution set.

The solution set is $\{x \mid -4 < x < \frac{3}{2}\}$, or equivalently in interval notation, $(-4, \frac{3}{2})$.

Skill Practice Solve the inequality.

3. $x^2 + x > 6$

Example 3 Solving a Polynomial Inequality by Using the Test Point Method

Solve the inequality by using the test point method. $x(x + 4)^2(x - 4) \geq 0$

Solution:

$x(x + 4)^2(x - 4) \geq 0$

$x(x + 4)^2(x - 4) = 0$ **Step 1:** Find the boundary points.

$x = 0 \quad x = -4 \quad x = 4$

Step 2: Plot the boundary points.

Step 3: Select a test point from each interval.

Test $x = -5$: $-5(-5 + 4)^2(-5 - 4) \stackrel{?}{\geq} 0$ $45 \stackrel{?}{\geq} 0$ True

Test $x = -1$: $-1(-1 + 4)^2(-1 - 4) \stackrel{?}{\geq} 0$ $45 \stackrel{?}{\geq} 0$ True

Test $x = 1$: $1(1 + 4)^2(1 - 4) \stackrel{?}{\geq} 0$ $-75 \stackrel{?}{\geq} 0$ False

Test $x = 5$: $5(5 + 4)^2(5 - 4) \stackrel{?}{\geq} 0$ $405 \stackrel{?}{\geq} 0$ True

Answer

3. $(-\infty, -3) \cup (2, \infty)$

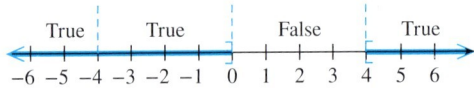

Step 4: The inequality symbol $\geq$ includes equality. Therefore, include the boundary points in the solution set.

The solution set is $\{x \mid x \leq 0 \text{ and } x \geq 4\}$, or equivalently in interval notation, $(-\infty, 0] \cup [4, \infty)$.

Skill Practice Solve the inequality.

4. $t(t-5)(t+2)^2 > 0$

TIP: In Example 3, one side of the inequality is factored, and the other side is zero. For inequalities written in this form, we can use a sign chart to determine the sign of each factor. Then the sign of the product (bottom row) is easily determined.

Sign of x	$-$	$-$	$+$	$+$
Sign of $(x+4)^2$	$+$	$+$	$+$	$+$
Sign of $(x-4)$	$-$	$-$	$-$	$+$
Sign of $x(x+4)^2(x-4)$	$+$	$+$	$-$	$+$

$$-4 \quad\quad 0 \quad\quad 4$$

The solution to the inequality $x(x+4)^2(x-4) \geq 0$ includes the intervals for which the product is positive (shown in blue).

The solution is $(-\infty, 0] \cup [4, \infty)$.

2. Solving Rational Inequalities

The test point method can be used to solve rational inequalities. A **rational inequality** is an inequality in which one or more terms is a rational expression. The solution set to a rational inequality must exclude all values of the variable that make the inequality undefined. That is, exclude all values that make the denominator equal to zero for any rational expression in the inequality.

Example 4 Solving a Rational Inequality by Using the Test Point Method

Solve the inequality. $\quad \dfrac{3}{x-1} > 0$

Solution:

$$\dfrac{3}{x-1} > 0$$

Step 1: Find the boundary points. Note that the inequality is undefined for $x = 1$, so $x = 1$ is a boundary point. To find any other boundary points, solve the related equation.

$$\dfrac{3}{x-1} = 0$$

$$(x-1) \cdot \left(\dfrac{3}{x-1}\right) = (x-1) \cdot 0 \quad\quad \text{Clear fractions.}$$

$$3 = 0 \quad\quad \text{There is no solution to the related equation.}$$

Answer

4. $(-\infty, -2) \cup (-2, 0) \cup (5, \infty)$

The only boundary point is $x = 1$.

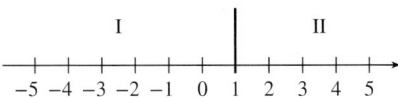

Test $x = 0$:

$\dfrac{3}{(0) - 1} \stackrel{?}{>} 0$

$\dfrac{3}{-1} \stackrel{?}{>} 0$ False

Test $x = 2$:

$\dfrac{3}{(2) - 1} \stackrel{?}{>} 0$

$\dfrac{3}{1} > 0$ True

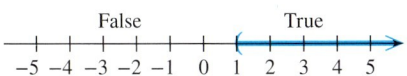

Step 2: Plot boundary points.

Step 3: Select test points.

Step 4: The boundary point $x = 1$ cannot be included in the solution set because it is undefined in the original inequality.

The solution is $\{x \mid x > 1\}$, or equivalently in interval notation, $(1, \infty)$.

Skill Practice Solve the inequality.

5. $\dfrac{-5}{y + 2} < 0$

TIP: Using a sign chart we see that the quotient of the factors 3 and $(x - 1)$ is positive on the interval $(1, \infty)$.

Therefore, the solution to the inequality $\dfrac{3}{x - 1} > 0$ is $(1, \infty)$.

Sign of 3	+	+
Sign of $(x - 1)$	−	+
Sign of $\dfrac{3}{x - 1}$	−	+

1
(undefined)

The solution to the inequality $\dfrac{3}{x - 1} > 0$ can be confirmed from the graph of the related rational function, $f(x) = \dfrac{3}{x - 1}$ (see Figure 17-4).

- The graph is above the x-axis where $f(x) = \dfrac{3}{x - 1} > 0$ for $x > 1$ (shaded red).
- Also note that $x = 1$ cannot be included in the solution set because 1 is not in the domain of f.

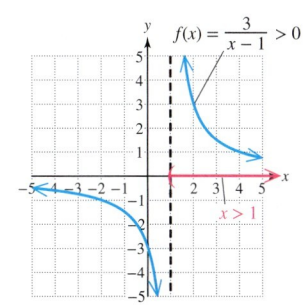

Figure 17-4

Answer

5. $(-2, \infty)$

Example 5: Solving a Rational Inequality by Using the Test Point Method

Solve the inequality by using the test point method. $\dfrac{x+2}{x-4} \leq 3$

Solution:

$$\dfrac{x+2}{x-4} \leq 3$$

$$\dfrac{x+2}{x-4} = 3$$

Step 1: Find the boundary points. Note that the inequality is undefined for $x = 4$. Therefore, $x = 4$ is automatically a boundary point. To find any other boundary points, solve the related equation.

$$(x-4)\left(\dfrac{x+2}{x-4}\right) = (x-4)(3) \qquad \text{Clear fractions.}$$

$$x + 2 = 3(x - 4) \qquad \text{Solve for } x.$$

$$x + 2 = 3x - 12$$

$$-2x = -14$$

$$x = 7$$

The solution to the related equation is $x = 7$, and the inequality is undefined for $x = 4$. Therefore, the boundary points are $x = 4$ and $x = 7$.

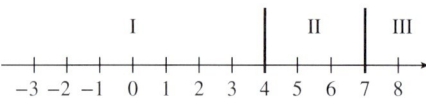

Step 2: Plot boundary points.

Step 3: Select test points.

Test $x = 0$

$$\dfrac{x+2}{x-4} \leq 3$$

$$\dfrac{0+2}{0-4} \overset{?}{\leq} 3$$

$$-\dfrac{1}{2} \overset{?}{\leq} 3 \quad \text{True}$$

Test $x = 5$

$$\dfrac{x+2}{x-4} \leq 3$$

$$\dfrac{5+2}{5-4} \overset{?}{\leq} 3$$

$$\dfrac{7}{1} \overset{?}{\leq} 3 \quad \text{False}$$

Test $x = 8$

$$\dfrac{x+2}{x-4} \leq 3$$

$$\dfrac{8+2}{8-4} \overset{?}{\leq} 3$$

$$\dfrac{10}{4} \overset{?}{\leq} 3$$

$$\dfrac{5}{2} \overset{?}{\leq} 3 \quad \text{True}$$

Test $x = 4$:

$$\dfrac{x+2}{x-4} \leq 3$$

$$\dfrac{4+2}{4-4} \overset{?}{\leq} 3$$

$$\dfrac{6}{0} \overset{?}{\leq} 3 \quad \text{Undefined}$$

Test $x = 7$:

$$\dfrac{x+2}{x-4} \leq 3$$

$$\dfrac{7+2}{7-4} \overset{?}{\leq} 3$$

$$\dfrac{9}{3} \overset{?}{\leq} 3 \quad \text{True}$$

Step 4: Test the boundary points.

The boundary point $x = 4$ cannot be included in the solution set because it is undefined in the inequality. The boundary point $x = 7$ makes the original inequality true and must be included in the solution set.

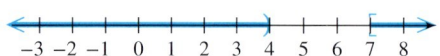

The solution is $\{x \mid x < 4 \text{ or } x \geq 7\}$, or equivalently in interval notation, $(-\infty, 4) \cup [7, \infty)$.

Skill Practice Solve the inequality.

6. $\dfrac{x-5}{x+4} \leq -1$

Answer

6. $\left(-4, \dfrac{1}{2}\right]$

Section 17.2 Activity

A.1. a. Consider the expression $x^2 - 2x - 3$. For different values of x, the expression may be zero, positive, or negative. To show this, evaluate the expression for $x = -1$, $x = 4$, and $x = 1$.

  b. Consider the function defined by $f(x) = x^2 - 2x - 3$. For different values of x in the domain, the function may be zero, positive, or negative. To show this, find $f(-1), f(4),$ and $f(1)$.

  c. The graph of $f(x) = x^2 - 2x - 3$ is shown. For what value(s) of x is the function equal to zero? (These are the values of x that are solutions to $x^2 - 2x - 3 = 0$).

  d. Refer to the graph of $f(x) = x^2 - 2x - 3$. For what values of x is the function negative (below the x-axis)? These are the values of x that are solutions to the inequality $x^2 - 2x - 3 < 0$. (*Hint:* There are infinitely many values of x that make the function negative. Express your answer in interval notation.)

  e. For what values of x is the function positive (above the x-axis)? These are the values of x that are solutions to the inequality $x^2 - 2x - 3 > 0$. (*Hint:* The function is positive over two intervals of x. Write the answer as the union of two intervals.)

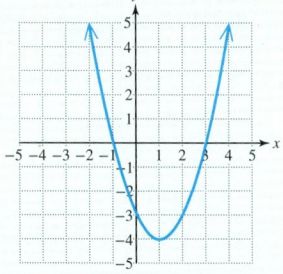

A.2. We will now walk through an algebraic approach to solve the quadratic inequality $x^2 - 2x - 3 > 0$.

  a. First solve the related *equation* $x^2 - 2x - 3 = 0$.

  b. Plot the solutions to the equation $x^2 - 2x - 3 = 0$ on the number line to divide the number line into intervals.

  c. Test an arbitrary point from each interval on the number line to determine if it makes the inequality $x^2 - 2x - 3 > 0$ true or false. Which intervals make the inequality true?

  d. Test the boundary points in the inequality $x^2 - 2x - 3 > 0$. In this case, the inequality is strict. Therefore, the boundary points (values for which $x^2 - 2x - 3$ *equals* zero) should be (choose one: included/excluded) from the solution set.

  e. Write the solution set in interval notation.

  f. Based on the outcomes of the test points in part (c), write the solution set to the inequality $x^2 - 2x - 3 < 0$.

  g. Compare the algebraic process to solve the inequality $x^2 - 2x - 3 > 0$ with the graphical approach taken in Exercise A.1.

A.3. Solve the rational inequality $\frac{x-3}{x-2} \geq 0$ algebraically by following these steps. As you work through the process, you can check your answers by referring to the graph of $f(x) = \frac{x-3}{x-2}$.

  a. First solve the related *equation* $\frac{x-3}{x-2} = 0$.

  b. Identify value(s) of x for which $\frac{x-3}{x-2}$ is undefined. Can this value be a solution to the original inequality?

  c. Plot the solutions to the equation $\frac{x-3}{x-2} = 0$ on the number line along with any values of x for which the expression is undefined. These are the boundary points that divide the number line into intervals.

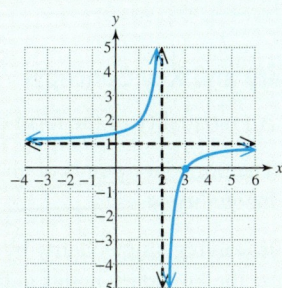

d. Test an arbitrary point from each interval on the number line to determine if it makes the inequality $\frac{x-3}{x-2} \geq 0$ true or false. Which interval(s) make the inequality true?

e. The expression $\frac{x-3}{x-2}$ equals zero for $x = 3$. Therefore, is 3 a solution to the inequality $\frac{x-3}{x-2} \geq 0$?

f. Write the solution set to the inequality $\frac{x-3}{x-2} \geq 0$ in interval notation.

g. How would the solution set change for the inequality $\frac{x-3}{x-2} > 0$?

h. Write the solution set to the inequality $\frac{x-3}{x-2} \leq 0$.

A.4. Consider the expression $x^2 - 6x + 9$.

a. Write the expression in factored form.

b. Write the inequality $x^2 - 6x + 9 < 0$ with the left side in factored form. Is it possible for a perfect square to be negative?

c. Write the solution set to the inequality $x^2 - 6x + 9 < 0$ and explain your answer.

d. How does the solution set to the inequality $x^2 - 6x + 9 \leq 0$ differ from the result of part (c)?

e. Write the solution set to $x^2 - 6x + 9 \geq 0$.

f. Write the solution set to $x^2 - 6x + 9 > 0$.

Section 17.2 Practice Exercises

Prerequisite Review

For Exercises R.1–R.8, solve the equation.

R.1. $x^2 + 9x + 14 = 0$

R.2. $y^2 - 19y + 18 = 0$

R.3. $4t^2 + 4t + 1 = 0$

R.4. $9p^2 - 30p + 25 = 0$

R.5. $2x(x - 6) = x - 6$

R.6. $3x(x + 3) = 2x + 6$

R.7. $\frac{x}{2x - 5} = 1$

R.8. $\frac{4y}{3y + 14} = 2$

Vocabulary and Key Concepts

1. a. An inequality of the form $ax^2 + bx + c > 0$ or $ax^2 + bx + c < 0$ is an example of a _____ inequality.

   b. The boundary points of an inequality consist of the real _____ to the related equation and the points where the inequality is _____.

   c. In solving an inequality by using the _____ _____ method, a point is selected from each interval formed by the boundary points and substituted into the original inequality.

   d. If a test point makes the original inequality _____, then that interval is part of the solution set.

2. a. The inequality $\frac{4}{x + 7} > 0$ is an example of a _____ inequality.

   b. The solution set to a rational inequality must exclude all values that make the denominator equal to _____ for any rational expression in the inequality.

Concept 1: Solving Quadratic and Polynomial Inequalities

For Exercises 3–6, estimate from the graph the intervals for which the inequality is true. (See Example 1.)

3.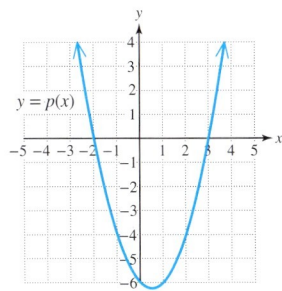

a. $p(x) > 0$ b. $p(x) < 0$
c. $p(x) \leq 0$ d. $p(x) \geq 0$

4.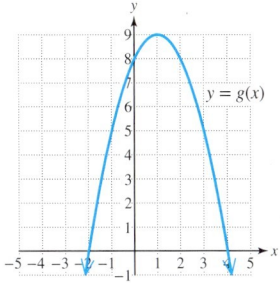

a. $g(x) > 0$ b. $g(x) < 0$
c. $g(x) \leq 0$ d. $g(x) \geq 0$

5.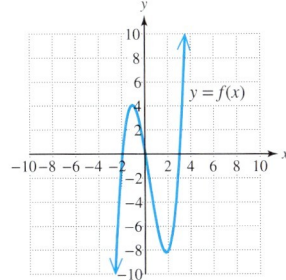

a. $f(x) > 0$ b. $f(x) < 0$
c. $f(x) \leq 0$ d. $f(x) \geq 0$

6.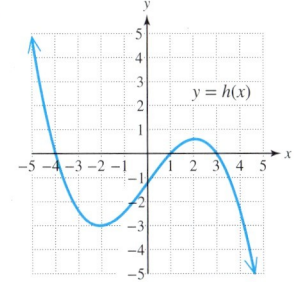

a. $h(x) > 0$ b. $h(x) < 0$
c. $h(x) \leq 0$ d. $h(x) \geq 0$

For Exercises 7–12, solve the equation and related inequalities. (See Examples 2–3.)

7. a. $3(4-x)(2x+1) = 0$
   b. $3(4-x)(2x+1) < 0$
   c. $3(4-x)(2x+1) > 0$

8. a. $5(y+6)(3-5y) = 0$
   b. $5(y+6)(3-5y) < 0$
   c. $5(y+6)(3-5y) > 0$

9. a. $x^2 + 7x = 30$
   b. $x^2 + 7x < 30$
   c. $x^2 + 7x > 30$

10. a. $q^2 - 4q = 5$
    b. $q^2 - 4q \leq 5$
    c. $q^2 - 4q \geq 5$

11. a. $2p(p-2) = p+3$
    b. $2p(p-2) \leq p+3$
    c. $2p(p-2) \geq p+3$

12. a. $3w(w+4) = 10-w$
    b. $3w(w+4) < 10-w$
    c. $3w(w+4) > 10-w$

For Exercises 13–30, solve the polynomial inequality. Write the answer in interval notation. (See Examples 2–3.)

13. $(t-7)(t-1) < 0$
14. $(p-4)(p-2) > 0$
15. $-6(4+2x)(5-x) > 0$
16. $-8(2t+5)(6-t) < 0$
17. $m(m+1)^2(m+5) \leq 0$
18. $w^2(3-w)(w+2) \geq 0$
19. $a^2 - 12a \leq -32$
20. $w^2 + 20w \geq -64$
21. $5x^2 - 4x - 1 > 0$
22. $x^2 + x \leq 6$
23. $b^2 - 121 < 0$
24. $c^2 - 25 < 0$

25. $3p(p-2) - 3 \geq 2p$ **26.** $2t(t+3) - t \leq 12$ **27.** $x^3 - x^2 \leq 12x$

28. $x^3 + 36 > 4x^2 + 9x$ **29.** $w^3 + w^2 > 4w + 4$ **30.** $2p^3 - 5p^2 \leq 3p$

Concept 2: Solving Rational Inequalities

For Exercises 31–34, estimate from the graph the intervals for which the inequality is true.

31.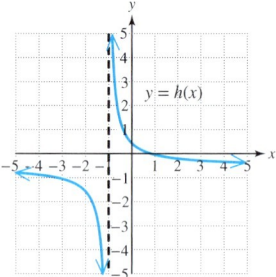

    a. $h(x) \geq 0$ **b.** $h(x) \leq 0$

    c. $h(x) < 0$ **d.** $h(x) > 0$

32.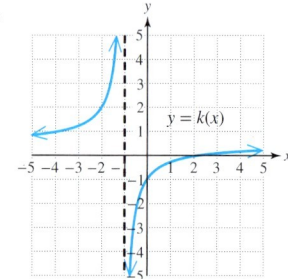

    a. $k(x) \leq 0$ **b.** $k(x) \geq 0$

    c. $k(x) > 0$ **d.** $k(x) < 0$

33.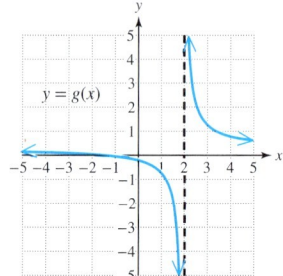

    a. $g(x) > 0$ **b.** $g(x) < 0$

    c. $g(x) \leq 0$ **d.** $g(x) \geq 0$

34.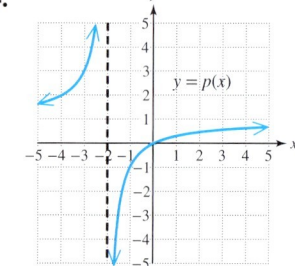

    a. $p(x) > 0$ **b.** $p(x) < 0$

    c. $p(x) \leq 0$ **d.** $p(x) \geq 0$

For Exercises 35–38, solve the equation and related inequalities. **(See Examples 4–5.)**

35. a. $\dfrac{10}{x-5} = 5$ **36. a.** $\dfrac{8}{a+1} = 4$ **37. a.** $\dfrac{z+2}{z-6} = -3$ **38. a.** $\dfrac{w-8}{w+6} = 2$

   b. $\dfrac{10}{x-5} < 5$ **b.** $\dfrac{8}{a+1} > 4$ **b.** $\dfrac{z+2}{z-6} \leq -3$ **b.** $\dfrac{w-8}{w+6} \leq 2$

   c. $\dfrac{10}{x-5} > 5$ **c.** $\dfrac{8}{a+1} < 4$ **c.** $\dfrac{z+2}{z-6} \geq -3$ **c.** $\dfrac{w-8}{w+6} \geq 2$

For Exercises 39–50, solve the rational inequality. Write the answer in interval notation. **(See Examples 4–5.)**

39. $\dfrac{2}{x-1} \geq 0$ **40.** $\dfrac{-3}{x+2} \leq 0$ **41.** $\dfrac{b+4}{b-4} > 0$ **42.** $\dfrac{a+1}{a-3} < 0$

43. $\dfrac{3}{2x-7} < -1$ **44.** $\dfrac{8}{4x+9} > 1$ **45.** $\dfrac{x+1}{x-5} \geq 4$ **46.** $\dfrac{x-2}{x+6} \leq 5$

47. $\dfrac{1}{x} \leq 2$ **48.** $\dfrac{1}{x} \geq 3$ **49.** $\dfrac{(x+2)^2}{x} > 0$ **50.** $\dfrac{(x-3)^2}{x} < 0$

Mixed Exercises

For Exercises 51–70, identify the inequality as one of the following types: linear, quadratic, rational, or polynomial (degree > 2). Then solve the inequality and write the answer in interval notation.

51. $2y^2 - 8 \leq 24$

52. $8p^2 - 18 > 0$

53. $(5x + 2)^2 > 4$

54. $(7x - 4)^2 < 9$

55. $4(x - 2) < 6x - 3$

56. $-7(3 - y) > 4 + 2y$

57. $\dfrac{2x + 3}{x + 1} \leq 2$

58. $\dfrac{5x - 1}{x + 3} \geq 5$

59. $4x^3 - 40x^2 + 100x > 0$

60. $2y^3 - 12y^2 + 18y < 0$

61. $2p^3 > 4p^2$

62. $w^3 \leq 5w^2$

63. $-3(x + 4)^2(x - 5) \geq 0$

64. $5x(x - 2)(x - 6)^2 \geq 0$

65. $x^2 - 4 < 0$

66. $y^2 - 9 > 0$

67. $\dfrac{a + 2}{a - 5} \geq 0$

68. $\dfrac{t + 1}{t - 2} \leq 0$

69. $2 \geq t - 3$

70. $-5p + 8 < p$

Technology Connections

71. To solve the inequality $\dfrac{x}{x - 2} > 0$, enter Y_1 as $x/(x - 2)$ and determine where the graph is above the x-axis. Write the solution in interval notation.

72. To solve the inequality $\dfrac{x}{x - 2} < 0$, enter Y_1 as $x/(x - 2)$ and determine where the graph is below the x-axis. Write the solution in interval notation.

73. To solve the inequality $x^2 - 1 < 0$, enter Y_1 as $x^2 - 1$ and determine where the graph is below the x-axis. Write the solution in interval notation.

74. To solve the inequality $x^2 - 1 > 0$, enter Y_1 as $x^2 - 1$ and determine where the graph is above the x-axis. Write the solution in interval notation.

For Exercises 75–78, determine the solution by graphing the inequalities.

75. $x^2 + 10x + 25 \leq 0$

76. $-x^2 + 10x - 25 \geq 0$

77. $\dfrac{8}{x^2 + 2} < 0$

78. $\dfrac{-6}{x^2 + 3} > 0$

Expanding Your Skills

The solution to an inequality is often one or more intervals on the real number line. Sometimes, however, the solution to an inequality may be a single point on the number line, the empty set, or the set of all real numbers. We call these "special case" solution sets.

For Exercises 79–94, solve the inequalities involving "special case" solution sets.

79. $x^2 + 10x + 25 \geq 0$

80. $x^2 + 6x + 9 < 0$

81. $x^2 + 2x + 1 < 0$

82. $x^2 + 8x + 16 \geq 0$

83. $x^4 + 3x^2 \leq 0$

84. $x^4 + 2x^2 \leq 0$

85. $x^2 + 12x + 36 < 0$

86. $x^2 + 12x + 36 \geq 0$

87. $x^2 + 3x + 5 < 0$

88. $2x^2 + 3x + 3 > 0$

89. $-5x^2 + x < 1$

90. $-3x^2 - x > 6$

91. $x^2 + 22x + 121 > 0$

92. $y^2 - 24y + 144 > 0$

93. $4t^2 - 12t \leq -9$

94. $9y^2 - 30y \leq -25$

Section 17.3 Absolute Value Equations

Concepts

1. Solving Absolute Value Equations
2. Solving Equations Containing Two Absolute Values

1. Solving Absolute Value Equations

An equation of the form $|x| = a$ is called an **absolute value equation**. The solution includes all real numbers whose absolute value equals a. For example, the solutions to the equation $|x| = 4$ are 4 as well as -4, because $|4| = 4$ and $|-4| = 4$. A geometric interpretation of the absolute value of a number is its distance from zero on the number line (Figure 17-5). Therefore, the solutions to the equation $|x| = 4$ are the values of x that are 4 units away from zero.

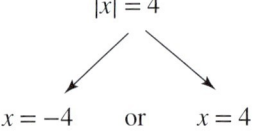
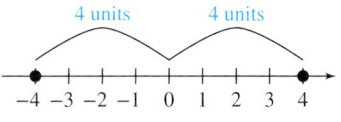

Figure 17-5

Solving Absolute Value Equations of the Form $|x| = a$

If a is a real number, then

- If $a \geq 0$, the solutions to the equation $|x| = a$ are $x = a$ and $x = -a$.
- If $a < 0$, there is no solution to the equation $|x| = a$.

To solve an absolute value equation of the form $|x| = a$ ($a \geq 0$), rewrite the equation as $x = a$ or $x = -a$.

Example 1 Solving Absolute Value Equations

Solve the absolute value equations.

a. $|x| = 5$ **b.** $|w| - 2 = 12$ **c.** $|p| = 0$ **d.** $|x| = -6$

Solution:

a. $|x| = 5$

$x = 5$ or $x = -5$ The equation is in the form $|x| = a$, where $a = 5$.

Rewrite the equation as $x = a$ or $x = -a$.

The solution set is $\{5, -5\}$.

b. $|w| - 2 = 12$ Isolate the absolute value to write the equation in the form $|x| = a$.

$|w| = 14$

$w = 14$ or $w = -14$ Rewrite the equation as $w = a$ or $w = -a$.

The solution set is $\{14, -14\}$.

c. $|p| = 0$

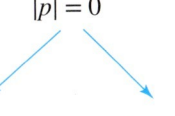

$p = 0$ or $p = -0$

The solution set is $\{0\}$.

Rewrite as two equations. Notice that the second equation $p = -0$ is the same as the first equation. Intuitively, $p = 0$ is the only number whose absolute value equals 0.

d. $|x| = -6$

No solution, $\{\ \}$

The equation is of the form $|x| = a$, but a is negative. There is no number whose absolute value is negative.

Skill Practice Solve the absolute value equations.

1. $|y| = 7$ **2.** $|v| + 6 = 10$ **3.** $|w| + 3 = 3$ **4.** $|z| = -12$

We have solved absolute value equations of the form $|x| = a$. Notice that x can represent any algebraic quantity. For example, to solve the equation $|2w - 3| = 5$, we still rewrite the absolute value equation as two equations. In this case, we set the quantity $2w - 3$ equal to 5 and to -5, respectively.

$|2w - 3| = 5$

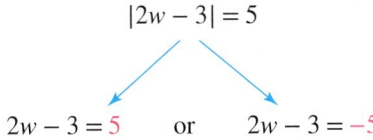

$2w - 3 = 5$ or $2w - 3 = -5$

Solving an Absolute Value Equation

Step 1 Isolate the absolute value. That is, write the equation in the form $|x| = a$, where a is a real number.
Step 2 If $a < 0$, there is no solution.
Step 3 Otherwise, if $a \geq 0$, rewrite the absolute value equation as $x = a$ or $x = -a$.
Step 4 Solve the individual equations from step 3.
Step 5 Check the answers in the original absolute value equation.

Example 2   Solving an Absolute Value Equation

Solve the equation. $|2w - 3| = 5$

Solution:

$|2w - 3| = 5$ The equation is already in the form $|x| = a$, where $x = 2w - 3$.

$2w - 3 = 5$ or $2w - 3 = -5$ Rewrite as two equations.

$2w = 8$ or $2w = -2$ Solve each equation.

$w = 4$ or $w = -1$

Answers
1. $\{7, -7\}$
2. $\{4, -4\}$
3. $\{0\}$ **4.** $\{\ \}$

Check: $w = 4$	Check: $w = -1$	Check the solutions in the original equation.
$\|2w - 3\| = 5$	$\|2w - 3\| = 5$	
$\|2(4) - 3\| \stackrel{?}{=} 5$	$\|2(-1) - 3\| \stackrel{?}{=} 5$	
$\|8 - 3\| \stackrel{?}{=} 5$	$\|-2 - 3\| \stackrel{?}{=} 5$	
$\|5\| \stackrel{?}{=} 5$ ✓	$\|-5\| \stackrel{?}{=} 5$ ✓	

The solution set is $\{4, -1\}$.

Skill Practice Solve the equation.

5. $|4x + 1| = 9$

Example 3 Solving an Absolute Value Equation

Solve the equation. $|2c - 5| + 6 = 2$

Solution:

$|2c - 5| + 6 = 2$

$|2c - 5| = -4$ Isolate the absolute value. The equation is in the form $|x| = a$, where $x = 2c - 5$ and $a = -4$. Because $a < 0$, there is no solution.

No solution, $\{\ \}$ There are no numbers c that will make an absolute value equal to a negative number.

Avoiding Mistakes

Always isolate the absolute value first. Otherwise you will get answers that do not check.

Skill Practice Solve the equation.

6. $|3z + 10| + 3 = 1$

Example 4 Solving an Absolute Value Equation

Solve the equation. $-2\left|\dfrac{2}{5}p + 3\right| - 7 = -19$

Solution:

$-2\left|\dfrac{2}{5}p + 3\right| - 7 = -19$

$-2\left|\dfrac{2}{5}p + 3\right| = -12$ Isolate the absolute value.

$\dfrac{-2\left|\dfrac{2}{5}p + 3\right|}{-2} = \dfrac{-12}{-2}$

$\left|\dfrac{2}{5}p + 3\right| = 6$

$\dfrac{2}{5}p + 3 = 6$ or $\dfrac{2}{5}p + 3 = -6$ Rewrite as two equations.

$2p + 15 = 30$ or $2p + 15 = -30$ Multiply by 5 to clear fractions.

$2p = 15$ or $2p = -45$

$p = \dfrac{15}{2}$ or $p = -\dfrac{45}{2}$ Both values check in the original equation.

The solution set is $\left\{\dfrac{15}{2}, -\dfrac{45}{2}\right\}$.

Answers

5. $\left\{2, -\dfrac{5}{2}\right\}$ 6. $\{\ \}$

Skill Practice Solve the equation.

7. $3\left|\dfrac{3}{2}a + 1\right| + 2 = 14$

Example 5   Solving an Absolute Value Equation

Solve the equation. $6.9 = |4.1 - p| + 6.9$

Solution:

$6.9 = |4.1 - p| + 6.9$

$|4.1 - p| + 6.9 = 6.9$ First write the absolute value on the left. Then subtract 6.9 from both sides to write the equation in the form $|x| = a$.

$|4.1 - p| = 0$ Isolate the absolute value.

$4.1 - p = 0$ or $4.1 - p = -0$ Rewrite as two equations. Notice that the equations are the same.

$-p = -4.1$ Subtract 4.1 from both sides.

$p = 4.1$

Check $p = 4.1$ in the original equation.

Check: $p = 4.1$

$|4.1 - p| + 6.9 = 6.9$

$|4.1 - 4.1| + 6.9 \stackrel{?}{=} 6.9$

$|0| + 6.9 \stackrel{?}{=} 6.9$

$6.9 \stackrel{?}{=} 6.9$ ✓

The solution set is $\{4.1\}$.

Skill Practice Solve the equation.

8. $-3.5 = |1.2 + x| - 3.5$

---

2. Solving Equations Containing Two Absolute Values

Some equations have two absolute values such as $|x| = |y|$. If two quantities have the same absolute value, then the quantities are equal or the quantities are opposites.

Equality of Absolute Values

$|x| = |y|$ implies that $x = y$ or $x = -y$.

Answers

7. $\left\{2, -\dfrac{10}{3}\right\}$ 8. $\{-1.2\}$

Example 6 — Solving an Equation Having Two Absolute Values

Solve the equation. $|2w - 3| = |5w + 1|$

Solution:

$$|2w - 3| = |5w + 1|$$

$2w - 3 = 5w + 1$ or $2w - 3 = -(5w + 1)$ Rewrite as two equations, $x = y$ or $x = -y$.

> **Avoiding Mistakes**
> To take the opposite of the quantity $5w + 1$, use parentheses and apply the distributive property.

$2w - 3 = 5w + 1$ or $2w - 3 = -5w - 1$ Solve for w.
$-3w - 3 = 1$ or $7w - 3 = -1$
$-3w = 4$ or $7w = 2$
$w = -\dfrac{4}{3}$ or $w = \dfrac{2}{7}$ Both values check in the original equation.

The solution set is $\left\{-\dfrac{4}{3}, \dfrac{2}{7}\right\}$.

Skill Practice Solve the equation.

9. $|3 - 2x| = |3x - 1|$

Example 7 — Solving an Equation Having Two Absolute Values

Solve the equation. $|x - 4| = |x + 8|$

Solution:

$|x - 4| = |x + 8|$

$x - 4 = x + 8$ or $x - 4 = -(x + 8)$ Rewrite as two equations, $x = y$ or $x = -y$.

$-4 = 8$ or $x - 4 = -x - 8$ Solve for x.
↑
contradiction
$2x - 4 = -8$
$2x = -4$
$x = -2$ $x = -2$ checks in the original equation.

> **FOR REVIEW**
> Recall that if an equation reduces to a contradiction such as $-4 = 8$, the equation has no solution. If an equation reduces to an identity such as $0 = 0$, the solution set is the set of real numbers.

The solution set is $\{-2\}$.

Skill Practice Solve the equation.

10. $|4t + 3| = |4t - 5|$

Answers

9. $\left\{\dfrac{4}{5}, -2\right\}$ 10. $\left\{\dfrac{1}{4}\right\}$

Section 17.3 Activity

A.1.
a. The expression $|x|$ represents the distance between the real number x and _____ on the number line.
b. Using this definition, solve the equation $|x| = 3$.
c. Solve the equation $|x| = 11$.
d. Solve the equation $|x| = 0$.
e. Explain why the equation $|x| = -2$ has no solution.

A.2.
a. The equation $|u| = 5$ is equivalent to $u = 5$ or $u = -5$. Write the equation $|x - 4| = 5$ as two equations joined by "or."
b. Solve $|x - 4| = 5$.

A.3.
a. What is the first step to solve the equation $4 = -6 + |3x + 1|$?
b. Solve the equation $4 = -6 + |3x + 1|$.

A.4.
a. Solve the equation $|5x - 6| + 8 = 6$.
b. Solve the equation $|5x - 6| + 8 = 8$.

A.5. Consider the expression $|x| = |y|$. If two quantities have the same absolute value, then the quantities are equal, or the quantities are opposites. For example: $|5| = |5|$ and $|5| = |-5|$. With this in mind, write the equation $|x| = |y|$ as two equations joined by "or."

A.6. Given the equation $|x - 5| = |2x - 1|$,
a. Write the equation as two equations joined by "or."
b. Solve the equation.

A.7. Solve the equation $|4x + 3| = |4x - 1|$.

Practice Exercises

Section 17.3

Prerequisite Review

For Exercises R.1–R.10, solve the equations.

R.1. a. $5x + 7 = 32$
b. $5x + 7 = -32$

R.2. a. $-4x - 3 = 17$
b. $-4x - 3 = -17$

R.3. a. $3w + 7 = 2w - 6$
b. $3w + 7 = -(2w - 6)$

R.4. a. $-4x + 5 = 2x - 1$
b. $-4x + 5 = -(2x - 1)$

R.5. $2(t - 3) + 4 = 2t - 3$

R.6. $3 - 4(p + 1) = -4p + 5$

R.7. $5x - 4 - x = 4(x - 1)$

R.8. $3(y + 1) - y = 2y + 3$

R.9. $\frac{1}{3}t - \frac{3}{4} = \frac{1}{6}t + \frac{1}{12}$

R.10. $\frac{2}{5}w + \frac{1}{3} = -\frac{1}{15}w - \frac{1}{5}$

Vocabulary and Key Concepts

1. **a.** An _____ value equation is an equation of the form $|x| = a$. If a is a positive real number then the solution set is _____.

   b. What is the first step to solve the absolute value equation $|x| + 5 = 7$?

   c. The absolute value equation $|x| = |y|$ implies that $x =$ _____ or $x =$ _____.

   d. The solution set to the equation $|x + 4| = -2$ is _____. The solution set to the equation $|x + 4| = 0$ is _____.

Concept 1: Solving Absolute Value Equations

For Exercises 2–6, solve the equation.

2. **a.** $|x| = 4$
   b. $|x| = -4$
   c. $|x| = 0$

3. **a.** $|t| = 5$
   b. $|t| = -5$
   c. $|t| = 0$

4. **a.** $|x + 3| = 2$
   b. $|x + 3| = -2$
   c. $|x + 3| = 0$

5. **a.** $|y - 4| = 3$
   b. $|y - 4| = -3$
   c. $|y - 4| = 0$

6. **a.** $|2m| = 12$
   b. $|2m| = -12$
   c. $|2m| = 0$

For Exercises 7–38, solve the equations. (See Examples 1–5.)

7. $|p| = 7$
8. $|q| = 10$
9. $|x| + 5 = 11$
10. $|x| - 3 = 20$

11. $|y| + 8 = 5$
12. $|x| + 12 = 6$
13. $|w| - 3 = -1$
14. $|z| - 14 = -10$

15. $|3q| = 0$
16. $|4p| = 0$
17. $|3x - 4| = 8$
18. $|4x + 1| = 6$

19. $5 = |2x - 4|$
20. $10 = |3x + 7|$
21. $\left|\dfrac{7z}{3} - \dfrac{1}{3}\right| + 3 = 6$
22. $\left|\dfrac{w}{2} + \dfrac{3}{2}\right| - 2 = 7$

23. $|0.2x - 3.5| = -5.6$
24. $|1.81 + 2x| = -2.2$
25. $1 = -4 + \left|2 - \dfrac{1}{4}w\right|$
26. $-12 = -6 - |6 - 2x|$

27. $10 = 4 + |2y + 1|$
28. $-1 = -|5x + 7|$
29. $-2|3b - 7| - 9 = -9$
30. $-3|5x + 1| + 4 = 4$

31. $-2|x + 3| = 5$
32. $-3|x - 5| = 7$
33. $0 = |6x - 9|$
34. $7 = |4k - 6| + 7$

35. $\left|-\dfrac{1}{5} - \dfrac{1}{2}k\right| = \dfrac{9}{5}$
36. $\left|-\dfrac{1}{6} - \dfrac{2}{9}h\right| = \dfrac{1}{2}$
37. $-3|2 - 6x| + 5 = -10$
38. $5|1 - 2x| - 7 = 3$

Concept 2: Solving Equations Containing Two Absolute Values

For Exercises 39–56, solve the absolute value equations. (See Examples 6–7.)

39. $|4x - 2| = |-8|$
40. $|3x + 5| = |-5|$
41. $|4w + 3| = |2w - 5|$

42. $|3y + 1| = |2y - 7|$
43. $|2y + 5| = |7 - 2y|$
44. $|9a + 5| = |9a - 1|$

45. $\left|\dfrac{4w - 1}{6}\right| = \left|\dfrac{2w}{3} + \dfrac{1}{4}\right|$
46. $\left|\dfrac{6p + 3}{8}\right| = \left|\dfrac{3}{4}p - 2\right|$
47. $|2h - 6| = |2h + 5|$

48. $|6n - 7| = |4 - 6n|$

49. $|3.5m - 1.2| = |8.5m + 6|$

50. $|11.2n + 9| = |7.2n - 2.1|$

51. $|4x - 3| = -|2x - 1|$

52. $-|3 - 6y| = |8 - 2y|$

53. $|8 - 7w| = |7w - 8|$

54. $|4 - 3z| = |3z - 4|$

55. $|x + 2| + |x - 4| = 0$

56. $|t + 6| + |t - 1| = 0$

Technology Connections

For Exercises 57–62, enter the left side of the equation as Y_1 and enter the right side of the equation as Y_2. Then use the *Intersect* feature to approximate the x values where the two graphs intersect (if they intersect).

57. $|4x - 3| = 5$

58. $|x - 4| = 3$

59. $|8x + 1| + 8 = 1$

60. $|3x - 2| + 4 = 2$

61. $|x - 3| = |x + 2|$

62. $|x + 4| = |x - 2|$

Expanding Your Skills

63. Write an absolute value equation whose solution is the set of real numbers 6 units from zero on the number line.

64. Write an absolute value equation whose solution is the set of real numbers $\frac{7}{2}$ units from zero on the number line.

65. Write an absolute value equation whose solution is the set of real numbers $\frac{4}{3}$ units from zero on the number line.

66. Write an absolute value equation whose solution is the set of real numbers 9 units from zero on the number line.

Absolute Value Inequalities

Section 17.4

1. Solving Absolute Value Inequalities by Definition

In this section, we will solve absolute value *inequalities*. An inequality in any of the forms $|x| < a$, $|x| \leq a$, $|x| > a$, or $|x| \geq a$ is called an **absolute value inequality**.

Recall that an absolute value represents distance from zero on the real number line. Consider the following absolute value equation and inequalities.

Concepts

1. Solving Absolute Value Inequalities by Definition
2. Solving Absolute Value Inequalities by the Test Point Method
3. Translating to an Absolute Value Expression

1. $|x| = 3$

   $x = 3$ or $x = -3$

Solution:

The set of all points 3 units from zero on the number line

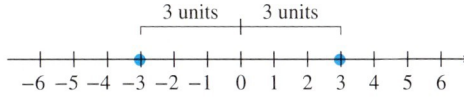

2. $|x| > 3$

   $x < -3$ or $x > 3$

Solution:

The set of all points more than 3 units from zero

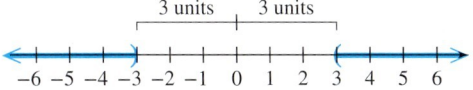

3. $|x| < 3$

   $-3 < x < 3$

Solution:

The set of all points less than 3 units from zero

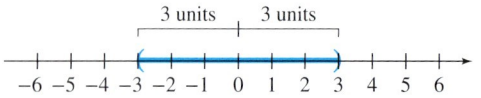

Solving Absolute Value Equations and Inequalities

Let a be a real number such that $a > 0$. Then

Equation/Inequality	Solution (Equivalent Form)	Graph
$\|x\| = a$	$x = -a$ or $x = a$	
$\|x\| > a$	$x < -a$ or $x > a$	
$\|x\| < a$	$-a < x < a$	

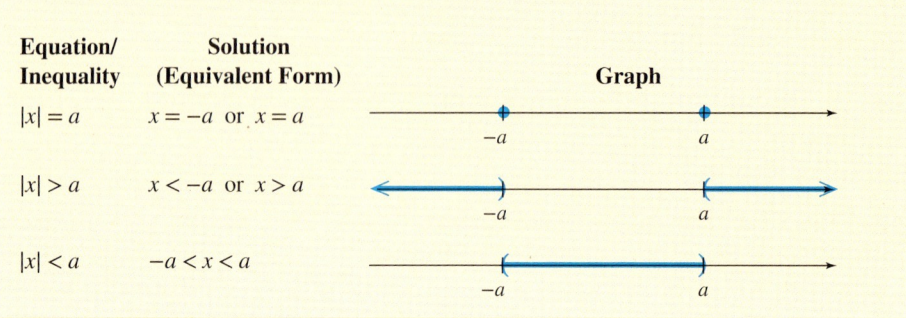

To solve an absolute value inequality, first isolate the absolute value and then rewrite the absolute value inequality in its equivalent form.

Example 1 Solving an Absolute Value Inequality

Solve the inequality. $|3w + 1| - 4 < 7$

Solution:

$|3w + 1| - 4 < 7$

$|3w + 1| < 11$ ← Isolate the absolute value first.

The inequality is in the form $|x| < a$, where $x = 3w + 1$.

$-11 < 3w + 1 < 11$ Rewrite in the equivalent form $-a < x < a$.

$-12 < 3w < 10$ Solve for w.

$-4 < w < \dfrac{10}{3}$

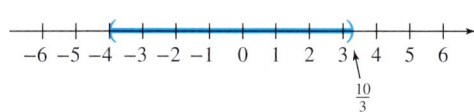

The solution is $\{w \mid -4 < w < \frac{10}{3}\}$, or equivalently in interval notation, $\left(-4, \frac{10}{3}\right)$.

Skill Practice Solve the inequality. Write the solution in interval notation.

1. $|2t + 5| + 2 \leq 11$

Answer

1. $[-7, 2]$

Section 17.4 Absolute Value Inequalities 1149

Example 2 — Solving an Absolute Value Inequality

Solve the inequality. $\quad 3 \leq 1 + \left|\frac{1}{2}t - 5\right|$

Solution:

$3 \leq 1 + \left|\frac{1}{2}t - 5\right|$

$1 + \left|\frac{1}{2}t - 5\right| \geq 3 \quad$ Write the inequality with the absolute value on the left.

$\left|\frac{1}{2}t - 5\right| \geq 2 \quad$ Isolate the absolute value.

The inequality is in the form $|x| \geq a$, where $x = \frac{1}{2}t - 5$.

$\frac{1}{2}t - 5 \leq -2 \quad$ or $\quad \frac{1}{2}t - 5 \geq 2 \quad$ Rewrite in the equivalent form $x \leq -a$ or $x \geq a$.

$\frac{1}{2}t \leq 3 \quad$ or $\quad \frac{1}{2}t \geq 7 \quad$ Solve the compound inequality.

$2\left(\frac{1}{2}t\right) \leq 2(3) \quad$ or $\quad 2\left(\frac{1}{2}t\right) \geq 2(7) \quad$ Clear fractions.

$t \leq 6 \quad$ or $\quad t \geq 14$

The solution is $\{t \mid t \leq 6 \text{ or } t \geq 14\}$, or equivalently in interval notation, $(-\infty, 6] \cup [14, \infty)$.

> **TIP:** It is generally easier to solve an absolute value inequality if the absolute value appears on the left-hand side of the inequality.

Skill Practice Solve the inequality. Write the solution in interval notation.

2. $5 < 1 + \left|\frac{1}{3}c - 1\right|$

By definition, the absolute value of a real number will always be nonnegative. Therefore, the absolute value of any expression will always be greater than a negative number. Similarly, an absolute value can never be less than a negative number. If a represents a positive real number, then

- The solution to the inequality $|x| > -a$ is all real numbers, $(-\infty, \infty)$.
- There is no solution to the inequality $|x| < -a$.

Example 3 — Solving Absolute Value Inequalities

Solve the inequalities.

a. $|3d - 5| + 7 < 4$ **b.** $|3d - 5| + 7 > 4$

Solution:

a. $|3d - 5| + 7 < 4 \quad$ Isolate the absolute value. An absolute value expression cannot be less than a negative number. Therefore, there is no solution.

$|3d - 5| < -3$

No solution, $\{\ \}$

Answer

2. $(-\infty, -9) \cup (15, \infty)$

b. $|3d - 5| + 7 > 4$

$|3d - 5| > -3$

Isolate the absolute value. The inequality is in the form $|x| > a$, where a is negative. An absolute value of any real number is greater than a negative number. Therefore, the solution is all real numbers.

All real numbers, $(-\infty, \infty)$

Skill Practice Solve the inequalities.

3. $|4p + 2| + 6 < 2$ **4.** $|4p + 2| + 6 > 2$

Example 4 **Solving Absolute Value Inequalities**

Solve the inequalities.

a. $|4x + 2| \geq 0$ **b.** $|4x + 2| > 0$ **c.** $|4x + 2| \leq 0$

Solution:

a. $|4x + 2| \geq 0$ ← The absolute value is already isolated.

The absolute value of any real number is nonnegative. Therefore, the solution is all real numbers, $(-\infty, \infty)$.

b. $|4x + 2| > 0$

An absolute value will be greater than zero at all points *except where it is equal to zero*. That is, the value(s) of x for which $|4x + 2| = 0$ must be excluded from the solution set.

$|4x + 2| = 0$

$4x + 2 = 0$ or $4x + 2 = -0$ The second equation is the same as the first.

$4x = -2$

$x = -\dfrac{1}{2}$ Therefore, exclude $x = -\dfrac{1}{2}$ from the solution.

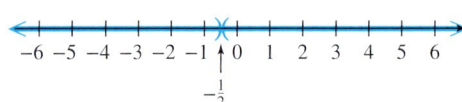

The solution is $\{x \mid x \neq -\tfrac{1}{2}\}$, or equivalently in interval notation, $(-\infty, -\tfrac{1}{2}) \cup (-\tfrac{1}{2}, \infty)$.

c. $|4x + 2| \leq 0$

An absolute value of a number cannot be less than zero. However, it can be *equal* to zero. Therefore, the only solutions to this inequality are the solutions to the related equation:

$|4x + 2| = 0$ From part (b), we see that the solution set is $\left\{-\dfrac{1}{2}\right\}$.

Skill Practice Solve the inequalities.

5. $|3x - 1| \geq 0$ **6.** $|3x - 1| > 0$ **7.** $|3x - 1| \leq 0$

Answers

3. { }
4. All real numbers; $(-\infty, \infty)$
5. $(-\infty, \infty)$
6. $\left\{x \mid x \neq \dfrac{1}{3}\right\}$ or $\left(-\infty, \dfrac{1}{3}\right) \cup \left(\dfrac{1}{3}, \infty\right)$
7. $\left\{\dfrac{1}{3}\right\}$

2. Solving Absolute Value Inequalities by the Test Point Method

In Examples 1 and 2, each absolute value inequality was converted to an equivalent compound inequality. However, sometimes students have difficulty setting up the appropriate compound inequality. To avoid this problem, you may want to use the test point method to solve absolute value inequalities.

Solving Inequalities by Using the Test Point Method

Step 1 Find the boundary points of the inequality. (Boundary points are the real solutions to the related equation and points where the inequality is undefined.)

Step 2 Plot the boundary points on the number line. This divides the number line into intervals.

Step 3 Select a test point from each interval and substitute it into the original inequality.
- If a test point makes the original inequality true, then that interval is part of the solution set.

Step 4 Test the boundary points in the original inequality.
- If the original inequality is strict ($<$ or $>$), do not include the boundary points in the solution set.
- If the original inequality is defined using $\leq$ or $\geq$, then include the boundary points that are defined within the inequality.

Note: Any boundary point that makes an expression within the inequality undefined must *always* be excluded from the solution set.

To demonstrate the use of the test point method, we will repeat the absolute value inequalities from Examples 1 and 2. Notice that regardless of the method used, the absolute value is always isolated *first* before any further action is taken.

Example 5 Solving an Absolute Value Inequality by the Test Point Method

Solve the inequality by using the test point method. $|3w + 1| - 4 < 7$

Solution:

$|3w + 1| - 4 < 7$

$|3w + 1| < 11$ ⟵ Isolate the absolute value.

$|3w + 1| = 11$ **Step 1:** Solve the related equation.

$3w + 1 = 11$ or $3w + 1 = -11$ Write as an equivalent system of two equations.

$3w = 10$ or $3w = -12$

$w = \dfrac{10}{3}$ or $w = -4$ These are the only boundary points.

Step 2: Plot the boundary points.

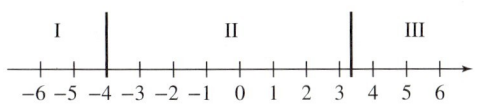

Step 3: Select a test point from each interval.

Test $w = -5$:	Test $w = 0$:	Test $w = 4$:
$\|3(-5)+1\|-4 \stackrel{?}{<} 7$	$\|3(0)+1\|-4 \stackrel{?}{<} 7$	$\|3(4)+1\|-4 \stackrel{?}{<} 7$
$\|-14\|-4 \stackrel{?}{<} 7$	$\|1\|-4 \stackrel{?}{<} 7$	$\|13\|-4 \stackrel{?}{<} 7$
$14-4 \stackrel{?}{<} 7$	$-3 \stackrel{?}{<} 7$ True	$13-4 \stackrel{?}{<} 7$
$10 \stackrel{?}{<} 7$ False		$9 \stackrel{?}{<} 7$ False

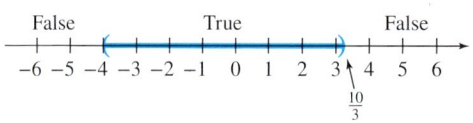

Step 4: Because the original inequality is a strict inequality, the boundary points (where equality occurs) are not included.

The solution is $\{w \mid -4 < w < \frac{10}{3}\}$, or equivalently in interval notation, $(-4, \frac{10}{3})$.

Skill Practice Solve the inequality.

8. $6 + |3t - 4| \leq 10$

Example 6 Solving an Absolute Value Inequality by the Test Point Method

Solve the inequality by using the test point method. $\quad 3 \leq 1 + \left|\frac{1}{2}t - 5\right|$

Solution:

$3 \leq 1 + \left|\frac{1}{2}t - 5\right|$

$1 + \left|\frac{1}{2}t - 5\right| \geq 3$ Write the inequality with the absolute value on the left.

$\left|\frac{1}{2}t - 5\right| \geq 2$ Isolate the absolute value.

$\left|\frac{1}{2}t - 5\right| = 2$ **Step 1:** Solve the related equation.

$\frac{1}{2}t - 5 = 2 \quad$ or $\quad \frac{1}{2}t - 5 = -2$ Write as an equivalent system of two equations.

$\frac{1}{2}t = 7 \quad$ or $\quad \frac{1}{2}t = 3$

$t = 14 \quad$ or $\quad t = 6$ These are the boundary points.

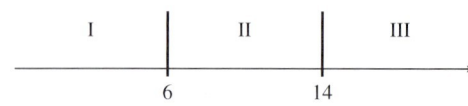

Step 2: Plot the boundary points.

Step 3: Select a test point from each interval.

Answer

8. $\left[0, \frac{8}{3}\right]$

Section 17.4 Absolute Value Inequalities   1153

Test $t = 0$:

$3 \stackrel{?}{\leq} 1 + \left|\frac{1}{2}(0) - 5\right|$

$3 \stackrel{?}{\leq} 1 + |0 - 5|$

$3 \stackrel{?}{\leq} 1 + |-5|$

$3 \stackrel{?}{\leq} 6$ True

Test $t = 10$:

$3 \stackrel{?}{\leq} 1 + \left|\frac{1}{2}(10) - 5\right|$

$3 \stackrel{?}{\leq} 1 + |5 - 5|$

$3 \stackrel{?}{\leq} 1 + |0|$

$3 \stackrel{?}{\leq} 1$ False

Test $t = 16$:

$3 \stackrel{?}{\leq} 1 + \left|\frac{1}{2}(16) - 5\right|$

$3 \stackrel{?}{\leq} 1 + |8 - 5|$

$3 \stackrel{?}{\leq} 1 + |3|$

$3 \stackrel{?}{\leq} 4$ True

Step 4: The original inequality uses the sign $\geq$. Therefore, the boundary points (where equality occurs) must be part of the solution set.

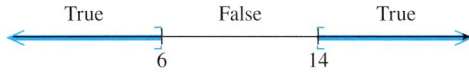

The solution is $\{t \mid t \leq 6 \text{ or } t \geq 14\}$, or equivalently in interval notation, $(-\infty, 6] \cup [14, \infty)$.

Skill Practice Solve the inequality.

9. $\left|\frac{1}{2}c + 4\right| + 1 > 6$

3. Translating to an Absolute Value Expression

Absolute value expressions can be used to describe distances. The distance between c and d is given by $|c - d|$. For example, the distance between -2 and 3 on the number line is $|(-2) - 3| = 5$ as expected.

Example 7 Expressing Distances With Absolute Value

Write an absolute value inequality to represent the following phrases.

  a. All real numbers x, whose distance from zero is greater than 5 units
  b. All real numbers x, whose distance from -7 is less than 3 units

Solution:

  a. All real numbers x, whose distance from zero is greater than 5 units

  $|x - 0| > 5$ or simply $|x| > 5$

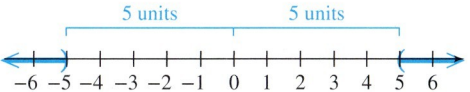

  b. All real numbers x, whose distance from -7 is less than 3 units

  $|x - (-7)| < 3$ or simply $|x + 7| < 3$

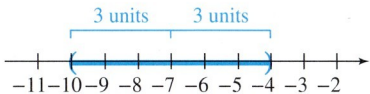

Skill Practice Write an absolute value inequality to represent the following phrases.
10. All real numbers whose distance from zero is greater than 10 units
11. All real numbers whose distance from 4 is less than 6 units

Answers
9. $(-\infty, -18) \cup (2, \infty)$
10. $|x| > 10$
11. $|x - 4| < 6$

Example 8 Expressing Measurement Error With Absolute Value

Latoya measured a certain compound on a scale in the chemistry lab at school. She measured 8 g of the compound, but the scale is only accurate to ±0.1 g. Write an absolute value inequality to express an interval for the true mass, x, of the compound she measured.

Solution:

Because the scale is only accurate to ±0.1 g, the true mass, x, of the compound may deviate by as much as 0.1 g above or below 8 g. This may be expressed as an absolute value inequality:

$|x - 8.0| \leq 0.1$ or equivalently $7.9 \leq x \leq 8.1$

Skill Practice

12. Vonzell molded a piece of metal in her machine shop. She measured the thickness at 12 mm. Her machine is accurate to ±0.05 mm. Write an absolute value inequality to express an interval for the true measurement of the thickness, t, of the metal.

Answer

12. $|t - 12| \leq 0.05$

Section 17.4 Activity

A.1.
  a. The inequality $|x| \geq 3$ is the set of real numbers that are 3 or more units from zero on the number line. Shade the region(s) on the number line that satisfy the inequality $|x| \geq 3$.
  b. Write the solution set to the inequality $|x| \geq 3$ in interval notation.
  c. Shade the region(s) on the number line that satisfy the inequality $|x| \leq 3$.
  d. Write the solution set to the inequality $|x| \leq 3$ in interval notation.

A.2. Given the inequality $|2x + 5| - 1 < 7$,
  a. What is the first step to solve the inequality?
  b. Write the inequality as an equivalent compound inequality without absolute value bars.
  c. Solve the inequality and write the solution set in interval notation.

A.3. Given the inequality $3 - |x - 9| < -7$,
  a. What is the first step to solve the inequality?
  b. Write the inequality as an equivalent compound inequality without absolute value bars.
  c. Solve the inequality and write the solution set in interval notation.

A.4. Solve each inequality by inspection and explain your answer.

   a. $|3x + 4| < -5$

   b. $|3x + 4| \leq -5$

   c. $|3x + 4| > -5$

   d. $|3x + 4| \geq -5$

A.5. Solve each inequality by inspection and explain your answer.

   a. $|2x - 6| < 0$

   b. $|2x - 6| \leq 0$

   c. $|2x - 6| \geq 0$

   d. $|2x - 6| > 0$

Practice Exercises   Section 17.4

Prerequisite Review

For Exercises R.1–R.4, solve the compound inequality. Write the solution set in interval notation.

R.1. $2x + 3 < -5$ or $2x + 3 > 5$

R.2. $5 - x \leq -1$ or $5 - x \geq 1$

R.3. $3 + y \geq -2$ and $3 + y \leq 2$

R.4. $4x + 1 > -9$ and $4x + 1 < 9$

For Exercises R.5–R.10, solve the equation.

R.5. $|3x - 4| + 1 = 9$

R.6. $|3 + 2x| - 8 = 11$

R.7. $|6x + 7| + 9 = 4$

R.8. $|-4x + 9| + 5 = 3$

R.9. $|-5x + 3| + 2 = 2$

R.10. $|-2x - 7| + 3 = 3$

Vocabulary and Key Concepts

1. **a.** If a is a positive real number, then the inequality $|x| < a$ is equivalent to _____ $< x <$ _____.

   b. If a is a positive real number, then the inequality $|x| > a$ is equivalent to $x <$ _____ or x _____ a.

   c. The solution set to the inequality $|x + 2| < -6$ is _____, whereas the solution set to the inequality $|x + 2| > -6$ is _____.

2. The solution set to the inequality $|x + 4| \leq 0$ (includes/excludes) -4, whereas the solution set to the inequality $|x + 4| < 0$ (includes/excludes) -4.

Concepts 1 and 2: Solving Absolute Value Inequalities

For Exercises 3–14, solve the equations and inequalities. For each inequality, graph the solution set and express the solution in interval notation. (See Examples 1–6.)

3. a. $|x| = 5$
   b. $|x| > 5$
   c. $|x| < 5$

4. a. $|a| = 4$
   b. $|a| > 4$
   c. $|a| < 4$

5. a. $|x - 3| = 7$
   b. $|x - 3| > 7$
   c. $|x - 3| < 7$

6. a. $|w + 2| = 6$
   b. $|w + 2| > 6$
   c. $|w + 2| < 6$

7. a. $|p| = -2$
   b. $|p| > -2$
   c. $|p| < -2$

8. a. $|x| = -14$
   b. $|x| > -14$
   c. $|x| < -14$

9. a. $|y + 1| = -6$
   b. $|y + 1| > -6$
   c. $|y + 1| < -6$

10. a. $|z - 4| = -3$
    b. $|z - 4| > -3$
    c. $|z - 4| < -3$

11. a. $|x| = 0$
    b. $|x| > 0$
    c. $|x| < 0$

12. a. $|p + 3| = 0$
    b. $|p + 3| > 0$
    c. $|p + 3| < 0$

13. a. $|k - 7| = 0$
    b. $|k - 7| > 0$
    c. $|k - 7| < 0$

14. a. $|2x + 4| + 3 = 2$
    b. $|2x + 4| + 3 > 2$
    c. $|2x + 4| + 3 < 2$

For Exercises 15–44, solve the absolute value inequality. Graph the solution set and write the solution in interval notation. (See Examples 1–6.)

15. $|x| > 6$

16. $|x| \leq 6$

17. $|t| \leq 3$

18. $|p| > 3$

19. $|y + 2| \geq 0$

20. $0 \leq |7n + 2|$

21. $5 \leq |2x - 1|$

22. $|x - 2| \geq 7$

23. $|k - 7| < -3$

24. $|h + 2| < -9$

25. $\left|\dfrac{w - 2}{3}\right| - 3 \leq 1$

26. $\left|\dfrac{x + 3}{2}\right| - 2 \geq 4$

27. $12 \leq |9 - 4y| - 2$

28. $5 > |2m - 7| + 4$

29. $4 > -1 + \left|\dfrac{2x + 1}{4}\right|$

30. $9 \geq 2 + \left|\dfrac{x-4}{5}\right|$

31. $8 < |4 - 3x| + 12$

32. $-16 < |5x - 1| - 1$

33. $5 - |2m + 1| > 5$

34. $3 - |5x + 3| > 3$

35. $|p + 5| \leq 0$

36. $|y + 1| - 4 \leq -4$

37. $|z - 6| + 5 > 5$

38. $|2c - 1| - 4 > -4$

39. $5|2y - 6| + 3 \geq 13$

40. $7|y + 1| - 3 \geq 11$

41. $-3|6 - t| + 1 > -5$

42. $-4|8 - x| + 2 > -14$

43. $|0.02x + 0.06| - 0.1 < 0.05$

44. $|0.05x - 0.04| - 0.01 < 0.11$

Concept 3: Translating to an Absolute Value Expression

For Exercises 45–48, write an absolute value inequality equivalent to the expression given. **(See Example 7.)**

45. All real numbers whose distance from 0 is greater than 7

46. All real numbers whose distance from −3 is less than 4

47. All real numbers whose distance from 2 is at most 13

48. All real numbers whose distance from 0 is at least 6

49. A 32-oz jug of orange juice may not contain exactly 32 oz of juice. The possibility of measurement error exists when the jug is filled in the factory. If the maximum measurement error is ±0.05 oz, write an absolute value inequality representing the range of volumes, x, in which the orange juice jug may be filled. **(See Example 8.)**

50. The length of a board is measured to be 32.3 in. The maximum measurement error is ±0.2 in. Write an absolute value inequality that represents the range for the length of the board, x.

51. A bag of potato chips states that its weight is $6\frac{3}{4}$ oz. The maximum measurement error is $\pm\frac{1}{8}$ oz. Write an absolute value inequality that represents the range for the weight, x, of the bag of chips.

52. A $\frac{7}{8}$-in. bolt varies in length by at most $\pm\frac{1}{16}$ in. Write an absolute value inequality that represents the range for the length, x, of the bolt.

53. The width, w, of a bolt is supposed to be 2 cm but may have a 0.01-cm margin of error. Solve $|w - 2| \leq 0.01$, and interpret the solution to the inequality in the context of this problem.

Andrew Bret Wallis/BananaStock/Getty Images

54. In a political poll, the front-runner was projected to receive 53% of the votes with a margin of error of 3%. Solve $|p - 0.53| \leq 0.03$ and interpret the solution in the context of this problem.

Technology Connections

To solve an absolute value inequality by using a graphing calculator, let Y_1 equal the left side of the inequality and let Y_2 equal the right side of the inequality. Graph both Y_1 and Y_2 on a standard viewing window and use an *Intersect* feature to approximate the intersection of the graphs. To solve $Y_1 > Y_2$, determine all x values where the graph of Y_1 is above the graph of Y_2. To solve $Y_1 < Y_2$, determine all x values where the graph of Y_1 is below the graph of Y_2.

For Exercises 55–64, solve the inequalities using a graphing calculator.

55. $|x + 2| > 4$

56. $|3 - x| > 6$

57. $\left|\dfrac{x+1}{3}\right| < 2$

58. $\left|\dfrac{x-1}{4}\right| < 1$

59. $|x - 5| < -3$

60. $|x + 2| < -2$

61. $|2x + 5| > -4$

62. $|1 - 2x| > -4$

63. $|6x + 1| \leq 0$

64. $|3x - 4| \leq 0$

Problem Recognition Exercises

Equations and Inequalities

For Exercises 1–24, identify the category for each equation or inequality (choose from the list here). Then solve the equation or inequality. Express the solution in interval notation where appropriate.

- Linear
- Quadratic
- Polynomial (degree greater than 2)
- Rational
- Absolute value

1. $z^2 + 10z + 9 > 0$

2. $5a - 2 = 6(a + 4)$

3. $\dfrac{x-4}{2x+4} \geq 1$

4. $4x^2 - 7x = 2$

5. $|3x - 1| + 4 < 6$

6. $2t^2 - t + 1 < 0$

7. $\dfrac{1}{2}p - \dfrac{2}{3} < \dfrac{1}{6}p - 4$

8. $p^2 + 3p \leq 4$

9. $3y^2 + y - 2 \geq 0$

10. $\dfrac{x+6}{x+4} = 7$

11. $3(2x - 4) \geq 1 - (x - 3)$

12. $|6x + 5| + 3 = 2$

13. $(x - 3)(2x + 1)(x + 5) \geq 0$

14. $-x^2 - x - 3 \leq 0$

15. $\dfrac{-6}{y-2} < 2$

16. $x^2 - 5 = 4x$

17. $\left|\dfrac{x}{4} + 2\right| + 6 > 6$

18. $x^2 - 36 < 0$

19. $3y^3 + 5y^2 - 12y - 20 = 0$

20. $|8 - 2x| = 16$

21. $5b - 10 - 3b = 3(b - 2) - 4$

22. $\dfrac{x}{x-5} \geq 0$

23. $|6x + 1| = |5 + 6x|$

24. $4x(x - 5)^2(2x + 3) < 0$

Linear Inequalities and Systems of Linear Inequalities in Two Variables

Section 17.5

Concepts
1. Graphing Linear Inequalities in Two Variables
2. Systems of Linear Inequalities in Two Variables
3. Graphing a Feasible Region

1. Graphing Linear Inequalities in Two Variables

A **linear inequality in two variables** x and y is an inequality that can be written in one of the following forms: $Ax + By < C$, $Ax + By > C$, $Ax + By \leq C$, or $Ax + By \geq C$, provided A and B are not both zero.

A solution to a linear inequality in two variables is an ordered pair that makes the inequality true. For example, solutions to the inequality $x + y < 6$ are ordered pairs (x, y) such that the sum of the x- and y-coordinates is less than 6. This inequality has an infinite number of solutions, and therefore it is convenient to express the solution set as a graph.

To graph a linear inequality in two variables, we will follow these steps.

Graphing a Linear Inequality in Two Variables

Step 1 Write the inequality with the y variable isolated, if possible.

Step 2 Graph the related equation. Draw a dashed line if the inequality is strict, $<$ or $>$. Otherwise, draw a solid line.

Step 3 Shade above or below the line as follows:
- Shade *above* the line if the inequality is of the form $y > mx + b$ or $y \geq mx + b$.
- Shade *below* the line if the inequality is of the form $y < mx + b$ or $y \leq mx + b$.

Note: A dashed line indicates that the line is *not* included in the solution set. A solid line implies that the line *is* included in the solution set.

This process is demonstrated in Example 1.

Example 1   Graphing a Linear Inequality in Two Variables

Graph the solution set. $-3x + y \leq 1$

Solution:

$-3x + y \leq 1$

$y \leq 3x + 1$ Solve for y.

Next graph the line defined by the related equation $y = 3x + 1$.

Because the inequality is of the form $y \leq mx + b$, the solution to the inequality is the region *below* the line $y = 3x + 1$. So to indicate the solution set, shade the region on and below the line. See Figure 17-6.

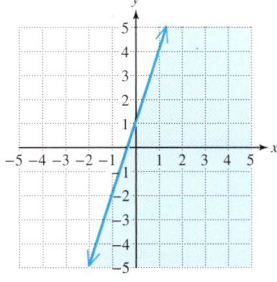

Figure 17-6

Skill Practice Graph the solution set.

1. $2x + y \geq -4$

Answer

1.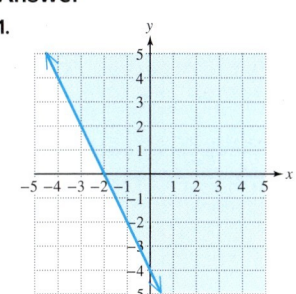

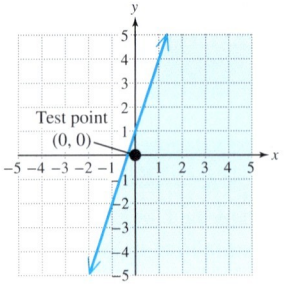

Figure 17-7

After graphing the solution to a linear inequality, we can verify that we have shaded the correct side of the line by using test points. In Example 1, we can pick an arbitrary ordered pair within the shaded region. Then substitute the x- and y-coordinates in the original inequality. If the result is a true statement, then that ordered pair is a solution to the inequality and suggests that other points from the same region are also solutions.

For example, the point $(0, 0)$ lies within the shaded region (Figure 17-7).

$-3x + y \leq 1$ Substitute $(0, 0)$ in the original inequality.

$-3(0) + (0) \stackrel{?}{\leq} 1$

$0 + 0 \stackrel{?}{\leq} 1$ ✓ True   The point $(0, 0)$ from the shaded region is a solution.

In Example 2, we will graph the solution set to a strict inequality. A strict inequality does not include equality and therefore, is expressed with the symbols $<$ or $>$. In such a case, the boundary line will be drawn as a dashed line. This indicates that the boundary itself is *not* part of the solution set.

Example 2   Graphing a Linear Inequality in Two Variables

Graph the solution set. $-4y < 5x$

Solution:

$-4y < 5x$

$\dfrac{-4y}{-4} > \dfrac{5x}{-4}$ Solve for y. Recall that when we divide both sides by a negative number, reverse the inequality sign.

$y > -\dfrac{5}{4}x$

Graph the line defined by the related equation, $y = -\dfrac{5}{4}x$. The boundary line is drawn as a dashed line because the inequality is strict. Also note that the line passes through the origin.

Because the inequality is of the form $y > mx + b$, the solution to the inequality is the region *above* the line. See Figure 17-8.

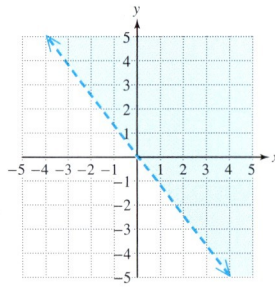

Figure 17-8

FOR REVIEW

Recall that to graph a linear equation in two variables, we can make a table of points and graph the line through the points. Alternatively, we can write the equation in slope-intercept form. Then graph the line using the y-intercept and slope.

Skill Practice Graph the solution set.

2. $-3y < x$

In Example 2, we cannot use the origin as a test point, because the point $(0, 0)$ is on the boundary line. Be sure to select a test point strictly within the shaded region. In this case, we choose $(2, 1)$. See Figure 17-9.

$-4y < 5x$ The test point $(2, 1)$ is indeed a solution to the original inequality.

$-4(1) \stackrel{?}{<} 5(2)$

$-4 \stackrel{?}{<} 10$ ✓ True

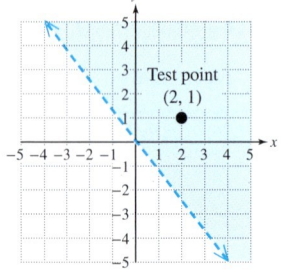

Figure 17-9

Answer

2.

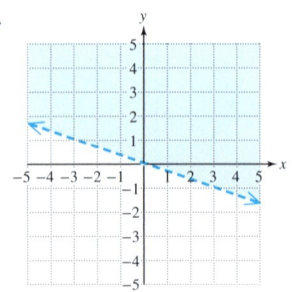

In Example 3, we encounter a situation in which we cannot solve for the *y* variable.

Example 3 Graphing a Linear Inequality in Two Variables

Graph the solution set. $4x \geq -12$

Solution:

$4x \geq -12$
$x \geq -3$

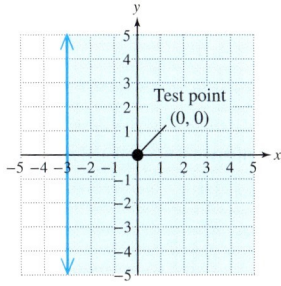

Figure 17-10

In this inequality, there is no *y* variable. However, we can simplify the inequality by solving for *x*.

Graph the related equation $x = -3$. This is a vertical line. The boundary is drawn as a solid line because the inequality is not strict, $\geq$.

To shade the appropriate region, refer to the inequality $x \geq -3$. The points for which *x* is greater than -3 are to the right of $x = -3$. Therefore, shade the region to the *right* of the line (Figure 17-10).

Selecting a test point such as (0, 0) from the shaded region indicates that we have shaded the correct side of the line.

$4x \geq -12$ Substitute $x = 0$.
$4(0) \stackrel{?}{\geq} -12$ ✓   True

FOR REVIEW

Recall that a vertical line is represented by an equation of the form $x = k$, such as $x = -3$. A horizontal line is represented by an equation of the form $y = k$, such as $y = 5$.

Skill Practice Graph the solution set.

3. $-2x \geq 2$

2. Systems of Linear Inequalities in Two Variables

Some applications require us to find the solutions to a system of linear inequalities.

Example 4 Graphing a System of Linear Inequalities

Graph the solution set. $y > \frac{1}{2}x + 1$
$x + y < 1$

Solution:

Solve each inequality for *y*.

First inequality

$y > \frac{1}{2}x + 1$

The inequality is of the form $y > mx + b$. Shade *above* the boundary line. See Figure 17-11.

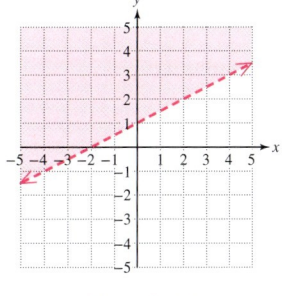

Figure 17-11

Answer

3.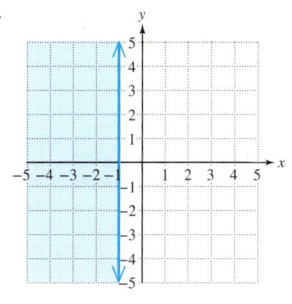

Second inequality

$x + y < 1$

$y < -x + 1$

The inequality is of the form $y < mx + b$. Shade *below* the boundary line. See Figure 17-12.

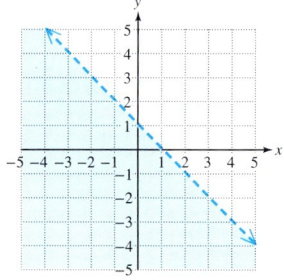

Figure 17-12

The region bounded by the inequalities is the region above the line $y = \frac{1}{2}x + 1$ and below the line $y = -x + 1$. This is the intersection or "overlap" of the two shaded regions (shown in purple in Figure 17-13).

The intersection is the solution set to the system of inequalities. See Figure 17-14.

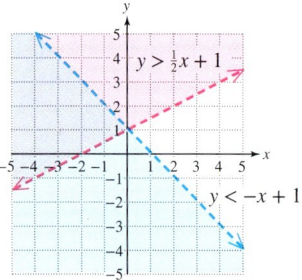

Figure 17-13

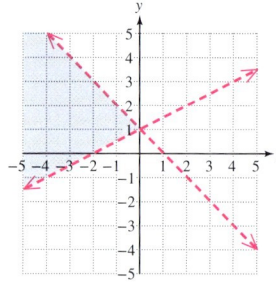

Figure 17-14

Skill Practice Graph the solution set.

4. $x - 3y > 3$
   $y < -2x + 4$

Example 5 — Graphing a System of Linear Inequalities

Graph the solution set. $3y \leq 6$
$y - x \leq 0$

Solution:

First inequality

$3y \leq 6$

$y \leq 2$

The graph of $y \leq 2$ is the region on and below the horizontal line $y = 2$. See Figure 17-15.

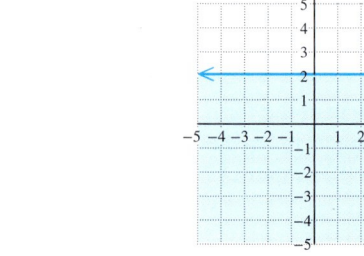

Figure 17-15

Answer

4.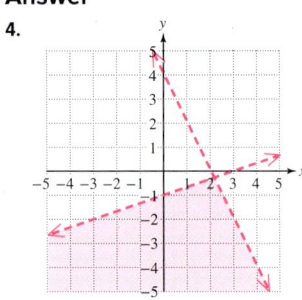

Second inequality

$y - x \leq 0$

$y \leq x$

The inequality $y \leq x$ is of the form $y \leq mx + b$. Graph a solid line and shade the region below the line. See Figure 17-16.

The solution to the system of inequalities is the intersection of the shaded regions. Notice that the portions of the lines not bounding the solution are dashed. See Figure 17-17.

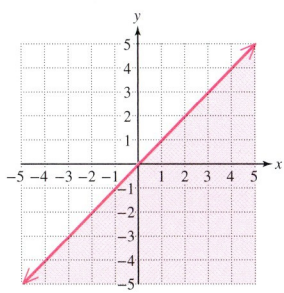

Figure 17-16

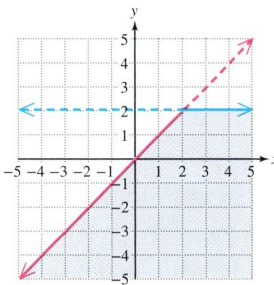

Figure 17-17

Skill Practice Graph the solution set.

5. $2y \leq 4$
   $y \leq x + 1$

Example 6 Graphing a System of Linear Inequalities

Describe the region of the plane defined by the system of inequalities.

$$x \leq 0$$
$$y \geq 0$$

Solution:

$x \leq 0$ $x \leq 0$ for points on the y-axis and in the second and third quadrants.

$y \geq 0$ $y \geq 0$ for points on the x-axis and in the first and second quadrants.

The intersection of these regions is the set of points in the second quadrant (with the boundaries included).

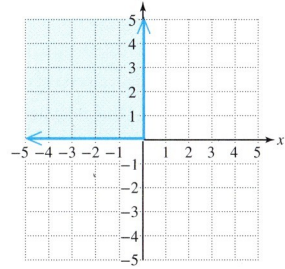

Skill Practice Graph the region defined by the system of inequalities.

6. $x \leq 0$
   $y \leq 0$

3. Graphing a Feasible Region

When two variables are related under certain constraints, a system of linear inequalities can be used to show a region of feasible values for the variables.

The feasible region represents the ordered pairs that are true for each inequality in the system.

Answers

5.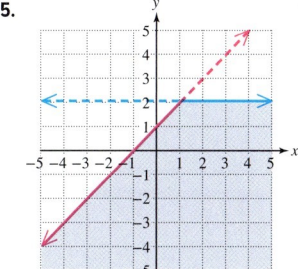

6.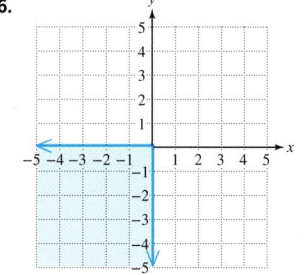

Example 7 — Graphing a Feasible Region

Susan has two tests on Friday: one in chemistry and one in psychology. Because the two classes meet in consecutive hours, she has no study time between tests. Susan estimates that she has a maximum of 12 hr of study time before the tests, and she must divide her time between chemistry and psychology.

Let x represent the number of hours Susan spends studying chemistry.

Let y represent the number of hours Susan spends studying psychology.

a. Find a set of inequalities to describe the constraints on Susan's study time.

b. Graph the constraints to find the feasible region defining Susan's study time.

Solution:

a. Because Susan cannot study chemistry or psychology for a negative period of time, we have $x \geq 0$ and $y \geq 0$. Furthermore, her total time studying cannot exceed 12 hr: $x + y \leq 12$.

A system of inequalities that defines the constraints on Susan's study time is:

$$x \geq 0$$
$$y \geq 0$$
$$x + y \leq 12$$

b. The first two conditions $x \geq 0$ and $y \geq 0$ represent the set of points in the first quadrant. The third condition $x + y \leq 12$ represents the set of points below and including the line $x + y = 12$ (Figure 17-18).

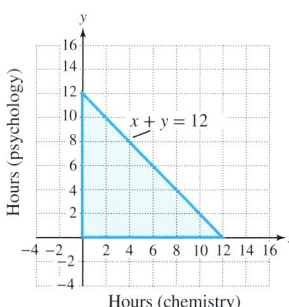

Figure 17-18

Discussion:

1. Refer to the feasible region drawn in Example 7(b). Is the ordered pair (8, 5) part of the feasible region?

    No. The ordered pair (8, 5) indicates that Susan spent 8 hr studying chemistry and 5 hr studying psychology. This is a total of 13 hr, which exceeds the constraint that Susan only had 12 hr to study. The point (8, 5) lies outside the feasible region, above the line $x + y = 12$ (Figure 17-19).

2. Is the ordered pair (7, 3) part of the feasible region?

    Yes. The ordered pair (7, 3) indicates that Susan spent 7 hr studying chemistry and 3 hr studying psychology.

    This point lies within the feasible region and satisfies all three constraints.

    $x \geq 0 \longrightarrow 7 \geq 0$ True
    $y \geq 0 \longrightarrow 3 \geq 0$ True
    $x + y \leq 12 \longrightarrow (7) + (3) \leq 12$ True

    Notice that the ordered pair (7, 3) corresponds to a point where Susan is not making full use of the 12 hr of study time.

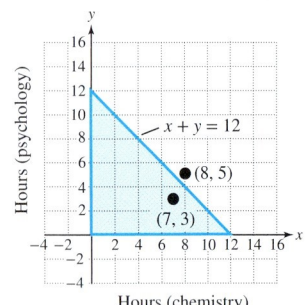

Figure 17-19

3. Suppose there was one additional constraint imposed on Susan's study time. She knows she needs to spend at least twice as much time studying chemistry as she does studying psychology. Graph the feasible region with this additional constraint.

Because the time studying chemistry must be at least twice the time studying psychology, we have $x \geq 2y$.

This inequality may also be written as $y \leq \frac{1}{2}x$.

Figure 17-20 shows the first quadrant with the constraint $y \leq \frac{1}{2}x$.

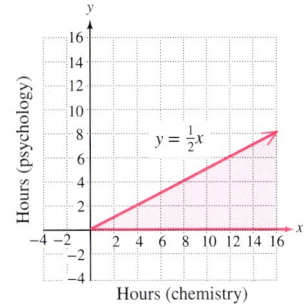

Figure 17-20

4. At what point in the feasible region is Susan making the most efficient use of her time for both classes?

First and foremost, Susan must make use of *all* 12 hr. This occurs for points along the line $x + y = 12$. Susan will also want to study for both classes with approximately twice as much time devoted to chemistry. Therefore, Susan will be deriving the maximum benefit at the point of intersection of the line $x + y = 12$ and the line $y = \frac{1}{2}x$.

Using the substitution method, replace $y = \frac{1}{2}x$ into the equation $x + y = 12$.

$x + \frac{1}{2}x = 12$

$2x + x = 24$ Clear fractions.

$3x = 24$

$x = 8$ Solve for x.

$y = \frac{(8)}{2}$ To solve for y, substitute $x = 8$ into the equation $y = \frac{1}{2}x$.

$y = 4$

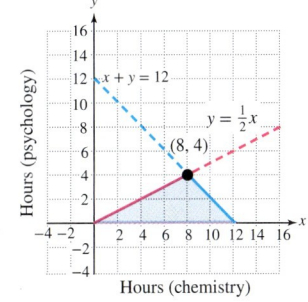

Therefore, Susan should spend 8 hr studying chemistry and 4 hr studying psychology.

Skill Practice

7. A local pet rescue group has a total of 30 cages that can be used to hold cats and dogs. Let x represent the number of cages used for cats, and let y represent the number used for dogs.
   a. Write a set of inequalities to express the fact that the number of cat and dog cages cannot be negative.
   b. Write an inequality to describe the constraint on the total number of cages for cats and dogs.
   c. Graph the system of inequalities to find the feasible region describing the available cages.

Answer

7. a. $x \geq 0$ and $y \geq 0$
   b. $x + y \leq 30$
   c.

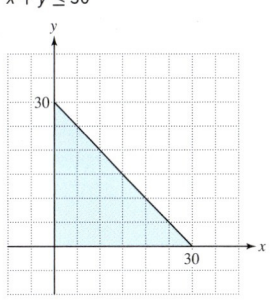

Section 17.5 Activity

A.1. Determine if the ordered pair is a solution to the inequality $-3x + 5y < 15$.
  a. $(2, 4)$ **b.** $(-4, 2)$ **c.** $(0, 3)$

A.2. a. Graph the line $-4x + 2y = 8$. **b.** Graph the inequality $-4x + 2y \leq 8$.

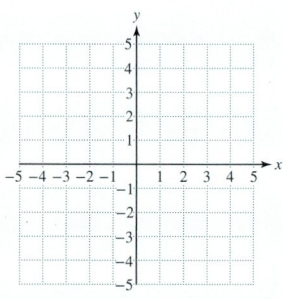

  c. Explain how the graph of part (b) would change for the inequality $-4x + 2y \geq 8$.
  d. Explain how the graph of part (b) would change for the inequality $-4x + 2y < 8$.

A.3. a. Graph the inequality $x < 0$. **b.** On the same coordinate system graph the inequality $y > 0$.

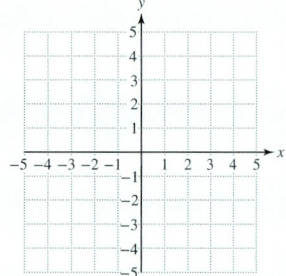
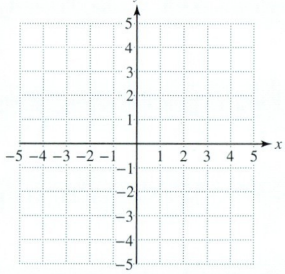

  c. Describe the area of overlap between the two inequalities.

  d. Now graph the inequality $x - y > -3$ on the same coordinate system. Note that the region of overlap of the three inequalities is the solution set to the system:

$$x < 0$$
$$y > 0$$
$$x - y > -3$$

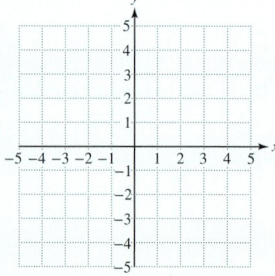

Practice Exercises

Section 17.5

Prerequisite Review

For Exercises R.1–R.4,

a. Find the *x*- and *y*-intercepts.
b. Graph the equation.

R.1. $4x - 3y = 12$

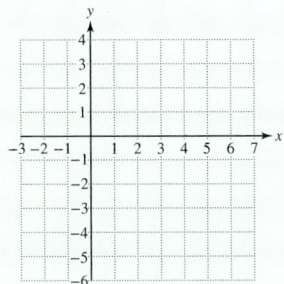

R.2. $3x = 4y$

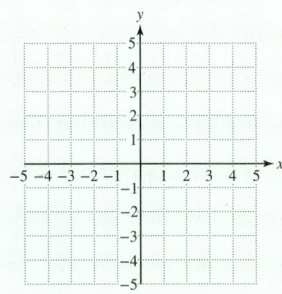

R.3. $4 - y = 2$

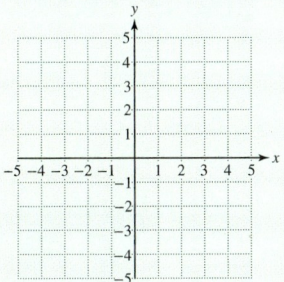

R.4. $-3x = 9$

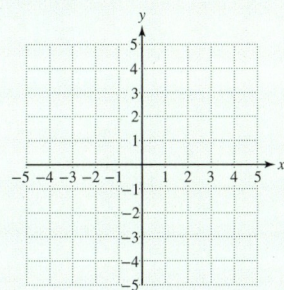

For Exercises R.5–R.6, determine whether the given value is a solution to the inequality.

R.5. $5x - 3 < 7$
  a. $x = 2$ b. $x = -2$ c. $x = 6$

R.6. $-4x - 8 \geq 8$
  a. $x = 4$ b. $x = -4$ c. $x = 0$

Vocabulary and Key Concepts

1. a. An inequality that can be written in the form $Ax + By > C$ is called a _____ inequality in two variables.

   b. Given the inequality $3x - 4y < 6$, the boundary line $3x - 4y = 6$ (is/is not) included in the solution set. However, given $3x - 4y \leq 6$, then the boundary line (is/is not) included in the solution set.

2. a. Given the inequality $5x + y > 5$, the boundary line $5x + y = 5$ should be drawn as a (solid/dashed) line to indicate that the line (is/is not) part of the solution set.

   b. Given the inequality $y \geq -x + 4$, the boundary line $y = -x + 4$ should be drawn as a (solid/dashed) line to indicate that the line (is/is not) part of the solution set.

3. The graph of $y < x + 2$ is the set of points below the line $y = x + 2$. How would the graph of $y > x + 2$ be different?

4. The graph of $y \geq 3x + 1$ is the set of points on and above the line $y = 3x + 1$. How would the graph of $y > 3x + 1$ be different?

5. The line $y = -2x$ passes through the origin. Does the graph of $y < -2x$ include the origin?

Concept 1: Graphing Linear Inequalities in Two Variables

For Exercises 6–9, determine if the given point is a solution to the inequality.

6. $2x - y > 8$
    a. $(3, -5)$
    b. $(-1, -10)$
    c. $(4, -2)$
    d. $(0, 0)$

7. $3y + x < 5$
    a. $(-1, 7)$
    b. $(5, 0)$
    c. $(0, 0)$
    d. $(2, -3)$

8. $y \leq -2$
    a. $(5, -3)$
    b. $(-4, -2)$
    c. $(0, 0)$
    d. $(3, 2)$

9. $x \geq 5$
    a. $(4, 5)$
    b. $(5, -1)$
    c. $(8, 8)$
    d. $(0, 0)$

10. When should you use a dashed line to graph the solution to a linear inequality?

For Exercises 11–16, decide which inequality symbol should be used ($<, >, \geq, \leq$) by looking at the graph.

11.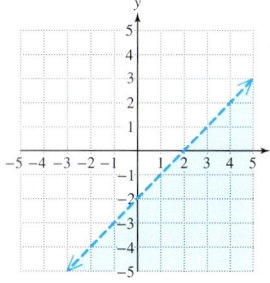
$x - y$ _____ 2

12.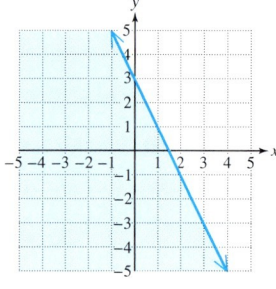
y _____ $-2x + 3$

13.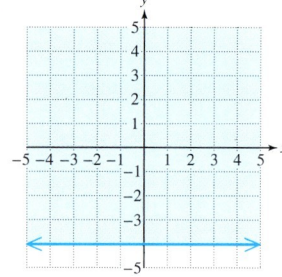
y _____ -4

14.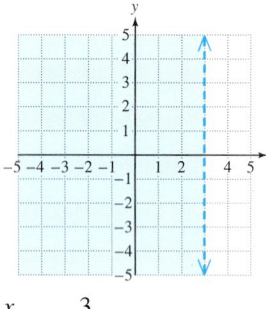
x _____ 3

15.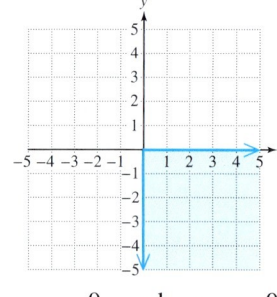
x _____ 0 and y _____ 0

16.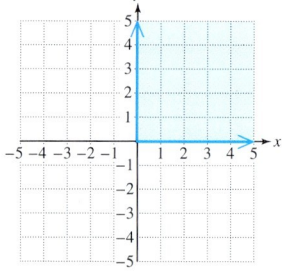
x _____ 0 and y _____ 0

For Exercises 17–40, graph the solution set. (See Examples 1–3.)

17. $x - 2y > 4$

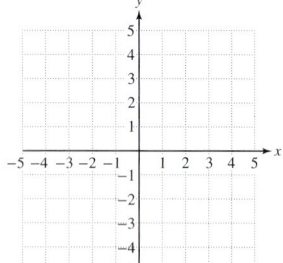

18. $x - 3y \geq 6$

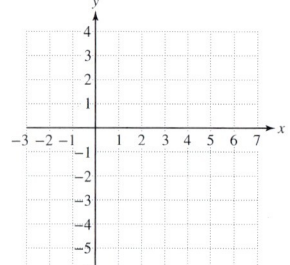

19. $5x - 2y < 10$

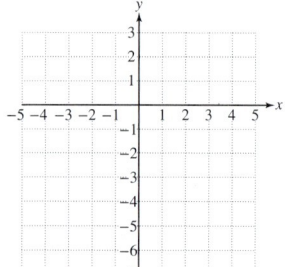

20. $x - 3y < 8$

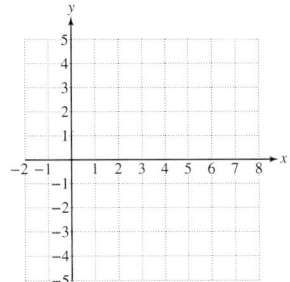

21. $2x \leq -6y + 12$

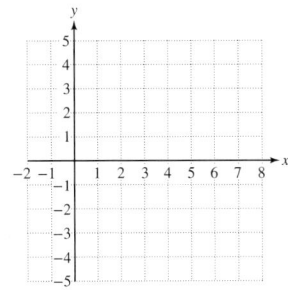

22. $4x < 3y + 12$

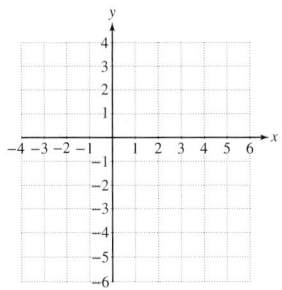

23. $2y \leq 4x$

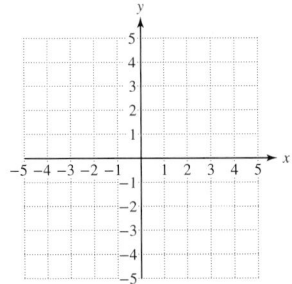

24. $-6x < 2y$

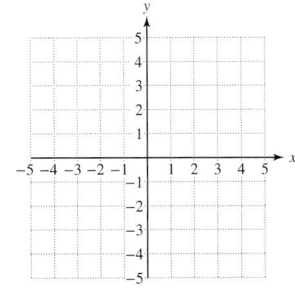

25. $y \geq -2$

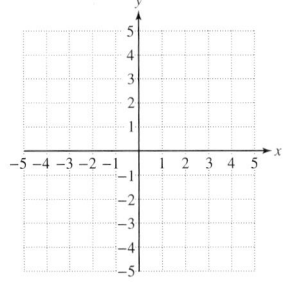

26. $y \geq 5$

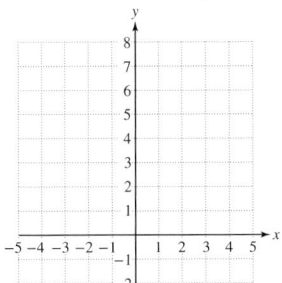

27. $4x < 5$

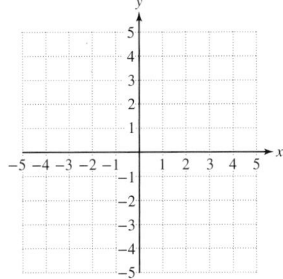

28. $x + 6 < 7$

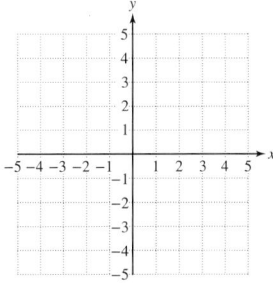

29. $y \geq \frac{2}{5}x - 4$

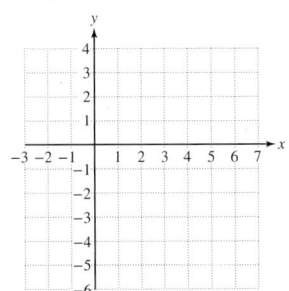

30. $y \geq -\frac{5}{2}x - 4$

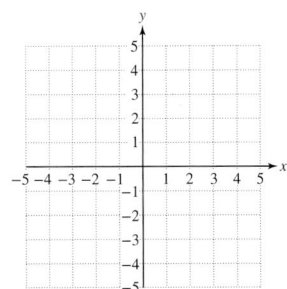

31. $y \leq \frac{1}{3}x + 6$

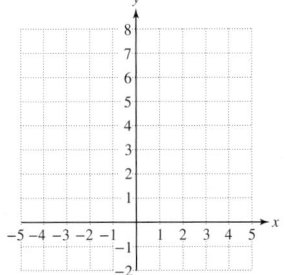

32. $y \leq -\dfrac{1}{4}x + 2$

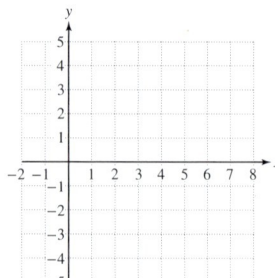

33. $y - 5x > 0$

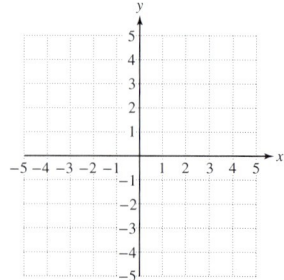

34. $y - \dfrac{1}{2}x > 0$

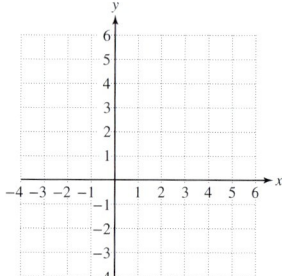

35. $\dfrac{x}{5} + \dfrac{y}{4} < 1$

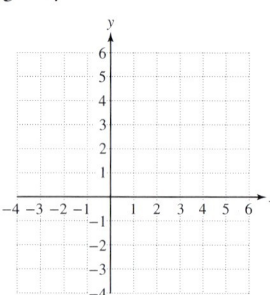

36. $x + \dfrac{y}{2} \geq 2$

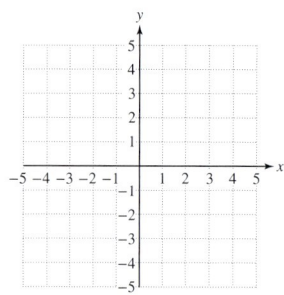

37. $0.1x + 0.2y \leq 0.6$

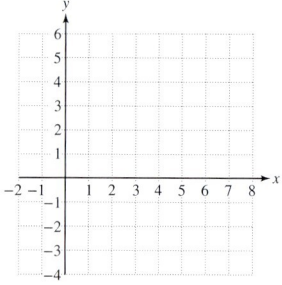

38. $0.3x - 0.2y < 0.6$

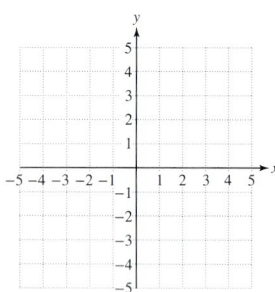

39. $x \leq -\dfrac{2}{3}y$

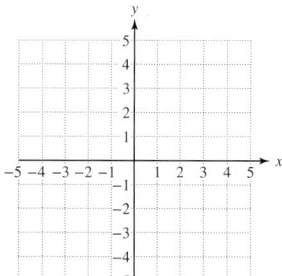

40. $x \geq -\dfrac{5}{4}y$

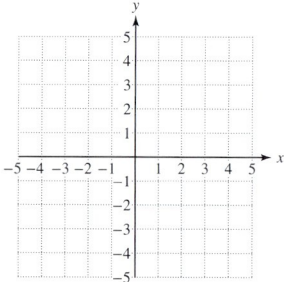

Concept 2: Systems of Linear Inequalities in Two Variables

For Exercises 41–55, graph the solution set. **(See Examples 4–6.)**

41. $y < 4$
$y > -x + 2$

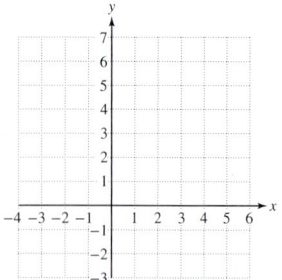

42. $y < 3$
$x + 2y < 6$

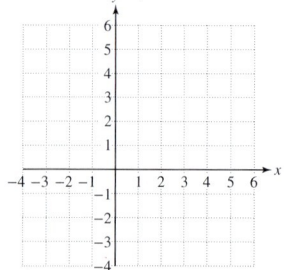

43. $2x + y \leq 5$
$x \geq 3$

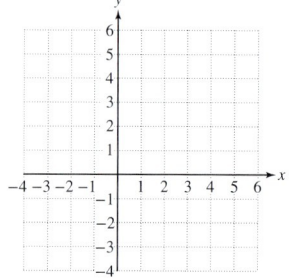

44. $x + 3y \geq 3$
$x \leq -2$

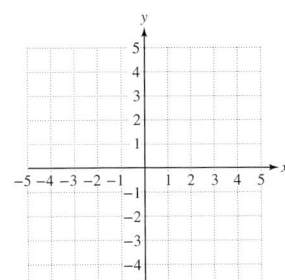

45. $x + y < 3$
$4x + y < 6$

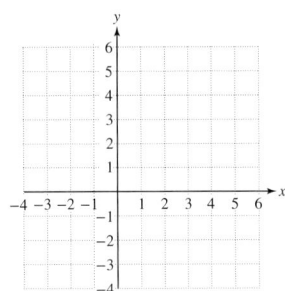

46. $x + y < 4$
$3x + y < 9$

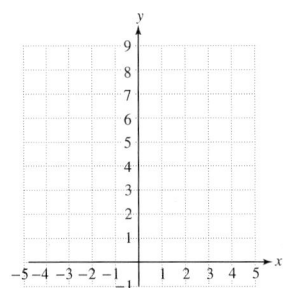

47. $2x - y \leq 2$
$2x + 3y \geq 6$

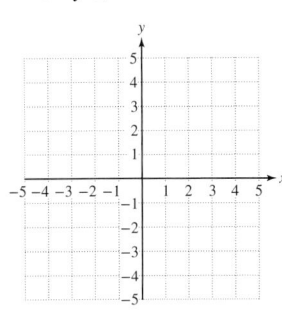

48. $3x + 2y \geq 4$
$x - y \leq 3$

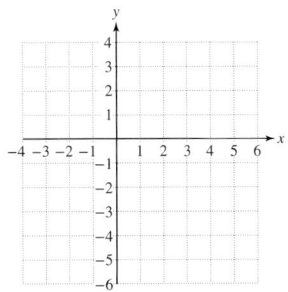

49. $x > 4$
$y < 2$

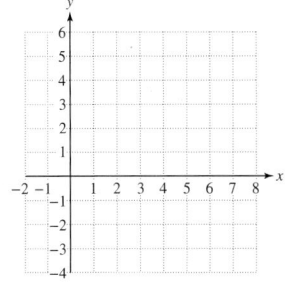

50. $x < 3$
$y > 4$

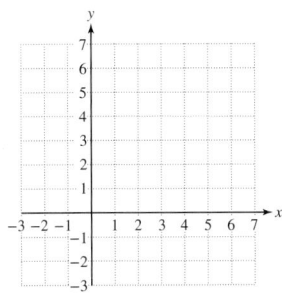

51. $x \leq -2$
$y \leq 0$

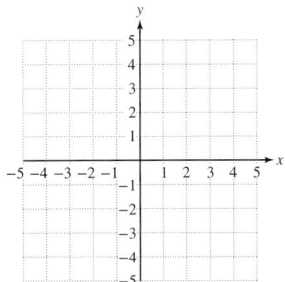

52. $x \geq 0$
$y \geq -3$

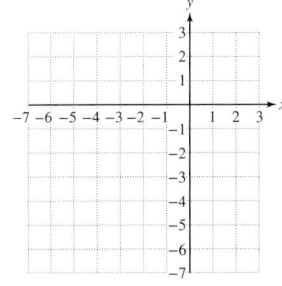

53. $x > 0$
$x + y < 6$

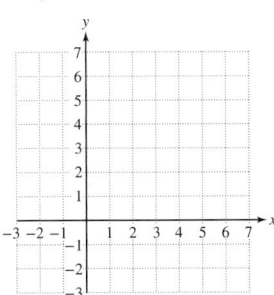

54. $x < 0$
$x + y < 2$

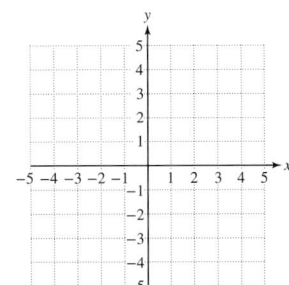

55. $y \geq 0$
$-2x + y \leq -4$

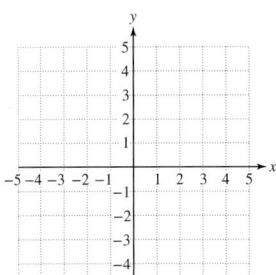

Concept 3: Graphing a Feasible Region

For Exercises 56–59, graph the feasible regions.

56. $x + y \leq 3$ and
$x \geq 0, y \geq 0$

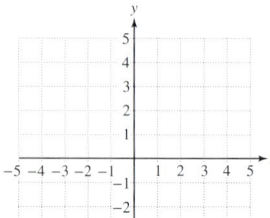

57. $x - y \leq 2$ and
$x \geq 0, y \geq 0$

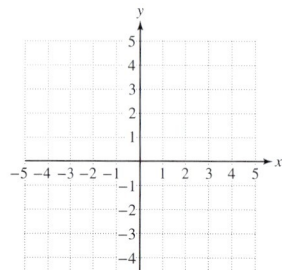

58. $x \geq 0, y \geq 0$
$x + y \leq 8$ and
$3x + 5y \leq 30$

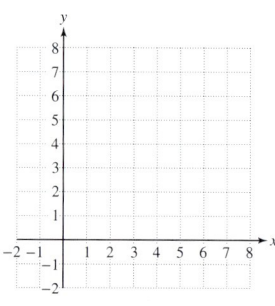

59. $x \geq 0, y \geq 0$
$x + y \leq 5$ and
$x + 2y \leq 6$

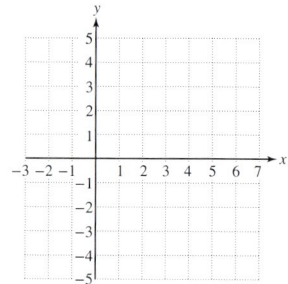

60. Suppose Sue has 50 ft of fencing with which she can build a rectangular dog run. Let x represent the length of the dog run and let y represent the width.

  a. Write an inequality representing the fact that the total perimeter of the dog run is at most 50 ft.

  b. Sketch part of the solution set for this inequality that represents all possible values for the length and width of the dog run. (*Hint:* Note that both the length and the width must be positive.)

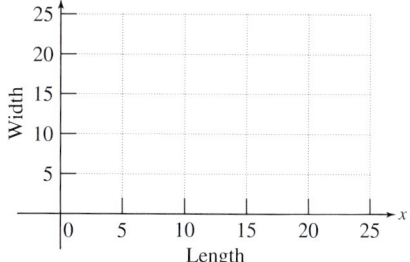

61. Suppose Rick has 40 ft of fencing with which he can build a rectangular garden. Let x represent the length of the garden and let y represent the width.

  a. Write an inequality representing the fact that the total perimeter of the garden is at most 40 ft.

  b. Sketch part of the solution set for this inequality that represents all possible values for the length and width of the garden. (*Hint:* Note that both the length and the width must be positive.)

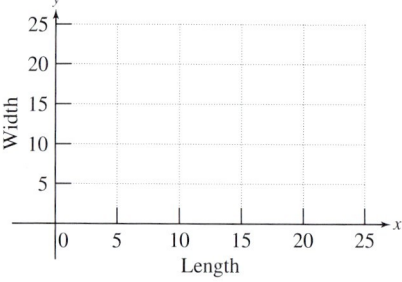

62. A manufacturer produces two models of desks. Model A requires 4 hr to stain and finish and 3 hr to assemble. Model B requires 3 hr to stain and finish and 1 hr to assemble. The total amount of time available for staining and finishing is 24 hr and for assembling is 12 hr. Let x represent the number of Model A desks produced, and let y represent the number of Model B desks produced.

  a. Write two inequalities that express the fact that the number of desks to be produced cannot be negative.

  b. Write an inequality in terms of the number of Model A and Model B desks that can be produced if the total time for staining and finishing is at most 24 hr.

  c. Write an inequality in terms of the number of Model A and Model B desks that can be produced if the total time for assembly is no more than 12 hr.

  d. Graph the feasible region formed by graphing the preceding inequalities.

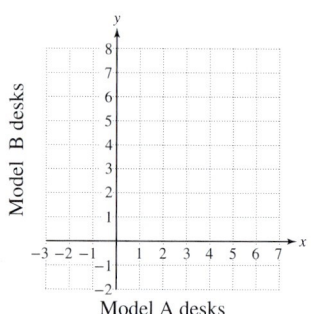

e. Is the point (3, 1) in the feasible region? What does the point (3, 1) represent in the context of this problem?

f. Is the point (5, 4) in the feasible region? What does the point (5, 4) represent in the context of this problem?

63. In scheduling two drivers for delivering pizza, James needs to have at least 65 hr scheduled this week. His two drivers, Karen and Todd, are not allowed to get overtime, so each one can work at most 40 hr. Let x represent the number of hours that Karen can be scheduled, and let y represent the number of hours Todd can be scheduled. (See Example 7.)

a. Write two inequalities that express the fact that Karen and Todd cannot work a negative number of hours.

b. Write two inequalities that express the fact that neither Karen nor Todd is allowed overtime (i.e., each driver can have at most 40 hr).

c. Write an inequality that expresses the fact that the total number of hours from both Karen and Todd needs to be at least 65 hr.

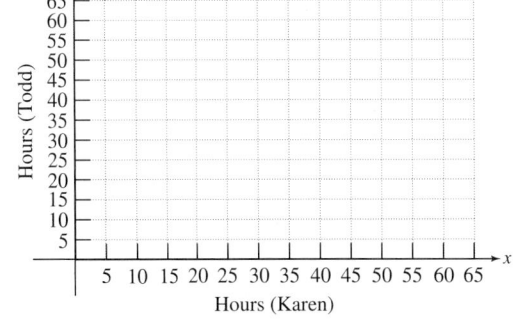

d. Graph the feasible region formed by graphing the inequalities.

e. Is the point (35, 40) in the feasible region? What does the point (35, 40) represent in the context of this problem?

f. Is the point (20, 40) in the feasible region? What does the point (20, 40) represent in the context of this problem?

Chapter 17 Review Exercises

Section 17.1

1. Explain the difference between the union and intersection of two sets. You may use the sets C and D in the following diagram to provide an example.

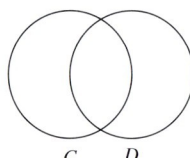

Let $X = \{x \mid x \geq -10\}$, $Y = \{x \mid x < 1\}$, $Z = \{x \mid x > -1\}$, and $W = \{x \mid x \leq -3\}$. For Exercises 2–7, find the intersection or union of the sets X, Y, Z, and W. Write the answers in interval notation.

2. $X \cap Y$
3. $X \cup Y$
4. $Y \cup Z$
5. $Y \cap Z$
6. $Z \cup W$
7. $Z \cap W$

For Exercises 8–13, solve the compound inequalities. Write the solutions in interval notation.

8. $4m > -11$ and $4m - 3 \leq 13$
9. $4n - 7 < 1$ and $7 + 3n \geq -8$
10. $\frac{2}{3}t - 3 \leq 1$ or $\frac{3}{4}t - 2 > 7$
11. $2(3x + 1) < -10$ or $3(2x - 4) \geq 0$
12. $2 \geq -(b - 2) - 5b \geq -6$
13. $-4 \leq \frac{1}{2}(x - 1) < -\frac{3}{2}$

14. Normal levels of total cholesterol vary according to age. For adults between 25 and 40 years old, the normal range is generally accepted to be between 140 and 225 mg/dL (milligrams per deciliter), inclusive. Let x represent cholesterol level.

    a. Write an inequality representing the normal range for total cholesterol for adults between 25 and 40 years old.

    b. Write a compound inequality representing abnormal ranges for total cholesterol for adults between 25 and 40 years old.

15. Normal levels of total cholesterol vary according to age. For adults younger than 25 years old, the normal range is generally accepted to be between 125 and 200 mg/dL, inclusive. Let x represent cholesterol level.

    a. Write an inequality representing the normal range for total cholesterol for adults younger than 25 years old.

    b. Write a compound inequality representing abnormal ranges for total cholesterol for adults younger than 25 years old.

Section 17.2

16. Solve the equation and inequalities. How do your answers to parts (a), (b), and (c) relate to the graph of $g(x) = x^2 - 4$?

    a. $x^2 - 4 = 0$
    b. $x^2 - 4 < 0$
    c. $x^2 - 4 > 0$

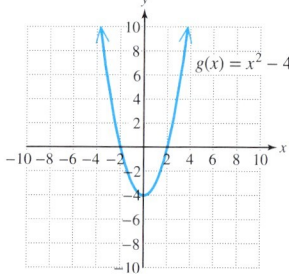

17. Solve the equation and inequalities. How do your answers to parts (a), (b), (c), and (d) relate to the graph of $k(x) = \dfrac{4x}{(x - 2)}$?

    a. $\dfrac{4x}{x - 2} = 0$
    b. For which values is $k(x)$ undefined?
    c. $\dfrac{4x}{x - 2} \geq 0$
    d. $\dfrac{4x}{x - 2} \leq 0$

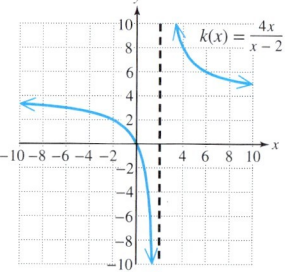

For Exercises 18–29, solve the inequalities. Write the answers in interval notation.

18. $w^2 - 4w - 12 < 0$
19. $t^2 + 6t + 9 \geq 0$
20. $\dfrac{12}{x+2} \leq 6$
21. $\dfrac{8}{p-1} \geq -4$
22. $3y(y-5)(y+2) > 0$
23. $-3c(c+2)(2c-5) < 0$
24. $-x^2 - 4x \geq 3$
25. $y^2 + 4y > 5$
26. $\dfrac{w+1}{w-3} > 1$
27. $\dfrac{2a}{a+3} \leq 2$
28. $t^2 + 10t + 25 \leq 0$
29. $-x^2 - 4x < 4$

Section 17.3

For Exercises 30–41, solve the absolute value equations.

30. $|x| = 10$
31. $|x| = 17$
32. $|8.7 - 2x| = 6.1$
33. $|5.25 - 5x| = 7.45$
34. $16 = |x+2| + 9$
35. $5 = |x-2| + 4$
36. $|4x - 1| + 6 = 4$
37. $|3x - 1| + 7 = 3$
38. $\left|\dfrac{7x-3}{5}\right| + 4 = 4$
39. $\left|\dfrac{4x+5}{-2}\right| - 3 = -3$
40. $|3x - 5| = |2x + 1|$
41. $|8x + 9| = |8x - 1|$

42. Which absolute value expression represents the distance between 3 and −2 on the number line?

$|3 - (-2)|$ $|-2 - 3|$

Section 17.4

43. Write the compound inequality $x < -5$ or $x > 5$ as an absolute value inequality.

44. Write the compound inequality $-4 < x < 4$ as an absolute value inequality.

For Exercises 45–46, write an absolute value inequality that represents the solution set graphed here.

45. (graph from −6 to 6, inclusive endpoints)

46. (graph with arrows outside $-\tfrac{2}{3}$ and $\tfrac{2}{3}$)

For Exercises 47–60, solve the absolute value inequalities. Graph each solution set and write the solution in interval notation.

47. $|x + 6| \geq 8$
48. $|x + 8| \leq 3$
49. $2|7x - 1| + 4 > 4$
50. $4|5x + 1| - 3 > -3$
51. $|3x + 4| - 6 \leq -4$
52. $|5x - 3| + 3 \leq 6$
53. $\left|\dfrac{x}{2} - 6\right| < 5$
54. $\left|\dfrac{x}{3} + 2\right| < 2$
55. $|4 - 2x| + 8 \geq 8$
56. $|9 + 3x| + 1 \geq 1$
57. $-2|5.2x - 7.8| < 13$
58. $-|2.5x + 15| < 7$
59. $|3x - 8| < -1$
60. $|x + 5| < -4$

61. State one possible situation in which an absolute value inequality will have no solution.

62. State one possible situation in which an absolute value inequality will have a solution of all real numbers.

63. The Neilsen ratings estimated that the percent, p, of the television viewing audience watching a popular reality show was 20% with a 3% margin of error. Solve the inequality $|p - 0.20| \leq 0.03$ and interpret the answer in the context of this problem.

64. The length, L, of a screw is supposed to be $3\frac{3}{8}$ in. Due to variation in the production equipment, there is a $\frac{1}{4}$-in. margin of error. Solve the inequality $|L - 3\frac{3}{8}| \leq \frac{1}{4}$ and interpret the answer in the context of this problem.

Section 17.5

For Exercises 65–72, solve the inequalities by graphing.

65. $2x > -y + 5$

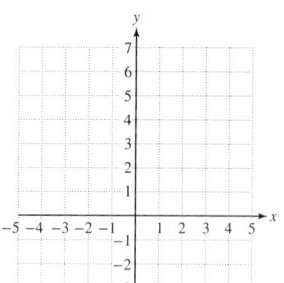

66. $2x \leq -8 - 3y$

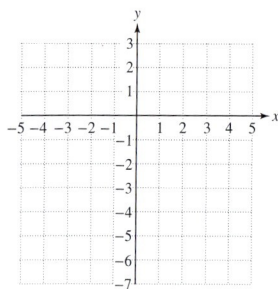

67. $y \geq -\frac{2}{3}x + 3$

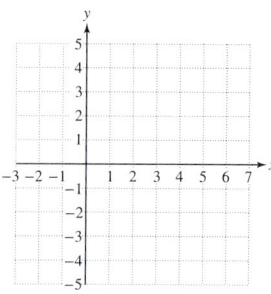

68. $y > \frac{3}{4}x - 2$

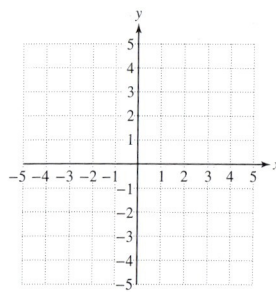

69. $x > -3$

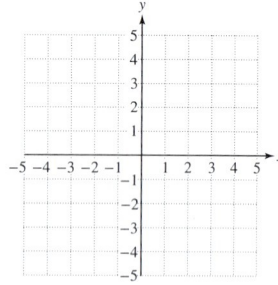

70. $x \leq 2$

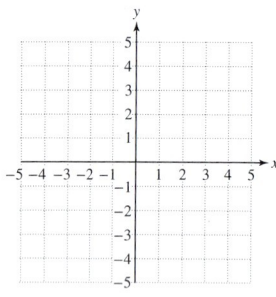

71. $x \geq \frac{1}{2}y$

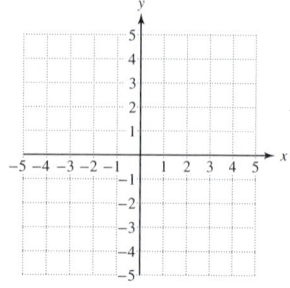

72. $x < \frac{2}{5}y$

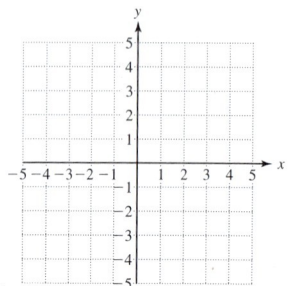

For Exercises 73–76 graph the system of inequalities.

73. $2x - y > -2$
$2x - y \leq 2$

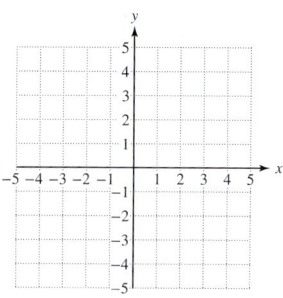

74. $3x + y < 6$
$-3x + y < -2$

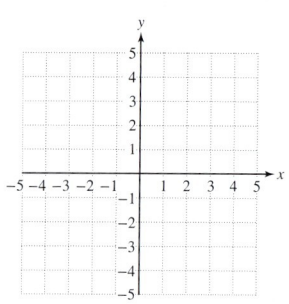

75. $y \geq x$
$y \leq -x$

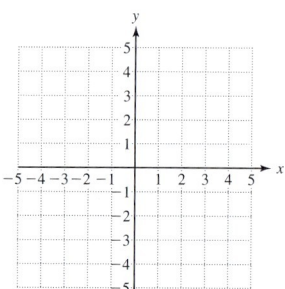

76. $x \geq 0$
$y \geq 0$
$y \leq -\frac{2}{3}x + 4$

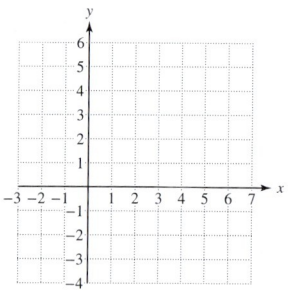

77. Suppose a farmer has 100 acres of land on which to grow oranges and grapefruit. Furthermore, because of demand from his customers, he wants to plant at least 4 times as many acres of orange trees as grapefruit trees.

Adalberto Rios Lanz/Sexto Sol/Photodisc/
Getty Images

Let x represent the number of acres of orange trees.

Let y represent the number of acres of grapefruit trees.

a. Write two inequalities that express the fact that the farmer cannot use a negative number of acres to plant orange and grapefruit trees.

b. Write an inequality that expresses the fact that the total number of acres used for growing orange and grapefruit trees is at most 100.

c. Write an inequality that expresses the fact that the farmer wants to plant at least 4 times as many orange trees as grapefruit trees.

d. Sketch the inequalities in parts (a)–(c) to find the feasible region for the farmer's distribution of orange and grapefruit trees.

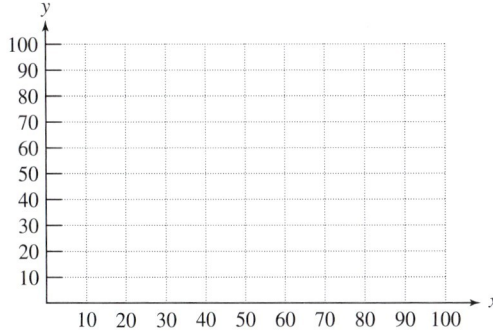

Radicals and Complex Numbers

18

CHAPTER OUTLINE

18.1 Definition of an *n*th Root   1180
18.2 Rational Exponents   1193
18.3 Simplifying Radical Expressions   1201
18.4 Addition and Subtraction of Radicals   1209
18.5 Multiplication of Radicals   1216
18.6 Division of Radicals and Rationalization   1225
   Problem Recognition Exercises: Operations on Radicals   1236
18.7 Solving Radical Equations   1237
18.8 Complex Numbers   1248
   Chapter 18 Review Exercises   1260

Mathematics in Architecture

The area A of a triangle can be found if the length of one side (the base b) and the corresponding height h of the triangle are known. $A = \frac{1}{2}bh$

However, if the base and height are not known, but the lengths of the three sides a, b, and c are known, then we can use Heron's formula to find the area of the triangle. Let s represent the semi-perimeter of the triangle.

That is, $s = \frac{1}{2}(a + b + c)$.

Then Heron's formula gives the area A as

$$A = \sqrt{s(s-a)(s-b)(s-c)}$$

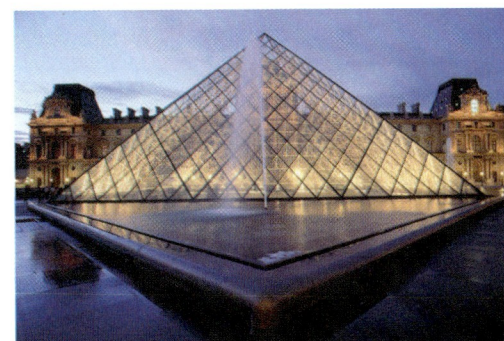

I. M. Pei (American architect, born 1917 in China)/ Photov.com/Pixtal/age fotostock

The Louvre pyramid, designed by architect I.M. Pei, is a glass structure that serves as the entrance to the Louvre Museum in Paris, France. Each triangular face is made of glass with sides of lengths 108.5 ft, 108.5 ft, and 116 ft. To determine the surface area of glass for each face, we can apply Heron's formula.

$$s = \frac{1}{2}(a + b + c) = \frac{1}{2}(108.5 + 108.5 + 116) = 166.5$$

$$A = \sqrt{s(s-a)(s-b)(s-c)} = \sqrt{166.5(166.5 - 108.5)(166.5 - 108.5)(166.5 - 116)}$$

$$\approx 5318 \text{ ft}^2 \quad \text{Each face is approximately 5318 ft}^2 \text{ of glass.}$$

Heron's formula is one application of radicals. In this chapter, we will simplify radical expressions and solve equations containing radicals.

Section 18.1 Definition of an *n*th Root

Concepts

1. Definition of a Square Root
2. Definition of an *n*th Root
3. Roots of Variable Expressions
4. Pythagorean Theorem
5. Radical Functions

1. Definition of a Square Root

The reverse operation to squaring a number is to find its square roots. For example, finding a square root of 36 is equivalent to asking, "when squared, what number equals 36?"

One obvious answer to this question is 6 because $(6)^2 = 36$, but -6 will also work, because $(-6)^2 = 36$.

> **Definition of a Square Root**
>
> b is a **square root** of a if $b^2 = a$.

Example 1   Identifying Square Roots

Identify the square roots of the real numbers.

    a. 25   **b.** 49   **c.** 0   **d.** -9

Solution:

a. 5 is a square root of 25 because $(5)^2 = 25$.

    -5 is a square root of 25 because $(-5)^2 = 25$.

b. 7 is a square root of 49 because $(7)^2 = 49$.

    -7 is a square root of 49 because $(-7)^2 = 49$.

c. 0 is a square root of 0 because $(0)^2 = 0$.

d. There are no real numbers that when squared will equal a negative number; therefore, there are no real-valued square roots of -9.

Skill Practice   Identify the square roots of the real numbers.

    1. 64   **2.** 16   **3.** 1   **4.** -100

> **TIP:**
> - All positive real numbers have two real-valued square roots (one positive and one negative).
> - Zero has only one square root, which is zero itself.
> - A negative number has no real-valued square roots.

Recall that the positive square root of a real number can be denoted with a **radical sign** $\sqrt{}$.

> **Positive and Negative Square Roots**
>
> Let a represent a positive real number. Then
>
> 1. $\sqrt{a}$ is the *positive* square root of a. The positive square root is also called the **principal square root**.
> 2. $-\sqrt{a}$ is the *negative* square root of a.
> 3. $\sqrt{0} = 0$

Answers

1. -8 and 8
2. -4 and 4
3. -1 and 1
4. No real-valued square roots

Section 18.1 Definition of an nth Root

Example 2 Simplifying Square Roots

Simplify the square roots.

a. $\sqrt{36}$ b. $\sqrt{\dfrac{4}{9}}$ c. $\sqrt{0.04}$

Solution:

a. $\sqrt{36}$ denotes the positive square root of 36.

$\sqrt{36} = 6$ because $(6)^2 = 36$

b. $\sqrt{\dfrac{4}{9}}$ denotes the positive square root of $\dfrac{4}{9}$.

$\sqrt{\dfrac{4}{9}} = \dfrac{2}{3}$ because $\left(\dfrac{2}{3}\right)^2 = \dfrac{4}{9}$

c. $\sqrt{0.04}$ denotes the positive square root of 0.04.

$\sqrt{0.04} = 0.2$ because $(0.2)^2 = 0.04$

Skill Practice Simplify the square roots.

5. $\sqrt{81}$ 6. $\sqrt{\dfrac{36}{49}}$ 7. $\sqrt{0.09}$

The numbers 36, $\dfrac{4}{9}$, and 0.04 are **perfect squares** because their square roots are rational numbers.

Radicals that cannot be simplified to rational numbers are irrational numbers. Recall that an irrational number cannot be written as a terminating or repeating decimal. For example, the symbol $\sqrt{13}$ is used to represent the *exact* value of the square root of 13. The symbol $\sqrt{42}$ is used to represent the *exact* value of the square root of 42. These values can be approximated by a rational number by using a calculator.

$$\sqrt{13} \approx 3.605551275 \qquad \sqrt{42} \approx 6.480740698$$

TIP: Before using a calculator to evaluate a square root, try estimating the value first.
$\sqrt{13}$ must be a number between 3 and 4 because $\sqrt{9} < \sqrt{13} < \sqrt{16}$.
$\sqrt{42}$ must be a number between 6 and 7 because $\sqrt{36} < \sqrt{42} < \sqrt{49}$.

A negative number cannot have a real number as a square root because no real number when squared is negative. For example, $\sqrt{-25}$ is *not* a real number because there is no real number b for which $(b)^2 = -25$.

Answers

5. 9 6. $\dfrac{6}{7}$ 7. 0.3

> **Example 3** Simplifying Square Roots
>
> Simplify the square roots, if possible.
>
> a. $\sqrt{-144}$ b. $-\sqrt{144}$ c. $\sqrt{-0.01}$ d. $-\sqrt{\dfrac{1}{9}}$
>
> **Solution:**
>
> a. $\sqrt{-144}$ is *not* a real number. No real number when squared equals -144.
>
> b. $-\sqrt{144}$
> $= -1 \cdot \sqrt{144}$
> $= -1 \cdot 12$
> $= -12$
>
> **TIP:** For the expression $-\sqrt{144}$, the factor of -1 is *outside* the radical.
>
> c. $\sqrt{-0.01}$ is *not* a real number. No real number when squared equals -0.01.
>
> d. $-\sqrt{\dfrac{1}{9}}$
> $= -1 \cdot \sqrt{\dfrac{1}{9}}$
> $= -1 \cdot \dfrac{1}{3}$
> $= -\dfrac{1}{3}$
>
> **Skill Practice** Simplify the square roots, if possible.
>
> 8. $\sqrt{-81}$ 9. $-\sqrt{64}$ 10. $-\sqrt{0.25}$ 11. $\sqrt{-\dfrac{1}{4}}$

2. Definition of an *n*th Root

Finding a square root of a number is the reverse process of squaring a number. This concept can be extended to finding a third root (called a cube root), a fourth root, and in general an **nth root**.

> **Definition of an *n*th Root**
>
> b is an *n*th root of a if $b^n = a$.
>
> Example: 2 is a square root of 4 because 2^2 is 4.
>
> Example: 2 is a third root of 8 because 2^3 is 8.
>
> Example: 2 is a fourth root of 16 because 2^4 is 16.

The radical sign $\sqrt{}$ is used to denote the principal square root of a number. The symbol $\sqrt[n]{}$ is used to denote the principal *n*th root of a number. In the expression $\sqrt[n]{a}$, n is called the **index** of the radical, and a is called the **radicand**. For a square root, the index is 2, but it is usually not written ($\sqrt[2]{a}$ is denoted simply as $\sqrt{a}$). A radical with an index of 3 is called a **cube root**, denoted by $\sqrt[3]{a}$.

Answers
8. Not a real number 9. -8
10. -0.5 11. Not a real number

Evaluating $\sqrt[n]{a}$

1. If $n > 1$ is an *even* integer and $a > 0$, then $\sqrt[n]{a}$ is the principal (positive) nth root of a. Example: $\sqrt[4]{625} = 5$
2. If $n > 1$ is an *odd* integer, then $\sqrt[n]{a}$ is the nth root of a.
   Example: $\sqrt[3]{8} = 2$, $\sqrt[3]{-8} = -2$
3. If $n > 1$ is an integer, then $\sqrt[n]{0} = 0$.

For the purpose of simplifying nth roots, it is helpful to know numbers that are perfect squares as well as the following common perfect cubes, fourth powers, and fifth powers.

Perfect Cubes	Perfect Fourth Powers	Perfect Fifth Powers
$1^3 = 1$	$1^4 = 1$	$1^5 = 1$
$2^3 = 8$	$2^4 = 16$	$2^5 = 32$
$3^3 = 27$	$3^4 = 81$	$3^5 = 243$
$4^3 = 64$	$4^4 = 256$	$4^5 = 1024$
$5^3 = 125$	$5^4 = 625$	$5^5 = 3125$

Example 4 Identifying the nth Root of a Real Number

Simplify the expressions, if possible.

a. $\sqrt{4}$ b. $\sqrt[3]{64}$ c. $\sqrt[5]{-32}$ d. $\sqrt[4]{81}$
e. $\sqrt[6]{1,000,000}$ f. $\sqrt{-100}$ g. $\sqrt[4]{-16}$

Solution:

a. $\sqrt{4} = 2$ because $(2)^2 = 4$
b. $\sqrt[3]{64} = 4$ because $(4)^3 = 64$
c. $\sqrt[5]{-32} = -2$ because $(-2)^5 = -32$
d. $\sqrt[4]{81} = 3$ because $(3)^4 = 81$
e. $\sqrt[6]{1,000,000} = 10$ because $(10)^6 = 1,000,000$
f. $\sqrt{-100}$ is not a real number.   No real number when squared equals -100.
g. $\sqrt[4]{-16}$ is not a real number.   No real number when raised to the fourth power equals -16.

Skill Practice Simplify if possible.

12. $\sqrt[4]{16}$ 13. $\sqrt[3]{1000}$ 14. $\sqrt[5]{-1}$ 15. $\sqrt[5]{32}$
16. $\sqrt[5]{100,000}$ 17. $\sqrt{-36}$ 18. $\sqrt[3]{-27}$

Examples 4(f) and 4(g) illustrate that an nth root of a negative quantity is not a real number if the index is even. This is because no real number raised to an even power is negative.

3. Roots of Variable Expressions

Finding an nth root of a variable expression is similar to finding an nth root of a numerical expression. For roots with an even index, however, particular care must be taken to obtain a nonnegative result.

Answers
12. 2 13. 10 14. -1
15. 2 16. 10
17. Not a real number 18. -3

Evaluating $\sqrt[n]{a^n}$

1. If n is a positive *odd* integer, then $\sqrt[n]{a^n} = a$.
2. If n is a positive *even* integer, then $\sqrt[n]{a^n} = |a|$.

The absolute value bars are necessary for roots with an even index because the variable a may represent a positive quantity or a negative quantity. By using absolute value bars, $\sqrt[n]{a^n} = |a|$ is nonnegative and represents the principal nth root of a^n.

Example 5 — Simplifying Expressions of the Form $\sqrt[n]{a^n}$

Simplify the expressions.

a. $\sqrt[4]{(-3)^4}$ b. $\sqrt[5]{(-3)^5}$ c. $\sqrt{(x+2)^2}$ d. $\sqrt[3]{(a+b)^3}$ e. $\sqrt{y^4}$

Solution:

a. $\sqrt[4]{(-3)^4} = |-3| = 3$ — Because this is an *even*-indexed root, absolute value bars are necessary to make the answer positive.

b. $\sqrt[5]{(-3)^5} = -3$ — This is an *odd*-indexed root, so absolute value bars are not necessary.

c. $\sqrt{(x+2)^2} = |x+2|$ — Because this is an *even*-indexed root, absolute value bars are necessary. The sign of the quantity $x + 2$ is unknown; however, $|x + 2| \geq 0$ regardless of the value of x.

d. $\sqrt[3]{(a+b)^3} = a + b$ — This is an *odd*-indexed root, so absolute value bars are not necessary.

e. $\sqrt{y^4} = \sqrt{(y^2)^2}$
$= |y^2|$ — Because this is an even-indexed root, use absolute value bars.
$= y^2$ — However, because y^2 is nonnegative, the absolute value bars are not necessary.

Skill Practice Simplify the expressions.

19. $\sqrt{(-4)^2}$ 20. $\sqrt[3]{(-4)^3}$ 21. $\sqrt{(y+9)^2}$ 22. $\sqrt[3]{(t+1)^3}$ 23. $\sqrt[4]{v^8}$

If n is an even integer, then $\sqrt[n]{a^n} = |a|$; however, if the variable a is assumed to be *nonnegative*, then the absolute value bars may be dropped. That is, $\sqrt[n]{a^n} = a$ provided $a \geq 0$. In many examples and exercises, we will make the assumption that the variables within a radical expression are positive real numbers. In such a case, the absolute value bars are not needed to evaluate $\sqrt[n]{a^n}$.

Take a minute to examine the following patterns associated with perfect squares and perfect cubes. In general, any expression raised to an even power is a perfect square. An expression raised to a power that is a multiple of 3 is a perfect cube.

Perfect Squares	Perfect Cubes
$(x^1)^2 = x^2$	$(x^1)^3 = x^3$
$(x^2)^2 = x^4$	$(x^2)^3 = x^6$
$(x^3)^2 = x^6$	$(x^3)^3 = x^9$
$(x^4)^2 = x^8$	$(x^4)^3 = x^{12}$

These patterns may be extended to higher powers.

Answers
19. 4   20. −4   21. $|y + 9|$
22. $t + 1$ 23. v^2

Example 6 Simplifying *n*th Roots

Simplify the expressions. Assume that all variables are positive real numbers.

a. $\sqrt{y^8}$ b. $\sqrt[3]{27a^3}$ c. $\sqrt[5]{\dfrac{a^5}{b^5}}$ d. $-\sqrt[4]{\dfrac{81x^4y^8}{16}}$

Solution:

a. $\sqrt{y^8} = \sqrt{(y^4)^2}$
   $= y^4$

b. $\sqrt[3]{27a^3} = \sqrt[3]{(3a)^3}$
   $= 3a$

c. $\sqrt[5]{\dfrac{a^5}{b^5}} = \sqrt[5]{\left(\dfrac{a}{b}\right)^5}$
   $= \dfrac{a}{b}$

d. $-\sqrt[4]{\dfrac{81x^4y^8}{16}} = -\sqrt[4]{\left(\dfrac{3xy^2}{2}\right)^4}$
   $= -\dfrac{3xy^2}{2}$

TIP: In Example 6, the variables are assumed to represent positive numbers. Therefore, absolute value bars are not necessary in the simplified form.

Skill Practice Simplify the expressions. Assume all variables represent positive real numbers.

24. $\sqrt{t^6}$ 25. $\sqrt[3]{64y^{12}}$ 26. $\sqrt[4]{\dfrac{x^4}{y^4}}$ 27. $-\sqrt[5]{\dfrac{32a^5}{b^{10}}}$

4. Pythagorean Theorem

Given a right triangle with legs of lengths a and b and hypotenuse of length c, the **Pythagorean theorem** can be stated as $a^2 + b^2 = c^2$. See Figure 18-1.

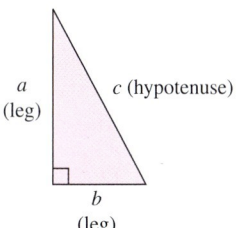

Figure 18-1

Example 7 Applying the Pythagorean Theorem

Use the Pythagorean theorem and the definition of the principal square root to find the length of the unknown side.

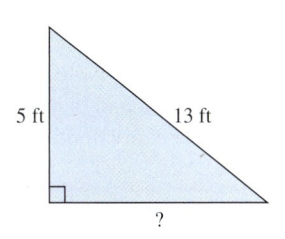

Solution:

Label the sides of the triangle.

$a^2 + b^2 = c^2$ Apply the Pythagorean theorem.
$(5)^2 + b^2 = (13)^2$
$25 + b^2 = 169$ Simplify.
$b^2 = 169 - 25$ Isolate b^2.
$b^2 = 144$ By definition, b must be one of the square roots of 144. Because b represents the length of a side of a triangle, choose the positive square root of 144.
$b = 12$

The third side is 12 ft long.

Skill Practice

28. Use the Pythagorean theorem and the definition of the principal square root to find the length of the unknown side of the right triangle.

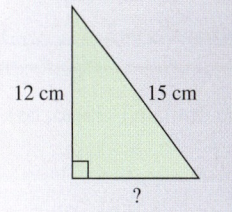

Answers

24. t^3 25. $4y^4$ 26. $\dfrac{x}{y}$

27. $-\dfrac{2a}{b^2}$ 28. 9 cm

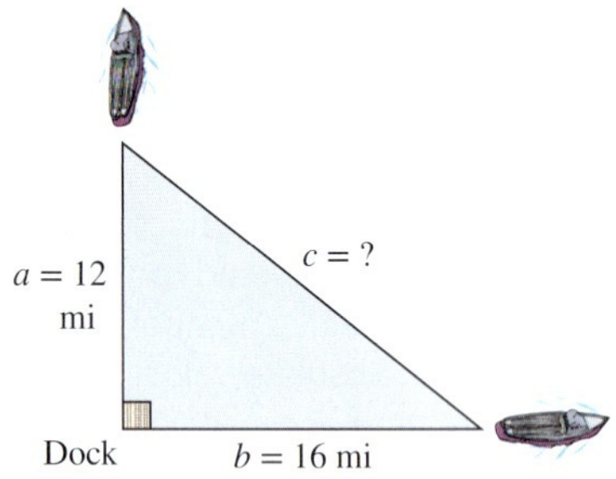

Example 8 Applying the Pythagorean Theorem

Two boats leave a dock at 12:00 noon. One travels due north at 6 mph, and the other travels due east at 8 mph (Figure 18-2). How far apart are the two boats after 2 hr?

Solution:

The boat traveling north travels a distance of (6 mph)(2 hr) = 12 mi. The boat traveling east travels a distance of (8 mph)(2 hr) = 16 mi. The course of the boats forms a right triangle where the hypotenuse represents the distance between them.

$$a^2 + b^2 = c^2$$
$$(12)^2 + (16)^2 = c^2 \quad \text{Apply the Pythagorean theorem.}$$
$$144 + 256 = c^2 \quad \text{Simplify.}$$
$$400 = c^2$$
$$\sqrt{400} = c \quad \text{By definition, } c \text{ must be one of the square roots of 400. Choose the positive square root of 400 to}$$
$$20 = c \quad \text{represent the distance between the two boats.}$$

The boats are 20 mi apart.

Figure 18-2
$a = 12$ mi
$b = 16$ mi
$c = ?$
Dock

Skill Practice

29. Two cars leave from the same place at the same time. One travels west at 40 mph, and the other travels north at 30 mph. How far apart are they after 2 hr?

5. Radical Functions

If n is an integer greater than 1, then a function written in the form $f(x) = \sqrt[n]{x}$ is called a **radical function**. Note that if n is an even integer, then the function will be a real number only if the radicand is nonnegative. Therefore, the domain is restricted to nonnegative real numbers, or equivalently, $[0, \infty)$. If n is an odd integer, then the domain is all real numbers.

Example 9 Determining the Domain of Radical Functions

For each function, write the domain in interval notation.

a. $g(t) = \sqrt[4]{t - 2}$ **b.** $h(a) = \sqrt[3]{a - 3}$ **c.** $k(x) = \sqrt{3 - 5x} + 2$

Solution:

a. $g(t) = \sqrt[4]{t - 2}$ The index is even. The radicand must be nonnegative.
$t - 2 \geq 0$ Set the radicand greater than or equal to zero.
$t \geq 2$ Solve for t.

The domain is $[2, \infty)$.

b. $h(a) = \sqrt[3]{a - 3}$ The index is odd; therefore, the domain is all real numbers.

The domain is $(-\infty, \infty)$.

FOR REVIEW

Recall that the domain of a function is the set of real numbers that will produce a real number when input into the function.

Answer

29. 100 mi

c. $k(x) = \sqrt{3 - 5x} + 2$ The index is even; therefore, the radicand must be nonnegative.

$3 - 5x \geq 0$ Set the radicand greater than or equal to zero.

$-5x \geq -3$ Solve for x.

$\dfrac{-5x}{-5} \leq \dfrac{-3}{-5}$ Reverse the inequality sign.

$x \leq \dfrac{3}{5}$

The domain is $\left(-\infty, \dfrac{3}{5}\right]$.

Skill Practice For each function, write the domain in interval notation.

30. $f(x) = \sqrt{x + 5}$ 31. $g(t) = \sqrt[3]{t - 9}$ 32. $h(a) = \sqrt{1 - 2a}$

FOR REVIEW

Recall that the square root of a negative real number is not a real number. For example, $\sqrt{-64}$ is not a real number.

The cube root of a negative real number is a real number. For example, $\sqrt[3]{-64} = -4$.

Example 10 Graphing a Radical Function

Given $f(x) = \sqrt{3 - x}$

a. Write the domain of f in interval notation.

b. Graph f by making a table of ordered pairs.

FOR REVIEW

When using interval notation, a parenthesis, (or), indicates that an endpoint is not included in the set. A square bracket, [or], indicates that an endpoint is included in the set.

Solution:

a. $f(x) = \sqrt{3 - x}$

$3 - x \geq 0$ The index is even. The radicand must be greater than or equal to zero.

$-x \geq -3$

$x \leq 3$ Reverse the inequality sign.

The domain is $(-\infty, 3]$.

b. Create a table of ordered pairs where x values are taken to be less than or equal to 3.

x	$f(x)$
3	0
2	1
−1	2
−6	3

$f(3) = \sqrt{3 - 3} = \sqrt{0} = 0$
$f(2) = \sqrt{3 - 2} = \sqrt{1} = 1$
$f(-1) = \sqrt{3 - (-1)} = \sqrt{4} = 2$
$f(-6) = \sqrt{3 - (-6)} = \sqrt{9} = 3$

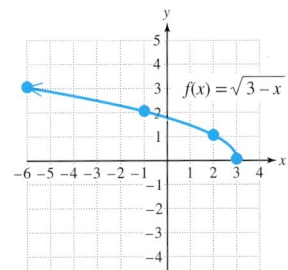

Answers

30. $[-5, \infty)$ 31. $(-\infty, \infty)$
32. $\left(-\infty, \dfrac{1}{2}\right]$
33. a. $[-4, \infty)$
    b.

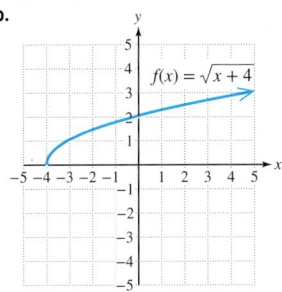

Skill Practice

33. Given $f(x) = \sqrt{x + 4}$
    a. Write the domain of f in interval notation.
    b. Graph f by making a table of ordered pairs.

Section 18.1 Activity

A.1. **a.** b is a **square root** of a if $b^\square = $ _____.
   b. 3 is a square root of 9 because $3^\square = $ _____.
   c. -3 is a square root of 9 because $(-3)^\square = $ _____.
   d. Parts (b) and (c) illustrate that there are two square roots of a positive real number. One square root is positive and the other is negative. The **principal square root** of a positive real number a is denoted by _____. By definition, a principal square root is (choose one: nonnegative/negative).
   e. Evaluate $\sqrt{9}$.

A.2. **a.** b is an ***n*th root** of a if $b^\square = $ _____.
   b. -4 is a cube root of -64 because $(-4)^\square = $ _____.
   c. 2 is a fifth root of 32 because $2^\square = $ _____.
   d. Given $\sqrt[3]{125}$, the index is _____ and the radicand is _____.
   e. Evaluate $\sqrt[3]{125}$.

A.3. **a.** Explain why $\sqrt{-4}$ is not a real number.
   b. Explain why $\sqrt[4]{-16}$ is not a real number.
   c. Evaluate $\sqrt[3]{-8}$ and explain why the answer *is* a real number.

A.4. **a.** The expression $\sqrt{a^2}$ simplifies as $|a|$. Explain why the absolute value bars are needed.

   b. The expression $\sqrt[3]{a^3} = a$. Explain why no absolute value bars are needed.

A.5. **a.** The radicand of a root with an (choose one: odd/even) index must be nonnegative.
   b. Given $f(x) = \sqrt{x}$, write the domain in interval notation.
   c. Given $g(x) = \sqrt{x-5}$, write the domain in interval notation.
   d. Given $h(x) = \sqrt{5-x}$, write the domain in interval notation.

A.6. **a.** The radicand of a root with an (choose one: odd/even) index can be positive, negative, or zero.
   b. Given $f(x) = \sqrt[3]{x}$, write the domain in interval notation.
   c. Given $g(x) = \sqrt[3]{x-5}$, write the domain in interval notation.
   d. Given $h(x) = \sqrt[3]{5-x}$, write the domain in interval notation.

Section 18.1 Practice Exercises

Prerequisite Review

For Exercises R.1–R.12, simplify the expression.

R.1. 10^2 **R.2.** 8^2 **R.3.** $(0.4)^2$

R.4. $(0.13)^2$ **R.5.** $\left(-\dfrac{3}{5}\right)^3$ **R.6.** $\left(-\dfrac{1}{10}\right)^5$

R.7. $\left(-\dfrac{1}{2}\right)^6$ **R.8.** $\left(-\dfrac{1}{9}\right)^2$ **R.9.** $(x^7)^2$

R.10. $(t^{10})^2$ **R.11.** $(3t^4)^4$ **R.12.** $(3p^2)^3$

Vocabulary and Key Concepts

1. **a.** If $b^2 = a$, then _____ is a square root of _____.

   b. Given $a > 0$, the symbol $\sqrt{a}$ denotes the positive or _____ square root of a.

   c. b is an nth root of a if _____ = _____.

   d. Given the symbol $\sqrt[n]{a}$, the value n is called the _____ and a is called the _____.

   e. The symbol $\sqrt[3]{a}$ denotes the _____ root of a.

   f. The expression $\sqrt{-4}$ (is/is not) a real number. The expression $-\sqrt{4}$ (is/is not) a real number.

   g. The expression $\sqrt[n]{a^n} = |a|$ if n is (even/odd). The expression $\sqrt[n]{a^n} = a$ if n is (even/odd).

   h. Given a right triangle with legs a and b and hypotenuse c, the _____ theorem is stated as $a^2 + b^2 =$ _____.

   i. In interval notation, the domain of $f(x) = \sqrt{x}$ is _____, whereas the domain of $g(x) = \sqrt[3]{x}$ is _____.

   j. Which of the following values of x are *not* in the domain of $h(x) = \sqrt{x+3}$: $x = -5, x = -4, x = -3, x = -2, x = -1, x = 0$?

Concept 1: Definition of a Square Root

2. Simplify the expression $\sqrt[3]{8}$. Explain how you can check your answer.

3. **a.** Find the square roots of 64. **(See Example 1.)**

   b. Find $\sqrt{64}$.

   c. Explain the difference between the answers in part (a) and part (b).

4. **a.** Find the square roots of 121.

   b. Find $\sqrt{121}$.

   c. Explain the difference between the answers in part (a) and part (b).

5. **a.** What is the principal square root of 81?

   b. What is the negative square root of 81?

6. **a.** What is the principal square root of 100?

   b. What is the negative square root of 100?

7. Using the definition of a square root, explain why $\sqrt{-36}$ is not a real number.

For Exercises 8–19, evaluate the roots without using a calculator. Identify those that are not real numbers. **(See Examples 2–3.)**

8. $\sqrt{25}$

9. $\sqrt{49}$

10. $-\sqrt{25}$

11. $-\sqrt{49}$

12. $\sqrt{-25}$

13. $\sqrt{-49}$

14. $\sqrt{\dfrac{100}{121}}$

15. $\sqrt{\dfrac{64}{9}}$

16. $\sqrt{0.64}$

17. $\sqrt{0.81}$

18. $-\sqrt{0.0144}$

19. $-\sqrt{0.16}$

Concept 2: Definition of an nth Root

20. Using the definition of an nth root, explain why $\sqrt[4]{-16}$ is not a real number.

For Exercises 21–38, evaluate the roots without using a calculator. Identify those that are not real numbers. **(See Example 4.)**

21. **a.** $\sqrt{64}$ **b.** $\sqrt[3]{64}$ **c.** $-\sqrt{64}$
    d. $-\sqrt[3]{64}$ **e.** $\sqrt{-64}$ **f.** $\sqrt[3]{-64}$

22. a. $\sqrt{16}$ b. $\sqrt[4]{16}$ c. $-\sqrt{16}$
    d. $-\sqrt[4]{16}$ e. $\sqrt{-16}$ f. $\sqrt[4]{-16}$

23. $\sqrt[3]{-27}$
24. $\sqrt[3]{-125}$
25. $\sqrt[3]{\dfrac{1}{8}}$
26. $\sqrt[5]{\dfrac{1}{32}}$

27. $\sqrt[5]{32}$
28. $\sqrt[4]{1}$
29. $\sqrt[3]{-\dfrac{125}{64}}$
30. $\sqrt[3]{-\dfrac{8}{27}}$

31. $\sqrt[4]{-1}$
32. $\sqrt[6]{-1}$
33. $\sqrt[6]{1{,}000{,}000}$
34. $\sqrt[4]{10{,}000}$

35. $-\sqrt[3]{0.008}$
36. $-\sqrt[4]{0.0016}$
37. $\sqrt[4]{0.0625}$
38. $\sqrt[3]{0.064}$

Concept 3: Roots of Variable Expressions

For Exercises 39–58, simplify the radical expressions. (See Example 5.)

39. $\sqrt{a^2}$
40. $\sqrt[4]{a^4}$
41. $\sqrt[3]{a^3}$
42. $\sqrt[5]{a^5}$

43. $\sqrt[6]{a^6}$
44. $\sqrt[7]{a^7}$
45. $\sqrt{(x+1)^2}$
46. $\sqrt[3]{(y+3)^3}$

47. $\sqrt{x^2 y^4}$
48. $\sqrt[3]{(u+v)^3}$
49. $-\sqrt[3]{\dfrac{x^3}{y^3}},\ y\ne 0$
50. $\sqrt[4]{\dfrac{a^4}{b^8}},\ b\ne 0$

51. $\dfrac{2}{\sqrt[4]{x^4}},\ x\ne 0$
52. $\sqrt{(-5)^2}$
53. $\sqrt[3]{(-92)^3}$
54. $\sqrt[6]{(50)^6}$

55. $\sqrt[10]{(-2)^{10}}$
56. $\sqrt[5]{(-2)^5}$
57. $\sqrt[7]{(-923)^7}$
58. $\sqrt[6]{(-417)^6}$

For Exercises 59–74, simplify the expressions. Assume all variables are positive real numbers. (See Example 6.)

59. $\sqrt{y^8}$
60. $\sqrt{x^4}$
61. $\sqrt{\dfrac{a^6}{b^2}}$
62. $\sqrt{\dfrac{w^2}{z^4}}$

63. $-\sqrt{\dfrac{25}{q^2}}$
64. $-\sqrt{\dfrac{p^6}{81}}$
65. $\sqrt{9x^2 y^4 z^2}$
66. $\sqrt{4a^4 b^2 c^6}$

67. $\sqrt{\dfrac{h^2 k^4}{16}}$
68. $\sqrt{\dfrac{4x^2}{y^8}}$
69. $-\sqrt[3]{\dfrac{t^3}{27}}$
70. $\sqrt[4]{\dfrac{16}{w^4}}$

71. $\sqrt[5]{32 y^{10}}$
72. $\sqrt[3]{64 x^6 y^3}$
73. $\sqrt[6]{64 p^{12} q^{18}}$
74. $\sqrt[4]{16 r^{12} s^8}$

Concept 4: Pythagorean Theorem

For Exercises 75–78, find the length of the third side of each triangle by using the Pythagorean theorem. (See Example 7.)

75.

12 cm, 15 cm, ?

76.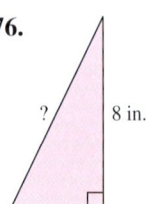

?, 8 in., 6 in.

77.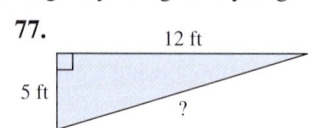

12 ft, 5 ft, ?

78.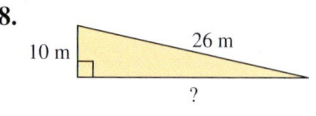

10 m, 26 m, ?

For Exercises 79–82, use the Pythagorean theorem.

79. Roberto and Sherona began running from the same place at the same time. They ran along two different paths that formed right angles with each other. Roberto ran 4 mi and stopped, while Sherona ran 3 mi and stopped. How far apart were they when they stopped? **(See Example 8.)**

80. Leine and Laura began hiking from their campground. Laura headed south while Leine headed east. Laura walked 12 mi and Leine walked 5 mi. How far apart were they when they stopped walking?

81. Two mountain bikers take off from the same place at the same time. One travels north at 4 mph, and the other travels east at 3 mph. How far apart are they after 5 hr?

82. Professor Ortiz leaves campus on her bike, heading west at 12 ft/sec. Professor Wilson leaves campus at the same time and walks south at 5 ft/sec. How far apart are they after 40 sec?

Concept 5: Radical Functions

For Exercises 83–86, evaluate the function for the given values of x. Then write the domain of the function in interval notation. **(See Example 9.)**

83. $h(x) = \sqrt{x-2}$

   a. $h(0)$

   b. $h(1)$

   c. $h(2)$

   d. $h(3)$

   e. $h(6)$

84. $k(x) = \sqrt{x+1}$

   a. $k(-3)$

   b. $k(-2)$

   c. $k(-1)$

   d. $k(0)$

   e. $k(3)$

85. $g(x) = \sqrt[3]{x-2}$

   a. $g(-6)$

   b. $g(1)$

   c. $g(2)$

   d. $g(3)$

86. $f(x) = \sqrt[3]{x+1}$

   a. $f(-9)$

   b. $f(-2)$

   c. $f(0)$

   d. $f(7)$

For each function defined in Exercises 87–94, write the domain in interval notation. **(See Example 9.)**

87. $f(x) = \sqrt{5-2x}$

88. $g(x) = \sqrt{3-4x}$

89. $k(x) = \sqrt[3]{4x-7}$

90. $R(x) = \sqrt[3]{x+1}$

91. $M(x) = \sqrt{x-5} + 3$

92. $N(x) = \sqrt{x+3} - 1$

93. $F(x) = \sqrt[3]{x+7} - 2$

94. $G(x) = \sqrt[3]{x-10} + 4$

For Exercises 95–102,

  a. Write the domain of f in interval notation.

  b. Graph f by making a table of ordered pairs. **(See Example 10.)**

95. $f(x) = \sqrt{1-x}$

96. $f(x) = \sqrt{2-x}$

97. $f(x) = \sqrt{x+3}$

98. $f(x) = \sqrt{x+1}$

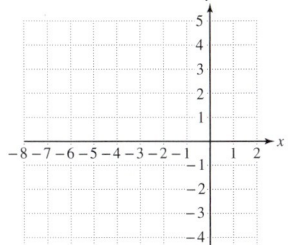 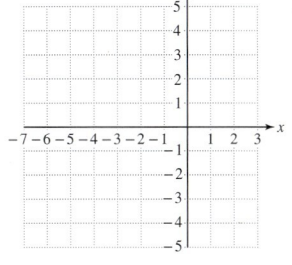 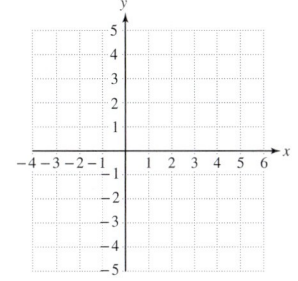 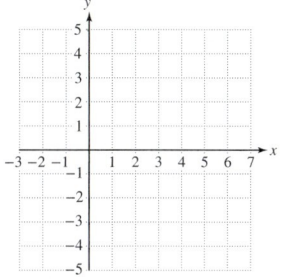

99. $f(x) = \sqrt{x} + 2$ 100. $f(x) = \sqrt{x} - 1$ 101. $f(x) = \sqrt[3]{x-1}$ 102. $f(x) = \sqrt[3]{x} + 2$

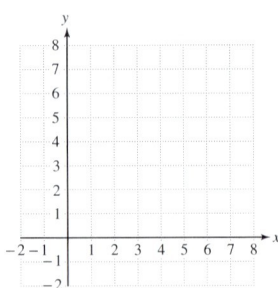
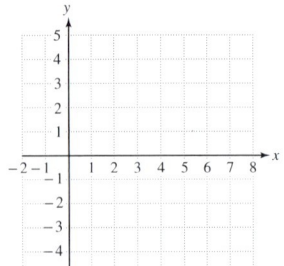
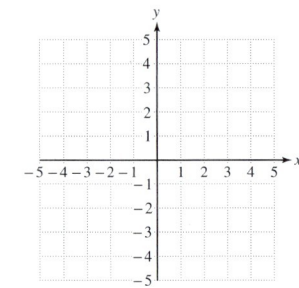
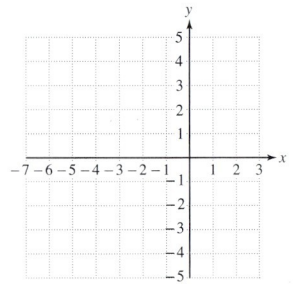

Mixed Exercises

For Exercises 103–106, write the English phrase as an algebraic expression.

103. The sum of q and the square of p

104. The product of 11 and the cube root of x

105. The quotient of 6 and the cube root of x

106. The difference of y and the principal square root of x

107. If a square has an area of 64 in.^{2}, then what are the lengths of the sides?

108. If a square has an area of 121 m^2, then what are the lengths of the sides?

Technology Connections

For Exercises 109–116, use a calculator to evaluate the expressions to four decimal places.

109. $\sqrt{69}$

110. $\sqrt{5798}$

111. $2 + \sqrt[3]{5}$

112. $3 - 2\sqrt[4]{10}$

113. $7\sqrt[4]{25}$

114. $-3\sqrt[3]{9}$

115. $\dfrac{3 - \sqrt{19}}{11}$

116. $\dfrac{5 + 2\sqrt{15}}{12}$

117. Graph $h(x) = \sqrt{x-2}$. Use the graph to confirm the domain found in Exercise 83.

118. Graph $k(x) = \sqrt{x+1}$. Use the graph to confirm the domain found in Exercise 84.

119. Graph $g(x) = \sqrt[3]{x-2}$. Use the graph to confirm the domain found in Exercise 85.

120. Graph $f(x) = \sqrt[3]{x+1}$. Use the graph to confirm the domain found in Exercise 86.

Rational Exponents

Section 18.2

Concepts

1. Definition of $a^{1/n}$ and $a^{m/n}$
2. Converting Between Rational Exponents and Radical Notation
3. Properties of Rational Exponents
4. Applications Involving Rational Exponents

1. Definition of $a^{1/n}$ and $a^{m/n}$

In this section, the properties for simplifying expressions with integer exponents are expanded to include expressions with rational exponents. We begin by defining expressions of the form $a^{1/n}$.

> **Definition of $a^{1/n}$**
>
> Let a be a real number, and let n be an integer such that $n > 1$. If $\sqrt[n]{a}$ is a real number, then
> $$a^{1/n} = \sqrt[n]{a}$$

Example 1 Evaluating Expressions of the Form $a^{1/n}$

Convert the expression to radical form and simplify, if possible.

a. $(-8)^{1/3}$ **b.** $81^{1/4}$ **c.** $-100^{1/2}$ **d.** $(-100)^{1/2}$ **e.** $16^{-1/2}$

Solution:

a. $(-8)^{1/3} = \sqrt[3]{-8} = -2$

b. $81^{1/4} = \sqrt[4]{81} = 3$

c. $-100^{1/2} = -1 \cdot 100^{1/2}$ The exponent applies only to the base of 100.

$\phantom{-100^{1/2}} = -1\sqrt{100}$

$\phantom{-100^{1/2}} = -10$

d. $(-100)^{1/2}$ is not a real number because $\sqrt{-100}$ is not a real number.

e. $16^{-1/2} = \dfrac{1}{16^{1/2}}$ Write the expression with a positive exponent. Recall that $b^{-n} = \dfrac{1}{b^n}$.

$\phantom{16^{-1/2}} = \dfrac{1}{\sqrt{16}}$

$\phantom{16^{-1/2}} = \dfrac{1}{4}$

Skill Practice Convert the expression to radical form and simplify, if possible.

1. $(-64)^{1/3}$ 2. $16^{1/4}$ 3. $-36^{1/2}$ 4. $(-36)^{1/2}$ 5. $64^{-1/3}$

If $\sqrt[n]{a}$ is a real number, then we can define an expression of the form $a^{m/n}$ in such a way that the multiplication property of exponents still holds true. For example:

$$16^{3/4} \begin{cases} (16^{1/4})^3 = (\sqrt[4]{16})^3 = (2)^3 = 8 \\ (16^3)^{1/4} = \sqrt[4]{16^3} = \sqrt[4]{4096} = 8 \end{cases}$$

Answers

1. -4 2. 2 3. -6
4. Not a real number 5. $\dfrac{1}{4}$

1194   Chapter 18   Radicals and Complex Numbers

> **Definition of $a^{m/n}$**
>
> Let a be a real number, and let m and n be positive integers such that m and n share no common factors and $n > 1$. If $\sqrt[n]{a}$ is a real number, then
>
> $$a^{m/n} = (a^{1/n})^m = (\sqrt[n]{a})^m \quad \text{and} \quad a^{m/n} = (a^m)^{1/n} = \sqrt[n]{a^m}$$

The rational exponent in the expression $a^{m/n}$ is essentially performing two operations. The numerator of the exponent raises the base to the mth power. The denominator takes the nth root.

Example 2   Evaluating Expressions of the Form $a^{m/n}$

Convert each expression to radical form and simplify.

a. $8^{2/3}$ **b.** $100^{5/2}$ **c.** $\left(\dfrac{1}{25}\right)^{3/2}$ **d.** $4^{-3/2}$ **e.** $(-81)^{3/4}$

Solution:

a. $8^{2/3} = (\sqrt[3]{8})^2$ Take the cube root of 8 and square the result.

$\phantom{8^{2/3}} = (2)^2$ Simplify.

$\phantom{8^{2/3}} = 4$

b. $100^{5/2} = (\sqrt{100})^5$ Take the square root of 100 and raise the result to the fifth power.

$\phantom{100^{5/2}} = (10)^5$ Simplify.

$\phantom{100^{5/2}} = 100{,}000$

c. $\left(\dfrac{1}{25}\right)^{3/2} = \left(\sqrt{\dfrac{1}{25}}\right)^3$ Take the square root of $\dfrac{1}{25}$ and cube the result.

$\phantom{\left(\dfrac{1}{25}\right)^{3/2}} = \left(\dfrac{1}{5}\right)^3$ Simplify.

$\phantom{\left(\dfrac{1}{25}\right)^{3/2}} = \dfrac{1}{125}$

d. $4^{-3/2} = \left(\dfrac{1}{4}\right)^{3/2} = \dfrac{1}{4^{3/2}}$ Write the expression with positive exponents.

$\phantom{4^{-3/2}} = \dfrac{1}{(\sqrt{4})^3}$ Take the square root of 4 and cube the result.

$\phantom{4^{-3/2}} = \dfrac{1}{2^3}$ Simplify.

$\phantom{4^{-3/2}} = \dfrac{1}{8}$

e. $(-81)^{3/4}$ is not a real number because $\sqrt[4]{-81}$ is not a real number.

Skill Practice Convert each expression to radical form and simplify.

6. $9^{3/2}$ 7. $8^{5/3}$ 8. $\left(\dfrac{1}{27}\right)^{4/3}$ 9. $32^{-4/5}$ 10. $(-4)^{3/2}$

2. Converting Between Rational Exponents and Radical Notation

Example 3 Using Radical Notation and Rational Exponents

Convert each expression to radical notation. Assume all variables represent positive real numbers.

a. $a^{3/5}$ b. $5^{1/3}x^{2/3}$ c. $3y^{1/4}$ d. $9z^{-3/4}$

Solution:

a. $a^{3/5} = \sqrt[5]{a^3}$ or $(\sqrt[5]{a})^3$

b. $5^{1/3}x^{2/3} = (5x^2)^{1/3} = \sqrt[3]{5x^2}$

c. $3y^{1/4} = 3\sqrt[4]{y}$ Note that the coefficient 3 is not raised to the $\frac{1}{4}$ power.

d. $9z^{-3/4} = 9 \cdot \dfrac{1}{z^{3/4}} = \dfrac{9}{\sqrt[4]{z^3}}$ Note that the coefficient 9 has an implied exponent of 1, not $-\frac{3}{4}$.

Skill Practice Convert each expression to radical notation. Assume all variables represent positive real numbers.

11. $t^{4/5}$ 12. $2^{1/4}y^{3/4}$ 13. $10p^{1/2}$ 14. $11q^{-2/3}$

Example 4 Using Radical Notation and Rational Exponents

Convert each expression to an equivalent expression by using rational exponents. Assume that all variables represent positive real numbers.

a. $\sqrt[4]{b^3}$ b. $\sqrt{7a}$ c. $7\sqrt{a}$

Solution:

a. $\sqrt[4]{b^3} = b^{3/4}$ b. $\sqrt{7a} = (7a)^{1/2}$ c. $7\sqrt{a} = 7a^{1/2}$

Skill Practice Convert to an equivalent expression using rational exponents. Assume all variables represent positive real numbers.

15. $\sqrt[3]{x^2}$ 16. $\sqrt{5y}$ 17. $5\sqrt{y}$

3. Properties of Rational Exponents

The properties and definitions for simplifying expressions with integer exponents also apply to rational exponents.

Answers

6. 27   7. 32   8. $\dfrac{1}{81}$
9. $\dfrac{1}{16}$ 10. Not a real number
11. $\sqrt[5]{t^4}$ 12. $\sqrt[4]{2y^3}$ 13. $10\sqrt{p}$
14. $\dfrac{11}{\sqrt[3]{q^2}}$ 15. $x^{2/3}$ 16. $(5y)^{1/2}$
17. $5y^{1/2}$

Definitions and Properties of Exponents

Let a and b be nonzero real numbers. Let m and n be rational numbers such that a^m, a^n, and b^m are real numbers.

Description	Property	Example
1. Multiplying like bases	$a^m a^n = a^{m+n}$	$x^{1/3} x^{4/3} = x^{5/3}$
2. Dividing like bases	$\dfrac{a^m}{a^n} = a^{m-n}$	$\dfrac{x^{3/5}}{x^{1/5}} = x^{2/5}$
3. The power rule	$(a^m)^n = a^{mn}$	$(2^{1/3})^{1/2} = 2^{1/6}$
4. Power of a product	$(ab)^m = a^m b^m$	$(9y)^{1/2} = 9^{1/2} y^{1/2} = 3y^{1/2}$
5. Power of a quotient	$\left(\dfrac{a}{b}\right)^m = \dfrac{a^m}{b^m}$	$\left(\dfrac{4}{25}\right)^{1/2} = \dfrac{4^{1/2}}{25^{1/2}} = \dfrac{2}{5}$

Description	Definition	Example
1. Negative exponents	$a^{-m} = \left(\dfrac{1}{a}\right)^m = \dfrac{1}{a^m}$	$(8)^{-1/3} = \left(\dfrac{1}{8}\right)^{1/3} = \dfrac{1}{2}$
2. Zero exponent	$a^0 = 1$	$5^0 = 1$

Example 5 Simplifying Expressions With Rational Exponents

Use the properties of exponents to simplify the expressions. Assume all variables represent positive real numbers.

a. $y^{2/5} y^{3/5}$ **b.** $\dfrac{6a^{-1/2}}{a^{3/2}}$ **c.** $\left(\dfrac{s^{1/2} t^{1/3}}{w^{3/4}}\right)^4$

Solution:

a. $y^{2/5} y^{3/5} = y^{(2/5)+(3/5)}$ Multiply like bases by adding exponents.

$\qquad\qquad = y^{5/5}$ Simplify.

$\qquad\qquad = y$

b. $\dfrac{6a^{-1/2}}{a^{3/2}}$

$\quad = 6a^{(-1/2)-(3/2)}$ Divide like bases by subtracting exponents.

$\quad = 6a^{-2}$ Simplify: $-\dfrac{1}{2} - \left(\dfrac{3}{2}\right) = -\dfrac{4}{2} = -2$

$\quad = \dfrac{6}{a^2}$ Simplify the negative exponent.

c. $\left(\dfrac{s^{1/2} t^{1/3}}{w^{3/4}}\right)^4 = \dfrac{s^{(1/2)\cdot 4} t^{(1/3)\cdot 4}}{w^{(3/4)\cdot 4}}$ Apply the power rule. Multiply exponents.

$\qquad\qquad = \dfrac{s^2 t^{4/3}}{w^3}$ Simplify.

Skill Practice Use the properties of exponents to simplify the expressions. Assume all variables represent positive real numbers.

18. $x^{1/2} \cdot x^{3/4}$ **19.** $\dfrac{4k^{-2/3}}{k^{1/3}}$ **20.** $\left(\dfrac{a^{1/3} b^{1/2}}{c^{5/8}}\right)^6$

Answers

18. $x^{5/4}$ **19.** $\dfrac{4}{k}$ **20.** $\dfrac{a^2 b^3}{c^{15/4}}$

4. Applications Involving Rational Exponents

Example 6   **Applying Rational Exponents**

Suppose P dollars in principal is invested in an account that earns interest annually. If after t years the investment grows to A dollars, then the annual rate of return r on the investment is given by

$$r = \left(\frac{A}{P}\right)^{1/t} - 1$$

Find the annual rate of return on \$5000 which grew to \$6894.21 after 6 years.

Solution:

$$r = \left(\frac{A}{P}\right)^{1/t} - 1$$

$$= \left(\frac{6894.21}{5000}\right)^{1/6} - 1 \quad \text{where } A = \$6894.21, P = \$5000, \text{ and } t = 6$$

$$\approx 0.055 \text{ or } 5.5\%$$

The annual rate of return is 5.5%.

Skill Practice

21. The formula for the radius of a sphere is

$$r = \left(\frac{3V}{4\pi}\right)^{1/3}$$

where V is the volume. Find the radius of a sphere whose volume is 113.04 in.^{3} (Use 3.14 for π.)

Answer

21. 3 in.

Section 18.2 Activity

A.1.
  a. Write the expression $a^{1/n}$ in radical notation.
  b. Write $9^{1/2}$ as a radical and simplify.
  c. Write $(-125)^{1/3}$ as a radical and simplify.
  d. Explain why $(-16)^{1/2}$ is undefined.

A.2.
  a. Write the expression $a^{m/n}$ in radical notation.
  b. Write $125^{2/3}$ as a radical and simplify.
  c. Write $16^{3/4}$ as a radical and simplify.

The properties of integer exponents can be extended to rational exponents, provided the corresponding radicals are real numbers. For Exercises A.3–A.9, simplify the expressions. Apply your knowledge of the properties of integer exponents to similar expressions involving rational exponents. Assume that all variables represent positive real numbers.

A.3.   **a.** $t^2 \cdot t^5$ **b.** $t^{1/2} \cdot t^{5/2}$

A.4.   **a.** $\dfrac{y^9}{y^5}$ **b.** $\dfrac{y^{6/5}}{y^{1/3}}$

A.5.   **a.** $(z^3)^5$ **b.** $(z^{2/3})^6$

A.6.   **a.** n^{-6} **b.** $n^{-1/2}$

A.7.   **a.** $\dfrac{2}{m^{-3}}$ **b.** $\dfrac{2}{m^{-1/3}}$

A.8.   **a.** $(a^2 b^4)^2$ **b.** $(a^2 b^4)^{1/2}$

A.9.   **a.** $\left(\dfrac{c^3}{d^4}\right)^3$ **b.** $\left(\dfrac{c^{1/3}}{d^{1/4}}\right)^{12}$

Section 18.2 Practice Exercises

Prerequisite Review

For Exercises R.1–R.10, simplify the expression.

R.1. $a^{-7} \cdot a^{12}$ **R.2.** $q^{-3} \cdot q^{10}$ **R.3.** $\dfrac{y^{17}}{y^{20}}$ **R.4.** $\dfrac{n^2}{n^{12}}$

R.5. $(n^4)^{-2}$ **R.6.** $(d^{-4})^3$ **R.7.** $\left(\dfrac{1}{5}a^2 b^3\right)^3$ **R.8.** $\left(\dfrac{2}{3}x^4 y\right)^2$

R.9. $w^4 \cdot w^{-4}$ **R.10.** $p^{-5} \cdot p^5$

For Exercises R.11–R.12,
   a. Identify the index.   b. Identify the radicand.   c. Simplify.

R.11. $\sqrt[4]{16}$ **R.12.** $\sqrt[3]{125}$

For Exercises R.13–R.16, simplify the radicals

R.13. $(\sqrt[4]{81})^3$ **R.14.** $(\sqrt[4]{16})^3$ **R.15.** $\sqrt[3]{(t+2)^3}$ **R.16.** $\sqrt[5]{(a^2+b)^5}$

For the exercises in this set, assume that all variables represent positive real numbers unless otherwise stated.

Vocabulary and Key Concepts

1. a. If n is an integer greater than 1, then radical notation for $a^{1/n}$ is _____.

   b. Assume that m and n are positive integers that share no common factors and $n > 1$. If $\sqrt[n]{a}$ exists, then in radical notation $a^{m/n} =$ _____.

2. a. The radical notation for $x^{-1/2}$ is _____.

   b. $8^{1/3} =$ _____ and $8^{-1/3} =$ _____.

Concept 1: Definition of $a^{1/n}$ and $a^{m/n}$

For Exercises 3–6, write the expression in radical form.

3. a. $49^{1/2}$
   b. $-49^{1/2}$
   c. $49^{-1/2}$

4. a. $121^{1/2}$
   b. $-121^{1/2}$
   c. $121^{-1/2}$

5. a. $(-64)^{1/3}$
   b. $-64^{1/3}$
   c. $64^{-1/3}$

6. a. $(-343)^{1/3}$
   b. $-343^{1/3}$
   c. $343^{-1/3}$

For Exercises 7–18, convert the expressions to radical form and simplify. (See Example 1.)

7. $144^{1/2}$ 8. $16^{1/4}$ 9. $-144^{1/2}$ 10. $-16^{1/4}$

11. $(-144)^{1/2}$ 12. $(-16)^{1/4}$ 13. $(-64)^{1/3}$ 14. $(-32)^{1/5}$

15. $25^{-1/2}$ 16. $27^{-1/3}$ 17. $-49^{-1/2}$ 18. $-64^{-1/2}$

19. Explain how to interpret the expression $a^{m/n}$ as a radical.

20. Explain why $(\sqrt[3]{8})^4$ is easier to evaluate than $\sqrt[3]{8^4}$.

For Exercises 21–24, simplify the expression, if possible. (See Example 2.)

21. a. $16^{3/4}$
    b. $-16^{3/4}$
    c. $(-16)^{3/4}$
    d. $16^{-3/4}$
    e. $-16^{-3/4}$
    f. $(-16)^{-3/4}$

22. a. $81^{3/4}$
    b. $-81^{3/4}$
    c. $(-81)^{3/4}$
    d. $81^{-3/4}$
    e. $-81^{-3/4}$
    f. $(-81)^{-3/4}$

23. a. $25^{3/2}$
    b. $-25^{3/2}$
    c. $(-25)^{3/2}$
    d. $25^{-3/2}$
    e. $-25^{-3/2}$
    f. $(-25)^{-3/2}$

24. a. $4^{3/2}$
    b. $-4^{3/2}$
    c. $(-4)^{3/2}$
    d. $4^{-3/2}$
    e. $-4^{-3/2}$
    f. $(-4)^{-3/2}$

For Exercises 25–48, simplify the expression. (See Example 2.)

25. $64^{-3/2}$
26. $81^{-3/2}$
27. $243^{3/5}$
28. $1^{5/3}$
29. $-27^{-4/3}$
30. $-16^{-5/4}$
31. $\left(\dfrac{100}{9}\right)^{-3/2}$
32. $\left(\dfrac{49}{100}\right)^{-1/2}$
33. $(-4)^{-3/2}$
34. $(-49)^{-3/2}$
35. $(-8)^{1/3}$
36. $(-9)^{1/2}$
37. $-8^{1/3}$
38. $-9^{1/2}$
39. $\dfrac{1}{36^{-1/2}}$
40. $\dfrac{1}{16^{-1/2}}$
41. $\dfrac{1}{1000^{-1/3}}$
42. $\dfrac{1}{81^{-3/4}}$
43. $\left(\dfrac{1}{8}\right)^{2/3} + \left(\dfrac{1}{4}\right)^{1/2}$
44. $\left(\dfrac{1}{8}\right)^{-2/3} + \left(\dfrac{1}{4}\right)^{-1/2}$
45. $\left(\dfrac{1}{16}\right)^{-3/4} - \left(\dfrac{1}{49}\right)^{-1/2}$
46. $\left(\dfrac{1}{16}\right)^{1/4} - \left(\dfrac{1}{49}\right)^{1/2}$
47. $\left(\dfrac{1}{4}\right)^{1/2} + \left(\dfrac{1}{64}\right)^{-1/3}$
48. $\left(\dfrac{1}{36}\right)^{1/2} + \left(\dfrac{1}{64}\right)^{-5/6}$

Concept 2: Converting Between Rational Exponents and Radical Notation

For Exercises 49–56, convert each expression to radical notation. (See Example 3.)

49. $q^{2/3}$
50. $t^{3/5}$
51. $6y^{3/4}$
52. $8b^{4/9}$
53. $x^{2/3}y^{1/3}$
54. $c^{2/5}d^{3/5}$
55. $6r^{-2/5}$
56. $7x^{-3/4}$

For Exercises 57–64, write each expression by using rational exponents rather than radical notation. (See Example 4.)

57. $\sqrt[3]{x}$
58. $\sqrt[4]{a}$
59. $10\sqrt{b}$
60. $-2\sqrt[3]{t}$
61. $\sqrt[3]{y^2}$
62. $\sqrt[6]{z^5}$
63. $\sqrt[4]{a^2b^3}$
64. $\sqrt{abc}$

Concept 3: Properties of Rational Exponents

For Exercises 65–88, simplify the expressions by using the properties of rational exponents. Write the final answers using positive exponents only. (See Example 5.)

65. $x^{1/4}x^{-5/4}$

66. $2^{2/3}2^{-5/3}$

67. $\dfrac{p^{5/3}}{p^{2/3}}$

68. $\dfrac{q^{5/4}}{q^{1/4}}$

69. $(y^{1/5})^{10}$

70. $(x^{1/2})^8$

71. $6^{-1/5}6^{3/5}$

72. $a^{-1/3}a^{2/3}$

73. $\dfrac{4t^{-1/3}}{t^{4/3}}$

74. $\dfrac{5s^{-1/3}}{s^{5/3}}$

75. $(a^{1/3}a^{1/4})^{12}$

76. $(x^{2/3}x^{1/2})^6$

77. $(5a^2c^{-1/2}d^{1/2})^2$

78. $(2x^{-1/3}y^2z^{5/3})^3$

79. $\left(\dfrac{x^{-2/3}}{y^{-3/4}}\right)^{12}$

80. $\left(\dfrac{m^{-1/4}}{n^{-1/2}}\right)^{-4}$

81. $\left(\dfrac{16w^{-2}z}{2wz^{-8}}\right)^{1/3}$

82. $\left(\dfrac{50p^{-1}q}{2pq^{-3}}\right)^{1/2}$

83. $(25x^2y^4z^6)^{1/2}$

84. $(8a^6b^3c^9)^{2/3}$

85. $(x^2y^{-1/3})^6(x^{1/2}yz^{2/3})^2$

86. $(a^{-1/3}b^{1/2})^4(a^{-1/2}b^{3/5})^{10}$

87. $\left(\dfrac{x^{3m}y^{2m}}{z^{5m}}\right)^{1/m}$

88. $\left(\dfrac{a^{4n}b^{3n}}{c^n}\right)^{1/n}$

Concept 4: Applications Involving Rational Exponents

89. If P dollars in principal grows to A dollars after t years with annual interest, then the interest rate r is given by $r = \left(\dfrac{A}{P}\right)^{1/t} - 1$. (See Example 6.)

    a. In one account, $10,000 grows to $16,802 after 5 years. Compute the interest rate. Round your answer to a tenth of a percent.

    b. In another account, $10,000 grows to $18,000 after 7 years. Compute the interest rate. Round your answer to a tenth of a percent.

    c. Which account produced a higher average yearly return?

90. If the area A of a square is known, then the length of its sides, s, can be computed by the formula $s = A^{1/2}$.

    a. Compute the length of the sides of a square having an area of 100 in.²

    b. Compute the length of the sides of a square having an area of 72 in.² Round your answer to the nearest 0.1 in.

91. The radius r of a sphere of volume V is given by $r = \left(\dfrac{3V}{4\pi}\right)^{1/3}$. Find the radius of a sphere having a volume of 85 in.³ Round your answer to the nearest 0.1 in.

92. Is $(a+b)^{1/2}$ the same as $a^{1/2} + b^{1/2}$? If not, give a counterexample.

Expanding Your Skills

For Exercises 93–104, write the expression using rational exponents. Then simplify and convert back to radical notation.

Example: $\sqrt[15]{x^{10}} \xrightarrow{\text{Rational exponents}} x^{10/15} \xrightarrow{\text{Simplify}} x^{2/3} \xrightarrow{\text{Radical notation}} \sqrt[3]{x^2}$

93. $\sqrt[6]{y^3}$
94. $\sqrt[4]{w^2}$
95. $\sqrt[12]{z^3}$
96. $\sqrt[18]{t^3}$

97. $\sqrt[9]{x^6}$
98. $\sqrt[12]{p^9}$
99. $\sqrt[6]{x^3 y^6}$
100. $\sqrt[8]{m^2 p^8}$

101. $\sqrt{16x^8 y^6}$
102. $\sqrt{81a^{12} b^{20}}$
103. $\sqrt[3]{8x^3 y^2 z}$
104. $\sqrt[3]{64 m^2 n^3 p}$

For Exercises 105–108, write the expression as a single radical.

105. $\sqrt{\sqrt[3]{x}}$
106. $\sqrt[3]{\sqrt{x}}$
107. $\sqrt[5]{\sqrt[3]{w}}$
108. $\sqrt[3]{\sqrt[4]{w}}$

For Exercises 109–116, use a calculator to approximate the expressions. Round to four decimal places, if necessary.

109. $9^{1/2}$
110. $125^{-1/3}$
111. $50^{-1/4}$
112. $(172)^{3/5}$

113. $\sqrt[3]{5^2}$
114. $\sqrt[4]{6^3}$
115. $\sqrt{10^3}$
116. $\sqrt[3]{16}$

Simplifying Radical Expressions — Section 18.3

1. Multiplication Property of Radicals

You may have already noticed certain properties of radicals involving a product or quotient.

Multiplication Property of Radicals

Let a and b represent real numbers such that $\sqrt[n]{a}$ and $\sqrt[n]{b}$ are both real. Then

$$\sqrt[n]{ab} = \sqrt[n]{a} \cdot \sqrt[n]{b}$$

Concepts

1. Multiplication Property of Radicals
2. Simplifying Radicals by Using the Multiplication Property of Radicals
3. Simplifying Radicals by Using the Order of Operations

The multiplication property of radicals follows from the property of rational exponents.

$$\sqrt[n]{ab} = (ab)^{1/n} = a^{1/n} b^{1/n} = \sqrt[n]{a} \cdot \sqrt[n]{b}$$

The multiplication property of radicals indicates that a product within a radicand can be written as a product of radicals, provided the roots are real numbers. For example:

$$\sqrt{144} = \sqrt{16} \cdot \sqrt{9}$$

The reverse process is also true. A product of radicals can be written as a single radical provided the roots are real numbers and they have the same indices.

$$\sqrt{3} \cdot \sqrt{12} = \sqrt{36}$$

2. Simplifying Radicals by Using the Multiplication Property of Radicals

In algebra, it is customary to simplify radical expressions.

> **Simplified Form of a Radical**
>
> Consider any radical expression where the radicand is written as a product of prime factors. The expression is in *simplified form* if all the following conditions are met:
>
> 1. The radicand has no factor raised to a power greater than or equal to the index.
> 2. The radicand does not contain a fraction.
> 3. There are no radicals in the denominator of a fraction.

For example, the following radicals are not simplified.

1. The expression $\sqrt[3]{x^5}$ fails condition 1.
2. The expression $\sqrt{\dfrac{1}{4}}$ fails condition 2.
3. The expression $\dfrac{1}{\sqrt[3]{8}}$ fails condition 3.

The expressions $\sqrt{x^2}$, $\sqrt{x^4}$, $\sqrt{x^6}$, and $\sqrt{x^8}$ are not simplified because they fail condition 1 (the exponents are not less than the index). However, each radicand is a perfect square and is easily simplified for $x \geq 0$.

$$\sqrt{x^2} = x$$
$$\sqrt{x^4} = x^2$$
$$\sqrt{x^6} = x^3$$
$$\sqrt{x^8} = x^4$$

However, how is an expression such as $\sqrt{x^9}$ simplified? This and many other radical expressions are simplified by using the multiplication property of radicals. We demonstrate the process in Examples 1–4.

Example 1   Using the Multiplication Property to Simplify a Radical Expression

Simplify the expression assuming that $x \geq 0$. $\sqrt{x^9}$

Solution:

The expression $\sqrt{x^9}$ is equivalent to $\sqrt{x^8 \cdot x}$.

$\sqrt{x^9} = \sqrt{x^8 \cdot x}$

$\phantom{\sqrt{x^9}} = \sqrt{x^8} \cdot \sqrt{x}$ Apply the multiplication property of radicals.

                 Note that x^8 is a perfect square because $x^8 = (x^4)^2$.

$\phantom{\sqrt{x^9}} = x^4 \sqrt{x}$ Simplify.

Skill Practice Simplify the expression. Assume $a \geq 0$.

1. $\sqrt{a^{11}}$

Answer

1. $a^5\sqrt{a}$

Section 18.3 Simplifying Radical Expressions

In Example 1, the expression x^9 is not a perfect square. Therefore, to simplify $\sqrt{x^9}$, it was necessary to write the expression as the product of the largest perfect square and a remaining or "left-over" factor: $\sqrt{x^9} = \sqrt{x^8 \cdot x}$. This process also applies to simplifying nth roots, as shown in Example 2.

Example 2 Using the Multiplication Property to Simplify Radical Expressions

Simplify each expression. Assume all variables represent positive real numbers.

a. $\sqrt[4]{b^7}$ b. $\sqrt[3]{w^7 z^9}$

Solution:

The goal is to rewrite each radicand as the product of the greatest perfect square (perfect cube, perfect fourth power, and so on) and a leftover factor.

a. $\sqrt[4]{b^7} = \sqrt[4]{b^4 \cdot b^3}$ b^4 is the greatest perfect fourth power in the radicand.

$= \sqrt[4]{b^4} \cdot \sqrt[4]{b^3}$ Apply the multiplication property of radicals.

$= b\sqrt[4]{b^3}$ Simplify.

b. $\sqrt[3]{w^7 z^9} = \sqrt[3]{(w^6 z^9) \cdot (w)}$ $w^6 z^9$ is the greatest perfect cube in the radicand.

$= \sqrt[3]{w^6 z^9} \cdot \sqrt[3]{w}$ Apply the multiplication property of radicals.

$= w^2 z^3 \sqrt[3]{w}$ Simplify.

Skill Practice Simplify the expressions. Assume all variables represent positive real numbers.

2. $\sqrt[4]{v^{25}}$ 3. $\sqrt[3]{p^{17} q^{10}}$

Each expression in Example 2 involves a radicand that is a product of variable factors. If a numerical factor is present, sometimes it is necessary to factor the coefficient before simplifying the radical.

Example 3 Using the Multiplication Property to Simplify a Radical

Simplify the expression. $\sqrt{56}$

Solution:

$\sqrt{56} = \sqrt{2^3 \cdot 7}$ Factor the radicand.

$= \sqrt{(2^2) \cdot (2 \cdot 7)}$ 2^2 is the greatest perfect square in the radicand.

$= \sqrt{2^2} \cdot \sqrt{2 \cdot 7}$ Apply the multiplication property of radicals.

$= 2\sqrt{14}$ Simplify.

2 | 56
2 | 28
2 | 14
   7

TIP: It may be easier to visualize the greatest perfect square factor within the radicand as follows:

$\sqrt{56} = \sqrt{4 \cdot 14}$
$= \sqrt{4} \cdot \sqrt{14}$
$= 2\sqrt{14}$

Skill Practice Simplify.

4. $\sqrt{24}$

Answers

2. $v^6 \sqrt[4]{v}$ 3. $p^5 q^3 \sqrt[3]{p^2 q}$ 4. $2\sqrt{6}$

TIP: The radical can also be simplified as:

$$6\sqrt{50} = 6\sqrt{25 \cdot 2}$$
$$= 6\sqrt{25} \cdot \sqrt{2}$$
$$= 6 \cdot 5\sqrt{2}$$
$$= 30\sqrt{2}$$

Example 4 — Using the Multiplication Property to Simplify Radicals

Simplify. $6\sqrt{50}$

Solution:

$$6\sqrt{50} = 6\sqrt{5^2 \cdot 2} \quad \text{Factor the radicand.}$$
$$= 6 \cdot \sqrt{5^2} \cdot \sqrt{2} \quad \text{Apply the multiplication property of radicals.}$$
$$= 6 \cdot 5 \cdot \sqrt{2} \quad \text{Simplify.}$$
$$= 30\sqrt{2} \quad \text{Simplify.}$$

Skill Practice Simplify.

5. $5\sqrt{18}$

TIP: In Example 5, the numerical coefficient within the radicand can be written $8 \cdot 5$ because 8 is the greatest perfect cube factor of 40:

$$\sqrt[3]{40x^3y^5z^7} = \sqrt[3]{8 \cdot 5x^3y^5z^7}$$
$$= \sqrt[3]{(8x^3y^3z^6)(5y^2z)}$$
$$= \sqrt[3]{8x^3y^3z^6} \cdot \sqrt[3]{5y^2z}$$
$$= 2xyz^2 \cdot \sqrt[3]{5y^2z}$$

Example 5 — Using the Multiplication Property to Simplify Radicals

Simplify the expression. Assume that x, y, and z represent positive real numbers.

$$\sqrt[3]{40x^3y^5z^7}$$

Solution:

$$\sqrt[3]{40x^3y^5z^7}$$
$$= \sqrt[3]{2^3 5 x^3 y^5 z^7} \quad \text{Factor the radicand.}$$
$$= \sqrt[3]{(2^3 x^3 y^3 z^6) \cdot (5y^2z)} \quad 2^3 x^3 y^3 z^6 \text{ is the greatest perfect cube.}$$
$$= \sqrt[3]{2^3 x^3 y^3 z^6} \cdot \sqrt[3]{5y^2z} \quad \text{Apply the multiplication property of radicals.}$$
$$= 2xyz^2 \sqrt[3]{5y^2z} \quad \text{Simplify.}$$

```
2 | 40
2 | 20
2 | 10
    5
```

Skill Practice Simplify. Assume that $a > 0$ and $b > 0$.

6. $\sqrt[4]{32a^{10}b^{19}}$

3. Simplifying Radicals by Using the Order of Operations

Often a radical can be simplified by applying the order of operations. In Example 6, the first step will be to simplify the expression within the radicand.

Example 6 — Using the Order of Operations to Simplify Radicals

Use the order of operations to simplify the expressions. Assume $a > 0$.

a. $\sqrt{\dfrac{a^7}{a^3}}$ b. $\sqrt[3]{\dfrac{3}{81}}$

Answers

5. $15\sqrt{2}$ 6. $2a^2b^4\sqrt[4]{2a^2b^3}$

Solution:

a. $\sqrt{\dfrac{a^7}{a^3}}$ The radicand contains a fraction. However, the fraction can be reduced to lowest terms.

$= \sqrt{a^4}$

$= a^2$ Simplify the radical.

b. $\sqrt[3]{\dfrac{3}{81}}$ The radical contains a fraction that can be simplified.

$= \sqrt[3]{\dfrac{1}{27}}$ Reduce to lowest terms.

$= \dfrac{1}{3}$ Simplify.

Skill Practice Use the order of operations to simplify the expressions. Assume $v > 0$.

7. $\sqrt{\dfrac{v^{21}}{v^5}}$ 8. $\sqrt[5]{\dfrac{64}{2}}$

Example 7   **Simplifying a Radical Expression**

Simplify. $\dfrac{7\sqrt{50}}{10}$

Solution:

$\dfrac{7\sqrt{50}}{10} = \dfrac{7\sqrt{25 \cdot 2}}{10}$ 25 is the greatest perfect square in the radicand.

$= \dfrac{7 \cdot 5\sqrt{2}}{10}$ Simplify the radical.

$= \dfrac{7 \cdot \overset{1}{5}\sqrt{2}}{\underset{2}{10}}$ Simplify the fraction to lowest terms.

$= \dfrac{7\sqrt{2}}{2}$

Avoiding Mistakes

The expression $\dfrac{7\sqrt{2}}{2}$ cannot be simplified further because one factor of 2 is in the radicand and the other is outside the radical.

Skill Practice Simplify.

9. $\dfrac{2\sqrt{300}}{30}$

Answers

7. v^8 8. 2   9. $\dfrac{2\sqrt{3}}{3}$

Section 18.3 Activity

A.1. **a.** If $\sqrt[n]{a}$ and $\sqrt[n]{b}$ are both real numbers, then $\sqrt[n]{a} \cdot \sqrt[n]{b} = $ _____.

b. The multiplication property of radicals can be used in "reverse" to simplify radical expressions. For example, $\sqrt{20} = \sqrt{4 \cdot 5} = \sqrt{4} \cdot \sqrt{5} = $ _____.

c. Simplify $\sqrt{50}$ by factoring 50 such that one factor is a perfect square.)

$$\sqrt{50} = \sqrt{\Box \cdot \Box} = \sqrt{\Box} \cdot \sqrt{\Box} = \underline{\qquad}$$

d. Simplify $\sqrt[3]{x^7}$ by factoring x^7 as the product of the largest perfect cube and a "leftover" factor.)

$$\sqrt[3]{x^7} = \sqrt[3]{\Box \cdot \Box} = \sqrt[3]{\Box} \cdot \sqrt[3]{\Box} = \underline{\qquad}$$

A.2. Simplify $\sqrt{48x^5y^6}$ by following these steps.

a. Factor 48 into a product of prime factors.

b. Write the radical with the radicand factored as a product of prime factors.

c. Write the radicand as the product of the largest perfect square factor and a "leftover" factor.

$$\sqrt{48x^5y^6} = \sqrt{\underbrace{(\Box \cdot x^\Box y^\Box)}_{\substack{\text{These factors}\\\text{should all be}\\\text{perfect squares.}}} \cdot (\underline{\quad})}$$

d. Simplify the radical.

A.3. Simplify $\sqrt{\dfrac{125x^5}{5x}}$ by first simplifying the radicand.

A.4. Simplify $\dfrac{5\sqrt{72}}{3}$. (*Hint:* Simplify the radical first.)

Section 18.3 Practice Exercises

Prerequisite Review

For Exercises R.1–R.2, find the prime factorization of the given number.

R.1. 882

R.2. 189

R.3. Which of the following are perfect squares?

4, 6, 9, 18, 25, 40, 100, 200, y^2, y^3, y^4, y^5, y^{10}, y^{15}

R.4. Which of the following are perfect cubes?

1, 4, 8, 16, 27, 30, 64, a^2, a^3, a^4, a^5, a^6, a^{10}, a^{15}

R.5. Which of the following are perfect fourth powers?

1, 4, 8, 16, 25, 40, 81, b^2, b^3, b^4, b^5, b^8, b^{10}

R.6. Which of the follow are perfect fifth powers?

1, 20, 25, 32, 100, 243, $10x$, x^5, x^{10}, x^{24}, x^{30}, x^{35}

For Exercises R.7–R.18, simplify the expression if possible. Assume that all variable expressions represent positive real numbers.

R.7. $\sqrt{\dfrac{81}{64}}$　　**R.8.** $\sqrt{\dfrac{121}{16}}$　　**R.9.** $\sqrt[3]{125}$　　**R.10.** $\sqrt[4]{81}$

R.11. $-\sqrt{36}$　　**R.12.** $\sqrt{-36}$　　**R.13.** $-\sqrt[3]{64}$　　**R.14.** $\sqrt[3]{-64}$

R.15. $\sqrt{b^{10}}$　　**R.16.** $\sqrt[4]{t^{12}}$　　**R.17.** $\sqrt{9c^4 d^6}$　　**R.18.** $\sqrt{16m^2 n^8}$

For the exercises in this set, assume that all variables represent positive real numbers unless otherwise stated.

Vocabulary and Key Concepts

1. **a.** The multiplication property of radicals indicates that if $\sqrt[n]{a}$ and $\sqrt[n]{b}$ are real numbers, then $\sqrt[n]{ab} = $ _____ · _____ .

    b. Explain why the following radical is not in simplified form. $\sqrt{x^3}$

    c. The radical expression $\sqrt[3]{x^{10}}$ (is/is not) in simplified form.

    d. The radical expression $\sqrt{18}$ simplifies to ___ $\sqrt{2}$.

    e. To simplify the radical expression $\sqrt[3]{t^{14}}$ the radicand is rewritten as $\sqrt[3]{\underline{} \cdot t^2}$.

2. On a calculator, $\sqrt{2}$ is given as 1.414213562. Is this decimal number the exact value of $\sqrt{2}$?

Concept 2: Simplifying Radicals by Using the Multiplication Property of Radicals

For Exercises 3–8, fill in the boxes inside the radicals. Then simplify the result.

3. $\sqrt{x^{11}} = \sqrt{x^{\square} \cdot x} = \sqrt{x^{\square}} \cdot \sqrt{x} = $ _____

4. $\sqrt{y^5} = \sqrt{y^{\square} \cdot y} = \sqrt{y^{\square}} \cdot \sqrt{y} = $ _____

5. $\sqrt[3]{x^{11}} = \sqrt[3]{x^{\square} \cdot x^2} = \sqrt[3]{x^{\square}} \cdot \sqrt[3]{x^2} = $ _____

6. $\sqrt[3]{y^5} = \sqrt[3]{y^{\square} \cdot y^2} = \sqrt[3]{y^{\square}} \cdot \sqrt[3]{y^2} = $ _____

7. $\sqrt[4]{x^{11}} = \sqrt[4]{x^{\square} \cdot x^3} = \sqrt[4]{x^{\square}} \cdot \sqrt[4]{x^3} = $ _____

8. $\sqrt[4]{y^5} = \sqrt[4]{y^{\square} \cdot y} = \sqrt[4]{y^{\square}} \cdot \sqrt[4]{y} = $ _____

For Exercises 9–32, simplify the radicals. **(See Examples 1–5.)**

9. $\sqrt{x^5}$　　**10.** $\sqrt{p^{15}}$　　**11.** $\sqrt[3]{q^7}$　　**12.** $\sqrt[3]{r^{17}}$

13. $\sqrt{a^5 b^4}$　　**14.** $\sqrt{c^9 d^6}$　　**15.** $-\sqrt[4]{x^8 y^{13}}$　　**16.** $-\sqrt[4]{p^{16} q^{17}}$

17. $\sqrt{28}$　　**18.** $\sqrt{63}$　　**19.** $\sqrt{20}$　　**20.** $\sqrt{50}$

21. $5\sqrt{18}$　　**22.** $2\sqrt{24}$　　**23.** $\sqrt[3]{54}$　　**24.** $\sqrt[3]{250}$

25. $\sqrt{25ab^3}$　　**26.** $\sqrt{64m^5 n^{20}}$　　**27.** $\sqrt[3]{40x^7}$　　**28.** $\sqrt[3]{81y^{17}}$

29. $\sqrt[3]{-16x^6 yz^3}$　　**30.** $\sqrt[3]{-192a^6 bc^2}$　　**31.** $\sqrt[4]{80w^4 z^7}$　　**32.** $\sqrt[4]{32p^8 qr^5}$

Chapter 18 Radicals and Complex Numbers

Concept 3: Simplifying Radicals by Using the Order of Operations

For Exercises 33–44, simplify the radical expressions. (See Examples 6–7.)

33. $\sqrt{\dfrac{x^3}{x}}$
34. $\sqrt{\dfrac{y^5}{y}}$
35. $\sqrt{\dfrac{p^7}{p^3}}$
36. $\sqrt{\dfrac{q^{11}}{q^5}}$

37. $\sqrt{\dfrac{50}{2}}$
38. $\sqrt{\dfrac{98}{2}}$
39. $\sqrt[3]{\dfrac{3}{24}}$
40. $\sqrt[3]{\dfrac{2}{250}}$

41. $\dfrac{5\sqrt[3]{16}}{6}$
42. $\dfrac{7\sqrt{18}}{9}$
43. $\dfrac{5\sqrt[3]{72}}{12}$
44. $\dfrac{3\sqrt[3]{250}}{10}$

Mixed Exercises

For Exercises 45–72, simplify the radical expressions.

45. $\sqrt{80}$
46. $\sqrt{108}$
47. $-6\sqrt{75}$
48. $-8\sqrt{8}$

49. $\sqrt{25x^4y^3}$
50. $\sqrt{125p^3q^2}$
51. $\sqrt[3]{27x^2y^3z^4}$
52. $\sqrt[3]{108a^3bc^2}$

53. $\sqrt{\dfrac{12w^5}{3w}}$
54. $\sqrt{\dfrac{64x^9}{4x^3}}$
55. $\sqrt{\dfrac{3y^3}{300y^{15}}}$
56. $\sqrt{\dfrac{4h}{100h^5}}$

57. $\sqrt[3]{\dfrac{16a^2b}{2a^2b^4}}$
58. $\sqrt[3]{\dfrac{-27a^4}{8a}}$
59. $\sqrt{2^3 a^{14} b^8 c^{31} d^{22}}$
60. $\sqrt{7^5 u^{12} v^{20} w^{65} x^{80}}$

61. $\sqrt[3]{54a^6b^4}$
62. $\sqrt[3]{72m^5n^3}$
63. $-5a\sqrt{12a^3b^4c}$
64. $-7y\sqrt{75xy^5z^6}$

65. $\sqrt[4]{7x^5y}$
66. $\sqrt[4]{10cd^7}$
67. $\sqrt{54a^4b^2}$
68. $\sqrt{48r^6s^2}$

69. $\dfrac{2\sqrt{27}}{3}$
70. $\dfrac{7\sqrt{24}}{2}$
71. $\dfrac{3\sqrt{125}}{20}$
72. $\dfrac{10\sqrt{63}}{12}$

For Exercises 73–76, write a mathematical expression for the English phrase and simplify.

73. The quotient of 1 and the cube root of w^6
74. The principal square root of the quotient of h^2 and 49
75. The principal square root of the quantity k raised to the third power
76. The cube root of $2x^4$

For Exercises 77–80, determine the length of the third side of the right triangle. Write the answer as a simplified radical.

77.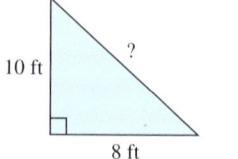

78.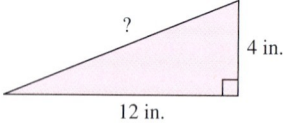

79.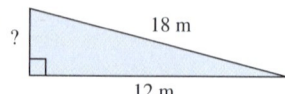

80.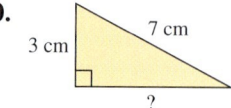

81. On a baseball diamond, the bases are 90 ft apart. Find the exact distance from home plate to second base. Then round to the nearest tenth of a foot.

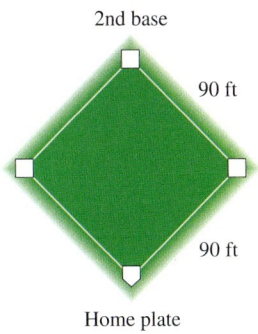

82. Linda is at the beach flying a kite. The kite is directly over a sand castle 60 ft away from Linda. If 100 ft of kite string is out (ignoring any sag in the string), how high is the kite? (Assume that Linda is 5 ft tall.) See figure.

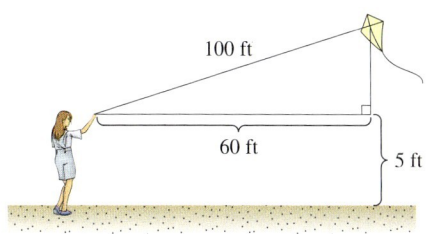

Expanding Your Skills

83. Tom has to travel from town A to town C across a small mountain range. He can travel one of two routes. He can travel on a four-lane highway from A to B and then from B to C at an average speed of 55 mph. Or he can travel on a two-lane road directly from town A to town C, but his average speed will be only 35 mph. If Tom is in a hurry, which route will take him to town C faster?

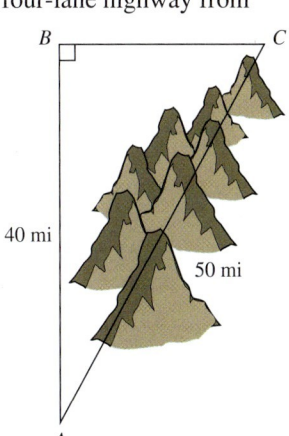

84. One side of a rectangular pasture is 80 ft in length. The diagonal distance is 110 yd. If fencing costs $3.29 per foot, how much will it cost to fence the pasture?

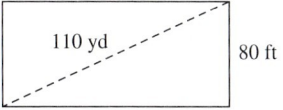

Addition and Subtraction of Radicals

Section 18.4

1. Addition and Subtraction of Radicals

Concept

1. Addition and Subtraction of Radicals

> **Definition of *Like* Radicals**
> Two radical terms are called ***like* radicals** if they have the same index and same radicand.

Like radicals can be added or subtracted by using the distributive property.

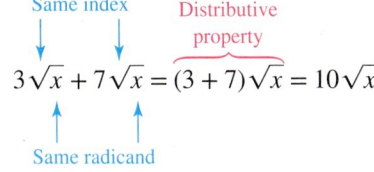

FOR REVIEW

When adding and subtracting *like* radicals, the distributive property is used to factor out the radical. Then the coefficients are added or subtracted as indicated.

$$6\sqrt{11} - 2\sqrt{11} = (6-2)\sqrt{11}$$
$$= 4\sqrt{11}$$

Example 1 — Adding and Subtracting Radicals

Add or subtract as indicated.

a. $6\sqrt{11} - 2\sqrt{11}$ **b.** $\sqrt{3} + \sqrt{3}$

Solution:

a. $6\sqrt{11} - 2\sqrt{11}$
$= (6-2)\sqrt{11}$ Apply the distributive property.
$= 4\sqrt{11}$ Simplify.

b. $\sqrt{3} + \sqrt{3}$
$= 1\sqrt{3} + 1\sqrt{3}$ Note that $\sqrt{3} = 1\sqrt{3}$.
$= (1+1)\sqrt{3}$ Apply the distributive property.
$= 2\sqrt{3}$ Simplify.

Avoiding Mistakes

The process of adding *like* radicals with the distributive property is similar to adding *like* terms. The end result is that the numerical coefficients are added and the radical factor is unchanged.

$$\sqrt{3} + \sqrt{3} = 1\sqrt{3} + 1\sqrt{3} = 2\sqrt{3}$$

Be careful: $\sqrt{3} + \sqrt{3} \neq \sqrt{6}$

In general: $\sqrt{x} + \sqrt{y} \neq \sqrt{x+y}$

Skill Practice Add or subtract as indicated.

1. $5\sqrt{6} - 8\sqrt{6}$ **2.** $\sqrt{10} + \sqrt{10}$

Example 2 — Adding and Subtracting Radicals

Add or subtract as indicated.

a. $-2\sqrt[3]{ab} + 7\sqrt[3]{ab} - \sqrt[3]{ab}$ **b.** $\frac{1}{4}x\sqrt{3y} - \frac{3}{2}x\sqrt{3y}$

Solution:

a. $-2\sqrt[3]{ab} + 7\sqrt[3]{ab} - \sqrt[3]{ab}$
$= (-2 + 7 - 1)\sqrt[3]{ab}$ Apply the distributive property.
$= 4\sqrt[3]{ab}$ Simplify.

b. $\frac{1}{4}x\sqrt{3y} - \frac{3}{2}x\sqrt{3y}$

$= \left(\frac{1}{4} - \frac{3}{2}\right)x\sqrt{3y}$ Apply the distributive property.

$= \left(\frac{1}{4} - \frac{6}{4}\right)x\sqrt{3y}$ Get a common denominator.

$= -\frac{5}{4}x\sqrt{3y}$ Simplify.

Answers
1. $-3\sqrt{6}$ **2.** $2\sqrt{10}$

Section 18.4 Addition and Subtraction of Radicals

Skill Practice Add or subtract as indicated.

3. $5\sqrt[3]{xy} - 3\sqrt[3]{xy} + 7\sqrt[3]{xy}$
4. $\dfrac{5}{6}y\sqrt{2} + \dfrac{1}{4}y\sqrt{2}$

Example 3 shows that it is often necessary to simplify radicals before adding or subtracting.

Example 3 Adding and Subtracting Radicals

Simplify the radicals and add or subtract as indicated. Assume all variables represent positive real numbers.

a. $3\sqrt{8} + \sqrt{2}$ b. $8\sqrt{x^3y^2} - 3y\sqrt{x^3}$

c. $\sqrt{50x^2y^5} - 13y\sqrt{2x^2y^3} + xy\sqrt{98y^3}$

Solution:

a. $3\sqrt{8} + \sqrt{2}$ The radicands are different. Try simplifying the radicals first.

$= 3 \cdot 2\sqrt{2} + \sqrt{2}$ Simplify: $\sqrt{8} = \sqrt{2^3} = 2\sqrt{2}$

$= 6\sqrt{2} + \sqrt{2}$

$= (6 + 1)\sqrt{2}$ Apply the distributive property.

$= 7\sqrt{2}$ Simplify.

b. $8\sqrt{x^3y^2} - 3y\sqrt{x^3}$ The radicands are different. Simplify the radicals first.

$= 8xy\sqrt{x} - 3xy\sqrt{x}$ Simplify $\sqrt{x^3y^2} = xy\sqrt{x}$ and $\sqrt{x^3} = x\sqrt{x}$.

$= (8 - 3)xy\sqrt{x}$ Apply the distributive property.

$= 5xy\sqrt{x}$ Simplify.

c. $\sqrt{50x^2y^5} - 13y\sqrt{2x^2y^3} + xy\sqrt{98y^3}$ Simplify each radical.

$\begin{cases} \sqrt{50x^2y^5} = \sqrt{25 \cdot 2x^2y^5} \\ \qquad = 5xy^2\sqrt{2y} \\ -13y\sqrt{2x^2y^3} = -13xy^2\sqrt{2y} \\ xy\sqrt{98y^3} = xy\sqrt{49 \cdot 2y^3} \\ \qquad = 7xy^2\sqrt{2y} \end{cases}$

$= 5xy^2\sqrt{2y} - 13xy^2\sqrt{2y} + 7xy^2\sqrt{2y}$

$= (5 - 13 + 7)xy^2\sqrt{2y}$ Apply the distributive property.

$= -xy^2\sqrt{2y}$

Skill Practice Simplify the radicals and add or subtract as indicated. Assume all variables represent positive real numbers.

5. $\sqrt{75} + 2\sqrt{3}$
6. $4\sqrt{a^2b} - 6a\sqrt{b}$
7. $-3\sqrt{2y^3} + 5y\sqrt{18y} - 2\sqrt{50y^3}$

Answers

3. $9\sqrt[3]{xy}$ 4. $\dfrac{13}{12}y\sqrt{2}$

5. $7\sqrt{3}$ 6. $-2a\sqrt{b}$

7. $2y\sqrt{2y}$

In some cases, when two radicals are added, the resulting sum is written in factored form. This is demonstrated in Example 4.

> **Example 4** **Adding Radical Expressions**
>
> Add the radicals. Assume that $x \geq 0$. $\quad 3\sqrt{2x^2} + \sqrt{8}$
>
> **Solution:**
>
> $3\sqrt{2x^2} + \sqrt{8}$
>
> $= 3x\sqrt{2} + 2\sqrt{2}$ Simplify each radical. Notice that the radicands are the same, but the terms are not *like* terms. The first term has a factor of x and the second does not.
>
> $= (3x + 2)\sqrt{2}$ Apply the distributive property. The expression cannot be simplified further because $3x$ and 2 are not *like* terms.
>
> **Skill Practice** Add the radicals. Assume that $y \geq 0$.
>
> 8. $4\sqrt{45} - \sqrt{5y^4}$

Answer
8. $(12 - y^2)\sqrt{5}$

Section 18.4 Activity

A.1. Two radical expressions are *like* radicals if they have the same _____ and same _____.

A.2. Determine whether the radicals are *like* radicals. If they are not *like* radicals, explain why.
  a. $\sqrt{7}$ and $\sqrt{3}$
  b. $\sqrt[3]{xy}$ and $\sqrt{xy}$
  c. $4\sqrt[3]{u}$ and $2\sqrt[3]{u}$
  d. $8\sqrt{5x}$ and $\sqrt{5x}$

A.3. Simplify the expressions.
  a. $4x + 7x - 2x$
  b. $4\sqrt{x} + 7\sqrt{x} - 2\sqrt{x}$
  c. $4\sqrt{2} + 7\sqrt{2} - 2\sqrt{2}$
  d. Discuss the similarity in the process to simplify each expression in parts (a)–(c).

A.4. Given the expression $16\sqrt{3t} - 12\sqrt{3t} + \sqrt{3t}$,
  a. Identify the coefficients of each term.
  b. Simplify the expression.

A.5. Simplify the expression by following these steps.
$3y\sqrt{5x^5} - 2xy\sqrt{20x^3} + 2x^2\sqrt{45xy^2}$
  a. Simplify $3y\sqrt{5x^5}$.
  b. Simplify $-2xy\sqrt{20x^3}$.
  c. Simplify $2x^2\sqrt{45xy^2}$.
  d. Write the original expression with the radicals in simplified form.
  e. Simplify the expression.

A.6. Consider the expression $3\sqrt{7x^2} + 5\sqrt{7}$.
  a. Write the expression with the radicals simplified.
  b. Add the radicals using the distributive property.

Practice Exercises

Section 18.4

Prerequisite Review

For Exercises R.1–R.2, determine whether the terms are *like* terms.

R.1. **a.** $3y$ and $\frac{1}{4}y$ **b.** $9a$ and $9b$ **c.** $2x^2y$ and $4x^2y$ **d.** $10x$ and 10

R.2. **a.** $-6p$ and -6 **b.** $\frac{3}{4}y^3$ and $6y^3$ **c.** $-11a^4b$ and $4ab^4$ **d.** $2.1x$ and $\frac{2}{5}x$

For Exercises R.3–R.6, combine *like* terms.

R.3. $4y - 6y + y$

R.4. $8p - p + 11p$

R.5. $\frac{3}{4}c^2d + \frac{1}{4}c^2d$

R.6. $\frac{2}{5}x^3y + \frac{8}{5}x^3y$

For Exercises R.7–R.10, simplify the expression. Assume that all variables represent positive real numbers.

R.7. $\sqrt{27}$

R.8. $\sqrt{84}$

R.9. $\sqrt[3]{250x^3y^4}$

R.10. $\sqrt[3]{16a^6b^8}$

For the exercises in this set, assume that all variables represent positive real numbers, unless otherwise stated.

Vocabulary and Key Concepts

1. **a.** Two radical terms are called *like* radicals if they have the same _____ and the same _____.
   b. The expression $\sqrt{3x} + \sqrt{3x}$ simplifies to _____.
   c. The expression $\sqrt{2} + \sqrt{3}$ (can/cannot) be simplified further, whereas the expression $\sqrt{2} \cdot \sqrt{3}$ (can/cannot) be simplified further.

2. **a.** The expression $\sqrt{18}$ simplifies to _____.
   b. The expression $\sqrt{2} + \sqrt{18}$ simplifies to _____.

Concept 1: Addition and Subtraction of Radicals

For Exercises 3–8, determine if the radical terms are *like*.

3. $\sqrt{2}$ and $\sqrt[3]{2}$

4. $7\sqrt[3]{x}$ and $\sqrt[3]{x}$

5. $\sqrt{2}$ and $3\sqrt{2}$

6. $\sqrt[3]{x}$ and $\sqrt[4]{x}$

7. $\sqrt{2}$ and $\sqrt{5}$

8. $2\sqrt[4]{x}$ and $x\sqrt[4]{2}$

9. Explain the similarities between the pairs of expressions.
   a. $7\sqrt{5} + 4\sqrt{5}$ and $7x + 4x$
   b. $-2\sqrt{6} - 9\sqrt{3}$ and $-2x - 9y$

10. Explain the similarities between the pairs of expressions.
    a. $-4\sqrt{3} + 5\sqrt{3}$ and $-4z + 5z$
    b. $13\sqrt{7} - 18$ and $13a - 18$

For Exercises 11–14, simplify the expressions.

11. **a.** $3x + 5x$
    b. $3\sqrt{x} + 5\sqrt{x}$

12. **a.** $-11m + 6m$
    b. $-11\sqrt{m} + 6\sqrt{m}$

13. **a.** $-8t + t$
    b. $-8\sqrt[3]{t} + \sqrt[3]{t}$

14. **a.** $-c + 7c$
    b. $-\sqrt[4]{c} + 7\sqrt[4]{c}$

For Exercises 15–32, add or subtract the radical expressions, if possible. (See Examples 1–2.)

15. $3\sqrt{5} + 6\sqrt{5}$

16. $5\sqrt{a} + 3\sqrt{a}$

17. $3\sqrt[3]{tw} - 2\sqrt[3]{tw} + \sqrt[3]{tw}$

18. $6\sqrt[3]{7} - 2\sqrt[3]{7} + \sqrt[3]{7}$

19. $6\sqrt{10} - \sqrt{10}$

20. $13\sqrt{11} - \sqrt{11}$

21. $\sqrt[4]{3} + 7\sqrt[4]{3} - \sqrt[4]{14}$

22. $2\sqrt{11} + 3\sqrt{13} + 5\sqrt{11}$

23. $8\sqrt{x} + 2\sqrt{y} - 6\sqrt{x}$

24. $10\sqrt{10} - 8\sqrt{10} + \sqrt{2}$

25. $\sqrt[3]{ab} + a\sqrt[3]{b}$

26. $x\sqrt[4]{y} - y\sqrt[4]{x}$

27. $\sqrt{2t} + \sqrt[3]{2t}$

28. $\sqrt[4]{5c} + \sqrt[3]{5c}$

29. $\frac{5}{6}z\sqrt[3]{6} + \frac{7}{9}z\sqrt[3]{6}$

30. $\frac{3}{4}a\sqrt[4]{b} + \frac{1}{6}a\sqrt[4]{b}$

31. $0.81x\sqrt{y} - 0.11x\sqrt{y}$

32. $7.5\sqrt{pq} - 6.3\sqrt{pq}$

33. Explain the process for adding the two radicals. Then find the sum. $3\sqrt{2} + 7\sqrt{50}$

34. Explain the process for subtracting two radicals. Then find the difference. $\sqrt{12x} - \sqrt{75x}$

For Exercises 35–64, add or subtract the radical expressions as indicated. (See Examples 3–4.)

35. $\sqrt{36} + \sqrt{81}$

36. $3\sqrt{80} - 5\sqrt{45}$

37. $2\sqrt{12} + \sqrt{48}$

38. $5\sqrt{32} + 2\sqrt{50}$

39. $4\sqrt{7} + \sqrt{63} - 2\sqrt{28}$

40. $8\sqrt{3} - 2\sqrt{27} + \sqrt{75}$

41. $5\sqrt{18} + \sqrt{27} - 4\sqrt{50}$

42. $7\sqrt{40} - \sqrt{8} + 4\sqrt{50}$

43. $\sqrt[3]{81} - \sqrt[3]{24}$

44. $17\sqrt[3]{81} - 2\sqrt[3]{24}$

45. $3\sqrt{2a} - \sqrt{8a} - \sqrt{72a}$

46. $\sqrt{12t} - \sqrt{27t} + 5\sqrt{3t}$

47. $2s^2\sqrt[3]{s^2 t^6} + 3t^2\sqrt[3]{8s^8}$

48. $4\sqrt[3]{x^4} - 2x\sqrt[3]{x}$

49. $7\sqrt[3]{x^4} - x\sqrt[3]{x}$

50. $6\sqrt[3]{y^{10}} - 3y^2\sqrt[3]{y^4}$

51. $5p\sqrt{20p^2} + p^2\sqrt{80}$

52. $2q\sqrt{48q^2} - \sqrt{27q^4}$

53. $\sqrt[3]{a^2 b} - \sqrt[3]{8a^2 b}$

54. $w\sqrt{80} - 3\sqrt{125w^2}$

55. $5x\sqrt{x} + 6\sqrt{x}$

56. $9y^2\sqrt{2} + 4\sqrt{2}$

57. $\sqrt{50x^2} - 3\sqrt{8}$

58. $\sqrt{9x^3} - \sqrt{25x}$

59. $11\sqrt[3]{54cd^3} - 2\sqrt[3]{2cd^3} + d\sqrt[3]{16c}$

60. $x\sqrt[3]{64x^5 y^2} - x^2\sqrt[3]{x^2 y^2} + 5\sqrt[3]{x^8 y^2}$

61. $\frac{3}{2}ab\sqrt{24a^3} + \frac{4}{3}\sqrt{54a^5 b^2} - a^2 b\sqrt{150a}$

62. $mn\sqrt{72n} + \frac{2}{3}n\sqrt{8m^2 n} - \frac{5}{6}\sqrt{50m^2 n^3}$

63. $x\sqrt[3]{16} - 2\sqrt[3]{27x} + \sqrt[3]{54x^3}$

64. $5\sqrt[4]{y^5} - 2y\sqrt[4]{y} + \sqrt[4]{16y^7}$

Mixed Exercises

For Exercises 65–72, answer true or false. If an answer is false, explain why or give a counter example.

65. $\sqrt{x} + \sqrt{y} = \sqrt{x+y}$
66. $\sqrt{x} + \sqrt{x} = 2\sqrt{x}$
67. $5\sqrt[3]{x} + 2\sqrt[3]{x} = 7\sqrt[3]{x}$
68. $6\sqrt{x} + 5\sqrt[3]{x} = 11\sqrt{x}$
69. $\sqrt{y} + \sqrt{y} = \sqrt{2y}$
70. $\sqrt{c^2 + d^2} = c + d$
71. $2w\sqrt{5} + 4w\sqrt{5} = 6w^2\sqrt{5}$
72. $7x\sqrt{3} - 2\sqrt{3} = (7x - 2)\sqrt{3}$

For Exercises 73–76, write the English phrase as an algebraic expression. Simplify each expression.

73. The sum of the principal square root of 48 and the principal square root of 12
74. The sum of the cube root of 16 and the cube root of 2
75. The difference of 5 times the cube root of x^6 and the square of x
76. The sum of the cube of y and the principal fourth root of y^{12}

For Exercises 77–80, write an English phrase that translates the mathematical expression. (Answers may vary.)

77. $\sqrt{18} - 5^2$
78. $4^3 - \sqrt[3]{4}$
79. $\sqrt[4]{x} + y^3$
80. $a^4 + \sqrt{a}$

For Exercises 81–82, find the exact value of the perimeter, and then approximate the value to one decimal place.

81.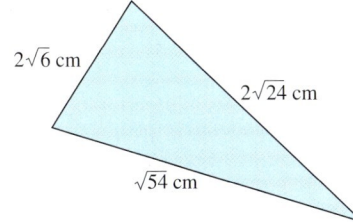

82.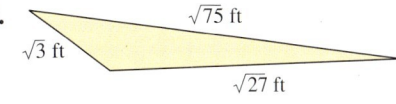

83. The figure has perimeter $14\sqrt{2}$ ft. Find the value of x.

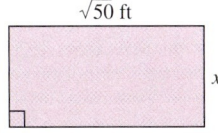

84. The figure has perimeter $12\sqrt{7}$ cm. Find the value of x.

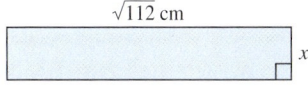

Expanding Your Skills

85. **a.** An irregularly shaped garden is shown in the figure. All distances are expressed in yards. Find the perimeter. (*Hint:* Use the Pythagorean theorem to find the length of each side.) Write the final answer in radical form.

   b. Approximate your answer to two decimal places.

   c. If edging costs $1.49 per foot and sales tax is 6%, find the total cost of edging the garden.

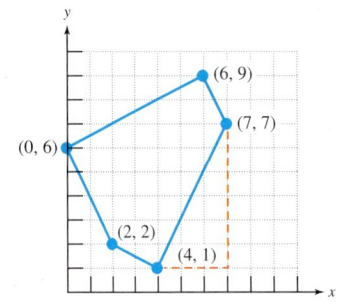

Section 18.5 Multiplication of Radicals

Concepts

1. Multiplication Property of Radicals
2. Expressions of the Form $(\sqrt[n]{a})^n$
3. Special Case Products
4. Multiplying Radicals With Different Indices

1. Multiplication Property of Radicals

In this section, we will learn how to multiply radicals by using the multiplication property of radicals.

> **The Multiplication Property of Radicals**
> Let a and b represent real numbers such that $\sqrt[n]{a}$ and $\sqrt[n]{b}$ are both real. Then
> $$\sqrt[n]{ab} = \sqrt[n]{a} \cdot \sqrt[n]{b}$$

To multiply two radical expressions, we use the multiplication property of radicals along with the commutative and associative properties of multiplication.

Example 1 — Multiplying Radical Expressions

Multiply each expression and simplify the result. Assume all variables represent positive real numbers.

a. $(3\sqrt{2})(5\sqrt{6})$ b. $(2x\sqrt{y})(-7\sqrt{xy})$ c. $(15c\sqrt[3]{cd})\left(\frac{1}{3}\sqrt[3]{cd^2}\right)$

Solution:

a. $(3\sqrt{2})(5\sqrt{6})$

$= (3 \cdot 5)(\sqrt{2} \cdot \sqrt{6})$ Commutative and associative properties of multiplication

$= 15\sqrt{12}$ Multiplication property of radicals

$= 15\sqrt{2^2 \cdot 3}$

$= 15 \cdot 2\sqrt{3}$ Simplify the radical.

$= 30\sqrt{3}$

b. $(2x\sqrt{y})(-7\sqrt{xy})$

$= (2x)(-7)(\sqrt{y} \cdot \sqrt{xy})$ Commutative and associative properties of multiplication

$= -14x\sqrt{xy^2}$ Multiplication property of radicals

$= -14xy\sqrt{x}$ Simplify the radical.

c. $(15c\sqrt[3]{cd})\left(\frac{1}{3}\sqrt[3]{cd^2}\right)$

$= \left(15c \cdot \frac{1}{3}\right)(\sqrt[3]{cd} \cdot \sqrt[3]{cd^2})$ Commutative and associative properties of multiplication

$= 5c\sqrt[3]{c^2 d^3}$ Multiplication property of radicals

$= 5cd\sqrt[3]{c^2}$ Simplify the radical.

Skill Practice Multiply the expressions and simplify the results. Assume all variables represent positive real numbers.

1. $(4\sqrt{6})(-2\sqrt{3})$
2. $(3ab\sqrt{b})(-2\sqrt{ab})$
3. $(2\sqrt[3]{4ab})(5\sqrt[3]{2a^2b})$

When multiplying radical expressions with more than one term, we use the distributive property.

Example 2 — Multiplying Radical Expressions

Multiply. $3\sqrt{11}(2+\sqrt{11})$

Solution:

$3\sqrt{11}(2+\sqrt{11})$

$= 3\sqrt{11} \cdot (2) + 3\sqrt{11} \cdot \sqrt{11}$ Apply the distributive property.

$= 6\sqrt{11} + 3\sqrt{11^2}$ Multiplication property of radicals

$= 6\sqrt{11} + 3 \cdot 11$ Simplify the radical.

$= 6\sqrt{11} + 33$

FOR REVIEW

Compare the multiplication of radicals to the multiplication of polynomials.

$3(x+2) = 3 \cdot x + 3 \cdot 2$

$3\sqrt{11}(2+\sqrt{11})$
$= 3\sqrt{11} \cdot 2 + 3\sqrt{11} \cdot \sqrt{11}$

Skill Practice Multiply.

4. $5\sqrt{5}(2\sqrt{5}-2)$

Example 3 — Multiplying Radical Expressions

Multiply.

a. $(\sqrt{5}+3\sqrt{2})(2\sqrt{5}-\sqrt{2})$ b. $(-10a\sqrt{b}+7b)(a\sqrt{b}+2b)$

Solution:

a. $(\sqrt{5}+3\sqrt{2})(2\sqrt{5}-\sqrt{2})$

$= 2\sqrt{5^2} - \sqrt{10} + 6\sqrt{10} - 3\sqrt{2^2}$ Apply the distributive property.

$= 2 \cdot 5 + 5\sqrt{10} - 3 \cdot 2$ Simplify radicals and combine *like* radicals.

$= 10 + 5\sqrt{10} - 6$

$= 4 + 5\sqrt{10}$ Combine *like* terms.

b. $(-10a\sqrt{b}+7b)(a\sqrt{b}+2b)$

$= -10a^2\sqrt{b^2} - 20ab\sqrt{b} + 7ab\sqrt{b} + 14b^2$ Apply the distributive property.

$= -10a^2b - 13ab\sqrt{b} + 14b^2$ Simplify and combine *like* terms.

Answers
1. $-24\sqrt{2}$ 2. $-6ab^2\sqrt{a}$
3. $20a\sqrt[3]{b^2}$ 4. $50 - 10\sqrt{5}$

Skill Practice Multiply.

5. $(2\sqrt{3} - 3\sqrt{10})(\sqrt{3} + 2\sqrt{10})$
6. $(x\sqrt{y} + y)(3x\sqrt{y} - 2y)$

Example 4 Multiplying Radical Expressions

Multiply. $(2\sqrt{x} + \sqrt{y})(6 - \sqrt{x} + 8\sqrt{y})$

Solution:

$(2\sqrt{x} + \sqrt{y})(6 - \sqrt{x} + 8\sqrt{y})$

$= 12\sqrt{x} - 2\sqrt{x^2} + 16\sqrt{xy} + 6\sqrt{y} - \sqrt{xy} + 8\sqrt{y^2}$ Apply the distributive property.

$= 12\sqrt{x} - 2x + 16\sqrt{xy} + 6\sqrt{y} - \sqrt{xy} + 8y$ Simplify the radicals.

$= 12\sqrt{x} - 2x + 15\sqrt{xy} + 6\sqrt{y} + 8y$ Combine *like* terms.

Skill Practice Multiply.

7. $(\sqrt{t} + 3\sqrt{w})(4 - \sqrt{t} - \sqrt{w})$

2. Expressions of the Form $(\sqrt[n]{a})^n$

The multiplication property of radicals can be used to simplify an expression of the form $(\sqrt{a})^2$, where $a \geq 0$.

$$(\sqrt{a})^2 = \sqrt{a} \cdot \sqrt{a} = \sqrt{a^2} = a \qquad \text{where } a \geq 0$$

This logic can be applied to *n*th roots.

- If $\sqrt[n]{a}$ is a real number, then $(\sqrt[n]{a})^n = a$.

Example 5 Simplifying Radical Expressions

Simplify the expressions. Assume all variables represent positive real numbers.

a. $(\sqrt{11})^2$ b. $(\sqrt[5]{z})^5$ c. $(\sqrt[3]{pq})^3$

Solution:

a. $(\sqrt{11})^2 = 11$ b. $(\sqrt[5]{z})^5 = z$ c. $(\sqrt[3]{pq})^3 = pq$

Skill Practice Simplify.

8. $(\sqrt{14})^2$ 9. $(\sqrt[7]{q})^7$ 10. $(\sqrt[5]{3z})^5$

Answers

5. $-54 + \sqrt{30}$
6. $3x^2y + xy\sqrt{y} - 2y^2$
7. $4\sqrt{t} - t - 4\sqrt{tw} + 12\sqrt{w} - 3w$
8. 14   9. q 10. $3z$

3. Special Case Products

From Examples 2–4, you may have noticed a similarity between multiplying radical expressions and multiplying polynomials.

Recall that the square of a binomial results in a perfect square trinomial:

$$(a+b)^2 = a^2 + 2ab + b^2$$
$$(a-b)^2 = a^2 - 2ab + b^2$$

The same patterns occur when squaring a radical expression with two terms.

Example 6 — Squaring a Two-Term Radical Expression

Square the radical expressions. Assume all variables represent positive real numbers.

a. $(\sqrt{d} + 3)^2$ **b.** $(5\sqrt{y} - \sqrt{2})^2$

Solution:

a. $(\sqrt{d} + 3)^2$ — This expression is in the form $(a+b)^2$, where $a = \sqrt{d}$ and $b = 3$.

$\qquad a^2 + 2ab + b^2$ — Apply the formula

$= (\sqrt{d})^2 + 2(\sqrt{d})(3) + (3)^2$ — $(a+b)^2 = a^2 + 2ab + b^2$.

$= d + 6\sqrt{d} + 9$ — Simplify.

TIP: The product $(\sqrt{d} + 3)^2$ can also be found by using the distributive property.

$$(\sqrt{d} + 3)^2 = (\sqrt{d} + 3)(\sqrt{d} + 3)$$

b. $(5\sqrt{y} - \sqrt{2})^2$ — This expression is in the form $(a-b)^2$, where $a = 5\sqrt{y}$ and $b = \sqrt{2}$.

$\qquad a^2 - 2ab + b^2$ — Apply the formula

$= (5\sqrt{y})^2 - 2(5\sqrt{y})(\sqrt{2}) + (\sqrt{2})^2$ — $(a-b)^2 = a^2 - 2ab + b^2$.

$= 25y - 10\sqrt{2y} + 2$ — Simplify.

Skill Practice Square the radical expressions. Assume all variables represent positive real numbers.

11. $(\sqrt{a} - 5)^2$ **12.** $(4\sqrt{x} + \sqrt{3})^2$

Recall that the product of two conjugate binomials results in a difference of squares.

$$(a+b)(a-b) = a^2 - b^2$$

The same pattern occurs when multiplying two conjugate radical expressions.

Example 7 — Multiplying Conjugate Radical Expressions

Multiply the radical expressions. Assume all variables represent positive real numbers.

a. $(\sqrt{3} + 2)(\sqrt{3} - 2)$ **b.** $\left(\dfrac{1}{3}\sqrt{s} - \dfrac{3}{4}\sqrt{t}\right)\left(\dfrac{1}{3}\sqrt{s} + \dfrac{3}{4}\sqrt{t}\right)$

Answers
11. $a - 10\sqrt{a} + 25$
12. $16x + 8\sqrt{3x} + 3$

TIP: The product $(\sqrt{3}+2)(\sqrt{3}-2)$ can also be found by using the distributive property.

$(\sqrt{3}+2)(\sqrt{3}-2)$

Solution:

a. $(\sqrt{3}+2)(\sqrt{3}-2)$ The expression is in the form $(a+b)(a-b)$, where $a=\sqrt{3}$ and $b=2$.

$\qquad\;\; a^2 - b^2$

$= (\sqrt{3})^2 - (2)^2$ Apply the formula $(a+b)(a-b) = a^2 - b^2$.

$= 3 - 4$ Simplify.

$= -1$

b. $\left(\dfrac{1}{3}\sqrt{s} - \dfrac{3}{4}\sqrt{t}\right)\left(\dfrac{1}{3}\sqrt{s} + \dfrac{3}{4}\sqrt{t}\right)$ This expression is in the form $(a-b)(a+b)$, where $a = \dfrac{1}{3}\sqrt{s}$ and $b = \dfrac{3}{4}\sqrt{t}$.

$\qquad\qquad a^2 - b^2$

$= \left(\dfrac{1}{3}\sqrt{s}\right)^2 - \left(\dfrac{3}{4}\sqrt{t}\right)^2$ Apply the formula $(a+b)(a-b) = a^2 - b^2$.

$= \dfrac{1}{9}s - \dfrac{9}{16}t$ Simplify.

Skill Practice Multiply the conjugates. Assume all variables represent positive real numbers.

13. $(\sqrt{5}+3)(\sqrt{5}-3)$ **14.** $\left(\dfrac{1}{2}\sqrt{a} + \dfrac{2}{5}\sqrt{b}\right) \cdot \left(\dfrac{1}{2}\sqrt{a} - \dfrac{2}{5}\sqrt{b}\right)$

4. Multiplying Radicals With Different Indices

The product of two radicals can be simplified provided the radicals have the same index. If the radicals have different indices, then we can use the properties of rational exponents to obtain a common index.

Example 8   **Multiplying Radicals With Different Indices**

Multiply the expressions. Write the answers in radical form.

a. $\sqrt[3]{5} \cdot \sqrt[4]{5}$ **b.** $\sqrt[3]{7} \cdot \sqrt{2}$

Solution:

a. $\sqrt[3]{5} \cdot \sqrt[4]{5}$

$= 5^{1/3} 5^{1/4}$ Rewrite each expression with rational exponents.

$= 5^{(1/3)+(1/4)}$ Because the bases are equal, we can add the exponents.

$= 5^{(4/12)+(3/12)}$ Write the fractions with a common denominator.

$= 5^{7/12}$ Simplify the exponent.

$= \sqrt[12]{5^7}$ Rewrite the expression as a radical.

Answers

13. -4 **14.** $\dfrac{1}{4}a - \dfrac{4}{25}b$

b. $\sqrt[3]{7} \cdot \sqrt{2}$

$\quad = 7^{1/3} 2^{1/2}$ Rewrite each expression with rational exponents.

$\quad = 7^{2/6} 2^{3/6}$ Write the rational exponents with a common denominator.

$\quad = (7^2 2^3)^{1/6}$ Apply the power rule of exponents.

$\quad = \sqrt[6]{7^2 2^3}$ Rewrite the expression as a single radical.

$\quad = \sqrt[6]{392}$ Simplify.

Skill Practice Multiply the expressions. Write the answers in radical form. Assume all variables represent positive real numbers.

15. $\sqrt{x} \cdot \sqrt[3]{x}$ 16. $\sqrt[4]{a^3} \cdot \sqrt[3]{b}$

Answers
15. $\sqrt[6]{x^5}$ 16. $\sqrt[12]{a^9 b^4}$

Section 18.5 Activity

A.1. Recall that the multiplication property of radicals indicates that $\sqrt[n]{a} \cdot \sqrt[n]{b} = $ _____, provided that the two radicals represent real numbers.

A.2. a. Multiply. $(5x) \cdot (2x)$
   b. Multiply. $(5\sqrt{x}) \cdot (2\sqrt{x})$
   c. Multiply. $(5\sqrt[3]{x}) \cdot (2\sqrt[3]{x})$
   d. Compare the process to multiply radicals in parts (b) and (c).

For Exercises A.3–A.6, multiply the expressions.

A.3. a. $\frac{1}{2}x(4x^2 + 8xy + 2y^2)$ b. $\frac{1}{2}\sqrt{x}(4\sqrt{x} + 8\sqrt{xy} + 2\sqrt{y})$

A.4. a. $(2x + 3)(5x - 6)$ b. $(2\sqrt{x} + 3)(5\sqrt{x} - 6)$
   c. $(2\sqrt{x} + \sqrt{3})(5\sqrt{x} - \sqrt{6})$

A.5. a. $(5x - 3)^2$ b. $(5\sqrt{x} - 3)^2$
   c. $(5\sqrt{x} - \sqrt{3})^2$

A.6. a. $(2y - 7)(2y + 7)$ b. $(2\sqrt{y} - 7)(2\sqrt{y} + 7)$
   c. $(2\sqrt{y} - \sqrt{7})(2\sqrt{y} + \sqrt{7})$

A.7. To multiply two radical expressions with different indices, write the radicals with rational exponents. Then write the rational exponents with a common denominator. To illustrate this process, simplify $\sqrt[3]{7} \cdot \sqrt[5]{2}$ by following these steps.
   a. Write $\sqrt[3]{7}$ using rational exponents.
   b. Write $\sqrt[5]{2}$ using rational exponents.
   c. Write the expression $\sqrt[3]{7} \cdot \sqrt[5]{2}$ using rational exponents.
   d. Write the expression from part (c) with a common denominator in the exponents.
   e. Convert the expression from part (d) as a product of radicals.
   f. Multiply the radicals. Leave the radicand in factored form.

Section 18.5 Practice Exercises

Prerequisite Review

For Exercises R.1–R.10, simplify the expression.

R.1. $(5x^2)(-8xy)$

R.2. $(-6t^2w)(-3t^3)$

R.3. $-\dfrac{1}{5}a(15a^2 + 25ab - 10b^2)$

R.4. $\dfrac{1}{4}m(4m^2 - 8mn + 24n^2)$

R.5. $(3c + 4)(c^2 - 2c + 5)$

R.6. $(5d - 1)(2d^2 + 3d + 7)$

R.7. $(2a - 5b)^2$

R.8. $(3m + 9n)^2$

R.9. $\left(\dfrac{1}{2}m + 4n\right)\left(\dfrac{1}{2}m - 4n\right)$

R.10. $\left(\dfrac{3}{5}x - y\right)\left(\dfrac{3}{5}x + y\right)$

For Exercises R.11–R.12, add or subtract as indicated.

R.11. $-2y\sqrt{28} + 4\sqrt{63y^2}$

R.12. $4\sqrt{8x^3} - x\sqrt{50x}$

For the exercises in this set, assume that all variables represent positive real numbers, unless otherwise stated.

Vocabulary and Key Concepts

1. **a.** If $\sqrt[n]{a}$ and $\sqrt[n]{b}$ are real numbers, then $\sqrt[n]{a} \cdot \sqrt[n]{b} = $ _____.

   b. If $x \geq 0$, then $\sqrt{x} \cdot \sqrt{x} = $ _____.

   c. If $\sqrt[n]{a}$ is a real number, then $(\sqrt[n]{a})^n = $ _____.

   d. Two binomials $(x + \sqrt{2})$ and $(x - \sqrt{2})$ are called _____ of each other, and their product is $(x)^2 - (\sqrt{2})^2$.

   e. If $m \geq 0$ and $n \geq 0$, then $(\sqrt{m} + \sqrt{n})(\sqrt{m} - \sqrt{n}) = $ _____.

2. Which is the correct simplification of $(\sqrt{c} + 4)^2$?

   $c + 16$ or $c + 8\sqrt{c} + 16$

Concept 1: Multiplication Property of Radicals

For Exercises 3–8, fill in the blanks and simplify the expressions. Assume that the variables represent positive real numbers.

3. **a.** $x \cdot x$

   b. $\sqrt{x} \cdot \sqrt{x} = \sqrt{x^\square} = $ _____

4. **a.** $y^2 \cdot y$

   b. $\sqrt[3]{y^2} \cdot \sqrt[3]{y} = \sqrt[3]{y^\square} = $ _____

5. **a.** $t \cdot t^5$

   b. $\sqrt[3]{t} \cdot \sqrt[3]{t^5} = \sqrt[3]{\square} = $ _____

6. **a.** $m^3 \cdot m^5$

   b. $\sqrt[4]{m^3} \cdot \sqrt[4]{m^5} = \sqrt[4]{\square} = $ _____

7. **a.** $(5x)(10x)$

   b. $\sqrt{5x} \cdot \sqrt{10x} = \sqrt{\square} = $ _____

8. **a.** $(6y)(3y)$

   b. $\sqrt{6y} \cdot \sqrt{3y} = \sqrt{\square} = $ _____

For Exercises 9–44, multiply the radical expressions. (See Examples 1–4.)

9. $\sqrt[3]{7} \cdot \sqrt[3]{3}$
10. $\sqrt[4]{6} \cdot \sqrt[4]{2}$
11. $\sqrt{2} \cdot \sqrt{10}$
12. $\sqrt[3]{4} \cdot \sqrt[3]{12}$
13. $\sqrt[4]{16} \cdot \sqrt[4]{64}$
14. $\sqrt{5x^3} \cdot \sqrt{10x^4}$
15. $(4\sqrt[3]{4})(2\sqrt[3]{5})$
16. $(2\sqrt{5})(3\sqrt{7})$
17. $(8a\sqrt{b})(-3\sqrt{ab})$
18. $(p\sqrt[4]{q^3})(\sqrt[4]{pq})$
19. $\sqrt{30} \cdot \sqrt{12}$
20. $\sqrt{20} \cdot \sqrt{54}$
21. $\sqrt{6x} \cdot \sqrt{12x}$
22. $(\sqrt{3ab^2})(\sqrt{21a^2b})$
23. $\sqrt{5a^3b^2} \cdot \sqrt{20a^3b^3}$
24. $\sqrt[3]{m^2n^2} \cdot \sqrt[3]{48m^4n^2}$
25. $(4\sqrt{3xy^3})(-2\sqrt{6x^3y^2})$
26. $(2\sqrt[4]{3x})(4\sqrt[4]{27x^6})$
27. $(\sqrt[3]{4a^2b})(\sqrt[3]{2ab^3})(\sqrt[3]{54a^2b})$
28. $(\sqrt[3]{9x^3y})(\sqrt[3]{6xy})(\sqrt[3]{8x^2y^5})$
29. $\sqrt{3}(4\sqrt{3} - 6)$
30. $3\sqrt{5}(2\sqrt{5} + 4)$
31. $\sqrt{2}(\sqrt{6} - \sqrt{3})$
32. $\sqrt{5}(\sqrt{3} + \sqrt{7})$
33. $-\dfrac{1}{3}\sqrt{x}(6\sqrt{x} + 7)$
34. $-\dfrac{1}{2}\sqrt{y}(8 - 3\sqrt{y})$
35. $(\sqrt{3} + 2\sqrt{10})(4\sqrt{3} - \sqrt{10})$
36. $(8\sqrt{7} - \sqrt{5})(\sqrt{7} + 3\sqrt{5})$
37. $(\sqrt{x} + 4)(\sqrt{x} - 9)$
38. $(\sqrt{w} - 2)(\sqrt{w} - 9)$
39. $(\sqrt[3]{y} + 2)(\sqrt[3]{y} - 3)$
40. $(4 + \sqrt[5]{p})(5 + \sqrt[5]{p})$
41. $(\sqrt{a} - 3\sqrt{b})(9\sqrt{a} - \sqrt{b})$
42. $(11\sqrt{m} + 4\sqrt{n})(\sqrt{m} + \sqrt{n})$
43. $(\sqrt{p} + 2\sqrt{q})(8 + 3\sqrt{p} - \sqrt{q})$
44. $(5\sqrt{s} - \sqrt{t})(\sqrt{s} + 5 + 6\sqrt{t})$

Concept 2: Expressions of the Form $(\sqrt[n]{a})^n$

For Exercises 45–52, simplify the expressions. Assume all variables represent positive real numbers. (See Example 5.)

45. $(\sqrt{15})^2$
46. $(\sqrt{58})^2$
47. $(\sqrt{3y})^2$
48. $(\sqrt{19yz})^2$
49. $(\sqrt[3]{6})^3$
50. $(\sqrt[5]{24})^5$
51. $\sqrt{709} \cdot \sqrt{709}$
52. $\sqrt{401} \cdot \sqrt{401}$

Concept 3: Special Case Products

53. a. Write the formula for the product of two conjugates. $(x + y)(x - y) =$
    b. Multiply. $(x + 5)(x - 5)$

54. a. Write the formula for squaring a binomial. $(x + y)^2 =$
    b. Multiply. $(x + 5)^2$

For Exercises 55–66, multiply the radical expressions. (See Examples 6–7.)

55. $(\sqrt{13} + 4)^2$
56. $(6 - \sqrt{11})^2$
57. $(\sqrt{p} - \sqrt{7})^2$
58. $(\sqrt{q} + \sqrt{2})^2$
59. $(\sqrt{2a} - 3\sqrt{b})^2$
60. $(\sqrt{3w} + 4\sqrt{z})^2$
61. $(\sqrt{3} + x)(\sqrt{3} - x)$
62. $(y + \sqrt{6})(y - \sqrt{6})$
63. $(\sqrt{6} + \sqrt{2})(\sqrt{6} - \sqrt{2})$
64. $(\sqrt{15} + \sqrt{5})(\sqrt{15} - \sqrt{5})$
65. $\left(\dfrac{2}{3}\sqrt{x} + \dfrac{1}{2}\sqrt{y}\right)\left(\dfrac{2}{3}\sqrt{x} - \dfrac{1}{2}\sqrt{y}\right)$
66. $\left(\dfrac{1}{4}\sqrt{s} + \dfrac{1}{5}\sqrt{t}\right)\left(\dfrac{1}{4}\sqrt{s} - \dfrac{1}{5}\sqrt{t}\right)$

Chapter 18 Radicals and Complex Numbers

For Exercises 67–68, multiply the expressions.

67.
a. $(\sqrt{3} + \sqrt{x})(\sqrt{3} - \sqrt{x})$
b. $(\sqrt{3} + \sqrt{x})(\sqrt{3} + \sqrt{x})$
c. $(\sqrt{3} - \sqrt{x})(\sqrt{3} - \sqrt{x})$

68.
a. $(\sqrt{5} + \sqrt{y})(\sqrt{5} - \sqrt{y})$
b. $(\sqrt{5} + \sqrt{y})(\sqrt{5} + \sqrt{y})$
c. $(\sqrt{5} - \sqrt{y})(\sqrt{5} - \sqrt{y})$

Mixed Exercises

For Exercises 69–76, identify each statement as true or false. If an answer is false, explain why.

69. $\sqrt{3} \cdot \sqrt{2} = \sqrt{6}$

70. $\sqrt{5} \cdot \sqrt[3]{2} = \sqrt{10}$

71. $(x - \sqrt{5})^2 = x - 5$

72. $3(2\sqrt{5x}) = 6\sqrt{5x}$

73. $5(3\sqrt{4x}) = 15\sqrt{20x}$

74. $\dfrac{\sqrt{5x}}{5} = \sqrt{x}$

75. $\dfrac{3\sqrt{x}}{3} = \sqrt{x}$

76. $(\sqrt{t} - 1)(\sqrt{t} + 1) = t - 1$

For Exercises 77–84, perform the indicated operations.

77. $(-\sqrt{6x})^2$

78. $(-\sqrt{8a})^2$

79. $(\sqrt{3x+1})^2$

80. $(\sqrt{x-1})^2$

81. $(\sqrt{x+3} - 4)^2$

82. $(\sqrt{x+1} + 3)^2$

83. $(\sqrt{2t-3} + 5)^2$

84. $(\sqrt{3w-2} - 4)^2$

For Exercises 85–88, find the exact area.

85.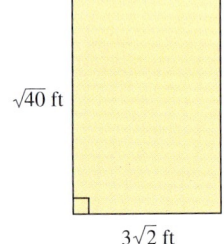
86.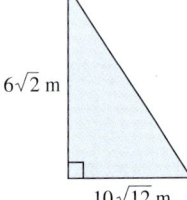
87.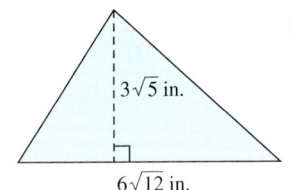
88.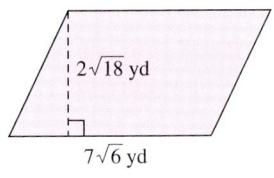

Concept 4: Multiplying Radicals With Different Indices

For Exercises 89–100, multiply or divide the radicals with different indices. Write the answers in radical form and simplify. (See Example 8.)

89. $\sqrt{x} \cdot \sqrt[4]{x}$

90. $\sqrt[3]{y} \cdot \sqrt{y}$

91. $\sqrt[5]{2z} \cdot \sqrt[3]{2z}$

92. $\sqrt[3]{5w} \cdot \sqrt[4]{5w}$

93. $\sqrt[3]{p^2} \cdot \sqrt{p^3}$

94. $\sqrt[4]{q^3} \cdot \sqrt[3]{q^2}$

95. $\dfrac{\sqrt{u^3}}{\sqrt[3]{u}}$

96. $\dfrac{\sqrt{v^5}}{\sqrt[4]{v}}$

97. $\sqrt[3]{x} \cdot \sqrt[6]{y}$

98. $\sqrt{a} \cdot \sqrt[6]{b}$

99. $\sqrt[4]{8} \cdot \sqrt[3]{3}$

100. $\sqrt{11} \cdot \sqrt[6]{2}$

Expanding Your Skills

For Exercises 101–106, multiply.

101. $\sqrt[3]{2xy} \cdot \sqrt[4]{5xy}$

102. $\sqrt{6ab} \cdot \sqrt[3]{7ab}$

103. $\sqrt[3]{4m^2n} \cdot \sqrt{6mn}$

104. $\sqrt[4]{5xy^3} \cdot \sqrt[3]{10x^2y}$

105. $(\sqrt[3]{a} + \sqrt[3]{b})(\sqrt[3]{a^2} - \sqrt[3]{ab} + \sqrt[3]{b^2})$

106. $(\sqrt[3]{a} - \sqrt[3]{b})(\sqrt[3]{a^2} + \sqrt[3]{ab} + \sqrt[3]{b^2})$

Division of Radicals and Rationalization

Section 18.6

Concepts

1. Simplified Form of a Radical
2. Division Property of Radicals
3. Rationalizing the Denominator—One Term
4. Rationalizing the Denominator—Two Terms

1. Simplified Form of a Radical

Recall that for a radical expression to be in simplified form the following three conditions must be met.

> **Simplified Form of a Radical**
>
> Consider any radical expression in which the radicand is written as a product of prime factors. The expression is in simplified form if all the following conditions are met:
>
> 1. The radicand has no factor raised to a power greater than or equal to the index.
> 2. The radicand does not contain a fraction.
> 3. No radicals are in the denominator of a fraction.

In the previous sections, we have concentrated on the first condition in the simplification process. Next, we will demonstrate how to satisfy the second and third conditions involving radicals and fractions.

2. Division Property of Radicals

The multiplication property of radicals makes it possible to write a product within a radical to be separated as a product of radicals. We now state a similar property for radicals involving quotients.

> **Division Property of Radicals**
>
> Let a and b represent real numbers such that $\sqrt[n]{a}$ and $\sqrt[n]{b}$ are both real. Then,
>
> $$\sqrt[n]{\frac{a}{b}} = \frac{\sqrt[n]{a}}{\sqrt[n]{b}} \quad b \neq 0$$

The division property of radicals indicates that a quotient within a radicand can be written as a quotient of radicals provided the roots are real numbers. For example:

$$\sqrt{\frac{25}{36}} = \frac{\sqrt{25}}{\sqrt{36}}$$

The reverse process is also true. A quotient of radicals can be written as a single radical provided that the roots are real numbers and that they have the same index.

$$\text{Same index} \longrightarrow \frac{\sqrt[3]{125}}{\sqrt[3]{8}} = \sqrt[3]{\frac{125}{8}}$$

In Examples 1 and 2, we will apply the division property of radicals to simplify radical expressions.

Example 1 — Using the Division Property to Simplify Radicals

Use the division property of radicals to simplify the expressions. Assume the variables represent positive real numbers.

a. $\sqrt{\dfrac{a^6}{b^4}}$ b. $\sqrt[3]{\dfrac{81y^5}{x^3}}$

Solution:

a. $\sqrt{\dfrac{a^6}{b^4}}$ The radicand contains an irreducible fraction.

$= \dfrac{\sqrt{a^6}}{\sqrt{b^4}}$ Apply the division property to rewrite as a quotient of radicals.

$= \dfrac{a^3}{b^2}$ Simplify the radicals.

b. $\sqrt[3]{\dfrac{81y^5}{x^3}}$ The radicand contains an irreducible fraction.

$= \dfrac{\sqrt[3]{81y^5}}{\sqrt[3]{x^3}}$ Apply the division property to rewrite as a quotient of radicals.

$= \dfrac{\sqrt[3]{3^4 \cdot y^5}}{\sqrt[3]{x^3}}$ Factor the radicand in the numerator to simplify the radical.

$= \dfrac{3y\sqrt[3]{3y^2}}{x}$ Simplify the radicals in the numerator and the denominator. The expression is simplified because it now satisfies all conditions.

Skill Practice Simplify the expressions.

1. $\sqrt{\dfrac{x^4}{y^{10}}}$ 2. $\sqrt[3]{\dfrac{w^7}{64}}$

Example 2 — Using the Division Property to Simplify a Radical

Use the division property of radicals to simplify the expressions. Assume the variables represent positive real numbers.

$$\dfrac{\sqrt[4]{8p^7}}{\sqrt[4]{p^3}}$$

Solution:

$\dfrac{\sqrt[4]{8p^7}}{\sqrt[4]{p^3}}$ There is a radical in the denominator of the fraction.

$= \sqrt[4]{\dfrac{8p^7}{p^3}}$ Apply the division property to write the quotient under a single radical.

$= \sqrt[4]{8p^4}$ Simplify the fraction.

$= p\sqrt[4]{8}$ Simplify the radical.

Skill Practice Simplify the expression.

3. $\dfrac{\sqrt[3]{16y^4}}{\sqrt[3]{y}}$

Answers

1. $\dfrac{x^2}{y^5}$ 2. $\dfrac{w^2\sqrt[3]{w}}{4}$ 3. $2y\sqrt[3]{2}$

3. Rationalizing the Denominator—One Term

The third condition restricts radicals from the denominator of an expression. The process to remove a radical from the denominator is called **rationalizing the denominator**. In many cases, rationalizing the denominator creates an expression that is computationally simpler. For example,

$$\frac{6}{\sqrt{3}} = 2\sqrt{3} \quad \text{and} \quad \frac{-2}{2 + \sqrt{6}} = 2 - \sqrt{6}$$

We will demonstrate the process to rationalize the denominator as two separate cases:

- Rationalizing the denominator (one term)
- Rationalizing the denominator (two terms involving square roots)

To begin the first case, recall that the *n*th root of a perfect *n*th power simplifies completely.

$$\sqrt{x^2} = x \qquad x \geq 0$$
$$\sqrt[3]{x^3} = x$$
$$\sqrt[4]{x^4} = x \qquad x \geq 0$$
$$\sqrt[5]{x^5} = x$$
$$\cdots$$

Therefore, to rationalize a radical expression, use the multiplication property of radicals to create an *n*th root of an *n*th power.

Example 3 Rationalizing Radical Expressions

Fill in the missing radicand to rationalize the radical expressions. Assume all variables represent positive real numbers.

a. $\sqrt{a} \cdot \sqrt{?} = \sqrt{a^2} = a$ b. $\sqrt[3]{y} \cdot \sqrt[3]{?} = \sqrt[3]{y^3} = y$
c. $\sqrt[4]{2z^3} \cdot \sqrt[4]{?} = \sqrt[4]{2^4 z^4} = 2z$

Solution:

a. $\sqrt{a} \cdot \sqrt{?} = \sqrt{a^2} = a$ What multiplied by $\sqrt{a}$ will equal $\sqrt{a^2}$?
$\sqrt{a} \cdot \sqrt{a} = \sqrt{a^2} = a$

b. $\sqrt[3]{y} \cdot \sqrt[3]{?} = \sqrt[3]{y^3} = y$ What multiplied by $\sqrt[3]{y}$ will equal $\sqrt[3]{y^3}$?
$\sqrt[3]{y} \cdot \sqrt[3]{y^2} = \sqrt[3]{y^3} = y$

c. $\sqrt[4]{2z^3} \cdot \sqrt[4]{?} = \sqrt[4]{2^4 z^4} = 2z$ What multiplied by $\sqrt[4]{2z^3}$ will equal $\sqrt[4]{2^4 z^4}$?
$\sqrt[4]{2z^3} \cdot \sqrt[4]{2^3 z} = \sqrt[4]{2^4 z^4} = 2z$

Skill Practice Fill in the missing radicand to rationalize the radical expression.

4. $\sqrt{7} \cdot \sqrt{?}$ 5. $\sqrt[5]{t^2} \cdot \sqrt[5]{?}$ 6. $\sqrt[3]{5x^2} \cdot \sqrt[3]{?}$

To rationalize the denominator of an expression with a single radical term in the denominator, we have the following strategy. Multiply the numerator and denominator by an appropriate expression to create an *n*th root of an *n*th power in the denominator.

Answers
4. 7   5. t^3 6. $5^2 x$

Chapter 18 Radicals and Complex Numbers

Example 4   Rationalizing the Denominator—One Term

Simplify the expression. $\quad \dfrac{5}{\sqrt[3]{a}} \quad a \neq 0$

Solution:

To remove the radical from the denominator, a cube root of a perfect cube is needed in the denominator. Multiply numerator and denominator by $\sqrt[3]{a^2}$ because $\sqrt[3]{a} \cdot \sqrt[3]{a^2} = \sqrt[3]{a^3} = a$.

$$\dfrac{5}{\sqrt[3]{a}} = \dfrac{5}{\sqrt[3]{a}} \cdot \dfrac{\sqrt[3]{a^2}}{\sqrt[3]{a^2}}$$

$$= \dfrac{5\sqrt[3]{a^2}}{\sqrt[3]{a^3}} \quad \text{Multiply the radicals.}$$

$$= \dfrac{5\sqrt[3]{a^2}}{a} \quad \text{Simplify.}$$

Skill Practice Simplify the expression. Assume $y > 0$.

7. $\dfrac{2}{\sqrt[4]{y}}$

Note that for $a \neq 0$, the expression $\dfrac{\sqrt[3]{a^2}}{\sqrt[3]{a^2}} = 1$. In Example 4, multiplying the expression $\dfrac{5}{\sqrt[3]{a}}$ by this ratio does not change its value.

Example 5   Rationalizing the Denominator—One Term

Simplify the expression. $\quad \sqrt{\dfrac{y^5}{7}}$

Solution:

$$\sqrt{\dfrac{y^5}{7}} \quad \text{The radical contains an irreducible fraction.}$$

$$= \dfrac{\sqrt{y^5}}{\sqrt{7}} \quad \text{Apply the division property of radicals.}$$

$$= \dfrac{y^2\sqrt{y}}{\sqrt{7}} \quad \text{Simplify the radical in the numerator.}$$

$$= \dfrac{y^2\sqrt{y}}{\sqrt{7}} \cdot \dfrac{\sqrt{7}}{\sqrt{7}} \quad \text{Rationalize the denominator.}$$
$$\quad\quad\quad\quad\quad\quad \text{Note: } \sqrt{7} \cdot \sqrt{7} = \sqrt{7^2} = 7$$

$$= \dfrac{y^2\sqrt{7y}}{\sqrt{7^2}}$$

$$= \dfrac{y^2\sqrt{7y}}{7} \quad \text{Simplify.}$$

Avoiding Mistakes

A factor within a radicand cannot be simplified with a factor outside the radicand. For example, $\dfrac{\sqrt{7y}}{7}$ cannot be simplified.

Answer

7. $\dfrac{2\sqrt[4]{y^3}}{y}$

Skill Practice Simplify the expression.

8. $\sqrt{\dfrac{8}{3}}$

Example 6 — Rationalizing the Denominator—One Term

Simplify the expression. $\dfrac{15}{\sqrt[3]{25s}}$

Solution:

$\dfrac{15}{\sqrt[3]{25s}}$

$= \dfrac{15}{\sqrt[3]{5^2 s}} \cdot \dfrac{\sqrt[3]{5s^2}}{\sqrt[3]{5s^2}}$ Because $25 = 5^2$, one additional factor of 5 is needed to form a perfect cube. Two additional factors of s are needed to make a perfect cube. Multiply numerator and denominator by $\sqrt[3]{5s^2}$.

$= \dfrac{15\sqrt[3]{5s^2}}{\sqrt[3]{5^3 s^3}}$

$= \dfrac{15\sqrt[3]{5s^2}}{5s}$ Simplify the perfect cube.

$= \dfrac{\overset{3}{\cancel{15}}\sqrt[3]{5s^2}}{\underset{1}{\cancel{5}s}}$ Reduce to lowest terms.

$= \dfrac{3\sqrt[3]{5s^2}}{s}$

TIP: In the expression $\dfrac{15\sqrt[3]{5s^2}}{5s}$, the factor of 15 and the factor of 5 may be reduced because both are outside the radical.

$\dfrac{15\sqrt[3]{5s^2}}{5s} = \dfrac{15}{5} \cdot \dfrac{\sqrt[3]{5s^2}}{s}$

$= \dfrac{3\sqrt[3]{5s^2}}{s}$

Skill Practice Simplify the expression.

9. $\dfrac{18}{\sqrt[3]{3y^2}}$

4. Rationalizing the Denominator—Two Terms

Example 7 demonstrates how to rationalize a two-term denominator involving square roots.
    First recall from the multiplication of polynomials that the product of two conjugates results in a difference of squares.

$$(a + b)(a - b) = a^2 - b^2$$

If either a or b has a square root factor, the expression will simplify without a radical. That is, the expression is *rationalized*. For example:

$$(2 + \sqrt{6})(2 - \sqrt{6}) = (2)^2 - (\sqrt{6})^2$$
$$= 4 - 6$$
$$= -2$$

Answers

8. $\dfrac{2\sqrt{6}}{3}$ 9. $\dfrac{6\sqrt[3]{9y}}{y}$

Example 7 Rationalizing the Denominator—Two Terms

Simplify the expression by rationalizing the denominator. $\dfrac{-2}{2 + \sqrt{6}}$

Solution:

$\dfrac{-2}{2 + \sqrt{6}}$

$= \dfrac{(-2)}{(2 + \sqrt{6})} \cdot \dfrac{(2 - \sqrt{6})}{(2 - \sqrt{6})}$ Multiply the numerator and denominator by the conjugate of the denominator.

conjugates

$= \dfrac{-2(2 - \sqrt{6})}{(2)^2 - (\sqrt{6})^2}$ In the denominator, apply the formula $(a + b)(a - b) = a^2 - b^2$.

$= \dfrac{-2(2 - \sqrt{6})}{4 - 6}$ Simplify.

$= \dfrac{-2(2 - \sqrt{6})}{-2}$

$= \dfrac{-\cancel{2}(2 - \sqrt{6})}{-\cancel{2}}$

$= 2 - \sqrt{6}$

Avoiding Mistakes

When constructing the conjugate of an expression, change only the sign between the terms (not the sign of the leading term).

Skill Practice Simplify by rationalizing the denominator.

10. $\dfrac{8}{3 + \sqrt{5}}$

Example 8 Rationalizing the Denominator—Two Terms

Rationalize the denominator. $\dfrac{4 + \sqrt{x}}{\sqrt{x} - 7}$

Solution:

$\dfrac{(4 + \sqrt{x})}{(\sqrt{x} - 7)} \cdot \dfrac{(\sqrt{x} + 7)}{(\sqrt{x} + 7)}$ Multiply numerator and denominator by the conjugate of the denominator.

$= \dfrac{(4)(\sqrt{x}) + (4)(7) + (\sqrt{x})(\sqrt{x}) + (\sqrt{x})(7)}{(\sqrt{x})^2 - (7)^2}$ Apply the distributive property.

$= \dfrac{4\sqrt{x} + 28 + x + 7\sqrt{x}}{x - 49}$ Multiply the radicals.

$= \dfrac{x + 11\sqrt{x} + 28}{x - 49}$ Simplify.

FOR REVIEW

Recall that the product of conjugates results in a difference of squares.

$(a - b)(a + b) = a^2 - b^2$

Skill Practice Rationalize the denominator.

11. $\dfrac{\sqrt{y} - 4}{8 - \sqrt{y}}$

Answers

10. $6 - 2\sqrt{5}$
11. $\dfrac{y + 4\sqrt{y} - 32}{64 - y}$

Section 18.6 Division of Radicals and Rationalization

Example 9 Rationalizing the Denominator—Two Terms

Rationalize the denominator. $\dfrac{3\sqrt{2} + 2\sqrt{5}}{\sqrt{2} - 4\sqrt{5}}$

Solution:

$\dfrac{(3\sqrt{2} + 2\sqrt{5})}{(\sqrt{2} - 4\sqrt{5})} \cdot \dfrac{(\sqrt{2} + 4\sqrt{5})}{(\sqrt{2} + 4\sqrt{5})}$ Multiply numerator and denominator by the conjugate of the denominator.

$= \dfrac{(3\sqrt{2}) \cdot (\sqrt{2}) + (3\sqrt{2})(4\sqrt{5}) + (2\sqrt{5})(\sqrt{2}) + (2\sqrt{5})(4\sqrt{5})}{(\sqrt{2})^2 - (4\sqrt{5})^2}$ Apply the distributive property.

$= \dfrac{6 + 12\sqrt{10} + 2\sqrt{10} + 40}{2 - 80}$ Multiply the radicals.

$= \dfrac{46 + 14\sqrt{10}}{-78}$ or $-\dfrac{46 + 14\sqrt{10}}{78}$

$= -\dfrac{2(23 + 7\sqrt{10})}{\underset{39}{78}}$ Factor and simplify.

$= -\dfrac{23 + 7\sqrt{10}}{39}$

Skill Practice Rationalize the denominator.

12. $\dfrac{5\sqrt{2} - 2\sqrt{5}}{\sqrt{2} - \sqrt{5}}$

Answer

12. $-\sqrt{10}$

Section 18.6 Activity

A.1. The division property of radicals indicates that $\sqrt[n]{\dfrac{a}{b}} = $ _____, provided that the radicals represent real numbers and that $b \neq 0$.

For Exercises A.2–A.3, apply the division property of radicals and simplify the expression. Assume that all variables represent positive real numbers.

A.2. $\sqrt[3]{\dfrac{40x^5}{y^3}}$

A.3. $\dfrac{\sqrt{50x^3}}{\sqrt{2x}}$ (*Hint:* Write the quotient as a single radical and simplify the fraction.)

A.4. The expression $\dfrac{6}{\sqrt[4]{x}}$ is not simplified because there is a radical in the denominator.

    a. Multiply the expression by a convenient ratio of 1 so that the product in the denominator is a perfect fourth power.

$\dfrac{6}{\sqrt[4]{x}} \cdot \dfrac{\sqrt[4]{\Box}}{\sqrt[4]{\Box}} = \dfrac{\Box}{\sqrt[4]{x^4}}$

    b. Simplify the expression.

A.5. The expression $\dfrac{14}{\sqrt[3]{49y}}$ is not simplified because there is a radical in the denominator.

   a. Factor the denominator as a product of prime factors.

   b. Multiply the expression by a convenient ratio of 1 so that the product in the denominator is a perfect square.

   $$\dfrac{14}{\sqrt[3]{7^2 y}} \cdot \dfrac{\sqrt[3]{\Box}}{\sqrt[3]{\Box}} = \dfrac{\Box}{\sqrt[3]{7^3 y^3}}$$

   c. Simplify the expression.

In Exercises A.6–A.7, we investigate how to rationalize the denominator when the denominator has two terms with one or more square roots.

A.6. Recall that the product of conjugates results in a difference of squares.

   a. Multiply. $(7x - 2)(7x + 2)$
   b. Multiply. $(\sqrt{7}x - 2)(\sqrt{7}x + 2)$
   c. Multiply. $(\sqrt{7} - 2)(\sqrt{7} + 2)$
   d. After multiplying the expressions in parts (b) and (c), do radicals still appear in the expressions?
   e. Now consider the expression $\dfrac{12}{\sqrt{7} - 2}$. To eliminate the radical in the denominator, we want to multiply the expression by a convenient ratio of 1. Fill in the blank so that in the resulting product, the radical in the denominator would be eliminated.

   $$\dfrac{12}{(\sqrt{7} - 2)} \cdot \dfrac{(\boxed{})}{(\boxed{})}$$

   f. Simplify the expression $\dfrac{12}{\sqrt{7} - 2}$ by using the result from part (e). Be sure to write the final answer in lowest terms.

A.7. Consider an expression with two terms in the denominator where one or more terms is a square root, such as $\dfrac{\sqrt{t} - 5}{\sqrt{t} + 4}$.

   a. To simplify the expression, we want to remove the radical(s) from the (choose one: numerator/denominator).
   b. Multiply numerator and denominator of the expression by the _____ of the denominator.
   c. Simplify. $\dfrac{\sqrt{t} - 5}{\sqrt{t} + 4}$

Section 18.6 Practice Exercises

Prerequisite Review

For Exercises R.1–R.4, simplify the expression. Assume that all variables represent positive real numbers.

R.1. $\sqrt[5]{x^5}$ **R.2.** $\sqrt[6]{y^6}$ **R.3.** $\sqrt{a^2 b^4}$ **R.4.** $\sqrt{p^8 q^6}$

For Exercises R.5–R.8, multiply. Assume that all variable expressions represent positive real numbers.

R.5. $(\sqrt{p} - 4)(\sqrt{p} + 4)$ **R.6.** $(\sqrt{w} + 5)(\sqrt{w} - 5)$

R.7. $(\sqrt{10} + \sqrt{7})(\sqrt{10} - \sqrt{7})$ **R.8.** $(\sqrt{11} - \sqrt{5})(\sqrt{11} + \sqrt{5})$

For the exercises in this set, assume that all variables represent positive real numbers unless otherwise stated.

Vocabulary and Key Concepts

1. **a.** In the simplified form of a radical expression, no _____ may appear in the denominator of a fraction.

   b. The division property of radicals indicates that if $\sqrt[n]{a}$ and $\sqrt[n]{b}$ are real numbers, then $\sqrt[n]{\dfrac{a}{b}} = \dfrac{\Box}{\Box}$ provided that $b \neq 0$.

   c. The simplified form of the expression $\sqrt[3]{\dfrac{64}{x^6}}$ is _____.

   d. The expression $\dfrac{\sqrt{3}}{3}$ (is/is not) in simplified form, whereas $\dfrac{3}{\sqrt{3}}$ (is/is not) in simplified form.

2. **a.** The process of removing a radical from the denominator of a fraction is called _____ the denominator.

   b. To rationalize the denominator for the expression $\dfrac{\sqrt{x}+3}{\sqrt{x}-2}$, multiply numerator and denominator by the conjugate of the _____.

Concept 2: Division Property of Radicals

For Exercises 3–14, simplify using the division property of radicals. Assume all variables represent positive real numbers. (See Examples 1–2.)

3. $\sqrt{\dfrac{49x^4}{y^6}}$

4. $\sqrt{\dfrac{100p^2}{q^8}}$

5. $\sqrt{\dfrac{8a^2}{x^6}}$

6. $\sqrt{\dfrac{4w^3}{25y^4}}$

7. $\sqrt[3]{\dfrac{-16j^3}{k^3}}$

8. $\sqrt[5]{\dfrac{32x}{y^{10}}}$

9. $\dfrac{\sqrt{72ab^5}}{\sqrt{8ab}}$

10. $\dfrac{\sqrt{6x^3}}{\sqrt{24x}}$

11. $\dfrac{\sqrt[4]{3b^3}}{\sqrt[4]{48b^{11}}}$

12. $\dfrac{\sqrt[3]{128wz^8}}{\sqrt[3]{2wz^2}}$

13. $\dfrac{\sqrt{3yz^2}}{\sqrt{w^4}}$

14. $\dfrac{\sqrt{50x^3z}}{\sqrt{9y^4}}$

Concept 3: Rationalizing the Denominator—One Term

The radical expressions in Exercises 15–22 have radicals in the denominator. Fill in the missing radicands to rationalize the denominators. (See Example 3.)

15. $\dfrac{x}{\sqrt{5}} = \dfrac{x}{\sqrt{5}} \cdot \dfrac{\sqrt{?}}{\sqrt{?}}$

16. $\dfrac{2}{\sqrt{x}} = \dfrac{2}{\sqrt{x}} \cdot \dfrac{\sqrt{?}}{\sqrt{?}}$

17. $\dfrac{7}{\sqrt[3]{x}} = \dfrac{7}{\sqrt[3]{x}} \cdot \dfrac{\sqrt[3]{?}}{\sqrt[3]{?}}$

18. $\dfrac{5}{\sqrt[4]{y}} = \dfrac{5}{\sqrt[4]{y}} \cdot \dfrac{\sqrt[4]{?}}{\sqrt[4]{?}}$

19. $\dfrac{8}{\sqrt{3z}} = \dfrac{8}{\sqrt{3z}} \cdot \dfrac{\sqrt{??}}{\sqrt{??}}$

20. $\dfrac{10}{\sqrt{7w}} = \dfrac{10}{\sqrt{7w}} \cdot \dfrac{\sqrt{??}}{\sqrt{??}}$

21. $\dfrac{1}{\sqrt[4]{8a^2}} = \dfrac{1}{\sqrt[4]{8a^2}} \cdot \dfrac{\sqrt[4]{??}}{\sqrt[4]{??}}$

22. $\dfrac{1}{\sqrt[3]{9b^2}} = \dfrac{1}{\sqrt[3]{9b^2}} \cdot \dfrac{\sqrt[3]{??}}{\sqrt[3]{??}}$

For Exercises 23–50, rationalize the denominator. (See Examples 4–6.)

23. $\dfrac{1}{\sqrt{3}}$

24. $\dfrac{1}{\sqrt{7}}$

25. $\sqrt{\dfrac{1}{x}}$

26. $\sqrt{\dfrac{1}{z}}$

27. $\dfrac{6}{\sqrt{2y}}$

28. $\dfrac{9}{\sqrt{3t}}$

29. $\sqrt{\dfrac{a^3}{2}}$

30. $\sqrt{\dfrac{b^3}{3}}$

31. $\dfrac{6}{\sqrt{8}}$
32. $\dfrac{2}{\sqrt{48}}$
33. $\dfrac{3}{\sqrt[3]{2}}$
34. $\dfrac{2}{\sqrt[3]{7}}$

35. $\dfrac{-6}{\sqrt[4]{x}}$
36. $\dfrac{-2}{\sqrt[5]{y}}$
37. $\dfrac{7}{\sqrt[3]{4}}$
38. $\dfrac{1}{\sqrt[3]{9}}$

39. $\sqrt[3]{\dfrac{4}{w^2}}$
40. $\sqrt[3]{\dfrac{5}{z^2}}$
41. $\sqrt[4]{\dfrac{16}{3}}$
42. $\sqrt[4]{\dfrac{81}{8}}$

43. $\dfrac{2}{\sqrt[3]{4x^2}}$
44. $\dfrac{6}{\sqrt[3]{3y^2}}$
45. $\dfrac{8}{7\sqrt{24}}$
46. $\dfrac{5}{3\sqrt{50}}$

47. $\dfrac{1}{\sqrt{x^7}}$
48. $\dfrac{1}{\sqrt{y^5}}$
49. $\dfrac{2}{\sqrt{8x^5}}$
50. $\dfrac{6}{\sqrt{27t^7}}$

Concept 4: Rationalizing the Denominator—Two Terms

51. What is the conjugate of $\sqrt{2} - \sqrt{6}$?
52. What is the conjugate of $\sqrt{11} + \sqrt{5}$?
53. What is the conjugate of $\sqrt{x} + 23$?
54. What is the conjugate of $17 - \sqrt{y}$?

For Exercises 55–74, rationalize the denominator. (See Examples 7–9.)

55. $\dfrac{4}{\sqrt{2} + 3}$
56. $\dfrac{6}{4 - \sqrt{3}}$
57. $\dfrac{8}{\sqrt{6} - 2}$
58. $\dfrac{-12}{\sqrt{5} - 3}$

59. $\dfrac{\sqrt{7}}{\sqrt{3} + 2}$
60. $\dfrac{\sqrt{8}}{\sqrt{3} + 1}$
61. $\dfrac{-1}{\sqrt{p} + \sqrt{q}}$
62. $\dfrac{6}{\sqrt{a} - \sqrt{b}}$

63. $\dfrac{x - 5}{\sqrt{x} + \sqrt{5}}$
64. $\dfrac{y - 2}{\sqrt{y} - \sqrt{2}}$
65. $\dfrac{\sqrt{w} + 2}{9 - \sqrt{w}}$
66. $\dfrac{10 - \sqrt{t}}{\sqrt{t} - 6}$

67. $\dfrac{3\sqrt{x} - \sqrt{y}}{\sqrt{y} + \sqrt{x}}$
68. $\dfrac{2\sqrt{a} + \sqrt{b}}{\sqrt{b} - \sqrt{a}}$
69. $\dfrac{3\sqrt{10}}{2 + \sqrt{10}}$
70. $\dfrac{4\sqrt{7}}{3 + \sqrt{7}}$

71. $\dfrac{2\sqrt{3} + \sqrt{7}}{3\sqrt{3} - \sqrt{7}}$
72. $\dfrac{5\sqrt{2} - \sqrt{5}}{5\sqrt{2} + \sqrt{5}}$
73. $\dfrac{\sqrt{5} + 4}{2 - \sqrt{5}}$
74. $\dfrac{3 + \sqrt{2}}{\sqrt{2} - 5}$

Mixed Exercises

For Exercises 75–78, write the English phrase as an algebraic expression. Then simplify the expression.

75. 16 divided by the cube root of 4

76. 21 divided by the principal fourth root of 27

77. 4 divided by the difference of x and the principal square root of 2

78. 8 divided by the sum of y and the principal square root of 3

79. The time $T(x)$ (in seconds) for a pendulum to make one complete swing back and forth is approximated by

$$T(x) = 2\pi\sqrt{\dfrac{x}{32}}$$

where x is the length of the pendulum in feet.

Determine the exact time required for one swing for a pendulum that is 1 ft long. Then approximate the time to the nearest hundredth of a second.

80. An object is dropped off a building x meters tall. The time $T(x)$ (in seconds) required for the object to hit the ground is given by

$$T(x) = \sqrt{\dfrac{10x}{49}}$$

Find the exact time required for the object to hit the ground if it is dropped off Three First National Plaza in Chicago, a height of 230 m. Then round the time to the nearest hundredth of a second.

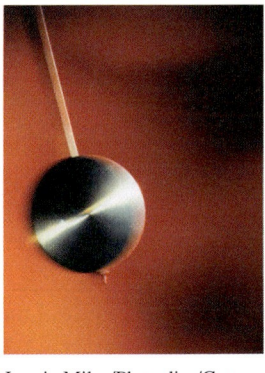

Jonnie Miles/Photodisc/Getty Images

For Exercises 81–84, rationalize the denominator.

81. a. $\dfrac{1}{\sqrt{2}}$ **b.** $\dfrac{1}{\sqrt[3]{2}}$

82. a. $\dfrac{1}{\sqrt[3]{x}}$ **b.** $\dfrac{1}{\sqrt[3]{x^2}}$

83. a. $\dfrac{1}{\sqrt{5a}}$ **b.** $\dfrac{1}{\sqrt{5}+a}$

84. a. $\dfrac{4}{\sqrt{2x}}$ **b.** $\dfrac{4}{2-\sqrt{x}}$

Expanding Your Skills

For Exercises 85–90, simplify each term of the expression. Then add or subtract as indicated.

85. $\dfrac{\sqrt{6}}{2}+\dfrac{1}{\sqrt{6}}$

86. $\dfrac{1}{\sqrt{7}}+\sqrt{7}$

87. $\sqrt{15}-\sqrt{\dfrac{3}{5}}+\sqrt{\dfrac{5}{3}}$

88. $\sqrt{\dfrac{6}{2}}-\sqrt{12}+\sqrt{\dfrac{2}{6}}$

89. $\sqrt[3]{25}+\dfrac{3}{\sqrt[3]{5}}$

90. $\dfrac{1}{\sqrt[3]{4}}+\sqrt[3]{54}$

For Exercises 91–98, rationalize the numerator by multiplying both numerator and denominator by the conjugate of the numerator.

91. $\dfrac{\sqrt{3}+6}{2}$

92. $\dfrac{\sqrt{7}-2}{5}$

93. $\dfrac{\sqrt{a}-\sqrt{b}}{\sqrt{a}+\sqrt{b}}$

94. $\dfrac{\sqrt{p}+\sqrt{q}}{\sqrt{p}-\sqrt{q}}$

95. $\dfrac{\sqrt{5+3h}-\sqrt{5}}{h}$

96. $\dfrac{\sqrt{7+2h}-\sqrt{7}}{h}$

97. $\dfrac{\sqrt{4+5h}-2}{h}$

98. $\dfrac{\sqrt{9+4h}-3}{h}$

Problem Recognition Exercises

Operations on Radicals

As you work through the following exercises, you will perform a variety of operations on radicals. For each exercise, if you are unsure what to do, try thinking of an analogy of the same operation on polynomials. For example, for each row of the table, compare the exercise on the left to the exercise on the right. Assume that all variables represent positive real numbers.

$(4x)(5x)$ Product of monomials $= (4 \cdot 5)(x \cdot x)$ $= 20x^2$	$(4\sqrt{x})(5\sqrt{x})$ $= (4 \cdot 5)(\sqrt{x} \cdot \sqrt{x})$ $= 20\sqrt{x^2}$ $= 20x$
$(2a - 5)(2a + 5)$ Product of conjugates $= (2a)^2 - (5)^2$ $= 4a^2 - 25$	$(2\sqrt{a} - 5)(2\sqrt{a} + 5)$ $= (2\sqrt{a})^2 - (5)^2$ $= 4\sqrt{a^2} - 25$ $= 4a - 25$
$(2 - 3x)^2$ Square of a binomial $= (2)^2 - 2(2)(3x) + (3x)^2$ $= 4 - 12x + 9x^2$	$(2 - 3\sqrt{x})^2$ $= (2)^2 - 2(2)(3\sqrt{x}) + (3\sqrt{x})^2$ $= 4 - 12\sqrt{x} + 9\sqrt{x^2}$ $= 4 - 12\sqrt{x} + 9x$
$(4c + 2)(3c + 5)$ Product of polynomials $= (4c + 2)(3c + 5)$ $= (4c)(3c) + (4c)(5) + (2)(3c) + (2)(5)$ $= 12c^2 + 20c + 6c + 10$ $= 12c^2 + 26c + 10$	$(4\sqrt{c} + 2)(3\sqrt{c} + 5)$ $= (4\sqrt{c} + 2)(3\sqrt{c} + 5)$ $= (4\sqrt{c})(3\sqrt{c}) + (4\sqrt{c})(5) + (2)(3\sqrt{c}) + (2)(5)$ $= 12\sqrt{c^2} + 20\sqrt{c} + 6\sqrt{c} + 10$ $= 12c + 26\sqrt{c} + 10$

For Exercises 1–30, simplify each expression. Assume that all variable expressions represent positive real numbers.

1. a. $\sqrt{24}$
   b. $\sqrt[3]{24}$

2. a. $\sqrt{54}$
   b. $\sqrt[3]{54}$

3. a. $\sqrt{200y^6}$
   b. $\sqrt[3]{200y^6}$

4. a. $\sqrt{32z^{15}}$
   b. $\sqrt[3]{32z^{15}}$

5. a. $\sqrt{80}$
   b. $\sqrt[3]{80}$
   c. $\sqrt[4]{80}$

6. a. $\sqrt{48}$
   b. $\sqrt[3]{48}$
   c. $\sqrt[4]{48}$

7. a. $\sqrt{x^5 y^6}$
   b. $\sqrt[3]{x^5 y^6}$
   c. $\sqrt[4]{x^5 y^6}$

8. a. $\sqrt{a^{10} b^9}$
   b. $\sqrt[3]{a^{10} b^9}$
   c. $\sqrt[4]{a^{10} b^9}$

9. a. $\sqrt[3]{32 s^5 t^6}$
   b. $\sqrt[4]{32 s^5 t^6}$
   c. $\sqrt[5]{32 s^5 t^6}$

10. a. $\sqrt[3]{96 v^7 w^{20}}$
    b. $\sqrt[4]{96 v^7 w^{20}}$
    c. $\sqrt[5]{96 v^7 w^{20}}$

11. a. $\sqrt{5} + \sqrt{5}$
    b. $\sqrt{5} \cdot \sqrt{5}$

12. a. $\sqrt{10} + \sqrt{10}$
    b. $\sqrt{10} \cdot \sqrt{10}$

13. a. $2\sqrt{6} - 5\sqrt{6}$
    b. $2\sqrt{6} \cdot 5\sqrt{6}$

14. a. $3\sqrt{7} - 10\sqrt{7}$
    b. $3\sqrt{7} \cdot 10\sqrt{7}$

15. a. $\sqrt{8} + \sqrt{2}$
    b. $\sqrt{8} \cdot \sqrt{2}$

16. a. $\sqrt{12} + \sqrt{3}$
    b. $\sqrt{12} \cdot \sqrt{3}$

17. a. $5\sqrt{18} - 4\sqrt{8}$
    b. $5\sqrt{18} \cdot 4\sqrt{8}$

18. a. $\sqrt{50} - \sqrt{72}$
    b. $\sqrt{50} \cdot \sqrt{72}$

19. a. $4\sqrt[3]{24} + 6\sqrt[3]{3}$
    b. $4\sqrt[3]{24} \cdot 6\sqrt[3]{3}$

20. a. $2\sqrt[3]{54} - 5\sqrt[3]{2}$
    b. $2\sqrt[3]{54} \cdot 5\sqrt[3]{2}$

21. a. $(\sqrt{3})(\sqrt{6})$ b. $\sqrt{3}+\sqrt{6}$ c. $\dfrac{\sqrt{6}}{\sqrt{3}}$

22. a. $\dfrac{\sqrt{14}}{\sqrt{2}}$ b. $(\sqrt{2})(\sqrt{14})$ c. $\sqrt{2}+\sqrt{14}$

23. a. $(3\sqrt{z})^2$ b. $(3+\sqrt{z})^2$ c. $(3+\sqrt{z})(3-\sqrt{z})$

24. a. $(4-\sqrt{x})^2$ b. $(4-\sqrt{x})(4+\sqrt{x})$ c. $(4\sqrt{x})^2$

25. a. $\dfrac{12}{\sqrt{2x}}$ b. $\sqrt{\dfrac{12}{2x}}$ c. $\dfrac{12}{\sqrt{2}+x}$

26. a. $\dfrac{15}{3-\sqrt{y}}$ b. $\dfrac{15}{\sqrt{3y}}$ c. $\sqrt{\dfrac{15}{3y}}$

27. a. $(2\sqrt{5}+1)+(\sqrt{5}-2)$ b. $(2\sqrt{5}+1)(\sqrt{5}-2)$ c. $2\sqrt{5}(\sqrt{5}-2)$

28. a. $(4\sqrt{3}-5)(\sqrt{3}+4)$ b. $4\sqrt{3}(\sqrt{3}+4)$ c. $(4\sqrt{3}-5)-(\sqrt{3}+4)$

29. a. $\sqrt{16a^{15}}$ b. $\sqrt[3]{16a^{15}}$

30. a. $\sqrt[3]{27y^9}$ b. $\sqrt{27y^9}$

Solving Radical Equations

Section 18.7

1. Solutions to Radical Equations

An equation with one or more radicals containing a variable is called a **radical equation**. For example, $\sqrt[3]{x}=5$ is a radical equation. Recall that $(\sqrt[n]{a})^n=a$, provided $\sqrt[n]{a}$ is a real number. The basis of solving a radical equation is to eliminate the radical by raising both sides of the equation to a power equal to the index of the radical.

To solve the equation $\sqrt[3]{x}=5$, cube both sides of the equation.

$$\sqrt[3]{x}=5$$
$$(\sqrt[3]{x})^3=(5)^3$$
$$x=125$$

By raising each side of a radical equation to a power equal to the index of the radical, a new equation is produced. Note, however, that some of (or all) the solutions to the new equation may *not* be solutions to the original radical equation. For this reason, it is necessary to *check all potential solutions* in the original equation. For example, consider the equation $\sqrt{x}=-7$. This equation has no solution because by definition, the principal square root of x must be a nonnegative number. However, if we square both sides of the equation, it appears as though a solution exists.

$$(\sqrt{x})^2=(-7)^2$$
$$x=49 \quad \text{The value 49 does not check in the original equation } \sqrt{x}=-7.$$
$$\text{Therefore, 49 is an } \textbf{extraneous solution}.$$

Concepts

1. Solutions to Radical Equations
2. Solving Radical Equations Involving One Radical
3. Solving Radical Equations Involving More than One Radical
4. Applications of Radical Equations and Functions

Solving a Radical Equation

Step 1 Isolate the radical. If an equation has more than one radical, choose one of the radicals to isolate.

Step 2 Raise each side of the equation to a power equal to the index of the radical.

Step 3 Solve the resulting equation. If the equation still has a radical, repeat steps 1 and 2.

*****Step 4** Check the potential solutions in the original equation.

*In solving radical equations, extraneous solutions *potentially occur* only when each side of the equation is raised to an even power.

2. Solving Radical Equations Involving One Radical

Example 1 Solving an Equation Containing One Radical

Solve the equation. $\quad \sqrt{p} + 5 = 9$

Solution:

$\sqrt{p} + 5 = 9$

$\sqrt{p} = 4 \qquad$ Isolate the radical.

$(\sqrt{p})^2 = 4^2 \qquad$ Because the index is 2, square both sides.

$p = 16$

Check: $p = 16 \qquad$ Check $p = 16$ as a potential solution.

$\sqrt{p} + 5 = 9$

$\sqrt{16} + 5 \stackrel{?}{=} 9$

$4 + 5 \stackrel{?}{=} 9 \checkmark \qquad$ True, 16 is a solution to the original equation.

The solution set is $\{16\}$.

Skill Practice Solve.

1. $\sqrt{x} - 3 = 2$

Example 2 Solving an Equation Containing One Radical

Solve the equation. $\quad \sqrt[3]{w - 1} - 2 = 2$

Solution:

$\sqrt[3]{w - 1} - 2 = 2$

$\sqrt[3]{w - 1} = 4 \qquad$ Isolate the radical.

$(\sqrt[3]{w - 1})^3 = (4)^3 \qquad$ Because the index is 3, cube both sides.

$w - 1 = 64 \qquad$ Simplify.

$w = 65$

Answer
1. $\{25\}$

Check: $w = 65$

$\sqrt[3]{65-1} - 2 \stackrel{?}{=} 2$ Check $w = 65$ as a potential solution.

$\sqrt[3]{64} - 2 \stackrel{?}{=} 2$

$4 - 2 \stackrel{?}{=} 2$ ✓   True, 65 is a solution to the original equation.

The solution set is $\{65\}$.

Skill Practice Solve.

2. $\sqrt[3]{t+2} + 5 = 3$

Example 3   Solving an Equation Containing One Radical

Solve the equation. $7 = (x+3)^{1/4} + 9$

Solution:

$7 = (x+3)^{1/4} + 9$

$7 = \sqrt[4]{x+3} + 9$ Note that $(x+3)^{1/4} = \sqrt[4]{x+3}$.

$-2 = \sqrt[4]{x+3}$ Isolate the radical.

$(-2)^4 = (\sqrt[4]{x+3})^4$ Because the index is 4, raise both sides to the fourth power.

$16 = x + 3$

$x = 13$ Solve for x.

Check: $x = 13$

$7 = \sqrt[4]{x+3} + 9$

$7 \stackrel{?}{=} \sqrt[4]{(13)+3} + 9$

$7 \stackrel{?}{=} \sqrt[4]{16} + 9$

$7 \stackrel{?}{=} 2 + 9$ (false)   13 is *not* a solution to the original equation.

The equation $7 = \sqrt[4]{x+3} + 9$ has no solution.

The solution set is the empty set, $\{\ \}$.

> **TIP:** After isolating the radical in Example 3, the equation shows a fourth root equated to a negative number:
> $$-2 = \sqrt[4]{x+3}$$
> By definition, a principal fourth root of any real number must be nonnegative. Therefore, there can be no real solution to this equation.

Skill Practice Solve.

3. $3 = 6 + (x-1)^{1/4}$

Example 4   Solving an Equation Containing One Radical

Solve the equation. $y + 2\sqrt{4y-3} = 3$

Solution:

$y + 2\sqrt{4y-3} = 3$

$2\sqrt{4y-3} = 3 - y$ Isolate the radical term.

$(2\sqrt{4y-3})^2 = (3-y)^2$ Because the index is 2, square both sides.

$4(4y-3) = 9 - 6y + y^2$ Note that $(2\sqrt{4y-3})^2 = 2^2(\sqrt{4y-3})^2$
                                               and $(3-y)^2 = (3-y)(3-y) = 9 - 6y + y^2$.

> **FOR REVIEW**
>
> Recall that the square of a binomial is a perfect square trinomial.
> $(a - b)^2 = a^2 - 2ab + b^2$
> $(3 - y)^2 = (3)^2 - 2(3)(y) + (y)^2$
> $= 9 - 6y + y^2$

Answers

2. $\{-10\}$
3. $\{\ \}$ (The value 82 does not check.)

$$16y - 12 = 9 - 6y + y^2$$

$$0 = y^2 - 22y + 21 \quad \text{The equation is quadratic. Set one side equal to zero. Write the other side in descending order.}$$

$$0 = (y - 21)(y - 1) \quad \text{Factor.}$$

$y - 21 = 0$ or $y - 1 = 0$ Set each factor equal to zero.

$y = 21$ or $y = 1$ Solve.

Check: $y = 21$

$$y + 2\sqrt{4y - 3} = 3$$
$$21 + 2\sqrt{4(21) - 3} \stackrel{?}{=} 3$$
$$21 + 2\sqrt{81} \stackrel{?}{=} 3$$
$$21 + 18 \stackrel{?}{=} 3$$
$$39 \stackrel{?}{=} 3 \text{ False}$$

Check: $y = 1$

$$y + 2\sqrt{4y - 3} = 3$$
$$1 + 2\sqrt{4(1) - 3} \stackrel{?}{=} 3$$
$$1 + 2\sqrt{1} \stackrel{?}{=} 3$$
$$1 + 2 \stackrel{?}{=} 3$$
$$3 \stackrel{?}{=} 3 \checkmark$$

The solution set is $\{1\}$. (The value 21 does not check.)

Skill Practice Solve.

4. $2\sqrt{m + 3} - m = 3$

3. Solving Radical Equations Involving More than One Radical

Example 5   Solving an Equation With Two Radicals

Solve the equation. $\sqrt[3]{2x - 4} = \sqrt[3]{1 - 8x}$

Solution:

$$\sqrt[3]{2x - 4} = \sqrt[3]{1 - 8x}$$
$$(\sqrt[3]{2x - 4})^3 = (\sqrt[3]{1 - 8x})^3 \quad \text{Because the index is 3, cube both sides.}$$
$$2x - 4 = 1 - 8x \quad \text{Simplify.}$$
$$10x - 4 = 1 \quad \text{Solve the resulting equation.}$$
$$10x = 5$$
$$x = \frac{1}{2} \quad \text{Solve for } x.$$

Check: $x = \frac{1}{2}$

$$\sqrt[3]{2x - 4} = \sqrt[3]{1 - 8x}$$
$$\sqrt[3]{2\left(\frac{1}{2}\right) - 4} \stackrel{?}{=} \sqrt[3]{1 - 8\left(\frac{1}{2}\right)}$$
$$\sqrt[3]{1 - 4} \stackrel{?}{=} \sqrt[3]{1 - 4}$$
$$\sqrt[3]{-3} \stackrel{?}{=} \sqrt[3]{-3} \checkmark \text{ (True)} \quad \text{Therefore, } \frac{1}{2} \text{ is a solution to the original equation.}$$

The solution set is $\left\{\frac{1}{2}\right\}$.

Answer

4. $\{-3, 1\}$

Skill Practice Solve.

5. $\sqrt[5]{2y-1} = \sqrt[5]{10y+3}$

Example 6   **Solving an Equation With Two Radicals**

Solve the equation. $\sqrt{3m+1} - \sqrt{m+4} = 1$

Solution:

$\sqrt{3m+1} - \sqrt{m+4} = 1$

$\sqrt{3m+1} = \sqrt{m+4} + 1$ Isolate one of the radicals.

$(\sqrt{3m+1})^2 = (\sqrt{m+4} + 1)^2$ Square both sides.

$3m + 1 = m + 4 + 2\sqrt{m+4} + 1$

Note: $(\sqrt{m+4} + 1)^2$
$= (\sqrt{m+4})^2 + 2(1)\sqrt{m+4} + (1)^2$
$= m + 4 + 2\sqrt{m+4} + 1$

$3m + 1 = m + 5 + 2\sqrt{m+4}$ Combine *like* terms.

$2m - 4 = 2\sqrt{m+4}$ Isolate the radical again.

$m - 2 = \sqrt{m+4}$ Divide both sides by 2.

$(m-2)^2 = (\sqrt{m+4})^2$ Square both sides again.

$m^2 - 4m + 4 = m + 4$ The resulting equation is quadratic.

$m^2 - 5m = 0$ Set one side equal to zero.

$m(m-5) = 0$ Factor.

$m = 0$ or $m = 5$

Check: $m = 0$

$\sqrt{3(0)+1} - \sqrt{(0)+4} \stackrel{?}{=} 1$

$\sqrt{1} - \sqrt{4} \stackrel{?}{=} 1$

$1 - 2 \stackrel{?}{=} 1$ (False)

Check: $m = 5$

$\sqrt{3(5)+1} - \sqrt{(5)+4} \stackrel{?}{=} 1$

$\sqrt{16} - \sqrt{9} \stackrel{?}{=} 1$

$4 - 3 \stackrel{?}{=} 1$ ✓

The solution set is {5}. (The value 0 does not check.)

TIP: In Example 6, we divided the equation by 2 because all coefficients were divisible by 2. This makes the coefficients smaller before we square both sides of the equation a second time.

Skill Practice Solve.

6. $\sqrt{3c+1} - \sqrt{c-1} = 2$

Answers

5. $\left\{-\dfrac{1}{2}\right\}$ 6. {1, 5}

Example 7 Solving an Equation With Two Radicals

Solve the equation. $\sqrt{5x+4} = 1 + \sqrt{5x-3}$

Solution:

$\sqrt{5x+4} = 1 + \sqrt{5x-3}$	One radical is already isolated.
$(\sqrt{5x+4})^2 = (1 + \sqrt{5x-3})^2$	Square both sides.
$5x + 4 = 1 + 2\sqrt{5x-3} + 5x - 3$	Note: $(1 + \sqrt{5x-3})^2 = (1)^2 + 2(1)\sqrt{5x-3} + (\sqrt{5x-3})^2$
$5x + 4 = 2\sqrt{5x-3} + 5x - 2$	Combine *like* terms.
$4 = 2\sqrt{5x-3} - 2$	Subtract $5x$ from both sides.
$6 = 2\sqrt{5x-3}$	Isolate the radical.
$3 = \sqrt{5x-3}$	Divide both sides by 2.
$(3)^2 = (\sqrt{5x-3})^2$	Square both sides again.
$9 = 5x - 3$	The resulting equation is linear.
$12 = 5x$	Solve for x.
$\dfrac{12}{5} = x$	Check: $x = \dfrac{12}{5}$

$$\sqrt{5\left(\frac{12}{5}\right) + 4} \stackrel{?}{=} 1 + \sqrt{5\left(\frac{12}{5}\right) - 3}$$

$$\sqrt{16} \stackrel{?}{=} 1 + \sqrt{9}$$

$$4 \stackrel{?}{=} 1 + 3 \quad \checkmark \text{ True}$$

The solution set is $\left\{\dfrac{12}{5}\right\}$.

Skill Practice Solve.

7. $\sqrt{4x-3} = 2 - \sqrt{4x+1}$

4. Applications of Radical Equations and Functions

Example 8 Applying a Radical Equation in Geometry

For a pyramid with a square base, the length of a side of the base b is given by

$$b = \sqrt{\dfrac{3V}{h}}$$

where V is the volume and h is the height.

The Pyramid of the Pharaoh Khufu (known as the Great Pyramid) at Giza has a square base (Figure 18-3). If the distance around the bottom of the pyramid is 921.6 m and the height is 146.6 m, what is the volume of the pyramid?

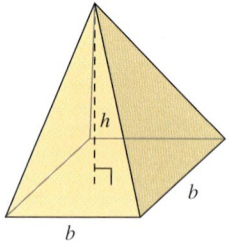

Figure 18-3

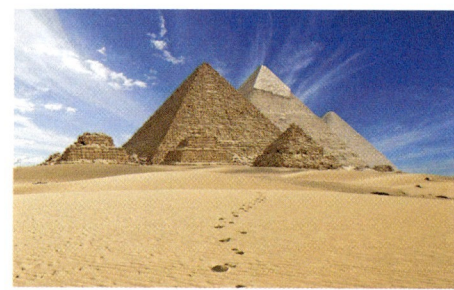

Waj/Shutterstock

Answer

7. $\left\{\dfrac{3}{4}\right\}$

Solution:

$$b = \sqrt{\frac{3V}{h}}$$

$$b^2 = \left(\sqrt{\frac{3V}{h}}\right)^2 \quad \text{Because the index is 2, square both sides.}$$

$$b^2 = \frac{3V}{h} \quad \text{Simplify.}$$

$$b^2 \cdot h = \frac{3V}{\not{h}} \cdot \not{h} \quad \text{Multiply both sides by } h.$$

$$b^2 h = 3V$$

$$\frac{b^2 h}{3} = \frac{3V}{3} \quad \text{Divide both sides by 3.}$$

$$\frac{b^2 h}{3} = V$$

The length of a side b (in meters) is given by $\frac{921.6}{4} = 230.4$ m.

$$\frac{(230.4)^2(146.6)}{3} = V \quad \text{Substitute } b = 230.4 \text{ and } h = 146.6.$$

$$2{,}594{,}046 \approx V$$

The volume of the Great Pyramid at Giza is approximately 2,594,046 m³.

Skill Practice

8. The length of the legs, s, of an isosceles right triangle is $s = \sqrt{2A}$, where A is the area. If the legs of the triangle are 9 in., find the area.

Example 9   **Applying a Radical Function**

On a certain surface, the speed $s(x)$ (in miles per hour) of a car before the brakes were applied can be approximated from the length of its skid marks x (in feet) by

$$s(x) = 3.8\sqrt{x} \quad x \geq 0 \quad \text{See Figure 18-4.}$$

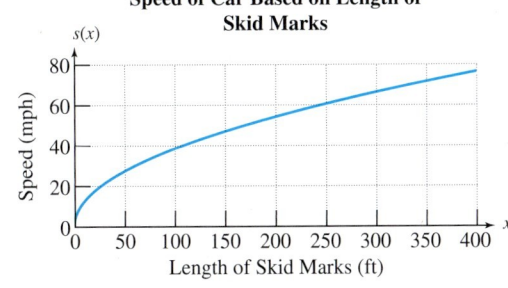

Figure 18-4

a. Find the speed of a car before the brakes were applied if its skid marks are 361 ft long.

b. How long would you expect the skid marks to be if the car had been traveling the speed limit of 50 mph? (Round to the nearest foot.)

Answer

8. 40.5 in.²

Solution:

a. $s(x) = 3.8\sqrt{x}$

$s(361) = 3.8\sqrt{361}$ Substitute $x = 361$.

$= 3.8(19)$

$= 72.2$

If the skid marks are 361 ft, the car was traveling approximately 72.2 mph before the brakes were applied.

b. $s(x) = 3.8\sqrt{x}$

$50 = 3.8\sqrt{x}$ Substitute $s(x) = 50$ and solve for x.

$\dfrac{50}{3.8} = \sqrt{x}$ Isolate the radical.

$\left(\dfrac{50}{3.8}\right)^2 = x$

$x \approx 173$

If the car had been going the speed limit (50 mph), then the length of the skid marks would have been approximately 173 ft.

Skill Practice When an object is dropped from a height of 64 ft, the time $t(x)$ (in seconds) it takes to reach a height x (in feet) is given by

$$t(x) = \dfrac{1}{4}\sqrt{64 - x}$$

9. Find the time to reach a height of 28 ft from the ground.
10. What is the height after 1 sec?

Answers

9. $\dfrac{3}{2}$ sec   10. 48 ft

Section 18.7 Activity

A.1. Consider the equations $x - 5 = 4$ and $3x = 6$.
   a. To isolate the variable, we perform the *inverse* operation to the operation acting on x. For example, to solve $x - 5 = 4$, we would add 5 to both sides of the equation. How would you solve $3x = 6$?
   b. By analogy, how would you solve the equation $\sqrt[3]{x} = 4$?

A.2. Consider the equation $\sqrt{x} = -5$.
   a. By inspection, the solution set is _____ because no real number x has a negative square root.
   b. However, if you were to square both sides of the equation $\sqrt{x} = -5$, the resulting equation would be _____. Does $x = 25$ check in the original equation $\sqrt{x} = -5$?
   c. Part (b) illustrates that if we square both sides of an equation, we must check the potential solution(s). Likewise, if we raise both sides of an equation to an (choose one: even/odd) power, we must check the potential solutions.

A.3. a. Consider the equation $2x - 3 = 5$. What is the first step to solve the equation? What is the second step to solve the equation?
   b. Consider the equation $2\sqrt{x} - 3 = 5$. What is the first step to solve the equation? What is the second step to solve the equation?
   c. Solve the equation. $2\sqrt{x} - 3 = 5$
   d. Solve the equation. $2\sqrt{3x + 1} - 3 = 5$

e. Solve the equation. $2\sqrt[3]{3x+1} - 3 = 5$

f. Solve the equation. $2(3x+1)^{1/4} - 3 = 5$

g. For which equations from parts (c)–(f) was it necessary to check the potential solution? Why?

A.4. a. Square the binomial. $(5-x)^2$
   b. Solve the equation. $\sqrt{x-3} - 5 = -x$

A.5. The equation $\sqrt{5x-3} = \sqrt{x+1}$ contains two square roots, and each square root is isolated on one side of the equation. Solve the equation and check the potential solution.

A.6. The equation $\sqrt{x+2} + \sqrt{2x+5} = 1$ contains two square roots, but because the equation has a third term, it is not possible to isolate both square roots simultaneously. This significantly complicates the process to solve the equation. Let's walk through the process using the following steps.

   a. First, we want to isolate one of the radicals. Write the equation with $\sqrt{x+2}$ isolated on the left side of the equation.
   b. Square both sides of the equation and write the new equation.
   c. The right side of the equation still has a radical. Isolate the term $-2\sqrt{2x+5}$ on the right side of the equation and write the new equation.
   d. To make the equation a bit simpler, multiply both sides by -1 and write the new equation.
   e. The term on the right side of the equation is $2\sqrt{2x+5}$. If we divided by 2 on both sides to isolate the radical, we would create fractional coefficients on the left. Instead, square both sides of the equation as is and write the new equation.
   f. The new equation is quadratic. Solve this equation, but do not write the final solution set yet.
   g. The solutions to the equation in part (f) are the *potential* solutions to the original equation $\sqrt{x+2} + \sqrt{2x+5} = 1$. Check the solutions from part (f) in the original equation and write the final solution set.

Practice Exercises

Section 18.7

Prerequisite Review

For Exercises R.1–R.6, square the expression as indicated. Assume all variable expressions represent positive real numbers.

R.1. $(t+8)^2$ **R.2.** $(y-10)^2$ **R.3.** $(\sqrt{3t+1})^2$

R.4. $(\sqrt{9-w})^2$ **R.5.** $(\sqrt{m}-7)^2$ **R.6.** $(3+\sqrt{x})^2$

For Exercises R.7–R.12, solve the equation.

R.7. $3y - 3 = 12$ **R.8.** $4w + 1 = 33$ **R.9.** $x^2 - 4x - 21 = 0$

R.10. $x^2 + 13x + 30 = 0$ **R.11.** $10x^2 - 7 = 33x$ **R.12.** $6x^2 - 4 = 5x$

Vocabulary and Key Concepts

1. a. The equation $\sqrt{x+5} + 7 = 11$ is an example of a _____ equation.
   b. The first step to solve the equation $\sqrt{x+5} + 7 = 11$ is to _____ the radical by subtracting _____ from both sides of the equation.
   c. When solving a radical equation, some potential solutions may not check in the original equation. These are called _____ solutions.

2. a. To solve the equation $\sqrt[3]{w-1} = 5$, raise both sides of the equation to the _____ power.
   b. To solve the equation $\sqrt[4]{m-3} = 2$, raise both sides of the equation to the _____ power.

Concept 2: Solving Radical Equations Involving One Radical

For Exercises 3–10, match the equation with the best first step used to solve the equation.

3. $\sqrt{x+1} = 5$
4. $2\sqrt{x+1} = 5$
5. $x + 1 = 5$
6. $(x+1)^{1/4} = 5$
7. $\sqrt{x+1} - 2 = 5$
8. $\sqrt[3]{x+1} = 5$
9. $\dfrac{\sqrt{x+1}}{2} = 5$
10. $\sqrt{x+1} + 2 = 5$

a. Raise both sides to the fourth power.
b. Divide both sides by 2.
c. Subtract 2 from both sides.
d. Square both sides.
e. Subtract 1 from both sides.
f. Multiply both sides by 2.
g. Cube both sides.
h. Add 2 to both sides.

For Exercises 11–30, solve the equations. (See Examples 1–3.)

11. $\sqrt{x} = 10$
12. $\sqrt{y} = 7$
13. $\sqrt{x} + 4 = 6$
14. $\sqrt{x} + 2 = 8$
15. $\sqrt{5y+1} = 4$
16. $\sqrt{9z-5} - 2 = 9$
17. $6 = \sqrt{2z-3} - 3$
18. $4 = \sqrt{8+3a} - 1$
19. $\sqrt{x^2+5} = x+1$
20. $\sqrt{y^2-8} = y-2$
21. $\sqrt[3]{x-2} - 1 = 2$
22. $\sqrt[3]{2x-5} - 1 = 1$
23. $(15-w)^{1/3} + 7 = 2$
24. $(k+18)^{1/3} + 5 = 3$
25. $3 + \sqrt{x-16} = 0$
26. $12 + \sqrt{2x+1} = 0$
27. $2\sqrt{6a+7} - 2a = 0$
28. $2\sqrt{3-w} - w = 0$
29. $(2x-5)^{1/4} = -1$
30. $(x+16)^{1/4} = -4$

For Exercises 31–34, assume all variables represent positive real numbers.

31. Solve for V: $r = \sqrt[3]{\dfrac{3V}{4\pi}}$
32. Solve for V: $r = \sqrt{\dfrac{V}{h\pi}}$
33. Solve for h^2: $r = \pi\sqrt{r^2 + h^2}$
34. Solve for d: $s = 1.3\sqrt{d}$

For Exercises 35–40, square the expression as indicated.

35. $(a+5)^2$
36. $(b+7)^2$
37. $(\sqrt{5a} - 3)^2$
38. $(2 + \sqrt{b})^2$
39. $(\sqrt{r-3} + 5)^2$
40. $(2 - \sqrt{2t-4})^2$

For Exercises 41–46, solve the radical equations, if possible. (See Example 4.)

41. $\sqrt{a^2 + 2a + 1} = a + 5$
42. $\sqrt{b^2 - 5b - 8} = b + 7$
43. $\sqrt{25w^2 - 2w - 3} = 5w - 4$
44. $\sqrt{4p^2 - 2p + 1} = 2p - 3$
45. $4\sqrt{p-2} - 2 = -p$
46. $x - 3\sqrt{x-5} = 5$

Concept 3: Solving Radical Equations Involving More than One Radical

For Exercises 47–70, solve the radical equations, if possible. (See Examples 5–7.)

47. $\sqrt[4]{h+4} = \sqrt[4]{2h-5}$
48. $\sqrt[3]{3b+6} = \sqrt[4]{7b-6}$
49. $\sqrt[3]{5a+3} - \sqrt[3]{a-13} = 0$
50. $\sqrt[3]{k-8} - \sqrt[3]{4k+1} = 0$
51. $\sqrt{5a-9} = \sqrt{5a} - 3$
52. $\sqrt{8+b} = 2 + \sqrt{b}$
53. $\sqrt{2h+5} - \sqrt{2h} = 1$
54. $\sqrt{3k-5} - \sqrt{3k} = -1$
55. $(t-9)^{1/2} - t^{1/2} = 3$
56. $(y-16)^{1/2} - y^{1/2} = 4$
57. $6 = \sqrt{x^2+3} - x$
58. $2 = \sqrt{y^2+5} - y$
59. $\sqrt{3t-7} = 2 - \sqrt{3t+1}$
60. $\sqrt{p-6} = \sqrt{p+2} - 4$
61. $\sqrt{z+1} + \sqrt{2z+3} = 1$

62. $\sqrt{2y+6} = \sqrt{7-2y} + 1$ **63.** $\sqrt{6m+7} - \sqrt{3m+3} = 1$ **64.** $\sqrt{5w+1} - \sqrt{3w} = 1$

65. $2 + 2\sqrt{2t+3} + 2\sqrt{3t-5} = 0$ **66.** $6 + 3\sqrt{3x+1} + 3\sqrt{x-1} = 0$ **67.** $3\sqrt{y-3} = \sqrt{4y+3}$

68. $\sqrt{5x-8} = 2\sqrt{x-1}$ **69.** $\sqrt{p+7} = \sqrt{2p} + 1$ **70.** $\sqrt{t} = \sqrt{t-12} + 2$

Concept 4: Applications of Radical Equations and Functions

71. If an object is dropped from an initial height h, its velocity at impact with the ground is given by

$$v = \sqrt{2gh}$$

where g is the acceleration due to gravity and h is the initial height. **(See Example 8.)**

    a. Find the initial height (in feet) of an object if its velocity at impact is 44 ft/sec. (Assume that the acceleration due to gravity is $g = 32$ ft/sec^2.)

    b. Find the initial height (in meters) of an object if its velocity at impact is 26 m/sec. (Assume that the acceleration due to gravity is $g = 9.8$ m/sec^2.) Round to the nearest tenth of a meter.

72. The time T (in seconds) required for a pendulum to make one complete swing back and forth is approximated by

$$T = 2\pi\sqrt{\frac{L}{g}}$$

where g is the acceleration due to gravity and L is the length of the pendulum (in feet).

    a. Find the length of a pendulum that requires 1.36 sec to make one complete swing back and forth. (Assume that the acceleration due to gravity is $g = 32$ ft/sec^2.) Round to the nearest tenth of a foot.

    b. Find the time required for a pendulum to complete one swing back and forth if the length of the pendulum is 4 ft. (Assume that the acceleration due to gravity is $g = 32$ ft/sec^2.) Round to the nearest tenth of a second.

73. The airline cost for x thousand passengers to travel round trip from New York to Atlanta is given by

$$C(x) = \sqrt{0.3x + 1}$$

where $C(x)$ is measured in millions of dollars and $x \geq 0$. **(See Example 9.)**

    a. Find the airline's cost for 10,000 passengers ($x = 10$) to travel from New York to Atlanta.

    b. If the airline charges $320 per passenger, find the profit made by the airline for flying 10,000 passengers from New York to Atlanta.

    c. Approximate the number of passengers who traveled from New York to Atlanta if the total cost for the airline was $4 million.

74. The time $t(d)$ in seconds it takes an object to drop d meters is given by

$$t(d) = \sqrt{\frac{d}{4.9}}$$

    a. Approximate the height of the JP Morgan Chase Tower in Houston if it takes an object 7.89 sec to drop from the top. Round to the nearest meter.

    b. Approximate the height of the Willis Tower in Chicago if it takes an object 9.51 sec to drop from the top. Round to the nearest meter.

David Forman/Image Source/Getty Images

Technology Connections

75. Graph Y_1 and Y_2 on a viewing window defined by $-10 \leq x \leq 40$ and $-5 \leq y \leq 10$.

$$Y_1 = \sqrt{2x} \quad \text{and} \quad Y_2 = 8$$

Use an *Intersect* feature to approximate the *x*-coordinate of the point of intersection of the two graphs. How does the point of intersection relate to the solution to the equation $\sqrt{2x} = 8$?

76. Graph Y_1 and Y_2 on a viewing window defined by $-10 \leq x \leq 20$ and $-5 \leq y \leq 10$.

$$Y_1 = \sqrt{4x} \quad \text{and} \quad Y_2 = 6$$

Use an *Intersect* feature to approximate the *x*-coordinate of the point of intersection of the two graphs. How does the point of intersection relate to the solution to the equation $\sqrt{4x} = 6$?

Expanding Your Skills

77. The number of hours needed to cook a turkey that weighs *x* pounds can be approximated by

$$t(x) = 0.90\sqrt[5]{x^3}$$

where $t(x)$ is the time in hours and *x* is the weight of the turkey in pounds.

a. Find the weight of a turkey that cooked for 4 hr. Round to the nearest pound.

b. Find $t(18)$ and interpret the result. Round to the nearest tenth of an hour.

For Exercises 78–81, use the Pythagorean theorem to find *a*, *b*, or *c*.

78. Find *b* when $a = 2$ and $c = y$.

79. Find *b* when $a = h$ and $c = 5$.

80. Find *a* when $b = x$ and $c = 8$.

81. Find *a* when $b = 14$ and $c = k$.

Section 18.8 Complex Numbers

Concepts

1. Definition of *i*
2. Powers of *i*
3. Definition of a Complex Number
4. Addition, Subtraction, and Multiplication of Complex Numbers
5. Division and Simplification of Complex Numbers

1. Definition of *i*

We have already learned that there are no real-valued square roots of a negative number. For example, $\sqrt{-9}$ is not a real number because no real number when squared equals -9. However, the square roots of a negative number are defined over another set of numbers called the **imaginary numbers**. The foundation of the set of imaginary numbers is the definition of the imaginary number *i*.

> **Definition of the Imaginary Number *i***
>
> $$i = \sqrt{-1}$$
>
> *Note:* From the definition of *i*, it follows that $i^2 = -1$.

Using the imaginary number *i*, we can define the square root of any negative real number.

> **Definition of $\sqrt{-b}$ for $b > 0$**
>
> Let *b* be a positive real number. Then $\sqrt{-b} = i\sqrt{b}$.

Example 1 — Simplifying Expressions in Terms of i

Simplify the expressions in terms of i.

a. $\sqrt{-64}$ b. $\sqrt{-50}$ c. $-\sqrt{-4}$ d. $\sqrt{-29}$

Solution:

a. $\sqrt{-64} = i\sqrt{64}$
$\phantom{\sqrt{-64}} = 8i$

b. $\sqrt{-50} = i\sqrt{50}$
$\phantom{\sqrt{-50}} = i\sqrt{5^2 \cdot 2}$
$\phantom{\sqrt{-50}} = 5i\sqrt{2}$

c. $-\sqrt{-4} = -1 \cdot \sqrt{-4}$
$\phantom{-\sqrt{-4}} = -1 \cdot i\sqrt{4}$
$\phantom{-\sqrt{-4}} = -1 \cdot 2i$
$\phantom{-\sqrt{-4}} = -2i$

d. $\sqrt{-29} = i\sqrt{29}$

Avoiding Mistakes

In an expression such as $i\sqrt{29}$, the i is often written in front of the square root. The expression $\sqrt{29}\,i$ is also correct, but may be misinterpreted as $\sqrt{29i}$ (with i incorrectly placed under the radical).

Skill Practice Simplify the expressions in terms of i.

1. $\sqrt{-81}$ 2. $\sqrt{-20}$ 3. $-\sqrt{-36}$ 4. $\sqrt{-7}$

If a and b represent real numbers such that $\sqrt[n]{a}$ and $\sqrt[n]{b}$ are both real, then

$$\sqrt[n]{ab} = \sqrt[n]{a} \cdot \sqrt[n]{b} \quad \text{and} \quad \sqrt[n]{\frac{a}{b}} = \frac{\sqrt[n]{a}}{\sqrt[n]{b}} \quad b \neq 0$$

The conditions that $\sqrt[n]{a}$ and $\sqrt[n]{b}$ must both be real numbers prevent us from applying the multiplication and division properties of radicals for square roots with a negative radicand. Therefore, to multiply or divide radicals with a negative radicand, first write the radical in terms of the imaginary number i. This is demonstrated in Example 2.

Example 2 — Simplifying a Product or Quotient in Terms of i

Simplify the expressions.

a. $\dfrac{\sqrt{-100}}{\sqrt{-25}}$ b. $\sqrt{-25} \cdot \sqrt{-9}$ c. $\sqrt{-5} \cdot \sqrt{-5}$

Answers

1. $9i$ 2. $2i\sqrt{5}$ 3. $-6i$ 4. $i\sqrt{7}$

Solution:

a. $\dfrac{\sqrt{-100}}{\sqrt{-25}}$

$= \dfrac{10i}{5i}$ Simplify each radical in terms of i *before* dividing.

$= 2$

b. $\sqrt{-25} \cdot \sqrt{-9}$

$= 5i \cdot 3i$ Simplify each radical in terms of i first *before* multiplying.

$= 15i^2$ Multiply.

$= 15(-1)$ Recall that $i^2 = -1$.

$= -15$ Simplify.

c. $\sqrt{-5} \cdot \sqrt{-5}$

$= i\sqrt{5} \cdot i\sqrt{5}$

$= i^2 \cdot (\sqrt{5})^2$

$= -1 \cdot 5$

$= -5$

Skill Practice Simplify the expressions.

5. $\dfrac{\sqrt{-36}}{\sqrt{-9}}$ 6. $\sqrt{-16} \cdot \sqrt{-49}$ 7. $\sqrt{-2} \cdot \sqrt{-2}$

Avoiding Mistakes

In Example 2, we wrote the radical expressions in terms of i first, before multiplying or dividing. If we had mistakenly applied the multiplication or division property first, we would have obtained an incorrect answer.

Correct: $\sqrt{-25} \cdot \sqrt{-9}$

$= (5i)(3i) = 15i^2$

$= 15(-1) = -15$ ↑ correct

Be careful: $\sqrt{-25} \cdot \sqrt{-9}$

$\neq \sqrt{225} = 15$
(incorrect answer)

$\sqrt{-25}$ and $\sqrt{-9}$ are not real numbers. Therefore, the multiplication property of radicals cannot be applied.

---

2. Powers of i

From the definition of $i = \sqrt{-1}$, it follows that

$i = i$

$i^2 = -1$

$i^3 = -i$ because $i^3 = i^2 \cdot i = (-1)i = -i$

$i^4 = 1$ because $i^4 = i^2 \cdot i^2 = (-1)(-1) = 1$

$i^5 = i$ because $i^5 = i^4 \cdot i = (1)i = i$

$i^6 = -1$ because $i^6 = i^4 \cdot i^2 = (1)(-1) = -1$

Answers
5. 2   6. -28 7. -2

This pattern of values $i, -1, -i, 1, i, -1, -i, 1, \ldots$ continues for all subsequent powers of i. Table 18-1 lists several powers of i.

Table 18-1 Powers of i

$i^1 = i$	$i^5 = i$	$i^9 = i$
$i^2 = -1$	$i^6 = -1$	$i^{10} = -1$
$i^3 = -i$	$i^7 = -i$	$i^{11} = -i$
$i^4 = 1$	$i^8 = 1$	$i^{12} = 1$

To simplify higher powers of i, we can decompose the expression into multiples of i^4 ($i^4 = 1$) and write the remaining factors as i, i^2, or i^3.

Example 3   Simplifying Powers of i

Simplify the powers of i.

a. i^{13} b. i^{18} c. i^{107} d. i^{32}

Solution:

a. $i^{13} = (i^{12}) \cdot (i)$ Write the exponent as a multiple of 4 and a remainder.
   $= (i^4)^3 \cdot (i)$
   $= (1)^3(i)$ Recall that $i^4 = 1$.
   $= i$ Simplify.

b. $i^{18} = (i^{16}) \cdot (i^2)$ Write the exponent as a multiple of 4 and a remainder.
   $= (i^4)^4 \cdot (i^2)$
   $= (1)^4 \cdot (-1)$ $i^4 = 1$ and $i^2 = -1$
   $= -1$ Simplify.

c. $i^{107} = (i^{104}) \cdot (i^3)$ Write the exponent as a multiple of 4 and a remainder.
   $= (i^4)^{26}(i^3)$
   $= (1)^{26}(-i)$ $i^4 = 1$ and $i^3 = -i$
   $= -i$ Simplify.

d. $i^{32} = (i^4)^8$
   $= (1)^8$ $i^4 = 1$
   $= 1$ Simplify.

Skill Practice Simplify the powers of i.
8. i^{45} 9. i^{22} 10. i^{31} 11. i^{80}

3. Definition of a Complex Number

We have already learned the definitions of the integers, rational numbers, irrational numbers, and real numbers. In this section, we define the complex numbers.

Answers
8. i 9. -1
10. $-i$ 11. 1

> **Definition of a Complex Number**
>
> A **complex number** is a number of the form $a + bi$, where a and b are real numbers and $i = \sqrt{-1}$.
>
> *Notes:*
>
> - If $b = 0$, then the complex number $a + bi$ is a real number.
> - If $b \neq 0$, then we say that $a + bi$ is an imaginary number.
> - The complex number $a + bi$ is said to be written in standard form. The quantities a and b are called the real and imaginary parts (respectively) of the complex number.
> - The complex numbers $a - bi$ and $a + bi$ are called **complex conjugates**.

From the definition of a complex number, it follows that all real numbers are complex numbers and all imaginary numbers are complex numbers. Figure 18-5 illustrates the relationship among the sets of numbers we have learned so far.

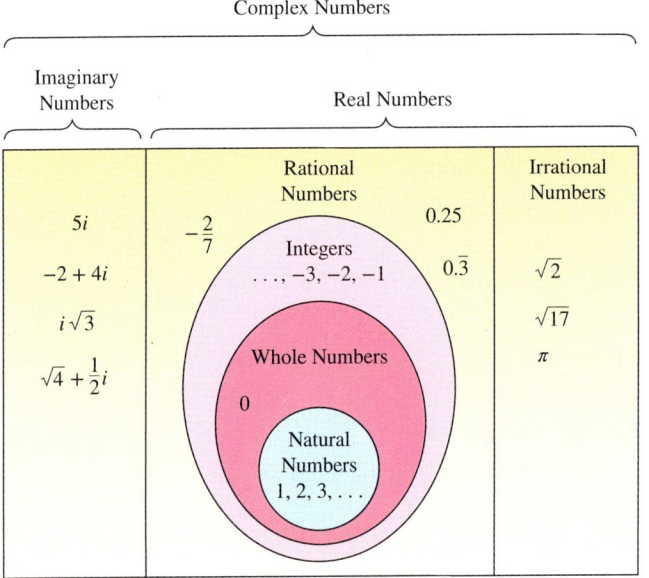

Figure 18-5

Example 4 — Identifying the Real and Imaginary Parts of a Complex Number

Identify the real and imaginary parts of the complex numbers.

a. $-8 + 2i$ **b.** $\dfrac{3}{2}$ **c.** $-1.75i$

Solution:

a. $-8 + 2i$ — -8 is the real part, and 2 is the imaginary part.

b. $\dfrac{3}{2} = \dfrac{3}{2} + 0i$ — Rewrite $\dfrac{3}{2}$ in the form $a + bi$. $\dfrac{3}{2}$ is the real part, and 0 is the imaginary part.

c. $-1.75i$
$= 0 + -1.75i$ — Rewrite $-1.75i$ in the form $a + bi$. 0 is the real part, and -1.75 is the imaginary part.

Skill Practice Identify the real and imaginary parts of the complex numbers.

12. $22 - 14i$ **13.** -50 **14.** $15i$

TIP: Example 4(b) illustrates that a real number is also a complex number.
$$\frac{3}{2} = \frac{3}{2} + 0i$$

Example 4(c) illustrates that an imaginary number is also a complex number.
$$-1.75i = 0 + -1.75i$$

---

4. Addition, Subtraction, and Multiplication of Complex Numbers

The operations of addition, subtraction, and multiplication of real numbers also apply to imaginary numbers. To add or subtract complex numbers, combine the real parts and combine the imaginary parts. The commutative, associative, and distributive properties that apply to real numbers also apply to complex numbers.

Example 5 — Adding and Subtracting Complex Numbers

Add or subtract as indicated. Write the answers in the form $a + bi$.

a. $(1 - 5i) + (-3 + 7i)$ **b.** $\left(-\frac{1}{4} + \frac{3}{5}i\right) - \left(\frac{1}{2} - \frac{1}{10}i\right)$

c. $\sqrt{-8} + \sqrt{-18}$

Solution:

a. $(1 - 5i) + (-3 + 7i) = (1 + -3) + (-5 + 7)i$ Add the real parts.
Add the imaginary parts.

$= -2 + 2i$ Simplify.

b. $\left(-\frac{1}{4} + \frac{3}{5}i\right) - \left(\frac{1}{2} - \frac{1}{10}i\right) = -\frac{1}{4} + \frac{3}{5}i - \frac{1}{2} + \frac{1}{10}i$ Apply the distributive property.

$= \left(-\frac{1}{4} - \frac{1}{2}\right) + \left(\frac{3}{5} + \frac{1}{10}\right)i$ Add real parts. Add imaginary parts.

$= \left(-\frac{1}{4} - \frac{2}{4}\right) + \left(\frac{6}{10} + \frac{1}{10}\right)i$ Get common denominators.

$= -\frac{3}{4} + \frac{7}{10}i$ Simplify.

c. $\sqrt{-8} + \sqrt{-18} = 2i\sqrt{2} + 3i\sqrt{2}$ Simplify each radical in terms of i.

$= 5i\sqrt{2}$ Combine *like* radicals.

Skill Practice Perform the indicated operations.

15. $\left(\frac{1}{2} - \frac{1}{4}i\right) + \left(\frac{3}{5} + \frac{2}{3}i\right)$ **16.** $(-6 + 11i) - (-9 - 12i)$ **17.** $\sqrt{-45} - \sqrt{-20}$

Answers

12. real: 22; imaginary: -14
13. real: -50; imaginary: 0
14. real: 0; imaginary: 15
15. $\frac{11}{10} + \frac{5}{12}i$ **16.** $3 + 23i$
17. $i\sqrt{5}$

> **Example 6** Multiplying Complex Numbers
>
> Multiply the complex numbers. Write the answers in the form $a + bi$.
>
> **a.** $(10 - 5i)(2 + 3i)$ **b.** $(1.2 + 0.5i)(1.2 - 0.5i)$
>
> **Solution:**
>
> **a.** $(10 - 5i)(2 + 3i)$
>
> $= (10)(2) + (10)(3i) + (-5i)(2) + (-5i)(3i)$ Apply the distributive property.
>
> $= 20 + 30i - 10i - 15i^2$
>
> $= 20 + 20i - (15)(-1)$ Recall $i^2 = -1$.
>
> $= 20 + 20i + 15$
>
> $= 35 + 20i$ Write in the form $a + bi$.
>
> **b.** $(1.2 + 0.5i)(1.2 - 0.5i)$
>
> The expressions $(1.2 + 0.5i)$ and $(1.2 - 0.5i)$ are complex conjugates. The product is a difference of squares.
>
> $(a + b)(a - b) = a^2 - b^2$ Apply the formula, where $a = 1.2$ and $b = 0.5i$.
>
> $(1.2 + 0.5i)(1.2 - 0.5i) = (1.2)^2 - (0.5i)^2$
>
> $= 1.44 - 0.25i^2$
>
> $= 1.44 - 0.25(-1)$ Recall $i^2 = -1$.
>
> $= 1.44 + 0.25$
>
> $= 1.69 + 0i$
>
> **Skill Practice** Multiply.
>
> **18.** $(4 - 6i)(2 - 3i)$ **19.** $(1.5 + 0.8i)(1.5 - 0.8i)$

5. Division and Simplification of Complex Numbers

The product of a complex number and its complex conjugate is a real number. For example:

$$(5 + 3i)(5 - 3i) = 25 - 9i^2$$
$$= 25 - 9(-1)$$
$$= 25 + 9$$
$$= 34$$

To divide by a complex number, multiply the numerator and denominator by the complex conjugate of the denominator. This produces a real number in the denominator so that the resulting expression can be written in the form $a + bi$.

Answers

18. $-10 - 24i$ **19.** $2.89 + 0i$

Example 7 Dividing by a Complex Number

Divide the complex numbers and write the answer in the form $a + bi$.

$$\frac{4 - 3i}{5 + 2i}$$

Solution:

$\dfrac{4 - 3i}{5 + 2i}$ Multiply the numerator and denominator by the complex conjugate of the denominator:

$$\frac{(4 - 3i)}{(5 + 2i)} \cdot \frac{(5 - 2i)}{(5 - 2i)} = \frac{(4)(5) + (4)(-2i) + (-3i)(5) + (-3i)(-2i)}{(5)^2 - (2i)^2}$$

$$= \frac{20 - 8i - 15i + 6i^2}{25 - 4i^2} \quad \text{Simplify numerator and denominator.}$$

$$= \frac{20 - 23i + 6(-1)}{25 - 4(-1)} \quad \text{Recall } i^2 = -1.$$

$$= \frac{20 - 23i - 6}{25 + 4}$$

$$= \frac{14 - 23i}{29} \quad \text{Simplify.}$$

$$= \frac{14}{29} - \frac{23}{29}i \quad \text{Write in the form } a + bi.$$

TIP: In Example 7, we asked for the answer in the form $a + bi$. However, if the second term is negative, we often leave an answer in terms of subtraction: $\frac{14}{29} - \frac{23}{29}i$. This is the same as $\frac{14}{29} + (-\frac{23}{29}i)$.

Skill Practice Divide the complex numbers. Write the answer in the form $a + bi$.

20. $\dfrac{2 + i}{3 - 2i}$

Example 8 Simplifying a Complex Number

Simplify the complex number. $\dfrac{6 + \sqrt{-18}}{9}$

Solution:

$$\frac{6 + \sqrt{-18}}{9} = \frac{6 + i\sqrt{18}}{9} \quad \text{Write the radical in terms of } i.$$

$$= \frac{6 + 3i\sqrt{2}}{9} \quad \text{Simplify } \sqrt{18} = 3\sqrt{2}.$$

$$= \frac{3(2 + i\sqrt{2})}{9} \quad \text{Factor the numerator.}$$

$$= \frac{\overset{1}{3}(2 + i\sqrt{2})}{\underset{3}{9}} \quad \text{Simplify.}$$

$$= \frac{2 + i\sqrt{2}}{3} \quad \text{or} \quad \frac{2}{3} + \frac{\sqrt{2}}{3}i \quad \text{Write in the form } a + bi.$$

TIP: As an alternative approach in Example 8, the expression $\dfrac{6 + i\sqrt{18}}{9}$ can be written in the form $a + bi$ and then simplified.

$$\frac{6 + i\sqrt{18}}{9} = \frac{6}{9} + \frac{3\sqrt{2}}{9}i$$

$$= \frac{2}{3} + \frac{\sqrt{2}}{3}i$$

Skill Practice Simplify the complex number.

21. $\dfrac{8 - \sqrt{-24}}{6}$

Answers

20. $\dfrac{4}{13} + \dfrac{7}{13}i$

21. $\dfrac{4 - i\sqrt{6}}{3}$ or $\dfrac{4}{3} - \dfrac{\sqrt{6}}{3}i$

Section 18.8 Activity

A.1. By definition, $\sqrt{-1} =$ _____ and $i^2 =$ _____.

A.2. **a.** By definition, if b is a positive real number, then $\sqrt{-b} =$ _____.
  b. Simplify. $\quad \sqrt{-13}$
  c. Simplify. $\quad \sqrt{-72}$
  d. Simplify. $\quad \dfrac{\sqrt{-50}}{\sqrt{25}}$
  e. Simplify. $\quad \dfrac{6 + \sqrt{-28}}{4}$

A.3. Recall that the product property of radicals, $\sqrt[n]{a} \cdot \sqrt[n]{b} = \sqrt[n]{ab}$, is true provided that the individual square roots $\sqrt[n]{a}$ and $\sqrt[n]{b}$ are *real* numbers.
  a. Therefore, what is the first step to simplify the expression $\sqrt{-4} \cdot \sqrt{-25}$?
  b. Simplify the expression $\sqrt{-4} \cdot \sqrt{-25}$.

A.4. Recall that the quotient property of radicals, $\dfrac{\sqrt[n]{a}}{\sqrt[n]{b}} = \sqrt[n]{\dfrac{a}{b}}$, is true provided that the individual square roots $\sqrt[n]{a}$ and $\sqrt[n]{b}$ are *real* numbers.
  a. Therefore, what is the first step to simplify the expression $\dfrac{\sqrt{-50}}{\sqrt{-2}}$?
  b. Simplify the expression $\dfrac{\sqrt{-50}}{\sqrt{-2}}$.

A.5. To simplify higher powers of i, we can use the fact that $i^2 = -1$. Decompose the higher power of i as a product of i^2 and another power.
  a. $i^3 = i^2 \cdot \square = -1 \cdot \square =$ _____
  b. $i^4 = i^2 \cdot \square = -1 \cdot \square =$ _____

A.6. From Exercise A.5, we see that $i^4 = 1$. Use the fact that $i^2 = -1$ and $i^4 = 1$ to decompose higher powers of i.
  a. $i^5 = i^4 \cdot \square = 1 \cdot \square =$ _____
  b. $i^6 = i^4 \cdot \square = 1 \cdot \square =$ _____
  c. $i^7 = i^4 \cdot i^2 \cdot \square =$ _____
  d. $i^8 = i^4 \cdot \square =$ _____

A.7. Simplify the powers of i.
  a. i^{43} **b.** i^{72} **c.** i^{50} **d.** i^{37}

A.8. Write the number in the form $a + bi$ and identify the real and imaginary parts.
  a. $3 - 7i$ **b.** $4 + \sqrt{-12}$ **c.** 9 **d.** $\dfrac{3}{4}i$

For Exercises A.9–A.14, simplify the expressions.

A.9. **a.** $(5 + 3x) - (4 - 7x)$ **b.** $(5 + 3i) - (4 - 7i)$ **A.10.** **a.** $(4 - 6x)(3 + 2x)$ **b.** $(4 - 6i)(3 + 2i)$

A.11. **a.** $(9 - 10x)(9 + 10x)$ **b.** $(9 - 10i)(9 + 10i)$ **A.12.** **a.** $(5 + 4x)^2$ **b.** $(5 + 4i)^2$

A.13. **a.** $\dfrac{5}{3 - 4\sqrt{x}}$ **b.** $\dfrac{5}{3 - 4i}$

A.14. $\dfrac{6 + \sqrt{-28}}{4}$

Practice Exercises

Section 18.8

Prerequisite Review

For Exercises R.1–R.10, perform the indicated operations.

R.1. $(3y + 4) + (-5y - 6)$ **R.2.** $(-7t + 8) + (9t - 11)$

R.3. $(-4x + 9) - (5x - 8)$ **R.4.** $(10n + 4) - (-5n - 3)$

R.5. $(2n - 7)(5n + 8)$ **R.6.** $(3z + 1)(z - 8)$

R.7. $(7d - 5)(7d + 5)$ **R.8.** $(11k + 4)(11k - 4)$

R.9. $(10b - 3)^2$ **R.10.** $(9x + 4)^2$

For Exercises R.11–R.14, simplify the expression.

R.11. $\sqrt{90}$ **R.12.** $\sqrt{112}$

R.13. $\dfrac{\sqrt{27}}{\sqrt{3}}$ **R.14.** $\dfrac{\sqrt{50}}{\sqrt{2}}$

For Exercises R.15–R.16, write the expression as a sum of two terms.

R.15. $\dfrac{10 + 5x}{15}$ **R.16.** $\dfrac{-2 - 4y}{6}$

Vocabulary and Key Concepts

1. **a.** A square root of a negative number is not a real number, but rather is an _____ number.
    b. $i =$ _____, and $i^2 =$ _____.
    c. For a positive number b, the value $\sqrt{-b} =$ _____.
    d. A complex number is a number in the form _____, where a and b are real numbers and $i =$ _____.
    e. Given a complex number $a + bi$, the value a is called the _____ part, and _____ is called the imaginary part.
    f. The complex conjugate of $a - bi$ is _____.

2. **a.** Answer true or false. All real numbers are complex numbers.
    b. Answer true or false. All imaginary numbers are complex numbers.

Concept 1: Definition of *i*

3. Simplify the expressions $\sqrt{-1}$ and $-\sqrt{1}$.

4. Simplify i^2.

For Exercises 5–30, simplify the expressions. **(See Examples 1–2.)**

5. $\sqrt{-49}$ **6.** $\sqrt{-121}$ **7.** $-\sqrt{49}$

8. $-\sqrt{121}$ **9.** $\sqrt{-\dfrac{1}{4}}$ **10.** $\sqrt{-\dfrac{9}{25}}$

11. $\sqrt{-144}$ 12. $\sqrt{-81}$ 13. $\sqrt{-3}$ 14. $\sqrt{-17}$

15. $-\sqrt{-20}$ 16. $-\sqrt{-75}$ 17. $(2\sqrt{-25})(3\sqrt{-4})$ 18. $(-4\sqrt{-9})(-3\sqrt{-1})$

19. $7\sqrt{-63} - 4\sqrt{-28}$ 20. $7\sqrt{-3} - 4\sqrt{-27}$ 21. $\sqrt{-7} \cdot \sqrt{-7}$ 22. $\sqrt{-11} \cdot \sqrt{-11}$

23. $\sqrt{-9} \cdot \sqrt{-16}$ 24. $\sqrt{-25} \cdot \sqrt{-36}$ 25. $\sqrt{-15} \cdot \sqrt{-6}$ 26. $\sqrt{-12} \cdot \sqrt{-50}$

27. $\dfrac{\sqrt{-50}}{\sqrt{25}}$ 28. $\dfrac{\sqrt{-27}}{\sqrt{9}}$ 29. $\dfrac{\sqrt{-90}}{\sqrt{-10}}$ 30. $\dfrac{\sqrt{-125}}{\sqrt{-45}}$

Concept 2: Powers of *i*

For Exercises 31–42, simplify the powers of *i*. **(See Example 3.)**

31. i^7 32. i^{38} 33. i^{64} 34. i^{75}

35. i^{41} 36. i^{25} 37. i^{52} 38. i^0

39. i^{23} 40. i^{103} 41. i^6 42. i^{82}

Concept 3: Definition of a Complex Number

43. What is the complex conjugate of a complex number $a + bi$?

44. True or false?
    a. Every real number is a complex number.
    b. Every complex number is a real number.

For Exercises 45–52, identify the real and imaginary parts of the complex number. **(See Example 4.)**

45. $-5 + 12i$ 46. $22 - 16i$ 47. $-6i$ 48. $10i$

49. 35 50. -1 51. $\dfrac{3}{5} + i$ 52. $-\dfrac{1}{2} - \dfrac{1}{4}i$

Concept 4: Addition, Subtraction, and Multiplication of Complex Numbers

For Exercises 53–76, perform the indicated operations. Write the answer in the form $a + bi$. **(See Examples 5–6.)**

53. $(2 - i) + (5 + 7i)$ 54. $(5 - 2i) + (3 + 4i)$ 55. $\left(\dfrac{1}{2} + \dfrac{2}{3}i\right) - \left(\dfrac{1}{5} - \dfrac{5}{6}i\right)$

56. $\left(\dfrac{11}{10} - \dfrac{7}{5}i\right) - \left(-\dfrac{2}{5} + \dfrac{3}{5}i\right)$ 57. $\sqrt{-98} - \sqrt{-8}$ 58. $\sqrt{-75} + \sqrt{-12}$

59. $(2 + 3i) - (1 - 4i) + (-2 + 3i)$ 60. $(2 + 5i) - (7 - 2i) + (-3 + 4i)$

61. $(8i)(3i)$ 62. $(2i)(4i)$ 63. $6i(1 - 3i)$ 64. $-i(3 + 4i)$

65. $(2 - 10i)(3 + 2i)$ 66. $(4 + 7i)(2 - 3i)$ 67. $(-5 + 2i)(5 + 2i)$ 68. $(4 - 11i)(4 + 11i)$

69. $(4 + 5i)^2$ **70.** $(3 - 2i)^2$ **71.** $(2 + i)(3 - 2i)(4 + 3i)$ **72.** $(3 - i)(3 + i)(4 - i)$

73. $(-4 - 6i)^2$ **74.** $(-3 - 5i)^2$ **75.** $\left(-\frac{1}{2} - \frac{3}{4}i\right)\left(-\frac{1}{2} + \frac{3}{4}i\right)$ **76.** $\left(-\frac{2}{3} + \frac{1}{6}i\right)\left(-\frac{2}{3} - \frac{1}{6}i\right)$

Concept 5: Division and Simplification of Complex Numbers

For Exercises 77–90, divide the complex numbers. Write the answer in the form $a + bi$. (See Example 7.)

77. $\dfrac{2}{1 + 3i}$ **78.** $\dfrac{-2}{3 + i}$ **79.** $\dfrac{-i}{4 - 3i}$ **80.** $\dfrac{3 - 3i}{1 - i}$

81. $\dfrac{5 + 2i}{5 - 2i}$ **82.** $\dfrac{7 + 3i}{4 - 2i}$ **83.** $\dfrac{3 + 7i}{-2 - 4i}$ **84.** $\dfrac{-2 + 9i}{-1 - 4i}$

85. $\dfrac{13i}{-5 - i}$ **86.** $\dfrac{15i}{-2 - i}$ **87.** $\dfrac{2 + 3i}{6i}$ (*Hint:* Consider multiplying numerator and denominator by i or by $-i$. This will make the denominator a real number.)

88. $\dfrac{4 - i}{2i}$ **89.** $\dfrac{-10 + i}{i}$ **90.** $\dfrac{-6 - i}{-i}$

For Exercises 91–98, simplify the complex numbers. Write the answer in the form $a + bi$. (See Example 8.)

91. $\dfrac{2 + \sqrt{-16}}{8}$ **92.** $\dfrac{6 - \sqrt{-4}}{4}$ **93.** $\dfrac{-6 + \sqrt{-72}}{6}$ **94.** $\dfrac{-20 + \sqrt{-500}}{10}$

95. $\dfrac{-8 - \sqrt{-48}}{4}$ **96.** $\dfrac{-18 - \sqrt{-72}}{3}$ **97.** $\dfrac{-5 + \sqrt{-50}}{10}$ **98.** $\dfrac{14 + \sqrt{-98}}{7}$

Expanding Your Skills

For Exercises 99–102, determine if the complex number is a solution to the equation.

99. $x^2 - 4x + 5 = 0$; $\quad 2 + i$ **100.** $x^2 - 6x + 25 = 0$; $\quad 3 - 4i$

101. $x^2 + 12 = 0$; $\quad -2i\sqrt{3}$ **102.** $x^2 + 18 = 0$; $\quad 3i\sqrt{2}$

Chapter 18 Review Exercises

For the exercises in this set, assume that all variables represent positive real numbers unless otherwise stated.

Section 18.1

1. True or false?
   a. The principal nth root of an even-indexed root is always positive
   b. The principal nth root of an odd-indexed root is always positive.

2. Explain why $\sqrt{(-3)^2} \neq -3$.

3. For $a > 0$ and $b > 0$, are the following statements true or false?
   a. $\sqrt{a^2 + b^2} = a + b$
   b. $\sqrt{(a+b)^2} = a + b$

For Exercises 4–6, simplify the radicals.

4. $\sqrt{\dfrac{50}{32}}$ 5. $\sqrt[4]{625}$ 6. $\sqrt{(-6)^2}$

7. Evaluate the function values for $f(x) = \sqrt{x-1}$.
   a. $f(10)$ b. $f(1)$ c. $f(8)$
   d. Write the domain of f in interval notation.

8. Evaluate the function values for $g(t) = \sqrt{5+t}$.
   a. $g(-5)$ b. $g(-4)$ c. $g(4)$
   d. Write the domain of g in interval notation.

9. Write the English expression as an algebraic expression: Four more than the quotient of the cube root of $2x$ and the principal fourth root of $2x$.

For Exercises 10–11, simplify the expression. Assume that x and y represent *any* real number.

10. a. $\sqrt{x^2}$ b. $\sqrt[3]{x^3}$
    c. $\sqrt[4]{x^4}$ d. $\sqrt[5]{(x+1)^5}$

11. a. $\sqrt{4y^2}$ b. $\sqrt[3]{27y^3}$
    c. $\sqrt[100]{y^{100}}$ d. $\sqrt[101]{y^{101}}$

12. Use the Pythagorean theorem to find the length of the third side of the triangle.

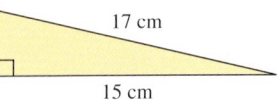

Section 18.2

13. Are the properties of exponents the same for rational exponents and integer exponents? Give an example. (Answers may vary.)

14. In the expression $x^{m/n}$ what does n represent?

15. Explain the process of eliminating a negative exponent from an algebraic expression.

For Exercises 16–21, simplify the expressions. Write the answers with positive exponents only.

16. $(-125)^{1/3}$ 17. $16^{-1/4}$

18. $\left(\dfrac{1}{16}\right)^{-3/4} - \left(\dfrac{1}{8}\right)^{-2/3}$ 19. $(b^{1/2} \cdot b^{1/3})^{12}$

20. $\left(\dfrac{x^{-1/4}y^{-1/3}z^{3/4}}{2^{1/3}x^{-1/3}y^{2/3}}\right)^{-12}$ 21. $\left(\dfrac{a^{12}b^{-4}c^7}{a^3b^2c^4}\right)^{1/3}$

For Exercises 22–23, rewrite the expressions by using rational exponents.

22. $\sqrt[4]{x^3}$ 23. $\sqrt[3]{2y^2}$

For Exercises 24–26, use a calculator to approximate the expressions to four decimal places.

24. $10^{1/3}$ 25. $17.8^{2/3}$ 26. $\sqrt[5]{147^4}$

Section 18.3

27. List the criteria for a radical expression to be simplified.

For Exercises 28–31, simplify the radicals.

28. $\sqrt{108}$ 29. $\sqrt[4]{x^5yz^4}$

30. $-2\sqrt[3]{250a^3b^{10}}$ 31. $\sqrt[3]{\dfrac{-16a^4}{2ab^3}}$

32. Write an English phrase that describes the following mathematical expressions: (Answers may vary.)

a. $\sqrt{\dfrac{2}{x}}$
b. $(x+1)^3$

33. An engineering firm made a mistake when building a $\frac{1}{4}$-mi bridge in the Florida Keys. The bridge was made without adequate expansion joints to prevent buckling during the heat of summer. During mid-June, the bridge expanded 1.5 ft, causing a vertical bulge in the middle. Calculate the height of the bulge h in feet. (*Note:* 1 mi = 5280 ft.) Round to the nearest foot.

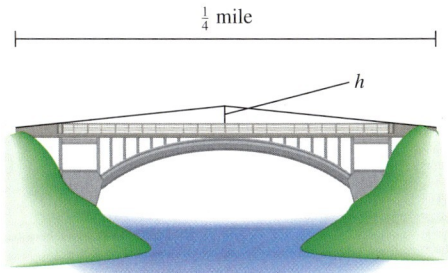

Section 18.4

34. Complete the following statement: Radicals may be added or subtracted if . . .

For Exercises 35–38, determine whether the radicals may be combined, and explain your answer.

35. $\sqrt[3]{2x} - 2\sqrt{2x}$
36. $2 + \sqrt{x}$
37. $\sqrt[4]{3xy} + 2\sqrt[4]{3xy}$
38. $-4\sqrt{32} + 7\sqrt{50}$

For Exercises 39–42, add or subtract as indicated.

39. $4\sqrt{7} - 2\sqrt{7} + 3\sqrt{7}$

40. $2\sqrt[3]{64} + 3\sqrt[3]{54} - 16$

41. $\sqrt{50} + 7\sqrt{2} - \sqrt{8}$

42. $x\sqrt[3]{16x^2} - 4\sqrt[3]{2x^5} + 5x\sqrt[3]{54x^2}$

For Exercises 43–44, answer true or false. If an answer is false, explain why. Assume all variables represent positive real numbers.

43. $5 + 3\sqrt{x} = 8\sqrt{x}$

44. $\sqrt{y} + \sqrt{y} = \sqrt{2y}$

Section 18.5

For Exercises 45–56, multiply the radicals and simplify the answer.

45. $\sqrt{3} \cdot \sqrt{12}$
46. $\sqrt[4]{4} \cdot \sqrt[4]{8}$

47. $-2\sqrt{3}(\sqrt{7} - 3\sqrt{11})$
48. $-3\sqrt{5}(2\sqrt{3} - \sqrt{5})$

49. $(2\sqrt{x} - 3)(2\sqrt{x} + 3)$
50. $(\sqrt{y} + 4)(\sqrt{y} - 4)$

51. $(\sqrt{7y} - \sqrt{3x})^2$
52. $(2\sqrt{3w} + 5)^2$

53. $(-\sqrt{z} - \sqrt{6})(2\sqrt{z} + 7\sqrt{6})$

54. $(3\sqrt{a} - \sqrt{5})(\sqrt{a} + 2\sqrt{5})$

55. $\sqrt[3]{u} \cdot \sqrt{u^5}$
56. $\sqrt{2} \cdot \sqrt[4]{w^3}$

Section 18.6

For Exercises 57–60, simplify the radicals.

57. $\sqrt{\dfrac{3y^5}{25x^6}}$
58. $\sqrt[3]{\dfrac{-16x^7y^6}{z^9}}$

59. $\dfrac{\sqrt{324w^7}}{\sqrt{4w^3}}$
60. $\dfrac{\sqrt[3]{3t^{14}}}{\sqrt[3]{192t^2}}$

For Exercises 61–68, rationalize the denominator.

61. $\sqrt{\dfrac{7}{2y}}$
62. $\sqrt{\dfrac{5}{3w}}$

63. $\dfrac{4}{\sqrt[3]{9p^2}}$
64. $\dfrac{-2}{\sqrt[3]{2x}}$

65. $\dfrac{-5}{\sqrt{15} + \sqrt{10}}$
66. $\dfrac{-6}{\sqrt{7} + \sqrt{5}}$

67. $\dfrac{t - 3}{\sqrt{t} - \sqrt{3}}$
68. $\dfrac{w - 7}{\sqrt{w} - \sqrt{7}}$

69. Write the mathematical expression as an English phrase. (Answers may vary.)

$$\dfrac{\sqrt{2}}{x^2}$$

Section 18.7

Solve the radical equations in Exercises 70–77, if possible.

70. $\sqrt{2y} = 7$
71. $\sqrt{a - 6} - 5 = 0$

72. $\sqrt[3]{2w - 3} + 5 = 2$

73. $\sqrt[4]{p+12} - \sqrt[4]{5p-16} = 0$

74. $\sqrt{t} + \sqrt{t-5} = 5$

75. $\sqrt{8x+1} = -\sqrt{x-13}$

76. $\sqrt{2m^2+4} - \sqrt{9m} = 0$

77. $\sqrt{x+2} = 1 - \sqrt{2x+5}$

78. A tower is supported by stabilizing wires. Find the exact length of each wire, and then round to the nearest tenth of a meter.

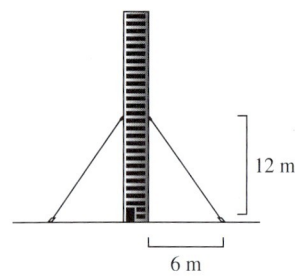

79. The velocity, $v(d)$, of an ocean wave depends on the water depth d as the wave approaches land.

$$v(d) = \sqrt{32d}$$

where $v(d)$ is in feet per second and d is in feet.

a. Find $v(20)$ and interpret its value. Round to one decimal place.

b. Find the depth of the water at a point where a wave is traveling at 16 ft/sec.

Section 18.8

80. Define a complex number.

81. Define an imaginary number.

82. Describe the first step in the process to simplify the expression.

$$\frac{3}{4+6i}$$

For Exercises 83–86, rewrite the expressions in terms of i.

83. $\sqrt{-16}$

84. $-\sqrt{-5}$

85. $\sqrt{-75} \cdot \sqrt{-3}$

86. $\dfrac{-\sqrt{-24}}{\sqrt{6}}$

For Exercises 87–90, simplify the powers of i.

87. i^{38}

88. i^{101}

89. i^{19}

90. $i^{1000} + i^{1002}$

For Exercises 91–94, perform the indicated operations. Write the final answer in the form $a + bi$.

91. $(-3 + i) - (2 - 4i)$

92. $(1 + 6i)(3 - i)$

93. $(4 - 3i)(4 + 3i)$

94. $(5 - i)^2$

For Exercises 95–96, write the expressions in the form $a + bi$, and determine the real and imaginary parts.

95. $\dfrac{17 - 4i}{-4}$

96. $\dfrac{-16 - 8i}{8}$

For Exercises 97–100, divide and simplify. Write the final answer in the form $a + bi$.

97. $\dfrac{2 - i}{3 + 2i}$

98. $\dfrac{10 + 5i}{2 - i}$

99. $\dfrac{5 + 3i}{-2i}$

100. $\dfrac{4i}{4 - i}$

For Exercises 101–102, simplify the expression.

101. $\dfrac{-8 + \sqrt{-40}}{12}$

102. $\dfrac{6 - \sqrt{-144}}{3}$

Quadratic Equations and Functions

19

CHAPTER OUTLINE

19.1 Square Root Property and Completing the Square 1264
19.2 Quadratic Formula 1276
19.3 Equations in Quadratic Form 1292
    Problem Recognition Exercises: Quadratic and Quadratic Type Equations 1299
19.4 Graphs of Quadratic Functions 1299
19.5 Vertex of a Parabola: Applications and Modeling 1314
    Chapter 19 Review Exercises 1326

Mathematics in Art

A **golden rectangle** is a rectangle in which the ratio of its length L to its width W is equal to the ratio of the sum of its length and width to its length.

$$\frac{L}{W} = \frac{L+W}{L}$$

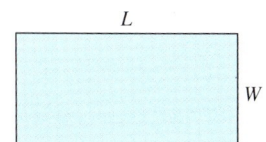

The values of L and W that meet this condition are said to be in the **golden ratio**. The golden ratio has been studied by artists and art historians for generations because the golden ratio represents an aesthetically pleasing ratio between the length and width of a figure. For example, the face of the Parthenon built in ancient Greece has the dimensions of a golden rectangle.

We can show that the length of a golden rectangle is approximately 1.62 times the width. Substituting 1 for the width, we have the proportion $\frac{L}{1} = \frac{L+1}{L}$.

Then, clearing fractions and writing the quadratic equation in standard form gives $L^2 - L - 1 = 0$. The expression on the left is not factorable, but fortunately in this chapter, we will learn two techniques to solve a quadratic equation when factoring fails. The positive solution for L in this equation is the golden ratio, $\frac{1+\sqrt{5}}{2} \approx 1.62$.

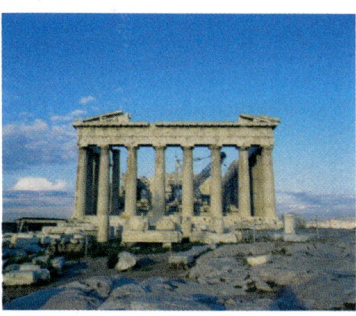

Purestock/SuperStock

Section 19.1 Square Root Property and Completing the Square

Concepts

1. Solving Quadratic Equations by Using the Square Root Property
2. Solving Quadratic Equations by Completing the Square
3. Literal Equations

1. Solving Quadratic Equations by Using the Square Root Property

We have already learned how to solve a quadratic equation by factoring and applying the zero product rule. For example:

$$x^2 = 81$$
$$x^2 - 81 = 0 \quad \text{Set one side equal to zero.}$$
$$(x - 9)(x + 9) = 0 \quad \text{Factor.}$$
$$x - 9 = 0 \quad \text{or} \quad x + 9 = 0 \quad \text{Set each factor equal to zero.}$$
$$x = 9 \quad \text{or} \quad x = -9$$

The solution set is $\{9, -9\}$.

It is important to note that the zero product rule can only be used if the equation is factorable. In this section, we will learn a method to solve quadratic equations containing expressions that are both factorable and nonfactorable.

Consider a quadratic equation of the form $x^2 = k$. The solutions are all numbers (real or imaginary) that when squared equal k, so for any nonzero constant k, there will be two solutions, $\sqrt{k}$ or $-\sqrt{k}$. For example:

$$x^2 = 25 \quad \text{The solutions are 5 and } -5.$$
$$x^2 = -25 \quad \text{The solutions are } 5i \text{ and } -5i.$$

This principle is stated formally as the **square root property**.

FOR REVIEW

The square root property states that if $x^2 = k$, then $x = \pm\sqrt{k}$. To understand why, recall that $\sqrt{x^2} = |x|$. Thus,

$$x^2 = 25$$
$$\sqrt{x^2} = \sqrt{25}$$
$$|x| = 5$$
$$x = 5 \text{ or } x = -5$$

> **The Square Root Property**
> For any real number k, if $x^2 = k$, then $x = \sqrt{k}$ or $x = -\sqrt{k}$.
>
> *Note:* The solution may also be written as $\pm\sqrt{k}$, read "plus or minus the square root of k."

Example 1 Solving a Quadratic Equation by Using the Square Root Property

Use the square root property to solve the equation. $\quad 4p^2 = 9$

Solution:

$$4p^2 = 9$$
$$p^2 = \frac{9}{4} \quad \text{Isolate } p^2 \text{ by dividing both sides by 4.}$$
$$p = \pm\sqrt{\frac{9}{4}} \quad \text{Apply the square root property.}$$
$$p = \pm\frac{3}{2} \quad \text{Simplify the radical.}$$

The solution set is $\left\{\frac{3}{2}, -\frac{3}{2}\right\}$.

Skill Practice Solve using the square root property.

1. $25a^2 = 16$

For a quadratic equation, $ax^2 + bx + c = 0$, if $b = 0$, then the equation is easily solved by using the square root property. This is demonstrated in Example 2.

Example 2 — Solving a Quadratic Equation by Using the Square Root Property

Use the square root property to solve the equation. $3x^2 + 75 = 0$

Solution:

$3x^2 + 75 = 0$ Rewrite the equation to fit the form $x^2 = k$.

$3x^2 = -75$

$x^2 = -25$ The equation is now in the form $x^2 = k$.

$x = \pm\sqrt{-25}$ Apply the square root property.

$= \pm 5i$

Check: $x = 5i$
$3x^2 + 75 = 0$
$3(5i)^2 + 75 \stackrel{?}{=} 0$
$3(25i^2) + 75 \stackrel{?}{=} 0$
$3(-25) + 75 \stackrel{?}{=} 0$
$-75 + 75 \stackrel{?}{=} 0$ ✓

Check: $x = -5i$
$3x^2 + 75 = 0$
$3(-5i)^2 + 75 \stackrel{?}{=} 0$
$3(25i^2) + 75 \stackrel{?}{=} 0$
$3(-25) + 75 \stackrel{?}{=} 0$
$-75 + 75 \stackrel{?}{=} 0$ ✓

The solution set is $\{\pm 5i\}$.

Avoiding Mistakes
A common mistake is to forget the ± symbol when solving the equation $x^2 = k$:
$$x = \pm\sqrt{k}$$

Skill Practice Solve using the square root property.

2. $8x^2 + 72 = 0$

Example 3 — Solving a Quadratic Equation by Using the Square Root Property

Use the square root property to solve the equation. $(w + 3)^2 = 20$

Solution:

$(w + 3)^2 = 20$ The equation is in the form $x^2 = k$, where $x = (w + 3)$.

$w + 3 = \pm\sqrt{20}$ Apply the square root property.

$w + 3 = \pm\sqrt{4 \cdot 5}$ Simplify the radical.

$w + 3 = \pm 2\sqrt{5}$

$w = -3 \pm 2\sqrt{5}$ Solve for w.

The solution set is $\{-3 \pm 2\sqrt{5}\}$.

TIP: Recall that $-3 \pm 2\sqrt{5}$ represents two solutions:
$-3 + 2\sqrt{5}$ and $-3 - 2\sqrt{5}$

Skill Practice Solve using the square root property.

3. $(t - 5)^2 = 18$

Answers

1. $\left\{\dfrac{4}{5}, -\dfrac{4}{5}\right\}$ 2. $\{\pm 3i\}$
3. $\{5 \pm 3\sqrt{2}\}$

2. Solving Quadratic Equations by Completing the Square

In Example 3 we used the square root property to solve an equation where the square of a binomial was equal to a constant.

$$\underbrace{(w+3)^2}_{\text{Square of a binomial}} = \underset{\underset{\text{Constant}}{\uparrow}}{20}$$

The square of a binomial is the factored form of a perfect square trinomial. For example:

Perfect Square Trinomial	Factored Form
$x^2 + 10x + 25$	$(x+5)^2$
$t^2 - 6t + 9$	$(t-3)^2$
$p^2 - 14p + 49$	$(p-7)^2$

For a perfect square trinomial with a leading coefficient of 1, the constant term is the square of one-half the linear term coefficient. For example:

$$x^2 + 10x + \underset{\underset{[\frac{1}{2}(10)]^2}{\uparrow}}{25}$$

In general, an expression of the form $x^2 + bx + n$ is a perfect square trinomial if $n = \left(\frac{1}{2}b\right)^2$. The process to create a perfect square trinomial is called **completing the square**.

Example 4 — Completing the Square

Determine the value of n that makes the polynomial a perfect square trinomial. Then factor the expression as the square of a binomial.

a. $x^2 + 12x + n$
b. $x^2 - 26x + n$
c. $x^2 + 11x + n$
d. $x^2 - \dfrac{4}{7}x + n$

Solution:

The expressions are in the form $x^2 + bx + n$. The value of n equals the square of one-half the linear term coefficient $\left(\frac{1}{2}b\right)^2$.

a. $x^2 + 12x + n$

$x^2 + 12x + 36 \qquad n = \left[\frac{1}{2}(12)\right]^2 = (6)^2 = 36$

$(x+6)^2 \qquad$ Factored form

b. $x^2 - 26x + n$

$x^2 - 26x + 169 \qquad n = \left[\frac{1}{2}(-26)\right]^2 = (-13)^2 = 169$

$(x-13)^2 \qquad$ Factored form

c. $x^2 + 11x + n$

$x^2 + 11x + \dfrac{121}{4} \qquad n = \left[\frac{1}{2}(11)\right]^2 = \left(\frac{11}{2}\right)^2 = \frac{121}{4}$

$\left(x + \dfrac{11}{2}\right)^2 \qquad$ Factored form

d. $x^2 - \dfrac{4}{7}x + n$

$x^2 - \dfrac{4}{7}x + \dfrac{4}{49}$ $n = \left[\dfrac{1}{2}\left(-\dfrac{4}{7}\right)\right]^2 = \left(-\dfrac{2}{7}\right)^2 = \dfrac{4}{49}$

$\left(x - \dfrac{2}{7}\right)^2$ Factored form

Skill Practice Determine the value of n that makes the polynomial a perfect square trinomial. Then factor.

4. $x^2 + 20x + n$ **5.** $y^2 - 16y + n$

6. $a^2 - 15a + n$ **7.** $w^2 + \dfrac{7}{3}w + n$

The process of completing the square can be used to write a quadratic equation $ax^2 + bx + c = 0$ $(a \neq 0)$ in the form $(x - h)^2 = k$. Then the square root property can be used to solve the equation. The following steps outline the procedure.

Solving a Quadratic Equation $ax^2 + bx + c = 0$ by Completing the Square and Applying the Square Root Property

Step 1 Divide both sides by a to make the leading coefficient 1.
Step 2 Isolate the variable terms on one side of the equation.
Step 3 Complete the square.
- Add the square of one-half the linear term coefficient to both sides, $\left(\dfrac{1}{2}b\right)^2$.
- Factor the resulting perfect square trinomial.

Step 4 Apply the square root property and solve for x.

Example 5 **Solving a Quadratic Equation by Completing the Square and Applying the Square Root Property**

Solve by completing the square and applying the square root property.

$$x^2 - 6x + 13 = 0$$

Solution:

$x^2 - 6x + 13 = 0$ **Step 1:** Since the leading coefficient a is equal to 1, we do not have to divide by a. We can proceed to step 2.

$x^2 - 6x = -13$ **Step 2:** Isolate the variable terms on one side.

$x^2 - 6x + 9 = -13 + 9$ **Step 3:** To complete the square, add $\left[\dfrac{1}{2}(-6)\right]^2 = 9$ to both sides of the equation.

$(x - 3)^2 = -4$ Factor the perfect square trinomial.

Answers
4. $n = 100$; $(x + 10)^2$
5. $n = 64$; $(y - 8)^2$
6. $n = \dfrac{225}{4}$; $\left(a - \dfrac{15}{2}\right)^2$
7. $n = \dfrac{49}{36}$; $\left(w + \dfrac{7}{6}\right)^2$

FOR REVIEW

Recall that a solution to an equation can be checked by substitution. Also recall that $i^2 = -1$.

$$x^2 - 6x + 13 = 0$$
$$(3 + 2i)^2 - 6(3 + 2i) + 13 \stackrel{?}{=} 0$$
$$9 + 12i + 4i^2 - 18 - 12i + 13 \stackrel{?}{=} 0$$
$$9 + 12i - 4 - 18 - 12i + 13 \stackrel{?}{=} 0$$
$$0 = 0 \checkmark$$

$$x - 3 = \pm\sqrt{-4}$$ **Step 4:** Apply the square root property.
$$x - 3 = \pm 2i$$ Simplify the radical.
$$x = 3 \pm 2i$$ Solve for x.

The solutions are imaginary numbers and can be written as $3 + 2i$ and $3 - 2i$.

The solution set is $\{3 \pm 2i\}$.

Skill Practice Solve by completing the square and applying the square root property.

8. $z^2 - 4z + 26 = 0$

Example 6 Solving a Quadratic Equation by Completing the Square and Applying the Square Root Property

Solve by completing the square and applying the square root property.

$$2m^2 + 10m = 3$$

Solution:

$$2m^2 + 10m = 3$$ The variable terms are already isolated on one side of the equation.

$$\frac{2m^2}{2} + \frac{10m}{2} = \frac{3}{2}$$ Divide by the leading coefficient, 2.

$$m^2 + 5m = \frac{3}{2}$$

$$m^2 + 5m + \frac{25}{4} = \frac{3}{2} + \frac{25}{4}$$ Add $\left[\frac{1}{2}(5)\right]^2 = \left(\frac{5}{2}\right)^2 = \frac{25}{4}$ to both sides.

$$\left(m + \frac{5}{2}\right)^2 = \frac{6}{4} + \frac{25}{4}$$ Factor the left side and write the terms on the right with a common denominator.

$$\left(m + \frac{5}{2}\right)^2 = \frac{31}{4}$$

$$m + \frac{5}{2} = \pm\sqrt{\frac{31}{4}}$$ Apply the square root property.

$$m = -\frac{5}{2} \pm \frac{\sqrt{31}}{2}$$ Subtract $\frac{5}{2}$ from both sides and simplify the radical.

The solution set is $\left\{-\frac{5}{2} \pm \frac{\sqrt{31}}{2}\right\}$. The solutions are irrational numbers.

TIP: The solutions to Example 6 can also be written as:

$$\frac{-5 \pm \sqrt{31}}{2}$$

Skill Practice Solve by completing the square and applying the square root property.

9. $4x^2 + 12x = 5$

Answers

8. $\{2 \pm i\sqrt{22}\}$
9. $\left\{-\frac{3}{2} \pm \frac{\sqrt{14}}{2}\right\}$

Example 7 Solving a Quadratic Equation by Completing the Square and Applying the Square Root Property

Solve by completing the square and applying the square root property.

$$2x(2x - 10) = -30 + 6x$$

Solution:

$2x(2x - 10) = -30 + 6x$

$4x^2 - 20x = -30 + 6x$ Clear parentheses.

$4x^2 - 26x + 30 = 0$ Write the equation in the form $ax^2 + bx + c = 0$.

$\dfrac{4x^2}{4} - \dfrac{26x}{4} + \dfrac{30}{4} = \dfrac{0}{4}$ **Step 1:** Divide both sides by the leading coefficient, 4.

$x^2 - \dfrac{13}{2}x + \dfrac{15}{2} = 0$

$x^2 - \dfrac{13}{2}x = -\dfrac{15}{2}$ **Step 2:** Isolate the variable terms on one side.

$x^2 - \dfrac{13}{2}x + \dfrac{169}{16} = -\dfrac{15}{2} + \dfrac{169}{16}$ **Step 3:** Add $\left[\tfrac{1}{2}(-\tfrac{13}{2})\right]^2 = \left(-\tfrac{13}{4}\right)^2 = \tfrac{169}{16}$ to both sides.

$\left(x - \dfrac{13}{4}\right)^2 = -\dfrac{120}{16} + \dfrac{169}{16}$ Factor the perfect square trinomial. Rewrite the right-hand side with a common denominator.

$\left(x - \dfrac{13}{4}\right)^2 = \dfrac{49}{16}$

$x - \dfrac{13}{4} = \pm\sqrt{\dfrac{49}{16}}$ **Step 4:** Apply the square root property.

$x - \dfrac{13}{4} = \pm\dfrac{7}{4}$ Simplify the radical.

$x = \dfrac{13}{4} \pm \dfrac{7}{4}$

$x = \dfrac{13}{4} + \dfrac{7}{4} = \dfrac{20}{4} = 5$

$x = \dfrac{13}{4} - \dfrac{7}{4} = \dfrac{6}{4} = \dfrac{3}{2}$

The solution set is $\left\{5, \dfrac{3}{2}\right\}$.   The solutions are rational numbers.

TIP: In general, if the solutions to a quadratic equation are rational numbers, the equation can be solved by factoring and using the zero product rule. Consider the equation from Example 7.

$2x(2x - 10) = -30 + 6x$
$4x^2 - 20x = -30 + 6x$
$4x^2 - 26x + 30 = 0$
$2(2x^2 - 13x + 15) = 0$
$2(x - 5)(2x - 3) = 0$
$x = 5 \text{ or } x = \dfrac{3}{2}$

Avoiding Mistakes
When the solutions are rational, combine the *like* terms. That is, do not leave the solution with the $\pm$ sign.

Skill Practice Solve by completing the square and applying the square root property.

10. $2y(y - 1) = 3 - y$

Answer

10. $\left\{\dfrac{3}{2}, -1\right\}$

3. Literal Equations

Example 8   Solving a Literal Equation

Ignoring air resistance, the distance d (in meters) that an object falls in t sec is given by the equation

$$d = 4.9t^2 \qquad \text{where } t \geq 0$$

a. Solve the equation for t. Do not rationalize the denominator.

b. Using the equation from part (a), determine the amount of time required for an object to fall 500 m. Round to the nearest second.

Solution:

a. $d = 4.9t^2$

$\dfrac{d}{4.9} = t^2$ Isolate the quadratic term. The equation is in the form $t^2 = k$.

$t = \pm\sqrt{\dfrac{d}{4.9}}$ Apply the square root property.

$= \sqrt{\dfrac{d}{4.9}}$ Because t represents time, $t \geq 0$. We reject the negative solution.

b. $t = \sqrt{\dfrac{d}{4.9}}$

$= \sqrt{\dfrac{500}{4.9}}$ Substitute $d = 500$.

$t \approx 10.1$

The object will require approximately 10.1 sec to fall 500 m.

Skill Practice

11. Given $x = 2yz^2$, solve for z where $z > 0$. Do not rationalize the denominator.

12. Use the equation from the previous exercise to find z when $x = 54$ and $y = 3$.

Answers

11. $z = \sqrt{\dfrac{x}{2y}}$

12. $z = 3$

Section 19.1 Activity

A.1. **a.** Consider the equation $x^2 = 49$. One method to solve this equation algebraically is to set one side equal to zero, factor the other side, and then apply the zero product rule. Solve the equation using this process.

b. Now consider a similar equation $x^2 = 16$. By inspection, the solutions to this equation are the real numbers that when squared equal 16. Write the solution set.

c. Now consider the equation $x^2 = -25$. There is no real number x such that the square of x is negative. However, if we solve the equation over the set of complex numbers, the solutions are $\sqrt{-25}$ and $-\sqrt{-25}$ or equivalently, $5i$ and $-5i$. To show this, fill in the blanks.
$(5i)^2 = $ _____ and $(-5i)^2 = $ _____

d. We can now generalize the results from parts (a)–(c) as the **square root property**. That is, for a real number k, the solutions to the equation $x^2 = k$ are _____ and _____.

For Exercises A.2–A.3, solve the equations.

A.2. **a.** $u^2 = 64$ **b.** $(x+3)^2 = 64$ **c.** $4(x+3)^2 = 64$

A.3. **a.** $u^2 = 20$ **b.** $(x-4)^2 = 20$ **c.** $(x-4)^2 + 26 = 20$

A.4. Consider the trinomial $x^2 + 10x + \boxed{?}$. Suppose we want to fill in the blank so that the expression is a perfect square trinomial. Thus,

$$x^2 + 10x + \boxed{?} = (x+a)^2.$$

The right side expands to $x^2 + 2ax + a^2$, which means that the middle term $10x$ on the left must equal the middle term $2ax$ on the right.

$$x^2 + \boxed{10x} + \boxed{?} = (x+a)^2$$
$$= x^2 + 2ax + a^2$$

a. If $10x = 2ax$, what is the value of a?

b. What is the value of a^2?

c. Fill in the blank to form a perfect square trinomial and then factor the result.

$$x^2 + 10x + \square = (x + \square)^2$$

d. Given the expression $x^2 + bx + n$, what is the value of n that would make the expression a perfect square trinomial?

A.5. Determine the value of n that would make the expression a perfect square trinomial. Then factor the result.

a. $y^2 + 18y + n$ **b.** $t^2 - 11t + n$ **c.** $x^2 + \dfrac{3}{5}x + n$

A.6. Consider the quadratic equation $2x^2 + 8x + 20 = 0$. The left side is not factorable using the techniques we have learned thus far. Therefore, we cannot apply the zero product property. Instead, we will solve the equation by completing the square and applying the square root property.

a. Divide both sides by the leading coefficient, 2, and write the resulting equation.

b. Using the equation from part (a), isolate the terms containing x on the left side of the equation. Write the new equation.

c. Write the left side of the equation as a perfect square trinomial by adding an appropriate constant. To balance the equation, be sure to add the same constant to the right side.

d. Write the left side in factored form and simplify on the right.

e. Solve the equation by using the square root property.

Practice Exercises

Section 19.1

Prerequisite Review

For Exercises R.1–R.4, solve the equation.

R.1. $4x - 3 = 17$

R.2. $-6t + 4 = 64$

R.3. $n - \dfrac{2}{3} = \dfrac{\sqrt{5}}{3}$

R.4. $p + \dfrac{1}{5} = \dfrac{\sqrt{7}}{5}$

R.5. Identify the square roots of 121.

R.6. Identify the square roots of 64.

For Exercises R.7–R.10, simplify the radical.

R.7. $\sqrt{72}$

R.8. $\sqrt{500}$

R.9. $\sqrt{\dfrac{37}{64}}$

R.10. $\sqrt{\dfrac{10}{49}}$

For Exercises R.11–R.14, square the binomial.

R.11. $(y-7)^2$

R.12. $(t+9)^2$

R.13. $(3w+4)^2$

R.14. $(2m-5)^2$

For Exercises R.15–R.18, factor completely.

R.15. $p^2 + 22p + 121$

R.16. $w^2 - 26w + 169$

R.17. $64x^2 - 144x + 81$

R.18. $4x^2 + 40x + 100$

Vocabulary and Key Concepts

1. **a.** The zero product rule states that if $ab = 0$, then $a = $ _____ or $b = $ _____.

   b. To apply the zero product rule, one side of the equation must be equal to _____ and the other side must be written in factored form.

   c. The square root property states that for any real number k, if $x^2 = k$, then $x = $ _____ or $x = $ _____.

   d. To apply the square root property to the equation $t^2 + 2 = 11$, first subtract _____ from both sides. The solution set is _____.

   e. The process to create a perfect square trinomial is called _____ the square.

   f. Fill in the blank to complete the square for the trinomial $x^2 + 20x + $ _____.

   g. To use completing the square to solve the equation $4x^2 + 3x + 5 = 0$, the first step is to divide by _____ so that the coefficient of the x^2 term is _____.

   h. Given the trinomial $y^2 + 8y + 16$, the coefficient of the linear term is _____.

Concept 1: Solving Quadratic Equations by Using the Square Root Property

For Exercises 2–21, solve the equations by using the square root property. Write imaginary solutions in the form $a + bi$.
(See Examples 1–3.)

2. $x^2 = 100$

3. $y^2 = 4$

4. $a^2 = 5$

5. $k^2 - 7 = 0$

6. $4t^2 = 81$

7. $36u^2 = 121$

8. $3v^2 + 33 = 0$

9. $-2m^2 = 50$

10. $(p-5)^2 = 9$

11. $(q+3)^2 = 4$

12. $(3x-2)^2 - 5 = 0$

13. $(2y+3)^2 - 7 = 0$

14. $(h-4)^2 = -8$

15. $(t+5)^2 = -18$

16. $6p^2 - 3 = 2$

17. $15 = 4 + 3w^2$

18. $\left(x - \dfrac{3}{2}\right)^2 + \dfrac{7}{4} = 0$

19. $\left(m + \dfrac{4}{5}\right)^2 + \dfrac{3}{25} = 0$

20. $-x^2 + 4 = 13$

21. $-y^2 - 2 = 14$

22. Given the equation $x^2 = k$, match the following statements.

   a. If $k > 0$, then _____
   b. If $k < 0$, then _____
   c. If $k = 0$, then _____

   i. there will be one real solution.
   ii. there will be two real solutions.
   iii. there will be two imaginary solutions.

23. State two methods that can be used to solve the equation $x^2 - 36 = 0$. Then solve the equation by using both methods.

24. Explain the difference between solving the equations: $x = \sqrt{16}$ and $x^2 = 16$.

For Exercises 25–26, solve the equations.

25. a. $\sqrt{x} = 4$
    b. $x^2 = 4$

26. a. $\sqrt{y} = 9$
    b. $y^2 = 9$

Concept 2: Solving Quadratic Equations by Completing the Square

For Exercises 27–38, find the value of n so that the expression is a perfect square trinomial. Then factor the trinomial. (See Example 4.)

27. $x^2 - 6x + n$

28. $x^2 + 24x + n$

29. $t^2 + 8t + n$

30. $v^2 - 18v + n$

31. $c^2 - c + n$

32. $x^2 + 9x + n$

33. $y^2 + 5y + n$

34. $a^2 - 7a + n$

35. $b^2 + \dfrac{2}{5}b + n$

36. $m^2 - \dfrac{2}{7}m + n$

37. $p^2 - \dfrac{2}{3}p + n$

38. $w^2 + \dfrac{3}{4}w + n$

39. Summarize the steps used in solving a quadratic equation of the form $ax^2 + bx + c = 0$ by completing the square and applying the square root property.

40. What types of quadratic equations can be solved by completing the square and applying the square root property?

For Exercises 41–60, solve the quadratic equation by completing the square and applying the square root property. Write imaginary solutions in the form $a + bi$. (See Examples 5–7.)

41. $t^2 + 8t + 15 = 0$

42. $m^2 + 6m + 8 = 0$

43. $x^2 + 6x = -16$

44. $x^2 - 4x = -15$

45. $p^2 + 4p + 6 = 0$

46. $q^2 + 2q + 2 = 0$

47. $-3y - 10 = -y^2$

48. $-24 = -2y^2 + 2y$

49. $2a^2 + 4a + 5 = 0$
50. $3a^2 + 6a - 7 = 0$
51. $9x^2 - 36x + 40 = 0$
52. $9y^2 - 12y + 5 = 0$

53. $25p^2 - 10p = 2$
54. $9n^2 - 6n = 1$
55. $(2w + 5)(w - 1) = 2$
56. $(3p - 5)(p + 1) = -3$

57. $n(n - 4) = 7$
58. $m(m + 10) = 2$
59. $2x(x + 6) = 14$
60. $3x(x - 2) = 24$

Concept 3: Literal Equations

61. The distance d (in feet) that an object falls in t sec is given by the equation $d = 16t^2$, where $t \geq 0$.

    a. Solve the equation for t. **(See Example 8.)**

    b. Using the equation from part (a), determine the amount of time required for an object to fall 1024 ft.

62. The volume V (in cubic inches) of a can that is 4 in. tall is given by the equation $V = 4\pi r^2$, where r is the radius of the can, measured in inches.

    a. Solve the equation for r. Do not rationalize the denominator.

    b. Using the equation from part (a), determine the radius of a can with a volume of 12.56 in.^{3} Use 3.14 for π.

For Exercises 63–68, solve for the indicated variable.

63. $A = \pi r^2$ for r $(r > 0)$
64. $E = mc^2$ for c $(c > 0)$

65. $a^2 + b^2 + c^2 = d^2$ for a $(a > 0)$
66. $a^2 + b^2 = c^2$ for b $(b > 0)$

67. $V = \frac{1}{3}\pi r^2 h$ for r $(r > 0)$
68. $V = \frac{1}{3}s^2 h$ for s $(s > 0)$

69. A corner shelf is to be made from a triangular piece of plywood, as shown in the diagram. Find the distance x that the shelf will extend along the walls. Assume that the walls are at right angles. Round the answer to a tenth of a foot.

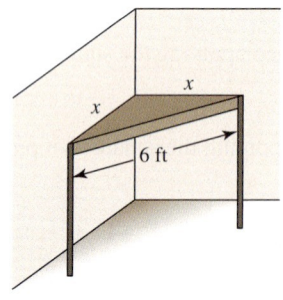

70. The volume $V(x)$ (in cubic inches) of a box with a square bottom and a height of 4 in. is given by $V(x) = 4x^2$, where x is the length (in inches) of the sides of the bottom of the box.

    a. If the volume of the box is 289 in.^{3}, find the dimensions of the box.

    b. Are there two possible answers to part (a)? Why or why not?

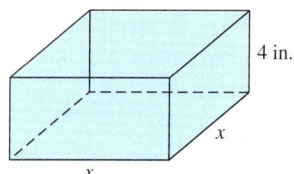

71. A square has an area of 50 in.² What are the lengths of the sides? (Round to one decimal place.)

72. The amount of money A in an account with an interest rate r compounded annually is given by

$$A = P(1 + r)^t$$

where P is the initial principal and t is the number of years the money is invested.

   a. If a $10,000 investment grows to $11,664 after 2 years, find the interest rate.

   b. If a $6000 investment grows to $7392.60 after 2 years, find the interest rate.

   c. Jamal wants to invest $5000. He wants the money to grow to at least $6500 in 2 years to cover the cost of his son's first year at college. What interest rate does Jamal need for his investment to grow to $6500 in 2 years? Round to the nearest hundredth of a percent.

73. A textbook company has discovered that the profit for selling its books is given by

$$P(x) = -\frac{1}{8}x^2 + 5x$$

where x is the number of textbooks produced (in thousands) and $P(x)$ is the corresponding profit (in thousands of dollars).

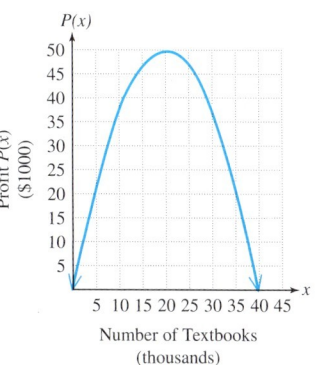

   a. Approximate the number of books required to make a profit of $20,000. [*Hint:* Let $P(x) = 20$. Then complete the square to solve for x.] Round to one decimal place.

   b. Why are there two answers to part (a)?

74. If we ignore air resistance, the distance $d(t)$ (in feet) that an object travels in free fall can be approximated by $d(t) = 16t^2$, where t is the time in seconds after the object was dropped.

   a. If the CN Tower in Toronto is 1815 ft high, how long will it take an object to fall from the top of the building? Round to one decimal place.

   b. If the Renaissance Tower in Dallas is 886 ft high, how long will it take an object to fall from the top of the building? Round to one decimal place.

Songquan Deng/Shutterstock

f11photo/Shutterstock

Section 19.2 Quadratic Formula

Concepts

1. Derivation of the Quadratic Formula
2. Solving Quadratic Equations by Using the Quadratic Formula
3. Using the Quadratic Formula in Applications
4. Discriminant
5. Mixed Review: Methods to Solve a Quadratic Equation

1. Derivation of the Quadratic Formula

If we solve a quadratic equation in standard form $ax^2 + bx + c = 0$, $a > 0$, by completing the square and using the square root property, the result is a formula that gives the solutions for x in terms of a, b, and c.

$ax^2 + bx + c = 0 \ (a \neq 0)$ — Begin with a quadratic equation in standard form.

$\dfrac{ax^2}{a} + \dfrac{bx}{a} + \dfrac{c}{a} = \dfrac{0}{a}$ — Divide by the leading coefficient.

$x^2 + \dfrac{b}{a}x + \dfrac{c}{a} = 0$

$x^2 + \dfrac{b}{a}x = -\dfrac{c}{a}$ — Isolate the terms containing x.

$x^2 + \dfrac{b}{a}x + \left(\dfrac{1}{2} \cdot \dfrac{b}{a}\right)^2 = \left(\dfrac{1}{2} \cdot \dfrac{b}{a}\right)^2 - \dfrac{c}{a}$ — Add the square of $\tfrac{1}{2}$ the linear term coefficient to both sides of the equation.

$\left(x + \dfrac{b}{2a}\right)^2 = \dfrac{b^2}{4a^2} - \dfrac{c}{a}$ — Factor the left side as a perfect square.

$\left(x + \dfrac{b}{2a}\right)^2 = \dfrac{b^2 - 4ac}{4a^2}$ — Combine fractions on the right side by getting a common denominator.

$x + \dfrac{b}{2a} = \pm\sqrt{\dfrac{b^2 - 4ac}{4a^2}}$ — Apply the square root property.

$x + \dfrac{b}{2a} = \dfrac{\pm\sqrt{b^2 - 4ac}}{2a}$ — Simplify the denominator.

$x = -\dfrac{b}{2a} \pm \dfrac{\sqrt{b^2 - 4ac}}{2a}$ — Subtract $\dfrac{b}{2a}$ from both sides.

$= \dfrac{-b \pm \sqrt{b^2 - 4ac}}{2a}$ — Combine fractions.

The solutions to the equation $ax^2 + bx + c = 0$ in terms of the coefficients a, b, and c are given by the **quadratic formula**.

The Quadratic Formula

For a quadratic equation of the form $ax^2 + bx + c = 0 \ (a \neq 0)$ the solutions are

$$x = \dfrac{-b \pm \sqrt{b^2 - 4ac}}{2a}$$

TIP: When applying the quadratic formula, note that a, b, and c are constants. The variable is x.

The quadratic formula gives us another technique to solve a quadratic equation. This method will work regardless of whether the equation is factorable or not factorable.

2. Solving Quadratic Equations by Using the Quadratic Formula

Example 1   Solving a Quadratic Equation by Using the Quadratic Formula

Solve the quadratic equation by using the quadratic formula. $2x^2 - 3x = 5$

Solution:

$2x^2 - 3x = 5$

$2x^2 - 3x - 5 = 0$ Write the equation in the form $ax^2 + bx + c = 0$.

$a = 2, \quad b = -3, \quad c = -5$ Identify a, b, and c.

$x = \dfrac{-b \pm \sqrt{b^2 - 4ac}}{2a}$ Apply the quadratic formula.

$= \dfrac{-(-3) \pm \sqrt{(-3)^2 - 4(2)(-5)}}{2(2)}$ Substitute $a = 2$, $b = -3$, and $c = -5$.

$= \dfrac{3 \pm \sqrt{9 + 40}}{4}$ Simplify.

$= \dfrac{3 \pm \sqrt{49}}{4}$

$= \dfrac{3 \pm 7}{4}$

$x = \dfrac{3 + 7}{4} = \dfrac{10}{4} = \dfrac{5}{2}$

$x = \dfrac{3 - 7}{4} = \dfrac{-4}{4} = -1$

The solution set is $\left\{\dfrac{5}{2}, -1\right\}$. Both solutions check in the original equation.

> **Avoiding Mistakes**
> - The term $-b$ represents the *opposite* of b.
> - Remember to write the *entire* numerator over $2a$.

Skill Practice   Solve the equation by using the quadratic formula.

1. $6x^2 - 5x = 4$

Example 2   Solving a Quadratic Equation by Using the Quadratic Formula

Solve the quadratic equation by using the quadratic formula. $-x(x - 6) = 11$

Solution:

$-x(x - 6) = 11$

$-x^2 + 6x - 11 = 0$ Write the equation in the form $ax^2 + bx + c = 0$.

$-1(-x^2 + 6x - 11) = -1(0)$ If the leading coefficient of the quadratic polynomial is negative, we suggest multiplying both sides of the equation by -1. Although this is not mandatory, it is generally easier to simplify the quadratic formula when the value of a is positive.

$x^2 - 6x + 11 = 0$

Answer

1. $\left\{-\dfrac{1}{2}, \dfrac{4}{3}\right\}$

$a = 1, b = -6,$ and $c = 11$ Identify a, b, and c.

$$x = \frac{-b \pm \sqrt{b^2 - 4ac}}{2a}$$ Apply the quadratic formula.

Avoiding Mistakes
When identifying a, b, and c, use the coefficients only, not the variable. For example, the value of b is -6, not $-6x$.

$$= \frac{-(-6) \pm \sqrt{(-6)^2 - 4(1)(11)}}{2(1)}$$ Substitute $a = 1, b = -6,$ and $c = 11$.

$$= \frac{6 \pm \sqrt{36 - 44}}{2}$$ Simplify.

$$= \frac{6 \pm \sqrt{-8}}{2}$$

Avoiding Mistakes
Always simplify the radical completely before trying to reduce the fraction to lowest terms.

$$= \frac{6 \pm 2i\sqrt{2}}{2}$$ Simplify the radical.

$$= \frac{2(3 \pm i\sqrt{2})}{2}$$ Factor the numerator.

$$= \frac{2(3 \pm i\sqrt{2})}{2}$$ Simplify the fraction to lowest terms.

$= 3 \pm i\sqrt{2}$ → $x = 3 + i\sqrt{2}$ The solutions are imaginary numbers.
   → $x = 3 - i\sqrt{2}$

The solution set is $\{3 \pm i\sqrt{2}\}$.

Skill Practice Solve the equation by using the quadratic formula.

2. $-y(y + 4) = 12$

3. Using the Quadratic Formula in Applications

Example 3 Using the Quadratic Formula in an Application

A delivery truck travels south from Hartselle, Alabama, to Birmingham, Alabama, along Interstate 65. The truck then heads east to Atlanta, Georgia, along Interstate 20. The distance from Birmingham to Atlanta is 8 mi less than twice the distance from Hartselle to Birmingham. If the straight-line distance from Hartselle to Atlanta is 165 mi, find the distance from Hartselle to Birmingham and from Birmingham to Atlanta. (Round the answers to the nearest mile.)

Solution:

The motorist travels due south and then due east. Therefore, the three cities form the vertices of a right triangle (Figure 19-1).

Let x represent the distance between Hartselle and Birmingham.

Then $2x - 8$ represents the distance between Birmingham and Atlanta.

Use the Pythagorean theorem to establish a relationship among the three sides of the triangle.

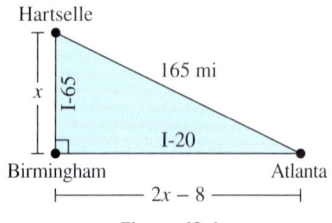

Figure 19-1

Answer

2. $\{-2 \pm 2i\sqrt{2}\}$

$$(x)^2 + (2x-8)^2 = (165)^2$$
$$x^2 + 4x^2 - 32x + 64 = 27{,}225$$
$$5x^2 - 32x - 27{,}161 = 0 \qquad \text{Write the equation in the form } ax^2 + bx + c = 0.$$

$$a = 5 \quad b = -32 \quad c = -27{,}161 \qquad \text{Identify } a, b, \text{ and } c.$$

$$x = \frac{-(-32) \pm \sqrt{(-32)^2 - 4(5)(-27{,}161)}}{2(5)} \qquad \text{Apply the quadratic formula.}$$

$$= \frac{32 \pm \sqrt{1024 + 543{,}220}}{10} \qquad \text{Simplify.}$$

$$= \frac{32 \pm \sqrt{544{,}244}}{10}$$

$$x = \frac{32 + \sqrt{544{,}244}}{10} \approx 76.97$$

$$x = \frac{32 - \sqrt{544{,}244}}{10} \approx -70.57$$

We reject the negative solution because distance cannot be negative. Rounding to the nearest whole unit, we have $x = 77$. Therefore, $2x - 8 = 2(77) - 8 = 146$.

The distance between Hartselle and Birmingham is 77 mi, and the distance between Birmingham and Atlanta is 146 mi.

Skill Practice

3. Steve and Tammy leave a campground, hiking on two different trails. Steve heads south and Tammy heads east. By lunchtime they are 9 mi apart. Steve walked 3 mi more than twice as many miles as Tammy. Find the distance each person hiked. (Round to the nearest tenth of a mile.)

Example 4 — Analyzing a Quadratic Function

A model rocket is launched straight upward from the side of a 144-ft cliff (Figure 19-2). The initial velocity is 112 ft/sec. The height of the rocket $h(t)$ is given by

$$h(t) = -16t^2 + 112t + 144$$

where $h(t)$ is measured in feet and t is the time in seconds after launch. Find the time(s) at which the rocket is 208 ft above the ground.

Solution:

$$h(t) = -16t^2 + 112t + 144$$
$$208 = -16t^2 + 112t + 144 \qquad \text{Substitute 208 for } h(t).$$
$$16t^2 - 112t + 64 = 0 \qquad \text{Write the equation in the form } at^2 + bt + c = 0.$$
$$\frac{16t^2}{16} - \frac{112t}{16} + \frac{64}{16} = \frac{0}{16} \qquad \text{Divide by 16. This makes the coefficients smaller, and it is less cumbersome to solve.}$$
$$t^2 - 7t + 4 = 0 \qquad \text{The equation is not factorable. Apply the quadratic formula.}$$

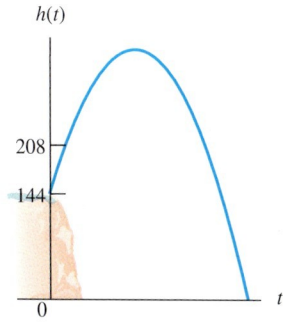

Figure 19-2

Answer

3. Tammy hiked 2.8 mi, and Steve hiked 8.6 mi.

$$t = \frac{-(-7) \pm \sqrt{(-7)^2 - 4(1)(4)}}{2(1)} \qquad \text{Let } a = 1, b = -7, \text{ and } c = 4.$$

$$= \frac{7 \pm \sqrt{33}}{2}$$

$$t = \frac{7 + \sqrt{33}}{2} \approx 6.37$$

$$t = \frac{7 - \sqrt{33}}{2} \approx 0.63$$

The rocket will reach a height of 208 ft after approximately 0.63 sec (on the way up) and after 6.37 sec (on the way down).

Skill Practice

4. A rocket is launched from the top of a 96-ft building with an initial velocity of 64 ft/sec. The height $h(t)$ of the rocket is given by $h(t) = -16t^2 + 64t + 96$. Find the time it takes for the rocket to hit the ground. [*Hint:* $h(t) = 0$ when the object hits the ground.]

4. Discriminant

The radicand within the quadratic formula is the expression $b^2 - 4ac$. This is called the **discriminant**. The discriminant can be used to determine the number of solutions to a quadratic equation as well as whether the solutions are rational, irrational, or imaginary numbers.

Using the Discriminant to Determine the Number and Type of Solutions to a Quadratic Equation

Consider the equation $ax^2 + bx + c = 0$, where a, b, and c are rational numbers and $a \neq 0$. The expression $b^2 - 4ac$ is called the *discriminant*. Furthermore,

- If $b^2 - 4ac > 0$, then there will be two real solutions.
  a. If $b^2 - 4ac$ is a perfect square, the solutions will be rational numbers.
  b. If $b^2 - 4ac$ is not a perfect square, the solutions will be irrational numbers.
- If $b^2 - 4ac < 0$, then there will be two imaginary solutions.
- If $b^2 - 4ac = 0$, then there will be one rational solution.

Example 5   Using the Discriminant

Use the discriminant to determine the type and number of solutions for each equation.

a. $2x^2 - 5x + 9 = 0$ b. $3x^2 = -x + 2$

c. $-2x(2x - 3) = -1$ d. $3.6x^2 = -1.2x - 0.1$

Solution:

For each equation, first write the equation in standard form $ax^2 + bx + c = 0$. Then determine the discriminant.

Equation	Discriminant	Solution Type and Number
a. $2x^2 - 5x + 9 = 0$	$b^2 - 4ac$ $= (-5)^2 - 4(2)(9)$ $= 25 - 72$ $= -47$	Because $-47 < 0$, there will be two imaginary solutions.

Answer

4. $2 + \sqrt{10} \approx 5.16$ sec

b. $3x^2 = -x + 2$

$3x^2 + x - 2 = 0$

$b^2 - 4ac$
$= (1)^2 - 4(3)(-2)$
$= 1 - (-24)$
$= 25$

$25 > 0$ and 25 is a perfect square. There will be two rational solutions.

c. $-2x(2x - 3) = -1$

$-4x^2 + 6x = -1$

$-4x^2 + 6x + 1 = 0$

$b^2 - 4ac$
$= (6)^2 - 4(-4)(1)$
$= 36 - (-16)$
$= 52$

$52 > 0$, but 52 is *not* a perfect square. There will be two irrational solutions.

d. $3.6x^2 = -1.2x - 0.1$

$3.6x^2 + 1.2x + 0.1 = 0$

$b^2 - 4ac$
$= (1.2)^2 - 4(3.6)(0.1)$
$= 1.44 - 1.44$
$= 0$

Because the discriminant equals 0, there will be only one rational solution.

Skill Practice Use the discriminant to determine the type and number of solutions for the equation.

5. $3y^2 + y + 3 = 0$ **6.** $4t^2 = 6t - 2$

7. $3t(t + 1) = 9$ **8.** $\dfrac{2}{3}x^2 - \dfrac{2}{3}x + \dfrac{1}{6} = 0$

With the discriminant we can determine the number of real-valued solutions to the equation $ax^2 + bx + c = 0$, and thus the number of x-intercepts to the graph of $f(x) = ax^2 + bx + c$. The following illustrations show the graphical interpretation of the three cases of the discriminant.

$f(x) = x^2 - 4x + 3$

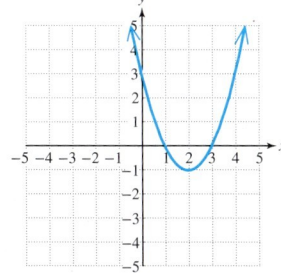

Use $x^2 - 4x + 3 = 0$ to find the value of the discriminant.

$$b^2 - 4ac = (-4)^2 - 4(1)(3)$$
$$= 4$$

Since the discriminant is positive, there are two real solutions to the quadratic equation. Therefore, there are two x-intercepts to the corresponding quadratic function, $(1, 0)$ and $(3, 0)$.

$f(x) = x^2 - x + 1$

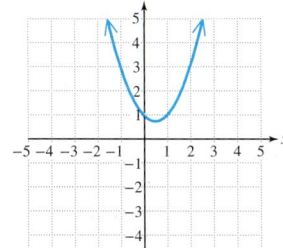

Use $x^2 - x + 1 = 0$ to find the value of the discriminant.

$$b^2 - 4ac = (-1)^2 - 4(1)(1)$$
$$= -3$$

Since the discriminant is negative, there are no real solutions to the quadratic equation. Therefore, there are no x-intercepts to the corresponding quadratic function.

Answers

5. -35; two imaginary solutions
6. 4; two rational solutions
7. 117; two irrational solutions
8. 0; one rational solution

$f(x) = x^2 - 2x + 1$

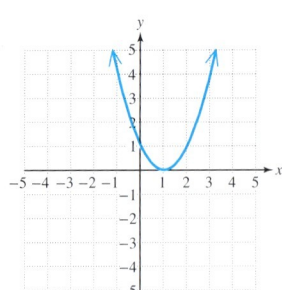

Use $x^2 - 2x + 1 = 0$ to find the value of the discriminant.

$$b^2 - 4ac = (-2)^2 - 4(1)(1)$$
$$= 0$$

Since the discriminant is zero, there is one real solution to the quadratic equation. Therefore, there is one x-intercept to the corresponding quadratic function, $(1, 0)$.

Example 6 Finding x- and y-Intercepts of a Quadratic Function

Given $f(x) = x^2 - 3x + 1$

a. Find the discriminant and use it to determine if there are any x-intercepts.
b. Find the x-intercept(s), if they exist.
c. Find the y-intercept.

Solution:

a. $a = 1$, $b = -3$, and $c = 1$.

The discriminant is $b^2 - 4ac = (-3)^2 - 4(1)(1)$
$$= 9 - 4$$
$$= 5$$

Since $5 > 0$, there are two x-intercepts.

TIP: Recall that an x-intercept is a point $(a, 0)$ where the graph of a function intersects the x-axis. A y-intercept is a point $(0, b)$ where the graph intersects the y-axis.

b. The x-intercepts are given by the real solutions to the equation $f(x) = 0$. In this case, we have

$f(x) = x^2 - 3x + 1 = 0$

$x^2 - 3x + 1 = 0$ The equation is in the form $ax^2 + bx + c = 0$.

$x = \dfrac{-(-3) \pm \sqrt{(-3)^2 - (4)(1)(1)}}{2(1)}$ Apply the quadratic formula.

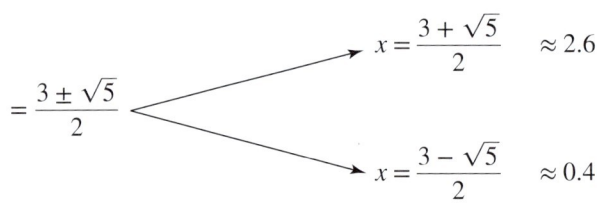

$= \dfrac{3 \pm \sqrt{5}}{2}$

$x = \dfrac{3 + \sqrt{5}}{2} \approx 2.6$

$x = \dfrac{3 - \sqrt{5}}{2} \approx 0.4$

The solutions are $\dfrac{3 + \sqrt{5}}{2}$ and $\dfrac{3 - \sqrt{5}}{2}$. Therefore, the x-intercepts are $\left(\dfrac{3 + \sqrt{5}}{2}, 0\right)$ and $\left(\dfrac{3 - \sqrt{5}}{2}, 0\right)$.

c. To find the y-intercept, evaluate $f(0)$.

$f(0) = (0)^2 - 3(0) + 1 = 1$

The y-intercept is located at $(0, 1)$.

The parabola is shown in the graph with the x- and y-intercepts labeled.

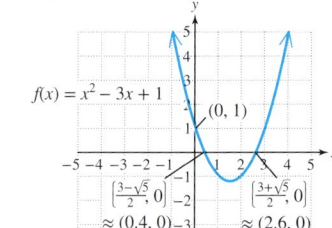

Skill Practice Given $f(x) = x^2 + 5x + 2$,

9. Find the discriminant and use it to determine if there are any x-intercepts.
10. Find the x- and y-intercepts.

Answers

9. Discriminant: 17; there are two x-intercepts.
10. x-intercepts: $\left(\dfrac{-5 + \sqrt{17}}{2}, 0\right)$, $\left(\dfrac{-5 - \sqrt{17}}{2}, 0\right)$

y-intercept: $(0, 2)$

Example 7 Finding x- and y-Intercepts of a Quadratic Function

Given $f(x) = x^2 + x + 2$

a. Find the discriminant and use it to determine if there are any x-intercepts.

b. Find the x-intercept(s), if they exist.

c. Find the y-intercept.

Solution:

a. $a = 1$, $b = 1$, and $c = 2$.

The discriminant is $b^2 - 4ac = (1)^2 - 4(1)(2)$
$$= 1 - 8$$
$$= -7$$

Since $-7 < 0$, there are no x-intercepts.

b. There are no x-intercepts.

c. $f(0) = (0)^2 + (0) + 2$
$$= 2$$

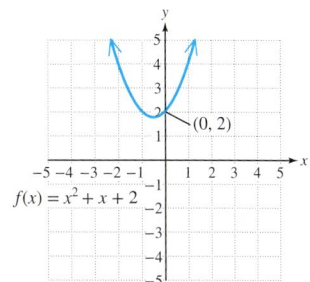

The y-intercept is located at (0, 2).

The parabola is shown. *Note:* The graph does not intersect the x-axis.

Skill Practice Given $f(x) = 2x^2 - 3x + 5$,

11. Find the discriminant and use it to determine if there are any x-intercepts.
12. Find the y-intercept.

5. Mixed Review: Methods to Solve a Quadratic Equation

Three methods have been presented to solve quadratic equations.

Methods to Solve a Quadratic Equation

Method	Example
Factor and use the zero product rule. • This method works well if you can factor the equation easily.	Example: $x^2 + 8x + 15 = 0$ factors as $(x + 3)(x + 5) = 0$
Use the square root property. Complete the square if necessary. This method is particularly good if • the equation is of the form $ax^2 + c = 0$ or • the equation is of the form $x^2 + bx + c = 0$, where b is even.	Examples: $4x^2 + 9 = 0$ $x^2 + 10x - 3 = 0$
Quadratic formula • This method works in all cases—just be sure to write the equation in the form $ax^2 + bx + c = 0$ before applying the formula.	Example: $7x^2 - 3x + 11 = 0$

Answers

11. Discriminant: -31; there are no x-intercepts.
12. (0, 5)

Before solving a quadratic equation, take a minute to analyze it first. Each problem must be evaluated individually before choosing the most efficient method to find its solutions.

Example 8 — Solving a Quadratic Equation by Using Any Method

Solve the equation by using any method. $(x + 3)^2 + x^2 - 9x = 8$

Solution:

$$(x + 3)^2 + x^2 - 9x = 8$$

$$x^2 + 6x + 9 + x^2 - 9x - 8 = 0 \quad \text{Clear parentheses and write the equation in the form } ax^2 + bx + c = 0.$$

$$2x^2 - 3x + 1 = 0 \quad \text{This equation is factorable.}$$

$$(2x - 1)(x - 1) = 0 \quad \text{Factor.}$$

$$2x - 1 = 0 \quad \text{or} \quad x - 1 = 0 \quad \text{Apply the zero product rule.}$$

$$x = \tfrac{1}{2} \quad \text{or} \quad x = 1 \quad \text{Solve for } x.$$

The solution set is $\{\tfrac{1}{2}, 1\}$.

This equation could have been solved by using any of the three methods, but factoring was the most efficient method.

Skill Practice Solve using any method.

13. $2t(t - 1) + t^2 = 5$

FOR REVIEW

Recall that a second-degree polynomial equation such as $2x^2 - 3x + 1 = 0$ is quadratic. A first-degree equation such as $2x - 1 = 0$ is linear.

Example 9 — Solving a Quadratic Equation by Using Any Method

Solve the equation by using any method. $x^2 + 5 = -2x$

Solution:

$$x^2 + 2x + 5 = 0 \quad \text{The equation does not factor.}$$

$$x^2 + 2x = -5 \quad \text{Because } a = 1 \text{ and } b \text{ is even, we can easily complete the square.}$$

$$x^2 + 2x + 1 = -5 + 1 \quad \text{Add } [\tfrac{1}{2}(2)]^2 = 1^2 = 1 \text{ to both sides.}$$

$$(x + 1)^2 = -4$$

$$x + 1 = \pm\sqrt{-4} \quad \text{Apply the square root property.}$$

$$x = -1 \pm 2i \quad \text{Solve for } x.$$

The solution set is $\{-1 \pm 2i\}$.

This equation could also have been solved by using the quadratic formula.

Skill Practice Solve using any method.

14. $x^2 - 4x = -7$

Answers

13. $\left\{-1, \dfrac{5}{3}\right\}$

14. $\{2 \pm i\sqrt{3}\}$

Example 10 Solving a Quadratic Equation by Using Any Method

Solve the equation by using any method. $\quad \dfrac{1}{4}x^2 - \dfrac{1}{2}x + \dfrac{1}{3} = 0$

Solution:

$12 \cdot \left(\dfrac{1}{4}x^2 - \dfrac{1}{2}x + \dfrac{1}{3} \right) = 12 \cdot (0)$ Clear fractions.

$3x^2 - 6x + 4 = 0$ The equation is in the form $ax^2 + bx + c = 0$.
The left-hand side does not factor.

$a = 3, b = -6,$ and $c = 4$

$x = \dfrac{-(-6) \pm \sqrt{(-6)^2 - 4(3)(4)}}{2(3)}$ Apply the quadratic formula.

$= \dfrac{6 \pm \sqrt{-12}}{6}$ Simplify.

$= \dfrac{6 \pm 2i\sqrt{3}}{6}$ Simplify the radical.

$= \dfrac{\overset{1}{2}(3 \pm i\sqrt{3})}{\underset{3}{6}}$ Factor and simplify.

$= \dfrac{3 \pm i\sqrt{3}}{3}$

$= \dfrac{3}{3} \pm \dfrac{\sqrt{3}}{3}i$ Write in the form $a \pm bi$.

$= 1 \pm \dfrac{\sqrt{3}}{3}i$ Simplify.

The solution set is $\left\{ 1 \pm \dfrac{\sqrt{3}}{3}i \right\}$.

Skill Practice Solve using any method.

15. $\dfrac{1}{5}x^2 - \dfrac{4}{5}x + \dfrac{1}{2} = 0$

Answer

15. $\left\{ \dfrac{4 \pm \sqrt{6}}{2} \right\}$

> **Example 11** **Solving a Quadratic Equation by Using Any Method**
>
> Solve the equation by using any method. $\quad 9p^2 - 11 = 0$
>
> **Solution:**
>
> $\quad 9p^2 - 11 = 0 \quad$ Because $b = 0$, use the square root property.
>
> $\quad 9p^2 = 11 \quad$ Isolate the variable term.
>
> $\quad p^2 = \dfrac{11}{9} \quad$ The equation is in the form $x^2 = k$.
>
> $\quad p = \pm\sqrt{\dfrac{11}{9}} \quad$ Apply the square root property.
>
> $\quad p = \pm\dfrac{\sqrt{11}}{3} \quad$ Simplify the radical.
>
> The solution set is $\left\{\pm\dfrac{\sqrt{11}}{3}\right\}$.
>
> **Skill Practice** Solve using any method.
>
> **16.** $4y^2 - 13 = 0$

Answer

16. $\left\{\pm\dfrac{\sqrt{13}}{2}\right\}$

Section 19.2 Activity

A.1. Given a quadratic equation $ax^2 + bx + c = 0$ ($a \neq 0$), the solutions are given by the **quadratic formula**:

$$x = \boxed{}$$

For Exercises A.2–A.4, an equation is given in the form $ax^2 + bx + c = 0$.
  a. Identify the values of a, b, and c.
  b. Solve the equation by using the quadratic formula.

A.2. $2x^2 - 12x + 16 = 0$ **A.3.** $x^2 - 4x + 5 = 0$ **A.4.** $-x^2 - 6x - 9 = 0$

A.5. Based on the results from Exercises A.2–A.4, the solutions to a quadratic equation $ax^2 + bx + c = 0$ may either be real numbers or imaginary numbers. One way to determine the type of solution is to analyze the radicand from the quadratic formula:

$$x = \dfrac{-b \pm \sqrt{b^2 - 4ac}}{2a}.$$

  a. If the radicand is negative, what type of solution does the equation have? (real or imaginary)
  b. If the radicand is nonnegative, what type of solution does the equation have? (real or imaginary)
  c. If the radicand is zero, *how many* solutions will the equation have? Otherwise, if the radicand is nonzero, how many solutions will there be?
  d. The radicand, $b^2 - 4ac$, is given a special name called the _____. It helps us determine the number and types of solutions to a quadratic equation.

For Exercises A.6–A.8, refer to the equations from Exercises A.2–A.4.
  a. Find the discriminant.
  b. Based on the discriminant, determine the number and types of solutions to the equation.
  c. Do your results from part (b) agree with the solutions you found in Exercises A.2–A.4?

A.6. $2x^2 - 12x + 16 = 0$ **A.7.** $x^2 - 4x + 5 = 0$ **A.8.** $-x^2 - 6x - 9 = 0$

A.9. Consider the graph of the quadratic function defined by $f(x) = 2x^2 - 12x + 16$ and the related quadratic equation $2x^2 - 12x + 16 = 0$ from Exercises A.2 and A.6.
  a. How are the solutions to the equation related to the graph?
  b. What does the discriminant to the equation tell you about the number of x-intercepts of the graph?

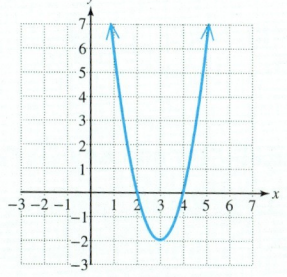

A.10. a. Based on the value of the discriminant from Exercise A.7, how many x-intercepts would the graph of $g(x) = x^2 - 4x + 5$ have?
  b. Based on the value of the discriminant from Exercise A.8, how many x-intercepts would the graph of $h(x) = -x^2 - 6x - 9$ have?
  c. Based on the results of parts (a) and (b), match $g(x) = x^2 - 4x + 5$ and $h(x) = -x^2 - 6x - 9$ with their respective graphs.

i.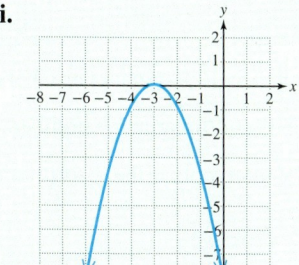

ii.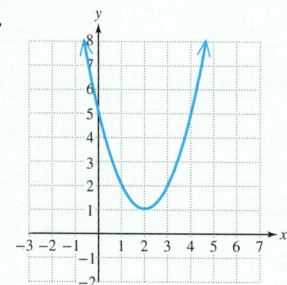

Practice Exercises

Section 19.2

Prerequisite Review

For Exercises R.1–R.6, simplify the expression.

R.1. $\sqrt{24}$

R.2. $\sqrt{128}$

R.3. $\dfrac{10 + \sqrt{125}}{5}$

R.4. $\dfrac{12 - \sqrt{48}}{4}$

R.5. $\sqrt{(-8)^2 - 4(1)(2)}$

R.6. $\sqrt{(-10)^2 - 4(4)(-3)}$

For Exercises R.7–R.10, solve the equation by first clearing fractions or decimals.

R.7. $\frac{1}{4}x - \frac{5}{12} = \frac{1}{2}x + 2$

R.8. $\frac{1}{5} - \frac{2}{15}y = 1 + \frac{2}{3}y$

R.9. $0.07x - 0.38 = 0.09x + 0.8$

R.10. $2.8w - 7.2 = 1.8w + 3.6$

For Exercises R.11–R.12, solve the equation by applying the square root property.

R.11. $(x + 7)^2 = 44$

R.12. $(y - 6)^2 = 50$

Vocabulary and Key Concepts

1. **a.** For the equation $ax^2 + bx + c = 0$ ($a \neq 0$), the _____ formula gives the solutions as $x =$ _____.

   b. To apply the quadratic formula, a quadratic equation must be written in the form _____ where $a \neq 0$.

   c. To apply the quadratic formula to solve the equation $8x^2 - 42x - 27 = 0$, the value of a is _____, the value of b is _____, and the value of c is _____.

   d. To apply the quadratic formula to solve the equation $3x^2 - 7x - 4 = 0$, the value of $-b$ is _____ and the value of the radicand is _____.

2. **a.** The radicand within the quadratic formula is _____ and is called the _____.

   b. If the discriminant is negative, then the solutions to a quadratic equation will be (real/imaginary) numbers.

   c. If the discriminant is positive, then the solutions to a quadratic equation will be (real/imaginary) numbers.

   d. Given a quadratic function $f(x) = ax^2 + bx + c = 0$, the function will have no x-intercepts if the discriminant is (less than, greater than, equal to) zero.

Concept 2: Solving Quadratic Equations by Using the Quadratic Formula

For Exercises 3–8, determine whether the equation is linear, quadratic, or neither.

3. $5x^2 - 3x = 8 + x$

4. $-4x + 5 = 2x^2 + 7$

5. $\frac{4}{x^2} + \frac{2}{x} + \frac{1}{3} = 0$

6. $x + 5\sqrt{x} + 4 = 0$

7. $2(x - 5) + x^2 = 3x(x - 4) - 2x^2$

8. $5x(x + 3) - 9 = 4x^2 - 3(x + 1)$

For Exercises 9–34, solve the equation by using the quadratic formula. Write imaginary solutions in the form $a + bi$.
(See Examples 1–2.)

9. $x^2 + 11x - 12 = 0$

10. $5x^2 - 14x - 3 = 0$

11. $9y^2 - 2y + 5 = 0$

12. $2t^2 + 3t - 7 = 0$

13. $12p^2 - 4p + 5 = 0$

14. $-5n^2 + 4n - 6 = 0$

15. $-z^2 = -2z - 35$

16. $12x^2 - 5x = 2$

17. $y^2 + 3y = 8$

18. $k^2 + 4 = 6k$

19. $25x^2 - 20x + 4 = 0$

20. $9y^2 = -12y - 4$

21. $w(w - 6) = -14$

22. $m(m + 6) = -11$

23. $(x + 2)(x - 3) = 1$

24. $3y(y + 1) - 7y(y + 2) = 6$

25. $-4a^2 - 2a + 3 = 0$

26. $-2m^2 - 5m + 3 = 0$

27. $\frac{1}{2}y^2 + \frac{2}{3} = -\frac{2}{3}y$
(*Hint:* Clear fractions first.)

28. $\frac{2}{3}p^2 - \frac{1}{6}p + \frac{1}{2} = 0$

29. $\frac{1}{5}h^2 + h + \frac{3}{5} = 0$

30. $\frac{1}{4}w^2 + \frac{7}{4}w + 1 = 0$

31. $0.01x^2 + 0.06x + 0.08 = 0$
(*Hint:* Clear decimals first.)

32. $0.5y^2 - 0.7y + 0.2 = 0$

33. $0.3t^2 + 0.7t - 0.5 = 0$

34. $0.01x^2 + 0.04x - 0.07 = 0$

For Exercises 35–38, factor the expression. Then use the zero product rule and the quadratic formula to solve the equation. There should be three solutions to each equation. Write imaginary solutions in the form $a + bi$.

35. a. Factor. $x^3 - 27$
   b. Solve. $x^3 - 27 = 0$

36. a. Factor. $64x^3 + 1$
   b. Solve. $64x^3 + 1 = 0$

37. a. Factor. $3x^3 - 6x^2 + 6x$
   b. Solve. $3x^3 - 6x^2 + 6x = 0$

38. a. Factor. $5x^3 + 5x^2 + 10x$
   b. Solve. $5x^3 + 5x^2 + 10x = 0$

Concept 3: Using the Quadratic Formula in Applications

39. The volume of a cube is 27 ft³. Find the lengths of the sides.

40. The volume of a rectangular box is 64 ft³. If the width is 3 times longer than the height, and the length is 9 times longer than the height, find the dimensions of the box.

41. The hypotenuse of a right triangle measures 4 in. One leg of the triangle is 2 in. longer than the other leg. Find the lengths of the legs of the triangle. Round to one decimal place. (**See Example 3.**)

42. The length of one leg of a right triangle is 1 cm more than twice the length of the other leg. The hypotenuse measures 6 cm. Find the lengths of the legs. Round to one decimal place.

43. The hypotenuse of a right triangle is 10.2 m long. One leg is 2.1 m shorter than the other leg. Find the lengths of the legs. Round to one decimal place.

44. The hypotenuse of a right triangle is 17 ft long. One leg is 3.4 ft longer than the other leg. Find the lengths of the legs.

45. The fatality rate (in fatalities per 100 million vehicle miles driven) can be approximated for drivers x years old according to the function $F(x) = 0.0036x^2 - 0.35x + 9.2$. (*Source:* U.S. Department of Transportation)

   a. Approximate the fatality rate for drivers 16 years old.

   b. Approximate the fatality rate for drivers 40 years old.

   c. Approximate the fatality rate for drivers 80 years old.

   d. For what age(s) is the fatality rate approximately 2.5?

46. The braking distance (in feet) of a car going v mph is given by
$$d(v) = \frac{v^2}{20} + v \qquad v \geq 0$$

   a. Find the speed for a braking distance of 150 ft. Round to the nearest mile per hour.

   b. Find the speed for a braking distance of 100 ft. Round to the nearest mile per hour.

47. Mitch throws a baseball straight up in the air from a cliff that is 48 ft high. The initial velocity is 48 ft/sec. The height (in feet) of the object after t sec is given by $h(t) = -16t^2 + 48t + 48$. Find the time at which the height of the object is 64 ft. **(See Example 4.)**

48. An astronaut on the moon throws a rock into the air from the deck of a spacecraft that is 8 m high. The initial velocity of the rock is 2.4 m/sec. The height (in meters) of the rock after t sec is given by $h(t) = -0.8t^2 + 2.4t + 8$. Find the time at which the height of the rock is 6 m.

Concept 4: Discriminant

For Exercises 49–56,

   a. Write the equation in the form $ax^2 + bx + c = 0$, $a > 0$.

   b. Find the value of the discriminant.

   c. Use the discriminant to determine the number and type of the solutions. Choose from imaginary, rational, and irrational. **(See Example 5.)**

49. $x^2 + 2x = -1$

50. $12y - 9 = 4y^2$

51. $19m^2 = 8m$

52. $2n - 5n^2 = 0$

53. $5p^2 - 21 = 0$

54. $3k^2 = 7$

55. $4n(n - 2) - 5n(n - 1) = 4$

56. $(2x + 1)(x - 3) = -9$

For Exercises 57–62, determine the discriminant. Then use the discriminant to determine the number of x-intercepts for the function.

57. $f(x) = x^2 - 6x + 5$

58. $g(x) = -x^2 - 4x - 3$

59. $h(x) = 4x^2 + 12x + 9$

60. $k(x) = 25x^2 - 10x + 1$

61. $p(x) = 2x^2 + 3x + 6$

62. $m(x) = 3x^2 + 4x + 7$

For Exercises 63–68, find the x- and y-intercepts of the quadratic function. **(See Examples 6–7.)**

63. $f(x) = x^2 - 5x + 3$

64. $g(x) = 2x^2 + 7x + 2$

65. $g(x) = -x^2 + x - 1$

66. $f(x) = 2x^2 + x + 5$

67. $p(x) = 2x^2 + 5x - 2$

68. $h(x) = 3x^2 + 2x - 2$

Concept 5: Mixed Review: Methods to Solve a Quadratic Equation

For Exercises 69–86, solve the quadratic equation by using any method. Write imaginary solutions in the form $a + bi$.
(See Examples 8–11.)

69. $a^2 + 2a + 10 = 0$

70. $4z^2 + 7z = 0$

71. $(x - 2)^2 + 2x^2 - 13x = 10$

72. $(x - 3)^2 + 3x^2 - 5x = 12$

73. $4y^2 - 20y + 43 = 0$

74. $k^2 + 18 = 4k$

75. $\left(x + \dfrac{1}{2}\right)^2 + 4 = 0$

76. $(2y + 3)^2 = 9$

77. $2y(y - 3) = -1$

78. $w(w - 5) = 4$

79. $(2t + 5)(t - 1) = (t - 3)(t + 8)$

80. $(b - 1)(b + 4) = (3b + 2)(b + 1)$

81. $\dfrac{1}{8}x^2 - \dfrac{1}{2}x + \dfrac{1}{4} = 0$

82. $\dfrac{1}{6}x^2 - \dfrac{1}{2}x + \dfrac{1}{4} = 0$

83. $32z^2 - 20z - 3 = 0$

84. $8k^2 - 14k + 3 = 0$

85. $4p^2 - 21 = 0$

86. $5h^2 - 120 = 0$

Sometimes students shy away from completing the square and using the square root property to solve a quadratic equation. However, sometimes this process leads to a simple solution. For Exercises 87–88, solve the equations two ways.

    a. Solve the equation by completing the square and applying the square root property.

    b. Solve the equation by applying the quadratic formula.

    c. Which technique was easier for you?

87. $x^2 + 6x = 5$

88. $x^2 - 10x = -27$

Technology Connections

89. Graph $Y_1 = x^3 - 27$. Compare the x-intercepts with the solutions to the equation $x^3 - 27 = 0$ found in Exercise 35.

90. Graph $Y_1 = 64x^3 + 1$. Compare the x-intercepts with the solutions to the equation $64x^3 + 1 = 0$ found in Exercise 36.

91. Graph $Y_1 = 3x^3 - 6x^2 + 6x$. Compare the x-intercepts with the solutions to the equation $3x^3 - 6x^2 + 6x = 0$ found in Exercise 37.

92. Graph $Y_1 = 5x^3 + 5x^2 + 10x$. Compare the x-intercepts with the solutions to the equation $5x^3 + 5x^2 + 10x = 0$ found in Exercise 38.

Section 19.3 Equations in Quadratic Form

Concepts

1. Solving Equations by Using Substitution
2. Solving Equations Reducible to a Quadratic

1. Solving Equations by Using Substitution

We have learned to solve a variety of different types of equations, including linear, radical, rational, and polynomial equations. Sometimes, however, it is necessary to use a quadratic equation as a tool to solve other types of equations.

In Example 1, we will solve the equation $(2x^2 - 5)^2 - 16(2x^2 - 5) + 39 = 0$. Notice that the terms in the equation are written in descending order by degree. Furthermore, the first two terms have the same base, $2x^2 - 5$, and the exponent on the first term is exactly double the exponent on the second term. The third term is a constant. An equation in this pattern is called **quadratic in form**.

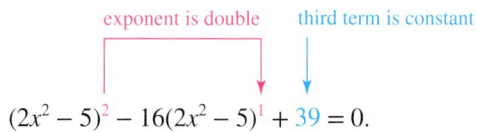

$$(2x^2 - 5)^2 - 16(2x^2 - 5)^1 + 39 = 0.$$

To solve this equation we will use substitution as demonstrated in Example 1.

Example 1   Solving an Equation in Quadratic Form

Solve the equation. $(2x^2 - 5)^2 - 16(2x^2 - 5) + 39 = 0$

Solution:

$(2x^2 - 5)^2 - 16(2x^2 - 5) + 39 = 0$ Once the substitution is made, the equation becomes quadratic in the variable u.

Substitute $u = (2x^2 - 5)$.

$u^2 - 16u + 39 = 0$ The equation is in the form $au^2 + bu + c = 0$.

$(u - 13)(u - 3) = 0$ Factor.

$u = 13$ or $u = 3$ Apply the zero product rule.

Reverse substitute.

$2x^2 - 5 = 13$ or $2x^2 - 5 = 3$ Reverse substitute.

$2x^2 = 18$ or $2x^2 = 8$

$x^2 = 9$ or $x^2 = 4$ Write the equations in the form $x^2 = k$.

$x = \pm\sqrt{9}$ or $x = \pm\sqrt{4}$ Apply the square root property.

$= \pm 3$ or $= \pm 2$

The solution set is $\{3, -3, 2, -2\}$.   All solutions check in the original equation.

Avoiding Mistakes

When using substitution, it is critical to reverse substitute to solve the equation in terms of the original variable.

Skill Practice Solve the equation.

1. $(3t^2 - 10)^2 + 5(3t^2 - 10) - 14 = 0$

Answer

1. $\{1, -1, 2, -2\}$

For an equation written in descending order, notice that u was set equal to the variable factor on the middle term. This is generally the case.

Example 2   **Solving an Equation in Quadratic Form**

Solve the equation. $p^{2/3} - 2p^{1/3} = 8$

Solution:

$$p^{2/3} - 2p^{1/3} = 8$$

$p^{2/3} - 2p^{1/3} - 8 = 0$ Set the equation equal to zero.

$(p^{1/3})^2 - 2(p^{1/3})^1 - 8 = 0$ Make the substitution $u = p^{1/3}$.

Substitute $u = p^{1/3}$.

$u^2 - 2u - 8 = 0$ Then the equation is in the form $au^2 + bu + c = 0$.

$(u - 4)(u + 2) = 0$ Factor.

$u = 4$ or $u = -2$ Apply the zero product rule.

Reverse substitute.

$p^{1/3} = 4$ or $p^{1/3} = -2$

$\sqrt[3]{p} = 4$ or $\sqrt[3]{p} = -2$ The equations are radical equations.

$(\sqrt[3]{p})^3 = (4)^3$ or $(\sqrt[3]{p})^3 = (-2)^3$ Cube both sides.

$p = 64$ or $p = -8$

The solution set is $\{64, -8\}$.   All solutions check in the original equation.

Skill Practice   Solve the equation.

2. $y^{2/3} - y^{1/3} = 12$

Example 3   **Solving an Equation in Quadratic Form**

Solve the equation. $x - \sqrt{x} - 12 = 0$

Solution:

The equation can be solved by first isolating the radical and then squaring both sides (this is saved for Exercise 25). However, this equation is also quadratic in form. By writing $\sqrt{x}$ as $x^{1/2}$, we see that the exponent on the first term is exactly double the exponent on the middle term.

$$x^1 - x^{1/2} - 12 = 0$$

$(x^{1/2})^2 - (x^{1/2})^1 - 12 = 0$ Let $u = x^{1/2}$.

$u^2 - u - 12 = 0$

$(u - 4)(u + 3) = 0$ Factor.

$u = 4$ or $u = -3$ Solve for u.

$x^{1/2} = 4$ or $x^{1/2} = -3$ Reverse substitute.

Answer

2. $\{64, -27\}$

Avoiding Mistakes

Recall that when each side of an equation is raised to an even power, we must check the potential solutions.

$\sqrt{x} = 4$ or $\sqrt{x} = -3$

$x = 16$

The solution set is $\{16\}$.

Solve each equation for x. Recall that the principal square root of a number cannot be negative.

The value 16 checks in the original equation.

Skill Practice Solve the equation.

3. $z - \sqrt{z} - 2 = 0$

2. Solving Equations Reducible to a Quadratic

Some equations are reducible to a quadratic equation. In Example 4, we solve a polynomial equation by factoring. The resulting factors are quadratic.

Example 4 Solving a Polynomial Equation

Solve the equation. $4x^4 + 7x^2 - 2 = 0$

Solution:

$4x^4 + 7x^2 - 2 = 0$ — This is a polynomial equation.

$(4x^2 - 1)(x^2 + 2) = 0$ — Factor.

$4x^2 - 1 = 0$ or $x^2 + 2 = 0$ — Set each factor equal to zero. Notice that the two equations are quadratic. Each can be solved by the square root property.

$x^2 = \dfrac{1}{4}$ or $x^2 = -2$

$x = \pm\sqrt{\dfrac{1}{4}}$ or $x = \pm\sqrt{-2}$ — Apply the square root property.

$x = \pm\dfrac{1}{2}$ or $x = \pm i\sqrt{2}$ — Simplify the radicals.

The solution set is $\left\{\dfrac{1}{2}, -\dfrac{1}{2}, i\sqrt{2}, -i\sqrt{2}\right\}$.

FOR REVIEW

If a quadratic equation cannot be solved by factoring and applying the zero product rule, then
- apply the square root property (sometimes complete the square first), or
- apply the quadratic formula.

Skill Practice Solve the equation.

4. $9x^4 + 35x^2 - 4 = 0$

Example 5 Solving a Rational Equation

Solve the equation. $\dfrac{3y}{y+2} - \dfrac{2}{y-1} = 1$

Solution:

$\dfrac{3y}{y+2} - \dfrac{2}{y-1} = 1$ — This is a rational equation. The LCD is $(y+2)(y-1)$. Also note that the restrictions on y are $y \neq -2$ and $y \neq 1$.

$\left(\dfrac{3y}{y+2} - \dfrac{2}{y-1}\right) \cdot (y+2)(y-1) = 1 \cdot (y+2)(y-1)$ — Multiply both sides by the LCD.

$\dfrac{3y}{y+2} \cdot (y+2)(y-1) - \dfrac{2}{y-1} \cdot (y+2)(y-1) = 1 \cdot (y+2)(y-1)$

Answers

3. $\{4\}$ (The value 1 does not check.)
4. $\left\{\pm\dfrac{1}{3}, \pm 2i\right\}$

$$3y(y-1) - 2(y+2) = (y+2)(y-1)$$ Clear fractions.
$$3y^2 - 3y - 2y - 4 = y^2 - y + 2y - 2$$ Apply the distributive property.
$$3y^2 - 5y - 4 = y^2 + y - 2$$ The equation is quadratic.
$$2y^2 - 6y - 2 = 0$$ Write the equation in descending order.

$$\frac{2y^2}{2} - \frac{6y}{2} - \frac{2}{2} = \frac{0}{2}$$

Each coefficient in the equation is divisible by 2. Therefore, if we divide both sides by 2, the coefficients in the equation are smaller. This will make it easier to apply the quadratic formula.

$$y^2 - 3y - 1 = 0$$

$$y = \frac{-(-3) \pm \sqrt{(-3)^2 - 4(1)(-1)}}{2(1)}$$ Apply the quadratic formula.

$$= \frac{3 \pm \sqrt{9 + 4}}{2}$$

$$= \frac{3 \pm \sqrt{13}}{2} \begin{cases} y = \dfrac{3 + \sqrt{13}}{2} \\ y = \dfrac{3 - \sqrt{13}}{2} \end{cases}$$

The solution set is $\left\{ \dfrac{3 + \sqrt{13}}{2}, \dfrac{3 - \sqrt{13}}{2} \right\}$.

Skill Practice Solve the equation.

5. $\dfrac{t}{2t - 1} - \dfrac{1}{t + 4} = 1$

Answer

5. $\left\{ \dfrac{-5 \pm 3\sqrt{5}}{2} \right\}$

Section 19.3 Activity

A.1. Consider the equation $(2x - 1)^2 - 13(2x - 1) + 36 = 0$.
   a. Notice that the factor $(2x - 1)$ appears in the first two terms. In the first term, $2x - 1$ is raised to the second power, and in the middle term, $2x - 1$ is raised to the first power. In such a case, the equation is said to be _____ in form.
   b. By making an appropriate substitution, this equation can be expressed as a quadratic equation in a new variable. Rewrite the equation by making the substitution $u = 2x - 1$.
   c. Find the values of u that are solutions to the equation from part (b).
   d. From part (c) you found that $u = 4$ and $u = 9$. In each equation, reverse the substitution. That is, replace u with $2x - 1$ and then solve each equation for x. Write the solution set to the original equation.

For Exercises A.2–A.5, determine an appropriate substitution to make the given equation quadratic in the variable u.

Equation in Quadratic Form	Substitution	New Equation
A.2. $(x^2 - 3)^2 - 9(x^2 - 3) - 52 = 0$	Let $u = $ _____	
A.3. $3x^{2/3} - x^{1/3} - 4 = 0$	Let $u = $ _____	
A.4. $x^4 - 29x^2 + 100 = 0$	Let $u = $ _____	
A.5. $\left(2 + \dfrac{3}{x}\right)^2 - \left(2 + \dfrac{3}{x}\right) - 12 = 0$	Let $u = $ _____	

For Exercises A.6–A.9, solve the equation.

A.6. $(x^2 - 3)^2 - 9(x^2 - 3) - 52 = 0$

A.7. $3x^{2/3} - x^{1/3} - 4 = 0$

A.8. $x^4 - 29x^2 + 100 = 0$

A.9. $\left(2 + \dfrac{3}{x}\right)^2 - \left(2 + \dfrac{3}{x}\right) - 12 = 0$

Section 19.3 Practice Exercises

Prerequisite Review

For Exercises R.1–R.4, write the expression in radical form. Assume that all variables represent positive real numbers.

R.1. $x^{1/2}$

R.2. $y^{1/3}$

R.3. $z^{2/3}$

R.4. $p^{3/4}$

For Exercises R.5–R.16, solve the equation.

R.5. $u^2 - 7u - 18 = 0$

R.6. $u^2 + 14u + 40 = 0$

R.7. $6u^2 + 2u = 6$

R.8. $2x^2 = 8x + 44$

R.9. $3x^2 - 7 = 20$

R.10. $4x^2 + 1 = 65$

R.11. $w^2 + 5 = 0$

R.12. $p^2 + 7 = 0$

R.13. $x^{1/3} = 5$

R.14. $y^{1/5} = 2$

R.15. $\dfrac{2}{x+1} - \dfrac{3}{x-2} = \dfrac{1}{x^2 - x - 2}$

R.16. $\dfrac{5}{t+3} - \dfrac{4}{t+2} = \dfrac{6}{t^2 + 5t + 6}$

Vocabulary and Key Concepts

1. **a.** An equation that can be written in the form $au^2 + bu + c = 0$, where u represents an algebraic expression, is said to be in _____ form.

   b. To use the method of substitution to solve the equation $(3x - 1)^2 + 2(3x - 1) - 8 = 0$, let $u = $ _____.

   c. To use the method of substitution to solve the equation $p^{2/3} - 2p^{1/3} - 15 = 0$, let $u = $ _____.

Concept 1: Solving Equations by Using Substitution

For Exercises 2–7, identify a substitution $u = $ _____ that would make the equation quadratic in the variable u. Then write the equation in terms of u.

2. $(2t - 3)^2 - 8(2t - 3) - 20 = 0$

3. $(3m + 4)^2 - 4(3m + 4) - 5 = 0$

4. $(x^2 - 3)^2 - 4(x^2 - 3) + 3 = 0$

5. $(y^2 + 1)^2 - 7(y^2 + 1) + 10 = 0$

6. $2\left(\dfrac{1}{x}\right)^2 - 5\left(\dfrac{1}{x}\right) - 3 = 0$

7. $\dfrac{3}{x^2} + \dfrac{10}{x} - 8 = 0$

8. a. Solve the quadratic equation by factoring. $u^2 - 2u - 24 = 0$
   b. Solve the equation by using substitution. $(x - 5)^2 - 2(x - 5) - 24 = 0$

9. a. Solve the quadratic equation by factoring. $u^2 + 10u + 24 = 0$
   b. Solve the equation by using substitution. $(y^2 + 5y)^2 + 10(y^2 + 5y) + 24 = 0$

10. a. Solve the quadratic equation by factoring. $u^2 - 2u - 35 = 0$
    b. Solve the equation by using substitution. $(w^2 - 6w)^2 - 2(w^2 - 6w) - 35 = 0$

For Exercises 11–24, solve the equation by using substitution. (See Examples 1–3.)

11. $(x^2 - 2x)^2 + 2(x^2 - 2x) = 3$

12. $(x^2 + x)^2 - 8(x^2 + x) = -12$

13. $(y^2 - 4y)^2 - (y^2 - 4y) = 20$

14. $(w^2 - 2w)^2 - 11(w^2 - 2w) = -24$

15. $m^{2/3} - m^{1/3} - 6 = 0$

16. $2n^{2/3} + 7n^{1/3} - 15 = 0$

17. $2t^{2/5} + 7t^{1/5} + 3 = 0$

18. $p^{2/5} + p^{1/5} - 2 = 0$

19. $y + 6\sqrt{y} = 16$

20. $p - 8\sqrt{p} = -15$

21. $2x + 3\sqrt{x} - 2 = 0$

22. $3t + 5\sqrt{t} - 2 = 0$

23. $16\left(\dfrac{x+6}{4}\right)^2 + 8\left(\dfrac{x+6}{4}\right) + 1 = 0$

24. $9\left(\dfrac{x+3}{2}\right)^2 - 6\left(\dfrac{x+3}{2}\right) + 1 = 0$

25. In Example 3, we solved the equation $x - \sqrt{x} - 12 = 0$ by using substitution. Now solve this equation by first isolating the radical and then squaring both sides. Don't forget to check the potential solutions in the original equation. Do you obtain the same solution as in Example 3?

Concept 2: Solving Equations Reducible to a Quadratic

For Exercises 26–36, solve the equations. (See Examples 4–5.)

26. $t^4 + t^2 - 12 = 0$

27. $w^4 + 4w^2 - 45 = 0$

28. $x^2(9x^2 + 7) = 2$

29. $y^2(4y^2 + 17) = 15$

30. $\dfrac{y}{10} - 1 = -\dfrac{12}{5y}$

31. $1 + \dfrac{5}{x} = -\dfrac{3}{x^2}$

32. $\dfrac{x+5}{x} + \dfrac{x}{2} = \dfrac{x+19}{4x}$

33. $\dfrac{3x}{x+1} - \dfrac{2}{x-3} = 1$

34. $\dfrac{2t}{t-3} - \dfrac{1}{t+4} = 1$

35. $\dfrac{x}{2x-1} = \dfrac{1}{x-2}$

36. $\dfrac{z}{3z+2} = \dfrac{2}{z+1}$

Mixed Exercises

For Exercises 37–60, solve the equations.

37. $x^4 - 16 = 0$

38. $t^4 - 625 = 0$

39. $(4x + 5)^2 + 3(4x + 5) + 2 = 0$

40. $2(5x+3)^2 - (5x+3) - 28 = 0$ 41. $4m^4 - 9m^2 + 2 = 0$ 42. $x^4 - 7x^2 + 12 = 0$

43. $x^6 - 9x^3 + 8 = 0$ 44. $x^6 - 26x^3 - 27 = 0$ 45. $\sqrt{x^2 + 20} = 3\sqrt{x}$

46. $\sqrt{x^2 + 60} = 4\sqrt{x}$ 47. $\sqrt{4t+1} = t+1$ 48. $\sqrt{t+10} = t+4$

49. $2\left(\dfrac{t-4}{3}\right)^2 - \left(\dfrac{t-4}{3}\right) - 3 = 0$ 50. $\left(\dfrac{x+1}{5}\right)^2 - 3\left(\dfrac{x+1}{5}\right) - 10 = 0$ 51. $x^{2/3} + x^{1/3} = 20$

52. $x^{2/5} - 3x^{1/5} = -2$ 53. $m^4 + 2m^2 - 8 = 0$ 54. $2c^4 + c^2 - 1 = 0$

55. $a^3 + 16a - a^2 - 16 = 0$ 56. $b^3 + 9b - b^2 - 9 = 0$ 57. $x^3 + 5x - 4x^2 - 20 = 0$
    (*Hint:* Factor by grouping first.)

58. $y^3 + 8y - 3y^2 - 24 = 0$ 59. $\left(\dfrac{2}{x-3}\right)^2 + 8\left(\dfrac{2}{x-3}\right) + 12 = 0$ 60. $\left(\dfrac{5}{x+1}\right)^2 - 6\left(\dfrac{5}{x+1}\right) - 16 = 0$

Technology Connections

61. **a.** Solve the equation $x^4 + 4x^2 + 4 = 0$.

    b. How many solutions are real and how many solutions are imaginary?

    c. How many x-intercepts do you anticipate for the function defined by $y = x^4 + 4x^2 + 4$?

    d. Graph $Y_1 = x^4 + 4x^2 + 4$ on a standard viewing window.

62. **a.** Solve the equation $x^4 - 2x^2 + 1 = 0$.

    b. How many solutions are real and how many solutions are imaginary?

    c. How many x-intercepts do you anticipate for the function defined by $y = x^4 - 2x^2 + 1$?

    d. Graph $Y_1 = x^4 - 2x^2 + 1$ on a standard viewing window.

63. **a.** Solve the equation $x^4 - x^3 - 6x^2 = 0$.

    b. How many solutions are real and how many solutions are imaginary?

    c. How many x-intercepts do you anticipate for the function defined by $y = x^4 - x^3 - 6x^2$?

    d. Graph $Y_1 = x^4 - x^3 - 6x^2$ on a standard viewing window.

64. **a.** Solve the equation $x^4 - 10x^2 + 9 = 0$.

    b. How many solutions are real and how many solutions are imaginary?

    c. How many x-intercepts do you anticipate for the function defined by $y = x^4 - 10x^2 + 9$?

    d. Graph $Y_1 = x^4 - 10x^2 + 9$ on a standard viewing window.

Problem Recognition Exercises

Quadratic and Quadratic Type Equations

For Exercises 1–4, solve each equation by

a. Completing the square and applying the square root property.

b. Using the quadratic formula.

1. $x^2 + 10x + 3 = 0$
2. $v^2 - 16v + 5 = 0$
3. $3t^2 + t + 4 = 0$
4. $4y^2 + 3y + 5 = 0$

In Exercises 5–24, we have presented all types of equations that you have learned up to this point. For each equation,

a. First determine the type of equation that is presented. Choose from: linear equation, quadratic equation, quadratic in form, rational equation, or radical equation.

b. Solve the equation by using a suitable method.

5. $t^2 + 5t - 14 = 0$
6. $a^2 - 9a + 20 = 0$
7. $a^4 - 10a^2 + 9 = 0$
8. $x^4 - 3x^2 - 4 = 0$
9. $x - 3x^{1/2} - 4 = 0$
10. $x^2 - 9x - 2 = 0$
11. $8b(b+1) + 2(3b-4) = 4b(2b+3)$
12. $6x(x+1) - 3(x+4) = 3x(2x+5)$
13. $5a(a+6) = 10(3a-1)$
14. $4x(x+3) = 6(2x-4)$
15. $\dfrac{t}{t+5} + \dfrac{3}{t-4} = \dfrac{17}{t^2+t-20}$
16. $\dfrac{v}{v+4} + \dfrac{12}{v^2+7v+12} = \dfrac{5}{v+3}$
17. $c^2 - 20c - 1 = 0$
18. $d^2 + 18d + 4 = 0$
19. $2u(u-3) = 4(2-u)$
20. $3y(y+2) = 9(y+1)$
21. $\sqrt{2b+3} = b$
22. $\sqrt{5t+6} = t$
23. $x^{2/3} + 2x^{1/3} - 15 = 0$
24. $y^{2/3} + 5y^{1/3} + 4 = 0$

Graphs of Quadratic Functions Section 19.4

A quadratic function is defined as a function of the form $f(x) = ax^2 + bx + c$ ($a \neq 0$). The graph of a quadratic function is a **parabola**. In this section, we will learn how to graph parabolas.

A parabola opens upward if $a > 0$ (Figure 19-3) and opens downward if $a < 0$ (Figure 19-4). If a parabola opens upward, the **vertex** is the lowest point on the graph. If a parabola opens downward, the **vertex** is the highest point on the graph. The **axis of symmetry** is the vertical line that passes through the vertex.

Concepts

1. Quadratic Functions of the Form $f(x) = x^2 + k$
2. Quadratic Functions of the Form $f(x) = (x - h)^2$
3. Quadratic Functions of the Form $f(x) = ax^2$
4. Quadratic Functions of the Form $f(x) = a(x - h)^2 + k$

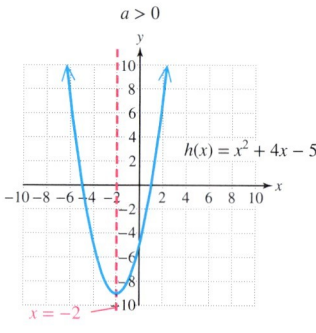

Figure 19-3

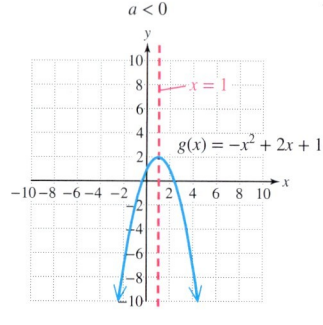

Figure 19-4

1. Quadratic Functions of the Form $f(x) = x^2 + k$

One technique for graphing a function is to plot a sufficient number of points on the graph until the general shape and defining characteristics can be determined. Then sketch a curve through the points.

Example 1 — Graphing Quadratic Functions of the Form $f(x) = x^2 + k$

Graph the functions f, g, and h on the same coordinate system.

$$f(x) = x^2 \qquad g(x) = x^2 + 1 \qquad h(x) = x^2 - 2$$

Solution:

Several function values for f, g, and h are shown in Table 19-1 for selected values of x. The corresponding graphs are pictured in Figure 19-5.

Table 19-1

x	$f(x) = x^2$	$g(x) = x^2 + 1$	$h(x) = x^2 - 2$
-3	9	10	7
-2	4	5	2
-1	1	2	-1
0	0	1	-2
1	1	2	-1
2	4	5	2
3	9	10	7

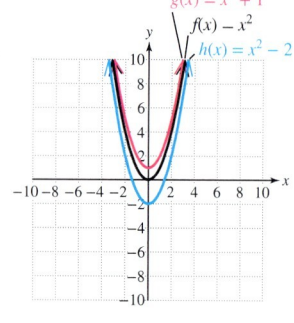

Figure 19-5

Skill Practice Refer to the graph of $f(x) = x^2 + k$ to determine the value of k.

1.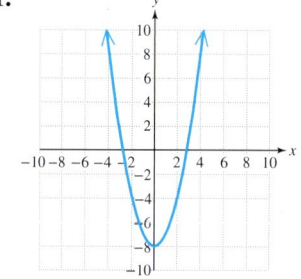

Answer

1. $k = -8$

Notice that the graphs of $g(x) = x^2 + 1$ and $h(x) = x^2 - 2$ take on the same shape as $f(x) = x^2$. However, the y values of g are 1 greater than the y values of f. Hence, the graph of $g(x) = x^2 + 1$ is the same as the graph of $f(x) = x^2$ shifted *up* 1 unit. Likewise the y values of h are 2 less than those of f. The graph of $h(x) = x^2 - 2$ is the same as the graph of $f(x) = x^2$ shifted *down* 2 units.

A function of the type $f(x) = x^2 + k$ represents a vertical shift of the graph of $f(x) = x^2$. The functions in Example 1 illustrate the following properties of quadratic functions of the form $f(x) = x^2 + k$.

Graphs of $f(x) = x^2 + k$

A function of the type $f(x) = x^2 + k$ represents a vertical shift of the graph of $f(x) = x^2$.

- If $k > 0$, then the graph of $f(x) = x^2 + k$ is the same as the graph of $f(x) = x^2$ shifted *up* k units.
- If $k < 0$, then the graph of $f(x) = x^2 + k$ is the same as the graph of $f(x) = x^2$ shifted *down* $|k|$ units.

Example 2 — Graphing Quadratic Functions of the Form $f(x) = x^2 + k$

Sketch the functions defined by

a. $m(x) = x^2 - 4$ b. $n(x) = x^2 + \dfrac{7}{2}$

Solution:

a. $m(x) = x^2 - 4$

$m(x) = x^2 + (-4)$

Because $k = -4$ (negative), the graph is obtained by shifting the graph of $f(x) = x^2$ down $|-4|$ units (Figure 19-6).

b. $n(x) = x^2 + \dfrac{7}{2}$

Because $k = \dfrac{7}{2}$ (positive), the graph is obtained by shifting the graph of $f(x) = x^2$ up $\dfrac{7}{2}$ units (Figure 19-7).

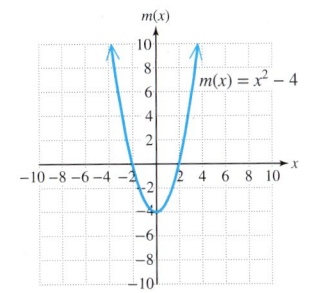

Figure 19-6

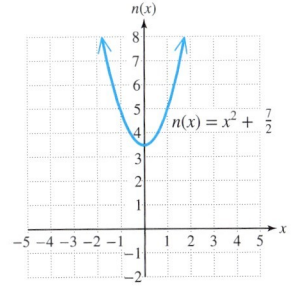

Figure 19-7

TIP: For more accuracy in the graph, plot one or two points near the vertex and use the symmetry of the curve to find additional points on the graph.

Answer

2.

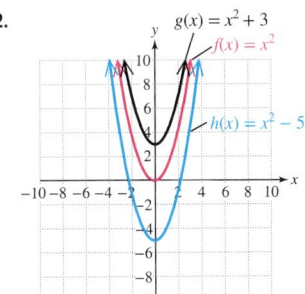

Skill Practice

2. Graph the functions f, g, and h on the same coordinate system.
   $f(x) = x^2$ $g(x) = x^2 + 3$ $h(x) = x^2 - 5$

2. Quadratic Functions of the Form $f(x) = (x - h)^2$

The graph of $f(x) = x^2 + k$ represents a vertical shift (up or down) of the graph of $f(x) = x^2$. Example 3 shows that functions of the form $f(x) = (x - h)^2$ represent a horizontal shift (left or right) of the graph of $f(x) = x^2$.

Example 3 — Graphing Quadratic Functions of the Form $f(x) = (x - h)^2$

Graph the functions f, g, and h on the same coordinate system.

$$f(x) = x^2 \qquad g(x) = (x + 1)^2 \qquad h(x) = (x - 2)^2$$

Solution:

Several function values for f, g, and h are shown in Table 19-2 for selected values of x. The corresponding graphs are pictured in Figure 19-8.

Table 19-2

x	$f(x) = x^2$	$g(x) = (x + 1)^2$	$h(x) = (x - 2)^2$
-4	16	9	36
-3	9	4	25
-2	4	1	16
-1	1	0	9
0	0	1	4
1	1	4	1
2	4	9	0
3	9	16	1
4	16	25	4
5	25	36	9

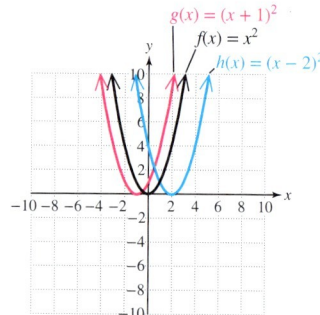

Figure 19-8

Skill Practice Refer to the graph of $f(x) = (x - h)^2$ to determine the value of h.

3.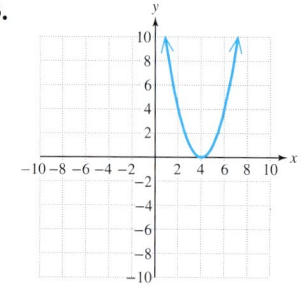

Example 3 illustrates the following properties of quadratic functions of the form $f(x) = (x - h)^2$.

Answer

3. $h = 4$

Graphs of $f(x) = (x - h)^2$

A function of the type $f(x) = (x - h)^2$ represents a horizontal shift of the graph of $f(x) = x^2$.

If $h > 0$, then the graph of $f(x) = (x - h)^2$ is the same as the graph of $f(x) = x^2$ shifted h units to the *right*.

If $h < 0$, then the graph of $f(x) = (x - h)^2$ is the same as the graph of $f(x) = x^2$ shifted $|h|$ units to the *left*.

From Example 3 we have

$$h(x) = (x - 2)^2 \quad \text{and} \quad g(x) = (x + 1)^2$$
$$g(x) = [x - (-1)]^2$$

$y = x^2$ shifted 2 units to the right

$y = x^2$ shifted $|-1|$ unit to the left

Example 4 — Graphing Functions of the Form $f(x) = (x - h)^2$

Sketch the functions p and q.

a. $p(x) = (x - 7)^2$ **b.** $q(x) = (x + 1.6)^2$

Solution:

a. $p(x) = (x - 7)^2$

Because $h = 7 > 0$, shift the graph of $f(x) = x^2$ to the *right* 7 units (Figure 19-9).

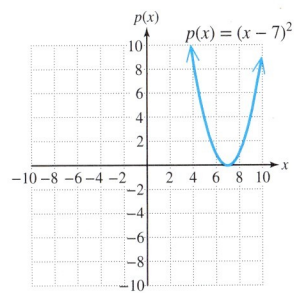

Figure 19-9

b. $q(x) = (x + 1.6)^2$

$q(x) = [x - (-1.6)]^2$

Because $h = -1.6 < 0$, shift the graph of $f(x) = x^2$ to the *left* 1.6 units (Figure 19-10).

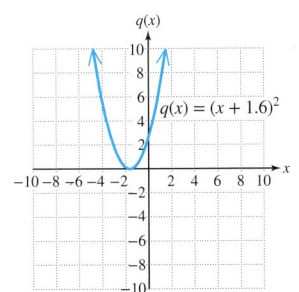

Figure 19-10

Answer

4.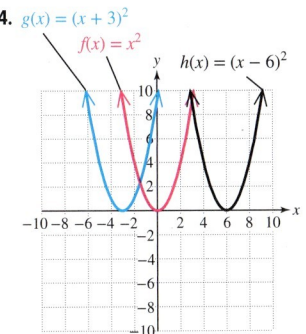

Skill Practice

4. Graph the functions f, g, and h on the same coordinate system.

$f(x) = x^2$

$g(x) = (x + 3)^2$

$h(x) = (x - 6)^2$

3. Quadratic Functions of the Form $f(x) = ax^2$

Examples 5 and 6 investigate functions of the form $f(x) = ax^2$ $(a \neq 0)$.

Example 5 — Graphing Functions of the Form $f(x) = ax^2$

Graph the functions f, g, and h on the same coordinate system.

$$f(x) = x^2 \qquad g(x) = 2x^2 \qquad h(x) = \frac{1}{2}x^2$$

Solution:

Several function values for f, g, and h are shown in Table 19-3 for selected values of x. The corresponding graphs are pictured in Figure 19-11.

Table 19-3

x	$f(x) = x^2$	$g(x) = 2x^2$	$h(x) = \frac{1}{2}x^2$
-3	9	18	$\frac{9}{2}$
-2	4	8	2
-1	1	2	$\frac{1}{2}$
0	0	0	0
1	1	2	$\frac{1}{2}$
2	4	8	2
3	9	18	$\frac{9}{2}$

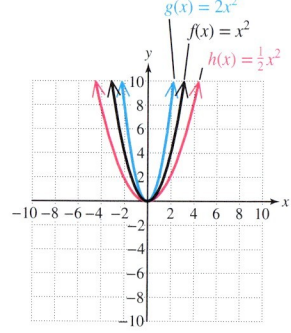

Figure 19-11

Skill Practice

5. Graph the functions f, g, and h on the same coordinate system.

   $f(x) = x^2$

   $g(x) = 3x^2$

   $h(x) = \frac{1}{3}x^2$

In Example 5, the function values defined by $g(x) = 2x^2$ are twice those of $f(x) = x^2$. The graph of $g(x) = 2x^2$ is the same as the graph of $f(x) = x^2$ **stretched vertically** by a factor of 2 (the graph appears narrower than $f(x) = x^2$).

In Example 5, the function values defined by $h(x) = \frac{1}{2}x^2$ are one-half those of $f(x) = x^2$. The graph of $h(x) = \frac{1}{2}x^2$ is the same as the graph of $f(x) = x^2$ **shrunk vertically** by a factor of $\frac{1}{2}$ (the graph appears wider than $f(x) = x^2$).

Example 6 — Graphing Functions of the Form $f(x) = ax^2$

Graph the functions f, g, and h on the same coordinate system.

$$f(x) = -x^2 \qquad g(x) = -3x^2 \qquad h(x) = -\frac{1}{3}x^2$$

Answer

5.

Solution:

Several function values for f, g, and h are shown in Table 19-4 for selected values of x. The corresponding graphs are pictured in Figure 19-12.

Table 19-4

x	$f(x) = -x^2$	$g(x) = -3x^2$	$h(x) = -\frac{1}{3}x^2$
-3	-9	-27	-3
-2	-4	-12	$-\frac{4}{3}$
-1	-1	-3	$-\frac{1}{3}$
0	0	0	0
1	-1	-3	$-\frac{1}{3}$
2	-4	-12	$-\frac{4}{3}$
3	-9	-27	-3

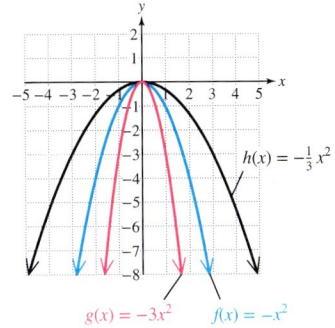

Figure 19-12

Skill Practice

6. Graph the functions f, g, and h on the same coordinate system.

$$f(x) = -x^2 \qquad g(x) = -2x^2 \qquad h(x) = -\frac{1}{2}x^2$$

Example 6 illustrates that if the coefficient of the squared term is negative, the parabola opens downward. The graph of $g(x) = -3x^2$ is the same as the graph of $f(x) = -x^2$ with a *vertical stretch* by a factor of $|-3|$. The graph of $h(x) = -\frac{1}{3}x^2$ is the same as the graph of $f(x) = -x^2$ with a *vertical shrink* by a factor of $|-\frac{1}{3}|$.

> ### Graphs of $f(x) = ax^2$
>
> 1. If $a > 0$, the parabola opens upward. Furthermore,
>   - If $0 < a < 1$, then the graph of $f(x) = ax^2$ is the same as the graph of $f(x) = x^2$ with a *vertical shrink* by a factor of a.
>   - If $a > 1$, then the graph of $f(x) = ax^2$ is the same as the graph of $f(x) = x^2$ with a *vertical stretch* by a factor of a.
> 2. If $a < 0$, the parabola opens downward. Furthermore,
>   - If $0 < |a| < 1$, then the graph of $f(x) = ax^2$ is the same as the graph of $f(x) = -x^2$ with a *vertical shrink* by a factor of $|a|$.
>   - If $|a| > 1$, then the graph of $f(x) = ax^2$ is the same as the graph of $f(x) = -x^2$ with a *vertical stretch* by a factor of $|a|$.

4. Quadratic Functions of the Form $f(x) = a(x - h)^2 + k$

We can summarize our findings from Examples 1–6 by graphing functions of the form $f(x) = a(x - h)^2 + k$ $(a \neq 0)$.

The graph of $f(x) = x^2$ has its vertex at the origin $(0, 0)$. The graph of $f(x) = a(x - h)^2 + k$ is the same as the graph of $f(x) = x^2$ shifted to the right or left h units and shifted up or down k units. Therefore, the vertex is shifted from $(0, 0)$ to (h, k). The axis of symmetry is the vertical line through the vertex. Therefore, the axis of symmetry must be the line $x = h$.

Answer

6.
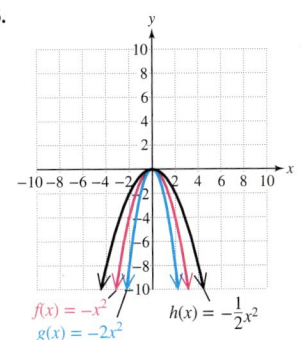

Graphs of $f(x) = a(x - h)^2 + k$

1. The vertex is (h, k).
2. The axis of symmetry is the line $x = h$.
3. If $a > 0$, the parabola opens upward, and k is the **minimum value** of the function.
4. If $a < 0$, the parabola opens downward, and k is the **maximum value** of the function.

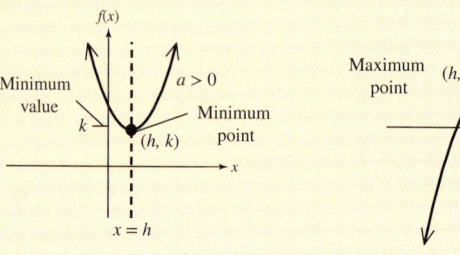

Example 7 — Graphing a Function of the Form $f(x) = a(x - h)^2 + k$

Given the function defined by $f(x) = 2(x - 3)^2 + 4$

a. Identify the vertex.
b. Sketch the function.
c. Identify the axis of symmetry.
d. Identify the maximum or minimum value of the function.
e. Use the graph to determine the domain and range of f in interval notation.

Solution:

a. The function is in the form $f(x) = a(x - h)^2 + k$, where $a = 2$, $h = 3$, and $k = 4$. Therefore, the vertex is $(3, 4)$.

b. The graph of f is the same as the graph of $f(x) = x^2$ shifted to the right 3 units, shifted up 4 units, and stretched vertically by a factor of 2 (Figure 19-13).

c. The axis of symmetry is the line $x = 3$.

d. Because $a > 0$, the function opens upward. Therefore, the minimum function value is 4. Notice that the minimum value is the minimum y value on the graph.

e. The domain is all real numbers: $(-\infty, \infty)$. The vertex is the lowest point on the parabola, and its y-coordinate is 4. Therefore, the range is $[4, \infty)$.

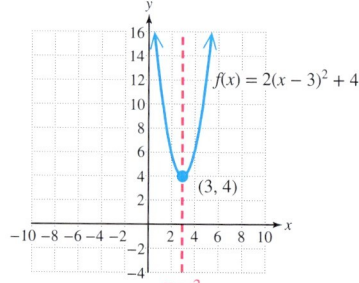

Figure 19-13

TIP: We can find additional points to the right or left of the vertex and use the symmetry of the parabola to refine the sketch of the curve.

x	f(x)
4	6
5	12

Skill Practice

7. Given the function defined by $g(x) = 3(x + 1)^2 - 3$
   a. Identify the vertex.
   b. Sketch the graph.
   c. Identify the axis of symmetry.
   d. Identify the maximum or minimum value of the function.
   e. Determine the domain and range in interval notation.

Answers

7. a. Vertex: $(-1, -3)$
   b.
   c. Axis of symmetry: $x = -1$
   d. Minimum value: -3
   e. Domain: $(-\infty, \infty)$; range: $[-3, \infty)$

Example 8 — Graphing a Function of the Form $f(x) = a(x - h)^2 + k$

Given the function defined by $g(x) = -(x + 2)^2 - \dfrac{7}{4}$

a. Identify the vertex.
b. Sketch the function.
c. Identify the axis of symmetry.
d. Identify the maximum or minimum value of the function.
e. Use the graph to determine the domain and range of f in interval notation.

Solution:

a. $g(x) = -(x + 2)^2 - \dfrac{7}{4}$

$= -1[x - (-2)]^2 + \left(-\dfrac{7}{4}\right)$

The function is in the form $g(x) = a(x - h)^2 + k$, where $a = -1$, $h = -2$, and $k = -\dfrac{7}{4}$. Therefore, the vertex is $\left(-2, -\dfrac{7}{4}\right)$.

b. The graph of g is the same as the graph of $f(x) = x^2$ shifted to the left 2 units, shifted down $\dfrac{7}{4}$ units, and opening downward (Figure 19-14).

c. The axis of symmetry is the line $x = -2$.

d. The parabola opens downward, so the maximum function value is $-\dfrac{7}{4}$.

e. The domain is all real numbers: $(-\infty, \infty)$.

The vertex is the highest point on the parabola, and its y-coordinate is $-\dfrac{7}{4}$. Therefore, the range is $\left(-\infty, -\dfrac{7}{4}\right]$.

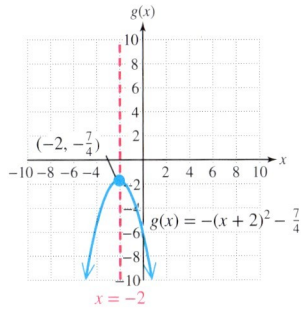

Figure 19-14

Skill Practice

8. Given the function defined by $h(x) = -\dfrac{1}{2}(x - 4)^2 + 2$

   a. Identify the vertex.
   b. Sketch the graph.
   c. Identify the axis of symmetry.
   d. Identify the maximum or minimum value of the function.
   e. Determine the domain and range in interval notation.

Answers

8. a. Vertex: (4, 2)
   b.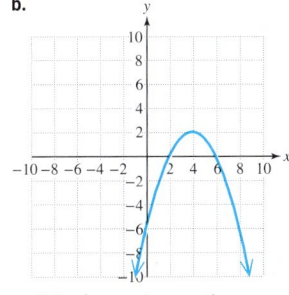
   c. Axis of symmetry: $x = 4$
   d. Maximum value: 2
   e. Domain: $(-\infty, \infty)$; range: $(-\infty, 2]$

Section 19.4 Activity

A.1. Graph the functions defined by $f(x) = x^2$, $g(x) = x^2 + 1$, and $h(x) = x^2 - 2$ by completing the table of points. Graph the functions on the same coordinate system for comparison.

x	$f(x) = x^2$	$g(x) = x^2 + 1$	$h(x) = x^2 - 2$
0			
1			
2			
3			
-1			
-2			
-3			

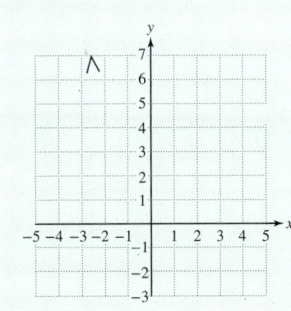

A.2. **a.** Based on the result of Exercise A.1, for $k > 0$, describe the graph of $y = x^2 + k$.

 b. Based on the result of Exercise A.1, for $k > 0$, describe the graph of $y = x^2 - k$.

A.3. Graph the functions defined by $f(x) = x^2$, $g(x) = (x + 1)^2$, and $h(x) = (x - 2)^2$ by completing the table of points. Graph the functions on the same coordinate system for comparison.

x	$f(x) = x^2$	$g(x) = (x + 1)^2$	$h(x) = (x - 2)^2$
0			
1			
2			
3			
−1			
−2			
−3			

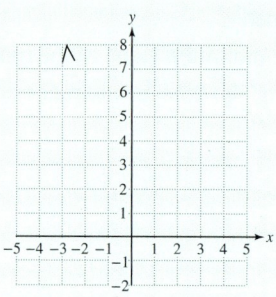

A.4. **a.** Based on the result of Exercise A.3, for $h > 0$, describe the graph of $y = (x + h)^2$.

 b. Based on the result of Exercise A.3, for $h > 0$, describe the graph of $y = (x - h)^2$.

A.5. Graph the functions defined by $f(x) = x^2$, $g(x) = 2x^2$, $h(x) = \frac{1}{2}x^2$, and $k(x) = -x^2$ by completing the table of points. Graph the functions on the same coordinate system for comparison.

x	$f(x) = x^2$	$g(x) = 2x^2$	$h(x) = \frac{1}{2}x^2$	$k(x) = -x^2$
0				
1				
2				
3				
−1				
−2				
−3				

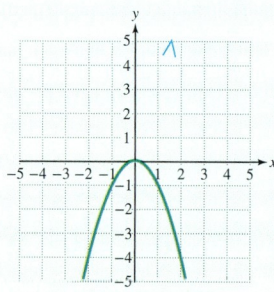

A.6. **a.** Based on the result of Exercise A.5, for $a > 1$, describe the graph of $y = ax^2$.

 b. Based on the result of Exercise A.5, for $0 < a < 1$, describe the graph of $y = ax^2$.

 c. Based on the result of Exercise A.5, for $a < 0$, describe the graph of $y = ax^2$.

A.7. Exercises A.1–A.6 illustrate *transformations* of graphs. These include shifting a graph right, left, upward, or downward; stretching or shrinking a graph vertically; and reflecting a graph over the x-axis. Given $y = a(x - h)^2 + k$, match the statement on the left with the conclusion on the right.

 a. The value of a _____ i. shrinks or stretches the graph of $y = x^2$ vertically and/or reflects the graph over the x-axis.

 b. The value of h _____ ii. shifts the graph vertically.

 c. The value of k _____ iii. shifts the graph horizontally.

A.8. **a.** Use transformations to graph $f(x) = -(x + 2)^2 + 4$.

 b. Identify the vertex of the parabola.

 c. Identify the axis of symmetry.

 d. Identify the maximum or minimum value of f.

 e. Use the graph to determine the domain and range of f.

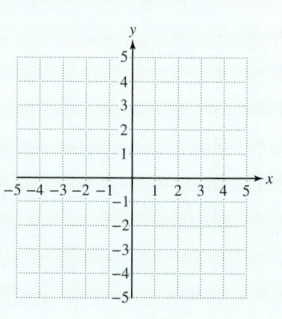

Practice Exercises

Section 19.4

Prerequisite Review

For Exercises R.1–R.4,
a. Write the domain in interval notation.
b. Write the range in interval notation.
c. Identify the *x*-intercepts if any exist.
d. Identify the *y*-intercpet.

R.1.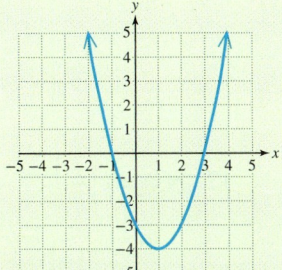

R.2.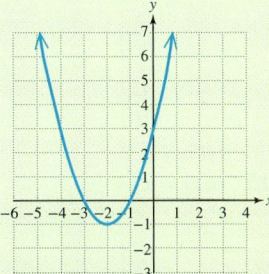

R.3.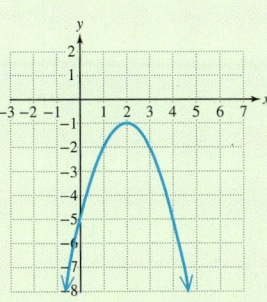

R.4.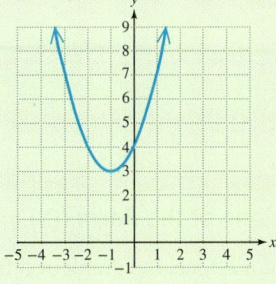

Vocabulary and Key Concepts

1. a. The graph of a quadratic function, $f(x) = ax^2 + bx + c$, is a _____.
   b. The parabola defined by $f(x) = ax^2 + bx + c$ ($a \neq 0$) will open upward if a _____ 0 and will open downward if a _____ 0.
   c. If a parabola opens upward, the vertex is the (highest/lowest) point on the graph. If a parabola opens downward, the vertex is the (highest/lowest) point on the graph.

2. a. Given $f(x) = a(x - h)^2 + k$ ($a \neq 0$), the vertex of the parabola is given by the ordered pair _____. If the parabola opens _____, the value of k is the minimum value of the function. If the parabola opens _____, the value of k is the maximum value of the function.
   b. Given $f(x) = a(x - h)^2 + k$ ($a \neq 0$), the axis of symmetry of the parabola is the vertical line that passes through the _____. An equation of the axis of symmetry is _____.

For Exercises 3–4, determine if the parabola opens upward or downward.

3. $f(x) = -2x^2 + 4$
4. $g(x) = x^2 - 9$

For Exercises 5–6, determine if the vertex of the parabola is a maximum point or a minimum point.

5. $f(x) = -2x^2 + 4$
6. $g(x) = x^2 - 9$

For Exercises 7–8, identify the vertex of the parabola and the axis of symmetry.

7. $h(x) = (x + 3)^2 + 1$
8. $k(x) = (x - 4)^2 - 2$

1310 Chapter 19 Quadratic Equations and Functions

Concept 1: Quadratic Functions of the Form $f(x) = x^2 + k$

9. Describe how the value of k affects the graph of a function defined by $f(x) = x^2 + k$.

For Exercises 10–17, graph the functions. (See Examples 1–2.)

10. $g(x) = x^2 + 1$

11. $f(x) = x^2 + 2$

12. $p(x) = x^2 - 3$

13. $q(x) = x^2 - 4$

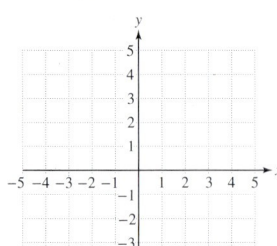 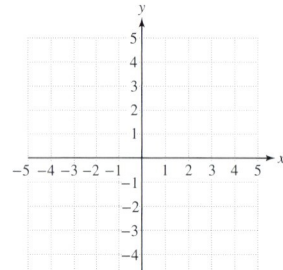 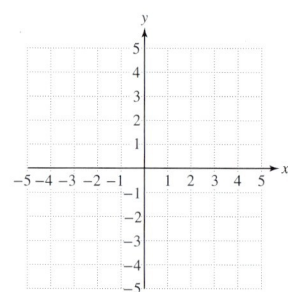 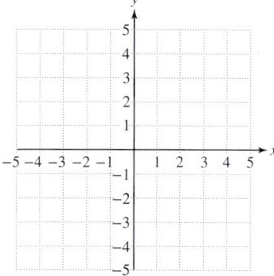

14. $T(x) = x^2 + \dfrac{3}{4}$

15. $S(x) = x^2 + \dfrac{3}{2}$

16. $M(x) = x^2 - \dfrac{5}{4}$

17. $n(x) = x^2 - \dfrac{1}{3}$

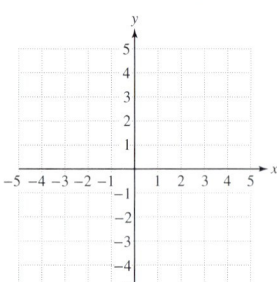 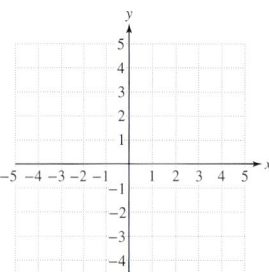 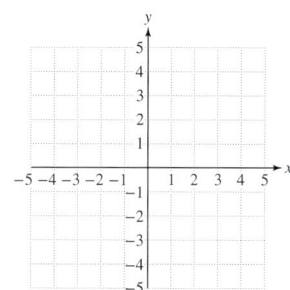 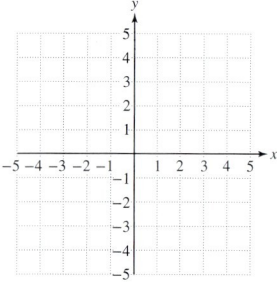

Concept 2: Quadratic Functions of the Form $f(x) = (x - h)^2$

18. Describe how the value of h affects the graph of a function defined by $f(x) = (x - h)^2$.

For Exercises 19–26, graph the functions. (See Examples 3–4.)

19. $r(x) = (x + 1)^2$

20. $h(x) = (x + 2)^2$

21. $k(x) = (x - 3)^2$

22. $L(x) = (x - 4)^2$

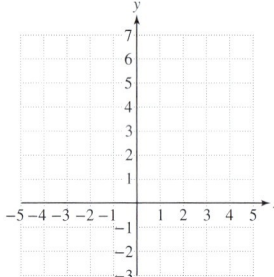 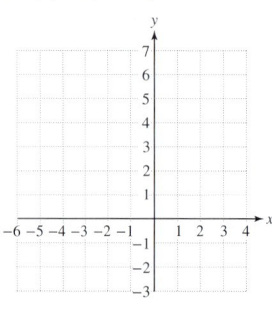 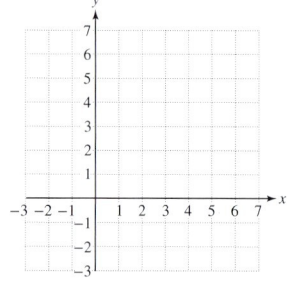 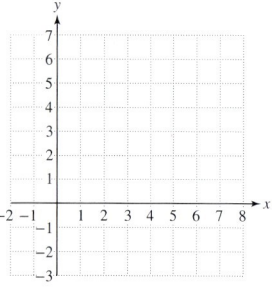

23. $A(x) = \left(x + \dfrac{3}{4}\right)^2$

24. $r(x) = \left(x + \dfrac{3}{2}\right)^2$

25. $W(x) = (x - 1.25)^2$

26. $V(x) = (x - 2.5)^2$

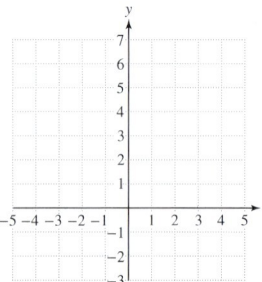 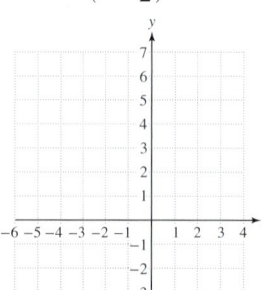 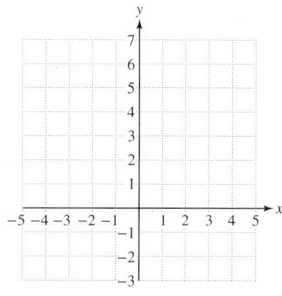 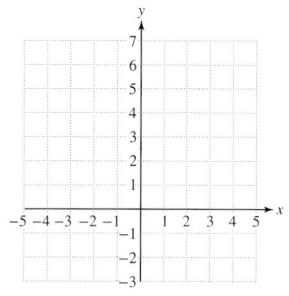

Concept 3: Quadratic Functions of the Form $f(x) = ax^2$

27. Describe how the value of a affects the graph of a function defined by $f(x) = ax^2$, where $a \neq 0$.

28. How do you determine whether the graph of a function defined by $h(x) = ax^2 + bx + c$ ($a \neq 0$) opens upward or downward?

For Exercises 29–36, graph the functions. **(See Examples 5–6.)**

29. $f(x) = 4x^2$

30. $g(x) = 3x^2$

31. $h(x) = \dfrac{1}{4}x^2$

32. $f(x) = \dfrac{1}{5}x^2$

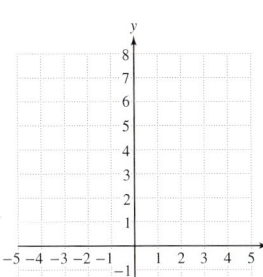
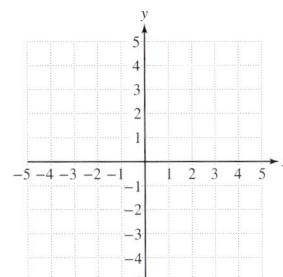
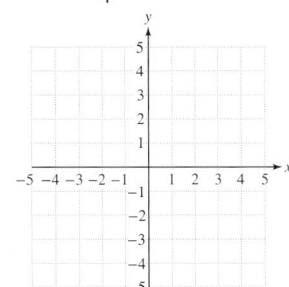
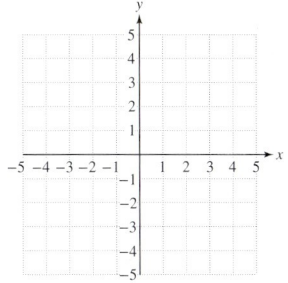

33. $c(x) = -x^2$

34. $g(x) = -4x^2$

35. $v(x) = -\dfrac{1}{5}x^2$

36. $f(x) = -\dfrac{1}{4}x^2$

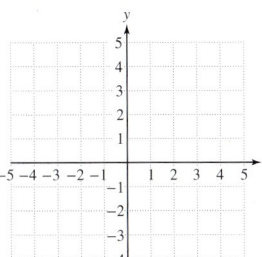
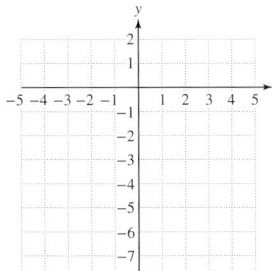
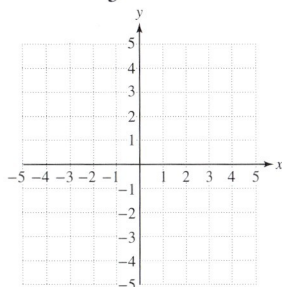
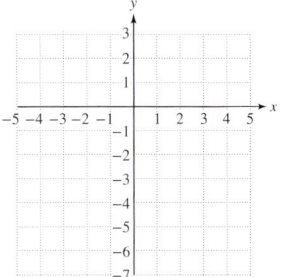

Concept 4: Quadratic Functions of the Form $f(x) = a(x - h)^2 + k$

For Exercises 37–44, match the function with its graph.

37. $f(x) = -\dfrac{1}{4}x^2$

38. $g(x) = (x + 3)^2$

39. $k(x) = (x - 3)^2$

40. $h(x) = \dfrac{1}{4}x^2$

41. $t(x) = x^2 + 2$

42. $m(x) = x^2 - 4$

43. $p(x) = (x + 1)^2 - 3$

44. $n(x) = -(x - 2)^2 + 3$

a.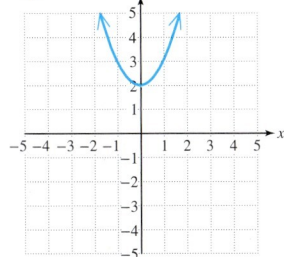
b.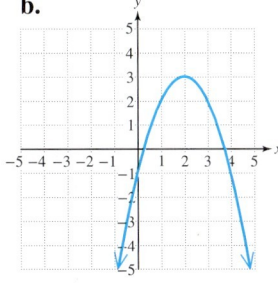
c.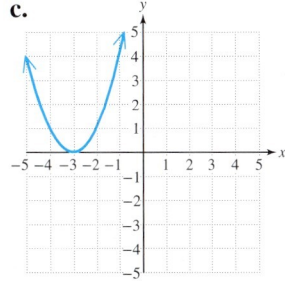
d.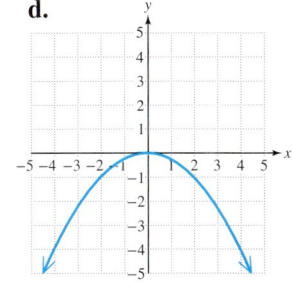

e.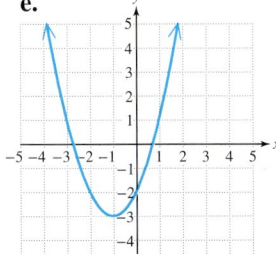
f.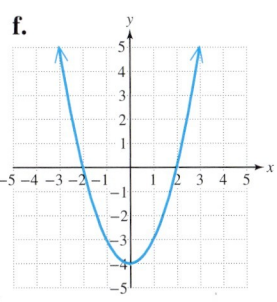
g.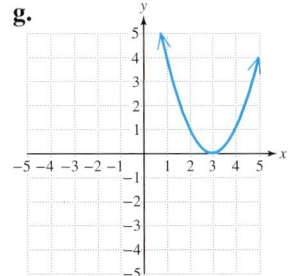
h.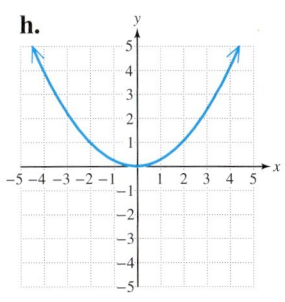

For Exercises 45–64, graph the parabola and the axis of symmetry. Label the coordinates of the vertex, and write the equation of the axis of symmetry. Use the graph to write the domain and range in interval notation. (See Examples 7–8.)

45. $f(x) = (x-3)^2 + 2$

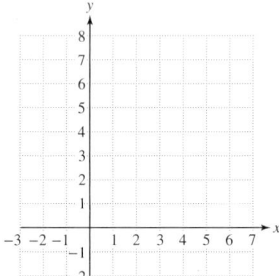

46. $f(x) = (x-2)^2 + 3$

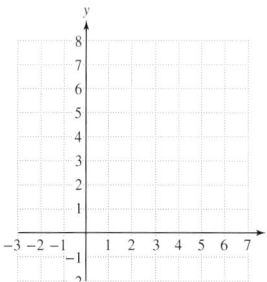

47. $f(x) = (x+1)^2 - 3$

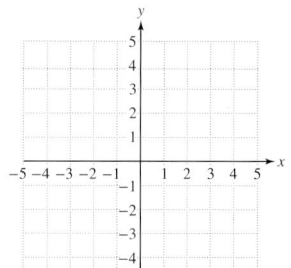

48. $f(x) = (x+3)^2 - 1$

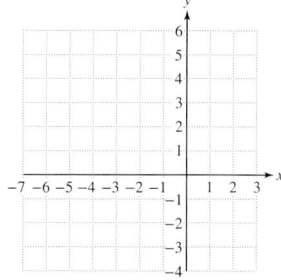

49. $f(x) = -(x-4)^2 - 2$

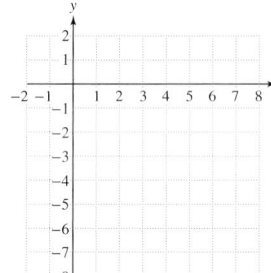

50. $f(x) = -(x-2)^2 - 4$

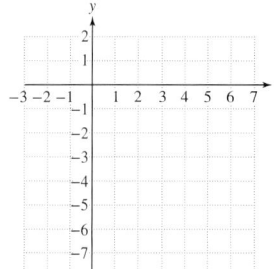

51. $f(x) = -(x+3)^2 + 3$

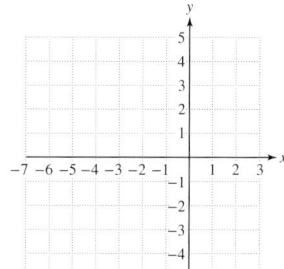

52. $f(x) = -(x+2)^2 + 2$

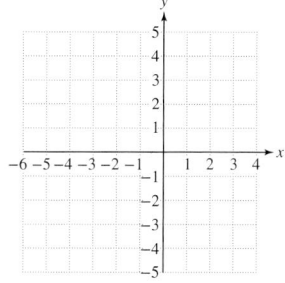

53. $f(x) = (x+1)^2 + 1$

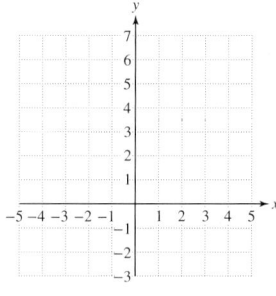

54. $f(x) = (x-4)^2 - 4$

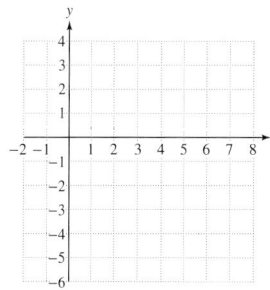

55. $f(x) = 3(x-1)^2$

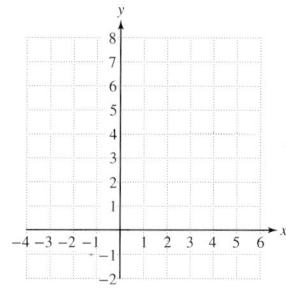

56. $f(x) = -3(x-1)^2$

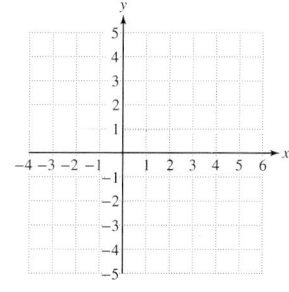

57. $f(x) = -4x^2 + 3$

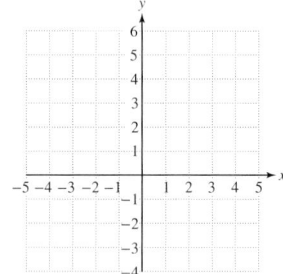

58. $f(x) = 4x^2 + 3$

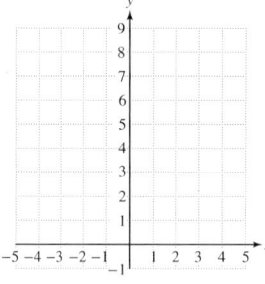

59. $f(x) = 2(x+3)^2 - 1$

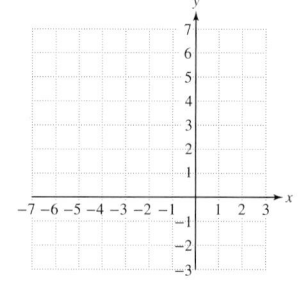

60. $f(x) = -2(x+3)^2 - 1$

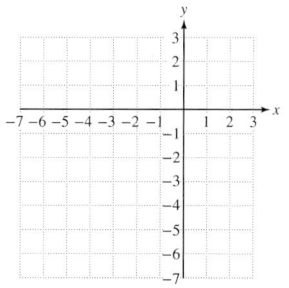

61. $f(x) = -\frac{1}{4}(x-1)^2 + 2$ **62.** $f(x) = \frac{1}{4}(x-1)^2 + 2$ **63.** $f(x) = \frac{1}{3}(x-2)^2 + 1$ **64.** $f(x) = -\frac{1}{3}(x-2)^2 + 1$

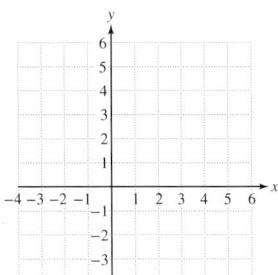
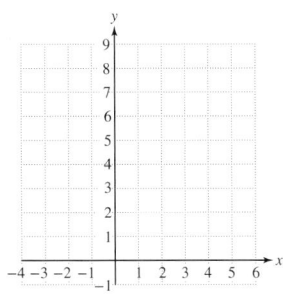
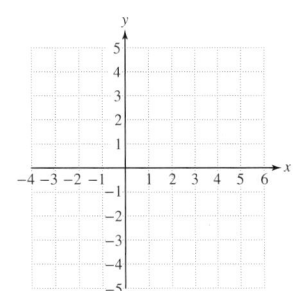
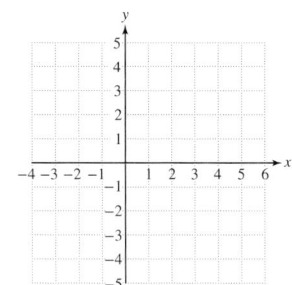

65. Compare the graphs of the following equations to the graph of $y = x^2$.

  a. $y = x^2 + 3$

  b. $y = (x+3)^2$

  c. $y = 3x^2$

66. Compare the graphs of the following equations to the graph of $y = x^2$.

  a. $y = (x-2)^2$

  b. $y = 2x^2$

  c. $y = x^2 - 2$

For Exercises 67–78, write the coordinates of the vertex and determine if the vertex is a maximum point or a minimum point. Then write the maximum or minimum value.

67. $f(x) = 4(x-6)^2 - 9$

68. $g(x) = 3(x-4)^2 - 7$

69. $p(x) = -\frac{2}{5}(x-2)^2 + 5$

70. $h(x) = -\frac{3}{7}(x-5)^2 + 10$

71. $k(x) = \frac{1}{2}(x+8)^2$

72. $m(x) = \frac{2}{9}(x+11)^2$

73. $n(x) = -6x^2 + \frac{21}{4}$

74. $q(x) = -4x^2 + \frac{1}{6}$

75. $A(x) = 2(x-7)^2 - \frac{3}{2}$

76. $B(x) = 5(x-3)^2 - \frac{1}{4}$

77. $F(x) = 7x^2$

78. $G(x) = -7x^2$

79. True or false: The function defined by $g(x) = -5x^2$ has a maximum value but no minimum value.

80. True or false: The function defined by $f(x) = 2(x-5)^2$ has a maximum value but no minimum value.

81. True or false: If the vertex $(-2, 8)$ represents a minimum point, then the minimum value is -2.

82. True or false: If the vertex $(-2, 8)$ represents a maximum point, then the maximum value is 8.

Expanding Your Skills

83. A suspension bridge is 120 ft long. Its supporting cable hangs in a shape that resembles a parabola. The function defined by $H(x) = \frac{1}{90}(x - 60)^2 + 30$ (where $0 \leq x \leq 120$) approximates the height $H(x)$ (in feet) of the supporting cable a distance of x ft from the end of the bridge (see figure).

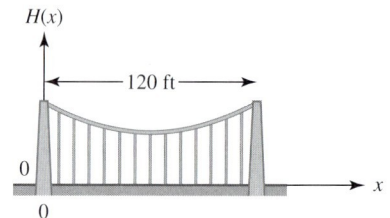

   a. What is the location of the vertex of the parabolic cable?

   b. What is the minimum height of the cable?

   c. How high are the towers at either end of the supporting cable?

84. A 50-m bridge over a crevasse is supported by a parabolic arch. The function defined by $f(x) = -0.16(x - 25)^2 + 100$ (where $0 \leq x \leq 50$) approximates the height $f(x)$ (in feet) of the supporting arch x meters from the end of the bridge (see figure).

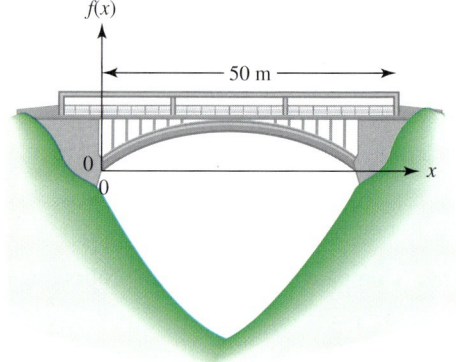

   a. What is the location of the vertex of the arch?

   b. What is the maximum height of the arch (relative to the origin)?

85. The staging platform for a fireworks display is 6 ft above ground, and the mortars leave the platform at 96 ft/sec. The height of the mortars $h(t)$ (in feet) can be modeled by $h(t) = -16(t - 3)^2 + 150$, where t is the time in seconds after launch.

   a. If the fuses are set for 3 sec after launch, at what height will the fireworks explode?

   b. Will the fireworks explode at their maximum height? Explain.

Section 19.5 Vertex of a Parabola: Applications and Modeling

Concepts

1. Writing a Quadratic Function in the Form $f(x) = a(x - h)^2 + k$
2. Vertex Formula
3. Determining the Vertex and Intercepts of a Quadratic Function
4. Applications and Modeling of Quadratic Functions

1. Writing a Quadratic Function in the Form $f(x) = a(x - h)^2 + k$

The graph of a quadratic function is a parabola, and if the function is written in the form $f(x) = a(x - h)^2 + k$ $(a \neq 0)$, then the vertex is (h, k). A quadratic function can be written in the form $f(x) = a(x - h)^2 + k$ $(a \neq 0)$ by completing the square. The process is similar to the steps to solve a quadratic equation by completing the square. However, in this context working with a function, we will perform all algebraic manipulations on the right side of the equation.

Section 19.5 Vertex of a Parabola: Applications and Modeling

Example 1 — Writing a Quadratic Function in the Form $f(x) = a(x - h)^2 + k$ $(a \neq 0)$

Given $f(x) = x^2 + 8x + 13$

a. Write the function in the form $f(x) = a(x - h)^2 + k$.

b. Identify the vertex, axis of symmetry, and minimum function value.

Solution:

a. $f(x) = x^2 + 8x + 13$

Rather than dividing by the leading coefficient on both sides, we will factor out the leading coefficient from the variable terms on the right-hand side.

$= 1(x^2 + 8x) + 13$

$= 1(x^2 + 8x \quad) + 13$

Next, complete the square on the expression within the parentheses: $[\frac{1}{2}(8)]^2 = 16$.

$= 1(x^2 + 8x + 16 - 16) + 13$

Rather than add 16 to both sides of the function, we *add and subtract* 16 within the parentheses on the right-hand side. This has the effect of adding 0 to the right-hand side.

Avoiding Mistakes

Do not factor out the leading coefficient from the constant term.

$= 1(x^2 + 8x + 16) - 16 + 13$

Use the associative property of addition to regroup terms and isolate the perfect square trinomial within the parentheses.

$= (x + 4)^2 - 3$

Factor and simplify.

b. $f(x) = (x + 4)^2 - 3$

The vertex is $(-4, -3)$.

The axis of symmetry is $x = -4$.

Because $a > 0$, the parabola opens upward.

The minimum value is -3 (Figure 19-15).

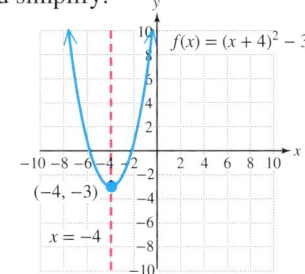

Figure 19-15

Skill Practice

1. Given: $f(x) = x^2 + 8x - 1$
   a. Write the function in the form $f(x) = a(x - h)^2 + k$.
   b. Identify the vertex, axis of symmetry, and minimum value of the function.

Answers

1. a. $f(x) = (x + 4)^2 - 17$
   b. Vertex: $(-4, -17)$; axis of symmetry: $x = -4$; minimum value: -17

> **Example 2**   **Analyzing a Quadratic Function**

Given $f(x) = -2x^2 + 12x - 16$

a. Write the function in the form $f(x) = a(x - h)^2 + k$.

b. Find the vertex, axis of symmetry, and maximum function value.

c. Find the x- and y-intercepts.

d. Sketch the graph of the function.

Solution:

a. $f(x) = -2x^2 + 12x - 16$ To find the vertex, write the function in the form $f(x) = a(x - h)^2 + k$.

$ = -2(x^2 - 6x) - 16$ If the leading coefficient is not 1, factor the coefficient from the variable terms.

$ = -2(x^2 - 6x + 9 - 9) - 16$ Add and subtract the quantity $[\frac{1}{2}(-6)]^2 = 9$ within the parentheses.

$ = -2(x^2 - 6x + 9) + (-2)(-9) - 16$ To remove the term -9 from the parentheses, we must first apply the distributive property. When -9 is removed from the parentheses, it carries with it a factor of -2.

$ = -2(x - 3)^2 + 18 - 16$ Factor and simplify.

$ = -2(x - 3)^2 + 2$

b. $f(x) = -2(x - 3)^2 + 2$

The vertex is $(3, 2)$. The axis of symmetry is $x = 3$. Because $a < 0$, the parabola opens downward and the maximum value is 2.

c. The y-intercept is given by $f(0) = -2(0)^2 + 12(0) - 16 = -16$.

The y-intercept is $(0, -16)$.

> **TIP:** In Example 2(c), we could have used the form of the equation found in part (a) to find the intercepts.

To find the x-intercept(s), find the real solutions to the equation $f(x) = 0$.

$f(x) = -2x^2 + 12x - 16$

$0 = -2x^2 + 12x - 16$ Substitute $f(x) = 0$.

$0 = -2(x^2 - 6x + 8)$ Factor.

$0 = -2(x - 4)(x - 2)$

$x = 4$ or $x = 2$

The x-intercepts are $(4, 0)$ and $(2, 0)$.

d. Using the information from parts (a)–(c), sketch the graph (Figure 19-16).

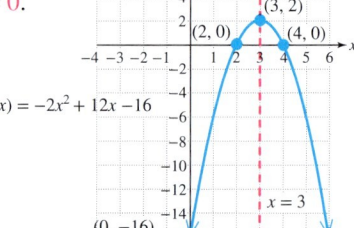

Figure 19-16

Skill Practice

2. Given $g(x) = -x^2 + 6x - 5$
   a. Write the function in the form $g(x) = a(x-h)^2 + k$.
   b. Identify the vertex, axis of symmetry, and maximum value of the function.
   c. Determine the x- and y-intercepts.
   d. Graph the function.

2. Vertex Formula

Completing the square and writing a quadratic function in the form $f(x) = a(x-h)^2 + k$ ($a \neq 0$) is one method to find the vertex of a parabola. Another method is to use the vertex formula. The **vertex formula** can be derived by completing the square on $f(x) = ax^2 + bx + c$ ($a \neq 0$).

$f(x) = ax^2 + bx + c \quad (a \neq 0)$

$= a\left(x^2 + \dfrac{b}{a}x \quad\quad\right) + c$ Factor a from the variable terms.

$= a\left(x^2 + \dfrac{b}{a}x + \dfrac{b^2}{4a^2} - \dfrac{b^2}{4a^2}\right) + c$ Add and subtract $\left[\tfrac{1}{2}(b/a)\right]^2 = b^2/(4a^2)$ within the parentheses.

$= a\left(x^2 + \dfrac{b}{a}x + \dfrac{b^2}{4a^2}\right) + (a)\left(-\dfrac{b^2}{4a^2}\right) + c$ Apply the distributive property and remove the term $-b^2/(4a^2)$ from the parentheses.

$= a\left(x + \dfrac{b}{2a}\right)^2 - \dfrac{b^2}{4a} + c$ Factor the trinomial and simplify.

$= a\left(x + \dfrac{b}{2a}\right)^2 + c - \dfrac{b^2}{4a}$ Apply the commutative property of addition to reverse the last two terms.

$= a\left(x + \dfrac{b}{2a}\right)^2 + \dfrac{4ac}{4a} - \dfrac{b^2}{4a}$ Obtain a common denominator.

$= a\left(x + \dfrac{b}{2a}\right)^2 + \dfrac{4ac - b^2}{4a}$

$= a\left[x - \left(-\dfrac{b}{2a}\right)\right]^2 + \dfrac{4ac - b^2}{4a}$

$\quad\quad\quad\quad\quad \downarrow \quad\quad\quad\quad\quad \downarrow$

$f(x) = a(x \quad - \quad h)^2 \; + \quad k$

The function is in the form $f(x) = a(x - h)^2 + k$, where

$$h = \dfrac{-b}{2a} \quad \text{and} \quad k = \dfrac{4ac - b^2}{4a}$$

Therefore, the vertex is $\left(\dfrac{-b}{2a}, \dfrac{4ac - b^2}{4a}\right)$.

Although the y-coordinate of the vertex is given by $\dfrac{4ac - b^2}{4a}$, it is usually easier to determine the x-coordinate of the vertex first and then find y by evaluating the function at $x = -b/(2a)$.

Answers

2. a. $g(x) = -(x-3)^2 + 4$
   b. Vertex: (3, 4); axis of symmetry: $x = 3$; maximum value: 4
   c. x-intercepts: (5, 0) and (1, 0); y-intercept: (0, −5)
   d.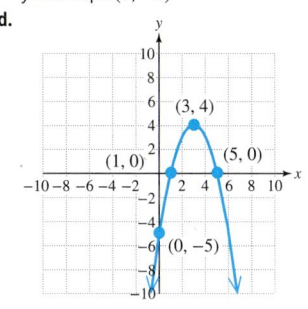

> **The Vertex Formula**
>
> For $f(x) = ax^2 + bx + c$ ($a \neq 0$), the vertex is given by
>
> $$\left(\frac{-b}{2a}, \frac{4ac - b^2}{4a}\right) \quad \text{or} \quad \left(\frac{-b}{2a}, f\left(\frac{-b}{2a}\right)\right)$$

3. Determining the Vertex and Intercepts of a Quadratic Function

Example 3   Determining the Vertex and Intercepts of a Quadratic Function

Given: $h(x) = x^2 - 2x + 5$

a. Use the vertex formula to find the vertex.

b. Find the x- and y-intercepts.

c. Sketch the function.

Solution:

a. $h(x) = x^2 - 2x + 5$

$a = 1 \quad b = -2 \quad c = 5 \qquad$ Identify a, b, and c.

The x-coordinate of the vertex is $\dfrac{-b}{2a} = \dfrac{-(-2)}{2(1)} = 1$.

The y-coordinate of the vertex is $h(1) = (1)^2 - 2(1) + 5 = 4$.

The vertex is $(1, 4)$.

b. The y-intercept is given by $h(0) = (0)^2 - 2(0) + 5 = 5$.

The y-intercept is $(0, 5)$.

To find the x-intercept(s), find the real solutions to the equation $h(x) = 0$.

$h(x) = x^2 - 2x + 5$

$0 = x^2 - 2x + 5 \qquad$ This quadratic equation is not factorable. Apply the quadratic formula: $a = 1$, $b = -2$, $c = 5$

$$x = \frac{-(-2) \pm \sqrt{(-2)^2 - 4(1)(5)}}{2(1)}$$

$$= \frac{2 \pm \sqrt{4 - 20}}{2(1)}$$

$$= \frac{2 \pm \sqrt{-16}}{2}$$

$$= \frac{2 \pm 4i}{2}$$

$$= 1 \pm 2i$$

The solutions to the equation $h(x) = 0$ are not real numbers. Therefore, there are no x-intercepts.

FOR REVIEW

Given a quadratic equation $ax^2 + bx + c = 0$, the quadratic formula gives the solutions for x.

$$x = \frac{-b \pm \sqrt{b^2 - 4ac}}{2a}$$

c.

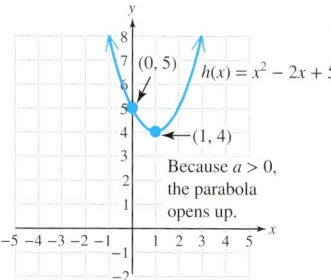

Because $a > 0$, the parabola opens up.

Figure 19-17

TIP: The location of the vertex and the direction that the parabola opens can be used to determine whether the function has any x-intercepts.

Given $h(x) = x^2 - 2x + 5$, the vertex (1, 4) is above the x-axis. Furthermore, because $a > 0$, the parabola opens upward. Therefore, it is not possible for the graph to cross the x-axis (Figure 19-17).

Skill Practice

3. Given: $f(x) = x^2 + 4x + 6$
   a. Use the vertex formula to find the vertex of the parabola.
   b. Determine the x- and y-intercepts.
   c. Sketch the graph.

4. Applications and Modeling of Quadratic Functions

Example 4 Applying a Quadratic Function

The crew from Extravaganza Entertainment launches fireworks at an angle of 60° from the horizontal. The height of one particular type of display can be approximated by

$$h(t) = -16t^2 + 128\sqrt{3}\,t$$

where $h(t)$ is measured in feet and t is measured in seconds.

a. How long will it take the fireworks to reach their maximum height? Round to the nearest second.

b. Find the maximum height. Round to the nearest foot.

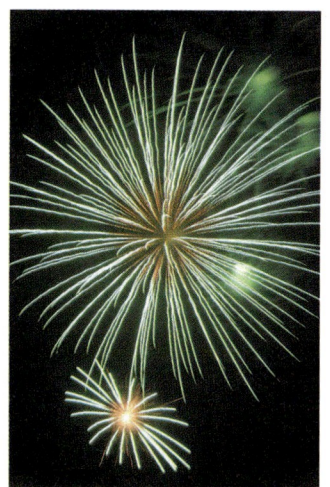

Lena Kofoed

Solution:

$h(t) = -16t^2 + 128\sqrt{3}\,t$ This parabola opens downward; therefore, the maximum height of the fireworks will occur at the vertex of the parabola.

$a = -16 \quad b = 128\sqrt{3} \quad c = 0$ Identify a, b, and c, and apply the vertex formula.

The x-coordinate of the vertex is

$$\frac{-b}{2a} = \frac{-128\sqrt{3}}{2(-16)} = \frac{-128\sqrt{3}}{-32} \approx 6.9$$

Answers

3. a. Vertex: (−2, 2)
   b. x-intercepts: none; y-intercept: (0, 6)
   c.

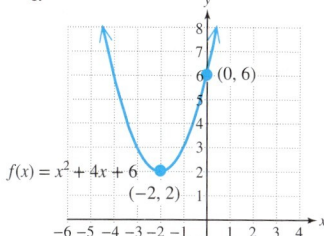

The y-coordinate of the vertex is approximately

$$h(6.9) = -16(6.9)^2 + 128\sqrt{3}(6.9) \approx 768$$

The vertex is (6.9, 768).

a. The fireworks will reach their maximum height in 6.9 sec.

b. The maximum height is 768 ft.

Skill Practice

4. An object is launched into the air with an initial velocity of 48 ft/sec from the top of a building 288 ft high. The height $h(t)$ of the object after t seconds is given by

$$h(t) = -16t^2 + 48t + 288$$

a. Find the time it takes for the object to reach its maximum height.
b. Find the maximum height.

In Example 5 we will learn how to write a quadratic model of a parabola given three points by using a system of equations and the standard form of a parabola: $y = ax^2 + bx + c$.

This process involves substituting the x- and y-coordinates of the three given points into the quadratic model. Then we solve the resulting system of three equations.

Example 5 Writing a Quadratic Model

Write an equation of a parabola that passes through the points (1, −1), (−1, −5), and (2, 4).

Solution:

Substitute (1, −1) into the equation
$y = ax^2 + bx + c$ ⟶ $(-1) = a(1)^2 + b(1) + c$
$-1 = a + b + c$
$a + b + c = -1$

Substitute (−1, −5) into the equation
$y = ax^2 + bx + c$ ⟶ $(-5) = a(-1)^2 + b(-1) + c$
$-5 = a - b + c$
$a - b + c = -5$

Substitute (2, 4) into the equation
$y = ax^2 + bx + c$ ⟶ $(4) = a(2)^2 + b(2) + c$
$4 = 4a + 2b + c$
$4a + 2b + c = 4$

Solve the system:
- A $a + b + c = -1$
- B $a - b + c = -5$
- C $4a + 2b + c = 4$

Answers

4. a. 1.5 sec   b. 324 ft

Notice that the c variables all have a coefficient of 1. Therefore, we choose to eliminate the c variable.

$$\boxed{A}\ a + b + c = -1 \longrightarrow a + b + c = -1$$
$$\boxed{B}\ a - b + c = -5 \xrightarrow{\text{Multiply by } -1.} -a + b - c = 5$$
$$2b = 4$$
$$b = 2\ \boxed{D}$$

$$\boxed{B}\ a - b + c = -5 \longrightarrow a - b + c = -5$$
$$\boxed{C}\ 4a + 2b + c = 4 \xrightarrow{\text{Multiply by } -1.} -4a - 2b - c = -4$$
$$-3a - 3b = -9\ \boxed{E}$$

Substitute b with 2 in equation $\boxed{E}$: $\quad -3a - 3(2) = -9$
$$-3a - 6 = -9$$
$$-3a = -3$$
$$a = 1$$

Substitute a and b in equation $\boxed{A}$ to solve for c: $\quad (1) + (2) + c = -1$
$$3 + c = -1$$
$$c = -4$$

Substitute $a = 1$, $b = 2$, and $c = -4$ in the standard form of the parabola for the final answer.

$$y = ax^2 + bx + c$$
$$y = (1)x^2 + (2)x + (-4) \longrightarrow y = x^2 + 2x - 4$$

A graph of $y = x^2 + 2x - 4$ is shown in Figure 19-18. Notice that the graph of the function passes through the points $(1, -1)$, $(-1, -5)$, and $(2, 4)$.

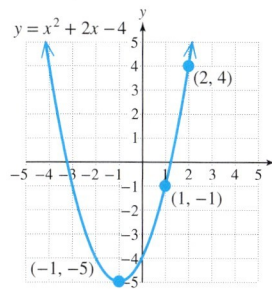

Figure 19-18

Skill Practice

5. Write an equation of the parabola that passes through the points $(1, 1)$, $(-2, 1)$, and $(3, -9)$.

Answer
5. $a = -1$, $b = -1$, $c = 3$;
$y = -x^2 - x + 3$

Section 19.5 Activity

A.1. Given a quadratic function defined by $f(x) = ax^2 + bx + c$, state two methods to find the vertex of the parabola.

A.2. Given $f(x) = ax^2 + bx + c$, explain how to use the vertex formula.

A.3. Consider the function defined by $f(x) = 2x^2 - 4x - 6$. One method to graph the function is to first write the function in vertex form $f(x) = a(x - h)^2 + k$. Follow these steps.
   a. Factor out the leading coefficient from the variable terms.
   $$f(x) = 2(\quad\quad) - 6$$
   b. What value would be added to the expression $x^2 - 2x$ to complete the square?
   c. Complete the square within the parentheses. However, rather than adding the constant from part (b) to both sides of the equation, *add* and *subtract* this value within the parentheses. This has the effect of adding 0 to the expression within the parentheses, which does not change the value of the expression.
   d. Next, remove the term -1 from inside the parentheses to combine it with the constant outside the parentheses. However, to remove -1 from inside the parentheses, first apply the distributive property. When -1 is removed from the parentheses, it carries with it a factor of 2.

e. Rewrite the function with the expression in parentheses factored.
f. The function should now be written in vertex form. Identify the vertex.
g. Does the graph of the function open upward or downward?
h. Identify the minimum or maximum value of the function.
i. Identify the axis of symmetry.
j. Find the x- and y-intercepts.
k. Graph the function.

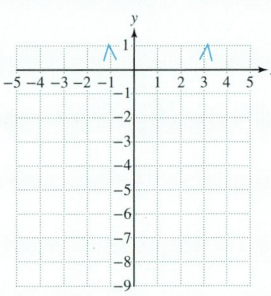

A.4. Refer to $f(x) = 2x^2 - 4x - 6$ from Exercise A.3. Use the vertex formula to find the vertex. Then compare the result to Exercise A.3(f).

A.5. Consider the function defined by $f(x) = -x^2 - 6x - 10$.
   a. Use the vertex formula to find the vertex of the parabola.
   b. Does the graph of the function open upward or downward?
   c. Identify the minimum or maximum value of the function.
   d. Identify the axis of symmetry.
   e. Find the x- and y-intercepts.
   f. Graph the function.

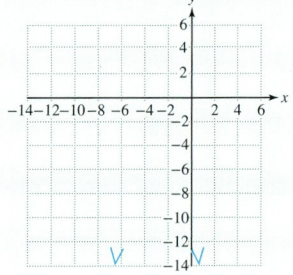

A.6. A rocket is fired upward from ground level with an initial velocity of 490 m/sec. The height of the rocket is given by $h(t) = -4.9t^2 + 490t$, where $h(t)$ is measured in meters and t is the time in seconds after launch.
   a. Function h is a quadratic function. Find the vertex.
   b. Interpret the meaning of the vertex in the context of this problem.

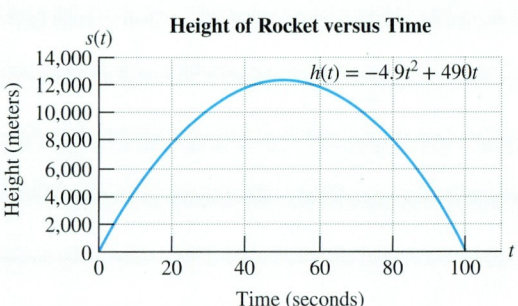

Height of Rocket versus Time

Section 19.5 Practice Exercises

Prerequisite Review

For Exercises R.1–R.2, simplify the expression.

R.1. $\dfrac{-(-6)}{2(4)}$ **R.2.** $\dfrac{-(-7)}{2(-14)}$

For Exercises R.3–R.4, find the x- and y-intercepts.

R.3. $f(x) = -4x - 8$ **R.4.** $f(x) = \dfrac{1}{2}x + 3$

For Exercises R.5–R.10, solve the equation.

R.5. $x^2 - 4x - 32 = 0$ **R.6.** $x^2 + 15x + 56 = 0$
R.7. $x^2 - 4 = 0$ **R.8.** $x^2 - 49 = 0$
R.9. $x^2 + 24x + 144 = 0$ **R.10.** $x^2 - 16x + 64 = 0$

For Exercises R.11–R.12, find the value of n so that the expression is a perfect square trinomial. Then factor the trinomial.

R.11. $x^2 - 14x + n$ **R.12.** $x^2 + 20x + n$

Vocabulary and Key Concepts

1. **a.** Given $f(x) = ax^2 + bx + c$ $(a \neq 0)$, the vertex formula gives the x-coordinate of the vertex as _____. The y-coordinate can be found by evaluating the function at $x = $ _____.
   b. Answer true or false. A parabola may have no x-intercepts.
   c. Answer true or false. A parabola may have one x-intercept.
   d. Answer true or false. A parabola may have two x-intercepts.
   e. Answer true or false. A parabola may have three x-intercepts.

2. Consider the quadratic function defined by $f(x) = x^2 - 6x + 10$.
   a. Identify the values of a and b that would be used to apply the vertex formula.
   b. Determine the x-coordinate of the vertex.
   c. Write the vertex as an ordered pair.

Concept 1: Writing a Quadratic Function in the Form $f(x) = a(x - h)^2 + k$

For Exercises 3–14, write the function in the form $f(x) = a(x - h)^2 + k$ by completing the square. Then identify the vertex. (See Examples 1–2.)

3. $g(x) = x^2 - 8x + 5$
4. $h(x) = x^2 + 4x + 5$
5. $n(x) = 2x^2 + 12x + 13$
6. $f(x) = 4x^2 + 16x + 19$
7. $p(x) = -3x^2 + 6x - 5$
8. $q(x) = -2x^2 + 12x - 11$
9. $k(x) = x^2 + 7x - 10$
10. $m(x) = x^2 - x - 8$
11. $F(x) = 5x^2 + 10x + 1$
12. $G(x) = 4x^2 + 4x - 7$
13. $P(x) = -2x^2 + x$
14. $Q(x) = -3x^2 + 12x$

Concept 2: Vertex Formula

For Exercises 15–26, find the vertex by using the vertex formula. (See Example 3.)

15. $Q(x) = x^2 - 4x + 7$
16. $T(x) = x^2 - 8x + 17$
17. $r(x) = -3x^2 - 6x - 5$
18. $s(x) = -2x^2 - 12x - 19$
19. $N(x) = x^2 + 8x + 1$
20. $M(x) = x^2 + 6x - 5$
21. $m(x) = \frac{1}{2}x^2 + x + \frac{5}{2}$
22. $n(x) = \frac{1}{2}x^2 + 2x + 3$
23. $k(x) = -x^2 + 2x + 2$
24. $h(x) = -x^2 + 4x - 3$
25. $A(x) = -\frac{1}{3}x^2 + x$
26. $B(x) = -\frac{2}{3}x^2 - 2x$

For Exercises 27–30, find the vertex two ways:
   a. by completing the square and writing in the form $f(x) = a(x - h)^2 + k$, and
   b. by using the vertex formula.

27. $p(x) = x^2 + 8x + 1$
28. $F(x) = x^2 + 4x - 2$
29. $f(x) = 2x^2 + 4x + 6$
30. $g(x) = 3x^2 + 12x + 9$

Concept 3: Determining the Vertex and Intercepts of a Quadratic Function

For Exercises 31–38
   a. Find the vertex.
   b. Find the y-intercept.
   c. Find the x-intercept(s), if they exist.
   d. Use this information to graph the function. (See Examples 2–3.)

31. $f(x) = x^2 + 2x - 3$
32. $f(x) = x^2 + 4x + 3$
33. $f(x) = 2x^2 - 2x + 4$
34. $f(x) = 2x^2 - 12x + 19$

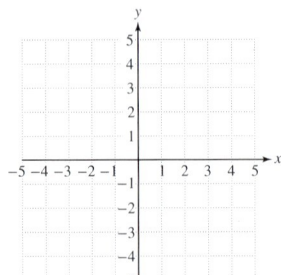
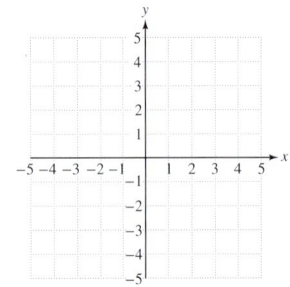
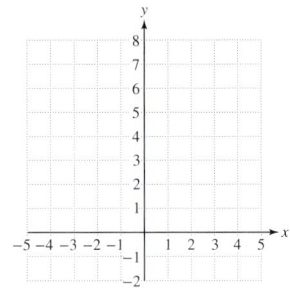
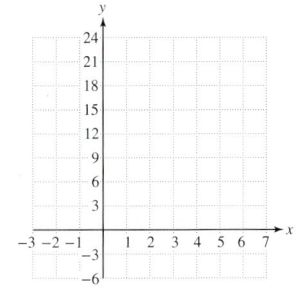

35. $f(x) = -x^2 + 3x - \dfrac{9}{4}$
36. $f(x) = -x^2 - \dfrac{3}{2}x - \dfrac{9}{16}$
37. $f(x) = -x^2 - 2x + 3$
38. $f(x) = -x^2 - 4x$

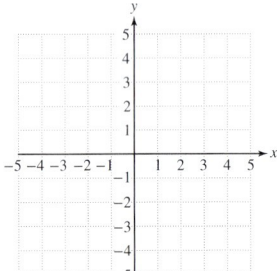
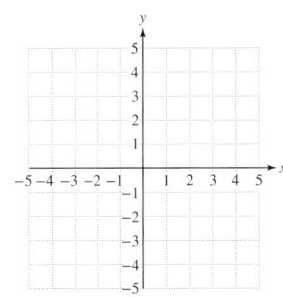
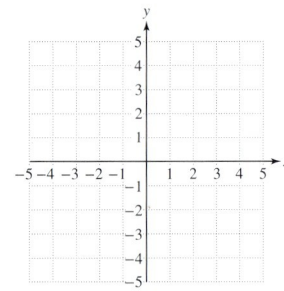
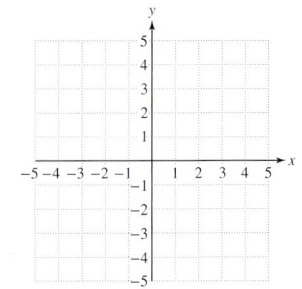

Concept 4: Applications and Modeling of Quadratic Functions

39. A set of fireworks mortar shells is launched from the staging platform at 100 ft/sec from an initial height of 8 ft above the ground. The height of the fireworks, $h(t)$, can be modeled by

    $$h(t) = -16t^2 + 100t + 8,$$ where t is the time in seconds after launch. **(See Example 4.)**

    a. What is the maximum height that the shells can reach before exploding?

    b. For how many seconds after launch would the fuses need to be set so that the mortar shells would in fact explode at the maximum height?

40. A baseball player throws a ball, and the height of the ball (in feet) can be approximated by

    $$y(x) = -0.011x^2 + 0.577x + 5$$

    where x is the horizontal position of the ball measured in feet from the origin.

    a. For what value of x will the ball reach its highest point? Round to the nearest foot.

    b. What is the maximum height of the ball? Round to the nearest tenth of a foot.

41. Gas mileage depends in part on the speed of the car. The gas mileage of a subcompact car is given by $m(x) = -0.04x^2 + 3.6x - 49$, where $20 \le x \le 70$ represents the speed in miles per hour and $m(x)$ is given in miles per gallon.

    a. At what speed will the car get the maximum gas mileage?

    b. What is the maximum gas mileage?

42. Gas mileage depends in part on the speed of the car. The gas mileage of a luxury car is given by $L(x) = -0.015x^2 + 1.44x - 21$, where $25 \le x \le 70$ represents the speed in miles per hour and $L(x)$ is given in miles per gallon.

    a. At what speed will the car get the maximum gas mileage?

    b. What is the maximum gas mileage? Round to the nearest whole unit.

43. The *Clostridium tetani* bacterium is cultured to produce tetanus toxin used in an inactive form for the tetanus vaccine. The amount of toxin produced per batch increases with time and then becomes unstable. The amount of toxin $b(t)$ (in grams) as a function of time t (in hours) can be approximated by:

$$b(t) = -\frac{1}{1152}t^2 + \frac{1}{12}t$$

a. How many hours will it take to produce the maximum yield?

b. What is the maximum yield?

44. The bacterium *Pseudomonas aeruginosa* is cultured with an initial population of 10^4 active organisms. The population of active bacteria increases up to a point, and then due to a limited food supply and an increase of waste products, the population of living organisms decreases. Over the first 48 hr, the population $P(t)$ can be approximated by:

$$P(t) = -1718.75t^2 + 82{,}500t + 10{,}000$$

where $0 \leq t \leq 48$

a. Find the time required for the population to reach its maximum value.

b. What is the maximum population?

For Exercises 45–50, use the standard form of a parabola given by $y = ax^2 + bx + c$ to write an equation of a parabola that passes through the given points. **(See Example 5.)**

45. $(0, 4), (1, 0)$ and $(-1, -10)$

46. $(0, 3), (3, 0), (-1, 8)$

47. $(2, 1), (-2, 5),$ and $(1, -4)$

48. $(1, 2), (-1, -6),$ and $(2, -3)$

49. $(2, -4), (1, 1),$ and $(-1, -7)$

50. $(1, 4), (-1, 6),$ and $(2, 18)$

Technology Connections

For Exercises 51–56, graph the functions in Exercises 31–36 on a graphing calculator. Use the *Max* or *Min* feature to approximate the vertex.

51. $Y_1 = x^2 + 2x - 3$ (Exercise 31)

52. $Y_1 = x^2 + 4x + 3$ (Exercise 32)

53. $Y_1 = 2x^2 - 2x + 4$ (Exercise 33)

54. $Y_1 = 2x^2 - 12x + 19$ (Exercise 34)

55. $Y_1 = -x^2 + 3x - \frac{9}{4}$ (Exercise 35)

56. $Y_1 = -x^2 - \frac{3}{2}x - \frac{9}{16}$ (Exercise 36)

Expanding Your Skills

57. A farmer wants to fence a rectangular corral adjacent to the side of a barn; however, she has only 200 ft of fencing and wants to enclose the largest possible area. See the figure.

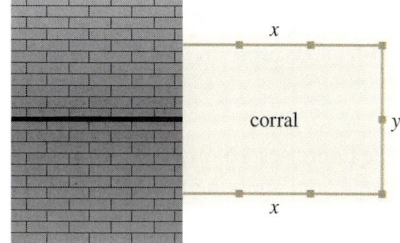

a. If x represents the length of the corral and y represents the width, explain why the dimensions of the corral are subject to the constraint $2x + y = 200$.

b. The area of the corral is given by $A = xy$. Use the constraint equation from part (a) to express A as a function of x, where $0 < x < 100$.

c. Use the function from part (b) to find the dimensions of the corral that will yield the maximum area. [*Hint:* Find the vertex of the function from part (b).]

58. A veterinarian wants to construct two equal-sized pens of maximum area out of 240 ft of fencing. See the figure.

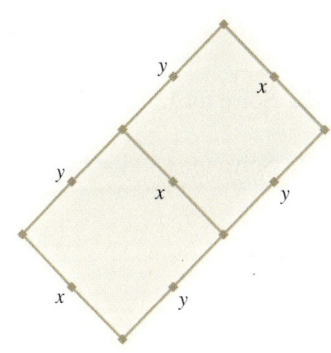

a. If x represents the length of each pen and y represents the width of each pen, explain why the dimensions of the pens are subject to the constraint $3x + 4y = 240$.

b. The area of each individual pen is given by $A = xy$. Use the constraint equation from part (a) to express A as a function of x, where $0 < x < 80$.

c. Use the function from part (b) to find the dimensions of an individual pen that will yield the maximum area. [*Hint:* Find the vertex of the function from part (b).]

Chapter 19 Review Exercises

Section 19.1

For Exercises 1–8, solve the equations by using the square root property.

1. $x^2 = 5$
2. $2y^2 = -8$
3. $a^2 = 81$
4. $3b^2 = -19$
5. $(x - 2)^2 = 72$
6. $(2x - 5)^2 = -9$
7. $(3y - 1)^2 = 3$
8. $3(m - 4)^2 = 15$

9. The length of each side of an equilateral triangle is 10 in. Find the exact height of the triangle. Then round to the nearest tenth of an inch.

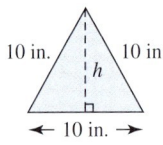

10. Use the square root property to find the length of the sides of a square whose area is 81 in^2.

11. Use the square root property to find the exact length of the sides of a square whose area is 150 in^2. Then round to the nearest tenth of an inch.

For Exercises 12–15, find the value of n so that the expression is a perfect square trinomial. Then factor the trinomial.

12. $x^2 + 16x + n$
13. $x^2 - 9x + n$
14. $y^2 + \frac{1}{2}y + n$
15. $z^2 - \frac{2}{5}z + n$

For Exercises 16–21, solve the equation by completing the square and applying the square root property.

16. $w^2 + 4w + 13 = 0$
17. $4y^2 - 8y - 20 = 0$
18. $2x^2 = 12x + 6$
19. $-t^2 + 8t - 25 = 0$
20. $3x^2 + 2x = 1$
21. $b^2 + \frac{7}{2}b = 2$

22. Solve for r. $V = \pi r^2 h$ $(r > 0)$

23. Solve for s. $A = 6s^2$ $(s > 0)$

Section 19.2

24. Describe the type and number of solutions to a quadratic equation whose discriminant is less than zero.

For Exercises 25–30, determine the type (rational, irrational, or imaginary) and number of solutions for the equations by using the discriminant.

25. $x^2 - 5x = -6$
26. $2y^2 = -3y$
27. $z^2 + 23 = 17z$
28. $a^2 + a + 1 = 0$
29. $10b + 1 = -25b^2$
30. $3x^2 + 15 = 0$

For Exercises 31–38, solve the equations by using the quadratic formula.

31. $y^2 - 4y + 1 = 0$
32. $m^2 - 5m + 25 = 0$
33. $6a(a - 1) = 10 + a$
34. $3x(x - 3) = x - 8$
35. $b^2 - \frac{4}{25} = \frac{3}{5}b$
36. $k^2 + 0.4k = 0.05$
37. $-32 + 4x - x^2 = 0$
38. $8y - y^2 = 10$

For Exercises 39–42, solve using any method.

39. $3x^2 - 4x = 6$
40. $\frac{w}{8} - \frac{2}{w} = \frac{3}{4}$
41. $y^2 + 14y = -46$
42. $(a + 1)^2 = 11$

43. The landing distance that a certain plane will travel on a runway is determined by the initial landing speed at the instant the plane touches down. The function D relates landing distance in feet to initial landing speed s:

$$D(s) = \frac{1}{10}s^2 - 3s + 22 \text{ for } s \geq 50$$

where s is in feet per second.
a. Find the landing distance for a plane traveling 150 ft/sec at touchdown.
b. If the landing speed is too fast, the pilot may run out of runway. If the speed is too slow, the plane may stall. Find the maximum initial landing speed of a plane for a runway that is 1000 ft long. Round to one decimal place.

44. The recent population of Kenya (in thousands) can be approximated by $P(t) = 4.62t^2 + 564.6t + 13{,}128$ where t is the number of years since 1974.

  a. If this trend continues, predict the number of people in Kenya for the year 2025.

  b. In what year after 1974 will the population of Kenya reach 50 million? (*Hint:* 50 million equals 50,000 thousand.)

45. A custom-built kitchen island is in the shape of a rectangle. The length is 1 ft more than twice the width. If the area is 22.32 ft², determine the dimensions of the island. Round the length and width to the nearest tenth of a foot.

Shutterstock

46. Lincoln, Nebraska, Kansas City, Missouri, and Omaha, Nebraska, form the vertices of a right triangle. The distance between Lincoln and Kansas City is 10 mi more than 3 times the distance between Lincoln and Omaha. If the distance from Omaha and Kansas City is 167 mi, find the distance between Lincoln and Omaha. Round to the nearest mile.

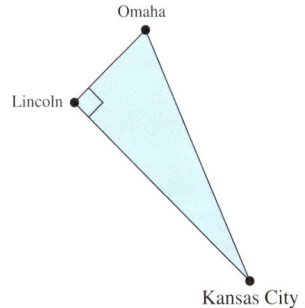

Section 19.3

For Exercises 47–56, solve the equations.

47. $x - 4\sqrt{x} - 21 = 0$

48. $n - 6\sqrt{n} + 8 = 0$

49. $y^4 - 11y^2 + 18 = 0$

50. $2m^4 - m^2 - 3 = 0$

51. $t^{2/5} + t^{1/5} - 6 = 0$

52. $p^{2/5} - 3p^{1/5} + 2 = 0$

53. $\dfrac{2t}{t+1} + \dfrac{-3}{t-2} = 1$

54. $\dfrac{1}{m-2} - \dfrac{m}{m+3} = 2$

55. $(x^2 + 5)^2 + 2(x^2 + 5) - 8 = 0$

56. $(x^2 - 3)^2 - 5(x^2 - 3) + 4 = 0$

Section 19.4

For Exercises 57–64, graph the function and write the domain and range in interval notation.

57. $g(x) = x^2 - 5$

58. $f(x) = x^2 + 3$

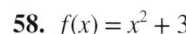
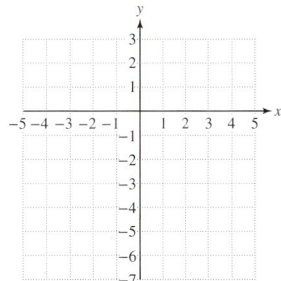
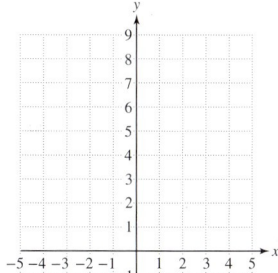

59. $h(x) = (x - 5)^2$

60. $k(x) = (x + 3)^2$

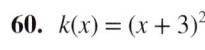
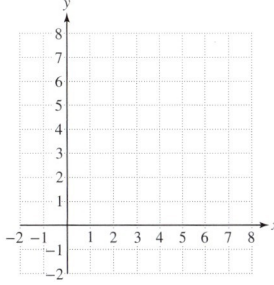
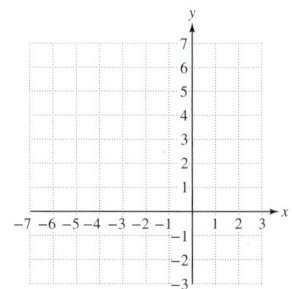

61. $m(x) = -2x^2$

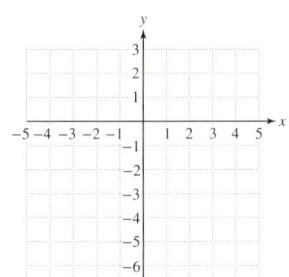

62. $n(x) = -4x^2$

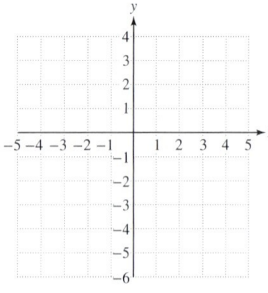

63. $p(x) = -2(x-5)^2 - 5$

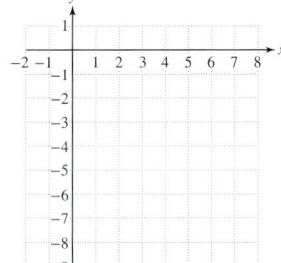

64. $q(x) = -4(x+3)^2 + 3$

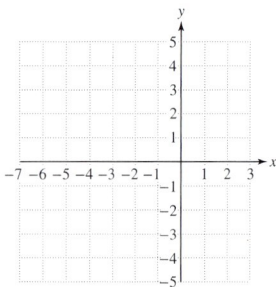

For Exercises 65–66, write the coordinates of the vertex of the parabola and determine if the vertex is a maximum point or a minimum point. Then write the maximum or the minimum value.

65. $t(x) = \frac{1}{3}(x-4)^2 + \frac{5}{3}$

66. $s(x) = -\frac{5}{7}(x-1)^2 - \frac{1}{7}$

For Exercises 67–68, write the equation of the axis of symmetry of the parabola.

67. $a(x) = -\frac{3}{2}\left(x + \frac{2}{11}\right)^2 - \frac{4}{13}$

68. $w(x) = -\frac{4}{3}\left(x - \frac{3}{16}\right)^2 + \frac{2}{9}$

Section 19.5

For Exercises 69–72, write the function in the form $f(x) = a(x-h)^2 + k$ by completing the square. Then write the coordinates of the vertex.

69. $z(x) = x^2 - 6x + 7$

70. $b(x) = x^2 - 4x - 44$

71. $p(x) = -5x^2 - 10x - 13$

72. $q(x) = -3x^2 - 24x - 54$

For Exercises 73–76, find the coordinates of the vertex of each parabola by using the vertex formula.

73. $f(x) = -2x^2 + 4x - 17$

74. $g(x) = -4x^2 - 8x + 3$

75. $m(x) = 3x^2 - 3x + 11$

76. $n(x) = 3x^2 + 2x - 7$

77. For the quadratic equation $y = \frac{3}{4}x^2 - 3x$,

a. Write the coordinates of the vertex.

b. Find the x- and y-intercepts.

c. Use this information to sketch a graph of the parabola.

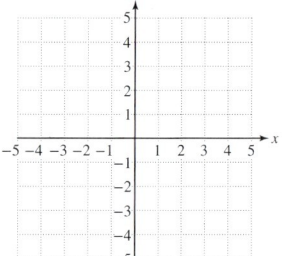

78. For the quadratic equation $y = -(x+2)^2 + 4$,

a. Write the coordinates of the vertex.

b. Find the x- and y-intercepts.

c. Use this information to sketch a graph of the parabola.

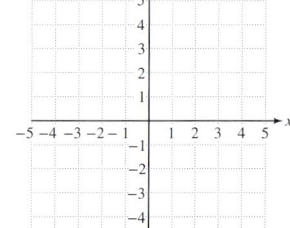

79. The height $h(t)$ (in feet) of a projectile fired vertically into the air from the ground is given by the equation $h(t) = -16t^2 + 96t$, where t represents the number of seconds after launch.

a. How long will it take the projectile to reach its maximum height?

b. What is the maximum height?

80. The weekly profit, $P(x)$ (in dollars), for a catering service is given by $P(x) = -0.053x^2 + 15.9x + 7.5$. In this context, x is the number of meals prepared.

a. Find the number of meals that should be prepared to obtain the maximum profit.

b. What is the maximum profit?

81. Write an equation of a parabola that passes through the points $(-3, -4)$, $(-2, -5)$, and $(1, 4)$.

82. Write an equation of a parabola that passes through the points $(4, 18)$, $(-2, 12)$, and $(-1, 8)$.

Exponential and Logarithmic Functions and Applications

20

CHAPTER OUTLINE

20.1 Inverse Functions   1330

20.2 Exponential Functions   1341

20.3 Logarithmic Functions   1352

   Problem Recognition Exercises: Identifying Graphs of Functions   1366

20.4 Properties of Logarithms   1367

20.5 The Irrational Number e and Change of Base   1376

   Problem Recognition Exercises: Logarithmic and Exponential Forms   1390

20.6 Logarithmic and Exponential Equations and Applications   1391

   Chapter 20   Review Exercises   1405

Mathematics and Population Growth

Over short periods of time, human population growth is proportional to the number of humans in the population at a given time. That is, the more people in a population, the faster the population will grow. For example, a country with 10 million people will typically have a greater yearly increase in population than a country with only 1 million people. However, other variables such as food supply, mortality rates, birth control, and immigration also affect population growth.

The graph shows the projected population growth for the United States, Mexico, and Japan based on growth rates of 0.73%, 1.27%, and −0.2%, respectively. Since Japan has a negative growth rate, its population is *decreasing*.

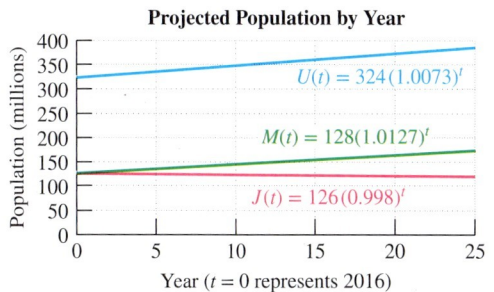

The functions U, M, and J that describe the population values versus time are called **exponential functions**. Exponential functions are characterized by having a constant base and variable exponent. In this chapter, we study exponential functions and their inverses, **logarithmic functions**. These functions have far-reaching applications, including the study of population growth, the growth of investments, and radioactive decay.

Section 20.1 Inverse Functions

Concepts

1. Introduction to Inverse Functions
2. Definition of a One-to-One Function
3. Definition of the Inverse of a Function
4. Finding an Equation of the Inverse of a Function

Avoiding Mistakes

f^{-1} denotes the inverse of a function. The -1 does not represent an exponent.

1. Introduction to Inverse Functions

A function is a set of ordered pairs (x, y) such that for every element x in the domain, there corresponds exactly one element y in the range. For example, the function f relates the weight of a package of deli meat x to its cost y.

$$f = \{(1, 4), (2.5, 10), (4, 16)\}$$

That is, 1 lb of meat sells for $4, 2.5 lb sells for $10, and 4 lb sells for $16. Now suppose we create a new function in which the values of x and y are interchanged. The new function, called the **inverse of f**, denoted f^{-1}, relates the price of meat x to its weight y.

$$f^{-1} = \{(4, 1), (10, 2.5), (16, 4)\}$$

Notice that interchanging the x and y values has the following outcome. The domain of f is the same as the range of f^{-1}, and the range of f is the domain of f^{-1}.

2. Definition of a One-to-One Function

A necessary condition for a function f to have an inverse function is that no two ordered pairs in f have different x-coordinates and the same y-coordinate. A function that satisfies this condition is called a **one-to-one function**. The function relating the weight of a package of meat to its price is a one-to-one function. However, consider the function g defined by

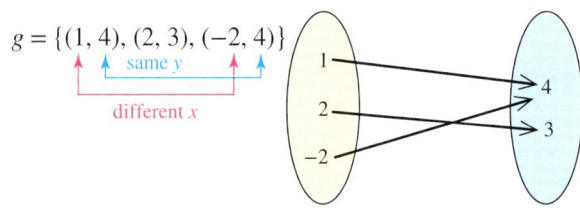

Not a one-to-one function because two different x values map to the same y value.

This function is not one-to-one because the range element 4 has two different x-coordinates, 1 and -2. Interchanging the x and y values produces a relation that violates the definition of a function.

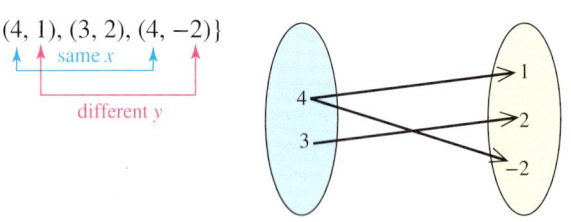

This relation is not a function because for $x = 4$, there are two different y values.

We have already learned that the vertical line test can be used to visually determine if a graph represents a function. Similarly, we use a **horizontal line test** to determine whether a function is one-to-one.

Section 20.1 Inverse Functions   1331

Using the Horizontal Line Test
Consider a function defined by a set of points (x, y) in a rectangular coordinate system. Then y is a *one-to-one* function of x if no horizontal line intersects the graph in more than one point.

FOR REVIEW

Recall that the vertical line test is used to determine if y is a function of x. If a vertical line crosses a graph in more than one point, the graph does not define y as a function of x.

To understand the horizontal line test, consider the functions f and g.

$f = \{(1, 4), (2.5, 10), (4, 16)\}$ $g = \{(1, 4), (2, 3), (-2, 4)\}$

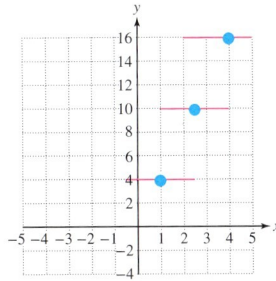
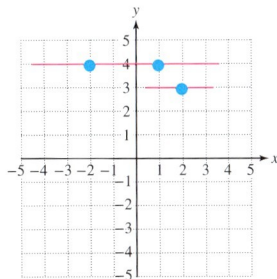

This function is one-to-one.
No horizontal line intersects more than once.

This function is *not* one-to-one.
A horizontal line intersects more than once.

Example 1 Identifying a One-to-One Function

For each function, determine if the function is one-to-one.

a.
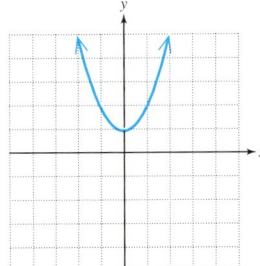

b.
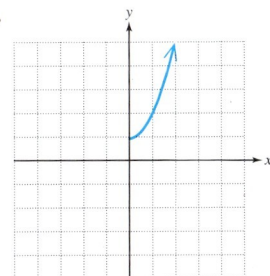

Solution:

a. Function is not one-to-one.
A horizontal line intersects in more than one point.

b. Function is one-to-one.
No horizontal line intersects more than once.

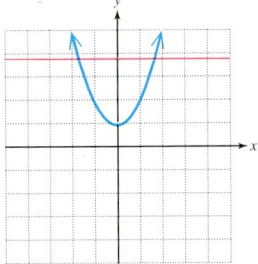
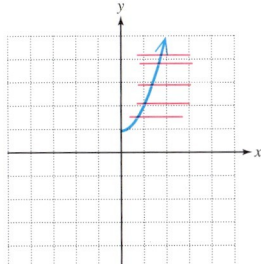

Skill Practice For each function determine if the function is one-to-one.

1.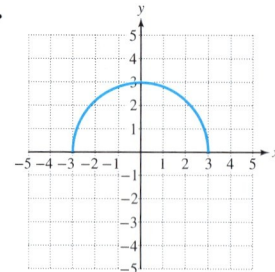

2.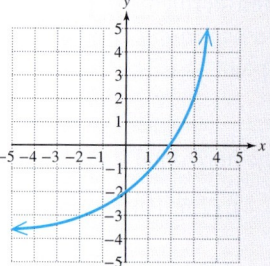

3. Definition of the Inverse of a Function

All one-to-one functions have an inverse function.

> **Definition of an Inverse Function**
>
> If f is a one-to-one function represented by ordered pairs of the form (x, y), then the inverse function, denoted by f^{-1}, is the set of ordered pairs given by (y, x).

Because the values of x and y are interchanged between a function f and its inverse f^{-1}, we have the following important relationship (Figure 20-1).

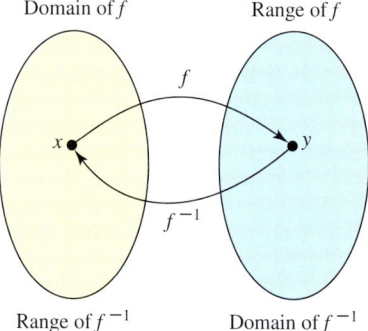

Figure 20-1

From Figure 20-1, we see that the operations performed by f are reversed by f^{-1}. This observation leads to the following important property.

> **Inverse Function Property**
>
> If f is a one-to-one function, then g is the inverse of f if and only if
>
> $$(f \circ g)(x) = x \quad \text{for all } x \text{ in the domain of } g$$
>
> and
>
> $$(g \circ f)(x) = x \quad \text{for all } x \text{ in the domain of } f.$$

Answers
1. Not one-to-one
2. One-to-one

Example 2 Composing a Function With Its Inverse

Show that the functions are inverses.

$$f(x) = 5x + 4 \qquad g(x) = \frac{x-4}{5}$$

Solution:

To show that the functions f and g are inverses, confirm that $(f \circ g)(x) = x$ and $(g \circ f)(x) = x$.

$$(f \circ g)(x) = f(g(x)) \qquad\qquad (g \circ f)(x) = g(f(x))$$
$$= f\left(\frac{x-4}{5}\right) \qquad\qquad\quad = g(5x+4)$$
$$= 5\left(\frac{x-4}{5}\right) + 4 \qquad\qquad = \frac{(5x+4)-4}{5}$$
$$= x - 4 + 4 \qquad\qquad\qquad\quad = \frac{5x}{5}$$
$$= x \checkmark \qquad\qquad\qquad\qquad\quad = x \checkmark$$

FOR REVIEW

Recall the composition of functions

$$(f \circ g)(x) = f(g(x))$$

means that function f is applied to the result of function g. That is, x is input into function g. Then the result, $g(x)$, is input into function f. Similarly,

$$(g \circ f)(x) = g(f(x))$$

means that function g is applied to the result of function f.

Skill Practice Show that the functions are inverses.

3. $f(x) = 3x - 2 \quad$ and $\quad g(x) = \dfrac{x+2}{3}$

4. Finding an Equation of the Inverse of a Function

For a one-to-one function defined by $y = f(x)$, the inverse is a function $y = f^{-1}(x)$ that performs the inverse operations in the reverse order. For example, the function defined by $f(x) = 2x + 1$ multiplies x by 2 and then adds 1. Therefore, the inverse function must *subtract* 1 from x and *divide* by 2. We have

$$f^{-1}(x) = \frac{x-1}{2} \qquad \text{The expression } f^{-1}(x) \text{ is read as ``}f\text{ inverse of }x\text{.''}$$

To facilitate the process of finding an equation of the inverse of a one-to-one function, we offer the following steps.

Finding an Equation of an Inverse of a Function

For a one-to-one function defined by $y = f(x)$, the equation of the inverse can be found as follows:

Step 1 Replace $f(x)$ with y.
Step 2 Interchange x and y.
Step 3 Solve for y.
Step 4 Replace y with $f^{-1}(x)$.

Answer

3. $(f \circ g)(x) = f(g(x))$
$$= 3\left(\frac{x+2}{3}\right) - 2 = x$$
$(g \circ f)(x) = g(f(x))$
$$= \frac{3x - 2 + 2}{3} = x$$

Example 3 Finding an Equation of the Inverse of a Function

Find the inverse. $f(x) = 2x + 1$

Solution:

Foremost, we know the graph of f is a nonvertical line. Therefore, $f(x) = 2x + 1$ defines a one-to-one function. To find the inverse we have

$y = 2x + 1$	**Step 1:** Replace $f(x)$ with y.
$x = 2y + 1$	**Step 2:** Interchange x and y.
$x - 1 = 2y$	**Step 3:** Solve for y. Subtract 1 from both sides.
$\dfrac{x-1}{2} = y$	Divide both sides by 2.
$f^{-1}(x) = \dfrac{x-1}{2}$	**Step 4:** Replace y with $f^{-1}(x)$.

Avoiding Mistakes

It is important to check that a function is one-to-one before finding the inverse function.

Skill Practice Find the inverse.

4. $f(x) = 4x + 6$

In Example 3, we can verify that $f(x) = 2x + 1$ and $f^{-1}(x) = \dfrac{x-1}{2}$ are inverses by using function composition.

$$(f \circ f^{-1})(x) = f(f^{-1}(x)) = 2\left(\dfrac{x-1}{2}\right) + 1 = x \checkmark \quad \text{and}$$

$$(f^{-1} \circ f)(x) = f^{-1}(f(x)) = \dfrac{(2x+1) - 1}{2} = x \checkmark$$

The key step in determining the equation of the inverse of a function is to interchange x and y. By so doing, a point (a, b) on f corresponds to a point (b, a) on f^{-1}. For this reason, the graphs of f and f^{-1} are symmetric with respect to the line $y = x$ (Figure 20-2). Notice that the point $(-3, -5)$ of the function f corresponds to the point $(-5, -3)$ of f^{-1}. Likewise, $(1, 3)$ of f corresponds to $(3, 1)$ of f^{-1}.

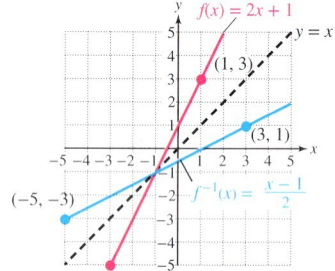

Figure 20-2

Example 4 Finding an Equation of the Inverse of a Function

Find the inverse of the one-to-one function. $g(x) = \sqrt[3]{5x} - 4$

Solution:

$y = \sqrt[3]{5x} - 4$	**Step 1:** Replace $g(x)$ with y.
$x = \sqrt[3]{5y} - 4$	**Step 2:** Interchange x and y.
$x + 4 = \sqrt[3]{5y}$	**Step 3:** Solve for y. Add 4 to both sides.
$(x+4)^3 = (\sqrt[3]{5y})^3$	To eliminate the cube root, cube both sides.
$(x+4)^3 = 5y$	Simplify the right side.
$\dfrac{(x+4)^3}{5} = y$	Divide both sides with 5.
$g^{-1}(x) = \dfrac{(x+4)^3}{5}$	**Step 4:** Replace y with $g^{-1}(x)$.

Answer

4. $f^{-1}(x) = \dfrac{x-6}{4}$

Skill Practice Find the inverse.

5. $h(x) = \sqrt[3]{2x-1}$

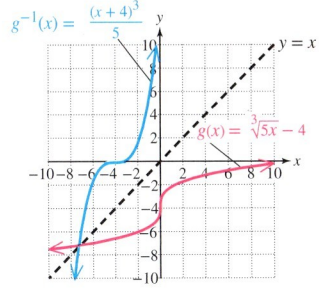

Figure 20-3

The graphs of g and g^{-1} from Example 4 are shown in Figure 20-3. Once again we see that the graphs of a function and its inverse are symmetric with respect to the line $y = x$.

For a function that is not one-to-one, sometimes we can restrict its domain to create a new function that is one-to-one. This is demonstrated in Example 5.

Example 5 Finding the Equation of an Inverse of a Function With a Restricted Domain

Given the function defined by $m(x) = x^2 + 4$ for $x \geq 0$, find an equation defining m^{-1}.

Solution:

We know that $y = x^2 + 4$ is a parabola with vertex at $(0, 4)$ (Figure 20-4). The graph represents a function that is not one-to-one. However, with the restriction on the domain $x \geq 0$, the graph of $m(x) = x^2 + 4$, $x \geq 0$, consists of only the "right" branch of the parabola (Figure 20-5). This *is* a one-to-one function.

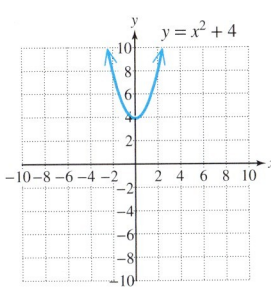

Figure 20-4   Figure 20-5

To find the inverse, we have

$y = x^2 + 4$ $\quad x \geq 0$	**Step 1:**	Replace $m(x)$ with y.
$x = y^2 + 4$ $\quad y \geq 0$	**Step 2:**	Interchange x and y. Notice that the restriction $x \geq 0$ becomes $y \geq 0$.
$x - 4 = y^2$ $\quad y \geq 0$	**Step 3:**	Solve for y. Subtract 4 from both sides.
$\sqrt{x-4} = y$ $\quad y \geq 0$		Apply the square root property. Notice that we obtain the *positive* square root of $x - 4$ because of the restriction $y \geq 0$.
$m^{-1}(x) = \sqrt{x-4}$	**Step 4:**	Replace y with $m^{-1}(x)$. Notice that the domain of m^{-1} has the same values as the range of m.

Figure 20-6 shows the graphs of m and m^{-1}.

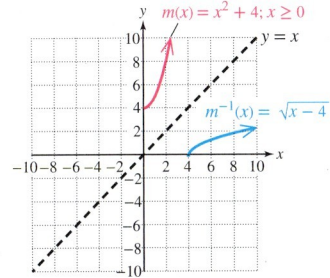

Figure 20-6

Skill Practice Find the inverse.

6. $g(x) = x^2 - 2 \quad x \geq 0$

Answers

5. $h^{-1}(x) = \dfrac{x^3 + 1}{2}$
6. $g^{-1}(x) = \sqrt{x+2}$

Section 20.1 Activity

A.1. If a function defined by $y = f(x)$ has the property that no two elements in the domain of f correspond to the same element in the range of f, then f is said to be _____.

A.2. To determine whether a function is one-to-one, use the _____ line test.

For Exercises A.3–A.4, determine whether the function is one-to-one. If the function is not one-to-one, explain why.

A.3. $\{(2, -4), (-3, 1), (0, -2), (-4, 3)\}$

A.4. $\{(4, 1), (5, -3), (-1, -3), (-2, 5)\}$

For Exercises A.5–A.6, consider the given function.
  a. Write each ordered pair with the x- and y-coordinates interchanged.
  b. Is the new relation a function? If not, explain why.

A.5. $\{(2, -4), (-3, 1), (0, -2), (-4, 3)\}$ (Refer to Exercise A.3.)

A.6. $\{(4, 1), (5, -3), (-1, -3), (-2, 5)\}$ (Refer to Exercise A.4.)

A.7. Given $f(x) = \dfrac{x - 4}{2}$,
  a. Describe the graph of the function.
  b. Is the function one-to-one? Does the function have an inverse function?

A.8. Find the inverse of $f(x) = \dfrac{x - 4}{2}$ by following these steps.
  a. Replace $f(x)$ with y.
  b. Interchange x and y in the equation. This is the key step to finding the inverse.
  c. Solve for y in the new equation.
  d. Replace y with $f^{-1}(x)$.
  e. To check your answer to part (d), find $(f \circ f^{-1})(x)$ and $(f^{-1} \circ f)(x)$.

A.9.
  a. Refer to Exercise A.8. Graph $f(x) = \dfrac{x - 4}{2}$ and $f^{-1}(x) = 2x + 4$ along with the line $y = x$ on the same coordinate system.
  b. Describe the symmetry between a function and its inverse.

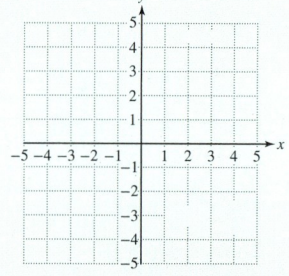

A.10.
  a. Graph $f(x) = x^2 + 2$ for $x \geq 0$ (right branch of the parabola only).
  b. Is the function graphed in part (a) one-to-one?
  c. Find the inverse of f.
  d. Graph f^{-1} on the same coordinate system as the graph in part (a). Also graph the line $y = x$. Are the graphs of the function and its inverse symmetric with respect to $y = x$?

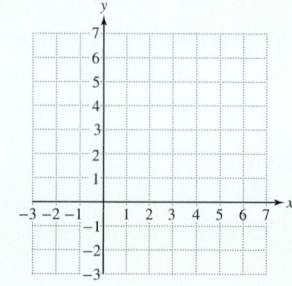
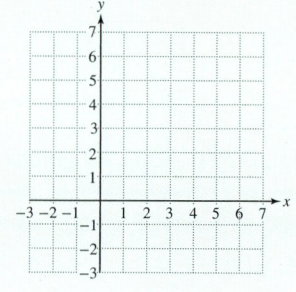

Practice Exercises

Section 20.1

Prerequisite Review

For Exercises R.1–R.2, use the vertical line test to determine if the graph defines y as a function of x.

R.1.
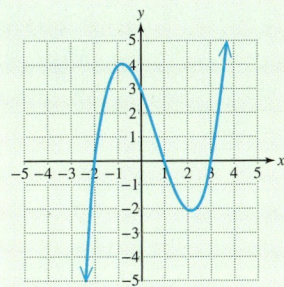

R.2.
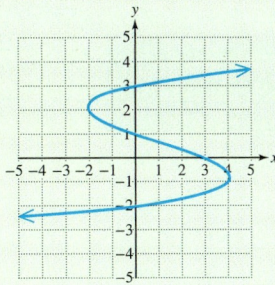

For Exercises R.3–R.6, solve for x.

R.3. $y = -3x - 4$

R.4. $y = 5x - 15$

R.5. $y = \sqrt[3]{x - 1}$

R.6. $y = (x - 4)^3$

R.7. Given $f(x) = 2x + 4$, find the function values.

    a. $f(1)$ **b.** $f(-2)$ **c.** $f(t)$ **d.** $f\left(\dfrac{x-4}{2}\right)$

R.8. Given $g(x) = \sqrt[3]{x + 5}$, find the function values.

    a. $g(3)$ **b.** $g(-6)$ **c.** $g(a)$ **d.** $g(x^3 - 5)$

For Exercises R.9–R.10, refer to the functions defined by $f(x) = 4x - 3$ and $g(x) = x^2 + 5$.

R.9. Find $(f \circ g)(x)$.

R.10. Find $(g \circ f)(x)$.

Vocabulary and Key Concepts

1. a. Given the function $f = \{(1, 2), (2, 3), (3, 4)\}$, write the set of ordered pairs representing f^{-1}.

   b. A necessary condition for a function f to be a _____-_____-_____ function is that no two ordered pairs in f have different x-coordinates and the same _____-coordinate.

   c. The function $f = \{(1, 5), (-2, 3), (-4, 2), (2, 5)\}$ (is/is not) a one-to-one function.

   d. A function defined by $y = f(x)$ (is/is not) a one-to-one function if no horizontal line intersects the graph of f in more than one point.

   e. The graph of a function and its inverse are symmetric with respect to the line _____.

2. a. Let f be a one-to-one function and let g be the inverse of f. Then $(f \circ g)(x) =$ _____ and $(g \circ f)(x) =$ _____.

   b. The notation _____ is often used to represent the inverse of a function f and not the reciprocal of f.

   c. If (a, b) is a point on the graph of a one-to-one function f, then the corresponding ordered pair _____ is a point on the graph of f^{-1}.

Concept 1: Introduction to Inverse Functions

For Exercises 3–6, write the inverse function for each function.

3. $g = \{(3, 5), (8, 1), (-3, 9), (0, 2)\}$

4. $f = \{(-6, 2), (-9, 0), (-2, -1), (3, 4)\}$

5. $r = \{(a, 3), (b, 6), (c, 9)\}$

6. $s = \{(-1, x), (-2, y), (-3, z)\}$

Concept 2: Definition of a One-to-One Function

7. The table relates a state x to the number of representatives in the House of Representatives y for a recent year. Does this relation define a one-to-one function? If so, write a function defining the inverse as a set of ordered pairs.

State x	Number of Representatives y
Colorado	7
California	53
Texas	32
Louisiana	7
Pennsylvania	19

8. The table relates a city x to its average January temperature y. Does this relation define a one-to-one function? If so, write a function defining the inverse as a set of ordered pairs.

City x	Temperature y (°C)
Gainesville, Florida	13.6
Keene, New Hampshire	−6.0
Wooster, Ohio	−4.0
Rock Springs, Wyoming	−6.0
Lafayette, Louisiana	10.9

For Exercises 9–14, determine if the function is one-to-one by using the horizontal line test. (See Example 1.)

9.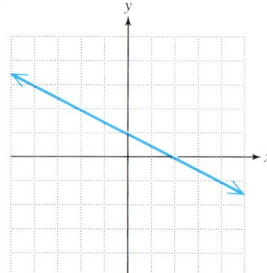

10.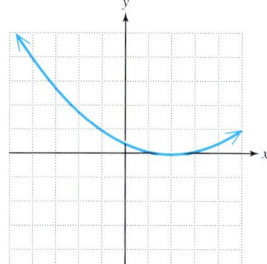

11.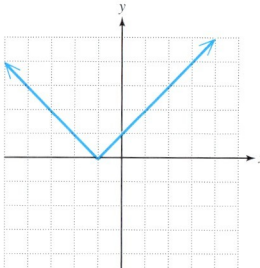

12.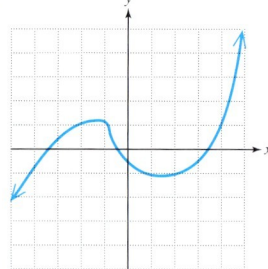

13.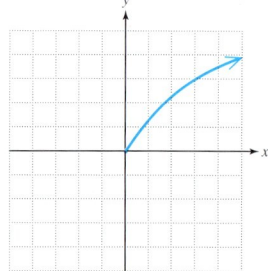

14.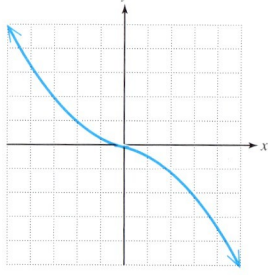

Concept 3: Definition of the Inverse of a Function

For Exercises 15–20, verify that f and g are inverse functions by showing that

    a. $(f \circ g)(x) = x$ **b.** $(g \circ f)(x) = x$ (See Example 2.)

15. $f(x) = 6x + 1$ and $g(x) = \dfrac{x-1}{6}$

16. $f(x) = 5x - 2$ and $g(x) = \dfrac{x+2}{5}$

17. $f(x) = \dfrac{\sqrt[3]{x}}{2}$ and $g(x) = 8x^3$

18. $f(x) = \dfrac{\sqrt[3]{x}}{3}$ and $g(x) = 27x^3$

19. $f(x) = x^2 + 1, x \geq 0$, and $g(x) = \sqrt{x - 1}, x \geq 1$

20. $f(x) = x^2 - 3, x \geq 0$, and $g(x) = \sqrt{x + 3}, x \geq -3$

Concept 4: Finding an Equation of the Inverse of a Function

21. If function f adds 4 to x, then f^{-1} _____ 4 from x. Function f is defined by $f(x) = x + 4$, and $f^{-1}(x) = $ _____.

22. If function g subtracts 10 from x, then g^{-1} _____ 10 to x. Function g is defined by $g(x) = x - 10$, and $g^{-1}(x) = $ _____.

23. If function h multiplies x by 5, then f^{-1} _____ x by 5. Function h is defined by $h(x) = 5x$, and $h^{-1}(x) = $ _____.

24. If function k multiplies x by $\frac{1}{3}$, then k^{-1} multiplies x by _____. Function k is defined by $k(x) = \frac{1}{3}x$, and $k^{-1}(x) = $ _____.

25. Suppose that function f multiplies x by 9 and subtracts 2 from the result.

  a. Write an equation for f.

  b. Write an equation for f^{-1}.

26. Suppose that function g cubes x and adds 1 to the result.

  a. Write an equation for g.

  b. Write an equation for g^{-1}.

For Exercises 27–42, write an equation of the inverse for each one-to-one function as defined. **(See Examples 3–5.)**

27. $h(x) = x + 4$

28. $k(x) = x - 3$

29. $m(x) = \frac{1}{3}x - 2$

30. $n(x) = 4x + 2$

31. $p(x) = -x + 10$

32. $q(x) = -x - \frac{2}{3}$

33. $n(x) = \frac{3x + 2}{5}$

34. $p(x) = \frac{2x - 7}{4}$

35. $h(x) = \frac{4x - 1}{3}$

36. $f(x) = \frac{6x + 3}{2}$

37. $f(x) = x^3 + 1$

38. $g(x) = \sqrt[3]{x} + 2$

39. $g(x) = \sqrt[3]{2x - 1}$

40. $f(x) = x^3 - 4$

41. $g(x) = x^2 + 9 \quad x \geq 0$

42. $m(x) = x^2 - 1 \quad x \geq 0$

43. The function defined by $f(x) = 0.3048x$ converts a length of x feet into $f(x)$ meters.

  a. Find the equivalent length in meters for a 4-ft board and a 50-ft wire.

  b. Find an equation defining $y = f^{-1}(x)$.

  c. Use the inverse function from part (b) to find the equivalent length in feet for a 1500-m race in track and field. Round to the nearest tenth of a foot.

44. The function defined by $s(x) = 1.47x$ converts a speed of x mph to $s(x)$ ft/sec.

  a. Find the equivalent speed in feet per second for a car traveling 60 mph.

  b. Find an equation defining $y = s^{-1}(x)$.

  c. Use the inverse function from part (b) to find the equivalent speed in miles per hour for a train traveling 132 ft/sec. Round to the nearest tenth.

For Exercises 45–51, answer true or false.

45. The function defined by $y = 2$ has an inverse function defined by $x = 2$.

46. The domain of any one-to-one function is the same as the domain of its inverse.

47. All linear functions with a nonzero slope have an inverse function.

48. The function defined by $g(x) = |x|$ is one-to-one.

49. The function defined by $k(x) = x^2$ is one-to-one.

50. The function defined by $h(x) = |x|$ for $x \geq 0$ is one-to-one.

51. The function defined by $L(x) = x^2$ for $x \geq 0$ is one-to-one.

52. Explain how the domain and range of a one-to-one function and its inverse are related.

53. If $(0, b)$ is the y-intercept of a one-to-one function, what is the x-intercept of its inverse?

54. If $(a, 0)$ is the x-intercept of a one-to-one function, what is the y-intercept of its inverse?

55. **a.** Find the domain and range of the function defined by $f(x) = \sqrt{x - 1}$.
    b. Find the domain and range of the function defined by $f^{-1}(x) = x^2 + 1, x \geq 0$.

56. **a.** Find the domain and range of the function defined by $g(x) = x^2 - 4, x \leq 0$.
    b. Find the domain and range of the function defined by $g^{-1}(x) = -\sqrt{x + 4}$.

For Exercises 57–60, the graph of $y = f(x)$ is given.
  a. State the domain of f.
  b. State the range of f.
  c. State the domain of f^{-1}.
  d. State the range of f^{-1}.
  e. Graph the function defined by $y = f^{-1}(x)$. The line $y = x$ is provided for your reference.

57.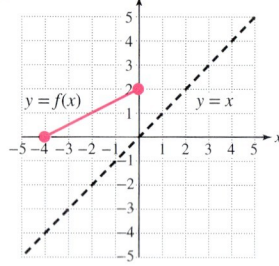

58.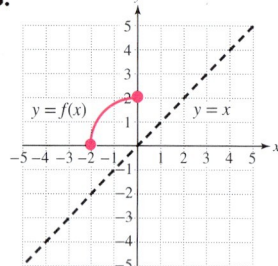

59.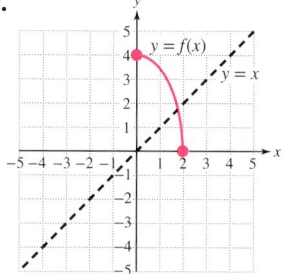

60.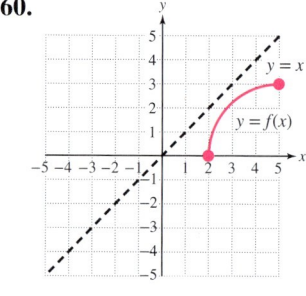

Technology Connections

For Exercises 61–64, use a graphing calculator to graph each function on the standard viewing window. Use the graph of the function to determine if the function is one-to-one on the interval $-10 \leq x \leq 10$. If the function is one-to-one, find its inverse and graph both functions on the standard viewing window.

61. $f(x) = \sqrt[3]{x+5}$

62. $k(x) = x^3 - 4$

63. $g(x) = 0.5x^3 - 2$

64. $m(x) = 3x - 4$

Expanding Your Skills

For Exercises 65–74, write an equation of the inverse of the one-to-one function.

65. $q(x) = \sqrt{x+4}$

66. $v(x) = \sqrt{x+16}$

67. $z(x) = -\sqrt{x+4}$

68. $u(x) = -\sqrt{x+16}$

69. $f(x) = \dfrac{x-1}{x+1}$

70. $p(x) = \dfrac{3-x}{x+3}$

71. $t(x) = \dfrac{2}{x-1}$

72. $w(x) = \dfrac{4}{x+2}$

73. $n(x) = x^2 + 9 \quad x \leq 0$

74. $g(x) = x^2 - 1 \quad x \leq 0$

Exponential Functions

Section 20.2

1. Definition of an Exponential Function

We have already learned to recognize several categories of functions, including constant, linear, rational, and quadratic functions. In this section and the next, we will define two new types of functions called exponential and logarithmic functions.

To introduce the concept of an exponential function, consider the following salary plans for a new job. Plan A pays $1 million for a month's work. Plan B starts with 2¢ on the first day, and every day thereafter the salary is doubled.

At first glance, the million-dollar plan appears to be more favorable. Look, however, at Table 20-1, which shows the daily payments for 30 days under plan B.

Concepts

1. Definition of an Exponential Function
2. Approximating Exponential Expressions With a Calculator
3. Graphs of Exponential Functions
4. Applications of Exponential Functions

Table 20-1

Day	Payment	Day	Payment	Day	Payment
1	2¢	11	$20.48	21	$20,971.52
2	4¢	12	$40.96	22	$41,943.04
3	8¢	13	$81.92	23	$83,886.08
4	16¢	14	$163.84	24	$167,772.16
5	32¢	15	$327.68	25	$335,544.32
6	64¢	16	$655.36	26	$671,088.64
7	$1.28	17	$1310.72	27	$1,342,177.28
8	$2.56	18	$2621.44	28	$2,684,354.56
9	$5.12	19	$5242.88	29	$5,368,709.12
10	$10.24	20	$10,485.76	30	$10,737,418.24

Notice that the salary on the 30th day for plan B is over $10 million. Taking the sum of the payments, we see the total salary for the 30-day period is $21,474,836.46.

The daily salary for plan B can be represented by the function $y = 2^x$, where x is the number of days on the job and y is the salary (in cents) for that day. An interesting feature of this function is that for every positive 1-unit change in x, the y value doubles. The function $y = 2^x$ is called an exponential function.

> **Definition of an Exponential Function**
>
> Let b be any real number such that $b > 0$ and $b \neq 1$. Then for any real number x, a function of the form $f(x) = b^x$ is called an **exponential function**.

An exponential function is recognized as a function with a constant base and exponent, x. Notice that the base of an exponential function must be a positive real number not equal to 1.

2. Approximating Exponential Expressions With a Calculator

Up to this point, we have evaluated exponential expressions with integer exponents and rational exponents, for example, $4^3 = 64$ and $4^{1/2} = \sqrt{4} = 2$. However, how do we evaluate an exponential expression with an irrational exponent such as 4^π? In such a case, the exponent is a nonterminating and nonrepeating decimal. The value of 4^π can be thought of as the limiting value of a sequence of approximations using rational exponents:

$$4^{3.14} \approx 77.7084726$$

$$4^{3.141} \approx 77.81627412$$

$$4^{3.1415} \approx 77.87023095$$

$$\vdots$$

$$4^\pi \approx 77.88023365$$

An exponential expression can be evaluated at all rational numbers and at all irrational numbers. Therefore, the domain of an exponential function is all real numbers.

Example 1 — **Approximating Exponential Expressions With a Calculator**

Approximate the expressions. Round the answers to four decimal places.

  a. $8^{\sqrt{3}}$ **b.** $5^{-\sqrt{17}}$ **c.** $\sqrt{10}^{\sqrt{2}}$

Solution:

On a calculator, use the ^, y^x, or x^y key to approximate an expression with an irrational exponent.

a. $8^{\sqrt{3}} \approx 36.6604$ **b.** $5^{-\sqrt{17}} \approx 0.0013$

c. $\sqrt{10}^{\sqrt{2}} \approx 5.0946$

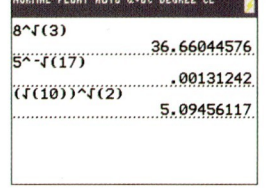

Skill Practice Approximate the value of the expressions. Round the answers to four decimal places.

1. 9^{π} 2. $15^{\sqrt{5}}$ 3. $\sqrt{7}^{\sqrt{3}}$

3. Graphs of Exponential Functions

The functions defined by $f(x) = 2^x$, $g(x) = \left(\frac{1}{2}\right)^x$, $h(x) = 3^x$, and $k(x) = 5^x$ are all examples of exponential functions. Example 2 illustrates the two general shapes of exponential functions.

Example 2 — Graphing Exponential Functions

Graph the functions f and g.

a. $f(x) = 2^x$ **b.** $g(x) = \left(\frac{1}{2}\right)^x$

Solution:

Table 20-2 shows several function values for $f(x)$ and $g(x)$ for both positive and negative values of x. The graphs are shown in Figure 20-7.

Table 20-2

x	$f(x) = 2^x$	$g(x) = \left(\frac{1}{2}\right)^x$
-4	$\frac{1}{16}$	16
-3	$\frac{1}{8}$	8
-2	$\frac{1}{4}$	4
-1	$\frac{1}{2}$	2
0	1	1
1	2	$\frac{1}{2}$
2	4	$\frac{1}{4}$
3	8	$\frac{1}{8}$
4	16	$\frac{1}{16}$

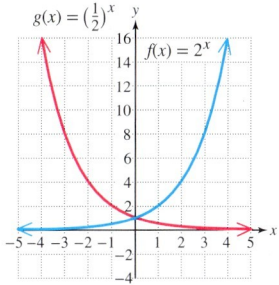

Figure 20-7

FOR REVIEW

Note the difference between the polynomial function $y = x^2$ and the exponential function $y = 2^x$. The polynomial function has a variable base and constant exponent. The exponential function has a constant base and variable exponent.

Skill Practice Graph the functions f and g on the same grid.

4. $f(x) = 3^x$ 5. $g(x) = \left(\frac{1}{3}\right)^x$

Answers

1. 995.0416   2. 426.4028
3. 5.3936
4–5.

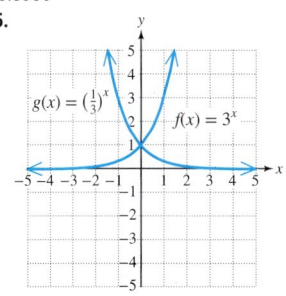

The graphs in Figure 20-7 illustrate several important features of exponential functions.

> ### Graphs of $f(x) = b^x$
>
> The graph of an exponential function defined by $f(x) = b^x$ ($b > 0$ and $b \neq 1$) has the following properties.
>
>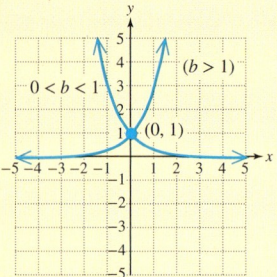
>
> 1. If $b > 1$, f is an *increasing* exponential function, sometimes called an **exponential growth function**.
>
>   If $0 < b < 1$, f is a *decreasing* exponential function, sometimes called an **exponential decay function**.
>
> 2. The domain is the set of all real numbers, $(-\infty, \infty)$.
>
> 3. The range is $(0, \infty)$.
>
> 4. The line $y = 0$ (x-axis) is a horizontal asymptote.
>
> 5. The function passes through the point $(0, 1)$ because $f(0) = b^0 = 1$.

These properties indicate that the graph of an exponential function is an increasing function if the base is greater than 1. Furthermore, the base affects the rate of increase. Consider the graphs of $f(x) = 2^x$, $h(x) = 3^x$, and $k(x) = 5^x$ (Figure 20-8). For every positive 1-unit change in x, $f(x) = 2^x$ increases by 2 times, $h(x) = 3^x$ increases by 3 times, and $k(x) = 5^x$ increases by 5 times (Table 20-3).

Table 20-3

x	$f(x) = 2^x$	$h(x) = 3^x$	$k(x) = 5^x$
-3	$\frac{1}{8}$	$\frac{1}{27}$	$\frac{1}{125}$
-2	$\frac{1}{4}$	$\frac{1}{9}$	$\frac{1}{25}$
-1	$\frac{1}{2}$	$\frac{1}{3}$	$\frac{1}{5}$
0	1	1	1
1	2	3	5
2	4	9	25
3	8	27	125

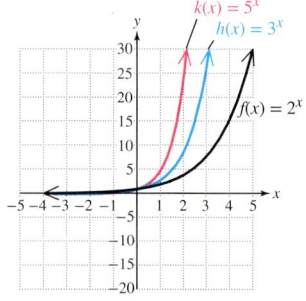

Figure 20-8

The graph of an exponential function is a *decreasing function* if the base is between 0 and 1. Consider the graphs of $g(x) = (\frac{1}{2})^x$, $m(x) = (\frac{1}{3})^x$, and $n(x) = (\frac{1}{5})^x$ (Table 20-4 and Figure 20-9).

Table 20-4

x	$g(x) = (\frac{1}{2})^x$	$m(x) = (\frac{1}{3})^x$	$n(x) = (\frac{1}{5})^x$
-3	8	27	125
-2	4	9	25
-1	2	3	5
0	1	1	1
1	$\frac{1}{2}$	$\frac{1}{3}$	$\frac{1}{5}$
2	$\frac{1}{4}$	$\frac{1}{9}$	$\frac{1}{25}$
3	$\frac{1}{8}$	$\frac{1}{27}$	$\frac{1}{125}$

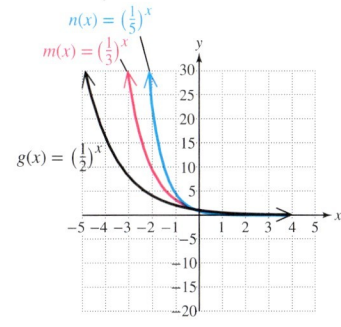

Figure 20-9

4. Applications of Exponential Functions

Exponential growth and decay can be found in a variety of real-world phenomena; for example,

- Short-term population growth can often be modeled by an exponential function.
- The growth of an investment under compound interest increases exponentially.
- The mass of a radioactive substance decreases exponentially with time.
- The temperature of a cup of coffee decreases exponentially as it approaches room temperature.

A substance that undergoes radioactive decay is said to be radioactive. The *half-life* of a radioactive substance is the amount of time it takes for one-half of the original amount of the substance to change into something else. That is, after each half-life the amount of the original substance decreases by one-half.

In 1898, Marie Curie discovered the highly radioactive element radium. She shared the 1903 Nobel Prize in Physics for her research on radioactivity and was awarded the 1911 Nobel Prize in Chemistry for her discovery of radium and polonium. Radium-226 (an isotope of radium) has a half-life of 1620 years and decays into radon-222 (a radioactive gas).

Marie and Pierre Curie
Everett Historical/Shutterstock

Example 3   Applying an Exponential Decay Function

In a sample originally composed of 1 g of radium-226, the amount of radium-226 present after t years is given by

$$A(t) = \left(\frac{1}{2}\right)^{t/1620}$$

where $A(t)$ is the amount of radium in grams and t is the time in years.

a. How much radium-226 will be present after 1620 years?

b. How much radium-226 will be present after 3240 years?

c. How much radium-226 will be present after 4860 years?

Solution:

$$A(t) = \left(\frac{1}{2}\right)^{t/1620}$$

a. $A(1620) = \left(\frac{1}{2}\right)^{1620/1620}$ Substitute $t = 1620$.

$= \left(\frac{1}{2}\right)^1$

$= 0.5$

After 1620 years (1 half-life), 0.5 g remains.

b. $A(3240) = \left(\frac{1}{2}\right)^{3240/1620}$ Substitute $t = 3240$.

$= \left(\frac{1}{2}\right)^2$

$= 0.25$

After 3240 years (2 half-lives), the amount of the original substance is reduced by one-half, 2 times. Thus, 0.25 g remains.

c. $A(4860) = \left(\frac{1}{2}\right)^{4860/1620}$ Substitute $t = 4860$.

$= \left(\frac{1}{2}\right)^3$

$= 0.125$

After 4860 years (3 half-lives), the amount of the original substance is reduced by one-half, 3 times. Thus, 0.125 g remains.

Skill Practice Cesium-137 is a radioactive metal with a short half-life of 30 years. In a sample originally composed of 1 g of cesium-137, the amount, $A(t)$ (in grams), of cesium-137 present after t years is given by

$$A(t) = \left(\frac{1}{2}\right)^{t/30}$$

6. How much cesium-137 will be present after 30 years?

7. How much cesium-137 will be present after 90 years?

Exponential functions are often used to model population growth. Suppose the initial value of a population at some time $t = 0$ is P_0. If the annual rate of increase of a population is r, then after t years, the population $P(t)$ is given by

$$P(t) = P_0(1 + r)^t$$

Answers

6. 0.5 g
7. 0.125 g

Example 4 — Applying an Exponential Growth Function

The population of the Bahamas in 2008 was estimated at 321,000 with an annual rate of increase of 1.39%.

a. Find a mathematical model that relates the population of the Bahamas as a function of the number of years since 2008.

b. If the annual rate of increase remains the same, use this model to estimate the population of the Bahamas in the year 2016. Round to the nearest thousand.

Solution:

a. The initial population is $P_0 = 321{,}000$, and the rate of increase is 1.39%.

$P(t) = P_0(1 + r)^t$ Substitute $P_0 = 321{,}000$ and $r = 0.0139$.

$ = 321{,}000(1 + 0.0139)^t$

$ = 321{,}000(1.0139)^t$ Here $t = 0$ corresponds to the year 2008.

b. Because the initial population ($t = 0$) corresponds to the year 2008, we use $t = 8$ to find the population in the year 2016.

$P(8) = 321{,}000(1.0139)^8$

$ \approx 358{,}000$ The population in the year 2016 was approximately 358,000.

Skill Practice The population of Colorado in 2000 was approximately 3,700,000 with an annual rate of increase of 2%.

8. Find a mathematical model that relates the population of Colorado as a function of the number of years since 2000.

9. Use this model to estimate the population of Colorado in 2016. Round to the nearest thousand.

Answers
8. $P(t) = 3{,}700{,}000(1.02)^t$
9. Approximately 5,079,000

Section 20.2 Activity

A.1. Given a positive real number b, such that $b \neq 1$, a function defined by $f(x) = b^x$ is called a(n) _____ function.

A.2. Which functions are exponential functions?

   a. $f(x) = x^2$ b. $g(x) = 6^x$ c. $h(x) = \left(\dfrac{1}{2}\right)^x$ d. $k(x) = 1^x$

A.3. Consider the functions defined by $f(x) = 3^x$ and $g(x) = \left(\dfrac{1}{3}\right)^x$.

x	$f(x) = 3^x$	$g(x) = \left(\dfrac{1}{3}\right)^x$
0		
1		
2		
−1		
−2		

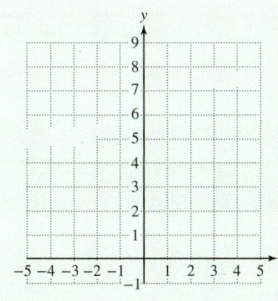

a. Complete the table.

b. Refer to the values of function f in the table. How do the function values change with each 1-unit increase in x?

c. Refer to the values of function g in the table. How do the function values change with each 1-unit increase in x?

d. Graph $f(x) = 3^x$ and $g(x) = \left(\dfrac{1}{3}\right)^x$ on the same coordinate system.

e. Is function f an exponential growth function or exponential decay function?

f. Is function g an exponential growth function or exponential decay function?

g. Does the graph of either function ever meet the x-axis?

h. Write the domain and range of f and g.

A.4. A radioactive form of iodine called iodine-131 (denoted by chemists as ^{131}I) is used to treat thyroid cancer. ^{131}I undergoes radioactive decay, meaning that it spontaneously emits particles and energy. Eventually the substance changes from one form (or *isotope*) of the element into another.

The half-life of ^{131}I is roughly 8 days. This means that due to radioactive decay, half of the original amount of the substance is present after 8 days.

Suppose that a patient is given 64 units of ^{131}I. Then the amount of substance $A(t)$ still present after t days can be modeled by $A(t) = 64 \cdot \left(\dfrac{1}{2}\right)^{t/8}$.

a. Evaluate $A(8)$ and interpret its meaning.

b. Evaluate $A(16)$ and interpret its meaning.

c. Evaluate $A(24)$ and interpret its meaning.

d. Compare the values from parts (a)–(c) and interpret the results.

Section 20.2 Practice Exercises

Prerequisite Review

For Exercises R.1–R.4, simplify the expressions.

R.1. a. 4^2 b. 4^0 c. 4^{-2} **R.2.** a. 2^3 b. 2^0 c. 2^{-3}

R.3. a. $\left(\dfrac{1}{4}\right)^2$ b. $\left(\dfrac{1}{4}\right)^0$ c. $\left(\dfrac{1}{4}\right)^{-2}$ **R.4.** a. $\left(\dfrac{1}{2}\right)^3$ b. $\left(\dfrac{1}{2}\right)^0$ c. $\left(\dfrac{1}{2}\right)^{-3}$

Vocabulary and Key Concepts

1. a. Given a real number b, where $b > 0$ and $b \neq 1$, a function of the form $f(x) =$ _____ is called an exponential function.

   b. The function defined by $f(x) = x^4$ (is/is not) an exponential function, whereas the function defined by $f(x) = 4^x$ (is/is not) an exponential function.

   c. The graph of $f(x) = \left(\dfrac{7}{2}\right)^x$ is (increasing/decreasing) over its domain.

   d. The graph of $f(x) = \left(\dfrac{2}{7}\right)^x$ is (increasing/decreasing) over its domain.

e. In interval notation, the domain of an exponential function $f(x) = b^x$ is _____ and the range is _____.

f. All exponential functions $f(x) = b^x$ pass through the point _____.

g. The horizontal asymptote of an exponential function $f(x) = b^x$ is the line _____.

2. The function defined by $f(x) = 1^x$ (is/is not) an exponential function.

Concept 2: Approximating Exponential Expressions With a Calculator

For Exercises 3–10, evaluate the expression by using a calculator. Round to four decimal places. (See Example 1.)

3. $5^{1.1}$

4. $2^{\sqrt{3}}$

5. 10^{π}

6. $3^{4.8}$

7. $36^{-\sqrt{2}}$

8. $27^{-0.5126}$

9. $16^{-0.04}$

10. $8^{-0.61}$

11. Solve for x.

   a. $3^x = 9$

   b. $3^x = 27$

   c. Between what two consecutive integers must the solution to $3^x = 11$ lie?

12. Solve for x.

   a. $5^x = 125$

   b. $5^x = 625$

   c. Between what two consecutive integers must the solution to $5^x = 130$ lie?

13. Solve for x.

   a. $2^x = 16$

   b. $2^x = 32$

   c. Between what two consecutive integers must the solution to $2^x = 30$ lie?

14. Solve for x.

   a. $4^x = 16$

   b. $4^x = 64$

   c. Between what two consecutive integers must the solution to $4^x = 20$ lie?

15. For $f(x) = \left(\frac{1}{5}\right)^x$ find $f(0), f(1), f(2), f(-1),$ and $f(-2)$.

16. For $g(x) = \left(\frac{2}{3}\right)^x$ find $g(0), g(1), g(2), g(-1),$ and $g(-2)$.

17. For $h(x) = 3^x$ use a calculator to find $h(0), h(1), h(-1), h(\sqrt{2}),$ and $h(\pi)$. Round to two decimal places if necessary.

18. For $k(x) = 5^x$ use a calculator to find $k(0), k(1), k(-1), k(\pi),$ and $k(\sqrt{2})$. Round to two decimal places if necessary.

Concept 3: Graphs of Exponential Functions

19. How do you determine whether the graph of $f(x) = b^x$ ($b > 0, b \neq 1$) is increasing or decreasing?

20. For $f(x) = b^x$ ($b > 0, b \neq 1$), find $f(0)$.

Graph the functions defined in Exercises 21–28. Plot at least three points for each function. (See Example 2.)

21. $f(x) = 4^x$

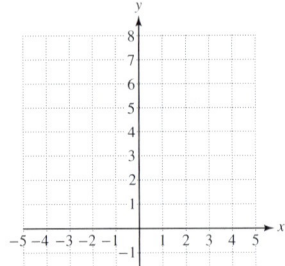

22. $g(x) = 6^x$

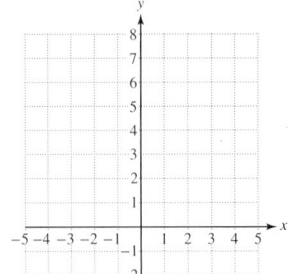

23. $m(x) = \left(\dfrac{1}{8}\right)^x$

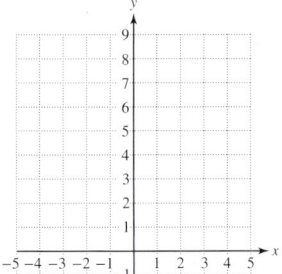

24. $n(x) = \left(\dfrac{1}{3}\right)^x$

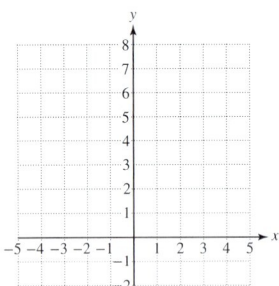

25. $h(x) = 2^{x+1}$

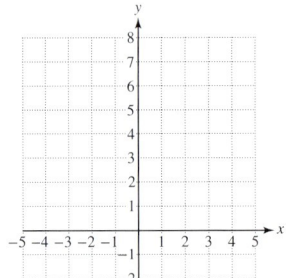

26. $k(x) = 5^{x-1}$

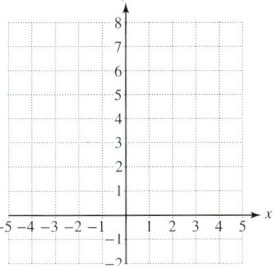

27. $g(x) = 5^{-x}$

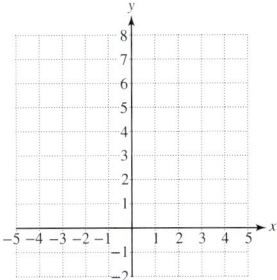

28. $f(x) = 2^{-x}$

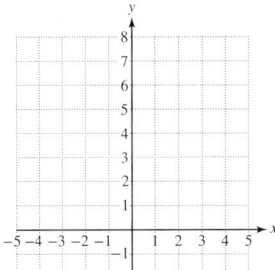

Concept 4: Applications of Exponential Functions

29. The half-life of the element radon (Rn-86) is 3.8 days. In a sample originally containing 1 g of radon, the amount left after t days is given by $A(t) = (0.5)^{t/3.8}$. (Round to two decimal places, if necessary.) (See Example 3.)

  a. How much radon will be present after 7.6 days?

  b. How much radon will be present after 10 days?

30. Nobelium, an element discovered in 1958, has a half-life of 10 min under certain conditions. In a sample containing 1 g of nobelium, the amount left after t min is given by $A(t) = (0.5)^{t/10}$. (Round to three decimal places.)

  a. How much nobelium is left after 5 min?

  b. How much nobelium is left after 1 hr?

31. Once an antibiotic is introduced to treat a bacterial infection, the number of bacteria decreases exponentially. For example, beginning with 1 million bacteria, the amount present t days from the time penicillin is introduced is given by $A(t) = 1{,}000{,}000(2)^{-t/5}$. Rounding to the nearest thousand, determine how many bacteria are present after

   a. 2 days **b.** 1 week **c.** 2 weeks

32. Once an antibiotic is introduced to treat a bacterial infection, the number of bacteria decreases exponentially. For example, beginning with 1 million bacteria, the amount present t days from the time streptomycin is introduced is given by $A(t) = 1{,}000{,}000(2)^{-t/10}$. Rounding to the nearest thousand, determine how many bacteria are present after

   a. 5 days **b.** 1 week **c.** 2 weeks

33. The population of Bangladesh was 153,000,000 in 2009 with an annual growth rate of 1.25%.
   (See Example 4.)

   a. Find a mathematical model that relates the population of Bangladesh as a function of the number of years after 2009.

   b. If the annual rate of increase remains the same, use this model to predict the population of Bangladesh in the year 2050. Round to the nearest million.

34. The population of Fiji was 908,000 in 2009 with an annual growth rate of 0.07%.

   a. Find a mathematical model that relates the population of Fiji as a function of the number of years after 2009.

   b. If the annual rate of increase remains the same, use this model to predict the population of Fiji in the year 2050. Round to the nearest thousand.

35. Suppose $1000 is initially invested in an account and the value of the account grows exponentially. If the investment doubles in 7 years, then the amount in the account t years after the initial investment is given by $A(t)$.

$$A(t) = 1000(2)^{t/7}$$

   a. Find the amount in the account after 5 years.

   b. Find the amount in the account after 10 years.

   c. Find $A(0)$ and $A(7)$ and interpret the answers in the context of this problem.

36. Suppose $1500 is initially invested in an account and the value of the account grows exponentially. If the investment doubles in 8 years, then the amount in the account t years after the initial investment is given by $A(t)$.

$$A(t) = 1500(2)^{t/8}$$

   a. Find the amount in the account after 5 years.

   b. Find the amount in the account after 10 years.

   c. Find $A(0)$ and $A(8)$ and interpret the answers in the context of this problem.

Technology Connections

For Exercises 37–44, graph the functions on your calculator to support your solutions to the indicated exercises.

37. $f(x) = 4^x$
(see Exercise 21)

38. $g(x) = 6^x$
(see Exercise 22)

39. $m(x) = (\frac{1}{8})^x$
(see Exercise 23)

40. $n(x) = (\frac{1}{3})^x$
(see Exercise 24)

41. $h(x) = 2^{x+1}$
(see Exercise 25)

42. $k(x) = 5^{x-1}$
(see Exercise 26)

43. $g(x) = 5^{-x}$
(see Exercise 27)

44. $f(x) = 2^{-x}$
(see Exercise 28)

Section 20.3 Logarithmic Functions

Concepts

1. Definition of a Logarithmic Function
2. Evaluating Logarithmic Expressions
3. The Common Logarithmic Function
4. Graphs of Logarithmic Functions
5. Applications of the Common Logarithmic Function

1. Definition of a Logarithmic Function

Consider the following equations in which the variable is located in the exponent of an expression. In some cases the solution can be found by inspection because the constant on the right-hand side of the equation is a perfect power of the base.

Equation	Solution
$5^x = 5$	$x = 1$
$5^x = 20$	$x = ?$
$5^x = 25$	$x = 2$
$5^x = 60$	$x = ?$
$5^x = 125$	$x = 3$

The equation $5^x = 20$ cannot be solved by inspection. However, we might suspect that x is between 1 and 2. Similarly, the solution to the equation $5^x = 60$ is between 2 and 3. To solve an exponential equation for an unknown exponent, we must use a new type of function called a logarithmic function.

> **Definition of a Logarithmic Function**
>
> If x and b are positive real numbers such that $b \neq 1$, then $y = \log_b x$ is called the **logarithmic function** with base b and
>
> $$y = \log_b x \quad \text{is equivalent to} \quad b^y = x$$
>
> *Note:* In the expression $y = \log_b x$, y is called the **logarithm**, b is called the **base**, and x is called the **argument**.

The expression $y = \log_b x$ is equivalent to $b^y = x$ and indicates that *the logarithm y is the exponent to which b must be raised to obtain x.* The expression $y = \log_b x$ is called the logarithmic form of the equation, and the expression $b^y = x$ is called the exponential form of the equation.

The definition of a logarithmic function suggests a close relationship with an exponential function of the same base. In fact, a logarithmic function is the inverse of the corresponding exponential function. For example, the following steps find the inverse of the exponential function defined by $f(x) = b^x$.

$f(x) = b^x$

$y = b^x$ Replace $f(x)$ with y.

$x = b^y$ Interchange x and y.

$y = \log_b x$ Solve for y using the definition of a logarithmic function.

$f^{-1}(x) = \log_b x$ Replace y with $f^{-1}(x)$.

The concept of a logarithm is new and unfamiliar. Therefore, it is often advantageous to rewrite a logarithm in its exponential form.

Example 1   Converting From Logarithmic Form to Exponential Form

Rewrite the logarithmic equations in exponential form.

   a. $\log_2 32 = 5$ **b.** $\log_{10}\left(\dfrac{1}{1000}\right) = -3$ **c.** $\log_5 1 = 0$

Solution:

Logarithmic Form		Exponential Form
a. $\log_2 32 = 5$	$\Leftrightarrow$	$2^5 = 32$
b. $\log_{10}\left(\dfrac{1}{1000}\right) = -3$	$\Leftrightarrow$	$10^{-3} = \dfrac{1}{1000}$
c. $\log_5 1 = 0$	$\Leftrightarrow$	$5^0 = 1$

Skill Practice Rewrite the logarithmic equations in exponential form.

1. $\log_3 9 = 2$ **2.** $\log_{10}\left(\dfrac{1}{100}\right) = -2$ **3.** $\log_8 1 = 0$

2. Evaluating Logarithmic Expressions

Example 2 Evaluating Logarithmic Expressions

Evaluate the logarithmic expressions.

a. $\log_8 64$ **b.** $\log_5\left(\dfrac{1}{125}\right)$ **c.** $\log_{1/2}\left(\dfrac{1}{8}\right)$

Solution:

a. $\log_8 64$ is the exponent to which 8 must be raised to obtain 64.

$y = \log_8 64$ Let y represent the value of the logarithm.

$8^y = 64$ Rewrite the expression in exponential form.

$8^y = 8^2$

$y = 2$ Therefore, $\log_8 64 = 2$.

b. $\log_5\left(\dfrac{1}{125}\right)$ is the exponent to which 5 must be raised to obtain $\dfrac{1}{125}$.

$y = \log_5\left(\dfrac{1}{125}\right)$ Let y represent the value of the logarithm.

$5^y = \dfrac{1}{125}$ Rewrite the expression in exponential form.

$5^y = \dfrac{1}{5^3} = 5^{-3}$

$y = -3$ Therefore, $\log_5\left(\dfrac{1}{125}\right) = -3$.

c. $\log_{1/2}\left(\dfrac{1}{8}\right)$ is the exponent to which $\dfrac{1}{2}$ must be raised to obtain $\dfrac{1}{8}$.

$y = \log_{1/2}\left(\dfrac{1}{8}\right)$ Let y represent the value of the logarithm.

$\left(\dfrac{1}{2}\right)^y = \dfrac{1}{8}$ Rewrite the expression in exponential form.

$\left(\dfrac{1}{2}\right)^y = \left(\dfrac{1}{2}\right)^3$

$y = 3$ Therefore, $\log_{1/2}\left(\dfrac{1}{8}\right) = 3$.

> **FOR REVIEW**
>
> Recall that for a nonzero real number b:
>
> • $b^{-n} = \left(\dfrac{1}{b}\right)^n = \dfrac{1}{b^n}$
>
> • $b^0 = 1$
>
> For example:
>
> $5^{-3} = \left(\dfrac{1}{5}\right)^3 = \dfrac{1}{5^3} = \dfrac{1}{125}$
>
> $5^0 = 1$

Skill Practice Evaluate the logarithmic expressions.

4. $\log_{10} 1000$ **5.** $\log_4\left(\dfrac{1}{16}\right)$ **6.** $\log_{1/3} 3$

Answers

1. $3^2 = 9$ **2.** $10^{-2} = \dfrac{1}{100}$
3. $8^0 = 1$ **4.** 3
5. -2 **6.** -1

Example 3 Evaluating Logarithmic Expressions

Evaluate the logarithmic expressions.

a. $\log_b b$ **b.** $\log_c c^7$ **c.** $\log_3 \sqrt[4]{3}$

Solution:

a. $\log_b b$ is the exponent to which b must be raised to obtain b.

$y = \log_b b$ Let y represent the value of the logarithm.

$b^y = b$ Rewrite the expression in exponential form.

$y = 1$ Therefore, $\log_b b = 1$.

b. $\log_c c^7$ is the exponent to which c must be raised to obtain c^7.

$y = \log_c c^7$ Let y represent the value of the logarithm.

$c^y = c^7$ Rewrite the expression in exponential form.

$y = 7$ Therefore, $\log_c c^7 = 7$.

c. $\log_3 \sqrt[4]{3} = \log_3 3^{1/4}$ is the exponent to which 3 must be raised to obtain $3^{1/4}$.

$y = \log_3 3^{1/4}$ Let y represent the value of the logarithm.

$3^y = 3^{1/4}$ Rewrite the expression in exponential form.

$y = \dfrac{1}{4}$ Therefore, $\log_3 \sqrt[4]{3} = \tfrac{1}{4}$.

FOR REVIEW

Recall that rational exponents can be used to represent radical expressions.

$x^{1/n} = \sqrt[n]{x}$

$x^{m/n} = \sqrt[n]{x^m}$ or $\left(\sqrt[n]{x}\right)^m$

provided that $\sqrt[n]{x}$ is a real number.

Skill Practice Evaluate the logarithmic expressions.

7. $\log_x x$ **8.** $\log_b b^{10}$ **9.** $\log_5 \sqrt[3]{5}$

3. The Common Logarithmic Function

The logarithmic function with base 10 is called the **common logarithmic function** and is denoted by $y = \log x$. Notice that the base is not explicitly written but is understood to be 10. That is, $y = \log_{10} x$ is written simply as $y = \log x$.

Example 4 Evaluating Common Logarithms

Evaluate the logarithmic expressions.

a. $\log 100{,}000$ **b.** $\log 0.01$

Solution:

a. $\log 100{,}000$ is the exponent to which 10 must be raised to obtain 100,000.

$y = \log 100{,}000$ Let y represent the value of the logarithm.

$10^y = 100{,}000$ Rewrite the expression in exponential form.

$10^y = 10^5$

$y = 5$ Therefore, $\log 100{,}000 = 5$.

Answers

7. 1   **8.** 10   **9.** $\dfrac{1}{3}$

b. log 0.01 is the exponent to which 10 must be raised to obtain 0.01 or $\frac{1}{100}$.

$y = \log 0.01$	Let y represent the value of the logarithm.
$10^y = 0.01$	Rewrite the expression in exponential form.
$10^y = 10^{-2}$	Note that $0.01 = \frac{1}{100}$ or 10^{-2}.
$y = -2$	Therefore, $\log 0.01 = -2$.

Skill Practice Evaluate the logarithmic expressions.

10. log 1,000,000   **11.** log 0.0001

On most calculators the **LOG** key is used to compute logarithms with base 10. This is demonstrated in Example 5.

Example 5   Evaluating Common Logarithms on a Calculator

Evaluate the common logarithms. Round the answers to four decimal places.

a. log 420   **b.** $\log(8.2 \times 10^9)$ **c.** log 0.0002

Solution:

a. $\log 420 \approx 2.6232$

b. $\log(8.2 \times 10^9) \approx 9.9138$

c. $\log 0.0002 \approx -3.6990$

TIP: On a scientific calculator, you may need to enter the logarithm and argument in reverse order. For example: 420 **log**.

Skill Practice Evaluate the common logarithms. Round answers to four decimal places.

12. log 1200   **13.** $\log(6.3 \times 10^5)$ **14.** log 0.00025

4. Graphs of Logarithmic Functions

Recall that $f(x) = \log_b x$ is the inverse of $g(x) = b^x$. Therefore, the graph of $y = f(x)$ is symmetric to the graph of $y = g(x)$ about the line $y = x$, as shown in Figures 20-10 and 20-11.

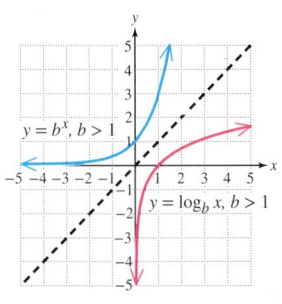

Figure 20-10

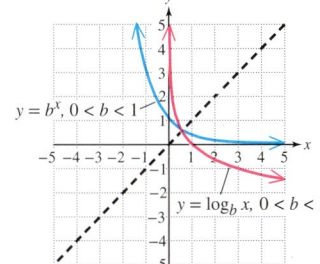

Figure 20-11

From Figures 20-10 and 20-11, we see that the range of $y = b^x$ is the set of positive real numbers. As expected, the domain of its inverse, the logarithmic function $y = \log_b x$, is also the set of positive real numbers. Therefore, the **domain of the logarithmic function** $y = \log_b x$ is the set of positive real numbers.

Answers
10. 6   **11.** −4
12. ≈ 3.0792   **13.** ≈ 5.7993
14. ≈ −3.6021

Example 6 — Graphing Logarithmic Functions

Graph the logarithmic functions.

a. $y = \log_2 x$ **b.** $y = \log_{1/4} x$

Solution:

A logarithmic function is the inverse of an exponential function with the same base. For this reason, we can reverse the coordinates of the ordered pairs on an exponential function to obtain ordered pairs on the related logarithmic function. For example, to graph the function $y = \log_2 x$, reverse the coordinates of the ordered pairs from its inverse function, $y = 2^x$.

a. Exponential function   Logarithmic function

x	$y = 2^x$
-3	$\frac{1}{8}$
-2	$\frac{1}{4}$
-1	$\frac{1}{2}$
0	1
1	2
2	4
3	8

x	$y = \log_2 x$
$\frac{1}{8}$	-3
$\frac{1}{4}$	-2
$\frac{1}{2}$	-1
1	0
2	1
4	2
8	3

Reverse the order.

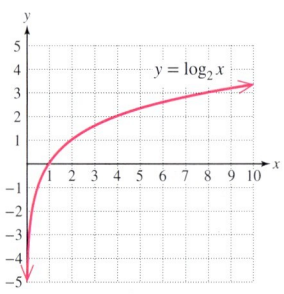

Figure 20-12

The graph of $y = \log_2 x$ is shown in Figure 20-12.

b. Exponential function   Logarithmic function

x	$y = \left(\frac{1}{4}\right)^x$
-3	64
-2	16
-1	4
0	1
1	$\frac{1}{4}$
2	$\frac{1}{16}$
3	$\frac{1}{64}$

x	$y = \log_{1/4} x$
64	-3
16	-2
4	-1
1	0
$\frac{1}{4}$	1
$\frac{1}{16}$	2
$\frac{1}{64}$	3

Reverse the order.

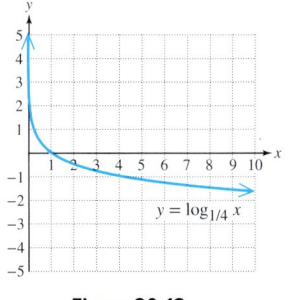

Figure 20-13

The graph of $y = \log_{1/4} x$ is shown in Figure 20-13.

Skill Practice Graph.

15. $y = \log_3 x$ **16.** $y = \log_{1/2} x$

Answers

15–16.

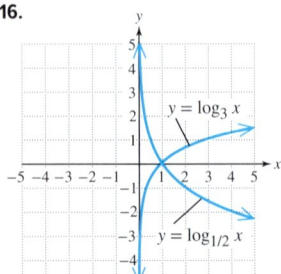

The general shape and important features of exponential and logarithmic functions are summarized as follows:

Graphs of Exponential and Logarithmic Functions

Exponential Functions

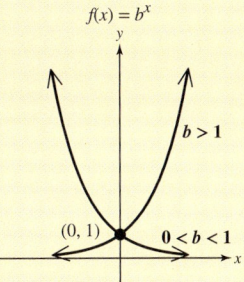

Logarithmic Functions

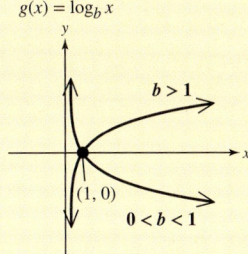

Exponential: $f(x) = b^x$
- Domain: $(-\infty, \infty)$
- Range: $(0, \infty)$
- Horizontal asymptote: $y = 0$
- Passes through $(0, 1)$
- If $b > 1$, the function is increasing.
- If $0 < b < 1$, the function is decreasing.

Logarithmic: $g(x) = \log_b x$
- Domain: $(0, \infty)$
- Range: $(-\infty, \infty)$
- Vertical asymptote: $x = 0$
- Passes through $(1, 0)$
- If $b > 1$, the function is increasing.
- If $0 < b < 1$, the function is decreasing.

Notice that the roles of x and y are interchanged for the functions $y = b^x$ and $b^y = x$. Therefore, it is not surprising that the domain and range are reversed between exponential and logarithmic functions. Moreover, an exponential function passes through $(0, 1)$, whereas a logarithmic function passes through $(1, 0)$. An exponential function has a horizontal asymptote at $y = 0$, whereas a logarithmic function has a vertical asymptote at $x = 0$.

When graphing a logarithmic equation, it is helpful to know its domain.

Example 7 Identifying the Domain of a Logarithmic Function

Find the domain of each function.

a. $f(x) = \log(4 - x)$ **b.** $g(x) = \log(2x + 6)$

Solution:

The domain of the function $y = \log_b x$ is the set of all positive real numbers. That is, the argument x must be greater than zero: $x > 0$.

a. $f(x) = \log(4 - x)$ The argument is $4 - x$.

$4 - x > 0$ The argument of the logarithm must be greater than zero.

$-x > -4$ Solve for x.

$x < 4$ Divide by -1 and reverse the inequality sign.

The domain is $(-\infty, 4)$.

b. $g(x) = \log(2x + 6)$ The argument is $2x + 6$.

$2x + 6 > 0$ The argument of the logarithm must be greater than zero.

$2x > -6$

$x > -3$

The domain is $(-3, \infty)$.

Skill Practice Find the domain of each function.

17. $f(x) = \log_3(x + 7)$ **18.** $g(x) = \log(4 - 8x)$

5. Applications of the Common Logarithmic Function

Example 8   Applying a Common Logarithm to Compute pH

The pH (hydrogen potential) of a solution is defined as

$$\text{pH} = -\log[\text{H}^+]$$

where $[\text{H}^+]$ represents the concentration of hydrogen ions in moles per liter (mol/L). The pH scale ranges from 0 to 14. The midpoint of this range, 7, represents a neutral solution. Values below 7 are progressively more acidic, and values above 7 are progressively more alkaline. Based on the equation $\text{pH} = -\log[\text{H}^+]$, a 1-unit change in pH means a 10-fold change in hydrogen ion concentration.

a. Normal rain has a pH of 5.6. However, in some areas of the northeastern United States the rainwater is more acidic. What is the pH of a rain sample for which the concentration of hydrogen ions is 0.0002 mol/L?

b. Find the pH of household ammonia if the concentration of hydrogen ions is 1.0×10^{-11} mol/L.

Solution:

a. $\text{pH} = -\log[\text{H}^+]$

$= -\log(0.0002)$ Substitute $[\text{H}+] = 0.0002$.

≈ 3.7 The pH of the rain sample is 3.7. (To compare this value with a familiar substance, note that the pH of orange juice is roughly 3.5.)

b. $\text{pH} = -\log[\text{H}^+]$

$= -\log(1.0 \times 10^{-11})$ Substitute $[\text{H}^+] = 1.0 \times 10^{-11}$.

$= -\log(10^{-11})$

$= -(-11)$

$= 11$ The pH of household ammonia is 11.

Skill Practice

19. A new all-natural shampoo on the market claims its hydrogen ion concentration is 5.88×10^{-7} mol/L. Use the formula $\text{pH} = -\log[\text{H}^+]$ to calculate the pH level of the shampoo.

Answers

17. Domain: $(-7, \infty)$

18. Domain: $\left(-\infty, \dfrac{1}{2}\right)$

19. pH ≈ 6.23

Example 9 Applying a Logarithmic Function to a Memory Model

One method of measuring a student's retention of material after taking a course is to retest the student at specified time intervals after the course has been completed. A student's score on a calculus test t months after completing a course in calculus is approximated by

$$S(t) = 85 - 25 \log(t + 1)$$

where t is the time in months after completing the course and $S(t)$ is the student's score as a percent.

a. What was the student's score at $t = 0$?
b. What was the student's score after 2 months?
c. What was the student's score after 1 year?

Solution:

a. $S(t) = 85 - 25 \log(t + 1)$

   $S(0) = 85 - 25 \log(0 + 1)$ Substitute $t = 0$.

   $= 85 - 25 \log 1$ $\log 1 = 0$ because $10^0 = 1$.

   $= 85 - 25(0)$

   $= 85 - 0$

   $= 85$ The student's score at the time the course was completed was 85%.

b. $S(t) = 85 - 25 \log(t + 1)$

   $S(2) = 85 - 25 \log(2 + 1)$

   $= 85 - 25 \log 3$ Use a calculator to approximate $\log 3$.

   ≈ 73.1 The student's score dropped to 73.1%.

c. $S(t) = 85 - 25 \log(t + 1)$

   $S(12) = 85 - 25 \log(12 + 1)$

   $= 85 - 25 \log 13$ Use a calculator to approximate $\log 13$.

   ≈ 57.2 The student's score dropped to 57.2%.

Skill Practice The memory model for a student's score on a statistics test t months after completion of the course in statistics is approximated by

$$S(t) = 92 - 28 \log(t + 1)$$

20. What was the student's score at the time the course was completed ($t = 0$)?

21. What was her score after 1 month?

22. What was the score after 2 months?

Answers
20. 92   **21.** 83.6   **22.** 78.6

Section 20.3 Activity

A.1. **a.** Consider an exponential function such as $y = 2^x$. If we input a value of x, such as 3, the y value is equal to the base, 2, raised to the third power.

$y = 2^3$ means that $y =$ _____.

b. The inverse of an exponential function reverses this process. The inverse of $y = 2^x$ is $x = 2^y$. Thus, if we input a value of x, such as 8, the y value is the exponent to which 2 is raised to make 8.

$8 = 2^y$ means that $y =$ _____.

c. The inverse of an exponential function is called a _____ function. Furthermore, $y = \log_b x$ is equivalent to _____.

For Exercises A.2–A.9, complete the table.

	Exponential Form	Logarithmic Form
A.2.		$\log_4 16 = 2$
A.3.	$3^4 = 81$	
A.4.		$\log_2 x = 7$
A.5.	$5^y = 125$	
A.6.		$\log_6 6 = 1$
A.7.	$7^0 = 1$	
A.8.		$\log_2 \frac{1}{16} = -4$
A.9.	$10^{-2} = \frac{1}{100}$	

For Exercises A.10–A.15, evaluate the logarithm.

A.10. $\log_2 32$ **A.11.** $\log 100$ **A.12.** $\log_3 \left(\frac{1}{9}\right)$

A.13. $\log_4 \left(\frac{1}{64}\right)$ **A.14.** $\log_5 1$ **A.15.** $\log_2 2$

A.16. **a.** Evaluate $\log 1000$.

b. Evaluate $\log 10{,}000$.

c. Approximate $\log 2500$ between two integer values.

For Exercises A.17–A.18,

a. Write the function in exponential form.

b. Complete the table and graph the function.

A.17. **a.** $y = \log_3 x$ _____ (exponential form)

b.

x	y
	-2
	-1
	0
	1
	2

A.18. **a.** $y = \log_{1/3} x$ _____ (exponential form)

b.

x	y
	-2
	-1
	0
	1
	2

For Exercises A.19–A.23, identify the similarities and differences between the graphs of $y = \log_b x$ with base b greater than 1 or with base b between 0 and 1. Refer to the graphs from Exercises A.17–A.18.

A.19. Write the domain of each function in interval notation.

A.20. Write the range of each function in interval notation.

A.21. Identify the asymptote of each function.

A.22. Identify the x-intercept of each function.

A.23. a. If $b > 1$, then the function (increases/decreases) from left to right.

  b. If $0 < b < 1$, then the function (increases/decreases) from left to right.

For Exercises A.24–A.27, write the domain in interval notation.

A.24. $f(x) = \log_b x$

A.25. $g(x) = \log_b(x - 4)$

A.26. $h(x) = \log_b(4 - x)$

A.27. $k(x) = \log_b(2x + 1)$

Practice Exercises

Section 20.3

Prerequisite Review

For Exercises R.1–R.6, simplify the expression.

R.1. $25^{1/2}$

R.2. $1000^{1/3}$

R.3. $64^{2/3}$

R.4. $32^{3/5}$

R.5. 8^{-2}

R.6. 2^{-3}

For Exercises R.7–R.14, fill in the blank to make a true statement.

R.7. $(3)^\square = 9$

R.8. $(2)^\square = 16$

R.9. $(3)^\square = \dfrac{1}{81}$

R.10. $(2)^\square = \dfrac{1}{8}$

R.11. $\left(\dfrac{1}{3}\right)^\square = 27$

R.12. $\left(\dfrac{1}{2}\right)^\square = 4$

R.13. $\left(\dfrac{3}{4}\right)^\square = \dfrac{16}{9}$

R.14. $\left(\dfrac{4}{5}\right)^\square = \dfrac{625}{256}$

For Exercises R.15–R.18, solve the inequality. Write the solution set in interval notation.

R.15. $2x - 6 > 0$

R.16. $3x + 12 > 0$

R.17. $6 - x > 0$

R.18. $-3 - x > 0$

Vocabulary and Key Concepts

1. a. The function defined by $y = \log_b x$ is called the _____ function with base _____. The values of x and b are positive, and $b \neq 1$.

  b. Given $y = \log_b x$, we call y the _____, we call b the _____, and we call x the _____.

  c. In interval notation, the domain of $y = \log_b x$ is _____ and the range is _____.

  d. A logarithmic function base b is the inverse of the _____ function base b.

  e. The logarithmic function base 10 is called the _____ logarithmic function and is denoted by $y = \log x$.

  f. The graph of a logarithmic function is (increasing/decreasing) if the base $b > 1$, whereas the graph is (increasing/decreasing) if the base is between 0 and 1.

  g. Which values of x are *not* in the domain of $y = \log(x - 4)$? $x = 2, x = 3, x = 4, x = 5, x = 6$

  h. The graphs of a logarithmic function base b and an exponential function base b are symmetric with respect to the line _____.

Concept 1: Definition of a Logarithmic Function

2. For the equation $y = \log_b x$, identify the base, the argument, and the logarithm.

3. Rewrite the equation in exponential form. $y = \log_b x$

4. Write the equation in logarithmic form. $b^y = x$

For Exercises 5–10, fill in the blanks.

5. a. $5^\square = 25$
   b. $\log_5 25 = \square$

6. a. $2^\square = 16$
   b. $\log_2 16 = \square$

7. a. $3^\square = 27$
   b. $\log_3 27 = \square$

8. a. $10^\square = 100{,}000$
   b. $\log_{10} 100{,}000 = \square$

9. a. $8^\square = 8$
   b. $\log_8 8 = \square$

10. a. $4^\square = 1$
    b. $\log_4 1 = \square$

For Exercises 11–22, write the equation in exponential form. (See Example 1.)

11. $\log_5 625 = 4$
12. $\log_{125} 25 = \dfrac{2}{3}$
13. $\log_{10}(0.0001) = -4$
14. $\log_{25}\left(\dfrac{1}{5}\right) = -\dfrac{1}{2}$

15. $\log_6 36 = 2$
16. $\log_2 128 = 7$
17. $\log_b 15 = x$
18. $\log_b 82 = y$

19. $\log_3 5 = x$
20. $\log_2 7 = x$
21. $\log_{1/4} x = 10$
22. $\log_{1/2} x = 6$

For Exercises 23–34, write the equation in logarithmic form.

23. $3^x = 81$
24. $10^3 = 1000$
25. $5^2 = 25$
26. $8^{1/3} = 2$

27. $7^{-1} = \dfrac{1}{7}$
28. $8^{-2} = \dfrac{1}{64}$
29. $b^x = y$
30. $b^y = x$

31. $e^x = y$
32. $e^y = x$
33. $\left(\dfrac{1}{3}\right)^{-2} = 9$
34. $\left(\dfrac{5}{2}\right)^{-1} = \dfrac{2}{5}$

Concept 2: Evaluating Logarithmic Expressions

For Exercises 35–50, evaluate the logarithm without the use of a calculator. (See Examples 2–3.)

35. $\log_7 49$
36. $\log_3 81$
37. $\log_{10} 0.1$
38. $\log_2\left(\dfrac{1}{16}\right)$

39. $\log_{16} 4$
40. $\log_8 2$
41. $\log_{7/2} 1$
42. $\log_{1/2} 2$

43. $\log_3 3^5$
44. $\log_9 9^3$
45. $\log_{10} 10$
46. $\log_7 1$

47. $\log_a a^3$
48. $\log_r r^4$
49. $\log_x \sqrt{x}$
50. $\log_y \sqrt[3]{y}$

Concept 3: The Common Logarithmic Function

For Exercises 51–58, evaluate the common logarithm without the use of a calculator. (See Example 4.)

51. $\log 10$
52. $\log 100$
53. $\log 1000$

54. $\log 10{,}000$
55. $\log(1.0 \times 10^6)$
56. $\log 0.1$

57. $\log 0.01$
58. $\log 0.001$

For Exercises 59–70, use a calculator to approximate the logarithms. Round to four decimal places. (See Example 5.)

59. $\log 6$ **60.** $\log 18$ **61.** $\log \pi$ **62.** $\log\left(\dfrac{1}{8}\right)$

63. $\log\left(\dfrac{1}{32}\right)$ **64.** $\log \sqrt{5}$ **65.** $\log 0.0054$ **66.** $\log 0.0000062$

67. $\log(3.4 \times 10^5)$ **68.** $\log(4.78 \times 10^9)$ **69.** $\log(3.8 \times 10^{-8})$ **70.** $\log(2.77 \times 10^{-4})$

71. Given: $\log 10 = 1$ and $\log 100 = 2$

    a. Estimate $\log 93$.

    b. Estimate $\log 12$.

    c. Evaluate the logarithms in parts (a) and (b) on a calculator and compare to your estimates.

72. Given: $\log\left(\dfrac{1}{10}\right) = -1$ and $\log 1 = 0$

    a. Estimate $\log\left(\dfrac{9}{10}\right)$.

    b. Estimate $\log\left(\dfrac{1}{5}\right)$.

    c. Evaluate the logarithms in parts (a) and (b) on a calculator and compare to your estimates.

Concept 4: Graphs of Logarithmic Functions

73. Let $f(x) = \log_4 x$.

    a. Find the values of $f\left(\dfrac{1}{64}\right)$, $f\left(\dfrac{1}{16}\right)$, $f\left(\dfrac{1}{4}\right)$, $f(1)$, $f(4)$, $f(16)$, and $f(64)$.

    b. Graph $y = f(x)$. (See Example 6.)

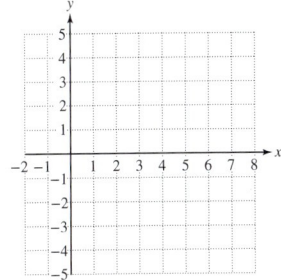

74. Let $g(x) = \log_2 x$.

    a. Find the values of $g\left(\dfrac{1}{8}\right)$, $g\left(\dfrac{1}{4}\right)$, $g\left(\dfrac{1}{2}\right)$, $g(1)$, $g(2)$, $g(4)$, and $g(8)$.

    b. Graph $y = g(x)$.

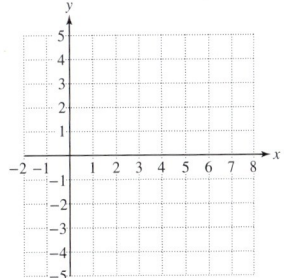

Graph the logarithmic functions in Exercises 75–78 by writing the function in exponential form and making a table of points. (See Example 6.)

75. $y = \log_3 x$

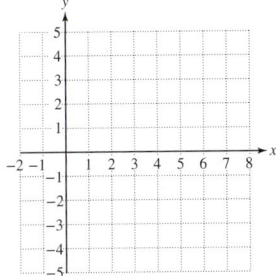

x	y

76. $y = \log_5 x$

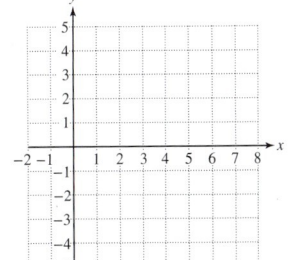

x	y

77. $y = \log_{1/2} x$ (See Example 6.)

78. $y = \log_{1/3} x$

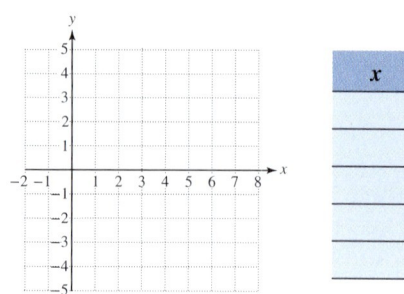
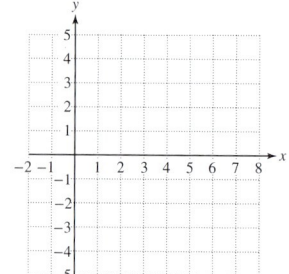

For Exercises 79–90, find the domain of the function and express the domain in interval notation. (See Example 7.)

79. $y = \log_7 (x - 5)$

80. $y = \log (x - 4)$

81. $g(x) = \log (2 - x)$

82. $f(x) = \log (1 - x)$

83. $y = \log (3x - 1)$

84. $y = \log_3 (2x + 1)$

85. $y = \log_3 (x + 1.2)$

86. $y = \log \left(x - \dfrac{1}{2} \right)$

87. $k(x) = \log (4 - 2x)$

88. $h(x) = \log (6 - 2x)$

89. $y = \log x^2$

90. $y = \log (x^2 + 1)$

Concept 5: Applications of the Common Logarithmic Function

For Exercises 91–92, use the formula pH = $-\log [H^+]$, where $[H^+]$ represents the concentration of hydrogen ions in moles per liter. Round to two decimal places. (See Example 8.)

91. Normally, the level of hydrogen ions in the blood is approximately 4.47×10^{-8} mol/L. Find the pH level of blood.

92. The normal pH level for streams and rivers is between 6.5 and 8.5. A high level of bacteria in a particular stream caused environmentalists to test the water. The level of hydrogen ions was found to be 0.006 mol/L. What is the pH level of the stream?

93. A graduate student in education is doing research to compare the effectiveness of two different teaching techniques designed to teach vocabulary to sixth-graders. The first group of students (group 1) was taught with method I, in which the students worked individually to complete the assignments in a workbook. The second group (group 2) was taught with method II, in which the students worked cooperatively in groups of four to complete the assignments in the same workbook.

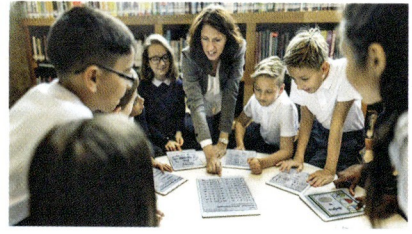
Rawpixel.com/Shutterstock

None of the students knew any of the vocabulary words before the study began. After completing the assignments in the workbook, the students were then tested on the vocabulary at 1-month intervals to assess how much material they had retained over time. The students' average scores t months after completing the assignments are given by the following functions:

Method I: $S_1(t) = 91 - 30 \log(t+1)$, where t is the time in months and $S_1(t)$ is the average score of students in group 1.

Method II: $S_2(t) = 88 - 15 \log(t+1)$, where t is the time in months and $S_2(t)$ is the average score of students in group 2.

a. Complete the table to find the average scores for each group of students after the indicated number of months. Round to one decimal place. (See Example 9.)

b. Based on the table of values, what were the average scores for each group immediately after completion of the assigned material ($t = 0$)?

t (months)	0	1	2	6	12	24
$S_1(t)$						
$S_2(t)$						

c. Based on the table of values, which teaching method helped students retain the material better for a long time?

94. Generally, the more money a company spends on advertising, the higher the sales. Let a represent the amount of money spent on advertising (in $100s). Then the amount of money in sales $S(a)$ (in $1000s) is given by

$$S(a) = 10 + 20 \log(a+1) \text{ where } a \geq 0$$

a. The value of $S(1) \approx 16.0$ means that if $100 is spent on advertising, $16,000 is returned in sales. Find the values of $S(11)$, $S(21)$, and $S(31)$. Round to one decimal place. Interpret the meaning of each function value in the context of this problem.

b. The graph of $y = S(a)$ is shown here. Use the graph and your answers from part (a) to explain why the money spent in advertising becomes less "efficient" as it is used in larger amounts.

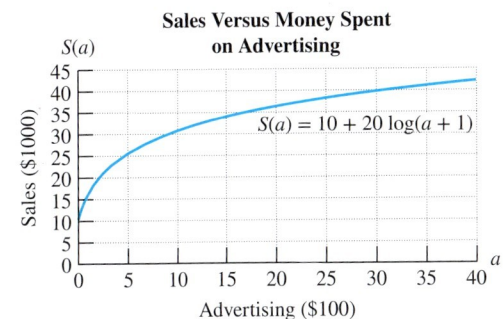

Sales Versus Money Spent on Advertising

Technology Connections

For Exercises 95–100, graph the function on an appropriate viewing window. From the graph, identify the domain of the function and the location of the vertical asymptote.

95. $y = \log(x + 6)$
96. $y = \log(2x + 4)$
97. $y = \log(0.5x - 1)$
98. $y = \log(x + 8)$
99. $y = \log(2 - x)$
100. $y = \log(3 - x)$

Problem Recognition Exercises

Identifying Graphs of Functions

Match the function with the appropriate graph. Do not use a calculator.

1. $g(x) = 3^x$
2. $f(x) = \log x$
3. $h(x) = x^2$
4. $k(x) = -2x - 3$
5. $L(x) = |x|$
6. $m(x) = \sqrt{x}$
7. $B(x) = 3$
8. $A(x) = -x^2$
9. $n(x) = \sqrt[3]{x}$
10. $p(x) = x^3$
11. $q(x) = \dfrac{1}{x}$
12. $r(x) = x$

a.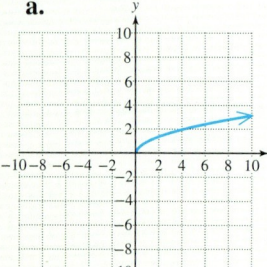

b.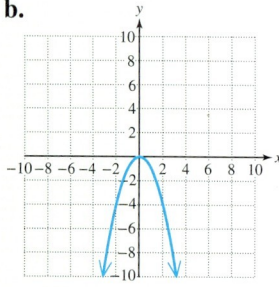

c.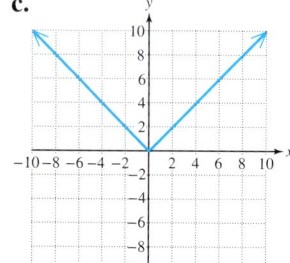

d.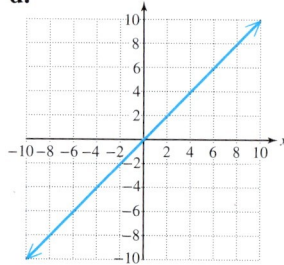

e.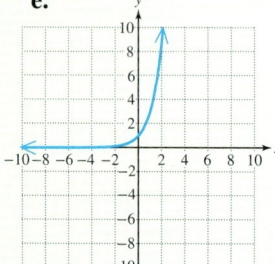

f.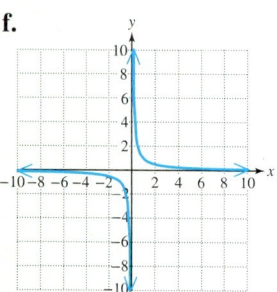

g.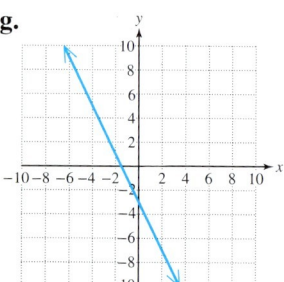

h.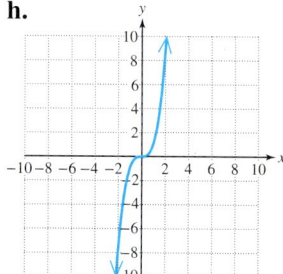

i.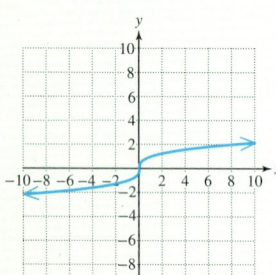

j.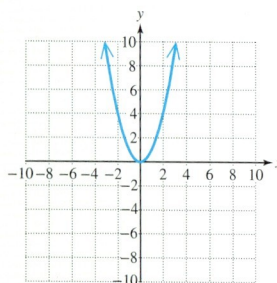

k.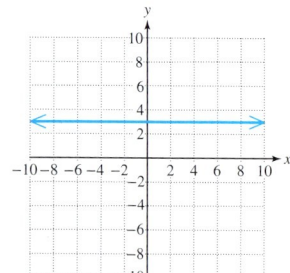

l.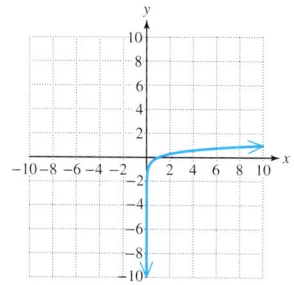

Properties of Logarithms

Section 20.4

Concepts

1. Properties of Logarithms
2. Expanded Logarithmic Expressions
3. Single Logarithmic Expressions

1. Properties of Logarithms

You have already been exposed to certain properties of logarithms that follow directly from the definition. Recall

$$y = \log_b x \quad \text{is equivalent to} \quad b^y = x \quad \text{for } x > 0, b > 0, \text{ and } b \neq 1$$

The following properties follow directly from the definition.

Property	Explanation
1. $\log_b 1 = 0$	Exponential form: $b^0 = 1$
2. $\log_b b = 1$	Exponential form: $b^1 = b$
3. $\log_b b^p = p$	Exponential form: $b^p = b^p$
4. $b^{\log_b x} = x$	Logarithmic form: $\log_b x = \log_b x$

Example 1 — Applying the Properties of Logarithms to Simplify Expressions

Use the properties of logarithms to simplify the expressions. Assume that all variable expressions within the logarithms represent positive real numbers.

a. $\log_8 8 + \log_8 1$ **b.** $10^{\log(x+2)}$ **c.** $\log_{1/2}\left(\dfrac{1}{2}\right)^x$

Solution:

a. $\log_8 8 + \log_8 1 = 1 + 0 = 1$ Properties 2 and 1

b. $10^{\log(x+2)} = x + 2$ Property 4

c. $\log_{1/2}\left(\dfrac{1}{2}\right)^x = x$ Property 3

Skill Practice Use the properties of logarithms to simplify the expressions.

1. $\log_5 1 + \log_5 5$ **2.** $15^{\log_{15} 7}$ **3.** $\log_{1/3}\left(\dfrac{1}{3}\right)^c$

Three additional properties are useful when simplifying logarithmic expressions. The first is the product property for logarithms.

> **Product Property of Logarithms**
>
> Let b, x, and y be positive real numbers where $b \neq 1$. Then
>
> $$\log_b (xy) = \log_b x + \log_b y$$
>
> The logarithm of a product equals the sum of the logarithms of the factors.

Answers
1. 1   **2.** 7   **3.** c

Proof:

Let $M = \log_b x$, which implies $b^M = x$.

Let $N = \log_b y$, which implies $b^N = y$.

Then $xy = b^M b^N = b^{M+N}$.

Writing the expression $xy = b^{M+N}$ in logarithmic form, we have,

$$\log_b (xy) = M + N$$

$$\log_b (xy) = \log_b x + \log_b y \checkmark$$

To demonstrate the product property for logarithms, simplify the following expressions by using the order of operations.

$$\log_3 (3 \cdot 9) \stackrel{?}{=} \log_3 3 + \log_3 9$$

$$\log_3 27 \stackrel{?}{=} 1 + 2$$

$$3 \stackrel{?}{=} 3 \checkmark \text{ True}$$

> **Quotient Property of Logarithms**
>
> Let b, x, and y be positive real numbers where $b \neq 1$. Then
>
> $$\log_b \left(\frac{x}{y}\right) = \log_b x - \log_b y$$
>
> The logarithm of a quotient equals the difference of the logarithms of the numerator and denominator.

The proof of the quotient property for logarithms is similar to the proof of the product property and is omitted here. To demonstrate the quotient property for logarithms, simplify the following expressions by using the order of operations.

$$\log \left(\frac{1{,}000{,}000}{100}\right) \stackrel{?}{=} \log (1{,}000{,}000) - \log (100)$$

$$\log (10{,}000) \stackrel{?}{=} 6 - 2$$

$$4 \stackrel{?}{=} 4 \checkmark \text{ True}$$

> **Power Property of Logarithms**
>
> Let b and x be positive real numbers where $b \neq 1$. Let p be any real number. Then
>
> $$\log_b x^p = p \log_b x$$

To demonstrate the power property for logarithms, simplify the following expressions by using the order of operations.

$$\log_4 4^2 \stackrel{?}{=} 2 \log_4 4$$

$$2 \stackrel{?}{=} 2 \cdot 1$$

$$2 \stackrel{?}{=} 2 \checkmark \text{ True}$$

The properties of logarithms are summarized in the box.

> **Summary of the Properties of Logarithms**
>
> Let b, x, and y be positive real numbers where $b \neq 1$, and let p be a real number. Then the following properties of logarithms are true.
>
> 1. $\log_b 1 = 0$
> 2. $\log_b b = 1$
> 3. $\log_b b^p = p$
> 4. $b^{\log_b x} = x$
> 5. $\log_b (xy) = \log_b x + \log_b y$ **Product property for logarithms**
> 6. $\log_b \left(\dfrac{x}{y}\right) = \log_b x - \log_b y$ **Quotient property for logarithms**
> 7. $\log_b x^p = p \log_b x$ **Power property for logarithms**

2. Expanded Logarithmic Expressions

In many applications, it is advantageous to expand a logarithm into a sum or difference of simpler logarithms.

Example 2   Writing a Logarithmic Expression in Expanded Form

Write the expression as the sum or difference of logarithms of x, y, and z. Assume all variables represent positive real numbers.

$$\log_3 \left(\dfrac{xy^3}{z^2}\right)$$

Solution:

$\log_3 \left(\dfrac{xy^3}{z^2}\right)$

$= \log_3 (xy^3) - \log_3 z^2$ Quotient property for logarithms (property 6)

$= [\log_3 x + \log_3 y^3] - \log_3 z^2$ Product property for logarithms (property 5)

$= \log_3 x + 3 \log_3 y - 2 \log_3 z$ Power property for logarithms (property 7)

Skill Practice Write the expression as the sum or difference of logarithms of a, b, and c. Assume all variables represent positive real numbers.

4. $\log_5 \left(\dfrac{a^2 b^3}{c}\right)$

Example 3   Writing a Logarithmic Expression in Expanded Form

Write the expression as the sum or difference of logarithms of x and y. Assume all variables represent positive real numbers.

$$\log \left(\dfrac{\sqrt{x+y}}{10}\right)$$

Answer

4. $2 \log_5 a + 3 \log_5 b - \log_5 c$

Solution:

$$\log\left(\frac{\sqrt{x+y}}{10}\right)$$

$= \log(\sqrt{x+y}) - \log 10$ Quotient property for logarithms (property 6)

$= \log(x+y)^{1/2} - 1$ Write $\sqrt{x+y}$ as $(x+y)^{1/2}$ and simplify $\log 10 = 1$.

$= \dfrac{1}{2}\log(x+y) - 1$ Power property for logarithms (property 7)

Skill Practice Write the expression as the sum or difference of logarithms of a and b. Assume $a > b$.

5. $\log\dfrac{(a-b)^2}{\sqrt{10}}$

Example 4   **Writing a Logarithmic Expression in Expanded Form**

Write the expression as the sum or difference of logarithms of x, y, and z. Assume all variables represent positive real numbers.

$$\log_b \sqrt[5]{\frac{x^4}{yz^3}}$$

Solution:

$\log_b \sqrt[5]{\dfrac{x^4}{yz^3}}$

$= \log_b \left(\dfrac{x^4}{yz^3}\right)^{1/5}$ Write $\sqrt[5]{\dfrac{x^4}{yz^3}}$ as $\left(\dfrac{x^4}{yz^3}\right)^{1/5}$.

$= \dfrac{1}{5}\log_b \left(\dfrac{x^4}{yz^3}\right)$ Power property for logarithms (property 7)

$= \dfrac{1}{5}[\log_b x^4 - \log_b(yz^3)]$ Quotient property for logarithms (property 6)

$= \dfrac{1}{5}[\log_b x^4 - (\log_b y + \log_b z^3)]$ Product property for logarithms (property 5)

$= \dfrac{1}{5}[\log_b x^4 - \log_b y - \log_b z^3]$ Distributive property

$= \dfrac{1}{5}[4\log_b x - \log_b y - 3\log_b z]$ Power property for logarithms (property 7)

or $\dfrac{4}{5}\log_b x - \dfrac{1}{5}\log_b y - \dfrac{3}{5}\log_b z$

Skill Practice Write the expression as the sum or difference of logarithms of a, b, and c. Assume all variables represent positive real numbers.

6. $\log_3 \sqrt[4]{\dfrac{a}{bc^5}}$

Answers

5. $2\log(a-b) - \dfrac{1}{2}$
6. $\dfrac{1}{4}\log_3 a - \dfrac{1}{4}\log_3 b - \dfrac{5}{4}\log_3 c$

3. Single Logarithmic Expressions

In some applications, it is necessary to write a sum or difference of logarithms as a single logarithm.

Example 5   Writing a Sum or Difference of Logarithms as a Single Logarithm

Rewrite the expression as a single logarithm, and simplify the result, if possible.

$$\log_2 560 - \log_2 7 - \log_2 5$$

Solution:

$\log_2 560 - \log_2 7 - \log_2 5$

$= \log_2 560 - (\log_2 7 + \log_2 5)$ Factor out -1 from the last two terms.

$= \log_2 560 - \log_2 (7 \cdot 5)$ Product property for logarithms (property 5)

$= \log_2 \left(\dfrac{560}{7 \cdot 5}\right)$ Quotient property for logarithms (property 6)

$= \log_2 16$ Simplify inside parentheses.

$= \log_2 2^4$ Write 16 in base 2. That is, $16 = 2^4$.

$= 4$ Property 3

Skill Practice   Write the expression as a single logarithm, and simplify the result, if possible.

7. $\log_3 54 + \log_3 10 - \log_3 20$

Example 6   Writing a Sum or Difference of Logarithms as a Single Logarithm

Rewrite the expression as a single logarithm, and simplify the result, if possible. Assume all variable expressions within the logarithms represent positive real numbers.

$$2 \log x - \dfrac{1}{2} \log y + 3 \log z$$

Solution:

$2 \log x - \dfrac{1}{2} \log y + 3 \log z$

$= \log x^2 - \log y^{1/2} + \log z^3$ Power property for logarithms (property 7)

$= \log x^2 + \log z^3 - \log y^{1/2}$ Group terms with positive coefficients.

$= \log (x^2 z^3) - \log y^{1/2}$ Product property for logarithms (property 5)

$= \log \left(\dfrac{x^2 z^3}{y^{1/2}}\right)$ or $\log \left(\dfrac{x^2 z^3}{\sqrt{y}}\right)$ Quotient property for logarithms (property 6)

Skill Practice   Write the expression as a single logarithm, and simplify, if possible.

8. $3 \log x + \dfrac{1}{3} \log y - 2 \log z$

Answers

7. 3 8. $\log \left(\dfrac{x^3 \sqrt[3]{y}}{z^2}\right)$

It is important to note that the properties of logarithms may be used to write a single logarithm as a sum or difference of logarithms. Furthermore, the properties may be used to write a sum or difference of logarithms as a single logarithm. In either case, these operations may change the domain.

For example, consider the function $y = \log_b x^2$. Using the power property for logarithms, we have $y = 2 \log_b x$. Consider the domain of each function:

$y = \log_b x^2$ Domain: $(-\infty, 0) \cup (0, \infty)$

$y = 2 \log_b x$ Domain: $(0, \infty)$

These two functions are equivalent only for values of x in the intersection of the two domains, that is, for $(0, \infty)$.

Section 20.4 Activity

For Exercises A.1–A.4, investigate the properties of logarithms by completing the table. Work across the table, doing one row at a time.

	Example	Example	Example	Property
A.1.	a. $\log_5 5 =$ ____	b. $\log_2 2 =$ ____	c. $\log_{1/2}\left(\dfrac{1}{2}\right) =$ ____	d. $\log_b b =$ ____
A.2.	a. $\log_5 1 =$ ____	b. $\log_2 1 =$ ____	c. $\log_{1/2} 1 =$ ____	d. $\log_b 1 =$ ____
A.3.	a. $\log_5 (5)^3 =$ ____	b. $\log_2 (2)^4 =$ ____	c. $\log_{1/2}\left(\dfrac{1}{2}\right)^5 =$ ____	d. $\log_b (b)^p =$ ____
A.4.	a. $5^{\log_5(25)} =$ ____	b. $2^{\log_2(16)} =$ ____	c. $\left(\dfrac{1}{2}\right)^{\log_{1/2}\left(\frac{1}{8}\right)} =$ ____	d. $b^{\log_b x} =$ ____

For Exercises A.5–A.7, investigate three more properties of logarithms. Work across the table, doing one row at a time.

	Example	Property
A.5.	$\log_2 8 = \log_2 4 + \log_2 2$ $\square = \square + \square$	**Product property of logarithms** $\log_b (x \cdot y) =$ ____
A.6.	$\log_5 \left(\dfrac{125}{5}\right) = \log_5 125 - \log_5 5$ $\square = \square - \square$	**Quotient property of logarithms** $\log_b \left(\dfrac{x}{y}\right) =$ ____
A.7.	$\log_4 (4)^3 = 3 \cdot \log_4 (4)$ $\square = \square \cdot \square$	**Power property of logarithms** $\log_b (x)^p =$ ____

A.8. Write the expression as a single logarithm. Assume that all variables represent positive real numbers.

$$5 \log_2 x - 2 \log_2 y + \frac{1}{2} \log_2 z - \frac{1}{3} \log_2 w$$

A.9. Write the expression as a sum or difference of $\log x$, $\log y$, and $\log z$. Assume that all variables represent positive real numbers.

$$\log\left(\frac{100\sqrt{x}\, y^3}{z}\right)$$

Practice Exercises

Section 20.4

Prerequisite Review

For Exercises R.1–R.10, use the properties of exponents to simplify the expression.

R.1. $x^{-4} \cdot x^{10}$ **R.2.** $p^{13} \cdot p^{-6}$ **R.3.** $\dfrac{w^{15}}{w^4}$ **R.4.** $\dfrac{m^{20}}{m^{15}}$

R.5. $(n^3)^4$ **R.6.** $(k^2)^5$ **R.7.** t^{-6} **R.8.** a^{-4}

R.9. $(3a^{-5}b^3)^2$ **R.10.** $\left(\dfrac{2c^4}{d^{-6}}\right)^3$

For Exercises R.11–R.16, simplify the expression.

R.11. $\log_4 64$ **R.12.** $\log_2 32$ **R.13.** $\log_2 \dfrac{1}{16}$ **R.14.** $\log_4 \dfrac{1}{16}$

R.15. $\log 1000$ **R.16.** $\log \dfrac{1}{1000}$

Vocabulary and Key Concepts

1. **a.** Fill in the blanks to complete the basic properties of logarithms.
   $\log_b b =$ _____, $\log_b 1 =$ _____, $\log_b b^x =$ _____, and $b^{\log_b x} =$ _____.

   b. If b, x, and y are positive real numbers and $b \neq 1$, then $\log_b(xy) =$ _____ and $\log_b\left(\dfrac{x}{y}\right) =$ _____.

   c. If b and x are positive real numbers and $b \neq 1$, then for any real number p, $\log_b x^p$ can be written as _____.

   d. Determine if the statement is true or false: $\log_b(xy) = (\log_b x)(\log_b y)$
   Use the expression $\log_2(4 \cdot 8)$ to help you answer.

   e. Determine if the statement is true or false: $\log_b\left(\dfrac{x}{y}\right) = \dfrac{\log_b x}{\log_b y}$
   Use the expression $\log_3\left(\dfrac{27}{9}\right)$ to help you answer.

   f. Determine if the statement is true or false: $\log_b(x)^p = (\log_b x)^p$
   Use the expression $\log(1000)^2$ to help you answer.

Concept 1: Properties of Logarithms

2. Select the values that are equivalent to $\log_5 5^2$.
   a. $\log_5 25$
   b. $2 \log_5 5$
   c. $\log_5 5 + \log_5 5$

3. Select the values that are equivalent to $\log_2 2^3$.
   a. $3 \log_2 2$
   b. $\log_2 8$
   c. 3

4. Select the values that are equivalent to $\log 10^4$.
   a. 4
   b. $4 \log 10$
   c. $\log 10{,}000$

For Exercises 5–28, evaluate each expression. (See Example 1.)

5. $\log_3 3$ **6.** $\log 10$ **7.** $\log_5 5^4$ **8.** $\log_4 4^5$

9. $6^{\log_6 11}$ **10.** $7^{\log_7 2}$ **11.** $\log 10^3$ **12.** $\log_6 6^3$

13. $\log_3 1$ **14.** $\log_8 1$ **15.** $10^{\log 9}$ **16.** $8^{\log_8 5}$

17. $\log_{1/2} 1$

18. $\log_{1/3}\left(\dfrac{1}{3}\right)$

19. $\log_2 1 + \log_2 2^3$

20. $\log 10^4 + \log 10$

21. $\log_4 4 + \log_2 1$

22. $\log_7 7 + \log_4 4^2$

23. $\log_{1/4}\left(\dfrac{1}{4}\right)^{2x}$

24. $\log_{2/3}\left(\dfrac{2}{3}\right)^p$

25. $\log_a a^4$

26. $\log_y y^2$

27. $\log 10^2 - \log_3 3^2$

28. $\log_6 6^4 - \log 10^4$

29. Compare the expressions by approximating their values on a calculator. Which two expressions are equivalent?
   a. $\log(3 \cdot 5)$
   b. $\log 3 \cdot \log 5$
   c. $\log 3 + \log 5$

30. Compare the expressions by approximating their values on a calculator. Which two expressions are equivalent?
   a. $\log\left(\dfrac{6}{5}\right)$
   b. $\dfrac{\log 6}{\log 5}$
   c. $\log 6 - \log 5$

31. Compare the expressions by approximating their values on a calculator. Which two expressions are equivalent?
   a. $\log 20^2$
   b. $[\log 20]^2$
   c. $2 \log 20$

32. Compare the expressions by approximating their values on a calculator. Which two expressions are equivalent?
   a. $\log \sqrt{4}$
   b. $\dfrac{1}{2} \log 4$
   c. $\sqrt{\log 4}$

Concept 2: Expanded Logarithmic Expressions

For Exercises 33–50, expand into sums and/or differences of logarithms. Assume all variables represent positive real numbers. (See Examples 2–4.)

33. $\log_3\left(\dfrac{x}{5}\right)$

34. $\log_2\left(\dfrac{y}{z}\right)$

35. $\log(2x)$

36. $\log_6(xyz)$

37. $\log_5 x^4$

38. $\log_7 z^{1/3}$

39. $\log_4\left(\dfrac{ab}{c}\right)$

40. $\log_2\left(\dfrac{x}{yz}\right)$

41. $\log_b\left(\dfrac{\sqrt{xy}}{z^3 w}\right)$

42. $\log\left(\dfrac{a\sqrt[3]{b}}{cd^2}\right)$

43. $\log_2\left(\dfrac{x+1}{y^2\sqrt{z}}\right)$

44. $\log\left(\dfrac{a+1}{b\sqrt[3]{c}}\right)$

45. $\log\left(\sqrt[3]{\dfrac{ab^2}{c}}\right)$

46. $\log_5\left(\sqrt[4]{\dfrac{w^3 z}{x^2}}\right)$

47. $\log\left(\dfrac{1}{w^5}\right)$

48. $\log_3\left(\dfrac{1}{z^4}\right)$

49. $\log_b\left(\dfrac{\sqrt{a}}{b^3 c}\right)$

50. $\log_x\left(\dfrac{x}{y\sqrt{z}}\right)$

Concept 3: Single Logarithmic Expressions

For Exercises 51–66, write the expressions as a single logarithm and simplify if possible. Assume all variable expressions represent positive real numbers. (See Examples 5–6.)

51. $\log_3 270 - \log_3 2 - \log_3 5$

52. $\log_5 8 + \log_5 50 - \log_5 16$

53. $\log_7 98 - \log_7 2$

54. $\log_6 24 - \log_6 4$

55. $2 \log_3 x - 3 \log_3 y + \log_3 z$

56. $\log C + \log A + \log B + \log I + \log N$

57. $2 \log_3 a - \dfrac{1}{4} \log_3 b + \log_3 c$

58. $\log_5 a - \dfrac{1}{2} \log_5 b - 3 \log_5 c$

59. $\log_b x - 3 \log_b x + 4 \log_b x$

60. $2 \log_3 z + \log_3 z - \dfrac{1}{2} \log_3 z$

61. $5 \log_8 a - \log_8 1 + \log_8 8$

62. $\log_2 2 + 2 \log_2 b - \log_2 1$

63. $2 \log (x + 6) + \dfrac{1}{3} \log y - 5 \log z$

64. $\dfrac{1}{4} \log (a + 1) - 2 \log b - 4 \log c$

65. $\log_b (x + 1) - \log_b (x^2 - 1)$

66. $\log_x (p^2 - 4) - \log_x (p - 2)$

Technology Connections

67. **a.** Graph $Y_1 = \log x^2$ and state its domain.

    b. Graph $Y_2 = 2 \log x$ and state its domain.

    c. For what values of x are the expressions $\log x^2$ and $2 \log x$ equivalent?

68. **a.** Graph $Y_1 = \log (x - 1)^2$ and state its domain.

    b. Graph $Y_2 = 2 \log (x - 1)$ and state its domain.

    c. For what values of x are the expressions $\log (x - 1)^2$ and $2 \log (x - 1)$ equivalent?

Expanding Your Skills

For Exercises 69–80, find the values of the logarithms given that $\log_b 2 \approx 0.693$, $\log_b 3 \approx 1.099$, and $\log_b 5 \approx 1.609$.

69. $\log_b 6$

70. $\log_b 4$

71. $\log_b 12$

72. $\log_b 25$

73. $\log_b 81$

74. $\log_b 30$

75. $\log_b \left(\dfrac{5}{2}\right)$

76. $\log_b \left(\dfrac{25}{3}\right)$

77. $\log_b 10^6$

78. $\log_b 15^3$

79. $\log_b 5^{10}$

80. $\log_b 2^{12}$

81. The intensity of sound waves is measured in decibels and is calculated by the formula

$$B = 10 \log \left(\dfrac{I}{I_0}\right)$$

where I_0 is the minimum detectable decibel level.

    a. Expand this formula by using the properties of logarithms.

    b. Let $I_0 = 10^{-16}$ W/cm² and simplify.

82. The Richter scale is used to measure the intensity of an earthquake and is calculated by the formula

$$R = \log \left(\dfrac{I}{I_0}\right)$$

where I_0 is the minimum level detectable by a seismograph.

    a. Expand this formula by using the properties of logarithms.

    b. Let $I_0 = 1$ and simplify.

Section 20.5 The Irrational Number e and Change of Base

Concepts

1. The Irrational Number e
2. Computing Compound Interest
3. The Natural Logarithmic Function
4. Change-of-Base Formula
5. Applications of the Natural Logarithmic Function

1. The Irrational Number e

The exponential function base 10 is particularly easy to work with because integral powers of 10 represent different place positions in the base-10 numbering system. In this section, we introduce another important exponential function whose base is an irrational number called e.

Consider the expression $(1 + \frac{1}{x})^x$. The value of the expression for increasingly large values of x approaches a constant (Table 20-5).

Table 20-5

x	$\left(1 + \dfrac{1}{x}\right)^x$
100	2.70481382942
1000	2.71692393224
10,000	2.71814592683
100,000	2.71826823717
1,000,000	2.71828046932
1,000,000,000	2.71828182710

As x approaches infinity, the expression $(1 + \frac{1}{x})^x$ approaches a constant value that we call **e**. From Table 20-5, this value is approximately 2.718281828.

$$e \approx 2.718281828$$

The value of e is an irrational number (a nonterminating, nonrepeating decimal) and like the number π, it is a universal constant.

Example 1   Graphing $f(x) = e^x$

Graph the function defined by $f(x) = e^x$.

Solution:

Because the base of the function is greater than 1 ($e \approx 2.718281828$), the graph is an increasing exponential function. We can use a calculator to evaluate $f(x) = e^x$ at several values of x.

Practice using your calculator by evaluating e^x for $x = 1$, $x = 2$, and $x = 3$.

If you are using your calculator correctly, your answers should match those found in Table 20-6. Values are rounded to three decimal places. The corresponding graph of $f(x) = e^x$ is shown in Figure 20-14.

TIP: On a scientific calculator, you may need to enter the base and exponent of an exponential expression in reverse order. For example, enter e^2 as:

2

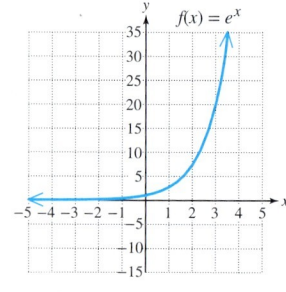

Table 20-6

x	$f(x) = e^x$
−3	0.050
−2	0.135
−1	0.368
0	1.000
1	2.718
2	7.389
3	20.086

Figure 20-14

Skill Practice

1. Graph $f(x) = e^x + 1$.

2. Computing Compound Interest

One particularly interesting application of exponential functions is in computing compound interest.

1. If the number of compounding periods per year is finite, then the amount $A(t)$ in an account is given by

$$A(t) = P\left(1 + \frac{r}{n}\right)^{nt}$$

where P is the initial principal, r is the annual interest rate, n is the number of times compounded per year, and t is the time in years that the money is invested.

2. If the number of compound periods per year is infinite, then interest is said to be **compounded continuously**. In such a case, the amount $A(t)$ in the account is given by

$$A(t) = Pe^{rt}$$

where P is the initial principal, r is the annual interest rate, and t is the time in years that the money is invested.

Example 2 — Computing the Balance on an Account

Suppose $5000 is invested in an account earning 6.5% interest. Find the balance in the account after 10 years under the following compounding options.

a. Compounded annually
b. Compounded quarterly
c. Compounded monthly
d. Compounded daily
e. Compounded continuously

Solution:

Compounding Option	n Value	Formula	Result
Annually	$n = 1$	$A(10) = 5000\left(1 + \frac{0.065}{1}\right)^{(1)(10)}$	$9385.69
Quarterly	$n = 4$	$A(10) = 5000\left(1 + \frac{0.065}{4}\right)^{(4)(10)}$	$9527.79
Monthly	$n = 12$	$A(10) = 5000\left(1 + \frac{0.065}{12}\right)^{(12)(10)}$	$9560.92
Daily	$n = 365$	$A(10) = 5000\left(1 + \frac{0.065}{365}\right)^{(365)(10)}$	$9577.15
Continuously	Not applicable	$A(10) = 5000e^{(0.065)(10)}$	$9577.70

Notice that there is a $191.46 difference in the account balance between annual compounding and daily compounding. However, the difference between compounding daily and compounding continuously is small, $0.55. As n gets infinitely large, the function defined by

$$A(t) = P\left(1 + \frac{r}{n}\right)^{nt} \qquad \text{converges to} \qquad A(t) = Pe^{rt}$$

Answer

1.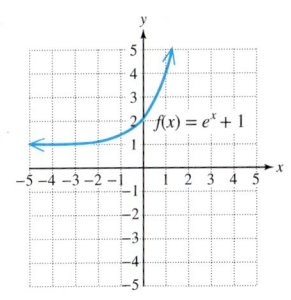

Skill Practice

2. Suppose $1000 is invested at 5%. Find the balance after 8 years under the following options.
   a. Compounded annually
   b. Compounded quarterly
   c. Compounded monthly
   d. Compounded daily
   e. Compounded continuously

3. The Natural Logarithmic Function

Recall that the common logarithmic function $y = \log x$ has a base of 10. Another important logarithmic function is called the **natural logarithmic function**. The natural logarithmic function has a base of e and is written as $y = \ln x$. That is,

$$y = \ln x = \log_e x$$

Example 3 — Graphing $y = \ln x$

Graph $y = \ln x$.

Solution:

Because the base of the function $y = \ln x$ is e and $e > 1$, the graph is an increasing logarithmic function. We can use a calculator to find specific points on the graph of $y = \ln x$ by pressing the **LN** key.

Practice using your calculator by evaluating $\ln x$ for the following values of x. If you are using your calculator correctly, your answers should match those found in Table 20-7. Values are rounded to three decimal places. The corresponding graph of $y = \ln x$ is shown in Figure 20-15.

TIP: On a scientific calculator, you may need to enter a logarithm and argument in reverse order. For example, enter ln 2 as

2 **LN**

Table 20-7

x	$\ln x$
0.25	−1.386
0.5	−0.693
1	0.000
2	0.693
3	1.099
4	1.386
5	1.609
6	1.792
7	1.946

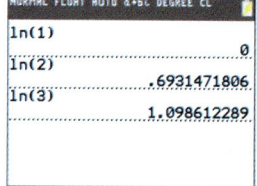

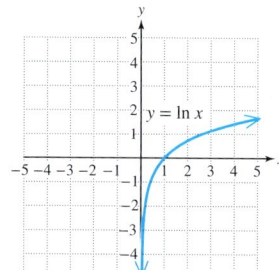

Figure 20-15

Skill Practice

3. Graph $y = \ln x + 1$.

Answers

2. a. $1477.46 b. $1488.13
   c. $1490.59 d. $1491.78
   e. $1491.82

3.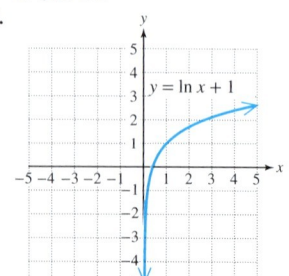

The properties of logarithms learned earlier in this chapter are also true for natural logarithms.

> **Properties of the Natural Logarithmic Function**
>
> Let x and y be positive real numbers, and let p be a real number. Then the following properties are true.
>
> 1. $\ln 1 = 0$
> 2. $\ln e = 1$
> 3. $\ln e^p = p$
> 4. $e^{\ln x} = x$
> 5. $\ln(xy) = \ln x + \ln y$ **Product property for logarithms**
> 6. $\ln\left(\dfrac{x}{y}\right) = \ln x - \ln y$ **Quotient property for logarithms**
> 7. $\ln x^p = p \ln x$ **Power property for logarithms**

Example 4 — Simplifying Expressions With Natural Logarithms

Simplify the expressions. Assume that all variable expressions within the logarithms represent positive real numbers.

a. $\ln e$ b. $\ln 1$ c. $\ln(e^{x+1})$ d. $e^{\ln(x+1)}$

Solution:

a. $\ln e = 1$ — Property 2

b. $\ln 1 = 0$ — Property 1

c. $\ln(e^{x+1}) = x + 1$ — Property 3

d. $e^{\ln(x+1)} = x + 1$ — Property 4

Skill Practice Simplify.

4. $\ln e^2$ 5. $-3 \ln 1$ 6. $\ln e^{(x+y)}$ 7. $e^{\ln(3x)}$

Example 5 — Writing a Sum or Difference of Natural Logarithms as a Single Logarithm

Write the expression as a single logarithm. Assume that all variable expressions within the logarithms represent positive real numbers.

$$\ln x - \frac{1}{3} \ln y - \ln z$$

Solution:

$\ln x - \dfrac{1}{3} \ln y - \ln z$

$= \ln x - \ln y^{1/3} - \ln z$ — Power property for logarithms (property 7)

$= \ln x - (\ln y^{1/3} + \ln z)$ — Factor out -1 from the last two terms.

$= \ln x - \ln(y^{1/3} \cdot z)$ — Product property for logarithms (property 5)

Answers

4. 2   5. 0   6. $x + y$ 7. $3x$

$$= \ln\left(\frac{x}{y^{1/3}z}\right) \quad \text{Quotient property for logarithms (property 6)}$$

$$= \ln\left(\frac{x}{\sqrt[3]{y}\,z}\right) \quad \text{Write the rational exponent as a radical.}$$

Skill Practice Write as a single logarithm.

8. $\dfrac{1}{4}\ln a - \ln b + \ln c$

Example 6 Writing a Logarithmic Expression in Expanded Form

Write the expression as a sum or difference of logarithms of x and y. Assume all variable expressions within the logarithm represent positive real numbers.

$$\ln\left(\frac{e}{x^2\sqrt{y}}\right)$$

Solution:

$$\ln\left(\frac{e}{x^2\sqrt{y}}\right)$$

$$= \ln e - \ln(x^2\sqrt{y}) \quad \text{Quotient property for logarithms (property 6)}$$

$$= \ln e - (\ln x^2 + \ln \sqrt{y}) \quad \text{Product property for logarithms (property 5)}$$

$$= 1 - \ln x^2 - \ln y^{1/2} \quad \text{Distributive property. Also simplify } \ln e = 1 \text{ (property 2).}$$

$$= 1 - 2\ln x - \frac{1}{2}\ln y \quad \text{Power property for logarithms (property 7)}$$

Skill Practice Write as a sum or difference of logarithms of x and y.

9. $\ln\left(\dfrac{x^3\sqrt{y}}{e^2}\right)$

4. Change-of-Base Formula

A calculator can be used to approximate the value of a logarithm with a base of 10 or a base of e by using the **LOG** key or the **LN** key, respectively. However, to use a calculator to evaluate a logarithmic expression with a base other than 10 or e, we must use the **change-of-base formula**.

> **Change-of-Base Formula**
>
> Let a and b be positive real numbers such that $a \neq 1$ and $b \neq 1$. Then for any positive real number x,
>
> $$\log_b x = \frac{\log_a x}{\log_a b}$$

Answers

8. $\ln\left(\dfrac{\sqrt[4]{a}\,c}{b}\right)$

9. $3\ln x + \dfrac{1}{2}\ln y - 2$

Proof:

Let $M = \log_b x$, which implies that $b^M = x$.

Take the logarithm, base a, on both sides: $\quad \log_a b^M = \log_a x$

Apply the power property for logarithms: $\quad M \cdot \log_a b = \log_a x$

Divide both sides by $\log_a b$: $\quad \dfrac{M \cdot \log_a b}{\log_a b} = \dfrac{\log_a x}{\log_a b}$

$$M = \dfrac{\log_a x}{\log_a b}$$

Because $M = \log_b x$, we have $\quad \log_b x = \dfrac{\log_a x}{\log_a b}$ ✓

The change-of-base formula converts a logarithm of one base to a ratio of logarithms of a different base. For the sake of using a calculator, we often apply the change-of-base formula with base 10 or base e.

Example 7 — Using the Change-of-Base Formula

a. Use the change-of-base formula to evaluate $\log_4 80$ by using base 10. (Round to three decimal places.)

b. Use the change-of-base formula to evaluate $\log_4 80$ by using base e. (Round to three decimal places.)

Solution:

a. $\log_4 80 = \dfrac{\log_{10} 80}{\log_{10} 4} = \dfrac{\log 80}{\log 4} \approx \dfrac{1.903089987}{0.6020599913} \approx 3.161$

b. $\log_4 80 = \dfrac{\log_e 80}{\log_e 4} = \dfrac{\ln 80}{\ln 4} \approx \dfrac{4.382026635}{1.386294361} \approx 3.161$

To check the result, we see that $4^{3.161} \approx 80$.

Skill Practice

10. Use the change-of-base formula to evaluate $\log_5 95$ by using base 10. Round to three decimal places.

11. Use the change-of-base formula to evaluate $\log_5 95$ by using base e. Round to three decimal places.

5. Applications of the Natural Logarithmic Function

Plant and animal tissue contains both carbon-12 and carbon-14. Carbon-12 is a stable form of carbon, whereas carbon-14 is a radioactive isotope with a half-life of approximately 5730 years. While a plant or animal is living, it takes in carbon from the atmosphere either through photosynthesis or through its food. The ratio of carbon-14 to carbon-12 in a living organism is constant and is the same as the ratio found in the atmosphere.

When a plant or animal dies, it no longer ingests carbon from the atmosphere. The amount of stable carbon-12 remains unchanged from the time of death, but the carbon-14 begins to decay. Because the rate of decay is constant, a tissue sample can be dated by comparing the percent of carbon-14 still present to the percentage of carbon-14 assumed to be in its original living state.

Answers

10. 2.829   11. 2.829

The age of a tissue sample is a function of the percent of carbon-14 still present in the organism according to the following model:

$$A(p) = \frac{\ln p}{-0.000121}$$

where $A(p)$ is the age in years and p is the percentage (in decimal form) of carbon-14 still present.

Example 8 — Applying the Natural Logarithmic Function to Radioactive Decay

Using the formula

$$A(p) = \frac{\ln p}{-0.000121}$$

a. Find the age of a bone that has 72% of its initial carbon-14.

b. Find the age of the Iceman, a body uncovered in the mountains of northern Italy in 1991. Samples of his hair revealed that 52.7% of the original carbon-14 was present after his death.

Solution:

a. $A(p) = \dfrac{\ln p}{-0.000121}$

$A(0.72) = \dfrac{\ln 0.72}{-0.000121}$ Substitute 0.72 for p.

≈ 2715 years

b. $A(p) = \dfrac{\ln p}{-0.000121}$

$A(0.527) = \dfrac{\ln 0.527}{-0.000121}$ Substitute 0.527 for p.

≈ 5300 years   The body of the Iceman is approximately 5300 years old.

Skill Practice

12. Use the formula

$$A(p) = \frac{\ln p}{-0.000121}$$

(where $A(p)$ is the age in years and p is the percent of carbon-14 still present) to determine the age of a human skull that has 90% of its initial carbon-14.

Answer

12. ≈ 871 years

Section 20.5 Activity

A.1. Evaluate the expression $\left(1 + \dfrac{1}{x}\right)^x$ for the following values of x. Round to five decimal places.

    a. 5000   **b.** 50,000   **c.** 500,000   **d.** 5,000,000

A.2. For larger and larger values of x, the expression $\left(1 + \dfrac{1}{x}\right)^x$ converges to a constant real number that mathematicians designate by e. The value of e is an irrational number (its decimal form is nonterminating and nonrepeating).

    a. Based on the results of Exercise A.1, write the value of e to five decimal places.

    b. Based on the value of e, is the function defined by $f(x) = e^x$ an exponential growth function or an exponential decay function?

A.3. The following three functions, A_1, A_2, and A_3, determine (in order) the amount of money in an account earning simple interest, interest compounded annually, and interest compounded continuously.

$$A_1(t) = P + Prt \qquad A_2(t) = P(1 + r)^t \qquad A_3(t) = Pe^{rt}$$

P represents the amount of principal invested, r represents the annual interest rate, and t represents the time in years of the investment. For this example, let $P = \$5000$ and $r = 0.08$.

    a. Evaluate $A_1(30)$ and interpret its meaning in context.

    b. Evaluate $A_2(30)$ and interpret its meaning in context.

    c. Evaluate $A_3(30)$ and interpret its meaning in context.

    d. From parts (a)–(c), which method of compounding yields the best return?

    e. How much more money is earned in 30 years for interest compounded continuously than for simple interest?

    f. Match the functions A_1, A_2, and A_3 with their graphs.

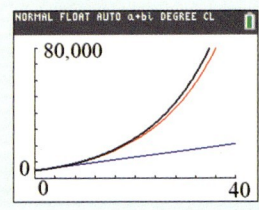

A.4. The inverse of the exponential function defined by $y = e^x$ is the _____ function, base e, denoted by _____.

A.5. **a.** Write the domain and range of $f(x) = e^x$ in interval notation.

    b. Write the domain and range of $g(x) = \ln x$ in interval notation.

    c. How are the domain and range of functions f and g related?

The properties of logarithms learned to this point also apply to the natural logarithm function. Exercises A.6–A.8 investigate these properties.

A.6. Evaluate the logarithmic expressions.

    a. $\ln 1$ **b.** $\ln e$ **c.** $\ln e^4$ **d.** $e^{\ln 6}$

A.7. Write the expression as a single logarithm. Assume that all variable expressions represent positive real numbers.

$$\ln(x^2 - y^2) - \ln(x - y) - 3\ln z$$

A.8. Write the expression as a sum or difference of $\ln a$, $\ln b$, and $\ln c$. Assume that all variables represent positive real numbers.

$$\ln\left(\dfrac{a^2}{b\sqrt[3]{c}}\right)$$

A.9. **a.** Write the domain of $f(x) = \ln(3x + 12)$ in interval notation.
   b. Write the domain of $g(x) = \ln(6 - x)$ in interval notation.

A.10. **a.** Complete the change-of-base formula by converting the expression into an equivalent expression using base a: $\log_b x = \boxed{}$
   b. Without using a calculator, evaluate $\log_2 16$.
   c. Without using a calculator, evaluate $\log_2 32$.
   d. Between which two integers is the value of $\log_2 20$?
   e. Use a calculator and the change-of-base formula with common logarithms to approximate the value of $\log_2 20$. Round to four decimal places.
   f. Use a calculator and the change-of-base formula with natural logarithms to approximate the value of $\log_2 20$. Round to four decimal places.
   g. Compare your answers to parts (e) and (f).

Section 20.5 Practice Exercises

Prerequisite Review

R.1. Evaluate the expression $\left(1 + \dfrac{1}{x}\right)^x$ for the given values of x. Round to six decimal places.
   a. $x = 2000$
   b. $x = 2{,}000{,}000$

R.2. Evaluate the expression $P\left(1 + \dfrac{r}{n}\right)^{nt}$ for the given values of the variables. Round to two decimal places.
   a. $P = 5000$, $r = 0.05$, $n = 12$, and $t = 10$
   b. $P = 8000$, $r = 0.06$, $n = 4$, and $t = 20$

For Exercises R.3–R.10, simplify the expression. Assume that all variables represent positive real numbers.

R.3. $\log 1$ **R.4.** $\log_5 1$ **R.5.** $\log_6 6$ **R.6.** $\log_8 8$

R.7. $\log_2(2^5)$ **R.8.** $\log(10^4)$ **R.9.** $10^{\log(3)}$ **R.10.** $5^{\log_5 7}$

For Exercises R.11–R.12, write the expression as a single logarithm. Assume that all variables represent positive real numbers.

R.11. $\dfrac{1}{3}\log_5 x + \log_5 y - 2\log_5 z$ **R.12.** $\log a - 5\log c + \dfrac{1}{2}\log d$

For Exercises R.13–R.14, write the expression as the sum or difference of logarithms of x, y, and z. Assume that all variables represent positive real numbers.

R.13. $\log\left(\dfrac{\sqrt{x}}{yz^4}\right)$ **R.14.** $\log_3\left(\dfrac{xy^5}{\sqrt[3]{z}}\right)$

Section 20.5 The Irrational Number e and Change of Base

Vocabulary and Key Concepts

1. a. As x becomes increasingly large, the value of $(1 + \frac{1}{x})^x$ approaches _____ ≈ 2.71828.

   b. The function $f(x) = e^x$ is the exponential function base _____.

   c. The logarithmic function base e is called the _____ logarithmic function and is denoted by $y =$ _____.

   d. If x is a positive real number, then $\ln 1 =$ _____, $\ln e =$ _____, $\ln e^p =$ _____, and $e^{\ln x} =$ _____.

   e. If x and y are positive real numbers, then $\ln(xy) =$ _____ and $\ln\left(\dfrac{x}{y}\right) =$ _____.

   f. If x is a positive real number, then for any real number p, $\ln x^p$ can be written as _____.

   g. The change-of-base formula states that $\log_b x = \dfrac{\log_a x}{\Box}$, where a is a positive real number and $a \neq 1$.

2. $\log(x)$ is a logarithmic expression with a base of _____.

3. $\ln(x)$ is a logarithmic expression with a base of _____.

4. $f(x) = e^x$ defines a(n) (increasing/decreasing) exponential function.

5. $g(x) = \ln x$ defines a(n) (increasing/decreasing) logarithmic function.

Concept 1: The Irrational Number e

6. From memory, write a decimal approximation of the number e, correct to three decimal places.

For Exercises 7–10, graph the equation by completing the table and plotting points. Identify the domain. Round to two decimal places when necessary. (See Example 1.)

7. $y = e^{x+1}$

x	y
−4	
−3	
−2	
−1	
0	
1	

8. $y = e^{x+2}$

x	y
−5	
−4	
−3	
−2	
−1	
0	

9. $y = e^x + 2$

x	y
−2	
−1	
0	
1	
2	
3	

10. $y = e^x - 1$

x	y
−4	
−3	
−2	
−1	
0	
1	

Concept 2: Computing Compound Interest

For Exercises 11–16, suppose that P dollars in principal is invested at an annual interest rate r. For interest compounded n times per year, the amount $A(t)$ in the account after t years is given by $A(t) = P\left(1 + \dfrac{r}{n}\right)^{nt}$. If interest is compounded continuously, the amount is given by $A(t) = Pe^{rt}$.

11. Suppose an investor deposits $10,000 in an account for 5 years for which the interest is compounded monthly. Find the total amount of money in the account for the following interest rates. Compare your answers and comment on the effect of interest rate on an investment. **(See Example 2.)**

    a. $r = 4.0\%$ b. $r = 6.0\%$ c. $r = 8.0\%$ d. $r = 9.5\%$

12. Suppose an investor deposits $5000 in an account for 8 years for which the interest is compounded quarterly. Find the total amount of money in the account for the following interest rates. Compare your answers and comment on the effect of interest rate on an investment.

    a. $r = 4.5\%$ b. $r = 5.5\%$ c. $r = 7.0\%$ d. $r = 9.0\%$

13. Suppose an investor deposits $8000 in an account for 10 years at 4.5% interest. Find the total amount of money in the account for the following compounding options. Compare your answers. How does the number of compound periods per year affect the total investment?

    a. Compounded annually
    b. Compounded quarterly
    c. Compounded monthly
    d. Compounded daily
    e. Compounded continuously

14. Suppose an investor deposits $15,000 in an account for 8 years at 5.0% interest. Find the total amount of money in the account for the following compounding options. Compare your answers. How does the number of compound periods per year affect the total investment?

    a. Compounded annually
    b. Compounded quarterly
    c. Compounded monthly
    d. Compounded daily
    e. Compounded continuously

15. Suppose an investor deposits $5000 in an account earning 6.5% interest compounded continuously. Find the total amount in the account for the following time periods. How does the length of time affect the amount of interest earned?

    a. 5 years b. 10 years c. 15 years d. 20 years e. 30 years

16. Suppose an investor deposits $10,000 in an account earning 6.0% interest compounded continuously. Find the total amount in the account for the following time periods. How does the length of time affect the amount of interest earned?

    a. 5 years b. 10 years c. 15 years d. 20 years e. 30 years

Concept 3: The Natural Logarithmic Function

For Exercises 17–20, graph the equation by completing the table and plotting the points. Identify the domain. Round to two decimal places when necessary. **(See Example 3.)**

17. $y = \ln(x - 2)$

x	y
2.25	
2.50	
2.75	
3	
4	
5	
6	

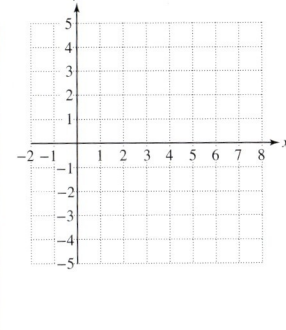

18. $y = \ln(x - 1)$

x	y
1.25	
1.50	
1.75	
2	
3	
4	
5	

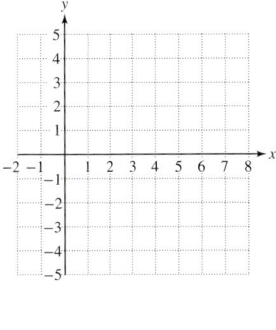

19. $y = \ln x - 1$

x	y
0.25	
0.5	
0.75	
1	
2	
3	
4	

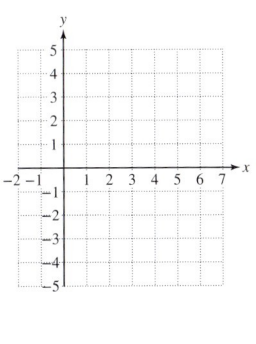

20. $y = \ln x + 2$

x	y
0.25	
0.5	
0.75	
1	
2	
3	
4	

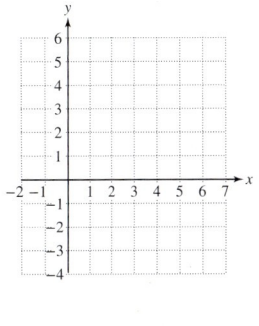

21. a. Graph $f(x) = 10^x$ and $g(x) = \log x$.

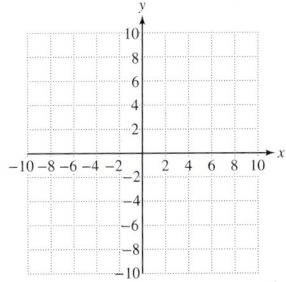

b. Identify the domain and range of f.

c. Identify the domain and range of g.

22. a. Graph $f(x) = e^x$ and $g(x) = \ln x$.

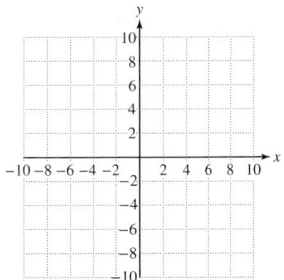

b. Identify the domain and range of f.

c. Identify the domain and range of g.

For Exercises 23–30, simplify the expressions. Assume all variables represent positive real numbers. **(See Example 4.)**

23. $\ln e$

24. $\ln e^2$

25. $\ln 1$

26. $e^{\ln x}$

27. $\ln e^{-6}$

28. $\ln e^{(5-3x)}$

29. $e^{\ln(2x+3)}$

30. $e^{\ln 4}$

For Exercises 31–38, write the expression as a single logarithm. Assume all variables represent positive real numbers. (See Example 5.)

31. $6 \ln p + \dfrac{1}{3} \ln q$

32. $2 \ln w + \ln z$

33. $\dfrac{1}{2}(\ln x - 3 \ln y)$

34. $\dfrac{1}{3}(4 \ln a - \ln b)$

35. $2 \ln a - \ln b - \dfrac{1}{3} \ln c$

36. $-\ln x + 3 \ln y - \ln z$

37. $4 \ln x - 3 \ln y - \ln z$

38. $\dfrac{1}{2} \ln c + \ln a - 2 \ln b$

For Exercises 39–46, write the expression as a sum and/or difference of $\ln a$, $\ln b$, and $\ln c$. Assume all variables represent positive real numbers. (See Example 6.)

39. $\ln \left(\dfrac{a}{b}\right)^2$

40. $\ln \sqrt[3]{\dfrac{a}{b}}$

41. $\ln(b^2 \cdot e)$

42. $\ln(\sqrt{c} \cdot e)$

43. $\ln\left(\dfrac{a^4 \sqrt{b}}{c}\right)$

44. $\ln\left(\dfrac{\sqrt{ab}}{c^3}\right)$

45. $\ln\left(\dfrac{ab}{c^2}\right)^{1/5}$

46. $\ln \sqrt{2ab}$

For Exercises 47–52, write the domain in interval notation.

47. $f(x) = \ln(x - 4)$

48. $g(x) = \ln(x + 3)$

49. $h(x) = \ln(2x + 5)$

50. $k(x) = \ln(4x - 1)$

51. $m(x) = \ln(7 - x)$

52. $n(x) = \ln(10 - x)$

Concept 4: Change-of-Base Formula

53. a. Evaluate $\log_6 200$ by computing $\dfrac{\log 200}{\log 6}$ to four decimal places.

   b. Evaluate $\log_6 200$ by computing $\dfrac{\ln 200}{\ln 6}$ to four decimal places.

   c. How do your answers to parts (a) and (b) compare?

54. a. Evaluate $\log_8 120$ by computing $\dfrac{\log 120}{\log 8}$ to four decimal places.

   b. Evaluate $\log_8 120$ by computing $\dfrac{\ln 120}{\ln 8}$ to four decimal places.

   c. How do your answers to parts (a) and (b) compare?

For Exercises 55–66, use the change-of-base formula to approximate the logarithms to four decimal places. Check your answers by using the exponential key on your calculator. (See Example 7.)

55. $\log_2 7$

56. $\log_3 5$

57. $\log_8 24$

58. $\log_4 17$

59. $\log_8 0.012$

60. $\log_7 0.251$

61. $\log_9 1$

62. $\log_2 \left(\dfrac{1}{5}\right)$

63. $\log_4 \left(\dfrac{1}{100}\right)$

64. $\log_5 0.0025$

65. $\log_7 0.0006$

66. $\log_2 0.24$

Concept 5: Applications of the Natural Logarithmic Function

Under continuous compounding, the amount of time t in years required for an investment to double is a function of the annual interest rate r according to the formula:

$$t = \frac{\ln 2}{r}$$

Use the formula for Exercises 67–70. **(See Example 8.)**

67. a. If you invest $5000, how long will it take the investment to reach $10,000 if the interest rate is 4.5%? Round to one decimal place.

    b. If you invest $5000, how long will it take the investment to reach $10,000 if the interest rate is 10%? Round to one decimal place.

    c. Using the doubling time found in part (b), how long would it take a $5000 investment to reach $20,000 if the interest rate is 10%?

68. a. If you invest $3000, how long will it take the investment to reach $6000 if the interest rate is 5.5%? Round to one decimal place.

    b. If you invest $3000, how long will it take the investment to reach $6000 if the interest rate is 8%? Round to one decimal place.

    c. Using the doubling time found in part (b), how long would it take a $3000 investment to reach $12,000 if the interest rate is 8%?

69. a. If you invest $4000, how long will it take the investment to reach $8000 if the interest rate is 3.5%? Round to one decimal place.

    b. If you invest $4000, how long will it take the investment to reach $8000 if the interest rate is 5%? Round to one decimal place.

    c. Using the doubling time found in part (b), how long would it take a $4000 investment to reach $16,000 if the interest rate is 5%?

70. On August 31, 1854, an epidemic of cholera was discovered in London, England, resulting from a contaminated community water pump at Broad Street. By the end of September more than 600 citizens who drank water from the pump had died.

    The cumulative number of deaths from cholera in the 1854 London epidemic can be approximated by

    $$D(t) = 91 + 160 \ln(t + 1)$$

    where t is the number of days after the start of the epidemic ($t = 0$ corresponds to September 1, 1854).

    a. Approximate the total number of deaths as of September 1 ($t = 0$).

    b. Approximate the total number of deaths as of September 5, September 10, and September 20.

Technology Connections

71. a. Graph the function defined by $f(x) = \log_3 x$ by graphing $Y_1 = \dfrac{\log x}{\log 3}$.

    b. Graph the function defined by $f(x) = \log_3 x$ by graphing $Y_2 = \dfrac{\ln x}{\ln 3}$.

    c. Does it appear that $Y_1 = Y_2$?

72. a. Graph the function defined by $f(x) = \log_7 x$ by graphing $Y_1 = \dfrac{\log x}{\log 7}$.

   b. Graph the function defined by $f(x) = \log_7 x$ by graphing $Y_2 = \dfrac{\ln x}{\ln 7}$.

   c. Does it appear that $Y_1 = Y_2$?

For Exercises 73–75, graph the functions on a graphing calculator.

73. Graph $s(x) = \log_{1/2} x$ **74.** Graph $y = e^{x-2}$ **75.** Graph $y = e^x - 4$

Problem Recognition Exercises

Logarithmic and Exponential Forms

Fill out the table by writing the exponential expressions in logarithmic form, and the logarithmic expressions in exponential form. Use the fact that:

$$y = \log_b x \quad \text{is equivalent to} \quad b^y = x$$

	Exponential Form	Logarithmic Form
1.	$2^5 = 32$	
2.		$\log_3 81 = 4$
3.	$z^y = x$	
4.		$\log_b a = c$
5.	$10^3 = 1000$	
6.		$\log 10 = 1$
7.	$e^a = b$	
8.		$\ln p = q$
9.	$\left(\frac{1}{2}\right)^2 = \frac{1}{4}$	
10.		$\log_{1/3} 9 = -2$
11.	$10^{-2} = 0.01$	
12.		$\log 4 = x$
13.	$e^0 = 1$	
14.		$\ln e = 1$
15.	$25^{1/2} = 5$	
16.		$\log_{16} 2 = \frac{1}{4}$
17.	$e^t = s$	
18.		$\ln w = r$
19.	$15^{-2} = \frac{1}{225}$	
20.		$\log_3 p = -1$

Logarithmic and Exponential Equations and Applications

Section 20.6

Concepts
1. Solving Logarithmic Equations
2. Solving Exponential Equations
3. Applications

1. Solving Logarithmic Equations

Equations containing one or more logarithms are called **logarithmic equations**. For example, the following are logarithmic equations.

$$\log_3(2x+5) = 1 \quad \text{and} \quad \log_4 x = 1 - \log_4(x-3)$$

To solve equations involving logarithms of first degree, we use the following guidelines.

Solving Logarithmic Equations

- **Step 1** Isolate the logarithms on one side of the equation.
- **Step 2** Write a sum or difference of logarithms as a single logarithm.
- **Step 3** Rewrite the equation in step 2 in exponential form.
- **Step 4** Solve the resulting equation from step 3.
- **Step 5** Check all solutions to verify that they are within the domain of the logarithmic expressions in the original equation.

Example 1 — Solving a Logarithmic Equation

Solve the equation. $\log_3(2x+5) = 1$

Solution:

$\log_3(2x+5) = 1$ — This equation has a single logarithm isolated on the left-hand side.

$2x + 5 = 3^1$ — Write the logarithm in its equivalent exponential form.

$2x + 5 = 3$ — Solve the resulting equation.

$2x = -2$

$x = -1$

Check: $x = -1$

$\log_3(2x+5) = 1$

$\log_3[2(-1) + 5] \stackrel{?}{=} 1$

$\log_3(3) \stackrel{?}{=} 1$ ✓

The solution set is $\{-1\}$.

Skill Practice Solve the equation.

1. $\log_6(5x - 4) = 2$

Answer

1. $\{8\}$

Example 2 Solving a Logarithmic Equation

Solve the equation. $\quad \log_4 x = 1 - \log_4 (x - 3)$

Solution:

$$\log_4 x = 1 - \log_4 (x - 3)$$

$\log_4 x + \log_4 (x - 3) = 1$	Isolate the logarithms on one side of the equation.
$\log_4 [x(x - 3)] = 1$	Write as a single logarithm.
$\log_4 (x^2 - 3x) = 1$	Simplify inside the parentheses.
$x^2 - 3x = 4^1$	Write the equation in exponential form.
$x^2 - 3x - 4 = 0$	The resulting equation is quadratic.
$(x - 4)(x + 1) = 0$	Factor.
$x = 4 \quad \text{or} \quad x = -1$	Apply the zero product rule.

Notice that -1 is *not* a solution because $\log_4 x$ is not defined at $x = -1$. However, $x = 4$ is defined in both expressions $\log_4 x$ and $\log_4 (x - 3)$. We can substitute $x = 4$ into the original equation to show that it checks.

Check: $x = 4$

$$\log_4 x = 1 - \log_4 (x - 3)$$
$$\log_4 4 \stackrel{?}{=} 1 - \log_4 (4 - 3)$$
$$1 \stackrel{?}{=} 1 - \log_4 1$$
$$1 \stackrel{?}{=} 1 - 0 \checkmark \text{ True}$$

The solution set is $\{4\}$. (The value -1 does not check.)

FOR REVIEW

Recall:
$$\log_b xy = \log_b x + \log_b y$$
$$\log_b \left(\frac{x}{y}\right) = \log_b x - \log_b y$$

Skill Practice Solve the equation.

2. $\log_3 (x - 8) = 2 - \log_3 x$

Example 3 Solving a Logarithmic Equation

Solve the equation. $\quad \log (x + 300) = 3.7$

Solution:

$\log (x + 300) = 3.7$	The equation has a single logarithm that is already isolated.
$10^{3.7} = x + 300$	Write the equation in exponential form.
$10^{3.7} - 300 = x$	Solve for x.
$x = 10^{3.7} - 300$	
≈ 4711.87	

Answer

2. $\{9\}$ (The value -1 does not check.)

Check: $x = 10^{3.7} - 300$ Check the exact value of x in the original equation.

$$\log(x + 300) = 3.7$$
$$\log[(10^{3.7} - 300) + 300] \stackrel{?}{=} 3.7$$
$$\log(10^{3.7} - 300 + 300) \stackrel{?}{=} 3.7$$
$$\log 10^{3.7} \stackrel{?}{=} 3.7$$
$$3.7 \stackrel{?}{=} 3.7 \checkmark \text{ True}$$

Property 3 of logarithms: $\log_b b^p = p$

The solution $10^{3.7} - 300$ checks.

The solution set is $\{10^{3.7} - 300\}$.

Skill Practice Solve the equation.

3. $\log(p + 6) = 1.3$

The following property is another useful tool to solve a logarithmic equation. This is demonstrated in Example 4.

Equivalence of Logarithmic Expressions

Let x, y, and b represent real numbers such that $x > 0$, $y > 0$, $b > 0$, and $b \neq 1$. Then

$$\log_b x = \log_b y \quad \text{implies} \quad x = y$$

The equivalence property of logarithmic expressions indicates that if two logarithms of the same base are equal, then their arguments are equal.

Example 4   Solving a Logarithmic Equation

Solve the equation.

$$\ln(x + 2) + \ln(x - 1) = \ln(9x - 17)$$

Solution:

$\ln(x + 2) + \ln(x - 1) = \ln(9x - 17)$

$\ln[(x + 2)(x - 1)] = \ln(9x - 17)$ Write the logarithms on the left-hand side as a single logarithm.

$\ln(x^2 + x - 2) = \ln(9x - 17)$ Simplify.

$x^2 + x - 2 = 9x - 17$ Use the equivalence property of logarithms. That is, equate the arguments of each log.

$x^2 - 8x + 15 = 0$ Solve the resulting quadratic equation.

$(x - 5)(x - 3) = 0$

$x = 5$ or $x = 3$

Answer

3. $\{10^{1.3} - 6\}$

The solutions 5 and 3 are both within the domain of the logarithmic functions in the original equation. Both solutions check.

The solution set is {5, 3}.

Skill Practice Solve the equation.

4. $\ln(t-3) + \ln(t-1) = \ln(2t-5)$

---

2. Solving Exponential Equations

An equation with one or more exponential expressions is called an **exponential equation**. The following property is often useful in solving exponential equations.

> **Equivalence of Exponential Expressions**
> Let x, y, and b be real numbers such that $b > 0$ and $b \neq 1$. Then
> $$b^x = b^y \quad \text{implies} \quad x = y$$

The equivalence property of exponential expressions indicates that if two exponential expressions of the same base are equal, then their exponents must be equal.

Example 5   Solving an Exponential Equation

Solve the equation. $4^{2x-9} = 64$

Solution:

$$4^{2x-9} = 64$$

$4^{2x-9} = 4^3$ Write both sides with a common base.

$2x - 9 = 3$ If $b^x = b^y$, then $x = y$.

$2x = 12$ Solve for x.

$x = 6$

To check, substitute $x = 6$ into the original equation.

$$4^{2(6)-9} \stackrel{?}{=} 64$$
$$4^{12-9} \stackrel{?}{=} 64$$
$$4^3 \stackrel{?}{=} 64 \checkmark \text{ True}$$

The solution set is {6}.

Skill Practice Solve the equation.

5. $2^{3x+1} = 8$

Answers

4. {4} (The value 2 does not check.)
5. $\left\{\dfrac{2}{3}\right\}$

Example 6 Solving an Exponential Equation

Solve the equation. $(2^x)^{x+3} = \dfrac{1}{4}$

Solution:

$(2^x)^{x+3} = \dfrac{1}{4}$

$2^{x^2+3x} = 2^{-2}$ Apply the multiplication property of exponents. Write both sides of the equation with a common base.

$x^2 + 3x = -2$ If $b^x = b^y$, then $x = y$.

$x^2 + 3x + 2 = 0$ The resulting equation is quadratic.

$(x+2)(x+1) = 0$ Solve for x.

$x = -2$ or $x = -1$ Both solutions check.

The solution set is $\{-2, -1\}$.

Skill Practice Solve the equation.

6. $(3^x)^{x-5} = \dfrac{1}{81}$

Example 7 Solving an Exponential Equation

Solve the equation. $4^x = 25$

Solution:

Because 25 cannot be written as a recognizable power of 4, we cannot easily use the property that if $b^x = b^y$, then $x = y$. Instead, we can take a logarithm of any base on both sides of the equation. Then by applying the power property of logarithms, the unknown exponent can be written as a factor.

$4^x = 25$

$\log 4^x = \log 25$ Take the common logarithm of both sides.

$x \log 4 = \log 25$ Apply the power property of logarithms to express the exponent as a factor. This is now a linear equation in x.

$\dfrac{x \log 4}{\log 4} = \dfrac{\log 25}{\log 4}$ Solve for x.

$x = \dfrac{\log 25}{\log 4}$ or approximately 2.322.

The solution set is $\left\{\dfrac{\log 25}{\log 4}\right\}$ or equivalently $\left\{\dfrac{\ln 25}{\ln 4}\right\}$.

FOR REVIEW

Recall:
$$\log_b x^p = p \log_b x$$

For example,
$$\log 4^x = x \cdot \log 4$$

Skill Practice Solve the equation.

7. $5^x = 32$

Answers

6. $\{1, 4\}$

7. $\left\{\dfrac{\log 32}{\log 5}\right\}$ or $\left\{\dfrac{\ln 32}{\ln 5}\right\}$

TIP: The equation from Example 7 is written in the form $b^x =$ constant. In this special case, we also have the option of rewriting the equation in its corresponding logarithmic form to solve for x.

$$4^x = 25$$
$$x = \log_4 25$$
$$= \frac{\ln 25}{\ln 4} \quad \text{(Change-of-base formula)}$$

Solving Exponential Equations

Step 1 Isolate one of the exponential expressions in the equation.

Step 2 Take a logarithm on both sides of the equation. (The natural logarithmic function or the common logarithmic function is often used so that the final answer can be approximated with a calculator.)

Step 3 Use the power property of logarithms (property 7) to write exponents as factors. Recall: $\log_b x^p = p \log_b x$.

Step 4 Solve the resulting equation from step 3.

Example 8 Solving an Exponential Equation by Taking a Logarithm on Both Sides

Solve the equation. $\quad e^{-3.6x} - 2 = 7.74$

Solution:

FOR REVIEW

Recall:
$$\log_b(b^x) = x$$

For example,
$$\ln(e^{-3.6x}) = -3.6x$$

This illustrates a function composed with its inverse.

$e^{-3.6x} - 2 = 7.74$

$e^{-3.6x} = 9.74 \qquad$ Add 2 to both sides to isolate the exponential expression.

$\ln e^{-3.6x} = \ln 9.74 \qquad$ The exponential expression has a base of e, so it is convenient to take the natural logarithm of both sides.

$(-3.6x) \ln e = \ln 9.74 \qquad$ Use the power property of logarithms.

$-3.6x = \ln 9.74 \qquad$ Simplify (recall that $\ln e = 1$).

$x = \dfrac{\ln 9.74}{-3.6} \approx -0.632 \qquad$ Divide both sides by -3.6.

The solution set is $\left\{-\dfrac{\ln 9.74}{3.6}\right\}$.

Skill Practice Solve the equation.

8. $e^{-0.2t} + 1 = 8.52$

Answer

8. $\left\{\dfrac{\ln 7.52}{-0.2}\right\}$

Section 20.6 Logarithmic and Exponential Equations and Applications

Example 9 Solving an Exponential Equation by Taking a Logarithm on Both Sides

Solve the equation. $\quad 2^{x+3} = 7^x$

Solution:

$2^{x+3} = 7^x$

$\ln 2^{(x+3)} = \ln 7^x \qquad$ Take the natural logarithm of both sides.

$(x+3) \ln 2 = x \ln 7 \qquad$ Use the power property of logarithms.

$x(\ln 2) + 3(\ln 2) = x \ln 7 \qquad$ Apply the distributive property.

$x(\ln 2) - x(\ln 7) = -3 \ln 2 \qquad$ Collect x terms on one side.

$x(\ln 2 - \ln 7) = -3 \ln 2 \qquad$ Factor out x.

$\dfrac{x(\ln 2 - \ln 7)}{(\ln 2 - \ln 7)} = \dfrac{-3 \ln 2}{\ln 2 - \ln 7} \qquad$ Solve for x.

$x = \dfrac{-3 \ln 2}{\ln 2 - \ln 7} \approx 1.66$

The solution set is $\left\{ -\dfrac{3 \ln 2}{\ln 2 - \ln 7} \right\}$.

> **TIP:** The exponential equation $2^{x+3} = 7^x$ could have been solved by taking a logarithm of *any* base on both sides of the equation.

> **TIP:** Using the properties of logarithms, we can write the solution to Example 9 in other forms, such as:
>
> $x = -\dfrac{\ln 8}{\ln 2 - \ln 7}$ and
>
> $x = \dfrac{\ln 8}{\ln 3.5}$

Skill Practice Solve the equation.

9. $3^x = 8^{x+2}$

---

3. Applications

Example 10 Applying an Exponential Function to World Population

The population of the world was estimated to have reached 6.5 billion in April 2006. The population growth rate for the world is estimated to be 1.4% (*source:* U.S. Census Bureau). The function defined by

$$P(t) = 6.5(1.014)^t$$

represents the world population $P(t)$ in billions as a function of the number of years after April 2006 ($t = 0$ represents April 2006).

a. Use the function to estimate the world population in April 2010.

b. Use the function to predict the amount of time after April 2006 required for the world population to reach 13 billion if this trend continues.

Solution:

a. $P(t) = 6.5(1.014)^t$

$P(4) = 6.5(1.014)^4 \qquad$ The year 2010 corresponds to $t = 4$.

$\approx 6.87 \qquad$ In 2010, the world's population was approximately 6.87 billion.

Answer

9. $\left\{ \dfrac{2 \ln 8}{\ln 3 - \ln 8} \right\}$

b. $P(t) = 6.5(1.014)^t$

$\quad\quad 13 = 6.5(1.014)^t$ Substitute $P(t) = 13$ and solve for t.

$\quad\quad \dfrac{13}{6.5} = \dfrac{\cancel{6.5}(1.014)^t}{\cancel{6.5}}$ Isolate the exponential expression on one side of the equation.

$\quad\quad 2 = 1.014^t$

$\quad\quad \ln 2 = \ln 1.014^t$ Take the natural logarithm of both sides.

$\quad\quad \ln 2 = t \ln 1.014$ Use the power property of logarithms.

$\quad\quad \dfrac{\ln 2}{\ln 1.014} = \dfrac{t \cancel{\ln 1.014}}{\cancel{\ln 1.014}}$ Solve for t.

$\quad\quad t = \dfrac{\ln 2}{\ln 1.014} \approx 50$ The population will reach 13 billion (double the April 2006 value) approximately 50 years after 2006.

Note: It has taken thousands of years for the world's population to reach 6.5 billion. However, with a growth rate of 1.4%, it will take only 50 years to gain an additional 6.5 billion, if this trend continues.

Skill Practice Use the population function from Example 10.

10. Predict the world population in April 2020.
11. Predict the year in which the world population will reach 9 billion.

On Friday, April 25, 1986, a nuclear accident occurred at the Chernobyl nuclear reactor, resulting in radioactive contaminates being released into the atmosphere. The most hazardous isotopes released in this accident were ^{137}Cs (cesium-137), ^{131}I (iodine-131), and ^{90}Sr (strontium-90). People living close to Chernobyl (in Ukraine) were at risk of radiation exposure from inhalation, from absorption through the skin, and from food contamination. Years after the incident, scientists have seen an increase in the incidence of thyroid disease among children living in the contaminated areas. Because iodine is readily absorbed in the thyroid gland, scientists suspect that radiation from iodine-131 is the cause.

Example 11 Applying an Exponential Equation to Radioactive Decay

The half-life of radioactive iodine, ^{131}I, is 8.04 days. If 10 g of iodine-131 is initially present, then the amount $A(t)$ (in grams) of radioactive iodine still present after t days is approximated by

$$A(t) = 10e^{-0.0862t}$$

where t is the time in days.

a. Use the model to approximate the amount of ^{131}I still present after 2 weeks. Round to the nearest 0.1 g.

b. How long will it take for the amount of ^{131}I to decay to 0.5 g? Round to the nearest 0.1 day.

Answers
10. Approximately 7.9 billion
11. The year 2029

Solution:

a. $A(t) = 10e^{-0.0862t}$

$A(14) = 10e^{-0.0862(14)}$ Substitute $t = 14$ (2 weeks).

≈ 3.0 g   There was 3.0 g still present after 2 weeks.

b. $A(t) = 10e^{-0.0862t}$

$0.5 = 10e^{-0.0862t}$ Substitute $A = 0.5$.

$\dfrac{0.5}{10} = \dfrac{10e^{-0.0862t}}{10}$ Isolate the exponential expression.

$0.05 = e^{-0.0862t}$

$\ln 0.05 = \ln e^{-0.0862t}$ Take the natural logarithm of both sides.

$\ln 0.05 = -0.0862t$ The resulting equation is linear.

$\dfrac{\ln 0.05}{-0.0862} = \dfrac{-0.0862t}{-0.0862}$ Solve for t.

$t = \dfrac{\ln 0.05}{-0.0862} \approx 34.8$ days   It will take 34.8 days for 10 g of ^{131}I to decay to 0.5 g.

Skill Practice Radioactive strontium-90 (^{90}Sr) has a half-life of 28 years. If 100 g of strontium-90 is initially present, the amount left after t years is approximated by

$$A(t) = 100e^{-0.0248t}$$

12. Find the amount of ^{90}Sr present after 85 years.
13. How long will it take for the amount of ^{90}Sr to decay to 40 g?

Answers

12. 12.1 g   13. 36.9 years

Section 20.6 Activity

A.1. a. The equivalence property of exponential expressions states that if $b^x = b^y$, then _____.
  b. Consider the equation $5^{x+1} = 125$. This can be written as $5^{x+1} = 5^{\square}$.
  c. Use the equivalence property of exponential expressions to solve $5^{x+1} = 125$.
  d. Solve the equation $3^{2x-6} = 81$.

A.2. If the two sides of an exponential equation cannot be easily written with a common base, we must use logarithms to solve the equation. Solve the equation $e^{2x+1} - 7 = 3$ by following these steps.
  a. Isolate the exponential expression on one side of the equation. In this case, isolate e^{2x+1}.
  b. Take a logarithm of the same base on each side of the equation. In this case, because one of the terms in the equation involves a base of e, take the natural logarithm of the expressions on each side.
  c. Apply the power property of logarithms to write the exponents as factors. (Recall $\log_b x^p = p \cdot \log_b x$.)
  d. Simplify both sides of the equation and solve for x. Write the exact solution and approximate the solution to four decimal places.

A.3. Solve the equation $3^{x-4} = 8^x$ by following these steps.
  a. Take a logarithm of each side (usually the common logarithm or the natural logarithm is used so that a calculator can be used to easily approximate the final answer). In this case, take the common logarithm of each side.
  b. Apply the power property of logarithms on each side.
  c. The equation in part (b) may look daunting, but it is a linear equation. Clear parentheses and collect all variable terms on one side of the equation. (Note that log 3 and log 8 are simply constants.) Solve the resulting equation.
  d. Factor out x from the variable terms and divide by the coefficient on x.
  e. Write the exact solution and approximate the solution to four decimal places.

A.4. a. The equivalence property of logarithmic expressions states that if $\log_b x = \log_b y$, then _____.
  b. Use the equivalence property of logarithms to solve the equation $\log_4(-3x) = \log_4(2x - 15)$ for x. Do not write the solution set yet.
  c. Check the potential solutions from part (b) and write the solution set.
  d. Explain why it is necessary to check the potential solutions to a logarithmic equation.

A.5. If the two sides of a logarithmic equation cannot be written easily with a common base, we must follow a different approach. Solve $\log_5 x = 3 - \log_5(x - 20)$ using these steps.
  a. Collect all logarithmic terms on one side of the equation.
  b. Write the logarithms as a single logarithm.
  c. Write the equation in exponential form.
  d. Solve the resulting equation from part (c).
  e. Check the potential solutions from part (d) and write the solution set to the original equation.

A.6. Radioactive iodine (^{131}I) is used to treat patients with thyroid cancer. Patients with this condition may have symptoms that include rapid weight loss, heart palpitations, and high blood pressure. Because iodine is readily absorbed in the thyroid gland, the radiation is localized and will reduce the size of the thyroid while minimizing damage to surrounding tissues. If a patient is given an initial dose of 2 µg (micrograms) of ^{131}I, the amount of ^{131}I remaining after t days is approximated by $A(t) = 2e^{-0.0866t}$.
  a. How much ^{131}I remains after 5 days? Round to one decimal place.
  b. How long will it take for the amount of ^{131}I to reach 0.2 µg? Round to the nearest day.

Section 20.6 Practice Exercises

Prerequisite Review

For Exercises R.1–R.4, write the expression as a single logarithm. Assume that all variable expressions represent positive real numbers.

R.1. $\log_2(x - 1) + \log_2(x + 2)$

R.2. $\log x + \log(2x + 3)$

R.3. $\log x - \log(1 - x)$

R.4. $\log_4(x + 2) - \log_4(3x - 5)$

For Exercises R.5–R.8, simplify the logarithmic expression.

R.5. $\log 10$

R.6. $\log \dfrac{1}{10}$

R.7. $\ln e^6$

R.8. $\ln \sqrt{e}$

For Exercises R.9–R.10, apply the power property of logarithms.

R.9. $\log 5^{2x+1}$

R.10. $\ln 2^{x-6}$

For Exercises R.11–R.14, simplify the expression.

R.11. $(4^2)^x$

R.12. $(5^4)^x$

R.13. $(2^5)^{x-3}$

R.14. $(3^2)^{4x+1}$

For Exercises R.15–R.20, solve the equation.

R.15. $3x - 4 = 5x + 6$

R.16. $4t + 7 = 3t + 1$

R.17. $\dfrac{2x-1}{x+5} = 3$

R.18. $\dfrac{x-7}{2x+1} = 3$

R.19. $x^2 + 8x = -12$

R.20. $x^2 - 7x = 18$

Vocabulary and Key Concepts

1. **a.** The equivalence property of logarithmic expressions states that if $\log_b x = \log_b y$, then _____.

   b. The equivalence property of exponential expressions states that if $b^x = b^y$, then _____.

Concept 1: Solving Logarithmic Equations

For Exercises 2–6, solve the equations.

2. a. $2^x = 8$
   b. $\log_2 x = 3$

3. a. $5^x = 25$
   b. $\log_5 x = 2$

4. a. $10^x = 1{,}000{,}000$
   b. $\log x = 6$

5. a. $10^x = \dfrac{1}{10}$
   b. $\log x = -1$

6. a. $6^x = \dfrac{1}{36}$
   b. $\log_6 x = -2$

For Exercises 7–38, solve the logarithmic equation. (See Examples 1–4.)

7. $\log_3 x = 2$

8. $\log_4 x = 9$

9. $\log p = 42$

10. $\log q = \dfrac{1}{2}$

11. $\ln x = 0.08$

12. $\ln x = 19$

13. $\log(x + 40) = -9.2$

14. $\log(z - 3) = -6.7$

15. $\log_x 25 = 2 \quad (x > 0)$

16. $\log_x 100 = 2 \quad (x > 0)$

17. $\log_b 10{,}000 = 4 \quad (b > 0)$

18. $\log_b e^3 = 3 \quad (b > 0)$

19. $\log_y 5 = \dfrac{1}{2} \quad (y > 0)$

20. $\log_b 8 = \dfrac{1}{2} \quad (b > 0)$

21. $\log_4 (c + 5) = 3$

22. $\log_5 (a - 4) = 2$

23. $\log_5 (4y + 1) = 1$

24. $\log_6 (5t - 2) = 1$

25. $\ln(1 - x) = 0$

26. $\log_4 (2 - x) = 1$

27. $\log_3 8 - \log_3 (x + 5) = 2$

28. $\log_2 (x + 3) - \log_2 (x + 2) = 1$

29. $\log_2 (h - 1) + \log_2 (h + 1) = 3$

30. $\log_3 k + \log_3 (2k + 3) = 2$

31. $\log(x + 2) = \log(3x - 6)$

32. $\log x = \log(1 - x)$

33. $\ln x - \ln(4x - 9) = 0$

34. $\ln(x + 5) - \ln x = \ln(4x)$

35. $\log_5 (3t + 2) - \log_5 t = \log_5 4$

36. $\log(6y - 7) + \log y = \log 5$

37. $\log(4m) = \log 2 + \log(m - 3)$

38. $\log(-h) + \log 3 = \log(2h - 15)$

Concept 2: Solving Exponential Equations

For Exercises 39–54, solve the exponential equation by using the property that $b^x = b^y$ implies $x = y$, for $b > 0$ and $b \neq 1$. (See Examples 5–6.)

39. $5^x = 625$
40. $3^x = 81$
41. $2^{-x} = 64$
42. $6^{-x} = 216$

43. $36^x = 6$
44. $343^x = 7$
45. $4^{2x-1} = 64$
46. $5^{3x-1} = 125$

47. $81^{3x-4} = \dfrac{1}{243}$
48. $4^{2x-7} = \dfrac{1}{128}$
49. $\left(\dfrac{2}{3}\right)^{-x+4} = \dfrac{8}{27}$
50. $\left(\dfrac{1}{4}\right)^{3x+2} = \dfrac{1}{64}$

51. $16^{-x+1} = 8^{5x}$
52. $27^{(1/3)x} = \left(\dfrac{1}{9}\right)^{2x-1}$
53. $(4^x)^{x+1} = 16$
54. $(3^x)^{x+2} = \dfrac{1}{3}$

For Exercises 55–74, solve the exponential equation by taking a logarithm of both sides. (See Examples 7–9.)

55. $8^a = 21$
56. $6^y = 39$
57. $e^x = 8.1254$
58. $e^x = 0.3151$

59. $10^t = 0.0138$
60. $10^p = 16.8125$
61. $e^{0.07h} - 6 = 9$
62. $e^{0.03k} + 7 = 11$

63. $e^{1.2t} = 3$
64. $e^{1.5t} = 7$
65. $3^{x+1} = 5^x$
66. $2^{x-1} = 7^x$

67. $2^{x+2} = 6^x$
68. $5^{x-2} = 3^x$
69. $32e^{0.04m} = 128$
70. $8e^{0.05n} = 160$

71. $6e^{x/3} = 125$
72. $8e^{x/5} = 155$
73. $5^{x-2} - 4 = 16$
74. $2^{x+4} + 17 = 50$

Concept 3: Applications

75. The population of China can be modeled by
$$P(t) = 1237(1.0095)^t$$
where $P(t)$ is in millions and t is the number of years since 1998. (See Example 10.)
    a. Using this model, what was the population in the year 2002?
    b. Estimate the population in the year 2016.
    c. If this growth rate continues, in what year will the population reach 2 billion people (2 billion is 2000 million)?

76. The population of Delhi, India, can be modeled by
$$P(t) = 9817(1.031)^t$$
where $P(t)$ is in thousands and t is the number of years since 2001.
    a. Using this model, estimate the population in the year 2018.
    b. In what year did the population reach 15 million (15 million is 15,000 thousand)?

77. The growth of certain bacteria in a culture is given by the model
$$A(t) = 500e^{0.0277t}$$
where $A(t)$ is the number of bacteria and t is time in minutes.
    a. What is the initial number of bacteria?
    b. What is the population after 10 min?
    c. How long will it take for the population to double (i.e., reach 1000)?

78. The population of the bacteria *Salmonella typhimurium* is given by the model

$$A(t) = 300e^{0.01733t}$$

where $A(t)$ is the number of bacteria and t is time in minutes.

   a. What is the initial number of bacteria?

   b. What is the population after 10 min?

   c. How long will it take for the population to double?

79. Suppose $5000 is invested at 7% interest compounded continuously. How long will it take for the investment to grow to $10,000? Use the model $A(t) = Pe^{rt}$ and round to the nearest tenth of a year.

80. Suppose $2000 is invested at 10% interest compounded continuously. How long will it take for the investment to triple? Use the model $A(t) = Pe^{rt}$ and round to the nearest year.

81. Suppose that $8000 is invested at 4.5% interest compounded monthly. How long will it take the investment to reach $10,000? Use the model $A(t) = P\left(1 + \dfrac{r}{n}\right)^{nt}$ and round to the nearest year.

82. Suppose that a couple invests $30,000 in a bond fund that pays 6.5% interest compounded quarterly. How long will it take the investment to reach $250,000? Use the model $A(t) = P\left(1 + \dfrac{r}{n}\right)^{nt}$ and round to the nearest year.

83. Phosphorus-32 (^{32}P) has a half-life of approximately 14 days. If 10 g of ^{32}P is present initially, then the amount $A(t)$ (in grams) of phosphorus-32 still present after t days is given by $A(t) = 10(0.5)^{t/14}$. **(See Example 11.)**

   a. Find the amount of phosphorus-32 still present after 5 days. Round to the nearest tenth of a gram.

   b. Find the amount of time necessary for the amount of ^{32}P to decay to 4 g. Round to the nearest tenth of a day.

84. Polonium-210 (^{210}Po) has a half-life of approximately 138.6 days. If 4 g of ^{210}Po is present initially, then the amount $A(t)$ (in grams) of polonium-210 still present after t days is given by $A(t) = 4e^{-0.005t}$.

   a. Find the amount of polonium-210 still present after 50 days. Round to the nearest tenth of a gram.

   b. Find the amount of time necessary for the amount of ^{210}P to decay to 0.5 g. Round to the nearest tenth of a day.

The decibel level of sound can be found by the equation $D = 10 \log\left(\dfrac{I}{I_0}\right)$, where I is the intensity of the sound and I_0 is the intensity of the least audible sound that an average person can hear. Generally I_0 is found as 10^{-12} watt per square meter (W/m^2). Use this information to answer Exercises 85–86.

85. Given that heavy traffic has a decibel level of 89.3 and that $I_0 = 10^{-12}$, find the intensity of the sound of heavy traffic.

86. Given that normal conversation has a decibel level of 65 and $I_0 = 10^{-12}$, find the intensity of the sound of normal conversation.

87. Suppose you save $10,000 from working an extra job. Rather than spending the money, you decide to save the money for retirement by investing in a mutual fund that averages 12% per year. How long will it take for this money to grow to $1,000,000? Use the model $A(t) = Pe^{rt}$ and round to the nearest tenth of a year.

88. The model $A = Pe^{rt}$ is used to compute the total amount of money in an account after t years at an interest rate r, compounded continuously. The value P is the initial principal. Find the amount of time required for the investment to double as a function of the interest rate. (*Hint:* Substitute $A = 2P$ and solve for t.)

Technology Connections

89. Graph $Y_1 = 8^x$ and $Y_2 = 21$ on a window where $0 \leq x \leq 5$ and $0 \leq y \leq 40$. Use the graph and an *Intersect* feature to support your answer to Exercise 55.

90. Graph $Y_1 = 6^x$ and $Y_2 = 39$ on a window where $0 \leq x \leq 5$ and $0 \leq y \leq 50$. Use the graph and an *Intersect* feature to support your answer to Exercise 56.

Expanding Your Skills

91. The isotope of plutonium of mass 238 (written ^{238}Pu) is used to make thermoelectric power sources for spacecraft. The heat and electric power derived from such units have made the Voyager, Galileo, and Cassini missions to the outer reaches of our solar system possible. The half-life of ^{238}Pu is 87.7 years.

    Suppose a space probe was launched in the year 2002 with 2.0 kg of ^{238}Pu. Then the amount of ^{238}Pu available to power the spacecraft decays over time according to

    $$P(t) = 2e^{-0.0079t}$$

    where $t \geq 0$ is the number of years since 2002 and $P(t)$ is the amount of plutonium still present (in kilograms).

    a. Suppose the space probe is due to arrive at Pluto in the year 2045. How much plutonium will remain when the spacecraft reaches Pluto? Round to two decimal places.

    b. If 1.5 kg of ^{238}Pu is required to power the spacecraft's data transmitter, will there be enough power in the year 2045 for us to receive close-up images of Pluto?

92. ^{99m}Tc is a radionuclide of technetium that is widely used in nuclear medicine. Although its half-life is only 6 hr, the isotope is continuously produced via the decay of its longer-lived parent ^{99}Mo (molybdenum-99), whose half-life is approximately 3 days. The ^{99}Mo generators (or "cows") are sold to hospitals in which the ^{99m}Tc can be "milked" as needed over a period of a few weeks. Once separated from its parent, the ^{99m}Tc may be chemically incorporated into a variety of imaging agents, each of which is designed to be taken up by a specific target organ within the body. Special cameras, sensitive to the gamma rays emitted by the technetium, are then used to record a "picture" (similar in appearance to an X-ray film) of the selected organ.

    Suppose a technician prepares a sample of ^{99m}Tc-pyrophosphate to image the heart of a patient suspected of having had a mild heart attack. If the injection contains 10 millicuries (mCi) of ^{99m}Tc at 1:00 P.M., then the amount of technetium still present is given by

    $$T(t) = 10e^{-0.1155t}$$

    where $t > 0$ represents the time in hours after 1:00 P.M. and $T(t)$ represents the amount of ^{99m}Tc (in millicuries) still present.

    a. How many millicuries of ^{99m}Tc will remain at 4:20 P.M. when the image is recorded? Round to the nearest tenth of a millicurie.

    b. How long will it take for the radioactive level of the ^{99m}Tc to reach 2 mCi? Round to the nearest tenth of an hour.

For Exercises 93–96, solve the equations.

93. $(\log x)^2 - 2 \log x - 15 = 0$
    (*Hint:* Let $u = \log x$.)

94. $(\log_2 z)^2 - 3 \log_2 z - 4 = 0$

95. $(\log_3 w)^2 + 5 \log_3 w + 6 = 0$

96. $(\ln x)^2 - 2 \ln x = 0$

Chapter 20 Review Exercises

Section 20.1

For Exercises 1–2, determine if the function is one-to-one by using the horizontal line test.

1.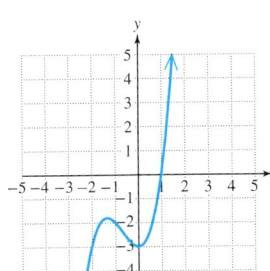

2.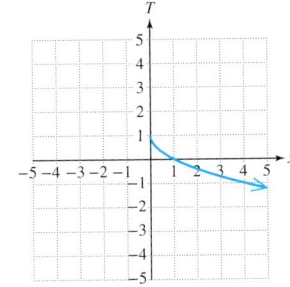

For Exercises 3–7, write the inverse for each one-to-one function.

3. $\{(3, 5), (2, 9), (0, -1), (4, 1)\}$

4. $q(x) = \dfrac{3}{4}x - 2$

5. $g(x) = \sqrt[5]{x} + 3$

6. $f(x) = (x - 1)^3$

7. $n(x) = \dfrac{4}{x - 2}$

8. Verify that the functions defined by $f(x) = 5x - 2$ and $g(x) = \frac{1}{5}x + \frac{2}{5}$ are inverses by showing that $(f \circ g)(x) = x$ and $(g \circ f)(x) = x$.

9. Graph the functions q and q^{-1} from Exercise 4 on the same grid. What can you say about the relationship between these two graphs?

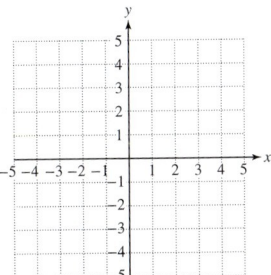

10. a. Find the domain and range of the function defined by $h(x) = \sqrt{x + 1}$.

    b. Find the domain and range of the function defined by $k(x) = x^2 - 1,\ x \geq 0$.

11. Determine the inverse of the function $p(x) = \sqrt{x} + 2$.

Section 20.2

For Exercises 12–19, evaluate the exponential expressions. Use a calculator and round to three decimal places, if necessary.

12. 4^5

13. 6^{-2}

14. $8^{1/3}$

15. $\left(\dfrac{1}{100}\right)^{-1/2}$

16. 2^{π}

17. $5^{\sqrt{3}}$

18. $(\sqrt{7})^{1/2}$

19. $\left(\dfrac{3}{4}\right)^{4/3}$

For Exercises 20–23, graph the functions.

20. $f(x) = 3^x$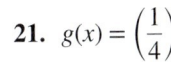

21. $g(x) = \left(\dfrac{1}{4}\right)^x$

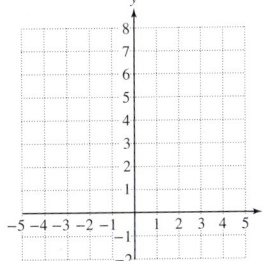
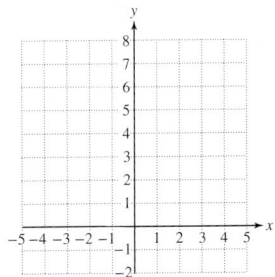

22. $h(x) = 5^{-x}$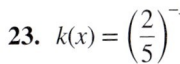

23. $k(x) = \left(\dfrac{2}{5}\right)^{-x}$

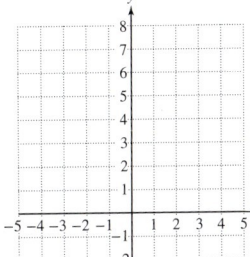
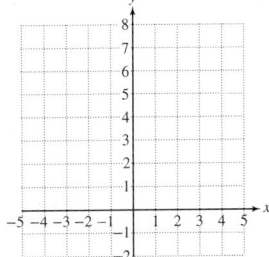

24. a. Does the graph of $y = b^x,\ b > 0,\ b \neq 1$, have a vertical or a horizontal asymptote?

    b. Write an equation of the asymptote.

25. Background radiation is radiation that we are exposed to from naturally occurring sources including the soil, the foods we eat, and the Sun. Background radiation varies depending on where we live. A typical background radiation level is 150 millirems (mrem) per year. (A rem is a measure of energy produced from radiation.) Suppose a substance emits 30,000 mrem per year and has a half-life of 5 years. The function defined by

$$A(t) = 30{,}000\left(\frac{1}{2}\right)^{t/5}$$

gives the radiation level (in millirems) of this substance after t years.

   a. What is the radiation level after 5 years?

   b. What is the radiation level after 15 years?

   c. Will the radiation level of this substance be below the background level of 150 mrem after 50 years?

Section 20.3

For Exercises 26–33, evaluate the logarithms without using a calculator.

26. $\log_3\left(\dfrac{1}{27}\right)$

27. $\log_5 1$

28. $\log_7 7$

29. $\log_2 2^8$

30. $\log_2 16$

31. $\log_3 81$

32. $\log 100{,}000$

33. $\log_8\left(\dfrac{1}{8}\right)$

For Exercises 34–35, graph the logarithmic functions.

34. $q(x) = \log_3 x$

35. $r(x) = \log_{1/2} x$

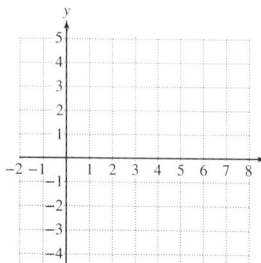
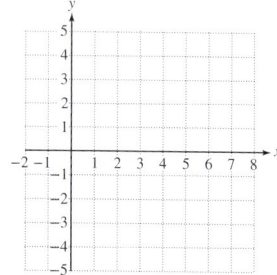

36. a. Does the graph of $y = \log_b x$ have a vertical or a horizontal asymptote?

    b. Write an equation of the asymptote.

37. Acidity of a substance is measured by its pH. The pH can be calculated by the formula pH = $-\log [H^+]$, where $[H^+]$ is the hydrogen ion concentration.

   a. What is the pH of a fruit with a hydrogen ion concentration of 0.00316 mol/L? Round to one decimal place.

   b. What is the pH of an antacid tablet with $[H^+] = 3.16 \times 10^{-10}$? Round to one decimal place.

Section 20.4

For Exercises 38–41, evaluate the logarithms without using a calculator.

38. $\log_8 8$

39. $\log_{11} 11^6$

40. $\log_{1/2} 1$

41. $12^{\log_{12} 7}$

42. Complete the properties Assume x, y, and b are positive real numbers such that $b \neq 1$.

   a. $\log_b (xy) =$

   b. $\log_b x - \log_b y =$

   c. $\log_b x^p =$

For Exercises 43–46, write the logarithmic expressions as a single logarithm and simplify if possible.

43. $\dfrac{1}{4}(\log_b y - 4\log_b z + 3\log_b x)$

44. $\dfrac{1}{2}\log_3 a + \dfrac{1}{2}\log_3 b - 2\log_3 c - 4\log_3 d$

45. $\log 540 - 3\log 3 - 2\log 2$

46. $-\log_4 18 + \log_4 6 + \log_4 3 - \log_4 1$

47. Which of the following is equivalent to

$$\dfrac{2\log 7}{\log 7 + \log 6}?$$

   a. $\dfrac{\log 7}{\log 6}$
   b. $\dfrac{\log 49}{\log 42}$
   c. $\log\left(\dfrac{7}{6}\right)$

48. Which of the following is equivalent to

$$\dfrac{\log 8^{-3}}{\log 2 + \log 4}?$$

   a. -3
   b. $-3\log\left(\dfrac{4}{3}\right)$
   c. $\dfrac{-3\log 4}{\log 3}$

Section 20.5

For Exercises 49–56, use a calculator to approximate the expressions to four decimal places.

49. e^5
50. $e^{\sqrt{7}}$
51. $32e^{0.008}$
52. $58e^{-0.0125}$
53. $\ln 6$
54. $\ln\left(\dfrac{1}{9}\right)$
55. $\log 22$
56. $\log e^3$

For Exercises 57–60, use the change-of-base formula to approximate the logarithms to four decimal places.

57. $\log_2 10$
58. $\log_9 80$
59. $\log_5 0.26$
60. $\log_4 0.0062$

61. An investor wants to deposit \$20,000 in an account for 10 years at 5.25% interest. Compare the amount $A(t)$ she would have if her money were invested with the following different compounding options. Use
$$A(t) = P\left(1 + \dfrac{r}{n}\right)^{nt}$$
for interest compounded n times per year and $A(t) = Pe^{rt}$ for interest compounded continuously.

   a. Compounded annually
   b. Compounded quarterly
   c. Compounded monthly
   d. Compounded continuously

62. To measure a student's retention of material at the end of a course, researchers give the student a test on the material every month for 24 months after the course is over. The student's average score t months after completing the course is given by
$$S(t) = 75e^{-0.5t} + 20$$
where $S(t)$ is the test score.

   a. Find $S(0)$ and interpret the result.
   b. Find $S(6)$ and interpret the result.
   c. Find $S(12)$ and interpret the result.

For Exercises 63–70, identify the domain. Write the answer in interval notation.

63. $f(x) = e^x$
64. $g(x) = e^{x+6}$
65. $h(x) = e^{x-3}$
66. $k(x) = \ln x$
67. $q(x) = \ln(x+5)$
68. $p(x) = \ln(x-7)$
69. $r(x) = \ln(3x-4)$
70. $w(x) = \ln(5-x)$

Section 20.6

For Exercises 71–88, solve the equations.

71. $\log_5 x = 3$
72. $\log_7 x = -2$
73. $\log_6 y = 3$
74. $\log_3 y = \dfrac{1}{12}$
75. $\log(2w-1) = 3$
76. $\log_2(3w+5) = 5$
77. $-1 + \log p = -\log(p-3)$
78. $\log_4(2+t) - 3 = \log_4(3-5t)$
79. $4^{3x+5} = 16$
80. $5^{7x} = 625$
81. $4^a = 21$
82. $5^a = 18$
83. $e^{-x} = 0.1$
84. $e^{-2x} = 0.06$
85. $10^{2n} = 1512$
86. $10^{-3m} = \dfrac{1}{821}$
87. $2^{x+3} = 7^x$
88. $14^{x-5} = 6^x$

89. Radioactive iodine (^{131}I) is used to treat patients with a hyperactive (overactive) thyroid. The half-life of radioactive iodine is 8.04 days. If a patient is given an initial dose of 2 μg, then the amount of iodine in the body after t days is approximated by
$$A(t) = 2e^{-0.0862t}$$
where t is the time in days and $A(t)$ is the amount (in micrograms) of ^{131}I remaining.

   a. How much radioactive iodine is present after a week? Round to two decimal places.
   b. How much radioactive iodine is present after 30 days? Round to two decimal places.
   c. How long will it take for the level of radioactive iodine to reach 0.5 μg?

90. The growth of certain bacteria in a culture is given by the model $A(t) = 150e^{0.007t}$, where $A(t)$ is the number of bacteria and t is time in minutes.

   a. What is the initial number of bacteria?
   b. What is the population after $\tfrac{1}{2}$ hr?
   c. How long will it take for the population to double?

91. The value of a car is depreciated with time according to

$$V(t) = 15{,}000e^{-0.15t}$$

where $V(t)$ is the value in dollars and t is the time in years after purchase.

a. Find $V(0)$ and interpret the result in the context of this problem.

b. Find $V(10)$ and interpret the result in the context of this problem. Round to the nearest dollar.

c. Find the time required for the value of the car to drop to $5000. Round to the nearest tenth of a year.

Conic Sections

CHAPTER OUTLINE

- **21.1** Distance Formula, Midpoint Formula, and Circles 1410
- **21.2** More on the Parabola 1422
- **21.3** The Ellipse and Hyperbola 1432
  - Problem Recognition Exercises: Formulas and Conic Sections 1441
- **21.4** Nonlinear Systems of Equations in Two Variables 1442
- **21.5** Nonlinear Inequalities and Systems of Inequalities in Two Variables 1449
  - Chapter 21 Review Exercises 1459

Mathematics in Engineering

In this chapter, we will revisit our study of parabolas as well as two new curves called *ellipses* and *hyperbolas*. These curves are called **conic sections**. Conic sections and their three-dimensional counterparts have numerous applications in engineering. For example, the mirror in a reflecting telescope has cross sections in the shape of a parabola. This is important because the curved surface of the mirror focuses light to a common point called the **focus**. This targeted focusing of light enables astronomers to view an image from a distant star with better clarity.

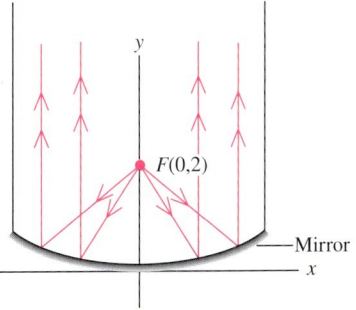

Aaron Roeth Photography

The cooling towers for nuclear power plants have cross sections in the shape of a hyperbola. One advantage of a hyperbolic cooling tower is that air accelerates as it rises inside, toward the narrow portion of the tower. This provides for an efficient, nonturbulent flow. Then, above the narrow part, the tower flares out for efficient dispersal of warm air.

The ellipse is an oval-shaped curve that also has numerous applications such as in architectural design and for describing the orbits of planets. For example, the Roman Coliseum is an elliptical stone and concrete amphitheater in the center of Rome, built between 70 A.D. and 80 A.D. The Coliseum seated approximately 50,000 spectators and was used for gladiatorial contests among other things.

Javier Larrea/Pixtal/age fotostock

Section 21.1 Distance Formula, Midpoint Formula, and Circles

Concepts

1. Distance Formula
2. Circles
3. Writing an Equation of a Circle
4. The Midpoint Formula

1. Distance Formula

Suppose we are given two points (x_1, y_1) and (x_2, y_2) in a rectangular coordinate system. The distance between the two points can be found by using the Pythagorean theorem (Figure 21-1).

First draw a right triangle with the distance d as the hypotenuse. The length of the horizontal leg a is $|x_2 - x_1|$, and the length of the vertical leg b is $|y_2 - y_1|$. From the Pythagorean theorem we have

$$d^2 = a^2 + b^2$$
$$= |x_2 - x_1|^2 + |y_2 - y_1|^2$$
$$= (x_2 - x_1)^2 + (y_2 - y_1)^2$$
$$d = \pm\sqrt{(x_2 - x_1)^2 + (y_2 - y_1)^2}$$
$$= \sqrt{(x_2 - x_1)^2 + (y_2 - y_1)^2}$$

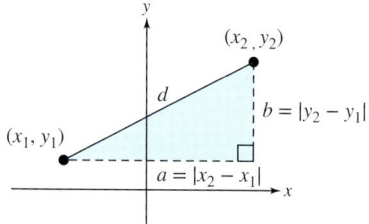

Figure 21-1

Because distance is positive, reject the negative value.

TIP: Squaring any real-valued quantity results in a nonnegative real number. Therefore, the absolute value bars can be dropped.

The Distance Formula

The distance d between the points (x_1, y_1) and (x_2, y_2) is

$$d = \sqrt{(x_2 - x_1)^2 + (y_2 - y_1)^2}$$

Example 1 Finding the Distance Between Two Points

Find the distance between the points $(-2, 3)$ and $(4, -1)$ (Figure 21-2).

Solution:

$\quad (-2, 3) \quad$ and $\quad (4, -1)$
$\quad (x_1, y_1) \quad\quad\quad\quad (x_2, y_2)$ Label the points.

$d = \sqrt{(x_2 - x_1)^2 + (y_2 - y_1)^2}$
$\quad = \sqrt{[(4) - (-2)]^2 + [(-1) - (3)]^2}$ Apply the distance formula.
$\quad = \sqrt{(6)^2 + (-4)^2}$
$\quad = \sqrt{36 + 16}$
$\quad = \sqrt{52}$
$\quad = \sqrt{4 \cdot 13}$
$\quad = 2\sqrt{13}$

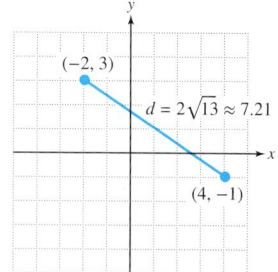

Figure 21-2

FOR REVIEW

Recall that to simplify $\sqrt{52}$, write the radicand as the product of the largest perfect square and a "leftover" factor. Then apply the product property of radicals.

$\sqrt{52} = \sqrt{2^2 \cdot 13}$
$\quad = \sqrt{2^2} \cdot \sqrt{13}$
$\quad = 2\sqrt{13}$

Skill Practice

1. Find the distance between the points $(-4, -2)$ and $(2, -5)$.

Answer

1. $3\sqrt{5}$

TIP: The order in which the points are labeled does not affect the result of the distance formula. For example, if the points in Example 1 had been labeled in reverse, the distance formula would still yield the same result:

$(-2, 3)$ and $(4, -1)$ $\quad d = \sqrt{(x_2 - x_1)^2 + (y_2 - y_1)^2}$
$(x_2, y_2) \quad\quad\quad (x_1, y_1)$
$\quad\quad\quad\quad\quad\quad\quad\quad\quad\quad = \sqrt{[(-2) - (4)]^2 + [3 - (-1)]^2}$
$\quad\quad\quad\quad\quad\quad\quad\quad\quad\quad = \sqrt{(-6)^2 + (4)^2}$
$\quad\quad\quad\quad\quad\quad\quad\quad\quad\quad = \sqrt{36 + 16}$
$\quad\quad\quad\quad\quad\quad\quad\quad\quad\quad = \sqrt{52}$
$\quad\quad\quad\quad\quad\quad\quad\quad\quad\quad = 2\sqrt{13}$

2. Circles

A **circle** is defined as the set of all points in a plane that are equidistant from a fixed point called the **center**. The fixed distance from the center is called the **radius** and is denoted by r, where $r > 0$.

Suppose a circle is centered at the point (h, k) and has radius r (Figure 21-3). The distance formula can be used to derive an equation of the circle.

Let (x, y) be any arbitrary point on the circle. Then, by definition, the distance between (h, k) and (x, y) must be r.

$$\sqrt{(x - h)^2 + (y - k)^2} = r$$
$$(x - h)^2 + (y - k)^2 = r^2 \quad \text{Square both sides.}$$

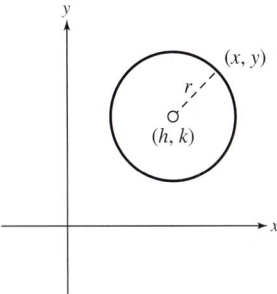

Figure 21-3

Standard Equation of a Circle

The **standard equation of a circle**, centered at (h, k) with radius r, is given by

$$(x - h)^2 + (y - k)^2 = r^2 \quad \text{where } r > 0.$$

Note: If a circle is centered at the origin $(0, 0)$, then $h = 0$ and $k = 0$, and the equation simplifies to $x^2 + y^2 = r^2$.

Example 2 — Graphing a Circle

Find the center and radius of the circle. Then graph the circle.

$$(x - 3)^2 + (y + 4)^2 = 36$$

Solution:

$(x - 3)^2 + (y + 4)^2 = 36$
$(x - 3)^2 + [y - (-4)]^2 = (6)^2$
$h = 3, k = -4,$ and $r = 6$

The equation is in standard form $(x - h)^2 + (y - k)^2 = r^2$. The center is $(3, -4)$, and the radius is $r = 6$. To graph the circle, first locate the center. From the center, mark points 6 units to the right, left, above, and below the center. Then sketch the circle through these four points (Figure 21-4).

Note that the center of the circle is not actually part of the circle. It is drawn as an open dot for reference.

Figure 21-4

Answer
2. Center: $(-1, 2)$; radius: 3

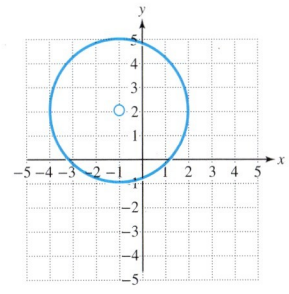

Skill Practice Find the center and radius of the circle. Then graph the circle.
2. $(x + 1)^2 + (y - 2)^2 = 9$

Example 3 — Graphing a Circle

Find the center and radius of each circle. Then graph the circle.

a. $x^2 + \left(y - \dfrac{10}{3}\right)^2 = \dfrac{25}{9}$ **b.** $x^2 + y^2 = 10$

Solution:

a. $x^2 + \left(y - \dfrac{10}{3}\right)^2 = \dfrac{25}{9}$

$(x - 0)^2 + \left(y - \dfrac{10}{3}\right)^2 = \left(\dfrac{5}{3}\right)^2$

The equation is now in standard form $(x - h)^2 + (y - k)^2 = r^2$.
The center is $\left(0, \dfrac{10}{3}\right)$ and the radius is $\dfrac{5}{3}$ (Figure 21-5).

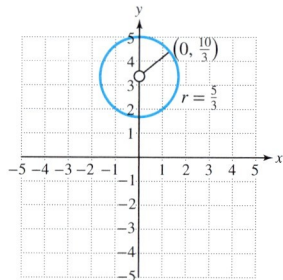

Figure 21-5

b. $x^2 + y^2 = 10$

$(x - 0)^2 + (y - 0)^2 = (\sqrt{10})^2$

$(x - h)^2 + (y - k)^2 = r^2$

The center is $(0, 0)$ and the radius is $\sqrt{10} \approx 3.16$ (Figure 21-6).

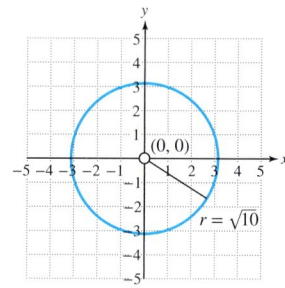

Figure 21-6

FOR REVIEW

From the vertical line test, we determine that a circle does not define y as a function of x.

Skill Practice Find the center and radius of the circle. Then graph the circle.

3. $\left(x + \dfrac{7}{2}\right)^2 + y^2 = \dfrac{9}{4}$

Sometimes it is necessary to complete the square to write an equation of a circle in standard form.

Example 4 — Writing an Equation of a Circle in the Form $(x - h)^2 + (y - k)^2 = r^2$

Identify the center and radius of the circle given by the equation.

$$x^2 + y^2 + 2x - 16y + 61 = 0$$

Solution:

$x^2 + y^2 + 2x - 16y + 61 = 0$ To identify the center and radius, write the equation in the form $(x - h)^2 + (y - k)^2 = r^2$.

$(x^2 + 2x \quad) + (y^2 - 16y \quad) = -61$ Group the x terms and group the y terms. Move the constant to the right-hand side.

Answer

3. Center: $\left(-\dfrac{7}{2}, 0\right)$; radius: $\dfrac{3}{2}$

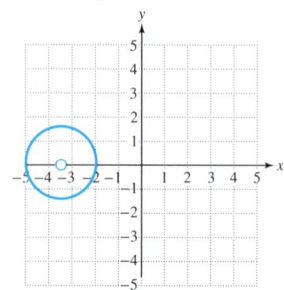

$(x^2 + 2x + 1) + (y^2 - 16y + 64) = -61 + 1 + 64$

- Complete the square on x. Add $[\frac{1}{2}(2)]^2 = 1$ to both sides of the equation.
- Complete the square on y. Add $[\frac{1}{2}(-16)]^2 = 64$ to both sides of the equation.

$(x + 1)^2 + (y - 8)^2 = 4$

$[x - (-1)]^2 + (y - 8)^2 = 2^2$

Factor and simplify.

Standard form: $(x - h)^2 + (y - k)^2 = r^2$

FOR REVIEW

Given $x^2 + bx + \square$, complete the square by adding the square of one-half the linear term coefficient. Thus,
$$x^2 + bx + \left(\frac{1}{2}b\right)^2.$$

The center is $(-1, 8)$ and the radius is 2.

Skill Practice Identify the center and radius of the circle given by the equation.

4. $x^2 + y^2 - 10x + 4y - 7 = 0$

3. Writing an Equation of a Circle

Example 5 Writing an Equation of a Circle

Write an equation of the circle shown in Figure 21-7.

Solution:

The center is $(-3, 2)$; therefore, $h = -3$ and $k = 2$. From the graph, $r = 2$.

$(x - h)^2 + (y - k)^2 = r^2$

$[x - (-3)]^2 + (y - 2)^2 = (2)^2$

$(x + 3)^2 + (y - 2)^2 = 4$

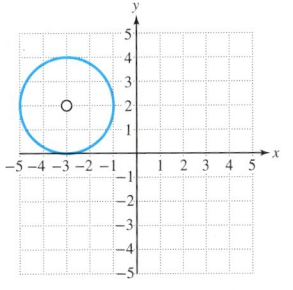

Figure 21-7

Avoiding Mistakes

Be sure that you *subtract* both coordinates of the center, when substituting the coordinates into the standard form of a circle.

Skill Practice

5. Write an equation for a circle whose center is $(6, -1)$ and whose radius is 8.

4. The Midpoint Formula

Consider two points in the coordinate plane and the line segment determined by the points. It is sometimes necessary to determine the point that is halfway between the endpoints of the segment. This point is called the *midpoint*.

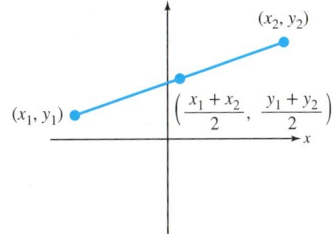

TIP: The midpoint of a line segment is found by taking the *average* of the x-coordinates and the *average* of the y-coordinates of the endpoints.

Midpoint Formula

Given two points (x_1, y_1) and (x_2, y_2), the midpoint of the line segment between the two points is given by

Midpoint: $\left(\dfrac{x_1 + x_2}{2}, \dfrac{y_1 + y_2}{2}\right)$

Answers

4. Center: $(5, -2)$; radius: 6
5. $(x - 6)^2 + (y + 1)^2 = 64$

Example 6 Finding the Midpoint of a Segment

Find the midpoint of the line segment with the given endpoints.

a. $(-4, 6)$ and $(8, 1)$ **b.** $(-1.2, -3.1)$ and $(-6.6, 1.2)$

Solution:

a. $(-4, 6)$ and $(8, 1)$
$(x_1, y_1) (x_2, y_2)$

$\left(\dfrac{-4+8}{2}, \dfrac{6+1}{2}\right)$ Apply the midpoint formula: $\left(\dfrac{x_1+x_2}{2}, \dfrac{y_1+y_2}{2}\right)$

$\left(2, \dfrac{7}{2}\right)$ Simplify.

The midpoint of the segment is $(2, \frac{7}{2})$.

b. $(-1.2, -3.1)$ and $(-6.6, 1.2)$
$(x_1, y_1) (x_2, y_2)$

$\left(\dfrac{-1.2+(-6.6)}{2}, \dfrac{-3.1+1.2}{2}\right)$ Apply the midpoint formula.

$(-3.9, -0.95)$ Simplify.

> **Avoiding Mistakes**
> Always remember that the midpoint of a line segment is a *point*. The answer should be an ordered pair, (x, y).

Skill Practice Find the midpoint of the line segment with the given endpoints.

6. $(5, 6)$ and $(-10, 4)$ **7.** $(-2.6, -6.3)$ and $(1.2, 4.1)$

Example 7 Applying the Midpoint Formula

Suppose that $(-2, 3)$ and $(4, 1)$ are endpoints of a diameter of a circle.

a. Find the center of the circle. **b.** Write an equation of the circle.

Solution:

a. Because the midpoint of a diameter of a circle is the center of the circle, apply the midpoint formula. See Figure 21-8.

$(-2, 3)$ and $(4, 1)$
$(x_1, y_1) (x_2, y_2)$

$\left(\dfrac{x_1+x_2}{2}, \dfrac{y_1+y_2}{2}\right)$

$\left(\dfrac{-2+4}{2}, \dfrac{3+1}{2}\right)$ Apply the midpoint formula.

$(1, 2)$ Simplify.

The center of the circle is $(1, 2)$.

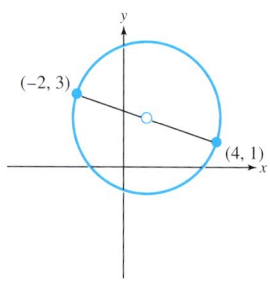

Figure 21-8

Answers

6. $\left(-\dfrac{5}{2}, 5\right)$ **7.** $(-0.7, -1.1)$

b. The radius can be determined by finding the distance between the center and either endpoint of the diameter.

$(1, 2)$ and $(4, 1)$
(x_1, y_1) (x_2, y_2)

$d = \sqrt{(x_2 - x_1)^2 + (y_2 - y_1)^2}$ Apply the distance formula.

$d = \sqrt{(4 - 1)^2 + (1 - 2)^2}$ Substitute $(1, 2)$ and $(4, 1)$ for (x_1, y_1) and (x_2, y_2).

$= \sqrt{(3)^2 + (-1)^2}$ Simplify.

$= \sqrt{10}$

The circle is represented by $(x - 1)^2 + (y - 2)^2 = (\sqrt{10})^2$
or $(x - 1)^2 + (y - 2)^2 = 10$.

Skill Practice

8. Suppose that $(3, 2)$ and $(7, 10)$ are endpoints of a diameter of a circle.
   a. Find the center of the circle.
   b. Write an equation of the circle.

Answer
8. a. $(5, 6)$
   b. $(x - 5)^2 + (y - 6)^2 = 20$

Section 21.1 Activity

A.1. Two bicyclists start out from the same point of origin. One rides east for $\frac{1}{4}$ hr and then north for $\frac{1}{2}$ hr at a rate of 16 mph. The other cyclist rides east for 1 hr and then south for $\frac{1}{3}$ hr at a rate of 12 mph.
   a. Find the ordered pairs representing the positions of the cyclists at their respective destinations. Use the point of origin as $(0, 0)$.
   b. Draw a right triangle using the distance between the cyclists as the hypotenuse.
   c. Find the horizontal distance between the two cyclists. Find the vertical distance between the two cyclists.
   d. Use the Pythagorean theorem to find the exact distance between the two cyclists at their destinations. Then approximate the distance to the nearest tenth of a mile.

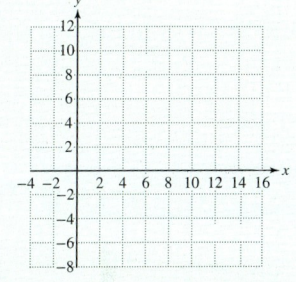

A.2. Refer to Exercise A.1. Label the point $(4, 8)$ as (x_1, y_1) and label the point $(12, -4)$ as (x_2, y_2).
   a. Write the generic distance formula to find the distance d between the points (x_1, y_1) and (x_2, y_2).
   b. Find the exact distance between the points $(4, 8)$ and $(12, -4)$.
   c. In the distance formula, the term $(x_2 - x_1)^2$ represents the square of the (choose one: horizontal/vertical) distance between the two points. The term $(y_2 - y_1)^2$ represents the square of the (choose one: horizontal/vertical) distance between the two points.
   d. Apply the distance formula with the point $(4, 8)$ labeled as (x_2, y_2) and the point $(12, -4)$ labeled as (x_1, y_1). Does the manner in which the points are labeled affect the result?

A.3. a. Explain how to find the average of two numbers.
   b. The midpoint of the line segment with endpoints (x_1, y_1) and (x_2, y_2) has coordinates given by $\left(\boxed{\text{Average of } x \text{ values}}, \boxed{\text{Average of } y \text{ values}} \right)$. Write the formula for the midpoint between the points (x_1, y_1) and (x_2, y_2).
   c. Refer to Exercise A.1. The bicyclists are at points $(4, 8)$ and $(12, -4)$ and plan to meet for lunch at the midpoint between them. Find the midpoint.

A.4. a. By definition, a circle is the set of points equidistant from a fixed point. The fixed point is called the _____ of the circle, and the common distance from the center to any point on the circle is called the _____.

b. Suppose that (x, y) is a point on a circle with center (h, k) and radius r. Write the standard equation for the circle.

For Exercises A.5–A.6, identify the center and radius of the circle.

A.5. $(x - 5)^2 + (y + 2)^2 = 4$

A.6. $\left(x + \dfrac{3}{2}\right)^2 + y^2 = 20$

A.7. The equation $x^2 + y^2 + 6x - 8y + 16 = 0$ is an equation of a circle in expanded form. To identify the center and radius, it is necessary to complete the square so that the equation is expressed as $(x - h)^2 + (y - k)^2 = r^2$. Follow these steps.

a. Group the x terms and group the y terms on the left side of the equation and move the constant term to the right. Fill in the blanks to complete the squares.

$$x^2 + 6x + \underline{\hspace{1cm}} + y^2 - 8y + \underline{\hspace{1cm}} = -16 + \underline{\hspace{1cm}} + \underline{\hspace{1cm}}$$

b. Factor on the left and simplify on the right.

c. Identify the center and radius.

d. Graph the circle.

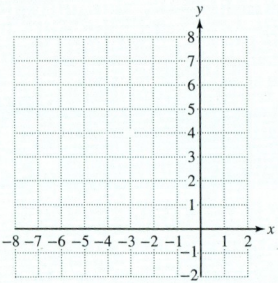

A.8. Write an equation of a circle with center $(0, -4)$ and radius 7.

Section 21.1 Practice Exercises

Prerequisite Review

For Exercises R.1–R.2, simplify the radical.

R.1. $\sqrt{50}$

R.2. $\sqrt{48}$

For Exercises R.3–R.6, simplify the expression.

R.3. $\dfrac{36}{48}$

R.4. $\dfrac{108}{72}$

R.5. $\dfrac{\dfrac{1}{2} + \dfrac{5}{4}}{2}$

R.6. $\dfrac{\dfrac{2}{3} + \dfrac{4}{5}}{2}$

For Exercises R.7–R.8, square the binomial.

R.7. $(x + 8)^2$

R.8. $(x - 9)^2$

For Exercises R.9–R.10, find the value of n that will make the trinomial a perfect square trinomial. Then factor.

R.9. $x^2 - 12x + n$

R.10. $x^2 + 20x + n$

Section 21.1 Distance Formula, Midpoint Formula, and Circles

Vocabulary and Key Concepts

1. **a.** The distance between two distinct points (x_1, y_1) and (x_2, y_2) is given by $d = $ _____.
   b. A _____ is the set of all points equidistant from a fixed point called the _____.
   c. The distance from the center of a circle to any point on the circle is called the _____ and is often denoted by r.
   d. The standard equation of a circle with center (h, k) and radius r is given by _____.
   e. The midpoint of the line segment with endpoints (x_1, y_1) and (x_2, y_2) is given by _____.

Concept 1: Distance Formula

For Exercises 2–16, use the distance formula to find the distance between the two points. **(See Example 1.)**

2. $(-2, 7)$ and $(4, -5)$
3. $(1, 10)$ and $(-2, 4)$
4. $(0, 5)$ and $(-3, 8)$
5. $(6, 7)$ and $(3, 2)$
6. $\left(\dfrac{2}{3}, \dfrac{1}{5}\right)$ and $\left(-\dfrac{5}{6}, \dfrac{3}{10}\right)$
7. $\left(-\dfrac{1}{2}, \dfrac{5}{8}\right)$ and $\left(-\dfrac{3}{2}, \dfrac{1}{4}\right)$
8. $(4, 13)$ and $(4, -6)$
9. $(-2, 5)$ and $(-2, 9)$
10. $(8, -6)$ and $(-2, -6)$
11. $(7, 2)$ and $(15, 2)$
12. $(-6, -2)$ and $(-3, -5)$
13. $(-1, -5)$ and $(-5, -9)$
14. $(3\sqrt{5}, 2\sqrt{7})$ and $(-\sqrt{5}, -3\sqrt{7})$
15. $(4\sqrt{6}, -2\sqrt{2})$ and $(2\sqrt{6}, \sqrt{2})$
16. $(6, 0)$ and $(0, -1)$

17. Explain how to find the distance between 5 and -7 on the y-axis.

18. Explain how to find the distance between 15 and -37 on the x-axis.

19. Find the values of y such that the distance between the points $(4, 7)$ and $(-4, y)$ is 10 units.

20. Find the values of x such that the distance between the points $(-4, -2)$ and $(x, 3)$ is 13 units.

21. Find the values of x such that the distance between the points $(x, 2)$ and $(4, -1)$ is 5 units.

22. Find the values of y such that the distance between the points $(-5, 2)$ and $(3, y)$ is 10 units.

For Exercises 23–26, determine if the three points define the vertices of a right triangle.

23. $(-3, 2), (-2, -4),$ and $(3, 3)$
24. $(1, -2), (-2, 4),$ and $(7, 1)$
25. $(-3, -2), (4, -3),$ and $(1, 5)$
26. $(1, 4), (5, 3),$ and $(2, 0)$

Concept 2: Circles

For Exercises 27–47, identify the center and radius of the circle and then graph the circle. Complete the square, if necessary. **(See Examples 2–4.)**

27. $(x - 4)^2 + (y + 2)^2 = 9$

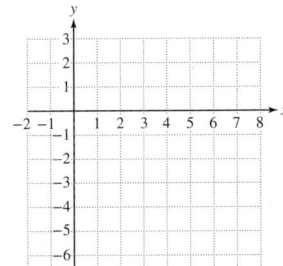

28. $(x - 3)^2 + (y + 1)^2 = 16$

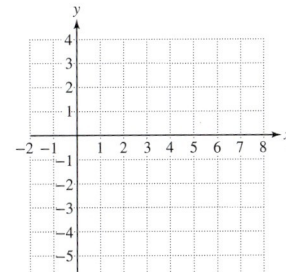

29. $(x + 1)^2 + (y + 1)^2 = 1$

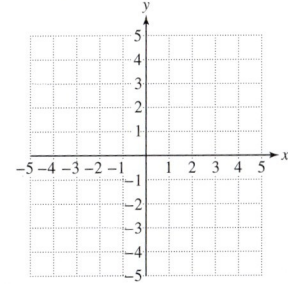

30. $(x-4)^2 + (y-4)^2 = 4$

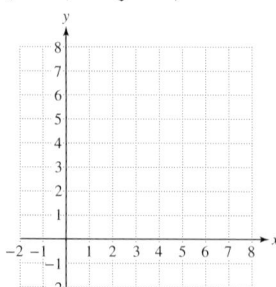

31. $x^2 + (y-2)^2 = 4$

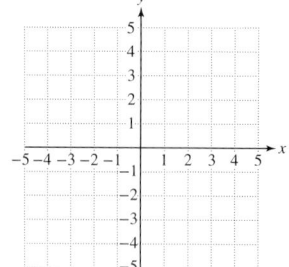

32. $(x+1)^2 + y^2 = 1$

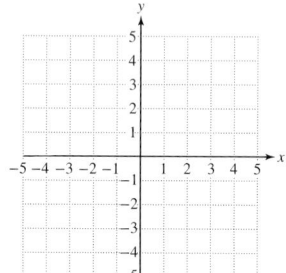

33. $(x-3)^2 + y^2 = 8$

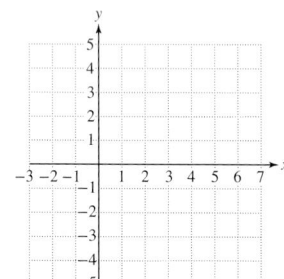

34. $x^2 + (y+2)^2 = 20$

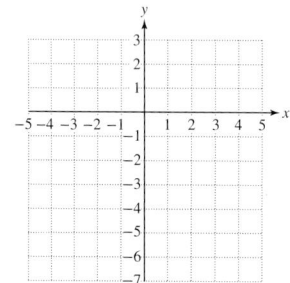

35. $x^2 + y^2 = 6$

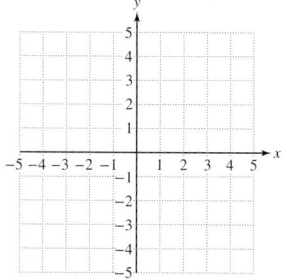

36. $x^2 + y^2 = 15$

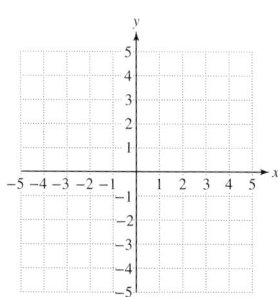

37. $\left(x + \dfrac{4}{5}\right)^2 + y^2 = \dfrac{64}{25}$

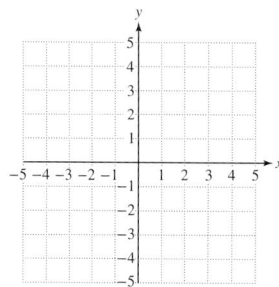

38. $x^2 + \left(y - \dfrac{5}{2}\right)^2 = \dfrac{9}{4}$

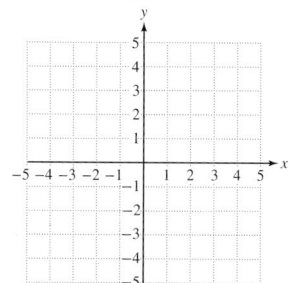

39. $x^2 + y^2 - 2x - 6y - 26 = 0$

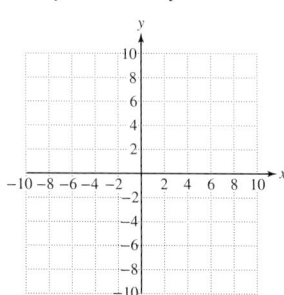

40. $x^2 + y^2 + 4x - 8y + 16 = 0$

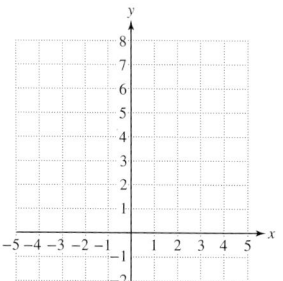

41. $x^2 + y^2 - 6y + 5 = 0$

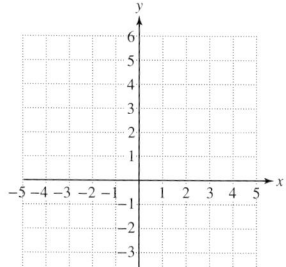

42. $x^2 + 2x + y^2 - 15 = 0$

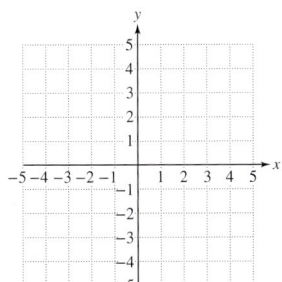

43. $x^2 + y^2 + 6y + \dfrac{65}{9} = 0$

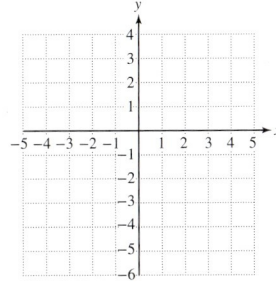

44. $x^2 + y^2 - 12x + 12y + 71 = 0$

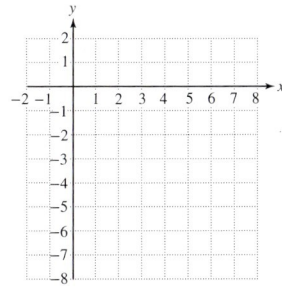

45. $x^2 + y^2 + 2x + 4y - 4 = 0$

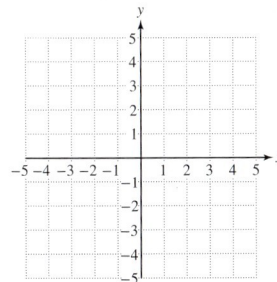

46. $2x^2 + 2y^2 = 32$

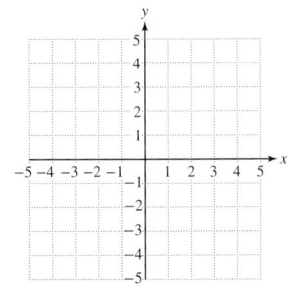

47. $3x^2 + 3y^2 = 3$

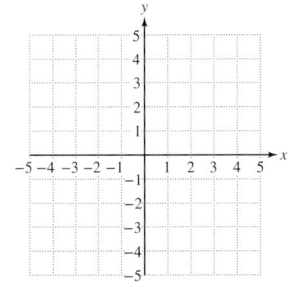

Concept 3: Writing an Equation of a Circle

48. If the diameter of a circle is 10 ft, what is the radius?

For Exercises 49–54, write an equation that represents the graph of the circle. (See Example 5.)

49.
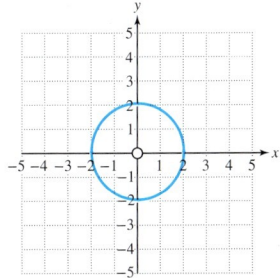

50.
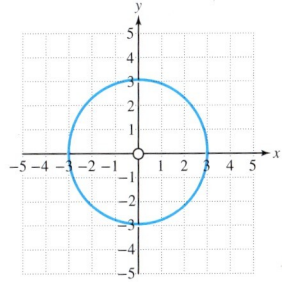

51.
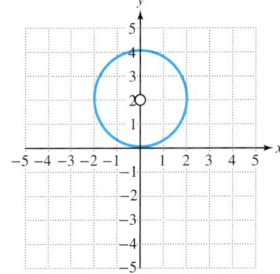

52.
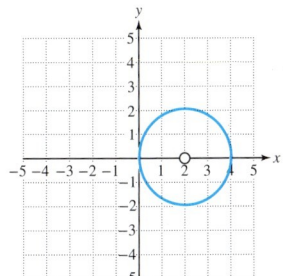

53.
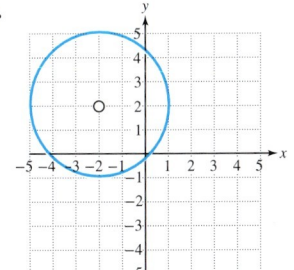

54.
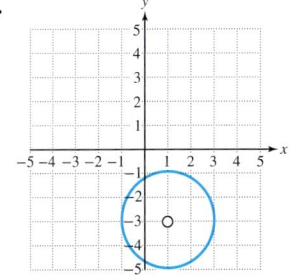

55. Write an equation of a circle centered at the origin with a radius of 7.

56. Write an equation of a circle centered at the origin with a radius of 12.

57. Write an equation of a circle centered at $(-3, -4)$ with a diameter of 12.

58. Write an equation of a circle centered at $(5, -1)$ with a diameter of 8.

59. A cell tower is a site where antennas, transmitters, and receivers are placed to create a cellular network. Suppose that a cell tower is located at a point $A(5, 3)$ on a map and its range is 1.5 mi. Write an equation that represents the boundary of the area that can receive a signal from the tower. Assume that all distances are in miles.

60. A radar transmitter on a ship has a range of 20 nautical miles. If the ship is located at a point $(-28, 32)$ on a map, write an equation for the boundary of the area within the range of the ship's radar. Assume that all distances on the map are represented in nautical miles.

Concept 4: The Midpoint Formula

For Exercises 61–64, find the midpoint of the line segment. Check your answers by plotting the midpoint on the graph.

61.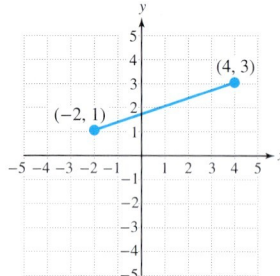

62.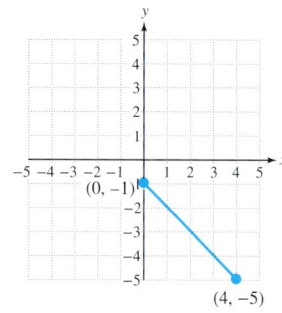

63.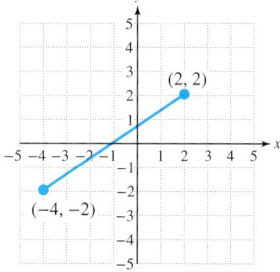

64.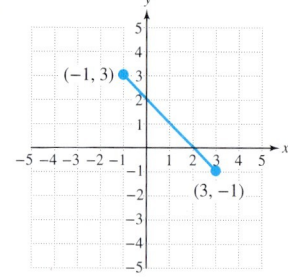

For Exercises 65–72, find the midpoint of the line segment between the two given points. (See Example 6.)

65. $(4, 0)$ and $(-6, 12)$

66. $(-7, 2)$ and $(-3, -2)$

67. $(-3, 8)$ and $(3, -2)$

68. $(0, 5)$ and $(4, -5)$

69. $(5, 2)$ and $(-6, 1)$

70. $(-9, 3)$ and $(0, -4)$

71. $(-2.4, -3.1)$ and $(1.6, 1.1)$

72. $(0.8, 5.3)$ and $(-4.2, 7.1)$

73. Two courier trucks leave the warehouse to make deliveries. One travels 20 mi north and 30 mi east. The other truck travels 5 mi south and 50 mi east. If the two drivers want to meet for lunch at a restaurant at a point halfway between them, where should they meet relative to the warehouse? (*Hint:* Label the warehouse as the origin, and find the coordinates of the restaurant. See the figure.)

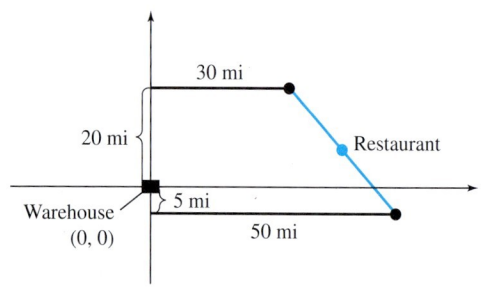

74. A map of a hiking area is drawn so that the Visitor Center is at the origin of a rectangular grid. Two hikers are located at positions $(-1, 1)$ and $(-3, -2)$ with respect to the Visitor Center where all units are in miles. A campground is located exactly halfway between the hikers. What are the coordinates of the campground?

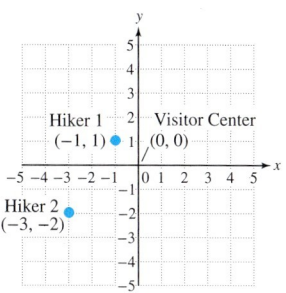

For Exercises 75–78, the two given points are endpoints of a diameter of a circle.

a. Find the center of the circle.

b. Write an equation of the circle. (See Example 7.)

75. $(-1, 2)$ and $(3, 4)$ 76. $(-3, 3)$ and $(7, -1)$ 77. $(-2, 3)$ and $(2, 3)$ 78. $(-1, 3)$ and $(-1, -3)$

Technology Connections

For Exercises 79–84, graph the circle from the indicated exercise on a square viewing window, and approximate the center and the radius from the graph.

79. $(x - 4)^2 + (y + 2)^2 = 9$ (Exercise 27)

80. $(x - 3)^2 + (y + 1)^2 = 16$ (Exercise 28)

81. $x^2 + (y - 2)^2 = 4$ (Exercise 31)

82. $(x + 1)^2 + y^2 = 1$ (Exercise 32)

83. $x^2 + y^2 = 6$ (Exercise 35)

84. $x^2 + y^2 = 15$ (Exercise 36)

Expanding Your Skills

85. Write an equation of a circle whose center is $(4, 4)$ and is tangent to the x- and y-axes. (*Hint:* Sketch the circle first.)

86. Write an equation of a circle whose center is $(-3, 3)$ and is tangent to the x- and y-axes. (*Hint:* Sketch the circle first.)

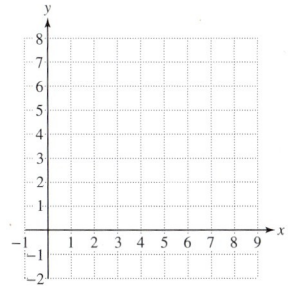
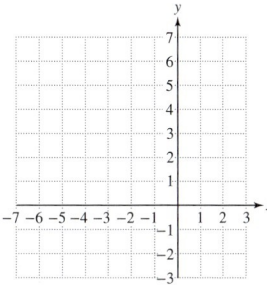

87. Write an equation of a circle whose center is $(1, 1)$ and that passes through the point $(-4, 3)$.

88. Write an equation of a circle whose center is $(-3, -1)$ and that passes through the point $(5, -2)$.

Section 21.2 More on the Parabola

Concepts

1. Introduction to Conic Sections
2. Parabola—Vertical Axis of Symmetry
3. Parabola—Horizontal Axis of Symmetry
4. Vertex Formula

1. Introduction to Conic Sections

The graph of a second-degree equation of the form $y = ax^2 + bx + c$ ($a \neq 0$) is a parabola and the graph of $(x - h)^2 + (y - k)^2 = r^2$ is a circle. These and two other types of figures called ellipses and hyperbolas are called **conic sections**. Conic sections derive their names because each is the intersection of a plane and a double-napped cone.

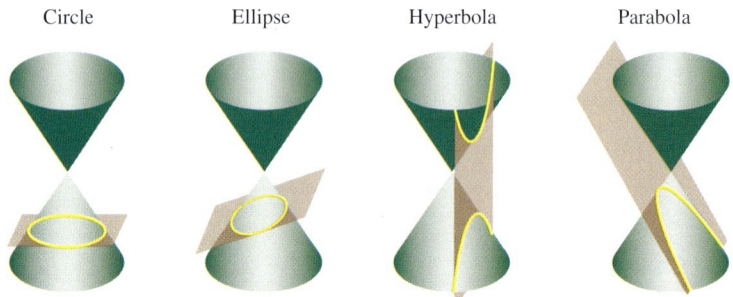

2. Parabola—Vertical Axis of Symmetry

A **parabola** is defined by a set of points in a plane that are equidistant from a fixed line (called the directrix) and a fixed point (called the focus) not on the directrix (Figure 21-9). Parabolas have numerous real-world applications. For example, a flashlight has a mirror with the cross section in the shape of a parabola. The bulb is located at the focus. The light hits the mirror and is reflected outward parallel to the sides of the cylinder. This forms a beam of light (Figure 21-10).

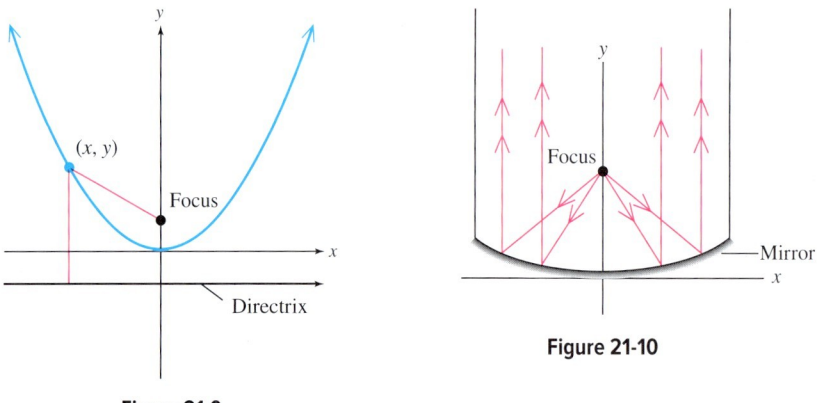

Figure 21-9

Figure 21-10

The graph of the solutions to the quadratic equation $y = ax^2 + bx + c$ is a parabola. Recall that the **vertex** is the highest or the lowest point of a parabola. The **axis of symmetry** of the parabola is a line that passes through the vertex and is perpendicular to the directrix (Figure 21-11).

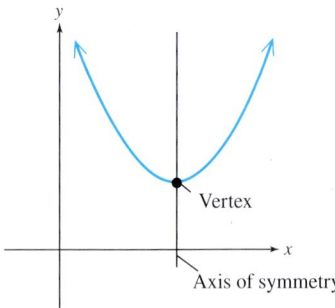

Figure 21-11

Standard Form of the Equation of a Parabola—Vertical Axis of Symmetry

The standard form of the equation of a parabola with vertex (h, k) and vertical axis of symmetry is

$$y = a(x - h)^2 + k \qquad \text{where } a \neq 0$$

If $a > 0$, the parabola opens upward; and if $a < 0$, the parabola opens downward.

The axis of symmetry is given by $x = h$.

It is sometimes necessary to complete the square to write an equation of a parabola in standard form.

Example 1 Graphing a Parabola by First Completing the Square

Given the equation of the parabola $y = -2x^2 + 4x + 1$,

a. Write the equation in standard form $y = a(x - h)^2 + k$.

b. Identify the vertex and axis of symmetry. Determine if the parabola opens upward or downward.

c. Graph the parabola.

Solution:

a. Complete the square to write the equation in the form $y = a(x - h)^2 + k$.

$y = -2x^2 + 4x + 1$

$y = -2(x^2 - 2x) + 1$ Factor out -2 from the variable terms.

$y = -2(x^2 - 2x + 1 - 1) + 1$ Add and subtract the quantity $[\frac{1}{2}(-2)]^2 = 1$.

$y = -2(x^2 - 2x + 1) + (-2)(-1) + 1$ Remove the -1 term from within the parentheses by first applying the distributive property. When -1 is removed from the parentheses, it carries with it the factor of -2 from outside the parentheses.

$y = -2(x - 1)^2 + 2 + 1$

$y = -2(x - 1)^2 + 3$

The equation is in the form $y = a(x - h)^2 + k$ where $a = -2$, $h = 1$, and $k = 3$.

b. The vertex is $(1, 3)$. Because a is negative, the parabola opens downward. The axis of symmetry is $x = 1$.

> **FOR REVIEW**
>
> Recall that a perfect square trinomial factors as the square of a binomial. For example:
>
> $x^2 - 2x + 1 = (x - 1)^2$

c. To graph the parabola, we know that its orientation is downward because the leading term is negative. Furthermore, we know the vertex is (1, 3). To find other points on the parabola, select several values of x and solve for y. Recall that the y-intercept is found by substituting $x = 0$ and solving for y.

x	y	
1	3	← Vertex
0	1	← y-intercept
−1	−5	
2	1	
3	−5	

↑ ↑
Choose x. Solve for y.

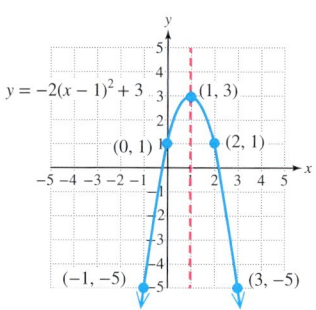

Notice that the points (2, 1) and (3, −5) are the mirror images of the points (0, 1) and (−1, −5) around the axis of symmetry, $x = 1$. For this reason, once we plot several points on one side of the axis of symmetry, the points on the other side automatically follow as a mirror image.

Skill Practice Given the equation of the parabola $y = 3x^2 + 6x + 4$,

1. Write the equation in standard form.
2. Identify the vertex and axis of symmetry.
3. Graph the parabola.

3. Parabola—Horizontal Axis of Symmetry

We have seen that the graph of a parabola $y = ax^2 + bx + c$ opens upward if $a > 0$ and downward if $a < 0$. A parabola can also open to the left or right. In such a case, the "roles" of x and y are essentially interchanged in the equation. Thus, the graph of $x = ay^2 + by + c$ opens to the right if $a > 0$ (Figure 21-12) and to the left if $a < 0$ (Figure 21-13).

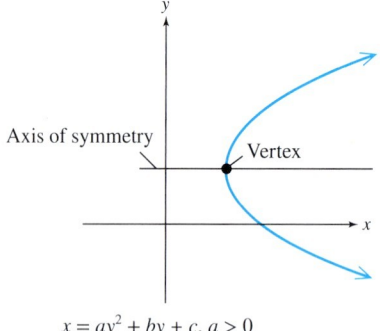

$x = ay^2 + by + c, a > 0$
Figure 21-12

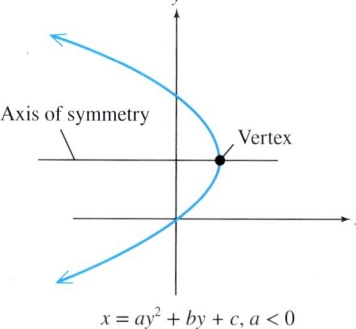

$x = ay^2 + by + c, a < 0$
Figure 21-13

Answers

1. $y = 3(x + 1)^2 + 1$
2. Vertex: (−1, 1); axis of symmetry: $x = −1$
3.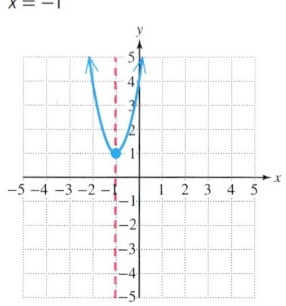

Standard Form of the Equation of a Parabola—Horizontal Axis of Symmetry

The standard form of an equation of a parabola with vertex (h, k) and horizontal axis of symmetry is

$$x = a(y - k)^2 + h \quad \text{where } a \neq 0$$

If $a > 0$, the parabola opens to the right and if $a < 0$, the parabola opens to the left.

The axis of symmetry is given by $y = k$.

Example 2 Graphing a Parabola With a Horizontal Axis of Symmetry

Given the equation of the parabola $x = 4y^2$,

a. Determine the coordinates of the vertex and the equation of the axis of symmetry.

b. Use the value of a to determine if the parabola opens to the right or left.

c. Plot several points and graph the parabola.

Solution:

a. The equation can be written in the form $x = a(y - k)^2 + h$:
$$x = 4(y - 0)^2 + 0$$

Therefore, $h = 0$ and $k = 0$.

The vertex is $(0, 0)$. The axis of symmetry is $y = 0$ (the x-axis).

b. Because a is positive, the parabola opens to the right.

c. The vertex of the parabola is $(0, 0)$. To find other points on the graph, select values for y and solve for x.

x	y
0	0 ← Vertex
4	1
4	−1
16	2
16	−2

Solve for x. Choose y.

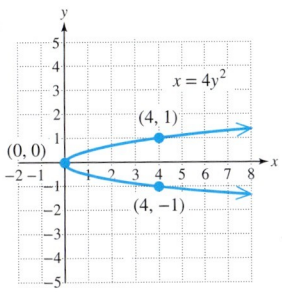

TIP: For Example 2, the x-axis is the axis of symmetry. Therefore, once we find points on the parabola above the axis of symmetry, their mirror image will be points on the parabola below the axis of symmetry.

Skill Practice Given the equation $x = -y^2 + 1$,

4. Identify the vertex and the axis of symmetry.
5. Determine if the parabola opens to the right or to the left.
6. Graph the parabola.

Answers
4. Vertex: $(1, 0)$; axis of symmetry: $y = 0$
5. Parabola opens left.
6.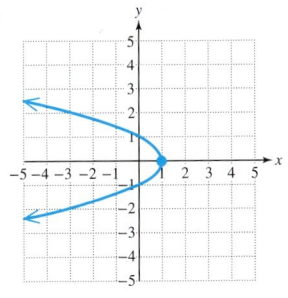

Example 3 — Graphing a Parabola by First Completing the Square

Given the equation of the parabola $x = -y^2 + 8y - 14$,

a. Write the equation in standard form $x = a(y - k)^2 + h$.

b. Identify the vertex and axis of symmetry. Determine if the parabola opens to the right or left.

c. Graph the parabola.

Solution:

a. Complete the square to write the equation in the form $x = a(y - k)^2 + h$.

$x = -y^2 + 8y - 14$

$x = -1(y^2 - 8y) - 14$ Factor out -1 from the variable terms.

$x = -1(y^2 - 8y + 16 - 16) - 14$ Add and subtract the quantity $\left[\tfrac{1}{2}(-8)\right]^2 = 16$.

$x = -1(y^2 - 8y + 16) + (-1)(-16) - 14$ Remove the -16 term from within the parentheses by first applying the distributive property. When -16 is removed from the parentheses, it carries with it the factor of -1 from outside the parentheses.

$x = -1(y - 4)^2 + 16 - 14$

$x = -1(y - 4)^2 + 2$

The equation is in the form $x = a(y - k)^2 + h$, where $a = -1$, $h = 2$, and $k = 4$.

b. The vertex is $(2, 4)$. Because a is negative, the parabola opens to the left. The axis of symmetry is $y = 4$.

c.

x	y	
2	4	← Vertex
1	5	
1	3	
-2	6	
-2	2	

↑ Solve for x.   ↑ Choose y.

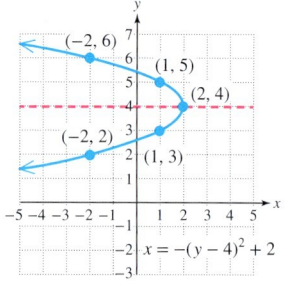

Skill Practice

Given the equation of the parabola $x = y^2 + 4y + 7$,

7. Identify the vertex and axis of symmetry.

8. Determine if the parabola opens to the right or to the left.

9. Graph the parabola.

Answers

7. Vertex: $(3, -2)$; Axis of symmetry: $y = -2$

8. Parabola opens right.

9.

4. Vertex Formula

The vertex formula can also be used to find the vertex of a parabola.

> **Vertex Formula**
>
> For a parabola defined by $y = ax^2 + bx + c$,
> - The x-coordinate of the vertex is given by $x = \frac{-b}{2a}$.
> - The y-coordinate of the vertex is found by substituting this value for x into the original equation and solving for y.
>
> For a parabola defined by $x = ay^2 + by + c$,
> - The y-coordinate of the vertex is given by $y = \frac{-b}{2a}$.
> - The x-coordinate of the vertex is found by substituting this value for y into the original equation and solving for x.

Example 4 Finding the Vertex of a Parabola by Using the Vertex Formula

Find the vertex by using the vertex formula. $x = y^2 + 4y + 5$

Solution:

$x = y^2 + 4y + 5$ The parabola is in the form $x = ay^2 + by + c$.

$a = 1 \quad b = 4 \quad c = 5$ Identify a, b, and c.

The y-coordinate of the vertex is given by $y = \dfrac{-b}{2a} = \dfrac{-(4)}{2(1)} = -2$.

The x-coordinate of the vertex is found by substitution:

$$x = (-2)^2 + 4(-2) + 5 = 1$$

The vertex is $(1, -2)$.

Skill Practice Find the vertex by using the vertex formula.

10. $x = -4y^2 + 12y$

Example 5 Find the Vertex of a Parabola by Using the Vertex Formula

Find the vertex by using the vertex formula. $y = \dfrac{1}{2}x^2 - 3x + \dfrac{5}{2}$

Solution:

$y = \dfrac{1}{2}x^2 - 3x + \dfrac{5}{2}$ The parabola is in the form $y = ax^2 + bx + c$.

$a = \dfrac{1}{2} \quad b = -3 \quad c = \dfrac{5}{2}$ Identify a, b, and c.

The x-coordinate of the vertex is given by $x = \dfrac{-b}{2a} = \dfrac{-(-3)}{2\left(\dfrac{1}{2}\right)} = 3$.

Answer

10. Vertex: $\left(9, \dfrac{3}{2}\right)$

The y-coordinate of the vertex is found by substitution.

$$y = \frac{1}{2}(3)^2 - 3(3) + \frac{5}{2}$$

$$= \frac{9}{2} - 9 + \frac{5}{2}$$

$$= \frac{14}{2} - 9$$

$$= 7 - 9$$

$$= -2$$

The vertex is $(3, -2)$.

Skill Practice Find the vertex by using the vertex formula.

11. $y = \frac{3}{4}x^2 + 3x + 5$

Answer

11. $(-2, 2)$

Section 21.2 Activity

A.1. a. Describe the graph of $y = a(x - h)^2 + k$.

b. Describe the graph of $x = a(y - k)^2 + h$.

A.2. Graph $y = -(x - 4)^2 + 1$.

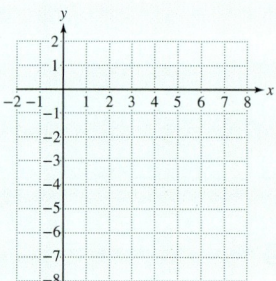

A.3. Refer to Exercise A.2. How would the graph change for the equation $x = -(y - 1)^2 + 4$?

A.4. The equation $x = y^2 + 4y + 1$ is of the form $x = ay^2 + by + c$.
  a. Identify the values of a, b, and c.
  b. Does the parabola open upward, downward, to the left, or to the right?
  c. Write the equation in the form $x = a(y - k)^2 + h$ by completing the square.
  d. Identify the vertex.
  e. Identify the axis of symmetry.

Practice Exercises

Section 21.2

Prerequisite Review

For Exercises R.1–R.4,

a. Identify the vertex of the parabola defined by the quadratic function.
b. Determine the y-intercept.
c. Determine the x-intercepts (if any exist).

R.1. $y = -x^2 + 2x + 3$

R.2. $y = x^2 + 6x + 8$

R.3. $y = x^2 + 4x + 4$

R.4. $y = x^2 - 10x + 25$

For Exercises R.5–R.6, evaluate the expression for the given values of the variables.

R.5. $\dfrac{-b}{2a}$ for $a = 6$ and $b = 24$

R.6. $\dfrac{-b}{2a}$ for $a = 5$ and $b = -12$

Vocabulary and Key Concepts

1. a. A circle, a parabola, an ellipse, and a hyperbola are curves that collectively are called _____ sections. These types of curves derive their names because each is the intersection of a _____ and a double-napped cone.

   b. A _____ is defined by a set of points in a plane that are equidistant from a fixed line (called the _____) and a fixed point (called the _____).

   c. Given a parabola defined by $y = a(x - h)^2 + k$ $(a \neq 0)$, the _____ is (h, k) and is the highest or lowest point on the graph. The axis of _____ is the vertical line through the vertex whose equation is $x = h$.

   d. Given a parabola defined by $x = a(y - k)^2 + h$ $(a \neq 0)$, the axis of symmetry is the horizontal line whose equation is _____.

2. Determine whether the parabola defined by the given function opens upward, downward, to the right, or to the left.

   a. $y = -x^2$
   b. $x = y^2$
   c. $x = -y^2$
   d. $y = x^2$

Concept 2: Parabola—Vertical Axis of Symmetry

For Exercises 3–10, use the equation of the parabola in standard form $y = a(x - h)^2 + k$ to determine the coordinates of the vertex and the equation of the axis of symmetry (complete the square if necessary). Then graph the parabola. (See Example 1.)

3. $y = (x + 2)^2 + 1$

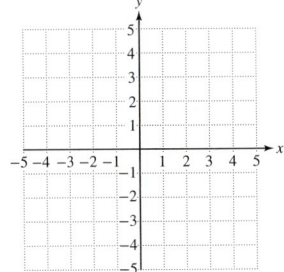

4. $y = (x - 1)^2 + 1$

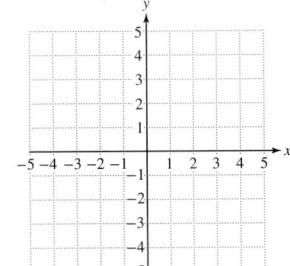

5. $y = x^2 - 4x + 3$

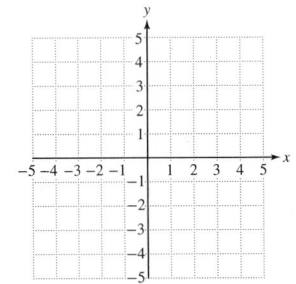

6. $y = x^2 + 6x + 5$

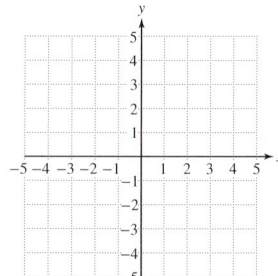

7. $y = -2x^2 + 8x$

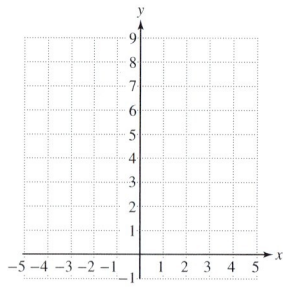

8. $y = -3x^2 - 6x$

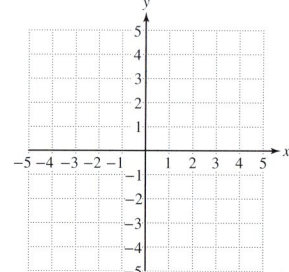

9. $y = -x^2 - 3x + 2$

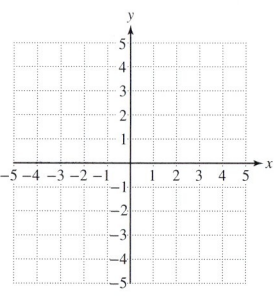

10. $y = -x^2 + x - 4$

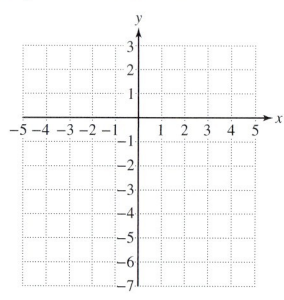

Concept 3: Parabola—Horizontal Axis of Symmetry

For Exercises 11–18, use the equation of the parabola in standard form $x = a(y - k)^2 + h$ to determine the coordinates of the vertex and the axis of symmetry (complete the square if necessary). Then graph the parabola. **(See Examples 2–3.)**

11. $x = y^2 - 3$

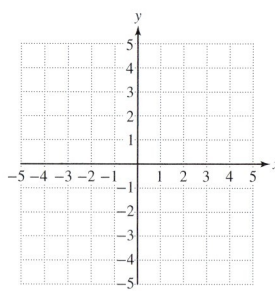

12. $x = y^2 + 1$

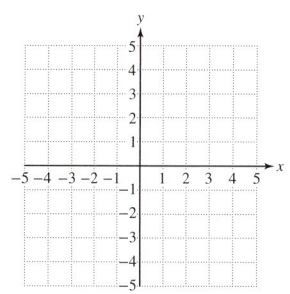

13. $x = -(y - 3)^2 - 3$

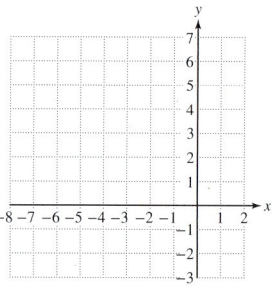

14. $x = -2(y + 2)^2 + 1$

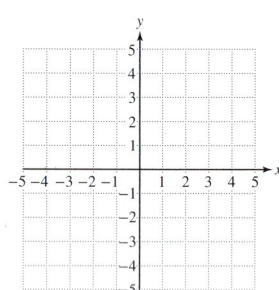

15. $x = -y^2 + 4y - 4$

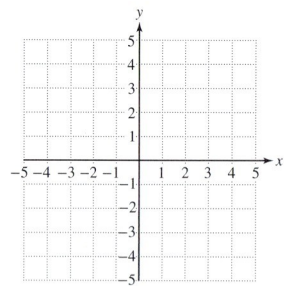

16. $x = -4y^2 - 4y - 2$

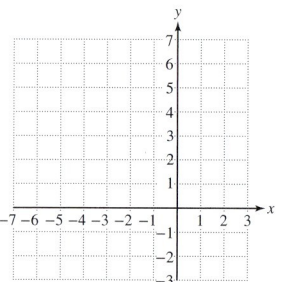

17. $x = y^2 - 2y + 2$

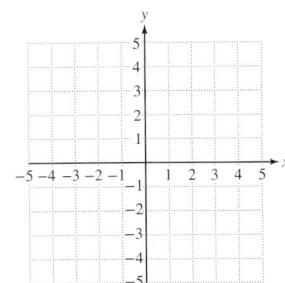

18. $x = y^2 + 4y + 1$

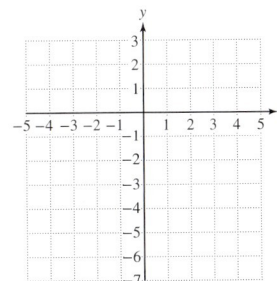

Concept 4: Vertex Formula

For Exercises 19–27, determine the vertex by using the vertex formula. **(See Examples 4–5.)**

19. $y = x^2 - 4x + 3$

20. $y = x^2 + 6x - 2$

21. $x = y^2 + 2y + 6$

22. $x = y^2 - 8y + 3$

23. $y = -\frac{1}{4}x^2 + x + \frac{3}{4}$

24. $y = -\frac{1}{2}x^2 - x + \frac{1}{2}$

25. $y = x^2 - 3x + 2$

26. $x = -2y^2 + 16y + 1$

27. $x = -3y^2 - 6y + 7$

28. Ricardo has a satellite dish for his television. The cross sections of the satellite dish are parabolic in shape. A cross section through the vertex can be described by the equation

$$d(x) = \frac{2}{25}x^2 - \frac{2}{5}x$$

where $d(x)$ is the depth of the dish (in feet) at a distance of x ft from the edge of the dish. How deep is the satellite dish?

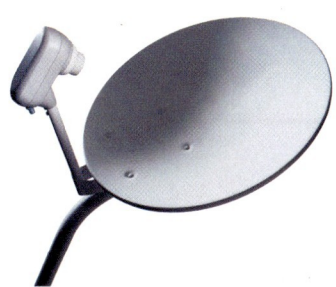

Ryan McVay/Photodisc/Getty Images

29. Water from an outdoor fountain is projected outward from a jet on the side wall of the fountain. The water follows a parabolic path and can be modeled by $h(x) = -x^2 + 10x - 3$. In this function, $h(x)$ represents the height of the water, x ft from the jet. Find the maximum height of the water.

Marcel Francois Loyau (French sculptor 1895-)/Rudolf Balasko/rudi1976/123RF

Mixed Exercises

30. Explain how to determine whether a parabola opens upward, downward, left, or right.

31. Explain how to determine whether a parabola has a vertical or horizontal axis of symmetry.

For Exercises 32–43, use the equation of the parabola first to determine whether the axis of symmetry is vertical or horizontal. Then determine if the parabola opens upward, downward, left, or right.

32. $y = (x - 2)^2 + 3$

33. $y = (x - 4)^2 + 2$

34. $y = -2(x + 1)^2 - 4$

35. $y = -3(x + 2)^2 - 1$

36. $x = y^2 + 4$

37. $x = y^2 - 2$

38. $x = -(y + 3)^2 + 2$

39. $x = -2(y - 1)^2 - 3$

40. $y = -2x^2 - 5$

41. $y = -x^2 + 3$

42. $x = 2y^2 + 3y - 2$

43. $x = y^2 - 5y + 1$

Section 21.3 The Ellipse and Hyperbola

Concepts

1. Standard Form of an Equation of an Ellipse
2. Standard Form of an Equation of a Hyperbola

1. Standard Form of an Equation of an Ellipse

In this section, we will study the two remaining conic sections: the ellipse and the hyperbola. An **ellipse** is the set of all points (x, y) such that the sum of the distances between (x, y) and two distinct points is a constant. The fixed points are called the foci (plural of *focus*) of the ellipse.

To visualize an ellipse, consider the following application. Suppose Sonya wants to cut an elliptical rug from a rectangular rug to avoid a stain made by the family dog. She places two tacks along the center horizontal line. Then she ties the ends of a slack piece of rope to each tack. With the rope pulled tight, she traces out a curve. This curve is an ellipse, and the tacks are located at the foci of the ellipse (Figure 21-14).

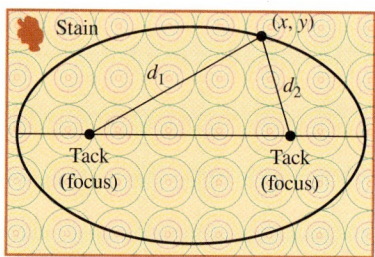

Figure 21-14

We will first graph ellipses that are centered at the origin.

> **Standard Form of an Equation of an Ellipse Centered at the Origin**
>
> An ellipse with the center at the origin has the equation $\dfrac{x^2}{a^2} + \dfrac{y^2}{b^2} = 1$, where a and b are positive real numbers. In the standard form of the equation, the right side must equal 1.

To graph an ellipse centered at the origin, find the x- and y-intercepts.

To find the x-intercepts, let $y = 0$.

$$\frac{x^2}{a^2} + \frac{0}{b^2} = 1$$

$$\frac{x^2}{a^2} = 1$$

$$x^2 = a^2$$

$$x = \pm\sqrt{a^2}$$

$$x = \pm a$$

The x-intercepts are $(a, 0)$ and $(-a, 0)$. See Figure 21-15.

To find the y-intercepts, let $x = 0$.

$$\frac{0}{a^2} + \frac{y^2}{b^2} = 1$$

$$\frac{y^2}{b^2} = 1$$

$$y^2 = b^2$$

$$y = \pm\sqrt{b^2}$$

$$y = \pm b$$

The y-intercepts are $(0, b)$ and $(0, -b)$. See Figure 21-15.

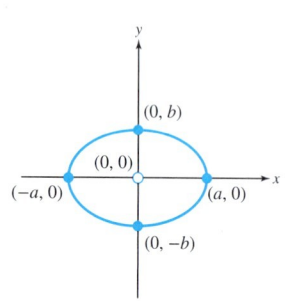

Figure 21-15

Section 21.3 The Ellipse and Hyperbola

Example 1 — Graphing an Ellipse

Graph the ellipse given by the equation. $\dfrac{x^2}{9} + \dfrac{y^2}{4} = 1$

Solution:

The equation can be written as $\dfrac{x^2}{3^2} + \dfrac{y^2}{2^2} = 1$.

The equation is in standard form $\dfrac{x^2}{a^2} + \dfrac{y^2}{b^2} = 1$, therefore, $a = 3$ and $b = 2$.

Graph the intercepts $(3, 0), (-3, 0), (0, 2), (0, -2)$ and sketch the ellipse.

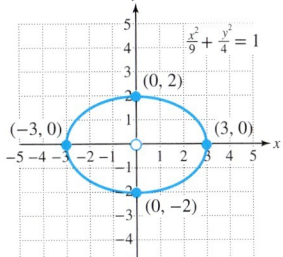

Skill Practice

1. Graph the ellipse given by the equation. $\dfrac{x^2}{4} + \dfrac{y^2}{16} = 1$

Example 2 — Graphing an Ellipse

Graph the ellipse given by the equation. $25x^2 + y^2 = 25$

Solution:

$25x^2 + y^2 = 25$ Divide both sides by 25.

$\dfrac{25x^2}{25} + \dfrac{y^2}{25} = \dfrac{25}{25}$

$x^2 + \dfrac{y^2}{25} = 1$

The equation can then be written as $\dfrac{x^2}{1^2} + \dfrac{y^2}{5^2} = 1$.

The equation is in standard form $\dfrac{x^2}{a^2} + \dfrac{y^2}{b^2} = 1$, therefore, $a = 1$ and $b = 5$.

Graph the intercepts $(1, 0), (-1, 0), (0, 5),$ and $(0, -5)$ and sketch the ellipse.

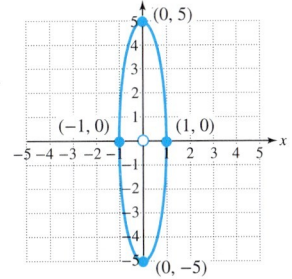

Skill Practice

2. Graph the ellipse given by the equation. $x^2 + 9y^2 = 9$

A circle is a special case of an ellipse where $a = b$. Therefore, it is not surprising that we graph an ellipse centered at the point (h, k) in much the same way we graph a circle.

Example 3 — Graphing an Ellipse Whose Center Is Not at the Origin

Graph the ellipse given by the equation. $\dfrac{(x-1)^2}{16} + \dfrac{(y+3)^2}{4} = 1$

Solution:

Just as we would find the center of a circle, we see that the center of the ellipse is $(1, -3)$. Now we can use the values of a and b to help us plot four strategic points to define the curve.

Answers

1.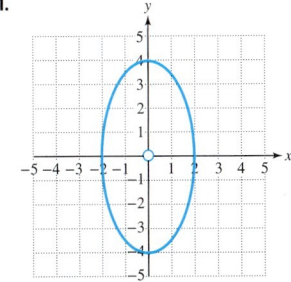

2.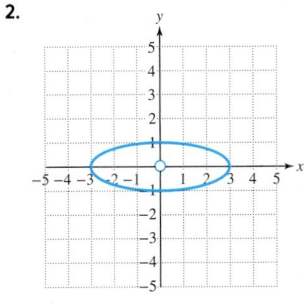

The equation can be written as

$$\frac{(x-1)^2}{(4)^2} + \frac{(y+3)^2}{(2)^2} = 1$$

From this, we have $a = 4$ and $b = 2$. To sketch the curve, locate the center at $(1, -3)$. Then move $a = 4$ units to the left and right of the center and plot two points. Similarly, move $b = 2$ units up and down from the center and plot two points.

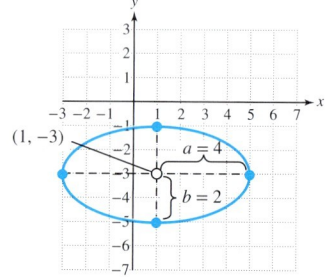

Skill Practice

3. Graph the ellipse. $\dfrac{(x+1)^2}{4} + \dfrac{(y-4)^2}{9} = 1$

2. Standard Form of an Equation of a Hyperbola

A **hyperbola** is the set of all points (x, y) such that the absolute value of the *difference* of the distances between (x, y) and two distinct points is a constant. The fixed points are called the foci of the hyperbola. The graph of a hyperbola has two parts, called branches. A hyperbola has two vertices that lie on an axis of symmetry called the **transverse axis**. For the hyperbolas studied here, the transverse axis is either horizontal or vertical.

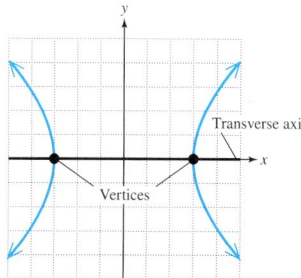 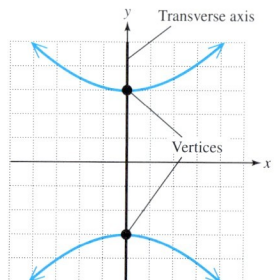

FOR REVIEW

From the vertical line test, neither the graph of an ellipse nor the graph of a hyperbola defines y as a function of x.

Standard Forms of an Equation of a Hyperbola With Center at the Origin

Let a and b represent positive real numbers.

Horizontal transverse axis:
The standard form of an equation of a hyperbola with a *horizontal transverse axis* and center at the origin is given by $\dfrac{x^2}{a^2} - \dfrac{y^2}{b^2} = 1$.

Note: The x term is positive. The branches of the hyperbola open left and right.

Vertical transverse axis:
The standard form of an equation of a hyperbola with a *vertical transverse axis* and center at the origin is given by $\dfrac{y^2}{b^2} - \dfrac{x^2}{a^2} = 1$.

Note: The y term is positive. The branches of the hyperbola open up and down.

In the standard forms of an equation of a hyperbola, the right side must equal 1.

Answer
3.
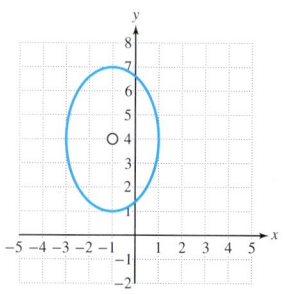

To graph a hyperbola centered at the origin, first construct a reference rectangle. Draw this rectangle by using the points (a, b), $(-a, b)$, $(a, -b)$, and $(-a, -b)$. Asymptotes lie on the diagonals of the rectangle. The branches of the hyperbola are drawn to approach the asymptotes.

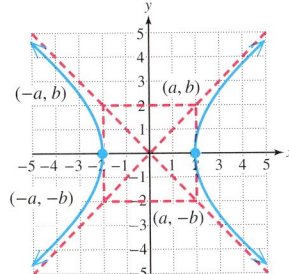 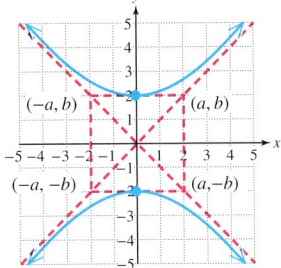

Example 4 Graphing a Hyperbola

Graph the hyperbola given by the equation. $\dfrac{x^2}{36} - \dfrac{y^2}{9} = 1$

a. Determine whether the transverse axis is horizontal or vertical.

b. Draw the reference rectangle and asymptotes.

c. Graph the hyperbola and label the vertices.

Solution:

a. Since the x term is positive, the transverse axis is horizontal.

b. The equation can be written $\dfrac{x^2}{6^2} - \dfrac{y^2}{3^2} = 1$; therefore, $a = 6$ and $b = 3$. Graph the reference rectangle from the points $(6, 3)$, $(6, -3)$, $(-6, 3)$, $(-6, -3)$.

c. The vertices are $(-6, 0)$ and $(6, 0)$.

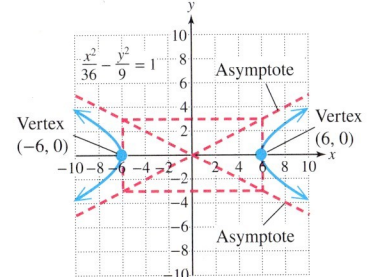

Skill Practice

4. Graph the hyperbola. $x^2 - \dfrac{y^2}{9} = 1$

Example 5 Graphing a Hyperbola

Graph the hyperbola given by the equation. $y^2 - 4x^2 - 16 = 0$

a. Write the equation in standard form to determine whether the transverse axis is horizontal or vertical.

b. Draw the reference rectangle and asymptotes.

c. Graph the hyperbola and label the vertices.

Answer

4.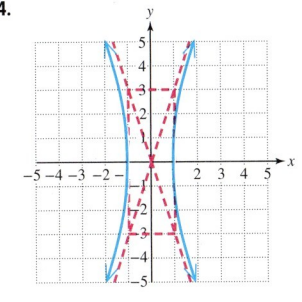

Solution:

a. $y^2 - 4x^2 - 16 = 0$

$y^2 - 4x^2 = 16$ Isolate the constant term on the right.

$\dfrac{y^2}{16} - \dfrac{4x^2}{16} = \dfrac{16}{16}$ Divide by 16 to make the constant term 1.

$\dfrac{y^2}{16} - \dfrac{x^2}{4} = 1$

Since the y term is positive, the transverse axis is vertical.

b. The equation can be written $\dfrac{y^2}{4^2} - \dfrac{x^2}{2^2} = 1$; therefore, $a = 2$ and $b = 4$. Graph the reference rectangle from the points $(2, 4)$, $(2, -4)$, $(-2, 4)$, $(-2, -4)$.

c. The vertices are $(0, 4)$ and $(0, -4)$.

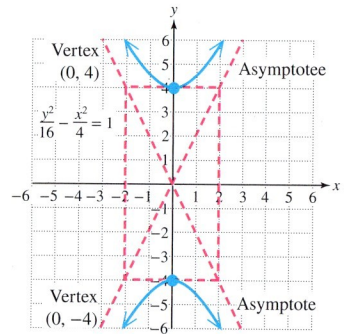

Skill Practice

5. Graph the hyperbola. $9y^2 - 4x^2 = 36$

The following summarizes the steps for graphing ellipses and hyperbolas.

Graphing an Ellipse With Center at the Origin

1. Write the equation in standard form:
$$\dfrac{x^2}{a^2} + \dfrac{y^2}{b^2} = 1$$

2. Plot the intercepts.
   x-intercepts: $(a, 0)$ and $(-a, 0)$
   y-intercepts: $(0, b)$ and $(0, -b)$

3. Sketch the ellipse through the intercepts.

Graphing a Hyperbola With Center at the Origin

1. Write the equation in standard form:
   - Horizontal transverse axis:
     $$\dfrac{x^2}{a^2} - \dfrac{y^2}{b^2} = 1 \quad \text{Branches open left and right}$$
   - Vertical transverse axis:
     $$\dfrac{y^2}{b^2} - \dfrac{x^2}{a^2} = 1 \quad \text{Branches open up and down}$$

2. Draw the reference rectangle through the points (a, b), $(a, -b)$, $(-a, b)$, and $(-a, -b)$.

3. Draw the asymptotes.

4. Plot the vertices and sketch the hyperbola.

Answer

5.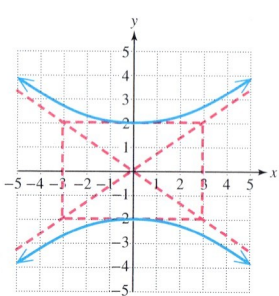

Section 21.3 Activity

A.1. a. Describe the graph of $x^2 + y^2 = 4$.

b. Describe the graph of $\dfrac{x^2}{9} + \dfrac{y^2}{4} = 1$.

c. Graph $\dfrac{x^2}{9} + \dfrac{y^2}{4} = 1$ from part (b).

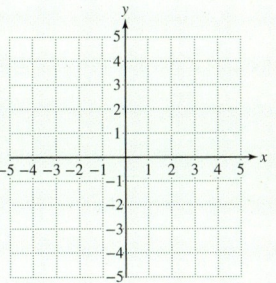

A.2. a. Refer to Exercise A.1(c). How is the graph of $\dfrac{(x-1)^2}{9} + \dfrac{(y+3)^2}{4} = 1$ different from the graph of $\dfrac{x^2}{9} + \dfrac{y^2}{4} = 1$?

A.3. a. Describe the graph of $\dfrac{x^2}{25} + \dfrac{y^2}{49} = 1$.

b. Describe the graph of $\dfrac{x^2}{25} - \dfrac{y^2}{49} = 1$, including the vertices of the reference rectangle.

c. Describe the graph of $\dfrac{y^2}{49} - \dfrac{x^2}{25} = 1$, including the vertices of the reference rectangle.

Practice Exercises — Section 21.3

Prerequisite Review

For Exercises R.1–R.4, determine the center and radius of the circle.

R.1. $x^2 + y^2 = 16$

R.2. $x^2 + y^2 = 36$

R.3. $(x-3)^2 + (y+2)^2 = 12$

R.4. $(x+5)^2 + (y-1)^2 = 20$

Vocabulary and Key Concepts

1. a. A(n) _____ is the set of all points (x, y) such that the sum of the distances between (x, y) and two distinct points (called _____) is constant.

b. The standard form of an equation of an _____ with center at the origin is $\dfrac{x^2}{a^2} + \dfrac{y^2}{b^2} = 1$.

c. A(n) _____ is the set of all points (x, y) such that the absolute value of the difference in distances between (x, y) and two distinct points (called _____) is a constant.

2. a. A hyperbola has two vertices that lie on an axis of symmetry called the _____ axis.

b. The standard form of an equation of a hyperbola with a *horizontal* transverse axis and center at $(0, 0)$, is given by _____.

c. The standard form of an equation of a hyperbola with a *vertical* transverse axis and center at $(0, 0)$, is given by _____.

Concept 1: Standard Form of an Equation of an Ellipse

For Exercises 3–10, find the *x*- and *y*-intercepts and graph the ellipse. **(See Examples 1–2.)**

3. $\dfrac{x^2}{4} + \dfrac{y^2}{9} = 1$

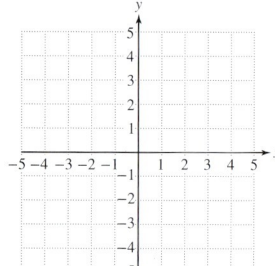

4. $\dfrac{x^2}{16} + \dfrac{y^2}{25} = 1$

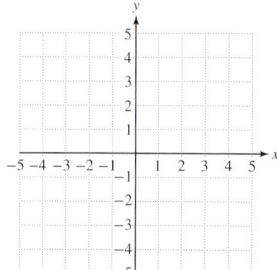

5. $\dfrac{x^2}{16} + \dfrac{y^2}{9} = 1$

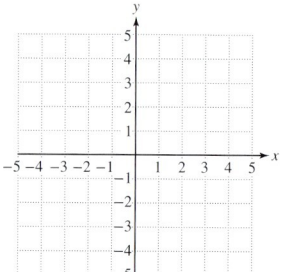

6. $\dfrac{x^2}{36} + \dfrac{y^2}{4} = 1$

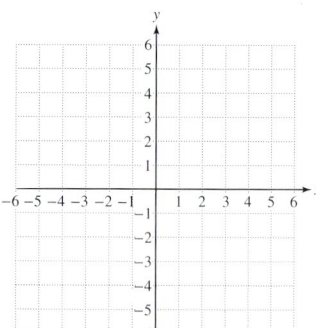

7. $4x^2 + y^2 = 4$

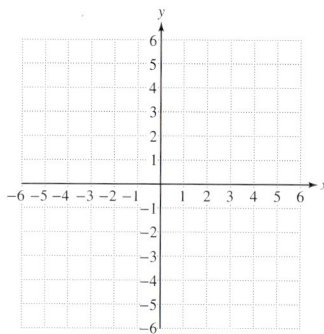

8. $9x^2 + y^2 = 36$

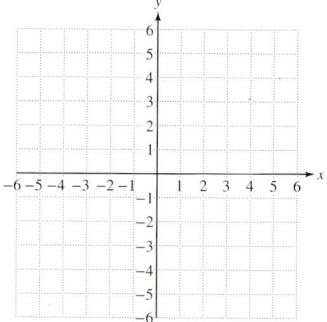

9. $x^2 + 25y^2 - 25 = 0$

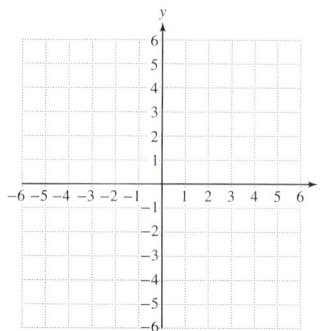

10. $4x^2 + 9y^2 = 144$

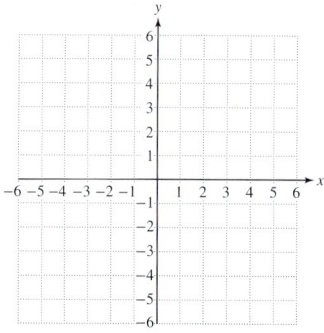

For Exercises 11–18, identify the center (h, k) of the ellipse. Then graph the ellipse. **(See Example 3.)**

11. $\dfrac{(x-4)^2}{4} + \dfrac{(y-5)^2}{9} = 1$

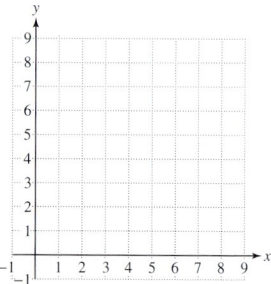

12. $\dfrac{(x+2)^2}{9} + \dfrac{(y-1)^2}{16} = 1$

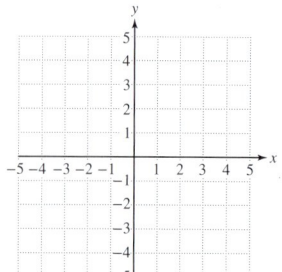

13. $\dfrac{(x+1)^2}{25} + \dfrac{(y-2)^2}{9} = 1$

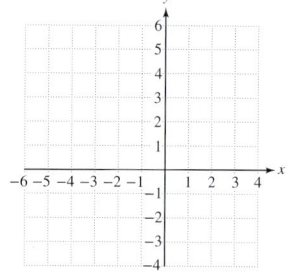

14. $\dfrac{(x-4)^2}{25} + \dfrac{(y+1)^2}{16} = 1$

15. $\dfrac{(x-2)^2}{9} + (y+3)^2 = 1$

16. $(x-5)^2 + \dfrac{(y-3)^2}{4} = 1$

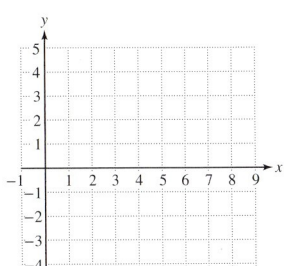 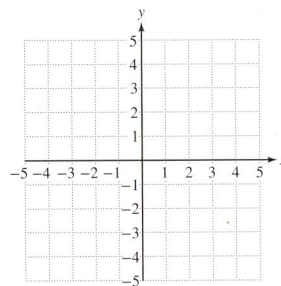 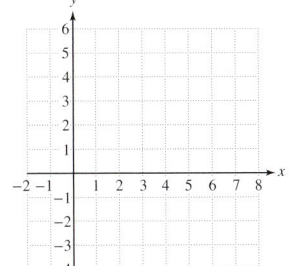

17. $\dfrac{x^2}{36} + \dfrac{(y-1)^2}{25} = 1$

18. $\dfrac{(x+5)^2}{4} + \dfrac{y^2}{16} = 1$

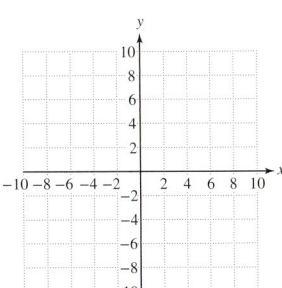 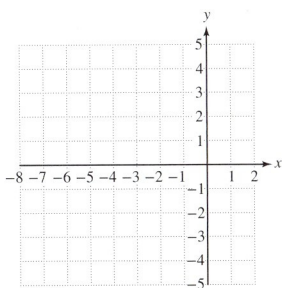

Concept 2: Standard Form of an Equation of a Hyperbola

For Exercises 19–26, determine whether the transverse axis is horizontal or vertical.

19. $\dfrac{y^2}{6} - \dfrac{x^2}{18} = 1$

20. $\dfrac{y^2}{10} - \dfrac{x^2}{4} = 1$

21. $\dfrac{x^2}{20} - \dfrac{y^2}{15} = 1$

22. $\dfrac{x^2}{12} - \dfrac{y^2}{9} = 1$

23. $x^2 - y^2 = 12$

24. $x^2 - y^2 = 15$

25. $x^2 - 3y^2 = -9$

26. $2x^2 - y^2 = -10$

For Exercises 27–34, use the equation in standard form to graph the hyperbola. Label the vertices of the hyperbola. (See Examples 4–5.)

27. $\dfrac{x^2}{25} - \dfrac{y^2}{16} = 1$

28. $\dfrac{x^2}{9} - \dfrac{y^2}{36} = 1$

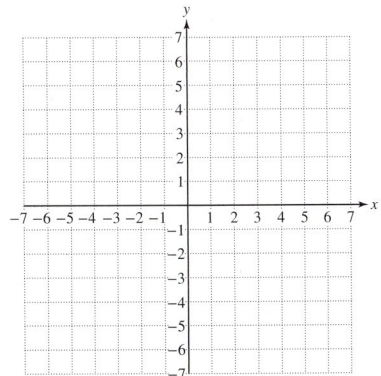 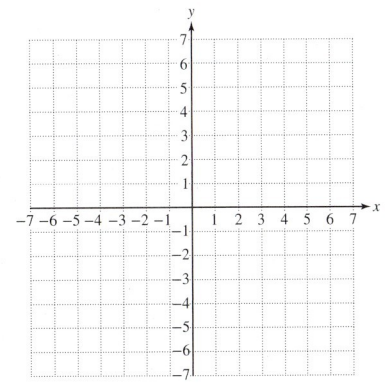

29. $\dfrac{y^2}{4} - \dfrac{x^2}{4} = 1$

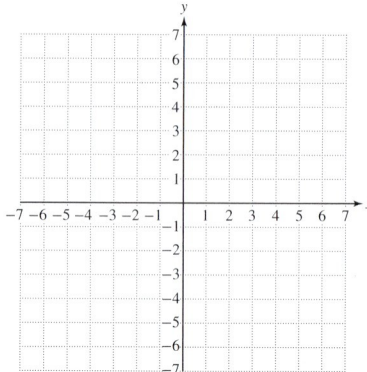

30. $\dfrac{y^2}{9} - \dfrac{x^2}{9} = 1$

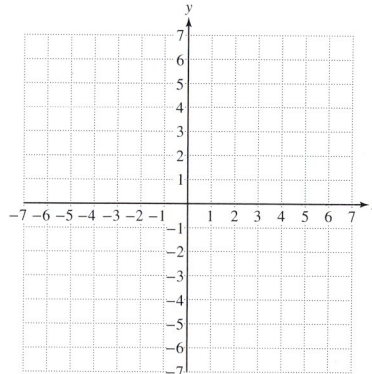

31. $36x^2 - y^2 = 36$

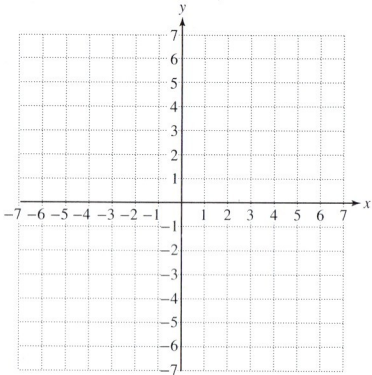

32. $x^2 - 25y^2 = 25$

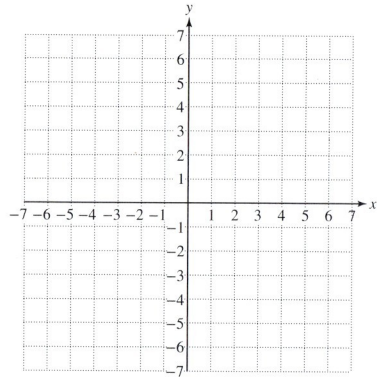

33. $y^2 - 4x^2 - 16 = 0$

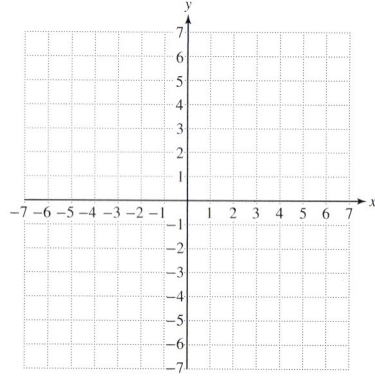

34. $y^2 - 4x^2 - 36 = 0$

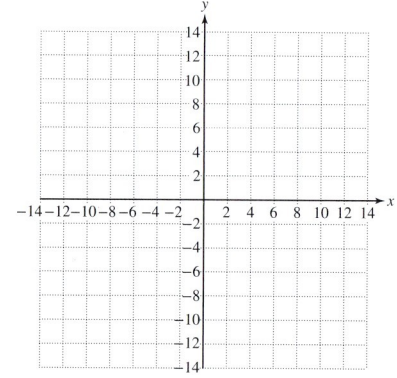

Mixed Exercises

For Exercises 35–46, determine if the equation represents an ellipse or a hyperbola.

35. $\dfrac{x^2}{6} - \dfrac{y^2}{10} = 1$

36. $\dfrac{x^2}{14} + \dfrac{y^2}{2} = 1$

37. $\dfrac{y^2}{4} + \dfrac{x^2}{16} = 1$

38. $\dfrac{x^2}{5} + \dfrac{y^2}{10} = 1$

39. $4x^2 + y^2 = 16$

40. $-3x^2 - 4y^2 = -36$

41. $-y^2 + 2x^2 = -10$

42. $x^2 - y^2 = -1$

43. $5x^2 + y^2 - 10 = 0$

44. $5x^2 - 3y^2 = 15$

45. $y^2 - 6x^2 = 6$

46. $3x^2 + 5y^2 = 15$

Expanding Your Skills

47. An arch for a tunnel is in the shape of a semiellipse. The distance between vertices is 120 ft, and the height to the top of the arch is 50 ft. Find the height of the arch 10 ft from the center. Round to the nearest foot.

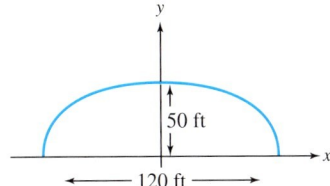

48. A bridge over a gorge is supported by an arch in the shape of a semiellipse. The length of the bridge is 400 ft, and the height is 100 ft. Find the height of the arch 50 ft from the center. Round to the nearest foot.

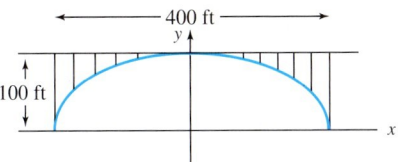

In Exercises 49–52, graph the hyperbola centered at (h, k) by following these guidelines. First locate the center. Then draw the reference rectangle relative to the center (h, k). Using the reference rectangle as a guide, sketch the hyperbola. Label the center and vertices.

49. $\dfrac{(x-1)^2}{9} - \dfrac{(y+2)^2}{4} = 1$

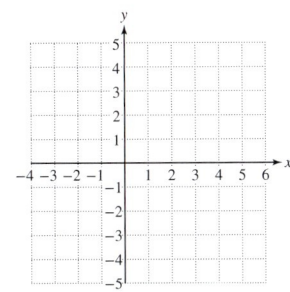

50. $\dfrac{(x+2)^2}{16} - \dfrac{(y-3)^2}{4} = 1$

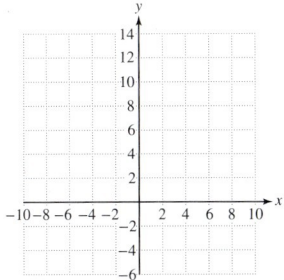

51. $\dfrac{(y-1)^2}{4} - (x+3)^2 = 1$

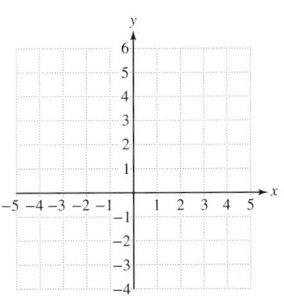

52. $(y+2)^2 - \dfrac{(x+3)^2}{4} = 1$

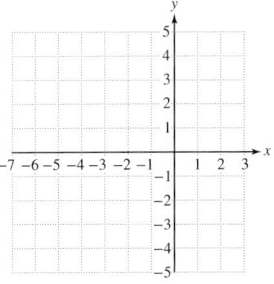

Problem Recognition Exercises

Formulas and Conic Sections

For Exercises 1–8, identify the formula.

1. $(x-h)^2 + (y-k)^2 = r^2$

2. $\dfrac{x^2}{a^2} + \dfrac{y^2}{b^2} = 1$

3. $\sqrt{(x_2-x_1)^2 + (y_2-y_1)^2}$

4. $\left(\dfrac{x_1+x_2}{2}, \dfrac{y_1+y_2}{2}\right)$

5. $y = a(x-h)^2 + k$

6. $\dfrac{x^2}{a^2} - \dfrac{y^2}{b^2} = 1$

7. $\dfrac{y^2}{b^2} - \dfrac{x^2}{a^2} = 1$

8. $x = a(y-k)^2 + h$

For Exercises 9–30, identify the equation as representing a circle, parabola, ellipse, hyperbola, or none of these.

9. $y = -2(x-3)^2 + 4$

10. $\dfrac{x^2}{4} - \dfrac{y^2}{1} = 1$

11. $(x+3)^2 + (y+2)^2 = 4$

12. $(x-2)^2 + (y-4)^2 = 9$

13. $\dfrac{x^2}{9} - \dfrac{y^2}{9} = 1$

14. $\dfrac{x^2}{9} + \dfrac{y^2}{16} = 1$

15. $\dfrac{x^2}{16} + \dfrac{y^2}{4} = 0$

16. $x^2 + y^2 - 2x + 4y - 4 = 0$

17. $y = \dfrac{1}{2}(x+2)^2 - 3$

18. $\dfrac{x^2}{4} - \dfrac{y^2}{2} = 1$

19. $x = (y+2)^2 - 4$

20. $x^2 + y^2 + 6x + 8 = 0$

21. $(x-1)^2 + (y+1)^2 = 0$

22. $x = -(y-2)^2 - 1$

23. $\dfrac{x^2}{25} + \dfrac{y^2}{4} = 1$

24. $x^2 + y^2 = 15$

25. $y = (x-6)^2 + 4$

26. $\dfrac{(x+1)^2}{2} + \dfrac{(y+1)^2}{5} = 1$

27. $\dfrac{y^2}{3} - \dfrac{x^2}{3} = 1$

28. $3x^2 + 3y^2 = 1$

29. $\dfrac{x^2}{9} + \dfrac{y^2}{12} = 1$

30. $x = (y+2)^2 - 5$

Section 21.4 Nonlinear Systems of Equations in Two Variables

Concepts

1. Solving Nonlinear Systems of Equations by the Substitution Method
2. Solving Nonlinear Systems of Equations by the Addition Method

1. Solving Nonlinear Systems of Equations by the Substitution Method

Recall that a linear equation in two variables x and y is an equation that can be written in the form $Ax + By = C$, where A and B are not both zero. We have solved systems of linear equations in two variables by using the graphing method, the substitution method, and the addition method. In this section, we will solve *nonlinear* systems of equations by using the same methods. A **nonlinear system of equations** is a system in which at least one of the equations is nonlinear.

Graphing the equations in a nonlinear system helps to determine the number of solutions and to approximate the coordinates of the solutions. The substitution method is used most often to solve a nonlinear system of equations analytically.

Example 1 Solving a Nonlinear System of Equations

Given the system
$$x - 7y = -25$$
$$x^2 + y^2 = 25$$

a. Solve the system by graphing.

b. Solve the system by the substitution method.

Solution:

a. $x - 7y = -25$ is a line (the slope-intercept form is $y = \tfrac{1}{7}x + \tfrac{25}{7}$).

$x^2 + y^2 = 25$ is a circle centered at the origin with radius 5.

From Figure 21-16, we appear to have two solutions $(-4, 3)$ and $(3, 4)$.

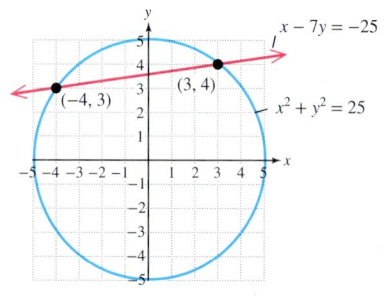

Figure 21-16

b. To use the substitution method, isolate one of the variables from one of the equations. We will solve for x in the first equation.

$$\boxed{A} \quad x - 7y = -25 \xrightarrow{\text{Solve for } x.} x = 7y - 25$$
$$\boxed{B} \quad x^2 + y^2 = 25$$

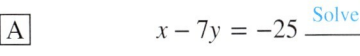

$\boxed{B}\quad (7y - 25)^2 + y^2 = 25$	Substitute $(7y - 25)$ for x in the second equation.
$49y^2 - 350y + 625 + y^2 = 25$	The resulting equation is quadratic in y.
$50y^2 - 350y + 600 = 0$	Set the equation equal to zero.
$50(y^2 - 7y + 12) = 0$	Factor.
$50(y - 3)(y - 4) = 0$	
$y = 3 \quad$ or $\quad y = 4$	

For each value of y, find the corresponding x value from the equation $x = 7y - 25$.

$y = 3$:	$x = 7(3) - 25 = -4$	The solution point is $(-4, 3)$.
$y = 4$:	$x = 7(4) - 25 = 3$	The solution point is $(3, 4)$. (See Figure 21-16.)

The solution set is $\{(-4, 3), (3, 4)\}$.

FOR REVIEW

Remember to use parentheses when substituting an expression for a variable. For example, using parentheses for the expression $(7y - 5)^2$ helps us remember to square the binomial, rather than the individual terms.

Skill Practice Given the system

$$2x + y = 5$$
$$x^2 + y^2 = 50$$

1. Solve the system by graphing.
2. Solve the system by the substitution method.

Example 2 Solving a Nonlinear System by the Substitution Method

Given the system
$$y = \sqrt{x}$$
$$x^2 + y^2 = 20$$

a. Sketch the graphs.

b. Solve the system by the substitution method.

Solution:

a. $y = \sqrt{x}$ is the basic square root function.

$x^2 + y^2 = 20$ is a circle centered at the origin with radius $\sqrt{20} \approx 4.5$.

From Figure 21-17, we see that there is one solution.

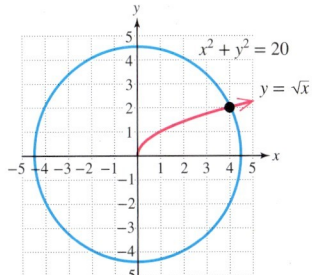

Figure 21-17

Answers

1.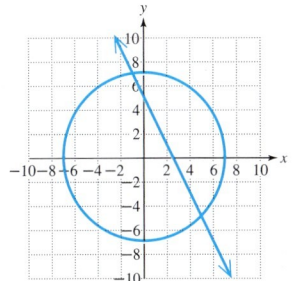

The solutions appear to be $(-1, 7)$ and $(5, -5)$.

2. $\{(-1, 7), (5, -5)\}$

b. To use the substitution method, we will substitute $y = \sqrt{x}$ into equation $\boxed{B}$.

$\boxed{A}$ $\quad y = \sqrt{x}$ $\qquad$ Substitute $y = \sqrt{x}$ into the second equation.
$\boxed{B}$ $\quad x^2 + y^2 = 20$

$\boxed{B}$ $\quad x^2 + (\sqrt{x})^2 = 20$

$\qquad x^2 + x = 20$

$\qquad x^2 + x - 20 = 0$ $\qquad$ Set the second equation equal to zero.

$\qquad (x + 5)(x - 4) = 0$ $\qquad$ Factor.

$\qquad \cancel{x = -5}$ or $x = 4$ $\qquad$ Reject $x = -5$ because it is not in the domain of $y = \sqrt{x}$.

Substitute $x = 4$ into the equation $y = \sqrt{x}$.

If $x = 4$, then $y = \sqrt{4} = 2$. $\qquad$ The solution set is $\{(4, 2)\}$.

Skill Practice Given the system

$$y = \sqrt{2x}$$
$$x^2 + y^2 = 8$$

3. Sketch the graphs.
4. Solve the system by using the substitution method.

Example 3 **Solving a Nonlinear System by the Substitution Method**

Solve the system by using the substitution method. $\quad y = \sqrt[3]{x}$
$\qquad\qquad\qquad\qquad\qquad\qquad\qquad\qquad\qquad\qquad y = x$

Solution:

$\boxed{A}$ $\quad y = \sqrt[3]{x}$
$\boxed{B}$ $\quad y = x$

$\qquad \sqrt[3]{x} = x$ $\qquad$ Because y is isolated in both equations, we can equate the expressions for y.

$\qquad (\sqrt[3]{x})^3 = (x)^3$ $\qquad$ To solve the radical equation, raise both sides to the third power.

$\qquad x = x^3$ $\qquad$ This is a third-degree polynomial equation.

$\qquad 0 = x^3 - x$ $\qquad$ Set the equation equal to zero.

$\qquad 0 = x(x^2 - 1)$ $\qquad$ Factor out the GCF.

$\qquad 0 = x(x + 1)(x - 1)$ $\qquad$ Factor completely.

$x = 0$ or $x = -1$ or $x = 1$

For each value of x, find the corresponding y value from either original equation. We will use equation $\boxed{B}$: $y = x$.

If $x = 0$, then $y = 0$. $\qquad$ The solution point is $(0, 0)$.

If $x = -1$, then $y = -1$. $\qquad$ The solution point is $(-1, -1)$.

If $x = 1$, then $y = 1$. $\qquad$ The solution point is $(1, 1)$.

The solution set is $\{(0, 0), (-1, -1), (1, 1)\}$. See Figure 21-18.

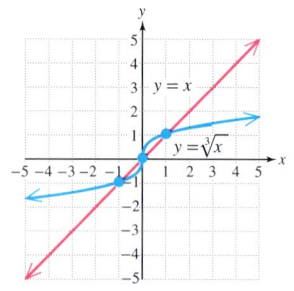

Figure 21-18

Answers

3.
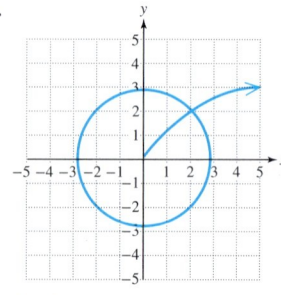

4. $\{(2, 2)\}$

Skill Practice Solve the system by using the substitution method.

5. $y = \sqrt[3]{9x}$
   $y = x$

2. Solving Nonlinear Systems of Equations by the Addition Method

The substitution method is used most often to solve a system of nonlinear equations. In some situations, however, the addition method offers an efficient means of finding a solution. Example 4 demonstrates that we can eliminate a variable from both equations provided the terms containing the corresponding variables are *like* terms.

Example 4 Solving a Nonlinear System of Equations by the Addition Method

Solve the system. $\quad 2x^2 + y^2 = 17$
$\qquad\qquad\qquad\quad x^2 + 2y^2 = 22$

Solution:

A $\quad 2x^2 + y^2 = 17 \qquad$ Notice that the y^2 terms are *like* in each equation.
B $\quad x^2 + 2y^2 = 22 \qquad$ To eliminate the y^2 terms, multiply the first equation by -2.

A $\quad 2x^2 + y^2 = 17 \quad\xrightarrow{\text{Multiply by }-2.}\quad -4x^2 - 2y^2 = -34$
B $\quad x^2 + 2y^2 = 22 \quad\longrightarrow\qquad\qquad\quad x^2 + 2y^2 = 22$
$\qquad\qquad\qquad\qquad\qquad\qquad\qquad\qquad \overline{-3x^2 \qquad\quad = -12}\quad$ Eliminate the y^2 term.

$$\frac{-3x^2}{-3} = \frac{-12}{-3}$$

$$x^2 = 4$$

$\qquad\qquad x = \pm 2 \quad$ Use the square root property.

Substitute each value of x into one of the original equations to solve for y. We will use equation A: $2x^2 + y^2 = 17$.

$x = 2$: $\quad$ A $\quad 2(2)^2 + y^2 = 17$
$\qquad\qquad\qquad 8 + y^2 = 17$
$\qquad\qquad\qquad\quad y^2 = 9$
$\qquad\qquad\qquad\quad y = \pm 3 \quad$ The solution points are $(2, 3)$ and $(2, -3)$.

$x = -2$: $\quad$ A $\quad 2(-2)^2 + y^2 = 17$
$\qquad\qquad\qquad 8 + y^2 = 17$
$\qquad\qquad\qquad\quad y^2 = 9$
$\qquad\qquad\qquad\quad y = \pm 3 \quad$ The solution points are $(-2, 3)$ and $(-2, -3)$.

The solution set is $\{(2, 3), (2, -3), (-2, 3), (-2, -3)\}$.

TIP: In Example 4, the x^2 terms are also *like* in both equations. We could have eliminated the x^2 terms by multiplying equation B by -2.

TIP: The two equations in Example 4 are ellipses, but an exact graph is difficult to render by hand.

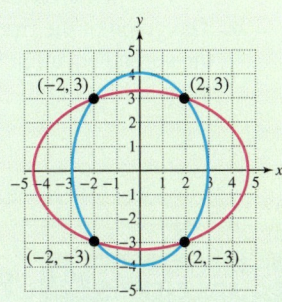

To check the solutions to the system, verify that each ordered pair satisfies both equations.

Skill Practice Solve the system by using the addition method.

6. $x^2 - y^2 = 24$
   $3x^2 + y^2 = 76$

Answers

5. $\{(0, 0), (3, 3), (-3, -3)\}$
6. $\{(5, 1), (5, -1), (-5, 1), (-5, -1)\}$

TIP: It is important to note that the addition method can be used only if two equations share a pair of *like* terms. The substitution method is effective in solving a wider range of systems of equations. The system in Example 4 could also have been solved by using substitution.

$$\boxed{A} \quad 2x^2 + y^2 = 17 \xrightarrow{\text{Solve for } y^2.} y^2 = 17 - 2x^2$$

$$\boxed{B} \quad x^2 + 2y^2 = 22$$

$$\boxed{B} \quad x^2 + 2(17 - 2x^2) = 22 \qquad x = 2: \quad y^2 = 17 - 2(2)^2$$

$$x^2 + 34 - 4x^2 = 22 \qquad\qquad\qquad y^2 = 9$$

$$-3x^2 = -12 \qquad\qquad\qquad y = \pm 3 \quad \text{The solutions are } (2, 3) \text{ and } (-2, -3).$$

$$x^2 = 4$$

$$x = \pm 2 \longrightarrow x = -2: \quad y^2 = 17 - 2(-2)^2$$

$$y^2 = 9$$

$$y = \pm 3 \quad \text{The solutions are } (-2, 3) \text{ and } (-2, -3).$$

Section 21.4 Activity

A.1. a. Graph the two equations on the same coordinate system.

$$(x - 2)^2 + y^2 = 9$$
$$y = x + 1$$

b. From the graph, determine the points of intersection.

c. Solve the system of nonlinear equations from part (a) using the substitution method.

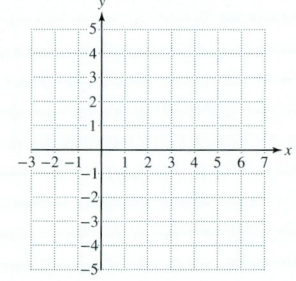

A.2. a. Use the addition method to solve the system of nonlinear equations.

$$x^2 + y^2 = 10$$
$$2x^2 + y^2 = 11$$

b. The graph of the first equation from part (a) is a circle, and the graph of the second equation is an ellipse. Verify your answers from part (a) from the graph.

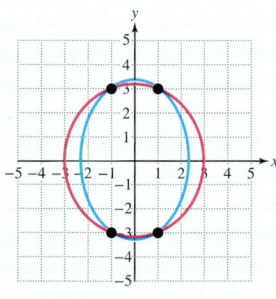

Section 21.4 Practice Exercises

Prerequisite Review

For Exercises R.1–R.2, solve the system using the addition method.

R.1. $2x - 3y = 7$
$-5x + 2y = -23$

R.2. $4x + 2y = 8$
$3x - 7y = 23$

For Exercises R.3–R.4, solve the system using the substitution method.

R.3. $7x - 2y = 3$
$x + 5y = 11$

R.4. $8x - y = -4$
$3x + 5y = -23$

For Exercises R.5–R.10, identify the type of curve defined by the equation. Choose from: line, parabola, circle, ellipse, or hyperbola.

R.5. $x^2 + (y - 4)^2 = 9$

R.6. $4x + 2y = 8$

R.7. $\dfrac{x^2}{9} + \dfrac{y^2}{16} = 1$

R.8. $\dfrac{x^2}{9} - \dfrac{y^2}{16} = 1$

R.9. $x = y^2 + 4$

R.10. $y = -x^2 - 6$

Vocabulary and Key Concepts

1. **a.** A _____ system of equations in two variables is a system in which at least one of the equations is nonlinear.

   b. Graphically, the solution set to a nonlinear system of equations is the set of ordered pairs representing the points of _____ of the graphs of the two equations.

Concept 1: Solving Nonlinear Systems of Equations by the Substitution Method

For Exercises 2–8, use sketches to explain.

2. How many points of intersection are possible between a line and a parabola?

3. How many points of intersection are possible between a line and a circle?

4. How many points of intersection are possible between two distinct circles?

5. How many points of intersection are possible between two distinct parabolas of the form $y = ax^2 + bx + c$, $a \neq 0$?

6. How many points of intersection are possible between a circle and a parabola?

7. How many points of intersection are possible between an ellipse and a hyperbola?

8. How many points of intersection are possible between an ellipse and a parabola?

For Exercises 9–14, sketch each system of equations. Then solve the system by the substitution method. **(See Example 1.)**

9. $y = x + 3$
$x^2 + y = 9$

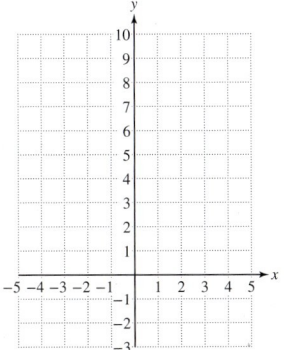

10. $y = x - 2$
$x^2 + y = 4$

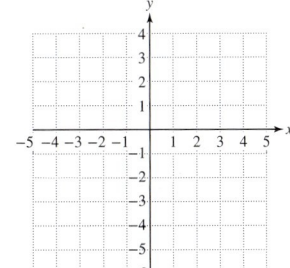

11. $x^2 + y^2 = 1$
$y = x + 1$

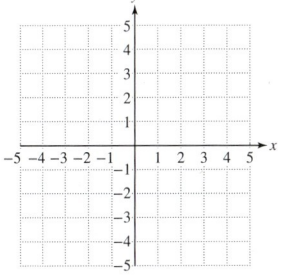

12. $x^2 + y^2 = 25$
    $y = 2x$

13. $x^2 + y^2 = 6$
    $y = x^2$
    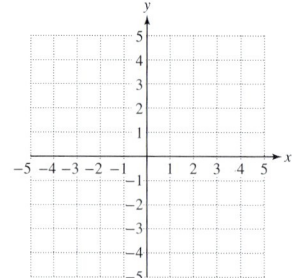

14. $x^2 + y^2 = 12$
    $y = x^2$
    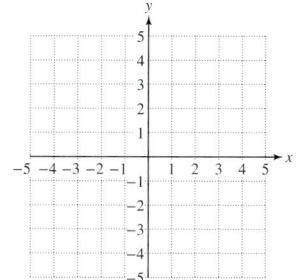

For Exercises 15–23, solve the system by the substitution method. (See Examples 2–3.)

15. $y = \sqrt{x}$
    $2x^2 - y^2 = 1$

16. $x^2 + y^2 = 30$
    $y = \sqrt{x}$

17. $y = x^2$
    $y = -\sqrt{x}$

18. $y = -x^2$
    $y = -\sqrt{x}$

19. $y = x^2$
    $y = (x - 3)^2$

20. $y = (x + 4)^2$
    $y = x^2$

21. $y = x^2 + 6x$
    $y = 4x$

22. $y = 3x^2 - 6x$
    $y = 3x$

23. $x^2 - 5x + y = 0$
    $y = 3x + 1$

Concept 2: Solving Nonlinear Systems of Equations by the Addition Method

For Exercises 24–40, solve the system of nonlinear equations by the addition method if appropriate. (See Example 4.)

24. $x^2 + y^2 = 13$
    $x^2 - y^2 = 5$

25. $4x^2 - y^2 = 4$
    $4x^2 + y^2 = 4$

26. $9x^2 + 4y^2 = 36$
    $x^2 + y^2 = 9$

27. $x^2 + y^2 = 4$
    $2x^2 + y^2 = 8$

28. $3x^2 + 4y^2 = 16$
    $2x^2 - 3y^2 = 5$

29. $2x^2 - 5y^2 = -2$
    $3x^2 + 2y^2 = 35$

30. $x^2 + y^2 = 169$
    $12x - 5y = 0$

31. $x^2 + y^2 = 100$
    $4y - 3x = 0$

32. $\dfrac{x^2}{4} + \dfrac{y^2}{9} = 1$
    $x^2 + y^2 = 4$

33. $\dfrac{x^2}{16} + \dfrac{y^2}{4} = 1$
    $x^2 + y^2 = 4$

34. $\dfrac{x^2}{10} + \dfrac{y^2}{10} = 1$
    $2x^2 + y^2 = 11$

35. $2y = -x + 2$
    $x^2 + 4y^2 = 4$

36. $2y = -3x - 6$
    $9x^2 + 4y^2 = 36$

37. $x^2 - xy = -4$
    $2x^2 - xy = 12$

38. $x^2 - xy = 3$
    $2x^2 + xy = 6$

39. $3x^2 + 2xy = 4$
    $x^2 - xy = 3$

40. $x^2 - 3xy = 8$
    $x^2 + xy = 4$

Technology Connections

For Exercises 41–44, use the *Intersect* feature to approximate the solutions to the system.

41. $y = x + 3$ (Exercise 9)
$x^2 + y = 9$

42. $y = x - 2$ (Exercise 10)
$x^2 + y = 4$

43. $y = x^2$ (Exercise 17)
$y = -\sqrt{x}$

44. $y = -x^2$ (Exercise 18)
$y = -\sqrt{x}$

For Exercises 45–46, graph the system on a square viewing window. What can be said about the solution to the system?

45. $x^2 + y^2 = 4$
$y = x^2 + 3$

46. $x^2 + y^2 = 16$
$y = -x^2 - 5$

Expanding Your Skills

47. The sum of two numbers is 7. The sum of the squares of the numbers is 25. Find the numbers.

48. The sum of the squares of two numbers is 100. The sum of the numbers is 2. Find the numbers.

49. The sum of the squares of two numbers is 32. The difference of the squares of the numbers is 18. Find the numbers.

50. The sum of the squares of two numbers is 24. The difference of the squares of the numbers is 8. Find the numbers.

Section 21.5 Nonlinear Inequalities and Systems of Inequalities in Two Variables

Concepts
1. Nonlinear Inequalities in Two Variables
2. Systems of Nonlinear Inequalities in Two Variables

1. Nonlinear Inequalities in Two Variables

In our earlier study of linear inequalities in two variables, we used the test point method to graph the solution set. For example, to graph the solution set to $y \leq 2x + 1$, first graph the related equation $y = 2x + 1$. This is the line shown in Figure 21-19. Then using test points, we see that points on and below the line make up the solution set to the inequality.

Test Point Above: (−2, 2)

$y \leq 2x + 1$

$2 \stackrel{?}{\leq} 2(-2) + 1$

$2 \stackrel{?}{\leq} -3$ False

Test Point Below: (0, 0)

$y \leq 2x + 1$

$0 \stackrel{?}{\leq} 2(0) + 1$

$0 \stackrel{?}{\leq} 1$ ✓ True

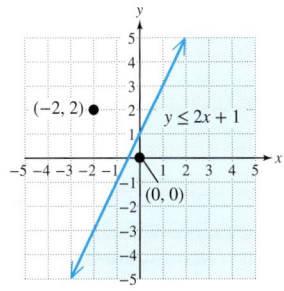

Figure 21-19

1450 Chapter 21 Conic Sections

Avoiding Mistakes

The first step in solving an inequality in two variables is to graph the related equation.

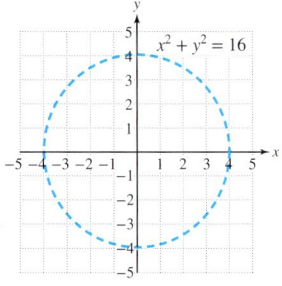

Figure 21-20

FOR REVIEW

Recall that when graphing an inequality in two variables, use a dashed curve to indicate that the curve or "boundary" is not part of the solution set. Use a solid curve to indicate that the curve is part of the solution set.

Example 1 — Graphing a Nonlinear Inequality in Two Variables

Graph the solution set of the inequality $x^2 + y^2 < 16$.

Solution:

The related equation $x^2 + y^2 = 16$ is a circle of radius 4, centered at the origin. Graph the related equation by using a dashed curve because the points satisfying the equation $x^2 + y^2 = 16$ are not part of the solution to the strict inequality $x^2 + y^2 < 16$. See Figure 21-20.

Notice that the dashed curve separates the xy-plane into two regions, one "inside" the circle, the other "outside" the circle. Select a test point from each region and test the point in the original inequality.

Test Point "Inside": (0, 0)

$x^2 + y^2 < 16$

$(0)^2 + (0)^2 \stackrel{?}{<} 16$

$0 \stackrel{?}{<} 16$ True

Test Point "Outside": (4, 4)

$x^2 + y^2 < 16$

$(4)^2 + (4)^2 \stackrel{?}{<} 16$

$32 \stackrel{?}{<} 16$ False

The inequality $x^2 + y^2 < 16$ is true at the test point (0, 0). Therefore, the solution set is the region "inside" the circle. See Figure 21-21.

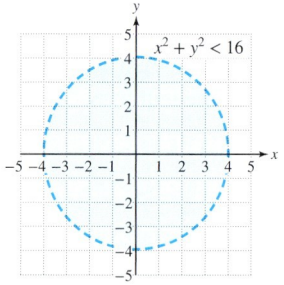

Figure 21-21

Skill Practice

1. Graph the solution set of the inequality. $x^2 + y^2 \geq 9$

Example 2 — Graphing a Nonlinear Inequality in Two Variables

Graph the solution set of the inequality $9y^2 \geq 36 + 4x^2$.

Solution:

First graph the related equation $9y^2 = 36 + 4x^2$. Notice that the equation can be written in the standard form of a hyperbola.

$9y^2 = 36 + 4x^2$

$9y^2 - 4x^2 = 36$ Subtract $4x^2$ from both sides.

$\dfrac{9y^2}{36} - \dfrac{4x^2}{36} = \dfrac{36}{36}$ Divide both sides by 36.

$\dfrac{y^2}{4} - \dfrac{x^2}{9} = 1$

Graph the hyperbola as a solid curve, because the original inequality includes equality. See Figure 21-22.

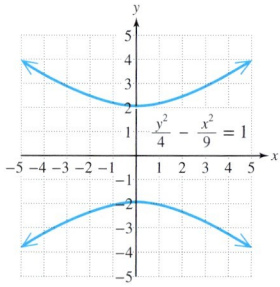

Figure 21-22

Answer

1.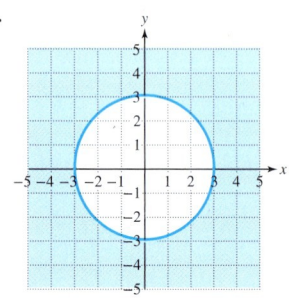

The hyperbola divides the xy-plane into three regions: a region above the upper branch, a region between the branches, and a region below the lower branch. Select a test point from each region.

$$9y^2 \geq 36 + 4x^2$$

Test: (0, 3)
$9(3)^2 \overset{?}{\geq} 36 + 4(0)^2$
$81 \overset{?}{\geq} 36$ True

Test: (0, 0)
$9(0)^2 \overset{?}{\geq} 36 + 4(0)^2$
$0 \overset{?}{\geq} 36$ False

Test: (0, −3)
$9(-3)^2 \overset{?}{\geq} 36 + 4(0)^2$
$81 \overset{?}{\geq} 36$ True

Shade the regions above the top branch and below the bottom branch of the hyperbola. See Figure 21-23.

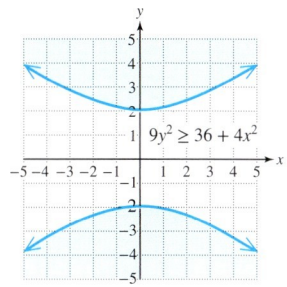

Figure 21-23

Skill Practice

2. Graph the solution set of the inequality. $9x^2 < 144 - 16y^2$

2. Systems of Nonlinear Inequalities in Two Variables

The solution set for a system of nonlinear equations in two variables is a set of ordered pairs that satisfy both equations simultaneously. We will now solve systems of nonlinear inequalities in two variables. Similarly, the solution set is the set of all ordered pairs that simultaneously satisfy each inequality. To solve a system of inequalities, we will graph the solution to each individual inequality and then take the intersection of the solution sets.

Example 3 Graphing a System of Nonlinear Inequalities in Two Variables

Graph the solution set. $y > \frac{1}{3}x^2$

$y < -x^2 + 2$

Solution:

The solution to $y > \frac{1}{3}x^2$ is the set of points above the parabola $y = \frac{1}{3}x^2$. See Figure 21-24. The solution to $y < -x^2 + 2$ is the set of points below the parabola $y = -x^2 + 2$. See Figure 21-25.

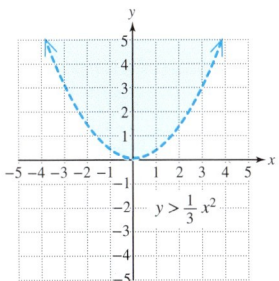

Figure 21-24

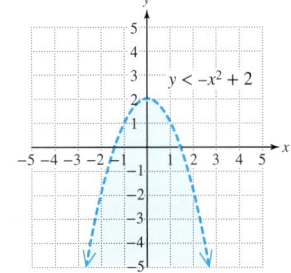

Figure 21-25

Answer

2.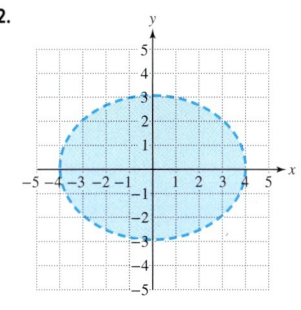

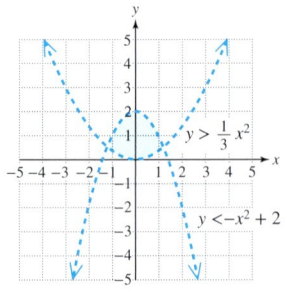

Figure 21-26

The solution to the system of inequalities is the intersection of the solution sets of the individual inequalities. See Figure 21-26.

Skill Practice

3. Graph the solution set.
$$\frac{x^2}{4} + \frac{y^2}{9} < 1$$
$$x > y^2$$

Example 4 Graphing a System of Nonlinear Inequalities in Two Variables

Graph the solution set. $y > e^x$
$y < -x^2 + 4$

Solution:

The solution to $y > e^x$ is the set of points above the curve $y = e^x$. See Figure 21-27. The solution to $y < -x^2 + 4$ is the set of points below the parabola $y = -x^2 + 4$. See Figure 21-28.

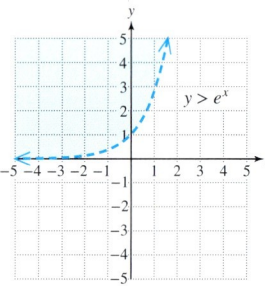

Figure 21-27

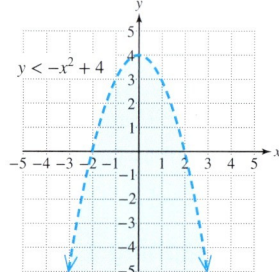

Figure 21-28

The solution to the system of inequalities is the intersection of the solution sets of the individual inequalities. See Figure 21-29.

Answers

3.

Figure 21-29

Skill Practice

4. Graph the solution set. $y > \frac{1}{2}x^2$
$x > y^2$

4.

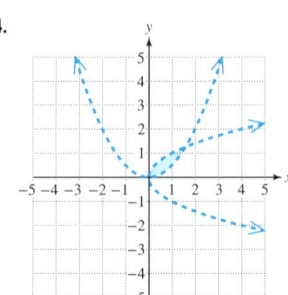

Section 21.5 Activity

A.1. a. Graph the equation $y = -x^2 + 4$.

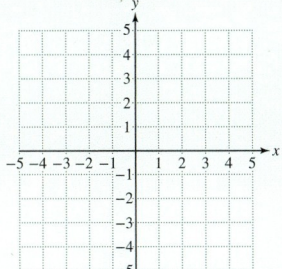

b. On the same graph as part (a), graph the inequality $y \leq -x^2 + 4$.

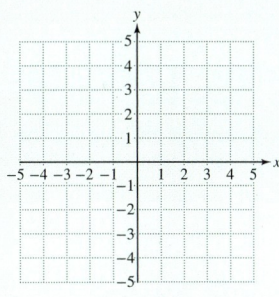

c. How would the graph from part (b) have changed if the inequality were $y < -x^2 + 4$?

d. How would the graph from part (b) have changed for the inequality $y \geq -x^2 + 4$?

A.2. Earlier in this text, we learned how to graph a system of linear inequalities. Use the same procedure to graph the system of *nonlinear* inequalities.

$$y \leq -x^2 + 4$$
$$y \geq x + 2$$

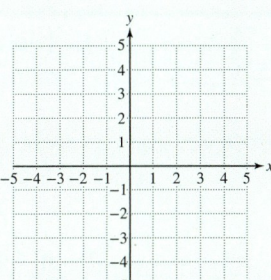

Practice Exercises

Section 21.5

Prerequisite Review

For Exercises R.1–R.2, graph the solution set to the inequality.

R.1. $3x + y \geq 2$

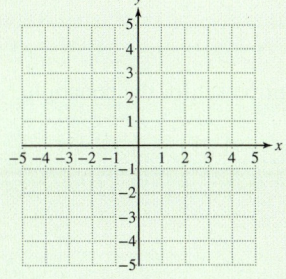

R.2. $4x - 2y < 6$

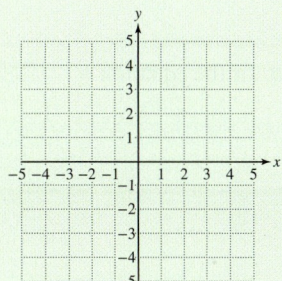

R.3. Refer to Exercise R.1.

    a. How would the graph of the solution set change for the inequality $3x + y > 2$?

    b. How would the graph of the solution set change for the inequality $3x + y \leq 2$?

R.4. Refer to Exercise R.2.

    a. How would the graph of the solution set change for the inequality $4x - 2y \leq 6$?

    b. How would the graph of the solution set change for the inequality $4x - 2y > 6$?

For Exercises R.5–R.10, identify the type of curve defined by the equation. Choose from: line, parabola, circle, ellipse, or hyperbola.

R.5. $x = y^2 - 1$

R.6. $\dfrac{x^2}{4} + y^2 = 4$

R.7. $3x + 2y = 8$

R.8. $(x - 4)^2 + (y + 2)^2 = 1$

R.9. $-\dfrac{x^2}{9} + \dfrac{y^2}{4} = 1$

R.10. $y = 3 - x^2$

Concept 1: Nonlinear Inequalities in Two Variables

1. True or false: The point $(2, 3)$ satisfies the inequality $-x^2 + y^3 > 1$.

2. True or false: The point $(4, -2)$ satisfies the inequality $4x^2 - 2x + 1 + y^2 < 3$.

3. True or false: The point $(5, 4)$ satisfies the system of inequalities.
$$\dfrac{x^2}{36} + \dfrac{y^2}{25} < 1$$
$$x^2 + y^2 \geq 4$$

4. True or false: The point $(1, -2)$ satisfies the system of inequalities.
$$y < x^2$$
$$y > x^2 - 4$$

5. a. Graph the solution set for $x^2 + y^2 \leq 9$.

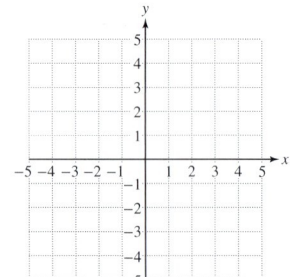

    b. Describe the solution set for the inequality $x^2 + y^2 \geq 9$.

    c. Describe the solution set for the equation $x^2 + y^2 = 9$.

6. a. Graph the solution set for $\dfrac{x^2}{4} + \dfrac{y^2}{9} \geq 1$.

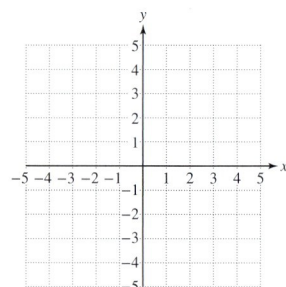

    b. Describe the solution set for the inequality $\dfrac{x^2}{4} + \dfrac{y^2}{9} \leq 1$.

    c. Describe the solution set for the equation $\dfrac{x^2}{4} + \dfrac{y^2}{9} = 1$.

7. **a.** Graph the solution set for $y \geq x^2 + 1$.

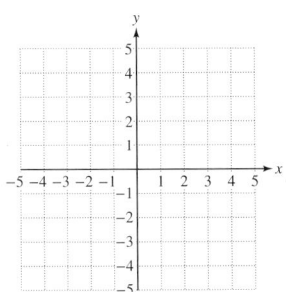

b. How would the solution change for the strict inequality $y > x^2 + 1$?

8. **a.** Graph the solution set for $\dfrac{x^2}{4} - \dfrac{y^2}{9} \leq 1$.

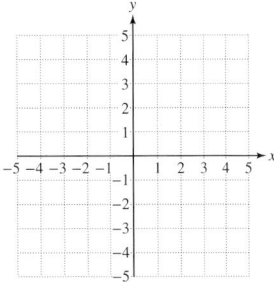

b. How would the solution change for the strict inequality $\dfrac{x^2}{4} - \dfrac{y^2}{9} < 1$?

9. A weak earthquake occurred in northern California roughly 4 mi south and 3 mi east of Sunol, California. The quake could be felt 25 mi away. Suppose the origin of a map is placed at the center of Sunol with the positive *x*-axis pointing east and the positive *y*-axis pointing north. Find an inequality that describes the points on the map for which the earthquake could be felt.

10. A coordinate system is placed at the center of a town with the positive *x*-axis pointing east and the positive *y*-axis pointing north. A cell tower is located 2 mi west and 4 mi north of the center of town. If the tower has a 30-mi range, write an inequality that represents the points on the map serviced by this tower.

For Exercises 11–25, graph the solution set. (See Examples 1–2.)

11. $2x + y \geq 1$

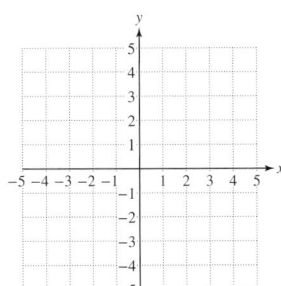

12. $3x + 2y \geq 6$

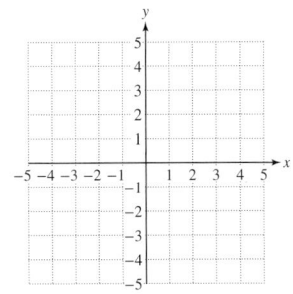

13. $x \leq y^2$

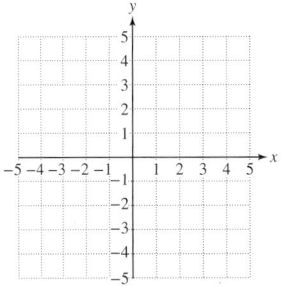

14. $y \leq -x^2$

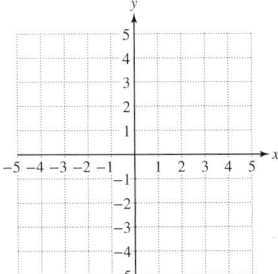

15. $(x-1)^2 + (y+2)^2 > 9$

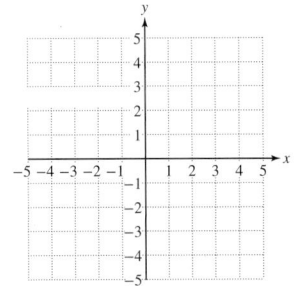

16. $(x+1)^2 + (y-4)^2 > 1$

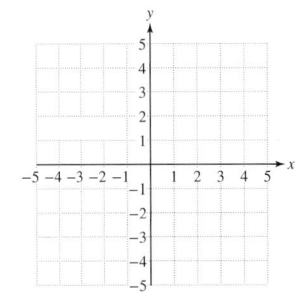

17. $x + y^2 \geq 4$

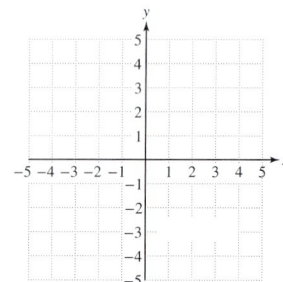

18. $x^2 + 2x + y - 1 \leq 0$

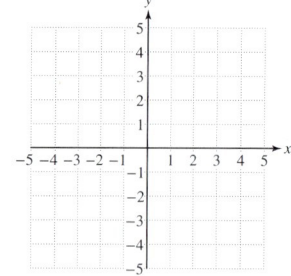

19. $9x^2 - y^2 > 9$

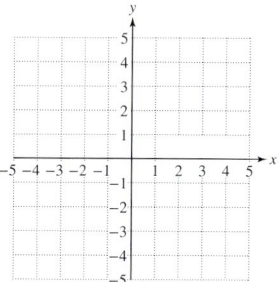

20. $y^2 - 4x^2 \leq 4$

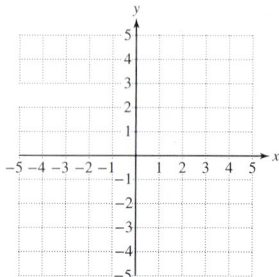

21. $x^2 + 16y^2 \leq 16$

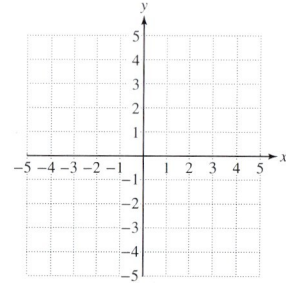

22. $4x^2 + y^2 \leq 4$

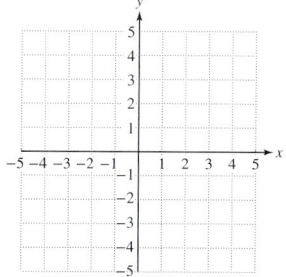

23. $y \leq \ln x$

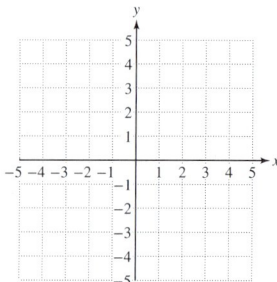

24. $y \leq \log x$

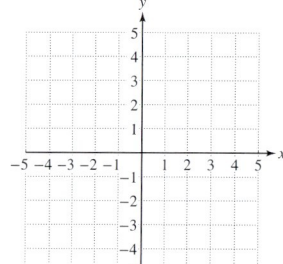

25. $y > 5^x$

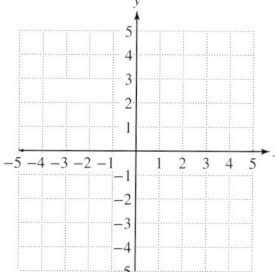

Concept 2: Systems of Nonlinear Inequalities in Two Variables

For Exercises 26–39, graph the solution set to the system of inequalities. **(See Examples 3–4.)**

26. $y < \sqrt{x}$
$x > 1$

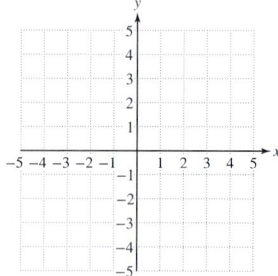

27. $y \geq \sqrt{x}$
$x \geq 0$

28. $\dfrac{x^2}{36} + \dfrac{y^2}{25} < 1$
$x^2 + y^2 \geq 4$

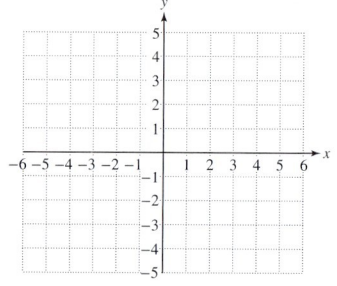

29. $x^2 - y^2 \geq 1$
$x \leq 0$

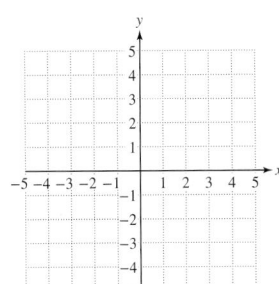

30. $y < x^2$
$y > x^2 - 4$

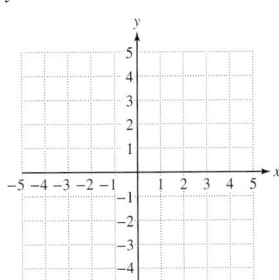

31. $y^2 - x^2 \geq 1$
$y \geq 0$

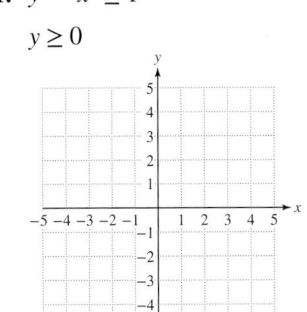

32. $y < \dfrac{1}{x}$
$y > 0$
$y < x$

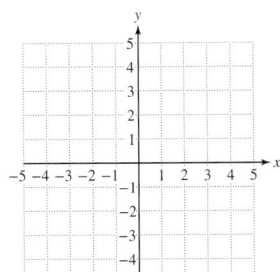

33. $y > x^3$
$y < 8$
$x > 0$

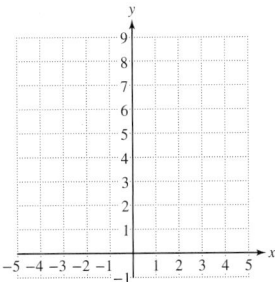

34. $x^2 + y^2 \geq 25$
$x^2 + y^2 \leq 9$

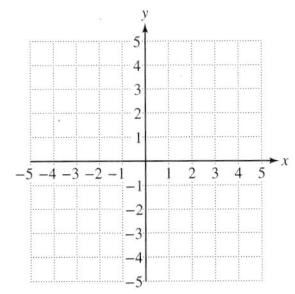

35. $\dfrac{x^2}{4} + \dfrac{y^2}{25} \geq 1$
$x^2 + \dfrac{y^2}{4} \leq 1$

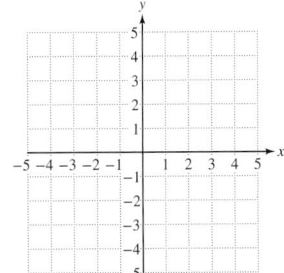

36. $x < -(y-1)^2 + 3$
$x + y > 2$

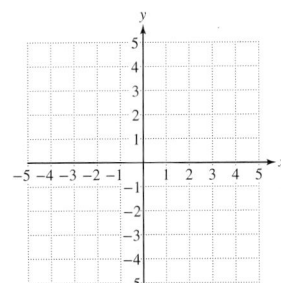

37. $x > (y-2)^2 + 1$
$x - y < 1$

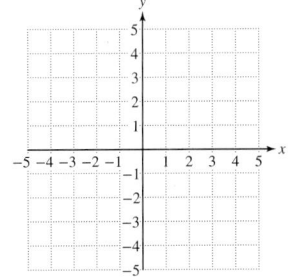

38. $x^2 + y^2 < 25$
$y < \dfrac{4}{3}x$
$y > -\dfrac{4}{3}x$

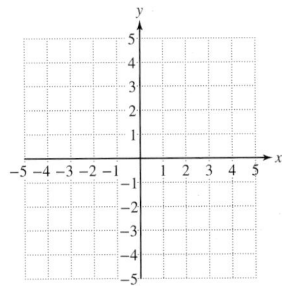

39. $y < e^x$
$y > 1$
$x < 2$

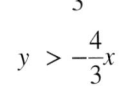

Expanding Your Skills

For Exercises 40–43, graph the compound inequalities.

40. $x^2 + y^2 \leq 36$ or $x + y \geq 0$

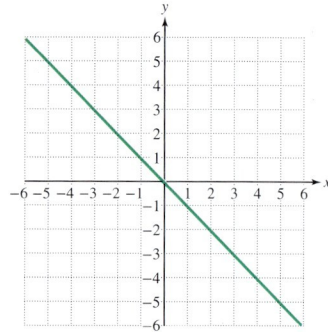

41. $y \leq -x^2 + 4$ or $y \geq x^2 - 4$

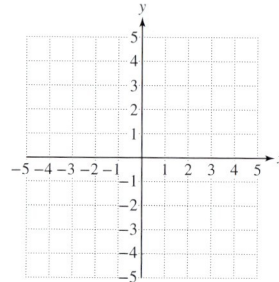

42. $y + 1 \geq x^2$ or $y + 1 \leq -x^2$

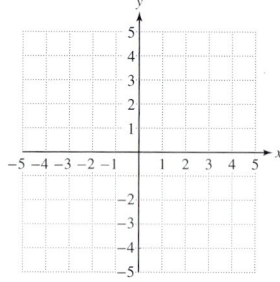

43. $(x+2)^2 + (y+3)^2 \leq 4$ or $x \geq y^2$

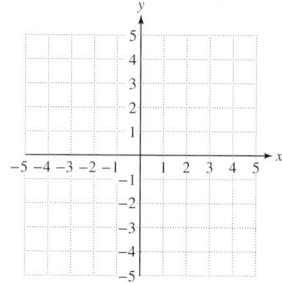

Chapter 21 Review Exercises

Section 21.1

For Exercises 1–2, find the distance between the two points by using the distance formula.

1. $(-6, 3)$ and $(0, 1)$
2. $(4, 13)$ and $(-1, 5)$

3. Find x such that $(x, 5)$ is 5 units from $(2, 9)$.

4. Find x such that $(-3, 4)$ is 3 units from $(x, 1)$.

For Exercises 5–8, find the center and the radius of the circle.

5. $(x - 12)^2 + (y - 3)^2 = 16$
6. $(x + 7)^2 + (y - 5)^2 = 81$
7. $(x + 3)^2 + (y + 8)^2 = 20$
8. $(x - 1)^2 + (y + 6)^2 = 32$

9. A stained glass window is in the shape of a circle with a 16-in. diameter. Find an equation of the circle relative to the origin for each of the following graphs.

a. 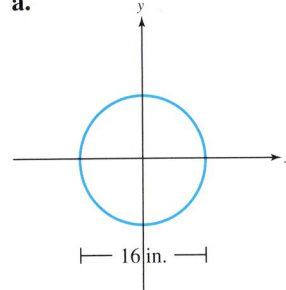   b.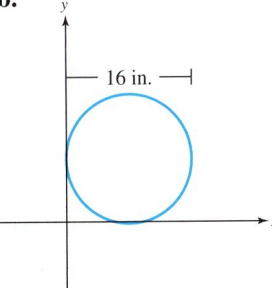

For Exercises 10–13, write the equation of the circle in standard form by completing the square.

10. $x^2 + y^2 + 12x - 10y + 51 = 0$
11. $x^2 + y^2 + 4x + 16y + 60 = 0$
12. $x^2 + y^2 - x - 4y + \dfrac{1}{4} = 0$
13. $x^2 + y^2 - 6x - \dfrac{2}{3}y + \dfrac{1}{9} = 0$

14. Write an equation of a circle with center at the origin and a diameter of 7 m.

15. Write an equation of a circle with center $(0, 2)$ and radius 3 m.

For Exercises 16–17, find the midpoint of the segment with the given endpoints.

16. $(-3, 1)$ and $(-5, -5)$
17. $(0, 9)$ and $(-2, 7)$

Section 21.2

For Exercises 18–21, determine whether the axis of symmetry is vertical or horizontal and if the parabola opens upward, downward, left, or right.

18. $y = -2(x - 3)^2 + 2$
19. $x = 3(y - 9)^2 + 1$
20. $x = -(y + 4)^2 - 8$
21. $y = (x + 3)^2 - 10$

For Exercises 22–25, determine the coordinates of the vertex and the equation of the axis of symmetry. Then use this information to graph the parabola.

22. $x = -(y - 1)^2$
23. $y = (x + 2)^2$

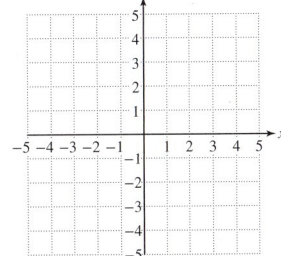   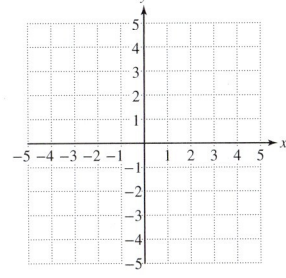

24. $y = -\dfrac{1}{4}x^2$
25. $x = 2y^2 - 1$

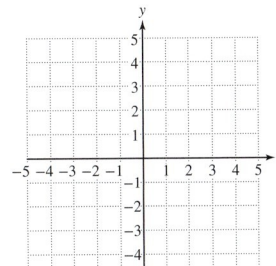   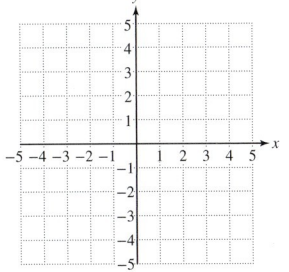

For Exercises 26–29, write the equation in standard form $y = a(x - h)^2 + k$ or $x = a(y - k)^2 + h$. Then identify the vertex and axis of symmetry.

26. $y = x^2 - 6x + 5$

27. $x = y^2 + 4y + 2$

28. $x = -4y^2 + 4y$

29. $y = -2x^2 - 2x$

Section 21.3

For Exercises 30–31, identify the x- and y-intercepts. Then graph the ellipse.

30. $\dfrac{x^2}{9} + \dfrac{y^2}{25} = 1$

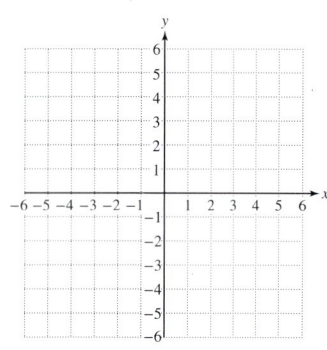

31. $x^2 + 4y^2 = 36$

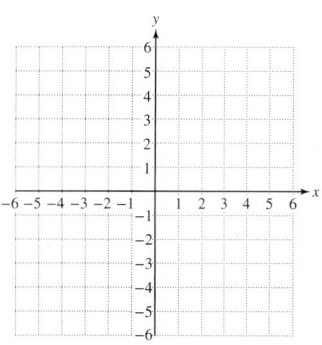

For Exercises 32–33, identify the center of the ellipse and graph the ellipse.

32. $\dfrac{(x-5)^2}{4} + \dfrac{(y+3)^2}{16} = 1$

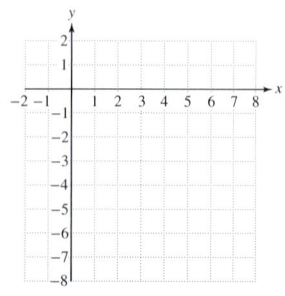

33. $\dfrac{x^2}{25} + \dfrac{(y-2)^2}{9} = 1$

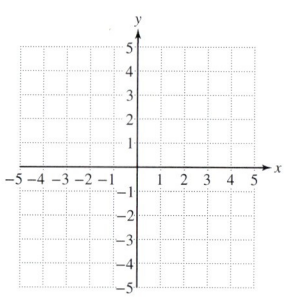

For Exercises 34–37, determine whether the transverse axis of the hyperbola is horizontal or vertical.

34. $\dfrac{x^2}{12} - \dfrac{y^2}{16} = 1$

35. $\dfrac{y^2}{9} - \dfrac{x^2}{9} = 1$

36. $y^2 - 8x^2 = 16$

37. $3x^2 - y^2 = 18$

For Exercises 38–39, graph the hyperbola by first drawing the reference rectangle and the asymptotes. Label the vertices.

38. $\dfrac{x^2}{4} - y^2 = 1$

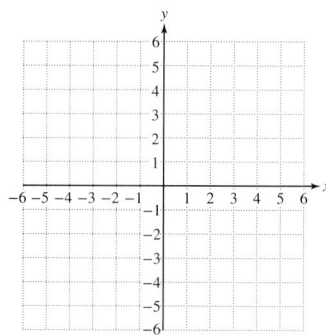

39. $y^2 - x^2 = 16$

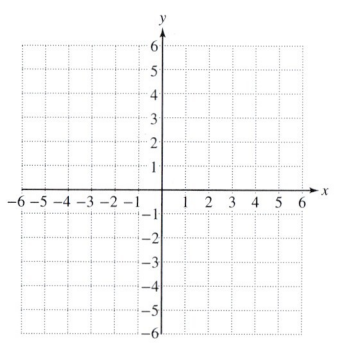

For Exercises 40–43, identify the equations as representing an ellipse or a hyperbola.

40. $\dfrac{x^2}{4} - \dfrac{y^2}{9} = 1$

41. $\dfrac{x^2}{16} + \dfrac{y^2}{9} = 1$

42. $\dfrac{x^2}{4} + \dfrac{y^2}{1} = 1$

43. $\dfrac{y^2}{1} - \dfrac{x^2}{16} = 1$

Section 21.4

For Exercises 44–47,

a. Identify each equation as representing a line, a parabola, a circle, an ellipse, or a hyperbola.

b. Graph both equations on the same coordinate system.

c. Solve the system analytically and verify the answers from the graph.

44. $3x + 2y = 10$

$y = x^2 - 5$

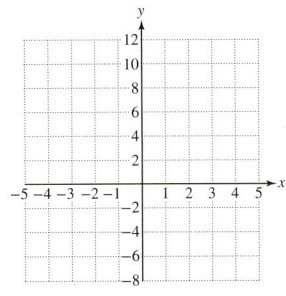

45. $4x + 2y = 10$

$y = x^2 - 10$

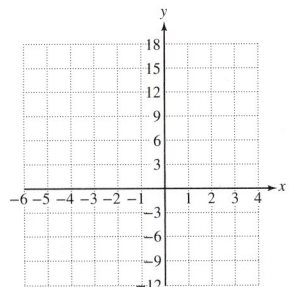

46. $x^2 + y^2 = 9$

$2x + y = 3$

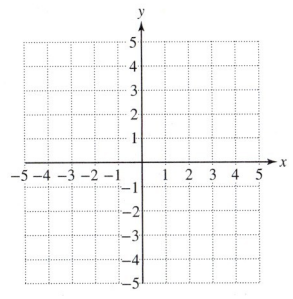

47. $x^2 + y^2 = 16$

$x - 2y = 8$

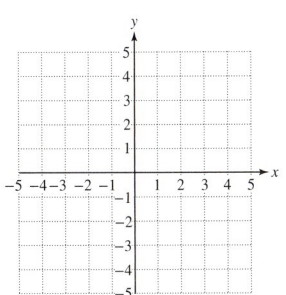

For Exercises 48–53, solve the system of nonlinear equations by using either the substitution method or the addition method.

48. $x^2 + 2y^2 = 8$

$2x - y = 2$

49. $x^2 + 4y^2 = 29$

$x - y = -4$

50. $x - y = 4$

$y^2 = 2x$

51. $y = x^2$

$6x^2 - y^2 = 8$

52. $x^2 + y^2 = 10$

$x^2 + 9y^2 = 18$

53. $x^2 + y^2 = 61$

$x^2 - y^2 = 11$

Section 21.5

For Exercises 54–59, graph the solution set to the inequality.

54. $\dfrac{x^2}{16} + \dfrac{y^2}{81} < 1$

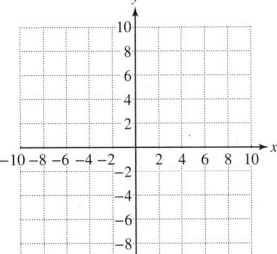

55. $\dfrac{x^2}{25} + \dfrac{y^2}{4} > 1$

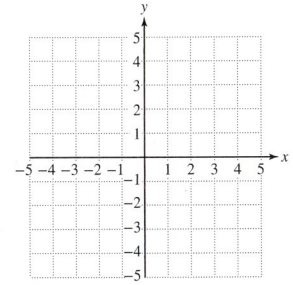

56. $(x-3)^2 + (y+1)^2 \geq 9$

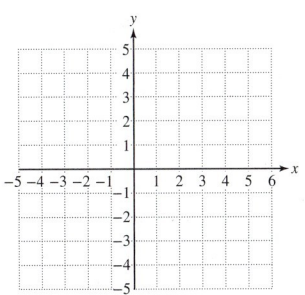

57. $(x+2)^2 + (y+1)^2 \leq 4$

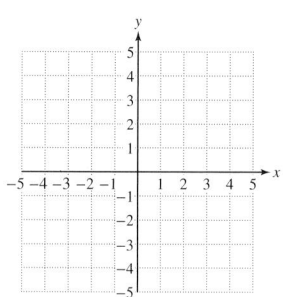

For Exercises 60–61, graph the solution set to the system of nonlinear inequalities.

60. $y > 2^x$
$x^2 + y^2 < 4$

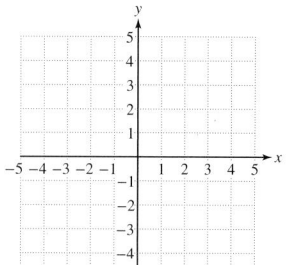

61. $y < x^2$
$x^2 + y^2 < 9$

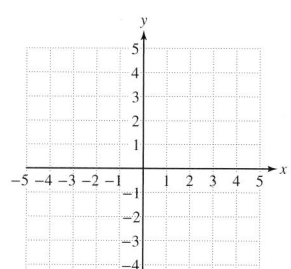

58. $y > (x-1)^2$

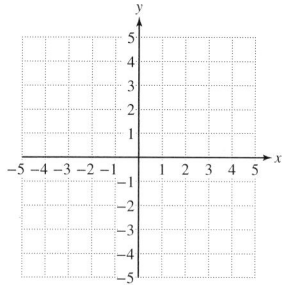

59. $x^2 - \dfrac{y^2}{4} \leq 1$

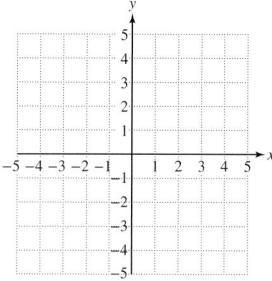

Binomial Expansions, Sequences, and Series

22

CHAPTER OUTLINE

22.1 Binomial Expansions   1464

22.2 Sequences and Series   1471

22.3 Arithmetic Sequences and Series   1481

22.4 Geometric Sequences and Series   1488

   Problem Recognition Exercises: Identifying Arithmetic and Geometric Sequences   1498

   Chapter 22 Review Exercises   1499

Mathematics in Business

In this chapter we present applications involving sequences and series. A **sequence** is a function whose domain is the set of positive integers. While infinitely many sequences can be defined, two specific categories are presented here: **arithmetic** and **geometric sequences**. An arithmetic sequence is a progression of numbers that increase or decrease by a fixed amount, whereas the numbers in a geometric sequence increase or decrease by a fixed ratio. A **series** is a sum of the terms in a sequence.

To illustrate the difference between an arithmetic and a geometric sequence, suppose that you have two different job offers. The first pays $60,000 per year with a $2000 raise at the start of each new year. The second pays $58,000 with a 4% raise at the start of each new year. The salaries for each job for each year can be written as the terms of a sequence. For example, the terms of the sequence of salaries for Job 1 are $60,000, 62,000, 64,000, and so on.

Notice that the yearly salary for Job 1 is initially more, but the salary increases by a fixed amount each year. The initial salary for Job 2 is less, but it grows exponentially and eventually overtakes the salary for Job 1. The sum of the salaries for the first 6 years can be added directly

Year	Job 1	Job 2	Total Job 1	Total Job 2
1	$60,000	$58,000	$60,000	$58,000
2	$62,000	$60,320	$122,000	$118,320
3	$64,000	$62,733	$186,000	$181,053
4	$66,000	$65,242	$252,000	$246,295
5	$68,000	$67,852	$320,000	$314,147
6	$70,000	$70,566	$390,000	$384,713
7	$72,000	$73,389	$462,000	$458,101
8	$74,000	$76,324	$536,000	$534,425
9	$76,000	$79,377	$612,000	$613,802
10	$78,000	$82,552	$690,000	$696,354

from the given data, or can be computed using formulas presented in this chapter. The total income over 6 years for Job 1 is $390,000, and the total income for Job 2 over 6 years is $384,713. However, because the yearly salary for Job 2 grows at a faster rate, by year 9 the cumulative income for Job 2 will be greater. The total income for Job 2 for the first 9 years will be $613,802, whereas for Job 1, the total income will be $612,000.

Our study of sequences and series in this chapter will help you learn to analyze numerical patterns and to make informed financial decisions.

Section 22.1 Binomial Expansions

Concepts

1. Binomial Expansions and Pascal's Triangle
2. Factorial Notation
3. The Binomial Theorem

1. Binomial Expansions and Pascal's Triangle

In this section, we will learn how to raise binomials to positive integer powers. First recall the formula to square a binomial.

$$(a + b)^2 = a^2 + 2ab + b^2$$

The expression $a^2 + 2ab + b^2$ is called the **binomial expansion** of $(a + b)^2$. To expand $(a + b)^3$, we can find the product $(a + b)(a + b)^2$.

$$(a + b)(a + b)^2 = (a + b)(a^2 + 2ab + b^2)$$

$$= a^3 + 2a^2b + ab^2 + a^2b + 2ab^2 + b^3$$
$$= a^3 + 3a^2b + 3ab^2 + b^3$$

Similarly, to expand $(a + b)^4$, we can multiply $(a + b)$ by $(a + b)^3$. Using this method, we can expand several powers of $(a + b)$ to find the following pattern:

$$(a + b)^0 = 1$$
$$(a + b)^1 = a + b$$
$$(a + b)^2 = a^2 + 2ab + b^2$$
$$(a + b)^3 = a^3 + 3a^2b + 3ab^2 + b^3$$
$$(a + b)^4 = a^4 + 4a^3b + 6a^2b^2 + 4ab^3 + b^4$$
$$(a + b)^5 = a^5 + 5a^4b + 10a^3b^2 + 10a^2b^3 + 5ab^4 + b^5$$

FOR REVIEW

Recall that for any nonzero real number x, we have $x^0 = 1$.

Notice that the exponents on a decrease from left to right, while the exponents on b increase from left to right. Also observe that for each term, the sum of the exponents is equal to the exponent to which $(a + b)$ is raised. Finally, notice that the number of terms in the expansion is exactly 1 more than the power to which $(a + b)$ is raised. For example, the expansion of $(a + b)^4$ has five terms, and the expansion of $(a + b)^5$ has six terms.

With these guidelines in mind, we know that $(a + b)^6$ will contain seven terms involving

$$a^6 \quad a^5b \quad a^4b^2 \quad a^3b^3 \quad a^2b^4 \quad ab^5 \quad b^6$$

We can complete the expansion of $(a + b)^6$ if we can determine the correct coefficients of each term.

If we write the coefficients for several expansions of $(a + b)^n$, where $n \geq 0$, we have a triangular array of numbers.

$$(a + b)^0 = 1 \qquad\qquad\qquad\qquad 1$$
$$(a + b)^1 = 1a + 1b \qquad\qquad\qquad 1 \quad 1$$
$$(a + b)^2 = 1a^2 + 2ab + 1b^2 \qquad\qquad 1 \quad 2 \quad 1$$
$$(a + b)^3 = 1a^3 + 3a^2b + 3ab^2 + 1b^3 \qquad 1 \quad 3 \quad 3 \quad 1$$
$$(a + b)^4 = 1a^4 + 4a^3b + 6a^2b^2 + 4ab^3 + 1b^4 \qquad 1 \quad 4 \quad 6 \quad 4 \quad 1$$
$$(a + b)^5 = 1a^5 + 5a^4b + 10a^3b^2 + 10a^2b^3 + 5ab^4 + 1b^5 \quad 1 \quad 5 \quad 10 \quad 10 \quad 5 \quad 1$$

Each row begins and ends with a 1, and each entry in between is the sum of the two entries from the line above. For example, in row six, $5 = 1 + 4$, $10 = 4 + 6$, and so on. This triangular array of coefficients for a binomial expansion is called **Pascal's triangle**, named after French mathematician Blaise Pascal (1623–1662).

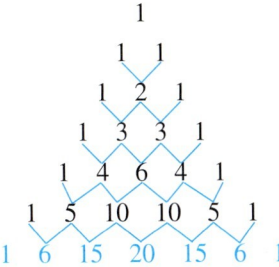

By using the pattern shown in Pascal's triangle, the coefficients corresponding to $(a + b)^6$ would be 1, 6, 15, 20, 15, 6, 1. By inserting the coefficients, the sum becomes

$$(a + b)^6 = 1a^6 + 6a^5b + 15a^4b^2 + 20a^3b^3 + 15a^2b^4 + 6ab^5 + 1b^6$$

2. Factorial Notation

Although Pascal's triangle provides an easy method to find the coefficients of $(a + b)^n$, it is impractical for large values of n. A more efficient method to find the coefficients of a binomial expansion involves **factorial notation**.

> ### Definition of n!
> Let n be a positive integer. Then $n!$ (read as "n factorial") is defined as the product of integers from 1 through n. That is,
> $$n! = n(n - 1)(n - 2) \cdots (2)(1)$$
> *Note:* We define $0! = 1$.

Example 1 Evaluating Factorial Notation

Evaluate the expressions.

a. $4!$ b. $10!$ c. $0!$

Solution:

a. $4! = 4 \cdot 3 \cdot 2 \cdot 1 = 24$

b. $10! = 10 \cdot 9 \cdot 8 \cdot 7 \cdot 6 \cdot 5 \cdot 4 \cdot 3 \cdot 2 \cdot 1 = 3{,}628{,}800$

c. $0! = 1$ by definition

Skill Practice Evaluate the expressions.

1. $3!$ 2. $8!$ 3. $1!$

Answers

1. 6   2. 40,320   3. 1

Sometimes factorial notation is used with other operations such as multiplication and division.

Example 2   **Evaluating Expressions with Factorials**

Evaluate the expressions.

a. $\dfrac{4!}{4! \cdot 0!}$ b. $\dfrac{4!}{3! \cdot 1!}$ c. $\dfrac{4!}{2! \cdot 2!}$ d. $\dfrac{4!}{1! \cdot 3!}$ e. $\dfrac{4!}{0! \cdot 4!}$

FOR REVIEW

Recall that to simplify a fraction, "cancel" common factors that form a ratio of 1. When working with factorial notation, sometimes it is easier to see the cancellation when the factors are listed. For example,

$$\dfrac{4!}{3! \cdot 1!} = \dfrac{4 \cdot 3 \cdot 2 \cdot 1}{(3 \cdot 2 \cdot 1) \cdot 1}$$
$$= \dfrac{4}{1} = 4$$

Solution:

a. $\dfrac{4!}{4! \cdot 0!} = \dfrac{4!}{4! \cdot 1} = 1$

b. $\dfrac{4!}{3! \cdot 1!} = \dfrac{4 \cdot 3!}{3! \cdot 1} = 4$ Note that $4! = 4 \cdot (3 \cdot 2 \cdot 1) = 4 \cdot 3!$

c. $\dfrac{4!}{2! \cdot 2!} = \dfrac{4 \cdot 3 \cdot 2!}{2! \cdot 2 \cdot 1} = \dfrac{12}{2} = 6$

d. $\dfrac{4!}{1! \cdot 3!} = \dfrac{4 \cdot 3!}{1 \cdot 3!} = 4$

e. $\dfrac{4!}{0! \cdot 4!} = \dfrac{4!}{1 \cdot 4!} = 1$

Skill Practice   Evaluate.

4. $\dfrac{5!}{5! \cdot 0!}$ 5. $\dfrac{5!}{4! \cdot 1!}$ 6. $\dfrac{5!}{3! \cdot 2!}$

3. The Binomial Theorem

Notice from Example 2 that the values of

$$\dfrac{4!}{4! \cdot 0!} \quad \dfrac{4!}{3! \cdot 1!} \quad \dfrac{4!}{2! \cdot 2!} \quad \dfrac{4!}{1! \cdot 3!} \quad \text{and} \quad \dfrac{4!}{0! \cdot 4!}$$

correspond to the values 1, 4, 6, 4, 1, which are the coefficients for the expansion of $(a + b)^4$. Generalizing this pattern, we see the coefficients for the terms in the expansion of $(a + b)^n$ are given by

$$\dfrac{n!}{r! \cdot (n - r)!}$$

where r corresponds to the exponent on the factor of a and $(n - r)$ corresponds to the exponent on the factor of b. Using this formula for the coefficients in a binomial expansion results in the **binomial theorem**.

Answers

4. 1   **5.** 5   **6.** 10

The Binomial Theorem

For any positive integer n,

$$(a+b)^n = \frac{n!}{n! \cdot 0!}a^n + \frac{n!}{(n-1)! \cdot 1!}a^{(n-1)}b + \frac{n!}{(n-2)! \cdot 2!}a^{(n-2)}b^2 + \cdots + \frac{n!}{0! \cdot n!}b^n$$

Example 3 — Applying the Binomial Theorem

Write out the first three terms of the expansion of $(a+b)^{10}$.

Solution:

The first three terms of $(a+b)^{10}$ are

$$\frac{10!}{10! \cdot 0!}a^{10} + \frac{10!}{9! \cdot 1!}a^9 b + \frac{10!}{8! \cdot 2!}a^8 b^2$$

$$= \frac{\cancel{10!}}{\cancel{10!} \cdot 1}a^{10} + \frac{10 \cdot \cancel{9!}}{\cancel{9!} \cdot 1!}a^9 b + \frac{10 \cdot 9 \cdot \cancel{8!}}{\cancel{8!} \cdot 2 \cdot 1}a^8 b^2$$

$$= a^{10} + 10a^9 b + 45a^8 b^2$$

TIP: When applying the binomial theorem, exponents on a and b must add up to n.

Skill Practice

7. Write out the first three terms of $(x+y)^5$.

Example 4 — Applying the Binomial Theorem

Use the binomial theorem to expand $(2m+7)^3$.

Solution:

This expression is of the form $(a+b)^3$, where $a = 2m$ and $b = 7$. Applying the binomial theorem, we have

$$(a+b)^3 = \frac{3!}{3! \cdot 0!}a^3 + \frac{3!}{2! \cdot 1!}a^2 b + \frac{3!}{1! \cdot 2!}ab^2 + \frac{3!}{0! \cdot 3!}b^3$$

$$= \frac{3!}{3! \cdot 0!}(2m)^3 + \frac{3!}{2! \cdot 1!}(2m)^2(7) + \frac{3!}{1! \cdot 2!}(2m)(7)^2 + \frac{3!}{0! \cdot 3!}(7)^3$$

$$= 1 \cdot (8m^3) + 3 \cdot (4m^2)(7) + 3 \cdot (2m)(49) + 1 \cdot (343)$$

$$= 8m^3 + 84m^2 + 294m + 343$$

TIP: The values of $\dfrac{n!}{r! \cdot (n-r)!}$ can also be found by using Pascal's triangle.

```
        1
       1 1
      1 2 1
     1 3 3 1
```

Skill Practice

8. Use the binomial theorem to expand $(3y+4)^3$.

Answers

7. $x^5 + 5x^4 y + 10x^3 y^2$
8. $27y^3 + 108y^2 + 144y + 64$

TIP: The values of $\dfrac{n!}{r! \cdot (n-r)!}$ can also be found by using Pascal's triangle.

```
        1
       1 1
      1 2 1
     1 3 3 1
    1 4 6 4 1
```

> **Example 5** Applying the Binomial Theorem
>
> Use the binomial theorem to expand $(3x^2 - 5y)^4$.
>
> **Solution:**
>
> Write $(3x^2 - 5y)^4$ as $[(3x^2) + (-5y)]^4$. In this case, the expressions $3x^2$ and $-5y$ may be substituted for a and b in the expansion of $(a+b)^4$.
>
> $$(a+b)^4 = \frac{4!}{4! \cdot 0!}a^4 + \frac{4!}{3! \cdot 1!}a^3 b + \frac{4!}{2! \cdot 2!}a^2 b^2 + \frac{4!}{1! \cdot 3!}ab^3 + \frac{4!}{0! \cdot 4!}b^4$$
>
> $$= \frac{4!}{4! \cdot 0!}(3x^2)^4 + \frac{4!}{3! \cdot 1!}(3x^2)^3(-5y) + \frac{4!}{2! \cdot 2!}(3x^2)^2(-5y)^2$$
>
> $$+ \frac{4!}{1! \cdot 3!}(3x^2)(-5y)^3 + \frac{4!}{0! \cdot 4!}(-5y)^4$$
>
> $$= 1 \cdot (81x^8) + 4 \cdot (27x^6)(-5y) + 6 \cdot (9x^4)(25y^2) + 4 \cdot (3x^2)(-125y^3)$$
> $$+ 1 \cdot (625y^4)$$
>
> $$= 81x^8 - 540x^6 y + 1350x^4 y^2 - 1500x^2 y^3 + 625y^4$$
>
> **Skill Practice**
>
> **9.** Use the binomial theorem to expand $(2a - 3b^2)^4$.

The binomial theorem may also be used to find a specific term in a binomial expansion.

> **Example 6** Finding a Specific Term in a Binomial Expansion
>
> Find the fourth term of the expansion $(a+b)^{13}$.
>
> **Solution:**
>
> There are 14 terms in the expansion of $(a+b)^{13}$. The first term will have variable factors $a^{13}b^0$. The second term will have variable factors $a^{12}b^1$. The third term will have $a^{11}b^2$, and the fourth term will have $a^{10}b^3$. Hence, the fourth term is
>
> $$\frac{13!}{10! \cdot 3!} a^{10} b^3 = 286 a^{10} b^3$$
>
> **Skill Practice**
>
> **10.** Find the fourth term of $(x+y)^8$.

From Example 6, we see that for the kth term in the expansion $(a+b)^n$, where k is an integer greater than zero, the variable factors are $a^{n-(k-1)}$ and b^{k-1}. Therefore, to find the kth term of $(a+b)^n$, we can make the following generalization.

> **Finding a Specific Term in a Binomial Expansion**
>
> Let n and k be positive integers such that $k \leq n$. Then the kth term in the expansion of $(a+b)^n$ is given by
>
> $$\frac{n!}{[n-(k-1)]! \cdot (k-1)!} \cdot a^{n-(k-1)} \cdot b^{k-1}$$

Answers

9. $16a^4 - 96a^3 b^2 + 216a^2 b^4 - 216ab^6 + 81b^8$

10. $56x^5 y^3$

Example 7 — Finding a Specific Term in a Binomial Expansion

Find the sixth term of $(p^3 + 2w)^8$.

Solution:

Apply the formula.

$$\frac{n!}{[n-(k-1)]! \cdot (k-1)!} \cdot a^{n-(k-1)} \cdot b^{k-1} \quad \text{with } n = 8, k = 6, a = p^3, \text{ and } b = 2w$$

$$\frac{8!}{[8-(6-1)]! \cdot (6-1)!} \cdot (p^3)^{8-(6-1)} \cdot (2w)^{6-1}$$

$$= \frac{8!}{(3)! \cdot (5)!} \cdot (p^3)^3 \cdot (2w)^5$$

$$= 56 \cdot (p^9) \cdot (32w^5)$$

$$= 1792 p^9 w^5$$

Skill Practice

11. Find the fifth term of $(t - m^2)^8$.

Answer

11. $70 t^4 m^8$

Section 22.1 Activity

For Exercises A.1–A.4, expand the binomial.

A.1. $(a+b)^0$ **A.2.** $(a+b)^1$ **A.3.** $(a+b)^2$

A.4. $(a+b)^3$ *Hint:* Multiply the result of Exercise A.3 by $(a+b)$.

A.5. a. Fill in the fifth row of Pascal's triangle.

  b. Write the expansion of $(a+b)^4$ by using the coefficients from Pascal's triangle and the pattern from Exercises A.1–A.4.

$$\begin{array}{c} 1 \\ 1 \quad 1 \\ 1 \quad 2 \quad 1 \\ 1 \quad 3 \quad 3 \quad 1 \\ \square \quad \square \quad \square \quad \square \quad \square \end{array} \qquad \begin{array}{c} 1 \\ a+b \\ a^2 + 2ab + b^2 \\ a^3 + 3a^2 b + 3ab^2 + b^3 \\ \square + \square + \square + \square + \square \end{array}$$

A.6. Evaluate the factorial expressions and compare the results to the fifth row of Pascal's triangle from Exercise A.5(a).

  a. $\dfrac{4!}{4! \cdot 0!}$ **b.** $\dfrac{4!}{3! \cdot 1!}$ **c.** $\dfrac{4!}{2! \cdot 2!}$ **d.** $\dfrac{4!}{1! \cdot 3!}$ **e.** $\dfrac{4!}{0! \cdot 4!}$

A.7. Simplify $(5x - y^2)^4$ by following these steps.

  a. Write the expression in parentheses as a sum.

  $$(5x - y^2)^4 = [5x + (\square)]^4$$

  b. Use the expansion of $(a+b)^4$ from Exercise A.5(b) to expand $(5x - y^2)^4$.

Section 22.1 Practice Exercises

Prerequisite Review

For Exercises R.1–R.2, simplify the expression.

R.1. $\dfrac{6 \cdot 5 \cdot 4 \cdot 3 \cdot 2 \cdot 1}{(4 \cdot 3 \cdot 2 \cdot 1)(2 \cdot 1)}$

R.2. $\dfrac{8 \cdot 7 \cdot 6 \cdot 5 \cdot 4 \cdot 3 \cdot 2 \cdot 1}{(3 \cdot 2 \cdot 1)(5 \cdot 4 \cdot 3 \cdot 2 \cdot 1)}$

For Exercises R.3–R.4, multiply the expression.

R.3. **a.** $(4x - 3)^2$ **b.** $(4x - 3)^3$

R.4. **a.** $(2x + 5)^2$ **b.** $(2x + 5)^3$

Vocabulary and Key Concepts

1. **a.** The expanded form of $(a + b)^2 = $ _____. The expanded form of $(a + b)^3 = $ _____. These are both called _____ expansions.

   b. Given a positive integer n, the value of $n!$ is _____. Furthermore, $n!$ is read as "n _____."

   c. $3! = $ _____, $2! = $ _____, $1! = $ _____, and $0! = $ _____.

   d. The _____ theorem states that for any positive integer n,

   $$(a + b)^n = \dfrac{n!}{n! \cdot 0!} a^n + \dfrac{n!}{(n-1)! \cdot 1!} a^{(n-1)} b + \dfrac{n!}{(n-2)! \cdot 2!} a^{(n-2)} b^2 + \cdots + \dfrac{n!}{0! \cdot n!} b^n.$$

   e. The coefficients of the binomial expansion of $(a + b)^n$ can also be found by using _____ triangle.

Concept 1: Binomial Expansions and Pascal's Triangle

For Exercises 2–7, expand the binomials. Use Pascal's triangle to find the coefficients.

2. $(x + y)^4$
3. $(a + b)^3$
4. $(4 + p)^3$
5. $(1 + g)^4$
6. $(a^2 + b)^6$
7. $(p + q^2)^7$

For Exercises 8–13, rewrite each binomial of the form $(a - b)^n$ as $[a + (-b)]^n$. Then expand the binomials. Use Pascal's triangle to find the coefficients.

8. $(t - 2)^5$
9. $(s - t)^5$
10. $(p^2 - w)^3$
11. $(5 - u^3)^4$
12. $(a - b^2)^4$
13. $(x^2 - 4)^3$

14. For $a > 0$ and $b > 0$, what happens to the signs of the terms when expanding the binomial $(a - b)^n$ compared with $(a + b)^n$?

Concept 2: Factorial Notation

For Exercises 15–18, evaluate the expression. (See Example 1.)

15. $5!$ **16.** $3!$ **17.** $0!$ **18.** $1!$

19. True or false: $0! \neq 1!$

20. True or false: $n!$ is defined for negative integers.

21. True or false: $n! = n$ for $n = 1$ and 2.

22. Show that $9! = 9 \cdot 8!$

23. Show that $6! = 6 \cdot 5!$

24. Show that $8! = 8 \cdot 7!$

For Exercises 25–32, evaluate the expression. (See Example 2.)

25. $\dfrac{8!}{4!}$ **26.** $\dfrac{7!}{5!}$ **27.** $\dfrac{3!}{0!}$ **28.** $\dfrac{4!}{0!}$

29. $\dfrac{8!}{3! \cdot 5!}$ **30.** $\dfrac{6!}{2! \cdot 4!}$ **31.** $\dfrac{4!}{0! \cdot 4!}$ **32.** $\dfrac{6!}{0! \cdot 6!}$

Concept 3: The Binomial Theorem

For Exercises 33–36, find the first three terms of the expansion. (See Example 3.)

33. $(m+n)^{11}$ **34.** $(p+q)^9$ **35.** $(u^2-v)^{12}$ **36.** $(r-s^2)^8$

37. How many terms are in the expansion of $(a+b)^8$?

38. How many terms are in the expansion of $(x+y)^{13}$?

For Exercises 39–50, use the binomial theorem to expand the binomials. (See Examples 4–5.)

39. $(s+t)^6$ **40.** $(h+k)^4$ **41.** $(b-3)^3$ **42.** $(c-2)^5$

43. $(2x+y)^4$ **44.** $(p+3q)^3$ **45.** $(c^2-d)^7$ **46.** $(u-v^3)^6$

47. $\left(\dfrac{a}{2}-b\right)^5$ **48.** $\left(\dfrac{s}{3}+t\right)^5$ **49.** $(x+4y)^4$ **50.** $(3y-w)^3$

For Exercises 51–56, find the indicated term of the binomial expansion. (See Examples 6–7.)

51. $(m-n)^{11}$; sixth term **52.** $(p-q)^9$; fourth term **53.** $(u^2-v)^{12}$; fifth term

54. $(2r-s^2)^8$; sixth term **55.** $(5f+g)^9$; 10th term **56.** $(4m+n)^{10}$; 11th term

Sequences and Series

Section 22.2

1. Finite and Infinite Sequences

In day-to-day life, we think of a sequence as a set of items with some order or pattern. In mathematics, a sequence is a list of terms that correspond to the set of positive integers.

For example, suppose Justine is allowed to choose between two salary plans. Plan A offers $48,000 as a starting salary with a 5% raise each year. Plan B offers

Concepts

1. Finite and Infinite Sequences
2. Series

$50,000 with a $2000 raise each year. Then the following sequences represent the salaries for each plan for the first 5 years.

| Plan A: | $48,000 | $50,400 | $52,920 | $55,566 | $58,344.30 |
| Plan B: | $50,000 | $52,000 | $54,000 | $56,000 | $58,000 |

By studying sequences we can see that Plan A yields a better salary if Justine plans to stay in the job for at least 5 years.

These sequences are called *finite sequences* because they have a finite number of terms. The sequence 1, 4, 9, 16, 25, . . . is called an *infinite sequence* because it continues indefinitely.

Because the terms in a sequence are related to the set of positive integers, we give a formal definition of finite and infinite sequences, using the language of functions.

> **Definition of Finite and Infinite Sequences**
>
> An **infinite sequence** is a function whose domain is the set of positive integers.
> A **finite sequence** is a function whose domain is the set of the first *n* positive integers.

For any positive integer n, the value of the sequence is denoted by a_n (read as "a sub n"). The values $a_1, a_2, a_3, \ldots$ are called the **terms of the sequence**. The expression a_n defines the **nth term** (or general term) **of the sequence**.

Example 1 Listing the Terms of a Sequence

List the terms of the following sequences.

a. $a_n = 3n^2 - 4, \quad 1 \leq n \leq 4$ **b.** $a_n = 3 \cdot 2^n$

Solution:

a. The domain is restricted to the first four positive integers, indicating that the sequence is finite.

n	a_n
1	$3(1)^2 - 4 = -1$
2	$3(2)^2 - 4 = 8$
3	$3(3)^2 - 4 = 23$
4	$3(4)^2 - 4 = 44$

The sequence is $-1, 8, 23, 44$.

b. The sequence $a_n = 3 \cdot 2^n$ has no restrictions on its domain; therefore, it is an infinite sequence.

n	a_n
1	$3 \cdot 2^1 = 6$
2	$3 \cdot 2^2 = 12$
3	$3 \cdot 2^3 = 24$
4	$3 \cdot 2^4 = 48$
. . .	

The sequence is 6, 12, 24, 48, . . .

Skill Practice List the terms of the sequences.

1. $a_n = n^3 - 2, \quad 1 \leq n \leq 3$ **2.** $a_n = \left(\dfrac{2}{3}\right)^n$

FOR REVIEW

Substituting values of n into a sequence defined by a_n is similar to substituting values of x into a function defined by $y = f(x)$.

Answers

1. $-1, 6, 25$ **2.** $\dfrac{2}{3}, \dfrac{4}{9}, \dfrac{8}{27}, \ldots$

Sometimes the terms of a sequence may have alternating signs. Such a sequence is called an **alternating sequence**.

Example 2   Listing the Terms of an Alternating Sequence

List the first four terms of each alternating sequence.

a. $a_n = (-1)^n \cdot \dfrac{1}{n}$ **b.** $a_n = (-1)^{n+1} \cdot \left(\dfrac{2}{3}\right)^n$

Solution:

a.

n	a_n
1	$(-1)^1 \cdot \dfrac{1}{1} = -1$
2	$(-1)^2 \cdot \dfrac{1}{2} = \dfrac{1}{2}$
3	$(-1)^3 \cdot \dfrac{1}{3} = -\dfrac{1}{3}$
4	$(-1)^4 \cdot \dfrac{1}{4} = \dfrac{1}{4}$

The first four terms are $-1, \tfrac{1}{2}, -\tfrac{1}{3}, \tfrac{1}{4}$.

TIP: Notice that the factor $(-1)^n$ makes the even-numbered terms positive and the odd-numbered terms negative.

b.

n	a_n
1	$(-1)^{1+1} \cdot \left(\dfrac{2}{3}\right)^1 = (-1)^2 \cdot \left(\dfrac{2}{3}\right) = \dfrac{2}{3}$
2	$(-1)^{2+1} \cdot \left(\dfrac{2}{3}\right)^2 = (-1)^3 \cdot \left(\dfrac{4}{9}\right) = -\dfrac{4}{9}$
3	$(-1)^{3+1} \cdot \left(\dfrac{2}{3}\right)^3 = (-1)^4 \cdot \left(\dfrac{8}{27}\right) = \dfrac{8}{27}$
4	$(-1)^{4+1} \cdot \left(\dfrac{2}{3}\right)^4 = (-1)^5 \cdot \left(\dfrac{16}{81}\right) = -\dfrac{16}{81}$

The first four terms are $\tfrac{2}{3}, -\tfrac{4}{9}, \tfrac{8}{27}, -\tfrac{16}{81}$.

TIP: Notice that the factor $(-1)^{n+1}$ makes the odd-numbered terms positive and the even-numbered terms negative.

Skill Practice List the first four terms of each alternating sequence.

3. $a_n = (-1)^n \left(\dfrac{1}{2}\right)^n$ **4.** $a_n = (-1)^{n+1} \left(\dfrac{2}{n}\right)$

In Examples 1 and 2, we were given the formula for the nth term of a sequence and asked to list several terms of the sequence. We now consider the reverse process. Given several terms of the sequence, we will find a formula for the nth term. To do so, look for a pattern that establishes each term as a function of the term number.

Answers

3. $-\dfrac{1}{2}, \dfrac{1}{4}, -\dfrac{1}{8}, \dfrac{1}{16}$

4. $2, -1, \dfrac{2}{3}, -\dfrac{1}{2}$

> **Example 3** Finding the nth Term of a Sequence
>
> Find a formula for the nth term of the sequence.
>
> **a.** $\dfrac{1}{2}, \dfrac{2}{3}, \dfrac{3}{4}, \dfrac{4}{5}, \ldots$ **b.** $-2, 4, -6, 8, -10, \ldots$ **c.** $\dfrac{1}{2}, \dfrac{1}{4}, \dfrac{1}{8}, \dfrac{1}{16}, \ldots$
>
> **Solution:**
>
> **a.** For each term in the sequence, the numerator is equal to the term number, and the denominator is equal to 1 more than the term number. Therefore, the nth term may be given by
>
> $$a_n = \dfrac{n}{n+1}$$
>
> **b.** The odd-numbered terms are negative, and the even-numbered terms are positive. The factor $(-1)^n$ will produce the required alternation of signs. The numbers 2, 4, 6, 8, 10, ... are equal to 2(1), 2(2), 2(3), 2(4), 2(5), ... Therefore, the nth term may be given by
>
> $$a_n = (-1)^n \cdot 2n$$
>
> **c.** The denominators are consecutive powers of 2. The sequence can be written as
>
> $$\dfrac{1}{2^1}, \dfrac{1}{2^2}, \dfrac{1}{2^3}, \dfrac{1}{2^4}, \ldots$$
>
> Therefore, the nth term may be given by
>
> $$a_n = \dfrac{1}{2^n}$$
>
> **Skill Practice** Find a formula for the nth term of the sequence.
>
> **5.** $\dfrac{1}{3}, \dfrac{1}{9}, \dfrac{1}{27}, \dfrac{1}{81}, \ldots$ **6.** $-1, 4, -9, 16, -25, \ldots$ **7.** $\dfrac{3}{2}, \dfrac{4}{3}, \dfrac{5}{4}, \dfrac{6}{5}, \dfrac{7}{6}, \ldots$

TIP: The first few terms of a sequence are not sufficient to define the sequence uniquely. For example, consider the sequence

$$\dfrac{1}{2}, \dfrac{1}{4}, \dfrac{1}{8}, \ldots$$

The following formulas both produce the first three terms, but differ at the fourth term:

$$a_n = \dfrac{1}{2^n} \quad \text{and} \quad b_n = \dfrac{1}{n^2 - n + 2}$$

$$a_4 = \dfrac{1}{16} \quad \text{whereas} \quad b_4 = \dfrac{1}{14}$$

To define a sequence uniquely, the nth term must be provided.

Answers

5. $a_n = \dfrac{1}{3^n}$

6. $a_n = (-1)^n n^2$

7. $a_n = \dfrac{n+2}{n+1}$

Example 4   **Using a Sequence in an Application**

A child drops a ball from a height of 4 ft. With each bounce, the ball rebounds to 50% of its height. Write a sequence whose terms represent the heights from which the ball falls (begin with the initial height from which the ball was dropped).

Solution:

The ball first drops 4 ft and then rebounds to a new height of 0.50(4 ft) = 2 ft. Similarly, it falls from 2 ft and rebounds 0.50(2 ft) = 1 ft. Repeating this process, we have

$$4, 2, 1, \frac{1}{2}, \frac{1}{4}, \frac{1}{8}, \ldots$$

The nth term can be represented by $a_n = 4 \cdot (0.50)^{n-1}$.

Skill Practice

8. The first swing of a pendulum measures 30°. If each swing that follows is 25% less, list the first three terms of this sequence.

2. Series

In many mathematical applications it is important to find the sum of the terms of a sequence. For example, suppose that the yearly interest earned in an account over a 4-year period is given by the sequence

$$\$250, \quad \$265, \quad \$278.25, \quad \$292.16$$

The sum of the terms gives the total interest earned

$$\$250 + \$265 + \$278.25 + \$292.16 = \$1085.41$$

By adding the terms of a sequence, we obtain a series.

Definition of a Series

The indicated sum of the terms of a sequence is called a **series**.

As with a sequence, a series may be a finite or an infinite sum of terms.

A convenient notation used to denote the sum of a set of terms is called **summation notation**, or sigma notation. The Greek letter Σ (sigma) is used to indicate sums. For example, the sum of the first four terms of the sequence defined by $a_n = n^3$ are denoted by

$$\sum_{n=1}^{4} n^3$$

This is read as "the sum from n equals 1 to 4 of n^3" and is simplified as

$$\sum_{n=1}^{4} n^3 = (1)^3 + (2)^3 + (3)^3 + (4)^3$$
$$= 1 + 8 + 27 + 64$$
$$= 100$$

In this example, the letter n is called the **index of summation**. Many times, the letters i, j, and k are also used for the index of summation.

Answer

8. 30°, 22.5°, 16.875°

Example 5 Finding a Sum from Summation Notation

Find the sum. $\sum_{i=1}^{4} 2^i$

Solution:

$$\sum_{i=1}^{4} 2^i = 2^1 + 2^2 + 2^3 + 2^4$$
$$= 2 + 4 + 8 + 16$$
$$= 30$$

Skill Practice Find the sum.

9. $\sum_{i=1}^{4} i(i+4)$

Example 6 Finding a Sum from Summation Notation

Find the sum. $\sum_{i=1}^{3} (-1)^{i+1} \cdot (3i+4)$

Avoiding Mistakes

The letter i is used as a variable for the index of summation. In this context, it does not represent an imaginary number.

Solution:

$$\sum_{i=1}^{3} (-1)^{i+1} \cdot (3i+4) = (-1)^{1+1} \cdot [3(1)+4] + (-1)^{2+1} \cdot [3(2)+4]$$
$$+ (-1)^{3+1} \cdot [3(3)+4]$$
$$= (-1)^2 \cdot (7) + (-1)^3 \cdot (10) + (-1)^4 \cdot (13)$$
$$= 7 - 10 + 13$$
$$= 10$$

Skill Practice Find the sum.

10. $\sum_{i=1}^{4} (-1)^i (4i-3)$

Example 7 Converting to Summation Notation

Write the series in summation notation.

a. $\dfrac{2}{3} + \dfrac{4}{9} + \dfrac{8}{27} + \dfrac{16}{81}$ Use n as the index of summation.

b. $1 - \dfrac{1}{2} + \dfrac{1}{3} - \dfrac{1}{4} + \dfrac{1}{5} - \dfrac{1}{6}$ Use j as the index of summation.

Answers

9. 70   10. 8

Solution:

a. The sum can be written as

$$\left(\frac{2}{3}\right)^1 + \left(\frac{2}{3}\right)^2 + \left(\frac{2}{3}\right)^3 + \left(\frac{2}{3}\right)^4$$

Taking n from 1 to 4, we have

$$\sum_{n=1}^{4} \left(\frac{2}{3}\right)^n$$

b. The even-numbered terms are negative. The factor $(-1)^{j+1}$ is negative for even values of j. Therefore, the series can be written as

$$\sum_{j=1}^{6} (-1)^{j+1} \cdot \frac{1}{j}$$

Skill Practice Write the series in summation notation.

11. $\dfrac{1}{4} + \dfrac{1}{16} + \dfrac{1}{64} + \dfrac{1}{256}$

12. $\dfrac{1}{2} - \dfrac{1}{4} + \dfrac{1}{6} - \dfrac{1}{8} + \dfrac{1}{10}$

Answers

11. $\sum_{i=1}^{4} \dfrac{1}{4^i}$

12. $\sum_{i=1}^{5} (-1)^{i+1} \dfrac{1}{2i}$

Section 22.2 Activity

A.1. List the terms of the sequence.

    a. $a_n = \dfrac{2}{n+1}$ for $1 \leq n \leq 5$ b. $b_n = (-1)^n \cdot \dfrac{2}{n+1}$ for $1 \leq n \leq 5$

    c. $c_n = (-1)^{n+1} \cdot \dfrac{2}{n+1}$ for $1 \leq n \leq 5$

A.2. Find the value of the sum.

    a. $\sum_{n=1}^{5} \dfrac{2}{n+1}$ b. $\sum_{n=1}^{5} (-1)^n \cdot \dfrac{2}{n+1}$ c. $\sum_{n=1}^{5} (-1)^{n+1} \cdot \dfrac{2}{n+1}$

A.3. Suppose that an Internet video channel has 100 subscribers in its first year. Every year thereafter, the number of subscriptions doubles.

    a. List the number of subscriptions for the first 5 years.

    b. Write a formula representing the nth term of the sequence representing the number of subscriptions s_n by year number, n.

    c. Suppose that each subscriber brings in an average of $10 in advertising revenue per year. Write a formula representing the revenue r_n brought in for year n.

    d. Write a series using summation notation that represents the total revenue brought in for 5 years. Use i as the index of summation.

    e. Write the sum from part (d) in expanded form.

    f. Determine the total revenue brought in for 5 years.

Section 22.2 Practice Exercises

Prerequisite Review

For Exercises R.1–R.6, evaluate the expression for the given value of n.

R.1. $(-1)^n$
  a. $n = 1$
  b. $n = 2$
  c. $n = 3$
  d. $n = 4$
  e. $n = 5$
  f. $n = 6$

R.2. $(-1)^{n+1}$
  a. $n = 1$
  b. $n = 2$
  c. $n = 3$
  d. $n = 4$
  e. $n = 5$
  f. $n = 6$

R.3. $-3 + (n - 1) \cdot 5$ for $n = 11$

R.4. $2 + 5n$ for $n = 21$

R.5. $10\left(\dfrac{2}{5}\right)^{n-1}$ for $n = 4$

R.6. $27\left(\dfrac{1}{3}\right)^{n-1}$ for $n = 5$

For Exercises R.7–R.10, simplify the expression.

R.7. $\dfrac{1}{2} - \dfrac{1}{3} + \dfrac{1}{4} - \dfrac{1}{5} + \dfrac{1}{6}$

R.8. $\dfrac{2}{3} + \dfrac{4}{9} + \dfrac{8}{27} + \dfrac{16}{81}$

R.9. $\dfrac{8!}{6!}$

R.10. $\dfrac{10!}{7!}$

Vocabulary and Key Concepts

1.
  a. A(n) (finite/infinite) sequence is a function whose domain is the set of positive integers. A(n) (finite/infinite) sequence is a function whose domain is the set of the first n positive integers.
  b. Given a sequence, the values a_1, a_2, a_3, and so on are called the _____ of the sequence. The expression a_n defines the _____ term (or general term) of the sequence.
  c. In an _____ sequence, consecutive terms alternate in sign.
  d. A sum of a sequence is called a _____.
  e. To represent a sum of terms, _____ notation is often used and is represented by the Greek letter Σ.
  f. Given the sum $\sum\limits_{n=1}^{5} n^2$, the letter n is called the _____ of summation. In expanded form, this series is $(1)^2 + (2)^2 + $ _____ $= 55$.

Concept 1: Finite and Infinite Sequences

For Exercises 2–6, look at the pattern shown in the sequence and fill in the next three terms.

2. $a_n = -1, 4, 9, 14, 19, \square, \square, \square$

3. $b_n = 7, 10, 13, 16, 19, \square, \square, \square$

4. $c_n = -2, 4, -6, 8, -10, \square, \square, \square$

5. $d_n = 3, -6, 9, -12, 15, \square, \square, \square$

6. $t_n = 1, 4, 9, 16, 25, \square, \square, \square$

For Exercises 7–18, list the terms of each sequence. (See Examples 1–2.)

7. $a_n = 3n + 1$, $1 \leq n \leq 5$

8. $a_n = -2n + 3$, $1 \leq n \leq 5$

9. $a_n = \sqrt{n+2}$, $1 \leq n \leq 4$

10. $a_n = \sqrt{n-1}$, $1 \leq n \leq 4$

11. $a_n = (-1)^n \dfrac{n+1}{n+2}$, $1 \leq n \leq 4$

12. $a_n = (-1)^n \dfrac{n-1}{n+2}$, $1 \leq n \leq 4$

13. $a_n = (-1)^{n+1}(n^2 - 1)$, $1 \leq n \leq 3$

14. $a_n = (-1)^{n+1}(n^2)$, $1 \leq n \leq 3$

15. $a_n = n^2 - n$, $1 \leq n \leq 6$

16. $a_n = n(n^2 - 1)$, $1 \leq n \leq 6$

17. $a_n = (-1)^n 3^n$, $1 \leq n \leq 4$

18. $a_n = (-1)^n n$, $1 \leq n \leq 4$

19. If the nth term of a sequence is $a_n = (-1)^n \cdot n^2$, which terms are positive and which are negative?

20. If the nth term of a sequence is $a_n = (-1)^{n-1} \cdot \frac{1}{n}$, which terms are positive and which are negative?

For Exercises 21–32, find a formula for the nth term of the sequence. Answers may vary. (See Example 3.)

21. 2, 4, 6, 8, . . .

22. 3, 6, 9, 12, . . .

23. 1, 3, 5, 7, . . .

24. 3, 5, 7, 9, . . .

25. $1, \dfrac{1}{4}, \dfrac{1}{9}, \dfrac{1}{16}, \ldots$

26. $\dfrac{1}{2}, \dfrac{2}{3}, \dfrac{3}{4}, \dfrac{4}{5}, \ldots$

27. 1, −1, 1, −1, . . .

28. −1, 1, −1, 1, . . .

29. −2, 4, −8, 16, . . .

30. 3, −9, 27, −81, . . .

31. $\dfrac{3}{5}, \dfrac{3}{25}, \dfrac{3}{125}, \dfrac{3}{625}, \ldots$

32. $\dfrac{1}{4}, \dfrac{1}{16}, \dfrac{1}{64}, \dfrac{1}{256}, \ldots$

33. Edmond borrowed $500. To pay off the loan, he agreed to pay 2% of the balance plus $50 each month. Write a sequence representing the amount Edmond will pay each month for the next 4 months. Round each term to the nearest cent.

34. Janice deposited $1000 in an account that pays 3% interest compounded annually. Write a sequence representing the amount Janice receives in interest each year for the first 4 years. Round each term to the nearest cent.

35. A certain bacteria culture doubles its size each day. If there are 25,000 bacteria on the first day, write a sequence representing the population each day for the first week (7 days). (See Example 4.)

36. A radioactive chemical decays by one-half of its amount each week. If there is 16 g of the chemical in week 1, write a sequence representing the amount present each week for 2 months (8 weeks).

Concept 2: Series

37. What is the difference between a sequence and a series?

38. a. Identify the index of summation for the series. $\sum_{k=1}^{7}(k^2 + 1)$

   b. How many terms does this series have?

For Exercises 39–54, find the sums. (See Examples 5–6.)

39. $\sum_{i=1}^{4}(3i)^2$

40. $\sum_{i=1}^{4}(2i^2)$

41. $\sum_{j=0}^{4}\left(\frac{1}{2}\right)^j$

42. $\sum_{j=0}^{4}\left(\frac{1}{3}\right)^j$

43. $\sum_{i=1}^{6}5$

44. $\sum_{i=1}^{7}3$

45. $\sum_{j=1}^{4}(-1)^j(5j)$

46. $\sum_{j=1}^{4}(-1)^j(4j)$

47. $\sum_{i=1}^{4}\frac{i+1}{i}$

48. $\sum_{i=2}^{5}\frac{i-1}{i}$

49. $\sum_{j=1}^{3}(j+1)(j+2)$

50. $\sum_{j=1}^{3}j(j+2)$

51. $\sum_{k=1}^{7}(-1)^k$

52. $\sum_{k=0}^{5}(-1)^{k+1}$

53. $\sum_{k=1}^{5}k^2$

54. $\sum_{k=1}^{5}2^k$

For Exercises 55–66, write the series in summation notation. (See Example 7.)

55. $1+2+3+4+5+6$

56. $1-2+3-4+5-6$

57. $4+4+4+4+4$

58. $8+8+8+8+8$

59. $4+8+12+16+20$

60. $3+6+9+12+15$

61. $\frac{1}{3}-\frac{1}{9}+\frac{1}{27}-\frac{1}{81}$

62. $\frac{1}{2}-\frac{1}{4}+\frac{1}{8}-\frac{1}{16}$

63. $\frac{5}{11}+\frac{6}{22}+\frac{7}{33}+\frac{8}{44}+\frac{9}{55}+\frac{10}{66}$

64. $-\frac{5}{7}-\frac{10}{8}-\frac{15}{9}-\frac{20}{10}-\frac{25}{11}-\frac{30}{12}-\frac{35}{13}$

65. $x+x^2+x^3+x^4+x^5$

66. $y^2+y^4+y^6+y^8+y^{10}$

Expanding Your Skills

Some sequences are defined by a recursive formula, which defines each term of a sequence in terms of one or more of its preceding terms. For example, if $a_1 = 5$ and $a_n = 2a_{n-1} + 1$ for $n > 1$, then the terms of the sequence are 5, 11, 23, 47, ... In this case, each term after the first is one more than twice the term before it.

For Exercises 67–70, list the first five terms of the sequence.

67. $a_1 = -3, a_n = a_{n-1} + 5$ for $n > 1$

68. $a_1 = 10, a_n = a_{n-1} - 3$ for $n > 1$

69. $a_1 = 5, a_n = 4a_{n-1} + 1$ for $n > 1$

70. $a_1 = -2, a_n = -3a_{n-1} + 4$ for $n > 1$

71. A famous sequence in mathematics is called the Fibonacci sequence, named after the Italian mathematician Leonardo Fibonacci of the thirteenth century. The Fibonacci sequence is defined by

$$a_1 = 1$$
$$a_2 = 1$$
$$a_n = a_{n-1} + a_{n-2} \quad \text{for } n > 2$$

This definition implies that beginning with the third term, each term is the sum of the preceding two terms. Write out the first 10 terms of the Fibonacci sequence.

Arithmetic Sequences and Series

Section 22.3

Concepts
1. Arithmetic Sequences
2. Arithmetic Series

1. Arithmetic Sequences

In this section, we present a special type of sequence called an arithmetic sequence, and then we will study its related sum called an arithmetic series. The sequence 4, 7, 10, 13, 16, ... is an example of an arithmetic sequence. Note the characteristic that each successive term after the first is a fixed value more than the previous term (in this case the terms differ by 3).

> **Definition of an Arithmetic Sequence**
>
> An **arithmetic sequence** is a sequence in which the difference between consecutive terms is constant.

The fixed difference between a term and its predecessor is called the **common difference** and is denoted by the letter d. That is,

$$d = a_{n+1} - a_n$$

Furthermore, if a_1 is the first term, then

$$a_1 = a_1 + 0d$$
$$a_2 = a_1 + 1d \quad \text{is the second term,}$$
$$a_3 = a_1 + 2d \quad \text{is the third term,}$$
$$a_4 = a_1 + 3d \quad \text{is the fourth term, and so on.}$$
$$\ldots$$

In general, $a_n = a_1 + (n-1)d$.

> **nth Term of an Arithmetic Sequence**
>
> The nth term of an arithmetic sequence is given by
>
> $$a_n = a_1 + (n-1)d$$
>
> where a_1 is the first term of the sequence and d is the common difference.

Example 1 — Writing the nth Term of an Arithmetic Sequence

Write the nth term of the sequence. $\quad 9, 2, -5, -12, \ldots$

Solution:

9, 2, −5, −12, ...
  −7 −7 −7

The common difference can be found by subtracting a term from its predecessor: $2 - 9 = -7$

> **TIP:** Notice that the terms of an arithmetic sequence increase or decrease linearly. In Example 1, the terms decrease by 7 for each subsequent term. This is similar to a line with a slope of −7.

FOR REVIEW

Notice the similarity between an arithmetic sequence such as $a_n = -7n + 16$ and a linear function such as $y = -7x + 16$. However, the domain of the sequence is the set of positive integers.

With $a_1 = 9$ and $d = -7$, we have

$$a_n = 9 + (n-1)(-7)$$
$$= 9 - 7n + 7$$
$$= -7n + 16$$

Skill Practice Write the nth term of the sequence.

1. $-3, -1, 1, 3, 5, \ldots$

In Example 1, the common difference between terms is -7. Accordingly, each term of the sequence *decreases* by 7.

The formula $a_n = a_1 + (n-1)d$ contains four variables: a_n, a_1, n, and d. Consequently, if we know the value of three of the four variables, we can solve for the fourth.

Example 2 Finding a Specified Term of an Arithmetic Sequence

Find the ninth term of the arithmetic sequence in which $a_1 = -4$ and $a_{22} = 164$.

Solution:

To find the value of the ninth term a_9, we need to determine the value of d. To find d, substitute $a_1 = -4$, $n = 22$, and $a_{22} = 164$ into the formula $a_n = a_1 + (n-1)d$.

$$a_n = a_1 + (n-1)d$$
$$164 = -4 + (22-1)d$$
$$164 = -4 + 21d$$
$$168 = 21d$$
$$d = 8$$

Therefore, $a_n = -4 + (n-1)(8)$

$$a_9 = -4 + (9-1)(8)$$
$$= -4 + (8)(8)$$
$$= 60$$

Skill Practice

2. Find the tenth term of the arithmetic sequence in which $a_1 = 14$ and $a_{15} = -28$. (*Hint:* First find d.)

Example 3 Finding the Number of Terms in an Arithmetic Sequence

Find the number of terms of the sequence $7, 3, -1, -5, \ldots, -113$.

Answers

1. $2n - 5$ 2. $d = -3$, $a_{10} = -13$

Solution:

To find the number of terms n, we can substitute $a_1 = 7$, $d = -4$, and $a_n = -113$ into the formula for the nth term.

$$a_n = a_1 + (n-1)d$$
$$-113 = 7 + (n-1)(-4)$$
$$-113 = 7 - 4n + 4$$
$$-113 = 11 - 4n$$
$$-124 = -4n$$
$$n = 31$$

Skill Practice Find the number of terms of the sequence.

3. $-15, -11, -7, -3, \ldots, 81$

2. Arithmetic Series

The indicated sum of an arithmetic sequence is called an **arithmetic series**. For example, the series

$$3 + 7 + 11 + 15 + 19 + 23$$

is an arithmetic series because the common difference between terms is constant (4). Adding the terms in a lengthy sum is cumbersome, so we offer the following "shortcut," which is developed here. Let S represent the sum of the terms in the series.

$S = 3 + 7 + 11 + 15 + 19 + 23$ Add the terms in ascending order.
$S = 23 + 19 + 15 + 11 + 7 + 3$ Add the terms in descending order.
$2S = 26 + 26 + 26 + 26 + 26 + 26$ Adding the two series produces six terms of 26.
$2S = 6 \cdot 26$
$S = \dfrac{6 \cdot 26}{2}$
$ = 78$

By adding the terms in ascending and descending order, we double the sum but create a pattern that is easily added. This is true in general. To find the sum, S_n, of the first n terms of the arithmetic series $a_1 + a_2 + a_3 + \cdots + a_n$, we have

$S_n = a_1 + (a_1 + d) + (a_1 + 2d) + \cdots + a_n$ Ascending order
$S_n = a_n + (a_n - d) + (a_n - 2d) + \cdots + a_1$ Descending order
$2S_n = (a_1 + a_n) + (a_1 + a_n) + (a_1 + a_n) + \cdots + (a_1 + a_n)$
$2S_n = n(a_1 + a_n)$
$S_n = \dfrac{n}{2}(a_1 + a_n)$

Answer

3. 25

Sum of an Arithmetic Series

The sum S_n of the first n terms of an arithmetic series is given by

$$S_n = \frac{n}{2}(a_1 + a_n)$$

where a_1 is the first term of the series and a_n is the nth term of the series.

Example 4 — Finding the Sum of an Arithmetic Series

Find the sum of the series. $\displaystyle\sum_{i=1}^{25}(2i+3)$

Solution:

In this series, $n = 25$. Furthermore, $a_1 = 2(1) + 3 = 5$ and $a_{25} = 2(25) + 3 = 53$. Therefore,

$$S_{25} = \frac{25}{2}(5 + 53) \qquad \text{Apply the formula } S_n = \frac{n}{2}(a_1 + a_n).$$

$$= \frac{25}{2}(58) \qquad \text{Simplify.}$$

$$= 725$$

Avoiding Mistakes

When we apply the sum formula $S_n = \dfrac{n}{2}(a_1 + a_n)$ to find the sum $\displaystyle\sum_{i=1}^{n} a_n$, the index of summation i must begin at 1.

Skill Practice Find the sum of the series.

4. $\displaystyle\sum_{i=1}^{20}(3i-2)$

Example 5 — Finding the Sum of an Arithmetic Series

Find the sum of the series.

$$-3 + (-5) + (-7) + \cdots + (-127)$$

Solution:

For this series, $a_1 = -3$ and $a_n = -127$. However, to determine the sum, we also need to find the value of n. The difference between the second term and its predecessor is $-5 - (-3) = -2$. Thus, $d = -2$. We have

$$-127 = -3 + (n-1)(-2) \qquad \text{Apply the formula } a_n = a_1 + (n-1)d.$$
$$-127 = -3 - 2n + 2 \qquad \text{Apply the distributive property.}$$
$$-127 = -1 - 2n \qquad \text{Combine } \textit{like} \text{ terms.}$$
$$-126 = -2n \qquad \text{Solve for } n.$$
$$n = 63$$

Using $n = 63$, $a_1 = -3$, and $a_{63} = -127$, we have

$$S_n = \frac{n}{2}(a_1 + a_n) = \frac{63}{2}[-3 + (-127)] = \frac{63}{2}(-130) = -4095$$

Skill Practice Find the sum of the series.

5. $8 + 3 + (-2) + \cdots + (-137)$

Answers

4. 590   5. -1935

Section 22.3 Activity

A.1. Joyce earns a salary of $55,000 during her first year of work. She receives a raise of $2000 per year thereafter.
   a. Fill in the table representing the sequence for Joyce's salary for year n.

Year	Base Salary Plus Raise(s)	Total Salary
Year 1	$55,000 + 0($2000)	$55,000
Year 2	$55,000 + 1($2000) (Joyce has had 1 raise.)	$57,000
Year 3	$55,000 + 2($2000) (Joyce has had 2 raises.)	$59,000
Year 4		
Year 5		
Year n	$55,000 +	

   b. Use the formula from the last row in the table to determine Joyce's salary in year 25.

A.2. a. The sequence from Exercise A.1 has the characteristic that each successive term differs by a constant, d. (In this case, the constant is the $2000 raise that Joyce earns each year.) This type of sequence is called a(n) _____ sequence.

   b. If a_1 is the first term of an arithmetic sequence, and d is the common difference, use the pattern below to write a formula for the nth term of the sequence.

$$a_1 = a_1 + 0d$$
$$a_2 = a_1 + 1d$$
$$a_3 = a_1 + 2d$$
$$\vdots$$
$$a_n = \underline{\qquad}$$

A.3. Consider the arithmetic sequence

$$17, 11, 5, -1, \ldots, -301$$

   a. What is the first term of the sequence?
   b. What is the common difference?
   c. Write a formula for the nth term of the sequence.
   d. How many terms are there in this sequence?

A.4. Refer to Exercise A.1.
   a. Write a series using summation notation that represents Joyce's total earnings over 25 years.
   b. Find Joyce's total earnings over 25 years.

Practice Exercises

Section 22.3

Prerequisite Review

For Exercises R.1–R.4, solve the equation.

R.1. $62 = 2 + (n - 1) \cdot 3$

R.2. $110 = 5 + (n - 1) \cdot 7$

R.3. $-95 = -15 + 40d$

R.4. $-142 = 14 + 52d$

For Exercises R.5–R.6, find the function values.

R.5. Find the function value for the function defined by $f(x) = 4x - 3$.
  a. $f(1)$ **b.** $f(2)$ **c.** $f(3)$ **d.** $f(4)$

R.6. Find the function value for the function defined by $g(x) = -2x + 5$.
  a. $g(1)$ **b.** $g(2)$ **c.** $g(3)$ **d.** $g(4)$

For Exercises R.7–R.8, write the sum in expanded form and then simplify.

R.7. $\sum_{n=1}^{5}(1 - 4n)$ **R.8.** $\sum_{n=1}^{6}(5n - 3)$

Vocabulary and Key Concepts

1. **a.** An _____ sequence is a sequence in which the difference between consecutive terms is constant.

   b. The common _____ between a term and its predecessor in an arithmetic sequence is often denoted by the letter d.

   c. The nth term of an arithmetic sequence is given by $a_n =$ _____, where a_n is the nth term, a_1 is the first term, n is the number of terms, and _____ is the common difference between terms.

   d. An indicated sum of an arithmetic sequence is called an arithmetic _____.

   e. The sum S_n of the first n terms of an arithmetic series is given by $S_n =$ _____, where a_1 is the first term and a_n is the nth term.

Concept 1: Arithmetic Sequences

2. Explain how to determine if a sequence is arithmetic.

For Exercises 3–6, write the first four terms of the sequence.

3. $a_n = -5 + 4n$

4. $b_n = 8 + (-5)n$

5. $c_n = \frac{1}{2} + (n-1)\frac{3}{2}$

6. $d_n = \frac{1}{4} + (n-1)\frac{3}{4}$

For Exercises 7–12, the first term of an arithmetic sequence is given along with its common difference. Write the first five terms of the sequence.

7. $a_1 = 3, d = 8$

8. $a_1 = -5, d = 2$

9. $a_1 = 80, d = -20$

10. $a_1 = -4, d = -5$

11. $a_1 = 3, d = \frac{3}{4}$

12. $a_1 = 1, d = \frac{2}{3}$

For Exercises 13–18, find the common difference d for each arithmetic sequence.

13. $1, 3, 5, 7, 9, \ldots$

14. $2, 8, 14, 20, 26, \ldots$

15. $6, 3, 0, -3, -6, \ldots$

16. $8, 3, -2, -7, -12, \ldots$

17. $-7, -9, -11, -13, -15, \ldots$

18. $-15, -11, -7, -3, 1, \ldots$

For Exercises 19–24, write the first five terms of the arithmetic sequence.

19. $a_1 = 3, d = 5$

20. $a_1 = -3, d = 2$

21. $a_1 = 2, d = \frac{1}{2}$

22. $a_1 = 0, d = \frac{1}{3}$

23. $a_1 = 2, d = -4$

24. $a_1 = 10, d = -6$

For Exercises 25–33, write the nth term of the sequence. (See Example 1.)

25. $0, 5, 10, 15, 20, \ldots$

26. $7, 12, 17, 22, 27, \ldots$

27. $-2, -4, -6, -8, -10, \ldots$

28. $1, -3, -7, -11, -15, \ldots$

29. $2, \dfrac{5}{2}, 3, \dfrac{7}{2}, 4, \ldots$

30. $1, \dfrac{4}{3}, \dfrac{5}{3}, 2, \dfrac{7}{3}, \ldots$

31. $21, 17, 13, 9, 5, \ldots$

32. $9, 6, 3, 0, -3, \ldots$

33. $-8, -2, 4, 10, 16, \ldots$

For Exercises 34–41, find the indicated term of each arithmetic sequence. (See Example 2.)

34. Find the eighth term given $a_1 = -6$ and $d = -3$.

35. Find the sixth term given $a_1 = -3$ and $d = 4$.

36. Find the 12th term given $a_1 = -8$ and $d = -2$.

37. Find the ninth term given $a_1 = -1$ and $d = 6$.

38. Find the tenth term given $a_1 = 1$ and $a_7 = 31$.

39. Find the seventh term given $a_1 = 0$ and $a_{10} = -45$.

40. Find the sixth term given $a_1 = 3$ and $a_{13} = 39$.

41. Find the 11th term given $a_1 = 12$ and $a_6 = -18$.

For Exercises 42–49, find the number of terms, n, of each arithmetic sequence. (See Example 3.)

42. $2, 0, -2, \ldots, -56$

43. $8, 13, 18, \ldots, 98$

44. $1, -3, -7, \ldots, -67$

45. $1, 5, 9, \ldots, 85$

46. $1, \dfrac{3}{4}, \dfrac{1}{2}, \ldots, -4$

47. $2, \dfrac{5}{2}, 3, \ldots, 13$

48. $-\dfrac{5}{3}, -1, -\dfrac{1}{3}, \ldots, \dfrac{37}{3}$

49. $\dfrac{13}{3}, \dfrac{19}{3}, \dfrac{25}{3}, \ldots, \dfrac{73}{3}$

50. If the third and fourth terms of an arithmetic sequence are 18 and 21, what are the first and second terms?

51. If the third and fourth terms of an arithmetic sequence are -8 and -11, what are the first and second terms?

Concept 2: Arithmetic Series

52. Explain the difference between an arithmetic sequence and an arithmetic series.

For Exercises 53–66, find the sum of the arithmetic series. (See Examples 4–5.)

53. $\displaystyle\sum_{i=1}^{20} (3i + 2)$

54. $\displaystyle\sum_{i=1}^{15} (2i - 3)$

55. $\displaystyle\sum_{i=1}^{20} (i + 4)$

56. $\displaystyle\sum_{i=1}^{25} (i - 3)$

57. $\displaystyle\sum_{j=1}^{10} (4 - j)$

58. $\displaystyle\sum_{j=1}^{10} (6 - j)$

59. $\displaystyle\sum_{j=1}^{15} \left(\dfrac{2}{3}j + 1\right)$

60. $\displaystyle\sum_{j=1}^{15} \left(\dfrac{1}{2}j - 2\right)$

61. $4 + 8 + 12 + \cdots + 84$

62. $4 + 9 + 14 + \cdots + 49$

63. $6 + 8 + 10 + \cdots + 34$

64. $-4 + (-3) + (-2) + \cdots + 12$

65. $-3 + (-7) + (-11) + \cdots + (-39)$

66. $2 + 5 + 8 + \cdots + 53$

67. Find the sum of the first 100 positive integers.

68. Find the sum of the first 50 positive even integers.

69. The seating in a certain theater is arranged so that there are 30 seats in row 1, 32 in row 2, 34 in row 3, and so on. If there are 20 rows, how many total seats are there? What is the total revenue if the average ticket price is $15 per seat and the theater is sold out?

70. A triangular array of dominoes has one domino in the first row, two dominoes in the second row, three dominoes in the third row, and so on. If there are 15 rows, how many dominoes are in the array?

Section 22.4 Geometric Sequences and Series

Concepts

1. Geometric Sequences
2. Geometric Series

1. Geometric Sequences

The sequence 2, 4, 8, 16, 32, . . . is not an arithmetic sequence because the difference between terms is not constant. However, a different pattern exists. Notice that each term after the first is 2 times the preceding term. This sequence is called a geometric sequence.

> **Definition of a Geometric Sequence**
>
> A **geometric sequence** is a sequence in which the ratio between each term and its predecessor is constant.

The constant quotient of a term and its predecessor is called the **common ratio** and is denoted by r. The common ratio is found by dividing a term by the preceding term. That is,

$$r = \frac{a_{n+1}}{a_n}$$

For the sequence 2, 4, 8, 16, 32, . . . we have $r = \frac{4}{2} = \frac{8}{4} = \frac{16}{8} = \frac{32}{16} = 2$.

If a_1 denotes the first term of a geometric sequence, then

$a_2 = a_1 r$ is the second term,

$a_3 = a_1 r^2$ is the third term,

$a_4 = a_1 r^3$ is the fourth term, and so on.

. . .

This pattern gives $a_n = a_1 r^{n-1}$.

> **nth Term of a Geometric Sequence**
>
> The nth term of a geometric sequence is given by
>
> $$a_n = a_1 r^{n-1}$$
>
> where a_1 is the first term and r is the common ratio.

Example 1   Finding the nth Term of a Geometric Sequence

Find the nth term of the sequence.

   a. $-1, 4, -16, 64, \ldots$ **b.** $12, 8, \frac{16}{3}, \frac{32}{9}, \ldots$

Solution:

a. The common ratio is found by dividing any term (after the first) by its predecessor.

$$r = \frac{4}{-1} = -4$$

With $r = -4$ and $a_1 = -1$, we have $a_n = -1(-4)^{n-1}$.

b. The common ratio is $r = \frac{8}{12} = \frac{2}{3}$. With $a_1 = 12$ and $r = \frac{2}{3}$, we have

$$a_n = 12\left(\frac{2}{3}\right)^{n-1}$$

Skill Practice Find the nth term of the geometric sequence.

1. $6, 12, 24, 48, \ldots$ 2. $6, 2, \frac{2}{3}, \frac{2}{9}, \ldots$

The formula $a_n = a_1 r^{n-1}$ contains the variables a_n, a_1, n, and r. If we know the value of three of the four variables, we can find the fourth.

Example 2   Finding a Specified Term of a Geometric Sequence

Given $a_n = 6\left(\frac{1}{2}\right)^{n-1}$, find a_5.

Solution:

$$a_n = 6\left(\frac{1}{2}\right)^{n-1}$$

$$a_5 = 6\left(\frac{1}{2}\right)^{5-1} = 6\left(\frac{1}{2}\right)^4 = 6\left(\frac{1}{16}\right) = \frac{3}{8}$$

Skill Practice

3. Given $a_n = 3(-2)^{n-1}$, find a_6.

Example 3   Finding a Specified Term of a Geometric Sequence

Find the first term of the geometric sequence where $a_5 = -162$ and $r = 3$.

Solution:

$-162 = a_1(3)^{5-1}$ Substitute $a_5 = -162$, $n = 5$, and $r = 3$ into the formula $a_n = a_1 r^{n-1}$.

$-162 = a_1(3)^4$ Simplify and solve for a_1.

$-162 = a_1(81)$

$a_1 = -2$

Skill Practice

4. Find the first term of the geometric sequence where $a_4 = \frac{25}{32}$ and $r = \frac{1}{4}$.

2. Geometric Series

The indicated sum of a geometric sequence is called a **geometric series**. For example,

$$1 + 3 + 9 + 27 + 81 + 243$$

is a geometric series. To find the sum, consider the following procedure. Let S represent the sum

$$S = 1 + 3 + 9 + 27 + 81 + 243$$

Answers

1. $a_n = 6(2)^{n-1}$ 2. $a_n = 6\left(\frac{1}{3}\right)^{n-1}$
3. -96 4. 50

Now multiply S by the common ratio, which in this case is 3.

$$3S = 3 + 9 + 27 + 81 + 243 + 729$$

Then

$$3S - S = (3 + 9 + 27 + 81 + 243 + 729) - (1 + 3 + 9 + 27 + 81 + 243)$$

$$2S = 3 + 9 + 27 + 81 + 243 + 729 - 1 - 3 - 9 - 27 - 81 - 243$$

$$2S = 729 - 1 \quad \text{The terms in red form a sum of zero.}$$

$$2S = 728$$

$$S = 364$$

A similar procedure can be used to find the sum S_n of the first n terms of any geometric series. Subtract rS_n from S_n.

$$S_n = a_1 + a_1 r + a_1 r^2 + \cdots + a_1 r^{n-1}$$

$$rS_n = a_1 r + a_1 r^2 + a_1 r^3 + \cdots + a_1 r^n$$

$$S_n - rS_n = (a_1 - a_1 r) + (a_1 r - a_1 r^2) + (a_1 r^2 - a_1 r^3) + \cdots + (a_1 r^{n-1} - a_1 r^n)$$

$$S_n - rS_n = a_1 - a_1 r^n \quad \text{The terms in red form a sum of zero.}$$

$$S_n(1 - r) = a_1(1 - r^n) \quad \text{Factor each side of the equation.}$$

$$S_n = \frac{a_1(1 - r^n)}{1 - r} \quad \text{Divide by } (1 - r).$$

Sum of a Geometric Series

The sum, S_n, of the first n terms of a geometric series $\sum_{i=1}^{n} a_1 r^{i-1}$ is given by

$$S_n = \frac{a_1(1 - r^n)}{1 - r}$$

where a_1 is the first term of the series, r is the common ratio, and $r \neq 1$.

Example 4 Finding the Sum of a Geometric Series

Find the sum of the series. $\sum_{i=1}^{6} 4\left(\frac{1}{2}\right)^{i-1}$

Solution:

By expanding the terms of this series, we see that the series is geometric.

$$\sum_{i=1}^{6} 4\left(\frac{1}{2}\right)^{i-1} = 4 + 2 + 1 + \frac{1}{2} + \frac{1}{4} + \frac{1}{8}$$

We have: $a_1 = 4$, $r = \frac{1}{2}$, and $n = 6$,

$$S_n = \frac{a_1(1 - r^n)}{1 - r} = \frac{4[1 - (\frac{1}{2})^6]}{1 - \frac{1}{2}} = \frac{4(1 - \frac{1}{64})}{\frac{1}{2}} = 8\left(\frac{63}{64}\right) = \frac{63}{8}$$

Skill Practice Find the sum of the geometric series.

5. $\frac{1}{9} + \frac{1}{3} + 1 + 3 + 9 + 27$

Answer

5. $\dfrac{364}{9}$

Example 5 Finding the Sum of a Geometric Series

Find the sum of the series $5 + 10 + 20 + \cdots + 5120$.

Solution:

The common ratio is 2 and $a_1 = 5$. The nth term of the sequence can be written as $a_n = 5(2)^{n-1}$. To find the value of n, substitute 5120 for a_n.

$$5120 = 5(2)^{n-1}$$

$$\frac{5120}{5} = \frac{5(2)^{n-1}}{5} \qquad \text{Divide both sides by 5.}$$

$$1024 = 2^{n-1} \qquad \text{To solve the exponential equation, write each side as}$$
$$2^{10} = 2^{n-1} \qquad \text{a power of 2.}$$

$$10 = n - 1 \qquad \text{Recall that if } b^x = b^y, \text{ then } x = y.$$

$$n = 11$$

With $a_1 = 5$, $r = 2$, and $n = 11$, we have

$$S_n = \frac{a_1(1 - r^n)}{1 - r} = \frac{5(1 - 2^{11})}{1 - 2} = \frac{5(1 - 2048)}{-1} = \frac{5(-2047)}{-1} = 10{,}235$$

Skill Practice Find the sum of the geometric series.

6. $3 + 6 + 12 + \cdots + 768$

Consider a geometric series where $|r| < 1$. For increasing values of n, r^n decreases. For example,

$$\left(\frac{1}{2}\right)^5 = 0.03125 \qquad \left(\frac{1}{2}\right)^{10} \approx 0.00097656 \qquad \left(\frac{1}{2}\right)^{15} \approx 0.00003052$$

For $|r| < 1$, r^n approaches 0 as n gets larger and larger. As n approaches infinity, the sum

$$S = \frac{a_1(1 - r^n)}{1 - r} \quad \text{approaches} \quad \frac{a_1(1 - 0)}{1 - r} = \frac{a_1}{1 - r}$$

Sum of an Infinite Geometric Series

Given an infinite geometric series $a_1 + a_1 r + a_1 r^2 + \cdots$, with $|r| < 1$, the sum S of all terms in the series is given by

$$S = \frac{a_1}{1 - r}$$

Note: If $|r| \geq 1$, then the sum does not exist.

Example 6 Finding the Sum of an Infinite Geometric Series

Find the sum of the series. $\sum_{i=1}^{\infty} \left(\frac{1}{3}\right)^{i-1}$

Solution:

In this example, the upper limit of summation is ∞. This indicates that we have an infinite series where the pattern for the sum of terms continues indefinitely.

$$\sum_{i=1}^{\infty} \left(\frac{1}{3}\right)^{i-1} = 1 + \frac{1}{3} + \frac{1}{9} + \frac{1}{27} + \frac{1}{81} + \cdots$$

Answer

6. 1533

The series is geometric with $a_1 = 1$ and $r = \frac{1}{3}$. Because $|r| = |\frac{1}{3}| < 1$, we have

$$S = \frac{a_1}{1-r} = \frac{1}{1-\frac{1}{3}} = \frac{1}{\frac{2}{3}} = \frac{3}{2}$$

The sum is $\frac{3}{2}$.

Skill Practice Find the sum of the series.

7. $\sum_{i=1}^{\infty} 4\left(\frac{3}{4}\right)^{i-1}$

Example 7 Using a Geometric Series in a Physics Application

A child drops a ball from a height of 4 ft. With each bounce, the ball rebounds to 50% of its original height. Determine the total distance traveled by the ball.

Solution:

The heights (in ft) from which the ball drops are given by the sequence

$$4, 2, 1, \frac{1}{2}, \frac{1}{4}, \ldots$$

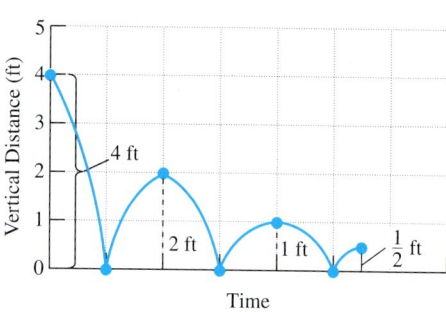

After the ball falls from its initial height of 4 ft, the distance traveled for every bounce thereafter is doubled (the ball travels up and down). Therefore, the total distance traveled is given by the series

$$4 + 2 \cdot 2 + 2 \cdot 1 + 2 \cdot \frac{1}{2} + 2 \cdot \frac{1}{4} + \cdots$$

or equivalently

$$4 + 4 + 2 + 1 + \frac{1}{2} + \cdots$$

The series $4 + 2 + 1 + \frac{1}{2} + \cdots$ is an infinite geometric series with $a_1 = 4$ and $r = \frac{1}{2}$.

Initial height from which the ball was dropped → 4

Infinite geometric series → $4 + 2 + 1 + \frac{1}{2} + \cdots$

$$= 4 + \frac{4}{1 - \frac{1}{2}}$$

$$= 4 + \frac{4}{\frac{1}{2}}$$

$$= 4 + 8 = 12$$

The ball traveled a total of 12 ft.

Skill Practice

8. A child drops a ball from a height of 3 ft. With each bounce, the ball rebounds to $\frac{1}{3}$ of its original height. Determine the total distance traveled by the ball.

Answers

7. 16   8. 6 ft

Section 22.4 Activity

A.1. Kevin earns a salary of $50,000 during his first year of work. He receives a raise of 4% per year thereafter.

a. Fill in the table representing the sequence for Kevin's salary for year n.

Year	Salary		Simplified
Year 1	$50,000(1.04)^0$		$50,000
Year 2	$50,000(1.04)^1$	(Kevin has 1 raise of 4% of $50,000 or equivalently 104% of his first year's salary.)	$52,000
Year 3	$50,000(1.04)^2$	(Kevin has 2 raises of 4%. This is 104% of 104% of his first year's salary.)	$54,080
Year 4	$50,000(1.04)^3$		
Year 5	$50,000(1.04)^4$		
Year n	$50,000(1.04)^{n-1}$		

b. Use the formula from the last row in the table to determine Kevin's salary in year 25.

c. Refer to Exercise A.1 of Section 22.3 Activity. Joyce received a salary of $55,000 in her first year and then a raise of $2000 each year thereafter. Kevin makes less money initially ($50,000), and his first raise is for $2000. But Kevin's yearly salary eventually overtakes Joyce's salary. Explain why.

d. Write a series using summation notation that represents Kevin's total earnings over 25 years.

e. The series in part (d) is a finite geometric series of the form $\sum_{i=1}^{n} a_1 r^{i-1}$. The sum can be evaluated by using the formula $\sum_{i=1}^{n} a_1 r^{i-1} = \frac{a_1(1 - r^n)}{1 - r}$. Find Kevin's total earnings over 25 years.

f. From Exercise A.4(b) of Section 22.3 Activity, you computed Joyce's total earnings over 25 years to be $\sum_{i=1}^{25}(53{,}000 + 2000i) = 1{,}975{,}000$. How much more money does Kevin make in 25 years?

For Exercises A.2–A.4, refer to the sequences a_n, b_n, and c_n.

$$a_n = 16, 8, 4, 2, \ldots, \tfrac{1}{16} \qquad b_n = 16, 8, 0, -8, \ldots, -360 \qquad c_n = 16, 8, 1, -5, \ldots, -20$$

A.2. a. Is sequence a_n arithmetic, geometric, or neither? Why?

b. Write a formula for the nth term of a_n.

c. Find the eighth term of a_n.

d. How many terms does the sequence a_n have?

A.3. a. Is sequence b_n arithmetic, geometric, or neither?

b. Write a formula for the nth term of b_n.

c. Find the 12th term of b_n.

d. How many terms does the sequence b_n have?

A.4. Is sequence c_n arithmetic, geometric, or neither?

For Exercises A.5–A.8, find the sum if possible. If the sum does not exist, explain why.

A.5. $\sum_{i=1}^{9} 16\left(\frac{1}{2}\right)^{i-1}$ **A.6.** $\sum_{i=1}^{\infty} 16\left(\frac{1}{2}\right)^{i-1}$ **A.7.** $\sum_{i=1}^{9} 16(2)^{i-1}$

A.8. $\sum_{i=1}^{\infty} 16(2)^{i-1}$

Section 22.4 Practice Exercises

Prerequisite Review

For Exercises R.1–R.2, find the function values.

R.1. Find the function value for the function defined by $f(x) = 50\left(\dfrac{2}{5}\right)^x$.

    a. $f(1)$ **b.** $f(2)$ **c.** $f(3)$ **d.** $f(4)$

R.2. Find the function value for the function defined by $g(x) = 8\left(\dfrac{1}{2}\right)^x$.

    a. $g(1)$ **b.** $g(2)$ **c.** $g(3)$ **d.** $g(4)$

For Exercises R.3–R.8, simplify the expression.

R.3. $\dfrac{16}{1 - \dfrac{1}{2}}$

R.4. $\dfrac{12}{1 - \dfrac{2}{3}}$

R.5. $\dfrac{10(1 - 2^4)}{1 - 2}$

R.6. $\dfrac{24(1 - 3^3)}{1 - 3}$

R.7. $\dfrac{16\left[1 - \left(\dfrac{1}{2}\right)^3\right]}{1 - \dfrac{1}{2}}$

R.8. $\dfrac{27\left[1 - \left(\dfrac{2}{3}\right)^3\right]}{1 - \dfrac{2}{3}}$

Vocabulary and Key Concepts

1. a. A _____ sequence is a sequence in which the ratio between each term and its predecessor is constant.

  b. The common _____ between a term and its predecessor in a geometric sequence is often denoted by r.

  c. The nth term of a geometric sequence is given by $a_n = $ _____, where a_n is the nth term in the sequence, a_1 is the first term, and r is the common ratio.

  d. The sum S_n of the first n terms of a geometric series $\sum_{i=1}^{n} a_1 r^{i-1}$ is given by $S_n = $ _____, where a_1 is the first term and r is the common ratio.

  e. Given an infinite geometric series $a_1 + a_2 r + a_1 r^2 + \cdots$ with $|r| < 1$, the sum of all terms in the sequence is given by $S = $ _____. If $|r| \geq $ _____, then the sum does not exist.

Concept 1: Geometric Sequences

2. Explain how to determine if a sequence is geometric.

For Exercises 3–6, write the first four terms of the sequence.

3. $a_n = 4 \cdot 2^{n-1}$

4. $b_n = \dfrac{1}{3} \cdot 3^{n-1}$

5. $c_n = (-5)^{n-1}$

6. $d_n = (-1)^{n+1} 4^{n-1}$

For Exercises 7–12, the first term of a geometric sequence is given along with the common ratio. Write the first four terms of the sequence.

7. $a_1 = 1$; $r = 10$

8. $a_1 = 2$; $r = 5$

9. $a_1 = 64$; $r = \dfrac{1}{2}$

10. $a_1 = 3$; $r = \dfrac{2}{3}$

11. $a_1 = 8$; $r = -\dfrac{1}{4}$

12. $a_1 = 9$; $r = -\dfrac{1}{3}$

For Exercises 13–18, determine the common ratio, r, for the geometric sequence.

13. 5, 10, 20, 40, ...

14. $-2, -1, -\frac{1}{2}, -\frac{1}{4}, \ldots$

15. $8, -2, \frac{1}{2}, -\frac{1}{8}, \ldots$

16. 4, −12, 36, −108, ...

17. 3, −6, 12, −24, ...

18. 1, 4, 16, 64, ...

For Exercises 19–24, write the first five terms of the geometric sequence.

19. $a_1 = -3, r = -2$

20. $a_1 = -4, r = -1$

21. $a_1 = 6, r = \frac{1}{2}$

22. $a_1 = 8, r = \frac{1}{4}$

23. $a_1 = -1, r = 6$

24. $a_1 = 2, r = -3$

For Exercises 25–30, find the nth term of each geometric sequence. **(See Example 1.)**

25. 3, 12, 48, 192, ...

26. 2, 6, 18, 54, ...

27. −5, 15, −45, 135, ...

28. −6, 12, −24, 48, ...

29. $\frac{1}{2}, 2, 8, 32, \ldots$

30. $\frac{16}{3}, 4, 3, \frac{9}{4}, \ldots$

For Exercises 31–40, find the indicated term of each geometric sequence. **(See Examples 2–3.)**

31. Given $a_n = 2(\frac{1}{2})^{n-1}$, find a_8.

32. Given $a_n = -3(\frac{1}{2})^{n-1}$, find a_8.

33. Given $a_n = 4(-\frac{3}{2})^{n-1}$, find a_6.

34. Given $a_n = 6(-\frac{1}{3})^{n-1}$, find a_6.

35. Given $a_n = -3(2)^{n-1}$, find a_5.

36. Given $a_n = 5(3)^{n-1}$, find a_4.

37. Given $a_5 = -\frac{16}{9}$ and $r = -\frac{2}{3}$, find a_1.

38. Given $a_6 = \frac{5}{16}$ and $r = -\frac{1}{2}$, find a_1.

39. Given $a_7 = 8$ and $r = 2$, find a_1.

40. Given $a_6 = 27$ and $r = 3$, find a_1.

41. If the second and third terms of a geometric sequence are 16 and 64, what is the first term?

42. If the second and third terms of a geometric sequence are $\frac{1}{3}$ and $\frac{1}{9}$, what is the first term?

Concept 2: Geometric Series

43. Explain the difference between a geometric sequence and a geometric series.

44. **a.** Write the series in expanded form. $\sum_{i=1}^{5} 4(2)^{i-1}$

  b. Find the sum. $\sum_{i=1}^{5} 4(2)^{i-1}$

45. **a.** Write the series in expanded form. $\sum_{i=1}^{4} 3(4)^{i-1}$

  b. Find the sum. $\sum_{i=1}^{4} 3(4)^{i-1}$

46. **a.** Write the series in expanded form. $\sum_{i=1}^{4} 6\left(\frac{1}{2}\right)^{i-1}$

  b. Find the sum. $\sum_{i=1}^{4} 6\left(\frac{1}{2}\right)^{i-1}$

For Exercises 47–56, find the sum of the geometric series. (See Examples 4–5.)

47. $10 + 2 + \dfrac{2}{5} + \dfrac{2}{25} + \dfrac{2}{125}$

48. $1 + 3 + 9 + 27 + 81 + 243$

49. $-2 + 1 + \left(-\dfrac{1}{2}\right) + \dfrac{1}{4} + \left(-\dfrac{1}{8}\right)$

50. $\dfrac{1}{4} + (-1) + 4 + (-16) + 64$

51. $12 + 16 + \dfrac{64}{3} + \dfrac{256}{9} + \dfrac{1024}{27}$

52. $9 + 6 + 4 + \dfrac{8}{3} + \dfrac{16}{9}$

53. $1 + \dfrac{2}{3} + \dfrac{4}{9} + \cdots + \dfrac{32}{243}$

54. $\dfrac{8}{3} + 2 + \dfrac{3}{2} + \cdots + \dfrac{243}{512}$

55. $-4 + 8 + (-16) + \cdots + (-256)$

56. $-\dfrac{7}{3} + 7 + (-21) + \cdots + 5103$

57. A deposit of $1000 is made in an account that earns 5% interest compounded annually. The balance in the account after n years is given by

$$a_n = 1000(1.05)^n \quad \text{for } n \geq 1$$

   a. List the first four terms of the sequence. Round to the nearest cent.

   b. Find the balance after 10 years, 20 years, and 40 years by computing a_{10}, a_{20}, and a_{40}. Round to the nearest cent.

58. A home purchased for $125,000 increases by 4% of its value each year. The value of the home after n years is given by

$$a_n = 125{,}000(1.04)^n \quad \text{for } n \geq 1$$

   a. List the first four terms of the sequence. Round to the nearest dollar.

   b. Find the value of the home after 5 years, 10 years, and 20 years by computing a_5, a_{10}, and a_{20}. Round to the nearest dollar.

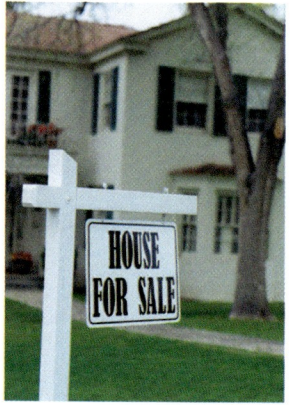

Steve Mason/Photodisc/Getty Images

For Exercises 59–64, first find the common ratio r. Then determine the sum of the infinite series, if it exists. (See Example 6.)

59. $1 + \dfrac{1}{6} + \dfrac{1}{36} + \dfrac{1}{216} + \cdots$

60. $-2 + \left(-\dfrac{1}{2}\right) + \left(-\dfrac{1}{8}\right) + \left(-\dfrac{1}{32}\right) + \cdots$

61. $\displaystyle\sum_{i=1}^{\infty} \left(-\dfrac{1}{4}\right)^{i-1}$

62. $\displaystyle\sum_{i=1}^{\infty} \left(-\dfrac{1}{5}\right)^{i-1}$

63. $\dfrac{2}{3} + (-1) + \dfrac{3}{2} + \left(-\dfrac{9}{4}\right) + \cdots$

64. $3 + 5 + \dfrac{25}{3} + \dfrac{125}{9} + \cdots$

65. Suppose $200 million is spent by tourists at a certain resort town. Further suppose that 75% of the revenue is respent in the community and then respent over and over, each time at a rate of 75%. The series

$$200 + 200(0.75) + 200(0.75)^2 + 200(0.75)^3 + \cdots$$

gives the total amount spent (and respent) in the community. Find the sum of the infinite series. **(See Example 7.)**

66. A bungee jumper jumps off a platform and stretches the cord 80 ft before rebounding upward. Each successive bounce stretches the cord 60% of its previous length. The total vertical distance traveled is given by

$$80 + 2(0.60)(80) + 2(0.60)^2(80) + 2(0.60)^3(80) + \cdots$$

After the first term, the series is an infinite geometric series. Compute the total vertical distance traveled.

67. A ball drops from a height of 4 ft. With each bounce, the ball rebounds to $\frac{3}{4}$ of its height. The total vertical distance traveled is given by

$$4 + 2\left(\frac{3}{4}\right)(4) + 2\left(\frac{3}{4}\right)^2(4) + 2\left(\frac{3}{4}\right)^3(4) + \cdots$$

After the first term, the series is an infinite geometric series. Compute the total vertical distance traveled.

68. The repeating decimal number $0.\overline{2}$ can be written as an infinite geometric series by

$$\frac{2}{10} + \frac{2}{100} + \frac{2}{1000} + \cdots$$

   a. What is a_1?   b. What is r?   c. Find the sum of the series.

69. The repeating decimal number $0.\overline{7}$ can be written as an infinite geometric series by

$$\frac{7}{10} + \frac{7}{100} + \frac{7}{1000} + \cdots$$

   a. What is a_1?   b. What is r?   c. Find the sum of the series.

Expanding Your Skills

70. The yearly salary for Job A is $40,000 initially with an annual raise of $3000 per year. The yearly salary for Job B is $38,000 initially with an annual raise of 6% per year.

    a. Find the total earnings for Job A over 20 years. Is this an arithmetic or geometric series?
    b. For Job B, what is the amount of the raise after 1 year?
    c. Find the total earnings for Job B over 20 years. Round to the nearest dollar. Is this an arithmetic or geometric series?
    d. What is the difference in total salary earned over 20 years between Job A and Job B?

71. a. Brook has a job that pays $48,000 the first year. She receives a 4% raise each year. Find the sum of her yearly salaries over a 20-year period. Round to the nearest dollar.
    b. Chamille has a job that pays $48,000 the first year. She receives a 4.5% raise each year. Find the sum of her yearly salaries over a 20-year period. Round to the nearest dollar.
    c. Chamille's raise each year was 0.5% higher than Brook's raise. How much more total income did Chamille receive than Brook over 20 years?

Problem Recognition Exercises

Identifying Arithmetic and Geometric Sequences

For Exercises 1–18, determine if the sequence is arithmetic, geometric, or neither. If the sequence is arithmetic, find d. If the sequence is geometric, find r.

1. $5, -\dfrac{5}{2}, \dfrac{5}{4}, -\dfrac{5}{8}, \ldots$

2. $\dfrac{1}{2}, 1, \dfrac{3}{2}, 2, \dfrac{5}{2}, \ldots$

3. $-2, -4, -8, -16, -32, \ldots$

4. $-2, -4, -7, -11, -16, \ldots$

5. $-\dfrac{1}{3}, \dfrac{1}{3}, 1, \dfrac{5}{3}, \ldots$

6. $1, -\dfrac{3}{2}, \dfrac{9}{4}, -\dfrac{27}{8}, \ldots$

7. $2, 6, 11, 17, 24, \ldots$

8. $2, -2, 2, -2, 2, \ldots$

9. $-2, -4, -6, -8, -10, \ldots$

10. $1, 3, 1, 3, 1, \ldots$

11. $0, 1, 0, 1, 0, 1, \ldots$

12. $2, 6, 10, 14, 18, 22, \ldots$

13. $\dfrac{1}{4}\pi, \dfrac{1}{2}\pi, \dfrac{3}{4}\pi, \pi, \dfrac{5}{4}\pi, \ldots$

14. $-1, 1, -1, 1, -1, \ldots$

15. $1, 4, 9, 16, 25, \ldots$

16. $\dfrac{1}{3}e, \dfrac{2}{3}e, e, \dfrac{4}{3}e, \dfrac{5}{3}e, \ldots$

17. $\dfrac{\pi}{2}, \dfrac{\pi}{3}, \dfrac{\pi}{4}, \dfrac{\pi}{5}, \dfrac{\pi}{6}, \ldots$

18. $2, 6, 18, 54, 162, \ldots$

Chapter 22 Review Exercises

Section 22.1

For Exercises 1–4, simplify the expression.

1. $8!$
2. $5!$
3. $\dfrac{12!}{10! \cdot 2!}$
4. $\dfrac{9!}{6! \cdot 3!}$

For Exercises 5–6, expand the binomial.

5. $(x^2 + 4)^5$

6. $(c - 3d)^4$

7. Find the first three terms of the binomial expansion. $(a - 2b)^{11}$

8. Find the fourth term of the binomial expansion. $(x^2 + 3y)^8$

9. Find the eighth term of the binomial expansion. $(5x - y^3)^{10}$

10. Find the middle term of the binomial expansion. $(a + 2b)^6$

Section 22.2

For Exercises 11–14, write the terms of the sequence.

11. $a_n = -3n + 4;\ 1 \le n \le 5$

12. $a_n = -2n^3;\ 1 \le n \le 3$

13. $a_n = (-1)^{n+1} \cdot \dfrac{n}{n+2};\ 1 \le n \le 4$

14. $a_n = \left(-\dfrac{2}{3}\right)^n;\ 1 \le n \le 4$

For Exercises 15–16, write a formula for the nth term of the sequence.

15. $\dfrac{1}{2}, \dfrac{2}{3}, \dfrac{3}{4}, \dfrac{4}{5}, \ldots$

16. $-\dfrac{3}{2}, \dfrac{3}{4}, -\dfrac{3}{8}, \dfrac{3}{16}, \ldots$

17. What is the index of summation for the series?

$$\sum_{k=1}^{7}(2n + 3)$$

18. How many terms are in the series? $\sum_{j=1}^{5} j^3$

For Exercises 19–20, find the sum of the series.

19. $\sum_{i=1}^{7} 5(-1)^{i-1}$

20. $\sum_{n=1}^{5} (-2)^n$

21. Write the sum in summation notation.

$$\dfrac{4}{1} + \dfrac{5}{2} + \dfrac{6}{3} + \dfrac{7}{4} + \dfrac{8}{5} + \dfrac{9}{6} + \dfrac{10}{7}$$

22. Study the figures and the corresponding expression for the area of each region. Write the nth term of a sequence representing the area of a region with base n.

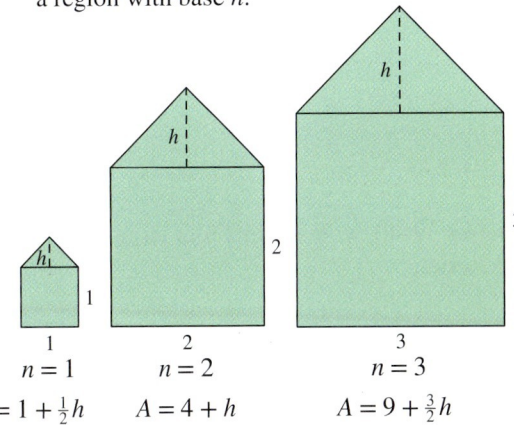

$n = 1$ $n = 2$ $n = 3$
$A = 1 + \tfrac{1}{2}h$ $A = 4 + h$ $A = 9 + \tfrac{3}{2}h$

Section 22.3

For Exercises 23–24, write the first five terms of the arithmetic sequence.

23. $a_1 = -12,\ d = 1.5$

24. $a_1 = \dfrac{2}{3},\ d = \dfrac{1}{3}$

For Exercises 25–26, write an expression for the nth term of the sequence.

25. $1, 11, 21, 31, \ldots$

26. $6, -8, -22, -36, \ldots$

For Exercises 27–28, find the indicated term of the arithmetic sequence.

27. $a_n = \dfrac{1}{2} + (n - 1) \cdot \dfrac{1}{4}$; 17th term

28. $a_1 = -11,\ d = -6$; 25th term

For Exercises 29–30, find the number of terms.

29. $3, 8, 13, 18, \ldots, 118$

30. $4, 1, -2, -5, \ldots, -50$

31. Find the common difference for an arithmetic sequence, given $a_1 = -40$ and $a_{19} = 140$.

32. Write an expression for the nth term of the arithmetic sequence, $14, 9, 4, -1, -6, \ldots$.

For Exercises 33–36, find the sum of the arithmetic series.

33. $\sum_{k=1}^{40}(5-k)$

34. $\sum_{i=1}^{20}(0.25i+2)$

35. $-6 + (-4) + (-2) + \cdots + 34$

36. $4 + 5 + 6 + \cdots + 109$

Section 22.4

For Exercises 37–38, find the common ratio.

37. $5, 15, 45, 135, \ldots$

38. $2, 2.4, 2.88, 3.456$

For Exercises 39–40, write the first four terms of the geometric series.

39. $a_1 = -1$ and $r = -\dfrac{1}{4}$

40. $a_n = 10(2)^{n-1}$

For Exercises 41–42, write an expression for the nth term of the geometric series.

41. $-4, -8, -16, -32, \ldots$

42. $6, -2, \dfrac{2}{3}, -\dfrac{2}{9}, \ldots$

43. Given $a_n = 4\left(\dfrac{2}{3}\right)^{n-1}$, find a_6.

44. Given $a_n = 10(3)^{n-1}$, find a_4.

45. Given $a_7 = \dfrac{1}{16}$ and $r = \dfrac{1}{2}$, find a_1.

46. Given $a_5 = 81$ and $r = 3$, find a_1.

47. Find the sum of the series. $\sum_{n=1}^{8} 5(2)^{n-1}$

48. Find the sum of the series.

$\dfrac{1}{3} + (-1) + 3 + (-9) + \cdots + (-81)$

49. Find the sum of the series.

$6 + 4 + \dfrac{8}{3} + \dfrac{16}{9} + \cdots$

50. Find the sum of the series.

$10 + 2 + \dfrac{2}{5} + \dfrac{2}{25} + \cdots$

51. An investment of \$10,000 is originally invested in a mutual fund. After 10 years, the yearly return averaged approximately 7%. The balance on the account is given by $b_n = 10{,}000(1.07)^n$.

  a. Find the first three terms of the sequence.

  b. Interpret the value of b_3.

  c. Find the amount in the account after 10 years.

Additional Topics Appendix

Additional Factoring

Section A.1

1. Factoring by Using Substitution

Concepts

1. Factoring by Using Substitution
2. Factoring 1 Term with 3 Terms
3. Additional Factoring Strategies

The trinomial $u^2 + 4u - 21$ has a leading coefficient 1 and factors as $(u + 7)(u - 3)$. Now compare the following polynomials to $u^2 + 4u - 21$. In each case, the variable expressions in the first and second terms are the same. Furthermore, the coefficients are 1, 4, and −21. By letting the single variable u represent the variable expression in parentheses, the polynomial is a quadratic trinomial in u.

Polynomial	Substitution	New form
$u^2 + 4u - 21$		
$(a + b)^2 + 4(a + b) - 21$	Let $u = a + b$	$u^2 + 4u - 21$
$(y - 3)^2 + 4(y - 3) - 21$	Let $u = y - 3$	$u^2 + 4u - 21$

With an appropriate substitution, the polynomials are easier to factor.

Example 1 Factoring by Using Substitution

Factor completely. $(a + b)^2 + 4(a + b) - 21$

Solution:

$(a + b)^2 + 4(a + b) - 21$ The polynomial is a quadratic trinomial with variable $(a + b)$.

$= u^2 + 4u - 21$ Let $u = a + b$.

$= (u + 7)(u - 3)$ Factor the trinomial.

$= [(a + b) + 7][(a + b) - 3]$ Back substitute.

$= (a + b + 7)(a + b - 3)$ Simplify.

Skill Practice Factor completely.

1. $(w - 6)^2 - 3(w - 6) - 10$

In Examples 2 and 3, we use substitution to factor a difference of squares and a sum of cubes.

Answer

1. $(w - 11)(w - 4)$

Example 2 — Factoring by Using Substitution

Factor completely. $9(4x - 1)^2 - 25$

Solution:

$9(4x - 1)^2 - 25$	The polynomial is a difference of squares.
$= 9u^2 - 25$	Let $u = 4x - 1$.
$= (3u - 5)(3u + 5)$	Factor as $a^2 - b^2 = (a - b)(a + b)$.
$= [3(4x - 1) - 5][3(4x - 1) + 5]$	Back substitute.
$= (12x - 3 - 5)(12x - 3 + 5)$	Apply the distributive property and simplify.
$= (12x - 8)(12x + 2)$	The remaining factors each have a common factor.
$= 4(3x - 2)2(6x + 1)$	Factor out 4 from the first binomial and 2 from the second binomial.
$= 8(3x - 2)(6x + 1)$	Simplify.

Skill Practice Factor completely.

2. $(2x + 1)^2 - 9$

Example 3 — Factoring by Using Substitution

Factor completely. $(x + 1)^3 + 8$

Solution:

$(x + 1)^3 + 8$	The polynomial is a sum of cubes.
$= u^3 + 8$	Let $u = x + 1$.
$= (u + 2)(u^2 - 2u + 4)$	Factor as $a^3 + b^3 = (a + b)(a^2 - ab + b^2)$.
$= [(x + 1) + 2][(x + 1)^2 - 2(x + 1) + 4]$	Back substitute.
$= (x + 3)[(x^2 + 2x + 1) - 2x - 2 + 4]$	Apply the distributive property.
$= (x + 3)(x^2 + 3)$	Simplify.

Skill Practice Factor completely.

3. $(y + 3)^3 + 64$

2. Factoring 1 Term with 3 Terms

When confronted with a 4-term polynomial, sometimes the standard method of grouping does not apply. For example, consider the polynomial $x^2 - y^2 - 6y - 9$. Notice that grouping 2 terms with 2 terms does not result in common binomial factors.

$$x^2 - y^2 \mid -6y - 9$$
$$= 1(x^2 - y^2) - 3(2x - 3)$$

Binomial factors do not match.

However, if a factor of -1 is removed from the last three terms, the last three terms result in a perfect square trinomial. This is called grouping 1 term with 3 terms.

Answers
2. $4(x - 1)(x + 2)$
3. $(y + 7)(y^2 + 2y + 13)$

Example 4 Factoring by Grouping 1 Term with 3 Terms

Factor completely. $x^2 - y^2 - 6y - 9$

Solution:

This polynomial has 4 terms. However, the standard grouping method does not work even if we try rearranging terms.

$x^2 - y^2 - 6y - 9$ We might try grouping one term with three terms. The reason is that after factoring out -1 from the last three terms we have a perfect square trinomial.

$= x^2 - 1(y^2 + 6y + 9)$

$= x^2 - (y + 3)^2$ ← difference of squares   The resulting expression is a difference of squares.

$= x^2 - u^2$ The difference of squares can be factored by using substitution. Let $u = y + 3$.

$= (x + u)(x - u)$ Factor.

$= [x + (y + 3)][x - (y + 3)]$ Back substitute.

$= (x + y + 3)(x - y - 3)$ Simplify.

Skill Practice Factor completely.

4. $n^2 - m^2 - 8m - 16$

The difference of squares $(x)^2 - (y + 3)^2$ can also be factored without using a substitution.

$$(x)^2 - (y + 3)^2$$
$$= [x + (y + 3)][x - (y + 3)]$$
$$= (x + y + 3)(x - y - 3)$$

3. Additional Factoring Strategies

Thus far, we have learned how to factor polynomials that fit the following patterns.

- Difference of squares
- Difference of cubes
- Sum of cubes
- Perfect square trinomials
- Nonperfect square trinomials
- Grouping 2 terms by 2 terms
- Grouping 3 terms by 1 term

Many polynomials do not fit any of these patterns, and many polynomials do not factor using these techniques. However, sometimes a combination of patterns exists and we can find the factored form. In Example 5, the polynomial has 6 terms. We might try grouping 3 terms with 3 terms or 4 terms with 2 terms, or even 2 terms, 2 terms and 2 terms. In any case, we should expect to try a variety of approaches before a workable strategy is identified.

Answer

4. $(n + m + 4)(n - m - 4)$

Example 5 Factoring a Polynomial

Factor completely. $2ac + 2a - bc - b - c - 1$

Solution:

$2ac + 2a - bc - b - c - 1$ The polynomial has 6 terms. After considerable trial and error, we might try grouping the first 2 terms, the second 2 terms, and the last 2 terms.

$= 2ac + 2a \mid -bc - b \mid -c - 1$

$= 2a(c + 1) - b(c + 1) - 1(c + 1)$ Factor out the GCF from each pair of terms.

$= (c + 1)(2a - b - 1)$ Factor out the common factor of $c + 1$.

Skill Practice Factor completely.

5. $xy + 2y + 3xz + 6z - x - 2$

Answer
5. $(x + 2)(y + 3z - 1)$

Section A.1 Practice Exercises

Vocabulary and Key Concepts

1. Given the expression $(3x - 2)^2 - (3x - 2) - 12$, what substitution could be used for u to make this a simpler quadratic trinomial in terms of u?

2. Given the expression $49(p - 3)^2 - 36$, what substitution could be used for u to make this a simpler difference of squares in terms of u?

3. What would be the first step to factor $x^2 - 4x + 4 - y^2$ by grouping 3 terms with 1 term?

4. What would be the first two steps to factor $9a^2 - 4b^2 - 20b - 25$ by grouping 1 term with 3 terms?

Concept 1: Factoring by Using Substitution

For Exercises 5–20, factor completely. (See Examples 1–3.)

5. $(4t - 1)^2 + 3(4t - 1) - 4$

6. $(2p + 5)^2 - 4(2p + 5) - 12$

7. $2(w - 2)^2 - 5(w - 2) - 3$

8. $3(y + 4)^2 + 11(y + 4) - 4$

9. $25(x + 4)^2 - y^2$

10. $36(y - 5)^2 - z^2$

11. $9(5a - 3)^2 - 4$

12. $25(3b + 7)^2 - 16$

13. $(t - 2)^2 - (2t + 1)^2$

14. $(w + 3)^2 - (3w - 5)^2$

15. $(a + b)^3 + c^3$

16. $(x + 4)^3 - y^3$

17. $27c^3 - (d + 2)^3$

18. $125v^3 + (w - 1)^3$

19. $4x^2 + 28x(2y + 1) + 49(2y + 1)^2$

20. $9a^2 - 12a(b + 1) + 4(b + 1)^2$

Concept 2: Factoring 1 Term with 3 Terms

For Exercises 21–28, factor completely. (See Example 4.)

21. $x^2 - 10x + 25 - y^2$

22. $m^2 - 22m + 121 - n^2$

23. $a^2 - x^2 - 18xy - 81y^2$

24. $c^2 - a^2 + 16ab - 64b^2$

25. $25a^2 + 20ab + 4b^2 - 100c^2$

26. $36x^2 + 60xy + 25y^2 - 121z^2$

27. $64w^2 - 9t^2 + 24tp - 16p^2$

28. $49k^2 - 25m^2 + 30mn - 9n^2$

Concept 3: Additional Strategies and Mixed Exercises

For Exercises 29–68, factor completely. (See Examples 1–5.)

29. $x^2 - 4y^2 - x - 2y$

30. $9v^2 - w^2 - 3v + w$

31. $w^7 + 8w^4 - w^3 - 8$

32. $t^5 - 125t^2 - 4t^3 + 500$

33. $xy - 4x + 2yz - 8z - 2y + 8$

34. $ax + 4a + 2bx + 8b - cx - 4c$

35. $x^2 + xy - 2y^2 + x - y$

36. $a^2 + 2ab - 3b^2 - b + a$

37. $n^6 - 63n^3 - 64$

38. $m^6 + 7m^3 - 8$

39. $(a + b)^2 - 6(a + b)(c - d) + 9(c - d)^2$

40. $(x + 1)^2 - 4(x + 1)(x - 2) + 4(x - 2)^2$

41. $4x^4 - 10x^3 - 36x^2 + 90x$

42. $15x^4 - 5x^3 - 60x^2 + 20x$

43. $a^2 + 2ab + b^2 - a - b$

44. $v^2 + 6vw + 9w^2 - v - 3w$

45. $x^2(x + y) - y^2(x + y)$

46. $u^2(u - v) - v^2(u - v)$

47. $(a + 3)^4 + 6(a + 3)^3$

48. $(4 - b)^4 - 2(4 - b)^3$

49. $24(3x + 5)^3 - 30(3x + 5)^2$

50. $10(2y + 3)^2 + 15(2y + 3)$

51. $\dfrac{1}{100}x^2 + \dfrac{1}{35}x + \dfrac{1}{49}$

52. $\dfrac{1}{25}a^2 + \dfrac{1}{15}a + \dfrac{1}{36}$

53. $(5x^2 - 1)^2 - 4(5x^2 - 1) - 5$

54. $(x^3 + 4)^2 - 10(x^3 + 4) + 24$

55. $16p^4 - q^4$

56. $s^4 t^4 - 81$

57. $y^3 + \dfrac{1}{64}$

58. $z^3 + \dfrac{1}{125}$

59. $6a^3 + a^2 b - 6ab^2 - b^3$

60. $4p^3 + 12p^2 q - pq^2 - 3q^3$

61. $\dfrac{1}{9}t^2 + \dfrac{1}{6}t + \dfrac{1}{16}$

62. $\dfrac{1}{25}y^2 + \dfrac{1}{5}y + \dfrac{1}{4}$

63. $6ax + 2bx - by - 3ay$

64. $5pq - 4q - 12 + 15p$

65. $x^8 - 1$

66. $y^8 - 256$

67. $25c^2 - 9d^2 + 5c - 3d$

68. $5wx^3 + 5wy^3 - 2zx^3 - 2zy^3$

Section A.2 Solving Systems of Linear Equations by Using Matrices

Concepts

1. Introduction to Matrices
2. Solving Systems of Linear Equations by Using the Gauss-Jordan Method

1. Introduction to Matrices

We have already learned how to solve systems of linear equations by using the substitution method and the addition method. We now present a third method called the Gauss-Jordan method that uses matrices to solve a linear system.

A **matrix** is a rectangular array of numbers (the plural of *matrix* is *matrices*). The rows of a matrix are read horizontally, and the columns of a matrix are read vertically. Every number or entry within a matrix is called an element of the matrix.

The **order of a matrix** is determined by the number of rows and number of columns. A matrix with m rows and n columns is an $m \times n$ (read as "m by n") matrix. Notice that with the order of a matrix, the number of rows is given first, followed by the number of columns.

Example 1   Determining the Order of a Matrix

Determine the order of each matrix.

a. $\begin{bmatrix} 2 & -4 & 1 \\ 5 & \pi & \sqrt{7} \end{bmatrix}$ b. $\begin{bmatrix} 1.9 \\ 0 \\ 7.2 \\ -6.1 \end{bmatrix}$ c. $\begin{bmatrix} 1 & 0 & 0 \\ 0 & 1 & 0 \\ 0 & 0 & 1 \end{bmatrix}$ d. $[a \ \ b \ \ c]$

Solution:

a. This matrix has two rows and three columns. Therefore, it is a 2×3 matrix.

b. This matrix has four rows and one column. Therefore, it is a 4×1 matrix. A matrix with one column is called a **column matrix**.

c. This matrix has three rows and three columns. Therefore, it is a 3×3 matrix. A matrix with the same number of rows and columns is called a **square matrix**.

d. This matrix has one row and three columns. Therefore, it is a 1×3 matrix. A matrix with one row is called a **row matrix**.

Skill Practice Determine the order of the matrix.

1. $\begin{bmatrix} -5 & 2 \\ 1 & 3 \\ 8 & 9 \end{bmatrix}$ 2. $[4 \ -8]$ 3. $\begin{bmatrix} 5 \\ 10 \\ 15 \end{bmatrix}$ 4. $\begin{bmatrix} 2 & -0.5 \\ -1 & 6 \end{bmatrix}$

A matrix can be used to represent a system of linear equations written in standard form. To do so, we extract the coefficients of the variable terms and the constants within the equation. For example, consider the system

$$2x - y = 5$$
$$x + 2y = -5$$

The matrix **A** is called the **coefficient matrix**.

$$\mathbf{A} = \begin{bmatrix} 2 & -1 \\ 1 & 2 \end{bmatrix}$$

If we extract both the coefficients and the constants from the equations, we can construct the **augmented matrix** of the system:

$$\begin{bmatrix} 2 & -1 & | & 5 \\ 1 & 2 & | & -5 \end{bmatrix}$$

A vertical bar is inserted into an augmented matrix to designate the position of the equal signs.

Answers
1. 3×2 2. 1×2
3. 3×1 4. 2×2

Example 2: Writing an Augmented Matrix for a System of Linear Equations

Write the augmented matrix for each linear system.

a. $-3x - 4y = 3$
$2x + 4y = 2$

b. $2x - 3z = 14$
$ 2y + z = 2$
$x + y = 4$

Solution:

a. $\begin{bmatrix} -3 & -4 & | & 3 \\ 2 & 4 & | & 2 \end{bmatrix}$

b. $\begin{bmatrix} 2 & 0 & -3 & | & 14 \\ 0 & 2 & 1 & | & 2 \\ 1 & 1 & 0 & | & 4 \end{bmatrix}$

TIP: Notice that zeros are inserted to denote the coefficient of each missing term.

Skill Practice Write the augmented matrix for each system.

5. $-x + y = 4$
$2x - y = 1$

6. $2x - y + z = 14$
$-3x + 4y = 8$
$x - y + 5z = 0$

Example 3: Writing a Linear System from an Augmented Matrix

Write a system of linear equations represented by each augmented matrix. Use x, y, and z as the variables.

a. $\begin{bmatrix} 2 & -5 & | & -8 \\ 4 & 1 & | & 6 \end{bmatrix}$

b. $\begin{bmatrix} 2 & -1 & 3 & | & 14 \\ 1 & 1 & -2 & | & -5 \\ 3 & 1 & -1 & | & 2 \end{bmatrix}$

c. $\begin{bmatrix} 1 & 0 & 0 & | & 4 \\ 0 & 1 & 0 & | & -1 \\ 0 & 0 & 1 & | & 0 \end{bmatrix}$

Solution:

a. $2x - 5y = -8$
$4x + y = 6$

b. $2x - y + 3z = 14$
$x + y - 2z = -5$
$3x + y - z = 2$

c. $x + 0y + 0z = 4$
$0x + y + 0z = -1$ or $x = 4$
$0x + 0y + z = 0$ $y = -1$
                       $z = 0$

Skill Practice Write a system of linear equations represented by each augmented matrix. Use x, y, and z as the variables.

7. $\begin{bmatrix} 2 & 3 & | & 5 \\ -1 & 8 & | & 1 \end{bmatrix}$

8. $\begin{bmatrix} -3 & 2 & 1 & | & 4 \\ 14 & 1 & 0 & | & 20 \\ -8 & 3 & 5 & | & 6 \end{bmatrix}$

9. $\begin{bmatrix} 1 & 0 & 0 & | & -8 \\ 0 & 1 & 0 & | & 2 \\ 0 & 0 & 1 & | & 15 \end{bmatrix}$

2. Solving Systems of Linear Equations by Using the Gauss-Jordan Method

We know that interchanging two equations results in an equivalent system of linear equations. Interchanging two rows in an augmented matrix results in an equivalent augmented matrix. Similarly, because each row in an augmented matrix represents a linear equation, we can perform the following elementary row operations that result in an equivalent augmented matrix.

Answers

5. $\begin{bmatrix} -1 & 1 & | & 4 \\ 2 & -1 & | & 1 \end{bmatrix}$

6. $\begin{bmatrix} 2 & -1 & 1 & | & 14 \\ -3 & 4 & 0 & | & 8 \\ 1 & -1 & 5 & | & 0 \end{bmatrix}$

7. $2x + 3y = 5$
$-x + 8y = 1$

8. $-3x + 2y + z = 4$
$14x + y = 20$
$-8x + 3y + 5z = 6$

9. $x = -8, y = 2, z = 15$

> **Elementary Row Operations**
>
> The following *elementary row operations* performed on an augmented matrix produce an equivalent augmented matrix:
>
> - Interchange two rows.
> - Multiply every element in a row by a nonzero real number.
> - Add a multiple of one row to another row.

When we are solving a system of linear equations by any method, the goal is to write a series of simpler but equivalent systems of equations until the solution is obvious. The *Gauss-Jordan method* uses a series of elementary row operations performed on the augmented matrix to produce a simpler augmented matrix. In particular, we want to produce an augmented matrix that has 1's along the diagonal of the matrix of coefficients and 0's for the remaining entries in the matrix of coefficients. A matrix written in this way is said to be written in **reduced row echelon form**. For example, the augmented matrix from Example 3(c) is written in reduced row echelon form.

$$\begin{bmatrix} 1 & 0 & 0 & | & 4 \\ 0 & 1 & 0 & | & -1 \\ 0 & 0 & 1 & | & 0 \end{bmatrix}$$

The solution to the corresponding system of equations is easily recognized as $x = 4$, $y = -1$, and $z = 0$.

Similarly, matrix **B** represents a solution of $x = a$ and $y = b$ to a system of two linear equations.

$$\mathbf{B} = \begin{bmatrix} 1 & 0 & | & a \\ 0 & 1 & | & b \end{bmatrix}$$

Example 4 Solving a System by Using the Gauss-Jordan Method

Solve by using the Gauss-Jordan method.

$$2x - y = 5$$
$$x + 2y = -5$$

Solution:

$\begin{bmatrix} 2 & -1 & | & 5 \\ 1 & 2 & | & -5 \end{bmatrix}$ Set up the augmented matrix.

$\xrightarrow{R_1 \Leftrightarrow R_2} \begin{bmatrix} 1 & 2 & | & -5 \\ 2 & -1 & | & 5 \end{bmatrix}$ Switch row 1 and row 2 to get a 1 in the upper left position.

$\xrightarrow{-2R_1 + R_2 \Rightarrow R_2} \begin{bmatrix} 1 & 2 & | & -5 \\ 0 & -5 & | & 15 \end{bmatrix}$ Multiply row 1 by -2 and add the result to row 2. This produces an entry of 0 below the upper left position.

$\xrightarrow{-\frac{1}{5}R_2 \Rightarrow R_2} \begin{bmatrix} 1 & 2 & | & -5 \\ 0 & 1 & | & -3 \end{bmatrix}$ Multiply row 2 by $-\frac{1}{5}$ to produce a 1 along the diagonal in the second row.

$$\xrightarrow{-2R_2 + R_1 \Rightarrow R_1} \begin{bmatrix} 1 & 0 & | & 1 \\ 0 & 1 & | & -3 \end{bmatrix}$$

Multiply row 2 by -2 and add the result to row 1. This produces a 0 in the first row, second column.

$$C = \begin{bmatrix} 1 & 0 & | & 1 \\ 0 & 1 & | & -3 \end{bmatrix}$$

The matrix **C** is in reduced row echelon form. From the augmented matrix, we have $x = 1$ and $y = -3$. The solution set is $\{(1, -3)\}$.

FOR REVIEW

Recall that to check a solution to a system of equations in two variables, test the ordered pair in both equations.

Skill Practice

10. Solve by using the Gauss-Jordan method.

$$x - 2y = -21$$
$$2x + y = -2$$

The order in which we manipulate the elements of an augmented matrix to produce reduced row echelon form was demonstrated in Example 4. In general, the order is as follows.

- First produce a 1 in the first row, first column. Then use the first row to obtain 0's in the first column below this element.
- Next, if possible, produce a 1 in the second row, second column. Use the second row to obtain 0's above and below this element.
- Next, if possible, produce a 1 in the third row, third column. Use the third row to obtain 0's above and below this element.
- The process continues until reduced row echelon form is obtained.

Example 5 Solving a System by Using the Gauss-Jordan Method

Solve by using the Gauss-Jordan method.

$$x = -y + 5$$
$$-2x + 2z = y - 10$$
$$3x + 6y + 7z = 14$$

Solution:

First write each equation in the system in standard form.

$$\begin{array}{lll} x = -y + 5 & \longrightarrow & x + y = 5 \\ -2x + 2z = y - 10 & \longrightarrow & -2x - y + 2z = -10 \\ 3x + 6y + 7z = 14 & \longrightarrow & 3x + 6y + 7z = 14 \end{array}$$

$$\begin{bmatrix} 1 & 1 & 0 & | & 5 \\ -2 & -1 & 2 & | & -10 \\ 3 & 6 & 7 & | & 14 \end{bmatrix}$$ Set up the augmented matrix.

Answer

10. $\{(-5, 8)\}$

$$2R_1 + R_2 \Rightarrow R_2 \atop -3R_1 + R_3 \Rightarrow R_3 \longrightarrow \begin{bmatrix} 1 & 1 & 0 & | & 5 \\ 0 & 1 & 2 & | & 0 \\ 0 & 3 & 7 & | & -1 \end{bmatrix}$$

Multiply row 1 by 2 and add the result to row 2.
Multiply row 1 by −3 and add the result to row 3.

$$-1R_2 + R_1 \Rightarrow R_1 \atop -3R_2 + R_3 \Rightarrow R_3 \longrightarrow \begin{bmatrix} 1 & 0 & -2 & | & 5 \\ 0 & 1 & 2 & | & 0 \\ 0 & 0 & 1 & | & -1 \end{bmatrix}$$

Multiply row 2 by −1 and add the result to row 1.
Multiply row 2 by −3 and add the result to row 3.

$$2R_3 + R_1 \Rightarrow R_1 \atop -2R_3 + R_2 \Rightarrow R_2 \longrightarrow \begin{bmatrix} 1 & 0 & 0 & | & 3 \\ 0 & 1 & 0 & | & 2 \\ 0 & 0 & 1 & | & -1 \end{bmatrix}$$

Multiply row 3 by 2 and add the result to row 1.
Multiply row 3 by −2 and add the result to row 2.

From the reduced row echelon form of the matrix, we have $x = 3$, $y = 2$, and $z = -1$. The solution set is $\{(3, 2, -1)\}$.

Skill Practice Solve by using the Gauss-Jordan method.

11. $x + y + z = 2$
    $x - y + z = 4$
    $x + 4y + 2z = 1$

It is particularly easy to recognize a system of dependent equations or an inconsistent system of equations from the reduced row echelon form of an augmented matrix. This is demonstrated in Examples 6 and 7.

Example 6 Solving a System of Dependent Equations by Using the Gauss-Jordan Method

Solve by using the Gauss-Jordan method.

$$x - 3y = 4$$
$$\frac{1}{2}x - \frac{3}{2}y = 2$$

Solution:

$$\begin{bmatrix} 1 & -3 & | & 4 \\ \frac{1}{2} & -\frac{3}{2} & | & 2 \end{bmatrix}$$ Set up the augmented matrix.

$$-\frac{1}{2}R_1 + R_2 \Rightarrow R_2 \longrightarrow \begin{bmatrix} 1 & -3 & | & 4 \\ 0 & 0 & | & 0 \end{bmatrix}$$ Multiply row 1 by $-\frac{1}{2}$ and add the result to row 2.

The second row of the augmented matrix represents the equation $0 = 0$. The equations are dependent. The solution set is $\{(x, y) | x - 3y = 4\}$.

Skill Practice Solve by using the Gauss-Jordan method.

12. $4x - 6y = 16$
    $6x - 9y = 24$

FOR REVIEW

Recall that if a system of equations reduces to an identity such as $0 = 0$, the system has infinitely many solutions. If the system reduces to a contradiction such as $0 = 7$, the system has no solution.

Answers
11. $\{(1, -1, 2)\}$
12. Infinitely many solutions; $\{(x, y) | 4x - 6y = 16\}$; dependent equations

Example 7 Solving an Inconsistent System by Using the Gauss-Jordan Method

Solve by using the Gauss-Jordan method.

$$x + 3y = 2$$
$$-3x - 9y = 1$$

Solution:

$$\begin{bmatrix} 1 & 3 & | & 2 \\ -3 & -9 & | & 1 \end{bmatrix}$$ Set up the augmented matrix.

$$3R_1 + R_2 \Rightarrow R_2 \quad \begin{bmatrix} 1 & 3 & | & 2 \\ 0 & 0 & | & 7 \end{bmatrix}$$ Multiply row 1 by 3 and add the result to row 2.

The second row of the augmented matrix represents the contradiction $0 = 7$. The system is inconsistent. There is no solution, { }.

Skill Practice Solve by using the Gauss-Jordan method.
13. $6x + 10y = 1$
    $15x + 25y = 3$

Answer
13. No solution; { }; inconsistent system

Practice Exercises   Section A.2

Vocabulary and Key Concepts

1. **a.** A _____ is a rectangular array of numbers. The order of a matrix is $m \times n$ where m is the number of _____ and n is the number of _____.

    b. A matrix that has exactly one column is called a _____ matrix. A matrix that has exactly _____ row is called a row matrix, and a matrix with the same number of rows and columns is called a _____ matrix.

    c. Given the system of equations shown, matrix **A** is called the _____ matrix. The matrix **B** is called the _____ matrix.

    $$\begin{array}{c} 3x - 2y = 6 \\ 4x + 5y = 9 \end{array} \quad \mathbf{A} = \begin{bmatrix} 3 & -2 \\ 4 & 5 \end{bmatrix} \quad \mathbf{B} = \begin{bmatrix} 3 & -2 & | & 6 \\ 4 & 5 & | & 9 \end{bmatrix}$$

    d. The matrix **A** is said to be written in reduced _____ form.

    $$\mathbf{A} = \begin{bmatrix} 1 & 0 & | & 8 \\ 0 & 1 & | & 2 \end{bmatrix}$$

Review Exercises

2. How much 50% acid solution should be mixed with pure acid to obtain 20 L of a mixture that is 70% acid?

For Exercises 3–5, solve the system by using any method.

3. $x - 6y = 9$
   $x + 2y = 13$

4. $x + y - z = 8$
   $x - 2y + z = 3$
   $x + 3y + 2z = 7$

5. $2x - y + z = -4$
   $-x + y + 3z = -7$
   $x + 3y - 4z = 22$

Concept 1: Introduction to Matrices

For Exercises 6–14, (**a**) determine the order of each matrix and (**b**) determine if the matrix is a row matrix, a column matrix, a square matrix, or none of these. (**See Example 1.**)

6. $\begin{bmatrix} 4 \\ 5 \\ -3 \\ 0 \end{bmatrix}$

7. $\begin{bmatrix} 5 \\ -1 \\ 2 \end{bmatrix}$

8. $\begin{bmatrix} -9 & 4 & 3 \\ -1 & -8 & 4 \\ 5 & 8 & 7 \end{bmatrix}$

9. $\begin{bmatrix} 3 & -9 \\ -1 & -3 \end{bmatrix}$

10. $[4 \ -7]$

11. $[0 \ -8 \ 11 \ 5]$

12. $\begin{bmatrix} 5 & -8.1 & 4.2 & 0 \\ 4.3 & -9 & 18 & 3 \end{bmatrix}$

13. $\begin{bmatrix} \frac{1}{3} & \frac{3}{4} & 6 \\ -2 & 1 & -\frac{7}{8} \end{bmatrix}$

14. $\begin{bmatrix} 5 & 1 \\ -1 & 2 \\ 0 & 7 \end{bmatrix}$

For Exercises 15–18, set up the augmented matrix. (**See Example 2.**)

15. $x - 2y = -1$
    $2x + y = -7$

16. $x - 3y = 3$
    $2x - 5y = 4$

17. $x - 2y = 5 - z$
    $2x + 6y + 3z = -2$
    $3x - y - 2z = 1$

18. $5x - 17 = -2z$
    $8x + 6z = 26 + y$
    $8x + 3y - 12z = 24$

For Exercises 19–22, write a system of linear equations represented by the augmented matrix. Use x, y, and z as the variables. (**See Example 3.**)

19. $\begin{bmatrix} 4 & 3 & | & 6 \\ 12 & 5 & | & -6 \end{bmatrix}$

20. $\begin{bmatrix} -2 & 5 & | & -15 \\ -7 & 15 & | & -45 \end{bmatrix}$

21. $\begin{bmatrix} 1 & 0 & 0 & | & 4 \\ 0 & 1 & 0 & | & -1 \\ 0 & 0 & 1 & | & 7 \end{bmatrix}$

22. $\begin{bmatrix} 1 & 0 & 0 & | & 0.5 \\ 0 & 1 & 0 & | & 6.1 \\ 0 & 0 & 1 & | & 3.9 \end{bmatrix}$

Concept 2: Solving Systems of Linear Equations by Using the Gauss-Jordan Method

23. Given the matrix **E**

$$\mathbf{E} = \begin{bmatrix} 3 & -2 & | & 8 \\ 9 & -1 & | & 7 \end{bmatrix}$$

   a. What is the element in the second row and third column?

   b. What is the element in the first row and second column?

24. Given the matrix **F**

$$\mathbf{F} = \begin{bmatrix} 1 & 8 & | & 0 \\ 12 & -13 & | & -2 \end{bmatrix}$$

   a. What is the element in the second row and second column?

   b. What is the element in the first row and third column?

25. Given the matrix **Z**

$$\mathbf{Z} = \begin{bmatrix} 2 & 1 & | & 11 \\ 2 & -1 & | & 1 \end{bmatrix}$$

   write the matrix obtained by multiplying the elements in the first row by $\frac{1}{2}$.

26. Given the matrix **J**

$$\mathbf{J} = \begin{bmatrix} 1 & 1 & | & 7 \\ 0 & 3 & | & -6 \end{bmatrix}$$

   write the matrix obtained by multiplying the elements in the second row by $\frac{1}{3}$.

27. Given the matrix **K**

$$\mathbf{K} = \begin{bmatrix} 5 & 2 & | & 1 \\ 1 & -4 & | & 3 \end{bmatrix}$$

   write the matrix obtained by interchanging rows 1 and 2.

28. Given the matrix **L**

$$\mathbf{L} = \begin{bmatrix} 9 & 6 & | & 13 \\ -7 & 2 & | & 19 \end{bmatrix}$$

   write the matrix obtained by interchanging rows 1 and 2.

29. Given the matrix **M**

$$M = \begin{bmatrix} 1 & 5 & | & 2 \\ -3 & -4 & | & -1 \end{bmatrix}$$

write the matrix obtained by multiplying the first row by 3 and adding the result to row 2.

30. Given the matrix **N**

$$N = \begin{bmatrix} 1 & 3 & | & -5 \\ -2 & 2 & | & 12 \end{bmatrix}$$

write the matrix obtained by multiplying the first row by 2 and adding the result to row 2.

31. Given the matrix **R**

$$R = \begin{bmatrix} 1 & 3 & 0 & | & -1 \\ 4 & 1 & -5 & | & 6 \\ -2 & 0 & -3 & | & 10 \end{bmatrix}$$

a. Write the matrix obtained by multiplying the first row by -4 and adding the result to row 2.

b. Using the matrix obtained from part (a), write the matrix obtained by multiplying the first row by 2 and adding the result to row 3.

32. Given the matrix **S**

$$S = \begin{bmatrix} 1 & 2 & 0 & | & 10 \\ 5 & 1 & -4 & | & 3 \\ -3 & 4 & 5 & | & 2 \end{bmatrix}$$

a. Write the matrix obtained by multiplying the first row by -5 and adding the result to row 2.

b. Using the matrix obtained from part (a), write the matrix obtained by multiplying the first row by 3 and adding the result to row 3.

For Exercises 33–36, use the augmented matrices **A**, **B**, **C**, and **D** to answer true or false.

$$A = \begin{bmatrix} 6 & -4 & | & 2 \\ 5 & -2 & | & 7 \end{bmatrix} \quad B = \begin{bmatrix} 5 & -2 & | & 7 \\ 6 & -4 & | & 2 \end{bmatrix} \quad C = \begin{bmatrix} 1 & -\frac{2}{3} & | & \frac{1}{3} \\ 5 & -2 & | & 7 \end{bmatrix} \quad D = \begin{bmatrix} 5 & -2 & | & 7 \\ -12 & 8 & | & -4 \end{bmatrix}$$

33. The matrix **A** is a 2×3 matrix.

34. Matrix **B** is equivalent to matrix **A**.

35. Matrix **A** is equivalent to matrix **C**.

36. Matrix **B** is equivalent to matrix **D**.

37. What does the notation $R_2 \Leftrightarrow R_1$ mean when one is performing the Gauss-Jordan method?

38. What does the notation $2R_3 \Rightarrow R_3$ mean when one is performing the Gauss-Jordan method?

39. What does the notation $-3R_1 + R_2 \Rightarrow R_2$ mean when one is performing the Gauss-Jordan method?

40. What does the notation $4R_2 + R_3 \Rightarrow R_3$ mean when one is performing the Gauss-Jordan method?

For Exercises 41–56, solve the system by using the Gauss-Jordan method. For systems that do not have one unique solution, also state the number of solutions and whether the system is inconsistent or the equations are dependent. **(See Example 4–7.)**

41. $x - 2y = -1$
$2x + y = -7$

42. $x - 3y = 3$
$2x - 5y = 4$

43. $x + 3y = 6$
$-4x - 9y = 3$

44. $2x - 3y = -2$
$x + 2y = 13$

45. $x + 3y = 3$
$4x + 12y = 12$

46. $2x + 5y = 1$
$-4x - 10y = -2$

47. $x - y = 4$
$2x + y = 5$

48. $2x - y = 0$
$x + y = 3$

49. $x + 3y = -1$
$-3x - 6y = 12$

50. $x + y = 4$
$2x - 4y = -4$

51. $3x + y = -4$
$-6x - 2y = 3$

52. $2x + y = 4$
$6x + 3y = -1$

53. $x + y + z = 6$
$x - y + z = 2$
$x + y - z = 0$

54. $2x - 3y - 2z = 11$
$x + 3y + 8z = 1$
$3x - y + 14z = -2$

55. $x - 2y = 5 - z$
$2x + 6y + 3z = -10$
$3x - y - 2z = 5$

56. $5x = 10z + 15$
$x - y + 6z = 23$
$x + 3y - 12z = 13$

Technology Connections

For Exercises 57–62, use the matrix features on a graphing calculator to express each augmented matrix in reduced row echelon form. Compare your results to the solution you obtained in the indicated exercise.

57. $\begin{bmatrix} 1 & -2 & | & -1 \\ 2 & 1 & | & -7 \end{bmatrix}$

Compare with Exercise 41.

58. $\begin{bmatrix} 1 & -3 & | & 3 \\ 2 & -5 & | & 4 \end{bmatrix}$

Compare with Exercise 42.

59. $\begin{bmatrix} 1 & 3 & | & 6 \\ -4 & -9 & | & 3 \end{bmatrix}$

Compare with Exercise 43.

60. $\begin{bmatrix} 2 & -3 & | & -2 \\ 1 & 2 & | & 13 \end{bmatrix}$

Compare with Exercise 44.

61. $\begin{bmatrix} 1 & 1 & 1 & | & 6 \\ 1 & -1 & 1 & | & 2 \\ 1 & 1 & -1 & | & 0 \end{bmatrix}$

Compare with Exercise 53.

62. $\begin{bmatrix} 2 & -3 & -2 & | & 11 \\ 1 & 3 & 8 & | & 1 \\ 3 & -1 & 14 & | & -2 \end{bmatrix}$

Compare with Exercise 54.

Section A.3 Determinants and Cramer's Rule

Concepts

1. Introduction to Determinants
2. Determinant of a 3 × 3 Matrix
3. Cramer's Rule

1. Introduction to Determinants

Associated with every square matrix is a real number called the **determinant** of the matrix. The determinant of a square matrix **A**, denoted det **A**, is written by enclosing the elements of the matrix within two vertical bars. For example,

If $\mathbf{A} = \begin{bmatrix} 2 & -1 \\ 6 & 0 \end{bmatrix}$ then $\det \mathbf{A} = \begin{vmatrix} 2 & -1 \\ 6 & 0 \end{vmatrix}$

If $\mathbf{B} = \begin{bmatrix} 0 & -5 & 1 \\ 4 & 0 & \frac{1}{2} \\ -2 & 10 & 1 \end{bmatrix}$ then $\det \mathbf{B} = \begin{vmatrix} 0 & -5 & 1 \\ 4 & 0 & \frac{1}{2} \\ -2 & 10 & 1 \end{vmatrix}$

Determinants have many applications in mathematics, including solving systems of linear equations, finding the area of a triangle, determining whether three points are collinear, and finding an equation of a line between two points.

The determinant of a 2 × 2 matrix is defined as follows.

Definition of a Determinant of a 2 × 2 Matrix

The determinant of the matrix $\begin{bmatrix} a & b \\ c & d \end{bmatrix}$ is the real number $ad - bc$. It is written as

$$\begin{vmatrix} a & b \\ c & d \end{vmatrix} = ad - bc$$

Example 1　Evaluating a 2 × 2 Determinant

Evaluate the determinants.

a. $\begin{vmatrix} 6 & -2 \\ 5 & \frac{1}{3} \end{vmatrix}$　　**b.** $\begin{vmatrix} 2 & -11 \\ 0 & 0 \end{vmatrix}$

Solution:

a. $\begin{vmatrix} 6 & -2 \\ 5 & \frac{1}{3} \end{vmatrix}$　　For this determinant, $a = 6$, $b = -2$, $c = 5$, and $d = \frac{1}{3}$.

$$ad - bc = (6)\left(\frac{1}{3}\right) - (-2)(5)$$
$$= 2 + 10$$
$$= 12$$

b. $\begin{vmatrix} 2 & -11 \\ 0 & 0 \end{vmatrix}$　　For this determinant, $a = 2$, $b = -11$, $c = 0$, $d = 0$.

$$ad - bc = (2)(0) - (-11)(0)$$
$$= 0 - 0$$
$$= 0$$

TIP: Example 1(b) illustrates that the value of a determinant having a row of all zeros is 0. The same is true for a determinant having a column of all zeros.

Skill Practice Evaluate the determinants.

1. $\begin{vmatrix} 2 & 8 \\ -1 & 5 \end{vmatrix}$　　**2.** $\begin{vmatrix} -6 & 0 \\ 4 & 0 \end{vmatrix}$

2. Determinant of a 3 × 3 Matrix

To find the determinant of a 3 × 3 matrix, we first need to define the **minor** of an element of the matrix. For any element of a 3 × 3 matrix, the minor of that element is the determinant of the 2 × 2 matrix obtained by deleting the row and column in which the element resides. For example, consider the matrix

$$\begin{bmatrix} 5 & -1 & 6 \\ 0 & -7 & 1 \\ 4 & 2 & 6 \end{bmatrix}$$

The minor of the element 5 is found by deleting the first row and first column and then evaluating the determinant of the remaining 2 × 2 matrix:

$$\begin{bmatrix} 5 & -1 & 6 \\ 0 & -7 & 1 \\ 4 & 2 & 6 \end{bmatrix}$$　Now evaluate the determinant:　$\begin{vmatrix} -7 & 1 \\ 2 & 6 \end{vmatrix} = (-7)(6) - (1)(2)$
$$= -44$$

For this matrix, the minor for the element 5 is -44.

To find the minor of the element -7, delete the second row and second column, and then evaluate the determinant of the remaining 2 × 2 matrix.

$$\begin{bmatrix} 5 & -1 & 6 \\ 0 & -7 & 1 \\ 4 & 2 & 6 \end{bmatrix}$$　Now evaluate the determinant:　$\begin{vmatrix} 5 & 6 \\ 4 & 6 \end{vmatrix} = (5)(6) - (6)(4) = 6$

For this matrix, the minor for the element -7 is 6.

Answers
1. 18　　**2.** 0

Example 2 — Determining the Minor for Elements in a 3 × 3 Matrix

Find the minor for each element in the first column of the matrix.

$$\begin{bmatrix} 3 & 4 & -1 \\ 2 & -4 & 5 \\ 0 & 1 & -6 \end{bmatrix}$$

Solution:

For 3: $\begin{bmatrix} 3 & 4 & -1 \\ 2 & -4 & 5 \\ 0 & 1 & -6 \end{bmatrix}$ The minor is: $\begin{vmatrix} -4 & 5 \\ 1 & -6 \end{vmatrix} = (-4)(-6) - (5)(1) = 19$

For 2: $\begin{bmatrix} 3 & 4 & -1 \\ 2 & -4 & 5 \\ 0 & 1 & -6 \end{bmatrix}$ The minor is: $\begin{vmatrix} 4 & -1 \\ 1 & -6 \end{vmatrix} = (4)(-6) - (-1)(1) = -23$

For 0: $\begin{bmatrix} 3 & 4 & -1 \\ 2 & -4 & 5 \\ 0 & 1 & -6 \end{bmatrix}$ The minor is: $\begin{vmatrix} 4 & -1 \\ -4 & 5 \end{vmatrix} = (4)(5) - (-1)(-4) = 16$

Skill Practice Find the minor for the element 3.

3. $\begin{bmatrix} -1 & 8 & -6 \\ \frac{1}{2} & 3 & 2 \\ 5 & 7 & 4 \end{bmatrix}$

The determinant of a 3 × 3 matrix is defined as follows.

Definition of a Determinant of a 3 × 3 Matrix

$$\begin{vmatrix} a_1 & b_1 & c_1 \\ a_2 & b_2 & c_2 \\ a_3 & b_3 & c_3 \end{vmatrix} = a_1 \cdot \begin{vmatrix} b_2 & c_2 \\ b_3 & c_3 \end{vmatrix} - a_2 \cdot \begin{vmatrix} b_1 & c_1 \\ b_3 & c_3 \end{vmatrix} + a_3 \cdot \begin{vmatrix} b_1 & c_1 \\ b_2 & c_2 \end{vmatrix}$$

From this definition, we see that the determinant of a 3 × 3 matrix can be written as

$$a_1 \cdot (\text{minor of } a_1) - a_2 \cdot (\text{minor of } a_2) + a_3 \cdot (\text{minor of } a_3)$$

Evaluating determinants in this way is called *expanding minors*.

Example 3 — Evaluating a 3 × 3 Determinant

Evaluate the determinant. $\begin{vmatrix} 2 & 4 & 2 \\ 1 & -3 & 0 \\ -5 & 5 & -1 \end{vmatrix}$

Solution:

$$\begin{vmatrix} 2 & 4 & 2 \\ 1 & -3 & 0 \\ -5 & 5 & -1 \end{vmatrix} = 2 \cdot \begin{vmatrix} -3 & 0 \\ 5 & -1 \end{vmatrix} - (1) \cdot \begin{vmatrix} 4 & 2 \\ 5 & -1 \end{vmatrix} + (-5) \cdot \begin{vmatrix} 4 & 2 \\ -3 & 0 \end{vmatrix}$$

$$= 2[(-3)(-1) - (0)(5)] - 1[(4)(-1) - (2)(5)] - 5[(4)(0) - (2)(-3)]$$

$$= 2(3) - 1(-14) - 5(6)$$

$$= 6 + 14 - 30$$

$$= -10$$

Answer

3. $\begin{vmatrix} -1 & -6 \\ 5 & 4 \end{vmatrix} = 26$

Skill Practice Evaluate the determinant.

4. $\begin{vmatrix} -2 & 4 & 9 \\ 5 & -1 & 2 \\ 1 & 1 & 6 \end{vmatrix}$

Although we defined the determinant of a matrix by expanding the minors of the elements in the first column, *any row or column can be used*. However, we must choose the correct sign to apply to each term in the expansion. The following array of signs is helpful.

$$\begin{matrix} + & - & + \\ - & + & - \\ + & - & + \end{matrix}$$

The signs alternate for each row and column, beginning with + in the first row, first column.

TIP: There is another method to determine the signs for each term of the expansion. For the a_{ij} element, multiply the term by $(-1)^{i+j}$.

Example 4 Evaluating a 3 × 3 Determinant

Evaluate the determinant by expanding minors about the elements in the second row.

$$\begin{vmatrix} 2 & 4 & 2 \\ 1 & -3 & 0 \\ -5 & 5 & -1 \end{vmatrix}$$

Solution:

Signs obtained from the array of signs

$\begin{vmatrix} 2 & 4 & 2 \\ 1 & -3 & 0 \\ -5 & 5 & -1 \end{vmatrix} = -(1) \cdot \begin{vmatrix} 4 & 2 \\ 5 & -1 \end{vmatrix} + (-3) \cdot \begin{vmatrix} 2 & 2 \\ -5 & -1 \end{vmatrix} - (0) \cdot \begin{vmatrix} 2 & 4 \\ -5 & 5 \end{vmatrix}$

$= -1[(4)(-1) - (2)(5)] - 3[(2)(-1) - (2)(-5)] - 0$

$= -1(-14) - 3(8)$

$= 14 - 24$

$= -10$ Notice that the value of the determinant is the same as the result obtained in Example 3.

Skill Practice Evaluate the determinant.

5. $\begin{vmatrix} 4 & -1 & 2 \\ 3 & 6 & -8 \\ 0 & \frac{1}{2} & 5 \end{vmatrix}$

In Example 4, the third term in the expansion of minors was zero because the element 0 when multiplied by its minor is zero. To simplify the arithmetic in evaluating a determinant of a 3 × 3 matrix, expand about the row or column that has the most 0 elements.

Answers

4. -42 5. 154

3. Cramer's Rule

In this section, we will learn another method to solve a system of linear equations. This method is called **Cramer's rule**.

> **Cramer's Rule for a 2 × 2 System of Linear Equations**
>
> The solution to the system $\quad a_1x + b_1y = c_1$
> $\qquad\qquad\qquad\qquad\qquad\quad a_2x + b_2y = c_2$
>
> is given by $\qquad x = \dfrac{\mathbf{D}_x}{\mathbf{D}} \quad$ and $\quad y = \dfrac{\mathbf{D}_y}{\mathbf{D}}$
>
> where $\mathbf{D} = \begin{vmatrix} a_1 & b_1 \\ a_2 & b_2 \end{vmatrix}$ (and $\mathbf{D} \neq 0$) $\quad \mathbf{D}_x = \begin{vmatrix} c_1 & b_1 \\ c_2 & b_2 \end{vmatrix} \quad \mathbf{D}_y = \begin{vmatrix} a_1 & c_1 \\ a_2 & c_2 \end{vmatrix}$

Example 5   Using Cramer's Rule to Solve a 2 × 2 System of Linear Equations

Solve the system by using Cramer's rule. $\quad 3x - 5y = 11$
$\qquad\qquad\qquad\qquad\qquad\qquad\qquad\quad -x + 3y = -5$

Solution:

For this system: $\quad a_1 = 3 \quad b_1 = -5 \quad c_1 = 11$
$\qquad\qquad\qquad\quad a_2 = -1 \quad b_2 = 3 \quad c_2 = -5$

$\mathbf{D} = \begin{vmatrix} 3 & -5 \\ -1 & 3 \end{vmatrix} = (3)(3) - (-5)(-1) = 9 - 5 = 4$

$\mathbf{D}_x = \begin{vmatrix} 11 & -5 \\ -5 & 3 \end{vmatrix} = (11)(3) - (-5)(-5) = 33 - 25 = 8$

$\mathbf{D}_y = \begin{vmatrix} 3 & 11 \\ -1 & -5 \end{vmatrix} = (3)(-5) - (11)(-1) = -15 + 11 = -4$

Therefore, $\qquad x = \dfrac{\mathbf{D}_x}{\mathbf{D}} = \dfrac{8}{4} = 2 \qquad y = \dfrac{\mathbf{D}_y}{\mathbf{D}} = \dfrac{-4}{4} = -1$

Check the ordered pair $(2, -1)$ in both original equations.

$\quad$ Check: $\quad 3x - 5y = 11 \longrightarrow 3(2) - 5(-1) \stackrel{?}{=} 11 \checkmark$
$\qquad\qquad\qquad -x + 3y = -5 \longrightarrow -(2) + 3(-1) \stackrel{?}{=} -5 \checkmark$

The solution set is $\{(2, -1)\}$.

Skill Practice Solve using Cramer's rule.

6. $2x + y = 5$
$\quad -x - 3y = 5$

Answer

6. $\{(4, -3)\}$

TIP: Here are some memory tips to help you remember Cramer's rule to solve:

$$a_1x + b_1y = c_1$$
$$a_2x + b_2y = c_2$$

1. The determinant **D** is the determinant of the coefficients of x and y.

   Coefficients of x-terms y-terms
   $$\mathbf{D} = \begin{vmatrix} a_1 & b_1 \\ a_2 & b_2 \end{vmatrix}$$

2. The determinant $\mathbf{D}_x$ has the column of x-term coefficients replaced by c_1 and c_2.

   x-coefficients replaced by c_1 and c_2
   $$\mathbf{D}_x = \begin{vmatrix} c_1 & b_1 \\ c_2 & b_2 \end{vmatrix}$$

3. The determinant $\mathbf{D}_y$ has the column of y-term coefficients replaced by c_1 and c_2.

   y-coefficients replaced by c_1 and c_2
   $$\mathbf{D}_y = \begin{vmatrix} a_1 & c_1 \\ a_2 & c_2 \end{vmatrix}$$

It is important to note that the linear equations must be written in standard form to apply Cramer's rule.

Example 6 — Using Cramer's Rule to Solve a 2 × 2 System of Linear Equations

Solve the system by using Cramer's rule.
$$-16y = -40x - 7$$
$$40y = 24x + 27$$

Solution:

$-16y = -40x - 7 \longrightarrow 40x - 16y = -7$ Rewrite each equation
$40y = 24x + 27 \longrightarrow -24x + 40y = 27$ in standard form.

For this system: $a_1 = 40$ $b_1 = -16$ $c_1 = -7$
                   $a_2 = -24$ $b_2 = 40$ $c_2 = 27$

$$\mathbf{D} = \begin{vmatrix} 40 & -16 \\ -24 & 40 \end{vmatrix} = (40)(40) - (-16)(-24) = 1216$$

$$\mathbf{D}_x = \begin{vmatrix} -7 & -16 \\ 27 & 40 \end{vmatrix} = (-7)(40) - (-16)(27) = 152$$

$$\mathbf{D}_y = \begin{vmatrix} 40 & -7 \\ -24 & 27 \end{vmatrix} = (40)(27) - (-7)(-24) = 912$$

Therefore, $x = \dfrac{\mathbf{D}_x}{\mathbf{D}} = \dfrac{152}{1216} = \dfrac{1}{8}$ $y = \dfrac{\mathbf{D}_y}{\mathbf{D}} = \dfrac{912}{1216} = \dfrac{3}{4}$

The ordered pair $\left(\dfrac{1}{8}, \dfrac{3}{4}\right)$ checks in both original equations.

The solution set is $\left\{\left(\dfrac{1}{8}, \dfrac{3}{4}\right)\right\}$.

Skill Practice Solve by using Cramer's rule.

7. $9x = 12y - 8$
   $30y = -18x - 7$

Answer

7. $\left\{\left(-\dfrac{2}{3}, \dfrac{1}{6}\right)\right\}$

Cramer's rule can be used to solve a 3 × 3 system of linear equations by using a similar pattern of determinants.

> **Cramer's Rule for a 3 × 3 System of Linear Equations**
>
> The solution to the system
> $$a_1x + b_1y + c_1z = d_1$$
> $$a_2x + b_2y + c_2z = d_2$$
> $$a_3x + b_3y + c_3z = d_3$$
>
> is given by $\quad x = \dfrac{D_x}{D} \quad y = \dfrac{D_y}{D} \quad$ and $\quad z = \dfrac{D_z}{D}$
>
> Where $D = \begin{vmatrix} a_1 & b_1 & c_1 \\ a_2 & b_2 & c_2 \\ a_3 & b_3 & c_3 \end{vmatrix}$ (and $D \neq 0$) $\quad D_x = \begin{vmatrix} d_1 & b_1 & c_1 \\ d_2 & b_2 & c_2 \\ d_3 & b_3 & c_3 \end{vmatrix}$
>
> $D_y = \begin{vmatrix} a_1 & d_1 & c_1 \\ a_2 & d_2 & c_2 \\ a_3 & d_3 & c_3 \end{vmatrix} \quad\quad D_z = \begin{vmatrix} a_1 & b_1 & d_1 \\ a_2 & b_2 & d_2 \\ a_3 & b_3 & d_3 \end{vmatrix}$

Example 7   **Using Cramer's Rule to Solve a 3 × 3 System of Linear Equations**

Solve the system by using Cramer's rule.
$$2x - 3y + 5z = 11$$
$$-5x + 7y - 2z = -6$$
$$9x - 2y + 3z = 4$$

Solution:

TIP: In Example 7, we expanded the determinants about the first column.

$$D = \begin{vmatrix} 2 & -3 & 5 \\ -5 & 7 & -2 \\ 9 & -2 & 3 \end{vmatrix} = 2 \cdot \begin{vmatrix} 7 & -2 \\ -2 & 3 \end{vmatrix} - (-5) \cdot \begin{vmatrix} -3 & 5 \\ -2 & 3 \end{vmatrix} + 9 \cdot \begin{vmatrix} -3 & 5 \\ 7 & -2 \end{vmatrix}$$
$$= 2(17) + 5(1) + 9(-29)$$
$$= -222$$

$$D_x = \begin{vmatrix} 11 & -3 & 5 \\ -6 & 7 & -2 \\ 4 & -2 & 3 \end{vmatrix} = 11 \cdot \begin{vmatrix} 7 & -2 \\ -2 & 3 \end{vmatrix} - (-6) \cdot \begin{vmatrix} -3 & 5 \\ -2 & 3 \end{vmatrix} + 4 \cdot \begin{vmatrix} -3 & 5 \\ 7 & -2 \end{vmatrix}$$
$$= 11(17) + 6(1) + 4(-29)$$
$$= 77$$

$$D_y = \begin{vmatrix} 2 & 11 & 5 \\ -5 & -6 & -2 \\ 9 & 4 & 3 \end{vmatrix} = 2 \cdot \begin{vmatrix} -6 & -2 \\ 4 & 3 \end{vmatrix} - (-5) \cdot \begin{vmatrix} 11 & 5 \\ 4 & 3 \end{vmatrix} + 9 \cdot \begin{vmatrix} 11 & 5 \\ -6 & -2 \end{vmatrix}$$
$$= 2(-10) + 5(13) + 9(8)$$
$$= 117$$

$$D_z = \begin{vmatrix} 2 & -3 & 11 \\ -5 & 7 & -6 \\ 9 & -2 & 4 \end{vmatrix} = 2 \cdot \begin{vmatrix} 7 & -6 \\ -2 & 4 \end{vmatrix} - (-5) \cdot \begin{vmatrix} -3 & 11 \\ -2 & 4 \end{vmatrix} + 9 \cdot \begin{vmatrix} -3 & 11 \\ 7 & -6 \end{vmatrix}$$

$$= 2(16) + 5(10) + 9(-59)$$
$$= -449$$

$$x = \frac{D_x}{D} = \frac{77}{-222} = -\frac{77}{222}$$

$$y = \frac{D_y}{D} = \frac{117}{-222} = -\frac{39}{74}$$

$$z = \frac{D_z}{D} = \frac{-449}{-222} = \frac{449}{222}$$

The solution $\left(-\frac{77}{222}, -\frac{39}{74}, \frac{449}{222}\right)$ checks in each of the original equations.

The solution set is $\left\{\left(-\frac{77}{222}, -\frac{39}{74}, \frac{449}{222}\right)\right\}$.

Skill Practice Solve by using Cramer's rule.
8. $x + 3y - 3z = -14$
   $x - 4y + z = 2$
   $x + y + 2z = 6$

Cramer's rule may seem cumbersome for solving a 3×3 system of linear equations. However, it provides convenient formulas that can be programmed into a computer or calculator to solve for x, y, and z. Cramer's rule can also be extended to solve a 4×4 system of linear equations, a 5×5 system of linear equations, and in general an $n \times n$ system of linear equations.

It is important to remember that Cramer's rule does not apply if $\mathbf{D} = 0$. In such a case, either the equations are dependent or the system is inconsistent, and another method may be needed to analyze the system.

Example 8 Analyzing a Dependent System of Equations

Solve the system. Use Cramer's rule if possible.
$$2x - 3y = 6$$
$$-6x + 9y = -18$$

Solution:

$$D = \begin{vmatrix} 2 & -3 \\ -6 & 9 \end{vmatrix} = (2)(9) - (-3)(-6) = 18 - 18 = 0$$

TIP: When Cramer's rule does not apply, that is, when $D = 0$, you may also use the substitution method or the Gauss-Jordan method to get a solution.

Because $\mathbf{D} = 0$, Cramer's rule does not apply. Using the addition method to solve the system, we have

$$2x - 3y = 6 \quad \xrightarrow{\text{Multiply by 3.}} \quad 6x - 9y = 18$$
$$-6x + 9y = -18 \quad \longrightarrow \quad \underline{-6x + 9y = -18}$$
$$0 = 0 \quad \text{The equations are dependent.}$$

The solution set is $\{(x, y) \mid 2x - 3y = 6\}$.

Answer
8. $\{(-2, 0, 4)\}$

Skill Practice Solve. Use Cramer's rule if possible.

9. $x - 6y = 2$
   $2x - 12y = -2$

TIP: In a 2 × 2 system of equations, if **D** = 0, then the equations are dependent or the system is inconsistent.
- If **D** = 0 and both $\mathbf{D}_x = 0$ and $\mathbf{D}_y = 0$, then the equations are dependent and the system has infinitely many solutions.
- If **D** = 0 and either $\mathbf{D}_x \neq 0$ or $\mathbf{D}_y \neq 0$, then the system is inconsistent and has no solution.

Answer

9. { }; inconsistent system

Section A.3 Practice Exercises

Vocabulary and Key Concepts

1. **a.** Given the matrix $\mathbf{A} = \begin{bmatrix} a & b \\ c & d \end{bmatrix}$, the _____ of **A** is denoted det **A** and is written as $\begin{vmatrix} a & b \\ c & d \end{vmatrix}$. The value of det **A** is the real number equal to _____.

   b. Given a 3 × 3 matrix, the _____ of an element in the matrix is the determinant of the 2 × 2 matrix formed by deleting the row and column in which the element resides.

   c. Complete the expression on the right to represent the value of the 3 × 3 determinant shown here.
   $$\begin{vmatrix} a_1 & b_1 & c_1 \\ a_2 & b_2 & c_2 \\ a_3 & b_3 & c_3 \end{vmatrix} = a_1 \begin{vmatrix} b_2 & c_2 \\ b_3 & c_3 \end{vmatrix} - \square \begin{vmatrix} b_1 & c_1 \\ b_3 & c_3 \end{vmatrix} + a_3 \begin{vmatrix} \square & \square \\ \square & \square \end{vmatrix}$$

Concept 1: Introduction to Determinants

For Exercises 2–7, evaluate the determinant of the 2 × 2 matrix. **(See Example 1.)**

2. $\begin{vmatrix} -3 & 1 \\ 5 & 2 \end{vmatrix}$

3. $\begin{vmatrix} 5 & 6 \\ 4 & 8 \end{vmatrix}$

4. $\begin{vmatrix} -2 & 2 \\ -3 & -5 \end{vmatrix}$

5. $\begin{vmatrix} 5 & -1 \\ 1 & 0 \end{vmatrix}$

6. $\begin{vmatrix} \frac{1}{2} & 3 \\ -2 & 4 \end{vmatrix}$

7. $\begin{vmatrix} -3 & \frac{1}{4} \\ 8 & -2 \end{vmatrix}$

Concept 2: Determinant of a 3 × 3 Matrix

For Exercises 8–11, evaluate the minor corresponding to the given element from matrix **A**. **(See Example 2.)**

$$\mathbf{A} = \begin{bmatrix} 4 & -1 & 8 \\ 2 & 6 & 0 \\ -7 & 5 & 3 \end{bmatrix}$$

8. 4

9. −1

10. 2

11. 3

For Exercises 12–15, evaluate the minor corresponding to the given element from matrix **B**.

$$\mathbf{B} = \begin{bmatrix} -2 & 6 & 0 \\ 4 & -2 & 1 \\ 5 & 9 & -1 \end{bmatrix}$$

12. 6   **13.** 5   **14.** 1   **15.** 0

16. Construct the sign array for a 3 × 3 matrix.

17. Evaluate the determinant of matrix **B**, using expansion by minors. (See Exercises 3–4.)

$$\mathbf{B} = \begin{bmatrix} 0 & 1 & 2 \\ 3 & -1 & 2 \\ 3 & 2 & -2 \end{bmatrix}$$

   a. About the first column

   b. About the second row

18. Evaluate the determinant of matrix **C**, using expansion by minors.

$$\mathbf{C} = \begin{bmatrix} 4 & 1 & 3 \\ 2 & -2 & 1 \\ 3 & 1 & 2 \end{bmatrix}$$

   a. About the first row

   b. About the second column

19. When evaluating the determinant of a 3 × 3 matrix, explain the advantage of being able to choose any row or column about which to expand minors.

For Exercises 20–25, evaluate the determinant. (See Examples 3–4.)

20. $\begin{vmatrix} 8 & 2 & -4 \\ 4 & 0 & 2 \\ 3 & 0 & -1 \end{vmatrix}$ **21.** $\begin{vmatrix} 5 & 2 & 1 \\ 3 & -6 & 0 \\ -2 & 8 & 0 \end{vmatrix}$ **22.** $\begin{vmatrix} -2 & 1 & 3 \\ 1 & 4 & 4 \\ 1 & 0 & 2 \end{vmatrix}$

23. $\begin{vmatrix} 3 & 2 & 1 \\ 1 & -1 & 2 \\ 1 & 0 & 4 \end{vmatrix}$ **24.** $\begin{vmatrix} -5 & 4 & 2 \\ 0 & 0 & 0 \\ 3 & -1 & 5 \end{vmatrix}$ **25.** $\begin{vmatrix} 0 & 5 & -8 \\ 0 & -4 & 1 \\ 0 & 3 & 6 \end{vmatrix}$

For Exercises 26–31, evaluate the determinant.

26. $\begin{vmatrix} x & 3 \\ y & -2 \end{vmatrix}$ **27.** $\begin{vmatrix} a & 2 \\ b & 8 \end{vmatrix}$ **28.** $\begin{vmatrix} a & 5 & -1 \\ b & -3 & 0 \\ c & 3 & 4 \end{vmatrix}$

29. $\begin{vmatrix} x & 0 & 3 \\ y & -2 & 6 \\ z & -1 & 1 \end{vmatrix}$ **30.** $\begin{vmatrix} p & 0 & q \\ r & 0 & s \\ t & 0 & u \end{vmatrix}$ **31.** $\begin{vmatrix} f & e & 0 \\ d & c & 0 \\ b & a & 0 \end{vmatrix}$

Concept 3: Cramer's Rule

For Exercises 32–34, evaluate the determinants represented by $\mathbf{D}$, $\mathbf{D}_x$, and $\mathbf{D}_y$.

32. $x - 4y = 2$
     $3x + 2y = 1$

33. $4x + 6y = 9$
     $-2x + y = 12$

34. $-3x + 8y = -10$
     $5x + 5y = -13$

For Exercises 35–40, solve the system by using Cramer's rule. (See Examples 5–6.)

35. $2x + y = 3$
     $x - 4y = 6$

36. $2x - y = -1$
     $3x + y = 6$

37. $4y = x - 8$
     $3x = -7y + 5$

38. $7x - 4 = -3y$
$5x = 4y + 9$

39. $4x - 3y = 5$
$2x + 5y = 7$

40. $2x + 3y = 4$
$6x - 12y = -5$

For Exercises 41–46, solve for the indicated variable by using Cramer's rule. **(See Example 7.)**

41. $2x - y + 3z = 9$
$x + 4y + 4z = 5$ for x
$3x + 2y + 2z = 5$

42. $x + 2y + 3z = 8$
$2x - 3y + z = 5$ for y
$3x - 4y + 2z = 9$

43. $3x - 2y + 2z = 5$
$6x + 3y - 4z = -1$ for z
$3x - y + 2z = 4$

44. $4x + 4y - 3z = 3$
$8x + 2y + 3y = 0$ for x
$4x - 4y + 6z = -3$

45. $5x + 6z = 5$
$-2x + y = -6$ for y
$3y - z = 3$

46. $8x + y = 1$
$7y + z = 0$ for y
$x - 3z = -2$

47. When does Cramer's rule not apply in solving a system of equations?

48. How can a system be solved if Cramer's rule does not apply?

For Exercises 49–58, solve the system by using Cramer's rule, if possible. Otherwise, use another method. If a system does not have a unique solution, determine the number of solutions and whether the system is inconsistent or the equations are dependent. **(See Example 8.)**

49. $4x - 2y = 3$
$-2x + y = 1$

50. $6x - 6y = 5$
$x - y = 8$

51. $4x + y = 0$
$x - 7y = 0$

52. $-3x - 2y = 0$
$-x + 5y = 0$

53. $x + 5y = 3$
$2x + 10y = 6$

54. $-2x - 10y = -4$
$x + 5y = 2$

55. $x = 3$
$-x + 3y = 3$
$y + 2z = 4$

56. $4x + z = 7$
$y = 2$
$x + z = 4$

57. $x + y + 8z = 3$
$2x + y + 11z = 4$
$x + 3z = 0$

58. $-8x + y + z = 6$
$2x - y + z = 3$
$3x - z = 0$

Expanding Your Skills

For Exercises 59–62, solve the equation.

59. $\begin{vmatrix} 6 & x \\ 2 & -4 \end{vmatrix} = 14$

60. $\begin{vmatrix} y & -2 \\ 8 & 7 \end{vmatrix} = 30$

61. $\begin{vmatrix} 3 & 1 & 0 \\ 0 & 4 & -2 \\ 1 & 0 & w \end{vmatrix} = 10$

62. $\begin{vmatrix} -1 & 0 & 2 \\ 4 & t & 0 \\ 0 & -5 & 3 \end{vmatrix} = -4$

For Exercises 63–64, evaluate the determinant by using expansion by minors about the first column.

63. $\begin{vmatrix} 1 & 0 & 3 & 0 \\ 0 & 1 & 2 & 4 \\ -2 & 0 & 0 & 1 \\ 4 & -1 & -2 & 0 \end{vmatrix}$

64. $\begin{vmatrix} 5 & 2 & 0 & 0 \\ 0 & 4 & -1 & 1 \\ -1 & 0 & 3 & 0 \\ 0 & -2 & 1 & 0 \end{vmatrix}$

For Exercises 65–66, refer to the following system of four variables.

$$x + y + z + w = 0$$
$$2x - z + w = 5$$
$$2x + y - w = 0$$
$$ y + z = -1$$

65. **a.** Evaluate the determinant **D**.

   b. Evaluate the determinant $\mathbf{D}_x$.

   c. Solve for x by computing $\dfrac{\mathbf{D}_x}{\mathbf{D}}$.

66. **a.** Evaluate the determinant $\mathbf{D}_y$.

   b. Solve for y by computing $\dfrac{\mathbf{D}_y}{\mathbf{D}}$.

67. Two angles are complementary. The measure of one angle is $\frac{5}{7}$ the measure of the other. Find the measures of the two angles.

68. Two angles are supplementary. The measure of the larger angle is $61°$ more than $\frac{3}{4}$ the measure of the smaller angle. Find the measures of the angles.

69. A theater charges $80 per ticket for seats in section A, $50 per ticket for seats in section B, and $30 per ticket for seats in section C. For one performance, 3000 tickets were sold for a total of $165,000 in revenue. If the total number of tickets in sections A and C is equal to the number of tickets sold in section B, how many tickets in each section were sold?

70. The measure of the largest angle in a triangle is $80°$ larger than the sum of the measures of the other two angles. The measure of the smallest angle is $22°$ less than the measure of the middle angle. Find the measure of each angle.

71. Suppose 1000 people were surveyed in southern California, and 445 said that they worked-out at least three times a week. If $\frac{1}{2}$ of the women and $\frac{3}{8}$ of the men said that they worked-out at least three times a week, how many men and how many women were in the survey?

72. During a 1-hr television program, there were 22 commercials. Some commercials were 15 sec long and some were 30 sec long. Find the number of 15-sec commercials and the number of 30-sec commercials if the total playing time for commercials was 9.5 min.

Section A.4 Introduction to Modeling

Concepts

1. Introduction to Modeling
2. Recognize Characteristics of Linear Behavior
3. Recognize Characteristics of Quadratic Behavior
4. Recognize Characteristics of Exponential Behavior

1. Introduction to Modeling

The natural world is a vastly complicated network of interactions among people, animals, and nature. Thus, when we study phenomena in the world around us, we often need to describe the relationships between two or more variables. For example,

- The dosage y of a medicine prescribed by a doctor depends on the patient's weight x.
- The perimeter P of a rectangle depends on the length l and the width w of the rectangle.
- The amount of interest I earned by an investment depends on the amount of time t that the principal is invested.

Defining the relationships among variables is called **mathematical modeling**. There are various ways to express one variable in terms of one or more other variables. For example, consider the recommended dosage of a pain reliever given to a child between 30 lb and 100 lb.

Verbal Model:
A doctor recommends that 5 mg of a pain reliever be given per pound of the child's weight.

Algebraic Model (equation):
Let x represent a child's weight in pounds.
Let y represent the recommended dosage of pain reliever.
Then $y = 5x$ for x between 30 lb and 100 lb.

Numerical Model:

Weight (lb) x	Dosage (mg) y
30	150
40	200
50	250
60	300
70	350
80	400
90	450
100	500

Graphical Model:

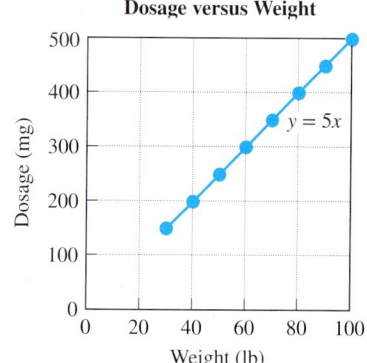

Figure A-1

In this example, the dosage y is the variable to be predicted or computed based on the patient's weight. Therefore, we call the dosage y the **dependent variable** because it *depends* on the weight x. The variable x is called the **independent** variable. In a scientific experiment, the independent variable is changed or controlled to test the effects on the dependent variable.

2. Recognize Characteristics of Linear Behavior

The graph of dosage versus weight in Figure A-1 is a line. This means that the dosage and weight are *linearly* related. The fundamental characteristic of a **linear model** is that the slope is constant and represents the rate of change of the dependent variable relative to the independent variable. Furthermore, linear models may have a positive slope, a negative slope, or a zero slope (Figures A-2 through A-4).

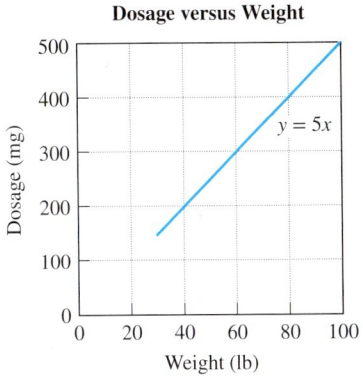

Figure A-2

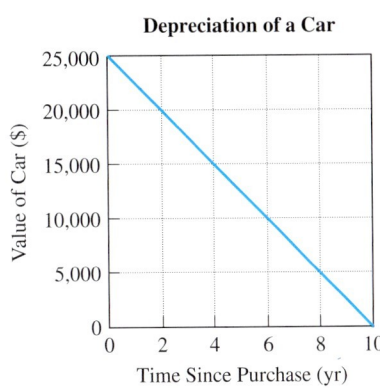

Figure A-3

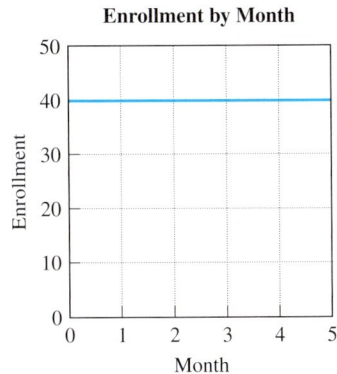

Figure A-4

In Figure A-2, the line rises upward from left to right. In such a case, we say that the graph is **increasing**. As a child's weight increases, so does the dosage.

The line in Figure A-3 represents the value y (in $) of a car x years after purchase. The value decreases and the line slopes downward from left to right. In such a case, we say that the graph is **decreasing**. The value of the car decreases with time.

The line in Figure A-4 represents enrollment in a first aid course offered monthly by the fire department. The course is very popular and is capped at 40 students. Thus, the enrollment is constant for the time period shown. The slope of the line is zero and we say that the graph is **constant**.

In Example 1 we develop a linear model based on a fixed value and a rate of change representing the slope.

Example 1 Writing a Linear Equation from a Verbal Model

A printing company charges $25 to design a business card and a printing fee of $0.06 per card.

a. Write an equation to model the cost y (in $) to print x cards.
b. Graph the equation from part (a).
c. Determine the values of x for which the graph is increasing, decreasing, or constant.
d. Use the model from part (a) to determine the cost to print 350 business cards.

Solution:

a. The cost to print x cards is made up of the printing cost and the design cost.

$$\begin{pmatrix} \text{Total} \\ \text{Cost} \end{pmatrix} = \begin{pmatrix} \text{Cost to} \\ \text{print } x \text{ cards} \end{pmatrix} + \begin{pmatrix} \text{Cost of the} \\ \text{design} \end{pmatrix}$$

$$y = 0.06x + 25 \quad \text{for } x > 0$$

b.

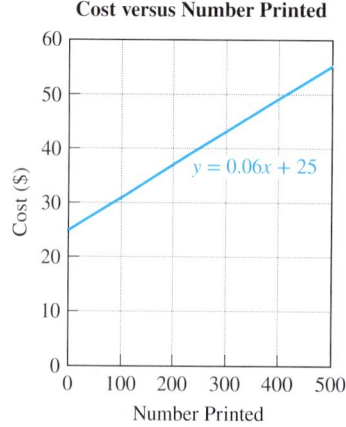

Cost versus Number Printed

c. The graph is increasing for $x > 0$.

d. $y = 0.06x + 25$

$y = 0.06(350) + 25$

$= 46$ The cost to print 350 cards is $46.00 before tax.

Skill Practice

1. A catering company charges $100 plus $35 per person to provide dinner at a wedding.

   a. Write an equation to model the cost y (in $) for a group of x people.

   b. Graph the equation from part (a).

   c. Determine the values of x for which the graph is increasing, decreasing, or constant.

   d. Use the model from part (a) to determine the cost of dinner for 45 people.

Example 2   Identifying a Linear Model from Observed Data

The data in the table give the depth y of a retention pond, x days after the start of a drought.

Day x	0	7	14	21	28
Depth (ft) y	6.00	5.25	4.50	3.75	3.0

a. Plot the points. The graph of a set of points representing observed data is called a **scatter diagram**.

b. Do x and y follow a linear relationship?

c. If the variables follow a linear relationship, what is the slope of the line?

d. Write an equation that models the depth y on day x.

e. Find the x- and y-intercepts and interpret their meaning in context.

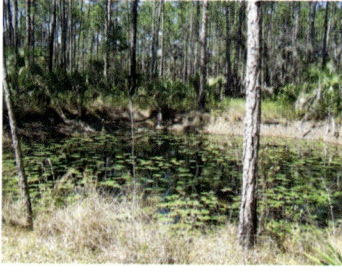

Courtesy of Julie Miller

Answers

1. a. $y = 35x + 100$
   b.

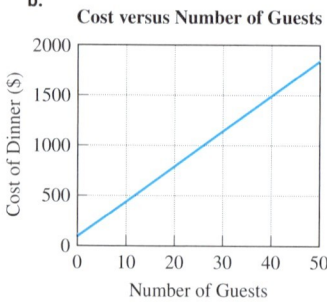

Cost versus Number of Guests

   c. The graph is increasing for $x > 0$.
   d. $1675

Solution:

a.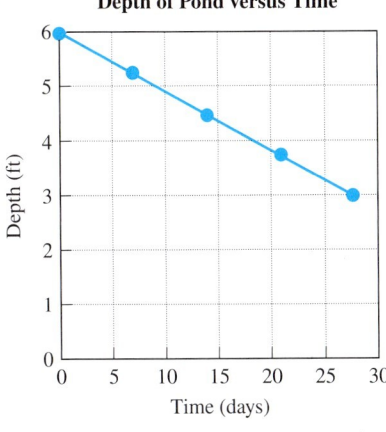
Depth of Pond versus Time

b. The relationship between x and y is linear because the data points are on the same line.

c. Choose any two points on the line from which to compute the slope. We will use $(0, 6)$ and $(28, 3)$ for (x_1, y_1) and (x_2, y_2), respectively.

$$m = \frac{y_2 - y_1}{x_2 - x_1} = \frac{3 - 6}{28 - 0} = -\frac{3}{28}$$

TIP: The slope of the line between any two points given in the table is the same.

d. The y-intercept is $(0, 6)$ and the slope is $m = -\frac{3}{28}$. Therefore, using slope-intercept form $y = mx + b$, we have $y = -\frac{3}{28}x + 6$.

e. The y-intercept $(0, 6)$ is given and means that the water level on day 0 (at the start of the drought) is 6 ft.

To find the x-intercept, substitute 0 for y and solve for x.

$$0 = -\frac{3}{28}x + 6 \qquad \text{Substitute 0 for } y.$$

$$\frac{3}{28}x = 6 \qquad \text{Add } \frac{3}{28}x \text{ to both sides.}$$

$$\frac{28}{3} \cdot \frac{3}{28}x = \frac{28}{3} \cdot 6 \qquad \text{Multiply both sides by the reciprocal of } \frac{3}{28}.$$

$$x = 56$$

The x-intercept is $(56, 0)$ and means that on the 56th day, the water level is 0 ft. The pond is dry.

Skill Practice

2. The data in the table give the depth of snowpack y (in inches) at a ski resort for x weeks at the end of winter.

Week x	0	1	2	3	4
Depth (in.) y	78	71.5	65	58.5	52

a. Plot the points.
b. Do x and y follow a linear relationship?
c. If the variables follow a linear relationship, what is the slope of the line? Interpret its meaning in context.
d. Write an equation that models the depth y at week x.
e. Find the x- and y-intercepts and interpret their meaning in context.

Answers

2. a. Depth of Snow versus Time
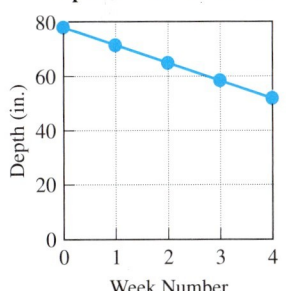

b. Yes
c. The slope is -6.5 and means that snow depth decreased by 6.5 in. per week during this period.
d. $y = -6.5x + 78$
e. The y-intercept is $(0, 78)$ and means that the snow depth was 78 in. at week 0. The x-intercept is $(12, 0)$ and means that at week 12, the depth of the snow was 0 in. The snow had all melted.

3. Recognize Characteristics of Quadratic Behavior

A linear relationship between two variables is the simplest form of mathematical model. However, many variables are not linearly related; that is, their relationship is **nonlinear**. We will next focus on two types of nonlinear models called quadratic models and exponential models.

A linear model can be expressed as an equation in slope-intercept form: $y = mx + b$. A **quadratic model** is expressed as $y = ax^2 + bx + c$ where $a \neq 0$.

The graph of a quadratic model is a **parabola**. A parabola resembles a "U" shape or an inverted "U" shape depending on whether the leading coefficient a is positive or negative. In Figure A-5, the graph of $y = x^2 - 2x - 3$ is shown. Notice that the leading coefficient (on the x^2 term) is positive 1 and the parabola opens upward in a "U" shape. In Figure A-6, we show the graph of $y = -2x^2 + 4$. The leading coefficient is -2 (negative), so the parabola opens downward in an inverted "U" shape.

TIP: Notice that a linear model has y expressed in terms of a first-degree polynomial in x. A quadratic model on the other hand expresses y in terms of a second-degree polynomial in x.

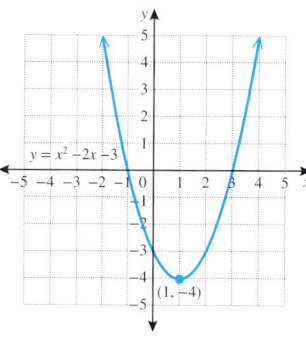

Figure A-5

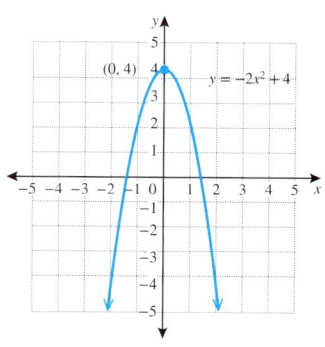

Figure A-6

From the graphs in Figures A-5 and A-6, we make the following observations about the graph of a quadratic model (parabola).

- The lowest point on a parabola opening upward is called the **vertex** of the parabola. Likewise, on a parabola opening downward, the highest point is the vertex. Think of the vertex as the point at the bottom of the "U" shape or at the top of an inverted "U" shape.
- The vertex of the graph in Figure A-5 is $(1, -4)$. Moving from left to right, notice that the graph decreases to the left of the vertex and increases to the right of the vertex. That is, the graph decreases for $x < 1$ and increases for $x > 1$.
- The vertex of the graph in Figure A-6 is $(0, 4)$. Notice that the graph increases to the left of the vertex and decreases to the right. That is, the graph increases for $x < 0$ and decreases for $x > 0$.

Example 3 Identifying Characteristics of a Quadratic Model

A golf ball is struck at ground level at an angle of 30° from the horizontal. The height y (in ft) of the ball, x seconds after being struck is given in the table.

Time (sec) x	0	0.5	1.0	1.5	2.0	2.5	3.0	3.5	4.0
Height (ft) y	0	28	48	60	64	60	48	28	0

a. Plot the points given in the table and sketch a curve through the points.
b. Describe the type of relationship between y and x.
c. Identify the vertex of the parabola and interpret its meaning in context.
d. Consider the parabola through the observed data points. Identify the values of x for which the graph is increasing.
e. Identify the values of x for which the graph is decreasing.

Solution:

a. Height of Golf Ball versus Time

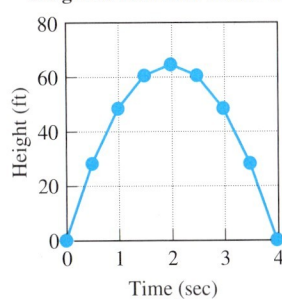

b. From the graph of the data, the relationship between y and x appears quadratic.

c. The vertex is (2, 64). This means that 2 sec after being struck, the ball reaches its maximum height of 64 ft.

d. The curve increases for $0 < x < 2$. This represents the times for which the ball is increasing in height.

e. The curve decreases for $2 < x < 4$. This represents the times for which the ball is decreasing in height.

Skill Practice

3. A child kicks a soccer ball from ground level at an angle of 45° from the horizontal. The height y (in ft) of the ball, x seconds after being kicked is given in the table. Repeat Example 3 using the given data.

Time (sec) x	0	0.5	1.0	1.5	2.0
Height (ft) y	0	12	16	12	0

4. Recognize Characteristics of Exponential Behavior

We now consider exponential models. These are nonlinear relationships whose equations have the form $y = b^x$ where b is a positive real number not equal to 1. In Figures A-7 and A-8, we show the graphs of $y = 2^x$ and $y = (\frac{1}{2})^x$, each based on a table of points.

From the graphs in Figures A-7 and A-8, we make the following observations about the graphs of exponential models.

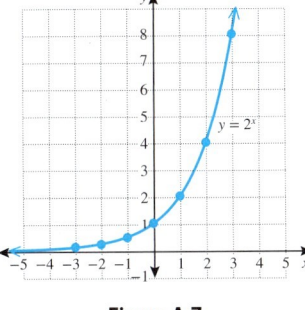

Figure A-7

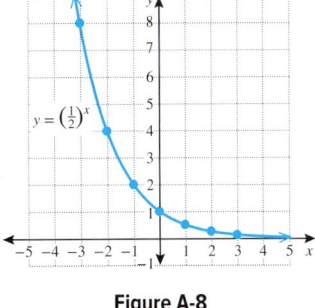

Figure A-8

If the base is greater than 1 ($b > 1$),
- The graph is always increasing ("rises" upward from left to right). This is called **exponential growth**.
- For large values of x, the y values are unbounded.
- For negative values of x taken far to the left of the origin, the y values get close to zero, so the graph gets very close to (but does not touch) the x-axis.

Answers

3. a. Height of Ball versus Time

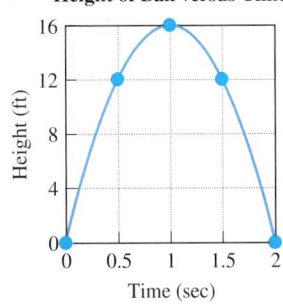

b. The relationship appears quadratic.
c. The vertex is (1, 16) and means that 1 sec after being kicked, the ball reaches its maximum height of 16 ft.
d. The curve increases for $0 < x < 1$. This represents the times for which the ball is increasing in height.
e. The curve decreases for $1 < x < 2$. This represents the times for which the ball is decreasing in height.

If the base is between 0 and 1 ($0 < b < 1$),

- The graph is always decreasing ("falls" downward from left to right). This is called **exponential decay**.
- For large values of x, the y values get close to zero, so the graph gets very close to (but does not touch) the x-axis.
- For negative values of x taken far to the left of the origin, the y values are unbounded.

Exponential growth is useful to model situations such as unbounded population growth. The population starts out growing slowly and then later, increases rapidly.

Exponential decay is useful to model a situation such as the temperature of a cake coming out of the oven. The temperature decreases rapidly at first and then levels off to room temperature. Likewise, scientists also use exponential decay to model the amount of a radioactive substance over time.

Example 4 Identifying Characteristics of an Exponential Model

A microbiologist starts out with a culture of Streptococcus bacteria with an estimated 1000 organisms. She also knows that Streptococcus grown in milk under optimal conditions, will double every 26 min. Based on this information, the population y at a time x minutes after the culture is started is modeled by the equation $y = 1000(2)^{x/26}$.

Time (min) x	Population $y = 1000(2)^{x/26}$
0	
26	
52	
78	
104	

a. Use the equation to complete the table.
b. Interpret the meaning of the first three ordered pairs in the table.

Solution:

a.

Time (min) x	Population $y = 1000(2)^{x/26}$
0	1000
26	2000
52	4000
78	8000
104	16,000

$y = 1000(2)^{(0)/26} = 1000(2)^0 = 1000(1) = 1000$
$y = 1000(2)^{(26)/26} = 1000(2)^1 = 1000(2) = 2000$
$y = 1000(2)^{(52)/26} = 1000(2)^2 = 1000(4) = 4000$
$y = 1000(2)^{(78)/26} = 1000(2)^3 = 1000(8) = 8000$
$y = 1000(2)^{(104)/26} = 1000(2)^4 = 1000(16) = 16{,}000$

b. The ordered pair (0, 1000) means that there were 1000 organisms initially.

The ordered pair (26, 2000) means that after 26 min, the projected population is 2000. This is reasonable, because 26 min is the doubling time for the population.

The ordered pair (52, 4000) means that after 52 min (two doubling times), the projected population is 4000. The population doubled twice: first from 1000 to 2000 organisms and then from 2000 to 4000 organisms.

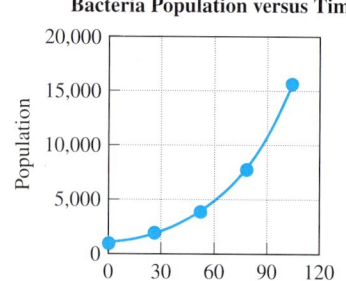

Bacteria Population versus Time

Skill Practice

4. The half-life of a widely used antibiotic is approximately 60 min. This means that half of the initial amount will remain after 60 min. Given a 500 mg dose, the amount of drug y (in mg) that remains in the body after x minutes is given by
$$y = 500\left(\frac{1}{2}\right)^{x/60}.$$
   a. Use the equation to complete the table.
   b. Interpret the meaning of the first three ordered pairs in the table.

Time (min) x	0	60	120	180
Amount of drug (mg) y $\quad y = 500\left(\frac{1}{2}\right)^{x/60}$				

It should also be noted that observed data that exhibit linear, quadratic, or exponential behavior usually do not follow these patterns "perfectly." (See Figures A-9 through A-11). For example, data values that show a linear trend such as those in Figure A-9 may not line up perfectly. However, using techniques presented in later courses in algebra and statistics, we can often find good models to represent the general trend and behavior.

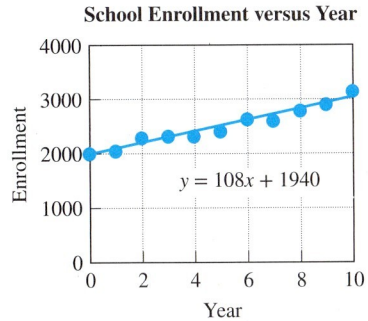

Figure A-9

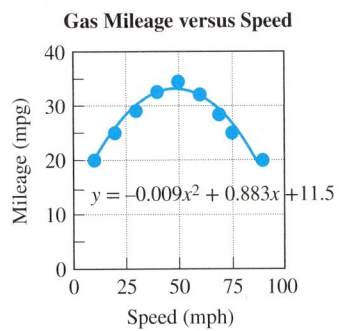

Figure A-10

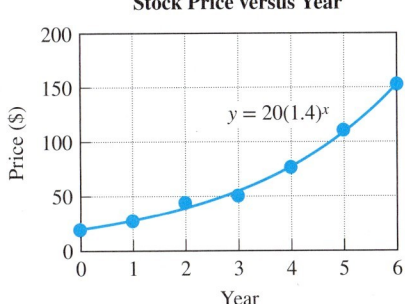

Figure A-11

Answers

4. a.

Time (min) x	Amount of drug (mg) y
0	500
60	250
120	125
180	62.5

b. (0, 500) means that initially, there was 500 mg of the drug in the body. (60, 250) means that after 60 min, 250 mg of the drug remains in the body. (120, 125) means that after 120 min, 125 mg of the drug remains in the body.

Section A.4 Practice Exercises

Vocabulary and Key Concepts

1. A line with a negative slope (increases/decreases) from left to right.
2. If a graph is constant over a given interval for x, what is the slope?
3. The graph of a quadratic model is in the shape of what letter? What is the formal name of the graph of a quadratic model?
4. For the graph of an exponential growth model, the graph (increases/decreases) for increasing values of x.

Concept 2: Recognize Characteristics of Linear Behavior

5. A two-bedroom apartment at an apartment complex near campus requires a deposit of $650 plus rent of $920 per month. **(See Example 1.)**

   a. Write an equation to model the cost y (in $) to rent the apartment for x months.
   b. Graph the equation from part (a).
   c. Determine the values of x for which the graph is increasing, decreasing, or constant.
   d. Use the model from part (a) to determine the total cost to rent the apartment for 15 months.

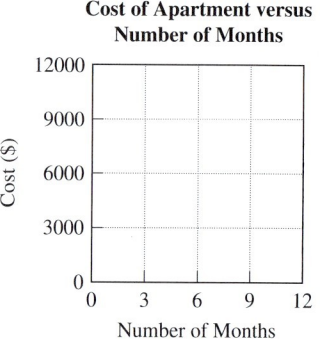

6. Jessica has $3250 in an account. She withdraws $125 per month to make payments on her student loan.

   a. Write an equation to model the amount y (in $) that remains in the account after x months.
   b. Graph the equation from part (a).
   c. Determine the values of x for which the graph is increasing, decreasing, or constant.
   d. Use the model from part (a) to determine the amount of money in the account after 1 yr.

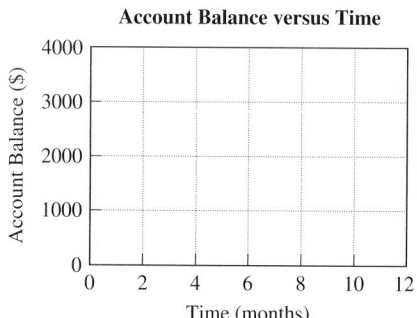

7. The data in the table give the value y of a motorcycle x years after it was purchased. **(See Example 2.)**

Years x	0	1	3	5	7
Value ($) y	14,520	13,200	10,560	7920	5280

   a. Plot the points and sketch a curve through the points.
   b. Do x and y follow a linear relationship?
   c. If the variables follow a linear relationship, what is the slope of the line? Interpret the meaning of the slope in context.
   d. Write an equation that models the value y for year x.
   e. Find the x- and y-intercepts and interpret their meaning in context.

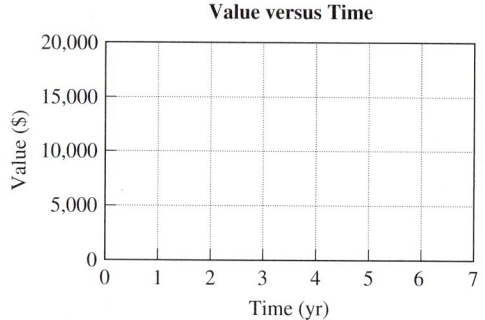

8. Jaden works out on a treadmill that displays the number of calories burned during his workout. The data in the table give the number of calories burned, y for a workout lasting x minutes.

Time (min) x	0	15	25	30	45
Calories y	0	153	255	306	459

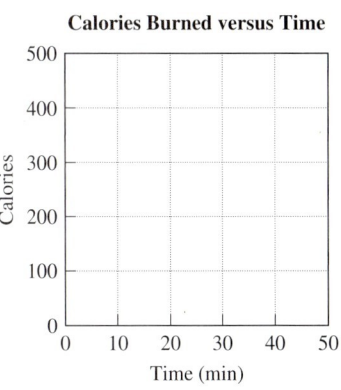

a. Plot the points and sketch a curve through the points.
b. Do x and y follow a linear relationship?
c. If the variables follow a linear relationship, what is the slope of the line? Interpret the meaning of the slope in context.
d. Write an equation that models the number of calories burned y for a workout lasting x minutes.
e. Find the x- and y-intercepts and interpret their meaning in context.

Concept 3: Recognize Characteristics of Quadratic Behavior

9. A pyrotechnic company launches fireworks straight into the air from the ground. The height y (in ft) of the fireworks, x seconds after being launched is given in the table. (See Example 3.)

Time (sec) x	0	1	2	3	4	5	6
Height (ft) y	0	80	128	144	128	80	0

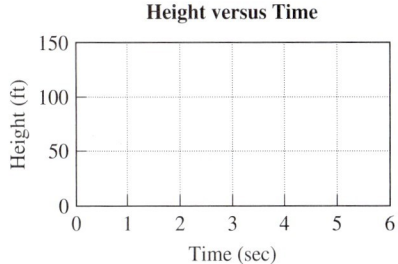

a. Plot the points given in the table and sketch a curve through the points.
b. Describe the type of relationship between y and x.
c. Identify the vertex of the parabola and interpret its meaning in context.
d. Consider the parabola through the observed data points. Identify the values of x for which the graph is increasing.
e. Identify the values of x for which the graph is decreasing.

10. The table given shows the average number of visits to doctor's offices per year, y, according to the age of an individual.

Age (yr) x	5	10	20	25	30	40	45
Number of Visits y	2.64	2.36	2.04	2	2.04	2.36	2.64

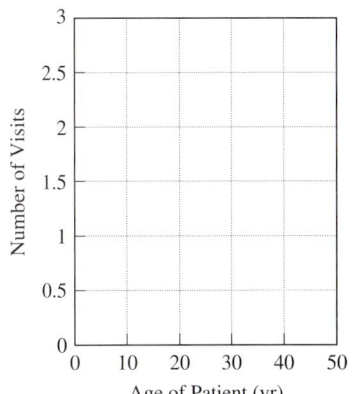

a. Plot the points given in the table and sketch a curve through the points.
b. Describe the type of relationship between y and x.
c. Identify the vertex of the parabola and interpret its meaning in context.
d. Consider the parabola through the observed data points. Identify the values of x for which the graph is decreasing.
e. Identify the values of x for which the graph is increasing.

Concept 4: Recognize Characteristics of Exponential Behavior

11. At one point during an Ebola outbreak in a densely populated area of west Africa, it was estimated that the number of cases of Ebola doubled every 3 weeks. At the time that researchers began the study, there were 200 cases of Ebola. The equation $y = 200(2)^{x/3}$ relates the number of cases of Ebola y to the time x in weeks after the study began. **(See Example 4.)**
    a. Use the equation to complete the table.
    b. What type of relationship exists between x and y?
    c. Interpret the meaning of the first three ordered pairs in the table.

Time (wks) x	Ebola Cases $y = 200(2)^{x/3}$
0	
3	
6	
12	
15	

12. Radon is a colorless, odorless gas that can seep up from the Earth and can cause health problems. The half-life of radon is 3.8 days. Given 100 grams of radon, the amount of radon y remaining after x days is given by $y = 100\left(\dfrac{1}{2}\right)^{x/3.8}$.
    a. Use the equation to complete the table.
    b. What type of relationship exists between x and y?
    c. Interpret the meaning of the first three ordered pairs in the table.

Time (days) x	Radon (gm) y $y = 100\left(\dfrac{1}{2}\right)^{x/3.8}$
0	
3.8	
7.6	
11.4	

Mixed Exercises

For Exercises 13–18, identify which type of model would best fit the scatter diagram. Choose from

a. Linear model with positive slope: $y = mx + b$ $(m > 0)$

b. Linear model with negative slope: $y = mx + b$ $(m < 0)$

c. Quadratic model: $y = ax^2 + bx + c$ $(a > 0)$ (parabola opening upward)

d. Quadratic model: $y = ax^2 + bx + c$ $(a < 0)$ (parabola opening downward)

e. Exponential growth model: $y = ab^x$ $(b > 1)$

f. Exponential decay model: $y = ab^x$ $(0 < b < 1)$

13.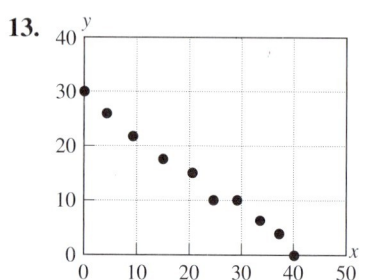

14.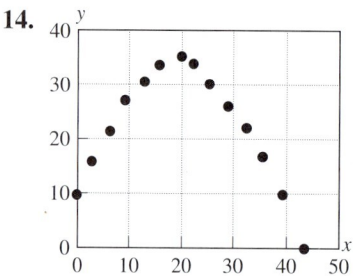

15.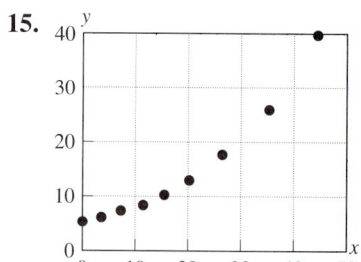

16. 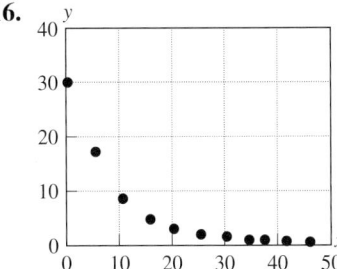 **17.** 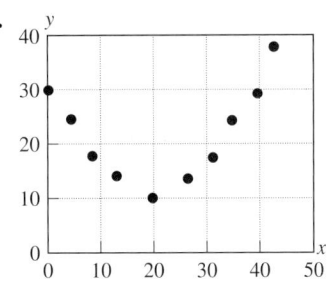 **18.**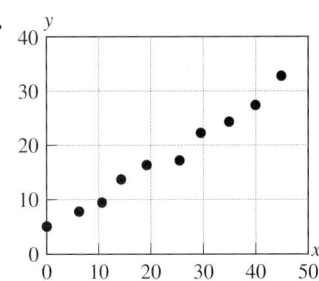

For Exercises 19–24, refer to the graphs in Exercises 13–18 as indicated and estimate the values of x for which the graph is increasing, decreasing, or constant. If the graph is a parabola, estimate the vertex.

19. Exercise 13

20. Exercise 14

21. Exercise 15

22. Exercise 16

23. Exercise 17

24. Exercise 18

For Exercises 25–30, based on the characteristics of graphs (a)–(f), match the equations and graphs.

25. $y = -1.6x + 8$

26. $y = 4(1.5)^x$

27. $y = 3\left(\dfrac{1}{4}\right)^x$

28. $y = -3x^2 + 6x + 2$

29. $y = x^2 + 4x + 1$

30. $y = \dfrac{1}{3}x + 2$

a.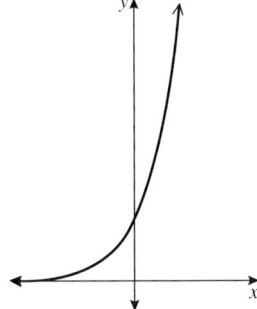

b.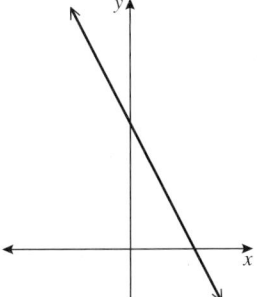

c.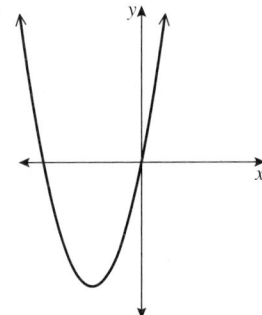

d.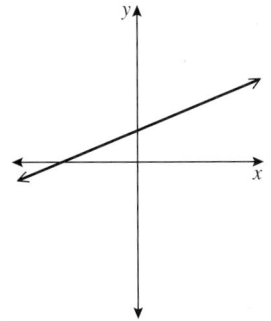

e.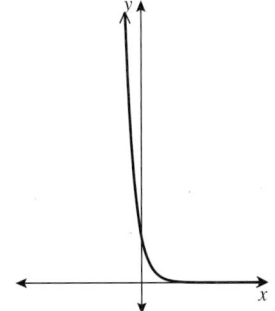

f.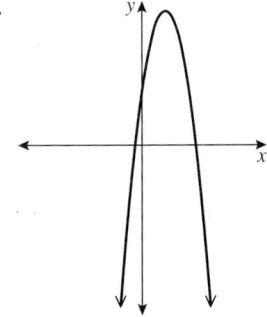

For Exercises 31–32, the data in the table follow a linear relationship. Fill in the missing table values.

31.

Time (hr) x	Distance (mi) y
2	96
4	192
6	?
8	384
10	?

32.

Time (hr) x	Value of Car ($) y
2	18,600
4	14,880
6	?
8	7440
10	?

For Exercises 33–44, determine if the statement suggests a linear, quadratic, or exponential model.

33. The median income for men and women in the United States has risen at a constant rate from the year 2000 to the year 2015.

34. The profit for a small printing company to produce x items rises to a point and then falls once the market becomes saturated.

35. The amount of money earned over time with compound interest rises slowly at first and then increases more rapidly.

36. The use of land line telephones has been dropping over time since the year 1990. The drop began slowly and has dropped more rapidly with time.

37. The number of vehicular accidents for drivers between 16 years old and 90 years old changes with age. The number of accidents is high for younger drivers and then drops as drivers approach middle age. The number of accidents then increases as drivers reach the age of senior citizens.

38. The consumption of beef has decreased at a steady rate during a period of a bad economy.

39. Alzheimer's disease usually occurs in individuals who are 60 years and older. After the age of 65, the probability of developing the disease doubles every 5 yr.

40. The kidneys filter the blood and remove caffeine and other drugs through urine. As a result, the biological half-life of caffeine is approximately 6 hr. This means that the amount of caffeine in the body is cut in half every 6 hr.

41. The temperature at which water boils decreases at a constant rate as altitude increases.

42. The cost of a speeding ticket in a rural western town is $80 plus $5 for each mph above the speed limit.

43. An object is tossed up in the air from the top of a building. The height of the object increases to a point and then decreases as it falls toward the ground.

44. After a toxic chemical leaked into a lake, the population of trout decreased by ¼ every month.

45. A shopping cart rolls across a parking lot at a constant speed and then crashes into a tree and stops. Which graph of speed versus time represents this scenario? Explain your answer.

a.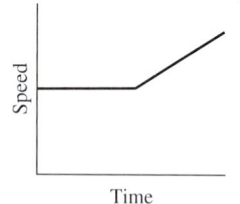

b.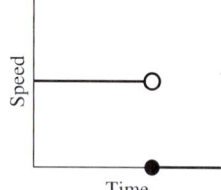

c.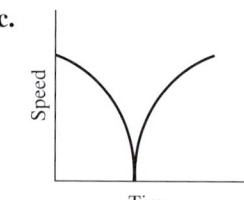

d.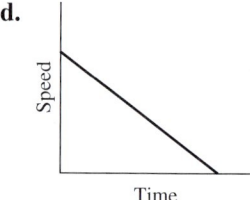

46. A student makes $16 per hour for the first 40 hr worked in a week. Then she makes time and half ($24/hr) for the work exceeding 40 hr. Which graph best represents her total salary versus time? Explain your reasoning.

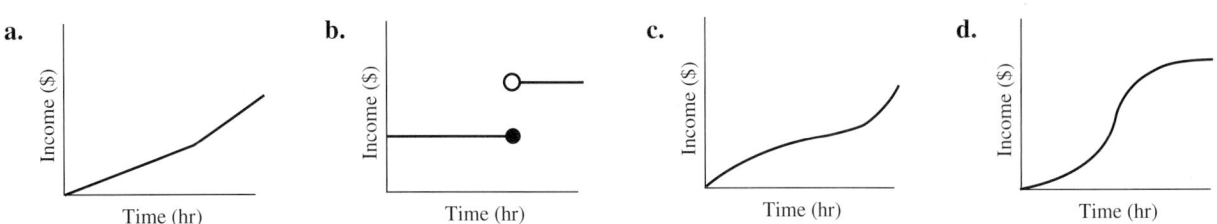

a.   b.   c.   d.

47. A linear model with a positive slope and an exponential growth model are both increasing for all values of x. How are the graphs different?

48. Describe the behavior of a quadratic model to the left and right of the vertex.

There are many other types of nonlinear models that describe everyday situations. For Exercises 49–50, apply the given model.

49. The concentration C (in ng/mL) of a drug in the bloodstream t hours after ingestion is shown in the graph and is modeled by
$$C = \frac{600t}{t^3 + 125}.$$

   a. Determine the concentration at 1 hr, 12 hr, 24 hr, and 48 hr after ingestion. Round to the nearest tenth of a ng/mL.

   b. What appears to be the limiting concentration for large values of t?

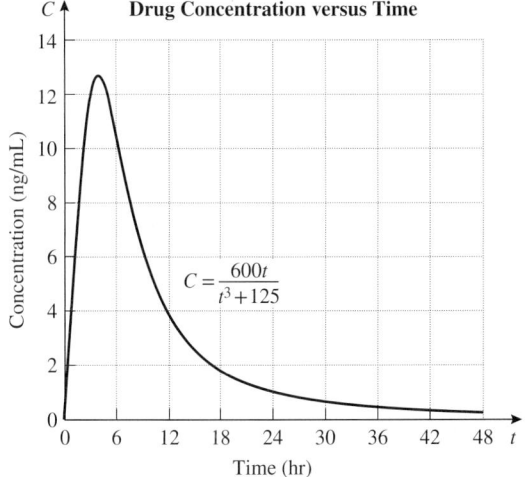

50. An object that is originally 35°C is placed in a freezer. The temperature T (in °C) of the object can be approximated by the model $T = \dfrac{350}{t^2 + 3t + 10}$, where t is the time in hours after the object is placed in the freezer.

   a. Determine the temperature at 2 hr, 4 hr, 12, hr, and 24 hr. Round to the nearest tenth of a degree Celsius.

   b. What appears to be the limiting temperature for large values of t?

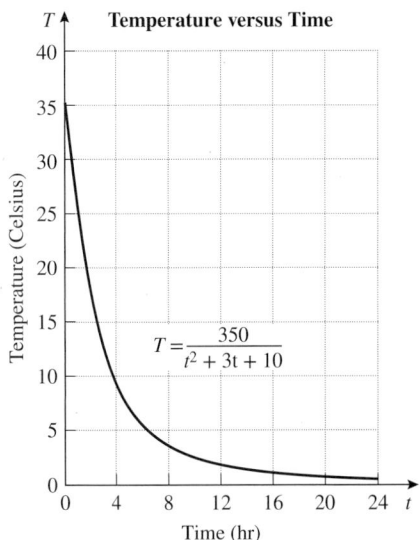

Technology Connections

For Exercises 51–56, use a graphing utility (graphing calculator or online graphing utility) to confirm the graphs in Exercises 25–30 on a window defined by $-10 \leq x \leq 10$ and $-10 \leq y \leq 10$.

51. $y = -1.6x + 8$

52. $y = 4(1.5)^x$

53. $y = 3\left(\dfrac{1}{4}\right)^x$

54. $y = -3x^2 + 6x + 2$

55. $y = x^2 + 4x + 1$

56. $y = \dfrac{1}{3}x + 2$

Expanding Your Skills

57. Suppose that a population y of bacteria begins at time $x = 0$ with 600 organisms and doubles every 30 min.

   a. Complete the table.

Time (min) x	Population y
30	
60	
90	
120	
150	

   b. Write an equation that represents y in terms of x. (*Hint:* See Exercise 11.)

58. A cake comes out of the oven at 160°C and is placed in a freezer at 0°C. Based on the physical properties of the cake, the temperature is cut in half approximately every 35 min.

   a. Complete the table.

Time (min) x	Temperature (°C) y
35	
70	
105	
140	
175	

   b. Write an equation represents y in terms of x. (*Hint:* See Exercise 12.)

59. The perimeter and area of a square both depend on the length of the sides x.

   a. Write an equation that expresses the area of the square A in terms of x. Is the relationship between A and x linear, quadratic, or exponential?

   b. Write an equation that expresses the perimeter of the square P in terms of x. Is the relationship between P and x linear, quadratic, or exponential?

   c. Complete the table.

Side (in.) x	Area (in.²) A	Perimeter (in.) P
2		
4		
6		
8		
10		

60. The circumference and area of a circle both depend on the radius r.

   a. Write an equation that expresses the circumference of the circle C in terms of r. Is the relationship between C and r linear, quadratic, or exponential?

   b. Write an equation that expresses the area of the circle A in terms of r. Is the relationship between A and r linear, quadratic, or exponential?

   c. Complete the table. Write the results in terms of π.

Radius (cm) x	Circumference (cm) C	Area (cm²) A
2		
4		
6		
8		
10		

Student Answer Appendix

Chapter 1

Section 1.1 Activity, p. 6

A.1. a. Hundred-thousands **b.** 1,000,000 + 900,000 + 2,000 + 500 + 5
c. One million, nine hundred two thousand, five hundred five
A.2. 1,896,531
A.3. a. 210 > 201 or 201 < 210
b. Two hundred ten is greater than two hundred one, or two hundred one is less than two hundred ten.
A.4. a. 2233 < 2323 or 2323 > 2233
b. Two thousand, two hundred thirty-three is less than two thousand, three hundred twenty-three, or two thousand three hundred twenty-three is greater than two thousand two hundred thirty-three.
A.5. a. 79 > 76 or 76 < 79
b. Seventy-nine is greater than seventy-six, or seventy-six is less than seventy-nine.
A.6. a. 614 < 641 or 641 > 614
b. Six hundred fourteen is less than six hundred forty-one, or six hundred forty-one is greater than six hundred fourteen.
A.7. a. 9,752 **b.** 2,579 **c.** Two thousand, five hundred seventy-nine.

Section 1.1 Practice Exercises, pp. 6–9

1. a. periods **b.** hundreds **c.** thousands
3. 7: ones; 5: tens; 4: hundreds; 3: thousands;
1: ten-thousands; 2: hundred-thousands; 8: millions
5. Tens **7.** Ones **9.** Hundreds **11.** Thousands
13. Hundred-thousands **15.** Billions
17. Ten-thousands **19.** Millions **21.** Ten-millions
23. Billions **25.** 5 tens + 8 ones; 5 × 10 + 8 × 1
27. 5 hundreds + 3 tens + 9 ones; 5 × 100 + 3 × 10 + 9 × 1
29. 5 thousands + 2 hundreds + 3 ones; 5 × 1,000 +
2 × 100 + 3 × 1 **31.** 1 ten-thousand + 2 hundreds + 4 tens +
1 one; 1 × 10,000 + 2 × 100 + 4 × 10 + 1 × 1 **33.** 524
35. 150 **37.** 1906 **39.** 85,007
41. Ones, thousands, millions, billions
43. Two hundred forty-one **45.** Six hundred three
47. Thirty-one thousand, five hundred thirty
49. One hundred thousand, two hundred thirty-four
51. Nine thousand, five hundred thirty-five
53. Twenty thousand, three hundred ten
55. Five hundred ninety thousand, seven hundred twelve
57. 6005 **59.** 672,000 **61.** 1,484,250
63.

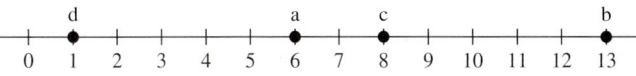

65. 10 **67.** 4
69. 8 is greater than 2, or 2 is less than 8
71. 3 is less than 7, or 7 is greater than 3
73. < **75.** > **77.** < **79.** > **81.** < **83.** <
85. False **87.** 99 **89.** There is no greatest whole number.
91. 7 **93.** 964

Section 1.2 Activity, pp. 19–20

A.1. a. The sum of 13 and 6. **b.** The total of 2, 8, and 10.
A.2. a. 60 **b.** 60 **c.** Yes **d.** The commutative property of addition.
A.3. a–c. Answers will vary. Some possible answers are minus, difference, decreased by, less than, subtracted from.
A.4. a. Lucas is correct.
b. The statement 12 minus 8 is equal to 4. The statement 12 subtracted from 8 is equal to −4. The answers are not equal because the order in which two numbers are subtracted affects the difference.
A.5. a. 941 ft **b.** 252 ft
A.6. 390 yd **A.7. a.** 20 ft **b.** 85 cm

Section 1.2 Practice Exercises, pp. 20–25

1. a. addends **b.** sum **c.** variable **d.** commutative **e.** a; a
f. $a + (b + c)$ **g.** minuend; subtrahend; difference **h.** polygon
i. perimeter
3. commutative **5.** Increase
7.

+	0	1	2	3	4	5	6	7	8	9
0	0	1	2	3	4	5	6	7	8	9
1	1	2	3	4	5	6	7	8	9	10
2	2	3	4	5	6	7	8	9	10	11
3	3	4	5	6	7	8	9	10	11	12
4	4	5	6	7	8	9	10	11	12	13
5	5	6	7	8	9	10	11	12	13	14
6	6	7	8	9	10	11	12	13	14	15
7	7	8	9	10	11	12	13	14	15	16
8	8	9	10	11	12	13	14	15	16	17
9	9	10	11	12	13	14	15	16	17	18

9. Addends: 1, 13, 4; sum: 18 **11.** 75 **13.** 59
15. 997 **17.** 119 **19.** 121 **21.** 111
23. 889 **25.** 701 **27.** 203 **29.** 15,203
31. 40,985 **33.** 44 + 101 **35.** $y + x$
37. 23 + (9 + 10) **39.** $(r + s) + t$
41. The commutative property changes the order of the addends, and the associative property changes the grouping.
43. Minuend: 12; subtrahend: 8; difference: 4
45. 18 + 9 = 27 **47.** 27 + 75 = 102 **49.** 5
51. 3 **53.** 1126 **55.** 1103 **57.** 17 **59.** 521
61. 4764 **63.** 1403 **65.** 2217 **67.** 713
69. 30,941 **71.** 5,662,119 **73.** The expression 7 − 4 means 7 minus 4, yielding a difference of 3. The expression 4 − 7 means 4 minus 7, which results in a difference of −3. (This is a mathematical skill we have not yet learned.)
75. 13 + 7; 20 **77.** 7 + 45; 52 **79.** 18 + 5; 23
81. 1523 + 90; 1613 **83.** 5 + 39 + 81; 125
85. 422 − 100; 322 **87.** 1090 − 72; 1018
89. 50 − 13; 37 **91.** 103 − 35; 68
93. a. 6010 ft **b.** 12,039 ft **95.** $200
97. Denali is 6064 ft higher than White Mountain Peak.
99. 7748 **101.** 195,489 **103.** 821,024 nonteachers

SA-1

105. 4256 ft   **107.** The total amount was $26,631.   **109.** 104 cm
111. 42 yd   **113.** 288 ft   **115.** 13 m   **117.** 192,780
118. 21,491,394   **119.** 5,257,179   **120.** 4,447,302
121. 897,058,513   **122.** 2,906,455   **123.** 49,408 mi²
124. 17,139 mi²   **125.** 96,572 mi²   **126.** 224,368 mi²

Section 1.3 Activity, p. 30

A.1. a. 4   **b.** 4
  c. Less than 5; keep it the same   **d.** 1400
  e.

  1,400   1,450   1,500

  f. Yes

A.2. a.

  490   495   500

  b. It is halfway between 490 and 500.   **c.** Round up   **d.** 500

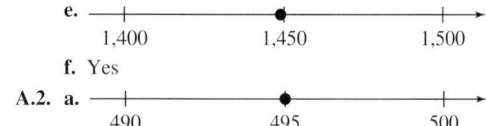

	Tens	Hundreds	Thousands
A.3. 3456	**a.** 3460	**b.** 3500	**c.** 3000
A.4. 9753	**a.** 9750	**b.** 9800	**c.** 10,000
A.5. 7999	**a.** 8000	**b.** 8000	**c.** 8000
A.6. 2199	**a.** 2200	**b.** 2200	**c.** 2000

A.7. a. Community college = 37,000; University = 27,000
  b. 10,000 students

Section 1.3 Practice Exercises, pp. 30–33

1. rounding   **3.** right
5. If the digit in the tens place is 0, 1, 2, 3, or 4, then change the tens and ones digits to 0. If the digit in the tens place is 5, 6, 7, 8, or 9, increase the digit in the hundreds place by 1 and change the tens and ones digits to 0.
7. Ten-thousands
9. 70

  70   80

11. 340   **13.** 730   **15.** 9400   **17.** 8500
19. 35,000   **21.** 3000   **23.** 10,000   **25.** 109,000
27. 490,000   **29.** $370,000   **31.** 239,000 mi
33. 160   **35.** 220   **37.** 1000   **39.** 2100
41. $151,000,000   **43.** 5,000,000 more votes
45. $10,000,000   **47. a.** Year 4; $3,500,000   **b.** Year 6; $2,000,000   **49.** Massachusetts; 79,000 students
51. 72,000 students   **53.** 10,000 mm   **55.** 440 in.

Section 1.4 Activity, p. 42

A.1. $5(6) = 30$ **A.2.** $5(2) + 5(4) = 10 + 20 = 30$
A.3. The results are the same; distributive property of multiplication over addition
A.4. 8; d   **A.5.** 2; a   **A.6.** $6 \cdot 9$; e   **A.7.** 0; b   **A.8.** 4; c
A.9. a. 32   **b.** 320   **c.** 3200   **d.** 32,000   **e.** Answers may vary. Possible answer: The products all contain the digits 32 followed by as many 0's as there are in the two numbers being multiplied.
A.10. $3120
A.11. a. 9000 yd²   **b.** $27,000

Section 1.4 Practice Exercises, pp. 42–46

1. a. factors; product   **b.** commutative   **c.** associative; $a \cdot (b \cdot c)$ **d.** 0; 0   **e.** a; a **f.** distributive; $a \cdot b + a \cdot c$ **g.** area   **h.** $l \cdot w$
3. Possible answers: $2 \cdot 6 = 12$, $2(6) = 12$, $(2)(6) = 12$
5. $4 \cdot 6$ **7.** 6×5; 30   **9.** 3×9; 27
11. Factors: 13, 42; product: 546
13. Factors: 3, 5, 2; product: 30

15. For example: 5×12; $5 \cdot 12$; $5(12)$
17. d   **19.** e   **21.** c   **23.** $8 \cdot 14$ **25.** $(6 \cdot 2) \cdot 10$
27. $(5 \cdot 7) + (5 \cdot 4)$ **29.** 144   **31.** 52   **33.** 655
35. 1376   **37.** 11,280   **39.** 23,184   **41.** 378,126
43. 448   **45.** 1632   **47.** 864   **49.** 2431   **51.** 6631
53. 19,177   **55.** 186,702   **57.** 21,241,448
59. 4,047,804   **61.** 24,000   **63.** 2,100,000
65. 72,000,000   **67.** 36,000,000   **69.** 60,000,000
71. 2,400,000,000   **73.** $1000   **75.** $1,370,000
77. 4000 min   **79.** $1665   **81.** 144 fl oz
83. 287,500 sheets   **85.** 372 mi   **87.** 276 ft²
89. 5329 cm²   **91.** 105,300 mi²   **93. a.** 2400 in.²
b. 42 windows   **c.** 100,800 in.²   **95.** 128 ft²

Section 1.5 Activity, p. 54

A.1. a. $50 \div 2$; $\dfrac{50}{2}$; $2\overline{)50}$
  b. Dividend = 50
    Divisor = 2
    Quotient = 25
  c. No. 50 is evenly divisible by 2.
  d. Multiplication; $2(25) = 50$
A.2. a. 0; Because $0(8) = 0$.
  b. Undefined; Because no number when multiplied by 0 equals 8.
  c. No. The answers are different because division is not commutative.
A.3. a. 1   **b.** 1   **c.** 1   **d.** Any nonzero number divided by itself is 1.
A.4. a. 15   **b.** 7   **c.** x **d.** Any number divided by 1 is the number itself.
A.5. a. $6\overline{)521}$ **b.** 86 R5   **c.** $6(86) + 5 = 521$
A.6. a. $400   **b.** $300   **c.** $240   **d.** The 5-year loan.

Section 1.5 Practice Exercises, pp. 54–58

1. dividend; divisor; quotient   **b.** 1   **c.** 5   **d.** 0   **e.** undefined   **f.** remainder
3. multiplication   **5.** No, $15 \div 5 \neq 5 \div 15$
7. Possible answers: divide, quotient, per, divides into, divided, or shared equally among
9. $18 \div 3 = 6$
11. 9; the dividend is 72; the divisor is 8; the quotient is 9
13. 8; the dividend is 64; the divisor is 8; the quotient is 8
15. 5; the dividend is 45; the divisor is 9; the quotient is 5
17. You cannot divide a number by zero (the quotient is undefined). If you divide zero by a number (other than zero), the quotient is always zero.
19. 15   **21.** 0   **23.** Undefined   **25.** 1
27. Undefined   **29.** 0   **31.** $2 \cdot 3 = 6$, $2 \cdot 6 \neq 3$
33. Multiply the quotient and the divisor to get the dividend.
35. 13   **37.** 41   **39.** 486   **41.** 409
43. 203   **45.** 822   **47.** Correct   **49.** Incorrect; 253 R2
51. Correct   **53.** Incorrect; 25 R3   **55.** 7 R5
57. 10 R2   **59.** 27 R1   **61.** 197 R2   **63.** 42 R4
65. 1557 R1   **67.** 751 R6   **69.** 835 R2   **71.** 479 R9
73. 43 R19   **75.** 308   **77.** 1259 R26   **79.** 22
81. 35 R1   **83.** 229 R96   **85.** 302   **87.** $497 \div 71$; 7
89. $877 \div 14$; 62 R9   **91.** $42 \div 6$; 7
93. 14 classrooms   **95.** 5 cases; 8 cans left over
97. There will be 120 classes of Beginning Algebra.
99. It will use 9 gal.
101. $1200 \div 20 = 60$; approximately 60 words per minute
103. Yes, they can all attend if they sit in the second balcony.
105. 7,665,000,000 bbl   **106.** 13,000 min
107. $888 billion   **108.** Each crate weighs 255 lb.

Chapter 1 Problem Recognition Exercises, p. 58

1. **a.** 120 **b.** 72 **c.** 2304 **d.** 4
2. **a.** 575 **b.** 525 **c.** 13,750 **d.** 22
3. **a.** 946 **b.** 612 **4. a.** 278 **b.** 612
5. **a.** 1201 **b.** 5500 **6. a.** 34,855 **b.** 22,718
7. **a.** 20,000 **b.** 400 **8. a.** 34,524 **b.** 548
9. **a.** 230 **b.** 5060 **10. a.** 15 **b.** 1875
11. **a.** 328 **b.** 4 **12. a.** 8 **b.** 547
13. **a.** 4180 **b.** 41,800 **c.** 418,000 **d.** 4,180,000
14. **a.** 35,000 **b.** 3500 **c.** 350 **d.** 35

Section 1.6 Activity, pp. 64–65

A.1. **a.** Base = 3 **b.** Exponent = 2 **c.** $3 \cdot 3$ **d.** 9
A.2. **a.** What positive number, when squared, equals 9?
  b. Finding the square root is reversing the process of squaring a number.
A.3. 8   A.4. 25   A.5. 64   A.6. 4   A.7. 8   A.8. 10
A.9. **a.** 11; you multiply or divide before you add. **b.** When expressions have multiple operations, you can obtain incorrect results by completing operations in different orders. **c.** 1. Perform operations inside parentheses or other grouping symbols. When multiple grouping symbols are present, simplify the innermost set first, then work your way out. 2. Simplify exponents or square roots. 3. Multiply or divide in the order from left to right. 4. Add or subtract in the order from left to right.
A.10. **a.** 19 **b.** 169 **c.** Yes, because in part (a) you are squaring 3 and then adding 10. In part (b) you are adding 10 and 3 and then squaring the resulting sum.
A.11. 3   A.12. 14   A.13. 65   A.14. 1
A.15. **a.** $10 - 5 - 2$ **b.** 3 **c.** No, the answer is numerical.
A.16. 7   A.17. 11   A.18. 1

Section 1.6 Practice Exercises, pp. 65–67

1. **a.** base; 4 **b.** powers **c.** square root; 81 **d.** order; operations **e.** variable; constants
3. squared   5. zeros   7. innermost   9. 9^4
11. 3^6 13. $4^4 \cdot 2^3$ 15. $8 \cdot 8 \cdot 8 \cdot 8$
17. $4 \cdot 4 \cdot 4 \cdot 4 \cdot 4 \cdot 4 \cdot 4$ 19. 8   21. 9
23. 27   25. 125   27. 32   29. 81
31. The number 1 raised to any power is 1.   33. 1000
35. 100,000   37. 2   39. 6   41. 10   43. 0
45. No, addition and subtraction should be performed in the order in which they appear from left to right.

47. 26   49. 1   51. 49   53. 3   55. 2
57. 53   59. 8   61. 45   63. 24   65. 4
67. 40   69. 5   71. 26   73. 4   75. 50
77. 2   79. 0   81. 5   83. 6   85. 3
87. 201   89. 6   91. 15   93. 10   95. 32
97. 24   99. 75   101. 400   103. 3   104. 24,336
105. 174,724   106. 248,832   107. 1,500,625   108. 79,507
109. 357,911   110. 8028   111. 293,834   112. 66,049
113. 1728   114. 35   115. 43

Section 1.7 Practice Exercises, pp. 71–74

R.1. $20 - 15$; 5   R.3. $71 + 14$; 85
R.5. $(10)(13)$; 130   R.7. $24 \div 6$; 4

1. mean   3. Jackson's monthly payment was $390.
5. Each trip will take 2 hr.   7. Perimeter
9. The cost will be $86.   11. The cost is $1020.
13. There will be $36 left in Gina's account.
15. The total bill was $154,032.
17. **a.** Latayne will receive $48. **b.** She can buy 6 CDs.
19. Michael Jordan scored 33,454 points with the Bulls.
21. **a.** One bottle will last for 30 days.
  b. The owner should order a refill no later than September 28.
23. **a.** The distance is 360 mi. **b.** 14 in. represents 840 mi.
25. 104 boxes will be filled completely with 2 books left over.
27. **a.** Marc needs five $20 bills. **b.** He will receive $16 in change.
29. He earned $520.   31. 19   33. 1154   35. 89
37. The average mileage rating was 34 mpg.
39. 16 in. per month

Chapter 1 Review Exercises, pp. 75–78

1. Ten-thousands   2. Hundred-thousands   3. 92,046
4. 503,160   5. 3 millions + 4 hundred-thousands + 8 hundreds + 2 tens; $3 \times 1,000,000 + 4 \times 100,000 + 8 \times 100 + 2 \times 10$
6. 3 ten-thousands + 5 hundreds + 5 tens + 4 ones; $3 \times 10,000 + 5 \times 100 + 5 \times 10 + 4 \times 1$
7. Two hundred forty-five
8. Thirty thousand, eight hundred sixty-one
9. 3602   10. 800,039
11. (number line with point at 7; 1–13)
12. (number line with point at 7; 1–13)
13. True   14. False   15. Addends: 105, 119; sum: 224
16. Addends: 53, 21; sum: 74   17. 71   18. 54
19. 17,410   20. 70,642   21. **a.** Commutative property **b.** Associative property **c.** Commutative property
22. Minuend: 14; subtrahend: 8; difference: 6
23. Minuend: 102; subtrahend: 78; difference: 24   24. 26
25. 20   26. 121   27. 1090   28. 31,019
29. 34,188   30. $403 + 79$; 482   31. $44 + 92$; 136
32. $38 - 31$; 7   33. $111 - 15$; 96   34. $36 + 7$; 43
35. $23 + 6$; 29   36. $251 - 42$; 209   37. $90 - 52$; 38
38. **a.** 96 cars **b.** 66 Fords   39. 45,797 thousand seniors
40. 7709 thousand   41. 45,096 thousand
42. 71,893,000 tons   43. $7,200,000   44. 177 m
45. 5,000,000   46. 9,330,000   47. 800,000
48. 1500   49. 13,000,000 people   50. 163,000 m^3
51. Factors: 33, 40; product: 1320
52. **a.** Yes **b.** Yes **c.** No   53. c   54. e   55. d
56. a   57. b   58. 6106   59. 52,224
60. 3,000,000   61. $429   62. 7714 lb
63. 7; divisor: 6, dividend: 42, quotient: 7
64. 13; divisor: 4, dividend: 52, quotient: 13   65. 3
66. 1   67. Undefined   68. 0
69. Multiply the quotient and the divisor to get the dividend.
70. Multiply the whole number part of the quotient and the divisor, and then add the remainder to get the dividend.
71. 58   72. 41 R7   73. 52 R3   74. $\frac{72}{4}$; 18
75. $9\overline{)108}$; 12   76. 26 photos with 1 left over
77. **a.** 4 T-shirts **b.** 5 hats   78. 8^5 79. $2^4 \cdot 5^3$
80. 125   81. 256   82. 1   83. 1,000,000   84. 8
85. 12   86. 7   87. 75   88. 90   89. 15   90. 12
91. 55   92. 3   93. 42   94. 0   95. 4   96. 100
97. **a.** 21 mi **b.** 840 mi   98. He received $19,600,000 per year.
99. **a.** She should purchase 48 plants. **b.** The plants will cost $144.
c. The fence will cost $80. **d.** Aletha's total cost will be $224.
100. 8   101. $89   102. 8 houses per month

Chapter 2

Section 2.1 Activity, pp. 83–84

A.1. a. Answers will vary. Some possible answers include a negative balance in a checking account, negative temperatures, and elevations below sea level.

A.2. a.

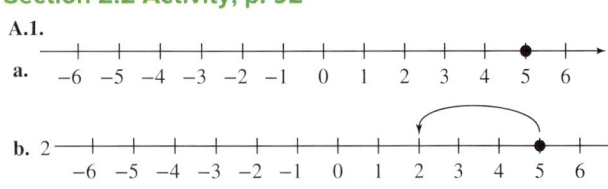

b. > **c.** < **d.** < **e.** 5 **f.** 5

g. Yes, because they are the same distance from zero on the number line. **h.** Yes, because they are the same distance from zero on the number line.

A.3. a. 7; This represents the opposite of −7, which is 7.
b. −7; This represents the opposite of |−7|. Since |−7| is equal to 7, the opposite is −7.

A.4. 2 **A.5.** 6 **A.6.** −15 **A.7.** −15
A.8. −30 **A.9.** 8 **A.10.** 0

Section 2.1 Practice Exercises, pp. 84–87

R.1. < **R.3.** >

1. a. positive; negative **b.** integers **c.** absolute **d.** opposites
3. −86 m **5.** $3800 **7.** −$500 **9.** −14 lb **11.** 1,400,000
13.

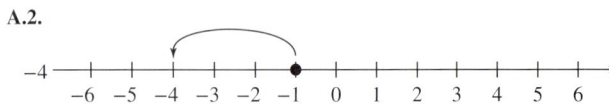

15. −2 **17.** > **19.** > **21.** < **23.** < **25.** 2 **27.** 2
29. 427 **31.** 100,000 **33. a.** −8 **b.** |−12| **35. a.** 7
b. |7| **37.** |−5| **39.** Neither, they are equal. **41.** −5
43. 12 **45.** 0 **47.** 1 **49.** 15 **51.** −15 **53.** −15
55. 15 **57.** 36 **59.** −107 **61. a.** 6 **b.** 6 **c.** −6
d. 6 **e.** −6 **63. a.** −8 **b.** 8 **c.** −8 **d.** 8 **e.** 8 **65.** −6
67. −(−2) **69.** |7| **71.** |−3| **73.** −|14| **75.** =
77. > **79.** > **81.** < **83.** < **85.** 25°F **87.** −22°F
89. −2°F **91.** 44°F **93.** September **95.** −60, −|−46|, |−12|, −(−24), 5² **97.** Positive **99.** Negative

Section 2.2 Activity, p. 92

A.1.
a.

b. 2

A.2.

−4

A.3. Subtract the smaller absolute value from the larger absolute value. Then apply the sign of the number having the larger absolute value.
A.4. Add their absolute values and apply the common sign.
A.5. 7 **A.6.** 8 **A.7.** −45 **A.8.** −239 **A.9.** 0
A.10. −6 **A.11.** 36 **A.12.** −110
A.13. a. −125 + (−110) + (−230) + (−86) + (−750) + 1100 + 955
b. $754 **A.14.** $x = 5$; Answers will vary.

Section 2.2 Practice Exercises, pp. 92–95

R.1. < **R.3.** > **R.5.** >

1. a. 0 **b.** negative; positive **3.** addition symbol **5.** 7
7. 0 **9.** 2 **11.** −2 **13.** −8 **15.** 6 **17.** −7
19. −4 **21.** To add two numbers with the same sign, add their absolute values and apply the common sign.
23. 15 **25.** −16 **27.** −124 **29.** 89
31. −3 **33.** 5 **35.** −24 **37.** 45 **39.** 0 **41.** 0
43. 9 **45.** −26 **47.** −41 **49.** −150 **51.** −17
53. −41 **55.** 529 **57.** 780 **59.** 2 **61.** −30 **63.** 10
65. −8 **67.** 0 **69.** −12 **71.** −191 **73.** −23 + 49; 26
75. 3 + (−10) + 5; −2 **77.** (−8 + 6) + (−5); −7 **79.** 7 in.; Marquette had above average snowfall. **81.** −16 **83.** 8°F
85. $333 **87.** $600 **89.** 0 **91.** −120 **92.** −566
93. −28,414 **94.** −24,325 **95.** 711 **96.** 339
97. For example: −12 + 2 **99.** For example: −1 + (−1)

Section 2.3 Activity, pp. 98–99

A.1. a. 13 **b.** 13 **c.** 4 **d.** 4
e. Answers will vary. Possible answer: Subtracting a negative number has the same result as adding a positive number.
f. When subtracting integers, change the subtraction sign to an addition sign, then add the opposite of the second number.

	English Phrase	Algebraic Expression	Simplified Result
A.2.	The difference of −13 and 4	−13 − 4	−17
A.3.	−21 decreased by 16	−21 − 16	−37
A.4.	−18 subtracted from 25	25 − (−18)	43
A.5.	4 less than −2	−2 − 4	−6

A.6. 180°F **A.7. a.** −11, −17, −23 **b.** −60, −74, −88

Section 2.3 Practice Exercises, pp. 99–102

R.1. 21 **R.3.** −21 **R.5.** −4

1. (−b) **3.** To subtract two integers, add the opposite of the second number to the first number.
5.

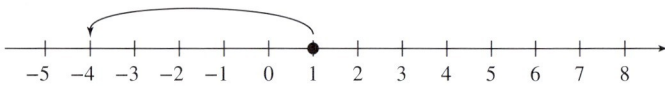

7. 9 + (−2); 7 **9.** 2 + (−9); −7 **11.** 4 + 3; 7
13. −3 + (−15); −18 **15.** −11 + 13; 2 **17.** 52 **19.** −33
21. −12 **23.** 8 **25.** 0 **27.** 161 **29.** −34 **31.** −22
33. −26 **35.** −1 **37.** 32 **39.** −15 **41.** −122
43. 1009 **45.** 1261 **47.** 0 **49.** −1 **51.** 52
53. minus, difference, decreased, less than, subtract from
55. 14 − 23; −9 **57.** 105 − 110; −5 **59.** 320 − (−20); 340
61. 5 − 12; −7 **63.** −34 − 21; −55 **65.** −35 − 24; −59
67. 6423°F **69.** His balance is −$375.
71. The balance is $18,084. **73.** 164 **75.** −112
77. The range is 3° − (−8°) = 11°. **79.** For example: 4 − 10
81. −11, −15, −19 **83.** −413 **84.** −433 **85.** −14,623
86. 19,906 **87.** 916,450 **88.** 129,777 **89.** 64,827 ft
90. 4478 m **91.** Positive **93.** Positive
95. Negative **97.** Negative

Section 2.4 Activity, pp. 107–108

A.1. a. 45 **b.** 45 **c.** 9 **d.** 9 **e.** The products and quotients are both positive. **f.** The product or quotient of two numbers with the same sign is positive.
A.2. a. −45 **b.** −45 **c.** −9 **d.** −9 **e.** The products and quotients are both negative. **f.** The product or quotient of two numbers with different signs is negative.
A.3. a. 1 **b.** −1 **c.** 1 **d.** −1 **e.** Positive **f.** Negative
g. Negative because 891 is an odd number.
A.4. a. 0 **b.** 90 **c.** −240 **d.** 100 **e.** 0 **f.** −82 **g.** −16 **h.** 21
A.5. a. −5 **b.** 2 **c.** (−5)(−5) **d.** 25 **e.** 25

A.6. a. 5 **b.** 2 **c.** $-1 \cdot (5)(5)$ **d.** -25 **e.** -25
A.7. -8 **A.8.** -8 **A.9.** 16   **A.10.** -16 **A.11.** Same
A.12. Opposite   **A.13.** Negative because the exponent is odd.

Section 2.4 Practice Exercises, pp. 109–111
R.1. a. 19   **b.** 9   **R.3.** 17

1. a. positive; negative **b.** positive; negative   **3.** positive; negative   **5.** -24 **7.** 0   **9.** -15 **11.** 40   **13.** -21
15. 48   **17.** -45 **19.** -72 **21.** 0   **23.** 95
25. $-3(-1); 3$ **27.** $-5 \cdot 3; -15$ **29.** $3(-5); -15$
31. 400   **33.** -88 **35.** 0   **37.** 1   **39.** 32
41. -100 **43.** 100   **45.** -1000 **47.** -1000 **49.** -625
51. 625   **53.** 1   **55.** -1 **57.** -20 **59.** 7   **61.** -3
63. 21   **65.** Undefined   **67.** 0   **69.** 4   **71.** -34
73. $-100 \div 20; -5$ **75.** $-64 \div (-32); 2$ **77.** $-52 \div 13; -4$
79. -6 ft   **81.** $-13°$F   **83.** $-\$235$ **85.** -18 ft
87. -108 **89.** -3 **91.** 108   **93.** 15   **95.** 0
97. Undefined   **99.** 40   **101.** 49   **103. a.** 7 **b.** -7
105. $-359{,}723$ **106.** $594{,}125$ **107.** 54
108. -629 **109.** $+1$ **111.** -2 **113.** Negative
115. Negative   **117.** Positive

Chapter 2 Problem Recognition Exercises, p. 112
1. a. 48 **b.** -22 **c.** -26 **d.** 12   **2. a.** -36 **b.** 15 **c.** 9 **d.** -4 **3.** $-5 + (-3); -8$ **4.** $9(-5); -45$
5. $-3 - (-7); 4$ **6.** $\dfrac{28}{-4}; -7$ **7.** $-23(-2); 46$
8. $-4 - 18; -22$ **9.** $\dfrac{42}{-2}; -21$ **10.** $-18 + (-13); -31$
11. $10 - (-12); 22$ **12.** $\dfrac{-21}{-7}; 3$ **13.** $-6(-9); 54$
14. $-7 + 4 + 8 + (-16) + (-5); -16$ **15. a.** 20 **b.** -75 **c.** 10 **d.** -3 **16. a.** 72 **b.** -34 **c.** 18 **d.** -38
17. a. -80 **b.** 80 **c.** -80 **d.** 80   **18. a.** 50 **b.** 50 **c.** 50 **d.** -50 **19. a.** -16 **b.** -90 **20. a.** -5 **b.** 25
21. a. 24 **b.** -24 **22. a.** 60 **b.** -60 **23.** 0
24. Undefined   **25.** 90   **26.** -593 **27.** -30 **28.** 40
29. 0   **30.** 0   **31.** 81   **32.** -32 **33.** -81 **34.** -32
35. Undefined   **36.** 0   **37.** $-10{,}149$ **38.** $-85{,}048$

Section 2.5 Activity, pp. 117–118

	Mathematical Steps	Description
A.1.	$-4 - 8(12 - 15)$ $= -4 - 8(-3)$ $= -4 - (-24)$ $= 20$	Perform operations within parentheses. Multiply (before subtracting). Subtract.
A.2.	$8^2 + 4^2 \div (-8 \div 4)$ $= 8^2 + 4^2 \div (-2)$ $= 64 + 16 \div (-2)$ $= 64 + (-8)$ $= 56$	Perform operations within parentheses. Perform operations involving exponents. Divide (before adding). Add.
A.3.	$(-10)^2 - (15)(-8)$ $= 100 - (15)(-8)$ $= 100 - (-120)$ $= 220$	Perform operations involving exponents. Multiply (before subtracting). Subtract.
A.4.	$\|3 - 8\| - 15 \div 3$ $= \|-5\| - 15 \div 3$ $= 5 - 15 \div 3$ $= 5 - (5)$ $= 0$	Perform operations within grouping symbols (absolute value bars). Simplify the absolute value. Divide (before subtracting). Subtract.

A.5. a. $(\ )^2 - (\ )^2$ **b.** $(6)^2 - (-4)^2$ **c.** 20 **d.** 52; No
e. The answer from part (c) is correct. The parentheses are necessary, otherwise the x^2 term will be incorrectly evaluated as -16.
A.6. -8 **A.7.** 8   **A.8.** -22 **A.9.** 98

Section 2.5 Practice Exercises, pp. 118–120
R.1. a. Undefined **b.** 0   **R.3. a.** 25 **b.** -96
R.5. a. 144 **b.** -144

1. does not   **3.** -14 **5.** 14   **7.** -14 **9.** -17
11. 120   **13.** 1   **15.** 16   **17.** -44 **19.** -150
21. 56   **23.** 110   **25.** -3 **27.** -4 **29.** 16   **31.** 29
33. 11   **35.** 2   **37.** -5 **39.** 8   **41.** -16 **43.** 6
45. -1 **47.** 1   **49.** 14   **51.** -5 **53.** 38   **55.** -18
57. $15x$ **59.** $t + 4$ **61.** $v - 6$ **63.** $2g$ **65.** $-12n$
67. $-9 - x$ **69.** $\dfrac{t}{-2}$ **71.** $y + (-14)$ **73.** $2(c + d)$
75. $x - (-8)$ **77.** -37 **79.** 17   **81.** 7   **83.** -48
85. 9   **87.** -4 **89.** 9   **91.** -9 **93.** -52 **95.** -5
97. $-2°$ **99.** -3

Chapter 2 Review Exercises, pp. 121–123
1. -4250 ft   **2.** $-\$3{,}000{,}000$
3–6. (number line from -6 to 6 with points at $-5, -3, -1, 3$)
7. Opposite: 4; absolute value: 4   **8.** Opposite: -6; absolute value: 6
9. 3   **10.** 1000   **11.** 74   **12.** 0   **13.** 9   **14.** 28
15. -20 **16.** -45 **17.** $<$ **18.** $<$ **19.** $>$ **20.** $=$
21. 4   **22.** 3   **23.** -5 **24.** -3 **25.** -6 **26.** -1
27. 13   **28.** -15 **29.** -70 **30.** -140 **31.** 3
32. -15 **33.** $23 + (-35); -12$ **34.** $57 + (-10); 47$
35. $-5 + (-13) + 20; 2$ **36.** $-42 + 12; -30$ **37.** $-12 + 3; -9$
38. $-89 + (-22); -111$ **39.** -2 in.; Caribou had below average snowfall.   **40.** -5 **41.** $-7 + (-5)$ **42.** 27   **43.** -25
44. 22   **45.** -419 **46.** -8 **47.** 40   **48.** -17
49. 100   **50. a.** $8 - 10; -2$ **b.** $10 - 8; 2$ **51.** For example: 14 subtracted from -2.   **52.** For example: Subtract -7 from -25.
53. The temperature rose $5°$F.   **54.** Sam's new balance is $\$92$.
55. The average is 1 above par.   **56.** 3450 ft   **57.** -18
58. -3 **59.** 15   **60.** 56   **61.** -4 **62.** -12 **63.** 96
64. -32 **65.** Undefined   **66.** 0   **67.** -125 **68.** -125
69. 36   **70.** -36 **71.** 1   **72.** -1 **73.** Negative
74. Positive   **75.** $-45 \div (-15); 3$ **76.** $-4 \cdot 19; -76$
77. $-3°$F   **78.** $-\$90$ **79.** 38   **80.** 57   **81.** -11
82. 8   **83.** -7 **84.** -2 **85.** -2 **86.** 3   **87.** 17
88. -13 **89.** $a + 8$ **90.** $3n$ **91.** $-5x$ **92.** $p - 12$
93. $(a + b) + 2$ **94.** $\dfrac{w}{4}$ **95.** $y - (-8)$ **96.** $2(5 + z)$
97. -23 **98.** -55 **99.** -18 **100.** -30 **101.** -2
102. -5 **103.** -10 **104.** 5

Chapter 3
Section 3.1 Activity, p. 132
A.1. a. $-8x^2, -x, 7$ **b.** $-8x^2, -x$ **c.** 7 **d.** -1
A.2. a. $-8x^2, -x, 7, 3x^2, -5x, 3$
  b. They each have the same variable and the variable is raised to the same power.
  c. $-x$ and $-5x$; 7 and 3.
  d. $-5x^2 - 6x + 10$
A.3. Mia is correct because of the commutative property of multiplication.

A.4. a. The first expression of $2(9 + y)$ contains a sum of two terms in the parentheses, whereas $2(9y)$ contains one term in the parentheses.
b. $18 + 2y$; distributive property of multiplication over addition.
c. $18y$; associative property of multiplication.
A.5. a. $5x - 5y + 15$ **b.** $-5x + 5y - 15$ **c.** The signs of all terms change.
A.6. $-15t$ **A.7.** $-24a + 35b - 46$ **A.8.** $-9x + 20$
A.9. $3w - 13$ **A.10.** $-10t + 13$

Section 3.1 Practice Exercises, pp. 132–135

R.1. -16 **R.3.** 14 **R.5.** -4

1. a. term   **b.** variable; constant   **c.** coefficient   **d.** like
e. $x + 6$; $6x$ **f.** associative   **g.** associative
h. $a \cdot b + a \cdot c$; $4x + 20$ **3.** addition; subtraction
5. The exponents on the variable x are different. Therefore, the terms are not *like* terms and cannot be combined.
7. a^2: variable term; 3: constant term
9. $2a$: variable term; $5b^2$: variable term; 6: constant term
11. 8: constant term; $9a$: variable term   **13.** $6, -4$
15. $-1, -12$ **17.** *Like* terms   **19.** Unlike terms; different variables   **21.** Unlike terms; different powers of y
23. Unlike terms; different variables   **25.** $w + 5$
27. $2r$ **29.** $-st$ **31.** $7 - p$ **33.** $(3 + 8) + t$; $11 + t$
35. $(-2 \cdot 6)b$; $-12b$ **37.** $(3 \cdot 6)x$; $18x$
39. $[-9 + (-12)] + h$; $-21 + h$ **41.** $4x + 32$
43. $4a + 16b - 4c$ **45.** $-2p - 8$ **47.** $-3x - 9 + 5y$
49. $-12 + 4n^2$ **51.** $15q + 6s + 9t$ **53.** $12x$
55. $12 + 6x$ **57.** $-4 - p$ **59.** $-32 + 8p$
61. $14r$ **63.** $7h$ **65.** $-2a^2b$ **67.** $6x - 15y + 9$
69. $-3k - 4$ **71.** $4uv + 6u$ **73.** $5t - 28$
75. $-6x - 16$ **77.** $6y - 14$ **79.** $7q$ **81.** $-2n - 4$
83. $4x + 23$ **85.** $2z - 11$ **87.** $-w - 4y + 9$
89. $8a - 9b$ **91.** $-5m + 6n - 10$ **93.** $12z^2 + 7$
95. a. 6   **b.** 6   **97. a.** 5   **b.** 5   **99. a.** -45 **b.** -45
101. a. 36   **b.** 36

Section 3.2 Activity, p. 140

A.1. Check -20 as the answer:
$3 + (-20) = -17$
$-17 = -17$
Solution.
Check -14 as the answer:
$3 + (-14) = -17$
$-11 \neq -17$
Not a solution.
A.2. Yes.
A.3. a. Add 7 to both sides of the equation. Adding 7 to both sides results in an equivalent equation, and since the sum of a number and its opposite is zero, also isolates the variable.   **b.** 9   **c.** 9; No
A.4. 7   **A.5.** -22 **A.6.** 12   **A.7.** 70   **A.8.** -110
A.9. a. There are variable terms on the left side and constant terms on the right that can be combined.   **b.** $x + 5 = 14$ **c.** 9
A.10. 26   **A.11.** 10   **A.12.** -5

Section 3.2 Practice Exercises, pp. 141–142

R.1. $-12a + 16b$ **R.3.** $-39y^2 + 15$
R.5. $x + 8$ **R.7.** 6

1. a. linear   **b.** solution   **c.** equivalent   **d.** addition   **e.** subtraction
3. Yes   **5.** No   **7.** Yes   **9.** No
11. Equation   **13.** Expression   **15.** Equation
17. 0   **19.** 7   **21.** 3   **23.** 37   **25.** 16   **27.** -15
29. 16   **31.** -78 **33.** 34   **35.** 52   **37.** 0
39. 100   **41.** -28 **43.** 3   **45.** -46 **47.** 61
49. -55 **51.** 42   **53.** -1 **55.** -6 **57.** 12
59. -5 **61.** -1 **63.** 11   **65.** -1 **67.** 10
69. 3   **71.** 28

Section 3.3 Activity, p. 147

A.1. a. 1   **b.** 1   **c.** 1
A.2. a. 1   **b.** 5   **c.** 3   **d.** -5 **e.** -3 **f.** No, because the numerical coefficient multiplied by x must be a *positive* 1.
A.3. 36   **A.4.** Answers will vary.
A.5. 4 (Explanations vary.)   **A.6.** 100 (Explanations vary.)
A.7. 15 (Explanations vary.)   **A.8.** 25 (Explanations vary.)
A.9. -6 **A.10.** 3   **A.11.** -1
A.12. -3 **A.13.** -34 **A.14.** 0

Section 3.3 Practice Exercises, pp. 147–149

R.1. 45   **R.3.** -24 **R.5.** 1

1. a. multiplication   **b.** division   **3.** No   **5.** Yes
7. 1   **9.** -15 **11.** 3   **13.** -7
15. 2   **17.** 13   **19.** -5 **21.** -21 **23.** 30
25. -28 **27.** 4   **29.** 0   **31.** 0   **33.** 6
35. division property   **37.** addition property
39. -16 **41.** -3 **43.** -8 **45.** -48 **47.** 2
49. 30   **51.** -15 **53.** -57 **55.** 0   **57.** 20
59. -31 **61.** 5   **63.** 13   **65.** 3   **67.** -2
69. 1   **71.** -5 **73.** -1

Section 3.4 Activity, p. 154

A.1. a. $4x - 3 + 3 = 13 + 3$ **b.** $\frac{4x}{4} = \frac{16}{4}$ **c.** 4; $4(4) - 3 \stackrel{?}{=} 13$; $13 = 13$ ✓
A.2. No, the result is not the same.
A.3. -7 (Explanations vary.)   **A.4.** -6 (Explanations vary.)
A.5. 21 (Explanations vary.)   **A.6.** -10 **A.7.** -10
A.8. The solution is the same; it does not matter.
A.9. 9   **A.10.** -3 **A.11.** 1
A.12. Step 1: Simplify both sides of the equation by clearing parentheses and combining *like* terms if necessary. Step 2: Use the addition or subtraction property of equality to collect the variable terms on one side of the equation. Step 3: Use the addition or subtraction property of equality to collect the constant terms on the other side of the equation. Step 4: Use the multiplication or division property of equality to make the coefficient of the variable term equal to 1. Step 5: Check the answer in the original equation.

Section 3.4 Practice Exercises, pp. 154–155

R.1. 3   **R.3.** -12 **R.5.** 4   **R.7.** 18

1. Add 6 to both sides first.   **3.** Yes   **5.** No
7. -2 **9.** 0   **11.** 6   **13.** 6   **15.** 5   **17.** 2
19. 9   **21.** -12 **23.** -60 **25.** -21 **27.** -3
29. 2   **31.** 1   **33.** 4   **35.** -3 **37.** -4
39. -13 **41.** 3   **43.** -12 **45.** -2 **47.** 0
49. 15   **51.** 6   **53.** 1   **55.** 5

Chapter 3 Problem Recognition Exercises, p. 156

1. Equation   **2.** Expression   **3.** Expression
4. Equation   **5.** Equation   **6.** Expression
7. Equation; 4   **8.** Equation; 6   **9.** Expression; $-15t$
10. Expression; $-30x$ **11.** Expression; $5w - 15$
12. Expression; $6x - 12$ **13.** Equation; 7
14. Equation; 8   **15.** Equation; 15   **16.** Equation; 30
17. Expression; $t + 25$ **18.** Expression; $x + 42$
19. Equation; -1 **20.** Equation; 9   **21.** Equation; 2
22. Equation; 3   **23.** Expression; $11u + 74$
24. Expression; $61w + 129$ **25.** Expression; $-3x + 26$
26. Expression; $-7x + 21$ **27.** Equation; 14
28. Equation; -7 **29.** Equation; 8   **30.** Equation; -34

Section 3.5 Activity, p. 162

A.1. b. $x =$ a number   **d.** $x + 8 = -4$ **e.** -12
f. -12 increased by 8 equals -4.
A.2. b. $x =$ a number   **d.** $-3x + 6 = 60$ **e.** -18
f. The sum of -3 times -18 and 6 is 60.
A.3. b. $x =$ a number   **d.** $-3(x + 6) = 60$ **e.** -26
f. -3 times the sum of -26 and 6 is 60.
A.4. b. One possibility: Let x represent the length of the short piece. Then the length of the longer piece is $3x - 16$.   **c.** Length of short piece + Length of long piece = Total length   **d.** $x + 3x - 16 = 136$ **e.** $x = 38$
f. The short piece is 38 ft and the long piece is 98 ft.
A.5. b. One possibility: Let x represent the cost of the used DVD player. Then the cost of the new DVD player is $x + 22$.   **c.** (Cost of the used DVD player) + (Cost of the new DVD player) = Total Cost
d. $x + x + 22 = 68$ **e.** $x = 23$ **f.** The cost of the used DVD player is $23, and the cost of the new DVD player is $45.
A.6. b. One possibility: Let x represent the length. Then the width is $x - 3$.
c. $P = 2l + 2w$ **d.** $42 = 2(x) + 2(x - 3)$ **e.** $x = 12$ **f.** The length of the screen is 12 in., and the width is 9 in.

Section 3.5 Practice Exercises, pp. 162–165

R.1. -3 **R.3.** -45 **R.5.** -2 **R.7.** 2

1. Step 3: Write an equation in words.
3. Step 4: Write a mathematical equation.
5. Step 2: Assign labels to unknown quantities.
7. division
9. a. $x + 6 = 19$ **b.** The number is 13.
11. a. $x - 14 = 20$ **b.** The number is 34.
13. a. $\frac{x}{3} = -8$ **b.** The number is -24.
15. a. $-6x = -60$ **b.** The number is 10.
17. a. $-2 - x = -14$ **b.** The number is 12.
19. a. $13 + x = -100$ **b.** The number is -113.
21. a. $60 = -5x$ **b.** The number is -12.
23. a. $3x + 9 = 15$ **b.** The number is 2.
25. a. $5x - 12 = -27$ **b.** The number is -3.
27. a. $\frac{x}{4} - 5 = -12$ **b.** The number is -28.
29. a. $8 - 3x = 5$ **b.** The number is 1.
31. a. $3(x + 4) = -24$ **b.** The number is -12.
33. a. $-4(3 - x) = -20$ **b.** The number is -2.
35. a. $-12x = x + 26$ **b.** The number is -2.
37. a. $10(x + 5) = 80$ **b.** The number is 3.
39. a. $3x = 2x - 10$ **b.** The number is -10.
41. $5x$ **43.** $6n$ **45.** $A - 30$ **47.** $p + 1481$
49. $20t$ **51.** $5x$ **53.** The pieces should be 2 ft and 6 ft.
55. The distance between Minneapolis and Madison is 240 mi.
57. The Beatles had 19, and Elvis had 10.
59. The Giants scored 17 points, and the Patriots scored 14 points.
61. Charlene's rent is $650 a month, and the security deposit is $300.
63. Stefan worked 6 hr of overtime.

Chapter 3 Review Exercises, pp. 166–167

1. $3, -5, 12$ **2.** $-6, -1, 2, 1$ **3.** Unlike terms
4. *Like* terms   **5.** *Like* terms   **6.** Unlike terms
7. $-5 + t$ **8.** $3h$ **9.** $(-4 \cdot 2)p; -8p$
10. $m + (10 - 12); m - 2$ **11.** $6b + 15$
12. $20x + 30y - 15z$ **13.** $-4c + 6d$
14. $4k - 8w + 12$ **15.** $2x$ **16.** $-9y$
17. $6x + 4y + 10$ **18.** $7a + 9b + 9$ **19.** $-2x + 26$
20. $-2z - 2$ **21.** $-u - 17v$ **22.** $p - 16$
23. -3 is a solution.   **24.** -3 is not a solution.
25. -35 **26.** -12 **27.** 14   **28.** 18   **29.** 13
30. 15   **31.** 2   **32.** -14 **33.** -7 **34.** 4
35. 26   **36.** 35   **37.** 6   **38.** -48 **39.** -1
40. 3   **41.** 8   **42.** -12 **43.** 32   **44.** 5
45. -28 **46.** -7 **47.** -4 **48.** 2   **49.** -1
50. 2   **51.** -3 **52.** -6 **53.** $x - 4 = 13$;
The number is 17.   **54.** $\frac{x}{-7} = -6$; The number is 42.
55. $-4x + 3 = -17$; The number is 5.   **56.** $3x - 7 = -22$;
The number is -5.   **57.** $2(x + 10) = 16$; The number is -2.
58. $3(4 - x) = -9$; The number is 7.   **59.** $9n$ **60.** $4x$
61. $x + 2$ **62.** $h + 6$ **63.** Monique drove 120 mi, and Michael drove 360 mi.   **64.** Angela ate 4 pieces, and Joel ate 8 pieces.
65. The longer piece is 35 ft, and the shorter piece is 30 ft.
66. Raul took 16 hours in the fall and 12 hours in the spring.

Chapter 4

Section 4.1 Practice Exercises, pp. 176–180

1. a. fractions   **b.** numerator; denominator
c. rational   **d.** proper   **e.** improper   **f.** mixed
3. Numerator: 8; denominator: 9
5. Numerator: $7p$; denominator: $9q$
7. $\frac{5}{9}$ **9.** $\frac{3}{8}$ **11.** $\frac{4}{7}$ **13.** $\frac{1}{8}$ **15.** $\frac{41}{103}$
17. $\frac{10}{21}$ **19.** -13 **21.** 1   **23.** 0   **25.** Undefined
27. $\frac{9}{10}$ **29.** b, c   **31.** a, b, c   **33.** Proper
35. Improper   **37.** Improper   **39.** $\frac{5}{2}$ **41.** $\frac{12}{4}$
43. $\frac{9}{8}$ in.   **45.** $\frac{7}{4}$; $1\frac{3}{4}$ **47.** $\frac{13}{8}$ in.; $1\frac{5}{8}$ in.   **49.** $\frac{7}{4}$
51. $-\frac{38}{9}$ **53.** $-\frac{24}{7}$ **55.** $\frac{27}{4}$ **57.** $\frac{137}{12}$ **59.** $-\frac{171}{8}$
61. 30   **63.** 19   **65.** 7   **67.** $4\frac{5}{8}$ **69.** $-7\frac{4}{5}$
71. $-2\frac{7}{10}$ **73.** $5\frac{7}{9}$ **75.** $12\frac{1}{11}$ **77.** $-3\frac{5}{6}$
79. $44\frac{1}{7}$ **81.** $1056\frac{1}{5}$ **83.** $810\frac{3}{11}$ **85.** $12\frac{7}{15}$
87. number line showing $\frac{3}{4}$ between 0 and 1
89. number line showing $\frac{1}{5}$ between 0 and 1
91. number line showing $-\frac{2}{3}$ between -1 and 0
93. number line showing $1\frac{1}{6}$ between 1 and 2
95. number line showing $-1\frac{1}{3}$ between -2 and 0
97. $\frac{3}{4}$ **99.** $\frac{1}{10}$ **101.** $-\frac{7}{3}$ **103.** $\frac{7}{3}$
105. False   **107.** True

Section 4.2 Activity, pp. 189–190

A.1. a. Answers will vary.   **b.** Answers will vary.
c. Answers will vary.   **d.** Answers will vary.

A.2. a. Divisible by 2 **b.** Divisible by 3 and 5 **c.** Divisible by 2, 3, 5, and 10 **d.** Not divisible by 2, 3, 5, or 10
A.3. a. $\dfrac{2}{19}, \dfrac{3}{23}, \dfrac{5}{29}, \dfrac{7}{\_\_}, \dfrac{11}{\_\_}, \dfrac{13}{\_\_}, \dfrac{17}{\_\_}$ **b.** Answers will vary.
A.4. a. 4 **b.** Answers will vary.
A.5. a. 1, 2, 4 **b.** 1, 2, 3, 4, 6, 8, 12, 24 **c.** 1, 47 **d.** 1, 2, 4, 5, 10, 20, 25, 50, 100
A.6. a. Answers will vary. **b.** Answers will vary. **c.** $2^3 \cdot 3 \cdot 5 \cdot 7$ **d.** Division by primes. **A.7.** $2 \cdot 3^2 \cdot 7$ **A.8.** $2^2 \cdot 5 \cdot 7$
A.9. $3^2 \cdot 7 \cdot 13$ **A.10.** $5 \cdot 7 \cdot 17$ **A.11. a.** 24 **b.** 24 **c.** Yes
A.12. = **A.13.** ≠ **14.** ≠ **A.15. a.** $16 = 2 \cdot 2 \cdot 2 \cdot 2$
b. $56 = 2 \cdot 2 \cdot 2 \cdot 7$ **c.** $\dfrac{16}{56} = \dfrac{\cancel{2}\cdot\cancel{2}\cdot\cancel{2}\cdot 2}{\cancel{2}\cdot\cancel{2}\cdot\cancel{2}\cdot 7} = \dfrac{2}{7}$
A.16. a. $16y^2 = 2 \cdot 2 \cdot 2 \cdot 2 \cdot y \cdot y$ **b.** $32y^3 = 2 \cdot 2 \cdot 2 \cdot 2 \cdot 2 \cdot y \cdot y \cdot y$
c. $\dfrac{16y^2}{32y^3} = \dfrac{\cancel{2}\cdot\cancel{2}\cdot\cancel{2}\cdot\cancel{2}\cdot\cancel{y}\cdot\cancel{y}}{2 \cdot \cancel{2}\cdot\cancel{2}\cdot\cancel{2}\cdot\cancel{2}\cdot\cancel{y}\cdot\cancel{y}\cdot y} = \dfrac{1}{2y}$
A.17. $-\dfrac{8}{5}$ **A.18.** $\dfrac{1}{3}$ **A.19.** $\dfrac{y}{3}$ **A.20.** $\dfrac{5z^3}{2v^3}$
A.21. There are $\dfrac{12}{17}$ faculty at the community college and $\dfrac{11}{17}$ faculty at the university.

Section 4.2 Practice Exercises, pp. 190–194

R.1. −12 **R.3.** Undefined **R.5.** $5\dfrac{3}{7}$ **R.7.** $\dfrac{44}{5}$

1. a. factor **b.** prime **c.** composite **d.** prime **e.** lowest
3. ones **5.** reducing **7.** For example: $2 \cdot 4$ and $1 \cdot 8$
9. For example: $4 \cdot 6$ and $2 \cdot 2 \cdot 2 \cdot 3$
11. A whole number is divisible by 2 if it is an even number.
13. A whole number is divisible by 3 if the sum of its digits is divisible by 3.
15. A whole number is divisible by 4 if the last two digits form a number that is divisible by 4.
17. 2, 3, 4, 6, 9 **19.** none **21.** 3, 5, 9 **23.** 2, 3, 6, 9
25. Yes **27.** Prime **29.** Composite **31.** Neither
33. Prime **35.** 2, 3, 5, 7, 11, 13, 17, 19, 23, 29, 31, 37, 41, 43, 47
37. False **39.** No, 9 is not a prime number. **41.** Yes
43. $2 \cdot 5 \cdot 7$ **45.** $2 \cdot 2 \cdot 5 \cdot 13$ or $2^2 \cdot 5 \cdot 13$
47. $3 \cdot 7 \cdot 7$ or $3 \cdot 7^2$ **49.** $2 \cdot 2 \cdot 2 \cdot 7 \cdot 11$ or $2^3 \cdot 7 \cdot 11$

51.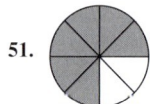
53. False **55.** ≠ **57.** = **59.** = **61.** ≠
63. 6 **65.** 7 **67.** 25 **69.** $3xy^3$
71. $\dfrac{1}{2}$ **73.** $\dfrac{1}{3}$ **75.** $-\dfrac{9}{5}$ **77.** $-\dfrac{5}{4}$ **79.** 1
81. $\dfrac{3}{4}$ **83.** −3 **85.** $\dfrac{7}{10}$ **87.** $\dfrac{77}{39}$ **89.** $\dfrac{2}{5}$ **91.** $\dfrac{3}{4}$
93. $-\dfrac{5}{3}$ **95.** $\dfrac{21}{11}$ **97.** $\dfrac{17}{100}$ **99.** $\dfrac{8b}{5}$ **101.** $2xy$
103. $\dfrac{x}{3}$ **105.** $-\dfrac{1}{2c^2}$ **107.** Heads: $\dfrac{5}{12}$; tails: $\dfrac{7}{12}$
109. a. $\dfrac{3}{13}$ **b.** $\dfrac{10}{13}$ **111. a.** Jonathan: $\dfrac{5}{7}$; Jared: $\dfrac{6}{7}$
b. Jared sold the greater fractional part. **113. a.** Raymond: $\dfrac{10}{11}$; Travis: $\dfrac{9}{11}$ **b.** Raymond read the greater fractional part.
115. $\dfrac{8}{9}$ **116.** $\dfrac{13}{14}$ **117.** $\dfrac{41}{51}$ **118.** $\dfrac{21}{10}$
119. $\dfrac{29}{30}$ **120.** $\dfrac{13}{7}$ **121.** $\dfrac{3}{2}$ **122.** $\dfrac{31}{19}$
123. For example: $\dfrac{6}{8}, \dfrac{9}{12}, \dfrac{12}{16}$ **125.** For example: $-\dfrac{6}{9}, -\dfrac{4}{6}, -\dfrac{2}{3}$

Section 4.3 Activity, pp. 204–205

A.1. a. $\dfrac{300}{210}$ **b.** $1\dfrac{3}{7}$
A.2. a. Divide 25 and 35 by the common factor of 5. Divide 12 and 6 by the common factor of 6.
b. $\dfrac{5}{1} \cdot \dfrac{2}{7} = \dfrac{10}{7}$ **c.** Yes, because the simplification occurred in part (a) before multiplying.
A.3. a. $\dfrac{2}{5}$ **b.** $\dfrac{x}{1} = x$ **c.** $\dfrac{2x}{5}$ **A.4.** $\dfrac{3}{20}$ **A.5.** $\dfrac{9}{2}$ **A.6.** $-\dfrac{36}{5}$
A.7. $\dfrac{2a^2}{b}$ **A.8.** $\dfrac{2}{3}$ km^2 **A.9.** $\dfrac{33}{40}$ yd^2 **A.10.** 35 ft^2
A.11. a. $\dfrac{4}{3}$ **b.** 1 **c.** 1 **d.** Yes, $\dfrac{1}{5}$ **A.12. a.** $\dfrac{9}{4}$ **b.** Process remains the same. The quotient would be negative if one of the fractions were negative and the other fraction were positive.
A.13. $\dfrac{5}{2}$ **A.14.** 12 **A.15.** $\dfrac{14}{3c^2}$ **A.16.** 30 electoral votes
A.17. 6 pieces

Section 4.3 Practice Exercises, pp. 205–209

R.1. a, c, d, e **R.3.** $\dfrac{3}{10}$ **R.5.** $\dfrac{4x}{5}$

1. a. $\dfrac{1}{2}bh$ **b.** reciprocals **3.** reciprocal **5.** positive

7. 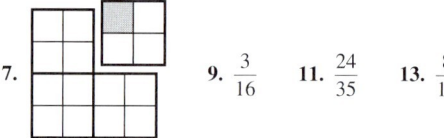 **9.** $\dfrac{3}{16}$ **11.** $\dfrac{24}{35}$ **13.** $\dfrac{8}{11}$
15. $-\dfrac{24}{5}$ **17.** $\dfrac{2}{15}$ **19.** $\dfrac{5}{8}$ **21.** $\dfrac{35}{4}$ **23.** $-\dfrac{8}{3}$
25. $\dfrac{4}{5}$ **27.** $\dfrac{30}{7}$ **29.** $\dfrac{3}{2}$ **31.** $-\dfrac{a}{10}$ **33.** $\dfrac{1}{20y}$

35. $4m^2$ **37.** 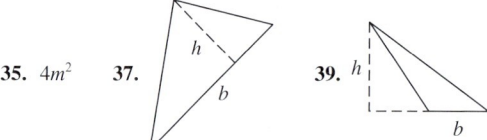 **39.** (figure)
41. 44 cm^2 **43.** 32 m^2 **45.** 4 yd^2 **47.** $\dfrac{49}{64}$ in.^{2}
49. $\dfrac{8}{7}$ **51.** $-\dfrac{9}{10}$ **53.** $-\dfrac{1}{4}$
55. No reciprocal exists. **57.** $\dfrac{1}{3}$ **59.** multiplying
61. $\dfrac{8}{25}$ **63.** $\dfrac{35}{26}$ **65.** $\dfrac{35}{9}$ **67.** −5 **69.** 1
71. $-\dfrac{21}{2}$ **73.** $\dfrac{3}{5}$ **75.** $-\dfrac{90}{13}$ **77.** $\dfrac{6y}{7}$ **79.** $-4c^2d$
81. $\dfrac{7}{2}$ **83.** $\dfrac{5}{36}$ **85.** −8 **87.** $\dfrac{2}{5}$ **89.** 2 **91.** $\dfrac{3}{2}$
93. $-\dfrac{xy}{2}$ **95.** $\dfrac{2}{d}$ **97.** Li wrapped 54 packages.
99. 24 cups of juice **101.** The stack will be 12 in. high.
103. a. 27 commercials in 1 hr **b.** 648 commercials in 1 day
105. a. Ricardo's mother will pay $16,000. **b.** Ricardo will have to pay $8000. **c.** He will have to finance $216,000.
107. Frankie mowed 960 yd^2. He has 480 yd^2 left to mow.
109. $\dfrac{1}{10}$ of the sample has O negative blood.
111. She can prepare 14 samples. **113.** Liu needs $1\dfrac{1}{4}$ gal.

115. 18   **117.** $\frac{2}{25}$ **119.** 12 ft, because $30 \div \frac{5}{2} = 12$.
121. $\frac{1}{32}$ **123.** They are the same.

Section 4.4 Activity, pp. 215–216

A.1. a. 9: 9, 18, 27, 36, 45, ⑤④, 63, 72
6: 6, 12, 18, 24, 30, 36, 42, 48, ⑤④, 60
27: 27, ⑤④, 81, 108
b. LCM is 54.

A.2. a. $9 = 3 \cdot 3$
$6 = ② \cdot 3$
$27 = ③ \cdot 3 \cdot 3$
b. LCM is 54.

A.3. 120   **A.4.** 105   **A.5.** $\frac{24}{30}$ **A.6.** $-\frac{10}{45}$ **A.7.** $\frac{6x}{14x}$

A.8. $\frac{4a}{a^2}$ **A.9. a.** 36   **b.** 36   **c.** $\frac{19}{36}$ and $\frac{20}{36}$ **d.** $\frac{19}{36}$

A.10. $\frac{1}{10}, \frac{37}{75}, \frac{13}{25}, \frac{17}{30}$

Section 4.4 Practice Exercises, pp. 216–219

R.1. $-\frac{26}{9}$ **R.3.** $\frac{3y}{2x^2}$ **R.5.** 132   **R.7.** $-\frac{4}{9}$

1. a. multiple   **b.** least common multiple
c. least common denominator
3. prime numbers   **5.** addition, subtraction   **7.** less than
9. a. 48, 72, 240   **b.** 4, 8, 12   **11. a.** 72, 360, 108   **b.** 6, 12, 9
13. 50   **15.** 48   **17.** 72   **19.** 60   **21.** 210
23. 120   **25.** 60   **27.** 240   **29.** 180   **31.** 180
33. The shortest length of floor space is 60 in. (5 ft).
35. It will take 120 hr (5 days) for the satellites to be lined up again.
37. $\frac{14}{21}$ **39.** $\frac{10}{16}$ **41.** $-\frac{12}{16}$ **43.** $\frac{-12}{15}$ **45.** $\frac{49}{42}$
47. $\frac{121}{99}$ **49.** $\frac{20}{4}$ **51.** $\frac{11,000}{4000}$ **53.** $-\frac{55}{15}$ **55.** $\frac{-15}{24}$
57. $\frac{16y}{28}$ **59.** $\frac{3y}{8y}$ **61.** $\frac{15p}{25p}$ **63.** $\frac{2x}{x^2}$ **65.** $\frac{8b^2}{ab^3}$
67. >   **69.** <   **71.** =   **73.** >   **75.** $\frac{7}{8}$
77. $\frac{2}{3}, \frac{3}{4}, \frac{7}{8}$ **79.** $-\frac{3}{8}, -\frac{5}{16}, -\frac{1}{4}$ **81.** $-\frac{4}{3}, \frac{13}{12}, \frac{17}{15}$
83. The longest cut is above the left eye. The shortest cut is on the right hand.   **85.** The least amount is $\frac{3}{4}$ lb of cheddar, and the greatest amount is $\frac{7}{8}$ lb of Swiss.   **87.** a and b
89. 336   **91.** 540

Section 4.5 Activity, p. 226

A.1. a. 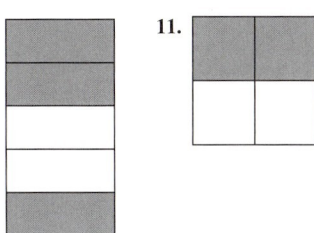 $\frac{2}{5} + \frac{1}{5} = \frac{3}{5}$
b. The denominator stays the same.

A.2. $\frac{17}{7}$ **A.3.** $\frac{6}{5}$ **A.4.** $-\frac{8}{21}$ **A.5.** $-\frac{11}{y}$

A.6. a. 8   **b.** $\frac{5}{8}$ and $\frac{6}{8}$ **c.** $\frac{11}{8}$ **A.7.** $\frac{1}{2}$ **A.8.** $-\frac{23}{24}$

A.9. $\frac{41}{60}$ **A.10.** $\frac{19}{24}$ **A.11.** $-\frac{49}{40}$ **A.12. a.** x^2 **b.** $\frac{6x}{x^2}$ and $\frac{5}{x^2}$
c. $\frac{6x + 5}{x^2}$ **A.13.** $\frac{2y - 15}{3y}$ **A.14.** $\frac{13}{3z}$ **A.15.** $\frac{1}{2}$ lb

A.16. $\frac{15}{8}$ in. or $1\frac{7}{8}$ in.

Section 4.5 Practice Exercises, pp. 226–229

R.1. $-\frac{9}{15}$ **R.3.** $\frac{96}{12}$ **R.5.** $\frac{6x}{15x^2}$

1. a. like   **b.** is   **c.** is not   **3.** unlike   **5.** like   **7.** unlike

9.   **11.**

13. $\frac{3}{2}$ **15.** $\frac{2}{3}$ **17.** $\frac{5}{2}$ **19.** $\frac{4}{5}$ **21.** $-\frac{5}{2}$
23. $-\frac{7}{4}$ **25.** $\frac{3y + 5}{2w}$ **27.** $-\frac{2x}{5y}$ **29.** $\frac{19}{16}$ **31.** $\frac{1}{6}$
33. $\frac{83}{42}$ **35.** $\frac{3}{8}$ **37.** $\frac{1}{3}$ **39.** $-\frac{55}{36}$ **41.** $-\frac{7}{8}$
43. $\frac{8}{3}$ **45.** $-\frac{2}{7}$ **47.** $-\frac{1}{100}$ **49.** $\frac{391}{1000}$ **51.** $\frac{9}{8}$
53. $-\frac{1}{8}$ **55.** $\frac{9}{16}$ **57.** $\frac{3x + 8}{4x}$ **59.** $\frac{10y + 7x}{xy}$
61. $\frac{10x - 2}{x^2}$ **63.** $\frac{5 - 2x}{3x}$ **65.** Inez added $\frac{9}{8}$ cups or $1\frac{1}{8}$ cups.
67. The storm delivered $\frac{5}{32}$ in. of rain.   **69. a.** $\frac{13}{36}$ **b.** $\frac{23}{36}$
71. $\frac{13}{5}$ m or $2\frac{3}{5}$ m   **73.** $a = \frac{3}{8}$ ft, $b = \frac{3}{8}$ ft; perimeter: 3 ft   **75.** b

Section 4.6 Practice Exercises, pp. 239–243

R.1. $\frac{24}{5}$ **R.3.** $\frac{13}{6}$ **R.5.** 12   **R.7.** $\frac{1}{9}$
R.9. $\frac{17}{5}$ **R.11.** $12\frac{5}{6}$

1. $7\frac{2}{5}$ **3.** $15\frac{2}{5}$ **5.** -38 **7.** $27\frac{2}{3}$
9. $72\frac{1}{2}$ **11.** 2   **13.** $-4\frac{5}{12}$ **15.** $2\frac{6}{17}$ **17.** $\frac{3}{5}$
19. $-2\frac{3}{4}$ **21.** $7\frac{4}{11}$ **23.** $15\frac{3}{7}$ **25.** $15\frac{9}{16}$ **27.** $10\frac{13}{15}$
29. 5   **31.** 2   **33.** $3\frac{1}{5}$ **35.** $8\frac{2}{3}$ **37.** $14\frac{1}{2}$
39. $23\frac{1}{8}$ **41.** $19\frac{17}{48}$ **43.** $9\frac{5}{12}$ **45.** $12\frac{19}{24}$ **47.** $9\frac{7}{8}$
49. $171\frac{1}{2}$ **51.** $11\frac{3}{5}$ **53.** $12\frac{1}{6}$ **55.** $11\frac{1}{2}$ **57.** $1\frac{3}{4}$
59. $7\frac{13}{14}$ **61.** $3\frac{1}{6}$ **63.** $2\frac{7}{9}$ **65.** $\frac{11}{16}$ **67.** $\frac{32}{35}$
69. $-7\frac{11}{14}$ **71.** $2\frac{7}{8}$ **73.** $5\frac{3}{4}$ **75.** $-2\frac{5}{6}$ **77.** $-\frac{36}{5}$
79. $\frac{7}{24}$ **81.** $\frac{50}{13}$ **83.** $\frac{61}{30}$ **85.** $7\frac{3}{4}$ in.
87. The index finger is longer.
89. Perimeter: $16\frac{1}{4}$ in.; area: $14\frac{7}{16}$ in.^{2}
91. $642\frac{1}{2}$ lb   **93.** The total is $16\frac{11}{12}$ hr.
95. a. 7 weeks old   **b.** $8\frac{1}{2}$ weeks old   **97. a.** 455 gal was required to transport the 130 gal.   **b.** A total of 585 gal was used.

99. $3\frac{5}{12}$ ft   **101.** The printing area width is 6 in.
103. a. $3\frac{3}{8}$ L   **b.** $\frac{5}{8}$ L   **105.** $318\frac{1}{4}$ **106.** $3\frac{1}{15}$ **107.** $17\frac{18}{19}$
108. $466\frac{1}{5}$ **109.** $1\frac{43}{168}$ **110.** $\frac{11}{30}$ **111.** $\frac{37}{132}$ **112.** $\frac{137}{391}$
113. $46\frac{25}{54}$ **114.** $25\frac{71}{84}$ **115.** $5\frac{17}{77}$ **116.** $3\frac{9}{68}$
117. $2\frac{2}{3}$ **119.** $2\frac{1}{6}$

Chapter 4 Problem Recognition Exercises, pp. 243–244

1. a. -1 **b.** $-\frac{14}{25}$ **c.** $-\frac{7}{2}$ or $-3\frac{1}{2}$ **d.** $-\frac{9}{5}$ or $-1\frac{4}{5}$
2. a. $\frac{10}{9}$ or $1\frac{1}{9}$ **b.** $\frac{8}{5}$ or $1\frac{3}{5}$ **c.** $\frac{13}{6}$ or $2\frac{1}{6}$ **d.** $\frac{1}{2}$
3. a. $\frac{5}{4}$ or $1\frac{1}{4}$ **b.** $\frac{17}{4}$ or $4\frac{1}{4}$ **c.** $-\frac{11}{6}$ or $-1\frac{5}{6}$
d. $-\frac{33}{8}$ or $-4\frac{1}{8}$ **4. a.** $\frac{221}{18}$ or $12\frac{5}{18}$ **b.** $\frac{26}{17}$ or $1\frac{9}{17}$
c. $\frac{3}{2}$ or $1\frac{1}{2}$ **d.** $\frac{43}{6}$ or $7\frac{1}{6}$ **5. a.** $-\frac{35}{8}$ or $-4\frac{3}{8}$
b. $-\frac{3}{2}$ or $-1\frac{1}{2}$ **c.** $-\frac{32}{3}$ or $-10\frac{2}{3}$ **d.** $-\frac{29}{8}$ or $-3\frac{5}{8}$
6. a. $\frac{11}{6}$ or $1\frac{5}{6}$ **b.** $\frac{5}{3}$ or $1\frac{2}{3}$ **c.** $\frac{17}{3}$ or $5\frac{2}{3}$ **d.** $\frac{22}{3}$ or $7\frac{1}{3}$
7. a. $-\frac{53}{15}$ or $-3\frac{8}{15}$ **b.** $-\frac{73}{15}$ or $-4\frac{13}{15}$ **c.** $\frac{14}{5}$ or $2\frac{4}{5}$
d. $\frac{63}{10}$ or $6\frac{3}{10}$ **8. a.** $\frac{25}{18}$ or $1\frac{7}{18}$ **b.** $\frac{50}{9}$ or $5\frac{5}{9}$ **c.** $\frac{7}{9}$
d. $\frac{43}{9}$ or $4\frac{7}{9}$ **9. a.** -1 **b.** $-\frac{56}{45}$ or $-1\frac{11}{45}$ **c.** $-\frac{81}{25}$ or $-3\frac{6}{25}$
d. $-\frac{106}{45}$ or $-2\frac{16}{45}$ **10. a.** 1   **b.** 1   **c.** 1   **d.** 1

Section 4.7 Activity, pp. 249–250

A.1. a. Multiplication, $\frac{1}{8}$ **b.** Addition, $-\frac{11}{40}$
A.2. a. In Exercise A.2 there are parentheses grouping the fractions that are being added.   **b.** Addition, $\frac{7}{20}$ **c.** Multiplication, $\frac{7}{120}$
A.3. $\frac{21}{2}$ **A.4.** -2 **A.5.** 5   **A.6.** $\frac{35}{8}$ or $4\frac{3}{8}$
A.7. a. $\left(\frac{1}{9} - \frac{2}{3}\right)$ **b.** $\left(\frac{1}{9} - \frac{2}{3}\right)\left(\frac{1}{9} - \frac{2}{3}\right)$ **c.** $\frac{25}{81}$
A.8. a. $\left(\frac{2}{3}\right)$ **b.** $\frac{1}{9} - \left(\frac{2}{3}\right)\left(\frac{2}{3}\right)$ **c.** $-\frac{1}{3}$ **A.9.** $\frac{9}{16}$
A.10. $-\frac{9}{16}$ **A.11. a.** $\frac{7}{6}$ **b.** $\frac{5}{12}$ **c.** $\frac{14}{5}$
A.12. a. 12   **b.** $8 + 6$ **c.** $9 - 4$ **d.** $\frac{14}{5}$ **A.13.** $-\frac{1}{3}$
A.14. $2w$ **A.15.** $-1x$ **A.16.** $-\frac{1}{5}x$
A.17. $-\frac{7}{20}x$ **A.18.** $-\frac{13}{18}x + \frac{13}{6}y$

Section 4.7 Practice Exercises, pp. 251–253

R.1. $\frac{23}{30}$ **R.3.** $-\frac{7}{15}$ **R.5.** $-\frac{12}{35}$ **R.7.** $6\frac{1}{12}$ **R.9.** $9\frac{5}{24}$

1. a. one-tenth   **b.** complex
3. $\frac{5}{8}$ **5.** $\frac{x}{5}$ **7.** simplify the exponent   **9.** $\frac{1}{81}$ **11.** $\frac{1}{81}$
13. $-\frac{27}{8}$ **15.** $-\frac{27}{8}$ **17.** $\frac{1}{1000}$ **19.** $\frac{1}{1,000,000}$

21. $-\frac{1}{1000}$ **23.** -27 **25.** $4\frac{1}{4}$ **27.** 42
29. -2 **31.** $\frac{2}{9}$ **33.** 3   **35.** $2\frac{3}{8}$ **37.** 0
39. $\frac{1}{36}$ **41.** $\frac{23}{24}$ **43.** $1\frac{3}{7}$ **45.** $\frac{25}{3}$ **47.** $5\frac{1}{4}$
49. -7 **51.** $7\frac{7}{9}$ **53.** $\frac{5}{6}$ **55.** $-\frac{7}{4}$ **57.** $\frac{x}{28}$
59. $-\frac{3}{5}$ **61.** $\frac{7}{4}$ **63.** $\frac{28}{11}$ **65.** -11 **67.** $\frac{2}{9}$
69. $2y$ **71.** $\frac{5}{8}a$ **73.** $\frac{17}{30}x$ **75.** $\frac{4}{3}y - \frac{7}{4}z$
77. a. $\frac{1}{36}$ **b.** $\frac{1}{6}$ **79.** $\frac{1}{5}$ **81.** $\frac{8}{9}$

Section 4.8 Activity, pp. 258–259

A.1. a. 6   **b.** Add 4 to both sides.   **c.** $\frac{6}{7}$ **d.** Add $\frac{4}{7}$ to both sides.
   e. The process is the same in both equations regardless of the fractions.
A.2. a. $\frac{4}{7}$ **b.** 1   **c.** $-\frac{8}{21}$ **d.** $\frac{7}{4}\left(-\frac{8}{21}\right) \stackrel{?}{=} -\frac{2}{3}; -\frac{2}{3} = -\frac{2}{3}$
A.3. $\frac{17}{4}$ **A.4.** -1 **A.5. a.** 12   **b.** $30x + 9 = -28$
c. $-\frac{37}{30}$ **d.** $\frac{5}{2}\left(-\frac{37}{30}\right) + \frac{3}{4} \stackrel{?}{=} -\frac{7}{3}; -\frac{7}{3} = -\frac{7}{3}$
A.6. LCD = 10; 23   **A.7.** LCD = 24; $-\frac{4}{3}$

Section 4.8 Practice Exercises, pp. 259–261

R.1. $-\frac{1}{12}$ **R.3.** $\frac{5}{4}$ **R.5.** $\frac{2}{15}$ **R.7.** -16

1. $\frac{7}{6}$ **3.** $-\frac{13}{10}$ **5.** $\frac{13}{12}$
7. $\frac{13}{8}$ **9.** 2   **11.** $\frac{7}{6}$ **13.** $-\frac{2}{5}$ **15.** -24 **17.** 21
19. -21 **21.** 30   **23.** 0   **25.** $\frac{2}{5}$ **27.** $\frac{1}{18}$ **29.** $\frac{35}{8}$
31. -5 **33.** $\frac{1}{4}$ **35.** 1   **37.** $-\frac{7}{2}$ **39.** $\frac{2}{3}$
41. $\frac{10}{9}$ **43.** 7   **45.** 5   **47.** -48 **49.** $\frac{1}{3}$
51. $-\frac{21}{10}$ **53.** $-\frac{1}{12}$ **55.** $\frac{7}{10}$ **57.** -12 **59.** $\frac{1}{3}$
61. $\frac{8}{5}$ **63.** $-\frac{44}{15}$ **65.** 6   **67.** $\frac{9}{5}$

Chapter 4 Problem Recognition Exercises, pp. 261–262

1. Equation; $\frac{1}{10}$ **2.** Equation; $\frac{6}{7}$ **3.** Expression; $\frac{3}{2}$
4. Expression; $\frac{1}{8}$ **5.** Expression; $\frac{7}{5}$ **6.** Expression; $\frac{9}{5}$
7. Equation; $-\frac{2}{5}$ **8.** Equation; $-\frac{4}{5}$ **9.** Equation; 4
10. Equation; $\frac{5}{3}$ **11.** Expression; $\frac{4}{9}$
12. Expression; 0   **13.** Equation; 6   **14.** Equation; $-\frac{5}{2}$
15. Expression; $6x - 24$ **16.** Expression; $2x + 30$
17. Equation; $\frac{8}{3}$ **18.** Equation; 7   **19.** Expression; $\frac{3}{2}$
20. Expression; $\frac{25}{7}$ **21.** Equation; $\frac{2}{7}$ **22.** Equation; $\frac{3}{4}$
23. Expression; $\frac{18}{7}c$ **24.** Expression; $\frac{21}{4}d$

Chapter 4 Review Exercises, pp. 262–265

1. $\frac{1}{2}$ 2. $\frac{4}{7}$ 3. a. $\frac{5}{3}$ b. Improper
4. a. $\frac{1}{6}$ b. Proper   5. a. $\frac{3}{8}$ b. $\frac{2}{3}$ c. $\frac{4}{9}$ 6. $\frac{23}{8}$ or $2\frac{7}{8}$
7. $\frac{7}{6}$ or $1\frac{1}{6}$ 8. $\frac{43}{7}$ 9. $\frac{57}{5}$ 10. $5\frac{2}{9}$ 11. $1\frac{2}{21}$
12., 13., 14., 15.

$-\frac{10}{5}$ $-1\frac{3}{8}$ $-\frac{7}{8}$ $\frac{13}{8}$

16. $134\frac{3}{7}$ 17. $60\frac{11}{13}$ 18. 21, 51, 1200
19. 55, 140, 260, 1200   20. 2, 53, 113
21. 12, 27, 51, 63, 130   22. $2\cdot3\cdot5\cdot11$
23. $2\cdot2\cdot3\cdot3\cdot5\cdot5$ or $2^2\cdot3^2\cdot5^2$ 24. $\neq$ 25. $=$
26. $\frac{1}{4}$ 27. $\frac{1}{5}$ 28. $-\frac{3}{2}$ 29. $-\frac{7}{3}$ 30. $\frac{2}{25}$
31. $\frac{7}{1000}$ 32. $\frac{2a}{5c}$ 33. $\frac{4t^2}{5}$ 34. $\frac{14}{15}; \frac{1}{15}$
35. a. $\frac{3}{5}$ b. $\frac{2}{5}$ 36. $-\frac{3}{7}$ 37. $-\frac{3}{2}$ 38. 63
39. 15   40. $\frac{4}{x}$ 41. $\frac{3y^2}{2}$ 42. $A=\frac{1}{2}bh$ 43. 51 ft^2
44. 1   45. 1   46. $\frac{2}{7}$ 47. $-\frac{1}{7}$ 48. $\frac{16}{9}$ 49. $\frac{7}{5}$
50. $-\frac{1}{21}$ 51. -14 52. $8a$ 53. $\frac{3y^2}{2}$
54. 36 bags of candy   55. Yes. $9\div\frac{3}{8}=24$ so he will have 24 pieces, which is more than enough for his class.

56. There are 900 African American students.
57. There are 300 Asian American students.
58. Amelia earned $576.
59. a. $2^2\cdot5^2$ b. $5\cdot13$ c. $2\cdot5\cdot7$
60. 420   61. 96   62. They will meet on the 12th day.
63. $\frac{15}{48}$ 64. $\frac{63}{35}$ 65. $\frac{35y}{60y}$ 66. $\frac{-28}{4x}$ 67. $<$
68. $>$ 69. $=$ 70. $-\frac{27}{35}, -\frac{7}{10}, -\frac{72}{105}, -\frac{8}{15}$
71. $\frac{3}{2}$ 72. $\frac{2}{3}$ 73. $\frac{29}{100}$ 74. $\frac{1}{25}$ 75. $-\frac{47}{11}$
76. $-\frac{117}{20}$ 77. $\frac{17}{40}$ 78. $\frac{12}{7}$ 79. $\frac{17}{5w}$
80. $\frac{11b+4a}{ab}$ 81. a. $\frac{35}{4}$ m or $8\frac{3}{4}$ m   b. $\frac{315}{128}$ m^2 or $2\frac{59}{128}$ m^2
82. a. $\frac{23}{3}$ yd or $7\frac{2}{3}$ yd   b. $\frac{7}{2}$ yd^2 or $3\frac{1}{2}$ yd^2
83. $23\frac{7}{15}$ 84. $23\frac{2}{3}$ 85. $\frac{10}{11}$ 86. $4\frac{1}{2}$ 87. $-2\frac{3}{11}$
88. $-\frac{3}{5}$ 89. 50; $50\frac{9}{40}$ 90. 23; $22\frac{71}{75}$ 91. $11\frac{11}{63}$
92. $14\frac{7}{16}$ 93. $2\frac{5}{8}$ 94. $1\frac{11}{12}$ 95. $3\frac{2}{5}$ 96. $3\frac{3}{14}$
97. $63\frac{15}{16}$ 98. $50\frac{1}{2}$ 99. $-2\frac{5}{6}$ 100. $-3\frac{7}{8}$
101. Corry drove a total of $8\frac{1}{6}$ hr.
102. Denise will have $\frac{7}{8}$ acre left.   103. It will take $3\frac{1}{8}$ gal.
104. There will be 10 pieces.   105. $\frac{9}{64}$ 106. $\frac{9}{64}$
107. $-\frac{1}{100,000}$ 108. $\frac{1}{10,000}$ 109. $-\frac{29}{10}$ 110. $\frac{31}{15}$
111. $\frac{1}{6}$ 112. $\frac{7}{16}$ 113. $\frac{14}{5}$ 114. $\frac{2x}{9}$ 115. $-\frac{3}{7}$
116. $\frac{1}{2}$ 117. $-\frac{4}{3}$ 118. $\frac{3}{5}$ 119. $\frac{5}{16}$ 120. $11\frac{2}{3}$
121. $-\frac{17}{12}x$ 122. $-\frac{13}{10}y$ 123. $\frac{2}{3}a+\frac{1}{6}c$ 124. $\frac{9}{10}w+\frac{5}{3}y$
125. $\frac{19}{15}$ 126. $-\frac{5}{14}$ 127. $\frac{10}{9}$ 128. $\frac{3}{8}$ 129. $\frac{3}{5}$
130. $\frac{3}{4}$ 131. -1 132. 2   133. -15 134. 12

Chapter 5
Section 5.1 Activity, p. 275

A.1. a. hundredths   b. ones   c. thousandths   d. tenths

	Decimal	Word Name
A.2.	3.006	Three and six thousandths
A.3.	0.13	Thirteen hundredths
A.4.	20.0012	Twenty and twelve ten-thousandths
A.5.	7.64	Seven and sixty-four hundredths
A.6.	205	Two hundred five
A.7.	0.065	Sixty-five thousandths

A.8. a. Fifteen hundredths   b. $\frac{15}{100}$ c. They are the same. There are two digits to the right of the decimal point and two zeros in the denominator.   d. $\frac{3}{20}$

A.9. a. 4   b. 0.15   c. $\frac{3}{20}$ d. $4\frac{3}{20}$ e. The whole number part of the mixed fraction would be 4238 but the fraction would remain $\frac{3}{20}$.

A.10. -0.64, -0.6, -0.568, 0.568, 0.59, 0.594

	Tenths	Hundredths	Thousandths
A.11. 1.8275	a. 1.8	b. 1.83	c. 1.828
A.12. 6.1529	a. 6.2	b. 6.15	c. 6.153
A.13. 30.3999	a. 30.4	b. 30.40	c. 30.400
A.14. 21.6959	a. 21.7	b. 21.70	c. 21.696

Section 5.1 Practice Exercises, pp. 276–278

1. a. decimal   b. tenths; hundredths; thousandths
3. 100   5. 10,000   7. $\frac{1}{100}$ 9. $\frac{1}{10,000}$
11. Tenths   13. Hundredths   15. Tens
17. Ten-thousandths   19. Thousandths   21. Ones
23. Nine-tenths   25. Twenty-three hundredths
27. Negative thirty-three thousandths
29. Four hundred seven ten-thousandths
31. Three and twenty-four hundredths
33. Negative five and nine-tenths
35. Fifty-two and three-tenths
37. Six and two hundred nineteen thousandths
39. -8472.014 41. 700.07   43. $-2,469,000.506$
45. $3\frac{7}{10}$ 47. $2\frac{4}{5}$ 49. $\frac{1}{4}$ 51. $-\frac{11}{20}$ 53. $20\frac{203}{250}$
55. $-15\frac{1}{2000}$ 57. $\frac{42}{5}$ 59. $\frac{157}{50}$ 61. $-\frac{47}{2}$
63. $\frac{1191}{100}$ 65. $<$ 67. $>$ 69. $<$ 71. $>$
73. a, b   75. 0.3444, 0.3493, 0.3558, 0.3585, 0.3664
77. These numbers are equivalent, but they represent different levels of accuracy.   79. 7.1   81. 49.9
83. 33.42   85. -9.096 87. 21.0   89. 7.000

91. 0.0079   **93.** 0.0036 mph

	Number	Hundreds	Tens	Tenths	Hundredths	Thousandths
95.	971.0948	1000	970	971.1	971.09	971.095
97.	21.9754	0	20	22.0	21.98	21.975

99. 0.972

Section 5.2 Activity, p. 284

A.1. a. Yes, the price is the same because $28 is equal to $28.00. The zeros after the decimal point do not change the numerical value of the number. **b.** Yes, it is to the right of the number 28. **c.** No
A.2. $119.45   **A.3.** $245.53
A.4. Answers will vary. Possible answer: Line up the numbers in a column with the decimal points and place values aligned. Add zeros as placeholders where necessary. Add or subtract the digits in columns from right to left. Place the decimal point in the sum or difference lined up with the decimal points from the original numbers.
A.5. 48.302   **A.6.** 1.0211   **A.7.** 3.088   **A.8.** 1.711
A.9. 11.11   **A.10.** 11.252   **A.11.** −11.78   **A.12.** 115.332
A.13. $2.3x$ **A.14.** $7.5y$ **A.15.** $-0.79c$

Section 5.2 Practice Exercises, pp. 284–288

R.1. b, d   **R.3.** a, c   **R.5.** 23.5   **R.7.** 26.000

1. decimal point   **3.** Estimation   **5.** deposits, payments
7. 32.24   **9.** 63.2   **11.** 8.951   **13.** 15.991   **15.** 79.8005
17. 31.0148   **19.** 62.6032   **21.** 100.414   **23.** 128.44
25. 82.063   **27.** 14.24   **29.** 3.68   **31.** 12.32   **33.** 5.677
35. 1.877   **37.** 21.6646   **39.** 14.765   **41.** 159.558
43. 0.9012   **45.** −422.94   **47.** −1.359   **49.** 50.979
51. −3.27   **53.** −4.432   **55.** 1.4   **57.** −111.2   **59.** 0.5346
61.

Check No.	Description	Payment	Deposit	Balance
				$245.62
2409	Electric bill	$52.48		193.14
2410	Groceries	72.44		120.70
2411	Department store	108.34		12.36
	Paycheck		$1084.90	1097.26
2412	Restaurant	23.87		1073.39
	Transfer from savings		200	1273.39

63. 1.35 million cells per microliter
65. a. The water is rising 1.7 in./hr.   **b.** At 1:00 P.M. the level will be 11 in.   **c.** At 3:00 P.M. the level will be 14.4 in.
67. The pile containing the two nickels and two pennies is higher.
69. $x = 8.9$ in.; $y = 15.4$ in.; the perimeter is 98.8 in.
71. $x = 2.075$ ft; $y = 2.59$ ft; the perimeter is 22.17 ft.
73. 27.2 mi   **75.** $3.87t$ **77.** $-13.2p$ **79.** $0.4v$
81. $0.845x + 0.52y$ **83.** $c - 5d$
85. IBM decreased by $1.99 per share.   **86.** FedEx increased by $6.56 per share.   **87.** Between March and April, FedEx increased the most, by $6.36 per share.   **88.** Between February and March, IBM increased the most, by $3.04 per share.
89. Between January and February, FedEx decreased the most, by $2.78 per share.
90. Between January and February, IBM decreased the most, by $6.92 per share.

Section 5.3 Practice Exercises, pp. 295–300

R.1. −0.40   **R.3.** >   **R.5.** 258,000

1. a. front   **b.** circle   **c.** radius   **d.** diameter   **e.** circumference   **f.** π **g.** 3.14; $\frac{22}{7}$ **h.** Both formulas can be used.   **i.** πr^2
3. left   **5.** 4   **7.** 3.6   **9.** 0.4   **11.** 3.6
13. 0.18   **15.** 17.904   **17.** 37.35   **19.** 4.176
21. −4.736   **23.** 2.891   **25.** 114.88   **27.** 2.248
29. −0.00144   **31. a.** 51   **b.** 510   **c.** 5100   **d.** 51,000
33. a. 0.51   **b.** 0.051   **c.** 0.0051   **d.** 0.00051   **35.** 3490
37. 96,590   **39.** −0.933   **41.** 0.05403
43. 2,600,000   **45.** 400,000   **47.** $20,549,000,000
49. a. 201.6 lb of gasoline   **b.** 640 lb of CO_2
51. The bill was $423.61.   **53.** $48.81 can be saved.
55. 0.00115 km^2 **57.** The area is 333 yd^2.   **59.** 0.16
61. 1.69   **63.** 0.001   **65.** −0.04   **67.** The length of a radius is one-half the length of a diameter.   **69.** 12.2 in.
71. 83 m   **73.** 62.8 cm   **75.** 15.7 km   **77.** 18.84 cm
79. 14.13 in.   **81.** 314 mm^2 **83.** 121 ft^2
85. 16.642 mi   **87.** 2826 ft^2
89. Area ≈ 517.1341 cm^2; circumference ≈ 80.6133 cm
90. Area ≈ 81.7128 ft^2; circumference ≈ 32.0442 ft
91. Area ≈ 70.8822 $in.^2$; circumference ≈ 29.8451 in.
92. Area ≈ 8371.1644 mm^2; circumference ≈ 324.3380 mm
93. a. 94.2 in.   **b.** It will roll 1507.2 in. Yes, it will reach the end of the driveway.   **95.** 69,080 in. or 5757 ft

Section 5.4 Activity, pp. 308–309

A.1. a. 648   **b.** 6.48   **c.** Answers will vary. One possibility: the numbers are the same but the decimal point is in a different place.   **d.** Multiplication; $648 \cdot 6 = 3888$
A.2. a. 0.233333...   **b.** The same values are obtained for each successive step.   **c.** $0.2\overline{3}$
A.3. 0.63   **A.4.** −103.2   **A.5.** 0.307
A.6. a. 10   **b.** 10   **c.** $\frac{5.7}{2.1} = \frac{57}{21}$ **d.** Yes
A.7. 0.059   **A.8.** 6.2   **A.9.** $2.\overline{2}$
A.10. a. $50.00   **b.** One place to the left   **c.** One, same as the number of places the decimal point moves.   **d.** $5.00
A.11. 50,000   **A.12.** 0.321038   **A.13.** 3,210.38
A.14. 3.21038   **A.15.** 321,038

Section 5.4 Practice Exercises, pp. 309–312

R.1. a. 17.06   **b.** 17.061   **R.3.** 42,300   **R.5.** 0.00423

1. a. repeating   **b.** terminating   **3.** left   **5.** quotient; divisor
7. 10   **9.** 1.8   **11.** 0.9   **13.** 0.18   **15.** 0.53
17. 21.1   **19.** 1.96   **21.** 0.035   **23.** 16.84   **25.** 0.12
27. −0.16   **29.** $5.\overline{3}$ **31.** $3.1\overline{6}$ **33.** $2.\overline{15}$ **35.** 503
37. 9.92   **39.** −56   **41.** 2.975   **43.** $208.\overline{3}$
45. 48.5   **47.** 1100   **49.** 42,060   **51.** The decimal point will move to the left two places.   **53.** 0.03923
55. −9.802   **57.** 0.00027   **59.** 0.00102
61. a. 2.4   **b.** 2.44   **c.** 2.444   **63. a.** 1.8   **b.** 1.79   **c.** 1.789
65. a. 3.6   **b.** 3.63   **c.** 3.626   **67.** 0.26   **69.** −14.8
71. 20.667   **73.** 35.67   **75.** 111.3   **77.** Unreasonable; $960
79. Unreasonable; $140,000   **81.** The monthly payment is $42.50.
83. a. 13 bulbs would be needed (rounded up to the nearest whole unit).   **b.** $9.75   **c.** The energy efficient fluorescent bulb would be more cost effective.   **85.** Babe Ruth's batting average was 0.342.
87. 2.2 mph   **89.** 1149686.166   **90.** 3411.4045   **91.** 1914.0625
92. 69568.83693   **93.** 95.6627907   **94.** 293.5070423

95. Answers will vary. **96.** Answers will vary.
97. a. $0.\overline{37}$ **b.** Yes, the claim is accurate. The decimal $0.\overline{37}$ is close to $0.\overline{3}$, which is equal to $\frac{1}{3}$. **98.** 239 people per square mile
99. a. 1,600,000 mi per day **b.** $66,666.\overline{6}$ mph
100. When we say that 1 year is 365 days, we are ignoring the 0.256 day each year. In 4 years, that amount is $4 \times 0.256 = 1.024$, which is another whole day. This is why we add one more day to the calendar every 4 years. **101.** -47.265 **103.** b, d

Chapter 5 Problem Recognition Exercises, pp. 313–314

1. a. 223.04 **b.** 12,304 **c.** 23.04 **d.** 1.2304 **e.** 123.05 **f.** 1.2304 **g.** 12,304 **h.** 123.03
2. a. 6078.3 **b.** 5,078,300 **c.** 4078.3 **d.** 5.0783 **e.** 5078.301 **f.** 5.0783 **g.** 5,078,300 **h.** 5078.299
3. a. -7.191 **b.** 7.191 **4. a.** -730.4634 **b.** 730.4634
5. a. 52.64 **b.** 52.64 **6. a.** 59.384 **b.** 59.384
7. a. 5.73 **b.** -12.67 **8. a.** 0.055 **b.** -0.139
9. a. -80 **b.** -448 **10. a.** -54 **b.** -496.8
11. 1 **12.** 1 **13.** 4000 **14.** 6,400,000
15. 200,000 **16.** 2700 **17.** 1,350,000,000
18. 1,700,000 **19.** 4.4001 **20.** 76.7001
21. 86.4 **22.** -5.4 **23.** 185 **24.** -46.25

Section 5.5 Practice Exercises, pp. 323–328

R.1. $\frac{3}{200}$ **R.3.** 0.0016 **R.5.** -1.06 **R.7.** 20 **R.9.** -21

1. a. rational **b.** repeating **c.** irrational **d.** real
3. 1000 **5.** $\frac{5}{10}$; 0.5 **7.** $\frac{84}{100}$; 0.84 **9.** $\frac{4}{10}$; 0.4
11. $\frac{98}{100}$; 0.98 **13.** 0.28 **15.** -0.632 **17.** -3.2
19. -5.25 **21.** 0.75 **23.** 7.45 **25.** $3.\overline{8}$
27. $0.52\overline{7}$ **29.** $-0.\overline{126}$ **31.** $1.1\overline{36}$ **33.** 0.9
35. 0.143 **37.** 0.08 **39.** -0.71
41. a. $0.\overline{1}$ **b.** $0.\overline{2}$ **c.** $0.\overline{4}$ **d.** $0.\overline{5}$
If we memorize that $\frac{1}{9} = 0.\overline{1}$, then $\frac{2}{9} = 2 \cdot \frac{1}{9} = 2 \cdot 0.\overline{1} = 0.\overline{2}$, and so on.

43.

	Decimal Form	Fraction Form
a.	0.45	$\frac{9}{20}$
b.	1.625	$1\frac{5}{8}$ or $\frac{13}{8}$
c.	$-0.\overline{7}$	$-\frac{7}{9}$
d.	$-0.\overline{45}$	$-\frac{5}{11}$

45.

	Decimal Form	Fraction Form
a.	$0.\overline{3}$	$\frac{1}{3}$
b.	-2.125	$-2\frac{1}{8}$ or $-\frac{17}{8}$
c.	$-0.8\overline{63}$	$\frac{19}{22}$
d.	1.68	$\frac{42}{25}$

47. Rational **49.** Rational **51.** Rational
53. Irrational **55.** Irrational **57.** Rational

59. = **61.** < **63.** > **65.** >
67. 1.75, $1.\overline{7}$, 1.8

69. $-\frac{1}{5}, -0.\overline{1}, -\frac{1}{10}$

71. 6.25 **73.** 10 **75.** 8.77 **77.** 25.75 **79.** -2
81. -8.58 **83.** 4 **85.** 12.1 **87.** 67.35 **89.** 25.05
91. 23.4 **93.** 1.28 **95.** -10.83 **97.** 2.84
99. 43.46 cm² **101.** 84.78 **103.** -78.4
105. a. 471 mi **b.** 62.8 mph **107.** Jorge will be charged $98.75. **109.** She has 24.3 g left for dinner.
111. Hannah should get $4.77 in change.
113. a. 237 shares **b.** $13.90 will be left.
114. a. Approximately 921,800 homes could be powered.
b. Approximately 342,678 additional homes could be powered.
115. a. The BMI is approximately 30.1. The person is at high risk.
b. The BMI is approximately 19.8. The person has a low risk.
116. a. Marty will have to finance $120,000. **b.** There are 360 months in 30 years. **c.** He will pay $287,409.60. **d.** He will pay $167,409.60 in interest.
117. Each person will get approximately $13,410.10.
118. a. 8.967 **b.** Yes **119.** 3.475 **121.** 0.52

Section 5.6 Activity, pp. 333–334

A.1. a. 315 **b.** 3.15 **c.** No, the same steps are followed. Subtract the constant from both sides of the equation.
A.2. a. Divide both sides by -52. **b.** Divide both sides by -5.2.
c. Yes **d.** Yes **e.** Each term was multiplied by 10.
A.3. a. 3.15 **b.** 3.15 **c.** Yes, because they are the same equations. The equation in part (a) was multiplied by 100 to clear the decimals, resulting in the equation in part (b).
A.4. 1.8 **A.5.** 2.44 **A.6.** -12 **A.7.** 2.5
A.8. b. x = a number **c.** The sum of x and 8.55 is 2.99.
d. $x + 8.55 = 2.99$ **e.** -5.56 **f.** -5.56 increased by 8.55 equals 2.99.
A.9. b. x = credit hours **c.** Cost = 35 + 92.50 (credit hours)
d. $1145 = 35 + 92.50x$ **e.** 12 **f.** Jenna is taking 12 credit hours, which cost $1145.

Section 5.6 Practice Exercises, pp. 334–336

R.1. 6 **R.3.** $-\frac{17}{10}$ **R.5.** 0

1. 0.86 **3.** -4.78 **5.** -49.9 **7.** 0.095 **9.** 42.78
11. -0.12 **13.** -2.9 **15.** 19 **17.** 5 **19.** 9
21. -24 **23.** 6.72 **25.** 50 **27.** -4 **29.** -600
31. 10 **33.** The number is 4.5. **35.** The number is 2.7.
37. The number is -9.4. **39.** The number is 7.6.
41. The sides are 4.6 yd, 7.7 yd, and 9.2 yd. **43.** Rafa made $65.25, Toni made $43.20, and Henri made $59.35 in tips. **45.** The painter rented the pressure cleaner for 3.5 hr. **47.** The previous balance is $104.75 and the new charges amount to $277.15. **49.** Madeline rode for 54.25 min and Kim rode for 62.5 min.

Chapter 5 Review Exercises, pp. 336–339

1. The 3 is in the tens place, 2 is in the ones place, 1 is in the tenths place, and 6 is in the hundredths place.

2. The 2 is in the ones place, 0 is in the tenths place, 7 is in the hundredths place, and 9 is the thousandths place.
3. Five and seven-tenths   4. Ten and twenty-one hundredths   5. Negative fifty-one and eight thousandths
6. Negative one hundred nine and one-hundredth
7. 33,015.047   8. −100.01   9. $-4\frac{4}{5}$ 10. $\frac{1}{40}$
11. $\frac{13}{10}$ 12. $\frac{27}{4}$ 13. <   14. <
15. .325, .330, .333, .338, .354   16. 89.92   17. 34.890
18. a. The amount in the box is less than the advertised amount.   b. The amount rounds to 12.5 oz.
19. a, b   20. b, c   21. 49.743   22. 273.22
23. 5.45   24. 1.902   25. −244.04   26. 29.007
27. 7.809   28. 82.265   29. $0.5y$ 30. $-13.5x + 6$
31. a. Between days 1 and 2, the increase was $0.194.
b. Between days 3 and 4, the decrease was $0.209.
32. 8.19   33. 74.113   34. −264.44   35. −346.5
36. 85,490   37. 100.34   38. 0.9201   39. 1.0422
40. 432,000   41. 33,800,000   42. a. Eight batteries cost $15.96 on sale.   b. A customer can save $2.03.
43. The yearly cost is $1493.88.   44. Area = 940 ft², perimeter = 127 ft   45. a. 7280 people   b. 19,260 people
46. 7.5 m   47. 13.6 ft   48. Area: 452.16 ft²; circumference: 75.36 ft   49. Area: 2826 yd²; circumference: 188.4 yd   50. 17.1   51. 42.8
52. $4.1\overline{3}$ 53. $8.7\overline{6}$ 54. −27   55. −0.03
56. 4.9393   57. 9.0234   58. 553,800   59. 260
60.

	$8.\overline{6}$	$52.\overline{52}$	$0.\overline{409}$
Tenths	8.7	52.5	0.4
Hundredths	8.67	52.53	0.41
Thousandths	8.667	52.525	0.409
Ten-thousandths	8.6667	52.5253	0.4094

61. −11.62   62. −11.97   63. a. $0.50 per roll
b. $0.57 per roll   c. The 12-pack is better.
64. 2.4   65. 3.52   66. −0.192   67. −0.4375
68. $0.58\overline{3}$ 69. $1.52\overline{7}$ 70. $-4.3\overline{18}$ 71. 0.29
72. 0.9   73. −3.667   74. a. Rational   b. Irrational
c. Irrational   d. Rational
75. $\frac{2}{9}$ 76. $3\frac{1}{3}$
77.

Stock	Closing Price ($) (Decimal)	Closing Price ($) (Mixed Number)
Ford	13.02	$13\frac{1}{50}$
Microsoft	30.50	$30\frac{1}{2}$
Citibank	29.37	$29\frac{37}{100}$

78. >   79. >   80. −5   81. −5.52   82. 59
83. 1.6   84. −3   85. −2   86. $89.90 will be saved by buying the combo package.   87. Marvin must drive 34 mi more.   88. −2.58   89. −1.53   90. 21.4
91. −11.32   92. −5.5   93. 2.5   94. −5.2
95. 110   96. −415   97. Multiply by 100; solution: −1.48
98. Multiply by 10; solution: 2.76   99. The number is 6.5.
100. The number is 2.4.   101. The number is 1600.
102. The number is 4.   103. The sides are 5.6 yd, 8 yd, and 11.2 yd.
104. The apartment costs $731.85 per week, and the car costs $129.50 per week.   105. Mykeshia can rent space for 6 months.

Chapter 6

Section 6.1 Activity, p. 346

A.1. a. $\frac{8}{11}$ b. $\frac{11}{8}$ c. $\frac{8}{19}$ A.2. $\frac{31}{44}$
A.3. No, there are 2 more infants scheduled than allowed by the 4:1 ratio.
A.4. $\frac{10}{1}$ A.5. $\frac{9}{13}$

Section 6.1 Practice Exercises, pp. 346–349

R.1. $\frac{3}{5}$ R.3. $\frac{23}{1}$
1. ratio   3. 5 : 6 and $\frac{5}{6}$ 5. 11 to 4 and $\frac{11}{4}$
7. 1 : 2 and 1 to 2   9. a. $\frac{5}{3}$ b. $\frac{3}{5}$ c. $\frac{3}{8}$
11. a. $\frac{2}{15}$ b. $\frac{2}{13}$ 13. a. $\frac{7}{18}$ b. $\frac{7}{25}$ 15. $\frac{2}{3}$
17. $\frac{1}{5}$ 19. $\frac{4}{1}$ 21. $\frac{11}{5}$ 23. $\frac{6}{5}$ 25. $\frac{1}{2}$ 27. $\frac{3}{2}$
29. $\frac{6}{7}$ 31. $\frac{8}{9}$ 33. $\frac{7}{1}$ 35. $\frac{1}{8}$ 37. $\frac{5}{4}$
39. a. 315 ft   b. $\frac{63}{1}$ 41. a. $\frac{2}{9}$ b. $\frac{2}{9}$ 43. $\frac{1}{11}$
45. $\frac{10}{1}$ 47. $\frac{15}{32}$ 49. $\frac{20}{61}$ 51. $\frac{2}{3}$ 53. $\frac{1}{4}$
55. 13 units   57. a. 1.5   b. $1.\overline{6}$ c. 1.6   d. 1.625; yes
59. Answers will vary.

Section 6.2 Activity, p. 354

A.1. 15 mi/gal   A.2. 55 mi/hr   A.3. $24.25/hr
A.4. a. $0.34, $0.56, $0.33   b. Plan C
A.5. a. 25 mi/gal   b. 21.4 mi/gal   c. Slower   d. 40 gal at 55 mph; 46.7 gal for 70 mph; Save 6.7 gal.   e. $14.41
A.6. The large pack of diapers is more cost effective at 0.35 cents per diaper vs. 0.353 cents per diaper for the jumbo box.

Section 6.2 Practice Exercises, pp. 354–358

R.1. $\frac{9}{11}$ R.3. $\frac{9}{17}$ R.5. $\frac{5}{1}$
1. a. rate   b. unit   3. divide   5. $\frac{33 \text{ mi}}{4 \text{ hr}}$
7. $\frac{\$32}{5 \text{ ft}^2}$ 9. $\frac{117 \text{ mi}}{2 \text{ hr}}$ 11. $\frac{\$29}{4 \text{ hr}}$ 13. $\frac{1 \text{ page}}{2 \text{ sec}}$
15. $\frac{65 \text{ calories}}{4 \text{ crackers}}$ 17. $-\frac{9°F}{4 \text{ hr}}$ 19. a, c, d
21. 113 mi/day   23. −400 m/hr   25. $55 per payment
27. $0.69/lb   29. $256,000 per person
31. 14.29 m/sec   33. $0.150 per oz   35. $0.995 per liter
37. $52.50 per tire   39. $5.417 per bodysuit
41. a. $0.334/oz   b. $0.334/oz   c. Both sizes cost the same amount per ounce.
43. The larger can is $0.123 per ounce. The smaller can is $0.164 per ounce. The larger can is the better buy.
45. Coca-Cola: 3.25 g/fl oz; MelloYello: 3.92 g/fl oz; Ginger Ale: 3 g/fl oz; MelloYello has the greatest amount per fluid oz.
47. Coca-Cola: 12 cal/fl oz; MelloYello: 14.2 cal/fl oz; Ginger Ale: 11.25 cal/fl oz; Ginger Ale has the least number of calories per fluid oz.

49. 295,000 vehicles/year **51. a.** 2.2 million per year **b.** 2.04 million per year **c.** Mexico **53.** Cheetah: 29 m/sec; antelope: 24 m/sec. The cheetah is faster. **54. a.** 9.9 wins/year **b.** 8.6 wins/year **c.** Shula **55. a.** 2.1 wins/loss **b.** 1.5 wins/loss **c.** Shula **56. a.** $0.38/oz **b.** $0.18/oz **c.** $0.19/oz; The best buy is the 8-bar pack. **57.** The unit prices are $0.181/oz, $0.280/oz, and $0.255/oz. The best buy is the 48-oz jar. **58. a.** $0.401/oz **b.** $0.167/oz **c.** $0.322/oz. The best buy is the 12-oz can. **59. a.** $0.062/oz **b.** $0.023/oz. The 12-pack of 12-oz cans for $3.33 is the better buy.

Section 6.3 Activity, p. 363

A.1. a. $\frac{4}{12} = \frac{16}{48}$ **b.** Yes, the ratios both reduce to $\frac{1}{3}$, and the cross products are equal.
A.2. Yes **A.3.** No **A.4.** No **A.5.** Yes **A.6.** 2
A.7. 2 **A.8.** Yes. Set the cross products equal to each other, then solve for the unknown. **A.9.** -9 **A.10.** -50
A.11. $3\frac{3}{4}$ or 3.75 **A.12. a.** $5x - 20$ **b.** -4

Section 6.3 Practice Exercises, pp. 363–365

R.1. $\frac{50}{17}$ **R.3.** $\frac{9 \text{ apples}}{2 \text{ pies}}$ **R.5.** 20.5 rev/sec

1. proportion **3.** cross, products **5.** are **7.** $\frac{2}{48} = \frac{1}{24}$
9. $\frac{4}{16} = \frac{5}{20}$ **11.** $\frac{-25}{15} = \frac{-10}{6}$ **13.** $\frac{2}{3} = \frac{4}{6}$
15. $\frac{-30}{-25} = \frac{12}{10}$ **17.** $\frac{\$6.25}{1 \text{ hr}} = \frac{\$187.50}{30 \text{ hr}}$
19. $\frac{1 \text{ in.}}{7 \text{ mi}} = \frac{5 \text{ in.}}{35 \text{ mi}}$ **21.** No **23.** Yes **25.** Yes
27. Yes **29.** Yes **31.** Yes **33.** No **35.** Yes
37. No **39.** 4 **41.** 3 **43.** -75 **45.** $\frac{3}{4}$
47. $-\frac{65}{4}$ or $-16\frac{1}{4}$ or -16.25 **49.** $\frac{15}{2}$ or $7\frac{1}{2}$ or 7.5
51. 3 **53.** 2.5 **55.** 4 **57.** $-\frac{1}{80}$ **59.** 36
61. 7.5 **63.** 30 **65.** $\frac{8}{7}$ **67.** -3 **69.** 12 **71.** -6

Chapter 6 Problem Recognition Exercises, p. 366

1. a. Proportion; $\frac{15}{2}$ **b.** Product of fractions; $\frac{15}{32}$
2. a. Product of fractions; $\frac{3}{25}$ **b.** Proportion; 4
3. a. Product of fractions; $\frac{3}{49}$ **b.** Proportion; 4
4. a. Proportion; 2 **b.** Product of fractions; $\frac{6}{25}$
5. a. Proportion; 9 **b.** Product of fractions; 32
6. a. Product of fractions; 8 **b.** Proportion; $\frac{98}{5}$
7. a. 14 **b.** $\frac{5}{2}$ **c.** $\frac{3}{5}$ **d.** $\frac{18}{245}$ **8. a.** $\frac{3}{25}$ **b.** $\frac{3}{5}$ **c.** $\frac{16}{3}$ **d.** $-\frac{88}{15}$
9. a. 4 **b.** $\frac{98}{5}$ **c.** $\frac{48}{35}$ **d.** $\frac{49}{25}$ **10. a.** 18 **b.** $\frac{29}{3}$ **c.** $\frac{11}{18}$ **d.** 22

Section 6.4 Activity, pp. 373–374

A.1. 13.2 min **A.2.** 108 dogs **A.3.** 1331 calories
A.4. 21.9 g **A.5.** 270 min or 4 hr, 30 min
A.6. 2.25 ft or $2\frac{1}{4}$ ft **A.7.** 14.25 in. or $14\frac{1}{4}$ in.

Section 6.4 Practice Exercises, pp. 374–378

R.1. $\frac{96}{5}$ or $19\frac{1}{5}$ or 19.2 **R.3.** 4.9 **R.5.** 6

1. a. similar **b.** similar **3.** Pam can drive 610 mi on 10 gal of gas.
5. 78 kg of crushed rock will be required.
7. The actual distance is about 80 mi.
9. There are 3800 male students.
11. Heads would come up about 315 times.
13. There would be approximately 3 earned runs for a 9-inning game.
15. Pierre can buy 720 Euros. **17.** 9 g
19. $\frac{2}{3}$ cup of water **21. a.** 195 emails **b.** 585 min or 9.75 hr
23. 0.98 megabytes or 980 kilobytes
25. 29.08 in. **27.** 252 mL of acid
29. There are approximately 357 bass in the lake.
31. There are approximately 4000 bison in the park.
33. $x = 24$ cm, $y = 36$ cm **35.** $x = 1$ yd, $y = 10.5$ yd
37. $x = 15$ cm, $y = 4$ in. **39.** The flagpole is 12 ft high.
41. The platform is 2.4 m high. **43.** $x = 17.5$ in.
45. $x = 6$ ft, $y = 8$ ft **47.** $x = 21$ ft; $y = 21$ ft; $z = 53.2$ ft
49. There were approximately 166,005 crimes committed.
50. The Washington Monument is approximately 555 ft tall.
51. Approximately 15,400 women would be expected to have breast cancer.
52. Approximately 295,000 men would be expected to have prostate disease.

Chapter 6 Review Exercises, pp. 378–381

1. 5 to 4 and $\frac{5}{4}$ **2.** 3 : 1 and $\frac{3}{1}$ **3.** 8 : 7 and 8 to 7
4. a. $\frac{2}{3}$ **b.** $\frac{3}{2}$ **c.** $\frac{3}{5}$ **5. a.** $\frac{4}{5}$ **b.** $\frac{5}{4}$ **c.** $\frac{5}{9}$
6. a. $\frac{12}{52}$ or $\frac{3}{13}$ **b.** $\frac{12}{40}$ or $\frac{3}{10}$ **7.** $\frac{4}{1}$ **8.** $\frac{7}{5}$ **9.** $\frac{2}{5}$
10. $\frac{1}{4}$ **11.** $\frac{9}{2}$ **12.** $\frac{4}{13}$ **13.** $\frac{4}{3}$ **14.** $\frac{170}{13}$
15. a. This year's enrollment is 1520 students. **b.** $\frac{4}{19}$
16. $\frac{19}{12}$ **17.** $\frac{1}{5}$ **18.** $\frac{24}{49}$ **19.** $\frac{4 \text{ hot dogs}}{9 \text{ min}}$
20. $\frac{2 \text{ mi}}{17 \text{ min}}$ **21.** $-\frac{\$1700}{3 \text{ months}}$ **22.** $-\frac{5 \text{ m}}{3 \text{ min}}$
23. All unit rates have a denominator of 1, and reduced rates may not.
24. 33 mph **25.** $-4°$ per hour **26.** 90 times/sec
27. 11 min/lawn **28.** $0.599 per ounce **29.** $6.667 per towel
30. a. $0.262/oz **b.** $0.280/oz The 32-oz bottle is the better buy.
31. a. $0.175/oz **b.** $0.159/oz The 44-oz jar is the better buy.
32. $0.499 per ounce
33. The difference is about 8¢ per roll or $0.08 per roll.
34. 0.6275 in./hr
35. a. There was an increase of 120,000 hybrid vehicles. **b.** There will be 10,000 additional hybrid vehicles per month.
36. a. There was an increase of 63 lb. **b.** Americans increased the amount of vegetables in their diet by 3.5 lb per year.
37. $\frac{16}{14} = \frac{12}{10\frac{1}{2}}$ **38.** $\frac{8}{20} = \frac{6}{15}$ **39.** $\frac{-5}{3} = \frac{-10}{6}$
40. $\frac{4}{-3} = \frac{20}{-15}$ **41.** $\frac{\$11}{1 \text{ hr}} = \frac{\$88}{8 \text{ hr}}$ **42.** $\frac{2 \text{ in.}}{5 \text{ mi}} = \frac{6 \text{ in.}}{15 \text{ mi}}$
43. No **44.** Yes **45.** Yes **46.** No
47. Yes **48.** No **49.** No **50.** Yes
51. 4 **52.** 27 **53.** 3 **54.** 2 **55.** -13.6 **56.** -0.9
57. The human equivalent is 84 years. **58.** He can buy 42,750 yen.
59. Alabama had approximately 4,600,000 people.
60. The tax would be $6.96. **61.** $x = 10$ in., $y = 62.1$ in.
62. The building is 8 m high. **63.** $x = 1.6$ yd, $y = 1.8$ yd
64. $x = 10.8$ cm, $y = 30$ cm

Chapter 7

Section 7.1 Activity, pp. 392–393

A.1. a. Division b. 100 c. Answers will vary. Possible answer: Divide by 100.

A.2.

	Number of Students	Fraction of Students	Percent of Students
Freshman	18	$\frac{18}{100}$	18%
Sophomores	30	$\frac{30}{100}$	30%
Juniors	40	$\frac{40}{100}$	40%
Seniors	12	$\frac{12}{100}$	12%

A.3. Answers will vary. Possible answer: 40% off is the better sale since 0.75% is less than 1%.
A.4. a. F b. D c. G d. A e. B f. H g. E h. C
A.5. a. 100 b. Moves two places to the left.
A.6. a. 100 b. Moves two places to the right.
A.7. a. $\frac{4}{5} \cdot \frac{100\%}{1}$ b. 80% **A.8.** a. 0.8 b. 80% c. Yes
  d. Answers will vary.
A.9. a. 0.53846153…; No b. 53.846153%; 53.8% c. 0.5; 50%
  d. No. The answer in part (b) was converted to a percent first then rounded, whereas the answer in part (c) was rounded first then converted to a percent. e. 22.2%

A.10.

Fraction	Decimal	Percent
$\frac{3}{8}$	0.375	37.5%
$\frac{9}{25}$	0.36	36%
$\frac{1}{250}$	0.004	0.4%
$1\frac{1}{2}$	1.5	150%
$\frac{1}{3}$	$0.\overline{3}$	$33\frac{1}{3}\%$
2	2	200%

A.11. Greater than, since 1.40 = 140% and 140% > 100%.
A.12. More than, since $\frac{7}{11} \approx 63.6\%$ and 63.6% > 50%.
A.13. Greater than, since 0.02 = 2% and 2% > 1%.
A.14. More than, since $\frac{19}{16} = 118.75\%$ and 118.75% > 100%.
A.15. More than, since 600% = 6 and 6 > 0.6.
A.16. Less than, since 0.057 = 5.7% and 5.7% < 50%.

Section 7.1 Practice Exercises, pp. 394–397

R.1. $\frac{3}{125}$ **R.3.** 0.48 **R.5.** $1.41\overline{6}$

1. percent **3.** 84% **5.** 10% **7.** 2% **9.** 70%
11. Replace the symbol % with $\times \frac{1}{100}$ (or ÷ 100). Then simplify the fraction to lowest terms.
13. $\frac{3}{100}$ **15.** $\frac{21}{25}$ **17.** $\frac{17}{500}$ **19.** $\frac{23}{20}$ or $1\frac{3}{20}$
21. $\frac{1}{200}$ **23.** $\frac{1}{400}$ **25.** $\frac{31}{600}$ **27.** $\frac{249}{200}$
29. Replace the % symbol with $\times 0.01$ (or ÷ 100).
31. 0.58 **33.** 0.085 **35.** 1.42 **37.** 0.0055
39. 0.264 **41.** 0.5505 **43.** 27% **45.** 19%
47. 175% **49.** 12.4% **51.** 0.6% **53.** 101.4%
55. 71% **57.** 87.5% or $87\frac{1}{2}\%$ **59.** $83.\overline{3}\%$ or $83\frac{1}{3}\%$
61. 175% **63.** $122.\overline{2}\%$ or $122\frac{2}{9}\%$ **65.** $166.\overline{6}\%$ or $166\frac{2}{3}\%$
67. 42.9% **69.** 7.7% **71.** 45.5% **73.** 86.7%
75. c **77.** e **79.** f **81.** e **83.** f **85.** a

87.

	Fraction	Decimal	Percent
a.	$\frac{1}{4}$	0.25	25%
b.	$\frac{23}{25}$	0.92	92%
c.	$\frac{3}{20}$	0.15	15%
d.	$\frac{8}{5}$ or $1\frac{3}{5}$	1.6	160%
e.	$\frac{1}{100}$	0.01	1%
f.	$\frac{1}{125}$	0.008	0.8%

89.

	Fraction	Decimal	Percent
a.	$\frac{7}{50}$	0.14	14%
b.	$\frac{87}{100}$	0.87	87%
c.	1	1	100%
d.	$\frac{1}{3}$	$0.\overline{3}$	$33.\overline{3}\%$ or $33\frac{1}{3}\%$
e.	$\frac{1}{500}$	0.002	0.2%
f.	$\frac{19}{20}$	0.95	95%

91. 25% **93.** 10% **95.** 0.096; $\frac{12}{125}$ **97.** 0.084; $\frac{21}{250}$
99. The fraction $\frac{1}{2} = 0.5$ and $\frac{1}{2}\% = 0.5\% = 0.005$.
101. 25% = 0.25 and 0.25% = 0.0025 **103.** a, c
105. a, c **107.** 1.4 > 100% **109.** 0.052 < 50%

Section 7.2 Activity, pp. 402–403

A.1. a. is b. % c. of **A.2.** a. 20% b. $1600 c. $8000
A.3. a. $\frac{x}{950} = \frac{38}{100}$ b. 361 **A.4.** a. $\frac{288}{x} = \frac{120}{100}$ b. 240
A.5. a. $\frac{5}{250} = \frac{p}{100}$ b. 2% **A.6.** a. $\frac{16}{40} = \frac{p}{100}$
  b. 40%. Since the amount of students who earned an "A" on the test was used in the proportion, this percent represents the number of students who received an "A" on the test. c. 60% earned a grade other than an "A".
A.7. a. 592 b. 29,008 **A.8.** 2196 **A.9.** $81 **A.10.** 29.5%

Section 7.2 Practice Exercises, pp. 403–407

R.1. 64 **R.3.** $\frac{25}{2}$ or $12\frac{1}{2}$ or 12.5 **R.5.** 2.34

1. a. percent b. cross **3.** total, whole amount
5. part **7.** base **9.** amount **11.** Yes
13. No **15.** Amount: 12; base: 20; $p = 60$
17. Amount: 99; base: 200; $p = 49.5$
19. Amount: 50; base: 40; $p = 125$ **21.** $\frac{10}{100} = \frac{12}{120}$
23. $\frac{80}{100} = \frac{72}{90}$ **25.** $\frac{104}{100} = \frac{21{,}684}{20{,}850}$ **27.** 108 employees
29. 0.2 **31.** 560 **33.** Pedro pays $20,160 in taxes.
35. Approximately 219 of the 304 teens were not wearing seat belts.
37. 36 **39.** 230 lb **41.** 1350

43. Albert makes $1600 per month. **45.** Aimee has a total of 35 emails. **47.** 35% **49.** 120% **51.** 87.5%
53. She answered 72.5% correctly. **55.** 20%
57. 26.7% **59.** 7.5 **61.** 40% **63.** 92 **65.** 77
67. 8800 **69.** 160% **71.** 70 mm of rain fell in August.
73. Approximately 27,826 students applied.
75. The hospital is filled to 84% occupancy.
77. a. 22 women would be expected to relapse. **b.** 465 would not be expected to relapse.
79. 73 were Chevys. **81.** There were 180 total vehicles.
83. $331.20 **85.** $11.60 **87.** $6.30

Section 7.3 Practice Exercises, pp. 412–415

R.1. 160 **R.3.** 700 **R.5.** 1.7

1. $x = (0.35)(700); x = 245$ **3.** $(0.0055)(900) = x; x = 4.95$
5. $x = (1.33)(600); x = 798$ **7.** 50% equals one-half of the number. So multiply the number by $\frac{1}{2}$. **9.** $2 \cdot 14 = 28$
11. $\frac{1}{2} \cdot 40 = 20$ **13.** There is 3.84 oz of sodium hypochlorite.
15. He completed approximately 5015 passes. **17.** $18 = 0.4x; x = 45$
19. $0.92x = 41.4; x = 45$ **21.** $3.09 = 1.03x; x = 3$ **23.** There were 1175 subjects tested. **25.** At that time, the population was about 308 million. **27.** $x \cdot 480 = 120; x = 25\%$ **29.** $666 = x \cdot 740; x = 90\%$
31. $x \cdot 300 = 400; x = 133.3\%$ **33.** 5% of American Peace Corps volunteers were over 50 years old. **35. a.** There are 80 total employees. **b.** 12.5% missed 3 days of work. **c.** 75% missed 1 to 5 days of work.
37. 27.9 **39.** 20% **41.** 150 **43.** 555 **45.** 600 **47.** 0.2%
49. There were 35 million total hospital stays that year.
51. Approximately 12.6% of Florida's panthers live in Everglades National Park. **53.** 416 parents would be expected to have started saving for their children's education. **55.** The total cost is $2400.
57. a. $40,200 **b.** $41,614 **59.** 12,240 accidents involved drivers 35–44 years old. **61.** There were 40,000 traffic fatalities.
63. a. 200 beats per minute **b.** Between 120 and 170 beats per minute

Chapter 7 Problem Recognition Exercises, p. 416

1. 8.2 **2.** 4.1 **3.** 16.4 **4.** 41 **5.** 164
6. 12.3 **7.** Greater than **8.** Less than
9. Greater than **10.** Greater than **11.** 3000
12. 24% **13.** 4.8 **14.** 15% **15.** 70 **16.** 36
17. 6.3 **18.** 250 **19.** 300% **20.** 0.7
21. 75,000 **22.** 37.5% **23.** 25 **24.** 135 **25.** 100
26. 6000 **27.** 0.8% **28.** 125% **29.** 2.6 **30.** 20
31. 75% **32.** 6.05 **33.** $133\frac{1}{3}\%$ **34.** 400%

Section 7.4 Activity, pp. 425–426

A.1. a. Amount of sales tax **b.** Tax rate **c.** Cost of merchandise
A.2. a. $7.20 **b.** Answers will vary. **c.** Answers will vary. **d.** Answers will vary.
A.3. 5.6% **A.4.** $14,175
A.5. Michael sells $160,000; Jorge sells $165,000; Dominique sells $172,500; Dominique receives the $2,000 bonus for the month.
A.6. a. $130 **b.** $780 **c.** $830.70
A.7. a. $119.60 **b.** $2.24 **c.** Retailer A
A.8. a. $8.00 **b.** 32% **A.9. a.** $135 **b.** $87.75 **c.** $10.53
A.10. a. Increase **b.** $1500 **c.** $13,500 **d.** 11.1%

A.11. a.

Movie	Amount of Decrease	Percent Decrease
Adventure	25M	47.2%
Sci-Fi	40M	42.1%

b. Sci-Fi **c.** Adventure; No

Section 7.4 Practice Exercises, pp. 426–430

R.1. 12 **R.3.** 200% **R.5.** 250

1. a. $\left(\begin{array}{c}\text{Sales}\\\text{tax}\end{array}\right) = \left(\begin{array}{c}\text{sales tax}\\\text{rate}\end{array}\right) \cdot \left(\begin{array}{c}\text{cost of}\\\text{merchandise}\end{array}\right)$
b. percent
3. a. $(\text{Discount}) = \left(\begin{array}{c}\text{discount}\\\text{rate}\end{array}\right) \cdot \left(\begin{array}{c}\text{original}\\\text{price}\end{array}\right)$
b. base
5. a. $\left(\begin{array}{c}\text{Percent}\\\text{increase}\end{array}\right) = \left(\frac{\text{amount of increase}}{\text{original value}}\right) \times 100\%$
b. subtracting
7. percent **9.** $x = (0.056)(1299)$ **11.** $11.72 = (0.062)(x)$

13.	Cost of Item	Sales Tax Rate	Amount of Tax	Total Cost
a.	$20	5%	$1.00	$21.00
b.	$12.50	4%	$0.50	$13.00
c.	$110	2.5%	$2.75	$112.75
d.	$55	6%	$3.30	$58.30

15. The total bill is $71.66. **17.** The tax rate is 7%.
19. The price is $44.50.

21.	Total Sales	Commission Rate	Amount of Commission
a.	$20,000	5%	$1000
b.	$125,000	8%	$10,000
c.	$5400	10%	$540

23. Zach made $3360 in commission. **25.** Rodney's commission rate is 15%. **27.** Her sales were $1,400,000.

29.	Original Price	Discount Rate	Amount of Discount	Sale Price
a.	$56	20%	$11.20	$44.80
b.	$900	$33\frac{1}{3}\%$	$300	$600
c.	$85	10%	$8.50	$76.50
d.	$76	50%	$38	$38

31.	Original Price	Markup Rate	Amount of Markup	Retail Price
a.	$92	5%	$4.60	$96.60
b.	$110	8%	$8.80	$118.80
c.	$325	30%	$97.50	$422.50
d.	$45	20%	$9	$54

33. The discounted lunch bill is $4.76. **35.** The discount rate is 25%.
37. a. The markup is $27.00. **b.** The retail price is $177.00.
c. The total price is $189.39. **39.** The markup rate is 25%.

41. The discount is $80.70 and the sale price is $188.30.
43. The markup rate is 54%.   **45.** The discount is $11.00, and the sale price is $98.99.   **47.** c   **49.** 100%   **51.** 10%   **53.** 7%
55. 29.5%   **57.** 5%   **59.** 68%   **61.** 15%

	Stock Fund	Price Jan. Year 1 ($ per share)	Price Jan. Year 3 ($ per share)	Change ($)	Percent Increase
63.	Healthcare	$16.96	$104.47	$87.51	516.0%
64.	Real Estate	$16.84	$23.90	$7.06	41.9%
65.	Foreign Markets	$6.06	$23.05	$16.99	280.4%
66.	Precious Metals	$28.66	$214.01	$185.35	646.7%
67.	Technology	$118.37	$132.45	$14.08	11.9%
68.	Industrial	$134.62	$212.70	$78.08	58.0%

69. a. $49 per ticket   **b.** 43.4%

Section 7.5 Activity, pp. 436–437

A.1. a. principal   **b.** principal and interest earned
A.2. a. I = amount of interest; P = amount of principal; r = interest rate; t = time   **b.** Decimal form   **c.** years
A.3. a. $1400   **b.** $6400   **c.** Demetri would have saved $400.
A.4. Liza's sister is correct. Answers may vary on why. Possible answer: Liza's sister is correct because she used the correct unit of time in the formula of 1.5 years vs. 18 months.   **A.5. a.** $420   **b.** $2420
A.6. a.

Compound Period	Principal (P)	Interest = P*r*t	Amount in the Account
1st Year	$2000 $2000	($2000)(0.07)(1) = $140 $140	$2000 + $140 = $2140 $2140
2nd Year	$2140 $2140	$149.80	$2289.80
3rd Year	$2289.80	$160.29	$2450.09

b. $450.09   **c.** Answers will vary. Possible answer: Compound interest account will earn more interest ($30.09) because interest is calculated on the principal and interest earned.
A.7. a. A = total amount in an account; P = amount of principal; r = interest rate; n = number of compounding periods per year; t = time
b. Decimal form   **c.** years   **d.** $2450.09
A.8. a. 1   **b.** 2   **c.** 4   **d.** 12   **e.** 365
A.9. Answers will vary. Possible answer: Compounding annually because you will be charged less interest than a loan that compounds monthly. Since interest is calculated less frequently, there will be a lower total amount due at the end of a loan.
A.10. a. $33,600   **b.** $6720   **c.** $26,880   **d.** $40,047.05
e. $13,167.05   **f.** $46,767.05

Section 7.5 Practice Exercises, pp. 437–440

R.1. 0.085   **R.3.** 1512   **R.5.** 8652.8

1. a. Simple; principal   **b.** $I = Prt$ **c.** Compound
d. $A = P \cdot \left(1 + \frac{r}{n}\right)^{n \cdot t}$
3. 12   **5.** 2; 4   **7.** $900, $6900   **9.** $1212, $6262
11. $2160, $14,160   **13.** $1890.00, $12,390.00   **15. a.** $350
b. $2850   **17. a.** $48   **b.** $448   **19.** $12,360
21. $5625   **23.** There are 6 total compounding periods.
25. There are 24 total compounding periods.   **27. a.** $560

b.

Year	Interest Earned	Total Amount in Account
1	$20.00	$520.00
2	20.80	540.80
3	21.63	**562.43**

29. a. $26,400   **b.**

Period	Interest Earned	Total Amount in Account
1st	$600	$24,600
2nd	615	25,215
3rd	630.38	25,845.38
4th	646.13	**26,491.51**

31. A = total amount in the account; P = principal; r = annual interest rate; n = number of compounding periods per year; t = time in years
33. $6230.91   **34.** $14,725.49   **35.** $6622.88
36. $4373.77   **37.** $10,934   **38.** $9941.60
39. $16,019.47   **40.** $10,555.99

Chapter 7 Review Exercises, pp. 440–443

1. 75%   **2.** 33%   **3.** 125%   **4.** 50%
5. b, c   **6.** c, d   **7.** $\frac{3}{10}$; 0.3   **8.** $\frac{19}{20}$; 0.95
9. $\frac{27}{20}$; 1.35   **10.** $\frac{53}{25}$; 2.12   **11.** $\frac{1}{500}$; 0.002
12. $\frac{3}{500}$; 0.006   **13.** $\frac{2}{3}$; $0.\overline{6}$ **14.** $\frac{1}{3}$; $0.\overline{3}$ **15.** 62.5%
16. 35%   **17.** 175%   **18.** 220%   **19.** 0.6%   **20.** 0.1%
21. 400%   **22.** 600%   **23.** 42.9%   **24.** 81.8%
25. $\frac{6}{8} = \frac{75}{100}$ **26.** $\frac{27}{180} = \frac{15}{100}$ **27.** $\frac{840}{420} = \frac{200}{100}$
28. $\frac{6}{2000} = \frac{0.3}{100}$ **29.** 6   **30.** 3.68   **31.** 12.5%
32. 0.32%   **33.** 39   **34.** 20   **35.** Approximately 11 people would be no-shows.   **36.** 850 people were surveyed.
37. Victoria spends 40% on rent.   **38.** There are 65 cars.
39. $0.18 \cdot 900 = x$; $x = 162$ **40.** $x = 0.29 \cdot 404$; $x = 117.16$
41. $18.90 = x \cdot 63$; $x = 30\%$ **42.** $x \cdot 250 = 86$; $x = 34.4\%$
43. $30 = 0.25 \cdot x$; $x = 120$ **44.** $26 = 1.30 \cdot x$; $x = 20$
45. The original price was $68.00.   **46.** Veronica read 55% of the novel.   **47.** Elaine can consume 720 fat calories.
48. a. 39,000,000   **b.** 80,800,000   **49.** The sales tax is $76.74.
50. The sales tax rate is 7%.   **51.** The cost before tax was $8.80. The cost per photo is $0.22.
52. The total amount for 4 nights will be $1053.00.
53. The commission rate was approximately 10.6%.
54. Andre earned $489 in commission.   **55.** Sela will earn $131 that day.   **56.** The house sold for $240,000.   **57.** The discount is $8.69. The sale price is $20.26.   **58.** The discount is $129.90. The final price is $1119.10.   **59.** The markup rate is 30%.   **60.** The baskets will sell for $59 each.   **61.** 118.3%   **62.** 35.4%   **63.** $1224, $11,424
64. $1400, $8400   **65.** Jean-Luc will have to pay $2687.50.
66. Steve will pay $840.
67.

Year	Interest	Total
1	$240.00	$6240.00
2	249.60	6489.60
3	259.58	**6749.18**

68.

Compound Periods	Interest	Total
Period 1 (end of first 6 months)	$150.00	$10,150.00
Period 2 (end of year 1)	152.25	10,302.25
Period 3 (end of 18 months)	154.53	10,456.78
Period 4 (end of year 2)	156.85	10,613.63

69. $995.91   **70.** $2624.17   **71.** $16,976.32   **72.** $9813.88

Chapter 8

Section 8.1 Activity, pp. 452–453

A.1. a. $\frac{2}{7}$; 5   **b.** 1   **c.** Itself   **A.2. a.** 12 in.   **b.** 1 ft = 12 in.
c. $\frac{1 \text{ ft}}{12 \text{ in.}}; \frac{12 \text{ in.}}{1 \text{ ft}}$ **d.** 1   **e.** Itself   **A.3. a.** foot   **b.** inches
A.4. a. $\frac{15 \text{ ft.}}{1}$ **b.** $\frac{15 \text{ ft.}}{1} \cdot \frac{1 \text{ yd}}{3 \text{ ft}}$ **c.** 5 yd   **d.** Numerator
e. Denominator
A.5. 2200 yd   **A.6.** 2640 ft   **A.7.** 0.6 hr   **A.8.** 1533 days
A.9. 81,500 lb   **A.10.** 3.375 lb   **A.11.** 0.875 qt
A.12. 96 c   **A.13.** 10 ft   **A.14.** 195 in.   **A.15.** 27.5 min
A.16. 223.13 sec   **A.17.** 1 lb 10 oz
A.18. No, because the recipe requires 3½ c which is equivalent to 28 fl oz, and the can holds 24.5 fl oz.

Section 8.1 Practice Exercises, pp. 454–458

R.1. 54   **R.3.** 128   **R.5.** 0.6 or $\frac{3}{5}$

1. conversion   **3.** 3 yd   **5.** 42 in.   **7.** $2\frac{1}{4}$ mi
9. $4\frac{2}{3}$ yd   **11.** 563,200 yd   **13.** $4\frac{3}{4}$ yd   **15.** 72 in.
17. 50,688 in.   **19.** $\frac{1}{2}$ mi   **21. a.** 76 in.   **b.** $6\frac{1}{3}$ ft
23. a. 8 ft   **b.** $2\frac{2}{3}$ yd   **25.** 6 ft   **27.** 3 ft 4 in.
29. 4′4″   **31.** 730 days   **33.** $1\frac{1}{2}$ hr   **35.** 3 min
37. 3 days   **39.** 1 hr   **41.** 1512 hr   **43.** 80.5 min
45. 175.25 min   **47.** 2 lb   **49.** 4000 lb   **51.** 6500 lb
53. 10 lb 8 oz   **55.** 8 lb 2 oz   **57.** 6 lb 8 oz   **59.** 2 c
61. 24 qt   **63.** 16 c   **65.** $\frac{1}{2}$ gal   **67.** 16 fl oz
69. 6 tsp   **71.** b   **73.** d   **75.** a   **77.** c   **79.** b
81. Yes, 3 c is 24 fl oz, so the 48-fl-oz jar will suffice.
83. The plumber used 7′2″ of pipe.   **85.** 7 ft is left over.
87. $5\frac{1}{2}$ ft   **89.** The unit price for the 24-fl-oz jar is about $0.112 per ounce, and the unit price for the 1-qt jar is about $0.103 per ounce; therefore, the 1-qt jar is the better buy.
91. The total length is 46′.   **93.** The total weight is 312 lb 8 oz.
95. 18 pieces of border are needed.   **97.** Gil ran for 5 hr 35 min.
99. The total time is 1 hr 34 min.   **101.** 6 yd²
103. 3 ft²   **105.** 720 in.²   **107.** 27 ft²

Section 8.2 Activity, pp. 466–467

A.1. Answers will vary.

A.2.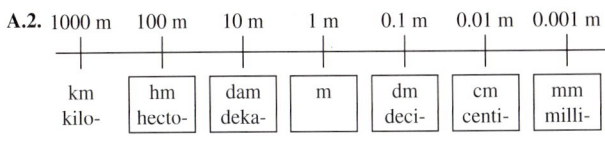

A.3. a. $\frac{144 \text{ m}}{1}$ **b.** $\frac{144 \text{ m}}{1} \cdot \frac{1 \text{ km}}{1000 \text{ m}}$ **c.** 0.144 m
A.4. a. Three places to the left.   **b.** 0.144   **c.** 0.144 km
d. Yes, because the metric system is based on powers of 10 the same as our numbering system. Thus, the decimal point moves the same direction and number of place positions as moved on the prefix line.
A.5. 640 m   **A.6.** 0.44 mm   **A.7.** 19 m   **A.8.** 0.0000192 dm
A.9. 1.82 dam   **A.10.** 315 pieces of pipe

A.11.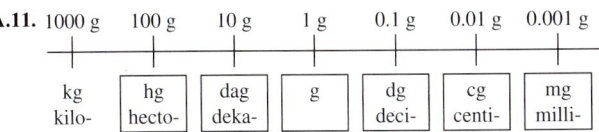

A.12. 3.82 kg   **A.13.** 0.00916 kg   **A.14.** 0.3 mg
A.15. 8000 dg   **A.16.** 1.87 g   **A.17.** 2 g

A.18.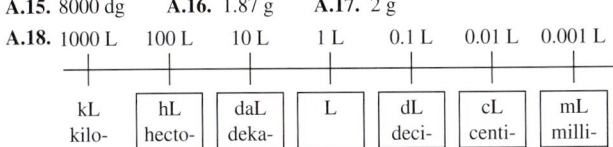

A.19. 60.2 kL   **A.20.** 240,000 mL   **A.21.** 100 mL
A.22. 0.15 L   **A.23.** 0.3 L   **A.24.** 0.008 kL
A.25. a. iv   **b.** ii   **c.** vii   **d.** i   **e.** vi   **f.** viii   **g.** iii   **h.** v

Section 8.2 Practice Exercises, pp. 468–472

R.1. 1.25 mi   **R.3.** 1.5 qt   **R.5.** 56 oz

1. a. metric   **b.** prefix   **c.** meter; m   **d.** gram; g   **e.** liter; L
f. cubic   **3.** right   **5.** 3; left   **7.** 2; left   **9.** b, f, g
11. 3.2 cm or 32 mm   **13.** a   **15.** d   **17.** d   **19.** 2.43 km
21. 50,000 mm   **23.** 4000 m   **25.** 43.1 mm   **27.** 0.3328 km
29. 300 m   **31.** 0.539 kg   **33.** 2500 g   **35.** 33.4 mg
37. 4.09 g   **39.** <   **41.** =   **43.** Cubic centimeter   **45.** 3.2 L
47. 700 cL   **49.** 0.42 dL   **51.** 64 mL   **53.** 40 cc
55. b, f   **57.** a, f

	Object	mm	cm	m	km
59.	Length of the Mississippi River	3,766,000,000	376,600,000	3,766,000	3766
61.	Diameter of a quarter	24.3	2.43	0.0243	0.0000243

	Object	mg	cg	g	kg
63.	Bag of rice	907,000	90,700	907	0.907
65.	Hockey puck	170,000	17,000	170	0.17

	Object	mL	cL	L	kL
67.	Bottle of vanilla extract	59	5.9	0.059	0.000059
69.	Capacity of a gasoline tank	75,700	7570	75.7	0.0757

71. 4,669,000 m   **73.** 305,000 mg   **75.** 250 mL
77. Cliff drives 17.4 km per week.

79. 8 cans hold 0.96 kL.   **81.** The bottle contains 49.5 cL.
83. 3.6 g   **85.** No, she needs 1.04 m of framing.
87. The difference is 4500 m.   **89.** 65 km²
91. 0.56 m²   **93.** 5.78 metric tons   **95.** 8500 kg

Section 8.3 Practice Exercises, pp. 478–481

R.1. 0.5 kg   **R.3.** 38,000 mL   **R.5.** 34,848 in.

1. a. Fahrenheit; 32; 212   **b.** Celsius; 0; 100
3. 1   **5.** denominator   **7.** $\frac{1 \text{ in.}}{2.54 \text{ cm}}$; $\frac{2.54 \text{ cm}}{1 \text{ in.}}$
9. $\frac{1 \text{ lb}}{0.45 \text{ kg}}$; $\frac{0.45 \text{ kg}}{1 \text{ lb}}$ **11.** b   **13.** a   **15.** 5.1 cm
17. 8.8 yd   **19.** 122 m   **21.** 1.1 m   **23.** 15.2 cm
25. 2.7 kg   **27.** 0.4 oz   **29.** 1.2 lb   **31.** 1980 kg
33. 5.7 L   **35.** 4 fl oz   **37.** 32 fl oz
39. The box of sugar costs $0.100 per ounce, and the packets cost $0.118 per ounce. The 2-lb box is the better buy.
41. 18 mi is about 28.98 km. Therefore, the 30-km race is longer than 18 mi.   **43.** 97 lb is approximately 43.65 kg.
45. The price is approximately $7.22 per gallon.
47. A hockey puck is 1 in. thick.   **49.** The football player weighs about 222 lb.   **51.** 45 cc is 1.5 fl oz.   **53.** 40.8 ft
55. 77°F   **57.** 20°C   **59.** 86°F
61. 7232°F   **63.** It is a hot day. The temperature is 95°F.
65. In Italy, the Celsius scale is used. Converting 25°C to Fahrenheit gives 77°F, which would be a warm day.
67. $F = \frac{9}{5}C + 32 = \frac{9}{5} \cdot 100 + 32 = 9(20) + 32 = 180 + 32 = 212$
69. 184 g   **71.** The large SUV weighs approximately 2.565 metric tons.
73. The average weight of the blue whale is approximately 240,000 lb.

Chapter 8 Problem Recognition Exercises, p. 481

1. 9 qt   **2.** 2.2 m   **3.** 12 oz   **4.** 300 mL
5. 4 yd   **6.** 6030 g   **7.** 4.5 m   **8.** $\frac{3}{4}$ ft
9. 2640 ft   **10.** 3 tons   **11.** 4 qt   **12.** $\frac{1}{2}$ T
13. 0.021 km   **14.** 6.8 cg   **15.** 36 cc   **16.** 4 lb
17. 4.322 kg   **18.** 5000 mm   **19.** 2.5 c   **20.** 8.5 min
21. 0.5 gal   **22.** 3.25 c   **23.** 5460 g   **24.** 902 cL
25. 16,016 yd   **26.** 3 lb   **27.** 3240 lb   **28.** 4600 m
29. 2.5 days   **30.** 8 mL   **31.** 512.4 min   **32.** 336 hr

Section 8.4 Practice Exercises, pp. 483–485

R.1. 9840 mm   **R.3.** 400,000 cL
R.5. 20.0 lb

1. microgram   **3.** 10 μg   **5.** 7.5 mg   **7.** 0.05 cg
9. 500 mcg   **11.** 200 mcg   **13.** 1 mg   **15. a.** 400 mg
b. 8000 mg or 8 g   **17.** 3 mL   **19.** 5.25 g of the drug would be given in 1 wk.   **21.** 2 mL   **23.** 500 people
25. 9.6 mg   **27.** 3.6 g

Section 8.5 Activity, pp. 490–492

A.1. a. v   **b.** iii   **c.** iv   **d.** ii   **e.** i
A.2. a. 40°   **b.** ∠BOA   **c.** acute
A.3. a. 88°   **b.** ∠COA   **c.** acute
A.4. a. 153°   **b.** ∠DOA   **c.** obtuse
A.5. a. 180°   **b.** ∠EOA   **c.** straight   **A.6.** O
A.7. The sum of their measures is 90°. The complement of 57° is 33°.
A.8. The sum of their measures is 180°. The supplement of 9° is 171°.
A.9. a. parallel   **b.**

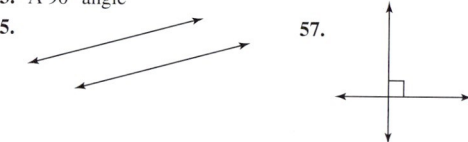

c. Answers will vary.

A.10. a. perpendicular
b. 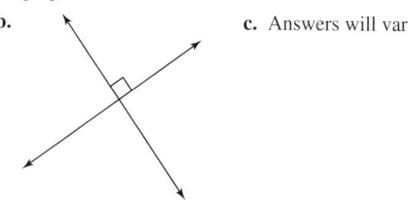   **c.** Answers will vary.

A.11. ∠a and ∠c are vertical angles. ∠b and ∠d are vertical angles. Vertical angles are equal in measure, thus: m(∠a) = m(∠c) and m(∠b) = m(∠d).
A.12. 62°   **A.13.** 62°   **A.14.** 118°   **A.15.** 118°
A.16. 62°   **A.17.** 62°   **A.18.** 118°

Section 8.5 Practice Exercises, pp. 492–496

1. a. point   **b.** line   **c.** segment   **d.** P; Q
e. angle; vertex   **f.** right; 180   **g.** protractor
h. acute; obtuse   **i.** complementary; supplementary
j. parallel   **k.** perpendicular
3. A line extends forever in both directions. A ray begins at an endpoint and extends forever in one direction.
5. Ray   **7.** Point   **9.** Line
11. For example:

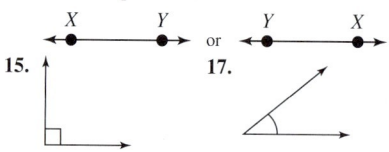

13. For example:

15.   **17.**

19. 20°   **21.** 90°   **23.** 148°   **25.** Right
27. Obtuse   **29.** Acute   **31.** Straight
33. 10°   **35.** 63°   **37.** 60.5°   **39.** 1°
41. 100°   **43.** 53°   **45.** 142.6°   **47.** 1°
49. No, because the sum of two angles that are both greater than 90° will be more than 180°.   **51.** Yes. For two angles to add to 90°, the angles themselves must both be less than 90°.
53. A 90° angle
55.   **57.**

59. m(∠a) = 41°; m(∠b) = 139°; m(∠c) = 139°
61. m(∠a) = 26°; m(∠b) = 112°; m(∠c) = 26°; m(∠d) = 42°
63. The two lines are perpendicular.   **65.** Vertical angles
67. a, c or b, h or e, g or f, d   **69.** a, e or f, b
71. m(∠a) = 55°; m(∠b) = 125°; m(∠c) = 55°; m(∠d) = 55°; m(∠e) = 125°; m(∠f) = 55°; m(∠g) = 125°
73. m(∠a) = 120°; m(∠b) = 60°; m(∠c) = 120°; m(∠d) = 120°; m(∠e) = 60°; m(∠f) = 120°; m(∠g) = 60°
75. True   **77.** True   **79.** False   **81.** True   **83.** True
85. 70°   **87.** 90°   **89. a.** 48°   **b.** 48°   **c.** 132°
91. 180°   **93.** 120°

Section 8.6 Activity, pp. 501–503

A.1. 180°   **A.2.** e, i, and iv   **A.3.** c, iii   **A.4.** f, ii   **A.5.** d, i
A.6. a, ii   **A.7.** b, iii, iv   **A.8.** 107°   **A.9.** 66°   **A.10.** 38°
A.11. 70°   **A.12.** 85°   **A.13.** 155°   **A.14.** 25°   **A.15.** 155°
A.16. a. 6   **b.** 7   **c.** 6.08   **d.** The square root is not a whole number. Also, it is in between the whole numbers of 6 and 7 since the $\sqrt{37}$ is between the $\sqrt{36}$ and the $\sqrt{49}$.
A.17. 12   **A.18. a.** Sides a and b   **b.** Side c
c. It is always the longest side of the triangle.   **d.** $a^2 + b^2 = c^2$

A.19. c. 10 ft **d.** Yes **A.20. c.** 10 ft
d. The answers are equal. Answers will vary. Possible answer: It does not matter which leg is labeled a or b since their squares are added in the Pythagorean theorem.
A.21. 12 in. **A.22.** 15 ft

Section 8.6 Practice Exercises, pp. 503–507

R.1. 146° **R.3.** 78° **R.5.** 9 **R.7.** 225

1. a. 180 **b.** acute; right; obtuse **c.** equilateral
d. isosceles **e.** scalene **f.** hypotenuse; legs
g. Pythagorean; c^2 **3.** angles **5.** perfect **7.** hypotenuse
9. $m(\angle a) = 54°$ **11.** $m(\angle b) = 78°$
13. $m(\angle a) = 60°, m(\angle b) = 80°$
15. $m(\angle a) = 40°, m(\angle b) = 72°$ **17.** c, f **19.** b, d
21. b, c, e **23.** 7 **25.** 49 **27.** 16 **29.** 4
31. 6 **33.** 36 **35.** 81 **37.** 9 **39.** $\frac{1}{4}$ **41.** 0.2
43. $c = 5$ m **45.** $b = 12$ yd **47.** Leg = 10 ft
49. Hypotenuse = 40 in. **51.** The brace is 20 in. long.
53. The height is 9 km. **55.** The car is 25 mi from the starting point.

	Square Root	Estimate	Calculator Approximation (Round to 3 Decimal Places)
	$\sqrt{50}$	is between 7 and 8	7.071
57.	$\sqrt{10}$	is between 3 and 4	3.162
58.	$\sqrt{90}$	is between 9 and 10	9.487
59.	$\sqrt{116}$	is between 10 and 11	10.770
60.	$\sqrt{65}$	is between 8 and 9	8.062
61.	$\sqrt{5}$	is between 2 and 3	2.236
62.	$\sqrt{48}$	is between 6 and 7	6.928

63. 20.682 **64.** 56.434 **65.** 1116.244 **66.** 7100.423
67. 0.7 **68.** 0.5 **69.** 0.748 **70.** 0.906
71. $b = 21$ ft **73.** Hypotenuse = 11.180 mi
75. Leg = 18.439 in. **77.** The diagonal length is 1.41 ft.
79. The length of the diagonal is 35.36 ft. **81.** 24 m **83.** 30 km

Section 8.7 Practice Exercises, pp. 514–519

R.1. 25 **R.3.** 50.41 **R.5.** 10 **R.7.** 75.36

1. a. perimeter; circumference **b.** area
3. quadrilateral **5.** rhombus **7.** perpendicular
9. a, b, c, d, e, h **11.** a, b, e **13.** a, b, e, h
15. 80 cm **17.** 260 mm **19.** 10.7 m **21.** 25.12 ft
23. 2.4 km or 2400 m **25.** 10 ft 6 in. **27.** 5 ft or 60 in.
29. $x = 550$ mm; $y = 3$ dm; perimeter = 26 dm or 2600 mm
31. 280 ft of rain gutters is needed. **33.** 576 yd^2
35. 23 in.^{2} **37.** 18.4 ft^2 **39.** 54 m^2
41. 656 in.^{2} **43.** 148.5 yd^2 **45.** 154 m^2
47. $346\frac{1}{2}$ cm^2 **49.** 113 mm^2 **51.** 5 ft^2 **53.** The area of the sign is 16.5 yd^2. **55.** $956.25 **57. a.** The area is 30,000 ft^2.
b. Lizette will need 6 bags of seed. **59.** 2.72 ft^2
61. 30.5 cm^2 **63.** $c = 5$ in.; perimeter = 28 in.

Chapter 8 Problem Recognition Exercises, p. 520

1. Area = 25 ft^2; perimeter = 20 ft
2. Area = 144 m^2; perimeter = 48 m
3. Area = 12 m^2 or 120,000 cm^2; perimeter = 14 m or 1400 cm
4. Area = 1 ft^2 or 144 in.^{2}; perimeter = 5 ft or 60 in.
5. Area = $\frac{1}{3}$ yd^2 or 3 ft^2; perimeter = 3 yd or 9 ft
6. Area = 0.473 km^2 or 473,000 m^2; perimeter = 3.24 km or 3240 m
7. Area = 6 yd^2; perimeter = 12 yd
8. Area = 30 cm^2; perimeter = 30 cm
9. Area = 44 m^2; perimeter = 32 m
10. Area = 88 in.^{2}; perimeter = 40 in.
11. Area $\approx$ 28.26 yd^2; circumference $\approx$ 18.84 yd
12. Area $\approx$ 1256 cm^2; circumference $\approx$ 125.6 cm
13. Area $\approx$ 2464 cm^2; circumference $\approx$ 176 cm
14. Area $\approx$ 616 ft^2; circumference $\approx$ 88 ft
15. a. perimeter **b.** 64 yd
16. a. circumference **b.** $\frac{11}{2}$ ft or $5\frac{1}{2}$ ft
17. a. area **b.** 125 ft^2 **18. a.** area **b.** 314 m^2

Section 8.8 Practice Exercises, pp. 525–529

R.1. 158 **R.3.** 452.16 **R.5.** 904.32

1. a. s^3 **3.** $\pi r^2 h$ **5.** surface area
7. 2.744 cm^3 **9.** 48 ft^3 **11.** 12.56 mm^3
13. 3052.08 yd^3 **15.** 235.5 cm^3 **17.** 452.16 ft^3
19. 289 in.^{3} **21.** 314 ft^3 **23.** 10 ft^3 **25. a.** 2575 ft^3
b. 19,260 gal **27.** 342 in.^{2} **29.** 13.5 cm^2
31. 113.0 in.^{2} **33.** 211.1 in.^{2} **35.** 492 ft^2
37. 96 cm^2 **39.** 1356.48 in.^{2} **41.** 1256 mm^2
43. $\frac{11}{36}$ ft^3 or 0.306 ft^3 or 528 in.^{3} **45.** 109.3 in.^{3}
47. 502.4 mm^3 **49.** 621.72 ft^3 **51.** 84.78 in.^{3}
53. 50,240 cm^3 **55.** 400 ft^3

Chapter 8 Review Exercises, pp. 530–534

1. 4 ft **2.** 39 in. **3.** 3520 yd **4.** $1\frac{1}{3}$ mi
5. 2640 ft **6.** 72 in. **7.** 9 ft 3 in. **8.** 6'4"
9. 2'10" **10.** 3 ft 9 in. **11.** $7\frac{1}{2}$ ft
12. There is 102 ft or 34 yd of wire left. **13.** 3 days
14. 360 sec **15.** 80 oz **16.** 168 hr **17.** $1\frac{1}{2}$ c
18. 500 lb **19.** $1\frac{3}{4}$ tons **20.** 16 pt **21.** $\frac{3}{4}$ lb
22. 4 gal **23.** 144.5 min **24.** The total weight was 11 lb 13 oz.
25. b **26.** d **27.** 520 mm **28.** 0.093 km
29. 3.4 m **30.** 0.21 dam **31.** 0.04 m
32. 1200 mm **33.** 610 cg **34.** 0.42 kg **35.** 3.212 g
36. 70 g **37.** 8.3 L **38.** 124 cc **39.** 22.5 cL
40. 490 L **41.** Perimeter: 6.5 m; area: 2.5 m^2
42. 5 glasses can be filled. **43.** The difference is 64.8 kg.
44. No, the board is 25 cm too short. **45.** 15.75 cm
46. 2.5 fl oz **47.** 5 oz **48.** 5.26 qt **49.** 1.04 m
50. 45 kg **51.** 74.53 mi **52.** 5.7 L **53.** 45 cc
54. 11,250 kg **55.** The difference in height is 38.2 cm.
56. There are approximately 6.72 servings.
57. The marathon is approximately 26.2 mi.
58. $C = \frac{5}{9}(F - 32)$ **59.** 82.2°C to 85°C
60. $F = \frac{9}{5}C + 32$ **61.** 46.4°F **62.** 450 µg
63. 1500 mcg **64.** 0.4 mg **65.** 0.5 cg
66. 2.5 mg/cc **67. a.** 3.2 mg **b.** 44.8 mg
68. The total amount of cough syrup is approximately 0.42 L.
69. There is 1.2 cc or 1.2 mL of fluid left. **70.** Clayton took 7.5 g.
71. d **72.** a **73.** c **74.** b
75. The measure of an acute angle is between 0° and 90°.
76. The measure of an obtuse angle is between 90° and 180°.
77. The measure of a right angle is 90°.
78. a. 57° **b.** 147° **79. a.** 70° **b.** 160°
80. 62° **81.** 118° **82.** 118° **83.** 62° **84.** 62°
85. $m(\angle x) = 40°$ **86.** $m(\angle x) = 80°$; $m(\angle y) = 32°$
87. An equilateral triangle has three sides of equal length and three angles of equal measure. **88.** An isosceles triangle has two sides of equal length and two angles of equal measure.

89. 5 **90.** 7 **91.** $b = 7$ cm **92.** $c = 20$ ft **93.** 13 m of string is extended. **94.** 90 cm **95.** 17.3 m **96.** 400 yd
97. 15.5 ft **98.** 62.8 cm **99.** 20 in.^{2} **100.** 51 ft^2
101. 7056 ft^2 **102.** 18 ft^2 **103.** 314 cm^2 **104.** 36 cm^2
105. Volume: 25,000 cm^3; SA: 5250 cm^2 **106.** Volume: 226.08 ft^3; SA: 207.24 ft^2 **107.** Volume: 14,130 in.^{3}; SA: 2826 in.^{2}
108. Volume: 512 mm^3; SA: 384 mm^2 **109.** 37.68 m^3
110. 249 in.^{3} **111.** 113 in.^{3} **112.** $2\frac{1}{3}$ ft^3 **113.** 335 cm^3
114. 28,500 in.^{3}

Chapter 9

Section 9.1 Practice Exercises, pp. 543–548

1. a. Statistics **b.** table; cells **c.** pictograph
3. 20° **5.** 77°
7. Asia **9.** 2514 ft **11.** 3.6 years
13. 2.8 years **15.** Men

17.

	Dog	Cat	Neither
Boy	4	1	3
Girl	3	4	5

19. a. The 18- to 29-year age group has the greatest percentage of Internet users.
b.

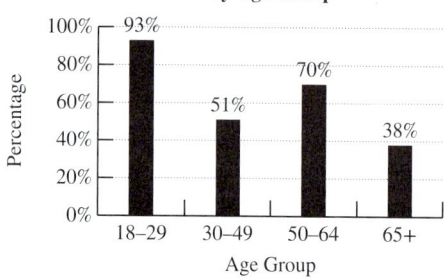

21.

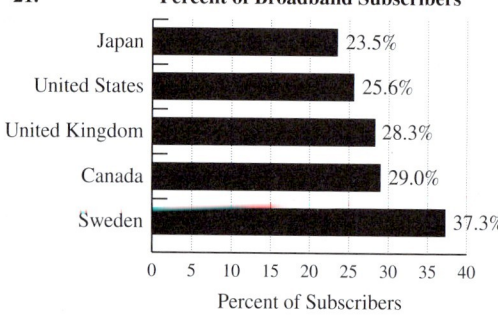

23. a. One icon represents 100 servings sold. **b.** About 450 servings
c. Sunday **25. a.** $65 million **b.** The superhero movie
c. Approximately $220 million **27.** 48.4% **29.** The trend for women over 65 in the labor force shows a slight increase. **31.** For example: 18%
33. The most cars were sold in year 2. 22,400 cars were sold.
35. 4800 cars **37.** The greatest increase was between year 1 and year 2.
39. a.

Average Height for Girls, Ages 2–9

b. ≈ 55 in.

41. There are 14 servings per container, which means that there is $8 \text{ g} \times 14 = 112$ g of fat in one container.
43. The daily value of fat is approximately 61.5 g.

Section 9.2 Activity, pp. 551–552

A.1.

Age Group	Tally	Frequency
18–21	|||||		6
22–26	|||||		6
27–30	|||	3
31–34	||||	4
35–38	||	2
39–42	|||	3

A.2.

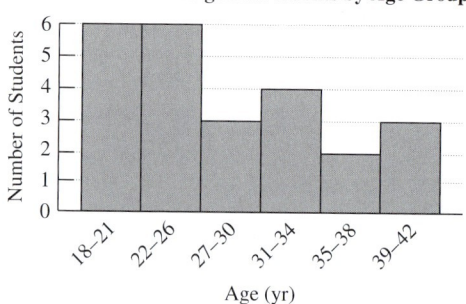

A.3. Both 18–21 and 22–26 are equally represented the most.
A.4. Students 35–38 are represented the least.
A.5. 29.2% **A.6.** 25%

A.7.

Age Group	Tally	Frequency
17–19	||||	4
20–22	||||| |||	8
23–25	||||| |||	8
26–28	||	2
29–31	||	2

A.8.

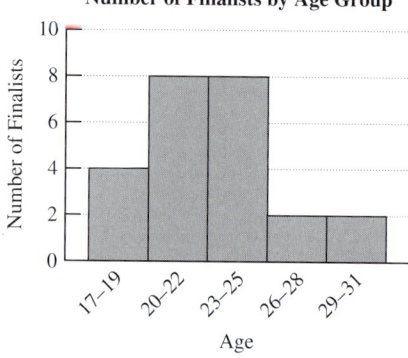

A.9. 20–22 and 23–25 year olds are the largest age groups represented with eight individuals in each group.
A.10. 83.3%

Section 9.2 Practice Exercises, pp. 552–556

1. a. frequency **b.** histogram
3. 30–39 **5.** 23.1%
7. There are 72 data values.
9. 9–12

11.

Class Intervals (Age in Years)	Tally	Frequency (Number of Professors)
56–58	||	2
59–61	|	1
62–64	|	1
65–67	卌 ||	7
68–70	卌	5
71–73	||||	4

a. The class of 65–67 years has the most values.
b. There are 20 values represented in the table.
c. Of the professors, 25% retire when they are 68 to 70 years old.

13.

Class Intervals (Amount in Gal)	Tally	Frequency (Number of Customers)
8.0–9.9	||||	4
10.0–11.9	|	1
12.0–13.9	卌	5
14.0–15.9	||||	4
16.0–17.9		0
18.0–19.9	||	2

a. The 12.0–13.9 gal class has the highest frequency.
b. There are 16 data values represented in the table.
c. Of the customers, 12.5% purchased 18 to 19.9 gal of gas.

15. The class widths are not the same. **17.** There are too few classes. **19.** The class intervals overlap. For example, it is unclear whether the data value 12 should be placed in the first class or the second class.

21.

Class Interval (Height, in.)	Frequency (Number of Students)
62–63	2
64–65	3
66–67	4
68–69	4
70–71	4
72–73	3

23.

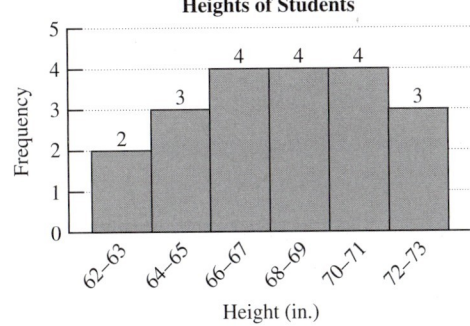

25.

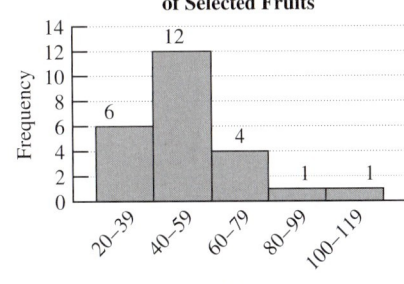

Section 9.3 Practice Exercises, pp. 561–564

R.1. 20%   **R.3.** 234

1. circle   **3.** fraction   **5.** 64,000   **7.** 640   **9.** 25%
11. 2.5 times   **13.** There are 200 bicycles represented.   **15.** There were 1.8 times as many road bikes.   **17.** 15% of the bicycles sold were touring bikes.   **19.** There are 960 Latina CDs.   **21.** There are 640 CDs that are classical or jazz.   **23.** 35,000 have private insurance.
25. 15,000 are uninsured.

27. **29.** **31.** **33.**

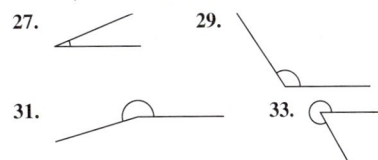

35.

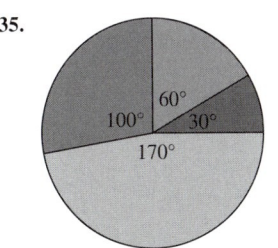

37.

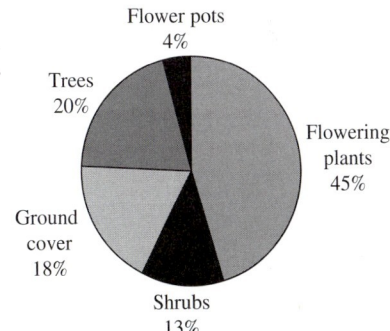

39. a.

	Expenses	Percent	Number of Degrees
Tuition	$9000	75%	270°
Books	600	5%	18°
Housing	2400	20%	72°

b.

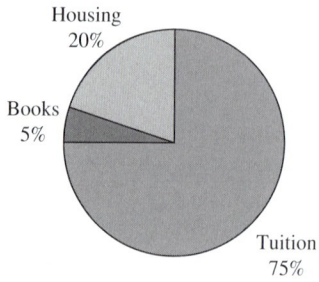

College Expenses for a Semester
Housing 20%
Books 5%
Tuition 75%

Section 9.4 Activity, pp. 568–569

A.1. a. $\{H, T\}$ b. 2 c. 1 d. $\frac{1}{2}$ or 50%

A.2. a. $\{1, 2, 3, 4, 5, 6, 7, 8, 9, 10\}$ b. 2 c. $\frac{1}{5}$

A.3. a. 10% b. 65% c. 35% d. 100% or 1 e. 1

A.4. a. 55.2% b. 6.4% c. 93.6% d. 0

Section 9.4 Practice Exercises, pp. 569–572

R.1. c   **R.3.** 25%

1. a. experiment b. sample c. probability d. complement e. 1 f. impossible g. 1

3. $\{H, T\}$ **5.** cannot

7. $\{1, 2, 3, 4, 5, 6, 7, 8, 9, 10\}$

9. $\{2, 3, 4, 5, 6, 7, 8, 9, 10, 11, 12\}$ **11.** In 3 ways

13. c, d, g, h   **15.** $\frac{2}{6} = \frac{1}{3}$ **17.** $\frac{3}{6} = \frac{1}{2}$ **19.** $\frac{5}{8}$

21. $\frac{1}{8}$ **23.** 1

25. An impossible event is one in which the probability is 0.

27. $\frac{12}{52} = \frac{3}{13}$ **29.** $\frac{12}{16} = \frac{3}{4}$

31. a. $\frac{18}{120} = \frac{3}{20}$ b. $\frac{27}{120} = \frac{9}{40}$ c. 30%

33. a. $\frac{21}{60} = \frac{7}{20}$ b. 50%

35. a. $\frac{7}{29}$ b. $\frac{11}{29}$ c. 62%

37. $1 - \frac{2}{11} = \frac{9}{11}$ **39.** $100\% - 1.2\% = 98.8\%$

Section 9.5 Activity, pp. 578–579

A.1. a. 1560 calories b. 5 days c. 312 calories
d. Mean = $\frac{\text{sum of the values}}{\text{number of values}}$

A.2. a. 18, 20, 21, 22, 23 b. 21 c. 14, 18, 20, 21, 22, 23 d. 20.5

A.3. a. Yes; 28 b. No, it is not the same. In this case the data are bimodal. There are two modes: 28 and 38.

A.4. Answers a–c.

Course	Grade	Number of Credit-Hours	Product: Number of Grade Points
Prealgebra	A = 4 points	3	4*3 = 12 pts.
College Writing	C = 2 points	3	2*3 = 6 pts.
Intro to Marketing	B = 3 points	4	3*4 = 12 pts.
		Total Hours: 10	Total Grade Pts: 30 pts.

d. Mean = $\frac{30}{10} = 3.0$. Hibeto's GPA for this term is 3.0.

A.5. a. 9.2 lb b. 9 lb c. 8 lb

Section 9.5 Practice Exercises, pp. 580–584

R.1. 30.6   **R.3.** 156

1. a. mean b. median c. mean d. mode e. weighted
3. 5   **5.** 6   **7.** −15.8   **9.** 5.8 hr
11. a. 397 Cal b. 386 Cal c. There is an 11-Cal difference in the means.
13. a. 86.5% b. 81% c. The low score of 59% decreased Zach's average by 5.5%.   **15.** 17   **17.** 110.5
19. −52.5   **21.** 3.93 deaths per 1000   **23.** 0
25. 51.7 million passengers   **27.** 4   **29.** −21 and −24
31. No mode   **33.** $300   **35.** 5.2%, 5.8%
37. Mean: 85.5%; median: 94.5%; The median gave Jonathan a better overall score.
39. Mean: $250; median: $256; mode: There is no mode.
41. Mean: $942,500; median: $848,500; mode: $850,000
43. 2.38   **45.** 2.77   **47.** 3.3; Elmer's GPA improved from 2.5 to 3.3.
49.

Number of Residents in Each House	Number of Houses	Product
1	3	3
2	9	18
3	10	30
4	9	36
5	6	30
Total:	37	117

The mean number of residents per house is approximately 3.2.

Chapter 9 Review Exercises, pp. 584–587

1. Godiva   **2.** Breyers   **3.** Blue Bell has 2 times more sodium than Edy's Grand.   **4.** There is a 10-g difference.   **5.** 1 icon represents 50 tornadoes.   **6.** 300   **7.** June   **8.** 75   **9.** 2015
10. 4900   **11.** Increasing   **12.** ≈ 7000

13.

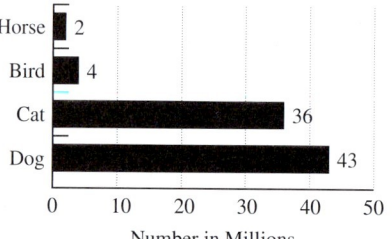

Number of Households with Pets
Horse 2
Bird 4
Cat 36
Dog 43
Number in Millions

14.

Class Intervals (Age)	Frequency
18–21	4
22–25	5
26–29	4
30–33	3
34–37	1
38–41	1
42–45	2

15.

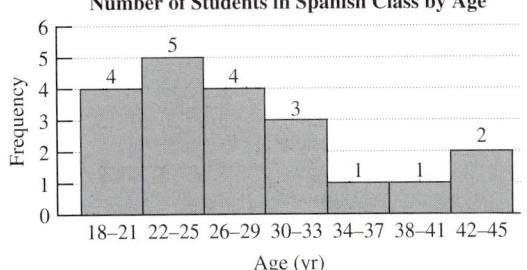

16. There are 24 types of subs.   **17.** $\frac{2}{3}$ of the subs contain beef.
18. $\frac{1}{3}$ of the subs do not contain beef.

19. a.

Education Level	Number of People	Percent	Number of Degrees
Grade school	10	5%	18°
High school	50	25%	90°
Some college	60	30%	108°
Four-year degree	40	20%	72°
Postgraduate	40	20%	72°

b.

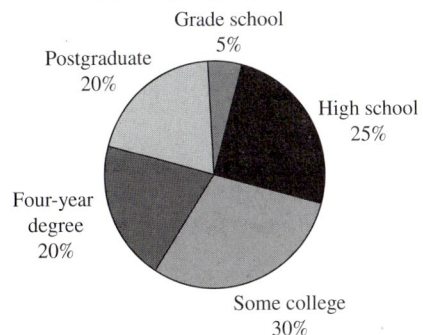

20. {blue, green, brown, black, gray, white}
21. $\frac{1}{6}$ **22.** $\frac{5}{6}$ **23.** a, c, d, e, g
24. a. $\frac{1}{2}$ **b.** $\frac{1}{2}$ **c.** 0   **25. a.** $\frac{1}{4}$ **b.** $\frac{3}{4}$
26. Mean: 17.5; median: 18; mode: 20
27. Mean: 1060 mg; median: 1000 mg; modes: 1000 mg and 1200 mg
28. The median is 20,562 seats.   **29.** 4   **30.** 3.0

Chapter 10

Section 10.1 Activity, p. 599

A.1. Substitute -4 for x in the equation. If the left side equals the right side, then -4 is a solution.
A.2. No   **A.3.** Yes
A.4. a. {25}; addition property of equality   **b.** {15}; subtraction property of equality   **c.** {4}; division property of equality   **d.** {100}; multiplication property of equality
A.5. a. {-10}; multiplication property of equality   **b.** $\left\{\frac{11}{5}\right\}$; addition property of equality   **c.** $\left\{\frac{9}{5}\right\}$; subtraction property of equality   **d.** $\left\{-\frac{2}{5}\right\}$; division property of equality

A.6. a. {-6}   **b.** {-6}   **c.** Answers will vary.
A.7. a. {3}   **b.** {3}   **c.** Answers will vary.
A.8. a. Multiply by the reciprocal of $-\frac{6}{5}$.   **b.** $\left\{\frac{1}{4}\right\}$

Section 10.1 Practice Exercises, pp. 600–602

R.1. $-y$ **R.3.** m **R.5.** $p+6$
R.7. $-\frac{3}{2}$ **R.9.** -22 **R.11.** $\frac{10}{3}$

1. a. equation   **b.** solution   **c.** linear   **d.** solution set
3. Expression   **5.** Equation   **7.** Substitute the value into the equation and determine if the right-hand side is equal to the left-hand side.   **9.** No   **11.** Yes   **13.** Yes   **15.** {-1}
17. {20}   **19.** {-17}   **21.** {-16}   **23.** {0}   **25.** {1.3}
27. $\left\{\frac{11}{2}\right\}$ or $\left\{5\frac{1}{2}\right\}$ **29.** {-2}   **31.** {-2.13}   **33.** {-3.2675}
35. {9}   **37.** {-4}   **39.** {0}   **41.** {-15}   **43.** $\left\{-\frac{4}{5}\right\}$
45. {-10}   **47.** {4}   **49.** {41}   **51.** {-127}
53. {-2.6}   **55.** $-8+x=42$; The number is 50.
57. $x-(-6)=18$; The number is 12.
59. $x \cdot 7 = -63$ or $7x=-63$; The number is -9.
61. $x-3.2=2.1$; The number is 5.3.
63. $\frac{x}{12}=\frac{1}{3}$; The number is 4.
65. $x+\frac{5}{8}=\frac{13}{8}$; The number is 1.   **67. a.** {10}   **b.** $\left\{-\frac{1}{9}\right\}$
69. a. {-12}   **b.** $\left\{\frac{22}{3}\right\}$ **71.** {-36}   **73.** {16}   **75.** {2}
77. $\left\{-\frac{7}{4}\right\}$ **79.** {11}   **81.** {-36}   **83.** $\left\{\frac{7}{2}\right\}$ **85.** {4}
87. {3.6}   **89.** {0.4084}   **91.** Yes   **93.** No   **95.** Yes
97. Yes   **99.** For example: $y+9=15$ **101.** For example: $2p=-8$ **103.** For example: $5a+5=5$ **105.** {-1}   **107.** {7}

Section 10.2 Activity, p. 609

A.1. Simplify both sides of the equation. Use the addition or subtraction property of equality to combine variable terms on one side of the equation and constant terms on the other side. Apply the multiplication or division properties of equality to isolate the variable. Check the solution in the original equation.
A.2. a. {5}   **b.** {5}   **c.** No
A.3. {2} (Explanations vary.)
A.4. {-2} (Explanations vary.)
A.5. $\left\{-\frac{3}{2}\right\}$ (Explanations vary.)
A.6. Subtracting x from both sides results in the statement $1=2$, which is a contradiction. The solution set is { }. Intuitively, if two different values are added to the same number x, then the resulting sums cannot be equal.
A.7. Simplifying both sides results in the statement $2x+6=2x+6$, which is an identity. The solution set is all real numbers.
A.8. 6; {6}

Section 10.2 Practice Exercises, pp. 610–612

R.1. $-5z+2$ **R.3.** $10p-10$ **R.5.** {19}
R.7. {-3}   **R.9.** $\left\{\frac{3}{2}\right\}$

1. a. conditional   **b.** contradiction   **c.** empty or null   **d.** identity
3. {2}   **5.** {6}   **7.** $\left\{\frac{5}{2}\right\}$ **9.** {-42}
11. $\left\{-\frac{3}{4}\right\}$ **13.** {5}   **15.** {-4}   **17.** {-26}   **19.** {10}
21. {-8}   **23.** $\left\{-\frac{7}{3}\right\}$ **25.** {0}   **27.** {-3}   **29.** {-2}

31. $\left\{\dfrac{9}{2}\right\}$ **33.** $\left\{-\dfrac{1}{3}\right\}$ **35.** $\{10\}$ **37.** $\{-6\}$ **39.** $\{0\}$
41. $\{-2\}$ **43.** $\left\{-\dfrac{25}{4}\right\}$ **45.** $\left\{\dfrac{10}{3}\right\}$ **47.** $\{-0.25\}$
49. $\{\ \}$; contradiction **51.** $\{-15\}$; conditional equation
53. The set of real numbers; identity **55.** One solution
57. Infinitely many solutions **59.** $\{7\}$ **61.** $\left\{\dfrac{1}{2}\right\}$ **63.** $\{0\}$
65. The set of real numbers **67.** $\{-46\}$ **69.** $\{2\}$
71. $\left\{\dfrac{13}{2}\right\}$ **73.** $\{-5\}$ **75.** $\{\ \}$ **77.** $\{2.205\}$ **79.** $\{10\}$
81. $\{-1\}$ **83.** $a = 15$ **85.** $a = 4$
87. For example: $5x + 2 = 2 + 5x$

Section 10.3 Activity, p. 617

A.1. a. 12, 24, 36, 48, 60 **b.** $4x$ **c.** -10 **d.** 24 **e.** $3x$ **f.** $\{34\}$
A.2. a. 100, 1000, 10,000, 100,000 **b.** -415 **c.** $-250x$ **d.** 1720
  e. $100x$ **f.** $\{-6.1\}$
A.3. $\{11\}$ (Explanations vary.)
A.4. $\{-66\}$ (Explanations vary.)
A.5. $\left\{\dfrac{7}{2}\right\}$ (Explanations vary.)

Section 10.3 Practice Exercises, pp. 617–619

R.1. 12 **R.3.** 1000 **R.5.** $\{0\}$ **R.7.** $\{\ \}$
R.9. The set of real numbers

1. a. clearing; fractions **b.** clearing; decimals **3.** $9w + 10$
5. $-2d - 7$ **7.** $25t - 40$ **9.** 18, 36 **11.** 100; 1000; 10,000
13. 30, 60 **15.** $\{4\}$ **17.** $\{-12\}$ **19.** $\left\{-\dfrac{15}{4}\right\}$
21. $\{8\}$ **23.** $\{3\}$ **25.** $\{15\}$ **27.** $\{\ \}$
29. The set of real numbers **31.** $\{5\}$ **33.** $\{2\}$ **35.** $\{-15\}$
37. $\{6\}$ **39.** $\{107\}$ **41.** The set of real numbers **43.** $\{67\}$
45. $\{90\}$ **47.** $\{4\}$ **49.** $\{4\}$ **51.** $\{\ \}$ **53.** $\{-0.25\}$
55. $\{-6\}$ **57.** $\left\{\dfrac{8}{3}\right\}$ or $\left\{2\dfrac{2}{3}\right\}$ **59.** $\{-11\}$ **61.** $\left\{\dfrac{1}{10}\right\}$
63. $\{-2\}$ **65.** $\{-1\}$ **67.** $\{2\}$

Chapter 10 Problem Recognition Exercises, p. 620

1. Expression; $-4b + 18$ **2.** Expression; $20p - 30$
3. Equation; $\{-8\}$ **4.** Equation; $\{-14\}$
5. Equation; $\left\{\dfrac{1}{3}\right\}$ **6.** Equation; $\left\{-\dfrac{4}{3}\right\}$
7. Expression; $6z - 23$ **8.** Expression; $-x - 9$
9. Equation; $\left\{\dfrac{7}{9}\right\}$ **10.** Equation; $\left\{-\dfrac{13}{10}\right\}$
11. Equation; $\{20\}$ **12.** Equation; $\{-3\}$
13. Equation; $\left\{\dfrac{1}{2}\right\}$ **14.** Equation; $\{-6\}$
15. Expression; $\dfrac{5}{8}x + \dfrac{7}{4}$ **16.** Expression; $-26t + 18$
17. Equation; $\{\ \}$ **18.** Equation; $\{\ \}$
19. Equation; $\left\{\dfrac{23}{12}\right\}$ **20.** Equation; $\left\{\dfrac{5}{8}\right\}$
21. Equation; The set of real numbers
22. Equation; The set of real numbers
23. Equation; $\left\{\dfrac{1}{2}\right\}$ **24.** Equation; $\{0\}$
25. Expression; 0 **26.** Expression; -1
27. Expression; $2a + 13$ **28.** Expression; $8q + 3$
29. Equation; $\{10\}$ **30.** Equation; $\left\{-\dfrac{1}{20}\right\}$
31. Expression; $-\dfrac{1}{6}y + \dfrac{1}{3}$ **32.** Expression; $\dfrac{7}{10}x + \dfrac{2}{5}$

Section 10.4 Activity, p. 627

A.1. a. 344 ft **b.** $400 - x$ ft
A.2. a. $30.96 **b.** $1.29p$
A.3. a. $x + 1, x + 2$ **b.** $x - 1, x - 2$
A.4. a. $x + 2, x + 4$ **b.** $x - 2, x - 4$
A.5. a. $3x + 21 = (x + 5) - 12$ **b.** $x = -14$ **c.** The number is -14.
A.6. b. One possibility: Let x represent the shortest side. Then $x + 2$ and $x + 4$ represent the lengths of the middle and longest sides, respectively.
  c. $\begin{pmatrix}\text{Length of}\\\text{shortest side}\end{pmatrix} + \begin{pmatrix}\text{length of}\\\text{middle side}\end{pmatrix} + \begin{pmatrix}\text{length of}\\\text{longest side}\end{pmatrix} = 54$
  d. $x + (x + 2) + (x + 4) = 54$ **e.** $x = 16$
  f. The sides are 16 in., 18 in., and 20 in.
A.7. b. Let x represent the number of people on one team. Then the other team has $2x - 6$ members.
  c. $\begin{pmatrix}\text{Number on}\\\text{one team}\end{pmatrix} + \begin{pmatrix}\text{number on}\\\text{the other team}\end{pmatrix} = 24$
  d. $x + (2x - 6) = 24$ **e.** $x = 10$
  f. One team has 10 people and the other has 14 people.

Section 10.4 Practice Exercises, pp. 628–631

R.1. $x + 16$ **R.3.** $x - 12$ **R.5.** $-23x$ **R.7.** $-4(x - 2)$

1. a. consecutive **b.** even **c.** odd **d.** 1 **e.** 2 **f.** 2
3. $x - 5,682,080$ **5.** $10x$ **7.** $3x - 20$ **9.** The number is -4.
11. The number is -3. **13.** The number is 5.
15. The number is -5. **17.** The number is 3. **19. a.** $x + 1, x + 2$
  b. $x - 1, x - 2$ **21.** The integers are -34 and -33.
23. The integers are 13 and 15. **25.** The sides are 14 in., 15 in., 16 in., 17 in., and 18 in. **27.** The integers are -54, -52, and -50.
29. The integers are 13, 15, and 17. **31.** The lengths of the pieces are 33 cm and 53 cm. **33.** Karen's music library has 23 playlists and Clarann's library has 35 playlists. **35.** There were 201 Republicans and 232 Democrats. **37.** There were 190 passenger cars and 230 SUVs sold. **39.** The Congo River is 4370 km long, and the Nile River is 6825 km. **41.** The area of Africa is 30,065,000 km². The area of Asia is 44,579,000 km².
43. They walked 12.3 mi on the first day and 8.2 mi on the second.
45. The pieces are 9 in., 17 in., and 22 in. **47.** The integers are 42, 43, and 44. **49.** The winner earned $14 million and the runner-up earned $5 million. **51.** The number is 11. **53.** The page numbers are 470 and 471. **55.** The number is 10. **57.** The deepest point in the Arctic Ocean is 5122 m.
59. The number is $\dfrac{7}{16}$. **61.** The number is 2.5.

Section 10.5 Activity, pp. 636–637

A.1. $x = 0.38(81)$; 30.78
A.2. $33.6 = x(28)$; 120%
A.3. $81 = 0.06x$; 1350
A.4. a. The $20 off coupon is better. **b.** The 15% off coupon is better.
A.5. 19%
A.6. Michael owes his father $1776.50.
A.7. b. Let x represent the original price.
  c. $\begin{pmatrix}\text{Original}\\\text{price}\end{pmatrix} + (\text{markup}) = 232.40$
  d. $x + 0.40x = 232.40$ **e.** $x = 166$
  f. The original price was $166.
A.8. b. Let x represent the retail price.
  c. $\begin{pmatrix}\text{Retail}\\\text{price}\end{pmatrix} + \begin{pmatrix}\text{sales}\\\text{tax}\end{pmatrix} = 65.10$
  d. $x + 0.05x = 65.10$ **e.** $x = 62$
  f. The shoes were $62.00 before tax.

A.9. b. Let x represent the amount of merchandise sold.

c. $\left(\begin{array}{c}\text{Base}\\\text{salary}\end{array}\right) + (\text{commission}) = 148$

d. $60 + 0.10x = 148$ **e.** $x = 880$

f. The employee sold $880 in merchandise.

Section 10.5 Practice Exercises, pp. 637–640

R.1. g **R.3.** b **R.5.** f **R.7.** a **R.9.** $1.20

R.11. 10% is $6 and half of that would be 5% which is $3. Together a 15% tip would be $9.

1. a. simple **b.** 100 **3.** $6
5. 12.5% **7.** 85% **9.** 0.75 **11.** 1050.8
13. 885 **15.** 2200 **17.** Molly will have to pay $106.99.
19. Approximately 231,000 cases **21.** 2% **23.** Javon's taxable income was $84,000. **25.** Aidan would earn $9 more in the CD.
27. Bob borrowed $1200. **29.** The rate is 6%.
31. Perry needs to invest $3302. **33. a.** $7.44 **b.** $54.56
35. The original price was $470.59. **37.** The discount rate is 12%.
39. The original dosage was 15 cc. **41.** The tax rate is 5%.
43. The original cost was $11.58 per pack.
45. The original price was $210,000. **47.** Alina made $4600 that month. **49.** Diane sold $645 over $200 worth of merchandise.

Section 10.6 Activity, p. 646

A.1. a. $d = \dfrac{C}{\pi}$ **b.** 7 in. **c.** Yes

d. Sometimes it is necessary to find the value of a different variable within a formula.

A.2. a. $\{7\}$ **b.** $y = \dfrac{w+z}{a}$

c. In each equation, combine the non-y terms on the right and divide by the coefficient of the y term.

A.3. a. $y = \dfrac{-3x+6}{2}$ or $y = -\dfrac{3}{2}x + 3$

b. Yes. The single fraction $y = \dfrac{-3x+6}{2}$ implies that the entire numerator is divided by 2. Therefore, the expression can be written as the sum of two fractions as $y = \dfrac{-3x}{2} + \dfrac{6}{2}$ or equivalently, $y = -\dfrac{3}{2}x + 3$.

A.4. The sum of their measures is 90°.
A.5. The sum of their measures is 180°.
A.6. 180°
A.7. $P = 2l + 2w$
A.8. They are vertical angles, and vertical angles are equal.
A.9. $x = 10$; The angles are 28°.
A.10. b. One possibility: Let x represent the length. Then the width is $x - 5$. **c.** $P = 2w + 2l$ **d.** $54 = 2(x - 5) + 2x$ **e.** $x = 16$
f. The length is 16 in., and the width is 11 in.
A.11. b. Let x represent the measure of one angle. Then the measure of the other angle is $5x - 6$.

c. $\left(\begin{array}{c}\text{Measure of}\\\text{one angle}\end{array}\right) + \left(\begin{array}{c}\text{measure of}\\\text{the other angle}\end{array}\right) = 90$

d. $x + (5x - 6) = 90$ **e.** $x = 16$
f. The angles are 16° and 74°.

A.12. b. One possibility: Let x represent the measure of the smallest angle. Then $2x + 4$ and $5x$ are the measures of the middle and largest angles, respectively.

c. $\left(\begin{array}{c}\text{Measure of}\\\text{smallest angle}\end{array}\right) + \left(\begin{array}{c}\text{measure of}\\\text{middle angle}\end{array}\right) + \left(\begin{array}{c}\text{measure of}\\\text{largest angle}\end{array}\right) = 180$

d. $x + (2x + 4) + (5x) = 180$ **e.** $x = 22$
f. The angles are 22°, 48°, and 110°.

Section 10.6 Practice Exercises, pp. 647–650

R.1. a. $P = 2l + 2w$ **b.** 20 cm **R.3. a.** 47° **b.** 137° **R.5.** 42°

1. $a = P - b - c$ **3.** $y = x + z$ **5.** $q = p - 250$
7. $b = \dfrac{A}{h}$ **9.** $t = \dfrac{PV}{nr}$ **11.** $x = 5 + y$ **13.** $y = -3x - 19$

15. $y = \dfrac{-2x+6}{3}$ or $y = -\dfrac{2}{3}x + 2$

17. $x = \dfrac{y+9}{-2}$ or $x = -\dfrac{1}{2}y - \dfrac{9}{2}$

19. $y = \dfrac{-4x+12}{-3}$ or $y = \dfrac{4}{3}x - 4$

21. $y = \dfrac{-ax+c}{b}$ or $y = -\dfrac{a}{b}x + \dfrac{c}{b}$

23. $t = \dfrac{A-P}{Pr}$ or $t = \dfrac{A}{Pr} - \dfrac{1}{r}$

25. $c = \dfrac{a-2b}{2}$ or $c = \dfrac{a}{2} - b$ **27.** $y = 2Q - x$

29. $a = MS$ **31.** $R = \dfrac{P}{I^2}$

33. The length is 7 ft and the width is 5 ft.
35. The length is 120 yd and the width is 30 yd.
37. The length is 195 m and the width is 100 m.
39. The sides are 22 m, 22 m, and 27 m.
41. "Adjacent supplementary angles form a straight angle." The words *Supplementary* and *Straight* both begin with the same letter.
43. The angles are 23.5° and 66.5°.
45. The angles are 34.8° and 145.2°.
47. $x = 20$; the vertical angles measure 37°.
49. The measures of the angles are 30°, 60°, and 90°.
51. The measures of the angles are 42°, 54°, and 84°.
53. $x = 17$; the measures of the angles are 34° and 56°.

55. a. $A = lw$ **b.** $w = \dfrac{A}{l}$ **c.** The width is 29.5 ft.

57. a. $P = 2l + 2w$ **b.** $l = \dfrac{P-2w}{2}$ **c.** The length is 103 m.

59. a. $C = 2\pi r$ **b.** $r = \dfrac{C}{2\pi}$ **c.** The radius is approximately 140 ft.

61. 140.056 **62.** 31.831 **63.** 1.273 **64.** 0.455

65. a. 415.48 m² **b.** 10,386.89 m³

Section 10.7 Activity, pp. 656–657

A.1. a. $16 - x$ **b.** $85x + 50(16 - x) = 975$
c. $x = 5$ **d.** The students have 5 seats near the stage and 11 seats in the balcony.
A.2. The solution has 2 L of acid and 8 L of water.
A.3. a. 0.3 oz **b.** 0.15x **c.** 0.1(x + 5)
A.4. a. Let x represent the amount of the 15% alcohol solution.

b.

	15% Alcohol Solution	6% Alcohol Solution	10% Alcohol Solution
Amount of solution (oz)	x	5	$x + 5$
Amount of pure alcohol (oz)	0.15x	0.06(5)	0.1(x + 5)

c. $0.15x + 0.06(5) = 0.1(x + 5)$ **d.** $x = 4$
e. 4 oz of 15% alcohol solution is needed.
A.5. $d = rt$
A.6. a. 2 mi **b.** 2 mi
A.7. a. One possibility: Let x represent Jennifer's speed without pushing the stroller. (Alternatively, we can let x represent Jennifer's speed while pushing the stroller. In such a case, her speed without pushing the stroller would be $x + 1$, and the table would be adjusted accordingly.)

b.

	Distance (mi)	Rate (mph)	Time (hr)
With the stroller	$\frac{2}{3}(x-1)$	$x-1$	$\frac{2}{3}$
Without the stroller	$\frac{1}{2}x$	x	$\frac{1}{2}$

c. $\frac{2}{3}(x-1) = \frac{1}{2}x$ **d.** $x = 4$

e. Jennifer walks 4 mph when not pushing the stroller and 3 mph when pushing the stroller.

Section 10.7 Practice Exercises, pp. 657–661

R.1. $\{-43\}$ **R.3. a.** $48 **b.** $12x$ dollars
R.5. a. 0.5 L **b.** $0.25x$ liters   **R.7. a.** 1600 mi **b.** $400t$ miles

1. $200 - t$ **3.** $100 - x$ **5.** $3000 - y$
7. 53 tickets were sold at $3 and 28 tickets were sold at $2.
9. There were 17 songs purchased at $1.29 and 8 songs purchased at $1.49.
11. Jolynn bought 21 books at $6.99 and 11 books at $9.99. **13.** $x + 7$
15. $d + 2000$ **17.** Mix 500 gal of 5% ethanol fuel mixture.
19. The pharmacist needs to use 21 mL of the 1% saline solution.
21. The contractor needs to mix 6.75 oz of 50% acid solution.
23. a. 300 mi **b.** $5x$ **c.** $5(x + 12)$ or $5x + 60$ **25.** She walks 4 mph to the lake. **27.** Bryan hiked 6 mi up the canyon.
29. The plane travels 600 mph in still air. **31.** The slower car travels 48 mph and the faster car travels 52 mph. **33.** The speeds of the vehicles are 40 mph and 50 mph. **35.** Sarah walks 3.5 mph and Jeanette runs 7 mph. **37. a.** 2 lb **b.** $0.10x$
c. $0.10(x + 3) = 0.10x + 0.30$ **39.** Mix 10 lb of coffee sold at $12 per pound and 40 lb of coffee sold at $8 per pound. **41.** The boats will meet in $\frac{3}{4}$ hr (45 min). **43.** Sam purchased 16 packages of wax and 5 bottles of sunscreen. **45.** 2.5 qt of 85% chlorine solution
47. 20 L of water must be added. **49.** The Japanese bullet train travels 300 km/hr and the Acela Express travels 240 km/hr.

Section 10.8 Activity, p. 672

A.1. $x \geq 18$ yr
A.2. $x < 2$ yr
A.3. $12.0 \leq x \leq 15.2$ g/dL
A.4. $x \leq 6$ yr
A.5. a. [graph: open circle at -2, arrow right] **b.** [graph: bracket at -2, arrow right]

c. Use a parenthesis when an "endpoint" is not included in the solution set. Use a bracket when an "endpoint" is included in the solution set.
d. $(-2, \infty)$ **e.** $[-2, \infty)$
f. Use a parenthesis when an "endpoint" is not included in the solution set and for ∞ and $-\infty$. Use a bracket when an "endpoint" is included in the solution set.
A.6. a. [graph: arrow left from 3] **b.** [graph: arrow left from 3]
c. The inequalities $x < 3$ and $3 > x$ are equivalent, which means that they have the same solution set.
d. $(-\infty, 3)$ **e.** $(-\infty, 3)$
A.7. a. $<$ **b.** $>$

A.8. a. [graph from 2 right] $(2, \infty)$
b. [graph from 5 right] $(5, \infty)$
c. [graph to -50 left] $(-\infty, -50)$
d. [graph to -2 left] $(-\infty, -2)$
e. [graph from 15 right] $(15, \infty)$

A.9. Parts (c) and (d). The inequality in part (c) required that we multiply both sides by a negative number. In part (d) we divided both sides by a negative number.
A.10. $(-5, \infty)$
A.11. $(-\infty, 1]$
A.12. $(3, \infty)$
A.13. $[-2, 6)$

Section 10.8 Practice Exercises, pp. 673–677

R.1. $<$ **R.3.** $>$ **R.5.** $\left\{\frac{20}{3}\right\}$ **R.7.** $\{\}$
R.9. The set of real numbers

1. a. linear inequality **b.** inequality **c.** set-builder; interval
3. b and d   **5.** [graph with open circle at 5]   **7.** [graph with bracket at $\frac{5}{2}$]
9. [arrow left from 13]   **11.** [bracket at 2, parenthesis at 6.5]   **13.** [0 to 4]
15. [bracket at 1 to bracket at 8]

Set-Builder Notation	Graph	Interval Notation
17. $\{x \mid x \geq 6\}$	[bracket at 6, →]	$[6, \infty)$
19. $\{x \mid x \leq 2.1\}$	[←, bracket at 2.1]	$(-\infty, 2.1]$
21. $\{x \mid -2 < x \leq 7\}$	[(at -2,] at 7]	$(-2, 7]$

Set-Builder Notation	Graph	Interval Notation
23. $\left\{x \mid x > \frac{3}{4}\right\}$	[(at $\frac{3}{4}$, →]	$\left(\frac{3}{4}, \infty\right)$
25. $\{x \mid -1 < x < 8\}$	[(at -1,) at 8]	$(-1, 8)$
27. $\{x \mid x \leq -14\}$	[←,] at -14]	$(-\infty, -14]$

Set-Builder Notation	Graph	Interval Notation
29. $\{x \mid x \geq 18\}$	[bracket at 18, →]	$[18, \infty)$
31. $\{x \mid x < -0.6\}$	[←,) at -0.6]	$(-\infty, -0.6)$
33. $\{x \mid -3.5 \leq x < 7.1\}$	[[at -3.5,) at 7.1]	$[-3.5, 7.1)$

35. a. $\{3\}$ **37. a.** $\{13\}$
 b. $\{x \mid x > 3\}$; $(3, \infty)$ **b.** $\{p \mid p \leq 13\}$; $(-\infty, 13]$
[graph from 3 right]   [graph to 13 left, bracket]

39. a. $\{-3\}$ **41. a.** $\left\{-\frac{3}{2}\right\}$
 b. $\{c \mid c < -3\}$; $(-\infty, -3)$ **b.** $\left\{z \mid z \geq -\frac{3}{2}\right\}$; $\left[-\frac{3}{2}, \infty\right)$
[graph to -3 left]   [graph from $-\frac{3}{2}$ right]

43. $(-1, 4]$ **45.** $(-3, 5)$
[graph from -1 to 4]   [graph from -3 to 5]

47. [2, 6]

49. a. $\{x|x \leq 1\}$ b. $(-\infty, 1]$

51. a. $\{q|q > 10\}$
    b. $(10, \infty)$

53. a. $\{x|x > 3\}$
    b. $(3, \infty)$

55. a. $\{c|c > 2\}$
    b. $(2, \infty)$

57. a. $\{c|c < -2\}$
    b. $(-\infty, -2)$

59. a. $\{h|h \geq 14\}$
    b. $[14, \infty)$

61. a. $\{x|x \geq -24\}$
    b. $[-24, \infty)$

63. a. $\{p|-3 \leq p < 3\}$
    b. $[-3, 3)$

65. a. $\{h|0 < h < \frac{5}{2}\}$
    b. $(0, \frac{5}{2})$

67. a. $\{x|10 < x < 12\}$
    b. $(10, 12)$

69. a. $\{x|-8 \leq x < 24\}$
    b. $[-8, 24)$

71. a. $\{y|y > -9\}$
    b. $(-9, \infty)$

73. a. $\{x|x \geq -\frac{15}{2}\}$
    b. $[-\frac{15}{2}, \infty)$

75. a. $\{x|x < -3\}$
    b. $(-\infty, -3)$

77. a. $\{b|b \geq -\frac{1}{3}\}$
    b. $[-\frac{1}{3}, \infty)$

79. a. $\{n|n > -3\}$
    b. $(-3, \infty)$

81. a. $\{x|x < 7\}$
    b. $(-\infty, 7)$

83. a. $\{x|x \leq -5\}$
    b. $(-\infty, -5]$

85. a. $\{z|z \leq -2\}$
    b. $(-\infty, -2]$

87. a. $\{a|a > -\frac{2}{3}\}$
    b. $(-\frac{2}{3}, \infty)$

89. a. $\{p|p \leq \frac{15}{4}\}$
    b. $(-\infty, \frac{15}{4}]$

91. a. $\{y|y < -9\}$
    b. $(-\infty, -9)$

93. a. $\{a|a > -3\}$
    b. $(-3, \infty)$

95. a. $\{x|x \leq 0\}$ b. $(-\infty, 0]$

97. No   99. Yes   101. $L \geq 10$ 103. $w > 75$
105. $t \leq 72$ 107. $L \geq 8$ 109. $2 < h < 5$

111. More than 10.2 in. of rain is needed.   113. Trevor needs at least 86 to get a "B" in the course.   115. a. $1539 b. 200 birdhouses cost $1440. It is cheaper to purchase 200 birdhouses because the discount is greater.   117. Company A is better if more than 400 flyers are printed.   119. Madison needs to babysit a minimum of 39.5 hr.

Chapter 10 Review Exercises, pp. 678–681

1. a. Equation b. Expression c. Equation d. Equation
2. A linear equation can be written in the form $ax + b = c, a \neq 0$.
3. a. No b. Yes c. No d. Yes   4. a. No b. Yes
5. $\{-8\}$ 6. $\{15\}$ 7. $\{\frac{21}{4}\}$ 8. $\{70\}$ 9. $\{-\frac{21}{5}\}$
10. $\{-60\}$ 11. $\{-\frac{10}{7}\}$ 12. $\{27\}$ 13. The number is 60.
14. The number is $\frac{7}{24}$.   15. The number is -8.
16. The number is -2.   17. $\{1\}$ 18. $\{-\frac{3}{5}\}$ 19. $\{2\}$
20. $\{-6\}$ 21. $\{-3\}$ 22. $\{18\}$ 23. $\{\frac{3}{4}\}$ 24. $\{-3\}$
25. $\{0\}$ 26. $\{\frac{1}{8}\}$ 27. $\{2\}$ 28. $\{6\}$ 29. A contradiction has no solution and an identity is true for all real numbers.
30. The set of real numbers; identity   31. $\{20\}$; conditional equation
32. $\{\ \}$; contradiction   33. The set of real numbers; identity
34. $\{\ \}$; contradiction   35. $\{2\}$; conditional equation
36. $\{6\}$ 37. $\{22\}$ 38. $\{13\}$ 39. $\{-27\}$ 40. $\{-10\}$
41. $\{-7\}$ 42. $\{\frac{5}{3}\}$ 43. $\{-\frac{9}{4}\}$ 44. $\{2.5\}$ 45. $\{-4\}$
46. $\{-4.2\}$ 47. $\{2.5\}$ 48. $\{-312\}$ 49. $\{200\}$ 50. $\{\ \}$
51. $\{\ \}$ 52. The set of real numbers   53. The set of real numbers   54. $\{0\}$ 55. $\{0\}$
56. The number is 30.   57. The number is 11.
58. The number is -7.   59. The number is -10.
60. The integers are 66, 68, and 70.
61. The integers are 27, 28, and 29.
62. The sides are 25 in., 26 in., and 27 in.
63. The sides are 36 cm, 37 cm, 38 cm, 39 cm, and 40 cm.
64. 23.8   65. 28.8   66. 12.5%   67. 95%   68. 160
69. 1750   70. The dinner was $40 before tax and tip.
71. a. $840 b. $3840   72. He invested $12,000.
73. The novel originally cost $29.50.   74. $K = C + 273$
75. $C = K - 273$ 76. $s = \frac{P}{4}$ 77. $s = \frac{P}{3}$
78. $x = \frac{y-b}{m}$ 79. $x = \frac{c-a}{b}$ 80. $y = \frac{-2x-2}{5}$
81. $b = \frac{Q-4a}{4}$ or $b = \frac{Q}{4} - a$ 82. The height is 7 m.
83. a. $h = \frac{3V}{\pi r^2}$ b. The height is 5.1 in.   84. The angles are 22°, 78°, and 80°.   85. The angles are 50° and 40°.
86. The length is 5 ft and the width is 4 ft.   87. $x = 20$. The angle measure is 65°.   88. The measure of angle y is 53°.
89. The truck travels 45 km/hr in bad weather and 60 km/hr in good weather.   90. Gus rides 15 mph and Winston rides 18 mph.
91. The cars will be 327.6 mi apart after 2.8 hr (2 hr and 48 min).
92. They meet in 2.25 hr (2 hr and 15 min).   93. 2 lb of 24% fat content beef is needed.   94. 20 lb of the 40% solder should be used.
95. $(-2, \infty)$
96. $(-\infty, \frac{1}{2}]$
97. $(-1, 4]$

98. a. $637 b. 300 plants cost $1410, and 295 plants cost $1416. 295 plants costs more.

99. $\{c|c < 17\}$; $(-\infty, 17)$

100. $\left\{w\middle|w > -\frac{1}{3}\right\}$; $\left(-\frac{1}{3}, \infty\right)$

101. $\{x|x \le -6\}$; $(-\infty, -6]$

102. $\left\{y\middle|y \le -\frac{14}{5}\right\}$; $\left(-\infty, -\frac{14}{5}\right]$

103. $\{a|a \ge 49\}$; $[49, \infty)$

104. $\{t|t < 34.5\}$; $(-\infty, 34.5)$

105. $\{k|k > 18\}$; $(18, \infty)$

106. $\left\{h\middle|h \le \frac{5}{2}\right\}$; $\left(-\infty, \frac{5}{2}\right]$

107. $\{x|x < -1\}$; $(-\infty, -1)$

108. $\{b| -3 < b \le 7\}$; $(-3, 7]$

109. $\{z| -6 \le z \le 5\}$; $[-6, 5]$

110. More than 2.5 in. is required.
111. Collette can have at most 18 wings.

Chapter 11

Section 11.1 Activity, p. 688

A.1. a.

Point	Ordered Pair	Location
A	$(-2, -3)$	Quadrant III
B	$(0, 1)$	y-axis
C	$(2, -3)$	Quadrant IV
D	$(-2, 0)$	x-axis
E	$(2, 0)$	x-axis

b.
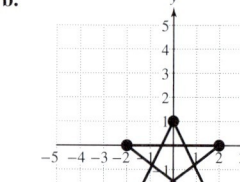

c. Video game programmers can use a rectangular coordinate system to position objects on the screen and to draw shapes by plotting and connecting points.

Section 11.1 Practice Exercises, pp. 689–694

R1.–R8.
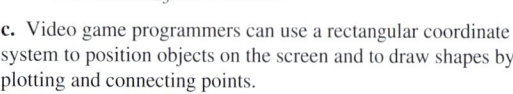

1. a. x; y-axis **b.** ordered **c.** origin; $(0, 0)$ **d.** quadrants **e.** negative **f.** III   **3. a.** Month 10 **b.** 30 **c.** Between months 3 and 5 and between months 10 and 12 **d.** Months 8 and 9 **e.** Month 3 **f.** 80   **5. a.** On day 1 the price per share was $89.25. **b.** $1.75 **c.** −$2.75

7.
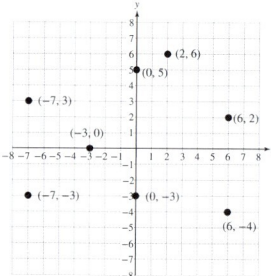

9.
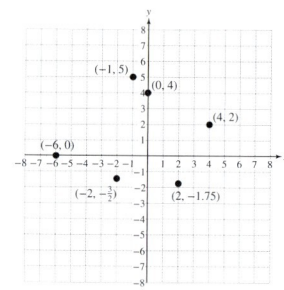

11. IV **13.** II **15.** III **17.** I
19. $(0, -5)$ lies on the y-axis.
21. $\left(\frac{7}{8}, 0\right)$ is located on the x-axis.
23. $A(-4, 2)$, $B(\frac{1}{2}, 4)$, $C(3, -4)$, $D(-3, -4)$, $E(0, -3)$, $F(5, 0)$
25. a. $A(400, 200)$, $B(200, -150)$, $C(-300, -200)$, $D(-300, 250)$, $E(0, 450)$ **b.** 450 m
27. a. $(250, 225)$, $(175, 193)$, $(315, 330)$, $(220, 209)$, $(450, 570)$, $(400, 480)$, $(190, 185)$; the ordered pair $(250, 225)$ means that 250 people produce $225 in popcorn sales.
b.
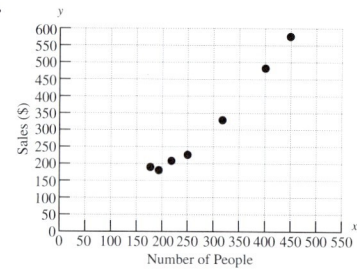

29. a. $(1, -10.2)$, $(2, -9.0)$, $(3, -2.5)$, $(4, 5.7)$, $(5, 13.0)$, $(6, 18.3)$, $(7, 20.9)$, $(8, 19.6)$, $(9, 14.8)$, $(10, 8.7)$, $(11, 2.0)$, $(12, -6.9)$
b.
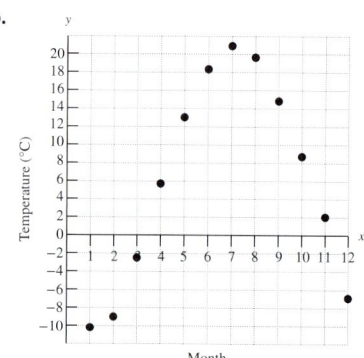

31. a.
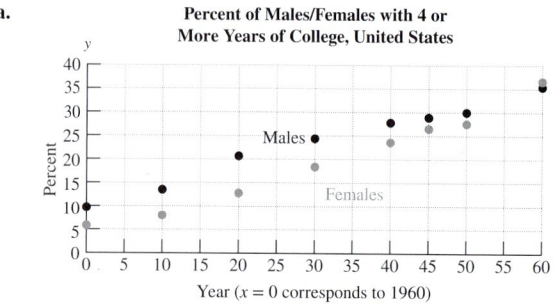

b. Increasing **c.** Increasing

Section 11.2 Activity, pp. 702–703

A.1. a.

x	y
-2	0
-4	4
0	-4
-3	2
1	-6

b.–c. The points all line up.

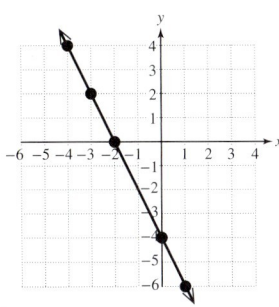

d. The equation $2x + y = -4$ is called a linear equation because the graph representing all solutions to the equation is a line.

A.2. a. Yes b. Yes c. No; in the term $3x^2$, the factor of x is raised to the second power rather than the first power. d. No; the variables x and y appear in the denominator rather than the numerator. e. Yes

A.3. An x-intercept is a point where the graph of an equation intersects the x-axis. A y-intercept is a point where the graph of an equation intersects the y-axis.

A.4. To find the x-intercept, substitute 0 for y and solve for x. To find the y-intercept, substitute 0 for x and solve for y.

A.5. a. $(-1, 0)$ b. $(0, 3)$ c. For example: $(-2, -3)$
d.

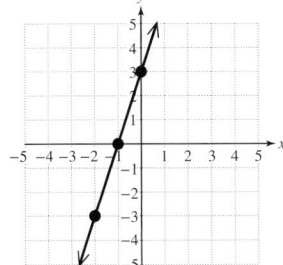

A.6. a. $(0, 0)$ b. $(0, 0)$ c. For example, $(2, 2)$
d.

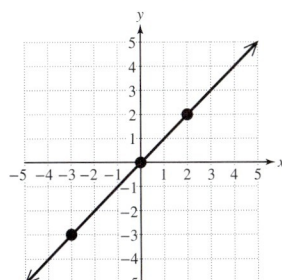

A.7. a. No x-intercept b. $(0, -2)$ c. For example, $(-3, -2)$
d.

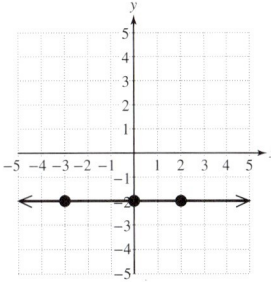

A.8. a. $(-1, 0)$ b. No y-intercept c. For example, $(-1, 2)$
d.

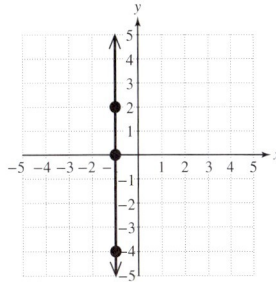

A.9. An equation represents a horizontal line if the equation has a y variable but no x variable. In such a case, the equation can be written in the form $y = k$, where k is a constant.

A.10. An equation represents a vertical line if the equation has an x variable but no y variable. In such a case, the equation can be written in the form $x = k$, where k is a constant.

A.11. An equation represents a slanted line passing through the origin if the equation has both x and y variables and the constant term is zero. In such a case, the equation can be written in the form $Ax + By = 0$, where neither A nor B is zero.

Section 11.2 Practice Exercises, pp. 703–710

R.1. {5} **R.3.** {2} **R.5.** 29 **R.7.** 14
R.9. (2, 4); Quadrant I **R.11.** (0, -1); y-axis
R.13. (3, -4); Quadrant IV

1. a. $Ax + By = C$ b. x-intercept c. y-intercept d. vertical
e. horizontal **3.** Linear **5.** Nonlinear **7.** Nonlinear **9.** Yes
11. Yes **13.** No **15.** No **17.** Yes

19.

x	y
1	-3
-2	0
-3	1
-4	2

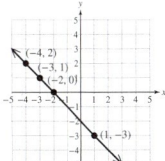

21.

x	y
-2	3
-1	0
-4	9

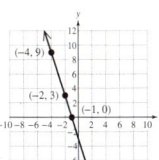

23.

x	y
0	4
2	0
3	-2

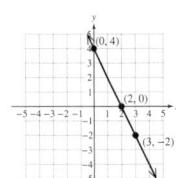

25.

x	y
0	−2
5	−5
10	−8

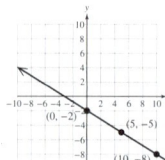

27.

x	y
0	−2
−3	−2
5	−2

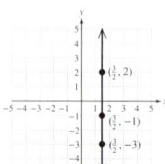

29.

x	y
3/2	−1
3/2	2
3/2	−3

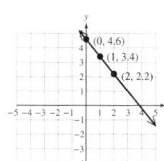

31.

x	y
0	4.6
1	3.4
2	2.2

33. 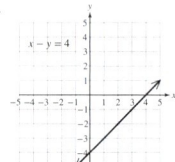 **35.** **37.**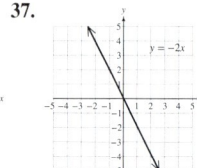

39. 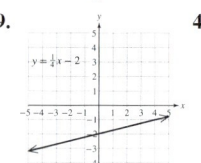 **41.** **43.**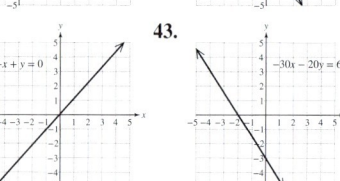

45. y-axis   **47.** x-intercept: (−1, 0); y-intercept: (0, −3)
49. x-intercept: (−4, 0); y-intercept: (0, 1)
51. x-intercept: $\left(-\frac{9}{4}, 0\right)$;   **53.** x-intercept: $\left(\frac{8}{3}, 0\right)$;
y-intercept: (0, 3)   y-intercept: (0, 2)

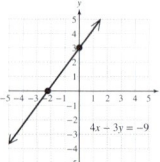 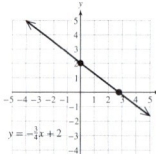

55. x-intercept: (−4, 0);   **57.** x-intercept: (0, 0);
y-intercept: (0, 8)   y-intercept: (0, 0)

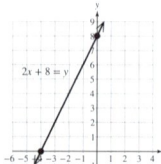 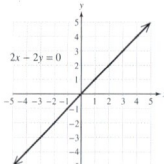

59. x-intercept: (10, 0);   **61.** x-intercept: (0, 0);
y-intercept: (0, 5)   y-intercept: (0, 0)

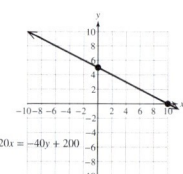 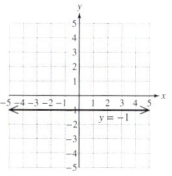

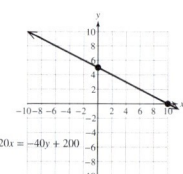

63. True   **65.** True
67. a. Horizontal line   **69. a.** Vertical line
   c. no x-intercept;   **c.** x-intercept: (4, 0);
   y-intercept: (0, −1)   no y-intercept

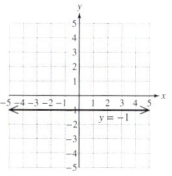 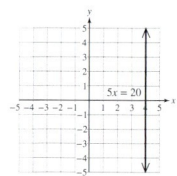

71. a. Horizontal line   **73. a.** Vertical line
   c. no x-intercept;   **c.** All points on the y-axis are
   y-intercept: (0, 3)   y-intercepts; x-intercept: (0, 0)

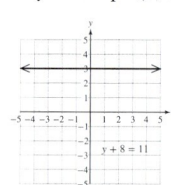 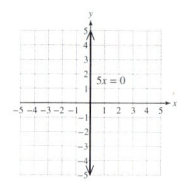

75. A horizontal line may not have an x-intercept. A vertical line may not have a y-intercept.   **77.** a, b, d

78. 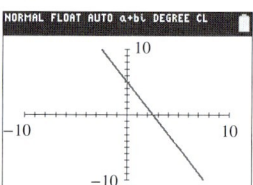 **79.**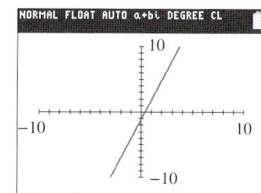

80. 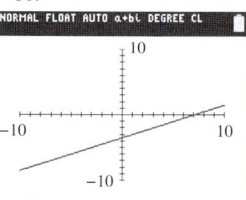 **81.**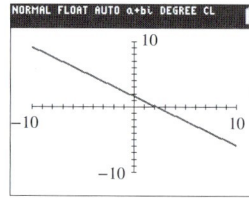

82. 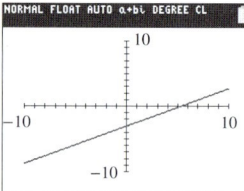 **83.**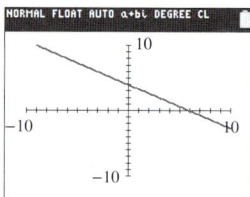

84. 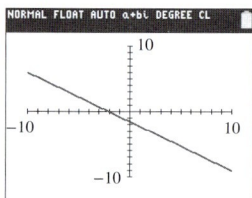 **85.**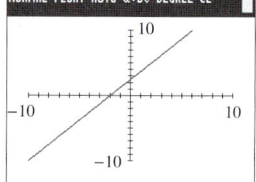

86. 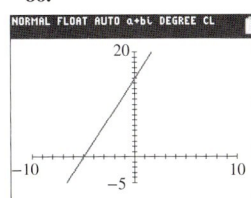   **87.**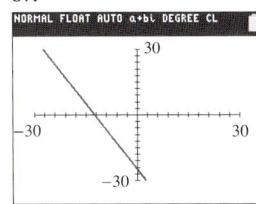

88. 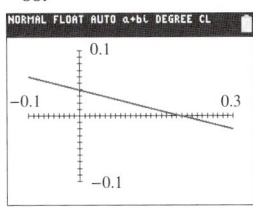   **89.**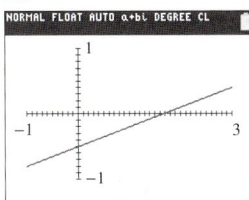

91. a. $y = 11{,}190$ **b.** $x = 3$ **c.** $(1, 11{,}190)$ One year after purchase the value of the car is $11,190. $(3, 9140)$ Three years after purchase the value of the car is $9140.

Section 11.3 Activity, pp. 719–720

A.1. Yes. The slope of the ramp is $\dfrac{3}{40}$ or 0.075. This is less than $\dfrac{1}{12}$, which is $0.08\overline{3}$.

A.2. $m = \dfrac{y_2 - y_1}{x_2 - x_1}$ **A.3.** $m = 0$; Graph iv

A.4. $m = 4$; Graph ii   **A.5.** Slope is undefined; Graph iii

A.6. $m = -\dfrac{1}{7}$; Graph i

A.7. a. Line l_1: $m = 4$, Line l_2: $m = -\dfrac{1}{4}$
b. Perpendicular
c.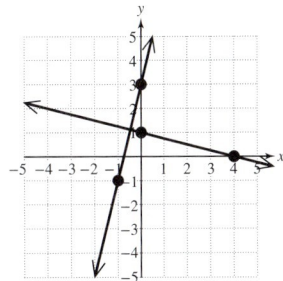

A.8. a. Line l_1: $m = 3$, Line l_2: $m = 3$ **b.** Parallel
c.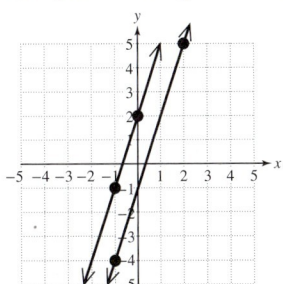

A.9. a. 21   **b.** The rise in out-of-pocket prescription drug cost is approximately $21 per year per capita.

Section 11.3 Practice Exercises, pp. 720–727

R.1. $-\dfrac{5}{3}$ **R.3.** 0   **R.5.** $\dfrac{9}{5}$ **R.7. a.** $\dfrac{3}{2}$ **b.** $-\dfrac{3}{2}$
R.9. $5.99/month

1. a. slope; $\dfrac{y_2 - y_1}{x_2 - x_1}$ **b.** parallel   **c.** right   **d.** -1
e. undefined; horizontal

3. $m = \dfrac{1}{3}$ **5.** $m = \dfrac{6}{11}$ **7.** undefined   **9.** positive

11. Negative   **13.** Zero   **15.** Undefined   **17.** Positive
19. Negative   **21.** $m = \dfrac{1}{2}$ **23.** $m = -3$ **25.** $m = 0$
27. The slope is undefined.   **29.** $\dfrac{1}{3}$ **31.** -3 **33.** $\dfrac{3}{5}$
35. Zero   **37.** Undefined   **39.** $\dfrac{28}{5}$ **41.** $-\dfrac{7}{8}$
43. -0.45 or $-\dfrac{9}{20}$ **45.** -0.15 or $-\dfrac{3}{20}$ **47. a.** -2 **b.** $\dfrac{1}{2}$
49. a. 0   **b.** undefined   **51. a.** $\dfrac{4}{5}$ **b.** $-\dfrac{5}{4}$ **53.** Perpendicular
55. Parallel   **57.** Neither   **59.** l_1: $m = 2$, l_2: $m = 2$; parallel
61. l_1: $m = 5$, l_2: $m = -\dfrac{1}{5}$; perpendicular
63. l_1: $m = \dfrac{1}{4}$, l_2: $m = 4$; neither   **65.** The average rate of change is $-$160 per year.   **67. a.** The ordered pair $(31, 3.833)$ means that on March 31, there were 3.833 thousand (3833) COVID-19 deaths in the United States.   **b.** The slope is 1.275, which means that the average number of COVID-19 deaths increased by 1.275 thousand (1275) per day from March 31 to April 7.   **c.** The death rate increased to approximately 1.861 thousand (1861), per day the following week.
69. a. Michael owes $11,550 after 5 months.   **b.** $m = -210$; The amount Michael owes his father decreases by $210 per month.
71. $m = \dfrac{3}{4}$ **73.** $m = 0$

75. For example: $(4, -4)$ and $(-2, 0)$
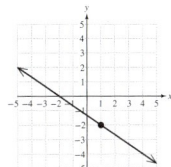

77. For example: $(3, 5)$ and $(1, -1)$
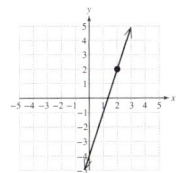

79. For example: $(-3, 1)$ and $(-3, 4)$
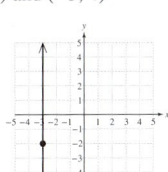

81.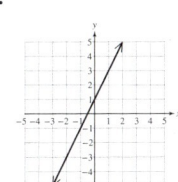

83.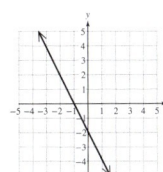

85.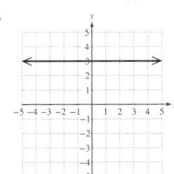

87. $\dfrac{3m - 3n}{2b}$ or $\dfrac{-3m + 3n}{-2b}$ **89.** $\left(\dfrac{c}{a}, 0\right)$

91. For example: $(7, 1)$

Section 11.4 Activity, pp. 732–733

A.1. a. $y = -2x + 4$ **b.** Slope: -2; y-intercept: $(0, 4)$
c.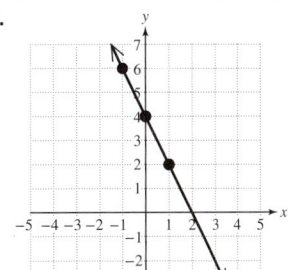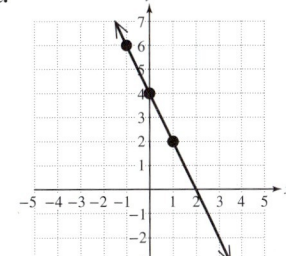

A.2. a. $y = -3x + 4$ **b.** -3 **c.** $y = \frac{1}{3}x + 1$ **d.** $\frac{1}{3}$
e. The lines are perpendicular because the slope of one line is the opposite of the reciprocal of the slope of the other line.
f.

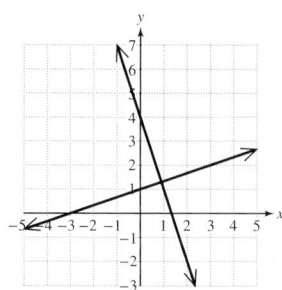

A.3. The first equation can be written in slope-intercept form as $y = 4x - 2$, and the second equation is $y = 4x + 1$. From the slope-intercept forms, we know that the slope of each line is 4 and that the y-intercepts are different. Two lines with the same slope and different y-intercepts are parallel.

A.4. $y = \frac{2}{3}x - 4$

A.5. a. $b = -2$ **b.** $y = -\frac{1}{2}x - 2$
c.

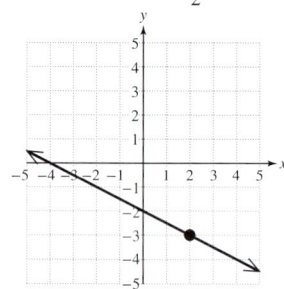

A.6. a. 42 in. **b.** 36 in. **c.** The slope is -0.5 in./hr, and this means that the depth of the pond decreases by 0.5 in. per hour.
d. The y-intercept is (0, 48), and this means that the pond was initially 48 in. deep. **e.** The x-intercept is (96, 0), and this means that after 96 hours, the pond will be empty (0 in. deep).

Section 11.4 Practice Exercises, pp. 733–738

R.1. $y = -2x + 4$ **R.3.** $y = 4x$ **R.5. a.** -4 **b.** $\frac{1}{4}$
R.7. a. $\frac{5}{7}$ **b.** $-\frac{7}{5}$ **R.9.** $\{7\}$

1. a. $y = mx + b$ **b.** standard
3. $m = -2$; y-intercept: (0, 3)
5. $m = 1$; y-intercept: (0, -2)
7. $m = -1$; y-intercept: (0, 0)
9. $m = \frac{3}{4}$; y-intercept: (0, -1) **11.** $m = \frac{2}{5}$; y-intercept: $\left(0, -\frac{4}{5}\right)$ **13.** $m = 3$; y-intercept: (0, -5)
15. $m = -1$; y-intercept: (0, 6)
17. Undefined slope; no y-intercept **19.** $m = 0$; y-intercept: $\left(0, -\frac{1}{4}\right)$
21. $m = \frac{2}{3}$; y-intercept: (0, 0)
23. 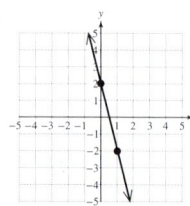 **25.**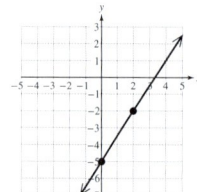

27. b **29.** e **31.** c

33. $y = -2x + 9$

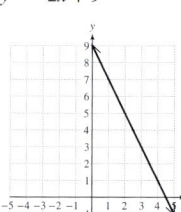

35. $y = \frac{1}{2}x - 3$

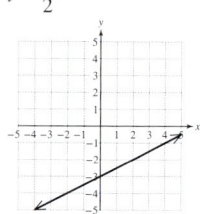

37. $y = -\frac{1}{2}x + \frac{3}{2}$

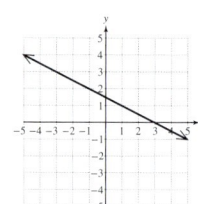

39. $y = -x$

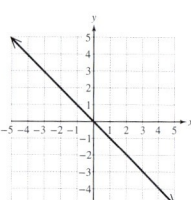

41. $y = \frac{4}{5}x$

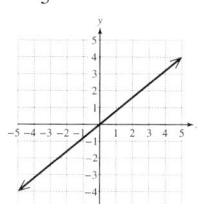

43. $y = -\frac{2}{3}$

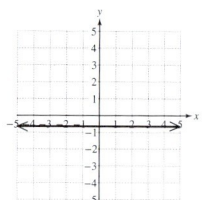

45. Perpendicular **47.** Neither **49.** Parallel
51. Perpendicular **53.** Parallel **55.** Perpendicular
57. Neither **59.** Parallel **61.** $y = -\frac{1}{3}x + 2$
63. $y = 10x - 19$ **65.** $y = 6x - 8$ **67.** $y = \frac{1}{2}x - 3$
69. $y = -11$ **71.** $y = 5x$
73. Perpendicular **74.** Parallel

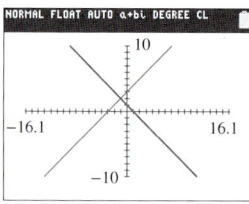 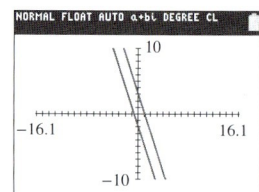

75. Neither

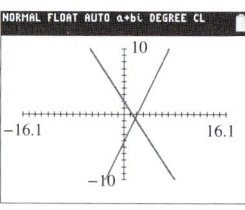

76. The lines may appear parallel; however, they are not parallel because the slopes are different.

77. The lines may appear to coincide on a graph; however, they are not the same line because the y-intercepts are different.

78. The line may appear to be horizontal, but it is not. The slope is 0.001 rather than 0.

79. $y = -2x + 3$ **81.** $y = -\frac{1}{3}x + 2$ **83.** $y = 4x - 1$

85. a. The slope is 3.50, which means that the yearly cost increases by $3.50 per gallon of propane purchased. **b.** The C-intercept is (0, 49) and represents the cost to rent the tank for one year but not purchase any propane. **c.** The cost to rent the tank and purchase 60 gal of propane during the year is $259. **d.** The homeowner purchased 42 gal.

87. $m = -\dfrac{2}{5}$ **89.** $m = \dfrac{4}{3}$

Chapter 11 Problem Recognition Exercises, pp. 738–739

1. a, c, d   **2.** b, f, h   **3.** a   **4.** f   **5.** b, f   **6.** c
7. c, d   **8.** f   **9.** e   **10.** g   **11.** b   **12.** h
13. g   **14.** e   **15.** c   **16.** b, h   **17.** e   **18.** e
19. b, f, h   **20.** c, d

Section 11.5 Activity, pp. 745–746

A.1. $y - y_1 = m(x - x_1)$
A.2. a. $y = -5x - 6$
  b.

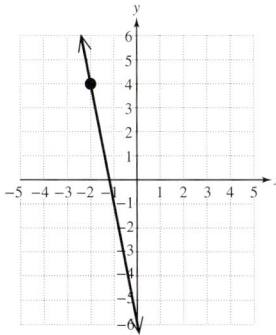

A.3. a. $y = -3x + 4$ **b.** $\dfrac{1}{3}$ **c.** $y = \dfrac{1}{3}x - 1$
  d. The line $y = \dfrac{1}{3}x - 1$ does indeed pass through $(3, 0)$ and is perpendicular to the line $3x + y = 4$.

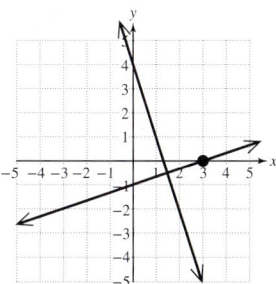

A.4. The slopes of the lines would each be -3 because the slopes of parallel lines are the same.
A.5. a. $m = 1$ **b.** $y = x - 1$ **c.** $y = x - 1$ **d.** The equations are the same. (This shows that either point on the line can be used with the slope of the line to find an equation of the line.)
A.6. a. Vertical line and the slope is undefined.
  b. Horizontal line and the slope is 0.
A.7. a.

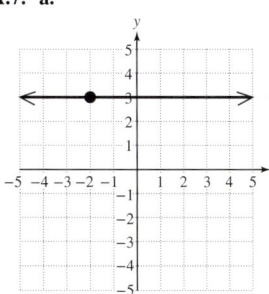

  b. $y = 3$

A.8. a.–b.

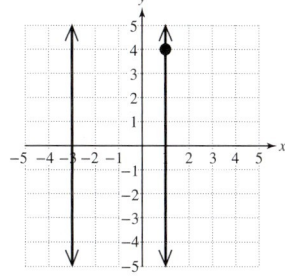

  c. $x = 1$

Section 11.5 Practice Exercises, pp. 746–750

R.1. Vertical   **R.3.** Slanted

R.5.   **R.7.**

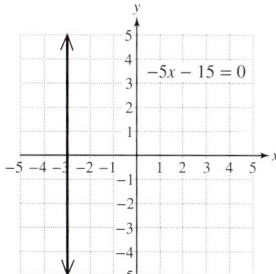

R.9. 9   **R.11.** 0   **R.13. a.** $-\dfrac{4}{3}$ **b.** $\dfrac{3}{4}$
R.15. a. 0   **b.** Undefined

1. a. $Ax + By = C$ **b.** horizontal   **c.** vertical   **d.** slope; y-intercept   **e.** $y - y_1 = m(x - x_1)$
3. $y = 3x + 7$ or $3x - y = -7$
5. $y = -4x - 14$ or $4x + y = -14$ **7.** $y = -\dfrac{1}{2}x - \dfrac{1}{2}$ or $x + 2y = -1$
9. $y = 2x - 2$ or $2x - y = 2$
11. $y = -x - 4$ or $x + y = -4$
13. $y = -0.2x - 2.86$ or $20x + 100y = -286$
15. $y = -2x + 1$ **17.** $y = 2x + 4$ **19.** $y = \dfrac{1}{2}x - 1$
21. $y = 4x + 13$ or $4x - y = -13$
23. $y = -\dfrac{3}{2}x + 6$ or $3x + 2y = 12$
25. $y = -2x - 8$ or $2x + y = -8$ **27.** $y = -\dfrac{1}{5}x - 6$ or $x + 5y = -30$ **29.** iv   **31.** vi   **33.** iii   **35.** $y = 1$
37. $x = 2$ **39.** $y = 2$ **41.** $y = \dfrac{1}{4}x + 8$ or $x - 4y = -32$
43. $y = 3x - 8$ or $3x - y = 8$ **45.** $y = 4.5x - 25.6$ or $45x - 10y = 256$
47. $x = -6$ **49.** $y = -2$ **51.** $x = -4$ **53.** $y = 2x - 4$
55. $y = -\dfrac{1}{2}x + 1$

Section 11.6 Activity, p. 754

A.1. a. y **b.** x **c.** 452 **d.** The slope is $25.5 per mile and means that the amount of money raised increases by $25.50 for each mile walked.   **e.** The y-intercept is $(0, 350)$ and means that the amount raised in donations and not based on miles walked is $350.
A.2. a. 9   **b.** The slope of 9 means that the average selling price of a house increased by $9000 per year during this time period.   **c.** $y = 9x + 150$ **d.** $330,000
A.3. a. $y = 25x + 60$ **b.** $210   **c.** $660   **d.** The slope is $25/month and represents the monthly fee.

Section 11.6 Practice Exercises, pp. 754–758

R.1. Slope: $\dfrac{3}{7}$; y-intercept: $(0, -2)$ **R.3.** $-\dfrac{1}{2}$
R.5. $y = -\dfrac{1}{3}x + 3$ **R.7.** $y = -3x - 11$

1. a. $2.73   **b.** 6.69/hr   **c.** The y-intercept is $(0, 1.6)$. This indicates that the minimum wage was $1.60 per hour in the year 1970.   **d.** The slope is 0.113. This indicates that the minimum wage has risen approximately $0.113/hr per year during this period.
3. a. $30.7°$ **b.** $13.4°$ **c.** $m = -2.333$. The average temperature in January decreases at a rate of $2.333°$ per $1°$ of latitude.   **d.** $(53.2, 0)$. At $53.2°$ latitude, the average temperature in January is $0°$.
5. a. $106.95   **b.** $201.95   **c.** $(0, 11.95)$. For 0 kilowatt-hours used, the cost consists of only the fixed monthly tax of $11.95.   **d.** $m = 0.095$. The cost increases at a rate of $0.095 for each kilowatt-hour used.

7. a. $m = -1.0$ **b.** $y = -1.0x + 1051$ **c.** The minimum pressure was approximately 921 mb.
9. a. $m = 21.5$ **b.** The slope means that the consumption of wind energy in the United States increased by 21.5 trillion Btu per year.
c. $y = 21.5x + 57$ **d.** 272 trillion Btu **11. a.** $y = 0.10x + 5000$
b. $6130 **13. a.** $y = 90x + 105$ **b.** $1185.00
15. a. $y = 0.8x + 100$ **b.** $260.00
17. 13.3 **18.** -42.3

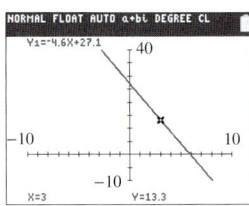 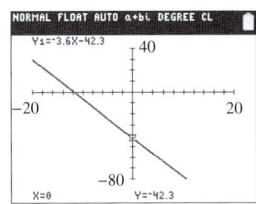

19. 345 **20.** 95

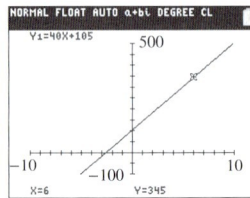 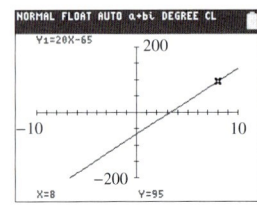

Chapter 11 Review Exercises, pp. 759–763

1.

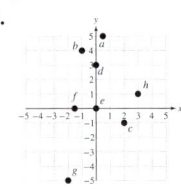

2. $A(4, -3)$; $B(-3, -2)$; $C(\frac{5}{2}, 5)$; $D(-4, 1)$; $E(-\frac{1}{2}, 0)$; $F(0, -5)$
3. III **4.** II **5.** IV **6.** I **7.** IV **8.** III **9.** x-axis
10. y-axis **11. a.** On day 1, the price was $26.25. **b.** Day 2
c. $2.25 **12. a.** (7, 67) means that a student had 7 absences and had a class average of 67. (1, 96) means that a student had 1 absence and had a class average of 96.
b.

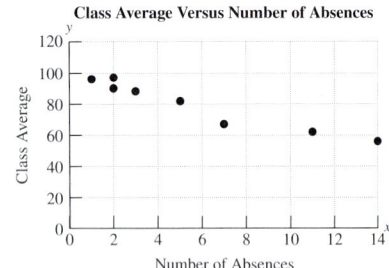

13. No **14.** No **15.** Yes **16.** Yes
17.

x	y
2	1
3	4
1	-2

18.

x	y
0	2
-2	7/3
-6	3

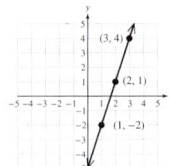 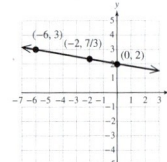

19.

x	y
0	-1
3	1
-6	-5

20.

x	y
0	-3
-3	3
1	-5

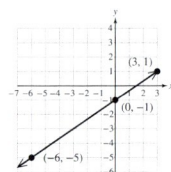 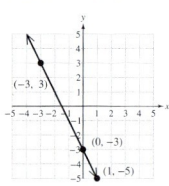

21. 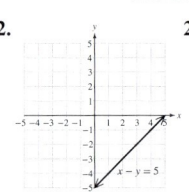 **22.** **23.**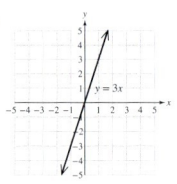

24. 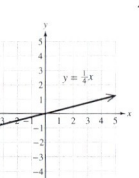 **25.** Vertical 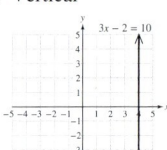 **26.** Vertical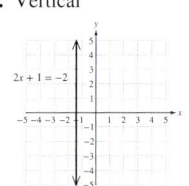

27. Horizontal **28.** Horizontal

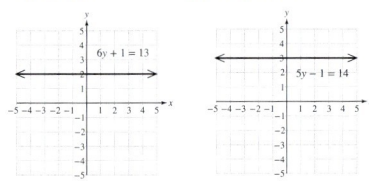

29. x-intercept: $(-3, 0)$; y-intercept: $\left(0, \frac{3}{2}\right)$
30. x-intercept: $(3, 0)$; y-intercept: $(0, 6)$ **31.** x-intercept: $(0, 0)$; y-intercept: $(0, 0)$ **32.** x-intercept: $(0, 0)$; y-intercept: $(0, 0)$
33. x-intercept: none; y-intercept: $(0, -4)$ **34.** x-intercept: none; y-intercept: $(0, 2)$ **35.** x-intercept: $\left(-\frac{5}{2}, 0\right)$; y-intercept: none
36. x-intercept: $\left(\frac{1}{3}, 0\right)$; y-intercept: none **37.** $m = \frac{12}{5}$
38. -2 **39.** $-\frac{2}{3}$ **40.** 8 **41.** Undefined **42.** 0
43. a. -5 **b.** $\frac{1}{5}$ **44. a.** 0 **b.** Undefined
45. $m_1 = \frac{2}{3}$; $m_2 = \frac{2}{3}$; parallel **46.** $m_1 = 8$; $m_2 = 8$; parallel
47. $m_1 = -\frac{5}{12}$; $m_2 = \frac{12}{5}$; perpendicular **48.** m_1 is undefined; $m_2 = 0$; perpendicular **49. a.** $m = 35$ **b.** The number of kilowatt-hours increased at a rate of 35 kilowatt-hours per day.
50. a. $m = -994$ **b.** The number of new cars sold decreased at a rate of 994 cars per year.
51. $y = \frac{5}{2}x - 5$; $m = \frac{5}{2}$; y-intercept: $(0, -5)$
52. $y = -\frac{3}{4}x + 3$; $m = -\frac{3}{4}$; y-intercept: $(0, 3)$
53. $y = \frac{1}{3}x$; $m = \frac{1}{3}$; y-intercept: $(0, 0)$ **54.** $y = \frac{12}{5}$; $m = 0$; y-intercept: $\left(0, \frac{12}{5}\right)$ **55.** $y = -\frac{5}{2}$; $m = 0$; y-intercept: $\left(0, -\frac{5}{2}\right)$
56. $y = x$; $m = 1$; y-intercept: $(0, 0)$ **57.** Neither
58. Perpendicular **59.** Parallel **60.** Parallel
61. Perpendicular **62.** Neither **63.** $y = -\frac{4}{3}x - 1$ or $4x + 3y = -3$ **64.** $y = 5x$ or $5x - y = 0$ **65.** $y = -\frac{4}{3}x - 6$ or $4x + 3y = -18$ **66.** $y = 5x - 3$ or $5x - y = 3$

67. For example: $y = 3x + 2$ 68. For example: $5x + 2y = -4$
69. $m = \dfrac{y_2 - y_1}{x_2 - x_1}$ 70. $y - y_1 = m(x - x_1)$
71. For example: $x = 6$ 72. For example: $y = -5$
73. $y = -6x + 2$ or $6x + y = 2$ 74. $y = \dfrac{2}{3}x + \dfrac{5}{3}$ or $2x - 3y = -5$
75. $y = \dfrac{1}{4}x - 4$ or $x - 4y = 16$ 76. $y = -5$
77. $y = \dfrac{6}{5}x + 6$ or $6x - 5y = -30$ 78. $y = 4x + 31$ or $4x - y = -31$ 79. **a.** 47.8 in. **b.** The slope is 2.4 and indicates that the average height for girls increases at a rate of 2.4 in. per year.
80. **a.** $m = 137$ **b.** The number of prescriptions increased by 137 million per year during this time period. **c.** $y = 137x + 2825$ **d.** 4880 million 81. **a.** $y = 20x + 55$ **b.** $235
82. **a.** $y = 8x + 700$ **b.** $1340

Chapter 12

Section 12.1 Activity, pp. 770–771

A.1. a. $y = -x + 2$ and $y = x - 4$ **b.** -1 and 1
c. Intersect at a single point.
d.

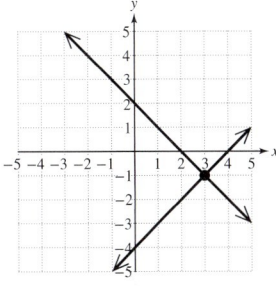

e. $(3, -1)$ **f.** The point $(3, -1)$ checks in both equations.

A.2. a. $y = 2x - 1$ and $y = 2x + 2$ **b.** 2 and 2
c. The lines are parallel.
d.

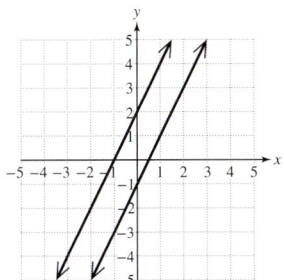

e. No solution; { } **f.** The system is inconsistent.

A.3. a. $y = \dfrac{1}{3}x + 2$ and $y = \dfrac{1}{3}x + 2$ **b.** $\dfrac{1}{3}$ and $\dfrac{1}{3}$
c. The equations represent the same line.
d.

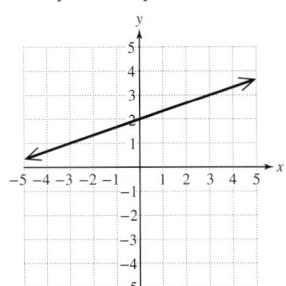

e. There are infinitely many solutions.
f. $\left\{ (x, y) \,\middle|\, y = \dfrac{1}{3}x + 2 \right\}$
g. The equations are dependent.

Section 12.1 Practice Exercises, pp. 771–777

R.1. Slope: $\dfrac{3}{5}$; x-intercept: (5, 0); y-intercept: (0, −3)
R.3. Slope: 0; x-intercept: none; y-intercept: (0, 3)
R.5.

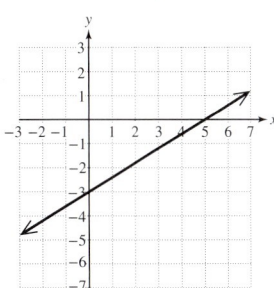

R.7.

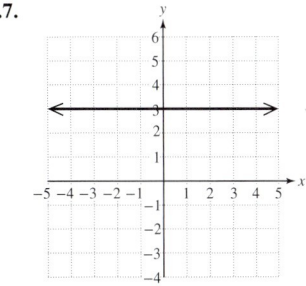

1. a. system **b.** solution **c.** intersect **d.** consistent **e.** { }
f. dependent **g.** independent **3.** Yes **5.** No **7.** Yes
9. No **11.** b **13.** d
15. a. **b.** **c.**

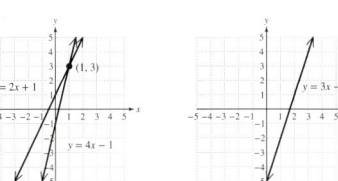

17. c **19.** a **21.** a **23.** b **25.** c
27. $\{(3, 1)\}$ **29.** $\{(1, -2)\}$ **31.** $\{(1, 4)\}$

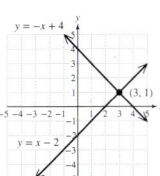 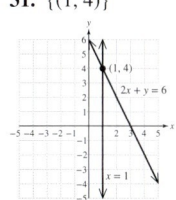

33. No solution; { }; inconsistent system **35.** Infinitely many solutions; $\{(x, y) \mid -2x + y = 3\}$; dependent equations

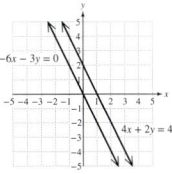 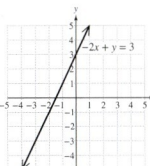

37. $\{(-3, 6)\}$ **39.** $\{(2, -2)\}$ **41.** No solution; { }; inconsistent system

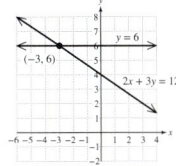 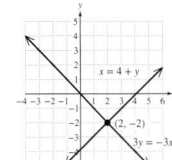 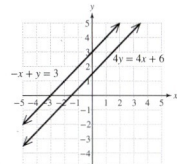

43. $\{(4, 2)\}$ **45.** $\{(1, 2)\}$

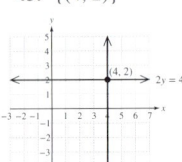
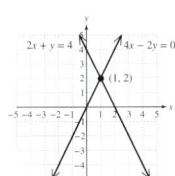

47. Infinitely many solutions; $\{(x, y) \mid y = 0.5x + 2\}$; dependent equations   **49.** $\{(-3, -1)\}$

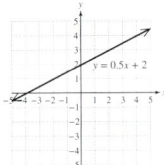
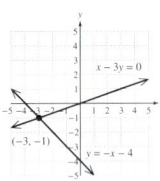

51. The same cost occurs when $2500 of merchandise is purchased.
53. The point of intersection is below the x-axis and cannot have a positive y-coordinate.
55. $\{(2, 1)\}$ **56.** $\{(6, -1)\}$

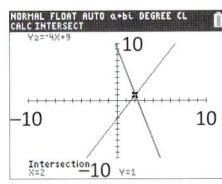
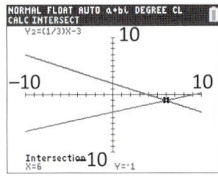

57. $\{(3, 1)\}$ **58.** $\{(-2, 0)\}$

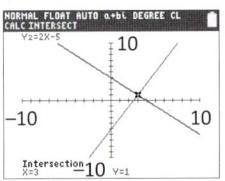
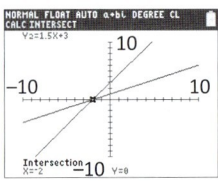

59. $\{\ \}$ **60.** $\left\{(x, y) \mid y = -\frac{1}{4}x + 2\right\}$

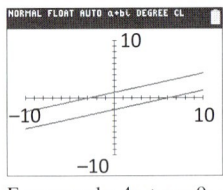
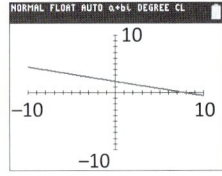

61. For example: $4x + y = 9; -2x - y = -5$
63. For example: $2x + 2y = 1$

Section 12.2 Activity, p. 785

A.1. a. $3x + 2(x + 6) = 2$ **b.** $x = -2$ **c.** $y = 4; \{(-2, 4)\}$
   d. The ordered pair $(-2, 4)$ checks in both equations.
A.2. a. The x variable in the second equation is easiest to isolate.
   b. $\{(-2, 2)\}$ **c.** The ordered pair $(-2, 2)$ checks in both equations.
A.3. a. The y variable in the first equation is easiest to isolate.
   b. $\{\ \}$ **c.** No solution   **d.** Inconsistent
A.4. a. The x variable in the second equation is easiest to isolate.
   b. $\left\{(x, y) \mid x = \frac{3}{5}y + 3\right\}$ **c.** Infinitely many
   d. Dependent
A.5. a. Let x represent one angle. Let y represent the other angle.
   b. $x + y = 90$ **c.** $\{(49, 41)\}$
   $x = y + 8$
   d. The two angles are $49°$ and $41°$.
   e. Yes; $49° + 41° = 90°$ and $49° = 41° + 8°$. ✓

Section 12.2 Practice Exercises, pp. 786–788

R.1. $y = 2x - 4$ **R.3.** $y = -\frac{2}{3}x + 2$
$y = 2x - 4$; coinciding lines $y = x - 5$; intersecting lines
R.5. $y = 4x - 4$ **R.7.** $x = 1$ **R.9.** $y = 9$
$y = 4x - 13$; parallel lines
R.11. a. Contradiction   **b.** $\{\ \}$
R.13. a. Identity   **b.** The set of real numbers
R.15. a. Conditional equation   **b.** $\{1\}$

1. $\{(3, -6)\}$ **3.** $\{(0, 4)\}$ **5. a.** y in the second equation is easiest to isolate because its coefficient is 1.   **b.** $\{(1, 5)\}$
7. $\{(5, 2)\}$ **9.** $\{(10, 5)\}$ **11.** $\left\{\left(\frac{1}{2}, 3\right)\right\}$
13. $\{(5, 3)\}$ **15.** $\{(1, 0)\}$ **17.** $\{(1, 4)\}$
19. No solution; $\{\ \}$; inconsistent system   **21.** Infinitely many solutions; $\{(x, y) \mid 2x - 6y = -2\}$; dependent equations   **23.** $\{(5, -7)\}$
25. $\left\{\left(-5, \frac{3}{2}\right)\right\}$ **27.** $\{(2, -5)\}$ **29.** $\{(-4, 6)\}$ **31.** $\{(0, 2)\}$
33. Infinitely many solutions; $\{(x, y) \mid y = 0.25x + 1\}$; dependent equations   **35.** $\{(1, 1)\}$ **37.** No solution; $\{\ \}$; inconsistent system
39. $\{(-1, 5)\}$ **41.** $\{(-6, -4)\}$ **43.** The numbers are 48 and 58.   **45.** The numbers are 13 and 39.   **47.** The angles are 165° and 15°.   **49.** The angles are 70° and 20°.   **51.** The angles are 42° and 48°.   **53.** For example: $(0, 3), (1, 5), (-1, 1)$

Section 12.3 Activity, p. 795

A.1. a. $7x + 0y = 7$ **b.** $-2x + 8y = 10$
   c. In part (a), the variable y was eliminated because the coefficients on the terms $-5y$ and $5y$ are opposites.
A.2. a. $\{(1, 2)\}$ **b.** The ordered pair $(1, 2)$ checks in both equations.
A.3. a. The x variable because multiplying the first equation by 2 will make the x term coefficients opposites. To eliminate the y terms, both equations would have to be multiplied by a constant to create opposite coefficients.   **b.** $\{(-1, 1)\}$
   c. The ordered pair $(-1, 1)$ checks in both equations.
A.4. a. $\{(3, 2)\}$ **b.** The ordered pair $(3, 2)$ checks in both equations.
A.5. a. 2   **b.** $\{(x, y) \mid -5x + 4y = -4\}$ or $\left\{(x, y) \mid y = \frac{5}{4}x - 1\right\}$
   c. Infinitely many

Section 12.3 Practice Exercises, pp. 795–799

R.1. No   **R.3.** $\{\ \}$ **R.5.** $\{-1\}$ **R.7.** The set of real numbers

1. Multiply the first equation by 2.
3. a. 6   **b.** Multiply the first equation by 3 and the second by 2.
5. a. 12   **b.** Multiply the first equation by -3 and the second equation by 4. (Or multiply the first equation by 3 and the second by -4.)
7. a. True   **b.** False, multiply the second equation by 5.
9. a. x would be easier.   **b.** $\{(0, -3)\}$
11. $\{(4, -1)\}$ **13.** $\{(4, 3)\}$ **15.** $\{(2, 3)\}$ **17.** $\{(1, -4)\}$
19. $\{(1, -1)\}$ **21.** $\{(-4, -6)\}$ **23.** $\left\{\left(\frac{7}{9}, \frac{5}{9}\right)\right\}$
25. The system will have no solution. The lines are parallel.
27. There are infinitely many solutions. The lines coincide.
29. The system will have one solution. The lines intersect at a point whose x-coordinate is 0.   **31.** No solution; $\{\ \}$; inconsistent system
33. Infinitely many solutions; $\{(x, y) \mid x + 2y = 2\}$; dependent equations   **35.** $\{(1, 4)\}$ **37.** $\{(-1, -2)\}$ **39.** $\{(2, 1)\}$
41. No solution; $\{\ \}$; inconsistent system   **43.** $\{(2, 3)\}$ **45.** $\{(3.5, 2.5)\}$
47. $\left\{\left(\frac{1}{3}, 2\right)\right\}$ **49.** $\{(-2, 5)\}$ **51.** $\left\{\left(-\frac{1}{2}, 1\right)\right\}$ **53.** $\{(0, 3)\}$
55. $\{\ \}$ **57.** $\{(1, 4)\}$ **59.** $\{(4, 0)\}$ **61.** Infinitely many solutions; $\{(a, b) \mid 9a - 2b = 8\}$ **63.** $\left\{\left(\frac{7}{16}, -\frac{7}{8}\right)\right\}$
65. The numbers are 17 and 19.   **67.** The numbers are -1 and 3.

69. The angles are 46° and 134°. **71.** $\{(1, 3)\}$ **73.** One line within the system of equations would have to "bend" for the system to have exactly two points of intersection. This is not possible. **75.** $A = -5, B = 2$

Chapter 12 Problem Recognition Exercises, pp. 799–801

1. Infinitely many solutions. The equations represent the same line.
2. No solution. The equations represent parallel lines.
3. One solution. The equations represent intersecting lines.
4. One solution. The equations represent intersecting lines.
5. No solution. The equations represent parallel lines.
6. Infinitely many solutions. The equations represent the same line.
7. The y variable will easily be eliminated by adding the equations. The solution set is $\{(-3, 1)\}$. **8.** The x variable will easily be eliminated by multiplying the second equation by 2 and then adding the equations. The solution set is $\{(3, 4)\}$. **9.** Because the variable x is already isolated in the first equation, the expression $-3y + 4$ can be easily substituted for x in the second equation. The solution set is $\{(-2, 2)\}$. **10.** Because the variable y is already isolated in the second equation, the expression $x - 8$ can be easily substituted for y in the first equation. The solution set is $\{(8, 0)\}$. **11.** $\{(5, 0)\}$
12. $\{(1, -7)\}$ **13.** $\{(4, -5)\}$ **14.** $\{(2, 3)\}$ **15.** $\{(2, 0)\}$
16. $\{(8, 10)\}$ **17.** $\left\{\left(2, -\dfrac{5}{7}\right)\right\}$ **18.** $\left\{\left(-\dfrac{14}{3}, -4\right)\right\}$
19. No solution; { }; inconsistent system **20.** No solution; { }; inconsistent system **21.** $\{(-1, 0)\}$ **22.** $\{(5, 0)\}$
23. Infinitely many solutions; $\{(x, y)\,|\,y = 2x - 14\}$; dependent equations **24.** Infinitely many solutions; $\{(x, y)\,|\,x = 5y - 9\}$; dependent equations **25.** $\{(2200, 1000)\}$
26. $\{(3300, 1200)\}$ **27.** $\{(5, -7)\}$ **28.** $\{(2, -1)\}$
29. $\left\{\left(\dfrac{2}{3}, \dfrac{1}{2}\right)\right\}$ **30.** $\left\{\left(\dfrac{1}{4}, -\dfrac{3}{2}\right)\right\}$

Section 12.4 Activity, pp. 807–809

A.1. a. $x + y = 248$ **b.** $\{(137, 111)\}$
   $x + 2y = 359$
   c. There were 137 one-dollar tickets sold and 111 two-dollar tickets sold. To check: $137 + 111 = 248$ and $137(\$1) + 111(\$2) = \$359$.
A.2. a. $I = Prt$ **b.** $\$210$ **c.** $0.03x$ **d.** $0.04y$
A.3. a. Let y represent the amount invested in the 4% account.
   b.

	3% Account	4% Account
Amount invested (\$)	x	y
Interest earned (\$)	$0.03x$	$0.04y$

   c. $y = x + 3000$ **d.** $\{(7000, 10{,}000)\}$
   $0.03x + 0.04y = 610$
   e. \$7000 was invested at 3% and \$10,000 was invested at 4%.
A.4. a. 1.92 L **b.** $0.12x$ **c.** $0.27y$
A.5. a. Let y represent the amount of the 27% solution.
   b.

	12% Mixture	27% Mixture	15% Mixture
Amount of solution (L)	x	y	20
Amount of pure acid (L)	$0.12x$	$0.27y$	$0.15(20)$

   c. $x + y = 20$ **d.** $\{(16, 4)\}$
   $0.12x + 0.27y = 0.15(20)$
   e. 16 L of the 12% mixture and 4 L of the 27% mixture should be used.
A.6. a. 12 mph; 24 mi **b.** 8 mph; 16 mi
   c. $(b + c)$; $(b + c) \cdot 2$ **d.** $(b - c)$; $(b - c) \cdot 3$

A.7. a. Let c represent the speed of the current.
   b.

	Distance (mi)	Rate (mph)	Time (hr)
With the current	48	$b + c$	2
Against the current	48	$b - c$	3

   c. $48 = (b + c) \cdot 2$ **d.** $\{(20, 4)\}$
   $48 = (b - c) \cdot 3$
   e. The speed of the boat is 20 mph in still water, and the speed of the current is 4 mph.

Section 12.4 Practice Exercises, pp. 809–813

R.1. $\{(-1, 4)\}$ **R.3.** $\left\{\left(\dfrac{5}{2}, 1\right)\right\}$ **R.5.** The numbers are 18 and 23.
R.7. The measures of the angles are 40° and 50°.
R.9. The numbers are 4 and 16. **R.11.** The angles are 80° and 10°.

1. A video game costs \$21.50 and a DVD costs \$15.
3. Technology stock costs \$16 per share, and the mutual fund costs \$11 per share. **5.** Armondo made 115 beef tacos and 85 pork tacos.
7. Shanelle invested \$3500 in the 10% account and \$6500 in the 7% account. **9.** \$9000 was borrowed at 6%, and \$3000 was borrowed at 9%.
11. Invest \$12,000 in the bond fund and \$18,000 in the stock fund.
13. 15 gal of the 50% mixture should be mixed with 10 gal of the 40% mixture. **15.** 12 gal of the 45% disinfectant solution should be mixed with 8 gal of the 30% disinfectant solution.
17. She should mix 20 mL of the 13% solution with 30 mL of the 18% solution. **19.** Chad needs 1 L of pure antifreeze and 5 L of the 40% solution. **21.** The speed of the boat in still water is 6 mph, and the speed of the current is 2 mph. **23.** The speed of the plane in still air is 300 mph, and the wind is 20 mph. **25.** The speed of the plane in still air is 525 mph and the speed of the wind is 75 mph. **27.** There are 17 dimes and 22 nickels. **29.** 2 qt of water should be mixed. **31. a.** 835 free throws and 1597 field goals
b. 4029 points **c.** Approximately 50 points per game
33. The speed of the plane in still air is 160 mph, and the wind is 40 mph. **35.** \$15,000 was invested in the 5.5% account, and \$45,000 was invested in the 6.5% account. **37.** 12 lb of candy should be mixed with 8 lb of nuts. **39.** Dallas scored 30 points, and Buffalo scored 13 points. **41.** There were 300 women and 200 men in the survey.

Section 12.5 Activity, pp. 819–820

A.1. a. Plane **b.** Pair; triple
A.2. a. No **b.** Yes
A.3. a. $-8x + 5y = 4$ **b.** $14x - 7y = 0$ **c.** $x = 2, y = 4$ **d.** $z = -3$
   e. $\{(2, 4, -3)\}$; The ordered triple checks in each original equation.
A.4. a. $3x + 7z = -29$ **b.** $x = 2, z = -5$ **c.** $y = -1$
   d. $\{(2, -1, -5)\}$; The ordered triple checks in each original equation.
A.5. a. $-4x - 6z = -6$ **b.** Identity; infinitely many solutions
A.6. a. $-15y + 12z = 6$ **b.** $-5y + 4z = 4$ **c.** Contradiction; no solution
   d. { }

Section 12.5 Practice Exercises, pp. 821–823

R.1. $\{(1, 1)\}$ **R.3.** { } **R.5.** $\{(x, y)\,|\,4x - 6y = 5\}$

1. **a.** linear **b.** ordered triples **3.** $(4, 0, 2)$ is a solution.
5. $(1, 1, 1)$ is a solution. **7.** $\{(1, -2, 4)\}$
9. $\{(-1, 2, 0)\}$ **11.** $\{(-2, -1, -3)\}$ **13.** $\{(1, -4, -2)\}$
15. $\{(1, 2, 3)\}$ **17.** $\{(-6, 1, 7)\}$ **19.** $\left\{\left(\dfrac{1}{2}, \dfrac{2}{3}, -\dfrac{5}{6}\right)\right\}$
21. $\left\{\left(\dfrac{1}{2}, 2, -\dfrac{1}{2}\right)\right\}$ **23.** $\{(6, -3, -2)\}$ **25.** $\{(0, 4, -4)\}$
27. $\{(1, 0, -1)\}$ **29.** $\{(-20, 8, 0)\}$
31. Dependent equations **33.** Inconsistent system
35. $\{(1, 3, 1)\}$ **37.** $\left\{\left(-\dfrac{1}{2}, \dfrac{1}{3}, \dfrac{1}{6}\right)\right\}$ **39.** Dependent equations
41. $\{(1, 0, 1)\}$ **43.** $\{(0, 0, 0)\}$ **45.** Dependent equations

Section 12.6 Activity, p. 825

A.1. a. One possibility: Let y represent Randall's second test score. Let z represent Randall's third test score.
b. $\dfrac{x+y+z}{3} = 92$ **c.** $x = y + 6$ **d.** $z = x - 3$
e. $\{(95, 89, 92)\}$ **f.** Randall's first test score is 95, his second is 89, and his third is 92.

A.2. a. One possibility: Let y represent the amount invested in the second fund. Let z represent the amount invested in the third fund.
b. $x + y + z = 14{,}500$ **c.** $z = x + y + 500$
d. $-0.02x + 0.03y + 0.05z = 485$ **e.** $\{(2000, 5000, 7500)\}$
f. The amount invested in the first fund is $2000, the amount invested in the second fund is $5000, and the amount invested in the third fund is $7500.

Section 12.6 Practice Exercises, pp. 826–827

R.1. $\{(-7, 0, 8)\}$ **R.3. a.** $6.45 **b.** $1.29y
R.5. a. $250 **b.** $0.025x$

1. 67°, 82°, 31° **3.** 25°, 56°, 99° **5.** 10 cm, 18 cm, 27 cm
7. 11 in., 17 in., 26 in. **9.** The fiber supplement has 3 g; the oatmeal has 4 g; and the cereal has 1.5 g. **11.** Kyoki invested $2000 in bonds, $6000 in mutual funds, and $2000 in the Treasury note.
13. $1500 at 5%, $2000 at 5.5%, $1000 at 4% **15.** He made six free throws, seven 2-point shots, and three 3-point shots. **17.** 24 oz peanuts, 8 oz pecans, 16 oz cashews **19.** 1500 seats in section A, 400 seats in section B, and 500 seats in section C were sold.

Chapter 12 Review Exercises, pp. 828–830

1. Yes **2.** No **3.** No **4.** Yes
5. Intersecting lines (The lines have different slopes.)
6. Intersecting lines (The lines have different slopes.)
7. Parallel lines (The lines have the same slope but different y-intercepts.)
8. Intersecting lines (The lines have different slopes.)
9. Coinciding lines (The lines have the same slope and same y-intercept.)
10. Intersecting lines (The lines have different slopes.)
11. $\{(0, -2)\}$ **12.** $\{(-2, 3)\}$

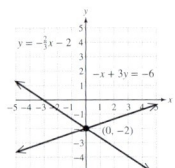

13. Infinitely many solutions; $\{(x, y) \mid 2x + y = 5\}$; dependent equations
14. Infinitely many solutions; $\{(x, y) \mid -x + 5y = -5\}$; dependent equations

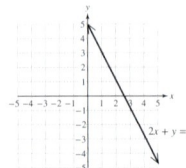

15. $\{(1, -1)\}$ **16.** $\{(-1, 2)\}$

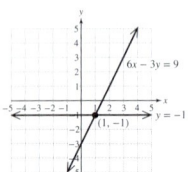 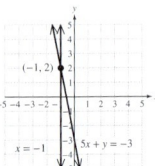

17. No solution; $\{\ \}$; inconsistent system
18. No solution; $\{\ \}$; inconsistent system

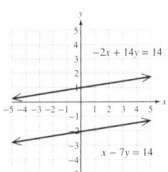 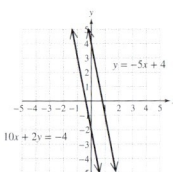

19. The two salespeople will make the same yearly compensation for $300,000 in sales.   **20.** $\left\{\left(\dfrac{2}{3}, -2\right)\right\}$ **21.** $\{(-4, 1)\}$
22. $\{\ \}$ **23.** Infinitely many solutions; $\{(x, y) \mid y = -2x + 2\}$
24. a. x in the first equation is easiest to isolate because its coefficient is 1. **b.** $\left\{\left(6, \dfrac{5}{2}\right)\right\}$ **25. a.** y in the second equation is easiest to isolate because its coefficient is 1. **b.** $\left\{\left(\dfrac{9}{2}, 3\right)\right\}$
26. $\{(5, -4)\}$ **27.** $\{(0, 4)\}$ **28.** Infinitely many solutions; $\{(x, y) \mid x - 3y = 9\}$ **29.** $\{\ \}$ **30.** The numbers are 50 and 8.
31. The angles are 41° and 49°.   **32.** The angles are $115\tfrac{1}{3}°$ and $64\tfrac{2}{3}°$.
33. b. $\{(-1, 4)\}$ **34. b.** $\{(-3, -2)\}$ **35. b.** $\{(2, 2)\}$
36. $\{(2, -1)\}$ **37.** $\{(-6, 2)\}$ **38.** $\left\{\left(-\dfrac{1}{2}, \dfrac{1}{3}\right)\right\}$
39. $\left\{\left(\dfrac{1}{4}, -\dfrac{2}{5}\right)\right\}$ **40.** Infinitely many solutions; $\{(x, y) \mid -4x - 6y = -2\}$ **41.** $\{\ \}$ **42.** $\{(-4, -2)\}$
43. $\{(1, 0)\}$ **44. a.** Use the substitution method because y is already isolated in the second equation. **b.** $\{(5, -3)\}$ **45. a.** Use the addition method. The substitution method would be cumbersome because isolating either variable from either equation would result in fractional coefficients. **b.** $\{(-2, -1)\}$
46. There were 8 adult tickets and 52 children's tickets sold.
47. He should invest $75,000 at 12% and $525,000 at 4%.
48. 20 gal of whole milk should be mixed with 40 gal of low-fat milk.
49. The speed of the boat is 18 mph, and that of the current is 2 mph.
50. The plane's speed in still air is 320 mph. The wind speed is 30 mph.
51. A hot dog costs $4.50 and a drink costs $3.50.
52. The score was 72 on the first round and 82 on the second round.
53. $\{(2, -1, 5)\}$ **54.** Inconsistent system
55. Dependent equations   **56.** $\{(-1, -2, 2)\}$
57. $\{(1, -3, -2)\}$ **58.** $\{(-1, 4, 0)\}$ **59.** 5 ft, 12 ft, and 13 ft
60. The pumps can drain 250, 300, and 400 gal/hr.
61. She invested $6000 in Fund A, $4000 in Fund B, and $2000 in Fund C.
62. The angles are 13°, 39°, and 128°.

Chapter 13

Section 13.1 Activity, pp. 838–839

A.1. a. $x \cdot x \cdot x$ **b.** $x \cdot x \cdot x \cdot x \cdot x$ **c.** $(x \cdot x \cdot x)(x \cdot x \cdot x \cdot x \cdot x)$
d. $x^3 \cdot x^5 = x^{\boxed{8}}$ **e.** $x^m \cdot x^n = x^{m+n}$
A.2. a. x^{19} **b.** 2^{19} **b.** y^{12}
A.3. a. $t \cdot t \cdot t \cdot t \cdot t \cdot t$ **b.** $t \cdot t$ **c.** $\dfrac{t \cdot t \cdot t \cdot t \cdot t \cdot t}{t \cdot t}$
d. $\dfrac{t^6}{t^2} = t^{\boxed{4}}$ **e.** $\dfrac{x^m}{x^n} = x^{m-n}$
A.4. a. p^8 **b.** 4^8 **c.** a^4
A.5. a. $\left(-10 \cdot \dfrac{1}{2}\right)(c^4 \cdot c^2)(d^7 \cdot d)$ **b.** $-5c^6 d^8$
A.6. a. $\left(\dfrac{-5}{-10}\right)\left(\dfrac{c^{10}}{c^4}\right)\left(\dfrac{d^8}{d^7}\right)$ **b.** $\dfrac{1}{2} c^6 d$

Section 13.1 Practice Exercises, pp. 839–842

R.1. 36 **R.3.** 36 **R.5.** 64 **R.7.** -64

1. a. exponent **b.** base; exponent **c.** 1 **d.** $I = Prt$
e. compound interest   **3.** Base: x; exponent: 4   **5.** Base: 3; exponent: 5   **7.** Base: -1; exponent: 4   **9.** Base: 13; exponent: 1
11. Base: 10; exponent: 3   **13.** Base: t; exponent: 6

15. v 17. 1   19. $(-6b)^2$ 21. $-6b^2$ 23. $(y+2)^4$
25. $\dfrac{-2}{t^3}$ 27. No; $-5^2 = -25$ and $(-5)^2 = 25$ 29. Yes; -32
31. Yes; $\dfrac{1}{8}$ 33. Yes; -16 35. 16   37. -1 39. $\dfrac{1}{9}$
41. $-\dfrac{4}{25}$ 43. 48   45. 4   47. 9   49. 50   51. -100
53. 400   55. 1   57. 1   59. 1000   61. -800
63. a. $(x \cdot x \cdot x \cdot x)(x \cdot x \cdot x) = x^7$ b. $(5 \cdot 5 \cdot 5 \cdot 5)(5 \cdot 5 \cdot 5) = 5^7$
65. z^8 67. a^9 69. 4^{14} 71. $\left(\dfrac{2}{3}\right)^4$ 73. c^{14} 75. x^{18}
77. a. $\dfrac{p \cdot p \cdot p \cdot p \cdot p \cdot p \cdot p \cdot p}{p \cdot p \cdot p} = p^5$ b. $\dfrac{8 \cdot 8 \cdot 8 \cdot 8 \cdot 8 \cdot 8 \cdot 8 \cdot 8}{8 \cdot 8 \cdot 8} = 8^5$
79. x^2 81. a^9 83. 7^7 85. 5^7 87. y 89. h^4
91. 7^7 93. 10^9 95. $6x^7$ 97. $40a^5b^5$ 99. s^9t^{16}
101. $-30v^8$ 103. $16m^{20}n^{10}$ 105. $2cd^4$ 107. z^4
109. $\dfrac{25hjk^4}{12}$ 111. $-8p^7q^9r^6$ 113. $-3stu$ 115. \$5724.50
117. \$4764.06   119. 201 in.²   121. 268 in.³
123. 1.338225578   124. 2.208039664   125. 6691.127888
126. 4416.079327   127. 3370.8   128. 1157.625
129. x^{2n+1} 131. p^{2m+3} 133. z 135. r^3

Section 13.2 Activity, p. 846

A.1. a.–e. b, d, e   A.2. a. p^{10} b. 5^{10}
A.3. a. q^{21} b. q^{21} c. In each case, we keep the base the same and multiply the exponents. The commutative property of multiplication guarantees that $3 \cdot 7 = 7 \cdot 3$.
A.4. a. m^{26} b. m^{26} A.5. a. k^{15} b. k^{15}
A.6. $125u^{12}v^6w^3$ A.7. $\dfrac{16m^{14}n^4}{p^2}$
A.8. a. $(3ab^4)(3a^2b)^2$ b. $(3ab^4)(9a^4b^2)$ c. $27a^5b^6$

Section 13.2 Practice Exercises, pp. 847–848

R.1. 4^9 R.3. a^{20} R.5. d^9 R.7. 7^6 R.9. a. $\dfrac{4}{9}$ b. $\dfrac{4}{9}$
1. 5^{12} 3. 12^6 5. y^{14} 7. w^{25} 9. a^{36} 11. y^{14}
13. They are both equal to 2^6.   15. $4^{3^2} = 4^9$; $(4^3)^2 = 4^6$; the expression 4^{3^2} is greater than $(4^3)^2$.   17. $25w^2$ 19. $s^4r^4t^4$
21. $\dfrac{16}{r^4}$ 23. $\dfrac{x^5}{y^5}$ 25. $81a^4$ 27. $-27a^3b^3c^3$ 29. $-\dfrac{64}{x^3}$
31. $\dfrac{a^2}{b^2}$ 33. $6^3u^6v^{12}$ or $216u^6v^{12}$ 35. $5x^8y^4$ 37. $-h^{28}$
39. m^{12} 41. $\dfrac{4^5}{r^5s^{20}}$ or $\dfrac{1024}{r^5s^{20}}$ 43. $\dfrac{3^5p^5}{q^{15}}$ or $\dfrac{243p^5}{q^{15}}$ 45. y^{14}
47. x^{31} 49. $200a^{14}b^9$ 51. $16p^8q^{16}$ 53. $-m^{35}n^{15}$
55. $25a^{18}b^6$ 57. $\dfrac{4c^2d^6}{9}$ 59. $\dfrac{c^{27}d^{31}}{2}$ 61. $-\dfrac{27a^9b^3}{c^6}$
63. $16b^{26}$ 65. x^{2m} 67. $125a^{6n}$ 69. $\dfrac{m^{2b}}{n^{3b}}$ 71. $\dfrac{3^na^{3n}}{5^nb^{4n}}$

Section 13.3 Activity, p. 856

A.1. a. 32   b. 16   c. 8   d. 4   e. 2
A.2. 1   A.3. a. $\dfrac{b \cdot b \cdot b \cdot b}{b \cdot b \cdot b \cdot b} = 1$ b. $b^0 = 1$
A.4. a. $\dfrac{b \cdot b}{b \cdot b \cdot b \cdot b \cdot b} = \dfrac{1}{b^3}$ b. $b^{-3} = \dfrac{1}{b^3}$ c. $b^{-n} = \left(\dfrac{1}{b}\right)^n = \dfrac{1}{b^n}$

A.5. a. 0.0625   b. 0.0625   c. They are the same.
d. Take the reciprocal of the base and take the opposite of the exponent.
A.6. $\dfrac{1}{x^4}$ A.7. x^4 A.8. a^6b^7 A.9. $\dfrac{1}{a^6b^7}$
A.10. a. Given $(-6)^0$, the base is -6 and the entire value is raised to the 0 power. The result is 1. The expression -6^0 is interpreted as the opposite of 6^0, or equivalently $-(6^0)$, which is -1.
b. The expression $(-3)^{-2}$ is interpreted as $\dfrac{1}{(-3)^2}$, which simplifies as $\dfrac{1}{9}$. The expression -3^{-2} simplifies as $-1 \cdot 3^{-2}$ or equivalently $-1 \cdot \dfrac{1}{3^2}$, which simplifies as $-\dfrac{1}{9}$.
c. The expression x^{-4} simplifies as $\dfrac{1}{x^4}$, whereas the expression $-x^{-4}$ is interpreted as $-1 \cdot x^{-4}$ or equivalently $-\dfrac{1}{x^4}$.
d. Given $\left(\dfrac{3}{4}\right)^0$, the base is $\dfrac{3}{4}$ and the entire fraction is raised to the zero power. Thus, $\left(\dfrac{3}{4}\right)^0 = 1$. Given $\dfrac{3^0}{4}$, the exponent zero is only applied to the numerator 3. Thus, $\dfrac{3^0}{4} = \dfrac{1}{4}$.
A.11. a. $\left(\dfrac{c^2d^6}{4}\right)^{-2}(cd^{-3})$ b. $\left(\dfrac{16}{c^4d^{12}}\right)(cd^{-3})$ c. $\dfrac{16cd^{-3}}{c^4d^{12}}$ d. $\dfrac{16}{c^3d^{15}}$

Section 13.3 Practice Exercises, pp. 857–859

R.1. a. 1   b. 1   c. 1   d. 1   R.3. 5   R.5. $\dfrac{1}{10}$ R.7. b^{11}
R.9. x^4 R.11. 9^{11} R.13. $6^5a^5b^{15}c^{10}$ or $7776a^5b^{15}c^{10}$
1. a. 1   b. $\left(\dfrac{1}{b}\right)^n$ or $\dfrac{1}{b^n}$ 3. a. 1   b. 1   5. 1   7. 1
9. -1 11. 1   13. 1   15. -7 17. a. $\dfrac{1}{t^5}$ b. $\dfrac{1}{t^5}$
19. $\dfrac{343}{8}$ 21. 25   23. $\dfrac{1}{a^3}$ 25. $\dfrac{1}{12}$ 27. $\dfrac{1}{16b^2}$
29. $\dfrac{6}{x^2}$ 31. $\dfrac{1}{64}$ 33. $-\dfrac{3}{y^4}$ 35. $-\dfrac{1}{t^3}$ 37. a^5
39. $\dfrac{x^4}{x^{-6}} = x^{4-(-6)} = x^{10}$ 41. $2a^{-3} = 2 \cdot \dfrac{1}{a^3} = \dfrac{2}{a^3}$ 43. $\dfrac{1}{x^4}$
45. 1   47. y^4 49. $\dfrac{n^{27}}{m^{18}}$ 51. $\dfrac{81k^{24}}{j^{20}}$ 53. $\dfrac{1}{p^6}$ 55. $\dfrac{1}{r^3}$
57. a^8 59. $\dfrac{1}{y^8}$ 61. $\dfrac{1}{7^7}$ 63. 1   65. $\dfrac{1}{a^4b^6}$ 67. $\dfrac{1}{w^{21}}$
69. $\dfrac{1}{27}$ 71. 1   73. $\dfrac{1}{64x^6}$ 75. $-\dfrac{16y^4}{z^2}$ 77. $-\dfrac{a^{12}}{6}$
79. $80c^{21}d^{24}$ 81. $\dfrac{9y^2}{8x^5}$ 83. $\dfrac{4d^{16}}{c^2}$ 85. $\dfrac{9y^2}{2x}$ 87. $\dfrac{9}{20}$
89. $\dfrac{9}{10}$ 91. $\dfrac{5}{4}$ 93. $\dfrac{10}{3}$ 95. 2   97. -11

Chapter 13 Problem Recognition Exercises, p. 859

1. t^8 2. 2^8 or 256   3. y^5 4. p^6 5. r^4s^8 6. $a^3b^9c^6$
7. w^6 8. $\dfrac{1}{m^{16}}$ 9. $\dfrac{x^4z^3}{y^7}$ 10. $\dfrac{a^3c^8}{b^6}$ 11. x^6 12. $\dfrac{1}{y^{10}}$
13. $\dfrac{1}{t^5}$ 14. w^7 15. p^{15} 16. p^{15} 17. $\dfrac{1}{v^2}$ 18. $c^{50}d^{40}$
19. 3   20. -4 21. $\dfrac{b^9}{2^{15}}$ 22. $\dfrac{81}{y^6}$ 23. $\dfrac{16y^4}{81x^4}$ 24. $\dfrac{25d^6}{36c^2}$
25. $3a^7b^5$ 26. $64x^7y^{11}$ 27. $\dfrac{y^4}{x^8}$ 28. $\dfrac{1}{a^{10}b^{10}}$ 29. $\dfrac{1}{t^2}$ 30. $\dfrac{1}{p^7}$
31. $\dfrac{8w^6x^9}{27}$ 32. $\dfrac{25b^8}{16c^6}$ 33. $\dfrac{q^3s}{r^2t^5}$ 34. $\dfrac{m^2p^3q}{n^3}$ 35. $\dfrac{1}{y^{13}}$
36. w^{10} 37. $-\dfrac{1}{8a^{18}b^6}$ 38. $\dfrac{4x^{18}}{9y^{10}}$ 39. $\dfrac{k^8}{5h^6}$ 40. $\dfrac{6n^{10}}{m^{12}}$

Section 13.4 Activity, p. 863

A.1.

Standard Form	Expanded Form	Scientific Notation
5000	5×1000	5×10^3
40,000,000	$4 \times 10,000,000$	4×10^7
800,000	$8 \times 100,000$	8×10^5
20	2×10	2×10^1

A.2.

Standard Form	Expanded Form	Scientific Notation
0.05	$5 \times \dfrac{1}{100}$ or $5 \times \dfrac{1}{10^2}$	5×10^{-2}
0.00002	$2 \times \dfrac{1}{100,000}$ or $2 \times \dfrac{1}{10^5}$	2×10^{-5}
0.7	$7 \times \dfrac{1}{10}$ or $7 \times \dfrac{1}{10^1}$	7×10^{-1}
0.00000006	$6 \times \dfrac{1}{100,000,000}$ or $6 \times \dfrac{1}{10^8}$	6×10^{-8}

A.3. Use a positive power of 10 for a number 10 or greater. Use a negative power of 10 for a number between 0 and 1.

A.4. a. Between the 5 and the 2. **b.** Place the decimal point to the right of the first nonzero digit. **c.** 5.249×10^8

A.5. a. 4.58×10^{-5} **b.** 8.57×10^4 **c.** 0.00749
d. 26,830,000,000,000

A.6. a. 4.22×10^8 mi **b.** 7.8×10^4 mph

Section 13.4 Practice Exercises, pp. 864–866

R.1. a. 1000 **b.** 10^5 **R.3. a.** $\dfrac{1}{1000}$ **b.** 10^{-5} **c.** 10^{-2}
R.5. $15t^6$ **R.7.** $\dfrac{2}{n^3}$ **R.9.** $\dfrac{1}{10^6}$ **R.11.** $c^{12}d^6$

1. scientific notation **3.** Move the decimal point between the 2 and 3 and multiply by 10^{-10}; 2.3×10^{-10}. **5.** 5×10^4
7. 2.08×10^5 **9.** 6.01×10^6 **11.** 8×10^{-6}
13. 1.25×10^{-4} **15.** 6.708×10^{-3} **17.** 1.7×10^{-24} g
19. $\$2.7 \times 10^{10}$ **21.** 6.8×10^7 gal; 1×10^2 mi
23. Move the decimal point nine places to the left; 0.000 000 0031.
25. 0.00005 **27.** 2800 **29.** 0.000603 **31.** 2,400,000
33. 0.019 **35.** 7032 **37.** 0.000 000 000 001 g
39. 1600 Cal and 2800 Cal **41.** 5×10^4 **43.** 3.6×10^{11}
45. 2.2×10^4 **47.** 2.25×10^{-13} **49.** 3.2×10^{14}
51. 2.432×10^{-10} **53.** 3×10^{13} **55.** 6×10^5 **57.** 1.38×10^1
59. 5×10^{-14} **61.** 3.75 in. **63.** $\$1.475 \times 10^7$
65. a. 6.5×10^7 **b.** 2.3725×10^{10} days **c.** 5.694×10^{11} hr
d. 2.04984×10^{15} sec
67. 23,920 **68.** 1.71696×10^{-11} **69.** 2×10^{-14}
70. 212,500 **71.** 19,200,000 **72.** 5×10^{-11}

Section 13.5 Activity, p. 872

A.1. a. $-5p^4$ **b.** $7y^2 - y - 2$

A.2. a. For example: 5 **b.** For example: $\dfrac{3}{4}x^3 + 2x$

A.3. a. $4x^2 - x + 5$ **b.** Add *like* terms by applying the distributive property.

A.4. a. $4w^3 - 2w^2 + 5$ **b.** Multiply each term by -1, which changes the sign of each term.

A.5. a. $5y^2 + y + 1$ **b.** $2a^2b + 3a^2 + 5b$ **c.** The difference of a and b is translated as $a - b$, whereas a subtracted from b is translated as $b - a$.

Section 13.5 Practice Exercises, pp. 873–876

R.1. $-4x$ **R.3.** $-\dfrac{1}{4}m$ **R.5.** $15x + 21y - 9$
R.7. $-x + 8$ **R.9. a.** 18 **b.** -18

1. a. polynomial **b.** coefficient; degree **c.** 1 **d.** one
e. binomial **f.** trinomial **g.** leading; leading coefficient
h. greatest **i.** zero **3. a.** $-7x^5 + 7x^2 + 9x + 6$ **b.** -7 **c.** 5
5. Binomial; 75 **7.** Monomial; 54 **9.** Binomial; -12
11. Trinomial; 27 **13.** Monomial; -192
15. The exponents on the *x*-factors are different. **17.** $35x^2y$
19. $8b^5d^2 - 9d$ **21.** $4y^2 + y - 9$ **23.** $-x^3 + 5x^2 + 9x - 10$
25. $10.9y$ **27.** $4a - 8c$ **29.** $a - \dfrac{1}{2}b - 2$
31. $\dfrac{4}{3}z^2 - \dfrac{5}{3}$ **33.** $7.9t^3 - 3.4t^2 + 6t - 4.2$
35. $-4h + 5$ **37.** $2.3m^2 - 3.1m + 1.5$
39. $-3v^3 - 5v^2 - 10v - 22$ **41.** $-8a^3b^2$ **43.** $-53x^3$
45. $-5a - 3$ **47.** $16k + 9$ **49.** $2m^4 - 14m$
51. $3s^2 - 4st - 3t^2$ **53.** $-2r - 3s + 3t$ **55.** $\dfrac{3}{4}x + \dfrac{1}{3}y - \dfrac{3}{10}$
57. $-\dfrac{2}{3}h^2 + \dfrac{3}{5}h - \dfrac{5}{2}$ **59.** $2.4x^4 - 3.1x^2 - 4.4x - 7.9$
61. $4b^3 + 12b^2 - 5b - 12$ **63.** $-\dfrac{7}{2}x^2 + 5x - 11$
65. $4y^3 + 2y^2 + 2$ **67.** $3a^2 - 3a + 5$
69. $9ab^2 - 3ab + 16a^2b$ **71.** $4z^5 + z^4 + 9z^3 - 3z - 2$
73. $2x^4 + 11x^3 - 3x^2 + 8x - 4$ **75.** $-w^3 + 0.2w^2 + 3.7w - 0.7$
77. $-p^2q - 4pq^2 + 3pq$ **79.** 0 **81.** $-5ab + 6ab^2$
83. $11y^2 - 10y - 4$ **85.** For example: $x^3 + 6$
87. For example: $8x^5$ **89.** For example: $-6x^2 + 2x + 5$

Section 13.6 Activity, p. 883

A.1. $9u^3v^5$ **A.2. a.** 2, 3, 6 **b.** $3t^3 - 83t^2 + 169t - 65$
A.3. a. conjugates **b.** $36p^2 + 6p - 6p - 1 = 36p^2 - 1$
c. $(6p)^2 - (1)^2 = 36p^2 - 1$
A.4. a. $4w^2 - 10w - 10w + 25 = 4w^2 - 20w + 25$
b. $(2w)^2 - 2(2w)(5) + (5)^2 = 4w^2 - 20w + 25$
A.5. $x^3 + 12x^2 + 48x + 64$

Section 13.6 Practice Exercises, pp. 883–887

R.1. $9x$ **R.3.** $20x^2$ **R.5.** $-7a^3b$ **R.7.** $10a^6b^2$
R.9. $3w^2 - 3w + 2$ **R.11.** $-9x + 21y + 24z$ **R.13.** $2x^2 - \dfrac{7}{4}x + \dfrac{3}{10}$

1. a. $5 - 2x$ **b.** squares; $a^2 - b^2$ **c.** perfect; $a^2 + 2ab + b^2$
3. $-12y$ **5.** $21p$ **7.** $12a^{14}b^8$ **9.** $-2c^{10}d^{12}$
11. $16p^2q^2 - 24p^2q + 40pq^2$ **13.** $-4k^3 + 52k^2 + 24k$
15. $-45p^3q - 15p^4q^3 + 30pq^2$ **17.** $y^2 + 19y + 90$
19. $m^2 - 14m + 24$ **21.** $12p^2 - 5p - 2$
23. $12w^2 - 32w + 16$ **25.** $p^2 - 14pw + 33w^2$
27. $12x^2 + 28x - 5$ **29.** $6a^2 - 21.5a + 18$
31. $9t^3 - 21t^2 + 3t - 7$ **33.** $9m^2 + 30mn + 24n^2$
35. $5s^3 + 8s^2 - 7s - 6$ **37.** $27w^3 - 8$
39. $p^4 + 5p^3 - 2p^2 - 21p + 5$ **41.** $6a^3 - 23a^2 + 38a - 45$
43. $8x^3 - 36x^2y + 54xy^2 - 27y^3$ **45.** $1.2x^2 + 7.6xy + 2.4y^2$
47. $y^2 - 36$ **49.** $9a^2 - 16b^2$ **51.** $81k^2 - 36$
53. $\dfrac{4}{9}t^2 - 9$ **55.** $u^6 - 25v^2$ **57.** $\dfrac{4}{9} - p^2$
59. $a^2 + 10a + 25$ **61.** $x^2 - 2xy + y^2$ **63.** $4c^2 + 20c + 25$
65. $9t^4 - 24st^2 + 16s^2$ **67.** $t^2 - 14t + 49$ **69.** $16q^2 + 24q + 9$

71. a. 36 b. 20 c. $(a+b)^2 \ne a^2 + b^2$ in general.
73. $4x^2 - 25$ 75. $16p^2 + 40p + 25$
77. $27p^3 - 135p^2 + 225p - 125$ 79. $15a^5 - 6a^2$
81. $49x^2 - y^2$ 83. $25s^2 + 30st + 9t^2$ 85. $21x^2 - 65xy + 24y^2$
87. $2t^2 + \dfrac{26}{3}t + 8$ 89. $-30x^2 - 35x + 25$
91. $6a^3 + 11a^2 - 7a - 2$ 93. $y^3 - 3y^2 - 6y - 20$
95. $\dfrac{1}{9}m^2 - \dfrac{2}{3}mn + n^2$ 97. $42w^3 - 84w^2$ 99. $16y^2 - 65.61$
101. $21c^4 + 4c^2 - 32$ 103. $9.61x^2 + 27.9x + 20.25$
105. $k^3 - 12k^2 + 48k - 64$ 107. $125x^3 + 225x^2 + 135x + 27$
109. $2y^4 + 3y^3 + 3y^2 + 5y + 3$ 111. $6a^3 + 22a^2 - 40a$
113. $2x^3 - 13x^2 + 17x + 12$ 115. $2x - 7$
117. $k = 10$ or -10 119. $k = 8$ or -8

Section 13.7 Activity, pp. 895–896

A.1. a. $\dfrac{12a^3}{-6a} + \dfrac{-6a^2}{-6a} + \dfrac{18a}{-6a}$ b. $-2a^2 + a - 3$
A.2. $6k^3 - k + 4 - \dfrac{3}{2k^2}$ A.3. Use long division when the divisor has two or more terms.   A.4. $1626\dfrac{11}{31}$
A.5. a. $2x^3 - 3x^2 + x - 4 + \dfrac{-2}{x - 3}$ b. $2x^3 - 3x^2 + x - 4$
   c. $x - 3$ d. $2x^4 - 9x^3 + 10x^2 - 7x + 10$ e. -2
   f. $2x^4 - 9x^3 + 10x^2 - 7x + 10 = (x - 3)(2x^3 - 3x^2 + x - 4) + (-2)$ ✓
A.6. a. $(27x^3 + 0x^2 + 0x + 8) \div (3x + 2)$ b. $9x^2 - 6x + 4$
   c. $9x^2 - 6x + 4$ d. $3x + 2$ e. $27x^3 + 8$ f. 0
   g. $27x^3 + 8 = (3x + 2)(9x^2 - 6x + 4)$ ✓
A.7. a. $\dfrac{3y^4 + 2y^3 + 0y^2 - 6y + 1}{y^2 + 0y - 2}$ b. $3y^2 + 2y + 6 + \dfrac{-2y + 13}{y^2 - 2}$
   c. $3y^4 + 2y^3 - 6y + 1 = (y^2 - 2)(3y^2 + 2y + 6) + (-2y + 13)$ ✓
A.8. a. $\underline{3\,|\ 2\ -9\ \ 10\ -7\ \ 10}$ b. $2x^3 - 3x^2 + x - 4$
      $\quad\quad\ \ 6\ -9\ \ 3\ -12$
      $\quad\ \ 2\ -3\ \ 1\ -4\ \underline{|-2}$
   c. -2 d. $2x^3 - 3x^2 + x - 4 + \dfrac{-2}{x - 3}$; Yes
   e. Answers will vary.
A.9. a. $t^2 + 5t + 25$ b. $t^3 - 125 = (t - 5)(t^2 + 5t + 25)$ ✓
A.10. Synthetic division can be used if the divisor is of the form $x - r$ where r is a constant.
A.11. Long division   A.12. Monomial division
A.13. Long division or synthetic division   A.14. Long division

Section 13.7 Practice Exercises, pp. 896–898

R.1. Dividend: 8945; divisor: 9; quotient: 993; remainder: 8
R.3. $8x^3 - 24x^2 + 16x - 8$ R.5. $16b^4 - 40b^3 + 96b^2$
R.7. $11x^2$ R.9. $2x^2 - 2x$

1. a. division; quotient; remainder b. Synthetic
3. p^5 5. $7a^4$ 7. $\dfrac{2x^2}{3y}$ 9. $-4t^3 + t - 5$
11. $12 + 8y + 2y^2$ 13. $x^2 + 3xy - y^2$ 15. $2y^2 + 3y - 8$
17. $-\dfrac{1}{2}p^3 + p^2 - \dfrac{1}{3}p + \dfrac{1}{6}$ 19. $a^2 + 5a + 1 - \dfrac{5}{a}$
21. $-3s^2 t + 4s - \dfrac{5}{t^2}$ 23. $8p^2q^6 - 9p^3q^5 - 11p - \dfrac{4}{p^2q}$
25. a. Divisor: $(x - 2)$; quotient: $(2x^2 - 3x - 1)$; remainder: (-3) b. Multiply the quotient and the divisor, then add the remainder. The result should equal the dividend.

27. $x + 7 + \dfrac{-9}{x + 4}$ 29. $3y^2 + 2y + 2 + \dfrac{9}{y - 3}$
31. $-4a + 11$ 33. $6y - 5$ 35. $6x^2 + 4x + 5 + \dfrac{22}{3x - 2}$
37. $4a^2 - 2a + 1$ 39. $x^2 - 2x + 2$
41. $x^2 + 2x + 5$ 43. $x^2 - 1 + \dfrac{8}{x^2 - 2}$
45. $n^3 + 2n^2 + 4n + 8$ 47. $3y^2 - 3 + \dfrac{2y + 6}{y^2 + 1}$
49. The divisor must be of the form $x - r$.
51. No. The divisor is not of the form $x - r$.
53. a. $x - 5$ b. $x^2 + 3x + 11$ c. 58
55. $x + 6$ 57. $t - 4$ 59. $5y + 10 + \dfrac{11}{y - 1}$
61. $3y^2 - 2y + 2 + \dfrac{-3}{y + 3}$ 63. $x^2 - x - 2$
65. $a^4 + 2a^3 + 4a^2 + 8a + 16$ 67. $x^2 + 6x + 36$
69. $4w^3 + 2w^2 + 6$ 71. $-x^2 - 4x + 13 + \dfrac{-54}{x + 4}$
73. $2x - 1 + \dfrac{3}{x}$ 75. $4y - 3 + \dfrac{4y + 5}{3y^2 - 2y + 5}$
77. $2x^2 + 3x - 1$ 79. $2k^3 - 4k^2 + 1 - \dfrac{5}{k^4}$
81. $x + \dfrac{9}{5} + \dfrac{2}{x}$

Chapter 13 Problem Recognition Exercises, p. 899

1. a. $8x^2$ b. $12x^4$ 2. a. $9y^3$ b. $8y^6$ 3. a. $16x^2 + 8xy + y^2$
b. $16x^2y^2$ 4. a. $4a^2 + 4ab + b^2$ b. $4a^2b^2$ 5. a. $6x + 1$
b. $8x^2 + 8x - 6$ 6. a. $6m^2 + m + 1$ b. $5m^4 + 5m^3 + m^2 + m$
7. a. $9z^2 + 12z + 4$ b. $9z^2 - 4$ 8. a. $36y^2 - 84y + 49$
   b. $36y^2 - 49$ 9. a. $2x^3 - 8x^2 + 14x - 12$ b. $x^2 - 1$
10. a. $-3y^4 - 20y^2 - 32$ b. $4y^2 + 12$ 11. a. $2x$ b. x^2
12. a. $4c$ b. $4c^2$ 13. $49m^2n^2$ 14. $64p^2q^2$ 15. $-x^2 - 3x + 4$
16. $5m^2 - 4m + 1$ 17. $-7m^2 - 16m$
18. $-4n^5 + n^4 + 6n^2 - 7n + 2$ 19. $8x^2 + 16x + 34 + \dfrac{74}{x - 2}$
20. $-4x^2 - 10x - 30 + \dfrac{-95}{x - 3}$ 21. $6x^3 + 5x^2y - 6xy^2 + y^3$
22. $6a^3 - a^2b + 5ab^2 + 2b^3$ 23. $x^3 + y^6$ 24. $m^6 + 1$
25. $4b$ 26. $-12z$ 27. $a^6 - 4b^2$ 28. $y^6 - 36z^2$
29. $4p + 4 + \dfrac{-2}{2p - 1}$ 30. $2v - 7 + \dfrac{29}{2v + 3}$ 31. $4x^2y^2$
32. $-9pq$ 33. $\dfrac{9}{49}x^2 - \dfrac{1}{4}$ 34. $\dfrac{4}{25}y^2 - \dfrac{16}{9}$
35. $-\dfrac{11}{9}x^3 + \dfrac{5}{9}x^2 - \dfrac{1}{2}x - 4$ 36. $\dfrac{13}{10}y^2 - \dfrac{9}{10}y + \dfrac{4}{15}$
37. $1.3x^2 - 0.3x - 0.5$ 38. $5w^3 - 4.1w^2 + 2.8w - 1.2$
39. $-6x^3y^6$ 40. $50a^4b^6$

Chapter 13 Review Exercises, pp. 900–902

1. Base: 5; exponent: 3   2. Base: x; exponent: 4
3. Base: -2; exponent: 0   4. Base: y; exponent: 1
5. a. 36   b. 36   c. -36 6. a. 64   b. -64 c. -64
7. 5^{13} 8. a^{11} 9. x^9 10. 6^9 11. 10^3 12. y^6
13. b^8 14. 7^7 15. k 16. 1   17. 2^8 18. q^6
19. Exponents are added only when multiplying factors with the same base. In such a case, the base does not change.
20. Exponents are subtracted only when dividing factors with the same base. In such a case, the base does not change.   21. $7146.10
22. $22,050   23. 7^{12} 24. c^{12} 25. p^{18} 26. 9^{28}

27. $\dfrac{a^2}{b^2}$ **28.** $\dfrac{1}{3^4}$ **29.** $\dfrac{5^2}{c^4 d^{10}}$ **30.** $-\dfrac{m^{10}}{4^5 n^{30}}$ **31.** $2^4 a^4 b^8$
32. $x^{14} y^2$ **33.** $-\dfrac{3^3 x^9}{5^3 y^6 z^3}$ **34.** $\dfrac{r^{15}}{s^{10} t^{30}}$ **35.** a^{11} **36.** 8^2
37. $4h^{14}$ **38.** $2p^{14}q^{13}$ **39.** $\dfrac{x^6 y^2}{4}$ **40.** $a^9 b^6$ **41.** 1
42. 1 **43.** -1 **44.** 1 **45.** 2 **46.** 1 **47.** $\dfrac{1}{z^5}$
48. $\dfrac{1}{10^4}$ **49.** $\dfrac{1}{36a^2}$ **50.** $\dfrac{6}{a^2}$ **51.** $\dfrac{17}{16}$ **52.** $\dfrac{10}{9}$
53. $\dfrac{1}{t^8}$ **54.** $\dfrac{1}{r}$ **55.** $\dfrac{2y^7}{x^6}$ **56.** $\dfrac{4a^6 bc}{5}$ **57.** $\dfrac{n^{16}}{16m^8}$
58. $\dfrac{u^{15}}{27v^6}$ **59.** $\dfrac{k^{21}}{5}$ **60.** $\dfrac{h^9}{9}$ **61.** $\dfrac{1}{2}$ **62.** $\dfrac{5}{4}$
63. a. 9.74×10^7 **b.** 4.2×10^{-3} in. **64. a.** 0.000 000 0001
  b. \$256,000 **65.** 9.43×10^5 **66.** 1.55×10^{10}
67. 2.5×10^8 **68.** 1.638×10^3 **69.** $\approx 9.5367 \times 10^{13}$.
This number has too many digits to fit on most calculator displays.
70. $\approx 1.1529 \times 10^{-12}$. This number has too many digits to fit on most calculator displays. **71. a.** $\approx 5.84 \times 10^8$ mi
b. $\approx 6.67 \times 10^4$ mph **72. a.** $\approx 2.26 \times 10^8$ mi
b. $\approx 1.07 \times 10^5$ mph **73. a.** Trinomial **b.** 4 **c.** 7
74. a. Binomial **b.** 7 **c.** -5 **75.** $7x - 3$
76. $-y^2 - 14y - 2$ **77.** $14a^2 - 2a - 6$
78. $\dfrac{15}{2}x^3 + \dfrac{1}{4}x^2 + \dfrac{1}{2}x + 2$ **79.** $10w^4 + 2w^3 - 7w + 4$
80. $0.01b^5 + b^4 - 0.1b^3 + 0.3b + 0.33$ **81.** $-2x^2 - 9x - 6$
82. $-5x^2 - 9x - 12$ **83.** For example, $-5x^2 + 2x - 4$
84. For example, $6x^5 + 8$ **85.** $6w + 6$ **86.** $-75x^6 y^4$
87. $18a^8 b^4$ **88.** $15c^4 - 35c^2 + 25c$ **89.** $-2x^3 - 10x^2 + 6x$
90. $5k^2 + k - 4$ **91.** $20t^2 + 3t - 2$ **92.** $6q^2 + 47q - 8$
93. $2a^2 + 4a - 30$ **94.** $49a^2 + 7a + \dfrac{1}{4}$ **95.** $b^2 - 8b + 16$
96. $8p^3 - 27$ **97.** $-2w^3 - 5w^2 - 5w + 4$
98. $4x^3 - 8x^2 + 11x - 4$ **99.** $12a^3 + 11a^2 - 13a - 10$
100. $b^2 - 16$ **101.** $\dfrac{1}{9}r^8 - s^4$ **102.** $49z^4 - 84z^2 + 36$
103. $2h^5 + h^4 - h^3 + h^2 - h + 3$ **104.** $2x^2 + 3x - 20$
105. $4y^2 - 2y$ **106.** $2a^2 b - a - 3b$ **107.** $-3x^2 + 2x - 1$
108. $-\dfrac{z^5 w^3}{2} + \dfrac{3zw}{4} + \dfrac{1}{z}$ **109.** $x + 2$ **110.** $2t + 5$
111. $p - 3 + \dfrac{5}{2p + 7}$ **112.** $a + 6 + \dfrac{-4}{5a - 3}$
113. $b^2 + 5b + 25$ **114.** $z + 4$ **115.** $y^2 - 4y + 2 + \dfrac{9y - 4}{y^2 + 3}$
116. $t^2 - 3t + 1 + \dfrac{-2t - 6}{3t^2 + t + 1}$ **117.** $w^2 + w - 1$

Chapter 14

Section 14.1 Activity, pp. 911–912

A.1. a. $10a^2 b^3 = 2 \cdot 5 \cdot a \cdot a \cdot b \cdot b \cdot b$
  $15a^3 b^4 = 3 \cdot 5 \cdot a \cdot a \cdot a \cdot b \cdot b \cdot b \cdot b$
  $-35a^4 b = -1 \cdot 5 \cdot 7 \cdot a \cdot a \cdot a \cdot a \cdot b$
  b. $5a^2 b$ **c.** $5a^2 b(2b^2 + 3ab^3 - 7a^2)$
A.2. a. $(c + d)$ **b.** $(c + d)(5x + 7y)$
A.3. a. $5m(-2m + 3)$ **b.** $-5m(2m - 3)$
  c. The sign of each term is changed.
A.4. $a^2 + 4a - ab - 4b$
A.5. a. $a(a - b) + 4(a - b)$ **b.** $(a - b)(a + 4)$
  c. The product equals the original polynomial $a^2 - ab + 4a - 4b$.
A.6. a. $(6p - q)(p - 5)$
  b. $(6p - q)(p - 5) = 6p^2 - 30p - pq + 5q$ ✓

Section 14.1 Practice Exercises, pp. 912–914

R.1. $2^3 \cdot 3 \cdot 5$ **R.3.** $-4x^2 + 5x - 3$ **R.5.** $-40c^3 + 50c^2$
R.7. $6x^4 y - 15x^3 y + 3x^2 y$ **R.9.** $8ax - 28a + 6bx - 21b$

1. a. product **b.** prime **c.** greatest common factor
  d. prime **e.** greatest common factor (GCF) **f.** grouping
3. 7 **5.** 6 **7.** y **9.** $4w^2 z$ **11.** $2xy^4 z^2$ **13.** $(x - y)$
15. a. $3x - 6y$ **b.** $3(x - 2y)$ **17.** $4(p + 3)$ **19.** $5(c^2 - 2c + 3)$
21. $x^3(x^2 + 1)$ **23.** $t(t^3 - 4 + 8t)$ **25.** $2ab(1 + 2a^2)$
27. $19x^2 y(2 - y^3)$ **29.** $6xy^5(x^2 - 3y^4 z)$ **31.** The expression is prime because it is not factorable. **33.** $7pq^2(6p^2 + 2 - p^3 q^2)$
35. $t^2(t^3 + 2rt - 3t^2 + 4r^2)$ **37. a.** $-2x(x^2 + 2x - 4)$
  b. $2x(-x^2 - 2x + 4)$ **39.** $-1(8t^2 + 9t + 2)$ **41.** $-15p^2(p + 2)$
43. $-3mn(m^3 n - 2m + 3n)$ **45.** $-1(7x + 6y + 2z)$
47. $(a + 6)(13 - 4b)$ **49.** $(w^2 - 2)(8v + 1)$ **51.** $7x(x + 3)^2$
53. $(2a - b)(4a + 3c)$ **55.** $(q + p)(3 + r)$ **57.** $(2x + 1)(3x + 2)$
59. $(t + 3)(2t - 1)$ **61.** $(3y - 1)(2y - 3)$ **63.** $(b + 1)(b^3 - 4)$
65. $(j^2 + 5)(3k + 1)$ **67.** $(2x^6 + 1)(7w^6 - 1)$
69. $(y + x)(a + b)$ **71.** $(vw + 1)(w - 3)$ **73.** $5x(x^2 + y^2)(3x + 2y)$
75. $4b(a - b)(x - 1)$ **77.** $6t(t - 3)(s - t^2)$ **79.** $P = 2(l + w)$
81. $S = 2\pi r(r + h)$ **83.** $\dfrac{1}{7}(x^2 + 3x - 5)$ **85.** $\dfrac{1}{4}(5w^2 + 3w + 9)$
87. For example: $6x^2 + 9x$ **89.** For example: $16p^4 q^2 + 8p^3 q - 4p^2 q$

Section 14.2 Activity, p. 919

A.1. a. 3 and 4 **b.** $(x + 3)(x + 4)$
  c. $(x + 3)(x + 4) = x^2 + 4x + 3x + 12 = x^2 + 7x + 12$ ✓
A.2. a. -12 and 4 **b.** $(y - 12)(y + 4)$
  c. $(y - 12)(y + 4) = y^2 + 4y - 12y - 48 = y^2 - 8y - 48$ ✓
A.3. a. Different **b.** $(t + 4)(t - 3)$
A.4. a. Both negative **b.** $(p - 7)(p - 2)$
A.5. a. Both positive **b.** $(w + 3)(w + 8)$
A.6. a. $(\boxed{x^2} \quad)(\boxed{x^2} \quad)$ **b.** $(x^2 + 9)(x^2 + 1)$
  c. $(x^2 + 9)(x^2 + 1) = x^4 + x^2 + 9x^2 + 9 = x^4 + 10x^2 + 9$ ✓
A.7. a. $5y^2$ **b.** Yes **c.** $5y^2(y^2 + 2y - 15)$ **d.** $5y^2(y + 5)(y - 3)$
  e. $5y^2(y + 5)(y - 3) = 5y^2(y^2 - 3y + 5y - 15) = 5y^2(y^2 + 2y - 15) = 5y^4 + 10y^3 - 75y^2$ ✓

Section 14.2 Practice Exercises, pp. 919–922

R.1. -9 and 5 **R.3.** $5y^2 + 15y + 10$ **R.5.** $5y(y^2 + 4y + 2)$
R.7. $x^2 - 8x - 48$ **R.9.** $m^2 + 16m + 64$
R.11. $4a^4 b + 4a^3 b - 120a^2 b$

1. a. positive **b.** different **3.** $(x + 8)(x + 2)$
5. $(z - 9)(z - 2)$ **7.** $(z - 6)(z + 3)$
9. $(p - 8)(p + 5)$ **11.** $(t + 10)(t - 4)$ **13.** Prime
15. $(n + 4)^2$ **17.** $(x^2 + 1)(x^2 + 9)$ **19.** $(w^2 + 5)(w^2 - 3)$
21. a **23.** c **25.** They are both correct because multiplication of polynomials is a commutative operation.
27. The expressions are equal and both are correct.
29. It should be written in descending order. **31.** $(x - 15)(x + 2)$
33. $(w - 13)(w - 5)$ **35.** $(t + 18)(t + 4)$ **37.** $3(x - 12)(x + 2)$
39. $8p(p - 1)(p - 4)$ **41.** $y^2 z^2(y - 6)(y - 6)$ or $y^2 z^2(y - 6)^2$
43. $-(x - 4)(x - 6)$ **45.** $-5(a - 3x)(a + 2x)$
47. $-2(c + 2)(c + 1)$ **49.** $xy^3(x - 4)(x - 15)$
51. $12(p - 7)(p - 1)$ **53.** $-2(m - 10)(m - 1)$
55. $(c + 5d)(c + d)$ **57.** $(a - 2b)(a - 7b)$ **59.** Prime
61. $(q - 7)(q + 9)$ **63.** $(x + 10)^2$ **65.** $(t + 20)(t - 2)$
67. The student forgot to factor out the GCF before factoring the trinomial further. The polynomial is not factored completely, because $(2x - 4)$ has a common factor of 2.
69. $x^2 + 9x - 52$
71. a. $3x^2 + 9x - 12$ **b.** $3(x + 4)(x - 1)$
73. 7, 5, -7, -5
75. For example: $c = -16$

Section 14.3 Activity, p. 929

A.1. a. 1 **b.** Yes **c.** 2 **d.** $2x^2$ **e.** $(2x+3)(x+5)$
f. $(2x+3)(x+5) = 2x^2 + 10x + 3x + 15 = 2x^2 + 13x + 15$ ✓

A.2. a. 1 **b.** Yes **c.** 5 **d.** $(5t\quad)(t\quad)$ **e.** $(5t-3)(t-3)$
f. $(5t-3)(t-3) = 5t^2 - 15t - 3t + 9 = 5t^2 - 18t + 9$ ✓

A.3. a. 1 **b.** All will produce a leading term of $4a^2$.
c. Both negative **d.** $(4a - 3b)(a - 2b)$
e. $(4a-3b)(a-2b) = 4a^2 - 8ab - 3ab + 6b^2 = 4a^2 - 11ab + 6b^2$ ✓

Section 14.3 Practice Exercises, pp. 930–932

R.1. $6y^4 - 11y^2 + 3$ **R.3.** $15p^2 + 23p - 28$
R.5. $9x^4 - 24x^2 + 16$ **R.7.** $8a^3 + 36a^2 - 72a$
R.9. $(n-1)(m-2)$ **R.11.** $6(a-7)(a+2)$

1. a. Both are correct. **b.** $2(3x-5)(x+1)$ **3.** a **5.** b
7. a. Same **b.** $(5x+3)(x+1)$ **9. a.** Different **b.** $(x+1)(2x-7)$
11. $(3n+1)(n+4)$ **13.** $(2y+1)(y-2)$ **15.** $(5x+1)(x-3)$
17. $(4c+1)(3c-2)$ **19.** $(10w-3)(w+4)$
21. $(3q+2)(2q-3)$ **23.** Prime **25.** $(5m+2)(5m-4)$
27. $(6y-5x)(y+4x)$ **29.** $2(m+4)(m-10)$
31. $y^3(2y+1)(y+6)$ **33.** $-(a+17)(a-2)$
35. $-10(4m+p)(2m-3p)$ **37.** $(x^2+1)(x^2+9)$
39. $(w^2+5)(w^2-3)$ **41.** $(2x^2+3)(x^2-5)$
43. $-2(z-9)(z-1)$ **45.** $(q-7)(q-6)$ **47.** $(2t+3)(3t-1)$
49. $(2m-5)^2$ **51.** Prime **53.** $(2x-5y)(3x-2y)$
55. $(4m+5n)(3m-n)$ **57.** $5(3r+2)(2r-1)$
59. Prime **61.** $(2t-5)(5t+1)$ **63.** $(7w-4)(2w+3)$
65. $(a-12)(a+2)$ **67.** $(x+5y)(x+4y)$
69. $(a+20b)(a+b)$ **71.** $(t-7)(t-3)$ **73.** $d(5d^2+3d-10)$
75. $4b(b-5)(b+4)$ **77.** $y^2(x-3)(x-10)$
79. $-2u(2u+5)(3u-2)$ **81.** $(2x^2+3)(4x^2+1)$
83. a. 36 ft **b.** $-4(4t+5)(t-2)$; Yes **85. a.** $(x-12)(x+2)$
b. $(x-6)(x-4)$ **87. a.** $(x-6)(x+1)$ **b.** $(x-2)(x-3)$

Section 14.4 Activity, p. 938

A.1. a. 1 **b.** Yes **c.** $a=2, b=7, c=5$ **d.** 10 **e.** 2 and 5
f. $2x^2 + 2x + 5x + 5$ **g.** $(x+1)(2x+5)$
h. $(2x+5)(x+1)$ is the same as $(x+1)(2x+5)$ by the commutative property of multiplication.

A.2. a. 1 **b.** Yes **c.** $a=6, b=-31, c=40$ **d.** -16 and -15
e. $= 6t^2 - 16t - 15t + 40$ **f.** $(3t-8)(2t-5)$
g. $(3t-8)(2t-5) = 6t^2 - 15t - 16t + 40 = 6t^2 - 31t + 40$ ✓

A.3. a. 4 **b.** $4(7a^2 + 20ab - 3b^2)$ **c.** $4(7a-b)(a+3b)$
d. $4(7a-b)(a+3b) = 4(7a^2 + 21ab - ab - 3b^2)$
$= 4(7a^2 + 20ab - 3b^2) = 28a^2 + 80ab - 12b^2$ ✓

Section 14.4 Practice Exercises, pp. 938–940

R.1. $7y(y-2)$ **R.3.** $(y+5)(9y+8)$ **R.5.** $-2x(x+5)$
R.7. $(3x+2)(2y-5)$ **R.9.** $6(a+4)(b-2)$

1. a. Both are correct. **b.** $3(4x+3)(x-2)$ **3.** 12, 1
5. $-8, -1$ **7.** -10 and -10 **9.** $5, -4$ **11.** $9, -2$
13. $(x+4)(3x+1)$ **15.** $(w-2)(4w-1)$ **17.** $(x+9)(x-2)$
19. $(m+3)(2m-1)$ **21.** $(4k+3)(2k-3)$ **23.** $(2k-5)^2$
25. Prime **27.** $(3z-5)(3z-2)$ **29.** $(6y-5z)(2y+3z)$
31. $2(7y+4)(y+3)$ **33.** $-(3w-5)(5w+1)$
35. $-4(x-y)(3x-2y)$ **37.** $6y(y+1)(3y+7)$
39. $(a^2+2)(a^2+3)$ **41.** $(3x^2-5)(2x^2+3)$
43. $(8p^2-3)(p^2+5)$ **45.** $(5p-1)(4p-3)$
47. $(3u-2v)(2u-5v)$ **49.** $(4a+5b)(3a-b)$
51. $(h+7k)(3h-2k)$ **53.** Prime **55.** $(2z-1)(8z-3)$
57. $(b-4)^2$ **59.** $-5(x-2)(x-3)$ **61.** $(t-3)(t+2)$
63. Prime **65.** $2(12x-1)(3x+1)$ **67.** $p(p+3)(p-9)$
69. $x(3x+7)(x+1)$ **71.** $2p(p-15)(p-4)$

73. $x^2(y+3)(y+11)$ **75.** $-(k+2)(k+5)$
77. $-3(n+6)(n-5)$ **79.** $(x^2-2)(x^2-5)$
81. No. $(2x+4)$ contains a common factor of 2.
83. a. 128 customers **b.** $-2(x-18)(x-2)$; 128; Yes
85. a. 42 **b.** $n(n+1)$; 42; Yes

Section 14.5 Activity, p. 946

A.1. a. 1, 4, 9, 16, 25, 36, 49, 64, 81, 100, 121
b. $x^2, x^4, x^6, x^8, x^{10}, x^{12}$ **c.** Even powers of n
A.2. a. $a^2 - b^2$ **b.** $(a-b)(a+b)$ **c.** $16x^2 - 9$ **d.** $(4x-3)(4x+3)$
e. The two terms have opposite signs and their absolute values are perfect squares.
A.3. $2y(y-7)(y+7)$ **A.4.** $2y(y^2+49)$
A.5. $(4t^2+9)(2t+3)(2t-3)$
A.6. a. $a^2 + 2ab + b^2$ **b.** $(a+b)^2$ **c.** $25w^2 + 60w + 36$ **d.** $(5w+6)^2$
e. The first and third terms are perfect squares. The middle term is twice the product of the square roots of the first and third terms.
A.7. $(9c-2d)^2$
A.8. a. Factor out the greatest common factor. **b.** $2(3x-2y)^2$
A.9. a. $(x+16)(x+1)$ **b.** $(x+4)^2$ **c.** The trinomial in part (b) is a perfect square trinomial.

Section 14.5 Practice Exercises, pp. 946–948

R.1. $4x^2 - 49$ **R.3.** $x^4 - 16$ **R.5.** $9x^2 - 12x + 4$
R.7. $36c^4 + 12c^2 + 1$ **R.9.** $(3x-5)(x+2)$ **R.11.** $3a^2b(2+a)$
R.13. $p(5p-3)(q+4)$ **R.15.** $(x-1)(x-5)$ **R.17.** Prime

1. a. difference; $(a+b)(a-b)$ **b.** sum **c.** is not
d. square **e.** $(a+b)^2$; $(a-b)^2$ **3.** $x^2 - 25$
5. $4p^2 - 9q^2$ **7.** $(x-6)(x+6)$ **9.** $3(w+10)(w-10)$
11. $(2a-11b)(2a+11b)$ **13.** $(7m-4n)(7m+4n)$
15. Prime **17.** $(y+2z)(y-2z)$ **19.** $(a-b^2)(a+b^2)$
21. $(5pq-1)(5pq+1)$ **23.** $\left(c - \frac{1}{5}\right)\left(c + \frac{1}{5}\right)$
25. $2(5-4t)(5+4t)$ **27.** $(x+4)(x-4)(x^2+16)$
29. $(2-z)(2+z)(4+z^2)$ **31.** $(x+5)(x+3)(x-3)$
33. $(c-1)(c+5)(c-5)$ **35.** $(x+3)(x-3)(2+y)$
37. $(y+3)(y-3)(x+2)(x-2)$ **39.** $9x^2 + 30x + 25$
41. a. $x^2 + 4x + 4$ is a perfect square trinomial.
b. $x^2 + 4x + 4 = (x+2)^2$; $x^2 + 5x + 4 = (x+1)(x+4)$
43. $(x+9)^2$ **45.** $(5z-2)^2$ **47.** $(7a+3b)^2$ **49.** $(y-1)^2$
51. $5(4z+3w)^2$ **53.** $(3y+25)(3y+1)$ **55.** $2(a-5)^2$
57. Prime **59.** $(2x+y)^2$ **61.** $3x(x-1)^2$ **63.** $y(y-6)$
65. $(2p-5)(2p+7)$ **67.** $(-t+2)(t+6)$ or $-(t-2)(t+6)$
69. $(2a-5+10b)(2a-5-10b)$ **71. a.** $a^2 - b^2$ **b.** $(a-b)(a+b)$

Section 14.6 Activity, p. 953

A.1. a. 1, 8, 27, 64, 125 **b.** x^3, x^6, x^9, x^{12} **c.** n must be a multiple of 3.
A.2. a. $a^3 - b^3$ **b.** $(a-b)(a^2 + ab + b^2)$
A.3. a. $(a+b)(a^2 - ab + b^2)$ **b.** $(2y)^3$ **c.** $(3)^3$ **d.** $a = 2y, b = 3$
e. $(2y+3)(4y^2 - 6y + 9)$
f. $(2y+3)(4y^2 - 6y + 9) = 8y^3 - 12y^2 + 18y + 12y^2 - 18y + 27$
$= 8y^3 + 27$ ✓
A.4. a. $(a-b)(a^2 + ab + b^2)$ **b.** $(p^2)^3$ **c.** $(4q)^3$ **d.** $a = p^2, b = 4q$
e. $(p^2 - 4q)(p^4 + 4p^2q + 16q^2)$
f. $(p^2 - 4q)(p^4 + 4p^2q + 16q^2)$
$= p^6 + 4p^4q + 16p^2q^2 - 4p^4q - 16p^2q^2 - 64q^3$
$= p^6 - 64q^3$ ✓
A.5. $3(p^2 + 16)$; Sum of squares; $3(p^2 + 16)$
A.6. $2(8x^3 - 27y^2)$; Difference of cubes; $2(2x - 3y^2)(4x^2 + 6xy^2 + 9y^4)$
A.7. $a(125a^3 + b^3)$; Sum of cubes; $a(5a+b)(25a^2 - 5ab + b^2)$
A.8. $625 - w^4$; Difference of squares; $(25 + w^2)(5+w)(5-w)$
A.9. $2y^2(8y-1)$; None of these; $2y^2(8y-1)$

Section 14.6 Practice Exercises, pp. 954–955

R.1. $y^3 - 125$ **R.3.** $3p$ **R.5.** $-5m^2$
R.7. $6(10-x)(10+x)$ **R.9.** $(a+b)(x+5)$
R.11. $(5y-2)(y+3)$ **R.13.** $-(c+5)^2$

1. **a.** sum; cubes **b.** difference; cubes **c.** $(a-b)(a^2+ab+b^2)$
   d. $(a+b)(a^2-ab+b^2)$ **3.** $x^3, 8, y^6, 27q^3, w^{12}, r^3s^6$
5. $t^3, -1, 27, a^3b^6, 125, y^6$ **7.** $(y-2)(y^2+2y+4)$
9. $(1-p)(1+p+p^2)$ **11.** $(w+4)(w^2-4w+16)$
13. $(x-10y)(x^2+10xy+100y^2)$ **15.** $(4t+1)(16t^2-4t+1)$
17. $(10a+3)(100a^2-30a+9)$ **19.** $\left(n-\frac{1}{2}\right)\left(n^2+\frac{1}{2}n+\frac{1}{4}\right)$
21. $(5x+2y)(25x^2-10xy+4y^2)$ **23.** $(x^2-2)(x^2+2)$
25. Prime **27.** $(t+4)(t^2-4t+16)$ **29.** Prime
31. $4(b+3)(b^2-3b+9)$ **33.** $5(p-5)(p+5)$
35. $\left(\frac{1}{4}-2h\right)\left(\frac{1}{16}+\frac{1}{2}h+4h^2\right)$ **37.** $(x-2)(x+2)(x^2+4)$
39. $\left(\frac{2}{3}x-w\right)\left(\frac{2}{3}x+w\right)$
41. $(q-2)(q^2+2q+4)(q+2)(q^2-2q+4)$
43. $(x^3+4y)(x^6-4x^3y+16y^2)$ **45.** $(2x+3)(x-1)(x+1)$
47. $(2x-y)(2x+y)(4x^2+y^2)$ **49.** $(3y-2)(3y+2)(9y^2+4)$
51. $(a+b^2)(a^2-ab^2+b^4)$ **53.** $(x^2+y^2)(x-y)(x+y)$
55. $(k+4)(k-3)(k+3)$ **57.** $2(t-5)(t-1)(t+1)$
59. $\left(\frac{4}{5}p-\frac{1}{2}q\right)\left(\frac{16}{25}p^2+\frac{2}{5}pq+\frac{1}{4}q^2\right)$ **61.** $(a^4+b^4)(a^8-a^4b^4+b^8)$
63. **a.** The quotient is x^2+2x+4. **b.** $(x-2)(x^2+2x+4)$
65. $x^2+4x+16$ **67.** $2x+1$

Chapter 14 Problem Recognition Exercises, pp. 956–957

1. A prime polynomial cannot be factored further.
2. Factor out the GCF. **3.** Look for a difference of squares: $a^2 - b^2$, a difference of cubes: $a^3 - b^3$, or a sum of cubes: $a^3 + b^3$. **4.** Grouping
5. **a.** Difference of squares **b.** $2(a-9)(a+9)$
6. **a.** Nonperfect square trinomial **b.** $(y+3)(y+1)$
7. **a.** None of these **b.** $6w(w-1)$
8. **a.** Difference of squares **b.** $(2z+3)(2z-3)(4z^2+9)$
9. **a.** Nonperfect square trinomial **b.** $(3t+1)(t+4)$
10. **a.** Sum of cubes **b.** $5(r+1)(r^2-r+1)$
11. **a.** Four terms-grouping **b.** $(3c+d)(a-b)$
12. **a.** Difference of cubes **b.** $(x-5)(x^2+5x+25)$
13. **a.** Sum of cubes **b.** $(y+2)(y^2-2y+4)$
14. **a.** Nonperfect square trinomial **b.** $(7p-1)(p-4)$
15. **a.** Nonperfect square trinomial **b.** $3(q-4)(q+1)$
16. **a.** Perfect square trinomial **b.** $-2(x-2)^2$
17. **a.** None of these **b.** $6a(3a+2)$
18. **a.** Difference of cubes **b.** $2(3-y)(9+3y+y^2)$
19. **a.** Difference of squares **b.** $4(t-5)(t+5)$
20. **a.** Nonperfect square trinomial **b.** $(4t+1)(t-8)$
21. **a.** Nonperfect square trinomial **b.** $10(c^2+c+1)$
22. **a.** Four terms-grouping **b.** $(w-5)(2x+3y)$
23. **a.** Sum of cubes **b.** $(x+0.1)(x^2-0.1x+0.01)$
24. **a.** Difference of squares **b.** $(2q-3)(2q+3)$
25. **a.** Perfect square trinomial **b.** $(8+k)^2$
26. **a.** Four terms-grouping **b.** $(t+6)(s^2+5)$
27. **a.** Four terms-grouping **b.** $(x+1)(2x-y)$
28. **a.** Sum of cubes **b.** $(w+y)(w^2-wy+y^2)$
29. **a.** Difference of cubes **b.** $(a-c)(a^2+ac+c^2)$
30. **a.** Nonperfect square trinomial **b.** Prime
31. **a.** Nonperfect square trinomial **b.** Prime
32. **a.** Perfect square trinomial **b.** $(a+1)^2$
33. **a.** Perfect square trinomial **b.** $(b+5)^2$
34. **a.** Nonperfect square trinomial **b.** $-(t+8)(t-4)$
35. **a.** Nonperfect square trinomial **b.** $-p(p+4)(p+1)$
36. **a.** Difference of squares **b.** $(xy-7)(xy+7)$
37. **a.** Nonperfect square trinomial **b.** $3(2x+3)(x-5)$
38. **a.** Nonperfect square trinomial **b.** $2(5y-1)(2y-1)$
39. **a.** None of these **b.** $abc^2(5ac-7)$
40. **a.** Difference of squares **b.** $2(2a-5)(2a+5)$
41. **a.** Nonperfect square trinomial **b.** $(t+9)(t-7)$
42. **a.** Nonperfect square trinomial **b.** $(b+10)(b-8)$
43. **a.** Four terms-grouping **b.** $(b+y)(a-b)$
44. **a.** None of these **b.** $3x^2y^4(2x+y)$
45. **a.** Nonperfect square trinomial **b.** $(7u-2v)(2u-v)$
46. **a.** Nonperfect square trinomial **b.** Prime
47. **a.** Nonperfect square trinomial **b.** $2(2q^2-4q-3)$
48. **a.** Nonperfect square trinomial **b.** $3(3w^2+w-5)$
49. **a.** Sum of squares **b.** Prime
50. **a.** Perfect square trinomial **b.** $5(b-3)^2$
51. **a.** Nonperfect square trinomial **b.** $(3r+1)(2r+3)$
52. **a.** Nonperfect square trinomial **b.** $(2s-3)(2s+5)$
53. **a.** Difference of squares **b.** $(2a-1)(2a+1)(4a^2+1)$
54. **a.** Four terms-grouping **b.** $(p+c)(p-3)(p+3)$
55. **a.** Perfect square trinomial **b.** $(9u-5v)^2$
56. **a.** Sum of squares **b.** $4(x^2+4)$
57. **a.** Nonperfect square trinomial **b.** $(x-6)(x+1)$
58. **a.** Nonperfect square trinomial **b.** Prime
59. **a.** Four terms-grouping **b.** $2(x-3y)(a+2b)$
60. **a.** Nonperfect square trinomial **b.** $m(4m+1)(2m-3)$
61. **a.** Nonperfect square trinomial **b.** $x^2y(3x+5)(7x+2)$
62. **a.** Difference of squares **b.** $2(m^2-8)(m^2+8)$
63. **a.** Four terms-grouping **b.** $(4v-3)(2u+3)$
64. **a.** Four terms-grouping **b.** $(t-5)(4t+s)$
65. **a.** Perfect square trinomial **b.** $3(2x-1)^2$
66. **a.** Perfect square trinomial **b.** $(p+q)^2$
67. **a.** Nonperfect square trinomial **b.** $n(2n-1)(3n+4)$
68. **a.** Nonperfect square trinomial **b.** $k(2k-1)(2k+3)$
69. **a.** Difference of squares **b.** $(8-y)(8+y)$
70. **a.** Difference of squares **b.** $b(6-b)(6+b)$
71. **a.** Nonperfect square trinomial **b.** Prime
72. **a.** Nonperfect square trinomial **b.** $(y+4)(y+2)$
73. **a.** Nonperfect square trinomial **b.** $(c^2-10)(c^2-2)$

Section 14.7 Activity, pp. 962–963

A.1. a. linear; 1 **b.** quadratic; 2
A.2. a. $a=0$ or $b=0$ **b.** $x+1$ or $x-4$ **c.** $\{-1, 4\}$
A.3. a. Yes **b.** $2y-5=0, y+7=0$ **c.** $\left\{\frac{5}{2}, -7\right\}$ **d.** Yes
A.4. a. $4x(x-3)=0$ **b.** $4x=0, x-3=0$ **c.** $\{0, 3\}$ **d.** Yes
A.5. a. $2(x-5)^2=0$ **b.** $2=0, x-5=0$ **c.** $\{5\}$ **d.** Yes
A.6. Quadratic; $\{-3, 4\}$ **A.7.** Linear; $\{6\}$ **A.8.** Linear; $\{4\}$
A.9. Quadratic; $\left\{-3, \frac{1}{5}\right\}$ **A.10.** Neither; $\left\{3, -\frac{1}{2}, 5\right\}$
A.11. Neither; $\left\{0, -\frac{1}{5}, \frac{1}{5}\right\}$

Section 14.7 Practice Exercises, pp. 963–965

R.1. $\{-9\}$ **R.3.** $\left\{\frac{7}{2}\right\}$ **R.5.** $(x-10)(x+7)$
R.7. $(2t-1)(t+5)$ **R.9.** $3h(h-5)(h+5)$

1. **a.** quadratic **b.** 0; 0 **3.** Neither **5.** Quadratic
7. Linear **9.** Linear **11.** Quadratic **13.** Neither
15. $\{-3, 1\}$ **17.** $\left\{\frac{7}{2}, -\frac{7}{2}\right\}$ **19.** $\{-5\}$ **21.** $\left\{0, \frac{1}{5}\right\}$
23. The polynomial must be factored completely.
25. $\{5, -3\}$ **27.** $\{-12, 2\}$ **29.** $\left\{4, -\frac{1}{2}\right\}$ **31.** $\left\{\frac{2}{3}, -\frac{2}{3}\right\}$
33. $\{6, 8\}$ **35.** $\left\{0, -\frac{3}{2}, 4\right\}$ **37.** $\left\{-\frac{1}{3}, 3, -6\right\}$ **39.** $\left\{0, 4, -\frac{3}{2}\right\}$
41. $\left\{0, -\frac{9}{2}, 11\right\}$ **43.** $\{0, 4, -4\}$ **45.** $\{-6, 0\}$
47. $\left\{\frac{3}{4}, -\frac{3}{4}\right\}$ **49.** $\{0, -5, -2\}$ **51.** $\left\{-\frac{14}{3}\right\}$ **53.** $\{5, 3\}$
55. $\{0, -2\}$ **57.** $\{-3\}$ **59.** $\{-3, 1\}$ **61.** $\left\{\frac{3}{2}\right\}$

63. $\{0, \frac{1}{3}\}$ 65. $\{0, 2\}$ 67. $\{3, -2, 2\}$ 69. $\{-5, 4\}$
71. $\{-5, -1\}$

Chapter 14 Problem Recognition Exercises, p. 965

1. a. $(x+7)(x-1)$ b. $\{-7, 1\}$ 2. a. $(c+6)(c+2)$
   b. $\{-6, -2\}$ 3. a. $(2y+1)(y+3)$ b. $\{-\frac{1}{2}, -3\}$
4. a. $(3x-5)(x-1)$ b. $\{\frac{5}{3}, 1\}$ 5. a. $\{\frac{4}{5}, -1\}$
   b. $(5q-4)(q+1)$ 6. a. $\{-\frac{1}{3}, \frac{3}{2}\}$ b. $(3a+1)(2a-3)$
7. a. $\{-8, 8\}$ b. $(a+8)(a-8)$ 8. a. $\{-10, 10\}$
   b. $(v+10)(v-10)$ 9. a. $(2b+9)(2b-9)$ b. $\{-\frac{9}{2}, \frac{9}{2}\}$
10. a. $(6t+7)(6t-7)$ b. $\{-\frac{7}{6}, \frac{7}{6}\}$
11. a. $\{-\frac{3}{2}, -\frac{1}{2}\}$ b. $2(2x+3)(2x+1)$
12. a. $\{-\frac{4}{3}, -2\}$ b. $4(3y+4)(y+2)$
13. a. $x(x-10)(x+2)$ b. $\{0, 10, -2\}$
14. a. $k(k+7)(k-2)$ b. $\{0, -7, 2\}$
15. a. $\{-1, 3, -3\}$ b. $(b+1)(b-3)(b+3)$
16. a. $\{8, -2, 2\}$ b. $(x-8)(x+2)(x-2)$
17. $(s-3)(2s+r)$ 18. $(2t+1)(3t+5u)$
19. $\{-\frac{1}{2}, 0, \frac{1}{2}\}$ 20. $\{-5, 0, 5\}$ 21. $\{0, 1\}$
22. $\{-3, 0\}$ 23. $\{\frac{1}{3}\}$ 24. $\{-\frac{7}{12}\}$
25. $(2w+3)(4w^2-6w+9)$ 26. $(10q-1)(100q^2+10q+1)$
27. $\{-\frac{7}{5}, 1\}$ 28. $\{-6, -\frac{1}{4}\}$ 29. $\{-\frac{2}{3}, -5\}$ 30. $\{-1, -\frac{1}{2}\}$
31. $\{3\}$ 32. $\{0\}$ 33. $\{-4, 4\}$ 34. $\{-\frac{2}{3}, \frac{2}{3}\}$
35. $\{1, 6\}$ 36. $\{-2, -12\}$

Section 14.8 Activity, pp. 970–971

A.1. a. $x+2$ b. $x(x+2)$ c. $x+(x+2)+23$
    d. $x(x+2) = x+(x+2)+23$ e. $x = 5, x = -5$
    f. The integers are 5 and 7 or −5 and −3.
A.2. a. $h+3$ b. $h(h+3) = 180$ c. $h = -15, h = 12$
    d. The negative solution is unreasonable to represent the height of a parallelogram.   e. The height is 12 m and the base is 15 m.
A.3. a. $a^2 + b^2 = c^2$ b. 15 ft
A.4. a.

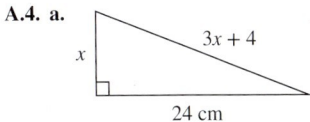

b. $(24)^2 + (x)^2 = (3x+4)^2$ c. $x = -10, x = 7$
d. The length of the unknown leg is 7 cm and the hypotenuse is 25 cm.
A.5. a. 0 ft   b. $t = 0, t = 8$ c. 8 sec

Section 14.8 Practice Exercises, pp. 971–973

R.1. $\{-\frac{1}{6}, 4\}$ R.3. $\{\frac{1}{2}, -\frac{1}{2}\}$ R.5. $\{-1, 6\}$
R.7. $\{-\frac{2}{3}, 4\}$ R.9. $x-8$
R.11. $t^2 + (-9)$ R.13. $6^2 - c^2$ or $36 - c^2$

1. a. $x+1$ b. $x+2$ c. $x+2$
3. The numbers are 7 and −7.   5. The numbers are 10 and −4.
7. The numbers are −9 and −7, or 7 and 9.   9. The numbers are 5 and 6, or −6 and −5.   11. The height of the painting is 11 ft

and the width is 9 ft.   13. a. The slab is 7 m by 4 m.   b. The perimeter is 22 m.   15. The base is 7 ft and the height is 4 ft.
17. The base is 5 cm and the height is 8 cm.
19. The ball hits the ground in 3 sec.
21. The object is at ground level at 0 sec and 1.5 sec.
23.

```
        hypotenuse
    a  _____
       |      /
  leg  |    / c
       |  /
       |/_____
            b
          leg
```

25. $c = 25$ cm   27. $a = 15$ in.
29. The brace is 20 in. long.   31. The kite is 19 yd high.
33. The bottom of the ladder is 8 ft from the house. The distance from the top of the ladder to the ground is 15 ft.   35. The hypotenuse is 10 m.

Chapter 14 Review Exercises, pp. 974–976

1. $3a^2b$ 2. $x+5$ 3. $2c(3c-5)$ 4. $-2yz$ or $2yz$
5. $2x(3x+x^3-4)$ 6. $11w^2y^3(w-4y^2)$
7. $t(-t+5)$ or $-t(t-5)$ 8. $u(-6u-1)$ or $-u(6u+1)$
9. $(b+2)(3b-7)$ 10. $2(5x+9)(1+4x)$
11. $(w+2)(7w+b)$ 12. $(b-2)(b+y)$
13. $3(4y-3)(5y-1)$ 14. $a(2-a)(3-b)$
15. $(x-3)(x-7)$ 16. $(y-8)(y-11)$
17. $(z-12)(z+6)$ 18. $(q-13)(q+3)$
19. $3w(p+10)(p+2)$ 20. $2m^2(m+8)(m+5)$
21. $-(t-8)(t-2)$ 22. $-(w-4)(w+5)$
23. $(a+b)(a+11b)$ 24. $(c-6d)(c+3d)$
25. Different   26. Both negative   27. Both positive
28. Different   29. $(2y+3)(y-4)$ 30. $(4w+3)(w-2)$
31. $(2z+5)(5z+2)$ 32. $(4z-3)(2z+3)$ 33. Prime
34. Prime   35. $10(w-9)(w+3)$ 36. $-3(y-8)(y+2)$
37. $(3c-5d)^2$ 38. $(x+6)^2$ 39. $(3g+2h)(2g+h)$
40. $(6m-n)(2m-5n)$ 41. $(v^2+1)(v^2-3)$
42. $(x^2+5)(x^2+2)$ 43. $5, -1$ 44. $-3, -5$
45. $(c-2)(3c+1)$ 46. $(y+3)(4y+1)$ 47. $(t+12w)(t+w)$
48. $(4x^2-3)(x^2+5)$ 49. $(w^2+5)(w^2+2)$
50. $(p-3q)(p-5q)$ 51. $-2(4v+3)(5v-1)$
52. $10(4s-5)(s+2)$ 53. $ab(a-6b)(a-4b)$
54. $2z^4(z+7)(z-3)$ 55. Prime   56. Prime   57. $(7x+10)^2$
58. $(3w-z)^2$ 59. $(a-b)(a+b)$ 60. Prime
61. $(a-7)(a+7)$ 62. $(d-8)(d+8)$ 63. $(10-9t)(10+9t)$
64. $(2-5k)(2+5k)$ 65. Prime   66. Prime   67. $(y+6)^2$
68. $(t+8)^2$ 69. $(3a-2)^2$ 70. $(5x-4)^2$ 71. $-3(v+2)^2$
72. $-2(x-5)^2$ 73. $2(c^2-3)(c^2+3)$ 74. $2(6x-y)(6x+y)$
75. $(p+3)(p-4)(p+4)$ 76. $(k-2)(2-k)(2+k)$ or $-(k-2)^2(2+k)$ 77. $(a+b)(a^2-ab+b^2)$
78. $(a-b)(a^2+ab+b^2)$ 79. $(4+a)(16-4a+a^2)$
80. $(5-b)(25+5b+b^2)$ 81. $(p^2+2)(p^4-2p^2+4)$
82. $\left(q^2 - \frac{1}{3}\right)\left(q^4 + \frac{1}{3}q^2 + \frac{1}{9}\right)$ 83. $6(x-2)(x^2+2x+4)$
84. $7(y+1)(y^2-y+1)$ 85. $x(x-6)(x+6)$
86. $q(q-4)(q^2+4q+16)$ 87. $4(2h^2+5)$ 88. $m(m-8)$
89. $(x+4)(x+1)(x-1)$ 90. $5q(p^2-2q)(p^2+2q)$
91. $n(2+n)(4-2n+n^2)$ 92. $14(m-1)(m^2+m+1)$
93. $(x-3)(2x+1) = 0$ can be solved directly by the zero product rule because it is a product of factors set equal to zero.
94. $\{\frac{1}{4}, -\frac{2}{3}\}$ 95. $\{9, \frac{1}{2}\}$ 96. $\{0, -3, -\frac{2}{5}\}$
97. $\{0, 7, \frac{9}{4}\}$ 98. $\{-\frac{5}{7}, 2\}$ 99. $\{-\frac{1}{4}, 6\}$
100. $\{12, -12\}$ 101. $\{5, -5\}$ 102. $\{0, \frac{1}{5}\}$ 103. $\{4, 2\}$
104. $\{-\frac{5}{6}\}$ 105. $\{-\frac{2}{3}\}$ 106. $\{\frac{2}{3}, 6\}$ 107. $\{\frac{11}{2}, -12\}$
108. $\{0, 7, 2\}$ 109. $\{0, 2, -2\}$ 110. The height is 6 ft, and the base is 13 ft.   111. The ball is at ground level at 0 and 1 sec.
112. The ramp is 13 ft long.   113. The legs are 6 ft and 8 ft; the hypotenuse is 10 ft.   114. The numbers are −8 and 8.
115. The numbers are 29 and 30, or −2 and −1.
116. The height is 4 m, and the base is 9 m.

Chapter 15

Section 15.1 Activity, p. 985

A.1. Substituting 5 for x would make the denominator equal to zero.

A.2. a. $\dfrac{5(d+3)}{(d+3)(d-4)}$ **b.** $d \neq -3, d \neq 4$

c. $\dfrac{5}{d-4}$ provided $d \neq -3, d \neq 4$

A.3. $\dfrac{3}{z^2}$ provided $z \neq 0$ **A.4.** $\dfrac{t-4}{t+3}$ provided $t \neq -3, t \neq -4$

A.5. a. $-x + 12$ or $12 - x$ **b.** -1 **A.6. a.** $2b + 5c$ **b.** -1

A.7. $-\dfrac{w+3}{2(w-3)}$ provided $w \neq 3$

Section 15.1 Practice Exercises, pp. 985–989

R.1. $2^2 \cdot 3^2 \cdot 7$ **R.3.** $\dfrac{7}{20}$ **R.5.** $-\dfrac{9}{2}$ **R.7.** -1 **R.9.** a^8
R.11. $\dfrac{1}{x}$ **R.13.** $21x^2(5x-3)$ **R.15.** $(5y-2)(5y+2)$
R.17. $(a-7)(a+3)$ **R.19.** $(5x-3)(2x-1)$
R.21. $(2a-5b)(3c+4)$ **R.23.** $(3x-5y)^2$

1. a. rational **b.** denominator; zero **c.** $\dfrac{p}{q}$ **d.** $1; -1$ **3.** 0
5. $-\dfrac{1}{8}$ **7.** $-\dfrac{1}{2}$ **9.** Undefined **11. a.** $3\tfrac{1}{5}$ hr or 3.2 hr
b. $1\tfrac{3}{4}$ hr or 1.75 hr **13.** $k = -2$ **15.** $x = \dfrac{5}{2}, x = -8$
17. $m = -2, m = -3$ **19.** There are no restricted values.
21. There are no restricted values. **23.** $t = 0$
25. For example: $\dfrac{1}{x-2}$ **27.** For example: $\dfrac{1}{(x+3)(x-7)}$
29. a. $\dfrac{2}{5}$ **b.** $\dfrac{2}{5}$ **31. a.** Undefined **b.** Undefined
33. a. $y = -2$ **b.** $\dfrac{1}{2}$ **35. a.** $t = -1$ **b.** $t - 1$
37. a. $w = 0, w = \dfrac{5}{3}$ **b.** $\dfrac{1}{3w-5}$ **39. a.** $x = -\dfrac{2}{3}$ **b.** $\dfrac{3x-2}{2}$
41. a. $a = -3, a = 2$ **b.** $\dfrac{a+5}{a+3}$ **43.** $\dfrac{b}{3}$ **45.** $-\dfrac{3xy}{z^2}$
47. $\dfrac{p-3}{p+4}$ **49.** $\dfrac{1}{4(m-11)}$ **51.** $\dfrac{2x+1}{4x^2}$ **53.** $\dfrac{1}{4a-5}$
55. $\dfrac{x-2}{3(y+2)}$ **57.** $\dfrac{2}{x-5}$ **59.** $a+7$ **61.** Cannot simplify
63. $\dfrac{y+3}{2y-5}$ **65.** $\dfrac{5}{(q+1)(q-1)}$ **67.** $\dfrac{c-d}{2c+d}$ **69.** $5x+4$
71. $\dfrac{x+y}{x-4y}$ **73.** They are opposites. **75.** -1 **77.** -1
79. $-\dfrac{1}{2}$ **81.** Cannot simplify **83.** $\dfrac{5x-6}{5x+6}$ **85.** $\dfrac{x+3}{4+x}$
87. $\dfrac{3x-2}{x+4}$ **89.** $-\dfrac{1}{2}$ **91.** $-\dfrac{w+2}{4}$ **93.** $\dfrac{3t^2}{2}$ **95.** $-\dfrac{2r-s}{s+2r}$
97. $-\dfrac{3}{2(x-y)}$ **99.** $\dfrac{1}{t(t-5)}$ **101.** $w-2$ **103.** $\dfrac{z+4}{z^2+4z+16}$

Section 15.2 Activity, p. 994

A.1. $\dfrac{1}{10}$ **A.2.** $-\dfrac{5(p-4)}{p(2p+1)}$

A.3. a. $\dfrac{4(t-6)(t+1)}{(3t-1)^2} \cdot \dfrac{(3t-1)(3t+1)}{2(3t+1)(t-6)}$ **b.** $\dfrac{2(t+1)}{3t-1}$

A.4. $-\dfrac{4}{3}$ **A.5. a.** $\dfrac{4-w}{2w-3} \cdot \dfrac{2(2w-3)}{w-4}$ **b.** -2

A.6. a. $\dfrac{100-y^2}{24y-3} \cdot \dfrac{12y+12}{y^2-9y-10}$ **b.** $\dfrac{(10-y)(10+y)}{3(8y-1)} \cdot \dfrac{12(y+1)}{(y-10)(y+1)}$

c. $-\dfrac{4(y+10)}{8y-1}$ **A.7.** $-\dfrac{x}{2(x-2)}$

Section 15.2 Practice Exercises, pp. 995–997

R.1. a. -2 **b.** $\dfrac{7}{3}$ **c.** $\dfrac{1}{8}$ **d.** $\dfrac{1}{x}$ **R.3.** $\dfrac{x^2+10x+9}{x-3}$
R.5. $\dfrac{3}{10}$ **R.7.** 2 **R.9.** $\dfrac{5}{2}$ **R.11.** $\dfrac{15}{4}$

1. $\dfrac{3}{2x}$ **3.** $3xy^4$ **5.** $\dfrac{x-6}{8}$ **7.** $\dfrac{2}{y}$ **9.** $-\dfrac{5}{8}$
11. $-\dfrac{b+a}{a-b}$ **13.** $\dfrac{y+1}{5}$ **15.** $\dfrac{2(x+6)}{2x+1}$ **17.** 6 **19.** $\dfrac{m^6}{n^2}$
21. $\dfrac{10}{9}$ **23.** $\dfrac{6}{7}$ **25.** $-m(m+n)$ **27.** $\dfrac{3p+4q}{4(p+2q)}$ **29.** $\dfrac{p}{p-1}$
31. $\dfrac{w}{2w-1}$ **33.** $\dfrac{4r}{2r+3}$ **35.** $\dfrac{5}{6}$ **37.** $\dfrac{1}{4}$ **39.** $\dfrac{y+9}{y-6}$ **41.** 1
43. $\dfrac{3t+8}{t+2}$ **45.** $\dfrac{x+4}{x+1}$ **47.** $-\dfrac{w-3}{2}$ **49.** $\dfrac{k+6}{k+3}$ **51.** $\dfrac{2}{a}$
53. $2y(y+1)$ **55.** $x+y$ **57.** 2 **59.** $\dfrac{1}{a-2}$ **61.** $\dfrac{p+q}{2}$

Section 15.3 Activity, pp. 1001–1002

A.1. a. 8, 16, 24, 32, 40, 48 **b.** 12, 24, 36, 48, 60, 72
c. 24, 48 **d.** 24

A.2. a. $8 = 2^3$ **b.** $12 = 2^2 \cdot 3$
c. The LCM of 8 and 12 is $2^3 \cdot 3$, which equals 24. This is the product of unique prime factors from 8 and 12, where each factor is raised to the highest power to which it appears.

d. 24 **e.** $\dfrac{3}{8} = \dfrac{9}{24}$ and $\dfrac{5}{12} = \dfrac{10}{24}$

A.3. a. $10p^2q^3$ **b.** $\dfrac{11}{5p^2q} = \dfrac{22q^2}{10p^2q^3}$ and $\dfrac{-1}{10pq^3} = \dfrac{-p}{10p^2q^3}$

A.4. a. $(2y+1)^2(y-7)$ **b.** $\dfrac{y^2-10y+21}{(2y+1)^2(y-7)}$ and $\dfrac{8y^2+4y}{(2y+1)^2(y-7)}$

A.5. a. $(x+3)(x-5)(x+5)$

b. $\dfrac{-x-5}{(x+3)(x-5)(x+5)}$ and $\dfrac{x^2+4x+3}{(x+3)(x-5)(x+5)}$

A.6. a. $\dfrac{-3}{y-5}$ **b.** $\dfrac{-x+2}{5-y}$

Section 15.3 Practice Exercises, pp. 1002–1004

R.1. $2^5 \cdot 3 \cdot 5$ **R.3.** 84 **R.5. a.** -3 **b.** $\dfrac{1}{3}$ **c.** Undefined
R.7. $x = 1, x = -1; \dfrac{3}{5(x-1)}$ **R.9.** $\dfrac{a+5}{a+7}$ **R.11.** $\dfrac{2}{3y}$
R.13. a, b, c, d

1. x^5 is the greatest power of x that appears in any denominator.
3. 45 **5.** 48 **7.** 63 **9.** $9x^2y^3$
11. w^2y **13.** $(p+3)(p-1)(p+2)$ **15.** $9t(t+1)^2$
17. $(y-2)(y+2)(y+3)$ **19.** $3-x$ or $x-3$
21. Because $(b-1)$ and $(1-b)$ are opposites, they differ by a factor of -1. **23.** $\dfrac{6}{5x^2}; \dfrac{5x}{5x^2}$ **25.** $\dfrac{24x}{30x^3}; \dfrac{5y}{30x^3}$
27. $\dfrac{10}{12a^2b}; \dfrac{a^3}{12a^2b}$ **29.** $\dfrac{6m-6}{(m+4)(m-1)}; \dfrac{3m+12}{(m+4)(m-1)}$
31. $\dfrac{6x+18}{(2x-5)(x+3)}; \dfrac{2x-5}{(2x-5)(x+3)}$
33. $\dfrac{6w+6}{(w+3)(w-8)(w+1)}; \dfrac{w^2+3w}{(w+3)(w-8)(w+1)}$
35. $\dfrac{6p^2+12p}{(p-2)(p+2)^2}; \dfrac{3p-6}{(p-2)(p+2)^2}$
37. $\dfrac{1}{a-4}; \dfrac{-a}{a-4}$ or $\dfrac{-1}{4-a}; \dfrac{a}{4-a}$

39. $\dfrac{8}{2(x-7)}$; $\dfrac{-y}{2(x-7)}$ or $\dfrac{-8}{2(7-x)}$; $\dfrac{y}{2(7-x)}$

41. $\dfrac{1}{a+b}$; $\dfrac{-6}{a+b}$ or $\dfrac{-1}{-a-b}$; $\dfrac{6}{-a-b}$

43. $\dfrac{-9}{24(3y+1)}$; $\dfrac{20}{24(3y+1)}$ 45. $\dfrac{3z+12}{5z(z+4)}$; $\dfrac{5z}{5z(z+4)}$

47. $\dfrac{z^2+3z}{(z+2)(z+7)(z+3)}$; $\dfrac{-3z^2-6z}{(z+2)(z+7)(z+3)}$; $\dfrac{5z+35}{(z+2)(z+7)(z+3)}$

49. $\dfrac{3p+6}{(p-2)(p^2+2p+4)(p+2)}$; $\dfrac{p^3+2p^2+4p}{(p-2)(p^2+2p+4)(p+2)}$; $\dfrac{5p^3-20p}{(p-2)(p^2+2p+4)(p+2)}$

Section 15.4 Activity, p. 1011

A.1. a. $\dfrac{6}{5x}+\dfrac{7}{2\cdot 5}$ b. $10x$ c. $\dfrac{12}{10x}+\dfrac{7x}{10x}$ d. $\dfrac{12+7x}{10x}$ e. $\dfrac{7x+12}{10x}$

A.2. a. $\dfrac{3}{2(x-2)}-\dfrac{6}{(x-2)(x+2)}$ b. $2(x-2)(x+2)$

c. $\dfrac{3(x+2)}{2(x-2)(x+2)}-\dfrac{6\cdot 2}{2(x-2)(x+2)}$

d. $\dfrac{3(x+2)-12}{2(x-2)(x+2)}$ e. $\dfrac{3}{2(x+2)}$

A.3. a. $\dfrac{1}{y+1}+\dfrac{1}{y-3}-\dfrac{4}{(y+1)(y-3)}$ b. $(y+1)(y-3)$

c. $\dfrac{1(y-3)}{(y+1)(y-3)}+\dfrac{1(y+1)}{(y+1)(y-3)}-\dfrac{4}{(y+1)(y-3)}$

d. $\dfrac{(y-3)+(y+1)-4}{(y+1)(y-3)}$ e. $\dfrac{2}{y+1}$

A.4. a. The expressions $t-5$ and $5-t$ differ only by a factor of -1.

b. $\dfrac{-6}{t-5}$ c. $\dfrac{6}{5-t}$

d. $\dfrac{-6}{t-5}=\left(\dfrac{-6}{t-5}\right)\cdot\left(\dfrac{-1}{-1}\right)=\dfrac{6}{-t+5}$ or equivalently $\dfrac{6}{5-t}$

Section 15.4 Practice Exercises, pp. 1011–1014

R.1. 90   R.3. $15a^2b^2$ R.5. $(x+2)^3(x-5)(x+1)$
R.7. $5t(t-3)^2$ R.9. $\dfrac{a^2}{2b}$; $a\neq 0, b\neq 0$ R.11. $\dfrac{2(x+4)}{x-4}$; $x\neq 4$
R.13. $\dfrac{4}{2w+1}$; $w\neq -\dfrac{1}{2}, w=\dfrac{7}{2}$ R.15. -2 R.17. $-3x^2+6x$
R.19. $8x^2-14x-15$

1. $\dfrac{5}{4}$ 3. $\dfrac{3}{8}$ 5. $-\dfrac{1}{x}$ 7. $\dfrac{t+5}{t+4}$ 9. 2   11. 5
13. $\dfrac{-2(t-2)}{t-8}$ 15. $3x+7$ 17. $m+5$
19. 2   21. $x-5$ 23. $\dfrac{1}{r+1}$ 25. $\dfrac{1}{y+7}$
27. $\dfrac{15x}{y}$ 29. $\dfrac{5a+6}{4a}$ 31. $\dfrac{2(6+x^2y)}{15xy^3}$ 33. $\dfrac{2s-3t^2}{s^4t^3}$
35. $-\dfrac{2}{3}$ 37. $\dfrac{19}{3(a+1)}$ 39. $\dfrac{-3(k+4)}{(k-3)(k+3)}$ 41. $\dfrac{a-4}{2a}$
43. $\dfrac{(x+6)(x-2)}{(x-4)(x+1)}$ 45. $\dfrac{x(4x+9)}{(x+3)^2(x+2)}$
47. $\dfrac{5p-1}{3}$ or $\dfrac{-5p+1}{-3}$ 49. $\dfrac{9}{x-3}$ or $\dfrac{-9}{3-x}$
51. $\dfrac{6n-1}{n-8}$ or $\dfrac{-6n+1}{8-n}$ 53. $\dfrac{2(4x+5)}{x(x+2)}$ 55. $\dfrac{3y}{2(2y+1)}$
57. $\dfrac{2(w-3)}{(w+3)(w-1)}$ 59. $\dfrac{4a-13}{(a-3)(a-4)}$ 61. $\dfrac{3x(x+7)}{(x+5)^2(x-1)}$
63. $\dfrac{-y(y+8)}{(2y+1)(y-1)(y-4)}$ 65. $\dfrac{1}{2p+1}$ 67. $\dfrac{-2mn+1}{(m+n)(m-n)}$

69. 0   71. $\dfrac{2(3x+7)}{(x+3)(x+2)}$ 73. $\dfrac{1}{n}$ 75. $\dfrac{5}{n+2}$
77. $n+\left(7\cdot\dfrac{1}{n}\right)$; $\dfrac{n^2+7}{n}$ 79. $\dfrac{1}{n}-\dfrac{2}{n}$; $-\dfrac{1}{n}$
81. $\dfrac{-w^2}{(w+3)(w-3)(w^2-3w+9)}$ 83. $\dfrac{p^2-2p+7}{(p+2)(p+3)(p-1)}$
85. $\dfrac{-m-21}{2(m+5)(m-2)}$ or $\dfrac{m+21}{2(m+5)(2-m)}$ 87. $\dfrac{3k+5}{4k+7}$ 89. $\dfrac{1}{a}$

Chapter 15 Problem Recognition Exercises, p. 1015

1. $\dfrac{-2x+9}{3x+1}$ 2. $\dfrac{1}{w-4}$ 3. $\dfrac{y-5}{2y-3}$ 4. $\dfrac{7}{(x+3)(2x-1)}$
5. $-\dfrac{1}{x}$ 6. $\dfrac{1}{3}$ 7. $\dfrac{c+3}{c}$ 8. $\dfrac{x+3}{5}$ 9. $\dfrac{a}{12b^4c}$
10. $\dfrac{2a-b}{a-b}$ 11. $\dfrac{p-q}{5}$ 12. 4   13. $\dfrac{10}{2x+1}$ 14. $\dfrac{w+2z}{w+z}$
15. $\dfrac{3}{2x+5}$ 16. $\dfrac{y+7}{x+a}$ 17. $\dfrac{1}{2(a+3)}$
18. $\dfrac{2(3y+10)}{(y-6)(y+6)(y+2)}$ 19. $(t+8)^2$ 20. $6b+5$

Section 15.5 Activity, pp. 1021–1022

A.1. a. $\dfrac{7}{6}$ b. $\dfrac{5}{12}$ c. $\dfrac{14}{5}$
A.2. a. $\dfrac{1+2b}{ab}$ b. $\dfrac{3a-1}{ab}$ c. $\dfrac{1+2b}{3a-1}$
A.3. a. 12   b. 14   c. 5   d. $\dfrac{14}{5}$
A.4. a. ab b. $1+2b$ c. $3a-1$ d. $\dfrac{1+2b}{3a-1}$ A.5. $\dfrac{w}{2w+1}$

Section 15.5 Practice Exercises, pp. 1023–1025

R.1. $9x-4$ R.3. $5w^2-4w-20$ R.5. $b-a$
R.7. $\dfrac{31}{24}$ R.9. $\dfrac{7}{18}$ R.11. $\dfrac{5}{t^2}$

1. complex   3. a. $\dfrac{7}{10}\cdot\dfrac{3}{14}$ b. $\dfrac{3}{20}$ 5. a. $\dfrac{2x}{15}\cdot\dfrac{25}{4x}$ b. $\dfrac{5}{6}$
7. $\dfrac{7}{4y}$ 9. $\dfrac{1}{2y}$ 11. $\dfrac{24b}{a^3}$ 13. $\dfrac{2r^5t^4}{s^6}$
15. $\dfrac{35}{2}$ 17. $k+h$ 19. $\dfrac{n+1}{2(n-3)}$ 21. $\dfrac{2x+1}{4x+1}$
23. $m-7$ 25. $\dfrac{2y(y-5)}{7y^2+10}$ 27. $-\dfrac{a+8}{a-2}$ or $\dfrac{a+8}{2-a}$
29. $\dfrac{t-2}{t-4}$ 31. $\dfrac{t+3}{t-5}$ 33. $\dfrac{1}{2}$ 35. $\dfrac{\frac{1}{2}+\frac{2}{3}}{5}$; $\dfrac{7}{30}$
37. $\dfrac{3}{\frac{2}{3}+\frac{3}{4}}$; $\dfrac{36}{17}$ 39. a. $\dfrac{6}{5}\Omega$ b. 6Ω 41. $\dfrac{xy}{y+x}$ 43. $\dfrac{y+4x}{2y}$
45. $\dfrac{1}{n^2+m^2}$ 47. $\dfrac{2z-5}{3(z+3)}$ 49. $-\dfrac{x+1}{x-1}$ or $\dfrac{x+1}{1-x}$ 51. $\dfrac{3}{2}$

Section 15.6 Activity, pp. 1033–1034

A.1. a. 6   b. $2y-3$ c. $5y+3$ d. $\{-2\}$
A.2. a. $\dfrac{2x}{x+2}=\dfrac{5}{(x+2)(x-3)}-\dfrac{1}{x-3}$ b. $x\neq -2, x\neq 3$
c. $(x+2)(x-3)$ d. $\left\{-\dfrac{1}{2}\right\}$; The value 3 does not check.
A.3. a. $c=\dfrac{Ka-b}{d}$ b. $a=\dfrac{cd+b}{K}$
A.4. a. Denominator   b. V_1V_2 c. $P_1V_1=P_2V_2$ d. $V_1=\dfrac{P_2V_2}{P_1}$

Section 15.6 Practice Exercises, pp. 1034–1036

R.1. $4a(a-5)$ **R.3.** $8(m+3)$ **R.5.** m^2 **R.7.** $19a - 35$
R.9. $8 - 2m$ **R.11.** $a \neq 0, a \neq 5$ **R.13.** $h = \dfrac{V}{lw}$
R.15. $y = \dfrac{5}{3}x - 2$

1. a. linear; quadratic **b.** rational **c.** denominator   **3.** $\left\{\dfrac{8}{5}\right\}$
5. $\left\{\dfrac{21}{4}\right\}$ **7.** $\{3\}$ **9.** $\left\{\dfrac{5}{11}\right\}$ **11.** $\left\{\dfrac{1}{3}\right\}$
13. $\{-8\}$ **15. a.** $z = 0$ **b.** $5z$ **c.** $\{5\}$ **17.** $\left\{-\dfrac{200}{19}\right\}$
19. $\{-7\}$ **21.** $\left\{\dfrac{47}{6}\right\}$ **23.** $\{3, -1\}$ **25.** $\{4\}$
27. $\{5\}$ (The value 0 does not check.)   **29.** $\{-5\}$
31. $\{\ \}$ (The value 4 does not check.)   **33.** $\{4\}$ **35.** $\{4, -3\}$
37. $\{-4\}$; (The value 1 does not check.)   **39.** $\{\ \}$ (The value -4 does not check.)   **41.** $\{4\}$ (The value -6 does not check.)
43. $\{-25\}$ **45.** $\{-2, 7\}$ **47.** The number is 8.
49. The number is -26.   **51.** $m = \dfrac{FK}{a}$ **53.** $E = \dfrac{IR}{K}$
55. $R = \dfrac{E-Ir}{I}$ or $R = \dfrac{E}{I} - r$ **57.** $B = \dfrac{2A-hb}{h}$ or $B = \dfrac{2A}{h} - b$
59. $h = \dfrac{V}{r^2\pi}$ **61.** $t = \dfrac{b}{x-a}$ or $t = \dfrac{-b}{a-x}$
63. $x = \dfrac{y}{1-yz}$ or $x = \dfrac{-y}{yz-1}$ **65.** $h = \dfrac{2A}{a+b}$ **67.** $R = \dfrac{R_1 R_2}{R_2 + R_1}$

Chapter 15 Problem Recognition Exercises, p. 1037

1. $\dfrac{y-2}{2y}$ **2.** $\{6\}$ **3.** $\{2\}$ **4.** $\dfrac{3a-17}{a-5}$ **5.** $\dfrac{4p+27}{18p^2}$
6. $\dfrac{b(b-5)}{(b-1)(b+1)}$ **7.** $\{5\}$ **8.** $\dfrac{2w+5}{(w+1)^2}$ **9.** $\{7\}$ **10.** $\{5\}$
11. $\dfrac{3x+14}{4(x+1)}$ **12.** $\left\{\dfrac{11}{3}\right\}$ **13.** $\left\{\dfrac{41}{10}\right\}$ **14.** $\dfrac{7-3t}{t(t-5)}$
15. $\dfrac{8a+1}{2a-1}$ **16.** $\{5\}$ (The value 2 does not check.)
17. $\{-1\}$ (The value 3 does not check.)   **18.** $\dfrac{11}{12k}$
19. $\dfrac{h+9}{(h-3)(h+3)}$ **20.** $\{7\}$

Section 15.7 Activity, pp. 1045–1046

A.1. a. 100   **b.** $\dfrac{1 \text{ in.}}{50 \text{ mi}} = \dfrac{2 \text{ in.}}{\boxed{100 \text{ mi}}}$
A.2. a. 48   **b.** $\dfrac{8 \text{ apples}}{\$4.00} = \dfrac{\boxed{48 \text{ apples}}}{\$24.00}$
A.3. a. Let x represent the amount of coffee required to make 300 cups.
  b. $\dfrac{2 \text{ lb}}{75 \text{ cups}} = \dfrac{x \text{ lb}}{300 \text{ cups}}$
  c. 8 lb; this means that 8 lb of ground coffee would be required to make 300 cups.
A.4. a. One possibility: Let x represent Sean's speed running.
  c. $t = \dfrac{d}{r}$
  b., d.

	Distance (mi)	Rate (mph)	Time (hr)
Sean running	12	x	$\dfrac{12}{x}$
Sean bicycling	72	$x + 20$	$\dfrac{72}{x+20}$

  e. $\dfrac{12}{x} = \dfrac{72}{x+20}$ **f.** $x = 4$
  g. Sean's speed running is 4 mph and his speed bicycling is 24 mph.

A.5. b. $\dfrac{1 \text{ job}}{4 \text{ hr}} + \dfrac{1 \text{ job}}{6 \text{ hr}} = \dfrac{1 \text{ job}}{x \text{ hr}}; \dfrac{1}{4} + \dfrac{1}{6} = \dfrac{1}{x}$ **c.** $x = \dfrac{12}{5}$
  d. It will take $\dfrac{12}{5}$ hr or equivalently 2.4 hr (2 hr, 24 min) to clean the pool if they work together.

Section 15.7 Practice Exercises, pp. 1046–1050

R.1. Equation; $\{44\}$ **R.3.** Expression; $\dfrac{m^2+m+2}{(m-1)(m+3)}$
R.5. Expression; $\dfrac{3}{10}$ **R.7.** Equation; $\{2\}$

1. proportional   **3.** $\{95\}$ **5.** $\{1\}$ **7.** $\left\{\dfrac{40}{3}\right\}$ **9.** $\{40\}$
11. $\{3\}$ **13.** $\{-1\}$ **15.** $\{1\}$ **17. a.** $V_f = \dfrac{V_i T_f}{T_i}$ **b.** $T_f = \dfrac{T_i V_f}{V_i}$
19. Toni can drive 297 mi on 9 gal of gas.   **21.** Mix 14.82 mL of plant food.   **23.** 5 oz contains 12 g of carbohydrate.
25. The minimum length is 20 ft.   **27.** $x = 4$ cm; $y = 5$ cm
29. $x = 3.75$ cm; $y = 4.5$ cm   **31.** The height of the pole is 7 m.
33. The light post is 24 ft high.   **35.** The speed of the boat is 20 mph.   **37.** The plane flies 210 mph in still air.   **39.** He runs 8 mph and bikes 16 mph.   **41.** Floyd walks 4 mph and Rachel walks 2 mph.   **43.** Sergio rode 12 mph and walked 3 mph.
45. $5\dfrac{5}{11} (5.\overline{45})$ min   **47.** $22\dfrac{2}{9} (22.\overline{2})$ min
49. 48 hr   **51.** $3\dfrac{1}{3} (3.\overline{3})$ days
53. There are 40 smokers and 140 nonsmokers.
55. There are 240 men and 200 women.

Chapter 15 Review Exercises pp. 1051–1053

1. a. $-\dfrac{2}{9}, -\dfrac{1}{10}, 0, -\dfrac{5}{6}$, undefined   **b.** $t = -9$
2. a. $-\dfrac{1}{5}, -\dfrac{1}{2}$, undefined, 0, $\dfrac{1}{7}$ **b.** $k = 5$ **3.** a, c, d
4. $x = \dfrac{5}{2}, x = 3; \dfrac{1}{2x-5}$ **5.** $h = -\dfrac{1}{3}, h = -7; \dfrac{1}{3h+1}$
6. $a = 2, a = -2; \dfrac{4a-1}{a-2}$ **7.** $w = 4, w = -4; \dfrac{2w+3}{w-4}$
8. $z = 4; -\dfrac{z}{2}$ **9.** $k = 0, k = 5; -\dfrac{3}{2k}$ **10.** $b = -3; \dfrac{b-1}{2}$
11. $m = -1; \dfrac{m-5}{3}$ **12.** $n = -3; \dfrac{1}{n+3}$ **13.** $p = -7; \dfrac{1}{p+7}$
14. y^2 **15.** $\dfrac{u^2}{2}$ **16.** $v + 2$ **17.** $\dfrac{3}{2(x-5)}$ **18.** $\dfrac{c(c+1)}{2(c+5)}$
19. $\dfrac{q-2}{4}$ **20.** $-2t(t-5)$ **21.** $4s(s-4)$ **22.** $\dfrac{1}{7}$
23. $\dfrac{1}{n-2}$ **24.** $-\dfrac{1}{6}$ **25.** $\dfrac{1}{m+3}$ **26.** $\dfrac{-1}{(x+3)(x+2)}$
27. $-\dfrac{2y-1}{y+1}$ **28.** LCD $= (n+3)(n-3)(n+2)$
29. LCD $= (m+4)(m-4)(m+3)$ **30.** $c - 2$ or $2 - c$
31. $3 - x$ or $x - 3$ **32.** $\dfrac{4b}{10ab}; \dfrac{3a}{10ab}$ **33.** $\dfrac{21y}{12xy}; \dfrac{22x}{12xy}$
34. $\dfrac{y}{x^2y^5}; \dfrac{3x}{x^2y^5}$ **35.** $\dfrac{5c^2}{ab^3c^2}; \dfrac{3b^3}{ab^3c^2}$
36. $\dfrac{5p-20}{(p+2)(p-4)}; \dfrac{p^2+2p}{(p+2)(p-4)}$ **37.** $\dfrac{6q+48}{q(q+8)}; \dfrac{q}{q(q+8)}$
38. 2   **39.** 2   **40.** $a + 5$ **41.** $x - 7$
42. $\dfrac{-y-18}{(y-9)(y+9)}$ or $\dfrac{y+18}{(9-y)(y+9)}$ **43.** $\dfrac{t^2+2t+3}{(2-t)(2+t)}$
44. $\dfrac{m+8}{3m(m+2)}$ **45.** $\dfrac{3(r-4)}{2r(r+6)}$ **46.** $\dfrac{p}{(p+4)(p+5)}$
47. $\dfrac{q}{(q+5)(q+4)}$ **48.** $\dfrac{1}{2}$ **49.** $\dfrac{1}{3}$ **50.** $\dfrac{a-4}{a-2}$

51. $\dfrac{3(z+5)}{z(z-5)}$ **52.** $\dfrac{w}{2}$ **53.** $\dfrac{8}{y}$ **54.** $y-x$ **55.** $-(b+a)$
56. $-\dfrac{2p+7}{2p}$ **57.** $-\dfrac{k+10}{k+4}$ **58.** $\{-8\}$ **59.** $\{-2\}$
60. $\{0\}$ **61.** $\{2\}$ **62.** $\{-2\}$ **63.** $\{-1\}$ (The value 3 does not check.)   **64.** $\{\ \}$ (The value 2 does not check.)   **65.** $\{-11, 1\}$
66. The number is 4.   **67.** $h = \dfrac{3V}{\pi r^2}$ **68.** $b = \dfrac{2A}{h}$
69. $\left\{\dfrac{6}{5}\right\}$ **70.** $\left\{\dfrac{96}{5}\right\}$ **71.** It contains 10 g of fat.
72. Ed travels 60 mph, and Bud travels 70 mph.   **73.** Together the pumps would fill the pool in 16.8 min.   **74.** $x = 11$; $b = 26$

Chapter 16

Section 16.1 Activity, p. 1061

A.1. a. x values   **b.** $\{-3, 0, 2, -1\}$ **c.** range   **d.** $\{1, 4, -5, -4\}$
A.2. a. $\{A, E, I, O, U\}$ **b.** $\{1, 9, 15, 21, 5\}$
A.3. a. $\{-4, -2, 0, 2, 3\}$ **b.** $\{-3, -2, 1, 2, 4\}$
A.4. a. $-4; 4$ **b.** $[-4, 4]$ **c.** $-2; 4$ **d.** $[-2, 4]$
A.5. a. The point $(1, -1)$ is *not* included in the graph.
  b. Infinitely far to the left; up to but not including $x = 1$ to the right.
  c. $(-\infty, 1)$ **d.** -2; no   **e.** $[-2, \infty)$

Section 16.1 Practice Exercises, pp. 1062–1065

R.1. $(-6, \infty)$ **R.3.** $\left(-\infty, \dfrac{3}{4}\right]$ **R.5.** $(0.5, 1.2]$
R.7. $(-4, 1)$ **R.9.** $(1, -4)$ **R.11.** $(4, 0)$

1. a. relation   **b.** domain   **c.** range
3. a. {(Northeast, 54.1), (Midwest, 65.6), (South, 110.7), (West, 70.7)}   **b.** Domain: {Northeast, Midwest, South, West}; Range: {54.1, 65.6, 110.7, 70.7}   **5. a.** {(USSR, 1961), (USA, 1962), (Poland, 1978), (Vietnam, 1980), (Cuba, 1980)}   **b.** Domain: {USSR, USA, Poland, Vietnam, Cuba}; range: {1961, 1962, 1978, 1980}
7. a. {(A, 1), (A, 2), (B, 2), (C, 3), (D, 5), (E, 4)}
  b. Domain: {A, B, C, D, E}; range: {1, 2, 3, 4, 5}
9. a. {(6, A), (6, B), (6, C)}   **b.** Domain: {6}; range: {A, B, C}
11. a. {(−4, 1), (0, 3), (1, −1), (3, 0)}
  b. Domain: {−4, 0, 1, 3}; range: {−1, 0, 1, 3}
13. a. {(−4, 4), (1, 1), (2, 1), (3, 1), (4, −2)}
  b. Domain: {−4, 1, 2, 3, 4}; range: {−2, 1, 4}
15. Domain: [0, 4]; range: [−2, 2]   **17.** Domain: [−5, 3]; range: [−2.1, 2.8]   **19.** Domain: (−∞, 2]; range: (−∞, ∞)
21. Domain: (−∞, ∞); range: (−∞, ∞)   **23.** Domain: {−3}; range: (−∞, ∞)   **25.** Domain: (−∞, 2); range: [−1.3, ∞)
27. Domain: {−3, −1, 1, 3}; range: {0, 1, 2, 3}
29. Domain: [−4, 5); range: {−2, 1, 3}   **31. a.** 2.85
  b. 9.33   **c.** Dec.   **d.** Nov.   **e.** 7.63   **f.** {Jan., Feb., Mar., Apr., May, June, July, Aug., Sept., Oct., Nov., Dec.}
33. a. 34.2 million or 34,200,000   **b.** The year 2019
35. a. For example: {(Julie, New York), (Peggy, Florida), (Stephen, Kansas), (Pat, New York)}   **b.** Domain: {Julie, Peggy, Stephen, Pat}; range: {New York, Florida, Kansas}
37. $y = 2x - 1$ **39.** $y = x^2$

Section 16.2 Activity, pp. 1072–1073

A.1. The relation defines y as a function of x if for each element x in the domain, there is exactly one value of y in the range.
A.2. a. Yes. The ordered pairs $(-5, 1)$ and $(-5, 2)$ have the same x value, but different y values.   **b.** No
A.3. a. No   **b.** Yes

A.4. a. Yes. The vertical line intersects the graph at $(4, 2)$ and $(4, -2)$.

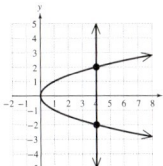

  b. No   **c.** Two points on a graph that are aligned vertically must have the same x-coordinate, but different y-coordinates. In such a case the graph fails to define y as a function of x.

A.5. a. 7   **b.** (3, 7)   **A.6. a.** $f(3) = 7$ **b.** (3, 7)
  c. The results are the same because evaluating a function for a given value of x indicates that we substitute the value for x into the function and solve for y.
A.7. a. $g(2) = 4$ **b.** $g(-3) = -1$ **c.** $g(0) = -4$
  d. Not possible. Substituting 1 for x would make the denominator zero.
A.8. a. $x = 1$ must be excluded because substituting 1 into the function would make the denominator equal to zero. Division by zero is undefined.
  b. $(-\infty, 1) \cup (1, \infty)$
A.9. a. $h(-1) = 1$ **b.** $h(2) = 2$ **c.** $h(7) = 3$
  d. Not possible. Substituting -6 for x would result in the square root of a negative number.
A.10. a. The radicand $x + 2$ must be greater than or equal to zero. That is, $x + 2 \geq 0$. This means that $x \geq -2$.   **b.** $[-2, \infty)$
A.11. a. $f(-2) = 14$ **b.** $f(t) = t^2 - 3t + 4$ **c.** $f(a+2) = a^2 + a + 2$

Section 16.2 Practice Exercises, pp. 1073–1078

R.1. -16 **R.3.** 17   **R.5.** $x = 5$ **R.7.** $x = -3, x = 3$
R.9. $[4, \infty)$ **R.11.** $(-6, -5]$ **R.13.** $\left[\dfrac{3}{2}, \infty\right)$ **R.15.** $(-\infty, 4]$

1. a. function   **b.** vertical   **c.** $2x + 1$ **d.** x values   **e.** y values   **f.** denominator   **g.** negative   **3.** h **5.** Function
7. Not a function   **9.** Function   **11.** Not a function
13. Function   **15.** Not a function   **17.** When x is 2, the function value y is 5.   **19.** $(0, -2)$ **21.** -11 **23.** 1
25. 2   **27.** $6t - 2$ **29.** 7   **31.** 4   **33.** 4
35. $6x + 4$ **37.** $-4x^2 - 8x + 1$ **39.** $-\pi^2 + 4\pi + 1$
41. 7   **43.** $-6a - 2$ **45.** $|-c - 2|$ **47.** 1   **49.** 7
51. -18.8 **53.** -7 **55.** 2π **57.** -5 **59.** 4
61. a. 3   **b.** 1   **c.** 1   **d.** $x = -3$ **e.** $x = 0, x = 2$
f. $(-\infty, 3]$ **g.** $(-\infty, 5]$ **63. a.** 3   **b.** $H(4)$ is not defined because 4 is not in the domain of H.   **c.** 4   **d.** $x = -3$ and $x = 2$ **e.** All x on the interval $[-2, 1]$ **f.** $[-4, 4)$ **g.** $[2, 5)$
65. a. -4 **b.** 0   **c.** -3 **d.** $x = -1$ **e.** There are no such values of x.   **f.** $(-\infty, \infty)$ **g.** $(-\infty, -3] \cup (-2, \infty)$
67. $\{-3, -7, -\tfrac{3}{2}, 1.2\}$ **69.** $\{6, 0\}$
71. $x = -3$ and $x = 1.2$ **73.** $x = 6$ and $x = 1$ **75.** -3
77. The domain is the set of all real numbers for which the denominator is not zero. Set the denominator equal to zero, and solve the resulting equation. The solution(s) to the equation must be excluded from the domain. In this case, setting $x - 2 = 0$ indicates that $x = 2$ must be excluded from the domain. The domain is $(-\infty, 2) \cup (2, \infty)$.   **79.** $(-\infty, -6) \cup (-6, \infty)$
81. $(-\infty, 0) \cup (0, \infty)$ **83.** $(-\infty, \infty)$ **85.** $[-7, \infty)$
87. $[3, \infty)$ **89.** $(-\infty, \tfrac{1}{2}]$ **91.** $(-\infty, \infty)$ **93.** $(-\infty, \infty)$
95. a. $h(1) = 64$ and $h(1.5) = 44$ **b.** $h(1) = 64$ means that after 1 sec, the height of the ball is 64 ft. $h(1.5) = 44$ means that after 1.5 sec, the height of the ball is 44 ft.   **97. a.** $d(1) = 11.5$ and $d(1.5) = 17.25$ **b.** $d(1) = 11.5$ means that after 1 hr, the distance traveled is 11.5 mi. $d(1.5) = 17.25$ means that after 1.5 hr, the distance traveled is 17.25 mi.
99. $f(x) = 2x + 3$ **101.** $f(x) = |x| - 10$

103. 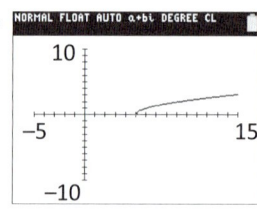 **104.**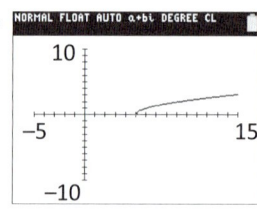

105. $(-2, \infty)$

Section 16.3 Activity, p. 1085

A.1. a. Slanted line **b.** $f(x) = -3x - 2$ **c.** Linear function
A.2. a. Horizontal line **b.** $f(x) = 5$ **c.** Constant function
A.3. a. Parabola **b.** $f(x) = x^2 - 4x + 3$ **c.** Quadratic function
A.4. a. Substitute 0 for $f(x)$ and solve the resulting equation for x.
  b. $\left(-\dfrac{2}{3}, 0\right)$
  c. Substitute 0 for x. Using function notation, this is equivalent to evaluating $f(0)$.
  d. $(0, -2)$
A.5. x-intercept: none; y-intercept: $(0, 5)$
A.6. x-intercepts: $(1, 0)$ and $(3, 0)$; y-intercept: $(0, 3)$

Section 16.3 Practice Exercises, 1085–1091

R.1. Vertical line   **R.3.** Slanted line
R.5. x-intercept: $(5, 0)$;   **R.7.** x-intercept: none;
  y-intercept: $(0, 3)$ y-intercept: $(0, -4)$

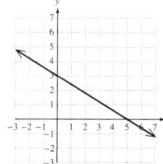 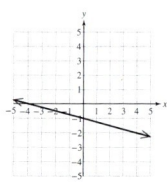

1. a. linear **b.** constant **c.** quadratic **d.** parabola **e.** 0 **f.** y
3. a. $g(1) = -3$ **b.** $g(2) = -3$ **c.** $g(3) = -3$
5. a. $k(1) = -3$ **b.** $k(2) = -6$ **c.** $k(3) = -9$ **7.** horizontal
9. Domain $(-\infty, \infty)$; range $\{2\}$

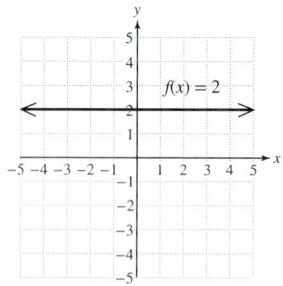

11.

x	$f(x)$	x	$f(x)$
-2	$-\frac{1}{2}$	$\frac{1}{4}$	4
-1	-1	$\frac{1}{2}$	2
$-\frac{1}{2}$	-2	1	1
$-\frac{1}{4}$	-4	2	$\frac{1}{2}$

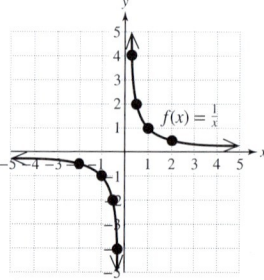

13.

x	$h(x)$
-2	-8
-1	-1
0	0
1	1
2	8

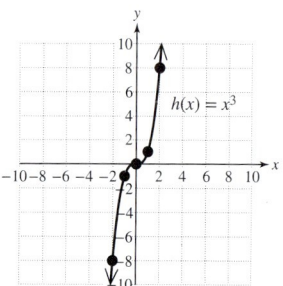

15.

x	$q(x)$
-2	4
-1	1
0	0
1	1
2	4

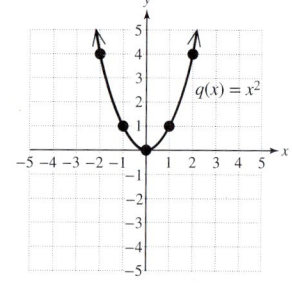

17. Quadratic   **19.** Linear   **21.** Constant
23. None of these   **25.** Linear   **27.** None of these
29. x-intercept: $(2, 0)$;   **31.** x-intercept: $\left(\dfrac{5}{6}, 0\right)$;
  y-intercept: $(0, -10)$ y-intercept: $(0, 5)$

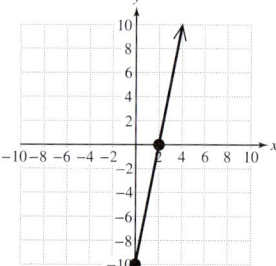 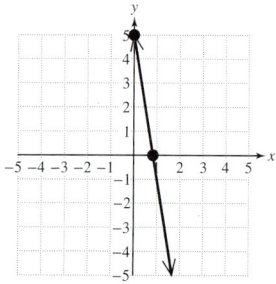

33. x-intercept: none;   **35.** x-intercept: $(-3, 0)$;
  y-intercept: $(0, 18)$ y-intercept: $(0, 2)$

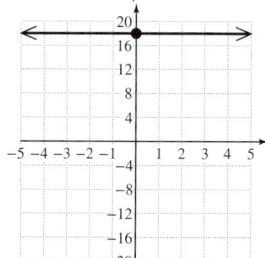 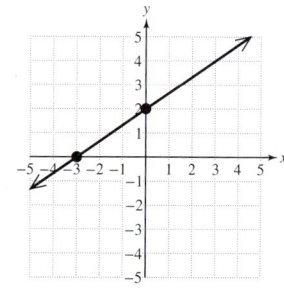

37. a. $x = -1$ **b.** $f(0) = 1$ **39. a.** $x = -2$ and $x = 2$
  b. $f(0) = -2$ **41. a.** None   **b.** $f(0) = 2$
43. a. $(-\infty, \infty)$ **b.** $(0, 0)$ **c.** vi   **45. a.** $(-\infty, \infty)$
  b. $(0, 1)$ **c.** viii   **47. a.** $[-1, \infty)$ **b.** $(0, 1)$ **c.** vii
49. a. $(-\infty, 3) \cup (3, \infty)$ **b.** $\left(0, -\dfrac{1}{3}\right)$ **c.** ii
51. a. $(-\infty, \infty)$ **b.** $(0, 2)$ **c.** iv   **53. a.** Linear
  b. $G(90) = 82.5$. This means that if the student gets a 90% score on her final exam, then her overall course average is 82.5%.
  c. $G(50) = 72.5$. This means that if the student gets a 50% score on her final exam, then her overall course average is 72.5%.

55. $f(x) = x^2 + 2$ **57.** $f(x) = 3$ **59.** $f(x) = \frac{1}{2}x - 2$

61. 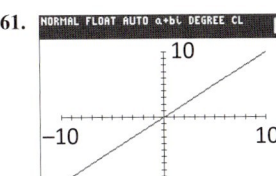   **62.**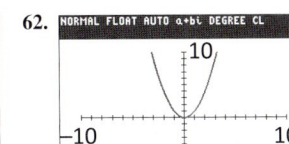

63. 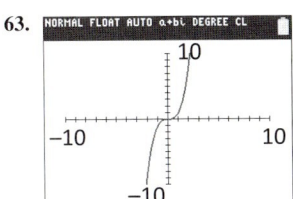   **64.**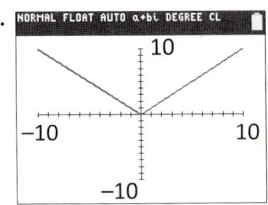

65. 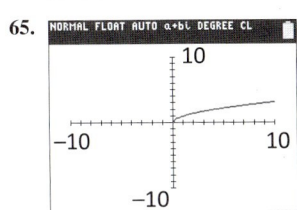   **66.**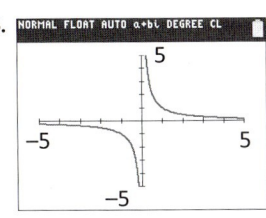

67. x-intercept: (8, 0); y-intercept: (0, 1)   **68.** x-intercept: (−6, 0); y-intercept: (0, −3)   **69.** x-intercept: (−5, 0); y-intercept: (0, 4)   **70.** x-intercept: (2, 0); y-intercept: (0, −7)

Chapter 16 Problem Recognition Exercises, p. 1091

1. a, c, d, f, g   **2.** a, b, d, f, h   **3.** 4   **4.** 2
5. $\{0, 1, \frac{1}{2}, -3, 2\}$ **6.** $\{4, 3, 6, 1, 10\}$ **7.** $[-2, 4]$
8. $[0, \infty)$ **9.** 0   **10.** $\left(\frac{9}{5}, 0\right)$ **11.** $(0, -1), (0, 1)$
12. 1   **13.** c   **14.** d   **15.** 3

Section 16.4 Activity, p. 1096

A.1. a. $f(0) = -4$ b. $g(0) = 2$ c. $f(0) + g(0) = -2$
   d. $(f + g)(1) = 1$ e. $(f + g)(-2) = -2$
A.2. a. $(f - g)(x) = f(x) - g(x)$ b. $(f \cdot g)(x) = f(x) \cdot g(x)$
   c. $\left(\frac{f}{g}\right)(x) = \frac{f(x)}{g(x)}$
A.3. a. Substitute the quantity $x - 4$ for all values of x in function f. Then simplify the result.
   b. $f(x - 4) = x^2 - 9x + 8$ c. $g(x^2 - x - 12) = x^2 - x - 16$
A.4. a. $(f \circ g)(x) = f(g(x))$; $(g \circ f)(x) = g(f(x))$
   b. $(f \circ g)(x) = x^2 - 9x + 8$ c. $(g \circ f)(x) = x^2 - x - 16$
A.5. a. $(h \circ k)(4) = h[k(4)] = h(-1) = -7$
   b. $(k \circ h)(0) = k[h(0)] = k(-4) = -\frac{1}{9}$
A.6. a. $(h \circ k)(6) = -1$
   b. $(h \circ k)(5)$ is undefined because $k(5)$ is undefined.
   c. $(k \circ h)(1) = -\frac{1}{6}$
   d. $(k \circ h)(3)$ is undefined because $h(3) = 5$, but 5 is not in the domain of k.
A.7. a. $(f \cdot g)(-1) = 0$ b. $(g - f)(-3) = -9$ c. $\left(\frac{f}{g}\right)(0) = -2$
   d. The value $f(-2) = 0$. Therefore, the quotient $\frac{g(-2)}{f(-2)}$ is undefined because division by 0 is undefined.
   e. $(f \circ g)(-1) = -4$ f. $(g \circ f)(-2) = 2$

Section 16.4 Practice Exercises, pp. 1096–1100

R.1. $9x^2 - 3x - 8$ **R.3.** $-2x^4 + 5x^2 - 6$ **R.5.** $\frac{4}{x-1}$
R.7. $-x$ **R.9.** $(-\infty, \infty)$ **R.11.** $\left(-\infty, -\frac{2}{3}\right) \cup \left(-\frac{2}{3}, \infty\right)$
R.13. $(-\infty, \infty)$ **R.15.** $(-\infty, 2]$ **R.17.** 8   **R.19.** -6

1. a. $f(x); g(x)$ b. $g(x); g(x)$ c. $f(g(x))$
3. $(f + g)(x) = 2x^2 + 5x + 4$
5. $(g - f)(x) = 2x^2 + 3x - 4$
7. $(f \cdot h)(x) = x^3 + 4x^2 + x + 4$
9. $(g \cdot f)(x) = 2x^3 + 12x^2 + 16x$
11. $\left(\frac{h}{f}\right)(x) = \frac{x^2+1}{x+4}, x \neq -4$
13. $\left(\frac{f}{g}\right)(x) = \frac{x+4}{2x^2+4x}, x \neq 0, x \neq -2$
15. $(f \circ g)(x) = 2x^2 + 4x + 4$
17. $(g \circ f)(x) = 2x^2 + 20x + 48$ **19.** $(k \circ h)(x) = \frac{1}{x^2+1}$
21. $(k \circ g)(x) = \frac{1}{2x^2+4x}, x \neq 0, x \neq -2$ **23.** No
25. $(f \circ g)(x) = 25x^2 - 15x + 1$; $(g \circ f)(x) = 5x^2 - 15x + 5$
27. $(f \circ g)(x) = |x^3 - 1|$; $(g \circ f)(x) = |x|^3 - 1$
29. $(h \circ h)(x) = 25x - 24$ **31.** 0   **33.** −64   **35.** 2
37. 1   **39.** $\frac{1}{64}$ **41.** 0   **43.** Not a real number
45. $\frac{4}{9}$ **47.** −2   **49.** 2   **51.** 0   **53.** 1   **55.** 0
57. Undefined   **59.** −1   **61.** 2   **63.** 2   **65.** −1
67. −2   **69.** 3   **71.** −6   **73.** −1   **75.** 4
77. 0   **79.** 2   **81.** a. $P(x) = 3.78x - 1$ b. $188
83. a. $F(t) = 0.2t + 6.4$; $F(t)$ represents the amount of child support (in billion dollars) not paid for year t.   b. $F(4) = 7.2$ means that in year 4, $7.2 billion of child support was not paid.
85. a. $(D \circ r)(t) = 560t$; This function represents the total distance Joe travels as a function of time that he rides.   b. 5600 ft

Section 16.5 Activity, pp. 1105–1106

A.1. The equation $z = kp$ represents a direct relationship between z and p, and $z = \frac{k}{p}$ represents an inverse relationship between z and p.
A.2. a. increase   b. decrease
A.3. a. $A = kw$ b. $k = \frac{1}{50}$; $A = \frac{1}{50}w$ c. more   d. 3.6 g
   e. 4.5 g   f. 2.4 g
A.4. a. $C = \frac{k}{n}$ b. $k = 2400$; $C = \frac{2400}{n}$ c. down   d. $0.40
   e. $0.30   f. $1.00
A.5. a. $y = kb\sqrt{c}$ b. $k = 3$ c. $y = 3b\sqrt{c}$ d. 126

Section 16.5 Practice Exercises, pp. 1106–1109

R.1. $\{5\}$ **R.3.** $\{0.64\}$ **R.5.** 48   **R.7.** 2

1. a. kx b. $\frac{k}{x}$ c. kxw **3.** Inversely   **5.** $T = kq$ **7.** $b = \frac{k}{c}$
9. $Q = \frac{kx}{y}$ **11.** $c = kst$ **13.** $L = kw\sqrt{v}$ **15.** $x = \frac{ky^2}{z}$
17. $k = \frac{9}{2}$ **19.** $k = 512$ **21.** $k = 1.75$ **23.** $x = 70$
25. $b = 6$ **27.** $Z = 56$ **29.** $Q = 9$ **31.** $L = 9$
33. $B = \frac{15}{2}$ **35.** a. The heart weighs 0.92 lb.   b. Answers will vary.
37. a. 5 turkeys   b. 10 turkeys   c. 50 turkeys
   d. 8 turkeys; Note that the answer was rounded up, because it would be better for the chef to have too much food than too little.
39. 355,000 tons   **41.** 42.6 ft   **43.** 300 W   **45.** 18.5 A
47. 20 lb   **49.** $3500

Chapter 16 Review Exercises, pp. 1110–1112

1. Domain: $\left\{\dfrac{1}{3}, 6, \dfrac{1}{4}, 7\right\}$; range: $\left\{10, -\dfrac{1}{2}, 4, \dfrac{2}{5}\right\}$
2. Domain: $\{-3, -1, 0, 2, 3\}$; range: $\left\{-2, 0, 1, \dfrac{5}{2}\right\}$
3. Domain: $[-3, 9]$; range: $[0, 60]$
4. Domain: $[-4, 4]$; range: $[1, 3]$
5.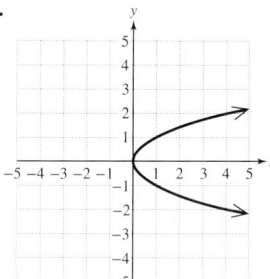
6.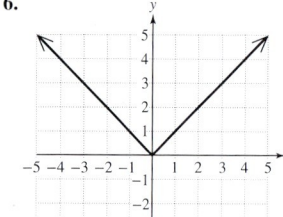
7. **a.** Not a function **b.** $[1, 3]$ **c.** $[-4, 4]$
8. **a.** Function **b.** $(-\infty, \infty)$ **c.** $(-\infty, 0.35]$
9. **a.** Function **b.** $\{1, 2, 3, 4\}$ **c.** $\{3\}$
10. **a.** Not a function **b.** $\{0, 4\}$ **c.** $\{2, 3, 4, 5\}$
11. **a.** Not a function **b.** $\{4, 9\}$ **c.** $\{2, -2, 3, -3\}$
12. **a.** Function **b.** $\{6, 7, 8, 9\}$ **c.** $\{9, 10, 11, 12\}$
13. -4 14. 2 15. 2 16. $6t^2 - 4$ 17. $6b^2 - 4$
18. $6\pi^2 - 4$ 19. $6a^2 - 4$ 20. 20 21. $(-\infty, \infty)$
22. $(-\infty, 11) \cup (11, \infty)$ 23. $[8, \infty)$ 24. $[-2, \infty)$
25. **a.** \$150 **b.** \$175 **c.** \$200
26.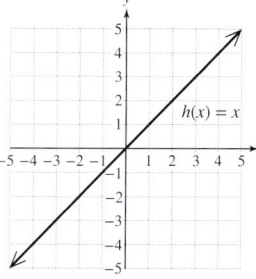
27.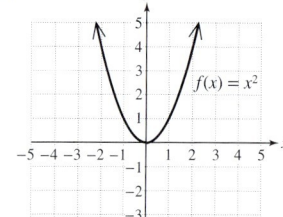
28.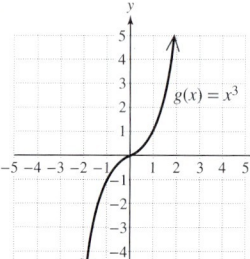
29.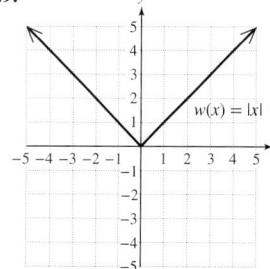
30.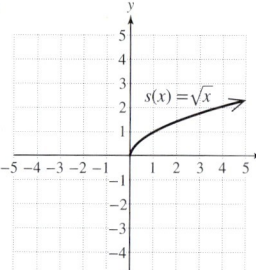
31.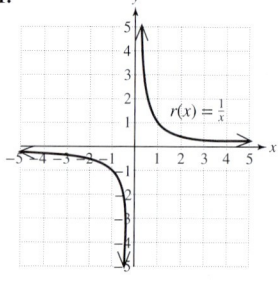

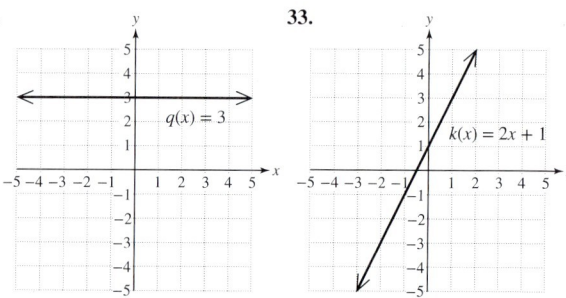

34. **a.** $s(4) = 4, s(-3) = 25, s(2) = 0, s(1) = 1, s(0) = 4$
    b. $(-\infty, \infty)$ 35. **a.** $r(2)$ is not a real number, $r(4) = 0, r(5) = 2, r(8) = 4$ **b.** $[4, \infty)$
36. **a.** $h(-3) = -\dfrac{1}{2}, h(0) = -1, h(2) = -3, h(5) = \dfrac{3}{2}$
    b. $(-\infty, 3) \cup (3, \infty)$
37. **a.** $k(-5) = -2, k(-4) = -1, k(-3) = 0,$
    $k(2) = -5$ **b.** $(-\infty, \infty)$ 38. x-intercept: $\left(\dfrac{7}{4}, 0\right)$; y-intercept: $(0, -7)$
39. x-intercept: $\left(\dfrac{9}{2}, 0\right)$; y-intercept: $(0, 9)$
40. **a.** $b(0) = 28.3$. In 2010 consumption was 28.3 gal of bottled water per capita. $b(5) = 36.5$. In 2015 consumption was 36.5 gal of bottled water per capita. **b.** $m = 1.64$. Consumption increased at a rate of 1.64 gal/year.
41. -1 42. 0 43. $x = 0$ and $x = 4$
44. There are no values of x for which $g(x) = 4$.
45. $(-4, \infty)$ 46. $(-\infty, 1]$ 47. $(f - g)(x) = 2x^3 + 9x - 7$
48. $(f + g)(x) = -2x^3 - 7x - 7$ 49. $(f \cdot n)(x) = \dfrac{x - 7}{x - 2}, x \neq 2$
50. $(f \cdot m)(x) = x^3 - 7x^2$ 51. $\left(\dfrac{f}{g}\right)(x) = \dfrac{x - 7}{-2x^3 - 8x}; x \neq 0$
52. $\left(\dfrac{g}{f}\right)(x) = \dfrac{-2x^3 - 8x}{x - 7}, x \neq 7$
53. $(m \circ f)(x) = (x - 7)^2$ or $x^2 - 14x + 49$
54. $(n \circ f)(x) = \dfrac{1}{x - 9}, x \neq 9$ 55. 100 56. $\dfrac{1}{8}$
57. -167 58. -10 59. **a.** $(2x + 1)^2$ or $4x^2 + 4x + 1$
    b. $2x^2 + 1$ **c.** No, $f \circ g \neq g \circ f$ 60. $\dfrac{1}{4}$ 61. -3
62. 0 63. -1 64. 4 65. 1
66. **a.** $F = kd$ **b.** $k = 3$ **c.** 12.6 lb
67. $y = \dfrac{8}{3}$ 68. $y = 12$ 69. 48 km

Chapter 17

Section 17.1 Activity, pp. 1122–1123

A.1.

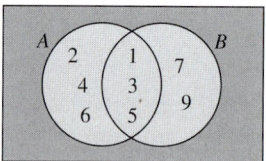

A.2. a. union; $A \cup B$ **b.** $A \cup B = \{1, 2, 3, 4, 5, 6, 7, 9\}$
A.3. a. intersection; $A \cap B$ **b.** $\{1, 3, 5\}$
A.4. a. ⟵———————[——————⟶ **b.** ———————]——————⟶
           2 -3
c. ⟵————————————————————⟶ **d.** $(-\infty, \infty)$
e. ⟵——[————————]——⟶ **f.** $[-3, 2)$
         -3 2

A.5. a. $(-5, \infty)$ **b.** $[-1, \infty)$ **c.** $[-1, \infty)$
   d. $(-5, \infty)$ **e.** intersection **f.** union
A.6. a. $-3 \leq \frac{1}{2}x - 5$ and $\frac{1}{2}x - 5 < 1$ **b.** $[4, 12)$
A.7. a. $4{,}200{,}000 \leq x \leq 5{,}900{,}000$ **b.** $x < 4{,}200{,}000$ or $x > 5{,}900{,}000$

Section 17.1 Practice Exercises, pp. 1123–1128

R.1. $(-\infty, 8)$ **R.3.** $(-9, -6]$ **R.5.** $[-6, \infty)$
R.7. $[-2, 6)$ **R.9.** $[18, \infty)$

1. a. union; $A \cup B$ **b.** intersection; $A \cap B$ **c.** intersection **d.** $a < x < b$ **e.** union **3.** $\{2, 4\}$ **5.** $\{2, 4, 5, 6, 7, 8, 10, 11\}$
7. a. $\{-3, -1\}$ **b.** $\{-4, -3, -2, -1, 0, 1, 3, 5\}$
9. $[-7, -4)$ **11.** $(-\infty, -4) \cup (2, \infty)$ **13.** $\{\ \}$ **15.** $[-7, \infty)$
17. $[0, 5)$ **19.** $[-7, \infty)$ **21. a.** $[-1, 5)$ **b.** $(-2, \infty)$
23. a. $(-1, 3)$ **b.** $\left(-\frac{5}{2}, \frac{9}{2}\right)$ **25. a.** $(0, 2]$ **b.** $(-4, 5]$
27. $[-2, 3]$
29. $(3, 6)$
31. $(-\infty, 2]$
33. $[8, \infty)$
35. $\{\ \}$ **37.** $-4 \leq t$ and $t < \frac{3}{4}$ **39.** The statement $6 < x < 2$ is equivalent to $6 < x$ and $x < 2$. However, no real number is greater than 6 and also less than 2. **41.** The statement $-5 > y > -2$ is equivalent to $-5 > y$ and $y > -2$. However, no real number is less than -5 and also greater than -2.
43. $\left[\frac{5}{2}, 7\right)$
45. $(-6, 6]$
47. $(2, 8)$
49. $\left[-\frac{10}{3}, -\frac{7}{3}\right]$
51. $\left[-\frac{1}{2}, \frac{3}{2}\right)$
53. $(-0.4, 0.2)$
55. $(-\infty, -2) \cup [2, \infty)$
57. $(-\infty, 2]$
59. $[-6, \infty)$
61. $(-\infty, \infty)$
63. $(-\infty, -4] \cup (4.5, \infty)$
65. a. $(-10, 8)$ **b.** $(-\infty, \infty)$ **67. a.** $(-\infty, -8] \cup [1.3, \infty)$ **b.** $\{\ \}$
69. $\left(-\frac{11}{2}, \frac{7}{2}\right]$ **71.** $\left(-\frac{13}{2}, \infty\right)$ **73.** $(-\infty, 1) \cup (5, \infty)$
75. a. $4800 \leq x \leq 10{,}800$ **b.** $x < 4800$ or $x > 10{,}800$
77. a. $44\% < x < 48\%$ **b.** $x \leq 44\%$ or $x \geq 48\%$
79. All real numbers between $-\frac{3}{2}$ and 6
81. All real numbers greater than 2 or less than -1
83. a. Amy would need 82% or better on her final exam.
   b. If Amy scores at least 32% and less than 82% on her final exam she will receive a "B" in the class.
85. $32° \leq F \leq 42.08°$

Section 17.2 Activity, pp. 1135–1136

A.1. a. $0, 5,$ and -4 **b.** $f(-1) = 0, f(4) = 5,$ and $f(1) = -4$
   c. -1 and 3 **d.** $(-1, 3)$ **e.** $(-\infty, -1) \cup (3, \infty)$
A.2. a. $\{-1, 3\}$ **b.**
   c. $(-\infty, -1)$ and $(3, \infty)$ **d.** excluded **e.** $(-\infty, -1) \cup (3, \infty)$
   f. $(-1, 3)$ **g.** The solutions to the related equation $x^2 - 2x - 3 = 0$ correspond to the x-intercepts of the quadratic function. The solutions to the inequality $x^2 - 2x - 3 > 0$ are the values of x for which the function is positive (above the x-axis). The solutions to the inequality $x^2 - 2x - 3 < 0$ are the values of x for which the function is negative (below the x-axis).
A.3. a. $\{3\}$ **b.** Undefined for $x = 2$; no
   c.
   d. $(-\infty, 2)$ and $[3, \infty)$ **e.** Yes **f.** $(-\infty, 2) \cup [3, \infty)$
   g. The solution set would not include the boundary $x = 3$. The solution set would be $(-\infty, 2) \cup (3, \infty)$ **h.** $(2, 3]$
A.4. a. $(x - 3)^2$ **b.** $(x - 3)^2 < 0$; no **c.** $\{\ \}$. The expression $x^2 - 6x + 9$ is a perfect square trinomial. Thus, the inequality is equivalent to $(x - 3)^2 < 0$. The square of any real number is greater than or equal to zero. This means that there are no real numbers that make the expression $(x - 3)^2$ negative. **d.** The inequality $x^2 - 6x + 9 \leq 0$ allows for the expression $x^2 - 6x + 9$, or equivalently, $(x - 3)^2$ to be equal to zero. This occurs for $x = 3$. The solution set is $\{3\}$. **e.** $(-\infty, \infty)$ **f.** $(-\infty, 3) \cup (3, \infty)$

Section 17.2 Practice Exercises, pp. 1136–1139

R.1. $\{-7, -2\}$ **R.3.** $\left\{-\frac{1}{2}\right\}$ **R.5.** $\left\{\frac{1}{2}, 6\right\}$ **R.7.** $\{5\}$

1. a. quadratic **b.** solutions; undefined **c.** test; point **d.** true
3. a. $(-\infty, -2) \cup (3, \infty)$ **b.** $(-2, 3)$ **c.** $[-2, 3]$ **d.** $(-\infty, -2] \cup [3, \infty)$
5. a. $(-2, 0) \cup (3, \infty)$ **b.** $(-\infty, -2) \cup (0, 3)$ **c.** $(-\infty, -2] \cup [0, 3]$
   d. $[-2, 0] \cup [3, \infty)$
7. a. $\{4, -\frac{1}{2}\}$ **b.** $(-\infty, -\frac{1}{2}) \cup (4, \infty)$ **c.** $(-\frac{1}{2}, 4)$
9. a. $\{-10, 3\}$ **b.** $(-10, 3)$ **c.** $(-\infty, -10) \cup (3, \infty)$
11. a. $\left\{-\frac{1}{2}, 3\right\}$ **b.** $\left[-\frac{1}{2}, 3\right]$ **c.** $\left(-\infty, -\frac{1}{2}\right] \cup [3, \infty)$
13. $(1, 7)$ **15.** $(-\infty, -2) \cup (5, \infty)$ **17.** $[-5, 0]$ **19.** $[4, 8]$
21. $\left(-\infty, -\frac{1}{5}\right) \cup (1, \infty)$ **23.** $(-11, 11)$ **25.** $\left(-\infty, -\frac{1}{3}\right] \cup [3, \infty)$
27. $(-\infty, -3] \cup [0, 4]$ **29.** $(-2, -1) \cup (2, \infty)$ **31. a.** $(-1, 1]$
b. $(-\infty, -1) \cup [1, \infty)$ **c.** $(-\infty, -1) \cup (1, \infty)$ **d.** $(-1, 1)$
33. a. $(-\infty, -1) \cup (2, \infty)$ **b.** $(-1, 2)$ **c.** $[-1, 2)$
d. $(-\infty, -1] \cup (2, \infty)$ **35. a.** $\{7\}$ **b.** $(-\infty, 5) \cup (7, \infty)$ **c.** $(5, 7)$
37. a. $\{4\}$ **b.** $[4, 6)$ **c.** $(-\infty, 4] \cup (6, \infty)$ **39.** $(1, \infty)$
41. $(-\infty, -4) \cup (4, \infty)$ **43.** $\left(2, \frac{7}{2}\right)$ **45.** $(5, 7]$
47. $(-\infty, 0) \cup \left[\frac{1}{2}, \infty\right)$ **49.** $(0, \infty)$ **51.** Quadratic; $[-4, 4]$
53. Quadratic; $\left(-\infty, -\frac{4}{5}\right) \cup (0, \infty)$ **55.** Linear; $\left(-\frac{5}{2}, \infty\right)$
57. Rational; $(-\infty, -1)$ **59.** Polynomial (degree > 2); $(0, 5) \cup (5, \infty)$ **61.** Polynomial (degree > 2); $(2, \infty)$
63. Polynomial (degree > 2); $(-\infty, 5]$ **65.** Quadratic; $(-2, 2)$
67. Rational; $(-\infty, -2] \cup (5, \infty)$ **69.** Linear; $(-\infty, 5]$

71. $(-\infty, 0) \cup (2, \infty)$ **72.** $(0, 2)$

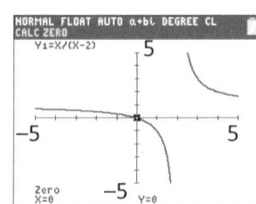 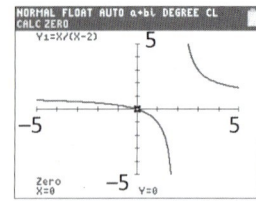

73. $(-1, 1)$ **74.** $(-\infty, -1) \cup (1, \infty)$

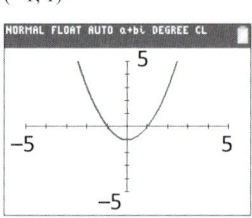 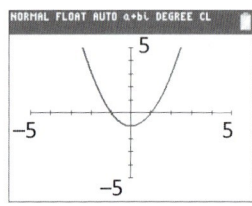

75. $\{-5\}$ **76.** $\{5\}$

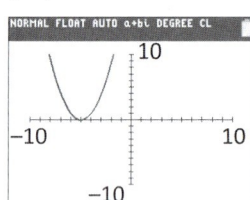 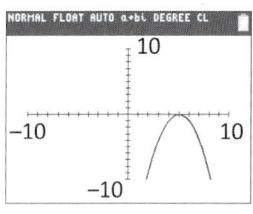

77. $\{\ \}$ **78.** $\{\ \}$

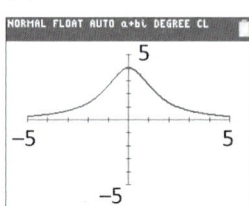 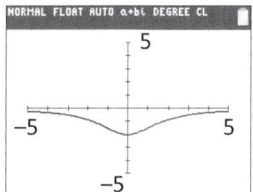

79. $(-\infty, \infty)$ **81.** $\{\ \}$ **83.** $\{0\}$ **85.** $\{\ \}$ **87.** $\{\ \}$
89. $(-\infty, \infty)$ **91.** $(-\infty, -11) \cup (-11, \infty)$ **93.** $\{\frac{3}{2}\}$

Section 17.3 Activity, p. 1145

A.1. a. zero **b.** $\{-3, 3\}$ **c.** $\{-11, 11\}$ **d.** $\{0\}$ **e.** There are no real numbers for which the absolute value is negative. The absolute value of any real number is greater than or equal to zero.

A.2. a. $x - 4 = 5$ or $x - 4 = -5$ **b.** $\{-1, 9\}$

A.3. a. Isolate the absolute value by adding 6 to both sides.
  b. $\left\{3, -\dfrac{11}{3}\right\}$

A.4. a. $\{\ \}$ **b.** $\left\{\dfrac{6}{5}\right\}$

A.5. $x = y$ or $x = -y$

A.6. a. $x - 5 = 2x - 1$ or $x - 5 = -(2x - 1)$ **b.** $\{-4, 2\}$

A.7. $\left\{-\dfrac{1}{4}\right\}$

Section 17.3 Practice Exercises, pp. 1145–1147

R.1. a. $\{5\}$ **b.** $\left\{-\dfrac{39}{5}\right\}$ **R.3. a.** $\{-13\}$ **b.** $\left\{-\dfrac{1}{5}\right\}$

R.5. $\{\ \}$ **R.7.** $\{x \mid x$ is a real number$\}$ **R.9.** $\{5\}$

1. a. absolute; $\{a, -a\}$ **b.** Subtract 5 from both sides.
  c. $y; -y$ **d.** $\{\ \}; \{-4\}$
3. a. $\{-5, 5\}$ **b.** $\{\ \}$ **c.** $\{0\}$ **5. a.** $\{1, 7\}$ **b.** $\{\ \}$ **c.** $\{4\}$
7. $\{7, -7\}$ **9.** $\{6, -6\}$ **11.** $\{\ \}$ **13.** $\{-2, 2\}$ **15.** $\{0\}$

17. $\left\{4, -\dfrac{4}{3}\right\}$ **19.** $\left\{\dfrac{9}{2}, -\dfrac{1}{2}\right\}$ **21.** $\left\{\dfrac{10}{7}, -\dfrac{8}{7}\right\}$ **23.** $\{\ \}$

25. $\{-12, 28\}$ **27.** $\left\{\dfrac{5}{2}, -\dfrac{7}{2}\right\}$ **29.** $\left\{\dfrac{7}{3}\right\}$ **31.** $\{\ \}$ **33.** $\left\{\dfrac{3}{2}\right\}$

35. $\left\{-4, \dfrac{16}{5}\right\}$ **37.** $\left\{-\dfrac{1}{2}, \dfrac{7}{6}\right\}$ **39.** $\left\{\dfrac{5}{2}, -\dfrac{3}{2}\right\}$ **41.** $\left\{-4, \dfrac{1}{3}\right\}$

43. $\left\{\dfrac{1}{2}\right\}$ **45.** $\left\{-\dfrac{1}{16}\right\}$ **47.** $\left\{\dfrac{1}{4}\right\}$ **49.** $\{-1.44, -0.4\}$

51. $\{\ \}$ **53.** $\{w \mid w$ is a real number$\}$ **55.** $\{\ \}$

57. $\left\{2, -\dfrac{1}{2}\right\}$

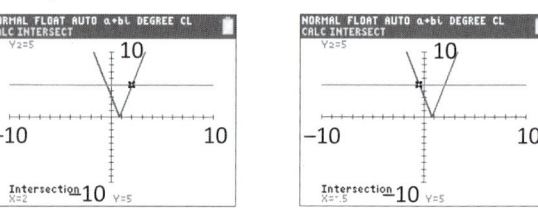

58. $\{7, 1\}$

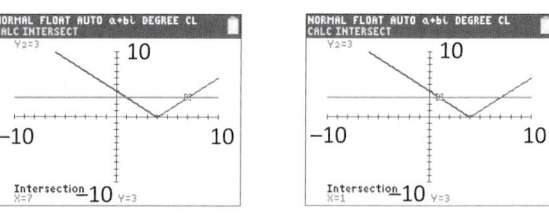

59. $\{\ \}$ **60.** $\{\ \}$

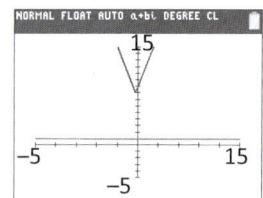 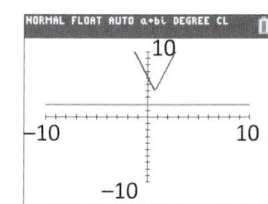

61. $\left\{\dfrac{1}{2}\right\}$ **62.** $\{-1\}$

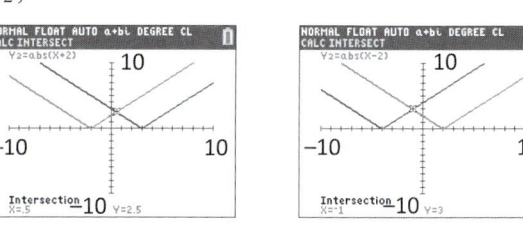

63. $|x| = 6$ **65.** $|x| = \dfrac{4}{3}$

Section 17.4 Activity, pp. 1154–1155

A.1. a. 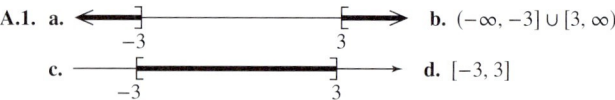 **b.** $(-\infty, -3] \cup [3, \infty)$
  c.   **d.** $[-3, 3]$

A.2. a. Isolate the absolute value by adding 1 to both sides.
  b. $-8 < 2x + 6 < 8$ **c.** $(-7, 1)$

A.3. a. Isolate the absolute value by subtracting 3 from both sides and then dividing both sides by -1. (Remember to reverse the direction of the inequality sign when dividing by -1.)
  b. $x - 9 < -10$ or $x - 9 > 10$ **c.** $(-\infty, -1) \cup (19, \infty)$

A.4. a. $\{\ \}$. By definition, the absolute value of a real number is greater than or equal to zero. Therefore, it cannot be less than a negative number.

b. { }. As with part (a), the absolute value of a real number cannot be less than or equal to a negative number.

   c. $(-\infty, \infty)$. Because the absolute value of any real number is greater than or equal to zero, it is automatically greater than a negative number.

   d. $(-\infty, \infty)$. Because the absolute value of any real number is greater than or equal to zero, it is automatically greater than or equal to a negative number.

A.5. a. { }. By definition, the absolute value of a real number is greater than or equal to zero. Therefore, it cannot be less than a negative number.

   b. {3}. Although the absolute value of a real-valued expression cannot be negative, it *can* be equal to zero. Therefore, the only solution is the value of x for which $2x - 6 = 0$.

   c. $(-\infty, \infty)$. By definition, the absolute value of any real number is greater than or equal to zero.

   d. $(-\infty, 3) \cup (3, \infty)$. The absolute value of any real number is greater than or equal to zero. Therefore, the solution set to the inequality $|2x - 6| > 0$ includes all real numbers, *except* those that make $2x - 6 = 0$. The value $x = 3$ must be excluded from the solution set.

Section 17.4 Practice Exercises, pp. 1155–1158

R.1. $(-\infty, -4) \cup (1, \infty)$ **R.3.** $[-5, -1]$ **R.5.** $\left\{-\frac{4}{3}, 4\right\}$

R.7. { }   **R.9.** $\left\{\frac{3}{5}\right\}$

1. a. $-a; a$ **b.** $-a; >$ **c.** { }; $(-\infty, \infty)$

3. a. $\{-5, 5\}$ **b.** $(-\infty, -5) \cup (5, \infty)$

c. $(-5, 5)$

5. a. $\{10, -4\}$ **b.** $(-\infty, -4) \cup (10, \infty)$

c. $(-4, 10)$

7. a. { }   **b.** $(-\infty, \infty)$ **c.** { }

9. a. { }   **b.** $(-\infty, \infty)$ **c.** { }

11. a. $\{0\}$ **b.** $(-\infty, 0) \cup (0, \infty)$ **c.** { }

13. a. $\{7\}$ **b.** $(-\infty, 7) \cup (7, \infty)$ **c.** { }

15. $(-\infty, -6) \cup (6, \infty)$

17. $[-3, 3]$

19. $(-\infty, \infty)$

21. $(-\infty, -2] \cup [3, \infty)$

23. { }   **25.** $[-10, 14]$

27. $\left(-\infty, -\frac{5}{4}\right] \cup \left[\frac{23}{4}, \infty\right)$

29. $\left(-\frac{21}{2}, \frac{19}{2}\right)$

31. $(-\infty, \infty)$

33. { }   **35.** $\{-5\}$

37. $(-\infty, 6) \cup (6, \infty)$

39. $(-\infty, 2] \cup [4, \infty)$

41. $(4, 8)$

43. $(-10.5, 4.5)$

45. $|x| > 7$

47. $|x - 2| \leq 13$

49. $|x - 32| \leq 0.05$

51. $\left|x - 6\frac{3}{4}\right| \leq \frac{1}{8}$

53. The solution set is $\{w \mid 1.99 \leq w \leq 2.01\}$, or equivalently in interval notation, $[1.99, 2.01]$. This means that the actual width of the bolt could be between 1.99 cm and 2.01 cm inclusive.

55. $(-\infty, -6) \cup (2, \infty)$

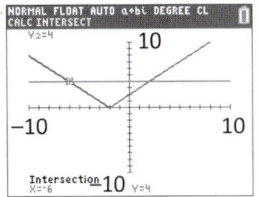 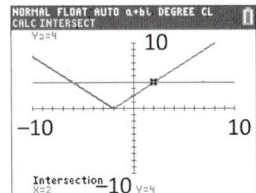

56. $(-\infty, -3) \cup (9, \infty)$

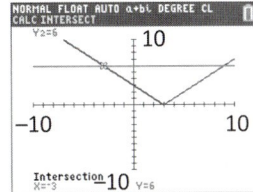 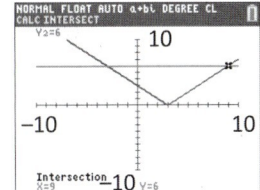

57. $(-7, 5)$

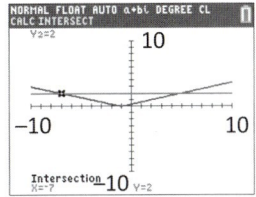 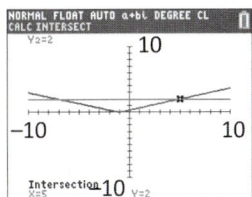

58. $(-3, 5)$

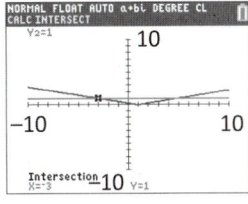 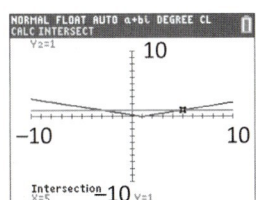

59. { }   **60.** { }

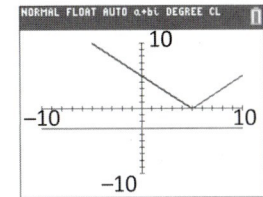 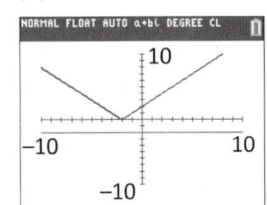

61. $(-\infty, \infty)$ 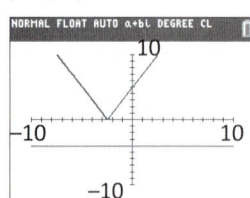 **62.** $(-\infty, \infty)$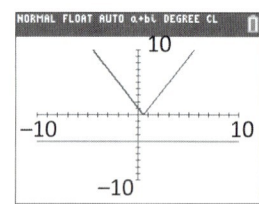

63. $\left\{-\dfrac{1}{6}\right\}$ 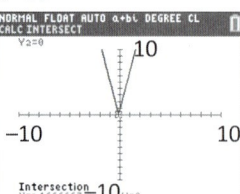 **64.** $\left\{\dfrac{4}{3}\right\}$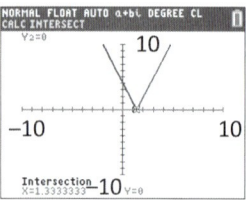

Chapter 17 Problem Recognition Exercises, p. 1158

1. Quadratic; $(-\infty, -9) \cup (-1, \infty)$ 2. Linear; $\{-26\}$
3. Rational; $[-8, -2)$ 4. Quadratic; $\{-\frac{1}{4}, 2\}$
5. Absolute value; $(-\frac{1}{3}, 1)$ 6. Quadratic; $\{\ \}$
7. Linear; $(-\infty, -10)$ 8. Quadratic; $[-4, 1]$
9. Quadratic; $(-\infty, -1] \cup [\frac{2}{3}, \infty)$ 10. Rational; $\{-\frac{11}{3}\}$
11. Linear; $\left[\dfrac{16}{7}, \infty\right)$ 12. Absolute value; $\{\ \}$
13. Polynomial; $[-5, -\frac{1}{2}] \cup [3, \infty)$ 14. Quadratic; $(-\infty, \infty)$
15. Rational; $(-\infty, -1) \cup (2, \infty)$ 16. Quadratic; $\{-1, 5\}$
17. Absolute value; $(-\infty, -8) \cup (-8, \infty)$
18. Quadratic; $(-6, 6)$ 19. Polynomial; $\{2, -2, -\frac{5}{3}\}$
20. Absolute value; $\{12, -4\}$ 21. Linear; $\{0\}$
22. Rational; $(-\infty, 0] \cup (5, \infty)$ 23. Absolute value; $\{-\frac{1}{2}\}$
24. Polynomial; $(-\frac{3}{2}, 0)$

Section 17.5 Activity, p. 1166

A.1. a. Yes **b.** No **c.** No

A.2. a. 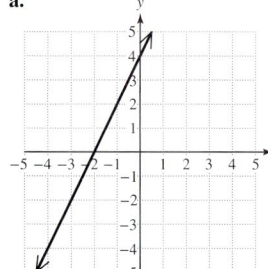 **b.**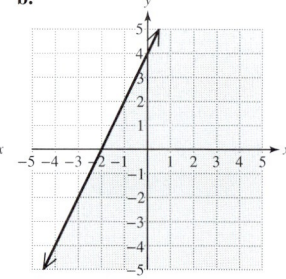

c. The solution set would be the region on and *above* the line.
d. The solution set would be the set of points *strictly below* the line. To show that the points on the line are *not* included in the solution set, the line would be drawn as a dashed line.

A.3. a.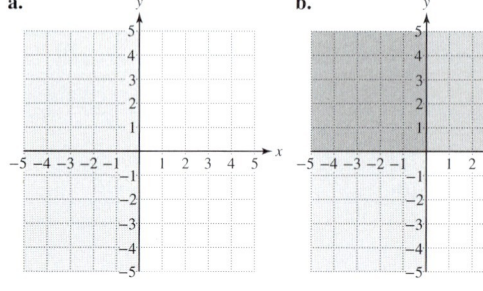

c. The area of "overlap" is the set of points in the second quadrant. That is, the solution set to the system of inequalities $x < 0$ and $y > 0$ is the set of points in Quadrant II.

d.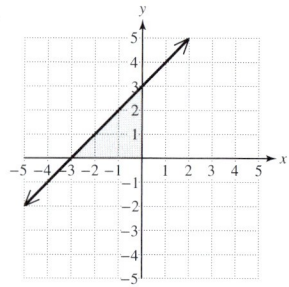

Section 17.5 Practice Exercises, pp. 1167–1173

R.1. x-intercept: $(3, 0)$; y-intercept: $(0, -4)$

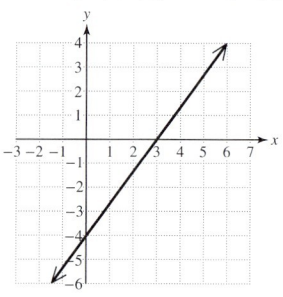

R.3. x-intercept: None; y-intercept: $(0, 2)$

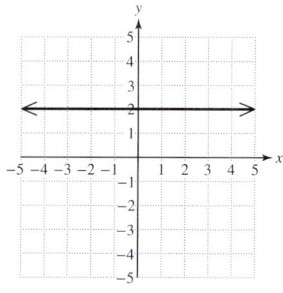

R.5. a. No **b.** Yes **c.** No

1. **a.** linear **b.** is not; is
3. The graph of $y > x + 2$ is the set of points *above* the line $y = x + 2$.
5. No.   7. **a.** No **b.** No **c.** Yes **d.** Yes
9. **a.** No **b.** Yes **c.** Yes **d.** No
11. $>$ 13. $\geq$ 15. $\geq, \leq$

17. 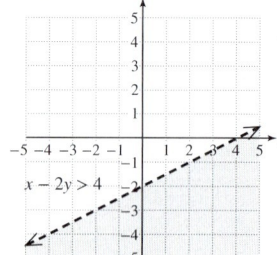   19.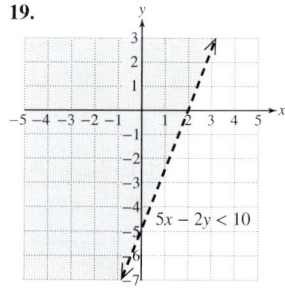

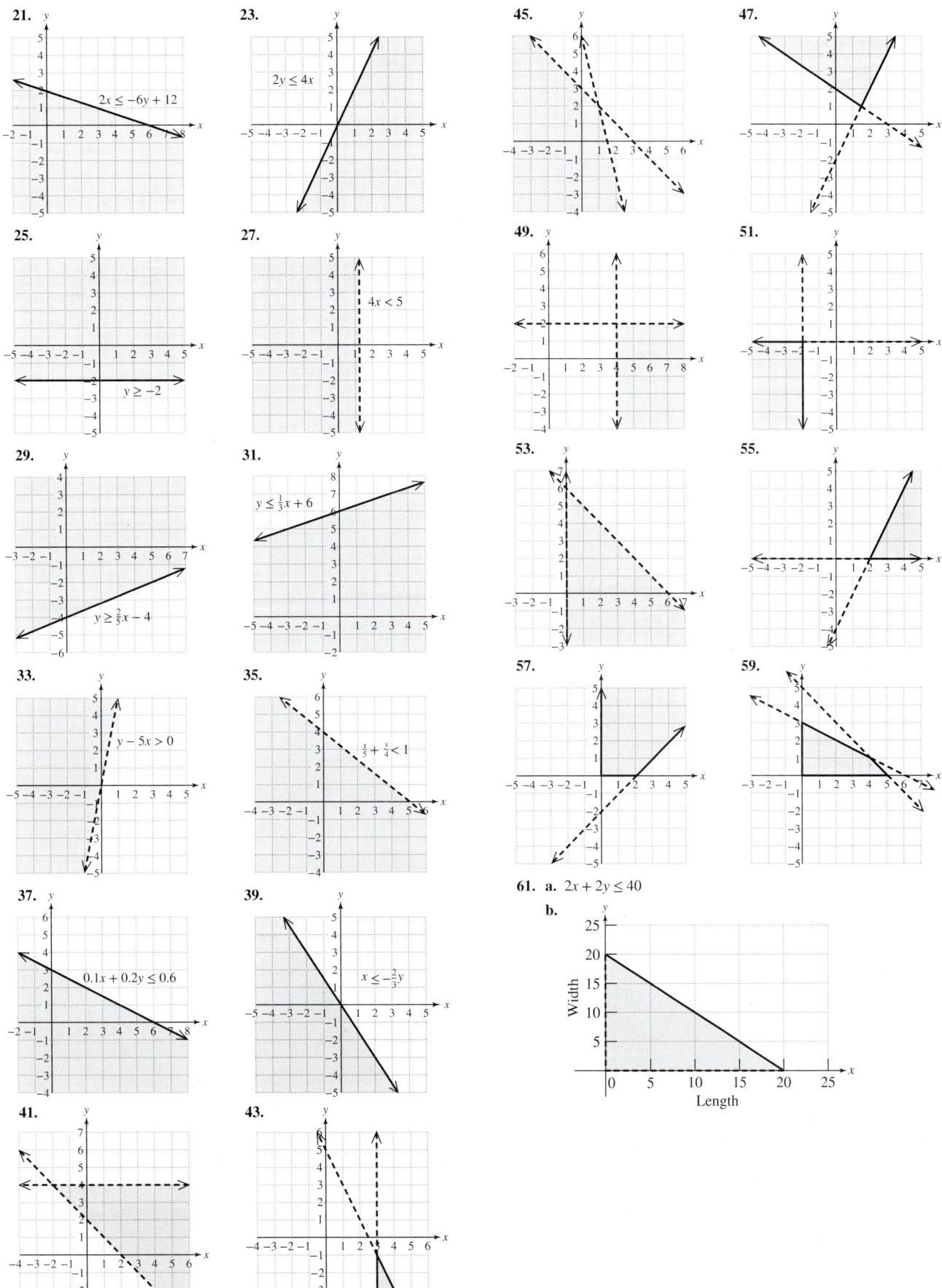

61. a. $2x + 2y \leq 40$

63. a. $x \geq 0, y \geq 0$ **b.** $x \leq 40, y \leq 40$ **c.** $x + y \geq 65$
d.

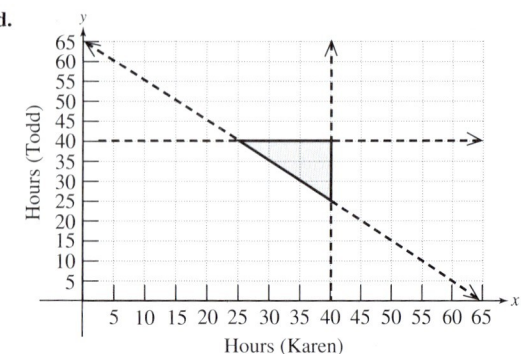

e. Yes. The point (35, 40) means that Karen works 35 hr and Todd works 40 hr. **f.** No. The point (20, 40) means that Karen works 20 hr and Todd works 40 hr. This does not satisfy the constraint that there must be at least 65 hr total.

Chapter 17 Review Exercises, pp. 1174–1177

1. $C \cap D$ is the set of elements common to both sets C and D. $C \cup D$ is the set of elements in either set C, set D, or both sets.

2. $[-10, 1)$ **3.** $(-\infty, \infty)$ **4.** $(-\infty, \infty)$ **5.** $(-1, 1)$
6. $(-\infty, -3] \cup (-1, \infty)$ **7.** $\{\}$ **8.** $\left(-\frac{11}{4}, 4\right]$
9. $[-5, 2)$ **10.** $(-\infty, 6] \cup (12, \infty)$
11. $(-\infty, -2) \cup [2, \infty)$ **12.** $\left[0, \frac{4}{3}\right]$ **13.** $[-7, -2)$
14. a. $140 \leq x \leq 225$ **b.** $x < 140$ or $x > 225$
15. a. $125 \leq x \leq 200$ **b.** $x < 125$ or $x > 200$
16. a. $(-2, 0)$ and $(2, 0)$ are the x-intercepts. **b.** On the interval $(-2, 2)$ the graph is below the x-axis. **c.** On the intervals $(-\infty, -2)$ and $(2, \infty)$ the graph is above the x-axis.
17. a. $x = 0$; $(0, 0)$ is the x-intercept. **b.** $x = 2$ is the vertical asymptote. **c.** On the intervals $(-\infty, 0]$ and $(2, \infty)$ the graph is on or above the x-axis. **d.** On the interval $[0, 2)$ the graph is on or below the x-axis. **18.** $(-2, 6)$ **19.** $(-\infty, \infty)$ **20.** $(-\infty, -2) \cup [0, \infty)$
21. $(-\infty, -1] \cup (1, \infty)$ **22.** $(-2, 0) \cup (5, \infty)$
23. $(-2, 0) \cup \left(\frac{5}{2}, \infty\right)$ **24.** $[-3, -1]$
25. $(-\infty, -5) \cup (1, \infty)$ **26.** $(3, \infty)$ **27.** $(-3, \infty)$
28. $\{-5\}$ **29.** $(-\infty, -2) \cup (-2, \infty)$ **30.** $\{10, -10\}$
31. $\{17, -17\}$ **32.** $\{1.3, 7.4\}$ **33.** $\{-0.44, 2.54\}$
34. $\{5, -9\}$ **35.** $\{3, 1\}$ **36.** $\{\}$
37. $\{\}$ **38.** $\left\{\frac{3}{7}\right\}$ **39.** $\left\{-\frac{5}{4}\right\}$ **40.** $\left\{6, \frac{4}{5}\right\}$
41. $\left\{-\frac{1}{2}\right\}$ **42.** Both expressions give the distance between 3 and -2.
43. $|x| > 5$ **44.** $|x| < 4$ **45.** $|x| < 6$ **46.** $|x| > \frac{2}{3}$
47. $(-\infty, -14] \cup [2, \infty)$
48. $[-11, -5]$
49. $\left(-\infty, \frac{1}{7}\right) \cup \left(\frac{1}{7}, \infty\right)$
50. $\left(-\infty, -\frac{1}{5}\right) \cup \left(-\frac{1}{5}, \infty\right)$

51. $\left[-2, -\frac{2}{3}\right]$
52. $\left[0, \frac{6}{5}\right]$
53. $(2, 22)$
54. $(-12, 0)$
55. $(-\infty, \infty)$
56. $(-\infty, \infty)$
57. $(-\infty, \infty)$
58. $(-\infty, \infty)$
59. $\{\}$ **60.** $\{\}$ **61.** If an absolute value is less than a negative number, there will be no solution. **62.** If an absolute value is greater than a negative number, then all real numbers are solutions. **63.** $0.17 \leq p \leq 0.23$, or equivalently in interval notation, $[0.17, 0.23]$. This means that the actual percentage of viewers is estimated to be between 17% and 23%, inclusive.
64. $3\frac{1}{8} \leq L \leq 3\frac{5}{8}$, or equivalently in interval notation, $[3\frac{1}{8}, 3\frac{5}{8}]$. This means that the actual length of the screw may be between $3\frac{1}{8}$ in. and $3\frac{5}{8}$ in., inclusive.

65. 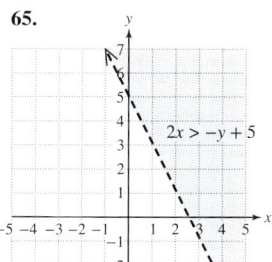 **66.**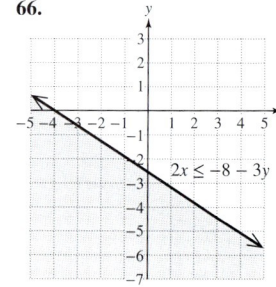

67. 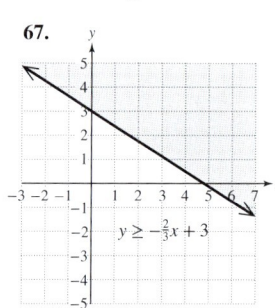 **68.**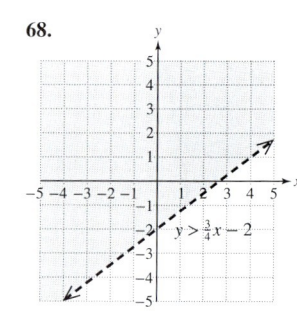

69. 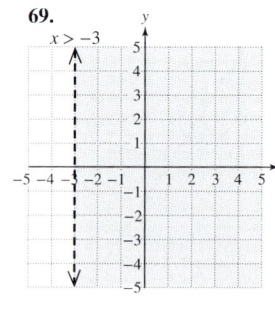 **70.**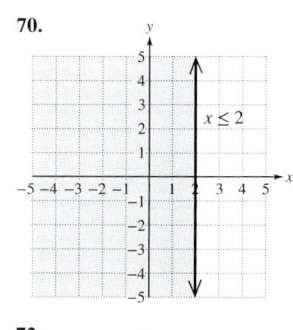

71. 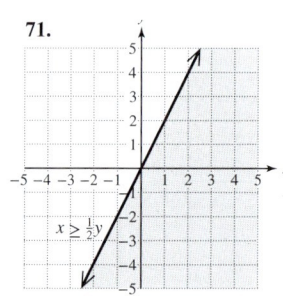 **72.**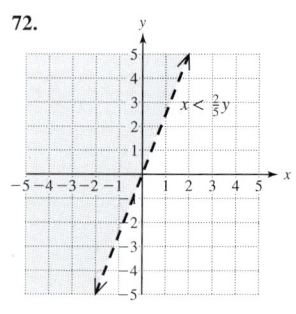

73. 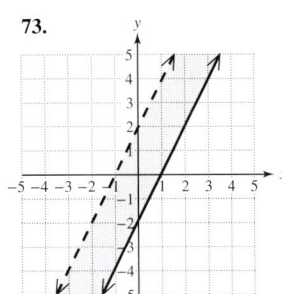   **74.**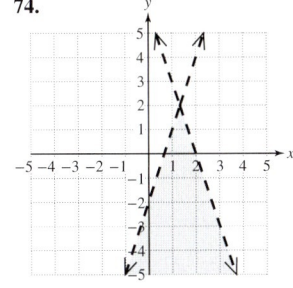

75. 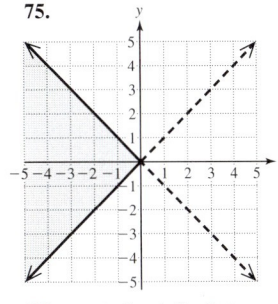   **76.**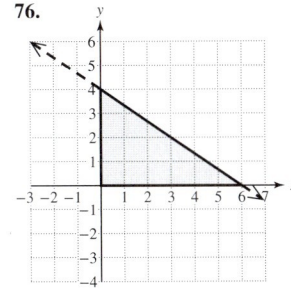

77. a. $x \geq 0, y \geq 0$ **b.** $x + y \leq 100$ **c.** $x \geq 4y$

d.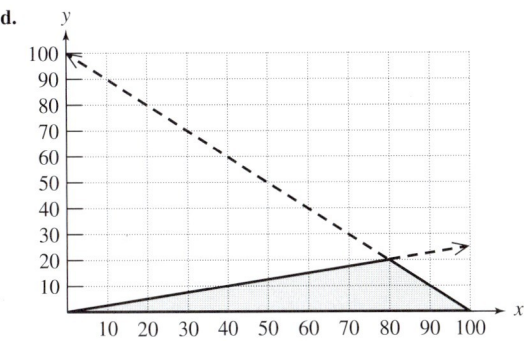

i. $[0, \infty); (-\infty, \infty)$ **j.** -5 and -4
3. a. $8, -8$ **b.** 8 **c.** There are two square roots for every positive number. $\sqrt{64}$ identifies the positive square root.
5. a. 9 **b.** -9 **7.** There is no real number b such that $b^2 = -36$.   **9.** 7 **11.** -7 **13.** Not a real number
15. $\dfrac{8}{3}$ **17.** 0.9 **19.** -0.4 **21. a.** 8 **b.** 4
c. -8 **d.** -4 **e.** Not a real number   **f.** -4
23. -3 **25.** $\dfrac{1}{2}$ **27.** 2 **29.** $-\dfrac{5}{4}$ **31.** Not a real number   **33.** 10 **35.** -0.2 **37.** 0.5 **39.** $|a|$
41. a **43.** $|a|$ **45.** $|x+1|$ **47.** $|x|y^2$ **49.** $-\dfrac{x}{y}$
51. $\dfrac{2}{|x|}$ **53.** -92 **55.** 2 **57.** -923 **59.** y^4
61. $\dfrac{a^3}{b}$ **63.** $-\dfrac{5}{q}$ **65.** $3xy^2z$ **67.** $\dfrac{hk^2}{4}$ **69.** $-\dfrac{t}{3}$
71. $2y^2$ **73.** $2p^2q^3$ **75.** 9 cm   **77.** 13 ft
79. They were 5 mi apart.   **81.** They are 25 mi apart.
83. a. Not a real number   **b.** Not a real number
c. 0 **d.** 1 **e.** 2; Domain: $[2, \infty)$ **85. a.** -2 **b.** -1
c. 0 **d.** 1; Domain: $(-\infty, \infty)$ **87.** $\left(-\infty, \dfrac{5}{2}\right]$
89. $(-\infty, \infty)$ **91.** $[5, \infty)$ **93.** $(-\infty, \infty)$
95. a. $(-\infty, 1]$ **97. a.** $[-3, \infty)$
b. 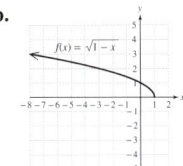   **b.**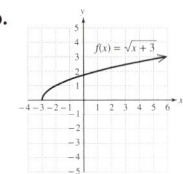

99. a. $[0, \infty)$ **101. a.** $(-\infty, \infty)$
b. 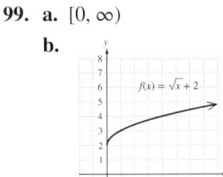   **b.**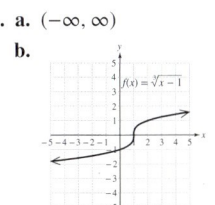

103. $q + p^2$ **105.** $\dfrac{6}{\sqrt[3]{x}}$ **107.** 8 in.   **109.** 8.3066 **110.** 76.1446
111. 3.7100 **112.** -0.5566 **113.** 15.6525 **114.** -6.2403
115. -0.1235 **116.** 1.0622
117.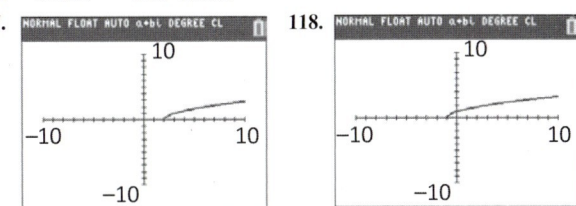

119. 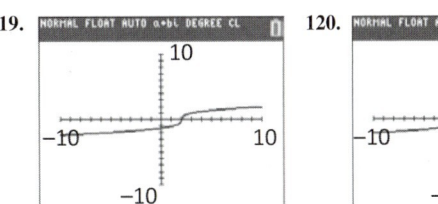   **120.**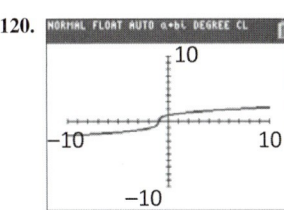

Chapter 18

Section 18.1 Activity, p. 1188

A.1. a. $b^2 = a$ **b.** $3^2 = 9$ **c.** $(-3)^2 = 9$ **d.** $\sqrt{a}$; nonnegative   **e.** 3
A.2. a. $b^n = a$ **b.** $(-4)^3 = -64$ **c.** $2^5 = 32$ **d.** $3; 125$ **e.** 5
A.3. a. No real number when squared equals a negative number.
  b. No real number when raised to the fourth power (or any even power) equals a negative number.
  c. $\sqrt[3]{-8} = -2$ because $(-2)^3 = -8$. In general, a negative real number when raised to an odd power is negative.
A.4. a. $\sqrt{a^2}$ is the principal square root of a^2. Therefore, the simplified form must not be negative. Because we do not know the value of a (which could be a positive or negative number), the absolute value bars ensure that the answer is nonnegative.
  b. For a root with an odd index, the absolute value bars are not needed because the value can be positive, negative, or zero.
A.5. a. even   **b.** $[0, \infty)$ **c.** $[5, \infty)$ **d.** $(-\infty, 5]$
A.6. a. odd   **b.** $(-\infty, \infty)$ **c.** $(-\infty, \infty)$ **d.** $(-\infty, \infty)$

Section 18.1 Practice Exercises, pp. 1188–1192

R.1. 100 **R.3.** 0.16 **R.5.** $-\dfrac{27}{125}$
R.7. $\dfrac{1}{64}$ **R.9.** x^{14} **R.11.** $81t^{16}$

1. a. $b; a$ **b.** principal   **c.** $b^n; a$ **d.** index; radicand
e. cube   **f.** is not; is   **g.** even; odd   **h.** Pythagorean; c^2

Section 18.2 Activity, p. 1197

A.1. a. $\sqrt[n]{a}$ **b.** $\sqrt{9} = 3$ **c.** $\sqrt[3]{-125} = -5$
  d. The definition $a^{1/n} = \sqrt[n]{a}$ is true only if the radical is a real number. The expression $\sqrt{-16}$ is not a real number.

A.2. a. $\sqrt[n]{a^m}$ or $(\sqrt[n]{a})^m$ **b.** $\sqrt[3]{125^2}$ or $(\sqrt[3]{125})^2 = 25$
c. $\sqrt[4]{16^3}$ or $(\sqrt[4]{16})^3 = 8$
A.3. a. t^7 **b.** t^3 **A.4. a.** y^4 **b.** $y^{13/15}$ **A.5. a.** z^{15} **b.** z^4
A.6. a. $\dfrac{1}{n^6}$ **b.** $\dfrac{1}{n^{1/2}}$ **A.7. a.** $2m^3$ **b.** $2m^{1/3}$
A.8. a. a^4b^8 **b.** ab^2 **A.9. a.** $\dfrac{c^9}{d^{12}}$ **b.** $\dfrac{c^4}{d^3}$

Section 18.2 Practice Exercises, pp. 1198–1201

R.1. a^5 **R.3.** $\dfrac{1}{y^3}$ **R.5.** $\dfrac{1}{n^8}$ **R.7.** $\dfrac{1}{125}a^6b^9$
R.9. 1 **R.11. a.** 4 **b.** 16 **c.** 2 **R.13.** 27 **R.15.** $t+2$

1. a. $\sqrt[n]{a}$ **b.** $(\sqrt[n]{a})^m$ or $\sqrt[n]{a^m}$
3. a. $\sqrt{49}$ **b.** $-\sqrt{49}$ **c.** $\dfrac{1}{\sqrt{49}}$
5. a. $\sqrt[3]{-64}$ **b.** $-\sqrt[3]{64}$ **c.** $\dfrac{1}{\sqrt[3]{64}}$ **7.** 12 **9.** -12
11. Not a real number **13.** -4 **15.** $\dfrac{1}{5}$ **17.** $-\dfrac{1}{7}$
19. $a^{m/n} = \sqrt[n]{a^m}$; The numerator of the exponent represents the power of the base. The denominator of the exponent represents the index of the radical. **21. a.** 8 **b.** -8 **c.** Not a real number
d. $\dfrac{1}{8}$ **e.** $-\dfrac{1}{8}$ **f.** Not a real number
23. a. 125 **b.** -125 **c.** Not a real number **d.** $\dfrac{1}{125}$
e. $-\dfrac{1}{125}$ **f.** Not a real number **25.** $\dfrac{1}{512}$ **27.** 27
29. $-\dfrac{1}{81}$ **31.** $\dfrac{27}{1000}$ **33.** Not a real number **35.** -2
37. -2 **39.** 6 **41.** 10 **43.** $\dfrac{3}{4}$ **45.** 1 **47.** $\dfrac{9}{2}$
49. $\sqrt[3]{q^2}$ **51.** $6\sqrt[4]{y^3}$ **53.** $\sqrt[3]{x^2y}$ **55.** $\dfrac{6}{\sqrt[5]{r^2}}$
57. $x^{1/3}$ **59.** $10b^{1/2}$ **61.** $y^{2/3}$ **63.** $(a^2b^3)^{1/4}$
65. $\dfrac{1}{x}$ **67.** p **69.** y^2 **71.** $6^{2/5}$ **73.** $\dfrac{4}{t^{5/3}}$
75. a^7 **77.** $\dfrac{25a^4d}{c}$ **79.** $\dfrac{y^9}{x^8}$ **81.** $\dfrac{2z^3}{w}$ **83.** $5xy^2z^3$
85. $x^{13}z^{4/3}$ **87.** $\dfrac{x^3y^2}{z^5}$ **89. a.** 10.9% **b.** 8.8%
c. The account in part (a). **91.** 2.7 in. **93.** $\sqrt{y}$
95. $\sqrt[4]{z}$ **97.** $\sqrt[3]{x^2}$ **99.** $y\sqrt{x}$ **101.** $4x^4y^3$
103. $2x\sqrt[3]{y^2z}$ **105.** $\sqrt[6]{x}$ **107.** $\sqrt[15]{w}$ **109.** 3
111. 0.3761 **113.** 2.9240 **115.** 31.6228

Section 18.3 Activity, p. 1206

A.1. a. $\sqrt[n]{ab}$ **b.** $2\sqrt{5}$ **c.** $\sqrt{50} = \sqrt{25 \cdot 2} = \sqrt{25} \cdot \sqrt{2} = 5\sqrt{2}$
d. $\sqrt[3]{x^7} = \sqrt[3]{x^6 \cdot x} = \sqrt[3]{x^6} \cdot \sqrt[3]{x} = x^2\sqrt[3]{x}$
A.2. a. $2^4 \cdot 3$ **b.** $\sqrt{2^4 \cdot 3x^5y^6}$
c. $\sqrt{(2^4x^4y^6) \cdot (3x)}$ or $\sqrt{(16x^4y^6) \cdot (3x)}$ **d.** $4x^2y^3\sqrt{3x}$
A.3. $5x^2$ **A.4.** $10\sqrt{2}$

Section 18.3 Practice Exercises, pp. 1206–1209

R.1. $2 \cdot 3^2 \cdot 7^2$ **R.3.** 4, 9, 25, 100, y^2, y^4, y^{10}
R.5. 1, 16, 81, b^4, b^8 **R.7.** $\dfrac{9}{8}$ **R.9.** 5
R.11. -6 **R.13.** -4 **R.15.** b^5 **R.17.** $3c^2d^3$

1. a. $\sqrt[n]{a}$; $\sqrt[n]{b}$ **b.** The exponent within the radicand is greater than the index. **c.** is not **d.** 3 **e.** t^{12}
3. $\sqrt{x^{11}} = \sqrt{x^{10} \cdot x} = \sqrt{x^{10}} \cdot \sqrt{x} = x^5\sqrt{x}$
5. $\sqrt[3]{x^{11}} = \sqrt[3]{x^9 \cdot x^2} = \sqrt[3]{x^9} \cdot \sqrt[3]{x^2} = x^3\sqrt[3]{x^2}$

7. $\sqrt[4]{x^{11}} = \sqrt[4]{x^8 \cdot x^3} = \sqrt[4]{x^8} \cdot \sqrt[4]{x^3} = x^2\sqrt[4]{x^3}$ **9.** $x^2\sqrt{x}$
11. $q^2\sqrt[3]{q}$ **13.** $a^2b^2\sqrt{a}$ **15.** $-x^2y^3\sqrt[4]{y}$ **17.** $2\sqrt{7}$ **19.** $2\sqrt{5}$
21. $15\sqrt{2}$ **23.** $3\sqrt[3]{2}$ **25.** $5b\sqrt{ab}$ **27.** $2x^2\sqrt[3]{5x}$
29. $-2x^2z\sqrt[3]{2y}$ **31.** $2wz\sqrt[4]{5z^3}$ **33.** x **35.** p^2 **37.** 5
39. $\dfrac{1}{2}$ **41.** $\dfrac{5\sqrt[3]{2}}{3}$ **43.** $\dfrac{5\sqrt[3]{9}}{6}$ **45.** $4\sqrt{5}$ **47.** $-30\sqrt{3}$
49. $5x^2y\sqrt{y}$ **51.** $3yz\sqrt[3]{x^2z}$ **53.** $2w^2$ **55.** $\dfrac{1}{10y^6}$ **57.** $\dfrac{2}{b}$
59. $2a^7b^4c^{15}d^{11}\sqrt{2c}$ **61.** $3a^2b\sqrt[3]{2b}$ **63.** $-10a^2b^2\sqrt{3ac}$
65. $x\sqrt[4]{7xy}$ **67.** $3a^2b\sqrt{6}$ **69.** $2\sqrt{3}$ **71.** $\dfrac{3\sqrt{5}}{4}$
73. $\dfrac{1}{\sqrt[3]{w^6}}$ simplifies to $\dfrac{1}{w^2}$ **75.** $\sqrt{k^3}$ simplifies to $k\sqrt{k}$
77. $2\sqrt{41}$ ft **79.** $6\sqrt{5}$ m **81.** The distance is $90\sqrt{2}$ ft or approximately 127.3 ft. **83.** The path from A to B and B to C is faster.

Section 18.4 Activity, p. 1212

A.1. index; radicand **A.2. a.** No; the radicands are different.
b. No; the indices are different. **c.** Yes **d.** Yes
A.3. a. $9x$ **b.** $9\sqrt{x}$ **c.** $9\sqrt{2}$
d. In each case, apply the distributive property to add *like* terms. For example, $4\sqrt{x} + 7\sqrt{x} - 2\sqrt{x} = \sqrt{x}(4 + 7 - 2) = \sqrt{x} \cdot (9)$ or $9\sqrt{x}$
A.4. a. 16, -12, 1 **b.** $5\sqrt{3t}$
A.5. a. $3x^2y\sqrt{5x}$ **b.** $-4x^2y\sqrt{5x}$ **c.** $6x^2y\sqrt{5x}$
d. $3x^2y\sqrt{5x} - 4x^2y\sqrt{5x} + 6x^2y\sqrt{5x}$ **e.** $5x^2y\sqrt{5x}$
A.6. a. $3x\sqrt{7} + 5\sqrt{7}$ **b.** $\sqrt{7}(3x+5)$ or $(3x+5)\sqrt{7}$

Section 18.4 Practice Exercises, pp. 1213–1215

R.1. a. Yes **b.** No **c.** Yes **d.** No **R.3.** $-y$ **R.5.** c^2d
R.7. $3\sqrt{3}$ **R.9.** $5xy\sqrt[3]{2y}$
1. a. index; radicand **b.** $2\sqrt{3x}$ **c.** cannot; can
3. Not *like* radicals **5.** *Like* radicals **7.** Not *like* radicals
9. a. Both expressions can be simplified by using the distributive property. **b.** Neither expression can be simplified because the terms do not contain *like* radicals or *like* terms.
11. a. $8x$ **b.** $8\sqrt{x}$ **13. a.** $-7t$ **b.** $-7\sqrt[3]{t}$ **15.** $9\sqrt{5}$
17. $2\sqrt[3]{tw}$ **19.** $5\sqrt{10}$ **21.** $8\sqrt[4]{3} - \sqrt[4]{14}$ **23.** $2\sqrt{x} + 2\sqrt{y}$
25. Cannot be simplified further
27. Cannot be simplified further **29.** $\dfrac{29}{18}z\sqrt[3]{6}$
31. $0.7x\sqrt{y}$ **33.** Simplify each radical: $3\sqrt{2} + 35\sqrt{2}$. Then add *like* radicals: $38\sqrt{2}$ **35.** 15 **37.** $8\sqrt{3}$
39. $3\sqrt{7}$ **41.** $-5\sqrt{2} + 3\sqrt{3}$ **43.** $\sqrt[3]{3}$ **45.** $-5\sqrt{2a}$
47. $8s^2t^2\sqrt[3]{s^2}$ **49.** $6x\sqrt[3]{x}$ **51.** $14p^2\sqrt{5}$
53. $-\sqrt[3]{a^2b}$ **55.** $(5x+6)\sqrt{x}$ **57.** $(5x-6)\sqrt{2}$
59. $33d\sqrt[3]{2c}$ **61.** $2a^2b\sqrt{6a}$ **63.** $5x\sqrt[3]{2} - 6\sqrt[3]{x}$
65. False; $\sqrt{9} + \sqrt{16} \ne \sqrt{9+16}$; $7 \ne 5$ **67.** True
69. False; $\sqrt{y} + \sqrt{y} = 2\sqrt{y} \ne \sqrt{2y}$
71. False; $2w\sqrt{5} + 4w\sqrt{5} = 6w\sqrt{5} \ne 6w^2\sqrt{5}$
73. $\sqrt{48} + \sqrt{12}$ simplifies to $6\sqrt{3}$
75. $5\sqrt[3]{x^6} - x^2$ simplifies to $4x^2$ **77.** The difference of the principal square root of 18 and the square of 5
79. The sum of the principal fourth root of x and the cube of y
81. $9\sqrt{6}$ cm ≈ 22.0 cm **83.** $x = 2\sqrt{2}$ ft
85. a. $10\sqrt{5}$ yd **b.** 22.36 yd **c.** $105.95

Section 18.5 Activity, p. 1221

A.1. $\sqrt[n]{ab}$ **A.2. a.** $10x^2$ **b.** $10x$ **c.** $10\sqrt[3]{x^2}$
d. Parts (b) and (c) illustrate that we multiply the coefficients of each term and multiply the radicals. In part (b), however, it was necessary to simplify the resulting radical.

A.3. **a.** $2x^3 + 4x^2y + xy^2$ **b.** $2x + 4x\sqrt{y} + \sqrt{xy}$
A.4. **a.** $10x^2 + 3x - 18$ **b.** $10x + 3\sqrt{x} - 18$
c. $10x - 2\sqrt{6x} + 5\sqrt{3x} - 3\sqrt{2}$
A.5. **a.** $25x^2 - 30x + 9$ **b.** $25x - 30\sqrt{x} + 9$ **c.** $25x - 10\sqrt{3x} + 3$
A.6. **a.** $4y^2 - 49$ **b.** $4y - 49$ **c.** $4y - 7$
A.7. **a.** $7^{1/3}$ **b.** $2^{1/5}$ **c.** $7^{1/3} \cdot 2^{1/5}$ **d.** $7^{5/15} \cdot 2^{3/15}$ **e.** $\sqrt[15]{7^5} \cdot \sqrt[15]{2^3}$
f. $\sqrt[15]{7^5 \cdot 2^3}$

Section 18.5 Practice Exercises, pp. 1222–1224

R.1. $-40x^3y$ **R.3.** $-3a^3 - 5a^2b + 2ab^2$ **R.5.** $3c^3 - 2c^2 + 7c + 20$
R.7. $4a^2 - 20ab + 25b^2$ **R.9.** $\frac{1}{4}m^2 - 16n^2$ **R.11.** $8y\sqrt{7}$

1. a. $\sqrt[n]{ab}$ **b.** x **c.** a **d.** conjugates **e.** $m - n$
3. a. x^2 **b.** $\sqrt{x} \cdot \sqrt{x} = \sqrt{x^{\boxed{2}}} = x$
5. a. t^6 **b.** $\sqrt[3]{t} \cdot \sqrt[3]{t^5} = \sqrt[3]{\boxed{t^6}} = t^2$
7. a. $50x^2$ **b.** $\sqrt{5x} \cdot \sqrt{10x} = \sqrt{50x^2} = 5x\sqrt{2}$
9. $\sqrt[3]{21}$ **11.** $2\sqrt{5}$ **13.** $4\sqrt[4]{4}$ **15.** $8\sqrt[3]{20}$
17. $-24ab\sqrt{a}$ **19.** $6\sqrt{10}$ **21.** $6x\sqrt{2}$ **23.** $10a^3b^2\sqrt{b}$
25. $-24x^2y^2\sqrt{2y}$ **27.** $6ab\sqrt[3]{2a^2b^2}$ **29.** $12 - 6\sqrt{3}$
31. $2\sqrt{3} - \sqrt{6}$ **33.** $-2x - \frac{7}{3}\sqrt{x}$ **35.** $-8 + 7\sqrt{30}$
37. $x - 5\sqrt{x} - 36$ **39.** $\sqrt[3]{y^2} - \sqrt[3]{y} - 6$
41. $9a - 28\sqrt{ab} + 3b$ **43.** $8\sqrt{p} + 3p + 5\sqrt{pq} + 16\sqrt{q} - 2q$
45. 15 **47.** $3y$ **49.** 6 **51.** 709 **53. a.** $x^2 - y^2$
b. $x^2 - 25$ **55.** $29 + 8\sqrt{13}$ **57.** $p - 2\sqrt{7p} + 7$
59. $2a - 6\sqrt{2ab} + 9b$ **61.** $3 - x^2$ **63.** 4
65. $\frac{4}{9}x - \frac{1}{4}y$ **67. a.** $3 - x$ **b.** $3 + 2\sqrt{3x} + x$
c. $3 - 2\sqrt{3x} + x$ **69.** True **71.** False;
$(x - \sqrt{5})^2 = x^2 - 2x\sqrt{5} + 5$ **73.** False; 5 is multiplied by 3 only.
75. True **77.** $6x$ **79.** $3x + 1$
81. $x + 19 - 8\sqrt{x+3}$ **83.** $2t + 10\sqrt{2t-3} + 22$
85. $12\sqrt{5}$ ft^2 **87.** $18\sqrt{15}$ in.^{2} **89.** $\sqrt[4]{x^3}$
91. $\sqrt[15]{(2z)^8}$ **93.** $p^2\sqrt[6]{p}$ **95.** $u\sqrt[6]{u}$
97. $\sqrt[6]{x^2y}$ **99.** $\sqrt[4]{2^3 \cdot 3^2}$ or $\sqrt[4]{72}$ **101.** $\sqrt[12]{2^4 5^3 x^7 y^7}$
103. $2m\sqrt[6]{3^3 2mn^5}$ **105.** $a + b$

Section 18.6 Activity, pp. 1231–1232

A.1. $\frac{\sqrt[n]{a}}{\sqrt[n]{b}}$ **A.2.** $\frac{2x\sqrt[3]{5x^2}}{y}$ **A.3.** $5x$
A.4. a. $\frac{6}{\sqrt[4]{x}} \cdot \frac{\sqrt[4]{x^3}}{\sqrt[4]{x^3}} = \frac{6\sqrt[4]{x^3}}{\sqrt[4]{x^4}}$ **b.** $\frac{6\sqrt[4]{x^3}}{x}$
A.5. a. $\frac{14}{\sqrt[3]{7^2y}}$ **b.** $\frac{14}{\sqrt[3]{7^2y}} \cdot \frac{\sqrt[3]{7y^2}}{\sqrt[3]{7y^2}} = \frac{14\sqrt[3]{7y^2}}{\sqrt[3]{7^3y^3}}$ **c.** $\frac{2\sqrt[3]{7y^2}}{y}$
A.6. a. $49x^2 - 4$ **b.** $7x^2 - 4$ **c.** 3 **d.** No
e. $\frac{12}{(\sqrt{7} - 2)} \cdot \frac{(\boxed{\sqrt{7} + 2})}{(\boxed{\sqrt{7} + 2})}$ **f.** $4(\sqrt{7} + 2)$ or $4\sqrt{7} + 8$
A.7. a. denominator **b.** conjugate **c.** $\frac{t - 9\sqrt{t} + 20}{t - 16}$

Section 18.6 Practice Exercises, pp. 1232–1235

R.1. x **R.3.** ab^2 **R.5.** $p - 16$ **R.7.** 3

1. a. radical **b.** $\frac{\sqrt[n]{a}}{\sqrt[n]{b}}$ **c.** $\frac{4}{x^2}$ **d.** is; is not **3.** $\frac{7x^2}{y^3}$ **5.** $\frac{2a\sqrt{2}}{x^3}$
7. $\frac{-2j\sqrt[3]{2}}{k}$ **9.** $3b^2$ **11.** $\frac{1}{2b^2}$ **13.** $\frac{z\sqrt{3y}}{w^2}$ **15.** $\frac{\sqrt{5}}{\sqrt{5}}$

17. $\frac{\sqrt[3]{x^2}}{\sqrt[3]{x^2}}$ **19.** $\frac{\sqrt{3z}}{\sqrt{3z}}$ **21.** $\frac{\sqrt[4]{2a^2}}{\sqrt[4]{2a^2}}$ **23.** $\frac{\sqrt{3}}{3}$
25. $\frac{\sqrt{x}}{x}$ **27.** $\frac{3\sqrt{2y}}{y}$ **29.** $\frac{a\sqrt{2a}}{2}$ **31.** $\frac{3\sqrt{2}}{2}$
33. $\frac{3\sqrt[3]{4}}{2}$ **35.** $\frac{-6\sqrt[4]{x^3}}{x}$ **37.** $\frac{7\sqrt[3]{2}}{2}$ **39.** $\frac{\sqrt[3]{4w}}{w}$
41. $\frac{2\sqrt[4]{27}}{3}$ **43.** $\frac{\sqrt[3]{2x}}{x}$ **45.** $\frac{2\sqrt{6}}{21}$ **47.** $\frac{\sqrt{x}}{x^4}$
49. $\frac{\sqrt{2x}}{2x^3}$ **51.** $\sqrt{2} + \sqrt{6}$ **53.** $\sqrt{x} - 23$
55. $\frac{4\sqrt{2} - 12}{-7}$ or $\frac{-4\sqrt{2} + 12}{7}$ **57.** $4\sqrt{6} + 8$
59. $-\sqrt{21} + 2\sqrt{7}$ **61.** $\frac{-\sqrt{p} + \sqrt{q}}{p - q}$ **63.** $\sqrt{x} - \sqrt{5}$
65. $\frac{w + 11\sqrt{w} + 18}{81 - w}$ **67.** $\frac{4\sqrt{xy} - 3x - y}{y - x}$
69. $5 - \sqrt{10}$ **71.** $\frac{5 + \sqrt{21}}{4}$ **73.** $-6\sqrt{5} - 13$
75. $\frac{16}{\sqrt[3]{4}}$ simplifies to $8\sqrt[3]{2}$ **77.** $\frac{4}{x - \sqrt{2}}$ simplifies to $\frac{4x + 4\sqrt{2}}{x^2 - 2}$
79. $\frac{\pi\sqrt{2}}{4}$ sec ≈ 1.11 sec **81. a.** $\frac{\sqrt{2}}{2}$ **b.** $\frac{\sqrt[3]{4}}{2}$
83. a. $\frac{\sqrt{5a}}{5a}$ **b.** $\frac{\sqrt{5} - a}{5 - a^2}$ **85.** $\frac{2\sqrt{6}}{3}$ **87.** $\frac{17\sqrt{15}}{15}$
89. $\frac{8\sqrt[3]{25}}{5}$ **91.** $\frac{-33}{2\sqrt{3} - 12}$ **93.** $\frac{a - b}{a + 2\sqrt{ab} + b}$
95. $\frac{3}{\sqrt{5 + 3h} + \sqrt{5}}$ **97.** $\frac{5}{\sqrt{4 + 5h} + 2}$

Chapter 18 Problem Recognition Exercises, pp. 1236–1237

1. a. $2\sqrt{6}$ **b.** $2\sqrt[3]{3}$ **2. a.** $3\sqrt{6}$ **b.** $3\sqrt[3]{2}$
3. a. $10y^3\sqrt{2}$ **b.** $2y^2\sqrt[3]{25}$ **4. a.** $4z^7\sqrt{2z}$ **b.** $2z^5\sqrt[3]{4}$
5. a. $4\sqrt{5}$ **b.** $2\sqrt[3]{10}$ **c.** $2\sqrt[5]{5}$ **6. a.** $4\sqrt{3}$ **b.** $2\sqrt[3]{6}$ **c.** $2\sqrt[4]{3}$
7. a. $x^2y^3\sqrt{x}$ **b.** $xy^2\sqrt[3]{x^2}$ **c.** $xy\sqrt[4]{xy^2}$
8. a. $a^5b^4\sqrt{b}$ **b.** $a^3b^3\sqrt[3]{a}$ **c.** $a^2b^2\sqrt[4]{a^2b}$ **9. a.** $2st^2\sqrt[3]{4s^2}$
b. $2st\sqrt[4]{2st^2}$ **c.** $2st\sqrt[5]{t}$ **10. a.** $2v^2w^6\sqrt[3]{12vw^2}$
b. $2vw^5\sqrt[4]{6v^3}$ **c.** $2vw^4\sqrt[5]{3v^2}$ **11. a.** $2\sqrt{5}$ **b.** 5
12. a. $2\sqrt{10}$ **b.** 10 **13. a.** $-3\sqrt{6}$ **b.** 60
14. a. $-7\sqrt{7}$ **b.** 210 **15. a.** $3\sqrt{2}$ **b.** 4
16. a. $3\sqrt{3}$ **b.** 6 **17. a.** $7\sqrt{2}$ **b.** 240
18. a. $-\sqrt{2}$ **b.** 60 **19. a.** $14\sqrt[3]{3}$ **b.** $48\sqrt[3]{9}$
20. a. $\sqrt[3]{2}$ **b.** $30\sqrt[3]{4}$
21. a. $3\sqrt{2}$ **b.** Cannot be simplified further. **c.** $\sqrt{2}$
22. a. $\sqrt{7}$ **b.** $2\sqrt{7}$ **c.** Cannot be simplified further.
23. a. $9z$ **b.** $9 + 6\sqrt{z} + z$ **c.** $9 - z$
24. a. $16 - 8\sqrt{x} + x$ **b.** $16 - x$ **c.** $16x$
25. a. $\frac{6\sqrt{2x}}{x}$ **b.** $\frac{\sqrt{6x}}{x}$ **c.** $\frac{12\sqrt{2} - 12x}{2 - x^2}$
26. a. $\frac{45 + 15\sqrt{y}}{9 - y}$ **b.** $\frac{5\sqrt{3y}}{y}$ **c.** $\frac{\sqrt{5y}}{y}$
27. a. $3\sqrt{5} - 1$ **b.** $8 - 3\sqrt{5}$ **c.** $10 - 4\sqrt{5}$
28. a. $-8 + 11\sqrt{3}$ **b.** $12 + 16\sqrt{3}$ **c.** $3\sqrt{3} - 9$
29. a. $4a^7\sqrt{a}$ **b.** $2a^5\sqrt[3]{2}$ **30. a.** $3y^3$ **b.** $3y^4\sqrt{3y}$

Section 18.7 Activity, pp. 1244–1245

A.1. a. Divide both sides by 3 or multiply both sides by $\frac{1}{3}$.
b. Cube both sides of the equation.
A.2. a. { } **b.** $x = 25$. No, because $\sqrt{25} \neq -5$. **c.** even

A.3. a. Add 3 to both sides. Divide by 2 (or multiply by $\frac{1}{2}$) on both sides.
b. Add 3 to both sides. Divide by 2 (or multiply by $\frac{1}{2}$) on both sides.
c. $\{16\}$ **d.** $\{5\}$ **e.** $\{21\}$ **f.** $\{85\}$
g. Parts (c), (d), and (f). In each part, it was necessary to check the potential solutions because we raised both sides of the equation to an *even* power.
A.4. a. $x^2 - 10x + 25$
b. $\{4\}$. (The value 7 does not check in the original equation.)
A.5. $\{1\}$
A.6. a. $\sqrt{x+2} = 1 - \sqrt{2x+5}$ **b.** $x + 2 = 2x + 6 - 2\sqrt{2x+5}$
c. $-x - 4 = -2\sqrt{2x+5}$ **d.** $x + 4 = 2\sqrt{2x+5}$
e. $x^2 + 8x + 16 = 4(2x+5)$ **f.** $x = 2, x = -2$
g. $\{-2\}$. (The value 2 does not check.)

Section 18.7 Practice Exercises, pp. 1245–1248

R.1. $t^2 + 16t + 64$ **R.3.** $3t + 1$ **R.5.** $m - 14\sqrt{m} + 49$
R.7. $\{5\}$ **R.9.** $\{-3, 7\}$ **R.11.** $\left\{-\frac{1}{5}, \frac{7}{2}\right\}$

1. a. radical **b.** isolate; 7 **c.** extraneous
3. d **5.** e **7.** h **9.** f **11.** $\{100\}$ **13.** $\{4\}$
15. $\{3\}$ **17.** $\{42\}$ **19.** $\{2\}$ **21.** $\{29\}$ **23.** $\{140\}$
25. $\{\ \}$ (The value 25 does not check.)
27. $\{7\}$ (The value -1 does not check.)
29. $\{\ \}$ (The value 3 does not check.) **31.** $V = \frac{4\pi r^3}{3}$
33. $h^2 = \frac{r^2 - \pi^2 r^2}{\pi^2}$ or $h^2 = \frac{r^2}{\pi^2} - r^2$ **35.** $a^2 + 10a + 25$
37. $5a - 6\sqrt{5a} + 9$ **39.** $r + 22 + 10\sqrt{r-3}$ **41.** $\{-3\}$
43. $\{\ \}$ (The value $\frac{1}{2}$ does not check.)
45. $\{2\}$ (The value 18 does not check.)
47. $\{9\}$ **49.** $\{-4\}$ **51.** $\left\{\frac{9}{5}\right\}$ **53.** $\{2\}$
55. $\{\ \}$ (The value 9 does not check.)
57. $\left\{-\frac{11}{4}\right\}$ **59.** $\{\ \}$ (The value $\frac{8}{3}$ does not check.)
61. $\{-1\}$ (The value 3 does not check.) **63.** $\left\{\frac{1}{3}, -1\right\}$
65. $\{\ \}$ (The values 3 and 23 do not check.) **67.** $\{6\}$
69. $\{2\}$ (The value 18 does not check.) **71. a.** 30.25 ft **b.** 34.5 m
73. a. $2 million **b.** $1.2 million **c.** 50,000 passengers
75. The x-coordinate of the point of intersection is the solution to the equation.

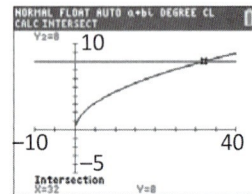

76. The x-coordinate of the point of intersection is the solution to the equation.

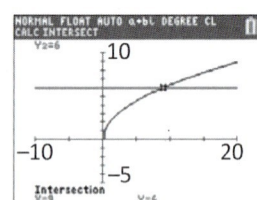

77. a. 12 lb **b.** $t(18) = 5.1$. An 18-lb turkey will take about 5.1 hr to cook. **79.** $b = \sqrt{25 - h^2}$ **81.** $a = \sqrt{k^2 - 196}$

Section 18.8 Activity, p. 1256

A.1. $i; -1$
A.2. a. $i\sqrt{6}$ **b.** $i\sqrt{13}$ **c.** $6i\sqrt{2}$ **d.** $i\sqrt{2}$ **e.** $\frac{3 \pm i\sqrt{7}}{2}$
A.3. a. Write the expression as $i\sqrt{4} \cdot i\sqrt{25}$. **b.** -10
A.4. a. Write the expression as $\frac{i\sqrt{50}}{i\sqrt{2}}$. **b.** 5
A.5. a. $i^3 = i^2 \cdot \boxed{i} = -1 \cdot \boxed{i} = -i$ **b.** $i^4 = i^2 \cdot \boxed{i^2} = -1 \cdot \boxed{-1} = 1$
A.6. a. $i^5 = i^4 \cdot \boxed{i} = 1 \cdot \boxed{i} = i$ **b.** $i^6 = i^4 \cdot \boxed{i^2} = 1 \cdot \boxed{-1} = -1$
c. $i^7 = i^4 \cdot i^2 \cdot \boxed{i} = -i$ **d.** $i^8 = i^4 \cdot \boxed{i^4} = 1$
A.7. a. $-i$ **b.** 1 **c.** -1 **d.** i
A.8. a. $3 + (-7)i$. Real part: 3; imaginary part: -7
b. $4 + 2\sqrt{3}i$. Real part: 4; imaginary part: $2\sqrt{3}$
c. $9 + 0i$. Real part: 9; imaginary part: 0
d. $0 + \frac{3}{4}i$. Real part: 0; imaginary part: $\frac{3}{4}$
A.9. a. $1 + 10x$ **b.** $1 + 10i$ **A.10. a.** $-12x^2 - 10x + 12$ **b.** $24 - 10i$
A.11. a. $81 - 100x^2$ **b.** 181 **A.12. a.** $16x^2 + 40x + 25$ **b.** $9 + 40i$
A.13. a. $\frac{5(3 + 4\sqrt{x})}{9 - 16x}$ or $\frac{15 + 20\sqrt{x}}{9 - 16x}$ **b.** $\frac{3}{5} + \frac{4}{5}i$ **A.14.** $\frac{3}{2} + \frac{\sqrt{7}}{2}i$

Section 18.8 Practice Exercises, pp. 1257–1259

R.1. $-2y - 2$ **R.3.** $-9x + 17$ **R.5.** $10n^2 - 19n - 56$
R.7. $49d^2 - 25$ **R.9.** $100b^2 - 60b + 9$ **R.11.** $3\sqrt{10}$
R.13. 3 **R.15.** $\frac{2}{3} + \frac{1}{3}x$

1. a. imaginary **b.** $\sqrt{-1}; -1$ **c.** $i\sqrt{b}$ **d.** $a + bi; \sqrt{-1}$
e. real; b **f.** $a + bi$ **3.** $\sqrt{-1} = i$ and $-\sqrt{1} = -1$ **5.** $7i$
7. -7 **9.** $\frac{1}{2}i$ **11.** $12i$ **13.** $i\sqrt{3}$ **15.** $-2i\sqrt{5}$
17. -60 **19.** $13i\sqrt{7}$ **21.** -7 **23.** -12
25. $-3\sqrt{10}$ **27.** $i\sqrt{2}$ **29.** 3 **31.** $-i$ **33.** 1
35. i **37.** 1 **39.** $-i$ **41.** -1 **43.** $a - bi$
45. Real: -5; imaginary: 12 **47.** Real: 0; imaginary: -6
49. Real: 35; imaginary: 0 **51.** Real: $\frac{3}{5}$; imaginary: 1
53. $7 + 6i$ **55.** $\frac{3}{10} + \frac{3}{2}i$ **57.** $5i\sqrt{2}$
59. $-1 + 10i$ **61.** $-24 + 0i$ **63.** $18 + 6i$
65. $26 - 26i$ **67.** $-29 + 0i$ **69.** $-9 + 40i$
71. $35 + 20i$ **73.** $-20 + 48i$ **75.** $\frac{13}{16} + 0i$
77. $\frac{1}{5} - \frac{3}{5}i$ **79.** $\frac{3}{25} - \frac{4}{25}i$ **81.** $\frac{21}{29} + \frac{20}{29}i$
83. $-\frac{17}{10} - \frac{1}{10}i$ **85.** $-\frac{1}{2} - \frac{5}{2}i$ **87.** $\frac{1}{2} - \frac{1}{3}i$
89. $1 + 10i$ **91.** $\frac{1}{4} + \frac{1}{2}i$ **93.** $-1 + i\sqrt{2}$
95. $-2 - i\sqrt{3}$ **97.** $-\frac{1}{2} + \frac{\sqrt{2}}{2}i$ **99.** Yes **101.** Yes

Chapter 18 Review Exercises, pp. 1260–1262

1. a. False; $\sqrt{0} = 0$ is not positive. **b.** False; $\sqrt[3]{-8} = -2$
2. $\sqrt{(-3)^2} = \sqrt{9} = 3$ **3. a.** False **b.** True
4. $\frac{5}{4}$ **5.** 5 **6.** 6 **7. a.** 3 **b.** 0 **c.** $\sqrt{7}$
d. $[1, \infty)$ **8. a.** 0 **b.** 1 **c.** 3 **d.** $[-5, \infty)$
9. $\frac{\sqrt[3]{2x}}{\sqrt[4]{2x}} + 4$ **10. a.** $|x|$ **b.** x **c.** $|x|$ **d.** $x + 1$
11. a. $2|y|$ **b.** $3y$ **c.** $|y|$ **d.** y **12.** 8 cm
13. Yes, provided the expressions are defined. For example: $x^5 \cdot x^3 = x^8$ and $x^{1/5} \cdot x^{2/5} = x^{3/5}$ **14.** n represents the root.

15. Take the reciprocal of the base and change the exponent to positive.
16. -5 **17.** $\dfrac{1}{2}$ **18.** 4   **19.** b^{10} **20.** $\dfrac{16y^{12}}{xz^9}$ **21.** $\dfrac{a^3c}{b^2}$
22. $x^{3/4}$ **23.** $(2y^2)^{1/3}$ **24.** 2.1544   **25.** 6.8173   **26.** 54.1819
27. a. The radicand has no factor raised to a power greater than or equal to the index.   **b.** The radicand does not contain a fraction.   **c.** There are no radicals in the denominator of a fraction.
28. $6\sqrt{3}$ **29.** $xz\sqrt[4]{xy}$ **30.** $-10ab^3\sqrt[3]{2b}$ **31.** $-\dfrac{2a}{b}$
32. a. The principal square root of the quotient of 2 and x **b.** The cube of the sum of x and 1   **33.** 31 ft
34. They are *like* radicals.   **35.** Cannot be combined; the indices are different.   **36.** Cannot be combined; one term has a radical, but the other does not.   **37.** Can be combined: $3\sqrt[4]{3xy}$
38. Can be added after simplifying: $19\sqrt{2}$
39. $5\sqrt{7}$ **40.** $9\sqrt[3]{2}-8$ **41.** $10\sqrt{2}$ **42.** $13x\sqrt[3]{2x^2}$
43. False; 5 and $3\sqrt{x}$ are not *like* radicals.
44. False; $\sqrt{y}+\sqrt{y}=2\sqrt{y}$ (add the coefficients).
45. 6   **46.** $2\sqrt{2}$ **47.** $-2\sqrt{21}+6\sqrt{33}$ **48.** $-6\sqrt{15}+15$
49. $4x-9$ **50.** $y-16$ **51.** $7y-2\sqrt{21xy}+3x$
52. $12w+20\sqrt{3w}+25$ **53.** $-2z-9\sqrt{6z}-42$
54. $3a+5\sqrt{5a}-10$ **55.** $u^2\sqrt[6]{u^5}$ **56.** $\sqrt[4]{4w^3}$
57. $\dfrac{y^2\sqrt{3y}}{5x^3}$ **58.** $\dfrac{-2x^2y^2\sqrt[3]{2x}}{z^3}$ **59.** $9w^2$ **60.** $\dfrac{t^4}{4}$
61. $\dfrac{\sqrt{14y}}{2y}$ **62.** $\dfrac{\sqrt{15w}}{3w}$ **63.** $\dfrac{4\sqrt[3]{3p}}{3p}$ **64.** $-\dfrac{\sqrt[3]{4x^2}}{x}$
65. $\sqrt{10}-\sqrt{15}$ **66.** $3\sqrt{5}-3\sqrt{7}$ **67.** $\sqrt{t}+\sqrt{3}$
68. $\sqrt{w}+\sqrt{7}$ **69.** The quotient of the principal square root of 2 and the square of x **70.** $\left\{\dfrac{49}{2}\right\}$ **71.** $\{31\}$
72. $\{-12\}$ **73.** $\{7\}$ **74.** $\{9\}$ **75.** $\{\ \}$ (The value -2 does not check.)   **76.** $\left\{\dfrac{1}{2},4\right\}$ **77.** $\{-2\}$ (The value 2 does not check.)
78. $6\sqrt{5}$ m ≈ 13.4 m   **79. a.** $v(20)\approx 25.3$ ft/sec. When the water depth is 20 ft, a wave travels about 25.3 ft/sec.   **b.** 8 ft
80. $a+bi$, where a and b are real numbers and $i=\sqrt{-1}$
81. $a+bi$, where $b\neq 0$ **82.** Multiply the numerator and denominator by $4-6i$, which is the complex conjugate of the denominator.
83. $4i$ **84.** $-i\sqrt{5}$ **85.** -15 **86.** $-2i$ **87.** -1 **88.** i
89. $-i$ **90.** 0   **91.** $-5+5i$ **92.** $9+17i$ **93.** $25+0i$
94. $24-10i$ **95.** $-\dfrac{17}{4}+i$; real part: $-\dfrac{17}{4}$; imaginary part: 1
96. $-2-i$; real part: -2; imaginary part: -1
97. $\dfrac{4}{13}-\dfrac{7}{13}i$ **98.** $3+4i$ **99.** $-\dfrac{3}{2}+\dfrac{5}{2}i$
100. $-\dfrac{4}{17}+\dfrac{16}{17}i$ **101.** $-\dfrac{2}{3}+\dfrac{\sqrt{10}}{6}i$ **102.** $2-4i$

Chapter 19
Section 19.1 Activity, pp. 1270–1271
A.1. a. $\{-7,7\}$ **b.** $\{-4,4\}$ **c.** $(5i)^2=-25$ and $(-5i)^2=-25$ **d.** $\sqrt{k}$ and $-\sqrt{k}$ or simply $\pm\sqrt{k}$
A.2. a. $\{\pm 8\}$ **b.** $\{-11,5\}$ **c.** $\{-7,1\}$
A.3. a. $\{\pm 2\sqrt{5}\}$ **b.** $\{4\pm 2\sqrt{5}\}$ **c.** $\{4\pm i\sqrt{6}\}$
A.4. a. 5   **b.** 25   **c.** $x^2+10x+25=(x+5)^2$ **d.** $\left(\dfrac{1}{2}b\right)^2$
A.5. a. $n=81;(y+9)^2$ **b.** $n=\dfrac{121}{4};\left(t-\dfrac{11}{2}\right)^2$ **c.** $n=\dfrac{9}{100};\left(x+\dfrac{3}{10}\right)^2$
A.6. a. $x^2+4x+10=0$ **b.** $x^2+4x=-10$ **c.** $x^2+4x+4=-10+4$ **d.** $(x+2)^2=-6$ **e.** $\{-2\pm i\sqrt{6}\}$

Section 19.1 Practice Exercises, pp. 1271–1275
R.1. $\{5\}$ **R.3.** $\left\{\dfrac{2}{3}+\dfrac{\sqrt{5}}{3}\right\}$ **R.5.** 11 and -11 **R.7.** $6\sqrt{2}$
R.9. $\dfrac{\sqrt{37}}{8}$ **R.11.** $y^2-14x+49$ **R.13.** $9w^2+24w+16$
R.15. $(p+11)^2$ **R.17.** $(8x-9)^2$

1. a. $0;0$ **b.** 0   **c.** $\sqrt{k};-\sqrt{k}$ **d.** $2;\{3,-3\}$ **e.** completing   **f.** 100   **g.** $4;1$ **h.** 8
3. $\{\pm 2\}$ **5.** $\{\pm\sqrt{7}\}$ **7.** $\left\{\dfrac{11}{6},-\dfrac{11}{6}\right\}$ **9.** $\{\pm 5i\}$
11. $\{-1,-5\}$ **13.** $\left\{\dfrac{-3\pm\sqrt{7}}{2}\right\}$ **15.** $\{-5\pm 3i\sqrt{2}\}$
17. $\left\{\pm\dfrac{\sqrt{33}}{3}\right\}$ **19.** $\left\{-\dfrac{4}{5}\pm\dfrac{\sqrt{3}}{5}i\right\}$ **21.** $\{4i,-4i\}$
23. 1. Factoring and applying the zero product rule.   2. Applying the square root property. $\{\pm 6\}$ **25. a.** $\{16\}$ **b.** $\{2,-2\}$
27. $n=9;(x-3)^2$ **29.** $n=16;(t+4)^2$ **31.** $n=\dfrac{1}{4};\left(c-\dfrac{1}{2}\right)^2$
33. $n=\dfrac{25}{4};\left(y+\dfrac{5}{2}\right)^2$ **35.** $n=\dfrac{1}{25};\left(b+\dfrac{1}{5}\right)^2$ **37.** $n=\dfrac{1}{9};\left(p-\dfrac{1}{3}\right)^2$
39. 1. Divide both sides by a to make the leading coefficient 1.   2. Isolate the variable terms on one side of the equation.   3. Complete the square.   4. Apply the square root property and solve for x.
41. $\{-3,-5\}$ **43.** $\{-3\pm i\sqrt{7}\}$ **45.** $\{-2\pm i\sqrt{2}\}$
47. $\{5,-2\}$ **49.** $\left\{-1\pm\dfrac{\sqrt{6}}{2}i\right\}$ **51.** $\left\{2\pm\dfrac{2}{3}i\right\}$
53. $\left\{\dfrac{1}{5}\pm\dfrac{\sqrt{3}}{5}\right\}$ **55.** $\left\{-\dfrac{3}{4}\pm\dfrac{\sqrt{65}}{4}\right\}$ **57.** $\{2\pm\sqrt{11}\}$
59. $\{1,-7\}$ **61. a.** $t=\dfrac{\sqrt{d}}{4}$ **b.** 8 sec
63. $r=\sqrt{\dfrac{A}{\pi}}$ or $r=\dfrac{\sqrt{A\pi}}{\pi}$ **65.** $a=\sqrt{d^2-b^2-c^2}$
67. $r=\sqrt{\dfrac{3V}{\pi h}}$ or $r=\dfrac{\sqrt{3V\pi h}}{\pi h}$ **69.** The shelf extends 4.2 ft.
71. The sides are 7.1 in.   **73. a.** 4.5 thousand textbooks or 35.5 thousand textbooks   **b.** Profit increases to a point as more books are produced. Beyond that point, the market is "flooded," and profit decreases. There are two points at which the profit is $20,000. Producing 4.5 thousand books makes the same profit with fewer resources as producing 35.5 thousand books.

Section 19.2 Activity, pp. 1286–1287
A.1. $x=\dfrac{-b\pm\sqrt{b^2-4ac}}{2a}$
A.2. a. $a=2,b=-12,c=16$ **b.** $\{2,4\}$
A.3. a. $a=1,b=-4,c=5$ **b.** $\{2\pm i\}$
A.4. a. $a=-1,b=-6,c=-9$ **b.** $\{-3\}$
A.5. a. Imaginary   **b.** Real   **c.** $1;2$ **d.** discriminant
A.6. a. 16   **b.** Two real solutions   **c.** Yes
A.7. a. -4 **b.** Two imaginary solutions   **c.** Yes
A.8. a. 0   **b.** One real solution   **c.** Yes
A.9. a. The solutions to the equation give the x-intercepts of the graph.
   b. The discriminant tells us the number of real solutions to the equation and thus the number of x-intercepts of the graph. In this case, the discriminant is 16, indicating that the equation $2x^2-12x+16=0$ has two real solutions, and the graph of $f(x)=2x^2-12x+16$ has two x-intercepts.
A.10. a. 0   **b.** 1
   c. Function g is shown in graph ii. Function h is shown in graph i.

Section 19.2 Practice Exercises, pp. 1287–1291
R.1. $2\sqrt{6}$ **R.3.** $2+\sqrt{5}$ **R.5.** $2\sqrt{14}$
R.7. $\left\{-\dfrac{29}{3}\right\}$ **R.9.** $\{-59\}$ **R.11.** $\{-7\pm 2\sqrt{11}\}$

1. **a.** quadratic; $\dfrac{-b \pm \sqrt{b^2 - 4ac}}{2a}$ **b.** $ax^2 + bx + c = 0$
   c. $8; -42; -27$ **d.** $7; 97$ **3.** Quadratic **5.** Neither
7. Linear **9.** $\{-12, 1\}$ **11.** $\left\{\dfrac{1}{9} \pm \dfrac{2\sqrt{11}}{9}i\right\}$
13. $\left\{\dfrac{1}{6} \pm \dfrac{\sqrt{14}}{6}i\right\}$ **15.** $\{7, -5\}$ **17.** $\left\{\dfrac{-3 \pm \sqrt{41}}{2}\right\}$
19. $\left\{\dfrac{2}{5}\right\}$ **21.** $\{3 \pm i\sqrt{5}\}$ **23.** $\left\{\dfrac{1 \pm \sqrt{29}}{2}\right\}$
25. $\left\{\dfrac{-1 \pm \sqrt{13}}{4}\right\}$ **27.** $\left\{-\dfrac{2}{3} \pm \dfrac{2\sqrt{2}}{3}i\right\}$ **29.** $\left\{\dfrac{-5 \pm \sqrt{13}}{2}\right\}$
31. $\{-2, -4\}$ **33.** $\left\{\dfrac{-7 \pm \sqrt{109}}{6}\right\}$
35. **a.** $(x-3)(x^2 + 3x + 9)$ **b.** $\left\{3, -\dfrac{3}{2} \pm \dfrac{3\sqrt{3}}{2}i\right\}$
37. **a.** $3x(x^2 - 2x + 2)$ **b.** $\{0, 1 \pm i\}$ **39.** The length of each side is 3 ft. **41.** The legs are approximately 1.6 in. and 3.6 in.
43. The legs are approximately 8.2 m and 6.1 m.
45. **a.** Approximately 4.5 fatalities per 100 million miles driven
    b. Approximately 1 fatality per 100 million miles driven
    c. Approximately 4.2 fatalities per 100 million miles driven
    d. For drivers 26 years old and 71 years old
47. The times are $\dfrac{3 + \sqrt{5}}{2}$ sec ≈ 2.62 sec or $\dfrac{3 - \sqrt{5}}{2}$ sec ≈ 0.38 sec.
49. **a.** $x^2 + 2x + 1 = 0$
    b. 0 **c.** 1 rational solution **51. a.** $19m^2 - 8m + 0 = 0$ **b.** 64
    c. 2 rational solutions **53. a.** $5p^2 + 0p - 21 = 0$ **b.** 420
    c. 2 irrational solutions **55. a.** $n^2 + 3n + 4 = 0$ **b.** -7
    c. 2 imaginary solutions **57.** Discriminant: 16; two x-intercepts
59. Discriminant: 0; one x-intercept
61. Discriminant: -39; no x-intercepts
63. x-intercepts: $\left(\dfrac{5 + \sqrt{13}}{2}, 0\right), \left(\dfrac{5 - \sqrt{13}}{2}, 0\right)$;
    y-intercept: $(0, 3)$ **65.** x-intercepts: none; y-intercept: $(0, -1)$
67. x-intercepts: $\left(\dfrac{-5 - \sqrt{41}}{4}, 0\right), \left(\dfrac{-5 + \sqrt{41}}{4}, 0\right)$;
    y-intercept: $(0, -2)$ **69.** $\{-1 \pm 3i\}$
71. $\left\{-\dfrac{1}{3}, 6\right\}$ **73.** $\left\{\dfrac{5}{2} \pm \dfrac{3\sqrt{2}}{2}i\right\}$ **75.** $\left\{-\dfrac{1}{2} \pm 2i\right\}$
77. $\left\{\dfrac{3 \pm \sqrt{7}}{2}\right\}$ **79.** $\{1 \pm 3i\sqrt{2}\}$ **81.** $\{2 \pm \sqrt{2}\}$
83. $\left\{-\dfrac{1}{8}, \dfrac{3}{4}\right\}$ **85.** $\left\{\pm \dfrac{\sqrt{21}}{2}\right\}$
87. **a.** $\{-3 \pm \sqrt{14}\}$ **b.** $\{-3 \pm \sqrt{14}\}$ **c.** Answers will vary

89.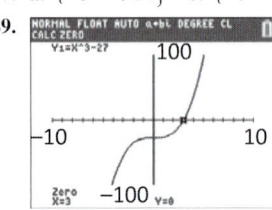

91.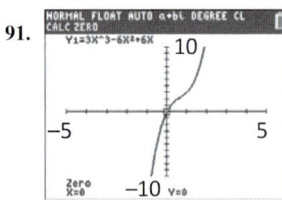

Section 19.3 Activity, pp. 1295–1296

A.1. a. quadratic **b.** $u^2 - 13u + 36 = 0$
 c. $u = 4$ and $u = 9$ **d.** $\left\{\dfrac{5}{2}, 5\right\}$

A.2. $u = x^2 - 3; u^2 - 9u - 52 = 0$ **A.3.** $u = x^{1/3}; 3u^2 - u - 4 = 0$
A.4. $u = x^2; u^2 - 29u + 100 = 0$ **A.5.** $u = 2 + \dfrac{3}{x}; u^2 - u - 12 = 0$
A.6. $\{\pm i, \pm 4\}$ **A.7.** $\left\{\dfrac{64}{27}, -1\right\}$ **A.8.** $\{\pm 5, \pm 2\}$ **A.9.** $\left\{\dfrac{3}{2}, -\dfrac{3}{5}\right\}$

Section 19.3 Practice Exercises, pp. 1296–1298

R.1. $\sqrt{x}$ **R.3.** $\sqrt[3]{z^2}$ or $\left(\sqrt[3]{z}\right)^2$ **R.5.** $\{-2, 9\}$
R.7. $\left\{\dfrac{-1 \pm \sqrt{37}}{6}\right\}$ **R.9.** $\{\pm 3\}$ **R.11.** $\{\pm \sqrt{5}\,i\}$
R.13. $\{125\}$ **R.15.** $\{-8\}$

1. **a.** quadratic **b.** $3x - 1$ **c.** $p^{1/3}$
3. Let $u = 3m + 4; u^2 - 4u - 5 = 0$
5. Let $u = y^2 + 1; u^2 - 7u + 10 = 0$
7. Let $u = \dfrac{1}{x}; 3u^2 + 10u - 8 = 0$
9. **a.** $\{-4, -6\}$ **b.** $\{-4, -1, -2, -3\}$
11. $\{1 \pm i\sqrt{2}, 1 \pm \sqrt{2}\}$ **13.** $\{5, -1, 2\}$ **15.** $\{27, -8\}$
17. $\left\{-\dfrac{1}{32}, -243\right\}$ **19.** $\{4\}$ (The value 64 does not check.)
21. $\left\{\dfrac{1}{4}\right\}$ (The value 4 does not check.)
23. $\{-7\}$ **25.** $\{16\}$; yes **27.** $\{\pm 3i, \pm \sqrt{5}\}$
29. $\left\{\pm \dfrac{\sqrt{3}}{2}, \pm i\sqrt{5}\right\}$ **31.** $\left\{\dfrac{-5 \pm \sqrt{13}}{2}\right\}$
33. $\left\{\dfrac{9 \pm \sqrt{73}}{4}\right\}$ **35.** $\{2 \pm \sqrt{3}\}$ **37.** $\{\pm 2, \pm 2i\}$
39. $\left\{-\dfrac{7}{4}, -\dfrac{3}{2}\right\}$ **41.** $\left\{\dfrac{1}{2}, -\dfrac{1}{2}, \sqrt{2}, -\sqrt{2}\right\}$
43. $\left\{2, 1, -1 \pm i\sqrt{3}, -\dfrac{1}{2} \pm \dfrac{\sqrt{3}}{2}i\right\}$ **45.** $\{5, 4\}$ **47.** $\{0, 2\}$
49. $\left\{1, \dfrac{17}{2}\right\}$ **51.** $\{64, -125\}$ **53.** $\{\pm \sqrt{2}, \pm 2i\}$
55. $\{\pm 4i, 1\}$ **57.** $\{4, \pm i\sqrt{5}\}$ **59.** $\left\{\dfrac{8}{3}, 2\right\}$
61. **a.** $\{\pm i\sqrt{2}\}$ **b.** Two imaginary solutions; no real solutions
    c. No x-intercepts
    d.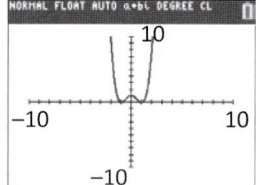

62. **a.** $\{1, -1\}$ **b.** Two real solutions; no imaginary solutions
    c. Two x-intercepts
    d.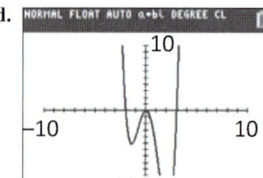

63. **a.** $\{0, 3, -2\}$ **b.** Three real solutions; no imaginary solutions
    c. Three x-intercepts
    d.

64. a. $\{\pm 1, \pm 3\}$ b. Four real solutions; no imaginary solutions
   c. Four x-intercepts
   d.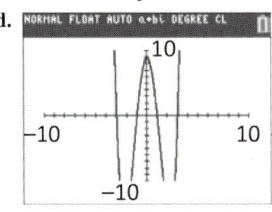

Chapter 19 Problem Recognition Exercises, p. 1299

1. $\{-5 \pm \sqrt{22}\}$ 2. $\{8 \pm \sqrt{59}\}$ 3. $\left\{-\dfrac{1}{6} \pm \dfrac{\sqrt{47}}{6} i\right\}$
4. $\left\{-\dfrac{3}{8} \pm \dfrac{\sqrt{71}}{8} i\right\}$ 5. a. Quadratic   b. $\{2, -7\}$
6. a. Quadratic   b. $\{4, 5\}$ 7. a. Quadratic form
   b. $\{-3, -1, 1, 3\}$ 8. a. Quadratic form   b. $\{2, -2, i, -i\}$
9. a. Quadratic form (or radical)   b. $\{16\}$ (The value 1 does not check.)   10. a. Quadratic   b. $\left\{\dfrac{9 \pm \sqrt{89}}{2}\right\}$
11. a. Linear   b. $\{4\}$ 12. a. Linear   b. $\{-1\}$
13. a. Quadratic   b. $\{\pm i\sqrt{2}\}$ 14. a. Quadratic
   b. $\{\pm i\sqrt{6}\}$ 15. a. Rational   b. $\{2, -1\}$
16. a. Rational   b. $\{-2, 4\}$
17. a. Quadratic   b. $\{10 \pm \sqrt{101}\}$
18. a. Quadratic   b. $\{-9 \pm \sqrt{77}\}$
19. a. Quadratic   b. $\left\{\dfrac{1 \pm \sqrt{17}}{2}\right\}$
20. a. Quadratic   b. $\left\{\dfrac{1 \pm \sqrt{13}}{2}\right\}$
21. a. Radical   b. $\{3\}$ (The value -1 does not check.)
22. a. Radical   b. $\{6\}$ (The value -1 does not check.)
23. a. Quadratic form (or radical)   b. $\{-125, 27\}$
24. a. Quadratic form (or radical)   b. $\{-64, -1\}$

Section 19.4 Activity, pp. 1307–1308

Instructor Note

Students can complete the tables by hand and plotting points or, if time is a consideration, consider having them use a graphing utility.

A.1.

x	$f(x) = x^2$	$g(x) = x^2 + 1$	$h(x) = x^2 - 2$
0	0	1	-2
1	1	2	-1
2	4	5	2
3	9	10	7
-1	1	2	-1
-2	4	5	2
-3	9	10	7

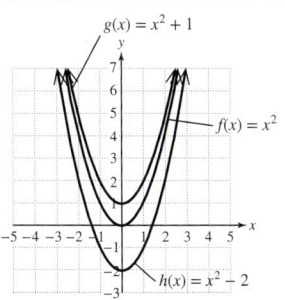

A.2. The graph of $y = x^2 + k$ is the graph of $y = x^2$ shifted up k units. The graph of $y = x^2 - k$ is the graph of $y = x^2$ shifted down k units.

A.3.

x	$f(x) = x^2$	$g(x) = (x+1)^2$	$h(x) = (x-2)^2$
0	0	1	4
1	1	4	1
2	4	9	0
3	9	16	1
-1	1	0	9
-2	4	1	16
-3	9	4	25

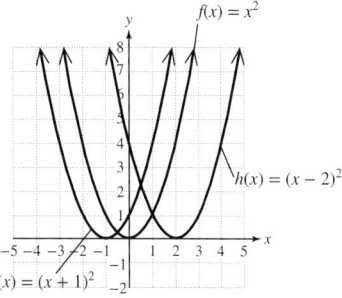

A.4. The graph of $y = (x + h)^2$ is the graph of $y = x^2$ shifted to the left h units. The graph of $y = (x - h)^2$ is the graph of $y = x^2$ shifted to the right h units.

A.5.

x	$f(x) = x^2$	$g(x) = 2x^2$	$h(x) = \frac{1}{2}x^2$	$k(x) = -x^2$
0	0	0	0	0
1	1	2	0.5	-1
2	4	8	2	-4
3	9	18	4.5	-9
-1	1	2	0.5	-1
-2	4	8	2	-4
-3	9	18	4.5	-9

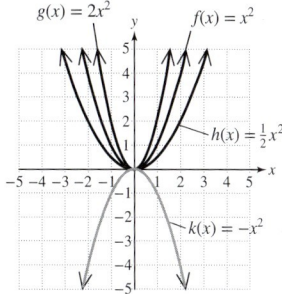

A.6. a. For $a > 1$, the graph of $y = ax^2$ is the graph of $y = x^2$ with a vertical stretch by a factor of a.
   b. For $0 < a < 1$, the graph of $y = ax^2$ is the graph of $y = x^2$ with a vertical shrink by a factor of a.
   c. For $a < 0$, the graph of $y = ax^2$ is the graph of $y = x^2$ reflected over the x-axis. There may also be a vertical shrink or stretch if $|a| \neq 1$.

A.7. a. i   b. iii   c. ii

A.8. a.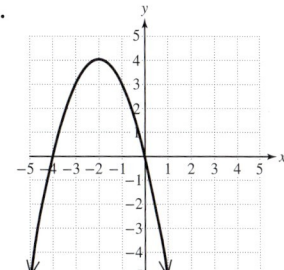

b. $(-2, 4)$ c. $x = -2$ d. Maximum value 4
e. Domain: $(-\infty, \infty)$; range: $(-\infty, 4]$

Section 19.4 Practice Exercises, pp. 1309–1314

R.1. a. $(-\infty, \infty)$ b. $[-4, \infty)$ c. x-intercepts: $(-1, 0)$ and $(3, 0)$
d. y-intercept: $(0, -3)$ **R.3.** a. $(-\infty, \infty)$ b. $(-\infty, -1]$
c. x-intercepts: None   d. y-intercept: $(0, -5)$

1. a. parabola   b. $>$; $<$ c. lowest; highest   3. Downward
5. Maximum   7. Vertex: $(-3, 1)$; axis of symmetry: $x = -3$
9. The value of k shifts the graph of $f(x) = x^2$ vertically.

11. 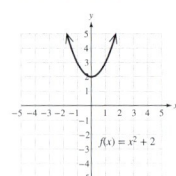   13.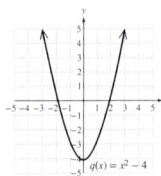

15. 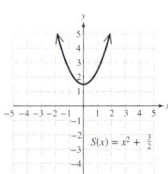   17.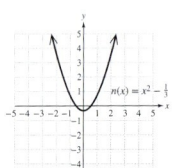

19. 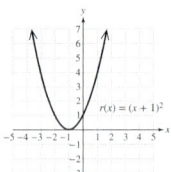   21.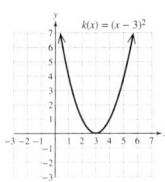

23. 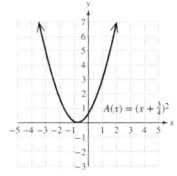   25.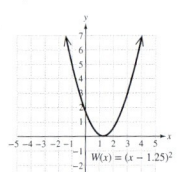

27. The value of a vertically stretches or shrinks the graph of $f(x) = x^2$.

29. 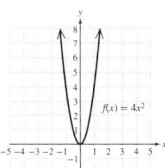   31.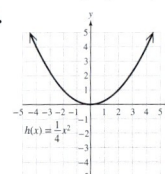

33. 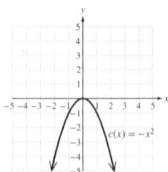   35.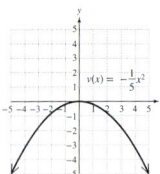

37. d   39. g   41. a   43. e

45. 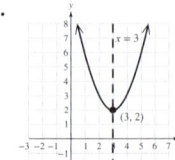   47.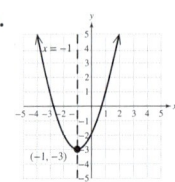

Domain: $(-\infty, \infty)$; range: $[2, \infty)$ Domain: $(-\infty, \infty)$; range: $[-3, \infty)$

49. 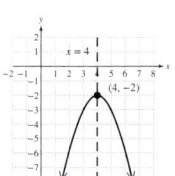   51.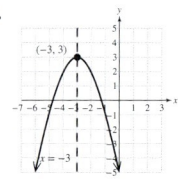

Domain: $(-\infty, \infty)$; range: $(-\infty, -2]$ Domain: $(-\infty, \infty)$; range: $(-\infty, 3]$

53. 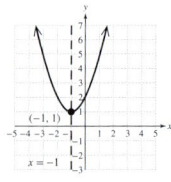   55.

Domain: $(-\infty, \infty)$; range: $[1, \infty)$ Domain: $(-\infty, \infty)$; range: $[0, \infty)$

57. 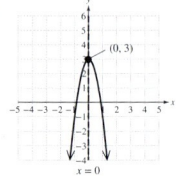   59.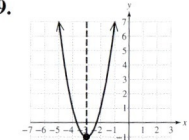

Domain: $(-\infty, \infty)$; range: $(-\infty, 3]$ Domain: $(-\infty, \infty)$; range: $[-1, \infty)$

61.    63.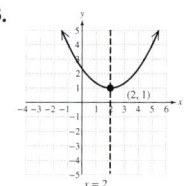

Domain: $(-\infty, \infty)$; range: $(-\infty, 2]$ Domain: $(-\infty, \infty)$; range: $[1, \infty)$

65. a. $y = x^2 + 3$ is $y = x^2$ shifted up 3 units.
b. $y = (x + 3)^2$ is $y = x^2$ shifted left 3 units.
c. $y = 3x^2$ is $y = x^2$ with a vertical stretch.
67. Vertex $(6, -9)$; minimum point; minimum value: -9
69. Vertex $(2, 5)$; maximum point; maximum value: 5
71. Vertex $(-8, 0)$; minimum point; minimum value: 0
73. Vertex $\left(0, \dfrac{21}{4}\right)$; maximum point; maximum value: $\dfrac{21}{4}$
75. Vertex $\left(7, -\dfrac{3}{2}\right)$; minimum point; minimum value: $-\dfrac{3}{2}$
77. Vertex $(0, 0)$; minimum point; minimum value: 0
79. True   81. False   83. a. $(60, 30)$ b. 30 ft   c. 70 ft
85. a. The fireworks will explode at a height of 150 ft.
b. Yes, because the ordered pair $(3, 150)$ is the vertex.

Section 19.5 Activity, pp. 1321–1322

A.1. Complete the square to write the function in vertex form $f(x) = a(x - h)^2 + k$, or apply the vertex formula.
A.2. The x-coordinate of the vertex is given by $\dfrac{-b}{2a}$. The y-coordinate is most easily found by substituting the x-coordinate of the vertex into the function.
A.3. a. $f(x) = 2(x^2 - 2x) - 6$ b. 1   c. $f(x) = 2(x^2 - 2x + 1 - 1) - 6$
d. $f(x) = 2(x^2 - 2x + 1) + 2(-1) - 6$ or equivalently
$f(x) = 2(x^2 - 2x + 1) - 8$
e. $f(x) = 2(x - 1)^2 - 8$ f. $(1, -8)$ g. Upward
h. Minimum value: -8 i. $x = 1$
j. x-intercepts: $(-1, 0)$ and $(3, 0)$; y-intercept: $(0, -6)$

k.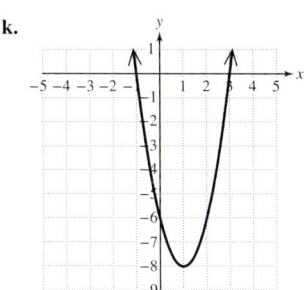

A.4. $(1, -8)$; They are the same.
A.5. a. $(-3, -1)$ b. Downward c. Maximum value: -1
d. $x = -3$ e. x-intercepts: None; y-intercept: $(0, -10)$
f.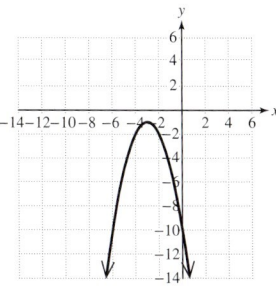

A.6. a. $(50, 12{,}250)$
b. The vertex means that 50 sec after launch, the rocket will reach its maximum height of 12,250 m.

Section 19.5 Practice Exercises, pp. 1322–1325

R.1. $\frac{3}{4}$ **R.3.** x-intercept: $(-2, 0)$; y-intercept: $(0, -8)$
R.5. $\{-4, 8\}$ **R.7.** $\{-2, 2\}$ **R.9.** $\{-12\}$ **R.11.** $n = 49$; $(x - 7)^2$

1. a. $\frac{-b}{2a}; \frac{-b}{2a}$ b. True   c. True   d. True   e. False
3. $g(x) = (x - 4)^2 - 11$; $(4, -11)$
5. $n(x) = 2(x + 3)^2 - 5$; $(-3, -5)$
7. $p(x) = -3(x - 1)^2 - 2$; $(1, -2)$
9. $k(x) = \left(x + \frac{7}{2}\right)^2 - \frac{89}{4}$; $\left(-\frac{7}{2}, -\frac{89}{4}\right)$
11. $F(x) = 5(x + 1)^2 - 4$; $(-1, -4)$
13. $P(x) = -2\left(x - \frac{1}{4}\right)^2 + \frac{1}{8}$; $\left(\frac{1}{4}, \frac{1}{8}\right)$ **15.** $(2, 3)$
17. $(-1, -2)$ **19.** $(-4, -15)$ **21.** $(-1, 2)$ **23.** $(1, 3)$
25. $\left(\frac{3}{2}, \frac{3}{4}\right)$ **27.** $(-4, -15)$ **29.** $(-1, 4)$
31. a. $(-1, -4)$ b. $(0, -3)$ **33.** a. $\left(\frac{1}{2}, \frac{7}{2}\right)$ b. $(0, 4)$
c. $(1, 0), (-3, 0)$ c. No x-intercepts
d. 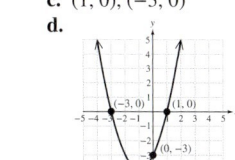   d.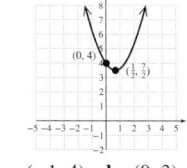

35. a. $\left(\frac{3}{2}, 0\right)$ b. $\left(0, -\frac{9}{4}\right)$ **37.** a. $(-1, 4)$ b. $(0, 3)$
c. $\left(\frac{3}{2}, 0\right)$ c. $(1, 0), (-3, 0)$
d. 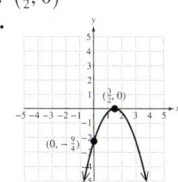   d.

39. a. 164.25 ft   b. 3.125 sec   **41.** a. 45 mph   b. 32 mpg
43. a. 48 hr   b. 2 g   **45.** $a = -9, b = 5, c = 4$; $y = -9x^2 + 5x + 4$
47. $a = 2, b = -1, c = -5$; $y = 2x^2 - x - 5$
49. $a = -3, b = 4, c = 0$; $y = -3x^2 + 4x$
51. 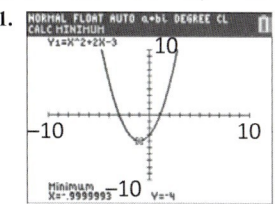   **52.**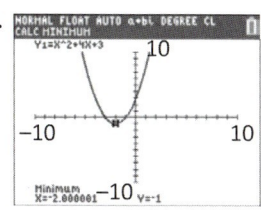
53. 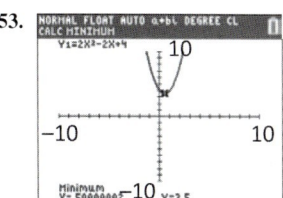   **54.**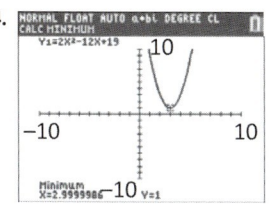
55. 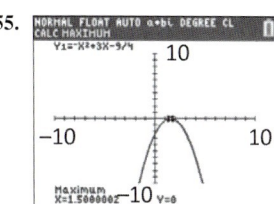   **56.**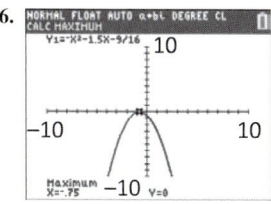

57. a. The sum of the sides must equal the total amount of fencing.
b. $A = x(200 - 2x)$ c. 50 ft by 100 ft

Chapter 19 Review Exercises, pp. 1326–1328

1. $\{\pm\sqrt{5}\}$ **2.** $\{\pm 2i\}$ **3.** $\{\pm 9\}$ **4.** $\left\{\pm\frac{\sqrt{57}}{3}i\right\}$
5. $\{2 \pm 6\sqrt{2}\}$ **6.** $\left\{\frac{5}{2} \pm \frac{3}{2}i\right\}$ **7.** $\left\{\frac{1 \pm \sqrt{3}}{3}\right\}$
8. $\{4 \pm \sqrt{5}\}$ **9.** $5\sqrt{3}$ in. ≈ 8.7 in.   **10.** 9 in.
11. $5\sqrt{6}$ in. ≈ 12.2 in.   **12.** $n = 64$; $(x + 8)^2$
13. $n = \frac{81}{4}$; $\left(x - \frac{9}{2}\right)^2$ **14.** $n = \frac{1}{16}$; $\left(y + \frac{1}{4}\right)^2$
15. $n = \frac{1}{25}$; $\left(z - \frac{1}{5}\right)^2$ **16.** $\{-2 \pm 3i\}$ **17.** $\{1 \pm \sqrt{6}\}$
18. $\{3 \pm 2\sqrt{3}\}$ **19.** $\{4 \pm 3i\}$ **20.** $\left\{\frac{1}{3}, -1\right\}$
21. $\left\{\frac{1}{2}, -4\right\}$ **22.** $r = \sqrt{\frac{V}{\pi h}}$ or $r = \frac{\sqrt{V \pi h}}{\pi h}$
23. $s = \sqrt{\frac{A}{6}}$ or $s = \frac{\sqrt{6A}}{6}$
24. There will be two imaginary solutions.
25. Two rational solutions   **26.** Two rational solutions
27. Two irrational solutions   **28.** Two imaginary solutions
29. One rational solution   **30.** Two imaginary solutions
31. $\{2 \pm \sqrt{3}\}$ **32.** $\left\{\frac{5}{2} \pm \frac{5\sqrt{3}}{2}i\right\}$ **33.** $\left\{2, -\frac{5}{6}\right\}$
34. $\left\{2, \frac{4}{3}\right\}$ **35.** $\left\{\frac{4}{5}, -\frac{1}{5}\right\}$ **36.** $\left\{\frac{1}{10}, -\frac{1}{2}\right\}$
37. $\{2 \pm 2i\sqrt{7}\}$ **38.** $\{4 \pm \sqrt{6}\}$ **39.** $\left\{\frac{2 \pm \sqrt{22}}{3}\right\}$
40. $\{8, -2\}$ **41.** $\{-7 \pm \sqrt{3}\}$ **42.** $\{-1 \pm \sqrt{11}\}$
43. a. 1822 ft   b. 115 ft/sec   **44.** a. $\approx 53{,}939$ thousand   b. 2021
45. The dimensions are approximately 3.1 ft by 7.2 ft.
46. The distance between Lincoln and Omaha is approximately 50 mi.
47. $\{49\}$ (The value 9 does not check.)
48. $\{4, 16\}$ **49.** $\{\pm 3, \pm\sqrt{2}\}$ **50.** $\left\{\pm\frac{\sqrt{6}}{2}, \pm i\right\}$

51. $\{-243, 32\}$ 52. $\{32, 1\}$ 53. $\{3 \pm \sqrt{10}\}$
54. $\left\{\dfrac{1 \pm \sqrt{181}}{6}\right\}$ 55. $\{\pm 3i, \pm i\sqrt{3}\}$ 56. $\{\pm\sqrt{7}, \pm 2\}$

57.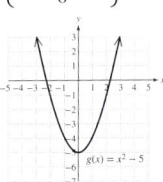
Domain: $(-\infty, \infty)$; range: $[-5, \infty)$

58.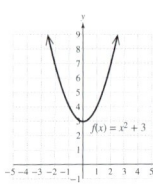
Domain: $(-\infty, \infty)$; range: $[3, \infty)$

59.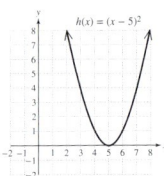
Domain: $(-\infty, \infty)$; range: $[0, \infty)$

60.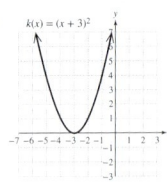
Domain: $(-\infty, \infty)$; range: $[0, \infty)$

61.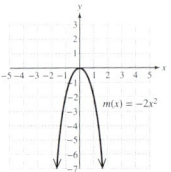
Domain: $(-\infty, \infty)$; range: $(-\infty, 0]$

62.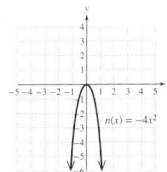
Domain: $(-\infty, \infty)$; range: $(-\infty, 0]$

63.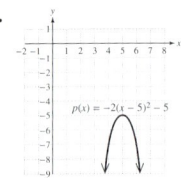
Domain: $(-\infty, \infty)$; range: $(-\infty, -5]$

64.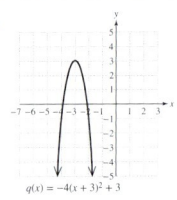
Domain: $(-\infty, \infty)$; range: $(-\infty, 3]$

65. $\left(4, \dfrac{5}{3}\right)$ is the minimum point. The minimum value is $\dfrac{5}{3}$.

66. $\left(1, -\dfrac{1}{7}\right)$ is the maximum point. The maximum value is $-\dfrac{1}{7}$.

67. $x = -\dfrac{2}{11}$ 68. $x = \dfrac{3}{16}$
69. $z(x) = (x-3)^2 - 2$; $(3, -2)$
70. $b(x) = (x-2)^2 - 48$; $(2, -48)$
71. $p(x) = -5(x+1)^2 - 8$; $(-1, -8)$
72. $q(x) = -3(x+4)^2 - 6$; $(-4, -6)$ 73. $(1, -15)$
74. $(-1, 7)$ 75. $\left(\dfrac{1}{2}, \dfrac{41}{4}\right)$ 76. $\left(-\dfrac{1}{3}, -\dfrac{22}{3}\right)$

77. **a.** $(2, -3)$ **b.** $(0, 0), (4, 0)$
c.

78. **a.** $(-2, 4)$ **b.** $(0, 0), (-4, 0)$
c.
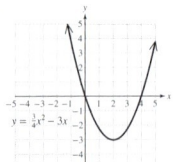

79. **a.** 3 sec **b.** 144 ft   80. **a.** 150 meals **b.** $1200
81. $a = 1, b = 4, c = -1$; $y = x^2 + 4x - 1$
82. $a = 1, b = -1, c = 6$; $y = x^2 - x + 6$

Chapter 20

Section 20.1 Activity, p. 1336

A.1. one-to-one
A.2. horizontal
A.3. Yes
A.4. No; the ordered pairs $(5, -3)$ and $(-1, -3)$ have different x-coordinates, but the same y-coordinate. These ordered pairs fail the *horizontal* line test.
A.5. **a.** $\{(-4, 2), (1, -3), (-2, 0), (3, -4)\}$ **b.** Yes
A.6. **a.** $\{(1, 4), (-3, 5), (-3, -1), (5, -2)\}$ **b.** No; the ordered pairs $(-3, 5)$ and $(-3, -1)$ have the same x-coordinate but different y-coordinates (these ordered pairs fail the *vertical* line test).
A.7. **a.** The graph of f is a line with slope $\dfrac{1}{2}$ and y-intercept $(0, -2)$.
b. Yes; yes
A.8. **a.** $y = \dfrac{x-4}{2}$ **b.** $x = \dfrac{y-4}{2}$ **c.** $y = 2x + 4$ **d.** $f^{-1}(x) = 2x + 4$
e. $(f \circ f^{-1})(x) = x$ and $(f^{-1} \circ f)(x) = x$

A.9. **a.**
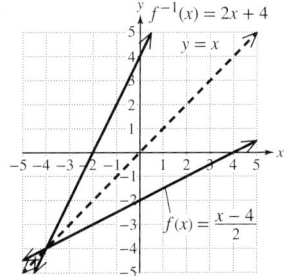
b. The graphs of a function and its inverse are symmetric with respect to the line $y = x$.

A.10. **a.**
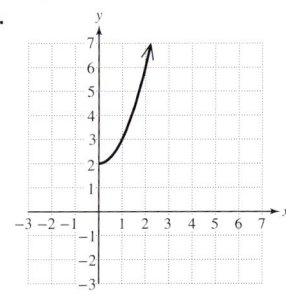
b. Yes **c.** $f^{-1}(x) = \sqrt{x-2}$
d. Yes

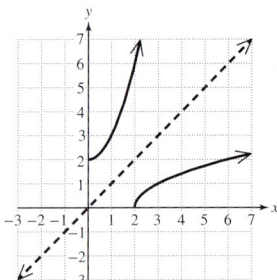

Section 20.1 Practice Exercises, pp. 1337–1341

R.1. Yes R.3. $x = -\dfrac{y+4}{3}$ or $x = -\dfrac{1}{3}y - \dfrac{4}{3}$ R.5. $x = y^3 + 1$
R.7. **a.** $f(1) = 6$ **b.** $f(-2) = 0$ **c.** $f(t) = 2t + 4$ **d.** $f\left(\dfrac{x-4}{2}\right) = x$
R.9. $(g \circ f)(x) = 4x^2 + 17$

1. a. {(2, 1), (3, 2), (4, 3)}   b. one-to-one; y
   c. is not   d. is   e. $y = x$
3. $g^{-1} = \{(5, 3), (1, 8), (9, -3), (2, 0)\}$ 5. $r^{-1} = \{(3, a), (6, b), (9, c)\}$
7. The function is not one-to-one.   9. Yes   11. No   13. Yes
15. a. $(f \circ g)(x) = 6\left(\dfrac{x-1}{6}\right) + 1 = x$ b. $(g \circ f)(x) = \dfrac{(6x+1)-1}{6} = x$
17. a. $(f \circ g)(x) = \dfrac{\sqrt[3]{8x^3}}{2} = x$ b. $(g \circ f)(x) = 8\left(\dfrac{\sqrt[3]{x}}{2}\right)^3 = x$
19. a. $(f \circ g)(x) = (\sqrt{x-1})^2 + 1 = x$ b. $(g \circ f)(x) = \sqrt{(x^2+1)-1} = x$
21. subtracts; $x - 4$ 23. divides; $\dfrac{x}{5}$
25. a. $f(x) = 9x - 2$ b. $f^{-1}(x) = \dfrac{x+2}{9}$ 27. $h^{-1}(x) = x - 4$
29. $m^{-1}(x) = 3(x+2)$ 31. $p^{-1}(x) = -x + 10$ 33. $n^{-1}(x) = \dfrac{5x-2}{3}$
35. $h^{-1}(x) = \dfrac{3x+1}{4}$ 37. $f^{-1}(x) = \sqrt[3]{x-1}$
39. $g^{-1}(x) = \dfrac{x^3+1}{2}$ 41. $g^{-1}(x) = \sqrt{x-9}$
43. a. 1.2192 m, 15.24 m   b. $f^{-1}(x) = \dfrac{x}{0.3048}$ c. 4921.3 ft
45. False   47. True   49. False   51. True   53. $(b, 0)$
55. a. Domain: $[1, \infty)$, range: $[0, \infty)$ b. Domain: $[0, \infty)$, range: $[1, \infty)$
57. a. $[-4, 0]$ b. $[0, 2]$ c. $[0, 2]$ d. $[-4, 0]$

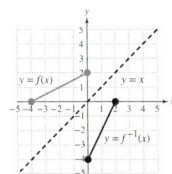

59. a. $[0, 2]$ b. $[0, 4]$ c. $[0, 4]$ d. $[0, 2]$

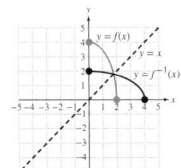

61. $f^{-1}(x) = x^3 - 5$ 62. $k^{-1}(x) = \sqrt[3]{x+4}$

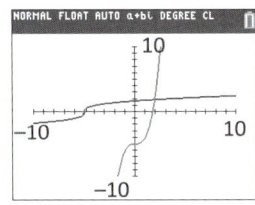 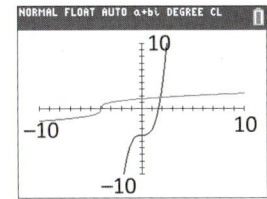

63. $g^{-1}(x) = \sqrt[3]{2x+4}$ 64. $m^{-1}(x) = \dfrac{x+4}{3}$

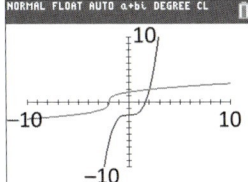 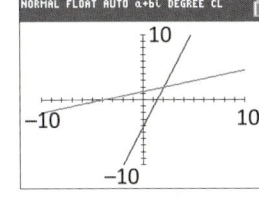

65. $q^{-1}(x) = x^2 - 4, x \geq 0$ 67. $z^{-1}(x) = x^2 - 4, x \leq 0$
69. $f^{-1}(x) = \dfrac{x+1}{1-x}$ 71. $t^{-1}(x) = \dfrac{x+2}{x}$
73. $n^{-1}(x) = -\sqrt{x-9}$

Section 20.2 Activity, pp. 1347–1348

A.1. exponential   A.2. b, c
A.3. a.

x	$f(x) = 3^x$	$g(x) = \left(\dfrac{1}{3}\right)^x$
0	1	1
1	3	$\dfrac{1}{3}$
2	9	$\dfrac{1}{9}$
−1	$\dfrac{1}{3}$	3
−2	$\dfrac{1}{9}$	9

b. For each 1-unit increase in x, the function values (y values) increase by a factor of 3.
c. For each 1-unit increase in x, the function values (y values) decrease by a factor of 3.
d.

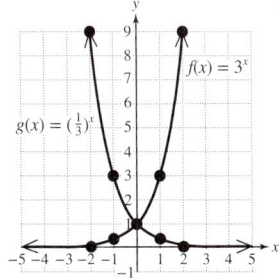

e. Exponential growth   f. Exponential decay   g. No
h. Domain: $(-\infty, \infty)$; range: $(0, \infty)$

A.4. a. $A(8) = 32$ and means that after 8 days, 32 units of ^{131}I is left.
   b. $A(16) = 16$ and means that after 16 days, 16 units of ^{131}I is left.
   c. $A(24) = 8$ and means that after 24 days, 8 units of ^{131}I is left.
   d. After each half-life (each increment of 8 days), the amount of substance is cut in half.

Section 20.2 Practice Exercises, pp. 1348–1351

R.1. a. 16   b. 1   c. $\dfrac{1}{16}$ R.3. a. $\dfrac{1}{16}$ b. 1   c. 16

1. a. b^x b. is not; is   c. increasing   d. decreasing
   e. $(-\infty, \infty); (0, \infty)$ f. $(0, 1)$ g. $y = 0$
3. 5.8731   5. 1385.4557   7. 0.0063   9. 0.8950
11. a. $x = 2$ b. $x = 3$ c. Between 2 and 3
13. a. $x = 4$ b. $x = 5$ c. Between 4 and 5
15. $f(0) = 1, f(1) = \dfrac{1}{5}, f(2) = \dfrac{1}{25}, f(-1) = 5, f(-2) = 25$
17. $h(0) = 1, h(1) = 3, h(-1) = \dfrac{1}{3}, h(\sqrt{2}) \approx 4.73, h(\pi) \approx 31.54$
19. If $b > 1$, the graph is increasing. If $0 < b < 1$, the graph is decreasing.

21. 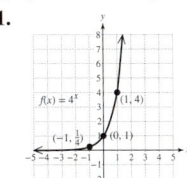   23.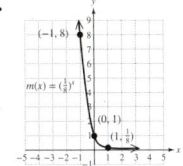

25. 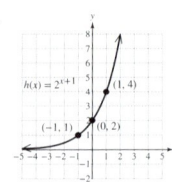 27.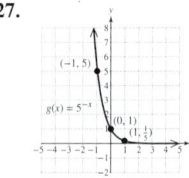

29. a. 0.25 g   b. ≈ 0.16 g   31. a. 758,000   b. 379,000
c. 144,000   33. a. $P(t) = 153,000,000(1.0125)^t$
b. $P(41) \approx 255,000,000$ 35. a. $1640.67   b. $2691.80
c. $A(0) = 1000$. The initial amount of the investment is $1000.
$A(7) = 2000$. The amount of the investment doubles in 7 years.

37. 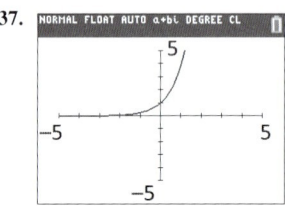 38.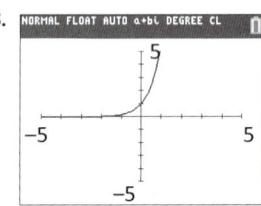

39. 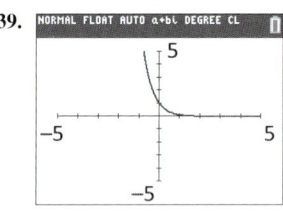 40.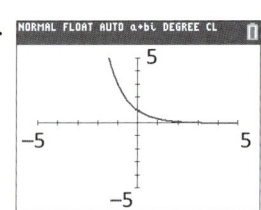

41. 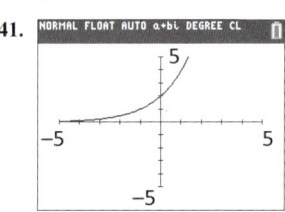 42.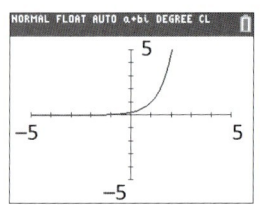

43. 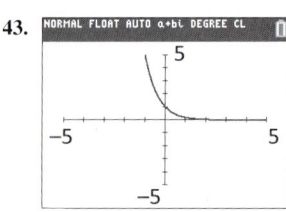 44.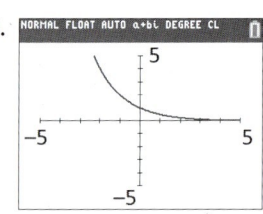

Section 20.3 Activity, pp. 1360–1361

A.1. a. 8   b. 3   c. logarithmic; $b^y = x$.

	Exponential Form	Logarithmic Form
A.2.	$4^2 = 16$	$\log_4 16 = 2$
A.3.	$3^4 = 81$	$\log_3 81 = 4$
A.4.	$2^7 = x$	$\log_2 x = 7$
A.5.	$5^y = 125$	$\log_5 125 = y$
A.6.	$6^1 = 6$	$\log_6 6 = 1$
A.7.	$7^0 = 1$	$\log_7 1 = 0$
A.8.	$2^{-4} = \frac{1}{16}$	$\log_2 \frac{1}{16} = -4$
A.9.	$10^{-2} = \frac{1}{100}$	$\log_{10} \frac{1}{100} = -2$

A.10. 5   A.11. 2   A.12. −2   A.13. −3   A.14. 0
A.15. 1   A.16. a. 3   b. 4   c. Between 3 and 4

A.17. a. $3^y = x$
b.

x	y
$\frac{1}{9}$	−2
$\frac{1}{3}$	−1
1	0
3	1
9	2

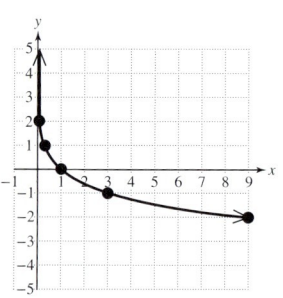

A.18. a. $\left(\frac{1}{3}\right)^y = x$
b.

x	y
9	−2
3	−1
1	0
$\frac{1}{3}$	1
$\frac{1}{9}$	2

A.19. $(0, \infty)$ A.20. $(-\infty, \infty)$ A.21. $x = 0$ (the y-axis)
A.22. $(1, 0)$ A.23. a. increases   b. decreases   A.24. $(0, \infty)$
A.25. $(4, \infty)$ A.26. $(-\infty, 4)$ A.27. $\left(-\frac{1}{2}, \infty\right)$

Section 20.3 Practice Exercises, pp. 1361–1365

R.1. 5   R.3. 16   R.5. $\frac{1}{64}$ R.7. 2   R.9. −4
R.11. −3   R.13. −2   R.15. $(3, \infty)$ R.17. $(-\infty, 6)$

1. a. logarithmic; b b. logarithm; base; argument
c. $(0, \infty); (-\infty, \infty)$ d. exponential   e. common
f. increasing; decreasing   g. 2, 3, and 4   h. $y = x$
3. $b^y = x$ 5. a. $5^{\boxed{2}} = 25$ b. $\log_5 25 = \boxed{2}$
7. a. $3^{\boxed{3}} = 27$ b. $\log_3 27 = \boxed{3}$ 9. a. $8^{\boxed{1}} = 8$ b. $\log_8 8 = \boxed{1}$
11. $5^4 = 625$ 13. $10^{-4} = 0.0001$ 15. $6^2 = 36$
17. $b^x = 15$ 19. $3^x = 5$ 21. $\left(\frac{1}{4}\right)^{10} = x$ 23. $\log_3 81 = x$
25. $\log_5 25 = 2$ 27. $\log_7\left(\frac{1}{7}\right) = -1$ 29. $\log_b y = x$
31. $\log_e y = x$ 33. $\log_{1/3} 9 = -2$ 35. 2   37. −1   39. $\frac{1}{2}$
41. 0   43. 5   45. 1   47. 3   49. $\frac{1}{2}$ 51. 1   53. 3
55. 6   57. −2   59. 0.7782   61. 0.4971   63. −1.5051
65. −2.2676   67. 5.5315   69. −7.4202
71. a. Slightly less than 2   b. Slightly more than 1
c. log 93 ≈ 1.9685, log 12 ≈ 1.0792
73. a. $f\left(\frac{1}{64}\right) = -3, f\left(\frac{1}{16}\right) = -2, f\left(\frac{1}{4}\right) = -1, f(1) = 0,$
$f(4) = 1, f(16) = 2, f(64) = 3$
b.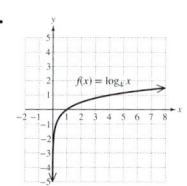

75. $3^y = x$

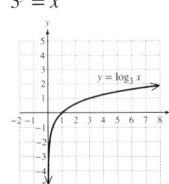

x	y
$\frac{1}{9}$	-2
$\frac{1}{3}$	-1
1	0
3	1
9	2

77. $\left(\frac{1}{2}\right)^y = x$

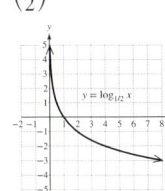

x	y
4	-2
2	-1
1	0
$\frac{1}{2}$	1
$\frac{1}{4}$	2

79. $(5, \infty)$ **81.** $(-\infty, 2)$ **83.** $\left(\frac{1}{3}, \infty\right)$ **85.** $(-1.2, \infty)$

87. $(-\infty, 2)$ **89.** $(-\infty, 0) \cup (0, \infty)$ **91.** ≈ 7.35

93. a.

t (months)	0	1	2	6	12	24
$S_1(t)$	91	82.0	76.7	65.6	57.6	49.1
$S_2(t)$	88	83.5	80.8	75.3	71.3	67.0

b. Group 1: 91; Group 2: 88 **c.** Method II

95. Domain: $(-6, \infty)$; asymptote: $x = -6$

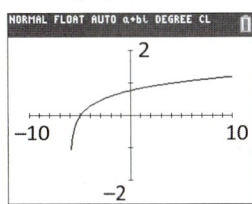

96. Domain: $(-2, \infty)$; asymptote: $x = -2$

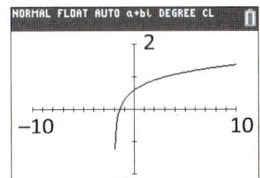

97. Domain: $(2, \infty)$; asymptote: $x = 2$

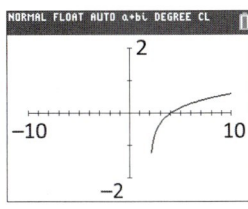

98. Domain: $(-8, \infty)$; asymptote: $x = -8$

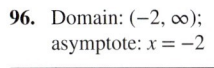

99. Domain: $(-\infty, 2)$; asymptote: $x = 2$

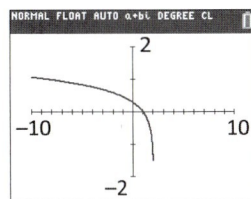

100. Domain: $(-\infty, 3)$; asymptote: $x = 3$

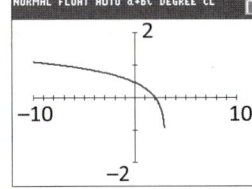

Chapter 20 Problem Recognition Exercises, p. 1366

1. e **2.** l **3.** j **4.** g **5.** c **6.** a
7. k **8.** b **9.** i **10.** h **11.** f **12.** d

Section 20.4 Activity, p. 1372

A.1. a. 1 **b.** 1 **c.** 1 **d.** 1 **A.2. a.** 0 **b.** 0 **c.** 0 **d.** 0
A.3. a. 3 **b.** 4 **c.** 5 **d.** p **A.4. a.** 25 **b.** 16 **c.** $\frac{1}{8}$ **d.** x

A.5. $\boxed{3} = \boxed{2} + \boxed{1}$; $\log_b(x \cdot y) = \log_b x + \log_b y$

A.6. $\boxed{2} = \boxed{3} - \boxed{1}$; $\log_b\left(\frac{x}{y}\right) = \log_b x - \log_b y$

A.7. $\boxed{3} = \boxed{3} \cdot \boxed{1}$; $\log_b(x)^p = p \cdot \log_b x$

A.8. $\log_2\left(\frac{x^5\sqrt{z}}{y^2\sqrt[3]{w}}\right)$ **A.9.** $2 + \frac{1}{2}\log x + 3\log y - \log z$

Section 20.4 Practice Exercises, pp. 1373–1375

R.1. x^6 **R.3.** w^{11} **R.5.** n^{12} **R.7.** $\frac{1}{t^6}$

R.9. $\frac{9b^6}{a^{10}}$ **R.11.** 3 **R.13.** -4 **R.15.** 3

1. a. 1; 0; x; x **b.** $\log_b x + \log_b y$; $\log_b x - \log_b y$ **c.** $p\log_b x$ **d.** False **e.** False **f.** False

3. a, b, c **5.** 1 **7.** 4 **9.** 11 **11.** 3 **13.** 0

15. 9 **17.** 0 **19.** 3 **21.** 1 **23.** $2x$ **25.** 4 **27.** 0

29. Expressions a and c are equivalent.

31. Expressions a and c are equivalent.

33. $\log_3 x - \log_3 5$ **35.** $\log 2 + \log x$ **37.** $4\log_5 x$

39. $\log_4 a + \log_4 b - \log_4 c$

41. $\frac{1}{2}\log_b x + \log_b y - 3\log_b z - \log_b w$

43. $\log_2(x+1) - 2\log_2 y - \frac{1}{2}\log_2 z$

45. $\frac{1}{3}\log a + \frac{2}{3}\log b - \frac{1}{3}\log c$ **47.** $-5\log w$

49. $\frac{1}{2}\log_b a - 3 - \log_b c$ **51.** 3 **53.** 2 **55.** $\log_3\left(\frac{x^2 z}{y^3}\right)$

57. $\log_3\left(\frac{a^2 c}{\sqrt[4]{b}}\right)$ **59.** $\log_b x^2$ **61.** $\log_8(8a^5)$ or $\log_8 a^5 + 1$

63. $\log\left[\frac{(x+6)^2\sqrt[3]{y}}{z^5}\right]$ **65.** $\log_b\left(\frac{1}{x-1}\right)$

67. a. Domain: $(-\infty, 0) \cup (0, \infty)$ **b.** Domain: $(0, \infty)$

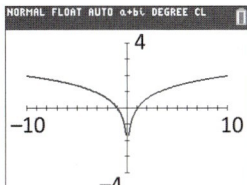 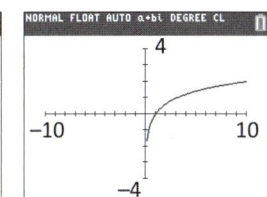

c. They are equivalent for all x in the intersection of their domains, $(0, \infty)$.

68. a. Domain: $(-\infty, 1) \cup (1, \infty)$ **b.** Domain: $(1, \infty)$

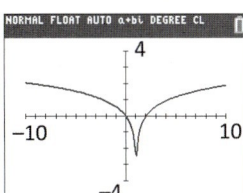 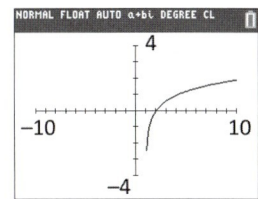

c. They are equivalent for all x in the intersection of their domains, $(1, \infty)$.

69. 1.792 **71.** 2.485 **73.** 4.396 **75.** 0.916 **77.** 13.812

79. 16.09 **81. a.** $B = 10\log I - 10\log I_0$ **b.** $10\log I + 160$

Section 20.5 Activity, pp. 1383–1384

A.1. a. 2.71801 **b.** 2.71825 **c.** 2.71828 **d.** 2.71828
A.2. a. 2.71828 **b.** Growth

A.3. a. $17,000; after 30 years of simple interest, $5000 will grow to $17,000.
 b. $50,313.28; after 30 years of interest compounded annually, $5000 will grow to $50,313.28.
 c. $55,115.88; after 30 years of continuous compounded interest, $5000 will grow to $55,115.88.
 d. Continuous compounding **e.** $38,115.88
 f.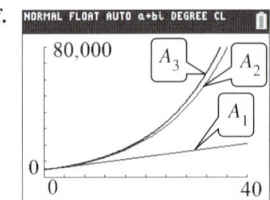

A.4. logarithmic; $y = \ln x$
A.5. a. Domain: $(-\infty, \infty)$; range: $(0, \infty)$
 b. Domain: $(0, \infty)$; range: $(-\infty, \infty)$
 c. The domain of f is the range of g, and the range of f is the domain of g.
A.6. a. 0 **b.** 1 **c.** 4 **d.** 6   **A.7.** $\ln\left(\dfrac{x+y}{z^3}\right)$
A.8. $2\ln a - \ln b - \dfrac{1}{3}\ln c$ **A.9. a.** $(-4, \infty)$ **b.** $(-\infty, 6)$
A.10. a. $\log_b x = \dfrac{\log_a x}{\log_a b}$ **b.** 4 **c.** 5 **d.** Between 4 and 5
 e. 4.3219 **f.** 4.3219 **g.** They are the same.

Section 20.5 Practice Exercises, pp. 1384–1390

R.1. a. 2.717603 **b.** 2.718281
R.3. 0   **R.5.** 1   **R.7.** 5   **R.9.** 3
R.11. $\log_5\left(\dfrac{\sqrt[3]{x}\cdot y}{z^2}\right)$ **R.13.** $\dfrac{1}{2}\log x - \log y - 4\log z$

1. a. e **b.** e **c.** natural; $\ln x$ **d.** 0; 1; p, x **e.** $\ln x + \ln y$; $\ln x - \ln y$ **f.** $p \ln x$ **g.** $\log_a b$ **3.** e **5.** increasing

7.

x	y
−4	0.05
−3	0.14
−2	0.37
−1	1
0	2.72
1	7.39

Domain: $(-\infty, \infty)$

9.

x	y
−2	2.14
−1	2.37
0	3
1	4.72
2	9.39
3	22.09

Domain: $(-\infty, \infty)$

11. a. $12,209.97 **b.** $13,488.50 **c.** $14,898.46 **d.** $16,050.09
An investment grows more rapidly at higher interest rates.
13. a. $12,423.76 **b.** $12,515.01 **c.** $12,535.94
 d. $12,546.15 **e.** $12,546.50
More money is earned at a greater number of compounding periods per year.
15. a. $6920.15 **b.** $9577.70 **c.** $13,255.84
 d. $18,346.48 **e.** $35,143.44
More money is earned over a longer period of time.

17.

x	y
2.25	−1.39
2.50	−0.69
2.75	−0.29
3	0
4	0.69
5	1.10
6	1.39

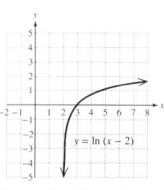

Domain: $(2, \infty)$

19.

x	y
0.25	−2.39
0.5	−1.69
0.75	−1.29
1	−1.00
2	−0.31
3	0.10
4	0.39

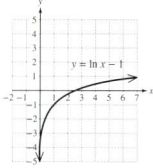

Domain: $(0, \infty)$

21. a. **b.** Domain: $(-\infty, \infty)$; range: $(0, \infty)$
 c. Domain: $(0, \infty)$; range: $(-\infty, \infty)$

23. 1   **25.** 0   **27.** −6   **29.** $2x + 3$ **31.** $\ln(p^6\sqrt[3]{q})$
33. $\ln\sqrt{\dfrac{x}{y^3}}$ **35.** $\ln\left(\dfrac{a^2}{b\sqrt[3]{c}}\right)$ **37.** $\ln\left(\dfrac{x^4}{y^3 z}\right)$
39. $2\ln a - 2\ln b$ **41.** $2\ln b + 1$ **43.** $4\ln a + \dfrac{1}{2}\ln b - \ln c$
45. $\dfrac{1}{5}\ln a + \dfrac{1}{5}\ln b - \dfrac{2}{5}\ln c$ **47.** $(4, \infty)$ **49.** $\left(-\dfrac{5}{2}, \infty\right)$
51. $(-\infty, 7)$ **53. a.** 2.9570 **b.** 2.9570 **c.** They are the same.
55. 2.8074   **57.** 1.5283   **59.** −2.1269   **61.** 0
63. −3.3219   **65.** −3.8124   **67. a.** 15.4 years **b.** 6.9 years
 c. 13.8 years   **69. a.** 19.8 years **b.** 13.9 years **c.** 27.8 years
71. a–b 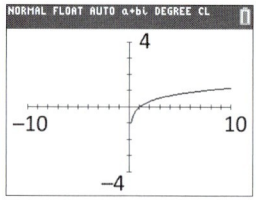 **c.** They appear to be the same.

72. a–b   **c.** They appear to be the same.

73.   **74.**

75.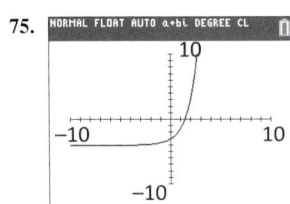

Chapter 20 Problem Recognition Exercises, p. 1390

	Exponential Form	Logarithmic Form
1.	$2^5 = 32$	$\log_2 32 = 5$
2.	$3^4 = 81$	$\log_3 81 = 4$
3.	$z^y = x$	$\log_z x = y$
4.	$b^c = a$	$\log_b a = c$
5.	$10^3 = 1000$	$\log 1000 = 3$
6.	$10^1 = 10$	$\log 10 = 1$
7.	$e^a = b$	$\ln b = a$
8.	$e^q = p$	$\ln p = q$
9.	$(\frac{1}{2})^2 = \frac{1}{4}$	$\log_{1/2}(\frac{1}{4}) = 2$
10.	$(\frac{1}{3})^{-2} = 9$	$\log_{1/3} 9 = -2$
11.	$10^{-2} = 0.01$	$\log 0.01 = -2$
12.	$10^x = 4$	$\log 4 = x$
13.	$e^0 = 1$	$\ln 1 = 0$
14.	$e^1 = e$	$\ln e = 1$
15.	$25^{1/2} = 5$	$\log_{25} 5 = \frac{1}{2}$
16.	$16^{1/4} = 2$	$\log_{16} 2 = \frac{1}{4}$
17.	$e^t = s$	$\ln s = t$
18.	$e^r = w$	$\ln w = r$
19.	$15^{-2} = \frac{1}{225}$	$\log_{15}(\frac{1}{225}) = -2$
20.	$3^{-1} = p$	$\log_3 p = -1$

Section 20.6 Activity, pp. 1399–1400

A.1. a. $x = y$ **b.** $5^{x+1} = 5^3$ **c.** $\{2\}$ **d.** $\{5\}$

A.2. a. $e^{2x+1} = 10$ **b.** $\ln e^{2x+1} = \ln 10$ **c.** $(2x+1)\ln e = \ln 10$

d. $\left\{\dfrac{-1 + \ln 10}{2}\right\}$; or approximately $\{0.6513\}$

A.3. a. $\log 3^{x-4} = \log 8^x$ **b.** $(x-4)\log 3 = x \log 8$

c. $x \log 3 - x \log 8 = 4 \log 3$ **d.** $x = \dfrac{4 \log 3}{\log 3 - \log 8}$

e. $\left\{\dfrac{4 \log 3}{\log 3 - \log 8}\right\}$; or approximately $\{-4.4803\}$

A.4. a. $x = y$ **b.** $x = 3$ **c.** $\{\ \}$; (The value 3 does not check.)
d. The domain of logarithmic functions is the set of positive real numbers. Therefore, any potential solution to a logarithmic equation that makes the argument to a logarithmic expression negative or zero must be discarded.

A.5. a. $\log_5 x + \log_5(x - 20) = 3$ **b.** $\log_5[x(x-20)] = 3$
c. $x(x - 20) = 5^3$ **d.** $x = 25, x = -5$
e. $\{25\}$; (The value -5 does not check.)

A.6. a. 1.3 µg **b.** 27 days

Section 20.6 Practice Exercises, pp. 1400–1404

R.1. $\log_2[(x-1)(x+2)]$ **R.3.** $\log\left(\dfrac{x}{1-x}\right)$ **R.5.** 1

R.7. 6 **R.9.** $(2x+1) \cdot \log 5$ **R.11.** 4^{2x} **R.13.** 2^{5x-15}

R.15. $\{-5\}$ **R.17.** $\{-16\}$ **R.19.** $\{-6, -2\}$

1. a. $x = y$ **b.** $x = y$ **3. a.** $\{2\}$ **b.** $\{25\}$

5. a. $\{-1\}$ **b.** $\left\{\dfrac{1}{10}\right\}$ **7.** $\{9\}$ **9.** $\{10^{42}\}$ **11.** $\{e^{0.08}\}$

13. $\{10^{-9.2} - 40\}$ **15.** $\{5\}$ **17.** $\{10\}$ **19.** $\{25\}$

21. $\{59\}$ **23.** $\{1\}$ **25.** $\{0\}$ **27.** $\left\{-\dfrac{37}{9}\right\}$

29. $\{3\}$ (The value -3 does not check.) **31.** $\{4\}$ **33.** $\{3\}$

35. $\{2\}$ **37.** $\{\ \}$ (The value -3 does not check.) **39.** $\{4\}$

41. $\{-6\}$ **43.** $\left\{\dfrac{1}{2}\right\}$ **45.** $\{2\}$ **47.** $\left\{\dfrac{11}{12}\right\}$ **49.** $\{1\}$

51. $\left\{\dfrac{4}{19}\right\}$ **53.** $\{-2, 1\}$ **55.** $\left\{\dfrac{\ln 21}{\ln 8}\right\}$ **57.** $\{\ln 8.1254\}$

59. $\{\log 0.0138\}$ **61.** $\left\{\dfrac{\ln 15}{0.07}\right\}$ **63.** $\left\{\dfrac{\ln 3}{1.2}\right\}$

65. $\left\{\dfrac{\ln 3}{\ln 5 - \ln 3}\right\}$ **67.** $\left\{\dfrac{2 \ln 2}{\ln 6 - \ln 2}\right\}$ **69.** $\left\{\dfrac{\ln 4}{0.04}\right\}$

71. $\left\{3 \ln\left(\dfrac{125}{6}\right)\right\}$ **73.** $\left\{\dfrac{\ln 20}{\ln 5} + 2\right\}$

75. a. ≈ 1285 million (or 1,285,000,000) people
b. ≈ 1466.5 million (or 1,466,500,000) people
c. The year 2049 ($t \approx 50.8$)
77. a. 500 bacteria **b.** ≈ 660 bacteria **c.** ≈ 25 min
79. It will take 9.9 years for the investment to double. **81.** 5 years
83. a. 7.8 g **b.** 18.5 days **85.** The intensity of sound of heavy traffic is $10^{-3.07}$ W/m². **87.** It will take 38.4 years.

89. 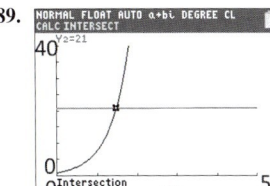 **90.**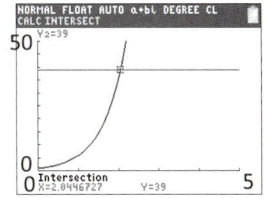

91. a. 1.42 kg **b.** No **93.** $\{10^5, 10^{-3}\}$ **95.** $\left\{\dfrac{1}{27}, \dfrac{1}{9}\right\}$

Chapter 20 Review Exercises, pp. 1405–1408

1. No **2.** Yes **3.** $\{(5, 3), (9, 2), (-1, 0), (1, 4)\}$

4. $q^{-1}(x) = \dfrac{4}{3}(x + 2)$ **5.** $g^{-1}(x) = (x - 3)^5$

6. $f^{-1}(x) = \sqrt[3]{x} + 1$ **7.** $n^{-1}(x) = \dfrac{2x + 4}{x}$

8. $(f \circ g)(x) = 5\left(\dfrac{1}{5}x + \dfrac{2}{5}\right) - 2 = x + 2 - 2 = x$

$(g \circ f)(x) = \dfrac{1}{5}(5x - 2) + \dfrac{2}{5} = x - \dfrac{2}{5} + \dfrac{2}{5} = x$

9. The graphs are symmetric about the line $y = x$.

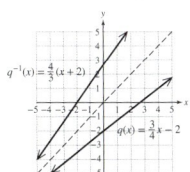

10. a. Domain: $[-1, \infty)$; range: $[0, \infty)$ **b.** Domain: $[0, \infty)$; range: $[-1, \infty)$ **11.** $p^{-1}(x) = (x - 2)^2, x \geq 2$

12. 1024 **13.** $\dfrac{1}{36} \approx 0.028$ **14.** 2 **15.** 10

16. 8.825 **17.** 16.242 **18.** 1.627 **19.** 0.681

20. **21.**

22. 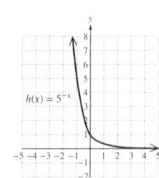   **23.**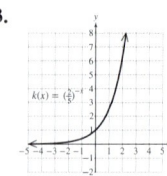

24. a. Horizontal   **b.** $y = 0$ **25. a.** 15,000 mrem
b. 3750 mrem   **c.** Yes   **26.** -3 **27.** 0   **28.** 1
29. 8   **30.** 4   **31.** 4   **32.** 5   **33.** -1

34. 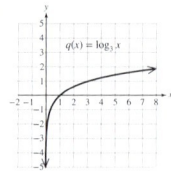   **35.**

36. a. Vertical asymptote   **b.** $x = 0$ **37. a.** 2.5   **b.** 9.5
38. 1   **39.** 6   **40.** 0   **41.** 7
42. a. $\log_b x + \log_b y$ **b.** $\log_b\left(\dfrac{x}{y}\right)$ **c.** $p \log_b x$
43. $\log_b\left(\dfrac{\sqrt[4]{x^3 y}}{z}\right)$ **44.** $\log_3\left(\dfrac{\sqrt{ab}}{c^2 d^4}\right)$ **45.** $\log 5$
46. 0   **47.** b   **48.** a   **49.** 148.4132   **50.** 14.0940
51. 32.2570   **52.** 57.2795   **53.** 1.7918   **54.** -2.1972
55. 1.3424   **56.** 1.3029   **57.** 3.3219   **58.** 1.9943
59. -0.8370 **60.** -3.6668 **61. a.** $33,361.92
b. $33,693.90   **c.** $33,770.48   **d.** $33,809.18
62. a. $S(0) = 95$; the student's score is 95 at the end of the course.
b. $S(6) \approx 23.7$; the student's score is 23.7 after 6 months.
c. $S(12) \approx 20.2$; the student's score is 20.2 after 1 year.
63. $(-\infty, \infty)$ **64.** $(-\infty, \infty)$ **65.** $(-\infty, \infty)$
66. $(0, \infty)$ **67.** $(-5, \infty)$ **68.** $(7, \infty)$ **69.** $\left(\dfrac{4}{3}, \infty\right)$
70. $(-\infty, 5)$ **71.** $\{125\}$ **72.** $\left\{\dfrac{1}{49}\right\}$ **73.** $\{216\}$ **74.** $\{3^{1/12}\}$
75. $\left\{\dfrac{1001}{2}\right\}$ **76.** $\{9\}$ **77.** $\{5\}$ (The value -2 does not check.)
78. $\left\{\dfrac{190}{321}\right\}$ **79.** $\{-1\}$ **80.** $\left\{\dfrac{4}{7}\right\}$ **81.** $\left\{\dfrac{\ln 21}{\ln 4}\right\}$
82. $\left\{\dfrac{\ln 18}{\ln 5}\right\}$ **83.** $\{-\ln 0.1\}$ **84.** $\left\{-\dfrac{\ln 0.06}{2}\right\}$
85. $\left\{\dfrac{\log 1512}{2}\right\}$ **86.** $\left\{\dfrac{\log 821}{3}\right\}$ **87.** $\left\{\dfrac{3 \ln 2}{\ln 7 - \ln 2}\right\}$
88. $\left\{\dfrac{5 \ln 14}{\ln 14 - \ln 6}\right\}$ **89. a.** 1.09 µg   **b.** 0.15 µg
c. 16.08 days   **90. a.** 150 bacteria   **b.** ≈ 185 bacteria
c. ≈ 99 min   **91. a.** $V(0) = 15,000$; the initial value of the car is
$15,000.   **b.** $V(10) = 3347$; the value of the car after 10 years is
$3347.   **c.** 7.3 years

Chapter 21
Section 21.1 Activity, pp. 1415–1416
A.1. a. Cyclist 1: (4, 8); Cyclist 2: (12, −4)
b.

c. Horizontal distance: 8 mi; vertical distance: 12 mi
d. $4\sqrt{13}$ mi ≈ 14.4 mi

A.2. a. $d = \sqrt{(x_2 - x_1)^2 + (y_2 - y_1)^2}$ **b.** $4\sqrt{13}$ mi
    c. horizontal; vertical   **d.** $4\sqrt{13}$ mi; No
A.3. a. Add the numbers and divide by 2.
    b. $\left(\dfrac{x_1 + x_2}{2}, \dfrac{y_1 + y_2}{2}\right)$ **c.** (8, 2)
A.4. a. center; radius   **b.** $(x - h)^2 + (y - k)^2 = r^2$
A.5. Center: $(5, -2)$; radius: 2
A.6. Center: $\left(-\dfrac{3}{2}, 0\right)$; radius: $2\sqrt{5}$
A.7. a. $x^2 + 6x + 9 + y^2 - 8y + 16 = -16 + 9 + 16$
    b. $(x + 3)^2 + (y - 4)^2 = 9$ **c.** Center: $(-3, 4)$; radius: 3
    d.

A.8. $x^2 + (y + 4)^2 = 49$

Section 21.1 Practice Exercises, pp. 1416–1421
R.1. $5\sqrt{2}$ **R.3.** $\dfrac{3}{4}$ **R.5.** $\dfrac{7}{8}$
R.7. $x^2 + 16x + 64$ **R.9.** $n = 36$; $(x - 6)^2$
1. a. $\sqrt{(x_2 - x_1)^2 + (y_2 - y_1)^2}$ **b.** circle; center   **c.** radius
   d. $(x - h)^2 + (y - k)^2 = r^2$ **e.** $\left(\dfrac{x_1 + x_2}{2}, \dfrac{y_1 + y_2}{2}\right)$
3. $3\sqrt{5}$ **5.** $\sqrt{34}$ **7.** $\dfrac{\sqrt{73}}{8}$ **9.** 4   **11.** 8
13. $4\sqrt{2}$ **15.** $\sqrt{42}$ **17.** Subtract 5 and -7. This becomes
$5 - (-7) = 12$.   **19.** $y = 13, y = 1$
21. $x = 0, x = 8$ **23.** Yes   **25.** No
27. Center $(4, -2)$; $r = 3$ **29.** Center $(-1, -1)$; $r = 1$

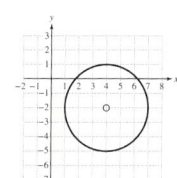   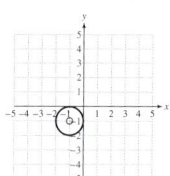

31. Center $(0, 2)$; $r = 2$ **33.** Center $(3, 0)$; $r = 2\sqrt{2}$

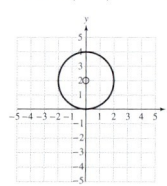   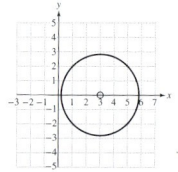

35. Center $(0, 0)$; $r = \sqrt{6}$ **37.** Center $\left(-\dfrac{4}{5}, 0\right)$; $r = \dfrac{8}{5}$

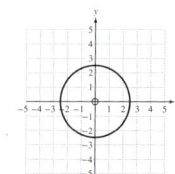   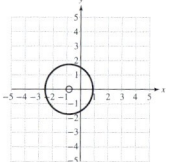

39. Center (1, 3); $r = 6$ **41.** Center: (0, 3); $r = 2$

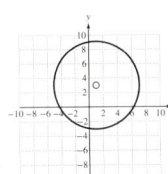 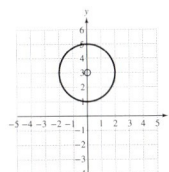

43. Center (0, −3); $r = \dfrac{4}{3}$ **45.** Center: (−1, −2); $r = 3$

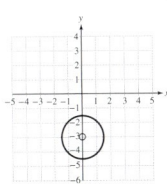 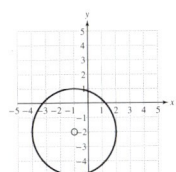

47. Center (0, 0); $r = 1$

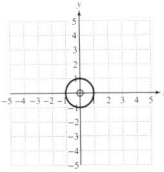

49. $x^2 + y^2 = 4$ **51.** $x^2 + (y − 2)^2 = 4$ **53.** $(x + 2)^2 + (y − 2)^2 = 9$
55. $x^2 + y^2 = 49$ **57.** $(x + 3)^2 + (y + 4)^2 = 36$
59. $(x − 5)^2 + (y − 3)^2 = 2.25$ **61.** (1, 2) **63.** (−1, 0)
65. (−1, 6) **67.** (0, 3) **69.** $\left(-\dfrac{1}{2}, \dfrac{3}{2}\right)$ **71.** (−0.4, −1)
73. $(40, 7\tfrac{1}{2})$; they should meet 40 mi east, $7\tfrac{1}{2}$ mi north of the warehouse.
75. a. (1, 3) **b.** $(x − 1)^2 + (y − 3)^2 = 5$
77. a. (0, 3) **b.** $x^2 + (y − 3)^2 = 4$

79.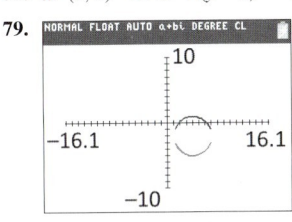

81.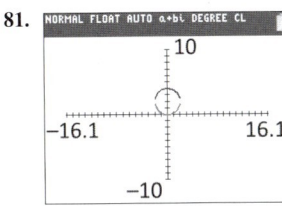

83.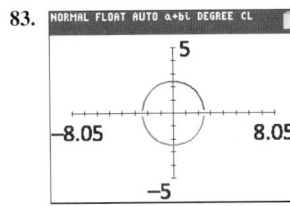

85. $(x − 4)^2 + (y − 4)^2 = 16$ **87.** $(x − 1)^2 + (y − 1)^2 = 29$

Section 21.2 Activity, p. 1428

A.1. a. The graph of $y = a(x − h)^2 + k$ is a parabola with vertex (h, k) and axis of symmetry $x = h$. The parabola opens upward if $a > 0$ and opens downward if $a < 0$.
  b. The graph of $x = a(y − k)^2 + h$ is a parabola with vertex (h, k) and axis of symmetry $y = k$. The parabola opens to the right if $a > 0$ and opens to the left if $a < 0$.

A.2.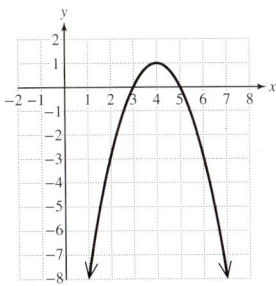

A.3. The graph would be a parabola with vertex (4, 1) but opening to the left.
A.4. a. $a = 1, b = 4, c = 1$ **b.** To the right
  c. $x = (y + 2)^2 − 3$ **d.** (−3, −2) **e.** $y = −2$

Section 21.2 Practice Exercises, pp. 1429–1431

R.1. a. (1, 4) **b.** (0, 3) **c.** (−1, 0) and (3, 0)
R.3. a. (−2, 0) **b.** (0, 4) **c.** (−2, 0) **R.5.** −2

1. a. conic; plane **b.** parabola; directrix; focus **c.** vertex; symmetry **d.** $y = k$

3. 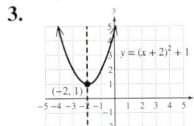   **5.**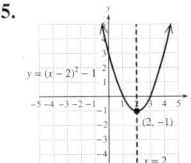

7. 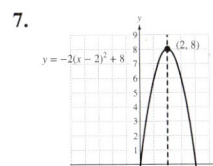   **9.**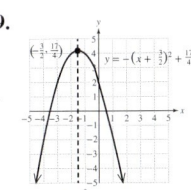

11. 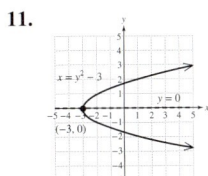   **13.**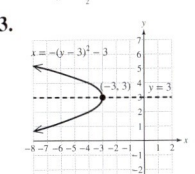

15. 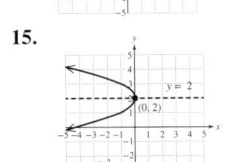   **17.**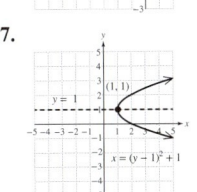

19. (2, −1) **21.** (5, −1) **23.** $\left(2, \dfrac{7}{4}\right)$ **25.** $\left(\dfrac{3}{2}, -\dfrac{1}{4}\right)$
27. (10, −1) **29.** The maximum height of the water is 22 ft.
31. A parabola whose equation is in the form $y = a(x − h)^2 + k$ has a vertical axis of symmetry. A parabola whose equation is in the form $x = a(y − k)^2 + h$ has a horizontal axis of symmetry.
33. Vertical axis of symmetry; opens upward
35. Vertical axis of symmetry; opens downward
37. Horizontal axis of symmetry; opens right
39. Horizontal axis of symmetry; opens left

41. Vertical axis of symmetry; opens downward
43. Horizontal axis of symmetry; opens right

Section 21.3 Activity, p. 1437

A.1. a. The graph of $x^2 + y^2 = 4$ is a circle centered at the origin with radius 2.

b. The graph of $\dfrac{x^2}{9} + \dfrac{y^2}{4} = 1$ is an ellipse centered at the origin with x-intercepts $(3, 0)$ and $(-3, 0)$ and y-intercepts $(0, 2)$ and $(0, -2)$.

c.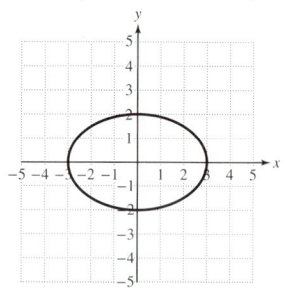

A.2. The graph of $\dfrac{(x-1)^2}{9} + \dfrac{(y+3)^2}{4} = 1$ is the same graph as $\dfrac{x^2}{9} + \dfrac{y^2}{4} = 1$ except that the center is $(1, -3)$.

A.3. a. The graph of $\dfrac{x^2}{25} + \dfrac{y^2}{49} = 1$ is an ellipse centered at the origin with x-intercepts $(5, 0)$ and $(-5, 0)$ and y-intercepts $(0, 7)$ and $(0, -7)$.

b. The graph of $\dfrac{x^2}{25} - \dfrac{y^2}{49} = 1$ is a hyperbola centered at the origin with a horizontal transverse axis (the branches of the hyperbola open to the left and right). The vertices of the reference rectangle are $(5, 7)$, $(5, -7)$, $(-5, 7)$, and $(-5, -7)$.

c. The graph of $\dfrac{y^2}{49} - \dfrac{x^2}{25} = 1$ is a hyperbola centered at the origin with a vertical transverse axis (the branches of the hyperbola open upward and downward). The vertices of the reference rectangle are $(5, 7)$, $(5, -7)$, $(-5, 7)$, and $(-5, -7)$.

Section 21.3 Practice Exercises, pp. 1437–1441

R.1. Center: $(0, 0)$; radius: 4   **R.3.** Center: $(3, -2)$; radius: $2\sqrt{3}$

1. a. ellipse; foci   **b.** ellipse   **c.** hyperbola; foci

3. 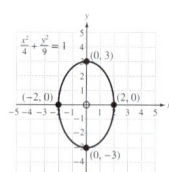   **5.**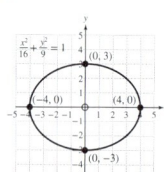

7. 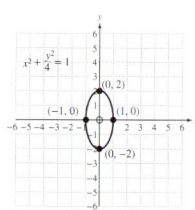   **9.**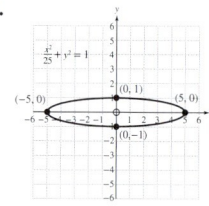

11. Center: $(4, 5)$ **13.** Center: $(-1, 2)$

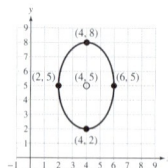   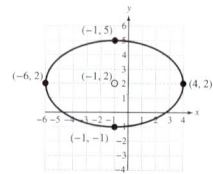

15. Center: $(2, -3)$ **17.** Center: $(0, 1)$

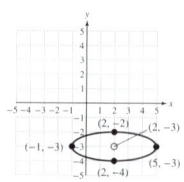   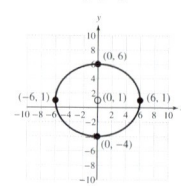

19. Vertical   **21.** Horizontal   **23.** Horizontal   **25.** Vertical

27. 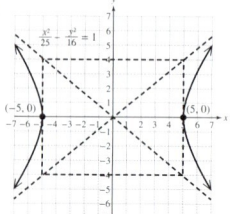   **29.**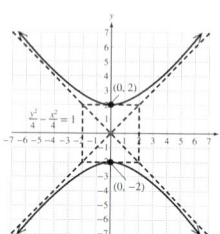

31. 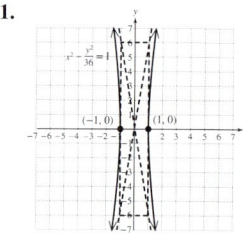   **33.**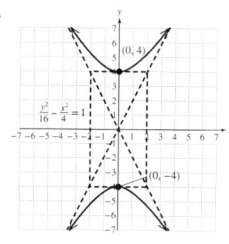

35. Hyperbola   **37.** Ellipse   **39.** Ellipse
41. Hyperbola   **43.** Ellipse   **45.** Hyperbola
47. The height 10 ft from the center is approximately 49 ft.

49. 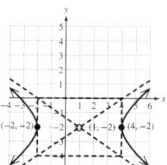   **51.**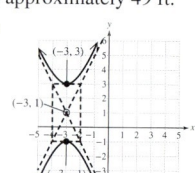

Section 21.3 Problem Recognition Exercises, pp. 1441–1442

1. Standard equation of a circle   **2.** Ellipse centered at the origin
3. Distance between two points   **4.** Midpoint between two points
5. Parabola with vertical axis of symmetry   **6.** Hyperbola with horizontal transverse axis   **7.** Hyperbola with vertical transverse axis
8. Parabola with horizontal axis of symmetry   **9.** Parabola
10. Hyperbola   **11.** Circle   **12.** Circle   **13.** Hyperbola
14. Ellipse   **15.** None of these   **16.** Circle   **17.** Parabola
18. Hyperbola   **19.** Parabola   **20.** Circle   **21.** None of these
22. Parabola   **23.** Ellipse   **24.** Circle   **25.** Parabola
26. Ellipse   **27.** Hyperbola   **28.** Circle   **29.** Ellipse
30. Parabola

Section 21.4 Activity, p. 1446

A.1. a. 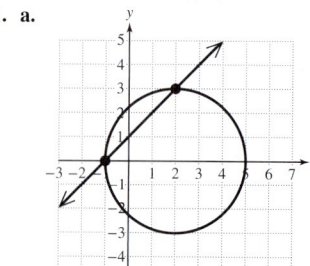   **b.** $(-1, 0)$ and $(2, 3)$
c. $\{(-1, 0), (2, 3)\}$

A.2. a. $\{(1, 3), (-1, 3), (1, -3), (-1, -3)\}$ **b.** The points of intersection of the graphs match the solutions found in part (a).

Section 21.4 Practice Exercises, pp. 1446–1449

R.1. $\{(5, 1)\}$ **R.3.** $\{(1, 2)\}$ **R.5.** Circle
R.7. Ellipse **R.9.** Parabola
1. a. nonlinear **b.** intersection **3.** Zero, one, or two
5. Zero, one, or two **7.** Zero, one, two, three, or four
9. $\{(-3, 0), (2, 5)\}$ **11.** $\{(0, 1), (-1, 0)\}$

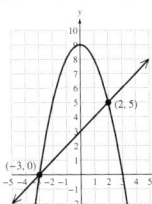 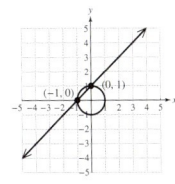

13. $\{(\sqrt{2}, 2), (-\sqrt{2}, 2)\}$

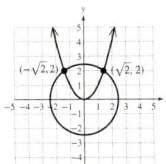

15. $\{(1, 1)\}$ **17.** $\{(0, 0)\}$ **19.** $\left\{\left(\dfrac{3}{2}, \dfrac{9}{4}\right)\right\}$
21. $\{(0, 0), (-2, -8)\}$ **23.** $\{(1, 4)\}$
25. $\{(1, 0), (-1, 0)\}$ **27.** $\{(2, 0), (-2, 0)\}$
29. $\{(3, 2), (-3, 2), (3, -2), (-3, -2)\}$
31. $\{(8, 6), (-8, -6)\}$ **33.** $\{(0, -2), (0, 2)\}$
35. $\{(2, 0), (0, 1)\}$ **37.** $\{(4, 5), (-4, -5)\}$
39. $\left\{\left(-\sqrt{2}, \dfrac{\sqrt{2}}{2}\right), \left(\sqrt{2}, -\dfrac{\sqrt{2}}{2}\right)\right\}$

41.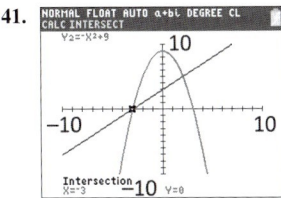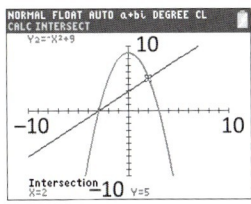

42.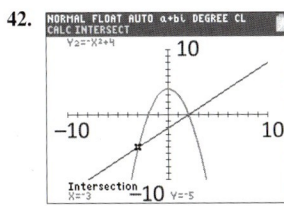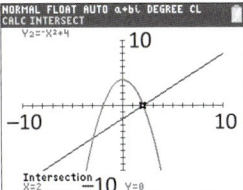

43.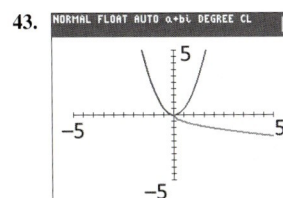

44.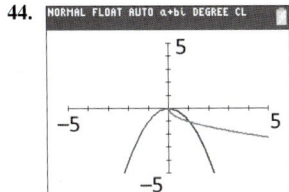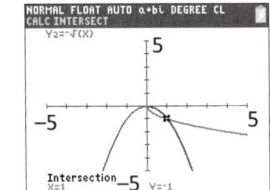

45. { } **46.** { }

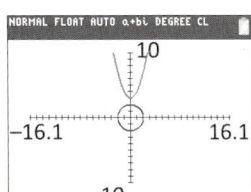 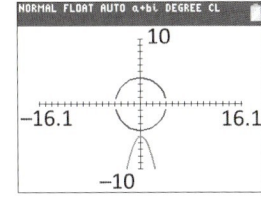

47. The numbers are 3 and 4. **49.** The numbers are 5 and $\sqrt{7}$, -5 and $\sqrt{7}$, 5 and $-\sqrt{7}$, or -5 and $-\sqrt{7}$.

Section 21.5 Activity, p. 1453

A.1. a. **b.**

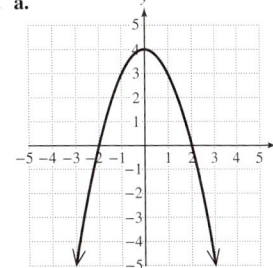 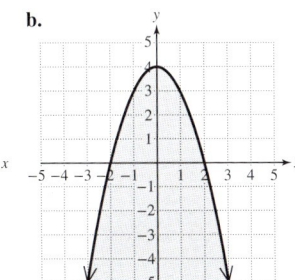

c. The solution set would not contain the boundary $y = -x^2 + 4$. To denote this, draw the parabola as a dashed curve.
d. The solution set would contain all points on and above the parabola.

A.2.

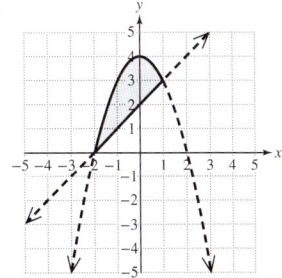

Section 21.5 Practice Exercises, pp. 1453–1458

R.1.

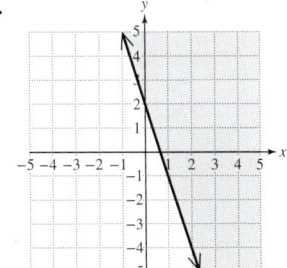

R.3. a. The solution set would be the same except that the bounding line would be dashed. **b.** The solution set would contain points on and *below* the line rather than above the line.
R.5. Parabola **R.7.** Line
R.9. Hyperbola
1. True **3.** False
5. a.

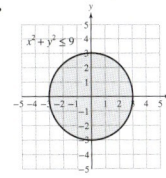

b. The set of points on and "outside" the circle $x^2 + y^2 = 9$
c. The set of points on the circle $x^2 + y^2 = 9$

7. a.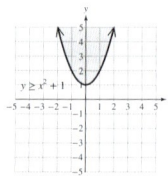

b. The parabola $y = x^2 + 1$ would be drawn as a dashed curve.

9. $(x - 3)^2 + (y + 4)^2 \leq 625$

11.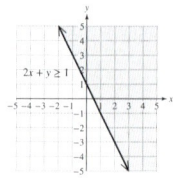

13.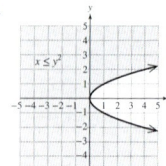

15.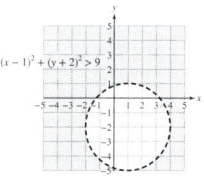

17.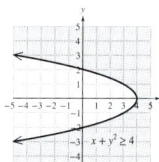

19.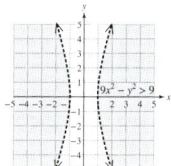

21.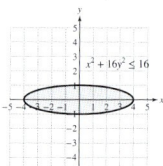

23.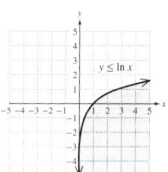

25.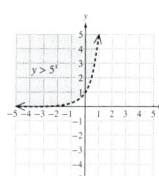

27.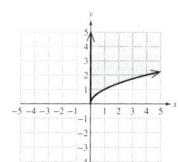

29.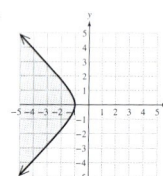

31.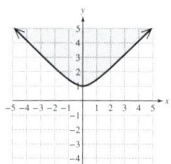

33.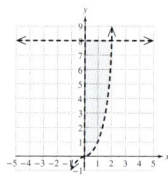

35. { }

37.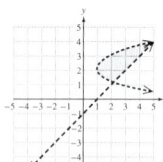

39.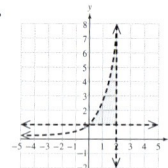

41.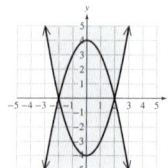

43.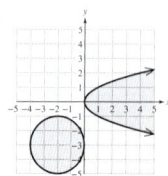

Chapter 21 Review Exercises, pp. 1459–1462

1. $2\sqrt{10}$ **2.** $\sqrt{89}$ **3.** $x = 5$ or $x = -1$
4. $x = -3$ **5.** Center $(12, 3)$; $r = 4$
6. Center $(-7, 5)$; $r = 9$ **7.** Center $(-3, -8)$; $r = 2\sqrt{5}$
8. Center $(1, -6)$; $r = 4\sqrt{2}$ **9. a.** $x^2 + y^2 = 64$
b. $(x - 8)^2 + (y - 8)^2 = 64$ **10.** $(x + 6)^2 + (y - 5)^2 = 10$
11. $(x + 2)^2 + (y + 8)^2 = 8$ **12.** $\left(x - \dfrac{1}{2}\right)^2 + (y - 2)^2 = 4$
13. $(x - 3)^2 + \left(y - \dfrac{1}{3}\right)^2 = 9$ **14.** $x^2 + y^2 = \dfrac{49}{4}$
15. $x^2 + (y - 2)^2 = 9$ **16.** $(-4, -2)$ **17.** $(-1, 8)$
18. Vertical axis of symmetry; parabola opens downward
19. Horizontal axis of symmetry; parabola opens right
20. Horizontal axis of symmetry; parabola opens left
21. Vertical axis of symmetry; parabola opens upward
22.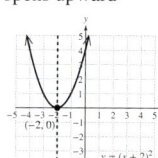
23.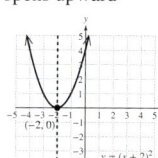

24. 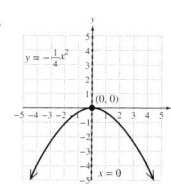 **25.**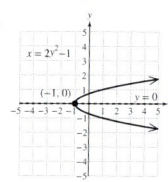

26. $y = (x - 3)^2 - 4$; vertex: $(3, -4)$; axis of symmetry: $x = 3$
27. $x = (y + 2)^2 - 2$; vertex: $(-2, -2)$; axis of symmetry: $y = -2$
28. $x = -4\left(y - \dfrac{1}{2}\right)^2 + 1$; vertex: $\left(1, \dfrac{1}{2}\right)$; axis of symmetry: $y = \dfrac{1}{2}$
29. $y = -2\left(x + \dfrac{1}{2}\right)^2 + \dfrac{1}{2}$; vertex: $\left(-\dfrac{1}{2}, \dfrac{1}{2}\right)$; axis of symmetry: $x = -\dfrac{1}{2}$

30. 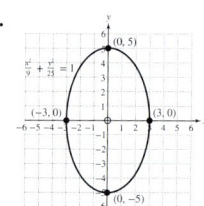 **31.**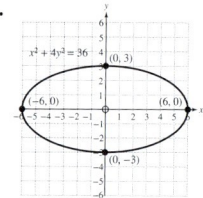

32. Center: $(5, -3)$ **33.** Center: $(0, 2)$

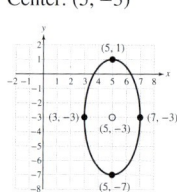 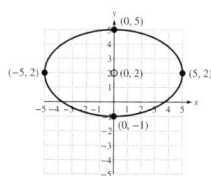

34. Horizontal **35.** Vertical **36.** Vertical **37.** Horizontal
38. 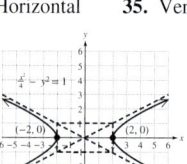 **39.**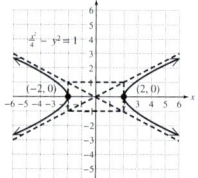

40. Hyperbola **41.** Ellipse **42.** Ellipse **43.** Hyperbola

44. a. Line and parabola
b.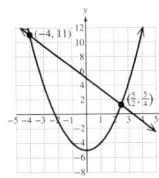
c. $\left\{\left(\dfrac{5}{2}, \dfrac{5}{4}\right), (-4, 11)\right\}$

45. a. Line and parabola
b.
c. $\{(-5, 15), (3, -1)\}$

46. a. Circle and line
b.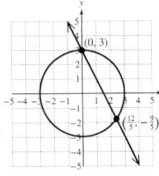
c. $\left\{(0, 3), \left(\dfrac{12}{5}, -\dfrac{9}{5}\right)\right\}$

47. a. Circle and line
b.
c. $\left\{(0, -4), \left(\dfrac{16}{5}, -\dfrac{12}{5}\right)\right\}$

48. $\left\{(0, -2), \left(\dfrac{16}{9}, \dfrac{14}{9}\right)\right\}$ **49.** $\left\{\left(-\dfrac{7}{5}, \dfrac{13}{5}\right), (-5, -1)\right\}$

50. $\{(8, 4), (2, -2)\}$ **51.** $\{(2, 4), (-2, 4), (\sqrt{2}, 2), (-\sqrt{2}, 2)\}$
52. $\{(3, 1), (3, -1), (-3, 1), (-3, -1)\}$
53. $\{(6, 5), (-6, 5), (6, -5), (-6, -5)\}$

54. 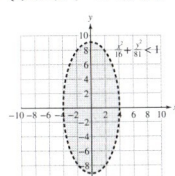   **55.**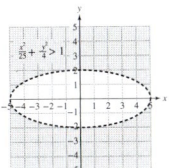

56. 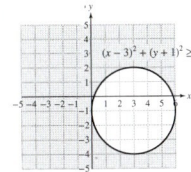   **57.**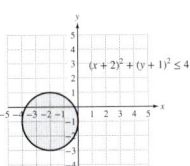

58. 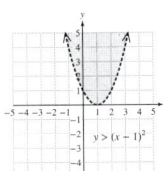   **59.**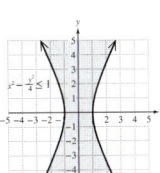

60. 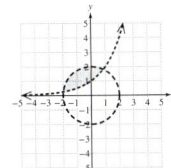 **61.**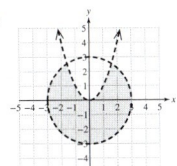

Chapter 22

Section 22.1 Activity, p. 1469
A.1. 1   **A.2.** $a + b$ **A.3.** $a^2 + 2ab + b^2$
A.4. $a^3 + 3a^2b + 3ab^2 + b^3$
A.5. a. 1 4 6 4 1 **b.** $a^4 + 4a^3b + 6a^2b^2 + 4ab^3 + b^4$
A.6. a. 1 **b.** 4 **c.** 6 **d.** 4 **e.** 1
A.7. a. $[5x + (-y^2)]^4$ **b.** $625x^4 - 500x^3y^2 + 150x^2y^4 - 20xy^6 + y^8$

Section 22.1 Practice Exercises, pp. 1470–1471
R.1. 15   **R.3. a.** $16x^2 - 24x + 9$ **b.** $64x^3 - 144x^2 + 108x - 27$

1. a. $a^2 + 2ab + b^2$; $a^3 + 3a^2b + 3ab^2 + b^3$; binomial
  b. $n(n-1)(n-2) \cdots (2)(1)$; factorial
  c. 6; 2; 1; 1 **d.** binomial **e.** Pascal's
3. $a^3 + 3a^2b + 3ab^2 + b^3$
5. $1 + 4g + 6g^2 + 4g^3 + g^4$
7. $p^7 + 7p^6q^2 + 21p^5q^4 + 35p^4q^6 + 35p^3q^8 + 21p^2q^{10} + 7pq^{12} + q^{14}$
9. $s^5 - 5s^4t + 10s^3t^2 - 10s^2t^3 + 5st^4 - t^5$
11. $625 - 500u^3 + 150u^6 - 20u^9 + u^{12}$
13. $x^6 - 12x^4 + 48x^2 - 64$ **15.** 120 **17.** 1 **19.** False
21. True **23.** $6! = 6 \cdot (5 \cdot 4 \cdot 3 \cdot 2 \cdot 1) = 6 \cdot 5!$ **25.** 1680
27. 6 **29.** 56 **31.** 1 **33.** $m^{11} + 11m^{10}n + 55m^9n^2$
35. $u^{24} - 12u^{22}v + 66u^{20}v^2$ **37.** 9 terms
39. $s^6 + 6s^5t + 15s^4t^2 + 20s^3t^3 + 15s^2t^4 + 6st^5 + t^6$
41. $b^3 - 9b^2 + 27b - 27$ **43.** $16x^4 + 32x^3y + 24x^2y^2 + 8xy^3 + y^4$
45. $c^{14} - 7c^{12}d + 21c^{10}d^2 - 35c^8d^3 + 35c^6d^4 - 21c^4d^5 + 7c^2d^6 - d^7$
47. $\dfrac{1}{32}a^5 - \dfrac{5}{16}a^4b + \dfrac{5}{4}a^3b^2 - \dfrac{5}{2}a^2b^3 + \dfrac{5}{2}ab^4 - b^5$
49. $x^4 + 16x^3y + 96x^2y^2 + 256xy^3 + 256y^4$
51. $-462m^6n^5$ **53.** $495u^{16}v^4$ **55.** g^9

Section 22.2 Activity, p. 1477
A.1. a. $1, \dfrac{2}{3}, \dfrac{1}{2}, \dfrac{2}{5}, \dfrac{1}{3}$ **b.** $-1, \dfrac{2}{3}, -\dfrac{1}{2}, \dfrac{2}{5}, -\dfrac{1}{3}$ **c.** $1, -\dfrac{2}{3}, \dfrac{1}{2}, -\dfrac{2}{5}, \dfrac{1}{3}$

A.2. a. $\dfrac{29}{10}$ **b.** $-\dfrac{23}{30}$ **c.** $\dfrac{23}{30}$

A.3. a. 100, 200, 400, 800, 1600 **b.** $s_n = 100 \cdot 2^{n-1}$
  c. $r_n = 10(100 \cdot 2^{n-1})$ or $r_n = 1000 \cdot 2^{n-1}$ **d.** $\sum_{i=1}^{5} 1000 \cdot 2^{i-1}$
  e. $1000 + 2000 + 4000 + 8000 + 16{,}000$ **f.** \$31,000

Section 22.2 Practice Exercises, pp. 1478–1480
R.1. a. -1 **b.** 1 **c.** -1 **d.** 1 **e.** -1 **f.** 1
R.3. 47 **R.5.** $\dfrac{16}{25}$ **R.7.** $\dfrac{23}{60}$ **R.9.** 56

1. a. infinite; finite **b.** terms; nth **c.** alternating **d.** series
  e. summation **f.** index; $(3)^2 + (4)^2 + (5)^2$
3. 22, 25, 28 **5.** $-18, 21, -24$ **7.** 4, 7, 10, 13, 16
9. $\sqrt{3}, 2, \sqrt{5}, \sqrt{6}$ **11.** $-\dfrac{2}{3}, \dfrac{3}{4}, -\dfrac{4}{5}, \dfrac{5}{6}$
13. $0, -3, 8$ **15.** 0, 2, 6, 12, 20, 30 **17.** $-3, 9, -27, 81$
19. When n is odd, the term is negative. When n is even, the term is positive. **21.** $a_n = 2n$ **23.** $a_n = 2n - 1$
25. $a_n = \dfrac{1}{n^2}$ **27.** $a_n = (-1)^{n+1}$ **29.** $a_n = (-1)^n 2^n$
31. $a_n = \dfrac{3}{5^n}$ **33.** \$60, \$58.80, \$57.62, \$56.47
35. 25,000; 50,000; 100,000; 200,000; 400,000; 800,000; 1,600,000
37. A sequence is an ordered list of terms. A series is the sum of the terms of a sequence.
39. 90 **41.** $\dfrac{31}{16}$ **43.** 30 **45.** 10 **47.** $\dfrac{73}{12}$
49. 38 **51.** -1 **53.** 55 **55.** $\sum_{n=1}^{6} n$
57. $\sum_{i=1}^{5} 4$ **59.** $\sum_{j=1}^{5} 4j$ **61.** $\sum_{k=1}^{4} (-1)^{k+1} \dfrac{1}{3^k}$
63. $\sum_{i=1}^{6} \dfrac{i+4}{11i}$ **65.** $\sum_{n=1}^{5} x^n$
67. $-3, 2, 7, 12, 17$ **69.** 5, 21, 85, 341, 1365
71. 1, 1, 2, 3, 5, 8, 13, 21, 34, 55

Section 22.3 Activity, p. 1485

A.1. a.

Year	Base Salary Plus Raise(s)	Total Salary
Year 1	$55,000 + 0($2000)	$55,000
Year 2	$55,000 + 1($2000) (Joyce has had 1 raise.)	$57,000
Year 3	$55,000 + 2($2000) (Joyce has had 2 raises.)	$59,000
Year 4	$55,000 + 3($2000) (Joyce has had 3 raises.)	$61,000
Year 5	$55,000 + 4($2000) (Joyce has had 4 raises.)	$63,000
Year n	$55,000 + (n − 1)($2000) (Joyce has had n − 1 raises.)	$53,000 + $2000n

b. $103,000   **A.2. a.** arithmetic   **b.** $a_n = a_1 + (n-1)d$
A.3. a. 17   **b.** −6   **c.** $a_n = 17 + (n-1)(-6)$ or $a_n = 23 - 6n$ **d.** 54
A.4. a. $\sum_{i=1}^{25}(53{,}000 + 2000i)$ **b.** $1,975,000

Section 22.3 Practice Exercises, pp. 1485–1487

R.1. {21}   **R.3.** {−2}   **R.5. a.** $f(1) = 1$ **b.** $f(2) = 5$
c. $f(3) = 9$ **d.** $f(4) = 13$
R.7. $-3 + (-7) + (-11) + (-15) + (-19)$; −55

1. a. arithmetic   **b.** difference   **c.** $a_1 + (n-1)d$; d
d. series   **e.** $\frac{n}{2}(a_1 + a_n)$ **3.** −1, 3, 7, 11   **5.** $\frac{1}{2}, 2, \frac{7}{2}, 5$
7. 3, 11, 19, 27, 35   **9.** 80, 60, 40, 20, 0   **11.** $3, \frac{15}{4}, \frac{9}{2}, \frac{21}{4}, 6$
13. 2   **15.** −3   **17.** −2   **19.** 3, 8, 13, 18, 23   **21.** $2, \frac{5}{2}, 3, \frac{7}{2}, 4$
23. 2, −2, −6, −10, −14   **25.** $a_n = -5 + 5n$
27. $a_n = -2n$ **29.** $a_n = \frac{3}{2} + \frac{1}{2}n$ **31.** $a_n = 25 - 4n$
33. $a_n = -14 + 6n$ **35.** $a_6 = 17$ **37.** $a_9 = 47$
39. $a_7 = -30$ **41.** $a_{11} = -48$ **43.** 19   **45.** 22
47. 23   **49.** 11   **51.** $a_1 = -2, a_2 = -5$ **53.** 670
55. 290   **57.** −15   **59.** 95   **61.** 924   **63.** 300
65. −210   **67.** 5050   **69.** 980 seats; $14,700

Section 22.4 Activity, p. 1493

A.1. a.

Year	Salary		Simplified
Year 1	$50,000(1.04)^0$		$50,000
Year 2	$50,000(1.04)^1$	(Kevin has 1 raise of 4% of $50,000 or equivalently 104% of his first year's salary.)	$52,000
Year 3	$50,000(1.04)^2$	(Kevin has 2 raises of 4%. This is 104% of 104% of his first year's salary.)	$54,080
Year 4	$50,000(1.04)^3$	(Kevin has 3 raises of 4%.)	$56,243.20
Year 5	$50,000(1.04)^4$	(Kevin has 4 raises of 4%.)	$58,492.93
Year n	$50,000(1.04)^{n-1}$	(Kevin has n − 1 raises of 4%.)	$50,000(1.04)^{n-1}$

b. $128,165.21
c. Kevin's salary increases each year, and his raise is a percentage of this increasing salary. Therefore, his raises also increase each year.
d. $\sum_{i=1}^{25} 50{,}000(1.04)^{i-1}$ **e.** $2,082,295.41   **f.** $107,295.41
A.2. a. Geometric; the ratio of two consecutive terms is a constant. In this case, each term is $\frac{1}{2}$ of its predecessor.
b. $a_n = 16\left(\frac{1}{2}\right)^{n-1}$ **c.** $\frac{1}{8}$ **d.** 9

A.3. a. Arithmetic; the difference between consecutive terms is constant. In this case, each term after the first is 8 less than its predecessor.
b. $b_n = 16 + (n-1)(-8)$ or $b_n = 24 - 8n$ **c.** −72   **d.** 48
A.4. Neither   **A.5.** $\frac{511}{16}$ or $31\frac{15}{16}$ **A.6.** 32   **A.7.** 8176
A.8. Does not exist; this is a geometric series with common ratio 2. Because the common ratio is greater than 1, the terms of the series increase without bound and the sum becomes infinitely large.

Section 22.4 Practice Exercises, pp. 1494–1497

R.1. a. $f(1) = 20$ **b.** $f(2) = 8$ **c.** $f(3) = \frac{16}{5}$ **d.** $f(4) = \frac{32}{25}$
R.3. 32   **R.5.** 150   **R.7.** 28

1. a. geometric   **b.** ratio   **c.** $a_1 r^{n-1}$ **d.** $\frac{a_1(1-r^n)}{1-r}$
e. $\frac{a_1}{1-r}$; 1   **3.** 4, 8, 16, 32   **5.** 1, −5, 25, −125
7. 1, 10, 100, 1000   **9.** 64, 32, 16, 8
11. $8, -2, \frac{1}{2}, -\frac{1}{8}$ **13.** 2   **15.** $-\frac{1}{4}$
17. −2   **19.** −3, 6, −12, 24, −48   **21.** $6, 3, \frac{3}{2}, \frac{3}{4}, \frac{3}{8}$
23. −1, −6, −36, −216, −1296   **25.** $a_n = 3(4)^{n-1}$
27. $a_n = -5(-3)^{n-1}$ **29.** $a_n = \frac{1}{2}(4)^{n-1}$ **31.** $\frac{1}{64}$
33. $-\frac{243}{8}$ **35.** −48   **37.** −9   **39.** $\frac{1}{8}$ **41.** 4
43. A geometric sequence is an ordered list of numbers in which the ratio between each term and its predecessor is constant. A geometric series is the sum of the terms of such a sequence.
45. a. 3 + 12 + 48 + 192   **b.** 255
47. $\frac{1562}{125}$ **49.** $-\frac{11}{8}$ **51.** $\frac{3124}{27}$ **53.** $\frac{665}{243}$
55. −172   **57. a.** $1050.00, $1102.50, $1157.63, $1215.51
b. $a_{10} = \$1628.89; a_{20} = \$2653.30; a_{40} = \$7039.99$
59. $r = \frac{1}{6}$; sum is $\frac{6}{5}$ **61.** $r = -\frac{1}{4}$; sum is $\frac{4}{5}$
63. $r = -\frac{3}{2}$; sum does not exist   **65.** The sum is $800 million.
67. The total vertical distance traveled is 28 ft.
69. a. $\frac{7}{10}$ **b.** $\frac{1}{10}$ **c.** $\frac{7}{9}$ **71. a.** $1,429,348
b. $1,505,828   **c.** $76,480

Section 22.4 Problem Recognition Exercises, p. 1498

1. Geometric, $r = -\frac{1}{2}$ **2.** Arithmetic, $d = \frac{1}{2}$
3. Geometric, $r = 2$ **4.** Neither   **5.** Arithmetic, $d = \frac{2}{3}$
6. Geometric, $r = -\frac{3}{2}$ **7.** Neither   **8.** Geometric, $r = -1$
9. Arithmetic, $d = -2$ **10.** Neither   **11.** Neither
12. Arithmetic, $d = 4$ **13.** Arithmetic, $d = \frac{1}{4}\pi$
14. Geometric, $r = -1$ **15.** Neither
16. Arithmetic, $d = \frac{1}{3}e$ **17.** Neither   **18.** Geometric, $r = 3$

Chapter 22 Review Exercises, pp. 1499–1500

1. 40,320   **2.** 120   **3.** 66   **4.** 84
5. $x^{10} + 20x^8 + 160x^6 + 640x^4 + 1280x^2 + 1024$
6. $c^4 - 12c^3d + 54c^2d^2 - 108cd^3 + 81d^4$
7. $a^{11} - 22a^{10}b + 220a^9b^2$ **8.** $1512x^{10}y^3$
9. $-15{,}000x^3y^{21}$ **10.** $160a^3b^3$ **11.** 1, −2, −5, −8, −11

12. $-2, -16, -54$ 13. $\frac{1}{3}, -\frac{1}{2}, \frac{3}{5}, -\frac{2}{3}$ 14. $-\frac{2}{3}, \frac{4}{9}, -\frac{8}{27}, \frac{16}{81}$
15. $a_n = \frac{n}{n+1}$ 16. $a_n = (-1)^n \cdot \frac{3}{2^n}$ 17. k 18. 5
19. 5 20. -22 21. $\sum_{i=1}^{7} \frac{i+3}{i}$ 22. $a_n = n^2 + \left(\frac{n}{2}\right)h$
23. $-12, -10.5, -9, -7.5, -6$ 24. $\frac{2}{3}, 1, \frac{4}{3}, \frac{5}{3}, 2$
25. $a_n = 10n - 9$ 26. $a_n = -14n + 20$ 27. $a_{17} = \frac{9}{2}$
28. $a_{25} = -155$ 29. 24 terms 30. 19 terms
31. $d = 10$ 32. $a_n = -5n + 19$ 33. -620
34. 92.5 35. 294 36. 5989 37. $r = 3$
38. $r = 1.2$ 39. $-1, \frac{1}{4}, -\frac{1}{16}, \frac{1}{64}$ 40. $10, 20, 40, 80$
41. $a_n = -4(2)^{n-1}$ 42. $a_n = 6\left(-\frac{1}{3}\right)^{n-1}$ 43. $a_6 = \frac{128}{243}$
44. $a_4 = 270$ 45. $a_1 = 4$ 46. $a_1 = 1$ 47. 1275
48. $-\frac{182}{3}$ 49. 18 50. $\frac{25}{2}$
51. a. $\$10{,}700, \$11{,}449, \$12{,}250.43$ b. The account is worth $\$12{,}250.43$ after 3 years. c. $\$19{,}671.51$

Additional Topics Appendix
Section A.1 Practice Exercises, pp. A-4–A-5
1. Let $u = 3x - 2$. 3. Factor the first three terms as $(x - 2)^2$.
5. $2(4t + 3)(2t - 1)$ 7. $(2w - 3)(w - 5)$
9. $(5x + 20 - y)(5x + 20 + y)$ 11. $(15a - 7)(15a - 11)$
13. $-1(t + 3)(3t - 1)$ 15. $(a + b + c)(a^2 + 2ab + b^2 - ac - bc + c^2)$
17. $(3c - d - 2)(9c^2 + 3cd + 6c + d^2 + 4d + 4)$
19. $(2x + 14y + 7)^2$ 21. $(x - 5 - y)(x - 5 + y)$
23. $(a - x - 9y)(a + x + 9y)$ 25. $(5a + 2b - 10c)(5a + 2b + 10c)$
27. $(8w - 3t + 4p)(8w + 3t - 4p)$ 29. $(x + 2y)(x - 2y - 1)$
31. $(w + 2)(w^2 - 2w + 4)(w^2 + 1)(w - 1)(w + 1)$
33. $(y - 4)(x + 2z - 2)$ 35. $(x - y)(x + 2y + 1)$
37. $(n - 4)(n^2 + 4n + 16)(n + 1)(n^2 - n + 1)$
39. $(a + b - 3c + 3d)^2$ 41. $2x(2x - 5)(x + 3)(x - 3)$
43. $(a + b)(a + b - 1)$ 45. $(x - y)(x + y)^2$
47. $(a + 3)^3(a + 9)$ 49. $18(3x + 5)^2(4x + 5)$
51. $\left(\frac{1}{10}x + \frac{1}{7}\right)^2$ 53. $5x^2(5x^2 - 6)$
55. $(4p^2 + q^2)(2p - q)(2p + q)$ 57. $\left(y + \frac{1}{4}\right)\left(y^2 - \frac{1}{4}y + \frac{1}{16}\right)$
59. $(a + b)(a - b)(6a + b)$ 61. $\left(\frac{1}{3}t + \frac{1}{4}\right)^2$
63. $(3a + b)(2x - y)$ 65. $(x^4 + 1)(x^2 + 1)(x + 1)(x - 1)$
67. $(5c - 3d)(5c + 3d + 1)$

Section A.2 Practice Exercises, pp. A-11–A-14
1. a. matrix; rows; columns b. column; one; square
c. coefficient; augmented d. row echelon
3. $\left\{\left(12, \frac{1}{2}\right)\right\}$ 5. $\{(1, 3, -3)\}$ 7. a. 3×1 b. column matrix
9. a. 2×2 b. square matrix 11. a. 1×4 b. row matrix
13. a. 2×3 b. none of these
15. $\begin{bmatrix} 1 & -2 & | & -1 \\ 2 & 1 & | & -7 \end{bmatrix}$ 17. $\begin{bmatrix} 1 & -2 & 1 & | & 5 \\ 2 & 6 & 3 & | & -2 \\ 3 & -1 & -2 & | & 1 \end{bmatrix}$
19. $4x + 3y = 6$
    $12x + 5y = -6$
21. $x = 4, y = -1, z = 7$

23. a. 7 b. -2 25. $\begin{bmatrix} 1 & \frac{1}{2} & | & \frac{11}{2} \\ 2 & -1 & | & 1 \end{bmatrix}$
27. $\begin{bmatrix} 1 & -4 & | & 3 \\ 5 & 2 & | & 1 \end{bmatrix}$ 29. $\begin{bmatrix} 1 & 5 & | & 2 \\ 0 & 11 & | & 5 \end{bmatrix}$
31. a. $\begin{bmatrix} 1 & 3 & 0 & | & -1 \\ 0 & -11 & -5 & | & 10 \\ -2 & 0 & -3 & | & 10 \end{bmatrix}$ b. $\begin{bmatrix} 1 & 3 & 0 & | & -1 \\ 0 & -11 & -5 & | & 10 \\ 0 & 6 & -3 & | & 8 \end{bmatrix}$
33. True 35. True 37. Interchange rows 1 and 2.
39. Multiply row 1 by -3 and add to row 2. Replace row 2 with the result. 41. $\{(-3, -1)\}$ 43. $\{(-21, 9)\}$ 45. Infinitely many solutions; $\{(x, y) | x + 3y = 3\}$; dependent equations
47. $\{(3, -1)\}$ 49. $\{(-10, 3)\}$ 51. No solution; $\{\ \}$; inconsistent system 53. $\{(1, 2, 3)\}$ 55. $\{(1, -2, 0)\}$

57. 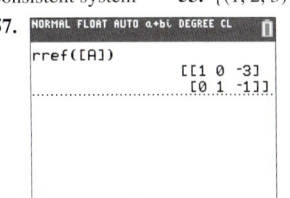 58.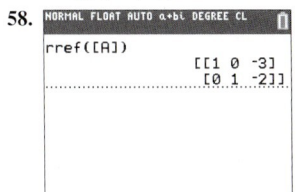

59. 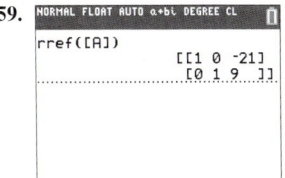 60.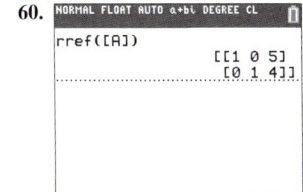

61. 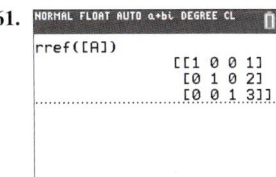 62.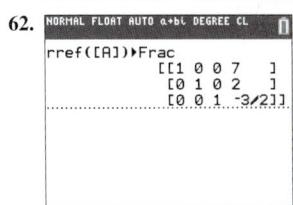

Section A.3 Practice Exercises, pp. A-22–A-25
1. a. determinant; $ad - bc$ b. minor c. $a_2; \begin{vmatrix} b_1 & c_1 \\ b_2 & c_2 \end{vmatrix}$
3. 16 5. 1 7. 4 9. 6 11. 26 13. 6
15. 46 17. a. 30 b. 30 19. Choosing the row or column with the most zero elements simplifies the arithmetic when evaluating a determinant.
21. 12 23. -15 25. 0 27. $8a - 2b$
29. $4x - 3y + 6z$ 31. 0
33. $D = 16; D_x = -63; D_y = 66$ 35. $\{(2, -1)\}$ 37. $\{(4, -1)\}$
39. $\left\{\left(\frac{23}{13}, \frac{9}{13}\right)\right\}$ 41. $x = 1$ 43. $z = \frac{1}{2}$ 45. $y = \frac{16}{41}$
47. Cramer's rule does not apply when the determinant $D = 0$.
49. No solution; $\{\ \}$; inconsistent system 51. $\{(0, 0)\}$
53. Infinitely many solutions; $\{(x, y) | x + 5y = 3\}$; dependent equations 55. $\{(3, 2, 1)\}$
57. No solution; $\{\ \}$; inconsistent system 59. $x = -19$
61. $w = 1$ 63. 36 65. a. 2 b. -2 c. $x = -1$
67. The measures are $37.5°$ and $52.5°$.
69. 900 tickets in section A, 1500 tickets in section B, and 600 tickets in section C were sold.
71. There were 560 women and 440 men in the survey.

Section A.4 Practice Exercises, pp. A-34–A-40

1. decreases 3. U; parabola
5. a. $y = 920x + 650$
b.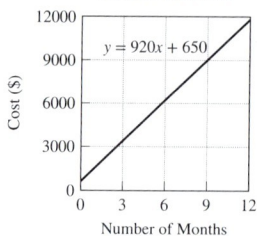
c. Increasing for $x > 0$ d. The cost for 15 months is $14,450.
7. a.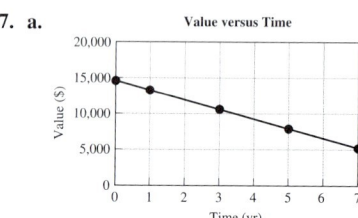
b. Yes c. The slope is -1320 and means that the value of the motorcycle decreases by $1320 per year. d. $y = -1320x + 14{,}520$
e. The y-intercept is $(0, 14{,}520)$ and means that the value of the motorcycle when new is $14,520. The x-intercept is $(11, 0)$ and means that the value of the motorcycle after 11 years is $0.
9. a.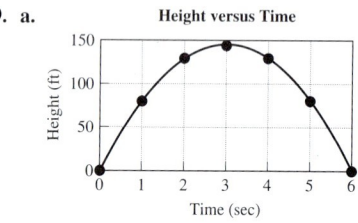
b. The relationship appears to be quadratic. c. The vertex is $(3, 144)$ which means that 3 sec after being launched the fireworks reach a maximum height of 144 ft. d. The curve increases for $0 < x < 3$. e. The curve decreases for $3 < x < 6$.
11. a.

Time (wks) x	Ebola Cases $y = 200(2)^{x/3}$
0	200
3	400
6	800
12	3200
15	6400

b. Exponential c. The ordered pair $(0, 200)$ means that there were initially 200 cases of Ebola. The ordered pair $(3, 400)$ means that after 3 weeks, the number of cases doubled to 400. The ordered pair $(6, 800)$ means that after 6 weeks, the number of cases was 800. (A time of 6 weeks is two doubling periods. The number of cases doubles from 200 to 400 and then from 400 to 800.)

13. b 15. e 17. c 19. Decreasing for $0 < x < 40$
21. Increasing for $0 < x < 45$
23. Decreasing for $0 < x < 20$ and increasing for $20 < x < 42$; The vertex is approximately $(20, 10)$. 25. b 27. e
29. c 31. 288; 480 33. linear 35. exponential
37. quadratic 39. exponential 41. linear
43. quadratic 45. Graph b; Constant speed is represented by a horizontal line. Then when the cart stops, the speed abruptly changes to 0.
47. A linear model has a constant slope and is thus characterized by a line. An exponential growth model shows increasing values of the slope for increasing values of x. The growth starts slowly and then increases more rapidly with increasing values of x.
49. a. At 1 hr: 4.8 ng/mL; At 12 hr: 3.9 ng/mL; At 24 hr: 1.0 ng/mL; At 48 hr: 0.3 ng/mL b. 0 ng/mL
51. 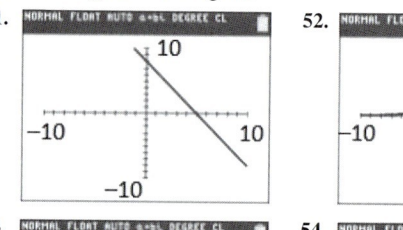 52.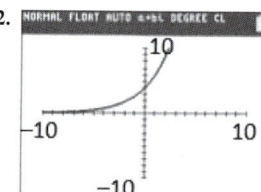
53. 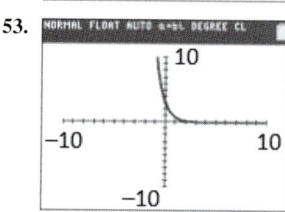 54.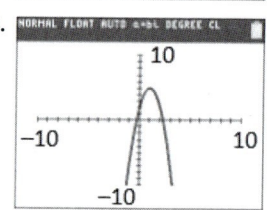
55. 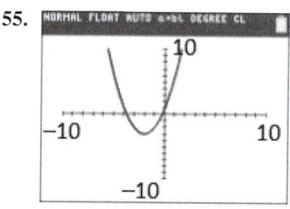 56.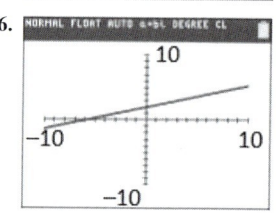

57. a.

Time (min) x	Population y
30	1200
60	2400
90	4800
120	9600
150	19,200

b. $y = 600(2)^{x/30}$

59. a. $A = x^2$; quadratic b. $P = 4x$; linear
c.

Side (in.) x	Area (in.^{2}) A	Perimeter (in.) P
2	4	8
4	16	16
6	36	24
8	64	32
10	100	40

Index

Note: Page numbers preceded by the letter *A* indicate Appendix pages.

A

Absolute value, 114
  adding integers and, 88–90
  applications of, 1153–1154
  explanation of, 81–82
  of fractions, 176
  method to find, 82, 176
  solving equations containing two, 1143–1144
Absolute value bars, 1184
Absolute value equations
  explanation of, 1140
  method to solve, 1140–1144, 1147
Absolute value expressions
  translations for, 1153–1154
Absolute value functions, 1079, 1082
Absolute value inequalities
  applications of, 1153–1154
  explanation of, 1147
  methods to solve, 1147–1153
  test point method to solve, 1151–1153
AC-method to factor trinomials, 933–937
Acute triangle, 497
Addends, 9, 90
Addition
  applications of, 17–18
  carrying in, 10, 233
  of complex numbers, 1253
  of decimals, 279–283
  distributive property of multiplication over, 36, 127, 129
  of functions, 1092–1093
  of integers, 87–91, 97, 103
  of like fractions, 220
  of like terms, 130, 831, 868, 869
  of mixed numbers, 232–234, 237
  of mixed units of measurement, 448–449
  on number line, 10, 87–88
  of opposites, 90
  of polynomials, 868–869
  properties of. *See* Addition properties
  of radicals, 1209–1212
  of rational expressions, 1004–1010
  of signed numbers, 88–90
  symbol for, 9, 10
  translating to/from words for, 15–16, 91, 157
  translations for, 15–16, 91
  of whole numbers, 9–12
Addition method
  explanation of, 788
  like terms and, 1446
  to solve nonlinear systems of equations, 1445–1446
  to solve systems of linear equations, 788–794
Addition properties
  associative, 12, 90, 127, 128
  commutative, 11, 12, 16, 90, 127, 128
  of equality, 137–140, 253, 254, 329, 591–593
  of inequality, 666–667
  of zero, 11
Additive inverse property, 90
Adjacent angles, 489

Algebraic expressions
  applications of, 115
  decimals in, 283
  English expressions translated to, 331–332
  evaluation of, 63–64, 116–117, 246, 322
  explanation of, 63
  simplification of, 249
  translating to, 115
Algebraic model, A–26. *See also* Equations
Alternate exterior angles, 489
Alternate interior angles, 489
Alternating sequence, 1473
Amount, of percent proportion, 398, 399
Angles
  adjacent, 489
  alternate exterior, 489
  alternate interior, 489
  complementary. *See* Complementary angles
  congruent, 487
  corresponding, 489
  explanation of, 486
  measurement of, 450, 486–487
  right, 486
  straight, 486
  supplementary. *See* Supplementary angles
  in triangle, 644–645
  of triangles, 496–497
  vertex of, 486
  vertical, 488
Applications. *See also* Word problems
  absolute value inequalities, 1153–1154
  addition of whole numbers, 17–18
  algebraic expressions, 115
  binomial theorem, 1467–1468
  of compound inequalities, 1121–1122
  consecutive integers, 624–625
  costs, 651–652, 802–803
  decimals, 281–283, 292–293, 307–308, 322–323, 331–333
  discount and markup, 635–636
  distance, rate, and time, 806–807, 1042–1044
  division, 52–53
  estimation, 29, 39
  of exponential equations, 1398–1399
  exponential functions, 1345–1347, 1397–1398
  exponents, 837–838
  fractions, 169, 188, 202–203, 225, 322, 323
  of geometric series, 1492
  geometry, 332, 513–514, 643–645, 784, 823–824, 871–872, 882, 966–967, 1242–1243
  integers, 91, 98
  least common multiple, 212
  linear equations in one variable, 159–161, 625–627
  linear equations in two variables, 750–753
  linear inequalities in, 670–671
  literal equations, 642, 1270
  of logarithmic functions, 1358–1359, 1381–1382
  mean, 71
  measurement, 445, 449, 450, 475–476, 482–483, 1154
  medical, 482–483
  mixed numbers, 238

mixtures, 652–653, 804–805, 824–825
multiple operations, 68–70
multiplication, 39–40
order of operations, 245–246
percent, 400–402, 417–424
percent equations, 410–412, 632–634
plotting points, 687–688
polynomials, 871–872
proportions, 367–369, 1039
Pythagorean theorem, 500–501, 968–969, 1185–1186
quadratic equations, 966–967, 969–970
quadratic formula, 1278–1280
quadratic functions, 1279–1280, 1319–1321
radical equations, 1242–1243
radical functions, 1243–1244
rational equations, 1038–1045
rational exponents, 1197
ratios, 345
relations, 1060
sequence, 1475
similar polygons, 373
similar triangles, 372, 1040–1041
simple interest, 634–635, 803–804
slope, 712, 717–718
substitution method, 783–784
subtraction of whole numbers, 17–18
systems of linear equations, 783–784, 802–807
uniform motion, 653–655
variation, 1102–1105
work, 1044–1045
Approximately equal to, 26
Approximation
  exponential expressions, 1342–1343
  with square roots, 1181
Area
  of circles, 295, 510, 512, 513
  explanation of, 510
  formulas for, 510
  method to find, 510–512
  of parallelograms, 510
  of rectangles, 40–41, 293, 510, 511
  of square, 510
  of trapezoids, 510, 512
  of triangles, 197–199, 322, 510, 511, 1179
Argument, of logarithmic expressions, 1352
Arithmetic sequence
  explanation of, 1463, 1481
  linear function and, 1482
  nth term of, 1481–1482
Arithmetic series
  explanation of, 1483
  finding sum of, 1484
Associative property
  of addition, 12, 90, 127, 128
  of multiplication, 35, 127, 128
Asymptotes, 1081
Augmented matrix, A–6 to A–7
Average. *See* Mean
Axis, transverse, 1433
Axis of symmetry
  explanation of, 1299
  of hyperbola, 1434
  of parabola, 1299, 1422–1426

I-1

B

b^0, 849–850
Bar graphs, 538–539, 550
Base
  explanation of, 59
  of exponential expressions, 832, 834–837, 1376
  of logarithmic expressions, 1352
  multiplication and division with like, 834–836
  of percent proportion, 398, 399
Bimodal, 575
Binomial(s). *See* Polynomial(s)
  explanation of, 866
  factoring, 908–909, 942–943, 950–953
  formulas for factoring, 950
  square of, 881, 969, 1094, 1464
  sum and difference of cubes to factor, 950–952
Binomial expansion
  binomial theorem and, 1466–1468
  explanation of, 1464–1465
  factorial notation to find coefficients of, 1465–1466
  finding specific term in, 1468–1469
  Pascal's triangle and, 1465
Binomial theorem
  applications of, 1467–1468
  explanation of, 1466–1467
b^{-v}, 850–852
Borrowing, 14, 15, 235, 236
Boundary points, 1130, 1151
Brackets, 662, 1072, 1114

C

Calculators
  approximating exponential expressions with, 1342–1343
  change-of-base formula on, 1380–1381
  logarithms on, 1355, 1380–1381
Capacity. *See also* Volume
  applying units of, 452
  converting units of, 451, 464–465, 475
  metric units of, 464–465
  U.S. Customary units of, 446, 451–452
Carrying
  in addition of mixed numbers, 233
  in addition of whole numbers, 10
  in multiplication of whole numbers, 37
Cells, 536
Celsius, 476, 477
Center, of circle, 1411
Centimeter, cubic, 464
Change-of-base formula, 1380–1381
Circle(s)
  area of, 295, 510, 512, 513
  center of, 1411
  circumference of, 294, 508
  diameter of, 293, 294
  explanation of, 293, 1411
  graphs of, 556–560, 1411–1412
  radius of, 293, 294, 1411
  standard equation of, 1411
  writing equation of, 1412–1413
Circle graphs
  construction of, 558–560
  explanation of, 556
  interpretation of, 556–557
  percents and, 557–558
Circumference
  application involving, 294, 645
  explanation of, 294, 508
  method to find, 294, 508, 509

Class intervals, 548, 549
Clearing decimals, 331, 615, 616
Clearing fractions, 257–258, 612–615, 1031–1033
Clearing parentheses, 131
Coefficient matrix, A–6
Coefficients
  explanation of, 126, 866
  leading, 866
  of polynomial, 866
Column matrix, A–6
Commission, 419–421
  formula, 419, 423
Common denominator
  explanation of, 220
  least. *See* Least common denominator (LCD)
Common difference, 1481
Common logarithmic functions
  applications of, 1358–1359
  evaluation of, 1354–1355
  explanation of, 1354
Common ratio, 1488
Commutative property
  of addition, 11, 12, 16, 90, 127, 128
  of multiplication, 34–35, 127, 128
Complementary angles
  explanation of, 487, 648, 784
  identifying, 488
  solving application involving, 644
Complementary events, 567–568
Completing the square
  equation of circle and, 1413
  explanation of, 1266
  to find vertex formula, 1318
  to graph parabola, 1423–1424, 1426
  to solve quadratic equations, 1266–1269, 1283
Complex conjugates, 1252
Complex fractions
  explanation of, 247
  method to simplify, 247–248, 1016–1021
Complex numbers
  addition of, 1253
  definition of i and, 1248–1250
  division of, 1254–1255
  explanation of, 1251–1253
  multiplication of, 1254
  powers of i and, 1250–1251
  real and imaginary parts of, 1252–1253
  simplification of, 1255
  subtraction of, 1253
Composite numbers, 182
Composition, of functions, 1093–1094
Compound inequalities
  applications of, 1121–1122
  explanation of, 663, 1114
  of form $a < x < b$, 1118–1119
  graphs of, 663
  involving union and intersection of sets, 1114–1116
  joined by word *and*, 1116–1118
  joined by word *or*, 1119–1121
  methods to solve, 670, 1116–1121
Compound interest
  application of, 838
  explanation of, 432, 837
  formula for, 434–435
  method to compute, 1377–1378
  simple interest *vs.*, 433
Conditional equations, 608–609
Cones, 523
Congruent angle, 487
Conic sections, 1409, 1422. *See also* Circles; Ellipse; Hyperbola; Parabola

Conjugates
  complex, 1252
  construction of, 1230
  explanation of, 880
  multiplication of, 880–881, 941, 1219–1220
Consecutive even integers, 623
Consecutive integers, 623, 624–625
Consecutive odd integers, 623
Consistent systems of linear equations, 767
Constant functions, 1079
Constant graphs, A–27
Constants, 63
Constant terms, 126
Contradictions, 608–609, 781, 961
Conversion. *See also* Measurement
  to/from scientific notation and standard form, 861–862
  to/from standard and expanded form, 3–4
  to/from U.S. Customary and metric units, 473–477
  units of capacity, 451–452, 464–465, 475
  units of length, 447, 448, 474
  units of mass, 463, 474
  units of temperature, 267, 477
  units of time, 449
  units of weight, 450, 474
Conversion factor
  explanation of, 446
  method to choose, 446
Corresponding angles, 489
Costs
  applications, 651–652, 802–803
  unit, 352–353
Cramer's rule, A–18 to A–22
  to analyze dependent system of equations, A–21 to A–22
  for 3×3 System of Linear Equations, A–20 to A–21
  for 2×2 System of Linear Equations, A–18 to A–19
Cube(s)
  factoring sum or difference of, 948–950
  perfect, 949, 1183, 1184
  surface area of, 524
  volume of, 521
Cube root, 1182, 1187
Cubic centimeter, 464
Cubic functions, 1079, 1081
Cylinders
  right circular, 521, 522
  surface area of, 524
  volume of, 522

D

Data
  explanation of, 536
  in table, 536–537
Decay functions, exponential, 1344–1346
Decimal fractions, 268, 288
Decimal notation, 268–270, 288
Decimal point, 267–270, 288, 461
Decimals
  addition of, 279–283
  in algebraic expressions, 283
  applications of, 281–283, 292–293, 307–308, 322–323
  clearing, 331, 615, 616
  converted to/from percents, 386–387, 390–391, 408
  division of, 301–309
  equations containing, 329–334
  explanation of, 267

linear equations with, 329–334, 615–616
method to read, 270
multiplication of, 288–293
on number line, 317–319
ordering, 272–273, 318
order of operations and, 319–321
repeating, 303–305, 315–316, 318
rounding, 273–274, 306–307, 316–317
subtraction of, 279–283
terminating, 304, 318
writing fractions as, 314–317
writing ratios of, 343–344
written as fractions, 270–272, 317
written as mixed numbers, 270–272
Decreasing functions, 1345
Decreasing graphs, A–27
Degree
  of angle, 486
  of polynomial, 866
  of term, 866, 917
Denominators
  common, 220
  explanation of, 170
  least common, 214–215. *See* Least common denominator (LCD)
  rational expressions with like, 1004–1006
  rational expressions with unlike, 1006–1010
  rationalizing, 1227–1231
Dependent systems of linear equations
  addition method to solve, 794
  Cramer's rule to analyze, A–21 to A–22
  explanation of, 767
  Gauss-Jordan method to solve, A–10
  graphing method to solve, 770
  substitution method to solve, 782
  in three variables, 818
Dependent variable, A–26
Derivation, of quadratic formula, 1276, 1280–1283
Descending order, polynomials in, 866, 917
Determinants
  explanation of, A–14
  of 3×3 Matrix, A–15 to A–17
  of 2×2 Matrix, A–14 to A–15
Diameter, 293, 294
Difference
  of cubes, 948–950
  estimating, 28, 29
  explanation of, 12
  of logarithms, 1371–1372
  of natural logarithms as single logarithm, 1379–1380
  of squares, 880, 941–943, 950, 951
Digits, 2, 3
Directrix, of parabola, 1422
Direct variation
  applications of, 1102–1103
  explanation of, 1100
Discount, 421, 635–636
  formula, 421, 423, 636
Discriminant
  explanation of, 1280
  use of, 1280–1283
Distance. *See also* Measurement
  with absolute value, 1153
  applications of, 368–369, 806–807, 1042–1044
  formula for, 1410, 1411
  rate, and time problems, 806–807, 1042–1044
Distributive property, 1209–1210
  application of, 129
  of multiplication over addition, 36, 127, 129
  to multiply polynomials, 878
  with radical expressions, 1217, 1218
  with rational expressions, 1006

Dividend, 47
Divisibility rules, 181, 182
Division
  applications of, 52–53
  of complex numbers, 1254–1255
  of decimals, 301–309
  explanation of, 47
  of fractions, 199–201
  of functions, 1092–1093
  of integers, 106–107
  of like bases, 834–836
  long, 49–50
  by many-digit divisor, 51
  of mixed numbers, 231–232
  of polynomials, 887–895
  properties. *See* Division properties
  of radicals, 1225–1226
  of rational expressions, 991–994
  in scientific notation, 862
  symbol for, 47
  synthetic, 892–895
  translating to/from words for, 52, 157
  undefined, 715
  of whole numbers, 47–53
  by zero, 715
Division algorithm, 889
Division properties
  of equality, 143, 144, 253, 329–330, 593–597
  explanation of, 48–49
  of inequality, 667–669
  of radicals, 1225–1226
Divisors
  dividing by many-digit, 51
  explanation of, 47
Domain
  of functions, 1071–1072, 1081–1082
  of inverse functions, 1332, 1335
  of logarithmic function, 1355, 1357–1358
  of radical functions, 1186–1187
  of relations, 1056–1059
  restricted, 1335

E

e (irrational number). *See* Irrational number *(e)*
Elementary row operations, A–8
Elimination method. *See* Addition method
Ellipse
  explanation of, 1409, 1432
  graphs of, 1433–1434, 1436, 1445
  standard form of equation of, 1432
Ellipses notation, 2
Empty set, 1114
Endpoints
  in angles, 486
  in rays, 485
English system of measurement. *See* U.S. Customary units
Equality
  addition properties of, 137–140, 253, 254, 329, 591–593
  comparing properties of, 145–146
  division properties of, 143, 144, 253, 593–597
  division property of, 329–330
  multiplication properties of, 143–145, 253, 255–256, 593–597, 612
  multiplication property of, 329–330
  solution set to, 591
  subtraction properties of, 137, 138–140, 253, 254–255, 329, 591–593
Equal sign (=), 359, 662

Equations. *See also* Linear equations in one variable; Linear equations in two variables; Nonlinear systems of equations; Systems of linear equations in three variables; Systems of linear equations in two variables; *Specific types of equations*
  absolute value. *See* Absolute value equations
  of circle, 1411, 1412–1413
  conditional, 608–609
  containing decimals, 329–334
  containing fractions, 253–256, 612–615, 1026
  contradictions as, 608–609
  dependent, 767
  of ellipse, 1432
  equivalent, 137
  explanation of, 125, 590
  exponential. *See* Exponential equations
  of hyperbola, 1434
  identities as, 608–609
  independent, 767
  of inverse functions, 1333–1335
  linear, 125, 1025
  literal, 640–642, 1270
  logarithmic, 1391–1394
  of parabola, 1423, 1425
  percent, 407–412
  polynomial, 961–962, 1294
  quadratic. *See* Quadratic equations
  in quadratic form, 1292–1294
  radical, 1237–1244
  rational. *See* Rational equations
  solutions to, 135–136, 138, 590
  standard form of, 727, 743, 744
  steps to solve, 152–153
  translating verbal statements into, 157–159
Equilateral triangle, 497
Equivalent equations, 137
Equivalent fractions
  explanation of, 184
  method to write, 212–214
Estimation. *See also* Rounding
  applications of, 29, 39
  explanation of, 28
  of probabilities, 566–567
  products by rounding, 38–39
Events
  complementary, 567–568
  explanation of, 565
  probability of, 565–566
Expanded form
  conversion to/from standard form, 3–4
  explanation of, 3
  logarithmic expressions in, 1369–1370, 1380
Expanding minors, A–16
Expansion. *See* Binomial expansion
Experiment, 564
Exponential equations
  applications of, 1398–1399
  explanation of, 1394
  method to solve, 1394–1397
Exponential expressions
  approximation with calculator, 1342–1343
  base of, 832, 834–837, 1376
  on calculators, 1376
  evaluation of, 59–60, 832–834
  explanation of, 105
  method to simplify, 105
Exponential form, 59, 1352–1353
Exponential functions
  applications of, 1345–1347, 1397–1398
  compound interest and, 1377–1378
  decay, 1344–1346
  explanation of, 1329, 1341–1342

Exponential functions—Cont.
graphs of, 1343–1345, 1357, 1376–1377
growth, 1344, 1347
Exponential models
characteristics of, A–31 to A–33
Exponential notation, 832
Exponents
applications of, 60, 837–838
on calculators, 1342–1343
evaluating expressions with, 832–834
explanation of, 59
integer, 852–855
negative, 850–852
power rule for, 843, 844
properties of, 843–846, 982, 1195–1196
rational, 1193–1197
simplifying expressions with, 245, 835–837, 844–846, 851–852, 853–855
zero, 849
Extraneous solution, 1237

F

Factorial notation, 1465–1466
Factoring out
binomials, 908–909
greatest common factor, 906–907, 925
negative factor, 908
Factorization
explanation of, 180
of numbers, 181
prime, 182–184, 904
Factors/factoring
additional strategies, A–3 to A–4
binomial, 908–909, 942–943, 950–953
difference of squares, 941–943
divisibility rules and, 181, 182
explanation of, 34, 180, 903
greatest common. *See* Greatest common factor (GCF)
by grouping, 909–911, 934
identification of, 34
negative, 908
1 term with 3 terms, A–2 to A–3
perfect square trinomials, 943–945
polynomial, 952
prime, 211
quadratic equations, 959–961, 1294
regrouping, 144, 145
steps for, 956
by substitution, A–1 to A–2
sum or difference of cubes, 948–950
switching order in, 910
terms *vs.*, 982
trinomials, 915–918, 922–928, 933–937, 1266
Factor tree, 183
Fahrenheit, 476, 477
Fibonacci, Leonardo, 1480
Fibonacci sequence, 1480
Finite sequence, 1472
First-degree equations. *See* Linear equations in one variable
Focus
of ellipse, 1432
explanation of, 1409
of hyperbola, 1434
of parabola, 1422
Foot/feet symbol, 446
Formulas. *See also specific formulas*
commission, 419, 423
for compound interest, 434–435
discount, 421, 423, 636

involving rational equations, 1031–1033
literal equations and, 640–642
markup, 422, 423
sales tax, 417, 423
simple interest, 431
temperature, 477
Fourth root, 1239
Fraction bar, 170
Fractions
applications of, 169, 188, 202–203, 225, 322–323
clearing, 257–258, 612–615, 1031–1033
complex, 247–248, 1016–1021
containing variables, 187, 197, 206
conversion to/from percents, 385–391
converted to repeating decimals, 315–316
decimal, 268, 288
determining absolute value and opposite of, 176
division of, 199–201
equivalent, 184, 212–214
explanation of, 170
fundamental principle of, 185
improper, 271–272. *See* Improper fractions
interpreted as division, 171
like, 220–221
linear equations containing, 612–615
method to simplify, 184–188, 196
method to write, 170–171
mixed numbers, 172–175
multiplication of, 195–197
on number line, 175–176
ordering, 214–215, 318–319
order of operations and, 244–246, 320–322
proper, 171
properties of, 171
reciprocal of, 199, 200
rewriting, 612
simplification of, 1466
simplified to lowest terms, 184–188
solving equations containing, 253–258, 612–615
unlike, 221–224
writing decimals as, 270–272, 317
written as decimals, 314–317
written as fractions, 317
Frequency distributions
explanation of, 548
method to create, 549
method to interpret, 549–550
Front-end rounding, 290
Function(s). *See also specific forms of functions*
absolute value, 1079, 1082
algebra of, 1092–1093
composition of, 1093–1094, 1333
constant, 1079
cubic, 1079, 1081
decreasing, 1345
domain of, 1071–1072, 1081–1082
evaluation of, 1068–1070
explanation of, 1055, 1066, 1330
exponential. *See* Exponential functions
graphs of, 1079–1085, 1095
identity, 1079, 1081
inverse. *See* Inverse functions
linear, 1079
logarithmic. *See* Logarithmic functions
maximum value of, 1306
method to combine, 1094–1095
minimum value of, 1306
one-to-one, 1330–1332
operations on, 1094–1095
quadratic. *See* Quadratic functions
radical, 1186–1187, 1242–1244
range of, 1071, 1080, 1081–1082

reciprocal, 1079, 1082
relations *vs.*, 1066–1067
square root, 1079, 1082
values, finding from graphs, 1070–1071
vertical line test for, 1067–1068
Function notation, 1068–1070
explanation of, 1068
finding intercepts using, 1083–1085
Fundamental principle of fractions, 185
Fundamental principle of rational expressions, 981

G

Gauss-Jordan method, A–7 to A–11
GCF. *See* Greatest common factor (GCF)
Geometric sequence
explanation of, 1463, 1488
finding specified term of, 1489
*n*th term of, 1488–1489
Geometric series
application of, 1492
explanation of, 1489
infinite, sum of, 1491–1492
sum of, 1490–1492
Geometry. *See also specific geometric shapes*
applications of, 332, 513–514, 643–645, 784, 823–824, 871–872, 882, 1242–1243
area, 510–514
circumference, 508, 509, 645
lines and angles, 485–490
perimeter, 508–509
perimeter and circumference, 294
surface area, 524–525
volume and, 521–523
Golden ratio, 349, 1263
Golden rectangle, 1263
Grams, 462, 463
Graphical model, A–26
Graphs. *See also* Rectangular coordinate system
bar, 538–539, 550
of circles, 1411–1412
of compound inequalities, 663
constant, A–27
decreasing, A–27
of ellipse, 1433–1434, 1436, 1445
of exponential functions, 1343–1345, 1357, 1376–1377
finding function values from, 1070–1071
of functions, 1079–1085, 1095
histograms, 550
of horizontal lines, 701
of hyperbola, 1435–1436
increasing, A–27
interpretation of, 684–685
of inverse functions, 1334, 1335
line, 540–542
of linear inequalities in one variable, 663
of linear inequalities in two variables, 1159–1161
of logarithmic function, 1355–1358
of natural logarithmic functions, 1378
of nonlinear inequalities, 1450–1451
of ordered pairs, 1069
of parabola, 1082, 1128, 1299–1307, 1423–1426
pictographs, 539–540
plotting points on, 685–686
of quadratic functions, 1299–1307
of quadratic inequalities, 1129–1131
of radical function, 1187
of slope-intercept form of linear equations, 728–729
of systems of linear equations, 767–770

of systems of linear inequalities in two variables, 1161–1163
of systems of nonlinear inequalities in two variables, 1451–1452
of vertical lines, 701–702
Greater than (>), 5
Greatest common factor (GCF)
explanation of, 186, 904
factoring out, 906–907, 925
method to identify, 186, 187, 904–906
removed from trinomials, 936
Great Pyramid application, 1242–1243
Grouping
factoring by, 909–911, 934
symbols, 11–12, 62
Growth functions, exponential, 1344, 1347

H

Heron's formula, 1179
Histograms, 550
Horizontal asymptote, 1081
Horizontal lines
explanation of, 700
graphs of, 701
linear equations and, 743
slope of, 715
Horizontal line test, 1330–1331
Horizontal transverse axis, 1434
Hyperbola
explanation of, 1409, 1434
focus of, 1434
graphs of, 1435–1436
standard forms of equation, 1434, 1436
Hypotenuse, 498, 499

I

Identities, 608–609
Identity functions, 1079, 1081
Identity property of multiplication, 983
Imaginary numbers (i)
explanation of, 1248
powers of, 1250–1251
simplifying product or quotient in terms of, 1249–1250
Improper fractions, 230. *See also* Fractions; Mixed numbers
adding mixed numbers by using, 233
converting to/from mixed numbers and, 173–174
explanation of, 171
method to write, 172
subtracting mixed numbers by using, 236
writing decimals as, 271–272
Inconsistent systems of linear equations
addition method to solve, 793
explanation of, 767
Gauss-Jordan method to solve, A–11
graphing method to solve, 769
substitution method to solve, 781
in three variables, 819
Increasing graphs, A–27
Independent systems of linear equations, 767
Independent variable, A–26
Index of summation, 1475, 1484
Index of the radical, 1220–1221, 1237
Inequalities. *See also* specific types of inequalities
absolute value. *See* Absolute value inequalities
addition properties of, 666–667
boundary points of, 1130, 1151
compound. *See* Compound inequalities
graphs of, 663

linear in one variable, 662–671
linear in two variables, 1113, 1159–1165
nonlinear, 1449–1451
on number line, 662–670
polynomial, 1131–1132
quadratic, 1128–1132
rational, 1132–1134
solution set of, 662, 663
solution to, 1133
subtraction properties of, 666–667
symbols, 662
systems of nonlinear, 1451–1452
Infinite geometric series, 1491–1492
Infinite sequence, 1472
Integer exponents, 852–855
Integers. *See also* Numbers
addition of, 87–91, 97
applications of, 91, 98
consecutive, 623, 624–625
division of, 106–107
division of decimals by, 302
explanation of, 80
method to write, 80
multiplication of, 103–105
on number line, 80–82, 87–88
opposites of, 83
subtraction of, 96–98
Interest
compound. *See* Compound interest
compounded continuously, 1377
simple. *See* Simple interest
Intersection, of sets, 1114–1116
Interval notation, 1187
explanation of, 664–665
union and intersection in, 1115–1116
use of, 665
Inverse functions, 1330–1335
definition of, 1332
domain of, 1332, 1335
equation of, 1333–1335
explanation of, 1332–1333
graphs of, 1334, 1335
one-to-one, 1330–1332
property, 1332
range of, 1332
Inverse variation
applications of, 1103–1104
explanation of, 1100
Irrational number (e), 1181, 1376–1377
explanation of, 318
Isolation of variables, 145, 150–152, 591, 605
Isosceles triangle, 497

J

Joint variation
applications of, 1104–1105
explanation of, 1101

L

LCD. *See* Least common denominator (LCD)
Leading coefficient, 866
Leading term, of polynomial, 866
Least common denominator (LCD)
to add and subtract unlike fractions, 222–225
clearing fractions and, 612, 614, 1032, 1033
explanation of, 214–215, 997
method to find, 997–999
simplifying complex fraction by multiplying by, 248
writing rational expressions with, 999–1001

Least common multiple (LCM), 614
applications of, 212
explanation of, 210
listing multiples to find, 210–211
Legs, 498, 499
Length
converting units of, 447, 474
metric units of, 459–462
multiple conversions of, 448
U.S. Customary units of, 446
Less than (<), 5
Like bases
division of, 834–836
multiplication of, 834–836
Like fractions
addition of, 220
explanation of, 220
subtraction of, 220–221
Like radicals, 1209, 1210
Like terms
addition method and, 1446
addition of, 130, 831, 868, 869
explanation of, 126
identification of, 126
method to combine, 130, 283
simplifying expressions with, 131
subtraction of, 130, 831, 868, 869
Line(s)
explanation of, 485
horizontal. *See* Horizontal lines
parallel. *See* Parallel lines
perpendicular. *See* Perpendicular lines
slope-intercept form of, 727–732
slope of, 711–715
vertical. *See* Vertical lines
x- and y-intercepts of, 698–700
Linear equations in one variable. *See also* Equations
addition property of equality to solve, 591–593
applications of, 159–161, 625–627
clearing decimals to solve, 331
clearing fractions to solve, 257–258, 612–615
conditional, 608–609
with consecutive integers, 623, 624–625
containing decimals, 329–334, 615–616
containing fractions, 612–615
contradictions as, 608–609
division property of equality to solve, 593–598
explanation of, 125, 136, 591, 957, 1025
graphs of, 663
identities as, 608–609
method to solve, 146, 149–153, 603–609
multiplication property of equality to solve, 593–598, 612
problem-solving flowchart, 156–157
problem-solving flowchart for, 621
steps to solve, 152–153, 257, 603–609
strategies to solve, 621
subtraction property of equality to solve, 591–593
translations to, 157–159, 598–599, 621–623
Linear equations in two variables
applications of, 750–753
explanation of, 694–695
forms of, 743–744
graphs of, 695–698
horizontal and vertical lines and, 700–702
interpretation of, 750–751
parallel and perpendicular lines and, 716–717
plotting points to graph, 695–698
point-slope formula and, 740–744
slope-intercept form and, 727–732
slope of line and, 711–715

Linear equations in two variables—*Cont.*
solutions to, 696
x- and *y*-intercepts of, 698–700
Linear functions, 1079, 1482
Linear inequalities in one variable
addition properties of, 666–667
applications of, 670–671
compound, 663, 670
division properties of, 667–669
explanation of, 662
multiplication properties of, 667–669
set-builder notation, 664, 665
solution to, 667
subtraction properties of, 666–667
test points for, 667
translations for, 671
Linear inequalities in two variables
explanation of, 1113
graphs of, 1159–1161
systems of, 1161–1163
Linear model, 752–753
characteristics of, A–27 to A–29
identifying, from observed data, A–28 to A–29
quadratic model *vs.,* A–30
Linear programming, 1113
Line graphs
construction of, 542
explanation of, 540
method to interpret, 541
Line segments, 485
Liter, 464
Literal equations. *See also* Quadratic formula
application of, 1270
applications of, 642
explanation of, 640–641
methods to solve, 641
Logarithmic equations
explanation of, 1391
method to solve, 1391–1394
Logarithmic expressions
equivalence of, 1393
evaluation of, 1353–1354
in expanded form, 1369–1370, 1380
as single logarithm, 1371–1372
Logarithmic functions
applications of, 1358–1359, 1381–1382
on calculators, 1355
common, 1354–1355, 1358–1359
converted to exponential form, 1352–1353
domain of, 1355, 1357–1358
explanation of, 1329, 1352
graphs of, 1355–1358
natural, 1378–1380
properties of, 1379–1380
Logarithms
on calculators, 1355, 1380–1381
change-of-base formula and, 1380–1381
explanation of, 1352
logarithmic expressions as single, 1371–1372
natural, simplifying expressions with, 1379
properties of, 1367–1369
sum or difference of natural logarithms as single, 1379–1380
Long division
explanation of, 49–50
of polynomials, 888–892
synthetic division *vs.,* 894
Louvre pyramid, 1179
Lowest terms
simplifying fractions to, 184–188
writing rates in, 350
writing ratios in, 343

M

Markup
explanation of, 422
formulas, 422, 423
Mass
explanation of, 462
units of, 462–463, 474, 482–483
Mathematical modeling, explanation of, A–26
Matrices
augmented, A–6 to A–7
coefficient, A–6
column, A–6
determinants of. *See* Determinants
elementary row operations, A–8
explanation of, A–6
order of, A–6
row, A–6
square, A–6
Maximum value, of function, 1306
Mean
computation of, 71
explanation of, 70, 572
method to find, 573
weighted, 577–578
Measurement
adding mixed units of, 448–449
of angles, 486–487, 490
applications of, 445, 449, 450, 475–476, 482–483, 1154
conversions between U.S. Customary and metric units of, 473–477
metric units of. *See* Metric units
subtracting mixed units of, 448–449
temperature, 267, 476–477
U.S. Customary units of. *See* U.S. Customary units
Median
explanation of, 573
method to find, 573–575
Medical applications, 482–483
Meter, 459
Metric prefix, 461, 462
Metric system. *See also* Measurement; Metric units
explanation of, 459
summary of, 465–466
Metric units. *See also* Measurement
of capacity, 464–465
conversion between U.S. Customary and, 473–477
explanation of, 459
of length, 459–462
of mass, 462–463, 482–483
prefixes for, 461, 462
summary of conversions in, 465–466
Micrograms, 482
Midpoint formula, 1413–1415
Minimum value, of function, 1306
Minor, of element of matrix, A–15, A–16
Minuend, 12, 14
Mixed numbers
addition of, 232–234
applications of, 238
converted to/from improper fractions and, 173–174
decimals written as, 270–272
division of, 231–232
explanation of, 172–173
multiplication of, 230–231
negative, 173, 237
quotients written as, 175
rounding, 232–234
subtraction of, 234–237
writing ratios of, 343–344

Mixture problems, 652–653, 804–805, 824–825
Mode
explanation of, 575
method to find, 576–577
Monomials. *See also* Polynomial(s)
division by, 887–888
explanation of, 866
multiplication of, 876–878
Motion, uniform, 653–655
Multiples
explanation of, 210
least common, 210–212
Multiplication. *See also* Products
applications of, 39–40
of complex numbers, 1254
of conjugates, 941, 1219–1220
of decimals, 288–293
explanation of, 34
of fractions, 195–197
of functions, 1092–1093
of integers, 103–105
of like bases, 834–836
of mixed numbers, 230–231
of polynomials, 916
properties. *See* Multiplication properties
of radicals, 1201–1204, 1216–1221
of rational expressions, 989–991, 993–994
in scientific notation, 862
symbol for, 34
translating to/from words for, 39, 157
of whole numbers, 33–40
Multiplication properties
associative, 35, 127, 128
commutative, 34–35, 127, 128
distributive property of multiplication over addition, 36, 127, 129
of equality, 143–145, 253, 255–256, 329–330, 593–597, 612
identity, 983
of inequality, 667–669
of radicals, 1201–1204, 1216–1221, 1225
of zero, 36

N

Natural logarithmic functions
applications of, 1381–1382
explanation of, 1378
graphs of, 1378
properties of, 1379–1380
Negative exponents, 850–852
Negative factors, 908
Negative numbers, 131, 237
adding and multiplying, 103
explanation of, 80
mixed, 173, 237
on number line, 80–82
opposites of, 82
ordering, 319
parentheses with, 607
square roots of, 1180, 1248
Negative square root, 1180, 1181
Nested parentheses, 62
Nonlinear inequalities in two variables
graphs of, 1450–1451
test point method for, 1449–1450
Nonlinear relationship, A–30
Nonlinear systems of equations
addition method to solve, 1445–1446
explanation of, 1442
substitution method to solve, 1442–1445

Notation/symbols
  addition, 9, 10
  angles, 486
  approximately equal to, 26
  decimal, 267–270, 288, 461
  division, 47
  ellipses, 2
  empty set, 1114
  equal sign, 359, 662
  expanded, 3–4
  exponential, 832
  factorial, 1465–1466
  foot/feet, 446
  fractions, 170
  function, 1068–1070
  grouping, 11–12, 62
  inequality, 662
  interval. *See* Interval notation
  multiplication, 34
  parentheses, 11–12, 62, 89, 607, 623, 1006
  percent, 384
  perpendicular lines, 489
  radical, 60, 498, 1180–1182, 1195
  scientific. *See* Scientific notation
  set-builder, 664, 665
  square bracket, 662
  standard, 3–4
  subtraction, 12
  summation/sigma, 1475–1477
  temperature, 476, 486
  union, 1082
  use of, 589
nth root
  evaluating, 1183
  explanation of, 1182–1183
  of real numbers, 1183
  simplification of, 1183, 1184–1185
  of variable expressions, 1183–1185
nth term of sequence
  explanation of, 1472
  formula for, 1474, 1475
  method to find, 1474, 1488–1489
  method to write, 1481–1482
Number(s). *See also* Integers
  complex. *See* Complex numbers
  composite, 182
  irrational, 318, 1181, 1376–1377
  negative, 319. *See* Negative numbers
  positive. *See* Positive numbers
  prime, 182, 192
  rational, 170, 1269
  reciprocal of, 199
  whole. *See* Whole numbers
Number line
  addition of integers by using, 87–88
  addition of whole numbers on, 10
  decimals on, 317–319
  explanation of, 5
  fractions on, 175–176
  inequalities on, 662–670
  integers on, 80–82, 87–88
  negative numbers on, 80–82
  positive numbers on, 80–82
  rounding on, 26
  subtraction on, 13, 96
Numerators, 170
Numerical model, A–26

O

Observed data
  constructing tables from, 537
  estimating probabilities from, 566–567

Obtuse triangle, 497
One, division by, 48
One-to-one function, 1330–1332
Opposites
  addition of, 90
  explanation of, 82
  of integers, 83
  subtraction of, 96
Ordered pairs
  explanation of, 683, 686
  graphs of, 1069
  in rectangular coordinate system, 686
  relation as set of, 1056–1058, 1066
  as solution to linear equation in two variables, 695–697
  as solution to system of linear equations, 766
Ordered triples, 814
Ordering
  of decimals, 272–273, 318–319
  of fractions, 214–215, 318–319
Order of a matrix, A–6
Order of operations
  application of, 61–63, 245–246
  decimals and, 319–321
  explanation of, 61
  fractions and, 244–246, 320–322
  method to apply, 113–114
  to simplify radicals, 1204–1205
Origin, 685

P

Parabola. *See also* Quadratic functions
  axis of symmetry of, 1299, 1422–1426
  explanation of, 1128, 1299, 1422, A–30
  graphs of, 1082, 1128, 1299–1307, 1315, 1423–1426
  standard form of equation, 1423, 1425
  vertex of, 1299, 1422, 1427–1428, A–30
Parallel lines
  explanation of, 488–489, 716
  point-slope formula and, 742
  in slope-intercept form, 730–731
  slope of, 716–717
Parallelograms
  area of, 510
  explanation of, 508
Parentheses, 136, 1070, 1072, 1115, 1187, 1443
  clearing, 131
  clearing fractions and, 615
  explanation of, 11
  as grouping symbol, 11–12
  guidelines for use of, 623
  with integers, 89
  with negative numbers, 607
  nested, 62
  with rational expressions, 1006
Pascal, Blaise, 1465
Pascal's triangle, 1465, 1467, 1468
Percent
  applications of, 400–402, 417–424
  circle graphs and, 557–558
  converted to/from decimals, 386–387, 390–391, 408
  converted to/from fractions, 385–391
  equations. *See* Percent equations
  explanation of, 383, 384
  list of common, 390
  proportions. *See* Percent proportions
  symbol, 384
  unknown, method to find, 633
Percent decrease, 423, 424

Percent equations
  applications of, 410–412, 632–634
  explanation of, 407–408
  method to solve, 408–410
Percent increase, 423–424
Percent proportions
  applications of, 400–402
  explanation of, 397–398, 407–408
  method to solve, 399–400, 408–410
  parts of, 398
Perfect cubes, 949, 1183, 1184
Perfect fifth powers, 1183
Perfect fourth powers, 1183
Perfect squares
  examples of, 941
  explanation of, 941, 1181
  patterns associated with, 1184
Perfect square trinomials
  checking for, 944
  explanation of, 880, 943, 969
  factoring, 943–945, 1266
Perimeter
  applications of, 643
  applying decimals to, 282–283
  explanation of, 18, 508
  formulas for, 508
  method to find, 18, 509
  of triangles, 508, 640–641
Periods, 2
Perpendicular lines
  explanation of, 489, 716
  point-slope formula and, 743
  in slope-intercept form, 730–731
  slope of, 716–717
  symbol, 489
Phrases. *See* Words/phrases
Pi (π), 294, 318
Pictographs, 539–540
Pie graphs. *See* Circle graphs
Placeholders, 870
Place value
  determination of, 2–3
  explanation of, 2, 268
  fractional, 268
  method to determine, 269
  rounding decimals to, 273–274
Place value chart, 268
Plotting points
  graphing linear equations in two variables by, 695–698
  in rectangular coordinate system, 685–686
Point, 485
Point-slope formula
  explanation of, 740
  writing equation of line using, 740–741, 742–743
Polygon(s). *See also* Triangles
  explanation of, 496
  similar, 372, 373
Polynomial(s). *See also* Binomials; Monomials; Trinomials
  addition of, 868–869
  applications of, 871–872
  degree of, 866
  in descending order, 866, 917
  division of, 887–895
  explanation of, 831, 866
  factoring, 952
    additional strategies, A–3 to A–4
    1 term with 3 terms, A–2 to A–3
    by substitution, A–1 to A–2
  identifying parts of, 867
  long division to divide, 888–892

Polynomial(s)—*Cont.*
  multiplication of, 876–879, 916
  prime, 907, 933, 980
  special case products of, 880–882
  subtraction of, 869–871
  synthetic division to divide, 892–895
Polynomial equations. *See also*
      Quadratic equations
  factoring to solve, 1294
  higher degree, 961–962
  second-degree, 957–962
Polynomial inequalities, 1131–1132
Positive numbers, 131
  explanation of, 80
  on number line, 80–82
  opposites of, 82
Power, 59
Power of 0.1, 290, 291
Power of product, 844
Power of quotient, 844
Power property of logarithms, 1368, 1369, 1370, 1379–1380
Power rule for exponents, 843, 844
Powers of i, 1250–1251
Powers of one-tenth, 245
Powers of 10
  division by, 306
  explanation of, 60
  metric system and, 461, 463
  multiplication by, 290, 291
Prefix line, 461–462
Prime factorization, 182–184, 904
Prime factors, 211
Prime numbers
  explanation of, 182
  sieve of Eratosthenes to find, 192
Prime polynomials, 907, 933, 980
Principal, 431
Principal square root
  explanation of, 1180, 1181
  Pythagorean theorem and, 1185
Probability
  of complementary events, 567–568
  estimated from observed data, 566–567
  of events, 565–566
  explanation of, 535, 564
Problem-solving flowchart, 156–157, 621
Product property of logarithms, 1367–1368, 1369, 1379–1380
Products. *See also* Multiplication
  of conjugates, 880–881
  explanation of, 34
  identification of, 34
  power of, 844
  special case, 880–882
Proper fractions
  explanation of, 171, 172
  writing decimals as, 271
Proportions
  applications of, 367–369, 1039
  explanation of, 358, 1038
  method to solve, 360–362
  pairs of numbers as, 360
  percent, 397–402
  rates and, 1038
  ratios and, 359, 1038
  writing, 359
Protractors, 486, 559
Pyramids, 1242–1243
Pythagorean theorem, 1410
  applications of, 500–501, 968–969, 1185–1186
  explanation of, 498–499, 1185

Q

Quadratic equations
  applications of, 966–967, 969–970
  completing the square to solve, 1266–1269, 1283
  equations reducible to, 1294–1295
  explanation of, 957–958, 1025
  factoring to solve, 959–961, 1294
  in one variable, 958
  quadratic formula to solve, 1277–1278, 1283
  square root property to solve, 1264–1265, 1267–1269, 1283
  summary of methods to solve, 1283–1286
  zero product rule to solve, 958–959, 1264, 1269, 1283
Quadratic formula
  applications of, 1278–1280
  derivation of, 1276, 1280–1283
  explanation of, 1276
  to solve quadratic equations, 1277–1278, 1283
Quadratic functions. *See also* Parabola
  analysis of, 1316–1317
  applications of, 1279–1280, 1319–1321
  explanation of, 1079, 1082–1083
  in form $f(x) = a(x - h)^2 + k$, 1314–1315
  graphs of, 1299–1307
  vertex and intercepts of, 1318–1319
  vertical shifts of, 1301
  x- and y-intercepts of, 1282, 1283
Quadratic inequalities
  explanation of, 1128
  graphs to solve, 1129–1131
  test point method to solve, 1130–1131
Quadratic in form, 1292–1294
Quadratic model, 1320–1321
  characteristics of, A–30 to A–31
  explanation of, A–30
  linear model *vs.*, A–30
Quadrilaterals, 508
Quotient property of logarithms, 1368, 1369, 1370, 1379–1380
Quotients
  explanation of, 47
  of nonzero number, 143, 595
  power of, 844
  as repeating decimals, 303
  rounding, 273–274, 306–307
  whole part of, 50
  written as mixed number, 175

R

Radical(s). *See also* Square roots
  conjugate, 1219–1220
  division property of, 1225–1226
  index of, 1182, 1220–1221, 1238
  like, 1209, 1210
  multiplication property of, 1201–1204, 1225, 1410
  rationalizing the denominator of, 1227–1231
  simplification of, 1181, 1202–1205, 1218, 1225–1226
Radical equations
  applications of, 1242–1243
  explanation of, 1237
  method to solve, 1237–1242
Radical functions
  applications of, 1243–1244
  domain of, 1186–1187
  explanation of, 1186
  graphs of, 1187
Radical sign, 60, 498, 1180–1182, 1195
Radicand, 1182
Radius, 293, 294, 1411
Range
  of functions, 1071, 1080, 1081–1082
  of inverse functions, 1332
  of relations, 1056–1059
  as set of ordered pairs, 1056–1058
Rate of change, 753
Rates. *See also* Ratio(s)
  applications of, 353
  conversion to unit rate, 351
  explanation of, 341, 350
  in lowest terms, 350
  proportions and, 1038
  unit, 341, 351–352
Ratio(s)
  of -1, 983
  applications of, 345
  explanation of, 341, 342, 1038
  golden, 349, 1263
  in lowest terms, 343
  of mixed numbers and decimals, 343–344
  proportions and, 358–362, 1038
  in similar triangles, 370
  writing, 342–343
Rational equations
  applications of, 1038–1045
  example of, 1026
  explanation of, 1026
  method to solve, 1027–1030, 1294–1295
  solving formulas involving, 1031–1033
  steps to solve, 1029
  translation to, 1031
Rational exponents
  applications of, 1197
  converting between radical notation and, 1195
  definition of $a^{1/n}$ and $a^{m/n}$, 1193–1195
  properties of, 1195–1196
Rational expressions
  addition of, 1004–1010
  division of, 991–994
  evaluation of, 978
  explanation of, 977, 978
  fundamental principle of, 981
  least common denominator of, 997–1001
  multiplication of, 989–991, 993–994
  restricted values of, 979–980
  simplification of, 980–985
  subtraction of, 1004–1010
  in translations, 1010
Rational inequalities
  explanation of, 1132
  test point method to solve, 1132–1134
Rationalizing the denominator
  explanation of, 1227
  one term, 1228–1229
  two terms, 1229–1231
Rational numbers, 170, 1269. *See also* Fractions
Rays, 485
Real number line, 318
Reciprocal functions, 1079, 1082
Reciprocals
  explanation of, 199, 200
  product of number and, 594
Rectangles
  area of, 40–41, 293, 510, 511
  explanation of, 40, 508
  golden, 1263
  perimeter of, 508
Rectangular coordinate system. *See also* Graphs
  explanation of, 685
  geometric relationships in, 767
  plotting points in, 686–688

Rectangular solids
  surface area of, 524
  volume of, 521, 522
Recursive formula, 1480
Regrouping factors, 144, 145
Relations
  applications of, 1060
  domain of, 1056–1059
  explanation of, 1056
  range of, 1056–1059
  as set of ordered pairs, 1056–1058, 1066
Remainder, 50
Repeating decimals
  converting fractions to, 315–316
  explanation of, 303–304, 318
  rounding, 306
Restricted values, of rational expressions, 979–980
Rhombus, 508
Richter scale, 1375
Right angles, 486
Right circular cones, 521
Right circular cylinders, 521
Right triangles
  explanation of, 497
  Pythagorean theorem and, 498–501, 968–969
Roots
  cube, 1182, 1187
  fourth, 1239
  nth, 1182–1185
  square. *See* Square roots
  of variable expressions, 1183–1185
Rounding. *See also* Estimation
  converting fractions to decimals with, 316–317
  decimal, 273–274, 306–307, 316–317
  to estimate products, 38–39
  explanation of, 26
  front-end, 290
  mixed numbers, 232–234
  usage of, 28
  of whole numbers, 26–28
Round-off error, 321
Row matrix, A–6

S

Sales tax, 417–419, 632–633
  formula, 417, 423
Sample space, 564
Scalene triangle, 497
Scientific notation
  converted to standard form, 861–862
  division in, 862
  explanation of, 860
  multiplication in, 862
  writing numbers in, 860–861
Sector, of circles, 556
Semicircles, 512
Sequence
  alternating, 1473
  applications of, 1475
  arithmetic. *See* Arithmetic sequence
  explanation of, 1463, 1471–1472
  Fibonacci, 1480
  finite, 1472
  geometric. *See* Geometric sequence
  infinite, 1472
  nth term of, 1472–1475
  terms of, 1472–1475, 1481–1483
Series
  arithmetic. *See* Arithmetic series
  explanation of, 1463, 1475
  geometric. *See* Geometric series
  summation notation and, 1475–1477

Sets
  empty, 1114
  of imaginary numbers, 1248
  solution, 591, 662, 663, 770
  union and intersection of, 1114–1116
Sieve of Eratosthenes, 192
Sigma notation. *See* Summation/sigma notation
Signed numbers. *See also* Integers
  addition of, 88–90
  division of, 106–107
  multiplication of, 103–105
  subtraction of, 96–98
Sign rules
  for factoring trinomials, 917
  for trial-and-error method, 824
Similar polygons, 372, 373
Similar triangles
  applications of, 372, 1040–1041
  explanation of, 370, 1039
  finding unknown sides in, 370–372
Simple interest
  application involving, 634–635, 803–804, 837
  compound interest *vs.*, 433
  explanation of, 431, 837
  formula for, 837
  formulas, 431
  method to compute, 431–432
Simplification
  of algebraic expressions, 249
  of complex fractions, 247–248, 1016–1021
  of complex numbers, 1255
  explanation of, 245
  of expressions with exponents, 245, 835–837, 844–846, 851–852, 853–855
  of fractions, 184–188, 196
  of like terms, 131
  of nth root, 1183, 1184–1185
  of radicals, 1181, 1202–1205, 1218, 1225–1226
  of rational exponents, 1196
  of rational expressions, 980–985
  solving equation *vs.*, 590
  of square roots, 1181–1182
  in terms of imaginary numbers, 1249
Slope
  applications of, 712, 717–718
  explanation of, 711–712
  formula, 712–714
  method to find, 713–715
  of parallel lines, 716–717, 730–731
  of perpendicular lines, 716–717, 730–731
Slope-intercept form
  explanation of, 727
  graphs of, 728–729
  systems of linear equations and, 768–769
  writing equation of line using, 731–732, 743, 744
Solutions
  to equations, 135–136, 138, 590
  extraneous, 1237
  to inequalities, 667, 1133
  to systems of linear equations in three variables, 814
  to systems of linear equations in two variables, 765–767
Solution set
  explanation of, 591
  for inequalities, 662, 663, 1132, 1450
  for system of dependent equations, 770
  for systems of nonlinear inequalities, 1451–1452
Special case products
  explanation of, 880–881
  in geometry application, 882
  radical expressions and, 1219–1220

Spheres
  surface area of, 524, 525
  volume of, 521, 523
Square(s). *See also* Completing the square
  of binomials, 881, 1464
  difference of, 880, 941–943, 950, 951
  evaluating, 498
  explanation of, 498
  perfect, 941, 1181, 1184
  sum of, 942
Square (shape)
  area of, 510
  explanation of, 508
  perimeter of, 508
Square brackets, 662, 1072, 1114
Square matrix, A–6
Square root(s). *See also* Radical(s)
  approximation of, 1181
  evaluation of, 60, 498
  explanation of, 60, 498, 1180
  method to simplify, 1181–1182
  negative, 1180, 1181
  of negative numbers, 1180, 1187, 1248
  positive, 1180, 1181
  symbol for, 60, 498, 1180
Square root functions, 1079, 1082
Square root property
  explanation of, 1264
  to solve quadratic equations, 1264–1265, 1267–1269, 1283
Square units, 40, 41
Standard equation of circle, 1411
Standard form
  converting scientific notation to, 861–862
  converting to/from expanded form, 3–4
  of equation of line, 727, 743, 744
  explanation of, 2, 727
  writing numbers in, 5
Standard form of equation
  of ellipse, 1432
  of hyperbola, 1434, 1436
  of parabola, 1423, 1425
Statistics
  explanation of, 535, 536
  percents in, 401
Straight angles, 486
Strike through method, 187
Substitution method
  applications of, 783–784
  explanation of, 777
  factoring by using, A–1 to A–2
  to solve dependent systems, 782
  to solve equations in quadratic form, 1292–1294
  to solve inconsistent systems, 781
  to solve nonlinear systems of equations, 1442–1445
  to solve systems of linear equations, 777–784, 794
Subtraction
  applications of, 17–18
  borrowing in, 14, 15, 235, 236
  of complex numbers, 1253
  of decimals, 279–283
  explanation of, 12
  of functions, 1092–1093
  of integers, 96–98
  of like fractions, 220–221
  of like terms, 130, 868
  of mixed numbers, 234–237
  of mixed units of measurement, 448–449
  on number line, 13
  of opposites, 96
  of polynomials, 869–871

Subtraction—*Cont.*
　of radicals, 1209–1211
　of rational expressions, 1004–1010
　of real numbers, 870
　symbol for, 12
　translating to/from words for, 16, 97–98, 157, 622
　of whole numbers, 12–15
　words and phrases for, 16
Subtraction properties
　of equality, 137–140, 253, 254–255, 329, 591–593
　of inequality, 666–667
Subtrahend, 12, 14
Sum
　of arithmetic series, 1484
　of cubes, 948–950
　estimating, 28, 29
　explanation of, 9
　of geometric series, 1490–1492
　of infinite geometric series, 1491–1492
　of logarithms, 1371–1372
　of natural logarithms as single logarithm, 1379–1380
　of squares, 942
Summation, index of, 1475, 1484
Summation/sigma notation
　conversion to, 1476–1477
　explanation of, 1475
　finding sum from, 1476
Supplementary angles
　explanation of, 488, 648, 784
　identifying, 488
Surface area, 524–525
　of cube, 524
　of cylinder, 524
　formulas, 524
　of rectangular solid, 524
　of sphere, 524, 525
Synthetic division
　explanation of, 892–893
　long division *vs.*, 894
　use of, 893–895
Systems of linear equations in three variables
　applications of, 823–825
　dependent, 818
　explanation of, 813–814
　inconsistent, 819
　methods to solve, 815–819
　ordered triples as solutions to, 814
　solutions to, 814
Systems of linear equations in two variables
　addition method to solve, 788–794
　applications of, 783–784, 802–807
　consistent, 767
　dependent, 767, 782
　explanation of, 765, 766
　graphing method to solve, 767–770
　inconsistent, 767, 769, 781
　independent, 767
　solutions to, 765–767
　substitution method to solve, 777–784, 794
　summary of methods to solve, 794
Systems of linear inequalities
　feasible regions and, 1163–1165
　graphs of, 1161–1163
Systems of nonlinear inequalities in two variables, 1451–1452

T

Tables
　data in, 536–537
　explanation of, 536
　frequency distribution, 548–550

Temperature
　converting units of, 267, 477
　formula, 477
　measurement of, 267
　symbol for, 476, 486
　units of, 476–477
Terminating decimals, 304, 318
Terms. *See also* Like terms; Lowest terms
　constant, 126
　degree of, 866
　explanation of, 126
　factors *vs.*, 982
　leading, 866
　of polynomials, 866
　of sequence, 1472–1475, 1481–1483, 1488–1489
　unlike, 126
　variable, 126
Test point method
　for absolute value inequalities, 1151–1153
　for linear inequalities, 667
　for nonlinear inequalities, 1449–1450
　for polynomial inequalities, 1131–1132
　for quadratic inequalities, 1130–1131
　for rational inequalities, 1132–1134
Time, U.S. Customary units of, 446, 449
Translations
　for addition, 15–16, 91, 157
　for algebraic expressions, 115
　compound inequalities, 1121–1122
　for division, 52, 157
　for English phrases to mathematical statements, 15–16
　for inequalities, 671
　for linear equations, 157–159, 598–599, 621–623
　for multiplication, 39, 157
　for rational equations, 1031
　rational expressions in, 1010
　for subtraction, 16, 97–98, 157, 622
　for variables, 114–115
　for variation, 1100–1102
Transverse axis, of hyperbola, 1433
Trapezoids
　area of, 510, 512
　explanation of, 508
Trial-and-error method
　factoring trinomials by, 822–828
　sign rules for, 824
　steps in, 823
Triangles
　acute, 497
　angles of, 496–497, 644–645, 823–824
　area of, 197–199, 322, 510, 511, 1179
　equilateral, 497
　explanation of, 496
　isosceles, 497
　method to find unknown side of, 1040
　obtuse, 497
　perimeter of, 508, 640–641
　Pythagorean theorem and, 498–501
　scalene, 497
　similar, 370–372, 1039–1041
Trinomials. *See also* Polynomial(s)
　AC-method to factor, 933–937
　explanation of, 866
　factoring, 915–918
　higher degree, 928, 937
　with leading coefficient of 1, 915–916
　perfect square. *See* Perfect square trinomials
　removing greatest common factor from, 936
　sign rules for factoring, 917
　trial-and-error method to factor, 822–828

U

Undefined division, 715
Uniform motion applications, 653–655
Union, of sets, 1114–1116
Unit costs, 352–353
Unit of measure, 446. *See* Measurement
Unit rates
　converting rate to, 351
　explanation of, 341, 351
　method to find, 351–352
Unlike fractions. *See also* Fractions
　addition of, 221–222, 223–224
　explanation of, 221
　subtraction of, 222–224
Unlike terms, 126
U.S. Customary units. *See also* Measurement
　of capacity, 446, 451–452
　conversions between metric units and, 473–477
　explanation of, 446
　of length, 446–449
　of time, 449
　of weight, 446, 450

V

Variables
　dependent, A–26
　explanation of, 11, 63, 114
　fractions containing, 187, 197, 206
　independent, A–26
　isolation of, 145, 150–152, 591, 605
　roots of, 1183–1185
　translations involving, 114–115
Variable terms, 126
Variation
　applications of, 1102–1105
　direct, 1100, 1102–1103
　inverse, 1100, 1103–1104
　joint, 1101, 1104–1105
　translations involving, 1100–1102
Verbal model, A–26
　linear equation from, A–27 to A–28
Vertex
　of angle, 486
　of hyperbola, 1434
　of parabola, 1299, 1318–1319, 1422, 1427–1428, A–30
　of quadratic function, 1318–1319
Vertex formula, 1318–1319, 1427–1428
Vertical angles, 488
Vertical asymptote, 1081
Vertical lines
　explanation of, 700
　graphs of, 701–702
　linear equations and, 743
　slope of, 715
Vertical line test, 1067–1068, 1331
Vertical transverse axis, 1434
Volume. *See also* Capacity
　explanation of, 521
　formulas for, 521
　method to find, 522–523

W

Weight
　converting units of, 450, 474
　U.S. Customary units of, 446, 450
Weighted mean, 577–578
Whole numbers
　addition of, 9–12
　algebraic expressions and, 63–64

computing mean and, 70–71
division of, 47–53
division of decimals by, 301
estimation of, 28, 29, 38–39
expanded notation for, 3–4
explanation of, 2
exponents and, 59–60
multiplication of, 33–40
on number line, 5, 13
order of operations and, 61–63
place value and, 2–3, 268
ratio of, 343–344
rounding of, 26–28
square roots and, 60
standard notation for, 3–4
subtraction of, 12–15
written in words, 4–5
Word problems flow chart, 157, 621. *See also* Applications
Words/phrases. *See also* Translations
for addition, 15–16, 91, 157
for division, 52, 157
for multiplication, 39, 157
for subtraction, 16, 97–98, 157
writing numbers in, 4–5

X

x-axis, 685
x-coordinates
explanation of, 686, 698
in one-to-one functions, 1330
x-intercepts
explanation of, 698
of functions, 1083, 1084–1085
method to find, 698–700
of quadratic functions, 1282, 1283, 1318, 1319
systems of linear equations and, 769

Y

y-axis, 685
y-coordinates
explanation of, 686, 698
in one-to-one functions, 1330
y-intercepts
explanation of, 698
of functions, 1083, 1084–1085
method to find, 698–700
of quadratic functions, 1282, 1283, 1318, 1319
slope-intercept form to find, 728–729
systems of linear equations and, 769

Z

Zero
addition property of, 11
division by, 48, 715
multiplication property of, 36
as place holder with decimals, 269
Zero exponents, 849
Zero product rule
explanation of, 958
to solve quadratic equations, 958–959, 1264, 1269, 1283

Perimeter and Circumference

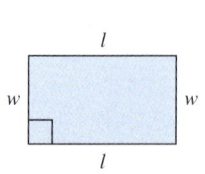
Rectangle
$P = 2l + 2w$

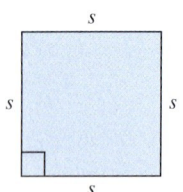
Square
$P = 4s$

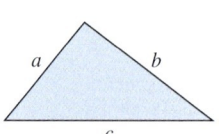
Triangle
$P = a + b + c$

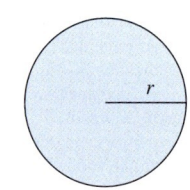
Circle
Circumference: $C = 2\pi r$

Area

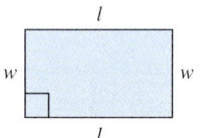
Rectangle
$A = lw$

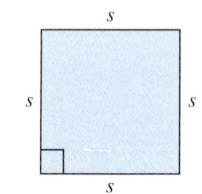
Square
$A = s^2$

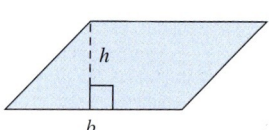
Parallelogram
$A = bh$

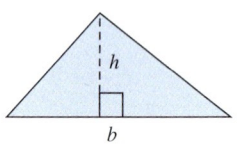
Triangle
$A = \frac{1}{2}bh$

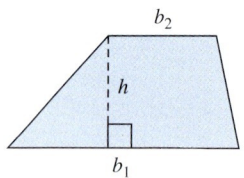
Trapezoid
$A = \frac{1}{2}(b_1 + b_2)h$

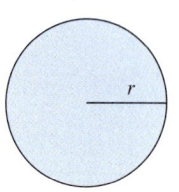
Circle
$A = \pi r^2$

Volume

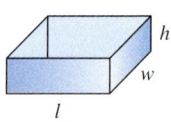
Rectangular Solid
$V = lwh$

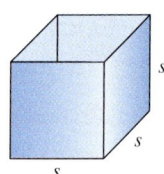
Cube
$V = s^3$

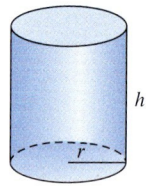
Right Circular Cylinder
$V = \pi r^2 h$

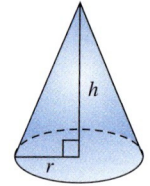
Right Circular Cone
$V = \frac{1}{3}\pi r^2 h$

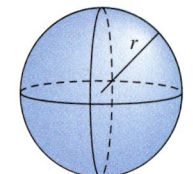
Sphere
$V = \frac{4}{3}\pi r^3$

Angles

- Two angles are **complementary** if the sum of their measures is 90°.

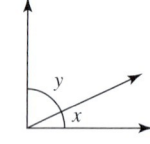

- Two angles are **supplementary** if the sum of their measures is 180°.

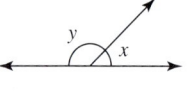

- $\angle a$ and $\angle c$ are vertical angles, and $\angle b$ and $\angle d$ are vertical angles. The measures of vertical angles are equal.

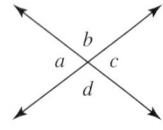

- The sum of the measures of the angles of a triangle is 180°.

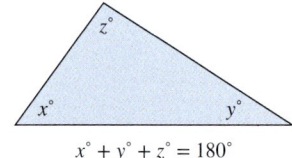

$x° + y° + z° = 180°$

Linear Equations and Slope

The slope, m, of a line between two distinct points (x_1, y_1) and (x_2, y_2):

$$m = \frac{y_2 - y_1}{x_2 - x_1}, \quad x_2 - x_1 \neq 0$$

Standard form: $Ax + By = C$ (A and B are not both zero.)

Horizontal line: $y = k$

Vertical line: $x = k$

Slope-intercept form: $y = mx + b$

Point-slope formula: $y - y_1 = m(x - x_1)$

Midpoint Formula

Given two points (x_1, y_1) and (x_2, y_2), the midpoint is

$$\left(\frac{x_1 + x_2}{2}, \frac{y_1 + y_2}{2}\right)$$

Properties and Definitions of Exponents

Let a and b ($b \neq 0$) represent real numbers and m and n represent positive integers.

$$b^m b^n = b^{m+n}; \quad \frac{b^m}{b^n} = b^{m-n}; \quad (b^m)^n = b^{mn};$$

$$(ab)^m = a^m b^m; \quad \left(\frac{a}{b}\right)^m = \frac{a^m}{b^m}; \quad b^0 = 1; \quad b^{-n} = \left(\frac{1}{b}\right)^n$$

The Quadratic Formula

The solutions to $ax^2 + bx + c = 0$ ($a \neq 0$) are given by

$$x = \frac{-b \pm \sqrt{b^2 - 4ac}}{2a}$$

The Vertex Formula

For $f(x) = ax^2 + bx + c$ ($a \neq 0$), the vertex is

$$\left(\frac{-b}{2a}, \frac{4ac - b^2}{4a}\right) \quad \text{or} \quad \left(\frac{-b}{2a}, f\left(\frac{-b}{2a}\right)\right)$$

The Distance Formula

The distance between two points (x_1, y_1) and (x_2, y_2) is

$d = \sqrt{(x_2 - x_1)^2 + (y_2 - y_1)^2}$

The Standard Form of a Circle

$(x - h)^2 + (y - k)^2 = r^2$ with center (h, k) and radius r

Properties of Logarithms

Let b, x, and y be positive real numbers where $b \neq 1$, and let p be a real number. Then the following properties are true.

1. $\log_b 1 = 0$
2. $\log_b b = 1$
3. $\log_b(b^p) = p$
4. $b^{\log_b(x)} = x$
5. $\log_b(xy) = \log_b x + \log_b y$
6. $\log_b\left(\frac{x}{y}\right) = \log_b x - \log_b y$
7. $\log_b(x^p) = p \log_b x$

Change-of-Base Formula:

$$\log_b x = \frac{\log_a x}{\log_a b} \quad a > 0, a \neq 1, b > 0, b \neq 1$$

Properties and Definitions of Radicals

Let a be a real number and n be an integer such that $n > 1$. If $\sqrt[n]{a}$ exists, then

$$a^{1/n} = \sqrt[n]{a}; \quad a^{m/n} = \left(\sqrt[n]{a}\right)^m = \sqrt[n]{a^m}$$

Let a and b represent real numbers such that $\sqrt[n]{a}$ and $\sqrt[n]{b}$ are both real. Then

$\sqrt[n]{ab} = \sqrt[n]{a} \cdot \sqrt[n]{b}$ Multiplication property

$\sqrt[n]{\frac{a}{b}} = \frac{\sqrt[n]{a}}{\sqrt[n]{b}}$ Division property

The Pythagorean Theorem:

$a^2 + b^2 = c^2$

Conversion of U.S. Customary Units of Length, Time, Weight, Capacity, Energy, and Power

Length	Time
1 foot (ft) = 12 inches (in.) 1 yard (yd) = 3 feet (ft) 1 mile (mi) = 5280 feet (ft) 1 mile (mi) = 1760 yard (yd)	1 year (yr) = 365 days 1 week (wk) = 7 days 1 day = 24 hours (hr) 1 hour (hr) = 60 minutes (min) 1 minute (min) = 60 seconds (sec)
Capacity	**Weight**
3 teaspoons (tsp) = 1 Tablespoon (Tbsp) 1 cup (c) = 8 fluid ounces (fl oz) 1 pint (pt) = 2 cups (c) 1 quart (qt) = 2 pints (pt) 1 gallon (gal) = 4 quarts (qt)	1 pound (lb) = 16 ounces (oz) 1 ton = 2000 pounds (lb)

Conversion of the Metric System (Prefix Line)

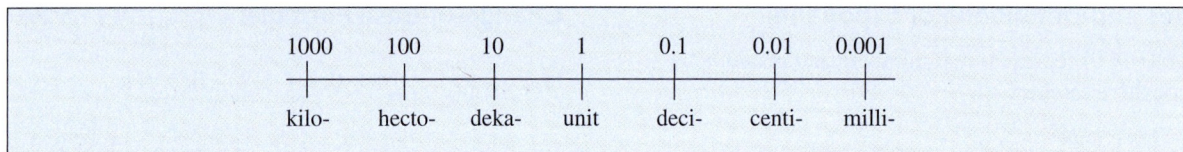

Conversion of U.S. Customary Measurement to Metric

Length	Capacity	Weight/Mass (on the Earth)
1 in. = 2.54 cm	1 qt ≈ 0.95 L	1 lb ≈ 0.45 kg
1 ft ≈ 0.305 m	1 fl oz ≈ 30 mL	1 oz ≈ 28 g
1 yd ≈ 0.914 m	1 mL = 1 cc	
1 mi. ≈ 1.61 km		